Handbook of Encapsulation and Controlled Release

Handbook of Encapsulation and Controlled Release

Edited by
Munmaya Mishra

CRC Press
Taylor & Francis Group
Boca Raton London New York

CRC Press is an imprint of the
Taylor & Francis Group, an **informa** business

CRC Press
Taylor & Francis Group
6000 Broken Sound Parkway NW, Suite 300
Boca Raton, FL 33487-2742

© 2016 by Taylor & Francis Group, LLC
CRC Press is an imprint of Taylor & Francis Group, an Informa business

No claim to original U.S. Government works

Printed in Canada on acid-free paper
Version Date: 20151019

International Standard Book Number-13: 978-1-4822-3232-5 (Hardback)

This book contains information obtained from authentic and highly regarded sources. Reasonable efforts have been made to publish reliable data and information, but the author and publisher cannot assume responsibility for the validity of all materials or the consequences of their use. The authors and publishers have attempted to trace the copyright holders of all material reproduced in this publication and apologize to copyright holders if permission to publish in this form has not been obtained. If any copyright material has not been acknowledged please write and let us know so we may rectify in any future reprint.

Except as permitted under U.S. Copyright Law, no part of this book may be reprinted, reproduced, transmitted, or utilized in any form by any electronic, mechanical, or other means, now known or hereafter invented, including photocopying, microfilming, and recording, or in any information storage or retrieval system, without written permission from the publishers.

For permission to photocopy or use material electronically from this work, please access www.copyright.com (http://www.copyright.com/) or contact the Copyright Clearance Center, Inc. (CCC), 222 Rosewood Drive, Danvers, MA 01923, 978-750-8400. CCC is a not-for-profit organization that provides licenses and registration for a variety of users. For organizations that have been granted a photocopy license by the CCC, a separate system of payment has been arranged.

Trademark Notice: Product or corporate names may be trademarks or registered trademarks, and are used only for identification and explanation without intent to infringe.

Library of Congress Cataloging-in-Publication Data

Handbook of encapsulation and controlled release / edited by Munmaya Mishra.
   pages cm
Includes bibliographical references and index.
   ISBN 978-1-4822-3232-5 (hardcover : alk. paper) 1. Controlled release technology--Handbooks, manuals, etc. 2. Microencapsulation--Handbooks, manuals, etc. I. Mishra, Munmaya K., editor.

TP156.C64H36 2016
660'.2--dc23   2015026119

Visit the Taylor & Francis Web site at
http://www.taylorandfrancis.com

and the CRC Press Web site at
http://www.crcpress.com

To my family

Also, to those who made and will make a difference through polymer research for improving the quality of life!

Contents

Preface xiii
Editor xv
Contributors xvii

SECTION I Fundamentals

Chapter 1 Overview of Encapsulation and Controlled Release 3

Munmaya K. Mishra

SECTION II Processes

Chapter 2 Process-Selection Criteria 23

James Oxley

Chapter 3 Microencapsulation by Spray Drying 35

Stephan Drusch and S. Diekmann

Chapter 4 Spray Drying and Its Application in Food Processing 47

Huang Li Xin and Arun S. Mujumdar

Chapter 5 Encapsulation via Spray Chilling/Cooling/Congealing 71

Carmen Sílvia Favaro-Trindade, Paula Kiyomi Okuro, and Fernando Eustáquio de Matos Jr.

Chapter 6 Encapsulation via Spinning Disk Technology 89

Aurélie Demont and Ian W. Marison

Chapter 7 Encapsulation via Fluidized Bed Coating Technology 111

Charles R. Frey

Chapter 8 Encapsulation via Pan-Coating 147

Charles R. Frey

Chapter 9 Microencapsulation by Dripping and Jet Break-Up 177

Aurélie Demont and Ian W. Marison

Chapter 10 Microencapsulation by Annular Jet Process 201
Thorsten Brandau

Chapter 11 Encapsulation via Hot-Melt Extrusion 213
Hemlata Patil, Roshan V. Tiwari, and Michael A. Repka

Chapter 12 Microencapsulation with Coacervation.. 235
Michael Yan

Chapter 13 Encapsulation via Microemulsion 247
Sushama Talegaonkar, Lalit Mohan Negi, and Harshita Sharma

Chapter 14 Ionotropic Gelation and Polyelectrolyte Complexation Technique: Novel Approach to Drug Encapsulation 273
J.S. Patil, S.C. Marapur, P.B. Gurav, and A.V. Banagar

Chapter 15 Microencapsulation via Interfacial Polymerization 297
Biao Duan

Chapter 16 Microencapsulation via *In Situ* Polymerization 307
Biao Duan

Chapter 17 Microencapsulation with Miniemulsion Technology... 315
Michael Yan

Chapter 18 Silica-Based Sol-Gel Microencapsulation and Applications 329
Rosaria Ciriminna and Mario Pagliaro

Chapter 19 Microencapsulation by Phase Inversion Precipitation 347
Ricard Garcia-Valls and Cinta Panisello

Chapter 20 Microfluidic Encapsulation Process 359
Fabrizio Sarghini

Chapter 21 Encapsulation Process in Granulation Technology.. 385
Himanshu K. Solanki

Chapter 22 Encapsulation via Electrohydrodynamic Atomization Spray Technology (Electrospray) 411
Milad Jafari-Nodoushan, Hamid Mobedi, and Jalal Barzin

Contents

Chapter 23 Encapsulation Process: Pulsed Combustion Spray Drying 439

Chilwin Tanamal and James A. Rehkopf

Chapter 24 Supercritical Fluid Technology for Encapsulation 447

Ángel Martín, Marta Fraile, Soraya Rodríguez-Rojo, and María José Cocero

Chapter 25 Melt-Dispersion Technique for Encapsulation 469

Verica Djordjević, Steva Lević, Thomas Koupantsis, Fani Mantzouridou, Adamantini Paraskevopoulou, Viktor Nedović, and Branko Bugarski

SECTION III Ingredients

Chapter 26 Materials of Natural Origin for Encapsulation 493

Munmaya K. Mishra

Chapter 27 Cellulose Ethers: Applications 517

Robert Schmitt, True Rogers, William Porter III, Oliver Petermann, and Britta Huebner-Keese

Chapter 28 Cellulose-Based Biopolymers: Formulation and Delivery Applications 535

J.D.N. Ogbonna, F.C. Kenechukwu, S.A. Chime, and A.A. Attama

Chapter 29 Starch-Based Polymeric Biomaterial: Drug Delivery 575

Akhilesh Vikram Singh and Ashok M. Raichur

Chapter 30 Biodegradable Polymers: Drug Delivery Applications 583

Satish Shilpi and Sanjay K. Jain

SECTION IV Characterization

Chapter 31 Encapsulation Field Polymers: Fourier Transform Infrared Spectroscopy (FTIR) 617

Oana Lelia Pop, Dan Cristian Vodnar, and Carmen Socaciu

SECTION V Applications

Chapter 32 Encapsulation Technologies for Modifying Food Performance 643

Maria Inês Ré, Maria Helena Andrade Santana, and Marcos Akira d'Ávila

Chapter 33 Microencapsulation: Probiotics 685
  Dan Cristian Vodnar, Oana Lelia Pop, and Carmen Socaciu

Chapter 34 Organogels as Food Delivery Systems 697
  Tarun Garg, Goutam Rath, and Amit K. Goyal

Chapter 35 β-Lactoglobulin: Bioactive Nutrients Delivery 729
  Li Liang and Muriel Subirade

Chapter 36 Encapsulation of Polyphenolics 741
  Florence Edwards-Lévy and Aude Munin-César

Chapter 37 Encapsulation of Bioactive Compounds 765
  Francesco Donsì, Mariarenata Sessa, and Giovanna Ferrari

Chapter 38 Encapsulation of Flavors, Nutraceuticals, and Antibacterials 801
  Stéphane Desobry and Frédéric Debeaufort

Chapter 39 Encapsulation of Aroma 833
  Christelle Turchiuli and Elisabeth Dumoulin

Chapter 40 Molecular (Cyclodextrin) Encapsulation of Volatiles and Essential Oils 867
  Paulo José Salústio, Maria Graça Miguel, and Helena Cabral-Marques

Chapter 41 Microencapsulation: Artificial Cells 907
  Thomas Ming Swi Chang

Chapter 42 Cell Encapsulation 917
  James Blanchette

Chapter 43 Cell Immobilization Technologies for Applications in Alcoholic Beverages 933
  Argyro Bekatorou, Stavros Plessas, and Athanasios Mallouchos

Chapter 44 Enzyme Immobilization in Biodegradable Polymers for Biomedical Applications 957
  S.A. Costa, Helena S. Azevedo, and Rui L. Reis

Chapter 45 Emulsion-Solvent Removal System for Drug Delivery 981
  Wasfy M. Obeidat

Chapter 46 Organogels in Controlled Drug Delivery 1035
  V.K. Singh, B. Behera, Sai S. Sagiri, Kunal Pal, Arfat Anis, and Mrinal K. Bhattacharya

Contents

Chapter 47 Microparticulate Drug Delivery Systems 1067

Hemant Kumar Singh Yadav, M. Navya, Abhay Raizaday,
V. Naga Sravan Kumar Varma, and H.G. Shivakumar

Chapter 48 Colloid Drug Delivery Systems...1111

Monzer Fanun

Chapter 49 Melt Extrusion: Pharmaceutical Applications 1127

James DiNunzio, Seth Forster, and Chad Brown

Chapter 50 Nanoparticles: Biomaterials for Drug Delivery 1151

Abhijit Gokhale, Thomas Williams, and Jason M. Vaughn

Chapter 51 Polymer Systems for Ophthalmic Drug Delivery 1167

Sepideh Khoee and Frazaneh Hashemi Nasr

Chapter 52 Drug Delivery Systems: Oral Mucosal 1225

Javier Octavio Morales

Chapter 53 Polymeric Biomaterials for Controlled Drug Delivery 1255

Sutapa Mondal Roy and Suban K. Sahoo

Chapter 54 Nanogels: Chemical Approaches to Preparation... 1271

Sepideh Khoee and Hamed Asadi

Chapter 55 Electrospinning Technology: Polymeric Nanofiber Drug Delivery........................1311

Narendra Pal Singh Chauhan, Kiran Meghwal, Priya Juneja, and Pinki B. Punjabi

Chapter 56 Polyelectrolyte Complexes: Drug Delivery Technology 1333

Lankalapalli Srinivas

Chapter 57 Polymeric Nano/Microparticles for Oral Delivery of Proteins and Peptides 1359

S. Sajeesh and Chandra P. Sharma

Chapter 58 Vegetable Oil–Based Formulations for Controlled Drug Delivery........................ 1381

V.K. Singh, Sai S. Sagiri, K. Pramanik, Arfat Anis, S.S. Ray,
I. Banerjee, and Kunal Pal

Chapter 59 Introduction to Commercial Microencapsulation 1413

George A. Stahler

Chapter 60 Stable Core-Shell Microcapsules for Industrial Applications 1423

Klaus Last

Chapter 61 Microencapsulation Applications in Food Packaging 1439

Artur Bartkowiak, Agnieszka Bednarczyk-Drag, Wioletta Krawczynska, Agnieszka Krudos, and Katarzyna Sobecka

Chapter 62 Microencapsulation of Phase Change Materials 1455

Jessica Giro-Paloma, Mònica Martínez, A. Inés Fernández, and Luisa F. Cabeza

Index 1483

Preface

The technology and applications of encapsulation (microencapsulation) are rapidly evolving. As a result, there is a clear need for a source of technical information that has broad coverage, is current, and is written at a level comprehensible to nonexperts. With these considerations in mind, the decision to publish the *Handbook of Encapsulation and Controlled Release* was apparent. My hope is to present the material in such a manner that it conveys important overviews of various processes and applications to help stimulate research for further advancements in the field. I have been working in the area for many years and am quite fascinated with the innovations that evolve around both science and art to some extent. I had the privilege of working with experts worldwide in preparing this book with its vast entries in the first edition spanning almost the entire field of encapsulation processes and many applications at that time.

This book is an authoritative and comprehensive reference on the broad subject of encapsulation (microencapsulation) and controlled release applications, which will enable readers to have an enriching experience in general and a targeted knowledge in this evolving arena. This groundbreaking work includes many chapters and offers a broad-based perspective on a variety of applications and processes, including research information, figures, tables, illustrations, and references. It provides the fundamentals, including chemical and physicochemical processes, and explores how to apply those processes for different applications in the industry. This book caters to engineers and scientists (polymer scientists, materials scientists, biomedical engineers, biochemists, and macromolecular chemists), researchers, pharmacists, doctors, and students, and general readers in academia, industry, and research institutions. It is envisioned that this book will serve as the most respected reference work on the process and application of encapsulation in various industries, such as those related to food, consumer products, pharmaceutical/medical, agriculture, nutraceuticals, dietary supplements, cosmetics, flavors, and fragrances.

I feel honored to undertake the important and challenging endeavor of developing the *Handbook of Encapsulation and Controlled Release*, which will cater to the needs of many who are working in the field or are somehow influenced by it. I would like to express my sincere gratitude and appreciation to the authors for their excellent professionalism and dedicated work. Needless to say, a book of this nature would never have existed if the expert authors had not devoted their valuable time for preparing the authoritative chapters. I thank the entire management team at Taylor & Francis Group (CRC Press) and, particularly, Barbara Glunn, who made this book possible.

I also express my sincere love and appreciation to my wife, Bidu Mishra, PhD, for her encouragement, sacrifice, and support during the weekends, early mornings, and holidays spent on this book. Without her help and support, this project would have never been started or completed.

Munmaya K. Mishra

Editor

Munmaya K. Mishra, PhD, is a polymer scientist who has worked in the industry for more than 25 years. He has been engaged in research, management, technology innovations, and product development and contributed immensely to multiple aspects of polymer applications, including encapsulation and controlled release technologies. He is the author/coauthor of hundreds of scientific articles and author/editor of seven books. He is the inventor of many technology innovations and holds over 40 U.S. patents, over 50 U.S. patent pending applications, and over 100 world patents. Dr. Mishra is the recipient of many recognitions and awards, including the Texaco Research Chairman's Award from the American Chemical Society's Mid-Hudson Section and the New York Research Award. He is currently the editor-in-chief of three renowned polymer journals published by the Taylor & Francis Group. He is the editor-in-chief of the recently published 11-volume *Encyclopedia of Biomedical Polymers and Polymeric Biomaterials*. He is also the founder of a scientific organization, the International Society of Biomedical Polymers and Polymeric Biomaterials. About 20 years ago, he founded and established a scientific meeting titled Advanced Polymers via Macromolecular Engineering, which has gained international recognition and is still being held under the sponsorship of the International Union of Pure and Applied Chemistry.

Contributors

Arfat Anis
Department of Chemical Engineering
SABIC Polymer Research Center
King Saud University
Riyadh, Saudi Arabia

Hamed Asadi
Department of Polymer Chemistry
School of Science
University of Tehran
Tehran, Iran

A.A. Attama
Faculty of Pharmaceutical Sciences
Department of Pharmaceutics
University of Nigeria
Nsukka, Nigeria

Helena S. Azevedo
Department of Polymer Engineering
University of Minho
Braga, Portugal

A.V. Banagar
VMVVS'S School of Pharmacy
Karnataka, India

I. Banerjee
Department of Biotechnology and Medical Engineering
National Institute of Technology
Odisha, India

Artur Bartkowiak
Center of Bioimmobilisation and Innovative Packaging Materials
West Pomeranian University of Technology
Szczecin, Poland

Jalal Barzin
Department of Biomaterials
Iran Polymer and Petrochemical Institute
Tehran, Iran

Agnieszka Bednarczyk-Drag
Center of Bioimmobilisation and Innovative Packaging Materials
West Pomeranian University of Technology
Szczecin, Poland

B. Behera
Department of Biotechnology and Medical Engineering
National Institute of Technology
Odisha, India

Argyro Bekatorou
Department of Chemistry
University of Patras
Patras, Greece

Mrinal K. Bhattacharya
Department of Botany and Biotechnology
Karimganj College
Assam, India

James Blanchette
University of South Carolina
Columbia, South Carolina

Thorsten Brandau
BRACE GmbH
Alzenau, Germany

Chad Brown
Merck & Co., Inc.
West Point, Pennsylvania

Branko Bugarski
Faculty of Technology and Metallurgy
University of Belgrade
Belgrade, Serbia

Luisa F. Cabeza
GREA Innovació Concurrent
Edifici Centre de Recerca en Economia Aplicada
Universitat de Lleida
Lleida, Spain

Helena Cabral-Marques
Research Institute for Medicine and
  Pharmaceutical Sciences
University of Lisbon
Lisbon, Portugal

Thomas Ming Swi Chang
Faculty of Medicine
Artificial Cells and Organs Research
  Center
McGill University
Montréal, Québec, Canada

Narendra Pal Singh Chauhan
Department of Polymer Science
University College of Science
Mohanlal Sukhadia University
Udaipur, India

S.A. Chime
Faculty of Pharmaceutical Sciences
Department of Pharmaceutics
University of Nigeria
Nsukka, Nigeria

Rosaria Ciriminna
Istituto per lo Studio dei Materiali
  Nanostrutturati
Palermo, Italy

María José Cocero
Department of Chemical Engineering and
  Environmental Technology
University of Valladolid
Valladolid, Spain

S.A. Costa
Department of Polymer Engineering
University of Minho
Braga, Portugal

Marcos Akira d'Ávila
School of Chemical Engineering
University of Campinas
Campinas, Brazil

Frédéric Debeaufort
Université de Bourgogne–EMMA EA 581
Institut Universitaire de Technologie
Dijon, France

Aurélie Demont
Laboratory of Integrated Bioprocessing
School of Biotechnology
Dublin City University
Dublin, Ireland

Stéphane Desobry
Laboratoire d'Ingenierie des Biomolecules
Nancy-Université-INPL-ENSAIA
Vandoeuvre, France

S. Diekmann
Institute of Food Technology and Food
  Chemistry
Technical University of Berlin
Berlin, Germany

James DiNunzio
Merck & Co., Inc.
Kenilworth, New Jersey

Verica Djordjević
Faculty of Technology and Metallurgy
University of Belgrade
Belgrade, Serbia

Francesco Donsì
Department of Industrial Engineering
University of Salerno
Fisciano, Italy

Stephan Drusch
Institute of Food Technology and Food
  Chemistry
Technical University of Berlin
Berlin, Germany

Biao Duan
Encapsys, a Division of Appvion, Inc.
Appleton Wisconsin

Elisabeth Dumoulin
AgroParisTech
Massy, France

Florence Edwards-Lévy
Faculty of Pharmacy
Institute of Molecular Chemistry of Reims
University of Reims Champagne-Ardenne
Reims, France

Contributors

Monzer Fanun
Colloids and Surfaces Research Center
Al-Quds University
East Jerusalem, Palestine

Carmen Sílvia Favaro-Trindade
College of Animal Science and Food Engineering
University of São Paulo
São Paulo, Brazil

A. Inés Fernández
Departament de Ciència dels Materials i Enginyeria Metal·lúrgica
Universitat de Barcelona
Barcelona, Spain

Giovanna Ferrari
Department of Industrial Engineering
University of Salerno
Fisciano, Italy

Seth Forster
Merck & Co., Inc.
West Point, Pennsylvania

Marta Fraile
Department of Chemical Engineering and Environmental Technology
University of Valladolid
Valladolid, Spain

Charles R. Frey
Coating Place, Inc.
Verona, Wisconsin

Ricard Garcia-Valls
Departament d'Enginyeria Química
Universitat Rovira i Virgili
and
Centre Tecnològic de la Química de Catalunya
Carrer Marcel·lí Domingo
Tarragona, Spain

Tarun Garg
Department of Pharmaceutics
ISF College of Pharmacy
Punjab, India

Jessica Giro-Paloma
Departament de Ciència dels Materials i Enginyeria Metal·lúrgica
Universitat de Barcelona
Barcelona, Spain

Abhijit Gokhale
Product Development Services
Patheon Pharmaceuticals, Inc.
Cincinnati, Ohio

Amit K. Goyal
Department of Pharmaceutics
ISF College of Pharmacy
Punjab, India

P.B. Gurav
SVERI's College of Pharmacy
Maharashtra, India

Frazaneh Hashemi Nasr
Department of Polymer Chemistry
School of Science
University of Tehran
Tehran, Iran

Britta Huebner-Keese
Dow Chemical Company
Midland, Michigan

Milad Jafari-Nodoushan
Department of Novel Drug Delivery Systems
Iran Polymer and Petrochemical Institute
Tehran, Iran

Sanjay K. Jain
Pharmaceutics Research Projects Laboratory
Department of Pharmaceutical Sciences
Dr. Hari Singh Gour University
Madhya Pradesh, India

Priya Juneja
Jubilant Life Sciences
Uttar Pradesh, India

F.C. Kenechukwu
Department of Pharmaceutics
Faculty of Pharmaceutical Sciences
University of Nigeria
Nsukka, Nigeria

Sepideh Khoee
Department of Polymer Chemistry
School of Science
University of Tehran
Tehran, Iran

Thomas Koupantsis
Laboratory of Food Chemistry and Technology
School of Chemistry
Aristotle University of Thessaloniki
Thessaloniki, Greece

Wioletta Krawczynska
Center of Bioimmobilisation and Innovative Packaging Materials
West Pomeranian University of Technology
Szczecin, Poland

Agnieszka Krudos
Center of Bioimmobilisation and Innovative Packaging Materials
West Pomeranian University of Technology
Szczecin, Poland

Klaus Last
Follmann Gmbh & Co. KG
Minden, Germany

Steva Lević
Faculty of Agriculture
Department of Food Technology and Biochemistry
University of Belgrade
Belgrade-Zemun, Serbia

Li Liang
Research Institute of Nutraceuticals and Functional Foods
Laval University
Quebec City, Quebec, Canada

and

State Key Laboratory of Food Science and Technology
School of Food Science and Technology
Jiangnan University
Wuxi, Jiangsu, People's Republic of China

Athanasios Mallouchos
Department of Food Science and Human Nutrition
Agricultural University of Athens
Athens, Greece

Fani Mantzouridou
Laboratory of Food Chemistry and Technology
School of Chemistry
Aristotle University of Thessaloniki
Thessaloniki, Greece

S.C. Marapur
BLDEA's College of Pharmacy
Karnataka, India

Ian W. Marison
School of Biotechnology
Dublin City University
Dublin, Ireland

Ángel Martín
Department of Chemical Engineering and Environmental Technology
University of Valladolid
Valladolid, Spain

Mònica Martínez
Departament de Ciència dels Materials i Enginyeria Metal·lúrgica
Universitat de Barcelona
Barcelona, Spain

Fernando Eustáquio de Matos Jr.
College of Animal Science and Food Engineering
University of São Paulo
São Paulo, Brazil

Kiran Meghwal
Department of Chemistry
University College of Science
Mohanlal Sukhadia University
Udaipur, India

Maria Graça Miguel
Faculdade de Ciências e Tecnologia
Universidade do Algarve
Faro, Portugal

Contributors

Munmaya K. Mishra
Altria Research Centre
Richmond, Virginia

Hamid Mobedi
Department of Novel Drug Delivery Systems
Iran Polymer and Petrochemical Institute
Tehran, Iran

Javier Octavio Morales
Department of Pharmaceutical Science and Technology
School of Chemical and Pharmaceutical Sciences
University of Chile
Santiago, Chile

Arun S. Mujumdar
Department of Mechanical Engineering
National University of Singapore
Singapore

Aude Munin-César
Faculty of Pharmacy
Institute of Molecular Chemistry of Reims
University of Reims Champagne-Ardenne
Reims, France

M. Navya
Department of Pharmaceutics
Chalapathi Institute of Pharmaceutical Sciences
Acharya Nagarjuna University
Andhra Pradesh, India

Viktor Nedović
Faculty of Agriculture
Department of Food Technology and Biochemistry
University of Belgrade
Belgrade-Zemun, Serbia

Lalit Mohan Negi
Faculty of Pharmacy
Department of Pharmaceutics
Jamia Hamdard University
New Delhi, India

Wasfy M. Obeidat
Department of Pharmaceutics
Jordan University of Science and Technology
Irbid, Jordan

J.D.N. Ogbonna
Faculty of Pharmaceutical Sciences
Department of Pharmaceutics
University of Nigeria
Nsukka, Nigeria

Paula Kiyomi Okuro
College of Animal Science and Food Engineering
University of São Paulo
São Paulo, Brazil

James Oxley
Department of Pharmaceuticals and Bioengineering
Southwest Research Institute
San Antonio, Texas

Mario Pagliaro
Istituto per lo Studio dei Materiali Nanostrutturati
Palermo, Italy

Kunal Pal
Department of Biotechnology and Medical Engineering
National Institute of Technology
Odisha, India

Cinta Panisello
Departament d'Enginyeria Química
Universitat Rovira i Virgili
and
Centre Tecnològic de la Química de Catalunya
Carrer Marcel·lí Domingo
Tarragona, Spain

Adamantini Paraskevopoulou
Laboratory of Food Chemistry and Technology
School of Chemistry
Aristotle University of Thessaloniki
Thessaloniki, Greece

Hemalata Patil
School of Pharmacy
University of Mississippi
University, Mississippi

J.S. Patil
VT'S Shivajirao S. Jondhale
College of Pharmacy
Maharashtra, India

Oliver Petermann
Dow Chemical Company
Midland, Michigan

Stavros Plessas
Faculty of Agricultural Development
Laboratory of Microbiology, Biotechnology
  and Hygiene
Democritus University of Thrace
Orestiada, Greece

Oana Lelia Pop
Department of Food Science and
  Technology
University of Agricultural Sciences and
  Veterinary Medicine
Cluj-Napoca, Romania

William Porter III
Dow Chemical Company
Midland, Michigan

K. Pramanik
Department of Biotechnology and Medical
  Engineering
National Institute of Technology
Odisha, India

Pinki B. Punjabi
Department of Chemistry
University College of Science
Mohanlal Sukhadia University
Rajasthan, India

Ashok M. Raichur
Department of Applied Chemistry
University of Johannesburg
Doornfontein, South Africa

and

Department of Materials Engineering
Indian Institute of Science
Karnataka, India

Abhay Raizaday
Department of Pharmaceutics
College of Pharmacy
Jagadguru Sri Shivarathreeswara University
Karnataka, India

Goutam Rath
Department of Pharmaceutics
ISF College of Pharmacy
Punjab, India

S.S. Ray
Department of Biotechnology and Medical
  Engineering
National Institute of Technology
Odisha, India

Maria Inês Ré
Research Center of Albi on Particulate Solids,
  Energy and Environment
Albi School of Mines
Albi, France

and

Center of Processes and Products Technology
Institute of Technological Research of the São
  Paulo State
São Paulo, Brazil

James A. Rehkopf
Pulse Holdings, LLC
Payson, Arizona

Rui L. Reis
Department of Polymer Engineering
University of Minho
Braga, Portugal

Michael A. Repka
School of Pharmacy
University of Mississippi
University, Mississippi

True Rogers
Dow Chemical Company
Midland, Michigan

Contributors

Soraya Rodríguez-Rojo
Department of Chemical Engineering and
  Environmental Technology
University of Valladolid
Valladolid, Spain

Sutapa Mondal Roy
Department of Applied Chemistry
Sardar Vallabhbhai National Institute of
  Technology, Surat
Gujarat, India

Sai S. Sagiri
Department of Biotechnology and Medical
  Engineering
National Institute of Technology
Odisha, India

Suban K. Sahoo
Department of Applied Chemistry
Sardar Vallabhbhai National Institute of
  Technology, Surat
Gujarat, India

S. Sajeesh
Division of Biosurface Technology
Sree Chitra Tirunal Institute for Medical
  Sciences and Technology
Kerala, India

Paulo José Salústio
Research Institute for Medicines and
  Pharmaceutical Sciences
University of Lisbon
Lisbon, Portugal

Maria Helena Andrade Santana
Department of Biotechnological Processes
University of Campinas
Campinas, Brazil

Fabrizio Sarghini
Department of Agriculture
University of Naples Federico II
Naples, Italy

Robert Schmitt
Dow Chemical Company
Midland, Michigan

Mariarenata Sessa
Department of Industrial Engineering
University of Salerno
Fisciano, Italy

Chandra P. Sharma
Division of Biosurface Technology
Sree Chitra Tirunal Institute for Medical
  Sciences and Technology
Kerala, India

Harshita Sharma
Department of Pharmaceutics
Jamia Hamdard University
New Delhi, India

Satish Shilpi
Pharmaceutics Research Projects Laboratory
Department of Pharmaceutical Sciences
Dr. Hari Singh Gour University
Madhya Pradesh, India

H.G. Shivakumar
Department of Pharmaceutics
College of Pharmacy
Jagadguru Sri Shivarathreeswara University
Karnataka, India

Akhilesh Vikram Singh
Department of Materials Engineering
Indian Institute of Science
Karnataka, India

V.K. Singh
Department of Biotechnology and Medical
  Engineering
National Institute of Technology
Odisha, India

Katarzyna Sobecka
Center of Bioimmobilisation and
  Innovative Packaging Materials
West Pomeranian University of Technology
Szczecin, Poland

Carmen Socaciu
Department of Food Science and
  Technology
University of Agricultural Sciences and
  Veterinary Medicine
Cluj-Napoca, Romania

Himanshu K. Solanki
Department of Pharmaceutics
SSR College of Pharmacy
Dadra and Nagar Haveli, India

Lankalapalli Srinivas
GITAM Institute of Pharmacy
GITAM University
Andhra Pradesh, India

George A. Stahler
Encapsys, a Division of Appvion, Inc.
Portage, Wisconsin

Muriel Subirade
Research Institute of Nutraceuticals and
  Functional Foods
Laval University
Québec City, Québec, Canada

Sushama Talegaonkar
Faculty of Pharmacy
Department of Pharmaceutics
Jamia Hamdard University
New Delhi, India

Chilwin Tanamal
Encapsys, a Division of Appvion, Inc.
Portage, Wisconsin

Roshan V. Tiwari
School of Pharmacy
University of Mississippi
University, Mississippi

Christelle Turchiuli
AgroParisTech Massy-Genial
Massy, France

V. Naga Sravan Kumar Varma
Department of Pharmaceutics
College of Pharmacy
Jagadguru Sri Shivarathreeswara University
Karnataka, India

Jason M. Vaughn
Patheon Pharmaceuticals, Inc.
Cincinnati, Ohio

Dan Cristian Vodnar
Department of Food Science and
  Technology
University of Agricultural Sciences and
  Veterinary Medicine
Cluj-Napoca, Romania

Thomas Williams
Product Development Services
Patheon Pharmaceuticals, Inc.
Cincinnati, Ohio

Huang Li Xin
Research Institute of Chemical Industry of
  Forestry Products
Nanjing, Chang, People's Republic of China

Hemant Kumar Singh Yadav
Department of Pharmaceutics
RAK College of Pharmaceutical Sciences
RAK Medical and Health Sciences University
Ras Al Khaimah, United Arab Emirates

Michael Yan
Encapsys, a Division of Appvion, Inc.
Appleton, Wisconsin

Section I

Fundamentals

1 Overview of Encapsulation and Controlled Release

Munmaya K. Mishra

CONTENTS

1.1 Definitions 4
1.2 Justification for Encapsulation 4
1.3 Classification 5
1.4 Techniques of Microencapsulation 6
1.5 Criteria for Selecting Encapsulation Technology 6
1.6 Annular Jet (Vibrational Nozzle) 7
1.7 Centrifugal Extrusion 8
1.8 Centrifugal Suspension Separation 8
1.9 Cocrystallization 9
1.10 Coacervation 9
1.11 Emulsification/Emulsion Polymerization 9
1.12 Fluid-Bed Coating 10
1.13 Interfacial Polymerization 10
1.14 Inclusion Complexation 10
1.15 Ionic Gelation (Hydrogel Microspheres) 11
1.16 Liposomes 11
1.17 Lyophilization 11
1.18 Melt Extrusion (Melt Injection) 12
1.19 Pan Coating 12
1.20 Phase Separation 12
1.21 Spray Drying 12
1.22 Spray Congealing 13
1.23 Spinning Disk (Centrifugal Suspension Separation) 13
1.24 Solvent Evaporation 13
1.25 Supercritical Fluid Assisted Encapsulation 13
1.26 Controlled Release 14
1.27 Release Mechanisms 14
1.28 Release Rates 14
1.29 Conclusions 15
References 15

Encapsulation or commonly referred to as microencapsulation involves the incorporation of actives (such as: flavors, drugs, enzymes, cells, or other materials) in small capsules. Capsules offer a means to protect sensitive components, transform liquids into easily handled solid ingredients, and incorporate controlled release attributes (such as time-release, targeted-release, or trigger-release mechanisms) into the product formulations. Various techniques are available to design capsules depending on the end use of the final product. A timely and targeted release improves the effectiveness of actives, broadens the application range of ingredients, and ensures optimal dosage, thereby improving the cost effectiveness for the manufacturer. Encapsulation techniques can facilitate product development through the use of carefully tuned controlled release attributes.

This chapter provides a short overview of encapsulation technologies commonly practiced to encapsulate active ingredients. The simplest way of looking at a microcapsule is that of imagining a grape or hen's egg reduced in size.[1] The shell has a number of names and can be referred to as a membrane, a wall, a covering, or a coating. Similarly the core also goes by a number of terms such as payload, encapsulant, fill, active ingredient, internal phase, or internal ingredient.

1.1 DEFINITIONS

Encapsulation is defined as a technology of casing solids, liquids, or gaseous materials in miniaturesealed capsules (which are nanometer to micrometer to millimeter range) that can release their contents at controlled rates under specific conditions.[2–7] This technique depends on the physical and chemical properties of the material to be encapsulated. The microencapsulation technology has been employed in a diverse range of industry such as chemicals, cosmetics, food, pharmaceuticals, printing, etc.

The development of early encapsulation technology and preparation of microcapsules dates back to 1950s when Green and coworkers[8,9] produced microencapsulated dyes by complex coacervation of gelatin and gum Arabic, for the manufacture of carbonless copying paper. The technologies developed for carbonless copy paper have led to the development of various microcapsule products in later years.

1.2 JUSTIFICATION FOR ENCAPSULATION

The capsule has the ability to preserve a substance in the finely divided state and to release it as needed.[10] The size of the capsules may range from submicrometer to several millimeters in size and have a multitude of different shapes, depending on the materials and methods used to prepare them. The encapsulations/entrapment of active ingredients are done for a variety of reasons:

- Protecting the core material from degradation by reducing its reactivity to the outside environment (such as UV light, heat, moisture, air oxidation, chemical attack, acids, bases, etc.).
- Reducing/retarding the evaporation or transfer rate of a volatile active ingredient (the core material) to the outside environment.
- Enhancing the visual aspect and marketing concept of the final encapsulated product.
- Modifying the physical characteristics of a material, making it easier to handle (e.g., converting liquid into solid form, improving the handling properties of a sticky material, etc.).
- Achieving controlled and/or targeted release of active ingredients. The product can be tailored to either release slowly over time or at a certain point. Improving shelf life by preventing degradative reactions (dehydration, oxidation, etc.).
- Masking of taste or odors of active ingredient(s).
- Handling highly valuable active ingredient (the core material can be diluted when only very small amounts are required, yet still achieve a uniform dispersion in the host material).
- Mixing incompatible compounds by separating components within a mixture that would otherwise react with one another.[11–16]
- Improved processing of materials (texture and less wastage of ingredients, control of hygroscopic attributes, enhance other attributes such as flowability, solubility, and dispersibility; dust-free powder).
- For safe handling of the toxic materials.

1.3 CLASSIFICATION

Microcapsules can be classified on the basis of their size or morphology, and they range in size from 1 μm to few millimeters. Some microcapsules whose diameter is in the nanometer range are referred to as nanocapsules to highlight their small size. Particles having diameter between 3 and 8 μm are called microparticles or microspheres or microcapsules. Particles larger than 1000 μm are called macroparticles. The microscopic size of microcapsules provides a huge surface area (e.g., the total surface area of 1 μm of hollow microcapsules having a diameter of 0.1 mm has been reported to be about 60 m^2) is available for sites of adsorption and desorption, chemical reactions, etc.[17]

The morphology of microcapsules depends mainly on the core material, how it is distributed within the system, and the deposition process of the shell. Similarly, the morphology of the internal structure of a microparticle depends largely on the selected shell materials and the microencapsulation methods that are employed. The microcapsules may be categorized into several arbitrary and overlapping classifications such as

- Mononuclear (also known as core–shell) microcapsules contain the shell around the core. This is also called a single-core or monocore capsule.
- Polynuclear capsules have many cores enclosed within the shell. This is also called polycore- or multicore-type capsule.
- Matrix encapsulation in which the core material is distributed homogeneously within the shell material.

In addition to these three basic morphologies, microcapsules can also be mononuclear with multiple shells (such as layering of shells), or they may form clusters of microcapsules (as noted in Figure 1.1).

Matrix encapsulation is the simplest structure, in which the active ingredient (core) is much more dispersed within the carrier/shell material either in the form of relatively small droplets or more homogenously distributed/embedded in a continuous matrix of wall material. The active ingredients in the matrix type morphology are also present at the surface unless there is additional coating applied.

The composition, mechanism of release, particle size, and final physical form of microcapsules can be changed to suit specific applications. The properties of the shell materials are extremely important for the stabilization of the core material. Importantly, it must be inert toward active ingredients.

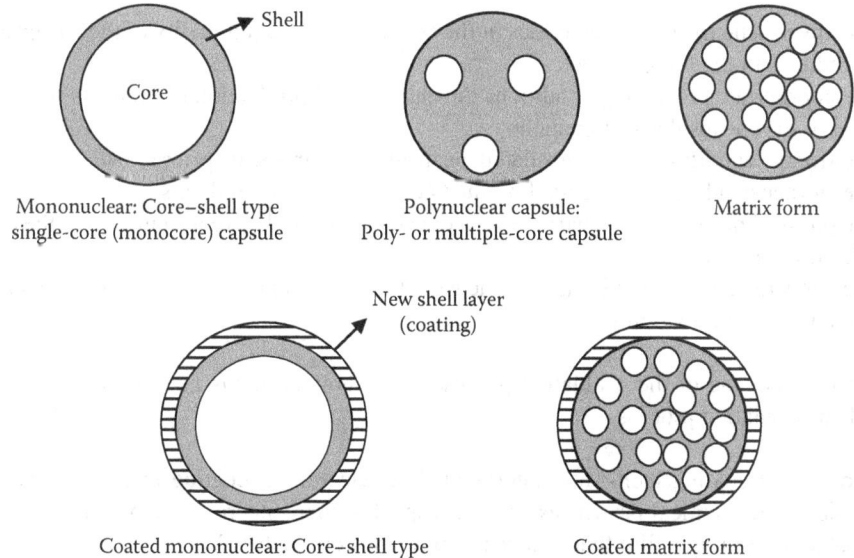

FIGURE 1.1 Schematics of types of microcapsules (spherical shaped shown; however, other forms possible).

It could be film-forming, pliable, tasteless, stable, nonhygroscopic, no high viscosity, economical, soluble in an aqueous media or solvent, or melting. In addition, the shell material can be flexible, brittle, hard, thin, etc. Microcapsule-based systems increase the life span of actives and control the release of said actives.

1.4 TECHNIQUES OF MICROENCAPSULATION

In a sense, the process of microencapsulation actually covers three separate processes on a time scale:

1. The first process consists of forming a shell wall around the core material.
2. The second process involves keeping the core materials intact inside the wall material so that it does not release.
3. The third process involves releasing the core material at the right time and at the right rate.

1.5 CRITERIA FOR SELECTING ENCAPSULATION TECHNOLOGY

The selection of the microencapsulation process is determined by the physical and chemical properties of core and shell/coating materials and the intended application. Various technologies and shell materials have been developed to design microcapsules with wide variety of functionalities. By using selective encapsulation techniques and shell materials, designed microcapsules with controlled and/or targeted release of the active encapsulated ingredients (by using triggers, such as pH change, mechanical stress, temperature, enzymatic activity, time, osmotic force, etc.) can be obtained.

It is not a matter of trial and error for selecting the proper encapsulation technology rather a well-thought process, which includes fulfilling certain requirements. The purpose of encapsulation must be clearly defined before considering the properties desired in encapsulated products. One of the first things to access is the type of benefits (such as desired properties, improving product/process, storage stability, etc.) one would like to accomplish through encapsulation technology. Various considerations come into play in designing the encapsulation process. The following items need to be considered for designing the encapsulation and selecting the proper encapsulation process:

- The physicochemical characteristics of the active and the functionality of the encapsulated ingredients in the final product.
- The appropriate processing conditions during the final production of products for the survival of the encapsulated ingredient.
- The storage conditions of encapsulatedencapsulated ingredients prior to use.
- The storage conditions of the final product containing the ingredients.
- The physical properties (the particle size, density, and stability requirements) of the encapsulated ingredient.
- The trigger(s) and mechanism(s) for releasing the active ingredient from microcapsules.
- The cost constraints if any.

Based on the outcome of the evaluation process, the following items need to be selected for the preferred encapsulation process.

- The type of coating/shell-wall material(s). The selection of coating material decides the physical and chemical properties of microcapsules/microspheres. The polymer should be capable of forming a film that is cohesive with the core material.
- The amount/percentage of core/loading for the microcapsule. The concentration of the active ingredient in the microcapsule.

TABLE 1.1
Various Encapsulation Processes

Annular Jet Process	Layer-by-Layer Deposition
Centrifugal suspension separation	Matrix encapsulation
Coacervation and phase separation	Melt-dispersion
Cocrystallization	Microfluidic encapsulation
Controlled precipitation	Molecular encapsulation
Dispersion/suspension polymerization	Organogels
Dripping and jet break up	Pan coating technology
Electrospraying	Phase inversion/precipitation
Emulsification/emulsion polymerization	Polyelectrolyte complexation
Emulsion-solvent removal	Self-assembly
Fluidized bed coating technology	Sol-gel process
Freeze-drying	Solvent evaporation
Garnulation	Spinning disk technology
Hot-melt extrusion process	Spray chilling/cooling/congealing
In situ polymerization	Spray drying
Interfacial cross-linking/reaction	Supercritical fluid technology
Interfacial polymerization	Templating
Ionotropic gelation	Vapor phase deposition

- The type of encapsulation technology/process.
- The legal issues, if any, to be considered concerning the technology.
- The intellectual property status and freedom to practice/use.
- The scale-up possibilities with no quality concerns.

Microencapsulation processes are usually categorized into two main groupings: chemical processes and mechanical or physical processes. These distinctions can be somewhat misleading as some processes classified as mechanical might involve or even rely upon a chemical reaction, and some chemical techniques rely solely on physical events. Some of the encapsulation processes are listed in Tables 1.1 and 1.2.

The subsequent sections will briefly describe the most widely used encapsulation processes however; the other chapters in the book will describe some of the techniques in detail.

1.6 ANNULAR JET (VIBRATIONAL NOZZLE)

The Annular Jet technology was developed by the Southwest Research Institute, USA. The technique involves two concentric jets, the inner containing the active ingredient (liquid core material) and the outer jet contains the liquid wall material, generally molten that solidifies when exiting the jet. Core–shell encapsulation can be done by using a laminar flow through a nozzle and an additional vibration of the nozzle or liquid. The vibrational nozzle also helps to control the droplet size giving a more uniform product with lower microcapsule sizes down to submicron diameters. The process involves the pumping of a dual fluid stream of liquid core and shell materials through concentric tubes/nozzles and forms droplets under the influence of vibration. The shell is then hardened by cooling (thermal gelation), chemical cross-linking, or solvent evaporation. Different types of extrusion nozzles have been developed in order to optimize the process. The vibration has to be done in resonance with Rayleigh instability and leads to the formation of very uniform droplets as the dual fluid stream naturally breaks.

TABLE 1.2
Microencapsulation Processes with Their Relative Particle Size Ranges

	Approximate Particle/Capsule Size (μm)	Approximate Payload%
Annular et	150–5000	30–90
Coacervation	2–1200	40–90
Electrospray	2–100	20–60
Emulsification	0.2–5000	1–100
Fluid-bed technology	5–2000	5–50
Freeze drying	20–5000	Varies
Granulation	Varies	Varies
In situ polymerization	1–1000	Varies
Inclusion complexation	0.001–0.01	5–20
Interfacial polymerization	1–1000	Varies
Ionic gelation (coextrusion)	200–5000	20–50
Jet cutting	100–3000	20–60
Liposome entrapment	1–10	5–50
Melt extrusion/hot melt	300–5000	5–40
Melt injection	200–2000	5–20
Microfluidics	2–1000	30–60
Pan coating	250–5000	Varies
Phase inversion	0.5–5	Varies
Polyelectrolyte multilayer	0.2–5000	1–90
Polymer–polymer incompatibility	0.5–1000	Varies
Solvent evaporation	0.5–1000	Varies
Spinning disc	5–1500	Varies
Spray-chilling/cooling	20–1000	10–20
Spray-drying	10–400	5–50
Supercritical fluid	10–400	20–50

1.7 CENTRIFUGAL EXTRUSION

This process is mostly used to encapsulate flavor oils. The liquid coextrusion process employs nozzles consisting of concentric orifices located on the outer circumference of a rotating cylinder. Simultaneously the liquid core material and the liquid wall material are fed through inner and outer orifice, respectively. The extruded rope of core material surrounded by wall material splits into round droplets directly after clearing the nozzle as the device rotates. By the action of surface tension, the coating material envelops the core material, thus accomplishing encapsulation. The shell wall of the droplets is solidified by cooling or gelling bath to produce capsules. The microcapsules thus formed can be collected on a moving bed of fine-grained starch, which cushions their impact and absorbs unwanted coating moisture. The capsule size depends on the rotational speed. Typical wall materials include starch, maltodextrins, gelatin, polyethylene glycol (PEG). Particles produced by this method have diameter ranging from 150 to 2000 mm.[18]

1.8 CENTRIFUGAL SUSPENSION SEPARATION

This is a continuous high-speed process that involves mixing the core and wall materials and then adding to a rotating disk. The core materials with a coating of residual liquid ejects out of the disk. The microcapsules are then dried or chilled after removal from the disk. Solids, liquids, or suspensions of 30 μm to 2 mm can be encapsulated in this manner.

1.9 COCRYSTALLIZATION

Cocrystallization encapsulation process utilizes sucrose as a matrix for the incorporation of core materials. The cocrystallization of sucrose/core ingredient is induced by adding a core material to supersaturated sucrose syrup (maintained at a high temperature to prevent crystallization) under vigorous mechanical agitation. The addition of core ingredient provides nucleation for the sucrose/core ingredient mixture to crystallize. A substantial heat is released as the syrup reaches the temperature at which transformation and crystallization begin. It is very important to properly control the rates of nucleation and crystallization as well as the thermal balance during the various phases.[19] By this process, core materials in a liquid form can be converted to a dry powdered form without additional drying.

1.10 COACERVATION

A coacervate is a tiny spherical droplet of assorted organic molecules, which is held together by hydrophobic forces and the sizes ranging from 1 to 100 µm across. The name derives from the Latin coacervare, meaning "to assemble together or cluster." There are two types of coacervation processes available, namely, simple and complex coacervation. In simple coacervation, a desolvation agent is added for phase separation. Whereas in complex coacervation, a complexation is involved between two oppositely charged polymers. *Complex coacervation* refers to the phase separation of a liquid precipitate, or phase, when solutions of two hydrophilic colloids are mixed under suitable conditions. As an example in a complex coacervation process, the core material (usually oil) is dispersed into a polymer solution (e.g., a cationic aqueous polymer, and gelatin) to which a second polymer (anionic, water soluble, and gum Arabic) solution is added. Deposition of the shell material onto the core particles occurs when the two polymers form a complex, which is generally triggered by the addition of salt or by changing the pH, temperature, or dilution of the medium. The microcapsules can be stabilized by cross linking, desolvation, or thermal treatment. The microcapsules are usually collected by filtration or centrifugation, washed with an appropriate solvent, and subsequently dried by standard techniques such as spray- or fluidized-bed drying to yield free-flowing, discrete particles.[20] In recent years, modified coacervation processes have also been developed to overcome some of the problems encountered during a typical gelatin/gum acacia complex coacervation process. For example, a room-temperature process for the encapsulation of heat-sensitive ingredients such as volatile flavor oils[21] by mixing the coating materials and inducing the phase separation (coacervation) by adjusting the pH. The formation of a multilayered coacervated microcapsule[22] can be achieved by repeating the coacervation stages in which an additional layer of wall material is applied to the microcapsule at each passage.

Complex coacervation can be used to produce microcapsules containing flavor, fragrant oils, liquid crystals, dyes, or inks as the core material. Capsules for carbonless paper applications were produced by complex coacervation process.

1.11 EMULSIFICATION/EMULSION POLYMERIZATION

Emulsions can be used as a delivery vehicle for either water soluble and/or lipophilic active ingredient.[23] In an emulsion polymerization, surfactant is dissolved in water until the critical micelle concentration is reached. The interior of the micelle provides the site necessary for polymerization. As the polymerization proceeds, these nuclei grow gradually and simultaneously entrap the core material to form the final microcapsules. The most common type of emulsion polymerization is an oil-in-water emulsion, in which droplets of monomer (the oil) are emulsified (with surfactants) in a continuous phase of water. Water-soluble active ingredients might be encapsulated in water-in-oil (w/o) emulsions or double emulsions of the type water-in-oil-in-water (w/o/w).[23] Similarly lipophilic active ingredients might be encapsulated via o/w emulsions.[23] Several technologies have been developed to produce highly uniform emulsion droplets,[24,25] including single-drop technologies like microfluidics. The release profile of entrapped active ingredients could be more defined and predictable with uniform/narrow dispersed emulsions than the polydispersed ones. This can be very important in pharmaceutical applications. The emulsions can be dried by spray drying or freeze drying to provide a dry powder version of the encapsulated product.

1.12 FLUID-BED COATING

Fluid-bed coating is a technique in which a coating is applied onto powder particles in a batch process or a continuous setup. Different types of fluid-bed coaters include top spray, bottom spray, and tangential spray and used for encapsulating solid or liquids absorbed into porous particles. The bottom spray version of the coating process is commonly known as Wurster coating and was developed in the 1950s and 1960s.[26] Wurster coater relies upon a bottom positioned nozzle spraying the wall material up into a fluidized bed of core particles. Particles of the active ingredient, spheres or granules, are suspended in an upward-moving stream of air and then covered with a spray of liquid coating material in a temperature and humidity-controlled chamber of high velocity air where the coating material is atomized.[27,28] The coating material must have an acceptable viscosity to enable pumping and atomizing and must be thermally stable and should be able to form a film over a particle surface. The rapid evaporation of the solvent helps in the formation of an outer layer on the particles. The particles to be coated by fluid bed should ideally be spherical and dense and should have a narrow particle size distribution and good flowability. The whole cycle of suspending, spraying, and cooling is continued and repeated as needed until the desired coating thickness and weight is obtained. The size of the core particle for this technique is usually large (~100 μm). This technique gives improved flexibility and control as compared to pan coating.

1.13 INTERFACIAL POLYMERIZATION

Most commercial processes use this type of polymerization to produce small uniform capsules in the range of 20–30 micron diameter; however, the process can be tuned to produce large microcapsules. The size of these microcapsules and the properties of the wall material/polymer matrix can be altered by using different monomers, utilizing additives, and adjusting reaction conditions. The encapsulation occurs by wall formation around the dispersed core material via the rapid polymerization of monomers at the surface of the droplets or particles. The solution of a multifunctional monomer in the core material is dispersed in an aqueous phase. The polymerization is commenced at the surfaces of the core droplets forming the capsule walls, by adding a reactant to the monomer dispersed in the aqueous phase.

1.14 INCLUSION COMPLEXATION

Encapsulation via inclusion complexation (host–guest complexation) occurs at a molecular level. This is also known as molecular encapsulation process or host–guest complexation. Cyclodextrin is one of the most useful host molecules for a variety of encapsulation applications from food to pharma industries. There are generally three types of cyclodextrins such as α-cyclodextrin (made up of six glucopyranose units), β-cyclodextrin (made up of seven glucopyranose units), and γ-cyclodextrin (made up of eight glucopyranose units). They are prepared from partially hydrolyzed starch (maltodextrin) by an enzymatic process. The external part of the cyclodextrin molecule is hydrophilic, whereas the internal part is hydrophobic. The guest molecules, which are apolar, can be entrapped into the apolar internal cavity through a hydrophobic interaction.[29] The size of the cavity varies depending on the cyclodextrin. Depending on the type of cyclodextrin one or more molecules can be entrapped in the interior of the molecule.[15] Cyclodextrin molecules form inclusion complexes with compounds that can fit dimensionally into their central cavity.

1.15 IONIC GELATION (HYDROGEL MICROSPHERES)

This is one of the simplest methods for making encapsulated material. Microspheres are microbeads composed of a biopolymer gel network entrapping an active. Calcium alginate gel is the best known gelling system used for the preparation of gel beads to encapsulate a wide variety of active agent, such as oil droplets containing aroma, cells, probiotics, yeast, enzymes, etc. Gelation of alginate in the presence of divalent or trivalent cations can be easily controlled. Microspheres made of gel-type polymers, such as alginate or pectinate are produced by dissolving the polymer in an aqueous solution then, suspending the active ingredient in the mixture followed by extruding through a precision device (the dripping tool can be simply a pipette, syringe, vibrating nozzle, spraying nozzle, jet cutter, atomizing disk, coaxial air-flow, or electric field) producing microdroplets. These microdroplets are hardened by cross-linking the wall material or polymer chain by using di- or multivalent metal ion (such as calcium chloride) aqueous solutions. There are several advantages of these methods that involve an all-aqueous system; the particle size of microspheres can be controlled by using various size extruders or by varying the polymer solution flow rates. Alternatively, the extrusion or dropping method can be used with a concentric nozzle (coextrusion), to prepare core–shell type of encapsulates with a lipophilic core and a shell of a gel network. Similarly, specially formulated emulsions can be used to make microspheres. Calcium chloride can be added to an emulsion of water droplets of an alginate solution and active ingredient in vegetable oil. This results in the "break-up" of the emulsion and microbeads are formed by the gelation of the alginate droplets. Alternatively, both alginate and calcium (in an insoluble form such as calcium carbonate) can already be present in the water phase of the emulsion. Upon addition of an oil-soluble acid (such as acetic acid) the pH decreases, liberating free calcium ions in the system and initiating the gel formation of alginate droplet with calcium.

1.16 LIPOSOMES

A liposome is a tiny vesicle generally made from phospholipids with diameter ranges from 25 nm to 10 μm capable of entrapping both hydrophobic and hydrophilic active ingredients. The liposomes consist of one or more layers of lipids and have been used for delivery of vaccines, hormones, enzymes, and vitamins.[20] The properties such as permeability, stability, surface activity, and affinity can be varied through size and lipid composition variations. The capsules can range from 25 nm to several microns in diameter, are easy to make, and can be stored by freeze-drying. Kirby and Gregoriadis have devised a method to encapsulate at high efficiency, which is easy to scale-up and uses mild conditions appropriate for enzymes.[30,31] The liposomes impart stability to water-soluble material in high water activity application. Another unique property of liposomes is the targeted delivery of the encapsulated ingredient their content at a specific and well-defined temperature in a specific location. The liposome bilayer is instantly broken down at the transition temperature of the phospholipids, typically around 50°C, at which temperature the content is immediately released.

1.17 LYOPHILIZATION

Lyophilization or freeze-drying is a simple technique that is suitable for the encapsulation of aromas, water-soluble essences, drugs, and importantly used for the dehydration of almost all heat-sensitive materials. It is a process that requires a long dehydration period. The retention of volatile compounds during lyophilization is dependent upon the chemical nature of the system.[32]

1.18 MELT EXTRUSION (MELT INJECTION)

Extrusion as applied for encapsulating flavor ingredient(s) is a relatively low-temperature entrapping method, which involves forcing a core material in a molten carbohydrate mass (composed of more than one ingredient, such as sucrose, maltodextrin, glucose syrup, glycerine, and glucose[33]) through a series of dies into a bath of dehydrating liquid. The process temperature and the pressure are around ~115°C or below and typically <100 psi, respectively.[34] The coating material hardens on contacting the liquids, forming an encapsulating matrix to entrap the core material, which can then be separated from the liquid bath, dried, and sized.[14] The polymer matrices and the plasticizers used can be modified to produce the capsules for controlled release application.[35]

1.19 PAN COATING

The pan coating process is amongst the oldest industrial techniques used particularly in the pharmaceutical industry for coating tablets or forming coated particles. The concept of the technology was initially patented by William E. Upjohn in the nineteenth century.[36] It generally required large core particles and produces the coated tablets that we are familiar with. The tablets/particles are tumbled in a pan or other device while the coating material is sprayed/applied slowly. In another aspect solid particles are mixed with a dry coating material and the temperature is raised so that the coating material melts and encloses the core particles, and then is solidified by cooling. On the other hand, the coating material can be gradually applied to core particles tumbling in a vessel rather than being wholly mixed with the core particles from the start of encapsulation.

1.20 PHASE SEPARATION

Phase separation process takes advantage of the phenomenon called polymer–polymer incompatibility. The process utilizes two polymers that are soluble in a common solvent; yet do not mix with one another in the solution. The polymers form two separate phases; one polymer intended to form the capsule walls, the other incompatible polymer meant to induce the separation of the two phases, but not meant to be part of the capsule wall material. This process is somewhat related to the complex coacervation process. The phase separation process is considered as the oldest true encapsulation technology first developed by the National Cash Register Company for carbonless copy-paper. Microencapsulation by coacervation involves the phase separation of one or more hydrocolloids from the initial solution, and the subsequent deposition of the newly formed coacervate phase around the active ingredient suspended or emulsified in the same reaction media. The size of the microcapsules formed may be in the range of 10–250 µm.

1.21 SPRAY DRYING

Microencapsulation by spray-drying is a low-cost commercial process, which is mostly used for the encapsulation of fragrances, oils, and flavors. Spray-drying encapsulation has been used in the food industry since the late 1950s to convert liquids to powders. The process is economical and flexible, in that it offers substantial variation in microencapsulation matrix, and produces particles of good quality. The flavor or ingredient to be encapsulated is added to the carrier and homogenized to create small droplets. The resultant emulsion is fed into the hot chamber of the spray dryer where it is atomized through a nozzle or spinning wheel. The size of the atomizing droplets depends on the surface tension and viscosity of the liquid, pressure drop across the nozzle, and the velocity of the spray. The size of the atomizing droplets also determines the drying time and particle size. Hot air contacts the atomized droplets; the shell material solidifies onto the core particles as the solvent evaporates, and leaving dried particles. It is immensely suitable for the continuous manufacture of dry solids as either powder, granulates, or agglomerates from liquid feeds. The microcapsules obtained are of polynuclear or matrix type.

Overview of Encapsulation and Controlled Release

1.22 SPRAY CONGEALING

The spray congealing is also known as spray chilling or spray cooling. In spray chilling and spray cooling, the core and wall mixtures are atomized into the cooled or chilled air, which causes the wall to solidify around the core. This technique can be accomplished with spray-drying equipment when the protective coating is applied as a melt. The process involves the dispersion of the core material in a coating material melt. Upon spraying, the hot mixture into a cool air stream the coating solidification (and microencapsulation) is accomplished. In spray cooling, the coating material can be some form of vegetable oil or its derivatives, fat (melting points of 45°C–122°C), as well as hard mono- and diacylglycerols (melting points of 45°C–65°C).[37] In spray chilling, the coating material is typically a fractionated or hydrogenated vegetable oil with a melting point in the range of 32°C–42°C.[38] Microcapsules prepared by spray chilling and spray cooling are insoluble in water due to the hydrophobic coating. Consequently, these techniques can be useful for encapsulating water-soluble core materials.[39]

1.23 SPINNING DISK (CENTRIFUGAL SUSPENSION SEPARATION)

This technology is rapid, cost-effective, and relatively simple and has high production efficiencies. The technology was developed by Robert E. Sparks in 1980s.[40] The suspensions of core particles in liquid shell material are poured into a rotating disk. Due to the spinning action of the disk, the core particles become coated with the shell material and the coated particles are then cast from the edge of the disc by centrifugal force. The shell material of the coated particles is solidified by external means (usually cooling). The microcapsules are then dried or chilled after removal from the disk, and the whole process can take between a few seconds to minutes. Solids, liquids, or suspensions of 30 μm to 2 mm can be encapsulated in this manner. Coatings can be 1–200 μm in thickness and include fats, PEG, diglycerides, and other meltable substances.

1.24 SOLVENT EVAPORATION

It is one of the most extensively used methods of microencapsulation. For example, the aqueous solution of the drug (may contain a viscosity building or stabilizing agent) is added to a water immiscible volatile organic phase consisting of the polymer solution in solvents like dichloromethane or chloroform with vigorous stirring to form the primary water in oil emulsion. The resulting emulsion is then added drop wise to a stirring aqueous solution containing an emulsifier/stabilizer like poly (vinyl alcohol) or poly (vinyl pyrrolidone) to form the multiple emulsion (w/o/w) forming small polymer droplets containing encapsulated material. Droplets hardened to produce the corresponding polymer microcapsules. This hardening process is accomplished by the removal of the solvent from the polymer droplets either by solvent evaporation (by heat or reduced pressure), or by solvent extraction. The microspheres can then be washed and dried.

1.25 SUPERCRITICAL FLUID ASSISTED ENCAPSULATION

Supercritical fluids are highly compressed gasses that possess several properties of both liquids and gases. The supercritical CO_2 or N_2O are normally used for this type of encapsulation process. The microcapsules are formed when the supercritical fluid under high pressure containing the active ingredient and the shell material are released through a small nozzle at atmospheric pressure. The sudden drop in pressure causes desolvation of the shell material, which is then deposited around the active ingredient (core) and forms a coating layer. Different core materials such as pesticides, pigments, vitamins, flavors, and dyes are encapsulated using this method. Although there are some advantages of this process, one of the requirements is that both the active ingredient and the shell material should be soluble in supercritical fluids.

1.26 CONTROLLED RELEASE

Controlled release is a term referring to the presentation or delivery of compounds in response to stimuli (such as pH, enzymes, light, magnetic fields, temperature, ultrasonic, osmosis, and more recently electronic control) or time. It may be defined as a method by which one or more active ingredients are made available at a desired site and time at a specific rate. Various applications include in areas such as pharmaceuticals, agriculture, cosmetics, food sciences, and personal care. This terminology commonly refers to time dependant release, sustained release, pulse release, delayed release in oral dose formulations. There are many advantages as well as challenges (such as: biocompatibility, the fate of controlled release system if not biodegradable, cost of formulation, etc.) associated with designing a controlled release system/formulation. The science of controlled release originates from the development of oral sustained-release products in the late 1940s and early 1950s, the development of controlled release of marine anti-foulants in the 1950s and controlled release fertilizer in the 1970s. Delivery of active ingredient is usually effected by dissolution, degradation, or disintegration of the shell material. Other encapsulation technologies including enteric coatings can further modify release profiles. Microencapsulation is a technology that can control dissolution profiles. The dissolution rates can be further controlled by further coating and layering the microcapsule/microsphere with insoluble substances.

1.27 RELEASE MECHANISMS

A variety of release mechanisms have been proposed for microcapsules. The pathways are as follows:

- A force breaks open the capsule by mechanical means/shear force.
- The shell wall is dissolved away, melt away from around the core.
- The core/active ingredient diffuses through the shell wall.

The mechanism can be further classified and elaborated as follows:

- *Degradation-controlled monolithic system*: The active ingredient is distributed uniformly throughout in a matrix. The active is released on degradation of the matrix. The diffusion of the active is slow as compared with degradation of the matrix.
- *Diffusion-controlled monolithic system*: The active ingredient is released by diffusion prior to or concurrent with the degradation of the polymer matrix. Rate of release also depend upon where the polymer degrades by homogeneous or heterogeneous mechanism.
- *Diffusion-controlled mononuclear (core–shell) system*: The active ingredient is encapsulated by a rate-controlling membrane through which the active diffuses and the membrane erodes only after its delivery is completed.
- *Erosion*: Erosion of the coating due to pH and enzymatic hydrolysis causes the release of active ingredients.

1.28 RELEASE RATES

The release rates that are achievable from a single microcapsule are generally "zero order," "half order," or "first order." Zero order occurs when the active ingredient/core is a pure material and releases through the wall of a mononuclear (core–shell) microcapsule as a pure material. Half-order release generally occurs with matrix particles. "First-order" release occurs when the core material is actually a solution. As the solute material releases from the capsule, the concentration of solute material in the solvent decreases and a first-order release is achieved. A mixture of microcapsules will include a distribution of capsules varying in size and wall thickness, and the release rate would be different because of the ensemble of microcapsules. It is therefore very desirable to carefully examine on an experimental basis the release rate from an ensemble of microcapsules and to recognize that the deviation from theory is due to the distribution in size and wall thickness.

1.29 CONCLUSIONS

The use of microencapsulated active ingredients is a promising avenue for variety controlled-release applications in many different areas and industries.[41–76] Lots of new techniques (hybrid processes) are being constantly developed for the encapsulation and controlled release formulations for active ingredients. The challenges are to select the appropriate microencapsulation technique and encapsulating material. Despite the wide range of encapsulated products that have been developed, manufactured, and successfully marketed in the pharmaceutical and cosmetic industries, microencapsulation has found a comparatively much smaller market in the food industry. Microencapsulation is gradually finding applications in many unconventional areas such as agriculture, energy storage and generation, catalysis, textiles, defense, etc.[77]

In agriculture, one of the applications of microencapsulated products is in the area of crop protection.[78–84] Another example is the utilization of encapsulated insect pheromones (instead of insecticides/pesticides) for controlling insect population.[82,84] In energy-generation applications[85] hollow multilayered (inner layer being the polystyrene followed by polyvinyl alcohol and the outer layer being a cross-linked polymer of 2-butene) microspheres loaded with gaseous deuterium are used to harness nuclear fusion for producing electrical energy. Transition metal–based catalytic processes are of vital importance to a variety of industries, including pharmaceuticals, agrochemicals, etc. Palladium- and Osmium-based metal catalysts have been encapsulated[86,87] in polyurea and used successfully as recoverable/reusable catalyst without much loss of activity. Similarly, the microencapsulation technology embedded in designing self-healing polymer and composites is finding applications in many advanced areas including defense.[88–94] Microencapsulation is also used for designing special fabrics for protecting against chemical warfare for military applications.[95] Some of the applications will be described in the subsequent chapters.

REFERENCES

1. Versic, R.J. Flavor Encapsulation—An Overview. 2014. http://www.rtdodge.com/fl-ovrvw.htm.
2. Chen, X.G.; Lee, C.M.; Park, H.J. O/W emulsification for the self-aggregation and nanoparticle formation of linolenic acid modified chitosan in the aqueous system. *Journal of Agricultural and Food Chemistry* 51, 3135–3139, 2003.
3. Kim, B.K.; Hwang, S.J.; Park, J.B.; Park, H.J. Preparation and characterization of drug-loaded microspheres by an emulsion solvent evaporation method. *Journal of Microencapsulation* 19, 811–822, 2002.
4. Lee, D.W.; Hwang, S.J.; Park, J.B.; Park, H.J. Preparation and release characteristics of polymer-coated and blended alginate microspheres. *Journal of Microencapsulation* 20, 179–192, 2003.
5. Ko, J.A.; Park, H.J.; Hwang, S.J.; Park, J.B.; Lee, J.S. Preparation and characterization of chitosan microparticles intended for controlled drug delivery. *International Journal of Pharmaceutics* 249, 165–174, 2002.
6. Lee, J.Y.; Park, H.J.; Lee, C.Y.; Choi, W.Y. Extending shelf life of minimally processed apples with edible coatings and antibrowning agents. *Lebensmittel-Wissenschaft und Technologie* 36, 323–329, 2003.
7. Cho, Y.H.; Shin, D.S.; Park, J. Optimization of emulsification and spray drying processes for the microencapsulation of flavor compounds. *Korean Journal of Food Science and Technology* 32, 132–139, 2000.
8. Green, B.K.; Schleicher, L. The National Cash Register Company, Dayton, Ohio. Oil containing microscopic capsules and method of making them. US Patent 2,800,457. 23 July 1957, 11.
9. Green, B.; B.K. The National Cash Register Company, Dayton, Ohio. Oil containing microscopic capsules and method of making them. US Patent 2,800,458. 23 July 1957, 7.
10. Andres, C. Encapsulation ingredients: I. *Food Processing* 38(12), 44–56, 1977.
11. Bakan, J.A. Microencapsulation of food and related products. *Food Technology* 27(11), 34–38, 1973.

12. Todd, R.D. Microencapsulation and food industry. *Flavor Industry* 1, 78–81, 1970.
13. Balsa, L.L.; Fanger, G.O. Microencapsulation in food industry. *Critical Reviews in Food Technology* 2, 245–249, 1971.
14. Shahidi, F.; Han, X.Q. Encapsulation of food ingredients. *Critical Reviews in Food Technology* 33(6), 501–504, 1993.
15. Dziezak, J.D. Microencapsulation and encapsulated food ingredients. *Food Technology* 42, 136–151, 1998.
16. Gibbs, B.F.; Kermasha, S.; Alli, I.; Mulligan, C.N. Encapsulation in food industry: A review. *International Journal of Food Science and Food Nutrition* 50, 213–234, 1999.
17. Gutcho, M.M. (Ed.). *Microcapsules and Microencapsulation Techniques*, Noyes Data Co., New Jersey, 1976; Arshady, R. (Ed.). *Microspheres, Microcapsules and Liposomes*, Citrus Books, London, U.K., 1999.
18. Schlameus, W. Centrifugal extrusion encapsulation. In: Risch, S.J.; Reineccius, G.A. (Eds.), *Encapsulation and Controlled Release of Food Ingredients*, American Chemical Society, Washington, DC, 1995.
19. Rizzuto, A.B.; Chen, A.C.; Veiga, M.F. Modification of the sucrose crystal structure to enhance pharmaceutical properties of excipient and drug substances. *Pharmaceutical Technology* 8(9), 32–35, 1984.
20. Kirby, C.J. Microencapsulation and controlled delivery of food ingredients. *Food Science and Technology Today* 5(2), 74–80, 1991.
21. Arneodo, C.J.F. Microencapsulation by complex coacervation at ambient temperature. FR 2732240 A1, 1996.
22. Ijichi, K.; Yoshizawa, H.; Uemura, Y.; Hatate, Y.; Kawano, Y. Multilayered gelatin = acacia microcapsules by complex coacervation method. *Journal of Chemical Engineering Japan* 30, 793–798, 1997.
23. Appelqvist, I.A.M.; Golding, M.; Vreeker, R.; Zuidam, N.J. Emulsions as delivery systems in foods. In: Lakkis, J.M. (Ed.) *Encapsulation and Controlled Release Technologies in Food Systems*, Blackwell Publishing, Ames, IA, 2007, pp. 41–81.
24. Link, D.R., Anna, S.L., Weitz, D.A., Stone, H.A. Geometrically mediated breakup of drops in microfluidic devices. *Physical Review Letters* 92(5), 054403-1–054403-4, 2004.
25. McClements, D.J. *Food Emulsions. Principles, Practices and Techniques*, CRC Press, Boca Raton, FL, 2005.
26. Wurster, D.E. U.S. Patent 2 799 241, 1957; 3 089 834, 1963; 3 117 027, 1964; 3 196 927, 1965; 3 207 824, 1965; 3 241 520, 1966; 3 253 944, 1966.
27. Balassa, L.L.; Fanger, G.O. Microencapsulation in the food industry. *CRC Reviews in Food Technology* 2, 245–263, 1971.
28. Zhao, L.; Pan, Y.; Li, J.; Chen, G.; Mujumdar, A.S. Drying of a dilute suspension in a revolving flow fluidized bed of inert particles. *Drying Technology* 22(1,2), 363–376, 2004.
29. Pagington, J.S. b-Cyclodextrin and its uses in the flavour industry. In: Birch, G.G., Lindley, M.G. (Eds.), *Developments in Food Flavours*, Elsevier Applied Science, London, U.K., 1986.
30. Gregoriadis, G. In *Liposome Technology*, vols. 1–3, CRC Press, Boca Raton, FL, 1984.
31. Kirby, C.J.; Gregoriadis, G. A simple procedure for preparing liposomes capable of high encapsulation efficiency under mild conditions. In: Gregoriadis, G. (Ed.), *Liposome Technology*, vol. 1, CRC Press, Boca Raton, FL, 1984.
32. Kopelman, I.J.; Meydav, S.; Wwilmersdorf, P. Storage studies of freezedried lemon crystals. *Journal of Food Technology* 12, 65–69, 1977.
33. Arshady, R. Microcapsules for food. *Journal of Microencapsulation* 10(4), 413–435, 1993.
34. Reineccius, G.A. Flavor encapsulation. *Food Reviews International* 5, 147–150, 1989.
35. Ubbink, J.; Schoonman, A. Flavour delivery systems. *Kirk-Othmer Encyclopedia of Chemical Technology*, John Wiley and Sons, New York, 2003.
36. W.E. Upjohn, U.S. Letters Patent Number 312,041 dated February 10, 1885.
37. Taylor, A.H. Encapsulation systems and their applications in the flavor industry. *Food Flavor Ingredients Packaging and Processing* 5(9), 48–51, 1983.
38. Blenford, D. Fully protected. *Food Flavor Ingredients Packaging and Processing* 8(7), 43–45, 1986.
39. Lamb, R. Spray chilling. *Food Flavor Ingredients Packaging and Processing* 9(12), 39–42, 1987.
40. Sparks, R.E.; Mason, N.S. Method for coating particles or liquid droplets. Patent US4675140, 1987.
41. Linko, P. Immobilized lactic acid bacteria. In: Larson, A. (Ed.), *Enzymes and Immobilized Cells in Biotechnology*, Benjamin Cummings, Meno Park, CA, 1985, pp. 25–36.
42. Seiss, W.; Divies, C. Microencapsulation. *Angewandte Chemie International Edition* 14, 539–550, 1975.

43. Kim, H.-H.Y.; Baianu, I.C. Novel liposome microencapsulation techniques for food applications. *Trends in Food Science and Technology* 2, 55–60, 1991.
44. Ono, F. New encapsulation technique with protein–carbohydrate matrix. *Journal of Japanese Food Science Technology* 27, 529–535, 1980.
45. Rosenberg, M.; Sheu, T.Y. Microencapsulation of volatiles by spray drying in whey protein based wall systems. *International Dairy Journal* 6, 273–284, 1996.
46. Shiga, H.; Yoshii, H.; Nishiyama, T.; Furuta, T.; Forssele, P.; Poutanen, K.; Linko, P. Flavor encapsulation and release characteristics of spray-dried powder by the blended encapsulant of cyclodextrin and gum arabic. *Drying Technology* 19(7), 1385–1395, 2001.
47. Liu, X.D.; Atarashi, T.; Furuta, T.; Yoshii, H.; Aishima, S.; Ohkawara, M.; Linko, P. Microencapsulation of emulsified hydrophobic flavors by spray drying. *Drying Technology* 19(7), 1361–1374, 2001.
48. Beristain, C.I.; Garcia, H.S.; Vernon-Carter, E.J. Mesquite gum (*Prosopis juliflora*) and maltodextrin blends as wall materials for spray-dried encapsulated orange peel oil. *Food Science and Technology International* 5, 353–356, 1999.
49. Beristain, C.I.; Garcia, H.S.; Vernon-Carter, E.J. Spray-dried encapsulation of cardamom (*Elettaria cardamomum*) essential oil with mesquite (*Prosopis juliflora*) gum. *Lebensmittel-Wissenschaft und Technologie* 34, 398–401, 2001.
50. Augustin, M.A.; Sanguansri, L.; Margetts, C.; Young, B. Microencapsulation of food ingredients. *Food Australia* 53, 220–223, 2001.
51. Millqvist-Fureby, A.; Malmsten, M.; Bergenstahl, B. An aqueous polymer two-phase system as carrier in the spray-drying of biological material. *Journal of Colloid and Interface Science* 225, 54–61, 2000.
52. Edris, A.; Benrgnstahl, B. Encapsulation of orange oil in a spray dried double emulsion. *Nahrung Food* 45, 133–137, 2001.
53. Re, M.I. Microencapsulation by spray drying. *Drying Technology* 16, 1195–1196, 1998.
54. Beristain, C.I.; Garcia, H.S.; Vernon-Carter, E.J. Mesquite gum (*Prosopis juliora*) and maltodextrin blends as wall materials for spray-dried encapsulated orange peel oil. *Food Science and Technology International* 5, 353–356, 1999.
55. Bhandari, B.R.; Dumoulin, H.M.J.; Richard, H.M.J. Flavor encapsulation of spray drying: Application to citral and linalyl acetate. *Journal of Food Science* 51, 1301–1306, 1992.
56. Kanawjia, S.K.; Pathania, V.; Singh, S. Microencapsulation of enzymes, micro-organisms and flavours and their applications in foods. *Indian Dairyman* 44, 280–287, 1992.
57. Dezarn, T.J. Food ingredient encapsulation. In: Risch, S.J., Reineccius, G.A. (Eds.), *Encapsulation and Controlled Release of Food Ingredients*, American Chemical Society, Washington, DC, 1995.
58. De Pauw, P.; Dewettinck, K.; Arnaut, F.; Huyghebaert, A. Microencapsulation improves the action of bakery ingredients. *Voedingsmiddelentechnologie* 29, 38–40, 1996.
59. Sparks, R.E. Microencapsulation. In: McKetta, J. (Ed.), *Encyclopedia of Chemical Process Technology*, Marcel Dekker, New York, 1989.
60. Beristain, C.; Vazquez, A.V.; Garcia, H.S.; Vernon-Carter, E.J. Encapsulation of orange peel oil by co-crystallization. *Lebensmittel-Wissenschaft und Technologie* 29, 645–647, 1996.
61. Desai, K.G.H.; Park, H.J. Recent developments in microencapsulation of food ingredients. *Drying Technology* 23, 1361–1394, 2005.
62. Gharsallaoui, A.; Roudaut, G.; Chambin, O.; Voilley, A.; Saurel, R. Applications of spray-drying in microencapsulation of food ingredients: An overview. *Food Research International* 40, 1107–1121, 2007.
63. Gouin, S. Microencapsulation: Industrial appraisal of existing technologies and trends. *Trends in Food Science and Technology* 15, 330–347, 2004.
64. Krasaekoopt, W; Bhandari, B.; Deeth, H. Evaluation of encapsulation techniques of probiotics for yoghurt. *International Dairy Journal* 13, 3–13, 2003.
65. Kosaraju, S.L.; Tran, C.; Lawrence, A. Liposomal delivery systems for encapsulation of ferrous sulfate: Preparation and characterization. *Journal of Liposome Research* 16, 347–358, 2006.
66. Martin Del Valle, E.M.; Galan, M.A. Supercritical fluid technique for particle engineering: Drug delivery applications. *Reviews in Chemical Engineering* 21(1), 33–69, 2005.
67. Mozafari, M.R.; Flanagan, J.; Matia-Merino, L.; Awati, A.; Omri, A.; Suntres, Z.E.; Singh, H. Recent trends in the lipid-based nanoencapsulation of antioxidants and their role in foods. *Journal of the Science of Food and Agriculture* 86(13), 2038–2045, 2006.
68. Porzio, M. Flavor encapsulation: A convergence of science and art. *Food Technology* 58(7), 40–47, 2004.

69. Reineccius, G.A. Multiple-core encapsulation: The spray drying of food ingredients. In: Vilstrup, P. (Ed.), *Microencapsulation of Food Ingredients*, Leatherhead Publishing, Surrey, U.K., 2001, pp. 151–185.
70. Reineccius, G.A. The spray drying of food flavors. *Drying Technology* 22(6), 1289–1324, 2004.
71. Semo, E., Kesselman, E., Danino, D., Livney, Y.D. Casein micelle as a natural nano-capsular vehicle for nutraceuticals. *Food Hydrocolloids* 21, 936–942, 2007.
72. Taylor, T.M.; Davidson, P.M.; Bruce, B.D.; Weiss, J. Liposomal nanocapsules in food science and agriculture. *Critical Reviews in Food Science and Nutrition* 45, 587–605, 2005.
73. Teunou, E.; Poncelet, D. Fluid-bed coating. In: Onwulata, C. (Ed.) *Encapsulated and Powdered Foods*, CRC Press, Boca Raton, FL, 2005a, pp. 197–212.
74. Teunou, E.; Poncelet, D. Dry coating. In: Onwulata C. (Ed.) *Encapsulated and Powdered Foods*, CRC Press, Boca Raton, FL, 2005b, pp. 179–195.
75. Thies, C. Microencapsulation of flavors by complex coacervation. In: Lakkis, J.M. (d) *Encapsulation and Controlled Release Technologies in Food Systems*, Blackwell Publishing, Ames, IA, 2007, pp. 149–170.
76. Uhlemann, J.; Schleifenbaum, B.; Bertram, H.J. Flavor encapsulation technologies: An overview including recent developments. *Perfumer and Flavorist* 27, 52–61, 2002.
77. Dubey, R.; Shami, T.C.; Bhasker Rao, K.U. Microencapsulation technology and applications. *Defence Science Journal* 59(1), 82–95, 2009.
78. Scher, H.B.; Rodson, M.; Lee, K.S. Microencapsulation of pesticides by interfacial polymerization utilizing isocyanate or aminoplast chemistry. *Pesticide Science* 54, 394–400, 1998.
79. Scher, H.B.; Groenwold, B.E.; Pereiro, F.; Purnell, T.J. Microencapsulated thiocarbamate herbicides. In *Proceedings of British Crop Protection Conference Weeds*, 1980, pp. 185–191.
80. Scher, H.B. Development of herbicide and insecticide microcapsule formulations. In *Proceedings of International Symposium on Controlled Release of Bioactive Materials*, 1985, pp. 110–111.
81. Bingham, G.; Gunning, R.V.; Gorman, K.; Field, L.M.; Moores, G.D. Temporal synergism by microencapsulation of piperonyl butoxide and-cypermethrin overcomes insecticide resistance in crop pests. *Pest Management Science* 63, 276–281, 2007.
82. Ilichev, A.L.; Stelinski, L.L.; Williams, D.G.; Gut, L.J. Sprayable microencapsulated sex pheromone formulation for mating disruption of oriental fruit moth (Lepidoptera: Tortricidae) in Australian peach and pear orchards. *Journal of Economic Entomology* 99(6), 2048–2054, 2006.
83. Mihou, A.P.; Michaelakis, A.; Krokos, F.D.; Majomenos, B.E.; Couladouros, E.A. Prolonged slow release of (Z)-11-hexadecenyl acetate employing polyuria microcapsules. *Journal of Applied Entomology* 131(2), 128–133, 2007.
84. Zengliang, C.; Yuling, F.; Zhongning, Z. Synthesis and assessment of attractiveness and mating disruption efficacy of sex pheromone microcapsules for the diamondback moth, *Plutella Xylostella* (L). *Chinese Science Bullettin* 57(10), 1365–1371, 2007.
85. Mishra, K.K.; Khardekar, R.K.; Singh, R.; Pant, H.C. Fabrication of polystyrene hollow microspheres as laser fusion targets by optimized density matched emulsion technique and characterization. *Pramana, Journal of Physics* 59(1), 113–131, 2002.
86. Ley, S.V.; Ramarao, C.; Lee, A.L.; Ostergaard, N.; Smith, S.C.; Shirley, I.M. Microencapsulation of osmium tetroxide in polyurea. *Organic Letters* 5(2), 185–187, 2003.
87. Ley, S.V.; Ramarao, C.; Gordon, R.S.; Holmes, A.B.; Morrison, A.J.; McConvey, I.F.; Shirley, I.M.; Smith, S.C.; Smith, M.D. Polyurea encapsulated palladium (II) acetate: A robust and recyclable catalyst for use in conventional and supercritical media. *Chemical Communication* 1134–1135, 2002.
88. Brown, E.N.; White, S.R.; Sottos, N.R. Microcapsule induced toughening in a self-healing polymer composite. *Journal of Materials Science* 39, 1703–710, 2004.
89. Li, Y.; Guozheng, L.; Jianqiang, X.; Lan, L.; Jing, G. Preparation and characterization of poly(ureaformaldehyde) microcapsules filled with epoxy resins. *Polymer* 47(15), 5338–5349, 2006.
90. Toohey, K.S.; Sottos, N.R.; Lewis, J.A.; Moore, J.S.; White, S.R. Self-healing materials with microvascular networks. *Nature Materials* 6(8), 581–585, 2007.
91. White, S.R.; Sottos, N.R.; Guebelle, P.H.; Moore, J.S.; Kessler, M.R.; Sriram, S.R.; Brown, E.N. Autonomic healing of polymer composites. *Nature* 409(6822), 794–797, 2001.
92. Yuan, L.; Liang, G.Z.; Xie, J.Q.; Li, L.; Guo, J. The permeability and stability of microencapsulated epoxy resins. *Journal of Materials Science* 42(12), 4390–4397, 2007.
93. Yuan, L.; Liang, G.Z.; Xie, J.Q.; He, S.B. Synthesis and characterisation of microencapsulated dicyclopentadiene with melamine-formaldehyde resins. *Colloid and Polymer Science* 285(7), 781–791, 2007.

94. Lee, J.K.; Liu, X., Yoon, S.H. Characterization of diene monomers as self-healing agent for polymer composite and its microcapsules, In *Conference Proceedings, Society of Plastics Engineers, Annual Technical Conference 2005, ANTEC* 2005, vol. 6, pp. 264–268.
95. Cowsar, D.R. The United States of America as represented by the Secretary of the Army, Washington, DC. Novel Fabric containing microcapsules of chemical decontaminants encapsulated within semipermeable polymers. U.S. Patent 4,201,822. 6 May 1980. 6.

Section II

Processes

2 Process-Selection Criteria

James Oxley

CONTENTS

2.1 Introduction 23
2.2 Criteria 24
    2.2.1 Size 25
    2.2.2 Morphology 26
        2.2.2.1 Microsphere 26
        2.2.2.2 Core–Shell 27
        2.2.2.3 Complex 27
    2.2.3 Payload 27
    2.2.4 Materials 28
        2.2.4.1 Shell Material 28
        2.2.4.2 Core Material 28
        2.2.4.3 Processing Aids 30
    2.2.5 Equipment 30
        2.2.5.1 Availability 30
        2.2.5.2 Cost 31
    2.2.6 Preprocessing 31
        2.2.6.1 Collection 31
    2.2.7 Scale 31
    2.2.8 Other Process Factors 32
2.3 Selection Process 32
2.4 Conclusions 33
References 33

2.1 INTRODUCTION

Two primary components of microencapsulation development are process and materials selection. All combinations of shell or matrix material and core material or active ingredient require the selection of a compatible process. With over two dozen encapsulation processes, it is important to know the strengths and weaknesses of each when making a selection. The variety of encapsulation processes can be divided into four general categories, listed in Table 2.1: atomization, spray coating, coextrusion (annular jet atomization), and emulsion-based processes. Additional processes are listed that, while less common, are either emerging as new leading technologies or have been established as suitable processes worth consideration. Within these general categories, additional variations are available to achieve capsules of different sizes, morphologies, or materials. This list and the contents of this chapter are offered as an initial guide to be combined with the information in subsequent chapters regarding individual processes.

TABLE 2.1
Encapsulation Processes

	Process	Morph[a]	Payload[b]	Core[c]	Shell[d]	Size[e] (µm)
Physical processes	*Atomization*					
	Spray drying	M	L–H	S/L	O/W/R	5–500
	Spray chilling	M	L–H	S/L	W	50–1000
	Spray congealing	M	L	S/L	M	50–1000
	Spinning disk	M/CS	L–H	S/L	O/W/M/R	5–1000
	Jet cutting	M	L–H	S/L	O/W/M/R	100–3000
	Electrospray	M	L–H	S/L	O/W/M/R	1–100
	Spray coating					
	Fluid bed/Wurster coating	CS	M–H	S	O/W/M	>75
	Pan coating	CS	M–H	S	O/W/M	>250
	Granulation	M	M–H	S	O/W/M	>5
	Coextrusion					
	Stationary nozzle	M/CS	M–H	L	O/W/M/R	500–5000
	Vibrating nozzle	M/CS	M–H	L	O/W/M/R	150–5000
	Centrifugal nozzle	M/CS	M–H	L	O/W/M/R	150–1500
	Submerged nozzle	M/CS	M–H	L	O/W/M/R	500–5000
	Flow focusing	M/CS	L–H	L	O/W/M/R	10–500
	Microfluidics	M/CS	L–H	L	O/W/M/R	1–1000
	Electroydrodynamics	M/CS	L–H	L	O/W/M/R	1–100
	Extrusion	M	L–H	S/L	O/W/M/R	10–5000
Emulsion/suspensions	*Emulsion*					
	Coacervation	M/CS	H	S/L	X	5–1000
	In situ polymerization	CS	H	S/L	X	1–100
	Interfacial polymerization	CS	H	L	R/X	1–500
	Solvent evaporation	M	L–H	S/L	O/W/R	<1–100
	Liposomes	CS/M	L–M	L	X	<1–10
	Sol-gel	CS/M	L–H	S/L	X	<1–100
	Layer-by-layer	CS	H	L	X	<1–500
	Molecular Complexation	M	L	L	X	n/a
	Yeast cells	CS	L–H	L	X	2–5

Source: Southwest Research Institute, San Antonio, TX.
[a] M, microsphere; CS, core–shell.
[b] L = <30%, M = 30%–60%, H = >60%.
[c] S, solid; L, liquid (including suspensions/solutions).
[d] O, organic solvent; W, water; M, melt; R, reactive; X, process specific.
[e] General size range, subject to variation based on formulation and process conditions.

2.2 CRITERIA

General criteria to consider when choosing an encapsulation process include capsule size, morphology, payload, core material, shell material, processing aids, pre/postprocessing, equipment, availability, scale, and cost. Many of these criteria are interdependent and all are process dependent. To complicate the selection process further, process limitations can vary when considering scale or availability of a process. For example, Table 2.2 lists some prominent variables for microcapsules that are commonly reported for most industrial processes or demonstrated in a research and development (R&D) laboratory. These discrepancies and limitations must be taken into account when reviewing research literature and selecting a process. Furthermore, there are exceptions to these

TABLE 2.2
General Encapsulation Parameter Limitations

Variable	Industry	R&D
Size	2 μm–5 mm	10 nm–10 mm
Shell thickness	0.5 nm–500 μm	1 nm–5 mm
Payload	30%–90%	1%–99%

general values. For example, molecular complexation with cyclodextrins is practiced on an industrial scale, and the corresponding size can be reported as the size of a cyclodextrin molecule when dispersed or dissolved into solution. However, when dry or agglomerated, the size is more often in the micron size range.

2.2.1 Size

Processes are available to produce microcapsules from tens of nanometers to tens of millimeters; however, no single process can achieve this range. The existence of multiple encapsulation processes is, in part, to achieve a broad size range demanded by various applications. For example, pharmaceutical intravascular injections may require the use of submicron encapsulated materials, while pesticides or fertilizers for agricultural use are often formulated into larger millimeter encapsulated granules.[1,2] Table 2.1 includes a list of common encapsulation processes and their general size ranges. The listed sizes are approximate values with some flexibility of the limits based on the process parameters and formulations. The table is arranged with physical encapsulation processes listed in the upper portion and are characterized by their use of drying, congealing, or gelling of an atomized droplet to form a matrix or shell. The bottom half of the table lists emulsion-based, or chemical-based, encapsulation processes. These processes rely on the use of emulsification to form droplets for encapsulation, followed by formation of a shell or matrix through chemical reaction or phase separation.

The physical processes are optimized for the preparation of larger microcapsules, from a few microns to millimeters. Spray drying occupies the lower end of the size scale for physical technologies, with most spray-drying processes and variants producing particles from a few microns to a few hundred microns. Spray drying of nanoparticles is feasible but only on a small scale.[3] The lower limit of size begins to increase as atomization moves from spray drying to spray chilling or spray congealing, which can be used to prepare microcapsules with diameters of approximately 50 μm up to millimeters. The ability to produce low micron or submicron droplets for spray drying is also a limitation for spray coating, where a common lower limit is 50–75 μm for fluid bed coating. Finally, coextrusion is further limited to capsule sizes of 150 μm and up on an industrial scale.[4] Recent advancements in this area have led to the ability to form smaller capsules using coextrusion processes, though production scale is limited.[5]

The emulsion-based processes, such as complex coacervation or interfacial polymerization, are best suited for preparing capsules from a few microns to a few hundred microns. Emulsification conditions are controlled to produce capsules of varying size. Surfactants can be introduced to adjust interfacial tension and further direct the emulsified droplet size. Paddle stirring, rotor-stator mixers, high-pressure homogenization, membrane emulsifiers, or other high shear processes can all be used to control droplet size.

In addition to the average size of the capsules produced, the size distribution will vary between encapsulation processes. While a monodisperse narrow size distribution is most often desired, most standard processes are limited to producing a wider unimodal particle size distribution. Atomized particle size distribution will vary with atomization technique.[6] The capsule size from spray coating processes will be directly related to the size distribution of the raw particles used for coating. Though, as coating thickness increases the size distribution will begin to narrow.[7] Particle size distribution resulting from the coextrusion methods will also vary depending on the method of coextrusion, such as stationary, centrifugal, submerged, or vibratory.[8] Vibratory coextrusion will yield the narrowest size distribution and is often sought after in cases where a monodisperse particle size distribution is required.[4] Droplet size distribution for the emulsion-based processes will vary significantly and is dependent on emulsification technique, interfacial tensions, and target droplet size. Techniques such as membrane emulsification or microfluidics can be employed to generate a monodisperse particle size distribution.[9,10]

2.2.2 Morphology

In addition to size, capsule morphology also dictates the encapsulation processes available for selection. Illustrated in Figure 2.1, the three general morphologies are microsphere, core–shell, and a more complex morphology combining the matrix encapsulation of a microsphere within a continuous shell. The atomization processes and some emulsion-based processes are used to prepare microspheres. Spray coating and coextrusion technologies are used to prepare core–shell capsules. A combination of techniques is typically required to achieve the complex morphologies of a microsphere within a continuous shell. Table 2.1 lists morphologies possible with each process. As with the other categories, the list is a generalization with exceptions possible for most processes.

2.2.2.1 Microsphere

Microsphere morphology particles are ubiquitous within the encapsulation industry due to the low cost of production. Microspheres are the product of spray atomization, which is generally the least expensive method of encapsulation. A microsphere morphology is recommended when the presence of some core material at or near the surface of a microsphere and a lower payload are acceptable. Depending upon the formulation, varying percentages of core material will not be completely enrobed within the encapsulating matrix. More protection and controlled release is offered to core material entrapped further into the microsphere. The amount of surface core material increases with increased payload, resulting in a common upper limit of 50% payload for atomized microspheres. Lower payloads are consequently needed to provide protection for the encapsulated core material.

Additional considerations for the choice of a microsphere morphology and its processes include release profile, release mechanism, core material properties, and capsule size. Presuming the matrix material remains intact, release from a microsphere morphology is typically mix of zero and first order.[11,12] Most release mechanisms can be achieved with a microsphere morphology, with the

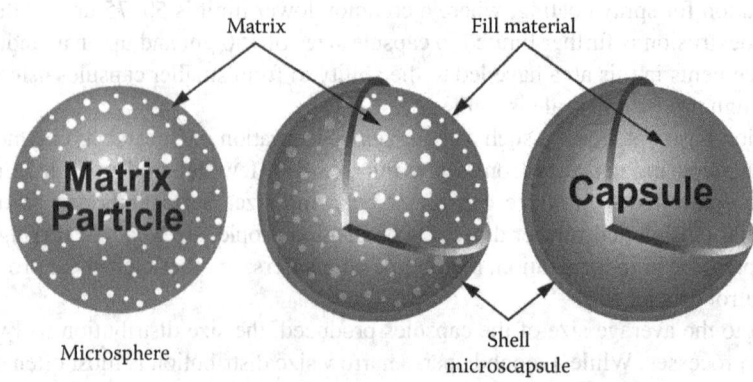

FIGURE 2.1 Encapsulation morphologies. (Courtesy of Southwest Research Institute, San Antonio, TX.)

Process-Selection Criteria

exception of mechanical burst release. Core material properties, such as size, viscosity, material state, solubility, thermal stability, surface tension, and cost will need to be considered for compatibility with a microsphere morphology and the respective encapsulation processes. Finally, size limitations of each process also impact selection of this morphology.

2.2.2.2 Core–Shell

The core–shell morphology is best associated with high payloads and increased protection of the core material. A continuous shell around the core material provides equally continuous protection with no core material at or near the surface of the capsule. Furthermore, the location of shell material exclusively around the perimeter of the capsule leaves more capacity within to host the core material, resulting in payloads over 90%.

Processes for preparing core–shell morphology capsules include spinning disk, spray coating, coextrusion, and most emulsion-based methods. In rare cases, spray drying can be used to prepare core–shell capsules, but is materials dependent.[13] Core–shell capsules with a solid core are limited to the use of spinning disk, spray coating processes, and some emulsion-based processes, while liquids are encapsulated using coextrusion and emulsion-based processes. Exceptions to these assignments can be made when a liquid core material is dispersed into a solid matrix or solid core material is dispersed into a liquid for encapsulation. This approach lowers the upper payload limit due to dilution of the core material into a carrier matrix prior to outer shell formation.

2.2.2.3 Complex

Complex morphologies are considered for additional core material protection, design of complex release mechanisms, adaptation of a core material for specific process, or for accommodation of multiple materials. The most common complex morphology is a core–shell morphology with a matrix particle as the core. The matrix core of a complex capsule provides controlled release of an active ingredient, while an outer shell applied to the microsphere surface provides a barrier against release until an initial trigger is applied. Additionally, the use of a continuous outer shell covers the small percentage of core material in the microsphere that may otherwise be present at or near the surface of the capsule. An alternative complex morphology is the application of multiple shells to a core–shell capsule, to provide unique barrier properties or unique protection and release combinations.

Some complex morphology capsules can be prepared with a single process; however, multiple processes are often needed. Spray-coating systems can be used to granulate core material into a matrix particle followed by use of the same equipment for overcoating. Additionally, spray coating is the most common category of technologies for applying multiple layers to the same core particle. However, other processes may need to be considered if the size restrictions or material restrictions of the spray coating processes prohibit use of one or both parts of a complex capsule. For example, atomization may be required to form the core matrix particle, followed by the use of spray coating or an emulsion-based method to apply the outer continuous shell.

2.2.3 Payload

The variation in payload limits for encapsulation processes, summarized in Table 2.1, is directly related to the morphologies. Core–shell morphologies offer a higher payload, typically 75%–90%. Higher payloads are possible at the expense of shell thickness, less protection for the core material, and reduced mechanical stability. Matrix capsules are most often prepared with payloads over 20% and less than 50%. Higher payloads are possible; however, the percentage of free core material at or near the surface increases with payload. As a result, higher payloads offer significantly less protection and quicker release than lower payload microspheres. Some processes, such as spray congealing, have physical limitations for payload.[14] Spray congealing is the atomization of a molten matrix material, such as a fat or wax, to form droplets that cool into spheres. Suspension of an active ingredient in the feed solution results in the formation of matrix particles from the atomized droplets. The viscosity of the molten feed material increases with solid content. Therefore, an upper payload limit is set by the amount of solid core material that can be added before the viscosity is too high to pump and atomize the slurry.

Some encapsulation processes have limited variability with regards to payload. For example, the payload capacity of molecular encapsulation or complexation in cyclodextrins is limited by affinity equilibrium of the active molecule to the host molecule.[15] Conversely, the payload for some of the common emulsion-based processes, such as interfacial polymerization, will remain high due to the inability to increase shell thickness set by the diffusion limits of the reactive monomers used to form the shell. While liposomes can also have a core–shell structure, their formation process and structure severely limit payload. Lipophilic active ingredients can be entrapped within the lipid bilayer of the liposome but are limited to low percentages to avoid disrupting the bilayer structure.[16–18] Hydrophilic active ingredients can be entrapped in the core of the liposome, but payload is again limited by their solubility or concentration in the inner aqueous environment.

2.2.4 Materials

The selection of materials is a separate exercise that can often have more variables than process selection. Materials can be selected and used to eliminate potential processes, or materials can be selected and evaluated based on the preceding identification of a process. To further complicate interdependency of processes and materials, processing aids are often required and should be considered along with the basic core and shell materials. Table 2.1 lists limitations for both core and shell materials.

2.2.4.1 Shell Material

Shell materials can be solvent-based, water-based, molten, reactive, or molecular. Variations of atomization, spray coating, and coextrusion are available to deposit shell or matrix materials from solvent, water, or as a molten material. For example, spray drying is suitable for encapsulating with solvent-based or water-based matrix materials, while spray congealing uses molten fats or waxes. Fewer shell material selections are available with the emulsion-based processes. For example, complex coacervation is most often associated with the use of gelatin as the shell, and the generation of polyurea or polymelamine formaldehyde shells is associated with in situ polymerization. Further limited examples include the use of cyclodextrins for molecular complexation or phospholipids for the formation of liposomes.

Additional process restrictions from materials selection include processing rates based on the concentration of a coating material in its respective solvent and safety of the material during processing. For example, atomization, spray coating, and coextrusion processes have viscosity limitations for the shell material solutions. Higher concentrations of coating material within a solvent system results in higher viscosity and requires selection of a process or processes that can accommodate the higher viscosity.

2.2.4.2 Core Material

Process selection around a core material is dependent upon core material properties such as thermal stability, viscosity if it is a liquid, particle size and shape if it is a solid, density, reactivity, and solubility. Processes such as spray chilling or spray coating can expose core materials to elevated temperatures for longer periods of time, compared to spray drying, as the mixture remains heated as it awaits pumping to an atomization nozzle. Other processes, such as coextrusion or some emulsion processes, can be carried out at or below room temperature. For example, ionic gelation coupled with coextrusion can be used to encapsulate oils or biological materials at or below room temperature.[4,19,20] Processes like solvent evaporation can be operated under vacuum to remove the matrix solvent rather than the use of temperature.[21]

Viscosity of the core material will impact emulsification or atomization processes, as the resulting droplet or capsule size is directly related to viscosity. The neat core material may be a viscous liquid or viscosity may increase when a solid core material is dissolved or suspended into a carrier fluid. If the viscosity is too high, processes such as atomization, coextrusion, or emulsion-based processes become impractical. Some process modification, such as elevated temperature, can be used to reduce viscosity during encapsulation.

If the core material is a solid, its particle size and shape will influence process selection. The processes and final capsule size will be highly dependent upon the raw core material size, so the list of available processes is reduced if the raw material is used as prepared or received. More processes can be considered if the core material size is modified. For example, milling can be used to reduce particle size, or granulation can be employed to agglomerate smaller particles into larger particles. Size distribution must also be considered and controlled. If encapsulated directly, the size distribution of the raw material will translate to a similar distribution with a spray-coating process. If a large size distribution of core material particles is used to form a suspension for coextrusion, larger particles may clog the nozzles. Finally, the shape of the particle can further influence process selection. Particles with large aspect ratios may require milling to produce a more spherical shape, or use of more viscous solutions (spinning disk coating) or longer coating times (fluid bed coating) to yield a uniform shell.

The solubility or miscibility of a core material will influence both shell material and process selection. A matrix morphology can be homogeneous or heterogeneous, so the core material can be miscible or immiscible with a matrix material used in an atomization process. For the spray-coating process, solubility of the core material in the liquid-coating solution can be tolerated with minimal leaching of the core into the shell. For the coextrusion processes, the core material can be miscible or soluble in the coating solution; however, the shell material must not be soluble in the liquid core material. Furthermore, the shell must form quickly enough to minimize mixing between the two systems. The emulsion-based processes have the greatest limitation, as they require the use of a two phase system for extended periods of time. Oil-in-water is the most common combination for emulsion-based encapsulation. Water-in-oil systems can be used in some cases, followed by double emulsion systems such as water-in-oil-in-water.[22,23] If the core material is a mixture, such as a fragrance or flavor, there is a risk of extracting water soluble components in an emulsion-based process.[23,24]

Density of the core material can negatively affect some encapsulation processes. For example, increasing differences in shell and core material density for a coextrusion process may result in incomplete encapsulation of the core material, or off-centered cores.[8] Density can also be a concern with emulsion-based processes, as a stable suspension is typically required during the formation of a shell or matrix. Density modification can be used to balance the core and shell systems to improve encapsulation efficiency.

Finally, reactive core materials must be considered when selecting a process. Atomization processes require the core and shell materials be mixed prior to atomization, providing some time for them to react. Spray drying requires brief exposure to heat, while spray chilling or congealing require the mixture to be prepared at elevated temperatures. Though the spray coating process also requires maintaining the core material at an elevated temperature for the duration of the coating process, the contact time with shell material solution is minimal. Similarly, the coextrusion process can also accommodate minimal contact time between the core material and liquid shell material. The emulsion-based processes have further restrictions, given that many of the processes require a chemical reaction to form the shell. Core materials can interfere with these chemical reactions and prevent shell formation or result in degradation of the core material. For example, the interfacial polymerization process typically uses multifunctional isocyanates or acyl chlorides in the core material to form the shell.[25] These chemicals are highly reactive toward alcohol or amine groups and can react with core material components that contain the same.

2.2.4.3 Processing Aids

Additional chemicals are often needed to facilitate the successful operation of an encapsulation process, including solvents, anticaking agents, drying aids, surfactants, plasticizers, or density modifiers. Organic solvents or water are required for spray drying or spray congealing. Spray drying quickly removes the solvent or water, while spray chilling may require postprocessing and drying to further remove the water from a hydrogel matrix. Similarly, the coextrusion processes also require removal of a carrier solvent or water following formation and collection of the capsules. The use of melts as a coating material with spray congealing or coextrusion requires no carrier fluid removal. When using the emulsion-based processes, a method for removing the carrier phase liquid must be used to recover dried capsules.

During the drying of capsules or collection of dried capsules, anticaking agents may be required to minimize agglomeration, accelerate any final drying that is required, and maintain a final flowable powder. The use of these additives can be avoided based on shell material selection or keeping the capsules in a liquid suspension. Surfactants may be required for liquid suspensions to improve the suspension stability and to prevent agglomeration.

2.2.5 Equipment

Many of the common encapsulation processes are developed around the use of specialized equipment. As a result, some processes are subject to availability, scale, and cost. Additional factors include collection requirements and the support structure for the process. Figure 2.2 begins to address some of the comparisons between process complexity, scale, and cost, where cost refers to the processing cost.

2.2.5.1 Availability

The processes discussed in this chapter and book are common; however, their availability for lab, pilot, or manufacturing scale can be varied or scarce. The atomization processes, particularly spray

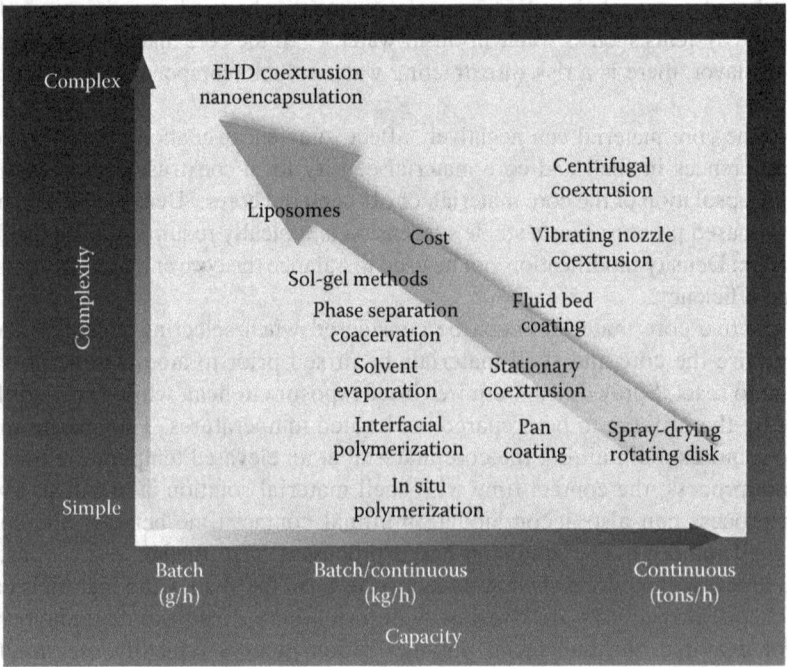

FIGURE 2.2 Comparison of basic encapsulation processes. (Courtesy of Southwest Research Institute, San Antonio, TX.)

Process-Selection Criteria

drying, are the most common. Both lab and production scale equipment are available from multiple suppliers such as BÜCHI Labortechnik AG, Pro-C-ept, Yamamoto, and GEA Niro. Similarly, fluid bed coating is also available in lab to production scale units from various suppliers, including Vector, FluidAir, and Glatt. The coextrusion processes are more limited with regards to availability. A few companies supply laboratory scale units, while even fewer have the capability to supply manufacturing scale facilities, such as Brace GmbH. An advantage of the emulsion based processes is that no specialized equipment is required. On a small scale, simple laboratory glassware, mixing, homogenization, temperature control, and pH monitoring are the general required components. This equipment is easily replicated and widely available on a large scale at multiple chemical facilities.

2.2.5.2 Cost

The economics of encapsulation is detailed in a previous chapter and should always be considered when selecting a process. The primary cost considerations are capital equipment and processing costs. In general, the atomization technologies are the least expensive to produce, followed by spray coating, coextrusion, and the emulsion-based technologies.

2.2.6 Preprocessing

When selecting a process and purchasing equipment, any preprocessing of the core material must be accounted for in the budget and planning. Preprocessing can include milling of a solid core material, granulation of a solid core material, melting of a core or shell material, blending and dissolution of core or shell materials, homogenization, sieving, filtration, or an encapsulation process in the case of complex morphologies. The melting of a core or shell material in large quantities can take time and is energy intensive. The dissolution of core or shell materials may also require heat or time to prepare homogeneous solutions. There are several methods and tools for homogenization, and will be dependent upon the desired droplet size. Raw materials may require sieving to remove fines or larger particles to avoid clogging atomizers or nozzles. Similarly, filtration may be required to also maintain a steady uninterrupted feed to an atomization, spray coating, or coextrusion system. Finally, all of these variables need to be considered again when using a second encapsulation process to prepare complex morphology capsules.

2.2.6.1 Collection

The collection of capsules can often be more costly and difficult than their preparation. Each process has its own set of common collection systems. Atomization techniques generally use cyclonic separation or bag houses, but additional methods include liquid collection, powdered bed collection, or misting chambers depending on the formulation. Fluid bed coating can yield a product that is ready for downstream processing after a coating batch is complete. However, when using a continuous fluid bed coater, a sieving operation may be required to remove over and undersize particles for destruction or recycling.[26] The collection methods of the coextrusion techniques vary with the formulation and employ collection methods similar to the atomization process. The emulsion-based techniques are self-contained batches and can often be used once the shell is formed with little or no post processing. However, if dry capsules are desired then the removal of the water or continuous phase is required. Isolation of capsules from a suspension can be achieved through the use of centrifugation, fluid bed drying, filtration, or lyophilization. It is generally observed that smaller particles and capsules are more difficult to concentrate, dry, and handle. As the size decreases, surface interactions become more dominant and lead to agglomeration, static interactions, and poor flowability.

2.2.7 Scale

The production capacity of an encapsulation process can often be the deciding factor for selection. As shown in Figure 2.2, the atomization processes yield the greatest production capacities.

The largest reported spray-drying capacity in the world is 30 tons/h.[27] The spray coating technologies, specifically fluid bed/Wurster coating, is available for 1 ton batches, or on a continuous platform at 1 ton/h.[28] The coextrusion processes can be operated over 10 tons/h.[4] The emulsion-based processes are mostly operated as batch processes and are only limited by the size and number of batch reactors. Production scale limitations become more prominent with specialized processes such as electrospray, microfluidics, flow focusing, or layer-by-layer processing.[5,29–31] While successfully demonstrated on lab and pilot scales, these processes currently have not demonstrated on production scales comparable to the other common encapsulation processes. Several recently established companies are working toward high production rates and larger scale processing.

Another consideration for the scale of a process is batch versus continuous. Batch processes limit production to the size and number of batches that can be produced in a given amount of time. Continuous production allows for uninterrupted operation, with the production rate controlled by the throughput of the equipment. The atomization and coextrusion technologies are the most common continuous encapsulation processes. Spray-coating processes are available as both batch and continuous. The emulsion-based processes are limited to batch processing, which is the basis for their increased operating costs over most of the physical processes.

2.2.8 Other Process Factors

Additional factors to consider during process selection are safety, labor, utilities, existing equipment, and industry-related regulations. Many of the common processes require the use of solvents or handling fine powders. Both have flammability or explosion hazard risks that require engineered solutions and protections. All processes will require labor to support operation and maintenance, and vary based on scale and process complexity. The availability and scope of utilities for a process will also vary between encapsulation techniques. For example, spray drying will require a significant heat source to dry the particles, or a large amount of water may be required for the emulsion-based processes. Some processes may be a better fit for existing utilities and equipment, resulting in the design of a capsule formulation around the process limitations. Finally, rules and guidelines from industry-specific regulatory organizations will have restrictions on materials that can be used, which directly translates to restrictions on processes that use those materials.

2.3 SELECTION PROCESS

The selection of an encapsulation process is not a simple endeavor. Given the number of process and material variables, more than one process can often achieve the same or similar results or no process may be immediately evident as a solution. The wide scope of the variables also increases the chance of overlooking a critical variable that can result in failure of the chosen encapsulation process. The first step for process selection is to list all of the critical variables and parameters, along with their respective tolerances. For example, what is the target capsule size along with lowest and highest acceptable sizes, or what is the lowest and highest permissible payload. These variables should then be ranked in order of importance as they relate to the encapsulation objectives. A list of some parameters is shown later in Table 2.3 and in no particular order.

Once these items are listed and ranked, along with associated values, an initial list of potential processes can be compiled. If this list has no processes, then some of the set values need to be expanded until some processes are identified. For example, particle size or distribution requirements may need to be expanded, or payload requirements raised or lowered. The consideration of alternative shell materials can also increase the list of potential processes. If this exercise still does not yield a viable list of techniques, then the feasibility of the objectives may need to be considered along with investigation into or development of a new encapsulation process. Alternatively, research into similar products or scenarios can yield examples of comparable products and their processes.

TABLE 2.3
Process Selection Variables

Core/shell material	Material particle size/distribution	Scale
Material state (liquid/solid)	Material solubility/miscibility	Capsule size/distribution
Material density	Dry/wet capsules	Thermal stability
Material MW	Collection	Suspension stability
Equipment availability	In house/contract manufacturing	Payload
Material reactivity	Material shelf-life	Existing utilities/equipment
Pre/postprocessing	Cost	Processing aids
Labor	Regulations	Safety

In many cases, multiple encapsulation processes are possible, each having their own set of advantages and disadvantages relative to a project objective. The process selection then requires laboratory research and development trials to prepare feasibility samples. The feasibility samples can provide insight into processing problems or differences in performance of the capsules prepared with different techniques. In lieu of purchasing multiple pieces of laboratory encapsulation processing equipment, many contract manufacturers and contract R&D organizations offer feasibility testing. The scope and complexity of the testing will dictate the cost and time, but with the benefit of providing the best evidence to support the selection of the best encapsulation process. The expense of paying for feasibility trials is typically offset, sometimes considerably, by the savings in time and money not spent on acquiring multiple pieces of equipment and in-depth operational knowledge of the processes.

2.4 CONCLUSIONS

With over two dozen encapsulation processes, selection of the best can be difficult. Each process has its own set of limitations, advantages, and disadvantages. Important considerations to evaluate are capsule size, capsule morphology, core material properties, shell materials and their properties, payload, process scale, process availability, material preprocessing, and material postprocessing. Careful collection, establishment, and review of the encapsulation objectives and variables will support identification of the most compatible process or processes. Additional work on a lab scale may be required to complete the investigation into the best combination of process and materials. Compromises on some objective limits, such as payload or particle size, may need to be made to accommodate a process that most closely matches a set of initial parameters. Maximizing the collected information about the raw materials, objectives, and flexibility of variables when entering into a process-selection decision will also maximize the chance of successfully identifying the most compatible process or processes.

REFERENCES

1. Rabinow, B.E. Nanosuspensions in drug delivery. *Nature Reviews Drug Discovery* 3(9) (2004): 785–796.
2. Scher, H.B., *Controlled-Release Delivery Systems for Pesticides*. Marcel Dekker, New York, 1999.
3. Li, X., Anton, N., Arpagaus, C., Belleteix, F., and Vandamme, T.F. Nanoparticles by spray drying using innovative new technology: The Büchi Nano Spray Dryer B-90. *Journal of Controlled Release* 147(2) (2010): 304–310.
4. Brandau, T. Preparation of monodisperse controlled release microcapsules. *International Journal of Pharmaceutics* 242(1–2) (2002): 179–184.
5. Berkland, C., Pollauf, E., Pack, D.W., and Kim, K. Uniform double-walled polymer microspheres of controllable shell thickness. *Journal of Controlled Release* 96(1) (2004): 101–111.
6. Masters, K., *Spray Drying Handbook*. Longman Scientific and Technical, Essex, U.K., 1991.

7. Iley, W.J. Effect of particle size and porosity on particle film coatings. *Powder Technology* 65(1–3) (1991): 441–445.
8. Oxley, J.D. Coextrusion for food ingredients and nutraceutical encapsulation: Principles and technology. In *Encapsulation Technologies and Delivery Systems for Food Ingredients and Nutraceuticals*, 640pp. Cambridge, U.K.: Woodhead Publishing Limited, 2012.
9. Shah, R.K., Shum, H.C., Rowat, A.C., Lee, D., Agresti, J.J., Utada, A.S., Chu, L.-Y., Kim, J.-W., Fernandez-Nieves, A., Martinez, C.J., and Weitz, D.A. Designer emulsions using microfluidics. *Materials Today* 11(4) (2008): 18–27.
10. Wagdare, N.A., Marcelis, A.T.M., Ho, O.B., Boom, R.M., and van Rijn, C.J.M. High throughput vegetable oil-in-water emulsification with a high porosity micro-engineered membrane. *Journal of Membrane Science* 347(1–2) (2010): 1–7.
11. Judy H. Senior and Michael L. Radomsky (eds.). *Sustained-Release Injectable Products*, Taylor & Francis, Boca Raton, FL, 2000.
12. Heger, R. Release kinetics/mechanism. In *Microencapsulation of Food Ingredients*, pp. 55–75. Leatherhead, Leatherhead, U.K., 2001.
13. Liu, W., Wu, W.D., Selomulya, C., and Chen, X.D. Facile spray-drying assembly of uniform microencapsulates with tunable core-shell structures and controlled release properties. *Langmuir* 27(21) (2011): 12910–12915.
14. Oxley, J.D. Spray cooling and spray chilling for food ingredient and nutraceutical encapsulation. In *Encapsulation Technologies and Delivery Systems for Food Ingredients and Nutraceuticals*, 640pp. Cambridge, U.K.: Woodhead Publishing Limited, 2012.
15. Del Valle, E.M.M. Cyclodextrins and their uses: A review. *Process Biochemistry* 39(9) (2004): 1033–1046.
16. Nii, T. and Ishii, F. Encapsulation efficiency of water-soluble and insoluble drugs in liposomes prepared by the microencapsulation vesicle method. *International Journal of Pharmaceutics* 298(1) (2005): 198–205.
17. Sharma, A. and Sharma, U.S. Liposomes in drug delivery: Progress and limitations. *International Journal of Pharmaceutics* 154(2) (1997): 123–140.
18. Gibbs, B.F., Kermasha, S., Alli, I., and Mulligan, C.N. Encapsulation in the food industry: A review. *International Journal of Food Sciences and Nutrition* 50(3) (1999): 213–224.
19. Reis, C.P., Neufeld, R.J., Vilela, S., Ribeiro, A.J., and Veiga, F. Review and current status of emulsion/dispersion technology using an internal gelation process for the design of alginate particles. *Journal of Microencapsulation* 23(3) (2006): 245–257.
20. Brinques, G.B. and Ayub, M.A.Z. Effect of microencapsulation on survival of *Lactobacillus plantarum* in simulated gastrointestinal conditions, refrigeration, and yogurt. *Journal of Food Engineering* 103(2) (2011): 123–128.
21. Freitas, S., Merkle, H.P., and Gander, B. Microencapsulation by solvent extraction/evaporation: Reviewing the state of the art of microsphere preparation process technology. *Journal of Controlled Release* 102(2) (2005): 313–332.
22. Chavez-Paez, M., Quezada, C.M., Ibarra-Bracamontes, L., Gonzalez-Ochoa, H.O., and Arauz-Lara, J.L. Coalescence in double emulsions. *Langmuir* 28(14) (2012): 5934–5939.
23. Rocha-Selmi, G.A., Theodoro, A.C., Thomazini, M., Bolini, H.M.A., and Favaro-Trindade, C.S. Double emulsion stage prior to complex coacervation process for microencapsulation of sweetener sucralose. *Journal of Food Engineering* 119(1) (2013): 28–32.
24. Porzio, M. Flavor encapsulation: A convergence of science and art. *Food Technology* 7(58) (2004): 40–47.
25. Kondo, A., *Microcapsule Processing and Technology*, Marcel Dekker, New York, 1979.
26. Teunou, E. and Poncelet, D. Batch and continuous fluid bed coating—Review and state of the art. *Journal of Food Engineering* 53(4) (2002): 325–340.
27. Niro, GEA, The world's largest spray dryer to be built by GEA Process Engineering, 2012. http://www.gea.com/global/en/stories/building-the-worlds-largest-dairy-spray-dryer.jsp.
28. Wysshaar, M., Use of ProCell®-Technology in Microencapsulation of Food Ingredients, in *4th Industrial Workshop on Microencapsulation*, Minneapolis, MN, 2012.
29. Loscertales, I.G., Barrero, A., Guerrero, I., Cortijo, R., Marquez, M., and Gañán-Calvo, A.M. Micro/nano encapsutation via electrified coaxial liquid jets. *Science* 295(5560) (2002): 1695–1698.
30. Martín-Banderas, L., Flores-Mosquera, M., Riesco-Chueca, P., Rodríguez-Gil, A., Cebolla, Á., Chávez, S., and Gañán-Calvo, A.M. Flow focusing: A versatile technology to produce size-controlled and specific-morphology microparticles. *Small* 1(7) (2005): 688–692.
31. Guzey, D. and McClements, D.J. Formation, stability and properties of multilayer emulsions for application in the food industry. *Advances in Colloid and Interface Science* 128(2006): 227–248.

3 Microencapsulation by Spray Drying

Stephan Drusch and S. Diekmann

CONTENTS

3.1 Introduction 35
3.2 General Description of the Process of Spray Drying 35
    3.2.1 Atomization 36
    3.2.2 Spray-Air Contact 38
    3.2.3 Evaporation and Particle Formation 39
3.3 Example: Microencapsulation of Nutritional Oils by Spray Drying 42
References 44

3.1 INTRODUCTION

Spray drying is a well-established process to encapsulate volatile, sensitive, and functional ingredients. The transformation from a liquid medium into a dry powder stabilizes the sensitive core materials in a closed matrix of the shell material. Since the core material is homogeneously distributed within the encapsulating matrix, the resulting type of capsule is usually referred to as matrix capsule. Air inclusion may occur in the particle in the form of a central void (Figure 3.1). With respect to applications in the food or pharmaceutical sector, spray drying of an emulsion is performed to encapsulate non–water-soluble core materials such as oils, vitamins, or flavors. Depending on the type of flavor, the process results in a high flavor retention above 95%,[1–5] and a similar retention for encapsulation of nutritional oils.[6–8]

Encapsulation by spray drying offers protection to adverse environmental conditions such as oxygen or light as well as protection against undesired reactions with other constituents or ingredients.[1,9,10] For this purpose, a carbohydrate-based matrix is usually used. A critical issue related to the process is to preserve the physical properties of the emulsion during drying. Since microcapsules are frequently redissolved before ingestion, the dried emulsions need to be physically stable after dissolution. To create a certain functionality, more sophisticated structures may be build up in the emulsion-like bilayer formation at the interface[11,12] or creation of fibrillar structures,[13] and these structures should not be negatively affected during spray drying.

The release mechanism of carbohydrate-based matrix capsules prepared by spray drying is solvent activated, which means that the powder readily dissolves upon contact with water. This release mechanism has limited the range of applications to dry products in the past. However, through selection of the matrix constituents and buildup of more sophisticated physical structures, a controlled release of encapsulated ingredients can be achieved. This is of particular importance with respect to the release of pharmaceutical or food ingredients, which need to be delivered to specific parts of the gastrointestinal system.[14–16]

3.2 GENERAL DESCRIPTION OF THE PROCESS OF SPRAY DRYING

Spray drying is a continuous process, which can be described in four stages. The first stage is the atomization where the liquid feed is transformed into droplets, the spray. By creating droplets, the surface of the fluid is greatly increased, which is important to ensure a fast and efficient evaporation

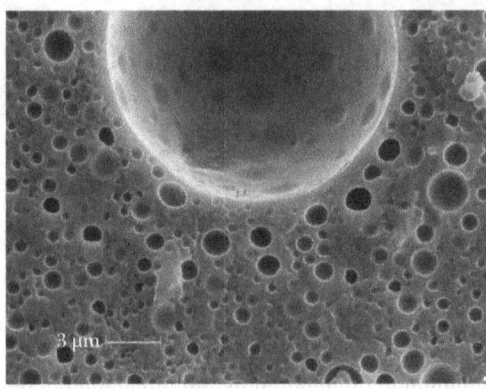

FIGURE 3.1 Scanning electron micrograph of the dross section of a matrix capsule prepared by spray drying.

TABLE 3.1
Impact of Droplet Size on the Total Surface Area of a Liquid Spray

Droplet Size	Number of Droplets	Surface Area (m²)
1 cm	1000	0.31
1 mm	1,000,000	3.14
100 μm	1,000,000,000	31.4
1 μm	1,000,000,000,000	3141

of the moisture from the drying droplet. The impact of droplet size reduction on the surface area is shown in Table 3.1.

The second stage is the spray-air contact when droplets meet the drying medium. Most commonly hot, dry air is used as drying medium. In case of very sensitive or reactive materials, nitrogen may be used. At this stage, the best conditions of contact between droplets and air must be achieved. Evaporation or drying, and thus particle formation within the drying chamber, is the third stage of the process followed by the product separation. The products can be directly recovered from the bottom of the drying chamber or in the case of fine particles through separation from the air with cyclones. Depending on the chemical and physical properties of the feed, the dryer design and operation, products can be powders, granules, or agglomerates.[17] The typical setup of a one-stage spray drier is given in Figure 3.2.

In a two-stage drying process, final drying of the product is performed in a fluidized bed. Since the last traces of residual moisture are difficult to remove, the particles are dried to a moisture content of 2%–5% above the final moisture content in the first stage. The remaining moisture is removed during fluidized bed drying, where the product is in contact with dry air at moderate temperature. Apart from a more efficient drying process, particle agglomeration can be achieved, since particle collisions occur in the fluidized bed with sufficient moisture available for agglomeration through interparticle bridges. Three-stage drying involves transfer of the second drying stage into the base of the spray drying chamber and having the final drying and cooling conducted in the third stage located outside the drying chamber.[18]

3.2.1 Atomization

Atomizer and nozzles are used to supply energy to the feed to form a spray. The feed can be a solution, emulsion, suspension, or paste, which needs to be pumpable. There are different types of

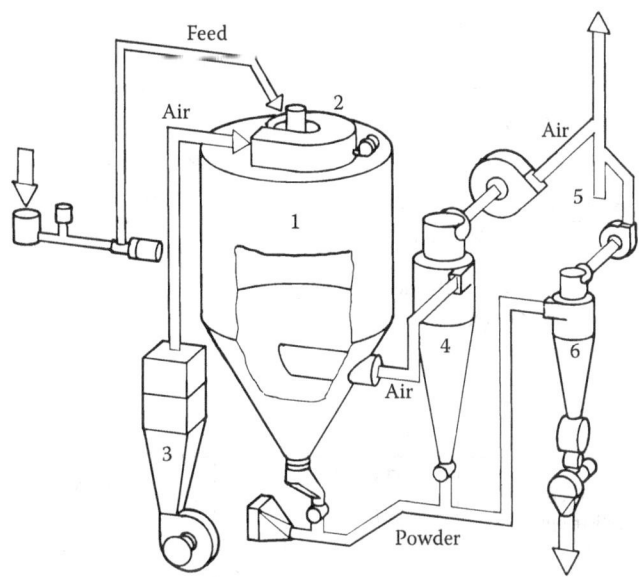

FIGURE 3.2 Typical setup of a one-stage spray drier (1: drying chamber, 2: atomizer, 3: air heater, 4: main cyclone, 5: fan/filter, and 6: transport system cyclone). (Reproduced with permission from Westergaard, V., *Milk Powder Technology: Evaporation and Spray-Drying*, Niro A/S, Soeborg, Denmark, 2004.)

atomizers; among them the most common types are pressure nozzles, two-fluid nozzles or pneumatic nozzles, and mechanical atomizers (Figure 3.3). The selection of an atomizer depends on the product requirements.[19] On one hand the physical properties of the feed need to be taken into consideration, while on the other hand the selection of an atomizer may affect the final particle properties. In general, a high viscosity, high total solids content, high surface tension, and high feed rate lead to an increase in droplet size. Finally, economic aspects play a crucial role in industrial applications. Mechanical atomization is less expensive than two-fluid nozzles since the supply of pressurized air in the latter system affects the costs in use.

Pressure nozzles are usually operated with a feed pressure of 30–200 bar and lead to a certain flow velocity at the orifice of the nozzle. The feed is forced into rotation in a swirl chamber within the nozzle resulting in a cone-shaped spray at the nozzle orifice. It readily integrates into a spray as it is unstable. An increase in the feed rate leads to a less homogeneous and coarse spray with an increase in the width of the droplet size distribution. The mean size of droplets is indirectly proportional to pressure up to 690 bar (680 atm) and directly proportional to feed rate and feed viscosity.[17] Working with pressure nozzles results in a particle size diameter between 50 and 500 μm.[20]

Two-fluid nozzles or pneumatic nozzles have pressurized air or steam as energy supply, whereas the feed has a comparably low velocity.[19] Operating pressure of two fluid nozzles is low, up to 7 bar. The feed and the pressurized air are separately conducted to the nozzle orifice. Two-fluid nozzles may show an internal or external mixing zone. In a two-fluid nozzle with internal mixing, the air stream is rotating and contacting the feed within the nozzle. Part of the energy of the pressurized air is consumed through the mixing process. The major part of the energy is used during disruption of the liquid by rapid gas expansion at the orifice. Recently, effervescent atomization has gained attention for application in spray drying of foods and microencapsulation of food ingredients. It is a special type of internal mixing pneumatic atomizer being distinct in the formation of a two-phase flow prior to the nozzle orifice outlet, in which gas is introduced into the feed stream through holes in the inner channel.[21] In nozzles with external mixing zone, the liquid emerges from the nozzle orifice and meets with the flowing gas. The atomization is achieved through frictional shearing forces between

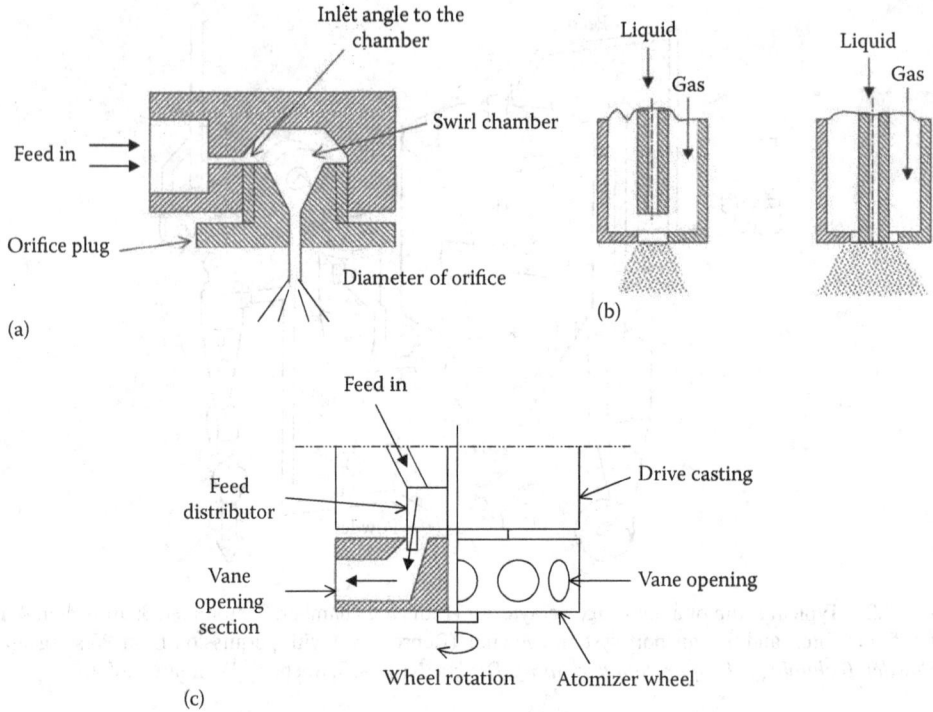

FIGURE 3.3 Examples of different atomization units. (a) Pressure nozzle, (b) two-fluid nozzle, and (c) mechanical atomizer (atomizing wheel). (Reproduced with permission from Barbosa-Cánovas, G. et al., *Food Powders. Physical Properties, Processing, and Functionality*, Kluwer Academic/Plenum Publishers, New York, 2005.)

the liquid surface and the air having a high velocity. A characteristic parameter affecting the droplet size is the air liquid ratio, which is the relation of air mass flow rate to feed flow rate. Two-fluid nozzles have much likelihood of occluded air content within the particles and a low production per drying units.[17] The resulting particles have a mean size with a diameter of ≤50 μm.[20]

Mechanical atomizer use mechanical oscillation or rotating discs or wheels to accelerate the feed.[19] They form a low-pressure system and work with centrifugal energy. The feed is supplied centrally to the wheel or onto the disc at rotating speed. In the case of a rotating disk, the liquid flows outwards over the surface, accelerates to the periphery, and disintegrates into a spray of droplets. To prevent liquid slippage over the surface, the friction between fluid and disc surface is increased. Atomizer wheels can be high or wide and can have straight or curved vanes or channels. In the latter case, the liquid exits the channel at the outer side of the wheel in the form of a liquid jet. Mechanical atomizers are frequently used in food industry and particularly in encapsulation by spray drying. They are operated at rotational speed up to 30,000 rotations per minute, produce a wide range of particle size, and are easy to scale up and produce low costs. It is a continuous process with mild conditions and no plugging.[22] The particle size is directly proportional to feed rate and viscosity and indirectly proportional to the wheel speed and wheel diameter. Rotary atomizer can handle high feed rates, fluctuating feed rates, abrasive feed stocks, or feed stocks with high viscosity. The mean size of particles is 30–200 μm.[17,20]

3.2.2 Spray-Air Contact

At the moment of atomization, the liquid droplets come into contact with the drying medium. The most common drying medium is hot air at an air inlet temperature between 150°C and 200°C.

Microencapsulation by Spray Drying

In some applications inert gas, namely nitrogen, is used. Inert gas prevents the risk of explosion when flammable organic or explosive solvents are involved as well as oxidation in the case of products, which are particularly sensitive to oxygen. In this case, a closed system is required, in which the inert gas is reused. In the case of hot air in an open system, the filtered air can be exhausted to the atmosphere.

The droplet-air contact determines drying kinetics and the properties of the resulting powder particles. In this context, the positioning of the atomizer unit and the air supply, and thus the air flow pattern within the drying chamber, are of major importance. Generally, there are three droplet-air flow patterns: cocurrent, countercurrent, and the mixed flow. In the cocurrent flow dryer, the spray and the hot drying medium enter the drying chambers at the same point, most frequently the top of the drying chamber. The liquid spray meets the drying medium at its highest temperature, and, therefore evaporation of moisture is rapid. Since the droplet is cooled by the evaporating air, the wet bulb temperature of the particle is comparably low. In countercurrent flow the air enters at the opposite end of drying chamber. This offers excellent heat utilization, since the almost dry powder comes into contact with the hot and dry air. As a consequence, a low-residual moisture content may be achieved, but the thermal impact on the encapsulated core material is higher than in cocurrent air flow. Furthermore, the upward stream of the air reduces the velocity of the large droplets and thus increases the residence time in the drying chamber. This system is not recommended for heat-sensitive products, but for the production of coarse products with high bulk density. They also have a low porosity because of the reduced tendency of the droplet to expand rapidly and fracture during drying.[17] Finally, a mixed flow pattern is a combination of cocurrent and countercurrent flow. The air enters at the top, and the spray is produced at the bottom of the drying chamber. The droplet is initially exposed to moderate temperature, meets the hottest air at an intermediate stage in the upper zone of the drying chamber, and is further dried on the way downward to the bottom of the drying chamber.

3.2.3 Evaporation and Particle Formation

Evaporation of water from the droplet or drying particle occurs on its surface. As mentioned earlier, the particle is cooled by the evaporating water. During the first stage of the drying process, the drying rate is constant. During this period, a solid structure develops in the interior of the drying droplet and transport of water through capillaries to the particle surface starts. At the end of this phase, moisture content at the surface of the particle has decreased to the moisture content that develops upon contact with saturated humidified air (maximum hygroscopic moisture content). In the second stage of the drying process, moisture content at the particle surface further decreases to an equilibrium state with the drying air. Heat conduction in the product takes place at this stage and negatively affects the drying rate. In this period of the drying process, moisture bound by sorption starts to be removed from the particle. Finally, in the third stage of the drying process drying rate further decreases due to a decrease in the difference of vapor pressure between the particle interior and the air.[23,24]

With respect to microencapsulation by spray drying, integrity of the particle is a prerequisite to achieve stabilization of the core material. If the drying process is too fast, particularly in the early stages of the drying process, particle ballooning may occur.[25] The phenomenon is well described in the literature, for example, for sucrose, maltodextrins, or emulsions.[26–28] Due to fast evaporation of the water from the particle surface, a crust develops. Through formation of the crust, transport of water to the particle surface is blocked. Since no evaporation occurs any longer, the drying particle heats up to the temperature of the air, which is above 100°C at that stage. Inside the particle, steam formation leads to inflation and particle ballooning occurs. If the internal pressure reaches a critical value, the crust bursts and steam is released from the inside. The particle may undergo this cycle for several times until the particle structure is completely fixed (Figures 3.4 and 3.5).

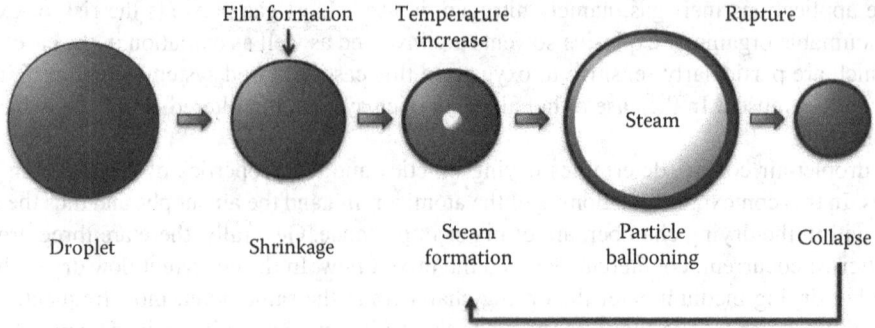

FIGURE 3.4 Scheme on the mechanism leading to particle ballooning during spray drying.

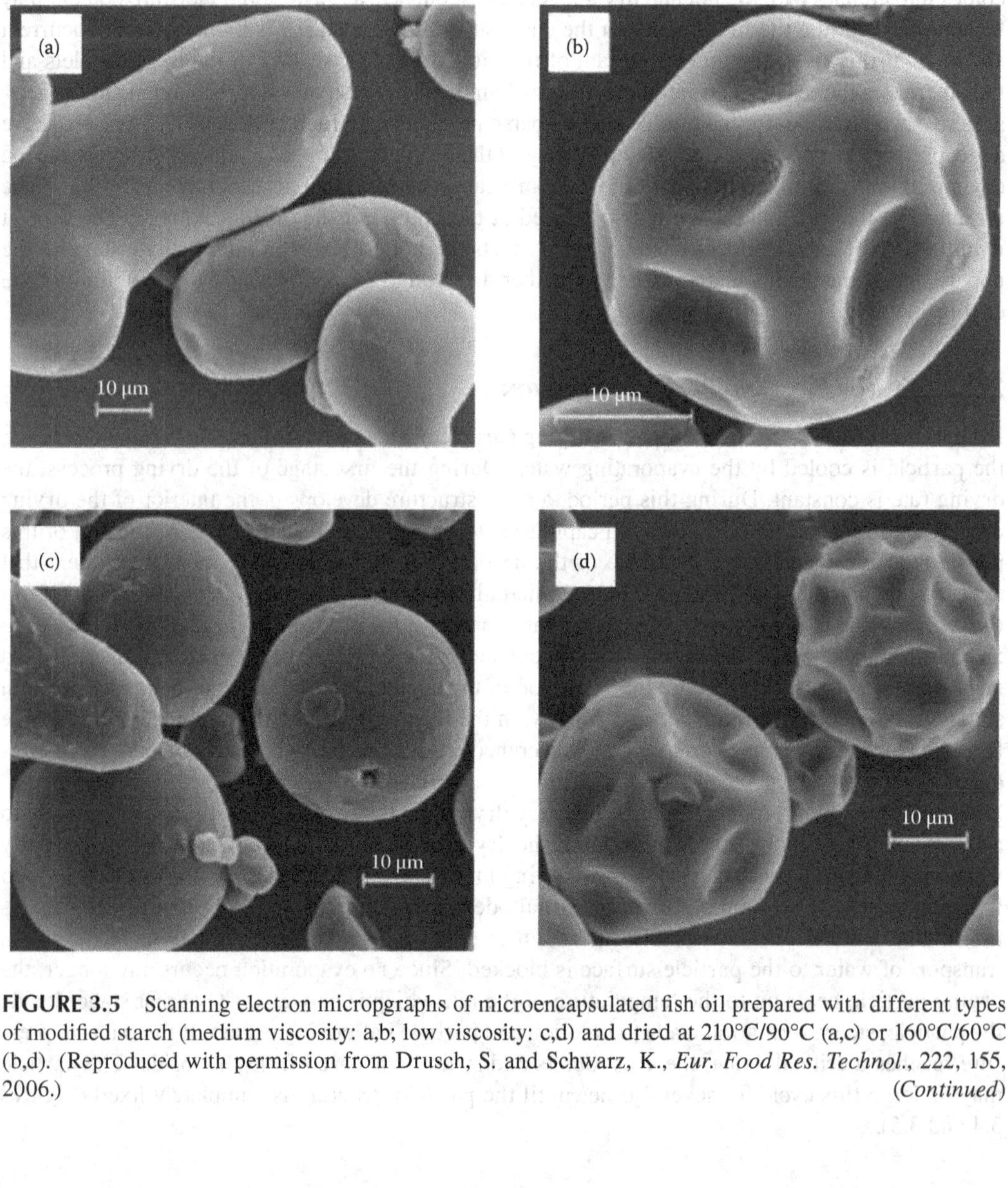

FIGURE 3.5 Scanning electron micropgraphs of microencapsulated fish oil prepared with different types of modified starch (medium viscosity: a,b; low viscosity: c,d) and dried at 210°C/90°C (a,c) or 160°C/60°C (b,d). (Reproduced with permission from Drusch, S. and Schwarz, K., *Eur. Food Res. Technol.*, 222, 155, 2006.) *(Continued)*

Microencapsulation by Spray Drying

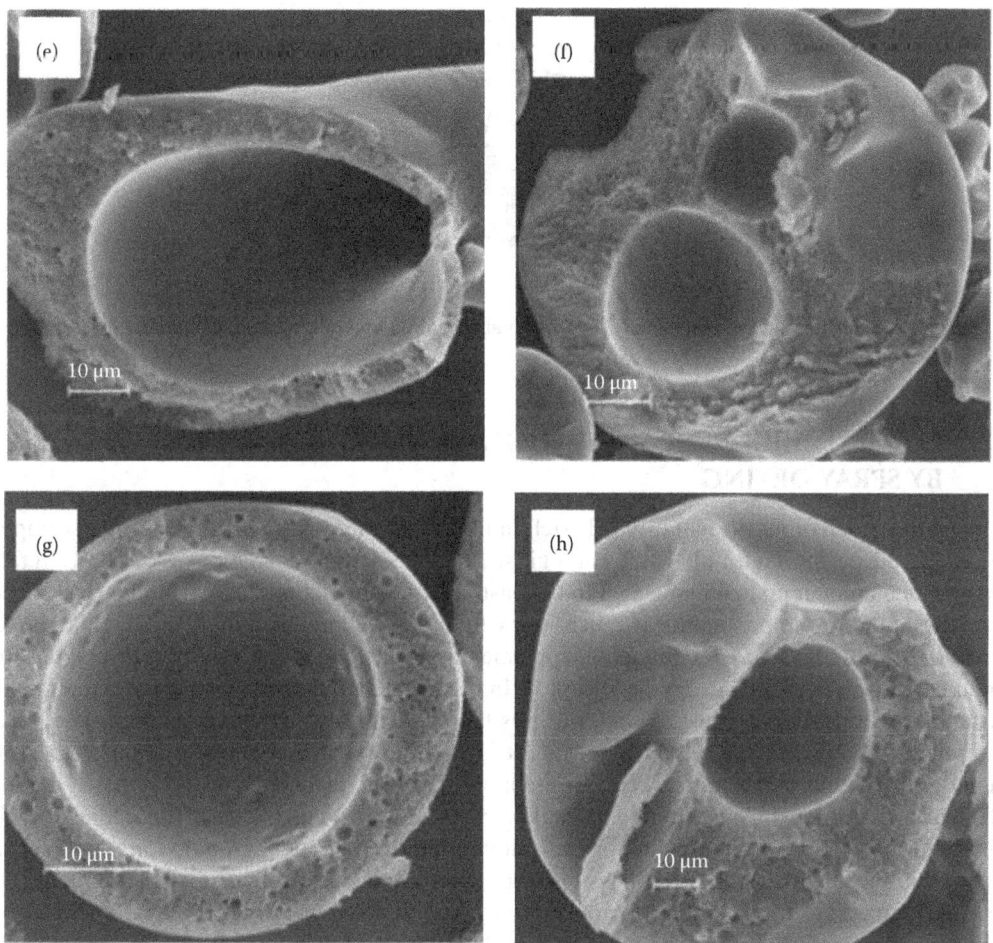

FIGURE 3.5 (*Continued*) Scanning electron micropgraphs of microencapsulated fish oil prepared with different types of modified starch (medium viscosity: e,f; low viscosity: g,h) and dried at 210°C/90°C (e,g) or 160°C/60°C (f,h). (Reproduced with permission from Drusch, S. and Schwarz, K., *Eur. Food Res. Technol.*, 222, 155, 2006.)

Another critical issue with respect to encapsulation is the fact that the particle undergoes a significant change in its composition during the process of particle formation. Approximately 40%–60% of the mass is lost since water is removed. On one hand, this means that the chemical environment constantly changes in ionic strength, pH, and content of other constituents dissolved in the aqueous phase. On the other hand, in case of dispersed systems, the emulsified or suspended core material droplets or particles come into close proximity. Stabilizing effects through steric hindrance or electrostatic repulsion are affected by the change in environmental conditions and aggregation or coalescence may occur (Figure 3.6), which in turn negatively affect encapsulation efficiency and physical stability of the redispersed system. Therefore, success in microencapsulation by spray drying can only be achieved, if the complex interplay between formulation, process design, functionality, and final application is analyzed. In the following section, critical aspects are described using the encapsulation of nutritional oils as an example.

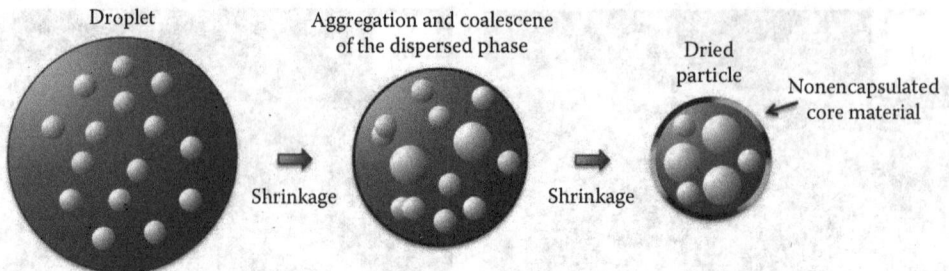

FIGURE 3.6 Scheme on the mechanism leading to aggregation and coalescence of the dispersed phase in spray-dried particles.

3.3 EXAMPLE: MICROENCAPSULATION OF NUTRITIONAL OILS BY SPRAY DRYING

Among the group of nutritional oils, oils rich in polyunsaturated fatty acids are of major importance. Based on the multitude of physiological functions recommendations for a daily intake exist, which amount up to 650 mg.[29-35] In the past, a range of epidemiological studies has shown that the intake in the population is much lower than the recommendation.[36,37] For this reason, nutritionist nowadays still recommend supplementation of foods with oils rich in polyunsaturated fatty acids to ensure an adequate supply. In this context, the sensitivity of long-chain polyunsaturated fatty acids against oxidation limits the applicability, and encapsulation is required to protect the oils. For more than a decade, it is well accepted that encapsulation of the core material in an amorphous matrix reduces the mobility of the reactants and slows down the rate of oxygen diffusion.[38] Thus, microencapsulation by spray drying is one of the suitable techniques for protection of oils rich in polyunsaturated fatty acids. Spray granulation is an alternative technique,[39] which offers the possibility to limit air inclusion, but process time and thus thermal impact are higher than in spray drying. Additional processing like agglomeration or coating may negatively affect the microencapsulation efficiency since rewetting of the particle surface occurs.[40,41]

A wide range of empirical studies on the performance of different matrix constituents in microencapsulation by spray drying are available.[42-49] However, the understanding of the complex interplay between formulation, process conditions, and functionality in terms of protection of the core material has significantly improved during the last decade. The influences of the emulsifier, other carrier matrix constituents, and the process parameters on the physical characteristics of spray-dried microcapsules have been investigated using octenylsuccinate-derivatized starch and proteins. Particle ballooning as described earlier has been described in addition to oxidative damage of the oil during drying of emulsions with high viscosity due to high thermal impact in this drying scenario.[50] Differences in the drying temperature resulted in significant differences in particle density in microcapsules based on octenylsuccinate-derivatized starch and maltodextrin as determined by mercury porosimetry and helium pycnometry. The impact of inlet air temperature on process-induced oxidation has been confirmed by Tonon et al.[51] Based on these results, it was hypothesized that structural differences in the pico- to micrometer range determine the gas diffusivity and thus the long-term stability of encapsulated nutritional oils.[28] The hypothesis was supported by the observation of Keogh et al.,[52] who described a relationship of vacuole volume and off-flavor development during storage of microencapsulated oil rich in polyunsaturated fatty acids.

In the following years, nanostructural differences in the particle morphology and thus differing gas diffusivity was identified as the cause of a different oxidative stability of microencapsulated nutritional oils. When using a octenylsuccinate-derivatized starch with a high proportion of low molecular weight disaccharides, a significant inhibition of autoxidation compared to a carrier system with higher molecular weight starch fragments was achieved.[44] Using positron annihilation lifetime spectroscopy, Townrow et al.[54] showed that nanometer-sized defects in amorphous maltodextrin systems exist, and that these defects are reduced with an increase in low molecular weight maltose in the system. The impact of these defects, the free volume elements, on the stability of encapsulated oils as shown in Figure 3.7 has been confirmed.[55] Furthermore, it was shown that also the protective effect of protein glycosylation on the oxidation of an encapsulated oil may not be attributed to an antioxidative effect of the conjugates, but rather a shift in the molecular weight profile in the carrier matrix.[52]

Furthermore, it is important to consider that the distribution of the core material in the matrix is very heterogeneous as it has been shown by confocal laser scanning microscopy.[9] The surface itself and pores close to the surface are filled with nonencapsulated or poorly encapsulated core material, which is easily oxidized. Microencapsulation efficiency is therefore without a doubt a critical factor. There is sufficient evidence that the surface area is inversely related to oxidative stability in oil-in-water emulsions.[56-59] In a study of Rusli et al.,[60] microcapsules with a low proportion of nonencapsulated oil had a lower stability compared to microcapsules with a high proportion of nonencapsulated oil. The authors hypothesized that a reduced oil droplet size in microcapsules with high microencapsulation efficiency was the reason for the decreased stability. A small oil droplet size is related to a high surface area and at a given amount of wall material subsequently with a thinner film of encapsulating material.[60] Furthermore, removal of water during rapid drying can result in conformational changes of high molecular weight emulsifying agents like proteins. Subsequently, destabilization of the emulsion and an increase in oil droplet size may occur. Low-molecular-weight carbohydrates like glucose syrup with a dextrose equivalent of 36 may compensate this effect by stabilizing the protein during dehydration.[61]

In this context, Let et al.[62] suggested that the composition of the interface may be more important than the total surface area itself. In recent years, research focused on modification of the oil-water-interface with the aim of physical stabilization of the interface during dehydration and/or modification of the release of the encapsulated core material. A well-described system is stabilization of the interface through bilayer formation using lecithin as emulsifier and chitosan as oppositely charged polymer for bilayer formation.[63-66] Due to the cationic surface of the droplet, surface repulsion

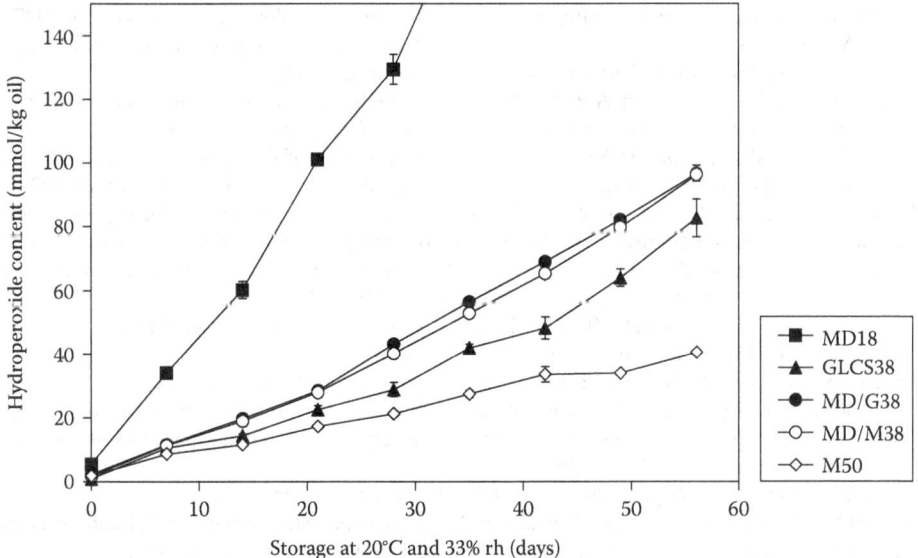

FIGURE 3.7 Development of the hydroperoxide content of fish oil microencapsulated into matrices with different molecular weight profile upon storage at 20°C and 33% relative humidity. (Reproduced with permission from Drusch, S. et al., *Food Biophys.*, 4, 42, 2009.)

of prooxidative metal ions leads to an increase in oxidative stability of nutritional oils in liquid emulsions with bilayer-covered interface. However, it has recently been shown that the bilayer is not stable during the process of spray drying due to the change in the composition of the aqueous phase. The instability is reflected in an increase in oil droplet size and a decrease in encapsulation efficiency and oxidative stability.[12] In contrast, bilayer formation using either pea protein or β-lactoglobulin and pectin is suitable for production of emulsions, which are stable during the process of spray drying.[12,67] Using interfacial dilatational rheology, Serfert et al.[11] showed that bilayer formation strongly increases the viscoelastic modulus of the film at the interface, and thus its resistance against mechanical stress and disruption. Finally, very promising results on the physical and chemical stabilization of nutritional oils have been achieved using β-lactoglobulin-based fibrils as emulsifying agent in spray-dried microparticles.[13]

REFERENCES

1. Charve J, Reineccius GA (2009) Encapsulation performance of proteins and traditional materials for spray dried flavors. *Journal of Agricultural and Food Chemistry* 57:2486–2492.
2. Liu X-D, Atarashi T, Furuta T et al. (2001) Microencapsulation of emulsified hydrophobic flavors by spray-drying. *Drying Technology* 19:1361–1374.
3. Porzio M (2004) Flavor encapsulation: A convergence of science and art. *Food Technology* 58:40–47.
4. Reineccius G (2010) Flavor deterioration during food storage. In: Skibsted LH, Risbo J, Andersen ML (eds.) *Chemical Deterioration and Physical Instability of Food and Beverages*. Woodhead Publishing, Oxford, U.K., pp. 95–112.
5. Uhlemann J, Reiß I (2010) Product design and process engineering using the example of flavors. *Chemical Engineering and Technology* 33:199–212. doi: 10.1002/ceat.200900508.
6. Drusch S, Regier M, Bruhn M (2012) Recent advances in the microencapsulation of oils high in polyunsaturated fatty acids. In: McElhatton A, Sobral PJ do A (eds.) *Novel Technologies in Food Science—Their Impact on Products, Consumer Trends and Environment*. Springer, New York, pp. 159–181.
7. Drusch S, Mannino S (2009) Patent-based review on industrial approaches for the microencapsulation of oils rich in polyunsaturated fatty acids. *Trends in Food Science and Technology* 20:237–244.
8. Hogan SA, McNamee BF, O'Riordan ED, O'Sullivan M (2001) Emulsification and microencapsulation properties of sodium caseinate/carbohydrate blends. *International Dairy Journal* 11:137–144.
9. Drusch S, Berg S (2008) Extractable oil in microcapsules prepared by spray-drying: Localisation, determination and impact on oxidative stability. *Food Chemistry* 109:17–24. doi: 10.1016/j.foodchem.2007.12.016.
10. Fuchs M, Turchiuli C, Bohin M et al. (2006) Encapsulation of oil in powder using spray drying and fluidised bed agglomeration. *Journal of Food Engineering* 75:27–35.
11. Serfert Y, Schröder J, Mescher A et al. (2013) Functionality and spray-drying behavior of emulsions with β-lactoglobulin/pectin -stabilised interface. *Food Hydrocolloids* 31:438–445.
12. Serfert Y, Schröder J, Mescher A et al. (2013) Characterisation of the spray-drying behavior of emulsions containing oil drop-lets with a structured interface. *Journal of Microencapsulation* 30:325–334.
13. Serfert Y, Lamprecht C, Tan C-P et al. (2014) Characterization and use of β-lactoglobulin fibrils for microencapsulation of lipophilic ingredients by spray drying. *Journal of Food Engineering* 143: 53–61.
14. Alcock R, Blair JA, O'Mahony DJ et al. (2002) Modifying the release of leuprolide from spray dried OED microparticles. *Journal of Controlled Release* 82:429–440.
15. Liu W, Selomulya C, Chen XD (2013) Design of polymeric microparticles for pH-responsive and time-sustained drug release. *Biochemical Engineering Journal* 81:177–186.
16. Park C-W, Li X, Vogt FG et al. (2013) Advanced spray-dried design, physicochemical characterization, and aerosol dispersion performance of vancomycin and clarithromycin multifunctional controlled release particles for targeted respiratory delivery as dry powder inhalation aerosols. *International Journal of Pharmaceutics* 455:374–392.
17. Masters K (1985) *Spray Drying Handbook*. John Wiley & Sons, New York.
18. Bylund G (1995) *Dairy Processing Handbook*. Tetra Pak Processing Systems AB, Lund, Sweden.
19. Richter T (2012) *Zerstäuben von Flüssigkeiten*. Expert Verlag, Renningen, Germany.
20. Walzel P (2011) Influence of the spray method on product quality and morphology in spray drying. *Chemical Engineering and Technology* 34:1039–1048. doi: 10.1002/ceat.201100051.
21. Schröder J, Kleinhans A, Serfert Y et al. (2012) Viscosity ratio: A key factor for control of oil droplet size distribution in effervescent atomization of oil-in-water emulsions. *Journal of Food Engineering* 111:265–271.

22. Teunou E, Poncelet D (2005) Rotary disc atomisation for microencapsulation applications—Prediction of the particle trajectories. *Journal of Food Engineering* 71:345–353.
23. Kessler H (2002) *Dairy Technology*. Verlag A. Kessler, München, Germany.
24. Ré MI (1998) Microencapsulation by spray-drying. *Drying Technology* 16:37–41.
25. Walton DE, Mumford CJ (1999) The morphology of spray-dried particles. The effect of process variables upon the morphology of spray-dried particles. *Trans IChemE* 77A:442–460.
26. Alamilla-Beltrán L, Chanona-Pérez JJ, Jimménez-Aparicio AR, Gutiérrez-Lopez GF (2005) Description of morphological changes of particles along spray drying. *Journal of Food Engineering* 67:179–184.
27. Hecht JP, King CJ (2000) Spray drying: Influence of developing drop morphology on drying rates and retention of volatile substances. 1. Single droplet experiments. *Industrial and Engineering Chemical Research* 39:1756–1765.
28. Drusch S, Schwarz K (2006) Microencapsulation properties of two different types of n-octenylsuccinate-derivatised starch. *European Food Research and Technology* 222:155–164.
29. Coste TC, Armand M, Lebacq J et al. (2007) An overview of monitoring and supplementation of omega 3 fatty acids in cystic fibrosis. *Clinical Biochemistry* 40:511–520.
30. Freemantle E, Vandal M, Tremblay-Mercier J et al. (2006) Omega-3 fatty acids, energy substrates, and brain function during aging. *Prostaglandins, Leukotrienes and Essential Fatty Acids* 75:213–220.
31. Garg ML, Wood LG, Singh H, Moughan PJ (2006) Means of delivering recommended levels of long chain n-3 polyunsaturated fatty acids in human diets. *Journal of Food Science* 71:R66–R71.
32. Kris-Etherton PM, Harris WS, Appel LJ, Association for the AH (2003) Omega-3 fatty acids and cardiovascular disease: New recommendations from the American Heart Association. *Arteriosclerosis, Thrombosis and Vascular Biology* 23:151–152.
33. Reisman J, Schachter HM, Dales RE et al. (2006) Treating asthma with omega-3 fatty acids: Where is the evidence? A systematic review. *BMC Complementary and Alternative Medicine* 6:26–34.
34. Sontrop J, Campbell MK (2006) W-3 polyunsaturated fatty acids and depression: A review of the evidence and a methodological critique. *Preventive Medicine* 42:4–13.
35. Szajewska H, Horvath A, Koletzko B (2006) Effect of n-3 long-chain polyunsaturated fatty acid supplementation of women with low-risk pregnancies on pregnancy outcomes and growth measures at birth: A meta-analysis of randomized controlled trials. *American Journal of Clinical Nutrition* 83:1337–1344.
36. Bauch A, Lindtner O, Mensink GBM, Niemann B (2006) Dietary intake and sources of long-chain n-3 PUFAs in German adults. *European Journal of Nutrition* 60:810–812.
37. Kris-Etherton PM, Taylor DS, Yu-Poth S et al. (2000) Polyunsaturated fatty acids in the food chain in the United States. *American Journal of Clinical Nutrition* 71:179S–188S.
38. Le Meste M, Champion D, Roudat G et al. (2002) Glass transition and food technology: A critical appraisal. *Journal of Food Science* 67:2444–2458.
39. Anwar SH, Kunz B (2011) The influence of drying methods on the stabilization of fish oil microcapsules: Comparison of spray granulation, spray drying, and freeze drying. *Journal of Food Engineering* 105:367–378.
40. Domian E, Sulek A, Cenkier J, Kerschke A (2014) Influence of agglomeration on physical characteristics and oxidative stability of spray-dried oil powder with milk protein and trehalose wall material. *Journal of Food Engineering* 125:34–43.
41. Anwar SH, Weissbrodt J, Kunz B (2010) Microencapsulation of fish oil by spray-granulation and fluid bed film coating. *Journal of Food Science* 75:E359–E371.
42. Keogh MK, O'Kennedy BT (1999) Milk fat microencapsulation using whey proteins. *International Dairy Journal* 9:657–663.
43. Hogan SA, O'Riordan ED, O'Sullivan M (2003) Microencapsulation and oxidative stability of spray-dried fish oil emulsions. *Journal of Microencapsulation* 20:675–688.
44. Drusch S, Serfert Y, Scampicchio M et al. (2007) Impact of physicochemical characteristics on the oxidative stability of fish oil microencapsulated by spray-drying. *Journal of Agricultural and Food Chemistry* 55:11044–11051.
45. Chung C, Sanguansri L, Augustin MA (2008) Effects of modification of encapsulant materials on the susceptibility of fish oil microcapsules to lipolysis. *Food Biophysics* 3:140–145.
46. Liao L, Luo Y, Zhao M, Wang Q (2012) Biointerfaces preparation and characterization of succinic acid deamidated wheat gluten microspheres for encapsulation of fish oil. *Colloids and Surfaces B: Biointerfaces* 92:305–314. doi: 10.1016/j.colsurfb.2011.12.003.
47. Wan Y, Bankston Jr DJ, Bechtel PJ, Sathivel S (2011) Microencapsulation of menhaden fish oil containing soluble rice bran fiber using spray drying technology. *Journal of Food Science* 76:E348–E356. doi: 10.1111/j.1750-3841.2011.02111.x.

48. Wang R, Tian Z, Chen L (2011) A novel process for microencapsulation of fish oil with barley protein. *Food Research International* 44:2735–2741.
49. Drusch S (2007) Sugar beet pectin: A novel emulsifying wall component for microencapsulation of lipophilic food ingredients by spray-drying. *Food Hydrocolloids* 21:1223–1228. doi: 10.1016/j.foodhyd.2006.08.007.
50. Drusch S, Serfert Y, Schwarz K (2006) Microencapsulation of fish oil with n-octenylsuccinate-derivatised starch: Flow properties and oxidative stability. *European Journal of Lipid Science and Technology* 108:501–512.
51. Tonon RV, Grosso CRF, Hubinger MD (2011) Influence of emulsion composition and inlet air temperature on the microencapsulation of flaxseed oil by spray drying. *Food Research International* 44:282–289. doi: 10.1016/j.foodres.2010.10.018.
52. Keogh MK, O'Kennedy BT, Kelly J, Auty MA, Kelly PM, Fureby A, Haahr A.-M. (2001) Stability of spray-dried fish oil powdermicroencapsulated using milk ingredients. *Journal of Food Science* 66:217–224.
53. Drusch S, Berg S, Scampicchio M et al. (2009) Role of glycated caseinate in stabilisation of microencapsulated lipophilic functional ingredients. *Food Hydrocolloids* 23:942–948. doi: 10.1016/j.foodhyd.2008.07.004.
54. Townrow S, Kilburn D, Alam A, Ubbink J (2007) Molecular packing in amorphous carbohydrate matrices. *Journal of Physical Chemistry B* 111:12643–12648.
55. Drusch S, Rätzke K, Shaikh MQ et al. (2009) Differences in free volume elements of the carrier matrix affect the stability of microencapsulated lipophilic food ingredients. *Food Biophysics* 4:42–48.
56. Lethuaut L, Métro F, Genot C (2002) Effect of droplet size on lipid oxidation rates of oil-in-water emulsions stabilized by protein. *Journal of the American Oil Chemists Society* 79:425–430.
57. Kiokias S, Dimakou C, Oreopoulou V (2007) Effect of heat treatment and droplet size on the oxidative stability of whey protein emulsions. *Food Chemistry* 105:94–100.
58. Osborn HT, Akoh CC (2004) Effect of emulsifier type, droplet size, and oil concentration on lipid oxidation in structured lipid-based oil-in-water emulsions. *Food Chemistry* 84:451–456.
59. Nakaya K, Ushio H, Matsuwaka S et al. (2005) Effect of oil droplet size on the oxidative stability of oil-in-water emulsions. *Lipids* 40:501–507.
60. Rusli JK, Sanguansri L, Augustin MA (2006) Stabilization of oils by microencapsulation with heated protein-glucose syrup mixtures. *Journal of the American Oil Chemists Society* 83:965–972.
61. Danviriyakul S, McClements DJ, Decker E et al. (2002) Physical stability of spray-dried milk fat emulsion as affected by emulsifiers and processing conditions. *Journal of Food Science* 67:2183–2189.
62. Let MB, Jacobsen C, Soerensen A-DM, Meyer AS (2007) Homogenization conditions affect the oxidative stability of fish oil enriched milk emulsions: Lipid oxidation. *Journal of Agricultural and Food Chemistry* 55:1773–1780.
63. Klinkesorn U, Sophanodora P, Chinachoti P et al. (2006) Characterization of spray-dried tuna oil emulsified in two-layered interfacial membranes prepared using electrostatic layer-by-layer deposition. *Food Research International* 39:449–457.
64. Klinkesorn U, Sophanodora P, Chinachoti P et al. (2005) Stability of spray-dried tuna oil emulsions encapsulated with two-layered interfacial membranes. *Journal of Agricultural and Food Chemistry* 53:8365–8371.
65. Klinkesorn U, Sophanodora P, Chinachoti P et al. (2005) Increasing the oxidative stability of liquid and dried tuna oil-in-water emulsions with electrostatic layer-by-layer deposition technology. *Journal of Agricultural and Food Chemistry* 53:4561–4566.
66. Shaw LA, Faraji H, Aoki T et al. (2008) Emulsion droplet interfacial engineering to deliver bioactive lipids into functional foods. In: Garti N (ed.) *Delivery and Controlled Release of Bioactives in Foods and Nutraceuticals*. Woodhead Publishing Limited, Cambridge, U.K., pp. 184–206.
67. Gharsallaoui A, Saurel R, Chambin O et al. (2010) Utilisation of pectin coating to enhance spray-dry stability of pea protein-stabilised oil-in-water emulsions. *Food Chemistry* 122:447–454.
68. Westergaard V (2004) *Milk Powder Technology: Evaporation and Spray-Drying*. Niro A/S, Soeborg, Denmark.
69. Barbosa-Cánovas G, Ortega-Rivas E, Juliano P, Yan H (2005) *Food Powders. Physical Properties, Processing, and Functionality*. Kluwer Academic/Plenum Publishers, New York.

4 Spray Drying and Its Application in Food Processing

Huang Li Xin and Arun S. Mujumdar

CONTENTS

4.1 Introduction 47
4.2 Principles of Spray Drying 48
    4.2.1 Spray-Drying Fundamentals 48
    4.2.2 Spray-Drying Components 51
        4.2.2.1 Atomization 51
        4.2.2.2 Air and Spray Contact and Droplet Drying 54
        4.2.2.3 Other Components of the Spray-Drying System 56
4.3 Modeling and Simulation of Spray Dryers Using Computational Fluid Dynamics 56
    4.3.1 Governing Equations for the Continuous Phase 58
    4.3.2 Governing Equations for the Particle 59
    4.3.3 Turbulence Models 60
    4.3.4 Heat- and Mass-Transfer Models 60
    4.3.5 Solver 62
    4.3.6 Typical Simulation Results Using CFD Model 62
4.4 Spray-Drying Applications in the Food Industry 63
    4.4.1 Spray-Drying of Milk 63
    4.4.2 Spray Drying of Tomato Juice 65
    4.4.3 Spray Drying of Tea Extracts 66
    4.4.4 Spray Drying of Coffee 66
    4.4.5 Spray Drying of Eggs 67
4.5 Summary 67
References 67

4.1 INTRODUCTION

Spray drying is a one-step continuous processing operation that can transform feed from a fluid state into a dried form by spraying the feed into a hot drying medium. The product can be a single particle or agglomerates. The feed can be a solution, a paste, or a suspension. This process has become one of the most important methods for drying liquid foods to powder form.

The main advantages of spray drying are the following:

- Product properties and quality are more effectively controlled.
- Heat-sensitive foods, biologic products, and pharmaceuticals can be dried at atmospheric pressure and low temperatures. Sometimes inert atmosphere is employed.
- Spray drying permits high-tonnage production in continuous operation and relatively simple equipment.
- The product comes into contact with the equipment surfaces in an anhydrous condition, thus simplifying corrosion problems and selection of materials of construction.

- Spray drying produces relatively uniform, spherical particles with nearly the same proportion of nonvolatile compounds as in the liquid feed.
- As the operating gas temperature may range from 150°C to 750°C, the efficiency is comparable to that of other types of direct dryers.

The principal disadvantages of spray drying are as follows:

- Spray drying generally fails if a high bulk density product is required.
- In general, it is not flexible. A unit designed for fine atomization may not be able to produce a coarse product, and vice versa.
- For a given capacity, evaporation rates larger than other types of dryers are generally required due to high liquid content requirement. The feed must be pumpable. Pumping power requirement is high.
- There is a high initial investment compared to other types of continuous dryers.
- Product recovery and dust collection increases the cost of drying.

The development of the process has been closely associated with the dairy industry. The use of spray drying in the dairy industry dates back to around 1800, but it was not until 1850 that it became possible to dry milk on industrial scale. Since then, this technology has been developed and expanded to cover a large food group which is now successfully spray dried. Over 20,000 spray dryers are estimated to be in use commercially, at present, to agro-chemical products, biotechnology products, fine and heavy chemicals, dairy products, foods, dyestuffs, mineral concentrates, and pharmaceuticals in evaporation capacities ranging from a few kg per hour to 50 tons/h (Mujumdar 2000).

4.2 PRINCIPLES OF SPRAY DRYING

4.2.1 Spray-Drying Fundamentals

A conventional spray dryer flow sheet in its most simplified form is shown in Figure 4.1. It consists of the following four essential components:

- Air heating system and hot-air distribution system
- Feed transportation and atomization
- Air and spray contacts and drying
- Dried particles collection system

From Figure 4.1, it is seen that the process operates in the following way.

The liquid is pumped from the product feed tank to the atomization device, e.g., a rotary disk atomizer, pressure nozzle, pneumatic nozzle, or ultrasonic nozzle, which is usually located in the air distributor at the top of the drying chamber. The drying air is drawn from the atmosphere through a filter by a supply fan and is passed through the air heater, e.g., oil furnace, electrical heater, and steam heater, to the air distributor. The droplets produced by the atomizer meet the hot air and the evaporation takes place, cooling the air in the meantime. After the drying of the droplets in the chamber, the majority of the dried product falls to the bottom of the chamber and entrains in the air. Then they pass through the bag filter for separation of the dried particles from air. The particles leave the bag filter at the bottom via a rotary valve and are collected or packed later. The exhausted air is discharged to the atmosphere via the exhaust fan.

This process shown in Figure 4.1 is the one generally used in industrial spray drying. It is called open-cycle process. Its main feature is that the air is drawn from the atmosphere, passed through the heating system and the drying chamber, and then exhausted to the atmosphere.

Some foodstuffs have to be prepared in organic solvents rather than water to prevent oxidation of one or more of the active ingredients. In these applications, a closed-cycle spray drying system

Spray Drying and Its Application in Food Processing

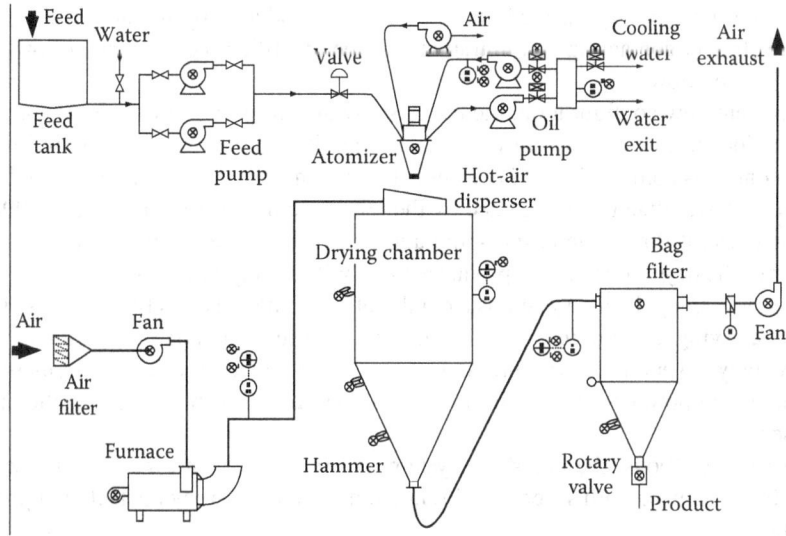

FIGURE 4.1 A typical basic spray dryer flow diagram.

using an inert gas, such as nitrogen, is typically used. This is a special system. Another application is the drying of flammable and toxic materials

In closed-cycle spray-drying plants, the atomized droplets are contacted by hot nitrogen in the spray-drying chamber and processed into a free-flowing powder like any other food formulation. Dried product is discharged from the drying chamber and the cyclone, but the spent drying gas must be introduced into a condensation system.

The solvent evaporated in the drying chamber has to be condensed and recovered. The off-gases from the condensation tower are then reheated in an indirect heater for being reused in the drying chamber. The process is shown in Figure 4.2.

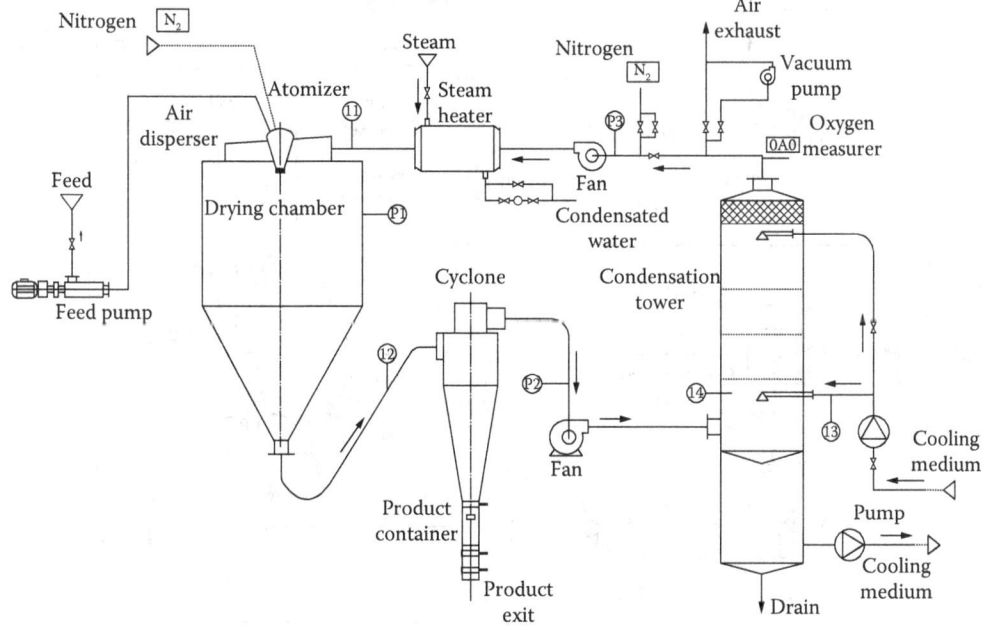

FIGURE 4.2 Layout of a closed-cycle spray-drying system.

Depending on how the drying medium and droplets produced by the atomizer are contacted, three basic air-droplet contacting configurations can be identified, i.e., cocurrent flow, countercurrent flow, and mixed flow.

In the cocurrent flow configuration, the liquid spray and air pass through the drying chamber in the same direction, although spray–air movement in reality is far from cocurrent in initial contact. This type of contact is commonly used in a centrifugal atomization spray dryer. It can lead to product temperatures lower than those obtained by the other two flow patterns (Masters 1991).

In a countercurrent flow system, the spray and hot drying air enter the drying chamber at the opposite ends of the dryer. It typically produces high bulk density powders.

The mixed-flow spray drying system is a combination of the previous two systems. The droplets may contact the drying medium in the same and opposite direction in one drying chamber. It is usually found in spray dryers fitted with pressure nozzle or pneumatic nozzle. It is sometimes used to obtain agglomerated powders in a small drying chamber. It should be noted that the product must be nonheat sensitive.

In spray drying of foods, the one-stage system like that shown in Figure 4.1 is the most common choice. It can be used with several different atomization, air-spray contact, and process layout arrangements.

Two-stage spray dryer systems are sometimes called Spray-Fluidizers, since a fluid bed is installed at the cone of the spray dryer chamber. Alternatively, a vibrated fluid bed (VFB) is installed at the bottom of the drying chamber. It can produce instantly soluble products, such as instant coffee, milk, cocoa, etc., by agglomeration of the product. It is ideal to handle heat-sensitive products.

The three-stage spray-dryer system includes a fluid bed and a VFB dryer or agglomerator together with the spray dryer. It is usually used to produce an agglomerated product by spraying the viscous feed at the beginning of VFB. The process is shown in Figure 4.3. Huang et al. (2001) reported that such a system can save about 20% energy compared to a single stage spray dryer. In this case, a spray dryer is used to evaporate the surface moisture in the drying chamber, and then the

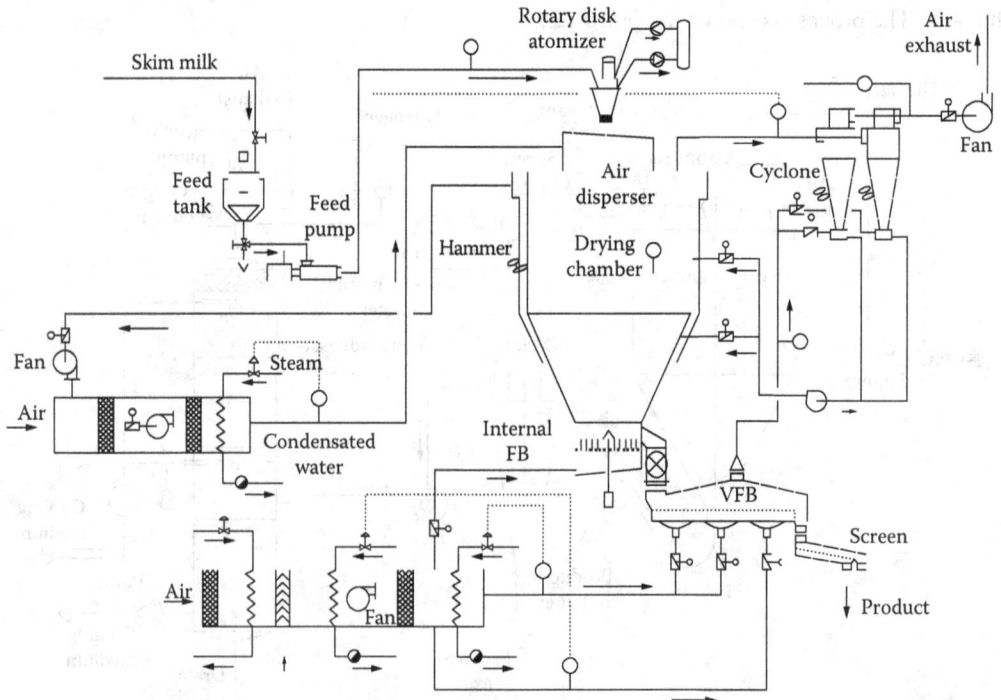

FIGURE 4.3 A schematic diagram of a typical three-stage spray-drying system.

moist powder is further dried in the fluid bed installed at the cone chamber of spray dryer. Finally, the powder leaves the integrated fluid bed to enter a VFB for final drying and cooling. In some applications, the VFB dryer is replaced by a belt dryer.

The main advantages of a three-stage system are:

- Improved agglomeration of particles
- Less thermal degradation of product
- Increased thermal efficiency
- Low product packaging temperature
- Easy to add new operations, such as, spray coating, agglomeration, etc.

4.2.2 Spray-Drying Components

From Figure 4.1, it is clear that spray drying consists of the following four essential stages, i.e., heating of the drying air and its distribution, feed transportation and atomization, contacting of hot air and spray for drying of spray, and recovery of dried products (final air cleaning and dried product handling).

Although the physical design, operation mode, handling of feedstock, and product requirements can be diverse, each stage must be carried out in all spray dryers. The formation of a spray and its effective contacting with the heated air are the key characteristic features of spray drying.

4.2.2.1 Atomization

Since the choice of the atomizer is very crucial, it is important to note the key advantages and limitations of different atomizers (centrifugal, pressure, and pneumatic atomizers). Other atomizers, e.g., ultrasonic atomizer, can also be used in spray dryers (Bittern and Kissel 1999) but they are expensive and have rather low capacity. Although different atomizers can be used to dry the same feedstock, the final product properties (bulk density, particle size, flowability, etc.) are quite different and hence a proper selection is necessary.

4.2.2.1.1 Centrifugal Atomizer

The centrifugal atomizer is sometimes called a rotary wheel or disk atomizer. This is a spinning disk assembly with radial or curved vanes, which rotate at high velocities (7,000–50,000 rpm) with wheel diameters of 5–50 cm. The feed is delivered near the center, spreads between the two plates, and is accelerated to high linear velocities before it is thrown off the disk in the form of thin sheets, ligaments, or elongated ellipsoids. However, the subdivided liquid immediately attains a spherical shape under the influence of surface tension.

The atomizing effect is dependent upon the centrifugal force generated by rotation of the disk; it also depends upon the frictional influence of the external air. The liquid is continuously accelerated to the disk rim by the centrifugal force produced by the disk rotation. Thus, the liquid is spread over the disk internal surface and discharged horizontally at a high speed from the periphery of the disk.

Masters (1991) noted that Equation 4.1 appears to be the most suitable one for the prediction of the mean droplet size generated by a rotating disc atomizer.

$$d_{32} = \frac{1.4 \times 10^4 (\dot{M}_1)^{0.24}}{(u_{atom} d_{atom})^{0.83} (n H_{atom})^{0.12}} \tag{4.1}$$

where
    d_{32} is the Sauter mean diameter (mm)
    $\dot{M}$ is the mass liquid feed rate (kg/h)
    u_{atom} is the atomizer wheel speed (rpm)

d_{atom} is the atomizer wheel diameter (m)
n is the number of channels in the wheel
H_{atom} is the atomizer wheel vane height (m)

Centrifugal atomizers have less tendency to become clogged, which is a great advantage. For this reason, they are preferred for spray drying of nonhomogeneous foods. Its advantages are summarized below:

- Handles large feed rates with single wheel or disk
- Suited for abrasive feeds with proper design
- Has negligible clogging tendency
- Change of wheel rotary speed to control the particle size distribution
- More flexible capacity (but with changes in powder properties)

The limitations associated with this type of atomizer are

- Higher energy consumption compared to pressure nozzles
- More expensive to be manufactured than the other nozzles
- Broad spray pattern requires large drying chamber diameter

4.2.2.1.2 Pressure Nozzle

High-pressure nozzles are alternative atomizing systems in which a fluid acquires a high-velocity tangential motion while being forced through the nozzle orifice. Orifice sizes are usually in the range of 0.5–3.0 mm. The fluid emerges with a swirling motion in a cone shaped sheet, which breaks up into droplets. Greater pressure drop across the orifice produces smaller droplets.

Keey (1991) and Masters (1991) introduced the following correlations to predict the mean droplet diameter produced by pressure nozzle for commercial conditions:

$$d_{32} = \frac{2774 Q_l^{0.25} \mu_l}{\Delta P^{0.5}} \qquad (4.2)$$

where
Q_l is the volumetric feed rate (mL/s)
μ_l is the feed viscosity (MPa s)
ΔP is the operating pressure (kPa)

Its advantages are

- Simple, compact, and cheap
- No moving parts
- Low energy consumption

Its limitations are

- Low capacity (feed rate for single nozzle)
- High tendency to clog
- Erosion can change spray characteristics

4.2.2.1.3 Pneumatic Nozzle

The pneumatic nozzle is an atomizer with internal or external mixing of gas and liquid. Here atomization is accomplished by the interaction of the liquid with a second fluid, usually compressed air.

Spray Drying and Its Application in Food Processing

Such a design permits air or steam to break up the stream of liquid into a mist of fine droplets. Neither the liquid nor the air requires very high pressure, with 200–450 kPa being typical. The particle size is controlled by varying the ratio of the compressed air flow to that of the liquid. The mean spray size produced by pneumatic nozzle atomization follows the relation (Masters 1991)

$$d_{32} = \frac{A_1}{(u_{rel}^2 \rho_g)^\alpha} + A_2 \left(\frac{\dot{M}_g}{\dot{M}_l}\right)^{-\beta} \tag{4.3}$$

where
The exponents α and β are function of nozzle design
A_1 and A_2 are constants involving nozzle design and liquid properties
u_{rel} is the relative velocity between gas and liquid (m/s)
$\dot{M}$ and $\dot{M}$ are mass flow rate of compressed air and feed, respectively (Masters 1991)

Its advantages are

- Simple, compact, and cheap
- No moving parts
- Handle the feedstocks with high viscosity
- Produce products with very small size particle

Its limitations are

- High energy consumption
- Low capacity (feed rate)
- High tendency to clog

4.2.2.1.4 Ultrasonic Nozzle

Ultrasonic nozzles are designed to specifically operate from a vibration energy source. In ultrasonic atomization, a liquid is subjected to a sufficiently high intensity of ultrasonic field that splits it into droplets, which are then ejected from the liquid-ultrasonic source interface into the surrounding air as a fine spray (Rajan and Pandit 2001). A number of basic ultrasonic atomizer types, like capillary wave, standing wave, bending wave, fountain, vibrating orifice, and whistle, etc., exist.

Rajan and Pandit (2001) assessed the impact of various physicochemical properties of liquid, its flow rate, the amplitude and frequency of ultrasonic, and the area and geometry of the vibrating surface on the droplet size distribution. A correlation was proposed to predict the droplet size formed using an ultrasonic atomizer taking into consideration the effect of liquid flow rate and viscosity. The droplet size distribution from an ultrasonic nozzle follows a log-normal distribution (Berger 1998).

Its advantages are

- Simple and compact
- No moving parts
- Droplets with narrow size distribution

Its limitations are

- Low capacity (feed rate)
- High tendency to clog

4.2.2.2 Air and Spray Contact and Droplet Drying

Spray and hot air contact determines the evaporation rate of volatiles in the droplet, droplet trajectory, droplet residence time in the drying chamber, and the deposit in the chamber wall. It also influences the morphology of particles and product quality. So, apart from the selection of atomizers, the drying chamber and air disperser selection are other important factors in spray drying. They determine the air flow pattern in the drying chamber. Several authors (e.g., Gauvin et al. 1975, Crowe 1980, Oakley 1994, Kieviet 1997, Langrish and Kockel 2001, Huang et al. 2003) worked in this area, but the amount of published data on spray–air contact is still limited and is mainly applicable to small-scale spray dryers. Experimental measurements are very difficult to make in an operating spray dryer.

4.2.2.2.1 Hot Air Distribution

Indeed, the hot air distribution is one of the crucial points in a spray drying system design. Today, there are three types of hot air distributors which can be found in spray drying systems in the food industry, i.e., rotating air flow distributor, plug air flow distributor, and central pipe air distributor. A schematic representation of each of these systems is shown in Figure 4.4.

The rotating air flow distributor (Figure 4.4a) is installed at the top of the drying chamber. At this condition, the hot air coming from the heater enters tangentially into a spiral-shaped house where it is distributed radially by the distributed guide vanes and led downward over the second set of guide vanes. The second set of guide vanes is used to make the distributed air rotate by the adjustment of the vanes. This type of distributor is usually used for the spray-drying system in which the rotary disc atomizer or nozzle atomizer is installed at the center of the distributor.

The plug flow air distributor (Figure 4.4b) is also installed at the top of the drying chamber. In this condition, the hot air enters radially from the distributor side and is distributed by air passing through the mesh or perforated plate. In order to make air distributed evenly, the mesh or plate is always arranged in two or three layers. Typically, the nozzle atomizer is used for this hot air distributor.

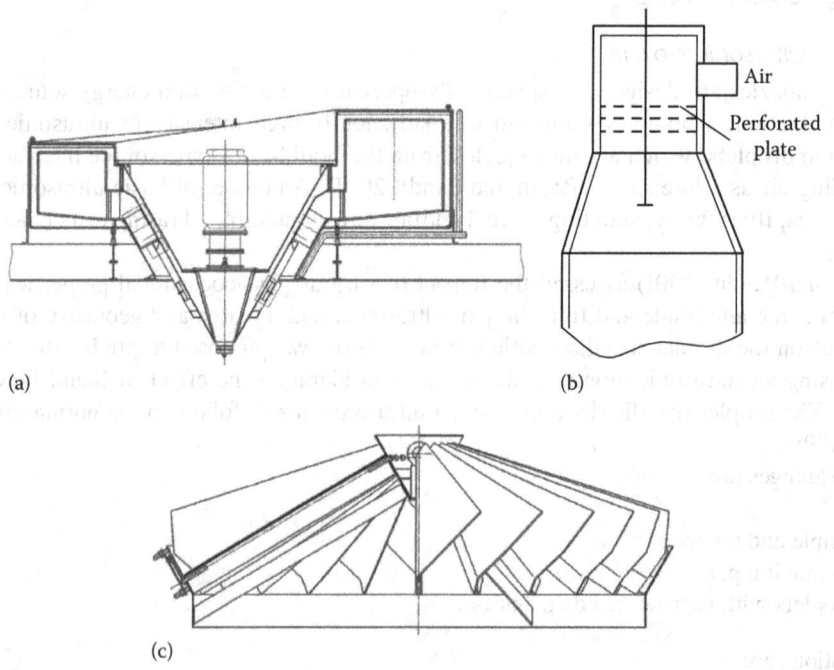

FIGURE 4.4 Hot air dispersers for spray drying in food products: (a) rotating distributor, (b) plug flow distributor; and (c) central pipe distributor.

Spray Drying and Its Application in Food Processing

The central pipe air distributor (Figure 4.4c) is installed at the hot air pipe located at the center of the drying chamber. Such a distributor is usually used for the spray dryer, which operates at high inlet temperatures.

4.2.2.2.2 Drying Chamber Selection and Design

Main heat and mass transfer between droplets and drying medium takes place in the drying chamber. The drying chamber designs are directly related to the results of droplets drying.

In spray-drying market, various designs of the drying chamber can be seen. The cylinder-on-cone chamber is commonly used. According to the product properties, the conical angle is adjusted within 40°–60°. Small angle will help the dried powder leave the chamber by gravity. But it has not necessarily been optimized. Huang et al. (2003) studied three new types of drying chamber designs, i.e., pure conical, lantern, and hourglass, comparing their performance to one of the cylinder-on-cone design under the same spray-drying operation conditions. They found that pure conical and a lantern geometry can also be used as the spray-dryer chamber.

Masters (2002) has suggested a simplified method to design the drying chamber. Huang et al. (1997) reported an empirical correlation for the volumetric drying intensity of the drying chamber in a centrifugal atomization spray dryer as follows:

$$\frac{\dot{M}_w}{V_{dryer}} = \frac{(T_{in} + 273)^{3.4287}}{(T_{out} + 273)^{3.34}} \quad (4.4)$$

where

$\frac{\dot{M}_w}{V_{dryer}}$ is the volumetric drying intensity (kg/m³), viz., evaporation rate per unit drying chamber volume

T_{in} and T_{out} are the inlet drying air temperature and outlet air temperature (°C)

Another important parameter for the drying chamber design is the droplets residence time in the drying chamber. The typical particle residence time in the drying chamber listed in Table 4.1 was suggested by Mujumdar (2000).

In the spray-drying market, the horizontal box type drying chambers are also found in food industries. In such a design, spray is generated horizontally in a box. The contacts and heat and mass transfer between spray and drying medium are within the box. At the bottom of the box, usually a forced powder removal system, e.g., screw, is installed. This is necessary for the system to remove the dried product so that the heat sensitive material, e.g., food, is not degraded. Huang (2005) suggested that a fluidized bed can be fitted at the bottom of such a box chamber. It may increase the drying capacity for the system, as well as remove the dried particles.

TABLE 4.1
Residence Time Requirements for Spray Drying of Various Products

Residence Time in Chamber	Recommended For
Short (10–20 s)	Fine, nonheat-sensitive products, surface moisture removal, nonhygroscopic
Medium (20–35 s)	Fine-to-coarse spray (d_{vs} = 180 mm), drying to low final moisture
Long (>35 s)	Large powder (200–300 mm); low final moisture, low-temperature operation for heat-sensitive products

Source: Mujumdar, A.S., Dryers for particulate solids, slurries, and sheet-form materials, in *Mujumdar's Practical Guide to Industrial Drying*, Devahastion, S., Ed., Exergex Corporation, Brossard, Quebec, Canada, 2000, pp. 37–71. With permission.

4.2.2.3 Other Components of the Spray-Drying System

4.2.2.3.1 Air-Heating System

There are two types of air heaters which can be used in a spray-drying system, e.g., direct air heaters and indirect heaters. Direct air heaters, such as direct gas or oil fired furnace, can be used whenever the contact between combustion gas and spray is acceptable. When products of combustion of fossil fuels cannot contact with the spray, an indirect heater, such as indirect steam air heater, indirect gas, or oil-fired heater, is recommended. Interested readers can find more details about it in the literature (Matsers 1991).

4.2.2.3.2 Dried-Particle-Collection System

The dried products that are entrained in the exhaust air from the drying chamber must be separated and collected. Generally, there are two main types of collectors, i.e., dry collectors and wet collectors. Dry collectors include cyclones, bag filters, and electrostatic precipitators, whereas wet collectors include wet scrubbers, wet cyclones, and spray towers.

Dry collectors are often used as the first-stage collector in a spray dryer. Due to the high cost and maintenance of electrostatic precipitator, it is less often used. Cyclones are first chosen due to their low cost and low maintenance requirements. But the relatively low collection efficiency (90%–98%) is not enough in some cases. Under this condition, a bag filter is used as the second collection stage or a wet collector follows. The limitations of a bag filter are its maintenance and cleaning difficulties. The bags rupture will lead to loss of products. Recently, Niro company in the United Kingdom brought out a new type of bag filter which could be cleaned-in-place (CIP) for better economy, quieter running, higher yields, and a greatly reduced chance of cross-contamination. Their new filter, the SANCIP, can be used with a cyclone in series or as the only means of powder separation and environmental control. SANCIP makes it possible to reduce water and chemical consumption and features a purge system that allows CIP of the entire air assembly.

Addition of wet collectors means additional cost. The scrubber liquid needs to be retreated. So the selection of collection method is dependent on the product value and environmental regulations. See Mujumdar (1995) for further details on product collection methods.

4.2.2.3.3 System Control

Spray dryers can be controlled manually or automatically. No matter what control method is used, the outlet temperature from the drying chamber is always controlled or monitored. It usually determines the residual moisture within the product. Based on the control of outlet temperature, two basic control systems (A and B) can be considered.

- System A: It maintains the outlet temperature by adjusting the feed rate. It is particularly suitable for centrifugal atomization spray dryers. This control system usually has another control loop, i.e., controlling the inlet temperature by regulating air heater.
- System B: It maintains the outlet temperature by regulating the air heater and maintaining the constant spray rate. This system can be particularly used for nozzle spray dryers because varying spray rate will result in the change of the droplet size distribution for pressure or pneumatic nozzle.

4.3 MODELING AND SIMULATION OF SPRAY DRYERS USING COMPUTATIONAL FLUID DYNAMICS

Although spray-drying systems are widely found and used in the industries, their design is still based on empirical methods and experience. Pilot tests are necessary for each spray-drying system. Therefore, more systematic studies must be carried out on spray formation and air flow and heat and mass transfer for optimizing and controlling the drying mechanisms to achieve the highest quality of the powder produced. This process–product association requires a more complex model, which must predict not only the material drying kinetics as a function of the spray drying (SD) operation variables, but also changes in the powder properties during drying in order to quantify the end-product quality. Such combination can be established by introducing into the SD operation, model empirical correlations for predicting the most important product quality requirements (statistical approach) or by describing mechanisms of changes in the material properties during drying (kinetic approach).

Fortunately, computer and software technology are under constant development, which makes the mathematical modeling of spray dryers possible. Such a model is used to predict the droplet or particle movement and the evaporation or drying of droplets in a spray dryer. Two kinds of numerical models, i.e., one-way coupling and two-way coupling, have appeared in the literature (Crowe et al. 1977). In the one-way coupling model, it is assumed that the condition of the drying medium is not affected by the spray or evaporating droplets, although the droplet or particle characteristics change due to the evaporation and drying process. In order to improve the model to simulate spray drying, taking into account heat and mass transfer between spray and drying medium, the two-way coupling approach was developed. This model considers the interaction, e.g., heat, mass, and momentum transfer, between the two phases, i.e., droplets and drying medium. Arnason and Crowe (1986) and Crowe et al. (1998) summarized this approach as shown in Figure 4.5.

On the other hand, the models can also be categorized in terms of the geometry, i.e., one-dimensional (1D), two-dimensional (2D), and three-dimensional (3D).

Crowe et al. (1977) proposed an axi-symmetric spray drying model called Particle-Source-In-Cell model (PSI-Cell model). This model includes two-way mass, momentum, and thermal coupling. In this model, the gas phase is regarded as a continuum (Eulerian approach) and is described by pressure, velocity, temperature, and humidity fields. The droplets or particles are treated as discrete phases which are characterized by velocity, temperature, composition, and the size along trajectories (Lagrangian approach). The model incorporates a finite difference scheme for both the continuum and discrete phases. The authors used this PSI-Cell model to simulate a cocurrent spray dryer. But no experimental data were compared with it. More details can be found in the work by Crowe et al. (1977).

Papadakis and King (1988a,b) used this PSI-Cell model to simulate a spray dryer and compare their predicted results with limited experimental results associated with a lab-scale spray dryer. They have shown that the measured air temperatures at various levels below the roof of the spray drying chamber were well predicted by the computational fluid dynamics (CFD) model. Negiz et al. (1995) developed a program to simulate a cocurrent spray dryer based on the PSI-Cell model. Straatsma et al. (1999) developed a drying model, named NIZO-DrySim, to simulate aspects of

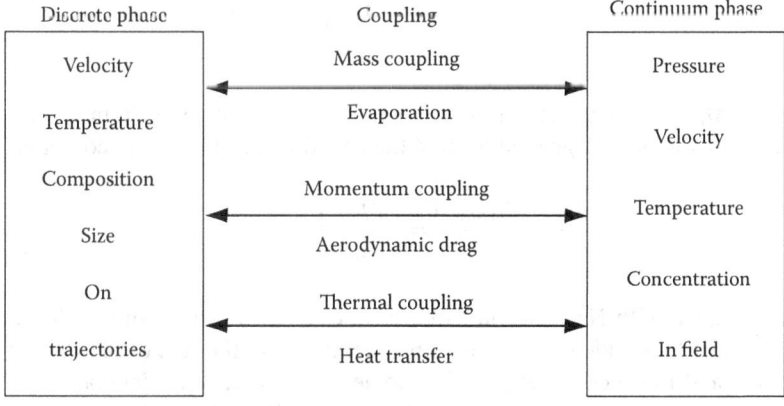

FIGURE 4.5 Two-way coupling between discrete and continuum phases.

drying processes in the food industry. It can simulate the gas flow in a 2D spray-dryer chamber and calculate the particle trajectories.

Livesley et al. (1992) and Oakley and Bahu (1990) found that numerical simulations using the k–e turbulence model are useful for simulating the measured particle sizes and mean axial velocities in industrial spray dryers.

Oakley and Bahu (1990) reported a 3D simulation using the CFD code FLOW3D which is an implementation of the PSI-Cell model. They proposed that additional research needs to be done to verify the performance of their model. This model was used by Goldberg (1987), who predicted the trajectories of typical small, medium, and large droplets of water in a spray dryer with a 0.76 m diameter chamber with 1.44 m height. But in the open literature, most of the studies were carried out in small scale spray dryers. For example, Langrish and Zbicinski (1994) carried out an experiment in a 0.779 m^3 spray dryer.

Kieviet (1997) carried out the measurement of air flow patterns and temperature profiles in a cocurrent pilot spray dryer (diameter 2.2 m). FLOW3D was used to model such a spray dryer and the results showed that experimental data agree qualitatively with the predicted ones, corroborating needs of improvements on measurement techniques and on turbulence model considerations.

Fletcher et al. (2003) used a commercial CFD software to simulate the full-scale spray dryers. Results obtained for the three scales (laboratory, pilot, and industrial) SD in two configurations (tall and short height to diameter chambers) showed the possibility of exploring the effect of the operational variables on flow stability, wall deposition, and product quality.

Cakaloz et al. (1997) studied a horizontal spray dryer to dry a-amylase. However, it is observed that in the flow pattern in Cakaloz et al. design is not optimal for spray drying since the main air inlet is located at a corner of the chamber. This arrangement makes the spray more likely to hit the top wall unless designed carefully.

Verdurmen et al. (2004) proposed an agglomeration model to be included into CFD models to predict agglomeration process in a spray dryer. However, this is still under investigation.

Huang and Mujumdar (2007) were the first to investigate a spray dryer fitted with a centrifugal atomizer using a CFD model. In their model, they model the rotary disk atomization into the disk side point injection which is the same as the holes in the disk.

4.3.1 Governing Equations for the Continuous Phase

For any fluid, its flow must obey the conservation of mass and momentum. These conservation equations can be found in standard fluid dynamic literature (for incompressible gas) (Bird et al. 1960).

The general form of the continuity equation for mass conservation is

$$\frac{\partial \rho}{\partial t} + \frac{\partial (\rho u_i)}{\partial x_i} = M_m \qquad (4.5)$$

The source term M_m is the mass added to the continuous phase, coming from the dispersed phase due to droplet evaporation. The general form of the equation for momentum conservation is

$$\frac{\partial (\rho u_i)}{\partial t} + \frac{\partial (\rho u_i u_j)}{\partial x_i} = -\frac{\partial P}{\partial x_i} + \frac{\partial \tau_{ij}}{\partial x_i} + \rho g_i + M_F \qquad (4.6)$$

This is in accordance with Newton's law (mass times acceleration = sum of forces) where the first term on the left-hand side of Equation 4.6 is dedicated to the rate of increase of momentum per unit volume and the second term is the momentum increase or decrease per unit volume due to convection. The first term on the right-hand side of Equation 4.6 is the pressure force on a fluid element per unit volume, the second term is the viscous force on a fluid element per unit

Spray Drying and Its Application in Food Processing

volume and the third term is the gravitational force on a fluid element per unit volume. The last term M_F is the momentum source term.

For Newtonian fluids, the components of the stress tensor τ_{ij} in Equation 4.6 can be written as

$$\tau_{ij} = \mu \left(\frac{\partial u_i}{\partial x_j} + \frac{\partial u_j}{\partial x_i} \right) - \frac{2}{3} \delta_{ij} \frac{\partial u_l}{\partial x_l}$$

with the fluid viscosity m and the volume dilation term with the "Kronecker" delta:

$$\delta_{ij} = \begin{cases} 1 & \text{for } i = j \\ 0 & \text{for } i \neq j \end{cases}$$

The general form of the energy equation is

$$\frac{\partial(\rho c_p T)}{\partial t} + \frac{\partial(\rho c_p u_i T)}{\partial x_i} = \frac{\partial}{\partial x_i} \left[\lambda \frac{\partial T}{\partial x_i} - \overline{\rho u_i' T'} \right] + M_h \quad (4.7)$$

where the first term on the left-hand side of Equation 4.7 is dedicated to the rate of increase of energy per unit volume and the second term is the energy increase/decrease per unit volume due to convection. The first term on the right-hand side of Equation 4.7 is energy on a fluid element per unit volume and the last term M_h is the energy source term.

4.3.2 Governing Equations for the Particle

Based on the solution obtained for the flow field of the continuous phase, using an Euler–Lagrangian approach, we can obtain the particle trajectories by solving the force balance for the particles taking into account the discrete phase inertia, aerodynamic drag, gravity g_i and further optional user-defined forces F_{xi}.

$$\frac{du_{pi}}{dt} = C_D \frac{18\mu}{\rho_p d_p^2} \frac{Re}{24} (u_i - u_{pi}) + g_i \frac{\rho_g - \rho}{\rho_g} + F_{xi} \quad (4.8)$$

with particle velocity u_{pi} and fluid velocity u_i in direction i, particle density ρ_p, gas density ρ_g, particle diameter d_p, and relative Reynolds number

$$Re = \frac{\rho d_p |u_p - u_g|}{\mu} \quad (4.9)$$

and drag coefficient

$$C_D = a_1 + \frac{a_2}{Re} + \frac{a_3}{Re} \quad (4.10)$$

where a_1 to a_3 are constants (FLUENT 2007).

Two-way coupling allows for interaction between both phases by including the effects of the particulate phase on the fluid phase. In order to simplify the model and computation, the particles are usually assumed to be fully dispersed, i.e., they are not interacting with each other. The particle trajectory is updated in fixed intervals (so-called length scales) along the particle path. Additionally, the particle trajectory is updated each time the particle enters a neighboring cell.

4.3.3 Turbulence Models

In turbulent flows, the instantaneous velocity component u is the sum of a time-averaged (mean) value u_i and a fluctuating component u_i' as shown in Equation 4.11.

$$u_i = \bar{u}_i + u_i' \tag{4.11}$$

These fluctuations need to be accounted for in the above illustrated Navier–Stokes equation

$$\frac{\partial}{\partial t}(\rho u_i) + \frac{\partial}{\partial x_j}(\rho u_i u_j) = -\frac{\partial P}{\partial x_i} + \frac{\partial}{\partial x_j}\left[\mu\left(\frac{\partial u_i}{\partial u_j} + \frac{\partial u_j}{\partial x_i} - \frac{2}{3}\delta_{ij}\frac{\partial u_l}{\partial x_l}\right)\right] + \frac{\partial}{\partial x_j}\left(-\overline{\rho u_i' u_j'}\right) \tag{4.12}$$

Compared with Equation 4.6, Equation 4.12 contains the term $-\overline{\rho u_i' u_j'}$, the so-called Reynolds stress, which represents the effect of turbulence and must be modeled by the CFD code. Limited computational resources restrict the direct simulation of these fluctuations, at least for the moment. Therefore the transport equations are commonly modified to account for the averaged fluctuating velocity components. Three commonly applied turbulence modeling approaches have been used in the CFD model of spray drying system, i.e., k–e model (Launder and Spalding 1972, 1974), RNG k–e model (Yakhot and Orszag 1986), and a Reynolds stress model (RSM) (Launder et al. 1975).

The standard k–e model focuses on mechanisms that affect the turbulent kinetic energy. Robustness, economy, and reasonable accuracy over a wide range of turbulent flows explain its popularity in industrial flow and heat transfer simulations. The RNG k–e model was derived using a rigorous statistical technique (called Re-Normalization Group theory). It is similar in form to the standard k–e model, but the effect of swirl on turbulence is included in the RNG mode enhancing the accuracy for swirling flows.

The RSM solves transport equations for all Reynolds stresses and the dissipation rate e and therefore does not rely on the isotropic turbulent viscosity m_t. This makes the RSM suitable to predict even swirling flows, however, the major drawback of this model is the computational effort needed to solve its equations. For 3D simulations, seven additional transport equations must be solved (six for the Reynolds stresses and one for e). However, the RSM is highly recommended if the expected flow field is characterized by anisotropy in the Reynolds stresses as is the case with swirling flows, e.g., cyclones or spray drying with tangential inlet ducts. Crowe (1980) emphasized that the k–e model is not suitable for swirl flow problems.

4.3.4 Heat- and Mass-Transfer Models

In general, there are two drying rate periods, i.e., constant drying rate period (CDRP) and falling drying rate period (FDRP) during droplet drying. CDRP is controlled by mass transfer between the drying medium and the droplet. But FDRP is controlled by the mass diffusion within the droplets or particles.

The heat transfer between the droplet and the hot gas is updated according to the heat-balance relationship given as follows:

$$M_p c_p \frac{dT_p}{dt} = hA_p(T_g - T_p) + \frac{dM_p}{dt}\Delta H^{vap} \tag{4.13}$$

where
  M_p is the mass of the particle (kg)
  c_p is heat capacity of the particle (J kg^{-1} K^{-1})

Spray Drying and Its Application in Food Processing

A_p is the surface area of the particle (m²)
T_g is the local temperature of the hot medium (K)
h is the convective heat-transfer coefficient (W m⁻² K⁻¹)
ΔH^{vap} is the latent heat (J kg⁻¹)
$\dfrac{dM_p}{dt}$ is the evaporation rate (kg s⁻¹)

The heat-transfer coefficient is evaluated using the correlation of Ranz and Marshall (1952a,b):

$$Nu = \frac{hd_p}{\lambda_g} = 2.0 + 0.6 Re^{1/2} Pr^{1/3} \tag{4.14}$$

where
  d_p is the particle diameter (m)
  λ_g is the thermal conductivity of the hot medium (Wm⁻¹ K)
  Re is the Reynolds number based on the particle diameter and the relative velocity (Equation 4.9)
  Pr is the Prandtl number of the hot medium ($Pr = c_p m_g / \lambda_g$) The mass transfer rate is given by

$$J_i^m = k(c_{i,sur} - c_{i,g}) \tag{4.15}$$

where
  J_i^m is the mass flux of vapor (kg m⁻² s⁻¹)
  k is the mass transfer coefficient (m s⁻¹)
  $c_{i,suri}$ is the vapor concentration at the droplet surface (kg m⁻³)
  $c_{i,g}$ is the vapor concentration in the bulk gas (kg m⁻³)

The concentrations $c_{w,sur}$ and $c_{w,g}$ are defined as:

$$c_{i,sur} = MW_i \frac{P_i^{sat}(T_p)}{RT_p} \tag{4.16}$$

$$c_{i,g} = MW_i y_i \frac{P_{op}}{RT_g} \tag{4.17}$$

where
  P_i^{sat} is the vapor pressure at the particle surface and corresponding temperature (Pa)
  R is the universal gas constant (8.314 J mol⁻¹ K⁻¹)
  y_i and MW_i are the local bulk mole fraction and the molecular weight of species i P_{op} is the operating pressure (Pa)

The mass-transfer coefficient in Equation 4.18 is calculated from the Sherwood correlation (Ranz and Marshall 1952a,b):

$$Sh = \frac{kd_p}{D_m} = 2.0 + 0.6 Re^{1/2} Sc^{1/3} \tag{4.18}$$

where
  D_m is the diffusion coefficient of vapor in the bulk (m² s⁻¹)
  Sc is the Schmidt number ($Sc = \mu/\rho D_m$)

4.3.5 Solver

In order to get the numerical results, the above equations must be solved in series. In general, the solution step of a CFD problem is carried out in two steps:

- Discretization: Integration of the governing equations for conservation of mass and momentum, and other scalars (e.g., turbulence parameters) on a cell (= control volume) yielding a set of mathematical expressions for the dependent variables, such as velocity, pressure, etc.
- Linearization: The above-obtained set of mathematical expressions has to be linearized and solved to update the dependent variables in the control volume (cells).

Starting with an initial guessed solution—provided by the user—this solution procedure is repeated until the preset convergence criteria are met and a final solution is obtained via an iterative process. The solution history can be monitored by plotting the sum of the residuals for each dependent variable at the end of each iteration. For a converged solution, the residuals should be a small value (so-called round off).

4.3.6 Typical Simulation Results Using CFD Model

Here is an example using CFD model by Huang (2005). Good agreement was obtained through comparison of its simulated results with the measurements. In the simulation, a pressure nozzle is used in this spray dryer. The feed is skim milk. The physical properties of skim milk are selected in constant values (Holman 1976). The properties of air in the simulation are varied with its temperature (Li et al. 1978). Different mesh sizes are designed and selected to get the mesh-independent results. In Table 4.2, the drying chamber dimension, main operating parameters, and values for simulation are summarized.

Figure 4.6 shows the typical results, e.g., air velocity vector, air streamline, temperature contour, and particles trajectories. From Figure 4.6a and b, it is seen that there is a strong recirculation zone in the drying chamber. It is also noted that there is a nonuniform velocity distribution in the core region of the chamber. The velocity is reduced as the air flows downwards in the chamber.

The temperature contours (Figure 4.6c) show that temperatures in the central core region vary significantly. This is due to the intense heat and mass transfer that occurs during the initial contact between the spray and drying air caused by the high relative velocity between these two phases and the large temperature driving force. Only a minor radial variation of air temperature is found in the remaining volume except for the region very near the chamber wall.

TABLE 4.2
Drying Chamber Dimension and Main Operating Parameters for Spray Drying

Drying Chamber			
Cylindrical Height (m)	Cylindrical Diameter (m)	Cone Height (m)	Cone Angle (°)
2.0	2.215	1.725	60
Main Operating Parameters				
Air Flow Rate (kg/s)	Air Temperature (°C)	Feed Rate (kg/h)	Droplet Size (μm)	Droplet Velocity (m/s)
0.336	195	50	10–138	59
Feed Solid Content (%)	Turbulent Kinetic Energy (m^2/s^2)		Energy-Dissipation Rate (m^2/s^3)	
42.5	0.027		0.3740	

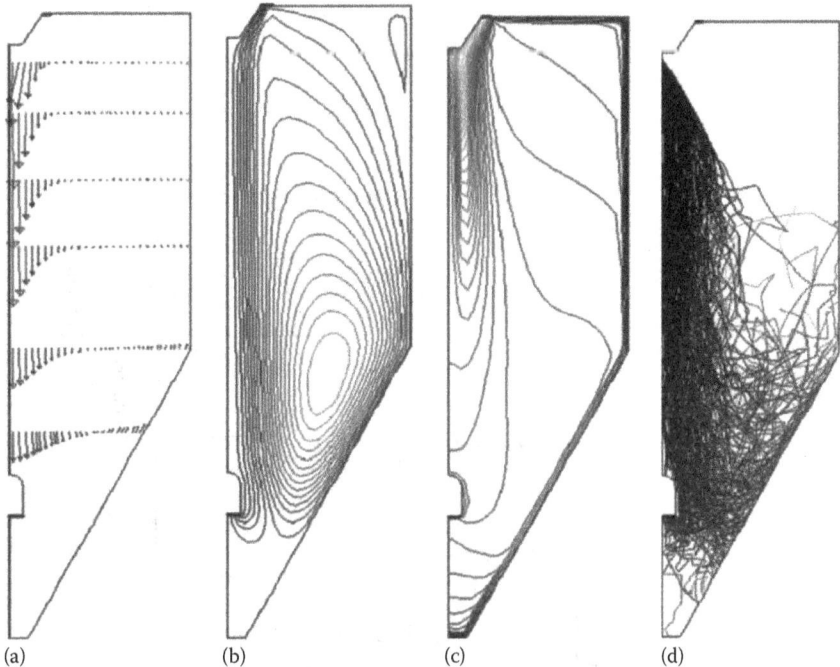

FIGURE 4.6 Simulation results using a CFD model. (a) Velocity vector, (b) streamline, (c) temperature contours, and (d) particle trajectory.

From Figure 4.6d of the particle trajectories, it is easy for the user to find the particle deposit position in the chamber. This information is useful for the designer as well as the spray dryer user to know the place of the deposit in the chamber.

4.4 SPRAY-DRYING APPLICATIONS IN THE FOOD INDUSTRY

Most food-processing companies use spray dryers to produce powdered products. Spray drying has the ability to handle heat-sensitive foods with maximum retention of their nutritive content. The flexibility of spray-dryer design enables powders to be produced in the various forms required by consumer and industry. This includes agglomerated and nonagglomerated powders having precise particles size distribution, residual moisture content, and bulk density. As examples, spray drying of milk, tomato juice, tea extracts, and coffee is discussed.

4.4.1 Spray-Drying of Milk

Milk is one of the most nutritious foods. It is rich in high quality protein providing all 10 essential amino acids. It contributes to total daily energy intake, as well as essential fatty acids, immunoglobulins, and other micronutrients. Commercially available milk can be classified into two major groups: liquid milk and dried or powdered milk. Due to long shelf life of the powdered milk, it is more popular in our daily life.

In general, there are two ways of spray drying milk, i.e., one-stage spray drying system with pneumatic conveying system (shown in Figure 4.7) and multistage spray drying system with or without inner static fluid bed (IFB) and external VFB dryer or cooler (shown in Figure 4.3). This is the basic dairy plant featuring a cocurrent drying chamber with either rotary disc or nozzle atomization. The drying chamber can be either a standard conical design or has a static fluid bed integrated into the chamber base.

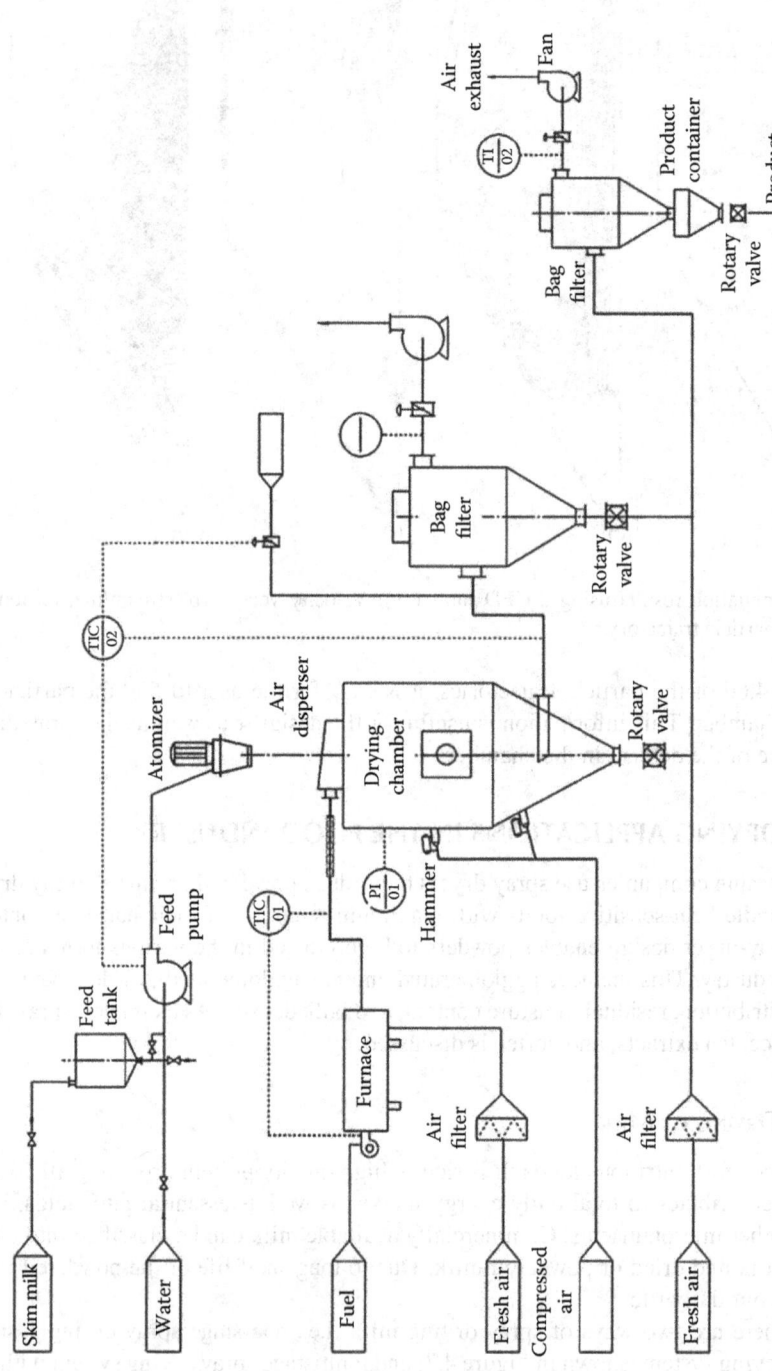

FIGURE 4.7 Conventional spray dryer with pneumatic conveying system.

Spray Drying and Its Application in Food Processing

The advantage of the multistage spray drying system can be summarized as follows: (1) higher capacity per unit drying air; (2) better economic performance, since low outlet temperature can be used; a (3) better product quality, e.g., good solubility, good flowability, and high bulk density of product. Here, the pneumatic conveying system is replaced by the internal fluid bed and VFB.

Both semi-instant skim milk and instant, agglomerated skim milk can be produced in MSD. Since the IFB can be used as the second drying stage and the main agglomeration device, lower outlet temperature can be obtained. The energy consumption is reduced significantly as well. At this process, the final stage (VFB) is usually used as a cooler. If the multistage system only includes SD and VFB, VFB is used as both dryer and cooler.

The performance of various designs for producing skim milk powders was compared by Masters (1991) and Westergaard (1994). Their results showed that MSD gives minimum energy consumption per unit product. The powders produced by MSD are agglomerated and free flowing and with low bulk density.

4.4.2 Spray Drying of Tomato Juice

Fresh, ripe tomatoes are soaked in a vat and then transported by a rolling conveyor to a spray-washing vat. Following washing, the tomatoes are manually sorted and crushed in a chopper to obtain the pulp. If the "hot break" is used, the tomatoes must be heated at 85°C–90°C prior to crushing. Seeds and skin need to be removed prior to refinement of particle size. Since tomato puree which contains skin and seeds produce a more dryable product, it is recommended that the seeds are grinded to 325 ìm in order to pass through a specific mesh screen. When a "cold break" is used, tomato pulp is held for a few seconds in order to obtain pectin decomposition, which will provide an easier-to-spread paste. However, the powder obtained from the hot break is more desirable. The juice is concentrated in a multieffect evaporator and this product goes to a feeding tank to be pumped to the spray drying (Masters 1976). The spray drying system is similar to that shown in Figure 4.3. The IFB and VFB system attached to the spray dryer plays an important role in the drying process of tomato juice. The cooling of drying chamber wall is sometimes necessary since it has some low melting-point ingredients in the tomato powder.

Cold break tomato pastes are spray dried at higher concentrations than hot break pastes. For moderate drying air temperatures, the intake of cool air is controlled in order to maintain a temperature between 40°C and 50°C. The tomato heavy paste goes to a rotary atomizer that contains several vanes. The paste is sprayed into a stream of hot air (140°C–150°C) and then cooled. Droplets of 120–250 ìm are desirable for tomatoes with 28% solids constituents. When low drying temperatures are used during spray drying, the system has slow evaporation rates and very long chambers are required. Usually, the chambers are tens of meter high to increase the droplet falling and drying time. Droplet expansion takes place very slowly because of the low drying temperature. However, the volatile compounds present in the paste will be retained as well as the quality of tomato solids. Tomato powder comes out of the drying chamber with 10% moisture content which is later reduced by using a fluidized bed attached to the base of the dryer chamber. The tomato powder is packaged in an air-conditioned packing room (Masters 1976).

Lumpiness decreases as cooling powder increases. However, the maximum powder temperature to obtain a lump-free product during storage depends upon the type of the tomato. If the product is utilized within few months and it is atmosphere packaged in dry air at low temperature, the product will suffice. In order to prevent lumpiness noncaking agents can be used. Since tomato powder cannot contain more than 2% moisture content, nitrogen or carbon dioxide atmosphere packaging is the most appropriated. For this reason, low-moisture content of dried fruit juices is required for storage. Food gel silica gel and other noncaking additives can be utilized to prevent caking.

4.4.3 Spray Drying of Tea Extracts

There are different kinds of tea, such as green tea, black tea, oolong tea, toasted tea, brown rice tea, etc. Here, we take green tea as an example. First of all, an extract solution from tea leaves is obtained by pouring water onto tea leaves, heat-treating, and filtering it. It is preferable to obtain this solution at a relatively low temperature, for example, in the range of 10°C–40°C. By a low-temperature extraction process, components for bitterness and astringency are not extracted from tea leaves to a great extent, whereas savory components can be extracted effectively. The separation of an extract solution and tea leaves may be carried out by pressing and centrifugal separation. Thus, the elution of components for bitterness and astringency can be prevented.

The extract solution of 3%–5% solid content obtained according to the aforementioned method is concentrated by means of multieffect tube-evaporators or a reverse osmosis membrane. This concentration by reverse osmosis has the advantages of the flavor not disappearing and the deterioration due to heat is small, since mild operating temperatures are employed. It is preferable to concentrate the solution at a temperature of about 25°C, since water flux can be too low at lower temperatures. The solids content in the extract solution is increased from 10% to 30%.

The tea extracts at a concentration of about 20%–25% are stored in a tank for spray drying. A pressure nozzle spray drying system is usually selected. The process is similar to that shown in Figure 4.2. When the tea extracts are transported into the pressure nozzle, it is separated into millions of fine droplets at an operating pressure of 2.7–3.5 MPa. The inlet air temperature is usually at 180°C–220°C. A cooled and dehumidified air must be used to transport and cool the powder products exiting from the drying chamber and cyclone base. Otherwise, too high packing temperature may degrade the tea powder quality.

4.4.4 Spray Drying of Coffee

Coffee is one of the world's most popular beverages. The production of coffee powder usually consists of roasting, grinding, extraction, spray drying, and agglomeration. The production of soluble coffee is a typical example that shows the need of this new drying process definition. Consumers are more demanding about the instant coffee quality requiring similar flavor and aroma of the regular coffee. To become more competitive in the international market, the coffee producers must enhance the quality of their soluble coffee. The traditional process can be categorized into three basic steps: (1) drying of green coffee beans; (2) roasting and grinding of these beans; and (3) extracting and drying of the coffee liquor. To improve the energy efficiency and to avoid environmental pollution, one more step has been added to this process corresponding to the treatment of the coffee sludge for reusing as a fuel (sludge must be dried and, if feasible, gasified). Roasting of green beans needs to be improved to enhance coffee flavor. Both, new gas-particle contactor equipment and grain propriety variable measurable on line (different from color) are required to save energy and to control precisely the roasting operation time, increasing the end-product quality. The spray-drying operation must be optimized as a function of the energy consumption and the powder quality. Changes in the solid material properties should be described together with the specific drying mechanisms observed in the spray chamber to overcome the loss of some volatile compounds responsible for the coffee flavor and aroma. This loss can be minimized by reducing the length of the spray formation region and decreasing faster water concentration into droplets until the critical value dictated by crust formation. Such considerations lead to work in a nozzle tower spray dryer (high height to diameter ratio with a pressure nozzle atomizer type) with a concurrent and a more concentrated coffee liquor. There is also the feasibility to use a secondary cooling air flow at the spray dryer base, as shown in Figure 4.3 (Huang and Mujumdar 2007).

Generally, the preconcentrated coffee extract (40%–50% solid content) is atomized by high pressure nozzles in a cocurrent flow drying chamber. A direct air heater is usually used. The drying operates at the inlet and outlet temperatures of 220°C–240°C and 105°C–115°C. Sometimes the second dehumidified and cool air is induced into the chamber cone to further cool the coffee powder. The exhaust air and fine coffee powder exit from the outlets in the middle of the chamber bustle. The upper chamber may be insulated and lower cone un-insulated. The powder from the drying chamber may be further dried and cooled in VFB dryer and cooler. The collected powder from the cyclone may be conveyed to the VFB or return to the atomization zone for agglomeration. The nonagglomerated coffee powder with 3% residual moisture normally has a particle size in the range 100–400 ìm.

4.4.5 Spray Drying of Eggs

Egg is the most nutritious natural product. Eggs are rich in protein, vitamins, and minerals. During the last three decades, the poultry industry in China has made remarkable progress and grown into an organized and highly productive industry. Dried egg powder can be stored and transported at room temperatures. It is quite stable and has long shelf life. The manufacture of egg powder is an important segment of egg consumption. Nowadays, there are some plants of eggs powder production with a suitable capacity across the world.

Manufacture of dried egg powder starts with breaking of eggs and removing egg-shells. After removal of shells, the mixture is filtered and stored in tanks at about 4°C. Before it is spray-dried, it is taken to a tubular heater wherein it is heated to about 65°C for 8–10 min. The conventional spray drying system with a pressure nozzle is usually selected. The centrifugal atomization is also found in some factories. The spray drying process is similar to that shown in Figure 4.1. The operating drying temperature is 180°C–200°C. Recently, at least two U.S. companies produce horizontal spray dryers, especially for heat-sensitive foods, such as eggs (FES 2008, Rogers 2008).

4.5 SUMMARY

In this chapter, a summary of the fundamentals of spray drying, selection of spray dryers, and the use of spray dryers in the food industry is provided. Spray dryers, both conventional and innovative, will continue to find increasing applications in various industries; almost all industries need or use or produce powders starting from liquid feedstocks. Therefore, although it is very difficult to generate rules for the selection of spray dryers in the food area because of numerous possible exceptions and new developments, it is important for the users of spray dryers to understand the typical and main characteristics of spray drying in the food industry. Despite advances in modeling of spray dryers, it is important to carry out careful pilot tests and evaluate the quality parameters that cannot yet be predicted with confidence. Further advances in mathematical modeling of sprays and spray dryers are needed before confident design and scale-up can be carried out using models.

REFERENCES

Arnason, G. and Crowe, C. T. 1986. Assessment of numerical models for spray drying. In *Drying'86*, Ed. A. S. Mujumdar. New York: Hemisphere Publishing Corporation.
Berger, H. L. 1998. *Ultrasonic Liquid Atomization: Theory and Application*. New York: Partridge Hill Publishers.
Bird, R. B., Stewart, W. E., and Lightfoot, E. N. 1960. *Transport Phenomena*. New York: John Wiley & Sons Inc.
Bittner, B. and Kissel, T. 1999. Ultrasonic atomization for spray drying: a versatile technique for the preparation of protein loaded biodegradable microspheres. *J. Microencapsul.* 16(3):325-341.
Cakaloz, T., Akbaba, H., Yesugey, E. T., and Periz, A. 1997. Drying model for a-amylase in a horizontal spray dryer. *J. Food Eng.* 31:499–510.
Crowe, C., Sommerfeld, M., and Tsuji, Y. 1998. *Multiphase Flows with Droplets and Particles*. Boca Raton, FL: CRC Press.

Crowe, C. T. 1980. Modeling spray air contact in spray drying systems. In *Advances in Drying*, Ed. A. S. Mujumdar, Vol. 1, pp. 63–99. New York: Hemisphere Publishing Corporation.

Crowe, C. T., Sharma, M. P., and Stock, D. E. 1977. The particle-source-in-cell (PSI-Cell) model for gas-droplet flows. *J. Fluid Eng.* 9:325–332.

FES company 2008. http://www.fesintl.com/htmfil.fld/sprydryh.htm.

Fletcher, D., Guo, B., Harvie, D. et al. 2003. What is important in the simulation of spray dryer performance and how do current CFD models perform? In *Proceedings of the Third International Conference on CFD in the Minerals and Process Industries*, Melbourne, Victoria, Australia, www.cfd.com.au.

FLUENT Manual 2007. http://www.fluent.com.

Gauvin, W. H., Katta, S., and Knelman, F. H. 1975. Drop trajectory predictions and their importance in the design of spray dryers. *Int. J. Multiphase Flow* 1:793–816.

Goldberg, J. E. 1987. Prediction of spray dryer performance, PhD thesis, University of Oxford, Oxford, U.K.

Holman, J. P. 1976. *Heat Transfer*. New York: McGraw-Hill.

Huang, L. X. 2005. Simulation of spray drying using computational fluid dynamics, PhD thesis, National University of Singapore, Kent Ridge, Singapore.

Huang, L. X., Kumar, K., and Mujumdar, A. S. 2003. Use of computational fluid dynamics to evaluate alternative spray chamber configurations. *Drying Technol.* 21:385–412.

Huang, L. X. and Mujumdar, A. S. 2007. Simulation of an industrial spray dryer and prediction of off-design performance. *Drying Technol.* 25:703–714.

Huang, L. X., Tang, J., and Wang, Z. 1997. Computer-aided design of centrifugal spray dryer, *J. Nanjing Forestry Univ.* 21(add.):68–71 (in Chinese).

Huang, L. X., Wang, Z., and Tang, J. 2001. Recent progress of spray drying in China, *Chem. Eng. (China)* 29:51–55 (in Chinese).

Keey, R. B. 1991. Private communication by Dr. Masters. In *Spray Drying Handbook*, 5th edn., p. 243. New York: John Wiley & Sons Inc.

Kieviet, F. G. 1997. Modeling quality in spray drying, PhD thesis, Eindhoven University of Technology, Eindhoven, the Netherlands.

Langrish, T. A. G. and Kockel, T. K. 2001. The assessment of a characteristic drying curve for milk powder for use in computational fluid dynamics modeling. *Chem. Eng. J.* 84:69–74.

Langrish, T. A. G. and Zbicinski, I. 1994. The effect of air inlet geometry and spray cone angle on the wall deposition rate in spray dryers. *Trans. I. Chem. E.* 72:420–430.

Launder, B. E., Reece, G. J., and Rodi, W. 1975. Progress in the development of a Reynolds-stress turbulence closure. *J. Fluid Mech.* 68:537–566.

Launder, B. E. and Spalding, D. B. 1972. *Lectures in Mathematical Models of Turbulence*. London, U.K.: Academic Press.

Launder, B. E. and Spalding, D. B. 1974. The numerical computation of turbulent flows. *Comput. Methods Appl. Mech. Eng.* 3:269–289.

Li, Y. K., Mujumdar, A. S., and Douglas, W. J. M. 1978. Coupled heat and mass transfer under a laminar impinging jet. In *Proceedings of the First International Symposium on Drying*, Ed. A. S. Mumudar, pp. 175–184. Montreal, Quebec, Canada: McGill University.

Livesley, D. M., Oakley, D. E., Gillespie, R. F. et al. 1992. Development and validation of a computational model for spray-gas mixing in spray dryers. In *Drying's 92*, Ed. A. S. Mujumdar, pp. 407–416. New York: Hemisphere Publishing Corporation.

Masters, K. 1976. *Spray Drying Handbook*, 1st edn. London, U.K.: George Godwin Ltd.

Masters, K. 1991. *Spray Drying Handbook*, 5th edn. New York: John Wiley & Sons.

Masters, K. 2002. *Spray Drying in Practice*. Charlottenlund, Denmark: SprayDryConsult International ApS.

Mujumdar, A. S. 1995. Superheated steam drying. In *Handbook of Industrial Drying*, 2nd edn., Ed. A. S. Mujumdar, pp. 1071–1086. New York: Marcel Dekker.

Mujumdar, A. S. 2000. Dryers for particulate solids, slurries and sheet-form materials. In *Mujumdar's Practical Guide to Industrial Drying*, Ed. S. Devahastion, pp. 37–71. Brossard, Quebec, Canada: Exergex Corporation.

Negiz, A., Lagergren, E. S., and Cinar, A. 1995. Mathematical models of cocurrent spray drying. *Ind. Eng. Chem. Res.* 34:3289–3302.

Oakley, D. 1994. Scale-up of spray dryers with the aid of computational fluid dynamics. *Drying Technol.*, 12:217–233.

Oakley, D. E. and Bahu, R. E. 1990. Spray/gas mixing behavior within spray dryers. In *Seventh International Symposium Drying*, in *Drying'91*, Ed. A. S. Mujumdar and I. Filkova, pp. 303–313. Amsterdam, the Netherlands: Elsevier.

Papadakis, S. E. and King, C. J. 1988a. Air temperature and humidity profiles in spray drying, part 1: Features predicted by the particle source in cell model. *Ind. Eng. Chem. Res.* 27:2111–2116.

Papadakis, S. E. and King, C. J. 1988b. Air temperature and humidity profiles in spray drying, part 2: Experimental measurements. *Ind. Eng. Chem. Res.* 27:2116–2123.

Rajan, R. and Pandit, A. B. 2001. Correlations to predict droplet size in ultrasonic atomization. *Ultrasonics* 39:235–255.

Ranz, W. E. and Marshall, W. R. 1952a. Evaporation from drops. *Chem. Eng. Prog.* 48:141–146.

Ranz, W. E. and Marshall, W. R. 1952b. Evaporation from drops. *Chem. Eng. Prog.* 48:173–180, Rogers company 2008. http: www.cerogers.com/html/horizontal_dryer.html.

Straatsma, J., Houwelingen, G. V., Steenbergen, A. E. et al. 1999. Spray drying of food products: 1. Simulation model. *J. Food Eng.* 42:67–72.

Verdurmen, R. E. M., Menn, P., Ritzert, J. et al. 2004. Simulation of agglomeration in spray drying installations: The EDECAD project. *Drying Technol.* 22:1403–1462.

Westergaard, V. 1994. *Milk Powder Technology: Evaporation and Spray Drying*. Søborg, Denmark: Niro A/S.

Yakhot, V. and Orszag, S. A. 1986. Renormalization group analysis of turbulence: I. Basic theory. *J. Sci. Comput.* 1:1–51.

5 Encapsulation via Spray Chilling/Cooling/Congealing

Carmen Sílvia Favaro-Trindade, Paula Kiyomi Okuro, and Fernando Eustáquio de Matos Jr.

CONTENTS

5.1 Introduction 71
5.2 The Encapsulation Platform 72
5.3 Encapsulation by Spray Chilling 73
5.4 Encapsulation of Probiotics 73
5.5 Encapsulation of Drugs/Cosmetics 76
5.6 Encapsulation of Vitamins.. 76
5.7 Characterization and Evaluation of Microspheres Produced by Spray Chilling........... 77
5.8 Size and Morphology of the Particle 78
5.9 Scanning Electron Microscopy (SEM)... 80
5.10 Differential Scanning Calorimetry... 81
5.11 Fourier-Transformed Infrared Spectroscopy 82
5.12 X-Ray Powder Diffraction 82
5.13 Determination of the Active Ingredient Content 83
5.14 Release Studies 84
5.15 Conclusion 85
References.. 85

5.1 INTRODUCTION

The allocation of an active ingredient in a matrix that is able to protect it from the adverse conditions of the surroundings, as well as to modulate its release in the gastrointestinal system, on the skin or in other sites have been created huge interest in researchers, in the industry and in the academy. In this context, the encapsulation technology by spray chilling has become promising.

In the literature, there are three names for this technique: spray chilling, spray cooling, and spray congealing. Besides, there are few studies that also call this technique "prilling" (Oxley, 2012; Dubey and Windhab, 2013). However, independent of the name, this technology is based on the atomization of a mixture (containing an active agent and a molten carrier) in a chamber whose temperature is below the melting point of the carrier. Thus, the carrier solidifies, and a powder made up of microparticles is obtained. Spray chilling is very similar to spray drying, however, a melt is atomized and cooled down in an airstream or a cool environment. Unlike spray drying, at spray chilling there is no mass transfer.

Microparticles formed by spray chilling are matrices type, that is, structures where the protected material is disperse all over the volume of the particle. The material may even be on the particle surface, and, because of this structure, these particles are called microspheres (Figure 5.1). Spray chilling typically produces massive and high-density microparticles, with regular spherical shape.

The most recommended carriers for the development of particles by spray chilling in food industries are fatty acids, alcohols, triacylglycerols, and waxes (Okuro et al., 2013b). For the food

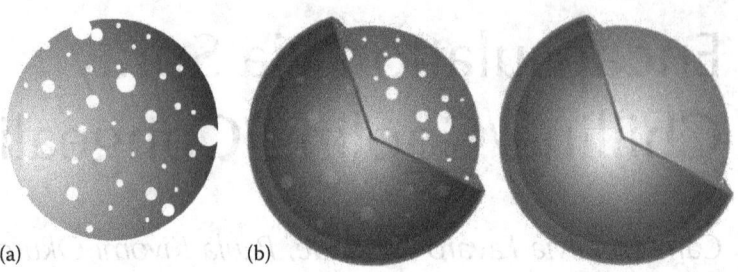

FIGURE 5.1 Structure of (a) a microsphere and (b) two microcapsules.

industry, the use of lipid carriers is interesting because these substances must be GRAS—*generally recognized as safe*—typically nonallergenic, and offer the possibility of being released in the gastrointestinal tract (lipid metabolism) after the carrier is digested. For chemical industries, some inorganic carriers may also be used. This encapsulation technique has received attention, especially in the pharmaceutical field, due to the possibility of using lipids as carriers in drug-delivery systems for protein and peptides. Besides these powdered lipid systems are biocompatible and can improve the bioavailability of active substances.

According to Cordeiro et al. (2013), compared with other "particle engineering" technologies, spray chilling is often faster because solvents are not used, and post-processing, such as coating and agglomeration, is not required typically. Besides, it is easy to scale the process up; it is a clean technology that does not demand the use of solvents and does not produce residues. Spray chilling enables rapid and continuous production of large quantities of product in a single-step operation, and the melts are much easier to handle—transportation, distribution, dosing, and storage—when available in powder form.

Nowadays, there are several publications in the specialized literature that describe the preparation of the particles by spray chilling for different uses. Among them, changing the dissolution profile of poor soluble drugs; extending the release of drugs with short half-lives; producing lipid carriers for topic use (cosmetology and dermopharmacy); masking odors and flavors; increasing the stability of photosensitive ingredients; and encapsulating bioactive proteins and peptides, adhesive materials sensitive to pressure, as well as probiotic microorganisms, vitamins, minerals, and flavors.

5.2 THE ENCAPSULATION PLATFORM

Spray chilling is a very simple technology that consists of first making a liquid feedstock and then atomizing it into a cooling chamber. The feedstock may consist of solutions, emulsions, or pumpable suspensions containing the active ingredient and the molten carrier. Generally, the chamber is maintained at low temperature with injections of cold air or liquid nitrogen. Under these conditions, the carrier solidifies instantly, forming matrix-type microspheres.

Spray chilling has a configuration similar to that of spray drying. However, instead of applying heat to promote solvent evaporation, heat is removed for the carrier solidification, so there is no mass transfer. The apparatus consists of two parts: a cooling chamber and an atomizer—a rotary wheel, a high pressure nozzle, a fountain nozzle, a vibrating nozzle, a double fluid, or a spinning disc atomizer.

According to Cordeiro et al. (2013), there are three rules to define the melting conditions for encapsulation by spray chilling:

1. The temperature of the melt should be 10°C above the melting temperature of the carrier(s).
2. It is recommended that feed viscosity is lower than 500 mPas. Higher viscosities may be possible, but only with appropriate nozzle design.
3. Drug loads for microencapsulation are usually limited to 50%.

The critical steps of the process are atomization of the melt and its solidification. The first one refers to the disintegration of the liquid melt into a spray of fine spherical droplets, whereas solidification is a coupled process of transformation of the melt into solid by cooling.

From an operational approach, insufficient cooling of the droplets can lead to their clumping and/or adhesion in the surface of the cooling chamber, an event that affects the particle morphology and impacts on the polymorphic behavior of the carrier and/or active ingredient, the process itself, and other properties of the microparticles.

For the atomization of the melt, the rotary and double-fluid nozzles are most commonly used. The use of different kinds of devices is associated with the viscosity of the molten mixture, the compatibility with the equipment configuration (size of the cooling chamber), and impacts in the size distribution and mean size of the microparticles produced and in the encapsulation efficiency (Cordeiro et al., 2013).

Albertini et al. (2008) described the difficulty in the atomization of mixtures containing drugs in a proportion greater than 30% (weight/weight). However, they reported the possibility of exceeding the feeding threshold in up to 50% (weight/weight) by changing the design of the atomizer.

Process conditions should be adjusted in order to avoid the particle disintegration and to prevent its adherence in the chamber. The time required to promoted the droplet solidification depends on the speed assumed by the droplet, the particle size and particle size distribution, specific heat of the material, temperature and speed of the cooling air (Cordeiro et al., 2013).

5.3 ENCAPSULATION BY SPRAY CHILLING

Countless uses have been studied for solid-lipid particles, such as masking odors and flavors (Akiyama et al., 1993; Yajima et al., 1999); protection of the filling material from adverse conditions, such as pH and enzyme action, moisture, oxygen, and light; optimization of the dissolution of poorly soluble drugs; modulation of the compound release; improvement of the flow properties; handling; appearance; and others (Ilić et al., 2009). With appropriate selection of polymers, microspheres produced by spray chilling can provide diffusion-controlled or erosion-controlled drug release, and can mask unpleasant and bitter taste of drugs, or can be used to modify their dissolution behavior (Cordeiro et al., 2013).

Table 5.1 shows some studies based on the use of spray chilling technique with several bioactive components.

Some uses of spray chilling technology to microencapsulate probiotics, drugs, some cosmetics, and vitamins will be described with details in the next topics.

5.4 ENCAPSULATION OF PROBIOTICS

Probiotics have been considered bioactive living cells, that is, microorganisms that can confer human health benefits and well-being when they are consumed alive and frequently. However, these microorganisms are very sensitive to many factors, because of that, they are inherently unstable. Therefore, microencapsulation has been used as an alternative to protect these microorganisms and to release them in their site of action (Fávaro-Trindade et al., 2011).

During the gap between processing and consuming, and from intake to reaching the site of action, these microorganisms have to be protected from variables such as high temperatures during processing; dissection if applied to a dry product; storage conditions; and interaction with the food matrix, packaging, and the environment (temperature, moisture, and presence of oxygen). Besides, after intake, probiotics must to survive the gastrointestinal tract, especially the acid environment in the stomach, the presence of bile salts in the small intestines (which have detergent properties that dissolve the microbial membrane), as well as the deleterious action of pancreatic enzymes (Manojlovic et al., 2010; Madureira et al., 2011; Senaka Ranadheera et al., 2012).

The ileum, a site that is responsible for a large part of the immune body response and where most of the Peyer patches are found in the mucous membrane, is considered an important site of probiotic

TABLE 5.1
Examples of the Use of Spray Chilling for the Microencapsulation of Different Compounds

Encapsulated Material	Carrier	Reference
Isoflavone	Medium chain triacylglyceride	Jeon et al. (2005)
Citric acid	Fats with different melting points	Nori (1996)
Clarithromycin	Glyceryl monostearate (GM) and aminoalkyl methacrylate copolymer E	Yajima et al. (1999)
Theophylline	Precirol(R) ATO 5	Albertini et al. (2004)
Diclofenac	Gelucire 50/13	Cavallari et al. (2005)
Iron, iodine, vitamin A	Fully hydrogenated palm oil	Wegmuller et al. (2006)
Propafenone hydrochloride and tocopheryl acetate	Lipophilic (carnauba wax, cetearyl, and stearyl alcohols) and hydrophilic (PEG 4000)	Albertini et al. (2008)
Glucose solution	Fatty acids (stearic and oleic) and vegetable hydrogenated fat	Leonel (2008)
Hydrophilic compounds	Lipid mixtures	Chambi et al. (2008)
Glucose solution	Combinations of: (1) stearic and oleic fatty acids, (2) fully hydrogenated soybean oil and oleic fatty acid, and (3) cetearyl alcohol and oleic fatty acid in different proportions	Ribeiro et al. (2010)
Liposoluble vitamins	Hydrogenated vegetable oils, such as almond, palm, cottonseed, and sesame seed oil, carnauba wax, and bee wax.	Zoet et al. (2011)
Bifidobacterium lactis and *Lactobacillus acidophilus*	Fat obtained by interesterification of fully hydrogenated palm and palm-kernel oil	Pedroso et al. (2012)
Lactobacillus acidophilus, polydextrose and inulin	Fat obtained by interesterification of fully hydrogenated palm and palm-kernel oil	Okuro et al. (2013a)
Bifidobacterium lactis and *Lactobacillus acidophilus*	Cocoa butter	Pedroso et al. (2013)

immunoregulatory activity (Kleerebezem and Vaughan, 2009). Therefore, the enteric delivery of these microorganisms is very important, as well as the development of a platform that enables the modulation of probiotic release in this site.

The purpose and rationale of probiotic microencapsulation is to confine these bacteria to an impermeable or semipermeable matrix that is able to protect them against external conditions and to form an internal microenvironment appropriate to their survival, with adequate communication between the external and internal environment. Thus, the synchrony associated with the fact that lipid digestion (the wall of the capsule) may effectively occur in the intestines, a site where probiotic may act, shows the potential use of the spray chilling technique.

In the microencapsulation of probiotics, matrices used in the formula cannot be cytotoxic or antimicrobial. When microparticles go through the stomach, bacteria have to continue retained and protected from the external acid environment, and they should be released only in the intestines, where immunological signaling will occur (Cook et al., 2012). Little is known about the effect of microencapsulation with lipid matrices on probiotic survival. However, it is known than fat/oils are barriers against oxygen and moisture.

Lahtinen et al. (2007) studied the immobilization of bifidobacteria (*B. longum*) in a lipid matrix that consisted of cocoa butter, in models that simulated fermented and nonfermented drinks. The study showed that the bacteria remained viable during storage, suggesting a positive result, as the lipid matrix probably protected the microorganism against water and H^+ ions.

Another study produced solid lipid microparticles with two probiotic strains by spray chilling: *Lactobacillus rhamnosus* IMC 501® and *Lactobacillus paracasei* IMC 502®, prebiotic components, and natural bioactive ingredients. Different types of vegetable fats were tested. It was observed that,

Encapsulation via Spray Chilling/Cooling/Congealing

for some fats, viable cells were preserved for at least 6 months of storage, showing the potential and applicability of the technique in probiotic microencapsulation (Cecchini et al., 2010).

Therefore, the possibility of encapsulating probiotics with lipid matrices is an interesting proposal, once lipids are digested by intestinal lipases, enabling the release of the microorganisms in the site where they are supposed to act, which is essential ensure their health benefits (Nori et al., 2009; Fávaro-Trindade et al., 2011; Pedroso et al. 2012, 2013; Okuro et al. 2013b), as shown in Figure 5.2.

Advantages such as the low toxicity level, biocompatibility, possibility of using GRAS formulations, easy processing, and low cost are issues that show the potential of the technique for the production of lipid microparticles.

Pedroso et al. (2012) produced solid lipid microparticles containing *B. animalis* subsp. *lactis* and *L. acidophilus* by spray chilling using an interesterified fat produced with fully hydrogenated palm oil and kernel oil. They observed that bacteria were highly resistant to the process. After analyzing the microparticles, they observed that encapsulated *L. acidophilus* were more resistant to simulated gastric and intestinal fluids when compared with free microorganisms, and that they were viable for 30–60 days of storage at 37°C and 7°C, respectively.

In line with the report above cited, another study produced symbiotic microparticles using as active materials two strains of probiotic microorganisms (*L. acidophilus*—LA and *L. rhamnosus*—LR), two prebiotics (inulin and polydextrose) and, as the carrier matrix a fat obtained by interesterification of fully hydrogenated palm oil and kernel oil. Bacteria were highly resistant to the process, as assessed by the little loss of viable cells during microparticle production. There was no interference due to the absence, presence, or type of prebiotic, either. Spherical microparticles with relatively uniform surface were obtained. Mean size was 62.4 ± 2.8 to 69.6 ± 5.1 μm. Microencapsulation favored survival in simulated gastric and intestinal fluids and enabled the maintenance of viable cells over 10^6 CFU/g up to 120 days of storage at controlled relative humidity for the formulation with *L. acidophilus* and polydextrose. The formula was influenced by the water activity of the particle and, in its turn, was affected by the incorporation of prebiotics (Okuro et al., 2013a).

There is a wide range of possible research areas on probiotic microencapsulation, especially by spray chilling. There still is no definitive report on the potential vs. performance for the binomium probiotic/spray chilling. From the studies carried out until today, promising results have been

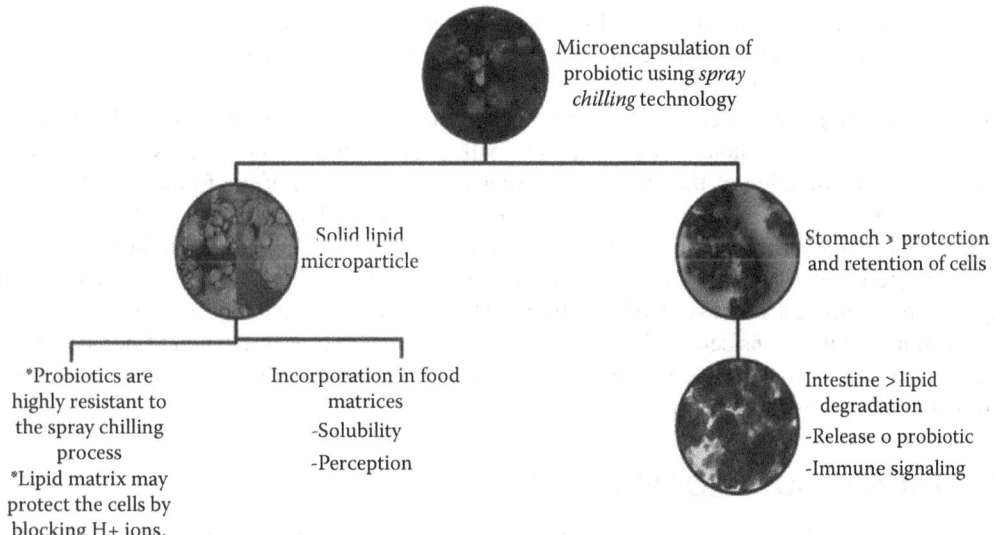

FIGURE 5.2 Microencapsulation of probiotics by spray chilling.

presented, and the use of lipid matrices for enteric delivery is important in terms of the benefit to the health of the consumer, because these microorganism are able to reach the site where they are supposed to act. In fact, it has been shown that the particles protect the viability of probiotic strains and their passage in simulated gastric and intestinal fluids. The study of different types of carriers with different melting points; new applications for lipid particles; the study of the interaction between the probiotic and the matrix, and the interaction with other ingredients, such as prebiotic components; the validation of methodologies to quantify viable cells in the particle system; and the conditions of the process to adjust the size and morphological characteristics of the particles are all perspectives in the study of microencapsulation of probiotic in lipid matrices.

5.5 ENCAPSULATION OF DRUGS/COSMETICS

The pharmaceutical area has the greatest number of studies involving spray chilling. Whereas in the food sector, microencapsulation basically focuses on improving the stability of the encapsulated material, in the pharmaceutical sector, drug encapsulation has different aims, such as improving dissolution profiles (Passerini et al., 2002; Qi et al., 2009; Dixit et al., 2010), controlling release (Rodriguez et al., 1999; Savolainen et al., 2002), masking flavors (Yajima et al., 1996), and maintaining the stability of the drug (Albertini et al., 2009).

The pharmaceutical industry is very interested in this technique because it does not use high temperatures, enabling the use of thermolabile drugs in the process. Besides, the absence of the use of organic solvents is very appropriate in a sector that has to comply with rigorous regulations on toxicity and safety.

Changes of the drug dissolution profile intends to improve its oral bioavailability. Studies involving spray chilling with different drugs have been successful. The microencapsulation of sodium diclofenac with gelucire 50/13, by Cavallari et al. (2007), showed an increase of 70% in the dissolution rate of the drug compared with its pure form. Positive results were also reported by Passerini et al. (2012), that microencapsulated carbamazepine and determined the dissolution profile of the microencapsulated form that was considerably greater than the free form.

Similar to the improved dissolution profile, careful attention should be given to the carrier used in the microencapsulation process to achieve extended release of a drug. Whereas hydrophilic carriers, especially low melting point polymers, are used for better solubility of the drug, hydrophobic carriers are used for extended release. Drugs for extended release chosen for microencapsulation by spray chilling are those with short half-lives, such as theophylline (Rodriguez et al., 1999; Albertini et al., 2004), fenbufen (Rodriguez et al., 1999), and felodipine (Savolainen et al., 2002).

Masking of flavors, one of the most common uses of microencapsulation, has already been tested with spray chilling. Yajima et al. (1996) microencapsulated clarithromycin, a highly bitter macrolide antibiotic. Glyceryl monostearate and aminoalkyl methacrilate copolymer were used as carriers. Microencapsulation of the antibiotic prevented the dissolution of the formula in the mouth, and it was immediately released in the gastrointestinal tract. Besides masking the bitter taste, the authors reported that the technique greatly improved the bioavailability of the antibiotic.

The objective of Albertini et al. (2009) was to improve the stability of an active sunscreen principle by spray chilling. These researchers observed that avobenzone microencapsulated in carnauba wax, had its stability considerably increased during simulated photodegradation. After 1 h of irradiation in a sun simulator, degradation was around 38% for the free form, and 15% for the spray chilling microencapsulated form.

5.6 ENCAPSULATION OF VITAMINS

Vitamins have been added to food products for a long time as a strategy to improve dietary intake of these micronutrients and to prevent nutritional deficiencies. Generally, supplementation is performed in foods that are easily available and in those that are habitually used by the population that that will be favored. With the progress in studies involving the action of vitamins in the organism, the addition of these organic compounds in foods started to be perceived at different way. Today, vitamins are not only added to comply with nutrient requirements, but also to act as functional ingredients capable of providing additional benefits to the organism besides their basic nutritional function. In this context, the interest in adding these micronutrients in different food matrices has been growing. However, the high sensitivity of these compounds to environmental factors, such as pH, presence of oxygen, temperature, and other specific compounds, is a challenge for the food industry. Besides having their functionality compromised, these vitamins may react with other food ingredients and change the flavor and color, for example, negatively affecting the sensory quality of the product. One alternative to overcome these problems is the use of microencapsulation.

There are several studies on vitamin encapsulation (Gonnet et al., 2010; Wang et al. 2011; Ziani et al., 2012; Comunian et al., 2013). More recently, some successful studies involving spray chilling have been reported. Spray chilling is interesting for the microencapsulation of vitamins, especially due to its advantages in relation to other techniques. In spray chilling high temperature is not required, as well as the use organic solvents, and the carriers are almost always lipids.

Due to its antioxidant properties, vitamin E has been extensively investigated as a bioactive compound, mainly to be used in food products. However, its instability when facing atmospheric oxygen, intensified by the presence of light and some metals, difficult vitamin E addition in foods as a bioactive compound. The first report on vitamin E microencapsulation using spray chilling was by Albertini et al. (2008). The vitamin in this study was used as a model of liquid ingredient for the production of lipid microparticles in a spray chiller that generated lipid microparticles with high concentrations of active material. The author used stearyl alcohol, carnauba wax, and low melting point polymers as carrier materials and tocopheryl acetate as the active ingredient in the microencapsulation process. The concentration of the active ingredient was 30% (w/w). Only the characteristics of the process and the microparticles were evaluated. The results were considered promising, owing to high yield of the particle, which was around 95%. The dissolution profile was satisfactory.

Gamboa et al. (2011) also microencapsulated vitamin E using spray chilling. However, their aim was to microencapsulate tocopherol using chemically interesterified low trans fat. Besides characterizing the microparticles, the authors used the tocopherol-loaded microparticles in a food matrix and analyzed the sensory acceptance of the product. It was observed that microparticles obtained by spray chilling presented satisfactory release profile. The use of microparticles loaded with tocopherol did not compromise the sensory characteristics of the product. In both studies (Albertini et al., 2008; Gamboa et al., 2011), there was no description of the oxidative stability of vitamin E after the encapsulation.

The techniques to evaluating the microparticles produced by spray chilling technology will be discussed in the following topics.

5.7 CHARACTERIZATION AND EVALUATION OF MICROSPHERES PRODUCED BY SPRAY CHILLING

One possible disadvantage of spray chilling is that the process makes the molten mixture solidify quickly, in which the crystalline structure of the carrier and/or bioactive ingredient can be change during process and storage, regarding drugs in the solid state (crystallinity, amorphous form) and the type of crystalline structure formed by the matrix. Based on this perspective, some analyses may be carried out, such as differential scanning calorimetry (DSC), hot stage microscopy (HTM), and x-ray diffraction (XRD) to detect possible modifications in the physical-chemical characteristics of the active compound and/or carrier, and the possible interaction between both of them (Passerini et al., 2003).

In the production of particles by spray chilling, it is important to study the system obtained both for the characterization of the solid particle and for the evaluation of the solid state of the

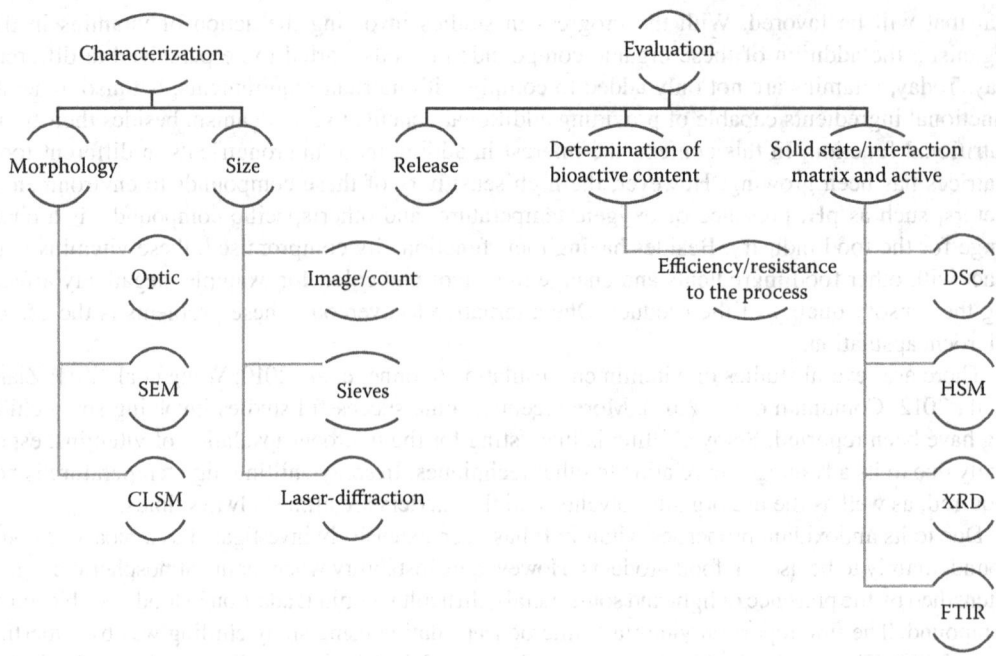

FIGURE 5.3 Analyses used in the characterization and evaluation of particles obtained by spray chilling.

active ingredient, and possible interactions between them. These factors may have an impact on the release profile of bioactive ingredients, and on the particle features. Figure 5.3 illustrates some of the analyses commonly performed in the evaluation and characterization of the particles obtained by spray chilling.

Calorimetric analysis by DSC provides information on the melting temperature, crystallization, enthalpy, polymorphic transition, and degree of crystallinity of the particles. XRD, on the other hand, also evaluates the crystallinity and polymorphic transition events that may eventually occur in microparticles.

In XRD, electrons are the main cause of dispersion. Molecular structures are analyzed mainly in samples in which the presence of crystals is recognized.

XRD is also used for polymorphic reorganization detection. As for lipid microparticles, polymorphic reorganization of the lipid to more thermodynamically organized and more energetically favorable levels is not desirable, once this event may lead to the expulsion of the incorporated active ingredient.

Some of these analyses used in the evaluation and characterization of the particles produced by spray chilling are presented below.

5.8 SIZE AND MORPHOLOGY OF THE PARTICLE

The characterization of the particles in terms of size and shape is essential and has a direct impact on the behavior of this particulate system. These data may be correlated with encapsulation efficiency, the form of release of the active ingredient from the matrix, and the efficiency in protecting the bioactive.

Different variables may have an impact on mean size and size distribution of the particles. *A priori,* the parameters used in the process, such as temperature and speed of the cooling air, atomization pressure, flow of the feeding mixture in the atomizer, and type and diameter of the atomizing nozzle directly influence the size of the particles. Lipid composition of the carrier matrix (which affects its viscosity), the presence and type of the surfactant in the mixture of active principle and molten matrix, the composition of the active ingredient, the proportion of the feeding mixture (bioactive/matrix), among other issues, are also important.

The microparticles size distribution, and mean size are important characteristic that may be determined according to the use of the microparticle, such as the time of release of the active ingredient. For example, smaller particles will release the bioactive compound more quickly than larger ones. Another example involves perception, in the case of incorporation in foods: a wide range in particle size may lead to different intervals in the release of the active ingredient and, therefore, an extended release, which can be useful for the application of sweetener encapsulated, for instance. On the other hand, large particles can negatively affect the texture of food products. Microparticles obtained by spray chilling are characterized by its spherical shape, relatively smooth surface, with some irregularities. Because of that, they provide fluidity to the formulation.

Some studies on spray chilling have been carried out in order to assess the influence of the parameters of the process on the particle size. Maschke et al. (2007) evaluated the influence of the atomization pressure and temperature parameters on the particle size and yield of the process. They observed that increasing the pressure from 5 to 6 bar, and temperature from 70°C to 80°C led to a reduction in the mean size of the microparticles from 315 to 182 μm, and that the increase in temperature led to a reduction in the viscosity of the carrier agent. This result was explained by the fact that the viscosity of the feeding mixture was reduced with the increase in the temperature, leading to a reduction in the microparticle size and greater yield of the process due to the lower rate of droplet collision against the cooling chamber.

Another study analyzed atomization pressure (1090, 1100, 1150, 1315, and 1590 bar) and the flow of the feeding mixture (6.0, 11.0, 16.0, and 23.4 mL/min), and observed that the particle size ranged from 58 to 278 μm, with total yield of the process ranging from 81% to 96%. Size was reduced with the increase of the atomization pressure. The greatest atomization pressures were associated with the greatest speeds of air flow from the atomizing nozzle, increasing the kinetic energy conferred to the liquid at the exit of the atomizer, increasing shear tension, and therefore, generating smaller particles.

The particle size was directly associated with the yield of the process. For a given feed flow and composition of the feeding mixture, the solidification capacity was fixed. The amount of energy necessary to be removed from the molten droplet is equal to the latent heat of solidification, and depends only on the feeding flow of the mixture. As smaller droplets show a larger area of contact with the cooled air of the cooling chamber, the heat transfer necessary to solidify them is theoretically lower (solidifies quickly), whereas larger droplets promote heat transfer more slowly, and then permanence in the chamber is extended. The increase in the feeding rate led to an increase in the size of the particle, and greater atomization pressure led to a reduction in the size of the particle. The dominant parameter, considering the impact on the particle size, was the atomization pressure (Ilić et al., 2009).

The analysis of the size distribution and particle size may be carried out by laser diffraction. This method evaluates the size distribution by measuring the angular variation in the intensity of diffused light when a laser beam interacts with the particles dispersed in the sample. The behavior in response to the intensity of the angular dispersion creates a dispersion pattern, so lipid particles must be suspended in the cell for analysis (Malvern, 2013). The solution used to suspend the microparticles must ensure that the particles remain dispersed and prevent clumping during measurement. Researchers have been using ethanol-water mixtures (64% v/v), aqueous solutions of surfactants, silicon oil, Triton X-100 solution in 1% deionized water (weight/volume), ethanol 96% and isobutanol. Besides, the sample is also sonicated (Al-Kassas et al., 2009; Passerini et al., 2012).

The distribution of size of the particles may also be analyzed with the use of vibratory sieves, with the sieve patterns determined according to the size range of the sample, which will go through progressively smaller meshes (Albertini et al., 2009). Ribeiro et al. (2012) suspended the samples in glycerol and manually measured the diameter of 500 particles in each sample. Two measurements were carried out (horizontal and vertical), and the mean was considered.

5.9 SCANNING ELECTRON MICROSCOPY (SEM)

The particle morphology is normally assessed by scanning electron microscopy (SEM). The sample preparation depends, in the first instance, on the type of equipment and normally requires the use of aluminum stubs with a carbon conductive tape and coating with gold–palladium layer. Sometimes the particles are placed on a double-sided carbon tape that is attached to aluminum stubs, without coating requirement. The analyses are carried out by applying to sample a difference of potential between 2 and 20 kV.

The SEM produces images due to the interaction of the electron beam and the sample, and the signals produced give information about sample. The interaction between the electron beam and the sample may produce heat, which can cause a negative effect in the case of solid-lipid microparticles. The sample can melt depending on the carrier matrix used, and this would compromise the information on the structure due to the transfer of energy from the electron beam to the sample.

In the case of SEM performed in organic samples, the most critical problems are probably heating, generation of superficial charges, and radiolysis (degradation of the compounds). The use of lower energy electrons reduces heating and the production of superficial charges. In this context, good preparation of the sample can prevents the damage if the contacts are efficient in dissipating the charges and the heat. On the other hand, radiolysis tends to increase when lower energy electrons are used, because it depends on inelastic shock section. It is also located on surface due to the lower interaction.

Figure 5.4 shows SEM images of particles obtained by spray chilling using different active ingredients and carriers. In the literature, it is recurrent to find that this technique produces

FIGURE 5.4 Solid-lipid microparticles (SLM) produced by spray chilling. (a) SLM loaded with lycopene, ×1.0 K, (b) SLM loaded with soybean protein hydrolysate, ×500, and (c) SLM loaded with *L. acidophilus* and polydextrose, 500×.

spherical microparticles of relatively smooth surfaces, which are shown in images obtained by SEM (Figure 5.4).

5.10 DIFFERENTIAL SCANNING CALORIMETRY

DSC and Fourier-transform infrared spectrophotometry (FTIR) are used to detect possible changes in the physical-chemical properties of the active ingredient and/or carrier, and possible interactions between the components used in the formulations, respectively.

The evaluation of the crystalline/amorphous nature of the core can be carried out by DSC, and the degree of impurities in the core at different time points can be evaluated by High Performance Liquid Chromatography (Cordeiro et al., 2013).

DSC measurement is based on the heat loss and absorption caused by changes in the sample as a function of the temperature programmed dynamics. The reference cell is an empty aluminum capsule (capsule without the sample). Therefore, when the sample undergoes any change, there is a difference between the temperature of the sample and that of the reference, and this difference is quantified and converted in thermograms. The response to the analysis is given by the flow of heat as a function of the temperature (Bunjes and Unruh, 2007).

Considering the peculiarities of the carriers and of the atomization/solidification process in spray chilling, the quick solidification of the droplets when they get in contact with the refrigerated environment may lead to changes in the original crystalline form, and in original polymorphic shape. Lipid carriers may be crystallized in three different polymorphic shapes, called α (hexagonal shape), β' (ortorhombic shape), and β (triclinic shape). The α shape is less thermodynamically stable and has the lower melting point. The β shape is more stable and has the higher melting point (Windbergs et al., 2009). Thermodynamic stability and melting point related to the type of crystal decrease in the following order: β, β', α (Mcclements and Decker, 2010). The lipid matrix may crystallize in different polymorphic shapes due to the different cooling rate.

The rationale of the analysis is based on the study of the raw materials individually. For example, the active ingredients and the carrier matrix separately, and in some cases, the surfactant agent and the physical mixture of the formulations before they are atomized. Then, the microparticles formed after the spray chilling process. Sometimes, the analysis of the same microparticles after a given storage time under controlled temperature and humidity is also carried out. The comparison of these thermograms may provide inferences on the condition of the active ingredient, the interaction between the components in the microparticles, the interaction between the bioactive compound and the carrier agent.

Passerini et al. (2003) used DSC to investigate the solid state of the drug and the carrier in particulate systems. Using the calorimetric analysis, they could draw conclusions on the racemic form of the drug due to an endothermic peak at 144.9°C. They studied the raw materials, individually, and the carrier agents microcrystalline wax and stearyl alcohol, and the surfactant, soy lecithin. Stearyl alcohol showed an endothermic peak and the curve at a lower temperature, indicating the possible presence of more than one crystalline state. They observed that soy lecithin did not interact with the other components (microcrystalline wax and the drug) given the similarity of the thermograms. Samples were differentiated only by the presence or absence of the surfactant. The thermogram for the formulations that used microcrystalline wax and stearyl alcohol as carriers showed to be very similar to each other, with an endothermic peak at 48°C corresponding to the fusion of carrier agent, and a little peak at 133°C, caused by the fusion of the drug. These results may be explained by the partial transformation of the drug into its amorphous state inside the microparticles. However, once the carrier melted at a temperature lower than that of the drug, the strong attenuation of the peak of the drug in the DSC curves for the microparticles was associated with a partial dissolution of the drug in the molten lipid carrier during the DSC analysis.

In terms of drug microencapsulation, DSC results alone are not enough to determine the physical state of the bioactive compound in the microparticles. Although the analysis may identify and quantify thermal events, it does not show the cause of the event. Then, in order to investigate these aspects in greater detail, HTM, and XRD are recommended (Passerini et al., 2003).

5.11 FOURIER-TRANSFORMED INFRARED SPECTROSCOPY

The molecules have vibrational and rotational energy due to the chemical bonds. These vibrations are associated with the infrared region (Skoog et al., 1992). Therefore, spectrophotometry in the infrared region determines the functional groups of a sample, and each group absorbs light in a characteristic frequency, according to the levels of energy of the molecule, which is given by its vibrational energy.

Vibrations of angular deformation, which correspond to specific energy levels of the molecules, are characterized by the variation in the angle between two bonds, and may be of four different types: scissoring or bending, rocking, wagging, and twisting. It is known that infrared absorption is different in the crystalline and amorphous states due to the presence of specific conformations in each one. Therefore, the analysis may differentiate these distinct states of the material.

The use of Fourier-transformed infrared spectroscopy (FTIR) in particles produced by spray chilling mainly predicts the detection of possible interactions between the active ingredient and the lipid carrier. Normally, the analysis individually studies the components (matrix and active compound, and other components, when available) that will form the capsules. Specific absorption bands will be observed in determined regions or in a specific wavelength. After that, the analysis of the microparticles that are already formed is carried out. The discussion involves changes in the absorption bands of ingredients and formed microparticles, with the comparison of the peaks observed in both cases.

The infrared region of the electromagnetic spectrum ranges from 1400 to 50 cm^{-1}, and this region is subdivided into three specific areas: proximal, medium, and thermal/distant. The medium infrared region ranges from 4000 to 400 cm^{-1}, and it is the most interesting region of the spectrum for the study of organic compounds, once the absorption bands are associated with the vibration of specific functional groups (Guillen and Cabo, 1997).

Considering that the most of the particles produced by spray chilling are fat-based structures, it is very important to understand the evaluation of fatty acids and triglycerols by FTIR.

Fatty acid chains are responsible for the absorption observed between 3000 and 2800 cm^{-1}, with ester bonds of triacylglycerol C–O (~1175 cm^{-1}), group C = O (~1750 cm^{-1}), and acyl chain C–H (3000–2800 cm^{-1}).

Albertini et al. (2010) encapsulated atenolol in a lipid excipient and characterized SLMs obtained by DSC, x-ray powder diffraction (XRPD), and FTIR. No modifications in the solid state or chemical interactions were detected in the microspheres after processing and storage.

Passerini et al. (2006) encapsulated praziquantel by spray congealing using an ultrasonic device and Gelucire 50/13 as carrier. The analysis of the spectra showed that those corresponding to physical mixtures and microparticles could be superposed, that is, there were no changes in the characteristic peaks of the drugs and carriers, suggesting that no significant changes or compounds interactions occurred.

5.12 X-RAY POWDER DIFFRACTION

The technique is responsible for microstructural characterization of crystalline materials. Simplistically, the technique is based on the fact that x-ray incidence on the crystal causes part of the energy to be absorbed and part to be reflected in different directions. Therefore, after the diffraction of the beam, the x-ray detector decodes this information, producing a diffractogram or diffraction profile. This response correlates the intensity of the radiation measured with the mirror angle (the incident beam and the reflected beam).

Ilic et al. (2009) studied the diffraction patterns for a pure bioactive compound (glimepiride), for the particles produced and carried with the active ingredient in all carriers analyzed (PEG6000, poloxamer 188, Gelucire R 50/13), and for the physical mixtures corresponding to each formulation of the microparticles produced by spray chilling. The pure drug was in the microcrystalline form, as demonstrated by the clear and intense diffraction peaks. The diffraction profile of the microparticles carried with the drug showed typical signals of the carriers and was according to the literature. The presence of low-intensity peaks in the diffractograms, both in the microparticles and in the physical mixture, was attributed to the drug, and the low intensity was proportional to the low concentration of glimepiride in the samples analyzed. Last, the analysis of the diffractograms of the microparticles with the respective physical mixtures of the components showed that they presented, similarly, patterns with low-intensity peaks, and that diffraction angles of the microparticles were substantially identical to those of the physical mixtures. The authors concluded that the crystallinity of the drug was not essentially reduced by the technological process used in the production of the microparticles.

In this analysis, considering that lipid particles will be evaluated, both the issue of crystallinity and storage should be explored. For example, studying the diffraction of recently prepared samples and those stored for a given period under controlled conditions. Diffractograms that are superposable, no matter the formulation or storage period, indicate that no recrystallization with change in the polymorphic shape occurred throughout storage. These changes could lead to the expulsion of the active ingredient. For lipid matrix it should be considered that it has a tendency for polymorphic reorganization to more energetically favorable levels, and this behavior may lead to an expulsion of the encapsulated material, once there would be in another crystalline arrangement (Jenning et al., 2000; Müller et al., 2002a,b; Gamboa et al., 2011).

Gamboa et al. (2011) produced solid lipid microparticles with tocopherol by spray chilling, and found main peaks corresponding to the angles 19.3°, 22.8°, and 23.1°. These peaks were associated to polymorphic shape β. Similarly, other studies carried out XRD analysis in lipid samples and observed the presence of a main peak around 19.5°. This peak was associated to crystalline form β, or the triclinic form, which is the most stable one and, therefore, has the higher melting point (Keller et al., 1996; Schenk et al., 2006).

A study conducted by Okuro et al. (2013a) used the technique to assess the behavior of lipid particles during storage. It was observed that, even in the absence of polymorphic changes during storage, diffractograms obtained in recently prepared samples and in those stored for 90 days were very similar. The same behavior during storage of lipid microparticles was observed by Gamboa et al. (2011), who studied tocopherol encapsulation in lipid matrices obtained by spray chilling. These authors observed similar diffraction patterns and did not find significant differences in the storage period of particles after 180 days of storage. Diffractograms were very similar between the tests, besides showing low percentage of crystallinity, possibly due to the presence of amorphous solids.

5.13 DETERMINATION OF THE ACTIVE INGREDIENT CONTENT

It is very important to establish an efficient and adequate protocol to determine the active ingredient loaded in the volume of the microparticle. As discussed before, the microparticles are dense and filled up by the matrix. The active ingredient is homogenously distributed all over the volume of the particle, including on the surface area.

Some studies in the pharmaceutical area used a common protocol, adapting it to the specific properties of the drug (solubility and wavelength). A certain quantity of the microparticles is suspended in a given volume of solution (distilled water, buffer pH 1.2, buffer pH 7.4, phosphate buffer pH 7.2, simulated vaginal fluid). The sample is then heated (in some case, 2°C above the melting point of the carrier) under stirring, for the carrier to melt (60°C; 65°C). After that, the sample is stirred for a given period (24 h; 1 h) at room temperature. In the end, it is filtered, and the content of the drug is determined by spectrophotometry (visible UV) (Passerini et al., 2002, 2003; Albertini et al., 2004, 2009; Di Sabatino et al., 2012).

In this same context, some studies that used solvents (methanol, acetonitrile, and chloroform) to determine the content of drug in the microparticle are described later.

Ilic et al. (2009) used a mixture of acetonitrile and water (80:20, v/v) as the solvent for the extraction of the drug from the microparticles. Then, the drug was quantified by high efficiency liquid chromatography—HPLC. In a study of insulin microencapsulation, the authors proposed that microparticles were initially stirred in a vortex with chloroform, and then insulin was extracted with HCl 0.01 M. After that, the solution was centrifuged and, in the end, insulin was quantified by HPLC (Maschke et al., 2007).

In the case of probiotic microencapsulation, particles were broken down in serial dilutions with warm dilutant (peptone water, sodium citrate solution, and sodium chloride solution), for the quantification of viable cells in the system. For this specific active ingredient, as they are live microorganisms, heating cannot be too high so as to damage or kill the bacteria. Therefore, it is necessary to choose a matrix that does not have a high melting point.

A characteristic of the particles produced by spray chilling is that the active ingredient may be allocated on the surface of the microparticle, besides being distributed inside its volume. Given this profile, some studies chose to quantify the active ingredient on the surface, although it is difficult to state that only the superficial material will be quantified. Ribeiro et al. (2012) quantified the load of superficial glucose of lipid particles produced by spray chilling. The particles were placed in a 0.1% Tween 80 solution and stirred for 5 min in shaker. Then, the content was filtered in Whatman qualitative paper, and glucose was determined in the filtrate using a specialized kit.

5.14 RELEASE STUDIES

Release studies are extremely important due to the possibility of modulating the releases of the active ingredient from microparticles produced by spray chilling. The type and geometry of the particles and, especially the encapsulating agent define the mechanism of release.

Release from solid-lipid microparticles may be caused by biodegradation or activated by temperature. In biodegradation, lipids are broken down by lipases during digestion. In relation to temperature activation, the bioactive is released in response to the increase in temperature. The carrier is melted and the material is released from the structure. In this context, there are two distinct concepts: sensitivity to temperature, related to material that shrink or expand when a critical temperature is reached; and activation by fusion, related to the melting of the material that forms the walls in response to a rise in temperature, as it is the case of walls made up of a modified lipid or wax.

Zaky et al. (2010) encapsulated a protein (BSA-albumin) using hydrogenated palm oil (S-154, Softisan® 154) by spray chilling and studied the release of the protein. Release studies were conducted by incubating the samples in phosphate buffer pH 7.4 supplemented with 0.02% sodium azide and kept at 37°C in a vibration water bath for up to 60 days. They observed that the size and morphology of the particles presented a significant effect on the kinetics of protein release. They proposed a mechanism for the release of the active ingredient loaded at lipid microparticles, initially suggesting that water entered into it and dissolved the incorporated protein, leading to the formation of pores that were filled with water, and enabling the diffusion of the protein to the outside of the matrix. The entry of the water and the release of the protein were strongly correlated.

In vitro studies of dissolution involving lipid particles are common. Normally, a given amount of particles are suspended in a buffer (borate pH 5.0; phosphate pH 6.8; and phosphate pH 7.4) added of sodium azide, and the mixture is incubated at 37°C for many days (Maschke et al., 2007; Al-Kassas et al., 2009; Ilic et al., 2009).

5.15 CONCLUSION

This chapter has focused on the encapsulation using spray cooling/chilling/congealing processes and has presented some the most used techniques to evaluate microparticles produced by these methods of encapsulation.

Despite the potential of the technology and wide range of encapsulated products that have been developed using it, these method of encapsulation has found a comparatively much smaller studies and industrial applications.

This technology is still far from being fully developed, understood and has yet to become a regular tool in the several industries for the many reasons reported in this chapter.

REFERENCES

Akiyama Y, Yoshioka A, Oribe H, Hirai S, Kitamori N, Toguchi H (1993) Novel oral controlled-release microspheres using polyglycerol esters of fatty acids. *Journal of Controlled Release* 26: 1–10.

Albertini B, Passerini N, Di Sabatino M, Vitali B, Brigidi P, Rodriguez L (2009) Polymer-lipid based mucoadhesive microspheres prepared by spray-congealing for the vaginal delivery of econazole nitrate. *European Journal of Pharmaceutical Sciences* 36: 591–601.

Albertini B, Passerini N, González-Rodríguez ML, Perissutti B, Rodriguez L (2004) Effect of Aerosil® on the properties of lipid controlled release microparticles. *Journal of Controlled Release* 100: 233–246.

Albertini B, Passerini N, Pattarino F, Rodriguez L (2008) New spray congealing atomizer for the microencapsulation of highly concentrated solid and liquid substances. *European Journal of Pharmaceutics and Biopharmaceutics* 69(1): 348–357.

Albertini B, Vitali B, Passerini N, Cruciani F, Di Sabatino M, Rodriguez L, Brigidi P (2010) Development of microparticulate systems for intestinal delivery of *Lactobacillus acidophilus* and *Bifidobacterium lactis*. *European Journal of Pharmaceutical Sciences* 40(4): 359–366.

Al-Kassas R, Donnelly RF, McCarron, PA (2009) Aminolevulinic acid-loaded Witepsol microparticles manufactured using a spray congealing procedure: Implications for topical photodynamic therapy. *Journal of Pharmacy and Pharmacology* 61: 1125–1135.

Bunjes H, Unruh T (2007) Characterization of lipid nanoparticles by differential scanning calorimetry, x-ray and neutron scattering. *Advances in Drug Delivery Review* 59: 379–402.

Cavallari C, Rodriguez L, Albertini B, Passerini N, Rosetti F, Fini A (2005) Thermal and fractal analysis of diclofenac/Gelucire 50/13 microparticles obtained by ultrasound-assisted atomization. *Journal of Pharmaceuticals Sciences* 94: 1124–1134.

Chambi HNM, Alvim ID, Barrera-Arellano D, Grosso CRF (2008) Solid lipid microparticles containing water-soluble compounds of different molecular mass: Production, characterization and release profiles. *Food Research International* 41: 229–236.

Comunian T, Thomazini M, Alves A, Matos-Jr, F, Balieiro J, Favaro-Trindade C (2013) Microencapsulation of ascorbic acid by complex coacervation: Protection and controlled. *Food Research International* 1: 373–379.

Cook MT, Tzortzis G, Charalampopoulos D, Khutoryanskiy VV (2012) Microencapsulation of probiotics for gastrointestinal delivery. *Journal of Controlled Release* 162: 56–67.

Cordeiro P, Temtem M, Winters C (2013) Spray congealing: Applications in the Pharmaceutical Industry. *Chimica Oggi-Chemistry Today* 31(5): 69–73.

Cecchini C, Verdenelli MC, Palmieri GF, Silvi S (2010) Evaluation of microgranulation of Lactobacillus rhamnosus IMC 501 and Lactobacillus paracasei IMC 502 with vegetable fats as an approach to prolonging viability during storage. *Journal of Biotechnology* 150: 325.

Di Sabatino M, Albertini B, Kett VL, Passerini N (2012) Spray congealed lipid microparticles with high protein loading: Preparation and solid state characterization. *European Journal of Pharmaceutical Sciences* 46: 346–356.

Dixit M, Kini A, Kulkarni P (2010) Preparation and characterization of microparticles of piroxicam by spray drying and spray chilling methods. *Research in Pharmaceutical Sciences* 5(2): 89–97.

Dubey NB, Windhab EJ (2013) Iron encapsulated microstructured emulsion-particle formation by prilling process and its release kinetics. *Journal of Food Engineering* 115: 198–206.

Fávaro-Trindade CS, Heinemann RJB, Pedroso DL (2011) Developments in probiotic encapsulation. *CAB Reviews: Perspectives in Agriculture, Veterinary Science, Nutrition and Natural Resources* 6(4): 1–8.

Gamboa OD, Gonçalves LG, Grosso CF (2011) Microencapsulation of tocopherols in lipid matrix by spray chilling method. *Procedia Food Science* 1: 1732–1739.

Gonnet M, Lethuaut L, Boury F (2010) New trends in encapsulation of liposoluble vitamins. *Journal of Controlled Release* 146(3): 276–290.

Guillen MD, Cabo N (1997) Characterization of edible oils and lard by Fourier transform infrared spectroscopy: Relationships between composition and frequency of concrete bands in the fingerprint region. *Journal of the American Oil Chemists' Society* 74: 1281–1286.

Heurtault B, Saulnier P, Pech B, Proust JE, Benoit JP (2003) Physico-chemical stability of colloidal lipid particles. *Biomaterials* 24: 4283–4300.

Ilić I, Dreu R, Burjak M, Homar M, Kerč J, Srčič S (2009) Microparticle size control and glimepiride microencapsulation using spray congealing technology. *International Journal of Pharmaceutics* 381: 176–183.

Jenning V, Thünemann AF, Gohla SH (2000) Characterisation of a novel solid lipid nanoparticle carrier system based on binary mixtures of liquid and solid lipids. *International Journal of Pharmaceutics* 199: 167–177.

Jeon BJ, Kim NC, Han EM, Kwak HS. Application of microencapsulated isoflavone into milk. *Archives of Pharmacal Research* 28(7): 859–865.

Keller G, Lavigne F, Loisel C, Ollivon M, Bourgaux C (1996) Investigation of the complex thermal behavior of fats: Combined DSC and x-ray diffraction techniques. *Journal of Thermal Analysis* 47: 1545–1565.

Kleerebezem M, Vaughan EE (2009) Probiotic and gut lactobacilli and bifidobacteria: Molecular approaches to study diversity and activity. *Annual Reviews of Microbiology* 63: 269–290.

Lahtinen SJ, Ouwehand AC, Salminen SJ, Forssell P, Myllärinen P (2007) Effect of starch and lipid based encapsulation on the culturability of two *Bifidobacterium longum* strains. *Letters in Applied Microbiology* 44: 500–505.

Madureira AR, Amorim M, Gomes AM, Pintado ME, Malcata FX (2011) Protective effect of whey cheese matrix on probiotic strains exposed to simulated gastrointestinal conditions. *Food Research International* 44(1): 465–470.

Manojlovic V, Nedovic VA, Kailasapathy K, Zuidam NJ (2010) Encapsulation of probiotics for use in food products. In: Nedovic, N.J.Z.a.V. (Ed.), *Encapsulation Technologies for Active Food Ingredients and Food Processing*. Springer, New York, pp. 269–302.

Maschke A, Becker C, Eyrich D, Kiermaier J, Blunk T, Göpferich A (2007) Development of a spray congealing process for the preparation of insulin-loaded lipid microparticles and characterization thereof. *European Journal of Pharmaceutics and Biopharmaceutics* 65(2): 175–187.

McClements DJ, Decker EA. Lipídeos (2010) In: Damodaran, S., Parhin, K.L., Fennema, O.R. *Química de Alimentos*, 4th edn. Artmed, Porto Alegre, Brazil, 900pp.

Mehnert W, Mäder K. Solid lipid nanoparticles—Production, characterization and applications. *Advanced Drug Delivery Reviews* 47: 165–196.

Müller RH, Radtke M, Wissing SA (2002a) Nanostructured lipid matrices for improved microencapsulation of drugs. *International Journal of Pharmaceutics* 242: 121–128.

Müller RH, Radtke M, Wissing SA (2002b) Solid lipid nanoparticles (SLN) and nanostructured lipid carriers (NLC) in cosmetic and dermatological preparations. *Advanced Drug Delivery Reviews* 54: S131–S155.

Nori MA (1996) Produção de microcápsulas de ácido cítrico para utilização em produtos cárneos. Dissertação (Mestrado em Ciências Farmacêuticas)- Universidade de São Paulo.

Nori M, Lopes CM, Favaro-Trindade CS, Souto EB (2009) Stability enhancement of Lactobacillus acidophilus and Bifidobacterium lactis in lipid microparticles produced by melt emulsification. *New Biotechnology* 25: S56–S57.

Okuro PK, Baliero JCC, Liberal RDCO, Favaro-Trindade CS (2013a) Co-encapsulation of Lactobacillus acidophilus with inulin or polydextrose in solid lipid microparticles provides protection and improves stability. *Food Research International* 53: 96–103.

Okuro PK, Matos Jr FE, Favaro-Trindade CS (2013b) Technological challenges for spray-chilling encapsulation of functional food ingredients. *Food Technology and Biotechnology* 51: 171–182.

Oxley JD (2012) Chapter 5—Spray cooling and spray chilling for food ingredient and nutraceutical encapsulation. In: Garti, N., McClements, D.J. *Encapsulation Technologies and Delivery Systems for Food Ingredients and Nutraceuticals*. pp. 110–130.

Passerini N, Albertini B, Perissutti B, Rodriguez L (2006) Evaluation of melt granulation and ultrasonic spray congealing as techniques to enhance the dissolution of praziquantel. *International Journal of Pharmaceutics* 318(2): 92–102.

Passerini N, Perissutti B, Albertini B, Franceschinis E, Lenaz D, Hasab D, Locatelli I, Voinovich D (2012) A new approach to enhance oral bioavailability of Silybum Marianum dry extract: Association of mechano-chemical activation and spray congealing. *Phytomedicine* 19: 160–168.

Passerini N, Perissutti B, Albertini B, Voinovich D, Moneghini M, Rodriguez L (2003) Controlled release of verapamil hydrochloride from waxy microparticles prepared by spray congealing. *Journal of Controlled Release* 88: 263–275.

Passerini N, Perissutti B, Moneghini M, Voinovich D, Albertini B, Cavallari C, Rodriguez L (2002) Characterization of carbamazepine-Gelucire 50/13 microparticles prepared by a spry-congealing process using ultrasounds. *Journal of Pharmaceutical Sciences* 91: 699–707.

Pedroso DL, Dogenski M, Thomazini M, Heinemann RJB, Favaro-Trindade CS (2013) Microencapsulation of *Bifidobacterium animalis* subsp. lactis. and *Lactobacillus acidophilus* in cocoa butter using spray chilling technology. *Brazilian Journal of Microbiology* 44(3): 777–783.

Pedroso DL, Thomazini M, Heinemann RJB, Favaro-Trindade CS (2012) Protection of Bifdobacterium lactis and Lactobacillus acidophilus by microencapsulation using spraychilling. *International Dairy Journal* 26: 127–132.

Qi S, Marchaud D, Craig D (2010) An investigation into the mechanism of dissolution rate enhancement of poorly water-soluble drugs from spray chilled gelucire 50/13 microspheres. *Journal of Pharmaceutical Sciences* 99: 262–274.

Ribeiro APB, Basso RC, Grimaldi R, Gioielli LA, Dos Santos AO, Cardoso LP, Guaraldo Gonçalves LA (2009) Influence of chemical interesterification on thermal behavior, microstructure, polymorphism and crystallization properties of canola oil and fully hydrogenated cottonseed oil blends. *Food Research International* 42(8): 1153–1162.

Ribeiro MDMM, Arellano DB, Grosso CRF (2012) The effect of adding oleic acid in the production of stearic acid lipid microparticles with a hydrophilic core by a spray-cooling process. *Food Research International* 47: 38–44.

Rodriguez L, Passerini N, Cavallari C, Cini M, Sancin P, Fini A (1999) Description and preliminary evaluation of a new ultrasonic atomiser for spray congealing processes. *International Journal of Pharmaceutics* 183: 133–143.

Savolainen M, Herder J, Khoo C, Lövqvist K, Dahlqvist C, Glad H, Juppo AM (2003) Evaluation of polar lipid-hydrophilic polymer microparticles. *International Journal of Pharmaceutics* 27; 262(1–2): 47–62.

Schenk H, Visser P, Peschar R (2006) Alternative precrystallization of cocoa butter. *The Manufacturing Confectioner* 86(1): 67–72.

Senaka Ranadheera C, Evans CA, Adams MC, Baines SK (2012) In vitro analysis of gastrointestinal tolerance and intestinal cell adhesion of probiotics in goat's milk ice cream and yogurt. *Food Research International* 49: 619–625.

Wegmuller R, Zimmermann MB, Buhr VG, Windhab EJ, Hurrel R (2006) Development, stability, and sensory testing of microcapsules containing iron, iodine and vitamin A for use in food fortification. *Journal of Food Science* 71: 181–187.

Windbergs M, Strachan CJ, Kleinebudde P (2009) Investigating the principles of recrystallization from glyceride melts. *AAPS PharmSciTech* 10(4): 1224–1233.

Yajima T, Nogata A, Demachi M, Umeki N, Itai S (1996) Particle design for taste-masking using a spray-congealing technique. *Chemical and Pharmaceutical Bulletin* 44: 187–191.

Zaky A, Elbakry A, Ehmer A, Breuning M, Goepferich A. (2010) The mechanism of protein release from triglyceride microspheres. *Journal of Controlled Release* 147: 202–210.

Ziani K, Fang Y, McClements D (2012) Encapsulation of functional lipophilic components in surfactant-based colloidal delivery systems: Vitamin E, vitamin D, and lemon oil. *Food Chemistry* 2: 1106–1112.

Zoet FD, Grandia J, Sibeijn N (2011) Encapsulated fat soluble vitamin. WO 2012/047098A1. 2011 October 2011.

6 Encapsulation via Spinning Disk Technology

Aurélie Demont and Ian W. Marison

CONTENTS

6.1 Introduction 89
6.2 Principles of Atomization on Spinning Disk 90
6.3 Theory 92
    6.3.1 Liquid Distribution 92
    6.3.2 Droplet Formation 95
    6.3.3 Droplet Collection and Bead Solidification for Encapsulation Purposes ... 98
6.4 Applications 102
6.5 Equipment 102
6.6 Advantages and Disadvantages 107
6.7 Conclusion 107
References 108

6.1 INTRODUCTION

The entrapment of various materials such as flavors, living cells, and pharmaceutical compounds within capsules for a wide range of applications is of considerable importance in the pharmaceutical, chemical, and food industries, as well as in agriculture, biotechnology, and medicine.[14,19] As a result microencapsulation has been extensively studied since the early to mid-twentieth century.[26,32]

Encapsulation is defined as the process involving the complete envelopment of a core material in a membrane, using different techniques.[14,34] The main aims of encapsulation are the immobilization, the protection, the stabilization, and the control of the release of the entrapped compound.[4]

The conventional methodology to produce capsules involves the reaction of a polymer dropping into a gelling agent solution, leading to the formation of a gel sphere.[2] Due to biocompatibility, cheap price, and the mild process conditions, alginate is the most widely used biopolymer for encapsulation.[4] It can form thermally stable hydrogels in the presence of divalent cations.[28] The prerequisites for industries such as cosmetics, agriculture, and food are usually high production rates and low cost; however, medical and biotechnological processes have more stringent criteria.[34] Methodologies must have the capacity to produce monodisperse, spherically shaped, homogeneous capsules, with a small size and a narrow size distribution. They must be performed under simple conditions with a short production time. Moreover, high efficiency and production rates should be possible with many different materials, including viscous solutions; it should allow the production of different sizes and the encapsulation of a variety of core materials, and enable, if required, to be performed under sterile condition.[33]

Many different techniques, such as emulsification, air jet, electrostatic dripping, mechanical jet cutting, vibrating nozzle, and spinning (rotating) disk atomization have been described in the literature[19,26,34] to produce small diameter capsules; however, all of them suffer from certain limitations, such as the possibility to scale-up production. Only mechanical jet cutting, spinning (rotating) disk

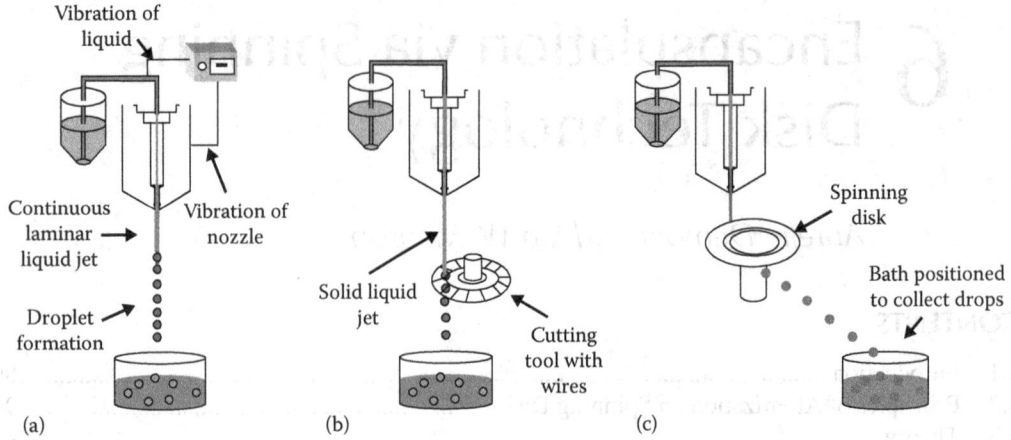

FIGURE 6.1 Techniques to produce microcapsules at high production rates: (a) vibrating jet, (b) jet cutting, and (c) spinning disk.

atomization, and vibrating-jet (nozzle) techniques have been reported to be capable of the high production rates, which are mandatory for most industrial applications[33] (Figure 6.1).

At present, the most widely used production technique is the vibrating-jet (nozzle) method, since it can produce small, monodisperse capsules and may be scaled-up by adding more nozzles to the system.[26,34] This method is based on the principle of controlling the breakup of a laminar jet by the application of a controlled vibrational frequency, with defined amplitude to the extruded jet. The frequency will cause the jet to break up into uniform droplets of equal size. This breakup, which is highly regular and reproducible, only takes place when the vibrational frequencies are near the natural frequency for the breakup of the jet itself.[34] An alternative to the vibrating-jet method to produce small diameter microcaspules with a high production rate is the spinning (rotating) disk technology, which will also be discussed here.

It was found by Johnson and Walton in 1947 that sprays of almost uniform drop size could be formed by means of a spinning disk sprayer in which the liquid is fed onto the center of a disk and centrifugal forces result in droplets forming at the edge of the disk.[32] The phenomenon was therefore studied in detail in order to develop an apparatus for the production of homogeneous droplet over as wide a size range as possible.[31] Such a spray source was discovered to be of great value for many fields of application as the drop size range can be from several millimeters to a few micrometers.[32] Nowadays, rotating disk atomization is well-known and is widely used in industrial processes such as air humidification, spray drying, spray cooling, fertilizer or pesticide sprinkling, oil burning, or encapsulation. This chapter will discuss the use of spinning disk atomization from a general standpoint followed by a more specific discussion about this for encapsulation.

6.2 PRINCIPLES OF ATOMIZATION ON SPINNING DISK

Disintegration of a liquid by a rotating wheel or disk is widely used for the production of fine particles of liquid in processes such as spray drying, cooling or freezing, humidifying, fertilizer or pesticide sprinkling,[7] and oil burning.[9] Atomization can be defined as the transformation of a liquid bulk into separated droplets.[22] Rotating disk atomization is a system where the rotation of the disk induces forces that will create the fluid breakup. The disk is circular and can take the form of a wheel, a plate, a bowl, or a cup, rotating around its axis and serving as a temporary support for a liquid feed.[22] The principle of the technique is to distribute a liquid onto a rotating disk, which due to centrifugal forces and wetting ability disperses into a thin film. The liquid then gains in velocity, is transported to the disk rim and is spread into droplets, which are thrown off tangentially from the disk and form a spray. The size of the drops depends essentially on the rotation speed, disk geometry, and liquid properties. The liquid is attached to the rim by interfacial and viscous forces, which are overcome by the rotation-induced forces. Once drops are detached, interfacial forces keep the atomized liquid as spherical shapes in order to minimize the specific surface area.[22] According to their size, each drop falls at a different distance from the disk. Large drops fly over a longer distance than smaller ones,[33] therefore allowing ballistic classification and monodisperse drop collection.[31] During the spreading process, a phenomenon called "surfing" or "slippage" can occur depending on the characteristics of the disk. This phenomenon is the result of the slippage of liquid over the disk that can reduce the energy transfer during atomization and cause the random formation of large and small drops with an uncontrollable size range called satellites[22] (Figure 6.2).

It is commonly accepted that there are three disintegration regimes for liquids discharged from a rotating disk, namely direct drop, ligament, and sheet formation[7,31] (Figure 6.3).

At low flow rates, in the direct drop regime, drops are formed near the disk rim and directly result from the liquid "bulges" created there. Since the liquid is supplied onto the disk surface, the liquid rotates and is carried outward by the centrifugal and centripetal forces. At the edge of the disk, a number of bulges are formed, due to surface tension, which grow into drops. Discrete drops of uniform size are shaped and regularly released from the disk. This regime is the most favorable in term of droplet formation as collisions and mutual perturbations are avoided.[22,24,31] In the ligament regime, at higher flow rates, the single drops converge to form thin ligaments or jets.[24] The drops are left attached to the rim by a thin string extending from the disk.[22] These ligaments, in metastable states, break into a chain of droplets, smaller in diameter than those in direct drop formation.[24]

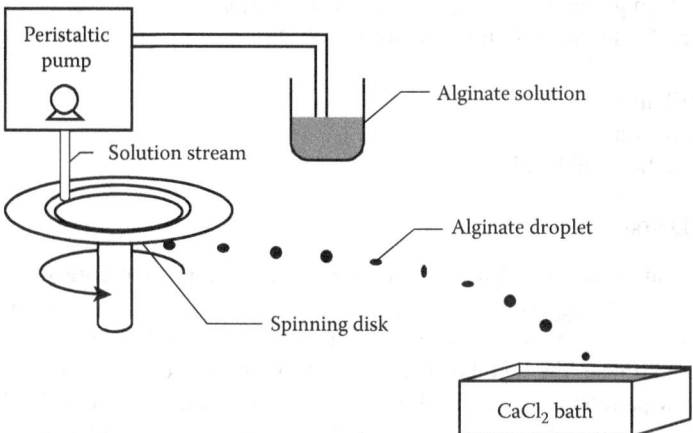

FIGURE 6.2 Schematic representation of an atomization system where the polymer solution is transported to the surface of the disk by a pump and atomized into droplets, which are collected upon landing at a defined distance from the disk. (Reproduced from Senuma, Y. et al., *Biotechnol. Bioeng.*, 67(5), 616, 2000. With permission.)

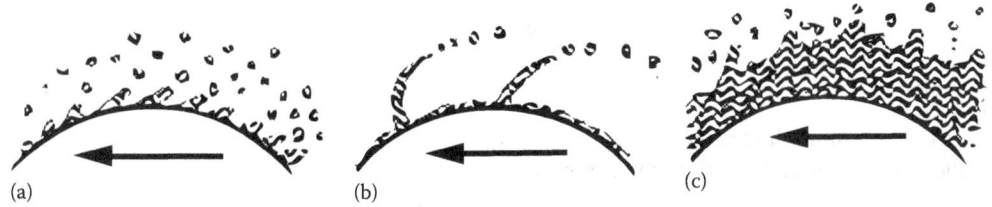

FIGURE 6.3 Mechanisms of drop formation by spinning disk atomization. (a) Direct formation, (b) ligament formation, and (c) sheet formation. (Reproduced from Teunou, E. and Poncelet, D., *J. Food Eng*, 71(4), 345, 2005. With permission.)

Moreover, ligaments often break up from singular disturbances, which results in the formation of droplets of unequal size. Necks and swells are created by disturbance waves and are considered to be the precursors to drops. Thin strings are also observed between large drops, which break into small drops leading to a wide drop-size distribution.[24] In this regime, the number of ligaments is practically independent of the flow rate.[22] Increasing the flow rate while operating in the ligament regime ultimately leads to the sheet formation regime. In this regime, the number of ligaments is insufficient to account for the liquid supply through either increasing ligament thickness or number. As a result the liquid forms a sheet, which will disintegrate into ligaments and clots of liquid in an irregular way, at a certain distance from the disk rim.[22] This regime is highly sensitive to perturbations and variations, making it very unfavorable for drop formation, particularly with respect to size distribution. A wide spectrum of drop sizes is created away from the disk rim, due to the variable condition at the region of disintegration.[24] The smallest size distribution is found in the direct drop regime.[24] As this shifts to the ligament or sheet regime, uniformity of drop formation is generally lost. Furthermore, disintegration of the liquid bulk tends to be delayed by higher surface tensions, thereby shifting drop formation to the most unfavorable regime.[24]

6.3 THEORY

Different studies have investigated the parameters involved in spinning disk atomization.[1,15,22–25,31] To produce drops using spinning disk technology, the liquid is first accelerated by the rotating disk and uniformly spread over the disk. Under proper condition, the liquid then leaves the edge of the disk to form isolated droplets that can be used in many applications such as spray combustion, spray cooling, or spray drying, fire extinguishing, and encapsulation.

The process can be divided into three main steps:[22,25]

1. Liquid distribution
2. Droplet formation
3. Droplet collection and bead solidification

6.3.1 LIQUID DISTRIBUTION

In spinning disk atomization, the disk provokes fluid breakup and is the critical step controlling the atomization process. It is suggested by Senuma[22] that the size and the size distribution of the droplets are strongly dependent on the liquid distribution and on the flow homogeneity, which take place during the residence time of the liquid on the disk.[25] It has been reported that liquid acceleration can occur along its flow path on the disk, until it is discharged as droplets at the rim. The fluid arrives from the feeding reservoir and is submitted to a radial acceleration, a_r, with respect to the disk, as it gains velocity (Equation 6.1).[22] The transfer of the liquid from the inner to the outer ring will then ensure wetting by increasing the specific contact area. Later, the fluid moves along the ring and finally reaches the edge of the disk where small teeth channel the flow to their tips as a precursor to atomization. Identical droplets are formed from each tooth.[25]

$$a_r = \frac{v_f^2}{r} \qquad (6.1)$$

where
v_f is the tangential fluid velocity
r the radial distance from the center of rotation

With a liquid continuously feeding the disk at a rate Q, the criteria for the transition between the different regimes is defined in terms of critical flow rate. As reported earlier, three modes of droplet formation exist and the liquid breaks up according to one of these modes. The transition from one mode to the other can occur when the flow rate of the liquid, disk diameter, and/or disk rotational speed are increased.[1] This transition phenomenon of liquid breakup in spinning disk technology has been investigated by many researchers (Table 6.1) such as Frost,[7,15] Kaniya and Kayano,[1] Tanasawa et al.,[30] Hinze and Milborn[15] or Matsumoto et al.,[15] and Fraser et al.[15] Frost developed criteria for: direct drop formation (Equation 6.2); for fully developed ligament formation (Equation 6.3); for the first appearance of sheet formation (Equation 6.4), while Kaniya and Kayano described the volume flow rate for fully developed sheet formation (Equation 6.5).[1] Tanasawa et al. proposed an experimental equation, which gives the condition of the transition between direct drop formation and ligament formation (Equation 6.6).[15,22]

TABLE 6.1
List of Previously Published[1,7,15,21,29] Criteria for the Transition between the Different Breakup Regimes

Author	Equation		Breakup Regime
Frost[7]	$Q \leq 1.52 \cdot \left(\frac{\sigma\rho D}{\mu^2}\right) \cdot \left(\frac{\mu D}{\rho}\right) / \left(\frac{\omega\rho D^2}{\mu}\right)^{0.95}$	(6.2)	Direct drop
Frost[7]	$Q > 0.46 \cdot \left(\frac{\sigma\rho D}{\mu^2}\right)^{0.9} \cdot \left(\frac{\mu D}{\rho}\right) / \left(\frac{\omega\rho D^2}{\mu}\right)^{0.63}$	(6.3)	Fully developed ligament
Frost[7]	$Q > 19.8 \cdot \left(\frac{\sigma\rho D}{\mu^2}\right)^{0.9} \cdot \left(\frac{\mu D}{\rho}\right) / \left(\frac{\omega\rho D^2}{\mu}\right)^{0.84}$	(6.4)	First appearance of the sheet
Kaniya and Kayano[1]	$Q > \left\{ 2700 / \left[\left(\frac{\rho\omega^2 D^3}{8\sigma}\right)\left(\frac{\rho}{\sigma D^3}\right)^{1.24} \right]^{0.403} \right\}$	(6.5)	Fully developed sheet
Tanasawa et al.[15,30]	$Q = 2.8 \dfrac{\left(\dfrac{D}{n}\right)^{2/3} \cdot \left(\dfrac{\sigma}{\rho}\right)}{1 + 10\left[\dfrac{\mu}{(\rho\sigma D)^{1/2}}\right]^{1/3}}$	(6.6)	Direct drop to ligament
Tanasawa et al.[15,30]	$Q = 5.3 \left(\dfrac{D}{n}\right)^{2/3} \cdot \left(\dfrac{\sigma}{\rho}\right) \cdot \left(\dfrac{\rho}{\mu}\right)^{1/3}$ for $D\rho/\mu < 30$	(6.7)	Ligament to sheet formation
Reported by Senuma[22]	$Q \cong 20 \cdot D^{1/2} \cdot \left(\dfrac{1}{n}\right)^{2/3} \cdot \left(\dfrac{\sigma}{\rho}\right)^{5/6}$ for $D\rho/\mu > 30$	(6.8)	Ligament to sheet formation
Hinze and Milborn[15]	$\left(\dfrac{\rho Q^2}{\sigma R^3}\right)\left(\dfrac{\rho R^3 \omega^2}{\sigma}\right)^{0.6}\left(\dfrac{\mu^2}{\rho\sigma R}\right)^{1/6} = 4.56$	(6.9)	Ligament to sheet formation
Fraser et al.[15]	$\left(\dfrac{\rho R^3 \omega^2}{\sigma}\right)\left(\dfrac{\bar{Q}}{R^3\omega}\right)\left(\dfrac{R\mu}{\rho\bar{Q}}\right)^{0.19} = 0.636$	(6.10)	Sheet to ligament

Note: Q the flow rate, D the diameter of the disk, n the rotational number and ρ, μ, and σ the density, viscosity, and surface tension of liquid, respectively, R the radius of the disk and ω the angular speed of the disk.

They also studied the transition from ligament to sheet formation and proposed a semi-empirical equation for viscous liquids (Equation 6.7).[15] A complementary equation was proposed by Tanasawa and reported by Senuma, for liquids with $D\rho/\mu > 30$ (where D is the diameter of the disk, ρ the density, and μ the viscosity of the polymer) (Equation 6.8).[22] Hinze and Milborn investigated the transition from ligament to sheet formation and suggested a semi-empirical equation assuming that film formation occurs when the feed rate of liquid exceeds a certain flow rate within which the liquid can be removed through ligaments (Equation 6.9).[15] Fraser et al. developed a semi empirical equation for the transition flow rate from sheet to ligament formation, assuming that the transition happens when the contracting velocity of the film is larger than the radial velocity of liquid leaving the disk (Equation 6.10).[15] The different equations are shown in Table 6.1.

In his work, Matsumoto determined experimentally the correlation of Re and We with the transition flow rate from direct drop to ligament, Q_1^+, and for the reverse process $\bar{Q}_1^+$. He showed that the conditions for the transition occur at a slightly smaller flow rate for the transition from ligament to direct drop $\bar{Q}_1^+$ than for direct drop to ligament (Q_1^+) (Equation 6.15). He also reported that the transition flow rate from sheet to ligament (Q_2^+) formation and the reverse can be correlated experimentally in terms of Re and We. They therefore rearranged Equations 6.7, 6.9, and 6.10 as shown in Table 6.2.[15] However, the authors concluded that Hinze and Milborn's equation applied to liquids of less than a few poises in viscosity, whereas more viscous liquids require the Tanasawa et al. equation to predict the transition flow rate.[15] These results can probably be explained by the difference in equipment, liquid, viscosity, and experiment devices used to develop the different transition flow rate equation. The sampling and analysis method may also be responsible for the differences observed in the conclusions of Matsumoto. It is therefore advisable to compare the different existing formulae with respect to the flow rate in order to determine which is the most applicable for the experimental conditions under study.

Senuma reported[22] that the critical flow rate decreases with increasing rotation speed or decreasing the diameter. The critical flow also decreases with increasing viscosity; however, in the sheet regime, Senuma reported that the decrease occurs only above a certain viscosity and then remains constant.[24] According to Senuma,[24] the homogeneity of the spray depends on the feed rate and the percentage of satellites can then be kept low if the flow rate is equally low. With atomization of uniform sprays,

TABLE 6.2
List of Previously Published[15] Correlation between Re, We, and the Different Transition Flow Rates

Author	Equation		Breakup Regime
Tanasawa et al.	$Q_2^+ = 0.297 \dfrac{Re^{6/5}}{We}$	(6.11)	Ligament to sheet
Hinze and Milborn	$Q_2^+ = 0.340 \dfrac{Re^{2/3}}{We^{0.883}}$	(6.12)	Ligament to sheet
Fraser et al.	$\bar{Q}_2^+ = 0.108 \dfrac{Re^{2/3}}{We^{0.875}}$	(6.13)	Sheet to ligament
Matsumoto et al.	$Q_1^+ = 0.096 \dfrac{Re^{0.95}}{We^{1.15}}$	(6.14)	Drop to ligament
Matsumoto et al.	$\bar{Q}_2^+ = 0.073 \dfrac{Re^{0.95}}{We^{1.15}}$	(6.15)	Ligament to drop

Note: Re = $\rho R^2 \omega/\mu$ the Reynolds number and We = $\rho R^3 \omega^2/\sigma$ the Weber number.

Encapsulation via Spinning Disk Technology

the drops land in a narrow band, centered around the disk, since the projection distance is the same for each drop. It was reported that increasing the flow rate does not affect the projection distance of the main drops, however, that of the satellites will be increased.[24] The projection distance changes with two opposing effects: the drop size and the disk speed. For the same initial speed, large drops fly further since inertia is larger; however, if the initial speed is increased, smaller drops are produced with the higher centrifugal force. The critical flow rate permitting a satisfactory separation of the main drops from satellites decreases hyperbolically with an increase of the disk speed. Using disks with larger diameters shifts this limit to higher values according to the work of Senuma.[22]

6.3.2 Droplet Formation

In spinning disk atomization, the factors governing the size of the droplets are the acceleration, a, or centrifugal forces, which tend to detach the drop, and the surface tension σ, which holds the drop in position. The acceleration is the result of the angular velocity of the disk (ω) and the radius (R)[25] (Equation 6.16).

$$a = \omega^2 R \qquad (6.16)$$

When the liquid is continuously feeding the drop with a rate Q, the drop reaches a critical volume at which it detaches, depending on the surface tension, σ and the density ρ.[25] In direct drop formation, the diameter of the main drop d can be given by Equation 6.17, according to the work of Walton and Prewett.[15]

$$d = 2.69 \cdot R \cdot \text{We}^{-1/2} \qquad (6.17)$$

with $\text{We} = \rho R^3 \omega^2 \sigma$ being the Weber number.[25]

The drop formation at the rim of the disk was investigated by Senuma and Hilborn.[24] They showed that the diameter of the main drop is not affected by the viscosity although increasing the viscosity broadens the size distribution and can alter the droplet formation at the teeth tips.[22]

They also reported that satellite drops are produced with each main drop and that the mass fraction of satellite drops in single drop regime can be expressed as follows (Equation 6.18):

$$\frac{m_{sat}}{m_{tot}} = \frac{1}{\left(\dfrac{d_{main}}{d_{sat}}\right)^3 + 1} \qquad (6.18)$$

where

m_{sat} is the mass of satellite droplets
m_{tot} the total mass of droplets
d_{main} is the diameter of the main drops
d_{sat} is the diameter of the satellite drops[24]

Senuma also reported[24] that in the ligament formation regime, drop size resulting from the breakup of the ligaments in disk atomization is, according to Weber's theory, given by Equation 6.19

$$\frac{d}{d_l} = 1.88 \left(1 + \frac{3}{\sqrt{\rho \sigma d_l}}\right)^{1/6} \qquad (6.19)$$

with

d_l the diameter of the ligaments
μ the viscosity
ρ the density
σ the surface tension of the liquid

It was also reported that this relationship can be applied to Newtonian liquids as well as power law liquids, since the second term of Equation 6.19 in parenthesis is much smaller than unity and therefore allows approximation to a linear relationship.[22]

$$d = 1.88 \cdot d_l \quad (6.20)$$

In his work, Senuma[22] stated that the maximum drop size varies with the flow rate Q and two sharp transitions corresponding to the change in liquid breakup regimes can be seen. The maximum drop size d_{max}^d was observed to abruptly decrease to a minimum value d_{max}^l when moving from direct drop to the ligament regime, and remains almost constant over the entire range of the ligament regime. If the flow rate is increased further, the breakup moves to the sheet regime and the maximum drop size d_{max}^s increases to a value higher than for direct drop. It was then observed that with further increase of the flow rate, the drop size slowly increases.[22] This observation was also reported by Ahmed and Youssef who observed a slight increase of the Sauter mean diameter of the drops, with increasing the flow rate over the range studied.[1] The maximum drop sizes reported by Senuma for each regime are given in Equations 6.21 through 6.23.

$$\text{Direct drop: } d_{max}^d = 43.2 \frac{1}{n \cdot D^{1/2}} \cdot \left(\frac{\sigma}{\rho}\right)^{1/2} \cdot \left(1 + 0.003 \frac{Q \cdot \rho}{\mu \cdot D}\right) \quad (6.21)$$

$$\text{Ligament: } d_{max}^l = 32 \frac{1}{n \cdot D^{1/2}} \cdot \left(\frac{\sigma}{\rho}\right)^{1/2} \cdot \left(\frac{\rho}{\mu}\right)^{0.1} \cdot Q^{0.1} \quad (6.22)$$

$$\text{Sheet: } d_{max}^s = 105 \frac{1}{n \cdot D^{0.8}} \cdot \left(\frac{\sigma}{\rho}\right)^{0.4} \cdot Q^{0.5} \quad (6.23)$$

where
μ the viscosity
ρ the density
σ the surface tension of the liquid
n the rotational number
D the diameter of the disk
Q the flow rate

Senuma[22] suggested that if the initial radial velocity and the resistance effects on the development of the ligament are neglected, the ligament is exactly an involute. The number of ligaments N_l increases with the flow rate to a maximum, which is smaller than the maximum number of bulges since some bulges do not transform into ligaments, and then it remains constant.[22] Satellite droplets are created in ligament mode. They result from the breakup of the thin ligaments between two drops, and the number is related to the drop velocity and a collision parameter.[22]

In the literature, different correlations have been reported for the mean size of droplets produced by spinning disk atomization. These correlations are summarized in Table 6.3.

The different correlations shown in Table 6.3 are inconsistent according to Ahmed and Youssef,[1] due to different causes such as the difference in breakup regime, the different ranges of variables, the difference in measuring and sampling techniques. The use of a single atomizer to develop these

TABLE 6.3
List of Previously Published[1,7,21,32] Correlations Reported for the Mean Droplet Size Produced by Spinning Disk Atomization

Author	Liquids	Equation		Breakup Regime
Walton and Prewett[32]	Water and methylsalicylate	$d_{av} = 3.8\left(\dfrac{\sigma}{\rho D \omega^2}\right)^{0.5}$	(6.24)	Direct drop, ligament and sheet
		$d = 2.69 \cdot R \cdot We^{-1/2}$	(6.25)	Direct drop
Boize and Dombrowski[1]	Oils	$d_{32} = 0.006 \omega^{-0.98}$	(6.26)	Direct drop and ligament
Ryley[21]	Water	$d_{32} = D^{-0.66} \omega^{-1.41} Q^{0.19} \mu^{-1.48} \sigma^{1.35} \rho^{-0.06}$	(6.27)	Ligament
Kayano and Kamiya[1]	Millet jelly solvent and soybean	$d_{vmd} = 2.0 \cdot D^{-0.69} \omega^{-0.79} Q^{0.32} \mu^{0.65} \sigma^{0.26} \rho^{-0.29}$	(6.28)	Ligament
Fraser and Eisenklam[1]	Water and oil emulsion	$d_{vmd} = K_1 \left(\dfrac{\sigma}{\rho D \omega^2}\right)^{0.5}$	(6.29)	Ligament
Frost[7]	Aqueous solutions of glycerol and proprietary wetting agent	$d_{vmd} = 1.87 \cdot D^{-0.8} \omega^{-0.75} Q^{0.44} \mu^{0.017} \sigma^{0.15} \rho^{-0.16}$	(6.30)	Ligament

Source: Ahmed, M. and Youssef, M.S., *J. Fluids Eng.*, 134(7), 071103, 2012. With permission.

Note: d_{32} the Sauter mean diameter defined as the arithmetic mean of several measurements of the Sauter diameter SD (SD = 6 V/A with V the volume and A the surface area of the particle), d_{vdm} the volume median diameter, which refers to the midpoint droplet size (mean), where half of the volume of spray is in droplets smaller and half of the volume is in droplets larger than the mean, ρ, μ, and σ, respectively the density, the viscosity, and the surface tension of liquid, D, the diameter of the disk, Q, the flow rate, and ω the angular speed of the disk.

equations may also be problematic as the variables are dimensional and depend on the atomizer characteristics. It is therefore not possible to extrapolate values for an industrial application. In order to overcome this, Ahmed and Youssef[1] investigated the Sauter mean diameter of the drops produced by spinning disk atomization. This showed that the Sauter mean diameter increases with the downstream distance, probably because the smaller droplets disappear. Moreover, the larger droplets travel a longer distance due to possessing a larger inertia. It was also reported that the Sauter mean diameter of the beads decreases with increasing disk diameter. Increasing the disk diameter leads to an increase of the downstream distance. This observation can be explained by the fact that increasing the disk diameter will increase the perimeter and therefore the peripheral velocity, more energy will then be imparted during the atomization process, which will cause a reduction of the thickness of the liquid sheet on the disk resulting in smaller drops. As a higher kinetic energy will be stored in the droplets, the droplets travel a longer distance after being discharged, which leads to an increase in the downstream distance. Ahmed and Youssef also reported[1] that increasing the rotational speed of the disk decreases the Sauter mean diameter of the droplets. Increasing the speed of the disk links this observation with the previously described observation, since, more energy will be attributed to the atomization process, resulting in a decrease in the liquid layer thickness and an increase in the angular displacement of the fluid. The liquid leaving the edge of the disk will then have a higher velocity and result in smaller droplets. Given the differences in the reported correlations, it is suggested that they should be tested in order to determine which equation best suits the experimental conditions employed.

6.3.3 Droplet Collection and Bead Solidification for Encapsulation Purposes

After the drops are discharged from the rotating disk, they need to be collected under buffered conditions, which maintain their structural integrity. Although the principles of spinning disk atomization have been investigated for many years, only a few studies have examined the distance needed by the particles to gel or to be collected without damage. The model developed by Teunou et al.[31] and the model of Sungkhaphaitoon et al.[29] are described in this chapter. The trajectory of the droplet from the disk is a very important aspect for the application of the spinning disk industrially. The trajectories of beads are parabolas, which increases with increasing bead size. Many environmental and technical factors influence the trajectory of a droplet leaving the disk. It is difficult to take in account all of the factors in one model consequently. Teunou et al. have made a number of assumptions in the model to simplify the analysis.[31] They assumed that

- The beads are produced in the direct drop formation regime
- The droplets are considered to be single particles (no interaction between droplets)
- The beads are spherical

According to the model of Teunou et al., if a spherical drop with a diameter d is leaving the rotating disk with a diameter D at a velocity V in the plane (x, y) (Figure 6.4), the two components of the droplet velocity can be written as[31]

$$\text{At } t = 0 \begin{cases} x = 0 \\ y = h_d \end{cases} \quad (6.31)$$

$$V = (V_x^2 + V_y^2)^{1/2} \quad (6.32)$$

where h_d is the position of the disk in the Y axis.

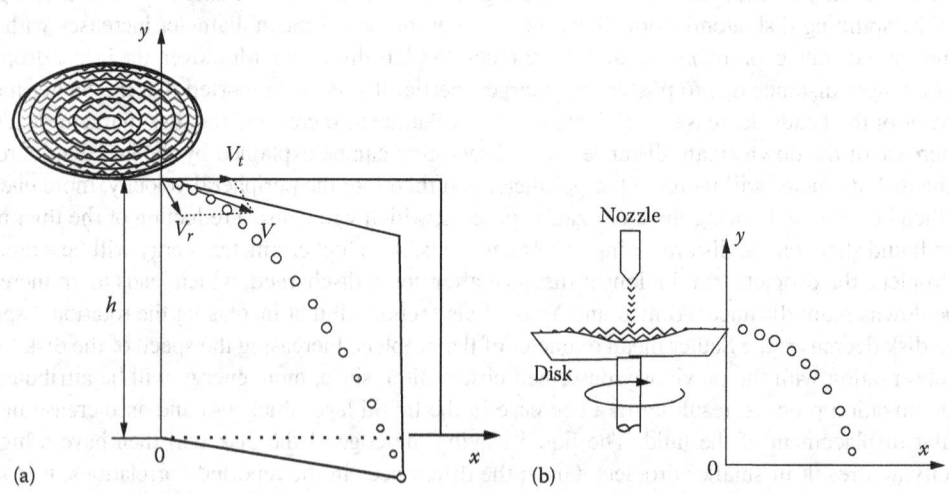

FIGURE 6.4 Projection of droplets from the disk edge. (a) Top view and (b) side view. (Reproduced from Teunou, E. and Poncelet, D., *J. Food Eng*, 71(4), 345, 2005. With permission.)

Encapsulation via Spinning Disk Technology

The air friction, which slows down the particles in their trajectory, and the slippage on the disk will have to be taken into account. The derivative of the velocity was given by the work of Teunou et al. and can be written as[31]

$$\begin{cases} \dfrac{dV_x}{dt} = -\dfrac{3}{4} \cdot \dfrac{\rho_a}{\rho_s} \cdot \dfrac{C_D}{d} \cdot V \cdot V_x \\ \dfrac{dV_y}{dt} = -g - \dfrac{3}{4} \cdot \dfrac{\rho_a}{\rho_s} \cdot \dfrac{C_D}{d} \cdot V \cdot V_y \end{cases} \quad (6.33)$$

$$\text{At } t = 0 \quad \begin{cases} V_x = V_t - V_s \\ V_y = 0 \end{cases}$$

where
  V is the velocity of the droplet as it leaves the disk (m/s)
  V_x the horizontal velocity of the droplet as it leaves the disk (m/s)
  V_y the vertical velocity of the droplet as it leaves the disk (m/s)
  ρ_a the air density (kg/m³)
  V_t the tangential velocity
  V_s the slipping velocity
  ρ_s the polymer density (kg/m³)
  g the standard acceleration of gravity (m/s²)
  d the diameter of the droplet (m)
  C_D is the drag coefficient, proportional to the total drag force exerted by air on the beads

Drag coefficient C_D is defined as the ratio of the total drag force per unit area to $\rho v_0^2/2$ can be calculated according to Equation 6.34.[8]

$$C_D = \dfrac{F_D/A_P}{\rho v_0^2/2} \quad (6.34)$$

where
  F_D is the total drag force
  A_P is the area of the particle in m²
  C_D is dimensionless
  v_0 is the free-stream velocity in m/s
  ρ is the density of the fluid in kg/m³

For a sphere, $A_P = \pi D_P^2/4$, where D_P is the sphere diameter.[8] A correlation of the Reynolds number and the drag coefficient can be observed, moreover, in the laminar region, for low Reynolds numbers (Re < 1):

$$F_D = 3\pi\mu D_P v_0 \quad (6.35)$$

By combining Equation 6.34 with Equation 6.35, the drag coefficient is[8]

$$C_D = \dfrac{24}{(D_P v_0 \rho)/\mu} = \dfrac{24}{\text{Re}} \quad (6.36)$$

with Re the Reynolds number calculated as follows:

$$\text{Re} = \dfrac{\rho_0 \cdot V \cdot d}{\mu_0} \quad (6.37)$$

If the Re > 1 in spinning disk technology, an empirical relation due to Klyachko[12] for C_D in terms of Reynolds is used by Teunou et al. in Equation 6.33. This relationship can be used when $1 < \text{Re} < 1000$; however, other empirical relations exist to determine C_D according to the value of Reynolds.[12]

$$C_D = \frac{24}{\text{Re}}\left(1 + \frac{\text{Re}^{2/3}}{6}\right) \quad (6.38)$$

The slipping velocity V_s used in Equation 6.33 can be calculated by iteration, applying the boundary layer theory for a laminar regime (Equation 6.39)[31]

$$V_s(r) = \frac{3.87 h_l \cdot \gamma_s^{1/2} \cdot \rho_a (\omega_0 - \omega)^{3/2} \cdot r^2}{2\pi \mu_s} \quad (6.39)$$

where
h_l is the thickness of the liquid on the disk
ρ_a is the air density (kg/m³)
ω the rotation speed of the disk (rpm)
ω_0 the initial rotating speed of the disk (rpm)
μ_s the liquid viscosity (Pa·s)
r the radius from the center of the disk (m)

However, to prevent slippage in commercial atomizers, radial vanes are used. The liquid is confined to the vane surface and at the periphery; the possible maximum liquid release velocity is attained.[31]

In this model, the value of the tangential velocity V_t in Equation 6.33 is calculated in the X–Y plane and by taking the radial velocity V_r into account (Equation 6.40).[31]

$$V_r = \frac{F_v}{2\pi R \cdot h_l} \quad (6.40)$$

where
F_v is the feed rate (m³/s)
R the radius of the disk
h_l is the thickness of the liquid layer sliding on the surface of the disk

Equation 6.40 is derived by assuming that the feed flow rate equals the rate of droplet formation (flow conversion). The value of the tangential velocity V_t in Equation 6.33 is then modified as follow[31]

$$\text{At } t = 0 \begin{cases} V_x = V_t - V_s = \sqrt{V_0^2 - V_r^2} - V_s \\ V_y = 0 \end{cases} \quad (6.41)$$

where
V_x the horizontal velocity of the droplet as it leaves the disk (m/s)
V_y the vertical velocity of the droplet as it leaves the disk (m/s)
V_t the tangential velocity
V_s the slipping velocity
V_r the radial velocity (m/s)
V_0 the initial horizontal velocity (m/s) (= the tangential velocity of the droplet as it leaves the disk)

The numerical resolution of these equations provides X and Y values of the droplet trajectories. The droplets have to be collected in the area corresponding to the minimum velocity in order to reduce percussions or impacts at the collecting point, which can affect the quality of the beads. It also has to be noticed that the model developed by Teunou et al. is calculated for a single and isolated drop, which is an approximation since a distribution of particles is produced. Teunou et al. noticed that the distribution can have an impact on the droplet trajectories, and the turbulence generated by different particles results in interactions between droplets, slowing them down.[30]

The model developed by Sungkhaphaitoon et al.[29] is based on a simple concept of projectile motion; however, it takes into account the effect of drag due to viscosity. As previously explained in Teunou model,[31] the velocity of the film on the disk has two components, the radial velocity V_r and the tangential velocity V_t. At the edge of the disk, the velocity of the liquid can then be estimated as shown in Equation 6.42,

$$V = V_t + V_r \tag{6.42}$$

where
  V_t is the tangential velocity in (m/s)
  V_r is the radial velocity in (m/s)

The drag force (F_D) that the air viscosity exerts on a flying droplet is calculated by Sungkhaphaitoon et al.[29] from Equation 6.43.

$$F_D = -\frac{1}{2} \cdot \rho_a \cdot V^2 \cdot A_p \cdot C_D \tag{6.43}$$

where
  ρ_a the air density (kg/m³)
  V the velocity of the melt droplet at the edge of the disk (m/s)
  A_p the area of a droplet particle (m²)
  C_D is expressed in Equation 6.38
  Re the Reynolds number

Sungkhaphaitoon et al.[29] then determined the equation of motion in the x- and y-directions of a single drop travelling in an air atmosphere (Equations 6.44 and 6.45),

$$\frac{d^2 x}{dt^2} = -\frac{\rho_a \left(\frac{dx}{dt}\right)^2 A_p \cdot C_D}{2 m_d} \tag{6.44}$$

$$\frac{d^2 y}{dt^2} = g - \frac{\rho_a \left(\frac{dy}{dt}\right)^2 A_p \cdot C_D}{2 m_d} \tag{6.45}$$

where
  A_p the area of the droplet particle (m²)
  m_d the mass of the droplet (kg)
  g the acceleration due to gravity (9.81 m/s²)

The productivity regarding the minimum surface required to produce beads with a given diameter was also calculated by Teunou et al.[31] for a given disk diameter D, rotation speed ω, and flow rate Q_v. This parameter allows estimation of the surface required for given production characteristics (Equation 6.46).[31]

$$\Pi_{(D,\omega,Q_v)} = \frac{S_m}{Q_v \cdot \rho_s} = \frac{\pi(D+2X)^2}{4Q_v \cdot \rho_s} \tag{6.46}$$

6.4 APPLICATIONS

Rotating disk atomization has been the focus of considerable research since the beginning of the twentieth century. This interest can be illustrated by the considerable amount of patents covering fields such as chemicals, resins, powder metallurgy, or agriculture, including pesticide spraying. The need and the importance of the drop size distribution was first noticed in the early 1900s with the influence of the droplet size spectrum on the flame of fuel burners, followed later by the discovery that the performance of turbine and rocket engines were also sensitive to droplet size.[9] Companies such as Shell applied for patents in the field of oil burners, concluding that sprays from rotating cups are much more uniform than with pressure atomizers. Another petrochemical company, Du Pont, also showed interests in rotary atomization and published papers about disk design on the spray characteristics over a wide range of operating parameters.[22] Since the late 1940s, the production of spray by rotating cups was also used for the application of pesticides and insecticides.[7,9] In 1947, Johnson and Walton used a disk atomizer to investigate the spraying of DDT solutions to kill mosquito larvae under natural conditions. Later, in 1948, Kennedy, Ainsworth and Toms used it for the experimental spraying of locusts, and in 1949, Parr and Busvine employed this method to prepare standardized deposits of insecticides for entomological tests.[32] Rotary atomizers were employed either as hand-held devices or as aircraft-mounted applicators, due to their ability to produce sprays with a narrow range of drop sizes.[32] More recently the possibility of using spinning disk atomizers on tractors was developed.[7] In the 1990s, new investigations on rotary atomization were performed in the field of powder metallurgy, however, problems relating to liquid "slippage" or "surfing" were reported.[22] This technique was first developed in the field of metallurgy by Pratt Whitney Aircraft Group for the production of rapidly solidified Ni-based superalloy powders and is still currently used for manufacturing powders of metallic materials such as An, Pb, Al, Zn, Mg, Ti, Mi, Co, and their alloys.[13] The technology is efficient to produce high quality powders[11] and to manufacture ring-shaped components with fine grain structures[35,36] The produced droplets of liquid metal either solidify in flight to form high quality powders or deposit onto a substrate to form microstructurally refined and chemically homogeneous preforms.[36] Rotary atomization can also find applications in the domain of polymer atomization with specific applications in encapsulation,[3,16,23,25] food,[10,27,31] or pharmaceuticals.[27] Rotating disk atomization also finds applications in spray drying.[6,18] Different types of equipment are commonly used to disperse the spray liquid in the drying chamber, including spinning disk atomizers. Centrifugal atomizers were first reported to be used in spray drying devices in 1938 by Fogler and Keinschmidt,[6] and are still used nowadays in such devices, as reported by Prabakaran and Hoti.[18] The advantages that are reported for the use of spinning disk atomization in spray drying device is that they can handle a large throughput and are less susceptible to clogging than a pressure nozzle. They are also better adapted to handle viscous solutions or heavy suspensions and can therefore be designed to disperse practically any material.[6,18]

6.5 EQUIPMENT

In spinning disk atomization, the construction of a rotating disk atomizer depends on the application and the liquid that will be used. However, the rotating disk atomizer will always require a pump or other device for feeding liquid onto a rotating disk from which the drops will be discharged.

A vessel for collecting and, if necessary, gelling the produced droplets should also be placed at a pre-determined distance from the disk rim to collect monodisperse drops[3,16,20,22] and were the velocity of the droplets is minimal in order to reduce percussions or impacts at the collecting point, which can affect the quality of the beads.[31] In centrifugal atomization, the properties of the final product are mostly determined by the size of the droplet and the velocity of the drops leaving the disk. To control these parameters, a good design of the atomizer is mandatory, however, only few theoretical considerations have been reported. Zhao[35] developed a model to facilitate the design of centrifugal atomizers, which was also reported by Sungkhaphaitoon et al.[29] In designing a spinning disk atomizer, the rotational speed of the atomizer should be considered. A powerful high speed electric motor is required to drive the atomizer; however, stability problems can occur with extremely high rotation speeds. When selecting an electrical motor, the power should be high enough to accelerate the liquid to a high velocity and to eject it into droplets. The creation of the surface area of the droplets only consumes a small amount of the input energy. The ratio between the surface energy of the droplet and the input energy is called the energy efficiency, which does not vary significantly with operating conditions. For centrifugal atomization, it is generally in the region of 0.5%. The power required to produce droplets of a certain size is therefore estimated by Zhao[35] as shown in Equation 6.48:

$$S = \frac{6Q}{D} \tag{6.47}$$

$$W = \frac{\gamma S}{\eta} = \frac{6\gamma Q}{\eta D} \tag{6.48}$$

where
Q is the volume flow rate of liquid
D the intended mean droplet size
S the surface area
γ the surface tension of the liquid
η the energy efficiency of centrifugal atomization (≈ 0.05)

In his model, Zhao[35] suggests that the majority of the input energy is consumed in accelerating the liquid to a high velocity at the edge of the atomizer. He approximated the power consumption by Equation 6.49:

$$W = \frac{1}{2}\rho Q v^2 \tag{6.49}$$

where
ρ is the density of the liquid
Q the flow rate of the liquid
v the velocity

In his model, Zhao[35] then estimated the velocity of the spray droplets at the edge of the atomizer as equal to the circumferential velocity of the atomizer. The higher velocity is therefore simply expressed as

$$v = \omega R \tag{6.50}$$

By combining Equations 6.49 and 6.50, the maximum radius of the disk can be established (Equation 6.51).

$$R_{max} = \frac{1}{\omega}\sqrt{\frac{2W}{\rho Q}} \qquad (6.51)$$

with

R the radius of the disk

ω the rotation speed in (radian/s)

To convert a rotation speed from rpm to radian/s, a factor of π/30 must be multiplied.

Zhao also warned that in designing an atomizer, one should also consider that the disk is subject to high tension due to the centrifugal force at high rotation speeds and that the material of the atomizer must be able to withstand the tensions without deformation or fracture.[35] To deter mine the minimum radius, Zhao had to take into account the liquid flow on a rotating disk or cup, which can be complex. An annular discontinuity in the flow called hydraulic jump often occurs on the atomizer. It manifests itself by an abrupt increase in the liquid thickness and a reduction in the radial velocity. Zhao reported that the flow after the hydraulic jump is insensitive to fluctuations in the liquid stream and controlled by the rotating atomizer. Zhao then suggested that it is therefore advantageous for the liquid to disintegrate after the hydraulic jump, which means that the minimal radius should be larger than the hydraulic jump radius. The minimal disk radius is then estimated as shown in Equation 6.52 by the Zhao model.

$$R_{min} = 0.55 \left(\frac{\rho Q^2}{\omega}\right)^{1/4} \qquad (6.52)$$

In rotary atomization, the role of the disk is essential as it determines the characteristics of the atomized drops.[22] Senuma reported that conventional rotating disks cannot avoid slippage of the fluid, however to avoid this phenomenon, various solutions using geometry-based approaches have been proposed. There are generally two categories of disk designs: vaned and vaneless disks.[22]

- Vaned disks: For vaned wheels, Senuma explained that the liquid is fed onto the disk and can only spread over the vertical vanes under the influence of the centrifugal force. The liquid is thereby confined to the vane surface, avoiding surfing. Thus, the liquid flow is, according to Senuma, entirely determined by the geometry of the vane and the liquid acquires maximum velocity at the disk rim and emerges from vertical surfaces.[22] Nevertheless, Senuma reported that some conditions are required for optimum size uniformity such as smooth vane surfaces, uniform liquid feed rate (varying feed rates change the drop size), complete wetting of the vane surfaces, uniform distribution of feed to the vaned wheel.[22] Different disk existing designs are mentioned by Senuma, all aimed at better size distribution, less maintenance, long durability, or higher yield capacity. Senuma explained that head design such as the curved vane disk and the nozzle wheel are adapted features conceived to reduce air-pumping effects.[22] Curved vane disks can increase the product bulk density at high feed rates compared to straight vane disks. The nozzle wheel is a tall disk with small orifices instead of channels, in which the feed rate is limited, because of the small cross sectional area of these orifices.[22] Senuma also warned that when suspensions of hard materials are used, abrasion of the disk is particularly important at the edge of the vanes, where the liquid leaves the disk; however, he suggested that wear-resistant bushings can increase the disk operating life.[22]

- Vaneless disks: In vaneless disks, Senuma suggests that the flow is modified to prevent surfing over the entire disk surface. The disk is designed to present an inverted bowl-like form, which increases the friction between the liquid and undesired rotating surface.[22] Inverted bowls, cups, and plates are classified as vaneless disk atomizers.[22] The liquid is fed from the reverse side to increase friction between the disk surface and the liquid, which is pressed against the surface as it flows to the edge.[22] According to Senuma, the difference between the three disk-type atomizers is their cone-angle, which defines the angle between the axis of rotation and the release surface.[22] Plate disks are flat, while bowl disks have a smaller cone angle and cup disks have the smallest angle.[22] The liquid flows out from the entire periphery of the disk. The liquid layer has, therefore, to be uniform over the entire disk rim, and the liquid should perfectly spread and accelerate to the disk speed prior to atomization.[22] The liquid leaves from a horizontal rim that is completely axisymmetrical when in vaned disk, liquid is pressed on a vertical plane due to centrifugation.[22] It is reported by Senuma that sprays produced with vaneless disks generate larger drops than those with vaned disks; however, the homogeneity of the spray is better with vaneless disks.[22] These properties make vane disks interesting for the production of powders without dust. Vaneless disks are therefore generally used for applications where large and uniform sprays are needed.[22]

Senuma reported that vaned disks are chosen for production of fine sprays, while inverted disks are more popular for applications requiring large particles.[22] However, two conditions are essential to produce the best monodisperse drops: the disk rotation must be vibrationless,[25,33] and the peripheral speed must be sufficient to reduce gravitational forces to negligible levels.[22] The liquid distribution velocity must be sufficiently low to avoid splashing, which will otherwise result in leaking distributors, deposits on the disk or drive surface, and undesired drop size distribution.[22]

Different adaptations to the rotating disk atomizer can be made depending on the application of the produced beads.[3,7,16,36] For encapsulation, the system developed for atomization for liquids can be used after some modifications. The first requirement is that the atomizer (disks and collection vessels) has to be housed within a larger, closed vessel, which allows the sterilization of the device. This larger vessel also acts as a water bath for temperature control during encapsulation. A container equipped with a mixing device should be provided for homogeneous mixing of the cells within the polymer solution during atomization.[3,16] In the case of polymer gelling at low temperatures, the container containing the liquid polymer should be temperature controlled in order to avoid gelling inside the container and the tubing.[16,22] The produced droplets should also fall directly in a collecting vessel containing a continuously stirred gelling agent solution.[3,17,31] An example of such an atomizer for cell encapsulation is the vortex-bowl disk atomizer system,[3] where the gelling bath stirring is produced by a rotating vessel, creating a vortex of the gelling agent solution. This principle is shown in Figure 6.5.

When molten liquids are used, different adaptations of the disk or of the vessel have to be made in order to keep the temperature of the disk constant. To ensure temperature stability, the atomization chamber can be insulated and the region of the drop break-off can be protected with plates confining the hot air flow through a narrow channel in order to reduce the temperature gradient. These modifications have been developed and tested to produce beads from molten polymers.[5,22] When the molten liquid gels by reducing the temperature, a drying tower as illustrated in Figure 6.6 can be used to create an appropriate gradient. Heated air is introduced at the level of the disk to ensure a uniform temperature around the disk and a liquid nitrogen stream is introduced at the bottom of the vessel, to chill the particles and allow them to gel. This system has been successfully developed and used in the production of bioresorbable microspheres.[23]

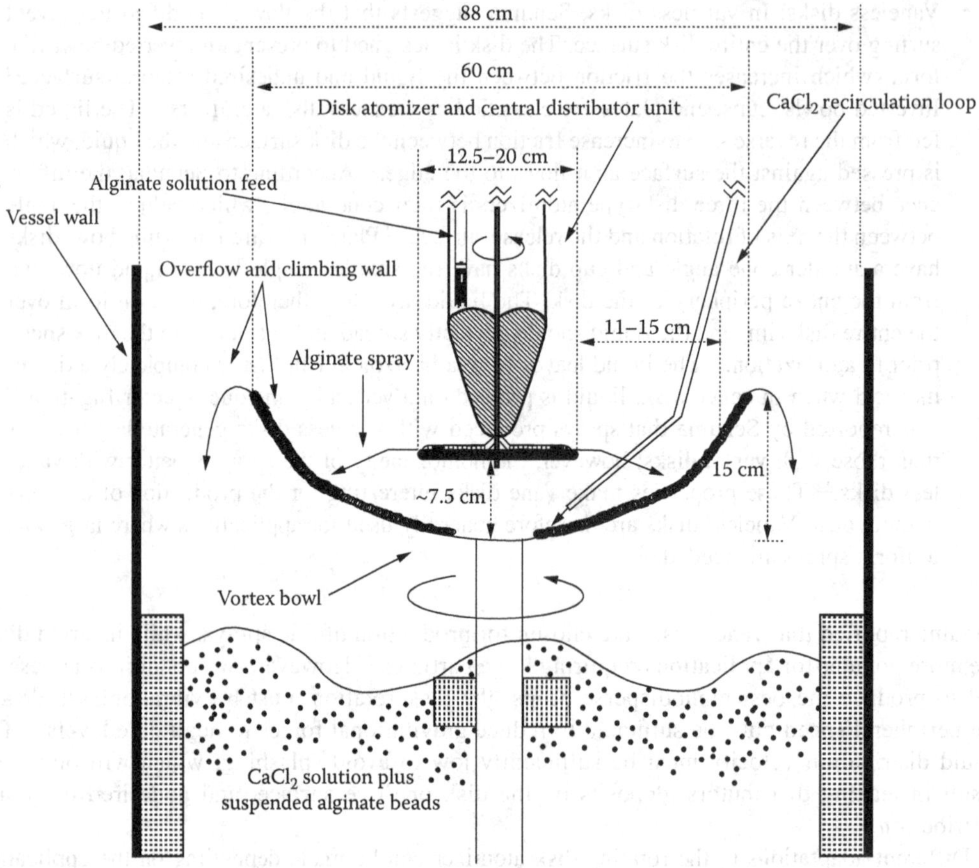

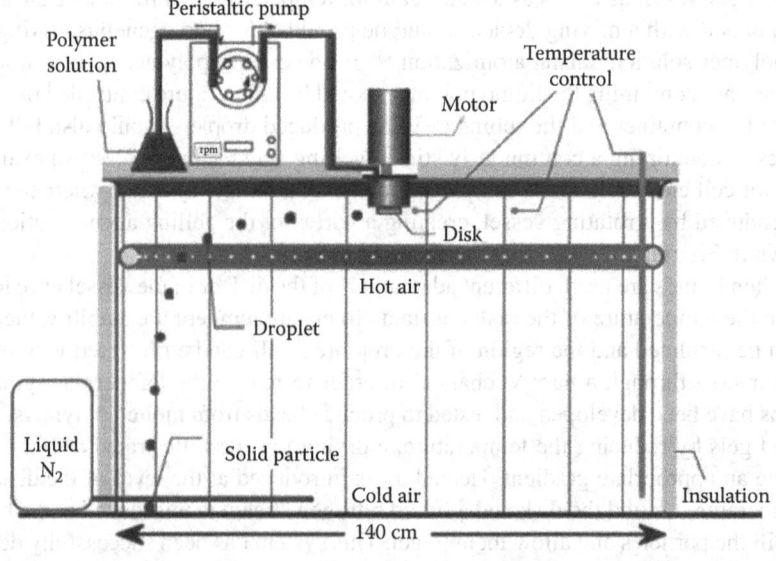

FIGURE 6.5 Vortex-bowl disk atomizer system for the production of alginate beads. (Reproduced from Champagne, C.P. et al., *Biotechnol. Bioeng.*, 68(6), 681, 2000. With permission.)

FIGURE 6.6 Schematic setup of the spinning disk atomizer equipment and heating tower. (Reproduced from Senuma, Y. et al., *Biotechnol. Bioeng.*, 67(5), 616, 2000. With permission.)

6.6 ADVANTAGES AND DISADVANTAGES

In encapsulation, the different techniques have to meet criteria such as producing monodisperse, homogeneous, and spherical beads, with a small size and a narrow size distribution. The device has to be relatively easy to set up and simple to operate, the time of production should be short, moreover the efficiency and the production rates should be high. The ability to extrude viscous solutions, to produce a range of different size droplets and to operate under sterile conditions are also required.[34] As encapsulation is one of the possible applications of rotary disk atomization, this technique should meet with the previous criteria. Spinning disk atomization is a technique that can produce monodisperse, homogeneous, and spherical beads, with a narrow size distribution; it can be used with different viscosities of liquid solutions[22] and operate under sterile conditions.[16,17] The system has the advantage to be able to produce a large quantity of beads in a short time and can be scaled-up easily by increasing the flow rate, the disk size, and the rotation speed of the disk.[31,33] However, the spinning disk processes require a large surface or volume to collect the produced beads without damaging them.[31] This large surface can also create problems with sterile operation in case of cell encapsulation. The process appears to be incapable of producing small particles of less than 100 μm in size. Despite this, a reduction in size could be achieved by increasing the rotational speed of the disk, rotating large disks at high speeds can cause the disk to vibrate and produce satellites drops.[22,25] The size can also be reduced by reducing the viscosity of the polymer, however if the viscosity is too low, the shape of the bead will seriously be affected.[22,33] Another disadvantage of the spinning disk is the loss of product caused by the production of satellite drops.[24,25,33] As the large particles travel further than the small particles, the small particles do not reach the collecting gelling bath.[31] This property can be used to select the size of the capsules and the size distribution,[31,33] however, all the drops that are not in the selected size range will be lost.[22] Increasing the collecting surface reduces the loss; however, it will also increase the size distribution.[33] Another possibility would be to recycle the lost polymer with a recirculating system. Spinning disk atomization is a technique that can be used in encapsulation, as it fulfills most of the criteria. However a few improvements or adaptations still need to be made in order to produce small size beads and to control the loss of polymer due to the satellite drops. It is also important to note that the technique has a high production rate and can be scaled-up easily; however, due to the principle of the process itself, it is more difficult to scale it down for small laboratory production.

6.7 CONCLUSION

The entrapment of various materials such as flavors, living cells, pharmaceutical compounds within beads for different purposes is of great importance in the pharmaceutical, chemical, and food industries, as well as in agriculture, biotechnology, and medicine and, diverse encapsulation techniques have been developed. The different techniques must reach some criteria in order to be employed in industry. These criteria can be separated in two main groups, the first group concerns the capsule characteristics and include the production of monodispersed, homogeneous, and spherical beads of small size and narrow size distribution, and the second group is more about the technique itself, which must be relatively easy to set up and simple to operate, with a high efficiency and a high production rate, the ability to use viscous solution, to produce a range of different-sized droplets, and to be sterilized; moreover, the production time must be short. Spinning disk atomization is a technique that has been widely used for diverse applications such as oil burners, pesticides spraying in the agriculture industry, powder production, and spray deposition in the metallurgy industry since the 1900s and was recently developed for encapsulation processes. The technique allows the production of monodisperse, homogeneous, and spherical beads, with a narrow size distribution, it can be used with different viscosities of liquid solutions and operate under sterile conditions. Moreover it can produce a large amount of beads in a short time and can be scaled-up very easily. Spinning disk atomization is an efficient technique for industrial microencapsulation using polymer solutions or molten liquids; however, the method requires a large surface to collect the beads without damaging them and a number of improvements or adaptations are required in order to produce small size beads and to control the loss of polymer due to satellite production.

REFERENCES

1. M. Ahmed and M. S. Youssef. Characteristics of mean droplet size produced by spinning disk atomizers. *Journal of Fluids Engineering*, 134(7):071103–071103, 2012.
2. A. V. Anilkumar, I. Lacik, and T. G. Wang. A novel reactor for making uniform capsules. *Biotechnology and Bioengineering*, 75(5):581–589, 2001.
3. C. P. Champagne, N. Blahuta, F. Brion, and C. Gagnon. A vortex-bowl disk atomizer system for the production of alginate beads in a 1500-liter fermentor. *Biotechnology and Bioengineering*, 68(6):681–688, 2000.
4. E.-S. Chan, B.-B. Lee, P. Ravindra, and D. Poncelet. Prediction models for shape and size of ca-alginate macrobeads produced through extrusion-dripping method. *Journal of Colloid and Interface Science*, 338(1):63–72, October 2009.
5. M. Eslamian, J. Rak, and N. Ashgriz. Preparation of aluminum/silicon carbide metal matrix composites using centrifugal atomization. *Powder Technology*, 184(1):11–20, 2008.
6. B. B. Fogler and R. V. Kleninschmidt. Spray drying. *Industrial and Engineering Chemistry*, 30(12):1372–1384, 1938.
7. A. Frost. Rotary atomization in the ligament formation mode. *Journal of Agricultural Engineering Research*, 26(1):63–78, January 1981.
8. C. J. Geankoplis. Principles of momentum transfer and applications. In *Transport Processes and Unit Operations*, 3rd edn. Allyn and Bacon, Boston, MA, 1993, pp. 114–213.
9. M. R. Gebhardt. Rotary disk atomization. *Weed Technology*, 2(1):106–113, January 1988.
10. S. Gouin. Microencapsulation: Industrial appraisal of existing technologies and trends. *Trends in Food Science and Technology*, 15(7–8):330–347, 2004.
11. K. Ho and Y. Zhao. Modelling thermal development of liquid metal flow on rotating disc in centrifugal atomisation. *Materials Science and Engineering: A*, 365(1–2):336–340, January 2004.
12. E. R. Lewis and S. E. Schwartz. Kinematics and dynamics of SSA particles. In *Sea Salt Aerosol Production: Mechanisms, Methods, Measurements, and Models—A Critical Review*. American Geophysical Union, January 2004, pp. 63–71.
13. H. Li and X. Deng. Prediction of powder particle size during centrifugal atomisation using a rotating disk. *Science and Technology of Advanced Materials*, 8(4):264, May 2007.
14. I. Marison, A. Peters, and C. Heinzen. Liquid core capsules for applications in biotechnology. In V. Nedović and R. Willaert, eds., *Fundamentals of Cell Immobilisation Biotechnology*, number 8A in Focus on Biotechnology. Springer, Dordrecht, the Netherlands, January 2004, pp. 185–204.
15. S. Matsumoto, K. Saito, and Y. Takashima. Phenomenal transition of liquid atomization from disk. *Journal of Chemical Engineering of Japan*, 7(1):13–19, 1974.
16. J. C. Ogbonna. Atomisation techniques for immobilisation of cells in micro gel beads. In V. Nedović and R. Willaert, editors, *Fundamentals of Cell Immobilisation Biotechnology*, number 8A in Focus on Biotechnology. Springer, Dordrecht, the Netherlands, January 2004, pp. 327–341.
17. J. C. Ogbonna, M. Matsumura, T. Yamagata, H. Sakuma, and H. Kataoka. Production of micro-gel beads by a rotating disk atomizer. *Journal of Fermentation and Bioengineering*, 68(1):40–48, 1989.
18. G. Prabakaran and S. Hoti. Optimization of spray-drying conditions for the large-scale preparation of bacillus thuringiensis var. israelensis after downstream processing. *Biotechnology and Bioengineering*, 100(1):103–107, 2008.
19. U. Pruesse, L. Bilancetti, M. Bucko, B. Bugarski, J. Bukowski, P. Gemeiner, D. Lewinska et al. Comparison of different technologies for alginate beads production. *Chemical Papers*, 62(4):364–374, August 2008.
20. G. A. Roth and G. E. Reins. Rotating disk apparatus for the production of droplets of uniform size. *Weeds*, 5(3):197–205, 1957.
21. D. J. Ryley. Analysis of a polydisperse aqueous spray from a high-speed spinning disk atomizer. *British Journal of Applied Physics*, 10(4):180, April 1959.
22. Y. Senuma. Generation of monodispersed polymer microspheres by spinning disk atomization. PhD thesis, EPFL, Lausanne, 1999.

23. Y. Senuma, S. Franceschin, J. Hilborn, P. Tissieres, I. Bisson, and P. Frey. Bioresorbable microspheres by spinning disk atomization as injectable cell carrier: From preparation to in vitro evaluation, *Biomaterials*, 21(11):1135–1144, 2000.
24. Y. Senuma and J. G. Hilborn. High speed imaging of drop formation from low viscosity liquids and polymer melts in spinning disk atomization. *Polymer Engineering and Science,* 42(5):969–982, 2002.
25. Y. Senuma, C. Lowe, Y. Zweifel, J. G. Hilborn, and I. Marison. Alginate hydrogel micro-spheres and microcapsules prepared by spinning disk atomization. *Biotechnology and Bioengineering*, 67(5):616–622, 2000.
26. D. Serp, E. Cantana, C. Heinzen, U. Von Stockar, and I. W. Marison. Characterization of an encapsulation device for the production of monodisperse alginate beads for cell immobilization. *Biotechnology and Bioengineering*, 70(1):41–53, 2000.
27. H. M. Shewan and J. R. Stokes. Review of techniques to manufacture micro-hydrogel particles for the food industry and their applications. *Journal of Food Engineering*, 119(4):781–792, 2013.
28. D. C. Sobeck and M. J. Higgins. Examination of three theories for mechanisms of cation-induced bioflocculation. *Water Research*, 36(3):527–538, 2002.
29. P. Sungkhaphaitoon, T. Plookphol, and S. Wisutmethangoon. Design and development of a centrifugal atomizer for producing zinc metal powder. *International Journal of Applied Physics and Mathematics*, 77–82, 2012.
30. Tansawa, Y., Miyasaka, Y. and Umehara, M., Atomization with Rotating Disk I-II-III, *Transactions of the Japan Society of Mechanical Engineers*, 25(156): 879, 888, 897, 1959.
31. E. Teunou and D. Poncelet. Rotary disc atomisation for microencapsulation applicationsprediction of the particle trajectories. *Journal of Food Engineering,* 71(4):345–353, 2005.
32. W. H. Walton and W. C. Prewett. The production of sprays and mists of uniform drop size by means of spinning disc type sprayers. *Proceedings of the Physical Society. Section B,* 62(6):341, June 1949.
33. M. Whelehan and I. W. Marison. Microencapsulation by dripping dand jet break up. *Bioen- capsulation innovations*, 4–10, September 2011.
34. M. Whelehan and I. W. Marison. Microencapsulation using vibrating technology. *Journal of Microencapsulation*, 28(8):669–688, December 2011.
35. Y. Zhao. Considerations in designing a centrifugal atomiser for metal powder production. *Materials and Design,* 27(9):745–750, 2006.
36. Y. Y. Zhao. Analysis of flow development in centrifugal atomization: Part I. Film thickness of a fully spreading melt. *Modelling and Simulation in Materials Science and Engineering*, 12(5):959, September 2004.
37. Y. Y. Zhao. Analysis of flow development in centrifugal atomization: Part II. Disintegration of a non-fully spreading melt. *Modelling and Simulation in Materials Science and Engineering,* 12(5):973, September 2004.

7 Encapsulation via Fluidized Bed Coating Technology

Charles R. Frey

CONTENTS

Abbreviations 111
Unit Conversions 112
7.1 Introduction 112
7.2 History 112
7.3 Process Descriptions 113
    7.3.1 Fluidization 113
    7.3.2 Wurster (Bottom Spray) Fluid Bed Coating 115
    7.3.3 Spouted Bed 118
    7.3.4 Top Spray 119
    7.3.5 Tangential Spray (Rotary) Granulation 119
7.4 Parameter Assessment 120
    7.4.1 Process Air 120
        7.4.1.1 Spouted Bed, Top Spray, and Tangential Spray Process Air 121
        7.4.1.2 Wurster Process Air 124
        7.4.1.3 Process Air Temperature and Volume Relationship 129
    7.4.2 Nozzles and Atomization 131
    7.4.3 Temperature 133
        7.4.3.1 Process Temperatures 133
        7.4.3.2 Hot Melt Temperatures 134
    7.4.4 Drying Capacity 135
    7.4.5 Tangential Spray Rotor Speed 140
7.5 Application 140
    7.5.1 Processing Obstacles 140
    7.5.2 Practical Considerations 141
    7.5.3 Commercial Applications 142
        7.5.3.1 Coating Ingredients 142
        7.5.3.2 Application Areas and Coat Functions 142
References 144

ABBREVIATIONS

atm	atmospheres
°C	degrees Centigrade
cfm	cubic feet per minute
cm	centimeters
°F	degrees Fahrenheit
ft	feet or foot
g	gram

°K   degrees Kelvin
min   minute
mg   milligram
mm   millimeter
psi   pounds per square inch
scfm   cubic feet per minute at standard pressure and temperature
s   seconds
Tg   glass transition temperature
μm   micrometer

UNIT CONVERSIONS

°C = (°F − 32)/1.8
1 in. = 2.54 cm
1 m = 3.28 ft

7.1 INTRODUCTION

Fluidized bed coating technology refers to processes where solid particulate material is fluidized in relation to coating spray from a nozzle or nozzles to apply a coating on the particulate. In the context of this publication, the coatings provide a controlled-release property to either the core particle or some element of the applied coating. The following processes are addressed to varying extents in this work:

Wurster (bottom spray) fluid bed coating
Spouted bed coating
Top spray granulation
Tangential spray (rotary) granulation

Other related processes exist and may have found some level of commercial success, but they are not addressed directly in this chapter. The principles that are addressed can generally be applied to any fluid bed coating process.

Applied coatings are film-forming materials or mixtures and must be deliverable as an atomized spray from a nozzle. In most applications, the coating is delivered with an organic solvent-based or water-based vehicle as a solution or suspension. The vehicle is removed by evaporation during processing. Molten materials such as waxes can be applied as coatings from the molten state without any vehicle. These options provide significant opportunity to tailor release properties through coat composition, coat layering, and coat thickness. Other than solvent vehicles employed in the coating process, coating materials and formulations are not discussed in this chapter.

This chapter provides process descriptions and supportive details to facilitate proper and safe application of fluid bed coating technology. This information provides readers with a practical knowledge that can be applied to the development and optimization of new applications or troubleshoot existing ones.

7.2 HISTORY

Fluidized bed coating technology emerged in the 1950s and 1960s. Many designs and developments have been disclosed in various patents extending from that time.[1-26] Some processes are designed to coat or agglomerate particles. Others are primarily used to agglomerate particles to form granulations. The Wurster (bottom spray) process has realized the greatest success and recognition as a means to uniformly apply a film coat to particulate materials.

Encapsulation via Fluidized Bed Coating Technology

The Wurster process was invented by Dr. Dale E. Wurster at the University of Wisconsin. This evolved from his interest in finding ways to coat tablets in the 1940s and 1950s. His patent disclosures in the 1960s[1–6] were focused on small particle coating and are the basis of systems and designs that remain in use today. Most, and likely all, commercially available Wurster equipment is linked to Dr. Wurster's work at the University of Wisconsin. Wurster technology was originally transferred either directly from the university or from Coating Place, Inc., a company built on the 1976 acquisition of all Wurster-related records and equipment from the university.

7.3 PROCESS DESCRIPTIONS

7.3.1 FLUIDIZATION

Fluidization is inherent in all fluid bed coating processes and is a condition realized when air is pushed or pulled upward through a bed of particulate material. A detailed treatise on fluidization is documented in a chapter authored by Dr. Wurster.[27] General fluidization states are illustrated in Figure 7.1. Briefly, below the point of fluidization, air moving through a particulate bed passes around the particles with minimal particle lift or movement. As air velocity is increased, there may be some decrease in bulk density, depending on particle physical characteristics such as size and shape and eventually, a point is reached where particles begin to lift and suspend in the moving air. When this happens, the particle bed takes on a fluid-like behavior as particles are relatively free to move about in all directions. This initial point of fluidization may also be referred as the point of "incipient" fluidization. As airflow is increased beyond this point of initial fluidization, the bed fluidizes and may bubble or "slug" as it takes on a turbulent character. Increased further, the air eventually transports particles pneumatically with it.

Incipient fluidization air velocity varies with particle physical characteristics such as size, shape, surface characteristics, and specific gravity; however, for a given material, it is not influenced by the amount of material in the bed. The incipient fluidization point may not be well defined, depending on particle characteristics and uniformity of air movement into and through the bed. If particles tend to interlock or "stick" to one another, increased air velocity to break up that interlocked

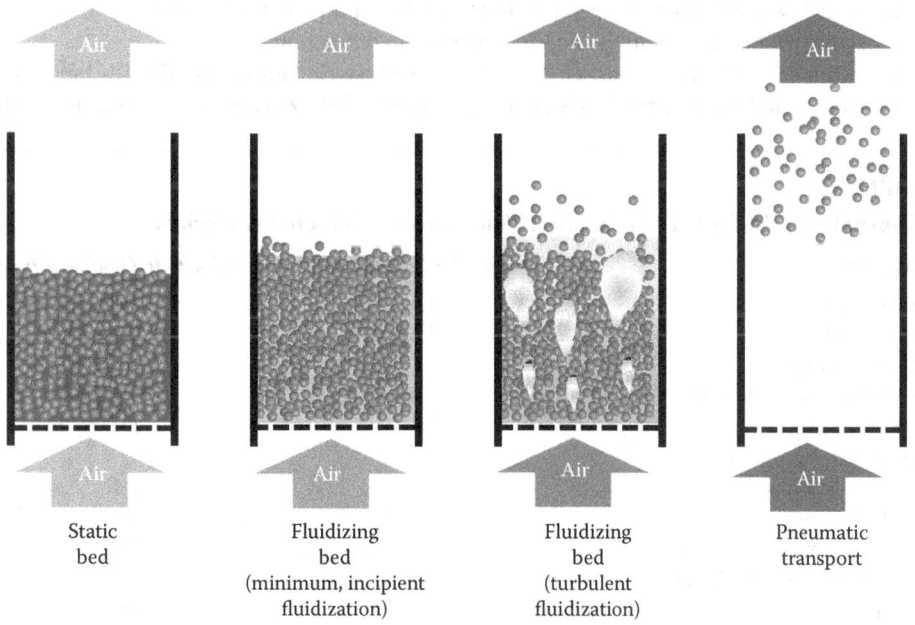

FIGURE 7.1 Fluidized states.

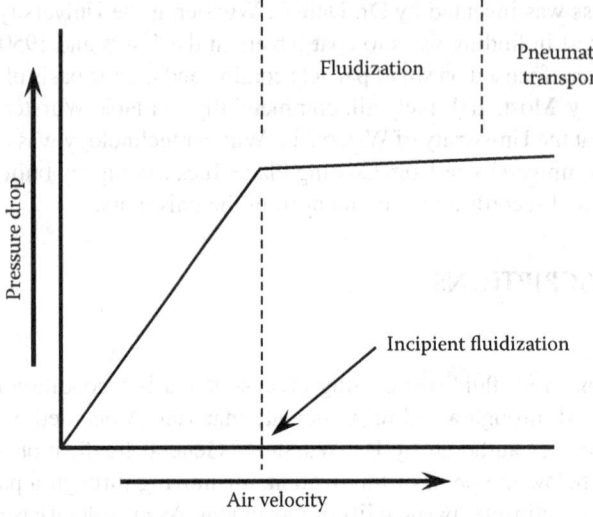

FIGURE 7.2 Pressure drop versus air velocity to estimate incipient fluidization velocity.

matrix or overcome the "sticking" forces may be required for adequate fluidization. The incipient fluidization velocity can be estimated by drawing air through a particle bed and visually observing when the fluidized state is realized. The airflow volume can be divided by the plate surface area to determine the air travel distance per unit time. For example, if 100 ft³/min airflow is the point where fluidization begins for a given particle bed with a cross-sectional area of 2 ft², the air velocity at insipient fluidization is ~50 ft/min (100/2 = 50).

Another means to estimate the incipient fluidization velocity is to use the pressure drop through the particle bed. This is illustrated in Figure 7.2. Pressure drop increases with air velocity at airflows below the incipient fluidization velocity. Above the incipient fluidization velocity, the pressure drop through the particle bed is relatively independent of airflow. The inflection point at the transition from airflow below the fluidization velocity to above the fluidization velocity in a graph of pressure drop versus airflow or air velocity is the ~incipient fluidization point.

Estimated incipient fluidization air velocities for various materials are summarized in Table 7.1. These values were estimated based either on visual observation of a fluidized bed with airflow measurements

TABLE 7.1
Approximate Incipient Fluidization Velocities for Selected Materials

Material	Specific Gravity (g/cc)	Air Velocity to Fluidize (ft/min)
550 mg tablet	~1.6	~458
9/16″ caplet	~1.6	~210
2 mm × 3 mm pellet	~0.8	~169
3 mm diameter gelatin sphere	~1.2	~158
~18 mesh salt	~3.34	~132
18/20 mesh sugar sphere	~1.6	~132
20 mesh sugar sphere	~1.6	~100
25/30 mesh sugar sphere	~1.6	~92
~1 mm × 1 mm aluminum flake	~2.55–3.6	~79
20/60 mesh potassium chloride	~1.98	~34
Bakers special sugar	~1.59	~25
100–200 μm microcrystalline cellulose sphere	~1.6	~15

Encapsulation via Fluidized Bed Coating Technology

provided by a calibrated airflow sensor or monitoring the pressure drop through the bed. These data provide a reasonable guide to approximate the incipient fluidization velocity for these or similar materials. Airflow requirements for material sized similarly to any one of the listed materials, but with alternative specific gravity or structure, could vary significantly from the table value.

Although particles are generally discrete entities with relatively low aspect ratios, materials with higher aspect ratios are applicable, provided they can be adequately fluidized. Elongated materials such as hard gelatin capsules, caplets, extrudates, and wires have been successfully fluidized for coating application. Some materials with uniformly sized and shaped particles can potentially align and channel air in a manner that "freezes" or "bridges" in a coater during fluidization. This concern is similar to the particle interlocking noted earlier. This can be overcome either by operating process air at suitable velocity or by introducing turbulence that prevents such alignment.

Airflow requirements for pneumatic transport vary with particle size, specific gravity, and structure.

7.3.2 Wurster (Bottom Spray) Fluid Bed Coating

The Wurster process is characterized by differential air velocity moving upward through two zones of the particle bed. Although a variety of configurations for the positioning of the zones were disclosed in the Wurster patents,[1–6] the preferred and optimal configuration was identified as a concentric design with the nozzle at the center and bottom of a cylindrical particle bed. Illustrations for the classic cylindrical Wurster design are shown in Figures 7.3 and 7.4. A high-velocity, "up-bed" region is positioned concentrically around the nozzle and a low-velocity, "down-bed" region is positioned concentrically around the up-bed. A physical barrier commonly referred to as a "partition" or

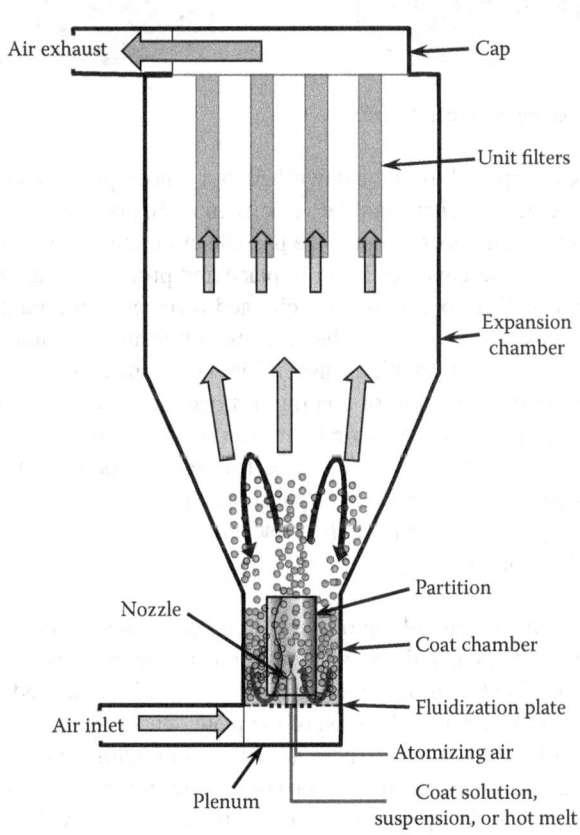

FIGURE 7.3 Wurster coater diagram.

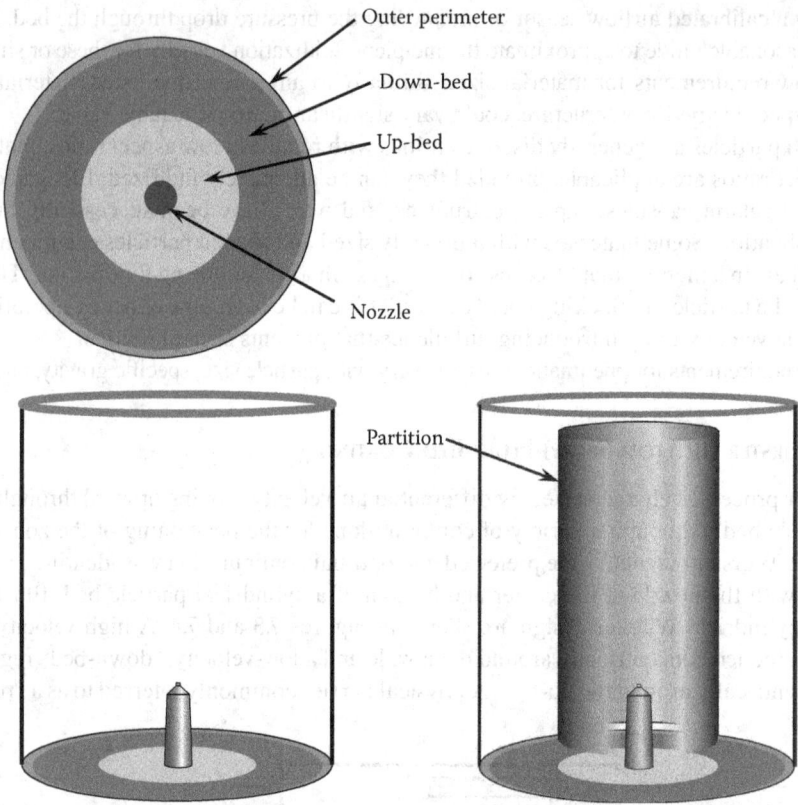

FIGURE 7.4 Wurster coating chamber diagram.

"column" may separate the up-bed from the down-bed, but is not a process requirement. A region of the down-bed with air velocity higher than the velocity in main portion of the down-bed is located at the outer perimeter of the chamber to optimize particle circulation during operation.

In a typical process, the coat chamber is put in place and process air and temperature are set to precondition the equipment. The coat chamber is charged with core material that is to be coated, the unit is reclosed, and process air is started. When process air establishes fluidization and conditions for the start of coating are reached, solution, suspension, or hot melt spray is started. Coating material is sprayed until the desired amount of coating has been applied. When spraying is complete, drying, curing, or cooling steps may be carried out prior to discharge.

Properly balanced fluidization is critical to successful Wurster coating. This fluidization includes critical aspects of the down-bed, up-bed, and expansion chamber regions. The dynamics of airflow and particle flow are complex and have been the subject of several studies.[28–32] The critical elements include the following three points:

1. The air velocity in the down-bed region should be near or above the incipient fluidization velocity for the particle bed. This velocity suspends particles in the down-bed region and allows them to move freely. The air entering the down-bed is composed of process air passing upward through the down-bed region of the plate. Once this air is through the plate, it follows the path of least resistance, which involves a split with a portion of the air passing upward through the down-bed and another portion sweeping across the plate toward the up-bed. The air sweep to the up-bed transfers fluidized particles with it. The rate of air and particle sweep to the up-bed is dependent on several relatively fixed and some dynamic factors including those noted in Table 7.2.

TABLE 7.2
Factors Affecting Particle Feed to the Up-Bed

Factor	Description
Plate physical dimensions and air distribution	Relative physical area and position of the up- and down-bed regions and the amount of air entering each region determine the relative velocity of the process air entering each region through the plate.
Partition length	Partition length influences the volume of the up-bed zone, which affects the particle mass in the up-bed.
Partition gap (distance between the lower edge of the partition and the fluidization plate)	The gap affects the feed area from the down-bed to the up-bed, which gates both air and particles to the up-bed from the down-bed. A smaller gap allows less air and particle entrance from the down-bed to the up-bed, which reduces the particle mass in the up-bed and promotes airflow to the up-bed; however, this is mitigated by reduced feed area under the partition. A larger gap allows more air and particle entrance and increases particle mass in the up-bed; however, this is mitigated by the added restriction created by adding particle mass to the up-bed.
Process air flow	If process air is insufficient to adequately fluidize in the down-bed, particle feed across the plate from the down-bed to the up-bed may be compromised. Excessive air may reduce feed to the up-bed by both reducing the particle mass differential between the up- and down-beds and reducing the particle density at the base of the down-bed.
Nozzle atomizing air volume	Atomizing air volume dilutes particle mass in the up-bed, which promotes air and particle feed from the down-bed to the up-bed.
Charge amount	Higher amounts of material in the unit add greater particle mass to the down-bed, which creates greater restriction to air flow upward through the down-bed and promotes air and particle feed to the up-bed. Smaller amounts reduce such restriction and allow more down-bed air to channel upward through the down-bed and reduce air and particle feed to the up-bed.

2. The air velocity in the up-bed region should be at or above the pneumatic transport velocity for the particles. Air in the up-bed is made up of process air passing upward through the up-bed region of the plate, process air traveling under the partition from the down-bed, and atomizing air from the nozzle. These combined airflows must be sufficient to pneumatically transport particles upward through the spray region in the up-bed with high efficiency. Any particles that do not clear the spray zone on a given pass may potentially be overwetted with spray and exposed to greater physical stresses from the atomizing air. Air entering the up-bed directly through the plate is typically the greatest contributor to air velocity in the up-bed. It also dilutes particles entering from the down-bed, which is required to provide proper air and particle feed from the down-bed to the up-bed.
3. The expansion chamber is located above the particle bed and partition. The cross-sectional area increases with the upward increase in chamber diameter. This increase expands the volume and reduces air and particle velocities. Velocity must be reduced sufficiently for particles to drop back to the top of the down-bed.

The coating capabilities of the Wurster process are realized due to the particle dilution in the up-bed and expansion chamber, the close proximity of particles to the spray as they pass through the coating zone, and the drying or congealing time in the expansion chamber prior to their fallback to the down-bed. The dilution aspect reduces particle-to-particle contact that potentially promotes particle agglomeration. The close proximity of particles to spray allows high coating efficiency with negligible spray drying or spray congealing prior to spray contact with the particles. The relatively extended time in the expansion chamber allows optimal drying or congealing of sprayed material prior to significant particle-to-particle contact in the down-bed.

Due to differing air velocity needs for smaller versus large particles (Table 7.1), the Wurster process is generally restricted to relatively narrow particle size distributions for the core material. A size range within ~1:5 (smallest to largest particles) in the distribution is best and up to ~1:10 is generally acceptable. As the distribution widens, smaller particles in the distribution have a tendency to accrete to larger particles, which can be acceptable depending on the goals of the process.

Wurster technology can also be used for agglomerating particles to form granulations. This typically involves configuration without a partition and an air distribution pattern that fluidizes in both the up- and down-bed regions, but provides only marginal pneumatic transport through the up-bed. Promotion of particle density in the coating zone and reduced time in the expansion chamber promotes accretion and agglomeration. Process conditions are also generally wetter than those used for coating to promote agglomeration.

7.3.3 Spouted Bed

A spouted bed process is similar to the Wurster process. The principal difference is that an air velocity differential needed to properly circulate particles through the nozzle spray zone is created by a tapered coat chamber design rather than an air distribution pattern created by the fluidization plate. An illustration of a spouted bed is provided in Figure 7.5. Process air entering through the fluidization plate passes upward past the nozzle, carrying particles with it as they feed in from the periphery of the plate and lower chamber. Spray is directed upward with the direction of particle flow. Vertical increase in cross-sectional area reduces the air and particle velocity to allow particles

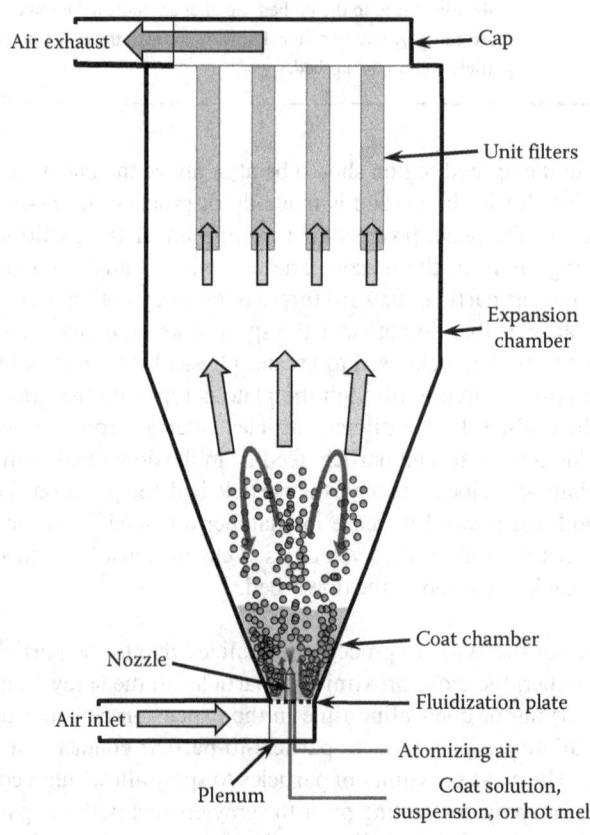

FIGURE 7.5 Spouted bed diagram.

to fall to the down-bed area. Particles are channeled downward and inward to the feed area near the fluidization plate by the tapered geometry of the lower chamber.

7.3.4 TOP SPRAY

The top spray process is illustrated in Figure 7.6. The top spray process is characterized by a spray nozzle or nozzles spraying coating material downward into a fluidizing particle bed. Random particle movement combined with the downward spray promotes particle interaction and agglomeration. In addition, the lower fluidizing air requirements of smaller particles in a particle distribution result in greater lift of those particles toward the nozzles compared to larger agglomerates as they form. This has a "leveling" effect as agglomeration of the finer particles is promoted.

The top spray process is used primarily to agglomerate particles and form granulations. Depending on vertical nozzle positioning and parameters employed, the process can achieve a marginal level of coating. Work comparing the top spray process to the Wurster process indicated significant coating uniformity and efficiency limitations with the top spray process.[33]

7.3.5 TANGENTIAL SPRAY (ROTARY) GRANULATION

A tangential spray configuration is illustrated in Figure 7.7. The base of the coating chamber is a solid rotating plate. This plate may have various surface topographies to promote desired product characteristics. Process air enters the chamber at the periphery of the rotating plate. Plate rotation forces particles outward to the periphery where they encounter the process air and are lifted upward and inward. The air velocity at the entrance to the product chamber is controlled by

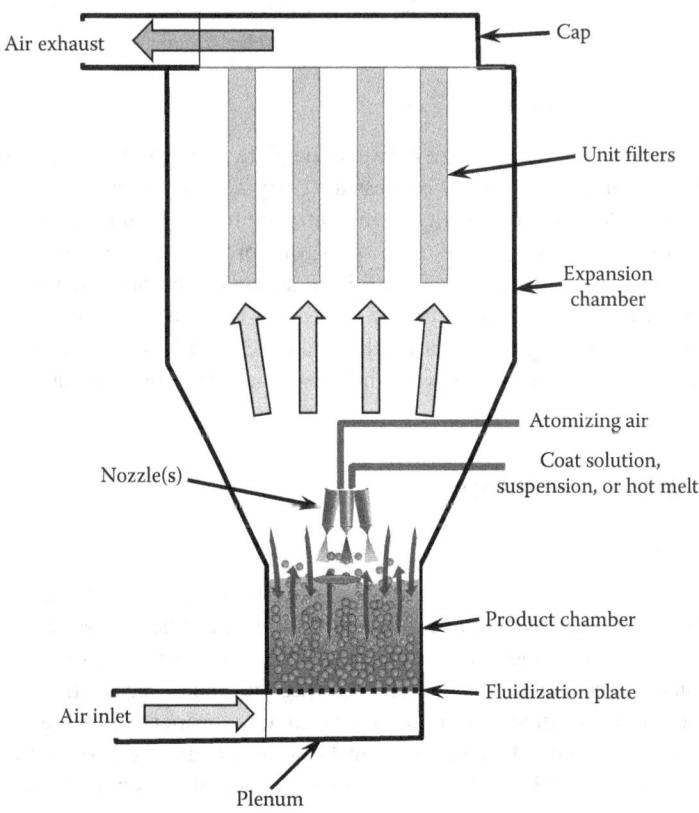

FIGURE 7.6 Top spray diagram.

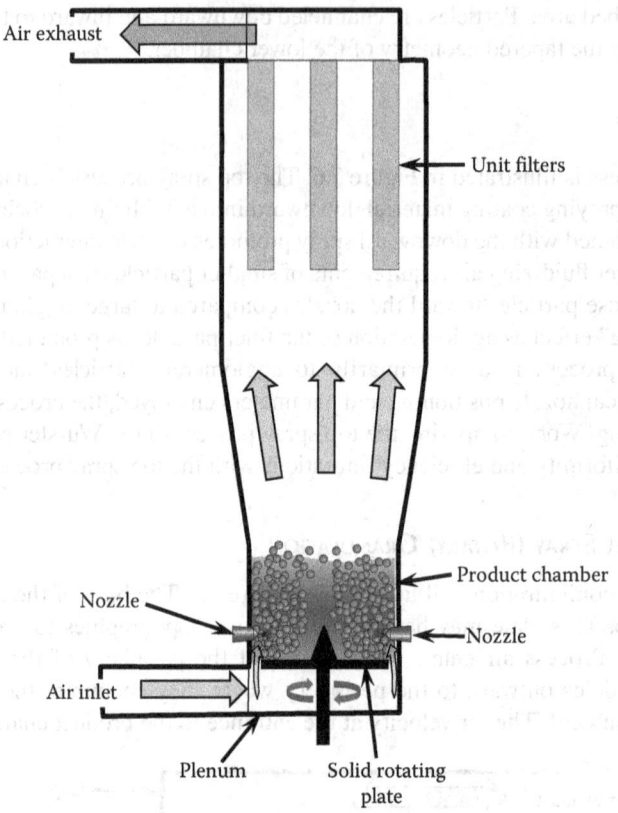

FIGURE 7.7 Tangential spray (rotary) diagram.

process air volume rate, and the gap size between the plate and chamber wall. Nozzles are located at the plate periphery near the point of process air entry and are directed *tangentially* to particle and plate movement. Process air velocity at the entrance to the product chamber is well into the pneumatic transport velocity range. Like top spray, this process is best suited to particle agglomeration and forming granules or pellets. The rolling action on the plate densifies and spheronizes the granules. The process can be used to powder coat, a process where granule-wetting agent or binder components and coating powder are delivered concurrently to build a coating of the powder on the granules. More detailed treatises on the tangential spray process can be found in several papers.[34–37]

7.4 PARAMETER ASSESSMENT

7.4.1 Process Air

Process air is the air used to fluidize and move material during processing. Since the process air requirement is dependent on material fluidization properties, it is generally best to set process air to meet those needs and adjust other parameters appropriately in relation to it. It is also important to consider the state of the material throughout the process, since it may be changing as coat level increases, particles agglomerate, or surface properties change, all of which may change the air velocity need. Process air can be adjusted during processing as the air need changes or be set adequately high throughout the process if it does not compromise fluidization or product quality.

7.4.1.1 Spouted Bed, Top Spray, and Tangential Spray Process Air

For processes that do not require a differential airflow such as spouted bed, top spray, and tangential spray, the process air need can be estimated from the air velocity requirement and the area of air introduction as described in the fluidization assessment in Section 7.2. Figure 7.8 (0–12,000 ft^3/min) and Figure 7.9 (0–2,000 ft^3/min) indicate the relationships between air volume and velocity for spouted bed and top stray processes at various scales. The air must be set to consistently provide required particle movement. The process air range is potentially from the point where any portion of the particle bed is not properly circulating to the point where particles are damaged or prevented from reaching the spray zone. The preferred range is likely more narrow than this and may depend on the goals of the process.

Tangential spray systems require suitable process air velocity and volume to move particles through the spray region and allow them to fall back toward the rotating plate that returns them outward to the spray zone. Velocity and volume are governed by the process air volume and slit size. Slit sizes generally range from ~0.2 to several mm. Air velocity can be determined from the process air volume and the passage area of the slit. The passage area can be accurately calculated from basic geometric calculations based on the plate diameter and slit width:

$$\text{Area of a circle} = \pi r^2$$

thus,

$$\text{Area of the slit} = \pi\left(d/2 + \text{slit width}\right)^2 - \pi\left(d/2\right)^2$$

where,
r = radius
d = plate diameter

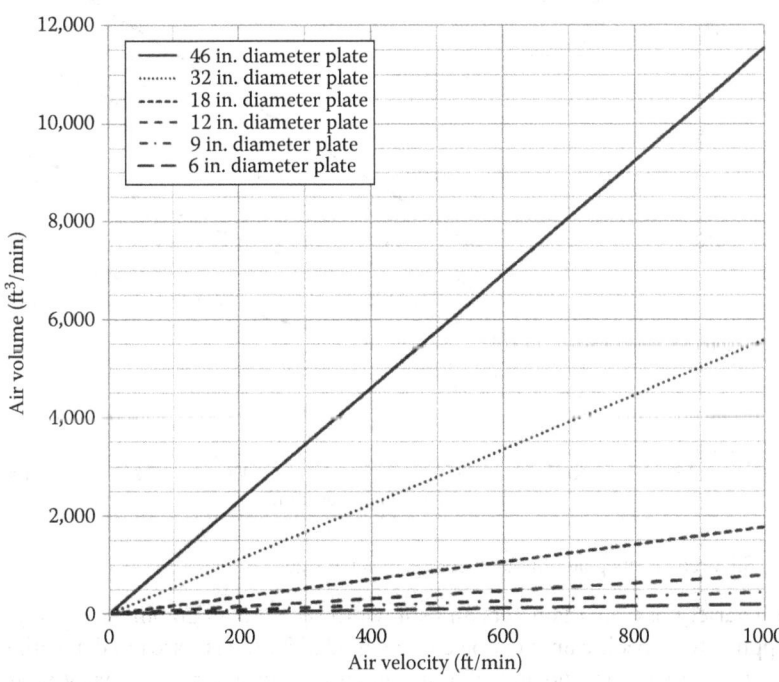

FIGURE 7.8 Process air volume versus process air velocity for various fluidization plate diameters.

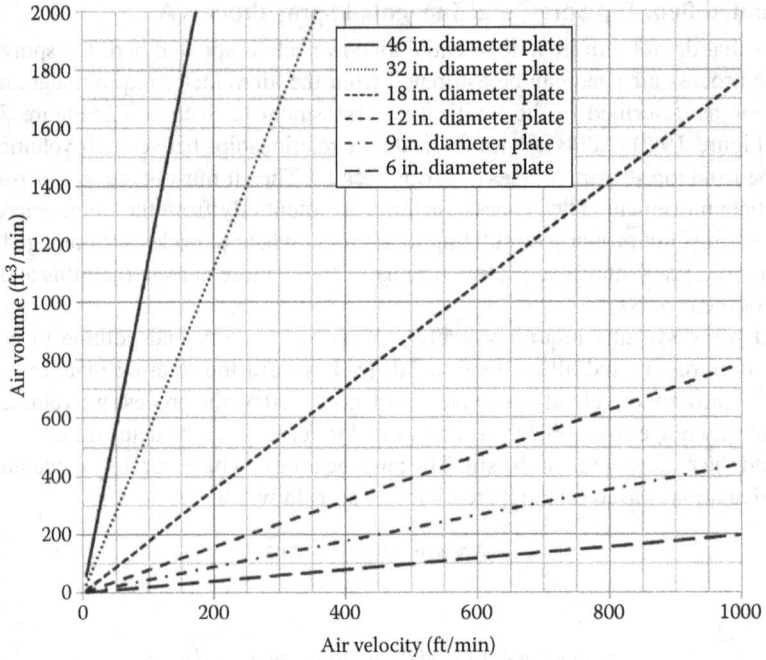

FIGURE 7.9 Process air volume versus process air velocity for various fluidization plate diameters.

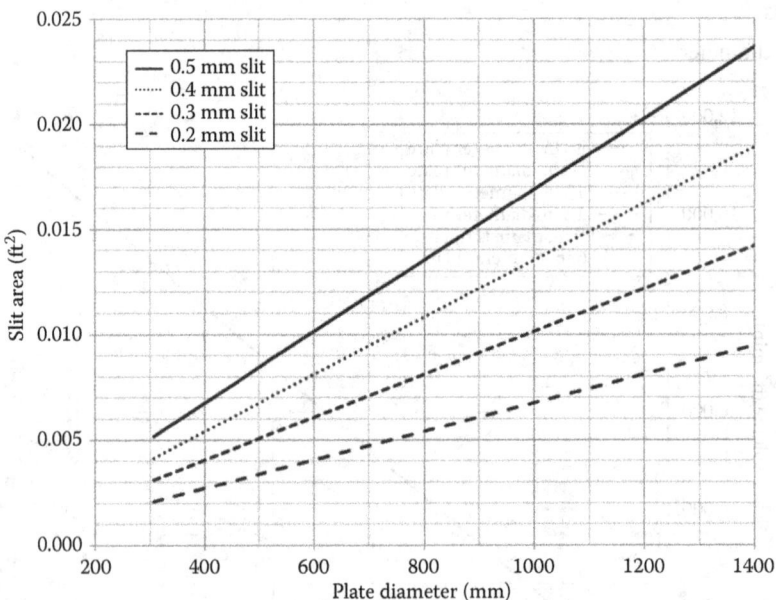

FIGURE 7.10 Slit area versus plate diameter for various slit sizes.

Slit areas for a variety of plate diameters and slit widths are shown graphically in Figure 7.10. The process air applied to a specific application can be divided by the slit area to determine the process air velocity at the slit (slit air velocity). Slit air velocity requirements may vary with a variety of formulation and process parameters such as granule or pellet size, slit width, and particle surface characteristics, but are likely below 5000 ft/min. A correlation of process air with slit velocity and slit area is provided in Table 7.3.

TABLE 7.3
Process Air Volume Correlation with Slit Air Velocity and Slit Area (Process Air Values in ft³/min)

		Slit Area (ft²)																						
		0.002	0.003	0.004	0.005	0.006	0.007	0.008	0.009	0.010	0.011	0.012	0.013	0.014	0.015	0.016	0.017	0.018	0.019	0.020	0.021	0.022	0.023	0.024
	500	1.0	1.5	2.0	2.5	3.0	3.5	4.0	4.5	5.0	5.5	6.0	6.5	7.0	7.5	8.0	8.5	9.0	9.5	10.0	10.5	11.0	11.5	12.0
Air Velocity (ft/min)	1,000	2.0	3.0	4.0	5.0	6.0	7.0	8.0	9.0	10.0	11.0	12.0	13.0	14.0	15.0	16.0	17.0	18.0	19.0	20.0	21.0	22.0	23.0	24.0
	1,500	3.0	4.5	6.0	7.5	9.0	10.5	12.0	13.5	15.0	16.5	18.0	19.5	21.0	22.5	24.0	25.5	27.0	28.5	30.0	31.5	33.0	34.5	36.0
	2,000	4.0	6.0	8.0	10.0	12.0	14.0	16.0	18.0	20.0	22.0	24.0	26.0	28.0	30.0	32.0	34.0	36.0	38.0	40.0	42.0	44.0	46.0	48.0
	2,500	5.0	7.5	10.0	12.5	15.0	17.5	20.0	22.5	25.0	27.5	30.0	32.5	35.0	37.5	40.0	42.5	45.0	47.5	50.0	52.5	55.0	57.5	60.0
	3,000	6.0	9.0	12.0	15.0	18.0	21.0	24.0	27.0	30.0	33.0	36.0	39.0	42.0	45.0	48.0	51.0	54.0	57.0	60.0	63.0	66.0	69.0	72.0
	3,500	7.0	10.5	14.0	17.5	21.0	24.5	28.0	31.5	35.0	38.5	42.0	45.5	49.0	52.5	56.0	59.5	63.0	66.5	70.0	73.5	77.0	80.5	84.0
	4,000	8.0	12.0	16.0	20.0	24.0	28.0	32.0	36.0	40.0	44.0	48.0	52.0	56.0	60.0	64.0	68.0	72.0	76.0	80.0	84.0	88.0	92.0	96.0
	4,500	9.0	13.5	18.0	22.5	27.0	31.5	36.0	40.5	45.0	49.5	54.0	58.5	63.0	67.5	72.0	76.5	81.0	85.5	90.0	94.5	99.0	103.5	108.0
	5,000	10.0	15.0	20.0	25.0	30.0	35.0	40.0	45.0	50.0	55.0	60.0	65.0	70.0	75.0	80.0	85.0	90.0	95.0	100.0	105.0	110.0	115.0	120.0
	5,500	11.0	16.5	22.0	27.5	33.0	38.5	44.0	49.5	55.0	60.5	66.0	71.5	77.0	82.5	88.0	93.5	99.0	104.5	110.0	115.5	121.0	126.5	132.0
	6,000	12.0	18.0	24.0	30.0	36.0	42.0	48.0	54.0	60.0	66.0	72.0	78.0	84.0	90.0	96.0	102.0	108.0	114.0	120.0	126.0	132.0	138.0	144.0
	6,500	13.0	19.5	26.0	32.5	39.0	45.5	52.0	58.5	65.0	71.5	78.0	84.5	91.0	97.5	104.0	110.5	117.0	123.5	130.0	136.5	143.0	149.5	156.0
	7,000	14.0	21.0	28.0	35.0	42.0	49.0	56.0	63.0	70.0	77.0	84.0	91.0	98.0	105.0	112.0	119.0	126.0	133.0	140.0	147.0	154.0	161.0	168.0
	7,500	15.0	22.5	30.0	37.5	45.0	52.5	60.0	67.5	75.0	82.5	90.0	97.5	105.0	112.5	120.0	127.5	135.0	142.5	150.0	157.5	165.0	172.5	180.0
	8,000	16.0	24.0	32.0	40.0	48.0	56.0	64.0	72.0	80.0	88.0	96.0	104.0	112.0	120.0	128.0	136.0	144.0	152.0	160.0	168.0	176.0	184.0	192.0
	8,500	17.0	25.5	34.0	42.5	51.0	59.5	68.0	76.5	85.0	93.5	102.0	110.5	119.0	127.5	136.0	144.5	153.0	161.5	170.0	178.5	187.0	195.5	204.0
	9,000	18.0	27.0	36.0	45.0	54.0	63.0	72.0	81.0	90.0	99.0	108.0	117.0	126.0	135.0	144.0	153.0	162.0	171.0	180.0	189.0	198.0	207.0	216.0
	9,500	19.0	28.5	38.0	47.5	57.0	66.5	76.0	85.5	95.0	104.5	114.0	123.5	133.0	142.5	152.0	161.5	171.0	180.5	190.0	199.5	209.0	218.5	228.0
	10,000	20.0	30.0	40.0	50.0	60.0	70.0	80.0	90.0	100.0	110.0	120.0	130.0	140.0	150.0	160.0	170.0	180.0	190.0	200.0	210.0	220.0	230.0	240.0

Figure 7.10 and Table 7.3 can be used for process air scaling. For example, from Figure 7.10, a process involving a 300 mm plate diameter and 0.2 mm slit has a slit area of ~0.002 ft². If process air for an application on this equipment is established at 7 ft³/min, the slit velocity is ~3500 ft/min, as indicated in Table 7.3. If this process is scaled up to a 1000 mm plate diameter and a 0.2 mm slit, the slit area increases to 0.068 ft², as in Figure 7.10. This 0.068 ft² slit area and 3500 ft/min slit velocity correlate to 23.8 ft³/min, as given in Table 7.3 (interpolated between values at 0.006 and 0.007 ft² slit area values). Note that this process air scale-up for the tangential spray process is almost linear with plate diameter for a given slit width (Area = ~π(diamter)(slit width)) versus the square function for the other processes (Area = π(radius)²). For example, doubling the plate diameter for tangential spray doubles the slit area, which doubles the process air need if slit velocity is maintained, while doubling the plate diameter for other processes increases area by a factor of four, which increases process air need by a factor of four. Tangential spray nozzle application rates are relatively slow compared to other fluid bed processes largely due to lower drying capacity resulting from more limited process air application.

Tangential spray batch size scales almost linearly with plate diameter; however, it can potentially be factored up more than this, depending on the height and thickness of the rotating material band at the various scales. It is also possible that a weight gain rate restriction may apply to some applications, which may limit scale-up to a linear function.

7.4.1.2 Wurster Process Air

7.4.1.2.1 Single Nozzle Wurster Configurations

The Wurster process introduces significant challenges to establishing optimal process airflow due to the differential air distribution and the dynamic nature of air flow, particle characteristics, and particle movement. As mentioned in Section 7.3, the two most critical concerns are fluidization in the down-bed and pneumatic transport in the up-bed. Many process variables affect these two concerns.

A theoretical breakdown of the flow dynamics in the Wurster process will help guide the setting of process airflow. Several hypothetical Wurster air distribution plates are described in Table 7.4; the first three (A–C) are 6 in. diameter plates, the next three (D–F) are 9 in. diameter plates, and the final three (G–I) are 18″ diameter plates. Each series presents variations in zone *area* distribution or zone *air* distribution. The upper portion of Table 7.4 contains details on the physical dimensions of the plates. Most of this information is related to the hypothetical design or can be calculated from those design parameters. Note that the % air entering the up-bed, down-bed, and rim differs from the % of open space in each of these zones. This difference results from the differing resistance to airflow through holes of smaller diameter compared to larger diameter. Air passing through a plate with two plate areas of equal open space, one open space created with larger holes and the other open space with smaller holes, will channel more air through the larger holed region.[38]

The midsection of Table 7.4 indicates a theoretical air distribution based on 100 ft/min velocity need entering the down-bed through the plate (a fluidization velocity appropriate for a ~25/30 mesh sugar sphere (Table 7.1)). The down-bed region referred to in this assessment is the critical velocity area of the down-bed between the up-bed and the outer rim area. Airflow in this area of the down-bed must be adequate to fluidize particles above it. This air need can be determined from the 100 ft/min velocity needed for fluidization:

$$\text{Down bed air flow} = (\text{Down bed air velocity})(\text{Down bed area})$$

For 6″ diameter plate "A" with 0.124 ft² in the critical area of the down-bed, ~12.4 ft³/min air is required in this region of the plate to provide 100 ft/min velocity (100 × 0.124 = 12.4). Since the air entering this area through the plate is only a portion of the overall process air, it must be translated

TABLE 7.4
Process Air Assessment for Hypothetical 6, 9, and 18 in. Diameter Wurster Air Distribution Plates

Plate	A	B	C	D	E	F	G	H	I
Plate Diameter (in.)	6	6	6	9	9	9	18	18	18
Up-bed area (ft^2)	0.0491	0.0491	0.0368	0.11	0.11	0.0828	0.442	0.442	0.331
Down-bed area (ft^2)	0.124	0.124	0.136	0.296	0.296	0.324	1.17	1.17	1.28
Rim area (ft^2)	0.0236	0.0236	0.0236	0.0353	0.0353	0.0353	0.159	0.159	0.159
Total plate area (ft^2)	0.196	0.196	0.196	0.442	0.442	0.442	1.77	1.77	1.77
Open area in up-bed (ft^2)	0.0118	0.00982	0.0118	0.0265	0.0221	0.0265	0.106	0.0884	0.106
Open area in down-bed (ft^2)	0.00599	0.00687	0.0055	0.0146	0.0168	0.0133	0.0601	0.0795	0.0583
Open area in rim (ft^2)	0.00187	0.00295	0.00236	0.00309	0.0053	0.00442	0.0106	0.00884	0.0124
Total open area (ft^2)	0.0196	0.0196	0.0196	0.0442	0.0442	0.0442	0.177	0.177	0.177
% Area in up-bed	25.00%	25.00%	18.75%	25.00%	25.00%	18.75%	25.00%	25.00%	18.75%
% Area in down-bed	63.00%	63.00%	69.25%	67.00%	67.00%	73.25%	66.00%	66.00%	72.25%
% Rim area	12.00%	12.00%	12.00%	8.00%	8.00%	8.00%	9.00%	9.00%	9.00%
Up-bed: % of open	60.00%	50.00%	60.00%	60.00%	50.00%	60.00%	60.00%	50.00%	60.00%
Down-bed: % of open	30.50%	35.00%	28.00%	33.00%	38.00%	30.00%	34.00%	45.00%	33.00%
Rim: % of open	9.50%	15.00%	12.00%	7.00%	12.00%	10.00%	6.00%	5.00%	7.00%
% Air in up-bed	68.00%	55.00%	70.00%	68.00%	55.00%	70.00%	68.00%	55.00%	70.00%
% Air in down-bed	20.00%	29.00%	16.00%	22.00%	30.00%	19.00%	25.00%	38.50%	22.00%
% Air in rim	12.00%	16.00%	14.00%	10.00%	15.00%	11.00%	7.00%	6.50%	8.00%
Assessment for 100 ft/min minimum fluidization velocity through plate in the down-bed									
Process air to reach fluidization velocity in down-bed (ft^3/min)	62.0	42.8	**85.0**	134.5	98.7	**170.5**	468.0	303.9	**581.8**
Velocity in up-bed (ft/min)	858.7	479.0	1616.8	831.7	493.3	1441.6	720.0	378.2	1230.4
Velocity in down-bed (ft/min)	100	100	100	100	100	100	100	100	100
Velocity in rim (ft/min)	315.3	289.9	504.2	381.1	419.3	531.4	206.0	124.2	292.7
Combined down-bed and rim velocity (ft/min)	134.4	130.4	159.8	130.0	134.0	142.4	112.7	102.9	121.3
Assessment for 1000 ft/min minimum air velocity through plate in the up-bed									
Process air for pneumatic transport velocity in the up-bed (ft^3/min)	**72.2**	**89.3**	52.6	**161.8**	**200.0**	118.3	**650.0**	**803.6**	472.9
Velocity in up-bed (ft/min)	1000	1000	1000	1000	1000	1000	1000	1000	1000
Velocity in down-bed (ft/min)	116.5	208.8	61.8	120.2	202.7	69.4	138.9	264.4	81.3
Velocity in rim (ft/min)	367.1	605.2	311.9	458.3	849.9	368.6	286.2	328.5	237.9
Combined down-bed and rim velocity (ft/min)	156.5	272.2	98.8	156.2	271.7	98.8	156.5	272.1	98.6
Theoretical process air to meet both up-bed and down-bed minimum velocity needs	72.2	89.3	85.0	161.8	200.0	170.5	650.0	803.6	581.8

Note: Bolded (largest) values from the above two assessments.

to the overall process air need by taking into account that this area of the down-bed receives only 20% of the total process air (Plate A, Table 7.4):

$$\text{Total process air} = \frac{\text{Down bed air flow}}{\text{Volume fraction of total process air in down bed}}$$

or

$$\text{Total process air} = \frac{(\text{Down bed air velocity})(\text{Down bed area})}{\text{Volume fraction of total process air in down bed}}$$

Thus, for hypothetical Plate A:

$$\text{Total process air} = \frac{(100 \text{ ft/min})(0.124 \text{ ft}^2)}{0.2} = 62.0 \text{ ft}^3/\text{min}$$

This process airflow need based on fluidization in the critical area of the down-bed has been calculated in the same way for all hypothetical plates in Table 7.4.

The second critical function of process air in a Wurster application is to provide sufficient air in the up-bed region to pneumatically transport particles through the spray region. Up-bed air is made up mostly of process air entering through the up-bed area of the plate, but it is supplemented with process air feeding from the base of the down-bed above the plate and atomization air delivered from the nozzle. Air feeding from the down-bed is dependent on relative particle mass in the up- and down-beds, which is controlled primarily by fluidization plate air distribution, particle load, partition gap and partition length, and atomizing air volume. A smaller particle load (smaller batch) will result in less particle load in the down-bed, which allows more down-bed air to pass upward through the down-bed and reduces particle feed to the up-bed. The partition gap physically regulates particle and air feed rate to the up-bed. Partition length affects overall up-bed volume, which affects particle mass in the up-bed. Atomizing air volume is dependent on atomizing air pressure and either the cross-sectional area of the concentric atomizing air gap at the nozzle tip or the internal design of the atomizing air pathways within the nozzle depending on which imparts the greatest restriction to air flow. Particle dilution in the up-bed must be adequate to pneumatically transport particles upward and maintain proper particle feed from the down-bed. Air entering through the up-bed region of the plate is a critical factor in that dilution.

Achieving adequate fluidization in the down-bed does not assure pneumatic transport in the up-bed. The 62.0 ft³/min total process air needed to achieve 100 ft/min air velocity in the Plate A down-bed can be converted to the air velocity it contributes directly to the up-bed:

$$\text{Air velocity in up bed} = \frac{(\text{Total process air})(\text{Air fraction in up bed})}{\text{Up bed area}}$$

$$= \frac{(62.0 \text{ ft}^3/\text{min})(0.68)}{0.0491} = 858.7 \text{ ft/min}$$

If 858.7 ft/min air velocity entering through the up-bed region of the plate, coupled with the air from the down-bed and atomizing air, is adequate to consistently dilute and pneumatically transport particles through the up-bed without excess velocity, 62.0 ft³/min would be a suitable process air setting for Plate A. If this velocity is too low for reliable pneumatic transport, process air can be raised to reach a suitable up-bed velocity, provided excess fluidization velocity in the down-bed is not realized. If suitable pneumatic transport velocity in the up-bed is realized only when there is excess fluidization velocity in the down-bed, an alternative air distribution plate or other adjustment may be needed.

This assessment can be worked in reverse from up-bed velocity target to the resulting down-bed velocity. If 1000 ft/min air velocity through the plate in the up-bed is needed, the following calculations can made for Plate A in Table 7.4:

$$\text{Total process air} = \frac{(1000 \text{ ft/min})(0.0491 \text{ ft}^2)}{(0.68)} = 72.2 \text{ ft}^3/\text{min}$$

$$\text{Air velocity in down-bed} = \frac{(72.2 \text{ ft}^3/\text{min})(0.20)}{0.124} = 116.5 \text{ ft/min}$$

Calculated air velocities and volumes for the up- and down-bed based on both 100 ft/min air velocity in the down-bed and 1000 ft/min in the up-bed are summarized in the middle and lower portions of Table 7.4, respectively. If these are the critical velocities for the up- and down-beds, process air that meets or exceeds each is needed. In general, fluidizing air entering the critical down-bed area can range from near the incipient fluidization velocity to three or more times that depending on material, process characteristics, and goals. It may be necessary to maintain a higher velocity in the down-bed to account for growing particle size due to coating or agglomeration during processing. Such increased velocity needs are generally required for smaller particles or materials with broad particle size distributions.

Typical fluidization plate designs include the area of higher air velocity at the outer rim or perimeter to help assure material in this region properly circulates to the coating zone during operation. This air is part of the down-bed air and is an area of higher velocity compared to the critical down-bed area between the rim and up-bed; thus, it may influence the air need in the critical fluidization area of the plate. Although not discussed in this text, the air velocity and volume entering this perimeter area through the plate can be calculated from the plate area and air fraction as performed above for the up- and down-bed regions.

This assessment is based on a theoretical model for the process. In actual practice, if all particles circulate and pneumatically transport through the up-bed, coating will be achieved. Even if coating is not done optimally, a functional product can be realized; however, it may be necessary to perform the same quality of coating in any scale-up or technical transfer with an equivalent level of optimization to realize consistent performance. Just like a poor job of painting a surface such as a wall might require three coats rather than two, a poor job of performing Wurster coating might take 15% coating rather than 10% to achieve a performance goal.

When an ideal air distribution plate is not available, it is possible to adjust partition gap, and partition length to promote proper circulation. Partition gap (if a partition is used) generally ranges from ~3/8 to ~3 in. Generalized effects of partition parameters were noted in Table 7.2 and include the following:

1. The gap is generally higher for large particles to physically allow adequate particle passage for suitable particle density in the spray region and ensure efficient coating.
2. Lowering a partition provides more restriction to air and particle flow from the down-bed to the up-bed, which reduces particle density (mass) in the up-bed. This reduced density also promotes airflow from the down-bed to the up-bed.
3. Raising a partition allows more air and particle flow from the down-bed to the up-bed, which increases particle density (mass) in the up-bed. This increased density also restricts airflow from the down-bed to the up-bed.
4. Increased partition length effectively increases particle mass in the up-bed, which will likely reduce air and particle feed from the down-bed to the up-bed and temper air velocity in the up-bed.

It is also possible to block fluidization plate holes with tape or other media to adjust the air distribution pattern for a given plate. If such "taping" is successful, the pattern and amount of taping can be applied consistently just like any other process parameter to assure process consistency.

7.4.1.2.2 Multiple Nozzle Configurations

Wurster process development efforts identified an 18 in. diameter coat chamber as an optimal maximum size.[38] This limit is based primarily on nozzle radial spray limits and particle horizontal transport limits. During operation, nozzle spray was found to extend radially up to ~4.5 in. from the nozzle tip due to particle density in the spray region. With ~100% of spray contacting particles within this 4.5 in. radius, there is no significant benefit and some potential detriments to extending the up-bed diameter beyond ~9 in. In addition, a down-bed region that extends radially more than ~4.5 in. from the up-bed region was found to come up against horizontal transport limits for moving particles from the outer edge of the down-bed inward to feed the up-bed region. This combined 4.5 in. up-bed radius and 4.5 in. particle feed distance resulted in the 18 in. optimal maximum chamber diameter. Scaling up beyond this size can be accomplished by extending upward with more vertical depth, provided proper fluidization can be maintained; however, spray rate will be subject to the same limits regardless of the bed depth.

Due to these single nozzle limitations, further scale-up is realized by merging multiple single nozzle configurations into a single process air stream. Some multiple nozzle configurations are shown in Figure 7.11. Each up-bed region is surrounded by the down-bed region and the down-bed region is continuous with no physical partitions between merged units. Each nozzle and up-bed zone is equivalent to that in a single nozzle process.

Process air for multinozzle units is scaled in the same way as single nozzle configurations with airflow delivered to provide fluidization in the down-bed and pneumatic transport in the up-bed. Since the "footprint" of the merged circular units does not match the footprint of the circular multinozzle housing, a greater horizontal particle transport distance is realized for regions of the down-bed that fall outside the original 18″ footprints. It is important for these regions to be addressed to assure proper particle transport to the coating zone.

It is critical that process air be uniformly distributed to each zone of a multinozzle unit. An air imbalance can result in inadequate fluidization and pneumatic transport air in some zones, excessive fluidization and pneumatic transport air in other zones, or complete collapse of one or more zones.

Some multinozzle units are linearly scaled from a single-nozzle configuration. For example, linear scale-up from the single-nozzle process would require three times or seven times the process air and load size for a three-nozzle or seven-nozzle unit, respectively. Nozzles would all spray at the

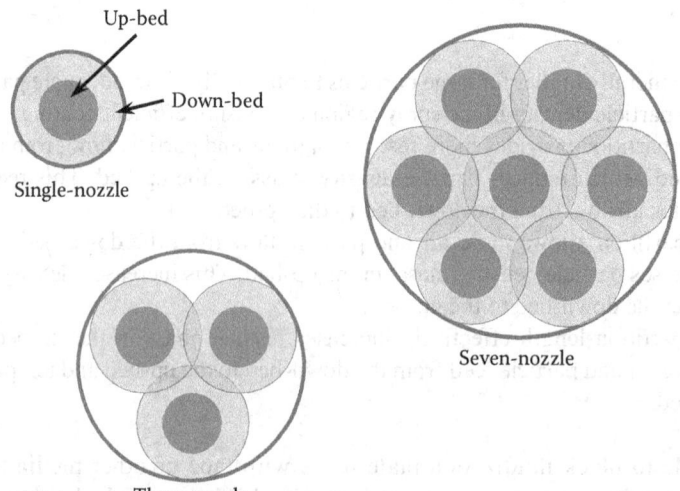

FIGURE 7.11 Multiple-nozzle configurations.

Encapsulation via Fluidized Bed Coating Technology

same rate as the single nozzle process. This linear scaling allows relatively easy scale-up or scale-down between these capacity options.

7.4.1.3 Process Air Temperature and Volume Relationship

Process air temperature should also be considered in process air assessments due to air volume expansion or contraction associated with temperature changes. Volume changes can be determined using the general gas equation:

$$V = \frac{nRT}{P}$$

where

V = Volume in ft^3

n = Number of moles of gas molecules

$R = 0.0028978 \frac{(\text{ft}^3)(\text{atmospheres})}{(\text{mole})(°K)}$

T = Temperature in °K = °C + 273.15

P = Pressure in atmospheres

Since R is a constant, n is constant through the isolated air path in the absence of any leaks, and pressure differences are negligible,

$$\frac{V_1}{T_1} = \frac{V_2}{T_2}$$

Thus,

$$V_2 = V_1 \frac{T_2}{T_1}$$

If the process air for a coating unit is measured at an inlet air temperature of 50°F (10°C) and the bed temperature of the process is 110°F (43.3°C), 500 ft^3/min at the inlet measurement point can be converted to the airflow in the bed:

$$\text{Volume in bed} = (500 \text{ ft}^3/\text{min}) \frac{(43.3°C + 273.15)}{(10°C + 273.15)} = 558.8 \text{ ft}^3/\text{min}$$

Table 7.5 correlates the volume of 1.000 ft^3 of air at temperature T_1 to its volume at temperature T_2. For example, 1.000 ft^3 of air at T_1 temperature 120°F will contract to 0.896 ft^3 at T_2 temperature 60°F and expand to 1.104 ft^3 at T_2 temperature 180°F. This conversion can be extended to process airflow. For example, 500 ft^3/min process air at 120°F would contract to 448 ft^3/min at 60°F (500 ft^3/min × 0.896 = 448 ft^3/min) and expand to 552 ft^3/min at 180°F (500 ft^3/min × 1.104 = 552 ft^3/min). Although the air volume units are ft^3 or ft^3/min in this example, this conversion will also apply to any other volume unit.

Process air is often measured and adjusted to standard temperature conditions to provide a consistent reference point. If so, recorded or specified values may be in units labeled scfm (ft^3/min at

TABLE 7.5
Volume of 1000 ft³ of Air at Temperature T₁ When Converted to Temperature T₂

T1 (°F)	T1 (°C)	T2 (°F): 0	10	20	30	40	50	60	70	80	90	100	110	120	130	140	150	160	170	180	190	200	210
	T2 (°C):	−17.8	−12.2	−6.7	−1.1	4.4	10.0	15.6	21.1	26.7	32.2	37.8	43.3	48.9	54.4	60.0	65.6	71.1	76.7	82.2	87.8	93.3	98.9
0	−17.8	1.000	1.022	1.044	1.065	1.087	1.109	1.131	1.152	1.174	1.196	1.218	1.239	1.261	1.283	1.305	1.326	1.348	1.370	1.392	1.413	1.435	1.457
10	−12.2	0.979	1.000	1.021	1.043	1.064	1.085	1.106	1.128	1.149	1.170	1.192	1.213	1.234	1.255	1.277	1.298	1.319	1.341	1.362	1.383	1.405	1.426
20	−6.7	0.958	0.979	1.000	1.021	1.042	1.063	1.083	1.104	1.125	1.146	1.167	1.188	1.208	1.229	1.250	1.271	1.292	1.313	1.334	1.354	1.375	1.396
30	−1.1	0.939	0.959	0.980	1.000	1.020	1.041	1.061	1.082	1.102	1.123	1.143	1.163	1.184	1.204	1.225	1.245	1.265	1.286	1.306	1.327	1.347	1.368
40	4.4	0.920	0.940	0.960	0.980	1.000	1.020	1.040	1.060	1.080	1.100	1.120	1.140	1.160	1.180	1.200	1.220	1.240	1.260	1.280	1.300	1.320	1.340
50	10.0	0.902	0.922	0.941	0.961	0.980	1.000	1.020	1.039	1.059	1.078	1.098	1.118	1.137	1.157	1.177	1.196	1.216	1.235	1.255	1.275	1.294	1.314
60	15.6	0.885	0.904	0.923	0.942	0.962	0.981	1.000	1.019	1.038	1.058	1.077	1.096	1.115	1.135	1.154	1.173	1.192	1.212	1.231	1.250	1.269	1.289
70	21.1	0.868	0.887	0.906	0.924	0.943	0.962	0.981	1.000	1.019	1.038	1.057	1.076	1.094	1.113	1.132	1.151	1.170	1.189	1.208	1.227	1.245	1.264
80	26.7	0.852	0.870	0.889	0.907	0.926	0.944	0.963	0.981	1.000	1.019	1.037	1.056	1.074	1.093	1.111	1.130	1.148	1.167	1.185	1.204	1.222	1.241
90	32.2	0.836	0.854	0.873	0.891	0.909	0.927	0.945	0.964	0.982	1.000	1.018	1.036	1.055	1.073	1.091	1.109	1.127	1.146	1.164	1.182	1.200	1.218
100	37.8	0.821	0.839	0.857	0.875	0.893	0.911	0.929	0.946	0.964	0.982	1.000	1.018	1.036	1.054	1.071	1.089	1.107	1.125	1.143	1.161	1.179	1.197
110	43.3	0.807	0.824	0.842	0.860	0.877	0.895	0.912	0.930	0.947	0.965	0.982	1.000	1.018	1.035	1.053	1.070	1.088	1.105	1.123	1.140	1.158	1.176
120	48.9	0.793	0.810	0.827	0.845	0.862	0.879	0.896	0.914	0.931	0.948	0.965	0.983	1.000	1.017	1.035	1.052	1.069	1.086	1.104	1.121	1.138	1.155
130	54.4	0.780	0.796	0.813	0.830	0.847	0.864	0.881	0.898	0.915	0.932	0.949	0.966	0.983	1.000	1.017	1.034	1.051	1.068	1.085	1.102	1.119	1.136
140	60.0	0.767	0.783	0.800	0.817	0.833	0.850	0.867	0.883	0.900	0.917	0.933	0.950	0.967	0.983	1.000	1.017	1.033	1.050	1.067	1.083	1.100	1.117
150	65.6	0.754	0.770	0.787	0.803	0.820	0.836	0.852	0.869	0.885	0.902	0.918	0.934	0.951	0.967	0.984	1.000	1.016	1.033	1.049	1.066	1.082	1.098
160	71.1	0.742	0.758	0.774	0.790	0.806	0.822	0.839	0.855	0.871	0.887	0.903	0.919	0.935	0.952	0.968	0.984	1.000	1.016	1.032	1.048	1.065	1.081
170	76.7	0.730	0.746	0.762	0.778	0.794	0.809	0.825	0.841	0.857	0.873	0.889	0.905	0.921	0.936	0.952	0.968	0.984	1.000	1.016	1.032	1.048	1.064
180	82.2	0.719	0.734	0.750	0.766	0.781	0.797	0.812	0.828	0.844	0.859	0.875	0.891	0.906	0.922	0.937	0.953	0.969	0.984	1.000	1.016	1.031	1.047
190	87.8	0.708	0.723	0.738	0.754	0.769	0.785	0.800	0.815	0.831	0.846	0.861	0.877	0.892	0.908	0.923	0.938	0.954	0.969	0.985	1.000	1.015	1.031
200	93.3	0.697	0.712	0.727	0.742	0.757	0.773	0.788	0.803	0.818	0.833	0.848	0.864	0.879	0.894	0.909	0.924	0.939	0.955	0.970	0.985	1.000	1.015
210	98.9	0.686	0.701	0.716	0.731	0.746	0.761	0.776	0.791	0.806	0.821	0.836	0.851	0.866	0.881	0.895	0.910	0.925	0.940	0.955	0.970	0.985	1.000

standard temperature and pressure) rather than cfm (ft³/min at the existing temperature and pressure). Standard pressure is generally 1 (one) atmosphere, which can be assumed for the purposes of fluid bed processing work performed without any special pressure controls. There are varying conventions for standard temperature; thus, it is important to know the convention used for a specified scfm value to accurately translate it. This can be an important factor depending on the extremes of temperature involved.

It is also important to recognize that airflow measurement is prone to error due to the compressible nature of air. Even in a properly configured system, up to 5% or 10% error in airflow measurement may be realized. Although this may result in little if any consequence for work performed on a given coating unit, it can be important consideration for process scale-up or technical transfer.

7.4.2 Nozzles and Atomization

Atomization of the coat solution, suspension, or hot melt (collectively referred to as solution in this discussion) is critical to fluid bed coating process performance. Atomization is most commonly provided by two fluid nozzles. An illustration of a nozzle is shown in Figure 7.12. These nozzles have a central nozzle tip for coat solution delivery and a cap that fits concentrically around the tip. One fluid is atomizing air that is introduced around the tip through the gap between the tip and cap. Atomizing air is delivered by pressure; a combination of atomizing air velocity and volume provides the atomization energy. Nozzles vary in outer and inner tip diameter, inner cap diameter, the amount of tip extension above the cap, and the internal structure of the atomizing air channel. The internal structure may be designed to optimize atomization or atomization air uniformity about the tip. The air volume is governed by both the applied atomizing air pressure and the limiting airflow restriction, which may be the cross-sectional passage area between the tip and cap or an internal restriction within the nozzle body.

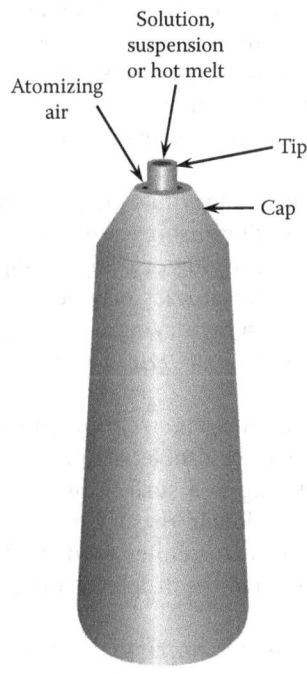

FIGURE 7.12 Two-fluid nozzle diagram.

Atomizing air pressures generally range from near 12 (~0.8 bar) to 80 psi (~5.5 bar). Based on volume data for airflow through selected two fluid nozzles and orifices, these pressures provide atomizing air velocities at the nozzle tip in the vicinity of 50,000–250,000 ft/min.

Nozzles generally produce liquid droplets with average sizes ranging from near 10 to over 50 μm. The droplet size from a given nozzle at a given atomizing air pressure is influenced by spray rate and solution viscosity. Higher spray rates and/or coat solution viscosity requires more energy to atomize than lower spray rate and/or viscosity; thus, a more viscous solution will require higher atomizing air to match spray droplet size of a less viscous solution and a higher spray rate will require higher atomizing air to match spray droplet size of the same solution at a lower spray rate. Atomizing air, solution viscosity, and spray rate can be adjusted to optimize atomization for a given nozzle and application. Solution viscosities ranging from below 5 to over 1000 centipoise have been successfully applied in coating applications.

Depending on the goals of a specific process, the droplet size distribution may also be critical. If large droplets are a concern, spray rate, viscosity, and atomizing air can be adjusted to minimize droplet size. Nevertheless, if the breadth of the droplet size distribution is wide, the larger droplets of the distribution could compromise the process or yield. For example, if every large droplet produces oversized agglomerates and 2% of the droplets are oversize, a significant amount of oversize will likely be created, depending on the amount of time in the process. A spray consisting of 15 μm average droplets with a distribution ranging from 10 to 25 μm may be preferable to one with a 12 μm average with a distribution from 7 to 40 μm if droplets over 30 μm create oversize. The droplet size distribution is significantly influenced by nozzle design.

For individual particle coating, droplet size is more critical when coating small particles compared to large ones. For standard fluid bed granulation work, larger droplets are often preferred, since they promote particle agglomeration and/or accretion regardless of the particle size. This phenomenon is often applied early in particle coating processes to "tie up" finer particles for more favorable coating logistics. Large droplets dry more slowly due to lower surface to volume ratio. Also, as droplet size approaches and exceeds that of the particles that are to be coated or granulated, multiple particles can attach to or penetrate droplets, which will promote agglomeration. Due to these concerns, particle coating in a Wurster process generally requires droplet sizes that are significantly smaller than the particles to be coated. This is one of the limiting factors for the applicable particle size range for Wurster coating. A lower droplet size limit near 10–15 μm results in a lower particle size limit near 50 μm. Even with 50 μm particles, coating is prone to significant levels of agglomeration depending on the "tackiness" characteristics of the coating system.

Alternative nozzle designs and technology has the potential to produce smaller droplets for coating. Nevertheless, there may be diminishing returns with smaller droplets. Atomization to an extreme will result in premature drying or spray drying, where the droplets dry prior to contacting particles. Application of such nozzles may be limited to certain coating systems or conditions that properly address any spray drying concerns. Some equipment manufacturers have introduced secondary means such as a nozzle shield to facilitate droplet formation prior to particle contact.[11,14]

A graph of surface area/mg versus droplet size for atomized water is shown in Figure 7.13. This relationship indicates that when the droplet size is halved for a given amount of liquid, the surface area doubles. Since drying is a surface phenomenon, this suggests that halving the droplet size has the potential to double the drying rate and allow faster application. In actual practice, drying is facilitated with smaller droplets, but it is also influenced by many other factors including transformation of the droplet shape as it wets or penetrates the particle surface.

It is important to be aware that high atomizing air pressures impart not only more energy into solution atomization, but also more energy into fluidizing particles. This energy can

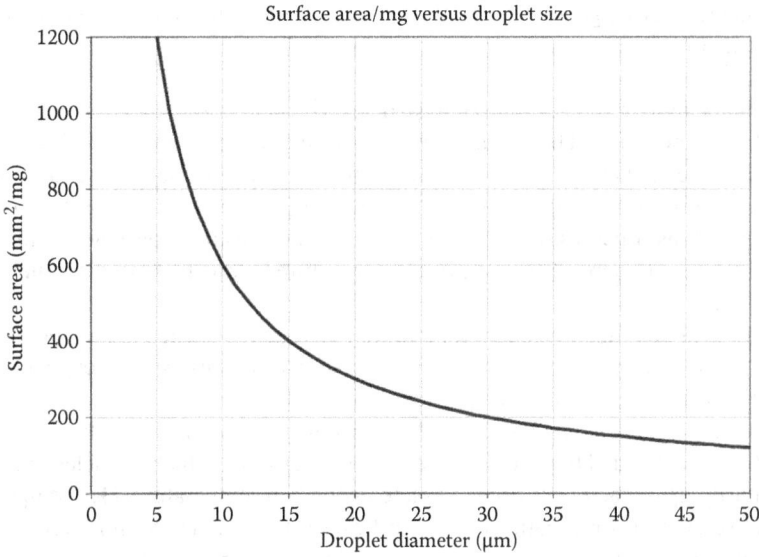

FIGURE 7.13 Surface area per milligram versus droplet diameter for water.

potentially fracture particles and cause particle attrition, both of which can compromise the coating process by creating smaller particles and exposing core material during processing. Several steps can be taken to potentially reduce or eliminate particle fracture and attrition caused by atomizing air including the following:

1. Reduce atomizing air to a minimum that does not cause excess agglomeration.
2. Ensure that spray rate is optimized with respect to coating requirements. Some formulations are prone to attrition when run too dry. Operating at cooler temperature conditions can reduce dryness.
3. Reduce spray rate to achieve required droplet size at lower atomizing air pressure.
4. Reduce solution viscosity to achieve required droplet size at lower atomizing air pressure.
5. Engineer a more robust core particle that can withstand the atomizing air forces.

Hot melt coating involves spraying molten coat material and congealing it on particle surfaces as they move through the coating zone. The melt tank, liquid lines, nozzle, and atomizing air are maintained at suitable temperatures to reliably maintain the melt in a liquid state until it is atomized and contacts particles. Good atomization is important to hot melt coating, since 100% solids are sprayed.

7.4.3 Temperature

7.4.3.1 Process Temperatures

Process temperatures are generally monitored or controlled at the inlet plenum, the particle bed, and the outlet from the coat chamber. Bed and outlet temperature are generally within a few degrees of one another. The difference between inlet temperature and outlet/bed temperatures is influenced by heat transfer/loss from the unit as air moves through the system and either evaporative cooling from coating vehicle evaporation or heat introduced by components of a hot melt application. Heat transfer/loss is significantly greater for small-scale work due to greater unit surface-to-volume ratio in small-scale systems.

Process temperatures are generally set within the limits of the formulation or equipment. Possible limits include the following:

1. Melt point of an ingredient: Fluidizing particles pass near the fluidization plate as they feed the up-bed of the coater. The temperature at the immediate top of the plate is approximately the same as the inlet temperature and quickly transitions to the bed temperature above that. Any material in the coating or core that has a melt point below the inlet temperature will soften or melt as it passes over the plate, which could result in particle agglomeration or sticking to the plate screen; thus, an inlet temperature below the ingredient melt point may be required.

    For hot melt coatings, the bed temperature must be maintained below the melt point of the coating to congeal it on particle, but not so low as to cause spray congealing (droplets congealing prior to contact with a particle.

2. Temperature stability of an ingredient: Process temperatures must be below the point where instability is realized. This could be the inlet temperature, which particles are exposed to momentarily upon each pass over the plate or the bed temperature. Bed temperatures for evaporative coating formulations are significantly lower than the inlet temperature, but may be the critical stability temperature consideration, since particles spend more time at that condition.
3. Drying condition or drying capacity concern: Some processes may be limited to a spray rate that provides a narrow range of relative humidity. This humidity level could be realized through a combination of spray rate, dew point, and temperature parameters.
4. Minimization of a volatile residual such as an organic solvent vehicle or water: Although aqueous or solvent vehicles are removed by evaporation during processing, a residual typically remains. The amount is typically related to the affinity or coating materials for the solvent. Higher process temperatures generally reduce the residual.
5. Film forming temperature restriction of a coating system: Latex systems have minimum film forming temperatures. Below this minimum, the film coating will not properly coalesce to form the film. Also, these coat systems typically become tacky when the temperature is too warm. As a result, process temperatures should be maintained between the extremes of the minimum film-forming temperature and the tackiness point if possible.
6. Glass transition temperature (Tg) of an ingredient or coating: The Tg is the temperature where a material transitions from a brittle or *glassy* state below the Tg to a more pliable or *rubbery* state above the Tg. Materials tend to be more tacky above their Tg, which can compromise particle movement and promote particle agglomeration. Solvent residual and plasticizer typically reduce the Tg of a polymer.

Beyond these limiting factors, process temperature restrictions may be imparted by solvent vehicle selection, drying capacity, and spray rate considerations. These concerns are discussed in Section 7.4.4.

7.4.3.2 Hot Melt Temperatures

Hot melt applications involve spraying molten coating material onto fluidizing particles. It is critical to maintain the melt in the liquid state in the melt tank and through the liquid lines and nozzle. Liquid lines are appropriately heat traced, the nozzle is heated, and the atomizing air may be heated. These temperatures are generally maintained sufficiently above the melt point to avoid congealing at a localized cool point in the system. In addition, liquid flow is typically maintained at all times to avoid congealing at localized cool spots in the lines. This includes recirculating the melt during idle time when material is not sprayed.

7.4.4 DRYING CAPACITY

Drying capacity is a measure of the process air capacity for evaporated solution vehicle. This capacity is dependent on the rate of air volume movement through the system, the temperature of that air, and the vapor pressure of the vehicle at that temperature. Figure 7.14 shows the relationship of air capacity and temperature for a variety of solvents. These data are based on vapor pressure data and the general gas equation noted previously in this chapter:

$$n = \frac{PV}{RT}$$

The solvent vapor capacity of 1.0 ft³ was determined by inserting the partial pressure at given temperatures from general literature resources. The mole amounts (n) can be multiplied by the molecular weight to arrive at the ~g/ft³ capacity. This information can be used to indicate the relationship of spray rate to drying capacity. For example, air capacity at 100°F is ~32 g/ft³ for acetone, ~8 g/ft³ for ethanol, or ~1.25 g/ft³ for water. At 500 ft³/min process air and 100°F, capacity would be reached at ~16,000 g/min acetone (32 × 500 = 16,000) spray, ~4,000 g/min ethanol spray (8 × 500 = 4,000), or ~625 g/min water spray (1.25 × 500 = 625). When spraying above these limits, sprayed solvent vapors cannot be vaporized and removed as fast as they are sprayed and sprayed solvent accumulates in the bed. These data are based on spraying straight solvent. When spraying actual coat solutions, the solvent is only a portion of the sprayed material and spray rate can be factored up appropriately. Be aware that other materials in coating solutions can potentially lower the vapor pressure as solvent evaporates or have a high affinity for the solvent and slow the drying process. In *coating* processes, spray rates are well below the capacity limit due to safety, formulation, and processing concerns. In aqueous *granulation* processes, spray rate is often above the capacity limit.

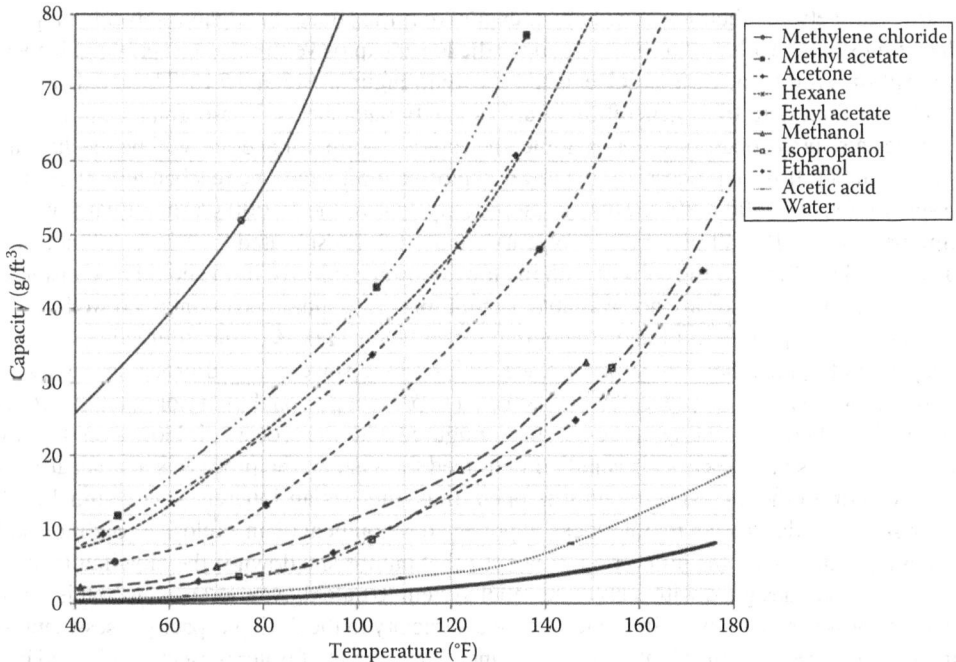

FIGURE 7.14 Air capacity versus temperature for selected solvents.

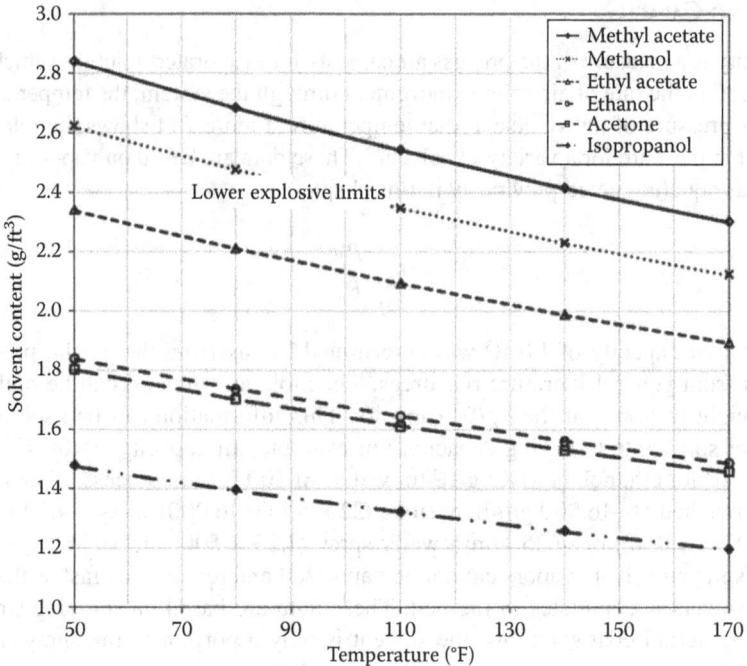

FIGURE 7.15 Lower explosive limits for selected organic solvents.

In applications using organic solvent vehicles, spray *must* be maintained below the lower explosive limit. Volatile organic materials such as organic solvents will create a flammable or explosive mixture in air between their lower and upper explosive limits; thus, the organic solvent composition in the air must *not* be allowed to increase above the lower explosive limit when spraying. Upper and lower explosive limits for various solvents are shown in Figures 7.15 and 7.16. These data are based on explosive limit data from general literature or safety data sheets. These graphs can be used to determine safe spray conditions. For example, the explosive limit for acetone will be reached at ~1.6 g/ft^3; thus, at 500 ft^3/min process air, the lower explosive limit would be reached at 800 g/min acetone spray rate (1.6 × 500 = 800). This amount is only ~5% of the 16,000 g/min air capacity at 500 ft^3/min process air. If multiple organic solvents are used, it is estimated that the combined solvent spray rate and the limit for the solvent with the lowest lower explosive limit should be considered in an assessment if actual data are not available. Actual spray rate should be maintained well below the calculated lower explosive limit and are commonly set below ½ or ¼ of the limit value.

Proper safety precautions must be in place with organic solvent–based processes to assure an explosive condition is either prevented or channeled to prevent personal injury or equipment/facility damage. Since fluid bed coating processes are configured with the process air blower on the exhaust side of the process and process air is pulled in to feed the system, the process is at a negative pressure in relation to the atmosphere and it is open; thus, there is an inherent vent channel built in. Some systems are designed with interlocks to prevent development of an explosive condition such as interlocking pumps with the process air blower so that pumps shutdown if the blow faults or stops. Other systems are designed with a pressure relief system to quickly vent if an explosion is realized.

The use of water as a solvent introduces some complexity to the drying capacity assessment, since water is always present in the air at varying amounts. The amount of water present is indicated by dew point. The air capacity versus temperature plot for water from Figure 7.14 is shown in Figures 7.17 and 7.18 along with curves indicating water content at various dew points. The saturation curve is the point at which air capacity for water is filled. For example, the air can hold ~0.965 g water vapor/ft^3 at

Encapsulation via Fluidized Bed Coating Technology

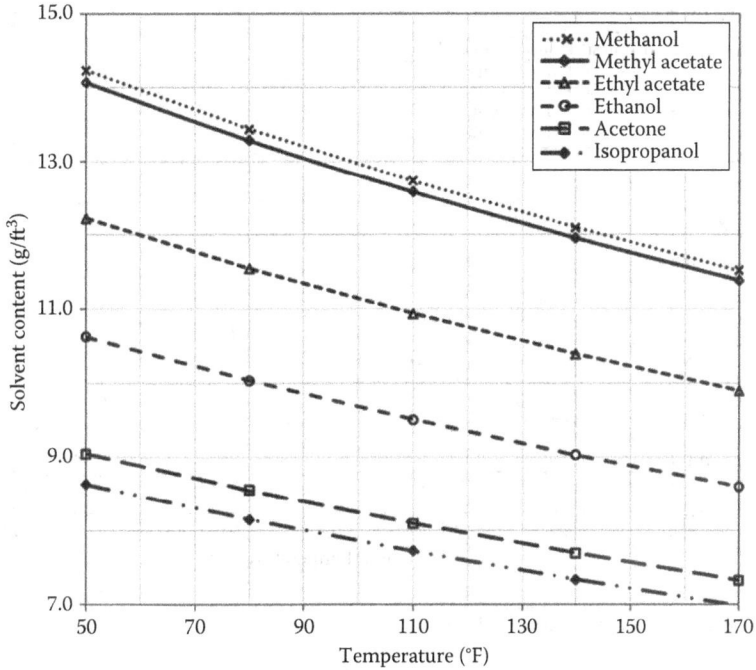

FIGURE 7.16 Upper explosive limits for selected organic solvents.

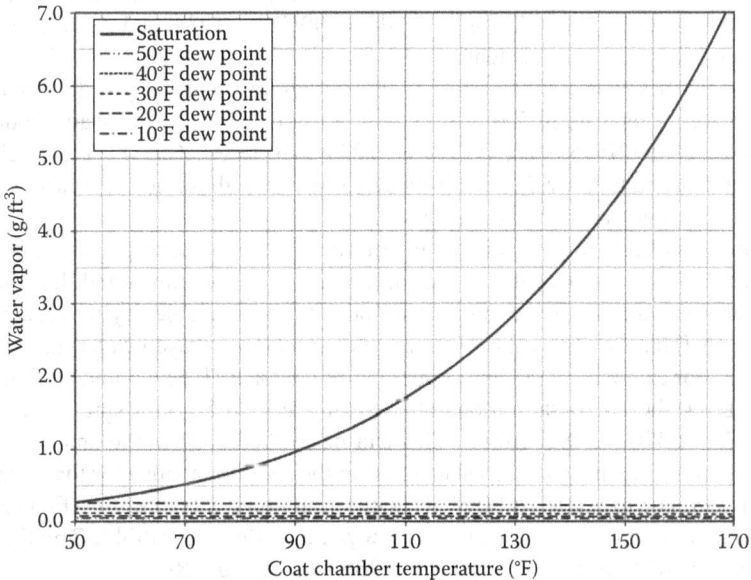

FIGURE 7.17 Water drying capacity (50°F–170°F).

90°F (Figure 7.18). Air at 90°F with a dew point of 50°F contains ~0.247 g water vapor/ft³ (Figure 7.18), which corresponds to ~25.6% relative humidity (0.247/0.965 × 100 = 25.6%); thus, the capacity for water vapor from sprayed water is ~0.718 g/ft³ (0.965 − 0.247 = 0.718). At 500 ft³/min process air under these conditions, water sprayed at 359 g/min (100% of its available water vapor capacity) would bring the air to almost saturation (100% relative humidity) (0.718 × 500 = 359). If this same 500 ft³/min process air

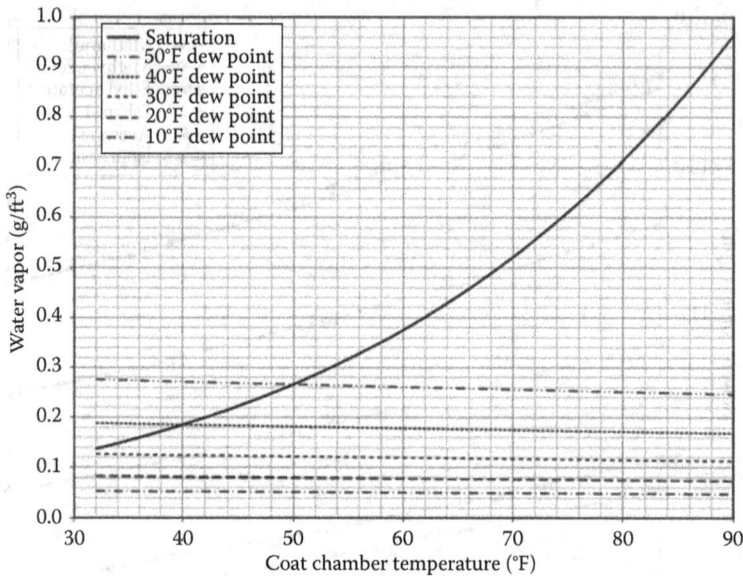

FIGURE 7.18 Water drying capacity (30°F–90°F).

at 90°F with a 50°F dew point were heated to 120°F, the air saturation point would shift to ~2.216 g water vapor/ft³ (Figure 7.17). The air would expand from 500 ft³/min at 90°F (305.3 K) to ~527 ft³/min at 120°F (322.0 K) (500 × 322.0/305.3 = 527 ft³/min). Due to this air expansion, the amount of water in the process air would shift to ~0.234 g/ft³ (0.247 × 500/527 = 0.234) and water applied at the same 359 g/min would bring the water content to ~0.915 g/ft³ (359/527 + 0.234 = 0.915), which is ~41.3% relative humidity (0.915/2.216 × 100 = 41.3%). The 359 g/min water spray shifts from being ~100% of the available drying capacity in 500 ft³/min air at 90°F to ~34.4% of the available drying capacity of the same air at 120°F ((0.915 − 0.234)/(2.216 − 0.234) × 100 = 34.4%). This is a rough estimate and does not take into account any volume adjustments other than those noted. It also assumes that all water sprayed is efficiently converted to vapor.

These examples indicate the dynamic nature of the drying relationships. To simplify this drying capacity assessment, the water content of air at a range of temperatures and dew points has been subtracted from the saturation water amount to indicate the available drying capacities and these data are shown in Table 7.6. The values indicate the amount of sprayed water that would bring the air to saturation. For example, at 90°F air temperature with a 40°F dew point, adding 0.797 g water vapor/ft³ will bring the air to saturation. This equates to 398.5 g/min spray rate at 500 ft³/min process air (0.797 × 500 = 398.5 g/min). This can be worked back to the amount of water in the process air by subtracting the value in Table 7.6 from the saturation value for that temperature from Figure 7.17 or 7.18. For example, at 90°F, the saturation point is 0.965 g/ft³ (Figure 7.18) and the water required to bring it to saturation at 40°F dew point is 0.797 g/ft³ (Table 7.6); thus, the air contains ~0.168 g/ft³ at 40°F dew point and 90°F (0.965 − 0.797 = 0.168).

Figure 7.17 can also be used to estimate the approximate dew point from relative humidity information. For example, 80°F air temperature has a ~0.714 g/ft³ saturation point (Figure 7.18). If the relative humidity of that air is 50%, then it only contains ~0.357 g/ft³ water vapor (0.714 × 0.5 = 0.357); 0.357 g/ft³ corresponds to ~59°F on the saturation curve (Figure 7.18), which is the estimated dew point. This estimate is a bit low, since it does not account for the volume change resulting from the temperature shift.

TABLE 7.6
Water Drying Capacity Correlation with Dew Point and Temperature (g/ft³)

		Air Temperature																
	°F	50	60	70	80	90	100	110	120	130	140	150	160	170	180	190	200	210
	°C	10.0	15.6	21.1	26.7	32.2	37.8	43.3	48.9	54.4	60.0	65.6	71.1	76.7	82.2	87.8	93.3	98.9
Dew Point °F	°C																	
0	−17.8	0.233	0.343	0.490	0.683	0.935	1.258	1.670	2.187	2.832	3.629	4.603	5.784	7.207	8.906	10.922	13.298	16.081
5	−15.0	0.225	0.335	0.482	0.675	0.927	1.250	1.662	2.180	2.825	3.621	4.595	5.777	7.200	8.899	10.916	13.292	16.074
10	−12.2	0.214	0.324	0.471	0.665	0.917	1.241	1.653	2.171	2.816	3.612	4.587	5.769	7.191	8.891	10.907	13.284	16.066
15	−9.4	0.201	0.312	0.459	0.653	0.905	1.229	1.641	2.159	2.805	3.601	4.576	5.758	7.181	8.881	10.897	13.274	16.056
20	−6.7	0.186	0.296	0.444	0.638	0.891	1.215	1.627	2.146	2.791	3.588	4.563	5.745	7.168	8.868	10.885	13.262	16.044
25	−3.9	0.167	0.278	0.426	0.620	0.873	1.198	1.610	2.129	2.775	3.572	4.547	5.729	7.153	8.853	10.870	13.247	16.030
30	−1.1	0.144	0.255	0.404	0.599	0.852	1.177	1.590	2.109	2.755	3.553	4.528	5.711	7.134	8.835	10.852	13.229	16.013
35	1.7	0.117	0.229	0.378	0.573	0.827	1.152	1.565	2.085	2.732	3.529	4.505	5.688	7.112	8.813	10.831	13.208	15.992
40	4.4	0.084	0.197	0.346	0.542	0.797	1.123	1.536	2.056	2.704	3.502	4.478	5.662	7.086	8.787	10.805	13.183	15.967
45	7.2	0.046	0.159	0.309	0.506	0.761	1.087	1.502	2.022	2.670	3.469	4.446	5.630	7.055	8.757	10.775	13.154	15.938
50	10.0	0.000	0.114	0.265	0.463	0.719	1.046	1.461	1.982	2.631	3.430	4.408	5.592	7.018	8.720	10.739	13.118	15.903
55	12.8	−0.054	0.062	0.214	0.412	0.669	0.997	1.413	1.935	2.584	3.385	4.363	5.548	6.975	8.678	10.697	13.077	15.862
60	15.6	−0.117	0.000	0.153	0.353	0.610	0.940	1.357	1.880	2.530	3.331	4.310	5.497	6.924	8.627	10.648	13.028	15.815

These water considerations are critical for aqueous work due to the effect on drying capacity. Processes can be developed with spray rate range to allow it to be set at a required humidity condition or percentage of available drying capacity. It is also possible to adjust temperature targets in relation to process air dew point to provide a required drying capacity. Some processes are run seasonally when dew point conditions are more consistently at a required level. Equipment can also be configured with dew point control systems that provide process air at a constant dew point regardless of the condition of the intake air.

Aqueous fluid bed granulation processes often required spray rate above the drying capacity. Spray rate, spray time, and temperature are balanced to reach a required level of water content prior to spray stoppage and drying. The overwetting is needed to promote the required granule build-up.

7.4.5 Tangential Spray Rotor Speed

Rotor rotation provides the means to circulate fluidized particles back to the spray region at the periphery of the rotor after they fall back to the rotor. Speed has an influence on the intensity of the feed and subsequently the quality of the pellets or granules produced. Rotor speeds may fall within ~1.5–5.0 m/s at the edge/slit, and the rotor speed is adjusted to the rpm speed needed for required edge speed. Since the rotor circumference varies linearly with diameter (circumference = pi × diameter), rpm scales linearly with diameter. For example, a 360 mm diameter rotor at 150 rpm is traveling at 2.826 m/s at the edge (0.36 × pi × 150 /60 = 2.826 m/s). For a 1000 mm rotor, the rpm would be reduced to ~54 rpm (2.826 × 60/pi/1.000 = 54 or 150 × 360/1000 = 54 rpm).

7.5 APPLICATION

7.5.1 Processing Obstacles

Most of the discussion in this chapter has focused on theoretical principles of operation. In actual practice, various other limiting concerns are realized. Fluidization plates may not always be ideal, and parameters may require adjustment to achieve coating under less favorable conditions. The most basic requirement in all cases is that all particles routinely circulate through the spray zone during processing so that they will be coated. Electrostatic forces between particles, or particles and equipment surfaces, are a common concern. These forces can promote agglomeration or prevent particles from circulating properly. Formulation or parameter adjustments are typically employed to overcome or minimize electrostatic concerns, but they are not always successful.

For individual particle coating, particle size limits have been mentioned in relation to airflow and droplet size. Individual particle coating by fluid bed processes is generally limited to particle sizes from ~50 μm to 2 or 3 cm. Near and below 50 μm, agglomeration, drying capacity, droplet size, and handling limits are realized. Above ~2 or 3 cm, blowers and equipment are not typically sized or configured to properly handle it. For particle sizes from ~2 mm to 2 or 3 cm and higher, fluid bed processes may not be applicable due to physical stresses of the fluidization process. Although tablets and the like can be coated in a fluid bed, they must be robust enough to withstand such fluidization stresses without chipping or fracturing. Alternative coating processes such as vented pan coating can be used to apply coatings on these larger particles with significantly less physical stress. This attrition concern can also be an issue for small, fragile particles.

For granulation work where the objective is to agglomerate particles to form granules, the lower particle size limit may be limited by the screens or filters that confine the particles in the process. If the plate screen does not adequately contain preagglomerated particles, they can be lost from the process. It is also possible for small particles to blind or pass through exhaust filters at the start of a granulation process prior to their incorporation into a granule.

7.5.2 Practical Considerations

The physical logistics of coating should also be considered in fluid bed coating processes. Since coating involves coverage of a surface, surface area in the bulk core material is a critical parameter. A bulk particulate material with a 53 μm area average size ($r = 26.5$ μm) has over 9.4 times as much surface area as the same bulk particulate material with a 500 μm diameter area average size ($r = 250$ μm). Clearly, the smaller 53 μm material would require significantly more coating and process time to achieve a coating thickness goal. Re-engineering a small particle to a larger size prior to coating can have significant cost advantage if the application can tolerate it.

The theoretical relationship between coat thickness and coating percentage can be useful in formulation work. Assuming a spherical particle, the following assessment can be made:

$$\text{Volume of a sphere} = \frac{4}{3}\pi r^3$$

where r = particle radius.

thus, for a particle with a radius r and a coat thickness t,

$$\% \text{ coating by volume} = \frac{\left(\frac{4}{3}\pi(r+t)^3 - \frac{4}{3}\pi r^3\right)}{\frac{4}{3}\pi(r+t)^3}(100)$$

or simplified

$$\% \text{ coating by volume} = \frac{\left((r+t)^3 - r^3\right)}{(r+t)^3}(100)$$

Incorporating the densities of the coating (d_{coat}) and core particle (d_{core}) provides the % coating by weight:

$$\% \text{ coating by weight} = \frac{\left((r+t)^3 - r^3\right)d_{coat}}{(r)^3 d_{core} + \left((r+t)^3 - r^3\right)d_{coat}}(100)$$

For a 100 μm core particle (1.6 g/cc core density) with a 10 μm thick coating (1.0 g/cc coat density), the coating would be 42.1% of the final particle by volume and 31.3% of the final particle by weight. These relationships can be rearranged to estimate the coat thickness for a given coat level and particle size.

This assessment would also work for a cubic particle by substituting r with the length of the cube side.

In actual practice, there are often various levels of attrition, agglomeration, and accretion that will affect particle morphology and coat thickness. In addition, there is variation in the circulation of small particles compared to larger particles of the particle size distribution and the surface-to-volume ratio varies with particle size; both of these concerns result in a theoretical size-dependent coat thickness variation within a given bulk particle size distribution. Although a theoretical assessment can be helpful as a guide to estimating coat level requirements, the extent to which these processing concerns are realized can significantly affect the actual coating thickness achieved.

7.5.3 COMMERCIAL APPLICATIONS

Fluid bed coating processes are used in a wide variety of industries to impart controlled-release properties to active ingredients in particles. The list of applications continues to grow as applications are developed to address new or unsolved controlled-release needs. A general list of common coating ingredients, application areas, and coating functions is provided in the following sections.

7.5.3.1 Coating Ingredients

Coating ingredients for fluid bed coating include an unlimited range of materials from an absolute perspective, since any material can be incorporated into a film coat as a dissolved or suspended component of a solution or hot melt. Even materials with a limited level of volatility can be included; however, a portion of the volatile material will be lost to evaporation during processing. It is critical that suspended materials are sized appropriately to both pass through the nozzle and remain incorporated in the film coat during application. It is generally best for suspended particles to be near 10 µm and smaller for optimal retention in the film; however, the binding properties of the film-forming components of the coating for the suspended particles will influence particle retention.

Film-forming ingredients are the key to coat formation, integrity, and adherence to particles. Film-forming ingredients are commonly polymers, but materials such as salts, sugars, waxes, and clays can also form films. A partial list of materials that have been successfully applied as coatings is provided in Table 7.7. More specific coating material lists and coating material trade names for oral pharmaceutical use can be found in other formulation resources.[39]

A simple means to assess film-forming properties is to cast a film of the material or coating formulation on a surface and observe whether it holds together. If it forms a flakey residue that is easily damaged with minimal force from a fingernail or other blunt edge, it will not likely form a good film on a particle. If it forms a film that is quite tacky, it might not be good for particle coating, but could be useful for granulation processes. Tackiness could be related to solvent retention in the cast film, which may be less of a concern with a sprayed film due to more efficient drying. Tackiness can also be mitigated with the use of suspended glidant particles in the coat formulation. Materials such as talc, magnesium stearate, or the like are commonly used as glidants at concentration ranging from ~10% to ~50% of the film coat. Be aware that a cast film may not match the properties of a sprayed on film coat if the film stratifies during drying; a sprayed-on film coat composed of many droplets of cast film may differ from that of a single cast film.

7.5.3.2 Application Areas and Coat Functions

A general list of coat function and release mechanisms commonly employed is provided in Table 7.8. In addition to coat material concerns noted in the previous section on coating ingredients, there are other practical considerations that should be taken into account when formulating.

1. Evaporative release of a contained ingredient that has an appreciable vapor pressure will be influenced by the vapor pressure, temperature, and affinity of other formulation ingredients for that ingredient.

TABLE 7.7
Example Fluid Bed Coating Materials

Acrylics	Gums, vegetable	Poly(ethylene glycol)
Acrylic aqueous dispersions	Hydrocarbon resins	Poly(vinyl acetate)
Cellulose acetate	Hydroxy propyl cellulose	Poly(vinyl pyrrolidone)
Cellulose acetate butyrate	Hydroxy propyl methyl cellulose	Poly(vinyl alcohol)
Cellulose acetate phthalate (enteric)	Hydroxy propyl methyl cellulose acetate succinate (enteric)	Poly(vinyl chloride)
Cellulose acetate phthalate aqueous dispersions (enteric)	Hydroxy propyl methyl cellulose phthalate (enteric)	Poly(vinylacetate phthalate) (enteric)
Caseinates	Fluoroplastics	Poly(vinylidene chloride)
Clay	Maltodextrins	Proteins
Coating butters	Methyl cellulose	Salts
Dextrins	Microcrystalline wax	Shellac
Ethylcellulose	Milk solids	Starches
Ethylcellulose aqueous dispersions	Molasses	Stearines
Ethylene vinyl acetate	Nylon	Sucrose
Fats	Paraffin wax	Surfactants
Fatty acids	Poly(lactides)	PTFE fluorocarbons
Gelatin	Poly(amino acids)	Vegetable waxes
Glycerides	Poly(ethylene)	Zein
	Poly(methacrylates)	

2. All insoluble coatings have some degree of permeability that is related to porosity and coat thickness. Although an insoluble coating delays or meters release of a core ingredient upon exposure, it does not eliminate it. The solubility of the core ingredient and osmolality of it and other excipients influence the release rate.
3. The porosity of a coating is influenced by the chemical structure of the insoluble components and can be modified by the addition of soluble "pore-forming" ingredients to the coating.
4. Release of a soluble ingredient from a particle coated with an insoluble coating will be influenced by the ingredient dissolution rate, coating permeability and thickness, and osmolality of the ingredient and other excipients. Osmotic pressures that develop within the particle upon exposure to an aqueous environment can be sufficient to fracture the coating and release ingredients.
5. Release of an ingredient from a particle coated with a soluble coating will be influenced by the coating dissolution rate.
6. Coating a pH basic ingredient with a neutral to base-soluble enteric polymer can trigger release from inside the particle as solvent penetrates and dissolves the ingredient to form a basic internal composition.
7. Coating an acidic ingredient with an acid-soluble reverse enteric polymer can trigger release from inside the particle as solvent penetrates and dissolves the ingredient to form an acidic internal composition.

An understanding of these and other concerns that influence ingredient release provides significant formulation insight. This insight coupled with a fundamental understanding of fluid bed coating principles will aid in the development, optimization, and troubleshooting of fluid bed coating processes and formulations.

TABLE 7.8
Fluid Bed Coating Is Commonly Used to Achieve a Controlled-Release Property for an Ingredient in the Core or Coating

Application	Description
Enteric delivery	Intestinal delivery using acidic polymer coatings that dissolve at near-neutral and basic conditions, but are relatively insoluble at acidic conditions. Coatings are applied to drug-containing particles, tablets, or capsules. There are regulatory limitations on the use of many enteric materials for food or nutritional work. There may also be daily intake considerations for some in pharmaceutical applications.
Sustained release	Sustained release may be achieved by coating an active core particle with a controlled-release layer. Various release mechanisms including the following can potentially be employed: 1. Slowly dissolving coat 2. Slowly eroding coat 3. Insoluble coat with controlled porosity and thickness. It is also possible to incorporate active into one of these coating approaches to realize a sustained-release profile from the coating.
Taste-masking or taste concealing	Taste masking often involves coating bad tasting ingredients to minimize exposure to taste receptors on the tongue in chewable or orally dissolving product. These coatings must also release appropriately in the gut. Depending on release needs, the coatings could have enteric or limited sustained-release properties, or could contain a *reverse* enteric, acid-soluble coating that is relatively insoluble in the mouth, but dissolves in the stomach.
Delayed or pulsatile release	Coating is designed to begin releasing active after a delay period. Delayed-release coatings include the following: 1. Enteric coatings that dissolve after the stomach as the pH increases 2. Taste-concealing coatings composed of acid-soluble polymers that release contents upon exposure to stomach acidity 3. Waxy coatings that fail when a desired temperature is reached 4. Insoluble brittle coatings that fracture as osmotic pressures build in a particle 5. Coatings that break down due to enzymatic cleavage in the gut or other enzymatic environment
Drug loading	Drug loading on nonpareil or other support particles offers a means to either a consistent particle size for application of controlled-release coatings or a uniform low dosing formulation need.
Stability	Coatings can shield formulation components from heat, oxygen, moisture, or other incompatible components.
Granulation or agglomeration	Particles can be accreted and join with binders to create larger particle masses for improved handling and/or uniformity.

Notes: Common coating properties or functions are listed above. Applications can potentially incorporate multiple delivery or release mechanisms to achieve a desired delivery profile. Although controlled-release fluid bed coatings are commonly used in oral pharmaceutical applications, the same release mechanisms can be employed in other applications including nutraceutical, nutritional, food, household, agricultural, industrial, and other product areas.

REFERENCES

1. Wurster, D. E. (1963). Granulating and coating process for uniform granules. US Patent No 3,089,824 to University of Wisconsin, Madison, WI.
2. Wurster, D. E. (1964). Apparatus for coating particles in a fluidized bed. US Patent No 3,117,027 to University of Wisconsin, Madison, WI.
3. Wurster, D. E. (1965). Apparatus for the encapsulation of discrete particles. US Patent No 3,196,827 to University of Wisconsin, Madison, WI.
4. Wurster, D. E. (1965). Process for preparing agglomerates. US Patent No 3,207,824 to University of Wisconsin, Madison, WI.
5. Wurster, D. E. (1966). Particle coating apparatus. US Patent No 3,241,520 to University of Wisconsin, Madison, WI.

6. Wurster, D. E. (1966). Particle coating process. US Patent No 3,253,944 to University of Wisconsin, Madison, WI.
7. Huttlin, H. (1990). Fluidized bed apparatus for the production and/or further treatment of granulate material. US Patent 4,970,804.
8. Imanidis, G., Leuenberger, H., Nowak R. Studer J. M., and Winzap S. (1990). Method and system for agglomerating particles and/or for coating particles. US Patent No 4,895,733 to Pharmatronic AG, Pratteln, Switzerland.
9. Huttlin, H. (1991). Fluidized bed apparatus, in particular for granulation of pulverulent substance. US Patent No 5,040,310.
10. Huttlin, H. (1992). Fluidized bed apparatus, in particular for granulation of pulverulent substance. US Patent No 5,085,170.
11. Jones, D. M. (1993). Fluidized bed with spray nozzle shielding. US Patent No 5,236,503 to Glatt Air Techniques, Inc., Ramsey, NJ.
12. Littman, H., Morgan, M. H., and Jovanovic S. D. (1993). Coating apparatus having opposed atomizing nozzles in a fluid bed column. US Patent No 5,254,168.
13. Hirschfeld, P. F. F. and Weh, M. (1994). Method and apparatus for coating particles agitated by a rotatable rotor. US Patent No 5,284,678 to Glatt GmbH, Binzen, Germany.
14. Jones, D. M. (1995). Fluidized bed with spray nozzle shielding. US Patent No 5,437,889 to Glatt Air Techniques, Inc., Ramsey, NJ.
15. Huttlin, H. (2000). Diffusing nozzle. US Patent No 6,045,061.
16. Walter, K. (2001). Apparatus and a method for treating particulate materials. US Patent No 6,270,801 to Aeromatic-Fielder AG, Bubendorf, Switzerland.
17. Huttlin, H. (2002). Device for treating particulate product. US Patent No 6,367,165.
18. Jones, D. M., Smith, R. A., Kennedy, J. P., Maurer, F., Tondar, M. G., Luy, B., and Baettig, M. J. (2004). Split plenum arrangement for an apparatus for coating tablets. US Patent No 6,692,571 to Glatt Air Techniques, Inc., Ramsey, NJ.
19. Huttlin, H. (2004). Method for treating particulate material with a coating medium and an apparatus for carrying out the method. US Patent No 6,740,162.
20. Huttlin, H. (2005). Device for treating particulate material. US Patent No 6,898,869.
21. Schneidereit, H. and Pritzke, H. (2006) Apparatus for the formation of coverings on surfaces of solid bodies in a coating chamber. US Patent No 7,083,683 to Glatt Ingenieurtechnik GmbH, Weimar, Germany.
22. Jacob, M., Rumpler, K., and Waskow, M. (2007). Fluidized bed apparatus for batch-by-batch or continuous process control and method for operating a fluidized bed apparatus. US Patent No 7,241,425 to Glatt Ingenieurtechnik GmbH, Weimar, Germany.
23. Mehta, A. M. and Natsuyama, S. (2007). Fluidized bed device. US Patent No 7,297,314 to Kabushiki Kaisha Powrex, Osaka, Japan.
24. Huttlin, H. (2009). Method and apparatus for treating particulate-shaped material, in particular for mixing, drying, graduating, pelletizing and/or coating the material. US Patent No 7,544,250.
25. Huttlin, H. (2010). Apparatus for treating particulate material. US Patent No 7,802,376.
26. Huttlin, H. (2010). Process and apparatus for treating a particulate material. US Patent No 7,798,092.
27. Wurster, D. E. (1990). Particle-coating methods. In *Pharmaceutical Dosage Forms: Tablets*, Vol. 3, Rev. 90, Lieberman, H.A. (Ed.). pp. 161–197. New York: Marcel Dekker.
28. Fries, L., Antonyuk, S., Heinrich, S., and Palzer, S. (2011). DEM-CFD modelling of a fluidized bed spray granulator. *Chem Eng Sci*, 66, 2340–2355.
29. Gryczka, O., Heinrich, S., Miteva, V., Deen, N. G., Kuipers, J. A. M., Jacob, M., and Mörl, L. (2008). Characterization of the pneumatic behavior of a novel spouted bed apparatus with two adjustable gas inlets. *Chem Eng Sci*, 63, 791–814.
30. Heng, P. W. S., Chan, L. W., and Tang, E. S. K. (2006). Comparative study of the fluidization dynamics of bottom spray fluid bed coaters. *AAPS PharmSciTech*, 7(2), E1–E9.
31. Mafadi, S. E., Hayert, M., and Poncelet, D. (2003). Fluidization control in the Wurster coating process. *Chem Ind*, 57(12), 641–644.
32. Palmer, S., Ingram, A., Fan, X., Fitzpatrick, S., and Seville, J. (2007). Investigation of the sources of variability in the Wurster coater: Analysis of particle cycle times using PEPT. In *The 12th International Conference on Fluidization—New Horizons in Fluidization Engineering*, Vancouver, British Columbia, Canada, ECI Symposium Series, Volume RP4 at http://dc.engconfintl.org/fluidization_xii/52.
33. Glatt International Times. (2006). No. 21.

34. Goodhart, F. W. (1989). Centrifugal equipment. In *Pharmaceutical Pelletization Technology*, GhebreSelassie, I. (Ed.), pp. 101–122. New York: Marcel Dekker.
35. Vertommen, J. and Kinget, R. (1997). The influence of five selected processing and formulation variables on the particle size, particle size distribution, and friability of pellets produced on a rotary processor. *Drug Dev Ind Pharm*, 23, 39–46.
36. Rashid, H. (2001). Centrifugal granulating process for preparing drug-layered pellets based on microcrystalline cellulose beads. PhD Dissertation, University of Helsinki, Helsinki, Finland.
37. Bouffard, J., Dumont, H., Bertrand, F., and Legros, R. (2007). Optimization and scale-up of a fluid bed tangential spray rotogranulation process. *Int J Pharm*, 335, 54–62.
38. University of Wisconsin Alumni Research Foundation. (circa 1963). Momentum transfer in Wurster air suspension coating equipment. (Out of print document archived at Coating Place, Inc., Verona, WI).
39. Skalsky, B. and Stegemann, S. (2011). Coated multiparticulates for controlling drug release. In *Controlled Release in Oral Drug Delivery*, Wilson, C.G. and Crowley, P.J. (Eds.), pp. 262–265. New York: Springer.

8 Encapsulation via Pan-Coating

Charles R. Frey

CONTENTS

Abbreviations 148
Unit Conversions 148
8.1 Introduction 148
8.2 History 149
8.3 Process Description 150
8.4 Conventional Pan-Coating 152
    8.4.1 Pan Design 152
    8.4.2 Ventilation 152
    8.4.3 Coating Delivery 152
    8.4.4 Charging and Discharging 153
8.5 Vented Pan-Coating 153
    8.5.1 Pan Design 153
    8.5.2 Ventilation 153
    8.5.3 Spray Guns 155
    8.5.4 Exhaust Filter 156
    8.5.5 Charging and Discharging 156
8.6 Critical Parameters 156
    8.6.1 Charge 156
    8.6.2 Pan Speed 157
    8.6.3 Process Air 158
    8.6.4 Temperature 158
    8.6.5 Gun Position 158
    8.6.6 Spray Rate 159
    8.6.7 Atomizing and Pattern Air 164
    8.6.8 Process Endpoint 165
8.7 Scale-Up 167
    8.7.1 Charge 167
    8.7.2 Pan Speed 168
    8.7.3 Process Air 168
    8.7.4 Spray Rate 169
    8.7.5 Atomizing and Pattern Air 170
8.8 Applications 172
    8.8.1 Coating Materials 172
    8.8.2 Conventional Pan-Coating Applications 173
    8.8.3 Vented Pan-Coating 173
8.9 Troubleshooting 173
8.10 Conclusions 175
References 175

ABBREVIATIONS

°C	degrees Centigrade
cm	centimeters
°F	degrees Fahrenheit
Ft	feet or foot
G	gravitational force
g	gram
in.	inch
kg	kilogram
L	liter
lb	pound
LEL	lower Explosive Limit
min	minute
mm	millimeter
psi	pounds per square inch
UEL	upper Explosive Limit

UNIT CONVERSIONS

°C = (°F − 32)/1.8
1 in. = 2.54 cm
1 m = 3.28 ft

8.1 INTRODUCTION

Pan-coating refers to processes where solid particulate materials or tablets generally ranging from near a few mm to several centimeters in diameter or size are placed in a rotating drum where they recirculate as coating material is applied from solution or suspension. Drying air is passed through or over the bed in some manner to evaporate the solvent vehicle and deposit coating on the particles. Pans are commonly fitted with angled "baffles" or "blades" and/or antislip "bars" on the inside surface to optimize particle circulation and coating uniformity.

Pan-coating is generally subdivided into two distinct subcategories:

1. Conventional pan-coating generally refers to configurations where the drum is solid and drying air is introduced above or into the bed and exhausted through the pan mouth or other exhaust channel. Coating material may be sprayed from a nozzle, which is commonly referred to as a spray "gun," onto the circulating particles, but it is ladled or poured in for many applications. The pan configuration may be similar to a cement mixer with an inclined axis (Figure 8.1) or the well-known Pelligrini design with a horizontal axis (Figure 8.2).
2. Vented pan-coating generally refers to configurations where the pan is perforated to allow drying air passage through the bed and pan wall (Figure 8.3). Drying air is blown through the particle bed as coating solution or suspension is sprayed downward onto the surface of the circulating bed. Vented pan-coating can be configured in various ways to optimize air flow through bed and is performed in fully or partially perforated pans.

Variations of these basic designs exist to optimize coating uniformity or application efficiency, minimize in-process particle stress, or offer some other proposed advantage. This chapter focuses on a critical assessment of the classic conventional and vented pan-coating configurations.

FIGURE 8.1 Conventional pan (inclined axis).

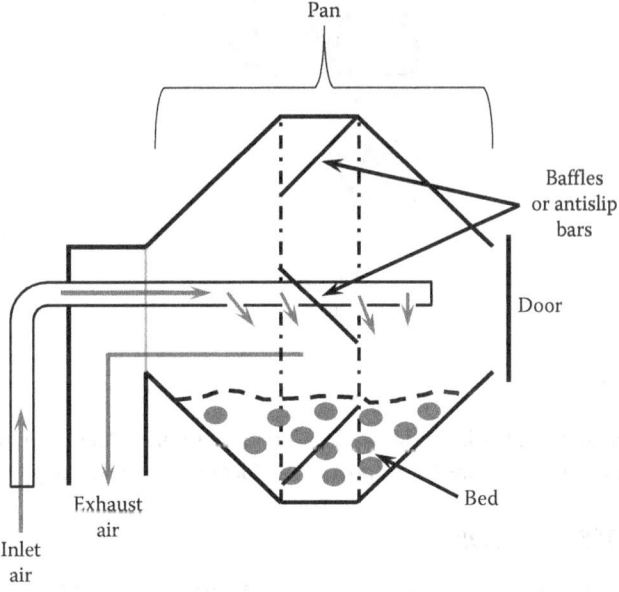

FIGURE 8.2 Conventional pan (Pellegrini design).

8.2 HISTORY

Pill coating was performed centuries ago either to overcome an unpleasant taste or odor with a pleasant-tasting overcoat or to impart an aesthetic improvement.[1] Pan processes dating back to the mid-1800s focused primarily on honey or sugar coating up until the 1950s when ventilation systems were introduced to improve drying processes and allowed the formation of more uniform film coats. This enabled a range of new applications involving attributes such as improved appearance,

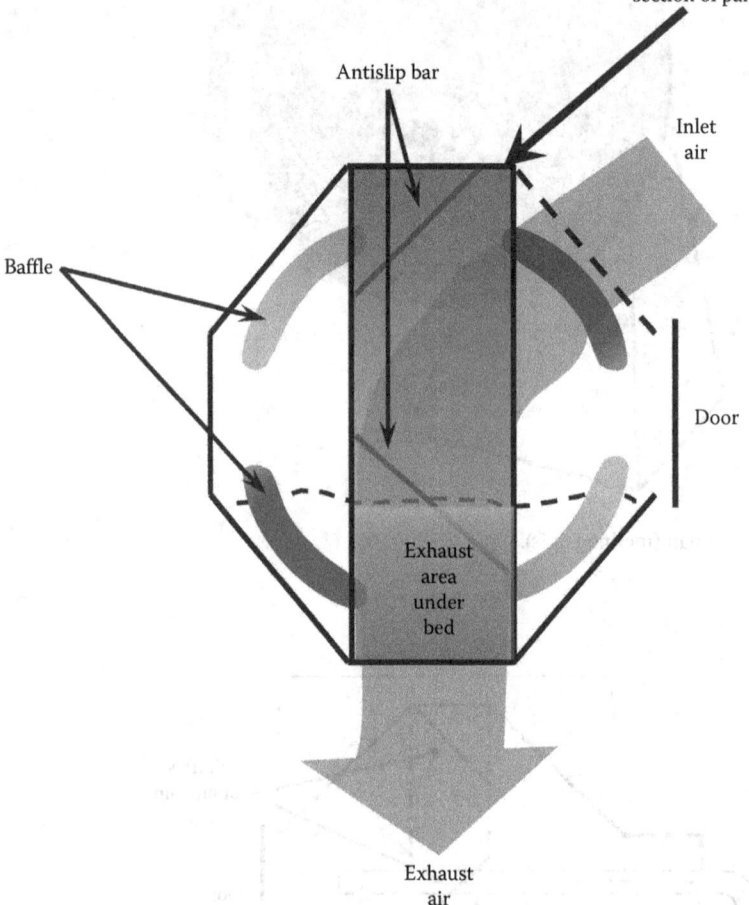

FIGURE 8.3 Vented pan.

improved shelf life, and controlled release. Pan-coating improvements in recent decades have involved primarily design variations for improved uniformity and efficiency.

8.3 PROCESS DESCRIPTION

Pan-coating involves applying or spraying a film coat on particles ranging from near a few millimeter to several centimeters in size. Film coat application on a particle by particle basis is not feasible due to time constraints and uniformity concerns. Although it is possible to coat such materials in a fluid bed coating process, the particles are often prone to fracture and breakage due to particle physical size and the intensity of particle contact with other particles and/or equipment surfaces when fluidizing. Pan-coating often employs a similar strategy to fluid bed coating with stationary spray delivery and particle movement in relation to it; however, particle movement is achieved by rotating a partially filled pan about an inclined or horizontal axis. The combination of particle lift created by the advancing surface of the rotating pan and gravity creates particle circulation including a cascading particle bed surface. Coating can be applied to the cascading surface via spray gun,

Encapsulation via Pan-Coating

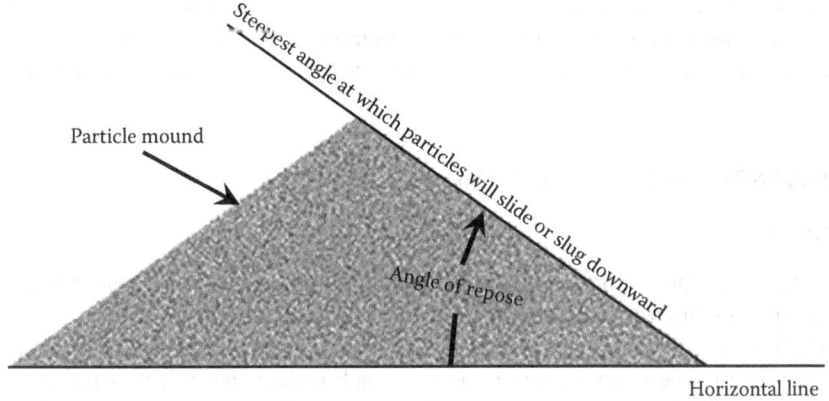

FIGURE 8.4 Angle of repose for a particulate material.

dripping, or ladle (some conventional pan processes). Peristaltic or gear pumps are generally used to deliver coating solution or suspension via spray gun or dripping. Air blowing over or through the pan and particle bed carries away solvent vehicle vapors to deposit coating on particles.

The cascading action in the rotating pan is realized once the particle bed surface exceeds the angle of repose for the particles as the pan rotates. The angle of repose is the steepest angle of descent in relation to horizontal for a particulate bed. It is commonly illustrated as shown in Figure 8.4, and the movement realized in a rotating pan is shown in Figure 8.5. Pan rotational speed is adjusted to maintain a continuous flow of recirculating particles and allow a gradual buildup of coating.

In addition to the classic inclined and horizontal axis configurations illustrated in Figures 8.1 through 8.3, some vertical axis designs have been patented.[1] These designs resemble an upright

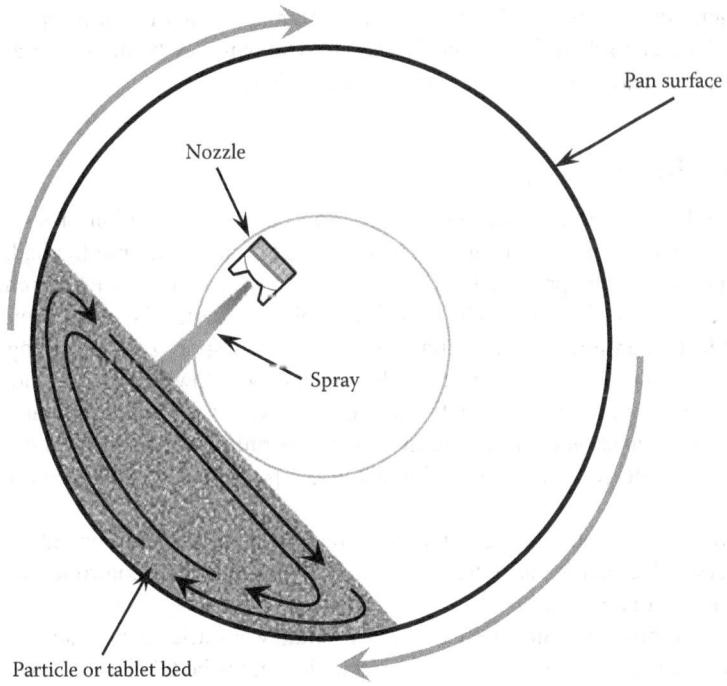

FIGURE 8.5 Generalized particle movement in a rotating pan.

spinning cup that forces core material outward and employs air and/or guide vanes to direct it upward and back toward the cup center and maintain circulation. The commercial viability of these processes has yet to be determined and these systems are not subject of the discussion in this chapter.

8.4 CONVENTIONAL PAN-COATING

8.4.1 Pan Design

Conventional pan-coating in this chapter refers to coating in nonperforated pans. The inclined axis design is shown in Figure 8.1. Upon pan rotation with a load of particles or tablets, material is lifted along the leading surface of the pan and eventually cascades downward as the angle of repose is reached or exceeded. The downward movement is back toward the trailing edge of the bed and toward the lower point of the pan axis in a manner similar to that shown in Figure 8.5. Baffles, fins, and/or antislip devices may be mounted on the interior pan surface to promote good bed circulation for optimal coating uniformity.

The Pellegrini pan design (Figure 8.2) typically maintains a horizontal pan axis. As the pan rotates, the material cascades from the leading edge of the bed toward the trailing edge as the angle of repose is reached and exceeded. The inclined/tapered pan design along with any baffles and/or anti-slip devices promote bed circulation to and from both the bed surface and pan edges to promote coating uniformity.

Conventional pans generally range from ~3 to ~2000 L in working capacity.

8.4.2 Ventilation

Drying air or other gas is drawn through the conventional pan in various ways to remove coating vehicle vapors. Air may be directed over the bed surface with inlet and outlet air ducts mounted at the mouth or back of the pan. Air may also be directed through the bed via inlet and/or outlet air ducts or diffusion "swords" inserted in the bed. Drying may be facilitated or optimized with heat and/and or humidity control applied to the drying air. The pan may be housed in an oven to allow the pan surface to be heated for optimal drying characteristics.

8.4.3 Coating Delivery

Coating is generally delivered to conventional coating pans from solution or suspension. Liquid components may be sprayed onto or into the bed, dripped or squirted onto the bed, or ladled onto the bed in various ways. The preferred delivery technique may be dependent on characteristics of the process, materials, or desired product attributes. When coating is sprayed on, uniformity is typically dependent on uniform particle circulation through the spray zone. Dripping or ladling processes generally rely on over wetting the particles/tablets and allowing particle and pan movement to distribute coating material to all particle surfaces as solvent vehicle is evaporated to deposit the coating. Ladled or dripped on coatings are generally less uniform than sprayed on coatings; thus, a higher coating percentage may be needed to achieve a performance specification compared to a sprayed of coating.

In powder coating applications, one or more coating components are delivered in powdered form during the process. The powder accretes to the particle surfaces in a controlled manner as liquid coating components are applied.

The vehicle is commonly water but can be any suitably volatile organic solvent or mixture of solvents. When working with organic solvents, precautions must be taken to vent vapors safely with respect to explosive concentration limits and within allowable/permitted evaporation quantity limits.

Encapsulation via Pan-Coating

8.4.4 CHARGING AND DISCHARGING

Charging and discharging of a conventional pan coater is done through the mouth of the pan or possibly though a hatch in the pan. Delivery or removal may be performed manually with a scoop or with any suitable system that can be adapted to the process.

8.5 VENTED PAN-COATING

8.5.1 PAN DESIGN

Vented pan-coating (sometimes referred to as perforated pan-coating) is similar to conventional pan-coating with the most significant difference being the addition of holes (perforations) through the pan. These holes allow drying air to pass directly through the bed to facilitate and accelerate drying. The perforation holes or slots are sized to retain product particles but allow air passage. The lower applicable particle size is limited primarily by hole or slot size, but may also be limited by formulation characteristics. Both fully perforated and partially perforated designs are offered. Fully perforated pans have a perforated zone that extends around the entire circumference of the pan, which allows uniform, continuous air passage over the entire region of air flow under the particle bed. Partially perforated pans have segments of perforated pan surface located at intervals around the pan circumference, which localizes airflow in those perforated regions when they pass under the particle bed. Perforated pan illustrations are provided in Figures 8.3 and 8.6.

The advantages of a fully perforated pan versus a partially perforated pan, if any, are not well established. A process performed in a partially perforated pan was found to require a significantly lower inlet temperature to realize a target exhaust temperature compared to the same for a fully perforated pan.[2] This was reasoned to be the result of less heat loss and/or more efficient heat transfer in the partially perforated pan.

Perforated and partially perforated pans are typically equipped with baffles, antislip ridges or bars, or other circulatory hardware to ensure optimal bed circulation. Antislip ridges or bars prevent the bed from sliding in the rotating pan and ensure a continuous cascading bed surface. Baffles are paddlelike devices mounted at an angle in the pan to promote particle/tablet circulation along the pan axis to optimize coating uniformity. Baffle designs can be critical due to their physical size in relation to bed volume and potential for particles/tablets to hang up or "wedge" into tight clearance spaces between the pan and baffle or baffle mount. Ideally baffles ride below the bed surface in the spray region to maintain a consistent bed surface flow and prevent sprayed coating buildup on an exposed portion of the baffle; however, they should not be so low that the center portion of a deep particle bed stagnates and cannot circulate adequately to the bed surface.

8.5.2 VENTILATION

Ventilation in perforated pan systems is provided by the combination of an inlet blower and an exhaust blower. The blowers work in combination to maintain the required process airflow and a suitable negative pan pressure in relation to the environment to confine mists, vapor, and spray-dried coating to the pan. Control of pan pressure may be maintained either by a feedback system that adjusts the inlet and exhaust blowers or by independent setting of an exhaust blower at a rate above that of the inlet blower.

Various terms have been applied to air flow configurations used in vented pans. All configurations are designed to pass drying air through the particle/tablet bed in some way. Although the bed is inclined in a rotating pan and air passes through the bed in a sideways or angled direction, air passage through the bed will be referred to as upward (from under the particle/tablet bed) or downward (from above the particle/tablet bed) as shown in Figure 8.6. Considerations for upward or downward air flow include the following:

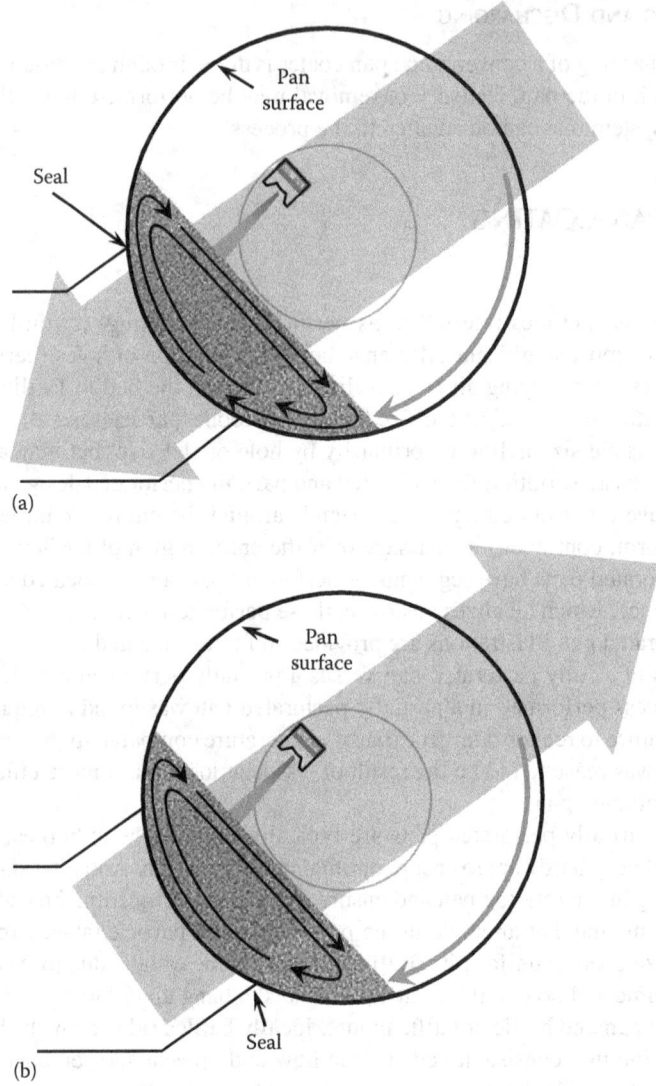

FIGURE 8.6 Air flow options in a vented pan include downward or upward air flow as indicated in illustrations (a) and (b). The exhaust/inlet area under the particle/tablet bed passes air only where pan perforations exist. Perforated pan area is continuous around a fully perforated pan. Perforation areas are staggered along the rotating pan surface in a partially perforated pan.

1. Upward air flow can reduce coating efficiency due to the countercurrent flow of air in relation to nozzle spray, which can interfere with gun spray pattern and promote premature drying.
2. Upward air flow must be restricted sufficiently to minimize particle/tablet fluidization that could result from excess upward air flow through the bed.
3. Upward air flow can minimize spray drying since hot inlet air is introduced under the bed away from the spray area. Evaporative cooling and solvent vaporization as the air passes through the bed reduces the drying capacity of the air in the spray zone.
4. Downward air flow introduces hot inlet air near the spray zone, which may promote spray drying.
5. Downward air flow may minimize spray pattern interference since air is introduced concurrently with spray.

Encapsulation via Pan-Coating

6. Downward air flow introduces potential concerns with particles being "pressed" to the pan surface by excess process air flow. Air flow should be set to minimize particle agglomeration and sticking to the pan that might be promoted by high air flow.

8.5.3 Spray Guns

In the classic perforated or partially perforated pan, guns are positioned with spray directed perpendicularly to the flowing bed surface as illustrated in Figure 8.6. The optimal bed contact region is where the bed surface is reasonably flat, particles/tablets are cascading at a consistent suitable rate, and baffles are not exposed. An optimal position assures that the bed surface velocity has accelerated to a suitable rate and maximizes the length of bed surface below the spray zone for optimal drying.

Guns are typically a two-fluid nozzle design with atomizing air delivered concentrically about the coat solution or suspension delivery tip. The resulting round pattern of the atomized spray is typically modified to a flat pattern with pattern air jets that are generally directed inward at the developing spray from opposite directions. Pattern air delivery is oriented in a variety of ways depending on the gun manufacturer. A classic gun design is illustrated in Figure 8.7. One or more guns are arranged along the axis of the bed to extend spray over the full expanse of the bed surface in the axial direction. In multiple gun systems, guns are positioned at intervals along the pan axis with minimal overlap of spray between adjacent guns at the bed surface. In larger pans, some systems may be configured with separate rows of guns to take advantage of the longer particle bed surface and either speed the coating process or balance process scaling requirements.

Optimal gun designs include an air-actuated, retractable shut off "pin" that allows the nozzle to be shut off as needed to perform weight gain checks or for other spray stoppages during the coating process. This shut off feature prevents nozzle drips, which would compromise the product. Solution recirculates in the solution tank during such in-process stoppages.

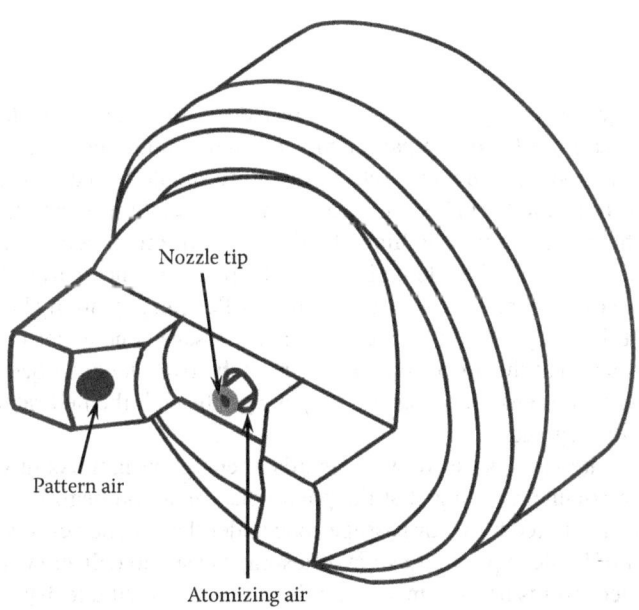

FIGURE 8.7 Illustration of a flat spray gun design commonly used in pancoating applications.

8.5.4 Exhaust Filter

Vented pan-coating systems are typically configured with exhaust filters to avoid discharge of fines and spray-dried coating to the environment. Due to the nature and amount of these fines, filter blinding occurs readily. To overcome this blinding, a pleated cartridge filter confined to properly designed housing with a back pressure pulsing mechanism is commonly employed to provide optimal filter surface area and a means to discharge blinding material. During processing, exhaust process air is directed to the outer surface of the pleats in the cylindrical cartridge, and when the pressure drop through the filter exceeds a predetermined level due to build up on the outer surface, a back pressure pulse is delivered to the inside cartridge/pleat surfaces to push accumulated powder from the outer surface. The housing that the filter cartridge resides in is constructed to allow released powder to fall into a "stagnant" accumulation area below the filter, where it will not be carried back to the filter surface and can easily be removed at an appropriate time after a batch, series of batches, or campaign.

8.5.5 Charging and Discharging

Charging of vented pan systems is generally done through the mouth of the pan or another accessible port built into the design. This may be performed manually with a scoop or with any suitable delivery system that can be adapted to the process. It may be important to run process air and pan rotation at minimal speed or with jogging (intermittent rotation on and off) prior to spraying to remove dust from the uncoated material with minimal physical stress.

Discharge can be done through the mouth of the pan manually or with a removable discharge chute or tube that channels material to a container as the pan rotates. Larger pans are designed with pan bottom discharges that can be opened at the conclusion of the process.

8.6 CRITICAL PARAMETERS

Critical pan-coating parameters are discussed in this section with respect primarily to classic perforated pan configurations.

8.6.1 Charge

The applicable charge range for a pan is generally from a minimum bed depth (deepest point from bed surface to pan) of ~1/8 of the pan diameter to a maximum bed depth of ~1/4 of the pan diameter. Actual limits specified in manufacturer literature should be maintained. The upper limit is generally referred to as the "brim" volume, and it is commonly near where the bed meets the mouth/port edge. The minimum may be limited either by a minimum bed depth or the air passage zone under the bed. A minimum bed depth limit will be realized when baffles become exposed to spray and coat during the coating process. Low-profile baffles can potentially be used to reduce this lower pan load limit, but only if bed coverage of the air passage zone under the bed is adequately maintained. It is critical that the air passage zone under the particle/tablet bed is entirely covered by the particle/tablet bed to ensure consistent air passage through the bed rather than around the bed through uncovered regions.

Conventional pans are less restrictive with regards to charge weight. As in vented pan-coating, an upper limit (brim volume) is realized at the point where material begins to spill from the pan. The lower limit is not restricted by an air passage zone under the bed; however, a lower limit may be realized at the point of baffle exposure to spray if a spray is used to deliver coating material. Also, if immersed perforated swords or an immersed air duct is used to facilitate drying, it may be critical for the perforations or duct opening to be immersed in the bed. If these concerns are not an issue, there is no significant physical limitation to a minimum batch size.

Encapsulation via Pan-Coating

8.6.2 Pan Speed

Particle/tablet movement in a rotating pan has been the subject of many studies. Critical considerations for pan speed are illustrated in Figure 8.8. With pan rotation, the angle of repose for a contained particulate material will eventually be reached and the material will begin to cascade downward along the inclined bed surface (Figure 8.8a). The point where this cascade is maintained without "slugging" or interruption of flow is a minimum limit for pan speed, which is generally measured in revolutions per minute (rpm). As the pan speed is further increased, the bed evolves from a sloped flat surface to an S-shaped profile (Figure 8.8b). Eventually, a point is reached where material in the upper bed cataracts as it extends beyond the bed surface into a downward free fall (Figure 8.8c). This point is above an upper preferred limit for pan speed. Optimal pan speed lies between these minimum (flat surface) and maximum (pre-cataract) limits.

Considerations for setting an ideal pan speed include the following:

1. Studies have indicated better coating uniformity at higher pan speeds.[4] This is likely realized due to improved bed mixing and increased frequency of particle/tablet passage through the spray zone (more total passages during the process time of the batch).
2. Higher pan speed shortens spray exposure during particle/tablet passage through the spray zone, which can mitigate over wetting that could be realized at slow pan speeds.
3. Higher pan speed shortens the initial surface dry time after passage through the spray zone due to increased bed surface velocity.

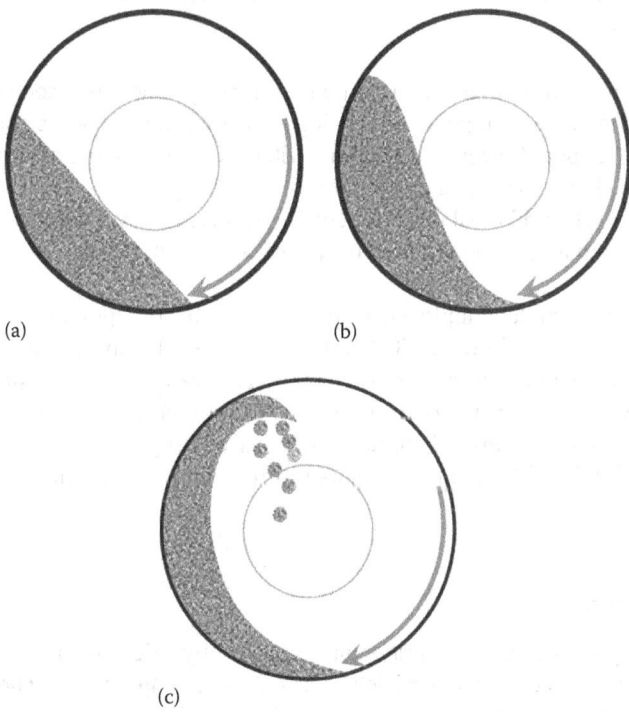

FIGURE 8.8 Bed character changes with pan speed. (a) Preferred flow with steady relatively flat bed surface. (b) S-shaped bed surface associated with high pan speed and higher physical stress to bed material. (c) Cataracting bed associated with excess pan speed that will result in significant physical stress to bed material and interfere with spray guns.

4. The cascading particle layer generally thickens as pan speed is increased. This can be critical depending on the depth to which spray droplets penetrate into the bed.
5. Higher pan speed creates more physical stress for particles/tablets, which could contribute to attrition and fracture depending on particle/tablet integrity.

Optimal pan speed may involve a balance between optimizing uniformity and minimizing attrition concerns.

8.6.3 Process Air

Optimal process air typically is below both the point where either the spray pattern or bed is excessively disrupted. Inlet and exhaust blowers are either set or adjusted by a feedback system to maintain a negative pan pressure in relation to the room environment at −0.1 to −0.5 in. of water while maintaining required process air flow. Equipment manufacturer support or literature should provide guidance on applicable process air ranges for their equipment. In general, process air is set as high as possible to provide maximum drying capacity. The upper limit will be dependent on various factors that have previously been mentioned including spray pattern and bed disruption.

8.6.4 Temperature

Process air temperature is typically controlled at the inlet, where it is adjusted to provide a required exhaust temperature. Control at the exhaust point would typically result in wide temperature swings due to the relatively slow temperature equilibration process in the pan.

The preferred exhaust temperature is at a point where several criteria are adequately addressed including the following:

1. The temperature is not so high that it introduces stability concerns associated with product degradation, a material melt point, or altering a release property of the core material.
2. The temperature should meet film-forming criteria for the coating system without introducing coat cracking, particle agglomeration, or twinning.
3. The temperature should be balanced for coating vehicle removal needs while minimizing spray drying and maintaining required coat surface characteristics.

Many organic solvent vehicles employed in pan-coating have vapor pressures/volatility suitable for exhaust temperatures in the 25°C (77°F) to 35°C (95°F) range, which provides suitable drying characteristics without excess spray drying. This range is also suitable for most latex coating formulas, which appropriately form films in this temperature range, but exhibit adverse tackiness at warmer temperatures. Nonlatex aqueous coating systems are generally applied at 38°C (100°F) to 50°C (122°F) exhaust temperatures to provide greater drying capacity and faster drying.

8.6.5 Gun Position

As discussed in the previous section, spray guns are typically oriented such that spray is directed perpendicularly to and directly at the particle bed as indicated in Figure 8.6. The target is optimally at a position in the upper area of the flowing particle bed within a range from 1/3 to 1/2 of the distance down the bed. One or more guns are positioned at the selected location and extend along the pan axis to uniformly cover the distance from the front to the back of the coating pan. The critical concerns of gun positioning include the following:

Encapsulation via Pan-Coating

1. Baffles must be below the bed surface at the spray point to avoid being coated.
2. Sufficient flowing bed surface below the spray point should be available for drying.
3. The bed surface should be suitably established and relatively flat with appropriate surface velocity and character.

Gun tip to bed distance is a critical parameter in pan-coating and is related to gun spray characteristics. For aqueous coating systems, gun to bed distance is typically 6–10 in. For organic solvent coating systems, gun to bed distance is typically 5–7 in. This difference is primarily related to the volatility of the coating vehicle. Water has relatively low volatility compared with commonly employed organic solvent vehicles. This allows spraying from a greater distance without excessive spray drying. The greater distance also reduces the potential to over wet any point on the bed surface by moving the spray impact point further out in the spray pattern. Spray drying is a greater concern with organic solvent vehicles due to higher volatility. Spray drying is minimized with guns positioned closer to the bed. In addition, the higher organic solvent vehicle volatility reduces the potential for localized over wetting at a short gun to bed distance.

8.6.6 SPRAY RATE

Spray rate is generally set in some relation to drying capacity. Drying capacity is a measure of how much vehicle vapor the process air can hold. Setting the spray rate at a consistent percentage of the available capacity is one strategy for maintaining a consistent process. Alternatively, there may be other considerations including a batch time concern or a coating gain rate requirement. If a coating process is too short, coating uniformity may be compromised as a result of insufficient passes of material through the spray zone. Longer time in the process provides more passes and improves the overall uniformity. Longer process time can be realized with slower spray rate or more dilute coat solution/suspension. A rough estimate of a minimum process time for good uniformity is near 1 h or more.

Drying capacities for various solvents are indicated in Figure 8.9. These curves indicate the amount of solvent vapor that could be carried by the air at the various temperature points. Although

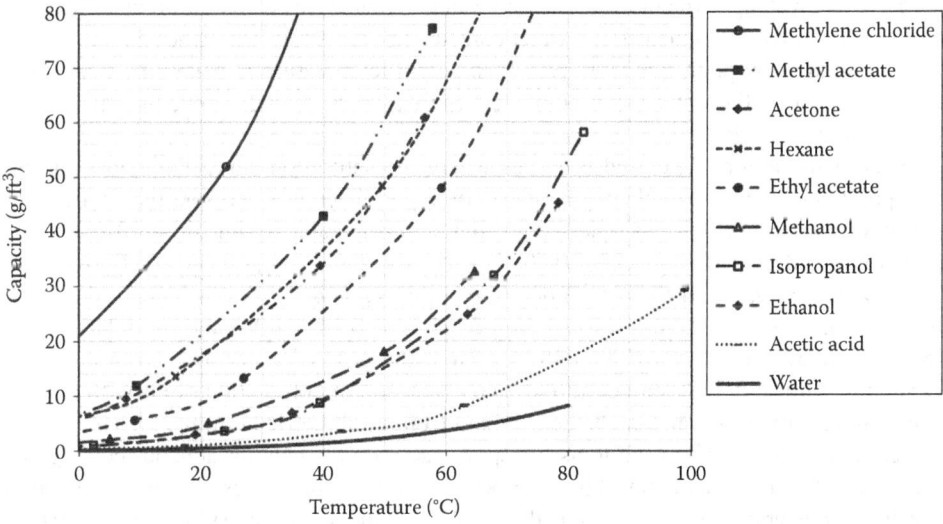

FIGURE 8.9 Capacity of air for solvent vapors vs. temperature. (*Note:* Data derived from referenced vapor pressure data.)

air is capable of holding this much vapor, the evaporation rate slows as vapor content increases. In addition, the drying process in a pan coater is relatively inefficient due to air movement variance through the particle bed. As a result, spray rate is much less than air capacity. Nevertheless, the capacity can be used as a guideline.

Use of organic solvent as a coating vehicle introduces safety concerns related to explosive limits. An explosive air composition is realized when the solvent content of the air is within a certain range specific to the solvent. The upper limit of that range is commonly referred to as the upper explosive limit (UEL) and the lower limit of that range as the lower explosive limit (LEL). To avoid a potentially explosive condition, the concentration should never be within the LEL to UEL range. Lower and upper explosive limits are shown graphically in Figure 8.10a and b, respectively. Spray rate should always remain below that which could bring the composition into the explosive range and preferably no more than 1/4 to potentially 1/2 of the LEL to account for non-uniformity of air flow and inherent inaccuracies in air flow measurement and delivery. For example, the LEL for acetone is ~1.6 g/ft^3 at 40°C. At 300 ft^3/min process air, the acetone spray rate at ¼ of the LEL would be 120 g/min (1.6 g/ft^3/4 × 300 ft^3/min = 120 g/min). This value is for spraying the pure solvent and does not account for other components of the sprayed coat formulation. If multiple organic solvent are collectively used as the vehicle, it is recommended that the solvent with the lowest LEL and the total content of sprayed organic solvents be used in this assessment. Although this calculation provides a safe upper limit for spray rate, actual spray rate may be limited further by the drying characteristics of the coating system.

If organic solvents are used, the pan coater should be configured with appropriate explosion proof safety features to minimize the potential ignition of any spilled, leaked, or built up vapors or arrest any event that might be realized during routine operation of the system. Such systems may include exhaust air and room air monitors to detect excess solvent vapor and trigger an alarm, explosion proofing to eliminate ignition sources, and pressure relief hatches or fire suppressant flooding systems that are triggered by a sudden pressure event. It is also possible to install interlocks that shut off spray if the process air is interrupted.

The drying capacity for water in aqueous coating formulations is significantly below that for organic solvents at comparable temperatures. This is caused by both a lower vapor pressure for water and water already present in the process air. The saturation curve for water in air is included in Figure 8.9 along with the selected organic solvents. This curve is expanded and shown in Figure 8.11a and b along with curves indicating water content at selected dew points. The dew point of the process air provides a measure of water content. The air is saturated when actual temperature is at or below the dew point temperature. When actual temperature is below the dew point temperature, a portion of the water vapor equivalent to that which extends above the saturation curve condenses and forms a fog.

In areas of the Figure 8.11a and b curves where the saturation curve lies above the process air dew point curve, the available drying capacity is the difference between the saturation curve and the dew point curve. For example, the available drying capacity for air at 30°C with a 10°C dew point is ~0.60 g/ft^3 (0.85 − 0.25 = 0.60). The relative humidity at those conditions is ~29.4% (0.25/0.85 × 100 = 29.4%). If the air temperature at that same dew point is increased to 37°C, the available drying capacity increases to ~1.06 g/ft^3 (1.30 − 0.24 = 1.06) and the relative humidity decreases to 18.5% (0.24/1.30 ×100 = 18.5%). Note the g/ft^3 of water at a given air dew point shifts slightly with actual air temperature due to expansion or contraction of air with temperature changes. This information can also be determined from cyclometric humidity charts.

These relationships can be used to assess spray rate requirements for aqueous coating systems. For example, 72 g/min water sprayed into a process at 300 ft^3/min process air, 10°C dew point, and 30°C exhaust temperature would be equivalent to 40% of the available drying capacity (0.60 g/min capacity × 0.4 × 300 = 72 g/min). A total spray rate of 90 g/min would apply if the coat suspension consists of only 80% water (0.72/0.8 = 90). If the coater were a two gun configuration, each gun would be set to deliver 45 g/min spray rate. Depending on the critical parameters of the process, a spray rate range can be set to allow spray adjustment to actual drying conditions across the applicable

Encapsulation via Pan-Coating

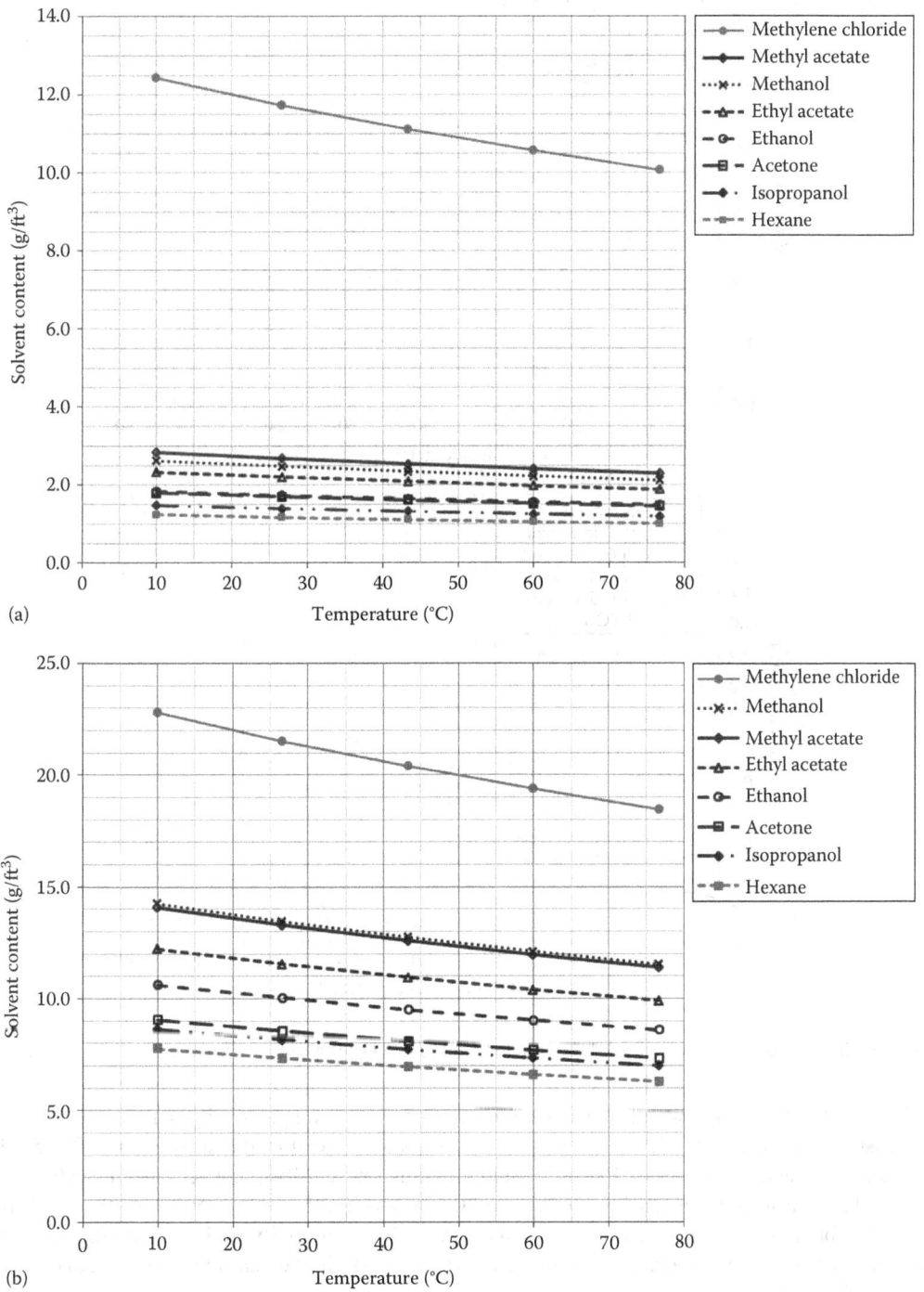

FIGURE 8.10 (a) Lower explosive limits for selected organic solvents. (b) Upper explosive limits for selected organic solvents. (*Note:* Data derived from explosive limit information documented in Material Safety Data Sheets and other sources and other sources.)

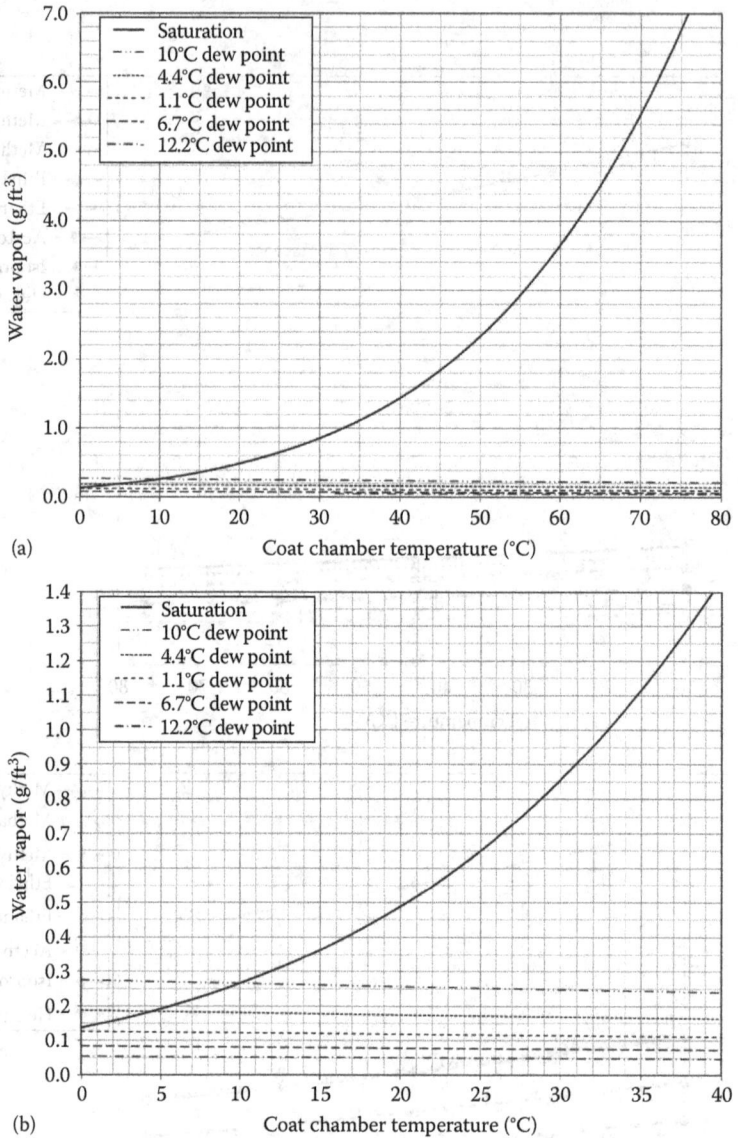

FIGURE 8.11 (a) Drying capacity for water at 0°C–80°C. (b) Drying capacity for water at 0°C–40°C.

range of allowed dew point and/or exhaust temperatures. For example for this same system at 30°C exhaust temperature and a dew point of −6.7°C, the water content of the air drops to ~0.075 g/ft³, the drying capacity increases to 0.775 g/ft³, and the spray rate at 40% of capacity would be 93 g/min water (116 g/min for a suspension containing 80% water), (0.85 − 0.075) × 0.4 × 300 = 93). It is also possible to target spray rate to achieve a desired relative humidity in the pan during spray, but this may not be possible depending on the dew point range, desired exhaust temperature, and associated performance factors.

A more comprehensive assessment of drying capacity in relation to dew point and exhaust temperature is provided in Table 8.1. The value at the intersection of dew point temperature and air temperature indicates the air capacity for water at those conditions. For example, 100°F air with a 40°F dew point has an available drying capacity of 1.123 g/ft³. If spray were targeted at 30% of this capacity for a coat solution/suspension containing 90% water using 300 ft³/min process air, the solution/suspension

TABLE 8.1
Available Air Capacity for Water at Selected Dew Points and Air Temperatures (g/ft^3)

Dew Point		\multicolumn{18}{c}{Air Temperature}
°F	°C	°F 50	60	70	80	90	100	110	120	130	140	150	160	170	180	190	200	210
		°C 10.0	15.6	21.1	26.7	32.2	37.8	43.3	48.9	54.4	60.0	65.6	71.1	76.7	82.2	87.8	93.3	98.9
0	−17.8	0.233	0.343	0.490	0.683	0.935	1.258	1.670	2.187	2.832	3.629	4.603	5.784	7.207	8.906	10.922	13.298	16.081
5	−15.0	0.225	0.335	0.482	0.675	0.927	1.250	1.662	2.180	2.825	3.621	4.595	5.777	7.200	8.899	10.916	13.292	16.074
10	−12.2	0.214	0.324	0.471	0.665	0.917	1.241	1.653	2.171	2.816	3.612	4.587	5.769	7.191	8.891	10.907	13.284	16.066
15	−9.4	0.201	0.312	0.459	0.653	0.905	1.229	1.641	2.159	2.805	3.601	4.576	5.758	7.181	8.881	10.897	13.274	16.056
20	−6.7	0.186	0.296	0.444	0.638	0.891	1.215	1.627	2.146	2.791	3.588	4.563	5.745	7.168	8.868	10.885	13.262	16.044
25	−3.9	0.167	0.278	0.426	0.620	0.873	1.198	1.610	2.129	2.775	3.572	4.547	5.729	7.153	8.853	10.870	13.247	16.030
30	−1.1	0.144	0.255	0.404	0.599	0.852	1.177	1.590	2.109	2.755	3.553	4.528	5.711	7.134	8.835	10.852	13.229	16.013
35	1.7	0.117	0.229	0.378	0.573	0.827	1.152	1.565	2.085	2.732	3.529	4.505	5.688	7.112	8.813	10.831	13.208	15.992
40	4.4	0.084	0.197	0.346	0.542	0.797	1.123	1.536	2.056	2.704	3.502	4.478	5.662	7.086	8.787	10.805	13.183	15.967
45	7.2	0.046	0.159	0.309	0.506	0.761	1.087	1.502	2.022	2.670	3.469	4.446	5.630	7.055	8.757	10.775	13.154	15.938
50	10.0	0.000	0.114	0.265	0.463	0.719	1.046	1.461	1.982	2.631	3.430	4.408	5.592	7.018	8.720	10.739	13.118	15.903
55	12.8	−0.054	0.062	0.214	0.412	0.669	0.997	1.413	1.935	2.584	3.385	4.363	5.548	6.975	8.678	10.697	13.077	15.862
60	15.6	−0.117	0.000	0.153	0.353	0.610	0.940	1.357	1.880	2.530	3.331	4.310	5.497	6.924	8.627	10.648	13.028	15.815

TABLE 8.2
Interpolation within Table 8.1 Water Drying Capacity Grid

Horizontal followed by vertical

Air Temp Dew point	37.8°C	40°C	43.3°C
4.4°C	1.123	1.288	1.536
5.0°C		1.28 g/ft^3	
7.2°C	1.087	1.253	1.502

Vertical followed by horizontal

Air Temp Dew point	37.8°C	40°C	43.3°C
4.4°C	1.123		1.536
5.0°C	1.115	1.28 g/ft^3	1.528
7.2°C	1.087		1.502

spray rate would be 112.3 g/min (1.123 × 0.3 × 300/0.9 = 112.3). For values that fall between displayed values, an interpolation between the adjacent values will provide a reasonable estimate. For example, the air capacity for water vapor at 5°C dew point and 40°C can be estimated filling in the interpolated values horizontally then vertically or vertically then horizontally as demonstrated in Table 8.2.

Spray rates can be limited for a variety of reasons but typically fall in the range of 80–220 g/min/gun in aqueous coating systems and 110–350 g/min/gun in organic solvent coating systems for commercial coating processes.

8.6.7 Atomizing and Pattern Air

A variety of gun spray techniques can be used in pan-coating including two fluid designs with internal or external mixing and hydraulic nozzles. An external mixing two fluid design is most commonly employed on a production scale. A generic drawing of a two-fluid external mix gun is provided in Figure 8.7. Atomizing air (one of the two fluids) is delivered concentrically around the coat solution/suspension delivery tip in the same manner as a typical two-fluid nozzle. Pattern air is delivered in various orientations on opposing sides of the spray pattern to "flatten" the round spray profile emerging from the nozzle tip. Atomizing air is the primary means of atomization, but the pattern air that flattens the pattern can marginally contribute to that atomization.

Figure 8.12 illustrates three potential spray pattern conditions from a two-fluid nozzle. Round spray patterns and the like do not adequately extend across the bed surface and tend to over wet at the localized spray positions. Addition of suitable pattern air widens and flattens the spray into a linear shape that extends over a wider swath of the bed surface. Excess pattern air disrupts the flat pattern to create an uneven spray and imparts unneeded energy that contributes to spray drying and errant spray droplets that can coat the pan or gun surfaces.

Atomizing and pattern air needs are dependent on the application. Spray guns should be selected based on their applicability at the required spray rate. The equipment manufacturer or equipment specifications can be consulted for guidance on nozzle selection. Optimum atomizing and pattern air settings take into account the following factors:

1. Nozzle design
2. Spray rate
3. The particle/tablet bed should not be unacceptably disturbed by atomizing and pattern air

Encapsulation via Pan-Coating

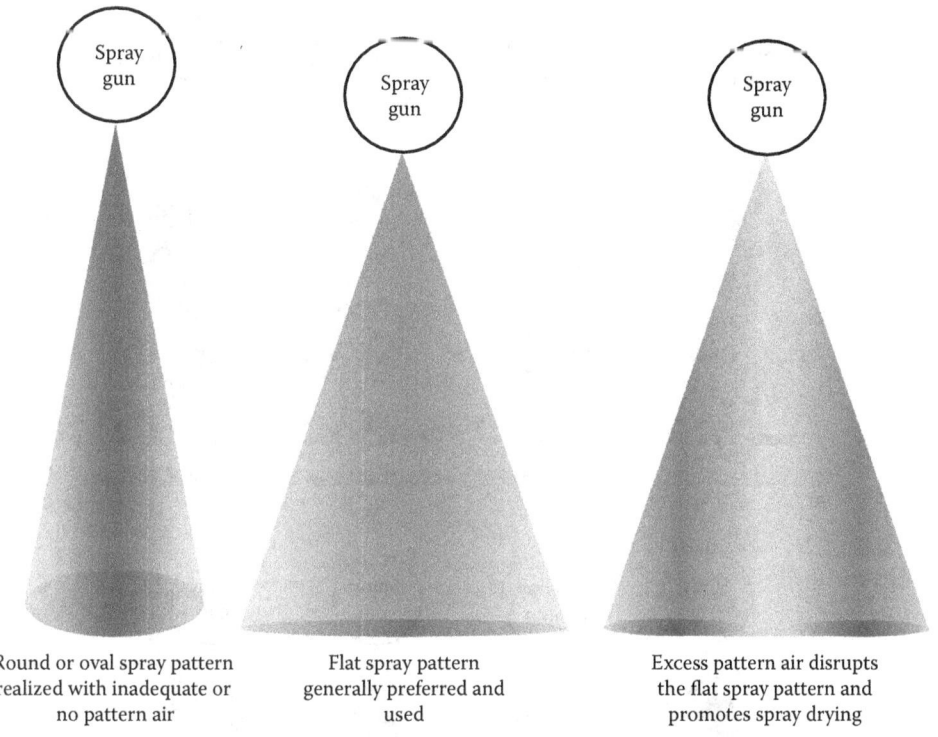

Round or oval spray pattern realized with inadequate or no pattern air

Flat spray pattern generally preferred and used

Excess pattern air disrupts the flat spray pattern and promotes spray drying

FIGURE 8.12 Potential gun spray patterns from two fluid pancoating spray guns.

Recommended atomizing and pattern air settings for the specific gun configuration, solution viscosity, and spray rate employed should also be available from the equipment manufacturer.

Spray characteristics influence nozzle placement. Figure 8.13a through c indicate gun positioning concerns for multinozzle configurations. Ideally, sprayed material extends uniformly across the particle/tablet bed as shown in Figure 8.13c. Undesired deviations from uniformity include the following:

1. Spray not extending to the edges of the bed.
2. Gaps between the individual gun spray zones (Figure 8.13a).
3. Overlap of spray zones for adjacent guns (Figure 8.13b).

These situations can be realized due to too many or too few guns, less than optimal atomizing and pattern air, and/or less than optimal gun to bed distance. Any of these situations can result in nonuniform application across the full width of the particle/tablet bed and localized areas of over wetting or under wetting that could reduce coating uniformity.

8.6.8 Process Endpoint

Spray drying is an inherent process concern due to gun to bed distance variance, process air drying capacity variance, and other contributing variables. In coating applications where these variances are within limits that consistently provide acceptable product, a process endpoint can be established when a fixed amount of coating solution has been applied.

In more precise coating applications where coating efficiency variations result in unacceptable variance in coat weight gain, coating solution is applied until the required weight gain is realized. Weight gain checks can be made during processing to determine the endpoint by stopping spray and weighing suitably sized subsets of tablets or particles to determine the per tablet/particle or per

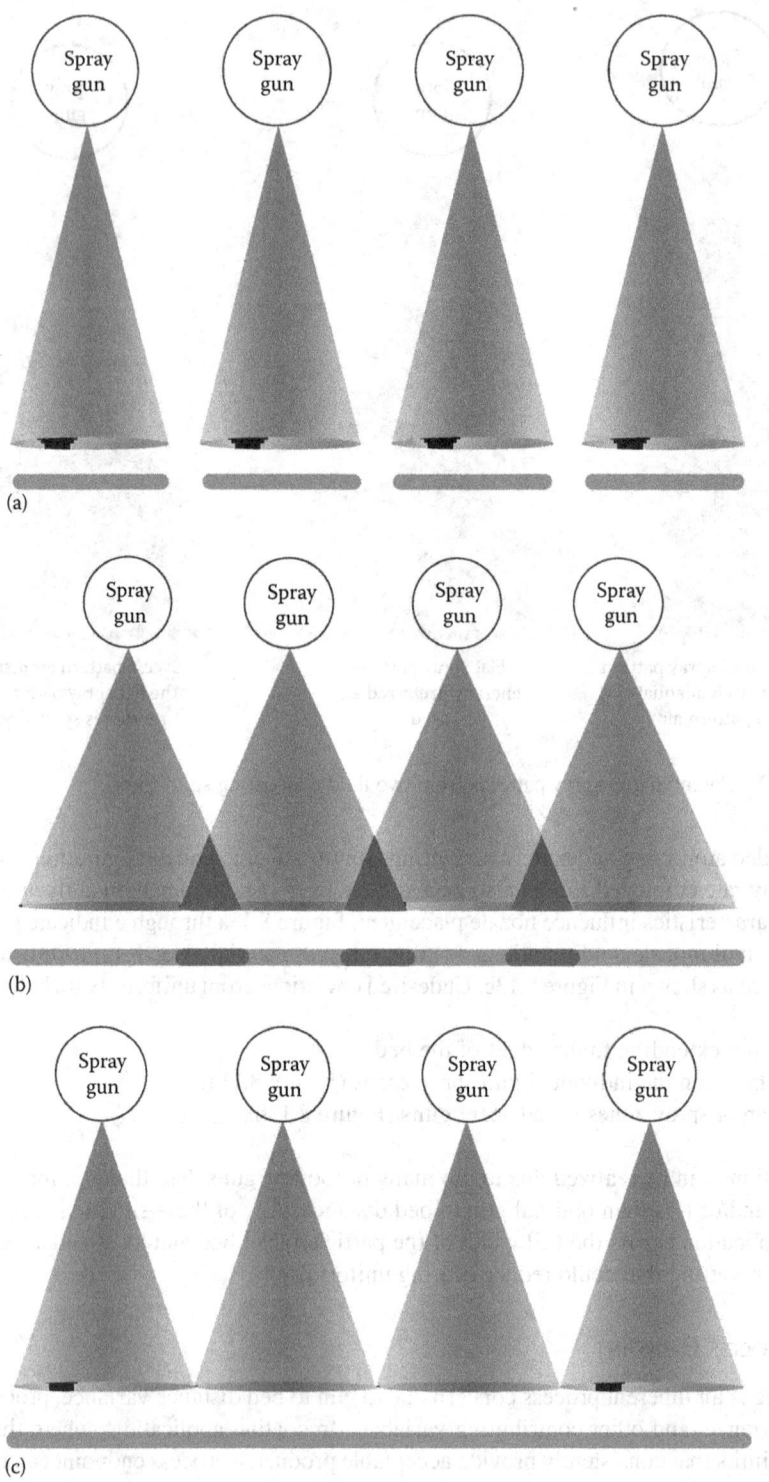

FIGURE 8.13 (a) Gaps between spray zones do not receive spray. Gun to bed distance is too short, guns are too far apart, or pattern air is inadequate. (b) Overlapped spray zones create a potential area of over wetting. Gun to bed distance is too long, guns are too close together, or excess pattern air. (c) Optimal positioning and conditions provide a relatively continuous spray zone across the particle.

TABLE 8.3
Example Weight Gain Check to Determine Process End Point

	Initial (Uncoated) Weight (g)	Coated Weight (g)	Weight Gain	Solution Applied (kg)	Cumulative Spray Time Spray Time (min)	Weight Gain Rate (%/kg)	Weight Gain Rate (%/min)	Weight Gain Target	Estimated Endpoint (Total kg of Solution)	Estimated Endpoint (Total Min Cumulative Spray Time)
Set #1 (50 ct)	17.502	19.381								
Set #2 (50 ct)	17.493	19.297								
Set #3 (50 ct)	17.517	19.405								
Avg. set weight	17.504	19.361	10.609%	21.000	90	0.505	0.118	12.000%	23.75	101.7

sample set weight. The coated weight and uncoated weight of such a set can then be used to establish weight gain. When a weight gain check is made, the weight gain and weight gain rate information can be used to estimate the amount of additional spray or spray time needed to reach the endpoint. It is generally better to spray to a point less than the calculated target to account for variance in the weight check result and avoid over coating; more coating can always be added, but it cannot be removed once it is applied. Using this approach, it is important that the count per sample set and/or number of sets weighed be sufficient to minimize individual tablet/particle variability. An example weight gain check from a hypothetical process is shown in Table 8.3.

If tablets are dusty prior to the start of coating, it may be important to charge the pan and start pan rotation (jog or minimal rate) and process air prior to sampling for the initial (uncoated) weight to remove dusts that could bias the initial weight and weight gain.

8.7 SCALE-UP

Critical concerns for scale-up will vary depending on process characteristics and goals. Some processes are robust and can tolerate significant application parameter variance. Others are less robust and require scaling parameters to match process attributes such as physical stresses, droplet size, and coating weight gain rate. Scale-up has been discussed in detail in various articles and book chapters including those referenced here.[3-5] The concepts of geometric, dynamic, and kinematic similarity date back to 1974 and offer guidance for matching a scaled process to the original process.[5] This section discusses considerations and options for pan-coating process scale-up and follows many of the principles outlined in the references.

Although this chapter is focused on a batch production strategy, continuous feed systems can potentially meet coating needs where uniformity needs are less critical.

8.7.1 Charge

Geometric similarity refers to the similarity of pan dimensions including the pan size, baffle size and position, and pan load. Pan size can be characterized with an aspect ratio (pan length to pan diameter ratio). This assures that as the pan diameter increased, the length is scaled proportionately. Pan aspect ratios can vary significantly from large diameters with relatively short pan lengths to smaller diameters with long pan lengths.

Pan baffle size should be scaled proportionately with the pan size increase to assure similar bed mixing. This size consideration includes the length, depth, and curvature of the baffle; all of which should mimic the baffle length in relation to the pan size, depth in relation to the bed depth, and pan curvature. Mounting angle should also be equivalent to the original pan.

Pan load to pan volume ratio should maintain a consistent h/D ratio, where h is the distance between pan center and bed surface, and D is the pan diameter.

8.7.2 Pan Speed

The determination of pan speed for a scale-up can be approached in various ways. The simplest approach is to scale to a consistent pan surface velocity. If a 24 in. diameter pan is rotated at 10 rpm, the pan surface speed is ~754 in./min (24 × pi × 10 = 754). If this is scaled to a 48 in. pan, the matching speed would be ~5 rpm (754/48/pi = 5).

Although this offers an equivalent fall rate for particle/tablet bed surface, it does not fully account for the ratio of forces throughout the pan, which would result in a dynamic difference from the original process. A generally accepted means to scale-up to the dynamic equivalent is to scale to a constant ratio of inertial to gravitational forces. This ratio is commonly referred to as the Froude Number (Fr), a dimensionless value that indicates the system dynamics. A generally accepted Froude Number expression for rotating drums is as follows:

$$\text{Fr} = \frac{(\text{rpm})^2 D}{G}$$

where
  D is the pan diameter
  G is the gravitational force

Since g is constant for both scales,

$$(\text{rpm})_1^2 D_1 = (\text{rpm})_2^2 D_2$$

and

$$(\text{rpm})_2 = \sqrt{\frac{(\text{rpm})_1^2 D_1}{D_2}}$$

Using this scaling option, the 10 rpm for the 24 in. pan translates to 7.1 rpm for the 48 in. pan.

The Froude Number scale-up provides a fundamental approach to a dynamic similarity, but there is some experiential evidence suggesting that the higher speed indicated by this approach (7.1 vs. 5.0 rpm) provides additional stress in the scale-up.[3]

8.7.3 Process Air

The equipment manufacturer or their literature should offer guidance on the range of applicable process air for a given coater design. Depending on the critical concerns of air delivery, it is possible that optimal process air may fall outside the recommended range. As previously indicated, primary concerns for process air include avoidance of excess air that would disrupt the spray or bed in an unacceptable way. In addition to factors surrounding air passage through the bed, the dynamics of the inlet air as it enters the pan can be a significant element of these concerns. Although a theoretical scale-up will provide guidance for a suitable setting, the above concerns should always be considered due to potential variance in the process air delivery system for the individual coater. Also, since it is generally preferred to maximize the process airflow for maximum drying capacity, there may be economic reasons to scale above a theoretical scale-up if the application can tolerate it.

Theoretically process air would scale to a consistent air velocity through the bed to maintain a similar air pressure on or air velocity through the bed. In a perforated pan, the velocity is related to the exhaust air passage area of the pan under the bed and the percentage of open space in the passage area (pan porosity). In addition, the depth of the bed will influence the pressure drop through the bed. Due to variation in the extent that these factors are a concern, an air assessment in this discussion will be limited to the air passage consideration only.

Since the air passage area will vary with pan coater design and perforated versus non-perforated configurations, the air passage area for each pan will be assumed to be equivalent to a square section of perforated pan surface with sides equal to 1/2 the pan diameter. A 24 in. diameter pan at 300 ft^3/min process air would pass air at a velocity of 300 ft/min ($300/(24/2/12)^2 = 300$). This air flow would translate to 1200 ft^3/min in a 48 in. diameter pan (300 ft/min × $(48/2/12)^2 = 1200$). Be aware that this is based on assumed air passage dimensions and actual dimensions would be applied in an actual scale-up.

8.7.4 SPRAY RATE

The requirements for spray rate scale-up are dependent on the critical parameters of the process. Some processes merely require coating application without twinning, picking, or other yield and appearance concerns. Other processes may be more dependent on equivalent drying conditions and weight gain rate during processing. In general, the coat solution/suspension is not typically altered in a scale-up; however, a change in the amount of solvent vehicle may potentially be required to match certain criteria. This assessment of spray rate scale-up will be made with the goal of matching critical elements of precision coating processes.

In fluid bed-coating processes, spray rate can be scaled linearly with process air due to a matching linear increase in drying capacity. The volume of material in the process is not significantly considered in this scale-up. In pan-coating processes, spray can also be scaled with process air for the same reason; however, drying kinetics can play a more prominent role. A critical phase of drying in a pan coater is the time of particle/tablet travel from the spray zone along bed surface to the lower edge of the bed. During that time, a majority of the solvent vehicle should be evaporated to avoid twinning, picking, and sticking to the pan that might be realized if too much solvent remains as it reaches the lower edge of the bed surface. As the pan size increases in a scale-up, the travel distance below the spray zone increases ~linearly with pan diameter; thus, a 48 in. diameter pan would have up to twice the travel distance as a 24 in. pan. This added distance ~doubles the available time for that initial drying phase. This potentially allows a significant spray rate increase per gun compared to the initial process.

In addition to the longer initial drying phase, the scaled up volume and pan size decreases the bed surface spray area to bed volume ratio; thus, individual particles/tablets pass through the spray zone less frequently than at the smaller scale. This effectively increases the drying time between passes; thus, a longer process time is required to achieve a similar number of passes through the spray zone.

The extent to which these scale-up coating factors match the original process is a measure of kinematic similarity with the original process.[4] Taking these factors into consideration, spray rate

TABLE 8.4
Hypothetical Scale Up of Process Air, Bed, Spray, and Guns

Pan diameter	24 in.	48 in.	60 in.
Process air	300	1200	1875
Bed volume	12 kg	120	240
Total spray rate (g/min)	160	800	1280
Gun count	2	4	5
Spray rate/gun (g/min)	80	200	256
Relative process time	1	2	2.5

can be scaled using the following relationship to provide an equivalent number of passes through the spray zone for each particle/tablet:

$$S_1 \frac{C_1}{V_1} = S_2 \frac{C_2}{V_2}$$

where
  S_1 = Spray rate process 1
  S_2 = Spray rate process 2
  C_1 = Gun count process 1
  C_2 = Gun count process 2
  V_1 = Bed volume process 1
  V_2 = Bed volume process 2

thus,

$$S_2 = S_1 \frac{C_1 V_2}{C_2 V_1}$$

A scale-up example from a hypothetical process in a 24 in. pan is shown in Table 8.4.

The changes in per nozzle spray rate will require alternative nozzle configurations and/or atomizing and pattern air settings to provide equivalent atomization.

8.7.5 ATOMIZING AND PATTERN AIR

Recommendations for spray gun parameters are often provided based on atomization performance mapping of parameters such as spray rate, atomizing and pattern air, and solution/suspension viscosity. There is often minimal theoretical guidance on these concerns and any such guidance may have limited value due to nozzle design variations and potentially significant effects that seemingly small differences can have on performance. Equipment manufacturer literature and equipment manufacturer support should be consulted for guidance on spray gun parameters. Nevertheless, this section is offered for general guidance with a strategy to approach equivalent performance at varying scales.

As discussed previously, atomizing and pattern air should ideally be set to provide suitable droplet size and spray pattern with minimal bed disturbance. Atomization involves an appropriate balance of air velocity and volume at the required spray rate. As spray is increased in a scale-up, an appropriate increase in atomizing air velocity and volume would be needed to maintain similar droplet characteristics from the same nozzle, which may disturb the bed. Potential nozzle configuration changes required for scale-up include use of larger nozzle tip and/or cap orifices. These will potentially allow scaled up spray with a nozzle tip liquid velocity similar to the original process and an appropriate increase in atomizing air volume at a comparable atomizing air velocity. Since air velocity is a significant contributor to potential bed disturbance, this should help address bed disturbance concerns.

Atomizing and pattern air are typically delivered by pressure. Units for pressure are weight per unit area such as lbs/in.^{2} (psi). For air delivered through an orifice at a given pressure, air volume scales ~linearly with area of the orifice. For example, 0.060 in. and 0.100 in. diameter orifices have 0.002826 in.^{2} and 0.00785 in.^{2} of open area, respectively. This is an open area ratio of 1: 2.78 and at a given pressure, the larger hole will have ~2.78 times more air volume passing through it than the smaller hole; however, the velocity of the air passing through each will be ~the same at that given pressure.

The atomizing air channel around the nozzle tip is similar to an orifice at a given pressure where atomizing air volume scales ~linearly with channel cross-sectional area and velocity remains ~constant. One approach to atomizing air scale-up is to match the atomizing air volume scale-up to the spray volume scale-up while maintaining a similar atomizing air velocity. The reasoning behind this is that the extent of atomization is related to the amount of energy delivered by the atomizing air and atomizing air volume need at a given velocity scales ~linearly with liquid volume. Since gun tip and atomizing air channels are fixed by gun design, a matched scale-up is not always possible and adjustment of the atomizing air volume/velocity may be needed to compensate for atomizing air volume underage or overage.

Table 8.5 may help clarify these concepts. The gun dimensions are hypothetical and used only to illustrate a potential scale-up strategy. The column labeled "Original Process" represents a process that is to be scaled and is assumed to be optimized with respect to spray requirements. The remaining columns indicate gun scale-up parameters with three different gun options: the same as the original gun (Gun #1) and two others (Gun #2 and Gun #3). The process air and spray rate scale-up parameters are carried over from the previous sections of this scale-up discussion. Each gun scale-up includes two potential gun spray rate increases that could be realized depending on the number of guns employed and the pan coater size options. Hypothetical tip and cap dimensions are indicated for each gun along with the tip and atomizing air area that those dimensions create. The atomizing air volume and spray rate for the original process are then translated to each nozzle option for the required scale-up spray rate and a comparable atomizing air velocity. Based on the assumption that atomizing air volume need will increase linearly with liquid volume, it is clear that using the same

TABLE 8.5
Spray Parameter Scale-Up Assessment

Process	Original Process	Scale Up Option 1		Scale Up Option 2		Scale Up Option 3	
Gun	Gun #1	Gun #1		Gun #2		Gun #3	
Tip ID (in.)	0.0472	0.0472		0.0591		0.0787	
Tip OD (in.)	0.100	0.100		0.100		0.115	
Cap ID (in.)	0.122	0.122		0.134		0.157	
Tip ID Area (in.^{2})	0.001749	0.001749		0.002742		0.004862	
Atomizing air gap area (in.^{2})	0.003834	0.003834		0.006245		0.008968	
Atomizing air volume (ft^3/min)	3.53	3.53		5.75		8.26	
Relative atomizing air volume	1	1		1.63		2.34	
Atomizing air velocity (ft/min)	132,582	132,582		132,586		132,632	
Spray rate (mL/min/gun)	80	200	256	200	256	200	256
Relative spray rate	1	2.5	3.2	2.5	3.2	2.5	3.2
Liquid velocity (ft/min)	233	582	744	371	475	209	268

nozzle in the scale-up as the original (Gun #1) will result in inadequate air volume at a comparable air velocity (same air volume as the original process with 2.5 and 3.2 factor spray volume increases for the two spray scale-up options).

For Scale-up Option 2/ Gun2, the atomizing air volume increases by a factor 1.63 to maintain the atomizing air velocity, which is less than the 2.5 and 3.2 factor spray rate scale-up. If this nozzle was employed in the scale-up, the atomizing air velocity would likely require adjustment upward to add air volume. The amount of adjustment should not exceed the volume increase needed to match the spray volume increase due to the higher velocity associated with such volume increase. An atomizing air volume increase to 8.82 and 11.3 ft^3/min with the resulting velocity increases to 203,375 and 260,560 for the two Gun #2 scale-up options would match liquid volume scale-up. Although the required increase will likely be below this estimated upper limit, the corresponding velocity increase would likely affect the particle/tablet bed surface significantly, which may prohibit use of this nozzle option in the scale-up.

Scale-up Option #3/Gun #3 would likely provide the best option for the scale-up. The atomizing air volume increases by a factor of 2.34 at the equivalent atomizing air velocity compared to the 2.5 and 3.2 factor spray rate increases. Since the spray increases are greater than the atomizing air factor, the atomizing air may potentially require scaling to 8.82 ft^3/min (~141,624 ft/min velocity) and 11.3 ft^3/min (~181,446 ft/min velocity) for the 200 and 256 mL/min/gun scale-up options, respectively, to match the spray rate increase. Although the atomizing air increases would not likely need to extend to these extremes, the effect of the actual atomizing and pattern air needs on the particle/tablet bed will be less than that of the Gun #2 option.

An additional consideration for the scale-up is liquid velocity at the gun tip. Table 8.5 includes the gun tip liquid velocities that would be realized at the various spray rates. Note that the Gun #3 options at 209 and 268 ft/min are comparable with the 233 ft/min in the original process. This further indicates that the gun #3 would be a preferred choice for the scale-up. Atomization performance decreases as spray rate increases for a given nozzle. This performance decrease is related to increased fluid volume and is likely contributed to significantly by velocity at high tip liquid velocities and a potential to "overshoot" the critical atomization zone.

Pattern air volume is generally set lower than atomizing air and will be dependent on gun design. Manufacturer data and recommendations should be consulted for guidance. Pattern air is primarily used to flatten the spray pattern. Inadequate pattern air will result in an oval spray pattern. Excess pattern air will contribute to a disrupted spray pattern, additional atomization, spray drying, and errant spray that may add to coating build up on pan and gun surfaces.

8.8 APPLICATIONS

Pan-coating can potentially be used to apply a film coat on any particulate material that can be confined to a coating pan and be coated. Beyond that, the applied coating system must be applicable to the critical process requirements such as particle size, coat solution viscosity, drying requirements, coat properties, and final product attributes. This section provides a general overview of applications and application considerations for pan-coating processes.

8.8.1 COATING MATERIALS

Coating materials are generally restricted to a list similar to that employed in fluid bed coating. Critical requirements include that the coat formulations has sufficient film forming character without excess brittleness or tackiness and that it adheres to the core particle.

The use of hot melt or wax coating is limited by the ability to deliver the wax in a molten state. Although this is physically possible, significant limitations are realized with application of molten spray and associated liquid line tracing and gun to bed spray distance. Application of a hot melt coat is feasible in a suitably temperature controlled conventional pan.

8.8.2 CONVENTIONAL PAN-COATING APPLICATIONS

Conventional pan-coating is generally used in sugar coating processes and some organic solvent vehicle-based processes. These processes can be used to provide a pleasant tasting top coat or a color coating for an improved appearance or brand defining color. Functional coatings for controlled release purposes such as enteric, delayed, or sustained release are possible, provided the required release profile can consistently be produced with the employed coating delivery system. If applicable, conventional pan application in these areas is, or may be, preferred due to low equipment cost, relatively low drying air need, and ease of cleaning. Processes that generally require high precision coating realize limited success in a conventional pan. Suitable care should be taken regarding safety concerns associated with LEL of any organic solvent systems.

Conventional pan-coating can be applied to powder coating processes. These processes are characterized by a combination of particle wetting and powder addition to a bed of core material to accrete powder to the core. This technique has realized significant interest for drug layering applications.

Conventional pan-coating may provide significant advantage over vented pan-coating in some applications. A conventional pan can be extended to smaller particles compared to vented pan-coating since the solid pan surface confines all pan material regardless of size to the process. If coating delivery is done via dripping or ladling, higher coating efficiencies are realized due to elimination of spray drying losses. These attributes can be critical to applications involving drug layering via solution delivery or powder coating techniques to eliminate spray drying losses of expensive APIs.

8.8.3 VENTED PAN-COATING

Vented pan processes offer better coating precision and drying efficiency compared to conventional pan-coating, both of which can contribute to improved economics. Better precision is generally realized due to more uniform spray configurations and more uniform drying, which allows application of uniform thin film coats. Better drying efficiency allows faster coat application, which can reduce overall process time and the added physical stress associated with that time.

Vented pans are applied most commonly to tablets for various purposes. Color coats are commonly used to provide an improved appearance or a brand defining color. Controlled release coats for enteric, delayed for sustained release are typically applied in vented pans. Coatings can be applied for improved surface characteristics such as a seal coat to minimize handling exposure to a core tablet, an oxygen or moisture barrier for improved stability, or to impart a slippery quality to a tablet when wetted to facilitate swallowing.

8.9 TROUBLESHOOTING

It is difficult to foresee all the potential problems that can be realized in a pan-coating process. Problems can be related to formulation shortcomings, process limitations, or suboptimal process parameters. Once a coat formulation has been established, there may not be opportunity to adjust composition depending on regulatory restrictions; thus, it may be prudent to optimize to a robust formulation before the adjustment period passes. The following characteristics should be included in such a formulation assessment:

1. Coating readily adheres to the core material
2. Coat solution/suspension readily "wets" the core surface for optimal drying
3. Coat solution/suspension has relatively low viscosity for optimal atomization and good particle surface wetting properties

4. Coating can be applied at conditions that are agreeable to both the core and coating
5. Coating is nontacky during drying and when dry
6. Coating is nonbrittle and forms an adequate film coat without splitting or cracking
7. Coating has low shrinkage during drying

This general wish list may not be fully met due to challenges associated with the application. Often, shortcomings of the coating system in any of these areas can be mitigated or overcome through formulation adjustments or optimized process parameters. For example, the plasticizer content can potentially be increased or decreased to address cracking or tackiness concerns. It may be necessary to apply a coating that is prone to cracking at relatively "wet" coat conditions to retain a higher coating vehicle residual early in the process to temporarily improve the film coat through plasticizing effects of the solvent residual. This could potentially be addressed with adjustments to temperature, spray rate, atomizing and/or pattern air, gun to bed distance, or pan speed. A coating prone to sticking may need to be processed at dryer conditions for more complete vehicle removal or cooler to a point below the glass transition temperature where tackiness is reduced or eliminated. Beyond temperature, these concerns might be addressed with adjustments to spray rate, atomizing and/or pattern air, gun to bed distance, or pan speed. If appearance is substandard with an undesired "orange peel" look, reduced atomizing air and/or increased spray could mitigate premature drying that leads to suboptimal particle/tablet surface wetting.

Several processing concerns can result in poor coating uniformity. Guns positioned or operated such that spray patterns overlap too much or too little will result in over wetted or under wetted areas of the spray line. Also, spray should extent all the way to near the edges of the bed. Any shortcomings in gun placement can affect coating uniformity depending on the extent that particle/tablet circulation overcomes these shortcomings.

Coat solution delivery imbalance to the guns has a high potential to negatively affect coating uniformity. In multigun systems, an imbalance where one or more guns spray faster than the others, those spray zones will receive more coating than the others. Although pans are designed to circulate particles throughout the bed, this circulation is not always adequate to fully overcome such an imbalance.

In some applications, a panload might involve dilution of a particle/tablet population with a "filler" particle/tablet to fill out the pan. These diluent particles are subsequently removed after the coating is complete to leave the coated product. Depending on the difference in size, weight, and shape of the diluent material, product may segregate within the bed in various ways such as to the outer edges, to the center, deeper into the bed below its surface, or to the bed surface. Depending on the primary location where the product particles/tablets congregate, the rate of coating weight gain may be highly dependent on gun position or pan speed.

In addition to coating weight gain, coating weight gain rate (% weight gain/kg solution or % weight gain /time) can be a critical concern for some processes. Weight gain rate is directly related to spray rate and particle/tablet panload amount but is also highly dependent on temperature, gun to bed distance, and atomizing/pattern air settings, which all influence the amount of spray drying that occurs during processing. Any of these parameters can be adjusted to meet weight gain rate requirements.

Picking, twinning, and related concerns occur when formulation and/or process conditions result in particles/tablets adhering to one another and then damaging the coating when they pull apart (picking) or adhere to one another and remain together. These conditions can be minimized by ensuring that spray conditions do not over wet the bed in any way and process temperature is appropriate. Possible parameter adjustments to overcome these concerns include gun positions, atomizing and pattern air, gun to bed distance, pan speed, spray rate, and temperature. It is also critical that the bed be sufficiently filled to create a uniform flowing bed surface. An inadequately filled bed can leave localized areas at the backside of baffles where the bed surface is not continuous and allows spray to extend to a slow moving area near the pan. Gun position adjustment or more fill in the bed can potentially overcome this concern. Particle/tablets sticking together or to the pan surface can be caused by excess process air, which presses particles/tablets more firmly together or to the pan surface in a vented pan system. Reduced process air could help reduce or eliminate this concern.

Blemishes and appearance concerns not directly related to picking, twinning, "orange peel" surface, and the like may be related to undissolved coating material (if solution/suspension is not filtered or properly prepared), coating build up on guns and pan surfaces that flakes off and accretes to particles/tablets, or gun dripping during spray stoppage for weight checks or other needs. Solutions to these concerns may include adjustments to atomizing and pattern air, process air, and gun position, all of which can contribute to coating build up on equipment surfaces. If coating build up on pan surfaces and guns is an inherent concern of the process, more frequent cleaning of the coating pan may be required. If accreting material is present in the uncoated particles/tablets, steps should be taken to remove it prior to coating; optimization of particle formation/tableting processes, sieving, and precoat dedusting offer potential means to remove such material.

Beyond these process and coat formulation concerns, there are some tablet designs or shapes that may be prone to coating problems. If particles/tablets erode readily or chip and fracture easily, minimizing pan speed may be important to reduce stresses. Also, applying coating as quickly as possible may be critical. Particle/tablet designs that minimize sharp edges or corners and are free of deep indented lines, figures, or logos are generally best for coating.

8.10 CONCLUSIONS

Pan-coating techniques originated centuries ago for coating tablets and the like and remain commercially viable today. These processes involve use of a rotating pan on an inclined or horizontal axis to circulate a particle/tablet bed. Coating is applied to the particles/tablets by introducing coating solution or suspension to the circulating bed via spray, drip, or ladle. Various ventilation strategies are applied to remove the coating vehicle and deposit coating on the particles.

Pan-coating processes are generally categorized as either conventional pan-coating characterized by solid pan surfaces or vented pan-coating characterized by perforated or partially perforated pan surfaces. Conventional pan-coating is associated with aqueous processes such as sugar coating and some solvent based processes. Perforated pans used in vented pan-coating significantly increase drying efficiency and allow application of relatively thin film coats with good uniformity. Vented pan-coating is commonly applied in controlled release applications where thin and/or highly uniform film coats are required.

Critical pan-coating parameters such as charge weight, process air, temperature, spray rate, and pan speed can be established and maintained based on sound scientific principles. These parameters can be scaled to commercial and economic requirements of a process.

REFERENCES

1. Behzadi SS, Toegel S, and Viernstein H. Innovations in coating technology. *Recent Patents on Drug Delivery and Formulation* (2008) 2, 209–230.
2. Liu L, Smith T, Sackett G, Poire E, and Sheskey P. Comparison of film coating process using fully and partially perforated coating pans. http://www.freund-vector.com/downloads/papers/2004/Comparison_of_Coating_Efficiency_AAPS_2004.pdf. Accessed June 30, 2015.
3. Turton R, Cheng XX. The scale-up of spray coating processes for granular solids and tablets. *Powder Technology* (2005) 150(2), 78–85.
4. Porter SC. Scale-up of film coating. Chapter in book *Pharmaceutical Process Scale-Up*, Levin M. (Ed.), Marcel Dekker, New York, (2011) pp. 259–310.
5. Pandey P, Turton R, Joshi N, Hammerman E, and Ergun J. Scale-up of a pan-coating process. *AAPS PhamScieTech* (2006) 7(4), E1–E8.

9 Microencapsulation by Dripping and Jet Break-Up

Aurélie Demont and Ian W. Marison

CONTENTS

9.1 Introduction 177
9.2 Formation of Beads/Capsules Using Techniques Based on Dripping and Jet Break-Up—Principles and Theoretical Aspects 178
    9.2.1 Extrusion of a Liquid through a Nozzle—Droplet Formation 179
    9.2.2 Production of Beads by Dripping and Jet Break-Up 179
        9.2.2.1 Dripping 181
        9.2.2.2 Flow Focusing 182
        9.2.2.3 Electrostatic Extrusion 183
        9.2.2.4 Vibrating-Jet 185
        9.2.2.5 Jet Cutter and Spinning Disk Atomization 187
    9.2.3 Effect of the Collecting Distance 188
    9.2.4 Droplets Impact on the Gelation Bath 191
9.3 Application 192
9.4 Equipment 193
    9.4.1 Simple Dripping, Electrostatic Extrusion, Flow Focusing, and Coaxial Flow 193
    9.4.2 Vibrating-Jet 194
9.5 Advantages and Disadvantages 195
9.6 Conclusion 196
References 197

9.1 INTRODUCTION

Encapsulation can be defined as a process that involves the complete envelopment of preselected core material(s) within a defined porous or impermeable membrane using various techniques.[27,55] The main aim of encapsulation could be the immobilization, the protection, the stabilization, and the control of the release of the entrapped compound.[9] Indeed, the entrapment of various materials such as flavor, living cells, pharmaceutical compounds inside beads for different purposes is of great importance in the pharmaceutical, chemical, and food industries, as well as in agriculture, biotechnology, or medicine.[27,39] The compound can be embedded within the polymer network or within an immiscible core phase in the microcapsules.[12,19,47] The beads/capsules can be manipulated in order to influence functionality. The properties that can be tuned include the degree of cross-linking, the size and size distribution, the polymer type, the surface charge, and the particle shape[19,47] as shown in Figure 9.1.

While the prerequisites for industries such as cosmetics, agriculture, and food are usually high production rates and low cost, medical and biotechnological processes have more stringent criteria.[55] Methodologies must have the capacity to produce mono-dispersed, spherically shaped, homogeneous beads, with a small size and a narrow size distribution. They must be produced under simple conditions and the production time has to be short. Moreover, high efficiency and production rates are to be obtained with many different materials, including viscous solutions; it should allow

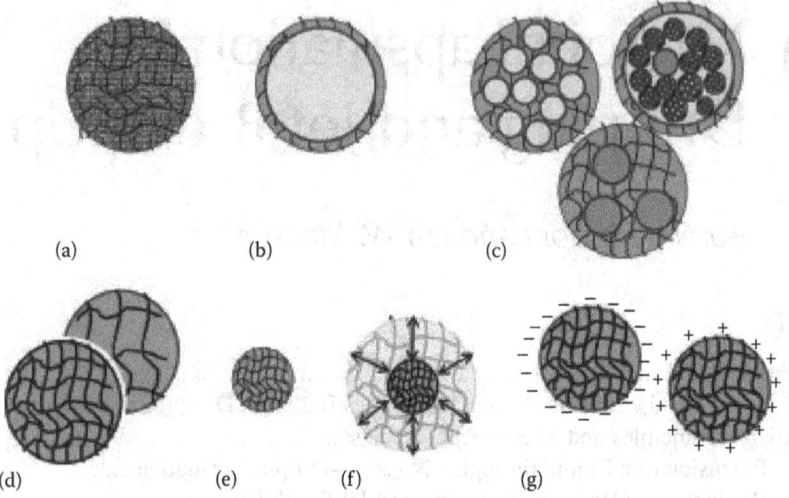

FIGURE 9.1 Schematic representation of the characteristics of capsules and the different types of encapsulation. (a) Dispersed, (b) core-shell, (c) droplets, (d) degree of cross-linking, (e) size, (f) swell or shrink, and (g) surface charge. (Reproduced from Shewan, H.M. and Stokes, J.R., *J. Food Eng.*, 119(4), 781, 2013. With permission.)

the production of different sizes and the encapsulation of a variety of core materials and finally enable, if required, to be performed under sterile conditions.[54]

Microcapsules can be produced by a large variety of methods that can often be combined. The main techniques used in encapsulation can therefore be classified into three main categories such as chemical, physicochemical, and mechanical processes.[57] Chemical and physicochemical processes include methods such as emulsion, interfacial polymerization, phase separation, liposome formation, and liquid membrane.[47,57] Chemical processes are those involving mainly polymerization and polycondensation and in which the capsule shell is made by *in situ* formed polymers. The disadvantage is the tendency to form very thin polymer membranes. Physico-chemical processes involve the formation of the capsule shell from a preformed polymer using processes such as solvent removal, gelatin, or coacervation. A subsequent cross-linking step can then be used to harden the coating.[57] Mechanical processes produce the polymer shell around the core using techniques such as spray drying and spray cooling, dripping, and jet break-up techniques.[47,57] Technologies based on dripping and jet break-up include methods such as simple dripping, electrostatic dispersion, flow focusing, vibrating nozzle, spinning disk atomization, and jet cutting.

This chapter will provide a detailed overview of the production of microbeads/capsules using dripping and jet break-up techniques. It will give a description of the principles involved, discuss the theoretical aspects behind droplet formation, describe the equipment needed and the different possible applications and finally compare the different methods, giving both their advantages and their disadvantages.

9.2 FORMATION OF BEADS/CAPSULES USING TECHNIQUES BASED ON DRIPPING AND JET BREAK-UP—PRINCIPLES AND THEORETICAL ASPECTS

Mechanical techniques are the most common type of processes used to produce microspheres.[55] These processes use a mechanical means to produce the desired particles instead of a physical or chemical phenomenon.[55,57] The droplets are generated from a polymer extruded through a nozzle, and mechanical means are used to increase the normal dripping process at the orifice of the liquid stream break up when it passes through the nozzle.[55] The droplets will then take a spherical shape during falling due to the surface tension of the liquid and will be solidified either by a physical or

Microencapsulation by Dripping and Jet Break-Up

chemical means.[55] The main mechanical techniques to produce microbeads are based on the dripping and jet break-up principle, which involve the formation of a droplet at an orifice or in a liquid jet. It is therefore mandatory to understand this phenomenon to understand each system.

9.2.1 Extrusion of a Liquid through a Nozzle—Droplet Formation

When a droplet is formed at the orifice of a nozzle, it can result from one of the five different formation processes occurring at the discharge point of the nozzle (Figure 9.2). The droplet formation is dependent on the velocity v of the extruded liquid, surface tension, gravitation, impulse, and friction forces.[17,55] At very low velocity ($v<$), single droplets are directly formed at the orifice of the nozzle. The extruded liquid sticks to the edge of the nozzle until the surface tension is overcome by gravitational force that results in the release of the droplet (mechanism 1).[17,54,55] The amount of droplets can then be increased by a small rise in velocity, whereas further augmentation will amplify droplet formation with a possible coalescence of the droplet occurring, thereby reducing the monodispersity (mechanism 2).[17,54,55] These two mechanisms are currently used at labscale in a process known as "dripping." Increasing the velocity further will cause the formation of an uninterrupted laminar jet, which breaks eventually by axial symmetrical vibration and the surface tension (mechanism 3). Another rise in the jet velocity will lead to a statistical distribution of the droplet size caused either by spiral symmetrical vibrations (mechanism 4) or by high friction forces present when the jet is sprayed (mechanism 5).[17,54,55] Mechanisms 1, 2, and 3 allow the controlled formation of monodisperse droplets and will be discussed further in this chapter.

9.2.2 Production of Beads by Dripping and Jet Break-Up

Mechanisms 1, 2, and 3 described in Figure 9.2 allow a controlled formation of monodisperse drop.[17,54,55] These mechanisms will therefore be used in bead production by dripping and jet break-up techniques. When an alginate solution is extruded through a nozzle, the droplet detachment at the dripping tip can be separated in two different mechanisms: dripping and jetting, which is influenced by the tip diameter, the flow rate, and the alginate solution itself.[22] It has been reported

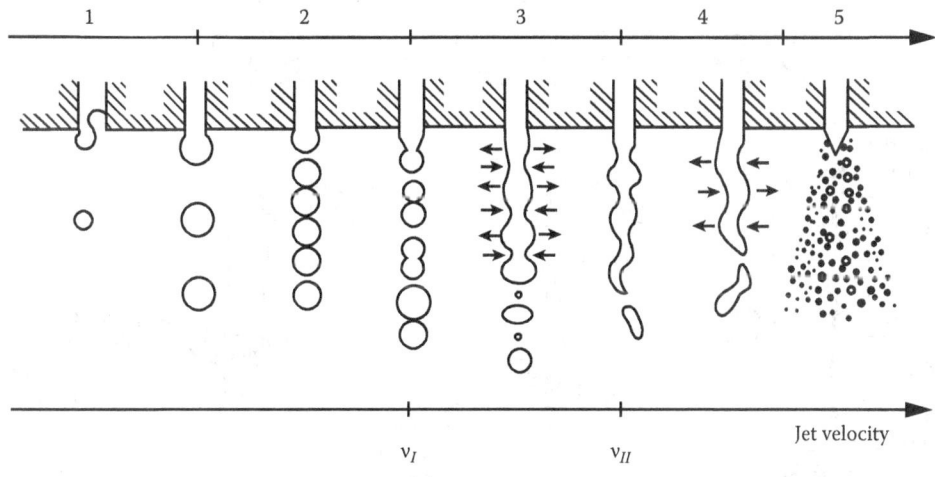

FIGURE 9.2 Different mechanisms of drop formation as a function of jet velocities. (With kind permission from Springer Science+Business Media: Heinzen, C. et al., Use of vibration technology for jet break-up for encapsulation of cells and liquids in monodisperse microcapsules, in: V. Nedovic and R. Willaert (Eds.), *Fundamentals of Cell Immobilisation Biotechnology*, number 8A in Focus on Biotechnology, Springer, Dordrecht, the Netherlands, January 2004, pp. 257–275.)

by Lee et al. that the influence of the parameter can be described by the dimensionless Ohnesorge number (Oh) and Reynolds number (Re).

$$Re = vd_d\rho/\eta \qquad (9.1)$$

$$Oh = \eta/(\rho d_d \sigma)^{1/2} \qquad (9.2)$$

with
- v the velocity of the alginate solution at the point of impact
- η the viscosity of the alginate solution
- ρ the density of the alginate solution
- d_d the diameter of the alginate solution droplet
- σ the surface tension of the alginate solution

An increase of flow rate (or Re) of an alginate solution at constant Oh will lead the droplet-detachment mechanism to a transition from dripping to jetting. Furthermore, the transition between dripping and jetting is strongly dependant on the viscosity (or Oh) of the alginate solution.[22] This finding reported by Lee et al. is also confirmed by Heinzen et al.[17,22] The most important methods in dripping and jet break-up will be discussed further in this chapter and include the simple dripping, the electrostatic extrusion, the flow focusing, the vibrating nozzle, the jet cutter, and the spinning disk atomization (Figure 9.3).[53,55]

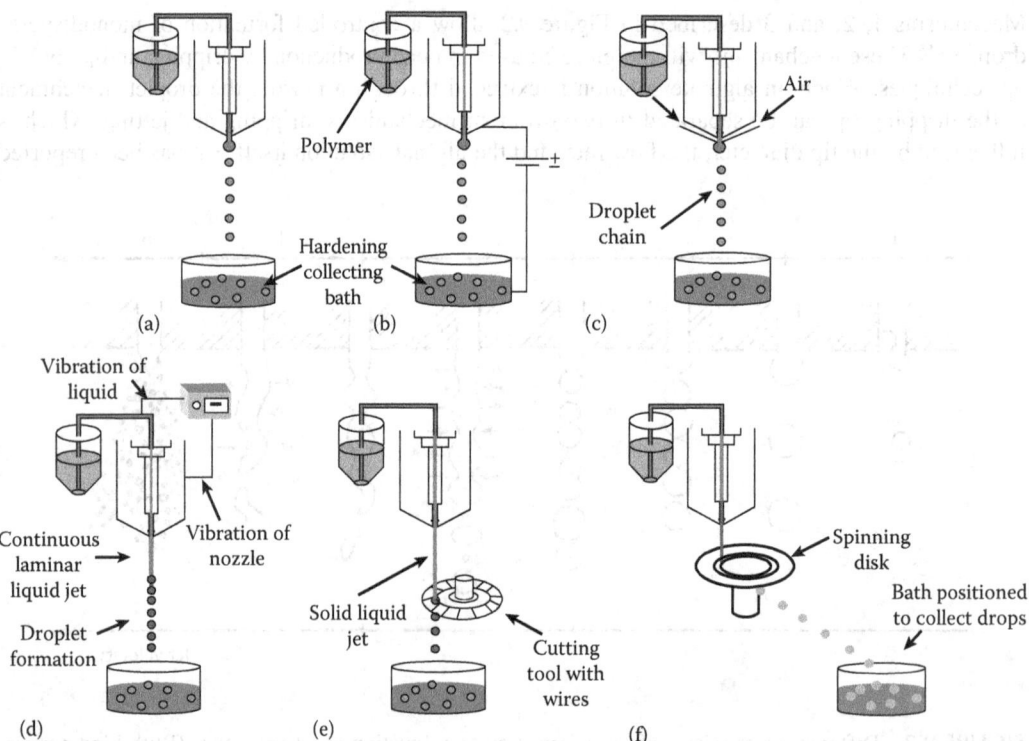

FIGURE 9.3 Different droplet production processes by dripping and jet break-up techniques. (a) Simple dripping, (b) electrostatic extrusion, (c) coaxial flow, (d) vibrating nozzle, (e) jet cutting, and (f) spinning disk atomization.

9.2.2.1 Dripping

During mechanisms 1 and 2 (Figure 9.2), the droplet diameter of a defined liquid only depends on the nozzle diameter.[17] These two mechanisms are currently used at lab-scale in a process known as "dripping" (Figure 9.3a).[17,54,55] The solution of alginate is delivered to the nozzle through a channel and a droplet is formed at the dripping tip, then the droplet grows and detaches as a single droplet from the dripping tip by the influence of gravity once it reaches its maximum volume (mechanism 1). Monodispersed alginate beads are therefore produced with this method, the bead diameter being approximately twice as big as the dripping tip diameter.[17,22,34] If the solution is pumped at a small flow rate (<1 mL min^{-1}), Lee et al. reported that the droplet grows relatively faster than that of natural extrusion; however, they also noticed that the droplet detaches prematurely from the tip with a greater volume than that by natural extrusion. The premature pendant droplet has the tendency to wet the external surface of the dripping tip, resulting in an increase in the bead diameter produced from the dripping tip as the flow rate is increased (mechanism 2).[22,34]

The diameter of the droplet, d_d formed during mechanisms 1 and 2 can be estimated from Equation 9.3. The balance between the two main force is approximated, the gravitational force is pulling the drop down when the surface tension is holding it to the tip at the instant of drop detachment.[9,17,34,55] Moreover, as seen in Equation 9.3, the size of the droplets produced with mechanisms 1 and 2 are mainly dependent on the nozzle diameter.[55]

$$d_d = \sqrt[3]{\frac{6d_n\sigma}{g\rho}} \tag{9.3}$$

with

d_n the diameter of the nozzle
σ the surface tension of the extruded liquid
g the acceleration due to gravity
ρ the density of the fluid

It is however reported by Chan et al.[9] and Whelehan et al.[55] that for systems producing alginate drops that will be gelled by dropping in a calcium chloride bath, the diameter of the produced calcium alginate bead d_{cab} after gelation is different than the diameter of the alginate drop d_d. In a study, Chan et al. extended Equation 9.3 by adding an overall size correction factor (K), which takes into account the liquid lost factor and the shrinkage factor.[9,22,55] The liquid lost factor is independent of the liquid properties, however it is dependent on the tip size. The shrinkage factor on the other hand was found to be affected by the M/G ratio of alginate.[9] The correction factor enables a good approximation of d_{cab} to be obtained. Chan et al. calculated that the overall correction factor K varies from 0.73 and 0.85 under the conditions examined, which correlated well with their experimental data with an average absolute deviation of less than 5%.[9]

$$d_{cab} = K \cdot \sqrt[3]{\frac{6d_n\sigma}{g\rho}} \tag{9.4}$$

with

d_n the diameter of the nozzle
σ the surface tension of the extruded liquid
g the acceleration due to gravity
ρ the density of the fluid
K the overall size correction factor ($K = k_{SF} \cdot k_{LF}$ with k_{SF} the shrinkage factor and k_{LF} the liquid lost factor)

It was reported by Lee et al. that the dripping tip length has no significant effect on the size and shape of the Ca-alginate beads; however, its length has a significant effect on the pressure drop of a bead production system.[22] It was also explained that reducing the tip length to half will increase the bead production rate to twofold.[22]

The diameter of the beads produced under the influence of the gravitational force is usually larger than 1000 µm and the production rate is less than 0.5 mL min^{-1}.[22] In order to improve the production rate and reduce the bead diameter, external forces such as an electrostatic potential, a coaxial flow or vibrations can be used to release the droplet before its maximum volume.[17,22,34,54,55]

9.2.2.2 Flow Focusing

Flow focusing is a process based on hydrodynamic focusing.[51] The polymer or dispersed phase is called "focused fluid" and flows into a central nozzle while an immiscible phase called the "focusing fluid" is delivered through two side channels and pressurizes the polymer.[28,51] The flow rate of the focusing fluid is several orders of magnitude higher than the flow rate of the focused fluid.[51] At low to moderate We and Re number, the focusing fluid is pressuring the inner fluid at the tip of the nozzle thereby forcing the inner stream through the orifice of the nozzle to form a thin stationary microjet much smaller than the orifice of the nozzle.[28,51] The jet eventually breaks-up into homogeneous droplets with a smaller size than the nozzle orifice and a narrow size distribution.[28,51] The focusing fluid can be either a gas or a liquid defining therefore the method which is either aerodynamic or hydrodynamic.[51] The particle diameter is reported to be generally 1/10 to 1/30 the diameter of the orifice.[51]

Gañán-Calvo showed that the breakup mode is axisymmetric and that the resulting droplets are monodispersed when the We number has a value below 40, with We defined as

$$We = \frac{d_j \Delta P_g}{\sigma} \quad (9.5)$$

where
d_j is the jet diameter
ΔP_g the pressure drop of the gas
σ the interfacial tension[14]

It, however, has to be noticed that this statement is valid provided that the fluctuations of the gas flow do not contribute to droplet coalescence, which can occur when the gas stream reaches a fully developed turbulent profile around the liquid jet breakup region.[14] Above the critical We value of 40, Gañán-Calvo pointed out that the sinuous non axisymmetric disturbances become apparent, coupled to the axisymmetric ones.[14] It is also mentioned that increasing again the We number will lead to a nonlinear growth rate of the sinous disturbances, which will overcome the axissymmetric ones and produce polydisperse drops.[14]

As mentioned by Martín-Banderas et al., the final microparticle diameter can be predicted with a remarkable accuracy.[29] The drop diameter can, according to their work, be calculated as

$$d_g = \left(\frac{3\pi}{2k}\right)^{1/3} \cdot d_j \quad (9.6)$$

where k is the wavenumber of the fastest growing perturbation on the jet (approximately $k \approx 0.5$ for most liquid–liquid combinations) depending on the viscosity and density ratio between the inner and outer liquids.[29]

These authors reported that the jet diameter depends on the nozzle geometry, mainly the orifice diameter D, and the operating conditions, mainly the liquid flow rates. The final droplet diameter can then be written as

$$d_g = \left(\frac{3\pi}{2k}\right)^{1/3} \cdot \left(\frac{Q_d}{Q_t}\right)^{1/2} \cdot D \qquad (9.7)$$

where

Q_d the inner fluid flow rates
Q_t the outer fluid flow rate

The particle diameter d_p can now be calculated with

$$d_p = \left(\frac{3\pi C}{2k\rho_p}\right)^{1/3} \cdot \left(\frac{Q_d}{Q_t}\right)^{1/2} \cdot D \qquad (9.8)$$

where

ρ_p is the polymer density
Q_d the inner fluid flow rates
Q_t the outer fluid flow rate
C the polymer concentration (mass of polymer per volume of solution)[29]

The diameter of the beads will decrease with increasing the focusing fluid flow rate, as reported in Tran et al. study.[51] The flow focusing is therefore a method able to generate particles with controlled size in the micron and submicron range.[28] The flow focusing method can also be performed in a dripping mode, which allows the reduction of the bead size compared to simple dripping. Beads produced in dripping mode have a diameter larger than 200 μm and are of uniform size and shape. Flow focusing performed in a dripping mode is usually called coaxial airflow or air-jet technique.[54]

9.2.2.3 Electrostatic Extrusion

Electrostatic extrusion is a process that uses electrical forces to overcome the surface tension force and pull the droplet off the orifice.[4,20,23,28,51,54] Two working methods can be used, the dripping mode and the jetting mode.[51] In the dripping mode, a low current is applied on the polymer solution flowing slowly through the nozzle. The electrical force then breaks-up the liquid and monodispersed microspheres between 500 and 1500 μm are obtained, with a production rate reported to be lower than 30 mL/h.[51] Smaller beads can be produced by increasing the current and using a smaller orifice, however an increase in the electric current will lead to a broader size distribution of the beads due to splitting of the polymer solution filament into numerous side filaments whithin the intense electric field.[4,34,36,51] An increase in velocity will lead to the formation of a stable jet, a higher electric current will then be necessary to break the jet into small droplet. This is the jetting mode, in which small monodispersed particle of 1–15 μm can be produced with a high productivity.[51]

In simple dripping mode, only the gravitational force is acting on the alginate drop, which grows at the tip of the nozzle until the surface tension force is overcome by the gravitational force thereby releasing the droplet. By applying a voltage on the nozzle, the solution reacts by accumulating the charges of the opposite sign on its surface. Coulombic repulsion forces results from this charges accumulation, creating a stress on the liquid surface.[13,15,51] The liquid at the nozzle tip is then distorted from a spherical shape into an inverted conical shape known as the Taylor cone.[13,15,49] Nedović et al. reported that a charge at the tip of the cone reduces the surface tension of the alginate solution[58] resulting in the formation of a neck-filament that will eventually break-up producing small droplets.[31] The meniscus then relaxes back to a spherical shape until the flow causes this phenomenon to start again.[31] Under appropriate process parameters, it was reported that this method allows a 6- to 10-fold reduction in bead diameter compared to the simple dripping mode.[31] Electrostatic extrusion can be used with a static of impulsed electrical potential.

In the dripping regime, the droplet diameter d continuously decreases with an increase of voltage U. The diameter will decrease from a value d_0 (Equation 9.9), which is the diameter of the drop when no electrical field is applied until the smallest diameter (Equation 9.10) is reached at the critical voltage U_c (Equation 9.11), which is the situation when surface tension becomes negligible.[23,34,36]

$$d_0 = \sqrt[3]{\frac{6d_s\sigma}{\rho g}} \tag{9.9}$$

where
  ρ is the liquid density
  g the gravitation constant
  d_0 the outer nozzle diameter
  σ the surface tension

$$d = d_0 \cdot \sqrt[3]{1 - \frac{U^2}{U_C^2}} \tag{9.10}$$

where
  d_0 is the droplet diameter when no electrical field is applied
  U is the electric voltage
  U_C the critical voltage value

$$U_C \cong \sqrt{\frac{d_s\sigma_0}{\varepsilon_0}} \tag{9.11}$$

where
  d_s is the outer nozzle diameter or the distance between the electrodes
  σ_0 the surface tension of the polymer at zero voltage
  ε_0 the air dielectric constant

When the critical voltage U_C is surpassed, the liquid flows as a jet.[23,34,36] The jet can be stable or unstable according to the voltages above the critical value and droplets are generated from the end of the jet. The detached droplets in the jetting regime are smaller than the inner nozzle diameter.[23]

Using electrostatic extrusion with an impulsed voltage can also offer a good control of the bead formation process in both electrostatic and dripping mode, as reported in Lewinska et al. study.[23] Lewinska et al. work mentioned that the droplet formed at the nozzle tip oscillates between the state of electrostatic dripping and the state of simple dripping when an impulsed electrical potential is applied.[23] The effective diameter of the droplet d_{eff} will then depend on the contribution of both states and could then be described as

$$d_{eff} = d_0 \left(1 - \alpha + \alpha \left(\sqrt[3]{1 - \frac{U^2}{U_C^2}}\right)\right) \tag{9.12}$$

with
  $\alpha = (f\tau) \times 10^{-3}$
  f the frequency of voltage impulses (Hz)
  τ the impulse duration time (ms)

Microencapsulation by Dripping and Jet Break-Up

U is the electric voltage

U_C the critical voltage value

α (in the range of 0–1) a coefficient describing a temporary contribution of the electrostatic dripping state to the entire process of droplet formation[23]

9.2.2.4 Vibrating-Jet

The vibrating-jet technique, also called vibrating nozzle or prilling[10,55] is one of the most widely used techniques to produce microspheres[45,55] The technology is based on the principle that the application of a vibration frequency with defined amplitude to the extruded laminar jet will break it into equally sized droplets.[17,18,55] Different methods can be used to apply the sinusoidal force such as vibrating the nozzle (vibrating nozzle technique), periodic changes of the nozzle/orifice diameter during extrusion or pulsating the polymer in a chamber before passing through the nozzle (vibrating chamber technique).[17,48,55,56] The choice of the technique applying the vibrational force is dependent on the system it is being applied to however these different methods to produce a sinusoidal force to the laminar jet will be termed as "vibrating-jet techniques" in this chapter[17,55] (Figure 9.4).

At the end of the nineteenth century, Lord Rayleigh investigated the instability of liquid jets and demonstrated that applying a permanent sinusoidal force at defined frequency on a jet allows to control the break-up of a laminar jet into uniform droplets of equal size.[40,41] As reported by Brandenberger et al., Rayleigh showed that for wavelengths smaller than the circumference of the jet, no break-up occurred, while for longer wavelengths the disturbances grow exponentially and cause break-up.[5,40,41] He therefore proposed an equation for inviscid

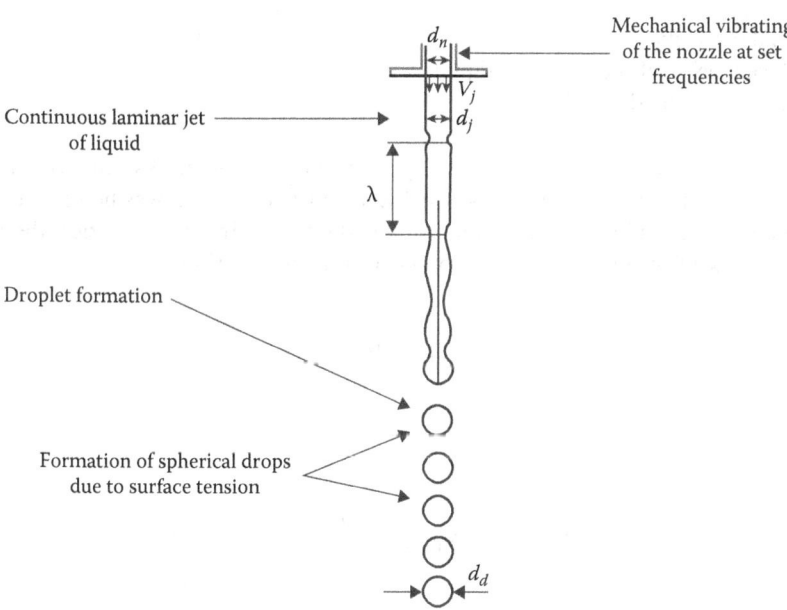

FIGURE 9.4 Controlled break-up of a laminar liquid jet into equal droplets by applying a sinusoidal force to the extruded jet by mechanically vibrating the nozzle at a defined frequency with a defined amplitude. (With kind permission from Springer Science+Business Media: Heinzen, C. et al., Use of vibration technology for jet break-up for encapsulation of cells and liquids in monodisperse microcapsules, in: V. Nedovic and R. Willaert (Eds.), *Fundamentals of Cell Immobilisation Biotechnology*, number 8A in Focus on Biotechnology, Springer, Dordrecht, the Netherlands, January 2004, pp. 257–275.)

and Newtonian fluids relating the optimal wavelength of the disturbance as a function of the jet diameter.[55,57]

$$\lambda_{opt} = \pi\sqrt{2} \cdot d_j \tag{9.13}$$

with λ_{opt} is the optimum wavelength for break-up of a jet of diameter d_j, knowing that the frequency is related to the wavelength and jet velocity by the following equation:[55]

$$\lambda = \frac{v_j}{f} \tag{9.14}$$

where
  v_j is the jet velocity
  f the frequency vibration

It is reported by Whelehan and Marison[55] and by Brandenberger et al. that Rayleigh's equation was extended further by Weber who included the effect of viscosity η into the analysis.[5,55]

$$\lambda_{opt} = \pi\sqrt{2} \cdot d_j \cdot \sqrt{1 + \frac{3 \cdot \eta}{\sqrt{\rho \sigma d_j}}} \tag{9.15}$$

where
  η is the fluid viscosity
  σ the surface tension of the extruded liquid
  ρ the density of the fluid

Different assumption such as $d_j = d_n$ proposed by Schneider and Heinricks[42] are used to determine the size of the jet diameter. As reported by Whelehan and Marison,[55] it was however demonstrated by Brandenberger and Widmer[6] that for precision-drilled sapphire stone nozzles, the relationship between d and d_n is a function of the Weber number of the nozzle (We_n).

$$\frac{d_j}{d_n} = 4.33 \cdot We_n^{-0.337} \tag{9.16}$$

where

$$We_n = \frac{v^2 \rho d_n}{\sigma} \tag{9.17}$$

with
  d_j is the jet diameter
  d_n the nozzle diameter
  v_n the velocity of the liquid in the nozzle
  ρ the density of the fluid
  σ the surface tension of the extruded liquid

Serp et al. and Whelehan and Marison reported that one droplet is generated by each hertz of vibration, thereby allowing calculation of the diameter of the drop d_d by a simple mass balance equation.[46,55,57]

$$d_d = \sqrt[3]{6\frac{F}{\pi f}} \qquad (9.18)$$

where F is the flow rate of the extruded liquid. Whelehan and Marison also reported that the vibrational frequency can be linked to the wavelength by the following equation:

$$f = \frac{F}{\lambda} \cdot \frac{4}{d_j^2 \cdot \pi} \qquad (9.19)$$

The droplet diameter d_d can therefore be expressed as a function of wavelength and jet diameter.[55,57]

$$d_d = \sqrt[3]{\frac{3}{2} d_j^2 \lambda_{opt}} \qquad (9.20)$$

It was reported by Heinzen et al. that the vibrational frequency and the jet velocity are the two main parameters to be determined in order to achieve optimal droplet formation with a given nozzle diameter.[17] It was also suggested by Stark et al. that uniform-sized droplets can be obtained with a range of frequencies around f_{opt}, which are dependant on the nozzle diameter, the rheology of the fluid and the surface tension.[48]

The vibration system is theoretically based on liquids with Newtonian fluid dynamics; however, it was reported by Serp et al. that it can be applied to non-Newtonian liquids such as alginate, to make uniform drops and therefore microspheres by ionotropic gelation.[46] The different equations can thereby be used as an approximation of the frequency and flow rates needed to break up the polymer into particles,[17] however, it was suggested by Serp et al. and by Whelehan and Marison to determine the required values for a given nozzle diameter with an empirical approach for each system in which the calculated value is used as a starting (reference) point.[46,55] It must also be noted that the size of the droplet does not necessarily equal the size of the produced bead/sphere.[55]

When droplets are formed by the break-up of a jet by the vibration method, they are often dispersed by a system based on electrostatic repulsion forces presented by Brandenberger et al.[5,17] As reported by Brandenberger et al., viscoelastic fluid jets break-up into monodisperse droplets; however, coalescence between droplets occurs due to irregularities in the filaments contraction after break-up, resulting in beads of double or triple volume.[5] Having the droplets flying through an electrostatic field after the break-up will charge them, therefore the droplets do not hit each other during the flight and are distributed over a large surface of the gelation bath, thereby resulting in monodisperse beads.[5,17] The electrode potential is, according to Brandenberger et al., in the range of 400–1500 V depending on the droplet diameter, the jet velocity, and the geometrical setup.[5]

9.2.2.5 Jet Cutter and Spinning Disk Atomization

The principle of the jet cutter is to form a solid jet by pressing the polymer out of a nozzle and then to cut it into uniform segments using a cutting tool formed of several wires.[38,39,54] The segments then form spherical beads due to surface tension, as they fall into the cool medium.[38]

The diameter of the resulting droplet is dependent on the number of cutting wires, the number of rotations of the cutting tool, the mass flow rate through the nozzle, and the mass flow depending both on the nozzle diameter and the velocity of the fluid.[38,39,54] In order to get a narrow size distribution of the beads, it is mandatory to get a steady flow through the nozzle and a uniform rotation speed of the cutting wire.[38] This method allows the production of narrowly distributed beads, with a size within the range of 200 μm up to several millimetres, even when using viscous fluids.[38,39,54] Moreover the productions rates are high. However, the main disadvantages of this method are the loss of material occurring during each cut of the liquid jet, known as the cutting loss, its inability to produce beads smaller than 200 μm and to produce high quantities of beads.[38,39,54]

The principle of spinning disk atomization is to distribute a liquid onto a rotating disk which, due to centrifugal forces and wetting ability, disperses into a thin film. The liquid then gains in velocity, is transported to the disk rim and is spread into droplets, which are thrown off tangentially from the disk and form a spray. The size of the drops depends essentially on the rotary speed, disk geometry, and liquid properties. The liquid holds to the rim by interfacial and viscous forces and the rotation induces forces tending to tear it off. Once drops are detached, interfacial forces keep the atomized liquid into a spherical shape to minimize the specific surface area.[43] According to their size, each drop falls at a different distance from the disk. Large drops fly over a longer distance than smaller ones,[54] therefore allowing ballistic classification and monodispersed drop collection.[50] Spinning disk atomization is a technique that can produce monodisperse, homogeneous, and spherical beads, with a narrow size distribution, it can be used with different viscosities of liquid solutions[43] and operate under sterile conditions.[8,32,33] The system has the advantage to be able to produce a large quantity of beads in a short time and can be scaled-up easily by increasing the flow rate, the disk size and the rotation speed of the disk.[50,54] However, spinning disk processes require a large surface or volume to collect the produced beads without damaging them.[50] This large surface can also create problems with sterilization in the case of cell encapsulation. The process seems incapable of producing small particles of less than 100 μm in size. Despite this a reduction in size can be achieved by increasing the rotational speed of the disk, rotating large disks at high speeds can cause the disk to vibrate and produce satellites drops.[43,45] The size can also be decreased by reducing the viscosity of the polymer; however, if the viscosity is too low, the shape of the bead will be seriously affected.[43,54] Another disadvantage of the spinning disk is the loss of product caused by the production of satellites drops.[44,45,54] As big particles travel further than small particles, the small particles will not reach the collecting gelling bath.[50] This property can be used to select the size of the capsules and the size distribution;[50,54] however, all the drops that are not in the selected size range will be lost.[43] Increasing the collecting surface will reduce the loss, however it will also increase the size distribution.[54] Another possibility would be to recycle the lost polymer with a recirculating system.

More informations about these jet break-up techniques can be found in later chapters of this book.

9.2.3 Effect of the Collecting Distance

Obtaining spherical sodium alginate droplets that will result in spherical beads after entering the gelation bath is critical in encapsulation processes. In the review of Lee et al., it was pointed out that spherical particles could only be obtained at a certain collecting distance, which is also confirmed in the study of Al-Hajry et al.[3,22] The determination of a suitable collecting distance is, according to Lee et al. review, strongly depending on the viscosity of the alginate solution. Droplets produced from a low-viscosity alginate solution are flattened upon impact on the gelation bath even at a short collection distance, while beads produced from 3% to 4% alginate solution are egg-shaped with a short distance. Lee et al. therefore reported that generally, small collection distances led to beads characterized by short tails while at large collection distances, a deformation of the particle is observed. Some study were reported in Lee et al. review, showing that a longer collecting distance increases the time to reshape the deformed alginate droplets into a spherical shape that is, the relaxation time of the alginate droplets, before curing in the gelation bath. However, in practice, the collecting distance is usually determined by adjusting the distance between the dripping tip and the gelation bath manually in a trial-and-error process. Different studies investigated the interrelationship effect of the collection distance and viscosity of the alginate solution on bead shape. These results are summarized in Lee et al. and are shown in Figure 9.5.[9,22]

In their work, Chan et al. developed a diagram (Figure 9.6) of the Oh–Re plot that reveals the operating region within which spherical Ca-alginate beads could be formed under the experimental condition. This diagram (Figure 9.6) was also reported in Lee et al. as an interesting means to estimate the collecting distance. The diagram describes the interinfluence of alginate solution physical properties, dripping tip diameter and collecting distance on the shape of Ca-alginate beads; moreover, it demonstrates that the inertia force, viscous force, and surface tension force of the falling alginate solution droplet are important in the formation of spherical

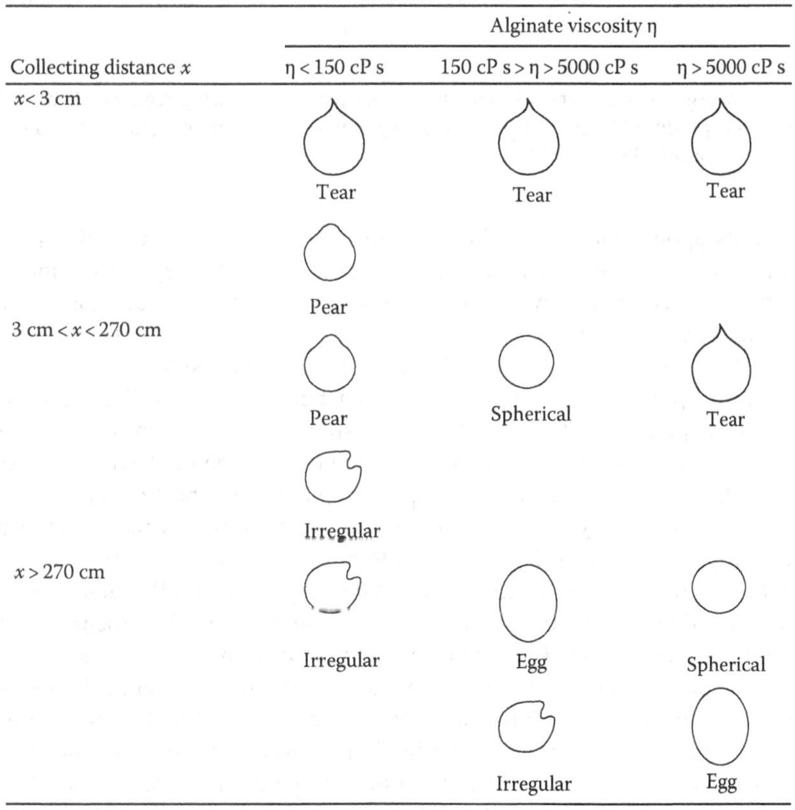

Collecting distance x	Alginate viscosity η		
	$\eta < 150$ cP s	150 cP s $> \eta > 5000$ cP s	$\eta > 5000$ cP s
$x < 3$ cm	Tear	Tear	Tear
3 cm $< x < 270$ cm	Pear / Pear	Spherical	Tear
$x > 270$ cm	Irregular / Irregular	Egg / Irregular	Spherical / Egg

FIGURE 9.5 Inter-relationship effect of the collection distance and viscosity of alginate solution on the shape of Ca-alginate beads. (Reproduced from Lee, B.-B. et al., *Chem. Eng. Technol.*, 36(10), 1627, 2013. With permission.)

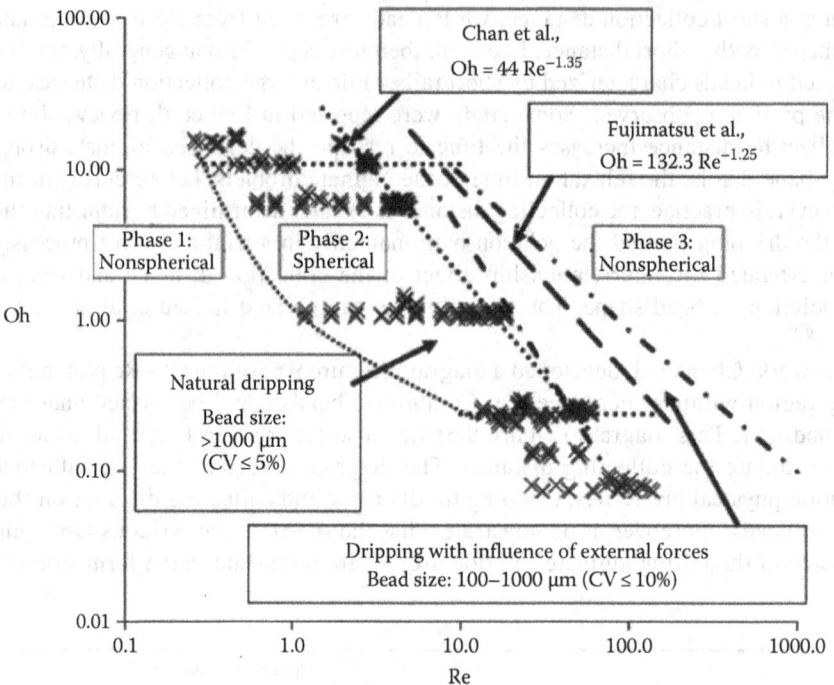

FIGURE 9.6 Ohnsorge number versus Reynolds number diagram showing regions where different shapes of Ca-alginate beads produced by the extrusion dripping method can be found. (Reproduced from Lee, B.-B. et al., *Chem. Eng. Technol.*, 36(10), 1627, 2013. With permission.)

Ca-alginate beads upon gelation. As illustrated in Figure 9.6, the lower and upper boundaries are dependent on the liquid properties and are independent of the Reynolds number. The lower boundary is given by Oh = 0.24, which means that spherical beads could not be formed below this value as the gel will be weak due to the diluted alginate solution.[9,22] The upper boundary is given by Oh = 11–13 and is limited by the viscosity of the alginate solution that can be processed using extrusion dripping.[9,22] The left boundary can be used to describe the transition point between phase I and phase II or the minimum collecting distance to form spherical beads. It was reported by Chan et al. that tailed-beads were formed outside (on the left) of this boundary due to the particle retainment of the tear-shape liquid drop at a very short collection distance. This left boundary is determined by the degree of impact or kinetic energy required to transform the less spherical droplet into a spherical Ca-alginate bead.[9,22] As reported by Chan et al and Lee et al., the right boundary shows the maximum collecting distance allowable to form spherical beads before shape deformation occurs. It reveals the influence of the process variables on the transition points between phase II and phase III and is given by Oh = 44 $Re^{-1.35}$.[9,22] Chan et al. and Lee et al. also pointed out that when the operating variables are over the boundary proposed by Fujimatsu et al., the alginate solution droplets may be disintegrated during the impact at the gelation bath surface.[9,22] The diagram shown in Figure 9.6 can be used to estimate suitable variables to produce spherical Ca-alginate beads within a targeted size range.[9,22] Beads produced by natural dripping (Re < 75) are generally > 1000 µm as the alginate droplet grows to its maximum volume at the dripping tip before being detached, thereby producing large beads. When an external force such as vibration, electrostatic potential, or coaxial flow is applied to separate the premature droplet from the dripping tip, smaller beads could be produced.[9,22] The diameter of the beads is therefore mainly governed by the diameter of the alginate droplet prior to gelation as suggested by Lee et al. and Chan et al.[9,22]

9.2.4 Droplets Impact on the Gelation Bath

After being grown at the dripping tip, the alginate droplet will reach a certain volume and be detached by gravitational or external forces, which are applied on the pendant droplet.[9,22] During its fall to the gelling bath, the droplet can have different shape transition depending on the distance between the dripping tip and the gelling bath.[9,22] The shape transition will follow the trend of a tear shape, egg shape, and spherical shape with increasing falling distance, regardless of the physical properties of the alginate solution or the corresponding Oh.[9,22] It was however shown by Chan et al. and reported in Lee et al. that the shape transition of the Ca-alginate beads with increasing collecting distance is due to the Oh of the alginate solution, that is, its physical properties.[9,22] Tear-shaped and pear-shape Ca-alginate beads are reported to be formed at short and long collecting distance when Oh < 0.24 (Figures 9.5 and 9.6). With Oh > 0.24, tear-shaped, spherical and egg-shaped Ca-alginate beads are formed with increasing of the collecting distance (Figures 9.5 and 9.6).[9,22]

The use of height speed cameras allowed to study the mechanism of impact and deformation of an alginate droplet on the gelation bath (Figure 9.7).[11,22,35,37] Figure 9.8 shows the different steps occurring after the impact on the gelation bath.[22] Deng et al. reported that the alginate solution droplet experiences the strongest deformation when hitting the surface of the gelation bath. According to Chan et al., if the Oh of the solution is below the critical value of 0.24, it could only form a flattened bead. Deng et al. using high speed imaging showed that after the strongest deformation, a detachment occurs at the surface of the gelation bath, where the droplet detaches from the liquid-air surface[11,22] and that the droplet is then deformed again due to the surface tension force and the gelation process.[11] The droplet/bead will finally regain its spherical shape, however this ability is depending on the physical properties of the droplet/bead such as surface tension and viscosity.[9,11,22,35,37] Some interesting soft landing approaches are reviewed in Lee et al.,[22] although they will not be presented in this chapter.

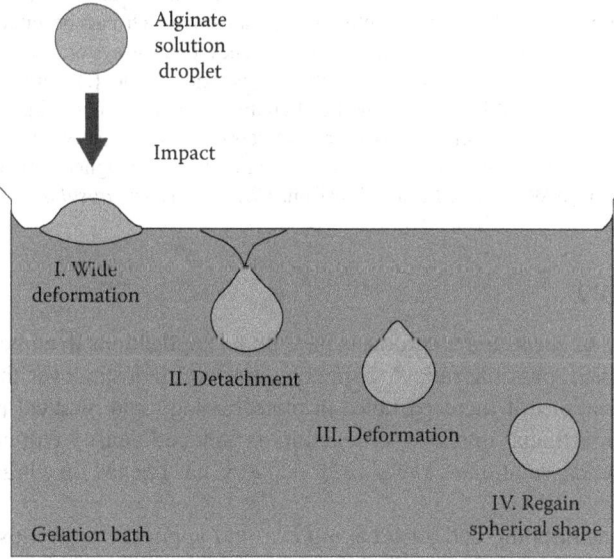

FIGURE 9.7 Impact and deformation of an alginate solution droplet at the interface of air-gelation bath. (Reproduced from Lee, B.-B. et al., *Chem. Eng. Technol.*, 36(10), 1627, 2013. With permission.)

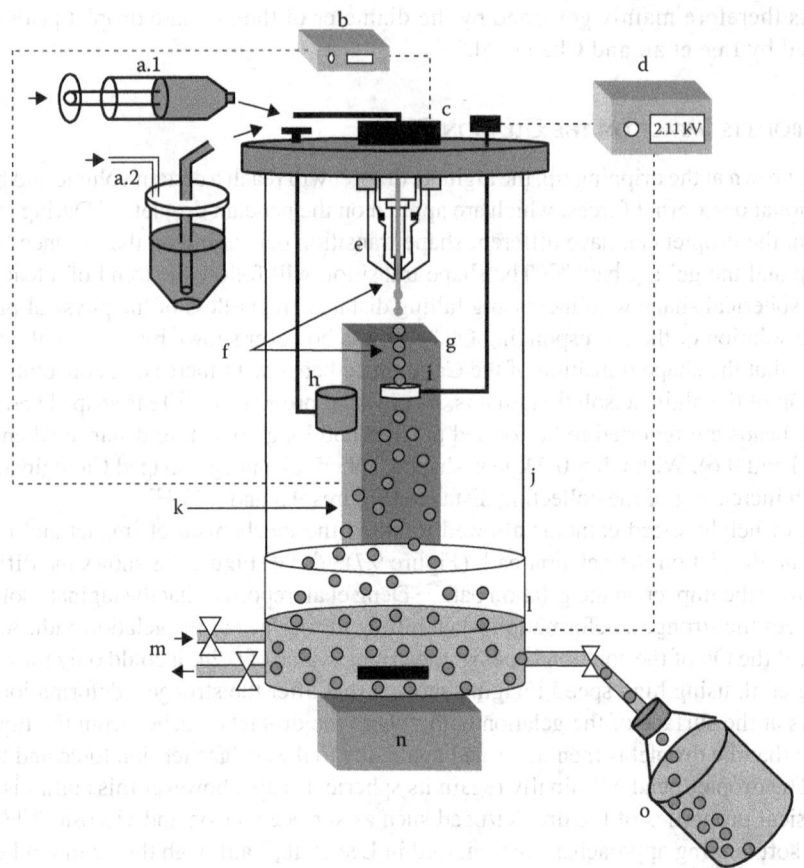

FIGURE 9.8 Representation of an encapsulation device based on the laminar jet break-up principle by vibrational frequencies with: (a) the product delivery mechanism which can be either a syringe pump (1) or a pressure regulation system (2), (b) a vibrational frequency control system, (c) a vibrating (pulsating) chamber, (d) an electrostatic charge generator, (e) a vibrating (single) nozzle, (f) a break-up of the liquid (optimal break-up of extruded jet resulting in a single bead chain), (g) a stoboscopic light, (h) a bypass system, (i) a electrode, (j) a reaction vessel casing, (k) dispersed droplets due to the negative charge applied, (l) a gelling bath, (m) the removal of gelling material for continuous operation, (n) a magnetic stirrer, and (o) the product collector. (Reproduced from Whelehan, M. and Marison, I.W., *J. Microencapsulation*, 28(8), 669, December 2011. With permission.)

9.3 APPLICATION

Microencapsulation and encapsulated products have found applications in numerous industries such as agriculture, chemical, pharmaceutical, cosmetic, and food industry over the last century.[12,54,55] More recently, applications of these particles in biotechnology and medical processes, including cell encapsulation for artificial implants, production of high cell density cultures and recombinant therapeutic proteins encapsulation as a means for delivery, has opened up a brand new field for this technology.[54,55]

Techniques such as dripping and jet break-up can find applications in most of the fields mentioned previously; however, they tend to find their most important application in cell encapsulation, mainly due to the relatively simplistic approach to produce the microcapsules, but also because they reach all or most of the criteria required for an application in medical and biotechnological processes.[54,55] Bacteria, yeast, plant cells, or mammalian cells have been widely encapsulated for different applications.[3,7,30,58] Encapsulating mammalian cells[52] can be used to produce bioartificial organs,[16] but also to get high cell density cultures to produce therapeutic recombinant proteins.[16] The first idea of immobilizing mammalian cells was suggested in 1933 by Bisceglie, who demonstrated that encapsulated insulin-producing cells remained viable.[24] The idea to use semi-permeable membrane was then introduced by Chang 30 years later, and Islets of Langerhans were then successfully immobilized and implanted in rats by Lim et al. in 1980.[24] The technology was shown to protect the encapsulated mammalian cells from mechanical stress[16] and the host immune system offering promising strategies to restore, maintain, and improve organ function and promote continuous drug delivery; however, encapsulation can also be a production method for the proliferation of mammalian cells.[21,24] Different studies investigated mammalian cell encapsulation, and it was shown that dripping and jet break-up techniques can be used for this purpose. Different cell lines such as Islet of Langerhans, hepatocytes, BHK, human parathyroid cells, etc. have been successfully encapsulated by dripping and jet break-up techniques.[15,21,23,25,26,58] Moreover, it was shown by Lewinska et al.[23] that electrostatic generation of droplets have no effect on the encapsulated cells. Bacteria,[7] yeast,[30] and plant cells[3] also were successfully encapsulated with dripping and jet break-up techniques.

Dripping and jet break-up techniques also have applications in agriculture,[12] food industry enabling the companies to incorporate minerals, vitamins, flavors, and other ingredients in their products,[12,47] in catalysis by enabling the recovery and re-use of catalytic material such as metals or enzymes,[12] in defense,[12] metal industry, and water treatment with capsular perstration enabling the recovery of pollutant substances.

Dripping and jet break-up techniques have indeed many possible applications; however, their actual development and main application is cell encapsulation as this field is growing fast in recent years.

9.4 EQUIPMENT

9.4.1 Simple Dripping, Electrostatic Extrusion, Flow Focusing, and Coaxial Flow

Dripping by gravity or simple dripping equipment has quite simple requirements. The device needs a tube to feed the polymer into a nozzle at a constant rate, the polymer will then go through the nozzle and drops will grow at the tip before getting separated from the stream and falling in the gelling solution. Such a system can be scaled-up by adding more nozzle for better productivity and are supplied by Nisco AG.[2]

Coaxial air flow and air dynamically driven devices are supplied by Nisco AG. Büchi Labortechnik also supplies an innovative nozzle that enables the use of encapsulators as an air flow system instead of a vibrating nozzle system, which can be an advantage for encapsulation of clusters or large particles.[1,25] A coaxial air flow or air-driven system must be equipped with a pump or mechanism to feed the polymer to the nozzle and two connections, the first to deliver the polymer and the second, which is concentric to the first, to deliver the air stream. The air stream is controlled by an air flow meter and a stirred gelling bath is mandatory to keep the beads separated from each other and avoid coalescence.

Devices based on electrostatic extrusion are supplied by Nisco AG. The device can be sterilized and consists of a pump or mechanism to feed the polymer to a nozzle, a high voltage power supply (7 kV) and a control system for fine tuning of the voltage magnitude, an autoclavable nozzle holder and a safety cage with a safety switch. A stirred gelling bath is mandatory to keep the beads separated from one another and avoid coalescence. Nozzles made of steel and of different diameters can be used, moreover, the device can be scaled-up by adding more nozzles, enabling production from 2400 capsules with single nozzle/min to approximately 24,000 capsules/min with a 10-nozzle head.[2]

9.4.2 Vibrating-Jet

Different devices based on vibrating technologies can be found on the market. Companies such as Nisco Engineering AG[2] or Büchi Labortechnik AG[1] are currently supplying devices, most of them consisting of several elements that can simply be assembled to make a lab scale encapsulator. As enumerated by Whelehan and Marison and shown in Figure 9.8, the main elements include

- A pump mechanism to feed the polymer(s) and/or core material to the nozzle(s);
- Nozzle(s) to create a laminar liquid jet;
- A vibration device and control system using signal frequency and amplitude to enable controlled break-up of the liquid jet;
- A stroboscopic light to allow visualization of droplet chain and tuning of frequency;
- An electrode and electrostatic charge generator to disperse the droplets with electrostatic repulsion forces;
- Agitated gelling bath to enable controlled gelification/polymerization of the droplet to form microspheres/microcapsules; and
- A collection device to recover the produced particles.

Most of the encapsulation devices can be used with a monocentric nozzle system or with a concentric nozzle system depending on the required capsules.[17,55]

The single flow nozzle system consists of a single orifice in which the polymer which is extruded passes through.[17,55] The nozzle can be made of different materials, the Inotech encapsulator uses a precision-drilled sapphire stone as the orifice on the tip of a stainless steel cone holder,[6,55] Nisco is producing nozzles with the orifice made of polished sapphire or polished ruby with a holder made of stainless steel,[2] while encapsulators from Büchi use nozzles made of stainless steel.[1] Nozzles diameter from 50 to 1000 µm are available depending on the supplier. Beads from 100 to 2000 µm can be produced with this method, and the production process can be carried out under sterile conditions by adding a glass casing around the apparatus followed by sterilization (autoclaving).[1,2,17,55] The single flow nozzle system can be used to produce microspheres in which solid particles or cells are encapsulated within a polymer matrix.[17,55] It can also be used to produce microcapsules by addition of an outer membrane followed by liquefaction of the inner core.[55]

Encapsulation devices produced by Inotech, Buchi Labortechnik AG and Nisco AG can also be used with a concentric nozzle system.[1,2,55] The concentric nozzle system allows the production of single liquid core microcapsules as the encapsulant is usually in the form of a liquid (if the encapsulation of a solid is required the solid can be suspended in a liquid which is then extruded as the core material).[17,55,56] The setup of the system is similar to the single flow nozzle however the single flow nozzle is replaced with a concentric system requiring two feeds, one for the core (inside) and one for the shell (outside).[55,56] The concentric nozzle is made from two nozzles, an internal nozzle placed within an external nozzle. Both can be used separately as single nozzle on a single flow system for the devices produced by Inotech and Büchi. Nozzles from 100 to 1000 µm are available for the outer nozzle and from 50 to 900 µm for the inner nozzle depending on the company supplying the device, enabling production of capsules from 200 to 2000 µm in diameter. The membrane thickness can be controlled by the difference in diameter between the inner and the outer nozzle and the average thickness can be obtained from the following equation:[55,56]

$$M_m = (d_m - d_c)/2 \qquad (9.21)$$

with
d_m the microcapsules diameter
d_c the core diameter

Microencapsulation by Dripping and Jet Break-Up

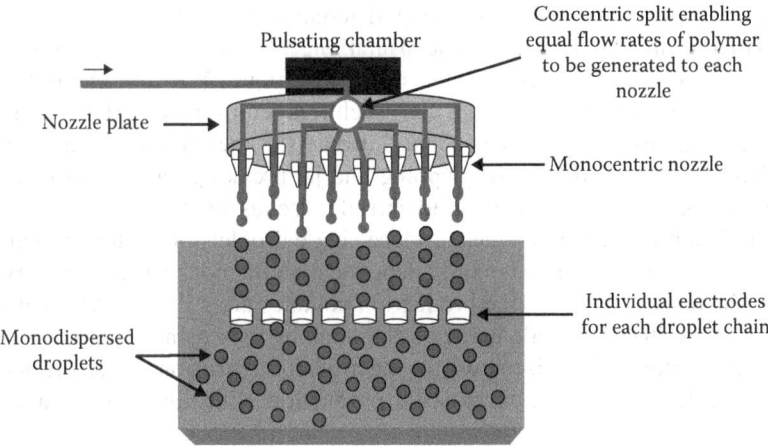

FIGURE 9.9 Multinozzle device such as that supplied by EncapBioSystems. (Reproduced from Whelehan, M. and Marison, I.W., *J. Microencapsulation*, 28(8), 669, December 2011. With permission.)

As the maximum flow rate per nozzle is limited in vibration technology, scale-up has to be done by increasing the number of nozzles (Figure 9.9). It has, however, to be taking into account that each nozzle of the multinozzle unit must show similar production conditions, meaning that the flow rate, the frequency, and the amplitude must be identical for each nozzle. Brandenberger and Widmer[6] increased microsphere output by adding more nozzles to the nozzle plate of their encapsulator and obtained a vibrating-jet monocentric multinozzle device.[6,55] Brandenberger and Widmer overcame the problem of keeping the flow rate constant of the multi-nozzle system by pumping the polymer through a concentric split, in this way they were able to obtain a relative flow difference of less than 2.0% between all the nozzles of the device.[6,55] The disturbance resulting in the jet break-up is transmitted in a vibrating chamber through which the polymer is flowing. Indeed, subjecting all liquid jets with the same sinusoidal force by vibrating each nozzle would be challenging from an engineering point of view. This method of creating the perturbation by a pulsating chamber allow a good size distribution of the droplets and any difference in droplets size is due to small difference in the diameters of the nozzles.[6,55] Such devices can be operated under sterile conditions with a mono or concentric nozzle system. Inotech and EncapBioSystems supplied large-scale devices operating under this principle. However, currently only Nisco AG are supplying such devices commercially.

9.5 ADVANTAGES AND DISADVANTAGES

The entrapment of various materials such as flavors, living cells, pharmaceutical compounds inside beads for different purposes is of great importance in the pharmaceutical, chemical, and food industries, as well as in agriculture, biotechnology or medicine, and diverse encapsulation techniques have been developed.[39,54,55] The different techniques must reach some criteria in order to be employed in industry. These criteria can be separated into two main groups, the first group concerns the capsule characteristics and includes the production of monodispersed, homogeneous and spherical beads of small size and narrow size distribution, the second group is more about the technique itself, which must be relatively easy to set up and simple to operate, with a high efficiency and a high production rate, the ability to use viscous solution, to produce a range of different sized droplets and to be sterilized, moreover, the production time must be short.[54,55] Dripping and jet break-up techniques are able to produce monodispersed homogeneous and spherical beads, with a narrow-size distribution. They have the advantage to be straightforward to setup and very simple to operate. They also have a short production time to produce droplets, a high efficiency, which can produce a range of different sized droplets and operate under sterile conditions, which is mandatory for cell encapsulation.[51,54] However producing small sized particles (<100 μm) can be a challenge for some of the techniques.[54] At present, only vibrating jet, electrostatic extrusion, and flow focusing can reach this requirement.[51] The inability of simple dripping and coaxial air flow to produce small droplets is therefore an inconvenience for those techniques. The main inconvenient of dripping and jet break-up techniques is, however, their inability to have high production rate and to extrude viscous solutions.[54]

Dripping and jet break-up techniques produce single droplets one after another at any given time and the production flow rate is mainly dependent on the nozzle diameter.[54] However, even by increasing the nozzle diameter, the production volumes are still low. As the maximum flow rate per nozzle is limited, the scale up has to be done by increasing the amount of nozzle.[51,54] Such devices are supplied by Nisco AG for electrostatic extrusion, vibrating jet, and simple dripping and enable to achieve a large scale production with vibrating jet and electrostatic extrusion.[2]

Another inconvenient of dripping and jet break-up techniques is the inability to extrude high viscosity polymers, leading to the use of low concentrations of polymer and therefore to the production of particles with a limited mechanical resistance.[54] This problem can be overcome by reducing the viscosity of the polymer solution by increasing its temperature during extrusion through the nozzle(s).[54] An apparatus to heat or maintain the polymer solution at controllable temperature as it passes through the pulsating chamber and the nozzle has being developed by EnCapBioSystems and is now supplied by Büchi.[1,54] This apparatus enabled work with alginate concentrations up to 4% (w/v) as reported by Whelehan and Marison and to successfully produce microcapsules, which showed considerable improvements in mechanical strength compared to lower alginates concentrations.[54]

Vibrating jet techniques also have the disadvantage that the produced bead size is related to the nozzle diameter, which can therefore be problematic when large particle or cell clusters have to be encapsulated. However concentric air-flow enables to encapsulate those large particles as the particle size can be more smaller than the nozzle size.[24]

9.6 CONCLUSION

For over a half century, encapsulation and microencapsulation have played an important role and found applications in many industries such as agriculture, chemical, pharmaceutical, cosmetic, food, medical, and biotechnology. The methodology to produce microcapsules for medical and biotechnological process must however meet some criteria such as producing monodisperse, homogeneous, and spherical beads, with a small size and a narrow size distribution. The device has to be relatively easy to set up and simple to operate, the time of production should be short; moreover, the efficiency and the production rates should be high. The ability to extrude viscous solutions, to produce a range of different sized droplets, and to operate under sterile conditions[55] is necessary. Dripping and jet break-up techniques are able to produce monodisperse homogeneous and spherical beads, with a narrow-size distribution. They have the advantage to be easy techniques to set up and very simple to operate. They also have a short production time to produce droplets, a high efficiency, can produce a range of different sized droplets and operate under sterile conditions, which is mandatory in cells encapsulation. Except simple dripping, all the techniques are scalable by adding more nozzles to the device, thereby enabling a large scale production. Polymers of higher viscosities can be extruded with the help of a heated nozzle apparatus, enabling the production of capsules with a higher mechanical strength. However, this solution cannot be used with heat sensitive encapsulants.

Taking these considerations into account, most of the methods discussed in this chapter are highly relevent, and the selection depends on the objective and requirements of the application.

REFERENCES

1. BÜCHI Labortechnik AG. Encapsulator B-390/B-395 Pro, Technical data sheet, http://www.buchi.com/en/content/spray-drying-encapsulation-solutions. Accessed December 2013.
2. Nisco AG. Encapsulation/Immobilisation systems. http://www.nisco.ch/units.htm. Accessed December 2013.
3. H. A. Al-Hajry, S. A. Al-Maskry, L. M. Al-Kharousi, O. El-Mardi, W. H. Shayya, and M. F. A. Goosen. Electrostatic encapsulation and growth of plant cell cultures in alginate. *Biotechnology Progress*, 15(4):768–774, 1999.
4. B. Amsden and M. Goosen. An examination of factors affecting the size, distribution and release characteristics of polymer microbeads made using electrostatics. *Journal of Controlled Release*, 43(23):183–196, January 1997.
5. H. Brandenberger, D. Nüssli, V. Pich, and F. Widmer. Monodisperse particle production: A method to prevent drop coalescence using electrostatic forces. *Journal of Electrostatics*, 45(3):227–238, January 1999.
6. H. Brandenberger and F. Widmer. A new multinozzle encapsulation/immobilisation system to produce uniform beads of alginate. *Journal of Biotechnology*, 63(1):73–80, 1998.
7. M. Bucko, A. Vikartovska, I. Lacik, G. Kollarikova, P. Gemeiner, V. Patoprsty, and M. Brygin. Immobilization of a whole-cell epoxide-hydrolyzing biocatalyst in sodium alginate cellulose sulfate poly(methylene-co-guanidine) capsules using a controlled encapsulation process. *Enzyme and Microbial Technology*, 36(1):118–126, January 2005.
8. C. P. Champagne, N. Blahuta, F. Brion, and C. Gagnon. A vortex-bowl disk atomizer system for the production of alginate beads in a 1500-liter fermentor. *Biotechnology and Bioengineering*, 68(6):681–688, 2000.
9. E.-S. Chan, B.-B. Lee, P. Ravindra, and D. Poncelet. Prediction models for shape and size of ca-alginate macrobeads produced through extrusion dripping method. *Journal of Colloid and Interface Science*, 338(1):63–72, October 2009.
10. P. Del Gaudio, P. Colombo, G. Colombo, P. Russo, and F. Sonvico. Mechanisms of formation and disintegration of alginate beads obtained by prilling. *International Journal of Pharmaceutics*, 302(12):1–9, September 2005.
11. Q. Deng, A. Anilkumar, and T. Wang. The phenomenon of bubble entrapment during capsule formation. *Journal of Colloid and Interface Science*, 333(2):523–532, 2009.
12. R. Dubey. Microencapsulation technology and applications. *Defence Science Journal*, 59(1):82–95, January 2009.
13. Y. Fukui, T. Maruyama, Y. Iwamatsu, A. Fujii, T. Tanaka, Y. Ohmukai, and H. Matsuyama. Preparation of monodispersed polyelectrolyte microcapsules with high encapsulation efficiency by an electrospray technique. *Colloids and Surfaces A: Physicochemical and Engineering Aspects*, 370(13):28–34, November 2010.
14. A. M. Ganan-Calvo. Generation of steady liquid microthreads and micron-sized monodisperse sprays in gas streams. *Physical Review Letters*, 80(2):285–288, January 1998.
15. L. Gasperini, D. Maniglio, and C. Migliaresi. Microencapsulation of cells in alginate through an electrohydrodynamic process. *Journal of Bioactive and Compatible Polymers*, 28(5):413–425, September 2013.
16. M. F. A. Goosen, A. S. Al-Ghafri, O. E. Mardi, M. I. J. Al-Belushi, H. A. Al-Hajri, E. S. E. Mahmoud, and E. C. Consolacion. Electrostatic droplet generation for encapsulation of somatic tissue. Assessment of high-voltage power supply. *Biotechnology Progress*, 13(4):497–502, 1997.
17. C. Heinzen, A. Berger, and I. Marison. Use of vibration technology for jet break-up for encapsulation of cells and liquids in monodisperse microcapsules. In V. Nedovic and R. Willaert, (Eds.), *Fundamentals of Cell Immobilisation Biotechnology*, number 8A in Focus on Biotechnology, Springer, Dordrecht, the Netherlands, January 2004, pp. 257–275.
18. C. Heinzen, I. Marison, A. Berger, and U. von Stockar. Use of vibration technology for jet break-up for encapsulation of cells, microbes and liquids in monodisperse microcapsules. *Landbauforschung Voelkenrode Sonderheft*, 241:19–25, 2002.
19. O. G. Jones and D. J. McClements. Functional biopolymer particles: Design, fabrication, and applications. *Comprehensive Reviews in Food Science and Food Safety*, 9(4):374–397, 2010.
20. T. I. Klokk and J. E. Melvik. Controlling the size of alginate gel beads by use of a high electrostatic potential. *Journal of Microencapsulation*, 19(4):415–424, 2002.
21. S. Koch, C. Schwinger, J. Kressler, C. Heinzen, and N. G. Rainov. Alginate encapsulation of genetically engineered mammalian cells: Comparison of production devices, methods and microcapsule characteristics. *Journal of Microencapsulation*, 20(3):303–316, 2003.

22. B.-B. Lee, P. Ravindra, and E.-S. Chan. Size and shape of calcium alginate beads produced by extrusion dripping. *Chemical Engineering and Technology*, 36(10):1627–1642, 2013.
23. D. Lewiska, S. Rosiski, and A. Weryski. Influence of process conditions during impulsed electrostatic droplet formation on size distribution of hydrogel beads. *Artificial Cells, Blood Substitutes and Biotechnology*, 32(1):41–53, 2004.
24. R. Mahou. *Hybrid Microspheres for Cell Encapsulation*. PhD thesis, EPFL, Lausanne, 2011.
25. R. Mahou, F. Borcard, R. Meier, C. Gonnelle-Gispert, L. Bühler, R. Plüss, M. Whelehan, and C. Wandrey. Engineered hydrogels for cell microencapsulation and subsequent transplantation. *XXI International Conference on Bioencapsulation*, 74–75, 2013.
26. V. Manojlovic, J. Djonlagic, B. Obradovic, V. Nedovic, and B. Bugarski. Immobilization of cells by electrostatic droplet generation: A model system for potential application in medicine. *International Journal of Nanomedicine*, 1(2):163–171, June 2006.
27. I. Marison, A. Peters, and C. Heinzen. Liquid core capsules for applications in biotechnology. In V. Nedovic and R. Willaert, (Eds.), *Fundamentals of Cell Immobilisation Biotechnology*, number 8A in Focus on Biotechnology, Springer, Dordrecht, the Netherlands, January 2004, pp. 185–204.
28. L. Martín-Banderas and A. M. Gañán Calvo. Making drops in microencapsulation processes. *Letters in Drug Design and Discovery*, 7(4):300–309, 2010.
29. G.-P. R. Martin-Banderas Lucía. Application of flow focusing to the break-up of a magnetite suspension jet for the production of paramagnetic microparticles. *Journal of Nanomaterials*, 2011(4), 2011.
30. V. A. Nedović, B. Obradović, I. Leskošek-Čukalovi, O. Trifunović, R. Pešić, and B. Bugarski. Electrostatic generation of alginate microbeads loaded with brewing yeast. *Process Biochemistry*, 37(1):17–22, September 2001.
31. V. A. Nedović, B. Obradović, D. Poncelet, M. F. Goosen, I. Leskošek-Čukalović, and B. Bugarski. Cell immobilisation by electrostatic droplet generation. *Landbauforschung Voelkenrode Sonderheft*, 241:11–17, 2002.
32. J. C. Ogbonna. Atomisation techniques for immobilisation of cells in micro gel beads. In V. Nedović and R. Willaert, (Eds.), *Fundamentals of Cell Immobilisation Biotechnology*, number 8A in Focus on Biotechnology, Springer, Dordrecht, the Netherlands, January 2004, pp. 327–341.
33. J. C. Ogbonna, M. Matsumura, T. Yamagata, H. Sakuma, and H. Kataoka. Production of micro-gel beads by a rotating disk atomizer. *Journal of Fermentation and Bioengineering*, 68(1):40–48, 1989.
34. D. Poncelet, V. Babak, R. Neufeld, M. Goosen, and B. Burgarski. Theory of electrostatic dispersion of polymer solutions in the production of microgel beads containing biocatalyst. *Advances in Colloid and Interface Science*, 79(23):213–228, 1999.
35. D. Poncelet, F. Davarci, M. Sayad, and S. Guessasma. How to observe a dripping process? *Bioencapsulation Innovations*, 18–19, September 2011.
36. D. Poncelet, R. J. Neufeld, M. F. A. Goosen, B. Burgarski, and V. Babak. Formation of microgel beads by electric dispersion of polymer solutions. *AIChE Journal*, 45(9):2018–2023, 1999.
37. S. Pregent, S. Adams, M. F. Butler, and T. A. Waigh. The impact and deformation of a viscoelastic drop at the air-liquid interface. *Journal of Colloid and Interface Science*, 331(1):163–173, March 2009.
38. U. Pruesse and K.D. Vorlop. The jetcutter technology. In V. Nedovic and R. Willaert, (Eds.), *Fundamentals of Cell Immobilisation Biotechnology*, number 8A in Focus on Biotechnology, Springer, Dordrecht, the Netherlands, January 2004, pp. 295–309.
39. U. Pruesse, L. Bilancetti, M. Bucko, B. Bugarski, J. Bukowski, P. Gemeiner, D. Lewinska et al. Comparison of different technologies for alginate beads production. *Chemical Papers*, 62(4):364–374, August 2008.
40. L. Rayleigh. On the instability of jets. *Proceedings of the London Mathematical Society*, s1–s10(1):4–13, November 1878.
41. L. Rayleigh. On the capillary phenomena of jets. *Proceedings of the Royal Society of London*, 29(196–199):71–97, January 1879.
42. J. M. Schneider and C. D. Hendricks. Source of Uniform Sized liquid droplets. *Review of Scientific Instruments*, 35(10):1349–1350, June 1964.
43. Y. Senuma. *Generation of monodispersed polymer microspheres by spinning disk atomization*. PhD thesis, EPFL, Lausanne, 1999.
44. Y. Senuma and J. G. Hilborn. High speed imaging of drop formation from low viscosity liquids and polymer melts in spinning disk atomization. *Polymer Engineering and Science*, 42(5):969–982, 2002.
45. Y. Senuma, C. Lowe, Y. Zweifel, J. G. Hilborn, and I. Marison. Alginate hydrogel microspheres and microcapsules prepared by spinning disk atomization. *Biotechnology and Bioengineering*, 67(5):616–622, 2000.

46. D. Serp, E. Cantana, C. Heinzen, U. Von Stockar, and I. W. Marison. Characterization of an encapsulation device for the production of monodisperse alginate beads for cell immobilization. *Biotechnology and Bioengineering*, 70(1):41–53, 2000.
47. H. M. Shewan and J. R. Stokes. Review of techniques to manufacture micro-hydrogel particles for the food industry and their applications. *Journal of Food Engineering*, 119(4):781–792, 2013.
48. D. Stark. *Extractive bioconversion of 2-phenylethanol from L-phenylalanine by Saccharomyces cerevisiae*. PhD thesis, EPFL, Lausanne, 2001.
49. G. Taylor. Disintegration of water drops in an electric field. *Proceedings of the Royal Society of London. Series A. Mathematical and Physical Sciences*, 280(1382):383–397, July 1964.
50. E. Teunou and D. Poncelet. Rotary disc atomisation for microencapsulation applications prediction of the particle trajectories. *Journal of Food Engineering*, 71(4):345–353, 2005.
51. V.-T. Tran, J.-P. Benot, and M.C. Venier-Julienne. Why and how to prepare biodegradable, monodispersed, polymeric microparticles in the field of pharmacy? *International Journal of Pharmaceutics*, 407(12):1–11, 2011.
52. H. Uludag, P. De Vos, and P. A. Tresco. Technology of mammalian cell encapsulation. *Advanced Drug Delivery Reviews*, 42(12):29–64, 2000.
53. M. Vemmer and A. V. Patel. Review of encapsulation methods suitable for microbial biological control agents. *Biological Control*, 67(3):380–389, December 2013.
54. M. Whelehan and I. W. Marison. Microencapsulation by dripping and jet break up. *Bioencapsulation innovations*, 4–10, September 2011.
55. M. Whelehan and I. W. Marison. Microencapsulation using vibrating technology. *Journal of Microencapsulation*, 28(8):669–688, December 2011.
56. A. Wyss, U. von Stockar, and I. Marison. Production and characterization of liquid-core capsules made from cross-linked acrylamide copolymers for biotechnological applications. *Biotechnology and Bioengineering*, 86(5):563–572, 2004.
57. A. Wyss-Peters. *Liquid core capsules as a tool in biotransformations*. PhD thesis, EPFL, Lausanne, 2005.
58. J. Xie and C.-H. Wang. Electrospray in the dripping mode for cell microencapsulation. *Journal of Colloid and Interface Science*, 312(2):247–255, 2007.

10 Microencapsulation by Annular Jet Process

Thorsten Brandau

CONTENTS

10.1 Introduction 201
10.2 Process Technologies 202
    10.2.1 Laminar Flow Breakup 202
    10.2.2 Vibrational Drip Casting 204
    10.2.3 Electrostatic Nozzle 206
    10.2.4 Submerged Nozzle 207
    10.2.5 Centrifugal Extrusion and Spinning Disk 208
    10.2.6 Flow Focusing 208
10.3 Conclusions 209
Summary 210
References 210

10.1 INTRODUCTION

Annular jet processes are used, when regular drop generation does not lead to a satisfactory result of the final properties of a particle. A normal particle, generated by a classical process using any jet, for example, atomization, vibrational drip casting, spinning disk, etc., consists of a more or less homogeneous mixture or a single material. The release properties of such particles are typically driven by a diffusion controlled process or by degradation. As, in case of an embedded active, the active is distributed over the total volume of the particle, it is partly unprotected. So, such actives at the outer parts of the particle can and will be subject to evaporation, oxidization, or degradation. In the best case, this means a loss of active. In the works, the final product is gone bad and has to be discarded—just think of a fruit yoghurt highly enriched with Ω_3-acid that smells strongly of rancid fish.

Even though certain processes can perform better here than others,[1] most of the applications need a different form of encapsulation.

A search for an alternative type of particle with superior properties was mainly driven by food technology, namely fermentation as initial industrialization efforts, but in the course of the development led to further benefits for many other fields of application.

In the field of further research, it was tried to use a multitude of concentric nozzles. The shell material was now used in the outermost nozzle, and the active was used in the core. This allowed not only for a much higher load of the particle of up to 90%, it also maintained many of the desired properties of the liquid flavors. Some of the early applications include tobacco products, fermentation of alcoholic beverages, and chewing gum. Nowadays, many products contain visible or invisible capsules made by the technologies described here to enhance product quality and protect the active agents (see Figure 10.1).

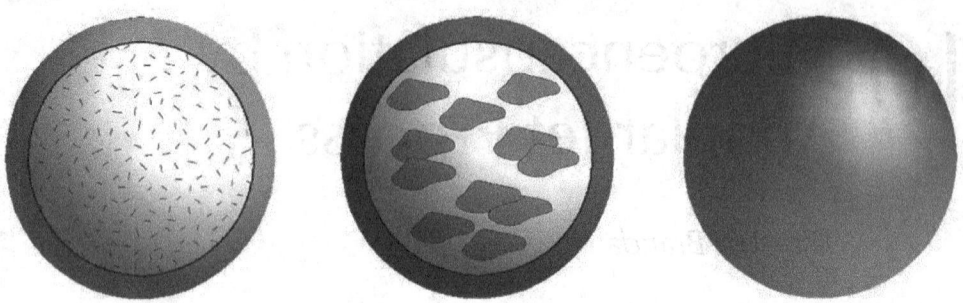

FIGURE 10.1 Different types of particles (from left to right): core–shell encapsulated particle with a solid or liquid core and a solid shell, core–shell encapsulated particle with a cell suspension inside, classical matrix encapsulated, respectively, granulated particle.

The use of flavors in food that were encapsulated goes back to the 1960s and 1970s where tobacco manufacturer tried to enhance cigarette filters with various devices that encapsulate liquid flavors.[2–4]

10.2 PROCESS TECHNOLOGIES

There are a number of process technologies used for annular jet processes. While there are some dominant ones, some smaller or niche processes should not be completely unmentioned. However, as the market is moving and evolving, it is not possible to give a complete overview about technologies, but the interested reader may inform himself by regularly visiting conferences and access the well-established journals to find out about newer developments.

At the current time, the annular jet processes most commonly used in food, pharma, cosmetic, and chemistry that produce particles of one or another kind are certainly laminar flow breakup processes. It has to be kept in mind, that the absolute majority of annular jet processes are used in other fields—as fuel engines—or to produce matrix particles, as two material nozzles used in atomization. Those, however, will not be subject to this chapter.

10.2.1 Laminar Flow Breakup

Laminar flow breakup is the general term for a number of processes that base on the same theoretical background, but lead to different apparatus and units, with different fields of applications and scalability.

In the beginning, the experiments of Savart[5,6] showed that a laminar flow of a liquid breaks up into droplets. Savart could verify that a circular flow of liquids separates into droplets (and satellite droplets). It was however not yet realized that the surface tension was the reason for the breakup, which was found later only by Plateau.[7] Later Rayleigh[6,8] was able to enhance this theory, which was further improved by Weber.[9] Nowadays, experimental aids as high-speed camera techniques made it possible to investigate and exploit the breakup of laminar flows in detail.

By extruding a liquid through an orifice with no more than gravitational force, droplets with diameters that are determined by viscosity and surface tension of the liquid are formed. The resulting droplets in that case (gravitational dripping) have a diameter x_t, which is only related to those material properties and the gravitational force, and not related to the diameter of the orifice.[10]

If the flow rate is accelerated, there is a resulting minimum flow rate for the transition of "dripping" to "flow," the latter meaning that a constant flow of liquid is extruded from the nozzle. This minimum flow or critical flow rate is a function of the surface tension, the viscosity, and the nozzle orifice.

Microencapsulation by Annular Jet Process

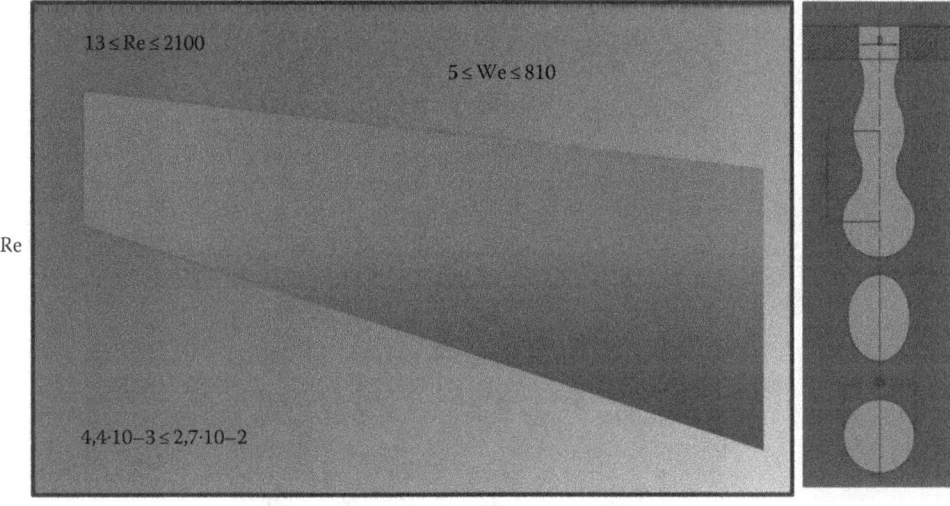

FIGURE 10.2 Ohnsorge–Reynolds plot.

If a liquid flow reaches the critical speed of v_{krit}, the surface tension tries to compress the liquid thread. An instability with a minimal wavelength of $x_t = 3.3\sqrt{(\sigma/\rho g)} \neq f(D)$ results from that.

This instability disrupts the thread and forms liquid cylinders that are forming (due to the surface tension) spherical droplets. Assuming the identical volume for droplet and cylinder, an ideal droplet diameter that is only corresponding to the nozzle diameter results ($x_T = 1.88\ D$[11]), always assuming that the flow is a laminar flow and is within the laminar flow breakup range of the Ohnsorge–Reynolds plot (Figure 10.2). Having said before that the flow rates are depending on material properties, resulting in a droplet size related to the nozzle diameter only, it is necessary to understand that the process is a two-step process. Initially, a laminar flow is generated—which properties are determined by the material—and secondly same-sized droplets are formed—which is determined by the nozzle diameter.

The disruption of the flow can lead to so-called satellite droplets, which have a diameter of $x_S \geq 0{,}1 x_T$. Their quantity is comparably small (0.1%–1%), but the total surface of the spray increases. Since droplets of different diameters have different falling speeds, the distance of the droplets is not constant anymore and droplets can remerge. This leads to the typical "noseforming," which can be found on unoptimized processes. The forming of the satellite droplets is determined mostly by the surface tension of the liquid. The viscosity has only a minor role (since $p_{stat} = (4\pi D\rho)/(D^2\pi) = 4\sigma/D$, $p_{dyn} = \rho v^2/2$), but for all practical purposes the viscosity can range between 10^{-3} and 10^4 [N/m]. Due to that, Oh can range over seven magnitudes. Therefore, the influence of the viscosity is in fact quite high. Materials that have a higher viscosity produce a longer wavelength and therefore larger droplets.

The maximum viscosity usable for the laminar flow break up processes is determined by the fact that a laminar flow has to be obtained. In practical terms upper limits (not absolute limits, only a guideline) of 5000 mPa/s are reasonable.

To avoid these problems, an external vibration is introduced in the flow. The vibration can be either in direction of the flow or rectangular to it. With that vibration the flow breakup is supported and leads to uniform droplets determined by the frequency of the external vibration and the reduction of satellites. When using optimized frequencies, the satellite forming is not existent anymore.

It is not important how the vibration is brought into the liquid (vibration of the nozzle or the liquid, vibration in the flow axis or perpendicular to it) but due to technical reasons, a vibrating nozzle with a vibration in the flow axis performs better (see Figure 10.3).

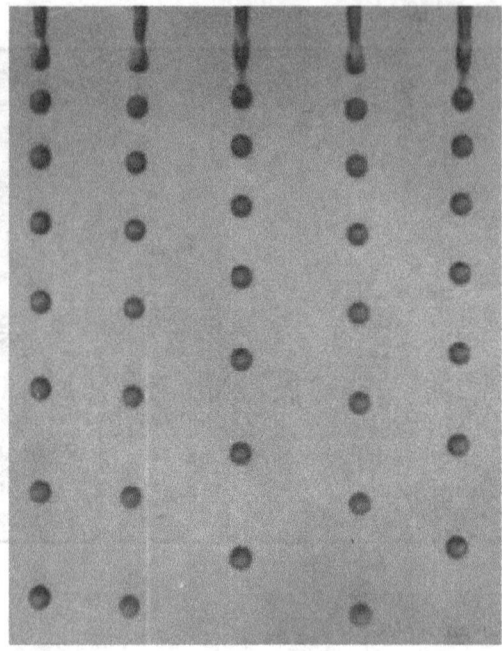

FIGURE 10.3 Drip casting with five nozzles with laminar flow drip casting and vibrating nozzle. (Courtesy of BRACE GmbH, Alzenau, Germany.)

10.2.2 Vibrational Drip Casting

Vibrational drip casting is one of the most scalable technologies to produce core–shell encapsulated microcapsules. There is a wide range of equipment sizes available, starting from small or very small units up to large scale equipment for the production of more than 20,000 kg products per hour. Of course, there are not many applications yet that use such large-scale equipment, but large products as encapsulated products can already reach several thousand tons per year.[12] Units making use of those processes can be obtained. For example, from BRACE GmbH, Germany, Büchi AG, Switzerland, or Nisco AG, Switzerland. The last two concentrate on laboratory scale machines, while the first one focuses on industry, with lab (Spherisator M, Figure 10.4) to large scale equipment.

FIGURE 10.4 BRACE Spherisator M. (Courtesy of BRACE GmbH, Alzenau, Germany.)

The principle of those technologies is long known but industrialization took a considerable time as there were besides the technical problems in scaling up also the need and acceptance of such products to be overcome. In fact, the first commercialization took place in nuclear fuel production in Germany for the high temperature reactor (Peeble-Bed-Reactor). With the politically decision to end nuclear technology in Germany, this technology was transferred to other fields of application, among them food technology. One of the early applications was the preparation of encapsulated yeast used to ferment Champagne.[13] Since then, many applications have been freshly developed, but also a lot of ideas have been newly reoccurred with this technology, as it was not economically possible to produce sufficient amounts with little effort.

Vibrational Drip Casting is a process working with a laminar flow breakup according to the Rayleigh flow break up with a vibration of the flow under resonance to give same sized droplets (see Figure 10.5[14]). In this process, the liquid feeds (one for the core and one for the shell material) are extruded through an annular gap nozzle. The flow is kept laminar so that it disrupts in droplets. By applying a vibration in resonance axial or lateral to the flow—either via the nozzle or the liquid directly—the resulting droplets have a very narrow monomodal size distribution. By adjusting the flow rates for core and shell nozzle separately, the load can be adjusted easily during processing.

These processes can easily be applied for room temperature or heated processes by installing a heating chamber around the nozzle head.

Another big advantage of those processes is that it can be scaled easily by increasing the nozzle types, up to exceedingly large number of nozzles is built, for example, by BRACE, Germany.

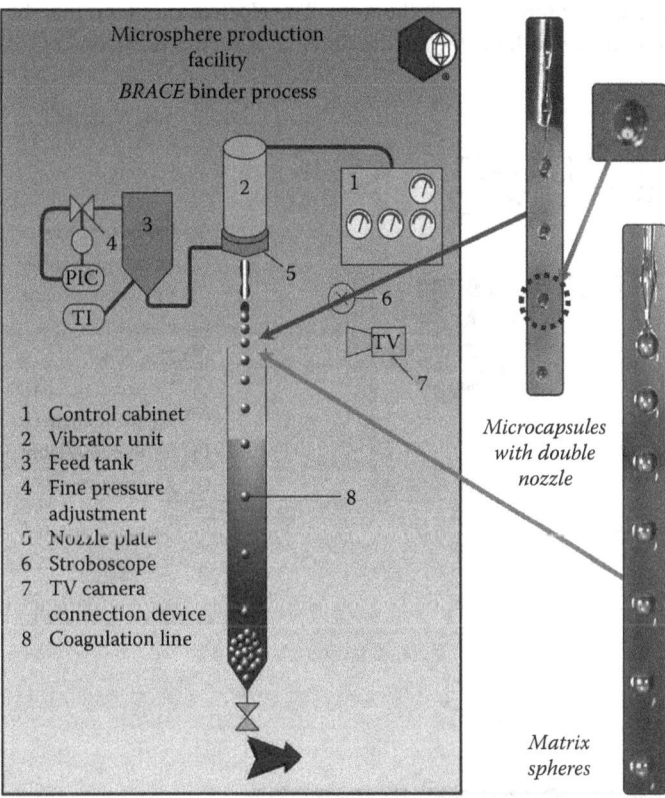

FIGURE 10.5 Schematic principle of annular jet nozzle processes with vibrating nozzles. Shown for room temperature processes of reactive mixtures, e.g. Alginate, Pectin, Cellulose, etc. Optional with heating and cooling towers in process drying or spray cooling is possible or other reactive processes. On the right side, pictures of microspheres (matrix encapsulation) and microcapsules (core–shell encapsulation) are shown during processing. (Courtesy of BRACE GmbH, Alzenau, Germany.)

By using multinozzle plates capacities of several hundred kg per hour can be reached. To further increase the capacity, the number of drip casting heads can be increased to reach capacities of 20 tons/h and more.

There are different number of nozzles used in the market like double nozzles—a core and a shell liquid—triple nozzles—a core, an intermediate, a shell—and even quadruple nozzles can be found. But with double nozzles the mechanical capabilities even of advanced tooling shops are already limited, so that the use of larger scale triple and quadruple nozzles is not widespread. As the tooling capabilities are bound to improve in the future there is a reasonable hope that triple and quadruple nozzle systems will be available for large scale production in the soon future.

10.2.3 Electrostatic Nozzle

The units are based on the vibrational drip casting with a focus on electrostatic dispersion. The electrostatic dispersion works by using the vibrational drip casting as source for generating identically sized droplets. Those are superficially charged with a high voltage load that repels the pellets from each other (Figure 10.6). So coagulation of the droplets during the free fall period is reduced if not eliminated. The drawback of this technology is that it only works well for a single nozzle and the space consumption of the reception bath is increased dramatically. For larger droplets, the electrostatic dispersion is seldom used, as those droplets sink into the reception liquid fast enough to avoid coagulation.

An advantage of this process is, that with the electrostatic field, it is possible to reduce the droplet size below the one given by the Rayleigh theorem. The electrostatic force hereby "pulls" the liquid out of the nozzle, having a similar effect as the flow focusing method (see the following section).

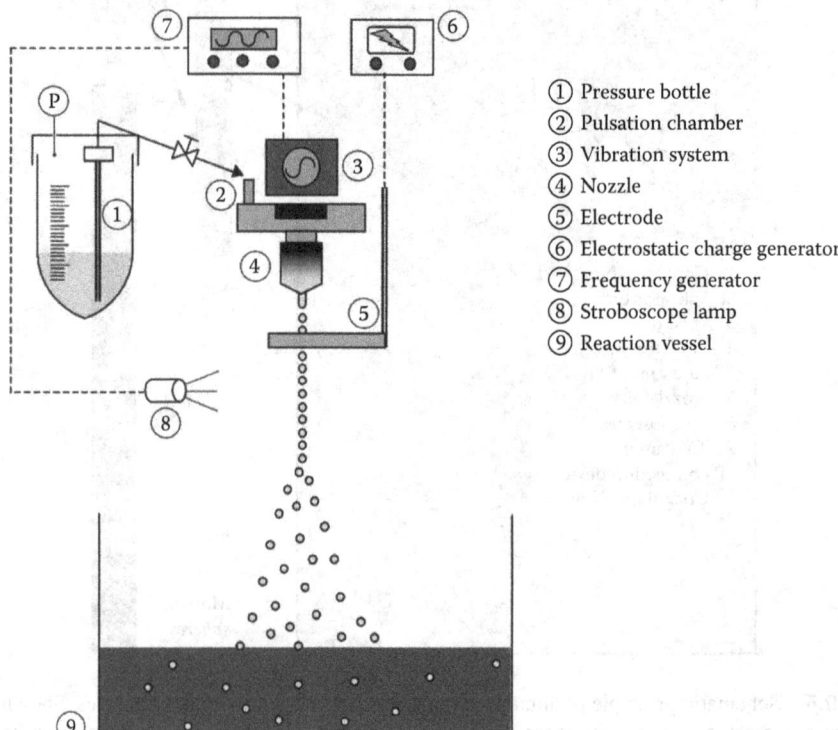

FIGURE 10.6 Process scheme Büchi Encapsulator. (From Neals, R,N. et al., Aromatic Filter, USPO 3,515,146, 1970, [Picture courtesy of Büchi], Switzerland.)

Microencapsulation by Annular Jet Process

This method works well with annular gap nozzles, as long as there is enough electrically potential in the liquids used in this process. Unfortunately the typically used high voltages tend to pose substantial problems when using inflammable liquids and during scale up. These problems can be addressed well by technical means, but might be a drawback to economical calculations.

10.2.4 Submerged Nozzle

The differences between submerged nozzle systems and vibrational drip casting systems are subtle, but essential with regard to the products produced.[15,16] A submerged nozzle system works by introducing the nozzle—inner and outer nozzle—into a transport medium[17] (see Figure 10.7). In most cases, it is oil. This transport medium flows codirectional to the nozzle and separates the droplets due to the external disruption introduced. The resulting droplets are of spherical shape due to the interfacial tension of the droplet shell material (usually gelatine) to the transport medium (usually oil). A vibration is in this case not necessary, but in some cases it is applied. The vibration applied here is in a low-frequency range and is typically little varied (around 50 Hz). There are also machines, especially on the Chinese market, that use a tube pump (or sometimes gear pumps) instead, to dose the liquid amounts into a solidification solution.

The resulting capsules range in the size of about 2 mm up to 10 mm. The scalability of this process is technically difficult, as the flow has to be applied evenly around each nozzle. Large-scale equipment with this process is not available on the market currently and the price for small scale equipment is typically quite high in comparison to vibrational drip casting processes. It is however not possible to produce such small particles as with vibrational drip casting.

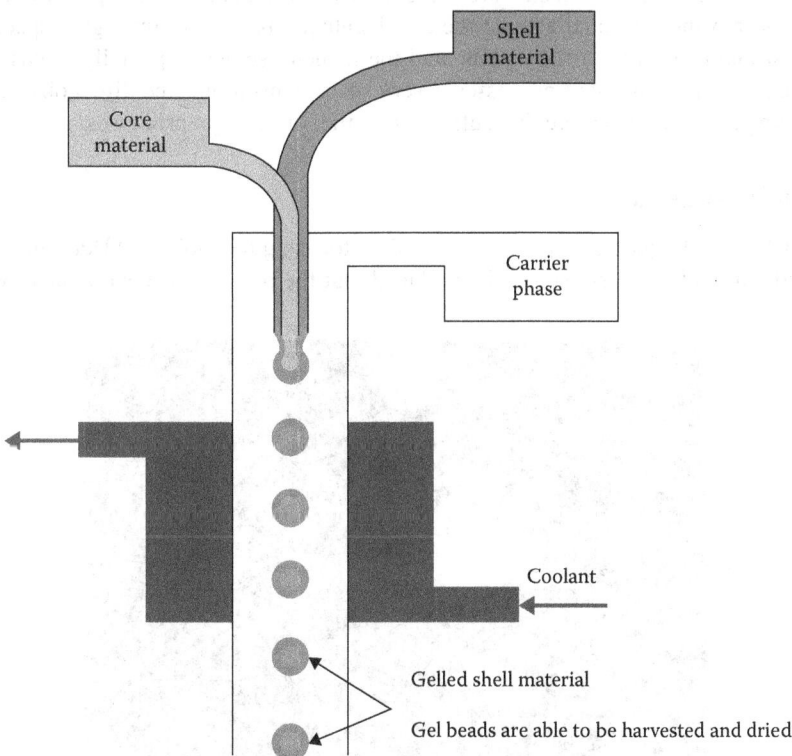

FIGURE 10.7 (See color insert.) Schematics of a submerged nozzle process. (From Hávarri, M., Marañón, I., and Villarán, M.C., Published in DOI: 10.5772/50046 under CC BY 3.0 license, Available from: http://dx.doi.org/10.5772/50046, © 2012.)

10.2.5 CENTRIFUGAL EXTRUSION AND SPINNING DISK

A spinning disk device—also called a rotary atomizer—is a device that forms on a rotary flat disk a liquid film that breaks up when leaving the disk into droplets (see Figure 10.8). A number of different physical shapes are common, as flat or vaned disk and cup or slotted wheel. The diameter of the rotary part vary between 25 and 450 mm with rotating speeds of 12,000–60,000 rpm^{-1}.[18–20] The droplet format can be assisted by vibration[21] or without vibration.

Typically, the materials are introduced from the top or into the disk and the disk throws them outwards, where they solidify by cooling, drying, or by collection in a reactive bath.

When instead of a flat disk annular gap nozzles are used and two materials are coextruded, the process is called centrifugal (co-)extrusion (see Figure 10.9).[22] The resulting capsules are typically solidified in a chemical reaction in a liquid bath, but can also be solidified, for example, by drying. With those rotary annular gap nozzles, the rotation speed is significantly lower than with a classical spinning disk, around 100–10,000 rpm^{-1}. The accessible capsules sizes are between 150 and 2000 µm. A single nozzle can hereby have a capacity of up to 10 kg/h, depending on material and particle size. Heads with up to 50 nozzles are known.

The theory behind those processes is very similarly to the processes described earlier. The main difference here is that the flow is generated by the centrifugal force and not by the pressure of the liquid. Simply speaking, the centrifugal force is manipulating the "gravity." The droplet diameter is here related to the thickness of the formed liquid sheet and the Weber number, that is, $D \propto \delta \cdot We^{-0.5}$.

The advantage therefore is that it is theoretically possible to produce smaller droplets as with processes where the liquid is extruded vertically. With the spinning disk also the nozzle is removed completely from the process, without that no nozzle blockage can occur and the challenges of producing many very precise nozzles are removed from the manufacturing of production lines. However, these advantages stand against the disadvantages as significant higher space consumption, the wear and tear on the rotary parts, and the difficult setups, especially in case of rotation and vibration. A scale up of those processes is very space consuming, and direct observation of the particles during production is more difficult as with other annular jet processes.

10.2.6 FLOW FOCUSING

A special niche annular jet-based process is the flow focusing technology.[23] Here, instead of using a liquid shell material, a matrix sphere is produced, but the outer nozzle is operated with an inert

FIGURE 10.8 Spinning Disk in operation. (Courtesy of the Southwest Research Institute, San Antonio, Texas.)

Microencapsulation by Annular Jet Process

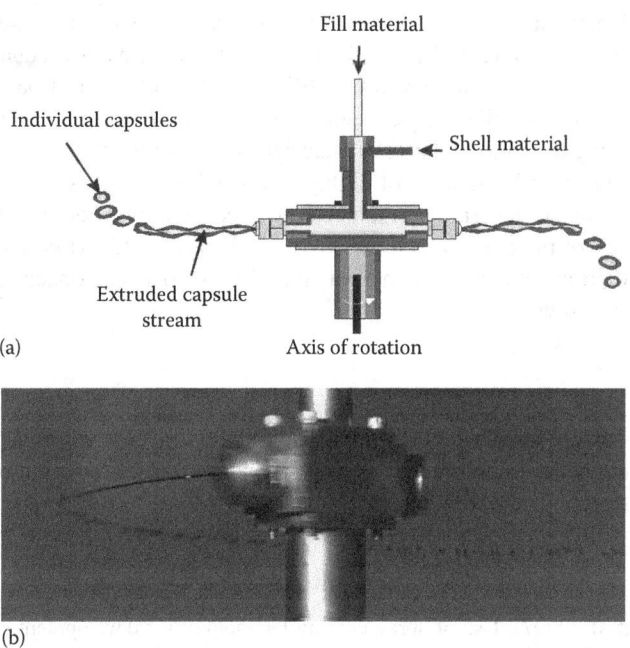

FIGURE 10.9 (a) Schematic of Centrifugal Extrusion with annular gap nozzles. (b) Picture of one annular gap nozzle in operation. (Courtesy of the Southwest Research Institute, San Antionio, Texas.)

medium (usually gas or a liquid).[24,25] In this case, a monomodal range of droplets can be produced, that have a mean diameter that is substantially smaller than the nozzle diameter. The size distribution is less narrow as with the vibrational drip casting method; however, with this technology, it is possible to produce very small droplets and even those which could not be processed in other nozzle based technologies, as the primary particle size in the feed mix is larger than the to-be-used-nozzle.

Flow focusing makes use of the effects of different velocities of the liquid jet and the surrounding medium. An increase in the relative velocity reduces the wavelength of the disturbance of the Rayleigh breakup. At 15 m/s for example λ_{opt} becomes 2.8 d (instead of 4.44 d in the case of a relative speed of 0), which reduces the drop size to 1.61 d. With higher speeds of the external fluid, the size of the particles can be reduced further to achieve down to about 1/10th of the nozzle diameter. The difficulty concerning the production of monodisperse droplets is that if the flow rate of the external medium is too high, a spray is formed. The minimum size achievable depends therefore a lot on the material properties. Additionally, scaling is very difficult, and it imposes significant technical and mechanical challenges, while the overall throughput is in any case comparatively low. Also, the flow focusing processes tend to be very hard to control, especially when non-Newtonian liquids are involved.

10.3 CONCLUSIONS

For the production of core–shell encapsulated products, annular jet processes offer a large number of advantages. Already a vast amount of products is produced using these processes, ranging from chewing gums, tobacco, medical products, supplements to electronic, and chemical applications.

Most of the processes are well scalable; although, sometimes technical issues—such as mechanical capabilities, control, or hazard control—have to be overcome. When it comes to scale, vibrational laminar flow break-up processes are the predominant technology in terms of flexibility, capacity, and costs. However, the process technology is not yet as far advanced as much older technologies as extrusion or spray drying, where basic knowledge about the processes can be assumed with any engineer, but they demand a good evaluation of materials and knowledge for successful implementation. As the processes are more widely used, the basic technology knowledge will improve, but a Microencapsulation with annular jet nozzles is still a system with a lot of parameters. To the outstanding witness, it might look like magic in some cases, what an experienced operator of a system can achieve by choosing the right and proper materials and combine them.

Certainly, the future will bring a lot of highly advanced products as knowledge increases and raw material makers start to understand the needs of encapsulation chemists, adopt their products so that a large amount of the current trial and error work is reduced, while equipment makers are already improving their machinery in such a way that the chemist can concentrate on his real work: To create a better tomorrow.

SUMMARY

Annular Jet processes are used widely in industries. A special interest and growing application are processes to produce Micrcapsules with annular gap nozzles. A selection of those processes is presented in this chapter.

REFERENCES

1. J.-P. Meunier et al. (2007). Use of spray cooling technology for development of micro-encapsulated *Capsicum Oleoresin* for the growing ping as an alternative to in-feed antibiotics. Study of Release using In Vitro Models. *J. Anim. Sci.*, October 2007; 85(10):2699–2710.
2. P.H. Leake et al. (1969). Tobacco Smoke Filter Element, USPO 3,428,049.
3. M.N. Boukair et al. (1969). Liquid-Containing Filter, USPO 3,420,242
4. R.N. Neals et al. (1970). Aromatic Filter, USPO 3,515,146.
5. F. Savart. (1833)., Mémoires sur la constitution des veines liquides. lancées par des orices circulaires enmince paroi. *Ann. Chim.*, 53:337–386.
6. J. Eggers. (2006). A brief history of drop formation. In *Nonsmooth Mechanics and Analysis* (P. Alart, O. Maisonneuve, and R.T. Rockafellar, Eds.), p. 163ff, Springer, Heidelberg, Germany.
7. J. Plateau. (1849), Statique Experimentale et Theorique des Liquides Soumis aux Seules Forces Moleculaires. *Acad. Sci. Bruxelles M´em.*, 5, 23.
8. L. Rayleigh. (1879). On the stability, or instability, of certain fluid motions.*Proc. London Math. Soc.*, 10.
9. C. Weber. (1931). Disintegration of liquid jets. *Z. Angewand. Math. Mech.*, 11(2):136–159
10. T. Brandau. (2002). Preparation of monodisperse controlled release microcapsules. *Int. J. Pharma.*, 242, 179–184.
11. A.H. Lefebvre. Atomization. http://www.thermopedia.com/content/573/. Accessed February 24, 2014. doi: 10.1615/AtoZ.a.atomization.
12. J.J. Burger et al. (1992). Chewing gum with improved flavor release. EPA 0528466.
13. F. Hill. (1991). Immobilization of yeast in alginate beads for production of alcoholic beverages. USPO 5,070,019.
14. Picture with kind curtesy of BRACE GmbH, Am Mittelberg 5, 63791 Karlstein, Germany, http://www.brace.de
15. M. Chávarri, I. Marañón and M.C. Villarán. (© 2012). Published in DOI: 10.5772/50046 under CC BY 3.0 license. Available from: http://dx.doi.org/10.5772/50046. Accessed February 2, 2014.
16. A. Christoffel et al. (1957). Method for producing seamless filled capsules. US Patent application US2799897.
17. A. Nakajima et al. (2003). Method of manufacturing seamless capsule. European Patent Application EP1310229A1.
18. K. Masters. (1976). *Spray Drying*, 2nd edn. John Wiley & Sons,New York.
19. Y. Pan et al. (2012). Effect of Flow and Operating Parameters on the Spreading of a Viscous Liquid on a Spinning Disc, *Ninth International Conference on CFD in the Minerals and Process Industries*, CSIRO, Melbourne, Australia, December 10–12, 2012.
20. H. Deng et al. (2011). Vibration of spinning discs and powder formation in centrifugal amtomization. *Proc. R. Soc.*, 467:361–380A.

21. J.M. Chicheportiche, Amtomiseur à disque torunant d'aérosols calibrés, EU patent application 91400621.8.
22. G.R. Somerville (Southwest Research Institute). (1967). Encapsulating method and apparatus. US Patent 3310612.
23. M. Ferrari ed. (2006). *BioMEMS and Biomedical Nanotechnology.* Vol. 1. p. 19 ff. Springer, Heidelberg, Germany.
24. http://en.wikipedia.org/wiki/Flow_focusing. Accessed February 24, 2014.
25. S. Turek et al. Numerische Simulation zur Herstellung monodisperser Tropfen in pneumatischen Ziehdüsen/Teilprojekt B5 im SPP Prozess-Spray (DFG-SPP 1423).

11 Encapsulation via Hot-Melt Extrusion

Hemlata Patil, Roshan V. Tiwari, and Michael A. Repka

CONTENTS

11.1 Introduction 213
11.2 HME Process and Equipment 214
    11.2.1 Equipment 214
    11.2.2 Hot-Melt Extrusion Process 216
11.3 Advantages and Disadvantages of Hot-Melt Extrusion 217
11.4 Materials Used in Hot-Melt Extrusion 218
11.5 Evaluation of Hot-Melt Extruded Products 221
    11.5.1 Thermal Analysis 221
    11.5.2 X-Ray Diffraction (XRD) 221
    11.5.3 Spectroscopic Techniques 222
    11.5.4 Microscopic Techniques 222
    11.5.5 Atomic Force Microscopy FM 222
    11.5.6 Water Vapor Sorption 222
11.6 Applications of Hot-Melt Extrusion Technology 222
    11.6.1 Solid Dispersions 222
    11.6.2 Oral Drug Delivery 224
    11.6.3 Films 224
    11.6.4 Implants 226
11.7 Encapsulation via Hot-Melt Extrusion 226
    11.7.1 Rationale for Encapsulation 226
    11.7.2 Techniques of Microencapsulation 227
11.8 Marketed Products 229
11.9 Conclusion 230
References 230

11.1 INTRODUCTION

Hot-melt extrusion (HME), a continuous pharmaceutical manufacturing process, has gained significant importance within the last few decades. It entails the process of pumping raw material at relatively high temperatures and pressures resulting in a product of uniform shape and density, many times to increase the dissolution profile of poorly water soluble drugs by converting it into an amorphous form.[1–3] Thus, HME has emerged as an alternative "platform technology" to other conventional techniques for production of pharmaceutical dosage forms, such as tablets, capsules, films, and implants for drug delivery via the oral, transdermal, and transmucosal routes.[2–7] This technology holds significant applications in several pharmaceutical areas, which remain yet unresolved by traditional pharmaceutical techniques.[4] To date, several research articles have been published describing the use of HME as a novel technique of choice to deal with day-to-day formulation challenges of new active pharmaceutical ingredients (API) by dispersing it with polymers and/or lipids at the molecular level, thus forming solid dispersions (or solutions).

It is well known that solid dispersions are commonly used for poorly water-soluble drugs because of its role in increasing the dissolution, absorption, and therapeutic efficacy of drugs. Moreover, several aspects of HME have been extensively cited.[1,4,8–12] Additionally, the total number of patents based on HME techniques has been on a steady rise worldwide in the last two decades. Thus, HME, being a proven manufacturing process, can meet the standards of the US Food and Drug Administration's process analytical technology measures for designing, analyzing, and controlling the manufacturing process through quality control measurements during the continuous extrusion process.[13]

11.2 HME PROCESS AND EQUIPMENT

11.2.1 Equipment

Pharmaceutical-class extruders have evolved and adapted to mix APIs with carriers, plasticizers, or with other processing aids to develop various solid dosage forms. The major difference between the pharmaceutical-class extruder and a "plastic extruder" are the contact parts, which are specifically configured to meet the current regulatory norms of manufacturing dosage forms.[8] The HME process is implemented using the equipment that mainly consists of an extruder, ancillary equipment for the extruder, downstream processing equipment, and other controlling tools used for performance and product quality evaluation. The extruder generally consists of a feeding hopper that is the opening through which material enters the barrel, a conveying section that is comprised of barrels and single or twin screws, the die for shaping the extrudates and a screw-driving unit. The ancillary equipment for the extruder consists of a heating/cooling device for the barrels, a conveyer belt to cool down the product and a solvent delivery pump for cooling, cutting, and/or collecting the finished product.[14,15] The controlling devices on the equipment include temperature gauges, a screw-speed controller, an extrusion torque monitor, and pressure gauges.

Different types of screw extruders are available such as single or multi-screw extruders. Multi-screw extruders can involve either two screws (twin screw design) or generally up to four.

Single screw extruders (SSE) are still utilized within the plastics field; however, their use in pharmaceuticals is limited. SSE consists of one rotating screw inside a stationary cylindrical barrel subdivided into three distinct zones: feeding zone, compressing zone, and metering zone. The barrel is generally manufactured in sections, which are either bolted or clamped with each other. The three primary functions of SSE are conveying, melting, and pumping polymers into a desired shape. The shape of the extruded product depends on the type of die connected to the end of the barrel.[1] Mechanical simplicity and a low cost of investment and minimal maintenance make SSE a good choice for the manufacturing of plastic extruded products. However, it has some limitations, such as poor mixing capability, and, therefore, it is not the preferred option for the manufacturing of most of the pharmaceutical formulations where high kneading and dispersing capacities are required. To overcome the stated limitations of SSE, multiscrew extruders were developed.

The first twin-screw extruder (TSE) was developed in the late 1930s in Italy.[8] The TSE has two agitator assemblies mounted on parallel shafts. Different types of TSE are available with each type having distinct operating principles and applications in processing. The two main types of TSE are either corotating (rotate in the same direction) or counte-rotating (the opposite direction). These two types of TSE can further be classified as either fully intermeshing or nonintermeshing. Schematic presentation of a TSE setup is provided in Figure 11.1.

A TSE is characterized by the following descriptive features:[1,8]

1. *Reduction in residence time*: In a typical extrusion process, the residence time ranges from 1–10 min based on the screw speed and feed rate.

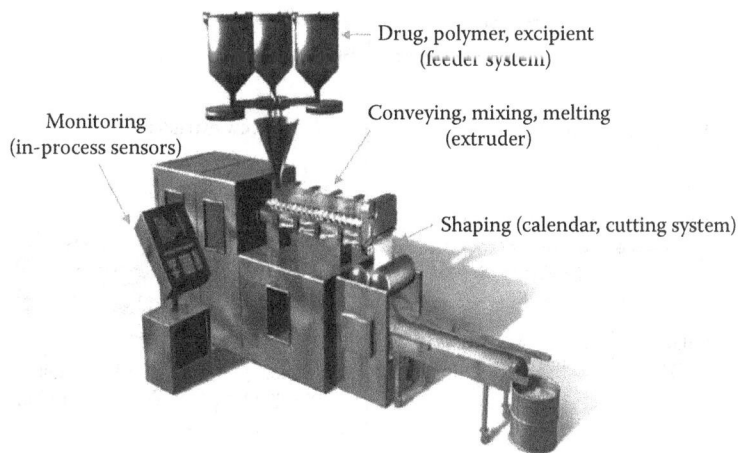

FIGURE 11.1 Schematic presentation of a twin-screw extruder setup. (Reprinted with permission from Breitenbach, J., *Eur. J. Pharm. Biopharm.*, 54, 107, 2002.)

2. *Self-cleaning screw profile*: Fully intermeshing type of screw design is self-cleaning, that is, the flight of one screw cleans the root of the adjacent screw ensuring complete emptying of the equipment and reduces product waste at the end of the production batch.
3. *Minimum supply*: Combining continuous operation of the equipment with the continuous feeding of the material helps in reducing the supply of work within the formulation batch.
4. *Flexibility*: Operating parameters can be altered easily and continuously to change extrusion rate or mixing functions. The segmented screw elements permits agitator designs to be easily optimized to work within the process application. Additionally die plates can also be easily exchanged to alter the extrudate diameter/shape. These attributes thus enables the processing of a variety of formulations on a single machine.
5. *Enhanced mixing*: The screws are designed in such a way that it provides two types of mixing, namely distributive mixing and dispersive mixing. In *distributive* mixing, the materials are evenly blended but not degraded. This type of mixing is implemented for heat and shear-sensitive APIs with minimal degradation. In the case of *dispersive* mixing, the droplet or solid domains are processed into fine morphologies using energy at or slightly higher than the threshold level needed. This mixing helps in efficient compounding of two or more API's in the TSE.

The difference between SSE and multiscrew extruders is their mode of operation. In an SSE, friction between the material and the rotating screw allows the material to rotate with the screw, and the friction between the rotating material and the barrel pushes the material forward, which results in the generation of heat. Therefore, for achieving high throughput in SSE by increasing screw speed results in the increase in frictional heat and therefore higher temperatures. Whereas, intermeshing screws of TSEs pushes the material forward with the relative motion of the flight of one screw inside the channel of the other. In TSEs, heat is therefore controlled independently from an outside source and is not influenced by screw speed. This difference in the operation becomes important when processing a thermolabile drug.[16]

When compared with SSE, intermeshing corotating TSEs generally provides better mixing to produce a homogeneous solid containing finely distributed and dispersed active compounds, better melt temperature control therefore less tendency to overheat, little dependence on the material's friction coefficients, and short transit time due to relatively fast melting.[17] A cross-section of a single and twin screw extruder barrel is shown in Figure 11.2.[18]

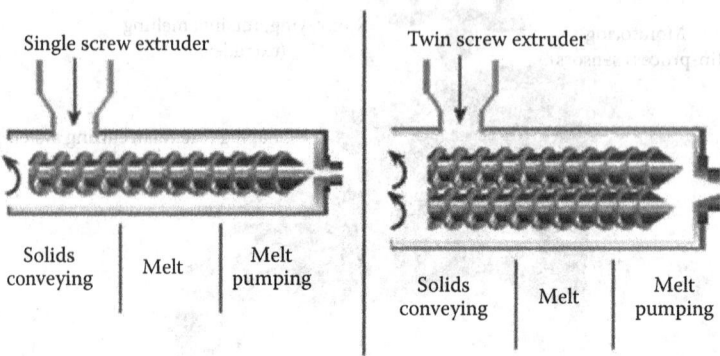

FIGURE 11.2 Single and twin screw extruder. (From Particle Sciences, *Technical Brief*, 3, 2011.)

11.2.2 HOT-MELT EXTRUSION PROCESS

Although HME is considered as a unit operation, for a theoretical understanding, the process can be categorized into the following subprocesses:[1,2,15–20]

1. Material feeding of the extruder with assistance of a hopper.
2. Conveying of mass.
3. Flow through the die.
4. Finally, extrudates from the die and subsequent downstream processing.
   1. *Material feeding*: Active compounds, carriers, plasticizers, or other processing excipients can be used as a preblend or can be introduced as an individual feed stream with the aid of a hopper(s) through feeding zone of the extruder by using either gravimetric (loss-in-weight) or volumetric feeders.[20] The preblend must have good flow. Feed rate, feed type, and oscillation in feeding rate affects the degree of fill, which influences the homogeneity, thermal, and mechanical energy input into the formulation.
   2. *Conveying of mass*: The material enters the feeder onto the rotating screws and is transported toward the die in the conveying section. This section includes several operations, such as melting, mixing, grinding, reduction in particle size, venting, and kneading. Therefore, the conveying process is mainly dependent upon screw speed and filling level, melting point of the individual components, particle size of the components, residence time, and screw configuration.[20] Modified screw designs are used when high shear is required to prepare an extrudate. Different types of screw elements to achieve desired results are shown in Figure 11.3.
   3. *Flow through the die*: At the end of barrel, a die is attached to the extruder in order to extrude the material. After the conveying section, the next step within the extrusion process is to pump the molten mass into the die. Different shapes of a die are available and are used depending on the desired end product shape.
   4. *Extrusion from the die and following downstream processing*: Just before the die, high-pressure is generally incurred, which allows nearly 100% screw fill level and ensures a constant melt flow through the die to provide an even shaping. Once the molten mass exit from the die, downstream processing plays a role into finishing, shaping, and analyzing the extruded product. The finishing process entails different downstream ancillary components are involved, such as conveyor belts that transfer the extruded product

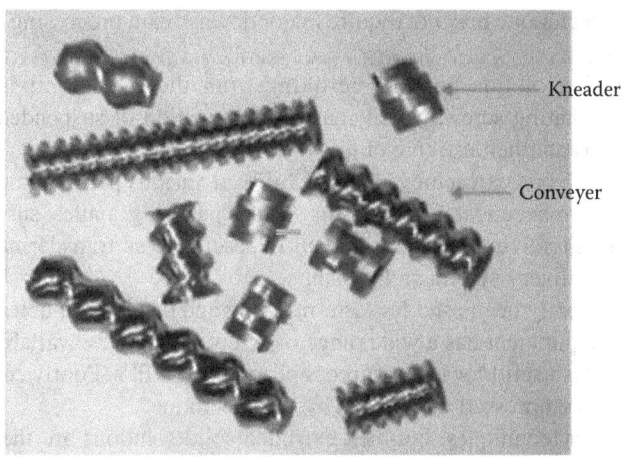

FIGURE 11.3 Modular screws design. (Reprinted with permission from Breitenbach, J., *Eur. J. Pharm. Biopharm.*, 54, 107, 2002.)

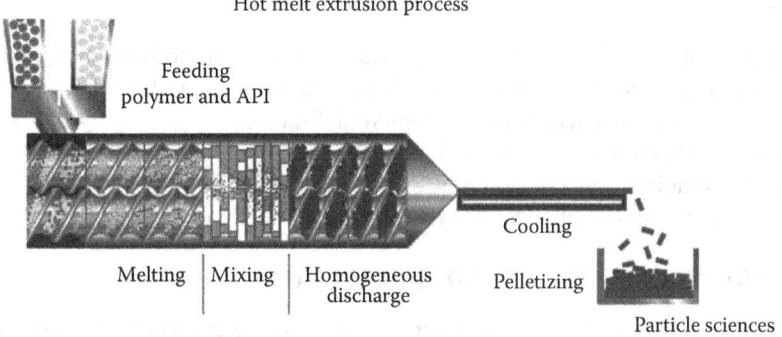

FIGURE 11.4 **(See color insert.)** Schematic presentation of HME process. (From Particle Sciences, *Technical Brief*, 3, 2011.)

from the die to the end of the processing line. The molten extrudates are often cooled by chill rolls and then either milled into powder or cut into rods by a strand-cutter. Sometimes pelletizers are also used to cut the extrudates into smaller units.

Schematic presentation of the HME process is provided in Figure 11.4.

11.3 ADVANTAGES AND DISADVANTAGES OF HOT-MELT EXTRUSION

HME offers several distinct advantages over conventional pharmaceutical formulation techniques:[2,4,8,21–23]

1. HME is a continuous and efficient process, which thus enables high throughput, online monitoring, minimal and controlled processing, and manageable process variables.
2. Solvent free techniques also reduce the number of processing steps and eliminate the time-consuming drying step. Also, there is no need to handle potentially dangerous solvents, resulting in the absence of residual solvents in the final formulation.
3. HME increases the solubility and potential bioavailability of poorly water-soluble and insoluble compounds, mainly by molecular dispersion of the drug within the final dosage form.

4. This processing technique may not require major downstream processing such as compression (tablets, etc.).
5. Uniform dispersion of fine particles resulting from the intense mixing and agitation imposed by the rotating screw, which cause deaggregation of suspended particles in the molten polymer, is another attribute of this technology.
6. Properly processed, one can achieve good stability at various pHs and moisture levels.
7. Wide array of pharmaceutical dosage forms such as pellets, granules, sustained/controlled release, oral fast dissolving system, targeted release such as transdermal, transmucosal, and transungual delivery system and implants.
8. HME has also recently been used for taste-masking and drug-abuse deterrent products.
9. Fortunately, the equipment has a wide range of screw geometries available for processing.
10. HME technology is useful for low compressibility index API's. Poorly compactable material can be easily compressed into tablets by this technique.
11. Compared to other techniques, hot-melt extruded solid solutions are thermodynamically stable if proper planning of the process is utilized.

Even though HME offers several advantages, like any pharmaceutical processing technique, it entails a few disadvantages, which includes the following. However, some of these can be managed and minimized:

1. HME may require high processing temperatures, which tend to limit its applicability in processing of thermolabile compounds. Thus, it is not suitable for relatively high heat-sensitive molecules such as proteins and microbial species.
2. There are a finite number of heat-stable polymers.
3. The process requires high energy input.
4. High flow properties of polymers and excipients are usually necessary.

11.4 MATERIALS USED IN HOT-MELT EXTRUSION

HME is used as a drug-delivery technology in which an active compound is embedded in a carrier system, which is usually comprised of one or more "meltable" substances and other functional excipients. Therefore, materials used in HME are mainly composed of APIs, meltable materials, which generally consist of low melting waxes and one or more thermoplastic polymers, plasticizers, and other processing aids. Molten polymers or waxes during extrusion process function as thermal binders, which upon cooling and solidification, act as drug depots and/or drug release retardants. The selection of API, meltable substances, and other functional excipients is one of the most critical steps during the HME process. As many drugs are heat sensitive, HME requires selection of appropriate polymers, which can be processed at low temperatures in order to prevent the thermal degradation of active compounds. Some of the carriers used to prepare hot-melt extruded dosage forms are listed in Table 11.1.[8]

In order to increase the physical and chemical stability, the majority of drugs available are prepared in a crystalline state, which is characterized by regular ordered lattice structures. This crystalline form of drug is mostly utilized in preparation of sustained/controlled release formulations. Crystalline products are generally thermodynamically stable, but it is still important to know possible polymorphic changes of the API during HME processing to ensure the stability of the product. To overcome poor dissolution limitations of crystalline products, the pharmaceutical delivery system may also be prepared in an amorphous form. One should be aware, however, that these amorphous systems might be thermodynamically unstable. Devitrification is a very common process that occurs with amorphous products, where upon storage, the amorphous form converts back into its more stable crystalline form. Therefore, for amorphous products glass transition temperature monitoring is very essential. Crowley and Repka (published in 2007) is a concise set of drug substances processed by HME techniques.[8] Table 11.2 presented in this chapter is a more updated and comprehensive list of drug substances used in HME techniques.

TABLE 11.1
Carrier Used to Prepare Hot-Melt Extruded Dosage Forms

Chemical Name	Trade Name
Ammonio methacrylate copolymer	Eudragit® RS/RL
Poly(dimethylaminoethylmethacrylate-comethacrylic esters)	Eudragit® E
Poly(methyl acrylate-co-methyl methacrylate-co-methacrylic acid) 7:3:1	Eudragit® 4135F
Poly(methacrylic acid-co-methyl methacrylate) 1:2	Eudragit® S
Hydroxypropyl cellulose	Klucel®
Ethyl cellulose	Ethocel®
Cellulose acetate butyrate	CAB 381-0.5
Cellulose acetate phthalate	—
Poly(ethylene oxide)	Polyox® WSR
Poly(ethylene glycol)	Carbowax®
Poly(vinyl pyrrolidone)	Kollidon®
Poly(vinyl acetate)	Sentry® plus
Hydroxypropyl methylcellulose phthalate	—
Polyvinylpyrrolidone-co-vinyl acetate	Kollidon® VA64
Hydroxypropyl methylcellulose	Methocel®
Hydroxypropyl methylcellulose acetate succinate	Aqoat-AS®
Poly(lactide-co-glycolide)	PLGA
Polyvinyl alcohol	Elvanol®
Chitosan lactate	Sea-cure®
Pectin	Obipektin®
Carbomer	Carbopol® 974P
Polycarbophil	Noveon®
Poly(ethylene-co-vinyl acetate)	Elvax® 40W
Polyethylene	—
Poly(vinyl acetate-co-methacrylic acid)	CIBA-I
Epoxy resin containing secondary amine	CIBA HI
Polycaprolactone	—
Carnauba wax	—
Ethylene-vinyl acetate copolymer	Evatane®
Glyceryl palmitostearate	Precirol® ATO 5
Hydrogenated castor and soybean oil	Sterotex® K
Microcrystalline wax	Lunacera® Paracera®
Corn starch	—
Maltodextrin	—
Pregelatinized starch	—
Isomalt	Palatinit®
Potato starch	—
Citric acid	—
Sodium bicarbonate	—
Methacrylic acid copolymer type C	Eudragit® L100-55
Chitosan	—
Xanthan gum	—
Agar	—
Povidone	Plasdone® S-30
Lactose	—
Microcrystalline cellulose	Avicel® PH 101
Dibasic calcium phophate	Emcompress®

Source: Crowley, M.M. et al., *Drug Dev. Ind. Pharm.*, 33, 909, 2007.

TABLE 11.2
Drug Substances Processed by Hot-Melt Extrusion Techniques

Drug	Tm (°C)	References
Glimpiride	207	[25]
Limaprost	97–100	[25]
Coumarin	71	[25]
Tamsulosin	226–228	[25]
Dexamethasone	262–264	[26,27]
Oxeglitazar	153	[28]
Paracetamol	169	[29,30]
Nifedipine	175	[21,33,34]
Indomethacin	162.7	[21,32,33]
Piroxicam	204.9	[32]
Tolbutamide	128.4	[21,32]
Lacidipine	184.8	[21,32,33]
Chlorpheniramine maleate	135	[35,36]
Theophylline	255	[37–40]
17β-estradiol hemihydrate	—	[41,42]
Oxprenolol hydrochloride	108	[43]
Fenoprofen calcium	—	[44]
Lidocaine	68.5	[45,46]
Hydrocortisone	220	[47,64]
Phenylpropanolamine hydrochloride	192	[48]
Hydrochlorothiazide	274	[49–51]
Carbamazepine	192	[52]
Ibuprofen	76	[53–55]
Melanotan-1	—	[56]
Diclofenac sodium	284	[57]
Acetaminophen	170	[58]
Nicardipine hydrochloride	180	[59]
Etonogestrel	200	[60]
Ethinyl estradiol	144	[60]
Acetylsalicylic acid	135	[61]
Diltiazem hydrochloride	210	[43]
5-Aminosalicylic acid	280	[62]
Itraconazole 166	166	[63]
Ketoconazole	148–152	[6]

Source: Crowley, M.M. et al., *Drug Dev. Ind. Pharm.*, 33, 909, 2007.

Additionally, functional excipients such as plasticizers are added into the formulation. Plasticizers are low molecular weight compounds capable of softening polymers to make them more flexible, may lower the processing temperature required for HME, which in turn can reduce the degradation of the active compound. Some of the common plasticizers are listed in Table 11.3.[8] Repka et al. in 2000 investigated the effectiveness of vitamin E TPGS as a plasticizer.[24] The study showed that vitamin E TPGS decreased the glass transition temperature of the extruded films and as the percent of TPGS increased the percent elongation of the film also were increased. Vitamin E TPGS was also found as an excellent processing aid, decreasing barrel pressure, drive amps, and torque as the TPGS percent increased.

TABLE 11.3
Common Plasticizers Used in Pharmaceutical Dosage Forms (8)

Type	Examples
Citrate esters	Triethyl citrate, tributyl citrate, acetyl triethyl citrate, acetyl tributyl citrate
Fatty acid esters	Butyl stearate, glycerol monostearate, stearyl alcohol
Sebacate esters	Dibutyl sebacate
Phthalate esters	Diethyl phthalate, dibutyl phthalate, dioctyl phosphate
Glycol derivatives	Polyethylene glycol, propylene glycol
Others	Triacetin, mineral oil, castor oil

The polymer degradation due to excessive temperature needed to process unplasticized, or under-plasticized polymers can be improved with the addition of other processing aids such as antioxidants, acid receptors, and/or light absorbers during HME. Incorporation of these processing aids in the HME process enhances its efficiency and overcomes many process limitations.

11.5 EVALUATION OF HOT-MELT EXTRUDED PRODUCTS

Several methods can be used to evaluate the hot-melt extruded products to distinguish among physical mixture of drug-carriers, solid solutions (molecularly dispersed), and solid dispersions, in which the drug is "partly" molecularly dispersed. An overview of the most common techniques used for the physicochemical characterization of hot-melt extruded formulations is provided in the following sections:

11.5.1 THERMAL ANALYSIS[12,88]

Thermal analytical methods examine the formulations as a function of temperature. As HME processes involve high processing temperatures, it is essential to determine the thermal stability of each individual component. Differential scanning calorimetry (DSC) one of the most sensitive thermal evaluation methods is generally used for the quantitative detection of transitions such as melting point and glass transition. DSC has been widely used for the assessment of drug crystallinity of hot-melt extruded products. Also, DSC is used as a preliminary screening method to study drug-polymer compatibility. The extrusion temperature of the polymer in the formulation should be higher (approx. 20°C–30°C) than the glass transition temperature (Tg) of the polymer to ensure good flow properties during HME, but it should be below the thermal degradation temperature of any of the ingredients.

Thermo gravimetric analysis (TGA) is another method of thermal analysis that measures the weight loss of a material as a function of increasing temperature.[77] TGA is commonly used to determine selected characteristics of materials that exhibit either weight loss or gain due to desolvation or decomposition.

11.5.2 X-RAY DIFFRACTION (XRD)[12,80]

X-ray diffraction (XRD) is commonly used to determine the crystalline nature of the drug. The principle of XRD is based on "Bragg's law of diffraction," in which the atomic planes of a crystal cause parallel incident beam of X-rays to interfere with one another as they leave the crystal lattice. Crystallinity is denoted by a featured fingerprint region in the diffraction order. The crystallinity of the drug and polymer following HME can be determined, if the fingerprints of drug and carrier do not overlay one another. Thus, X-ray can be used to differentiate between solid solutions (API is amorphous) and solid dispersions (API is at minimum present in the crystalline form, regardless of whether the carrier is amorphous or crystalline).

11.5.3 Spectroscopic Techniques

Spectroscopic techniques can be used to detect drug crystallinity, drug-polymer compatibility, or to measure the concentration of the drug present in the HME formulations. Commonly used spectroscopic methods are UV-visible, infrared, Raman, and near-infrared. FTIR is also used to detect the phase separation within the HME samples.[30]

11.5.4 Microscopic Techniques

Hot-stage microscopy (HSM) and scanning electron microscopy (SEM) are used as microscopic techniques to evaluate hot-melt extrudates. HSM can be used to detect the changes in a hot melt extruded formulation as a function of temperature. SEM is used to study the surface morphology of the hot melt extruded formulations by evaluating physical appearance and extent of drug aggregation of particles before and after extrusion.[63] SEM can be used to identify the differences of the crystal growth in the bulk and at the surface of the dosage form.

11.5.5 Atomic Force Microscopy FM

Atomic force microscopy (AFM) is used to determine surface microstructure of hot-melt extrudates, can be used to identify phase separation or non-homogeneity of HME samples. The AFM-based assessment of API:excipient combinations is a robust method to rapidly identify miscible and stable solid dispersions in a routine manner. It is a novel analytical method used for the optimization of HME processes.[31]

11.5.6 Water Vapor Sorption

Water vapor sorption (WVS) is used to assess the hygroscopicity of the hot melt extruded formulations.[65] WVS is used in quantification of amorphous content in pharmaceutical solids. Prodduturi et al. investigated the use of rapid dynamic vapor sorption techniques for analysis of films containing hydroxypropyl cellulose and clotrimazole.[96]

11.6 APPLICATIONS OF HOT-MELT EXTRUSION TECHNOLOGY

Recently HME has found its place in the array of technologies used to produce pharmaceutical dosage forms comprising oral, parental, and topical routes of administration. The primary application of HME technology is to enhance the solubility and bioavailability of lipophilic compounds. HME has demonstrated to provide sustained,[66,67] modified, and targeted drug delivery resulting in an improved bioavailability.

11.6.1 Solid Dispersions

Solid dispersion is defined as the dispersion of drug molecules in a solid matrix consisting of either a small molecule or polymer.[68,69] It is an established and well-explored solubilization technique for poorly water-soluble drugs.[70] Solid dispersions are mainly used to enhance the solubility and bioavailability of poorly water-soluble drugs.

There are several common technologies that are being used to manufacture solid dispersions such as by melting excipients or fusion technology,[71] embedding a drug in a carrier system by either roll-mixing/comilling,[72,73] coevaporation, coprecipitation,[74] freeze drying,[75] or spray drying.[76]

Hulsmann et al. in 2000 investigated the use of HME techniques to increase the solubility and dissolution rate of a poorly water-soluble drug, 17-estradiol hemihydrate by preparing solid dispersions of the drug into different compositions of excipients such as PEG 4000, PVP K 30, Kollidon, Sucroester®

WE15, or Gelucire® 44/14. When compared to the pure drug and physical mixture of drug and excipients, the solid dispersion demonstrated a significant increase in the dissolution rate (Figure 11.5). Researchers have also processed solid dispersions into tablets and studied dissolution behavior for those dosage forms. The enhancement in dissolution was also maintained with the tablets (Figure 11.6).[41]

Djuris et al. in 2013 utilized HME technology to prepare carbamazepine-Soluplus® solid dispersions and also studied the drug-polymer miscibility by thermodynamic model fitting.[77] Soluplus®

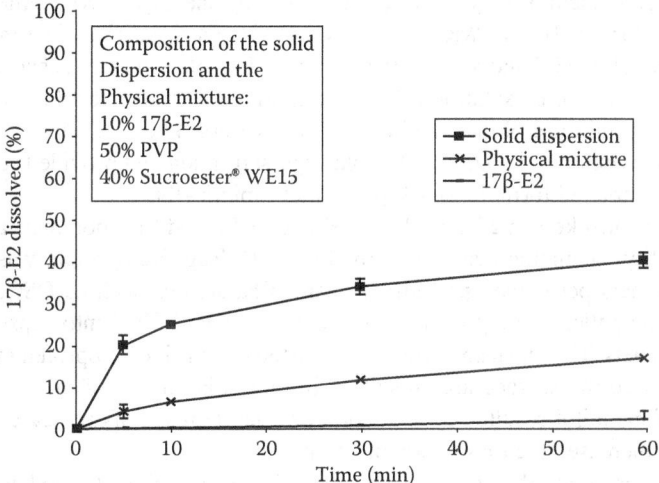

FIGURE 11.5 Comparison of a melt extruded solid dispersion with a physical mixture and pure 17β-E2 (dissolution media 0.1 N HCl). (Reprinted with permission from Hulsmann, S. et al., *Eur. J. Pharm, Biopharm.*, 49(3), 237, 2000.)

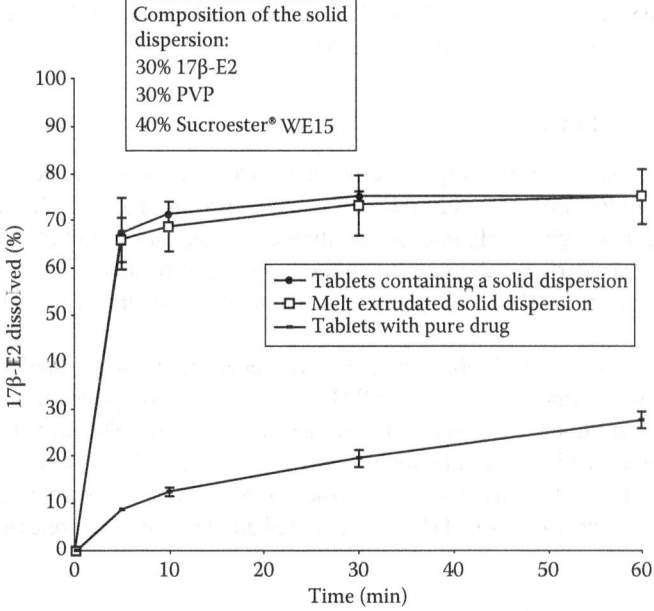

FIGURE 11.6 Drug release from solid dispersions and tablets containing pure drug or a solid dispersion (dissolution media: 0.3% aqueous SDS-solution). (Reprinted with permission from Hulsmann, S. et al., *Eur. J. Pharm. Biopharm.*, 49(3), 237, 2000.)

is a polyethylene-glycol-polyvinyl caprolactum-polyvinyl acetate grafted copolymer. Solid dispersions prepared by HME were characterized by TGA, DSC, attenuated total reflectance infrared (ATR-FTIR) spectroscopy, and hot stage microscopy. DSC and FTIR results have showed that molecular dispersions are formed when the drug concentration does not exceed 5% w/w, whereas higher drug concentrations result in microcrystalline dispersions instead of molecular dispersions.

Kalivoda et al. in 2012 studied the use of blends of different polymeric carriers to enhance the dissolution rate of the poorly water-soluble model drug fenofibrate via HME technology.[78] Copovidone (COP), polyvinyl caprolactam–polyvinyl acetate–polyethylene glycol copolymer (PVCL–PVAc–PEG) and hypromellose 2910/5 (HPMC) were used as a polymeric carriers. Hot-melt extruded fenofibrate was evaluated by DSC and x-ray diffractometry. Dissolution studies showed that there was a significant increase in the dissolution rate of the hot-melt extruded fenofibrate compared to the pure drug or physical mixture of drug and polymeric carriers. These researchers reported that the fenofibrate extrudates exhibited a 7.6- to 12.1-fold supersaturation and provided superior dissolution performance than marketed formulations Lipidil® and Lipidil®-Ter.

Mohammad and coworkers in 2012 evaluated Klucel™ EF and ELF polymers as an HME matrix former and solubility-enhancing agent. The BCS class II drug, Ketoprofen, was used as a model drug.[79] The researchers performed preliminary screening studies such as DSC, hot-stage microcopy, and TGA. The pellets were prepared from extrudates and filled into capsules or milled and compressed into tablets. The extrusion process converted crystalline ketoprofen into the amorphous form, which increased the surface area resulting in rapid dissolution of milled ketoprofen extrudates. Addition of mannitol resulted in formation of micropores. This increase in porosity of the extrudates further increased the dissolution of ketoprofen.

Albers et al. demonstrated the drug release mechanism from polymethacrylate-based extrudates prepared by HME.[80] These researchers reported a correlation between the nature of solid state, dissolution, and stability of the system. In order to increase the dissolution rate of poorly soluble drugs, the investigators emphasized the importance of knowing the mechanism of drug release from solid dispersions. Therefore, they studied the influence of mechanical stress on the solid dispersions. It was determined that solid-state properties determined the dissolution mechanism of solid dispersions, such as glassy solid solutions, followed carrier controlled mechanisms, whereas crystalline glass suspensions possess a drug-controlled mechanism.

11.6.2 Oral Drug Delivery

Young et al. in 2002 successfully prepared controlled release spherical pellets of theophylline by HME and spheronization process.[81] A Randcastle extruder was used to extrude the mixture of anhydrous theophylline, Eudragit 4135F, microcrystalline cellulose, and PEG 8000. Hot-melt extruded rods were cut into symmetrical pellets and then these pellets were spheronized at an elevated temperature. The dissolution profile of the spherical pellets was dependent on the matrix polymer solubility in the media.

Gryczke and coworkers used HME technology to prepare Ibuprofen granules, which were then compressed into orally disintegrating tablets (ODT).[82] In this study, Ibuprofen was encapsulated into a methacrylate copolymer Eudragit® EPO to produce a solid dispersion. The investigators' in-vivo taste masking study showed that HME techniques can be used to mask the bitter taste of active compounds. Also, the author mentioned that in comparison to the commercial tablet, Nurofen®, hot-melt extruded Ibuprofen granules compressed into ODTs demonstrated an increased drug release rate.

11.6.3 Films

The current methods of preparation of films are mainly based on solvent-casting approaches, which entail numerous disadvantages. To overcome solvent casting method limitations, HME technology was employed to prepare polymeric films.

Repka et al. in 1999 investigated the effect of plasticizers and drugs on the relationship of physical-mechanical properties of hot melt extruded hydrophilic films.[47] Hydrocortisone (1%) or chlorpheniramine maleate (1%) was used as a model drug. Researchers prepared the hydroxypropyl cellulose (Klucel, HPC) films containing drug and plasticizing agents such as polyethylene glycol 8000, triethyl citrate, acetyltributyl citrate, and polyethylene glycol 400. The prepared films were evaluated for tensile strength, percentage elongation, and Young's modulus. These investigators concluded that HME was a feasible process to prepare thin, flexible, and stable HPC films. However, without a plasticizer, or some processing aid, the HPC films could not be produced due to the high stress that was exhibited in the extruder.

Low et al. recently investigated the effect of the type and ratio of solubilizing agents on mechanical properties and release rate of hot-melt extruded orodispersible films.[83] Chlorpheniramine and indomethacin were used as model drugs. These researchers reported that the drug and a two-way interaction between the drug and solubilizing polymer were influencing the mechanical properties of the films. Dissolution depended primarily with the drug, solubilizing polymer, and the two-way interaction between the solubilizing polymer and ratio of solubilization to the film-forming polymer.

In 2005, Repka and coworkers prepared an oral mucoadhesive lidocaine-containing film using the combination of two cellulosic polymers, HPC, and hydroxypropyl methyl cellulose (HPMC) by HME processes.[46] Films prepared with only an HPC polymer was compared with the one prepared by combination of HPC:HPMC (80:20). Bioadhesive testing of both films was performed using a TA.XT2i texture analyzer. These results demonstrated that the hot-melt extruded HPC:HPMC films had a greater area under the curve (work of adhesion) and a higher peak adhesive force than that of the HPC-only films. The peak force and work of adhesion results for the HPC and HPC:HPMC films with an applied force of 3.5 N as a function of time are depicted in Figure 11.7.

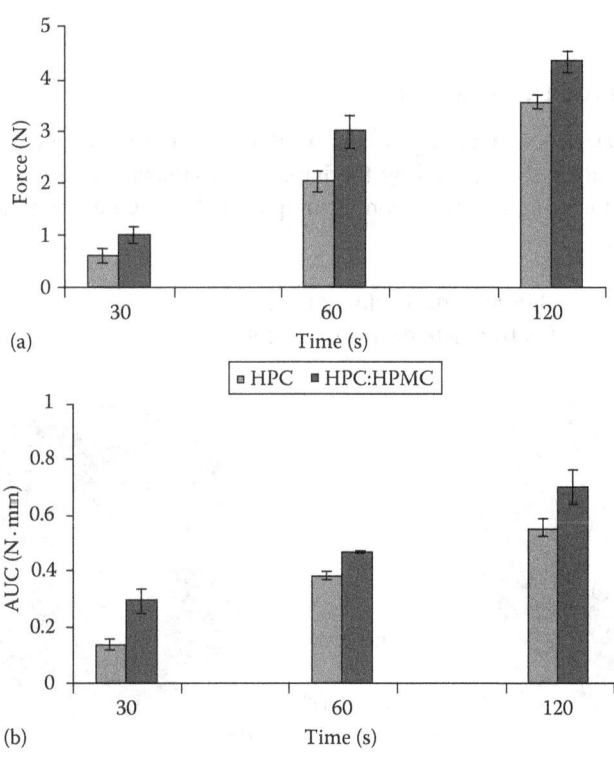

FIGURE 11.7 (a) Peak force (adhesive strength) and (b) work of adhesion of HPC and HPC:HPMC films measured using Texture Analyzer and rabbit intestinal mucosa as a substrate ($n = 5$). (Reprinted with permission from Repka, M.A. et al., *Eur. J. Pharm. Biopharm.*, 59, 189, 2005.)

11.6.4 IMPLANTS

Rothen-Weinhold et al. investigated the use of HME technology to prepare the long-acting poly (lactic acid) implants containing vapreotide, a somatostatin analogue.[84] The peptide degraded during processing with the formation of a lactoyl-vapreotide conjugate. These investigators discovered that the residual lactide in the polylactic acid significantly affected the formation of the peptide impurity, which illustrates that carrier purity is an essential parameter to be considered in order to develop a quality dosage form.

Li and coworkers prepared Dexamethasone-loaded implants by poly(D,L-lactic acid) (PLA) and poly(ethylene glycol)-block-poly(propylene glycol)-block-poly(ethylene glycol) copolymer (PEG-PPG-PEG, Pluronic® F68) using HME techniques.[26] Drug-implant compatibility was studied utilizing DSC. Mass loss and SEM techniques were used to confirm the degradation behavior of these dosage forms. Drug loading and encapsulation efficiency for the implants was reported to be up to 48.9% and 97.9%, respectively. The in vitro results of the implants demonstrated controlled drug release for up to 120 days.

11.7 ENCAPSULATION VIA HOT-MELT EXTRUSION

Encapsulation is a formulation technique in which an active compound can be enclosed inside a second material, such as polymeric or nonpolymeric carrier. The product obtained by this process is referred to as microcapsules. These microcapsules are obtainable in two types of morphology—either matrix type or core–shell (Figure 11.8).[85] Matrix-type microcapsules are in which the API is homogeneously distributed within the carrier system. In contrast, core–shell microcapsules consist of a solid shell made up of polymeric materials surrounding a core-forming space, which entraps the API.

11.7.1 RATIONALE FOR ENCAPSULATION[86]

Encapsulation of materials is a strategy to ensure that the enclosed material reaches the targeted area without being adversely affected by the harsh environment through which it passes. There are several reasons for using encapsulation techniques within the pharmaceutical industry. These include the following:

1. Masking of odor and taste of the enclosed material.
2. Protection of the APIs from a harsh environment.

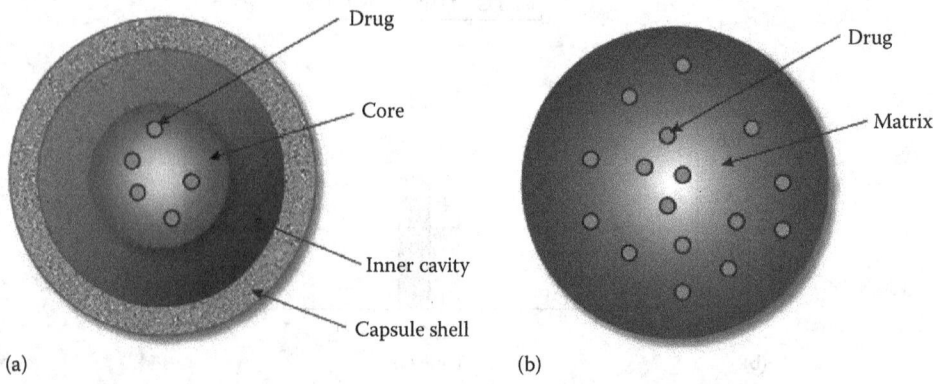

FIGURE 11.8 Schematic morphologies of the two types of microcapsules. (a) Core–shell microcapsule or reservoir and (b) matrix type microcapsule. (From Lembo, D. and Cavalli, R., *Antivir. Chem. Chemother.*, 21, 53, 2010.)

3. Targeted release of the encapsulated material.
4. Increasing stability by preventing oxidation or degradation of the encapsulated material by an adverse environment.
5. Solving incompatibility problems by separating APIs from each other or carrier materials.
6. Controlled release of the active compound (sustained release or delayed release).

11.7.2 Techniques of Microencapsulation

Drug encapsulation techniques are important for the delivery of poorly water soluble, fragile, or toxic compounds. Increasing a drug's encapsulation efficiency in drug-carrier particles can increase therapeutic benefits while minimizing adverse effects. Thus, by using new materials and several types of drug-carrier interactions, new encapsulation methods can be developed.[87] To date, several chemical and physical/mechanical methods have been used for microencapsulation. While considering the reasons or benefits of encapsulation, it is very crucial to consider the limitations of these available methods such as follows:

1. Increased costs.
2. Complexity of the production.
3. Stability of the active component while using solvents in the manufacturing of microcapsules.

To overcome these limitations, HME technology offers some distinct advantages over traditional methods of encapsulation. Notably, HME is generally a solvent free (or minimal amount of solvent) technique, is cost efficient, entails a small equipment "footprint," is a continuous (melting, mixing, and shaping) process, and is suitable for numerous matrix materials and encapsulants.

Loreti and coworkers in 2013 investigated HPC as a thermoplastic carrier for the monolithic matrix systems intended for oral sustained release.[88] In this study, they compared matrices prepared by HME with directly compressed matrices. Model drugs used in the study were theophylline or ketoprofen (5%–70% or 5%–40% range, respectively). The effect of the method of preparation on hydration, swelling, and erosion rate of the polymer was studied. Promising results were obtained, which demonstrated the use of HME technology as a continuous manufacturing process in the preparation of drug encapsulated HPC matrices, sustaining the release of the model drugs (Figure 11.9).

Schulz and Winter in 2009 utilized HME technology as an alternative technique for the preparation of lipid extrudates as novel sustained release systems for pharmaceutical proteins.[89] The focus of this research entailed a thorough investigation of lipid modification and the protein stability after extrusion. Since HME technology involves high temperatures, maintenance of the protein structure during manufacturing was the most critical parameter in this study. The investigators utilized lipid blends of low and high melting lipids such that extrusion was performed at moderate temperatures. Interferon-α was used as the model protein. Protein integrity was analyzed by SDS-PAGE western blot analysis after extraction. It was determined that changing the implant diameter or modulating the PEG content of the system could control the release rate of the protein from the prepared extrudates.

HME technology has been used for taste masking of bitter APIs. Maniruzzaman and coworkers investigated the potential application of HME technology for masking the bitter taste of APIs by incorporating and selected them into different polymer formulations.[90] Cetrizine HCl and verapamil HCl were used as model drugs. Anionic polymers such as Eudragit® L100 and Eudragit L100-55 were used as matrix components. The extrudates characterized by SEM, X-ray, and DSC demonstrated that the API was molecularly dispersed within the anionic polymer matrices. The researchers concluded that by enhancing drug-polymer interaction HME technology could be successfully used to mask the bitter taste of the tested APIs.

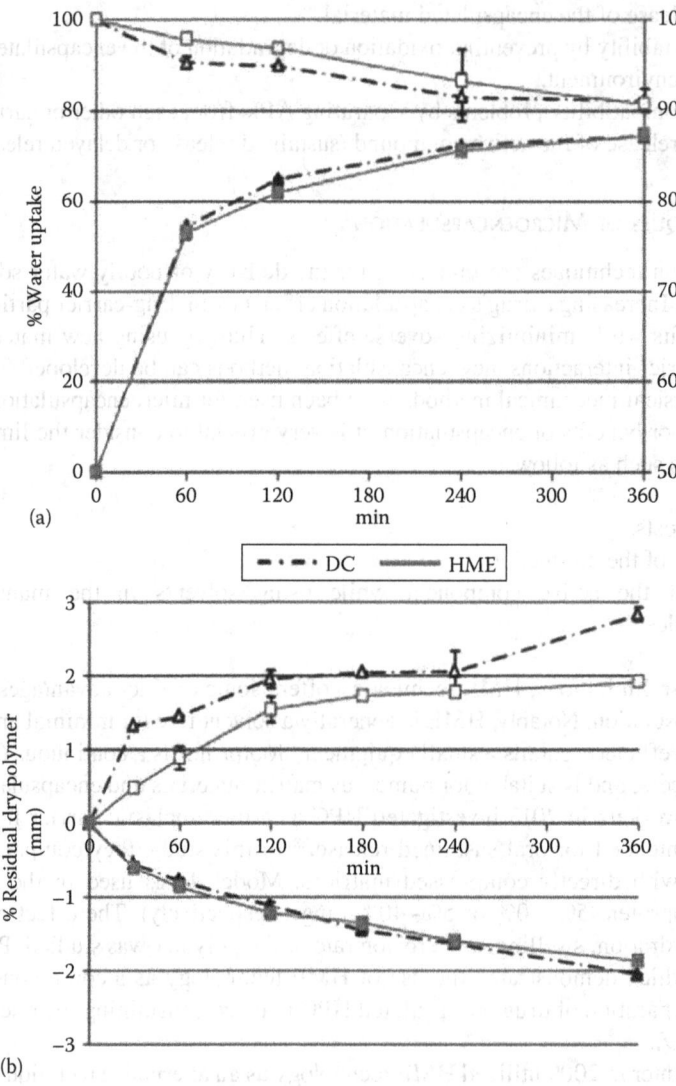

FIGURE 11.9 (a) Water uptake (■, ▲) and residual dry polymer (□, △) profiles and (b) swelling (■, ▲) and erosion (□, △) front profiles of HPC matrices manufactured by direct compression (dotted lines) and HME (solid lines). (Reprinted with permission from Loreti, G. et al., *Eur. J. Pharm. Sci.*, 52, 77, 2014.)

Remon et al. developed a sustained release system consisting of hot-melt extruded ethylcellulose cylinders containing an HPMC-Gelucire core.[91] This study represents as an example of a core-shell-type microcapsule system prepared by HME technology. Different types and viscosity grades of HPMC were used. All systems exhibited only 40% drug release over a 24 h period. An increase in drug release beyond 24 h in the system was observed by shortening the length of the ethylcellulose cylinder, whereas a change in the diameter did not affect the drug release rate. In addition, drug solubility had no effect on the drug-release rate or the release mechanism.

Vervaet et al. investigated another example of a core–shell microcapsule system via HME technology in 2005.[92] These researchers used HME as an alternative technique for enteric delivery. Polyvinyl acetate phthalate and hydroxypropyl methyl cellulose acetate succinate (HPMC AS) enteric coating polymers were premixed with plasticizers and extruded into hollow cylinders. A model drug was then filled into these hollow cylinders and both open ends were sealed. Dissolution

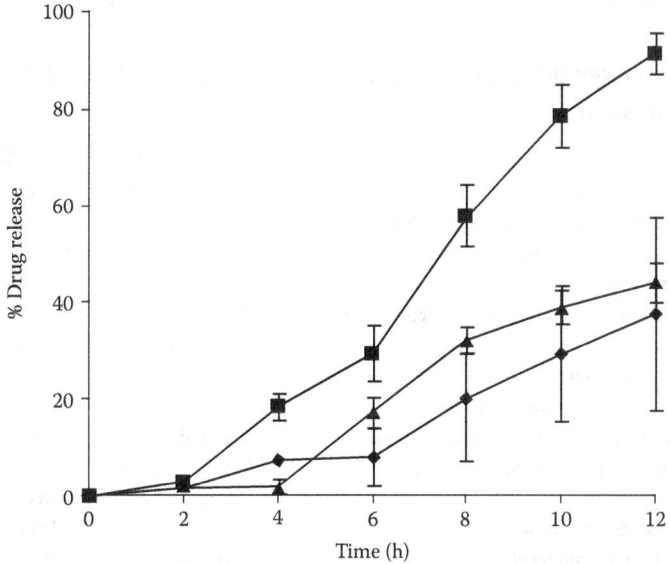

FIGURE 11.10 Influence of TEC concentration and pre-plasticization on the drug release rate of hot-melt extruded tablets containing 25% w/w 5-ASA. (▲) Formulation A2, preplasticized 12% w/w TEC; (♦) Formulation A1, no preplasticization 12% w/w TEC; (■) Formulation B, pre-plasticized 23% w/w TEC; Dissolution media consisting of 0.1 N HCl, pH 1.2 from 0 to 2 h; 50 mM phosphate buffer, pH 6.8 from 2 to 6 h and pH 7.4 from 6 to 12 h, 37°C, 100 rpm, apparatus 2 ($n = 3$). (Reprinted with permission from Bruce, L.D. et al., *Eur. J. Pharm. Biopharm.*, 59(1), 85, 2005.)

profiles demonstrated that there was no drug release after 2 h in 0.1N HCl. The investigated enteric coating acted as an excellent system for gastro-resistance.

Bruce and coworkers investigated the application of HME technology for targeted drug delivery. The researchers studied the properties of tablets prepared by HME technology for targeting delivery of 5-Aminosalicylicacid (ASA) to the colon.[93] Eudragit S 100 was used as a polymeric carrier to target delivery of 5-ASA. HME extrudates of 5-ASA were characterized by modulated DSC, SEM, and XRD methods. The results showed that 5-ASA remained in its crystalline form and was homogeneously dispersed into the polymer matrix. The investigators observed that the addition of citric acid monohydrate in the formulation may act as a plasticizer and addition of TEC in the formulation not only reduced the processing temperature, but also influenced the drug release rates from the extruded tablets due to leaching of the TEC during dissolution testing (Figure 11.10). Citric acid delayed the release of 5-ASA in phosphate buffer solution due to a lowering of the micro-environmental pH of the tablet matrix, which suppressed ionization of the Eudragit S100.

11.8 MARKETED PRODUCTS

HME technology is an increasingly attractive process for the manufacturing of pharmaceutical dosage forms. Pharmaceutical products manufactured via HME have been approved in the United States, Europe, and Asia. Over the last three decades, patents related to pharmaceutical systems prepared by HMT have steadily increased. The United States and Germany hold approximately 56% of all issued patents for HME worldwide.[94]

Several commercialized HME pharmaceutical products either currently marketed or under development are shown in Table 11.4. Due to the production feasibility and scaling ability for commercialization of HME techniques, many multinational companies have been performing extensive research in HME as continued discovery of new, poorly soluble drug candidates demand the advancement of novel formulation development technologies.

TABLE 11.4
Currently Marketed and Developed Drug Products Produced Utilizing Hot-Melt Extrusion Technology

Product	Indication	Company
Lacrisert® (opthalmic insert)	Dry eye syndrome	Merck
Zoladex™ (goserelin acetate injectable implant)	Prostate cancer	AstraZeneca
Implanon® (etonogestrel implant)	Contraceptive	Organon
Gris-PEG (griseofulvin)	Anti-fungal	Pedinol Pharmacal Inc.
NuvaRing® (etonogestrel, ethinyl estradiol depot system)	Contraceptive	Merck
Norvir® (ritonavir)	Antiviral (HIV)	Abbott Laboratories
Kaletra® (ritonavir/lopinavir)	Antiviral (HIV)	Abbott Laboratories
Eucreas® (vildagliptin/metformin HCl)	Diabetes	Novartis
Zithromax® (Azythromycin enteric-coated multiparticulates)	Antibiotic	Pfizer
Orzurdex® (dexamethasone implantable device)	Macular edema	Allergan
Fenoglide™ (fenofibrate)	Dyslipidemia	LifeCycle Pharma
Anacetrapib (under development)	Atherosclerosis	Merck
Posaconazole (under development)	Antifungal	Merck

Source: With kind permission from Springer Science+Business Media: DiNunzio, J.C. et al., Melt extrusion, in: Williams III, R.O., Watts, A.B., and Miller, D.A. (Eds.), *Formulating Poorly Water Soluble Drugs*, Springer, New York, 2012, pp. 311–362.

11.9 CONCLUSION

It is well documented that HME is a solvent-free, robust, and cost-effective manufacturing process for the production of a wide variety of pharmaceutical dosage forms. HME technology is now viewed as an advanced technique that represents an efficient pathway for the manufacture of various delivery systems, including oral dosage forms, topical films, implants, and ophthalmic inserts. Depending on one's targeted release profile, the physical state of the extrudates may be modified by considering instrument engineering design (process) and/or by using different types of polymeric carriers (formulation). A drug can be present in a crystalline form to provide sustained release dosage forms or dissolved in a polymer to improve the dissolution profile of poorly water-soluble drugs.

Increasing work in the field of HME and published literature reveals the innovative aspects of this technology. These include, but are not limited to, in situ salt formation, quick dispersing systems with foam like structures, coextrusion to prepare extrudates in the form of laminar structures with multiple layers, nanoparticles released from molecular dispersions manufactured by melt extrusion, and twin-screw melt granulation, which can provide continuous manufacturing of granules yielding consistent product quality attributes.

In summary, HME as a continuous manufacturing process entails an interesting future within the pharmaceutical industry—a versatile technology as a conduit for novel drug-delivery systems.

REFERENCES

1. Breitenbach, J., *Eur. J. Pharm. Biopharm.*, 54, 107–117 (2002).
2. Maniruzzaman, M., Boatend, J. S. Snowden, M. J., and Douroumis, D., *ISRN Pharmaceutics*, 2012, pp. 1–9.
3. Keen, J. M., McGinity, J. W., and Williams III, R. O., *Int. J. Pharm.*, 450, 185–196 (2013).
4. Shah, S. and Reka, M. A. Melt extrusion in drug delivery: Three decades of progress. In: Repka, M. A., Langley, N., and DiNunzio, J., eds. *Melt Extrusion*. Springer, New York, 2013, pp. 3–46.
5. Prodduturi, S., Manek, R. V., Kolling, W. M., Stodghill, S. P., and Repka, M. A., *J. Pharm. Sci.*, 94(10), 2232–2245 (2005).

6. Mididoddi, P. K., Prodduturi, S., and Repka, M. A., *Drug Dev. Ind. Pharm.*, 32, 1059–1066 (2006).
7. Repka, M. A. and McGinity, J. W., *J. Contr. Release*, 70(3), 341–351 (2001).
8. Crowley, M. M., Zhang, F., Repka. M. A., Thumma, S., Upadhye, S. B., Battu S. K., McGinity J. W., and Martin, C., *Drug Dev. Ind. Pharm.*, 33, 909–926 (2007).
9. Repka, M. A., Battu, S. K., Upadhye, S. B., Thumma, S., Crowley, M. M., Zhang, F., Martin, C., and McGinity, J. W., *Drug Dev. Ind. Pharm.*, 33, 1043–1057 (2007).
10. Repka, M. A., Majumdar, S., Battu, S. K., Srirangam, R., and Upadhye, S. B., *Expert Opin. Drug Deliv.*, 5(12), 1357–1376 (2008).
11. Repka, M. A., Shah, S., Lu, J., Morott, J., Patwardhan, K., and Mohammed, N. N., *Expert Opin. Drug Deliv.*, 9(1), 105–125 (2012).
12. Shah, S., Maddineni, S., Lu, J., and Repka, M. A., *Int. J. Pharm.*, 453, 233–252 (2013).
13. Charlie, M., *Pharmaceut. Technol.*, 32(10), 76–86 (2008).
14. Kruder, G. A. Extrusion. In: H. F. Mark, N. M. Bikales, C. G. Overberger, and G. Menges. (eds.), *Encyclopedia of Polymer Science and Engineering*, Vol. 1, 2nd edn. John Wiley & Sons Inc., New York, 1985, pp. 571–631.
15. Choksi, R. and Zia, I. *J. Pharm. Res.* 3, 3–16 (2004).
16. Cheremisinoff, N. P., Multiple screw extruders. In: Cheremisinoff, N. P. (ed.), *Polymer Mixing and Extrusion Technology*. Marcel Dekker, Inc. New York, 1987.
17. Carneiro, O. S., Caldeira, G., and Covas, J. A., *J. Mater. Process. Tech.*, 92–93, 309–315 (1999).
18. Particle Sciences. *Technical Brief*, 3 (2011).
19. Dreiblatt, A. Process design. In: Ghebre-Sellassie, I. and Martin, C. (eds.), *Pharmaceutical Extrusion Technology*. Marcel Dekker, Inc., New York, 2003, pp. 153–169.
20. Leister, D., Geilen, T., and Geissler, T. Twin-screw extruders for pharmaceutical hot-melt extrusion: Technology, techniques and practices. In: Douroumis, D. (ed.), *Hot-Melt Extrusion: Pharmaceutical Applications*, 1st edn. Wiley, Chichester, U.K., 2012, pp. 23–42.
21. Forster, A., Hempenstall, J., and Rades, T., *J. Pharm. Pharmacol.*, 53(3), 303–315 (2001).
22. Ndindayino, F., Vervaet, C., VandenMooter, G., and Remon, J. P., *Int. J. Pharm.*, 246(1–2), 199–202 (2002).
23. Breitenbach, J. and Magerlein, M. Melt extruded molecular dispersions. In Ghebre-Sellassie, I. and Martin, C. (eds.), *Pharmaceutical Extrusion Technology*, Vol. 133. Marcel Dekker Inc., New York, 2003.
24. Repka, M. A. and McGinity, J. W., *Int. J. Pharm.*, 202, 63–70 (2000).
25. Park, J. B., Kang, C. Y., Kang, W. S., Choi, H. G., and Han, H. K., *Int. J. Pharm.*, 458, 245–253 (2013).
26. Li, D., Guo, G., Fan, R., Liang, J., Deng, X., Luo, F., and Qian, Z., *Int. J. Pharm.*, 441, 365–372 (2013).
27. Liu, J., Cao, F., Zhang, C., and Ping, Q., *Acta Pharm. Sinic. B*, 3(4), 263–272 (2013).
28. Kalivoda, A., Fischbach, M., and Kleinebudde, P., *Int. J. Pharm.*, 439, 145–156 (2012).
29. Maniruzzaman, M., Boateng, J. S., Bonnefille, M., Aranyos, A., Mitchell, J. C., and Douroumis, D., *Eur. J. Pharm. Biopharm.*, 80, 433–442 (2012).
30. Qi, S., Gryczke, A., Belton, P., and Craig, D. Q. M., *Int. J. Pharm.*, 354(1–2), 158–167 (2008).
31. Lauer, M. E., Siam, M., Tardio, J., Page, S., Kindt, J. H., and Grassmann, O., *Pharm. Res.*, 30(8), 2010–2022 (2013).
32. Forster, A., Hempenstall, J., Tucker, I., and Rades, T., *Drug Dev. Ind. Pharm.*, 27(6), 549–560 (2001).
33. Forster, A., Hempenstall, J., Tucker, I., and Rades, T., *Int. J. Pharm.*, 226, 147–161 (2001).
34. Nakamichi, K., Yasuura, H., Fukui, H., Oka, M., and Izumi, S., *Int. J. Pharm.*, 218(1–2), 103–112 (2001).
35. Crowley, M. M., Zhang, F., Koleng, J. J., and McGinity, J. W., *Biomaterials*, 23(21), 4241–4248 (2002).
36. Fukuda, M., Peppas, N. A., and McGinity, J. W. *J. Contr. Release*, 115(2), 121–129 (2006).
37. Henrist, D., Lefebvre, R. A., and Remon, J. P., *Int. J. Pharm.*, 187(2), 185–191 (1999).
38. Henrist, D. and Remon, J. P. *Int. J. Pharm.*, 188(1), 111–119 (1999).
39. Henrist, D. and Remon, J. P. *Int. J. Pharm.*, 189(1), 7–17 (1999).
40. Sprockel, O. L., Sen, M., Shivanand, P., and Prapaitrakul, W., *Int. J. Pharm.*, 155(2), 191–199 (1997).
41. Hulsmann, S., Backensfeld, T., Keitel, S., and Bodmeier, R., *Eur. J. Pharm. Biopharm.*, 49(3), 237–242 (2000).
42. Hulsmann, S., Backensfeld, T., and Bodmeier, R., *Pharm. Dev. Technol.*, 6(2), 223–229 (2001).
43. Follonier, N., Doelker, E., and Cole, E. T., *Drug Dev. Ind. Pharm.*, 20(8), 1323–1339 (1994).
44. Cuff, G. and Raouf, F., *Pharmaceut. Tech.*, 22, 96–106 (1998).
45. Aitken-Nichol, C., Zhang, F., and McGinity, J. W., *Pharm. Res.*, 13(5), 804–808 (1996).
46. Repka, M. A., Gutta, K., Prodduturi, S., Munjala, M., and Stodghill, S. P., *Eur. J. Pharm. Biopharm.*, 59, 189–196 (2005).

47. Repka, M. A., Gerding, T. G., Repka, S. L., and McGinity, J. W., *Drug Dev. Ind. Pharm.*, 25(5), 625–633 (1999).
48. Liu, J., Zhang, F., and McGinity, J. W., *Eur. J. Pharm. Biopharm.*, 52, 181–190 (2001).
49. Keleb, E. I., Vermeire, A., Vervaet, C., and Remon, J. P., *Eur. J. Pharm. Biopharm.*, 52(3), 359–368 (2001).
50. Ndindayino, F., Vervaet, C., Van Den Mooter, G., and Remon, J., *Int. J. Pharm.*, 246(1–2), 199–202 (2002).
51. Ndindayino, F., Vervaet, C., Van Den Mooter, G., and Remon, J. P., *Int. J. Pharm.*, 235(1–2), 159–168 (2002).
52. Perissutti, B., Newton, J. M., Podczeck, F., and Rubessa, F., *Eur. J. Pharm. Biopharm.*, 53(1), 125–132 (2002).
53. De Brabander, C., Vervaet, C., Fiermans, L., and Remon, J. P., *Int. J. Pharm.*, 199(2), 195–203 (2000).
54. De Brabander, C., Van Den Mooter, G., Vervaet, C., and Remon, J. P., *J. Pharm. Sci.*, 91(7), 1678–1685 (2002).
55. Kidokoro, M., Shah, N. H., Malick, A. W., Infeld, M. H., and McGinity, J. W., *Pharm. Dev. Technol.*, 6(2), 263–275 (2001).
56. Bhardwaj, R. and Blanchard, J., *J. Contr. Release*, 45(1), 49–55 (1997).
57. Lyons. J. G., Kennedy, D. D. M., Geever, J. E., O'sullivan, L. M., and Higginbotham, C. L. *Eur. J. Pharm. Biopharm.*, 64(1), 75–81 (2006).
58. Ndindayino, F., Henrist, D., Kiekens, F., Van Den Mooter, G., Vervaet, C., and Remon, J. P., *Int. J. Pharm.*, 235(1–2), 149–157 (2002).
59. Nakamichi, K., Yasuura, H., Fukui, H., Oka, M., and Izumi, S., *Int. J. Pharm.*, 218(1–2), 103–112 (2001).
60. Van Laarhoven, J. A. H., Kruft, M. A. B., and Vromans, H., *Int. J. Pharm.*, 232(1–2), 163–173 (2002).
61. Stepto, R. F. T., *Polym. Int.*, 43(2), 155–158 (1997).
62. Bruce, L. D., Shah, N. H., Malick, A. W., Infeld, M. H., and McGinity, J. W., *Eur. J. Pharm. Biopharm.*, 59(1), 85–97 (2005).
63. Miller, D. A., McConville, J. T., Yank, W., Williams III, R. O., and Mcginity, J. W., *J. Pharm. Sci.*, 96, 361–376 (2007).
64. DiNunzio, J. C., Brough, C., Hughey, J. R., Miller, D. A., Williams III, R. O., and McGinity, J. W., *Eur. J. Pharm. Biopharm.*, 74, 340–351 (2010).
65. Dong, Z., Chatterji, A., Sandhu, H., Choi, D. S., Chokshi, H., and Shah, N., *Int. J. Pharm.*, 355, 141–149 (2008).
66. Vithani, K., Maniruzzaman, M., Slipper, I. J., Mostafa, S., Miolane, C., Cuppok, Y., Marchaud, D., and Douroumis, D., *Colloids Surf. B*, 110, 403–410 (2013).
67. Fukuda, M., Peppas, N. A., and McGinity, J. W., *Int. J. Pharm.*, 310 (1–2), 90–100 (2006).
68. Chiou, W. L. and Riegelman, S., *J. Pharm. Sci.*, 60(9), 1281–1302 (1971).
69. Hughey, J. R., DiNunzio, J. C., Bennett, R. C., Brough, C., Miller, D. A., Ma, H., Williams III, R. O., and McGinity, J. W., *AAPS PharmSciTech*, 11(2), 760–774 (2010).
70. Huang, Y. and Dai, W. G., *Acta Pharm. Sinic. B*, 4(1), 18–25 (2014).
71. Dittgen, M., Fricke, S., Gerecke, H., and Osterwals, H., *Pharmazie*, 50, 225–226 (1995).
72. Nozawa, Y., Mizumoto, T., and Higashide, F., *Pharm. Acta Hely.*, 60, 175–177 (1985).
73. Nozawa, Y., Mizumoto, T., and Higashide, F., *Pharm. Ind.*, 8, 967–969 (1986).
74. Sekikawa, H., Arita, T., and Nakano, M., *Chem. Pharm. Bull.*, 26, 118–126 (1978).
75. Sekikawa, H., Fukyda, W., Takada, M., Ohtani, K., Arita, T., and Nakano, M., *Chem. Pharm. Bull.*, 31, 1350–1356 (1983).
76. Jung, J., Yoo, S., Lee, S., Kim, K., Yoon, D., and Lee, K., *Int. J. Pharm.*, 187, 209–218 (1999).
77. Djuris, J., Nikolakakis, I., Ibric, S., Djuric, Z., and Kachrimanis, K., *Eur. J. Pharm. Biopharm.*, 84(1), 228–237 (2013).
78. Kalivoda, A., Fischbach, M., and Kleinebudde, P., *Int. J. Pharm.*, 429, 58–68 (2012).
79. Mohammed, N. N., Majumdar, S., Singh, A., Deng, W., Murthy, N. S., Pinto, E., Tewari, D., Durig, T., and Repka, M. A., *AAPS PharmSciTech*, 13(4), 1158–1169 (2012).
80. Albers, J., Alles, R., Matthee, K., Knop, K., Nahrup, J. S., and Kleinebudde, P., *Eur. J. Pharm. Biopharm.*, 71, 387–394 (2009).
81. Young, C. R., Koleng, J. J., and McGinity, J. W., *Int. J. Pharm.*, 242 (1–2), 87–92 (2002).
82. Gryczke, A., Schminke, S., Maniruzzaman, M., Beck, J., and Douroumis, D., *Colloids Surf. B*, 86 (2), 275–284 (2011).
83. Low, A. Q. J., Parmentiera, J., Khonga, Y. M., Chai, C. C. E., Tun, T. Y., Berania, J. E., Liu, X., Gokhale, R., and Chan, S. Y., *Int. J. Pharm.*, 455, 138–147 (2013).

84. Rothen-Weinhold, A., Oudry, N., Schwach-Abdellaoui, K., Frutiger-Hughes, S., Hughes, G. J., Jeannerat, D., Burger, U., Besseghir, K., and Gurny, R., *Eur. J. Pharm. Biopharm.*, 49, 253–257 (2000).
85. Lembo, D. and Cavalli, R., *Antivir. Chem. Chemother.*, 21, 53–70 (2010).
86. Dubey, R., Shami, T. C., and Bhasker Rao, K. U., *Defence Sci. J.*, 59(1), 82–95 (2009).
87. Kita, K. and Dittrich, C. *Expert Opin. Drug Deliv.*, 8(3), 329–42 (2011).
88. Loreti, G., Maroni, A., Curto, M. D. D., Melocchi, A., Gazzaniga, A., and Zema, L., *Eur. J. Pharm. Sci.*, 52, 77–85 (2014).
89. Schulze, S. and Winter, G., *J. Control Release*, 134, 177–185 (2009).
90. Maniruzzaman, M., Bonnefille, M., Aranyos, A., Snowden, M. J., and Douroumis, D., *J. Pharm. Pharmacol.*, 66(2), 323–337 (2013).
91. Mehuys, E., Vervaet, C., and Remon, J. P., *J. Control Release*, 94, 273–280 (2004).
92. Mehuys, E., Remon, J. P., and Vervaet, C., *Eur. J. Pharm. Biopharm.*, 24, 207–212 (2005).
93. Bruce, L. D., Shah, N. H., Waseem, M. A., Infeld, M. H. and McGinity, J. W., *Eur. J. Pharm. Biopharm.*, 59(1), 85–97 (2005).
94. Wilson, M., Williams, M. A., Jones, D. S., and Andrews, G. P., *Ther. Deliv.*, 3(6), 787–797 (2012).
95. DiNunzio, J. C., Martin, Z. F. C., and Mcginity, J. W. Melt extrusion. In: Williams III, R. O., Watts, A. B., and Miller, D. A. (eds.), *Formulating Poorly Water Soluble Drugs*. Springer, New York, 2012, pp. 311–362.
96. Prodduturi, S., Urman, K. L., Otaigbe, J. U., and Repka, M. A., *AAPS PharmSciTech*, 8(2), E152–E161 (2007).

84. Rotheα-Witthohl, A., Ourjeith, Schwarab, Abolhosny, K. Flinger, Hug, Bess, Hughes, O., Insinger, D., Burger, U., Beresschi, K., and Gnarig, K., Eur. J. Pharm. Biopharm., 58, 253–257 (2004).
85. Lamba, D. and Cavalli, R., Asthma Genet. J. Genet. Syst., 21, 53–70 (2010).
86. Dobe, Z., Sharif, T. G., and Shackerlag, L. J., Pharma. Sci. J. Sci.(1), 82–97 (2009).
87. Kiba, R., Fard-ruth, C., Expert Opin. Drug Deliv., 8(2), 125–141 (2011).
88. Longo, O., Marano, A., Chen, M., D., Meriotti, A., Chiaroffia, A., and Zampa, L., Ann. N. Y. Acad. Sci., 53(2), 85–OOH.
89. Scholze, S. and Winter, G., Chem. Rev., 101, 129–165 (2440).
90. Manivasagam, M., Brasiliar, M. A., Reim, A., Snarron, M. J., and Petrocomhe, D., J. Pharm. Pharmacol., 60(2), 325–337 (2012).
91. M. Jaya, B., Vernon, G., and Berry, V. R., J. Pharm. Res., 1, 11, 275–290 (2008).
92. Mehive, B., Ruaaso, J. D., and Vaccari, C., Eur. J. Pharm. Biopharm. 73, 309–312 (2005).
93. Snoeck, D., Szok, A. I., Waesen, M. A., Dridi, D. H., and Vierbol, W., Ad. Drug Deliver. Rev., 57, 98–112 (2005).
94. Wilson, M., Whitemill, L., Jones, D. S., et al., J. Coll. Pharmacol., 776–797 (2005).
95. Dicostanzio, J. C., Harris, Z. S., J., and Maginya, J. V., Manufacturing Technology, HDP, Wiley, A. B., and Miller, D., eds, Informa Healthcare Press, Huggel-Springer, New York, 2012, p. 311–362.
96. Bhaduri, S., Truman, K. L., D'alchesio, F., Andolighan, M. A., AAPS Pharm. Tech., 8(2), E102–E107 (2007).

12 Microencapsulation with Coacervation

Michael Yan

CONTENTS

12.1 Introduction 235
12.2 Simple Coacervation..236
12.3 Complex Coacervation...237
    12.3.1 Principles 237
    12.3.2 Encapsulation Process 238
    12.3.3 Wall Materials 239
    12.3.4 Gelatin Replacement..240
12.4 Conclusions and Future Outlook 242
References..243

12.1 INTRODUCTION

Polymer coacervation is one of the earliest microencapsulation techniques. The first commercial application of encapsulation with coacervation was to the development of carbonless copy paper by the National Cash Register Company, which resulted from patents by Barrett K. Green and Lowell Schleicher in the late 1950s.[1,2] Since then, coacervation has been used for various applications.

Coacervation, as defined by International Union of Pure and Applied Chemistry (IUPAC), is the separation into two liquid phases in colloidal systems. The phase more concentrated in colloid component is the coacervate, and the other phase is the equilibrium solution.[3] It is to be distinguished from precipitation which is observed in the form of coagulum or flocs in colloidally unstable systems. The term coacervation was coined in 1929 by Bungenberg de Jong and Kruyt to describe a process in which aqueous colloid solutions separate, upon alteration of the thermodynamic condition of state, into two immiscible liquid layers, one rich in colloid, i.e., the *coacervate*, and the other containing little colloid.[4]

Polymer coacervation can occur in either aqueous or organic liquids. Coacervation in aqueous liquids and the related processes are mainly used to encapsulate water-immiscible liquids or water-insoluble solid particles. On the other hand, coacervation in organic liquids, or sometimes called phase separation in organic liquids, is used to encapsulate core materials that are not miscible or soluble in the organic liquids. It may be induced by the addition of a nonsolvent to the polymer solution or by the addition of an incompatible polymer based on polymer–polymer incompatibility. This chapter will only discuss the coacervation in aqueous liquids.

Based on phase-separation mechanisms, coacervation systems can be classified into two general types: simple coacervation and complex coacervation. When only one polymer is involved, the process is referred to as simple coacervation, and when two or more polymers with opposite charges are involved, it is referred to as complex coacervation. In both cases, the coacervation takes place just before precipitation from solution. This separated phase in the form of amorphous, liquid droplets constituted the coacervate which is the polymer-rich solution. Deposition of this coacervate around the individual insoluble oil droplets or solid particles dispersed in the equilibrium liquid forms the embryonic capsules, and subsequent gelling of the deposited coacervate results in microcapsules.

The unique property of coacervation system is the fact that the solvent components of the two phases (polymer-rich coacervate and the polymer-poor aqueous phase) are the same, that is, water. This is the basic distinguish of a coacervate as compared with the two-phase system involving two immiscible liquid.

Hydrophilic polymers have been used for coacervation in an aqueous phase. They are large molecules which are dispersible or soluble in aqueous solution. They can be of natural origin or can be synthesized. Examples of the natural polymers include gelatin, alginate, gellan, carrageenans, gum arabic, albumin, casein, whey protein isolate, soy protein isolate, pea protein isolate, chitosan, pectin and starch, etc. Some of the materials are not obtained directly from nature, but have undergone some modifications. One of the examples is carboxyl methyl cellulose which is a natural material gone through chemical modifications. Other polymer materials are synthesized outright, such as polyphosphate, polyacrylic acid, polyacrylamide, etc.

12.2 SIMPLE COACERVATION

This process involves only one polymer (e.g., gelatin, polyvinyl alcohol, pectin, agar, or carboxylmethylcellulose, etc.) in its aqueous solution, and the polymer phase separation can be induced by adding to the solution a water-miscible nonsolvent for the polymer. The term nonsolvent is used here for all poor solvents for the polymer to be coacervated. Examples of the water-miscible nonsolvent include ethanol, acetone, dioxane, isopropanol, or propanol, etc., depending on the polymers to be coacervated.[5–9]

Simple coacervation can also be induced by adding an inorganic salt for polymer desolvation, and the phenomenon is sometimes called salting-out.[2,9–11] The ability of inorganic cations to induce coacervation is expressed by the Hofmeister or lyotropic series, which arranges ions in the order of decreasing salting out capacity for the polymers:

$$Mg^{+2} > Ca^{+2} > Si^{+2} > Ba^{+2} > Li^+ > Na^+ > K^+ > NH_4^+ > Rb^+ > Cs^+$$

Several anions of Hofmeister series in decreasing order of inducing coacervation are citrate, tartrate, sulfate, acetate, chloride, nitrate, bromide, and iodide.

During the early development of encapsulation technology, two parallel approaches were pursued. These were simple coacervation by addition of salt and complex coacervation. However, the complex coacervation system was more readily converted to a commercial process, and was more fully developed in preference to the simple coacervation. As encapsulation became more diversified, the disadvantages of the simple coacervation were minimized because of its insensitivity to water-soluble additives, as well as to a wide pH range in the system. This is of importance in the encapsulation of pH-sensitive materials. The major disadvantages of coacervation with salt are the critical entry into the coacervation zone and the necessity of almost complete salt removal from the encapsulated product.

An example of gelatin coacervation with the addition of sodium sulfate is shown in the phase diagram of Figure 12.1. Starting with a 10% gelatin solution (point X) and gradually diluting with 20% Na_2SO4 solution, the concentration changes would follow the line XY. At point P, phase separation occurs, although the phase boundary is not sharp and the higher molecular weight gelatin separates first. From point P to point A, coacervate phase separates out as tiny spheres which are visible under microscope. As more salt is added, and point B is reached, solid particle will appear; and with further salt addition, gelatin precipitation becomes complete.

The phase diagram shown in Figure 12.1 is temperature dependent. With higher temperature, the phase boundary shifts toward higher salt concentration, and the reverse as temperature is decreased. Therefore the best method of inducing phase separation is to add sufficient salt to bring the system to the point P in Figure 12.1, and to complete the coacervation process by allowing the temperature to decrease slowly to room temperature.

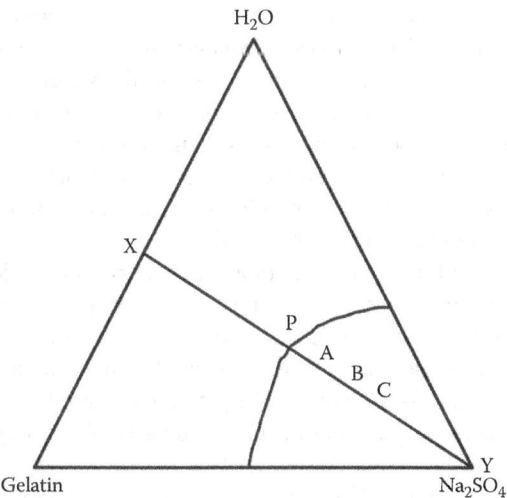

FIGURE 12.1 Schematic phase diagram of simple coacervation with salt addition.

12.3 COMPLEX COACERVATION

12.3.1 Principles

Complex coacervation is based on the ability of cationic and anionic water-soluble polymers to interact in water to form a polymer-rich phase called complex coacervate. Gelatin and gum arabic are typical polymers for complex coacervation. In this system, gelatin is an amphoteric polymer, carrying both carboxyl and amino groups. The IEP of Type A gelatin (acid processed) is typically from pH 7 to 9. When pH of the gelatin solution is below the IEP, it becomes positively charged. Gum arabic on the other hand is always negatively charged as illustrated in Figure 12.2. When pH of the gelatin-gum arabic solution is in the shaded area of Figure 12.2, it is possible for gelatin and gum arabic to form a coacervate under proper temperature, ionic strength, concentration and ratio

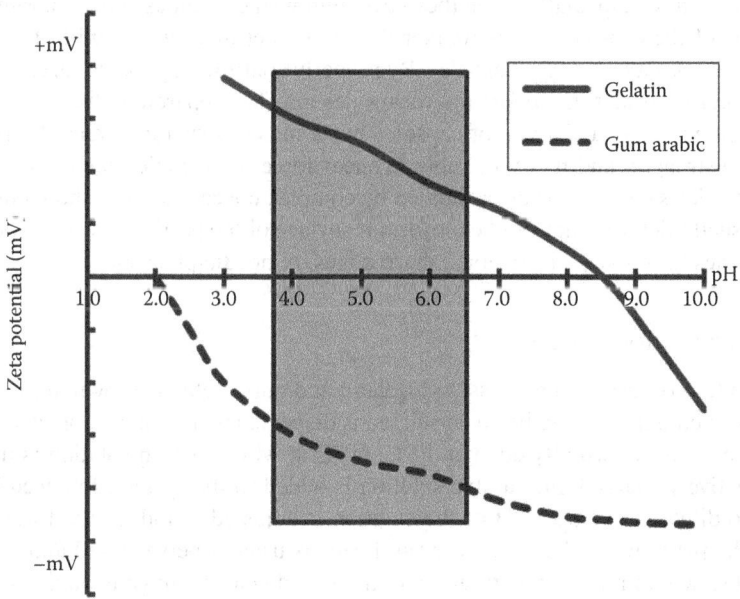

FIGURE 12.2 Zeta potential chart.

of the two polymers. During coacervation process, gelatin and gum arabic interact with each other to form a polymer-rich phase called a *complex coacervate* which is in equilibrium with a dilute solution called the *supernatant*. In this two-phase system, the supernatant acts like a continuous phase whereas the complex coacervate acts as a dispersed phase. When a water-insoluble core material to be encapsulated is dispersed into the system, and if the coacervate wets the dispersed core material, the coacervate will be deposited onto the core material surface, accompanied by spreading and fusion into a membrane to form microcapsules. The microcapsules can be solidified or hardened, and isolated from water if it is required.

The interaction between polymers during complex coacervation is electrostatic in nature. The charge density that each polymer carries is the most important factor affecting the complex coacervation. At low charge density, complex coacervation is suppressed, and at high charge density, precipitation and/or gelation of the polymers may occur. The maximum coacervation yield is usually obtained for a ratio of the two polymers at a pH value where they carry equal and opposite charges. Therefore, any factor that can alter the charge density of the polymers can affect the complex coacervation.

In addition to solution pH which affects the charge of the polymers, the formation of complex coacervate is also affected by many other variables such as temperature, ions present in the system, ionic strength, polymer chain structure and molecular weight, charge density, concentration and ratio of the polymers involved. Complex coacervation process is favored by dilution and suppressed by ions (salts) present in the system. For a given system, coacervation only occurs in a limited pH range. A number of researchers have worked on optimizing coacervate formation.[12–16] Although this information provides very valuable guide for developing microencapsulation process with complex coacervation, it is important to recognize that optimized conditions for the coacervate formation in a specific system may not be directly transferred to conditions for preparing microcapsules. The process conditions may have to be optimized to accommodate core materials with different properties, such as solubility, polarity, functional groups, interfacial tension, wettability, etc. These optimized conditions may be shifted away from those for coacervate formation.

Microcapsules prepared with complex coacervation may have single core or multicore morphologies as shown in Figure 12.3. The single core microcapsules usually have an oval shape with non-uniform wall thickness, especially when they have high wall content as shown in Figure 12.3a. The thinner portion of the capsule wall is weaker than the other area, which will affect the strength, stability or barrier properties of the capsules. By properly controlling process variables to agglomerate tiny single core capsules, multicore microcapsules may be prepared with complex coacervation as shown in Figure 12.3b. This type of capsules has a matrix structure, generally provides stronger mechanical strength, and more tolerable to shear force than single core capsules. Non water-soluble solid particles can also be encapsulated by complex coacervation as shown in Figure 12.3c. However, the shell thickness may not be uniform if surface of the particles is not smooth, especially when the particles have sharp protrusions, sharp edges, or needle/plate shapes.

12.3.2 ENCAPSULATION PROCESS

A typical complex coacervation process with gelatin and gum arabic is shown in Figure 12.4. First, a water-insoluble core material to be encapsulated is dispersed to a desired droplet size into a warm gelatin solution. This is normally done at 40°C–60°C at which gelatin solution is melted and liquid. Then negatively charged gum arabic solution is added to the system, followed by addition of warm water to dilute the system. pH of the solution is adjusted to induce the formation of liquid coacervate. The pH at which liquid coacervate forms is usually between 4.0 and 5.0. The system is then cooled slowly to room temperature, and the liquid coacervate generated is coated onto the core material surface to form embryo capsules. The system is then further cooled to 10°C, thereby gelling the liquid coacervate capsule shell.

Microencapsulation with Coacervation

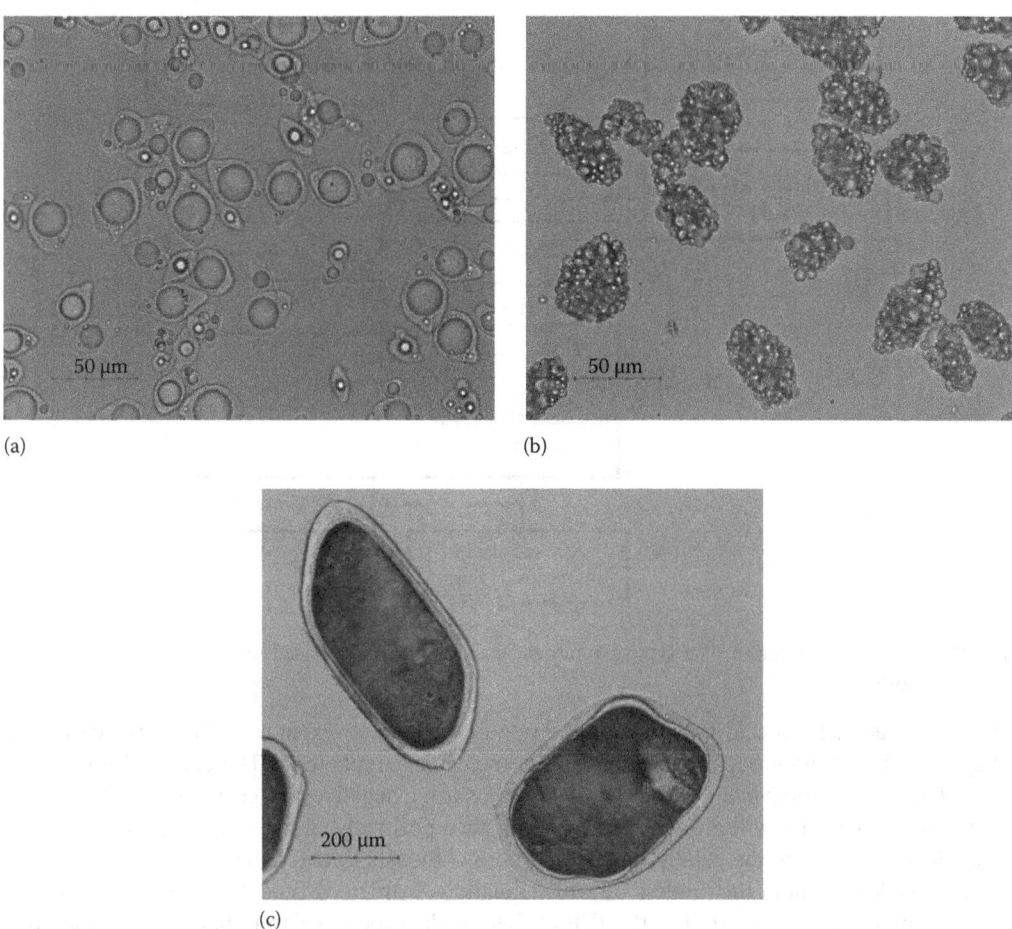

FIGURE 12.3 (a) Single core, (b) multicore, and (c) solid core.

At this stage, the liquid coacervate capsule shell is highly water swollen. If heated, it may swell, dissolve and eventually disappear thereby releasing the core materials. This is undesirable in their applications, or in the post process to isolate or to dry the capsules. In order to make the capsule shell not thermally reversible, and also to increase shell strength, the capsules are normally cross-linked with chemicals.

The chemicals that have been used to crosslink gelatin are formaldehyde, glutaraldehyde, tannins, genipin, phenolics and alum.[17–21] Although glutaraldehyde is effective in cross-linking gelatin, and the capsules treated with it have been approved for specific flavor use, better alternatives have been proposed. Other alternatives are tannins, plant phenolics and transglutaminase (TG). TG is a commercially available enzyme that cross-links between protein molecules, and has been used successfully in commercial scale microencapsulation with complex coacervation.[22]

12.3.3 Wall Materials

There are many complex coacervation systems suitable for microencapsulation. Most of them use gelatin as the cationic polymer to interact with a wide range of anionic polymers to form coacervates.

Gelatin is derived from pork or bovines skin, or from fish skin and scales. The gelatin molecule is made up of amino acids joined together by peptide linkages in a long molecular chain. The linkages involving the carboxylic groups and the α-amino groups form the backbone of the molecule.

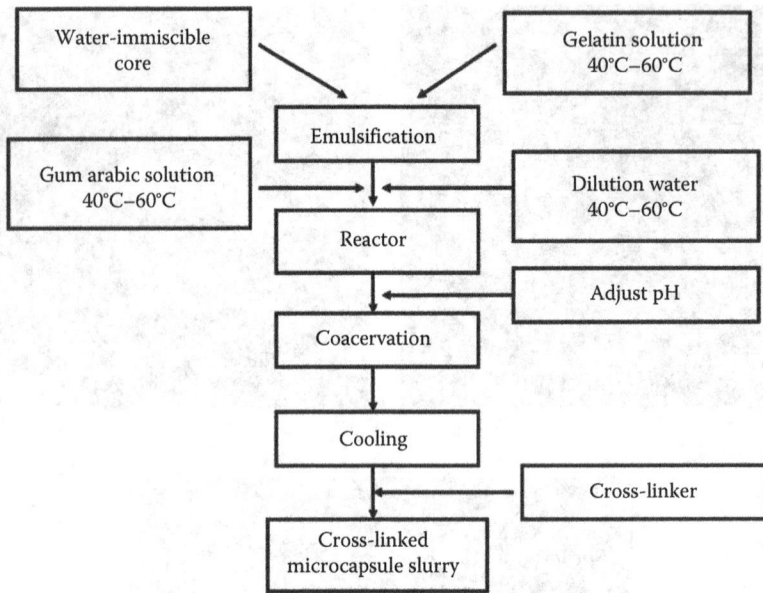

FIGURE 12.4 Flow diagram of a typical microencapsulation process based on complex coacervation of gelatin and gum arabic.

The rest of each amino acid is known as the side chain. The co-existence of both the carboxylic groups and the amino groups makes the gelatin molecules amphoteric: The carboxyl groups will be ionized to the carboxylate ion and thereby give gelatin a negative charge. On the other hand, an amino group on gelatin will attract a proton and form a positively charged NH_3^+-group. The pH at which the net charge on the gelatin molecule is zero is the isoelectric point (IEP). The IEP values of gelatin is dependent of the process by which gelatin is extracted. Type A gelatin obtained by an acid extraction process usually has the IEP of 7–9, while Type B gelatin obtained by an alkaline extraction process usually has the IEP of 4–5.

In addition to gelatin–gum arabic complex coacervation system, many other systems have been studied and reported, and Type A gelatin is most often used as the cationic polymer. Table 12.1 lists some of the examples of gelatin-based complex coacervation systems. It is interesting to note that by using Type A gelatin as the cationic polymer and Type B gelatin as the anionic polymer, complex coacervation process can yield coacervates composed of only gelatin.[25] As mentioned before, the IEP of Type A gelatin is typically between 7 and 9, while the IEP of Type B gelatin is typically between 4 and 5. At an appropriate pH between these two IEPs, Type A gelatin is positively charged, while Type B gelatin is negatively charged, thereby forming complex coacervates.

Another example of using type B gelatin as a negatively charged polymer is also listed in Table 12.1.[33] In the coacervate system involving chitosan and type B gelatin, chitosan is always positively charged in solution while the charge of gelatin molecules depends on pH. Coacervation between chitosan and gelatin should be restricted to a narrow pH range where both molecules carry opposite charges. The investigated pH range was 4.5–6.5, which was above the IEP of the type B gelatin and below the pH of precipitation for chitosan. The reported optimum pH for the maximum coacervate formation was between 5.25 and 5.50. The coacervate yield dropped at pH value below or above this pH range.

12.3.4 Gelatin Replacement

Although gelatin has been successfully used in complex coacervation for many years and in various applications, efforts have been made in recent years to find its replacement to meet the growing demand for polymers suitable for use in complex coacervation while still meet the requirement for

TABLE 12.1
Examples of Gelatin-Based Complex Coacervation Systems

Polycation	Polyanion	Core	Coacervation pH	Reference
Gelatin, Type A	Polyphosphate	Fish oil	4.5	[22]
Gelatin, Type A	Gum arabic	Paraffin oil	4.5	[23]
Gelatin, Type A	Gum arabic	Garlic oils	4.5	[24]
Gelatin, Type B	Gum arabic	Garlic oils	3.5	[24]
Gelatin, Type A	Gelatin, Type B	Naproxen	5.4	[25]
Gelatin, Type A	Pectin	Sulfamerazine	3.2–4.6	[26]
Gelatin, Type A	Gantraz	Nitrofurantoin		[27]
Gelatin, Type A	CMC	Triclosan	4.4	[28]
Gelatin, Type A	Sodium alginate	Olive oil	3.5–3.8	[29]
Gelatin, Type A	Carrageenan	Sunflower oil, paraffin	3.5–5.0	[30]
Gelatin, Type A	Gellan gum	Sun flower oil, paraffin	3.5–5.0	[31]
Fish gelatin, Type A	Gum arabic	Sunflower oil	2.7–4.5	[32]
Chitosan	Gelatin, Type B	Clofibrate, piroxicam, sulfamethoxazole	5.25–5.50	[33]

specific applications, such as biocompatibility, biodegradability, nontoxicity, physiological inertness, antibacterial properties, heavy metal ions chelation, and regulatory requirements, such as GRAS, kosher and halal for food and dietary supplement applications.[34,35]

Gelatin has traditionally been isolated from pork skin or beef hide and bone by acid or alkaline extraction process. The appearance of Bovine Spongiform Encephalopathy (BSE, mad cow disease) has stimulated efforts to develop alternate gelatin sources. Although gelatin suppliers have worked hard to demonstrate that prions associated with BSE have been destroyed during the extraction process, concerns about BSE and other unknown potential for future emergence of diseases harmful to human still remain. Kosher and halal dietary restrictions are also reasons to avoid gelatin derived from animals. Fish Gelatin is a promising alternative.[36] It meets the kosher and halal requirements, and has been successfully adapted to the existing complex coacervation protocols,[33,37] however, cost and commercial availability to meet the growing demand remain issues that are to be resolved. The potential health threats associated with gelatin from animals and a growing demand in vegetarian diet have prompted the development of polymers derived from other sources, such as plants and dairy products.

There have been excellent reviews published in recent years as related to gelatin replacement for microencapsulation.[35,38,39] Table 12.2 lists some of the examples of non-gelatin based complex coacervation systems. Proteins derived from plants, such as pea protein,[40–42] wheat protein,[40,41] soy protein,[42–46] whey protein from dairy products,[42,47,48] chitosan from marine crustaceans,[49,50] and collagen[51] have been used to replace gelatin in complex coacervation.

It should be noted that making a successful change of capsule wall material needs considerable effort. Firstly, the polymers should be able to form a coacervate within a certain convenient range of operational variables, such as pH, temperature and concentration. Secondly, the rheology of the formed coacervate is of critical importance during encapsulation process. In order for a coacervate to deposit on a core material to form a microcapsule, the coacervate must have certain viscosity and can attach to as well as spread around the core material. If a coacervate material is too viscous, it may not be able to coat on the core materials to be encapsulated. Furthermore, wall materials derived from natural products may have lot to lot variation in properties that might be difficult to quantify, but have a significant impact on the formation of complex coacervate and also on its deposition onto the core materials, thereby affects the reproducibility of the encapsulation process. Zhang et al. characterized four types of whey protein isolate and five types of gum arabic (three types were from Senegal species, and two were from Seyal species) for the encapsulation of

TABLE 12.2
Examples of Non Gelatin-Based Complex Coacervation Systems

Polycation	Polyanion	Core	Coacervation pH	References
Pea globulin	Gum arabic	Triglyceride	3.5	[40,41]
Alpha-gliadin	Gum arabic	Triglyceride	3.0	[40,41]
Pea protein isolate	Caseinate	Fish oil	5.0	[42]
Soy protein isolate	Caseinate	Fish oil	5.0	[42]
Soy protein isolate	Gellan gum	Fish oil	5.0	[42]
Soy protein isolate	Pectin	Casein hydrolysate	4.4	[43]
Soy protein isolate	Pectin	Propolis	4.0	[44]
Soy protein isolate	Gum arabic	Orange oil	4.0	[45]
Soy glycinin	Glycinin-sodium dodecyl sulfate	Hexadecane	4.0	[46]
Whey protein isolate	Gum arabic	Sunflower oil	3.0–4.5	[47]
Whey protein isolate	Gum arabic	Fish oil	3.9	[42]
Whey protein isolate	Na caseinate	Fish oil	4.7	[42]
Whey protein isolate	Pectin	Fish oil	3.1	[42]
Whey protein isolate	Gum arabic	Fish oil	Various	[48]
Chitosan	Alginate	Cells		[49]
Chitosan	Silk fibroin	n-Eicosane	5.2	[50]
Collagen	Chondroitin sulfate	Albumin	2–6	[51]

omega-3 lipids.[48] Their study demonstrated that material properties of whey protein isolates and gum arabic affect the complex coacervation process significantly. For whey proteins, the coacervation capability can be correlated with their degree of denaturation and the content of divalent metal ions. For gum arabic, both molecular weight/species and the content of divalent metal ions affect their coacervation capability. Senegal gums produced a low level of soluble solid and a high level of coacervate content in comparison with Seyal gums.

One of the advantages of using plant- or milk-based proteins as the capsule wall materials is that the proteins can be denatured by heating to 80°C–95°C prior to spray drying.[42] In contrast, the traditional gelatin-based complex coacervate capsule wall needs to be cross-linked with some type of chemicals, such as glutaraldehyde, before the capsules to be isolated by a drying step with heat. The chemicals used to cross-link gelatin cannot meet the regulatory requirements for some applications, such as food ingredients, dietary supplements, and pharmaceuticals.

12.4 CONCLUSIONS AND FUTURE OUTLOOK

Since the first commercial application to carbonless copy paper in the 1950s, microencapsulation with coacervation technology has been successfully applied to many other areas, such as food, pharmaceuticals, cosmetics, biotechnology, and agrochemicals. The capsules provide such functions as controlled release, taste masking, improved heat and oxidative stability, reduced volatility/flammability/toxicity, separation of reactive incompatibles, improved shelf-life, conversion of liquids to solids, and improved flowability as well as material handling. Among various coacervation processes, complex coacervation is most prevalent.

Microencapsulation with complex coacervation has many advantages. It can produce capsules with a payload as high as 95%. The wall of the microcapsules is non–water soluble when it is either cross-linked with chemicals or treated with heat. This is a significant advantage over the microcapsules prepared with other technologies, such as spray drying or fluid bed coating by which the microcapsule wall produced is often water soluble. The microcapsules produced have excellent oxidation stability at low relative humidity, and core release can be initiated by different mechanisms, such as diffusion, pH change, temperature, osmotic pressure, dissolution, wall degradation, and shear. The commercial encapsulation process with complex coacervation is well established, and does not require any elaborate manufacturing equipment.

Although many successful complex coacervation systems have been developed, they have a number of limitations. It is difficult to encapsulate a core material containing water-soluble or polar components. For a given system, coacervation occurs at a narrow range of pH, which may lead to difficulties in process operation and control. As coacervation only occurs at low concentration of the capsule wall-forming materials, the solids content from these processes is usually low, which significantly affect the process economics. Most of the single core capsules produced from complex coacervation process are not spherical but have an oval-shaped morphology, resulting in non-uniform capsule wall thickness as shown in Figure 12.3. The thinner portion of the capsule wall is weaker than the other area, which will affect the strength, stability and barrier properties of the capsules.

In spite of these limitations, the development of microcapsules by complex coacervation remains a viable area of study today from both a fundamental understanding and a commercial application point of view. Polymers derived from plant or milk that are adaptable to the complex coacervation encapsulation protocols have been used to replace gelatin isolated from animals. Interest in capsules produced by complex coacervation for food-related applications remains high because a variety of GRAS polymers can be used to produce commercially viable capsules accepted by the food industry. New technically and commercially viable polymers are being developed to meet the increasing demand from various applications and to fulfill the regulatory requirements.

REFERENCES

1. Green, B. K. and L. Schleicher, U.S. Patents 2,730,456 (1956) and 2,730,457 (1957).
2. Green, B. K., U.S. Patents 2,730,458 (1957).
3. IUPAC, *Compendium of Chemical Terminology*, 2nd edn., Compiled by A. D. McNaught and A. Wilkinson. Blackwell Scientific Publications, Oxford, U.K., 1997.
4. Bungenberg de Jong, H. G., Crystallisation-coacervation-flocculation. In *Colloid Science II*, Kruyt, H.R., Ed., Elsevier, Amsterdam, the Netherlands, 1949, Vol. VIII, pp. 232–258.
5. Mohanty, B. and H. B. Bohidar, Systematic of alcohol-induced simple coacervation in aqueous gelatin solutions. *Biomacromolecules*, 4(4) (2003): 1080–108.
6. Okada, J., A. Kusai, and S. Ueda, Factors affecting microencapsulability in simple gelatin coacervation method. *J. Microencapsul.*, 2(3) (1985): 163–173.
7. Okada, J., A. Kusai, and S. Ueda, Core treatment for improving microencapsulability in simple gelatin coacervation method. *J. Microencapsul.*, 2(3) (1985): 175–182.
8. Khaledi, M. G., S. I. Jenkins, and S. Liang, Perfluorinated alcohols and acids induce coacervation in aqueous solutions of amphiphiles. *Langmuir*, 29(8) (2013): 2458–2464.
9. Khalil, S. A., J. R. Nixon, and J. E. Carless, Role of pH in the coacervation of the systems: Gelatin-water-ethanol and gelatin-water-sodium sulphate. *J. Pharm. Pharmacol.*, 20 (1968): 215–225.
10. Phares, R. E. and G. J. Sperandio, Coating pharmaceuticals by coacervation. *J. Pharm. Sci.*, 53(5) (1964): 515–518.
11. Siddiqui, O. and H. Taylor, Physical factors affecting microencapsulation by simple coacervation of gelatin. *J. Pharm. Pharmacol.*, 35(2) (1983): 70–73.
12. Thies, C., Microencapsulation of flavors by complex coacervation. In *Encapsulation and Controlled Release Technologies in Food Systems*, Lakkis, J. M., Ed., Blackwell Publishing, Ames, IA, 2007, pp. 149–170.
13. Burgess, D. J., Practical analysis of complex coacervate systems. *J. Colloid Interf. Sci.*, 140 (1990): 227–238.
14. Schmitt, C. and S. L. Turgeon, Protein/polysaccharide complexes and coacervates in food systems. *Adv. Colloid Interf. Sci.*, 167(1–2) (2011): 63–70.
15. Schmitt, C., C. Sanchez, S. Desobry-Banon, and J. Hardy, Structure and technofunctional properties of protein-polysaccharide complexes: A review. *Crit. Rev. Food Sci. Nutr.*, 38(8) (1998): 689–753.
16. Veis, A. A review of the early development of the thermodynamics of the complex coacervation phase separation. *Adv. Colloid Interf. Sci.*, 167(1–2) (2011): 2–11.

17. Thies, C., The reaction of gelatin-gum arabic coacervate gels with glutaraldehyde. *J. Colloid Interf. Sci.*, 44(1) (1973): 133–141.
18. Xing, F., G. Cheng, B. Yang, and L. Ma, Microencapsulation of capsaicin by the complex coacervation of gelatin, acacia and tannins. *J. Applied Polym. Sci.*, 91(4) (2004): 2669–2675.
19. Hussian, M. R. and T. K. Maji, Preparation of genipin cross-linked chitosan-gelatin microcapsules for encapsulation of *Zanthoxylum limonella* oil (ZLO) using salting-out method. *J. Microencapsul.*, 25(6) (2008): 414–420.
20. Soper, J. C. and M. T. Thomas, Enzymatically protein-encapsulating oil particles by complex coacervation. US Patent 6,039,901 (March 21, 2000).
21. Straussa, G. and S. M. Gibsonb, Plant phenolics as cross-linkers of gelatin gels and gelatin-based coacervates for use as food ingredients. *Food Hydrocolloids*, 18(1) (2004): 81–89.
22. Yan, N., Encapsulated agglomeration of microcapsules and method for the preparation thereof. US Patent 6,974,592 (December 13, 2005).
23. Bhattatharyya, A. and J.-F. Argillier, Microencapsulation by complex coacervation: Effect of cationic surfactants. *J. Surf. Sci. Technol.*, 21(3–4) (2005): 161–168.
24. Siow, L.-F. and C.-S. Ong, Effect of pH on garlic oil encapsulation by complex coacervation. *J. Food Process. Technol.*, 4(1) (2013): 1–5.
25. Burgess, D. J. and J. E. Carless, Manufacture of gelatin/gelatin coacervate microcapsules. *Intern. J. Pharma.*, 27 (1985): 61–70.
26. McMullen, J. N., D. W. Newton, and C. H. Becker, Pectin-gelatin complex coacervates II: Effect of microencapsulated sulfamerazine on size, morphology, recovery, and extraction of water-dispersible microglobules. *J. Pharm. Sci.*, 73(12) (1984): 1799–1803.
27. Mortada, S. A. M., A. M. El Egaky, A. M. Motawi, and K. A. El Khodery, Preparation of microcapsules from complex coacervation of gantrez-gelatin. II. In vitro dissolution of nitrofurantoin microcapsules. *J. Microencapsul.*, 4(1) (1987): 23–37.
28. Kim, J.-C., M.-E. Song, E.-J. Lee, S.-K. Park, M.-J. Rang, and H.-J. Ahn, Preparation and characterization of triclosan-containing microcapsules by complex coacervation. *J. Dispersion Sci. Technol.*, 22(6) (2001): 591–596.
29. Devi, N., D. Hazarika, C. Deka, and D. K. Kakati, Study of complex coacervation of gelatin A and sodium alginate for microencapsulation of olive oil. *J. Macromol. Sci. Part A Pure Appl. Chem.*, 49(11) (2012): 936–945.
30. Devi, N. and T. K. Maji, Genipin crosslinked microcapsules of gelatin A and κ-carrageenan polyelectrolyte complex for encapsulation of Neem (*Azadirachta indica* A.Juss.) seed oil. *Polym. Bull.*, 65 (2010): 347–362.
31. Chilvers, G. R. and V. J. Morris, Coacervation of gelatin-gellan gum mixtures and their use in microencapsulation. *Carbohydr. Polym.*, 7(2) (1987): 111–120.
32. Piacentini, E., L. Giorno, M. M. Dragosavac, G. T. Vlasdisavljevic, and R. G. Holdich, Microencapsulation of oil droplets using cold water fish gelatine/gum arabic complex coacervation by membrane emulsification. *Food Res. Int.*, 53(1) (2013): 362–372.
33. Remunan-Lopez, C. and R. Bodmeier, Effect of formation and process variables on the formation of chitosan-gelatin coacervates. *Int. J. Pharmaceut.*, 135(1–2) (1996): 63–72.
34. Karim, A. A. and R. Bhat, Gelatin alternatives for food industry: Recent development, challenges and prospects. *Trends Food Sci. Technol.*, 19 (2008): 644–655.
35. Thies, C., Microencapsulation methods based on biopolymer phase separation and gelation phenomena in aqueous media. In *Encapsulation Technologies and Delivery Systems for Food Ingredients and Neutraceuticals*, Garti, N. and D. J. McClements, Eds., Woodhead Publishing Limited, Philadelphia, PA, 2012, pp. 177–207.
36. Karim, A. A. and R. Bhat, Fish gelatin: Properties, challenges, and prospects as an alternative to mammalian gelatins. *Food Hydrocolloids*, 23 (2009): 563–576.
37. Soper, J. C., Method of encapsulating food or flavor particles using warm water fish gelatin and capsules produced therefrom. US patent 5,603,952 (February 18, 1997).
38. Thies, C., Biopolymers and complex coacervation encapsulation procedures. *Agro Food Industry Hi Tech*, 24(4) (2013): 50–52.
39. Nesterenko, A., I. Alric, F. Silvestre, and V. Durrieu, Vegetable proteins in microencapsulation: A review of recent interventions and their effectiveness. *Ind. Crop Prod.*, 42 (2013): 469–479.
40. Ducel, V., J. Richard, P. Saulnier, Y. Popineau, and F. Boury, Evidence and characterization of complex coacervates containing plant proteins: Application to the microencapsulation of oil droplets. *Colloid Surf. A Physicochem. Eng. Aspects*, 232(2–3) (2004): 239–247.

41. Ducel, V., J. Richard, Y. Popineau, and F. Boury, Rheological interfacial properties of plant protein-arabic gum coacervates at the oil-water interface. *Biomacromolecules*, 6(2) (2005): 790–796.
42. Yan, C., W. Zhang, Y. Jin, L. A. Webber, and C. Barrow, Vegetarian microcapsules. WO/2008085997 (2008).
43. Mendanha, D. V., S. E. M. Ortiz, C. S. Favaro-Trindade, A. Mauri, E. S. Monterrey-Quintero, and M. Thomazini, Microencapsulation of casein hydrolysate by complex coacervation with SPI/pectin. *Food Res. Int.*, 42 (2009): 1099–1104.
44. Nori, M. P., C. S. Favaro-Trindade, S. M. Alencar, S. M. Thomazini, and J. C. C. Balieiro, Microencapsulation of propolis extract by complex coacervation. *Food Sci. Technol.*, 44 (2010): 429–435.
45. Xiao, J.-X., H.-Y. Yu, and J. Yang, Microencapsulation of sweet orange oil by complex coacervation with soybean protein isolate/gum arabic. *Food Chem.*, 125 (2011): 1267–1272.
46. Lazko, J., Y. Popineau, D. Renard, and J. Legrand, Microcapsules based on glycinin glycinin-sodium dodecyl sulfate complex coacervation. *J. Microencapsul.*, 21 (2004): 59–70.
47. Weinbreck, F., M. Minor, and C. G. De Kruif, Microencapsulation of oils using whey protein/gum arabic coacervates. *J. Microencapsul.*, 21(6) (2004): 667–679.
48. Zhang, W., C. Yan, J. May, and C. J. Barrow, Whey protein and gum arabic encapsulated omega-3 lipids—The effect of material properties on coacervation. *Agro Food Industry Hi Tech*, 20(4) (2009): 20–24.
49. Baruch, L. and M. Machluf, Alginate–chitosan complex coacervation for cell encapsulation: Effect on mechanical properties and on long-term viability. *Biopolymers*, 82(6) (2006): 570–579.
50. Deveci, S. S. and G. Basal, Preparation of PCM microcapsules by complex coacervation of silk fibroin and chitosan. *Colloid Polym. Sci.*, 287(12) (2009): 1455–1467.
51. Shao, W. and K. W. Leong, Microcapsules obtained from complex coacervation of collagen and chondroitin sulfate. *J. Biomater. Sci. Polym. Ed.*, 7(5) (1995): 389–399.

13 Encapsulation via Microemulsion

Sushama Talegaonkar, Lalit Mohan Negi, and Harshita Sharma

CONTENTS

13.1 Introduction 247
    13.1.1 Difference between Macroemulsion and Microemulsion 248
    13.1.2 Advantages of Microemulsion 248
13.2 Structure and Types of Microemulsion 249
    13.2.1 Phase Diagrams 250
    13.2.2 Components of Microemulsion 251
        13.2.2.1 Oil Phase 252
        13.2.2.2 Surfactants 252
        13.2.2.3 Classification of Surfactants 253
        13.2.2.4 Cosurfactants 255
        13.2.2.5 Cosolvents 256
13.3 Methods of Microemulsion Preparation 256
    13.3.1 Phase Titration Method (Water Titration Method) 256
    13.3.2 Phase Inversion Method...256
        13.3.2.1 Transitional Phase Inversion 257
        13.3.2.2 Catastrophic Phase Inversion 257
13.4 Microemulsion in Microencapsulation Technique 258
    13.4.1 Oil-in-Water Microemulsion 258
    13.4.2 Water-in-Oil Microemulsion 261
    13.4.3 Bicontinuous Microemulsion 262
13.5 Conclusion 263
References 268

13.1 INTRODUCTION

An emulsion is a system containing two immiscible phases, water and oil, stabilized by a surfactant. One phase is the dispersed phase as droplets (internal phase) and the other is the continuous phase (external phase). Depending on the droplet size of the dispersed phase, emulsions can be divided into three categories: (1) macroemulsion, (2) microemulsion, and (3) nanoemulsion. Macroemulsion is a conventional emulsion of oil and water, which is kinetically stabilized where droplets of the dispersed phase have a diameter greater than 100 nm. Macroemulsions appear milky, because they scatter light effectively, as their droplets are greater than the wavelength of light.[1] Microemulsion, having droplet size in the range 1–100 nm, is thermodynamically stable, is isotropically clear dispersion of oil and water, and is stabilized by an interfacial film of surfactant molecules (frequently in combination with a cosurfactant), having very low oil/water interfacial tension.[2–6] Nanoemulsion is defined as a dispersion of oil, surfactant, and aqueous phase, which is a single optically isotropic as well as thermodynamically stable liquid system and usually having droplet diameter within the range 1–100 nm. Though both microemulsions and nanoemulsions can be in the 100 nm size range, the two systems are very different, since nanoemulsions are formed by mechanical shear and microemulsion phases are formed by self-assembly.[7]

Microemulsions are currently of great interest to the pharmaceutical industry because of their significant potential of microencapsulating drug and to act as drug delivery vehicles, which can be incorporated with a wide range of drug molecules. They offer numerous advantages: some of them are spontaneous formation, ease of manufacturing and scale-up, thermodynamic stability, improved drug solubilization, and enhanced bioavailability.[8] The IUPAC defines microemulsion as "dispersion made of water, oil, and surfactant(s) that is anisotropic and thermodynamically stable system with dispersed domain diameter varying approximately from 1 to 100 nm, usually 10–50 nm."[9] Microemulsions are transparent: the droplet size is less than 25% of the wavelength of visible light. These systems are formed readily and spontaneously, generally without the input of high energy. Microemulsion is a ternary system where water and oil (two immiscible phases) are present with a surfactant, in which the surfactant molecules form a monolayer at the interface of oil and water. The hydrophobic tails of the surfactant molecules orient toward the oil phase and the hydrophilic head groups toward the aqueous phase. The aqueous phase may contain salt(s) and/or other ingredients, and the "oil" may be a mixture of different hydrocarbons and olefins. Unlike ordinary emulsions, microemulsions form by simple mixing of the components and do not require the high shear conditions, which are generally used in the formation of ordinary emulsions. Sometimes, a cosurfactant or cosolvent needs to be used in addition to the basic three components (surfactant, the oil phase, and the water phase).[10]

The production of microemulsions is comparatively simple and cost-effective, and thus, they have attracted a great interest as drug-delivery vehicles. Microemulsions have the capability of transporting lipophilic substances through an aqueous medium, and can also carry hydrophilic substances across lipoidal medium. Based on this attribute, potential of microemulsions has been explored for oral, transdermal, parenteral, topical, and pulmonary administration of lipophilic and hydrophilic drugs. In the last decade, microemulsions have also been explored for their potential as vehicles for topical ocular drug delivery.[11–14]

13.1.1 Difference between Macroemulsion and Microemulsion

The main difference between emulsions and microemulsions lies in the size and shape of the droplets of dispersed phase, which causes the differences in the thermodynamic stability of the two systems.[15] Emulsions allow the drug to be administered as a dispersed oil solution and thus are kinetically stable but thermodynamically unstable. After storage or aging, droplets will coalesce and the two phases separate. Unlike emulsions, microemulsions are thermodynamically stable and phases do not separate on storage.[13–14,16,17] Another important difference between the two systems is their appearance; emulsions have a cloudy appearance, while microemulsions are transparent because of the lower dispersed phase size than macroemulsions.

In addition to the aforementioned differences, their methods of preparation also differ distinctly. During preparation of macroemulsions, a large input of energy is required, whereas preparation of microemulsions does not require energy. In contrast to ordinary emulsions, microemulsions form upon simple mixing of the components and do not require the high shear conditions generally used in the formation of ordinary emulsions.[10]

13.1.2 Advantages of Microemulsion

Microemulsions offer a number of advantages, including

- Microemulsion increases the rate of absorption.
- It provides an aqueous dosage form for water-insoluble drugs.
- It enhances the bioavailability of drugs.
- Microemulsion can deliver drug through various routes like topical, oral, intravenous, etc.
- Rapid and efficient penetration of the drug moiety is achieved.
- Loading of drug into microemulsion is helpful in taste masking.

- Microemulsion provides protection from hydrolysis and oxidation, as drug in the dispersed phase of the microemulsion is not exposed to attack by water and air.
- Preparation of microemulsion requires less amount of energy and is spontaneously formed.
- Use of microemulsion eliminates intersubject and intrasubject variabilities in absorption.
- As microemulsion is a thermodynamically stable system, it can remain stable for a long time without any type of aggregation or creaming.
- Microemulsion minimizes first-pass metabolism.
- Microemulsion has a wide application in colloidal drug delivery systems for the purpose of drug targeting and controlled release.[8,18]

13.2 STRUCTURE AND TYPES OF MICROEMULSION

Structurally, microemulsions consist of microdroplets of oil- or water-entrapped pockets, which is the dispersed phase, stabilized by a layer of surfactant (or a mixture or surfactant and cosurfactant) on the surface, similar to conventional emulsions (Figure 13.1).[16]

The interface of water and oil in microemulsions is stabilized by a suitable combination of surfactants and/or cosurfactants. A wide variety of structures and phases can be formed by the mixture of water, oil, and surfactant/cosurfactant. The type of the structure and phase depends upon the proportions of the components. In this regard, flexibility of the surfactant film plays an important role. A surfactant film, flexible enough, results in the existence of several different structures like droplets, aggregates, and bicontinuous structures. This broadens the range of microemulsion existence. A rigid film of surfactant does not allow the existence of bicontinuous structures, which reduces the range of existence. Conventional emulsions, anisotropic crystalline hexagonal or cubic phases, and lamellar structures can also exist besides microemulsion, depending on the proportion of the components. The internal structure of a microemulsion vehicle is very important for the diffusivity of the phases, and thereby also for the diffusion of a drug in the respective phases.[2,18]

Depending on structure and composition of components, microemulsion can be any of the following types:

- *Oil-in-water microemulsions*: In o/w microemulsions, droplets of oil are dispersed in the continuous aqueous phase.
- *Water-in-oil microemulsions*: In w/o microemulsions, droplets of water are dispersed in the continuous oil phase.
- *Bicontinuous microemulsions*: In bicontinuous microemulsions, microdomains of oil and water are interdispersed within the system. A bicontinuous microemulsion is formed in case where the amounts of water and oil are more or less similar.[13–14,17]

Three types of emulsions are represented in Figure 13.2.

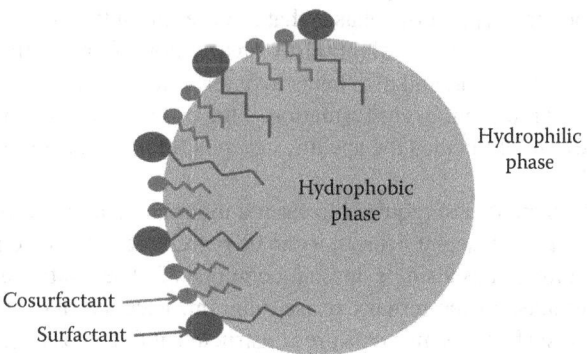

FIGURE 13.1 Surfactant and cosurfactant molecules oriented at the surface of the oil microdroplet.

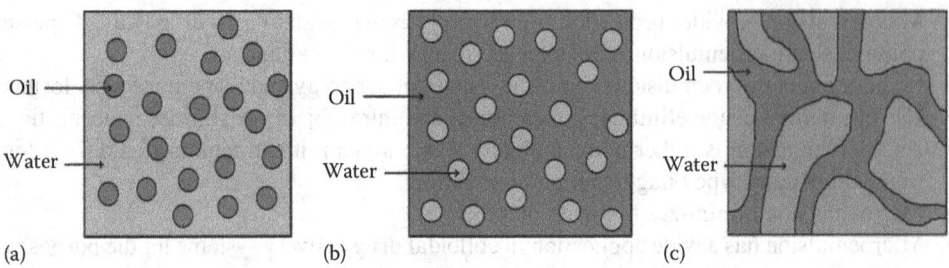

FIGURE 13.2 (a) Oil-in-water microemulsion, (b) water-in-oil microemulsion, and (c) bicontinuous microemulsion.

Winsor has reported the relationship between the phase behavior of surfactants, oil, water, and nature of the different components of ternary system. According to Winsor, there are four types of microemulsion phases, which exist in equilibria. These phases are referred to as Winsor phases. They are

- *Winsor I*: The microemulsion composition corresponding to Winsor I is characterized by two phases, the lower oil/water (O/W) microemulsion phase in equilibrium with excess oil.
- *Winsor II*: The microemulsion composition corresponding to Winsor II is characterized by very low interfacial tension and maximal solubilization of oil and water for a fixed quantity of surfactant. Since, in this phase, microemulsion coexists with both excess phases, the dispersed phase cannot be distinguished from the continuous phase.
- *Winsor III*: The microemulsion composition corresponding to Winsor III comprises of three phases, middle microemulsion phase (O/W plus W/O, called bicontinuous) in equilibrium with upper excess oil and lower water.
- *Winsor IV*: The microemulsion composition corresponding to Winsor IV comprises of single phase, with oil, water, and surfactant homogenously mixed.[13,17,19–21]

13.2.1 PHASE DIAGRAMS

Microemulsion domains are characterized by constructing ternary phase diagrams. A number of variables need to be considered for the preparation of an isotropic homogeneous, stable, nontoxic, transparent microemulsion. Mixing of an appropriate amount of the corresponding components is an essential requirement for the preparation of the thermodynamically stable microemulsions. For this, mixtures with different compositions of the components should be prepared, and checked for the type and number of phases present in the system. These resulting diagrams, which show the number and type of the phases that are present in the system associated with each specific composition, are called phase diagrams. Construction of phase diagrams helps in reducing the number of trials and labor, and also helps to determine the minimum amount of surfactant that will be required for microemulsion formation. This process is of great importance from the industrial point of view and indicates the specific compositions of the components giving a stable microemulsion.[19,22,23]

Basically, three components are required to form a microemulsion: two immiscible liquids and a surfactant. In case, if a cosurfactant is used, it can be represented at a fixed ratio to surfactant as a single component, and treated as a single "pseudocomponent." The relative amounts of these three components are represented in the ternary phase diagram. Pseudoternary phase diagrams of oil, water, and cosurfactant/surfactants mixtures are constructed at fixed cosurfactant/surfactant weight ratios.[10] Gibbs phase diagrams can be used to show the influence of changes in the volume fractions of the different phases on the phase behavior of the system. (Figure 13.3)

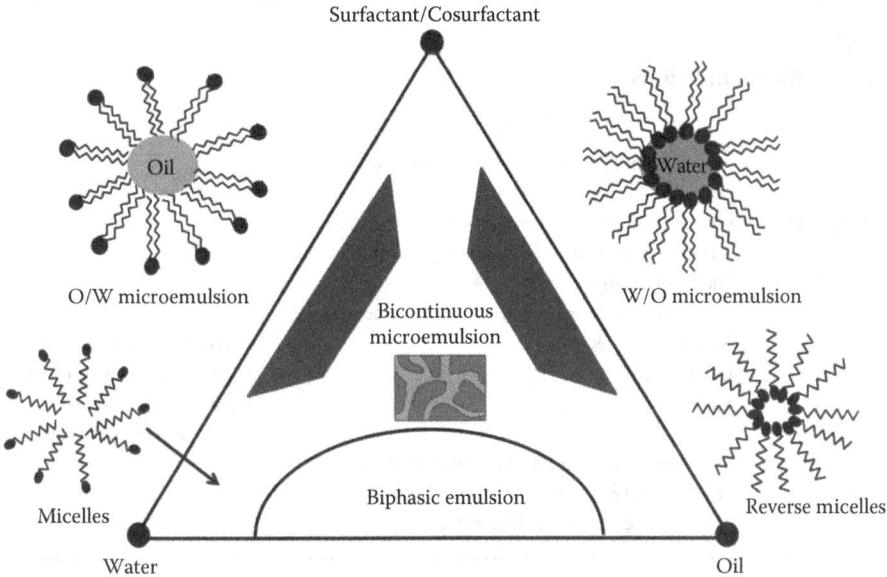

FIGURE 13.3 Phase diagram.

The three components that compose the microemulsion system are kept at an apex of the triangle. Each apex represents 100% volume fraction of the corresponding component. While moving away from the apex, the volume fraction of that specific component reduces and the volume fraction of one or both of the two other components increases. Each point in the triangle represents a possible composition of the mixture of the three components, which may consist of one, two, or all the three phases. These points form the regions, which represent "phase behavior" of the system at constant temperature and pressure. Phase diagrams are constructed by mixing the ingredients. Ingredients shall be preweighed and titrated with water. Formation of monophasic or biphasic system is confirmed by inspecting the system visually. The sample shall be considered as biphasic when it is turbid and gradually separates into different phases. The sample is monophasic when it appears clear and transparent after stirring. These samples shall be marked as points in the phase diagram. The area covered by points of monophasic systems is defined as the microemulsion region. The phase behavior of surfactants, which form microemulsion in the absence of cosurfactant, can be completely represented by ternary diagram.[10,19]

13.2.2 COMPONENTS OF MICROEMULSION

The selection of the components to be used in the microemulsion is a very critical step. The pharmaceutical acceptability of the components and their toxicity issues must be considered. A large number of oils and surfactants are available, but their use in the microemulsion formulation is restricted due to their toxicity, incompatibility, and stability. Components that are used for the formulation of microemulsion should be biocompatible, nontoxic, and clinically acceptable. Emulsifiers should be used in an appropriate concentration range that will result in mild and nonaggressive microemulsion. The selection of generally regarded as safe (GRAS) excipients should always be emphasised.[4,24–26]

A microemulsion has the following components:

1. Oil phase
2. Surfactant
3. Cosurfactant
4. Cosolvent

TABLE 13.1
Oils Used in Microemulsions

Type	Examples	
Natural oils	Soya bean oil, peanut oil, castor oil, olive oil, corn oil, safflower oil	
Mono/di/tri glycosides	Glyceryl caprylate, glycerol mono-oleate, glyceryl monostearate, glyceryl distearate, glycerol triacetace	
Polyoxyl glycerides	Oleyl polyoxylglycerides, linoleyl polyoxylglycerides, lauroyl macrogolglycerides	
Fatty acids	Saturated fatty acids	Lauric acid, myristic acid, caproic acid
	Unsaturated fatty acids	Oleic acid, linoleic acid, linolenic acid
	Fatty acid esters	Methyl or ethyl esters of lauric acid, myristic acid, oleic acid
Propylene glycol esters	Propylene glycol monolaurate, propylene glycol monocaprylate, propylene glycol dicaprylic ester, propylene glycol dicaprylocaprate	

13.2.2.1 Oil Phase

The oil represents one of the most important components in the formulation of the microemulsion. It can solubilize the required dose of the lipophilic drug and can also increase the fraction of lipophilic drug transported via the intestinal lymphatic system, thereby increasing absorption from the GI tract, depending on the molecular nature of the triglyceride.[26]

Oil influences curvature of the surfactant film by its tendency to swell the tail group region of the monolayer of surfactant. Short chain oils penetrate the tail group region to a greater extent (as compared to long chain alkanes) and hence swell this region to a greater extent, which results in increased negative curvature and reduced effective HLB. Saturated and unsaturated fatty acids have their own penetration-enhancing property and have been extensively studied. Fatty acid esters have also been used as the oil phase. Lipophilic drugs have a tendency to solubilize in the microdroplets of oil dispersed in the continuous aqueous phase (i.e., o/w microemulsions). Selection of the oil phase mainly depends on the criterion that the drug should be highly soluble in it. This minimizes the volume of the formulation, which will be required to deliver the therapeutic dose of the drug in an encapsulated form.[8,18]

Table 13.1 enlists the different oils mainly used for the formulation of microemulsion.

13.2.2.2 Surfactants

Selecting a suitable surfactant is a crucial step for the preparation of microemulsions, because formulation of a microemulsion requires the use of moderate to high concentration of surfactant.[27] Surfactants are those compounds, which lower the interfacial tension between two liquids or between a liquid and a solid. Surfactants are usually amphiphilic compounds, that is, the molecule of a surfactant has two parts: one part is soluble in polar solvents, which is hydrophilic, and the other part is insoluble in the polar solvent, which is hydrophobic.[22,28] The polar part of the surfactant molecule is referred as head, and the nonpolar part of the molecule is referred as tail. This amphiphilic nature confers unique capabilities to the surfactant molecules: tendency to adsorb at the surfaces and interfaces, which causes the decrease of the surface tension, and also formation of the aggregates inside the solutions, resulting in the formation of the microemulsions.[22,29–31] Surfactants will diffuse in water and adsorb at interfaces between air and water or at the interface between oil and water, in the case where water is mixed with oil. Surfactant in solution below its

CMC value (critical micellization concentration) increases drug solubility by providing regions for hydrophobic drug interactions in solution. Above the CMC value, surfactant molecules self-aggregate in definite orientation to form micelles having a hydrophilic surface and a hydrophobic core. The hydrophobic core improves the entrapment of drug, hence increasing its solubility. In the presence of a significant amount of oil, surfactant molecules concentrate at the interface of the oil and water, forming oil-in-water emulsion. The drug is solubilized in the internal oil phase. When the oil content is low, minute oil-entrapped surfactant globules are produced, which are known as swollen micelles or microemulsions. The predominant location of drug solubilization depends on its hydrophobic or hydrophilic nature and interactions with the surfactant and cosurfactant.[16] In the bulk aqueous phase, surfactants form aggregates, such as micelles, where the hydrophobic tails form the core of the aggregate and the hydrophilic heads are in contact with the surrounding liquid. Other types of aggregates such as spherical or cylindrical micelles or bilayers can be formed. The shape of the aggregates depends on the chemical structure of the surfactants, depending on the balance of the sizes of the hydrophobic tail and hydrophilic head. This is known as the hydrophilic–lipophilic balance (HLB). Surfactants having low HLB value (such as sorbitan monoesters) are lipophilic and are preferably used for water-in-oil microemulsions, whereas surfactants having high HLB values (such as polysorbates 80 or 20) are hydrophilic and are preferred for oil-in-water microemulsion. Sometimes, a mixture of lipophilic (low HLB) and hydrophilic (high HLB) surfactants is considered as useful.[2,10,32]

13.2.2.3 Classification of Surfactants

Surfactants are most commonly classified according to their polar head group. Surfactants, according to the composition of their head, are classified into following four groups:

1. Anionic
2. Cationic
3. Nonionic
4. Zwitterionic

The ionic surfactants should not be used for pharmaceutical formulations because of their toxicity, such as skin irritation and membrane perturbation.[33] Zwitterionic and nonionic surfactants are commonly used in formulating microemulsion systems, because they have lower toxicity than ionic surfactants, and greater stability toward change in ionic strength and pH, which is likely to be encountered after *in vivo* administration.[34,35] Nonionic surfactants are usually considered acceptable for oral intake.[16]

Table 13.2 gives a brief classification of the types of surfactants, which are used in the formulation of microemulsions.

TABLE 13.2
Classification of Surfactants

Type	Examples
Anionic surfactants	Sodium lauryl sulfate, sodium laureth sulfate, dioctyl sodium sulfosuccinate, sodium stearate, perfluro-octanoate
Cationic surfactants	Octenidine dihydrochloride, cetyl trimethylammonium bromide, benzalkonium chloride, cetrimonium bromide
Zwitterionic surfactants	Cocamidopropyl betaine, lecithin
Nonionic surfactants	Octaethylene glycol monododecyl ether, decyl glucoside, triton X-100, glyceryl laurate, Spans

13.2.2.3.1 Anionic Surfactants

Anionic surfactants are those which contain anionic functional groups at their head, such as sulfate, sulfonate, and carboxylates. Their relatively lower cost makes them the most sought-after surfactant in the industry.[29,36,37] Table 13.3 mentions the anionic surfactants used in the formulation of microemulsions.

13.2.2.3.2 Cationic Surfactants

The surface-active part of cationic surfactants bears positive charge. Cationic surfactants are nonbiodegradable and are used as bactericides. Due to the tendency to be adsorbed at negatively charged surfaces, they are anticorrosive and antistatic agents.[36] Examples of cationic surfactants are enlisted in Table 13.4.

13.2.2.3.3 Zwitterionic Surfactants

Zwitterionic surfactants, also known as amphoteric surfactants, have both cationic and anionic centers present in the same molecule. Generally, their properties are highly dependent on the pH of the solution. The cationic part is based on primary, secondary, or tertiary amines or quaternary ammonium cations. The anionic part can be more variable and include sulfonates, as in CHAPS (3-[(3-Cholamidopropyl)dimethylammonio]-1-propanesulfonate). Other anionic groups are sultaines illustrated by cocamidopropyl hydroxysultaine. Examples of zwitterionic surfactants include betaines (such as cocamidopropyl betaine, dodecyl betaine, lauramidopropyl betaine and cocoamido-2-hydroxypropyl sulfobetaine) and phosphates (such as lecithin).

13.2.2.3.4 Nonionic Surfactants

Nonionic surfactants apparently bear no ionic charge but are polar. The use of nonionic surfactants has been widely accepted, since they can be used over a broad range of pH values, in addition to

TABLE 13.3
Anionic Surfactants

Type	Examples	
Sulfate	Alkyl sulfates	Lauryl sulfate, sodium lauryl sulfate, sodium dodecyl sulfate
	Related alkyl-ether sulfates	Sodium laureth sulfate, sodium myreth sulfate
Sulfonates	Perfluoro-octanesulfonate, perfluorobutanesulfonate, linear alkylbenzene sulfonates	
Docusates	Dioctyl sodium sulfosuccinate,	
Carboxylates	Alkyl carboxylates	Sodium stearate
	Carboxylate-based fluorosurfactants	Perfluorononanoate, Perfluorooctanoate

TABLE 13.4
Cationic Surfactants

Types	Examples
pH-dependent primary, secondary, or tertiary amines	Octenidine dihydrochloride
Permanently charged quaternary ammonium cation	Alkyltrimethylammonium salts (cetyl trimethylammonium bromide, cetyl trimethylammonium chloride), cetylpyridinium chloride, benzalkonium chloride, benzethonium chloride, 5-bromo-5-nitro-1,3-dioxane, dimethyldioctadecylammonium chloride, cetrimonium bromide, dioctadecyldimethylammonium bromide

TABLE 13.5
Nonionic Surfactants

Type	Examples
Polyoxyethylene glycol alkyl ethers	Octaethylene glycol monododecyl ether, Pentaethylene glycol monododecyl ether
Glucoside alkyl ethers	Decyl glucoside, lauryl glucoside, octyl glucoside
Polyoxyethylene glycol octylphenol ethers	Triton X-100
Polyoxyethylene glycol alkylphenol ethers	Nonoxylon-9
Glycerol alkyl esters	Glyceryl laurate
Polyoxyethylene glycol sorbitan alkyl esters	Polysorbate
Sorbitan alkyl esters	Spans
Block copolymers of polyethylene glycol and polypropylene glycol	Poloxamers

their biocompatible nature.[34,38] Two important classes of nonionic surfactants are the ones based on ethylene oxide, referred to as ethoxylated surfactants and the multihydroxy products such as glycerols and sucrose esters. Prominent among these are the fatty alcohols, cetyl alcohol, stearyl alcohol, cetostearyl alcohol (consisting predominantly of cetyl and stearyl alcohols), and oleyl alcohol. Table 13.5 enlists the some more types of nonionic surfactants.

13.2.2.4 Cosurfactants

In most of the emulsions, surfactants alone are not able to sufficiently reduce the interfacial tension between oil and water. Cosurfactants further reduce the interfacial tension and increase the fluidity of the interfacial film. The use of cosurfactants imparts sufficient flexibility to the interfacial film to take up different curvatures, which may be required to form microemulsion over a wide range of proportions of the components. The main role of cosurfactant is to destroy liquid crystalline or gel structures that form in place of a microemulsion phase. Typically used cosurfactants are short chain alcohols (C3–C8), glycols such as propylene glycol, medium chain alcohols, amines, or acids.[27,38,39] Cosurfactants are mainly used in microemulsion formulation for the following reasons:

- Cosurfactants allow the interfacial film sufficient flexibility to take up different curvatures required to form microemulsion over a wide range of composition.
- They reduce the interfacial tension and increase the fluidity of the interface.
- They destroy liquid crystalline or gel structure, which would prevent the formation of microemulsion.
- Surfactants having HLB greater than 20 often require the presence of cosurfactant to reduce their effective HLB to a value within the range required for microemulsion formulation.[8,18,39]

A major disadvantage of microemulsions that contain cosurfactant is their instability on dilution with aqueous biological fluids, which generally occurs after *in vivo* administration through most routes of administration. Use of cosurfactants may also cause irritancy. Microemulsion systems devoid of cosurfactants could be prepared by using double alkyl chain surfactants and nonionic surfactants.[34,40,41]

Different cosurfactants mainly used for microemulsion include

- Ethanol
- Isopropyl alcohol
- Ethylene glycol

- Sorbitan mono-oleate
- Sorbitan monostearate
- Polyglyceryl oleate
- Propylene glycol monocaprylate, etc.

13.2.2.5 Cosolvents

Cosolvents are often used in the formulation of microemulsions to increase the solubility of drug by cosolvency and to stabilize the dispersed phase. In addition to making the environment more hydrophobic by reducing the dielectric constant of water, cosolvents also increase the amount of molecularly dispersed surfactant in the aqueous phase. Availability of free surfactant aids in drug solubilization by creating pockets of hydrophobic regions within the aqueous phase.[16,42]

Examples of cosolvents include

- Propylene glycol
- Glycerine
- Alcohol, etc.

13.3 METHODS OF MICROEMULSION PREPARATION

Microemulsions are spontaneously formed on mixing the components. Formulation of microemulsions does not require the input of high amount of energy; it is a low-energy process. Commonly used methods for the preparation of microemulsions are

1. Phase titration method
2. Phase inversion method

13.3.1 Phase Titration Method (Water Titration Method)

Microemulsions are prepared by the spontaneous emulsification method (phase titration method) and can be depicted with the help of phase diagrams. Construction of phase diagram is a useful approach to study the complex series of interactions that can occur when different components are mixed. Microemulsions are formed along with various association structures (including emulsion, micelles, lamellar, hexagonal, cubic, and various gels and oily dispersion), depending on the chemical composition and concentration of each component.

The understanding of their phase equilibria and demarcation of the phase boundaries are essential aspects of the study. As quaternary phase diagram (four-component system) is time consuming and difficult to interpret; pseudoternary phase diagram is often constructed to find the different zones including microemulsion zone, in which each corner of the diagram represents 100% of the particular component. The region can be separated into w/o or o/w microemulsion by simply considering the composition, that is, whether it is oil rich or water rich. Observations should be made carefully so that the metastable systems are not included.[2,17,43]

13.3.2 Phase Inversion Method

Phase inversion of microemulsions occurs upon addition of excess of the dispersed phase or in response to temperature. During phase inversion, drastic physical changes occur including changes in particle size that can affect drug release both *in vivo* and *in vitro*. These methods make use of changing the spontaneous curvature of the surfactant. For nonionic surfactants, this can be achieved by changing the temperature of the system, forcing a transition from an o/w microemulsion at low temperatures to a w/o microemulsion at higher temperatures (transitional phase inversion [PIT]). Other parameters such as salt concentration or pH value may be considered as well instead of the

Encapsulation via Microemulsion

temperature alone. Additionally, a transition in the spontaneous radius of curvature can be obtained by changing the water volume fraction. Short-chain surfactants form flexible monolayers at the o/w interface, resulting in a bicontinuous microemulsion at the inversion point.

Microemulsions can be prepared by controlled addition of lower alkanols (butanol, pentanol, and hexanol) to milky emulsions to produce transparent solutions comprising dispersions of either water-in-oil (w/o) or oil-in-water (o/w) in nanometer or colloidal dispersions (~100 nm). The alkanols (called cosurfactants) lower the interfacial tension between oil and water sufficiently for almost spontaneous formation. The miscibility of oil, water, and amphiphile (surfactant plus cosurfactant) depends on the overall composition, which is system specific.[8,17,18] Phase inversion method is further divided into two types:

1. Transitional phase inversion (PIT)
2. Catastrophic phase inversion (PIC)

13.3.2.1 Transitional Phase Inversion

PIT method, which was introduced by Shinoda, is an emulsification technique for preparation of microemulsions with low energy. Advantages like being low cost have made the PIT method more attractive in recent years. The PIT concept is based on one type of phase inversion in emulsions (transitional inversion) induced by changing temperature, which affects the HLB of the system.

When the affinity of the surfactant for the water phase equilibrates its affinity for the oil phase, PIT occurs. The variation in the affinity or HLB of the surfactant can be performed by changing temperature. It is based on the changes in affinity of polyoxyethylene-type nonionic surfactants with temperature. At low temperature, the surfactant monolayer becomes hydrophilic and hence, O/W emulsion is produced. At high temperatures, because of dehydration of hydrophilic surfactant, the surfactant becomes lipophilic and, therefore, W/O emulsion is produced. At intermediate temperatures (the HLB temperature), a microemulsion (bicontinuous) or lamellar liquid crystalline phase region appears. The interfacial tension is very low at the HLB temperature, and hence, very small-sized emulsions are produced. It is possible that these nanodrops coalesce to form a macroemulsion with drops in the 1–10 mm range and beyond. If there exists a large amount of surfactant and if some liquid crystal structure is formed near HLB of 0, then the droplets do not coalesce immediately and while the HLB is moved away from the unstable region, the droplet size remains in the range. If the emulsion prepared at a temperature near the PIT is rapidly cooled or heated, kinetically stable emulsions with small droplet size and narrow size distribution can be produced. If the emulsion is rapidly cooled, the W/O emulsion inverts to an O/W emulsion, and if the emulsion is rapidly heated, the O/W emulsion inverts to W/O emulsion occurs. If the cooling or heating process is not fast, coarse emulsions are formed. The characterization tools for determination of the PIT point include: conductivity, viscosity, dynamic light scattering, cryo-TEM, light microscopy, turbidity, and optical microscopy. By monitoring the changes of the conductivity and turbidity with temperatures in emulsion system with nonionic surfactant for preparation of W/O microemulsion, there will be a sudden drop in the conductivity curve and sudden increase in turbidity curve (inverse relation): with increasing temperature, O/W emulsion changes to an intermediate phase and then changes to W/O emulsion. For preparation of O/W nanoemulsions, the result of this experiment is reversed. An average temperature between the temperatures at the maximum and minimum conductivity and turbidity value is considered as PIT point.[2,4]

13.3.2.2 Catastrophic Phase Inversion

In this technique, a transition in the affinity is obtained by changing the water volume fraction, instead of changing the temperature. By successively adding water into oil, initially water droplets are formed in a continuous oil phase. Increasing the water volume fraction changes the spontaneous curvature of the surfactant from initially stabilizing a w/o microemulsion to an o/w microemulsion at the inversion locus. This transition is referred to as PIC. PIC method of emulsification involves addition of continuous phase components gradually, over the dispersed phase components. As a result, in some places along the emulsification path, a phase inversion occurs in which bicontinuous phase appears. Although emulsification is a spontaneous process, the driving forces are small and reaching equilibrium system needs long time. Emulsions are prepared using PIC methods in five steps: (1) first surfactant and cosurfactant (Smix) are mixed, then oil and Smix are mixed together to form oil phase; (2) water is added drop wise to the surfactant and oil mixture. The addition rate is very important and should be adjusted to ensure it is slow enough such that the bicontinuous phase or oil-in water phase is formed; if this rate is very slow, the droplet size increases due to emulsion destabilization. Water droplets are produced in a continuous oil phase by adding water into oil (W/O microemulsion); (3) as the water volume fraction increased, the droplets start to become bigger; (4) an increase in water volume fraction causes droplets to merge together and bicontinuous or lamellar structures are formed; at this time, emulsion inversion point is reached, in which surfactant curvature changes and minimal interfacial tensions are achieved and, therefore, the emulsions with small droplet are formed. The emulsion inversion point can be determined by conductivity measurement. If in the inversion zone, W/O emulsion converts to O/W emulsion, a sharp increase in conductivity will be observed; in contrast if O/W emulsion converts to W/O emulsion, a sharp decrease in conductivity will be observed; (5) by further increasing the water content, the emulsion inversion point is passed and the bicontinuous structure decomposes into smaller oil droplets; from this stage, increasing the water does not change the droplet size. For a complete solubilization of the oil near the emulsion inversion point, a high surfactant concentration is required.

The PIC emulsification method for preparation of microemulsions enjoys many advantages, such as low preparation cost, absence of organic solvents, good production feasibility, long stability, and thermodynamic stability.[4]

13.4 MICROEMULSION IN MICROENCAPSULATION TECHNIQUE

Microemulsion is a suitable candidate for encapsulating the drugs, as the drug gets entrapped in the microdroplets of oil or water present as the dispersed phase in the microemulsion. Microemulsion is also one of the most important techniques for preparation of microcapsules and microspheres. Both hydrophobic and hydrophilic drugs can be encapsulated, in o/w microemulsions and w/o microemulsions, respectively. Encapsulating the drug into the dispersed phase of microemulsions may serve a number of purposes and offers several advantages, for example, improving the transdermal absorption of the drug, easy administration of a poorly water-soluble drug, increasing the bioavailability of a less permeable drug, maintaining the drug stability in the formulation, developing controlled release dosage forms, etc. These advantages and applications of microemulsions have been explored by a number of scientists and researchers, and microemulsions have been extensively worked upon in the last decade.

13.4.1 OIL-IN-WATER MICROEMULSION

Oil-in-water microemulsion has been widely studied and used for transdermal delivery of the drugs. Apart from its advantages as topical formulation, oil-in-water emulsion also offers numerous advantages such as bioavailability enhancement via transdermal, oral, ocular, and vaginal routes.

The skin permeation enhancement of many kinds of drugs by microemulsions has been widely known. In a study, the influence of microemulsions types on *in vitro* skin permeation of model hydrophobic drugs and their hydrophilic salts was investigated. The microemulsion systems were composed of isopropyl palmitate, water, a 2:1 w/w mixture of Aerosol OT and 1-butanol, and a model drug. Transdermal flux of lidocaine, tetracaine, dibucaine, and their respective hydrochloride salts from the drug-loaded microemulsions through heat-separated human epidermis was investigated. The o/w microemulsions resulted in the highest fluxes of the model drugs in base form as compared with the other formulations within the same group of drugs.[44] Very recently, a novel oil-in-water microemulsion of adapalene was developed for transfollicular delivery. A pseudoternary phase diagram was developed for microemulsion consisting of oleic acid as oil phase, Tween 20 as surfactant, Transcutol as cosurfactant, and deionized water. Penetration of microemulsion through hair follicles was studied. The drug penetration in the hair follicles was found to be increased significantly, as the microstructure of microemulsion shifted from oil-in-water to bicontinuous, with increase in water content of microemulsion. Results of this study suggested that microemulsion penetrated through hair follicles and is promising for transfollicular drug delivery.[45] In another experiment, nicotinic acid was encapsulated in the o/w microemulsion to achieve its transdermal delivery. Nicotinic acid has a lipid-regulating effect and is used clinically, but its utility is strongly limited by several disadvantages such as extensive hepatic metabolism and flushing. Therefore, transdermal delivery of nicotinic acid may help in reducing side effects associated with oral administration, and in maintaining constant therapeutic blood levels for longer duration. The aim of this experiment was to deliver nicotinic acid through skin without causing vasodilatation and flushing and optimizing its delivery to the blood stream. The microemulsion system was composed of isopropyl myristate, water, and a 4:1 (w/w) mixture of Labrasol and Peceol where a pseudoternary phase diagram was constructed. According to the investigations, microemulsion formulations that formed were of oil-in-water type. The transdermal permeability of nicotinic acid and its prodrugs was evaluated. The selected formulations were found to be promising for developing a transdermal drug-delivery system of nicotinic acid from dodecyl nicotinate that would offer advantages like possible controlled drug release, reduced flushing, increased drug stability, and ease of large-scale production.[46] Transdermal delivery of indomethacin was also attempted to improve drug permeability by encapsulating the drug into the microdroplets of microemulsion. Formulations were based on the oil/water microemulsion region of pseudoternary phase diagrams. *In vitro* permeation studies were performed using Franz diffusion cells. Permeation through skin and skin retention of indomethacin microemulsions and ointment were tested. The cumulative amount of permeated indomethacin and its skin retention were found significantly higher in microemulsion formulations compared with ointment, and thus, microemulsions represented a novel transdermal delivery vehicle for increasing the solubility and permeability of indomethacin.[47] In one more study, transdermal delivery of silymarin, a standardized extract from *Silybum marianum* seeds known for its many skin benefits such as antioxidant, anti-inflammatory, and immunomodulatory properties, was achieved using o/w microemulsion. In this study, the potential of o/w water emulsion for dermal delivery of silymarin was evaluated, which was prepared using glyceryl monooleate, oleic acid, ethyl oleate, or isopropyl myristate as the oily phase; a mixture of Tween 20, Labrasol, or Span 20 with HCO-40 as surfactants; and Transcutol as a cosurfactant. *In vitro* release studies showed prolonged release for microemulsions when compared to silymarin solution.[48]

Several other studies have been performed in which o/w microemulsions are used as transdermal delivery system for delivery of many drugs, such as ketoconazole, acyclovir, terbinafine, tretinoin, etc. In all these experiments, o/w microemulsions were proved to be a promising vehicle for topical delivery of these drugs.[49–52] Oil-in-water microemulsion has also been explored for intravenous delivery of drugs. In one such study, poorly soluble anesthetic drug propofol was incorporated in the o/w microemulsion, composed of pluronic F68, propylene glycol, and saline, which solubilized poorly soluble anesthetic drug propofol for intravenous administration. The ternary diagram was constructed to identify the regions of microemulsions, and the optimal composition of microemulsion was determined. Hemolysis percent of propofol microemulsions was found to be lower than that of commercial lipid emulsion. No significant difference in time for unconsciousness and recovery of righting reflex was observed between the prepared microemulsions and commercial lipid emulsion. Thus, it was concluded that microemulsion would be a promising intravenous delivery system for propofol.[53] Peptide rhPTH1–34 is clinically used for osteoporosis treatment. However, this peptide drug has no oral bioavailability because of proteolysis and low membrane permeability in gastrointestinal gut. Guo L and his coworkers explored the possibility of absorption enhancement of rhPTH1–34 through the oral delivery of the microemulsion. The microemulsion consisted of

Labrasol, Crodamol GTCC, Solutol HS 15, d-α-tocopheryl acetate, and saline water. The microemulsion showed high drug-loading efficiency and permeability, and significantly higher resistance to proteolysis. Also, the proximal tibia bone mineral content and density in oral rats was significantly increased compared to the control rats, and the proximal tibia microstructure of oral rats was improved greatly. These findings revealed that oral microemulsion might represent an effective oral delivery system for rhPTH1–34.[54]

Fluconazole, a synthetic triazole antifungal agent, is the most prescribed drug used in treating *Candida albicans*. Because of its poor water solubility and the emergence of resistant strains against this antimycotic drug, a unique microemulsion drug delivery system for fluconazole was devised against candidiasis. A clear o/w microemulsion system, consisting of clove oil as oil phase, Tween 20 as surfactant, and water as aqueous phase, was developed using a ternary phase diagram. The efficacy of fluconazole was found to be greatly improved when compared with its conventional bulk form. The optimized microemulsion exhibited significantly higher antifungal activity. Thus, this report disclosed an excellent oral drug-delivery system.[55]

Dixit RP[56] along with his coworkers developed solid microemulsions for improved delivery of simvastatin. Pseudoternary phase diagrams were constructed and microemulsions were optimized for oil and drug content. Solid microemulsions were prepared using colloidal silicon dioxide to adsorb the liquid microemulsion. Dissolution studies revealed remarkable increase in dissolution of the drug as compared to plain drug. All the formulations provided significant reduction in the total cholesterol levels in hyperlipidemic rats with reference to rats of control group. Thus, the proposed solid microemulsion has potential to deliver water-insoluble drugs like simvastatin by oral route for better efficacy.[56] Oil-in-water microemulsion has also been used for encapsulation of drugs. Following this approach, Wilk KA[57] encapsulated the cyanine-type photosensitizer IR-780 in poly(n-butyl cyanoacrylate) (PBCA) nanocapsules, which can be used for photodynamic therapy, which enables a photosensitizer to be selectively delivered to tumor cells with enhanced bioavailability and diminished dark cytotoxicity. Nanocapsules were fabricated by interfacial polymerization in o/w microemulsions formed by dicephalic and gemini saccharide-derived surfactants. The effects of encapsulated IR-780 were compared with those of native photosensitizer. Cyanine IR-780 delivered in nanocapsules to MCF-7/WT cells retained its sensitivity upon photoirradiation and was regularly distributed in the cell cytoplasm. The intensity of the photosensitizer-generated oxidative stress depended on IR-780 release from the effective uptake of polymeric nanocapsules and seemed to remain dependent upon the surfactant structure in o/w microemulsion-based templates applied to nanocapsule fabrication.[57] In a similar experiment, encapsulation of cyanine IR-768 in o/w microemulsion was performed for generation of photodynamic agent suitable for photodynamic therapy (PDT). Oil-cored PBCA nanocapsules were prepared by interfacial polymerization in o/w microemulsions formed by the nonionics Tween 80 and Brij 96. Iso-propyl myristate, ethyl oleate, iso-octane, and oleic acid were used as the oil phases and iso-propanol and propylene glycol as the cosurfactants. Droplets of o/w microemulsion, also containing hydrophobic IR-768 in the oil phase, were applied in the interfacial polymerization of n-butyl cyanoacrylate. The results of *in vitro* erythrocyte hemolysis and the cell viability of breast cancer MCF-7 cells proved the nanocapsules to be a safe carrier of IR-768 in the circulation, having a very low hemolytic potential and being nontoxic to the studied cells. Photoirradiation of the cancer cells with entrapped photosensitizer decreased cell viability, demonstrating that this effect may be utilized in photodynamic therapy.[58]

In another study, radiolabeled and fluorescent lipid nanocapsules were synthesized by using a phase inversion process that followed the formation of an o/w microemulsion containing triglycerides, lecithins, and a nonionic surfactant. Results of the experiment revealed that lipid nanocapsules were rapidly accumulated within cells through active and saturating mechanisms. Nanocapsules could bypass the endo-lysosomal compartment with only 10% of the cell-internalized fraction found in isolated lysosomes. When nanocapsules were loaded with paclitaxel, smallest lipid nano capsules (LNCs) also were found to trigger the best cell death activity.[59]

13.4.2 WATER-IN-OIL MICROEMULSION

Water-in-oil microemulsion, having water microdroplets dispersed in the continuous oil phase, has the ability to encapsulate water-soluble drugs in the water microdroplets and has emerged as a successful dosage form for the delivery of hydrophilic drug.

In a recent study, skin permeation of diclofenac has been enhanced using w/o microemulsion. The w/o microemulsion formulations were selected based on constructed pseudoternary phase diagrams depending on water solubilization capacity and thermodynamic stability. Permeation of diclofenac across rat skin from selected w/o microemulsion formulations was evaluated and compared with control formulations Microemulsion formulations exhibited significantly increased diclofenac permeation compared to conventional formulations.[60] Another topical w/o microemulsion formulation has been developed and studied by Xiao YY et al. This microemulsion contained NaCl and fluorouracil (5-Fu) as a model drug to investigate the transdermal characteristics and skin irritation of the microemulsion *in vitro*. Other components were isopropylmyristate acting as oil phase, Aerosol-OT as surfactant, and Tween 85 as cosurfactant. The cumulative amount of fluorouracil permeated was found to be many folds higher than the fluorouracil aqueous solution and fluorouracil cream. Microemulsion exhibited some irritation, but could be reversed after drug withdrawal. The addition of NaCl significantly increased the content of water and the drug loading in microemulsion. The microemulsion system promoted the permeation of fluorouracil greatly, which is a promising vehicle for the transdermal delivery of fluorouracil and other hydrophilic drug.[61] An important experiment explored the potential of w/o microemulsions as an effective means of topical delivery of peptides and proteins of all sizes, and in high doses, as these formulations are a cheap, stable, pain-free means of delivery of peptides and proteins to the skin. The results using insulin, IGF-I, and GHRP-6 given topically were found to be particularly intriguing.[62] Another study attempted to improve the efficiency of the intradermal delivery of genistein and other two isoflavones (daidzein and biochanin A) using microemulsion consisting of isopropyl myristate, NaCl solution, Tween 80, and ethanol as a vehicle. The solubility of all the isoflavones markedly increased and significant amounts of isoflavones were delivered to the skin. The effect of w/o microemulsion was larger than that of o/w microemulsion. Among three isoflavones tested, genistein showed significant enhancement in both solubility as well as skin accumulation. Genistein retained in the skin significantly inhibited lipid peroxidation. Furthermore, pretreatment of guinea pig dorsal skin with genistein containing w/o microemulsion prevented UV irradiation-induced erythema formation. These findings indicated the potential use of w/o microemulsion for the delivery of genistein to protect skin against UV-induced oxidative damage.[63]

Peptide and protein drugs have become the new generation of therapeutics, yet most of them are only available as injections, and reports on oral local intestinal delivery of peptides and proteins are quite limited. Thus, a w/o microemulsion system was developed and evaluated for local intestinal delivery of water soluble peptides after oral administration. Water-in-oil microemulsions consisting of Miglyol 812, Capmul MCM, Tween 80, and water were developed. A fluorescent labeled peptide, 5-(and-6)-carboxytetramethylrhodamine labeled HIV transactivator protein, TAMRA-TAT was loaded in the microemulsion. The half-life of TAMRA-TAT in microemulsion was enhanced nearly three-fold compared to that in the water solution. The *in vitro* and *in vivo* studies both suggested TAMRA-TAT was better protected in the w/o microemulsion in an enzyme-containing environment, suggesting that the w/o microemulsions developed in this study may serve as a potential delivery vehicle for local intestinal delivery of peptides or proteins after oral administration.[64] Another work performed by GundogduE includes development of novel loaded w/o microemulsion formulation to improve the bioavailability of fexofenadine as compared to fexofenadine syrup. Microemulsion composed of span 80, lutrol F 68, oleic acid, isopropyl alcohol, and water. This microemulsion increased the oral bioavailability of fexofenadine, which has high water solubility but low permeability. These results suggested that novel w/o microemulsions plays an important role in enhancing oral bioavailability of low permeability drugs.[65] Another w/o microemulsion was developed as oral formulation using propylene glycol dicaprylocaprate, Cremophor, and water, and investigated as a system for enhancing the oral bioavailability of Biopharmaceutic Classification System (BCS) III drugs. The microemulsion increased the oral bioavailability of hydroxysafflor yellow A, which was highly water soluble but very poorly permeable. *In vitro* lipolysis showed that the microemulsion was digested very quickly by pancreatic lipase. These results suggested that digestion of the microemulsion by pancreatic lipase plays an important role in enhancing oral bioavailability of water-soluble drugs.[66] Earthworm fibrinolytic enzyme (EFE-d, Mr 24177), a water-soluble protein, is clinically used for the management of cardiovascular diseases. However, this protein drug has a very low oral bioavailability because of its low membrane permeability. Cheng MB explored the possibility of absorption and efficacy enhancement for EFE-d through the delivery of the w/o microemulsions. The w/o microemulsion consisting of Labrafac CC, Labrasol, Plurol Oleique CC 497, and saline was developed and characterized. The w/o microemulsion possessed higher intestinal membrane permeability *in vitro* as well as a higher absorption and efficacy *in vivo*, when compared to control solution. The intraduodenal bioavailability of EFE-d for microemulsions was 208-fold higher than that of control solution. These findings indicated that the w/o microemulsion may represent a safe and effective oral delivery system for hydrophilic bioactivity macromolecules.[67]

Absorption of the drug released from w/o microemulsion through nasal route has also been studied. Novel w/o microemulsion system was developed and loaded with insulin. Nasal absorption of insulin from the w/o microemulsion spray was investigated. The bioavailability of insulin via the nasal route using w/o microemulsion was found to reach 21.5% relative to subcutaneous administration. The profile of plasma glucose levels obtained after nasal spray application of the microemulsion was similar to the subcutaneous profile and resulted in a 30%–40% drop in glucose levels.[68]

Water-in-oil microemulsion has also been used to develop anticancer formulations. In an instance, anticancer drug imatinib-loaded w/o microemulsion was formulated as an alternative formulation for cancer therapy and evaluated for permeability and cytotoxic effect. Microemulsion had a significant cytotoxic effect on MCF-7 cells. The permeability studies of imatinib across Caco-2 cells also showed a higher permeability value. Thus, in conclusion, the imatinib-loaded microemulsion formulation may be used as an effective alternative breast cancer therapy for oral delivery of imatinib.[69]

13.4.3 Bicontinuous Microemulsion

Bicontinuous microemulsions have also been explored for their potential as dosage form to encapsulate and deliver drugs. In a recent study, a microemulsion system was formulated as a topical delivery system of naproxen for relief of symptoms of rheumatoid arthritis, osteoarthritis, and treatment of dysmenorrhea. Microemulsion was prepared by mixing the appropriate amounts of surfactant, co-surfactant and oil phase i.e., Tween 80-Span 80 mix, propylene glycol (PG) and Labrafac PG–transcutol P mix, respectively. SEM photographs of the microemulsion revealed the existence of hexagonal and bicontinuous structures. Drug release profile showed that 26.15% of the drug was released in the first 24 h of experiment. Microemulsion was found to be preferable as topical naproxen formulation, based on the results of characterization, physicochemical properties, and *in vitro* release studies.[70] Salimi A and coworkers designed a topical microemulsion of vitamin B12 and studied the correlation between internal structure and physicochemical properties of the microemulsions. They investigated the phase behavior and microstructure of traditional and novel microemulsions of vitamin B12. Water-in-oil and bicontinuous microemulsion with different microstructures were found in novel and traditional formulations. The results showed that both microemulsions provided good solubility of vitamin B12 with a wide range of internal structure.[71] In another experiment, cationic microemulsions containing a protein transduction domain (penetratin) were developed for optimizing paclitaxel localization within the skin. Microemulsions were prepared by mixing a surfactant blend (BRIJ:ethanol:propylene glycol) with monocaprylin (oil phase), and adding water at 30% (ME-30), 43% (ME-43), and 50% (ME-50).

Electrical conductivity and viscosity measurements indicated that ME-30 was most likely a bicontinuous system, whereas ME-43 and ME-50 were water continuous. Their irritation potential was found to decrease as aqueous content increased. ME-43 was selected for penetratin incorporation, as ME-50 was not stable in the presence of paclitaxel. The microemulsion containing penetratin (ME-P) displayed a 1.8-fold increase in paclitaxel cutaneous (but not transdermal) delivery compared with the plain ME-43, suggesting that penetratin addition increases the barrier-disrupting and penetration-enhancing effects of microemulsions. The results suggested potential of ME-P for drug localization within cutaneous tumor lesions.[72] A successful colloidal soft nanocarrier viz. microemulsion system, for the transdermal delivery of an angiotensin II receptor blocker: olmesartan medoxomil, was developed and characterized in an experiment. Different microemulsion formulations were prepared. The physicochemical and spectroscopic methods revealed the presence of water-in-oil and bicontinuous structures. Olmesartan medoxomil was delivered successfully across the skin. Higher bioavailability compared to commercial oral tablets with a more sustained behavior was achieved.[73] Topical delivery of Cyclosporin A (CysA) is of great interest for the treatment of autoimmune skin disorders. Consequently, microemulsion systems possessing a potentially improved skin bioavailability of CysA were designed using AOT, Tween85, and isopropyl myristate. The results of diffusion-ordered NMR spectroscopy measurements indicated the presence of bicontinuous and w/o microemulsions depending on microemulsion composition. Therapeutic advantage of dermal administration of CysA was also evaluated in rat model. In case of bicontinuous microemulsion containing CysA, the deposition of the drug into skin and subcutaneous fat was respectively almost 30- and 15-fold higher than the concentrations compared with oral administration. Systemic distribution in blood, liver, and kidney was much lower following topical administration than that of following oral administration. It was thus concluded that topical microemulsion vehicle loaded with CysA delivered maximal therapeutic effect to local tissue while avoiding side effects seen with systemic therapy and is a safe vehicle for topical drug delivery of CysA.[74]

Table 13.6 enlists some more studies performed on microemulsions, which proved its potential as an effective drug delivery system to overcome a number of limitations of other dosage forms, administered through different routes of administration.

13.5 CONCLUSION

Microemulsion is one of the most suitable techniques for encapsulation of drugs, which in turn may offer a number of advantages like improvement in bioavailability of the drug, solving solubility problems, avoiding stability issues, increasing absorption of a less permeable drug, and also aids in developing a controlled release formulation. Potential of microemulsion technique for microencapsulation of drugs has been realized through the success achieved in several studies and experiments performed by researchers and scientists. Oil-in-water microemulsion, the most suitable of all the three types, has been most widely used in encapsulating drugs and developing novel formulations. It has been most commonly used to develop transdermal formulations, solving problems of poor absorption, bioavailability, and permeability; but it has also been successfully used in developing oral, subcutaneous, ocular, intragastric, and even intravaginal formulations, offering many advantages over conventional alternatives. Water-in-oil microemulsion has also been emerged as a suitable alternative of conventional formulations, giving a number of successful transdermal formulations solving bioavailability, permeability, toxicity, and other common issues. It has been explored for few other routes of administration also, for example, intragastric and intraduodenal. Bicontinuous microemulsion, though not as successful as o/w and w/o, has been used to develop transdermal and oral formulations, and positive results are obtained. All the results indicate the huge potential that microencapsulation technique holds for the future of microencapsulation of wide variety of drugs (hydrophilic, hydrophobic, and amphipathic), in order to produce better and more promising formulations.

TABLE 13.6
Studies Performed on Microemulsions as Drug Delivery System

Route of Administration	Drug	Category/Class/ Use	Limitation	Inference	References
Oil-in-water microemulsion					
Transdermal	Ondansetron	Anti emetic	Poor absorption and bioavailability	Pharmacokinetic profile was compared with ondansetron conventional gel and oral marketed syrup. Bioavailability from microemulsion was found to be 6.03 folds more than oral conventional syrup and 9.66 times more than OCG gel.	[75]
	Minoxidil	Anti alopecia	Variable levels of success	Microemulsion containing minoxidil, diclofenac, and tea tree oil achieved significantly superior mean hair count, mean hair weight, and mean hair thickness. Also, it showed no appreciable side effects.	[76]
	Griseofulvin	Anti fungal	Poor skin permeability	Microemulsion produced many folds increase in drug permeation. Treatment resulted in a complete clinical and mycological cure in 7 days. Formulation was observed to be nonsensitizing and histopathologically safe.	[77]
	Clobetasol propionate	Antivitiligo	Poor permeability and retention in skin	Microemulsion-based gel showed greater retention of drug into skin layers than microemulsion and market preparation. Microemulsion-based gel was found to be significantly less irritating. Results indicated interaction of microemulsion components with skin, resulting in permeation enhancement and retention of drug into skin layers.	[78]
	Oxymatrine	Antiviral	Low oral bioavailability, low membrane permeability	Microemulsion of OMT–PLC (oxymatrine–phospholipid complex) improves lipid solubility and effectiveness of oxymatrine) provided better skin permeability and higher retention ratio of oxymatrine in skin. Microemulsion also significantly enhanced antiproliferative activity of oxymatrine on scar fibroblasts.	[79]
	Vitamin C and Vitamin E	Vitamins	Poor skin permeability, difficult coadministration	Microemulsion was found to be most appropriate for effective skin delivery of both vitamins, followed by gel-like microemulsion and by microemulsion carbomer. Cytotoxicity studies revealed good cell viability after exposure to microemulsion and confirmed all tested microemulsions as nonirritants.	[80]
	Babchi oil and Psoralen	Anti psoriasis	Poor skin permeability	Topical delivery of babichi oil and psoralen was found to be improved from microemulsion gel.	[81]
	Celecoxib	Anti-inflammatory	Poor skin permeability	Microemulsion showed higher permeation rate and significant anti-inflammatory activity.	[82]

(Continued)

TABLE 13.6 (Continued)
Studies Performed on Microemulsions as Drug Delivery System

Route of Administration	Drug	Category/Class/ Use	Limitation	Inference	References
	Vinpocetine	Neuroprotective	Poor solubility and skin permeability	Solubility and *in vitro* percutaneous permeation flux of vinpocetin was improved significantly and irritation study showed that microemulsion was safe.	[83]
	Diclofenac	Anti inflammatory	Poor skin permeability	Percutaneous absorption of diclofenac from microemulsions was enhanced with increasing lauryl alcohol and water contents, and with decreasing Labrasol:ethanol mixing ratio in the formulation.	[84]
	Ascorbyl palmitate	Anti oxidant	Poor delivery	O/w microemulsions delivered ascorbyl palmitate to the skin significantly better than w/o microemulsions. In both types of microemulsions, the effectiveness increased at higher concentrations of ascorbyl palmitate.	[85]
Subcutaneous	Zolmitriptan	Anti migraine	High pain recurrence due to rapid drug elimination	Microemulsion provided differential release of zolmitriptan and diclofenac acid both *in vitro* and *in vivo* that may be potentially beneficial to migraine patients.	[86]
Oral	T-OA (3β-hydroxyol-2a-12-en-28-oic acid-3, 5, 6-trimethylpyrazin-2-methyl ester)	Anti tumor	Water insolubility	Solubility of T-OA was enhanced. AUC of T-OA was also significantly increased. Relative bioavailability of T-OA microemulsion was found to be 57-fold higher than the pure drug.	[87]
	5-F (ent-11α-hydroxy-15-oxo-kaur-16-en-19-oic-acid)	Anti tumor	Poor bioavailability	*In vivo* bioavailability of 5-F was markedly improved. Toxicity tests showed that microemulsion had no hepatotoxicity in mice.	[88]
	Andrographolide	Anti-inflammatory	Low aqueous solubility and oral bioavailability	Microemulsion had much better anti-inflammatory effect and a higher biological availability than andrographolide tablets. Also, it showed a very low acute oral toxicity.	[89]
Ocular	Chloramphenicol	Antibiotic	Tendency to hydrolize in commercial eye drops	Results of HPLC revealed that contents of glycols in microemulsion were much lower than that in the commercial eye drops. It implied that stability of chloramphenicol in microemulsion was increased remarkably.	[90]

(*Continued*)

TABLE 13.6 (Continued)
Studies Performed on Microemulsions as Drug Delivery System

Route of Administration	Drug	Category/Class/Use	Limitation	Inference	References
	Dexamethasone	Glucocorticoid steroid	Poor solubility and absorption	Microemulsion showed greater penetration of dexamethasone in anterior segment of eye and also release of drug for a longer time. AUC was more than twofold higher than that of the conventional preparation.	[91]
Intragastric	Epidermal growth factor	Anti ulcer	Incomplete healing, no reduction in basal acid secretion	Microemulsion significantly reduced basal gastric secretion. Mean ulcer score was reduced with microemulsion in 7 days and was almost completely healed in four of the animals. Mucus levels increased significantly.	[92]
Intravaginal	Spermicide	Contraceptive	Less effective, toxic	Studies confirmed potent contraceptive activity. Vaginal irritation test was not associated with any local, systemic, or reproductive toxicity.	[93]
Water-in-oil microemulsion					
Transdermal	Iodide ions		Poor skin permeability	Microemulsion formulations were stable and compatible with iodide ions. In vitro human skin permeation studies revealed that microemulsion improved iodide ion diffusion significantly.	[94]
	Alkaloidal extract of Tabernaemontana divaricata	Anti-acetylcholinestrase	Poor skin absorption	Microemulsion showed significantly higher acetylcholinesterase inhibition. Microemulsion significantly increased transdermal delivery of extract within 24 h.	[95]
	Nicotinamide	Vitamin	Lower release rate	Release kinetics of microemulsion-based gel was best fitted to zero order model. In vitro release of nicotinamide from microemulsion-based gel was found to be highest followed by microemulsion, as compared to commercial cream.	[96]
	Metronidazole	Anti infective	Uncontrolled release, incomplete relief from symptoms of rosacea	Release rate of metronidazole from microemulsion was found to be lower than that of commercial gel. Microemulsion resulted in complete relief in 38% of patients with telangiectasia, while commercial product did not provide any relief from telangiectasia symptoms.	[97]
	Quercetin	Photoprotective	Toxicity issues	Microemulsion reduced incidence of histological skin alterations, mainly the connective-tissue damage, induced by exposure to UVB irradiation. Study also demonstrated no toxicity of the microemulsion, by evaluating the cytotoxic effect on L929 cells and histological aspects.	[98]

(Continued)

Encapsulation via Microemulsion 267

TABLE 13.6 (Continued)
Studies Performed on Microemulsions as Drug Delivery System

Route of Administration	Drug	Category/Class/Use	Limitation	Inference	References
	Fluorouracil	Anti tumor	Poor transdermal delivery ability	Cumulative amount permeated was 19.1- and 7-fold more than fluorouracil aqueous solution and fluorouracil cream, respectively.	[99]
	Desmopressin	Anti diuretic/anti diabetic	Poor skin penetration	Greater fraction of applied dose reached acceptor compartment from microemulsion as compared to commercial cream. Concentration of drug that penetrated into upper layers of skin was higher from cream than from microemulsion, whereas higher amount of drug was found in deeper skin layers and in acceptor compartment from microemulsion.	[100]
Intragastric	Insulin	Anti diabetic	Insufficient reduction in blood glucose levels	Insulin loaded nanocapsules dispersed in biocompatible microemulsion resulted in significantly greater reduction in blood glucose levels than aqueous insulin solution. This demonstrated that formulation of peptides within nanocapsules administered dispersed in microemulsion can facilitate oral absorption of encapsulated peptide. Such system can be prepared in situ by the interfacial polymerization of a w/o microemulsion.	[101]
Intraduodenal	Ginsenoside RB1	Steroid glycoside	Poor intestinal absorption	Results indicated that components of microemulsion can facilitate membrane permeability of drug. W/o microemulsion enhanced intestinal absorption of drug. This is attributed to its enhancement on membrane fluidity.	[102]
Bicontinuous microemulsions					
Transdermal	Betahistine hydrochloride	Anti vertigo	Poor skin permeation	High permeation fluxes were obtained with microemulsion. Confocal laser scanning microscopy was used to confirm the permeation enhancement and to reveal the penetration pathways.	[103]
	Ascorbic acid	Vitamin	Unfavorable transdermal absorption	Ascorbic acid was found to be mostly located in the epidermis where the decomposition of melanin occurred. This suggested that microemulsion could be considered as a suitable carrier system for application of ascorbic acid as a whitening agent.	[104]
	Diclofenac sodium	Analgesic/anti inflammatory	Limited skin penetration	Microemulsion increased in vivo transdermal penetration rate of diclofenac by two fold, compared to commercial macroemulsion cream.	[105]
Oral	Berberine	Antifungal	Poor bioavailability	Bioavailability of oral berberine-loaded microemulsion was found to be 6.47 times greater than that of berberine tablet suspensions.	[106]

REFERENCES

1. Shah DO. (1985). *Macro-and Microemulsions: Theory and Applications: Based on a Symposium Sponsored by the Division of Industrial and Engineering Chemistry at the 186th Meeting* of the American Chemical Society, Washington, DC, August 28–September 2, 1983. Washington, DC, American Chemical Society, pp. 1–13.
2. Talegaonkar S, Azeem A, Ahmad FJ, Khar RK, Pathan SA, and Khan ZI. (2008). Microemulsions: A novel approach to enhanced drug delivery. *Recent Patents on Drug Delivery and Formulation*, 2, 238–257.
3. Kumar P and Mittal KL. (1999). *Handbook of Microemulsion Science and Technology*, 1st edn.; CRC Press, New York, p. 1.
4. Maali A and Hamed Mosavian MT. (2013). Preparation and application of nanoemulsions in the last decade (2000–2010). *Journal of Dispersion Science and Technology*, 34, 92–105.
5. Jafari SM, Assadpoor E, He Y, and Bhandari B. (2008). Re-coale scence of emulsion droplets during high-energy emulsification. *Food Hydrocolloid*, 22, 1191–1202.
6. Becher P. (2001). *Emulsions: Theory and Practice*; Oxford University Press, Oxford, U.K.
7. Mason TG, Wilking JN, Meleson K, Chang CB, and Graves SM. (2006). Nanoemulsions: Formation, structure, and physical properties. *Journal of Physics: Condensed Matter*, 18, R635–R666.
8. Jha SK, Dey S, and Karki R. (2011). Microemulsions-potential carrier for improved drug delivery. *Asian Journal of Biomedical and Pharmaceutical Sciences*, 1, 5–9.
9. Slomkowski S, Alemán JV, Gilbert RG, Hess M, Horie K, Jones RG, Kubisa P et al. (2011). Terminology of polymers and polymerization processes in dispersed systems (IUPAC Recommendations 2011). *Pure and Applied Chemistry*, 83, 2229–2259.
10. Chandra A. (2008). Microemulsions: An overview. http://www.pharmainfo.net/reviews/microemulsions-overview. Accessed January 6, 2014.
11. Vandamme TF. (2002). Microemulsions as ocular drug delivery systems: Recent developments and future challenges. *Progress in Retinal and Eye Research*, 21, 15–34.
12. Tenjarla S. (1999). Microemulsions: An overview and pharmaceutical applications. *Therapeutic Drug Carrier Systems*, 16, 461.
13. Aboofazeli R and Lawrence MJ. (1993). Pseudo-ternary phase diagrams of systems containing water-lecithin-alcohol-isopropyl myristate. *International Journal of Pharmaceutics*, 93, 161–175.
14. Pattarino F, Marengo E, Gasco MR, and Carpignano R. (1993). Experimental design and partial least squares in the study of complex mixtures: Microemulsions as drug carriers. *International Journal of Pharmaceutics*, 91, 157–165.
15. Kreilgaard M. (2002). Influence of microemulsions on cutaneous drug delivery. *Bulletin Technique Gattefossé*, 95, 79–100.
16. Narang AS, Delmarre D, and Gao D. (2007). Stable drug encapsulation in micelles and microemulsions. *International Journal of Pharmaceutics*, 345, 9–25.
17. Singh V, Bushettii SS, Raju SA, Ahmad R, Singh M, and Bisht A. (2011). Microemulsions as promising delivery systems: A review. *Indian Journal of Pharmaceutical Education and Research*, 45, 392–401.
18. Naimish AS, Mayur AN, Vipul PP, Samir AA, and Thusarbindu RD. Emerging trend of microemulsion in formulation and research. *International Bulletin of Drug Research*, 1, 54–83.
19. Fanun M. (2008). *Microemulsions: Properties and Applications (Surfactant Science)*, 1st edn.; CRC Press, p. 1.
20. Winsor PA. (1954). *Solvent Properties of Amphiphilic Compounds*; Butterworth Scientific, London, U.K.
21. Hasse A and Keipert S. (1997). Development and characterization of microemulsions for ocular application. *European Journal of Pharmaceutics and Biopharmaceutics*, 430, 179–183.
22. Reza N. (Ed.) (2012). *Microemulsions—A Brief Introduction, Microemulsions—An Introduction to Properties and Applications*, InTech, Available from:http://www.intechopen.com/books/microemulsions-an-introduction-to-properties-andapplications/microemulsions-a-brief-introduction.
23. Ekwall P, Danielsson I, and Mandell L. (1960). Assoziations- und Phasengleichgewichte bei der Einwirkung von Paraffinkettenalkoholen an wässrigen Lösungen von Assoziations-kolloiden. *Kolloid-Zeitschrift*, 169, 113.
24. Kyatanwar AU, Jadhav KR, and Kadam VJ. (2010). Self micro-emulsifying drug delivery system (SMEDDS): Review. *Journal of Pharmacy Research*, 3, 75–83.
25. Pouton CW and Porter CJ. (2008). Formulation of lipid-based delivery systems for oral administration: Materials, methods and strategies. *Advanced Drug Delivery Review*, 60, 625–637.

26. Kimura M, Shizuki M, Miyoshi K, Sakai T, Hidaka H, Takamura H, and Matoba T. (1994). Relationship between the molecular structures and emulsification properties of edible oils. *Bioscience, Biotechnology and Biochemistry*, 58, 1258–1261.
27. Lawrence MJ. (1996). Microemulsions as drug delivery vehicles. *Current Opinion Colloid Interface Science*, 1, 826–832.
28. Holmberg K, Jönsson B, Kronberg B, and Lindman B. (2002). *Surfactants and Polymers in Aqueous Solution*; John Wiley & Sons, Ltd., West Sussex, England.
29. Tadros TF. (2005). *Applied Surfactants Principles and Applications*; Wiley-VCH Verlag GmbH & Co. KGaA, Weinheim, Germany, p. 3.
30. Rosen MJ and Dahanayake M. (2000). *Industrial Utilization of Surfactants: Principles and Practice*; MCS Press, Urbana, IL.
31. van Os NM. (1998). *Nonionic Surfactants: Organic Chemistry*; CRC Press/Marcel Dekker, Inc., New York.
32. Constantinides P, Welzel P, Ellenns G, Smith H, Sturgis P, Li S, and Yiv SH. (1996). Water-in-oil microemulsions containing medium-chain fatty acids/salts: Formulation and intestinal absorption enhancement evaluation. *Pharmaceutical Research*, 13, 210–215.
33. Florence AT and Attwood D. (1998). *Physicochemical Principles of Pharmacy*; Macmillan, London, U.K.
34. Gupta S. (2011). Biocompatible microemulsion systems for drug encapsulation and delivery. *Current Science*, 101, 174–188.
35. Tenjarala, S. (1999). Microemulsions: An overview and pharmaceutical applications. *Critical Review in Therapeutic Drug Carrier Systems*, 16, 461–521.
36. Harsha Mohan P. (2012). Microemulsion: Prediction of the phase diagram with a modified Helfrich free energy: A thesis submitted for the degree of Doctor of Philosophy in Natural Sciences. Max Planck Institute for Polymer Research Mainz, Germany and Johannes Gutenberg University, Mainz, Germany.
37. Goodman JF, Corkill JM, and Harrold SP. (1964). Thermodynamics of micellization of non-ionic detergents. *Transactions of the Faraday Society*, 60, 202–207.
38. Lawrence MJ. (1994). Surfactant systems: Microemulsions and vesicles as vehicles for drug delivery. *European Journal of Drug Metabolism and Pharmacokinetics*, 3, 257–269.
39. Roux D and Coulon C. (1986). Modelling interactions in microemulsion phases. *Journal of Physique*, 47, 1257–1264.
40. Cho YH, Kim S, Bae EK, Mok CK, and Park J. (2008). Formation of a co-surfactant free o/w microemulsion using non-ionic surfactant minures. *Journal of Food Science*, 73, E115–E121.
41. Rataj VN, Caron L, Borde C, and Aubry JM. (2008). Oxidation in three lipid phase microemulsion systems using balanced 'catalytic surfactants. *Journal of American Chemical Society*, 130, 14914–14915.
42. Strickley RG. (2004). Solubilizing excipients in oral and injectable formulations. *Pharmaceutical Research*, 21, 201–230.
43. Gi HJ, Chen SN, Hwang JS, Tien C, and Kuo MT. (1992). Studies of formation and interface of oil-water microemulsion. *Chinese Journal of Physics*, 30, 665–678.
44. Junyaprasert VB, Boonme P, Wurster DE, and Rades T. (2008). Aerosol OT microemulsions as carriers for transdermal delivery of hydrophobic and hydrophilic local anesthetics. *Drug Delivery*, 15, 323–330.
45. Bhatia G, Zhou Y, and Banga AK. (2013). Adapalene microemulsion for transfollicular drug delivery. *Journal of Pharmaceutical Science*, 102, 2622–2631.
46. Tashtoush BM, Bennamani AN, and Al-Taani BM. (2013). Preparation and characterization of microemulsion formulations of nicotinic acid and its prodrugs for transdermal delivery. *Pharmaceutical Development and Technology*, 18, 834–843.
47. Chen L, Tan F, Wang J, and Liu F. (2012). Microemulsion: A novel transdermal delivery system to facilitate skin penetration of indomethacin. *Pharmazie*, 67, 319–323.
48. Panapisal V, Charoensri S, and Tantituvanont A. (2012). Formulation of microemulsion systems for dermal delivery of silymarin. *AAPS PharmSciTech*, 13, 389–399.
49. Patel MR, Patel RB, Parikh JR, Solanki AB, and Patel BG. (2011). Investigating effect of microemulsion components: In vitro permeation of ketoconazole. *Pharmaceutical Development and Technology*, 16, 250–258.
50. Shishu, Rajan S, Kamalpreet. (2009). Development of novel microemulsion-based topical formulations of acyclovir for the treatment of cutaneous herpetic infections. *AAPS PharmSciTech*, 10, 559–565.
51. Baboota S, Al-Azaki A, Kohli K, Ali J, Dixit N, and Shakeel F. (2007). Development and evaluation of a microemulsion formulation for transdermal delivery of terbinafine. *PDAJ PharmSciTechnol*, 61, 276–285.

52. Moghimipour E, Salimi A, and Eftekhari S. (2013). Design and characterization of microemulsion systems for naproxen. *Advanced Pharmaceutical Bulletin*, 3, 63–71.
53. Li G, Fan Y, Li X, Wang X, Li Y, Liu Y, and Li M. (2012). In vitro and in vivo evaluation of a simple microemulsion formulation for propofol. *International Journal of Pharmaceutics*, 425, 53–61.
54. Guo L, Ma E, Zhao H, Long Y, Zheng C, and Duan M. Preliminary evaluation of a novel oral delivery system for rhPTH1–34: In vitro and in vivo. *International Journal of Pharmaceutics*, 420, 172–179.
55. Nirmala MJ, Mukherjee A, and Chandrasekaran N. (2013). Design and formulation technique of a novel drug delivery system for azithromycin and its anti-bacterial activity against *Staphylococcus aureus*. *AAPS PharmSciTech*, 14, 1045–1054.
56. Dixit RP and Nagarsenker MS. (2010). Optimized microemulsions and solid microemulsion systems of simvastatin: Characterization and in vivo evaluation. *Journal of Pharmaceuticals Science*, 99, 4892–4902.
57. Wilk KA, Zielińska K, Pietkiewicz J, Skołucka N, Choromańska A, Rossowska J, Garbiec A, and Saczko J. (2012). Photo-oxidative action in MCF-7 cancer cells induced by hydrophobic cyanines loaded in biodegradablemicroemulsion-templated nanocapsules. *International Journal of Oncology*, 41, 105–116.
58. Pietkiewicz J, Zielińska K, Saczko J, Kulbacka J, Majkowski M, and Wilk KA. (2010). New approach to hydrophobic cyanine-type photosensitizer delivery using polymeric oil-cored nanocarriers: Hemolytic activity, in vitro cytotoxicity and localization in cancer cells. *European Journal of Pharmaceuticals Science*, 39, 322–335.
59. Paillard A, Hindré F, Vignes-Colombeix C, Benoit JP, and Garcion E. (2010). The importance of endo-lysosomal escape with lipid nanocapsules for drug subcellular bioavailability. *Biomaterials*, 31, 7542–7554.
60. Thakkar PJ, Madan P, and Lin S. (2014). Transdermal delivery of diclofenac using water-in-oil microemulsion: Formulation and mechanistic approach ofdrug skin permeation. *Pharmaceutical Development and Technology*, 19, 373–384.
61. Xiao YY, Liu F, Chen ZP, and Ping QN. (2011). Water in oil microemulsions containing NaCl for transdermal delivery of fluorouracil. *Yao Xue Xue Bao*, 46, 720–726.
62. Russell-Jones G and Himes R. (2011). Water-in-oil microemulsions for effective transdermal delivery of proteins. *Expert Opinion in Drug Delivery*, 8, 537–546.
63. Kitagawa S, Inoue K, Teraoka R, and Morita SY. (2010). Enhanced skin delivery of genistein and other two isoflavones by microemulsion and prevention against UV irradiation-induced erythema formation. *Chemical and Pharmaceutical Bulletin*(Tokyo), 58, 398–401.
64. Liu D, Kobayashi T, Russo S, Li F, Plevy SE, Gambling TM, Carson JL, and Mumper RJ. (2013). In vitro and in vivo evaluation of a water-in-oil microemulsion system for enhanced peptide intestinal delivery. *AAPSJ*, 15, 288–298.
65. Gundogdu E, Alvarez IG, and Karasulu E. (2011). Improvement of effect of water-in-oil microemulsion as an oral delivery system for fexofenadine: In vitro and in vivo studies. *International Journal of Nanomedicine*, 6, 1631–1640.
66. Qi J, Zhuang J, Wu W, Lu Y, Song Y, Zhang Z, Jia J, and Ping Q. (2011). Enhanced effect and mechanism of water-in-oil microemulsion as an oral delivery system of hydroxysafflor yellow A. *International Journal of Nanomedicine*, 6, 985–991.
67. Cheng MB, Wang JC, Li YH, Liu XY, Zhang X, Chen DW, Zhou SF, and Zhang Q. (2008). Characterization of water-in-oil microemulsion for oral delivery of earthworm fibrinolytic enzyme. *Journal of Controlled Release*, 129, 41–48.
68. Sintov AC, Levy HV, and Botner S. (2010). Systemic delivery of insulin via the nasal route using a new microemulsion system: In vitro and in vivo studies. *Journal of Controlled Release*, 148, 168–176.
69. Gundogdu E, Karasulu HY, Koksal C, and Karasulu E. (2013). The novel oral imatinib microemulsions: Physical properties, cytotoxicity activities and improved Caco-2 cell permeability. *Journal of Microencapsulation*, 30, 132–142.
70. Moghimipour E, Salimi A, and Eftekhari S. (2013). Design and characterization of microemulsion systems for naproxen. *Advanced Pharmaceutical Bulletin*, 3, 63–71.
71. Salimi A, SharifMakhmalZadeh B, and Moghimipour E. (2013). Preparation and characterization of cyanocobalamin (vit B12) microemulsion properties and structure for topical and transdermal application. *Iranian Journal of Basic Medical Science*, 16, 865–872.
72. Pepe D, McCall M, Zheng H, and Lopes LB. (2013). Protein transduction domain-containing microemulsions as cutaneous delivery systems for an anticancer agent. *Journal of Pharmaceutical Science*, 102, 1476–1487.

73. Hathout RM and Elshafeey AH. (2012). Development and characterization of colloidal soft nanocarriers for transdermal delivery and bioavailability enhancement of an angiotensin II receptor blocker. *European Journal of Pharmaceutical Science*, 82, 230–240.
74. Liu H, Wang Y, Lang Y, Yao H, Dong Y, and Li S. (2009). Bicontinuous cyclosporin a loaded water-AOT/Tween 85-isopropylmyristate microemulsion: Structural characterization and dermal pharmacokinetics in vivo. *Journal of Pharmaceutical Science*, 98, 1167–1176.
75. AlAbood RM, Talegaonkar S, Tariq M, and Ahmad FJ. (2013). Microemulsion as a tool for the transdermal delivery of ondansetron for the treatment of chemotherapy induced nausea and vomiting. *Colloids Surface B: Biointerfaces*, 101, 143–151.
76. Sakr FM, Gado AM, Mohammed HR, and Adam AN. (2013). Preparation and evaluation of a multimodal minoxidil microemulsion versus minoxidil alone in the treatment of androgenic alopecia of mixed etiology: A pilot study. *Drug Design Development and Therapy*, 7, 413–423.
77. Aggarwal N, Goindi S, and Khurana R. (2013). Formulation, characterization and evaluation of an optimized microemulsion formulation of griseofulvin for topical application. *Colloids Surface B: Biointerfaces*, 105, 158–166.
78. Patel HK, Barot BS, Parejiya PB, Shelat PK, and Shukla A. (2013). Topical delivery of clobetasol propionate loaded microemulsion based gel for effective treatment of vitiligo: Ex vivo permeation and skin irritation studies. *Colloids Surface B: Biointerfaces*, 102, 86–94.
79. Cao FH, OuYang WQ, Wang YP, Yue PF, and Li SP. (2011). A combination of a microemulsion and a phospholipid complex for topical delivery of oxymatrine. *Archives of Pharmacal Research*, 34, 551–562.
80. Rozman B, Zvonar A, Falson F, and Gasperlin M. (2009). Temperature-sensitive microemulsion gel: An effective topical delivery system for simultaneous delivery of vitamins C and E. *AAPS PharmSciTech*, 10, 54–61.
81. Ali J, Akhtar N, Sultana Y, Baboota S, and Ahuja A. (2008). Antipsoriatic microemulsion gel formulations for topical drug delivery of babchi oil (Psoralea corylifolia). *Methods and Findings in Experimental and Clinical Pharmacology*, 30, 277–285.
82. Subramanian N, Ghosal SK, and Moulik SP. (2004). Topical delivery of celecoxib using microemulsion. *Acta Poloniae Pharmaceutical*, 61, 335–341.
83. Li H, Pan WS, Wu Z, Li JY, and Xia LX. (2004). Optimization of microemulsion containing vinpocetine and its physicochemical properties. *Yao Xue Xue Bao*, 39, 681–685.
84. Kweon JH, Chi SC, and Park ES. (2004). Transdermal delivery of diclofenac using microemulsions. *Archieves of Pharmaceutical Research*, 27, 351–356.
85. Jurkovic P, Sentjurc M, Gasperlin M, Kristl J, and Pecar S. (2003). Skin protection against ultraviolet induced free radicals with ascorbyl palmitate in microemulsions. *European Journal of Pharmaceutical and Biopharmaceutical*, 56, 59–66.
86. Dubey R, Martini LG, and Christie M. (2013). Duel-acting subcutaneous microemulsion formulation for improved migraine treatment with Zolmitriptan and Diclofenac: Formulation and in vitro-in vivo characterization. *AAPSJ*, 16, 214–220.
87. Hou P, Cao S, Ni J, Zhang T, Cai Z, Liu J, Wang Y, Wang P, Lei H, and Liu Y. (2013). In-vitro and in-vivo comparison of T-OA microemulsions and solid dispersions based on EPDC. *Drug Development and Industrial Pharmacy*, 41, 263–271.
88. Lu Y, Wu K, Li L, He Y, Cui L, Liang N, and Mu B. (2013). Characterization and evaluation of an oral microemulsion containing the antitumor diterpenoid compound ent-11alpha-hydroxy-15-oxo-kaur-16-en-19-oic-acid. *International Journal of Nanomedicine*, 8, 1879–1886.
89. Du H, Yang X, Li H, Han L, Li X, Dong X, Zhu O, Ye M, Feng O, and Niu X. (2012). Preparation and evaluation of andrographolide-loaded microemulsion. *Journal of Microencapsulation*, 29, 657–665.
90. Lv FF, Zheng LQ, and Tung CH. (2005). Phase behavior of the microemulsions and the stability of the chloramphenicol in the microemulsion-based ocular drug delivery system. *International Journal of Pharmaceuticals*, 301, 237–246.
91. Fialho SL and daSilva-Cunha A. (2004). New vehicle based on a microemulsion for topical ocular administration of dexamethasone. *Clinical and Experimental Ophthalmology*, 32, 626–632.
92. Celebi N, Türkyilmaz A, Gönül B, and Ozogul C. (2002). Effects of epidermal growth factor microemulsion formulation on the healing of stress-induced gastric ulcers in rats. *Journal of Controlled Release*, 83, 197–210.
93. D'Cruz OJ and Uckun FM. (2001). Gel-microemulsions as vaginal spermicides and intravaginal drug delivery vehicles. *Contraception*, 64, 113–123.
94. Lou H, Qiu N, Crill C, Helms R, and Almoazen H. (2013). Development of w/o microemulsion for transdermal delivery of iodide ions. *AAPSPharmSciTech*, 14, 168–176.

95. Chaiyana W, Rades T, and Okonogi S. (2013). Characterization and in vitro permeation study of microemulsions and liquid crystalline systems containing the anticholinesterase alkaloidal extract from Tabernaemontana divaricata. *International Journal of Pharmaceuticals*, 452, 201–210.
96. Boonme P, Suksawad N, and Songkro S. (2012). Characterization and release kinetics of nicotinamide microemulsion-based gels. *Journal of Cosmetic Science*, 63, 397–406.
97. Tirnaksiz F, Kayiş A, Çelebi N, Adişen E, and Erel A. (2012). Preparation and evaluation of topical microemulsion system containing metronidazole for remission in rosacea. *Chemical and Pharmaceutical Bulletin*(Tokyo), 60, 583–592.
98. Vicentini FT, Fonseca YM, Pitol DL, Iyomasa MM, Bentley MV, and Fonseca MJ. (2010). Evaluation of protective effect of a water-in-oil microemulsion incorporating quercetin against UVB-induced damage in hairless mice skin. *Journal of Pharmacy and Pharmaceutical Science*, 13, 274–285.
99. Liu F, Xiao YY, Ping QN, and Yang C. (2009). Water in oil microemulsions for transdermal delivery of fluorouracil. *Yao Xue Xue Bao*, 44, 540–547.
100. Getie M, Wohlrab J, and Neubert RH. (2005). Dermal delivery of desmopressin acetate using colloidal carrier systems. *Journal of Pharmacy and Pharmaceutical Science*, 57, 423–427.
101. Watnasirichaikul S, Rades T, Tucker IG, and Davies NM. (2002). In-vitro release and oral bioactivity of insulin in diabetic rats using nanocapsules dispersed in biocompatible microemulsion. *Journal of Pharmacy and Pharmaceutical Science*. 54, 473–480.
102. Han M, Fu S, and Fang XL. (2007). Screening of Panax notoginsenoside water in oil microemulsion formulations and their evaluation in vitro and in vivo. *Yao Xue Xue Bao*, 42, 780–786.
103. Hathout RM and Nasr M. (2013). Transdermal delivery of betahistine hydrochloride using microemulsions: Physical characterization, biophysical assessment, confocal imaging and permeation studies. *Colloids Surface B Biointerfaces*, 110, 254–260.
104. Pakpayat N, Nielloud F, Fortuné R, Tourne-Peteilh C, Villarreal A, Grillo I, and Bataille B. (2009). Formulation of ascorbic acid microemulsions with alkyl polyglycosides. *European Journal of Pharmaceutics and Biopharmaceutics*, 72, 444–452.
105. Shevachman M, Garti N, Shani A, and Sintov AC. (2008). Enhanced percutaneous permeability of diclofenac using a new U-type dilutable microemulsion. *Drug Development and Industrial Pharmacy*, 34, 403–412.
106. Gui SY, Wu L, Peng DY, Liu QY, Yin BP, and Shen JZ. (2008). Preparation and evaluation of a microemulsion for oral delivery of berberine. *Pharmazie*, 63, 516–519.

14 Ionotropic Gelation and Polyelectrolyte Complexation Technique
Novel Approach to Drug Encapsulation

J.S. Patil, S.C. Marapur, P.B. Gurav, and A.V. Banagar

CONTENTS

14.1 Introduction	274
14.2 Ionotropic Gelation Technique	275
14.2.1 Discussion	276
14.2.2 Concept of Internal Gelation	276
14.2.3 Biopolymers Used in Ionotropic Gelation Technique	277
14.2.3.1 Alginate	277
14.2.3.2 Chitosan	278
14.2.3.3 Gelan Gum	279
14.2.3.4 Carboxymethylcellulose	280
14.2.3.5 K-Carrageenan	280
14.2.4 Effect of Divalent Cation Characteristics on Gelation Process	280
14.2.5 Factors That Affect Ionotropic Gelation	281
14.2.5.1 Concentration of Polymer and Cross-Linking Electrolyte	281
14.2.5.2 Temperature of Gelation Process	281
14.2.5.3 Cross-Linking Solution and Its pH	281
14.2.5.4 Concentration of Encapsulating Agent and Polymer	281
14.2.5.5 Concentration of Gas Generating Agent	281
14.2.6 Types of Ionotropic Gelation	282
14.2.6.1 Ionotropic Pregelation/Polyelectrolyte Complexation	282
14.2.6.2 Modified Ionotropic Gelation under a High-Voltage Electrostatic Field	282
14.2.6.3 Emulsification-Internal Gelation	282
14.2.6.4 Ionotropic Gelation Followed by Coacervation	282
14.2.6.5 Electrostatic Interaction and Chemical Reaction	282
14.2.6.6 Multipolyelectrolyte Gelispheres	283
14.2.6.7 Ionotropic Gelation Followed by Compression	283
14.3 Polyelectrolyte Complexation	283
14.3.1 PECs with Chitosan: Applications in Drug Encapsulation	284
14.3.2 Chitosan-Reinforced Nanoparticles	284
14.3.3 Classification of PECs	285
14.3.3.1 PEC Formed between Biopolymers	285
14.3.3.2 PEC Formed between Natural and Synthetic Polymer	285

    14.3.4 Mechanism of Gelation...285
        14.3.4.1 Physical or Reversible Gelation 285
        14.3.4.2 Heating or Cooling a Polymer Solution...285
        14.3.4.3 Gelation through Ionic Interaction 286
        14.3.4.4 Complex Coacervation Gelation..286
        14.3.4.5 Gelation through Hydrogen Bonding 286
        14.3.4.6 Heat-Induced Aggregation or Maturation Gelation................................286
        14.3.4.7 Gelation through Freeze–Thawing..286
        14.3.4.8 Chemical or Irreversible Gelation 286
        14.3.4.9 Gelation through Use of Chemical Gelling Agents................................286
        14.3.4.10 Gelation through Grafting...287
    14.3.5 Mechanism of Complex Formation 287
    14.3.6 Merits of Ionotropic Gelation and Polyelectrolyte Complexation Technique 287
    14.3.7 Demerits of Ionotropic Gelation and Polyelectrolyte Complexation Technique 288
14.4 Common Laboratory Method of Encapsulation via Ionotropic Gelation............................288
14.5 Conclusion 290
References..291

14.1 INTRODUCTION

Continuous efforts are under progress in the field of advanced drug-delivery research. Pharmaceutical scientists, along with bioengineers, biophysicists, and biomedical scientists, are actively being involved in the interdisciplinary research activities for the formulation and optimization of advanced drug-delivery system. A well-designed controlled drug-delivery system is always capable of delivering the agent to the target tissue in the optimal amount in the specified period of time there by causing little toxicity and minimal side effects with maximum therapeutic efficacy. Recently, many efforts are being made on the fabrication of polymeric microparticles, also sometimes termed as microcapsules or microspheres. A typical microsphere and microcapsule are given in Figure 14.1.

Exhaustive research was carried out to evaluate physicochemical characteristics of microparticles, particularly release behavior under *in vitro* and *in vivo* conditions. Microspheres are characteristically free-flowing particulate systems consisting of proteins or synthetic or natural polymers, which are biodegradable in nature and ideally having a particle size less than 200 μm. Encapsulation has attracted considerable interest as a technology and its advancement will not only stimulate the exploration of new drug delivery systems, but it will also lead to engineering revolutions and, as a consequence, become a driving force for the therapy and diagnosis of numerous diseases in the current century. Biodegradable polymeric particles, especially nanoparticles, have attracted considerable attention as potential drug controlled delivery devices.[1] Microencapsulation is a process by which very minute droplets or particles of liquid or solid material are surrounded or coated by a continuous film/layer of polymeric material. This technique had its origin in the late 1930s as a cleaner substitute for carbon paper and carbon ribbons as sought by the business machines industry. The ultimate development in the 1950s of reproduction paper and ribbons that contained dyes in minute gelatin capsules released on impact by a typewriter key or the pressure of a pen or pencil was the stimulus for the development of a host of microencapsulated materials, including drugs.[2,3] Microencapsulation also includes restricted entrapment of a biologically active substance such as DNA to entire cell or group of cells. This is done in order generally to improve its performance and/or enhance its shelf-life.[4,5] Microencapsulation technique offers the methodology to convert the liquids to solids and alter the colloidal and surface properties. This process also provides an environmental protection to the entrapped substance and controls the release characteristics or availability of coated materials. Encapsulation of cell or drug substances offers many advantages. For instance, oxygen, moisture, or light-sensitive agents can be stabilized, incompatibility among the drugs can be prevented, and vaporization of volatile drugs can be avoided. GI irritation and toxicity have been reduced for many drugs including ferrous sulfate and

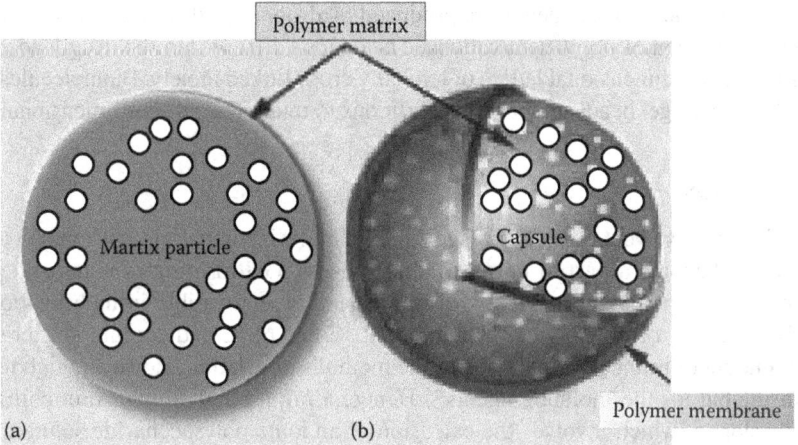

FIGURE 14.1 Picture showing typical (a) microsphere and (b) microcapsule.

KCl by microencapsulation. In this chapter, a comprehensive attempt has beenmade to discuss the different aspects of ionotropic gelation and polyelectrolyte complexation approaches by considering these techniques as great promising tools for the development of encapsulation process.

14.2 IONOTROPIC GELATION TECHNIQUE

The trend of utilizing natural and chemically modified polysaccharides as a part of drug development has increased in the past two decades. Great attention has also been paid on development of biopolymer-based hydrogels for use as potential carriers in controlled drug delivery.[6,7] Hydrogels beads are three-dimensional, hydrophilic networks capable of imbibing large amount of water or biological fluids, mimics biological tissues.[8] Due to the uniqueness of this nature, great attention was devoted to these systems for biomedical applications. Indeed, these networks can be made suitable as the modulated drug delivery devices by tuning the physicochemical properties of the hydrogels with varying the degree of crosslinking either by physical or chemical or both physical–chemical means. The polymeric gel beads are prepared by using number of natural, biodegradable polymers. Beads can provide sustained release properties and a more uniform distribution of drugs, including inside the gastrointestinal tract.[9] Although the encapsulation techniques have become popular, all are mainly based on use of organic solvents. There may be a possibility of toxicity in chronic dosing due to presence of even traces of organic solvents in the final drug products, the flammability, the environmental pollution associated with stringent governmental regulations that restricts their use, all these have raised question about its long-term viability.[10] To overcome such demerits of the popularized encapsulation techniques, consequently, much research efforts have been concentrated on the development of natural polymer-based hydrogel beads that do not require organic solvents, are easily available, and qualify for a number of chemical modifications.[11] Drug-loaded hydrogel beads offer an inert environment within the matrix and encapsulation is usually achieved in a media free of organic solvents. The concept of cross-linking of hydrogels in the presence of polyelectrolyte as counterions is termed as ionotropic gelation. Ionotropic gelation is based on the ability of polyelectrolytes to cross-link in the presence of counterions to form hydrogels. Since the use of alginates, gellan gum, chitosan, and carboxymethylcellulose (CMC) for the encapsulation of drug and even cells, ionotropic gelation technique has been widely used for this purpose.[12] The natural based polyelectrolytes in spite of, having a property of coating over the particulate surface of drug core and works as release rate retardants contains certain anions on their chemical structure. These anions form meshwork structure when exposed with the polyvalent cations and induce gelation by binding mainly to the anion blocks. The hydrogel beads are produced by dropping a drug-loaded polymeric solution into the aqueous solution of polyvalent cations. The cations diffuse into the drug-loaded polymeric drops, forming a three-dimensional lattice of ionically cross-linked moiety. Biomolecules can also be loaded into these hydrogel beads under mild conditions to retain their three-dimensional structure.

14.2.1 Discussion

Utilization of organic solvents or other reagents, which are incompatible with many of biological encapsulants, has become a part of encapsulation process. Gel-forming proteins and polysaccharides can be used alternatively to synthesize milder and biocompatible immobilization materials and methods.[13] Heating of the polymer solution at 40°C and 60°C and cooling it after the addition of sufficient amount of immobilizant is one of the methods of gelation. However, high temperatures are often unsuitable for thermolabile agents.[14] Hence, a milder and simpler immobilization technique was developed, which involves the extrusion of an ionic polysaccharide solution (immobilizant) drop-wise into a solution of a divalent cation usually calcium. Mixing of the content is done during gel formation; otherwise, aggregation of microparticles may result.[15] The calcium cation diffuses rapidly inward, forming a calcium-polysaccharide gel.[16] Extrusion method suffers from many demerits such as limited efficiency of size reduction due to nozzle diameter as well as the viscosity of the solution, and the method is not suitable for industrial scale-up for producing microparticles on a large scale, which requires a large number of nozzles to be operated simultaneously. Finally, microparticles tend to be teardrop-shaped due to drag forces following impact with the gelation bath. In order to solve these problems, several techniques have been developed such as the use of multiple needles, electrostatics, vibration, droplet propulsion from the needle tip by concentric airflow, and liquid jet cutters. To produce the smaller microspheres, atomizing spray techniques have also been tested, but due to higher rates, shearing effects in such a system could be harmful to many biological encapsulants.[14] Encapsulation devices have emerged commercially and became popular among droplet extrusion technologies. To produce hydrogel beads in the millimeter size range, high production rates are feasible and multiple needles permit small-scale industrial production, but once again further large scale-up and size reduction could be limited. Emulsion/gelation or polymerization methods can be used to solve some of the problems associated with droplet extrusion technologies. For example, warm carrageenan/oil emulsions were chilled in cold water,[17] which involves use of elevated temperatures that again, may not be suitable for heat-sensitive material and oil/alginic acid emulsions were extruded drop-wise into calcium chloride solution[12] where particle size cannot be easily controlled and particles tend to aggregate before hardening properly.[18]

14.2.2 Concept of Internal Gelation

Although several industrial methods based on *in situ* gelation of alginate existed, a method to produce alginate gel slabs was first proposed[19] in a procedure termed "internal gelation." For production of gel microbeads, another innovative procedure termed emulsification/internal gelation was developed based on the concept of internal gelation. In this method, the gelation reaction is based on the use of insoluble calcium microcrystals dispersed in polysaccharide aqueous solution. This mixture is emulsified into an oil phase containing surfactant as an emulsifying agent. To release the calcium from calcium complex, the pH of the mixture is reduced from 7.5 to 6.5 that triggers gelation to form Ca-polysaccharide.[16] This innovative technique is fundamentally based on cross-linking of the polysaccharide residues and makes possible the immobilization of the labile biological agents such as DNA and proteins.[20–23] Formulation parameters including the nature, composition, and structure of polysaccharide used and form and concentration of the divalent cation vector are significant to the process. Also, pH range, type of oil, emulsifying agents, and the type of acid represent key parameters to the development of this method. This method has been only applied so far to produce microspheres and extrapolation of the method to produce nanospheres is described. The material to be encapsulated is dissolved or entrapped within or attached to a nanoparticle matrix and, depending upon the method of preparation, nanospheres or nanocapsules can be obtained.[24]

14.2.3 BIOPOLYMERS USED IN IONOTROPIC GELATION TECHNIQUE

Over the last few years, there has been a growing interest by medical and pharmaceutical industries regarding the use of natural polymers as drug carriers in ionotropic gelation technique. This may be mainly due to their biocompatibility and biodegradability. Utilization of polymers produced from renewable natural resources has become increasingly important because of their low cost, easy availability, aqueous solubility, and gel-forming ability. Applications of agar as conventional solidifying agent used in the field of microbiology and biotechnology is limited because of its instability and raising costs due to the shortage of resources of the species of red algae from which it is isolated.[25] Due to this factor, a considerable investigation has been done in an attempt to find new polymers and gelling systems that can act as agar substitutes. The natural polymers such as alginates, gellan gum, chitosan, and pectin are widely use for the entrapment of drug by this technique. These natural polyelectrolytes contain certain anions/cations that leads to formation of meshwork structure by combining with the counterions and induce gelation by cross-linking. These natural polymers additionally act as release rate retardant, in spite of having a property of coating on the drug core.

14.2.3.1 Alginate

Alginate has become the most widely used encapsulation matrix, for biological materials including plant cells,[26] mammalian cells,[27] yeasts,[28] bacteria,[29] insulin,[30] and food products. Alginate forms stable reversible gels in the presence of multivalent cations under gentle preparation conditions at room temperature. Alginate is cheap and widely available in different grades. Alginate has attracted most of the formulation scientists for its several unique properties that have enabled its use as a matrix for entrapment and/or delivery of a variety of proteins and cells.[27] During the last two decades, number of quality alginate suppliers also increased considerably and alginates are now being sold partially or fully characterized in terms of its physicochemical properties.

14.2.3.1.1 Characteristics of Sodium Alginate

Alginate is a nontoxic, biodegradable, naturally occurring polysaccharide obtained from all species of brown algae and certain species of bacteria. Chemically sodium alginate is a sodium salt of alginic acid, a natural polysaccharide, and a linear polymer composed of 1,4-linked β-D-mannuronic acid (M) and α-D-glucuronic acid (G) residues in varying proportions and arrangements. The homopolymer regions composed of M-blocks and G-blocks are interspersed with M G heteropolymeric regions known as "egg box junction."[31] The solubility of alginate in water depends on the associated cations and pH. Sodium alginate is soluble in water and forms a reticulated structure, which can be cross linked with divalent or polyvalent cations to form insoluble meshwork. Calcium and zinc cations have been reported for cross-linking of acid groups of alginate. Composition and concentration of alginate are the important parameters in alginate particle formation. There will be a seasonal variation in monomer content and viscosity of alginates in brown seaweed, usually increasing from the lowest at the youngest stage of the plant to the maximum at a mature stage. This is an important property, since sodium and calcium percentages will strongly influence the swelling and healing properties of the alginate.[32] The average chain length of glucuronic blocks decides the reactivity with calcium and the subsequent gel formation capacity. Initially, the divalent calcium cation begins to fit into the glucuronate structures like eggs in an "egg box junction." Consequently, this binds the alginate chains together by forming junction zones, sequentially leading to gelling of the solution mixture and bead formation. When aqueous solution of sodium alginate is added drop-wise to an aqueous solution of divalent metal ions such as calcium, barium, and stannous, the rapid ion binding and formation of a polymeric network produces an inwardly moving gelling zone. In fact, alginate moves from the gel core toward this gelling zone, leading to the deletion of alginate within the core. Polymer gradient is essentially governed by the relative diffusion rate between the soluble alginate molecules and the gel forming ions[33] finally leads to the formation of a spherical gel with regular shape and size, also known as an "alginate bead." Alginate beads have the advantages of being nontoxic orally, high biocompatibility, inability to reswell in acidic environment, and easily reswell in an alkaline environment, thereby protecting the acid-sensitive drugs from gastric environment.[34] Viscosity of alginate solution typically changes in proportion to the glucuronic acid content. Alginate gelling takes place in a highly cooperative manner by preferential binding of divalent cations to glucuronic acid blocks and the size of the cooperative unit is reported to be more than 20 monomers.[35] A stronger and stable gel is formed by higher interaction between alginate and calcium due to high glucuronic content and homopolymer blocks. However, larger beads with larger dispersions[36] and more porous gels are formed in the emulsification step due to high glucuronic content, which gives premature gelation. On the other hand, more elastic weaker gels with good freeze–thaw behavior are formed with a high mannuronic content. Almost no spherical particles are formed below 1.0% sodium alginate concentration; this may be probably due to the lack of enough carboxyl groups for gelation. Viscosity of the alginate aqueous solutions proportionally increases with increase in sodium alginate concentration, resultings in larger droplets with a wide distribution.[37] In order to control the particle size, shape, and distribution of intended alginate particles for a given application, utmost priority is necessary to optimize the concentration of alginate. Thus, sodium alginate appeared to be highly promising polymer utilized in gelation process owing to its nontoxic, biodegradable, and biocompatible nature. Its unique property of forming water-insoluble calcium alginate gel through ionotropic gelation with calcium ions is a simple, mild, and eco-friendly condition that has made possible the encapsulation of macromolecular bioactive agents like cell, enzyme, protein, and vaccine.[38–41] Recently, much research efforts have been concentrated to develop calcium alginate beads loaded with various low molecular weight therapeutic agents such as stavudine.[42] In various studies, alginate beads have been used as excellent vehicles. Another important property of alginate beads is their reswelling ability. This property is sensitive to the environment pH. Hence, acid-sensitive drugs incorporated within the beads can be protected from the gastric juice.[43]

14.2.3.2 Chitosan

Chitosan is a linear randomly distributed, binary heteropolysaccharide prepared by deacetylation of chitin, a linear polymer of β(1–4) linked N-acetyl-D-glucosamine units composed of mucopolysaccharides and amino sugars. It is a natural poly-(aminosaccharide), having structural characteristics similar to glycosaminoglycans.

14.2.3.2.1 Characteristics of Chitosan

Chitosan is broadly classified as chitosan oligomer which is composed of about 12 monomer units and chitosan polymer with more than 12 monomer units. It is further subdivided into three different types, namely, low molecular weight chitosan of less than 150 kDa, high molecular weight chitosan of 700–1000 kDa, and medium molecular weight chitosan.[44] The degree of deacetylation of typical commercial chitosan usually ranges between 66% and 95%.[45,46] Chitosan in its solid state exists in a semicrystalline form, which exhibits polymorphism.[47] Chitosan is readily soluble in aqueous acidic solution after the removal of acetyl moieties that are present in the amine functional groups. Protonation of amino functional groups on the C-2 position of D-glucosamine residues of chitosan whereby polysaccharide is converted into polycation in acidic media and solubilization of chitosan takes place thereby. Chitosan has a low solubility at physiological or higher pH as it is a weak base with pK_a values ranging from 6.2 to 7.[48,49] Adjusting solution pH to approximately 7.5 induces flocculation due to deprotonation and insolubility of this polysaccharide.[47] When the degree of deacetylation of chitosan exceeds 50%, its solubility in acidic media is enhanced. Whereas with

40%–50% degree of deacetylation, it has been claimed to soluble in an aqueous medium with pH up to 9 and in neutral solution respectively.[47,48,50] It has been suggested that water solubility of chitosan results from random distribution of N-acetyl groups.[51] In the presence of both organic and inorganic acids, chitosan forms salt.[48,49] But salting-out of chitosan can be observed when chitosan forms chitosan chlorohydrate in the presence of excessive hydrochloric acid.[47] Addition of electrolyte to the aqueous solution of chitosan greatly enhances its solubility.[48,50] Properties such as molecular weight, ionic strength, pH, and temperature of chitosan solution are largely responsible for the viscosity of chitosan solution. The viscosity of chitosan solution increases as the molecular weight, concentration, and degree of deacetylation of chitosan oligomers increase[49] but decreases with a decrease in solution temperature and pH.[48,50] Chitosan is a biopolymer, which could be used for the preparation of various polyelectrolyte complex products with natural polyanions such as xanthan, alginate, and carrangeenan. Chitosan–polyanions complexes have been widely investigated for the applications like drug and protein delivery, cell transplantation, and enzyme immobilization. Among these complexes, chitosan–alginate hydrogel complex system might be the most important drug delivery system. The chitosan–alginate complex forms due to the strong electrostatic interaction of amine groups of chitosan with the carboxyl groups of alginate. Due to the protonation of amino group on chitosan and the ionization of carboxylic acid group on alginate, the stability of chitosan is influenced by environmental parameters such as pH and ionic strength. It was found that the macromolecular chitosan rapidly binds onto the surface of alginate droplet, but there is limited scope to diffuse into the inner core. In the gelation of alginate, calcium chloride can be used in order to increase the stability of chitosan–alginate complex. As the concentration of calcium chloride increases, the rate and extent of chitosan binding process also proportionally increases.[43]

14.2.3.3 Gelan Gum

Gelan gum is a bacterial exopolysaccharide prepared commercially by aerobic submerged fermentation of *Sphingomones elodea*. Deacetylated gelan gum is obtained by alkali treatment of the native polysaccharide. Both native and deacetylated gellun gum are capable of physical gelation. A concentrated water solution of gellan gum warmed up first preliminary to induce the gelan gelation. When the temperature is decreased, the chains undergo a conformational transition from random coils to double helices (coil–helix transition). Then rearrangement of a double helices occurs, leading to the formation of ordered junction zones (sol–gel transition), thus giving a thermo-reversible hydrogel. Much stronger physical thermoreversible hydrogels are also obtained by the addition of monovalent and divalent cations to gel solutions. The physical gelation ability of this polysaccharide makes it suitable as structuring and gelling agent in foods and toothpastes, binder, and as a sustained release matrix. The gellan gum is also utilized in the fields of modified release of bioactive molecules. Aqueous solutions of gelan are used as *in situ* gelling systems, mainly for ophthalmic preparations, and for oral drug delivery. Physical gellan hydrogels, prepared with mono- or divalent cations, are used also for the preparation of tablets, beads, or microspheres. Interpenetrating polymer networks or cocross linked polymer networks based on gelan and other polysaccharide systems have also been developed as drug delivery matrices.[43]

14.2.3.3.1 Characteristics of Gelan Gum

Temperature-dependent and cation-induced gelation is the characteristic property of gelan gum. The gelation process involves the formation of double helical junction zones followed by a gelan gum gelation of the double helical segments to form a three-dimensional network by complexation with cations and hydrogen bonding with water.[52] There are two chemical forms of gelan gum; these are natural forms, which has high acyl contents, and low or deacetylated form. Both forms have a similar linear structure. The natural form of gelan is composed of the linear structure of a repeating tetrasaccharide unit of glucose, glucuronic acid, and rhamnose in a molar ratio of 2:1:1. It is partially acetylated with acetyl and L-glyceryl groups located on the glucose residues.[53]

The native form contains two acyl substituents, namely, acetate and glycerate, both located on the same glucose residue and on the average, there is one glycerate and a half acetate group per every tetrasaccharide repeating unit. This difference in substitutions leads to a difference in the gelling potential. However, the presence of acetyl groups interferes with the ion bonding ability. Because of the presence of free carboxylate groups in gelan gum, it becomes anionic in nature and thus, would possess the characteristic property of undergoing ionic gelation in the presence of mono- and divalent cations.[53] However, the affinity for divalent cations is much stronger than the monovalent cations.[54] Swelling ability of this gum is one of the most remarkable and useful features, as its swelling can be stimulated by changing the surrounding environment of the delivery system. Based on the polymer nature, the change in environmental may be pH change, temperature, or ionic strength. So the systems can either shrink or swell. The drug release is achieved only when the polymer swells and because gelan gum is potentially most useful pH-sensitive polymer that swells at high pH values and collapse at low pH values. Thus, gellan gum is considered as most ideal natural polymer for oral delivery systems, in which the drug is not released at low pH values in the stomach, but rather at high pH values in the upper small intestine.[55–59]

14.2.3.4 Carboxymethylcellulose

Naturally occurring biopolymers can be suitably tuned by chemical modification to change the properties in order to make them useful in the development of sustained, modulated drug delivery systems. CMC is a chemically modified semisynthetic polymer that can be prepared by carboxymethylation process from a plant product called cellulose. CMC is a chemically modified biopolymer that was first prepared in 1918 and was produced commercially in the early 1920s. Today CMC is available in different quality and is employed in many areas of industry and human life. The interactions of the carboxylic groups of the CMC with multivalent metal ions can be used to form so-called ionotropic gels, which are predominantly stabilized by the electrostatic interactions. In addition, interactions between the hydroxy groups of the polymer and the metal ions contribute to the stability and the water insolubility of these polymeric aggregates. The CMC can be cross-linked with ferric/aluminum salt to get biodegradable hydrogel beads. Controlled release pattern can also be improved by coating these hydrogels with chitosan/gelation and by cross-linking.[43]

14.2.3.5 K-Carrageenan

Carrageenan is a naturally available polysaccharide that is extracted from marine macroalgae. This polysaccharide can be dissolved in aqueous medium at elevated temperatures between 60°C and 80°C at the concentrations ranging from 2% to 5% w/v. Temperature plays an important role in the gelation of this polymer. Gelation of this gum is generally dependent on a change in temperature. The cell slurry is added to the heat-sterilized carrageenan solution at 40°C–45°C and gelation occurs by cooling to room temperature. The beads are formed after dropping the mixture of polymer and cells into a potassium chloride (KCl) solution. K-carrageenan beads have found applications in encapsulating probiotics[60] and this natural polysaccharide is commonly used as a food additive.

The ionotropic gelation technique utilizes other polymers to a lesser extent; these are proteins such as gelatin and egg whites, semisynthetic polymers like hydroxyl propyl methyl cellulose, and synthetic polymers like polyacrylamides.

14.2.4 EFFECT OF DIVALENT CATION CHARACTERISTICS ON GELATION PROCESS

Alginate gels in the presence of divalent cations is converted into insoluble hydrogels. Such gels can be heat treated without melting, although they may degrade at the end. Control of cation addition is an important criterion in the gelling process to produce homogeneous gels. Due to the additional merits of calcium ions such as clinical safety, easy accessibility, and economic affordability, it is being used as a main cation even though alginate particles can be produced using zinc ions. Structurally, calcium ions are located in electronegative cavities, being described in the literature as an egg-box model. Ionic interactions between glucuronate blocks and calcium ions cause the formation of a strong thermostable gel. Properties of final gels produced largely depend on the characteristics of the polymer used and the preparation method adopted. The desired range of pH values decides the type of suitable calcium vector to be selected for internal gelation of alginate. Over the pH range of interest, the concentration of free calcium must be very low initially with rapid release of calcium while reducing pH. Gelation varied significantly with the pH range. Different calcium salts such as oxalate, tartrate, phosphate, carbonate, and citrate may be used. Hydrogel beads were not formed with calcium oxalate or with calcium tartrate with no release of calcium from the complexes within a suitable pH range was found, which has been proved by some studies.[61] The large grain size of calcium phosphate limits its use as a source of calcium, additionally incomplete gelation resulted in bead clumping with the formation of salt agglomerates at the centre of the beads.[61] More spherical and stable beads with a smaller unimodal size distribution can be typically obtained when calcium citrate is used as a source of calcium. Calcium carbonate-alginate suspension remained stable in a wide pH range.[61] To obtain the hydrogel particles in few micron sizes, preferably use of calcium carbonate microcrystals is convenient. The resulting particles are more spherical and less likely to aggregate.[62]

14.2.5 Factors That Affect Ionotropic Gelation[63]

Ultimate properties of hydrogel beads prepared by ionotropic gelation depend mainly upon different factors, which affect the ionic gelation process. Some of the factors which affect ionotropic gelation are discussed in the following.

14.2.5.1 Concentration of Polymer and Cross-Linking Electrolyte

Concentration of both polymer and cross-linking electrolyte play a major role in deciding the properties of intended hydrogel beads. Concentration of both polymer and electrolyte is calculated based on the ratio of number of cross-linking units present. Drug encapsulation efficiency of the prepared hydrogel beads depends mainly on type and concentration of electrolytes used in the process.

14.2.5.2 Temperature of Gelation Process

Size of the particulate hydrogel beads depends on temperature at which ionotropic gelation process takes place. Curing time or time required for cross-linking of the beads also depends on temperature condition.

14.2.5.3 Cross-Linking Solution and Its pH

Rate of ionotropic gelation reaction, shape, and size of particulate beads depend on pH of cross-linking solution.

14.2.5.4 Concentration of Encapsulating Agent and Polymer

It is very important to select a proper ratio of substance to be encapsulated and a polymer used to encapsulate. The ratio of ensapsulant to encapsulating agent greatly affects the preliminary characteristics such as entrapment efficiency, size, and shape of the hydrogel beads.

14.2.5.5 Concentration of Gas Generating Agent

Electrolytes such as calcium carbonate, sodium bicarbonate which upon contact with water generates gas are added in to the formulation which tremendously affects the size and shape of the particulate product. Presence of such gas-generating agent produces porous gelispheres, breaks the lining of gelispheres, and results into the irregular surface.

14.2.6 Types of Ionotropic Gelation

14.2.6.1 Ionotropic Pregelation/Polyelectrolyte Complexation

Hydrogel beads prepared by ionotropic gelation method can be further coated by suitable oppositely charged cations through polyelectrolyte complexation technique in order to improve the quality of hydrogel beads. The mechanical strength and permeability barrier of hydrogels can be improved by the addition of oppositely charged another polyelectrolyte to the ionotropically gelated gelispheres. For instance, addition of polycations allows a membrane of polyelectrolyte complex to form on the surface of alginate gelispheres. Polyelectrolyte beads of ampicillin prepared by ionotropic gelation method using alginate and chitosan for complexation are reported with enhancement in encapsulation efficiency and improved properties of controlled release of formed multilayer ampicillin.[64]

14.2.6.2 Modified Ionotropic Gelation under a High-Voltage Electrostatic Field

A modified ionotropic gelation method by combining it with a high voltage electrostatic field was reported to prepare protein-loaded chitosan microspheres. This is a new method for sustained delivery of bovine serum albumin by encapsulating in chitosan that exhibited good sphericity and dispersibility. The results from the literature survey suggest that ionotropic gelation method combined with a high voltage electrostatic field is an effective method for sustained delivery of protein by hydrogel beads.[65]

14.2.6.3 Emulsification-Internal Gelation

Ionotropic gelation technique can be improved by the incorporation of oily phase and emulsifier. Emulsification-internal gelation has been suggested as an alternative to extrusion/external gelation in the encapsulation of several compounds including sensitive biologicals such as protein drugs. Protein-loaded microparticles offer an inert environment within the matrix and encapsulation is conducted at room temperature in a media free of organic solvents. Recently, the concept of internal gelation has been applied to formulating nanoparticles as drug delivery systems. Emulsification/internal gelation technologies available for microparticles preparation, particularly involving alginate polymer, as well as recent advances towards applications in nanotechnology are described. These methods show great promise as a tool for the development of encapsulation processes, especially for the new field of nanotechnology using natural polymers.[66]

14.2.6.4 Ionotropic Gelation Followed by Coacervation

Ionotropic gelation technique is further modified to improve the stability and physicochemical properties of beads by coacervation after ionotropic gelation. This successfully developed a new encapsulation method involving two polymers (alginate and chitosan) and using methods of functionalization (acylation) and ionotropic gelation followed by coacervation. Beads were reported to form by ionotropic gelation via calcium cross-linking and by alginate–chitosan complex coacervation. The main difference between native and functionalized beads is the presence of fatty acid chains in the core (palmitoylated alginate) and external layer (palmitoylated chitosan) of beads. Hence, alginate cross-links improved insolubility of beads by ionotropic gelation and alginate–chitosan coacervation, which led to polyionic links between the core bead and the external layer. Functionalization increases hydrophobic interactions in the polymeric matrix involving structural changes that also improves the polymers barrier property by decreasing water uptake and water vapour pressure.[67]

14.2.6.5 Electrostatic Interaction and Chemical Reaction

Alginate-poly (ethylene glycol) hybrid gelispheres can be prepared by a new approach, which involves the combination of electrostatic interaction of calcium ions with sodium alginate and the chemical reaction of vinyl sulfone-terminated poly (ethylene glycol) (PEG-VS) with threo-1,4-dimercapto-2,3-butanediol. One-step extrusion process under physiological conditions yields calcium alginate-poly (ethylene glycol) hybrid microspheres, an interpenetrating network with well-controllable physical properties. It was reported that the permeability of the hydrogel can be tailored by adequate choice of the arm length of PEG-VS, while the swelling degree can be tuned by varying the PEG-VS concentration and/or by liquefaction of calcium-alginate. Overall, important physical properties of the hydrogel spheres are obtained in the range desired for biotechnological, biomedical, and pharmaceutical applications.[63]

14.2.6.6 Multipolyelectrolyte Gelispheres

In this technique, the polymer solution contains both sodium alginate and pectin for cross-linking and added to cellulose acetophthalate solution. To this multicomponent solution, a cross-linking acidified solution of calcium chloride is incorporated. Gelispheres are formed by titration of the polymer suspension at 2 mL/min with the cross-linking solution using flat-tip 19-guage opening. The gelispheres formed are allowed to cure for period of 24 h at 218°C, then the cross-linking solution is decanted and gelispheres are washed and dried for 48 h at 218°C under extractor.[68]

14.2.6.7 Ionotropic Gelation Followed by Compression

A new method for alginate-hydroxyethylcellulose gelispheres for controlled intrastriatal nicotine release in Parkinson's disease was developed. Hydroxyethylcellulose was incorporated as a reinforcing protective colloidal polymer to induce interactions between the free carboxyl groups of alginate with hydroxyethylcellulose monomers. Further to prolong the release of nicotine, gelispheres were compressed within an external poly (lactic-co-glycolic acid) (PLGA) matrix.[69]

14.3 POLYELECTROLYTE COMPLEXATION

Macromolecular compounds, which when are dissolved in a suitable polar solvent, generally water, spontaneously acquire or can be made to acquire a large number of elementary charges distributed along the macromolecular chain that are generally denoted by a term polyelectrolyte. In its uncharged state, a polyelectrolyte behaves like any other macromolecule, but the dissociation of even a small fraction of its ionic groups leads to dramatic changes of its properties. The quality of hydrogel beads prepared by ionotropic gelation method can also be further improved by polyelectrolyte complexation (PEC) technique. The mechanical strength and permeability barrier of hydrogels can be improved by the addition of oppositely charged another polyelectrolyte to the ionotropically gelated hydrogel beads. For instance, addition of polycations allows a membrane of polyelectrolyte complex to form on the surface of alginate beads.[70] When two macromolecules in liquid form are mixed together, interchain interaction takes place, which finally leads to polymer–polymer complexation.[71,72] The complexation reaction may be driven by Coulombic attraction or other interactions, such as hydrogen bonding, dipole–dipole interaction, charge-transfer interaction, and the hydrophobic effects. This phenomenon of polyelectrolyte complexation is being effectively and practically applied in the fields of microencapsulation of water-insoluble molecules, such as pigments, drugs, or proteins,[73,74] protein separation and purification[75] models for understanding protein–nucleic acid interactions in cells, and gene therapy.[76] Polyelectrolyte complexes are generally formed by electrostatic interaction between two oppositely charged polyelectrolyte solutions.[77–79] These complexes exhibit unique physical and chemical properties, as the electrostatic interactions within the polyelectrolyte complex gels are considerably stronger than most secondary binding interactions. The use of polyelectrolyte complex of chitosan and polyanions for pharmaceutical applications has been reported by many researchers.[80,81] Chitosan is a naturally occurring cationic polysaccharide derived from the N-deacetylation of chitin and at acidic pH ranges, the ionizable amino groups in chitosan molecules are protonated. Different studies have reported the formation of PEC with polyanionic molecules.[82,83] The properties of PEC depend on various factors such as position of the ionic groups, charge density, concentration of anionic and cationic polymers, proportion of opposite charges, molecular weight of the macromolecules, and condition of synthesis.[84-89] The polymer such as alginate has very high hydrophilicity due to presence of both its carboxylic and hydroxyl groups in the solution forms inter polymer complex with chitosan was investigated in many researches.[90-93] The drug release from *in situ* polyionic complexes has been proved to show comparatively more sustained effect than the single polymer. Polymer solutions can interact to give a superior quality complex for the dosage form.

14.3.1 PECs with Chitosan: Applications in Drug Encapsulation

Among different polymers, chitosan, the N-deacetylated product of the polysaccharide chitin, is gaining increasing importance in medical and pharmaceutical applications due to its good mucoadhesion and absorption enhancing ability. Moreover, chitosan shows the ability to form hydrogels that are able to control the rate of drug release from the delivery system as well as protect the drug from chemical and enzymatic degradation in the administration site. In particular, when chitosan is cross-linked or complexed with an oppositely charged polyelectrolyte, a three-dimensional network is formed in which the drug can be incorporated in order to control its release. PECs with chitosan have been developed for local or systemic administration of drugs and biodrugs. Precise drug delivery requires that the stimuli-responsive mechanism in the delivery system only responds to the physiological or pathological conditions particular to the treatment site. Hence, continuous efforts are being focused by many research groups on designing delivery systems with improved site specificity and versatile drug release kinetics to accommodate different therapeutic needs. Today's research concerns are the development of innovative drug delivery systems based on new natural polymeric materials and innovative preparative technologies. Application of chitosan has been extensively investigated in various fields, such as biotechnology,[94] oral dosage forms,[95] wastewater treatment,[96] cosmetics,[97] and food science.[98] These broad fields of applications are due to its unique set of properties. Favorable biological properties, low toxicity, biodegradability, proper mucoadhesive properties, absorption enhancer across intestinal epithelium and mucosal sites for drugs, peptides and proteins, metal complexation, antimicrobial activity, good homeostatic properties, and acceptable anticancerous properties[99-107] are the most important set of properties. Among all applications of chitosan, drug delivery systems are widely recognized as a promising research field.[108]

14.3.2 Chitosan-Reinforced Nanoparticles

In the past 20 years, chitosan nanoparticles have been extensively studied, because they can control drug release rate, prolong the duration of action, and drug delivery to the specific site.[109] Owing to their small and quantum size effect, chitosan nanoparticles exhibit superior activity.[110] Chitosan nanoparticles are commonly prepared by ionotropic gelation, because it is a very simple and mild method. This process can be completed either by chemical or physical cross-linking. The abundant amine groups of chitosan are protonated to form $-NH_3+$ in acidic solution. These positively charged groups can be chemically cross-linked with dialdehydes such as glutaraldehyde[111] or physically cross-linked with multivalent anions derived from sodium tripolyphosphate (TPP),[112] citrate,[113] and sulphate.[114] Glutaraldehyde is toxic and can cause irritation to mucosal membranes.[115] Physically cross-linked chitosan gels have been used in drug delivery systems due to their enhanced biocompatibility over chemically cross-linked chitosan.[116] Due to the nontoxic, and quick gelling ability, TPP is considered as a favorable cross-linking agent for ionic gelation of chitosan[107] and in addition, it has been also recognized as an acceptable food additive by the US Food and Drug Administration.[117] Moreover, the chitosan-ionotropic gelation process with TPP as a cross-linking agent is feasible for the scale-up of entrapment in a particle processing operation.[118] Further, reinforced chitosan nanoparticles with TPP are homogeneous, and possess positive surface charges that make them suitable for mucosal adhesion applications.[107]

14.3.3 Classification of PECs[119]

14.3.3.1 PEC Formed between Biopolymers

Natural polyanions such as alginic acid, dextran sulfate, heparin, carrageenan, pectin, and xanthane are abundantly used to prepare PEC with chitosan. It has been reported that positively and negatively charged proteins interact with many macromolecular and increases the functional properties including foaming and aggregation phenomena or gelation. The intensity of interactions not only depends on the concentration of each protein, but also on the ionic strength and pH of the solution. When soya protein is mixed with sodium alginate, both the polymers interact with each other to form an electrostatic complex that finally leads to improvement in the solubility and emulsifying activity.

14.3.3.2 PEC Formed between Natural and Synthetic Polymer

Phase separation of complex coacrvate or a solid precipitate is an evidence for the formation of biological polymeric complex between natural polymer proteins with synthetic polyelectrolytes through intermolecular interactions. Similarly, a potassium poly(vinyl alcohol sulfate) forms complex with carboxyhemoglobin in the presence of poly(dimethyldialylammonium chloride). The interaction between proteins and synthetic polyelectrolytes can be investigated using turbidity and quasielastic light-scattering techniques.

14.3.4 Mechanism of Gelation

Linking of macromolecular chains together which initially leads to progressively larger branched yet soluble polymers, is usually referred as gelation. The mixture of such polydisperse soluble branched polymer is called "sol." Linking process of chains continues and eventually results in increasing the size of the branched polymer with decreasing solubility. This "infinite polymer" is called the "gel" or "network" and is permeated with finite branched polymers. The transition from a system with finite branched polymer to infinite molecules is called sol-gel transition or gelation and the critical point where gel first appears is called the "gel point."[120] Gelatin can take place through two different mechanisms either by physical linking called physical gelation or by chemical linking called as chemical gelation. Physical gels are further classified as strong physical gels and weak gels. Later type of gels has strong physical bonds between polymer chains and is effectively permanent at a given set of experimental conditions and former gels have reversible links formed from temporary associations between chains. These associations have finite lifetimes, breaking and reforming continuously. Hence, strong physical gels are analogous to chemical gels. Examples of strong physical bonds are lamellar microcrystals, glassy nodules, or double and triple helices. Examples of weak physical bonds are hydrogen bond, block copolymer micelles, and ionic associations. On the other hand, chemical gelation involves formation of covalent bonds and always results in a strong gel. The three main chemical gelation processes include condensation, vulcanization, and addition polymerization.[121]

14.3.4.1 Physical or Reversible Gelation

Due to the relative ease of production and without using cross-linking agents, it has been observed that there is an increased interest in physical or reversible gels as cross-linking agents affect the integrity of substances such as cells and proteins to be entrapped. The various mechanisms reported in literature for gelation process are

14.3.4.2 Heating or Cooling a Polymer Solution

Cooling of hot solutions of polymers such as gelatin or carrageenan physical cross-linking takes place to produce hydrogels due to helix formation, association of the helices, and forming junction zones.[122] Polymeric structure of carrageenan in hot solution above the melting transition temperature is present as random coil conformation. Upon cooling, it converts to rigid helical rods. In the presence of salts such as potassium and sodium, due to screening of repulsion of sulphonic group, double helices further aggregate to form stable gels.

14.3.4.3 Gelation through Ionic Interaction

Ionic polymers can be cross-linked in the presence of di- or tri-valent counterions. This mechanism of gelation follows the principle of gelling polyelectrolyte solutions such as sodium alginate with multivalent ions of opposite charges like calcium chloride.

14.3.4.4 Complex Coacervation Gelation

When two polymers with opposite charges are mixed together, polyanion, and polycation stick together and form soluble and insoluble complex coacervate gels depending on the concentration and pH of the respective solutions. This concept is called as complex coacervation gelation mechanism. One such example is: proteins below its isoelectric point are positively charged and likely to associate with anionic hydrocolloids and form polyion complex coacervate hydrogel.[123]

14.3.4.5 Gelation through Hydrogen Bonding

When the pH of aqueous solution of polymers carrying carboxyl groups is lowered, hydrogen bonded hydrogel can be obtained. Examples of such hydrogel are a hydrogen-bound CMC network formed by dispersing it in 0.1 M HCl.[124] The mechanism involves replacing the sodium in CMC with hydrogen in the acid solution to promote hydrogen bonding. The hydrogen bonds induce a decrease of polymer solubility in water and result in the formation of an elastic hydrogel.

14.3.4.6 Heat-Induced Aggregation or Maturation Gelation

Proteinaceous components of the gum Arabic are aggregated upon heat treatment due to increase in the molecular weight and subsequently produces hydrogels with enhanced mechanical properties and water-binding capability. The molecular changes which accompany the maturation process demonstrate that gelation can takes place with precisely structured molecular dimensions in a hydrogel. The controlling feature is the agglomeration of the proteinaceous components within the molecularly disperse system that is present in the naturally occurring gum. Maturing of the gum leads to transfer of the protein associated with the lower molecular weight components to give larger concentrations of high molecular weight fraction.[125]

14.3.4.7 Gelation through Freeze–Thawing

Physical gelation of a polymer to form its hydrogel network can also be achieved by using freeze–thaw cycles that involve the formation of microcrystal in the structure due to repeated freezing and thawing. Examples of this type of gelation are freeze-thawed gels of polyvinyl alcohol and xanthan.[126]

14.3.4.8 Chemical or Irreversible Gelation

Chemical gelation mechanism normally involves grafting of monomers on the backbone of the polymers or the utilization of a cross-linking agent to link two polymer chains. The gelation of natural and synthetic polymers can be achieved through the reaction of their functional groups with cross-linking agents such as aldehydes. There are a number of methods reported in literature to obtain chemically cross-linked irreversible hydrogels.

14.3.4.9 Gelation through Use of Chemical Gelling Agents

Chemical gelling agents such as glutaraldehyde have been widely used to produce hydrogel network of various synthetic and natural polymers. The mechanism mainly involved in this technique is introduction of new molecules between the polymeric chains to produce cross-linked chains. One such example is hydrogel prepared by cross-linking of corn starch and polyvinyl alcohol using glutaraldehyde as a cross-linker. The prepared hydrogel membrane could be used as artificial skin and at the same time, various nutrients/healing factors and medicaments can be delivered to the site of action.

14.3.4.10 Gelation through Grafting

Polymerization of preformed polymer on its backbone is called grafting or gelation through grafting. The mechanism involved in grafting is activation of polymer chains by the action of chemical reagents, or high-energy radiation treatment. The growth of functional monomers on activated macroradicals leads to branching and further to cross-linking. The method can be divided as chemical and physical grafting. In chemical grafting, macromolecular backbones are activated by the action of a chemical reagent. Starch grafted with acrylic acid by using N-vinyl-2-pyrrolidone is an example of this kind of process.[127] Such hydrogels show an excellent stimuli responsive, pH-dependent swelling behavior and possess ideal characteristic features to be used as drug and vitamin delivery device in the small intestine. Whereas, in radiation grafting, high-energy radiations such as gamma and electron beam are utilized to initiate the gelation process. In some of the reported works, CMC was grafted with acrylic acid in the presence of electron beam irradiation, in aqueous solution. Electron beam was used to initiate the free radical polymerization of acrylic acid on the backbone of polymer.[128]

14.3.5 MECHANISM OF COMPLEX FORMATION

Two oppositely charged polyelectrolytes in an aqueous solution react to yield PECs. Network which is formed by ionic interaction is characterized by a hydrophilic microenvironment with a high water content and electrical charge density. Electrostatic attraction between the cationic amino groups of chitosan and the anionic groups of the other polyelectrolyte leads to the formation of the PEC. Moreover, additional secondary interactions such as those between crystalline domains of xylan or hydrogen and amide bonds can occur between chitosan and the additional polymer. Along with chitosan, a polyanionic polymer is required to form the PEC. The PEC formation mechanism does not require any more auxiliary molecules such as catalysts or initiators in aqueous solution. This leads to main advantage over covalently cross-linked networks and thus favors biocompatibility and avoids purification before administration. The most commonly used polyanions are polysaccharides bearing carboxylic groups such as alginate, pectin, or xanthenes. PEC can also be formed by positively charged chitosan derivatives such as glycol-chitosan or N-dodecylated chitosan. PEC can be reinforced by additional covalent cross-linking of chitosan. However, the addition of covalent cross-linking agent may decrease the biocompatibility. PEC can also be reinforced by the addition of ions, inducing the formation of ionically cross-linked systems. Calcium ions can be added with alginate or pectin; aluminum ions can be added with CMC and potassium with carrageenan. The properties of final hydrogels not only depend on cross-linking density, but also on the degree of interaction between the polymers. In addition, there are secondary factors such as flexibility of polymers, molecular weight and degree of deacetylation of chitosan, the substitution degree of other polyelectrolyte and the nature of the solvent also decides the properties of final product.[119]

14.3.6 MERITS OF IONOTROPIC GELATION AND POLYELECTROLYTE COMPLEXATION TECHNIQUE

Although the existing encapsulation techniques have become popular, all are mainly based on use of organic solvents, which may lead to a possibility of toxicity in chronic dosing due to presence of even traces of organic solvents in the final drug products. Flammability, environmental pollution associated use of organic solvents presents stringent governmental regulations that restrict their use, whereas, drug encapsulation via ionotropic gelation and polyelectrolyte complexation technique is simple, fast, and cost effective. Gelation process usually proceedes under very mild formulation conditions that lead to maintaining cell viability and integrity while encapsulating the protein and peptide types of agents. Since this method avoids the use of organic solvents and elevated temperature conditions, only water-soluble polyelectrolytes can be effectively utilized for encapsulation of protein materials. Thus, damage of integrity of protein and other thermolabile agents can be successfully prevented.

14.3.7 Demerits of Ionotropic Gelation and Polyelectrolyte Complexation Technique

Ionotropic gelation process presents numerous advantages along with few major demerits for controlled protein delivery, where the matrix and membrane formed is not capable of controlling the release rate for a long period of time. It does not matter in cell encapsulation, where the membrane should provide sufficient permeability for the cell products such as therapeutic protein. However, for controlled delivery of protein entities, biodegradable polymers that can produce dense membrane, preferably capable of controlling the release rate, should be incorporated or alternative combinations of polyelectrolytes should be explored to control the permeability of membrane, so that the release rate may be controlled over the desired span.[129]

In case of alginate-poly-L-lysine system, all parameters are tied to a single chemical complex; therefore, it is not easy to optimize the capsule condition. For this reason, new combinations of polyelectrolytes were proposed to allow independent modification of capsule parameters including size, wall thickness, mechanical strength, permeability, and surface characteristics.[130] Another issue in IG/PEC is that some polyelectrolytes might have biocompatibility problem. There have been contradictory arguments concerning the biocompatibility of alginates.[131–133] A host foreign-body reaction (fibrosis) occurred on account of the implanted alginate-based particles and mannuronic acid block was found to be responsible for the fibrotic response, as a potent stimulator of IL-1 and TNF-α production. Use of alginates of high glucuronic acid content was suggested to minimize the cytokine response. On the other hand, another study indicated that alginate particles with high glucuronic acid contents provoked stronger response than the high mannuronic acid alginate particles.[133] In general, polyionic hydrogels obtained by IG/PEC have rather low mechanical strength. A few papers have suggested ways to overcome this problem by exploring alternative combinations of polyanions and polycations,[130,134] introducing a different process,[135] or applying additional coating.[136,137]

14.4 COMMON LABORATORY METHOD OF ENCAPSULATION VIA IONOTROPIC GELATION

A routine laboratory method of microencapsulation is ionotropic gelation and polyelectrolyte complexation. Alginate is one of the most widely and commonly used polyanions for microencapsulation. The cell- or drug-loaded alginate solution is slowly dropped into the aqueous calcium chloride solution, which leads to gelation followed by formation of microspheric or macrospheric particles. Gelation takes place by diffusion of calcium ions into the alginate drops, forming a three-dimensional lattice of ionically cross-linked alginate. To increase the mechanical strength and/or to provide a permeability barrier of the hydrogel particles thus formed, polyelectrolyte complexation is achieved by the addition of oppositely charged polyelectrolytes. For instance, addition of polycations such as chitosan or poly-L-lysine allows a membrane of polyelectrolyte complex to form on the surface of alginate beads.[138] Lot of work has been reported on ionotropic gelation and polyelectrolyte complexation technique to prepare sustained and modulated release drug-delivery systems using various natural, chemically modified, and synthetic polymers. Few examples for such reported work are summarized in Table 14.1.

In laboratory scale, encapsulated hydrogel particulate system can be prepared by syringe droplet method, which involves dropping of polyanion and encapsulant solution through a needle into an aqueous solution of divalent or trivalent cation. Figure 14.2 shows the diagrammatic conventional procedure of drug encapsulation.

TABLE 14.1
Examples of Microencapsulating Substances by Ionotropic Gelation and Polyelectrolyte Complexation

Biopolymer	Cations Used for Gelation	Polyelectrolyte	Encapsulant	References
Sodium alginate	Calcium chloride	Chitosan, poly-L-lysine	DNA	[139,140]
Gelan gum	Calcium chloride	Chitosan	Stavudine,	[141]
Chitosan	Trypolyphosphate	—	Pyrimethamine	[142]
CMC	Aluminum chloride	Chitosan	Valsartan	[136,143]
K-Carrageenan	Potassium chloride	Chitosan	alpha-Amylase	[144]
Pectin	Calcium chloride	—	Theophylline	[145]

Polymer–drug solution [alginate (−)/Gellun gum (−)/CMC (−) + Drug]

↓

Add drop-wise under magnetic stirring by 21 G needle into

↓

Calcium chloride solution (+): Cross-linking Agent + Chitosan solution (+): PEC Agent

↓

Ionotropically gelled and chitosan-reinforced hydrogel beads

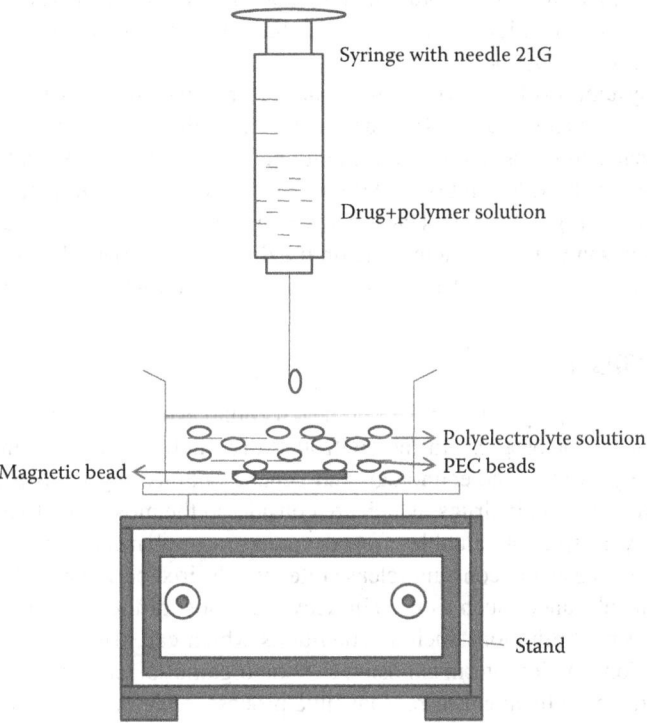

FIGURE 14.2 Schematic diagram showing conventional preparation of hydrogel microparticles.

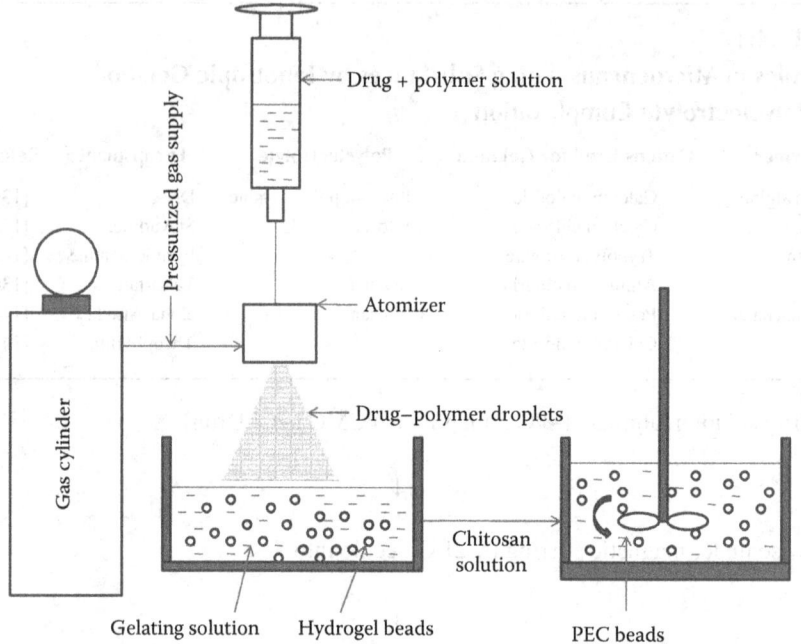

FIGURE 14.3 Schematic diagram showing spray-atomization provision for the preparation of hydrogel microparticles.

The particles formed by conventional syringe droplet method are relatively large, because most of the droplets are made with syringe needles. To control the particle size, it is essential to produce relatively smaller droplets, for this a vibration system or air atomization provisions are to be made to extrude the alginate solution. Figure 14.3 shows the diagrammatic procedure of drug encapsulation process using size controlling options.

This vibration system involves a Turbotak air-atomizer in which a pressurized air is injected to mix with the sodium alginate solution. Pressurized air forces the minute liquid droplets out through the orifice of a nozzle and is sprayed into the calcium chloride solution. When these droplets come in contact with calcium chloride solution, calcium ions diffuses into the droplets of sodium alginate, which leads to form microgel droplets. Hydrogel particles thus formed can be further reinforced by poly-L-lysine or chitosan to form a membrane on the droplets. The particle size of hydrogel beads can be obtained in the range of 5 and 15 μm by the use of air atomization technique.[146]

14.5 CONCLUSION

Significant exploitation of polymer chemistry and conceptual evolution of microencapsulation technique contributed to major progress in the controlled drug delivery system since its introduction in 1970s. Explosive growth of biotechnology and genetic engineering fields presented the easiest way for encapsulation of protein drugs, which are considered the most attractive candidates of controlled drug delivery. Despite of valuable effort devoted by our pharmaceutical scientist to develop a microcapsule system ensuring constant release rate over desired period and structural integrity of certain encapsulant substances such as protein. However, most of the current encapsulation methods suffer from several limitations such as conditions which can affect the drug entity, difficult methods of production, and long-term exposure to toxic organic solvents. To minimize such limitations by ensuring the establishment of reproducible process, which can be easily scaled up to the commercial scale. Due to abundant availability of numerous novel protein drugs from the genome projects, new microencapsulation systems that overcome unfavorable conditions will be invaluable tools in the future. One among such encapsulation technique presently available is ionotropic gelation and polyelectrolyte complexation. Ionotropic gelation is a promising tool in the development of biocompatible novel sustained and targeted controlled drug delivery systems, which are capable to encapsulate large number of microtherapeutic and macrotherapeutic molecules in their hydrogel meshwork structure. The larger utilization of expensive and toxic organic solvents in the microencapsulation process has been drastically reduced due to evolution of ionotropic gelation and hence provides an ecofriendly pharmaceutical product development process in the preparation of hydrogel beads. To overcome the limitations of ionotropic gelation technique, continuous efforts of our scientists helped to bring out certain advances in the conventional gelation method. Sensitive macromolecules such as proteins and peptides can be successfully encapsulated into hydrogel meshwork through the modified gelation techniques, which yield more acceptable drug delivery pattern by retaining their structural integrity. Ionotropic gelation and polyelectrolyte complexation technique provide tremendous opportunities for designing new controlled release oral formulations and stimuli responsive hydrogel formulations, thus extending the frontier of future pharmaceutical product development.

REFERENCES

1. Friese, A., Seiller, E., Quack, G., Lorenz, B., and Kreuter, J. Increase of the duration of the anticonvulsive activity of a novel NMDA receptor antagonist using poly (butylcyanoacrylate) nanoparticles as a parental controlled release system. *Eur. J. Pharm. Biopharm.* 49 (2000): 103–109.
2. Alagusundaram, M., Chetty, M. S., and Umashankari, C. Microspheres as a Novel drug delivery system—A review. *Int. J. Chem. Tech.* 12 (2009): 526–534.
3. Allen, L. V., Popovich, N. G., and Ansel, H. C. *Pharmaceutical Dosage Forms and Drug Delivery Systems*. Delhi, India: BI Publication, 2005, p. 265.
4. Banker, G. S. and Rhodes, C. T. *Modern Pharmaceutics*. In Parma Publication, New York, Basil, 2002, pp. 501–527.
5. Bungenburg de Jong, H. G. *Proc. Acad. Sci. Amsterdam* 41 (1938): 646.
6. Van Tomme, S. R., Storm, G., Hennink, W. E. In situ gelling hydrogels for pharmaceutical and biomedical applications. *Int. J. Pharm.* 355(1–2) (2008): 1–18.
7. Coviello, T., Matricardi, P., Marianecci, C., and Alhaique, F. Polysaccharide hydrogels for Modified release formulations. *J. Control Release* 119 (2007): 5–24.
8. Peppas, N. A., Bures, P., Leobandung, W., and Ichikawa, H. Hydrogels in pharmaceutical formulations, *Eur. J. Pharm. Biopharm.* 50 (2000): 27–46.
9. Follonier, N. and Doelkar, E. Biopharmaceutical comparison of oral multiple-unit and single-unit sustained release dosage forms. *STP Pharm. Sci.* 2 (1992): 141–158.
10. Harris, M. R. and Ghabre-Sellassie, I. *Aqueous Polymeric Coating for Modified Release Pellets*. In: Mchinity, J. W. (Ed.), Duleieu, Dawei, 1989, p. 64.
11. Vyas, S. P. and Khar, R. P. *Control Drug Delivery Concepts and Advances*. Delhi, India: Vallab Prakashan, 2002, p. 102.
12. Lim, F. and Sun, A. M. Microencapsulated islets as bioartificial endocrine pancreas. *Science* 210 (1980): 908–910.
13. Poncelet, D., Lencki, R., Beaulieu, C., Halle, J. P., Neufeld, R. J., and Fournier, A. Production of alginate beads by emulsification/internal gelation. I. Methodology. *Appl. Microbiol. Biotechnol.* 38 (1992): 39–45.
14. Lencki, R. W. J., Neufeld, R. J., and Spinney, T. Method of producing microspheres. 1989, US Patent 4822534.
15. Chan, L. W., Lee, H. Y., and Heng, P. W. S. Production of alginate microspheres by internal gelation using emulsification method. *Int. J. Pharm.* 242 (2002): 259–262.
16. Liu, X. D., Yu, W. Y., Zhang, Y., Xue, W. M., Yu, W. T., Xiong, Y., Ma, X. J., Chen, Y., and Yuan, Q. Characterization of structure and diffusion behavior of Ca-alginate beads prepared with external or internal calcium sources. *J. Microencapsul.* 19 (2002): 775–782.
17. Lacroix, C., Paquin, C., and Arnaud, J. P. Batch fermentation with entrapped growing cells of lactobacillus casei. Optimization of rheological properties of the entrapment gel matrix. *Appl. Microbiol. Biotechnol.* 32 (1990): 403–408.

18. Poncelet, D., Babak, V., Dulieu, C., and Picot, A. A physico-chemical approach to production of alginate beads by emulsification-internal ionotropic gelation. *Coll. Surf. A: Physiochem. Eng. Aspects* 155 (1999): 171–176.
19. Pelaez, C. and Karel, M. Improved method for preparation of fruit-simulating alginate gels. *J. Food Process. Preserv.* 5 (1981): 63–81.
20. Alexakis, T., Boadi, D. K., Quong, D., Groboillot, A., O'Neill, I. K., Poncelet, D., and Neufeld, R. J. Microencapsulation of DNA within alginate microspheres and crosslinked chitosan membranes for in vivo application. *Appl. Biochem. Biotechnol.* 50 (1995): 93–106.
21. Quong, D., O'Neil, I. K., Poncelet, D., and Neufeld, R. J. Gastrointestinal protection of cellular component DNA within an artificial cell system for environment carcinogen biomonitoring. In: Wijffels, R. J., Buitelaar, R. M., Bucke, C., Tramper, J. (Eds.) *Immobilized Cells: Basics and Applications.* Amsterdam, the Netherland: Elsevier Science, 1996, pp. 814–820.
22. Vandenberg, G. W. and Noue, J. D. L. Evaluation of protein release from chitosan-alginate microcapsules produced using external or internal gelation. *J. Microencapsul.* 18 (2001): 433–441.
23. Liu, X. D., Bao, D. C., Xue, W. M., Xiong, Y., Yu, W. T., Yu, X. J., Ma, X. J., and Yuan, Q. Preparation of uniform calcium alginate gel beads by membrane emulsification coupled with internal gelation. *J. Appl. Polym. Sci.* 87 (2002): 848–852.
24. Reis, C. P., Ribeiro, A. J., Neufeld, R. J., and Veiga, F. Insulin-loaded alginate nanoparticles obtained by emulsification/internal gelation. XII International workshop on bioencapsulation; 24–26 September 2004; Vitoria, Spain: Servicio editorial Universidad del País Vasco, 2004, p. 251.
25. McLachlan, J. Macroalgae (seaweeds): Industrial resources and their utilization. *Plant Soil* 89 (1985): 137–157.
26. Redenbaugh, K. P. B., Nichol, J. W., Kossler, M. E., Viss, P. R., and Ka, W. Somatic seeds: Encapsulation of asexual plant embryos. *Biotechnol. Technol.* 4 (1986): 797–801.
27. Lim, F. and Sun, A. M. Microencapsulated islets as bioartificial endocrine pancreas. *Science* 210 (1980): 908–910.
28. Shiotani, T. and Yamane, T. A horizontal packed-bed bioreactor to reduce carbon dioxide gas hold-up in the continuous production of ethanol in immobilized yeast cells. *Eur. J. Appl. Microbiol. Biotechnol.* 13 (1981): 96–101.
29. Provost, H., Divies, C., and Rousseau, E. Continuous production with *Lactobacillus bulgarius* and *Streptococcus termophillus* entrapped in calcium alginate. *Biotechnol. Lett.* 7 (1985): 247–252.
30. Lim, F. Microcapsules containing viable tissue cells. 1983, US Patent 4391909.
31. Rastogi, R., Sultana, Y., Aquiol, M., Ali, A., Kumar, S., Chuttani, K., and Mishra, A. K. Alginate microspheres of isoniazid for oral sustained drug delivery. *Int. J. Pharm.* 334 (1–2) (2007): 71–77.
32. Sartori, C., Finch, D. S., and Ralph, A. B. Determination of the cation content of alginate thin films by FTIR spectroscopy. *Polymer* 38 (1996): 43–51.
33. Thu, B., Skjak-Braek, G., Micali, F., Vittur, F., and Rizzo, R. The spatial distribution of calcium in alginate gel beads analyzed by synchron-radiation induced x-ray emission (SRIXE). *Carbohydr. Res.* 297 (1997): 101–105.
34. Zonghua, L., Yanpeng, J., and Ziyong, Z. Calcium-carboxymethyl chitosan hydrogel beads for protein drug delivery system. *J. Appl. Polym. Sci.* 103 (2007): 3164–3168.
35. Walsh, P. K., Isdell, F. V., Noone, S. M., O'Donovan, M. G., and Malone, D. M. Growth patterns of *Saccharomyces cerevisiae* microcolonies in alginate and carrageenan gel particles: Effect of physical and chemical properties of gels. *Enzyme Microbiol. Technol.* 18 (1996): 366–372.
36. Poncelet, D. Production of alginate beads by emulsification/internal gelation. *Ann. N.Y. Acad. Sci.* 944 (2001): 74–82.
37. Liu, X., Ma, Z., Xing, J., and Liu, H. Preparation and characterization of amino-silane modified superparamagnetic silica nanospheres. *J. Magnetism Magnet. Mater.* 270 (2004): 1–6.
38. Lamberti, F. V., Sefton, M. V. Microencapsulation of erythrocytes in eudragit-RL-coated calcium alginate. *Biochem. Biophys. Acta* 795(1983) 81–91.
39. Burns, M. A. and Kvesitadze, G. I. Dried calcium alginate/magnetite spheres: A new support for chromatographic separations and enzyme immobilization. *Biotechnol. Bioeng.* 27 (1985): 137.
40. Bowersock, T. L., HogenEsch, H., Suckow, M., Guimond, P., Martin, S., Borie, D., Torregrosa, S., Park, H., and Park, K. Oral vaccination of animals with antigens encapsulated in alginate microspheres. *Vaccine* 17(13–14) (1999): 1804–1811.
41. Polk, A., Amsden, B., Yao, K., Peng, T., and Goosen, M. F. Controlled release of albumin from chitosan-alginate microcapsules. *J. Pharm. Sci.* 83 (1994): 78–185.

42. Patil, J. S., Kamalapur, M. V., Marapur, S. C., Shiralshetti, S. S., and Kadam, D. V. Ionotropically gelled chitosan-alginate complex hydrogel beads: Preparation, characterization and *in-vitro* evaluation. *Indian J. Pharm. Edu. Res.* 46(3) (2012): 248–252.
43. Patil, J. S., Kamalapur, M. V., Marapur, S. C., and Kadam, D. V. Ionotropic gelation and polyelectrolyte complexation; the novel techniques to design hydrogel particulate, sustained, modulated drug delivery system: A review. *Digest J. Nanomat. Biostruct.* 5(1) (2010): 241–248.
44. Jon, S., Lee, E., Lee, J. J., and Lee, I.-H. (2007). A conjugate for transmucosal delivery comprising a pharmacologically active substance covalently bound via a linker to chitosan or its derivative. US Patents 2007/0292,387.
45. Agnihotri, S. A., Mallikarjuna, N. N., and Aminabhavi. T. M. Recent advances on chitosan-based micro- and nanoparticles in drug delivery. *J. Control Release* 100 (2004): 5–28.
46. Sinha, V. R., Singla, A. K., Wadhawan, S., Kaushik, R., Kumria, R., and Bansal, K. Chitosan microspheres as a potential carrier for drugs. *Int. J. Pharm.* 274(1–2) (2004): 1–33.
47. Rinaudo, M. Chitin and chitosan: Properties and applications. *Prog. Polym. Sci.* 31 (2006): 603–632.
48. Hejazi, R. and Amiji, M. Chitosan-based gastrointestinal delivery systems. *J. Control. Rel.* 89 (2003): 151–165.
49. Issa, M. M., Köping-Höggård, M., and Artursson, P. Chitosan and the mucosal delivery of biotechnology drugs. *Drug Discov. Today.* 2(1) (2005): 1–6.
50. Singla, A. K. and Chawla, M. Chitosan: Some pharmaceutical and biological aspects-an update. *J. Pharm. Pharmacol.* 53 (2001): 1047–1067.
51. Kubota, N., Tatsumoto, N., Sano, T., and Toya, K. A simple preparation of half N-acetylated chitosan highly soluble in water and aqueous organic solvents. *Carbohydr. Res.* 324 (2000): 268–274.
52. Kubo, W., Miyazaki, S., and Attwood, D. Oral sustained delivery of paracetamol from in situ-gelling gellan and sodium alginate formulations. *Int. J. Pharm.* 258 (2003): 55–64.
53. Jansson, P. E., Lindberg, B., and Sandford, P. A. Structural studies of gellan gum, an extracellular polysaccharide elaborated by *Pseudomonas elodea. Carbohydr. Res.* 124 (1983): 135–139.
54. Sanderson, G. R. and Clark, R. C. Gellan gum. *Food Technol.* 37 (1983): 62–70.
55. Miyoshi, E., Takaya, T., Nishinari, K. Rheological and thermal studies of gel-sol transition in Gellan gum aqueous solutions. *Carbohydr. Polym.* 30 (1996): 109–119.
56. Grasdalen, H. and Smidsroed, O. Gelation of gellan gum. *Carbohydr. Polym.* 7 (1987): 371–393.
57. Yuguchi, M. Structural characteristics of gellan in aqueous solution. *Food Hydrocoll.* 7 (1993): 373–385.
58. Mao, M., Tang, J., and Swanson, B. G. Texture properties of high and low acyl mixed gellan gels. *Carbohydr. Polym.* 41 (2000): 331–338.
59. Huang, Y. Gelling temperatures of high acyl gellan as affected by monovalent and divalent cations with dynamic rheological analysis. *Carbohydr. Polym.* 56 (2004): 27–33.
60. Anil Kumar, A. and Harjinder, S. Recent advances in Microencapsulation of probiotics for industrial applications and targeted delivery. *Trends Food Sci. Technol.* 18 (2007): 240–251.
61. Poncelet, D., Smet, B. P. D., Beaulieu, C., Huguet, M. L., Fournier, A., and Neufeld, R. J. Production of alginate beads by emulsification/internal gelation. II. Physicochemistry. *Appl. Microbiol. Biotechnol.* 43 (1995): 644–650.
62. Quong, D., O'Neil, I. K., Poncelet, D., and Neufeld, R. J. 1996. Gastrointestinal protection of cellular component DNA within an artificial cell system for environment carcinogen biomonitoring. In: Wijffels, R. J., Buitelaar, R. M., Bucke, C., Tramper, J., (Eds.). *Immobilized Cells: Basics and Applications.* Amsterdam, the Netherlands: Elsevier Science B.V., pp. 814–820.
63. Patil, P., Chavanke, D., and Wagh, M. A review on ionotropic gelation method: Novel approach for controlled gastroretentive gelispheres. *Int. J. Pharm. Pharm. Sci.* 4 (2012): 27–32.
64. Anal, A. K. and Stevens, W. F. Chitosan–alginate multilayer beads for controlled release of ampicillin. *Int. J. Pharm.* 290 (2005): 45–54.
65. Lihua, M. and Changsheng, L. Preparation of chitosan microspheres by ionotropic gelation under a high voltage electrostatic field for protein delivery. *Coll. Surf. B: Biointerfaces* 75(2) (2010): 448–453.
66. Catarina, P. R., Ronald, J. N., Sandra, V., Antonio, J. R., and Francisco, V. Review and current status of emulsion/dispersion technology using an internal gelation process for the design of alginate particles. *J. Microencapsul.* 23(3) (2006): 245–257.
67. Han, J., Guenier, A. S., Salmieri, S., and Lacroix, M. Alginate and chitosan functionalization for micronutrient encapsulation. *J. Agricult. Food Chem.* 56(7) (2008): 2528–2535.
68. Pillay, V. and Danckwerts, M. P. Textural profiling and statistical optimization of crosslinked calcium-alginate-pectinate-cellulose acetophthalate gelisphere matrices. *J. Pharm. Sci.* 91 (2002): 2559–2570.

69. Choonara, Y. E., Pillay, V., Khan, R. A., Singh, N., and Toit, L. C. Mechanistic evaluation of alginate-HEC gelisphere compacts for controlled intrastriatal nicotine release in parkinson's disease. *J. Pharm. Sci.* 98 (2009): 2059–2072.
70. Badarinath, A. V., Ravi kumar, reddy, J., Mallikarjuna rao, K., Alagusundaram, M, Gnanaprakash, K., and Madhu sudhana chetty, C. Formulation and characterization of alginate microbeads of Flurbiprofen by ionotropic gelation technique. *Int. J. ChemTech Res.* 2(1) (2010): 361–367.
71. Tsuchida, E. and Abe, K. Interactions between macromolecules in solution and intermacromolecular complexes. *Advances in Polymer Science* 45. New York: Springer-Verlag, 1982.
72. Petrak, K. In *Polyelectrolytes, Science and Technology*; Hara, M. (Ed.). New York: Marcel Dekker, 1993, p. 265.
73. Deasy, P. B. *Microencapsulation and Related Drug Process*. Basel, Switzerland: Marcel Dekker, 1984.
74. Guiot, P. and Couvreur, P. *Polymeric Nanoparticles and Microspheres*. Boca Raton, FL: CRC Press, 1986.
75. Albertsson, P. A. *Partition of Cell Particles and Macromolecules*, 2nd edn. New York: Wiley Intersciences, 1971.
76. Felgner, P. L., Particulate systems and polymers for in vitro and in vivo delivery of polynucleotides. *Adv. Drug Delivery Rev.* 5 (1990): 163–187.
77. Fuoss, R. M. and Sadek, H. Mutual interaction of polyelectrolytes. *Science* 110 (1949): 552–573.
78. Tsuchida, E., Osada, Y., and Abe, K. Formation of polyion complexes between polycarboxylic acids and polycations carrying charges in the chain backbone. *Makromol. Chem.* 175 (1974): 583–592.
79. Tsuchida, E. Formation of interpolymer complexes. *J. Macromol. Sci.* 17 (1980): (Part B): 683–714.
80. Vert, M. Ionizable polymers for temporary use in medicine. In: *IUPAC 28th Macromolecular Symposium Proceedings*, Amherst, Massachusetts, 1982, 377.
81. Vert, M. Polyvalent polymeric drug carriers. *Crit. Rev., Ther. Drug Carr. Syst.* 2 (1986): 291–327.
82. Shinya, Y., Tsurushima, H., Tsurumi, T., Kajiuchi, T., Kam, and Leong, W. Polyelectrolyte complex films derived from polyethyleneoxide–maleic acid copolymer and chitosan: Preparation and characterization. *Macromol. Biosci.* (2004): 526–531.
83. Tapia, C., Escobar, Z., Costa, E., Sapag-Hagar, J., Valenzuela, F., Basualto, C., Gai, M. N., and Yazdani-Pedram, M. Comparative studies on polyelectrolyte complexes and mixtures of chitosan alginate and chitosan-carrageenan as prolonged diltiazem clorhydrate release systems. *Eur. J. Pharm. Biopharm.* 57 (2004): 65–75.
84. Tsuchida, E. and Abe, K. Interactions between macromolecules in solution and intermacromolecular complex. *Adv. Polym. Sci.* 45 (1982): 1–119.
85. Tsuchida, E. Formation of polyelectrolyte complexes and their structures. *J. Macromol. Sci.* 31 (1994): (Part A): 1–15.
86. Kim, H. J., Lee, H. C., Oh, J. S., Shin, B, A., Oh, C. S., Park, R, D., Yang, K. S., and Cho, C. S. Polyelectrolyte complex composed of chitosan and sodium alginate for wound dressing application. *J. Biomat. Sci., Polym. Ed.* 10 (1999): 543–556.
87. Kabanov, A. V. and Kabanov, V. A. DNA complexes with polycations for the delivery of genetic material into cells. *Bioconjug. Chem.* 6 (1995): 7–20.
88. Luo, D. and Saltzman, W. M. Synthetic DNA delivery systems. *Nat. Biotechnol.* 18 (2000): 33–37.
89. Schaffer, D. V., Fidelman, N. A., Dan, N., and Lauffenburger, D. A. Vector unpacking as a potential barrier for receptor-mediated polyplex gene delivery. *Biotechnol. Bioeng.* 67 (2000): 598–606.
90. Miyazaki, S., Nakayama, A., Oda, M., Takada, M., and Attwood, D. Chitosan and sodium alginate based bioadhesive tablets for intraoral drug delivery. *Biol. Pharm. Bull.* 17 (1994): 745–747.
91. Fukuda, H. and Kikuchi, Y. Polyelectrolyte complexes of sodium carboxymethylcellulose with chitosan. *Makromol. Chem.* 180 (1979): 1631–1633.
92. Lin, S. Y. and Lin, P. C. Effect of acid type, acetic acid and sodium carboxymethylcellulose concentrations on the formation, micromeretic, dissolution and floating properties of theophylline chitosan microcapsules. *Chem. Pharm. Bull.* 40 (1992): 2491–2497.
93. Meshali, M. M. and Gabr, K. E. Effect of interpolymer complex formation of chitosan with pectin or acacia on the release behaviour of chlorpromazine HCl. *Int. J. Pharm.* 89 (1993): 177–181.
94. Liu, X., Yang, F., Song, T., Zeng, A., Wang, Q., Sun, Z., and Shen, J. Effects of chitosan, O-carboxymethyl chitosan and N-[(2-hydroxy-3-N,N-dimethylhexadecyl ammonium) propyl] chitosan chloride on lipid metabolism enzymes and low-densitylipoprotein receptor in a murine diet-induced obesity. *Carbohydr. Polym.* 85 (2011): 334–340.
95. Baldrick, P. The safety of chitosan as a pharmaceutical excipient. *Regul. Toxicol. Pharmacol.* 56 (2010): 290–299.
96. Gerente, C., Andres, Y., McKay, G., and Le Cloirec, P. Removal of arsenic(V) onto chitosan: From sorption mechanism explanation to dynamic water treatment process. *Chem. Eng. J.* 158 (2010): 593–598.

97. Harris, R., Lecumberri, E., Mateos-Aparicio, I., Mengibar, M., and Heras, A. Release characteristics of vitamin E incorporated chitosan microspheres and *in vitro-in vivo* evaluation for topical application. *Carbohydr. Polym.* 84 (2011): 803–806.
98. Fernandez-Saiz, P., Ocio, M. J., and Lagaron, J. M. Antibacterial chitosan-based blends with ethylenevinyl alcohol copolymer. *Carbohydr. Polym.* 80 (2010): 874–884.
99. Alves, N. M. and Mano, J. F. Chitosan derivatives obtained by chemical modifications for biomedical and environmental applications. *Int. J. Biolog. Macromol.* 43 (2008): 401–414.
100. Gao, J.-Q., Zhao, Q.-Q., Lv, T.-F., Shuai, W.-P., Zhou, J., Tang, G.-P., and Liang, W.-Q., Tabata, Y., and Hu, Y.-L. Genecarried chitosan-linked-PEI induced high gene transfection efficiency with low toxicity and significant tumor-suppressive activity. *Int. J. Pharm.* 387 (2010): 286–294.
101. Bagheri-Khoulenjani, S., Taghizadeh, S. M., and Mirzadeh, H. An investigation on the short-term biodegradability of chitosan with various molecular weights and degrees of deacetylation. *Carbohydr. Polym.* 87 (2009): 773–778.
102. Patil, S. B. and Sawant, K. K. Chitosan microspheres as a delivery system for nasal insufflations. *Coll. Surf. B: Biointerfaces* 84 (2011): 384–389.
103. Van der Lubben, I. M., Verhoef, J. C., Borchard, G., and Junginger, H. E. Chitosan for mucosal vaccination. *Adv. Drug Delivery Rev.* 52 (2001): 139–144.
104. Trimukhe, K. D. and Varma, A. J. A morphological study of heavy metal complexes of chitosan and crosslinked chitosans by SEM and WAXRD. *Carbohydr. Polym.* 71 (2008): 698–702.
105. Kong, M., Chen, X, G., Xing, K., and Park, H. J. Antimicrobial properties of chitosan and mode of action: a state of the art review. *Int. J. Food Microbiol.* 144 (2010): 51–63.
106. Ong, S. Y., Wu, J., Moochhala, S. M., Tan, M. H., and Lu, J. Development of a chitosan-based wound dressing with improved hemostatic and antimicrobial properties. *Biomaterials* 29 (2008): 4323–4332.
107. Gan, Q. and Wang, T. Chitosan nanoparticle as protein delivery carrier-systematic examination of fabrication conditions for efficient loading and release. *Coll. Surf. B: Biointerfaces* 59 (2007): 24–34.
108. Rao, K. S. V. K, Rao, K. M., Kumar, P. V. N., and Chung, I. D. Novel chitosan-based pH sensitive micronetworks for the controlled release of 5-fluorouracil. *Iran Polym. J.* 19 (2010): 265–276.
109. Papadimitriou, S., Bikiaris, D., Avgoustakis, K., Karavas, E., and Georgarakis, M. Chitosan nanoparticles loaded with dorzolamide and pramipexole. *Carbohydr. Polym.* 73 (2008): 44–54.
110. Qi, L., Xu, Z., Jiang, X., Hu, C., and Zou, X. Preparation and antibacterial activity of chitosan nanoparticles. *Carbohydr. Res.* 339 (2004): 2693–2700.
111. Xiong, W. W., Wang, W, F., Zhao, L., Song, Q., and Yuan, L. M. Chiral separation of (R,S)-2-phenyl-1-propanol through glutaraldehyde-crosslinked chitosan membranes. *J. Membr. Sci.* (2009): 328, 268–272.
112. Liu, H. and Gao, C. Preparation and properties of ionically cross-linked chitosan nanoparticles. *Polym. Adv. Technol.* 20 (2009): 613–619.
113. Chen, S., Liu, M., Jin, S., and Wang, B. Preparation of ionic-crosslinked chitosan-based gel beads and effect of reaction conditions on drug release behaviors. *Int. J. Pharm.* 349 (2008): 180–187.
114. Zhong, Z., Li, P., Xing, R., and Liu, S. Antimicrobial activity of hydroxylbenzenesulfonailides derivatives of chitosan, chitosan sulfates and carboxymethyl chitosan. *Int. J. Biolog. Macromol.* 45 (2009): 163–168.
115. Gupta, K. C. and Jabrail, F. H. Ontrolled-release formulations for hydroxyl urea and rifampicin using polyphosphate-anion-crosslinked chitosan microspheres. *J. Appl. Polym. Sci.* 104 (2007): 1942–1956.
116. Rayment, P. and Butler, M. F. Investigation of ionically crosslinked chitosan and chitosan-bovine serum albumin beads for novel gastrointestinal functionality. *J. Appl. Polym. Sci.* 108 (2008): 2876–2885.
117. Lin, Y. H., Sonaje, K., Lin, K. M., Juang, J. H., Mi, F. L., Yang, H. W., and Sung, H. W. Multi-ion-crosslinked nanoparticles with pH-responsive characteristics for oral delivery of protein drugs. *J. Control Release* 132 (2008): 141–149.
118. Stulzer, H. K., Tagliari, M. P., Parize, A. L., Silva, M. A. S., and Laranjeira, M. C. M. Evaluation of cross-linked chitosan microparticles containing acyclovir obtained by spray-drying. *Mater. Sci. Eng. C* 29 (2009): 387–392.
119. Gubbala, S. K. and Nimisha, C. Polyelectrolyte complex: A pharmaceutical review. *Int. J. Pharm. Biolog. Sci.* 2 (3) (2012): 399–407.
120. Rubinstein, M. and Colby, R. H. (2003) *Polymer Physics*. Oxford, New York: Oxford University Press.
121. Syed, K. H., Gulrez, S. A., and C Phillips, G. O. Hydrogels: Methods of preparation, characterisation and applications. *Progress in Molecular and Environmental Bioengineering—From Analysis and Modeling to Technology Applications*. www.intechopen.com., pp. 117–150.
122. Funami, T., Hiroe, M., Noda, S., Asai, I., Ikeda, S., and Nishimari, K. (2007) Influence of molecular structure imaged with atomic force microscopy on the rheological behavior of carrageenan aqueous systems in the presence or absence of cations. *Food Hydrocolloids* 21: 617–629.

123. Magnin, D., Lefebvre, J., Chornet, E., and Dumitriu, S. Physicochemical and structural characterization of a polyionic matrix of interest in biotechnology, in the pharmaceutical and biomedical fields. *Carbohyd. Polym.* 55 (2004): 437–453.
124. Takigami, M., Amada, H., Nagasawa, N., Yagi, T., Kasahara, T., Takigami, S., and Tamada, M. Preparation and properties of CMC gel. *Trans. Mater. Res. Soc. Jpn.* 32(3) (2007): 713–716.
125. Aoki, H., Katayama, T., Ogasawara, T., Sasaki, Y., Al-Assaf, S., and Phillips, G. O., Characterization and properties of Acacia senegal (L.) Willd. var. Senegal with enhanced properties (Acacia (sen) SUPER GUM(TM)): Part 5. Factors affecting the emulsification of Acacia senegal and Acacia (sen) SUPER GUM(TM). *Food Hydrocolloids* 21 (2007): 353–358.
126. Giannouli, P. and Morris, E. R. Cryogelation of xanthan. *Food Hydrocolloids.* 17 (2003): 495–501.
127. Spinelli, L. S., Aquino, A. S., Lucas, E., d'Almeida, A. R., Leal, R., and Martins, A. L. Adsorption of polymers used in drilling fluids on the inner surfaces of carbon steel pipes. *Polym. Eng. Sci.* 48 (2008): 1885–1891.
128. Said, H. M., Alla, S. G. A., and El-Naggar, A. W. M. Synthesis and characterization of novel gels based on carboxymethyl cellulose/acrylic acid prepared by electron beam irradiation. *React. Funct. Polym.* 61 (2004): 397–404.
129. Peters, M. C., Isenberg, B. C., Rowley, J. A., and Mooney, D. J. Release from alginate enhances the biological activity of vascular endothelial growth factor. *J. Biomater. Sci., Polym. Ed.* 9 (1998): 1267–1278.
130. Wang, T. I., Lacik, M., Brissova, A. V., Anikumar, A., Prokop, D. Hunkeler, R., Green, K., and Shahrokhi, A. C. (1997). Powers an encapsulation system for the immuno-isolation of pancreatic islets. *Nat. Biotechnol.* 15: 358–362.
131. Yang, H., James, R., and Wright, J. (1999). Calcium alginate. In: W. M. Kuhtreiber, R. P. Lanza, and W. L. Chick (eds.). *Cell Encapsulation Technology and Therapeutics.* Birkhauser, Boston, MA, pp. 79–89.
132. Soon-Shiong, P., Otterlie, M., Skjak-Braek, G., Smidsrod, O., Heintz, R., Lanza, R. P., and Espevik, T. An immunologic basis for the fibrotic reaction to implanted microcapsules. *Transplant. Proc.* 23 (1991): 758–759.
133. Clayton, H. A., London, N. J. M., Colloby, P. S., Bell, P. R. F, and James, R. F. L. The effect of capsule composition on the biocompatibility of alginate-poly-l-lysine capsules. *J. Microencapsulation* 8 (1991): 221–233.
134. Brissova, M., Lacik, I., Powers, A. C., Anilkumar, A. V., and Wang, T. Control and measurement of permeability for design of microcapsule cell delivery system. *J. Biomed. Mater. Res.* 39 (1998): 61–70.
135. Shu, X. Z. and Zhu, K. J. A novel approach to prepare tripolyphosphate/chitosan complex beads for controlled release drug delivery. *Int. J. Pharm.* 201: (2000), 51–58.
136. Long, D. D. Chitosan-carboxymethylcellulose hydrogels as supports for cell immobilization. *J. M. S. Pure Appl. Chem.* A 33 (1996): 1875–1884.
137. Hearn, E. and R. J. Neufeld poly (methylene coguanidine) coated alginate as an encapsulation matrix for urease. *Process Biochem.* 35 (2000): 1253–1260.
138. Dulieu, C., Poncelet, D., and Neufeld, R. J. (1999) Encapsulation and immobilization techniques. In: Kuhtreiber, W. M., Lanza, R. P., and W. L. Chick (eds.). *Cell Encapsulation Technology and Therapeutics.* Boston, MA: Birkhauser, pp. 3–17.
139. Takka, S. and Acarturk, F. Calcium alginate microparticles for oral administration: I: Effect of sodium alginate type on drug release and drug entrapment efficiency. *J. Microencapsul.* 16 (1999): 275–290.
140. Thu, B., Bruheim., P., Espevik., T., Smidsrod., O., Skjak-Braek., G., Soon-Shiong., P. Alginate polycation microcapsules. I. Interaction between alginate and polycation. *Biomaterials* 17 (1996): 1031–1040.
141. Patil, J. S., Kamalapur, M. V., Marapur, S. C., and Shiralshetti, S. S. Ionotropically gelled novel hydrogel beads: Preparation, characterization and *in-vitro* evaluation. *Indian J. Pharm. Sci.* 73(5) (2011): 504–509.
142. Emmanuel, C. I., Cristina, T. A., Cristina, M., Bianca, B., Damian, C. O., Felipe, F. D., and Ionically, L. Cross-linked chitosan/tripolyphosphate microparticles for the controlled delivery of pyrimethamine. *Ibnosina J. Med. BS.* 3(2011): 77–88.
143. Saleem, M. A., Murli, T. D., Naheem, M. D., Patel, J., and Malvania, D. Preparation and evaluation of valsartan hydrogel beads. *Int. Res. J. Pharm.* 3(6) (2012): 80–85.
144. Sankalia, M. G., Mashru, R. C., Sankalia, J. M., and Sutariya, V. B. Stability improvement of alpha-amylase entrapped in kappa-carrageenan beads: Physicochemical characterization and optimization using composite index. *Int. J. Pharm.* 7 (2006): 1–14.
145. Maestrelli, F., Cirri, M., Corti, G., Mennini, N., and Mura, P. Development of enteric-coated calcium pectinate microspheres intended for colonic drug delivery. *Eur. J. Pharm. Biopharm.* 69(2) (2008): 508–518.
146. Kwok, K. K., Groves, M. J., and Burgess, D. J. Production of 5–15 μm diameter alginate-polylysine microcapsules by an air-atomization technique. *Pharm. Res.* 8 (1991): 341–344.

15 Microencapsulation via Interfacial Polymerization

Biao Duan

CONTENTS

15.1 Introduction 297
15.2 Microcapsule Wall Formation 298
15.3 Factors on Microcapsule Formation via Interfacial Polymerization 298
    15.3.1 Chemical Structure of the Reactants 299
    15.3.2 Concentration and Temperature 299
    15.3.3 Emulsifier 299
    15.3.4 Core/Wall Ratio 299
    15.3.5 A Process Example 300
15.4 Development and Future 300
References 301

15.1 INTRODUCTION

Polymerization plays a key role in chemical microencapsulation. The basic mechanism of this method is to put a polymer wall (can be multilayer) through polymerization on a core material, which is in a form of small liquid droplets, solid particles, or even gas bubbles; or to embed the core material in a polymer matrix through polymerization. Interfacial polymerization is one of the most important methods that have been extensively developed and industrialized for microencapsulation. According to Thies[1] and Salaun,[2] interfacial polymerization includes five types of processes represented by the methods of emulsion polymerization, suspension polymerization, dispersion polymerization, interfacial polycondensation/polyaddition, and in situ polymerization. This chapter is only focused on interfacial polycondensation and polyaddition in a narrow sense of interfacial polymerization.

Historically, interfacial polymerization was considered as a gentle alternative to bulk polymerization, which would require high temperature. Morgan[3] and his coworkers published a series of works in this field since 1959, and this process was employed to prepare nylon microcapsules by Chang in 1964.[4] Many patents and studies were published since then (refer Table 15.1).

Both interfacial polycondensation and polyaddition involve two reactants dissolved in a pair of immiscible liquids, one of which is preferably water, which is normally the continuous phase, and the other one is the dispersed phase, which is normally called the oil phase. The polymerization takes place at the interface and controlled by reactant diffusion. Researches[5–8] indicate that the polymer film occurs and grows toward the organic phase, and this was visually observed by Yuan et al.[9] In most cases, oil-in-water systems are employed to make microcapsules, but water-in-oil systems are also common for the encapsulation of hydrophilic compounds.[10–13] Even oil-in-oil systems were applied to prepare polyurethane and polyurea microcapsules.[14]

Equations 15.1 through 15.5 illustrate the main polymers and reactant systems of microcapsule making through interfacial polycondensation and polyaddition.[15] The thus-prepared microcapsules

TABLE 15.1
Major Applications of the Microcapsules with Various Active Cores through Interfacial Polymerization

Application Field	Polymer Wall	Active Core
Pharmaceutical	Polyamide[4,16–25]	Enzyme, antibody, amino acid, thyroxine, hemolysate, liposome, yeast cell, etc.
Biomedical	Polyurea[26]	
Food	Polycarbonate[27]	
Agrochemical	Polyamide[28,29] Polyurea[30–35]	Insecticide, pesticide, etc.
Personal care	Polyurea/polyurethane[36,37,39,54] Polyamide[40]	Essential oil, Jojoba oil, etc.
Graphic arts: carbonless paper, thermal paper and toners, etc.	Polyurea[41–56] Polyurethane[44,51,54] Polyamide[51,57–59]	Dye, pigments, toner, color precursor, lightsensitive, thermosensitive and photosensitive compounds, etc.
Other Industry	Polyamide[60–65] Polyurea[62,66–78] Polyurethane[76,79,80]	Flame retardant, phase change material, catalyst, self-healing agent, cooling agent, demulsifier, etc.

have many applications in various fields such as pharmaceutical, agrochemical, graphic art, cosmetic, and adhesive industries. Some typical examples are listed in Table 15.1.

$$\text{(Di or poly)chlorides} + \text{(Di or poly)amines} \rightarrow \text{Polyamides} \tag{15.1}$$

$$\text{(Di or poly)isocyanates} + \text{(Di or poly)amines} \rightarrow \text{Polyureas} \tag{15.2}$$

$$\text{(Di or poly)isocyanates} + \text{(Di or poly)ols} \rightarrow \text{Polyurethane} \tag{15.3}$$

$$\text{(Di or poly)acyl Chlorides} + \text{(Di or poly)ols} \rightarrow \text{Polyesters} \tag{15.4}$$

$$\text{Bischloroformates} + \text{(Di or poly)amines} \rightarrow \text{Polycarbonates} \tag{15.5}$$

15.2 MICROCAPSULE WALL FORMATION

The formation of a microcapsule wall through interfacial polycondensation/addition takes place in two steps. First step is the deposit of the oligomer (initial wall) at the oil droplet, and the second step is the wall thickness builds up. As described earlier, the polymerization occurs in oil phase, and the formed initial wall can limit the diffusion of the reactants. This reduces the polymerization rate that has great impact on the surface morphology and thickness of the microcapsule wall.[5–7,81,82] Polycondensation by which polyamide, polyester, and polycarbonate microcapsules are prepared can generate acid byproduct during the process; therefore, a base is needed to neutralize the acid and drive the reaction to complete.[83]

15.3 FACTORS ON MICROCAPSULE FORMATION VIA INTERFACIAL POLYMERIZATION[28]

In order to prepare the microcapsule with expected structure, the chosen polymer wall material should match with the core material. The affinity interaction between the polymer wall and the core material can determine the structure of the microcapsule. Highly cross-linked polyurea and polyurethane can form so-called "compact" capsules with 2-methylbenzothiazole homogeneously distributed in the polymer matrix, while in the case of polyamide core-shell structure capsules were found. The "compact" form is due to the high affinity of the core to the polymers, and the core-shell form is resulted from the low affinity.[84]

Like other chemical reactions, the chemical structure of the reactants, the concentrations, reaction temperature, reaction time, emulsification, and stirring are major factors that can affect this process and thus the formation and quality of the microcapsule. Furthermore, reactants should be compatible with the core material without any adverse reactions.

15.3.1 CHEMICAL STRUCTURE OF THE REACTANTS

Aliphatic or aromatic structure, as well as liner or branched structure of the reactants, can give the microcapsule shell different porosity and permeability, which can greatly influence the release performances.[85–87] Multifunctional reactants can help to achieve more thermal mechanical stable microcapsules since the wall is a three-dimensional cross-linked polymer network.[85] Experiments[15] have shown that dichlorides with less than eight carbon atoms do not produce quality polyamide microcapsules. The reason behind this is the competition between interfacial condensation and the hydrolysis reaction of dichlorides. More hydrophobic dichlorides can favor the polymerization and slow the hydrolysis. Similarly, for polyurethane and polyurea type microcapsules, polymeric isocynates are preferred because they might favor the formation of less permeable microcapsules for the hydrolysis of isocynate groups are limited, which consequently reduced the CO_2 release that contribute to the porosity increase of the polymer wall.[4]

15.3.2 CONCENTRATION AND TEMPERATURE

Higher concentration and higher temperature can accelerate the formation of the polymer wall,[6,82,88] but the reaction rate is also controlled by the reactants diffusion.[6,7,81,82,73] Some organic solvents like chloroform[83] and toluene[88] were employed to improve the diffusion. But for industry practice, Norpar, vegetable oil, etc., are more environmental friendly and cost effective.[89,90] Each oil droplet dispersed in the water phase can be treated as a tiny reactor since the interfacial polymerization only happens at the interface between the water and oil. Therefore, the smaller oil droplet size means higher reaction rate.[81]

15.3.3 EMULSIFIER

Emulsifiers (sometimes with protective colloid) and proper stirring are required to produce stable emulsions or suspensions not only in the first step but also prevent microcapsules from agglomeration during curing process. Normally, higher emulsifier concentration or higher protective colloid concentration generates smaller microcapsules with narrower size distribution.[91,92] Higher homogenization rate helps to reduce the microcapsule size but results in broader size distribution.[93] It is of no surprise that at a given condition such as a given emulsifier concentration and stirring speed, higher portion of the organic phase generates larger microcapsules.[92] One thing to notice is that some surface active impurities such as fatty acid in the system, especially in the core material, can greatly affect the polymer layer deposition and even cause the failure of the capsule formation.[94]

15.3.4 CORE/WALL RATIO

Core/wall ratio cannot only decide the wall thickness of a microcapsule but also the effectiveness of the microencapsulation.[11,73,75,92] The work[11] on the microencapsulation of xylitol by poly (urethane-urea), which was performed in a water-in-oil system, indicates that the most proper core/wall ratio is 77/23 in terms of high encapsulation yield and xylitol loading content. Another work[92] on the microencapsulation of perfume by polyurea, which was performed in a oil-in-water system, found that with the decrease of the core/wall ratio, the size of the formed microcapsule increased even the original droplet size is roughly the same. An outward diffusion mechanism was proposed to explain the phenomenon. Based on this mechanism, isocynates migrate outward through the formed wall and part of them react with water to form amines, which then react with additional isocynates and accumulate polyurea on the wall surface. In this way, as the process continues, the wall becomes thicker and thicker and the microcapsule size become larger and larger. Contrary to most other models proposed in the reports,[6,8,81] this mechanism attributes the microcapsule wall growth to the outward diffusion of the isocynates through the wall, not the inward diffusion of water.

Of course, the preparation of a quality microcapsule is not only determined by a single factor but the collective interactions of a large number of the factors. Besides what described above, other factors like reactant feeding speed,[95] polymerization catalyst,[96] etc. can also affect the polymerization process, which consequently affect the properties of the microcapsule.

15.3.5 A Process Example

A process[89] is schemed in Figure 15.1 to show the industry art on how to prepare a microcapsule by interfacial polymerization.

This unique process introduced a reaction period at elevated temperature between emulsification and polyamine addition, and allows microcapsules of 10 μm or less average diameter to be made at greater than 40% by weight solids without agglomeration or resultant excess viscosity.

Figure 15.2 is a microscope image of the polyurea microcapsule prepared through a similar process.

15.4 DEVELOPMENT AND FUTURE

Many efforts have been devoted to the microcapsules with tailored structure and the fabrication methods thereof. Examples like double-shell microcapsule of polyurea/polyurethane[75] show improved thermal mechanical property and ethanol resistance, poly(acrylonitrile-divinylbenzene-styrene)/polyamide two-layer microcapsule was prepared to encapsulate water,[13] and self-bursting microcapsules[34] may have potential application in agricultural field because of its unique release profile. Additionally, monodispersed microcapsules based on microfludic processes like SPG (Shirasu

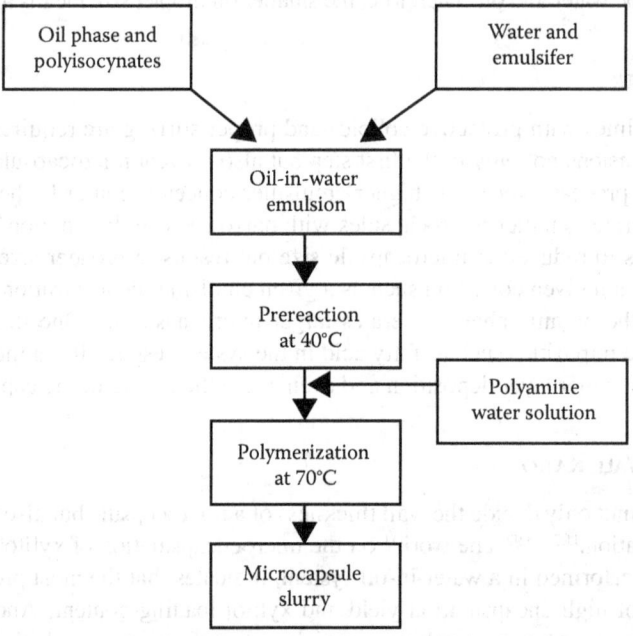

FIGURE 15.1 Flow diagram of an interfacial polymerization process to make high solids aqueous slurry of microcapsules.

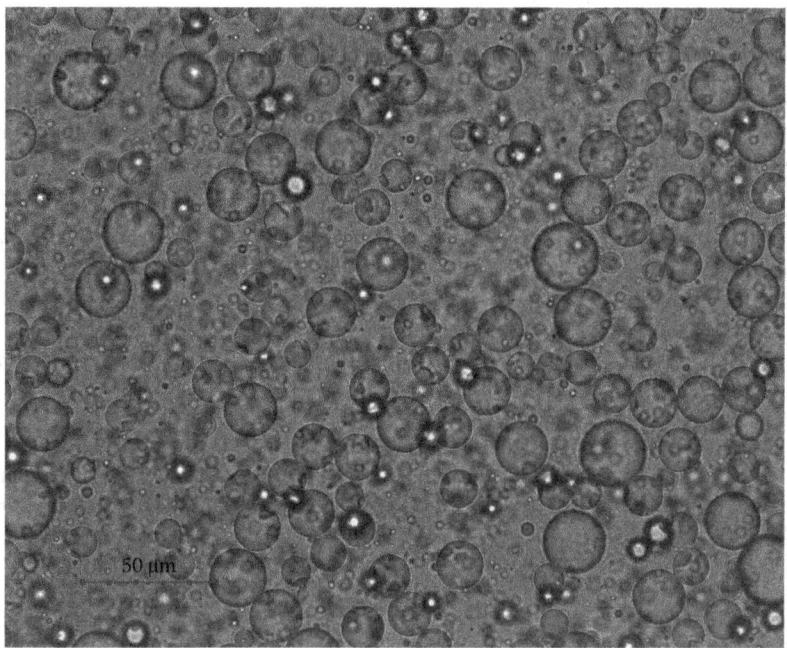

FIGURE 15.2 Microscope image of the polyurea microcapsules.

porous glass) membrane emulsification[96] and axisymmetric flow-focusing microfluidic device[97] also attract many science and industry attention even though the industry use of both the techniques are limited because of the scaling up challenge.[98] As a versatile technology, microencapsulation through interfacial polymerization has been developing steadily with the developments of other cutting-edge technologies, and gains more and more applications in different industry areas.

REFERENCES

1. Thies, C. Microencapsulation. *Encyclopedia of Chemical Technology*, 4th edn, Kroschwitz, J.I. and Howe-Grant, M., Eds., Hoboken, NJ: John Wiley & Sons, Vol. 16, 1995, pp. 628–651.
2. Salaun, F. Microencapsulation by interfacial polymerization. *Encapsulation Nanotechnologies*, Mittal, V., Ed., Hoboken, NJ: John Wiley & Sons, 2013, pp. 137–173.
3. a. Emerson, L. W.; Morgan, P. W. Interfacial polycondensation. I. *Journal of Polymer Science* (1959), 40(137), 289–297; b. Morgan, P. W.; Kwolek, S. L. Interfacial polycondensation. II. Fundamentals of polymer formation at liquid interfaces. *Journal of Polymer Science* (1959), 40(137), 299–327; c. Beaman, R. G.; Morgan, P. W.; Koller, C. R.; Wittbecker, E. L.; Magat, E. E. Interfacial polycondensation. III. Polyamides. *Journal of Polymer Science* (1959), 40(137), 329–336.
4. Chang, T. M. S. Semipermeable microcapsules. *Science* (1964), 146(3643), 524–525.
5. Janssen, L. J. J. M.; Te Nijenhuis, K. Encapsulation by interfacial polycondensation. II. The membrane wall structure and the rate of the wall growth. *Journal of Membrane Science* (1992), 65(1–2), 69–75.
6. Janssen, L. J. J. M.; Te Nijenhuis, K. Encapsulation by interfacial polycondensation. I. The capsule production and a model for wall growth. *Journal of Membrane Science* (1992), 65(1–2), 59–68.
7. Yadav, S. K.; Khilar, K. C.; Suresh, A. K. Microencapsulation in polyurea shell: Kinetics and film structure. *AIChE Journal* (1996), 42(9), 2616–2626.
8. Ji, J.; Childs, R. F.; Mehta, M. Mathematical model for encapsulation by interfacial polymerization. *Journal of Membrane Science* (2001), 192(1–2), 55–70.
9. Yuan, F.; Wang, Z.; Yu, X.; Wei, Z.; Li, S.; Wang, J.; Wang, S. Visualization of the formation of interfacially polymerized film by an optical contact angle measuring device. *Journal of Physical Chemistry C* (2012), 116(21), 11496–11506.

10. Li, J.; Hughes, A.; Kalantar, T.; Drake, I.; Tucker, C.; Moore, J. S. Efficient encapsulation of hydrophilic actives including aliphatic amines. Abstracts of Papers, *246th ACS National Meeting and Exposition*, Indianapolis, IN, September 8–12, 2013, PMSE-257.
11. Salauen, F.; Bedek, G.; Devaux, E.; Dupont, D.; Gengembre, L. Microencapsulation of a cooling agent by interfacial polymerization: Influence of the parameters of encapsulation on poly(urethane-urea) microparticles characteristics. *Journal of Membrane Science* (2011), 370(1–2), 23–33.
12. Pense, A. M.; Vauthier, C.; Benoit, J. P. Preparation of microcapsules containing water-soluble amphiphiles by interfacial polycondensation. EP 407257, 1991.
13. Hatate, Y.; Imafuku, T. Two-layer water-containing microcapsules having good retention of contents. JP 02258052, 1990.
14. Kobaslija, M.; McQuade, D. T. Polyurea microcapsules from oil-in-oil emulsions via interfacial polymerization. *Macromolecules* (2006), 39(19), 6371–6375.
15. Madan, P. L. Microencapsulation. II. Interfacial reactions. *Drug Development and Industrial Pharmacy* (1978), 4(3), 289–304.
16. Chibata, I.; Tosa, T.; Sato, T.; Mori, T.; Matsuo, Y. Microencapsulation of enzymes. JP 47035191, 1972.
17. Lim, F.; Moss, R. D. Process for preparing semi-permeable microcapsules. US 4251387, 1981.
18. O'Grady, P.; Joyce, P. Microencapsulation of bovine liver arginase: Characterization and in vivo evaluation of its effect on the growth of the L1210 murine leukemia. *Enzyme and Microbial Technology* (1981), 3(2), 149–152.
19. Aruna, G.; Sastri, N. V. S. Enzyme immobilization by encapsulation: A kinetic study. *Indian Chemical Engineer* (1989), 31(1), 76–79.
20. Wallace, A. M.; Wood, D. A. Development of a simple procedure for the preparation of semipermeable antibody-containing microcapsules and their analytical performance in a radioimmunoassay for 17-hydroxyprogesterone. *Clinica Chimica Acta* (1984), 140(2), 203–212.
21. Suenaga, E.; Sato, N. Microcapsules. JP 48020705, 1973.
22. Lim, F.; Moss, R. D. Microcapsules with walls of which the upper limit of permeability is in a chosen range. Belg. BE 878084, 1979.
23. Arakawa, M.; Kondo, T. Preparation of hemolysate-loaded poly(Nα,Nε-L-lysinediylterephthaloyl) nanocapsules. *Journal of Pharmaceutical Sciences* (1981), 70(4), 354–357.
24. Yeung, V. W.; Nixon, J. R. Preparation of microencapsulated liposomes. *Journal of Microencapsulation* (1988), 5(4), 331–337.
25. Green, K. D.; Gill, I. S.; Khan, J. A.; Vulfson, E. N. Microencapsulation of yeast cells and their use as a biocatalyst in organic solvents. *Biotechnology and Bioengineering* (1996), 49(5), 535–543.
26. Hoshino, K.; Muramatsu, N.; Kondo, T. A study on the thermostability of microencapsulated glucose oxidase. *Journal of Microencapsulation* (1989), 6(2), 205–211.
27. Lukashcheva, E. V. Properties of alpha-chymotrypsin enclosed into polycarbonate microcapsules. Estimation of the diffusion effect. *Biokhimiia* (1977), 42(11), 2013–2019.
28. Heinrich, R.; Frensch, H.; Albrecht, K. Pressure-resistant microcapsules with a polyamide outer shell and an inner composition structured by polyurethane-polyurea and their use. DE 3020781, 1981.
29. Choi, K. Y.; Min, K. S.; Chang, T Microencapsulation of pesticides by interfacial polymerization. 2. Polyamide microcapsules containing water-soluble drug. *Polymer* (1991), 15(5), 548–55.
30. Scher, H. B.; Rodson, M.; Lee, K.-S. Microencapsulation of pesticides by interfacial polymerization utilizing isocyanate or aminoplast chemistry. *Pesticide Science* (1998), 54(4), 394–400.
31. Hirech, K.; Payan, S.; Carnelle, G.; Brujes, L.; Legrand, J. Microencapsulation of an insecticide by interfacial polymerization. *Powder Technology* (2003), 130(1–3), 324–330.
32. Fu, G.; Zhong, B.; Chen, J.; Wang, H.; Zhang, D.; Wang, G. Pesticide microcapsules prepared by an interfacial polymerization method. *Nongyao* (2005), 44(2), 66–68, 73.
33. Zhu, L.; Wang, Z.; Zhang, S.; Long, X. Fast microencapsulation of chlorpyrifos and bioassay. *Journal of Pesticide Science* (2010), 35(3), 339–343.
34. Tsuda, N.; Ohtsubo, T.; Fuji, M. Study on the breaking behavior of self-bursting microcapsules. *Advanced Powder Technology* (2012), 23(6), 845–849.
35. Bristow, J. T. Agrochemical composition, method for its preparation and use thereof. GB 2496330, 2013.
36. Markus, A.; Linder, C. Polyurea/urethane encapsulated essential oils for controlled release. WO 2004098767, 2004.
37. Scarfato, P.; Avallone, E.; Iannelli, P.; De Feo, V.; Acierno, D. Synthesis and characterization of polyurea microcapsules containing essential oils with antigerminative activity. *Journal of Applied Polymer Science* (2007), 105(6), 3568–3577.

38. Fletcher, R. B.; Malotky, D. L.; Zhang, X. Encapsulated hydrophobic actives via interfacial polymerization. WO 2009091726, 2009.
39. Zhang, B.; Fei, X.; Fu, Y.; Yu, M.; Zhao, H. Self-bonded double-walled essence microcapsule preparation and application. CN 103230766, 2013.
40. Persico, P.; Carfagna, C.; Danicher, L.; Frere, Y. Polyamide microcapsules containing jojoba oil prepared by inter-facial polymerization. *Journal of Microencapsulation* (2005), 22(5), 471–486.
41. Kobayashi, M.; Iwasaki, H. Manufacture of microcapsule encapsulating color developer by interfacial polymerization suitable for pressure-sensitive copying paper. JP 11290675, 1999.
42. Takasu, M.; Kawaguchi, H. Superfine microcapsules, their manufacture, and recording liquids using them. JP 2006021164, 2006.
43. Suda, H. Microcapsules containing isocyanates, their manufacture, and coating compositions, adhesive compositions, and plastic modifiers containing them. JP 2006061802, 2006.
44. Kobayashi, T.; Yagi, H. Encapsulated pigments. JP 49015719, 1974.
45. Horiike, T.; Okimoto, T.; Shiozaki, T. Microcapsules by interfacial polymerization of a polyisocyanate. FR 2476100, 1981.
46. Riecke, K. Solvent mixture for microcapsules. DE 3346601, 1985.
47. Hosoi, N.; Hatakeyama, A. Electrostatographic encapsulated toner. GB 178182, 1987.
48. Kori, S.; Machida, J. Non-magnetic microencapsulated toner containing composite metal oxide colorant. JP 62283347, 1987.
49. Iida, Y.; Satsuta, K. Powder capsule toners with wall formed by interfacial polymerization. JP 01057268, 1989.
50. Miyatake, M.; Tenmaya, E.; Ohashi, T.; Kawasaki, M. Water-based microencapsulated mixed pigment dispersions for inks and coatings. JP 01115976, 1989.
51. Nuyken, O.; Dauth, J.; Pekruhn, W. Thermosensitive microcapsules. II. New azomonomers, homo- and cocondensation, microencapsulation, release measurements, size distribution and thermo-printing. *Angewandte Makromolekulare Chemie* (1991), 190, 81–98.
52. Tan, H. S.; Ng, T. H.; Mahabadi, H. K. Processes for encapsulated toners. US 5114824, 1992.
53. Kao, S. V.; Allison, G. R.; Hawkins, M. S.; Mahabadi, H. K. Process for preparing encapsulated electrophotographic toner. US 5153092, 1992.
54. Imai, T.; Agata, T. Electrophotographic developer encapsulated toner and its manufacture. JP 06317924, 1994.
55. Lai, W.; Li, X.; Feng, H.; Fu, G. Photo-polymerization property of the photosensitive core-shell structured microcapsule material. *Journal of Photopolymer Science and Technology* (2008), 21(6), 761–765.
56. Li, X.-Z.; Li, X.-W.; Lai, W.-D.; Bai, B.; Feng, H.-G. Influence of monomer curing characteristic on the developing density of light-thermal sensitive microcapsule. *Gongneng Cailiao* (2011), 42(3), 404–406, 410.
57. Moffat, K. A.; Breton, M. P.; Martin, T. I.; Gerroir, P. J. Process for controlling electrical characteristics of toners. US 4937167, 1990.
58. Hsieh, B. R.; Gruber, R. J.; Dalal, E. N. Processes for the preparation of encapsulated toner compositions. US 5108863, 1992.
59. Parrish, C. F. Water fire extinguishers microencapsulated with flame-retardant polymers. WO 2002096519, 2002.
60. Chao, H. Y. Microencapsulated adhesive prepared by polymerization of encapsulated monomers. GB 2272446, 1994.
61. Durand, G.; Poirier, J.-E.; Chappat, M. Aqueous bitumen emulsions containing encapsulated demulsification agents for controlled bitumen addition to paving materials. EP 864611, 1998.
62. Valea, A.; Miguez, J. C.; Juanes, F. J.; Gonzalez, M. L. Microencapsulation of paraffins as phase change materials by interfacial polycondensation of a polyamide. *Comunicaciones presentadas a las Jornadas del Comite Espanol de la Detergencia* (2007), 37, 335–346.
63. Hatade, Y.; Yoshida, M. Manufacture of microcapsules capable of encapsulating latent heat storage materials in high concentrations. JP 2007244935, 2007.
64. Yan, L.; Jiang, J.; Zhang, Y.; Liu, J. Preparation and characterization of large-size halloysite nanotubes particles by a combined technique of interfacial polymerization and condensation polymerization. *Journal of Nanoparticle Research* (2011), 13(12), 6555–6561.
65. Liu, Y.-Q.; Zhao, G.-Z. Microencapsulation of chlorocyclophosphazene by interfacial polymerization. *Journal of China Ordnance* (2007), 3(1), 71–73.
66. Hosokawa, T.; Matsushita, T.; Hata, M. Encapsulated cross-linking agents or accelerators, epoxy resins containing the microcapsules, curing process of the resins, and cured products. JP 08337633, 1996.

67. Ramarao, C.; Ley, S. V.; Smith, S. C.; Shirley, I. M.; DeAlmeida, N. Encapsulation of palladium in polyurea microcapsules. *Chemical Communications* (2002), (10), 1132–1133.
68. Reed, J. L.; McFarland, B. H. Synthesis of microcapsules containing dialkylanilines and their use in frontal polymerization. Abstracts of Papers, *235th ACS National Meeting*, New Orleans, LA, April 6–10, 2008, CHED-687.
69. McFarland, B. H.; Pojman, J. A. Preparation and analysis of peroxide microcapsules. *Polymer Preprints (American Chemical Society, Division of Polymer Chemistry)* (2004), 45(1), 1–2.
70. Cho, J.-S.; Kwon, A.; Cho, C.-G. Microencapsulation of octadecane as a phase-change material by interfacial polymerization in an emulsion system. *Colloid and Polymer Science* (2002), 280(3), 260–266.
71. Xing, J.; Li, Y.; Newton, E.; Yeung, K.-W. Method for encapsulating phase-transitional paraffin compound and their microcapsules. US 20030222378, 2003.
72. Periyasamy, S.; Palanikkumaran, M.; Agrawal, A. K.; Kotresh, T. M. Microencapsulation of N-octadecane using interfacial polymerization. Jassal, M.; Agrawal, A. K., Eds., From Emerging Trends in Polymers and Textiles, *Proceedings of [the] International Conference*, New Delhi, India, January 7–8, 2005, pp. 139–145.
73. Lin, Y.-H.; Wei, C.-S. Composition and method for fabricating microcapsules encapsulating phase-change material. US 20080157415, 2008.
74. Lu, S.; Xing, J.; Zhang, Z.; Jia, G. Preparation and characterization of polyurea/polyurethane double-shell microcapsules containing butyl stearate through interfacial polymerization. *Journal of Applied Polymer Science* (2011), 121(6), 3377–3383.
75. Iwasawa, A. Manufacture of resin dispersions for coatings with good storage stability. JP 2010082527, 2010.
76. Iwasawa, A.; Mizuno, T. Aqueous dispersions containing microcapsules and method for curing resins using them. JP 2011031147, 2011.
77. Zhou, Y. F.; Xie, S.; Chen, C. H. Pyrolytic polyurea encapsulated natural graphite as anode material for lithium ion batteries. *Electrochimica Acta* (2005), 50(24), 4728–4735.
78. Arai, H.; Takagi, N. Manufacture of capsules for heat storage and capsule dispersions. JP 2011111512, 2011.
79. Brochu, A. B. W.; Chyan, W. J.; Reichert, W. M. Microencapsulation of 2-octylcyanoacrylate tissue adhesive for self-healing acrylic bone cement. *Journal of Biomedical Materials Research, Part B: Applied Biomaterials* (2012), 100B(7), 1764–1772.
80. Yadav, S. K.; Suresh, A. K.; Khilar, K. C. Microencapsulation in polyurea shell by interfacial polycondensation. *AIChE Journal* (1990), 36(3), 431–438.
81. Janssen, L. J. J. M.; Boersma, A.; te Nijenhuis, K. Encapsulation by interfacial polycondensation. III. Microencapsulation; the influence of process conditions on wall permeability. *Journal of Membrane Science* (1993), 79(1), 11–26.
82. Zydowicz, N.; Chaumont, P.; Soto-Portas, M. L. Formation of aqueous core polyamide microcapsules obtained via interfacial polycondensation—Optimization of the membrane formation through pH control. *Journal of Membrane Science* (2001), 189(1), 41–58.
83. Latnikova, A.; Grigoriev, D. O.; Moehwald, H.; Shchukin, D. G. Capsules made of cross-linked polymers and liquid core: Possible morphologies and their estimation on the basis of Hansen solubility parameters. *Journal of Physical Chemistry C* (2012), 116(14), 8181–8187.
84. Mathiowitz, E.; Cohen, M. D. Polyamide microcapsules for controlled release. I. Characterization of the membranes. *Journal of Membrane Science* (1989), 40(1), 1–26.
85. Mathiowitz, E.; Cohen, M. D. Polyamide microcapsules for controlled release. II. Release characteristics of the microcapsules. *Journal of Membrane Science* (1989), 40(1), 27–41.
86. Stefanescu, E. A.; Stefanescu, C.; Huvard, G. S.; McHugh, M. A. Influence of co-monomers architecture on the structure of polyurea microcapsules. *Polymer Preprints (American Chemical Society, Division of Polymer Chemistry)* (2009), 50(2), 506–507.
87. Frere, Y.; Danicher, L; Gramain, P. Preparation of polyurethane microcapsules by interfacial polycondensation. *European Polymer* (1998), 34(2), 193–199.
88. Kalishek, R. J.; Hayford, D. E. Process for microencapsulation without capsule agglomeration. US 5164126, 1992.
89. Mulqueen, P. J.; Waller, A.; Shirley, I. M.; Chavant, M. Sustained-release microcapsules. EP 1965638, 2011.
90. Meyers, P. A. Particle size distribution control during the microencapsulation of a dinitroaniline herbicide. Book of Abstracts, *213th ACS National Meeting*, San Francisco, CA, April 13–17, 1997, COLL-026.

91. Li, Z.; Chen, S.; Zhou, S. Factors affecting the particle size and size distribution of polyurea microcapsules by interfacial polymerization of polyisocyanates. *International Journal of Polymeric Materials* (2004), 53(1), 21–31.
92. Tan, H. S.; Ng, T. H.; Mahabadi, H. K. Interfacial polymerization encapsulation of a viscous pigment mix: Emulsification conditions and particle size distribution. *Journal of Microencapsulation* (1991), 8(4), 525–536.
93. Duan, B. Lab study observations.
94. Madan, P. L.; Chareonboonsit, P. Nylon microcapsules. II. Effect of selected variables on theophylline release. *Pharmaceutical Research* (1989), 6(8), 714–718.
95. Zhuo, L.; Chen, S. Effects of catalyst and core materials on the morphology and particle size of microcapsules. *International Journal of Polymeric Materials* (2004), 53(5), 385–393.
96. Xie, R.; Chu, L.-Y.; Chen, W.-M.; Zhao, Y.; Xiao, X.-C.; Wang, S. Preparation of monodispersed porous microcapsule membranes with SPG membrane emulsification and interfacial polymerization. *Gaoxiao Huaxue Gongcheng Xuebao* (2003), 17(4), 400–405.
97. Takeuchi, S.; Garstecki, P.; Weibel, D. B.; Whitesides, G. M. An axisymmetric flow-focusing microfluidic device. *Advanced Materials* (2005), 17(8), 1067–1072.
98. Molly, K. M.; Jonathan, P. R. Scale-up and control of droplet production in coupled microfluidic flow-focusing geometries. *Microfluidics and Nanofluidics* (2012), 13(1), 65–73.

16 Microencapsulation via *In Situ* Polymerization

Biao Duan

CONTENTS

16.1 Introduction 307
16.2 Process and Related Chemistry 308
    16.2.1 Process 308
    16.2.2 Wall Formation and Related Chemistry 309
16.3 Influence Factors 310
    16.3.1 Emulsifier 310
    16.3.2 pH Value 310
    16.3.3 Salt 310
    16.3.4 Temperature and Stirring 310
    16.3.5 Formaldehyde Scavenger 310
16.4 Conclusion 311
References 311

16.1 INTRODUCTION

In situ polymerization is a very developed and commercialized chemical process for microencapsulation.[1-4] In this process the reactants, either monomers or oligomer, are in a single phase. The polymerization occurs in the continuous phase in most cases, or in the dispersed phase, for example, *in situ* water-in-oil microencapsulation.[5-7] The formed polymer migrates and deposits on the dispersed phase to generate a solid wall thus forming a microcapsule. In a wide sense, emulsion polymerization, suspension polymerization, and some other processes fall into this category.[8] This chapter is only focused on *in situ* microencapsulation based on amino resins, especially melamine-formaldehyde (MF) and urea-formaldehyde (UF) systems.

Amino resins are thermosetting polymers made by the reaction of an aldehyde with an amino containing compound. UF and MF resins are predominant in this area.[9,10] The main advantages of this method are as follows: inexpensive and easy-to-get materials, relatively simple and controllable process, highly cross-linked impermeable wall with superior thermal and mechanical properties, high loading core (up to 95%), high resistance to harsh chemical environments (e.g., in detergents, softeners), and easy to large industry scale-up.[4,8,10] By this method, many microcapsules with different active core materials such as color precursor, fragrance, phase change material (PCM), insecticide, etc., were prepared. Some typical examples are listed in Table 16.1.

However, there are two major disadvantages of this process, which may affect the microcapsule's properties made from and its future development. The first one is the same wall thicknesses apply to all the microcapsules with different sizes in a batch.[54] That means larger microcapsules have thinner wall relative to their sizes and thus are weaker, while smaller microcapsules have thicker wall and stronger. This will impact the microcapsules release profile and consequently the applications. The second disadvantage of this method is this process is formaldehyde involved. For environmental and health considerations, formaldehyde is regulated. Efforts have been devoted to minimize the residual formaldehyde to satisfy technical requirements.[10,55-60]

TABLE 16.1
Materials Microencapsulated by *In Situ* Polymerization and Their Functionalities/Applications

Polymer Wall	Active Core	Functionality/Application
MF	Essence,[11] Fragrant oil[12,13]	Consumer product
	Expandable Graphite,[14] Cage Phosphate,[15] Ammonium polyphosphate,[16,17] Caged bicyclic phosphate[18]	Flame retardant
	n-Dodecane,[19] Dodecanol,[20] n-Octadecane,[21] Other PCM[22]	Phase change material
	Emamectin benzoate[23]	Insecticide
	Blue E 4R[24]	Dye
	Light Fast Yellow G[25]	Pigment
	TiO_2[26]	Electrophoretic ink
	Color precursor,[27] Color developer,[28] Color former[29]	Carbonless paper, thermal-sensitive paper, pressure-sensitive recording
	Ag-80,[30] Styrene/epoxydiacrylate,[31] Dicyclopentadiene,[32] Polythiol[33]	Self-healing agent
UF	Polyethylene glycol,[7] Paraffin[34]	Phase change material
	Chlorpyrifos[35]	Insecticide
	3-Iodo-2-propynylbutyl carbamate,[36] 2,3,5,6-tetrachloro-4-methylsulfonylpyridine and tetrachloroisophthalonitrile[37]	Fungicide
	Isocyanate [38], Single-walled nanotube[39]	Functional material
	Red 5R,[40] Red pigment[41]	Electronic ink
	Dyestuff precursor[42]	Carbonless paper
	Epoxy resin/butyl glycidyl ether,[43] Tung oil,[44] Epoxy resin,[45,46] Ethylene silicon oil,[47] Polythiol[33]	Self-healing agent
	Poly-alpha-olefin[48]	Drag reduce
	Fragrant oil[49]	Consumer product
Phenol-formaldehyde resin	Red phosphorus[50]	Flame retardant
	Calcium hydroxide[51]	Additive for polymer
Melamine-urea-formaldehyde	Dodecanol[52]	Phase change material
	Essence oil[53]	Consumer product

16.2 PROCESS AND RELATED CHEMISTRY

16.2.1 PROCESS

With different starting materials and formulas,[10,61] the process might be different. But in general, to encapsulate water-immiscible oil, the key steps should include the following:

- Disperse oil-in-water phase that contains the reactants (i.e., MF or UF resin prepolymer) and create a stable emulsion with the expected oil droplet size
- Polymerize from the water phase and deposit monomer/oligomer on the oil phase
- Build wall thickness and cure

Figure 16.1 illustrates a process of making a MF perfume microcapsule.[62]

Figure 16.2 is the microscope and scanning electron microscope (SEM) images of the MF microcapsules made following the aforementioned process.

Microencapsulation via *In Situ* Polymerization

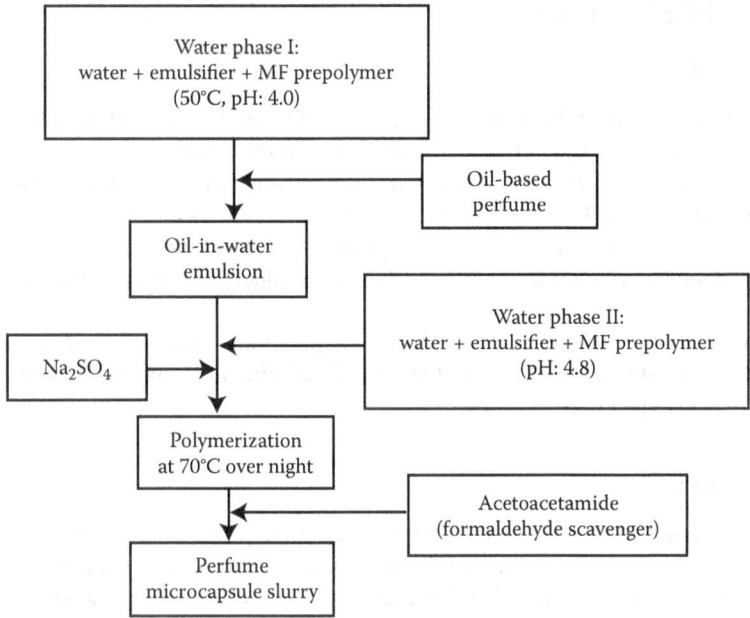

FIGURE 16.1 Example of perfume MF microcapsule making.

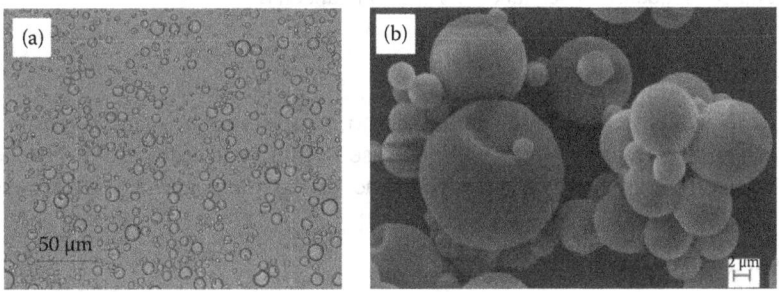

FIGURE 16.2 The MF microcapsules: (a) microscope image of the aqueous slurry and (b) SEM image of the spray dried powder.

16.2.2 Wall Formation and Related Chemistry

The wall formation is a chemical and physical process.[8,12,63,64] Under acidic condition, the amino resin prepolymer, which is methylol compound and water soluble, becomes more hydrophobic because of the etherification and methylene bridge formation. The hydrophilic and hydrophobic interactions drive the oligomer concentrated on the interface and create a primary wall on the oil droplet. And finally with the further polycondensation, the wall is cross-linked and the wall thickness is built up. The related chemistry is presented in Equations 16.1 through 16.4.[8,9]

$$R\text{-}NH_2 + HCHO \rightarrow RNH\text{-}CH_2OH \tag{16.1}$$

$$RNH\text{-}CH_2OH + H_2NR \rightarrow RNH\text{-}CH_2\text{-}NHR + H_2O \tag{16.2}$$

$$2RNH\text{-}CH_2OH \rightleftharpoons RNHCH_2\text{-}O\text{-}CH_2NHR + H_2O \tag{16.3}$$

$$RNHCH_2\text{-}O\text{-}CH_2NHR \rightarrow RNH\text{-}CH_2\text{-}NHR + HCHO \tag{16.4}$$

16.3 INFLUENCE FACTORS

16.3.1 Emulsifier

In most cases, anionic water-soluble polymers such as poly(styrene-maleic anhydride), polyacrylic acid, etc., are applied.[4,10,65] These kinds of emulsifiers can influence the microcapsule preparation, mean particle size, and particle size distribution. By emulsification, an electric double layer generates on the dispersed phase. Then the electrostatic interactions between the protonated amino resin prepolymer and the negatively charged organic phase can act as a driving force, which enable the wall material polycondensate on the surface of the oil droplets but not throughout the whole water phase.[53,61,66]

Nonionic and small molecular emulsifiers such as poly(vinyl alcohol),[67] polyoxyethylene nonyl phenyl ether,[68] sodium dodecyl benzene sulfonate, and sodium lauryl sulfate[69] can also be employed to make amino resin-type microcapsules.

16.3.2 pH Value

In situ polymerization is an acid catalyzed polycondensation process[9]; pH value cannot only affect the polycondensation rate, but also the surface activity of the amino resin prepolymer. Low pH process promotes the formation of methylene bridges affecting the amino resin solubility and therefore obtains a rough surface.[8] However, for MF microcapsule making, at pH values below 3.0, the fraction of protonated melamine resin prepolymer is increased, and the MF prepolymer becomes more hydrophilic and consequently loses its encapsulation capacity.[63]

16.3.3 Salt

Salts like KCl, Na2SO4, KH2PO4, etc., can help to reduce the processing viscosity and achieve high solid content microcapsule slurry with increased wall impermeability.[62,65] NaCl can increase the encapsulation efficiency[34] and has positive impact on the microcapsules' surface morphology.[70]

16.3.4 Temperature and Stirring

Higher stirring rate reduces the microcapsule size and narrows the size distribution. The core content of the microcapsules decreases as the stirring rate increases since the core weight fraction in the microcapsule reduced for smaller microcapsules.[31] Ultra fine microcapsules have higher surface energy and are thermodynamically unstable and have a strong tendency to agglomerate, higher content of emulsifier can help to reduce the agglomeration.[19] When low reaction temperature and stirring rate are applied to prepare UF microcapsules, the encapsulation yield decreases. The possible cause may be that under such a condition, the oil phase is not well dispersed and the polycondensation occurs mainly in the aqueous phase rather than at the oil/water interface, which leads to a decreased encapsulation yield.[46]

16.3.5 Formaldehyde Scavenger

Various formaldehyde scavengers are known. Some common ones are urea,[71] ammonia,[58,72] melamine,[73,74] ammonium chloride,[60] and acetoacetamide.[62,75] A linear correlation exists between the added ammonia amount and the reduction of residual formaldehyde in a microcapsule slurry. Furthermore, ammonia can improve the microcapsules' wall impermeability/durability and thermal resistance.[58] A similar finding was also reported by using melamine as the formaldehyde scavenger.[73]

Besides the factors described earlier, other factors such as monomer ratio,[76] addition of resorcinol,[46,77] oil-water phase ratio, and reactant adding sequence have direct impact on the *in situ* polymerization process and thus the formation of the microcapsule.[78] A successful process is a synergistic effect of all the factors with the right core material and the right wall choice.

16.4 CONCLUSION

In situ polymerization has been industrialized for several decades. The large-scale amino resin microcapsules are widely applied for carbonless paper, fabric care, and PCM, etc. Even though free formaldehyde is a big concern for its development, as a conventional and economic process, and especially through which excellent thermally and mechanically stable microcapsules can be made, *in situ* polymerization continues as one of the leading industry methods of microencapsulation.

REFERENCES

1. Foris, P. L.; Brown, R. W.; Phillips, P. S., Jr. Polymer microencapsulation. DE 2757528, 1978.
2. Foris, P. L.; Brown, R. W.; Phillips, P. S., Jr. Micro-capsules. AU 529814, 1983.
3. Hawkins, S.; Wolf, M.; Guyard, G.; Greenberg, S.; Dayan, N. Microcapsules as a delivery system. *Delivery System Handbook for Personal Care and Cosmetic Products*, Rosen, M. R. Ed., Norwich, NY: William Andrew Pub. 2005, pp. 191–213.
4. Boh, B.; Sumiga, B. In situ polymerization microcapsules. *Bioencapsulation Innovations* March 2013, pp. 4–6.
5. Rodson, M.; Scher, H. B. Water-in-oil microencapsulation process and microcapsules produced thereby. US 6113935, 2000.
6. Yan, N. UV curable coating material of encapsulated water dispersed core material. US7629394, 2009.
7. Ghosh, S.; Bhatkhande, P. Encapsulation of PCM for thermo-regulating fabric application. *International Journal of Organic Chemistry* 2012, 2 (4), 366–370.
8. Salaun, F. Microencapsulation by interfacial polymerization. *Encapsulation Nanotechnologies*, Mittal, V. Ed, New Jersy, Wiley-Scrivener. 2013, pp. 137–173.
9. Williams, L. L. Amino resin and plastics. *Encyclopedia of Chemical Technology*, 4th edn., Kroschwitz, J.I., Howe-Grant, M. Eds., New York: John Wiley and Sons. 1992, pp. 604–637.
10. Dietrich, K.; Herma, H.; Nastke, R.; Bonatz, E.; Teige, W. Amino resin microcapsules. *Acta Polymerica* 1989, 40(4), 243–251.
11. Zhang, B.; Fei, X.; Fu, Y.; Yu, M.; Zhao, H. Self-bonded double-walled essence microcapsule preparation and application. CN 103230766, 2013.
12. Lee, H. Y.; Lee, S. J.; Cheong, I. W.; Kim, J. H. Microencapsulation of fragrant oil via in situ polymerization: Effects of pH and melamine-formaldehyde molar ratio. *Journal of Microencapsulation* (2002), 19(5), 559–569.
13. Hong, K.; Park, S. Melamine resin microcapsules containing fragrant oil: Synthesis and characterization. *Materials Chemistry and Physics* (1999), 58 (25), 128–131.
14. Ou, K.; Duan, H.; Zhang, W.; Tang, J.; Li, Z. Synthesis of core-shell expandable graphite/melamine-formaldehyde particles and their flame retardant applications. *Gaofenzi Cailiao Kexue Yu Gongcheng* (2013), 29(1), 22–26.
15. Fang, X.; Xu, Y.; Ding, T. Synthesis and characterization of cage phosphate microparticles coated with melamine resin based on composite properties of materials. *Advanced Materials Research (Durnten-Zurich, Switzerland)* (2012), 583 (*Advanced Composite Materials and Manufacturing Engineering*), 236–239.
16. Feng, S.; Xu, J.; Guo, B.; Li, X. Preparation of encapsulated ammonium polyphosphate and its applications in SEBS. *Zhongguo Suliao* (2011), 25(8), 81–85.
17. Geng, Y.; Tao, J.; Cui, Y.; Zeng, Z. Preparation of microcapsulated ammonium polyphosphate. *Boligang/Fuhe Cailiao* (2006), 188(3), 39–41.
18. Chang, H.; Lu, T.; Xu, Q.; Sun, Q.; Han, G. Preparation and characterization of microencapsulated red phosphorus with phenol-formaldehyde resin as shell material. *Zhongguo Suliao* (2010), 24(6), 72–75.
19. Zhu, K.; Qi, H.; Wang, S.; Zhou, J.; Zhao, Y.; Su, J.; Yuan, X. Preparation and characterization of melamine-formaldehyde resin micro- and nanocapsules filled with n-dodecane. *Journal of Macromolecular Science, Part B: Physics* (2012), 51(10), 1976–1990.
20. Lu, R.; Wang, J.; Li, Y.; Xu, K.; Zheng, X. Influencing factors on encapsulation rate of melamine-formaldehyde resin microcapsule phase change material. *Huaxue Gongcheng* (2011), 39(12), 26–30, 39.

21. Su, J.; Wang, L.; Ren, L. Fabrication and thermal properties of microPCMs: Used melamine-formaldehyde resin as shell material. *Journal of Applied Polymer Science* (2006), 101(3), 1522–1528.
22. Wang, L.-X.; Su, J.-F.; Ren, L. Preparation of thermal energy storage microcapsule by phase change. *Gaofenzi Cailiao Kexue Yu Gongcheng* (2005), 21(1), 276–279.
23. Li, W.; Lu, F.-S.; Guo, W.-T.; Li, H. Preparation and characterization of emamectin benzoate microcapsules. *Yingyong Huaxue* (2010), 27(12), 1381–1385.
24. Luo, Y.; Chen, S.; Wang, Z. Method for preparing disperse dye microcapsules. CN 1443808, 2003.
25. Xun, Y.; Liu, Y.; Shu, W.; Xiong, L. In-situ polymerization for preparation of encapsulated Light Fast Yellow G pigment. *Tuliao Gongye* (2003), 33(7), 15–18.
26. Song, J. K.; Myoung, H. J.; Kim, K.; Sung, J. H.; Choi, H. J.; Chin, I.-J. Preparation of TiO_2 based microcapsules for electrophoretic ink. Abstracts of Papers, *229th ACS National Meeting*, San Diego, CA, March 13–17, 2005.
27. Lu, J.; Pan, J.; Chen, S.; Hua, Z. Synthesis and properties of pressure-sensitive color—Developing microcapsule—Effects of the dosage of system modifier and the pH of microcapsule slurry on the properties of microcapsule. *Zhongguo Fangzhi Daxue Xuebao* (1998), 24(1), 5–8.
28. Goto, H.; Kawamura, E. Reversibly thermochromic microcapsule and its preparation. JP 04193583, 1992.
29. Miyamoto, A. Pressure-sensitive microcapsular recording material. GB 2101648, 1983.
30. Tian, R.; Fu, X.; Zheng, Y.; Liang, X.; Wang, Q.; Ling, Y.; Hou, B. The preparation and characterization of double-layer microcapsules used for the self-healing of resin matrix composites. *Journal of Materials Chemistry* (2012), 22(48), 25437–25446.
31. Wang, H. Microencapsulation of styrene/epoxydiacrylate via in situ polymerization of melamine-formaldehyde. *Advanced Materials Research* (2012), 393–395(Pt. 3, *Biotechnology, Chemical and Materials Engineering*), 1279–1282.
32. Hu, J.; Xia, Z.; Situ, Y.; Chen, H. Mechanical properties of microcapsules encapsulating DCPD with MF. *Journal of Chemical industry and Engineering*. (2010), 61, 2738–2742.
33. Yuan, Y.; Zhang, M.; Rong, M. Polythiol microcapsule with improved mechanical strength and its preparation method. CN 101116806, 2008.
34. Yan, Y.; Liu, J.; Zhang, H. Effects of NaCl and dispersants on preparation of microencapsulated phase change paraffin. *Huagong Xinxing Cailiao* (2009), 37(1), 56–59.
35. Zhao, D.; Liu, F.; Mu, W.; Ma, C.; Chen, Z. Preparation of chlorpyrifos aqueous capsule suspension and optimization of encapsulation conditions. *Nongyaoxue Xuebao* (2006), 8(1), 77–82.
36. Ibrahim, W. A.; Seman, A. S. M.; Nasir, N. M.; Sudin, R. Performance of microencapsulated fungicide in exterior latex paint on wood substrate. *Pertanika* (1989), 12(3), 409–412.
37. Noren, G. K.; Clifton, M. F.; Migdal, A. H. Investigation of microencapsulated fungicides for use in exterior trade sales paints. *Journal of Coatings Technology* (1986), 58(734), 31–39.
38. Wang, W.; Xu, L.; Liu, F.; Li, X.; Xing, L. Synthesis of isocyanate microcapsules and micromechanical behavior improvement of microcapsule shells by oxygen plasma treated carbon nanotubes. *Journal of Materials Chemistry A: Materials for Energy and Sustainability* (2013), 1(3), 776–782.
39. Caruso, M. M.; Schelkopf, S. R.; Jackson, A. C.; Landry, A. M.; Braun, P. V.; Moore, J. S. Microcapsules containing suspensions of carbon nanotubes. *Journal of Materials Chemistry* (2009), 19(34), 6093–6096.
40. Guo, H.; Li, H.; Fan, H.; Zhao, X. Preparation of red encapsulated electrophoretic ink by in situ polymerization. *Gongneng Cailiao* (2006), 37(4), 559–561.
41. Guo, H. L.; Zhao, X. P. Preparation of a kind of red encapsulated electrophoretic ink. *Optical Materials* (2004), 26(3), 297–300.
42. Michael, E. S. Melamine formaldehyde microencapsulation in aqueous solutions containing high concentrations of organic solvent. US 5401577, 1995.
43. Li, Y.; Li, R.; He, J.; Wang, Z.; Wang, H. Study on biological modeling self-healing microcapsules. *Zhongguo Suliao* (2011), 25(7), 58–62.
44. Samadzadeh, M.; Boura, S. H.; Peikari, M.; Ashrafi, A.; Kasiriha, M. Tung oil: An autonomous repairing agent for self-healing epoxy coatings. *Progress in Organic Coatings* (2011), 70(4), 383–387.
45. Liao, L. P.; Zhang, W.; Xin, Y.; Wang, H. M.; Zhao, Y.; Li, W. J. Preparation and characterization of microcapsule containing epoxy resin and its self-healing performance of anticorrosion covering material. *Chinese Science Bulletin* (2011), 56(4–5), 439–443.
46. Cosco, S.; Ambrogi, V.; Musto, P.; Carfagna, C. Urea-formaldehyde microcapsules containing an epoxy resin: Influence of reaction parameters on the encapsulation yield. *Macromolecular Symposia* (2006), 234 (*Trends and Perspectives in Polymer Science and Technology*), 184–192.

47. Ai, Q.-S.; Zhang, Q.-Y.; Xing, R.-Y.; Zhang, J.-P. Direct in-situ polymerization for preparing microcapsules based on reactive ethylene silicone oil encapsulated by poly(urea-formaldehyde). *Zhongguo Jiaonianji* (2010), 19(4), 13–17.
48. Xiao, H.; Jiang, J.; Tang, Y. Study on encapsulation of poly-alpha-olefin with PVA-modified urea-formaldehyde resin. *Huaxue Yanjiu Yu Yingyong* (2008), 20(6), 696–699.
49. Huang, G.; Xiao, J. Study on in-situ encapsulated flavor. *Huagong Shikan* (2009), 23(2), 1–3.
50. Chang, H.; Lu, T.; Xu, Q.; Sun, Q.; Han, G. Preparation and characterization of microencapsulated red phosphorus with phenol-formaldehyde resin as shell material. *Zhongguo Suliao* (2010), 24(6), 72–75.
51. Liu, X.-P.; Xu, J.; Xiao, L.-R.; Huang, B.-Q.; Chen, Q.-H. Synthesis of capsule calcium hydroxide by in-situ polymerization. *Huadong Ligong Daxue Xuebao, Ziran Kexueban* (2006), 32(10), 1221–1225.
52. Yang, B.; Li, Y.; Cui, J.; Zhou, Y.; Guo, J.; Ma, Y.; Liu, Y.; Li, J. In-situ polymerization method for manufacturing dodecanol phase change microcapsules. CN 102504765, 2012.
53. Zhen, Z. Effects of protective colloid on the MF-UF double wall micro-encapsulation containing essence oil. *Advanced Materials Research* (2011), 236–238(Pt. 2, *Application of Chemical Engineering*), 1169–1173.
54. Blythe, D., Churchill, D.; Glanz, K.; Stutz, J. 1999. Microcapsules for carbonless copy paper. In R. Arshady, ed. *Microcapsules and Liposomes*, Vol. 1: Preparation and Chemical Applications. London: Citus 1999, pp. 421–439.
55. U. S. Environmental Protection Agency, Office of Air and Radiation. *Report to Congress on Indoor Air Quality*, Volume II: Assessment and Control of Indoor Air Pollution, Washington, DC, 1989.
56. International Agency for Research on Cancer (June 2004). IARC monographs on the evaluation of carcinogenic risks to humans. Volume 88 (2006): Formaldehyde, 2-Butoxyethanol and 1-tert-Butoxypropan-2-ol.
57. National Toxicology Program (June 2011). *Report on Carcinogens*, 12th edn. Research Triangle Park, NC: Department of Health and Human Services, Public Health Service, National Toxicology Program.
58. Sumiga, B.; Knez, E.; Vrtacnik, M. et al. Production of melamine-formaldehyde PCM microcapsules with ammonia scavenger used for residual formaldehyde reduction. *Acta Chim. Slov.* (2011), 58(1), 14–25.
59. Bone, S.; Vauttrin, C.; Barbesant, V. et al. Microencapsulated fragrance in melamine formaldehyde. *Chimia* (2011), 65(3), 177–181.
60. Li, W.; Wang, J.; Wang, X.; Wu, S.; Zhang, X. Effects of ammonium chloride and heat treatment on residual formaldehyde contents of melamine-formaldehyde microcapsules. *Colloid and Polymer Science* (2007), 285(15), 1691–1697.
61. Dietrich, K.; Herma, H.; Nastke, R.; Bonatz, E.; Teige, W. Amino resin microcapsules, II Preparation and morphology. *Acta Polymerica* (1989), 40(5), 325–331.
62. Smets, J.; Dihora, J. O.; Pintens, A.; Guinebretiere, S. J.; Druckrey, A. K.; Sands, P. D.; Yan, N. Benefit agent containing delivery particle. US 8551935, 2013.
63. Dietrich, K.; Herma, H.; Nastke, R.; Bonatz, E.; Teige, W. Amino resin microcapsules, IV. Surface tension of the resins and mechanism of capsule formation. *Acta Polymerica* (1990), 41(2), 91–95.
64. Rochmadi, A. P.; Hasokowati, W. Mechanism of microencapsulation with urea-formaldehyde polymer. *American Journal of Applied Science* (2010) 7(6), 9–745.
65. Donald, E. H. Capsule manufacture. US4444699, 1984.
66. Zhen, Z.-H.; Chen, Z.-H. Effect of emulsifier on the encapsulation of melamine formaldehyde resin microcapsule. *Zhongguo Zaozhi Xuebao* (2006), 21(1), 47–51.
67. Hwang, J.; Kim, J.; Wee, Y. et al. Preparation and characterization of melamine-formaldehyde resin microcapsules containing fragrant oil. *Biotechnology and bioprocess engineering* (2006), 11, 332–336.
68. Liu, G.; Xie, B.; Fu, D.; Wang, Y.; Fu, Q.; Wang, D. Preparation of nearly monodisperse microcapsules with controlled morphology by *in situ* polymerization of a shell layer. *Journal of Materials Chemistry* (2009), 19(36), 6605–6609.
69. Zhu, G.; Lu, L.; Tang, J.; Dong, B.; Han, N.; Xing, F. Prepararion of mono-sized epoxy/MF microcapsules in the appearance of polyvinyl alcohol as co-emulsifier. *Proceedings of the Fourth International Conference on Self-Healing Materials*, Ghent, 2013, pp. 225–228.
70. Zhao, D.; Liu, F.; Mu, W.; Han, Z.-R. Factors affecting morphology and encapsulation ratio of chlorpyrifos microcapsules with UF-resin during preparation. *Yingyong Huaxue* (2007), 24(5), 589–592.
71. Frank, G.; Biastoch, R. Low-formaldehyde dispersion of microcapsules of melamine-formaldehyde resins. US6224795, 2001.
72. Scher, H. B.; Rodson, M.; Davis, R. W.; Baker, D. R.; Kezerian, C. Process for formaldehyde content reduction in microcapsule formulations. EP 0623052,1996.

73. Li, W.; Zhang, X.-X.; Wang, X.-C.; Niu, J.-J. Preparation and characterization of microencapsulated phase change material with low remnant formaldehyde content. *Materials Chemistry and Physics* (2007), 106(2–3), 437–442.
74. Hoffman, D.; Eisermann, H. Low-viscosity melamine-formaldehyde resin microcapsule dispersions with reduced formaldehyde content. US6759931, 2004.
75. Greene, J. T. Formaldehyde scavenging process useful in manufacturing durable press finished fabric. US 5112652, 1992.
76. Then, S.; Neon, G. S.; Kasim, N. H. A. Optimization of microencapsulation process for self-healing polymeric materials. *Sains Malaysiana* (2011), 40(7), 795–802.
77. Zhang, H.; Wang, X. Fabrication and performances of microencapsulated phase change materials based on n-octadecane core and resorcinol-modified melamine–formaldehyde shell. *Colloids and Surfaces A: Physicochemical and Engineering Aspects* (2009), 332(2–3), 129–138.
78. Duan, B. Lab study works.

17 Microencapsulation with Miniemulsion Technology

Michael Yan

CONTENTS

17.1 Introduction 315
17.2 Emulsion Polymerization and Miniemulsion Polymerization 315
17.3 Miniemulsion Formulation 316
    17.3.1 Monomers 316
    17.3.2 Surfactants 317
    17.3.3 Costabilizers 317
    17.3.4 Initiators 317
17.4 Method of Preparation 318
17.5 Polymer Materials 318
17.6 Applications 319
    17.6.1 Encapsulation of Hydrophobic Oil with Radical Polymerization 319
    17.6.2 Encapsulation of Hydrophobic Oil with Interfacial Polymerization 320
    17.6.3 Encapsulation of Hydrophilic Core with Interfacial Polymerization 321
    17.6.4 Encapsulation of Hydrophilic Core with Nanoprecipitation 321
    17.6.5 Encapsulation of Organic Particles 321
    17.6.6 Encapsulation of Inorganic Nanoparticles in Direct Miniemulsion 323
    17.6.7 Encapsulation of Inorganic Nanoparticles in Inverse Miniemulsion 323
17.7 Conclusions 324
References 325

17.1 INTRODUCTION

Microencapsulation with miniemulsion polymerization involves the preparation of miniemulsion and subsequent polymerization to form nanoparticles or nanocapsules. Miniemulsion is a special class of emulsion that is stabilized against coalescence by a surfactant and suppressing Ostwald ripening by an osmotic pressure agent, or costabilizer. The miniemulsion is produced by high-energy homogenization and usually yields stable and narrowly distributed droplets with a size ranging from 50 to 500 nm. Nonpolar droplets can be dispersed into a polar liquid to give a direct miniemulsion, or the oil-in-water miniemulsion, whereas polar droplets dispersed into a nonpolar liquid leads to inverse miniemulsion, or the water-in-oil miniemulsion. During the preparation of miniemulsion, the materials to be encapsulated are normally included in the emulsion droplets, and subsequent polymerization of the system allows for the formation of nanocapsules. The materials that can be encapsulated with this technique may be a hydrophilic or a hydrophobic, a liquid or a solid, an inorganic or an organic material.

17.2 EMULSION POLYMERIZATION AND MINIEMULSION POLYMERIZATION

The mechanisms of conventional emulsion polymerization and miniemulsion polymerization appear to be similar, but in some ways are significantly different. In order to adequately discuss miniemulsion polymerization, it is necessary to review the mechanism of conventional emulsion polymerization.

A conventional emulsion polymerization can be divided into three intervals. Particle nucleation occurs during Interval I and is usually completed at low monomer conversion (2%–10%) when most of the monomer is located in relatively large monomer droplets (1–10 μm). Particle nucleation is believed to take place when radicals formed in the aqueous phase grow via propagation and then enter into micelles or become large enough in the continuous phase to precipitate and form primary particles, which may undergo limited flocculation until a stable particle population is obtained. Significant nucleation of particles from monomer droplets is discounted because of the small total surface area of the large droplets. Interval II involves polymerization within the monomer-swollen polymer particles with monomer supplied by diffusion from the droplets. Interval III begins when the droplets disappear, or at least reach a polymer fraction similar to that of the particles of around 100 nm, and continues to the end of the reaction. From this mechanism, monomer has to transfer from monomer droplets, through aqueous phase, to monomer-swollen micelle or monomer-swollen polymer particles until the monomer droplet is depleted. Mass transport from the droplets (Interval II) can be a problem, however, for strongly hydrophobic monomers with very low water solubility.

The significant differences of miniemulsion polymerization to emulsion polymerization are the much smaller droplet size obtained by using high shear force during miniemulsion preparation and the use of costabilizer to suppress the Ostwald ripening. The monomer droplet size for miniemulsion polymerization is typically around 50–300 nm, much smaller than that of the monomer droplets in emulsion polymerization. Hence, the droplet surface area in miniemulsion systems is very large compared to emulsion. The surfactant concentration in the formulation is so controlled that little free surfactant is present in the form of micelles as most of the surfactant is adsorbed at the droplet surface. Because of the large surface area in miniemulsion polymerization systems, particle nucleation is primarily via radical entry into monomer droplets, which are the polymerization loci in miniemulsion polymerization. There is no monomer transfer between the droplets, and each droplet acts like a nanoreactor itself and stays almost unchanged during polymerization. This allows the use of very hydrophobic monomers in the miniemulsion polymerization, which is not accessible to the conventional emulsion polymerization.

When a miniemulsion is created by the application of high shear force, a distribution of droplet sizes is resulted. Monomers will, over time, diffuse from the smaller monomer droplets into the larger ones even in the presence of surfactant to prevent droplet coalescence. This process of small droplet degradation is referred to as Ostwald ripening, and is driven by the reduction in interfacial area (energy). If Ostwald ripening is allowed to continue, creaming of monomer droplets will occur as the result of larger droplet sizes according to Stokes law. A costabilizer is added to suppress Ostwald ripening by retarding monomer diffusion from the smaller droplets to the larger ones. Costabilizers should be highly insoluble in the aqueous phase and highly soluble in the monomer droplets so that they will not diffuse out of the droplets.

17.3 MINIEMULSION FORMULATION

Monomer miniemulsion suitable for miniemulsion polymerization is submicron monomer-in-water dispersions stabilized against both coalescence by a surfactant and Ostwald ripening by an osmotic pressure agent, or costabilizer. The key issues in the preparation of stable monomer miniemulsion are the formulation and the method of preparation. There have been a few review articles published over the years.[1–4]

A typical miniemulsion useful for encapsulation includes water, monomer(s), core materials to be encapsulated, surfactant(s), a costabilizer, and an initiator system.

17.3.1 Monomers

In principle, the process of forming a miniemulsion allows for almost all types of monomers to be used, including those not miscible with the continuous phase. In the case of prevailing droplet nucleation or the start of the polymer reaction in the droplet phase, each miniemulsion droplet can be treated as a nanoreactor. This enables a whole variety of monomers to be used, such as methyl methacrylate (MMA), dodecyl methacrylate, stearyl methacrylate, butyl acrylate, styrene, vinyl chloride, vinyl acetate, vinyl hexanoate, etc. In addition, miniemulsion with multiple monomers has been reported, some even including certain amount of completely water-soluble monomers (comonomers) such as methacrylic acid or acrylic acid to promote the capsule wall formation as will be discussed in Section 17.6.1.

17.3.2 SURFACTANTS

The majority of miniemulsions reported in the literature have been stabilized with anionic surfactants such as sodium dodecyl sulfate (SDS), probably because of the widespread use of anionic surfactants in conventional emulsion polymerization, and due to their compatibility with neutral or anionic monomers and anionic initiators. Other types of surfactant have also been used, such as cationic, nonionic, and mixed nonionic/anionic, reactive as well as nonreactive surfactants. The surfactants useful for miniemulsion polymerization should meet the same requirements as in the conventional emulsion polymerization.

17.3.3 COSTABILIZERS

Emulsion droplets obtained after homogenization are rather unstable and can undergo growing by Ostwald ripening, which can efficiently be suppressed by the addition of a costabilizer, or sometimes called the hydrophobic agent, to the dispersed phase to counteract the Laplace pressure of the droplets. The hydrophobic agent cannot diffuse from one droplet to the other and is trapped inside the droplets, thereby providing an osmotic pressure inside the droplets to counteract the Laplace pressure. To be effective in suppressing Ostwald ripening, the costabilizer should have high solubility in monomer, low solubility in water, and low molecular weight. The high monomer solubility and low water solubility will ensure that the distribution coefficient for the costabilizer strongly favors the monomer droplets. Low molecular weight will give high mole ratio of costabilizer to monomer in the droplets. Under these conditions, diffusion of monomer out of the droplets would increase concentration of the costabilizer resulting in increased free energy, which can balance the reduced interfacial energy to limit Ostwald ripening. It should be noted that even with the costabilizer in the emulsion droplets, Ostwald ripening will still proceed, but on a much longer time scale, which is unimportant, since the time scale of polymerization is usually in hours.

Various costabilizers have been used in the miniemulsion polymerization. Cetyl alcohol and hexadecane are most often used since the original discovery of miniemulsion polymerization by Ugelstad et al.[5] Other researchers have used initiators, polymers, chain transfer agents, or monomers as costabilizers.[2]

Reimers and Schork used lauroyl peroxide as the initiator and also as the costabilizer,[6] and found that diffusion instability was reduced to the point where nucleation in the monomer droplets and polymerization could be carried out before significant diffusional degradation of the droplets took place. Alduncin et al. compared different oil-soluble initiators in conjunction with hexadecane as the costabilizer, and found that only lauroyl peroxide was water-insoluble enough to stabilize the monomer droplets against degradation by molecular diffusion.[7]

17.3.4 INITIATORS

Most of the miniemulsion polymerization processes have been carried out by using water-soluble initiators following the conventional emulsion polymerization, although a number of researchers have looked at the possibilities of using oil-soluble initiators.[6–9] Reimers and Schork used lauroyl peroxide as both the initiator and also as the costabilizer.[6] Alduncin et al. used other oil-soluble initiators, such as lauroyl peroxide, benzoyl peroxide, and azobis (isobutyronitrile), for miniemulsion polymerization of styrene.[7]

Combination of both water-soluble and oil-soluble initiators has also been used in miniemulsion polymerization. Choi et al.[10] successfully used both water-soluble potassium persulfate and oil-soluble 2,2'-azobis-(2-methyl butyronitrile) initiators in the miniemulsion polymerization of styrene. Ghazaly et al.[11] used both water-soluble and oil-soluble initiators in the copolymerization of n-butyl methacrylate with cross-linking monomers. Variations in the particle morphologies were found between the water-soluble and oil-soluble initiators, depending on the hydrophobicity of the cross-linking monomer.

17.4 METHOD OF PREPARATION

Figure 17.1 shows a schematic diagram most often used for the preparation of oil-containing nanocapsules with miniemulsion technique. First, an oil-in-water emulsion is prepared by dispersing an oil phase containing monomer, costabilizer, the core material to be encapsulated, and optionally an initiator, into an aqueous phase containing a surfactant. This step is achieved by using a stirrer. Then the preformed emulsion is subject to a high shear processing to form a miniemulsion. The high shear may be provided by a sonicator or by a mechanical homogenizer. A sonicator is convenient for laboratory use, but may not be practical for large-scale industrial applications. Two types of homogenizer have been used: One is a fine clearance valve homogenizer, and the other is a rotor-stator homogenizer. The shear intensity is adjusted in order to obtain required droplet size and distribution in the miniemulsion. In a well-designed formula, the surfactant in the system should keep the droplets from coalescing, creaming, or settling, and the costabilizer should prevent/suppress the Ostwald ripening from happening within the required time scale. Polymerization reactions are then initiated in the miniemulsion system to form the nanocapsules.

17.5 POLYMER MATERIALS

A variety of materials can be used as capsule wall materials with miniemulsion polymerization technique. In the conventional emulsion polymerization, diffusion of monomers in the aqueous phase is required, and monomers should have certain solubility in water to be useful in the process.

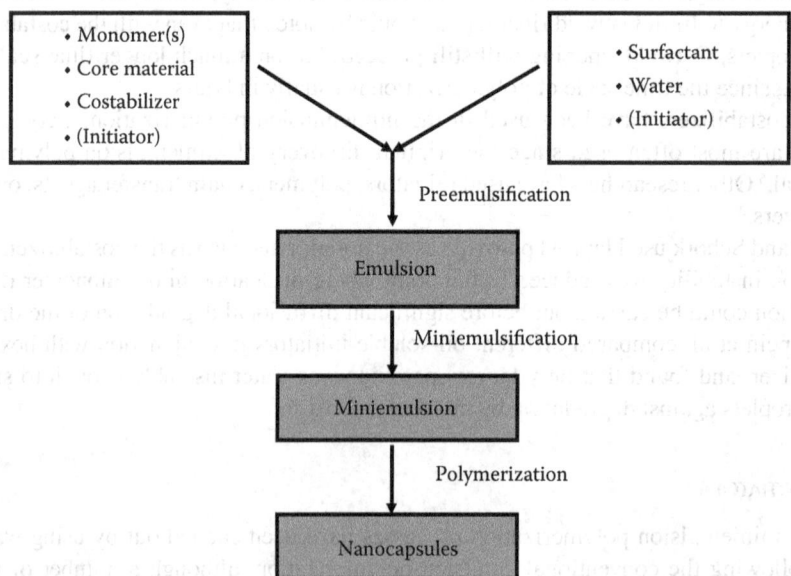

FIGURE 17.1 Encapsulation process for oily core–based miniemulsion polymerization.

In comparison, the polymerization reaction in miniemulsion to form capsules is performed in the miniemulsion droplets, and no monomer diffusion is required. This allows for all types of monomers and polymerization reactions to be used in the formation of nanocapsules. Following is a list of various types of polymerization reactions that has been used in miniemulsion polymerization.[12]

- Anionic polymerization in nonaqueous miniemulsion leads to polyamide nanoparticles.[13]
- Anionic polymerization in aqueous phase results in polybutylcyanoacrylate nanoparticles.[14]
- Cationic polymerization allows the formation of poly-p-methoxystyrene particles.[15,16]
- Catalytic polymerization enables the synthesis of polyolefine[17] or polyketone particles.[18]
- Ring-opening metathesis polymerization leads to polynorbonene nanoparticles.[19]
- Step-growth acylic diene metathesis polymerization forms oligo (phenylene vinylenes) particles.[20]
- Polyaddition in miniemulsion allows the formation of polyepoxides[21] or polyurethanes particles.[22]
- Polycondensation processes yield stable polyester nanoparticles.[23]
- Enzymatic polymerization of lactone miniemulsions leads to biodegradable polyester nanoparticles with a molecular weight of about 100,000 g/mol.[24]
- Oxidative polymerization of aniline generates polyaniline nanoparticles using both inverse and direct miniemulsion polymerization techniques.[25]

17.6 APPLICATIONS

Miniemulsion technology is versatile and can be used to encapsulate a variety of materials with different properties, such as hydrophobic or hydrophilic, liquid or solid, and inorganic or organic.

17.6.1 ENCAPSULATION OF HYDROPHOBIC OIL WITH RADICAL POLYMERIZATION

In this process, a hydrophobic oil material inert to polymerization reaction is encapsulated to form nanocapsules with miniemulsion polymerization. The monomer and the oily core material to be encapsulated, together with costabilizer, are initially miscible and are dispersed as miniemulsion droplets in an aqueous phase during emulsification. During polymerization, polymers are gradually phase separated from the core material, and subsequently serve as a locus for polymerization. The phase separation may result in three different scenarios: The polymers completely engulf, partially engulf, or nonengulf the core material, depending on the spreading coefficients among the three phases, that is, the core, the polymer, and dispersion medium.[26] Nanocapsules could be obtained only when the polymers completely engulf the core material. The morphology of the nanocapsule is determined by many factors, such as the type of surfactant chosen, the polarity of the monomers and comonomers, the interaction between polymers and core materials, and the choice of costabilizer and initiators.

Nanocapsules can be prepared in a miniemulsion process by a variety of monomers in the presence of large amount of hydrophobic oil. Tiarks et al. used poly (methyl methacrylate) (PMMA) to encapsulate hexadecane as a model oil.[27] The miniemulsion was obtained by mixing monomer MMA and hexadecane together with oil-soluble initiator 2,2′-azo-bis-isobutyronitrile (AIBN) and then miniemulsifying the mixture in an aqueous solution of SDS. In the miniemulsion, monomer MMA is miscible with hexadecane, but phase separation occurs during polymerization due to immiscibility of PMMA in hexadecane. PMMA is regarded as polar, whereas hexadecane is very hydrophobic so that the polymer is separated from hexadecane and migrates to the oil–water interface to form a PMMA shell around hexadecane droplets, resulting in nanocapsules. With increasing hexadecane content, a decrease of the shell thickness was detected. Shell stability of the nanocapsules can be improved by using up to 10 wt% of a difunctional monomer ethylene glycol dimethacrylate (EGDMA) as a cross-linking agent.

These authors concluded that the differences in the hydrophobicity of the oil and the polymer turned out to be the driving force for the formation of nanocapsules. Due to the pronounced difference of polarity of PMMA and hexadecane, the system was very well suited for the formation of nanocapsules. With more hydrophobic monomers such as styrene, however, it was more difficult to create nanocapsules as the cohesion energy density of the polymer phase was close to that of the oil, and adjustment of parameters to influence the interfacial tensions and spreading coefficients became critical in order to form nanocapsules. The parameters studied were monomer concentration, type and amount of surfactant and initiators, and the addition of functional comonomers. For example, addition of 10 wt% acrylic acid as a comonomer in the miniemulsion leads to an increase in the number of close-to-perfect nanocapsules.

Luo et al. reported nanoencapsulation of hydrophobic compounds by miniemulsion polymerization.[28] Paraffin was encapsulated with styrene as the monomer, methyl acrylic acid as the comonomer, and EGDMA as the cross-linker. Both thermodynamic and kinetic factors as well as nucleation modes have great influence on the morphology of capsules. The same author also used amphiphilic oligomers of styrene and maleic anhydride for nanoencapsulation of n-nonadecane via interfacially confined controlled/living radical miniemulsion polymerization.[29] The oligomers are synthesized by bulk reversible addition fragmentation transfer polymerization, and can self-assemble at the interface of water/droplets once miniemulsion is formed, allowing the polymerization reaction to take place at the interface.

Theisinger et al. encapsulated hydrophobic fragrance 1,2-dimethyl-1-phenyl-butyramide in PMMA, polystyrene, or acrylic copolymer nanoparticles using a one-step miniemulsion process.[30] It was shown that this hydrophobic fragrance compound directly influenced the kinetics, the molecular weight, and the morphology of the nanocapsules. The release behavior could be tuned by the temperature in relation to the glass transition temperature (T_g) of the polymer, which makes these nanocapsules interesting candidates for temperature-sensitive delivery systems.

17.6.2 Encapsulation of Hydrophobic Oil with Interfacial Polymerization

When an oil-soluble monomer and a water-soluble monomer reside in their respective oil and aqueous phase, respectively, they react at the miniemulsion droplet surface, thereby forming the nanocapsules. Torini et al. have shown that interfacial polycondensation of isophorone diisocyanate (IPDI) and 1,6-hexanediol in cyclohexane-water miniemulsions stabilized with SDS resulted in stable polyurethane nanocapsules with a diameter of around 200 nm and narrow size distribution.[31] A critical SDS concentration higher than 1 wt%, relative to the organic dispersed phase, is required to ensure the stability of the miniemulsion and nanocapsules. Hexadecane was used as costabilizer, and the surfactant-to-costabilizer ratio, varied between 1:1 and 1:3, has no influence on the size and size distribution of the miniemulsion and nanocapsules. The nanocapsules obtained in such conditions are stable over 18 months, without any increase in the z-average diameter and the *zeta*-potential.

Groison et al. reported nanocapsules with diameter of around 100 nm and the shell is made of a supramolecular polymer in which repeating units were held together by reversible interactions rather than covalent bonds.[32] These nanocapsules were prepared in direct miniemulsion through interfacial addition reaction of a diisocyanate (IPDI) and a monoamine (iBA), forming low-molecular weight bis-ureas moieties, which are strong self-complementary interacting molecules through hydrogen bonding. The authors claimed that this was the first example of the direct preparation of nanocapsules through interfacial reaction involving supramolecular interactions from low-molecular weight moieties. Such work dealing with stimuli-responsive systems opens the scope of an exciting field on which temperature-sensitive drug-delivery system or self-healing capsules can be directly obtained by controlling the chemical structures, kinetics, and thermodynamics.

17.6.3 ENCAPSULATION OF HYDROPHILIC CORE WITH INTERFACIAL POLYMERIZATION

Hydrophilic materials can be encapsulated with the inverse miniemulsions by using interfacial polymerization such as polyaddition and polycondensation, radical, or anionic polymerization.[33,34] Crespy et al. reported that silver nitrate was encapsulated and subsequently reduced to give silver nanoparticles inside the nanocapsules.[33] The miniemulsions were prepared by emulsifying a solution of amines or alcohols in a polar solvent with cyclohexane as the nonpolar continuous phase. The addition of suitable hydrophobic diisocyanate or diisothiocyanate monomers to the continuous phase allows the polycondensation or the cross-linking reactions to occur at the interface of the droplets. By using different monomers, polyurea, polythiourea, or polyurethane nanocapsules can be formed. The wall thickness of the capsules can be directly tuned by the quantity of the reactants. The nature of the monomers and the continuous phase are the critical factors for the formation of the hollow capsules, which is explained by the interfacial properties of the system. The resulting polymer nanocapsules could be subsequently dispersed in water.

Jagielski et al. reported the encapsulation of hydrophilic contrast agents Magnevist® and Gadovist® for use in magnetic resonance imaging (MRI) with inverse miniemulsion process.[35] Nanodroplets of 100–550 nm were formed, which were subsequently encapsulated by an interfacial polyaddition in polymeric shells of polyurethane, polyurea, and cross-linked dextran. The capsules were subsequently transferred to the water phase. Their experimental results clearly showed the potential of using nanocapsules as new contrast agent material for MRI.

17.6.4 ENCAPSULATION OF HYDROPHILIC CORE WITH NANOPRECIPITATION

This method is based on the precipitation of preformed polymers onto the aqueous nanodroplets in the miniemulsions by addition of a nonsolvent for the polymer.[36,37] Paiphansiri et al. prepared well-defined nanocapsules with this method whose core was composed of an antiseptic agent, and capsule wall of polymethylmethacrylate.[36] The stable nanodroplets were obtained by inverse miniemulsions with an aqueous antiseptic solution dispersed in an organic medium of solvent/nonsolvent mixture containing an oil-soluble surfactant and the polymer for the shell formation. By varying the ratio of solvent/nonsolvent mixture of dichloromethane/cyclohexane, polymers precipitated in the organic continuous phase and deposited onto the interface of the aqueous miniemulsion droplets to form nanocapsules. The monodisperse polymer nanocapsules with the size range of 80–240 nm were achieved as a function of the amount of surfactant. The nanocapsules could be easily transferred into an aqueous phase resulting in aqueous dispersions with nanocapsules containing antiseptic agent. The encapsulated amount of the antiseptic agent was evaluated to indicate the durability of the nanocapsule's wall. In addition, the use of different types of polymers having glass transition temperatures (T_g) ranging from 10 °C to 100 °C in this process has been also successful.

17.6.5 ENCAPSULATION OF ORGANIC PARTICLES

Organic pigments have many applications, such as coatings, paints, and printing inks. As the pigments are usually in submicron size, they tend to aggregate due to their high surface area. A polymer coating on the pigment particles may provide benefits such as reducing agglomeration; providing better storage stability, color stability, and durability; or protecting the particles from unwanted environmental influences such as UV radiation, moisture, or pH change.

Organic pigments such as carbon black are hydrophobic. It is possible to directly disperse the pigment powder in the monomer phase prior to emulsification, and full encapsulation of the nonagglomerated carbon particles can be provided by choosing an appropriate hydrophobe as reported by Bechthold et al.[38] It was shown that for full encapsulation of nonagglomerated carbon black particles, the type of hydrophobe is essential. In this case, the hydrophobe not only acts as a stabilizing agent against Oswald ripening for the miniemulsion process, but also mediates to the monomer phase by partial adsorption, which prevents the formation of rigid aggregates. The main drawback of this method of dispersing carbon black in the monomer is that the carbon is still highly agglomerated in the monomer, and only relatively low carbon black content (<10%) can be obtained in the resulting nanocomposites. Furthermore, a drastic increase in viscosity of the monomer phase would occur if the carbon black content in the monomer exceeds by about 10 wt%, making it difficult to disperse this monomer phase into the aqueous phase.[38,39] In order to increase the amount of carbon black that could be encapsulated in the nanocomposite particles, Tiarks et al. presented a new process called cosonication to completely encapsulate carbon black pigments with minimal particle size.[40] The process steps are as follows, and are shown schematically in Figure 17.2.

- *Preparation of monomer miniemulsion*: Monomer containing a costabilizer and oil-soluble initiator is dispersed in aqueous surfactant solution to form a miniemulsion.
- *Preparation of aqueous carbon black dispersion*: Carbon black particles are dispersed in water with appropriate surfactant to form a dispersion.
- *Sonication*: The monomer miniemulsion is mixed with the particle dispersion, and a fusion/fission process triggered by ultrasonication leads to an encapsulation of the particles into the monomer droplets. During the incorporation of the particles into the monomer droplet, surfactant desorbs from the pigment surface as monitored by surface tension measurement, and monomers adsorbs onto the surface of the particles.
- *Initiation of polymerization*: Polymerization reaction in the adsorbed monomer layer is initiated to allow for the formation of encapsulated nanoparticles.

With this technique, the amount of carbon black that could be encapsulated in the nanocomposite increases to 80%. Although this technique was initially developed for carbon black, it has been successfully applied to other pigments, such as organic pigment[41] and hydrophilic magnetite.[42,43]

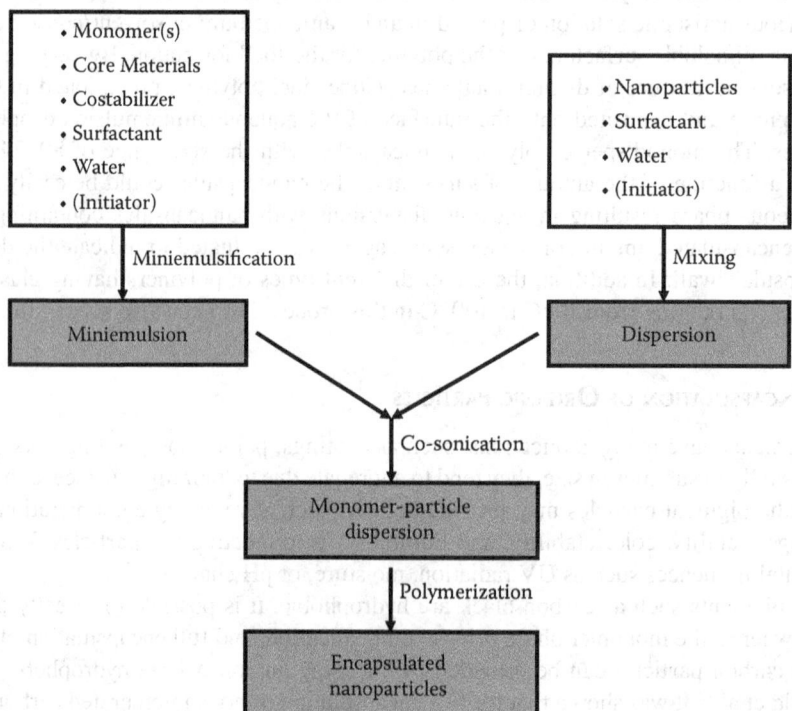

FIGURE 17.2 Encapsulation of nanoparticles with miniemulsion polymerization using a cosonication.

It should be noted that in order to encapsulate hydrophilic particles with this technique, the particles should be pretreated with surfactants to render them hydrophobic as discussed in the next section.

17.6.6 Encapsulation of Inorganic Nanoparticles in Direct Miniemulsion

Inorganic particles encapsulated in a polymer shell result in hybrid materials. These materials combine the excellent properties of inorganic materials such as thermal, mechanical, optical, electrical, fire-retardant, and catalytic properties, with the superior properties of organic materials such as elasticity, easy processing, and film formation. For pigment particles, such as titanium dioxide, that provide opacity in paper or paints/coating, the polymer shell on each pigment particle provides a spacer effectively separating the pigment particles, thereby improving the light scattering efficiency and thus the opacity of the products such as paper, coating, and paints. In addition, the polymer shell may protect the particles from detrimental effects from the environments, modulate the release rate, prevent the particles from aggregation, or modify surface properties for the particles to be compatible with their surrounding materials.

In order to encapsulate inorganic particles with direct miniemulsion polymerization, the particles have to be dispersed in the monomer, and stay in the monomer droplet during miniemulsification and subsequent polymerization steps. Generally, inorganic particles are difficult to be dispersed in a typical organic monomer phase due to their hydrophilic character. They will come out of the monomer droplet during emulsification step if untreated. Therefore, it is crucial to treat the particles with a suitable hydrophobilizing agent so that they are compatible with the monomers.

Erdem et al. described the encapsulation of titanium dioxide (TiO_2) nanoparticles in polystyrene via miniemulsion polymerization.[44–46] The TiO_2 nanoparticles were first surface-modified with polybutene–succinimide pentamine (OLOA 370), and then a dispersion containing 5 wt% of the hydrophobilized TiO_2 in styrene was prepared prior to a miniemulsification. The dispersion was then emulsified into an aqueous phase containing SDS to form stable miniemulsion droplets, which were polymerized in a stirred reactor at 70°C for 2–4 h by addition of a water-soluble initiator potassium persulfate. The authors reported that not all the TiO_2 particles in the styrene monomer were encapsulated, and to achieve the greatest encapsulation efficiencies (83 wt% TiO_2 and 73 wt% polystyrene), the TiO_2 particles should be as small and as stable as possible in the styrene monomer.

Other inorganic nanoparticles have been encapsulated with miniemulsion polymerization, and a hydrophobilizing agent was used to render the particles hydrophobic prior to miniemulsification. For example, calcium carbonate was pretreated with stearic acid prior to being dispersed into the monomer phase.[38] Alumina[47] and magnetite[42] were pretreated with oleic acid, laponite was pretreated with a cetyltrimethylammonium bromide,[48] and silica was pretreated with cetyltrimethylammonium chloride[49] or methacryloxy(propyl)trimethoxysilane.[50]

Ramírez and Landfester reported that magnetite particles were treated with oleic acid, and then were encapsulated with polystyrene with the cosonication technique.[42] It was shown that full encapsulation of the magnetite particles was obtained, and small magnetite particles were well separated, indicating that each particle was presumably completely coated with a thin layer of oleic acid and then the entire aggregate was covered with a layer of polymer as shown in Figure 17.3.

17.6.7 Encapsulation of Inorganic Nanoparticles in Inverse Miniemulsion

Magnetic nanoparticles have potential application in many fields. For biomedical applications, the iron oxide nanoparticles have to be brought into contact the bloodstream of a target and must be shielded from the aqueous environment to protect them from being degraded or metabolized. The encapsulation of superparamagnetic iron oxide nanoparticles in hydrophilic shells has been studied by using an inverse miniemulsion process.[51–53] Xu et al. reported the encapsulation of a magnetic fluid in polyacrylamide cross-linked with N,N'-methylene bis acrylamide using an inverse miniemulsion polymerization process.[51] A representative transmission electron

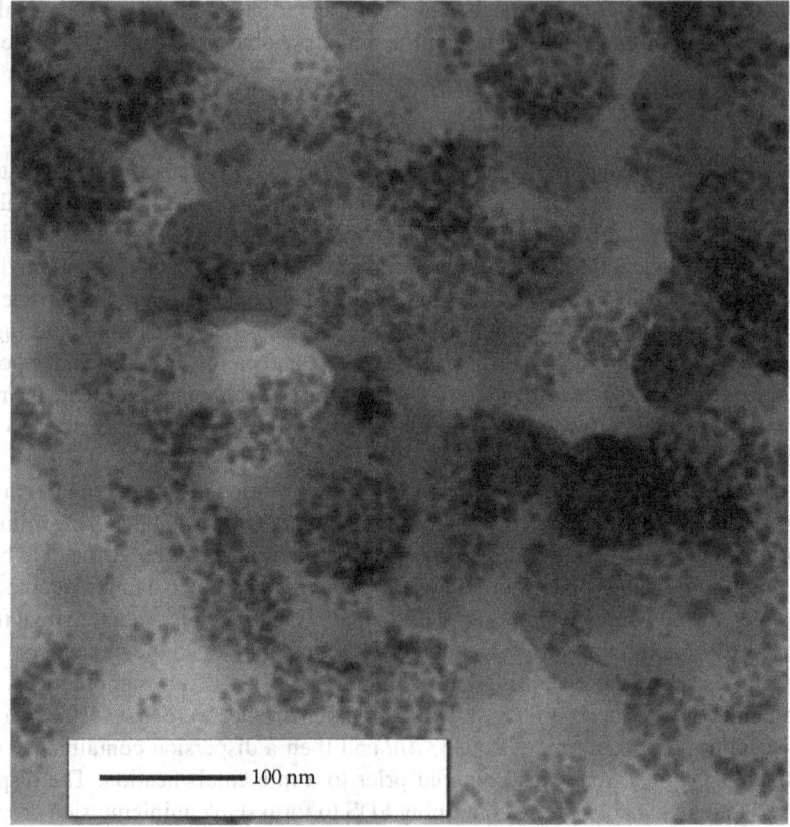

FIGURE 17.3 Transmission electron micrograph (TEM) for magnetite polystyrene particles. (From Ramírez, L.P. and Landfester, K., *Macromol. Chem. Phys.*, 204, 22, 2003.)

microscope image of the magnetic polymeric particles is shown in Figure 17.4. Their experimental results demonstrate that the inverse miniemulsion polymerization is an effective way to synthesize magnetic polymeric particles. The magnetic polymeric particles are spherical and their size ranges from 60 to 160 nm depending on the reaction parameters. The nanosized iron oxide particles can be well encapsulated in polyacrylamide particles and the magnetic polymeric particles are superparamagnetic. These cross-linked magnetic polyacrylamide particles can be easily dispersed into water.

17.7 CONCLUSIONS

Miniemulsion is a special class of emulsion that is stabilized against coalescence by a surfactant and Ostwald ripening by an osmotic pressure agent, or costabilizer. Compared with conventional emulsion polymerization process, the miniemulsion polymerization process allows all types of monomers to be used in the formation of nanoparticles or nanocapsules, including those not miscible with the continuous phase. Each miniemulsion droplet can indeed be treated as a nanoreactor, and the colloidal stability of the miniemulsion ensures a perfect copy from the droplets to the final product. The versatility of polymerization process makes it possible to prepare nanocapsules with various types of core materials, such as hydrophilic or hydrophobic, liquid or solid, organic or inorganic materials. Different techniques can be used to initiate the capsule wall formation, such as radical, ionic polymerization, polyaddition, polycondensation, or phase separation from preformed polymers.

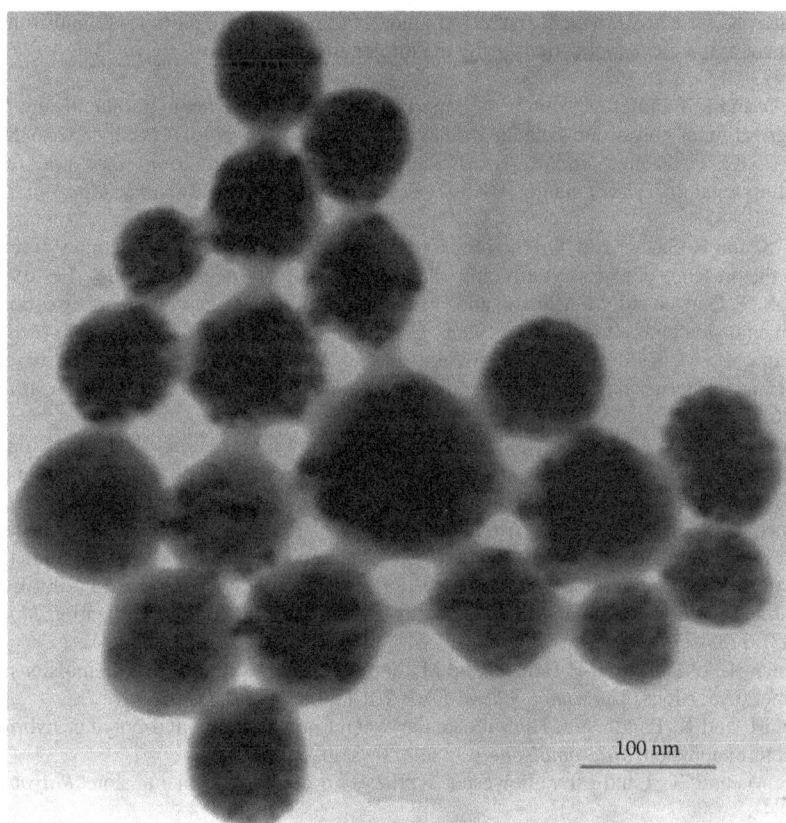

FIGURE 17.4 TEM image of magnetic Fe3O4/polyacrylamide particles. (From Xu, Z. et al., *J. Magn. Magn. Mater.*, 277, 136, 2004.)

REFERENCES

1. Asua, J. M. Miniemulsion polymerization. *Prog. Polym. Sci.* 27 (2002): 1283–1346.
2. Schork, F. J., Y. Luo, W. Smulders, J. Russum, A. Butte, and K. Fontenot, Miniemulsion polymerization. *Adv. Polym. Sci.* 175 (2005): 129–255.
3. Landfester, K. Miniemulsion polymerization and the structure of polymer and hybrid nanoparticles. *Angew. Chem. Int. Ed.* 48 (2009): 4488–4507.
4. Clemens, W. and K. Landfester, Miniemulsion polymerization as a means to encapsulate organic and inorganic materials. *Adv. Polym. Sci.* 233 (2010). 185–236.
5. Ugelstad, J., M. S. El-Aasser, and J. W. Vanderhoff, Emulsion polymerization: Initiation of polymerization in monomer droplets. *J. Polym. Sci. C Polym. Lett.* 11 (1973): 503–513.
6. Reimers, J. L. and F. J. Schork, Lauroyl peroxide as a cosurfactant in miniemulsion polymerization. *Ind. Eng. Chem. Res.* 36(4) (1997): 1085–1087.
7. Alduncin, J. A., J. Forcada, and J. M. Asua, Miniemulsion polymerization using oil-soluble initiators. *Macromolecules* 27(8) (1994): 2256–2261.
8. Alduncin, J. A. and J. M. Asua, Molecular-weight distributions in the miniemulsion polymerization of styrene initiated by oil-soluble initiators. *Polymer* 35(17) (1994): 3758–3765.
9. Blythe, P. J., A. Klein, J. A. Phillips, E. D. Sudol, and M. S. El-Aasser, Miniemulsion polymerization of styrene using the oil-soluble initiator AMBN. *J. Polym. Sci. A Polym. Chem.* 37 (1999): 4449–4457.
10. Choi, Y. T., M. S. El-Aasser, E. D. Sudol, and J. W. Vanderhoff, Polymerization of styrene miniemulsions. *J. Polym. Sci. A Polym. Chem.* 23 (1985): 2973–2987.
11. Ghazaly, H. M., E. S. Daniels, V. L. Dimonie, A. Klein, and M. S. El-Aasser, Miniemulsion copolymerization of *n*-butyl methacrylate with cross-linking monomers. *J. Appl. Polym. Sci.* 81 (2001): 1721–1730.

12. Landfester, K., A. Musyanovych, and V. Mailander, From polymeric particles to multifunctional nanocapsules for biomedical applications using the miniemulsion process. *J. Polym. Sci. A Polym. Chem.* 48 (2010): 493–515.
13. Crespy, D. and K. Landfester, Anionic polymerization of ε-caprolactam in miniemulsion: Synthesis and characterization of polyamide-6 nanoparticles. *Macromolecules* 38(16) (2005): 6882–6887.
14. Weiss, C. K., U. Ziener, and K. Landfester, A route to nonfunctionalized and functionalized poly(n-butylcyanoacrylate) nanoparticles: Preparation in miniemulsion. *Macromolecules* 40(4) (2007): 928–938.
15. Cauvin, S. and F. Ganachaud, On the preparation and polymerization of p-methoxystyrene miniemulsions in the presence of excess ytterbium triflate. *Macromol. Symp.* 215(1) (2004): 179–190.
16. Cauvin, S., F. Ganachaud, M. Moreau, and P. Hémery, High molar mass polymers by cationic polymerisation in emulsion and miniemulsion. *Chem. Commun.* 21 (2005): 2713–2715.
17. Wehrmann, P., M. Zuideveld, R. Thomann, and S. Mecking, Copolymerization of ethylene with 1-butene and norbornene to higher molecular weight copolymers in aqueous emulsion. *Macromolecules* 39 (2006): 5995–6002.
18. Held, A., I. Kolb, M. A. Zuideveld, R. Thomann, S. Mecking, M. Schmid, R. Pietruschka, E. Lindner, M. Khanfar, and M. Sunjuk, Aqueous polyketone latices prepared with water-insoluble palladium(II) catalysts. *Macromolecules* 35 (2002): 3342–3347.
19. Quemener, D., V. Heroguez, and Y. Gnanou, Design of PEO-based ruthenium carbene for aqueous metathesis polymerization. Synthesis by the "macromonomer method" and application in the miniemulsion metathesis polymerization of norbornene. *J. Polym. Sci. A Polym. Chem.* 44 (2006): 2784–2793.
20. Pecher, J. and S. Mecking, Nanoparticles from step-growth coordination polymerization. *Macromolecules* 40 (2007): 7733–7735.
21. Landfester, K., F. Tiarks, H.-P. Hentze, and M. Antonietti, Polyaddition in miniemulsions: A new route to polymer dispersions. *Macromol. Chem. Phys.* 201 (2000): 1–5.
22. Barrere, M. and K. Landfester, High molecular weight polyurethane and polymer hybrid particles in aqueous miniemulsion. *Macromolecules* 36 (2003): 5119–5125.
23. Barrere, M. and K. Landfester, Polyester synthesis in aqueous miniemulsion. *Polymer* 44 (2003): 2833–2841.
24. Taden, A., M. Antonietti, M., and K. Landfester, Enzymatic polymerization towards biodegradable polyester nanoparticles. *Macromol. Rapid Commun.* 24 (2003): 512–516.
25. Marie, E., R. Rothe, M. Antonietti, and K. Landfester, Synthesis of polyaniline particles via inverse and direct miniemulsion. *Macromolecules* 36 (2003): 3967–3973.
26. Torza, S. and S. G. Mason, Three-phase interaction in shear and electric fields. *J. Colloid Interf. Sci.* 33(1) (1970): 67–83.
27. Tiarks, F., K. Landfester, and M. Antonietti, Preparation of polymeric nanocapsules by miniemulsion polymerization. *Langmuir* 17(3) (2001): 908–918.
28. Luo, Y. and X. Zhou, Nanoencapsulation of a hydrophobic compound by a miniemulsion polymerization process. *J. Polym. Sci. A Polym. Chem.* 42 (2004): 2145–2154.
29. Luo, Y. and H. Gu, Nanoencapsulation via interfacially confined reversible addition fragmentation transfer (RAFT) miniemulsion polymerization. *Polymer* 48 (2007): 3262–3272.
30. Theisinger, S., K. Schoeller, B. Osborn, M. Sarkar, and K. Landfester, Encapsulation of a fragrance via miniemulsion polymerization for temperature-controlled release. *Macromol. Chem. Phys.* 210 (2009): 411–420.
31. Torini, L., J. F. Argillier, and N. Zydowicz, Interfacial polycondensation encapsulation in miniemulsion. *Macromolecules*, 38 (2005): 225–3236.
32. Groison, E., S. Adjili, A. Ferrand, F. Lortie, D. Portinha, and N. Zydowicz, 'All-supramolecular' nanocapsules from low-molecular weight ureas through interfacial addition reaction in miniemulsion. *Macromol. Rapid Commun.* 32 (2011): 491–496.
33. Crespy, D., M. Stark, C. Hoffmann-Richter, U. Ziener, and K. Landfester, Polymeric nanoreactors for hydrophilic reagents synthesized by interfacial polycondensation on miniemulsion droplets. *Macromolecules*, 40 (2007): 3122–3135.
34. Musyanovych, A. and K. Landfester, Synthesis of poly(butylcyanoacrylate) nanocapsules by interfacial polymerization in miniemulsions for the delivery of DNA molecules. *Prog. Colloid Polym. Sci.* 134 (2008): 120–127.
35. Jagielski, N., S. Sharma, V. Hombach, V. Mailander, V. Rasche, and K. Landfester, Nanocapsules synthesized by miniemulsion technique for application as new contrast agent materials. *Macromol. Chem. Phys.* 208 (2007): 2229–2241.

36. Paiphansiri, U., P. Tangboriboonrat, and K. Landfester, Polymeric nanocapsules containing an antiseptic agent obtained by controlled nanoprecipitation onto water-in-oil miniemulsion droplets. *Macromol. Biosci.* 6(1) (2006): 33–40.
37. De Faria, T. J., A. M. de Campos, and E. L. Senna, Preparation and characterization of poly(D,L-lactide) (PLA) and poly(D,L-lactide)-poly(ethylene glycol) (PLA-PEG) nanocapsules containing antitumoral agent methotrexate. *Macromol. Symp.* 229 (2005): 228–233.
38. Bechthold, N., F. Tiarks, M. Willert, K. Landfester, and M. Antonietti. Miniemulsion polymerization: Applications and new materials. *Macromol. Symp.* 151 (2000): 549–555.
39. Lelu, S., C. Novat, C. Graillat, A. Guyot, and E. Bourgeat-Lami, Encapsulation of an organic phthalocyanine blue pigment into polystyrene latex particles using a miniemulsion polymerization. *Polym. Intern.* 52 (2003): 542–547.
40. Tiarks, F., K. Landfester, and M. Antonietti, Encapsulation of carbon black by miniemulsion polymerization. *Macromol. Chem. Phys.* 202 (2001): 51–60.
41. Steiert, N. and K. Landfester, Encapsulation of organic pigment particles via miniemulsion polymerization. *Macromol. Mater. Eng.* 292 (2007): 1111–1125.
42. Ramírez, L. P. and K. Landfester, Magnetic polystyrene nanoparticles with a high magnetite content obtained by miniemulsion processes. *Macromol. Chem. Phys.* 204 (2003): 22–31.
43. Landfester, K. and L. P. Ramírez, Encapsulated magnetite particles for biomedical application. *J. Phys. Condens. Mater.* 15 (2003): S1345–S1361.
44. Erdem, B., E. D. Sudol, V. L. Dimonie, and M. S. El-Aasser, Encapsulation of inorganic particles via miniemulsion polymerization. I. Dispersion of titanium dioxide particles in organic media using OLOA 370 as stabilizer. *J. Polym. Sci. A Polym. Chem.* 38 (2000): 4419–4430.
45. Erdem, B., E. D. Sudol, V. L. Dimonie, and M. S. El-Aasser, Encapsulation of inorganic particles via miniemulsion polymerization. II. Preparation and characterization of styrene miniemulsion droplets containing TiO_2 particles. *J. Polym. Sci. A Polym. Chem.* 38 (2000): 4431–4440.
46. Erdem, B., E. D. Sudol, V. L. Dimonie, and M. S. El-Aasser, Encapsulation of inorganic particles via miniemulsion polymerization. III. Characterization of encapsulation. *J. Polym. Sci. A Polym. Chem.* 38 (2000): 4441–4450.
47. Mahdavian, A. R., Y. Sarrafi, and M. Shabankareh, Nanocomposite particles with core–shell morphology III: Preparation and characterization of nano Al_2O_3–poly(styrene–methyl methacrylate) particles via miniemulsion polymerization. *Polym. Bull.* 63 (2009): 329–340.
48. Sun, Q., Y. Deng, Z. L. Wang, Synthesis and characterization of polystyrene-encapsulated laponite composites via miniemulsion polymerization. *Macromol. Mater. Eng.* 289 (2004): 288–295.
49. Tiarks, F., K. Landfester, and M. Antonietti, Silica nanoparticles as surfactants and fillers for latexes made by miniemulsion polymerization. *Langmuir* 17 (2001): 5775–5780.
50. Zhang, S., S.-X. Zhou, Y.-M. Weng, and L.-M. Wu, Synthesis of SiO_2/polystyrene nanocomposite particles via miniemulsion polymerization. *Langmuir* 21 (2005): 2124–2128.
51. Xu, Z., C. Wang, W. Yang, Y. Deng, and S. Fu, Encapsulation of nanosized magnetic iron oxide by polyacrylamide via inverse miniemulsion polymerization. *J. Magn. Magn. Mater.* 277 (2004): 136–143.
52. Wormuth, K., Superparamagnetic latex via inverse emulsion polymerization. *J. Colloid Interf. Sci.* 241 (2001): 366–377.
53. Xu, Z., C. Wang, W. Yang, and S. Fu, Synthesis of superparamagnetic Fe_3O_4/SiO_2 composite particles via sol-gel process based on inverse miniemulsion. *J. Mater. Sci.* 40 (2005): 4667–4669.

18 Silica-Based Sol-Gel Microencapsulation and Applications

Rosaria Ciriminna and Mario Pagliaro

CONTENTS

18.1 Introduction 329
18.2 A Green, Advanced Nanochemistry Technology 330
18.3 Advantages and Limitations of Sol-Gel Microencapsulation 335
18.4 Controlled Release from Sol-Gel Microparticles 337
18.5 Advantages and Limitations Of Sol-Gel Microencapsulation 339
18.6 Emerging Applications 340
18.7 Outlook and Perspectives 342
Acknowledgments..345
References..345

18.1 INTRODUCTION

Microencapsulation, i.e. the process of enclosing micron-sized particles of solids or droplets of liquids or gasses in an inert shell which in turn isolates and protects them from the external environment,[1] is one of the main research topics in contemporary chemical research, as shown by the large body of scientific publications and scientific symposia devoted to the subject.* Heterogenization in micron-sized particles is useful for eliminating a processing step and uses more efficiently costly ingredients by enhancing the precision and effectiveness of action through controlled release, or for isolating molecules that would otherwise react with each other.

Chemical companies are interested in the development of innovative "system solutions,"[2] that is, new functional materials in which known molecules are integrated to show new effects. Flavor and fragrance companies are interested to protect and precisely release their valued molecules (the "payload"). The polymer industry demands encapsulated curing agents and encapsulated reactants. These are just three examples, out of many, that show that microencapsulation is mainly used for the purpose of protection and controlled release. Since at least a decade, therefore, traditional low-volume markets for microcapsule-based products are expanding to include fine chemicals, adhesives, inks, fragrances, toners, sealants, and detergent manufacturers.[3]

Microencapsulation companies, including chemical makers, flavor and fragrance houses, and specialist firms, use competing technologies to manufacture encapsulated functional materials using either chemical or physical techniques.

A 2004 study[4] aiming to identify trends in microencapsulation technologies (since 1955) found that liposome entrapment and spinning-disk were the dominant approaches; with nanoencapsulation growing but still far from the mainstream methods that include chemical (*in situ* processes such

---

* See, for example, The 42nd Annual Meeting and Exposition of the Controlled Release Society, Edinburgh, Scotland, July 26–29, 2015; or 23rd International Conference on Bioencapsulation, Delft, Netherlands, September 2–4, 2015.

as emulsion, suspension, precipitation or dispersion polymerization, and interfacial polycondensation) and physical (spinning-disk and spray-drying) methodologies.

The sol-gel microencapsulation in silica-based materials is an emerging and powerful nanochemistry technology in which the active ingredients are protected (stabilized) in silica-based particles.[5]

In 2004, Barbé and coworkers reviewed sol-gel microencapsulation of bioactive molecules for drug-delivery.[6] Four years later, the same team published a first summary of their studies on silica-based microparticles doped with hydrophilic molecules obtained from water-in-oil (W/O) microemulsions. The Australian scientists showed[7] how both the particle size and the release rate of silica-based microparticles can be finely and easily tailored in a wide range by controlling the conditions affecting the sol-gel process. In 2006, van Driessche and Hoste published the first account[8] on the topic in a book addressing microencapsulation techniques, mainly covering the findings of Barbé and coworkers.

The first comprehensive review on sol-gel microencapsulation covering all methods was published by Chemical Reviews in 2011.[5] The method was found to be still a relatively new solution for controlled release formulations. Yet, following an ample section dealing with economic and environmental arguments, we were concluding that "the sol-gel microencapsulation will become one of the most relevant chemical technologies with applications in numerous industrial sectors." Four years later, indeed, we find a number of new applications that have reached the marketplace, while research in the field has boomed.

As of late 2013, a quick Boolean search in Google Scholar with the query "sol-gel" and "microencapsulation" returns about 4550 articles, patents, and scholar reports. Most of the technical and economic problems that limited the practical application of sol-gel derived silica-based functional materials have been addressed and resolved. Numerous new materials, formulations, and devices that use functional sol-gel materials have been developed and marketed. Even though the global financial crisis started in 2008, the global sol-gel market has not stopped to grow, reaching $1.5 billion in 2012, with a projected annual growth rate of 7.9% for the subsequent 5 years.[9] Research in the field is flourishing on a true international scale, with academic groups operating in countries as distant as Australia, India, China, Japan, Israel, Italy, Hong Kong, the United States, Iran, Malaysia, Germany, Korea, Canada, and France.

The aforementioned literature references cited provide a fine, detailed description of the chemistry of sol-gel microencapsulation process. Hence, in the following, instead of reporting again the basic chemistry of the sol-gel microencapsulation, we address a number of relevant and often overlooked practical issues such as methods to break the capsules, health and safety, practical advantages and limitations, as well as some economic aspects. The message of this chapter is that the sol-gel encapsulation in micron-sized silica and organosilica particles is an important chemical technology with a large applicative potential that must be understood.

The outcome is a report that will hopefully be truly useful to heads of chemical research as well as to managers in the fine chemical, fragrance and flavor, pharmaceutical, and polymer industries for years to come.

18.2 A GREEN, ADVANCED NANOCHEMISTRY TECHNOLOGY

We have shown elsewhere[10] that sol-gel microparticle formation is one case in which soft matter is used to template the resulting porous material in a basic chemical strategy for making functional nanomaterials using lyotropic mesophases, foams and emulsion, that has been named by Ozin "nanochemistry."[11]

In 1998, Avnir and coworkers patented the interfacial polymerization process in which the mild sol-gel production of silica glass is combined with emulsion chemistry to form sol-gel spherical microparticles in place of irregular granules.[12] In brief, the emulsion droplets provide a microreactor environment for the hydrolysis and condensation reactions of Si alkoxides. All sorts of molecules can be entrapped and stabilized in similar ceramic microparticles with broad control over the release rate for a range of applications (such as drug delivery, release of specialty chemicals and cosmeceutical/nutraceutical, and beyond, Figure 18.1), with no need for reformulation for different molecules.

Silica-Based Sol-Gel Microencapsulation and Applications

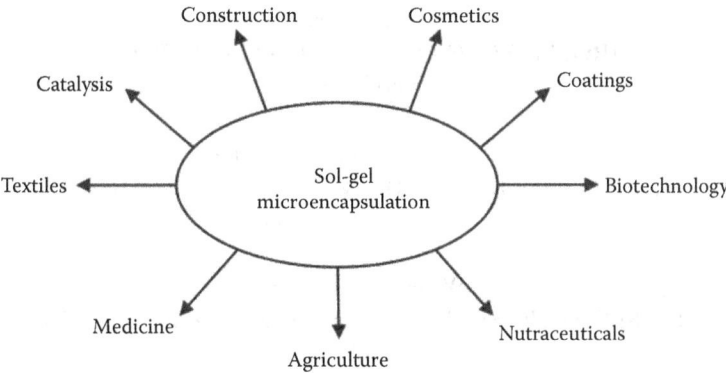

FIGURE 18.1 All sort of molecules can be entrapped and stabilized in silica-based microparticles with broad control over the release rate, hence opening the route to a wide range of applications.

We have dubbed similar doped silica gels "system materials" in the same sense for which in systems theory, the properties of the system include and go beyond the properties of the parts comprising the system. For example, once encapsulated in SiO_2 microparticles, hydrophobic molecules can be incorporated into the aqueous phase (Scheme 18.1). For example, chemists can now formulate perfumes in water or deliver poorly soluble drugs through encapsulation (Scheme 18.1) in hydrophilic sol-gel silica, whose correct chemical formula is not SiO_2, but rather (as emphasized by Avnir)[13] $(SiO_mH_n)_p$, in which p approaches the Avogadro's number and where n is always $\neq 0$ due to densely coated silanol groups at the surface of the highly porous amorphous silica gel.

Again, summarizing what has been extensively described elsewhere,[5] the two main techniques to prepare sol-gel microparticles start either from W/O or from oil-in-water (O/W) emulsions to entrap, respectively, hydrophilic or lipophilic ingredients inside the spherical shell of amorphous silica or organosilica.

In general, hydrolysis inside the water droplets of a W/O microemulsion occurs more than an order of magnitude faster than in typical single-phase sol-gel solution at the same concentration, as a result of the high local water concentration.

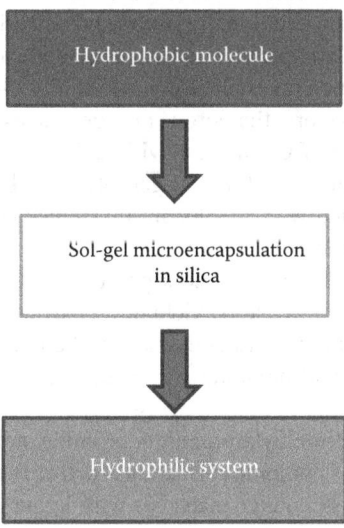

SCHEME 18.1 (See color insert.) Once entrapped in hydrophilic silica microparticles, hydrophobic organic molecules can be formulated and incorporated into the aqueous phase.

$$\text{(RO)}_3\text{Si-OR} + \text{H}_2\text{O} \underset{\text{Esterification}}{\overset{\text{Hydrolysis}}{\rightleftarrows}} \text{RO}_3\text{Si-OH} + \text{ROH} \quad (18.1)$$

$$\text{(RO)}_3\text{Si-OH} + \text{RO-Si(OR)}_3 \underset{\text{Alcoholsis}}{\overset{\text{Alcohol condensation}}{\rightleftarrows}} \text{RO}_3\text{Si-OH} + \text{ROH} \quad (18.2)$$

$$\text{(RO)}_3\text{Si-OH} + \text{HO-Si(OR)}_3 \underset{\text{Hydrolysis}}{\overset{\text{Water condensation}}{\rightleftarrows}} \text{(RO)}_3\text{Si-O-Si(OR)}_3 + \text{H}_2\text{O} \quad (18.3)$$

The rate of hydrolysis of Si(OR)_4 (Equation 18.1) exhibits a minimum at pH 7, and increases exponentially at either lower or higher pH. In contrast, the rate of silane condensation (Equations 18.2 and 18.3) exhibits a minimum at pH 2, and a maximum around pH 7 where SiO_2 solubility and dissolution rates are maximized.

At low pH, acid catalysis promotes hydrolysis but hinders both condensation and dissolution reactions,[14] leading to small and homogeneous particles.[15] Base catalysis of sol-gel hydrolysis and condensation reactions, in contrast, promotes fast condensation and dissolution. This leads to the production of an inhomogeneous system due to rapid condensation of the hydrolyzed precursor monomers and to dense silica particles formed by the ripening of aggregates during the collision of droplets. As a result, the microparticles show essentially *no* porosity, with the particles being stabilized by a water/surfactant layer on the particle surface that prevents particle precipitation.[24]

In other words, base-catalyzed microparticle synthesized in W/O microemulsions results in the formation of relatively large matrix particles of low and even negligible porosity. Thus, in order to obtain porous microcapsules under base-catalyzed conditions, the "two-step" sol-gel polycondensation process must be employed, in which hydrolysis is first conducted under acidic conditions followed by condensation catalyzed by base. When this approach is adopted, mesoporous microspheres can be synthesized also in basic W/O emulsions. In contrast, smaller particles synthesized in acid exhibit a strong microporous component (Figure 18.2).

The process has been extensively explored by Barbé and coworkers in Australia (where the same scientists established the spin-off company Ceramisphere), and can be viewed as an emulsification of a sol-gel solution in which gelation takes place concomitantly. Depending on the order of addition of the different chemicals, furthermore, the porous microparticles prepared from interfacial hydrolysis and condensation of TEOS in W/O emulsion will be full porous matrix particles or core–shell capsules. Normally, if the emulsification of the sol-gel solution takes place concomitantly with gelation, full microparticles are formed with the dopant molecules homogeneously distributed within the inner huge porosity of the particles (Figure 18.3).

Encapsulation occurs as the silicon precursors polymerize to build an oxide cage around the polar droplets, which, acting as microreactors, yield microparticles with size comparable to the size of the droplets. The larger the amount of water employed, the larger are the drops and thus the final microparticles. In general, ionic and nonionic surfactants are used as structuring agent to make these inorganic microparticles.

Ionic detergents, such as cetyltrimethylammonium bromide or sodium dodecyl sulfate, give pore sizes between 2 and 4 nm, whereas nonionic surfactants such as the Pluronics and the Tween surfactant series (ethoxylated sorbitan esters) give materials with larger pores (around 10 nm) and thicker walls (Table 18.1).

The technology best suited to simply encapsulate hydrophobic molecules such as essential oils, flavors, vitamins, proteins (including enzymes), and many other biomolecules makes use of an O/W

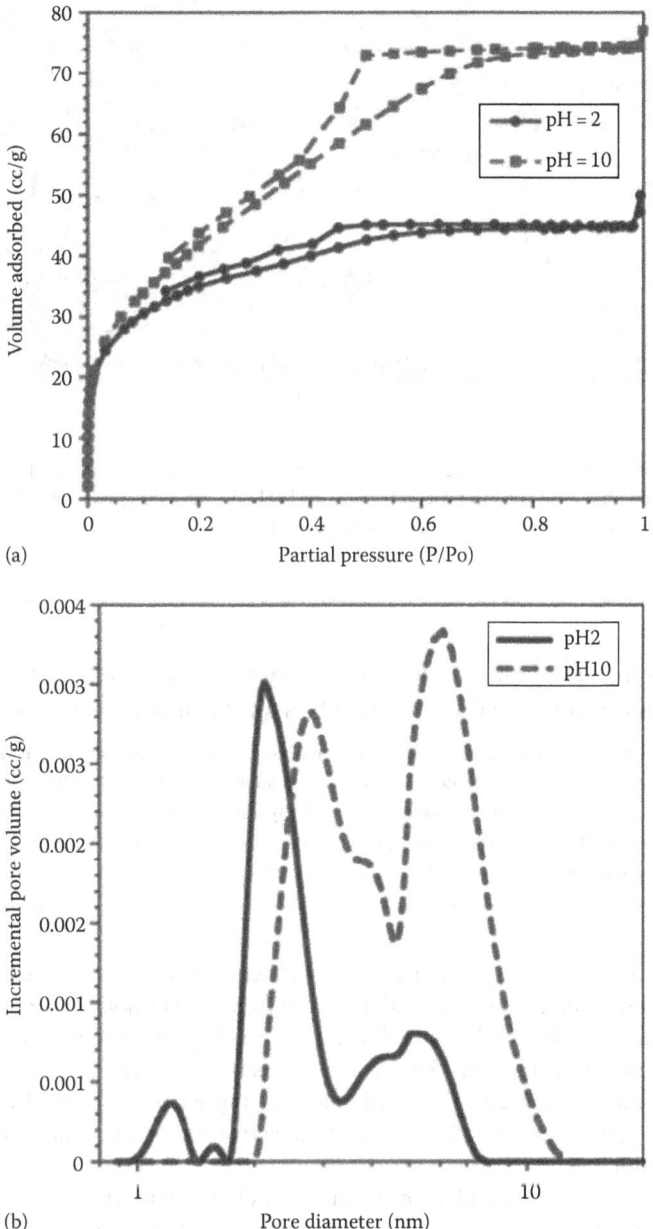

FIGURE 18.2 Adsorption isotherms and pore size distribution for silica microparticles synthesized from W/O emulsions at acid and basic pH. (a) after isotherms and (b) after distribution. (Reproduced from Barbé, C.J. et al., *J. Sol-Gel Sci. Technol.*, 46, 393, 2008. With kind permission.)

emulsion. The actives are thus solubilized in the aqueous droplets containing also prehydrolyzed monomers such as tetraethoxysilane (TEOS), dispersed in a nonpolar solvent and stabilized by a surfactant. Further hydrolysis of the Si alkoxides and condensation of the hydrolyzed species to silica or organosilica occurs at the oil–water interface, normally with formation of core–shell capsules. In detail, a mixture containing the dopant, the water-insoluble Si alkoxides, and a surfactant is emulsified by stirring it in water. In the second step, the microcapsules are obtained by hydrolytic polycondensation in water phase containing a surfactant catalyzed by base or acid, typically at temperatures in the

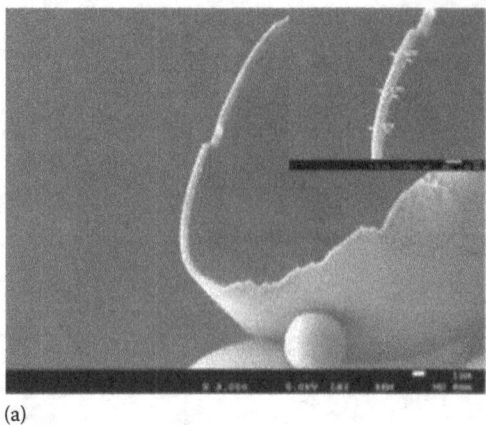 

(a)   (b)

FIGURE 18.3 SEM images of shell thickness of broken silica microcapsules doped with aqueous glycerol, obtained from [water + glycerol]/TEOS ratio = 10/1 (a) and 1/1 (b). Aqueous glycerol is homogeneously microencapsulated in the capsules core. (Reproduced from Galgali, G. et al., *Mater. Res. Bullet.*, 46, 2445, 2011.)

TABLE 18.1
The HLB Number in Surfactant Amphiphilic Molecules (A Value between 0 and 60), Defining the Affinity of a Surfactant for Water or Oil,[45] Has a Crucial Influence on the Particle Size

By changing the solvent–surfactant combination, namely, the emulsion parameters, the particle size can be varied in the wide 10 nm–100 μm size range. In detail, HLB numbers >10 point to an affinity for water (hydrophilic), whereas HLB values <10 indicate affinity for oil (lipophilic). Hence, surfactants with HLB (hydrophile–lipophile balance) lesser than 10 (such as Span 80, with HLB = 4.3; or Span 20, with HLB = 8.6) are preferred for microsphere synthesis, whereas surfactants with mid range HLB (between 10 and 15) are normally selected for nanoparticle synthesis.

50°C–70°C range, normally using a high water-to-surfactant molar ratio as well as fast agitation to ensure formation of oil droplets small enough to afford formation of homogeneous microparticles.

The resulting particles typically have size values ranging between 0.3 and 3 μm and a characteristic core–shell structure, with the dopant enclosed within the silica shell. This approach enables to incorporate into a preparation a high load of ingredient (even 90% of the weight of the particles) and thus achieve sustained delivery of the active ingredients under defined mechanical or chemical conditions.

These SiO_2 microcapsules doped with organic sunscreen molecules were the first large-scale commercial application of sol-gel-doped microparticles labeled "UV-Pearls." In cosmetics and medicine, water-based products or emulsions with external water phase are preferred over oil-based products (ointments) or W/O emulsions.

For example, advanced sunscreens are supplied as freely dispersed suspension in water, and hydrophilic silica microcapsules enhance the bioavailability of the entrapped active ingredients. Moreover, silica microcapsules from O/W emulsions are clear and smooth to the touch, ideally suited for cosmetic formulations. Developed by Sol-Gel Technologies, the "UV-Pearls" know-how and production line were sold by Sol-Gel Technologies to Merck, which started to market them in 2002. These suspensions ("Eusolex UV-Pearls") are nowadays used by several cosmetic companies to formulate high sun protection factors lotions (Figure 18.4) with improved safety profile.

Many of the molecules used as sunscreens such as 4-methylbenzylidene camphor and octyl methoxycinnamate, in fact, show estrogenic activity and are potential endocrine disrupters.[16] The product is normally supplied as aqueous white liquid dispersion containing approximately

FIGURE 18.4 Silica-microencapsulated sunscreens UV-Pearls are used by several cosmetic companies to formulate high sun protection factors lotions with improved safety profile. (Image courtesy of Isomers, kindly reproduced from www.isomers.ca.)

35% (w/w) of the UV absorber at pH between 3.8 and 4.2.[17] The capsules are about 1.0 μm in diameter, and thus sufficiently small to be transparent when applied to the skin and to give a pleasant feeling. Penetration of the encapsulated sunscreens into the body is at least half reduced compared to penetration from nonencapsulated sunscreens. In practice, the encapsulated UV filters predominantly remain on the surface of the skin, because the microparticles are small enough to give a pleasant feel, but large enough so as to avoid penetrating the epidermis.

More recently, the same approach has been extended by Sol-Gel Technologies Ltd. to microcapsules made of sol-gel-entrapped benzoyl peroxide (BPO) for effective, nonirritating treatment of acne. The inert SiO_2 core shells now serve as a safe protective barrier, preventing direct contact between the BPO and the skin, thereby significantly reducing side effects. In this case, the amount of skin lipids controls the rate at which the BPO is released as the release mechanism involves migration of the skin's natural oily secretions through the silica pores into the capsule. The oils dissolve the BPO crystals and carry the dissolved BPO molecules to the sebaceous follicles, but avoiding the irritation, redness, and dryness caused by freely available BPO.

18.3 ADVANTAGES AND LIMITATIONS OF SOL-GEL MICROENCAPSULATION

The three main advantages of encapsulating active ingredients in silica-based ceramic particles are: (1) the versatility of the sol-gel process, enabling unprecedented control of the particle structure; (2) the chemical, biological, and physical stability of the doped glassy particles; and (3) the negligible toxicity and biocompatibility of amorphous silica.

The immense versatility of the sol-gel process in liquid phase enables control on the size, porosity, surface area, morphology (full or core–shell particle), and hydrophilic–lipophilic balance of the encapsulating silica cage. Not only the conditions of the process can be controlled by many parameters including acid–base catalysis, surfactants, temperature, and mixing conditions, but also the nature of the silicon alkoxides can be varied to include organically modified silanes, opening the route to organically modified silica microparticles that share the versatility of organic polymers.

The second intrinsic advantage of the glassy silica particles over competing polymeric formulations lies in their high mechanical resistance, thermal stability, and chemical and biological inertness. The silica matrix offers far better tightness and resistance to extraction forces than polymers or waxes, as well as chemical protection of the entrapped dopant from extreme acid or base, from enzymatic degradation, and from detergents. The chemical and biological inertness of silica makes it fully compatible with most formulations of interest to medicine, cosmetics, and food. Furthermore, the sol-gel encapsulation results in pronounced protection deriving from the caging and isolation of the dopant species from its surroundings.

The third major advantage derives from silica full biocompatibility. The SiO_2 synthesized by the sol-gel process is biologically inert, and at the same time completely biodegradable. Porous silica particles, indeed, dissolve at physiological pH in a dilute environment. Suspended in water in open configuration in physiological buffers at pH 7.4, which closely models in vivo conditions, SiO_2 microparticles dissolve in 6 h.[18] A ^{29}Si tracer study revealed that the dissolved silicon diffuses through the blood stream or lymph and is excreted in the urine.

It is also relevant that silica microspheres being readily biodegradable and of no toxicity, they will not contaminate the environment as the current commercial trend in their use continues,[19] production of increasing quantities of sol-gel-derived silica-based functional materials will inevitably result in the introduction of granules and microparticles made of amorphous silica (or organosilica) in environmental matrices and in the biosphere (Table 18.2).

It is perhaps then no surprising that, following the original patents dating back to 1998, tens of patent applications have been filed in the last 10 years to disclose and protect specific processes to make functional microcapsules and use thereof. Table 18.3 lists a selection of said patents. Other relevant patents exist and applications are filed on a regular basis. They can easily be accessed online using a patent database.

Table 18.3 clearly shows that, following the first patents from academic researchers and their spin-off companies, in the last 5 years, several large companies entered the field showing the manifest interest of mainstream chemical, flavor and fragrance, food, cosmetic, specialty chemicals, and even aerospace industry. It is also relevant here that practical innovation will not be stopped by patent disputes, too.

Despite the size and relevance of industrial organizations now involved in sol-gel-doped materials, the story of the patent conflict concerning sol-gel-encapsulated sunscreens encourages innovative companies willing to adapt the technology to enter the field.[20] Indeed, large chemical and cosmetic companies tried to invalidate the original patent used by Sol-Gel Technologies to

TABLE 18.2

Advantages of Silica-Based Ceramic Particles over Polymeric Encapsulants

- A very wide range of active molecules can be encapsulated and released.
- Encapsulation at room temperature (or lower) for temperature-sensitive materials.
- Silica provides pronounced protection of the payload.
- No alteration of the physical and chemical properties of the payload.
- Large control over the release rate by control of the microparticle structure.
- Homogenous distribution of the payload throughout the particle.
- Silica is chemically and biologically inert.
- Amorphous silica is fully biodegradable.

TABLE 18.3
Selected Relevant Patents on Sol-Gel Microencapsulation

Priority Year	Inventors/Company	Title and Number
1998	D. Dupuis et al./Rhodia Chimie	Method for preparing capsules consisting of a liquid active material core enclosed in a mineral coating US 6537583 B1
1998	M. Yoshioka et al./Seiwa Kasei Company	Microcapsule containing core material and method for producing the same US 6337089 B1
1999	D. Avnir et al./Sol-Gel Technologies	Method for the preparation of oxide microcapsules loaded with functional molecules and the products obtained thereof WO 2000009652 A3
1999	D. Avnir et al./Sol-Gel Technologies	Sunscreen composition containing sol-gel microcapsules—CA 2370364 A1
2000	C. J. Barbé et al./ANSTO & Ceramisphere	Controlled release ceramic particles, compositions thereof, processes of preparation, and methods of use US 7354603 B2
2003	S. Il Seok et al./Unitech Co. and Korea Research Institute of Chemical Technology	Process for preparing silica microcapsules US 6855335 B2
2006	C. J. Barbé et al./Ceramisphere	Particles having hydrophobic material therein—WO2006/133518
2009	A. de Schrijver et al./Altachem	Leach-proof microcapsules, the method for preparation and use of leach-proof microcapsules—WO 2011003805 A2
2009	L. Kong et al./European Aeronautic Defence and Space Company	Coating compositions including ceramic particles WO 2010146001 A1
2010	J. Dreher et al./BASF	In situ sol-gel encapsulation of fragrances, perfumes, or flavors—WO 2011124706 A1
2011	S. Bone/Givaudan Sa	Process - WO2013083760A2
2011	X. Huang et al./International Flavors & Fragrances	Microcapsules produced from blended sol-gel precursors and method for producing the same EP 2500087 A2
2012	G. Dardelle et al./Firmenich	Core–Shell Capsules—PCT/EP2012/060339

manufacture the first encapsulated UV filters. Yet, despite different legal strategies employed, no claims of previous priority were acknowledged, establishing a clear legal precedent. When novelty is granted to a specific class of microencapsulated actives and to a specific process to make them, each patent has its own legitimacy.

18.4 CONTROLLED RELEASE FROM SOL-GEL MICROPARTICLES

As recently emphasized by a cosmetic industry's practitioner,[21] the way the active entrapped compounds are released (or are not released and thus retained) is at least as important as efficient compartmentalization for industry.

For example, the transparent "UV-Pearls" that are the basic ingredients for advanced sunscreens need to be totally leach-proof in order to retain the encapsulated UV absorber and ensure prolonged action of the lotion applied to the skin. Hence, the porosity of the silica capsules must be negligible such as in the case of capsules obtained from accelerated basic sol-gel polycondensation. On the other hand, the "Cool Pearls" doped with BPO to treat acne need to be porous to ensure controlled

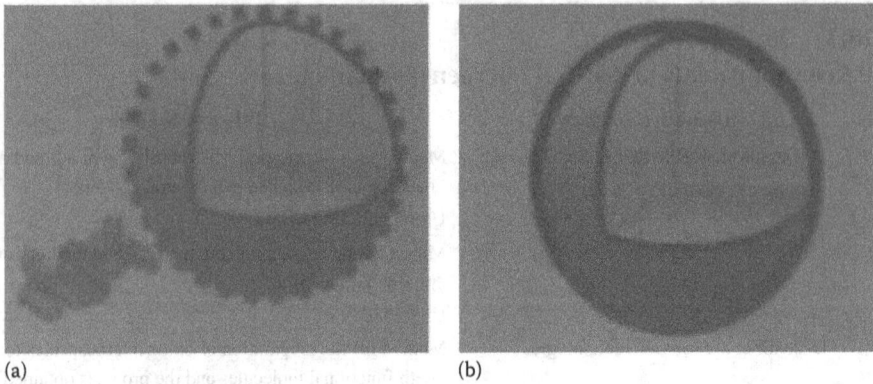

FIGURE 18.5 In certain cases, application requires porous particles (a) to ensure controlled release of the encapsulated active. In other cases, encapsulation in closed, totally leach-proof particles (b) is required. The sol-gel microencapsulation technology allows to meet both requirements by making different capsules. (Image courtesy of Prof. D. Avnir.)

release of the encapsulated active. The sol-gel microencapsulation technology allows to fully meet both requirements (Figure 18.5).

In general, compared to irregular xerogel matrix particles, microparticles encapsulate a far higher load of active ingredient (up to 90% in weight of the final materials), and afford a wide control over the release rate (from hours to months and up to unlimited retention of the entrapped ingredient), thanks to control of the microstructure and thus by the initial sol-gel chemistry.

Dopant molecules encapsulated in the inner core of silica microcapsules can only diffuse through the pores in the shells. The trigger for release is usually the presence of a liquid in which the active is soluble. The proportion of active in contact with the external fluid (the diffusion front) in the microspheres is much larger than in sol-gel granules produced by grinding and sieving. When the concentration of entrapped drugs in the confined space is halved, the diffusion driving force declines and so does the release rate. As a result, doped silica microcapsules usually give place to a two-step release, namely, a diffusion-controlled process whose release profile is best described by the Baker–Lonsdale model for diffusion from matrices of spherical shape.[22]

Dopant molecules uniformly distributed throughout matrix particles diffuse through the huge microparticle inner porosity. Also in this case, the release velocity can be varied from mg/hours to mg/month by changing parameters such as the water-to-alkoxide ratio, pH, alkoxide concentration, and matrix hydrophobicity.

For example, increasing the hydrophobicity of the organosilica matrix drastically reduces the release rate as shown by Figure 18.6 profiling release for hydrophilic dyes entrapped in full microparticles.[7] Clearly, with increasing amount of methyltrimethoxysilane (MTMS) in the TMOS/MTMS precursor sol, the release rate drastically decreases.

As mentioned earlier and claimed also by Marteaux,[23] a silicone industry's practitioner, the industrial interest in the sol-gel encapsulation technology not only depends on its ability to protect and control the delivery of actives but also on the different triggers one can use to release them, with silica and organically modified silica showing many advantages compared to organic shell materials.

A unique means to release the active species from hard silica spheres, for example, is the drying of the microcapsules. The simple desiccation of the microcapsule suspension destabilizes the capsules, thanks to the Laplace pressure, that is, the pressure difference across a curved interface, that occurs between the microcapsules upon the evaporation of the solvent phase (Equation 18.4). For a typical microcapsule size of 3 μm, considering an amorphous

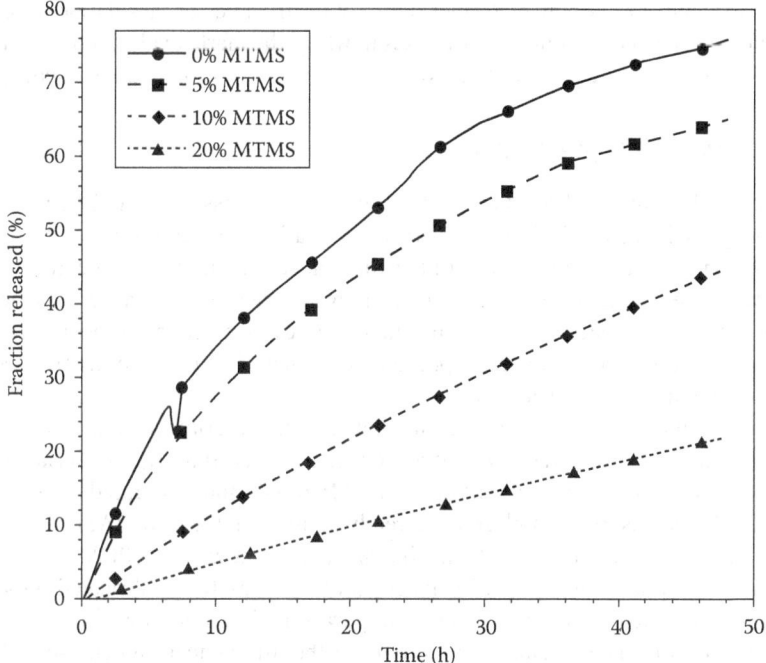

FIGURE 18.6 Increasing the hydrophobicity of the organosilica matrix reduces the release rate of a hydrophilic dye in water from microparticles. (Reproduced from Barbé, C.J. et al., *J. Sol-Gel Sci. Technol.*, 46, 393, 2008. With kind permission.)

silica shell surface tension of 330 mN/m,[24] the Laplace pressure difference existing between both sides of the shell (Equation 18.4) is 4.3 atmospheres.

$$\Delta P = 2\gamma/r \qquad (18.4)$$

Finally, adding to a good solvent for the active can result in capsule burst. In colloidal systems at equilibrium such as microcapsule suspensions, chemical potentials tend to become equal. The largely different chemical composition difference between each side of the microcapsule shells implies that the overall chemical potential must be compensated by the osmotic pressure. The latter can be stronger than the mechanical resistance of the shell, breaking the microcapsules.

18.5 ADVANTAGES AND LIMITATIONS OF SOL-GEL MICROENCAPSULATION

Silica-based microparticles obtained by the sol-gel process also have limitations that must be known and, when possible, faced. One problem arises with the tendency to gel of the colloidal silica particles formed as a side product in the tetra-alkoxysilane polymerization reaction. In this case, the addition of a sequestering agent or the reduction in the amount of colloidal silicate particles in the suspension leads to aqueous suspensions of silica shell microcapsules having improved storage stability.[25]

Another issue posed by silica capsules has to do with the brittle nature of the glass shell that often requires (see the following text for encapsulated waxes) the use of organically modified silanes in the precursor mixture along or in place of TEOS.

Finally, Si alkoxides used as precursors are expensive as well as toxic. Toxicity and high flammability are addressed by proper storage and handling at the manufacturing site. For example, one of us visiting the Sol-Gel Technologies industrial premises in Israel in late 2001 recalls that the TEOS storage tank was located outside the building including the tank reactor. Obviously, the high cost of alkoxysilanes compared to the low cost of the oil-derived chemicals (urea, formaldehyde, melamine, etc.)

normally used to synthesize polymer capsules does not justify silica encapsulation in place of microencapsulation routes of lower cost in all those cases in which the market value of the final product does allow to recover the higher incurred cost (a problem, e.g., with encapsulated phase change materials).

18.6 EMERGING APPLICATIONS

At the last (2013) International Sol-Gel Workshop, the new session established to highlight the potential of sol-gel science for different industrial applications hosted only industry's communications[26] focusing on sol-gel microparticles, rather than on sol-gel granules. According to this trend, numerous new sol-gel products based on silica microparticles will soon be launched to address relevant issues of global economic and environmental relevance. In detail, we forecast herein that sol-gel-encapsulated anticorrosion agents, phase change materials (PCMs), fragrances, and curing agents will shortly reach the marketplace.

PCM are important, because they are latent heat storage materials suitable for the construction industry as low-energy alternatives to air-conditioning systems. Indeed, a material series made of wax entrapped into microcapsules of acrylic glass (Micronal) and integrated in conventional construction materials such as gypsum plaster is already commercialized by BASF.*

In 2010, Wang's and Fang's teams in China independently showed that PCM with enhanced thermal conductivity and phase-change performance can be successfully fabricated by sol-gel microencapsulation of wax such as n-octadecane[27] and paraffin[28] in silica microspheres obtained from TEOS polycondensation. The thermal conductivity of the microencapsulated n-octadecane is also significantly enhanced due to the presence of the high thermally conductive silica shell. However, the silica microcapsules prepared from TEOS only have poor mechanical properties, with the brittle shell of the microencapsulated PCM easily cracking.

Recently, thus, the mechanical properties of the capsules doped with paraffin were considerably improved by another Chinese research team using MTMS in place of TEOS.[29] The sol-gel encapsulation of paraffin in methylsilica takes place with both high encapsulation ratio (82%) and efficacy (no leaching of the melted paraffin) of the silica capsules, whereas the thermal stability and lowered flammability of the silica-microencapsulated paraffin is retained.

Smart anticorrosion coating is another technology that will soon be commercialized. For example, Ceramisphere and aerospace company EADS researchers make pigment-sized solid porous particles (trademarked "Inhibispheres") doped with anticorrosion agents.[30] Many effective anticorrosion chemicals are incompatible with most coating formulations. The sol-gel microencapsulated agents instead have size similar to that of the pigment particles used in paint.

For corrosion protection, the trigger is the presence of water in the coating, which corresponds also to the onset of corrosion. With active chemicals trapped inside the particles and with the particles incorporated into the coating, the chemicals can be readily released (in response to a stimulus such as water) to provide protection for sustained periods of time (Figure 18.7).

Another relevant forthcoming commercial application will be sol-gel microencapsulated odorant and flavored silicas.[31] Fragrance chemicals are added to an ample variety of consumer products including perfumes, laundry detergents, fabric softeners, soaps, detergents, and personal care products, whereas aroma chemicals are widely employed by the food industry to enhance taste. Accordingly, the flavor and fragrance industry is an important sector of the chemical industry (21.8 billion $ sales in 2011).[32] Today this industry uses an extensive range of encapsulation technologies, as microencapsulated flavor and fragrances of improved efficacy are demanded by a wide range of industries for applications including cosmetics (perfumes), personal care (hand and body wash, toothpaste, etc.), food (flavors), and home care (laundry and detergent).[33]

---

* According to company's information, a layer of PCM plaster approximately 1.5 cm thick has the same heat capacity as a concrete or brick wall meaning that one can reap the benefits of lightweight design while still storing heat. They have been incorporated in numerous buildings, including the Haus der Gegenwart in Munich. See at the URL: www.micronal.de.

Silica-Based Sol-Gel Microencapsulation and Applications

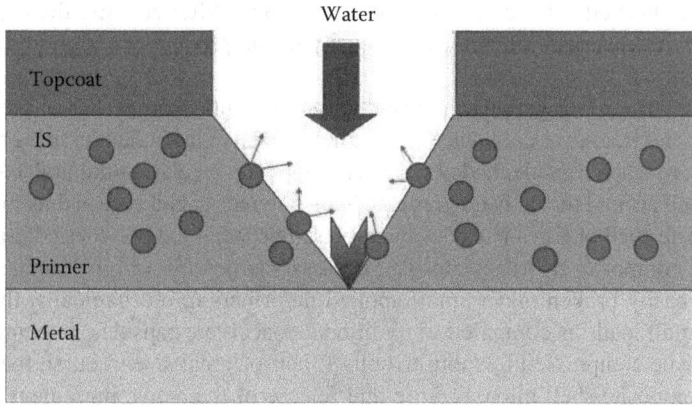

FIGURE 18.7 Biocides are microencapsulated in silica microparticles and the particles incorporated into the coating. When water diffuses toward the metal surface to be protected, the biocide chemicals are released, providing protection for sustained periods of time. (Image courtesy of Ceramisphere, kindly reproduced from www.ceramisphere.com.)

In cosmetic and personal care products, for instance, the objective is to extend the life and improve the delivery of the highly volatile fragrances (that mixed to the cleaning products may escape before the time of use); the interest is also high to infuse scent into everyday materials such as fabrics; or to turn liquid flavors into free-flowing powders so as to increase their shelf life in food products.

Significant health and environmental advantages, too, may originate from the efficient encapsulation of fragrances. For example, most of the synthetic nitro- and polycyclic musks used in perfumes, deodorants, and detergents are toxic and nonbiodegradable. As a result, they tend to accumulate in the environment,[34] and in human mother's milk.[35] Traditional microencapsulation of fragrance material in a polymeric coating is often not sufficient to improve fragrance performance in consumer products, as organic polymers suffer from limited stability, and do not allow proper controlled release.

The sol-gel encapsulation protects the fragrance against oxidation, while the nontoxic and biocompatible nature of amorphous silica (in the United States, it has the GRAS status and is approved as flavor carrier with up to 2% in weight in the final food) makes spherical silica microcapsules particularly well suited for cosmetic applications: for example, in personal care and in cleaning product compositions, such as body wash, liquid soap, skin cream, and shampoos. Accordingly, several patent applications describe the benefits of entrapping fragrances, perfumes, aromas, and flavors as core in a shell of SiO_2, made via emulsification of fragrance oils containing TEOS.

For example, data in Table 18.4 from one patent[36] application of a flavor and fragrance company illustrate how once applied through an antiperspirant (AP) roll-on base, silica capsules doped with

TABLE 18.4
Panelist Evaluation of Fragrance in SiO_2 Microcapsules

	Sample		
Fragrance Intensity Immediately after Application, I_0	Fragrance Intensity 5 h after Application with Rubbing, I_{5hr}	I_0/I_{5hr}	
Neat fragrance	16.5	9.8	59
SiO_2 capsules	14.5	13.5	93

Source: Reproduced from US 20100143422, with kind permission.

a fragrance retain their odor far better than neat fragrance. After rubbing the doped capsule, the intensity was almost unvaried, whereas the intensity of the product containing the neat fragrance was almost halved.

Further progress toward sol-gel capsules of enhanced mechanical properties was recently described by another flavor and fragrance company.[37] Now, the oily active ingredient is first entrapped in a traditional gelatin core–shell capsule made by coacervation with a gelatin/gum arabic capable of forming a hydrogel shell around the active ingredient. This polymer-caged core is thus mixed with TEOS to form a composite shell of silica particles interspersed between the polymeric lattice.

The resulting composite hydrogel/capsules possess shells with a high Young's modulus: even though they are easily broken under small applied deformation, mechanically, they are far more resistant under small loads as compared to traditional coacervate capsules. In contrast to SiO_2 capsules that need to be compressed by more than half their original size to burst, the capsules with a polymer/silica composite shell allow rupture and release of the active ingredient at a small much smaller deformation of the capsule. Again, the researchers carried out sensory evaluation tests, for example, with encapsulated peppermint oil in a toothpaste, or encapsulated fragrance applied in a skin cream. In each case, the panels judged the encapsulated products to be superior to the nonencapsulated counterparts.

Further progress can be easily envisaged. For example, attempts to create perfumes based on suspended capsules go back to the early 1970s.[38] Now formulation in water of sol-gel-entrapped perfumes of tunable scent to avoid skin irritation becomes possible.* Stabilized natural fragrances replacing toxic synthetic musks are also forecasted herein. In particular, sol-gel microencapsulated essential oils will replace nonbiodegradable musks as a consequence of the chemical and physical stabilization of nice fragrances that thus far could not be widely commercialized due to well-known poor chemical stability.

Finally, another forthcoming applications of the sol-gel microencapsulation technology concerns encapsulated curing agents for curing multicomponent polymer formulations and composite mixtures that are widely used as functional coatings, molding compounds, adhesives, and sealant. Sprayed from a pressurized aerosol can, for example, some 500 million cans containing one component foam (OCF) polyurethane formulations are sold worldwide each year (a figure growing at 6% rate) posing numerous health and safety problems. Now, the sol-gel microencapsulation of free-radical initiator BPO[39] (Figure 18.8) or of aqueous glycerol[40] in methyl-modified silica microcapsules of core–shell geometry allows to cure OCF formulations without the need to compartmentalize reactants from the curing agent.

Combined with the thermosetting resin precursors, the organosilica microcapsules are smoothly dispersed in the resin composition without swelling, while the silica-based capsule wall provides excellent chemical and physical stability, that is ideally suited for prolonging the shelf-life of these highly reactive formulations. Release of the entrapped initiator takes place upon spraying the capsules from cans that are typically pressurized at 4–6 bars under hydrocarbon atmosphere in the presence of the polymer precursors, without leaching of the encapsulated curing agent in the reactant mixture.

The capsules are prepared from O/W emulsions via ammonia-catalyzed condensation of TEOS and MTMS assisted by a surfactant. The synthesis is fully reproducible and can be easily scaled up.

18.7 OUTLOOK AND PERSPECTIVES

The sol-gel process to make doped silica-based materials has evolved from encapsulates in irregular SiO_2 xerogel particles, to sophisticate core–shell particles capable to encapsulate high amounts of functional organic species, and effectively release the entrapped species under small load. Sol-gel microcapsule and microparticle delivery systems will soon be introduced by numerous industries,

---

* Yissum Research Development Company of the Hebrew University of Jerusalem, «Controlled Release of Complex Fragrances and Aromas», Technology Opportunities. code: 9–2011–2594.

Silica-Based Sol-Gel Microencapsulation and Applications 343

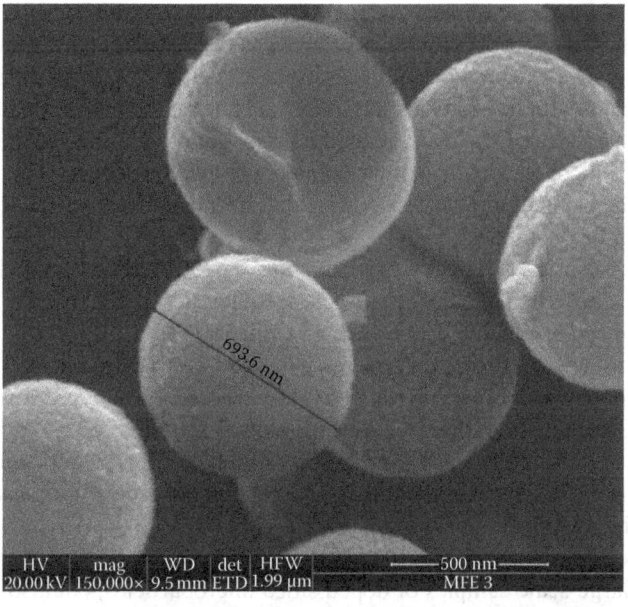

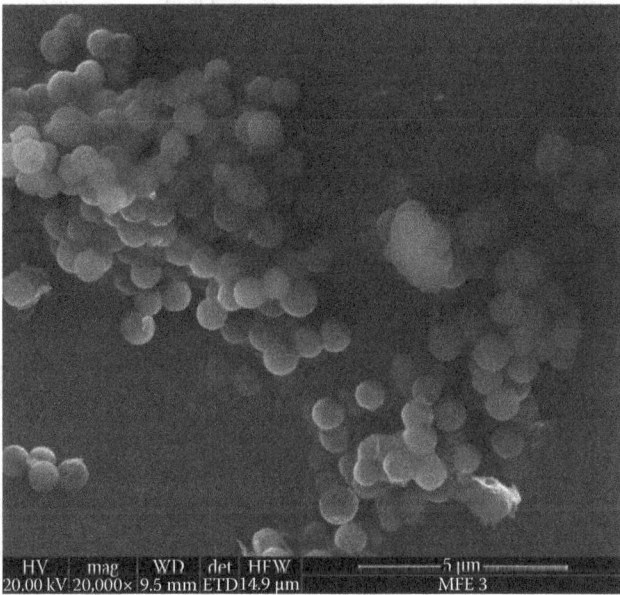

FIGURE 18.8 SEM pictures of methylsilica capsules doped with free-radical initiator BPO. These capsules, with a size between 700 and 900 nm, are leach-proof curing agents for polyurethane foams and polyester resins. (Reproduced from Ciriminna, R. et al., *ACS Sustainable Chem. Eng.*, 1, 2013, 1572 with kind permission.)

starting with the cosmetic and flavor and fragrance companies, especially now that the innovation can be made clearly visible for consumers. In the case of toxic sunscreens, for example, the side effects of organics absorbed by the skin were eliminated. Hence, "sunglasses for the skin" was (and is) the term aptly used by Merck to market ethylhexyl methoxycinnamate and related UV filters entrapped in sol-gel SiO_2 capsules. In the early 2000s, Merck asked Sol-Gel Technologies, then a start-up specializing in sol-gel materials, to conduct the scale-up of their technology to manufacture sol-gel-entrapped sunscreens. When the technology was proven successful, and following rapid growth in customer's demand, the company internalized the process, bought the manufacturing unit in Israel, and expanded production of the UV Pearls at another site in Germany. Today a number of commercial cosmetic preparations use these leach-proof capsules as effective and safe sunscreens providing the benefits of UVA and UVB filters, while reducing undesired side effects due, for example, to photodegradation of the sunscreen molecules with free radicals formation and interaction with body tissues.

Following the emergence of these and related new sol-gel products, as well as the booming production of elemental silicon due to the world's boom of photovoltaic energy, the production of TEOS expanded by two orders of magnitude. TEOS price thus dropped to first to 2 $/kg and then to current 1–1.5 $/kg range (for bulk samples).

In the late 1990s, the price of TEOS was so high that the IRB Italy's start-up biotechnology company established by Carturan and coworkers to encapsulate living cells opted to build its own small TEOS manufacturing plant (starting from $SiCl_4$ and ethanol).

New products and new exporting companies will soon be based on sol-gel microencapsulated actives. Sol-gel technology, however, is relatively new. Hence, when asked to identify the main challenges of commercialization faced by Australia-based Ceramisphere, the first problem identified[41] by the company's researchers and managers was finding versatile personnel with right mix of technical and managerial skills. Another issue was assessing the cost of industrial manufacturing of competitively priced products.

The company then recently built a 1000 L pilot plant (Figure 18.9) not only to enable the production and testing of large-scale samples of doped silica microparticles, but also to generate production cost data. Commenting on sol-gel market innovation, Abraham in 2006 was writing that[42]

> It is difficult to convince end users to switch from well-established technologies to new ones, despite property advantages, especially if they are not convinced that the process will cost the same or if they are unfamiliar with the technology.
>
> End users also often do not know how to apply the technology since they usually are not experts in sol gel or do not understand cost/performance benefits. Those who are experts are often ignorant of the end user's needs.

FIGURE 18.9 The 1000 L pilot plant built at Ceramisphere premises in 2013 enables the production and testing of large-scale samples of doped silica microparticles, as well as to generate production cost data. (Reproduced from Finnie, K.S. et al., Challenges of commercialising sol-gel technology. Ceramisphere experience, *XVII International Sol-Gel Conference*, Madrid, August 25–30, 2013, with kind permission.)

Eight years later, this aspect is still the largest obstacle. We have detailed elsewhere[43] why and how, in place of ultraspecialized researchers exclusively dedicated to the production of new knowledge, we need to educate and shape *scholars*, namely, scientists who are researchers producing new knowledge as well as teachers and communicators who share and valorize knowledge. Arguments provided in this chapter will help said scholars (and managers) to achieve all these objectives in the case of sol-gel-templated microspheres that will actually emerge as one eminent class of those "instruments of sustainability"[44] invoked by Wiesner for a variety of chemistry-enabled nanotechnology applications.

ACKNOWLEDGMENTS

Thanks to Dr Kim Finnie and Dr Chris Barbé, Ceramisphere, for sharing valued insight on industrial sol-gel microencapsulation. We are indebted to our friends and coworkers in this field of our research, David Avnir, The Hebrew University of Jerusalem, Laura M. Ilharco, Alexandra Fidalgo and João M. Bordado, Instituto Superior Técnico, Lisboa, François Béland and Delphine Desplantier-Giscard, SiliCycle, Ana Marques and Aster de Schrijver, Greenseal Research and Greenseal Chemicals.

REFERENCES

1. S.K. Ghosh, Functional coatings and microencapsulation: A general perspective. In *Functional Coatings*, Ghosh, S.K., (Ed.) Wiley-VCH, Weinheim, Germany, 2006.
2. A. Kreimeyer, BASF keeps R&D spending at high level, 28 January 2010, http://www.basf.ca/group/corporate/ca/en_GB/news-and-media-relations/news-releases/P-10-129.
3. M. McCoy, Encapsulating a new business, *Chem. Eng. News.* **2008**, *86* (12), 26.
4. S. Gouin, *Trends Food Sci. Technol.* **2004**, *15*, 330.
5. R. Ciriminna, M. Sciortino, G. Alonzo, A. de Schrijver, M. Pagliaro, *Chem. Rev.* **2011**, *111*, 765.
6. C. Barbé, J. Bartlett, L. Kong, V. Finnie, H.Q. Lin, M. Larkin, S. Calleja, A. Bush, G. Calleja, *Adv. Mater.* **2004**, *16*, 1959.
7. C.J. Barbé, L. Kong, K.S. Finnie, S. Calleja, J.V. Hanna, E. Drabarek, D.T. Cassidy, M.G. Blackford, *J. Sol-Gel Sci. Technol.* **2008**, *46*, 393.
8. I. van Driessche, S. Hoste, Encapsulations through the sol-gel technique and their applications in functional coatings. In *Functional Coatings*, Ghosh, S.K., (Ed.) Wiley-VCH, Weinheim, Germany, 2006.
9. BCC Research, *Sol-Gel Processing of Ceramics and Glass*, Wellesley, MA, 2012.
10. M. Pagliaro, R. Ciriminna, G. Palmisano, *Chem. Rec.* **2010**, *10*, 17.
11. A. Arsenault, G.A. Ozin, *Nanochemistry: A Chemical Approach to Nanomaterials*, RSC Publishing, Cambridge, 2005.
12. WO2000009652 A3.
13. D. Avnir, *Acc. Chem. Res.* **1998**, *28*, 328.
14. D.L. Meixner, P.N. Dyer, *J. Sol-Gel Sci. Technol.* **1999**, *14*, 223.
15. K.S. Finnie, J.R. Bartlett, C.J.A. Barbe, L. Kong, *Langmuir* **2007**, *23*, 3017.
16. M. Schlumpf, B. Cotton, M. Conscience, V. Haller, B. Steinmann, W. Lichtensteiger, *Environ. Health Perspect.* **2001**, *109*, 239.
17. S. Fireman, O. Toledano, K. Neimann, N. Looboda, N. Dayan, *Dermatol. Ther.* **2011**, *24*, 477.
18. K.S. Finnie, D.J. Waller, F.L. Perret, A.M. Krause-Heuer, H.Q. Lin, J.V. Hanna, C.J. Barbé, *J. Sol-Gel Sci. Technol.* **2008**, *49*, 12.
19. C.J. Murphy, *J. Mater. Chem.* **2008**, *18*, 2173.
20. D. Avnir, The long route from basic science to an exporting company, Asian Science Camp 2012, The Hebrew University of Jerusalem, 26–30 August 2012. The lecture can be visualized at the URL: http://www.youtube.com/watch?v=PsCGY8zAjLY.
21. E. Perrier, Encapsulation in cosmetics or not, that is the question!, *Bioencapsulation Innovation*, Newsletter of the Bioencapsulation Research Group, March 2012, pp. 1–2.
22. R.W. Baker, H.S. Lonsdale, Controlled release: mechanisms and rates. In *Controlled Release of Biologically Active Agents*, Taquary, A.C., Lacey, R.E. (Eds.), Plenum Press, New York, 1974, pp. 15–71.
23. F. Galeone, B.L. Zimmerman, A. Marteaux, Inorganic microencapsulation, *Bioencapsulation Innovations*, March 2013, pp. 12–14.

24. A. Roder, W. Kob, K. Binder, *Chem. Phys.* **2001**, *114*, 7602.
25. EP 2386353 A1 (2011).
26. *XVII International Sol-Gel Conference*, Madrid, August 25–30, 2013. https://www.isgs.org/index.php/isgs-sol-gel-2013/industrial-session.
27. H. Zhang, X. Wang, D. Wu, *J Colloid Interface Sci.* **2010**, *343*, 246.
28. G. Y. Fang, Z. Chen, H. Li, *Chem. Eng. J.* **2010**, *163*, 154.
29. Z. Chen, L. Cao, G. Fang, F. Shan, *Nanosc. Microsc. Therm.* **2013**, *17*, 112.
30. C. G. da Silva, C. Barbé, E. Campazzi, P.-J. Lathière, N. Pébère, E. Rumeau, L. Tran, M. Villatte, LEIS to study corrosion protection of AA 2024 by smart coatings containing encapsulated inhibitors, *EMCR-2012—10th Symposium on Electrochemical Methods in Corrosion Research*, Maragogi-Al, Brazil, 18–23 November, 2012.
31. R. Ciriminna, M. Pagliaro, *Chem. Soc. Rev.* **2013**, *42*, 9243.
32. Leffingwell and Associates, 2010–2014 Flavor and Fragrance Industry Leaders, http://www.leffingwell.com/top_10.htm.
33. J.J.G. van Soest, Encapsulation of fragrances and flavours as a way to control odour and aroma in consumer products. In *Flavours and Fragrances: Chemistry, Bioprocessing and Sustainability*, R.G. Berger (Ed), Springer, Berlin, Germany, 2007.
34. A.M. Peck, E.K. Linebaugh, K.C. Hornbuckle, *Environ. Sci. Technol.* **2006**, *40*, 5629.
35. J.L. Reiner, M.C. Wong, K.F. Arcaro, K. Kannan, *Environ. Sci. Technol.* **2007**, *41*, 3815.
36. US 20100143422.
37. PCT/EP2012/060339.
38. US 4428869 A.
39. R. Ciriminna, M. Sciortino, A. De Schrijver, D. Desplantier-Giscard, F. Béland, M. Pagliaro, *ACS Sustainable Chem. Eng.* **2013**, *1*, 1572.
40. R. Ciriminna, A. Alterman, V. Loddo, A. De Schrijver, M. Pagliaro, *ACS Sustainable Chem. Eng.* **2014**, *2*, 506.
41. K.S. Finnie, C.J. Barbé, K. Nagy, Challenges of commercialising sol-gel technology. The ceramisphere experience, *XVII International Sol-Gel Conference*, Madrid, Spain, August 25–30, 2013.
42. T. Abraham, *Sol-Gel Processing of Ceramics and Glass*, BCC Research, Norwalk, CT, 2006.
43. M. Pagliaro, *Nano-Age*, Wiley-VCH, Weinheim, Germany, 2010, Chapter 9.
44. M.R. Wiesner, M.R. Lecoanet, M. Cortalezzi, Nanomaterials, sustainability, and risk minimization, *IWA International Conference on Nano and Microparticles in Water and Wastewater Treatment*, Zurich, Switzerland, 22–24 September 2003.
45. K. Holmberg, B. Jönsson, B Kronberg, B. Lindman, *Surfactants and Polymers in Aqueous Solution*, 2nd edn., Wiley, New York, 2002.

19 Microencapsulation by Phase Inversion Precipitation

Ricard Garcia-Valls and Cinta Panisello

CONTENTS

19.1 Introduction 347
19.2 Principles 347
19.3 Process Technology 352
    19.3.1 Phase Inversion by Immersion in a Liquid Nonsolvent 352
    19.3.2 Phase Inversion Induced by Nonsolvent Vapor 353
19.4 Practical Example 353
    19.4.1 Introduction 353
    19.4.2 Materials and Methods 353
        19.4.2.1 Materials 353
        19.4.2.2 Microcapsules Production 354
        19.4.2.3 Microcapsules Characterization 354
        19.4.2.4 Release Experiments 354
    19.4.3 Results and Discussion 354
        19.4.3.1 Microcapsules Mean Size and Size Distribution 354
        19.4.3.2 Microcapsules Morphology 355
        19.4.3.3 Vanillin Release 356
19.5 Conclusions 356
References 356

19.1 INTRODUCTION

One of the possible methods for microcapsules production is phase inversion precipitation. This technique requires the preparation of microdroplets of the solved polymer containing the core materials. The microdroplets are then precipitated due to the presence of a nonsolvent for the polymer. In this chapter, we will describe the technique by using as example the production of polysulfone microcapsules by sprays and phase inversion precipitation in two modes, liquid and vapor non solvent media.

19.2 PRINCIPLES

Phase inversion precipitation technique is widely used for the preparation of both flat membranes[1] and microcapsules.[2]

This technique is based on the interaction of at least three compounds: the chosen polymer and two other compounds, one in which the polymer is soluble (solvent) and another in which the polymer is insoluble (nonsolvent). The solvent and the nonsolvent should be miscible with each other.

Broadly, when a polymeric solution gets in contact with a nonsolvent (whether solvent being in liquid or in vapor phase) changes happen in the composition of the solution and a solid structure is obtained. Herein, the fundamentals of the technique are described, first in general and later adding the required clarification depending on if you use the nonsolvent in liquid or in vapor state.

In general, in the process of preparing flat membranes or microcapsules by phase inversion, the first step is to prepare a polymeric solution. The polymeric solution is prepared by dissolving the polymer in a suitable solvent. The polymer concentration usually ranges between 10 and 30 wt.%.

This solution is stirred the time required for the polymer to dissolve completely. In this work, the solutions were prepared at least 24 h before its use. In addition, the bottles should be hermetically closed, because if not water vapor could penetrate (from environmental moisture), which could cause the precipitation of the polymer.

In a second step, the polymer solution must get in contact with the nonsolvent, whether it is by immersion in a bath containing nonsolvent liquid as by exposure to this nonsolvent vapors.

When this contact takes place, a mass-transfer process begins, which causes an exchange between solvent and nonsolvent in the solution. After a short time (small, of the order of a second), the polymeric solution becomes thermodynamically unstable, because of the increase of nonsolvent concentration and the decrease of solvent concentration.

When the solution becomes unstable it splits into two phases, which are immiscible between them. In one phase, the polymer concentration is high (polymer-rich phase) and in the other it is low (polymer-poor phase). Phase separation process is known as liquid–liquid demixing.

The nuclei of the poor phase are responsible for the formation of pores in the membrane while the polymer-rich phase precipitates resulting in the solid membrane. At this point, the morphology of the membrane is completely defined and stationary.

Consequently, the final morphology of the membrane is determined by the rate of liquid–liquid demixing and, in this rate, there are two main influencing factors: thermodynamic interactions between the compounds and kinetics of the mass transfer process.[3,4]

For a better understanding of the process graphical aids are available, which are known as ternary phase diagrams. In these diagrams, all possible concentrations of the three components are represented in a triangle. Figure 19.1 shows a generic ternary diagram.

The points located on the vertices of the triangle represent the pure components, the points on the sides of the triangle show the three possible binary combinations and a point inside the triangle represents a ternary mixture. The shadowed area corresponds to the zone where the solution is not stable, and therefore it decomposes into two immiscible phases, the composition of each of the phases can be determined from the equilibrium line. The area of the diagram that is not shaded is the region where the three components coexist in a single phase. The line separating these two areas is named binodal. Ternary diagrams can also show other interesting parameters such as spinodal and/or critical point, which are not on the scope of this work and therefore have not been represented in the diagram of the example. Thus, the present explanation is limited to the parameters herein detailed.[4,5]

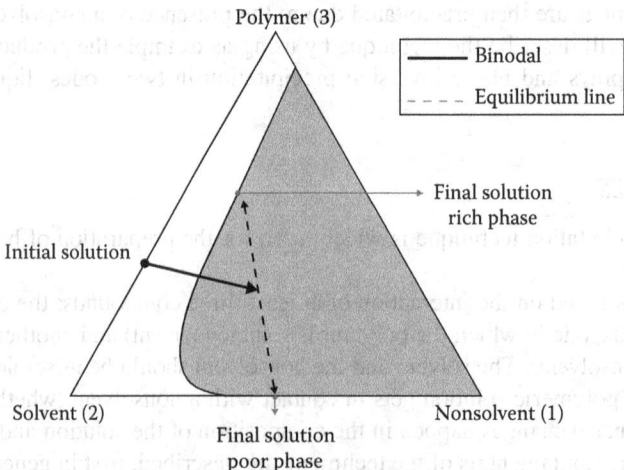

FIGURE 19.1 Representation of a ternary system in a triangular diagram.

Microencapsulation by Phase Inversion Precipitation

Figure 19.1 shows a representation of the phase inversion process. The process starts with a polymeric solution thermodynamically stable, the composition of which is usually located at the axis polymer–solvent, being a binary mixture of these two components. The process begins when this solution gets in contact with the nonsolvent, and so, the fraction of nonsolvent in the solution increases (ternary mixture).

The binodal is the limit between the stable and the unstable (biphasic) regions. When composition crosses the binodal, liquid–liquid demixing begins, leading to the formation of two immiscible phases. When the composition in the polymer-rich phase is high enough the polymer precipitates.

According to Flory–Huggins theory, both the location and size of the different regions depend on the Flory–Huggins binary interaction parameters, which are parameters that characterize the interaction between pairs of compounds. Thus, we have the solvent–polymer parameter (χ_{23}), the nonsolvent-Polymer (χ_{13}), and solvent–nonsolvent (χ_{12}). The subscripts, as shown also in the figure, refer to the nonsolvent,[1] solvent,[2] and polymer.[3,6]

Flory–Huggins interaction parameters indicate the affinity between substances. If the parameter of interaction for a pair of compounds is small, it indicates that the compounds have high affinity. However, if the parameter is high, the affinity between the compounds is low.[5]

Moreover, the theory defines the Gibbs free energy of a mixture on the basis of these parameters, as shown in expression 19.1.

$$\Delta G_m = RT(n_1 \ln \phi_1 + n_2 \ln \phi_2 + n_3 \ln \phi_3 + \chi_{12} n_1 \phi_2 + \chi_{13} n_1 \phi_3 + \chi_{23} n_2 \phi_3) \tag{19.1}$$

where
R is the universal constant for gases
T the temperature
n the number of moles
ϕ the volume fraction[7]

To calculate the binodal Equation 19.1 is used, taking into account that the sum of the mole fractions of the components in each phase must be 1 and that the chemical potential of a compound in two phases in equilibrium is the same in both of them (equilibrium condition).

The development of these considerations gives rise to a complex mathematical system, which resolution allows binodal calculation for any system with known interaction parameters and molar volumes. This calculation has been described more in depth in previous works.[3,7,8]

It is also essential to know that the decomposition process can take place in two ways: instantaneous or delayed. Figure 19.2 shows a diagram of generic processes.

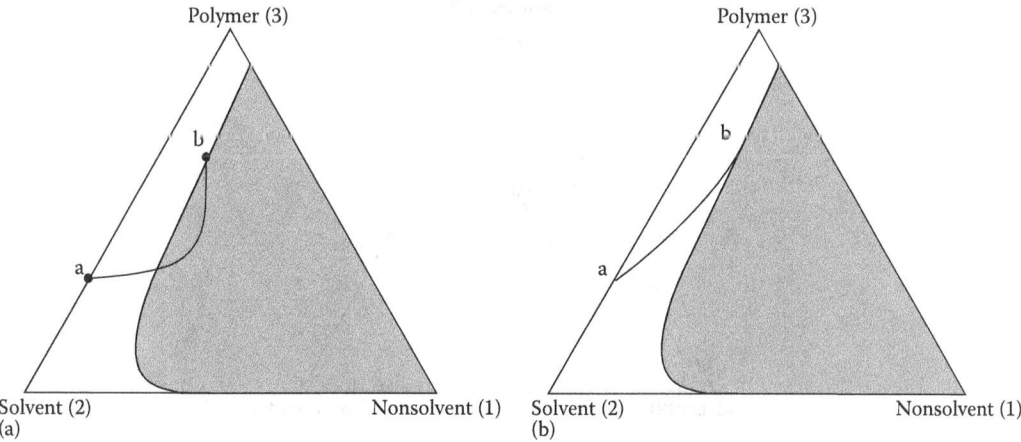

FIGURE 19.2 Composition profile in the liquid–liquid demixing process: (a) instantaneous and (b) delayed.

When liquid–liquid demixing takes place, first effects happen on the first area that is directly in contact with the nonsolvent (surface). As time passes, the nonsolvent penetrates to the most remote areas from the interface, progressing from the outside toward the inside of the film or drop in the microcapsules case.

Thus, at the beginning of the process, the point "b" of the graph corresponds to the composition in the outermost part of the sphere, while "a" corresponds to the composition in the inner part. The line connecting the two points represents intermediate points in the section of the film or sphere.

When an instantaneous process begins, immediately after immersion some points inside the external layer have already crossed the binodal. The demixing starts immediately after the immersion. However in delayed demixing, areas below the outer boundary layer are still in the stable region, which means that the decomposition does not start until after a certain time has passed.

The fact of being the decomposition instantaneous or delayed depends of the mass transfer flows and, as it can be observed, of the binodal position. So, it is important for the development of this work to know the influence of the different interaction parameters in the binodal position. Binodal plays an important role in how the demixing takes place, and thus, in the final morphology of the capsules. Figure 19.3 shows how the position of the binodal changes according to variation on the different interaction parameters. The graphic is based on calculations from previous works.[8]

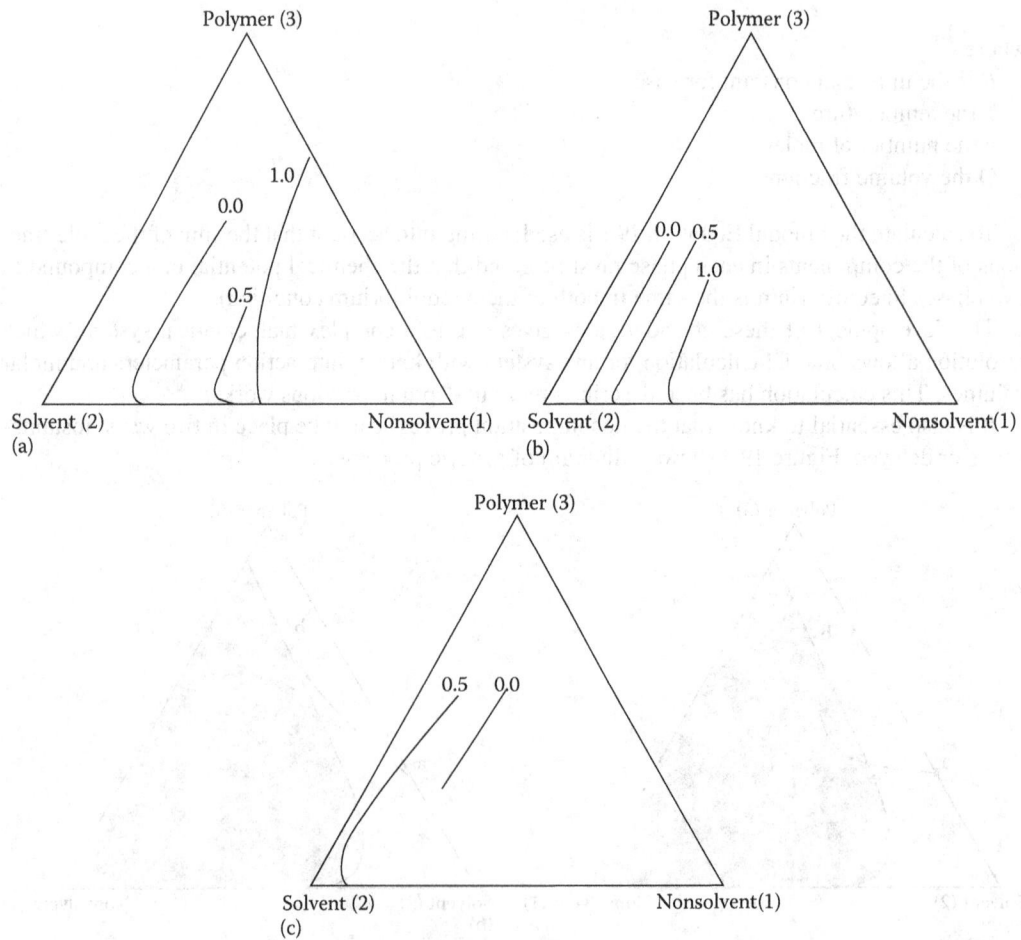

FIGURE 19.3 (a-c) Effect of changes in the interaction parameters on the position of the binodal.

Microencapsulation by Phase Inversion Precipitation

In Figure 19.3 we can observe how the position of the binodal changes when one of the interaction parameters is varied, while maintaining constant the other two.

- In the case of Figure 19.3a the variable parameter is χ_{12}, two different values are given to it (values appear indicated over the binodal). χ_{23} and χ_{13} are constant and equal to 1.
- Figure 19.2 through 19.3b has increased the value of χ_{13} to 1.5, maintaining χ_{23} constant and representing the same χ_{12} than in 3a, indicated over the corresponding binodal.
- Finally, in Figure 19.3c binodals are represented for different values of χ_{23} (indicated on the corresponding curve) for the case of χ_{13} constant and equal to 1 and χ_{12} constant and equal to 0.

From the observation of the figure, it is concluded that

- The smaller the parameter of interaction nonsolvent-solvent (χ_{12}), which is to say more affinity between solvent and nonsolvent, the binodal is closer to the axis polymer—nonsolvent (Figure 19.3a). This means that less nonsolvent is needed to produce separation and thus it begins earlier.

Other studies show that if χ_{12} increases the time it takes to start the separation also increases.[5] Both observations are interrelated and consistent. High affinity between solvent and nonsolvent favors instantaneous demixing.

- The nonsolvent–polymer interaction parameter (χ_{13}) also plays an important role in determining the surface area of the biphasic region, as it can be observed by comparison of Figure 19.3a and b. If χ_{13} increases the binodal is closer to the axis polymer—nonsolvent. Thus, as more different are the polymer and the nonsolvent, less of the latter is required to cause the separation. Experimental observations have shown that, in this case, time required to initiate decomposition is lower. High χ_{13} parameters favor instantaneous demixing.
- Finally in Figure 19.3c, it is observed that, if the solvent–polymer interaction parameter (χ_{23}) increases, the binodal gets closer to the polymer–solvent axis. When reducing the interaction, binodal moves away from that axis. Therefore, as closer are the polymer and the solvent, more nonsolvent would be required to cause the demixing.

Thus, broadly speaking, we can understand that the factors that promote instantaneous demixing are the high affinity between solvent and nonsolvent (lower χ_{12}) and the low affinity of the polymer for nonsolvent (high χ_{13}). The low affinity between polymer and solvent (high χ_{23}) also favors instantaneous demixing, but keep in mind that the range of this setting is limited, because if they are not sufficiently similar the solvent would not be suitable to dissolve the polymer.

The time required to start and the rate at which decomposition occurs are crucial in the morphology of the membrane as during demixing, nuclei of the polymer poor phase originate the pores.

Once a nucleus has been initiated, it draws solvent into it, in order to achieve the thermodynamic equilibrium of the phase. If affinity between solvent and nonsolvent is low, the driving force for this diffusion process will also be low, and pores growth would be restricted. On the other hand, if affinity is high, pores will grow fast.

The growth of pores takes place (1) while there is solvent available (2) until the concentration of polymer in the polymer-rich phase is sufficiently high so the polymer precipitates. However, it is logical that if many cores are growing at the same time, there will be less solvent available for their growth, because all of them are consuming solvent. Competition for the solvent, together with physical space available, is factors that limit the growth of the pores.

When the affinity between solvent and nonsolvent is high, the solvent diffuses rapidly to the precipitation bath, and a polymer concentrated layer is formed in the surface. This dense layer restricts the diffusion to the bottom layers. So, few nuclei are initiated under the surface layer, but

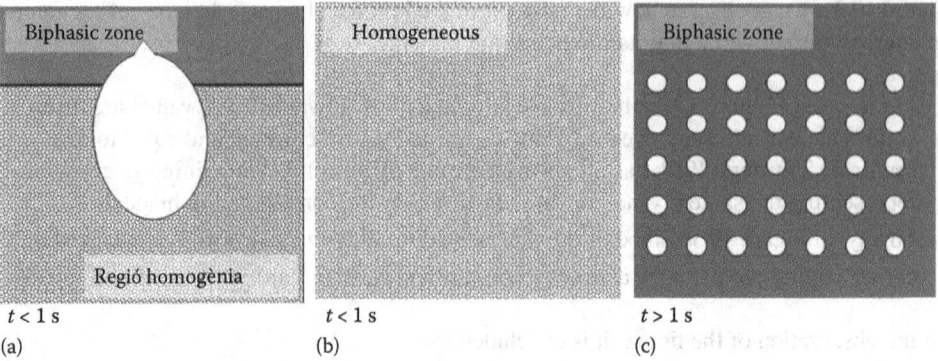

FIGURE 19.4 Comparison instantaneous versus delayed demixing: (a) instantaneous demixing, (b) delayed onset of demixing, and (c) delayed demixing by the time it starts.

little amount of nonsolvent in a nuclei is enough to drive solvent toward it, in order to maintain phase equilibrium. Thus, this few nuclei grow rapidly and in addition, more solvent is available for each of them. These conditions favor the formation of large pores.

In addition, the high affinity between solvent and nonsolvent (low χ_{12}) causes the instantaneous demixing. In the case of an instantaneous process, the governing factor is the concentration gradient, which diverges in the direction in which diffusion occurs and causing asymmetric structures and macrovoids. Macrovoids consist in porus that could have sizes similar to the thickness of the membrane.

In the case of delayed demixing, as the decomposition does not start immediately, solution is still homogeneous throughout the section at the start time of the phenomenon. This causes the formation of nuclei of equal size throughout the thickness of the membrane. The growth of these nuclei is limited by the low driving force for diffusion, due to the low affinity between solvent and nonsolvent.

In addition, many nuclei are growing at the same time so there will be less solvent available for their growth, as all of them are consuming it. Competition between nuclei for the solvent and the limitations of physical space are factors that restrain the growth of the pores and do not allow the formation of macrovoids in these systems. Figure 19.4 compares these two phenomena.

Finally it has to be mentioned that macrovoids cause weaknesses in the membrane and therefore, normally they are undesirable.[4]

19.3 PROCESS TECHNOLOGY

Two different processes can be done by sending the sprays to the nonsolvent, depending on the aggregation state of the nonsolvent (liquid or vapor).

19.3.1 Phase Inversion by Immersion in a Liquid Nonsolvent

One of the ways to generate contact between polymer solution and nonsolvent is immersion polymeric solution (casted or microdroplet) in a precipitation bath containing the nonsolvent.

This method is called phase inversion precipitation induced by immersion in nonsolvent. In this method, diffusion processes are very important. The nonsolvent diffuses from the bath to the polymeric solution, while the solvent migrates from the solution to the bath.

The speed in which this exchange occurs determines the final morphology of the membrane or microcapsule.

The exchange is faster when the affinity between solvent and nonsolvent is high. In this case, most likely, as already mentioned, the demixing of the solution into two phases is instantaneous and macrovoids are obtained.

However, the addition of solvent in the precipitation bath causes a reduction in the speed of diffusion. By adding solvent to the bath, the chemical potential gradient between the polymeric solution and the bath decreases. This potential reduction causes a delay on the onset of liquid–liquid demixing. When the separation begins, a certain amount of nonsolvent has already diffused throughout the cross section of the structure, and therefore many poor polymer phase nuclei have been initiated simultaneously. The growth of these cores is slower because the driving force has been reduced. In addition growth of the pores is limited by pores that surround them, as all of them are consuming solvent.

Under these conditions it is very rare to obtain macrovoids (typical from instantaneous demixing). Therefore, adding solvent to the precipitation bath limits the formation of a macrovoids in a system in which the thermodynamic relations between its components would produce instantaneous demixing, being likely to generate these macrovoids.

19.3.2 Phase Inversion Induced by Nonsolvent Vapor

If instead of a nonsolvent liquid, the nonsolvent is used in vapor state, the mechanism of precipitation is called phase inversion induced by nonsolvent vapor. Vapor nonsolvent is drawn into the polymer solution due to the affinity between solvent and nonsolvent.[9]

This method has been used to prepare flat membranes, both from polysulfone,[9-11] and from other polymers.[12-15]

In the previous works, performed with membranes, different membrane morphologies have been obtained depending on the relative humidity and time of exposure of the polymeric solution to the vapor. Under none of the conditions assessed was there evidence of macrovoids formation when separation is induced by vapor.[9,10,13-17]

This is an advantage of the technique, because as noted above, these macrovoids can cause weaknesses in the membrane.

Recently, the technique has been used, as far as we know for the first time, for the production of polysulfone microcapsules.[18]

19.4 PRACTICAL EXAMPLE

An example of the methodologies we are presenting is the case of producing polysulfone microcapsules by phase inversion by immersion precipitation.

19.4.1 Introduction

In this experimental section, the production of polysulfone microcapsules containing vanillin is described. Our method is based on the phase inversion by immersion precipitation technique. The herein described process has been successfully employed for the encapsulation of vanillin into polysulfone microcapsules and, in addition, vanillin release from those capsules has been characterized.[2,19-21] Polysulfone/vanillin microcapsules could have an application in laundry industry and also in medical applications.[22]

19.4.2 Materials and Methods

19.4.2.1 Materials

The materials used were Polysulfone (PSf, Sigma–Aldrich, Spain, transparent pellets of Mw = 16,000) as polymer, dimethylformamide (DMF, Scharlab, reagent grade ACS-ISO) as solvent, vanillin (Acros organics, 99% pure) as core material and distilled water as nonsolvent.

19.4.2.2 Microcapsules Production

An air atomizing nozzle (nozzle diameter = 0.8 mm) was installed over a beaker containing the precipitation bath (pure water). Air pressure was set to 2.5 bar. The air flow at this pressure was 250 L/h. Polymeric solution (15 wt.% PSf, 10 wt.% vanillin, both solved in DMF) was prepared 24 h before microcapsules production and it was kept in a closed, amber bottle in order to avoid its contact with humid atmospheric air that could cause polymer precipitation and to prevent vanillin degradation by U.V. radiation. The polymeric solution microdroplets were projected onto the precipitation bath and immediately, microcapsules were formed. Finally, capsules were collected by filtration and kept into a desiccator.

19.4.2.3 Microcapsules Characterization

Morphology of the products was studied by using Scanning Electron Microscopy (SEM) by using a Jeol JSM-6400 Scanning Microscopy, working with a voltage between 15 and 20 kV. Microcapsules samples were sputtered with gold, at 30 mA for 180 s, and afterward, their surface features were investigated. Moreover, with the aim to elucidate the main differences in the wall structure, samples were cut by cryogenic breaking,[23] and their cross-sections were also sputtered with gold at the conditions aforementioned, and observed by SEM. The procedure for obtaining cross sections of polysulfone microcapsules was described in detail in previous work.[21]

SEM images were analyzed by using ImageJ, which is a free program for image processing.

19.4.2.4 Release Experiments

The method used to follow the release of vanillin and DMF has been reported in a previous work.[20] The release medium was composed of 100 mL of distilled water, in which 1 g of microcapsules was added. The preparation was stirred (SBS multipoint magnetic stirrer, Spain) at 700 rpm for 96 h. Samples were taken from the release medium periodically and hermetically stored until analyzed.

19.4.3 RESULTS AND DISCUSSION

19.4.3.1 Microcapsules Mean Size and Size Distribution

As it can be observed in Figure 19.5, microcapsules were obtained by using the described equipment and procedure.

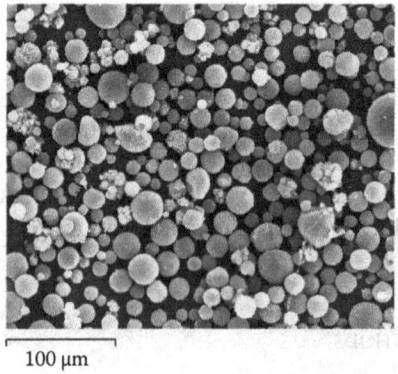

FIGURE 19.5 Polysulfone/vanillin microcapsules produced by phase inversion precipitation.

Microencapsulation by Phase Inversion Precipitation

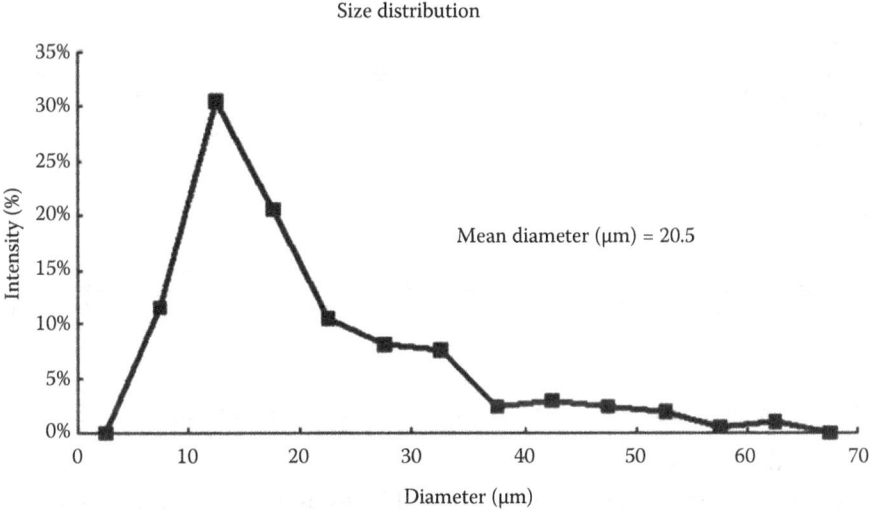

FIGURE 19.6 Mean size and size distribution of polysulfone/vanillin microcapsules.

Mean size and size distribution of the product were determined analyzing several micrographies by using ImageJ. For the results herein presented 300 capsules were measured. Figure 19.6 shows mean size and size distribution of the product.

Although the technology used produced a wide size distribution, this could be improved by using more precise nozzles. However, this was not in the scope of the work.

19.4.3.2 Microcapsules Morphology

Surface and cross section features of the capsules can be observed in Figure 19.7.

Capsules were porous on its surface (Figure 19.7a). On the other hand, cross-section of the microcapsules with vanillin showed that microcapsules prepared by using pure water as precipitation bath had macrovoids in their wall. This feature can be observed in Figure 19.7b). Fundamentals of the precipitation technique provide a suitable explanation for this phenomenon, as it has been explained before. When affinity between solvent and nonsolvent is high, as it is the case of DMF and water,[3] a fast demixing of the solution will take place. Fast demixing has been observed to favor the apparition of macrovoids.

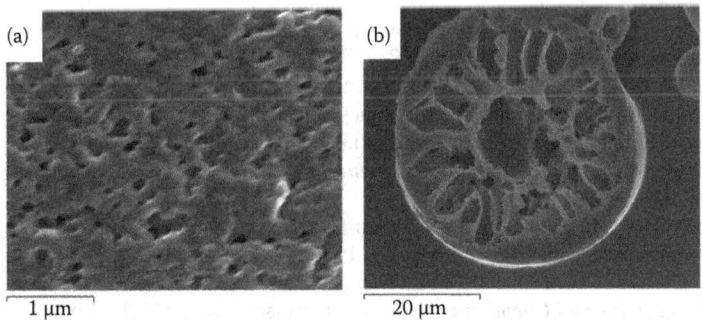

FIGURE 19.7 (a) Surface and (b) cross-section images of polysulfone/vanillin microcapsules.

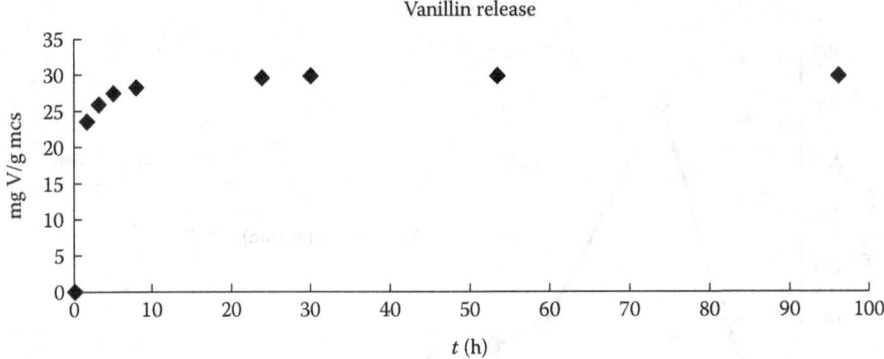

FIGURE 19.8 Vanillin release versus time for polysulfone/vanillin microcapsules.

19.4.3.3 Vanillin Release

Vanillin release was determined for a 96 h period by using HPLC. The analysis showed that the vanillin release rate was fast during the first 8 h of experiment, and it decreased over that period of time until it reached a plateau in the first 24 h. This tendency can be observed in Figure 19.8. Similar release tendencies were also observed in previous works with PSf macro and microcapsules.[19,20,24] Maximum vanillin release after 96 h was 29.90 29.90 ± 0.58 mg vanillin/g capsules.

19.5 CONCLUSIONS

The production of microcapsules by phase inversion precipitation is an advisable technology. In this production process, the polymer that will constitute the microcapsule already exists before the actual production of the capsules. It is solved and then only precipitated due to the presence of a non-solvent phase that makes the microcapsule precipitate from the droplet with the desired properties. As a remarkable capacity of the method, the final capsules can be designed as desired just by playing with the experimental conditions. In that way, the engineer can decide the size and distribution, the porosity, the internal morphology, and the core materials. On the other hand, the production is fast and even in the "so called" delayed process, the precipitation takes place within seconds, compared to other technologies that require longer times (e.g., interfacial polymerization) or less energy that in the case that evaporation has to occur rapidly.

REFERENCES

1. Guillen G, Pan Y, Li M. Preparation and characterization of membranes formed by nonsolvent induced phase separation: A review. *Ind Eng Chem Res*. 2011;50(7):3798–3817.
2. Peña B, Gumí T. State of the art of polysulfone microcapsules. *Curr Org Chem*. 2013;17(1):22-29.
3. Torras C. Obtenció de membranes polimèriques selectives [Internet]. Tarragona, Spain: Universitat Rovira i Virgili; 2005. Available from: http://hdl.handle.net/10803/8526.
4. Mulder M. *Basic Principles of Membrane Technology*. Dordrecht, the Netherlands: Kluwer Academic Publishers; 2003.
5. VandeWitte P, Dijkstra P, VandenBerg J, Feijen J. Phase separation processes in polymer solutions in relation to membrane formation. *J Membr Sci*. 1996;117(1–2):1–31.
6. Ullmann F, Gerhartz W, Yamamoto YS, Campbell FT, Pfefferkorn R, Rounsaville JF. *Ullmann's Encyclopedia of Industrial Chemistry*. Weinheim, Germany : Wiley-VCH; 1985.
7. Kim J, Lee H, Baik K, Kim S. Liquid-liquid phase separation in polysulfone/solvent/water systems. *J Appl Polym Sci*. 1997;65(13):2643–2653.
8. Altena FW, Smolders CA. Calculation of liquid liquid—Phase separation in ternary-system of a polymer in a mixture of a solvent and a nonsolvent. *Macromolecules*. 1982;15(6):1491–1497.

9. Su YS, Kuo CY, Wang DM, Lai JY, Deratani A, Pochat C et al. Interplay of mass transfer, phase separation, and membrane morphology in vapor-induced phase separation. *J Membr Sci.* 2009;338(1–2):17–28.
10. Park H, Kim Y, Kim H, Kang Y. Membrane formation by water vapor induced phase inversion. *J Membr Sci.* 1999;156(2):169–178.
11. Tsai J, Su Y, Wang D, Kuo J, Lai J, Deratani A. Retainment of pore connectivity in membranes prepared with vapor-induced phase separation. *J Membr Sci.* 2010;362(1–2):360–373.
12. Ripoche A, Menut P, Dupuy C, Caquineau H, Deratani A. Poly(ether imide) membrane formation by water vapour induced phase inversion. Macromolecular symposia. Weinheim, Fed. Rep. of Germany: Wiley-VCH Verlag GmbH; 2002. p. 37–48.
13. Sun H, Liu S, Ge B, Xing L, Chen H. Cellulose nitrate membrane formation via phase separation induced by penetration of nonsolvent from vapor phase. *J Membr Sci.* 2007;295(1–2):2–10.
14. Caquineau H, Menut P, Deratani A, Dupuy C. Influence of the relative humidity on film formation by vapor induced phase separation. *Polym Eng Sci.* 2003;43(4):798–808.
15. Matsuyama H, Teramoto M, Nakatani R, Maki T. Membrane formation via phase separation induced by penetration of nonsolvent from vapor phase. I. Phase diagram and mass transfer process. *J Appl Polym Sci.* 1999;74(1):159–170.
16. Khare V, Greenberg A, Krantz W. Vapor-induced phase separation—Effect of the humid air exposure step on membrane morphology Part I. Insights from mathematical modeling. *J Membr Sci.* 2005;258(1–2):140–156.
17. Matsuyama H, Teramoto M, Nakatani R, Maki T. Membrane formation via phase separation induced by penetration of nonsolvent from vapor phase. II. Membrane morphology. *J Appl Polym Sci.* 1999;74(1):171–178.
18. Panisello C, Garcia-Valls R. Polysulfone/vanillin microcapsules production based on vapor induced phase inversion precipitation. *Ind Eng Chem Res.* 2012;51(47):15509–15516.
19. Peña B, de Menorval L-C, Garcia Valls R, Gumi T. Characterization of polysulfone and polysulfone/vanillin microcapsules by (1)H NMR spectroscopy, solid-state (13)C CP/MAS-NMR spectroscopy, and N(2) adsorption desorption analyses. *ACS Appl Mater Interf.* 2011;3(11):4420–4430.
20. Peña B, Panisello C, Aresté G, Garcia-Valls R, Gumí T. Preparation and characterization of polysulfone microcapsules for perfume release. *Chem. Eng. J.* 2012;179:394–403.
21. Panisello C, Peña B, Gumi T, Garcia Valls R. Polysulfone microcapsules with different wall morphology. *J Appl Polym Sci.* 2013;129:1625–1636.
22. Panisello C, Peña B, Gilabert Oriol G, Constantí M, Gumí T, Garcia-Valls R. Polysulfone/vanillin microcapsules for antibacterial and aromatic finishing of fabrics. *Ind Eng Chem Res.* 2013;52:9995–10003.
23. Torras C, Pitol L, Garcia Valls R. Two methods for morphological characterization of internal microcapsule structures. *J Membr Sci.* 2007;305(1–2):1–4.
24. Peña B, Casals M, Torras C, Gumi T, Garcia-Valls R. Vanillin release from polysulfone macrocapsules. *Ind Eng Chem Res.* Feb 4, 2009;48(3):1562–1565.

20 Microfluidic Encapsulation Process

Fabrizio Sarghini

CONTENTS

20.1 Introduction 359
20.2 Physics of Microfluidic 360
20.3 Basic of Microfluidic Devices: Cross Flowing Streams, Flow Focusing, and Coflow Focusing...363
    20.3.1 T-Junctions..364
    20.3.2 Focusing and Coflow-Focusing Devices...366
    20.3.3 Size, Mixing, and Shape Control...370
    20.3.4 Shell Thickness and Loading Efficiency Control..371
        20.3.4.1 Drug-Release Rate Control 372
20.4 Device Materials and Construction Techniques 372
    20.4.1 Photolithography-Based Microfabrication..372
    20.4.2 Methods Based on Replication-Based Methods 374
        20.4.2.1 Soft Lithography 374
        20.4.2.2 Hot Embossing...375
        20.4.2.3 Injection Molding...376
20.5 Perspectives and Challenges 378
References 378

20.1 INTRODUCTION

Conventional nanoparticles (NPs) or encapsulated active compounds fabrication techniques are often characterized by polydispersity and batch-to-batch variations.

The technological framework is plenty of emulsification approaches, mostly involving mixing two liquids in bulk processes, and many of them using turbulence to enhance drop breakup.

In these "top-down" approaches to emulsification, little control over the formation of individual droplets is available and, as a consequence, a broad distribution of sizes is typically obtained, together with a limited loading efficiency control.

This heterogeneity remains a significant obstacle in bulk preparation of drug delivery systems or encapsulation of active compounds for functional foods. Innovations in microfluidics, the science of manipulating fluids in nano-/picoliter scale, introduce several opportunities to improve the fabrication process.

Several benefits of conducting reaction in microfluidic devices can be mentioned: rapid mixing of reagents at microscales, homogeneous reaction environment, possibility of multistep reaction design, enhanced processing accuracy and efficiency, improved heat transfer due to high surface-to-volume ratio, miniaturization, parallelization for mass production, and cost savings from reduced consumption of reagents.[1]

However, all these advantages raise a new set of fluid dynamical problems related to the deformable interface of the droplets: the need to take into account interfacial tension and its variations and the complexity of singular events such as merging or splitting of drops.

Microfluidic devices can be used both in NP fabrication and for core–shell and matrix encapsulation structures.

In these processes, the active compounds are driven to the area of adsorption by a carrier enclosed in matrix or core–shell configuration. Important parameters to increase the bioavailability are, for example, the size of the global compound, the encapsulated quantities (loading efficiency), the thickness of the external shell in case of time-driven controlled release, and also the shape of the particle.

The carrier can be organic (polylactic acid [PLA][2,3] and poly(lactic-co-glycolic acid) [PLGA][4,5]) or inorganic (chitosan;[6,7] poly(N-isopropylacrylamide);[8–11] hydrogelator;[12] silk protein;[13] pectin;[14] hydrazide- and aldehyde-functionalized carbohydrates;[15] dextran hydroxyethyl methacrylate;[16] and silica[17]).

20.2 PHYSICS OF MICROFLUIDIC

The technological advances in micro- and nanodevices fabrication techniques introduced significant changes in the ability to manipulate very small volumes of fluid, including micro- and nanoquantities contained therein. Several applications of these small devices had been developed in sensors, separation and analysis, biological characterization, cell capture and counting, micropumps, actuators, high-throughput design and parallelization, and system integration, to name just a few areas.

Because biological and chemical applications are typically concerned with molecules and bioparticles with small dimensions, the tools used to manipulate these objects are usually fabricated on a similar scale, and the developments in micro- and nanofabrication in recent decades has brought engineering tools to a scale that easily matches these objects.

To understand the forces involved in microfluidic systems, we can assert in general that dynamic properties $p(A)$, function of the area of interaction A, decrease more slowly than properties $p(V)$ that depend on the volume V, and the ratio between such properties can be expressed by the "square-cube" law:

$$\frac{p(A)}{p(V)} \propto \frac{L^2}{L^3}$$

and given the scales involved in microfluidic devices, this ratio is order of 10^6.

As a consequence, in liquid-phase devices, the inertial forces tend to be quite small, and surface effects dominate the behavior of these small systems. In most biological and biochemical applications, if we shrink the length scales, then interfacial and electrokinetic phenomena become more important, reducing the influence of traditional driving forces like gravity and pressure.

Also some general certainty like the no-slip boundary condition does not always fit, and this shifted the interest of researchers from how to solve equations to which equations is better to apply.

The main differences between fluid mechanics at microscales and at macroscales can be generally classified into four classes of effects:

- Noncontinuum effects
- Surface-dominated effects
- Low-Reynolds-number effects
- Multiscale and multiphysics effects

Although the low Reynolds number characteristic of most of these flows eliminates the challenges of nonlinearity in the convective term and the associated difficulty in modeling turbulent flows (which is actually not true in some gas–liquid devices), we are instead forced to face the nonlinearity of the source term in the Poisson–Boltzmann equation, the nonlinearity of the coupling of electrodynamics with fluid flow, and the uncertainty in predicting electroosmotic boundary conditions.

Microfluidic Encapsulation Process

Microscale flows can be considered typically laminar, due to the short length scales involved, but they can have large mass transfer Peclet numbers, owing to the low diffusivity of macromolecules of interest.

These flows can be driven with pressure, but sometimes applied electric fields are often used to actuate these systems.

In any case, even if not applied, intrinsic electric fields exist at interfaces in all cases, driven usually by chemical reaction.

At the scales involved, electrodynamics, chemistry, and fluid mechanics are inextricably intertwined: electric fields can create fluid flow, and fluid flow can create electric fields, with the surface chemistry driving the degree of coupling. The flow coupling effect can be described by electrostatic source terms in the Navier–Stokes equations or particle transport equations. Boundary conditions become an issue in microsystems, due to high surface area–volume ratios. Boundary conditions that are taken for granted at the macroscale (e.g., the no-slip condition) can often fail in these systems.

Multiphase implementations, designed to optimize certain aspects of transport, lead to additional interfacial concerns. Last but not least, we often need to work in non-Newtonian systems, which require a modification of the constitutive relation used in the Navier–Stokes equations.

From a physical point of view, the main modification that droplets bring, with respect to single-phase microfluidic flows, is the introduction of interfacial tension. This new physical ingredient can be thought in two complementary ways. Interfacial tension is a force per unit of length, which pulls the interface with a magnitude γ (N m^{-1}): any spatial imbalance in the value of γ generates a flow along the interface from the low to the high interfacial tension regions, a phenomenon known as Marangoni flow.

Since the value of the surface tension varies with temperature and with the contamination of the interface by the possible presence of surfactant molecules, each change of these parameters can lead to a Marangoni flow, which is referred, respectively, as thermocapillary or solute-capillary flow.

Interfacial tension can also be described as an energy per unit area (J m^{-2}), which acts to minimize the total surface area and reduce to the minimum the free energy associated with the interface, and the minimum area for a given volume is a sphere, which is not by chance the shape of an isolated droplet or bubble.

This is because when two immiscible fluids are brought in contact, it is evident that they develop an interface between them, and the presence of this interface introduces a cost to the free energy of the system, the interfacial energy $E\gamma$ proportional to the surface area A of the interface and to the interfacial tension γ:

$$E\gamma = A\gamma \tag{20.1}$$

The expression (Equation 20.1) relates the surface of the drop to the force exercised when a deformation take places.

If we consider a small change of the shape of the interface dr, the resulting force will be

$$F = dE\gamma/dr = \gamma dA/dr \tag{20.2}$$

If we are interested in describing the physics in terms of pressure (force per unit of area), in case of a sphere, we can introduce $A = 4\pi r^2$ and obtain

$$P\gamma = (1/A)(dE\gamma/dr) = \gamma(1/A)(dA/dr) = 2\gamma(1/r). \tag{20.3}$$

This means that the pressure inside a droplet in equilibrium is larger by the value of $P\gamma$, also known as the Laplace pressure, in respect to the pressure with its surrounding immiscible fluid. For an arbitrary surface, the two principal radii of curvature may be pointing in different directions (e.g., confined drops must adapt their shape to the presence of boundary walls deforming their interface),

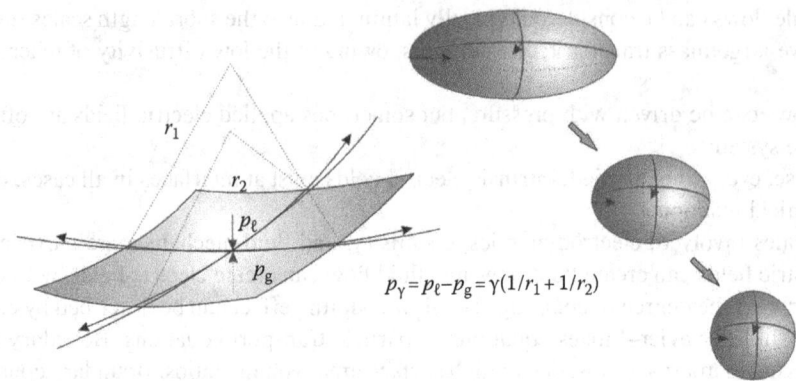

FIGURE 20.1 (See color insert.) The Laplace pressure exerted by curved interfaces is proportional to the sum of the principal curvatures. The blue arrows signify the direction and magnitude of the interfacial forces, explaining why a deformed droplet restores its spherical shape.

and then the Laplace pressure is given by the difference of their magnitudes, oriented into the direction of the smaller radii of curvature, that is, into the direction where concavity is more pronounced (see Figure 20.1):

$$\Delta P_\gamma = \gamma \left(\frac{1}{r_1} + \frac{1}{r_2} \right) \tag{20.4}$$

where r_1 and r_2 are the two principal radii of curvature of the interface.

These supplementary pressure variations play a major role in determining the flow conditions.

The presence of droplets also introduces new kinematic and dynamic boundary conditions on the fluid flow. Since the immiscible fluids cannot cross the interface, boundary condition states that the local normal component of the velocities in each fluid must be equal to the interface velocity, the velocity tangents to the interface must be also equal inside and outside the droplet, and the tangential shear stresses must be balanced at the interface when it is clean of surfactants.

This means that the variation of the tangential velocity $u\tau$ in respect to the normal direction r, inside and outside the drop, must balance

$$\mu_{in} \left. \frac{\partial u_\tau}{\partial r} \right|_{in} = \mu_{out} \left. \frac{\partial u_\tau}{\partial r} \right|_{out} \tag{20.5}$$

where μ_{in} and μ_{out} are the viscosities of internal and external fluid, and the viscosity ratio μ_{in}/μ_{out} plays an important role inside and outside a moving drop or bubble.

In presence of surfactant molecules, the local surface coverage affects the value of interfacial tension. Sometimes, these molecules are often added on purpose, in order to facilitate the creation and transport of drops, but other times, they can also be present in the fluids as by-products of chemical reactions or as impurities, so that the spatial value of interfacial tension could be different if the surface concentration displays variations.

As a consequence, it is necessary to introduce in Equation 20.5 a tangential stress $\nabla_\tau \gamma$ along the tangent to the interface at every point (Marangoni stress).

$$\mu_{in} \left. \frac{\partial u_\tau}{\partial r} \right|_{in} = \mu_{out} \left. \frac{\partial u_\tau}{\partial r} \right|_{out} + \nabla_\tau \gamma \tag{20.6}$$

The relation between γ and the local surfactant concentration is nonlinear, and it was modeled in some cases using Langmuir model.[18]

Surfactant transport plays another important role, as these molecules can be transported either by advection of the main flow or through molecular diffusion either in the bulk or along the interface.[19,20]

Additional effects can be introduced by variation of surfactant characteristics, like the adsorption and desorption rates on the interface, by measuring the chemical kinetics and the partition coefficient, and by measuring the relative bulk and surface concentrations at equilibrium.

As any change in the shape of a drop induces a local contraction or expansion of the interface and as a consequence an increase or a decrease of surface concentration, it is clear that in practical microfluidics applications, the presence of a surfactant involves a complex interplay between several parameters.

Generally speaking, the fluid behavior can be parameterized using the values taken by some important dimensionless numbers comparing different physical quantities.

In microfluidic devices, generally but not always (e.g., inertial effects can be significant in case of high-speed flows, for high production rates or droplet breakup situations), inertial effects are minimal, meaning that they work in small Reynolds number regimes.

The Weber number $We = \rho U^2 l / \gamma$, where U is a characteristic velocity scale and l is a spatial scale, compares inertial effects to interfacial tension and is usually small as well. As the effect of gravity can be neglected, the only two remaining players are viscosity and interfacial tension.

The relative strength of the two can be expressed using the capillary number $Ca = \mu U/\gamma$, where μ is generally the larger and most significant viscosity working in the system.

Low values of Ca mean that the stresses due to interfacial tension are prevalent if compared to viscous stresses, and drops in low Ca regimes tend to minimize their surface area by producing spherical shapes. On the contrary, in regimes described by high values of Ca, viscous effects dominate, and we can observe large deformations of the drops and asymmetric shapes.

The geometry of the device of course plays a major role and, for example, in case of an expansion or a contraction, the velocity varies over a length scale quite different from the radius of the drop itself.

In this case, we need to consider a new capillary number, based on the characteristic magnitude of the shear stress in a specific spatial direction s, to the flow $\mu dU/ds$, where s represents a spatial direction.

This stress is still compensated by the Laplace pressure, introducing a directional capillary number, which describes the magnitude of deformation observed on a drop due to variations in velocity of the flow field[21] when a drop enters a bifurcating microchannel.[22,23]

20.3 BASIC OF MICROFLUIDIC DEVICES: CROSS FLOWING STREAMS, FLOW FOCUSING, AND COFLOW FOCUSING

The majority of microfluidic methods produce droplet using passive devices generating a uniform, evenly spaced, continuous stream of droplet,[24] whose volume ranges from femtoliters to nanoliters. Their operational modes take advantage of the characteristics of the flow field to deform the interface and promote the natural growth of interfacial instabilities, avoiding in this way the necessity of any local external actuation. Droplet polydispersity, defined as the ratio between the standard deviation of the size distribution and the mean droplet size, can be as small as 1%–3%.

From a fluid dynamics point of view, three main approaches, based on different physical mechanisms of droplet formation and breakup, have emerged, and they can be characterized considering the flow field topology in the vicinity of the drop production zone:

1. Breakup in cross-flowing streams (Figure 20.2a)
2. Breakup in coflowing streams (Figure 20.2b)
3. Breakup in elongational flows (Figure 20.2c)

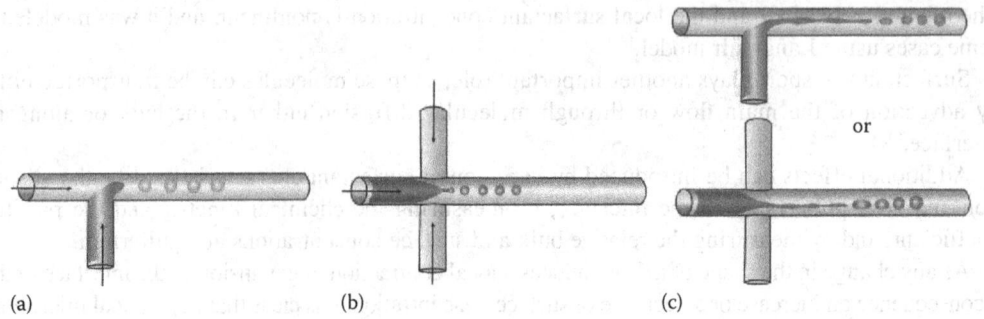

FIGURE 20.2 (See color insert.) Droplet breakup (a) in cross flowing streams, (b) in coflowing streams, and (c) in elongational flows.

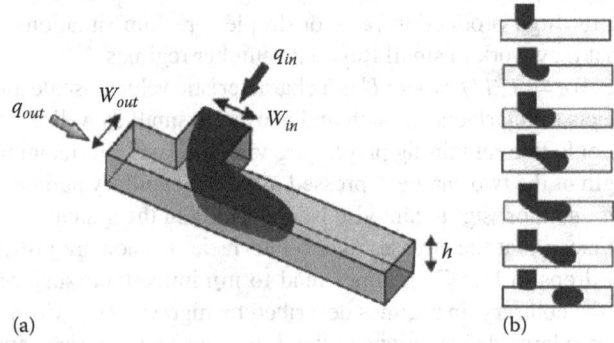

FIGURE 20.3 (a) A typical T-junction with characteristics dimensions (a) and (b) breakup mechanism.

In all three cases, the phase to be dispersed (from hereafter defined dispersed or inner fluid) is driven by a pump or by a fixed pressure boundary condition into a microchannel, where it encounters the immiscible external fluid (continuous or outer fluid).

The geometry of zone where the two fluids meet (the junction), together with the physical properties of the fluids such as viscosities and interfacial tension, and the flow rates of both fluids set the local flow field, where the fluid–fluid interface deforms and eventually leads to drop or bubble production depending on the phases involved. The size of the droplet is set by a competition between the already mentioned players, the pressure due to the external flow and viscous shear stresses, and the capillary Laplace pressure resisting deformation of the interface.

Among all dimensionless numbers, the most important is the capillary number Ca based on the mean continuous phase velocity, usually ranging between 10^{-3} and 10^1 in most microfluidic devices, while additional dimensionless parameters are the ratio between flow rates $Q_r = q_{in}/q_{out}$ and between viscosities $\lambda = \mu_{in}/\mu_{out}$, and the geometric dimensional ratio typically given by the ratio of channel widths $wr = w_{in}/w_{out}$.

T-junctions belong to the first class of flow topology, while focusing and cofocusing device to class (2) and (3) (see Figures 20.3 and 20.4).

The latter can be further classified in mostly 2D devices (planar or nonplanar) and 3D coaxial injectors whose working principle is related to the Rayleigh–Plateau instability described in following sections.

20.3.1 T-Junctions

T-junctions are the most common and simple devices developed to generate droplets in a controlled way.

Figure 20.3a illustrates schematically the geometry of a T-junction, where two channels merge at a right angle.

Microfluidic Encapsulation Process

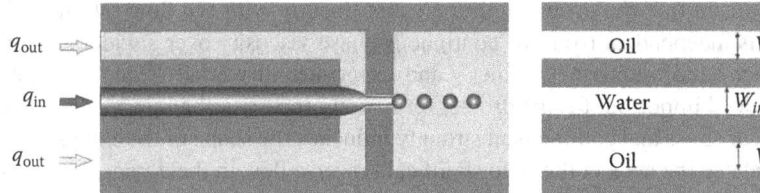

FIGURE 20.4 A typical flow-focusing device for oil/water emulsions with characteristics dimensions.

In the main channel flows the continuous fluid, while the orthogonal channel introduces the fluid that will be dispersed forming droplets. If both channels have rectangular cross sections, geometry can be defined by three parameters configuring topology of the T-junction: the width w_{out} of the main channel, the width w_{in} of the channel supplying the secondary fluid, and the height h of the two channels; this elementary configuration could be modified in order to use, for example, different channel heights or different channel connections.

The usual process of formation of droplets in the T-junction geometries can be described as follows: the two immiscible fluids form an interface at the junction of the main channel and the orthogonal inlet.

The stream of the discontinuous phase penetrates into the main channel flow, and a droplet begins to grow, while the pressure gradient and the flow in the main channel start to warp the droplet in the downstream direction.

At this point, the interface on the upstream part of the droplet moves downstream, and when the interface approaches the downstream edge of the lateral inlet, the little connection of the inlet channel with the droplet breaks.

The disconnected liquid plug flows downstream in the main channel starting the regularization process, while the tip of the stream of the discontinuous phase retracts to the end of the inlet, and the process repeats.

Following the early work by Thorsen et al.,[25] focused on the formation of monodisperse aqueous droplets in an organic carrier fluid performed on a microfluidic chip, and then followed by others works,[26–29] the breakup mechanism responsible of droplet formation was later analyzed by Garstecki et al.[30] showing that when Q_r is order of 1 the dominant contribution to the dynamics of breakup at low capillary numbers is not dominated by shear stresses, but it is driven by the pressure drop across the emerging droplet.

Three operational regimes can be identified: the squeezing, the dripping, and the jetting regimes.

At low values of the capillary number (typically $Ca < 10^{-2}$), formation of droplets follows the squeezing model, at intermediate values of Ca the device operates in the dripping mode in which the viscous effects become more important and, at highest flow rates, the system develops a long jet and droplets are sheared off due to fluid dynamics instabilities effects.[31]

The dynamics of droplet breakup in squeezing regime can be qualitatively described in the following way. as the tip of the dispersed phase enters the main channel, it starts to fill the main cross section, while the hydraulic resistance to flow in the thin films, between the dispersed phase interface and the walls of the obstructed microchannel, creates an additional pressure drop along the growing droplet.

This pressure drop has a strong influence on the dynamics of breakup, as once the main channel is obstructed by the growing droplet, the upstream interface of the droplet is pushed downstream by the continuous fluid. Once the droplet interface is pushed against the downstream wall of the channel, the neck connecting the orthogonal stream of dispersed fluid with the droplet narrows and eventually breaks, and the shaped droplet is released.

When the dispersed phase flow rate becomes larger than the continuous phase flow rate, the squeezing regime further evolves into the formation of stable parallel flowing streams, starting the dripping regime.

Garstecki et al. predicted that the drop length increases linearly with the flow rate ratio[30] and that the droplet length is independent from the continuous phase viscosity over a wide range of oil viscosities. On the other hand, numerical studies[32] and experimental works[33,34] demonstrate that the viscosity ratio is indeed important for the droplet formation process in the intermediate regime ($Q_r < 1$), where both shear stress and confinement strongly influence the shape of the emerging droplet. Van Steijn et al.[35] related the neck collapse to significant reverse flow in the corners between the phase to be dispersed and the channel walls.

Transition from squeezing to dripping regime was analyzed by De Menech et al.[36] and Christopher et al.[34]

Although shear stress effects do play a role in both squeezing and dripping regimes, results showed that these effects are secondary in the squeezing regime, and the squeezing model approximates the volume of the droplets quite well. Both the shear stress exerted by the continuous fluid on the growing droplet and the pressure drop previously described are important in the dripping regime, and the scaling of the volume of the droplet exhibits in this case a dependence on the value of the capillary number.

In T-junctions, the strong effects of confinement exerted by the presence of walls in the microchannels, coupled with the importance of the evolution of the pressure field during the process of formation of a droplet, confers a quasistatic character of the collapse of the dispersed streams, while the separation of time scales between the slow evolution of the interface during breakup and fast equilibration of the shape of the interface itself via capillary waves, and of the pressure field in the fluids via acoustic waves, are the basis of the observed monodispersity of the droplets and bubbles formed in microfluidic systems at low values of the capillary number.

20.3.2 FOCUSING AND COFLOW-FOCUSING DEVICES

Flow-focusing systems are a class of microfluidics devices working in a different way in respect to the T-junctions and, qualitatively, their operation can be described in the following way: two immiscible phases (e.g., a gas and a liquid, or two immiscible liquids like water and oil) are driven via their inlet channels to the flow-focusing junction (Figure 20.4).

In this junction, the central channel delivering the fluid to be dispersed ends upstream of an orifice. From the two sides of the central channel (or the annulus around the central capillary in the case of coaxial glass capillaries configuration[37]), additional streams deliver the continuous fluid.

The boundary conditions for the fluids can be assigned either using constant flow rate or fixed pressure applied to the inlet. In both cases,[38–40] the pressure gradient along the central axis of the device pushes the two immiscible fluids through the orifice: the tip of the inner phase enters the orifice and starts to inflate a bubble in case of gas (or fill a liquid droplet) growing downstream of the orifice.

At the same time, in the orifice, the inner stream begins to thin, and a neck is formed, due to the restriction generated by the continuous fluid passing by, and eventually breaks, releasing a bubble or a droplet and retracting upstream to its original position, where the process starts again. The range of pressures (or flow rates) applied to the boundaries generating a periodic dynamics depends on the particular device, coupling the dynamics of flow in the orifice and the flow in the outlet channel.[41]

The droplets produced using focusing flows are strongly monodispersed, presenting a standard deviation of their diameters usually below of 5% of the mean value.

Squeezing, dripping, and jetting are the basic regimes involved also in flow-focusing microfluidic device.[165]

In the case of a gas–liquid system, squeezing of a gas bubble confined in a fluid can be described using a quasistatic model.

When the tip of the gaseous phase enters the orifice, most of the cross section of the microchannel is filled.

Microfluidic Encapsulation Process

The interfacial forces dominate the shear stress, and the capillary number is low so that the tip assumes a compact shape minimizing the surface, restricting the flow of the continuous liquid to thin layers between the interface and the walls of the orifice.

The flow in these thin layers is subject to an increased viscous dissipation and resistance, opposing the liquid inflowing from the inlet channels to pass through the orifice.

At this point, the pressure upstream of the orifice rises, and the liquid squeezes the neck of the gas stream. Setting a fixed flow rate Q_r for the continuous liquid, this squeezing proceeds at a rate proportional to Q_r and independent of pressure, viscosity of the liquid, and the value of interfacial tension. The quasistatic model has been confirmed in detailed experiments by Dollet et al.,[22] and it can explain the observed monodispersity of the bubbles as collapse of bubbles and perturbations occur at different time scales: if the collapsing speed of the bubbles is much smaller than the capillary speeds and the speed of sound in the liquid, any perturbations in flow given by variation of pressure or shape of the interface are balanced on time scales much shorter than the interval required for the formation of a single bubble.

The quasistatic model of breakup can be extended also to the liquid–liquid systems, where the framework is little more complicated because the viscosities of both of the liquids play an important role in the process, and the shear stresses can be transferred between them (Figure 20.5a[41]).

Further, the quasistatic model will work only in the regimes where the capillary speed is larger than the characteristic time for breakup, which is related to the time needed for filling the volume of the orifice with the continuous fluid.

Depending on the flow rates of the two immiscible liquids and on the rheological properties of the dispersed phase, emulsification occurred in two different modes also in dripping regime as it is strongly influenced by the geometry of the channels, and we can distinguish different submodes of operation.

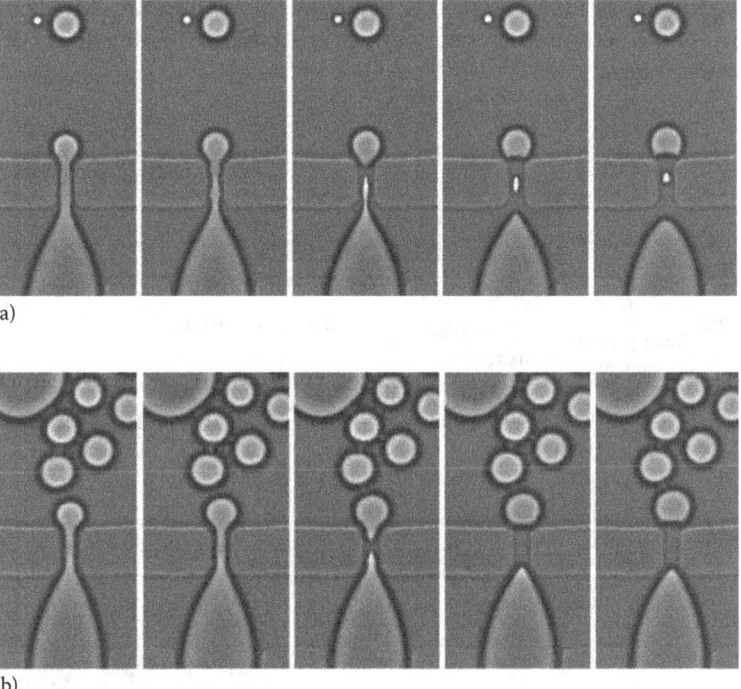

FIGURE 20.5 The quasistatic model of droplet breakup for liquid–liquid systems: (a) squeezing mode and (b) dripping mode. (Reproduced with permission from Anna, S.L. et al., Formation of dispersions using "flow focusing" in microchannels, Carnegie Mellon University, Pittsburgh, PA, 2003.)

At lower flow rates of the inner phase, we distinguish a mode in which the thread broke in the orifice, and after each breakup, the tip of the discontinuous stream retracted upstream of the orifice, to the end of its inlet channel (squeeze), and the thread typically occupies most of the cross-section of the orifice. At higher flow rates of dispersed phase, the breakup occurs either in the orifice or slightly upstream of it, but the tip of the discontinuous stream did not retract to its inlet, which remains at the inlet to the orifice, feeding immediately the orifice for a successive breakup (see Figure 20.5b).

In this mode, the inner phase did not always occupy completely the cross-section of the orifice, and further increasing the flow rates is focused in the orifice by the continuous flow.[42]

In the squeezing mode, the breakup obeys the quasistatic model, with only a slight dependence of the diameter of the droplets on the value of the capillary number Ca.[43]

A particular class of focusing device is represented by coaxial injectors (Figure 20.6), first implemented using microcapillaries in the context of microfluidics by Cramer et al.[44]

They showed that the breakup of the liquid stream into droplets could be separated into two distinct regimes: dripping, in which droplets pinch off near the capillary's tip, and jetting in which droplets pinch off from an extended jet downstream of the tube tip. The transition from dripping to jetting is regulated by a critical value U^* of the continuous phase velocity, which decreases as the flow rate of the dispersed phase increases.

This critical value was also found to depend on the viscosities ratio λ, as well as on the interfacial tension γ.

The work of Cramer was confirmed simultaneously by Utada et al.[45,46] and Guillot et al.[47,48] through stability analyses of viscous stream confined within a viscous outer liquid in a microchannel.

Both groups reached the conclusion that the transition from dripping to jetting is a transition from an absolute to a convective instability, a terminology that refers to the ability of perturbations to grow and to withstand the mean advection.

While absolute instabilities grow faster than they are advected and contaminate the whole domain yielding to a self-sustained oscillation, convective instabilities are characterized by a dominating advection of the perturbations, and they can be considered as amplifiers of the noise that already exist in the system.[49]

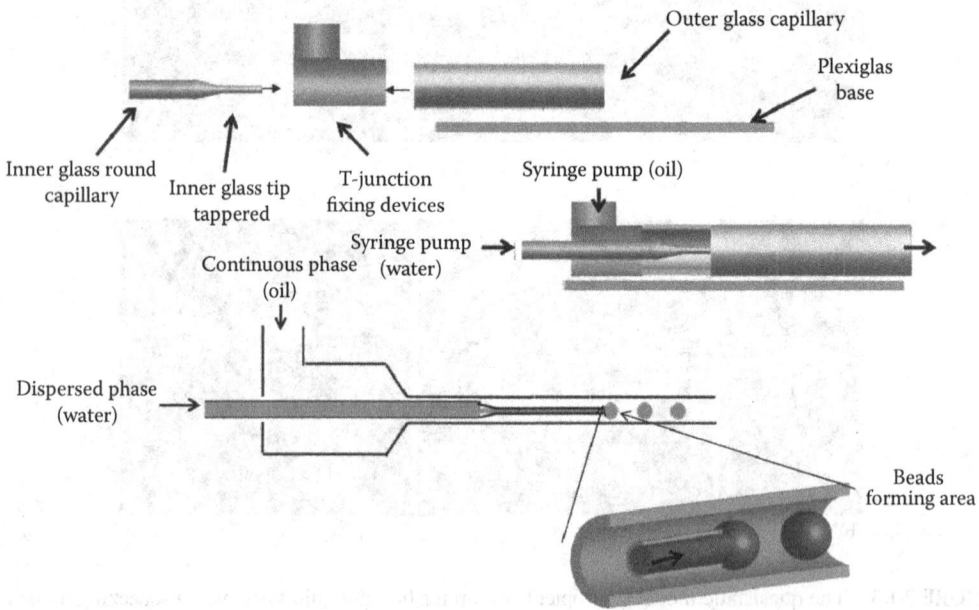

FIGURE 20.6 A typical coaxial capillary assembly for focusing flows.

Microfluidic Encapsulation Process

In the case of axisymmetric devices, the control of instability generating the transition from dripping to jetting regime is the basis of operation of this type of device.

As a matter of fact, a cylindrical jet of fluid is always unstable, because when the cylindrical jet breaks into drops, the surface energy decreases toward a minimum.

Any infinitesimal undulation of the shape of axial profile of the cylinder can be decomposed—via a series expansion—into sinusoidal perturbations, and all of these perturbations, characterized by a length larger than a certain critical value, spontaneously amplify and, as a consequence, the cylinder breaks into droplets.

The speed of this spontaneous breakup can be estimated using dimensional analysis: when the viscous terms dominate the dynamics, then the surface of the cylinder will be imploding with a speed that can be approximated as $u = \gamma/\mu$, where μ is the viscosity of the fluid and γ is the surface tension. Most of the common techniques of emulsification and atomization are based on the Rayleigh–Plateau instability (Figure 20.7): it is enough to deform the immiscible fluid into elongated morphologies, and this deformation will trigger the onset of the Rayleigh–Plateau instability.

The principle behind this drop formation in jetting regime can be easily understood following the aforementioned theory.[50,51] Any perturbation in the jet will result in a slightly thinner region,

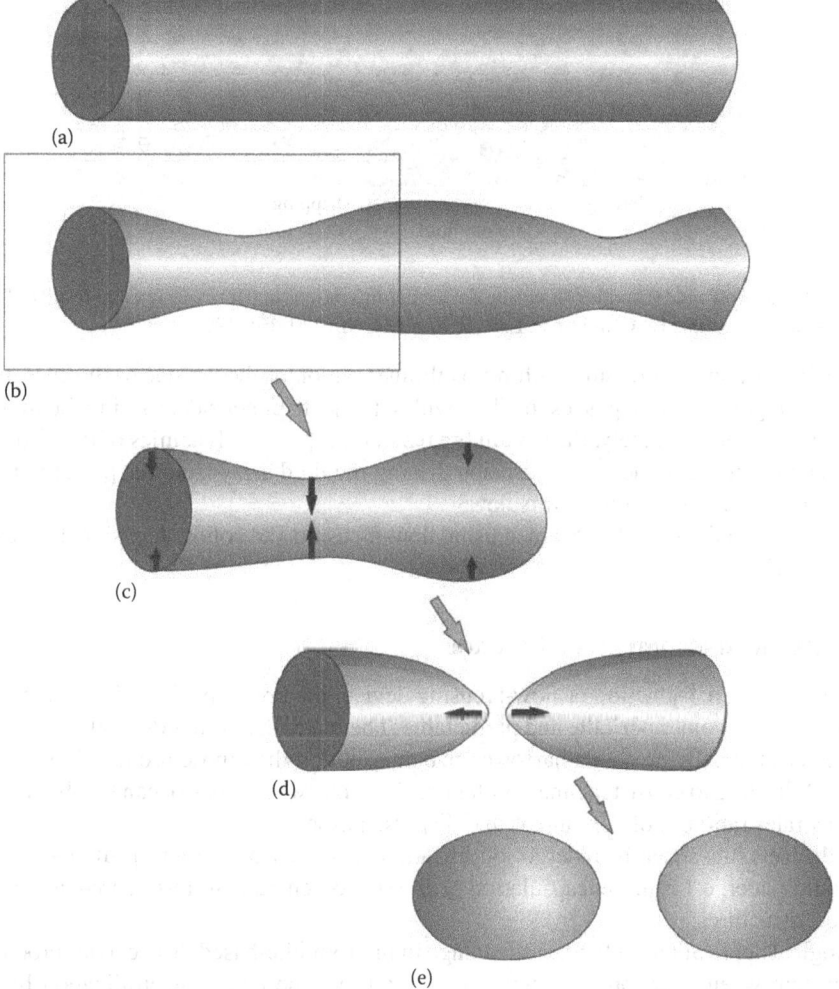

FIGURE 20.7 The physical mechanism of Rayleigh–Plateau instability: the undeformed jet (a) start to deform due to external pressure oscillations (b) surface deformations are amplified by Laplace pressure effects (c) until a jet breakup occurs (d). The new drops eventually recover their spherical shape (e).

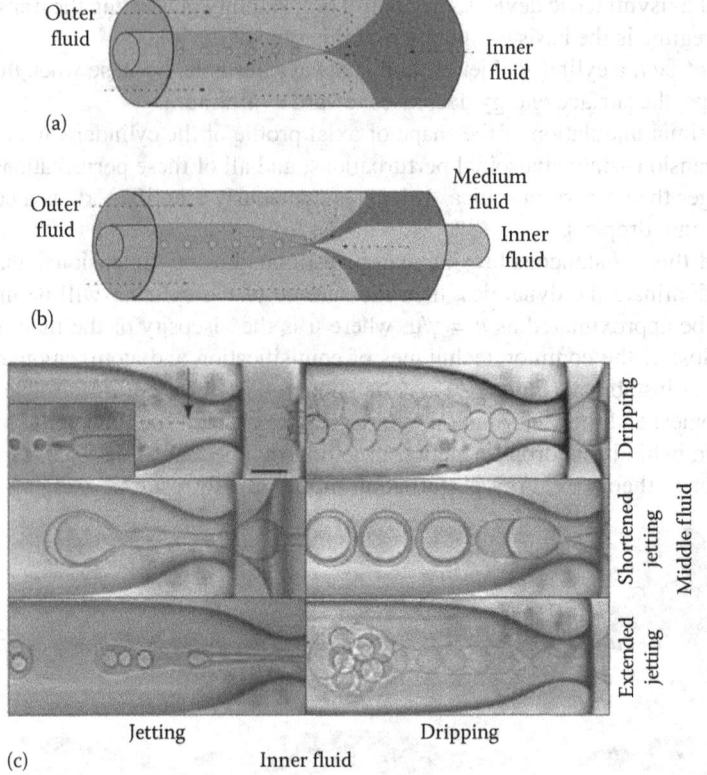

FIGURE 20.8 (a) Flow- and (b) coflow-focusing device configurations and (c) experimental results. (Reproduced with permission from Stone, H.A. and Leal, L.G., *J. Fluid Mech.*, 220, 161, 1990.)

with the rising of Laplace pressure within this thinner region of the jet, due to the curvature of the interface; this higher pressure pushes the fluid within the jet to either side, causing the thin region to become even thinner; ultimately, the stream breaks into drops. The dynamics of this drop pinch off reflects a balance between the surface tension driving the fluid away from the perturbation and the viscous drag of the fluid that resists this flow.

Several configurations can be obtained using flow-focusing and coflow-focusing devices, as illustrated in Figure 20.8.

20.3.3 Size, Mixing, and Shape Control

It is clear that using a T-junction or flow-focusing device, the breakup of the disperse phase by the continuous phase becomes periodic and predictable. The micro- or even NPs produced using microfluidics devices typically present a narrower size distribution than those produced by conventional methods,[52–57] leading to consistent and regular droplets size, where control can be obtained by altering the flow rates ratio Qr_r of continuous and disperse phases.

These droplets subsequently undergo solidification to obtain a matrix encapsulation microsphere, or they could undergo further encapsulation processes to generate uniform core–shell with single or multicore structures.

In biological applications, a further advantage in microfluidic-based drug carrier production systems is present when nanocomplex synthesis is required; the charge neutralization between the negatively charged nucleic acid and the cationic gene carrier is supposed to be more complete by confining the two components in a picoliter droplet, obtaining more uniform and compact nanocomplexes while exhausting any unreacted cationic gene carrier, often cytotoxic.

Microfluidic Encapsulation Process

The efficiency of mixing in microfluidic devices is another important feature to consider, as in hydrodynamic focusing, the fluid stream to be mixed globally described as dispersed phase, flowing along the central microfluidics channel, is confined into a narrow stream delimited by the two adjacent streams at higher flow rates.

For example, diblock copolymers or lipids dissolved in a solvent self-assemble as solubility decreases via solvent dilution, obtained through mixing with buffer in adjacent streams.

This process takes place in micron scale, but as a matter of fact, the efficiency of mixing by diffusion is length dependent, and mixing time drops significantly as diffusion distance falls below 100 µm,[1] while in reactors with a characteristic length scale larger than 1 mm, mixing by diffusion is not very efficient.

For this reason, microfluidic channels are usually designed with width of 100 µm or less, representing an efficient reactor for rapid mixing of fluids, transforming the stages of carrier fabrication into a self-assembly (nucleation, growth by aggregation, and stabilization after a characteristic time scale) in a well-controlled automatic process.

Another important feature to consider in order to improve the effectiveness of controlled release is the shape control, and some recent works showed the effect of particle shape on drug carrier or functional food performance because of its influence on tissue biodistribution and endocytic uptake.[58,59]

Although conventional soft lithography techniques can be used to fabricate nonspherical particles, the production rate is limited by the mold size, limiting mass production due to the increased complexity in parallel configuration setup, and a possible solution involves the combined use of a microfluidic device and microscope projection lithography, so that the fabrication process for nonspherical objects could be transformed into a continuous.[60]

In this approach, a mask with desired geometry is inserted in the microscope to generate a mask-defined UV light beam projection on the monomer stream flowing in a polydimethylsiloxane (PDMS) microfluidics device described earlier.

Particles with desired shape are polymerized and transported along with the unpolymerized monomer stream, and in this way, particles with a variety of shapes can be efficiently generated.

20.3.4 Shell Thickness and Loading Efficiency Control

Another important target in microcapsule production is the shell thickness control, as the shell represents a selective diffusion barrier to control the sustained release from the capsule of the active principle contained inside. If the shell thickness is irregular, then the drug release profile associated with microcapsules would be inhomogeneous, an undesirable behavior. In the double emulsion approach, the thickness of the capsule shell can be easily controlled via optimizing the design of microfluidics device or simply altering the flow rate adopted in the encapsulation process. The use of microfluidic platform, for example, allowed to produce PLA microcapsules with a shell thickness as thin as 80 nm, a thickness that is otherwise not achievable in bulk preparation process.[61] Moreover, the precise control of the capsule shell thickness allowed by microfluidics enables fine tuning of the drug release profile of microcapsules in different controlled drug-release applications.

In conventional fabrication methods for microspheres, the drug encapsulation efficiency is ranging from 50% to 90%,[56] while several studies investigating the drug-loading efficiency associated with a microfluidics production technique reported a consistent encapsulation efficiency of 95% or greater.[7] The result is linked to generation of isolated single emulsion droplets during microparticle or microcapsule synthesis, where no drug is lost from the droplets. In the case of nanodrug delivery systems synthesis, one study reported an encapsulation efficiency of 21%–45% for PLGA–polyethylene glycol (PEG) NPs fabricated in bulk (by emulsification solvent diffusion), while a microfluidics approach achieved an encapsulation efficiency of 28%–51%.[54] The effect of different mixing rates can explain the difference: when the mixing rate is slow, the time scale of NPs assembly is smaller than the time scale of solvent diffusion, meaning that there is not enough time for the solvent to work. Thus, in bulk mixing, NPs start assembling when the solvent concentration is still high and results in drugs escaping the encapsulation process. In the microfluidics approach, the time scales of the two processes are comparable, and hence more drugs can be encapsulated within the carrier, providing a significant advantage dealing with expensive drugs.

20.3.4.1 Drug-Release Rate Control
While most of the work already done focused on the successful demonstration of microfluidics fabrication of different types of delivery vehicles, little effort (for the complexity required) has been devoted to characterization of the drug release kinetics profile of the particles generated from a microfluidics versus the conventional approaches. Xu et al. compared the kinetics of drug release of PLGA microparticles generated from the two approaches.[62] Results showed that the monodisperse particles prepared using microfluidics devices release drug more slowly and have a smaller initial burst effect than those observed in conventional polydispersed particles. The effect to more uniform drug distribution in the microfluidics-produced particles could be the reason of the observed behavior; hence, drug release by degradation of the former would be more gradual. Similarly, Karnik et al. reported the microfluidic-aided self-assembled PLGA–PEG NPs had slower and smaller initial burst of drug release than those fabricated by conventional approaches.[54] Again, rapid mixing feature pertaining to microfluidics approach may lead to a more uniform drug distribution inside the particles so that drug release rate is more stable given by uniform degradation of the assembly. Once again, these studies suggest the superiority of the use of microfluidics production techniques over conventional approaches in controlling the drug release rate. For a review about microfluidic application in drug delivery systems, see Reference 63.

20.4 DEVICE MATERIALS AND CONSTRUCTION TECHNIQUES
Nowadays, most of the microfluidic devices are fabricated in either glass,[13,64–76] providing some advantages in particular when aggressive chemical reaction are involved by the presence of solvents, or PDMS,[2,4,10,12,15,17,52–55,56,62,77–85] but the use of silicon,[86–88] perfluoropolyether,[89] poly(methyl methacrylate) (PMMA),[5,6] polyurethane, polycarbonate,[89] and stainless steel[14] microfluidic devices is also reported.

The microfabrication techniques used in construction of microfluidic devices can be broadly classified into two groups: a class of techniques based on photolithography and another group where devices are produced starting from a mold and that we can define replication based.

In the first group, the active principle is the light used to define patterns on a photosensitive material, and the resolution that can be achieved depends on the light wavelength. The final result, however, also strongly depends on the limitation of optical components and on material properties such as numerical aperture and the polarity of photoresist as well.

In some cases, the photosensitive material itself can be used as a structural component of the device, while in other approaches, it can be used to transfer the pattern onto another structural material.

In the replication method, a master mold, which could be of any material, is made using either the photolithographic process or traditional micromachining processes, strong enough to resist the operating conditions of the fabrication process, and then it is used to replicate the pattern onto another softer material by direct physical contact. The choice between the two fabrication methods depends on various factors such as the size, the cost, the desired substrate, the chemical properties of the flows, the speed, the smoothness, and the required section geometry.

20.4.1 PHOTOLITHOGRAPHY-BASED MICROFABRICATION
Photolithography is a technique consisting in the use of light to define features on a photosensitive material. This technique, first develop in semiconductors applications along with manufacturing process of thin film deposition and etching, was later transferred to produce microfluidic devices.

Initially adapted to produce microfluidic devices on silicon,[91,92] it was later used also on a glass substrate.[93,94]

In the same years, new methods to fabricate open microchannel structures in silicon or glass substrate using various etching techniques were developed, such as reactive ion etching, hydrofluoric acid, or potassium hydroxide wet etching.

Assembly of the microchannels was obtained using bonding techniques such as hydrophobic silicon bonding,[95] electromagnetic induction heating,[96] fusion bonding,[97] and anodic bonding.[98]

For a review of these early microfabrication techniques, see Stokes and Palmer[99] and others before.[100–107]

The major drawback of these techniques was the high production cost of the substrates and the requirement of microfabrication in a clean room facility, not affordable for disposable microfluidic devices, and this consideration led the researchers to investigate the possibility of using of photodefinable polymers, converting conventional photoresists that have been already used for patterning, in the microelectronics industry into structural elements of microfluidic devices.

Burns et al. applied these conventional photoresists to create manifolds for microfluidic devices,[108] although channel height was limited to $h < 3$ μm.

X-rays were also used earlier in 1986[109] to define shapes on a photoresist using a process known as LIGA (*lithographie, galvanoformung, abformung*), obtaining microstructures with channel height $h > 350$ μm and an aspect ratio of >100:1.

Later, with the introduction of the photoresist SU-8, microstructures could be produced using the standard photolithography process,[110–112] obtaining channel height of 100 μm and an aspect ratio of >10:1. Several typologies of SU-8 with different viscosities can be used to obtain different channel heights.

Since then, various methods have been adopted for fabrication of photoresist-based microfluidic devices. The first method shown in Figure 20.9a begins with a spin coating of photoresist onto a substrate and patterning with a photomask.[113,114] Once the open microchannels are created, a sacrificial material is filled into the space of the microchannel. Subsequently, a second layer of photoresist is spin coated and patterned on top to define the access holes for inlet and outlet. Finally, the sacrificial layer is dissolved to create the closed microchannels. The major disadvantage in this process is the slow dissolution, therefore only short microchannels are applicable.

The second method shown in Figure 20.9b laminates a dry SU-8 or Kapton film on top of the open microchannels.[115] Although this process is relatively simple, the alignment and the bonding strength of lamination could be a challenging problem.

The third method shown in Figure 20.9c uses a double exposure at different wavelengths to create the embedded microchannels in SU-8.[116] The first exposure (365 nm) defines the sidewalls of the microchannels, while the second exposure (254 nm) creates the encapsulation layer for the microchannels, due to its shallower absorption depth.

The exposure with two different wavelengths could be inconvenient in certain circumstances, and a slow dissolution problem similar to that in the first method could appear.

Stereolithography[117,118] and laser ablation are also used as alternative techniques for the microfabrication for microfluidic devices.

The former method is an additive manufacturing process that employs liquid resins and high-intensity light beams to build 3D microstructures, a technique widely used in rapid prototyping. The photo-induced cross-linking happens upon the exposure of the liquid resin.

The process time depends on the complexity of the 3D microstructures; therefore, it is commonly used for prototyping. An extensive review on stereolithography can be found in Reference 119. The latter method is a subtractive manufacturing process that uses a focused high-intensity laser beam to evaporate the material from the surface.[120,121] Laser ablation is mostly used to fabricate microchannels in thermosetting polymers such as polyimide due to its physical properties.[122,123] Microstructures of nanometer scale have been demonstrated,[124] but the surface roughness and properties using laser

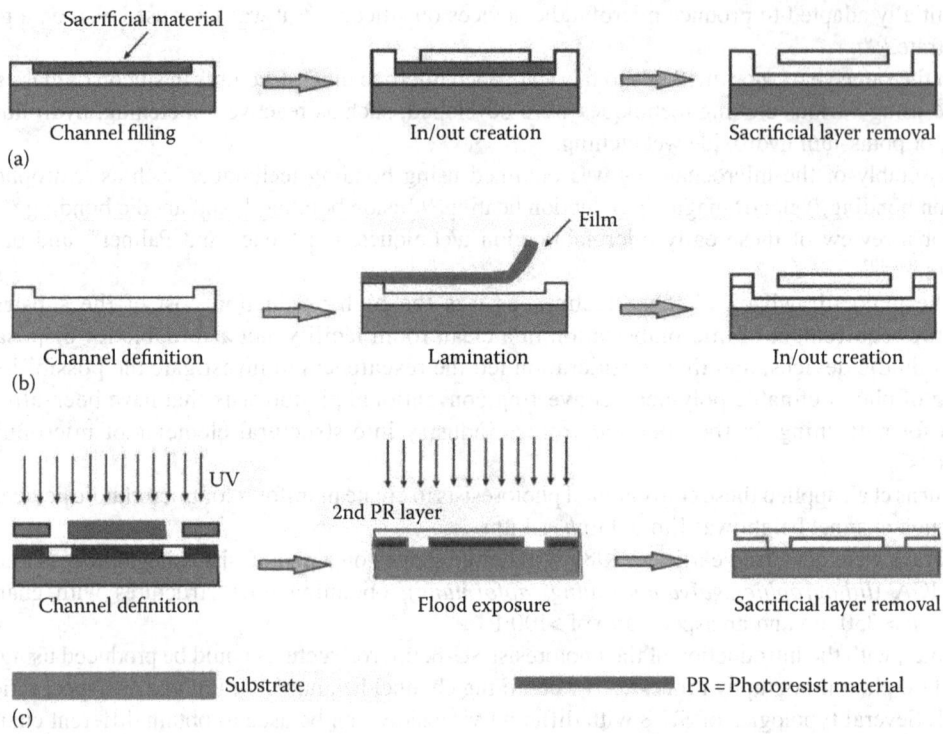

FIGURE 20.9 Photolithography-based microfabrication techniques: (a) sacrificial layer, (b) lamination, and (c) multiwave exposure.

ablation are difficult to control and highly dependent on the manufacturing parameters. An extensive review on laser micromachining can be found in References 135 through 137.

20.4.2 Methods Based on Replication-Based Methods

One of the major advantages of using polymers in fabrication of microfluidic devices is largely linked to low cost and to the possibility to use high-volume-replication methods such as soft lithography, hot embossing, and injection molding. The first step in these methods consists in the preparation of a master (also called a mold), which is an essential part in all replication methods, and it can be fabricated through conventional photolithography, laser ablation, silicon etching, LIGA, or microelectrode discharge machining. The correct choice of mold fabrication method depends on several parameters, like the available material, resolution, aspect ratio, and the processing conditions.

20.4.2.1 Soft Lithography

Soft lithography process, due to its simple process, excellent material properties, low manufacturing cost, and high replicating accuracy, has become one of the main rapid prototyping methods[128] used in microfluidics. Commonly used materials, such as Sylgard 184 by Dow Corning and RTV 615 by Elastosil, are two components mixture, a base elastomer and a curing agent. The mixture is degassed before use to prevent the formation of air bubbles and then cast on the mold as shown in Figure 20.10.

After curing, the elastomer is peeled off from the mold and bonded to a glass slide or other polymer sheets to create closed microchannels. Various methods have been developed to increase the bonding strength for each polymer. For example, a treatment with oxygen or air plasma can lead to a permanent bond between two PDMS layers or PDMS and glass. The mold can be used to produce numerous devices, and a replication accuracy of 10 nm feature has been demonstrated using this

Microfluidic Encapsulation Process

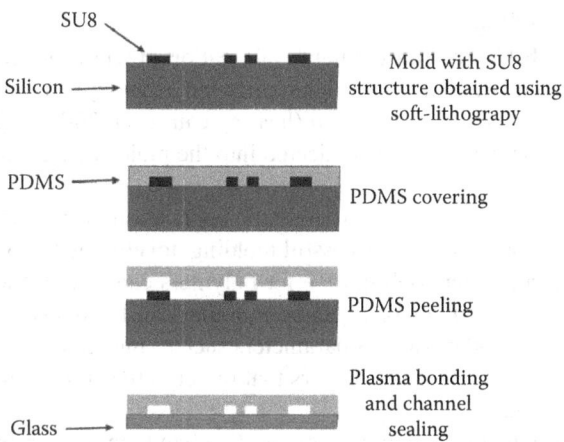

FIGURE 20.10 Soft lithography replication method using polydimethylsiloxane process flow.

method.[129] Other polymers, such as polyurethane and polyester, can also be used in casting microfluidic structures since they can be cured through temperature or UV exposure.[130] An extensive review on soft lithography can be found here.[131,132]

20.4.2.2 Hot Embossing

Hot embossing technique had been commonly used for the microstructuring of polymers in industry. Before, it was adapted for microfluidic applications due to its relatively simple process, wide selection of materials, and availability of facility.

As shown in Figure 20.11, the microstructures are transferred from the master to the polymer by stamping the master into the polymer, which is previously softened by heating above its glass transition temperature. This method is limited to thermoplastic polymers, and the technique has been used successfully on a variety of polymers, including polycarbonate,[133] polyimide,[134] cyclic olefin copolymer,[135] and PMMA.[136] The main parameters to control are the surface quality, temperature uniformity, and chemical compatibility of the master.

Using hot embossing, a replication accuracy of a few tens of nanometers can be achieved.[137–139] An implementation of hot embossing is known as nanoimprinting, where the feature size of the device ranges from a few tens to hundreds of nanometers. An extensive review of hot embossing and nanoimprint lithography can be found in Matthias[140] and Worgull.[141,142]

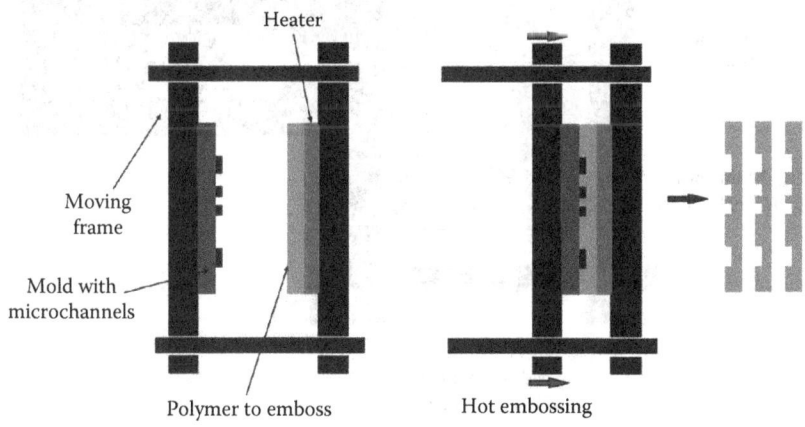

FIGURE 20.11 Hot embossing process flow.

20.4.2.3 Injection Molding

Injection molding is probably the most suitable fabrication process for polymers in high-volume production, due to its fast process time and high replication accuracy. As shown in Figure 20.12, the polymer is fed into a heated screw and melted (heating can reach 200°C–350°C depending on the melting temperature of polymer). It is then injected into the mold cavity, usually made in steel, and cooled to form a replica.

Due to the dimension involved, in the microfluidic application, filling without trapping air bubbles is one of the key requirements for successful molding, together with cooling time.

This is because an insufficient cooling time for polymers can lead to thermal stress and defect formation, although the cycle times for injection molding can be as fast as several seconds. The process control involves also other various parameters such as injection pressure, molding temperatures and their duration, and cycle times, factors that increase the complexity of application of this technology to a micron scale.

This technology presents pro and cons: a major disadvantage is the relatively high cost of the mold material, usually a special steel requiring sophisticated micromachinery work, to ensure it is capable of withstanding the high-temperature injection process: any mistake requires a new mold from scratch. On the contrary, the major advantage of injection molding over other replication methods, once the system is set, is its ability to form 3D microstructures without any geometrical constraint, including circular channels, using both thermosetting and thermoplastic polymers. Moreover, additional microfluidic components required for the device, such as interconnections and connectors, can also be integrated together in one piece,[143,144] reducing the effort of bonding parts together. An extensive review on injection molding can be found in References 145–147.

The fabrication techniques previously described are the most commonly used methods for fabrication of microfluidic devices but, in addition, also other methods, such as microthermoforming,

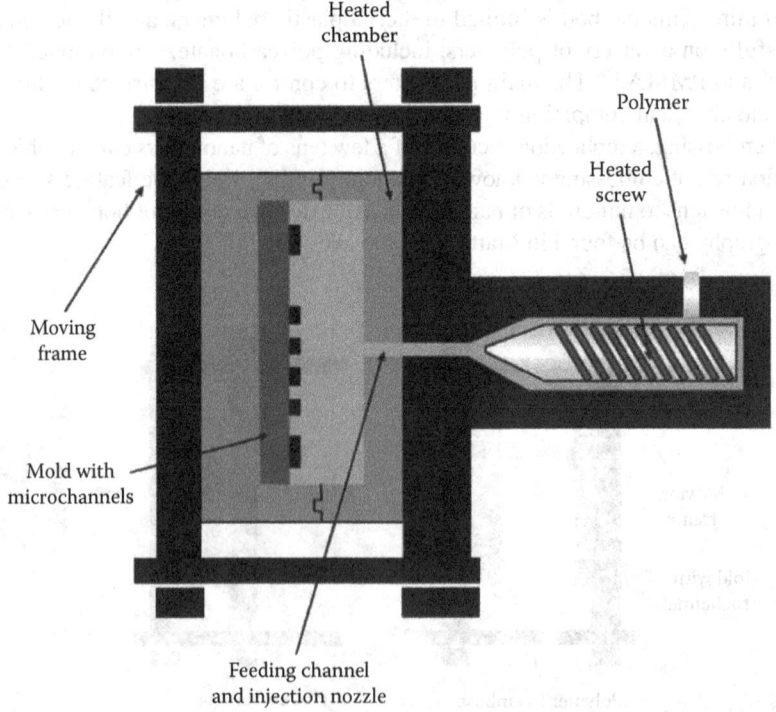

FIGURE 20.12 Injection molding device production process flow.

Microfluidic Encapsulation Process

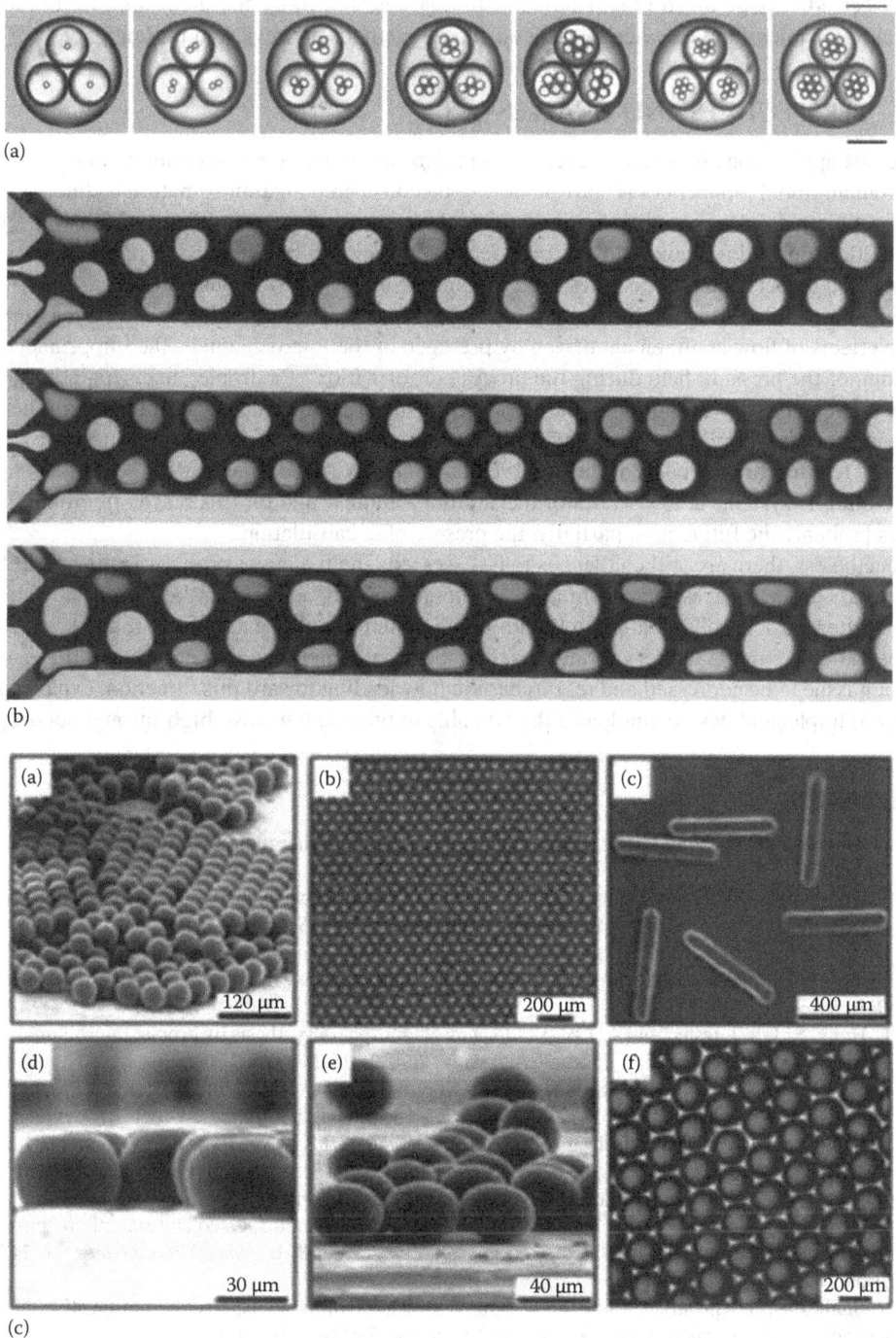

FIGURE 20.13 Examples of multiple emulsions formed in microfluidic systems: (a) multiple shells–multiple cores configurations of monodisperse triple emulsions made with cascaded microcapillary devices results. (Reproduced with permission from Utada, A.S. et al., *Bull. MRS*, 32(09), 702, 2007.), (b) composite emulsion formed by droplets of different composition and different volumes. (Reproduced with permission from Hashimoto, M. et al., *Small*, 3(10), 1792, 2007.), and (c) examples of anisotropic particles formed by either polymerization (spheres and disks, rods) of droplets of monomer or thermal setting of droplets. (Reproduced with permission from Xu, S. et al., *Angew. Chem. Int. Ed. Engl.*, 44(5), 724, 2005.)

microelectrodischarge machining, LIGA machining, micromilling, and precision machining have been used for microfluidic device fabrication. These methods are reviewed in References 148–152.

20.5 PERSPECTIVES AND CHALLENGES

Numerous applications have been developed in microfluidic-based encapsulation, and the technology from an initial pioneer era is starting now to be considered an almost mature technology.

A graphic picture of possible configurations obtained using microfluidic devices is reported in Figure 20.13, including liquid and solid particles, and they can be extended to production of more complicated objects and architectures such as multiple emulsions or Janus particles.

Parameters influencing the droplet production have been focused, and they can be reassumed in strong effects of flow confinement driven by presence of the microchannels, the importance of the evolution of the pressure field during the process of formation of a droplet, and separation of time scales between the slow evolution of the interface during breakup and fast equilibration of the shape of the interface via capillary waves and of the pressure field in the fluids via acoustic waves.

These features are the basis of the observed high-quality monodispersity of the droplets formed in microfluidic systems at low values of the capillary number and for this reason microfluidic represents probably the future and, partially, the present of encapsulation.

Nonetheless, there are still challenges to be assessed, and the major problem is the possibility to obtain mass production, as some techniques are not feasible for an easy parallelization, and cleaning process after particle solidification in liquid–liquid systems could be a problem too.

The scale up from laboratory to industrial applications is not straightforward, and this is one of the main issue to be addressed and researches are now leading toward this direction, exploring possibility to implement new technologies that are able to provide low-cost, high-throughput droplets.

REFERENCES

1. A. De Mello, Control and detection of chemical reactions in microfluidic systems, *Nature*, 442, 394–402, 2006.
2. T. Watanabe and Y. Kimura, Continuous fabrication of monodisperse polylactide microspheres by droplet—Particle technology using microfluidic emulsification and, *Soft Matter*, 7, 9894–9897, 2011.
3. D. Jagadeesan, I. Nasimova, I. Gourevich, S. Starodubtsev, and E. Kumacheva, Microgels for the encapsulation and stimulus-responsive release of molecules with distinct polarities, *Macromolecular Bioscience*, 11, 889–896, 2011.
4. L. Hung, S. Teh, J. Jester, and A. Lee, PLGA micro/nanosphere synthesis by droplet microfluidic solvent evaporation and extraction approaches, *Lab on a Chip*, 10, 1820–1825, 2010.
5. Q. Xu, M. Hashimoto, T. Dang, T. Hoare, D. Kohane, G. Whitesides, and R. Langer, Preparation of monodisperse biodegradable polymer microparticles using a microfluidic flow-focusing device for controlled drug delivery, *Small*, 5, 1575–1581, 2009.
6. J. Xu, Z. H., W. Lan, and G. Luo, A novel microfluidic approach for monodispersed chitosan microspheres with controllable structures, *Advanced Healthcare Materials*, 1, 106–111, 2012.
7. J. Xu, S. Li, C. Tostado, W. Lan, and G. Luo, Preparation of monodispersed chitosan microspheres and in situ encapsulation of BSA in a co-axial microfluidic device, *Biomedical Microdevices*, 11, 243–249, 2009.
8. R. Shah, J. K., J. Agresti, D. Weitz, and L. Chu, Fabrication of monodisperse thermosensitive microgels and gel capsules in microfluidic devices, *Soft Matter*, 4, 2303–2309, 2008.
9. J. Kim, A. Utada, A. Fernandez-Nieves, Z. Hu, and D. Weitz, Fabrication of monodisperse gel shells and functional microgels in microfluidic devices, *Angewandte Chemie International Edition in English*, 46, 1819–1822, 2007.
10. S. D. Seiffert, Controlled fabrication of polymer microgels by polymer analogous gelation in droplet microfluidics, *Soft Matter*, 6, 3184–3190, 2010.
11. B. L. J. Q. S. Huang, Microfluidic synthesis of tunable poly-(N-isopropylacrylamide) microparticles via PEG adjustment, *Electrophoresis*, 32, 3364–3370, 2011.

12. W. Chen, Y. Yang, C. Rinadi, D. Zhou, and A. Shen, Formation of supramolecular hydrogel microspheres via microfluidics, *Lab on a Chip*, 9, 2947–2951, 2009.
13. D. Breslauer, S. Muller, and L. Lee, Generation of monodisperse silk microspheres prepared with microfluidics, *Biomacromolecules*, 11, 643–647, 2010.
14. D. Ogonczyk, M. Siek, and P. Garstecki, Microfluidic formulation of pectin microbeads for encapsulation and controlled release of nanoparticles, *Biomicrofluidics*, 5, 13405, 2011.
15. L. Kesselman, S. Shinwary, P. Selvaganapathy, and T. Hoare, Synthesis of monodisperse, covalently cross-linked, degradable "smart" microgels using microfluidics, *Small*, 8, 1092–1098, 2010.
16. B. De Geest, J. Urbanski, T. Thorsen, J. Demeester, and S. De Smedt, Synthesis of monodisperse biodegradable microgels in microfluidic devices, *Langmuir*, 21, 10275–10279, 2005.
17. I. Lee, Y. Y., Z. Cheng, and H. Jeong, Generation of monodisperse mesoporous silica microspheres with controllable size and surface morphology in a microfluidic device, *Advanced Functional Materials*, 18, 4014–4021, 2008.
18. V. G. Levich, *Physicochemical Hydrodynamics*, Prentice Hall, Englewood Cliffs, NJ, 1962.
19. C. D. Eggleton, T.-M. Tsai, and K. J. Stebe, Tip streaming from a drop in the presence of surfactants, *Physical Review Letters*, 87(4), 48302, 2001.
20. H. A. Stone and L. G. Leal, The effects of surfactants on drop deformation and breakup, *Journal of Fluid Mechanics*, 220, 161–186, 1990.
21. G. I. Taylor, The formation of emulsions in definable fields of flow, *Proceedings of the Royal Society of London, Series A*, 146, 501–523, 1934.
22. B. Dollet, W. van Hoeve, J. P. Raven, P. Marmottant, and M. Versluis, Role of the channel geometry on the bubble pinch-off in flow-focusing devices, *Physical Review Letters*, 100, 2008.
23. L. Ménéetrier-Demble and P. Tabeling, Droplet breakup in microfluidic junctions of arbitrary angles, *Physical Review E*, 74, 035303, 2006.
24. G. F. Christopher and S. L. Anna, Microfluidic methods for generating continuous droplet streams, *Journal of Physics D: Applied Physics*, 40, 319–336, 2007.
25. T. Thorsen, R. W. Roberts, F. H. Arnold, and S. Quake, Dynamic pattern formation in a vesicle-generating microfluidic device, *Physical Review Letters*, 86, 4163–4166, 2001.
26. J. D. Tice, H. Song, A. D. Lyon, and R. F. Ismagilov, Formation of droplets and mixing in multiphase microfluidics at low values of the Reynolds and the capillary numbers, *Langmuir*, 19, 9127–9133, 2003.
27. S. Okushima, T. Nisisako, T. T., and T. Higuchi, Controlled production of monodisperse double emulsions by two-step droplet breakup in microfluidic devices, *Langmuir*, 20, 9905–9908, 2004.
28. A. Gunther, S. A. Khan, M. Thalmann, F. Trachsel, and K. F. Jensen, Transport and reaction in microscale segmented gas–liquid flow, *Lab on a Chip*, 4, 278–286, 2004.
29. B. Zheng and R. F. Ismagilov, A microfluidic approach for screening submicroliter volumes against multiple reagents by using preformed arrays of nanoliter plugs in a three-phase liquid/liquid/gas flow, *Angewandte Chemie International Edition*, 117, 2576–2579, 2005.
30. P. Garstecki, M. J. Fuerstman, H. Stone, and G. Whitesides, Formation of droplets and bubbles in a microfluidic T-junction—Scaling and mechanism of break-up, *Lab on a Chip*, 6, 437–446, 2006.
31. B. Zheng, L. Roach, and R. Ismagilov, Screening of protein crystallization conditions on a microfluidic chip using nanoliter-size droplets, *Journal of the American Chemical Society*, 125, 11170–11171, 2003.
32. M. De Menech, Modeling of droplet breakup in a microfluidic t-shaped junction with a phase-field model, *Physical Review E*, 73, 031505, 2006.
33. V. van Steijn, M. T. Kreutzer, and C. R. Kleijn, m-PIV study of the formation of segmented flow in microfluidic T-junctions, *Chemical Engineering Science*, 62(24), 7505–7514, 2007.
34. G. Christopher, N. Noharuddin, J. Taylor, and S. Anna, Experimental observations of the squeezing-to-dripping transition in T-shaped microfluidic junctions, *Physical Review E*, 78(3), 36317, 2008.
35. V. van Steijn, C. R. Kleijn, and M. T. Kreutzer, Flows around confined bubbles and their importance in triggering pinch-off, *Physical Review Letters*, 103(21), 214501, 2009.
36. M. De Menech, P. Garstecki, F. Jousse, and H. Stone, Transition from squeezing to dripping in a microfluidic T-shaped junction, *Journal of Fluid Mechanics*, 595, 141–161, 2008.
37. A. Utada, E. Lorenceau, D. Link, P. Kaplan, H. Stone, and D. Weitz, Monodisperse double emulsions generated from a microcapillary device, *Science*, 308(5721), 537–541, 2005.
38. T. Ward, M. Faivre, M. Abkarian, and H. A. Stone, Microfluidic flow focusing: Drop size and scaling in pressure versus flow-rate-driven pumping, *Electrophoresis*, 26, 3716–3724, 2005.
39. J. Raven and P. Marmottant, Microfluidic crystals: Dynamic interplay between rearrangement waves and flow, *Physical Review Letters*, 102, 084501, 2009.

40. M. Sullivan and H. Stone, The role of feedback in microfluidic flow focusing devices, *Philosophical Transactions of the Royal Society*, 366, 2131–2143, 2008.
41. S. L. Anna, N. Bontoux, and H. A. and Stone, Formation of dispersions using "flow focusing" in microchannels, Carnegie Mellon University, , Pittsburgh, PA, 2003.
42. Z. Nie, M. Seo, S. Xu, P. Lewis, M. K. Mok, G. Whitesides, P. Garstecki, and H. Stone, Emulsification in a microfluidic flow-focusing device: Effect of the viscosities of the liquids, *Microfluidics and Nanofluidics*, 5, 585–594, 2008.
43. W. Lee, L. Walker, and S. Anna, Role of geometry and fluid properties in droplet and thread formation processes in planar flow focusing, *Physics of Fluids*, 21(3), 032103, 2008.
44. C. Cramer, P. Fischer, and E. J. Windhab, Drop formation in a coflowing ambient fluid, *Chemical Engineering Science*, 59(15), 3045–3058, 2004.
45. A. S. Utada, A. Fernandez-Nieves, H. A. Stone, and D. A. Weitz, Dripping to jetting transitions in coflowing liquid streams, *Physical Review Letters*, 99, 094502, 2007.
46. A. S. Utada, A. Fernandez-Nieves, J. M. Gordillo, and D. A. Weitz, Absolute instability of a liquid jet in a coflowing stream, *Physical Review Letters*, 100, 014502, 2008.
47. P. Guillot, A. Colin, A. S. Utada, and A. Ajdari, Stability of a jet in confined pressure-driven biphasic flows at low Reynolds numbers, *Physical Review Letters*, 99, 104502, 2007.
48. P. Guillot, A. Colin, and A. Ajdari, Stability of a jet in confined pressure-driven biphasic flows at low Reynolds number in various geometries, *Physical Review E*, 99, 104502, 2007.
49. P. Huerre and P. A. Monkewitz, Local and global instabilities in spatially developing flows, *Annual Review of Fluid Mechanics*, 22(1), 473–537, 1990.
50. J. Plateau. *Acad. Sci. Bruxelles Mem.*, XXIII, 5(1849).
51. L. Rayleigh, On the instability of jets. *Proceedings of the London Mathematical Society*, 10(4), 4–13, 1879.
52. R. Karnik, F. Gu, P. Basto, C. Cannizzaro, L. Dean, W. Kyei-Manu, R. Langer, and O. Farokhzad, Microfluidic platform for controlled synthesis of polymeric nanoparticles, *Nano Letters*, 8, 2906–2912, 2008.
53. N. Kolishetti, S. Dhar, P. Valencia, L. Lin, R. Karnik, S. Lippard, R. Langer, and O. Farokhzad, Engineering of self-assembled nanoparticle platform for precisely controlled combination drug therapy, *Proceedings of the National Academy of Sciences of the United States of America*, 107, 17939–17944, 2010.
54. P. Valencia, P. Basto, L. Zhang, M. Rhee, R. Langer, O. Farokhzad, and R. Karnik, Single-step assembly of homogenous lipid-polymeric and lipid-quantum dot nanoparticles enabled by microfluidic rapid mixing, *ACS Nano*, 4, 1671–1679, 2010.
55. X. Gong, S. Peng, W. Wen, P. Sheng, and W. Li, Design and fabrication of magnetically functionalized core/shell microspheres for smart drug delivery, *Advanced Functional Materials*, 18, 1–6, 2008.
56. T. He, Q. Liang, K. Zhang, X. Mu, T. Luo, Y. Wang, and G. Luo, A modified microfluidic chip for fabrication of paclitaxel-loaded poly(L-lactic acid) microspheres, *Microfluidics and Nanofluidics*, 10, 1289–1298, 2011.
57. T. Y. K. T. Watanabe, Continuous fabrication of monodisperse polylactide spheres by droplet—Particle technology using microfluidic emulsification and emulsion–solvent diffusion, *Soft Matter*, 7, 9894–9897, 2011.
58. S. Gratton, P. Ropp, P. Pohlhaus, J. Luft, V. Madden, M. Napier, and J. De Simone, The effect of particle design on cellular internalization pathways, *Proceedings of the National Academy of Sciences of the United States of America*, 105, 11613–11618, 2008.
59. J. Champion, Y. Katare, and S. Mitragotri, Particle shape: A new design parameter for micro- and nanoscale drug delivery carriers, *Journal of Controlled Release*, 121, 3–9, 2007.
60. D. Dendukuri, D. Pregibon, J. Collins, T. Hatton, and P. Doyle, Continuous-flow lithography for high-throughput microparticle synthesis, *Nature Materials*, 5, 365–369, 2006.
61. S. Kim, J. Kim, J. Cho, and Weitz, D. A., Double-emulsion drops with ultra-thin shells for capsule templates, *Lab on a Chip*, 11, 3162–3166, 2011.
62. Q. Xu, M. Hashimoto, T. Dang, T. Hoare, D. Kohane, G. Whitesides, R. Langer, and D. Anderson, Preparation of monodisperse biodegradable polymer microparticles using a microfluidic flow-focusing device for controlled drug delivery, *Small*, 5, 1575–1581, 2009.
63. Y. Zhang, H. Chan, and K. W. Leon, Advanced materials and processing for drug delivery: The past and the future, *Advanced Drug Delivery Reviews*, 65(1), 104–120, 2013.
64. W. Duncanson, T. Lin, A. Abate, S. Seiffert, and R. W. Shah, Microfluidic synthesis of advanced microparticles for encapsulation and controlled release, *Lab on a Chip*, 12, 2135–2145, 2012.

65. D. Lensen, K. van Breukelen, D. Vriezema, and J. van Hest, Preparation of biodegradable liquid core PLLA microcapsules and hollow PLLA microcapsules using microfluidics, *Macromolecular Bioscience*, 10, 475–480, 2010.
66. H. Okochi and M. Nakano, Preparation and evaluation of w/o/w type emulsions containing vancomycin, *Advanced Drug Delivery Reviews*, 45, 5–26, 2000.
67. A. Utada, E. Lorenceau, D. Link, P. Kaplan, H. Stone, and D. Weitz, Monodisperse double emulsions generated from a microcapillary device, *Science*, 308(5721), 537–541, 2005.
68. T. Endres, M. Zheng, M. Beck-Broichsitter, O. Samsonova, H. Debus, and T. Kissel, Optimising the self-assembly of siRNA loaded PEG-PCL-lPEI nano-carriers employing different preparation techniques, *Journal of Controlled Release*, 160, 583–591, 2012.
69. J. Kim, A. Utada, A. Fernandez-Nieves, Z. Hu, and D. Weitz, Fabrication of monodisperse gel shells and functional microgels in microfluidic devices, *Angewandte Chemie International Edition in English*, 46, 1819–1822, 2007.
70. R. Shah, J. W. K., J. Agresti, D. Weitz, and L. Chu, Fabrication of monodisperse thermosensitive microgels and gel capsules in microfluidic devices, *Soft Matter*, 4, 2303–2309, 2008.
71. L. Chu, J. Y. K., R. Shah, and D. Weitz, Monodisperse thermoresponsive microgels with tunable volume-phase transition kinetics, *Advanced Functional Materials*, 17, 3499–3504, 2007.
72. W. Wang, R. Xie, X. Ju, T. Luo, L. Liu, and D. Weitz, Controllable microfluidic production of multicomponent multiple emulsions, *Lab on a Chip*, 11, 1587–1592, 2011.
73. P. Ren, X. Ju, R. Xie, and L. Chu, Monodisperse alginate microcapsules with oil core generated from a microfluidic device, *Journal of Colloid and Interface Science*, 343, 392–395, 2010.
74. R. Shah, J. Kim, and D. Weitz, Monodisperse stimuli-responsive colloidosomes by self-assembly of microgels in droplets, *Langmuir*, 26, 1561–1565, 2010.
75. L. Liu, J. Yang, X. Ju, R. Xie, Y. Liu, W. Wang, J. Zhang, C. Niu, and L. Chu, Monodisperse core-shell chitosan microcapsules for pH-responsive burst release of hydrophobic drugs, *Soft Matter*, 7, 4821–4827, 2011.
76. M. Romanowsky, A. Abate, A. Rotem, C. Holtze, and D. Weitz, High throughput production of single core double emulsions in a parallelized microfluidic device, *Lab on a Chip*, 12, 802–807, 2012.
77. A. Hsieh, N. Hori, R. Massoudi, P. Pan, H. Sasaki, Y. Lin, and A. Lee, On viral gene vector formation in monodispersed picolitre incubator for consistent gene delivery, *Lab on a Chip*, 9, 2638–2643, 2009.
78. Y. Ho, C. Grigsby, F. Zhao, and K. Leong, Tuning physical properties of nanocomplexes through microfluidics-assisted confinement, *Nano Letters*, 11, 2178–2182, 2011.
79. S. Huang, B. Lin, and J. Qin, Microfluidic synthesis of tunable poly-(N-isopropylacrylamide) microparticles via PEG adjustment, *Electrophoresis*, 32, 3364–3370, 2011.
80. M. Marimuthu, S. Kim, and J. An, Amphiphilic triblock copolymer and a microfluidic device for porous microfiber fabrication, *Soft Matter*, 6, 2200–2207, 2010.
81. K. Hettiarachchi, S. Zhang, S. Feingold, A. Lee, and P. Dayton, Controllable microfluidic synthesis of multiphase drug-carrying liposheres for site-targeted therapy, *Biotechnology Progress*, 25, 938–945, 2009.
82. S. Shin, J. Park, J. Lee, H. Park, Y. Park, K. Lee, C. Whang, and S. Lee, "On the fly" continuous generation of alginate fibers using a microfluidic device, *Langmuir*, 23, 9104–9108, 2007.
83. J. Su, Y. Zheng, and H. Wu, Generation of alginate microfibers with a roller-assisted microfluidic system, *Lab on a Chip*, 9, 996–1001, 2009.
84. C. Choi, H. Yi, S. Hwang, D. Weitz, and C. Lee, Microfluidic fabrication of complex shaped microfibers by liquid template-aided multiphase microflow, *Lab on a Chip*, 11, 1477–1483, 2011.
85. C. Choi, J. Jung, D. Kim, Y. Chung, and C. Lee, Monodisperse thermosensitive hollow microcapsules in a microfluidic system, *Lab on a Chip*, 8, 1544–1551, 2008.
86. A. Jahn, W. Vreeland, M. Gaitan, and L. Locascio, Controlled vesicle self-assembly in microfluidic channels with hydrodynamic focusing, *Journal of the American Chemical Society*, 126, 2674–2675, 2004.
87. A. Jahn, W. Vreeland, D. DeVoe, L. Locascio, and M. Gaitan, Microfluidic directed formation of liposomes of controlled size, *Langmuir*, 23, 6289–6293, 2007.
88. A. Jahn, S. Stavis, J. Hong, W. Vreeland, D. DeVoe, and M. Gaitan, Microfluidic mixing and the formation of nanoscale lipid vesicles, *ACS Nano*, 4, 2077–2087, 2010.
89. T. Tran, C. Nguyen, D. Kim, Y. Lee, and K. Huh, Microfluidic approach for highly efficient synthesis of heparin-based bioconjugates for drug delivery, *Lab on a Chip*, 12, 589–594, 2012.

90. Z. Nie, W. Li, M. Seo, S. Xu, and E. Kumacheva, Janus and ternary particles generated by microfluidic synthesis: Design, synthesis, and self-assembly, *Journal of the American Chemical Society*, 128, 9408–9412, 2006.
91. A. Manz, D. J. Harrison, E. M. J. Verpoorte, J. C. Fettinger, H. Lüdi, and H. M. Widmer, Miniaturization of chemical analysis systems: A look into next century's technology or just a fashionable craze, *Chemia*, 45(40, 103–105, 1991.
92. S. Terry, J. Jerman, and J. Angell, A gas chromatographic air analyzer fabricated on a silicon wafer, *IEEE Transactions on Electron Devices*, 26(12), 1880–1886, 1979.
93. Z. Fan, H. Ludi, and H. Widmers, Capillary electrophoresis and sample injection systems integrated on a planar glass chip, *Analytical Chemistry*, 64(17), 1926–1932, 1992.
94. S. C. Jacobson, R. Hergenroder, L. B. Koutny, and J. M. Ramsey, High-Speed Separations on a Microchip, *Analytical Chemistry*, 66(7), 1114–1118, 1994.
95. Q. Tong, E. Schmidt, and U. Gselea, Hydrophobic silicon wafer bonding, *Applied Physics Letters*, 64(5), 625, 1994.
96. K. Thompson and Y. B. Gianchandani, Direct silicon–silicon bonding by electromagnetic induction heating, *J. Microelectromechanical Systems*, 11(4), 285–292, 2002.
97. C. Harendt, H. G. Graf, B. Hofflinger, and E. Penteker, Silicon fusion bonding and its characterization, *Journal of Micromechanical and Microengineering*, 2(3), 113, 1992.
98. V. Kutchoukov, F. Laugerea, W. van der Vlist, L. Pakulaa, Y. Garinib, and A. Bossche, Fabrication of nanofluidic devices using glass-to-glass anodic bonding, *Sensors and Actuators A, Physical*, 114, 521–527, 2004.
99. C. Stokes and P. Palmer, 3D micro-fabrication processes: A review, *Proceedings of the Seminar on MEMS Sensors and Actuators*, Institution of Engineering and Technology, London, U.K., pp. 289–298, 2006.
100. J. Bustillo, R. Howe, and R. Muller, Surface micromachining for microelectromechanical systems, *Proceedings of the IEEE*, 86(8), 1552–1574, 1998.
101. M. Gad-El-Hak, *The Mems Handbook*, CRC Press, Boca Raton, FL, 2002.
102. M. Hoffmann and E. Voges, Bulk silicon micromachining for MEMS in optical communication systems, *Journal of Micromechanics and Microengineering*, 12(4), 349–360, 2002.
103. J. Judy, Microelectromechanical systems (MEMS): Fabrication, design and applications, *Smart Materials and Structures*, 10(6), 1115–1134, 2001.
104. W. Lang, Silicon microstructuring technology, *Materials Science and Engineering: Reports*, 17(1), 1–55, 1996.
105. N. Maluf, An introduction to microelectromechanical systems engineering, *Measurement Science and Technology*, 13(2), 229, 2002.
106. N. Miki, Wafer bonding techniques for MEMS, *Sensor Letters*, 3(4), 11, 2005.
107. K. Petersen, Silicon as a mechanical material, *Proceedings of the IEEE*, 70(5), 420–457, 1982.
108. M. Burns et al., An integrated nanoliter DNA analysis device, *Science*, 282(5388), 484–487, 1998.
109. E. Becker, W. Ehrfeld, P. Hagmann, A. Maner, and D. Münchmeyer, Fabrication of microstructures with high aspect ratios and great structural heights by synchrotron radiation lithography galvanoforming, and plastic moulding (LIGA process), *Microelectronic Engineering*, 4(1), 35–56, 1986.
110. C. Lin, G. Lee, B. Chang, and G. Chang, A new fabrication process for ultra-thick microfluidic microstructures utilizing SU-8 photoresist, *Journal of Micromechanics and Microengineering*, 12(5), 590–597, 2002.
111. T. Sikanen, S. Tuomikoski, R. Ketola, R. Kostiainen, S. Franssila, and T. Kotiaho, Characterization of SU-8 for electrokinetic microfluidic applications, *Lab on a Chip*, 5(8), 888–896, 2005.
112. L. Yang, Fabrication of SU-8 embedded microchannels with circular cross-section, *International Journal of Machine Tools and Manufacture*, 44(10), 1109–1114, 2004.
113. S. Metz, S. Jiguet, A. Bertsch, and P. Renaud, Polyimide and SU-8 microfluidic devices manufactured by heat-depolymerizable sacrificial material technique, *Lab on a Chip*, 4(2), 114–120, 2004.
114. F. E. H. Tay, J. A. van Kan, F. Watt, and W. O. Choong, A novel micro-machining method for the fabrication of thick-film SU-8 embedded microchannels, *Journal of Micromechanics and Microengineering*, 11(1), 27–32, 2001.
115. M. Agirregabiria, F. Blanco, J. Berganzo, M. Arroyo, A. Fullaondo, K. Mayora, and J. Ruano-López, Fabrication of SU-8 multilayer microstructures based on successive CMOS compatible adhesive bonding and releasing steps, *Lab on a Chip*, 5(5), 545–552, 2005.

116. J. Dykes, D. K. Poon, J. Wang, D. Sameoto, J. T. K. Tsui,. C. Choo, G. H. Chapman, A. M. Parameswaren, and B. L. Gray, Creation of embedded structures in SU-8, *Proceedings of SPIE*, 6465, 64650, 2007.
117. Y. Morimoto, W.-H. Tan, and S. Takeuchi, Three-dimensional axisymmetric flow-focusing device using stereolithography, *Biomedical Microdevices*, 11(2), 367–377, 2009.
118. L. A. Tse, P. J. Hesketh, D. W. Rosen, and J. L. Gole, Stereolithography on silicon for micro fluidics and microsensor packaging, *Microsystem Technologie*, 9(5), 319–23, 2003.
119. F. Melchels, J. Feijen, and D. Grijpma, A review on stereolithography and its applications in biomedical engineering, *Biomaterials*, 31(24), 6121–6130, 2010.
120. M. F. Jensen, J. E. McCormack, and B. Helbo, Rapid prototyping of polymer microsystems via excimer laser ablation of polymeric mould, *Lab on a Chip*, 4(4), 391–395, 2004.
121. M. Khan, Laser processing for bio-microfluidics applications (part I), *Analytical and Bioanalytical Chemistry*, 385(8), 1362–1369, 2006.
122. S. Metz, B. A. D. Bertrand, and P. Renaud, Flexible polyimide probes with microelectrodes and embedded microfluidic channels for simultaneous drug delivery and multi-channel monitoring of bioelectric activity, *Biosensors and Bioelectronics*, 19(10), 1309–1318, 2004.
123. H. Yin, K. Killeen, R. Brennen, D. Sobek, M. Werlich, and T. van de Goor, Microfluidic chip for peptide analysis with an integrated HPLC column, sample enrichment column, and nanoelectrospray tip, *Analytical Chemistry*, 77, 527–533, 2005.
124. T. N. Kim, K. Campbell, A. Groisman, and D. S. B. Kleinfeld, Femtosecond laser-drilled capillary integrated into a microfluidic device, *Applied Physics Letters*, 86(20), 201106, 2005.
125. N. Rizvi, Femtosecond laser micromachining: Current status and applications, *Riken Review*, 50, 107–112, 2003.
126. A. Dubey and V. Yadava, Laser beam machining—A review, *International Journal of Machine Tools and Manufacture*, 48(6), 609–628, 2008.
127. R. Gattass and E. Mazur, Femtosecond laser micromachining in transparent materials, *Nature Photonics*, 2(4), 219–225, 2008.
128. Y. Xia and G. Whitesides, Soft lithography, *Angewandte Chemie International Edition*, 37(5), 550–575, 1998.
129. F. Hua, A. Gaur, Y. Sun, M. Word, N. Jin, I. Adesida, M. Shim, A. Shim, and J. A. Rogers, Processing dependent behavior of soft imprint lithography on the 1-10-nm scale, *IEEE Transactions on Nanotechnology*, 5(3), 301–308, 2006.
130. G. Y. M. Fiorini, G. Jeffries, P. Schiro, S. Mutch, R. Lorenz, and D. Chiu, Fabrication improvements for thermoset polyester (TPE) microfluidic devices, *Lab on a Chip*, 7(7), 923–926, 2007.
131. J. Rogers and R. Nuzzo, Recent progress in soft lithograph, *Materials Today*, 8(2), 50–56, 2005.
132. D. Qin, Y. Xia, and G. Whitesides, Soft lithography for micro- and nanoscale patterning, *Nature Protocols*, 5(3), 491–502, 2010.
133. M. Svedberg, F. Nikolajeff, and G. Thornell, Fabrication of a paraffin actuator using hot embossing of polycarbonate, *Sensors and Actuators A Physical*, 103(3), 307–316, 2003.
134. S. Youn, T. Noguchi, M. Takahashi, and R. Maeda, Dynamic mechanical thermal analysis, forming and mold fabrication studies for hot-embossing of a polyimide microfluidic platform, *Journal of Micromechanics and Microengineering*, 18(4), 45025, 2008.
135. J. S. Jeon, S. Chung, R. D. Kamm, and J. L. Charest, Hot embossing for fabrication of a microfluidic 3D cell culture platform, *Biomedical Microdevices*, 13(2), 325–333, 2011.
136. H. Becker, Hot embossing as a method for the fabrication of polymer high aspect ratio structures, *Sensors and Actuators A: Physical*, 83, 130–135, 2000.
137. A. Kolew, D. Münch, K. Sikora, and M. Worgull, Hot embossing of micro and sub-micro structured inserts for polymer replication, *Microsystem Technologies*, 17(4), 609–618, 2010.
138. N. Roos, H. Schulz, L. Bendfeldt, M. Fink, K. Pfeiffer, and H.-C. Scheer, First and second generation purely thermoset stamps for hot embossing, *Engineering Conference*, 61, 399–405, 2002.
139. H. H. Schift, C. David, J. Gabriel, J. Gohrecht, L. Heydemlan, W. Kaiser, S. Kiippelrl, and L. Scandella, Nanoreplication in polymers using hot embossing and injection molding, *Microelectronic Engineering*, 53, 171–174, 2000.
140. M. Worgull, M. Heckele, and W. Schomburg, Large-scale hot embossing, *Microsystem Technologies*, 12, 110–115, 2005.
141. L. Guo, Nanoimprint lithography: Methods and material requirements, *Advanced Materials*, 19(4), 495–513, 2007.

142. M. Worgull, *Hot Embossing: Theory and Technology of Microreplication*, William Andrew, Oxford, U.K., 2009.
143. C. Gartner, R. Klemm, and H. Becker, Methods and instruments for continuous flow PCR on a chip, *Proc. SPIE 6465, Microfluidics, BioMEMS, and Medical Microsystems*, 6465, 646502, 2007.
144. D. A. Mair, E. Geiger, A. P. Pisano, J. M. Frechetacd, and F. Svec, Injection molded microfluidic chips featuring integrated interconnects, *Lab on a Chip*, 6(10), 1346–1400, 2006.
145. U. Attia and J. Alcock, A review of micro-powder injection moulding as a microfabrication technique, *Journal of Micromechanics and Microengineering*, 21(4), 043001, 2011.
146. U. Attia, S. Marson, and J. Alcock, Micro-injection moulding of polymer microfluidic devices, *Microfluidics and Nanofluidics*, 7(1), 1–28, 2009.
147. M. Heckele and W. Schomburg, Review on micro molding of thermoplastic polymers, *Journal of Micromechanics and Microengineering*, 14(2), 1–14, 2004.
148. M. L. Hupert, W. J. Guy, S. D. Llopis, H. Shadpour, S. Rani, D. E. Nikitopoulos, and S. A. Soper, Evaluation of micromilled metal mold masters for the replication of microchip electrophoresis devices, *Microfluidics and Nanofluidics*, 3(1), 1–11, 2006.
149. C. Malek and V. Saile, Applications of LIGA technology to precision manufacturing of high-aspect-ratio micro-components and -systems: A review, *Microelectronics Journal*, 35(1), 131–143, 2004.
150. S. Nikumb, Q. Chen, C. Li, H. Reshef, H. Zheng, H. Qiu, and D. Low, Precision glass machining, drilling and profile cutting by short pulse lasers, *Thin Solid Films*, 477, 216–221, 2005.
151. R. Truckenmüller, Z. Rummler, T. Schaller, and W. K. Schomburg, Low-cost thermoforming of micro fluidic analysis chips, *Journal of Micromechanics and Microengineering*, 12(4), 375–379, 2009.
152. W. Ehrfeld, H. Lehr, F. Michel, A. Wolf, H. Gruber, and A. Bertholds, Microelectro discharge machining as a technology in micromachining, *Proc. SPIE 2879, Micromachining and Microfabrication Process Technology II*, pp. 332–337, 1996.
153. G. F. Christopher, N. N. Noharuddin, J. A. Taylor, and J. A. S. L. Anna, Experimental observations of the squeezing-to-dripping transition in t-shaped microfluidic junctions, *Physical Review E.*, 78(3), 036317, 2008.
154. S. Xu, Z. Nie, M. Seo, P. Lewis, E. Kumacheva, H. Stone, P. Garstecki, D. Weibel, I. Gitlin, and G. Whitesides, Generation of monodisperse particles by using microfluidics: Control over size, shape, and composition, *Angewandte Chemie International Edition in English*, 44(5), 724–728, 2005.
155. A. S. Utada, L.-Y. Chu, A. Fernandez-Nieves, D. R. Link, C. Holtze, and D. A. Weitz, Dripping, jetting, drops, and wetting: The magic of microfluidics, *Bulletin of MRS*, 32(09), 702–708, 2007.
156. M. Hashimoto, P. Garstecki, and G. Whitesides, Synthesis of composite emulsions and complex foams with the use of microfluidic flow-focusing devices, *Small*, 3(10), 1792–1802, 2007.

21 Encapsulation Process in Granulation Technology

Himanshu K. Solanki

CONTENTS

21.1 Introduction 386
    21.1.1 Objective 386
    21.1.2 Various Techniques of Granulation on Large-Scale Production 386
    21.1.3 Encapsulation Process 386
21.2 Principles and Process of Encapsulation in Granulation Technology 388
    21.2.1 Encapsulation by Wet Granulation 388
    21.2.2 Encapsulation by Spray Granulation 389
        21.2.2.1 Principle 389
        21.2.2.2 Atomization 390
        21.2.2.3 Rotary Atomizers: Atomization by Centrifugal Energy 391
        21.2.2.4 Pressure Nozzles: Atomization by Pressure Energy 392
        21.2.2.5 Two-Fluid or Pneumatic Nozzles: Atomization by Kinetic Energy 392
        21.2.2.6 Ultrasonic Atomization 393
        21.2.2.7 Mixing and Drying 393
        21.2.2.8 Drying Chamber 394
        21.2.2.9 Powder Separation 394
        21.2.2.10 Parameter to Be Controlled 394
        21.2.2.11 Advantages of Spray Dry Granulation 395
        21.2.2.12 Disadvantages 395
    21.2.3 Encapsulation by Fluid Bed Granulation or Air Suspension Techniques (Wurster Process) 395
        21.2.3.1 Advantages 396
        21.2.3.2 Disadvantages 397
    21.2.4 Encapsulation by Fluid Bed Spray Granulation 397
    21.2.5 Encapsulation by Fluid Bed Rotor Granulation 397
    21.2.6 Encapsulation by Extrusion–Spheronization 397
        21.2.6.1 Advantages 398
        21.2.6.2 Disadvantages 398
21.3 Novel Approaches in Encapsulation Processes in Granulation Technology 399
    21.3.1 Encapsulation by Vibrational Nozzle Technology 399
    21.3.2 Encapsulation by Pneumatic Dry Granulation 400
        21.3.2.1 Granulate Any API 400
        21.3.2.2 Pneumatic Dry Granulation Replaces Wet Granulation 400
        21.3.2.3 Advantages of PDG Technology 400
        21.3.2.4 Benefits to Pharmaceutical Companies 401
    21.3.3 Encapsulation by Freeze Granulation Technology 401
    21.3.4 Encapsulation by Foamed Binder Technologies 402
        21.3.4.1 How Foam Binder Granulation Works 403
        21.3.4.2 Extremely Efficient Binder Delivery and Particle Coverage 403

	21.3.5	Encapsulation by Melt Granulation Technology	403
		21.3.5.1 Principle of Melt Granulation	403
		21.3.5.2 Requirements of Melt Granulation	404
		21.3.5.3 Meltable Binders	405
		21.3.5.4 Advantage of Melt Granulation	405
		21.3.5.5 Disadvantages	405
	21.3.6	Encapsulation by Steam Granulation	406
		21.3.6.1 Advantages	406
		21.3.6.2 Disadvantages	406
	21.3.7	Encapsulation by Moisture-Activated Dry Granulation	406
		21.3.7.1 Advantages	407
	21.3.8	Encapsulation by Granurex® Technology	408
		21.3.8.1 Key Features	408
	21.3.9	Encapsulation by Thermal Adhesion Granulation Process	408
21.4	Conclusion		409
References			409

21.1 INTRODUCTION

Encapsulation by granulation of particulate materials is a fundamental operation widely practiced in a variety of chemical industries including pharmaceuticals, food, fertilizer, cosmetics, biomedical, and nuclear.

21.1.1 OBJECTIVE

Generally, the encapsulation by granulation process is performed to achieve one or several of the following objectives:

- To protect powders from oxygen, humidity, light, or any other incompatible element
- To delay and/or control the release of active agents involved in core particles
- To confer desired interfacial properties to the particles making them more proper for the final target applications (e.g., dispersion in plastics, electrostatic pulverization)
- To reduce the affinity of powders with respect to aqueous or organic solvents
- To avoid caking phenomena during storage and transport
- To improve appearance, taste, or odors of products
- To conserve nutrients contained in food products
- To functionalize powders (catalysts, enzyme-coated detergents, etc.)
- To increase the particle size

Granulation may be defined as a size-enlargement process that converts fine or coarse particles into physically stronger and larger agglomerates having good flow property, better compression characteristics, and uniformity[1] (Figure 21.1).

The art and science for the processing and production of granules by encapsulation is known as encapsulation by granulation technology.

21.1.2 VARIOUS TECHNIQUES OF GRANULATION ON LARGE-SCALE PRODUCTION[2]

Various granulation technologies are presented in Figure 21.2.

21.1.3 ENCAPSULATION PROCESS

Through the process of encapsulation, solid or liquid active ingredients are homogeneously embedded into a carrier material, producing a stable product with defined characteristics (Figure 21.3).

Encapsulation Process in Granulation Technology

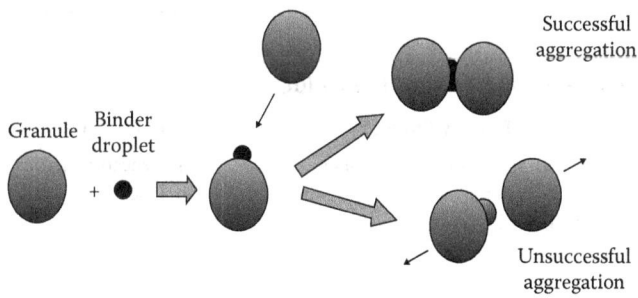

FIGURE 21.1 Schematic of aggregation process in granulation.

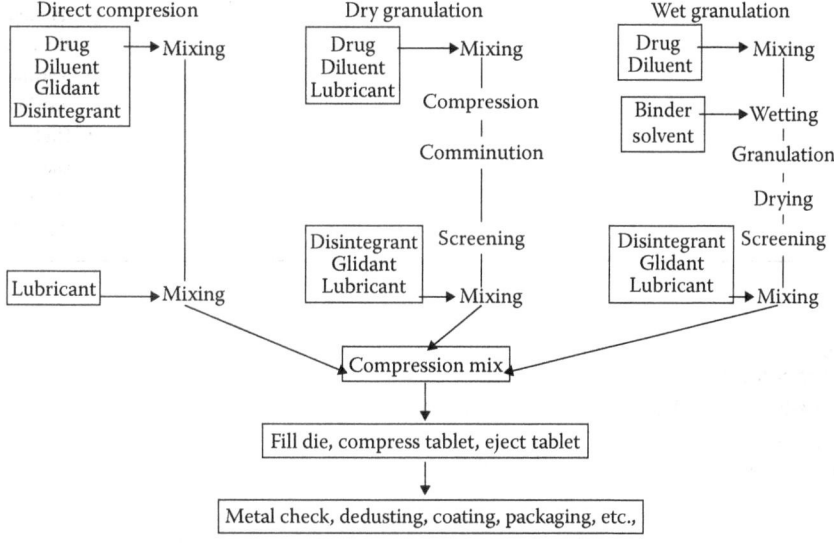

FIGURE 21.2 Schematic representation of various granulation technologies.

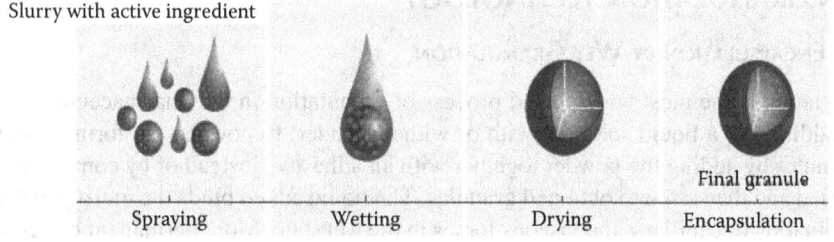

FIGURE 21.3 Principle of encapsulation by granulation.

This process is especially useful for controlled and targeted release of active pharmaceutical ingredients (APIs), food additives, flavors, etc.[3,4] (Table 21.1).

Examples of product formulated by encapsulation process are as follows:

- Scent oils
- Aromas
- Perfumes
- Vitamins
- Lactic acid bacillus

TABLE 21.1
Overview of Microencapsulated Granule Product

Process	Primary Products	End Products	Product Benefits
Micro encapsulation	Start granules (seeds from original or secondary material) + suspensions, emulsions, or solutions with active agents	Microencapsulated granules	• Shelf life • Encapsulation of volatile components • Protection against chemical reactions • Protection against oxygen or light • Protection against humidity • Depot effect • Narrow particle size distribution • High bulk density • Building of solid particles from liquid products

- Enzymes
- Starter cultures
- Omega-3 fatty acids, fish oils
- Pigments
- Amino acids
- Minerals

21.2 PRINCIPLES AND PROCESS OF ENCAPSULATION IN GRANULATION TECHNOLOGY

21.2.1 Encapsulation by Wet Granulation

Wet granulation is the most widely used process of granulation in the pharmaceutical industry.[5] It involves addition of a liquid solution (with or without binder) to powders, to form a wet mass or it forms granules by adding the powder together with an adhesive, instead of by compaction. The wet mass is dried and then sized to obtained granules. The liquid added binds the moist powder particles by a combination of capillary and viscous forces in the wet state. More permanent bonds are formed during subsequent drying, which leads to the formation of agglomerates.[6]

The basic reasons for wet granulation are as follows:

- Render material free-flowing
- Density materials
- Prepare uniform mixes that do not segregate
- Improve the compression characteristics of a drug
- Control the rate of drug release
- Facilitate volume dispensing
- Reduce dust
- Improve the appearance of a product
- Reduce variations in different batches of raw materials (Figure 21.4)

Encapsulation Process in Granulation Technology

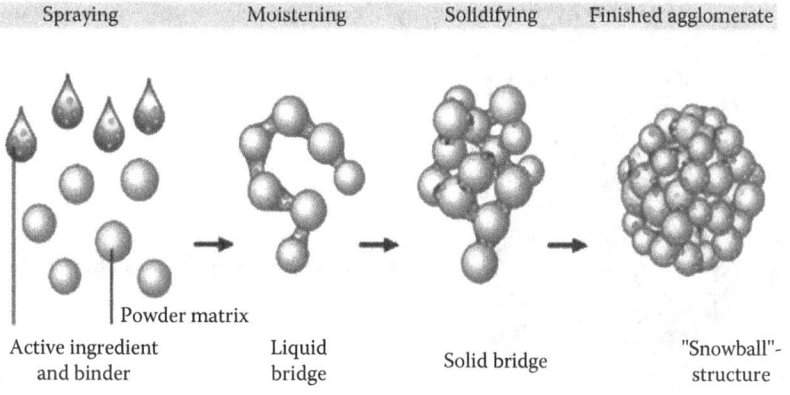

FIGURE 21.4 Principle of the encapsulation process as matrix pellets.

TABLE 21.2
Merits and Demerits of Wet Granulation Process

Merits	Demerits
• It improves flow property and compression characteristics.	• Process is expensive because of labor, space, time, special equipment, and energy requirement.
• Increases density of granules.	• Multiple processing steps involved in the process add complexity.
• Better distribution of color and soluble drugs if added in the binding solution.	• Loss of material during various stages of processing.
• It reduces dust hazards.	• Moisture-sensitive and thermolabile drugs are poor candidates.
• Prevents segregation of powders.	• Any incompatibility between the formulation components is aggravated during the processing.
• Makes hydrophobic surfaces more hydrophilic.	

Although the process is most widely used in the pharmaceutical industry, the conventional wet granulation process has the following merits and demerits[1] (Table 21.2).

21.2.2 Encapsulation by Spray Granulation

21.2.2.1 Principle[7,8]

There are three fundamental steps (Figure 21.5) involved in spray drying:

1. Atomization of a liquid feed into fine droplets
2. Mixing of these sprays droplets with a heated gas stream, allowing the liquid to evaporate and leave dried solids
3. Separation and collection of dried powder from the gas stream

Spray drying involves the atomization of a liquid feedstock into a spray of droplets and contacting the droplets with hot air in a drying chamber (Figure 21.6).

The sprays are produced by either rotary (wheel) or nozzle atomizers. Evaporation of moisture from the droplets and formation of dry particles proceed under controlled temperature and

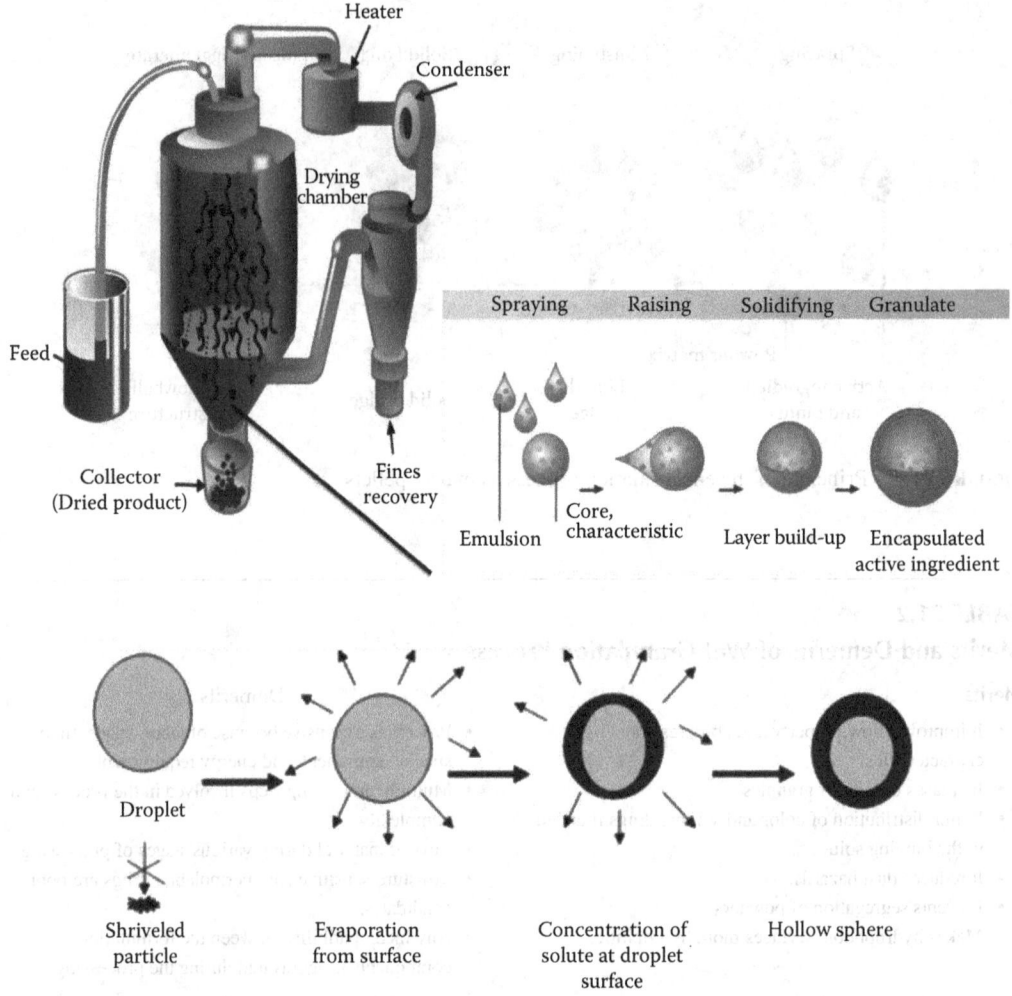

FIGURE 21.5 Principle of encapsulation process by spray granulation.

airflow conditions. Powder is discharged continuously from the drying chamber. Operating conditions and dryer design are selected according to the drying characteristics of the product and powder specification.

21.2.2.2 Atomization[9,10]

The atomizing device, which forms the spray, is the "heart" of the spray-drying process.

Atomizer: Equipment that breaks bulk liquid into small droplets, forming a spray.

The prime functions of atomization are:

1. A high surface-to-mass ratio resulting in high evaporation rates
2. Production of particles of the desired shape, size, and density

The aim of atomizing the concentrate is to provide a very large surface, from which the evaporation can take place. The smaller droplets, the bigger surface, the easier evaporation, and a better thermal efficiency of the dryer are obtained.

Encapsulation Process in Granulation Technology

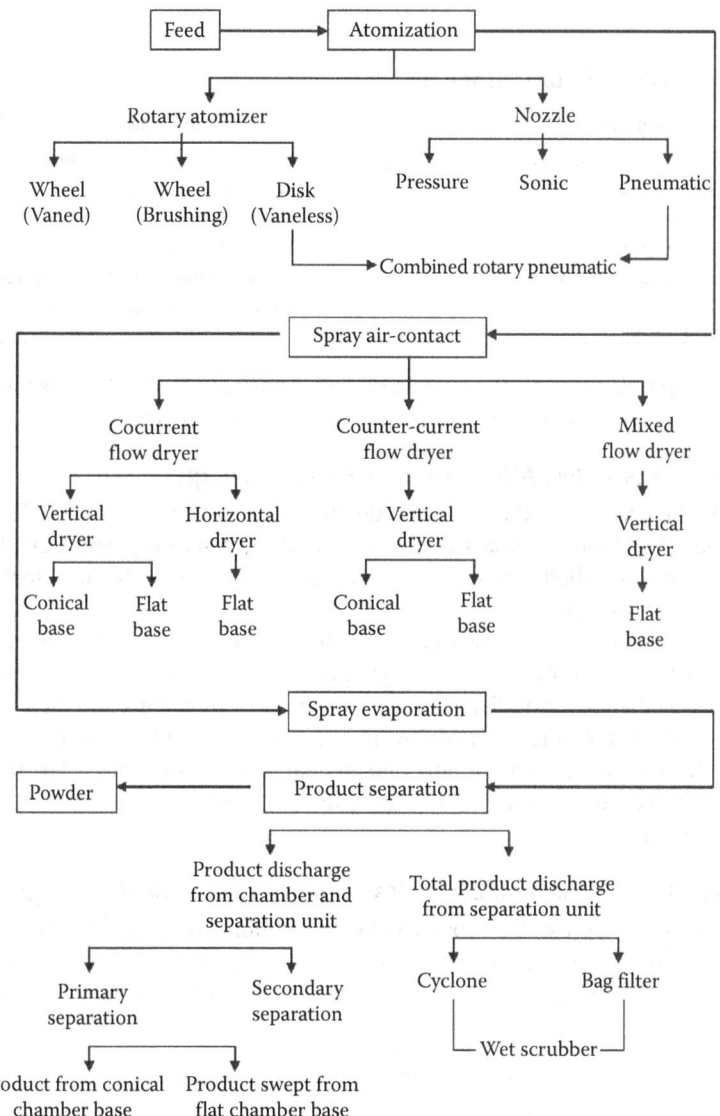

FIGURE 21.6 Schematic of spray-drying process.

The ideal from a drying point of view would be a spray of drops of same size, which would mean that the drying time for all particles would be the same for obtaining equal moisture content.

In order to produce top-quality products in the most economical manner, it is crucial to select the right atomizer. Three basic types of atomizers are used commercially:

1. Rotary atomizer (atomization by centrifugal energy)
2. Pressure nozzle (atomization by pressure energy)
3. Two-fluid nozzle (atomization by kinetic energy)

21.2.2.3 Rotary Atomizers: Atomization by Centrifugal Energy

Rotary atomizer uses the energy of a high-speed rotating wheel to divide bulk liquid into droplets. Feedstock is introduced at the center of the wheel, flows over the surface to the periphery, and disintegrates into droplets when it leaves the wheel (Table 21.3).

TABLE 21.3
Merits and Demerits of Rotary Atomizer

Advantages of Rotary Atomizers	Disadvantages of Rotary Atomizers
• Great flexibility and ease of operation	• Produce large quantities of fine particles, which can result in pollution control problems
• Low-pressure feed system	
• No blockage problems	• High capital cost
• Handling of abrasive feeds	• Very expensive to maintain
• Ease of droplet size control through wheel speed adjustment	• Cannot be used in horizontal dryers
	• Difficult to use with highly viscous materials

Because of the problems and costs associated with rotary atomizers, there is interest within segments of the spray dry industry in replacing rotary atomizers with spray nozzles.

21.2.2.4 Pressure Nozzles: Atomization by Pressure Energy

Pressure nozzle (Figure 21.7) is the most commonly used atomizer for spray drying.

Nozzles generally produce coarse, free-flowing powders than rotary atomizers. Pressure nozzles used in spray drying are called "vortex" nozzles because they contain features that cause the liquid passing through them to rotate.

The rotating fluid allows the nozzle to convert the potential energy of liquid under pressure into kinetic energy at the orifice by forming a thin, high-speed film at the exit of the nozzle. As the unstable film leaves the nozzle, it disintegrates, forming first ligaments and then droplets. Pressure nozzles can be used over a large range of flow rates and can be combined in multiple-nozzle installations to give them a great amount of flow rate and particle size flexibility. The range of operating pressure range for pressure nozzles used in spray drying is from about 250 PSI (17.4 bar) to about 10,000 PSI (690 bar).

21.2.2.5 Two-Fluid or Pneumatic Nozzles: Atomization by Kinetic Energy

Liquid feedstock and compressed air (or steams) are combined in a two-fluid nozzle (Figure 21.8).

The design utilizes the energy of compressed gas to atomize the liquid. Two advantages of the two-fluid nozzle are its ability to produce very fine particles and to atomize highly viscous feeds.

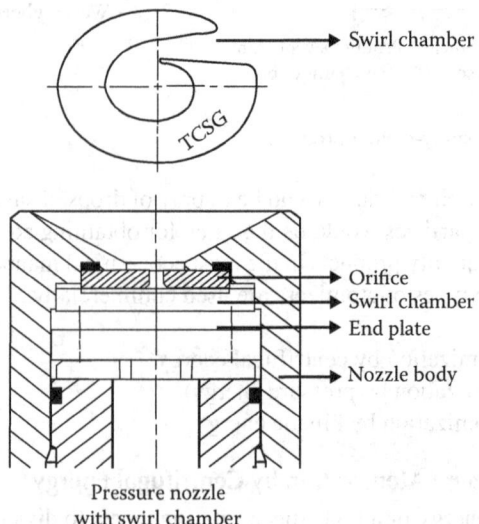

FIGURE 21.7 Pressure nozzle.

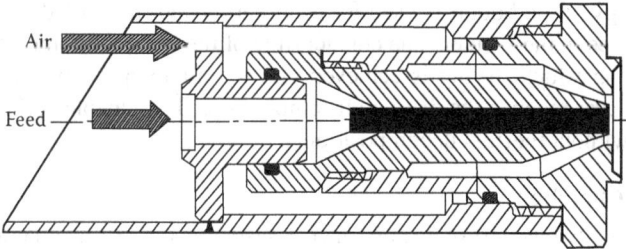

FIGURE 21.8 Two-fluid or pneumatic nozzles.

However, two-fluid nozzles are expensive to operate because of the high cost of compressed air. Two-fluid nozzles are often used in laboratory and pilot plant spray dry applications because of their ability to produce a wide range of flow rates and droplet sizes. The range of operating pressure range for pressure nozzles used in spray drying is from about 250 PSI (17.4 bar) to about 10,000 PSI (690 bar).

21.2.2.6 Ultrasonic Atomization

Recently ultrasonic energy has been used in place of pressure or centrifugal force to form droplets. In this method, a liquid is placed on a rapidly vibrating surface at ultrasonic frequencies. At sufficiently high amplitude, the liquid spreads, becomes unstable and collapses, resulting in the formation of very fine droplets. These devices are excellent for droplets below 50 μm.

21.2.2.7 Mixing and Drying

Once the liquid is atomized it must be brought into intimate contact with the heated gas for evaporation to take place equally from the surface of all droplets within the drying chamber. The heated gas is introduced into the chamber by an *air disperser*, which ensures that the gas flows equally to all parts of the chamber.

21.2.2.7.1 Air Disperser

The air disperser uses perforated plates or vane channels through which the gas is directed, creating a pressure drop and, thereby, equalizing the flow in all directions. It is critical that the gas entering the air disperser is well mixed and has no temperature gradient across the duct leading into it. As a result, it is important that any type of heater used inherently produces a well-mixed gas stream or that a mixing section is placed between the heater and the air disperser. The air disperser is normally built into the roof of the drying chamber, and the atomization device is placed in or adjacent to the air disperser. This arrangement allows instant and complete mixing of the heated drying gas with atomized cloud of droplets. To fully understand the characteristics of spray-dried powders, one needs to examine the mechanism for drying within a single droplet. Typically, there are many very small particles suspended in a sphere of liquid. When the droplet is first exposed to hot gas, rapid evaporation takes place. Material dissolved in the liquid will tend to form a thin shell at the surface of the sphere. Although the evaporation has kept the particle itself quite cool, as the liquid concentration decreases, the particle will begin to heat. Evaporation then takes only as quickly as the liquid can diffuse to the surface of the sphere. This phase of the drying process is called first-order drying or is said to be diffusion rate limited.

Fortunately, this phase occurs in the cooler part of the dryer where the drying gas is at or near the outlet temperature of the dryer. As a result the solids in each particle are never heated above the outlet temperature of the dryer, even though the dryer inlet may be considerably higher. The final dried powder will be at a temperature approximately 20°C lower than the air outlet temperature.

The thermal energy of the hot air is used for evaporation and the cooled air pneumatically conveys the dried particles in the system. The contact time of the hot air and the spray droplets is only a few seconds, during which drying is achieved and the air temperature drops instantaneously. The dried particle never reaches the drying air temperature. This enables efficient drying of heat-sensitive materials without thermal decomposition.

21.2.2.8 Drying Chamber

The largest and most obvious part of a spray-drying system is the drying chamber. This vessel can be taller and slander or have large diameter with a short cylinder height.

Selecting these dimensions is based on two process criteria that must be met.

- First, the vessel must be of adequate volume to provide enough contact time between the atomized cloud and the heated glass.
- The second criterion is that all droplets must be sufficiently dried before they contact a surface.

This is where the vessel shape comes into play. Centrifugal atomizer requires larger diameter and less cylinder height. Nozzles are just the opposite. Most spray dryer manufacturers can estimate, a given powder's mean particle size, what dimensions are needed to prevent wet deposits on the drying chamber walls.

Drying chambers are usually constructed of stainless steel sheet metal, with stiffeners for structural support and vessel integrity. Sheet steel finish and weld polish can be specified to meet any requirement. Insulation is usually applied to the outside of the vessel, and stainless steel wrapping is seam welded over the entire vessel. This provides a thermally efficient and safe system that is easy to clean and has no crevice areas that might become contaminated.

21.2.2.9 Powder Separation

In almost every case, spray-drying chambers have cone bottoms to facilitate the collection of the dried powder. When the coarse powder is to be collected, they are usually discharged directly from the bottom of the cone through a suitable airlock, such as a rotary valve. The gas stream, now cool and containing all the evaporate moisture, is drawn from the center of the cone above the cone bottom and discharged through a side outlet. In effect, the chamber bottom is acting as a cyclone separator. Because of the relatively low efficienctly of collection, some fines are always carried with the gas stream. This must be separated in high-efficiency cyclones, followed by a wet scrubber or in a fabric filter (bag collector). Fines are collected in the dry state (bag collector) are often added to the larger powder stream or recycled.

21.2.2.10 Parameter to Be Controlled[11]

The pharmaceutical spray-dried products have important properties:

- Uniform particle size
- Nearly spherical regular particle shape
- Excellent flowability
- Improved compressibility
- Low bulk density
- Better solubility
- Reduced moisture content
- Increased thermal stability and suitability for further applications

Such product characteristics majorly depend on the physical properties of feed, equipment components, and processing parameters (Table 21.4).

TABLE 21.4
Parameters to Be Controlled

Sr.No.	Physical Properties of Feed	Equipment and Process Parameters	Spray-Dried Product Characteristics
1	Feed concentration increased or high viscosity of polymer	Two-fluid nozzle: 10–200 μm Pressure nozzle: 20–200 μm	Particle size increased with low bulk density
2	Low surface tension with smaller droplet size	Increased energy of atomization and higher drop velocity	Reduced particle size and better compressibility
3	Increased temperature of the spray solution with correct drop formation	Increased spray flow rate (to optimum level)	Better flow characteristics
4	Low concentration (<0.5% w/w) of hydrophilic polymers such as NaCMC and HPMC	—	Improved release rate of a poorly water-soluble drug
5	—	Increased dryer outlet temperature	Lower final product moisture content
6	—	Multistage and energy recovery units	Suitability for processing heat-sensitive pharmaceuticals

21.2.2.11 Advantages of Spray Dry Granulation[12]

- Drying is very rapid and it completes within 3–30 s.
- Labor costs are low as it performs the functions of evaporator, crystallizer, dryer, size reduction unit, and classifier.
- By using a suitable atomizer, the product of uniform and controllable size can be obtained.
- Fine droplets formed provide a large surface area for heat and mass transfer.
- Product shows excellent solubility.
- Either solution or suspension or thin paste can be dried in a single step to get the final product ready for package.
- It is suitable for drying of sterile products.
- Globules of emulsion can be dried.

21.2.2.12 Disadvantages[12]

- It is very bulky and expensive.
- Labor and running costs are high.
- Thermal efficiency is low.

21.2.3 ENCAPSULATION BY FLUID BED GRANULATION OR AIR SUSPENSION TECHNIQUES (WURSTER PROCESS)

Fluidization is the operation by which fine solids are transformed into a fluid-like state through contact with a gas. At certain gas velocity, the fluid will support the particles giving them free mobility without entrapment.[13,14]

Fluid bed granulation is a process by which granules are produced in single equipment by spraying a binder solution onto a fluidized powder bed. The materials processed by fluid bed granulation are finer, free-flowing, and homogeneous. The system involves the heating of air and then directing it through the material to be processed. Later, the same air exits through the voids of the product.[15]

Fluid bed processing of pharmaceuticals was first reported by Wurster,[16] by using air suspension technique to coat tablets later used this technique in granulating and drying of pharmaceuticals, for

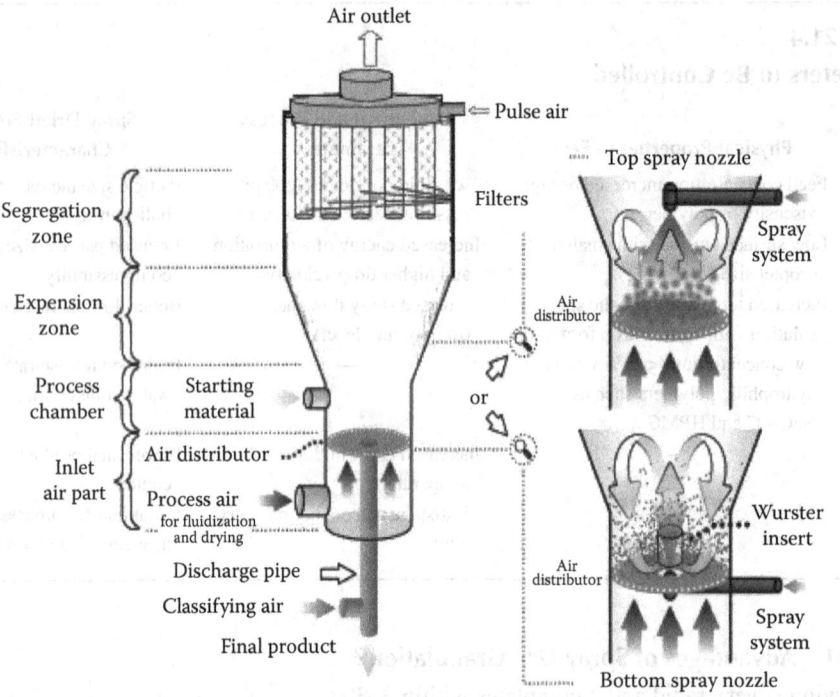

FIGURE 21.9 Schematics of a fluid bed system (Glatt type).

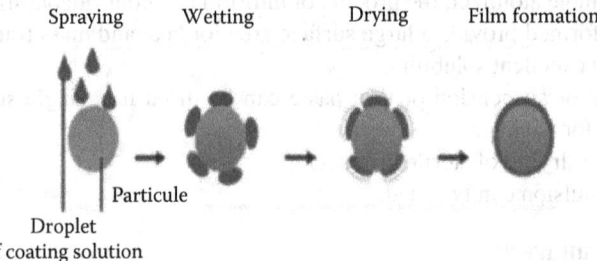

FIGURE 21.10 Mechanism of film formation.

the preparation of compressed tablets.[17] Fluidized bed system contains various components such as (Figures 21.9 and 21.10)

- Air-handling unit
- Product container and air distributor
- Spray nozzle
- Disengagement area and process filters
- Exhaust blower or fan
- Control system
- Solution delivery system

21.2.3.1 Advantages
1. It reduces dust formation during processing, thus improving housekeeping.
2. It reduces product loss.
3. It improves worker safety.

Encapsulation Process in Granulation Technology

21.2.3.2 Disadvantages
1. The fluid bed cleaning is labor intensive and time consuming.
2. Difficulty of assuring reproducibility.

21.2.4 ENCAPSULATION BY FLUID BED SPRAY GRANULATION[18,19]

Through the process of fluid bed spray granulation, spherical and compact granules with extraordinary physical properties can be obtained. The solid-containing liquids (e.g., suspensions or emulsions) are atomized in the fluid bed over starter seed granules. As the atomized droplets fall, the liquid evaporates, and the solid is drawn to the granule seed, forming a stable coating. This mechanism is repeated steadily in the fluid bed, resulting in shell-like granules. Process parameters such as particle size, residual moisture, and solid content can be adjusted. The process is suitable for all applications, where a homogeneous, dust-free granule with high bulk density is desired (Figure 21.11).

21.2.5 ENCAPSULATION BY FLUID BED ROTOR GRANULATION[10,17]

- Forces on the powder are balanced by the airflow (drag force and buoyancy) and the centrifugal force. It can uniformly fluidize much finer powders than conventional fluidized beds.
- A novel rotating fluidized bed system has been developed for fluidizing, granulating, and coating cohesive fine powders to tailor their properties and functionalities (Figure 21.12).

21.2.6 ENCAPSULATION BY EXTRUSION–SPHERONIZATION[20]

Extruders are thermomechanical mixers that consist of one or more screws in a barrel. Extrusion technology was initially applied in the plastics processing area, and after years of development and application, it has become a well-elaborated tool with technical solutions available for other fields like the pharmaceutical industry (for the production of controlled-release formulations).

The basic idea behind encapsulation using extrusion is to create a molten mass in which the active agents (either liquids or solids) are dispersed or dissolved. Upon cooling, this mass will solidify, thereby entrapping the active components.

It is a multiple-step process involving at least five steps capable of making uniform-sized spherical particles:

1. Dry mixing of materials to achieve homogeneous dispersion
2. Wet granulation of the resulted mixture to form wet mass
3. Extrusion of wet mass to form rod-shaped particles
4. Rounding off (in spheronizer)
5. Drying

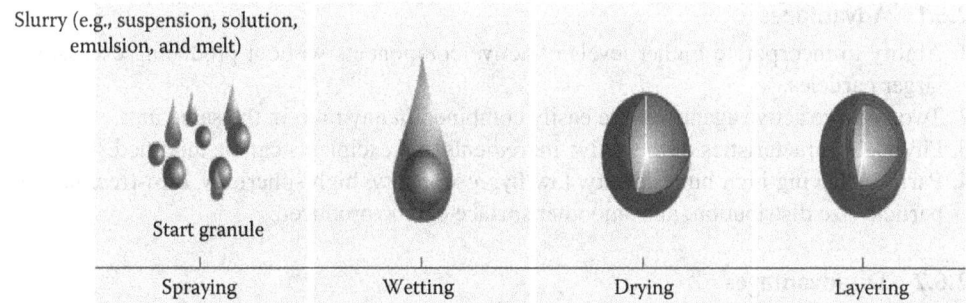

FIGURE 21.11 Fluid bed spray granulation.

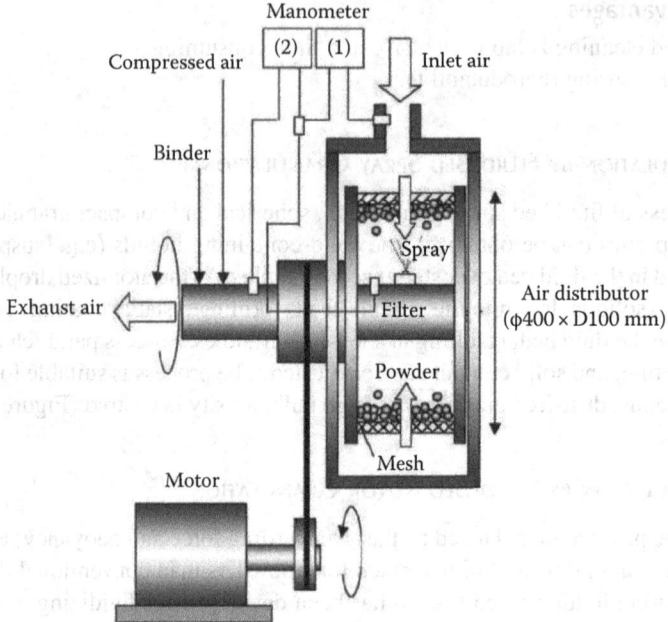

FIGURE 21.12 Fluid bed rotor granulation.

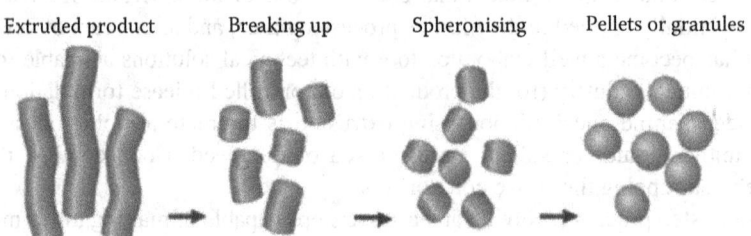

FIGURE 21.13 Different steps involved in the extrusion–spheronization process.

These dried, rounded particles can be optionally screened to achieve a targeted mean size distribution that describes schematically the steps involved in the extrusion–spheronization process (Figure 21.13).

21.2.6.1 Advantages
1. Ability to incorporate higher levels of active components without producing excessively larger particles.
2. Two or more active agents can be easily combined in any ratio in the same unit.
3. Physical characteristics of the active ingredients and excipients can be modified.
4. Particles having high bulk density, low hygroscopicity, high sphericity, dust-free, narrow particle size distribution, and smoother surface can be produced.

21.2.6.2 Disadvantages
This process is more labor and time intensive than other commonly used granulation techniques.

21.3 NOVEL APPROACHES IN ENCAPSULATION PROCESSES IN GRANULATION TECHNOLOGY

Over a period of time, due to technological advancements and in an urge to improve commercial output, various, newer granulation technologies have been evolved such as the following:

21.3.1 ENCAPSULATION BY VIBRATIONAL NOZZLE TECHNOLOGY[4,21]

- Core–shell encapsulation or microgranulation (matrix encapsulation) can be done using a laminar flow through a nozzle and an additional vibration of the nozzle or the liquid (Figure 21.14).
- The vibration has to be done in resonance of the Rayleigh instability and leads to very uniform droplets.
- The liquid can consists of any liquids with limited viscosities (0–10,000 mPa·s has been shown to work), for example, solutions, emulsions, suspensions, and melts.
- The solidification can be done according to the used gelation system with an internal (e.g., sol–gel processing, melt) or an external (additional binder system, e.g., in a slurry) gelation.
- The process works very well for generating droplets between 100 and 5000 μm (3.9–200 miles); applications for smaller and larger droplets are known.
- The units are deployed in industries and research mostly with capacities of 1–10,000 kg/h (2–22,000 lb/h) at working temperatures of 20°C–1500°C (68°F–2732°F) (room temperature up to molten silicon).
- Nozzles heads are available from one up to several hundred thousand.

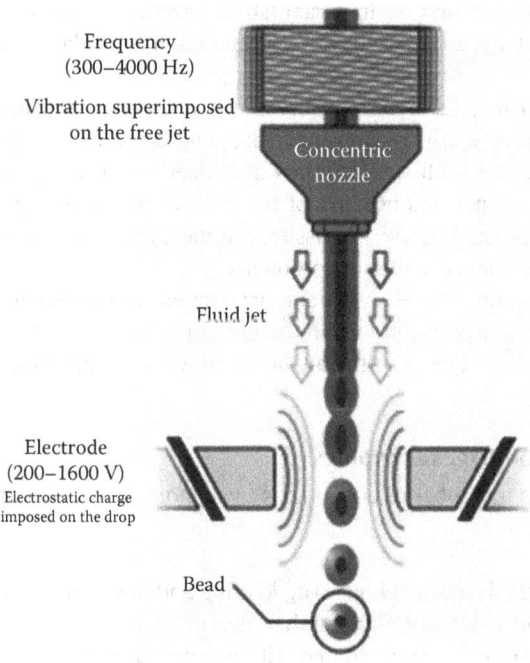

FIGURE 21.14 Droplet formation based on nozzle vibration technology.

21.3.2 Encapsulation by Pneumatic Dry Granulation[22,23]

The pneumatic dry granulation (PDG) Technology

- Is based on a PDG process, a novel dry method for automatic or semiautomatic production of granules
- Enables flexible modification of drug load, disintegration time, and tablet hardness
- Can achieve the following:
  - High drug loading, even with "difficult" APIs and combinations
  - Taste masking
  - Excellent stability
- Is compatible with other technologies, such as sustained release, fast release, and coating
- Is suitable for heat-labile and moisture-sensitive drugs
- Is the subject of a number of patent applications

The PDG Technology™ produces porous granules with excellent compressibility and flowability characteristics.

21.3.2.1 Granulate Any API

The PDG process can granulate virtually any pharmaceutical solid dosage ingredient. The granulated material has exceptionally good flowability and compressibility properties.

PDG Technology has been used with superior results in developing fast-release, controlled-release, fixed-dose, and orally disintegrating tablets. The technology is applicable to practically any solid dosage pharmaceutical product.

21.3.2.2 Pneumatic Dry Granulation Replaces Wet Granulation

Today, wet granulation is the most commonly used granulation method. Formulation teams will usually target a direct compression or dry granulation formulation where possible but in approximately 80% of the cases they end up with a wet granulation formulation due to processing issues (Figure 21.15).

Wet granulation is also unsuitable for moisture-sensitive and heat-sensitive drugs, is more expensive than dry granulation, is relatively labor intensive, and can take a long time. There are a large number of process steps, and each step requires qualification, cleaning, and cleaning validation; high material losses can be incurred because of the transfer between stages; and there is the need for long drying times. Scale-up is usually an issue, and there are considerable capital requirements. PDG Technology solves the aforementioned problems.

PDG Technology granules have excellent properties compared to wet granulation, dry granulation, and direct compression. At the same time, the granules show both high compressibility and flowability. The results can be archived without using exotic and expensive excipients.

21.3.2.3 Advantages of PDG Technology

The PDG Technology has a number of advantages to support the aforementioned claims including the following:

- Good granulation results even at high drug loading and has been achieved even with materials known to be historically difficult to handle.
- Faster speed of manufacturing compared with wet granulation.
- Lower cost of manufacturing compared with wet granulation.
- The system is closed offering safety advantages due to low dust levels and potential for sterile production or handling of toxic materials.

Encapsulation Process in Granulation Technology

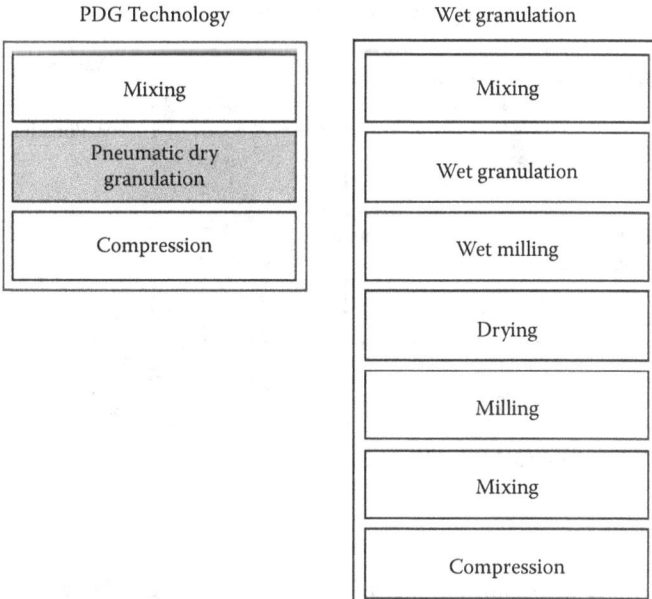

FIGURE 21.15 PDG Technology and wet granulation comparison.

- The end products are very stable; shelf life may be enhanced.
- Little or no waste of material.
- Scale-up is straightforward.
- The granules and tablets produced show fast disintegration properties, offering the potential for fast-release dosage forms.
- Release time can be tailored to requirements.

21.3.2.4 Benefits to Pharmaceutical Companies

PDG Technology is the key solution to challenges faced by pharmaceutical companies in development of solid oral dosage forms. The technology replaces existing solid dosage form development and manufacturing technologies, offering more rapid development and better quality. The unique capabilities of the technology have been demonstrated in number of evaluation studies with top-tier pharmaceutical companies.

21.3.3 ENCAPSULATION BY FREEZE GRANULATION TECHNOLOGY[23–25]

The Swedish Ceramic Institute has adopted and developed an alternative technique, freeze granulation (FG), which enables preservation of the homogeneity from suspension to dry granules by spraying a powder suspension into liquid nitrogen; the drops (granules) are instantaneously frozen. In a subsequent freeze-drying, the granules are dried by sublimation of the ice without any segregation effects as in the case of conventional drying in air. The result will be spherical, free-flowing granules, with optimal homogeneity (Figure 21.16).

FG provides optimized condition for the subsequent processing of the granules, for example, easy crushing to homogeneous and dense powder compacts in a pressing operation. High degree of compact homogeneity will then support the following sintering with minimal risks for granule defects.

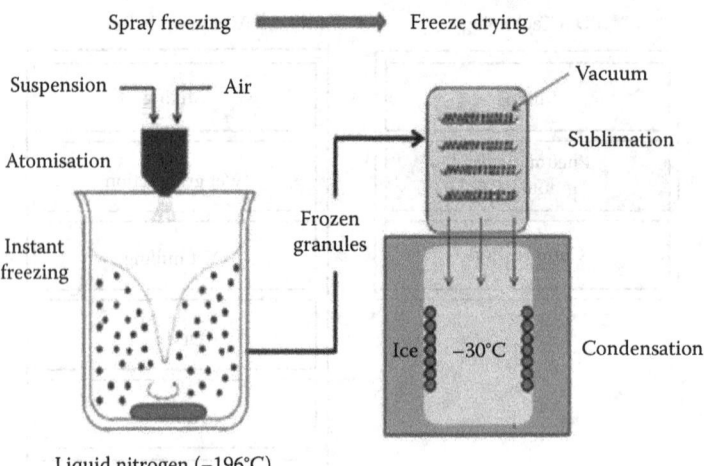

FIGURE 21.16 Freeze granulation.

Besides a high degree of granule homogeneity, FG offers several other advantages:

- Control of granule density by the solids content of the suspension.
- Mild drying prevents serious oxidation of nonoxides and metals.
- No cavities in the granules.
- Low material waste (high yield).
- Small (50–100 mL suspension) as well as large granule quantities can be produced to equal quality.
- Easy clean of the equipment (latex binder can be used).
- Possibility to recycle organic solvents.

The granule size distribution will be controlled by the suspension rheology (flow properties) and the process parameters (pump speed and air pressure). Normally, a certain size distribution width is achieved with an average size, typically around 100–200 μm.

Several companies and research labs around the world have applied the FG process with the support of PowderPro. Typical ceramic powders are oxides (Al_2O_3, ZrO_2, and SiO_2), nitrides (Si_3N_4), and carbides (SiC), but also nanopowders, diamonds, and pharmaceuticals like proteins and enzymes.

21.3.4 Encapsulation by Foamed Binder Technologies[23]

Foamed binder technology (FBT) from the Dow Chemical Company can help you achieve faster, simpler, and safer wet granulation processing.[26] Using familiar, proven METHOCEL polymers, this technology greatly improves binder distribution in the formulation mix and yields a remarkable array of processing advantages.[27]

Compared to conventional spray processing, FBT can shorten processing times by reducing water requirements. It can improve reproducibility through more uniform binder distribution. Moreover, it eliminates spray nozzles and their many variables in granulation processing equipment. Foam processing also offers better end-point determinations and reduced equipment clean-up time.

While foamed binder processing offers many advantages, this technology doesn't demand new equipment or radical changes in processing techniques. You can very easily use it with familiar high-shear, low-shear, or fluid bed granulation equipment, in both laboratory- and production-scale settings.[28] Our evaluations also show it yields familiar metrics for particle size distributions, solid dose physical properties, and dissolution profiles.

21.3.4.1 How Foam Binder Granulation Works

Foam granulation takes advantage of the tremendous increase in the liquid surface area and volume of polymeric binder foams to improve the distribution of the water/binder system throughout the powder bed of a solid dose pharmaceutical formulation.

A simple foam generation apparatus is used to incorporate air into a conventional water-soluble polymeric excipients binder such as METHOCEL hypromellose (hydroxypropyl methylcellulose). The resulting foam has a consistency like shaving cream. Hypromellose polymers are ideal candidates for this technology because they are excellent film formers and create exceptionally stable foams.

In a small-scale laboratory setting or in a full-scale production setting, the foam generator can be connected directly to high-shear, low-shear, or fluid bed granulation equipment.[29]

21.3.4.2 Extremely Efficient Binder Delivery and Particle Coverage

The key to the effectiveness of foam binder performance is rapid and extremely efficient particle coverage. Compared to sprayed liquid binders, foamed binders offer much higher surface area, and they spread very rapidly and evenly over powder surfaces. The foamed binders and the powder particles show excellent mutual flow through one another.

The foam binder also shows a low soak: spread ratio, so particle surfaces are quickly and completely covered. By contrast, spraying is a cumulative process that begins with small liquid droplets "dappling" particle surfaces until enough binder liquid accumulates to initiate particle agglomeration. Spraying requires considerably more water and processing time than a foamed binder to achieve particle agglomeration.

The foam binder technology also eliminates the need for spray nozzles and all of their attendant variables, such as nozzle configuration, distance from the moving powder bed, spray patterns, clogging, droplet size, and droplet distribution. The dilute binder solutions are easy to handle in processing. Overall, foam binder processing is easier and faster and allows safer handling of potent drug compounds.[29]

21.3.5 ENCAPSULATION BY MELT GRANULATION TECHNOLOGY[30]

Melt granulation is a process by which granules are obtained through the addition of either a molten binder or a solid binder that melts during the process. This process is also called "melt agglomeration" and "thermoplastic granulation."

21.3.5.1 Principle of Melt Granulation

The process of granulation consists of a combination of three phases:

1. Wetting and nucleation
2. Coalescence step
3. Attrition and breakage

21.3.5.1.1 Wetting and Nucleation Step
- During the nucleation step, the binder comes into contact with the powder bed and some liquid bridges are formed, leading to the formation of small agglomerates.

Two nucleation mechanisms are proposed by Schafer and Mathiesen:

1. Immersion
2. Distribution

21.3.5.1.1.1 Immersion
- Nucleation by immersion occurs when the size of the molten binder droplets is greater than that of the fine solid particles.
- Immersion proceeds by the deposition of fine solid particles onto the surfaces of molten binder droplets (Figure 21.17).

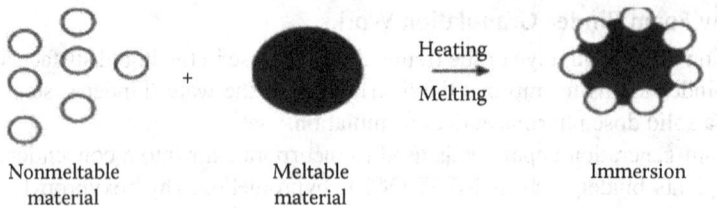

FIGURE 21.17 Modes of melt agglomeration: immersion.

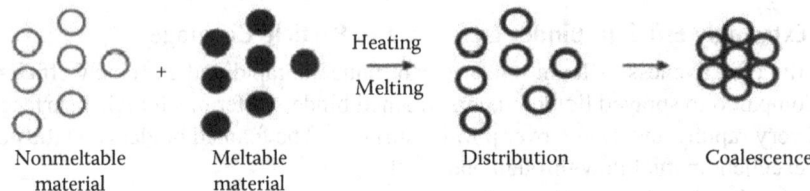

FIGURE 21.18 Modes of melt agglomeration: distribution.

21.3.5.1.1.2 Distribution
- In the distribution method, a molten binding liquid is distributed onto the surfaces of fine solid particles.
- The nuclei are formed by the collision between the wetted particles.
- Generally, small binder droplet size, low binder viscosity, and high shearing forces are favorable conditions for nucleation by the distribution method (Figure 21.18).

21.3.5.1.2 Coalescence Step
- It involves nuclei that have residual surface liquid to promote successful fusion of nuclei.
- The surface liquid imparts plasticity to the nuclei and is essential for enabling the deformation of nuclei surface for coalescence as well as promoting the rounding of granulation.[31,32]

21.3.5.1.3 Attrition–Breakage Step
- Attrition and breakage refer to the phenomenon of granulation fragmentation in that are solidified by tray cooling to ambient temperature without the need for drying by a tumbling process.
- Consequently, breakage is known to have a more essential role in affecting the resultant properties of the melt granulation during the granulation phase.

21.3.5.2 Requirements of Melt Granulation
- Generally, an amount of 10%–30% w/w of meltable binder, with respect to that of fine solid particles, is used.
- A meltable binder suitable for melt a granulation has a melting point typically within the range of 50°C–100°C.
- Hydrophilic meltable binders are used to prepare immediate-release dosage forms, while the hydrophobic meltable binders are preferred for prolonged-release formulations.
- The melting point of fine solid particles should be at least 20°C higher than that of the maximum processing temperature.

TABLE 21.5
Hydrophilic Meltable Binders Used in the Melt Granulation Technique

Hydrophilic Meltable Binder	Typical Melting Range (°C)
Gelucire 50/13	44–50
Poloxamer 188	50.9
Polyethylene glycols:	
PEG 2000	42–53
PEG 3000	48–63
PEG 6000	49–63
PEG 8000	54–63

TABLE 21.6
Hydrophobic Meltable Binders Used in the Melt Granulation Technique

Hydrophobic Meltable Binder	Typical Melting Range (°C)
Bees wax	56–60
Carnauba wax	75–83
Cetyl palmitate	47–50
Glyceryl stearate	54–63
Hydrogenated castor oil	62–86
Microcrystalline wax	58–72
Paraffin wax	47–65

21.3.5.3 Meltable Binders
- It must be solid at room temperature and melts between 40°C and 80°C.
- It has physical and chemical stability.
- It has hydrophilic–lipophilic balance to ensure the correct release of the active substance.
- There are two types of meltable binder:
  - Hydrophilic meltable binders (Table 21.5)
  - Hydrophobic meltable binder (Table 21.6)

21.3.5.4 Advantage of Melt Granulation
- Neither solvent nor water is used.
- Fewer processing steps are needed; thus time-consuming drying steps are eliminated.
- Uniform dispersion of fine particle occurs.
- Good stability at varying pH and moisture levels.
- Safe application in humans due to their nonswellable and water-insoluble nature.

The melt granulation process carries several advantages over conventional pharmaceutical granulation methods, as the process does not require the use of solvents. A further significant advantage of melt granulation is that judicious choice of the granulation excipient may enable the formulator to manipulate the drug dissolution rate from the corresponding dosage form.

21.3.5.5 Disadvantages
- Heat-sensitive materials are poor candidates.
- Binders having melting point in the specific range can only be utilized in the process.

The melt granulation process uses substances that melt at relatively low temperature (i.e., 50°C–80°C). These substances can be added to the molten form over the substrate or to a solid form, which is then heated above its melting points by hot air or by a heating jacket. In both cases, the substance acts like a liquid binder after it melts. Thus, melt granulation does not require the organic or aqueous solvents. Moreover, the drying step is not necessary in melt granulation; thus the process is less time-consuming and more energy efficient than wet granulation.[33]

After selecting a suitable binder, one can use melt granulation to prepare controlled-release or improved-release granules. Polyoxyl stearates may be considered as potentially useful hydrophilic binders in melt granulation. When water-soluble binders are needed, polyethylene glycol (PEG) is used as melting binders.[34] When water-insoluble binders are needed, stearic acid, cetyl or stearyl alcohol, various waxes, and mono-, di-, and triglycerides are used as melting binders.

21.3.6 Encapsulation by Steam Granulation[35,36]

- It is a modification of wet granulation. Here, steam is used as a binder instead of water.
- This method of granulating particles involves the injection of the required amount of liquid in the form of steam.
- This steam injection method, which employs steam at a temperature of about 150°C, tends to produce local overheating and excessive wetting of the particles in the vicinity of the steam nozzles, thereby causing the formation of lumps in the granulated product.

21.3.6.1 Advantages
- Higher distribution uniformity.
- Higher diffusion rate into powders.
- Steam granules are more spherical.
- Have large surface area, hence increased dissolution rate of the drug from granules.
- Processing time is shorter; therefore, more number of tablets are produced per batch.
- Compared to the use of an organic solvent, water vapor is environmentally friendly.
- Lowers dissolution rate so it can be used for the preparation of taste-masked granules without modifying availability of the drug.

21.3.6.2 Disadvantages
- Requires special equipment for steam generation and transportation.
- Requires high energy inputs.
- Thermolabile materials are poor candidates.
- More safety measures required.
- Not suitable for all the binders.

21.3.7 Encapsulation by Moisture-Activated Dry Granulation

- In this method, moisture is used to activate the granules formation, but the granules' drying step is not necessary due to moisture-absorbing materials such as microcrystalline cellulose (MCC) add.[37,38]
- The moisture-activated dry granulation (MADG) process consists of two steps, wet agglomeration of the powder mixture followed by moisture absorption stages.
- A small amount of water (1%–4%) is added first to agglomerate the mixture of the API, a binder, and excipients. Moisture-absorbing material such as MCC and potato starch is then added to absorb any excessive moisture.[39,40]
- After mixing with a lubricant, the resulting mixture can then be compressed directly into tablets. Hence, this process offers the advantage of wet granulation that eliminates the need for a drying step (Figure 21.19).

Encapsulation Process in Granulation Technology

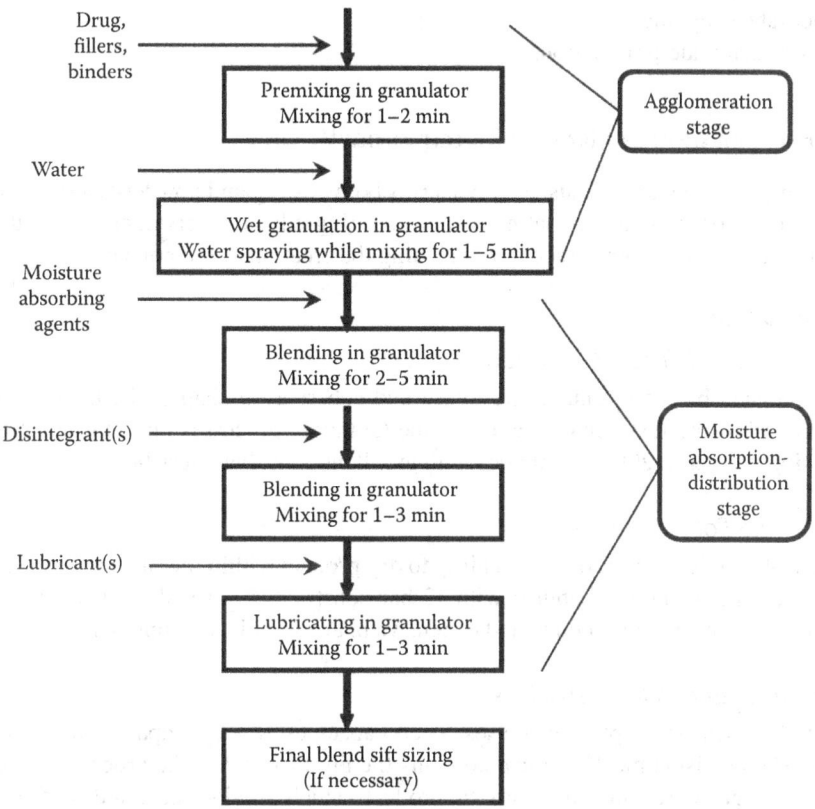

FIGURE 21.19 Flow diagram of moisture-activated dry granulation process.

FMC Biopolymer has introduced two new excipient products to the pharma market, Avicel HFE-102 and Avicel PH-200 LM, which are based on already existing excipients but have been generated to produce a different entities with improved benefits.

Avicel PH-200 LM, based on MCC, has been formulated to reduce the amount of water added to the granulation process. Avicel PH-200 LM is a step up from FMC Biopolymer's Avicel PH-200, which had a moisture level of 5%. The new product has a moisture level of no more than 1.5% and can absorb approximately three to four times as much water from the granule. This advantage, along with enabling the use of MADG, meant the use of Avicel PH-200 LM could eliminate the extra steps of milling, drying, and screening, thereby reducing manufacturing costs and energy used. The process also produced a larger particle size for optimal flow. This increases efficiencies to the manufacturing process. It takes aspects of wet granulation but eliminates the drawbacks of it. Also it is useful for the use of APIs that were sensitive to moisture.[40,41]

Avicel HFE-102 is a new, proprietary cospray-dried MCC/mannitol high functionality binding excipient for direct compression. The cospray drying added extra benefits to the excipient as it changed its properties combining the high compactibility of MCC and the low lubricant sensitivity of mannitol. The outcome was a harder, less friable, and faster disintegrating tablet.

21.3.7.1 Advantages
- It utilizes very little granulating fluid.
- It decreases drying time and produces granules with excellent flow ability.
- Single production equipment (high shear granulator).
- No equipment change.

- Lower tablet capping.
- No over- and undergranulation.

21.3.8 Encapsulation by Granurex® Technology[42,43]

The Granurex® precisely and consistently performs both coating and powder-layering processes. In the pictures to the right, multiple coating and powder (ingredient) layers demonstrate the accuracy and control of a Granurex rotor processor, including the creation of the nonpareil.

21.3.8.1 Key Features

21.3.8.1.1 Unique, Efficient Granulation Processes

Granules produced by the Granurex are dense and spherical in shape. The pictures and graphs shown in the following text demonstrate how the Granurex processes ciprofloxacin from a 7 μm poorly flowing powder to 200 μm granules with excellent flow characteristics.

21.3.8.1.2 One-Pot Processing

A patented feature of the Granurex is its ability to dry product within the same processing chamber. This unique drying method, combined with 12 bar construction, provides a true one-pot system, ideal for manufacturing highly potent and expensive pharmaceutical compounds.

21.3.8.1.3 Increased Batch Capacities

The patented conical rotor plate increases batch capacities when compared to traditional rotor processors. The precision machined gap contains the product within the processing area, and the peripheral spray guns are embedded into the product, which provides accurate coating with minimal spraying defects.

21.3.8.1.4 Maximum Process Flexibility

Using micronized acetaminophen as the base material, the Granurex produced both a 100 μm granulation and a 200 μm spherical bead. In both of the examples in the following text, the APAP core material had the same initial mean particle size (X50) of 40 μm.

21.3.9 Encapsulation by Thermal Adhesion Granulation Process[23,44]

It is applicable for preparing direct tabletting formulations. The thermal adhesion granulation (TAG) process is performed under low moisture content or low content of pharmaceutically acceptable solvent by subjecting a mixture containing one or more diluents and/or active ingredients, a binder, and optionally a disintegrant to heating at a temperature in the range from about 30°C to about 130°C in a closed system under mixing by tumble rotation until the formation of granules. This method utilizes less water or solvent than traditional wet granulation method. It provides granules with good flow properties and binding capacity to form tablets of low friability, adequate hardness and have a high uptake capacity for active substances whose tabletting is poor.

In TAG, granules are formed during mixing of the moist powder under continuous tumble rotation, as the heated powder mass flows within the container and agglomerates with the aid of the binder. Drying and milling to form the desired granules are unnecessary in the present invention due to the low amount of moisture introduced to the tabletting mixture.

Another major advantage of granulating pharmaceutical products in a closed system is that it helps to minimize the generation of dust during powder processing. This technique serves to contain fine-powder active ingredients whose spread or loss from the system is not desirable due to their cost or biological activity.

21.4 CONCLUSION

This article provides a review on various granulation techniques that are accessible in the pharmaceutical industry. A judicial assortment of appropriate technology for carrying out the granulation process is the key to achieve a targeted granulation and final product parameters. In-depth knowledge of the processing techniques and their merits and demerits is required to adopt during the development stage of product. A systematic approach should be followed for selecting the suitable granulation process. Which method is chosen depends on the ingredients individual characteristics and ability to properly flow, compresses, eject, and disintegrate. Choosing a method requires thorough investigation of each ingredient in the formula, the combination of ingredients, and how they work with each other. Then the proper granulation process can be applied. A systematic approach should be followed for selecting the suitable granulation process. The review article aims to provide comprehensive information in this regard, which will be useful for the researchers and scientists involved at the product development stage.

REFERENCES

1. Parikh DM. Theory of granulation, in *Handbook of Pharmaceutical Granulation Technology*. Marcel Dekker Inc., New York, 1997, pp. 7–13.
2. Lachman L, Liberman HA, Kanig JL. Tablets, in *The Theory and Practice of Industrial Pharmacy*, 3rd edn. Varghese Publishing House, Bombay, India, 1991, pp. 320–321.
3. Allen LV, Popovich NG, Ansel HC. *Pharmaceutical Dosage Forms and Drug Delivery Systems*. BI Publication, Delhi, India, 2005, Vol. 8, p. 265.
4. Kreitz M, Brannon-Peppas L, Mathiowitz E. *Microencapsulation Encyclopedia of Controlled Drug Delivery*. John Wiley & Sons Publishers, 1999, pp. 493–553.
5. Kristensen HG, Schaefer T. Granulation, a review on wet granulation, *Drug Dev. Ind. Pharm.* (1987); 13(4&5): 803–872.
6. Iveson SM, Litster JD, Hapgood K, Ennis BJ. Nucleation, growth and breakage phenomena in agitated wet granulation processes: A review. *Powder Technol.* (2001); 117: 3–39.
7. Ré M.-I. Formulating drug delivery systems by spray drying. *Drying Technol.* (2006); 24: 433–446.
8. Dilip MP. Spray drying as a granulation technique, in *Handbook of Pharmaceutical Granulation Technology, Drugs and the Pharmaceutical Sciences*. Marcel Dekker, New York, 1997, pp. 75–96.
9. Swarbrick J, Boylan J. Spray drying and spray congealing of pharmaceuticals, in *Encyclopedia of Pharmaceutical Technology*, Marcel Dekker, New York, 1992, pp. 207–221.
10. Aulton M. Drying, in *Pharmaceutics—The Science of Dosage Form Design*. Churchill Livingstone, Edinburgh, U.K., 2002, p. 390.
11. Oral solid dosage forms, in *Remington—The Science and Practice of Pharmacy*, 1995, 4: 1627–1628.
12. Mujumdar S. *Handbook of Industrial Drying*. CRC Press, Boca Raton, FL, 2007, p. 710.
13. Thiel W. J. The theory of fluidization and application to the industrial processing of pharmaceutical products. *Int. J. Pharm. Tech. Prod. Manuf.* (1981); 2(5):5–8.
14. Lipsanen T, Antikainen O, Räikkönen H, Airaksinen S, Yliruusi J. Effect of fluidization activity on endpoint detection of a fluid bed drying process. *Int. J. Pharm.* (2008); 357: 37–43.
15. Gore AY, McFarland DW, Datuyios NII. Fluid bed granulation. Factors affecting the process in laboratory development and production scale-up. *Pharm. Technol.* (1985); 9(9): 114.
16. Wurster DE. Air-suspension technique of coating drug particles. *J. Am. Pharm. Assoc.* (1959); 48: 451–454.
17. Banks M, Aulton ME. Fluidised-bed granulation: A chronology. *Drug Dev. Ind. Pharm.* (1991); 17: 1437–1463.
18. Rubino OR. Fluid-bed technology; Overview and criteria for process selection. *Pharm. Technol.* (1999); 6: 104–113.
19. www.niroinc.com/pharma_systems/spray_fluid_bed_granulation.asp.
20. Parikh DM. Extrusion spheronization as a granulation technique, in *Handbook of Pharmaceutical Granulation Technology*, Marcel Dekker Inc., New York, 1997, pp. 334–337.
21. Bansode SS, Banarjee SK, Gaikwad DD, Jadhav SL. Microencapsulation: A review. *Int. J. Pharm. Sci. Rev. Res.* (2010); 1: 38–43.

22. Sandler N, Lammens RF. Pneumatic dry granulation: Potential to improve roller compaction technology in drug manufacture. *Exp. Opin. Drug Deliv.* (2011) Feb; 8(2): 225–36. doi: 10.1517/17425247.2011.548382.
23. Himanshu KS, Basuri T, Thakkar JH, Patel CA. Recent advances in granulation technology. *Int. J. Pharm. Sci. Rev. Res.* (2010); 5(3): 008.
24. Rundgren K, Lyckfeldt O, Sjöstedt M., Improving powders with freeze granulation, *Ceram. Ind.* (2003); pp. 40–44.
25. www.powderpro.se/uploads/media/Freeze_Granulation_of_Nano_MaterialsLondon_June_2010.pdf.
26. Paul J, Shesky R, Colin K, New foam binder technology from Dow improves granulation process. *Pharmaceutical Canada*, June 2006; pp. 19–22.
27. Sheskey P. et al. Foam technology: the development of a novel technique for the delivery of aqueous binder systems in high-shear and fluid-bed wet-granulation applications, in Poster Presented at *AAPS Annual Meeting and Exposition*, Salt Lake City, UT, Oct. 2003; pp. 26–30.
28. Sheskey P. et al. Scale-up trials of foam granulation technology—High shear," *Pharm. Technol.* (2007); 31 (4): 94–108.
29. Keary CM, Sheskey PJ. Preliminary report of the discovery of a new pharmaceutical granulation process using foamed aqueous binders. *Drug Dev. Ind. Pharm.* (2004); 30(8): 831–845.
30. Chaudhari PD. Melt granulation technique: A review. *Pharmainfo.net.* (2006); 4 (1).
31. Breitenbach J. Melt extrusion: From process to drug delivery technology. *Eur. J. Pharm. Biopharm.* (2002); 54: 107–117.
32. Chokshi R, Zia H. Hot melt extrusion technique: A review. *Iran. J. Pharm. Res.* (2004); 3: 3–16.
33. Heng WS, Wong TW. Melt processes for oral solid dosage forms. *Pharm. Technol.* (2003); 1–6.
34. Kidokoro M, Sasaki K, Haramiishi Y, Matahira N. Effect of crystallization behavior of polyethylene glycol 6000 on the properties of granule prepared by fluidized hot melt granulation (FHMG). *Chem. Pharm. Bull.* (2003); 51 (5): 487–493.
35. United States Patent 4489504. Steam granulation apparatus and method.
36. http://www.pharmaprdia.com/Tablet:manufacturing method/Granulation.
37. Ismat U, Jennifer W. Moisture-activated dry granulation: The one pot process. *Pharm. Technol. Eur.* (2010); 22(3).
38. Ullah I, Wang J, Chang S-Y, Wiley GJ, Jain NB, Kiang S. Moisture-activated dry granulation—Part I: A guide to excipient and equipment selection and formulation development. *Pharmaceut. Tech.* (2009); 33(11): 62–70.
39. Ullah I, Wang J, Chang S-Y, Guo H, Kiang S, Jain NB. Moisture-activated dry granulation—Part II: The effects of formulation ingredients and manufacturing-process variables on granulation quality attributes. *Pharm. Technol.* (2009);33(12),42–51
40. Gerhardt AH. Moisture effects on solid dosage forms formulation, processing and stability. *J. GXP Compliance* (Winter 2009); 13(1); 58–66.
41. www.vectorcorporation.com/news/papers.asp, Optimization of Binder Level in Moisture Activated Dry Granulation (MADG) using absorbent starch to distribute moisture.
42. www.vectorcorporation.com/technology/granurex.asp.
43. www.freund-vector.com/downloads/.../Granurex%20Technology.pdf.
44. Hong-Liang LIN, Hsiu HO, Chi-Chia C, Ta-Shuong Y, Ming-Thau S. Process and formulation characterizations of the thermal adhesion granulation (TAG) process for improving granular properties. *Int. J. Pharm.* (2008); 357(1–2): 206–212.

22 Encapsulation via Electrohydrodynamic Atomization Spray Technology (Electrospray)

*Milad Jafari-Nodoushan, Hamid Mobedi, and Jalal Barzin**

CONTENTS

22.1 Introduction 411
22.2 Principles 412
22.3 Different Configurations 416
    22.3.1 Spray Head 417
    22.3.2 Collection Chambers 418
    22.3.3 Other Configurations 420
22.4 Advantages and Disadvantages 420
    22.4.1 Advantages 420
    22.4.2 Disadvantages 422
22.5 Applications 422
22.6 Controlling Characteristics 424
    22.6.1 Process Parameters 424
    22.6.2 Solution Parameters 430
    22.6.3 Drug Release 431
22.7 Conclusion 433
References 433

22.1 INTRODUCTION

Encapsulation is the process of protecting, packing, entrapping, or immobilizing an active material such as a food ingredients, flavorings, drug molecules, and chemical substances, in a solid, liquid, or gas state within a shielding material released at a controlled rate or immediately under specific conditions.[1] Encapsulation is used in industries such as agriculture and food and pharmaceutical technologies. Encapsulation improves the stability of an incorporated substance against the environment, masking undesirable tastes, flavors and colors, determining dosage, controlling the handling and release of active ingredients at specified sites.[2] Several studies have proposed techniques to fabricate micro- and nanoparticles that act on their own or encapsulate active substances. The most extensively used technique to produce microspheres is an emulsion/evaporation-based method.[3]

Micro-/nanoparticles are preferred to produce in a continuous process instead of in batches; this increases their output and quality and decreases the end product cost. Electrohydrodynamic atomization (EHDA; electrospraying) is a useful technique for continuous production of fine particles.

---

* Lead author.

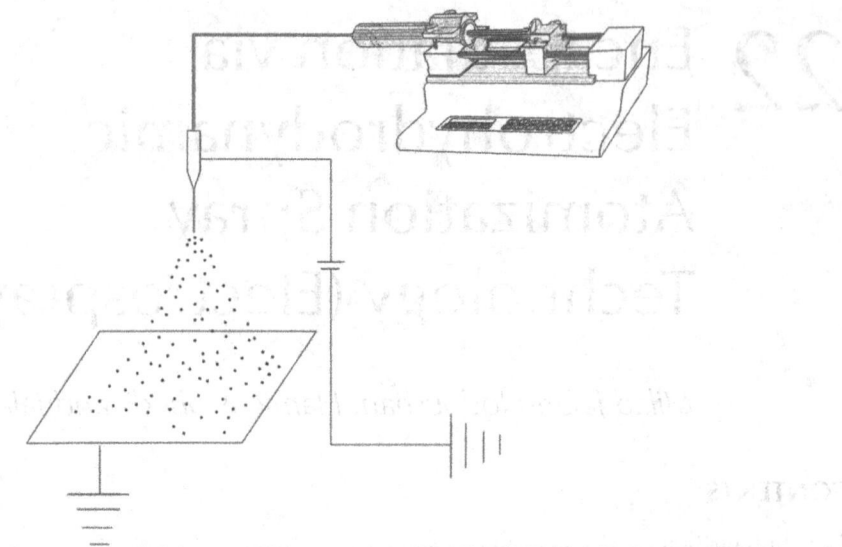

FIGURE 22.1 A simple electrospray setup. From Jafari-Nodoushan, M. et al., Design of a combined morphine delivery system (morphine microsphere-embedded in-situ forming), in *Iranian Controlled Release Society*, Tehran, Iran, 2014.

In electrospraying, high-voltage electrical current is applied to a polymer solution flowing through a nozzle, which causes the surface of the solution meniscus at the nozzle opening to become highly charged (Figure 22.1).[4]

The induced electrical tension at the meniscus surface pushes it away from the nozzle tip, contrasting the surface tension. Increasing the voltage allows electrical forces to overcome the surface tension that minimizes the meniscus surface area; a conical shape develops and a very thin capillary jet is ejected from its apex.[5] The jet breaks up and forms micro- and nanoscale droplets. Depending on the type of collector, collecting media, material used, process parameters, and equipment configuration, a series of applications ranging from nanoparticle drug transporters[6] to deposition of micro-/nanoscale particle film[7] can be achieved.

Zeleny was a pioneer of EDHA[8]; however, Rayleigh studied the stability of jets more than 20 years before Zeleny[9] in 1879. After Zeleny, little improvement was achieved in the understanding of the process until 1952, when Vonnegut and Neubauer investigated the monodispersity of droplets from electrostatic spray.[10] In 1964, Taylor explained the conical shape of the jet and the balance between electrical and surface tension that oppose each other.[11] Subsequent researchers investigated the use of electrostatic spraying for different applications[3] and have continued to improve the process.[12]

22.2 PRINCIPLES

The formation of electrically induced liquid spray is achieved by applying high electrical potential on the scale of a few kilovolts to the solution. The electrical potential is usually applied to a metal needle carrying the solution to extrude it into the atmosphere. The flow rate of the fluid is usually controlled by a precision pump. When no voltage is applied, the liquid drips from the capillary at a critical volume, depending on its surface tension.[13] The surface tension gives a spherical shape to falling droplets because that shape has smallest area for a given volume.

Under relatively low voltage, the electrical forces act with gravitational forces and push the droplets, decreasing its volume to what is known as the dripping mode. A further increase in voltage accumulates a charge on the liquid surface that causes the electrostatic pressure to oppose the capillary pressure, resulting in a decrease in apparent surface tension of the liquid and droplet volume (microdripping).[14]

A gradual increase in voltage increases the applied potential past the instability limit, producing instability in the fluid meniscus. In this condition, liquid is extruded in small droplets or long filaments. Each time a filament breaks off, the liquid meniscus relaxes back to its initial shape[13]; this cycle is repeated frequently, causing a pulsating mode. As the applied electrical field increases, the frequency of pulsation increases until the electric field reaches a sufficiently high value to form a jet of conical shape at the exit.

The jet of liquid then breaks into fine monodisperse droplets. This is the preferred mode of atomization because it continuously produces stable uniform particles.[15] Studies have shown that this mode is the preferred operating regime for particle production because it produces particles with a narrow size distribution.[16–19] This type of cone is known as the Taylor cone after the researcher who theoretically demonstrated the existence of conical menisci.[11]

A persistent increase in the electric field shifts the jet laterally. The shape of the jet becomes asymmetric, the jet splits into two emissive cups, and several jets develop at the end of the capillary[20,21] (Figures 22.2 and 22.3).

Researchers agree that the most important feature of the spray system is the mode of spraying.[22] The most important parameters in this system are (1) the physical properties of the liquid, especially its electrical conductivity, surface tension, and viscosity; (2) the electrical field; (3) the liquid flow rate; (4) the dielectric strength of the ambient atmosphere; and (5) the system configuration. Changes in parameter values produce different spraying modes, giving rise to aerosols with highly varied characteristics and unequal practical values.[23]

Conical jets can form different shapes under varying conditions. Figure 22.4 shows possible shapes for a conical jet to illustrate the process. Liquids with relatively high conductivities produce a cone with a virtually straight generatrix (Figure 22.4a) or may be shaped as shown in Figure 22.4b. The acceleration zone extends further toward the base of the cone as the liquid conductivity decreases

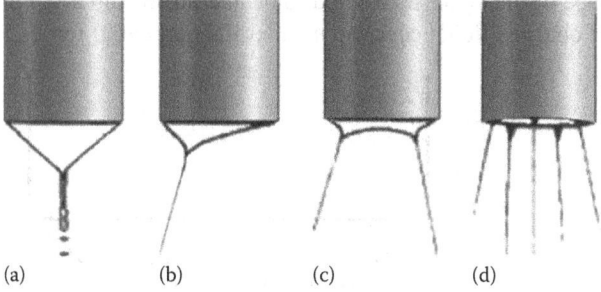

FIGURE 22.2 (a, b) Meniscus and jet during cone-jet mode and (c, d) variants of this mode. (Reprinted from *J. Electrostat.*, 22(2), Cloupeau, M. and Prunet-Foch, B., Electrostatic spraying of liquids in cone-jet mode, 135–159. Copyright 1989, with permission from Elsevier.)

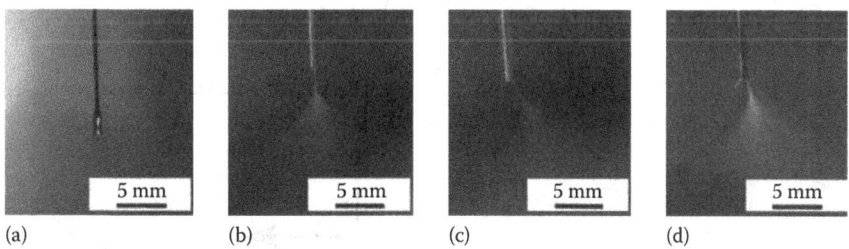

FIGURE 22.3 Different spraying modes: (a) dripping, (b) single cone jet, (c) asymmetric single cone jet, and (d) multijet. (Park, C.H. and Lee, J.: Electrosprayed polymer particles: Effect of the solvent properties. *J. Appl. Polym. Sci.* 2009. 114(1). 430–437. Copyright Wiley-VCH Verlag GmbH & Co. KGaA. Reproduced with permission.)

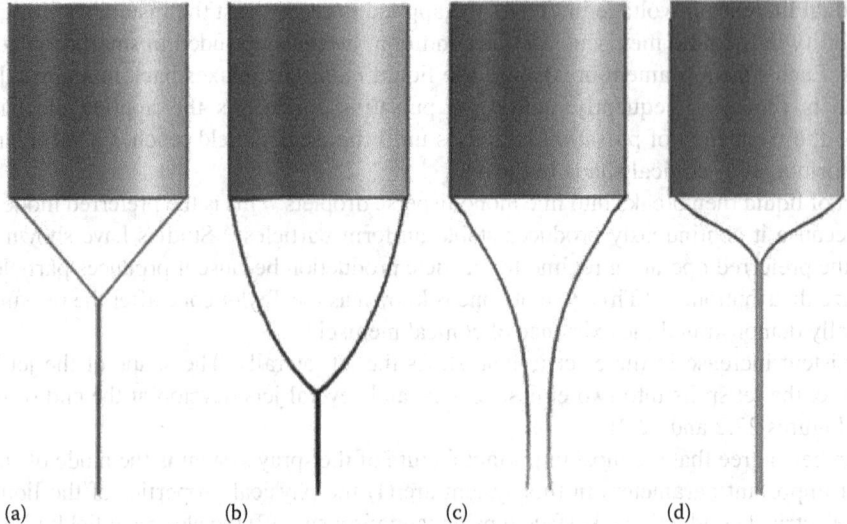

FIGURE 22.4 (a–d) Different forms of meniscus in cone-jet mode. (Reprinted from *J. Electrostat.*, 22(2), Cloupeau, M. and Prunet-Foch, B., Electrostatic spraying of liquids in cone-jet mode, 135–159. Copyright 1989, with permission from Elsevier.)

(Figures 22.4c and d).[20] Generally, liquids with high conductivity produce fine particles in cone-jet mode at low flow rates; liquids with low conductivity have the opposite effect.[14] Investigations have shown that a minimum solution conductivity of 0.01 μSm^{-1} is desired for EDHA processing.[24]

Figure 22.5 shows the acting force in formation of the conical jet. In Taylor cone-jet mode, the force on the surface of the cone is assumed to be at equilibrium at all points except near the apex, where the jet of charged fluid accelerates under the tangential forces.[25] In this mode, three

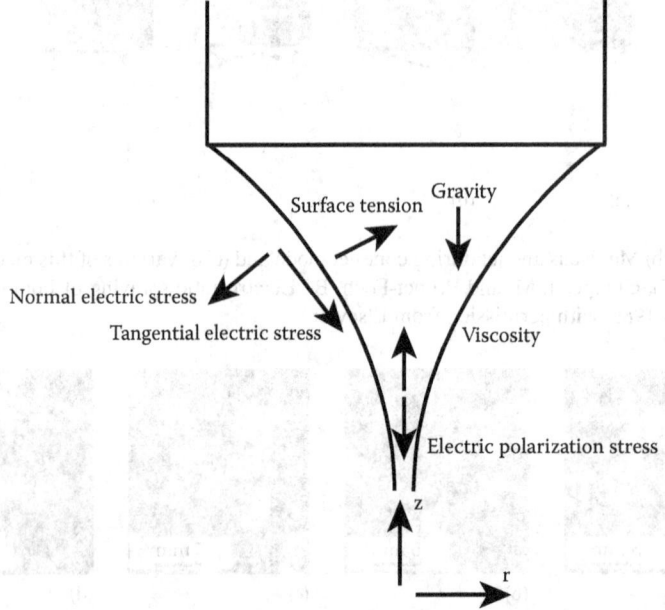

FIGURE 22.5 Forces acting on the Taylor cone in EHDA process. (Reprinted from *J. Aerosol Sci.*, 30(7), Hartman, R.P.A., Brunner, D.J., Camelot, D.M.A., Marijnissen, J.C.M., and Scarlett, B., Electrohydrodynamic atomization in the cone–jet mode physical modeling of the liquid cone and jet, 823–849. Copyright 1999, with permission from Elsevier.)

possible mechanisms exist for particle formation in response to droplet surface charge density. These mechanisms depend on the definition described by Rayleigh.

The Rayleigh limit is defined as the amount of charge on a drop at which the repulsive electric force overcomes the surface tension, leading to drop instability.[27] This limit is expressed as

$$q^2 = 64\pi^2 \varepsilon_0 \gamma \alpha^3 \qquad (22.1)$$

where
α is the drop radius
ε_0 is the permittivity of the atmosphere surrounding the jet
γ is the surface tension of the liquid[28]

If the surface charge density is too low, the Rayleigh limit will not be attained, even after evaporation of the solvent and the consequent shrinkage of the particles produced; thus, no Coulomb fission will occur. Higher surface charge densities also do not reach the Rayleigh limit, but the decreasing droplet diameter caused by solvent evaporation increases the surface charge density past the limit, resulting in Coulomb fission. This fission causes disintegration of the particles into smaller particles and changes the size distribution of the particles. If the charge is even higher, a third mechanism occurs in which the droplets break up rapidly into finer particles with a narrow size distribution[29] (Figure 22.6). Droplets produced by electrospraying, however, usually carry a charge equal to about half the Rayleigh limit.[30]

Different shapes and morphologies can be produced by varying the process and solvent parameters of the three main mechanisms of particle formation. Figure 22.7 shows possible morphologies for particles produced by EHDA, as demonstrated by Enayati et al.[31] If the time

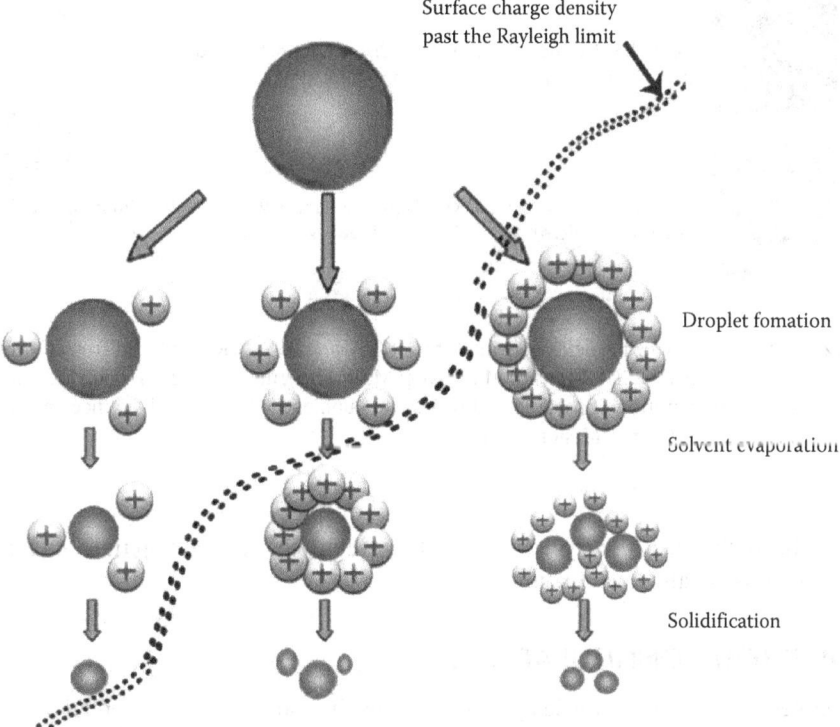

FIGURE 22.6 Effect of charge density on Coulomb fission of solidifying particles produced by EHDA process.

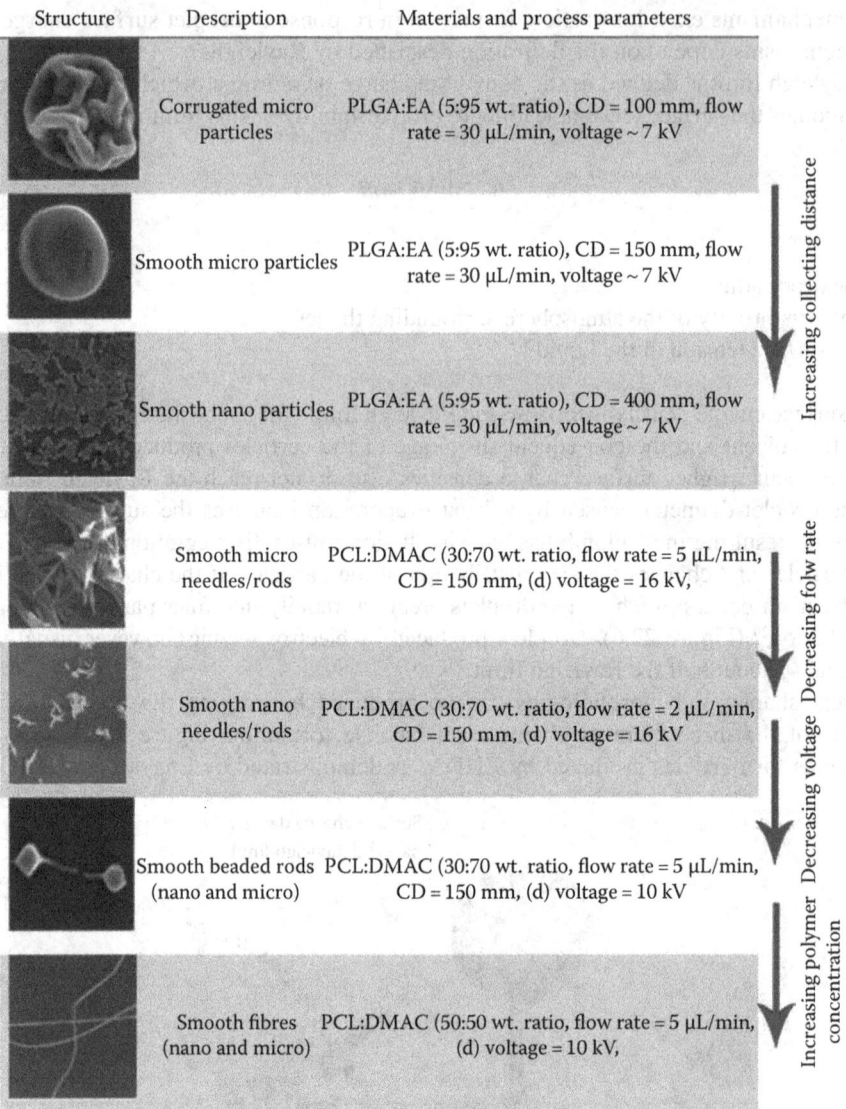

FIGURE 22.7 Morphology of particles generated by EHDA processes. (Reproduced from *Colloid Surface A Physicochem. Eng. Aspect*, 382(1), Enayati, M., Chang, M.-W., Bragman, F., Edirisinghe, M., and Stride, E., Electrohydrodynamic preparation of particles, capsules and bubbles for biomedical engineering applications, 154–164. Copyright 2011, with permission from Elsevier.)

needed to solidify the electrosprayed liquid is shorter than the jet breakup time, nanofibers will form. This process is called electrospinning.[24]

22.3 DIFFERENT CONFIGURATIONS

EHDA operates by means of a simple apparatus. A few thousands volts of potential difference in a needle supplied with liquid are all that is required under a defined flow rate (Figure 22.1). Several electrospray setups have been investigated. The present study classifies the research by type of spray head and method of particle collection.

22.3.1 Spray Head

The spray head configuration is an interesting aspect of EHDA. The nozzle varies from a simple hypodermic needle of different diameters to highly technologically advanced multiplex nozzles (Figure 22.8). The coaxial needle shown increases the number of centered nozzles to four and is mainly used to fabricate multilayer spheres. In coaxial setups, the most commonly used is the coaxial two-capillary nozzle used to fabricate core–shell spheres (capsules). The inner needle is supplied by a solution of active substance and the outer needle is filled with shell material.[32] In this process, parameters are used as adjusting switches to control the diameter of the capsules, thickness of the shell, and number of inner cores.[32] The two solvents used in the coaxial setup are immiscible and wettable, and the inner one has a higher surface tension.[33]

Making capsules is the most common application for the coaxial system, but, for example, the outer needle could supply a surfactant[34] to modify the surface properties of the particles. Previous studies have used a coaxial setup to modify the cone-jet shape to prevent nozzle head blockage by a solidified polymer solution of volatile solvent by applying gas saturated from the solvent vapor[35] (Figure 22.9) or to compensate for the low conductivity of the solution by supplying the outer needle with an appropriate liquid.[36]

The coaxial system used two concentric needles until a three-capillary coaxial needle was developed by Labbaf et al.[37] They have also used a four-capillary coaxial system to produce fibers and particles in a layered structure. Figure 22.10 shows the TEM of a particle produced and the schematic configuration.[38] The results confirm the ability to produce multilayered structures using coaxial configurations with up to four capillaries.

Flow-limited field injection electrostatic spraying is a method in which a smooth glass nozzle covering a sharp needle is used to spray the liquid (Figure 22.8). The nozzle minimizes imperfection and allows spraying of low-conductivity liquid.[39]

A multiplex nozzle is a novel approach for increasing the spray source and, subsequently, increasing the rate of production. Several multihead spray systems have been designed. Some arrange a number of parallel nozzles in a linear array,[24] and some use a polycarbonate line fabricated by CNC[40] (Figure 22.8) on a silicon wafer with multiple nozzles produced by deep reactive ion technology.[36]

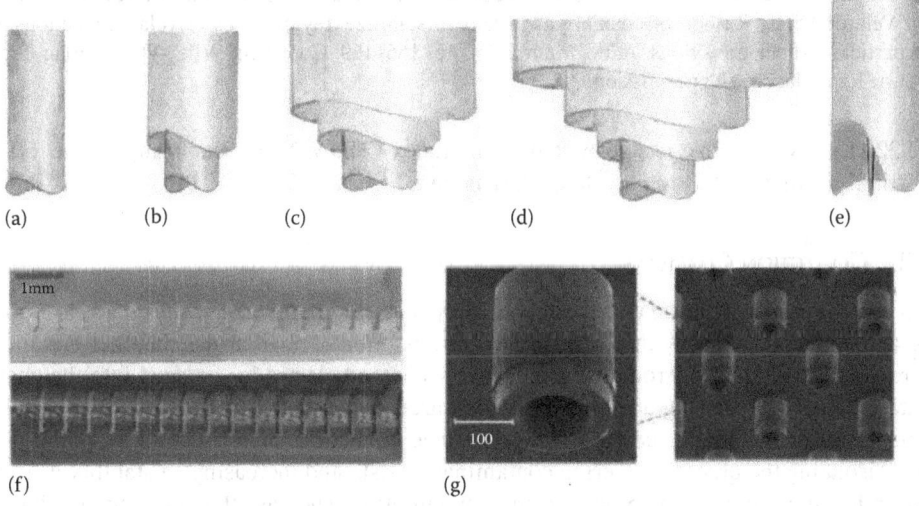

FIGURE 22.8 Different needle configurations in EHDA process. (a) Single axial, (b) two-capillary coaxial, (c) tri-capillary coaxial, (d) four capillary coaxial, (e) FFESS, (f) linear multiplex, and (g) sheet multiplex. (Adapted from Lojewski, B. et al., *Aerosol Sci. Technol.*, 47(2), 146, 2012; Reprinted from *J. Aerosol Sci.*, 37(6), Deng, W., Klemic, J.F., Li, X., Reed, M.A., and Gomez, A., Increase of electrospray throughput using multiplexed microfabricated sources for the scalable generation of monodisperse droplets, 696–714, Copyright 2006, with permission from Elsevier.)

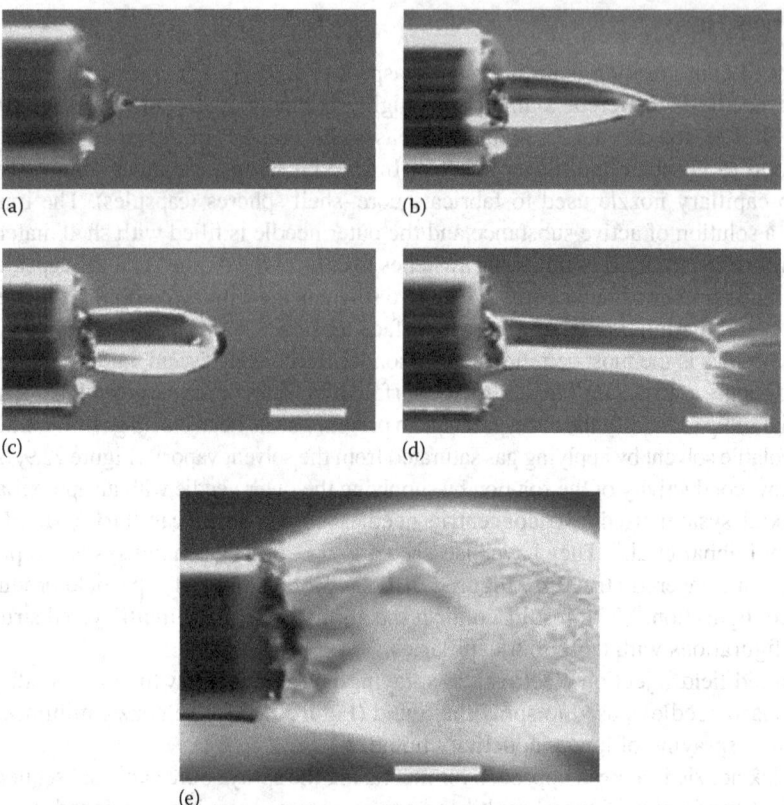

FIGURE 22.9 Taylor cone structure at the tip of the nozzle. Applied voltages are as follows: (a) 1.2, (b) 0.8, (c) 0.6 kV (with DCM saturated N_2 gas in outer nozzle), (d) 1.2, and (e) 0.8 kV (without N_2 gas flow in outer nozzle). For all cases a solution of 10% PLA in DCM was applied at a rate of 0.5 mL/h. (Larsen, G., Spretz, R., and Velarde-Ortiz, R.: Use of coaxial gas jackets to stabilize Taylor cones of volatile solutions and to induce particle-to-fiber transitions. *Adv. Mater.* 2004. 16. 166–169. Copyright Wiley-VCH Verlag GmbH & Co. KGaA. Reproduced with permission.)

Figure 22.11 shows the cone jets generated by multiplex EHDA. Many jets form, but the total flow rate is still 5 mL/h, which is too low for industrial use.

22.3.2 Collection Chambers

Researchers have also investigated the use of a cylindrical chamber, usually glass, to cover the nozzle and collector. A gas with a defined flow rate is often applied at the chamber entrance and exits from the other side. A grounded or negatively charged electrode is placed into the chamber, opposite the high-voltage ring and needle, to discharge the particle[17,18,41] (Figure 22.12). This configuration has advantages including effective particle collection, better solvent evaporation from particles, isolating the process, lowering contamination risk, and increasing jet stability.[41] Attempts have been made to improve the process using configurations such as attaching a high-voltage ring or plate above the nozzle.[29]

Lin et al. selected liquid nitrogen as the collection media in their configuration. The generated droplet fell into the liquid nitrogen and froze and then was freeze-dried to produce a highly porous microsphere. Freezing the solvent and sublimating it efficiently prevented loss of drug during solidification and increased loading efficiency.[42] Studies on other types of collection setups have suggested use of a charged tube[43] or a vessel grounded by metal mesh.[44]

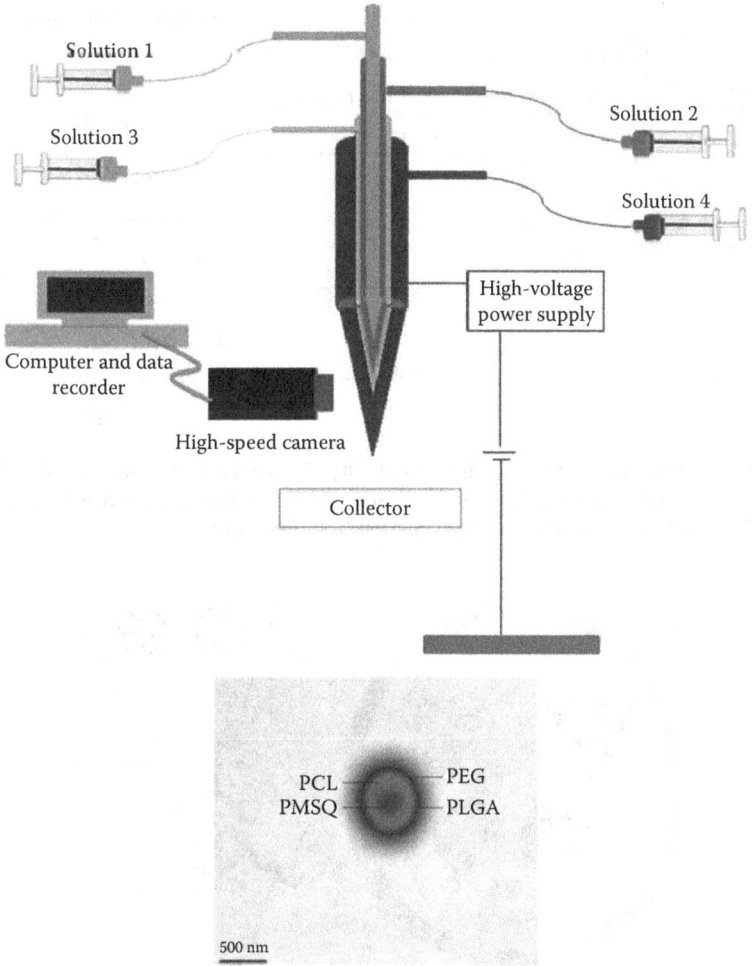

FIGURE 22.10 Four-capillary coaxial EHDA configuration and resulting particle (in this case solutions prepared from (1) PMSQ, (2) PCL, (3) PLGA, and (4) PEG). (Labbaf, S., Ghanbar, H., Stride, E., and Edirisinghe, M.: Preparation of multilayered polymeric structures using a novel four-needle coaxial electrohydrodynamic device. *Macromol. Rapid Commun.* 2014. 35. 618–623. Copyright Wiley-VCH Verlag GmbH & Co. KGaA. Reproduced with permission.)

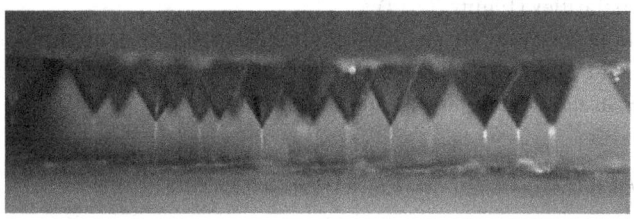

FIGURE 22.11 Several cone jets are achieved by means of multiplex nozzle. (Reprinted from *J. Aerosol Sci.*, 36, Bocanegra, R., Galán, D., Márquez, M., Loscertales, I.G., and Barrero, A., Multiple electrosprays emitted from an array of holes, 1387–1399, Copyright 2005, with permission from Elsevier.)

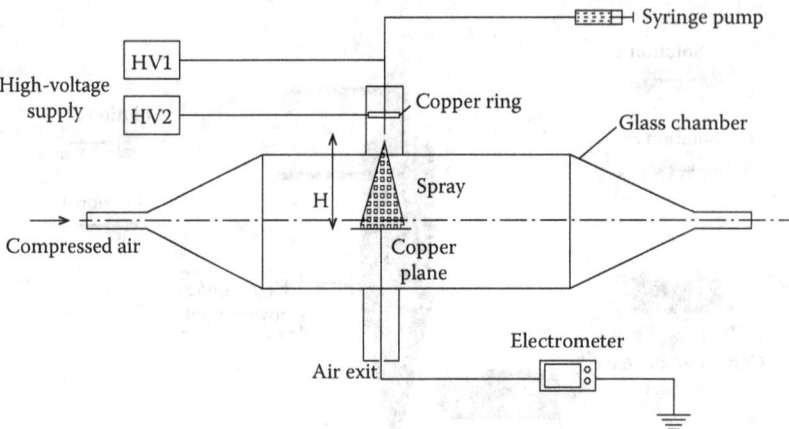

FIGURE 22.12 Electrospray source with chamber facility. (Reprinted from *J. Aerosol Sci.*, 39, Yao, J., Lim, L.K., Xie, J., Hua, J., and Wang, C.-H., Characterization of electrospraying process for polymeric particle fabrication, 987–1002, Copyright 2008, with permission from Elsevier.)

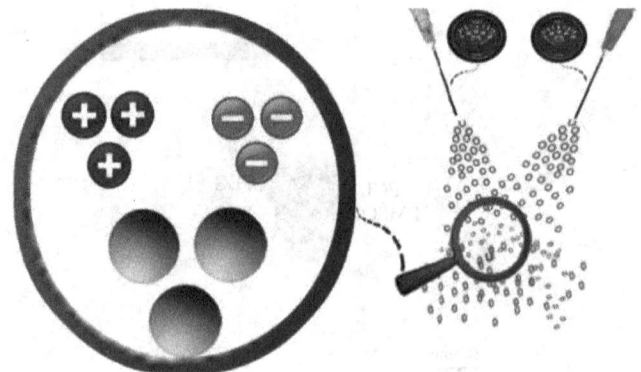

FIGURE 22.13 (See color insert.) Bipolar electrospray setup (reactive electrospray).

22.3.3 OTHER CONFIGURATIONS

One configuration employs two oppositely charged needles directed toward each other for droplets with opposing charges that attract each other and coagulate.[22] Figure 22.13 is a schematic of this bipolar configuration.

Electrospinning fabricates nanofibers, primarily for biomedical use.[45–47] The electrospinning/spraying configuration is similar to the reactive electrospray, where two nozzles are subjected to a grounded or negatively charged rotating collector. Each nozzle is connected to a separate voltage supplier with same polarities (Figure 22.14).

Other configurations are used for coatings. Figure 22.15 shows configurations for coating glass surfaces with TiO_2. The major difference in the deposition setup for this configuration is the heater placed under the substrate to facilitate solvent evaporation.[48]

22.4 ADVANTAGES AND DISADVANTAGES

22.4.1 ADVANTAGES

The main advantage of EHDA over conventional methods of micro-/nanoparticle production like emulsion/solvent evaporation is that EHDA eliminates or strongly decreases the need for surfactants.

Encapsulation via Electrohydrodynamic Atomization Spray Technology (Electrospray)

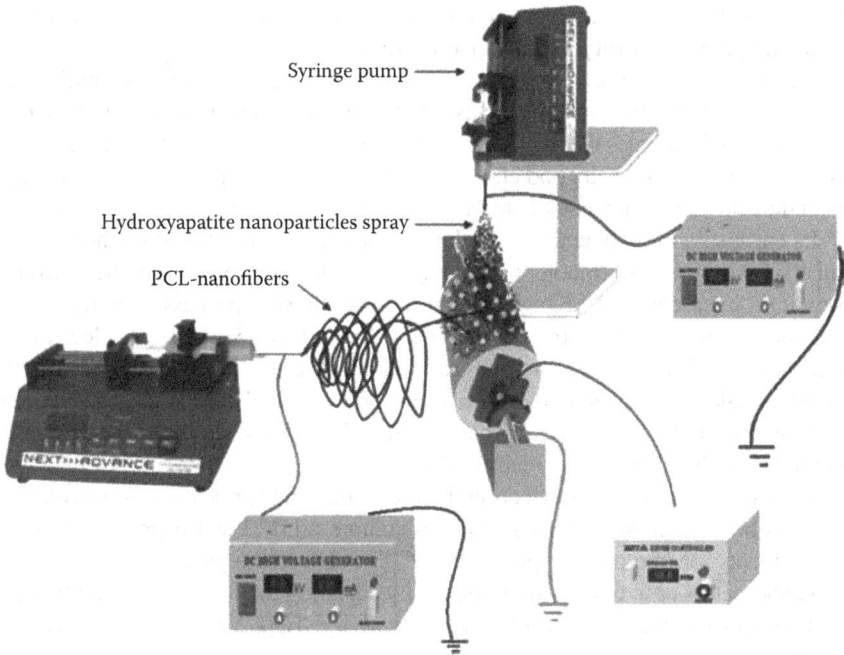

FIGURE 22.14 Schematic diagram of EHDA system has capability of 3D structure fabrication. (From Venugopal, J. et al., *J. Biomater. Sci. Polymer Ed.*, 24(2), 170–184, 2012. With permission. 2013 Taylor & Francis.)

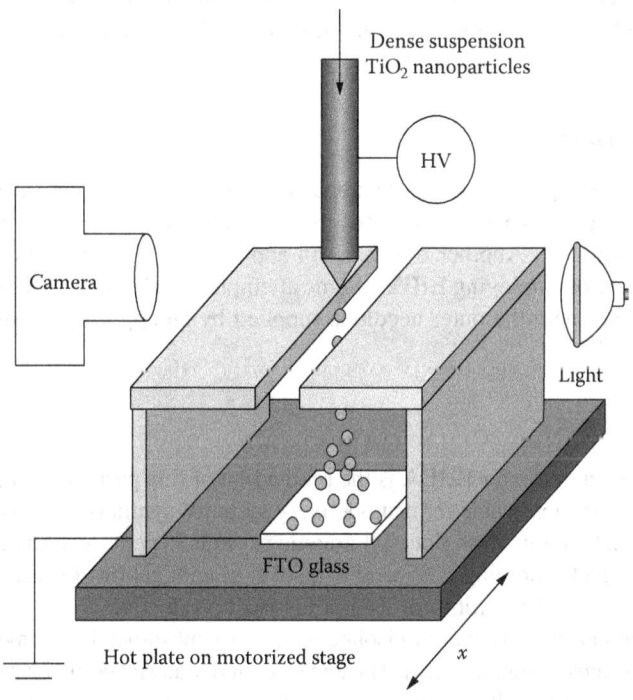

FIGURE 22.15 Schematic diagram of configuration that is used for coating the TiO_2 photoelectrode for application in solar cell plate. (Adapted from Zhu, T. et al., *Aerosol Sci. Technol.*, 47(12), 1302, 2013. With permission. 2013 Taylor & Francis.)

The process easily operates at ambient temperature and pressure without application of shear stress,[38] achieving greater stability in the active substance.

In a coaxial setup, two or more liquids are applied simultaneously to form a multilayer particle system.[31] Multilayer micro-/nanospheres are useful where controlled or targeted release of the active substance is required. Examples of this are sustained-release drug delivery and enteric-coated drug formulation.[49] A coaxial setup is a good choice for sensitive active ingredient pharmaceuticals such as proteins and peptides that are not stable in organic solvent or denatured in an organic/inorganic interface.[50] Its encapsulation efficiency is much higher than in conventional methods.[51]

For industrial purposes, continuous processing in the fewest steps is preferred over multistep batch processing. Since electrospraying is a continuous one-step process, it is used in a number of industries that benefit from its attractive features. One unique feature is the electrostatic charge of the particles. In applications such as pharmaceuticals, the use of microspheres for parenterals or aerosols for inhalation prevents coalescence and allows good dispersion after reconstitution. Although charged particles may seem difficult to handle, the trajectory of the charged particles allows them to be easily guided by an electric field.[52]

Size is the most important factor affecting the properties and application of nano-/microparticle systems and can be determined by the surface-to-volume ratio.[5] Electrospraying allows the most sensitive adjustments for fine particle production.[53] The size of the particle can be fine-tuned to a few hundred nanometers or to tens of micrometers by adjusting the process and solution parameters (liquid flow rate, applied electric field, solution conductivity). EHDA allows fine control for size and produces particles within a narrow size distribution.[52]

Electrospraying also effectively transfers biomaterial into cells. The advantage of EHDA over conventional bombardment techniques is the charge-space effect that is the driving force for biological substances; bombardment techniques employ external forces that cause the biomaterial to penetrate the cell.[54] Coating biomaterials with biocompatible agents like hydroxyl apatite by EHDA has two advantages over ink-jet printing; the larger 4× needle diameter allows use of concentrated ceramic solution without needle blockage and the generation of 10× finer particles.[55]

22.4.2 Disadvantages

The low flow rate used to produce micro-/nanoparticles strongly decreases throughput. This problem is addressed by multiplexing the spray source in applications where the value added compensates for the increased cost. Another drawback in applications using organic and nonconductive buffers is the difficulty of employing EHDA. Some attempts have been made to modify this process using coaxial needles where the outer needle is supplied by an appropriate liquid to draw out the nonconductive liquid.[36,56]

22.5 APPLICATIONS

Perhaps the most common use for EHDA is the production of fine particles in narrow size distributions. These particles are applicable for drug-delivery systems, synthesis of nanoparticles, thin film deposits, superconductors, quantum dots, photoionic crystals,[36] paint technology, and agricultural applications like pesticide spraying.[57] Several examples of applications for electrospraying in published studies demonstrate the abilities of EHDA for mass production.

Common uses for encapsulation are to isolate a sensitive substance from environmental damage and to deliver compounds to specific sites. The encapsulated material could be a gas, liquid, or solid; the encapsulating material usually is a polymer in solution or in a molten state that will solidify during the process. The controlled release of the micro-/nanocapsules and its coating characteristics[56] are major attractions of the method.

The electrospraying of biomaterials into cells is used in gene therapy and gene transfection. The biomaterial penetrates the cell by means of the intrinsic droplet charge that develops during electrospraying.[54]

Nanoparticles produced by EHDA are used in the field of nanotechnology. For example, the production of zinc sulfide particles under 50 nm in diameter has been reported using electrospray pyrolysis[58,59] (Figure 22.16c). Nanodevices, such as in nanoxerography, require highly charged aerosols to eliminate noise deposition from random Brownian motion. EHDA achieves this objective in the processing of colloidal suspensions of gold.[60] TiO_2, SnO_2,[7] metal nanoparticles,[61] hydroxyapatite,[62] and composites of hydroxyapatite and polylactic acid[63] have been used to coat substrate to provide properties such as biocompatibility. The treating substrate is usually heated to facilitate solvent evaporation from the particles.[7] Figure 22.16a shows the SEM cross section of deposited SnO_2 thin film on a glass substrate.

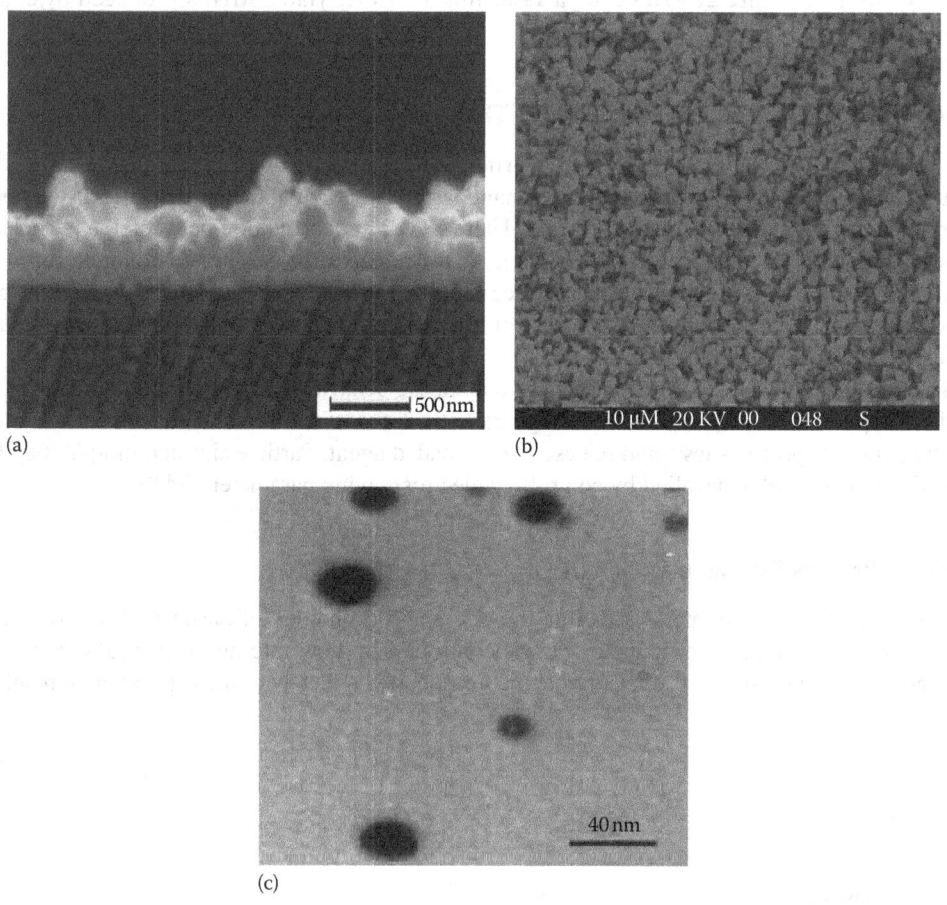

FIGURE 22.16 Application of EHDA in the field of nanotechnology: (a) SEM image of deposited SnO_2 thin film (cross section), (b) SEM image of deposited SiC thin film (surface), and (c) TEM photograph of ZnS particles prepared by electrospray pyrolysis at 600°C. (Reprinted from *Sens. Actuators B Chem.*, 96(1), Matsushima, Y., Nemoto, Y., Yamazaki, T., Maeda, K., and Suzuki, T., Fabrication of SnO_2 particle-layer on the glass substrate using electrospray pyrolysis method and the gas sensitivity for H_2, 133–138, 2003, Copyright 2014, with permission from Elsevier; Reprinted from *J. Electrostat.*, 50(4), Balachandran, W., Miao, P., and Xiao, P., Electrospray of fine droplets of ceramic suspensions for thin-film preparation, 249–263. Copyright 2001, with permission from Elsevier; Reprinted from *Chem. Eng. Sci.*, 58(3), Okuyama, K. and Lenggoro, I.W., Preparation of nanoparticles via spray route, 537–547, Copyright 2003, with permission from Elsevier.)

Another application is uniform deposition of a thin film of ceramic materials onto a substrate. Thin film formation is achieved by electrospraying a ceramic solution. This method is more economical than chemical/physical vapor deposition methods. Figure 22.16b shows the SEM of SiC ceramic thin film deposited on a substrate.[64]

Several studies have used electrospraying in conjunction with other production methods to fabricate composite structures. An example of this is the electrospraying/spinning method. Venugopal et al. electrosprayed hydroxyapatite while electrospinning polycaprolactone as a scaffold with a morphology suitable for cell attachment.[46] The same method has been used to deposit nanoparticles of PLGA on fibers[65] (Figure 22.17a) that provides a 3D structural construct for better cell attachment that is suitable as a drug release scaffold.[46] For PEGylation of magnetic nanoparticles (MNPs), a coaxial electrospray is used, the outer needle of which is supplied with polysorbate-80. The droplets fall into distilled water and are stirred to complete synthesis.[34] Figure 22.17b shows a TEM image of PEGylated MNPs produced by coaxial EHDA setup.

22.6 CONTROLLING CHARACTERISTICS

The size and morphology of a micro-/nanoparticle affects its properties and final applications. For encapsulation of an active drug into a micro-/nanoparticle, the size of the particle affects the release characteristics and particle degradation rate. There is a direct correlation between the particle and its surface-to-volume ratio; increasing particle size decreases the surface-to-volume area. For a given mass of particles, increasing the surface area increases the release rate; the penetration of water into the particle also increases because of its smaller diameter.[5] Controlling the size and morphology of the particle is important.

The morphology of the particle influences the release behavior and degradation process. In biomedical applications, shape and morphology can affect processes like cell attachment, clearance, and targeting of specific sites[66] and release rate of loaded agent. Particle size and morphology have been shown to be well controlled by controlling electrospraying parameters.[39,41,67]

22.6.1 PROCESS PARAMETERS

Flow rate is the main parameter affecting the size and morphology of particles.[68] Several studies have shown that the particle diameter can be controlled by flow rate and electrical conductivity. Ganan-Calvo proposed a theoretical model that reasonably agrees with the experimental results:

$$d_d = \alpha \left(\frac{Q^3 \varepsilon \rho}{\pi^4 \sigma \gamma} \right)^{1/6} \tag{22.2}$$

where

α is a constant
Q is the liquid flow rate
ε is the dielectric constant in a vacuum
ρ is the liquid density
γ is the conductivity of the liquid
σ is its surface tension[69]

As shown, the size of the droplet is proportional to the root square of Q; if the particle diameters are plotted against $Q^{0.5}$, the result is linear with a good R^2.

Jafari-Nodoushan et al. plotted particle size by the root square of the flow rate for the particles produced by electrospraying a PLGA solution in dichloromethane. The results show good

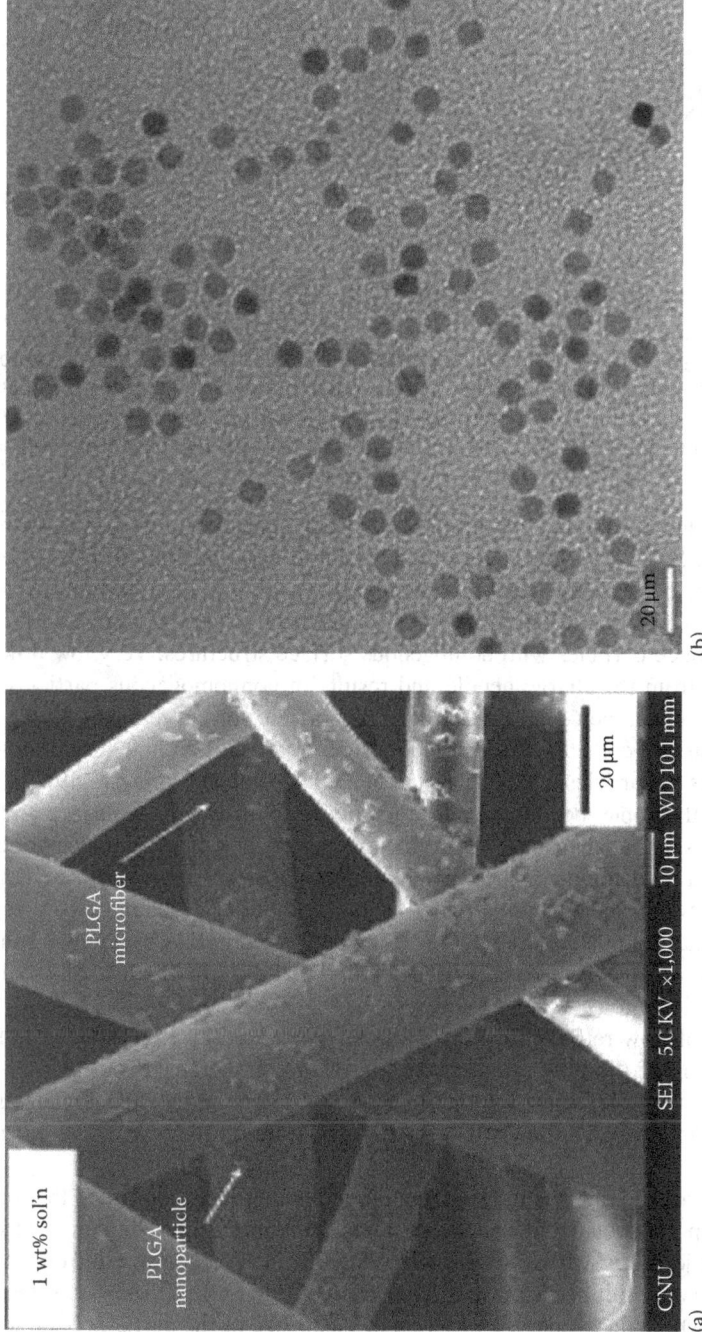

FIGURE 22.17 Electrospray in conjunction with other fabrication method: (a) SEM image of electrosprayed PLGA particles deposited on electrospinned PLGA fibers and (b) TEM image of PEGylated MNPs produced by coaxial electrospray setup. (From Kim, S.J., Jang, D.A., Park, W.H., and Min, B.-M.: Fabrication and characterization of 3-dimensional PLGA nanofiber/microfiber composite scaffolds. *Polymer*. 2010. 51(6). 1320–1327. Copyright Wiley-VCH Verlag GmbH & Co. KGaA. Reproduced with permission; From Kim, S.-Y., Yang, J., Kim, B., Park, J., Suh, J.-S., Huh, Y.-M., Haam, S., and Hwang, J.: Continuous coaxial electrohydrodynamic atomization system for water-stable wrapping of magnetic nanoparticles. *Small*. 2013. 9(13). 2325–2330. Copyright Wiley-VCH Verlag GmbH & Co. KGaA. Reproduced with permission.)

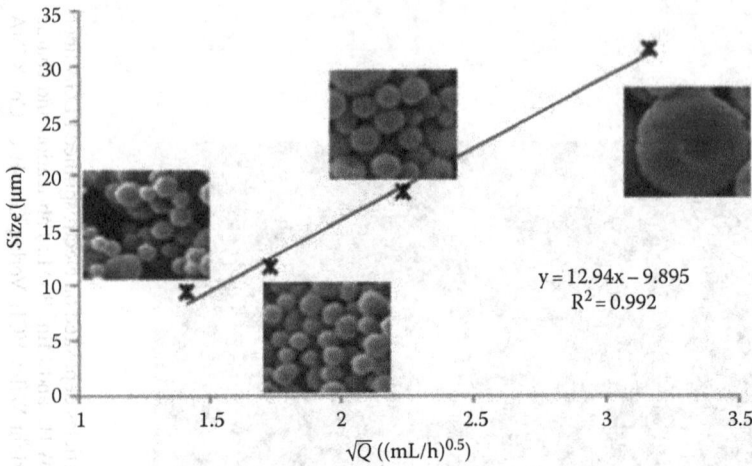

FIGURE 22.18 Effect of polymeric solution flow rate on mean particle size of PLGA microspheres produced by EHDA. (From Jafari-Nodoushan, M. et al., Monitoring the effect of flow rate on size and morphology of microspheres produced by the means of electrospraying method, in *International Seminar of Polymer Science and Technology*, Tehran, Iran, 2012. With permission. International Seminar on Polymer Science and Technology.)

agreement with the suggested scaling law (Figure 22.18).[70] Bock et al. generated polymeric microspheres by electrospraying 5%–10% (w/v) solution of polycaprolactone in chloroform. They found that increasing the flow rate increase the particle diameter.[71]

The flow rate of liquid in EHDA is not usually very high and is on a scale of mL/h. The flow rate should be adjusted to produce particles with homogenous surface structures. Very low flow rates prevent quick dissipation from the charge needle and results in nonhomogenous particles. Very high flow rates can cause very large drops and result in incomplete solvent removal, which alters the particle shape as it hits the collector.[39] Jafari-Nodoushan et al. showed how a high flow rate changes spherical particles to particles of irregular and coagulated shape (Figure 22.19).[72]

Ruala stated that a small droplet has more chance to solidify with a smooth surface and solid core and a large particle has higher chance of solidifying into a particle with a wrinkled surface and porous core.[73] Heptane is a model low-conductivity liquid that has been subjected to EHDA. The results in Figure 22.20 show that the range of sizes vary up to two orders of magnitude in response to variations in the flow rate from 0.25 to 28 mL/h.[52] Hong et al. used EHDA to produce spherical colloidal crystals using AC voltage instead of DC. They found that the size of silica colloidal crystals produced were proportional to $Q^{0.33}$.[74]

A study on the effect of flow rate on chitosan particles generated by EHDA found a linear correlation between the log of the flow rate and the log of the diameter.[15] From this observation and others, it can be concluded that flow rate has a direct effect on the size of the particle, regardless of its power. Figure 22.21 shows the effect of flow rate on the size of the fabricated particles.[17]

Variation in the electrical field dictates the spraying mode. Studies have confirmed that the size of the particle is reversely proportional to the applied potential (Figure 22.20). This can be observed at low voltages in the cone-jet mode, where voltage had no significant effect on size and only a slight decrease in size was observed with an increase in voltage.[2,53,68] Figure 22.22 shows the results of a study by Enayati et al., where they reported variation of mean size of the particles obtained as a function of applied voltage and flow rate.[75]

The preferred mode of electrospraying is cone jet, where, for each flow rate, a minimum and maximum limit exists in which the cone is stable. This limit varies in accordance with solution properties such as conductivity and surface tension of liquid. Figure 22.23 shows a typical stable

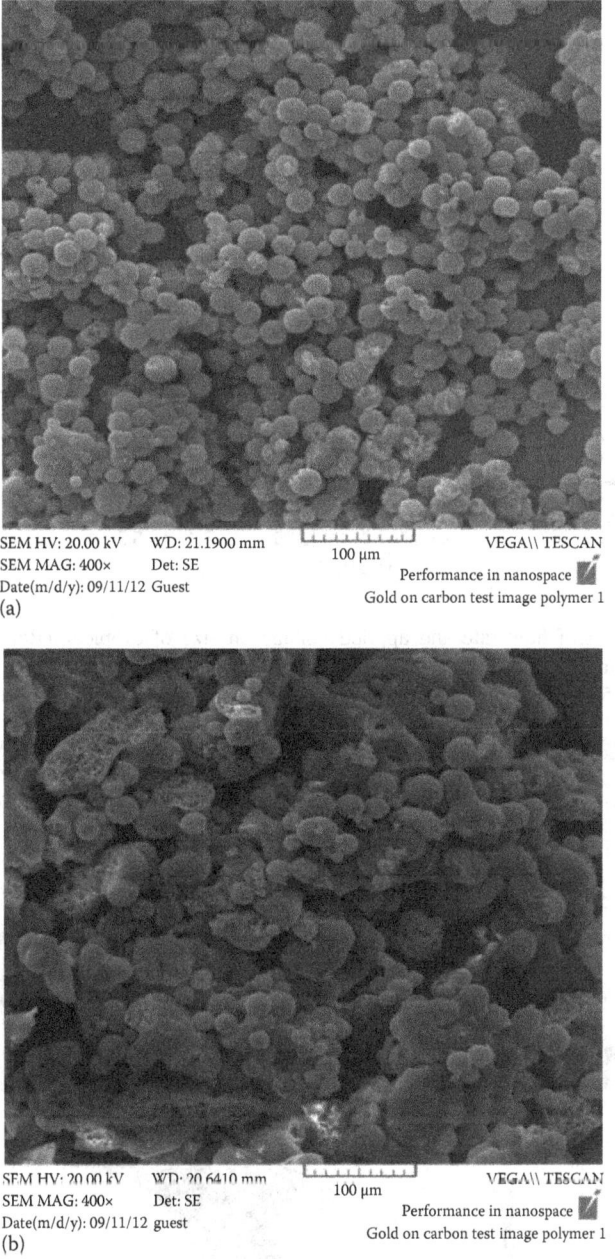

FIGURE 22.19 SEM images of PLGA microparticles produced by electrospray with different flow rates: (a) 2 mL/h and (b) 3 mL/h. (From Jafari-Nodoushan, M., Barzin, J., and Mobedi, H.: Size and morphology controlling of PLGA microparticles produced by electro hydrodynamic atomization. *Poly. Adv. Technol.* 2015. 26(5). 502–513. Copyright Wiley-VCH Verlag GmbH & Co. KGaA. Reproduced with permission.)

range as a function of applied potential and liquid flow rate with the other parameter held constant; the area between the two lines is the stable range.[76] A spherical particle is usually produced at a lower range and elongated particles and fibers at higher voltages.[68]

The strength of the electrical field is defined as the applied voltage divided by the distance between the tip and collector.[77] Changing the collecting distance changes the strength of the electrical field. The collecting distance is proportional to the flying time of the droplet to the collector and

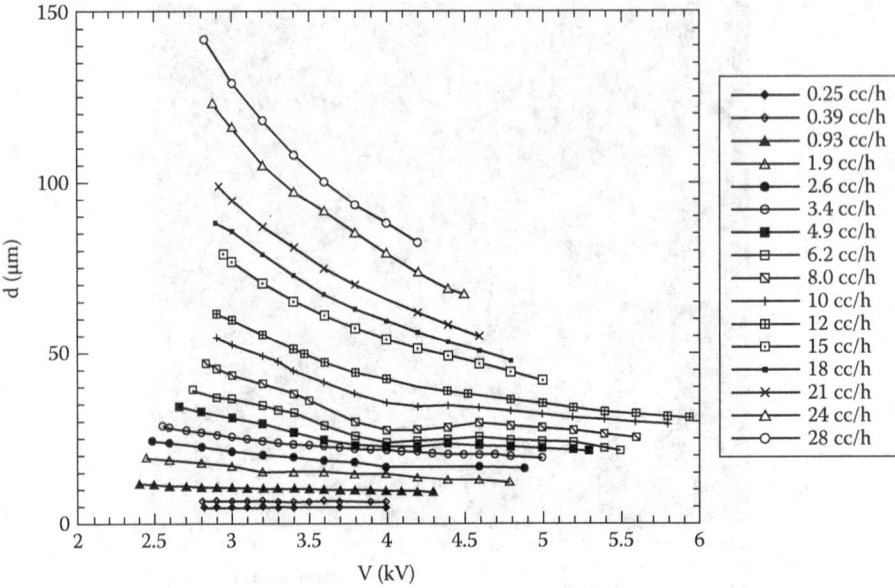

FIGURE 22.20 Effect of flow rate and applied voltage on size of droplets. (Reprinted from *J. Colloid Interface Sci.*, 184, Tang, K. and Gomez, A., Monodisperse electrosprays of low electric conductivity liquids in the cone-jet mode, 500–511, Copyright 1996, with permission from Elsevier.)

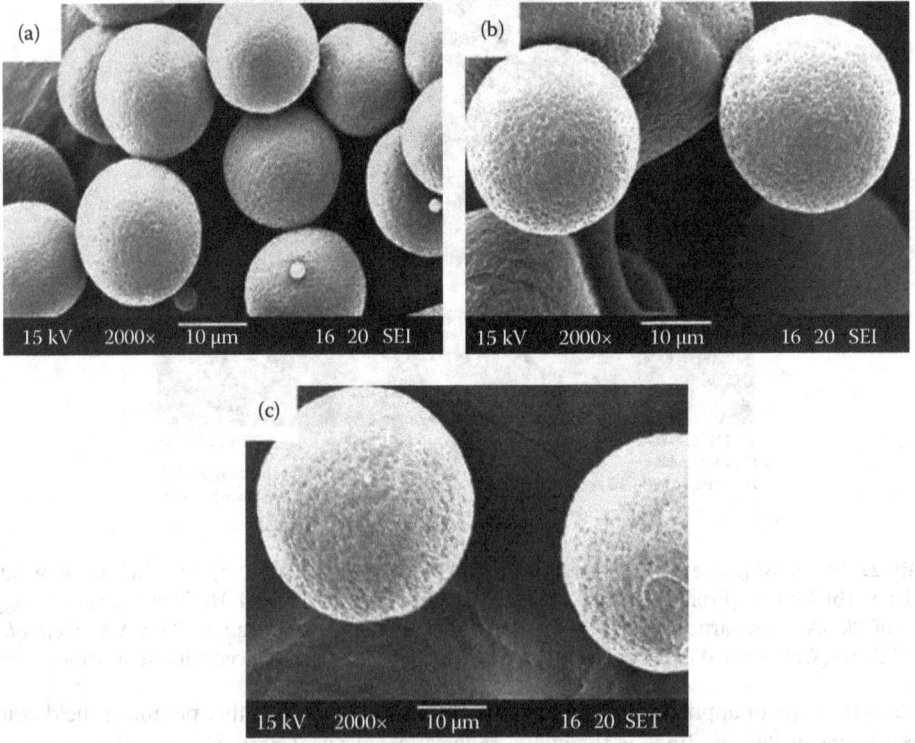

FIGURE 22.21 SEM images of PCL microparticles fabricated under different polymer solution flow rates: (a) 3 mL/h, (b) 10 mL/h, and (c) 15 mL/h. (Reprinted from *Biomaterials*, 27, Xie, J., Marijnissen, J.C.M., and Wang, C.-H., Microparticles developed by electrohydrodynamic atomization for the local delivery of anticancer drug to treat C6 glioma in vitro, 3321–3332, Copyright 2006, with permission from Elsevier.)

Encapsulation via Electrohydrodynamic Atomization Spray Technology (Electrospray)

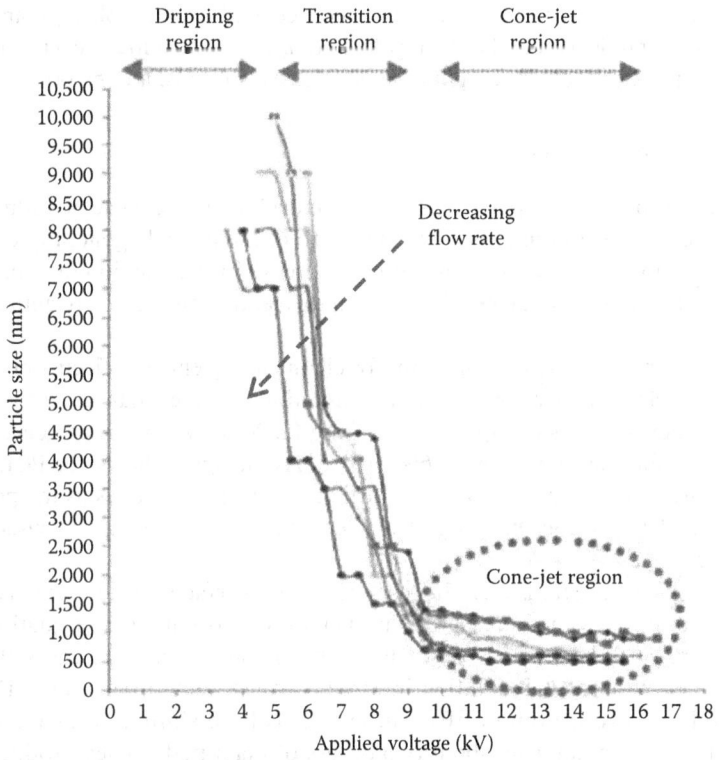

FIGURE 22.22 Variation of mean size of the particles obtained as a function of applied voltage and flow rate. (Adapted from Enayati, M., Ahmad, Z., Stride, E., and Edirisinghe, M., Size mapping of electric field-assisted production of polycaprolactone particles, *J. R. Soc. Interface*, 7(Suppl. 4), S398, figure 4, 2010, The Royal Society.)

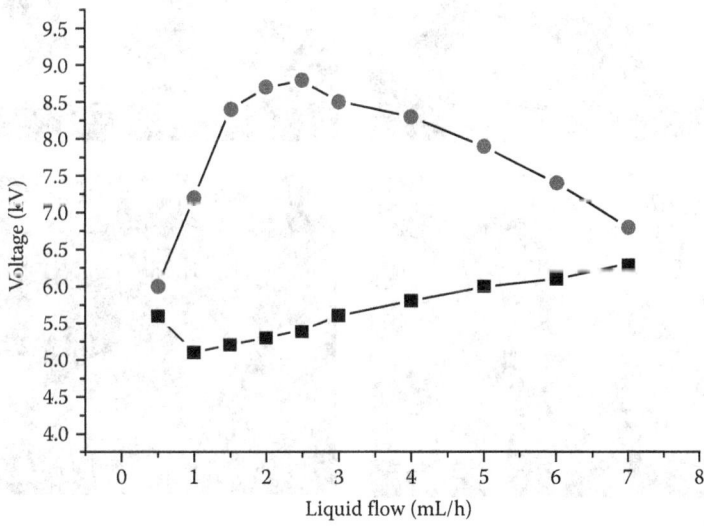

FIGURE 22.23 Typical stability map of the EHDA, 1% paracetamol in isopropyl alcohol. (Reprinted from *Int. J. Pharmaceut.*, 324, Ciach, T., Microencapsulation of drugs by electro-hydro-dynamic atomization, 51–55, Copyright 2006, with permission from Elsevier.)

affects solvent evaporation from the droplet and, consequently, its morphology and size distribution.[68,71] Studies have produced conflicting results in this area, which may be a result of differences in configuration and rate of solvent evaporation of different solvents, among others.

22.6.2 Solution Parameters

Polymer concentration affects particle size and morphology because it influences polymer chain entanglements and solution viscosity. The size of the particle generally decreases as the polymer concentration decreases because of the decrease in the fraction of solid material for a defined volume of solution. A further decrease in concentration creates irregular or biconcave particle shapes.

An increase in concentration past the limit for chain entanglements (electrospinning) produces a fiber-shaped particle; polymer concentrations three times greater than the critical overlap concentration are sufficient for electrospinning.[78] Figure 22.24 shows particles generated by EHDA from PCL at decreasing concentrations of 6%–0.5%.[17] As the figure shows, the PCL microparticles changed from spherical to smaller biconcave particles to irregular shapes as the polymer concentration decreased. Polymer concentration also affects viscosity. A change in viscosity varies the particle size (Figure 22.25).

The type of solvent affects EHDA in response to its surface tension, electrical conductivity, and boiling point or vapor pressure. The Ganan-Calvo equation (Equation 22.2) shows that increasing the surface tension or conductivity of the solvent decreases the diameter of the particle. Xie et al. used PCL solution in three solvents, dichloromethane (DCM), acetonitrile (ACN), and tetrahydrofuran (THF). These liquids have similar surface tensions, but the conductivity of ACN is greater by about two orders of magnitude. They produced particles of

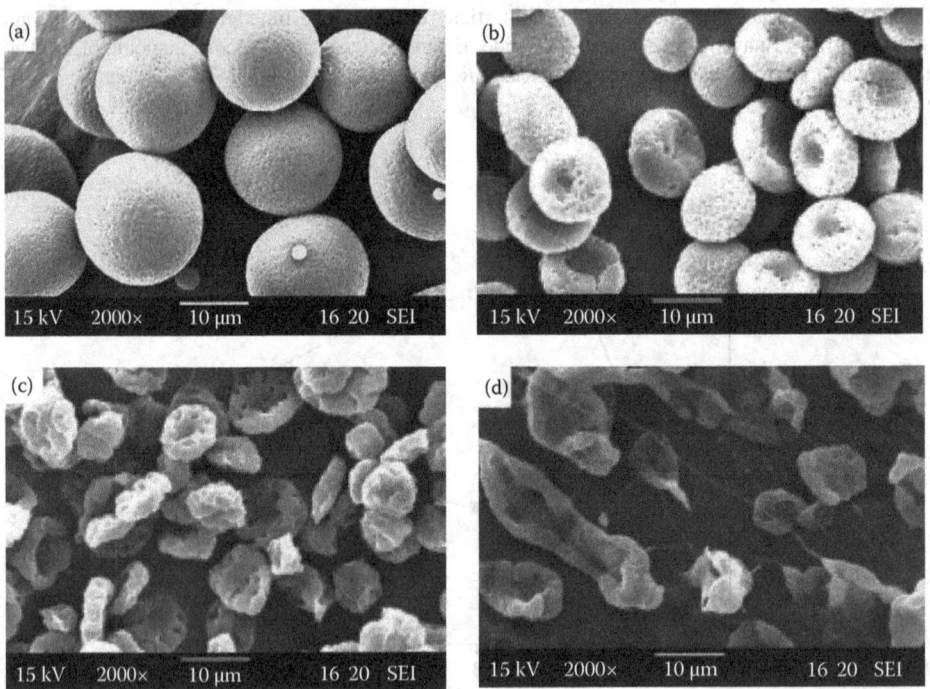

FIGURE 22.24 SEM images of PCL microparticles fabricated under different polymer concentrations: (a) 6%, (b) 3%, (c) 1%, and (d) 0.5%. (Reprinted from *Biomaterials*, 27, Xie, J., Marijnissen, J.C.M., and Wang, C.-H., Microparticles developed by electrohydrodynamic atomization for the local delivery of anticancer drug to treat C6 glioma in vitro, 3321–3332, Copyright 2006, with permission from Elsevier.)

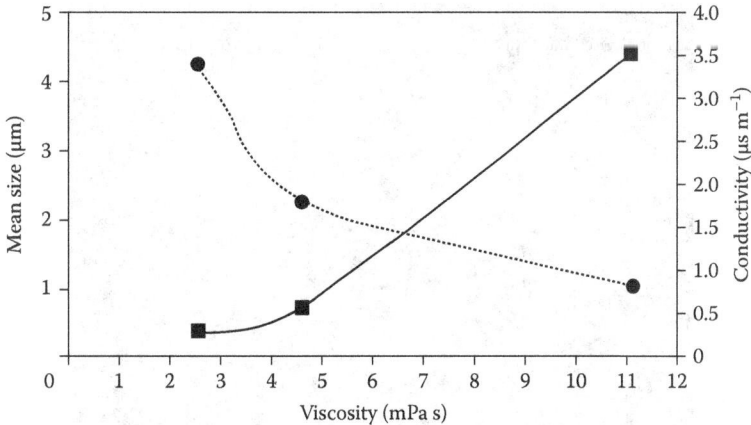

FIGURE 22.25 Relationship between mean particle size with the electrical conductivity and the viscosity of the PCL solutions used (flow rate, 10 μL min^{-1}; applied voltage, 10 kV). v, mean size; λ, conductivity. (Adapted from Enayati, M., Ahmad, Z., Stride, E., and Edirisinghe, M., Size mapping of electric field-assisted production of polycaprolactone particles, *J. R. Soc. Interface*, 7(Suppl. 4), S398, figure 5, 2010, The Royal Society.)

similar size for THF and DCM but of smaller size for ACN in response to the different conductivities.[17] Researchers agree that the diameter of a droplet decreases as the liquid conductivity increases.[21,53,79,80]

The boiling point of solvent influences the morphology and size of the particles. When a droplet is ejected from a jet, the fast evaporation of a solvent with a low boiling point causes a skin to form; if the skin is strong enough to resist contraction pressure, a hollow particle is generated. If the strength of the skin is low, it collapses into cuplike or irregular shapes. A solvent with a high boiling point usually produces droplets that are not fully dry and that deposits on the collector or flatten upon hitting the surface. Park et al. demonstrated this effect in cases where two solvents were used.[21]

The formation of a porous structure results from phase separation (or phase inversion) mechanisms that are not limited to electrospraying. It is the process that controls membrane formation, as the solvent exchanges with a nonsolvent, polymer solution solidifies and polymeric device forms. The phase separation is fully investigated in fabrication of flat or hollow fiber membranes[81–85] or in situ forming drug delivery systems.[86–89] Usually, quick evaporation of the solvent produces particles with porous or golf ball–shaped surfaces[90] (Figure 22.26).

22.6.3 Drug Release

The main application of nano-/microparticles generated by EHDA is as a drug carrier. Ciach generated PLGA microparticles using a single-axial configuration. He investigated the effect of solvent type and evaporation rate on the morphology of the particles. Figure 22.27 shows the release of Taxol from particles with different morphologies. The change in solvent evaporation rate altered the internal structure of the particles and affected drug release.[76] The hollow microparticles showed faster drug release because of their porous structure that filled with water. The next fastest release was from PLGA microspheres containing 10% polyethylene glycol (PEG) and the slowest was from PLGA microspheres. The difference between the two slower patterns (PLGA and PLGA with PEG) may be a result of low molecular weight of PEG that plasticized the PLGA matrix with a high molecular weight.

The type of polymer also affects the release rate of drugs from microspheres. For example, a study on the release of Taxol from PLGA and PCL microparticles indicated that drug-loaded PLGA microparticles released the drug faster than the PCL microparticles and that the biconcave particles released the drug faster than the spherical particles.[17]

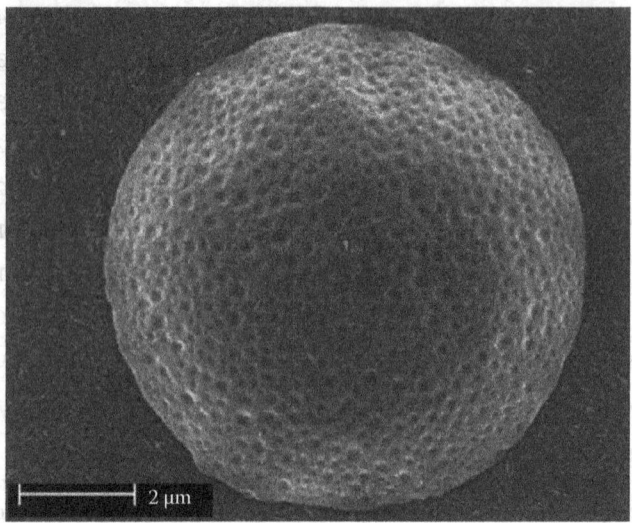

FIGURE 22.26 SEM image of particle obtained by EHDA from a solution of PCL in chloroform, as a fast evaporating solvent (golf-shaped surface). (Reprinted from *J. Colloid Interface Sci.*, 310, Wu, Y. and Clark, R.L., Controllable porous polymer particles generated by electrospraying, 529–535, Copyright 2007, with permission from Elsevier.)

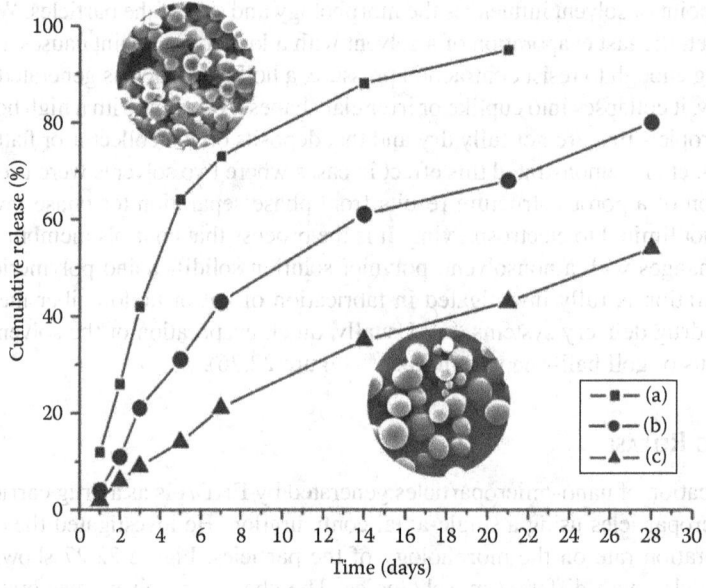

FIGURE 22.27 Drug release from the following particles: (a) PLA with 10% Taxol, hollow sphere; (b) PLA with 10% Taxol and 10% PEG, solid sphere; and (c) PLA with 10% Taxol, solid sphere. (Reprinted from *Int. J. Pharmaceut.*, 324(1), Ciach, T., Microencapsulation of drugs by electro-hydro-dynamic atomization, 51–55, Copyright 2006, with permission from Elsevier.)

Another parameter affecting EHDA is polymer concentration. Enayati et al. investigated the effect of polymer concentration on the drug release rate of microparticles. They found that the rate of release could be effectively controlled by this parameter. Figure 22.28 shows that as concentration decreased, the size of the particles decreased and the surface-to-volume area increased, which increased the rate of drug release.[75]

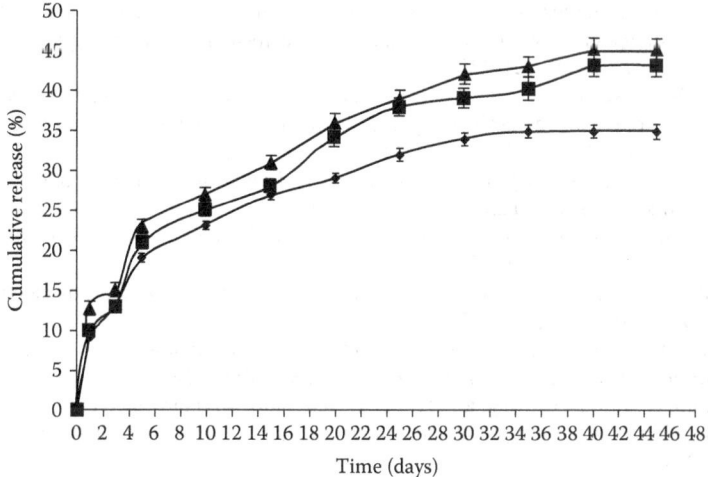

FIGURE 22.28 β-Estradiol release profile of PCL particles prepared from different concentrations: τ, PCL 10 wt%; ν, PCL 5 wt%; and ▲, PCL 2 wt%. (Adapted from Enayati, M., Ahmad, Z., Stride, E., and Edirisinghe, M., Size mapping of electric field-assisted production of polycaprolactone particles, *J. R. Soc. Interface*, 7(Suppl. 4), S409, figure 9, 2010, The Royal Society.)

In general, it was found that the morphology and size of the particles are responsible for the release patterns of the particulate system. Each parameter that can vary these characteristics can affect the release behavior of these structures.

22.7 CONCLUSION

EDHA has many applications in food, agriculture, coating, and pharmaceutical industries. It has the ability to fully control the size of the particles to within a narrow size distribution. Its continuous one-step method operates under different conditions depending on its use. The present study reviewed the principles, applications, advantages, disadvantages, and configuration of the process. The parameters affecting the characteristics of the particles have been highlighted and the potential uses of EHDA reported. Much effort has been expended to overcome difficulties associated with the process to make it suitable for mass production. It is anticipated that this technique will be a standard form of production for many industries in the near future.

REFERENCES

1. Fang, Z. and B. Bhesh. Spray drying, freeze drying and related processes for food ingredient and nutraceutical encapsulation, in *Encapsulation Technologies and Delivery Systems for Food Ingredients and Nutraceuticals*, N. Garti and D.J. McClements, Eds. WP Woodhead Publishing: Oxford, U.K., pp. 73–109, 2012.
2. Gomez-Estaca, J., M.P. Balaguer, R. Gavara, and P. Hernandez-Munoz. Formation of zein nanoparticles by electrohydrodynamic atomization: Effect of the main processing variables and suitability for encapsulating the food coloring and active ingredient curcumin. *Food Hydrocolloids* 28(1) (2012): 82–91.
3. Bock, N., T.R. Dargaville, and M.A. Woodruff. Electrospraying of polymers with therapeutic molecules: State of the art. *Progress in Polymer Science* 37(11) (2012): 1510–1551.
4. Jafari-Nodoushan, M., J. Barzin, and H. Mobedi. Design of a combined morphine delivery system (morphine microsphere-embedded in-situ forming), in *Iranian Controlled Release Society*, Tehran, Iran, 2014.
5. Kim, K.K. and D.W. Pack. Microspheres for drug delivery, in *BioMEMS and Biomedical Nanotechnology*, M. Ferrari, Ed. Springer: New York, pp. 19–50, 2006.

6. Almería, B. and A. Gomez. Electrospray synthesis of monodisperse polymer particles in a broad (60 nm–2 μm) diameter range: Guiding principles and formulation recipes. *Journal of Colloid and Interface Science* 417 (2014): 121–130.
7. Matsushima, Y., Y. Nemoto, T. Yamazaki, K. Maeda, and T. Suzuki. Fabrication of SnO_2 particle-layer on the glass substrate using electrospray pyrolysis method and the gas sensitivity for H_2. *Sensors and Actuators B: Chemical* 96(1) (2003): 133–138.
8. Zeleny, J. The electrical discharge from liquid points, and a hydrostatic method of measuring the electric intensity at their surfaces. *Physical Review* 3(2) (1914): 69.
9. Rayleigh, L. On the stability, or instability, of certain fluid motions. *Proceedings of the London Mathematical Society* 1(1) (1879): 57–72.
10. Vonnegut, B. and R.L Neubauer. Production of monodisperse liquid particles by electrical atomization. *Journal of Colloid Science* 7(6) (1952): 616–622.
11. Taylor, G. Disintegration of water drops in an electric field. *Proceedings of the Royal Society of London. Series A. Mathematical and Physical Sciences* 280(1382) (1964): 383–397.
12. Cao, L., J. Luo, K. Tu, L.-Q. Wang, and H. Jiang. Generation of nano-sized core–shell particles using a coaxial tri-capillary electrospray-template removal method. *Colloids and Surfaces B: Biointerfaces* 115 (2014): 212–218.
13. Hayati, I., A. Bailey, and T.F. Tadros. Investigations into the mechanism of electrohydrodynamic spraying of liquids: II. Mechanism of stable jet formation and electrical forces acting on a liquid cone. *Journal of Colloid and Interface Science* 117(1) (1987): 222–230.
14. Cloupeau, M. and B. Prunet-Foch. Electrohydrodynamic spraying functioning modes: A critical review *Journal of Aerosol Science* 25(6) (1994): 1021–1036.
15. Zhang, S and K. Kawakami. One-step preparation of chitosan solid nanoparticles by electrospray deposition. *International Journal of Pharmaceutics* 397(1) (2010): 211–221.
16. Xie, J. and C.-H. Wang. Encapsulation of proteins in biodegradable polymeric microparticles using electrospray in the Taylor cone-jet mode. *Biotechnology and Bioengineering* 97(5) (2007): 1278–1290.
17. Xie, J., J.C.M. Marijnissen, and C.-H. Wang. Microparticles developed by electrohydrodynamic atomization for the local delivery of anticancer drug to treat C6 glioma in vitro. *Biomaterials* 27(17) (2006): 3321–3332.
18. Ding, L., T. Lee, and C.-H. Wang. Fabrication of monodispersed Taxol-loaded particles using electrohydrodynamic atomization. *Journal of Controlled Release* 102(2) (2005): 395–413.
19. Li, J. On the stability of electrohydrodynamic spraying in the cone-jet mode. *Journal of Electrostatics* 65(4) (2007): 251–255.
20. Cloupeau, M. and B. Prunet-Foch. Electrostatic spraying of liquids in cone-jet mode. *Journal of Electrostatics* 22(2) (1989): 135–159.
21. Park, C.H. and J. Lee. Electrosprayed polymer particles: Effect of the solvent properties. *Journal of Applied Polymer Science* 114(1) (2009): 430–437.
22. Jaworek, A. Micro-and nanoparticle production by electrospraying. *Powder Technology* 176(1) (2007): 18–35.
23. Cloupeau, M. and B. Prunet-Foch. Electrostatic spraying of liquids: Main functioning modes. *Journal of Electrostatics* 25(2) (1990): 165–184.
24. Bocanegra, R., D. Galán, M. Márquez, I.G. Loscertales, and A. Barrero. Multiple electrosprays emitted from an array of holes. *Journal of Aerosol Science* 36(12) (2005): 1387–1399.
25. Scholten, E., H. Dhamankar, L. Bromberg, G.C. Rutledge, and T Alan Hatton. Electrospray as a tool for drug micro-and nanoparticle patterning. *Langmuir* 27(11) (2011): 6683–6688.
26. Hartman, R.P.A., D.J. Brunner, D.M.A. Camelot, J.C.M. Marijnissen, and B. Scarlett. Electrohydrodynamic atomization in the cone–jet mode physical modeling of the liquid cone and jet. *Journal of Aerosol Science* 30(7) (1999): 823–849.
27. Rayleigh, L. XX. On the equilibrium of liquid conducting masses charged with electricity. *The London, Edinburgh, and Dublin Philosophical Magazine and Journal of Science* 14(87) (1882): 184–186.
28. Taflin, D.C., T.L. Ward, and E. James Davis. Electrified droplet fission and the Rayleigh limit. *Langmuir* 5(2) (1989): 376–384.
29. Zarrabi, A., M. Vossoughi, I. Alemzadeh, and M.R. Chitsazi. Monodispersed polymeric nanoparticles fabrication by electrospray atomization. *International Journal of Polymeric Materials and Polymeric Biomaterials* 61(8) (2012): 611–626.
30. Jaworek, A. Electrostatic micro- and nanoencapsulation and electroemulsification: A brief review. *Journal of Microencapsulation* 25(7) (2008): 443–468.

31. Enayati, M., M.-W. Chang, F. Bragman, M. Edirisinghe, and E. Stride. Electrohydrodynamic preparation of particles, capsules and bubbles for biomedical engineering applications. *Colloids and Surfaces A: Physicochemical and Engineering Aspects* 382(1) (2011): 154–164.
32. Bocanegra, R., A.G. Gaonkar, A. Barrero, I.G. Loscertales, D. Pechack, and M. Marquez. Production of cocoa butter microcapsules using an electrospray process. *Journal of Food Science* 70(8) (2005): e492–e497.
33. Langer, G. and G. Yamate. Encapsulation of liquid and solid aerosol particles to form dry powders. *Journal of Colloid and Interface Science* 29(3) (1969): 450–455.
34. Kim, S.-Y., J. Yang, B. Kim, J. Park, J.-S. Suh, Y.-M. Huh, S. Haam, and J. Hwang. Continuous coaxial electrohydrodynamic atomization system for water-stable wrapping of magnetic nanoparticles. *Small* 9(13) (2013): 2325–2330.
35. Larsen, G., R. Spretz, and R. Velarde-Ortiz. Use of coaxial gas jackets to stabilize Taylor cones of volatile solutions and to induce particle-to-fiber transitions. *Advanced Materials* 16(2) (2004): 166–169.
36. Deng, W., J.F. Klemic, X.Li, M.A. Reed, and A. Gomez. Increase of electrospray throughput using multiplexed microfabricated sources for the scalable generation of monodisperse droplets. *Journal of Aerosol Science* 37(6) (2006): 696–714.
37. Labbaf, S., S. Deb, G. Cama, E. Stride, and M. Edirisinghe. Preparation of multicompartment submicron particles using a triple-needle electrohydrodynamic device. *Journal of Colloid and Interface Science* 409 (2013): 245–254.
38. Labbaf, S., H. Ghanbar, E. Stride, and M. Edirisinghe. Preparation of multilayered polymeric structures using a novel four-needle coaxial electrohydrodynamic device. *Macromolecular Rapid Communications* 35(6) (2014): 618–623.
39. Berkland, C., D.W. Pack, and K. Kim. Controlling surface nano-structure using flow-limited field-injection electrostatic spraying (FFESS) of poly(d,l-lactide-co-glycolide). *Biomaterials* 25(25) (2004): 5649–5658.
40. Lojewski, B., W. Yang, H. Duan, C. Xu, and W. Deng. Design, fabrication, and characterization of linear multiplexed electrospray atomizers micro-machined from metal and polymers. *Aerosol Science and Technology* 47(2) (2012): 146–152.
41. Yao, J., L.K. Lim, J. Xie, J. Hua, and C.-H. Wang. Characterization of electrospraying process for polymeric particle fabrication. *Journal of Aerosol Science* 39(11) (2008): 987–1002.
42. Liu, G., X. Miao, W. Fan, R. Crawford, and Y. Xiao. Porous PLGA microspheres effectively loaded with BSA protein by electrospraying combined with phase separation in liquid nitrogen. *Journal of Biomimetics, Biomaterials, and Tissue Engineering* 6 (2010): 1–18.
43. Reyderman, L. and S. Stavchansky. Electrostatic spraying and its use in drug delivery—Cholesterol microspheres. *International Journal of Pharmaceutics* 124(1) (1995): 75–85.
44. Fantini, D., M. Zanetti, and L. Costa. Polystyrene microspheres and nanospheres produced by electrospray. *Macromolecular Rapid Communications* 27(23) (2006): 2038–2042.
45. Mota, A., A.S. Lotfi, J. Barzin, M. Massumi, M. Hatam, and B. Adibi. Study of hBMSC adhesion and proliferation on RGD-modified polycaprolactone/gelatin nanofibrous scaffold. *Modares Journal of Medical Sciences: Pathobiology* 16(1) (2013): 75–87.
46. Venugopal, J., R. Rajeswari, M. Shayanti, S. Low, A. Bongso, V.R. Giri Dev, G. Deepika, A.T. Choon, and S. Ramakrishna. Electrosprayed hydroxyapatite on polymer nanofibers to differentiate mesenchymal stem cells to osteogenesis. *Journal of Biomaterials Science, Polymer Edition* 24(2) (2012): 170–184.
47. Mota, A., A.S. Lotfi, J. Barzin, M. Hatam, B. Adibi, Z. Khalaj, and M. Massumi. Human bone marrow mesenchymal stem cell behaviors on PCL/gelatin nanofibrous scaffolds modified with A collagen IV-derived RGD-containing peptide. *Cell Journal (Yakhteh)* 16(1) (2014): 1.
48. Zhu, T., C. Li, W. Yang, X. Zhao, X. Wang, C. Tang, B. Mi, Z. Gao, W. Huang, and W. Deng. Electrospray dense suspensions of TiO_2 nanoparticles for dye sensitized solar cells. *Aerosol Science and Technology* 47(12) (2013): 1302–1309.
49. Anal, A.K. and W.F. Stevens. Chitosan–alginate multilayer beads for controlled release of ampicillin. *International Journal of Pharmaceutics* 290(1) (2005): 45–54.
50. Xie, J., W.J. Ng, L.Y. Lee, and C.-H. Wang. Encapsulation of protein drugs in biodegradable microparticles by co-axial electrospray. *Journal of Colloid and Interface Science* 317(2) (2008): 469–476.
51. Xie, J., R.S. Tan, and C.-H. Wang. Biodegradable microparticles and fiber fabrics for sustained delivery of cisplatin to treat C6 glioma in vitro. *Journal of Biomedical Materials Research Part A* 85(4) (2008): 897–908.
52. Tang, K. and A. Gomez. Monodisperse electrosprays of low electric conductivity liquids in the cone-jet mode. *Journal of Colloid and Interface Science* 184(2) (1996): 500–511.

53. Gañán-Calvo, A.M., J. Dávila, and A. Barrero. Current and droplet size in the electrospraying of liquids. Scaling laws. *Journal of Aerosol Science* 28(2) (1997): 249–275.
54. Chen, D.-R., C.H. Wendt, and D.Y.H. Pui. A novel approach for introducing bio-materials into cells. *Journal of Nanoparticle Research* 2(2) (2000): 133–139.
55. Edirisinghe, M.J. and S.N. Jayasinghe. Electrohydrodynamic atomization of a concentrated nano-suspension. *International Journal of Applied Ceramic Technology* 1(2) (2004): 140–145.
56. Loscertales, I.G., A. Barrero, I. Guerrero, R. Cortijo, M. Marquez, and A.M. Ganan-Calvo. Micro/nano encapsulation via electrified coaxial liquid jets. *Science* 295(5560) (2002): 1695–1698.
57. Asano, K. Electrostatic spraying of liquid pesticide. *Journal of Electrostatics* 18(1) (1986): 63–81.
58. Lenggoro, I.W., K. Okuyama, J.F. de la Mora, and N. Tohge. Preparation of ZnS nanoparticles by electrospray pyrolysis. *Journal of Aerosol Science* 31(1) (2000): 121–136.
59. Okuyama, K. and I.W. Lenggoro. Preparation of nanoparticles via spray route. *Chemical Engineering Science* 58(3) (2003): 537–547.
60. Suh, J., B. Han, K. Okuyama, and M. Choi. Highly charging of nanoparticles through electrospray of nanoparticle suspension. *Journal of Colloid and Interface Science* 287(1) (2005): 135–140.
61. Kaelin, M., H. Zogg, A.N. Tiwari, O. Wilhelm, S.E. Pratsinis, T. Meyer, and A. Meyer. Electrosprayed and selenized Cu/In metal particle films. *Thin Solid Films* 457(2) (2004): 391–396.
62. Huang, J., S.N. Jayasinghe, S.M. Best, M.J. Edirisinghe, R.A. Brooks, and W. Bonfield. Electrospraying of a nano-hydroxyapatite suspension. *Journal of Materials Science* 39(3) (2004): 1029–1032.
63. Zhou, H. and S.B. Bhaduri. Deposition of PLA/CDHA composite coating via electrospraying. *Journal of Biomaterials Science, Polymer Edition* 24(7) (2012): 784–796.
64. Balachandran, W., P. Miao, and P. Xiao. Electrospray of fine droplets of ceramic suspensions for thin-film preparation. *Journal of Electrostatics* 50(4) (2001): 249–263.
65. Kim, S.J., D.H. Jang, W.H. Park, and B.-M. Min. Fabrication and characterization of 3-dimensional PLGA nanofiber/microfiber composite scaffolds. *Polymer* 51(6) (2010): 1320–1327.
66. Almería, B., W. Deng, T.M. Fahmy, and A. Gomez. Controlling the morphology of electrospray-generated PLGA microparticles for drug delivery. *Journal of Colloid and Interface Science* 343(1) (2010): 125–133.
67. Wu, Y., J.A. MacKay, J.R. McDaniel, A. Chilkoti, and R.L. Clark. Fabrication of elastin-like polypeptide nanoparticles for drug delivery by electrospraying. *Biomacromolecules* 10(1) (2008): 19–24.
68. Enayati, M., Z. Ahmad, E. Stride, and M. Edirisinghe. Preparation of polymeric carriers for drug delivery with different shape and size using an electric jet. *Current Pharmaceutical Biotechnology* 10(6) (2009): 600–608.
69. Ganan-Calvo, A.M. Cone-Jet analytical extension of Taylor's electrostatic solution. The asymptotic universal scaling laws in electrospraying of liquids, in *APS Division of Fluid Dynamics Meeting Abstracts*, San Francisco, CA, 1997.
70. Jafari-Nodoushan, M., H. Mobedi, and J. Barzin. Monitoring the effect of flow rate on size and morphology of microspheres produced by the means of electrospraying method, in *International Seminar of Polymer Science and Technology*, Tehran, Iran, 2012.
71. Bock, N., M.A. Woodruff, D.W. Hutmacher, and T.R. Dargaville. Electrospraying, a reproducible method for production of polymeric microspheres for biomedical applications. *Polymers* 3(1) (2011): 131–149.
72. Jafari-Nodoushan, M., J. Barzin, and H. Mobedi. Size and morphology controlling of PLGA microparticles produced by electro hydrodynamic atomization. *Polymers for Advanced Technologies* 26(5) (2015): 502–513.
73. Raula, J., H. Eerikäinen, and E.I. Kauppinen. Influence of the solvent composition on the aerosol synthesis of pharmaceutical polymer nanoparticles. *International Journal of Pharmaceutics* 284(1) (2004): 13–21.
74. Hong, S.-H., J.H. Moon, J.-M. Lim, S.-H. Kim, and S.-M. Yang. Fabrication of spherical colloidal crystals using electrospray. *Langmuir* 21(23) (2005): 10416–10421.
75. Enayati, M., Z. Ahmad, E. Stride, and M. Edirisinghe. Size mapping of electric field-assisted production of polycaprolactone particles. *Journal of the Royal Society, Interface/the Royal Society* 7 (Suppl 4) (2010): S393–402.
76. Ciach, T. Microencapsulation of drugs by electro-hydro-dynamic atomization. *International Journal of Pharmaceutics* 324(1) (2006): 51–55.
77. Hu, J.-F., S.-F. Li, G.R. Nair, and W.-T. Wu. Predicting chitosan particle size produced by electrohydrodynamic atomization. *Chemical Engineering Science* 82 (2012): 159–165.
78. Gupta, P., C. Elkins, T.E. Long, and G.L. Wilkes. Electrospinning of linear homopolymers of poly (methyl methacrylate): exploring relationships between fiber formation, viscosity, molecular weight and concentration in a good solvent. *Polymer* 46(13) (2005): 4799–4810.

79. Yeo, L.Y., Z. Gagnon, and H.-C. Chang. AC electrospray biomaterials synthesis. *Biomaterials* 26(31) (2005): 6122–6128.
80. Ganan-Calvo, A.M., J. Davila, and A. Barrero. Current and droplet size in the electrospraying of liquids. Scaling laws. *Journal of Aerosol Science* 28(2) (1997): 249–275.
81. Barzin, J., C. Feng, K.C. Khulbe, T. Matsuura, S.S. Madaeni, and H.Mirzadeh. Characterization of polyethersulfone hemodialysis membrane by ultrafiltration and atomic force microscopy. *Journal of Membrane Science* 237(1–2) (2004): 77–85.
82. Barzin, J., S.S. Madaeni, and S. Pourmoghadasi. Hemodialysis membranes prepared from poly (vinyl alcohol): Effects of the preparation conditions on the morphology and performance. *Journal of Applied Polymer Science* 104(4) (2007): 2490–2497.
83. Barzin, J. and B. Sadatnia. Correlation between macrovoid formation and the ternary phase diagram for polyethersulfone membranes prepared from two nearly similar solvents. *Journal of Membrane Science* 325(1) (2008): 92–97.
84. Barzin, J. and B. Sadatnia. Theoretical phase diagram calculation and membrane morphology evaluation for water/solvent/polyethersulfone systems. *Polymer* 48(6) (2007): 1620–1631.
85. Barzin, J., S.S. Madaeni, H. Mirzadeh, and M. Mehrabzadeh. Effect of polyvinylpyrrolidone on morphology and performance of hemodialysis membranes prepared from polyether sulfone. *Journal of Applied Polymer Science* 92(6) (2004): 3804–3813.
86. Zare, M., H. Mobedi, J. Barzin, H. Mivehchi, A. Jamshidi, and R. Mashayekhi. Effect of additives on release profile of leuprolide acetate in an in situ forming controlled-release system: In vitro study. *Journal of Applied Polymer Science* 107(6) (2008): 3781–3787.
87. Astaneh, R., M. Erfan, J. Barzin, H. Mobedi, and H.R. Moghimi. Effects of ethyl benzoate on performance, morphology, and erosion of PLGA implants formed in situ. *Advances in Polymer Technology* 27(1) (2008): 17–26.
88. Farrokhzad, H., H. Mobedi, J. Barzin, and A. Poorkhalil. Evaluation of polymer concentration effect on doxycycline hyclate drug release from in situ forming system based on poly (lactide-co-glycolide). *Science and Technology* 22(6) (2010): 495–505.
89. Jafari-Nodoushan, M., T. Ebrahimi, J. Barzin, and H. Mobedi, Evaluation of naltroxone release kinetic from in situ-forming drug delivery system (investigation the effect of polymer end group functionality), in *International Seminar of Polymer Science and Technology*, Tehran, Iran, 2012.
90. Wu, Y. and R.L. Clark. Controllable porous polymer particles generated by electrospraying. *Journal of Colloid and Interface Science* 310(2) (2007): 529–535.

23 Encapsulation Process
Pulsed Combustion Spray Drying

Chilwin Tanamal and James A. Rehkopf

CONTENTS

23.1 Introduction 439
23.2 Principles of Pulsed Combustion Spray Drying 439
23.3 Equipment 441
23.4 Advantages/Disadvantages of Pulsed Combustion Spray Drying 441
23.5 Application/Examples (Microcapsule Consideration) 442
23.6 Ingredients/Materials 442
23.7 Slurry Properties 442
23.8 Experiment Procedure 442
23.9 Findings/Observations 442
23.10 Conclusion 445
References 445

23.1 INTRODUCTION

Pulsed combustion spray drying is a form of spray drying that is well suited for the drying of slurries of encapsulated materials. The atomization method employed in pulsed combustion drying is the key to good results on encapsulates. Scientists from Appvion Encapsys group recognized this feature and conducted the initial drying trials in the fall of 2006. These results were promising and provided the foundation of a long relationship between Pulse Combustion Systems (PCS) and Encapsys.

23.2 PRINCIPLES OF PULSED COMBUSTION SPRAY DRYING

A pulse combustion spray dryer is identical to a conventional tall-form spray dryer in the drying chamber and all components downstream, including the ducting, cyclone, baghouse/scrubber, and exhaust fan. Upstream of the drying chamber, however, things are different. In a conventional spray dryer, atomization is accomplished by pressurizing the liquid feed in a high-pressure pump (at up to 3000 psi) and forcing it through a tiny nozzle (located at the top center of the dryer) to create a fan-shaped spray of tiny droplets. As drying air is introduced into the drying chamber, the warm air evaporates the water from the atomized droplets, leaving behind a dry powder after 15–30 s of residence time in the dryer. This atomization method tends to apply high mechanical shear forces on the liquid feed, which can shred the capsule wall of encapsulated particles.

In the pulse dryer, the atomization method is reversed. Instead of accelerating the liquid feed, the pulse dryer accelerates the drying gas to about 300 miles per hour in a resonating pulse combustor, as indicated in Figure 23.1, and directs it downward into the drying chamber from the top center. The liquid feed is introduced through an open pipe at low pressure and velocity into the center of the high-velocity exhaust of the pulse combustor. When the liquid exits the feed pipe into the pulsating, high-velocity exhaust gases, it is quickly atomized by a process called "gas dynamic atomization." The atomization zone is turbulent, providing good mixing between the atomized droplets and the pulsing hot air exiting the pulse combustor. This pulsating and turbulent zone produces

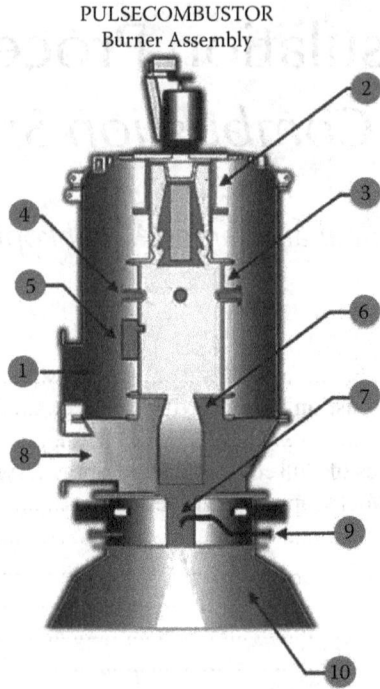

FIGURE 23.1 **(See color insert.)** Pulse spray drying basic process operation diagram. Here's how pulse combustion spray driers work: Air is (1) pumped into the pulse combustor's outer shell at low pressure, where it flows through the patented unidirectional air valve. (2) The air enters a tuned combustion chamber (a *Helmholtz resonator*) (3) where fuel is added (4) to the spray dryer. The air valve closes. The fuel–air mixture is ignited by a pilot (5) and explodes, creating hot air pressurized to about 3 psi above combustion fan pressure. The hot gases rush down the spray dryer's tailpipe (6) toward the atomizer (7). The air valve reopens (2) and allows the next air charge to enter. The fuel valve admits fuel, and the mixture explodes in the hot chamber. This spray drying cycle is controllable from about 80 to 110 Hz. Just above the spray dryer's atomizer, quench air (8) is blended in to achieve the desired product contact temperature. The exclusive PCS atomizer releases the liquid (9) into a carefully balanced gas flow, which dynamically controls atomization, drying, and particle trajectory. The atomized liquid enters a conventional tall-form drying chamber (10). Downstream of the spray-drying process, the suspended powder is retrieved using standard collection equipment, such as cyclones and baghouses.

heat transfer rates up to five times that of conventional drying.[1] The high heat transfer rate, in turn, results in rapid drying—on the order of one-half of a second.

Because the atomization in a pulse dryer is achieved with little mechanical shear, the encapsulating sheath is largely unaffected, leaving the core intact. As described in the following, a measure for encapsulate integrity is "free core"—the lower, the better. Pulsed drying can produce low levels of free core.

Two other features of pulsed combustion drying were not optimized in the Encapsys trials, but are worthy of mention for other users. First, pulse combustion dryers are 12%–20% more thermally efficient than conventional spray dryers[1] and can handle higher solids loadings in the feed as well as higher viscosities. Both features reduce drying costs. Second, pulse drying is gentle, causing little scorching or degradation. If an encapsulate shell were made of a delicate protein, as an example, the shell could be expected to exit the pulse-drying process in a better shape than when conventionally spray dried. Based on this low-degradation feature, PCS obtained U.S. Patent 8,517,723, methods for drying a viral vaccine.[2]

FIGURE 23.2 A demonstration pulse combustion dryer installed in the Arizona plant of PCS.

23.3 EQUIPMENT

The equipment used in a pulse dryer is the same as that used in a conventional tall-form, pressure-nozzle spray dryer, with the following equipment replacements:

1. A pulse combustor for a conventional atmospheric burner (e.g., Maxon and Eclipse) and ducting for the hot air from the burner to the drying chamber.
2. A 6 psi air pump for a low-pressure burner fan.
3. A 1 horsepower (HP) peristaltic pump for a high HP, high-pressure, atomizing feed pump.
4. An open-pipe *atomizer* (impervious to wear) for an atomizing nozzle.

A photograph of a demonstration pulse combustion dryer installed in the Arizona plant of PCS is shown in Figure 23.2.

Pulse dryers in operation in industry range in evaporative capacity between 30 pounds of water per hour and 2000 pounds per hour.

23.4 ADVANTAGES/DISADVANTAGES OF PULSED COMBUSTION SPRAY DRYING

The *zero-shear* and rapid-drying environment of a pulse dryer tend to produce a superior powder to what can be expected from a conventional spray dryer in many cases. Past and present customers of PCS report little degradation in the drying process, improved powder *surface characteristics*, and more uniform particle size distributions. In addition, pulse dryers have a larger operating envelope (inlet and outlet temperatures, atomization energy, feed solids, etc.), and this envelope was explored in the Encapsys trials, detailed below. Finally, the high *delta-T* that is natural for pulse dryers produces improved thermal efficiencies, as low as 1250 Btu per pound of water.[1] The thermal efficiencies observed while drying Encapsys' slurries ranged from 2000 to 3000 Btu per pound water, which is still yet to be optimized. Pulse dryer capital costs are competitive with conventional spray dryers, and maintenance costs are less, mainly because there is no high-pressure pump and no spray nozzles.

A disadvantage with pulse drying is that there are few pulse dryers in operation. As a result, some of the customer-led innovations, such as in-vessel agglomeration, have not yet been developed for pulse dryers. Also, there are few used dryers available and few technicians trained in the art. Finally, although pulse dryers are more efficient and more flexible than conventional spray dryers, they are not a good choice for feeds that are elastomeric, high in sugars, or high in free fats.

23.5 APPLICATION/EXAMPLES (MICROCAPSULE CONSIDERATION)

As noted earlier, one important factor in spray drying microcapsules is the ability to keep individual capsules intact. However, there are other factors, such as moisture, particle size, and flowability, to consider during spray drying of microcapsule slurry. In addition to these quality considerations, processing factors such as thermal efficiency and yield are also important. When all these considerations are properly addressed, spray drying of microcapsules becomes a complex process. The application section of this chapter provides an overview of microcapsule spray drying through the pulsed combustion technology utilization and offers points to consider for such application.

23.6 INGREDIENTS/MATERIALS

Results from microcapsule spray drying runs on PCS' pulsed combustion dryer will serve as the basis for this discussion. The rationale behind showing the different data sets is simply to illustrate the many variables to consider while drying microcapsule slurries with the pulsed combustion dryer technology. These data sets also show that different products tend to behave differently. The encapsulated material presented in this chapter is primarily fragrance based.

23.7 SLURRY PROPERTIES

The microcapsule slurries used to generate the data have volume-weighted median particle size ranges from 5 to 25 µm. There are also a few different capsule chemistries for the slurry to show the different data curve generated among them. This chemistry difference results from the type of polymer and cross-linking agent used to form the shell of the capsule. Also, the densities of these slurries are lighter than water. If the slurries are not regularly agitated, there is an observable two-layer phase separation. Therefore, the slurry needs to be continuously agitated during the drying process to ensure consistency.

23.8 EXPERIMENT PROCEDURE

The data, chart, and discussion points presented in this application section are based on several trials conducted at the PCS plant in Payson, AZ, utilizing their pulsed combustion spray dryer technology. The trials were not specifically designed for this documentation, so some definitive protocols cannot be provided. However, we investigated how several parameters such as contact and exit temperatures (components of the dryer operating envelope), flow aid level, and percentage of feed solids affected capsule integrity as measured analytically and presented as % free core.

23.9 FINDINGS/OBSERVATIONS

During multiple trials, we observed the impact of contact temperature, flow aid level, and feed solid content on capsule integrity, depicted as percentage of core that is not contained within the capsule. Figures 23.3 through 23.7 display various trends as each variable changes. Because these trials were not conducted for the purpose of this chapter, each figure does not necessarily depict trends for the same product. In Figure 23.3, we observed that an increase in contact temperature could lead to an increase in capsule breakage during the drying process for one product, while for another product, the difference in the amount of capsule breakage level was insignificant with an increase in contact temperature. Figure 23.4 shows that the amount of flowing

Encapsulation Process

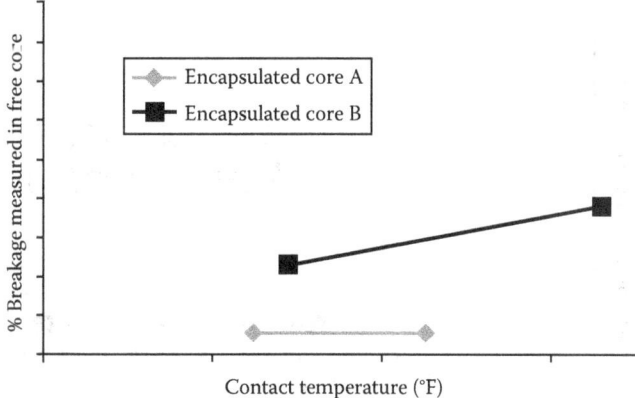

FIGURE 23.3 Percent free core versus contact temperature.

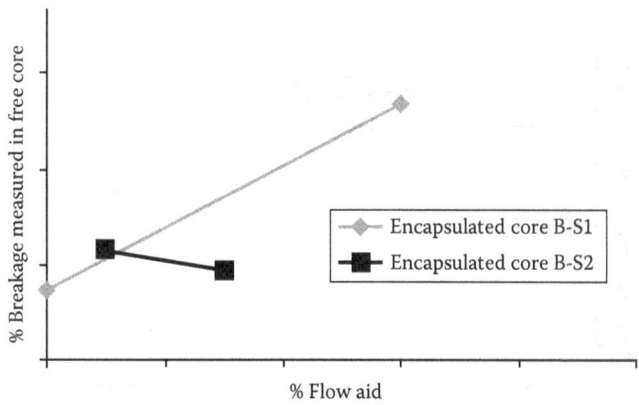

FIGURE 23.4 Percent free core versus percent flow aid.

agent used to improve powder flowability characteristics may impact capsule breakage unfavorably, but depending on the product, it may also not impact capsule breakage. Another variable that could be important in the level of capsule breakage observed is feed solids. As depicted in Figure 23.5, an increase in feed solids could significantly result in higher level of capsule breakage. This is caused by the tendency for higher solid slurries to also have higher viscosity, which requires higher atomization energy. Similar to the contact temperature and flow aid loading, certain products could also have higher atomization tolerance in which an increase in feed solid may not affect the level of breakage significantly. Figure 23.6 shows that no significant impact was observed on capsule integrity based on exit temperature variation, which is evidence of the gentle drying mentioned earlier. Increasing exit temperature simply decreases residual moisture, as expected and as shown in Figure 23.7.

Thermal efficiency is one advantage claimed by the pulsed combustion technology. For all the slurry runs that Encapsys has performed, while the lowest attainable efficiency has never been reached, the efficiency value is fairly consistent throughout a long run. Figure 23.8 displays the consistency of thermal efficiency for a trial run that spanned a 2-day period.

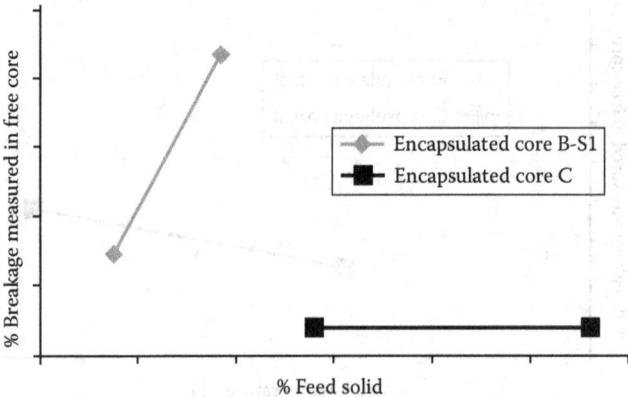

FIGURE 23.5 Percent free core versus percent feed solid.

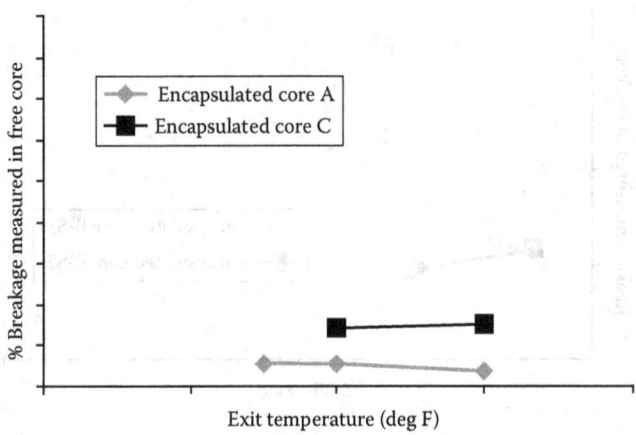

FIGURE 23.6 Percent free core versus exit temperature.

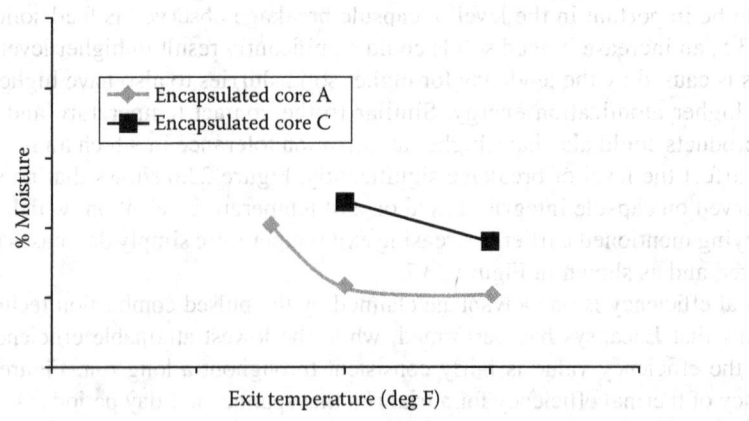

FIGURE 23.7 Percent moisture versus exit temperature.

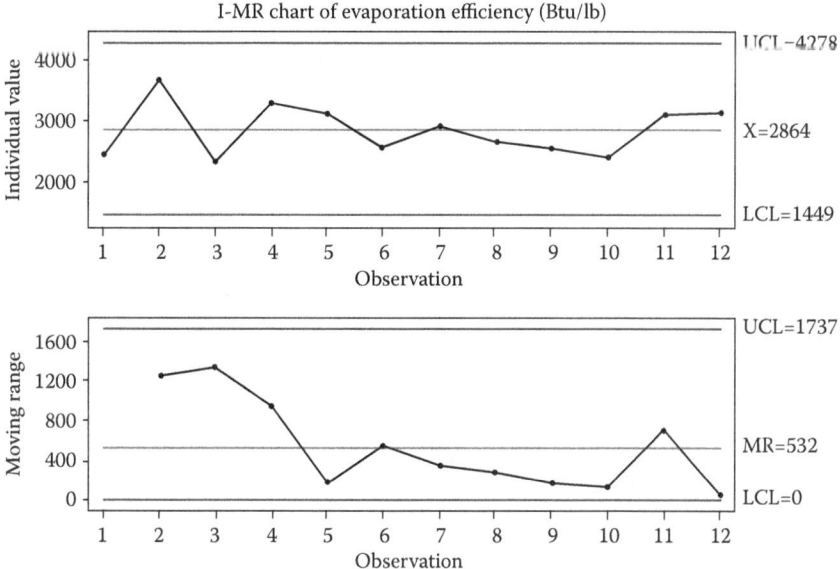

FIGURE 23.8 Thermal efficiency calculated for a trial run spanned over 2 days.

23.10 CONCLUSION

Pulsed combustion spray drying can be an effective and efficient method for spray drying of encapsulated slurries. The key feature of pulsed drying as related to encapsulated materials is an atomization method that applies minimal mechanical shear on the core/shell structure of the individual encapsulates. The atomization method keeps the shell in place throughout the drying process and results in a powder consisting of mostly intact encapsulates. However, certain parameter changes could impact the amount of breakage seen on a given formula. As a result, a robust formula is also needed in conjunction with the right process parameters and technology.

Additionally, there are other benefits associated with the pulse combustion drying technology associated with less degradation, improved surface characteristics, and more uniform particle size distribution. These possible benefits were not explored during the trial run conducted by Encapsys, but may be worth exploring for specific products.

REFERENCES

1. Corliss, J.M. and Putnam, A.A. (1986). Heat-transfer enhancement by pulse combustion in industrial processes, *American Society of Metals*, 39–48.
2. Tate, J.L., Mirko, D.A., and Rehkopf, J.A. (2013). US Patent No. 8,517,723. Washington, DC: U.S. Patent and Trademark Office.

FIGURE 22.5. Trend of 16 hour reduced CO_2 fermenting peas over 2 days.

22.10 CONCLUSION

Postharvest modified atmosphere can be an effective method for prolonging storage of non-carbohydrate foods. In the case of fruit storage, it is limited to certain fruit that is a major portion of world fruit supply, apples, pears, etc. This is useful, thus also of the fruits and vegetables. The storage and the modern egg plant is taking place to improve the dry life process and results in a powder of essentially all-dry peas. It appears, however, that major changes you contemplate the amount of modified atmosphere which is exist, in most complex applications used in CONA with a view in the right proportions yielding a future key.

Vitality in the air of the farmers, and with the peas combined with the reduced energy input from degradable biopolymer. The atmosphere also needs to be under the better developed. These packaging techniques are not expected at the moment as carried out by European factories for the world and entire for apes of produce.

REFERENCES

1.

2.

24 Supercritical Fluid Technology for Encapsulation

Ángel Martín, Marta Fraile, Soraya Rodríguez-Rojo, and María José Cocero

CONTENTS

24.1 Properties of Supercritical Fluids 447
24.2 Supercritical Carbon Dioxide as Media for Precipitation Processes 448
    24.2.1 Rapid Expansion of Supercritical Solutions 449
    24.2.2 Batch Gas Antisolvent Process 449
    24.2.3 Semicontinuous Supercritical Antisolvent (SAS) Process and Its Variants 451
    24.2.4 Supercritical Fluid Extraction of Emulsions (SFEE) 452
    24.2.5 Particles from Gas-Saturated Solutions Process 455
24.3 Encapsulation Processes with Supercritical Fluids: Interactions between Supercritical Carbon Dioxide and Common Polymeric Carrier Materials 457
24.4 Case Study: Encapsulation of Quercetin in Pluronic® F127 Poloxamer by SAS Process 460
References 464

24.1 PROPERTIES OF SUPERCRITICAL FLUIDS

A pure component is considered to be in supercritical conditions when its temperature and pressure are higher than its critical temperature and pressure (Figure 24.1).[1,2] For example, in the case of CO_2, one of the most commonly used fluids in supercritical encapsulation processes, the critical conditions are approximately 7.4 MPa and 31.1°C.

The properties of a supercritical fluid (SCF) are often described as being intermediate between those of a liquid and a gas. This is because the supercritical region includes the transition from liquid to gas regions of the phase diagram that at subcritical conditions is represented by the vapor–liquid saturation line. Compared to the normal vapor–liquid phase change, in the supercritical region, this transition takes place through a monotonous (yet very pronounced) variation in the physical properties of the fluid, and not through a sharp phase change at the saturation point (Figure 24.2). In the gas to liquid transition region, SCFs present a combination of thermodynamic and transport properties, like high densities, low viscosities, and high diffusivities, which makes them suitable for the development of new processes that cannot be carried out with conventional liquid or gaseous solvents.

Besides the unique combination of physical properties of SCFs, the most useful characteristic of these fluids is the possibility of modifying their properties with changes in their state parameters (pressure and temperature), thanks to the smooth variation of these properties in the supercritical region. This allows to tune the properties of the supercritical solvent for a specific application. In particular, in the near-critical region, the density of the fluid is very sensitive to changes in pressure. Since many solvent properties are directly related to the bulk density, they will also have a strong pressure dependence. For instance, it has been found that the solvent power of the SCF is proportional to the density of the fluid.[1,2] This allows using SCFs for highly selective extraction processes.

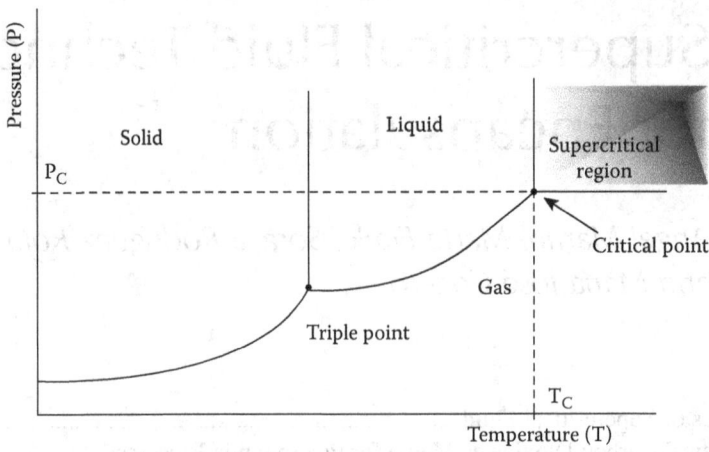

FIGURE 24.1 Definition of the supercritical fluid region in the phase diagram of a pure substance.

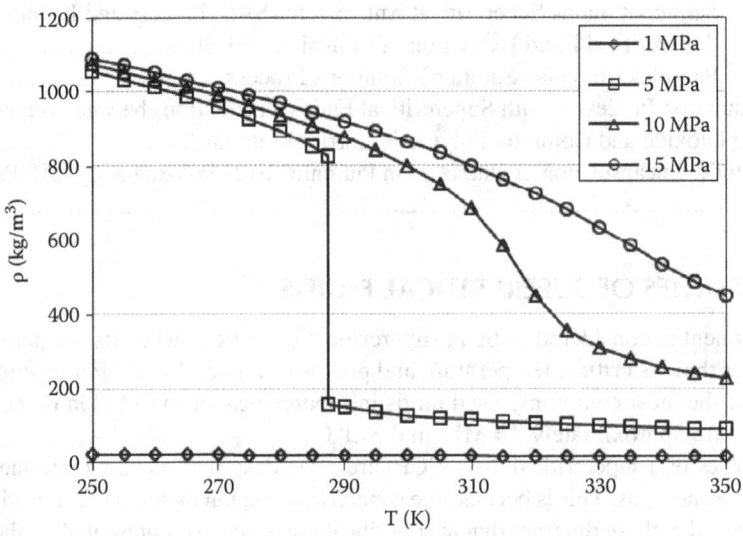

FIGURE 24.2 Variation of the density of CO_2 with temperature at subcritical and supercritical pressures.

24.2 SUPERCRITICAL CARBON DIOXIDE AS MEDIA FOR PRECIPITATION PROCESSES

The application of SCFs in precipitation processes arises directly from their characteristic physical properties, and in particular from the possibility of varying these properties with changes in the operating conditions. The benefits claimed for the application of SCF technologies instead of conventional precipitation processes are the possibility of obtaining particles with a smaller particle size and controlled particle size distributions (PSDs), and the simplicity of the solvent removal from the product, which can be achieved with a pressure reduction, in contrast with the complex purification processes often required when conventional liquid solvents are used.

The development of precipitation processes is one of the most active research topics in the area of SCFs, which started in a large scale in the mid-1980s,[3] and continues today. In this period, several different precipitation techniques with SCFs have been proposed.

Supercritical Fluid Technology for Encapsulation

In these processes, the most frequently used supercritical solvent is supercritical carbon dioxide (SC-CO_2). The reason for this are the favorable properties of SC-CO_2, including: no toxicity, no flammability, low critical temperature that allows to carry out the precipitation with mild operating conditions, and phase equilibrium properties relatively favorable for many precipitation processes. Supercritical precipitation processes with CO_2 have been used to process a large variety of materials, including natural substances, pharmaceuticals, polymers, explosives, and inorganic substances.[3–6]

Different precipitation processes based on supercritical carbon dioxide have been proposed in which CO_2 performs different functions: as solvent, in the rapid expansion of supercritical solutions (RESS) process, as antisolvent, in the supercritical antisolvent (SAS) process, or as solute, in the particles from gas saturated solutions (PGSS) process.

24.2.1 RAPID EXPANSION OF SUPERCRITICAL SOLUTIONS

This process takes advantage of the large variations in the solvent power of the SCF with changes in pressure. The process consists of a first extraction step in which the SCF is saturated with the substrates of interest, followed by a sudden depressurization in a nozzle, which produces a large decrease in the solvent power and the temperature of the fluid, therefore, causing the precipitation of the solute (Figure 24.3).[7]

This process has been successfully applied to obtain nano- and microparticles of different substances, with narrow PSDs. It is also suitable for the production of films. More recently, it has also been used for coprecipitation processes. Usually, the process parameters with a larger influence on product characteristics are the nozzle design, and the preexpansion pressure and temperature. The main limitation of this process is that it can only be applied which substances which have a relatively large solubility in SC-CO_2, since its application to substances with low solubilities requires large amounts of CO_2, which makes the process uneconomical. This limitation precludes the application of the RESS process to high-molecular-weight or to polar substances, since these materials typically have very low solubilities in SC-CO_2.

24.2.2 BATCH GAS ANTISOLVENT PROCESS

In this process, the SCF is not used as a solvent, but as an antisolvent, which causes the precipitation of a solute initially dissolved in an organic liquid. This process typically operates at moderate pressures (5–8 MPa). In this range, CO_2 has a large solubility in most organic solvents, while at the

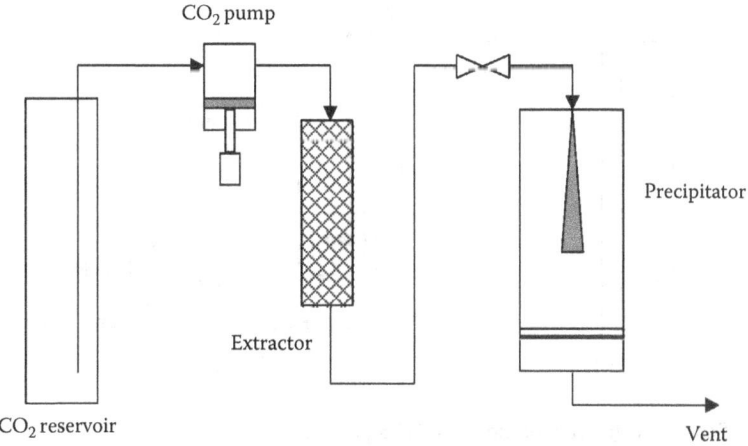

FIGURE 24.3 Schematic diagram of the RESS process.

same time, the solvent power of CO_2 for relatively high-molecular-weight substrates is very low. Therefore, the saturation of the organic solvent with CO_2 causes a decrease in the solvent power of the liquid mixture and the precipitation of the solute. A schematic diagram of a gas antisolvent (GAS) process is presented in Figure 24.4. This process consists of a vessel that initially is partially filled with the organic solution. CO_2 is pumped into this vessel until the desired operating pressure is achieved. After this, the solution is drained, and the CO_2 flow is maintained isobarically during a period long enough for the complete removal of the solvent. One peculiarity of this process is that when relatively large volumes of solution are processed, it is necessary to attach a stirrer to the vessel for improving the mixing between the solution and the CO_2.

As the driving force for the precipitation in the GAS process is the antisolvent effect caused by the solubilization of CO_2 in the liquid phase, and the solvent power of a liquid is often proportional to its density; it has been found that it is possible to select the optimum thermodynamic conditions for this process by studying the volumetric expansion in the solvent caused by CO_2. De la Fuente et al.[8] presented a definition of the volumetric expansion based on the variation of the partial molar volume of the solvent, which allows an optimum selection of the solvent and the operating pressure and temperature for a particular application. These authors also showed that a study of the solubility of the solute in the solvent–CO_2 mixtures allows predicting if the GAS process can yield satisfactory results: in systems in which there is a sharp decrease in solubility at some concentration of CO_2, the performance of the GAS process will be optimum, since in this case, the precipitation will take place very quickly and homogenously upon reaching this region; systems that show a slow decrease in solubility as the CO_2 amount increases are likely to yield worse results, since in this case, the precipitation will take place continuously and relatively slowly as CO_2 is fed to the precipitator.

In addition to the thermodynamic factors (pressure, temperature, concentration of the solute, and organic solvent), there are a number of other factors that have influence on the results of the precipitation, as the velocity of addition of CO_2, and the stirrer speed, which affects the secondary nucleation as in other crystallization processes. The main advantage of this process over the RESS process is that with a proper selection of the solvent, it is possible to micronize a very wide range of products, obtaining particle sizes typically in the range 1–10 μm, with a narrow PSD. The main disadvantages of this process are an increase in the mechanical complexity of the equipment due to the necessity of using a stirrer, the use of an organic solvent, and the relatively small production capacity of this process due to the operation in batch.

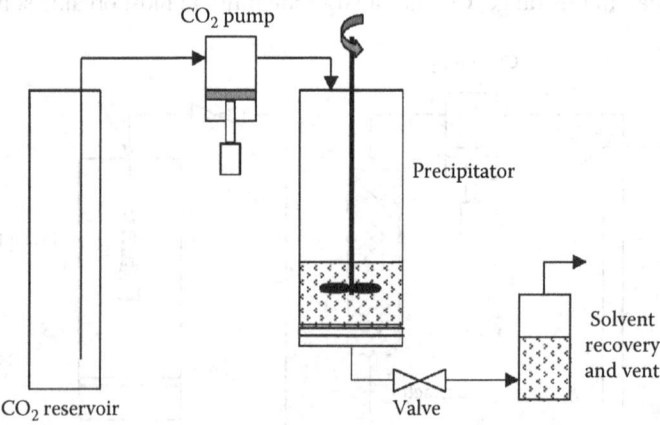

FIGURE 24.4 Schematic diagram of the batch GAS process.

24.2.3 SEMICONTINUOUS SUPERCRITICAL ANTISOLVENT (SAS) PROCESS AND ITS VARIANTS

The semicontinuous SAS processes were developed in order to overcome the limitations in the production capacity of the batch GAS processes. In these precipitation methods, CO_2 and a solution of the substance of interest are continuously fed to a precipitator. As in the GAS process, the mixing between these streams causes the precipitation of the solute due to the antisolvent effect of CO_2 (Figure 24.5). In order to obtain a good mixing between the two streams without the aid of a mechanical device, the semicontinuous processes typically operate at higher pressures than in the GAS process, usually in the range 9–15 MPa. In this range, CO_2 is completely miscible with many organic solvents (Figure 24.6), and therefore, the mixing can take place without interface limitations. Additionally, the characteristic point of many of the variants of the semicontinuous anti solvent process (PGA, SEDS, ASES, etc.) is the design of the mixer for these two streams.

The influence of the position of the operating point with respect to the P-xy diagrams shown in Figure 24.6 has been the subject of several studies. Reverchon et al.[9] found that when operating at pressures above the mixture critical point, the parameters that influence the mixing between the two streams, as the design of the nozzle and the precipitator, or the Reynolds number in the nozzle, have a small effect on the precipitation. This indicates that, in these conditions, the mixing between solution and CO_2 occurs faster than the precipitation, and thus, mixing parameters cannot influence the precipitation. However, at pressures below the mixture critical point, there is a significant effect of these parameters. Moreover, a change in particle morphology is frequently observed when pressure is reduced below this point, and sometimes, the particles are agglomerated, forming empty spheres, which could indicate that the atomization and precipitation occurs through the formation of droplets during the atomization of the organic solvent in CO_2. This difference in the precipitation mechanism depending on pressure is a further reason for operating above the mixture critical point, since the mixer parameters that have a critical influence at lower pressures are more difficult to scale up than pure thermodynamic parameters such as pressure, temperature, or solute concentration.

The semicontinuous SAS processes share the wide applicability and the performance of the batch processes in terms of particle size, and improve the production capacity. Its main

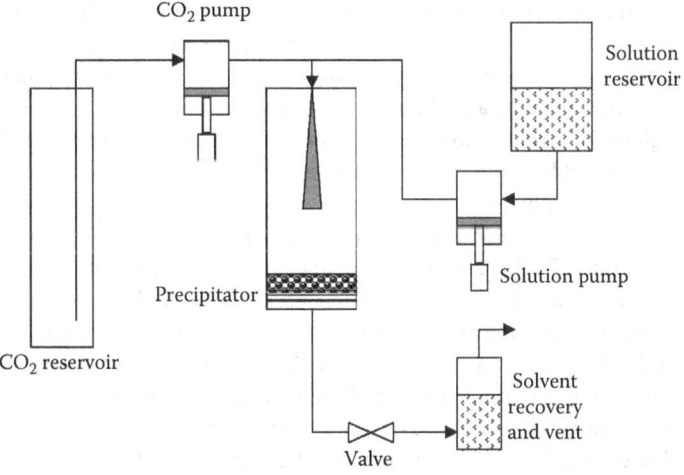

FIGURE 24.5 Schematic diagram of the semicontinuous SAS process.

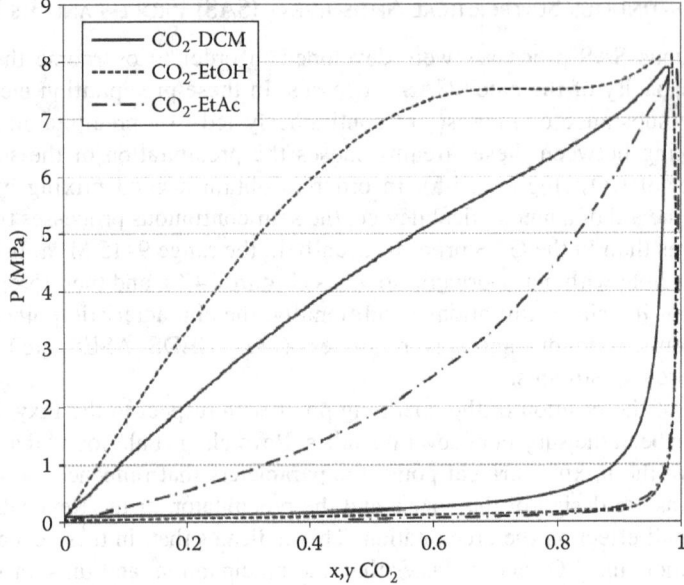

FIGURE 24.6 P-xy diagrams of CO_2-dichloromethane (DCM), CO_2-ethanol (EtOH), and CO_2-ethyl acetate (EtAc) at 40°C, calculated with the Peng–Robinson EOS.

disadvantage is again the use of organic solvents, and the large number of operating and design parameters that influence the process (pressure, temperature, concentrations, flow rate, and nozzle design), which makes difficult the analysis of experimental results and the scale-up of the process. A brief summary of some of the most relevant recent research contributions employing SAS is presented in Table 24.1.

24.2.4 SUPERCRITICAL FLUID EXTRACTION OF EMULSIONS (SFEE)

Supercritical fluid extraction of emulsions (SFEE) is a combination of the conventional emulsion precipitation process with the SAS processes. This process emerges as a solution for several of the drawbacks of each separated technology. Conventional emulsion precipitation methods involve large quantities of organic solvents and their removal requires additional separation techniques that commonly need high temperatures and high cost. On the other hand, as shown in Table 24.1 by SAS process, often it is not possible to obtain particle sizes within the nanometric scale and the products can present agglomeration problems.

In the SFEE process, an organic solvent in water emulsion is prepared and put into contact with carbon dioxide. Figure 24.7 presents micrographs of the evolution of an organic solvent droplet immersed in water during this process. In the first stages of the process, emulsion droplets are saturated by CO_2, thus causing the precipitation of solutes by antisolvent effect. In this way, each emulsion droplet behaves as a miniature gas antisolvent precipitator. By confining particle growth within the droplet, it is possible to control the final particle size by modification of the initial emulsion droplet size, which in some cases allows reducing particle sizes to the nanometer scale. Furthermore, agglomeration is reduced, thanks to the surfactants forming the emulsion. As the process continues, CO_2 extracts the organic solvent from the emulsion, thus leading to a solvent-free product.[42]

The experimental setup of these processes can be performed in a batch or continuous form, and the equipments involved in the precipitation process are practically the same as that of the batch

TABLE 24.1
Recent Applications of the SAS Process

Active Compound	Carrier Material	Solvent	T (K)	P (MPa)	Solution/CO_2 Flow	Product Characteristics	Reference
Quercetin	—	EtOH	309	7.9		120–450 nm	Kakran (2013)[10]
Quercetin dihydrate		EA	313.15	10	(1.4 mg/mL) were 0.6 kg/h CO_2 and 0.2 mL/min	2 µm	Santos (2013)[11]
β-Carotene		DCM	313.15	0.8	(8 mg/mL) were 1.5 kg/h and 1 mL/min	16 µm.	
Bixin-rich extract	PEG	DCM	313.15	10	1 mL/min 0.6 and 1.5 kg/h	33.2 µm	Sacchetin (2013)[12]
17α-Methyltestosterone (MT)	PLA	DCM	313	8–16	0.5–2.5 mL/min	5.4–20.5 µm	
Lutein		DCM/EtOH	328	8–12	0.25 mL/min	902 nm–1.58 µm	Boonnoun (2013)[13]
Apigenin (AP)		DMSO	308	14.5	0.5 mL/min	400–800 nm	Zhang (2013)[14]
Amoxicillin		DCM and DMSO	308–323	10–15	2 mL/min 11 g/min	0.5–5 µm	Montes (2013)[5]
Naproxen	Ethyl cellulose	EtOH/ACE	313	14	4 mL/min 11 g/min	0.08–1.3 µm	Montes (2013)[6]
Silica		EtOH	313	12	4 mL/min 11 g/min	1–2.7 µm	Montes (2013)[7]
	L-PLA	DCM	308–323	—	1–7.5 mL/min 30–60 g/min	0.62 µm	Bakhbakhi (2013)[18]
Rifampicin	Lactose	MetOH	313	12.4	1 mL/min 20 g/min	1 µm	Ober (2013)[19]
10-Hydroxycamptothecin (HCPT)	L-PLA	DCM and EtOH	303–313	7.5–14	0.5–1.7 mL/min 1–9 g/L	794.5 nm	Wang (2013)[20]
Ibuprofen sodium		EtOH	308–323		30–75 g/min	0.5–17.7 µm	Bakhbakhi (2013)[21]
Lignin		ACE	308	30	6.5 mL/min	(0.144 ± 0.03 µm)	Lu (2012)[22]
Vinblastine		N-methyl-2-pyrrolidone	333	25	6.7 mg/min	121 ± 5.3 nm	Zhang (2012)[23]
Vitexin		DMSO	313–343	15–30	8.5 kg/h 3.3–8.4 mL/min 1–2.5 mg/mL	126 ± 18.5 nm	Zu (2012)[24]
Furosemide	Crospovidone	MetOH, EtOH, or ACE	313	10–20	1 L/min	3–6 µm	De Zordi (2012)[25]

(Continued)

TABLE 24.1 (Continued)
Recent Applications of the SAS Process

Active Compound	Carrier Material	Solvent	T (K)	P (MPa)	Solution/CO$_2$ Flow	Product Characteristics	Reference
Paclitaxel (PTX)	L-PLA	DCM/EtOH	308	10–12	0.5 mL/min		Li Wenfeng (2012)[26]
Taxol		EtOH	330	20	2.5 mg/mL 6.6 mL/min	150.5 nm	Zhao (2012)[27]
Ursolic acid		EtOH	308–338	12.5–20	5–8 mL/min 1.25–5 mg/mL	139.2 ± 19.7 to 1039.8 ± 65.2 nm	Yang (2012)[28]
Caffeine		DCM	303–333	8–12	4 kg/h 4 mL/min	2.5 μm	Weber Brun (2012)[29]
Rosemary antioxidants	Pluronic® F88 or Pluronic® F127	EtOH	298–323	8–12	0.7 kg/min 1 mL/min 0.0067–0.08 g/mL	<1 μm	Visentin (2012)[30]
Glycyrrhizic acid (GA)		EtOH	338	25	15 mL/min 8 mL/min 20 mg/mL	95–174 nm	Sui (2012)[31]
Polymeric procyanidins (PPC)		EtOH	353	25	4 mL/min 25 mg/mL 8.5 kg/h	111 ± 5 nm	Yang (2011)[32]
Itraconazole (ITZ) microflakes	Sodium dodecyl sulfate (SDS) and polaxamer 407 (PLX)	DCM	313	11.6	20 g/min 1 mL/min	0.7–1 μm	Sathigari (2011)[33]
Lecithin		EtOH	308	9–13	22.8 mL/h	0.1–10 μm	Lesoin (2011)[34]
Green tea	Poly-caprolactone, MW: 25,000		283–307	8–12	0.15–007 kg/h 2–4 kg/h	3–5 μm	Sosa (2011)[35]
Cholesterol	PLA	DCM	318	9	0.6 kg/h 48 kg/h	1.7–8 μm	Guha (2011)[36]
Cefuroxime axetil	PVP	MetOH	308–323	7–20	0.85–2.5 mL/min	1.88–3.97 μm	Uzun (2011)[37]
Sulfathiazole		EtOH/ACE	308–328	10–14	1–2 mL/min	2.1 ± 0.8 im	Chen (2010)[38]
Erlotinib hydrochloride/Fulvestrant		MetOH and EA	308–328	10–18	0.25–2 mL/min	2 im	Tien (2010)[39]
Andrographolide		EtOH	308–328	5–24	0.5–1.5 g/m	3–228 nm	Imsanguan (2010)[40]
Indomethacin (IDMC)	PVP	80 and 20 vol.% of ACE DCM	308	8.5	1 mL/min 60 g/min		Lim (2010)[41]

Supercritical Fluid Technology for Encapsulation

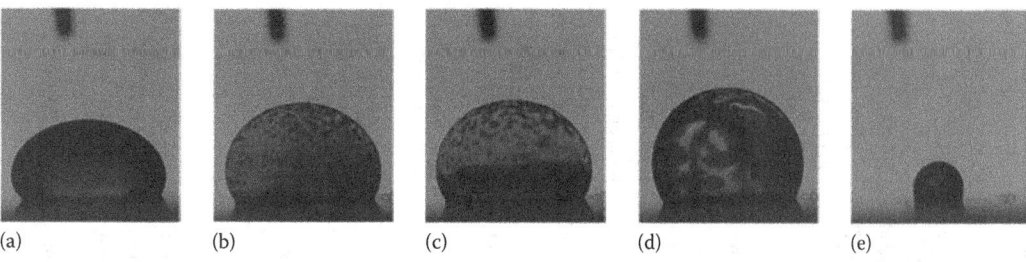

(a)   (b)   (c)   (d)   (e)

FIGURE 24.7 Behavior of an organic solvent drop during the supercritical extraction of emulsions.

GAS and semicontinuous SAS process, respectively. The differences from GAS and SAS processes and the SFEE are: (1) an emulsion containing the desired substance to be precipitated dissolved in its dispersed phase is injected instead of injecting a simple solution of the substances; (2) a liquid product is formed, allowing the continuous extraction of the product but requiring additional steps to produce a powdery product; (3) the preparation of the initial materials is more complex, involving the use of surfactants and high-energy dispersion techniques; and (4) there is an additional controlling parameter besides the usual parameters of GAS and SAS processes (pressure, temperature, flow rates, and concentrations), that is, the emulsion droplet size distribution. On the other hand, some researchers employ a different process design based on a countercurrent packed column for putting into contact emulsion and CO_2.

To optimize this technology, it is important to understand the variation of the properties of the emulsion during the saturation with the SCF, considering parameters as the changes in the droplet size of the dispersed phase due to the dissolution of SC-CO_2, or the possible reduction of stability of the emulsion caused by CO_2 and the relationship between the kinetics of deemulsification and crystallization process.

Table 24.2 presents a review of recent research studies working with SCF extraction of emulsions.

24.2.5 Particles from Gas-Saturated Solutions Process

The PGSS process takes advantage of the high solubility of carbon dioxide in several food products such as fats, phospholipids, oils, or many polymers of interest for application as carrier materials, as well as of the intense cooling produced by Joule–Thomson effect when CO_2 is depressurized from supercritical conditions to ambient pressure.[59] In the PGSS process (Figure 24.8), CO_2 is dissolved in a melted solid, and the mixture is expanded through a nozzle. The expansion causes the vaporization of the dissolved CO_2, which has an intense cooling effect on the melt. This process is specially suitable for substances such as polymers, in which CO_2 has a large solubility, and moreover, it has a melting-point-depression effect, as in this systems, the increase in the melting point caused by the release of CO_2 from the melt strengthens the cooling effect. The PGSS process has also been used to micronize suspensions of different substances in polymer melts, in order to obtain composite materials. In a modification of the PGSS process denominated PGSS drying, an aqueous solution of the active compounds is saturated with CO_2 and expanded.[60] By PGSS drying, aqueous solutions can be successfully dried and micronized at temperatures below 80°C, which makes this process an interesting alternative of conventional spray drying for thermally sensitive materials. The main advantages of PGSS processes are their wide applicability, the reduced consumption of CO_2 with respect to the previous processes, and their simplicity. For these reasons, PGSS processes are already in operation at large scales.[61] Table 24.3 presents some recent studies dealing with the PGSS process.

TABLE 24.2
Recent Applications of SFEE

Active compound	Encapsulant (Coating Material)	Solvent(o/w)	T (K)	P (MPa)	Elimination solvent (ppm)	Encapsulation	Product Characteristisd 0.5 Morphology	Research
Ibuprofen	PLGA	EA	308–338	8–20	Not considered	60%	100–300 nm	Lin (2013)[43]
Crystallization		EA	318–333	8–10	—	—	3–6 μm	Kluge (2013)[44]
Phenanthrene								
	PLGA and PCL	ACE	311	8–10	74–2700	—	170–200 nm	Campardelli (2012)[45]
		ACE/EtOH		10–14	31–112	—	160–170 nm	
Insulin	PLGA	EA	311	8	20000–1300 600	70%	1.8–4.8 μm	Falco (2012)[46]
Carotenoid fraction of pink shrimp residue	(OSA)-modified starch	DCM	313	10	—	97%	0.7 μm	Mezzomo (2012)[47]
Lactobacillus acidophilus	(OSA)-modified starch	EA	310	9	Less than 50	80%	20 μm	Della Porta (2012)[48]
β-carotene and lycopene	PLGA	DCM	323	9–13	1000	34–89%	344–366 nm	Santos (2012)[49]
Vitamin A complex	PLGA	ACE/glycerol	309	8	—	80–90%	3.2–3.5 μm	Della Porta (2011)[50]
Lysozyme	PLGA	EA	318	8	—	50%	100–1000 nm	Kluge (2009)[51,52]
Ketoprofen			—	—	—	—	100–200 nm	
Piroxicam (PX) and Diclophenac sodium (DF)	PLGA	EA	311	8	10 ppm	88–97%	1–3 μm	Della Porta (2008)[53]
Ketoprofen (KET)	PLGA	EA	311	8	100–300 ppm	—	1–3 μm	Chattopadhyay (2006, 2007)[54,55]
		EA	318	8	—	—	—	
Indomethacin (IN) Ketoprofen (KET)	Tripalmitin Tristearin Gelucire 50/13 Lecithin	Chloroform	308	8	20 ppm	>90%	50 nm	
	PLGA/Eudragit RS	EA	308–323	8–15	Below 50 ppm	98%	200–1000 nm	Furlan (2010)[56]
Magnetite Fe3O4	PLGA	DCM			—	—	230–140 nm	
Cholesterol acetate (CA), Griseofulvin (GF), and Megestrol acetate (MA)	(PVA, Pluronics, lecithin, Span 80, or Tween 80)	EA/toluene/ DCM		8–20	9–55 ppm	—	200–1000 nm	Shekunov (2005)[57]
Plasmid DNA (pDNA)	PLGA	EA	318	8	Less than 25 ppm	98%	150–320 nm	Mayo (2010)[58]

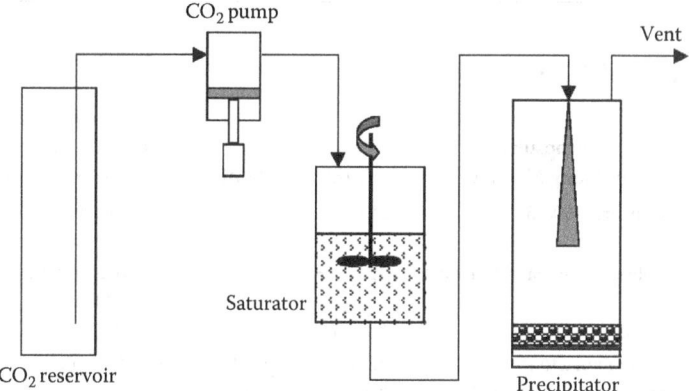

FIGURE 24.8 Schematic diagram of the PGSS process.

24.3 ENCAPSULATION PROCESSES WITH SUPERCRITICAL FLUIDS: INTERACTIONS BETWEEN SUPERCRITICAL CARBON DIOXIDE AND COMMON POLYMERIC CARRIER MATERIALS

In general, the different techniques presented in Section 23.2 can be used for the precipitation of pure compounds as well as for the encapsulation of these compounds in different carrier materials. The application of the techniques for encapsulation often requires additional considerations or adjustments in the processing conditions that are frequently related to the specific interactions between SCFs, active compounds, and carrier materials.

In particular, the complex phase behavior of mixtures of polymers, pharmaceuticals, and SCFs has a crucial influence on the performance of the precipitation processes. Frequently, polymers+SCF systems exhibit type III phase behavior,[90] which is characterized by a region of liquid–liquid immiscibility between the polymer and the SCF that is extended up to any arbitrarily high pressure without being limited by a mixture critical point.

Although the solubility of polymers in SC-CO_2 could be reasonably expected to be extremely low considering the high molecular weight of most polymers, it is remarkable that in certain cases, it is possible to observe certain polymer solubility, particularly in the case of fluorine-substituted polymers such as perfluoroalkyl ethers or acrylates. However, the solubility of SC-CO_2 in polymers can be very high (e.g., up to 30 wt% in PEG 4000),[91] a property that is exploited in PGSS processes. Regarding the variation of the solubility with pressure, normally two distinct regions are observed: at low pressures, the solubility of SC-CO_2 increases almost linearly with pressure. At a certain pressure, the polymer appears to be totally "saturated" with the CO_2 and for increasing CO_2 solubility to a small extent over this saturation point, it is necessary to apply a large increase in pressure, a feature that as previously explained is characteristic of type III phase diagrams. This behavior may be related to the total occupation of the free intermolecular volume of the polymer by CO_2 at the saturation pressure.

The dissolution of CO_2 in polymeric carriers can cause important variations in their physical properties.[92] Two properties of particular importance for precipitation and encapsulation processes are the glass transition temperature and the melting temperature. The dissolution of SC-CO_2 in the polymer can reduce the glass transition temperature of amorphous polymers by as much as 4°C–30°C/MPa,[92] an effect that has been related to the intermolecular interactions between dissolved CO_2 and polymer. Significant reductions of melting temperature can be observed as well. As these effects are caused by the dissolved CO_2, the trend of variation of these properties with pressure can be related to the trend of variation of solubility: at

TABLE 24.3
Recent Applications of PGSS

Active Compound/Core Material	Encapsulant (Coating Material)	T (K)	P (MPa)	Product Characteristics d0.5 Morphology	Research
C_3F_8 or C_4F_8	Gelucire ® 50/13	353	8.5	Particles with holes	Rodriguez-Rojo (2013)[62]
Lavandin Essential Oil	Poly-(epsilon-caprolactones)	323–343	5–11	Spherical particles 100–700 µm	Varona (2013)[63]
	PEG	343–363	5–9	30–100 µm	Varona (2010)[64]
β-carotene	Polycaprolactones	323–343	11–15	100–600 µm	De Paz (2012)[65]
Water	Hydrogenated castor oil	359–369	5.9–15.2	3–22 µm	Hanu (2012)[66]
Chitosan	PLGA mPEG	313	13.8	90–100 µm	Casettari (2011)[67]
Flufenamic acid	Triacetyl-β-cyclodextrin	308	25	—	Nunes (2010)[68]
Ribonuclease A	PEGylated, tristearin, phosphatidylcholine	323–338	13–14	4–15 µm	Vezzu (2010)[69]
	PEG 5000 Triestearin (TS) Phosphatidylcholine (PC)	318–338	10–17	4–15 µm	Sinha (2004)[70]
Cydia pomonella granulovirus	Palm oil–based fat	338	10	23 µm Spheres	Pemsel (2010)[71]
	Polybutyleneterephthalate/ zinc oxide and Bentonite as additive	503–593	5–41	100–300 µm spherical particles	Pollak (2010)[72]
Caffeine Glutathione Ketoprofen	Glyceryl monostearate (Lumulse), waxy triglyceride (Cutina® HR), Silanized TiO2	345	13	—	García González (2010)[73]
YNS3107	PEG4000, PEG400, poloxamer 407	335	17.7	30 µm	Brion (2009)[74]
Insulin	Tristearin, Tween-80, phosphatidylcholine, PEG, dioctyl sulfosuccinate	318	15	200–400 nm	Salmaso (2009)[75]
Human growth hormone (hgH)	PLGA/PLA	305	7.6	60–100 µm	Jordan (2009)[76]
Insulin Human growth hormone (rh-GH)	Tristearin/ phosphatidylcholine/PEG mixtures	313	15	200 nm	Salmasso (2009)[77]
Trans-chalcone	Precito® atos Gelucire® 50/13	315–325	12	2–7 µm	de Sousa (2009)[78]
	Monostearate Tristearate	343–353 333–343	6–21	10–40 µm	Mandzuka (2008)[79]
Caffeine	GMS	335	13	5.5 µm Nedlee aggregates	de Sousa (2007)[80]
	Rapessed 70(RP70)	333–373	7–18	15–20 µm Spheres/ Aggregates	Munuklu (2007)[81]
Cyclosporine		298–318	16–20	150 nm	Tandya (2006)[82]
Cocoa butter		293–373	20–32	Agglomeration of smaller particles.	Letourneau (2005)[83]

(Continued)

TABLE 24.3 (Continued)
Recent Applications of PGSS

Active Compound/Core Material	Encapsulant (Coating Material)	T (K)	P (MPa)	Product Characteristics d0.5 Morphology	Research
Theophylline	Hydrogenated palm oil	359	12–18	2–3 μm spheres/ Needles	Rodrigues (2004)[84]
	Hydrogenated palm oil	333–373	9–22	2–3 μm Spheres and crystals	Li (2005)[85]
Lysozyme Ribonuclease Insulin	P(DLLA)	308	32	150 μm Irregular/ porous	Whitaker (2005)[86]
	P(DLLA)	308	30.6	10–20 μm fibres	Hao (2004)[87]
Nifedipine Felodipine Fenobiate	PEG 4000	323–343 423 338–353	10–20 20 19	15–30 μm 42 μm Irregular/ porous 32 μm	Kerc (1999)[88]
	PEG (MW 1500/4000/8000/35000)	318–343	10–25	150–400 μm spheres	Weidner. (1996)[89]

low pressures, melting temperature decreases almost linearly with pressure. Upon reaching the maximum CO_2 solubility, the melting temperature reaches its minimum value, and then increases as CO_2 pressure increases due to the effect of hydrostatic pressure. As an example, Figure 24.9 presents the evolution of the melting temperature of Pluronic® F127 poloxamers with CO_2 pressure.[93]

The dissolution of CO_2 also causes certain modifications in the structure of the polymer: a decrease in the crystallinity[94] as well as a significant swelling (of up to 34% in PEG1500–CO_2 systems[95]). Although these effects are exploited in different supercritical encapsulation technologies, it also has to be taken into account that if after the precipitation, the particles stay in contact with pressurized CO_2 for a prolonged time, the decrease of glass transition and melting temperatures can promote agglomeration and cause other variations of the properties of particles.

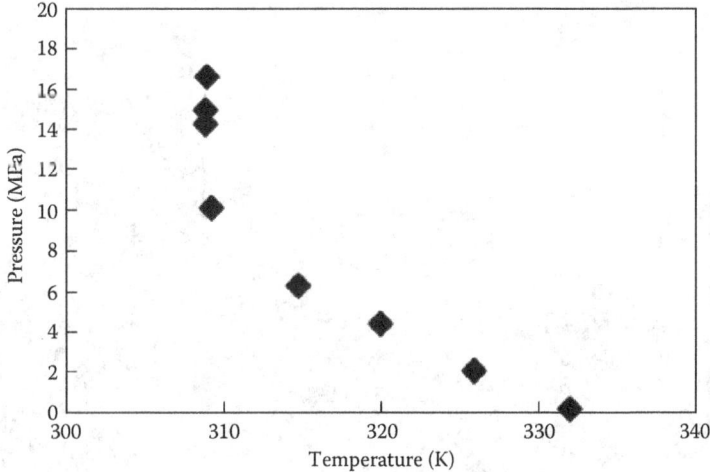

FIGURE 24.9 Melting temperature of Pluronic F127 as a function of CO_2 pressure.

24.4 CASE STUDY: ENCAPSULATION OF QUERCETIN IN PLURONIC® F127 POLOXAMER BY SAS PROCESS

Quercetin ($C_{15}H_{10}O_7$) is a flavonoid with strong antioxidant properties, and other important biological activities, including antibacterial, anti-inflammatory, and antihistaminic effects. In human diet, some benefits of moderate ingestions of quercetin for the treatment of different chronic diseases have been claimed.[96]

These applications of quercetin are limited by the low bioavailability of this compound, which is related to its low solubility in gastrointestinal fluids and its sensitivity to degradation. A way of overcoming these limitations is the encapsulation of quercetin in a carrier material suitable for protecting the compound from degradation, as well as for enhancing its aqueous solubility.

Pluronic® poloxamers can be suitable carrier materials for this purpose due to their capacity to enhance the absorption of water-insoluble compounds by formation of micelles in aqueous environment that can host such hydrophobic compounds.[97] Different authors have described the formation of quercetin-loaded Pluronic micelles by thin-film hydration methods[98] and have shown that the resulting micelles enhanced the solubilization of the active compound.[99] Ghanem et al. described the encapsulation of quercetin in Pluronic F127 via spray drying.[100]

In this case study, the encapsulation of quercetin in Pluronic F127 from acetone solutions by SAS process is described. Pressure and temperature operating conditions, 10 MPa and 40°C, were chosen in order to operate in the single phase region of the solvent (acetone)–CO_2 system.[101] Solution flow rate and antisolvent flow rate were 2 mL/min and 2 kg/h. Different solution concentrations and carrier/quercetin ratios were tested in order to optimize the particle production process.

Figure 24.10 presents a schematic diagram of the SAS pilot plant employed in encapsulation experiments. CO_2 is taken from a gas bottle and cooled down using ethylene glycol as refrigerant in order to maintain it in the liquid state, before compressing it with a DOSAPRO diaphragm pump, and preheating it to the desired operating temperature. On the other hand, the liquid solution of quercetin and Pluronic in acetone is pumped with a Jasco chromatographic pump. Both solutions are continuously injected into the precipitation vessel through a concentric tube nozzle, consisting of an inner 1/16 "tube used to inject the solution, and an outer 1/4" tube used to introduce CO_2.

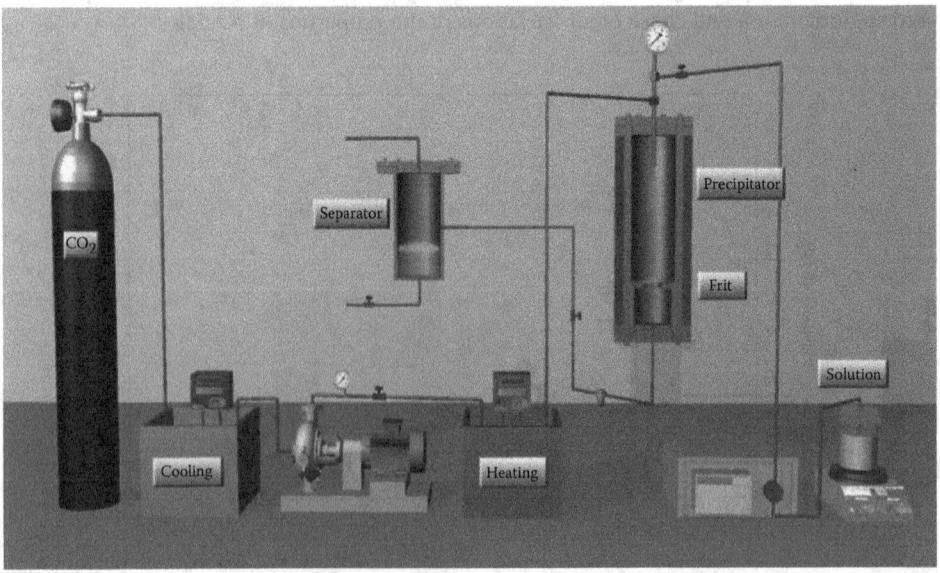

FIGURE 24.10 Schematic representation of SAS plant.

Supercritical Fluid Technology for Encapsulation

The precipitator is a jacketed vessel of 2.5 L of inner volume. The formed particles are collected in a frit at the bottom of the precipitator vessel, which consists of a metallic porous disk used to support a polymeric filter with a pore size of 0.1 μm. Afterward, the effluent from the precipitator is depressurized using a GO back-pressure valve, the organic solvent condensed after the depressurization is collected in a flash vessel, and gaseous CO_2 is vented.

An SAS experiment consists of three steps. Firstly, the SCF is introduced in the precipitation vessel until the desired pressure and temperature conditions are reached and are maintained constant. Then the solution is injected with the desired flow rate until an amount of at least 100 mL of solution has been processed. Finally, the SCF flow is maintained constant during at least 30 min to eliminate the remaining organic solvent from the particles. After this, the precipitator is depressurized and particles are collected from the filter.

Figure 24.11 shows SEM micrographs of particles obtained in SAS experiments with pure quercetin. Unprocessed quercetin presented prismatic crystalline morphology, with particle sizes in the range of 10–20 μm. As shown in this figure, the prismatic morphology of particles was preserved by SAS processing, but particle size was drastically reduced. Moreover, a significant reduction of particle size was observed as the concentration of quercetin in the initial acetone solution was reduced. With the lowest concentration of 0.005 g/mL, submicrometric or nanometric quercetin particles, highly agglomerated in flocks of about 1 μm, were obtained.

FIGURE 24.11 SEM micrographs of particles obtained by SAS coprecipitation of quercetin. (a) Raw quercetin without processing, (b) SAS-processed quercetin, initial concentration in acetone solution: 0.02 g/mL, (c) initial concentration: 0.01 g/mL, and (d) initial concentration: 0.005 g/mL. All micrographs are presented with the same magnification ratio (5000×), and a size bar of 5 μm.

A frequent challenge in encapsulation experiments is to achieve an effective encapsulation of the active compound in the polymer instead of a segregated crystallization of polymer and active compound. In this study, a successful encapsulation was achieved exploiting the phase behavior of Pluronic–CO_2 mixtures presented in Figure 24.9. As shown in this figure, CO_2 pressure drastically reduces the melting temperature of the polymer. At operating pressures in the range of 10 MPa, the melting temperature is slightly below the temperature of 40°C employed in SAS experiments. Thus, during the atomization, a polymer melt is obtained rather than solid polymer particles.

The results of working in these conditions can be observed in Figure 24.12. As presented in this figure, SAS micronization experiments with pure Pluronic are, of course, unsuccessful, since in the conditions employed in experiments, the polymer is melted and forms a film rather than microparticles. However, in quercetin-Pluronic coprecipitation experiments, this polymer film is formed over quercetin crystals, thus effectively encapsulating them and restraining their growth.

SAS processing also produces changes in the crystalline structure of the particles. As presented in x-ray diffractograms in Figure 24.13, the polymer is amorphized by SAS processing, while the crystallinity of quercetin particles appears to be reduced.

FIGURE 24.12 SEM micrographs of particles obtained by SAS coprecipitation of Pluronic F127+quercetin. (a) Pure F127, (b) quercetin + Pluronic F127 in mass proportion 2:1, (c) quercetin + Pluronic F127 in mass proportion 1:1, (d) quercetin + Pluronic F127 in mass proportion 1:1.7. Micrographs (b), (c), and (d) are presented with a magnification ratio of 5000× and a size bar of 5 μm, while micrograph (a) is presented with a magnification ratio of 100× and a size bar of 200 μm.

Supercritical Fluid Technology for Encapsulation

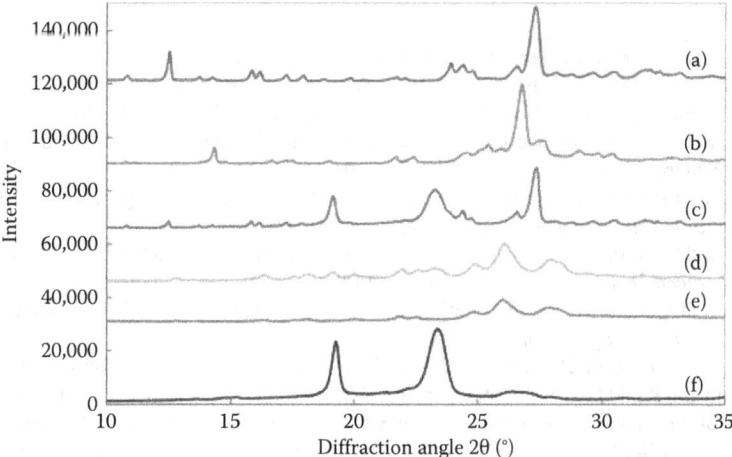

FIGURE 24.13 (See color insert.) X-ray diffractograms of (a) unprocessed quercetin, (b) SAS-processed quercetin, (c) physical mixture of quercetin-Pluronic® F127 (mass ratio 1 g:1 g), (d) SAS-processed mixture of quercetin-Pluronic® F127 (mass ratio 2 g:1 g), (e) SAS-processed mixture of quercetin-Pluronic® F127 (mass ratio 1 g:1 g), and (f) unprocessed Pluronic® F127. For clarity, all diffractograms were vertically displaced by arbitrary amounts.

By SAS encapsulation of quercetin in Pluronic F127, their solubility and dissolution rate in gastrointestinal fluids are enhanced. Results of quercetin release experiments in simulated intestinal fluid are presented in Figure 24.14. As presented in this figure, a significant difference was observed between the results with unprocessed and SAS processed pure quercetin, both in the dissolution rate and in the final solubility. Similarly, SAS-processed mixtures of quercetin and Pluronic showed a faster solubilization and a higher final solubility than a simple physical mixture of these two compounds. This result indicates that the properties of the product obtained by SAS processing previously described, including the reduction of particle size and the production of a homogeneous dispersion of the active compound in an amorphous polymer matrix, have a positive impact on the dissolution behavior of the formulation.

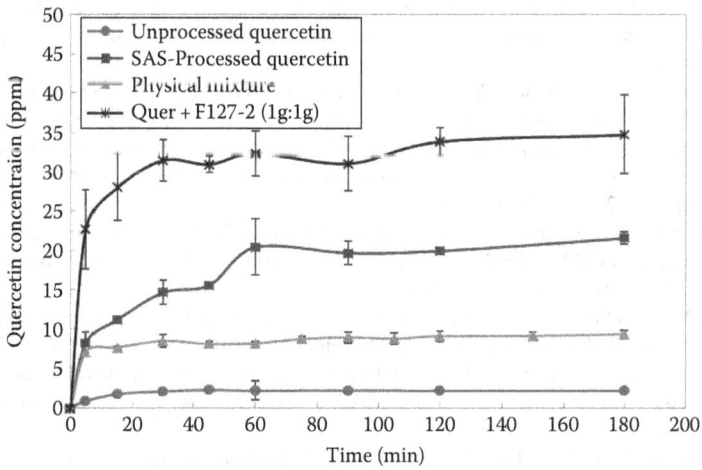

FIGURE 24.14 Release tests of quercetin encapsulated in Pluronic F127 in simulated intestinal fluid.

REFERENCES

1. Brunner, G. *Gas Extraction*. Springer, Berlin, Germany, 1994.
2. Bertucco, A. and Vetter, G. (Eds.), *High Pressure Process Technology: Fundamentals and Applications*. Industrial Chemistry Library, Vol. 9. Elsevier, Amsterdam, the Netherlands, 2001.
3. Jung, J. and Perrut, M. Particle design using supercritical fluids: Literature and patent survey, *The Journal of Supercritical Fluids* 20 (2001): 179–219.
4. Shariati, A. and Peters, C.J. Recent developments in particle design using supercritical fluids. *Current Opinion in Solid State and Materials Science* 7 (2003): 371–383.
5. Martín, A. and Cocero, M.J. Micronization processes with supercritical fluids: fundamentals and mechanisms. *Advanced Drug Delivery Reviews* 60 (2008): 339–350.
6. Martín, Á. and Cocero, M.J. Precipitation processes with supercritical fluids: patents review. *Recent Patents in Chemical Engineering* 2 (2008): 9–20.
7. Türk, M., Hils, P., Helfgen, B., Schaber, K., Martin, H.J., and Wahl, M.A. Micronization of pharmaceutical substances by the Rapid Expansion of Supercritical Solutions (RESS): A promising method to improve bioavailability of poorly soluble pharmaceutical agents. *The Journal of Supercritical Fluids* 22 (2002): 75–84.
8. De la Fuente Badilla, J.C., Peters, C.J., and de Swaan Arons, J. Volume expansion in relation to the gas-antisolvent process. *The Journal of Supercritical Fluids* 17 (2000): 13–23.
9. Reverchon, E., Caputo, G., and De Marco, I. Role of phase behavior and atomization in the supercritical antisolvent precipitation. *Industrial and Engineering Chemistry Research*. 42 (2003): 6406–6414.
10. Kakran, M., Sahoo, N.G., Antipina, M.N., and Li, L. Modified supercritical antisolvent method with enhanced mass transfer to fabricate drug nanoparticles. *Materials Science and Engineering: C* 33 (2013): 2864–2870.
11. Santos, D.T. and Meireles, M.A.A. Micronization and encapsulation of functional pigments using supercritical carbon dioxide. *Journal of Food Process Engineering* 36 (2013): 36–49.
12. Sacchetin, P.S.C., Morales, A.R., Moraes, Â.M., and Rosa, Paulo de Tarso Vieira. Formation of PLA particles incorporating 17α-methyltestosterone by supercritical fluid technology. *The Journal of Supercritical Fluids* 77 (2013): 52–62.
13. Boonnoun, P., Nerome, H., Machmudah, S., Goto, M., and Shotipruk, A. Supercritical anti-solvent micronization of marigold-derived lutein dissolved in dichloromethane and ethanol. *The Journal of Supercritical Fluids* 77 (2013) 103–109.
14. Zhang, J., Huang, Y., Liu, D., Gao, Y., and Qian, S. Preparation of apigenin nanocrystals using supercritical antisolvent process for dissolution and bioavailability enhancement. *European Journal of Pharmaceutical Sciences* 48 (2013): 740–747.
15. Montes, A., Nunes, A., Gordillo, M.D., Pereyra, C., Duarte, C.M.M., and Martínez de la Ossa, E.J. Amoxicillin and ethyl cellulose precipitation by two supercritical antisolvent processes. *Chemical Engineering and Technology* 36 (2013): 665–672.
16. Montes, A., Bendel, A., Kürti, R., Gordillo, M.D., Pereyra, C., and Martínez de la Ossa, E.J. Processing naproxen with supercritical CO_2. *The Journal of Supercritical Fluids* 75 (2013): 21–29.
17. Montes, A., Gordillo, M.D., Pereyra, C., de la Rosa-Fox, N., and Martínez de la Ossa, E.J. Silica microparticles precipitation by two processes using supercritical fluids. *The Journal of Supercritical Fluids* 75 (2013): 88–93.
18. Bakhbakhi, Y., Asif, M., Chafidz, A., and Ajbar, A. Formation of biodegradable polymeric fine particles by supercritical antisolvent precipitation process. *Polymer Engineering and Science* 53 (2013): 564–570.
19. Ober, C.A., Kalombo, L., Swai, H., and Gupta, R.B. Preparation of rifampicin/lactose microparticle composites by a supercritical antisolvent-drug excipient mixing technique for inhalation delivery. *Powder Technology* 236 (2013): 132–138.
20. Wang, W., Liu, G., Wu, J., and Jiang, Y. Co-precipitation of 10-hydroxycamptothecin and poly (l-lactic acid) by supercritical CO2 anti-solvent process using dichloromethane/ethanol co-solvent. *The Journal of Supercritical Fluids* 74 (2013): 137–144.
21. Bakhbakhi, Y., Alfadul, S., and Ajbar, A. Precipitation of ibuprofen sodium using compressed carbon dioxide as antisolvent. *European Journal of Pharmaceutical Sciences* 48 (2013): 30–39.
22. Lu, Q., Zhu, M., Zu, Y., Liu, W., Yang, L., and Zhang, Y. Comparative antioxidant activity of nanoscale lignin prepared by a supercritical antisolvent (SAS) process with non-nanoscale lignin. *Food Chemistry* 135 (2012): 63–67.
23. Zhang, J., Huang, Y., Liu, D., Gao, Y., and Qian, S. Preparation of apigenin nanocrystals using supercritical antisolvent process for dissolution and bioavailability enhancement. *European Journal of Pharmaceutical Sciences* 48 (2013). 740–747.

24. Zu, Y., Zhang, Q., Zhao, X., Wang, D., Li, W., Sui, X. Preparation and characterization of vitexin powder micronized by a supercritical antisolvent (SAS) process. *Powder Technology* 228 (2012): 47–55.
25. De Zordi, N., Moneghini, M., Kikic, I., Grassi, M., Del Rio Castillo, A.E., and Solinas, D. Applications of supercritical fluids to enhance the dissolution behaviors of furosemide by generation of microparticles and solid dispersions. *European Journal of Pharmaceutics and Biopharmaceutics* 81 (2012): 131–141.
26. Li, W., Liu, G., Li, L., Wu, J., Lu, Y., Jiang, Y. Effect of process parameters on co-precipitation of paclitaxel and poly(L-lactic acid) by supercritical antisolvent process. *Chinese Journal of Chemical Engineering* 20 (2012): 803–813.
27. Zhao, X., Chen, X., Zu, Y., Jiang, R., and Zhao, D. Recrystallization and micronization of taxol using the supercritical antisolvent (SAS) process. *Industrial and Engineering Chemistry Research* 51 (2012): 9591–9597.
28. Yang, L., Sun, Z., Zu, Y., Zhao, C., Sun, X., and Zhang, Z. Physicochemical properties and oral bioavailability of ursolic acid nanoparticles using supercritical anti-solvent (SAS) process. *Food Chemistry* 132 (2012): 319–325.
29. Weber Brun, G., Martin, A., Cassel, E., Figueiro Vargas, R.M., and Cocero, M.J. Crystallization of caffeine by supercritical antisolvent (SAS) process: Analysis of process parameters and control of polymorphism." *Crystal Growth and Design* 12 (2012): 1943–1951.
30. Visentin, A., Rodríguez-Rojo, S., Navarrete, A., Maestri, D., and Cocero, M.J. Precipitation and encapsulation of rosemary antioxidants by supercritical antisolvent process. *Journal of Food Engineering* 109 (2012): 9–15.
31. Sui, X., Wei, W., Yang, L., Zu, Y., Zhao, C., and Zhang, L. Preparation, characterization and in vivo assessment of the bioavailability of glycyrrhizic acid microparticles by supercritical anti-solvent process. *International Journal of Pharmaceutics* 423 (2012): 471–479.
32. Yang, L., Huang, J., Zu, Y., Ma, C., Wang, H., and Sun, X. Preparation and radical scavenging activities of polymeric procyanidins nanoparticles by a supercritical antisolvent (SAS) process. *Food Chemistry* 128 (2011): 1152–1159.
33. Sathigari, S.K., Ober, C.A., Sanganwar, G.P., Gupta, R.B., and Babu, R.J. Single-step preparation and deagglomeration of itraconazole microflakes by supercritical antisolvent method for dissolution enhancement. *Journal of Pharmaceutical Sciences* 100 (2011): 2952–2965.
34. Lesoin, L., Crampon, C., Boutin, O., and Badens, E. Preparation of liposomes using the supercritical anti-solvent (SAS) process and comparison with a conventional method. *The Journal of Supercritical Fluids* 57 (2011): 162–174.
35. Sosa, M.V., Rodríguez-Rojo, S., Mattea, F., Cismondi, M., and Cocero, M.J. Green tea encapsulation by means of high pressure antisolvent coprecipitation. *The Journal of Supercritical Fluids* 56 (2011): 304–311.
36. Guha, R., Vinjamur, M., and Mukhopadhyay, M. Demonstration of mechanisms for coprecipitation and encapsulation by supercritical antisolvent process. *Industrial and Engineering Chemistry Research*, 50 (2011): 1079–1088.
37. Uzun, İ.N., Sipahigil, O., and Dinçer, S. Coprecipitation of cefuroxime Axetil–PVP composite microparticles by batch supercritical antisolvent process. *The Journal of Supercritical Fluids* 55 (2011): 1059–1069.
38. Chen, Y., Tang, M., and Chen, Y. Recrystallization and micronization of sulfathiazole by applying the supercritical antisolvent technology. *Chemical Engineering Journal* 165 (2010): 358–364.
39. Tien, Y., Su, C., Lien, L., Chen, Y. Recrystallization of erlotinib hydrochloride and fulvestrant using supercritical antisolvent process. *The Journal of Supercritical Fluids* 55 (2010): 292–299.
40. Imsanguan, P., Pongamphai, S., Douglas, S., Teppaitoon, W., and Douglas, P.L. Supercritical antisolvent precipitation of andrographolide from andrographis paniculata extracts: Effect of pressure, temperature and CO2 flow rate. *Powder Technology* 200 (2010): 246–253.
41. Lim, R.T.Y., Ng, W.K., Tan, R.B.H. Amorphization of pharmaceutical compound by co-precipitation using supercritical anti-solvent (SAS) process (part I). *The Journal of Supercritical Fluids*, 53 (2010): 179–184.
42. F. Mattea, Á. Martín, C. Schulz, P. Jaeger, R. Eggers, and M.J. Cocero. Behavior of an organic solvent drop during the supercritical extraction of emulsions. *AIChE J.* 56 (2010): 1184–1195.
43. Lin, C.S., Xu, J.J., Ng, K.M., Wibowo, C., and Luo, K.Q. Encapsulation of a low aqueous solubility substance in a biodegradable polymer using supercritical fluid extraction of emulsion. *Industrial and Engineering Chemistry Research* 52 (2013): 134–141.
44. Kluge, J., Mazzotti, M., and Muhrer, G. Solubility of ketoprofen in colloidal PLGA. *International Journal of Pharmaceutics* 399 (2010): 163–172.

45. Campardelli, R., Della Porta, G., Reverchon, E. Solvent elimination from polymer nanoparticle suspensions by continuous supercritical extraction. *Journal of Supercritical Fluids* 70 (2012): 100–105.
46. Falco, N., Reverchon, E., and Della Porta, G. Continuous supercritical emulsions extraction: Packed tower characterization and application to poly(lactic-co-glycolic acid) plus insulin microspheres production. *Industrial and Engineering Chemistry Research* 51 (2012) 8616–8623.
47. Mezzomo, N., Paz, E.d., Maraschin, M., Martín, Á., Cocero, M.J., and Ferreira, S.R.S. Supercritical anti-solvent precipitation of carotenoid fraction from pink shrimp residue: Effect of operational conditions on encapsulation efficiency. *The Journal of Supercritical Fluids*, 66 (2012): 342–349.
48. Della Porta, G., Castaldo, F., Scognamiglio, M., Paciello, L., Parascandola, P., Reverchon, E. Bacteria microencapsulation in PLGA microdevices by supercritical emulsion extraction. *The Journal of Supercritical Fluids* 63 (2012): 1–7.
49. Santos, D.T., Martín, Á., Meireles, M.A.A., and Cocero, M.J. Production of stabilized sub-micrometric particles of carotenoids using supercritical fluid extraction of emulsions. *The Journal of Supercritical Fluids* 61 (2012): 167–174.
50. Della Porta, G., Campardelli, R., Falco, N., and Reverchon, E. PLGA microdevices for retinoids sustained release produced by supercritical emulsion extraction: Continuous versus batch operation layouts. *Journal of Pharmaceutical Sciences* 100 (2011): 4357–4367.
51. Kluge, J., Fusaro, F., Casas, N., Mazzotti, M., and Muhrer, G. Production of PLGA micro- and nanocomposites by supercritical fluid extraction of emulsions: I. Encapsulation of lysozyme. *The Journal of Supercritical Fluids* 50 (2009): 327–335.
52. Kluge, J., Fusaro, F., Mazzotti, M., and Muhrer, G. Production of PLGA micro- and nanocomposites by supercritical fluid extraction of emulsions: II. encapsulation of ketoprofen. *The Journal of Supercritical Fluids* 50 (2009): 336–343.
53. Della Porta, G. and Reverchon, E. Nanostructured microspheres produced by supercritical fluid extraction of emulsions. *Biotechnology and Bioengineering* 100 (2008): 1020–1033.
54. Chattopadhyay, P., Huff, R., and Shekunov, B. Drug encapsulation using supercritical fluid extraction of emulsions. *Journal of Pharmaceutical Sciences* 95 (2006): 667–679.
55. Chattopadhyay, P., Shekunov, B.Y., Yim, D., Cipolla, D., Boyd, B., and Farr, S. Production of solid lipid nanoparticle suspensions using supercritical fluid extraction of emulsions (SFEE) for pulmonary delivery using the AERx system. *Advanced Drug Delivery Reviews* 59 (2007): 444–453.
56. Furlan, M., Kluge, J., Mazzotti, M., and Lattuada, M. Preparation of biocompatible magnetite–PLGA composite nanoparticles using supercritical fluid extraction of emulsions. *The Journal of Supercritical Fluids* 54 (2010): 348–356.
57. Shekunov, B., Chattopadhyay, P., Seitzinger, J., and Huff, R. Nanoparticles of poorly water-soluble drugs prepared by supercritical fluid extraction of emulsions. *Pharmaceutical Research* 23 (2006): 196–204.
58. Mayo, A.S., Ambati, B.K., and Kompella, U.B. Gene delivery nanoparticles fabricated by supercritical fluid extraction of emulsions. *International Journal of Pharmaceutics* 387 (2010): 278–285.
59. Yeo, S.D. and EKiran, E. Formation of polymer particles with supercritical fluids: A review. *The Journal of Supercritical Fluids* 34 (2005): 287–3008.
60. Martín, Á. and Weidner, E. PGSS-drying: Mechanisms and modelling. *The Journal of Supercritical Fluids* 55 (2010): 271–281.
61. Weidner, E."High-pressure micronization for food applications. *The Journal of Supercritical Fluid* 47 (2009): 556–565.
62. Rodríguez-Rojo, S., Lopes, D.D., Alexandre, A.M.R.C., Pereira, H., Nogueira, I.D., and Duarte, C.M.M. Encapsulation of perfluorocarbon gases into lipid-based carrier by PGSS. *The Journal of Supercritical Fluids* 81 (2013): 226–235.
63. Varona, S., Martin, A., Jose Cocero, M., and Duarte, C.M.M. Encapsulation of lavandin essential oil in poly-(epsilon-caprolactones) by PGSS process. *Chemical Engineering and Technology* 36 (2013): 1187–1192.
64. Varona, S., Kareth, S., Martin, A., and Cocero, M.J. Formulation of lavandin essential oil with biopolymers by PGSS for application as biocide in ecological agriculture. *The Journal of Supercritical Fluids* 54 (2010): 369–377.
65. de Paz, E., Martin, A., Duarte, C.M.M., and Cocero, M.J. Formulation of beta-carotene with poly-(epsilon-caprolactones) by PGSS process. *Powder Technology* 217 (2012): 77–83.
66. Hanu, L.G., Alessi, P., Kilzer, A., and Kareth, S. Manufacturing and characterization of water filled micro-composites. *The Journal of Supercritical Fluids* 66 (2012): 274–281.
67. Casettari, L., Castagnino, E., Stolnik, S., Lewis, A., Howdle, S.M., and Illum, L. Surface characterisation of bioadhesive PLGA/Chitosan microparticles produced by supercritical fluid technology. *Pharmaceutical Research* 28 (2011): 1668–1682.

68. Nunes, A.V., Rodriguez-Rojo, S., Almeida, A.P., Matias, A.A., Rego, D., and Simplicio, A.L. Supercritical fluids strategies to produce hybrid structures for drug delivery. *Journal of Controlled Release*, 148 (2010): 11–12.
69. Vezzu, K., Borin, D., Bertucco, A., Bersani, S., Salmaso, S., Caliceti, P. Production of lipid microparticles containing bioactive molecules functionalized with PEG. *Journal of Supercritical Fluids* 54 (2010): 328–334.
70. Sinha, V., Bansal, K., Kaushik, R., Kumria, R., and Trehan, A. Poly-epsilon-caprolactone microspheres and nanospheres: An overview. *International Journal of Pharmaceutics* 278 (2004): 1–23.
71. Pemsel, M., Schwab, S., Scheurer, A., Freitag, D., Schatz, R., and Schluecker, E. Advanced PGSS process for the encapsulation of the biopesticide cydia pomonella granulovirus. *The Journal of Supercritical Fluids* 53 (2010): 174–178.
72. Pollak, S., Petermann, M., Kareth, S., and Kilzer, A. Manufacturing of pulverised nanocomposites-dosing and dispersion of additives by the use of supercritical carbon dioxide. *The Journal of Supercritical Fluids* 53 (2010): 137–141.
73. Garcia-Gonzalez, C.A., Argemi, A., Sampaio de Sousa, A.R., Duarte, C.M.M., Saurina, J., and Domingo, C. Encapsulation efficiency of solid lipid hybrid particles prepared using the PGSS (R) technique and loaded with different polarity active agents. *The Journal of Supercritical Fluids* 54 (2010): 342–347.
74. Brion, M., Jaspart, S., Perrone, L., Piel, G., and Evrard, B. The supercritical micronization of solid dispersions by particles from gas saturated solutions using experimental design. *The Journal of Supercritical Fluids* 51 (2009): 50–56.
75. Salmaso, S., Elvassore, N., Bertucco, A., and Caliceti, P. Production of solid lipid submicron particles for protein delivery using a novel supercritical gas-assisted melting atomization process. *Journal of Pharmaceutical Sciences* 98 (2009): 640–650.
76. Jordan, F., Naylor, A., Kelly, C.A., Howdle, S.M., Lewis, A., and Illum, L. Sustained release hGH microsphere formulation produced by a novel supercritical fluid technology: in vivo studies. *Journal of Controlled Release* 141 (2010): 153–160.
77. Salmaso, S., Bersani, S., Elvassore, N., Bertucco, A., and Caliceti, P. Biopharmaceutical characterisation of insulin and recombinant human growth hormone loaded lipid submicron particles produced by supercritical gas micro-atomisation. *International Journal of Pharmaceutics* 379 (2009): 51–58.
78. de Sousa, A.R.S., Silva, R., Tay, F.H., Simplicio, A.L., Kazarian, S.G., and Duarte, C.M.M. Solubility enhancement of trans-chalcone using lipid carriers and supercritical CO2 processing. *The Journal of Supercritical Fluids* 48 (2009): 120–125.
79. Mandzuka, Z. and Knez, Z. Influence of temperature and pressure during PGSS (TM) micronization and storage time on degree of crystallinity and crystal forms of monostearate and tristearate. *The Journal of Supercritical Fluids* 45 (2008): 102–111.
80. de Sousa, A.R.S., Simplicio, A.L., de Sousa, H.C., and Duarte, C.M.M. Preparation of glyceryl mono stearate-based particles by PGSS((R))—Application to caffeine. *The Journal of Supercritical Fluids* 43 (2007): 120–125.
81. Munuklu, P. and Jansens, P.J. Particle formation of an edible fat (rapeseed 70) using the supercritical melt micronization (ScMM) process. *The Journal of Supercritical Fluids*, 40 (2007): 433–442.
82. Tandya, A., Dehghani, F., and Foster, N. Micronization of cyclosporine using dense gas techniques. *The Journal of Supercritical Fluids* 37 (2006): 272–278.
83. Letourneau, J.J., Vigneau, S., Gonus, P., and Fages, J. Micronized cocoa butter particles produced by a supercritical process. *Chemical Engineering and Processing* 44 (2005): 201–207.
84. Rodrigues, M., Peirico, N., Matos, H., de Azevedo, E., Lobato, M., and Almeida, A. Microcomposites theophylline/hydrogenated palm oil from a PGSS process for controlled drug delivery systems. *The Journal of Supercritical Fluids* 29 (2004): 175–184.
85. Li, J., Rodrigues, M., Paiva, A., Matos, H., and de Azevedo, E. Modeling of the PGSS process by crystallization and atomization. *AIChE Journal* 51 (2005): 2343–2357.
86. Whitaker, M., Hao, J., Davies, O., Serhatkulu, G., Stolnik-Trenkic, S., and Howdle, S. The production of protein-loaded microparticles by supercritical fluid enhanced mixing and spraying. *Journal of Controlled Release* 101 (2005): 85–92.
87. Hao, J., Whitaker, M.J., Wong, B., Serhatkulu, G., Shakesheff, K.M., and Howdle, S.M. Plasticization and spraying of poly (DL-lactic acid) using supercritical carbon dioxide: Control of particle size. *Journal of Pharmaceutical Sciences* 93 (2004): 1083–1090.
88. Kerc, J., Srcic, S., Knez, Z., and Sencar-Bozic, P. Micronization of drugs using supercritical carbon dioxide. *International Journal of Pharmaceutics*, 182 (1999): 33–39.
89. Weidner, E., Steiner, R., and Knez, Z. Powder generation from polyethyleneglycols with compressible fluids. *High Pressure Chemical Engineering* 12 (1996): 223–228.

90. Kirby, C.F. and McHugh, M.A. Phase behaviour of polymers in supercritical fluid solvents. *Chemical Reviews* 99 (1999): 565–602.
91. Weidner, E., Wiesmet, V., Knez, Z., and Skerget, M. Phase equilibrium (solid-liquid-gas) in polyethyleneglycol-carbon dioxide systems. *The Journal of Supercritical Fluids* 10 (1997): 139–147.
92. Tomasko, D.L., Li, H., Lui, D., Han, X., Wingert, M.J., Lee, L.J., and Koelling, K.W. A review of CO_2 applications in the processing of polymers. *Industrial and Engineering Chemistry Research* 42 (2003): 6431–6456.
93. Fraile, M., Martín, Á., Deodato, D., Rodríguez-Rojo, S., Nogueira, I.D., Simplício, A.L., Cocero, M.J., and Duarte, C.M.M. Production of new hybrid systems for drug delivery by PGSS process. *The Journal of Supercritical Fluids* 81 (2013): 226–235.
94. Lei, Z., Ohyabu, H., Sato, Y., Inomata, H., and Smith Jr, R.L., Solubility, swelling degree and crystallinity of carbon dioxide-polypropilene system. *The Journal of Supercritical Fluids* 40 (2007): 452–461.
95. Pasquali, I., Comi, L., Pucciarelli, F., and Bettini, R. Swelling, melting point reduction and solubility of PEG 1500 in supercritical CO_2. *International Journal of Pharmaceutics* 356 (2008): 76–81.
96. Formica, J.V. Review of the biology of quercetin and related bioflavonoids. *Food and Chemical Toxicology* 33 (1995): 1061–1080.
97. Batrakova, E.V. and Kabanov, A.V. Pluronic block copolymers: Evolution of drug delivery concept from inert nanocarriers to biological response modifiers. *Journal of Controlled Release* 130 (2008): 98–106.
98. Zhao, L., Shi, Y., Zou, S., Sun, M., Li, N., and Zhai, G. Formulation and in vitro evaluation of quercetin loaded polymeric micelles composed of pluronic P123 and D-a-tocopheryl polyethylene glycol succinates. *Journal of Biomedical Nanotechnology* 7 (2011): 358–365.
99. Parmar, A., Singh, K., Bahadur, A., Marangoni, G., and Bahadur, P. Interaction and solubilization of some phenolic antioxidants in Pluronic micelles. *Colloid and Surfaces B* 86 (2011): 319–326.
100. Ghanem, A.S., Mohamed Ali, H.S., Mostafa El-Shanawany, S., and Ali Ibrahim, E.S. Solubility and dissolution enhancement of quercetin via preparation of spray dried microstructured solid dispersion. *Thai Journal of Pharmaceutical Science* 37 (2013): 12–24.
101. Day, C.Y., Chang, C.J., and Chen, C.Y. Phase equilibrium of ethanol + CO2 and acetone + CO2 at elevated pressures. *Journal of Chemical and Engineering Data* 41 (1996): 839–843.

25 Melt-Dispersion Technique for Encapsulation

*Verica Djordjević, Steva Lević, Thomas Koupantsis,
Fani Mantzouridou, Adamantini Paraskevopoulou,
Viktor Nedović, and Branko Bugarski*

CONTENTS

25.1 Introduction 469
25.2 Basic Principle of Melt Dispersion 470
25.3 Melt Spraying 470
25.4 Melt Emulsification/Melt Homogenization 472
25.5 Materials for Encapsulation by Applying Melt-Dispersion Technique 474
     25.5.1 Fatty Acids and Fatty Alcohols 480
     25.5.2 Glycerides and Polyglycolyzed Glycerides 481
     25.5.3 Natural and Synthetic Waxes 481
     25.5.4 Other Carriers 482
25.6 Applications 482
25.7 Properties of Encapsulates 483
25.8 Limitations of Melt-Dispersion Technology and Trials to Overcome Them 485
25.9 Summary and Future Perspectives 486
25.10 Experiment—An Example: Production of Carnauba Wax Microparticles
Encapsulating Ethyl Vanilline 487
     25.10.1 Equipment and Reagents 487
     25.10.2 Method 487
Acknowledgments 487
References 487

25.1 INTRODUCTION

Encapsulation involves the coating or entrapment of a pure material or a mixture into another material. The coated or entrapped material, usually a liquid, is known as the "core" or "active" material, while the coating material is known as the "wall" material.[1] At the end of any applicable technique for encapsulation, the final products called particles (micro- or nanoparticle depending on the size) can be dried or not.[2] Considering the aforementioned facts, a number of technologies have been used in the preparation of encapsulates, such as spray-drying, fluidized-bed coating, spray-cooling, extrusion technologies, emulsification, inclusion encapsulation, coacervation, nanoencapsulation, and liposome entrapment. There are a number of excellent recent reviews summarizing all encapsulation processes.[2–8] Although the principle of dispersing of a molten matrix has been frequently employed for production of encapsulates, there are not many, if any, papers overviewing the processes and equipments utilizing this principle. The aim of this chapter is to describe technologies utilizing melt dispersion, melt spraying, melt emulsification, and melt homogenization. It also surveys applications of melt dispersion, describes its advantages and limitations, and emphasizes trends and innovations.

25.2 BASIC PRINCIPLE OF MELT DISPERSION

The basic principle of melt-dispersion technique involves the atomizing of a molten matrix with a low melting point (32°C–85°C) containing the bioactive compounds in finely dispersed microdroplets, which then solidify to give solid microparticles. This is the same principle as in spray-cooling or spray-chilling technology, alternatively called spray-congealing technology. This technology is the most convenient method of transforming melted feedstocks into free-flowing particulates of controlled particle size, which has been used in chemical, pharmaceutical, and food industries. It also enables encapsulation of various bioactive compounds (both hydrophilic and lipophilic), even heat-sensitive compounds, in matrices with low melting points (usually fats). Although melt-dispersion, spray-cooling, spray-chilling, and spray-congealing techniques, all refer to the same concept, and in literature, they are mentioned in an interchangeable fashion, there are some delicate disparities between their exact meanings. Both spray cooling and spray chilling are assigned to encapsulation technologies, but they differ in melting point range of the used matrix material (34°C–42°C for spray chilling and higher for spray-cooling[5]). Spray congealing is not strictly related to encapsulation. It is generally referred to any system of two compounds forming a mixture with a single eutectic temperature at which liquid and solid phases can coexist. The mixture congeals in a solid dispersion upon rapid cooling. In this chapter, spray cooling, spray chilling, and spray congealing are going to be discussed under the umbrella of melt-dispersion technique, since they all rely on the same method—dispersing of a melt.

25.3 MELT SPRAYING

A typical system for spray cooling is presented in Figure 25.1. The matrix and active material are mixed and melted together in a separate feed tank and then pumped toward the cooling chamber. Herein the melt is atomized into droplets, which solidify in particles forming a powder. The apparatus employs centrifugal, kinetic, pressure, or ultrasonic energy to atomize a melt into fine microdroplets. The equipment for atomizing molten matrix can exist in two different configurations: pneumatic spraying system (as that one presented in Figure 25.1) and rotary (centrifugal) atomizing system. Pneumatic spraying device includes a system of air nozzles, airless nozzles, or two-fluid nozzles. In a rotary (centrifugal) atomizer, a melt is dispersed onto a disk rotating at high speeds. This type of atomization is often employed as it can process even viscous molten mixtures and the

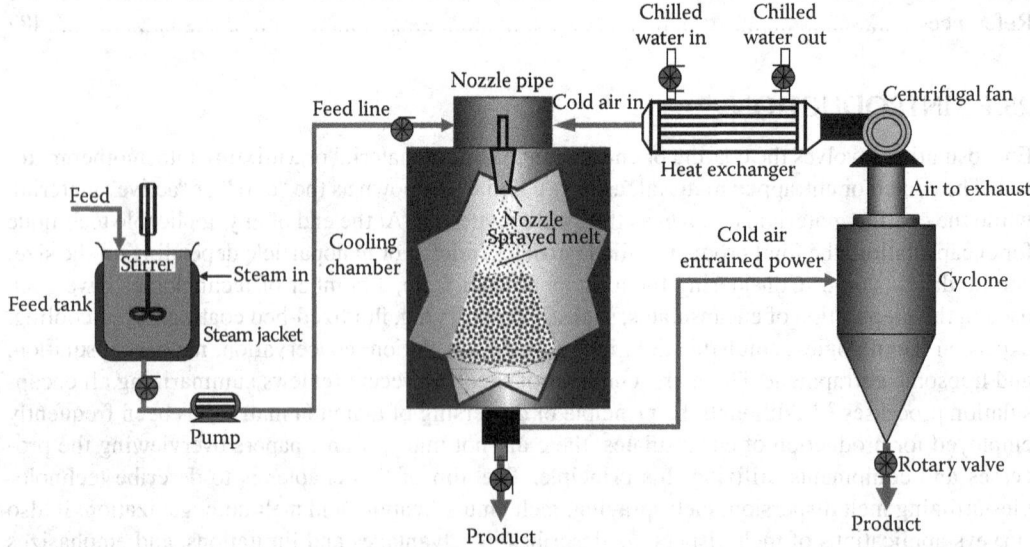

FIGURE 25.1 Schematic presentation of an encapsulation system employing spray cooling.

obtained microparticles are relatively uniform. However, the spray pattern is wide and short, so the cooling chambers have to be wide.[9] Air nozzles have been frequently employed in contrast to airless nozzles. The problem with airless nozzles appears when high viscous melt has to be processed; in this case, the obtained microparticles have nonuniform size. Equipments of different construction characteristics are commercially available, depending on the targeted size of microparticles. Thus, fine particles with a mean particle size of 50–150 μm are often congealed in cocurrent spray-cooling chamber with high-speed rotary atomization. Grainer particles with a mean particle size of 150–500 μm that solidify slowly are frequently congealed in a spray-cooling chamber with a fountain nozzle. For extra coarse particles with a mean particle size of 500–2000 μm (also called "prills"), a low-speed rotary atomizer with a specially designed prilling wheel is generally used.

The atomized droplets solidify upon their exposure to an ambient airflow, carbon-dioxide ice bath, or by cold air injected into the vessel (as that presented in Figure 25.1). The produced microspheres are cooled in a controllable manner and at the same cooling rate. The spray-cooling chamber can be designed with cocurrent, mix-current, or countercurrent air flow. The bottom of the chamber is either conical or flat (Figure 25.1). Novel apparatus consists of a chamber with integrated bag filters for cooling and bag filters for direct separation of products in one unit.[9,10] Otherwise, the final separation of the product can be achieved in a separate cyclone unit. The obtained microparticles are in the range of 50–2000 μm, depending on processing conditions and type of matrix, that is, more specifically the viscosity of the matrix material.

Apart from the standard setup of spray coolers described here, nowadays, process engineers are developing novel systems for either or both of two main reasons: (1) in order to reduce energy consumption during processing, and (2) in order to produce particles with specific structural or morphological characteristics. One of the strategies for more energy-efficient processing is to achieve higher heat transfer coefficients with lower mass flow rates.[11] The second reason inspired, for example, Mejia et al.,[12] who aimed at fabrication of wax microparticles, which should be disk shaped. For this purpose, they developed a novel electrospray device in which high electric field is applied (via a high voltage amplifier—voltage range from 2.6 to 2.9 kV), which forces a melt (α-eicosene above its melting point (26°C) with an addition of Stadis-450 to increase electrical conductivity) through a metallic needle to form sprayed droplets. The wax droplets were then collected in an ethanol/water mixture containing a surfactant in order to obtain an emulsion containing monodisperse wax droplets with a diameter ranging from 0.5 to 5 μm, depending on the flow rate of the dispersed phase. After ethanol evaporation, in order to reduce solubility of α-eicosene, the wax emulsion underwent phase changing during storage (at 4°C for one to several days depending on the droplets' size) and the wax droplets were transformed into monodisperse discotic particles.

The process parameters influencing particles' characteristics are: heating temperature, pump (liquid feed) flow rate, compressed air pressure (or wheel speed), solidifying air flow rate, and the fraction of an active compound in the mixture, but also the relationship between all these variables is an important factor. There is only limited literature data about the specific impact of each of the process variables on the product features and the results are often nonconsistent between different reports. The atomizing device system has the major impact on the particles characteristics and the particles will be bigger if the nozzle orifice is larger.[13] An increase in the energy used for atomization (pressure or kinetic energy) led to a decrease of the resulting particle size for pressure nozzles as well as for two-fluid nozzles and centrifugal wheels.[14,15] Ilic et al.[14] reported a decrease in the total product yield after increasing the rate of feed pumping explained by insufficient cooling of the molten droplets before reaching the device walls. Gavory et al.[13] applied a screening design methodology involving six parameters related to the process in order to evaluate how they affect the production yield and some particle properties (i.e., the volume-surface mean diameter, the residual humidity, the ratio of the fusion enthalpies of the polymorphs, and the normalized peeling force). Surprisingly, the statistical analysis of the results showed a negligible impact of the parameters related to the process, while the heating temperature and the active compound (an industrial water-based pressure-sensitive adhesive) fraction were the most significant factors. The increase of the mean particle size with the pump flow rate was consistent with other studies.[14] As regards the influence of heating temperature on the size of the microparticles, opposite findings have been reported in the literature. In general, an increase in heating temperature leads to a decrease in viscosity of fusible fats, but this tendency reflected on the particle size as a decrease in some studies,[13,15] while in some others as *vice versa*.[16,17]

25.4 MELT EMULSIFICATION/MELT HOMOGENIZATION

The term emulsification refers to the technique that involves mixing of two immiscible materials (usually liquids) in order to produce a homogeneous system. Usually, one of the two materials has an oily nature and the second is the water. In the emulsion, the liquid present in the larger proportion is called "continuous phase," while the liquid in the smaller proportion, which disperses, is called "dispersed phase." Depending on the dispersed phase, there are different types of emulsions, that is, "oil-in-water" (o/w) wherein oil is the dispersed phase, "water-in-oil" (w/o) wherein water is the dispersed phase, as well as multiple emulsions, like "oil-in-water-in-oil" (o/w/o), in which there are continuous layers of the two immiscible materials. Emulsification process includes the use of an emulsifying agent, which will be adsorbed on the interface of the two immiscible materials, in order to achieve the miscibility of them. The structure of these agents contains a polar and nonpolar part, and such molecules may be proteins, phospholipids, etc. Emulsification can also include the use of other colloidal macromolecules, which are able to form multilayer films on the interface, in order to achieve a better kinetic stabilization of the system.[18,19]

There is a type of melt-dispersion technology, which actually represents a combination of melt-dispersion and emulsification technique. Namely, a melt (dispersed phase) is fed into a vessel containing water (continuous phase) wherein mechanical stirring is used for dispersing it; thus, mixing is performed simultaneously with cooling (Figure 25.2).

In the literature, there are a number of papers that describe the critical points of the process aiming at its optimization. These are the following:

- The use of emulsifiers (different types and concentrations)
- The chronological order of the events—melting, addition of an active compound, emulsification, and cooling

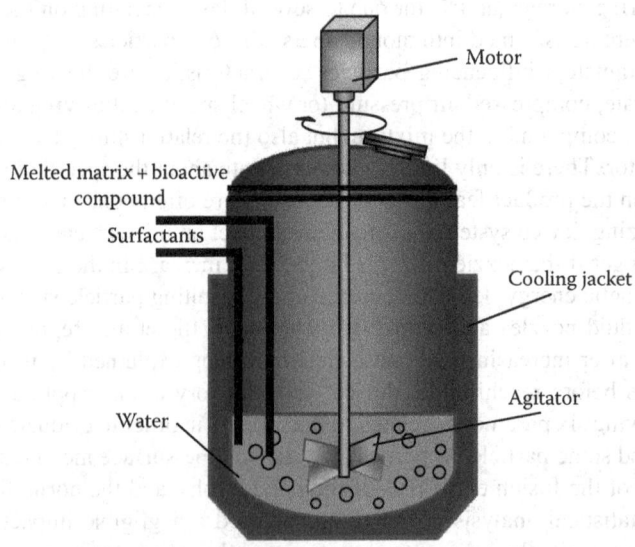

FIGURE 25.2 Schematic presentation of melt emulsification for encapsulation.

- The agitation conditions during the process
- Melt and water temperature
- The cooling rate of water after completion of homogenization

Optionally, emulsifiers can be supplemented in order to obtain microparticles of desired and controlled size. The choice of the emulsifier depends on the nature of the emulsion, that is, o/w or w/o, and the characteristic parameter, which helps to select the appropriate one, is its hydrophilic–lipophilic balance (HLB).[20] Different types of Tween (polyethoxylated sorbitan esters) and Span (sorbitan esters) have been mostly used as model emulsifiers.[16] HLB values of Tween are larger than 10 (more water soluble), while HLB values of Span are smaller than 10 (more oil soluble). Additionally, there is an optimum combination between the amount of core material and the amount of wall materials that have to be used, mainly affecting the thickness of wall materials and thus the decrease of losses.[21] In general, the existence of holes on wall material's surface or when capsules' surface is not smooth, but rough, results in increased losses of core material. Stirring can be performed either by a simple mixing apparatus (a stir bar or impeller) or by using high shear homogenization units. Simple mixing apparatus usually results in negligible encapsulation loadings. Furthermore, sonication, another commonly used homogenization method, resulted in the formation of fine emulsion droplets and high encapsulation loadings.[21]

The melt droplets rapidly solidify in contact with cold water to give solid particles. The water is cooled either gradually or rapidly, depending on the system matrix/active compound. An active compound is either mixed with a matrix prior to dispersing, or, alternatively, it is inserted directly into water simultaneously with the addition of a melt. In the last case, the particular condition is that the active material is hydrophobic; otherwise, it is likely going to be diluted in water instead of being embraced by the matrix droplets. Also, irregular crystallization may happen, which would consequently reflect on subsequent variability in dissolution rate. When the active compound is firstly dissolved in the liquid carrier, prior to cooling and congealing the mixture, the resulting solid will contain active's particles dispersed at the molecular or near-molecular level within the solidified matrix carrier (commonly termed amorphous solid dispersion). Solid dispersions may also contain multiple amorphous phases or a fraction of the active material in an amorphous phase and a fraction as a separate crystalline phase. In most cases, the solubility of the active compound in the external phase is important parameter and affects encapsulation efficiency and loading during the emulsification process.[21]

In some reports, phase inversion methodology of emulsification is proposed in order to avoid loss of the active compound during the preparation process. Briefly, preheated water containing a surfactant is added to the molten lipid phase, in which the active compound has been previously dispersed. For encapsulation of shear- and heat-sensitive compounds (e.g., DNA and some proteins), Schubert et al.[22] proposed the so-called postloading methodology, in which at first empty microparticles should be prepared by melt emulsification and then followed by adsorptive loading with the respective active compound. However, there are two main issues one should take into consideration before accepting this approach: (1) an active often ends up rather at the surface of particles than really incorporated within a matrix; (2) certain surface modifications of microparticles are required to assure efficient adsorptive loading. In the study of Schubert et al.,[22] the amphiphilic lipid lecithin was incorporated into the lipid matrix (made from hydrogenated palm oil and Phospholipon® 90G) and the surface properties of the obtained solid lipid nanoparticles were additionally manipulated by variation of the nonionic emulsifier concentration in the aqueous phase.

As previously mentioned, in melt-emulsification technique, the process variables are agitation speed, melt and water temperatures, as well as the type and concentration of the surfactant. In addition, critical parameters are different for different carrier materials and the type of lipid plays an important role, that is, the rate of lipid crystallization, the lipid hydrophilicity, and the surface of lipid crystals. The literature dealing with the specific influence of the process parameters is limited and one of the studies was done by our research group.[23–26] In these works, ethyl vanilline was encapsulated as a model aroma compound in carnauba wax microparticles by using a mechanical stirrer with two blade impellers rotating at low rates (1000–1500 min^{-1}). The effect of process variables such as wax fraction (1%–10%), stirring time (4–15 min), stirring rate (1000–1500 min^{-1}), and concentration of the emulsifiers (0%–1% of Tween 20/Span 40 or Tween 20/Span 60) on the mean particle diameter was assessed. The important outcome is that without any surfactant, it is possible to produce spherical microsized beads with bimodal size distribution with the main fraction having diameter in the range 210–360 μm. The obtained wax aroma formulations were evaluated as cost-effective as they were produced with very few consumables by using a simple apparatus and with low energy input. They can be considered as food additives in animal feeds, while for human usage, it is needed to downscale particles. This can be achieved by melt-emulsification process, which usually requires the usage of surfactants; otherwise, molten droplets will coalesce upon their collisions. Herein, one should be careful with selection of the suitable emulsifier, since the emulsifiers or additives required for droplet stabilization have to be thermally stable at elevated emulsification temperatures. Melt emulsification involving high-energy inputs (usually via an Ultra-Turrax mixer) is required in order to produce submicron-sized particles, since any lipid/emulsifier blend has considerable viscosity. As a temperature of a melt is higher, the size of particles obtained is lower due to decreased viscosity of the lipid phase. However, high temperatures may also increase the degradation rate of an active compound. For production of nanoscale particles, high-pressure homogenization or probe sonication is often employed subsequent to melt-emulsification step, as can be seen in Table 25.1, which presents some of the recent applications of melt dispersion. The homogenization step can be repeated several times (usually 3–5 cycles), but one should have in mind that a particle coalescence may occur resulting in enlargement of particles due to high kinetic energy of the particles.[27] The resultant particles are often referred in literature as solid lipid nanoparticles. Recently, a new device has been developed named "Simultaneous Homogenizing and Mixing" nozzle with the acronym SHM.[28] It combines a simple homogenizing nozzle with a mixing unit in a single T-shaped micromixer. SHM-valve was tested on different waxes with melting points between 50°C and 62°C, and it was found that at pressures between 100 and 1000 bar, molten droplets were homogenized without the aid of emulsifiers or other additives, which are absolutely required in conventional melt-emulsification processes. This was realized by a controlled and quick dilution and cooling down of molten fat globules directly after their disruption in the nozzle itself. The obtained microparticles exhibited monomodal particle distribution with mean value $d_{3,50}$ in the order of 1 μm. Moreover, the device is suitable for large feed capacities.

25.5 MATERIALS FOR ENCAPSULATION BY APPLYING MELT-DISPERSION TECHNIQUE

Materials used for encapsulation by melt-dispersion technique should have some specific characteristics, mainly related to their thermal behavior. In the recent review of Turton et al.,[53] ideal properties of carriers suitable to be processed by melt congealing are described and summarized. The melting point range should be narrow, ideally 75°C–80°C, and they should not undergo softening before melting in order to avoid agglomeration during processing. They should be stable in the range of 30°C–200°C and their thermal behavior should be independent of their thermal history, which also means that they have to be able to sustain thermal cycling. Moreover, in this temperature range, they should not experience crystal modifications upon heating. Viscosity of their melts should be low to facilitate flow and spray formation. In addition, the matrix material should have other attributes mandatory for any encapsulation material, such as to be nontoxic, inert, that is, not to show tendency for reaction with an active compound. The matrix material may act as a solvent for the active compound, resulting in improved dissolution characteristics.

Fatty acids and fatty alcohols, glycerides, and waxes have been commonly used for spray-congealing. These are summarized in Table 25.2. Apart from pure compounds, mixtures have beenWW often

TABLE 25.1
Application of Melt Dispersion: Examples Published Since 2000

Carrier	Active Compound	Particle Size	Active Loading/Encapsulation Efficiency (EE)/Production Yield	Specifics	References
Hydrogenated palm oil	An industrial water-based pressure-sensitive adhesive	the particle volume-surface average diameter 11–28 μm (average diameter <100 μm)	Product yield: 70%	—	[13]
Cutina®HR	Zidovudine	175 nm	EE: 49%	Melt emulsification followed by probe sonication	[29]
Compritol®888ATO/Miglyol®812N mixture	Rhodamine B Coumarin-6	70 nm	—	Melt emulsification followed by probe sonication	[30]
Erythrytol (anhydrous sugar alcohol)	Flavors: limonene, nicotine, methyl salicylate, cinnamic aldehyde and Neobee® M5	40–300 μm (produced by pressure nozzle) 500 μm–2 mm (produced by ejecting through dispensing needle)	Active loading: 35 wt%	Erythrytol is a lipophobic crystallizing carrier and immiscible with lipophilic flavors; the effects—superior retention of flavors	[21]
Tristearin	Quercetin	Volume median diameter <6.7 μm	Active loading: 11.8 wt%	—	[31]
Glycerylbehenate Poloxamer188 Stearicacid Cetylpalmitate			EE: 72.4%	Melt emulsification followed by a sonication step	
Compritol®888 ATO Gelucire 44/14 Miglyol 812N	Genistein	90 nm (for the optimal encapsulate form)	EE: 91.1%	Melt emulsification combined with ultrasonication technique	[32]

(Continued)

TABLE 25.1 (Continued)
Application of Melt Dispersion: Examples Published Since 2000

Carrier	Active Compound	Particle Size	Active Loading/Encapsulation Efficiency (EE)/Production Yield	Specifics	References
Beeswax/Phospholipon®90H/PEG-4000 mixture	Gentamicin	271–306 μm	Active loading: 39.2–44.7 wt% EE: 74.7–86.3%	Melt homogenization	[33]
Gelucire® 50/13	*Silybum marianum* dry extract	355–500 μm size fraction, representing the 85%	Product yield: 94–98 wt% EE: 87%–99%	Mechanochemical activation coupled with spray congealing	[34]
Stearic acid/Oleic acid mixture	Glucose solution	1–350 μm mean diameter	Incorporated active load: 75.1–96.9 wt% Superficial active load: 2.6–23.7 wt%	Spray congealing of the w/o emulsions (prepared from the lipid mixture (oil phase) and glucose solution (water phase) with an emulsifier)	[35]
Glyceryl palmitostearate Tristearin	Albumin	Volume mean diameter 167–291 μm	Active loading: 10.6–20.1 wt% Product yield: 60%–80%	Spray congealing using wide pneumatic nozzle from Albertini et al. (2008)	[36]
EUDRAGIT® L 100/Carnauba wax and Stearic acid	Mesalazine	500–700 μm	Active loading: 11.9–15.9 wt%	2-step spray congealing followed by dry coating by tumbling with mannitol/lecithin, the result–layered microcapsules for mesalazine pH-dependent delayed release	[10]

(*Continued*)

TABLE 25.1 (Continued)
Application of Melt Dispersion: Examples Published Since 2000

Carrier	Active Compound	Particle Size	Active Loading/Encapsulation Efficiency (EE)/Production Yield	Specifics	References
Poloxamer 407 (Lutrol® or Pluronic®F127 poly(oxyethylene)–poly(oxypropylene)–poly(oxyethylene))/Gelucire® 50/13 mixture	Atenolol	<50 μm 6.5 wt% 50–100 μm 72.7 wt% 100–355 μm 20.8 wt%	Active loading: 10.0–16.9 wt%	Spray congealing using wide pneumatic nozzle from Albertini et al, 2008	[37,38]
Carnauba wax/Bees wax mixture	Ketoprofen	65–100 nm	Active loading: 0.5% w/v EE: 96%	Melt emulsification	[39]
Carnauba wax	Ethyl vanilline	Bimodal—the main fraction: 210–360 μm	Active loading: 10% wt% EE: 86%	Melt homogenization	[23]
Gelucire® 50/13/Polymer mixture (polymer: chitosan, sodium carboxymethylcellulose or poloxamers)	Econazole nitrate	100–355 μm prevailing fraction	Active loading: 21.0–35.2 wt%	Wide pneumatic nozzle from Albertini et al. (2008)	[40]
Gelucire® 50/13 Poloxamer 188 PEG 6000	Glimepiride EE% 100	$d_{(0.5)}$ 84–140 μm	Active loading: 1.7 wt%	Spray congealing	[14]
Cysteine-polyethylene glycol stearate conjugate	CyclosporineA	60–67 nm	Active loading: 3.6 mg/mL	Melt emulsification followed by probe sonication	[41]
Carnauba wax Cetearyl alcohol Stearyl alcohol PEG 4000	Propafenone hydrochloride Vitamin E	75–500 μm	Active loading: 50% for propafenone hydrochloride 30% for vitamin E Product yield: 95%	Unique device–wide pneumatic nozzle	[9]
Dynasan116® (Glycerol tripalmitate)	Insulin	$d_{(0.5)}$ 182–315 μm	Active loading: 0.5–2.7 wt% Product yield: 79%–95%	Spray-congealing using customized spray-congealing apparatus	[15]

(*Continued*)

TABLE 25.1 (Continued)
Application of Melt Dispersion: Examples Published Since 2000

Carrier	Active Compound	Particle Size	Active Loading/Encapsulation Efficiency (EE)/Production Yield	Specifics	References
Monostearin (MS)/stearic acid (SA) mixture monostearin (MS)/PEG mixture	Progesterone	322–485 nm	Active loading: 2.7–11.6 wt% EE: 55%–77%	Melt emulsification followed by ultrasonic pulverization	[42]
Gelucire® 50/13	Praziquantel	75–355 μm size fraction representing the 80%	Active loading: 6.1–34.1 wt%	Ultrasonic spray-congealing	[43]
Carnauba wax–decyl oleate mixture	Pigments: barium sulfate, strontium carbonate, and titanium dioxide	247–562 nm (loaded with barium sulfate and strontium carbonate) 367–556 nm (loaded with titanium dioxide)	Active loading: 2–6 wt% EE: 88–98%	Melt emulsification followed by high-pressure homogenization	[44]
Softisan® 154 (hydrogenated palm oil)/Phospholipon® 90G/Lecithin mixture	Albumin	98–262 nm	Active loading: 3–15 wt%	Melt emulsification followed by high-pressure homogenization and adsorptive loading with albumin	[22]
Precirol® ATO 5 (mixture of mono-, di- and triglycerides of stearic and palmitic acids); Aerosil® 90, 200 and 300 (colloidal silicon dioxide).	Theophylline	50–1000 μm	Active loading: 11–29 wt%	Ultrasonic spray-congealing method. Aerosil® was used as thickening and suspending agent.	[45]
Cetyl alcohol/Palmitic acid mixture	Ibuprofen	—	Product yield: 84–93 wt% Active loading: 77.3–86.6 wt%	Melt solidification technique	[46]
Cetyl alcohol	Flurbiprofen	250–1400 μm	Product yield: 94.7–98.7 wt%. Active loading: 72.8–81.2 wt%	Melt solidification technique	[47]

(Continued)

TABLE 25.1 (Continued)
Application of Melt Dispersion: Examples Published Since 2000

Carrier	Active Compound	Particle Size	Active Loading/Encapsulation Efficiency (EE)/Production Yield	Specifics	References
Microcrystalline wax/Stearyl alcohol/Soya lecithin	Verapamil hydrochloride	75–250 μm size fraction representing >80%	EE: 49%–100%	Ultrasonic spray-congealing technique	[48]
Carnauba wax Cetanol Precirol® ATO 5 Stearic acid Compritol®888 ATO Glycerol monosterate Precirol®WL2155 Hydrophilic excipients: Pluronic® F127 PEG 4000	Felodipine	Median 20.9–34.6 μm	—	Spray chilling for drug encapsulation followed by compression for tablets production, the carriers tested are mixtures of lipophilic matrix and hydrophilic polymers.	[49]
Carnauba wax Cetanol Cutina®HR Precirol® ATO 5 Stearic acid	Felodipine	Median: 26.5–30.3μm	—	Spray chilling was used for drug encapsulation followed by compression for tablets production.	[50]
Dynasan 116® (Glyceryl tripalmitate)	Somatostatin acetate	Medium diameter: 92.8 μm	Product yield: 85%–90%. EE: ~90%	Melt-dispersion method	[51]
Imwitor 900 (glyceryl monostearate), Compritol 888 ATO (glyceryl behenate) Dynasan 116 (tripalmitate) Cutina CP (cetyl palmitate) Beeswax	Retinol	0.1–1 μm (cetyl palmitate and beeswax solid- lipid microparticles) 0.1–5 μm (tripalmitate) 0.1–100 μm bimodal size distribution (glyceryl behenate)	—	Melt-emulsification technique	[52]

TABLE 25.2
Common Materials for Encapsulation by Melt-Dispersion Technique

Materials	Melting Point (°C)
Fatty acids	
Lauric acid $CH_3(CH_2)_{10}COOH$	+44
Myristic acid $CH_3(CH_2)_{12}COOH$	+55
Palmitic acid $CH_3(CH_2)_{14}COOH$	+63
Stearic acid $CH_3(CH_2)_{16}COOH$	+70
Behenic acid $CH_3(CH_2)_{20}COOH$	+74–78
Arachidic acid $CH_3(CH_2)_{18}COOH$	+76
Fatty alcohols	
Tetradecanol (Myristyl alcohol) $CH_3(CH_2)_{13}OH$	+38
Pentadecanol $CH_3(CH_2)_{14}OH$	+42
Hexadecanol (Cetyl alcohol) $CH_3(CH_2)_{15}OH$	+49
Heptadecanol (Margaryl alcohol) $CH_3(CH_2)_{16}OH$	+53
Octadecanol (Stearyl alcohol) $CH_3(CH_2)_{17}OH$	+59
Nonadecanol $CH_3(CH_2)_{18}OH$	+62
Eicosanol (Arachidyl alcohol) $CH_3(CH_2)_{19}OH$	+65
Henicosanol $CH_3(CH_2)_{20}OH$	+68
Docosanol (Behenyl alcohol) $CH_3(CH_2)_{21}OH$	+70
Tricosanol $CH_3(CH_2)_{22}OH$	+72
Tetracosanol (Lignoceryl alcohol) $CH_3(CH_2)_{23}OH$	+72
Pentacosanol $CH_3(CH_2)_{24}OH$	+75
Hexacosanol $CH_3(CH_2)_{25}OH$	+73
Heptacosanol $CH_3(CH_2)_{26}OH$	+80
Octacosanol $CH_3(CH_2)_{27}OH$	+81
Nonacosanol $CH_3(CH_2)_{28}OH$	+83.5
Glycerides	
Myverol™ 18–06 (Glycerol monostearate)	+55
Compritol 888 ATO® (Glyceryl behenate)	+70
Cutina HR® (Hydrogenated Castor Oil)	+85–88
Hydrogenated Palm Oil	+40–62
Gelucire® 44/14 (polyethylene glycol (PEG) and PEG esters)	+44
Waxes	
Beeswax	+62–64
Lanolin	+38
Bayberry wax	+45
Japan wax	+51
Candelilla wax	+68.5–72.5
Rice bran wax	+77–86
Carnauba wax	+78–86
Castor wax	+80
Ouricury wax	+81–84
Paraffin wax	+46–68

employed (see Table 25.1). Some studies show that in a matrix mixture, the liquid lipid modifies the crystallization kinetics of the matrix, by modifying the organization of the crystalline network; consequently, encapsulation capacity is increased, while expulsion of the core is decreased.[22]

25.5.1 FATTY ACIDS AND FATTY ALCOHOLS

Fatty acids comprise a large group of chemical compounds divided into saturated and unsaturated acids. Saturated fatty acids are linear monocarboxylic acids, while unsaturated fatty acids have one or more double bonds in their chain. Fatty acids are classified, according to their chain length, as short-chain fatty acids (having a number of carbons less than 8 in the aliphatic tail), medium-chain fatty acids (8–14 carbons), and long-chain fatty acids (more than 16 carbons). Unsaturated fatty acids occur in different chain configurations, namely, *cis*- and *trans*-. In fatty alcohols, a carboxyl group is replaced by hydroxyl group. The water solubility of fatty acids rapidly decreases with increasing chain length. The melting points of fatty acids vary over a wide range, and unsaturated fatty acids have a much higher melting point. The main drawback of these materials is their poor stability as they are prone to auto-oxidation at room temperatures. Fatty acids have been used as nonionic surfactants and have remarkable emulsifying properties.[54] They are produced by the hydrolysis of the ester linkage of naturally occurring oils and fats, which are in general triglycerides.

25.5.2 Glycerides and Polyglycolyzed Glycerides

Glycerides consist of one, two, or three fatty acid chains covalently bonded to a glycerol molecule by ester linkages and thus, they are categorized as monoglycerides, diglycerides, and triglycerides, respectively. Diglycerides and triglycerides (also called fats) may consist of saturated or unsaturated fatty acid chains. The melting points strongly depend on the chemical nature, but also on the symmetry of the fatty acid residues and their distribution along the carbon skeleton of glycerol residue of the molecule.[54] Hydrocarbon chains of saturated fatty acids lie parallel with strong dispersion forces between their chains; they pack into well-ordered, compact crystalline forms and melt above room temperature. In contrast, unsaturated fatty acids, because of the *cis*-configuration of the double bonds, have hydrocarbon chains of less ordered structure; these triglycerides have melting points below room temperature. Also, melting point increases as the number of carbons in the hydrocarbon chains increases. Glycerides and polyglycolyzed glycerides have been commonly used in the preparation of sustained-release matrix formulations. Among them, triglyceride vegetable oils are most commonly used as lipid matrices in drug delivery as these lipids are fully digested and absorbed.[55] According to the length of fatty acids chain, they are divided into the following categories: long-chain triglycerides, medium-chain triglycerides, and short-chain triglycerides. Medium-chain triglycerides have a higher solvent capacity for lipophilic compounds than long-chain triglycerides, and they are less prone to oxidation.[55] Vegetable oils differ in proportion of fatty acids. Pure triglycerides can be found in refined vegetable oils, while mixed glycerides are obtained by partial hydrolysis of vegetable oils where the extent of hydrolysis determines the chemical composition of the mixed glycerides product. There are several commercial products on the market, such as Cutina HR® (hydrogenated castor oil), Compritol 888 ATO® (glyceryl behenate), Myverol™ (distilled monoglycerides), Myveplex 600, and Myverol™ 18-06 (glycerol monostearate). Glycerides crystallize in different subcell arrangements: hexagonal, orthorhombic, and triclinic. They exhibit marked polymorphism with three and often more individual forms.[52]

25.5.3 Natural and Synthetic Waxes

The use of waxes in pharmaceutical and food industry is a common practice. Their main use, as described previously, is to play the role of wall material in melt-dispersion technique and widely in melt techniques. The extent of this use stems from the advantages that waxes have as components intended for human use. These advantages include good chemical and physical stability at varying pH values and moisture levels, well-established safe use in humans due to their nonswelling and water-insoluble nature, and minimal effect on food.[46]

The major constituents of waxes are esters of fatty acids, isolated from animals and plants. Thus, most of natural waxes are basically edible, such as those summarized in Table 25.2, but often indigestible. Among them, carnauba, candelilla, and beeswax have been frequently used for encapsulation by melt dispersion. Waxes crystallize mainly in an orthorhombic subcell arrangement and the polymorphic transition rate is low.[52]

Carnauba wax (also called Brazil wax or palm wax) is isolated from the leaves of the palm *Copernicia prunifera* (carnauba palm), a plant native to northeastern part of Brazil. It has an appearance of yellow flakes. Carnauba consists mostly of aliphatic esters (40.0 wt%), diesters of 4-hydroxycinnamic acid (21.0 wt%), ω-hydroxycarboxylic acids (13.0 wt%), and fatty acid alcohols (12.0 wt%). These compounds are mainly derived from acids and alcohols in the C26-C30 range. Distinctive for carnauba wax is the high content of methoxycinnamic acid as well as diesters. This type of wax is one of the hardest natural waxes with the highest melting point between 78°C and 86°C, but most often the melting-point range is 82°C–85°C.[56] Like all other waxes, it is insoluble in water, but soluble on heating in ethyl acetate and in xylene, and practically insoluble in ethyl alcohol (European Pharmacopeia 6.0).

Beeswax is secreted by young honeybees to build the honeycomb. It is mainly esters of fatty acids and various long chain alcohols. However, the monoesters in beeswax are poorly hydrolyzed in the guts of humans and mammals and are, therefore, of no significant food value. Bees wax has a melting-point range 62°C–64°C.[54]

Candelilla wax is derived from the leaves of the small Candelilla shrub, which grows in northern Mexico. It is yellowish-brown, hard, brittle, and aromatic. Candelilla wax mainly consists of hydrocarbons (about 50%, chains with 29–33 carbons), as well as esters of high molecular weight (20%–29%), free fatty acids (7%–9%), and resins (12%–14%, mainly triterpenoid esters). It is insoluble in water, but soluble in many organic solvents such as acetone, chloroform, and benzene. Candelilla wax melts in the range 68.5°C–72.5°C.[57]

25.5.4 OTHER CARRIERS

Common carriers for spray melt-dispersion encapsulation listed earlier are lipophilic materials. Such materials are often miscible with the active ingredients having the same nature and hence poor barrier efficiency. There have been many efforts, even from early days, to use a lipophobic crystallizing carrier, which should also have narrow melting point range. Some typical examples include mannitol[20,58] and meso-erythritol.[21] Both of them are sugar alcohols with high melting points of 164°C–169°C and 121°C for mannitol and erythritol, respectively. In order to quench melts with such high melting points and assure their rapid crystallization, supercooling or supersaturated conditions needs to be provided.[59] Erythritol has been used mainly in pharmaceutical formulations[60–62] since it has good thermal stability, low hygroscopicity, sweet taste, low toxicity, and good compatibility with drugs.[21]

25.6 APPLICATIONS

The melt-congealing technique has been practiced in the manufacture of wax-based suppositories for over 100 years.[53] The basic concept of melt-congealing process was employed by Sekiguchi et al.[63]

These researchers melted a sulphathiazole–urea mixture of eutectic composition at above its eutectic temperature, solidified the dispersion in an ice bath, and pulverized it into a powder. New applications have evolved quickly in the last 40 years to meet the increasing need for improving the solubility and consequent bioavailability of poorly water-soluble drugs and for production of encapsulates with controlled-release properties. A large number of pharmaceutical dosage forms have been fabricated using melt congealing, since lipid-based carriers improve the solubility and bioavailability of drugs with poor water solubility.[55] Namely, it has been determined that lipids administrated together with drugs (either *via* meals or as carriers for drugs) enhance oral bioavailability, affect gastrointestinal physiology, and maximize drug transfer into the systemic circulation. Hence, lipid-based formulations can be used to reduce the dose of a drug, while simultaneously enhancing its oral bioavailability.[55] In addition to encapsulation of pharmaceuticals, melt dispersion has been used for flavors, enzymes, and functional and textural ingredients such as minerals and proteins in the food sector. This technology can be applied even to heat-sensitive compounds, since during processing, an active ingredient is in contact with a hot melt just for a short time. Thus, about 5%–10% of all aroma encapsulates are produced by melt-dispersion technique.[5] With this technology, the incorporation of lipophilic and hydrophilic drugs is feasible, the carriers are safe (matrices are made from physiological lipids, which decreases a risk of acute and chronic toxicity), the usage of organic solvents is avoided, and large-scale production is possible.

25.7 PROPERTIES OF ENCAPSULATES

It should be pointed out that the properties of both the active compound and the carrier, their interactions, the rate of cooling and congealing as well as the conditions of storing the final dosage form all play an important role in the performance of the end product. Chemical, physical, and crystallographic differences between different carriers will reflect on their encapsulation properties. In Table 25.1, one can see that most of the lipids used represent mixtures of several chemical compounds, the exact portions of which vary from different suppliers, and might even be different for different batches of the same supplier. Even small differences in the mixture composition may reflect on the quality of encapsulates. Section 25.7 gives an overview of the encapsulates' characteristics frequently discussed in scientific reports, where some general conclusions are pointed out with the few examples confirming these rules. But it should be emphasized that one can easily extract from literature the opposite conclusions, as well.

The active loadings of encapsulates produced by spray congealing are rather low, as can be seen in Table 25.1. Generally, with fat crystalline matrices and by applying this technique, it is difficult to achieve high drug loadings (usually below 20%). Furthermore, a lipophilic active material with low molecular weight might easily diffuse out from fat crystals. The losses happen even during spray-chilling processing, especially with very volatile and/or carrier mixable actives. Therefore, spray-chilled encapsulates are usually not recommended for applications that require long shelf-lives.[5] Some studies showed that glyceride-based matrices showed better encapsulation of lipophilic actives compared to wax-based particles, but poorer physical stability and these differences were attributed in part to different crystal packing.[52]

As far as stability is concerned, during storage, a rearrangement of the crystal lattice may occur in favor of thermodynamically stable configuration and this is often connected with expulsion of the active molecules.[27] Upon dissolution of an active compound in a low molecular melt, conditions must be thermodynamically favorable during cooling to assure stability of the composition. Otherwise, recrystallization of an active is inevitable, which may alter the dissolution characteristics on storage. Given that the kinetics of recrystallization and the active solubility in carriers are both temperature dependent, the rate of cooling after incorporation may alter the morphology of an active ingredient.[64]

One of the main disadvantages of particles prepared by melt emulsification is that they are often nonuniform in size (Figure 25.3) as a consequence of inhomogeneous distribution of the power density. Thus, particles localized in different volumes of the sample experience forces of different magnitudes and, therefore, the degree of particle disruption diverges within the sample volume. Furthermore, in case of waxy particles surface morphology, these may have an appearance from completely smooth to quite rough (Figure 25.4a). In addition, particles may have nonhomogenous internal structure; thus, they can gain more/less compact matrix type of interior structure or hollow-shell morphology (Figure 25.4b). On the other hand, lipids may form spherical or platelet forms during homogenization and they develop a number of layers.[27] The surfactants have impact on the final properties of encapsulates, in particular on particle size; in general, particle size decreases with an increase in surfactant content due to reduction of surface tension and improved particle partition during homogenization, although there are many exceptions to this rule. Actually, during emulsification, two processes compete with each other: the covering of the particles surfaces with surfactant molecules and the agglomeration of uncovered lipid

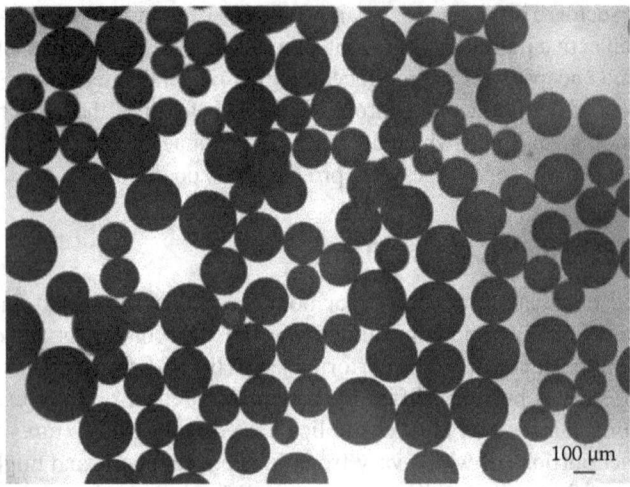

FIGURE 25.3 Microscopic images of carnauba wax microparticles produced by spray congealing under following conditions: melt temperature 115°C, pneumatic spraying system with air pressure 1 bar, air nozzle internal diameter 1 mm (unpublished data).

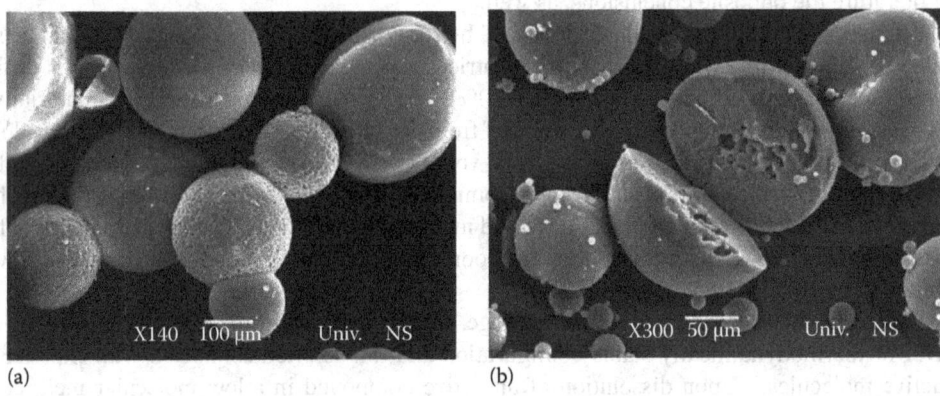

FIGURE 25.4 SEM images of carnauba wax microparticles entrapping ethyl vanilline (unpublished micrographs) produced by melt emulsification using mechanical stirrer with two-blade impeller under conditions from [23]: stirring rate 1200 rpm, stirring time 4 min, carnauba wax/water weight ratio 0.08:1, surfactant (Tween 20/Span 40 in weight ratio 0.53:0.47) concentration 1% wt. (a) low magnification and (b) high magnification.

surfaces.[27] The kinetics of the two processes determines time scale of the redistribution processes of surfactant molecules between particles, and it varies from system to system, depending on the power density and power distribution during dispersing process as well as on the molecular weight of the surfactants. The choice of surfactant is an important issue when producing encapsulates by melt homogenization, since surfactant type can affect the particle size and also stability of particles during periods of storage. Thus, mixed surfactants often reduce interfacial tension more than single-surfactant formulations.[39] Surfactants provide either electrostatic or steric stabilization of the particles; that is, long-chain surfactant molecules, after interpenetration, limit freedom of the particles and prevent them from associating with each another.

The encapsulates produced by melt dispersion often release the active material easily, especially the water-soluble one, when brought into contact with a medium of high water activity.[1] One of the reasons is that a significant portion of an active compound is actually located at the surface of microparticles and has direct access to the environment. In a large number of publications, *in vitro* release studies of encapsulates aimed at oral administration and were performed mimicking the gastric and intestinal pH conditions. Generally, wax-based matrices release active compounds without responding to the pH of the release media. The two most important mechanisms of drug release from waxy materials are erosion (hydrolytic degradation) of the waxy matrix and diffusion of the active compound through the matrix, whereas this last step has been considered to be the rate-controlling step. Other release mechanisms may also be involved, such as diffusion of water through the shell defects, osmotic forces, and mechanical disruption of the particles.[1] Waxy solid-lipid particles degrade in the presence of water by an ester hydrolysis reaction that is acid-catalyzed and reversible.[39] Long chain esters degrade via hydrolysis of free fatty acids and hydroxyl groups of wax structures to carboxylic acids with short chains. Increase of lipid matrix porosity and mass loss will affect the drug release profiles. In addition, active release is slower from more lipophilic matrices. For example, carnauba wax is a more lipophilic matrix compared to bees wax and also contains 5% of resins that constrain water to penetrate into the pores of the lipid structure. In addition, carnauba wax contains lower percentages of free fatty acids and hydroxyl groups (acid value: 2–7) than beeswax (acid value: 17–24), which also contributes to slower release of poorly water-soluble substances from carnauba wax encapsulates compared to bees wax.

Thermal characteristics of encapsulates are important, especially for aroma active compounds, from the viewpoint of their release in thermally processed food. In one of the recent publications,[65] the thermal release of vanillin encapsulated in Carnauba wax microcapsules was studied by isothermal thermogravimetric analysis at a temperature range of 170°C–210°C. Kinetic studies revealed that the release is not a single-step reaction but a complex kinetic process that can satisfactorily be described by the Avrami–Erofe'ev kinetic model A3. More importantly, thermal release of vanillin encapsulated into Carnauba wax proceeded with an activation energy lower than 40 kJ mol^{-1}, indicating that the Carnauba wax microcapsules release vanillin relatively easily and thus suggesting that the Carnauba wax can be suitably used as a carrier for aromas especially in the food industry.

Mechanical properties of fatty and wax encapsulates have been less explored compared to others mentioned earlier. The rupture forces as well as adhesion forces increased with particles diameter and the structure of the particles could have been weakened by the surfactant in case of particles made from hydrogenated palm oil matrix.[13]

25.8 LIMITATIONS OF MELT-DISPERSION TECHNOLOGY AND TRIALS TO OVERCOME THEM

One of the disadvantages of melt-dispersion technology is the high-energy input required for processing. First, the matrix material has to be heated in order to get the melt; then, in many cases, the fluid (usually water), in which solidification of the melt droplets occurs, has to be cooled. In addition, in order to atomize highly viscous fluids, it is necessary to provide very high pressures up to 70 bar.[9] The cost increases more with the scale-up, since heating and mixing time increase due to surface area/volume efficiencies and this can be problematic for heat-sensitive compounds. In some cases, the cooling step is followed by drying of microparticles for several hours,[15] which also requires a substantial amount of energy. Furthermore, considerable capital costs should also be taken into consideration as the equipment is complex and often large, for example, when cooling chambers are ought to be spacious. For example, for products that solidify slowly, the spray-cooling chamber can be equipped with an integrated fluid bed and/or even with an external unit for final cooling and solidification. In order to reduce the extra energy consumption involved, industrial common practice recycles the process air (as presented in Figure 25.1), thereby minimizing the load on the dehumidification/cooling unit.

Processing conditions, such as pressure, temperature, and rotation speed of a disk or mechanical stirrer, must be optimized in order to maintain an adequate temperature profile, which should consequently lead to harmonized sequence of events: melting of a matrix material, melt dispersing, and cooling. Then, processing parameters have to be carefully controlled in order to avoid undesired accidents, for example, obstructions of nozzle orifices.

A breakthrough in this field was done by Rodriguez et al.[66] who proposed a novel technological solution where ultrasonic energy is employed instead of traditionally used centrifugal or pressure energy in order to atomize a liquid matrix material. The invented device was tested on different matrices: stearic acid, carnauba wax, Cutina HR®, and Compritol 888 ATO®. The particle sizes depended on the amount of drug present (theophylline and fenbufen were used as model drugs) and in each case, the maximum size value of the distribution frequency was found to be 375 μm. Later, the same research group developed a new device, called Wide Pneumatic Nozzle, which utilizes Venturi effect and, in this way, high kinetic energy is created at the expense of the pressure drop caused by melt flow through a constricted section of a pipe. Wide pneumatic nozzle requires moderate air consumption (1–3 bar) and the heating of nozzles is provided by two resistors connected to an inverter instead by hot air. With this innovative technological concept, very good results were obtained regarding product yield (95% w/w), flexibility of the processing (an active compound can be either solid or liquid), and product quality (spherical, nonaggregated microparticles (75–500 μm) with high drug content (50%)). In addition, the authors proved that the new design was able to nebulize very viscous systems (up to 500 mPa s) that are usually not processed by conventional apparatus.

25.9 SUMMARY AND FUTURE PERSPECTIVES

In this chapter, the scientific literature on the utilization of melt-dispersion technique employed in melt-spraying, melt-emulsification, and melt-homogenization processes is reviewed. The principles behind these technologies are explained and matrix materials used for encapsulation are discussed. The basic equipment required for processing is described with a critical outlook on innovation in the field of process engineering, while the inclusion of some common properties of encapsulates was attempted. In the market, there is a increased need for encapsulates, which fulfill many demands: delayed release, stability, thermal protection, suitable sensorial profile, and others. It is almost impossible to fulfill all these criteria in a single encapsulation step by any known method and matrix material. Therefore, there is a need for multiencapsulated formulations, which can be obtained only if more than one technique is employed and through several encapsulation steps.

One of the strategies that should be explored more for an improvement of the release kinetic could be the modification of the crystalline structure of the shell material. For example, glyceride esters of fatty acids have various crystalline forms, where one form may stabilize an active compound more than the others, but the knowledge on how to assure the preferable crystal form is still lacking.

Additionally, the low loading capacity limits the application of melt-dispersion technologies, especially at industrial scale, since the cost-effectiveness of the process is in line. Another consequence of such low loads is that a substantial amount of a matrix material has to be consumed in order to achieve an adequate dosage of an active compound. Although considered as safe, fats and waxes are not welcomed in high amounts in some products. For these reasons, the scientists are trying to achieve maximum loads by seeking for a suitable matrix composition for each particular active compound. Without a deep knowledge in chemistry, thermodynamics, and physics and a multidisciplinary approach, it is not possible to do it.

The melt-dispersion technique has been employed for various chemicals (as can be seen in Table 25.1), but is not yet as widely used as some other techniques. However, this does not necessarily mean that melt-dispersion technique is not suitable for other compounds and it is likely to expect that in the future, its applications will increase, especially on heat-sensitive compounds, such as polyphenols.

25.10 EXPERIMENT—AN EXAMPLE: PRODUCTION OF CARNAUBA WAX MICROPARTICLES ENCAPSULATING ETHYL VANILLINE

25.10.1 EQUIPMENT AND REAGENTS

Carnauba wax Carl Roth GmbH (Germany)

Tween 20
Span 40
Distilled water
Water bath
Thermometer
Mechanical stirrer with two blade impellers (IKAWerke RW, Germany)
Vacuum filtration unit
Ethyl vanilline

25.10.2 METHOD

Place a vessel (internal volume of 600 mL) in a water bath and fill it up to 250 mL with water. Add emulsifiers (mixture Tween 20/Span 40 0.53:0.47 w/w) in a weight ratio emulsifier mixture:water 1:100. Place a mechanical stirrer centrally in the vessel and turn it on at 1200 r min^{-1}. Place the thermometer in the vessel close to the walls of the vessel. Turn on the thermostat of the water bath at 95°C to start with heating. Place another vessel into the same water bath and put 20 g of Carnauba wax together with 2 g of ethyl vanilline. After Carnauba wax is completely melted, pour the melt (wax/aroma mixture) to the vessel containing water and emulsifiers. After 4 min of stirring, stop the heating, and rapidly pour 250 mL of cold water (2°C–5°C) to the resulting dispersion in order to cool it down and to enable solidification of wax droplets. Turn off the stirring, remove the stirrer from the vessel, and leave the dispersion to cool down spontaneously up to room temperature. Filtrate the dispersion using vacuum filtration unit in order to collect solid wax microparticles. Wash them thoroughly with 250 mL of water and dry in an oven at a temperature of 50°C up to a constant weight.

ACKNOWLEDGMENTS

This work was supported by the Ministry of Education, Science and Technological Development, Republic of Serbia (Projects No. III46010 and No III46001) and the COST Action FA0907 "Yeast flavour production – New biocatalysts and novel molecular mechanisms (BIOFLAVOUR)"

REFERENCES

1. Gouin, S. Micro-encapsulation: Industrial appraisal of existing technologies and trends. *Trends Food Sci Technol*, 15 (2004): 330–347.
2. Madene, A., Jacquot, M., Scher, J., and S. Desobry. Flavour encapsulation and controlled release—A review. *Int J Food Sci Technol*, 41 (2006): 1–21.
3. Nedovic, V., Kalusevic, A., Manojlovic, V., Petrovic, T., and B. Bugarski. Encapsulation systems in the food industry. In: Yanniotis, S., Taoukis, P., Stoforos, N. and Karathanos, V.T. Eds., *Advances in Food Process Engineering Research and Applications*, pp. 229–255. New York: Springer (2013).
4. Shewan, H. and J. Stokes. Review of techniques to manufacture micro-hydrogel particles for the food industry and their applications. *J Food Eng*, 119 (2013): 781–792.
5. Zuidam, N.J. and E. Heinrich. Encapsulation of aroma. In: Zuidam, N.J. and Nedovic, V. Eds., *Encapsulation Technologies for Active Food Ingredients and Food Processing*, pp. 127–161. New York: Springer (2010).
6. de Vos, P., Faas, M., Spasojevic, M., and J. Sikkema. Encapsulation for preservation of functionality and targeted delivery of bioactive food components. *Int Dairy J*, 20 (2010): 292–302.

7. Fang, Z. and B. Bhandari. Encapsulation of polyphenols—A review. *Trends Food Sci Technol*, 21 (2010): 510–523.
8. Jyothi, V.N., Prasanna, M., Sakarkar, S.N., Prabha, S., Ramaiah, S., and G.Y. Srawan. Microencapsulation techniques, factors influencing encapsulation efficiency. *J Microencapsul*, 27(3) (2010): 187–197.
9. Albertini, B., Passerini, N., Pattarino, F., and L. Rodriguez. New spray congealing atomizer for the microencapsulation of highly concentrated solid and liquid substances. *Eur J Pharm Biopharm*, 69 (2008): 348–357.
10. Balducci, A., Colombo, G., Corace, G., Cavallari, C., Rodriguez, L., Buttini, F., Colombo, P., and A. Rossi. Layered lipid microcapsules for mesalazine delayed-release in children. *Int J Pharm*, 421 (2011): 293–300.
11. Chunqiang, S., Shuangquan, S., Changqing, T., and X. Hongbo. Development and experimental investigation of a novel spray cooling system integrated in refrigeration circuit. *Appl Therm Eng*, 33–34 (2012): 246–252.
12. Mejia, A., He, P., Luo, D., Marquez, M., and Z. Cheng. Uniform discotic wax particles via electrospray emulsification. *J Colloid Interf Sci*, 334 (2009): 22–28.
13. Gavory, C., Abderrahmen, R., Bordes, C., Chaussy, D., Belgacem, M.N., Fessi, H., and S. Briançon. Encapsulation of a pressure sensitive adhesive by spray-cooling: Optimum formulation and processing conditions. *Adv Powder Technol*, 25 (2014): 292–300.
14. Ilic, I., Dreu, R., Burjak, M., Homar, M., Kerc, J., and S. Srcic. Microparticle size control and glimepiride microencapsulation using spray congealing technology. *Int J Pharm*, 381 (2009): 176–183.
15. Maschke, A., Becker, C., Eyrich, D., Kiermaier, J., Blunk, T., and A. Göpferich. Development of a spray congealing process for the preparation of insulin loaded lipid microparticles and characterization thereof. *Eur J Pharm Biopharm*, 65 (2007): 175–187.
16. Scott, M., Robinson, M., Pauls, J., and R. Lantz. Spray congealing: Particle size relationships using a centrifugal wheel atomizer. *J Pharm Sci*, 53 (1964): 670–675.
17. Cusimano, A. and C. Becker. Spray-congealed formulations of sulfaethylthiadiazole (SETD) and waxes for prolonged-release medication—Effect of wax. *J Pharm Sci*, 57 (1968): 1104–1112.
18. McClements, J.D. Advances in fabrication of emulsions with enhanced functionality using structural design principles. *Curr Opin Colloid Interface Sci*, 17 (2012): 235–245.
19. Matalanis, A., Owen, G.J., and J.D. McClements. Structured biopolymer-based delivery systems for encapsulation, protection and release of lipophilic compounds. *Food Hydrocolloid*, 25 (2011): 1865–1880.
20. Kanig, J. Properties of fused mannitol in compressed tablets. *J Pharm Sci*, 53 (1964): 188–192.
21. Sillick, M. and C.M. Gregson. Spray chill encapsulation of flavors within anhydrous erythritol crystals. *LWT—Food Sci Technol*, 48 (2012): 107–113.
22. Schubert, M.A., Müller-Goymann, C.C., and C. Christel. Characterisation of surface-modified solid lipid nanoparticles (SLN): Influence of lecithin and nonionic emulsifier. *Eur J Pharm Biopharm*, 61 (2005): 77–86.
23. Milanovic, J., Manojlovic, V., Levic, S., Rajic, N., Nedovic, V., and B. Bugarski. Microencapsulation of flavors in Carnauba wax. *Sensor*, 10(1) (2010): 901–912.
24. Milanovic, J., Levic, S., Manojlovic, V., Nedovic, V., and B. Bugarski. Carnauba wax microparticles produced by melt dispersion technique. *Chem Pap*, 65(2) (2011): 213–220.
25. Bugarski, B., Levic, S., Milanovic, J., Manojlovic, V., and V. Nedovic. Microencapsulation of flavours in Carnauba wax. *XVIIth International Conference on Bioencapsulation*, Groningen, Netherlands, (September 24–26, 2009) pp. 276–277.
26. Levic, S., Nedovic, V., Milivojevic, M., Manojlovic, V., and B. Bugarski. Microencapulation of flavours in Carnauba wax. In: Daran, J.-M. Ed., *Second European Yeast Falvour Workshop* [Cost Action FA0907], p. 5. Delft, The Netherlands: Programme and Abstract Book (May 26–27, 2011).
27. Mehnert, W. and K. Mader. Solid lipid nanoparticles production, characterization and applications. *Adv Drug Deliver Rev*, 47 (2001): 165–196.
28. Köhler, K., Hensel, A., Kraut, M., and H. Schuchmann. Melt emulsification—Is there a chance to produce particles without additives? *Particuology*, 9(5) (2011): 506–509.
29. Kumbhar, D. and V. Pokharkar. Physicochemical investigations on an engineered lipid–polymer hybrid nanoparticle containing a model hydrophilic active, zidovudine. *Colloid Surface A*, 436 (2013): 714–725.
30. Tian, B.-C., Zhang, W.-J., Xu, H.-M., Hao, M.-X., Liu, Y.-B., Yang, X.-G., Pan, W.-S., and X.-H. Liu. Further investigation of nanostructured lipid carriers as an ocular delivery system: In vivo transcorneal mechanism and in vitro release study. *Colloid Surface B*, 102 (2013): 251–256.

31. Scalia, S., Trotta, V., Traini, D., Young, P., Sticozzi, C., Cervellati, F., and G. Valacchi. Incorporation of quercetinin respirable lipid microparticles: Effect on stability and cellular uptake on A549 pulmonary alveolar epithelial cells. *Colloid Surface B*, 112 (2013): 322–329.
32. Zhang, W., Li, X., Ye, T., Chen, F., Sun, X., Kong, J., Yang, X., Pan, W., and S. Li. Design, characterization, and in vitro cellular inhibition and uptake of optimized genistein-loaded NLC for the prevention of posterior capsular opacification using response surface methodology. *Int J Pharm*, 454 (2013): 354–366.
33. Momoh, M.A. and C. Okechukwu Esimone. Phospholipon 90H (P90H)-based PEGylated microscopic lipospheres delivery system for gentamicin: An antibiotic evaluation. In: Momoh, M.A. and Esimone, C.O. Eds., *Asian Pac J Trop. Biomed*, 2(11) (2012): 889–894.
34. Passerini, N., Perissutti, B., Albertini, B., Franceschinis, E., Lenaz, D., Hasa, D., Locatelli, I., and D. Voinovich. A new approach to enhance oral bioavailability of Silybum Marianum dry extract: Association of mechanochemical activation and spray congealing. *Phytomedicine*, 19 (2012): 160–168.
35. Ribeiro, M., Morselli, A., Daniel, B., and C.F. Grosso. The effect of adding oleic acid in the production of stearic acid lipid microparticles with a hydrophilic core by a spray-cooling process. *Food Res Int*, 47 (2012): 38–44.
36. Di Sabatino, M., Albertini, B., Kett, V., and N. Passerini. Spray congealed lipid microparticles with high protein loading: Preparation and solid state characterisation. *Eur J Pharm Sci*, 46 (2012): 346–356.
37. Albertini, B., Passerini, N., Di Sabatino, M., Monti, D., Burgalassi, S., Chetoni, P., and L. Rodriguez. Poloxamer 407 microspheres for orotransmucosal drug delivery. Part I: Formulation, manufacturing and characterization. *Int J Pharm*, 399 (2010): 71–79.
38. Monti, D., Burgalassi, S., Rossato, M.S., Albertini, B., Passerini, N., Rodriguez, L., and P. Chetoni. Poloxamer 407 microspheres for orotransmucosal drug delivery. Part II: In vitro/in vivo evaluation. *Int J Pharm*, 400 (2010): 32–36.
39. Kheradmandnia, S., Vasheghani-Farahani, E., Nosrati, M., and F. Atyabi. Preparation and characterization of ketoprofen-loaded solid lipid nanoparticles made from beeswax and carnauba wax. *Nanomed Nanotechnol Biol Med*, 6 (2010): 753–759.
40. Albertini, B., Passerini, N., Sabatino, M., Di Franco, P., Vitali, B., Brigidi, P., and L. Rodriguez. Polymer–lipid based mucoadhesive microspheres prepared by spray-congealing for the vaginal delivery of econazole nitrate. *Eur J Pharm Sci*, 36 (2009): 591–601.
41. Shen, J., Wang, Y., Ping, Q., Xiao, Y., and X. Huang. Mucoadhesive effect of thiolated PEG stearate and its modified NLC for ocular drug delivery. *J Control Release*, 137 (2009): 217–223.
42. Yuan, H., Wang, L.-L., Du, Y.-Z., You, J., Hu, F.-Q., and S. Zeng. Preparation and characteristics of nanostructured lipid carriers for control-releasing progesterone by melt-emulsification. *Colloid Surface B*, 60 (2007): 174–179.
43. Passerini, N., Albertini, B., Perissutti, B., and L. Rodriguez. Evaluation of melt granulation and ultrasonic spray congealing as techniques to enhance the dissolution of praziquantel. *Int J Pharm*, 318 (2006): 92–102.
44. Villalobos-Hernández, J.R. and C.C. Müller-Goymann. Physical stability, centrifugation tests, and entrapment efficiency studies of carnauba wax–decyl oleate nanoparticles used for the dispersion of inorganic sunscreens in aqueous media. *Eur J Pharm Biopharm*, 63 (2006): 115–127.
45. Albertini, B., Passerini, N., González-Rodríguez, M.L., Perissutti, B., and L. Rodriguez. Effect of Aerosil® on the properties of lipid controlled release microparticles. *J Control Release*, 100 (2004): 233–246.
46. Kamble, R., Maheshwari, M., Paradkar, A., and S. Kadam. Melt solidification technique: Incorporation of higher wax content in ibuprofen beads. *AAPS Pharm Sci Technol*, 5(4) (2004): 75–83.
47. Paradkar, A., Maheshwari, M., Amit Kumar, T., Chauhan, B., and S.S. Kadam. Preparation and characterization of flurbiprofen beads by melt solidification technique. *AAPS Pharm Sci Technol*, 4(4) (2003): 514–522.
48. Passerini, N., Perissutti, B., Albertini, B., Voinovich, D., Moneghini, M., and L. Rodriguez. Controlled release of verapamil hydrochloride from waxy microparticles prepared by spray congealing. *J Control Release*, 88 (2003): 263–275.
49. Savolainen, M., Herder, J., Khoo, C., Lövqvist, K., Dahlqvist, C., Glad, H., and A.M. Juppo. Evaluation of polar lipid-hydrophilic polymer microparticles. *Int J Pharm*, 262 (2003): 47–62.
50. Savolainen, M., Khoo, C., Glad, H., Dahlqvist, C., and A.M. Juppo. Evaluation of controlled-release polar lipid microparticles. *Int J Pharm*, 244 (2002): 151–161.
51. Reithmeier, H., Herrmann, J., and A. Göpferich. Development and characterization of lipid microparticles as a drug carrier for somatostatin. *Int J Pharm*, 218 (2001): 133–143.

52. Jenning, V. and S. Gohla. Comparison of wax and glyceride solid lipid nanoparticles (SLN®). *Int J Pharm*, 196 (2000): 219–222.
53. Turton, R. and X.X. Cheng. Cooling processes and congealing. In: Swarbrick, J. Ed., *Encyclopedia of Pharmaceutical Technology*, Vol. 2, pp. 61–774. New York: Informa Healthcare USA Inc. (2007).
54. Wandrey, C., Bartkowiak, A., and S.E. Harding. Materials for encapsulation. In: Zuidam, N.J. and Viktor, A. Nedovic, Eds., *Encapsulation Technologies for Active Food Ingredients and Food Processing*, pp. 31–101. New York: Springer (2010).
55. Kalepun, S., Manthina, M., and V. Padavala. Oral lipid-based drug delivery systems—An overview. *Acta Pharm Sinic B*, 3(6) (2013): 361–372.
56. Hwang, K.T., Cuppett, S.L., Weller, C.L., and M.A. Hanna. Properties, composition, and analysis of grain sorghum wax. *J Am Oil Chem Soc*, 79(6) (2002): 521–527.
57. European Food Safety Authority (EFSA). Scientific opinion on the re-evaluation of candelilla wax (E 902) as a food additive. *EFSA J*, 10(11) (2012): 2946.
58. Zajc, N., Obreza, A., Bele, M., and S. Srcic. Physical properties and dissolution behaviour of nifedipine/mannitol solid dispersions prepared by hot melt method. *Int J Pharm*, 291 (2005): 51–58.
59. Lopes, J., Nunes, S.C., Ramos, S., Matos, B., and J.S. Redinha. Erythritol: Crystal growth from the melt. *Int J Pharm*, 388 (2010): 129–135.
60. Ohmori, S., Ohno, Y., Makino, T., and T. Kashihara. Characteristics of erythritol and formulation of a novel coating with erythritol termed thin-layer sugarless coating. *Int J Pharm*, 278 (2004): 447–457.
61. Endo, K., Amikawa, S., Matsumoto, A., Sahashi, N., and S. Onoue. Erythritol-based dry powder of glucagon for pulmonary administration. *Int J Pharm*, 290 (2005): 63–71.
62. Gonnissen, Y., Remon, J.P., and C. Vervaet. Development of directly compressible powders via co-spray drying. *Eur J Pharm Biopharm*, 67 (2007): 220–226.
63. Sekiguchi, K. and N. Obi. Studies on absorption of eutectic mixtures. I. A comparison of the behaviour of eutectic mixture of sulfathiazole and that of ordinary sulfathiazoleinman. *Chem Pharm Bull*, 9 (1961): 866–872.
64. Keen, J., McGinity, J., and R. Williams. Enhancing bioavailability through thermal processing. *Int J Pharm*, 450 (2013): 185–196.
65. Stojakovic, D., Bugarski, B., and N. Rajic. A kinetic study of the release of vanillin encapsulated in Carnauba wax microcapsules. *J Food Eng*, 109 (2012): 640–642.
66. Rodriguez, L., Passerini, N., Cavallari, C., Cini, M., Sancin, P., and A. Fini. Description and preliminary evaluation of a new ultrasonic atomizer for spray-congealing processes. *Int J Pharm*, 183(2) (1999): 133–143.

Section III

Ingredients

Section III

Ingredients

26 Materials of Natural Origin for Encapsulation

Munmaya K. Mishra

CONTENTS

26.1 Carbohydrate Polymers 493
26.2 Proteins 503
26.3 Lipids 506
26.4 Strategies for the Selection of Materials..508
26.5 Conclusions..508
References..509

As the encapsulation technologies advancing rapidly, it will be a daunting task to include a comprehensive list of materials, which can be used to entrap, coat, or encapsulate solids, liquids, or gases. Indeed, there are a multitude of substances (of different types, origins, and properties), including natural or synthetic origin available, which can be used for the purpose. However, only a limited number thereof have been approved for food, drug, and cosmetics applications (the regulations for food additives are much stricter than for pharmaceuticals or cosmetics). In other words, the materials suitable for drug encapsulation may not be applicable for food industry and *vice versa*. This chapter hopes to provide a brief introduction of different class of materials (mostly natural origin) that can be considered for encapsulating active ingredients.

Although there are various materials available for encapsulation and so as technologies, the challenges do exist concerning the selection of appropriate microencapsulation technique and encapsulation material. The cost consideration of materials for food applications need to be taken into account unlike the pharmaceutical industry, which can tolerate high costs. The majority of materials used for microencapsulation in the food sector are bio-based materials such as carbohydrate polymers (polysaccharides), proteins, lipids, etc.

Polysaccharides studied as a matrix for microencapsulation are starches,[1,2] maltodextrins,[3–5] gums arabic,[6,7] pectins,[8,9] chitosans,[10,11] and alginates[12–14] (Figure 26.1). The major advantages of these biopolymers are their good solubility in water. Often carbohydrates are mixed with proteins[15–21] to improve the emulsifying and film-forming properties during microencapsulation.

26.1 CARBOHYDRATE POLYMERS

These natural homo- and copolymers (such as starch and cellulose) are composed of sugar residues and/or their derivatives. The names carbohydrate polymer and polysaccharide refer to the chemical structure. Carbohydrate polymers are also designated as gum or hydrocolloid, which refers to the property that these polysaccharides hydrate in hot or cold water to form viscous solutions or dispersions at low concentration. The gums/hydrocolloids may be harvested from nature or obtained by the chemical modification of native polysaccharides.

Examples of carbohydrate polymers from natural sources and the structural characteristics are presented in Schemes 26.1 and 26.2. Also, the chemical structures of few carbohydrates are presented as well.

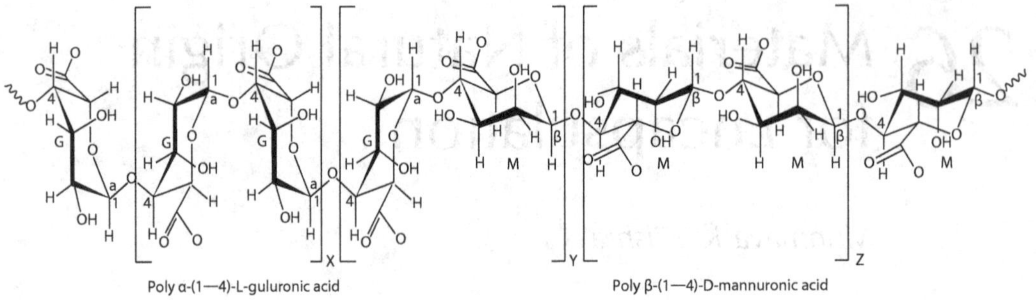

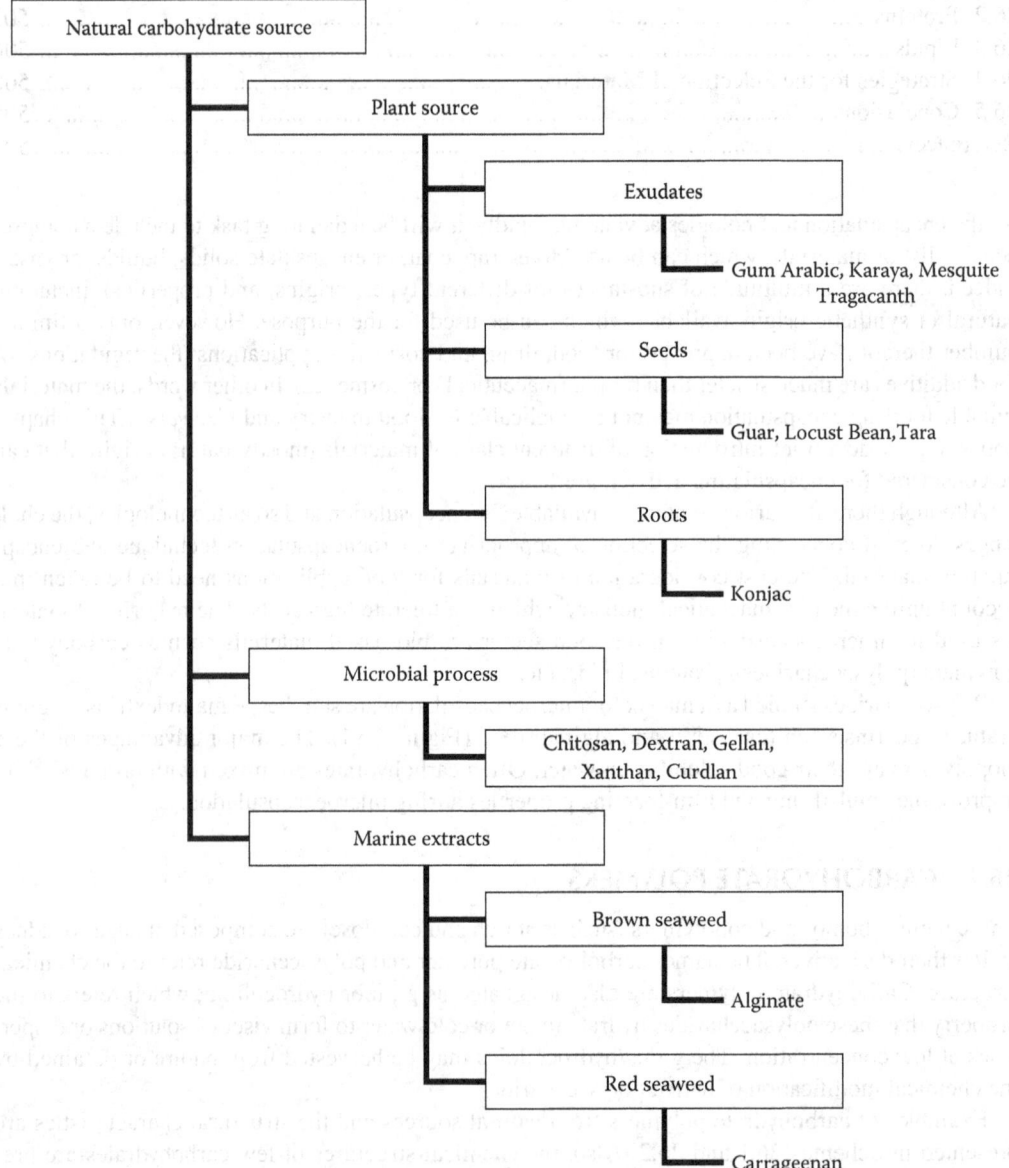

FIGURE 26.1 Typical structure of Alginate polymer.

SCHEME 26.1 Example of carbohydrate polymers from natural sources.

Materials of Natural Origin for Encapsulation

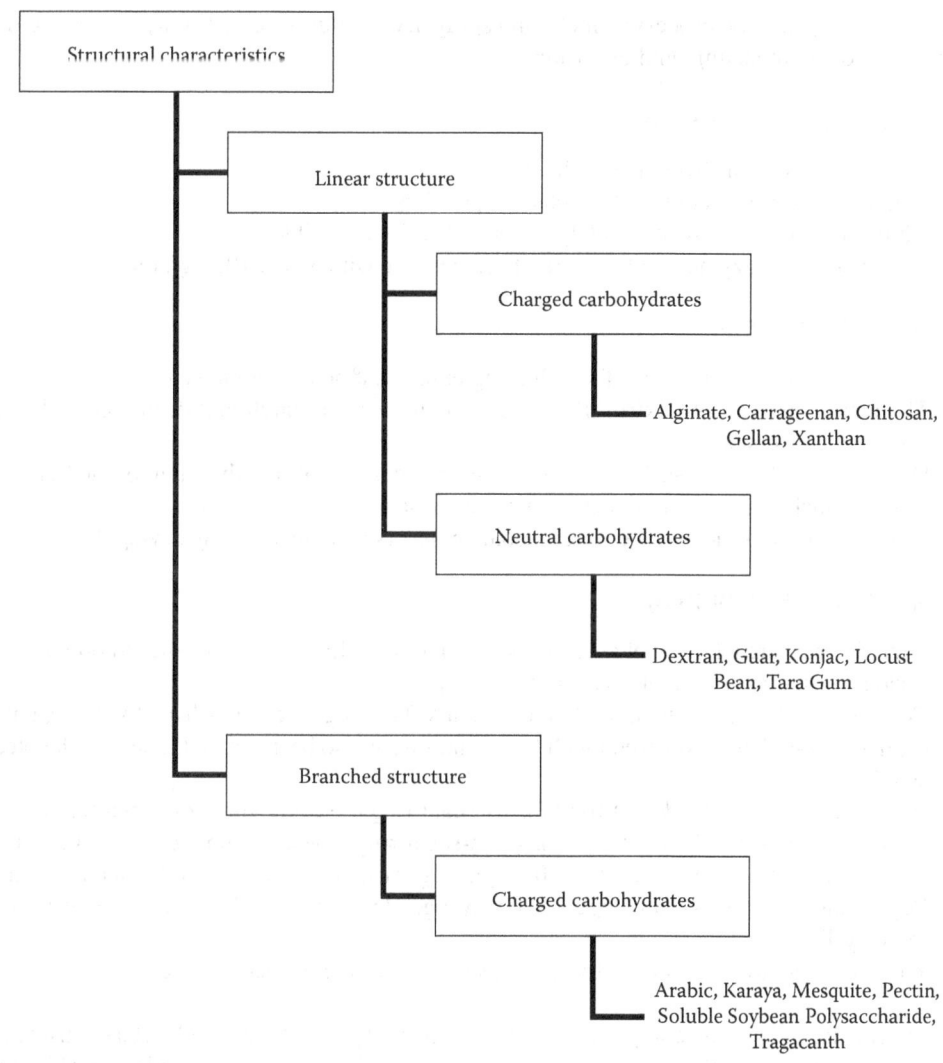

SCHEME 26.2 Structural characteristics of carbohydrate polymers.

Starch and Starch Derivatives: Starch is normally a white powder (mostly odorless and tasteless), insoluble in cold water, ethanol, and most common solvents. The starch is a polymer of α-d-glucose with the general chemical composition $(C_6H_{10}O_5)n$ and consists of two architecturally different polysaccharide molecules (the linear amylose and the highly branched amylopectin) but having similar molecular composition. The starch granules swell, lose crystallinity, and leach amylose as they absorb water. The higher the amylose content, the lower is the swelling power and the lower is the gel strength for the same starch concentration.[22,23] Both the amylose and amylopectin are complex macromolecules that form ordered structures[24] by intra- and intermolecular interactions. Starch can be modified[25,26] by chemical (acid hydrolysis),[27,28] biochemical (using enzymes), and physical means to produce functional derivatives, including cross-linked, oxidized, acetylated, hydroxypropylated, partially hydrolyzed molecules, etc. The resulting derivatives are assigned a Dextrose Equivalent (DE) value, which represents the degree of hydrolysis. The modification process alters the structure and affects the hydrogen bonding in order to enhance and extend the applicability for various uses. Modified starch and maltodextrin have been used for coating and encapsulation processes, including spray drying, fluidized bed spray drying, fluidized bed granulation, compacting, and extrusion.[29]

Maltodextrins (CAS# 9050-36-6):

- Maltodextrins are hydrolysates with a DE < 20.
- Creamy white hygroscopic polysaccharide powders.
- Almost tasteless or only moderately sweet and easily digestible.
- It is an acid- or enzyme-catalyzed starch hydrolysate with $Mw < 4000$ g/mol.

Dextrins (CAS# 9004-53-9):

- Produced by the degradation of starch using either heat or acid or enzyme.
- The tensile strength of dextrin film is lower than that for starch and decreases with the degree of conversion.
- Dextrin formulations can be prepared at higher concentrations than unmodified starch yielding thicker films with higher proportions of solids.
- Dextrin is used to encapsulate water-insoluble flavors and oils by spray drying.[26]

Cyclodextrin (CAS# 12619-70-4):

- Cyclodextrins are a family of cyclic oligosaccharides, which are non-branched oligomeric cycloamyloses composed of α-(1→4) d-glucopyranoside units.
- Typical cyclodextrins contain 6–8 glucose units. The rings are cone-shaped with a cavity depth of 0.7–0.8 nm. However, much larger rings have also been derived. They are denoted as follows:[30]
  - α-cyclodextrin (CAS# 10016-20-3): six-sugar ring molecule, inner diameter 0.5 nm
  - β-cyclodextrin (CAS# 7585-39-9): seven-sugar ring molecule, inner diameter 0.6 nm
  - γ-cyclodextrin (CAS# 17465-86-0): eight-sugar ring molecule; inner diameter 0.8 nm
- Degradation of amylose, amylopectin, or glycogen by diluted acids or by enzymes yields these cyclic dextrins.
- The ring structure results from the helical segments of the polysaccharide.

Cyclodextrins form inclusion complexes,[31] and their solubility in water at 25°C differs considerably, with γ-cyclodextrin being the best soluble (23.2 g/100 mL) followed by α-cyclodextrin (14.5 g/100 mL) and β-cyclodextrin (1.85 g/100 mL).[28]

Cellulose and Cellulose Derivatives (CAS# 9004-39-1): Cellulose is insoluble in water and other ordinary solvents. Cellulose is much more crystalline than starch, and it is a stiff polymer with an extended rod-like conformation. Cellulose becomes amorphous in water around 320°C and 25 MPa, whereas starch undergoes a crystalline to amorphous transition at 60°C–70°C in water. Cellulose[32,33] is a polymer of β-d-glucose and the chain units are linked by β-(1→4)-glycosidic bonds. Microcrystalline cellulose is widely used for matrix encapsulation and is composed of 100–300 glucose units. The hydroxyl groups of cellulose can be partially or completely modified by reacting with various reagents to produce derivatives with desired properties for various applications. The properties of modified cellulose depend on the substituent type and distribution along the polymer backbone, the molecular weight, and distribution, etc. In general, aqueous solutions of modified celluloses are odorless, colorless, and clear. Cellulose ethers are the most important cellulose derivatives for food, drug, and coating applications.

Methylcellulose (MC):

- It is a hydrophilic white powder soluble in cold water forming a clear viscous solution. The solution of MC exhibit pseudoplastic non-thixotropic flow properties.

- The solubility decreases with higher degree of substitution (DS) and the degree of polymerization.
- The viscosity of MC solutions is reasonably stable over a wide pH range (pH 3–11).
- Three-dimensional gel formation occurs on heating above 50°C (MC of higher DS has a lower gelation temperature).
- MC has good film-forming properties, which is very useful for coating.

Hydroxypropyl Methyl Cellulose:

- It is very similar to MC, a white to off-white powder or granules, which swell and dissolve in water. The DS influences the solubility.
- Typically, hydroxypropyl methyl cellulose (HPMC) has a molar mass higher than 10,000 g/mol, $T_m = 220°C$, and a density of 1.6 g/mL.[34]
- The solutions of HPMC exhibit pseudoplastic non-thixotropic flow properties.
- HPMC is also soluble in most polar solvents.
- Aqueous solutions are surface active and form films upon drying.
- HPMC undergoes reversible thermal gelation (transformation from sol to gel upon heating and cooling). The gel texture also changes with increasing hydroxypropyl substitution.

Hydroxypropyl Cellulose:

- This nonionic cellulose ether is soluble in cold water, ethanol, and mixtures of ethanol and water.
- Hydroxypropyl cellulose (HPC) possesses good film-forming properties, and high surface activity compared to most other hydrocolloids.[35] The films are flexible, glossy, and nontacky.[36]
- HPC is compatible with a number of high molecular weight, high-boiling waxes, and oils and can, therefore, be used to modify the properties of these materials.
- HPC is compatible with most water-soluble gums and resins, and yields homogeneous solutions with MC, hydroxyethyl cellulose (HEC), carboxymethyl cellulose (CMC), guar, alginate, and locust bean gum, gelatin, sodium caseinate, poly(ethylene oxide), carbowax, etc. (Figure 26.2)
- A synergistic viscosity increase is observed with a solution containing HPC and anionic polymers such as CMC and alginate. On the other hand a lower than expected viscosity is observed[36] with a solution containing HPC and nonionic polymers such as HEC, MC, and guar gum.

Ethylcellulose and Ethyl Methylcellulose:

- Ethylcellulose (EC) ($T_m = 135°C$) is water insoluble used for coating and controlled release applications.
- Ethyl methylcellulose (EMC) behaves like MC and HPMC and is soluble in cold water and forms gels on heating; however, the gels are comparably weak.

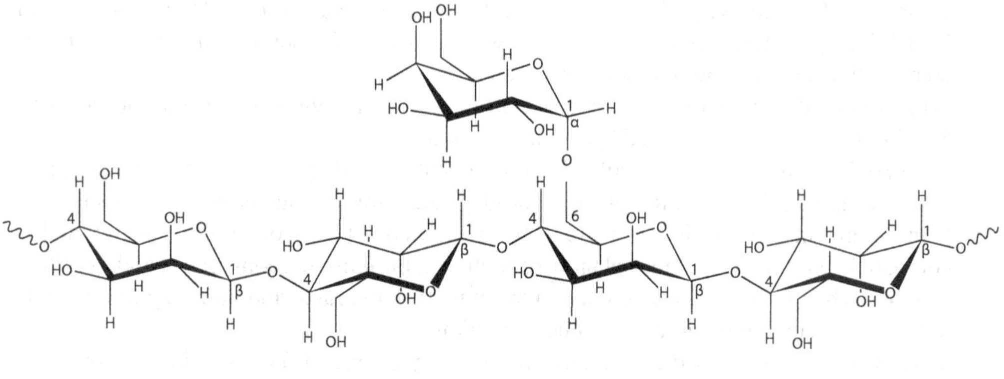

FIGURE 26.2 Typical structure of Locust Bean Gum.

Sodium Carboxymethyl Cellulose:

- The purified product is a tasteless, odorless free-flowing powder with white to off-white color.
- CMC is an anionic linear polyelectrolyte, soluble in both cold and hot water, and the molecular conformation in aqueous solution strongly depends on the concentration, ionic strength, and pH. The solution viscosity decreases during heating. Nevertheless, the solution behavior strongly depends on the molar mass and the DS.
- At high pH and low concentrations, the CMC polymer chains are most extended. However, the coiling of CMC polymer chains occurs with increasing concentration and decreasing pH.
- Thermo-reversible gels are formed in water at concentrations above the overlap concentration.
- Gelation and precipitation occurs with multivalent cations and with some polycations. However, solubilization was observed with proteins (soy protein, caseinate)[37] in the isoelectric range. At low pH, CMC may form cross-links through lactonization between carboxylic acid and free hydroxyl groups.[38]
- The lower substituted CMCs are hydrophobic and show thixotropic behavior; however, the higher substituted ones are more hydrophilic and exhibit pseudoplastic behavior.

Plant Exudates: Polysaccharide plant exudates (products of gummosis or gums from plant) are complex macromolecular substances consisting mixtures of oligomers and polymers of different chemical structure and/or chain architecture. Gums are produced in response to stress conditions such as injury of the plant. Only a few plant species (Leguminosae family) are cultivated at present to provide gums for the encapsulation and food use. Some examples of gums from plants are gum arabic (gum senegal), gum tragacanth, guar gum, and locust bean gum.

Gum Arabic (CAS# 9000-01-5):

- Gum arabic (also known as gum acacia) is a natural plant gum, a dried exudate obtained from the stems and branches of *Acacia senegal* or *Acacia seyal* or other *Acacia* species.[39-41]
- Gum arabic is odorless, colorless, and tasteless and is highly soluble in water and dissolves in both cold and hot water with concentrations up to 50 wt%. The solutions exhibit Newtonian behavior at concentrations up to 40 wt% and become pseudoplastic at higher concentration.
- Gum arabic is a complex mixture of arabinogalactan oligosaccharides, polysaccharides, and glycoproteins. The chemical composition of the mixture can vary with the source, climate, and other factors.
- The backbone consists of β-(1→3)-linked d-galactopyranosyl units. The side chains are composed of two to five β-(1→3)-linked d-galactopyranosyl units, joined to the main chain by 1,6-linkages. Because it is a mixture, which varies with source, the exact chemical composition and molecular structures are unclear.
- The gum Arabic type, pH, and ionic strength influence the viscosity of solution in water. Maximum viscosity is achieved between pH 6 and 7.
- Gum Arabic acts as protective colloid and excellent emulsifier and the molecular aggregation can cause both shear thinning and time-dependent thickening behavior at low shear.[42]
- Gum Arabic has the ability to create a strong protective film around oil droplets[43] and is compatible with the most other plant hydrocolloids, proteins, carbohydrates, and modified starches. The viscosity of a solution of a mixture of gum arabic and gum tragacanth tends to be lower than that of either constituent solution.
- When Gum Arabic is used as a fixative in the spray drying of flavors, it forms a thin and impenetrable film around the flavor particle.[44,45]

Gum Tragacanth (CAS# 9000-65-1):

- Gum tragacanth is a dried exudate obtained from the stems and branches of *Astragalus gummifer* and other Asiatic species of *Astragalus*.[46]
- Gum tragacanth is one of the most acid-resistant gums. It acts as an emulsifier for stabilizing oil-in-water emulsions and since it increases the viscosity of the aqueous phase, it is considered as a bifunctional emulsifier.
- Gum tragacanth forms viscous aqueous solutions even at low concentration, which exhibits pseudoplastic flow properties. It forms a protective film around oil droplets.[47]
- Gum tragacanth is a complex mixture of highly branched, heterogeneous polysaccharides.[48] It occurs naturally as slightly acidic calcium, magnesium, or potassium salt.
- Gum tragacanth consists of two components:[49] (1) a water-swellable component and (2) a water-soluble component, a colloidal hydrosol. The ratio of the water swellable to the water-soluble fraction varies depending on the source.
- The main chain is formed by (1→4)-linked d-galactose residues with side chains of d-xylose units attached to the main chain by (1→3)-linkages.
- Gum tragacanth forms viscous aqueous solutions even at low concentration, which exhibits pseudoplastic flow properties. It forms a protective film around oil droplets.[47]

Gum Karaya (CAS# 9000-36-6):

- Gum Karaya (also known as Sterculia gum) is a dried exudate.[50]
- Powdered gum karaya is white to grayish white and one of the least soluble of the exudate gums (only ~10% of the native gum solubilizes in cold water, increasing to 30% in hot water. After deacetylation, 90% dissolves in water.[41]
- Gum Karaya[51] is a complex, partially acetylated polysaccharide with a branched structure obtained as a calcium and magnesium salt.
- The backbone consists of α-d-galacturonic acid and α-l-rhamnose residues. Side chains are attached by (1→2)-linkage of β-d-galactose or by (1→3)-linkage of β-d-glucuronic acid to the galacturonic acid of the main chain. Furthermore, half of the rhamnose residues of the main chain are (1→4)-linked to β-d-galactose units.[52]
- Gum karaya is compatible with other plant hydrocolloids, proteins, and carbohydrates.

Mesquite Gum:

- This gum is obtained from the mesquite tree or shrub.[53]
- Mesquite gum has a good level of solubility and solutions of up to 50% can be prepared (the color of the solution is brown).
- Mesquite gum is the neutral salt of a complex acidic branched polysaccharide. Its backbone is formed by (1→3)-linked β-d-galactose residues with (1→6)-linked branches, bearing l-arabinose (pyranose and furanose ring forms), l-rhamnose, β-d-glucuronate, and 4-*O*-methyl β-d-glucuronate as single sugar or oligosaccharide side chains.[54-59]
- Mesquite gum has good film formation properties.[60]

Galactomannans:

- Guar gum (CAS# 9000-30-0, 9000-30-3, 9066-07-3), Locust bean gum (LBG)[61] (CAS# 9000-40-2) (also known as carob bean gum), and tara gum (CAS# 39300-88-4) belongs to the family of galactomannans.
- Guar gum, LBG, and tara gum are isolated from guar plant, the endosperm of the seeds of the carob tree, the tara shrub, respectively.

- They consist of linearly (1→4)-linked β-d-mannopyranosyl units with single α- d-galactopyranosyl units connected by (1→6) linkages as side branches.
- The ratio of d-mannosyl to d-galactosyl varies for three gum types.
- The three gum types differ in their solubility and the solubility increases with increasing number of side units.
- The solubility at ambient temperature for three gums is in the order guar gum > tara gum > LBG. Under identical conditions guar gum is fully water soluble at ambient temperature whereas LBG has very limited solubility and Tara gum is only 70% soluble. However, LBG and tara gum is completely soluble above 60°C and 70°C respectively.[62,63]
- The solutions of all three gum types exhibit pseudoplastic behavior.

Pectins (CAS# 9000-69-5):

- Pectins are isolated from a variety of sources, including the citrus peel, orange peel, and residues from the extraction of citrus juice, apple juice, etc.
- Pectins[64,65] are high molecular weight polysaccharides with at least 65 wt% of α-(1→4)-linked d-galacturonic acid-based units, which may exist as free acid, salt (calcium, sodium, potassium, ammonium, etc.), naturally esterified with methanol, or as acid amide in amidated pectins.
- Pectins can differ by the degree of esterification (in the range of 20%–80%) of the carboxyl groups of the galacturonic acid. Pectins with less than 50% ester groups are designated as low-esterified pectins (low methoxylated) whereas more than 50% esterification are designated as high-esterified pectins (HM, high methoxylated). The gelation depends on the degree of esterification.
- Pectins are insoluble in most organic solvents but, soluble in water up to 12% concentrations depending on the pectin type. The rheological behavior is close to Newtonian behavior at low concentrations but became pseudoplastic at higher concentration.[66]
- Pectins are negatively charged macromolecules and behave as polyelectrolytes. The solution properties (gelation behavior) depend on the degree of esterification and are sensitive to pH and the presence of cations in the solution. Pectin solutions are most stable at pH 3–4.[67]
- HM pectins gel only in the presence of sugars or other cosolutes in the low pH range, where the acid groups are not completely ionized. The gel strength increases with decreasing pH.
- Pectins are widely used as shell materials for encapsulation.

Soluble Soybean Polysaccharide:

- Soluble soybean polysaccharide (SSP) is extracted from a variety of sources including soybean.[68]
- SSP is considered as an anionic polyelectrolyte and soluble in both cold and hot water.
- SSP is mainly composed of galactose, arabinose, and galacturonic acid, but also contains many sugars at low levels.[69]
- SSP solutions do not gel and the solution viscosity is not very much affected by heat, salt, or by acidic pH variation.[70]
- SSP has the ability to stabilize protein particles at low pH without increasing the viscosity, and it can be used as an emulsifier/stabilizer for emulsions.
- The adhesive and film-forming properties of SSP are excellent. The films are generally colorless, transparent, and suitable for coating applications.

Marine Extracts: Seaweed provides a number of polysaccharides for many applications, including coating, encapsulation, etc.

Carrageenans (CAS# 9000-07-1):

- Carrageenans are anionic polyelectrolytes prepared from red seaweed. There are three types k- (kappa), i- (iota), and l- (lambda) carrageenan commercially available (Figure 26.3).
- Carrageenans[71–74] are a family of high molecular weight sulfated polysaccharides. The structures strongly depend on the source and conditions during extraction and purification[75] process.
- The polymer chains comprise alternating (1→3)-linked β-d-galactopyranosyl and (1→4)-linked α-d-galactopyranosyl units.
- Carrageenans are highly polydisperse[72] and exhibit a wide spectrum of the rheological behavior.[76] They can form viscous solutions and also thermally reversible gels with a texture varying from soft and elastic to firm and brittle. l-Carrageenan is non-gelling but is the type with the highest anionic charge density, which causes the best solubility, extended chain conformation, and strong electrostatic interaction in solution.[77,78]
- k- and i-carrageenans have the ability to form elastic gels in the presence of certain cations such as K^+ and Ca^{2+}.
- Synergistic effects are obtained by combining carrageenans with other gum types such as brittle k-carrageenan gels may be softened with locust bean gum.

Alginate (CAS 9005-38-3 Na):

- Alginates are quite abundant in nature and sodium alginates are generally produced from marine brown algae. They may also be synthesized as an exocellular material by some bacteria.[79,80]
- Alginate degrades even as a dry powder and, in particular as a solution. However, sodium alginate may be kept in a freezer for several years without significant degradation.
- Alginate[81–84] is a family of linear anionic polysaccharides, which can be considered as copolymers of (1→4) linked α-l-glucuronic acid (G) and β-d-mannuronic acid (M) residues.
- The functional properties of alginates are dependent on composition, structure, and molecular weight.
- Sodium alginate is an example of a water-soluble alginate, and the solubility is related to the rate of dissociation and the type of the counterion.

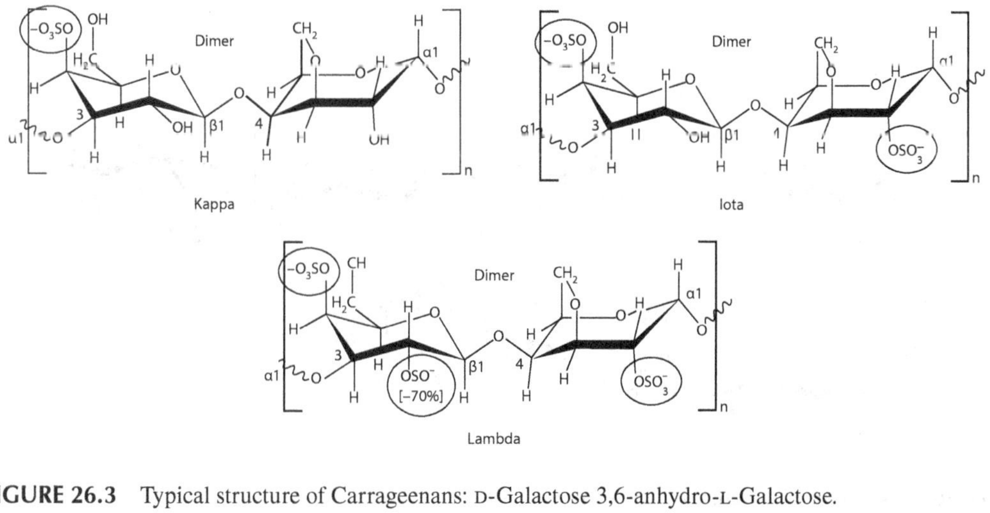

FIGURE 26.3 Typical structure of Carrageenans: d-Galactose 3,6-anhydro-l-Galactose.

- The rheological behavior of sodium alginate solutions depends on the concentration and the shear rate. The Newtonian behavior and pseudoplastic behavior is observed at low shear rate and high sear rate, respectively.[81] Higher concentrated solutions exhibit pseudoplastic flow even at low shear rates.
- This high viscosity observed at relatively low alginate concentration is due to the intramolecular electrostatic repulsion between the neighboring negative charges of each monomer unit that forces alginate molecules into an extended random coil conformation.[85]
- Alginates form ionotropic gels via selective binding of cations, and it forms gel-like networks with the polycation chitosan.

Polysaccharides via Microbial Process: Polysaccharides produced by bacteria using biotechnology approach are biopolymers with novel and unique functional properties (such as xanthan, gellan, and curdlan).

Xanthan (CAS# 11138-66-2):

- Xanthan is nongelling, hydrates rapidly, and soluble in cold water (Figure 26.4). The solutions show a very pronounced pseudoplastic behavior. The viscosity decreases under shear stress but recovers after removal of the shear stress. The solution viscosity is normally stable over a wide range of pH (2–12) and temperature.
- Xanthan is a high molar weight anionic polyelectrolyte occurs as a mixed salt of sodium, potassium, and calcium. Its backbone consists of β-(1→4)-d-glucopyranosyl units with every second unit having a tri-saccharide side chain attached at the C-3 position, one d-glucuronosyl unit between two d-mannosyl units.
- Xanthan undergoes cryogelation.[86]
- Interaction of xanthan with guar gum in solution leads to enhanced viscosity, whereas elastic thermo-reversible gels are obtained with LBG and konjac mannan soft[87] (Figure 26.5). Since xanthan may contain cellulases, it cannot be used with cellulose derivatives.

Gellan (CAS# 71010-52-1):

- Gellan[88] is a high molecular weight anionic polyelectrolyte. The solubility and solution properties depend on the DS and the type and concentration of ions present in the solution.
- Thermo-reversible gelation takes place upon cooling of solutions in the presence and absence of gelling cations for low acyl and high acyl types, respectively.

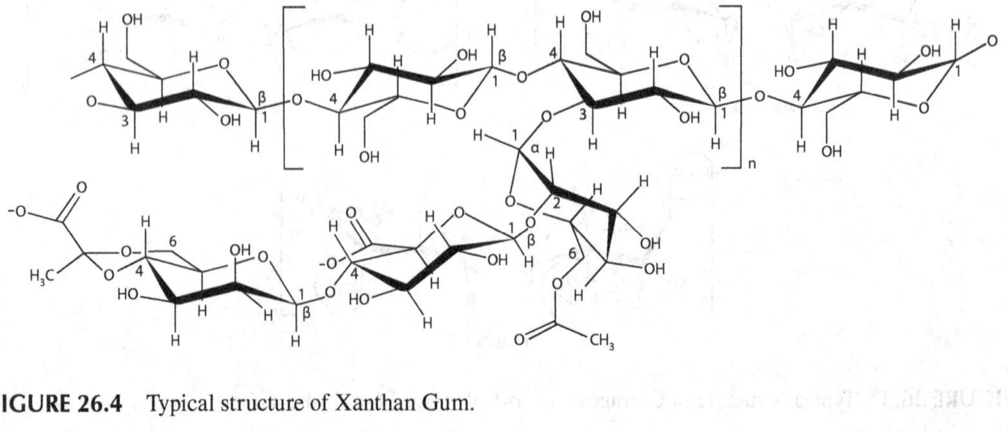

FIGURE 26.4 Typical structure of Xanthan Gum.

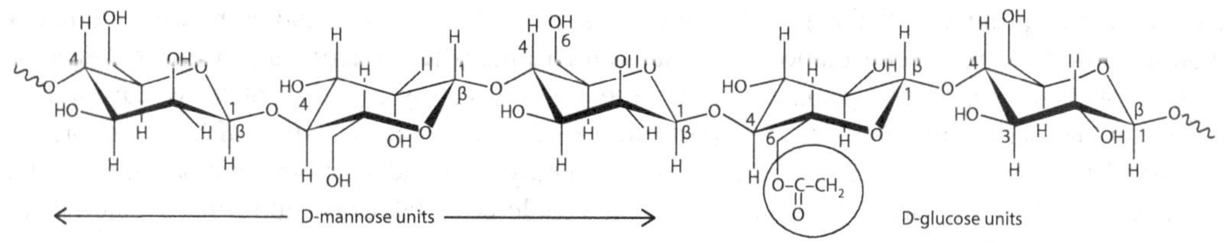

Materials of Natural Origin for Encapsulation

FIGURE 26.5 Typical structure of Konjac.

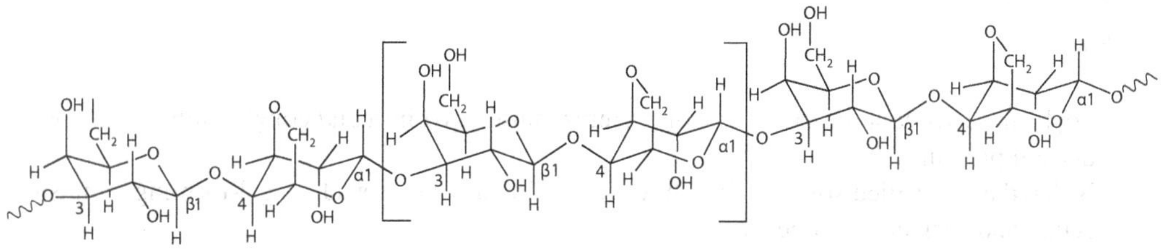

FIGURE 26.6 Typical structure of Agar.

- The resulting gel texture depends on the degree of acetylation. Low acyl types form hard, nonelastic brittle gels, whereas high acyl types yield soft, elastic, transparent, and flexible gels (Figure 26.6).

Dextran (CAS# 9004-54-0):

- Microbial fermentation processes of sucrose yield dextrans, which are mainly linear neutral polymers of α-d-glucose linked by α-(1→6) glycosidic bonds (may include variable amounts of α-(1→3) branches).
- Dextrans are soluble in water with non-soluble portions result from higher degree of branching.

Chitosan (CAS# 9012-76-4):

- Chitosan[89–91] is a linear polysaccharide, which can be considered as a copolymer consisting of randomly distributed β(1→4) linked d-glucosamine and N-acetyl-d-glucosamine.
- Chitosan is a nonpermanently charged cationic polyelectrolyte and soluble in acidic to neutral medium.
- Chitosan forms gels with tripolyphosphate and alginate.
- It has a very good film-forming ability.[92]

26.2 PROTEINS

Proteins are natural macromolecules composed of linear chains of amino acids and can be used in variety of applications including encapsulation. There are several types of proteins available. Proteins extracted from animal derived products (whey proteins, gelatin, and casein) and from vegetables (soy proteins, pea proteins, and cereal proteins) are widely used for encapsulation of active substances. Proteins present several advantages such as biocompatibility, biodegradability, good water solubility, emulsifying, and foaming capacity. The use of vegetable proteins as wall-forming materials in microencapsulation is a growing trend in the pharmaceutical, cosmetics, and food industries. The vegetable sourced proteins are known to be less allergenic compared to animal derived proteins.[93,94] The proteins of cereals (oat, wheat, barley, and corn) were studied as wall material for microencapsulation[16,95,96] due to their interesting functional properties. Vegetable proteins widely used as encapsulants are pea protein isolate, soy protein isolate, wheat gliadins, corn zein and barley protein.[97–100] They perform well for microencapsulation of hydrophobic and hydrophilic compounds alone, as well as mixed with polysaccharides or synthetic polymers. Other proteins extracted from rice, oat, or sunflower seeds could be suitable as wall forming materials for microencapsulation.

Examples of microencapsulation using proteins are provided in Table 26.1.

Vegetable Sourced Protein

Soy Proteins:

- Soybean seeds contain 35%–40% of proteins mainly glycinin and conglycinin (50%–90% of total proteins).[116]
- Isolated and purified soy proteins show interesting attributes such as gel-forming, emulsifying, and surfactant properties.[117]
- The solubility of proteins is strongly dependent on pH, heat treatment, and the presence and concentration of salts or other ingredients (oil, carbohydrate, and surfactant).
- Soybean proteins can be used as wall forming materials for masking the undesirable taste of some nutritional additives (bioactive compounds) such as casein hydrolysate)[17,102,104,118] or to protect components sensitive to oxidation and/or volatile aromas (orange oil).[6,119,120]

TABLE 26.1
Examples of Microencapsulation Using Proteins

Process	Encapsulate	Active-Ingredient	References
Spray drying	SPI	Orange oil	[6]
Spray drying	SPI/maltodextrin	Phospholipid	[21]
Spray drying	SPI	Flavors	[101]
Spray drying	SPI	Casein hydrolysate	[102]
Spray drying	SPI	Paprika oleoresin	[103]
Spray drying	SPI/gelatin	Casein hydrolysate	[104]
Spray drying	SPI	α-tocopherol	[105]
Spray drying	Pea proteins/Maltodextrin	Ascorbic acid	[18,19]
Spray drying	Pea proteins/Maltodextrin	α-tocopherol	[20]
Spray drying	Barley Protein	Fish oil	[96,98]
Coacervation	SPI	Fish oil	[106]
Coacervation	Soy glycinin	Hexadecane	[107]
Coacervation	SPI/Pectin	Casein Hydrolysate	[17]
Coacervation	SPI/Pectin	Propolis	[108]
Coacervation	SPI/Gum Arabic	Orange oil	[109]
Coacervation	Pea globulins/gum Arabic/CMC/Sodium alginate	Triglyceride	[16,110]
Coacervation	Gluten/Casein	Pyrrolnitrin	[111]
Coacervation	Gliadin	Hexadecane	[112]
Coacervation	α-Gliadin/Arabic gum	Vaselin oil	[16,95,110]
Gelation	SPI	Riboflavin	[113]
Solvent evaporation	Gliadin	Retinoic acid	[114,115]
Phase separation	Corn Zein	Essential oils	[97]
Precipitation	Corn Zein	Quercetin	[100]

- Soy protein isolate (SPI) can be used as an individual coating material or as mixed system with polysaccharides[15,21,121] for application in microencapsulation. The latter as a carrier material favors better protection, oxidative stability, and drying properties.

Pea Proteins:

- Pea seeds contain 20%–30% proteins mainly globulins (65%–80% of total proteins) and two minor fractions of albumins and glutelins.[122]
- Pea proteins possess interesting gel-forming[123] and emulsifying[124] properties.
- Pea proteins in combination with polysaccharides have been widely used in microencapsulation[9,16,18–20,125]
- Pea protein/polysaccharide interactions provide interesting attributes for creating stable emulsions toward good particle size distribution and thus improving the overall microencapsulation process.[9,10]
- Pea proteins show good properties for their potential application, in particular for the production of adhesives, bio plastics, emulsifiers, and wall forming materials for microencapsulation.[126]

Wheat Proteins (Gluten):

- Wheat contains a specific protein, gluten (~80% of wheat seed proteins), obtained as a by-product during starch isolation from wheat flour.[127]
- The low water solubility and viscoelasticity of wheat protein provide various interesting physico-chemical characteristics, such as gel- and film-forming properties.[128]
- Gluten alone, or in combination with polysaccharides are widely used for encapsulating active core materials using various techniques.[16,95,111,112,115,129]

Rice Proteins:

- Rice and rice bran contains 12%–20% proteins.[130,131]
- Rice proteins form complex with polysaccharides (alginate and carrageenan).[132]
- The foaming properties of rice protein are similar to those of albumin from egg white.[133]
- The physico-chemical properties of rice proteins are similar to those of casein.[134] The main amino acid content of rice proteins is similar to that of casein and soy proteins. The rice proteins have favorable characteristics for wall material in microencapsulation.

Gluten (CAS# 8002-80-0):

- Gluten is a complex mixture of gliadins (monomeric gluten proteins) and glutenins (polymeric gluten proteins) comprise about 80% of the proteins present in wheat grain
- Glutenins show very low solubility in water due to high content of nonpolar amino acids and glutamine.[135].
- Gliadins are soluble in distilled water but aggregate in salt solution.[136]
- Gluten film is elastic in nature.[137]

Milk Proteins: Bovine milk contains about 3.0–3.6 wt% proteins comprising caseins and whey proteins.[138–142]

Caseins (CAS#9000-71-9):

- Caseins are the most predominant phosphoproteins found in milk.
- Caseins can vary in their net charge, hydrophilicity, and metal binding.

- Caseins are extremely heat-stable proteins, and thus do not coagulate by heat.
- They are insoluble at their isoelectric point, at about pH 4.6.
- Calcium caseinate is the only milk protein system reported to exhibit reversible thermal gelation.

Whey Proteins:

- Whey is a by-product of cheese or casein production and has several uses including encapsulation.
- Whey proteins are soluble in their native forms in the ionic environment of milk, almost independent of pH, but they become insoluble at their isoelectric point (pH about 5).
- Whey proteins denature at temperatures above 70°C and become insoluble and form thermally irreversible gels.
- Whey proteins exhibit Newtonian flow at concentrations in the range of 4–12 wt% but become pseudoplastic in the range of 18–29 wt% concentration.
- Films of whey proteins (formed by thermally induced disulfide cross-linking) are flavorless, transparent to translucent depending on the protein source,[143,144] excellent gas barriers,[145] and possess good tensile strengths similar to synthetic films.
- Whey proteins are globular proteins[146] and gelation of globular whey protein[146,147] can be achieved by adding calcium to a preheated mixture.
- Whey proteins rapidly adsorb on the emulsion interface, where they self-aggregate and form continuous and homogeneous membranes around oil droplets[146,148] Various procedures can yield microcapsules from whey proteins[149]

Gelatin (CAS# 9000-70-8):

- Gelatins[150] vary widely in their size and charge distribution. Gelatins in their pure form are solid, translucent, brittle, which are colorless or slightly yellow.
- Gelatins are obtained from collagens (cartilage, sinews, skin, etc.) by complex, a process that yields different types of gelatin: type A gelatins by acid treatment (pH 1.5–3.0) and type B by alkaline treatment (pH 12) of collagen.
- Upon heating gelatins melt but solidify when cooled. Mammalian gelatins form high viscosity aqueous solution, which gel on cooling below 35°C–40°C. For a given molecular weight the type A gelatins usually have lower intrinsic viscosities than type B gelatins.
- Gelatin is soluble in most polar solvents and the aqueous solutions show viscoelastic behavior.
- Gelatins are amphiphilic and depending on the pH it can behave as polyelectrolytes in aqueous solution.
- The isoelectric points of type A gelatins and type B gelatins are in the range of
- pH 7–9.4 and pH 4.8–5.5, respectively.
- Gelatin films with higher triple-helix content swell less in water and are therefore much stronger.[151]
- At low pH gelatin (becomes positively charged) forms coacervates with negatively charged gellan gum[145]
- Fish gelatins[152] can be advantageous for encapsulation applications to reduce water loss at low temperatures.

26.3 LIPIDS

Lipids are hydrophobic (generally insoluble in water) and exist in different forms such as oils, fats, waxes, and phospholipids, etc.

Fatty Acids and Fatty Alcohols:

- There are two subgroups of fatty acids and fatty alcohols such as saturated and unsaturated versions.
- Unsaturated fatty acids or fatty alcohols possess one or more double bonds in their chain.
- Short chain aliphatic acids are miscible in water; however, the water-solubility rapidly decreases with increasing chain length.
- Fatty alcohols behave as nonionic surfactants and have emulsifying properties.

Glycerides:

- The glycerides family includes triglyceride, diglyceride, and monoglyceride. Triglyceride is the main constituent in animal fats and plant/vegetable oils. Natural fats may contain a mixture of different triglycerides.
- Depending on the type of glycerides, one, two, or three fatty acid chains (the chain length may be same or different, saturated or unsaturated) are covalently bonded to a glycerol molecule by ester linkages.
- Mono- and diglycerides have emulsifying properties, and the glycerides are not soluble in water.
- The melting points of the glycerides strongly depend on various factors, including chemical structure, the distribution of double bonds, mixture of glycerides present, etc.

Waxes (Beeswax, Carnauba Wax, and Candelilla Wax)—(CAS# 8012-89-3):

- Waxes are esters of fatty acids, isolated from animal and plant products, and are insoluble in water.[153]
- The beeswax melts in the range of 62°C–64°C and is[154] compatible with most other waxes and oils, fatty acids, glycerides, and hydrocarbons.
- The melting point of Carnauba wax (CAS# 8015-86-9)[155,156] is in the range 78°C–85°C, and it is compatible with most other waxes and oils, fatty acids, glycerides, and hydrocarbons as well.
- Candelilla wax (CAS# 8006-44-8)[157,158] is soluble in many organic solvents, melts in the range of 67°C–79°C, and is compatible with all vegetable and animal waxes, fatty acids, a large variety of natural and synthetic resins, glycerides, etc.

Phospholipids—Liposomes:

- Phospholipids contain two long chain fatty acids, and the fatty acid components may be saturated or unsaturated (for example palmitic, stearic, oleic, linoleic, or linolenic acid).
- Phospholipids are present in all animal and plant cells (isolated from the egg yolk, soybean oil, and also milk).[159,160]
- Phospholipids are ionic amphiphiles, and due to the amphiphilic character they function well as emulsifying and dispersing agents.[161]
- Phospholipids when mixed with water aggregate or self-assemble into well-organized and defined structures and bilayers.
- The liposomes are formed from bilayers by applying energy during the mixing process.[162]
- In liposomes the core, an aqueous interior is separated by one or more phospholipid bilayers from the aqueous exterior.[161] The size of the liposomes depends upon the manufacturing technique, intensity of mixing, etc.[163] and is stable for a defined period of time.[31] Liposomal properties and functionality depend on external parameters, such as pH and ionic strength of the medium.[162]

Paraffin (CAS# 8002-74-2):

- Paraffin's[164–166] are linear hydrocarbons C_nH_{2n+2} (the solid paraffin wax, the solid form, has $n > 20$) generally derived from petroleum source.
- Food grade paraffin wax is a white, odorless, tasteless material with melting points ranging from 48°C to 95°C.
- Good for coating and encapsulation applications.

26.4 STRATEGIES FOR THE SELECTION OF MATERIALS

Many published encapsulation strategies includes several materials and combinations on a trial-and-error basis in order to identify the most suitable one. The strategies for the selection of a matrix/wall material are affected by various aspects/goals, such as: improvement of an existing product, protection of a known ingredient, improvement of an existing process, replacement of an existing technology, etc. There are several publications available on the microencapsulation of food ingredients, including appropriate material recommendations.[145,146,162,167–179]

The reasons for encapsulation play a major role on the selection of a shell/wall/coating material. As described in the Introduction Chapter, the encapsulations/entrapment of actives are done for a variety of reasons such as improving shelf life, masking of taste or odors of actives, simplification of handling, controlling the release profile of active, enhancing visual aspects of the products, etc. Thus, the selection process of shell/coating material is dependent on the selection of the microencapsulation process or more appropriately the selection of coating material and encapsulation method is interdependent. The selection process of shell/coating material is achieved by reviewing various project criteria.

- The functionality of the encapsulated ingredients in the final product
- The processing conditions to be used during the final production of products
- The storage conditions of the final product containing the encapsulates
- The trigger(s) and mechanism(s) for releasing the active ingredient
- The cost constraints, if any
- The film-forming ability of shell/coating material and compatibility with the core material
- The intellectual property status and freedom to practice/use

The shell/coating materials can be selected from a wide variety of natural or synthetic polymers, depending on the characteristics desired in the final microcapsules. Coating materials may be employed in combinations since a single material may not be able to meet all of the desired attributes for selecting a microencapsulation process or the final product. Although the performance of the coating in the final application is crucial, the matching of the material to the process technology and process conditions is likely to be of equal importance.[167] An example of a quantitative method for selecting the most suitable biopolymer has been given by Pérez-Alonso et al. proposing blends.[180]

An ideal coating material may possess various desired attributes including the following:

- Desired rheological properties at a reasonable high concentration
- Possess the ability to disperse/emulsify the active ingredient and stabilize the emulsified system
- No reactivity toward the active ingredient during processing and prolonged storage
- The ability to seal and hold the active material within its structure during processing or storage

26.5 CONCLUSIONS

This chapter provides a brief overview of a variety of biopolymers of natural origin available for the use as shell or coating materials in microencapsulation of actives. However, a detailed knowledge of the chemical and physical properties of the encapsulation material is required prior to use as a coating/film/matrix during the encapsulation process. The interaction between encapsulation material and actives needs careful consideration. Also, the economic considerations will remain crucial for the selection of the most appropriate encapsulant materials.

REFERENCES

1. Jeon, Y.J., Vasanthan, T., Temelli, F., Song, B.K., 2003. The suitability of barley and corn starches in their native and chemically modified forms for volatile meat flavor encapsulation. *Food Res. Int.* 36, 349–355.
2. Murúa-Pagola, B., Beristain-Guevara, C.I., Martínez-Bustos, F., 2009. Preparation of starch derivatives using reactive extrusion and evaluation of modified starches as shell materials for encapsulation of flavoring agents by spray drying. *J. Food Eng.* 91, 380–386.
3. Krishnan, S., Bhosale, R., Singhal, R.S., 2005. Microencapsulation of cardamom oleoresin: Evaluation of blends of gum arabic, maltodextrin and a modified starch as wall materials. *Carbohydr. Polym.* 61, 95–102.
4. Saénz, C., Tapia, S., Chávez, J., Robert, P., 2009. Microencapsulation by spray drying of bioactive compounds from cactus pear (*Opuntia ficus-indica*). *Food Chem.* 114, 616–622.
5. Semyonov, D., Ramon, O., Kaplun, Z., Levin-Brener, L., Gurevich, N., Shimoni, E., 2010. Microencapsulation of *Lactobacillus paracasei* by spray freeze drying. *Food Res. Int.* 43, 193–202.
6. Kim, Y.D., Morr, C.V., Schenz, T.W., 1996. Microencapsulation properties of gum Arabic and several food proteins: Spray-dried orange oil emulsion particles. *J. Agric. Food Chem.* 44, 1314–1320.
7. Shaikh, J., Bhosale, R., Singhal, R., 2006. Microencapsulation of black pepper oleoresin. *Food Chem.* 94, 105–110.
8. Drusch, S., 2007. Sugar beet pectin: A novel emulsifying wall component for microencapsulation of lipophilic food ingredients by spray-drying. *Food Hydrocolloids* 21, 1223–1228.
9. Gharsallaoui, A., Saurel, R., Chambin, O., Cases, E., Voilley, A., Cayot, P., 2010. Utilisation of pectin coating to enhance spray-dry stability of pea protein-stabilised oil-in-water emulsions. *Food Chem.* 122, 447–454.
10. Higuera-Ciapara, I., Felix-Valenzuela, L., Goycoolea, F.M., Arguelles-Monal, W., 2004. Microencapsulation of astaxanthin in a chitosan matrix. *Carbohydr. Polym.* 56, 41–45.
11. Pedro, A.S., Cabral-Albuquerque, E., Ferreira, D., Sarmento, B., 2009. Chitosan: An option for development of essential oil delivery systems for oral cavity care? *Carbohydr. Polym.* 76, 501–508.
12. Yoo, S.H., Song, Y.B., Chang, P.S., Lee, H.G., 2006, Microencapsulation of α-tocopherol using sodium alginate and its controlled release properties. *Int. J. Biol. Macromol.* 38, 25–30.
13. Huang, S.B., Wu, M.H., Lee, G.B., 2010. Microfluidic device utilizing pneumatic microvibrators to generate alginate microbeads for microencapsulation of cells. *Sensor. Actuator.* 147, 755–764.
14. Wikstrom, J., Elomaa, M., Syvajarvi, H., Kuokkanen, J., Yliperttula, M., Honkakoski, P., Urtti, A., 2008. Alginate-based microencapsulation of retinal pigment epithelial cell line for cell therapy. *Biomaterial* 29, 869–876.
15. Augustin, M.A., Sanguansri, L., Bode, O., 2006. Maillard reaction products as encapsulants for fish oil powders. *J. Food Sci.* 71, 25–32.
16. Ducel, V., Richard, J., Saulnier, P., Popineau, Y., Boury, F., 2004. Evidence and characterization of complex coacervates containing plant proteins: Application to the microencapsulation of oil droplets. *Colloid Surf.* 232, 239–247.
17. Mendanha, D.V., Ortiz, S.E.M., Favaro-Trindade, C.S., Mauri, A., Monterrey-Quintero, E.S., Thomazini, M., 2009. Microencapsulation of casein hydrolysate by complex coacervation with SPI/pectin. *Food Res. Int.* 42, 1099–1104.
18. Pereira, H.V.R., Saraiva, K.P., Carvalho, L.M.J., Andrade, L.R., Pedrosa, C., Pierucci, A.P.T.R., 2009. Legumes seeds protein isolates in the production of ascorbic acid microparticles. *Food Res. Int.* 42, 115–121.
19. Pierucci, A.P.T.R., Andrade, L.R., Baptista, E.B., Volpato, N.M., Rocha-Leao, M.H.M., 2006. New microencapsulation system for ascorbic acid using pea protein concentrate as coat protector. *J. Microencapsul.* 23, 654–662.
20. Pierucci, A.P.T.R., Andrade, L.R., Farina, M., Pedrosa, C., Rocha-Leao, M.H.M., 2007. Comparison of α-tocopherol microparticles produced with different wall materials: Pea protein a new interesting alternative. *J. Microencapsul.* 24, 201–213.
21. Yu, C., Wang, W., Yao, H., Liu, H., 2007. Preparation of phospholipid microcapsules by spray drying. *Dry. Technol.* 25, 695–702.
22. Li, J.-Y., Yeh, A.-I., 2001. Relationships between thermal, rheological characteristics and swelling power for various starches. *J. Food Eng.* 50, 141–148.
23. Singh, N., Singh, J., Kaur, L., Singh Sodhi, N., Singh Gill, B., 2003. Morphological, thermal and rheological properties of starches from different botanical sources. *Food Chem.* 81, 219–231.
24. Parker, R., Ring, S.G., 2001. Aspects of the physical chemistry of starch. *J. Cereal Sci.* 34, 1–17.

25. Wurzburg, O.B., 1995. Modified starches. In: Stephen, A.M. (ed.) *Food Polysaccharides and Their Applications*. Marcel Dekker, Inc, New York, pp. 67–97.
26. Wurzburg, O.B., 2006. Modified starches. In: Stephen, A.M., Phillips, G.O., Williams, P.A. (eds.) *Food Polysaccharides and Their Applications*, 2nd edn. Taylor & Francis, Boca Raton, FL, pp. 87–118.
27. Blanchard, P.H., Katz, F.R., 1995. Starch hydrolysates. In: Stephen, A.M. (ed.) *Food Polysaccharides and Their Applications*. Marcel Dekker, Inc, New York, pp. 99–122.
28. Blanchard, P.H., Katz, F.R., 2006. Starch hydrolysates. In: Stephen, A.M., Phillips, G.O., Williams, P.A. (eds.) *Food Polysaccharides and Their Applications*, 2nd edn. Taylor & Francis, Boca Raton, FL, pp. 119–145.
29. Shahidi, F.R., Regg, R.B., 1991. Encapsulation of the pre-formed cooked cured-meat pigment. *J. Food Sci.* 56, 1500.
30. Szejtli, J., 1998. Introduction and general overview of cyclodextrin chemistry. *Chem. Rev.* 98, 1743.
31. Reineccius, T.A., Reineccius, G.A., Peppard, T.L., 2005. The effect of solvent interactions on alpha-, beta-, and gamma-cyclodextrin flavor molecular inclusion complexes. *J. Agric. Food Chem.* 53, 388–392.
32. Coffey, D.G., Bell, D.A., Henderson, A., 1995. Cellulose and cellulose derivatives. In: Stephen, A.M. (ed.) *Food Polysaccharides and Their Applications*, 2nd edn. Marcel Dekker, Inc., New York, pp. 123–153.
33. Coffey, D.G., Bell, D.A., Henderson, A., 2006. Cellulose and cellulose derivatives. In: Stephen, A.M., Phillips, G.O., Williams, P.A. (eds.) *Food Polysaccharides and Their Applications*, 2nd edn. Taylor & Francis, Boca Raton, FL, pp. 147–179.
34. Greminger, G.K. Jr, Krumel, K.L., 1980. Alkyl and hydroxyalkylalkylcellulose. In: Davidson, R.L. (ed.) *Handbook of Water-Soluble Gums and Resins*. McGraw-Hill, New York, pp. 3–1–3–25.
35. Murray, J.C.F., 2000. Cellulosics. In: Phillips, G.O., Williams, P.A. (eds.) *Handbook of Hydrocolloids*. Woodhead Publishing Limited, Cambridge, England, pp. 219–230.
36. Butler, R.W., Klug, E.D., 1980. Hydroxypropylcellulose. In: Davidson, R.L. (ed.) *Handbook of Watersoluble Gums and Resins*. McGraw-Hill, New York, pp. 13–1–13–6.
37. Stelzer, G.I., Klug, E.D., 1980. Carboxymethylcellulose. In: Davidson, R.L. (ed.) *Handbook of Watersoluble Gums and Resins*. McGraw-Hill, New York, pp. 4–1–4–28.
38. Kästner, U., Hoffmann, H., Dönges, R., Hilbig, J., 1997. Structure and solution properties of sodium carboxymethyl cellulose. *Colloids Surf. A: Physicochem. Eng. Asp.* 123, 307–328.
39. FAO, 1995. *Gums, Resins and Latexes of Plant Origin*. (Non-wood forest products 6). FAO, Rome, Italy.
40. FAO, 1999. *Gum Arabic*. (Food and nutrition paper 52, addendum 7). FAO, Rome, Italy.
41. Verbeken, D., Dierckx, S., Dewttlinck, K., 2003. Exudate gums: Occurrence, production, and applications. *Appl. Microbiol. Biotechnol.* 63, 10–21.
42. Sanchez, C., Renard, D., Robert, P., Schmitt, C., Lefebvre, J., 2002. Structure and rheological properties of acacia gum dispersions. *Food Hydrocolloids* 16, 257–267.
43. Krishnan, S., Kshirsagar, A.C., Singhal, R.S., 2005. The use of gum arabic and modified starch in the microencapsulation of a food flavoring agent. *Carbohydr. Polym.* 62(4), 309–315.
44. Meer, W., 1980a. Gum arabic. In: Davidson, R.L. (ed.) *Handbook of Water-Soluble Gums and Resins*. McGraw-Hill, New York, pp. 8–1–8–21.
45. McNamee, B.F., O'Riordan, E.D., O'Sullivan, M., 1998. Emulsification and microencapsulation properties of gum arabic. *J. Agric. Food Chem.* 46, 4551–4555.
46. FAO, 1992. *Tragacanth Gum*. (Food and nutrition paper 53). FAO, Rome, Italy.
47. Mohammadifar, M.A., Musavi, S.M., Kiumarsi, A., Williams, P.A., 2006. Solution properties of targacanthin (water-soluble part of gum tragacanth exudate from *Astragalus gossypinus*). *Int. J. Biol. Macromol.* 38, 31–39.
48. Stauffer, K.R., 1980. Gum tragacanth. In: Davidson, R.L. (ed.) *Handbook of Water-Soluble Gums and Resins*. McGraw-Hill, New York, pp. 11–1–11–30.
49. Elias, H.-G., 1992. *Makromoleküle. Band 2, Technologie*. Hüthig & Wepf Verlag, Basel, Heidelberg, New York.
50. FAO, 1992. *Karaya Gum*. (Food and nutrition paper 52). FAO, Rome, Italy.
51. Meer, W., 1980b. Gum Karaya. In: Davidson, R.L. (ed.) *Handbook of Water-Soluble Gums and Resins*. McGraw-Hill, New York, pp. 10–1–10–13.
52. Weiping, W., 2000. Tragacanth and karaya. In: Phillips, G.O., Williams, P.A. (eds.) *Handbook of Hydrocolloids*. Woodhead Publishing Limited, Cambridge, England, pp. 231–246.
53. Orozco-Villafuerte, J., Buendía-González, L., Cruz-Sosa, F., Vernon-Carter, E.J., 2005. Increased mesquite gum formation in nodal explants cultures after treatment with a microbial biomass preparation. *Plant Physiol. Biochem.* 43, 802–807.

54. Orozco-Villafuerte, J., Cruz-Sosa, F., Ponce-Alquicira, E., Vernon-Carter, E.J., 2003. Mesquite gum: Fractionation and characterization of the gum exuded from *Prosopis laevigata* obtained from plant tissue culture and from wild trees. *Carbohydr. Polym.* 54, 327–333.

55. Anderson, D.M.W., Farquhar, J.G.K., 1982. Gum exudates from the genus *Prosopis*. *Int. Tree Crops J.* 2, 15–24.

56. Anderson, D.M.W., Weiping, W., 1989. The characterization of proteinaceous *Prosopis* (mesquite) gums which are not permitted food additives. *Food Hydrocolloids J.* 3, 235–242.

57. Goycoolea, F.M., Calderón de la Barca, A.M., Balderrama, J.R., Valenzuela, J.R., 1997. Immunological and functional properties of the exudate gum from northwestern Mexican mesquite (Prosopis spp.) in comparison with gum arabic. *Int. J. Biol. Macromol.* 21, 29–36.

58. Vernon-Carter, E.J., Beristain, C.I., Pedroza-Islas, R., 2000. Mesquite gum (Prosopis gum). In: Doxastakis, G., Kiosseoglou, V. (eds.) *Novel Macromolecules in Food Systems Development in Food Science 41*. Elsevier, Amsterdam, the Netherlands, pp. 217–238.

59. Islam, A.M., Phillips, G.O., Sljivo, A., Snowden, M.J., Williams, P.A., 1997. A review of recent developments on the regulatory, structural and functional aspects of gum arabic. *Food Hydrocolloids* 11(4), 493–505.

60. Diaz-Sobac, R., Garcia, H., Beristain, C.I., Vernon-Carter, E.J., 2002. Morphology and water vapor permeability of emulsion films based on mesquite gum. *J. Food Process Preserv.* 26(2), 129–141.

61. Gidley, M.J., Reid, J.S.G., 2006. Galactomannans and other cell wall storage polysaccharides in seeds. In: Stephen, A.M., Phillips, G.O., Williams, P.A. (eds.) *Food Polysaccharides and Their Applications*, 2nd edn. Taylor & Francis, Boca Raton, FL, pp. 181–215.

62. Hoefler, A.C., 2004. *Hydrocolloids*. Eagan Press, St. Paul, MN.

63. Seaman, J.K., 1980. Locust bean gum. In: Davidson, R.L. (ed.) *Handbook of Water-Soluble Gums and Resins*. McGraw-Hill, New York, pp. 14–1–14–16.

64. May, C.D., 2000. Pectins. In: Phillips, G.O., Williams, P.A. (eds.) *Handbook of Hydrocolloids*. Woodhead Publishing Limited, Cambridge, England, pp. 169–188.

65. Lopez da Silva, J.A., Rao, M.A., 2006. Pectins: Structure, functionality, and uses. In: Stephen, A.M., Phillips, G.O., Williams, P.A. (eds.) *Food Polysaccharides and Their Applications*, 2nd edn. Taylor & Francis, Boca Raton, FL, pp. 353–411.

66. Pedersen, J.K., 1980. Pectins. In: Davidson, R.L. (ed.) *Handbook of Water-Soluble Gums and Resins*. McGraw-Hill, New York, pp. 15–1–15–21.

67. Voragen, A.G.J., Pilnik, W., Thibault, J.-F., Axelos, M.A.V., Renard, C.M.G.C., 1995. Pectins. In: Stephen, A.M. (ed.) *Food Polysaccharides and Their Applications*. Marcel Dekker, Inc, New York, pp. 287–339.

68. Maeda, H., 2000. Soluble soybean polysaccharide. In: Phillips, G.O., Williams, P.A. (eds.) *Handbook of Hydrocolloids*. Woodhead Publishing Limited, Cambridge, England, pp. 309–320.

69. Nakamura, A., Furuta, H., Maeda, H., Nagamatsu, Y., Yoshimoto, A., 2000. The structure of soluble soybean polysaccharide. In: Nishinari, V. (ed.) *Hydrocolloids*, Part 1. Elsevier Science, Amsterdam, the Netherlands, pp. 235–241.

70. Furuta, H., Maeda, H., 1999. Rheological properties of water-soluble soybean polysaccharides extacted under weak acidic conditions. *Food Hydrocolloids* 13, 267–274.

71. Guiseley, K.B., Stanley, N.F., Whitehouse, P.A., 1980. Carrageenan. In: Davidson, R.L. (ed.) *Handbook of Water-Soluble Gums and Resins*. McGraw-Hill, New York, pp. 5–1–5–30.

72. Piculell, L., 1995. Gelling Carrageenans. In: Stephen, A.M. (ed.) *Food Polysaccharides and Their Applications*. Marcel Dekker, Inc, New York, pp. 205–244.

73. Piculell, L., 2006. Gelling Carrageenans. In: Stephen, A.M., Phillips, G.O., Williams, P.A. (eds.) *Food Polysaccharides and Their Applications*, 2nd edn. Taylor & Francis, Boca Raton, FL, pp. 239–287.

74. Imeson, A., 2007. Carrageenan. In: Phillips, G.O., Williams, P.A. (eds.) *Handbook of Hydrocolloids*. Woodhead Publishing Limited, Cambridge, England, pp. 87–102.

75. Falshaw, R., Bixler, H.J., Johndro, K., 2001. Structure and performance of commercial kappa-2 carrageenan extracts I. Struct Anal, *Food Hydrocolloids* 15, 441–452.

76. Mangione, M.R., Giacomazza, D., Bulone, D., Martorana, V., San Biagio, P.L., 2003. Thermoreversible gelation of k-carrageenan: Relation between conformational transition and aggregation. *Biophys. Chem.* 104, 95–105.

77. Janaswamy, S., Chandrasekaran, R., 2002. Effect of calcium ions on the organization of iota-carragenan helices: An X-ray investigation. *Carbohydr. Res.* 337, 523–535.

78. Kara, S., Arda, E., Kavzak, B., Pekcan, Ö., 2006. Phase transitions of k-carrageenan gels in various types of salts. *J. Appl. Polym. Sci.* 102, 3008–3016.

79. Chapman, D.J., 1980. *Seaweeds and Their Uses*. Chapman & Hall, London, U.K.
80. Valla, S., Ertesvåg, H., Skjåk-Bræk, G., 1996. Genetics and biosynthesis of alginates. *Carbohydr. Eur.* 14, 14–18.
81. Draget, K.I., 2000. Alginates. In: Phillips, G.O., Williams, P.A. (eds.) *Handbook of Hydrocolloids*. Woodhead Publishing Limited, Cambridge, England, pp. 379–395.
82. Cottrell, I.W., Kovacs, P., 1980b. Alginates. In: Davidson, R.L. (ed.) *Handbook of Water-Soluble Gums and Resins*. McGraw-Hill, New York, pp. 2–1–2–43.
83. Moe, S.T., Draget, K.I., Skjåk-Bræk, G., Smidsrød, O., 1995. Alginates. In: Stephen, A.M. (ed.) *Food Polysaccharides and Their Applications*. Marcel Dekker, Inc, New York, pp. 245–286.
84. Draget, K.I., Moe, S.T., Skjåk-Bræk, G., Smidsrød, O., 2006. Alginates. In: Stephen, A.M., Phillips, G.O., Williams, P.A. (eds.) *Food Polysaccharides and Their Applications*, 2nd edn. Taylor & Francis, Boca Raton, FL, pp. 289–334.
85. Smidsrød, O., Hang, A., 1968. Dependence upon uronic acid composition of some ion.exchange properties of Alginates. *Acta Chem. Scand.* 22(6), 1989–1997.
86. Giannouli, P., Morris, E.R., 2003. Cryogelation of xanthan. *Food Hydrocolloids* 17, 495–501.
87. Sworn, G., 2000. Xanthan gum. In: Phillips, G.O., Williams, P.A. (eds.) *Handbook of Hydrocolloids*. Woodhead Publishing Limited, Cambridge, England, pp. 103–115.
88. Sworn, G., 2000. Gellan gum. In: Phillips, G.O., Williams, P.A. (eds.) *Handbook of Hydrocolloids*. Woodhead Publishing Limited, Cambridge, England, pp. 117–135.
89. Winterowd, J.G., Sandford, P.A., 1995. Chitin and Chitosan. In: Stephen, A.M. (ed.) *Food Polysaccharides and Their Applications*. Marcel Dekker, Inc, New York, pp. 441–462.
90. Vårum, K.M., Smidsrød, O., 2006. Chitosans. In: Stephen, A.M., Phillips, G.O., Williams, P.A. (eds.) *Food Polysaccharides and Their Applications*, 2nd edn. Taylor & Francis, Boca Raton, FL, pp. 497–520.
91. Inoue, Y., 1997. NMR determination of the degree of acetylation. In: Muzzarelli, R.A.A., Peter, M.G. (eds.) *Chitin Handbook*. Atec, Grottamare, Italy, pp. 133–136.
92. Domard, A., Domard, M., 2002. Chitosan: Structure-properties relationship and biomedical application. In: Dumitriu, S. (ed.) *Polymeric Biomaterials*. Marcel Dekker, New York, pp. 187–212.
93. Jenkins, J.A., Breiteneder, H., Mills, E.N., 2007. Evolutionary distance from human homologs reflects allergenicity of animal food proteins. *J. Allergy Clin. Immunol.* 120, 1399–1405.
94. Li, H., Zhu, K., Zhou, H., Peng, W., 2012. Effects of high hydrostatic pressure treatment on allergenicity and structural properties of soybean protein isolate for infant formula. *Food Chem.* 132, 808–814.
95. Ducel, V., Richard, J., Popineau, Y., Boury, F., 2005. Rheological interfacial properties of plant protein-arabic gum coacervates at the oil-water interface. *Biomacromolecules* 6, 790–796.
96. Wang, R., Tian, Z., Chen, L., 2011. A novel process for microencapsulation of fish oil with barley protein. *Food Res. Int.* 44, 2735–2741.
97. Parris, N., Cooke, P.H., Hicks, K.B., 2005. Encapsulation of essential oils in Zein nanospherical particles. *J. Agric. Food Chem.* 53, 4788–4792.
98. Wang, R., Tian, Z., Chen, L., 2011. Nano-encapsulations liberated from barley protein microparticles for oral delivery of bioactive compounds. *Int. J. Pharm.* 406, 153–162.
99. Zhong, Q., Jin, M., Davidson, P.M., Zivanovic, S., 2009. Sustained release of lysozyme from zein microcapsules produced by a supercritical anti-solvent process. *Food Chem.* 115, 697–700.
100. Patel, A.R., Heussen, P.C.M., Hazekamp, J., Dorst, E., Velikov, K.P., 2012. Quercetin loaded biopolymeric colloidal particles prepared by simultaneous precipitation of quercetin with hydrophobic protein in aqueous medium. *Food Chem.* 133, 423–429.
101. Charve, J., Reineccius, G.A., 2009. Encapsulation performance of proteins and traditional materials for spray dried flavors. *J. Agric. Food Chem.* 57, 2486–2492.
102. Ortiz, S.E.M., Mauri, A., Monterrey-Quintero, E.S., Trindade, M.A., 2009. Production and properties of casein hydrolysate microencapsulated by spray drying with soybean protein isolate. *Food Sci. Technol.* 42, 919–923.
103. Rascon, M.P., Beristain, C.I., Garcie, H.S., Salgado, M.A., 2010. Carotenoid retention and storage stability of spray-dried paprika oleoresin using gum arabic and soy protein isolate as wall materials. *Food Sci. Technol.* 44, 549–557.
104. Favaro-Trindade, C.S., Santana, A.S., Monterrey-Quintero, E.S., Trindade, M.A., Netto, F.M., 2010. The use of spray drying technology to reduce bitter taste of casein hydrolysate. *Food Hydrocolloids* 24, 336–340.
105. Nesterenko, A., Alric, I., Silvestre, F., Durrieu, V., 2012. Influence of soy protein's structural modifications on their microencapsulation properties: α-tocopherol microparticles preparation. *Food Res. Int.* 48, 387–396.

106. Gan, C.Y., Cheng, L.H., Easa, A.M., 2008. Evaluation of microbial transglutaminase and ribose cross-linked soy protein isolate-based microcapsules containing fish oil. *Innov. Food Sci. Emerg. Technol.* 9, 563–569.
107. Lazako, J., Popineau, Y., Legrand, J., 2004. Soy glycinin microcapsules by simple coacervation method. *Colloid Surf.* 37, 1–8.
108. Nori, M.P., Favaro-Trindade, C.S., Alencar, S.M., Thomazini, S.M., Balieiro, J.C.C., 2010. Microencapsulation of propolis extract by complex coacervation. *Food Sci. Technol.* 44, 429–435.
109. Jun-xia, X., Hai-yan, Y., Jian, Y., 2011. Microencapsulation of sweet orange oil by complex coacervation with soybean protein isolate/gum Arabic. *Food Chem.* 125, 1267–1272.
110. Ducel, V., Richard, J., Popineau, Y., Boury, F., 2004a. Adsorption kinetics and rheological interfacial properties of plant proteins at the oil-water interface. *Biomacromolecules* 5, 2088–2093.
111. Yu, J.Y., Lee, W.C., 1997. Microencapsulation of pyrrolnitrin from Pseudomonas cepacia using gluten and casein. *J. Ferment. Bioeng.* 84, 444–448.
112. Mauguet, M.C., Legrand, J., Brujes, L., Carnelle, G., Larre, C., Popineau, Y., 2002. Gliadin matrices for microencapsulation processes by simple coacervation method. *J. Microencapsul.* 19(3), 377–384.
113. Chen, L., Subirade, M., 2009. Elaboration and characterization of soy/zein protein microspheres for controlled nutraceutical delivery. *Biomacromolecules* 10, 3327–3334.
114. Ezpeleta, I., Irache, J.M., Stainmesse, S., Chabenat, C., Gueguen, J., Orecchioni, A.M., 1996. Preparation of lectin-vicilin nanoparticle conjugates using the carbodiimide coupling technique. *Int. J. Pharm.* 142, 227–233.
115. Ezpeleta, I., Irache, J.M., Stainmesse, S., Chabenat, C., Gueguen, J., Popineau, Y., Orecchioni, A.M., 1996. Gliadin nanoparticles for the controlled release of all-transretinoic acid. *Int. J. Pharm.* 131, 191–200.
116. Ruiz-Henestrosa, V.P., Sanchez, C.C., Escobar, M.M.Y., Jimenez, J.J.P., Rodrıguez, F.M., Patino, J.M.R., 2007. Interfacial and foaming characteristics of soy globulins as a function of pH and ionic strength. *Colloid Surf.* 309, 202–215.
117. Gu, X., Campbell, L.J., Euston, S.R., 2009. Effects of different oils on the properties of soy protein isolate emulsions and gels. *Food Res. Int.* 42, 925–932.
118. Sun-Waterhouse, D., Wadhwa, S.S., 2013. Industry-relevant approaches for minimising the bitterness of bioactive compounds in functional foods: A review. *Food Bioprocess Technol.* 6(8), 607–627.
119. Gharsallaoui, A., Roudaut, G., Chambin, O., Voilley, A., Saurel, R., 2007. Applications of spray-drying in microencapsulation of food ingredients: An overview. *Food Res. Int.* 40, 1107–1121.
120. Xiao, J. X., Yu, H. Y., Yang, J., 2011. Microencapsulation of sweet orange oil by complex coacervation with soyabean protein ispolate/gum Arabic. *Food Chem.* 125, 1267–1272.
121. Rusli, J.K., Sanguansri, L., Augustin, M.A., 2006. Stabilization of oils by microencapsulation with heated protein-glucose syrup mixtures. *JAOCS* 83, 965–971.
122. Koyoro, H., Powers, J.R., 1987. Functional properties of pea globulin fractions. *Cereal Chem.* 64, 97–101.
123. Akintayo, E.T., Oshodi, A.A., Esuoso, K.O., 1999. Effects of NaCl, ionic strength and pH on the foaming and gelation of pigeon pea (Cajanus cajan) protein concentrates. *Food Chem.* 66, 51–56.
124. Raymundo, A., Gouveia, L., Batista, A.P., Empis, J., Sousa, I., 2005. Fat mimetic capacity of Chlorella vulgaris biomass in oil-in-water food emulsions stabilized by pea protein. *Food Res. Int.* 38, 961–965.
125. Liu, S., Elmer, C., Low, N.H., Nickerson, M.T., 2010. Effect of pH on the functional behaviour of pea protein isolate-gum arabic complexes. *Food Res. Int.* 43, 489–495.
126. De Graaf, L.A., Harmsen, P.F.H., Vereijken, J.M., Monikes, M., 2001. Requirements for non-food applications of pea proteins. A review. *Nahrung/Food.* 45, 408–411.
127. Day, L., Augustin, M.A., Batey, I.L., Wrigley, C.W., 2006. Wheat-gluten uses and industry needs. *Trends Food Sci. Technol.* 17, 82–90.
128. Sun, S., Song, Y., Zheng, Q., 2009. Rheological behavior of heat-induced wheat gliadin gel. *Food Hydrocolloids* 23, 1054–1056.
129. Iwami, K., Hattori, M., Nakatani, S., Ibuki, F., 1987. Spray-dried gliadin powders inclusive of linoleic acid (microcapsules): Their preservability, digestibility and application to bread making. *Agric. Biol. Chem.* 51, 3301–3307.
130. Hamada, J.S., 2000. Characterization and functional properties of rice bran proteins modified by commercial exoproteases and endoproteases. *J. Food Sci.* 65, 305–310.
131. Bienvenido, O.J., 1994. *Le riz dans la nutrition humaine*. Organisation des nations unies pour l'alimentation et l'agriculture, Rome, Italy.

132. Fabian, C.B., Huynh, L.H., Ju, Y.H., 2010. Precipitation of rice bran protein using carrageenan and alginate. *Food Sci. Technol.* 43, 375–379.
133. Wang, M., Hettiarachchy, N.S., Qi, M., Burks, W., Siebenmorgen, T., 1999. Preparation and functional properties of rice bran protein isolate. *J. Agric. Food Chem.* 47, 411–416.
134. Chandi, G.K., Sogi, D.S., 2007. Functional properties of rice bran protein concentrates. *J. Food Eng.* 79, 592–597.
135. Singh, H., MacRitchie, F., 2001. Application of polymer science to properties of gluten. *J. Cereal Sci.* 33, 231–243.
136. Van Vliet, T., Martin, A.H., Bos, M.A., 2002. Gelation and interfacial behavoiur of vegetable proteins. *Curr. Opin. Colloid Interface Sci.* 7, 462–468.
137. Kokelaar, J.J., Prins, A., de Gee, M., 1991. A new method for measuring the surface dilational modules of a liquid. *J. Colloid Interface Sci.* 146, 507–511.
138. Ennis, M.P., Mulvihill, D.M., 2000. Milk proteins. In: Phillips, G.O., Williams, P.A. (eds.) *Handbook of Hydrocolloids*. Woodhead Publishing Limited, Cambridge, England, pp. 189–217.
139. Fox, P.F., 1990. The milk protein system. In: Fox, P.F. (ed.) *Developments in Dairy Chemistry*, Elsevier Applied Science Book, New York. p. 4.
140. Fox, P.F., 1992. *Advanced Dairy Chemistry*, Vol. 1. Elsevier Applied Science Publishers, London, U.K.
141. Holt, C., 1992. Structure and stability of bovine casein micelles. *Adv. Protein Chem.* 43, 63–151.
142. Creamer, L.K., MacGibbon, A.K.H., 1996. Some recent advances in the basic chemistry of milk proteins and lipids. *Int. Dairy J.* 6, 539–568.
143. Chen, H., 1995. Functional properties and applications of edible films made of milk proteins. *J. Diary Sci.* 78, 2563–2583.
144. Lee, S.J., Rosenberg, M., 2000. Whey protein-based microcapsules prepared by double emulsification and heat gelation. *Food Sci. Technol.—Lebenson. Wiss. Technol.* 33(2), 80–88.
145. Madene, A., Jacquot, M., Scher, J., Desobry, S., 2006. Flavour encapsulation and controlled release—A review. *Int. J. Food Sci. Technol.* 41, 1–21.
146. Chen, L.Y., Remondetto, G.E., Subirade, M., 2006. Food protein-based materials as nutraceutical delivery systems. *Trends Food Sci. Technol.* 17, 272–283.
147. Barbut, S., Foegeding, E.A., 1993. Ca^{2+}-induced gelation of pre-heated whey protein isolate. *J. Food Sci.* 58, 867–871.
148. Lefèvre, T., Subirade, M., 2003. Formation of intermolecular β-sheet structures: A phenomenon relevant to protein film structure at oil-water interfaces of emulsions. *J. Colloid Interface Sci.* 263, 59–67.
149. Rosenberg, M., Lee, S.J., 2004. Water-insoluble, whey protein-based microspheres prepared by an all-aqueous process. *J. Food Sci.* 69, E50–E58.
150. Ledward, D.A., 2000. Gelatin. In: Phillips, G.O., Williams, P.A. (eds.) *Handbook of Hydrocolloids*. Woodhead Publishing Limited, Cambridge, England, pp. 67–86.
151. Bigi, A., Panzavolta, S., Rubini, K., 2004. Relationship between triple-helix content and mechanical properties of gelatin films. *Biomaterials*, 25, 5675–5680.
152. Avena-Bustillos, R.J., Olsen, C.W., Olson, D.A., Chiou, B., Yee, E., Bechtel, P.J., McHugh, T.H., 2006. Water vapor permeability of mammalian and fish gelatin films. *J. Food Sci.* 71, E202–E207.
153. Parish, E.J., Boos, T.L., Li, S., 2002. The chemistry of waxes and sterols. In: Akoh, C.C., Min, D.B. (eds.) *Food Lipids: Chemistry, Nutrition, and Biotechnology*, 2nd edn. Marcel Dekker, New York, pp. 103–132.
154. Ross Waxes, Frank B. Ross Co. Inc., 2008. Data tables for Beeswax: http://www.frankbross.com/beeswax.htm. Accessed on 2014.
155. Ross Waxes, Frank B. Ross Co. Inc., 2008. Data tables for Carnauba Wax: http://www.frankbross.com/carnauba_wax.htm. Accessed on 2014.
156. Strahl & Pitsch. Quality waxes, 2008. Data tables Carnauba Wax: http://www.spwax.com/spcarnau.htm. Accessed on 2014.
157. Ross Waxes, Frank B. Ross Co. Inc., 2008. Data tables for Candelilla Wax: http://www.frankbross.com/candelilla_wax.htm. Accessed on 2014.
158. Strahl & Pitsch. Quality waxes, 2008. Data tables Candelilla Wax: http://www.spwax.com/spcandel.htm. Accessed on 2014.
159. Hsieh, Y.F., Chen, T.L., Wang, Y.T., Chang, J.H., Chang, H.M., 2002. Properties of liposomes prepared with various lipids. *J. Food Sci.* 67(8), 2808–2813.
160. Thompson, A.K., Singh, H., 2006. Preparation of liposomes from milk fat globule membrane phospholipids using a Microfluidizer. *J. Dairy Sci.* 89, 410–419.
161. Weiner, A.L., 2002. Lipid excipients in phamaceutical dosage forms. In: Swarbrick, J., Boylan, J.C. (eds.) *Encyclopedia of Pharmaceutical Technology*. Marcel Dekker, New York, pp. 1659–1672.

162. Taylor, T.M., Davidson, P.M., Bruce, B.D., Weiss, J., 2005. Liposomal nanocapsules in food science and agriculture. *Crit. Rev. Food Sci. Nutr.* 45, 587–605.
163. Winterhalter, M., Lasis, D.D., 1993. Liposome stability and formation: Experimental parameters and theories on the size distribution. *Chem. Phys. Lipids*, 64, 35–43.
164. Freund, M., Mózes, G., Jakab, E. (trans), 1982. *Paraffin Products: Properties, Technologies, Applications*. Elsevier, Amsterdam, the Netherlands, p. 121.
165. Ross Waxes, Frank B. Ross Co. Inc., 2008. Data tables for Paraffin Wax: http://www.frankbross.com/paraffin_wax.htm. Accessed on 2014.
166. Strahl and Pitsch. Quality waxes, 2008. Data tables Paraffin and Microcrystalline Wax: http://www.spwax.com/spparaff.htm. Accessed on 2014.
167. Werner, S.R.L., Jones, J.R., Paterson, A.H.J., Archer, R.H., Pearce, D.L., 2007. Air-suspension particle coating in the food industry: Part I—State of the art. *Powder Technol.* 171(1), 25–33.
168. Lopez-Rubio, A., Gavara, R., Lagaron, J.A., 2006. Bioactive packaging: Turning foods into healthier foods through biomaterials. *Trends Food Sci. Technol.* 17(10), 567–575.
169. Ubbink, J., Krüger, J., 2006. Physical approaches for the delivery of active ingredients in foods. *Trends Food Sci. Technol.* 17, 244–254.
170. Desai, K.G.H., Park, H.J., 2005a. Recent developments in microencapsulation of food ingredients. *Dry. Technol.* 23, 1361–1394.
171. Kruckeberg, S., Kunz, B., Weissbrodt, J., 2003. Chances and limits of microencapsulation in progressive food industry. *Chem. Ingenieur Techn.* 75(11), 1733–1774.
172. Gouin, S., 2004. Microencapsulation: Industrial appraisal of existing technologies and trends. *Trends Food Sci. Technol.* 15(7–8), 330–347.
173. Szente, L., Szejtli, J., 2004. Cyclodextrins as food ingredients. *Trends Food Sci. Technol.* 15, 137–142.
174. Mattila-Sandholm, T., Myllarinen, P., Crittenden, R., Mogensen, G., Fonden, R., Saarela, M., 2002. Technological challenges for future probiotic foods. *Int. Dairy J.* 12(2–3), 173–182.
175. Gibbs, B.F., Kermasha, S., Alli, I., Mulligan, C.N., 1999. Encapsulation in the food industry: A review. *Int. J. Food Sci. Nutr.* 50, 213–224.
176. Lasic, D.D., 1998. Novel applications of liposomes. *Trends Biotechnol.* 16, 307–321.
177. Pothakamury, U.R., Barbosa-Canovas, G.V., 1995. Fundamental aspects of controlled release in foods. *Trends Food Sci. Technol.* 6(12), 397–406.
178. Jackson, L.S., Lee, K., 1991. Microencapsulation and the Food Industry. *Food Sci. Technol.—Lebenson. Wiss. Technol.* 24, 289–297.
179. Wandrey, C., Bartkowiak, A., Harding, S.E., 2010. In: Zudiam, N.J. and Nedovic, V.A. (eds.) *Encapsulation Technologies for Active Food Ingredients and Food Processing*, Springer Science, Springer-Verlag, New York, pp. 31–100.
180. Perez-Alonso, C., Baez-Gonzalez, J.G., Beristain, C.I., Vernon-Carter, E.J., Vizcarra-Mendoza, M.G., 2003. Estimation of the activation energy of carbohydrate polymers blends as selection criteria for their use as wall material for spray-dried microcapsules. *Carbohydr. Polym.* 53(2), 197–203.

27 Cellulose Ethers
Applications

*Robert Schmitt, True Rogers, William Porter III,
Oliver Petermann, and Britta Huebner-Keese*

CONTENTS

27.1 Introduction 517
27.2 Chemical Structure and Manufacture 517
27.3 Important Commercial Cellulose Ethers and Ether Esters 519
    27.3.1 Methylcellulose 519
    27.3.2 Ethylcellulose 519
    27.3.3 Hydroxypropyl Methylcellulose 519
    27.3.4 Hydroxypropyl Methylcellulose Acetate Succinate 519
27.4 Oral Drug Delivery 520
    27.4.1 Immediate Release 521
    27.4.2 IR Coatings 521
    27.4.3 Hard-Shell Capsules 521
    27.4.4 Solubility Enhancement 522
27.5 Modified Release 524
27.6 Health Enhancement 526
27.7 Dietary Fiber 526
27.8 Replacement of Gluten in Bakery Products 527
27.9 Reduction of Fat Uptake in Foods 527
27.10 Healthier Fats 527
27.11 Reduction of Blood Cholesterol 528
27.12 Glycemic Response 528
27.13 Effects on Lipid Metabolism: Reduction of Body Weight Gain 528
    27.13.1 Satiety 529
27.14 Conclusions 529
References 530

27.1 INTRODUCTION

Cellulose derivatives have found industrial applications for over 150 years. They have been used in such a wide-ranging application adhesives, building materials, inks, and munitions. Within the health and wellness area, cellulose derivatives have a long history of use as binders, coatings, viscosity modifiers, and release modifiers. This entry will present the chemistry of cellulose ethers a key class of cellulose derivatives and the clinical benefits they bring to food and pharmaceutical products.

27.2 CHEMICAL STRUCTURE AND MANUFACTURE

Cellulose derivatives are widely used in biomedical and pharmaceutical applications. Most significant from a commercial and technical point of view are cellulose ether and cellulose ether esters.

The common structural element of all cellulose derivatives is the polymeric backbone of cellulose, which contains as the basic repeating structure a β-D-anhydroglucose unit with three hydroxyl groups at the C–2, C–3, and C–6 positions available for substitution.

The nature of the substituent and the number of substituted hydroxyl groups are mainly responsible for the individual properties of the different cellulose derivatives like solubility, gelation, or water retention.

Important substituents are low molecular mass alkyl (ethers) and aliphatic or aromatic carboxylate groups (esters), which are listed in Table 27.1.

The amount of substituent groups on the anhydroglucose unit is usually designated by the degree of substitution (DS), the average number of substituted hydroxyl groups with a maximum value of three or by weight percent.

Hydroxyalkyl substituents are characterized by weight percent or by the molar degree of substitution (MS), which is the average number of alkylene oxide reacted with each anhydroglucose unit. The MS includes also side chain formation due to further etherification of hydroxyl groups of the hydroxyalkyl substituent and may take on any value.

Other characteristics of cellulose derivatives that influence its properties are molecular weight (MW), MW distribution, and substituent distribution in and over the polymer chains.[1]

The cellulose ether production process is heterogeneous in nature and consists of two major steps: an initial mercerization step and the final etherification step. Both steps may be conducted simultaneously or subsequently. In the mercerization step, the cellulose is being suspended and

TABLE 27.1
Important Commercial Cellulose Ether and Ether Ester

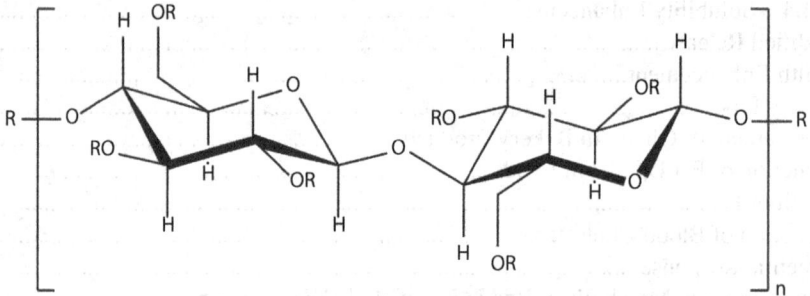

Cellulose and Derivatives	Substituent R
Cellulose	–H
Methylcellulose	–OCH$_3$, –H
Ethylcellulose	–OCH$_2$CH$_3$, –H
Hydroxypropyl methylcellulose	–OCH$_3$, –(CH$_2$CH(CH$_3$)O)$_n$, –H
	–(CH$_2$CH(CH$_3$)O–)$_n$–H, –H
Hydroxypropyl cellulose	–(CH$_2$CH(CH$_3$)O–)$_n$–H, –H
Sodium carboxymethylcellulose	–CH$_2$COO–Na$^+$, H
Hydroxypropymethylcelluloseacetatesuccinate	–OCH$_3$, –(CH$_2$CH(CH$_3$)O)$_n$, –H
	–(CH$_2$CH(CH$_3$)O–)$_n$–CH$_3$,
	–COCH$_3$, –COCH$_2$CH$_2$COOH,
	–(CH$_2$CH(CH$_3$)O–)$_n$–COCH$_3$,
	–(CH$_2$CH(CH$_3$)O–)$_n$–COCH$_2$CH$_2$COOH, –H
Hydroxypropyl methylcellulose phthalate	–OCH$_3$, –(CH$_2$CH(CH$_3$)O–)$_n$–H,
	–(CH$_2$CH(CH$_3$)O–)$_n$–CH$_3$,
	–COC$_6$H$_4$COOH, –H
Cellulose acetate phthalate	–COCH$_3$, COC$_6$H$_4$COOH, –H

pre-activated in aqeous caustic soda and then reacted with alkyl chlorides (Williamson etherification) and/or epoxides (alkali-catalyzed oxalkylation) as etherifying reagents (etherification step).[2] Purification is performed by hot water washing or by treatment with mixtures of water and organic solvents for cellulose ethers without flocculation point.

For the synthesis of cellulose ether esters at industrial scale, purified cellulose ether is reacted with carboxylic anhydrides in acetic acid as solvent system and sodium acetate as catalyst. The dissolved cellulose ether ester is being obtained then by precipitation after addition of water. Purification is performed by washing with water.

27.3 IMPORTANT COMMERCIAL CELLULOSE ETHERS AND ETHER ESTERS

27.3.1 METHYLCELLULOSE

Methylcellulose (MC) is a mono methyl-substituted cellulose ether. MC with a DS > 1.4 is soluble in cold water and shows a reversible thermal gelation behavior in aqueous media. Above the gelation temperature of about 50°C, a strong elastic gel is formed caused by hydrophobic polymer-to-polymer association. By cooling the gel under the gelation temperature, this gel effect can be reversed and repeated afterward if desired. The gelation temperature and gel strength strongly depend on the substituent distribution in and over the polymer chains,[3,4] but is also influenced by concentration and additives.

27.3.2 ETHYLCELLULOSE

Ethylcellulose (EC) is a cellulose ether substituted with ethyl groups and more hydrophobic than MC. EC with a DS of 0.7–1.7 is soluble in water and with DS > 1.7 is soluble in organic solvents.[5] Commercial grades usually have DS > 2.2 and excellent organic solubility with high clarity of solution but are not water-soluble anymore. EC possesses superb thermoplasticity and is an exceptional film former. The softening point depends on the DS and is for commercial unplasticized grades from 135°C to 160°C. Films made from EC are tough, with high tensile strength and an unusual degree of flexibility even at low temperatures.

27.3.3 HYDROXYPROPYL METHYLCELLULOSE

HPMC is a mixed cellulose ether with hydroxypropyl and methyl substitution on the anhydroglucose unit. In general, the introduction of a second substituent leads to clearer solutions, higher gelation temperature, and thermoplastic properties at higher MS compared to MC. Due to the secondary substitution, gels from HPMC show much lower gel strength and a softer gel texture than gels from MC. Adjusting the ratio of methyl to hydroxypropyl allows to tailor many of the properties of HPMC. HPMC with high MS have a good solubility in hot organic solvents while maintaining water solubility, whereas HPMC with lower MS are less thermoplastic and soluble in organic solvents but showing excellent water solubility and water retention. Another important attribute is the MW. The MW ranges from very low for coating solutions with high solid content to high for controlled release applications. Polymer hydration, gel formation, and polymer erosion[6] are primarily controlled by MW, substitution, and morphology and are mainly responsible for controlled release performance of HPMC in hydrophilic matrix systems.

27.3.4 HYDROXYPROPYL METHYLCELLULOSE ACETATE SUCCINATE

Hydroxypropyl methylcellulose acetate succinate (HPMCAS), a cellulose ether ester derived from HPMC esterified with acetate and succinate groups, is an enteric polymer soluble only in water of high pH or in polar organic solvents like acetone or tetrahydrofuran (THF). The enteric performance is due to the presence of carboxylic groups. At low pH, HPMCAS exists in its protonated form and is insoluble in water, whereas at high pH HPMCAS is being deprotonated and therefore soluble in water. The dissolution pH is controlled by the acetate/succinate ratio, which is also being used to tailor the amphiphilic properties of HPMCAS, and therefore its solubility in organic solvents and interactions with active pharmaceutical drugs (APIs).

27.4 ORAL DRUG DELIVERY

According to the online resource, Pharmacircle,[7] 53% of all commercialized pharmaceutical dosage forms are oral drug-delivery systems. Oral drug delivery is the most common route of administration due to the fact that it is more convenient, less invasive, and less painful compared to other routes of administration. Of the various oral drug-delivery systems shown in Figures 27.1 and 27.2, 68% are tablets and 20% are capsules.[7] In particular, tablets are relatively easy to manufacture and are generally well-accepted by patients.

Oral drug delivery systems can be designed to deliver active pharmaceutical ingredients (APIs) in either immediate or modified fashion, and there are two major application spaces for which immediate- and modified-release (MR) oral dosage forms are utilized: (1) systemic delivery and (2) localized delivery. The majority of this section is focused upon systemic delivery.

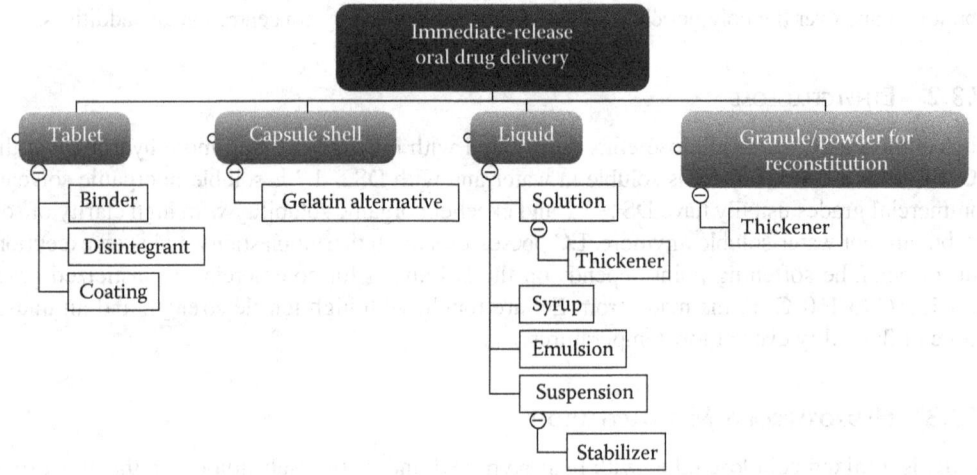

FIGURE 27.1 IR application into which cellulose ethers are commonly used.

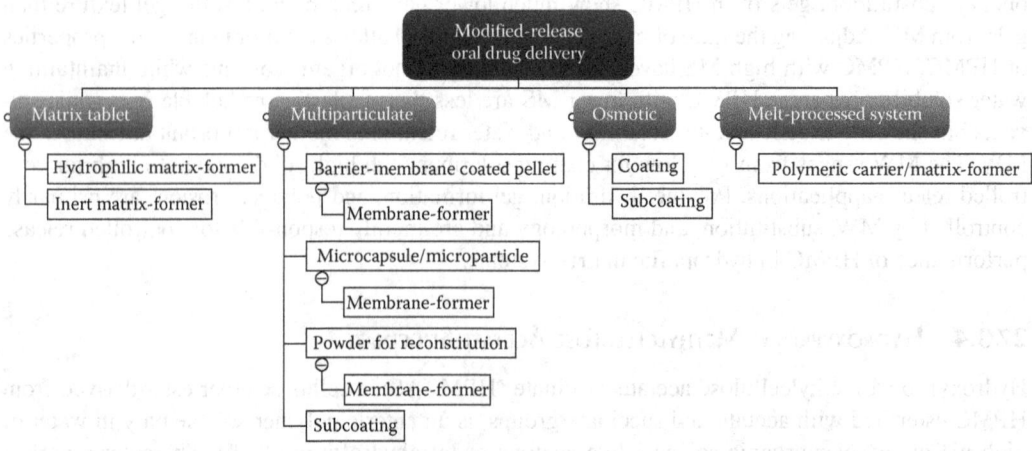

FIGURE 27.2 MR applications into which cellulose ethers are commonly used.

An immediate-release (IR) dosage form is designed not to impede API dissolution, so the entire dosage would be available for dissolution in the gastric media. IR dosage forms can also be utilized to deliver poorly soluble APIs by placing the compound in a more easily absorbed state. Some absorption might occur across the gastric lining, but most would occur once the dosage enters the small intestine, where the presence of villi and microvilli significantly increase the surface area for systemic absorption. In addition, the small intestine offers at least an equivalent transit time compared to the stomach (2.5–5 h).[8] An MR dosage form is designed to gradually release API as the dosage traverses through the stomach, small intestine, and even large intestine. It can be designed to release API over any interval within typically a 24 h time period.

27.4.1 Immediate Release

Cellulose ethers are very commonly used in the production of IR dosage forms, serving such roles as binder, coating agent, API solubilizer/crystallization inhibitor, suspension stabilizer, thickener, and gelatin substitute (see Figure 27.1). The most widely used applications are for producing tablet coatings and hard-shell capsules, both of which will be elaborated upon.

27.4.2 IR Coatings

IR coatings are typically applied onto tablets in order to improve aesthetics. HPMC was first reported as an IR coating in 1965,[9] with HPMC 2910 being the most commonly utilized grade in this application. HPMC is typically dissolved in water, and the coating is subsequently applied. A plasticizer, like polyethylene glycol 400, is included in the aqueous coating formulation in order to lower the T_g (~170°C) of HPMC, render the film more ductile, and ensure a continuous film is applied across the substrate surface. Pigments, such as titanium dioxide and aluminum lakes, can also be included in the coating formulation. The aqueous coating formulation is atomized onto the substrates, most typically as the substrates are tumbled in a perforated coating pan; and the aesthetic coating coalesces upon evaporation of water. The coating formulation viscosity is typically less than 500 mPa s in order to ensure sufficient atomization.[10] Hence, a low-viscosity grade of HPMC 2910 is most commonly employed. Refer to the work of Khot et al. for more details on the perforated pan coating process.[10]

HPMC can also be dissolved in a layering formulation (in which API may be dissolved or suspended) and subsequently coated onto a tablet or nonpareil pellet.[11–14] In this application, HPMC binds the API onto the surface of the substrate. HPMC can also be applied as either a subcoating or top-coating. Subcoatings may serve to separate layers within a multilayer dosage form due to ingredient incompatibilities or to smooth the substrate surface for subsequent deposition of an MR barrier coating.[15,16] Top coatings may be used to protect underlying ingredients from attrition during downstream processing or to minimize fracture of underlying coatings upon compression of multiparticulates to tablets.[17]

27.4.3 Hard-Shell Capsules

Dietary preferences, religious beliefs, and nutritional awareness impact patient decisions and compliance. Many patients opt not to consume dosage forms containing gelatin, which is a mixture of peptides and proteins derived from collagen-containing animal tissues, such as skin, white connective tissues, and bones.[18] HPMC is derived from renewable cellulose resources, such as sustainable timber forests and is a viable alternative to gelatin in the manufacture of hard-shell capsules.

HPMC capsules are commonly made using low-viscosity grades of HPMC 2910 or 2906. HPMC and other materials are suspended in hot water, and subsequent decrease in water temperature then allows the polymer to dissolve. Heated metal pins are then submerged into the concentrated solution. The heated pins cause local gelling of the polymer solution, leaving them coated with a thick film upon removal. Film coalescence occurs upon evaporation of water, and the resulting capsule half-shells are removed from the pins. Refer the book by Podczeck and Jones for more details on manufacturing of capsules.[19]

The capsule shells are, in essence, delivery vehicles designed to contain mainly powders or multiparticulates. Materials contained within the capsules may dissolve immediately but may also be coated to impart MR performance. In addition to the dietary advantages of HPMC capsules, they typically contain less residual moisture; so, the materials contained within will be exposed to less moisture.[20–23] Furthermore, the capsule is designed to quickly dissolve and thus release its materials within a matter of minutes. Gelatin cross-links if exposed to aldehydes, which hinders dissolution of the gelatin capsule.[24,25] HPMC capsules are much less susceptible to cross-linking.

27.4.4 Solubility Enhancement

The rapid advancement in combinatorial chemistry and high-throughput screening techniques has resulted in a myriad of potential APIs that have extremely low aqueous solubility making their development into a viable dosage form challenging.[1,2] These compounds fall into the biopharmaceutics classification system as class 2 if they have high permeability and class 4 if they have low permeability. A key area of research is focused on the formation of amorphous solid dispersions (ASDs), which place the API in a high energy state, which facilitates dissolution and hence bioavailability.

Spray drying and hot melt extrusion are the two methods that are most commercially viable for the formation of ASDs. During the formation of ASDs, several considerations such as the relative glass transition temperature of the polymer and API and the miscibility of the API in the polymer must be evaluated.[28] Other methods of note include Kinetisol® Dispersing.

Most cellulosics that are approved for pharmaceutical use have been used in the formation of ASD for solubility enhancement. By far, the most common cellulosic utilized in this capacity is hypromellose. As demonstrated in Table 27.2, hypromellose has been utilized in a wide range of studies including model API compounds and in new drug development. Hypromellose is advantageous because it comes in a wide range of chemistries and viscosities that can be selected to control

TABLE 27.2
API Studies Utilizing HPMC as a Solubilization Enhancing Polymer

API	Polymer (When Disclosed)	Preparation Method	References
Felodipine	HPMC 2910	Spin coating	[29]
Itraconazole	HPMC 2910	Hot-melt extrusion (HME)	[30]
	HPMC 2910	HME	[31]
	HPMC 2910	Solvent evaporation	[32]
AMG 517	HPMC 2910	Spray drying	[33]
ER-34122	HPMC	Solvent evaporation	[34]
Indomethacin	HPMC	Common solvent	[35]
Itraconazole	HPMC 2910	HME, Kinetisol® dispersing	[36]
	HPMC 2910	Supercritical antisolvent	[37]
MFB-1041	HPMC 2910	Spray drying	[38]
Nimodipine	HPMC 2910	HME	[39,40]
R103757	HPMC 2910	HME	[41]
Tacrolimus	HPMC	Solvent evaporation	[42]
		Swelled polymer	
Torcetrapib	HPMC 2910	Spray drying	[43]
Griseofulvin	HPMC 2910	Melt quench	[44]
Danazol	HPMC 2910	Melt quench	[44]
	HPMC 2910	SEEDS	[45]
NVS981	HPMC	HME	[46]
Etravirine	HPMC 2910	Spray drying	[47]

the release profile and compatibility with the API. Despite these advantages, hypromellose has low organic solubility, making spray drying challenging, and does not extrude well without the addition of a plasticizer.

Hypromellose acetate succinate (HPMCAS) is also frequently leveraged for use in solubilization methodologies with examples demonstrated in Table 27.3. HPMCAS is unique in that it is enteric in nature and contains both hydrophilic and hydrophobic substituents. The enteric nature is advantageous in that the ASD does not dissolve until it encounters the higher pH values associated with the upper intestine and thus release the API in an environment more suitable for maintaining supersaturation of the drug compound.

Although hypromellose and HPMCAS are the dominant cellulosics utilized in the solubilization of poorly soluble drug compounds, most cellulosics accepted as an excipient for pharmaceutical formulation have been studied. These polymers, shown in Table 27.4, include MC, hypromellose phthalate (HPMC-P), EC, carboxymethyl ethylcellulose, hydroxypropyl cellulose, and sodium carboxymethylcellulose.

TABLE 27.3
API Studies Utilizing HPMCAS as a Solubilization Enhancing Polymer

API	Polymer (When Disclosed)	Preparation Method	References
Felodipine	HPMCAS-MF	Spin coating	[29]
Itraconazole	HPMCAS-HG	Spray drying	[48]
AMG 517	HPMCAS-MF	Spray drying	[33]
Torcetrapib	HPMCAS-MG	Spray drying	[43]
Griseofulvin	HPMCAS-HF	Melt quench	[44]
Danazol	HPMCAS-HF	Melt quench	[44]
NVS981	HPMCAS	HME	[46]
Celecoxib	HPMCAS-L	Spray drying	[49]
BMS-A	HPMCAS-MF	Spray drying	[50]
Telaprevir	HPMCAS	Spray drying	[51]
Ivacaftor	HPMCAS	Spray drying	[52]

TABLE 27.4
API Studies Utilizing Cellulosics Other than HPMC and HPMCAS as a Solubilization Enhancing Polymer

API	Polymer (When Disclosed)	Preparation Method	References
ER-34122	MC	Solvent evaporation	[34]
	HPMC-P	Solvent evaporation	
Indomethacin	HPC-SL	Common solvent	[35]
	HEC	Common solvent	
Itraconazole	HPMC-P grade 55	Spray drying	[48]
MFB-1041	HPMC-P grade 55	Spray drying	[38]
	CMEC®	Spray drying	
NVS981	HPMC-P (HP-50)	HME	[46]
KCA-098	HPC-SL	Co-grinding	[53]
Celecoxib	EC	Nanoparticle	[49]
Flurbiprofen	Na-CMC	SEEDS	[54]

27.5 MODIFIED RELEASE

Many chronic conditions are most effectively treated via MR oral drug delivery, mainly due to the fact that the number of required daily doses can be reduced to one or two, thus facilitating convenience and compliance. APIs exhibiting narrow therapeutic indices can be formulated into oral MR dosage forms in order to prevent API concentration spikes in the blood above that of the minimum toxic concentration (see Figure 27.3).

Cellulose ethers serve as rate-modifying polymers in various oral drug delivery systems. HPMC is commonly used as the rate-modifying polymer in hydrophilic matrices,[55–63] while hydroxypropylcellulose (HPC) can be used as a sub- or laminating coating in osmotic delivery systems.[64,65] EC, a water-insoluble cellulose ether, is used as the barrier-forming polymer in coated multiparticulates and microcapsules[11,66–71] and is also used as the matrix-forming polymer in inert matrices.[72–75] Thermoplastic cellulose ethers, like HPC and EC, can be used to produce melt-processed dosage forms.[76–89] Two of the more commonly utilized systems, hydrophilic matrices and barrier-coated multiparticulates, are elaborated upon in greater detail.

The most common MR oral delivery system is the hydrophilic matrix tablet. This dosage form consists of a high-MW water-soluble polymeric matrix containing the API. Additional additives, such as microcrystalline cellulose, are typically added to tailor the physical properties of the tablet. HPMC 2208 is the most widely utilized polymer in hydrophilic matrices. It is the most hydrophilic of the HPMC chemistries, typically exhibiting a relatively high powder dissolution temperature (51°C–53°C).[55] Thus, HPMC 2208 more readily wets and begins to dissolve upon introduction of the matrix tablet into aqueous media equilibrated to body temperature. A partially dissolved, swollen HPMC layer forms around the matrix tablet (see Figure 27.4). API release is modulated by both diffusional transport through, and dilution/erosion of, the swollen HPMC layer. Hydrophilic matrices typically enable first-order release kinetics (see Figure 27.5) over a time interval ranging anywhere from 8 to 24 h.[30,90–93]

Hydrophilic matrices are the most commonly utilized MR oral dosage forms because they are more easily developed, require fewer production steps, and are less expensive to manufacture.[94–96] Hydrophilic matrices are typically made by granulating the API with other ingredients, such as filler and binder. Granulation occurs in a fluidized bed, high-shear granulator, or roller compactor; and HPMC can be included in the formulation prior to (intragranular) or following granulation (extragranular). A glidant and lubricant may be blended extragranularly shortly before compression to tablets.

The second most common MR dosage form is the coated multiparticulate. Although not always the case, the barrier-coated multiparticulate is typically a multilayered entity with a nonpareil

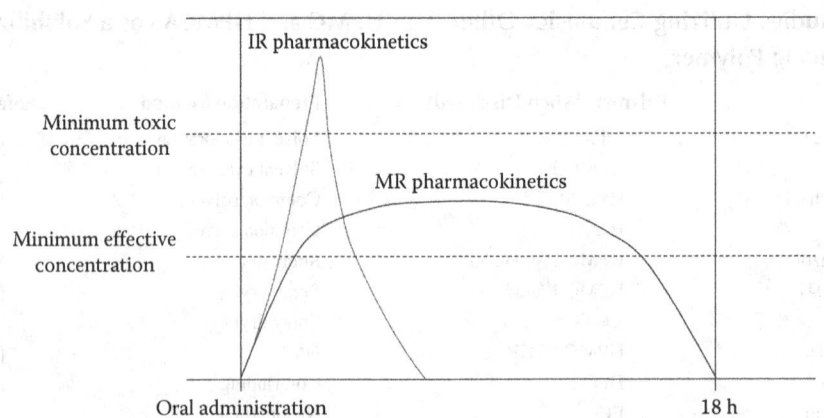

FIGURE 27.3 Illustration of how MR dosage forms can more safely deliver narrow therapeutic index APIs.

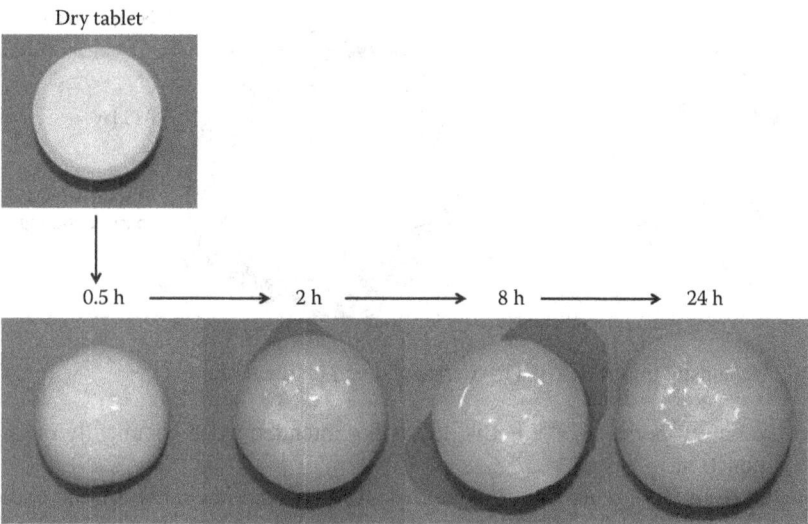

FIGURE 27.4 Visual observation of a hydrophilic matrix following introduction into aqueous media.

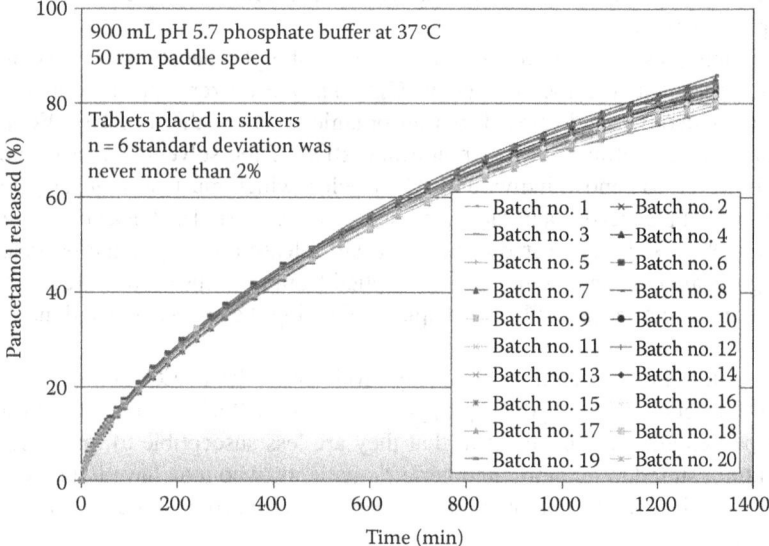

FIGURE 27.5 Typical first-order release kinetics from hydrophilic matrices.

sucrose- or microcrystalline cellulose–based core, which is coated with an API-hypromellose 2910 layer. Upon the API layer can be applied an MR barrier layer (see Figure 27.6), where EC can serve as both the film-forming and rate-modifying polymer because of the fact that EC is a water-insoluble cellulose ether across the physiological pH range. Barrier-coated multiparticulates are commonly produced in a fluidized bed fitted with a Würster insert (bottom spray configuration).

EC is soluble in many organic solvents and insoluble in water, so the polymer can be applied to the substrate from: (1) an organic solution, typically using a low MW alcohol, acetone, or a mixture thereof; or (2) an aqueous pseudolatex dispersion. The term "pseudolatex" is used because the dispersed nanoparticles in an aqueous latex are considered to be produced from a polymerization reaction. Aqueous EC pseudolatexes, however, are produced utilizing either melt-processing

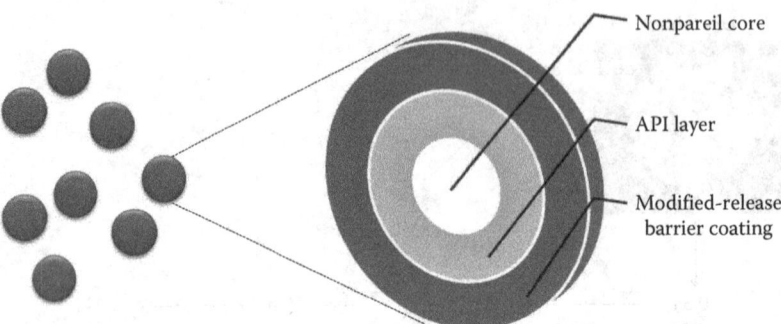

FIGURE 27.6 Schematic illustration of a barrier-coated multiparticulate.

or emulsion-solvent evaporation,[97,98] that is, EC is not intended to be chemically modified during pseudolatex production.

When coated from an organic solution, the dissolved polymeric chains overlap and entangle to render a highly uniform and durable film across the substrate surface. The higher the MW of EC, the more durable the applied film. However, a balance must be met between a sufficiently high MW and avoiding a solution viscosity that is too high for suitable atomization.[99] Hence, EC grades ranging from 50,000 to 120,000 g/mol (polystyrene-equivalent weight-average MW) are typically used for organic spray coating.

EC is commonly plasticized with materials, such as dibutyl sebacate or triethyl citrate, in order to reduce its glass transition temperature (~130°C).[100] The plasticizer is particularly necessary when coating with an aqueous pseudolatex. From an organic solution, the dissolved EC polymer molecules coalesce into a continuous film upon evaporation of the solvent. Water evaporates from a pseudolatex, however, to deposit individual EC particles, which must subsequently coalesce into a film. The presence of plasticizer facilitates coalescence of the particles, but a curing step is typically required as well.[101,102] Stability-related changes in API release due to gradual particle coalescence represent a significant challenge with coatings applied from aqueous pseudolatex,[102,103] that is, particle coalescence may not necessarily be complete following the curing step and may continue during storage.

Multiparticulates can be coated with varying levels or grades of EC, and may include varying levels of pore formers, such as hydroxypropylcellulose or HPMC, in order to adjust MR performance. Multiparticulates are beneficial in that they are less susceptible to food effects[104] and are more easily administered to pediatric and geriatric patients, who may have issues swallowing tablets.[105,107] Although they can be compressed to tablets,[109] multiparticulates are typically delivered in hard-shell capsules.[109,110]

27.6 HEALTH ENHANCEMENT

Cellulose ethers have been used in foods for several decades as functional additives. During this time, most of the uses focused on the ability of these polymers to tailor the rheology of the food. However, within the last decade, a new focus has emerged due to their ability to impart broad health benefits when included in the diet. Cellulose ethers can be used as fibers to substitute allergenic food ingredients, tshift fat content to healthier oils, to help control body weight, and blunt postprandial insulin levels.

27.7 DIETARY FIBER

Dietary fibers impact all aspects of gut physiology and are a vital part of healthy diet. There is a need for development of novel fiber-rich foods that are both acceptable to the consumer and have proven health benefits. However, many fibers (particularly those that are rapidly fermented) result in gas production, bloating, and diarrhea.[112] Several cellulose ethers are recognized as fibers (e.g., HPMC is a nonfermentable soluble dietary fiber), which have been used in the manufacturing of many foods for several decades. It has a long safety record and GRAS status up to 20 g/day in the United States.[113] Due to HPMC's unique value as a dietary fiber, the Association of Official Analytical Chemists (AOACs) published a special test method (AOAC 2006.08) to analyze the soluble fiber content provided by HPMC, MC, and CMC.

27.8 REPLACEMENT OF GLUTEN IN BAKERY PRODUCTS

Celiac disease (CD), an immune-mediated enteropathy, is one of the most common lifelong disorders on a worldwide basis. At present, the only available treatment for CD is strict adherence to a gluten-free diet, which means a permanent withdrawal of gluten from daily food. The majority of leavened cereal-based products are made of wheat flour or other cereal flours containing gluten. Gluten is an essential structure-building component in bread and other bakery products, too: its removal impairs the dough's capacity to properly develop during kneading, leavening, and baking. Hydrocolloids and mixtures of hydrocolloids can mimic the structuring role of gluten due to their physical properties.[114] The usage of HPMC or the combination of HPMC with CMC as a gluten replacement in baked goods has been in practice since at least 1985.[115] Figure 27.7 demonstrates the ability of HPMC to build a structural network inside the product ensuring that cakes and breads retain the desired shape and volume.

27.9 REDUCTION OF FAT UPTAKE IN FOODS

An interesting approach to reduce the health risks of consumer foods is to reduce the overall fat content of fried food. Cellulose derivatives such as CMC, MC, and HPMC are used in the batter or breadcrumbs to reduce oil absorption during frying (e.g., in doughnuts, fried dough products, structured, extruded, and coated products).[116] The following example highlights the ability of HPMC to reduce the fat content of French fries.

27.10 HEALTHIER FATS

Although trans- and saturated fats have beneficial attributes from the standpoint of food formulation (firmness, reduction of oil migration, and leakage), they have also been linked to detrimental health effects. As a result, the World Health Organization recommends that fat consumption should be shifted toward unsaturated fatty acids as opposed to saturated and trans-fats. However, fat sources higher in unsaturated fatty acids, such as edible oils, lack structure at room temperature. As a consequence, they can produce adverse effects in food products, which often result in a reduction in

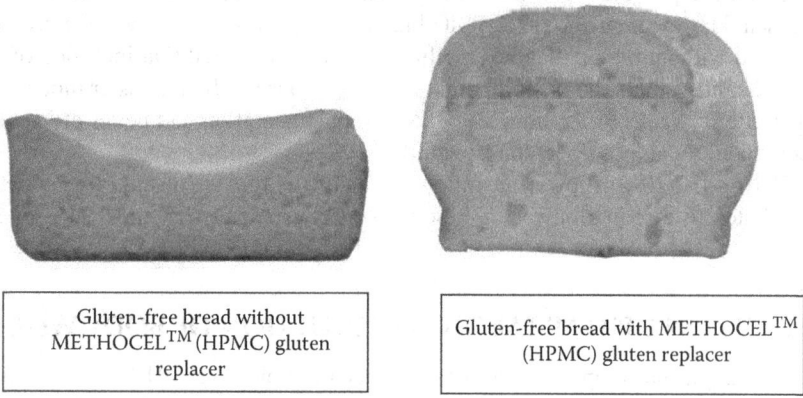

FIGURE 27.7 Impact of METHOCEL HPMC on the structure of gluten-free bread.

product quality when used as a direct substitute for solid fats. An emerging strategy is to entrap the liquid oil within a gel network (e.g., with EC). The incorporation of edible oil-based EC oleogels into food systems is now an active area of research and applications currently under investigation include finely comminuted meat products, cream fillings, and chocolate.[116]

27.11 REDUCTION OF BLOOD CHOLESTEROL

Elevated levels of total cholesterol (TC) and low-density lipoprotein cholesterol (LDL-C) are associated with an increased risk for coronary heart disease. In several human clinical trials, HPMC was found to reduce total blood cholesterol (TC) and LDL cholesterol while having only a minor effect on high-density lipoprotein-cholesterol (HDL-C).[113] For example, Maki et al. found that HPMC with varying dose and viscosity combinations showed LDL-C reductions ranging from 6.1% to 13.3% compared to a nonsignificant reduction (1.9%) in the control group. Changes in total and nonhigh density lipoprotein cholesterol paralleled those for LDL-C. Concentrations of HDL-C were not significant altered.[117] A recent study demonstrates that HPMC is an effective adjunct to statin therapy for further lowering atherogenic lipids and lipoproteins in humans with primary hypercholesterolemia.[118] In the European Union, the commission regulation (EU) No. 432/2012 allows the following health claim: "HPMC contributes to the maintenance of normal blood cholesterol levels."

In a diet-induced obesity mouse model, significant decreases in concentrations of plasma cholesterol were seen by feeding cationic hydroxylethyl cellulose (cHEC): Plasma TC were 19.7%, and 25.3% lower in mice fed the 2% or 4% cHEC-supplemented diet.[119]

27.12 GLYCEMIC RESPONSE

Diabetes is associated with numerous adverse health outcomes, including cardiovascular disease. Dietary fibers that produce viscous solutions in the digestive system blunt postprandial glucose and insulin excursions.[120] The degree of viscosity appears to be inversely related to glycemic response, with the more viscous dietary fibers producing greater effects. The fibers form viscous solutions when mixed with the gastrointestinal tract contents, slowing gastric emptying and thickening small intestine contents. This may reduce contact between food and digestive enzymes and interfere with diffusion of nutrients to absorptive surfaces, thus slowing the rate at which glucose molecules become available for absorption at the small intestine brush border. It has been suggested that lowering dietary glycemic load may be advantageous for individuals at risk for type 2 diabetes, coronary heart disease, and obesity.[118,121]

Several studies are suggesting that consumption of HPMC has potential therapeutic values in the management of risk factors for type 2 diabetes and cardiovascular disease.[121]

Consumption of high viscosity (HV) and ultra high viscosity (UHV) HPMC with a meal significantly blunted postprandial insulin excursions and was well tolerated in overweight and obese men and women. However, only UHV HPMC blunted the peak glucose level.[121] Further studies in subjects with and without type 2 diabetes mellitus have demonstrated that inclusion of HV-HPMC in a meal significantly blunts the postprandial glucose response.[118] In a hamster model, it was demonstrated that after 8 weeks of feeding, 8% HPMC supplementation had better efficacy in glucose reduction compared to natural fibers (e.g., pectin).[113]

In the EU No. 432/2012 allows the following health claim: "Consumption of HPMC with a meal contributes to a reduction in the blood glucose rise after that meal (4 g of HPMC per portion)."

27.13 EFFECTS ON LIPID METABOLISM: REDUCTION OF BODY WEIGHT GAIN

HPMC modulates plasma lipoprotein profiles and hepatic lipid levels. HPMC is not absorbed by the body, but its presence in the intestinal lumen increases fecal fat, sterol, and bile acid excretion, which results indirectly in changes in hepatic lipid metabolism. In recent studies of hepatic gene expression levels in the hamster, upregulation of bile acid synthesis and export were observed.[122] A first study suggests that HPMC may be facilitating fat excretion in a biased manner with preferential fecal excretion of both trans- and saturated fat in hamsters fed fast food diets.[123]

In preliminary studies, maturing hamsters on a high-fat diet supplemented with HPMC put on significantly less body weight than control animals, due primarily to reduced deposition of abdominal fat tissue and fat accumulations in liver and skeletal muscle. In obese B6 mice, 4% and 8% HPMC supplementation in a high-fat diet leads to significant weight loss. Reductions in plasma cholesterol, glucose, and insulin levels were seen, which are strongly correlated with reduced leptin concentrations. Moreover, increases in fecal secretion of total bile acids, sterols, and fats indicated altered fat absorption when HPMC is incorporated in the diet. The data indicate that HPMC not only reduces body weight but also normalizes the metabolic abnormalities associated with obesity and suggest that the effect of HPMC on glucose and lipid homeostasis in B6 mice is mediated through improvement in leptin sensitivity resulting from reduced fat absorption.[115] Hydroxy ethyl methyl cellulose (HEMC) was shown to be similarly effective in improving the lipid metabolism under high-fat diet condition.[124]

27.13.1 SATIETY

Overweight and obesity, as well as consequent cardiovascular disease, are primarily driven by over availability of food and an increasingly sedentary lifestyle. One approach to treatment is to manipulate appetite and reduce food intake through control of satiety. Dietary fibers are thought to impact on satiety (inhibition of hunger as a result of having eaten) because of their property of producing viscosity. Viscous soluble fibers may be useful because they prolong the intestinal phase of nutrient digestion and absorption. Not all dietary fibers have an impact on satiety.[125]

Materials incorporated into the diet that form a gel mass in the stomach, such as alginate and pectin, have been shown to enhance satiety.[126] This effect is thought to be caused by distending of the stomach wall.

A novel, food grade, MC (Satisfit™LTG) was developed, which gels at temperatures below body temperature but which is unaffected by pH.

It was shown in vivo by MRI trials that Satisfit™LTG forms a gel mass, which persists for at least 2 h. In contrast, the conventional MC not gelling at body temperature clears the stomach rapidly. A clinical trial with healthy human volunteers demonstrates clear perception of greater satiety. Analysis of results from the Visual Analogue Scale indicates that appetite recovery is slower with novel gelling MC than with control products. A significant reduction in energy intake was seen, which might be explained by the gelation of the MC in the stomach. The satiety effect has been shown to last for at least 2 h after ingestion of the product.[127]

27.14 CONCLUSIONS

This entry has clearly established the enabling role cellulose ethers play in our everyday health and wellness. Employed in tablets, capsules, liquids, and films, these versatile polymers have found continuously expanding uses over the past 50 plus years.

In the pharmaceutical market, polymers such and HPMC and EC provide the controlling function to modified release oral tablets, which improve both the efficacy and patient compliance of many marketed drug products. HPMC and HPMC-AS enable the formation of solubility-enhanced formulations of poorly soluble active ingredients, which comprise an ever-increasing percentage of new APIs. HPMC has expanded the use of hard shell capsules to new customer bases not open to the consumption of gelatin. Finally, HPMC is the most widely used tablet coating polymer providing patients with a desirable product with ease of recognition.

Within the food market, MC and HPMC have long been used to improve the texture of processed foods. In addition, they are now being utilized to bring health enhancements to our everyday foods.

The ability to improve the structure of gluten-free baked products allowing this growing population to obtain desirable products, which meet their unique dietary needs. HPMC has been recognized as a dietary fiber, which allows food producers to increase daily fiber without the negative effects of fermentable fibers. Finally, as the world struggles to address obesity and metabolic syndrome, cellulose ethers have been demonstrated to reduce oil uptake in common foods, enhance the utility of health oils, and increased satiety when consumed in a normal diet.

REFERENCES

1. Doeker, E. Cellulose derivatives. *Adv. Polym. Sci.* 1993, 107, 199–265.
2. Donges, R. Non-ionic cellulose ethers. *Br. Polym. J.* 1990, 23(4), 315–326.
3. Hirrien, M., Chevillard, C., Desbrieres, J., Axelos, M.A.V., Rinaudo, M. Thermal gelation of methyl celluloses: New evidence for understanding the gelation mechanism. *Polymer* 1998, 39(25), 6251–6259.
4. Reibert, K.C., Conklin, J.R. Enhanced gel strength methycellulose, EP1171471, December 17, 2003.
5. Brandt, L. Cellulose ethers. In *Ullmann's Encyclopedia of Industrial Chemistry: Cancer Chemotherapy to Ceramic Colorants*, 5th edn., Gerhartz, W., Yamamoto, Y.S., Campbell, F.T., Pfefferkorn, R., Rounsaville, J.F. Eds., VCH: Wienheim, Germany, 1986, Vol. A5, pp. 461–488.
6. Reynolds, T.D., Gehrke, S.H., Hussain, A.S., Shenouda, S.L. Polymer erosion and drug release, characterization of hydroxypropyl methylcellulose matrices. *J. Pharm. Sci.* 1998, 87(9), 1115–1123.
7. Pharmacircle. Available from www.pharmacircle.com (accessed July 2012).
8. Bowen, R.A., Austgen, L., Rouge, M. Pathophysiology of the digestive system. 2006 7/5/2006 09/22/2012.
9. Anon. *Coating of Tablets*, Merck & Co., Inc.: NL, 1965, Vol. 6502190, 16pp.
10. Khot, S.N., Hwang, J., Anderson, J., Coppens, K., Faham, A., Sheskey, P. How very-low-viscosity coatings reduce the total cost of ownership of coating operations. *Tablet Capsule* 2012, 4–9.
11. Liang, L., Wang, H., Bhatt, P.P., Veiira, M.L. Sustained-release formulations of topiramate for oral administration, WO 2008061226 A3, 2008, 58pp.
12. Feng, H. Compressible mixture for coated compositions, US 20060263429 A1, 2006, 18pp.
13. Trehan, A., Arora, V.K., Madan, S., Malik, R. Pharmaceutical dosage forms of biguanide-sulfonylurea combinations, WO 2004045622 A1, 2004, p. 38.
14. Trehan, A., Arora, V.K., Madan, S., Malik, R. Pharmaceutical compositions containing a biguanide-glitazone combination, WO 2004045608 A9, June 2, 2004, 47pp.
15. Yoneyama, S., Bando, H. Drug composition having active ingredient adhered at high concentration to spherical core, WO 2004080439 A1, September 23, 2004, 237pp.
16. Ikemoto, K. et al. Enteric-coated stable omeprazole tablets and their manufacture, JP 2004175768 A, 2004, 5pp.
17. Nguyen, C., Christensen, J.M., Ayres James, W. Novel mesalamine-loaded beads in tablets for delayed release of drug to the colon. *Pharm. Dev. Technol.* 2012, 17(1), 73–83.
18. USP 35—NF 30. The United States Pharmacopeial Convention, 2012.
19. Podczeck, F., Jones, B.E. *Pharmaceutical Capsules*, Pharmaceutical Press: London, U.K., 2004.
20. Chiba, T., Muto, H., Tanioka, S., Nishayama, Y., Hoshi, N., Onda, Y. Cellulose ether composition and a hard medicinal capsule prepared Therefrom, EP 180287 B1, 1986, 21pp.
21. Matsuura, S., Yamamoto, T. New hard capsules prepared from water-soluble cellulose derivative. *Yakuzaigaku* 1993, 53(2), 135–140.
22. Nagata, S. Cellulose capsules—An alternative to gelatin: Structural, functional and community aspects. *Biomed. Polym. Polym. Ther.* 2001, 3, 53–62.
23. Ogura, T., Furuya, Y., Matsuura, S. HPMC capsules. An alternative to gelatin. *Pharm. Technol. Eur.* 1998, 10(11), 32, 34, 36, 40, 42.
24. Tengroth, C., Gasslander, U., Andersson, F.O., Jacobsson, S.P. Cross-linking of gelatin capsules with formaldehyde and other aldehydes: An FTIR spectroscopy study. *Pharm. Dev. Technol.* 2005, 10(3), 405–412.
25. Yong, C.-S., Li, D.X., Oh, D.H., Kim, J.A., Yoo, B.K., Woo, J.S., Rhee, J.D., Choi, H.G. Retarded dissolution of ibuprofen in gelatin microcapsule by cross-linking with glutaradehyde. *Arch. Pharm. Res.* 2006, 29(6), 520–524.
26. Lipinski, C.A. Drug-like properties and the causes of poor solubility and poor permeability. *J. Pharmacol. Toxicol. Methods* 2000, 44(1), 235–249.

27. Lipinski, C.A., Lombardo, F., Dominy, B.W., Feeney, P.J. Experimental and computational approaches to estimate solubility and permeability in drug discovery and development settings. *Adv. Drug Deliv. Rev.* 2001, 46(1–3), 3–36.
28. Robert O. Williams III, Alan B. Watts, Dave A. Miller. *Formulating Poorly Water Soluble Drugs*, Springer: New York, 2012.
29. Konno, H., Taylor, L.S. Influence of different polymers on the crystallization tendency of molecularly dispersed amorphous felodipine. *J. Pharm. Sci.* 2006, 95(12), 2692–2705.
30. Six, K., Berghmans, H., Leuner, C., Dressman, J., Van Werde, K., Mullens, J., Benoist, L. et al. Characterization of solid dispersions of itraconazole and hydroxypropylmethylcellulose prepared by melt extrusion—Part II. *Pharm. Res.* 2003, 20(7), 1047–1054.
31. Verreck, G., Six, K., Van den Mooter, G., Baert, L., Peeters, J., Brewster, M.E. Characterization of solid dispersions of itraconazole and hydroxypropylmethylcellulose prepared by melt extrusion—Part I. *Int. J. Pharm.* 2003, 251(1–2), 165–174.
32. Matteucci, M.E., Paguio, J.C., Miller, M.A., Williams, R.O., Johnston, K.P. Highly supersaturated solutions from dissolution of amorphous itraconazole microparticles at pH 6.8. *Mol. Pharm.* 2009, 6(2), 375–385.
33. Kennedy, M., Hu, J., Gao, P., Li, L., Ali-Reynolds, A., Chai, B., Gupta, V. et al. Enhanced bioavailability of a poorly soluble VR1 antagonist using an amorphous solid dispersion approach: A case study. *Mol. Pharm.* 2008, 5(6), 981–993.
34. Kushida, I., Ichikawa, M., Asakawa, N. Improvement of dissolution and oral absorption of ER-34122, a poorly water-soluble dual 5-lipoxygenase/cyclooxygenase inhibitor with anti-inflammatory activity by preparing solid dispersion. *J. Pharm. Sci.* 2002, 91(1), 258–266.
35. Chowdary, K.P.R., Babu, K. Dissolution, bioavailability and ulcerogenic studies on solid dispersions of indomethacin in water-soluble cellulose polymers. *Drug Dev. Ind. Pharm.* 1994, 20(5), 799–813.
36. DiNunzio, J.C., Brough, C., Miller, D.A., Williams, R.O., McGinity, J.W. Fusion processing of itraconazole solid dispersions by kinetisol (R) dispersing: A comparative study to hot melt extrusion. *J. Pharm. Sci.* 2010, 99(3), 1239–1253.
37. Lee, S., Nam, K., Kim, M.S., Jun, S.W., Park, J.S., Woo, J.S., Hwang, S.J. Preparation and characterization of solid dispersions of itraconazole by using aerosol solvent extraction system for improvement in drug solubility and bioavailability. *Arch. Pharm. Res.* 2005, 28(7), 866–874.
38. Kai, T., Akiyama, Y., Nomura, S., Sato, M. Oral absorption improvement of poorly soluble drug using solid dispersion technique. *Chem. Pharm. Bull.* 1996, 44(3), 568–571.
39. Zheng, X., Yang, R., Tang, X., Zheng, L.Y. Part I: Characterization of solid dispersions of nimodipine prepared by hot-melt extrusion. *Drug Dev. Ind. Pharm.* 2007, 33(7), 791–802.
40. Zheng, X., Yang, R., Zhang, Y., Wang, Z.J., Tang, X., Zheng, L.Y. Part II: Bioavailability in beagle dogs of nimodipine solid dispersions prepared by hot-melt extrusion. *Drug Dev. Ind. Pharm.* 2007, 33(7), 783–789.
41. Verreck, G., Vandecruys, R., De Conde, V., Baert, L., Peeters, J., Brewster, M.E. The use of three different solid dispersion formulations—Melt extrusion, film-coated beads, and a glass thermoplastic system—To improve the bioavailability of a novel microsomal triglyceride transfer protein inhibitor. *J. Pharm. Sci.* 2004, 93(5), 1217–1228.
42. Yamashita, K., Nakate, T., Okimoto, K., Ohike, A., Tokunaga, Y., Ibuki, R., Higaki, K., Kimura, T. Establishment of new preparation method for solid dispersion formulation of tacrolimus. *Int. J. Pharm.* 2003, 267(1–2), 79–91.
43. Friesen, D.T., Shanker, R., Crew, M., Smithey, D.T., Curatolo, W.J., Nightingale, J.A.S. Hydroxypropyl methylcellulose acetate succinate-based spray-dried dispersions: An overview. *Mol. Pharm.* 2008, 5(6), 1003–1019.
44. Murdande, S.B., Pikal, M.J., Shanker, R.M., Bogner, R.H. Solubility advantage of amorphous pharmaceuticals, part 3: Is maximum solubility advantage experimentally attainable and sustainable? *J. Pharm. Sci.* 2011, 100(10), 4349–4356.
45. Anby, M.U., Williams, H.D., McIntosh, M., Benameur, H., Edwards, G.A., Pouton, C.W., Porter, C.J.H. Lipid digestion as a trigger for supersaturation: Evaluation of the impact of supersaturation stabilization on the in vitro and in vivo performance of self-emulsifying drug delivery systems. *Mol. Pharm.* 2012, 9(1), 2063–2079.
46. Ghosh, I., Snyder, J., Vippagunta, R., Alvine, M., Vakil, R., Tong, W.Q., Vippagunta, S. Comparison of HPMC based polymers performance as carriers for manufacture of solid dispersions using the melt extruder. *Int. J. Pharm.* 2011, 419(1–2), 12–19.
47. Kiekens, F.R.I., Voorspoels, J.F.M., Baert, L.E.C. Process for preparing spray dried formulations of Etravirine, WO2007141308A1, June 06, 2007.
48. Engers, D., Teng, J., Jimenez-Novoa, J., Gent, P., Hossack, S., Campbell, C., Thomson, J. et al. A Solid-state approach to enable early development compounds: Selection and animal bioavailability studies of an itraconazole amorphous solid dispersion. *J. Pharm. Sci.* 2010, 99(9), 3901–3922.

49. Morgen, M., Bloom, C., Beyerinck, R., Bello, A., Song, W., Wilkinson, K., Steenwyk, R., Shamblin, S. Polymeric nanoparticles for increased oral bioavailability and rapid absorption using celecoxib as a model of a low-solubility, high-permeability drug. *Pharm. Res.* 2012, 29(2), 427–440.
50. Qian, F., Wang, J., Hartley, R., Tao, J., Haddadin, R., Mathias, N., Hussain, M. Solution behavior of PVP-VA and HPMC-AS-based amorphous solid dispersions and their bioavailability implications. *Pharm. Res.* 2012, 29(10), 2766–2776.
51. Murphy, M., Dinehart, K., Hurter, P., Connelly, P., Cui, Y. Pharmaceutical compositions containing amorphous VX-950, preparation thereof, and method for treating hepatitis C infection, WO2005123076A2, December 29, 2005.
52. Hurter, P. Solid forms of N-[2,4-bis(1,1-dimethylethyl)-5-hydroxyphenyl]-1,4-dihydro-4-oxoquinoline-3-carboxamide, WO 2007079139A2, July 12, 2007.
53. Yamada, T., Saito, N., Imai, T., Otagiri, M. Effect of grinding with hydroxypropyl cellulose on the dissolution and particle size of a poorly water-soluble drug. *Chem. Pharm. Bull.* 1999, 47(9), 1311–1313.
54. Kang, J.H., Oh, D.N., Oh, Y.K., Yong, C.S., Choi, H.G. Effects of solid carriers on the crystalline properties, dissolution and bioavailability of flurbiprofen in solid self-nanoemulsifying drug delivery system (solid SNEDDS). *Eur. J. Pharm. Biopharm.* 2012, 80(2), 289–297.
55. Rogers, T.L. et al. *Investigation and Rank-Ordering of Hypromellose 2208 Properties Impacting Modified Release Performance of a Hydrophilic Matrix Tablet*, American Association of Pharmaceutical Scientists: Washington, DC, 2011.
56. Chen, B., Joshi, S.C., Lam, Y.C. Bio-fluid uptake and release of Indomethacin of direct-compressed HPMC tablets. *Carbohydr. Polym.* 2009, 75(2), 282–286.
57. Jaya, A., Gour, S., Bhavesh, S., Rajneesh, S. Stabilized controlled release dosage form of gliclazide comprising release controlling polymer and free from saccharides, WO 2008062470 A3, January 29, 2008, 19pp.
58. Lin, J.L.-Y., Wong, D., Chow, S.-L. Sustained-release pharmaceutical compositions with zero order release profile, US 7063862 B2, 2004, 10pp.
59. Ellstroem, K. et al. Extended release oral dosage forms containing dicyclobutylamino(fluoro)dihydro-benzopyrancarboxamide, EP 1423098 A1, June 2, 2003, 40pp.
60. Haque, S.F. et al., Studies on sustained release tablets of flurbiprofen. *Pak. J. Pharmacol.* 2002, 19(1), 37–40.
61. Nellore, R.V., Singh Rekhi, G., Hussain, A.S., Tillman, L.G., Augsburger, L.L. Development of metoprolol tartrate extended-release matrix tablet formulations for regulatory policy consideration. *J. Control. Release* 1998, 50(1–3), 247–256.
62. Kim, C.-J. Controlled-release tablets containing water-swellable polymers, EP 910349 A2, 1997, 70pp.
63. Infeld, M.H., Malick, A.W., Phuapradit, W., Shah, N.H. Pharmaceutical compositions with constant erosion volume for zero-order controlled release, US 5393765 A, 1995, 21pp.
64. Evenstad, K.L., Malhotra, K.R., O'Neill, V.A. Controlled-release tablets containing a water-soluble medicament, US 5268181 A, 1992, 8pp, Cont of U S Ser No 337,460, abandoned.
65. Edgren, D., Jao, F., Li, S., Skluzacek, R., Wong, P. Volume efficient controlled release dosage form with cellulose-based bilayer membrane, US 20030185888 A1, October 2, 2003, 30pp.
66. Yam, N.V. et al. Stepwise delivery of Topiramate over prolonged period of time by formulations including polyethylene oxide and polyvinylpyrrolidone, WO 2005020957 A3, 2003, 66pp.
67. Venkatesh, G.M., Stevens, P.J., Lai, J.-W. Compositions comprising melperone and controlled-release dosage forms, US 20100151015 A1, September 23, 2010, 16pp.
68. Danagher, H.K., Hayes, G.G., Mohammad, H., Whitehouse, M.W.J.O. Pharmaceutical spheroids comprising glyceryl monostearate and a polymer binder. EP 2001445 A1, December 17, 2007, 43pp.
69. Venkatesh, G.M., Lai, J.-W., Vyas, N.H. Drug delivery systems comprising weakly basic selective serotonin 5-HT3 blocking agent and organic acids, EP 1976491 A2, October 8, 2007, 52pp.
70. Rogers, T.L., Wallick, D. Reviewing the use of ethylcellulose, methylcellulose and hypromellose in microencapsulation. Part 2: Techniques used to make microcapsules. *Drug Dev. Ind. Pharm.* 2011, 37(11), 1259–1271.
71. Rogers, T.L., Wallick, D. Reviewing the use of ethylcellulose, methylcellulose and hypromellose in microencapsulation. Part 3: Applications for microcapsules. *Drug Dev. Ind. Pharm.* 2012, 38(5), 521–539.
72. Rogers, T.L., Wallick, D. Reviewing the use of ethylcellulose, methylcellulose and hypromellose in microencapsulation. Part 1: Materials used to formulate microcapsules. *Drug Dev. Ind. Pharm.* 2012, 38(2), 129–157.
73. Pollock, D.K., Sheskey, P.J. Opportunities in direct-compression controlled-release tablets. *Pharm. Technol.* 1996, 20(9), 120, 122, 124, 126, 128, 130.
74. Pollock, D.K., Sheskey, P.J. Micronized ethyl cellulose: Opportunities in direct-compression controlled-release tablets. *Pharm. Technol. Eur.* 1997, 9(1), 26, 28, 30, 32, 34, 36.

75. Pollock, D.K. Influence of filler/binder composition on the properties of a inert matrix CR system. *Proc. Int. Symp. Control. Release Bioact. Mater.* 1997, 24, 481–482.
76. Nakamoto, A., Ogawa, K., Ukigaya, T. Sustained-release medicinal composition, US 3773920 A, November 20, 1973, 3pp.
77. Fischer, G., Slot, L., Bar-Shalom, D., Anderson, C., Lademann, A.-M. Controlled release pharmaceutical compositions containing polymers, EP 1929998 A3, November 26, 2003, 105pp.
78. Fischer, G., Bar-Shalom, D., Lademann, A.-M., Jenser, G. Controlled release solid dispersions containing Carvedilol, EP 1429734 B1, December 26, 2003, 110pp.
79. De Brabander, C., Vervaet, C., Remon, J.P. Development and evaluation of sustained release mini-matrices prepared via hot melt extrusion. *J. Control. Release* 2003, 89(2), 235–247.
80. De Brabander, C., Vervaet, C., Bortel, L.V., Remon, J.P. Bioavailability of ibuprofen from hot-melt extruded mini-matrices. *Int. J. Pharm.* 2004, 271(1–2), 77–84.
81. Dhanraj, A.P. et al. Sustained release extrudates using melt extrusion technology, IN 2006MU01439 A, 2011, 34pp.
82. Katz, I.M. Hydroxypropyl cellulose-containing preparations for the treatment of keratoconjuctivitis sicca (dry eye syndrome), GB 1543189 A, 1977, 19pp.
83. Miller Dave, A., McConville, J.T., Yang, W., Williams, R.O. 3rd, McGinity, J.W. Hot-melt extrusion for enhanced delivery of drug particles. *J. Pharm. Sci.* 2007, 96(2), 361–376.
84. Prodduturi, S., Manek, R.V., Kolling, W.M., Stodghill, S.P., Repka, M.A. Water vapor sorption of hot-melt extruded hydroxypropyl cellulose films: Effect on physico-mechanical properties, release characteristics, and stability. *J. Pharm. Sci.* 2004, 93(12), 3047–3056.
85. Read, M. et al. Thermal and rheological evaluation of pharmaceutical excipients for hot melt extrusion. *Annu. Tech. Conf. Soc. Plast. Eng.* 2005, 63, 3100–3104.
86. Repka, M.A., Gerding, T.G., Repka, S.L., McGinity, J.W. Influence of plasticizers and drugs on the physical-mechanical properties of hydroxypropylcellulose films prepared by hot melt extrusion. *Drug Dev. Ind. Pharm.* 1999, 25(5), 625–633.
87. Repka, M.A., Gutta, K., Prodduturi, S., Munjal, M., Stodghill, S.P. Characterization of cellulosic hot-melt extruded films containing lidocaine. *Eur. J. Pharm. Biopharm.* 2005, 59(1), 189–196.
88. Repka, M.A., McGinity, J.W. Bioadhesive properties of hydroxypropyl cellulose topical films produced by hot-melt extrusion. *J. Control. Release* 2001, 70(3), 341–351.
89. Roth, W., Burst, A., Zietsch, M. Abuse resistant melt extruded formulation having reduced alcohol interaction, EP 2389172 A1, November 30, 2010, 130pp.
90. Baki, G., Bajdik, J., Kelemen, A., Pintye-Hódi, K. Formulation of a solid intravaginal matrix system to prolong the pH-decreasing effect of lactic acid. *J. Drug Deliv. Sci. Technol.* 2009, 19(2), 133–137.
91. Chatterji, A., Huang, J., Koennings, S., Lindenstruth, K., Sandhu, H.K., Shah, N.H. Pharmaceuticals of metabotropic glutamate 5 receptor (mGlu5) antagonists, WO 2012019989 A3, August 2, 2012, 50pp.
92. Doshi, M.M., Joshi, M.D., Mody, S.B. Controlled release formulations of nimesulide, EP 1147767 A1, October 24, 2001, 9pp.
93. Jani, R.H. et al. Matrix release pharmaceutical compositions and process for the preparation thereof, IN 212955 A1, 2006, 33pp.
94. Modi, V.C., Seth, A.K. Formulation and evaluation of diltiazem sustained release tablets. *Int. J. Pharm. Biol. Sci.* 2010, 1(3), 1–10.
95. Adden, R. et al. Innovative renewable cellulose-based materials for modified-release drug delivery. Abstracts of Papers, *244th ACS National Meeting and Exposition*, Philadelphia, PA, August 19–23, 2012, p. CELL-25.
96. Kim, H., Fassihi, R. Matrix for controlled delivery of highly soluble srugs, US 6337091 B1, 1999, 31pp.
97. Varma, M.V.S., Kaushal, A.M., Garg, A., Garg, S. Factors affecting mechanism and kinetics of drug release from matrix-based oral controlled drug delivery systems. *Am. J. Drug Deliv.* 2004, 2(1), 43–57.
98. Leng, D.E., Sigelko, W.L., Sounders, F.L. Aqueous dispersions of plasticized polymer particles, EP 0113443 B1, May 15, 1983, p. 27.
99. Banker, G.S. Food and pharmaceutical coating composition, method of preparation and products so coated, WO 80/00659, April 17, 1980, p. 27.
100. Sheskey, P.J., Keary, C.M. In situ, liquid-activated film-coated tablets and a process for making the same, EP 2164343 A1, March 24, 2009, 3pp.
101. Coppens, K.A., Hall, M.J., Mitchell, S.A., Read, M.D. Hypromellose, ethyl cellulose, and polyethylene oxide use in hot melt extrusion. *Pharm. Technol.* 2006, 30(1), 62, 64, 66, 68, 70.
102. Muschert, S., Siepmann, F., Leclercq, B., Siepmann, J. Dynamic and static curing of ethylcellulose: PVA-PEG graft copolymer film coatings. *Eur. J. Pharm. Biopharm.* 2011, 78(3), 455–461.

103. Siepmann, F., Hoffmann, A., Leclercq, B., Carlin, B., Siepmann, J. How to adjust desired drug release patterns from ethylcellulose-coated dosage forms. *J. Control. Release* 2007, 119(2), 182–189.
104. Koerber, M., Hoffart, V., Walther, M., Macrae, R.J., Bodmeier, R. Effect of unconventional curing conditions and storage on pellets coated with Aquacoat ECD. *Drug Dev. Ind. Pharm.* 2010, 36(2), 190–199.
105. Muschert, S., Siepmann, F., Leclercq, B., Carlin, B., Siepmann, J. Simulated food effects on drug release from ethylcellulose: PVA-PEG graft copolymer-coated pellets. *Drug Dev. Ind. Pharm.* 2010, 36(2), 173–179.
106. El-Gazayerly, O.N., Rakkanka, V., Ayres, J.W. Novel chewable sustained-release tablet containing verapamil hydrochloride. *Pharm. Dev. Technol.* 2004, 9(2), 181–188.
107. Kumar, A., Sharma, P.K., Kumar, P. Formulation of mouth dissolving tablets of glipizide by microencapsulation technique using various polymers. *Biosci. Biotechnol. Res. Asia* 2009, 6(1), 281–284.
108. Percel, P.J., Venkatesh, G.M., Vishnupad, K.S. Functional coating of linezolid microcapsules for taste-masking and associated formulation for oral administration, EP 1248616 B1, March 21, 2001, 12pp.
109. Hsiao, C., Chou, C.T.K. Controlled release potassium chloride, EP 211946 B1, 1986, 24pp.
110. Lippmann, I., Bell, L.G., Miller, L.G., Popli, S.D. Controlled release potassium dosage form, US 4259315 A, March 31, 1981, 5pp.
111. Sherman, D.M. Extended release formulation containing Venlafaxine, EP 797991 B1, 1997, 9pp.
112. Brownlee, I.A. The physiological roles of dietary fibre. *Food Hydrocolloids* 2011, 25(2), 238–250.
113. Hung, S.C., Anderson, W.H.K., Alvers, D.R., Langhorst, M.L., Young S.A. Effect of hydroxypropyl methylcellulose on obesity and glucose metabolism in a diet-induced obesity mouse model. *J. Diab.* 2011, 3(2), 158–167.
114. Mariotti, M., Pagani, M.A., Lucisano, M. The role of buckwheat and HPMC on the breadmaking properties of some commercial gluten-free bread mixtures. *Food Hydrocolloids* 2013, 30(1), 393–400.
115. Emeson, A. ed. *Thickening and Gelling Agents for Foods*, 2nd edn., Aspen Publication, Inc.: Gaithersburg, 1999.
116. Gravelle, A.J., Barbut, S., Marangoni, A.G. Ethylcellulose oleogels: Manufacturing considerations and effects of oil oxidation. *Food Res. Int.* 2012, 48(2), 578–583.
117. Maki, K.C et al. Lipid-altering effects of different formulations of hydroxypropylmethylcellulose. *J. Clin. Lipidol.* 2009, 3(3), 159–166.
118. Maki, K.C., Carson, M.L., Miller, M.P., Anderson, W.H.K., Turowski, M., Reeves, M.S., Kaden, V., Dicklin, M.R. Hydroxypropylmethylcellulose lowers cholesterol in statin-treated men and women with primary hypercholesterolemia. *Eur. J. Clin. Nutr.* 2009, 63(8), 1001–1007.
119. Young, S.A., Hung, S.-C., Anderson, W.H.K., Albers, D.R., Langhorst, M.L., Yokoyama, W. Effects of cationic hydroxyethyl cellulose on glucose metabolism and obesity in a diet-induced obesity mouse model. *J. Diab.* 2012, 4(1), 8594.
120. Maki, K.C., Reeve, M.S., Carson, M.L., Miller, M.P., Turowski, M., Rains, T.M., Anderson, K., Papanikoleaeou, Y., Wilder, D.M. Dose-response characteristics of high-viscosity hydroxypropylmethylcellulose in subjects at risk for the development of type 2 diabetes mellitus. *Diab. Technol. Therap.* 2009, 11(2), 119–125.
121. Maki, C.K., Carson, M.L., Miller, M.P., Turowski, M., Bell, M., Wilder, D.M., Rains, T.M., Reeves, M.S. Hydroxyproplmethylcellulose and methylcellulose consumption reduce postprandial insulinemia in overweight and obese men and women. *J. Nutr.* 2008, 138(2), 292–296.
122. Bartley, G.E., Yokoyama, W., Young, S.A., Anderson, W.H.K., Hung, S.C., Albers, D.R., Langhorst, M.L., Turowski, M., Hyunsook, K.H. Hypocolesterolemic effects of hydroxypropylmethylcellulose are mediated by altered gene expression in hepatic bile and cholesterol pathways of male hamsters. *J. Nutr.* 2010, 140(7), 1255–1260.
123. Yokoyama, W., Anderson, W.H.K., Albers, D.R., Hong, Y.J., Langhorst, M.L., Hung, S.C., Lin, J.T., Young, S.A. Dietary HPMC increases excretion of saturated and trans fats by hamsters fed fast food diets. *J. Agric. Food Chem.* 2011, 50(20), 11249–11254.
124. Ban, S.J., Rico, C.W., Um, I.C., Kang, M.Y. Comparative evaluation of the hypolipidemic effects of HEMC and HPMC in high fat-fed mice. *Food Chem. Toxicol.* 2012, 50(2), 130–134.
125. Slavin, J., Green, H. Dietary fibre and satiety. *Nutr. Bull.* 2007, 32(Suppl. 1), 32–42.
126. Hoad, C.L., Rayment, P., Spiller, R.C., Marciani, B., de Celis, A.B., Traynor, C., Mela, D.J., Peters, H.P.F., Gowland, P.A. In-vivo imaging of intragastric gelation and its effect on satiety in humans. *J. Nutr.* 2004, 134(9), 2293–2300.
127. Re, R. *Presentation at Vitafoods*, Geneva, Switzerland, May 2011.

28 Cellulose-Based Biopolymers
Formulation and Delivery Applications

*J.D.N. Ogbonna, F.C. Kenechukwu,
S.A. Chime, and A.A. Attama*

CONTENTS

- 28.1 Introduction 536
- 28.2 The Assembly of Biopolymers 536
- 28.3 Cellulose 537
  - 28.3.1 Cellulose-Based Biopolymers Used in Formulation and Delivery of Bioactive Compounds 538
    - 28.3.1.1 Oxycellulose 538
    - 28.3.1.2 Microcrystalline Cellulose 538
    - 28.3.1.3 Cellulose Ethers 538
    - 28.3.1.4 Cellulose Esters 540
    - 28.3.1.5 Hemicellulose 540
- 28.4 Functional Properties of Cellulose-Based Biopolymers in Drug Formulation and Delivery 541
  - 28.4.1 Particle Properties 541
  - 28.4.2 Swelling 542
  - 28.4.3 Bioadhesion 542
  - 28.4.4 Crystallinity and Solid State Characteristics 543
- 28.5 Modification of Cellulose-Based Biopolymers 543
  - 28.5.1 Physical Modifications 544
  - 28.5.2 Silylation, Mercerization, and Other Surface Chemical Modifications 544
  - 28.5.3 Polymer Grafting 545
  - 28.5.4 Bacterial Modification 545
  - 28.5.5 Chemical Modifications 545
  - 28.5.6 Preparation of Cellulose Ethers 546
  - 28.5.7 William Etherification 547
  - 28.5.8 Alkaline-Catalyzed Oxalkylation 547
  - 28.5.9 Enzymic Method of Chemical Modification 547
  - 28.5.10 Cellulose-Hydrolyzing Enzymes 547
  - 28.5.11 Endo-1,4-β-Glucanase 547
  - 28.5.12 β-Glucosidase 548
  - 28.5.13 Cellobiohydrolase 548
  - 28.5.14 Other Derivatization Processes 548
    - 28.5.14.1 Cellulose-Based Hydrogels and Cross-Linking Strategies 548
  - 28.5.15 Conventional Methods for Cellulose Dissolution and Modifications 549
  - 28.5.16 Ionic Liquids for Cellulose Dissolution and Modification 549
  - 28.5.17 Ionic Liquid as Media for Modification of Cellulose by Grafting Copolymerization and Blends (Composites) 551
  - 28.5.18 Cellulose Blends (Composites) in ILs Media 551

28.6 Application of Cellulose-Based Biopolymers in Dosage Form Design and Novel Drug Delivery Systems 551
    28.6.1 Immediate-Release Dosage Forms 552
    28.6.2 Sustained/Modified/Controlled-Release 553
        28.6.2.1 Dosage Forms 553
        28.6.2.2 Specialized Bioactive Carriers 557
        28.6.2.3 Cellulose-Based Biopolymer Microparticles for Drug Delivery 559
    28.6.3 Application of Cellulose-Based Biopolymer 560
        28.6.3.1 Nanoparticles in Drug and Bioactive Delivery 560
        28.6.3.2 Cellulose-Based Biopolymers in Protein and Gene Delivery 563
        28.6.3.3 Cellulose-Based Biopolymers in Wound Healing 564
28.7 Conclusions 565
References 565

28.1 INTRODUCTION

Biopolymers are polymers produced by living organisms. They contain monomeric units that are covalently bonded to form macromolecules. There are three main classes of biopolymers, classified according to the monomeric units used and the structure of the biopolymer formed[1]: polynucleotides (RNA and DNA), which are long polymers composed of 13 or more nucleotide monomers; polypeptides, which are short polymers of amino acids; and polysaccharides, which are often linear, bonded, polymeric carbohydrate structures. Polysaccharides include cellulose, starch, gums, glycogen, etc.

Cellulose and cellulose derivatives have long been used in the pharmaceutical industry as excipients in many drug device formulations. Devices other than swelling tablets have been developed for controlled drug delivery using these biopolymers. The most recent advances aim not only at the sustained release of a bioactive molecule over a long time period, ranging from hours to weeks, but also at a space-controlled delivery, directly at the site of interest. The need to encapsulate bioactive molecules into a cellulose hydrogel matrix or other delivery devices (e.g., microspheres and nanospheres) is also related to the short half-life displayed by many biomolecules in vivo. Smart cellulose-based hydrogels are particularly useful to control the time- and space-release profile of drugs, as swelling–deswelling transitions, which modify the mesh size of the hydrogel network, occur upon changes of physiologically relevant variables, such as pH, temperature, and ionic strength. Controlled release through oral drug delivery is usually based on the strong pH variations encountered when transiting from the stomach to the intestine. Cellulose-based polyelectrolyte hydrogels (e.g., hydrogels containing sodium carboxymethylcellulose [SCMC]) are particularly suitable for this application. The use of cellulose and its derivatives as biomaterials for the design of tissue engineering scaffolds has received increasing attention, due to the excellent biocompatibility of cellulose and its good mechanical properties, and in the form of sponges or fabrics, have been applied for the treatment of severe skin burns, and in studies on the regeneration of cardiac, vascular, neural, cartilage, and bone tissues.[2] This entry focuses on the current design and use of cellulose-based biopolymer hydrogels, which usually couple their biodegradability with a smart stimuli-sensitive behavior. These features, together with the large availability of cellulose in nature and the low cost of cellulose derivatives, make cellulose-based biopolymers particularly attractive. Both well-established and innovative applications of cellulose-based biopolymers are discussed.

28.2 THE ASSEMBLY OF BIOPOLYMERS

Polysaccharides (e.g., cellulose, starch, and glycogen), proteins, and nucleic acids are polymers composed of subunits (monomers) linked to one another in linear sequence by a particular type of bond. The sequence of monomer units in nucleic acids and proteins is always strictly linear; a few polysaccharides (e.g., glycogen) are branched molecules, but such branching is a secondary biosynthetic event, superimposed on a primary linear arrangement of the subunits. Some polysaccharides are homopolymers, consisting of a single, chemically identical repeating subunit. However, many polysaccharides, all proteins and all nucleic acids are heteropolymers, consisting of chemically similar but nonidentical subunits. The polysaccharides that contain more than one kind of subunit show a regular arrangement of the subunits, whereas in proteins and nucleic acids, the sequences of subunits are irregular. The bonds that link together the subunits in some biological polymers such as proteins and polysaccharides are amide and glycosidic bonds, respectively. A characteristic feature of polysaccharide synthesis is the requirement of a primer—a short segment of the polysaccharide in question to act as an acceptor for the monomer units. For instance, in the synthesis of glycogen, it has been found that the primer must contain more than four sugar units to function effectively. Special branching enzymes bring about the branching in these polymers.

Nearly all synthetic polymers and naturally occurring biopolymers possess a range of molecular weights—except proteins and polypeptides. The molecular weight determined is thus an average molecular weight, the value of which depends on the method of measurement: chemical analysis, osmotic pressure, light scattering, viscosimetric method, etc. Small molecules and many biopolymers are monodisperse; i.e., all molecules of a given pure compound have the same molecular weight. In synthetic polymerization reactions, no two chains grow equally fast or for the same length of time. The resultant biopolymers are heterodisperse; i.e., they have different chain lengths and a range of molecular weights, which can be described by an average molecular weight and by a molecular weight distribution. A number of functional properties are attributed to biopolymers for drug delivery.

28.3 CELLULOSE

Cellulose is an essential structural component of cell walls in higher plants and remains the most abundant renewable organic polymer on earth. Cellulose is a homo polymer and its monomer is glucose. It consists of long chains of anhydro-D-glucopyranose units (AGUs) with each cellulose molecule having three hydroxyl groups per AGU, with the exception of the terminal ends (Figure 28.1). The fundamental unit of cellulose is the microfibril, constituting of a bundle of β-1,4-glucane that are formed by intra- and intermolecular hydrogen bonding of glucan chains, forming a crystalline array. These microfibrils are about 10–38 nm diameter and 30–100 cellulose molecules in extended chains, depending on their origin.[3] The glucose units in cellulose are held together by 1,4-β-glucosidic linkages, which account for the high crystallinity of cellulose and its insolubility in water and other common solvents. However, modification products of cellulose have shown different degrees of solubility in different solvents. Cellulose derivatives constitute a large class of biopolymers with diverse physicochemical properties and large functional versatility in food, cosmetics, and pharmaceutical applications. Cellulose derivatives have been used as diluents, binders

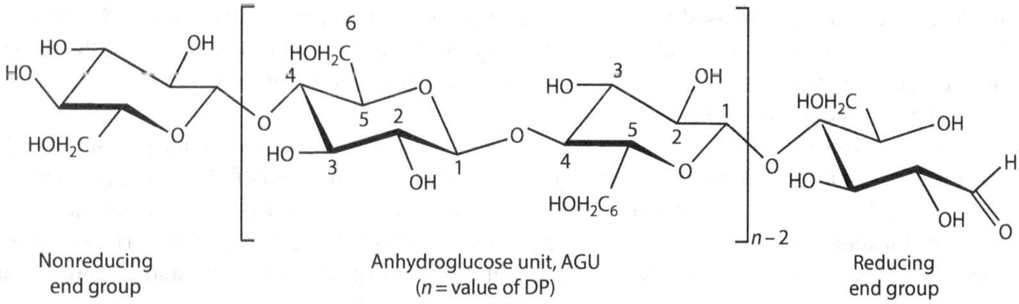

FIGURE 28.1 Molecular structure of cellulose. (From Builders, P.F. and Attama, A.A., Functional properties of biopolymers for drug delivery applications, in *Biodegradable Materials*, Johnson, B.M. and Berkel, Z.E., Eds., Nova Science Publishers Inc., New York. Copyright 2011, with permission from Nova Science Publishers, Inc.)

in direct compression and wet granulation processes, film coating, viscosity enhancers, and drug-release retardants in controlled-release matrix systems for microparticles and nanoparticles.[4] The excellent biocompatibility of cellulose, cellulosics, and cellulase-mediated degradation products has prompted the large use of cellulose-based devices in biomedical applications.

Both natural and synthetic polymers have been employed for drug delivery purposes. However, cellulose-based biopolymers have continued to gain greater popularity in pharmaceutical manufacture, especially for drug delivery applications because of their diverse and numerous physicochemical and functional properties. In general, the major attraction to the use cellulose-based biopolymers for drug delivery application includes ready availability, low cost, relatively low toxicity, ease of modification, biodegradability, biocompatibility, etc.[5]

28.3.1 Cellulose-Based Biopolymers Used in Formulation and Delivery of Bioactive Compounds

28.3.1.1 Oxycellulose

Oxidized cellulose (oxycellulose) is cellulose in which some of the terminal primary alcohol groups of the glucose residues have been converted to carboxyl groups.

28.3.1.2 Microcrystalline Cellulose

Since its introduction in the 1960s, microcrystalline cellulose (MCC) has offered great advantages in the formulation of solid dosage forms, but some characteristics have limited its application, such as relatively low bulk density, moderate flowability, loss of compactibility after wet granulation, and sensitivity to lubricants. MCC is purified partially depolymerized cellulose. In many conventional pharmaceutical formulations, it is primarily used as binder and diluents in oral tablets and capsules in both wet granulation and direct compression processes. It is also used as a lubricant and disintegrant in tablets. Because of its desirable physicochemical and functional properties, MCC has been coprocessed with other excipients.[6] It is nonirritant and nontoxic, and is widely used in oral pharmaceutical formulations and food products.[7] Silicification of MCC improves the functionality of MCC with properties such as enhanced density, low moisture content, flowability, lubricity, larger particle size, compactibility, and compressibility. Silicified MCC (SMCC) is manufactured by codrying a suspension of MCC particles and colloidal silicon dioxide such that the dried finished product contains 2% colloidal silicon dioxide.[8] Silicon dioxide simply adheres to the surface of MCC and occurs mainly on the surface of MCC particles; only a small amount was detected in the internal regions of the particles. So, SMCC shows higher bulk density than the common types of MCC.[9]

28.3.1.3 Cellulose Ethers

Cellulose ethers are widely used as important excipients for designing matrix tablets. On contact with water, the cellulose ether swells and forms a hydrogel layer around the dry core of the tablet. The hydrogel presents a diffusional barrier for water molecules penetrating into the polymer matrix and the drug molecules being released.[10–14]

Sodium carboxymethyl cellulose: SCMC is a low-cost, soluble, and polyanionic polysaccharide derivative of cellulose that has been employed in medicine as an emulsifying agent in pharmaceuticals and in cosmetics.[15] The many important functions provided by this polymer make it a preferred thickener, suspending aid, stabilizer, binder, and film-former in a wide variety of uses. In biomedicine, it has been employed for preventing postsurgical soft tissue and epidural scar adhesions.

Methylcellulose: Methylcellulose is neutral, odorless, tasteless, and inert. It swells in water to produce a clear to opalescent, viscous, colloidal solution and it is insoluble in most of the common organic solvents. However, aqueous solutions of methylcelluose can be diluted with ethanol. Methylcellulose solutions are stable over a wide range of pH (2–12) with no apparent change in viscosity. They can be used as bulk laxatives, so it can be used to treat constipation, and in nose drops, ophthalmic preparations, burn preparations ointments, etc. Although methylcellulose, when used as a bulk laxative, takes up water quite uniformly, tablets of methylcellulose have caused fecal impaction and intestinal obstruction. Methylcellulose attracts large amounts of water into the colon, producing a softer and bulkier stool so it is used to treat constipation, diverticulosis, hemorrhoids, and irritable bowel syndrome.[16] Methylcellulose dissolves in cold water, but those with higher degree of substitution (DS) result in lower solubility, because the polar hydroxyl groups are masked.

Ethylcellulose: It is the nonionic, pH-insensitive cellulose ether, insoluble in water but soluble in many polar organic solvents. It is used as a nonswellable, insoluble component in matrix or coating systems. When water-soluble binders cannot be used in dosage processing because of water sensitivity of the active ingredient, ethylcellulose is often used. It can also be used to coat one or more active ingredients of a tablet to prevent them from reacting with other materials or with one another. Ethylcellulose can also be used on its own or in combination with water-soluble polymers to prepare sustained-release film coatings that are frequently used for the coating of microparticles, pellets, and tablets.

In addition to ethylcellulose, hydroxyethylcellulose (HEC) is also nonionic, water-soluble, cellulose ether, easily dispersed in cold or hot water to give solutions of varying viscosities and desired properties, yet it is insoluble in organic solvents. It is used as a modified-release tablet matrix, a film former, and a thickener, stabilizer, and suspending agent for oral and topical applications when a nonionic material is desired. Many researchers such as Mura et al.[17] Friedman and Golomb,[18] and Soskolne et al.[19] have demonstrated the ability of ethylcellulose to sustain release of drugs. Hydroxypropyl Cellulose. Hydroxypropyl cellulose (HPC) is nonionic, water-soluble, and pH-insensitive cellulose ether. It can be used as thickening agent, tablet binder, modified-release polymer, and in film coating. By using solid dispersions containing a polymer blend, such as HPC and ethylcellulose, it is possible to precisely control the rate of release of an extremely water-soluble drug, such as oxprenolol hydrochloride.[20–24] In this case, the water-soluble HPC swells in water and is trapped in the water-insoluble EC so that the release of the drug is slowed. These studies have shown that there is a linear relationship between the rate of release of the water-insoluble drug and its interaction with the polymer.[25–27]

Hydroxypropylmethyl cellulose: Hydroxypropylmethyl cellulose (HPMC) (Figure 28.2) is water-soluble cellulose ether and it can be used as hydrophilic polymer for the preparation of controlled-release tablets. Water penetrates the matrix and hydrates the polymer chains, which eventually disentangle from the matrix. Drug release from HPMC matrices follows two mechanisms, drug diffusion through the swelling gel layer and release by matrix erosion of the swollen layer.[28–31]

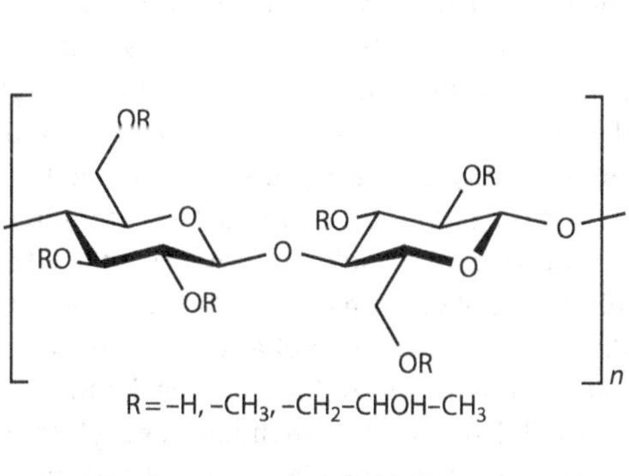

$R = -H, -CH_3, -CH_2-CHOH-CH_3$

FIGURE 28.2 Structural formula of hydroxypropylmethylcellulose. (From Builders, P.F. and Attama, A.A., Functional properties of biopolymers for drug delivery applications, in *Biodegradable Materials*, Johnson, B.M. and Berkel, Z.E., Eds., Nova Science Publishers Inc., New York. Copyright 2011, with permission from Nova Science Publishers, Inc.)

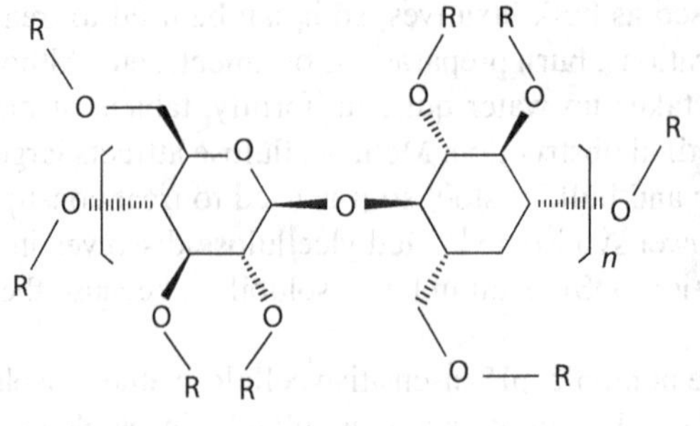

$C_6H_7O_2(OH)_x(OCH_3)_y[OCH_2CH(CH_3)OH]_z(OCOC_6H_5COOH)_u$

FIGURE 28.3 Structural formula of cellulose acetate phthalate. (From Builders, P.F. and Attama, A.A., Functional properties of biopolymers for drug delivery applications, in *Biodegradable Materials*, Johnson, B.M. and Berkel, Z.E., Eds., Nova Science Publishers Inc., New York. Copyright 2011, with permission from Nova Science Publishers, Inc.)

28.3.1.4 Cellulose Esters

Cellulose esters are part of a large family of cellulose derivatives that are produced by the esterification of the hydroxyl groups of each anhydroglucose monomer. The physical properties of cellulose esters depend on the cellulose chain length and on the type and amount of ester groups attached to the polymer chain. Cellulose acetate phthalate (CAP, Figure 28.3) is a partial acetate ester of cellulose that has been reacted with phthalic anhydride. One carboxyl of the phthalic acid is esterified with the cellulose acetate. The finished product contains about 20% acetyl groups and about 35% phthalyl groups. In the acid form, it is soluble in organic solvents and insoluble in water. The nonenteric esters do not show pH-dependent solubility characteristics with the exception of cellulose acetate with low levels of acetyl group. Most nonenteric esters are insoluble in water. The salt form is readily soluble in water. This combination of properties makes it useful in enteric coating of tablets because it is resistant to the acid condition of the stomach but soluble in the more alkaline environment of the intestinal tract.[32]

28.3.1.5 Hemicellulose

In nature, hemicelluloses are found in the cell walls of woody and annual plants, together with cellulose and lignin.[33] Hemicellulose is made up of a group of complex low-molecular-weight polysaccharides that are bound to the surface of cellulose microfibrils, but their structure prevents them from forming microfibrils by themselves.

The sugar monomers in hemicellulose may include xylose, mannose, galactose, rhamnose, and arabinose. Hemicelluloses contain most of the D-pentose sugars, and occasionally small amounts of l-sugars as well. Their backbones consist mainly of β-1,4-linked D-glycans. Xylose is always the sugar monomer present in the largest amount, but mannuronic acid and galacturonic acid may also be present. Thus, the hemicellulose polysaccharides consist of mannans (Figure 28.4), xylan (Figure 28.5), and xyloglucan (Figure 28.6). These can be extracted from the plant cell wall with a strong alkali. Most of the hemicellulose fraction is soluble in water after alkaline extraction. Xyloglucan has a similar backbone as cellulose, but contains xylose branches on three out of every four glucose monomers and the β(1,4)linked D-xylan backbone of arabinoxylan contains arabinose branches.[34,35]

Glucomannan is the most commonly used form of the hemicelluloses. The chain structure consists of a linear backbone of (1,4)-linked β-D-mannopyranosyl units to which (1,6)-linked α-D-galactopyranosyl units are substituted. The Man:Gal ratio in the chemical structure of glucomannan shown in Figure 28.4 is 2:1. Glucomannan is a water-soluble polysaccharide that is considered a dietary fiber. It is commonly used as a food additive, as an emulsifier, and a thickener. It is also used in drug delivery applications.[36,37] Glucomannans have been specifically derived from softwoods,

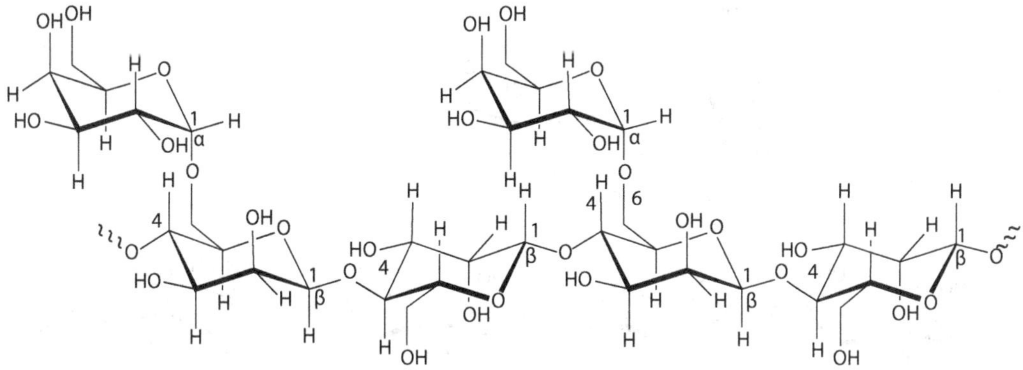

Cellulose-Based Biopolymers

FIGURE 28.4 The chemical structure of a typical galactomannan. (From Builders, P.F. and Attama, A.A., Functional properties of biopolymers for drug delivery applications, in *Biodegradable Materials*, Johnson, B.M. and Berkel, Z.E., Eds., Nova Science Publishers Inc., New York. Copyright 2011, with permission from Nova Science Publishers, Inc.)

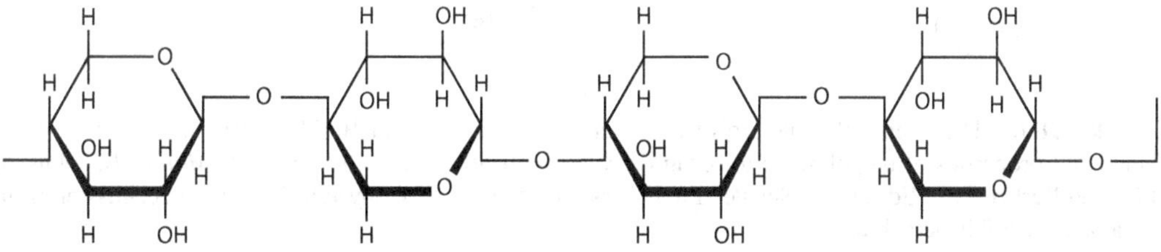

FIGURE 28.5 The chemical structure of a typical xylan. (From Builders, P.F. and Attama, A.A., Functional properties of biopolymers for drug delivery applications, in *Biodegradable Materials*, Johnson, B.M. and Berkel, Z.E., Eds., Nova Science Publishers Inc., New York. Copyright 2011, with permission from Nova Science Publishers, Inc.)

roots, tubers, and bulbs. Their use in foods and drug delivery applications are primarily based on their viscosity-imparting, gelation, and swelling potentials. Majority have been used as matrices in oral controlled-release formulations. There has been an increase in the number of commercial glucomannan processed and standardized to meet pharmaceutical grade. The most commonly used types of glucomannan include konjac glucomannan, guar gum, and locust bean gum.[37–39]

28.4 FUNCTIONAL PROPERTIES OF CELLULOSE-BASED BIOPOLYMERS IN DRUG FORMULATION AND DELIVERY

A clear understanding of the physical and chemical (physicochemical) properties of biopolymers, especially those to be used for drug delivery application, is very critical to their use for both conventional and modified drug delivery systems. These properties aid in identification and can as well help in modeling the polymer to give a clear understanding how the materials will behave under various conditions. The physical properties of polymers essentially refer to those properties that do not change their chemical nature. The chemical properties refer to those properties that change their chemical nature. To determine these properties, a number of parameters listed here among others are usually evaluated.[5]

28.4.1 Particle Properties

Drug molecules and polymers for formulation consists multiparticulate powders, which are heterogeneous in shape, size, and size distributions. The particle properties of these polymers are critical

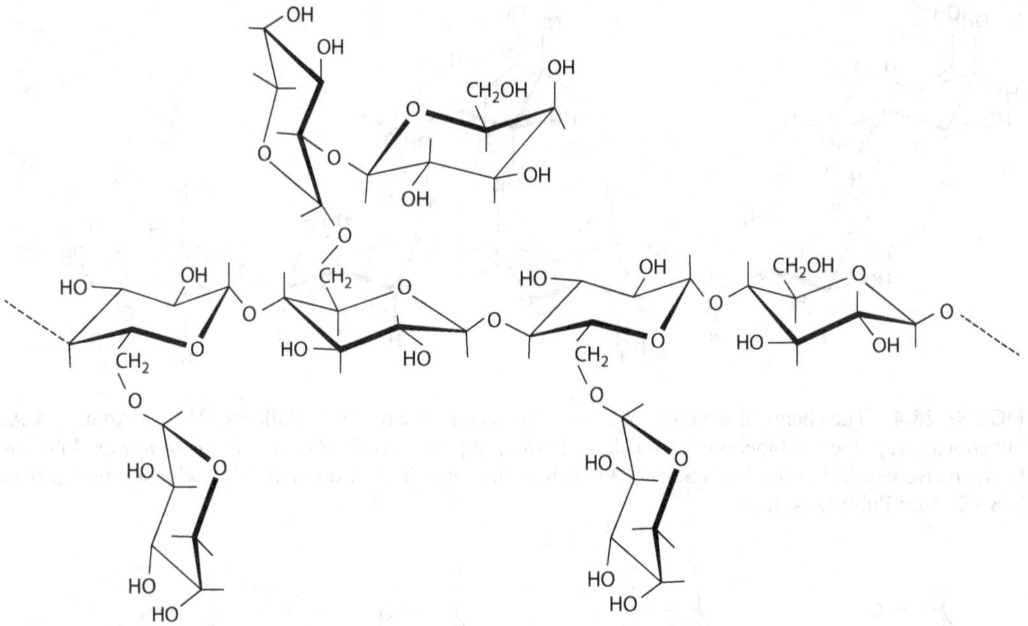

FIGURE 28.6 The chemical structure of a typical xyloglucan. (From Builders, P.F. and Attama, A.A., Functional properties of biopolymers for drug delivery applications, in *Biodegradable Materials*, Johnson, B.M. and Berkel, Z.E., Eds., Nova Science Publishers Inc., New York. Copyright 2011, with permission from Nova Science Publishers, Inc.)

as many of their physico-functional properties (both molecular and particulate) as well as the biopharmaceutical properties and the performance of finished products are controlled by their particle characteristics. In the production and processing of cellulose powder into its derivative for direct compression technology, the materials' particle size and size distribution are critical as these affect powder bulkiness, flow, and compaction characteristics. Some particle properties measured during the assessment of solid particulate delivery systems and powdered polymeric excipients include particle size and size distribution, fractal dimension, particle density, and particle strength. These control the bulk powder properties such as flow properties, moisture content, bulk density, water sorption isotherms, compression properties, critical relative humidity, thermal properties, and moisture diffusivity.

28.4.2 Swelling

Swelling is one of the functional properties used to characterize biopolymers required for modified or controlled drug delivery systems. The biopolymers employed for modified drug delivery applications are hydrogels that form three-dimensional polymeric networks when they come into contact with water, they absorb many times their weight of water but they do not dissolve. Swelling of the cellulosics in different biorelevant media must be established.

28.4.3 Bioadhesion

In drug delivery applications, biopolymers with mucoadhesive properties serve to increase the residence time of the bioactive agents at the site of absorption, resulting in a steep concentration gradient to favor drug absorption and localization in specified regions, thus, improving the bioavailability of the drug. Mucoadhesive polymers are often characterized by certain specific intrinsic properties that have been related to the muco/bioadhesive: These include the presence of strong hydrogen bond–forming group(s) such as carboxylate and hydroxyl groups, presence of a strong anionic charge, high molecular weight, high viscosity, high hydration capacity, sufficient chain flexibility, and high surface energy that favors spreading onto the mucus.

28.4.4 CRYSTALLINITY AND SOLID STATE CHARACTERISTICS

Polymer crystallinity refers to the amount of crystalline region in a polymer with respect to its amorphous content. Characterization of polymer crystallinity is based on the molecular chain arrangement. The crystallinity of cellulose partly results from hydrogen bonding between the cellulosic chains, but some hydrogen bonding also occurs in the amorphous phase, although its organization is low. The structural integrity and stability have been monitored using differential scanning calorimetry (DSC). X-ray diffraction has been used to measure the degree of crystallinity of a biopolymer. In the solid state, a polymer may be completely amorphous, perfectly crystalline, or semi-crystalline.

Semi-crystalline polymers are composed of a combination of an amorphous and a crystalline region. Biopolymers are predominantly semi-crystalline as they are composed of varying ratios of crystalline and amorphous regions. The presence of crystalline and amorphous structures forms a material with superior properties in terms of strength and stiffness. The crystallinity of a biopolymer can be used to predict properties such as hardness, modulus, tensile strength, stiffness, moisture sorption, swelling, and melting characteristics. Highly crystalline biopolymers are rigid, high melting, and less affected by solvent penetration. Amorphous polymers are usually less rigid, weaker, and more easily deformed.

Cellulose is stable to thermal degradation until about 370°C, and then decomposes almost completely over a very short temperature range. At 300°C the cellulose molecule is highly flexible and undergoes depolymerization by transglycosylation to create products such as anhydromonosaccharides, which include levoglucosan (1,6-anhydro-β-D-glyucopyranose) and (1,6-anhydro-β-D-glyucofuranose). Cellulose also produces combustible volatiles such as acetaldehyde, propenal, methanol, butanedione, and acetic acid. These are converted into low-molecular-weight products, randomly linked oligosaccharides and polysaccharides, which lead to carbonized products. Modification of the particle morphology of a biopolymer such as cellulose has been achieved by changing its glass transition temperature.

28.5 MODIFICATION OF CELLULOSE-BASED BIOPOLYMERS

The crystalline nature of cellulose originates from intermolecular forces between neighboring cellulose chains over long lengths. All native celluloses show the same crystal lattice structure, called cellulose I. However, various modifications of native cellulose can alter the lattice structure to yield other types of crystals.[40–42] The intermolecular forces in the crystalline domains are mainly hydrogen bonds between adjacent cellulose chains in the same lattice plane, which results in a sheet-like structure of packed cellulose chains. In addition, the sheets are probably connected to one another by hydrogen bonds and/or van der Waal's forces. The organization of cellulose molecules into parallel arrangements is responsible for the formation of crystallites. The length of an elementary crystallite ranges from 12 to 20 nm (≈24–40 glucose units) and the width from 2.5 to 4 nm.[40–42] The crystalline parts of cellulose are rather resistant to degradation by enzymes and a system of several synergistically acting enzymes is necessary to obtain any significant hydrolysis.[43] However, the reactivity of cellulose can be greatly enhanced by various forms of treatment, such as swelling, degradation, or mechanical grinding, which break down the fibrillar aggregations.[40]

To increase their use and to fulfill the various demands for functionality of different cellulose products, they are often modified by physical, chemical, enzymic, or genetic means. Modification leads to changes in the properties and behavior of the polymer and, consequently, improvement of the positive attributes and reduction of the negative characteristics.[42,44,45] The properties of a modified polysaccharide depend on several factors, such as the modification reaction, the nature of the substitution group, the DS, and the distribution of the substitution groups. To direct a modification reaction toward a certain product with the desired properties, it is of importance to have knowledge of the correlations that exist between the modification process, chemical structure, and functional properties of the final product. However, the relationships, if any, between these parameters are still far from fully understood, largely due to difficulties in the elucidation of the modified polymer structure, including the distribution of substituent.[42]

Chemical modification can be used to tune some other cellulose properties, such as hydrophobic or hydrophilic character, elasticity, adsorption, microbial resistance, and heat and mechanical resistance.[46,47] The chemical modification of a surface of an organic polysaccharide follows the same principles as those established for other media, such as for silica gel. However, the hydroxyl groups of cellulose are less reactive and the beginning of the chemical modification takes place in primary hydroxyl found in carbon 6, which may occur for several different routes, highlighting that it is yet to occur also in the secondary hydroxyl groups present on carbons 2 and 3. Cellulose is not modified thermally through melting process,[48] because the decomposition temperature is less than melting temperature.[49] The major modifications of cellulose occur through halogenation, oxidation, etherification, and esterification.[46,50,51] Chemical modification implies the substitution of free hydroxyl groups in the polymer with functional groups, yielding different cellulose derivatives.[42] Chemical modification of cellulose is performed to improve processability and to produce cellulose derivatives (cellulosics) that can be tailored for specific industrial applications.[52,53] Large-scale commercial cellulose ethers include carboxymethyl cellulose (CMC), methyl cellulose, HEC, HPMC, HPC, ethyl hydroxyethyl cellulose (EHEC), and methyl hydroxyethyl cellulose (MHEC).[53]

Pretreatments of the cellulose fiber surface or physical modifications can clean the fiber surface, chemically modify the surface, stop the moisture absorption process, and increase the surface roughness.[54–57]

28.5.1 Physical Modifications

Native cellulose are commonly modified by physical, chemical, enzymic, or genetic means in order to obtain specific functional properties,[42,44,58,59] and to improve some of the inherent properties that limit their utility in certain application. Physical/surface modification of cellulose are performed in order to clean the fiber surface, chemically modify the surface, stop the moisture absorption process, and increase the surface roughness.[54–56] Among the various pretreatment techniques, silylation, mercerization, peroxide, benzoylation, graft copolymerization, and bacterial cellulose treatment are the best methods for surface modification of natural fibers.

28.5.2 Silylation, Mercerization, and Other Surface Chemical Modifications

Silane-coupling agents usually improve the degree of cross-linking in the interface region and offer a perfect bonding. Among the various coupling agents, silane-coupling agents were found to be effective in modifying the natural fiber–matrix interface and may reduce the number of cellulose hydroxyl groups in the fiber–matrix interface. In the presence of moisture, hydrolyzable alkoxy group leads to the formation of silanols. The silanol then reacts with the hydroxyl group of the fiber, forming stable covalent bonds to the cell wall that are chemisorbed onto the fiber surface.[60] Therefore, the hydrocarbon chains provided by the application of silane restrain the swelling of the fiber by creating a cross-linked network due to covalent bonding between the matrix and the fiber.[54] Cellulose fiber treatment with toluene dissocyanate and triethoxyvinyl silane could improve the interfacial properties. Silanes after hydrolysis undergo condensation and bond-formation stage and can form polysiloxane structures by reacting with hydroxyl group of the fibers.[54,56,61]

Mercerization is the common method to produce high-quality fibers.[62] Mercerization leads to fibrillation, which causes the breaking down of the composite fiber bundle into smaller fibers and

Cellulose-Based Biopolymers

reduces fiber diameter, thereby increases the aspect ratio, which leads to the development of a rough surface topography that results in better fiber–matrix interface adhesion and an increase in mechanical properties.[56,63] Moreover, mercerization increases the number of possible reactive sites and allows better fiber wetting. Mercerization has an effect on the chemical composition, degree of polymerization, and molecular orientation of the cellulose crystallites due to cementing substances like lignin and hemicellulose, which are removed during the mercerization process.

Peroxide treatment of cellulose fiber has attracted the attention of various researchers due to easy processability and improvement in mechanical properties. Organic peroxides tend to decompose easily to free radicals, which further react with the hydrogen group of the matrix and cellulose fibers.[56]

28.5.3 Polymer Grafting

Desirable and targeted properties can be imparted to the cellulose fibers through graft copolymerization in order to meet the requirement of specialized applications. Graft copolymerization is one of the best methods for modifying the properties of cellulose fibers. Different binary vinyl monomers and their mixtures have been graft-copolymerized onto cellulosic material for modifying the properties of numerous polymer backbones.[54,64] During past decades, several methods have been suggested for the preparation of graft copolymers by conventional chemical techniques. Creation of an active site on the preexisting polymeric backbone is the common feature of most methods for the synthesis of graft copolymers. The active site may be either a free radical or a chemical group that may get involved in an ionic polymerization or in a condensation process. Polymerization of an appropriate monomer onto this activated backbone polymer leads to the formation of a graft copolymer. Ionic polymerization has to be carried out in the presence of anhydrous medium and/or in the presence of considerable quantity of alkali metal hydroxide. Another disadvantage of the ionic grafting is that low-molecular-weight graft copolymers are obtained, while in case of free-radical grafting high-molecular-weight polymers can be prepared. The molecular weight affects the drug delivery applicability of these cellulosics.

28.5.4 Bacterial Modification

The coating of bacterial cellulose onto cellulose fibers provides new means of controlling the interaction between fibers and polymer matrices. Coating of fibers with bacterial cellulose not only facilitates good distribution of bacterial cellulose within the matrix, but also results in an improved interfacial adhesion between the fibers and the matrix. This enhances the interaction between the fibers and the polymer matrix through mechanical interlocking.[55,56,65] Surface modification of cellulose fibers using bacterial cellulose is one of the best methods for greener surface treatment of fibers. Bacterial cellulose has gained attention in the research area for the encouraging properties it possesses; such as its significant mechanical properties in both dry and wet states, porosity, water absorbency, moldability, biodegradability, and excellent biological affinity.[66] Because of these properties, bacteria cellulose has a wide range of potential applications in drug delivery. Acetobacterxylinum (or Gluconacetobacter xylinus) is the most efficient producer of bacterial cellulose. Bacteria cellulose is secreted as a ribbon-shaped fibril, less than 100 nm wide, which is composed of much finer 2–4 nm nanofibrils.

28.5.5 Chemical Modifications

Chemical modification is based on reactions of the free hydroxyl groups in the anhydroglucose monomers, resulting in changes in the chemical structure of the glucose units and, ultimately, the production of cellulose derivatives.[42] Chemical modification implies the substitution of free hydroxyl groups in the polymer with functional groups, yielding different cellulose derivatives.[42] Usually, these modifications

TABLE 28.1
Some Commercially Marketed Cellulose Esters and Ethers

Cellulose Esters			
Nitrate	1.5–3.0	MeOH, PhNO$_2$, ethanol-ether	Films, fibers, explosives
Acetate	1.0–3.0	Acetone	Films, fibers, coatings, heat- and rot-resistant fabrics
Cellulose Ethers			
Methyl	1.5–2.4	Hot H$_2$O	Food additives, films, cosmetics, greaseproof paper
Carboxymethyl	0.5–1.2	H$_2$O	Food additives, fibers, coatings, oil-well drilling muds, paper size, paints, detergents
Ethyl	2.3–2.6	Organic solvents	Plastics, lacquers
Hydroxyethyl	Low DS	H$_2$O	Films
Hydroxypropyl	1.5–2.0	H$_2$O	Paints
Hydroxypropylmethyl	1.5–2.0	H$_2$O	Paints
Cyanoethyl	2.0	Organic solvents	Products with high dielectric constants, fabrics with heat and rot resistances

Sources: Reprinted from *Cellulose Fibers: Bio- and Nano-Polymer Composites*, Varshney, V.K., and Naithani, S., Chemical functionalization of cellulose derived from nonconventional sources, Kalia, S., Ed., Springer-Verlag, Berlin, Germany, pp. 43–58. Copyright 2011, with permission from Elsevier; Reprinted from *Comprehensive Polymer Science*, Allen, G. and Bevington, J.C., Eds., Pergamon, Oxford, U.K., pp. 26, 681. Copyright 1986, with permission from Elsevier.

involve esterification or etherification reactions of the hydroxyl groups. Each AGU is available for up to three sites of substitution; the hydroxyl group on C-2, C-3 or C-6. The DS describes the average number of substituted hydroxyl groups per AGU and ranges from 0 to 3. The term DS applies to derivatives in which the substitution group terminates the reactive hydroxyl sites. Substitution by chemical groups that generate new free hydroxyl groups for further substitution is quantified by the molar substitution (MS). This value is defined as the average number of moles of substituent added per AGU.[67] The MS has no theoretical upper limit. Cellulose derivatives can be characterized by a number of factors, such as type and nature of substitution group, DS, MS, average chain length, and DP. These factors influence the functional properties of the derivative in various ways.[42,67,68]

Chemically modified celluloses were developed primarily in order to overcome the insoluble nature of cellulose, thus extending the range of applications of the polymer. Commercial cellulose derivatives are usually ethers or esters that are soluble in water and/or organic solvents. They are produced by reacting the free hydroxyl groups in the AGUs with various chemical substitution groups. The introduction of substituent disturbs the inter- and intramolecular hydrogen bonds in cellulose, which leads to liberation of the hydrophilic character of the numerous hydroxyl groups and restriction of the chains.[42,68,69] However, substitution with alkyl groups reduces the number of free hydroxyl groups. Cellulose derivatives are used in a wide range of industrial fields and their availability, economic efficiency, easy handling, and low toxicity are reasons for a continuously expanding worldwide market. Common industrial derivatives, properties, and their fields of application are summarized in Table 28.1. Cellulose ethers are the most widely used derivatives, although there are also some commercial esters, as shown in Table 28.1.

28.5.6 Preparation of Cellulose Ethers

Etherification of cellulose proceeds under alkaline conditions, generally in aqueous NaOH solutions. Treatment of native cellulose with NaOH causes the cellulose to swell, which makes it more readily accessible to the modification reagent.[42,68] Thorough mixing and stirring are of vital importance to ensure uniform swelling and alkali distribution, which are the most important conditions for homogeneous etherification. Uneven distribution of the substituents causes severe loss in solubility due to the unetherified regions in the final product. Two types of reactions dominate cellulose etherification.[42,70]

28.5.7 WILLIAM ETHERIFICATION

An organic halide is used as the etherification reagent and alkali in amounts that are stoichiometrically equivalent to the reagent are consumed. Unreacted alkali must be washed out of the final product as a salt.[42]

28.5.8 ALKALINE-CATALYZED OXALKYLATION

In this reaction, an epoxide is added to the swollen alkali cellulose. Only catalytic amounts of alkali are required; thus, in principle, no alkali is consumed. The reaction may proceed further as new hydroxyl groups are generated during this reaction.

28.5.9 ENZYMIC METHOD OF CHEMICAL MODIFICATION

Enzymic methods for the determination of the substituent distribution in cellulose derivatives are based on the selective degradation of the modified polymer; nonsubstituted regions are easily hydrolyzed compared with low-substituted regions, whereas highly substituted areas are not hydrolysed at all and remain intact. However, one limitation of this enzymic approach is that the hydrolysis of a certain glucosidic linkage does not simply result in cleavage or not, but the hydrolysis rate may differ by some orders of magnitude. The rate of enzymic hydrolysis seems to depend not only on several factors, such as the DS, the nature of the substituent, and position of the substituents in the neighboring glucose unit, but also on the substituent distribution further along the chain, as an oligomeric sequence is usually involved in the formation of the enzyme–substrate complex.[42,43] In addition, the accessibility to enzymic hydrolysis is dependent on the physical features of the polymer, such as crystallinity, degree of swelling, and solubility. In many cases, chemical modification makes the polymer less crystalline and enhances its water solubility, thus increasing the susceptibility of cellulose derivatives to enzymic attack.[42,43]

28.5.10 CELLULOSE-HYDROLYZING ENZYMES

Cellulose-hydrolyzing enzymes are known as cellulases and they catalyze the hydrolysis of β-D-glucosidic linkages in cellulose and other β-D-glucans. Cellulases are produced by microorganisms, including bacteria and fungi, for example, the fungi Trichoderma reesei and Humicola insolens.[73]

28.5.11 ENDO-1,4-β-GLUCANASE

Endo-1,4-β-glucanase [1,4-(1,3; 1,4)-β-D-glucan 4-glucanohydrolase], often called cellulase, is an endo-enzyme that catalyzes the hydrolysis of (1,4)-β-D-glucosidic linkages in cellulose, lichenin, and cereal β-glucans.[42,73] Endo-1,4-β-glucanase shows low activity on crystalline cellulose, and it is generally believed that the enzyme acts through random hydrolysis of the internal bonds in the amorphous regions of the polymer. Thus, new chain ends that become available for further enzymic hydrolysis are formed.[43] The major hydrolysis products are cellobiose and cellotriose, and also longer cello-oligosaccharides with different chain lengths depending on the source of the enzyme.[42,43]

28.5.12 β-Glucosidase

β-Glucosidase (β-D-glucoside glucohydrolase) is an exoenzyme that hydrolyzes m-terminal nonreducing β-D-glucose residues in β-D-oligosaccharides with the release of β-glucose. The enzyme is also called cellobiase, as it readily hydrolyzes cellobiose to glucose.[42,73]

28.5.13 Cellobiohydrolase

Cellobiohydrolase (1,4-D-glucan cellobiohydrolase) is an exo-glucanase that catalyzes the hydrolysis of (1,4)-β-D-glucosidic linkages in cellulose with the release of cellobiose. Various forms of this enzyme are thought to hydrolyze cellulose chains either from the reducing or from the nonreducing end.[42,43]

28.5.14 Other Derivatization Processes

28.5.14.1 Cellulose-Based Hydrogels and Cross-Linking Strategies

Cellulose-based hydrogels can be obtained via physical or chemical stabilization of aqueous solutions of cellulosics. Additional natural and/or synthetic polymers might be combined with cellulose to obtain composite hydrogels with specific properties.[2,74,75] Physical, thermoreversible gels are usually prepared from water solutions of methylcellulose and/or hydroxypropyl methylcellulose (in a concentration of 1%–10% by weight).[76] The gelation mechanism involves hydrophobic associations among the macromolecules possessing the methoxy group. At low temperatures, polymer chains in solution are hydrated and simply entangled with one another. As temperature increases, macromolecules gradually lose their water of hydration, until polymer–polymer hydrophobic associations take place, thus forming the hydrogel network. The sol–gel transition temperature depends on the DS of the cellulose ethers as well as on the addition of salts. A higher DS of the cellulose derivatives provides them a more hydrophobic character, thus lowering the transition temperature at which hydrophobic associations take place. A similar effect is obtained by adding salts to the polymer solution, since salts reduce the hydration level of macromolecules by recalling the presence of water molecules around themselves. Both the DS and the salt concentration can be properly adjusted to obtain specific formulations gelling at 37°C and thus potentially useful for biomedical applications.[2,77–79] Liquid formulations, either mixed with therapeutic agents or not, are envisaged to be injected in vivo and their cross-linking reaction triggered by the only physiological environment. However, physically cross-linked hydrogels are reversible,[80] thus might flow under given conditions (e.g., mechanical loading) and might degrade in an uncontrollable manner. Owing to such drawbacks, physical hydrogels based on methylcellulose and HPMC are not recommended for use in vivo. In vitro, methylcellulose hydrogels have been recently proposed as novel cell sheet harvest systems.[78] As opposed to physical hydrogels that show flow properties, stable and stiff networks of cellulose can be prepared by inducing the formation of chemical, irreversible cross-links among the cellulose chains. Either chemical agents or physical treatments (i.e., high-energy radiation) can be used to form stable cellulose-based networks. The degree of cross-linking, defined as the number of cross-linking sites per unit volume of the polymer network, affects the diffusive, mechanical, and degradation properties of the hydrogel and can be controlled to a certain extent during the synthesis. Specific chemical modifications of the cellulose backbone might be performed before cross-linking, in order to obtain stable hydrogels with given properties. For instance, silylated HPMC has been developed, which cross-links through condensation reactions upon a decrease of the pH in water solutions. Such hydrogels show potential for the in vivo delivery of chondrocytes in cartilage tissue engineering.[81,82] As a further example, tyramine-modified SCMC has been synthesized to obtain enzymatically gellable formulations for cell delivery.[83] Photocross-linking of water solutions of cellulose derivatives is achievable following proper functionalization of cellulose. Depending on the cellulose derivatives used, a number of cross-linking agents and catalysts can be employed to form hydrogels. Epichlorhydrin, aldehydes and aldehyde-based reagents, urea derivatives, carbodiimides, and multifunctional carboxylic acids are the most widely used cross linkers for cellulose. However, some reagents, such as aldehydes, are highly toxic in their unreacted state. Although unreacted chemicals are usually eliminated after cross-linking through extensive washing in distilled water, as a rule, toxic cross-linkers should be avoided to preserve the biocompatibility of the final hydrogel, as well as to ensure an environmentally sustainable production process. The cross-linking reactions among the cellulose chains activated by chemical agents might take place in water solution, organic solvents, or even in the dry state (e.g., polycarboxylic acids can cross-link cellulose macromolecules via condensation reactions, which are favored at high temperature and in the absence of water).[2,84–87] Novel superabsorbent cellulose-based hydrogels crosslinked with citric acid have been recently reported, which combine good swelling properties with biodegradability and absolute safety of the production process.[2,87] In light of environmental and health safety concerns, radiation cross-linking of polymers, based on gamma radiation or electron beams, has been receiving increasing attention in the past years as it does not involve additional chemical reagents, is easily controllable, and, in cases of biomedical applications, allows the simultaneous sterilization of the product. High-energy radiation usually leads to chain scission of the polymer and this has been shown also for cellulose.[88] However, several cellulosics can be cross-linked under relatively mild radiation, both in aqueous solutions and in solid form, because cross-linking prevails over degradation.[89–91] The cross-linking reaction is affected by the irradiation dose as well as by the cellulose concentration in solution.[2]

28.5.15 CONVENTIONAL METHODS FOR CELLULOSE DISSOLUTION AND MODIFICATIONS

Cellulose in aqueous phase are modified by introducing functional groups into the cellulose macromolecules to substitute the free hydroxyl groups, either in the heterogeneous phase or in the homogeneous phase.[49,92] There are many solvents used to dissolve and modify cellulose, such as N,N-dimethylacetamide/lithium chloride, dimethylsulfoxide (DMSO)/tetrabutylammonium fluoride, N-methylmorpholine-N-oxide and other molten salt hydrates, such as $(LiClO_4)3H_2O$.[93,94] Commercially, two major solvent systems were used for dissolving and regenerating cellulose materials; one is carbon disulfide solvent used in viscose process and the other is N-methylmorpholine-N-oxide (NMMO) solvent used in Lyocell process.[49]

28.5.16 IONIC LIQUIDS FOR CELLULOSE DISSOLUTION AND MODIFICATION

Recently, certain ionic liquids (ILs) have been applied as green solvents that would dissolve cellulose[95–98] and function as inert and homogeneous reaction media. During the past decade, many researches and studies were done on the application of ionic liquid in cellulose dissolution and modification.[96,97,99–104] ILs have been attracting interest because of their wonderful properties such as high thermal stability, lack of inflammability, low volatility, chemical stability, and excellent solubility with many organic compounds.[105,106] Many kinds of room-temperature ILs,[107] with a variety of structures, have shown a good ability to dissolve cellulose, such as halogen-based ILs, e.g., 1-butyl-3-methyl- and 1-allyl-3methyl-imidazolium, which contain chloride. Also some imidazolium-based ILs containing phosphate, formate, and acetate anion have shown high ability for dissolving cellulose.[105,106] ILs have been regarded as low-melting (<100°C) salts, and hence, it forms liquids that consist of pure cations and anions.[49] In 1934, Graenacher discovered that molten N-ethylpyridinium chloride, in the presence of nitrogen-containing bases, could be used to dissolve cellulose.[98] It was reported that cellulose could be dissolved in ILs without formation of any derivative—a great progress in cellulose dissolution and cellulose modification had been established and interesting results have been reported by various pioneers in the optimization and applications of ILs in cellulose dissolution and functionalization.[49,95,97,107] Table 28.2 lists some ILs used for cellulose dissolution, while Figure 28.7 shows the possible ways of modifying cellulose with other reagents by use of ILs as a medium for cellulose modification.

TABLE 28.2
Different Types of ILs Used for Cellulose Dissolution

Cellulose	Ionic Liquids	Conditions		Solubility (%)
		Temperature (°C)	Time (min)	
MCC pulp cotton (linter)	[Amim]Cl	100–130	40–240	5–14.5
Avicel (MCC)	[Amim]Cl	90	720	5
Pulp (DP = 1000)	[Bmim]Cl	100	—	10
Pulp (DP = 286)	[Bmim]Cl	83	720	18
Avicel (MCC)	[Bmim]Cl	100	120	20
Avicel (MCC)	[Bmim]Cl	100	60	20
Pulp	[Bmim][Cl]	85	—	13.6
Pulp	[Emim]Cl	85	—	15.8
Avicel (MCC)	[Emim]Cl	100	60	10–14
Avicel (MCC)	[Emim]Cl	90	—	5
Pulp	[Emim]OAc	85	—	13.5
Avicel (MCC)	[Emim]OAc	110	—	15

Source: Magdi, E.G. et al., *Int. J. Eng. Sci. Technol.*, 4(7), 3556, 2012, with permission from the International Journal of Engineering Science and Technology.

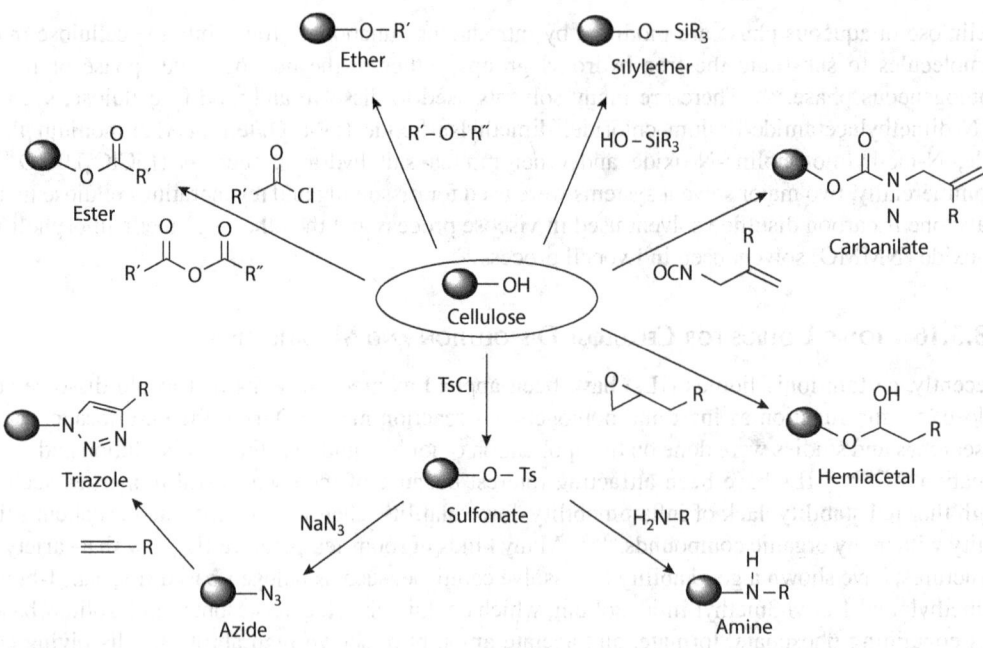

FIGURE 28.7 Functionalization routes for cellulose in ILs medium. (From Magdi, E.G. et al., *Int. J. Eng. Sci. Technol.*, 4(7), 3556, 2012, with permission from the International Journal of Engineering Science and Technology.)

Cyclic anhydrides like acetic, phthalic, and succinic anhydride have been widely used in cellulose modifications to produce different cellulose derivatives such as cellulose acetate, cellulose butyrate, cellulose benzoate, cellulose phthalate, and cellulose with or without catalyst,[49,96] which have several applications such as water absorbents for soil in agriculture, drug delivery system, and as thermoplastic.

Due to the wide application of cellulose acetate, cellulose acetylation in ILs has taken significant attention. Several investigations of cellulose acetylation with acetic anhydride or acetyl chloride in ILs have been done, and cellulose acetates with a high DS have been obtained.[95,108–110] The acetylation of cellulose in an ionic liquid has been introduced as a novel process by Zhang and coworkers.[107] They reported that acetylation of cellulose occurs under homogeneous conditions and in the absence of any catalyst by using allylimidazolium-based ionic liquid. Cellulose solution (5%) was prepared by dissolving cellulose with degree of polymerization of 650 into [Amim]Cl ILs at 100°C for 15 min, followed by the addition of acetic anhydrate to acetylate cellulose under a nitrogen atmosphere and temp 80°C–100°C for 25 min–23 h. Different degrees of substitution in the range 0.94–2.74 were obtained.[49]

28.5.17 IONIC LIQUID AS MEDIA FOR MODIFICATION OF CELLULOSE BY GRAFTING COPOLYMERIZATION AND BLENDS (COMPOSITES)

Cellulose has the ability to blend with other biopolymers such as silk, wool, chitin, chitosan, elastin, collagen, keratin, and polyhydroxyalkanoate after dissolution in ILs.[49,98,111] Hence, this part of entry has been directed to overview the cellulose derivatives and their modification by blending graft copolymerization in ILs. Grafting copolymerization is usually achieved by modifying the cellulose molecules through the creation of branches (grafts) of synthetic polymers that impart specific properties onto the cellulose substrate, without destroying its intrinsic properties.[112] Chemical modification of cellulose through graft copolymerization has also been demonstrated to be a promising method for the preparation of new materials, enabling the introduction of special properties into cellulose without destroying their intrinsic characteristics and enlarging their scope for potential applications.[113] Ring-opening polymerization is a way of grafting copolymers. Generally, it has been accomplished by using cyclic monomer such as oxiranes (epoxides), lactons, amino acid N-carboxy anhydrides (NCAs), and 2-alkyl oxazolines,[114] with cellulose/IL solution. But the formation of homopolymer during the grafting process is unwanted, which should be removed by suitable solvents in extraction process. Zhang et al.[113,115] grafted cellulose in opening ring polymerization. The grafted materials were characterized by FTIR (confirmed the introduction of the side chain into the cellulose backbone via graft copolymerization) and SEM optical microscopy (showed good uniformity). Lin et al.[116] applied microwave heating and obtained similar results.

28.5.18 CELLULOSE BLENDS (COMPOSITES) IN ILs MEDIA

Kadokawa, Masa-Aki, and Akihiko-Kaneko[117] successfully obtained a gel by dissolving a mixture of cellulose–starch in [Bmim]Cl at 100+°C for 24 h and thereafter, cooling the composites solution to room temperature and kept between two glass plates for 5 days. ILs were removed by ethanol or acetone. The composite materials were characterized by XRD and SEM, which showed good compatibility with the thermal analysis and TGA exhibiting higher weight loss at higher temperatures.[118] A novel heparin- and cellulose-based biocomposite has been prepared by dissolution of cellulose and heparin in ILs. This study was investigated by Murugesan et al.[119] They obtained a membrane film of cellulose/heparin composite. Cellulose was dissolved in [Bmim]Cl at 70°C at a concentration of 10%wt with microwave for 4–5 s and heparin was dissolved in [Emim]BA at 35°C for 20 min.[49]

28.6 APPLICATION OF CELLULOSE-BASED BIOPOLYMERS IN DOSAGE FORM DESIGN AND NOVEL DRUG DELIVERY SYSTEMS

Pharmaceutical formulation development involves various components in addition to the active pharmaceutical ingredients. In recent years, excipient development has become a core area of research in pharmaceutical drug delivery because it influences the formulation development and drug delivery process in various ways. Biopolymers are choice of research as excipient because of their low toxicity, biodegradability, stability, and renewable nature. Cellulose-based biopolymers are naturally obtainable macromolecules and they play an important role in biomedicine with applications in tissue engineering, regenerative medicine, drug-delivery systems, and biosensors. The inherent recyclability, reproducibility, cost-effectiveness, and availability in a wide variety of forms, biocompatibility and biodegradability of these materials make them particularly useful in biomedical applications. Thus, cellulose-based biopolymers are employed pharmaceutically in the formulation and delivery of bioactive compounds as immediate-release and/or sustained/prolonged/extended/modified/controlled-release dosage forms, specialized bioactive carriers as well as nanoparticulate drug delivery systems.

28.6.1 IMMEDIATE-RELEASE DOSAGE FORMS

Immediate release refers to the instantaneous availability of drug for absorption or pharmacologic action in which drug products allow drugs to dissolve with no intention of delaying or prolonging dissolution or absorption of the drug. Cellulose-based excipients have been well explored and used in pharmaceutical formulation development of immediate-release dosage forms. Cellulose-based biopolymers are often used as tablet binding, thickening and rheology control agents, for film formation, water retention, improving adhesive strength, and as suspending and emulsifying agents.[8] Cellulose ethers are widely used as important excipients for designing matrix tablets. Large-scale commercial cellulose ethers including CMC, methyl cellulose, HEC, HPMC, HPC, EHEC, and MHEC have been exploited as potential raw materials in the design of immediate-release dosage forms.[48] For instance, methylcellulose has been employed in the formulation of bulk laxatives, as well as solid dispersion granules[120]; MCC has been employed as suspending agent in zinc oxide and sulfadimidine suspensions and is used as diluent and disintegrating agent for immediate-release oral solid dosage forms[121,122]; SCMC is employed as suspending agent in chalk suspensions and could be used in the formulation of pharmaceutical suspensions for extemporaneous use.[123] HEC and HPC are used in hydrophilic matrix systems, while ethylcellulose can be used in hydrophobic matrix systems.[124] HPC has been employed to formulate buccal tablets containing lidocaine. These tablets were more effective in reducing pain and decreasing the healing time than both experimental and plain tablets.[125] HEC is used as a thickner, stabilizer, and suspending agent for oral and topical applications when a nonionic material is desired. Ethylcellulose can be used to coat one or more active ingredients of a tablet to prevent them from reacting with other materials or with one another, can prevent discoloration of easily oxidizable substances such as ascorbic acid, and allows granulations for easily compressed tablets and other dosage forms.[17] Also, liquid and semi-solid pharmaceutical dosage forms are important physicochemical systems for medical treatment, which require rheological control and stabilizing excipients as essential additives. CMC can be used to adjust the viscosity of syrups.[124] SCMC is a low-cost, soluble, and polyanionic polysaccharide derivative of cellulose that has been employed as an emulsifying agent in pharmaceuticals. The very many important functions provided by this polymer make it a preferred thickening, gelling agent, protective colloid, and film-former in jellies; stabilizer, thickner, and film-former in ointments, creams, lotions, and emulsions; thickner and suspending aid in syrups and suspensions; tablet binder, granulating aid, and tablet-coating film-former.[126] HPMC, a water-soluble cellulose ether, can be used as a hydrophilic polymer for the preparation of immediate-release dosage forms. For instance, immediate-release HPMC matrix tablets of pseudoephedrine, atenolol, and naproxen have been assessed by researchers.[124,127,128]

Immediate-release oral dosage forms (such as tablets and capsules) are the most widely used drug-delivery systems available. These products are designed to disintegrate in the stomach followed by their dissolution in the fluids of the gastrointestinal tract.[129] Dissolution of the drug substance, under physiological conditions, is essential for its systemic absorption. For this reason, dissolution testing is typically performed on solid dosage forms to measure the drug release from the drug product as a test for product quality assurance/product performance and to determine the compliance with the dissolution requirements.[130] Thus, cellulose-based biopolymers, in addition to their usefulness in the design and formulation of immediate-release dosage forms, could also be important excipients for establishing the potential usefulness of the disintegration test as drug product acceptance. A group of researchers has previously assessed verapamil hydrochloride tablets formulated using lactose monohydrate (LMH) as filler, HPMC as binder, SCMC as disintegrating agent, and established that only one formulation (of these tablets) might be suitable for using the disintegration test instead of the dissolution test as the drug product acceptance criteria.[130] These researchers highlighted the need for systemic studies before using the disintegration test, instead of the dissolution test as the drug acceptance criterion.

28.6.2 Sustained/Modified/Controlled-Release

28.6.2.1 Dosage Forms

Cellulose-based biopolymers are an important class of materials for pharmaceutical and biotechnological applications. There is a great potential in utilizing cellulose-based biopolymers as pharmaceutical adjuvants and in modified/sustained/controlled-release drug delivery systems. Modified-release dosage forms include both delayed and extended-release drug products. Delayed release is defined as the release of a drug at a time other than immediately following administration, while extended-release products are formulated to make the drug available over an extended period after administration. Controlled release includes extended-release and pulsatile-release products. Pulsatile release involves the release of finite amounts (or pulses) of drug at distinct intervals that are programmed into the drug product. Modified-release technologies utilize polymers such as cellulose-based biopolymers to alter the site or time of drug release within the gastrointestinal tract.[124] The need for modified-release technologies arose from an understanding that disintegration of a dosage form in the stomach, resulting in immediate release of drug, is not always desirable. In recent years, there has been an increasing tendency to deliver drug entities as modified-release formulations, and even though it can be appreciated that the unit cost of a modified-release formulation will be greater than the equivalent immediate-release variety, the former version may confer a reduction in overall healthcare costs. This may be in terms of a reduction in the number of doses to be taken to achieve the desired therapeutic effect, therefore reducing overall medication costs, or the subsequent improvement in compliance, negating the implications of ineffective therapy, or the need for further medication required to treat drug-induced side effects of the original treatment.[131]

Different cellulose-based biopolymers can be used either singly or in combination as matrix formers for sustained/modified/controlled drug delivery in different dosage forms ranging from oral to topical drug delivery systems. Hydrophilic cellulose-based biopolymer matrix systems are widely used for designing oral controlled drug delivery dosage forms because of their flexibility to provide a desirable drug-release profile, cost-effectiveness, and broad regulatory acceptance. But owing to rapid diffusion of the dissolved highly water-soluble drugs through the hydrophilic gel network, hydrophobic polymers are usually included in the matrix system to extend the release of highly water-soluble drugs.[132,133] These hydrophobic polymers may or may not be cellulose-based.

Controlled drug delivery remains a research focus for public health to enhance patient compliance and drug efficiency and to reduce the side effects of drugs. Cellulose-based biopolymers are employed in the controlled release of drugs via solid dispersions, a novel drug-delivery system in which compounds are dispersed into water-soluble carriers. This has been generally used to improve the dissolution properties and the bioavailability of drugs that are poorly water soluble.[120,134] Methylcellulose has the hydroxyl group in a structure and is interactive with the carboxylic acid of carboxyvinyl polymer, as well as poly(ethylene oxide) (PEO). Ozeki et al.[134] studied the release of phenacetin from the solid dispersion granules containing different ratios of methylcellulose and carboxyvinyl polymer. They found out that it is feasible to control phenacetin release from methylcellulose–carboxyvinyl polymer solid dispersions by controlling the complex formation between methylcellulose and carboxyvinyl polymer, which can be accomplished by varying the methylcellulose–carboxyvinyl polymer ratio and the molecular weight of the methylcellulose. In addition, by using solid dispersions containing a polymer blend, such as HPC and ethylcellulose, it is possible to precisely control the rate of release of an extremely water-soluble drug, such as oxprenolol hydrochloride.[23] In this case, the water-soluble HPC swells in water and is trapped in the water-insoluble ethylcellulose so that the release of the drug is slowed. These studies have shown that there is a linear relationship between the rate of release of the water-insoluble drug and its interaction with the polymer.[27] Furthermore, HPC offers interesting characteristics as controlled-release matrices. Gon et al.[135] observed that graft copolymers could stand alone as an effective matrix for tablets designed for drug-delivery systems. Similarly, buccal delivery formulations containing HPC and polyacrylic acid have been in use for many years,[136,137] with various ratios of the two polymers. More so, mucoadhesive delivery systems based on HPC have been reported for different drugs.[125,138] Furthermore, the ability of ethylcellulose and HEC to sustain the release of drugs has been demonstrated.[126] Additionally, SCMC can be used in the preparation of semi-interpenetrating polymer network microspheres by using glutaraldehyde as a cross-linker. Ketorolac tromethamine, an anti-inflammatory and analgesic agent, was successfully encapsulated into these microspheres and drug encapsulation of up to 67% was achieved. The diffusion coefficients decreased with increasing cross-linking as well as increasing content of SCMC in the matrix, and in vitro release studies indicated a dependence of release rate on both the extent of cross-linking and the amount of SCMC used to produce microspheres.[139] Controlled-release preparations of indomethacin could be employed to increase patient compliance and to reduce adverse effects, fluctuation in plasma concentration, and dosing frequency. Waree and Garnpimol[140] prepared a complex of chitosan and CMC and cross-linked by glutaraldehyde to control the release of indomethacin from microcapsules. The membrane of the microcapsules was formed by electrostatic interaction between positively charged amine on the chitosan chain and the negatively charged hydroxyl group on the CMC chain, and the concentration of CMC affected the formability of chitosan–CMC microcapsules.[141]

Moreover, HPMC, a water-soluble cellulose ether, can be used as hydrophilic polymer for the preparation of controlled-release tablets. Khanvilkar et al.[142] investigated the effects of a mixture of two different grades of HPMC and the apparent viscosity on drug-release profiles of extended-release matrix tablets. The study indicated that lower and higher viscosity grades of HPMC can be mixed uniformly in definite proportions to get the desired apparent viscosity. Also, incorporating a low viscosity grade of HPMC in the formulation would lead to a significantly shorter lag time. However, it imposes minimal impact on the overall dissolution profile. The study showed that the drug release from an HPMC matrix tablet prepared by dry blend and direct compression approach is independent of tablet hardness, is diffusion-controlled, and depends mostly on the viscosity of the gel layer formed.[142] In addition, Ye et al.[143] studied the effect of manufacturing process on the dissolution characteristics of HPMC matrix tablets. These researchers reported that when HPMC matrix tablets were prepared by wet-granulation approach, the tablet hardness, distribution of HPMC within the tablet (intergranular and intragranular), and the amount of water added in the wet granulation step all have a significant impact on dissolution. The results also indicated that incorporating partial amount of HPMC inter-granularly in the dry-blend step, drug-release profiles could be made much less sensitive to the manufacturing process. In a related study, Liu et al.[144] employed alginate as the gelling agent in combination with HPMC (viscosity-enhancing agent) in controlling the release of gatifloxacin. The study showed that the alginate/HPMC solution retained the drug better than the alginate or HPMC solutions alone, indicating that the alginate/HPMC mixture can be used as an in situ gelling vehicle to enhance ocular bioavailability and patient compliance.

Furthermore, owing to the hydration and gel-forming properties of HPMC, it can be used to prolong the release of bioactives. The yahom (a well-known traditional remedy/medicine for treatment of nausea, vomiting, flatulence, and unconsciousness in Thailand) buccal tablet possessed antimicrobial activities that could be able to cure the oral microbial infection and aid in wound healing, but the addition of polyvinyl pyrrolidone (PVP) combined with HPMC could promote the bioadhesivity of yahom tablet.[145] The research by Chantana et al [146] indicated that PVP had higher water sorption and erosion, whereas HPMC could prolong the erosion of yahom buccal tablet, and that the tablet containing 50% yahom, which had the polymer mixture of PVP: HPMC 1:2 was suitable for use as buccal tablet. In addition to the cellulose derivatives, cross-linked high amylose starch (CLA) has been successfully used as a controlled-release excipient for the preparation of solid dosage forms.[147] Rahmouni et al.[148] characterized the gel matrix properties of binary mixtures of CLA/HPMC, and studied the effect of incorporated HPMC on the release kinetics of three model drugs of different solubilities such as pseudoephedrine sulfate (very soluble), diclofenac sodium (sparingly soluble), and prednisone (very slightly soluble). These researchers found out that the swelling characteristics and erosion of granulated cross-linked high amylose starch (CLAgr)/HPMC tablets increased with HPMC concentration and incubation time, and that the presence of HPMC in CLA tablets at concentration 10% protected CLA against α-amylase hydrolysis and reduced the release rate of poorly and moderately water-soluble drugs. However, the release of the highly water-soluble model drug, which occurred predominantly by diffusion, was rapid both in the presence or in the absence of HPMC. In another study, the effect of the concentration of HPMC on naproxen release rate was evaluated. The result showed that an increased amount of HPMC resulted in reduced drug release. The inclusion of buffers to increase the dissolution and to decrease the gastric irritation of weak acid drugs, such as naproxen in the HPMC matrix tablets, enhanced naproxen release. The inclusion of sodium bicarbonate and calcium carbonate in the HPMC matrix improved the naproxen dissolution; however, including sodium citrate did not produce any effect on naproxen dissolution.[149]

Many cellulose-based biopolymer blends have been used in the formulation of bioadhesive/mucoadhesive drug-delivery systems in the form of microparticles, tablets, patches, hydrogels or films. Builders et al.[7] prepared and evaluated mucinated cellulose microparticles for controlled drug-delivery application by mixing of colloidal dispersions of porcine mucin and MCC. The hybrid biopolymer was recovered by precipitating at controlled temperature and pH conditions using acetone. The mucoadhesive property of the new polymer was similar to that of mucin. Scanning electron micrographs (SEMs) showed that the microparticles generated from the hybidization were similar to those of MCC, but with larger and denser particles. The FT-IR spectrum and DSC thermogram of the hybrid polymer were characteristically different from mucin and MCC. The presence of new peaks in the FT-IR spectrum and distinct cold crystallization exotherm, which were absent in both mucin and MCC, confirmed the formation of a new polymer type with synergistic physicochemical and functional properties.[36] The use of admixtures of Carbopols (Carbopols 940 and 941) and SCMC in the formulation of bioadhesive metronidazole tablets has been studied by Ibezim et al.[150] In a related research, ternary cellulose-based biopolymer blends consisting of SCMC, acacia, and Veegum were evaluated for bioadhesive delivery of metronidazole comparing them with the performances of the single biopolymers.[151] Bioadhesive characteristics of tablets prepared with SCMC alone was highest, but blending improved the bioadhesive strengths of acacia and Veegum tablets. Drug release from the single polymers and the ternary biopolymer blends was prolonged, indicating their suitability for the delivery of metronidazole by bioadhesive controlled-release mechanism. Similarly, hydrogel bead-delivery systems of hydrochlorothiazide were studied by Attama and Adikwu.[152] The hydrogel beads consisted of blends of tacca starch and SCMC and Carbopols 940 and 941. The admixtures studied showed improvement on the bioadhesive properties of tacca starch, and confirmed they could be used as bioadhesive motifs for drug delivery into the gastrointestinal tract. More so, the buccoadhesiveness, swelling characteristics, and release profile of hydrochlorothiazide from patches formulated with ethylcellulose and HPMC interpolymer complexes of different ratio were studied to evaluate their applicability in sustained drug delivery.[153] The study indicated that, although the in vitro release of hydrochlorothiazide from the patches (prepared by casting) was not appreciably prolonged, the blends with low area swelling ratio are more suitable for the formulation of buccoadhesive drug-delivery systems. In addition, release of diclofenac sodium from bioadhesive hydrophilic matrix tablets composed of polyvinyl pyrrolidone (PVP) and SCMC was also studied.[154] Tablets that satisfied pharmacopoeial standards were obtained, and prolonged release of diclofenac sodium was achieved.

Furthermore, Proddurituri et al.[155] studied a method of improving the physical stability of clotrimazole (CT) and the polymer contained within hot-melt extrusion (HME) films using polymer blends of HPC and poly(ethylene oxide) (PEO). Films containing HPC:PEO:CT in the ratio of 55:35:10 demonstrated optimum physico-mechanical, bioadhesive, and release properties. These researchers concluded that polymer blends of HPC and PEO could be used successfully to tailor the drug release, mechanical and bioadhesive properties, and stability of the HME films, and that the glass transition temperature of the polymers played an important role in determining the physical stability of the solubilized drug. In a study comparing the performance of CMC, HPC, and their admixtures in tableting, tablets containing CMC alone had poor compression characteristics.[156] The hardness values for HPC-containing tablets were the same as or slightly less than those results seen with the HPC/CMC polymer blends. Dissolution test results clearly demonstrate the effects of polymer blending on release rate modification. The tablets that contained 100% CMC reached a T_{80} in 88 min, while 100% HPC tablets had a T_{80} of 224 min. When the tablets that contained the polymer mixture of HPC/CMC (75/25) were tested, the T_{80} value increased to 339 min. Also, combination of HPC/carrageenan (75/25) was tableted and tested. The hardness data showed that the blend is equivalent to or better than the individual polymers demonstrating the advantage of polymer blending in tablets for sustained drug delivery. On the same lines, Kuksal et al.[157] prepared and characterized extended-release matrix tablets of zidovudine (AZT) using hydrophilic Eudragit® RLPO and Eudragit® PSPO alone or their combination with hydrophobic ethylcellulose. Results of the study demonstrated that a combination of both hydrophilic and hydrophobic polymers could be successfully employed for formulating sustained-release matrix tablets of AZT. The investigated sustained-release matrix tablet was capable of maintaining constant plasma AZT concentration through 12 h, with good correlation between the dissolution profiles and bioavailability. Mukherjee et al.[158] developed novel transdermal drug delivery system (TDDS) of matrix type containing dexamethasone using blends of two different polymeric combinations, PVP and ethylcellulose, and Eudragit® with PVP. All the formulations were found to be suitable for formulating TDDS in terms of physicochemical characteristics and there was no significant interaction noticed between the drug and polymers used. However, PVP–ethylcellulose polymers performed better than PVP-Eudragit as TDDS for dexamethasone. Yerri-Swamy et al.[159] evaluated interpenetrating polymer network microspheres of HPMC/poly(vinyl alcohol) for controlled release of ciprofloxacin hydrochloride. The HPMC and poly(vinyl alcohol) blend microspheres were prepared by water-in-oil emulsion method and ciprofloxacin hydrochloride was loaded into the interpenetrating polymer network microspheres that were cross-linked with glutaraldehyde. in vitro dissolution experiments performed in pH 7.4 buffer medium at 35°C indicates a sustained and controlled release of ciprofloxacin hydrochloride from the interpenetrating polymer network microspheres up to 10 h.[159]

Iqual et al.[160] studied bacterial cellulose as a promising biopolymer for controlled drug delivery applications. Model tablets were film-coated with bacterial cellulose, using a spray-coating technique, and in vitro drug release studies of these tablets were investigated. They concluded that bacterial cellulose could be used as novel aqueous film-coating agent with lower cost and better-film forming properties than existing film-coating agents.[160] Moreover, enteric coating is another important application of cellulose-based biopolymer blends in controlled delivery of bioactives. In this perspective, cellulose acetate phthalate, hydroxypropyl methylcellulose phthalate, and ethylcellulose in blends with other polymers are commonly used for enteric coating.[161–163] Using ethylcellulose as an example, amylose, a plant polysaccharide from starch, can be combined with ethylcellulose to produce a film coating capable of effecting colon-specific drug release from a dosage form through bacterial fermentation of the amylose component. Ethylcellulose is present in the system as a structuring agent in the form of the aqueous dispersion Surelease® grade EA-7100.

28.6.2.2 Specialized Bioactive Carriers

Cellulose based biopolymers could be useful in designing specialized drug-delivery devices, such as hydrogels, xerogels/aerogels, osmotic pumps, dual-drug dosage forms with improved separation of drugs as well as dosage forms combining both immediate-release and prolonged-release modes of drug delivery. Cellulose-based hydrogels are biocompatible and biodegradable materials, which show promise for a number of industrial uses, especially in cases where environmental issues are concerned, as well as biomedical applications. Apart from swelling tablets, more sophisticated hydrogel-based devices have been developed for controlled drug delivery.

The advances in hydrogel-based devices aim not only at the sustained release of a bioactive molecule over a long period of time, ranging from hours to weeks, but also at a space-controlled delivery, directly at the site of interest.[164] Several water-soluble cellulose derivatives can be used, singularly or in combination, to form hydrogel networks possessing specific properties in terms of swelling capability and sensitivity to external stimuli. The trend in the design of cellulose hydrogels is related to the use of nontoxic cross-linking agents or cross-linking treatments to further improve the safety of both the final product and the manufacturing process. Controlled release through oral drug delivery is usually based on the strong pH variations encountered when transiting from the stomach to the intestine. Cellulose-based polyelectrolyte hydrogels (e.g., hydrogels containing SCMC) are particularly suitable for this application. For instance, anionic hydrogels based on CMC have been investigated for colon-targeted drug delivery.[165] The advances in controlled release through a hydrogel matrix deal with the delivery of proteins, growth factors, and genes to specific sites, the need for which has been prompted by tissue engineering strategies. While hydrogel formulations for oral and transdermal delivery can be nondegradable, the direct delivery of drugs or proteins to different body sites requires the hydrogel biodegradation to avoid foreign body reactions and further surgical removal. Injectable hydrogel formulations are particularly appealing and currently under investigation. The cross-linking reaction has to be performed under mild conditions to avoid denaturing the loaded molecule. The microenvironment resulting from degradation of the polymer should be mild as well for the same purpose. With particular regard to cellulose-based hydrogels, injectable formulations, based on HPMC, have been developed to deliver both biomolecules and exogenous cells in vivo.[84,85,166] The need to encapsulate bioactive molecules into a hydrogel matrix or other delivery devices (e.g., microspheres) is also related to the short half-life displayed by many biomolecules in vivo. When using hydrogels to modulate the drug release, the loading of the drug is performed either after cross-linking or simultaneously during network formation.[164] Moreover, the bioactive molecule can be covalently or physically linked to the polymer network to further tune the release rate. The smart behavior of some cellulose derivatives (e.g., SCMC and HPMC) in response to physiologically relevant variables (i.e., pH, ionic strength, and temperature) makes the resulting hydrogels particularly appealing for in vivo applications. Smart hydrogels are particularly useful to control the time- and space-release profile of the drug as swelling–deswelling transitions, which modify the mesh size of the hydrogel network, occur upon changes of physiologically relevant variables, such as pH, temperature, and ionic strength.[167]

In addition, cellulose-based hydrogel devices could be useful as stomach bulking agents. Novel bulking agents, effective in promoting weight loss, could be developed with cellulose-based superabsorbent hydrogels, since not only can their swelling capacity be properly designed by controlling their chemical composition and physical microstructure, but it can also be modulated by changing the environmental conditions (e.g., pH, ionic strength, and temperature). Here, the notion is that a xerogel-based pill is administered orally before each meal, and that the xerogel powder swells once in the stomach. By so doing, the space available for food intake is reduced, giving a feeling of fullness. Subsequently, the swollen hydrogel is eliminated from the body via the feces. In this direction, the hydrogel is envisaged to pass through the gastrointestinal tract, thus it is supposed to encounter the different pH environments of the stomach and the intestine. Along with superporous acrylate-based hydrogels, which swell very rapidly in aqueous solutions,[168] novel cellulose-based hydrogels, obtained by cross-linking aqueous mixtures of SCMC and HEC, have been shown to be appealing for the production of dietary bulking agents.[169,170] Indeed, such hydrogels possess high biocompatibility, with respect to intestinal tissues, and a high, pH-sensitive water-retention capacity.[170] Although the polyanionic nature of the SCMC network provides higher swelling capabilities at neutral pHs rather than at acid ones, the swelling ratio obtained at acid pHs might still be significant for use of the hydrogel as stomach filler. In particular, cellulose-based hydrogels obtained from nontoxic cross-linking agents are particularly attractive for this kind of application.[90,169,170] Moreover, shaped/silicized cellulosic aerogels have been developed. Reinforced shaped cellulosic aerogels consisting of two interpenetrating networks of cellulose and silica were prepared from shaped cellulose solutions by regenerating (reprecipitating) cellulose with ethanol; subjecting the obtained shaped alcogels to sol–gel condensation with tetraethoxysilane as the principal network-forming compound; and by drying the reinforced cellulose bodies with supercritical carbon dioxide. The influence of different types of cellulose and sol–gel forming parameters on porosity, cellulose integrity, and silica content were studied. The results showed improved functional and physico-chemical properties over the normal aerogels in cellulose aerogel applications.[171]

Furthermore, cellulose-based hydrogels hold promise as devices for the removal of excess water from the body (body water retainers) in the treatment of some pathological conditions, such as renal failure and diuretic-resistant edemas. The hydrogel in powder form is envisaged to be administered orally and absorb water in its passage through the intestine, where the pH is about 6–7, without previously swelling in the acid environment of the stomach. The hydrogel is then expelled through the feces, thus performing its function without interfering with body functions. As sensitivity to pH is required, polyelectrolyte cellulose hydrogels based on SCMC and HEC have been investigated for such applications.[172–174] Also, the use of hydrogels in combination with diuretic therapies might be useful in substituting some drugs and in using an intestinal pathway, instead of the systemic one, to remove water from the body.[173]

Additionally, osmotic pump, a specialized drug delivery device developed about 30 years ago, remains an excellent example of the use of polymeric properties to control the delivery of bioactives. The drug-release mechanism in this device is driven by a difference in osmotic pressure between the drug solution and the environment outside the formulation. It represents a family of technologies that have been developed for the extended, optimally zero-order release of pharmaceutical actives.[175] These technologies rely upon the encapsulation of the pharmaceutical active within a membrane (cellulose diacetate is by far the most commonly used membrane polymer), which is highly permeable to water, but is impermeable to salts and to many organics. Upon ingestion of the pill or capsule, water permeates through the membrane and dissolves water-soluble ingredients inside. If the active itself develops sufficient osmotic pressure, it may be used without other osmogents; otherwise inert water-soluble agents (e.g., sodium chloride or other salts) are added to help develop osmotic within the dosage form. That osmotic pressure is very high in comparison to that in the surrounding gut, and powers the ejection of an aqueous drug solution at a constant rate through a small (usually laser-drilled) orifice. Cellulose acetate was one of the first materials used for manufacturing semipermeable membranes in elementary osmotic pumps developed by ALZA Corporation®. These membranes continue to be used in commercial OROS® products, where the semipermeable membrane controls drug-release rate.[175] The release rate of drugs from an OROS is controlled by semipermeable membranes composed typically of cellulose acetate with various flux enhancers. Cellulose acetate butyrate (CAB) was identified as a viable alternative. The CAB membrane matched the cellulose acetate membrane in robustness but had superior drying properties, offering particular advantages for thermolabile formulations.

Many review works has been published on osmotic delivery over the past 40 years, including some recent ones.[176–179] In addition, a research group at the University of Mumbai[180] described a system in which a cellulose diacetate membrane is used for osmotic delivery of pseudoephedrine, with plasticizers used as film dopants in an attempt to create the pores. Good results were obtained with diethylphthalate (DEP), in some cases containing PEG (both of which have significant miscibility with cellulose acetate). Release rates could be controlled by film thickness and PEG content, and near-zero-order rates were seen in some cases. Reasonable control over release rates was also seen with more hydrophobic (and incompatible with cellulose acetate) plasticizers such as dibutyl sebacate (DBS). It has also been reported that acetaminophen could be microencapsulated mixed with sodium chloride osmogent, using a cellulose acetate coating.[181] The drug, osmogent, and other inert ingredients were extruded, then solution-coated with cellulose acetate. The authors demonstrated zero-order release under certain conditions, and showed that much faster release was obtained in the presence of NaCl than in its absence. There were no holes drilled, nor were pore-forming agents included; the authors speculate that the osmotic pressure tore holes in the cellulose acetate coating of these small particles. One area of more recent concentration has been the development of simpler forms of osmotic delivery systems. In this perspective, Catellani et al.[182] devised a method that avoids laser-drilling entirely, and makes membrane failure unlikely. They created formulations in which a polymer/drug core tablet is dipped partially in cellulose ester solution, creating a cellulose ester coating on some portion but not the entire pill. For these experiments they used HPMC as the matrix and buflomedil pyridoxal phosphate as the drug. They found out that these systems had an osmotic component to the release rate by virtue of the permeation of water into the system partly via the cellulose ester coating. They could release slowly by using CAP as the coating, or speed it by using increasing amounts of PEG plasticizer in a cellulose acetate coating. The study is interesting but is of unclear practical portent.

In furtherance to that, an interesting application of cellulose-based biopolymers as specialized bioactive carriers has been described.[183] The method involves manufacturing a pharmaceutical tablet for oral administration, the tablet combining both immediate-release and prolonged-release modes of drug delivery and using immediate-release drug that is either insoluble in water or only sparingly soluble and is present in a very small amount compared with the prolonged-release drug. The method involves the use of particles of the immediate-release drug (e.g., glimepiride) that is equal to or less than 10 µm in diameter, applied as a layer or coating over a core of the prolonged-release drug (e.g., metformin hydrochloride), the layer or coating being either the drug particles themselves, applied as an aqueous suspension (e.g., by using HPMC as a suspending agent), or a solid mixture containing the drug, in admixture with a material that disintegrates rapidly in gastric fluid (e.g., lactose, MCC, and combinations of lactose and MCC). The result in both cases is a high degree of uniformity in the proportions of the immediate-release and prolonged-release drugs, uniformity that is otherwise difficult to achieve in view of the insolubility of the immediate-release drug and its relatively small amount compared with the prolonged-released drug.

Another specialized bioactive carrier based on cellulose biopolymers could be seen in dual-drug dosage forms with improved separation of drugs.[184] Drug tablets that include a prolonged-release core and an immediate-release layer or shell are prepared with a thin barrier layer of drug-free polymer between the prolonged-release and immediate-release portions of the tablet. The barrier layer is penetrable by gastrointestinal fluid, thereby providing full access of the gastrointestinal fluid to the prolonged-release core, but remains intact during the application of the immediate-release layer, substantially reducing or eliminating any penetration of the immediate-release drug into the prolonged-release portion. This has been demonstrated using a solid matrix prepared with materials selected from the group consisting of poly(ethylene oxide), HPMC, and combinations of poly(ethylene oxide) and HPMC. It has also been established using a second solid matrix prepared with materials selected from the group consisting of lactose, MCC, and combinations of lactose and MCC.[184]

28.6.2.3 Cellulose-Based Biopolymer Microparticles for Drug Delivery

By hybridizing mucin and MCC, a novel polymer with a combination of the physicochemical and functional properties characteristic of the two-component polymers was obtained. The new polymer was directly compressible and it possessed mucus membrane protectant. Thus, to produce a novel excipient with membrane-protective, mucoadhesive, and direct compression properties for therapeutics and drug-delivery purposes, a mucin and MCC hybrid was prepared by regenerating colloidal mixtures of mucin and MCC at controlled pH and temperature conditions. Excipients with multiple functional properties confer many advantages, such as reduction of cost of production and the number of steps used during production. MCC–maize starch composite was generated by mixing colloidal dispersions of MCC and chemically gelatinized maize starch at controlled temperature conditions, and this process of polymer composite formation was termed compatibilized reactive polymer blending.[185] This led to direct compression efficiency of the novel polymer formed.

The study of the release kinetic profiles of naproxen from microcapsule compressed as well as matrix tablets using a combination of water-insoluble materials (like beeswax, cetyl alcohol, and stearic acid) with hydrophilic polymers was investigated. The ethylcellulose/HPMC combinations, contributing to an increase in hydrophilic part of blend system, rationally increased the release rate, kinetic constant, and diffusion coefficient thereby, whereas HPMC/beeswax, HPMC/cetyl alcohol, and HPMC/stearic acid combinations, contributing an increase in hydrophobic part of the blend system, caused a substantial reduction of release.[186]

In several investigations the feasibility of development of a sustained-release form for diclofenac sodium was studied. Matrix-type formulation was designed, which appears to be a very attractive approach from process development and scale up points of view. HPMC is the most important hydrophilic polymer used for the preparation of oral controlled-release drug-delivery systems.[187,188] One of the most important characteristics of HPMC is the high swellability, which has a considerable effect on the release kinetics of the incorporated drug.

28.6.3 Application of Cellulose-Based Biopolymer

28.6.3.1 Nanoparticles in Drug and Bioactive Delivery

Nanotechnology has applications across most economic sectors and allows the development of new enabling science with broad commercial potential. Cellulose-based nanoparticles have received considerable attention in recent years as one of the most promising nanoparticulate drug-delivery systems owing to their unique potentials. Nanoparticle drug-delivery systems are defined as particulate dispersions or solid particles with a size in the range of 10–1000 nm and with various morphologies, including nanospheres, nanocapsules, nanomicelles, nanoliposomes, and nanodrugs. The drug is dissolved, entrapped, encapsulated, or attached to a nanoparticle matrix.[189,190] The nanoparticles take on novel properties and functions such as small size, modified surface, improved solubility, and multifunctionality.[191] Drug-delivery systems of nanoparticles have several advantages, such as high drug-encapsulation efficiency, efficient drug protection against chemical or enzymatic degradation, unique ability to create a controlled release, cell internalization as well as the ability to reverse the multidrug resistance of tumor cells.[192] The use of cellulose-based biopolymer nanoparticles is receiving a significant amount of attention due to the impressive mechanical properties, reinforcing capability, abundance, low weight, low filler load requirements, and biodegradable nature of nanoparticles. These properties make it an ideal candidate for the development of green polymer nanocomposites, especially for drug-delivery applications. Cellulose-based nanoparticles are usually identified with different terminologies such as cellulose nanowhiskers,[193] cellulose nanocrystals,[194] nanocrystalline cellulose (NCC),[195] nanofibrillated cellulose,[196] cellulose nanofibrils,[197] and crystalline nanocellulose.[198] They can be prepared by solvent evaporation method, spontaneous emulsification or solvent diffusion method, self-assembly of hydrophobically modified, and dialysis method.[48,199,200] Application of cellulose nanoparticles in drug delivery is a relatively new research area. Cellulose and lignocellulose have great potential as nanomaterials because they are abundant, renewable, have a nanofibrillar structure, can be made multifunctional, and can self-assemble into well-defined architectures.[201] Thus, the various applications of cellulose-based biopolymer nanoparticles in drug and bioactive delivery cannot be overemphasized.

Cellulose-based biopolymers possess several potential advantages as drug-delivery excipients.[202] They are employed in advanced pelleting systems whereby the rate of tablet disintegration and drug release may be controlled by microparticle inclusion, excipient layering, or tablet coating.[203,204] The very large surface area and negative charge of crystalline nanocellulose suggest that large amounts of drugs might be bound to the surface of this material with the potential for high payloads and optimal control of dosing. The established biocompatibility of cellulose supports the pharmaceutical use of nanocellulose in the controlled delivery of drugs and bioactives. The abundant surface hydroxyl groups on crystalline nanocellulose provide a site for the surface modification of the material with a range of chemical groups by a variety of methods.[205] Surface modification may be used to modulate the loading and release of drugs that would not normally bind to nanocellulose, such as nonionized and hydrophobic drugs. For example, Lönnberg et al.[206] suggested that poly(caprolactone) chains might be conjugated onto NCC for such purpose. Additionally, since crystalline nanocellulose is a low-cost, readily abundant material from a renewable and sustainable resource, its use provides a substantial environmental advantage compared with other nanomaterials.[202]

Recently, the hydrophobically modified cellulose-based biopolymers have received increasing attention because they can form self-assembled nanoparticles for biomedical uses.[207] In the aqueous phase, the hydrophobic cores of polymeric nanoparticles are surrounded by hydrophilic outer shells. Thus, the inner core can serve as a nanocontainer for hydrophobic drugs. For instance, although the successful uses of cellulose-based biopolymers such as CMC for nanoparticle technologies are quite limited due to toxicity issues, Uglea et al.[208] conjugated benzocaine to CMC and oxidized CMC, and tested the effects of these polymers on subcutaneous sarcoma tumors in rat models, and reported some antitumor effect from a single intraperitoneal injection. Also, Sievens-Figueroa et al.[209] prepared and evaluated HPMC films containing stable BCS class II drug nanoparticles for pharmaceutical applications. The ultimate aim of these reserachers was to enhance the dissolution rate of poorly water-soluble drugs. Nanosuspensions produced from wet stirred media milling (WSMM) were transformed into polymer films containing drug nanoparticles by mixing with a low-molecular-weight HPMC (E15V) solution containing glycerin, followed by film casting and drying. Three different BCS class II drugs, naproxen, fenofibrate, and griseofulvin, were studied. The study demonstrated the enhancement in drug dissolution rate of films due to the large surface area and smaller drug particle size.[209] Similarly, cellulose-based biopolymers have been engineered to deliver a hydrophobic anticancer drug, docetaxel. Here, a compound comprising an acetylated carboxymethylcellulose (CMC-Ac) covalently linked to at least one PEG and at least one hydrophobic drug (docetaxel). The compound was transformed into a functional polymeric self-assembling nanoparticle with the following attributes: it dissolves or transports a hydrophobic drug in an aqueous environment; the drug is protected from metabolism by the particle; the particle is protected from reticulo-endothelial system elimination by PEG or other suitable chemistry; the polymer self-assembles into a suitably scaled nanoparticle due to a balance in hydrophobic and hydrophilic elements; the particle accumulates in the targeted disease compartment through passive accumulation; the link between the drug and particle is reversible, so that the drug can be released; the polymer is biocompatible; and the particle contains an agent to provide imaging contrast or detection in the physiological system.[210]

In addition, a group of researchers has developed methods for the synthesis of cellulose nanocrystals with optimal properties for applications in targeted drug delivery; covalent attachment of fluorescein isothiocyanate molecules to the surface of cellulose nanocrystals for fluorescent labeling; covalent attachment of folic acid molecules to the surface of fluorescently labeled cellulose nanocrystals for cancer targeting; and covalent attachment of doxorubicin molecules to the surface of folic acid-conjugated cellulose nanocrystals for cancer therapy. They assessed the toxicity of cellulose nanocrystals to a variety of human, mouse, and rat cell lines; the cellular uptake of fluorescently labeled cellulose nanocrystals; the cellular uptake of folic acid-conjugated, fluorescently labeled cellulose nanocrystals; and the in vitro efficacy of the targeted drug nanoconjugates on mouth epidermal carcinoma cells (KB cells) as a cancer model. According to the cytotoxicity studies, the cellulose nanocrystals have no cytotoxic effects. The demonstrated lack of cytotoxicity is a necessary prerequisite for the use of cellulose nanocrystals in targeted drug-delivery applications. The cellular uptake studies have demonstrated that targeting of cellulose nanocrystals through folic acid conjugation leads to uptake of cellulose nanocrystals by cancer cells and that nonspecific cellular uptake of cellulose nanocrystals is minimal. These results confirm that cellulose nanocrystals can be targeted for selective uptake by specific cells. The efficacy studies with doxorubicin-conjugated, folate receptor-targeted cellulose nanocrystals have shown that the targeted drug nanoconjugates are more effective in eradicating cancer cells than free doxorubicin. The studies have also shown that doxorubicin-conjugated cellulose nanocrystals that were not targeted to the folate receptor are less toxic to cancer cells than free doxorubicin. The lower toxicity of doxorubicin-conjugated cellulose nanocrystals, compared with that of free doxorubicin, indicates that targeting of the drug nanoconjugates to the folate receptor is crucial for its high efficacy. The findings by these researchers confirmed that cellulose nanocrystals are promising nanoparticles for targeted drug-delivery applications. In addition, a chemical method was developed for covalently attaching molecules of fluorescein isothiocyanate, one of the most widely used fluorescent labels, to the surface of cellulose nanocrystals. The method for fluorescent labeling of cellulose nanocrystals, developed by these researchers, enables the use of fluorescence techniques, such as spectrofluorometry, fluorescence microscopy, and flow cytometry, to study the interaction of cellulose nanocrystals with cells and the biodistribution of cellulose nanocrystals in vivo.[211–219]

In a related study, Aswathy et al.[220] developed multifunctional nanoparticles based on CMC. In the study, folate group was attached to nanoparticle for specific recognition of cancerous cells and 5-fluorouracil was encapsulated for delivering cytotoxicity. The whole system was able to be tracked by the semiconductor quantum dots that were attached to the nanoparticle. The multifunctional nanoparticle was characterized by spectroscopic techniques such as ultraviolet-visible spectra and FTIR and microsocopic techniques such as transmission electron microscopy and scanning electron microscopy (SEM) and was targeted to human breast cancer cell, MCF7. The biocompatibility of nanoparticle without drug and cytotoxicity rendered by nanoparticle with drug were studied with MCF7 and L929 cell lines. The epifluorescent images suggest that the folate-conjugated nanoparticles were more internalized by folate receptor positive cell line, MCF7, than the non-cancerous L929 cells.

Another area of application of cellulose-based biopolymer nanoparticles is in the formulation and delivery of bioactive compounds such as curcumin (CUR), a natural diphenol used in the treatment of tumors. Yallapu et al.[221] designed curcumin-loaded cellulose nanoparticles for prostrate cancer. They evaluated the comparative cellular uptake and cytotoxicity of β-cyclodextrin, HPMC, poly(lactic-co-glycolic acid) (PLGA), magnetic nanoparticles, and dendrimer-based CUR nanoformulations in prostate cancer cells. The study showed that curcumin-loaded cellulose nanoparticles (cellulose-CUR) formulation exhibited the highest cellular uptake and caused maximum ultrastructural changes related to apoptosis (presence of vacuoles) in prostate cancer cells. Secondly, the anticancer potential of the cellulose-CUR formulation was evaluated in cell culture models using cell proliferation, colony formation, and apoptosis (7-AAD staining) assays. In these assays, the cellulose-CUR formulation showed improved anticancer efficacy compared with free curcumin.[221]

Moreover, novel cellulose-based drug-delivery systems were recently developed by Neha,[222] who designed novel cellulosic nanoparticles with potential pharmaceutical and personal care applications. The study involved the synthesis and characterization of polyampholyte nanoparticles composed of chitosan and CMC, a cellulosic ether. 1-Ethyl-3-(3-dimethylaminopropyl)carbodiimide (EDC) chemistry and inverse microemulsion technique was used to produce cross-linked nanoparticles. Chitosan and CMC provided amine and carboxylic acid functionality to the nanoparticles, thereby making them pH responsive. Chitosan and CMC also make the nanoparticles biodegradable and biocompatible, making them suitable candidates for pharmaceutical applications. The synthesis was then extended to chitosan and modified methylcellulose microgel system. The prime reason for using methylcellulose was to introduce thermo-responsive characteristics to the microgel system. Methylcellulose was modified by carboxymethylation to introduce carboxylic acid functionality, and the chitosan-modified methylcellulose microgel system was found to be responsive to pH as well as temperature. Several techniques were used to characterize the two microgel systems. For both systems, polyampholytic behavior was observed in a pH range of 4–9. The microgels showed swelling at low and high pH values and deswelling at isoelectric point (IEP). Zeta potential values confirmed the presence of positive charges on the microgel at low pH, negative charges at high pH, and neutral charge at the IEP. For chitosan-modified methylcellulose microgel system, temperature-dependent behavior was observed with dynamic light scattering. The second study undertaken by the same researcher involved the study of binding interaction between NCC and an oppositely charged surfactant tetradecyl trimethyl ammonium bromide (TTAB). NCC is a crystalline form of cellulose obtained from natural sources like wood, cotton, or animal sources. These rodlike nanocrystals prepared by acid hydrolysis of native cellulose possess negatively charged surface. The interaction between negatively charged NCC and cationic TTAB surfactant was examined, and it was observed that in the presence of TTAB, aqueous suspensions of NCC became unstable and phase separated. A study of this kind is imperative since NCC suspensions are proposed to be used in personal care applications (such as shampoos and conditioners), which also consist of surfactant formulations. Therefore, NCC suspensions would not be useful for applications that employ an oppositely charged surfactant. To prevent destabilization, poly(ethylene glycol) methacrylate (PEGMA) chains were grafted on the NCC surface to prevent the phase separation in presence of a cationic surfactant. Grafting was carried out using the free radical approach. The NCC–TTAB polymer surfactant interactions were studied via isothermal titration calorimetry (ITC), surface tensiometry, conductivity measurements, phase separation, and zeta potential measurements. Grafting of PEGMA on the NCC surface was confirmed using FTIR and ITC experiments. In phase separation experiments, NCC-g-PEGMA samples showed greater stability in the presence of TTAB compared with unmodified NCC. By comparing ITC and phase separation results, an optimum grafting ratio (PEGMA:NCC) for steric stabilization was also proposed.[222]

Applications of cellulose-based biopolymers in formulation and delivery of bioactive compounds discussed here are in no way exhaustive of the works done in the use of cellulose-based biopolymers in the formulation and delivery of bioactive compounds.

28.6.3.2 Cellulose-Based Biopolymers in Protein and Gene Delivery

Cellulose ethers have long been used in the pharmaceutical industry as excipients in many drug device formulations.[223] Their use in solid tablets allows a swelling-driven release of the drug as physiological fluids come into contact with the tablet itself. The cellulose ether on the tablet surface (e.g., HPMC) starts to swell, forming chain entanglements and physical hydrogel. More sophisticated hydrogel-based devices other than swelling tablets have been developed for controlled drug delivery. Smart hydrogels are particularly useful to control the time- and space-release profile of the drug. The most recent advances aim not only at the sustained release of a bioactive molecule over a long time period, ranging from hours to weeks, but also at a space-controlled delivery, directly at the site of interest. The most recent advances in controlled release through a hydrogel matrix deal with the delivery of proteins, growth factors, and genes to specific sites, the need for which has been prompted by tissue engineering strategies.

Both chitosan and a chitosan oligomer could complex CMC to form stable cationic nanoparticles for subsequent plasmid DNA coating.[224] Chitosan–CMC was subsequently coated with plasmid DNA for genetic immunization.[225] Microcapsules modified with talc and MCC have been shown to exhibit high protein retention in the core in different pH media and could facilitate targeting of protein to the colon.[226,227]

Microparticles containing mixtures of proteins in powder form have been coated with cellulose acetate phthalate using simple preparation techniques based on single emulsion/solvent evaporation. Using aprotinin as a model drug, it was found that these procedures were effective in microencapsulating protein in the solid form without affecting its biological activity. The particles showed adequate in vitro release patterns for application to the intestine.[228]

In recent years, a number of polymeric drug/gene-loaded nanoparticles have been developed as drug delivery carriers and their mechanism of circulation in human bodies has been extensively investigated.[229,230] When drug- or gene-loaded nanoparticles are injected into the body, they cross epithelial barriers and circulate in the blood vessels before reaching the target site. Gene therapy has been applied in many different diseases such as cancer, acquired immune deficiency syndrome, and cardiovascular diseases, and is based on the concept that human disease may be treated by the transfer of genetic materials into specific cells of a patient to supply defective genes responsible for disease development.[231] To transfer the genes to the specific site, genes must escape the processes that affect the disposition of macromolecules. Furthermore, the degradation of gene by serum nucleases needs to be avoided. Thus, encapsulation of genes in delivery carrier is necessary to protect the gene until it reaches its target. The delivery carriers must be small enough to internalize into cells and passage to the nucleus. They also need to be capable of escaping endosome–lysosome processing following endocytosis.[231] While both viral and nonviral vectors have been developed for the delivery of genes, nonviral vectors have been studied more actively due to their low immunogenicity and ease of control of their properties.[232,233] Thus, the cationic polymers have a potential for DNA complexation as nonviral vectors for gene therapy applications. To introduce the specificity into the nanoparticle surfaces, the conjugation of cell-specific ligands to the surface of nanoparticles allows for targeted transgene expression. For example, the nanoparticles of genes and the cationic polymers can be modified with proteins (knob, transferrin, or antibodies/antigens) to allow for cell-specific targeting and enhanced gene transfer.[227,234]

Well-defined comb-shaped cationic copolymers composed of long biocompatible HPC backbones and short poly[(2-dimethyl amino) ethyl methacrylate] [P(DMAEMA)] side chains were prepared as gene vectors via atom transfer radical polymerization from the bromoisobutyryl-terminated HPC biopolymers. The P(DMAEMA) side chains of HPDs was further partially quaternized to produce the quaternary ammonium HPDs (QHPDs). HPDs and QHPDs were assessed in vitro for nonviral gene delivery. HPDs exhibit much lower cytotoxicity and better gene transfection yield than high-molecular-weight P(DMAEMA) homopolymers. QHPDs exhibit a stronger ability to complex pDNA, due to increased surface cationic charges. Thus, the approach to well-defined comb-shaped cationic copolymers provided versatile means for tailoring the functional structure of nonviral gene vectors to meet the requirements of strong DNA-condensing ability and high transfection capability.[235]

28.6.3.3 Cellulose-Based Biopolymers in Wound Healing

Traditional plant-originated cellulose and cellulose-based materials, usually in the form of woven cotton gauze dressings, have been used in medical applications for many years and are mainly utilized to stop bleeding. Even though this conventional dressing is not ideal, its use continues to be widespread. These cotton gauzes consist of an oxidized form of regenerated plant cellulose. In addition, several studies described the implantation of regenerated cellulose hydrogels and revealed their biocompatibility with connective tissue formation and long-term stability. Other in vitro studies showed that regenerated cellulose hydrogels promote bone cell attachment and proliferation and are very promising materials for orthopedic applications. Although chemically identical to plant cellulose, the cellulose synthesized by Acetobacter is characterized by a unique fibrillar nanostructure that determines its extraordinary physical and mechanical properties, characteristics that are quite promising for modern medicine and biomedical research. The nonwoven ribbons of microbial cellulose microfibrils closely resemble the structure of native extracellullar matrices, suggesting that it could function as a scaffold for the production of many tissue-engineered constructs. In addition, microbial cellulose membranes, having a unique nanostructure, could have many other uses in wound healing and regenerative medicine, such as guided tissue regeneration, periodontal treatments, or as a replacement for dura mater (a membrane that surrounds brain tissue).[236] Microbial cellulose could function as a scaffold material for the regeneration of a wide variety of tissues, showing that it could eventually become an excellent platform technology for medicine.

28.7 CONCLUSIONS

Cellulose derivatives constitute a large class of biopolymers with diverse physicochemical properties and large functional versatility in pharmaceutical application. Cellulose derivatives have been used as diluents, binders in direct compression and wet granulation processes, film coating, viscosity enhancers, drug-release retardants in controlled-release matrix systems for microparticles and nanoparticles. By modifying the naturally occurring inexpensive renewable resources, both durable and environmentally acceptable materials can be developed. The most recent advances in utilization of cellulosics is in the delivery of proteins, growth factors, and genes to specific sites, the need for which has been prompted by tissue engineering strategies. Microbial could be used in many biomedical and biotechnological applications, such as tissue engineering, drug delivery, wound dressings, and medical implants. Different cellulose-based biopolymers can be combined with non-cellulose-based materials to obtain novel polymer composites with unique properties for applications in drug delivery and tissue engineering. Due to the excellent biocompatibility of cellulose and its good mechanical properties, they form good replacement for synthetic materials used as biopolymers. To fully realize the potential of the newly developed celluose-based biopolymers in near future would require coordinated and dedicated multidisciplinary research.

REFERENCES

1. Biopolymer, http://en.wikipedia.org/wiki/Biopolymer (accessed March 2013).
2. Sannino, A.; Demitri, C.; Madaghiele, M. Biodegradable cellulose-based hydrogels: Design and applications. *Materials* 2009, 2 (2), 353–373.
3. Baruah, S.D. Biodegradable polymer: The promises and the problems. *Sci. Cult.* November, December 2011, 77, 466–470.
4. Attama, A.A.; Builders, P.F. Particulate drug delivery: Recent applications of batural biopolymers. In *Biopolymers in Drug Delivery: Recent Advances and Challenges*; Adikwu, M.U., Esimone, C.O., Eds.; Bentham Science Publishers: Saif Zone, Sharjah, 2009; pp. 63–94.
5. Builders, P.F.; Attama, A.A. Functional properties of biopolymers for drug delivery applications. In *Biodegradable Materials*; Johnson, B.M., Berkel, Z.E., Eds.; Nova Science Publishers Inc.: New York, 2011.
6. Rowe, R.C.; Shesky, P.J.; Weoller, P.J. *Handbook of Pharmaceutical Excipients*, 2nd edn.; Pharmaceutical Press: London, U.K., 2003; pp. 120–122, 544–545.
7. Builders, P.F.; Ibekwe, N.; Okpako, L.C.; Attama, A.A.; Kunle, O.O. Preparation and characterization of mucinated cellulose microparticles for therapeutic and drug delivery purposes. *Eur. J. Pharm. Biopharm.* 2009, 72 (1), 34–41.
8. Kibbe, A.H. *Handbook of Pharmaceutical Excipients: Cellulose, Silicified Microcrystalline*; American Public Health Association: Washington, DC, 2000.
9. Luukkonen, P.; Schaefer, T.; Hellen, J.; Juppo, A.M.; Yliruusi, J. Rheological characterization of microcrystalline cellulose and silicified microcrystalline cellulose wet masses using a mixer torque rheometer. *Int. J. Pharm.* 1999, 188 (2), 181–192.
10. Siepmann, J.; Kranz, H.; Bodmeier, R.; Peppas, N.A. HPMC-matrices for controlled drug delivery: A new model combining diffusion, swelling, and dissolution mechanisms and predicting the release kinetics. *Pharm. Res.* 1999, 16 (11), 1748–1756.
11. Colombo, P.; Bettini, R.; Peppas, N.A. Observation of swelling process and diffusion front position during swelling in hydroxypropylmethyl cellulose (HPMC) matrices containing a soluble drug. *J. Control. Release* 1999, 61 (1–2), 83–91.
12. Lowman, A.M.; Peppas, N.A. Hydrogels. In *Encyclopedia of Controlled Drug Delivery*; Mathiowitz, E., Ed.; Wiley: New York, 2000; pp. 397–417.
13. Le Neel, T.; Morlet-Renaud, C.; Lipart, C.; Gouyette, A.; Truchaud, A.; Merle, C. Image analysis as a new technique for the study of water uptake in tablets. *STP Pharma. Sci.* 1997, 7 (2), 117–122.
14. Baumgartner, S.; Šmid-Korbar, J.; Vreèer, F.; Kristl, J. Physical and technological parameters influencing floating properties of matrix tablets based on cellulose ethers. *STP Pharma. Sci.* 1998, 8 (5), 182–187.

15. Arion, H. Carboxymethyl cellulose hydrogel-filled breast implants. Our experience in 15 years. *Ann. Chir. Plast. Esthet.* 2001, 46 (1), 55–59.
16. Methyl Cellulose, http://en.wikipedia.org/wiki/Methyl_cellulose (accessed March 2013).
17. Mura, P.; Faucci, M.T.; Manderioli, A.; Bramanti, G.; Parrini, P. Thermal behavior and dissolution properties of naproxen from binary and ternary solid dispersion. *Drug Dev. Ind. Pharm.* 1999, 25 (3), 257–264.
18. Friedman, M.; Golomb, G. New sustained release dosage form of chlorhexidine for dental use. *J. Periodontal Res.* 1982, 17 (3), 323–328.
19. Soskolne, W.A.; Golomb, G.; Friedman, M.; Sela, M.N. New sustained release dosage form of chlorhexidine for dental use. *J. Periodontal Res.* 1983, 18 (3), 330–336.
20. Yuasa, H.; Ozeki, T.; Kanaya, Y.; Oishi, K.; Oyake, T. Application of the solid dispersion method to the controlled release of medicine. I. Controlled release of water soluble medicine by using solid dispersion. *Chem. Pharm. Bull.* 1991, 39 (2), 465–467.
21. Yuasa, H.; Ozeki, T; Kanaya, Y.; Oishi, K. Application of the solid dispersion method to the controlled release of medicine. II. Sustained release tablet using solid dispersion granule and the medicine release mechanism. *Chem. Pharm. Bull.* 1992, 40 (6), 1592–1596.
22. Ozeki, T.; Yuasa, H.; Kanaya, Y.; Oishi, K. Application of the solid dispersion method to the controlled release of medicine. V. Suppression mechanism of the medicine release rate in the three-component solid dispersion system. *Chem. Pharm. Bull.* 1994, 42 (2), 337–343.
23. Ozeki, T.; Yuasa, H.; Kanaya, Y.; Oishi, K. Application of the solid dispersion method to the controlled release of medicine. VII. Release mechanism of a highly water-soluble medicine from solid dispersion with different molecular weight of polymer. *Chem. Pharm. Bull.* 1995, 43 (4), 660–665.
24. Ozeki, T.; Yuasa, H.; Kanaya, Y.; Oishi, K. Application of the solid dispersion method to the controlled release of medicine. VIII. Medicine release and viscosity of the hydrogel of a water-soluble polymer in a three-component solid dispersion system. *Chem. Pharm. Bull.* 1995, 43 (9), 1574–1579.
25. Yuasa, H.; Takahashi, H.; Ozeki, T.; Kanaya, Y.; Ueno, M. Application of the solid dispersion method to the controlled release of medicine. III. Control of the release rate of slightly water soluble medicine from solid dispersion granules. *Chem. Pharm. Bull.* 1993, 41 (2), 397–399.
26. Yuasa, H.; Ozeki, T.; Takahashi, H.; Kanaya, Y.; Ueno, M. Application of the solid dispersion method to the controlled release of medicine. VI. Release mechanism of slightly water soluble medicine and interaction between flurbiprofen and hydroxypropyl cellulose in solid dispersion. *Chem. Pharm. Bull.* 1994, 42 (2), 354–358.
27. Ozeki, T.; Yuasa, H.; Kanaya, Y. Application of the solid dispersion method to the controlled release of medicine. IX. Difference in the release of flurbiprofen from solid dispersions with poly(ethylene oxide) and hydroxypropylcellulose and interaction between medicine and polymers. *Int. J. Pharm.* 1997, 115 (2), 209–217.
28. Tahara, K.; Yamamoto, K.; Nishihata, T. Overall mechanism behind matrix sustained release (SR) tablets prepared with hydroxypropylmethyl cellulose. *J. Control. Release* 1995, 35 (1), 59–66.
29. Skoug, J.W.; Mikelsons, M.V.; Vigneron, C.N.; Stemm, N.L. Qualitative evaluation of the mechanism of release of matrix sustained release dosage forms by measurement of polymer release. *J. Control. Release* 1993, 27 (3), 227–245.
30. Ford, J.L.; Rubinstein, M.H.; McCaul, F.; Hogan, J.E.; Edgar, P.J. Importance of drug type, tablet shape and added diluents on drug release kinetics from hydroxypropylmethyl cellulose matrix tablets. *Int. J. Pharm.* 1987, 40 (3), 223–234.
31. Ranga-Rao, K.V.; Padmalatha, D.; Buri, P. Influence of molecular size and water solubility of the solute on its release from swelling and erosion controlled polymeric matrices. *J. Control. Release* 1990, 12 (2), 133–141.
32. Delgado, J.N.; William, A. *Wilson and Gisvold's Textbook of Organic Medicinal and Pharmaceutical Chemistry*; Lippincott-Raven Publishers: Wickford, U.K., 1998.
33. Karaaslan, A.M.; Tshabalala, M.A.; Buschle-Diller, G. Wood hemicellulose/chitosan-based semi-interpenetrating network hydrogels: Mechanical, swelling and controlled drug release properties. *BioRes* 2010, 5 (2), 1036–1054.
34. Lerouxel, O.; Cavalier, D.M.; Liepman, A.H.; Keegstra, K. Biosynthesis of plant cell wall polysaccharides—A complex process. *Curr. Opin. Plant Biol.* 2006, 9 (6), 621–630.
35. Chaa, L.; Joly, N.; Lequart, V.; Faugeron, C.; Mollet, J.; Martin, P.; Morvan, H. Isolation, characterization and valorization of hemicelluloses from Aristida pungens leaves as biomaterial. *Carbohydr. Polym.* 2008, 74 (3), 597–602
36. Petkowicz, C.L.O.; Reicher, F.; Mazeau, K. Conformational analysis of galactomannans. *Carbohydr. Polym.* 1998, 37 (1), 25–39.

37. Chourasia, M.K.; Jain, S.K. Polysaccharides for colon targeted drug delivery. *Drug Deliv.* 2004, 11 (2), 129–148.
38. Vendruscolo, C.W.; Andreazza, I.F; Ganter, J.L.M.S.; Ferrero, C.; Bresolin, T.M.B. Xanthan and galactomannan (from M. scabrella) matrix tablets for oral controlled delivery of theophylline. *Int. J. Pharm.* 2005, 296 (1), 1–11.
39. Beneke, C.E.; Viljoen, A.M.; Hamman, J.H. Polymeric plant-derived excipients in drug delivery. *Molecules* 2009, 14 (7), 2602–2620.
40. Krässing, H.; Schurz, J.; Steadman, R.G.; Schliefer, K.; Albrecht, K.X. Cellulose. In *Ullmann's Encyclopedia of Industrial Chemistry*; Campbell, F.T., Pfefferkorn, R., Rounsaville, J.F., Eds.; VCH Verlagsgesellschaft: Weinheim, Germany, 1986; p. 375.
41. Fengel, D.; Wegener, G. *Wood: Chemistry, Ultrastructure, Reactions*; Walter de Gruyter: Berlin, Germany, 1989; pp. 26–226.
42. Richardson S.; Gorton L. Characterisation of the substituent distribution in starch and cellulose derivatives. *Anal. Chim. Acta* 2003, 497 (1), 27–65.
43. Saake, B.; Horner, S.; Puls, J. Progress in the enzymatic hydrolysis of cellulose derivatives. In *Cellulose Derivatives: Modification, Characterization and Nanostructures*; Heinze, T., Glasser G., Eds.; American Chemical Society: Washington, DC, 1998; p. 201.
44. Guilbot, A.; Mercier, C. Starch. In *The Polysaccharides*; Aspinall, G., Ed.; Academic Press: New York, 1985; p. 209.
45. BeMiller, J.N. Starch modification: Challenges and prospects. *Starch/Stärke* 1997, 49 (7–8), 127.
46. Edson, C.S.F.; Luciano, C.B.L.; Kaline, S.S; Maria, G.F.; Francisco, A.R.P. Calorimetry studies for interaction in solid/liquid interface between the modified cellulose and divalent cation. *J. Therm. Anal. Calorim.* 2013, 114 (1), 57–66.
47. McDowall, D.J.; Gupta, B.S.; Stannett, V.T. Grafting of vinyl monomers to cellulose by ceric ion initiation. *Prog. Polym. Sci.* 1984, 31 (12), 1–50.
48. Klemm, D.H.; Heublein, B.; Fink, H.P.; Bohn, A. Cellulose: Fascinating biopolymer and sustainable raw material. *Angew. Chem. Int. Ed.* 2005, 44 (22), 3358–3393.
49. Magdi, E.G.; Zhang, Y.; Li, X.; Li, H.; Zhong, X.; Li, H.F.; Yu, M. Current status of applications of ionic liquids for cellulose dissolution and modifications: Review. *Int. J. Eng. Sci. Technol.* 2012, 4 (7), 3556–3571.
50. Da SilvaFilho, E.C.; De Melo, J.C.P.; Airoldi, C. Preparation of ethylenediamine-anchored cellulose and determination of thermochemical data for the interaction between cations and basic centers at the solid/liquid interface. *Carbohydr. Res.* 2006, 341 (17), 2842–2850.
51. De Melo, J.C.P.; Da SilvaFilho, E.C.; Santana, S.A.A.; Airoldi, C. Exploring the favorable ion-exchange ability of phthalylated cellulose biopolymer using thermodynamic data. *Carbohydr. Res.* 2010, 345 (13), 1914–1921.
52. Akira, I. Chemical modification of cellulose. In *Wood and Cellulosic Chemistry*; Hon, D.N.S., Shiraishi, N., Eds.; Marcel Dekker: New York, 2001; pp. 599–626.
53. Swati, A.; Achhrish, G.; Sandeep, S. Drug delivery: Special emphasis given on biodegradable polymers. *Adv. Polym. Sci. Technol.* 2012, 2 (1), 1–15.
54. Kalia, S.; Kaith, B.S.; Kaur, I. Pretreatments of natural fibers and their application as reinforcing material in polymer composites—A review. *Polym. Eng. Sci.* 2009, 49 (7), 1253–1272.
55. Kalia, S.; Kaith, B.S.; Sharma, S.; Bhardwaj, B. Mechanical properties of flax-g-poly(methyl acrylate) reinforced phenolic composites. *Fibers Polym.* 2008, 9 (4), 416–422.
56. Kalia, S.; Dufresne, A.; Cherian, B.M.; Kaith, B.S.; Luc, A.; Njuguna, J.; Nassiopoulos, E. Cellulose-based bio- and nanocomposites: A review. *Int. J. Poly. Sci.* 2011, Article ID 837875, 35, doi:10.1155/2011/837875.
57. Belgacem, M.N.; Salon-Brochier, M.C.; Krouit, M.; Bras, J. Recent advances in surface chemical modification of cellulose fibres. *J. Adh. Sci. Tech.* 2011, 25 (6–7), 661–684.
58. Kennedy, J.F.; Phillips, G.O.; Wedlock, D.J.; Williams, P.A. *Cellulose and Its Derivatives: Chemistry, Biochemistry and Applications*; Ellis Horwood: Chichester, U.K., 1985; p. 551.
59. Whistler, R.L.; BeMiller, J.N.; Paschall, E.F. *Starch: Chemistry and Technology*, 2nd edn.; Academic Press: London, U.K., 1984; pp. 26–86.
60. Agrawal, R.; Saxena, N.S.; Sharma, K.B.; Thomas, S.; Sreekala, M.S. Activation energy and crystallization kinetics of untreated and treated oil palm fibre reinforced phenol formaldehyde composites. *Mater. Sci. Eng.* 2000, 277 (1–2), 77–82.
61. Sreekala, M.S.; Kumaran, M.G.; Joseph, S.; Jacob, M.; Thomas, S. Oil palm fibre reinforced phenol formaldehyde composites: Influence of fibre surface modifications on the mechanical performance. *App. Comp. Mater.* 2007, 7 (5–6), 295–329.

62. Ray, D.; Sarkar, B.K.; Rana, A.K.; Bose, N.R. Effect of alkali treated jute fibres on composite properties. *Bull. Mater. Sci.* 2001, 24 (2), 129–135.
63. Joseph, K.; Mattoso, L.H.C.; Toledo, R.D.; Thomas, S.; de Carvalho, L.H. Natural fiber reinforced thermoplastic composites. In *Natural Polymers and Agrofibers Composites*; Frollini, E.; Leão, A.L.; Mattoso, L.H.C.; Eds.; Embrapa: Sãn Carlos, Brazil, 2000; 159–201,
64. Kaith, B.S.; Singha, A.S.; Kumar, S.; Misra, B.N. FASH2O2 initiated graft copolymerization of methylmethacrylate onto flax and evaluation of some physical and chemical properties. *J. Polym. Mater.* 2005, 4 (22), 425–432.
65. Pommet, M.; Juntaro, J.; Heng, J.Y.Y.; Athanasios, M.; Adam F.L.; Karen, W.; Gerhard, K.; Milo, S.P.S.; Bismarck, A. Surface modification of natural fibers using bacteria: Depositing bacterial cellulose onto natural fibers to create hierarchical fiber reinforced nanocomposites. *Biomacromolecules* 2008, 9 (6), 1643–1651.
66. Shoda M.; Sugano, Y. Recent advances in bacterial cellulose production. *Biotechnol. Bioprocess. Eng.* 2005, 10 (1), 1–8.
67. Rutenberg, M.W.; Solarek, D. Starch derivatives: Production and uses. In *Starch Chemistry and Technology*; Whistler, R.L., BeMiller, J.N., Paschall, E.F., Eds.; Academic Press: London, U.K., 1984; p. 311.
68. Brandt, L. Cellulose ethers. In *Ullmann's Encyclopedia of Industrial Chemistry*; Campbell, F.T., Pfefferkorn, R., Rounsaville, J.F., Eds.; VCH Verlagsgesellschaft: Weinheim, Germany, 1986; p. 461.
69. Marchessault, R.H.; Sundararajan, P.R. Cellulose. In *The Polysaccharides*; Aspinall, G.O., Ed.; Academic Press: New York, 1983; p. 11.
70. Mondt, J.L. The use of cellulose derivatives in the paint and building industries. In *Cellulose Sources Exploitation*; Kennedy, J.F., Phillips, G.O., Williams, P.A., Eds.; Ellis Horwood: London, U.K., 1983; p. 269.
71. Varshney, V.K.; Naithani, S. Chemical functionalization of cellulose derived from nonconventional sources. In *Cellulose Fibers: Bio- and Nano-Polymer Composites*; Kalia, S., Ed.; Springer-Verlag: Berlin, Germany, 2011; pp. 43–58.
72. Arthur, J.C., Jr. Technology and engineering cellulose chemical technology. In *Comprehensive Polymer Science*; Allen G., Bevington, J.C., Eds.; Pergamon: Oxford, U.K., 1986; pp. 26, 681.
73. Schomburg, D.; Salzmann, M. *Enzyme Handbook*; Springer: Berlin, Germany, 1991; p. 3.2.1.3.18.
74. Chen, H.; Fan, M. Novel thermally sensitive pH-dependent chitosan/carboxymethyl cellulose hydrogels. *J. Bioact. Compat. Polym.* 2008, 23 (1), 38–48.
75. Chang, C.; Lue, A.; Zhang, L. Effects of crosslinking methods on structure and properties of cellulose/PVA hydrogels. *Macromol. Chem. Phys.* 2008, 209 (12), 1266–1273.
76. Sarkar, N. Thermal gelation properties of methyl and hydroxypropyl methylcellulose. *J. Appl. Polym. Sci.* 1979, 24 (4), 1073–1087.
77. Tate, M.C.; Shear, D.A.; Hoffman, S.W.; Stein, D.G.; LaPlaca, M.C. Biocompatibility of methylcellulose-based constructs designed for intracerebral gelation following experimental traumatic brain injury. *Biomaterials* 2001, 22 (10), 1113–1123.
78. Chen, C.; Tsai, C.; Chen, W.; Mi, F.; Liang, H.; Chen, S.; Sung, H. Novel living cell sheet harvest system composed of thermoreversible methylcellulose hydrogels. *Biomacromolecules* 2006, 7 (3), 736–743.
79. Stabenfeldt, S.E.; Garcia, A.J.; LaPlaca, M.C. Thermoreversible laminin-functionalized hydrogel for neural tissue engineering. *J. Biomed. Mater. Res. A* 2006, 77 (4), 718–725.
80. Te Nijenhuis, K. On the nature of crosslinks in thermoreversible gels. *Polym. Bull.* 2007, 58 (1), 27–42.
81. Vinatier, C.; Magne, D.; Weiss, P.; Trojani, C.; Rochet, N.; Carle, G.F.; Vignes-Colombeix, C. et al. A silanized hydroxypropyl methylcellulose hydrogel for the three-dimensional culture of chondrocytes. *Biomaterials* 2005, 26 (33), 6643–6651.
82. Vinatier, C.; Magne, D.; Moreau, A.; Gauthier, O.; Malard, O.; Vignes-Colombeix, C.; Daculsi, G.; Weiss, P.; Guicheux, J. Engineering cartilage with human nasal chondrocytes and a silanized hydroxypropyl methylcellulose hydrogel. *J. Biomed. Mater. Res. A* 2007, 80 (1), 66–74.
83. Ogushi, Y.; Sakai, S.; Kawakami, K. Synthesis of enzymatically-gellable carboxymethylcellulose for biomedical applications. *J. Biosci. Bioeng.* 2007, 104 (1), 30–33.
84. Wang, C.; Chen, C. Physical properties of the crosslinked cellulose catalyzed with nanotitanium dioxide under UV irradiation and electronic field. *Appl. Catal. A* 2005, 293 (2B), 171–179.
85. Coma, V.; Sebti, I.; Pardon, P.; Pichavant, F.H.; Deschamps, A. Film properties from crosslinking of cellulosic derivatives with a polyfunctional carboxylic acid. *Carbohydr. Polym.* 2003, 51 (3), 265–271.
86. Xie, X.; Liu, Q.; Cui, S.W. Studies on the granular structure of resistant starches (type 4) from normal, high amylose and waxy corn starch citrates. *Food Res. Int.* 2006, 39 (3), 332–341.

87. Demitri, C.; Del Sole, R.; Scalera, F.; Sannino, A.; Vasapollo, G.; Maffezzoli, A.; Ambrosio, L.; Nicolais, L. Novel superabsorbent cellulose-based hydrogels crosslinked with citric acid. *J. Appl. Polym. Sci.* 2008, 110 (4), 2453–2460.
88. Charlesby, A. The degradation of cellulose by ionizing radiation. *J. Polym. Sci.* 1955, 15 (79), 263–270.
89. Wach, R.A.; Mitomo, H.; Nagasawa, N.; Yoshii, F. Radiation crosslinking of methylcellulose and hydroxyethylcellulose in concentrated aqueous solutions. *Nucl. Instrum. Methods Phys. Res. Sect. B* 2003, 211 (4), 533–544.
90. Pekel, N.; Yoshii, F.; Kume, T.; Guven, O. Radiation crosslinking of biodegradable hydroxypropylmethylcellulose. *Carbohydr. Polym.* 2004, 55 (2), 139–147.
91. Liu, P.; Peng, J.; Li, J.; Wu, J. Radiation crosslinking of CMC-Na at low dose and its application as substitute for hydrogels. *Rad. Phys. Chem.* 2005, 72 (5), 635–638.
92. Li, W.Y.J.; Liu, A.X.; Sun, C.F.; Zhang, R.C.; Kennedy, J.F. Homogeneous modification of cellulose with succinic anhydride in ionic liquid using 4-dimethylaminopyridine as a catalyst. *Carbohydr. Polym.* 2009, 78 (3), 389–395.
93. Gericke, M.L.; TimHeinze, T. Solvent for cellulose chemistry. *Nachrichten Aus. Der. Chemie.* 2011, 59 (4), 405–409.
94. Schobitz, M.M.; F.Heinze, T. Unconventional reactivity of cellulose dissolved in ionic liquids. *Macromol. Symp.* 2009, 280 (1), 102–111.
95. Heinze, T.D.; Susann, S.; Michael, L.; Tim, K.; Sarah, M.F. Interactions of ionic liquids with polysaccharides—2: Cellulose. *Macromol. Symp.* 2008, 262 (1), 8–22.
96. Pinkert, A.M.; Kenneth, N.P.; Shusheng, S.; Mark, P. Ionic liquids and their interaction with cellulose. *Chem. Rev.* 2009, 109 (12), 6712–6728.
97. Swatloski, R.P.S.; Holbrey, J.S.K.; Rogers, R.D. Dissolution of cellulose with ionic liquids. *J. Am. Chem. Soc.* 2002, 124 (18), 4974–4975.
98. Zhu, S.; Wu, Y.; Chen, Q.; Yu, Z.; Wang, C.; Jin, S.; Ding, Y.; Wu, G. Dissolution of cellulose with ionic liquids and its application: A mini-review. *Green Chem.* 2006, 8 (4), 325–327.
99. Cao, Y.W.; Jin, Z.; Jun, L.; Huiquan, Z.; Yi, H.J. Room temperature ionic liquids (RTILs): A new and versatile platform for cellulose processing and derivatization. *Chem. Eng. J.* 2009, 147 (1), 13–21.
100. El Seoud, O.A.K.; A.Fidale, L.; C.Dorn, S.; Heinze, T. Applications of ionic liquids in carbohydrate chemistry: A window of opportunities. *Biomacromolecules* 2007, 8 (9), 2629–2647.
101. Zhu, J.; Wang, W.T.; Wang, X.L.; Li, B.; Wang, Y.Z. Green synthesis of a novel biodegradable copolymer base on cellulose and poly(p-dioxanone) in ionic liquid. *Carbohydr. Polym.* 2009, 76 (1), 139–144.
102. Przemys, A.; Aw, K. Ionic liquids as solvents for polymerization processes Progress and challenges. *Prog. Polym. Sci.* 2009, 34 (12), 1333–1347.
103. Wilpiszewska, K.S.A.T. Ionic liquids: Media for starch dissolution, plasticization and modification. *Carbohydr. Polym.* 2011, 86 (2), 424–428.
104. Feng, L.C.; Zhong, I. Research progress on dissolution and functional modification of cellulose in ionic liquids. *J. Mol. Liq.* 2008, 142 (1–3), 1–5.
105. Kim, K.W.; Song, B.; Choi, M.Y.; Kim, M.J. Biocatalysis in ionic liquids: Markedly enhanced enantioselectivity of lipase. *Org. Lett.* 2001, 3 (10), 1507–1509.
106. Ren, J.P.; Xinwen, P.; FengSun, R. *Ionic Liquid as Solvent for a Biopolymer: Acetylation of Hemicelluloses*; Sun, R., Fu, S., Eds.; Research Progress in Paper Industry and Biorefinery, South China University of China: Guangzhou, China, 2010; pp. 57–60.
107. Wu, J.; Zhang, J.; Zhang, H.; He, J.; Ren, Q.; Guo, M. Homogeneous acetylation of cellulose in a new ionic liquid. *Biomacromolecules* 2004, 5 (2), 266–268.
108. Koehler, S.L.; Tim, S.; Michael, S.; Jens, M.; Wolfgang, G.F.; Heinze, T. Interactions of ionic liquids with polysaccharides 1. Unexpected acetylation of cellulose with 1-ethyl-3-methylimidazolium acetate. *Macromol. Rapid Commun.* 2007, 28 (24), 2311–2317.
109. Cao, Y.L.; Zhang, H.Q.J. Homogeneous synthesis and characterization of cellulose acetate butyrate (CAB) in 1-allyl-3- methylimidazolium chloride (AmimCl) ionic liquid. *Ind. Eng. Chem. Res.* 2004, 50 (13), 7808–7814.
110. Abbott, A.P.B.; Handa, S.T.J.; Stoddart, B. O-acetylation of cellulose and monosaccharides using a zinc based ionic liquid. *Green Chem.* 2005, 7 (10), 705–707.
111. Turner, M.B.S.; Holbrey, J.D.S.K.; Rogers, R.D. Production of bioactive cellulose films reconstituted from ionic liquids. *Biomacromolecules* 2004, 5 (4), 1379–1384.
112. Semsarilar, M.L.; Vincent-Perrier, S. Synthesis of a cellulose supported chain transfer agent and its application to RAFT polymerization. *J. Polym. Sci. A* 2010, 48 (19), 4361–4365.

113. Lin, C.X.Z.; Zhang, H.-Y.; Liu, M.-H.; Fu, S.-Y.; Lucia, L.A. Novel preparation and characterization of cellulose microparticles functionalized in ionic liquids. *Langmuir* 2009, 25 (17), 10116–10120.
114. Tomasik, P.A. Chemical modifications of polysaccharides. In *Chemical and Functional Properties of Food Saccharides*; CRC Press: New York, 2003; pp. 217–229.
115. Yan, C.Z.; Zhang, J.; Lv, Y.; Lu, J.; Wu, J.; Zhang, J.; He, J. Thermoplastic cellulose-graft-poly(L-lactide) copolymers homogeneously synthesized in an ionic liquid with 4 dimethylaminopyridine catalyst. *Biomacromolecules* 2009, 10 (8), 2013–2018.
116. Lin, C.X.Z.; Huai-yu, L.; Ming-hua, F.; Shi-yu, H.L. Rapid homogeneous preparation of cellulose graft copolymer in BMIMCL under microwave irradiation. *J. Appl. Polym. Sci.* 2010, 118 (1), 399–404.
117. Kadokawa, J.I.M.; Masa-Aki, T.; Akihiko-Kaneko, Y. Preparation of cellulose-starch composite gel and fibrous material from a mixture of the polysaccharides in ionic liquid. *Carbohydr. Polym.* 2009, 75 (1), 180–183.
118. Hameed, N.G.; Qipeng, T.; Feng, H.K.; Sergei, G. Blends of cellulose and poly(3-hydroxybutyrate-co-3-hydroxyvalerate) prepared from the ionic liquid 1-butyl-3-methylimidazolium chloride. *Carbohydr. Polym.* 2011, 86 (1), 94–104.
119. Murugesan, S.M.; Shaker, V.; Aravind, A.; Pulickel, M.; Linhardt, R.J. Ionic liquid-derived bloodcompatible composite membranes for kidney dialysis. *Appl. Biomater.* 2006, 79B (2), 298–304.
120. Suzuki, H.; Sunada, H. Influence of water-soluble polymers on the dissolution of nifedipine solid dispersions with combined carriers. *Chem. Pharm. Bull.* 1998, 46 (3), 482–487.
121. Ofoefule, S.I.; Chukwu, A. Application of blends of microcrystalline cellulose-cissus gum in the formulation of aqueous suspensions. *Boll. Chim. Farm.* 1999, 138 (5), 217–222.
122. Hwang, R.-C.; Peck, G.R. A systematic evaluation of the compression and tablets characteristics of various types of microcrystalline cellulose. *Pharm. Technol.* 2001, 25 (3), 112–132.
123. Attama, A.A.; Adikwu, M.U.; Esimone, C.O. Sedimentation studies on chalk suspensions containing blends of Veegum and detarium gum as suspending agents. *Boll. Chim. Farm.* 1999, 138 (10), 521–525.
124. Kamel, S.; Ali, N.; Jahangir, K.; Shah, S.M.; El-Gendy, A.A. Pharmaceutical significance of cellulose: A review. *Expr. Polym. Lett.* 2008, 2 (11), 758–778.
125. Okamoto, H.; Nakamori, T.; Arakawa, Y.; Iida, K.; Danjo, K. Development of polymer film dosage forms of lidocaine for buccal administration. II. Comparison of preparation methods. *J. Pharm. Sci.* 2002, 91 (11), 2424–2432.
126. Hercules Incorporated, Aqualon Division, http://www.aqualon.com (accessed July 2008).
127. Katzhendler, I.; Mader, K.; Friedman, M. Structure and hydration properties of hydroxypropyl methylcellulose matrices containing naproxen and naproxen sodium. *Int. J. Pharm.* 2000, 200 (2), 161–179.
128. Vazquez, M.-J.; Casalderrey, M.; Duro, R.; Gómez-Amoza, J.-L.; Martinez-Pacheco, R.; Souto, C.; Concheiro, A. Atenolol release from hydrophilic matrix tablets with hydroxypropylmethylcellulose (HPMC) mixtures as gelling agent: Effects of the viscosity of the HPMC mixture. *Eur. J. Pharm. Sci.* 1996, 4 (1), 39–48.
129. Gupta, A.; Hunt, R.L.; Shah, R.B.; Sayeed, V.A.; Khah, M.A. Disintegration of highly soluble immediate release tablets: A surrogate for dissolution. *AAPS PharmSciTech* 2009, 10 (2), 495–499.
130. Vogelpoel, H.; Welink, J.; Amidon, G.L.; Junginger, H.E.; Midha, K.K.; Moller, H. Biowaiver monographs for immediate release solid oral dosage forms based on Biopharmaceutics Classification System (BCS) literature data: Verapamil hydrochloride, propranolol hydrochloride, and atenolol. *J. Pharm. Sci.* 2004, 93 (8), 1945–1956.
131. Kendall, R.A.; Basit, A.W. The role of polymers in solid oral dosage forms. In *Polymers in Drug Delivery*; Uchegbu, I.F., Ed.; CRC Press, Taylor & Francis: Boca Raton, FL, 2006; p. 280.
132. Alderman, D.A. A review of cellulose ethers in hydrophilic matrices for oral controlled-release dosage forms. *Int. J. Pharm. Tech. Prod. Mfr.* 1984, 5 (1), 1–9.
133. Liu, J.; Zhang, F.; McGinity, J.W. Properties of lipophilic matrix tablets containing phenylpropanolamine hydrochloride prepared by hot-melt extrusion. *Eur. J. Pharm. Biopharm.* 2001, 52 (2), 181–190.
134. Ozeki, T.; Yuasa, H.; Okada, H. Controlled release of drug via methylcellulose-carboxyvinylpolymer interpolymer complex solid dispersion. *AAPS PharmSciTech* 2005, 6 (2), E231–E236.
135. Gon, M.C.; Ferrero, R.M.; Jimenez, C.; Gurruchaga, M. Synthesis of hydroxypropyl methacrylate/polysaccharide graft copolymers as matrices for controlled release tablets. *Drug Dev. Ind. Pharm.* 2002, 28 (9), 1101–1115.
136. Han, R.-Y.; Fang, J.-Y.; Sung, K.C.; Hu, O.Y.P. Mucoadhesive buccal disks for novel nalbuphine prodrug controlled delivery: Effect of formulation variables on drug release and mucoadhesive performance. *Int. J. Pharm.* 1999, 177 (2), 201–209.
137. Park, C.R.; Munday, D.L. Development and evaluation of a biphasic buccal adhesive tablet for nicotine replacement therapy. *Int. J. Pharm.* 2002, 237 (1), 215–226.

138. Senel, S.; Hincal, A.A. Drug permeation enhancement via buccal route: Possibilities and limitations. *J. Control. Release* 2001, 72 (1), 133–144.
139. Rokhade, A.P.; Agnihotri, S.A.; Patil, S.A.; Mallikarjuna, N.N.; Kulkarni, P.V.; Aminabhavi, T.M. Semi-interpenetrating polymer network microspheres of gelatin and sodium carboxymethyl cellulose for controlled release of ketorolac tromethamine. *Carbohyd. Polym.* 2006, 65 (3), 243–252.
140. Waree, T.; Garnpimol, C.R. Development of indomethacin sustained release microcapsules using chitosan-carboxymethyl cellulose complex coacervation. *Songklanakarin J. Sci. Technol.* 2003, 25 (2), 245–254.
141. Ritthidej, G.C.; Tiyaboonchai, W. Formulation and drug entrapment of microcapsules prepared from chitosan-carboxymethylcellulose complex coacervation. *Thai. J. Pharm. Sci.* 1997, 21 (1), 137–144.
142. Khanvilkar, K.H.; Ye, H.; Moore, A.D. Influence of hydroxypropyl methylcellulose mixture, apparent viscosity, and tablet hardness on drug release using a 23 full factorial design. *Drug Dev. Ind. Pharm.* 2002, 28 (5), 601–608.
143. Ye, H.; Khanvilkar, K.H.; Moore, A.D.; Hilliard-Lott, M. Effects of manufacturing process variables on in vitro dissolution characteristics of extended-release tablets formulated with hydroxypropylmethyl cellulose. *Drug Dev. Ind. Pharm.* 2003, 29 (1), 79–88.
144. Liu, Z.; Li, J.; Nie, S.; Liu, H.; Ding, P.; Pan, W. Study of an alginate/HPMC-based in situ gelling ophthalmic delivery system for gatifloxacin. *Int. J. Pharm.* 2006, 315 (1), 12–17.
145. Phaechamu, T.; Vesapun, C.; Kraisit, P. Yahom in dosage form of buccal tablet. In *Proceedings of the 10th World Congress on Clinical Nutrition*, Phuket, Thailand, 2004, pp. 193–198.
146. Chantana, V.; Juree, C.; Thawatchaj, P. Effect of hydroxypropylmethyl cellulose and polyvinyl pyrrolidone on physical properties of Yahom buccal tablets and the antimicrobial activity of Yahom. In *The Fourth Thailand Materials Science Technology Conference*, Khong Wan: Bangkok, Thailand, PP01/1–PP01/3, 2006.
147. Lenaerts, V.; Dumoulin, Y.; Mateescu, M.A. Controlled release of theophylline from cross-linked amylose tablets. *J. Control. Release* 1991, 15 (1), 39–46.
148. Rahmouni, M.; Lenaerts, V.; Massuelle, D.; Doelker, E.; Johnson, M.; Leroux, J.-C. Characterization of binary mixtures consisting of cross-linked high amylase starch and hydroxypropylmethyl cellulose used in the preparation of controlled release tablets. *Pharm. Dev. Technol.* 2003, 8 (4), 335–348.
149. Amaral, M.H.; Sousa-Lobo, J.M.; Ferreira, D.C. Effect of hydroxypropylmethyl cellulose and hydrogenated castor oil on naproxen release from sustained-release tablets. *APS PharmSciTech* 2001, 2 (2), 1–8.
150. Ibezim, E.C.; Attama, A.A.; Dimgba, I.C.; Ofoefule, S.I. Use of carbopols-sodium carboxymethylcellulose admixtures in the formulation of bioadhesive metronidazole tablets. *Acta Pharma.* 2000, 50 (2), 121–130.
151. Ibezim, E.C.; Ofoefule, S.I. in vitro evaluation of the bioadhesive properties of sodium carboxymethylcellulose, acacia, Veegum and their admixtures. *Afr. J. Pharm. Res. Dev.* 2006, 2 (1), 67–72.
152. Attama, A.A.; Adikwu, M.U. Bioadhesive delivery of hydrochlorothiazide using tacca starch/SCMC and Carbopols 940 and 941 admixtures. *Boll. Chim. Farm.* 1999, 138 (7), 329–336.
153. Attama, A.A.; Akpa, P.A.; Onugwu, L.E.; Igwilo, G. Novel buccoadhesive delivery system of hydrochlorothiazide formulated with ethyl cellulose-hydroxy propyl methyl cuelolose interpolymer complex. *Sci. Res. Essays* 2008, 3 (6), 343–347.
154. Attama, A.A.; Nnamani, P.O.; Adikwu, M.U. Diclofenac release from bioadhesive hydrophilic matrix tablets formulated with polyvinyl pyrrolidone-sodium carboxymethyl cellulose copolymer. *J. Pharm. Appl. Sci.* 2003, 1 (1), 1–7.
155. Proddurituri, S.; Urman, K.L.; Otaigbe, J.U.; Rekpa, M.A. Stabilization of hot-melt extrusion formulations containing solid solutions using polymer blends. *AAPS PharmSciTech* 2007, 8 (2), E1–E10.
156. Hercules Pharmaceutical Technology Report (PTR-016). In Polymer blend matrix for oral sustained drug delivery. In *25th International Symposium on Controlled Release of Bioadhesive Materials*, Las Vegas, NV, June 21–26, 1998.
157. Kuksal, A.; Tiwary, A.K.; Jain, N.K.; Jain, S. Formulation and in vitro-in vivo evaluation of extended-release matrix tablet of zidovudine: Influence of combination of hydrophilic and hydrophobic matrix formers. *AAPS PharmSciTech* 2006, 7 (1), E1–E9.
158. Mukherjee, B.; Mahapatra, S.; Gupta, R.; Patra, B.; Tiwari, A.; Arora, P. A comparison betwen povidone-ethylcellulose and povidone-Eudragit transdermal dexamethasone matrix patches based on in vitro skin permeation. *Eur. J. Pharm. Biopharm.* 2005, 59 (3), 475–483.
159. Yerri-Swamy, B.; Prasad, C.V.; Reedy, C.L.N.; Mallikarjuma, B.; Rao, K.C.; Subha, M.C.S. Interpenetrating polymer network microspheres of hydroxy propyl methyl cellulose/poly(vinyl alcohol) for control release of ciprofloxacin hydrochloride. *Cellulose* 2011, 18 (2), 349–357.

160. Iqual, M.C.; Amin, M.; Abadi, A.G.; Ahmad, N.; Jamia, K.H.; Jamal, A. Bacteria cellulose film caoting as drug delivery system: Thermal and drug release properties. *Sains Malays.* 2012, 41 (5), 561–568.
161. Siepmann, F.; Siepmann, J.; Walther, M.; MacRae, R.J.; Bodmeier, R. Blends of aqueous polymer dispersions used for pellet coating: Importance of the particle size. *J. Control. Release* 2005, 105 (3), 226–239.
162. Tezuka, Y.; Imai, K.; Oshima, M.; Ito, K. 13C-n.m.r. structural study on an enteric pharmaceutical coating cellulose derivative having ether and ester substituents. *Carbohydr. Res.* 1991, 222 (1), 255–259.
163. Wu, S.; Wyatt, D.; Adams, M. Chemistry and application of cellulosic polymers for enteric coatings. In *Aqueous Polymeric Coatings for Pharmaceutical Dosage Forms*; McGinity, J., Ed.; Dekker: New York, 1997; pp. 385–418.
164. Drury, J.L.; Mooney, D.J. Hydrogels for tissue engineering: Scaffold design variables and applications. *Biomaterials* 2003, 24 (24), 4337–4351.
165. El-Hag Ali, A.; Abd El-Rehim, H.; Kamal, H.; Hegazy, D. Synthesis of carboxymethyl cellulose based drug carrier hydrogel using ionizing radiation for possible use as specific delivery system. *J. Macromol. Sci. Pure Appl. Chem.* 2008, 45 (8), 628–634.
166. Trojani, C.; Weiss, P.; Michiels, J.F.; Vinatier, C.; Guicheux, J.; Daculsi, G.; Gaudray, P.; Carle, G.F.; Rochet, N. Three-dimensional culture and differentiation of human osteogenic cells in an injectable hydroxypropylmethylcellulose hydrogel. *Biomaterials* 2005, 26 (27), 5509–5517.
167. Peppas, N.A. Hydrogels and drug delivery. *Curr. Opin. Colloid Interface Sci.* 1997, 2 (5), 531–537.
168. Chen, J.; Park, H.; Park, K. Synthesis of superporous hydrogels: Hydrogels with fast swelling and superabsorbent properties. *J. Biomed. Mater. Res.* 1999, 44 (1), 53–62.
169. Sannino, A.; Pappadà, S.; Madaghiele, M.; Maffezzoli, A.; Ambrosio, L.; Nicolais, L. Crosslinking of cellulose derivatives and hyaluronic acid with water-soluble carbodiimide. *Polymer* 2005, 46 (25), 11206–11212.
170. Sannino, A.; Madaghiele, M.; Lionetto, M.G.; Schettino, T.; Maffezzoli, A. A cellulose-based hydrogel as a potential bulking agent for hypocaloric diets: An in vitro biocompatibility study on rat intestine. *J. Appl. Polym. Sci.* 2006, 102 (2), 1524–1530.
171. Litschauer, M.; Neouze, M.-A.; Haimar, E.; Henniges, U.; Potthast, A.; Rosenau, T.; Liebner, F. Silica modified cellulosic aerogels. *Cellulose* 2011, 18 (1), 143–149.
172. Sannino, A.; Esposito, A.; Nicolais, L.; Del Nobile, M.A.; Giovane, A.; Balestrieri, C.; Esposito, R.; Agresti, M. Cellulose-based hydrogels as body water retainers. *J. Mater. Sci.* 2000, 11 (4), 247–253.
173. Sannino, A.; Esposito, A.; De Rosa, A.; Cozzolino, A.; Ambrosio, L.; Nicolais, L. Biomedical application of a superabsorbent hydrogel for body water elimination in the treatment of edemas. *J. Biomed. Mater. Res. A* 2003, 67 (3), 1016–1024.
174. Esposito, A.; Sannino, A.; Cozzolino, A.; Quintiliano, S.N.; Lamberti, M.; Ambrosio, L.; Nicolais, L. Response of intestinal cells and macrophages to an orally administered cellulose-PEG based polymer as a potential treatment for intractable edemas. *Biomaterials* 2005, 26 (19), 4101–4110.
175. Shanbhag, A.; Barclay, B.; Koziana, J.; Shivanand, P. Application of cellulose acetate butyrate-based membrane for osmotic drug delivery. *Cellulose* 2007, 14 (1), 65–71.
176. Verma, R.K.; Mishra, B.; Garg, S. Osmotically controlled oral drug delivery. *Drug Dev. Ind. Pharm.* 2000, 26 (7), 695–708.
177. Theeuwes, F. Oros-osmotic system development. *Drug Dev. Ind. Pharm.* 1983, 9 (7), 1331–1357.
178. Santus, G.; Baker, R.W. Osmotic drug delivery: A review of the patent literature. *J. Control. Release* 1995, 35 (1), 1–21.
179. Theeuwes, F. Elementary osmotic pump. *J. Pharm. Sci.* 1975, 64 (12), 1987–1991.
180. Makhija, S.N.; Vavia, P.R. Controlled porosity osmotic pump-based controlled release systems of pseudoephedrine. I. Cellulose acetate as a semipermeable membrane. *J. Control. Release* 2003, 89 (1), 5–18.
181. Schultz, P.; Kleinebudde, P. A new multiparticulate delayed release system. Part I: Dissolution properties and release mechanism. *J. Control. Release* 1997, 47 (2), 181–189.
182. Catellani, P.L.; Colombo, P.; Peppas, N.A.; Santi, P.; Bettini, R. Partial permselective coating adds an osmotic contribution to drug release from swellable matrixes. *J. Pharm. Sci.* 1998, 87 (6), 726–731.
183. Lim, S. Manufacture of oral dosage forms delivering both immediate-release and sustained-release drugs. US Patent 0795324; 0598309; 1330839; WO00/23045. 6682759. January 27, 2004.
184. Lee, M. Controlled release of dual drug-loaded hydroxypropylmethylcellulose matrix tablet using drug-containing polymeric coatings. *Int. J. Pharm.* 1999, 188 (1), 71–80.
185. Builders, P.F.; Agbo, M.B.; Adelakun, T.; Okpako, L.C.; Attama, A.A. Novel multifunctional pharmaceutical excipients derived from microcrystalline cellulose-starch microparticulate composites prepared by compatibilized reactive polymer blending. *Int. J. Pharm.* 2010, 388 (1–2), 159–167.

186. Molla, M.A.K.; Shaheen, S.M.; Rashid, M.; Hossain, A.K.M.M. Rate controlled release of naproxen from HPMC Based sustained release dosage form, I. Microcapsule compressed tablet and matrices. *Dhaka Univ. J. Pharm. Sci.* 2005, 4 (1), 588–599.
187. Siepmann, J.; Peppas, N.A. Modeling of drug release from delivery systems based on hydroxypropyl methylcellulose (HPMC). *Adv. Drug Deliv. Rev.* 2001, 48 (2–3), 139–157.
188. Siepmann, J.; Streubel, A.; Peppas, N.A. Understanding and predicting drug delivery from hydrophilic matrix tablets using the "sequential layer" model. *Pharm. Res.* 2002, 19 (3), 306–314.
189. Kommareddy, S.; Tiwari, S.; Amiji, M. Long-circulating polymeric nanovectors for tumor-selective gene delivery. *Technol. Cancer Res. Treat.* 2005, 4 (6), 615–625.
190. Lee, M.; Kim, S. Polyethylene glycol-conjugated copolymers for plasmid DNA delivery. *Pharm. Res.* 2005, 22 (1), 1–10.
191. Hamidi, M.; Azadi, A.; Rafiei, P. Hydrogel nanoparticles in drug delivery. *Adv. Drug Deliv. Rev.* 2008, 60 (15), 1638–1649.
192. Soma, C.E.; Dubernet, C.; Barratt, G.; Nemati, F.; Appel, M.; Benita, S.; Couvreur, P. Ability of doxorubicin-loaded nanoparticles to overcome multidrug resistance of tumour cells after their capture by macrophages. *Pharm. Res.* 1999, 16 (11), 1710–1716.
193. Habibi, Y.; Goffin, A.L.; Schiltz, N.; Duquesne, E.; Dubois, P.; Dufresne, A. Bionanocomposites based on poly(epsilon-caprolactone)-grafted cellulose nanocrystals by ring-opening polymerization. *J. Mater. Chem.* 2008, 18 (1), 5002–5010.
194. Paralikar, S.A.; Simonsen, J.; Lombardi, J. Poly(vinyl alcohol)/cellulose nanocrystals barrier membranes. *J. Membr. Sci.* 2008, 320 (1), 248–258.
195. Bai, W.; Holbery, J.; Li, K. A techinique for production of nanocrystalline cellulose with a narrow size distribution. *Cellulose* 2009, 16 (3), 455–465.
196. Mörseburg, K.; Chinga-Carrasco, G. Assessing the combined benefits of clay and nanofibrillated cellulose in layered TMP-based sheets. *Cellulose* 2009, 16 (5), 795–806.
197. Chinga-Carrasco, G.; Syverud, K. Computer-assisted quantification of the multi-scale structure of films made of nanofibrillated cellulose. *J. Nanopart. Res.* 2010, 12 (3), 841–851.
198. Moran, J.I.; Alvarez, V.A.; Cyras, V.P.; Vazquez, A. Extraction of cellulose and preparation of nanocellulose from sisal fibers. *Cellulose* 2008, 15 (1), 149–159.
199. Aumelas, A.; Serrero, A.; Durand, A.; Dellacherie, E.; Leonard, M. Nanoparticles of hydrophobically modified dextrans as potential drug carrier systems. *Colloids Surf. B* 2007, 59 (1), 74–80.
200. Couvreur, P.; Lemarchand, C.; Gref, R. Polysaccharide-decorated nanoparticles. *Eur. J. Pharm. Biopharm.* 2004, 58 (2), 327–341.
201. Wegnar, T.H.; Jones, P.E. Advancing cellulose-based nanotechnology. *Cellulose* 2006, 13 (2), 115–118.
202. Edgar, K.J. Cellulose esters in drug delivery. *Cellulose* 2007, 14 (1), 49–64.
203. Baumann, M.D.; Kang, C.E.; Stanwick, J.C.; Wang, Y.; Kim, H.; Lapitsky, Y.; Shoichet, M.S. An injectable drug delivery platform for sustained combination therapy. *J. Control. Release* 2009, 138 (3), 205–213.
204. Watanabe, Y.; Mukai, B.; Kawamura, K.I.; Ishikawa, T.; Namiki, M.; Utoguchi, N.; Fujji, M. Preparation and evaluation of press-coated aminophylline tablet using crystalline cellulose and polyethylene glycol in the outer shell for timed-release dosage forms. *Yakugaku Zasshi* 2002, 122 (2), 157–162.
205. Gilberto, S.; Julien, B.; Alain, D. Cellulosic bionanocomposite: A review of preparation, properties and applications. *Polymers* 2010, 2 (4), 728–765.
206. Lönnberg, H.; Fogelström, L.; Samir, M.A.S.A.; Berglund, L.; Malmström, E.; Hult, A. Surface grafting of microfibrillated cellulose with poly(ε-caprolactone)—Synthesis and characterization. *Eur. Polym. J.* 2008, 44 (9), 2991–2997.
207. Hassan, N.; Farzaneh, F.; Abolfazl, H. Nanoparticles based on modified polysaccharides. In *The Delivery of Nanoparticles*; Hashim, A.A., Ed.; InTech: Rijeka, Croatia, 2012; pp. 149–184.
208. Uglea, C.V.; Pary, A.; Corjan, M.; Dumitriu, A.D.; Ottenbrite, R.M. Biodistribution and antitumor activity induced by carboxymethylcellulose conjugates. *J. Bioact. Comp. Polym.* 2005, 20 (6), 571–583.
209. Sievens-Figueroa, L.; Bhakay, A.; Jerez-Rozo, J.T.; Pandya, N.; Romanach, R.J.; Michniak-Kohn, B.; Iqbal, Z.; Bilgili, E.; Dave, R.N. Preparation and characterization of hydroxypropylmethyl cellulose films containing stable BCS Class II drug nanoparticles for pharmaceutical applications. *Int. J. Pharm.* 2012, 423 (2), 496–508.
210. Ernsting, M.J. Cellulose-based nanoparticles for drug delivery. US Patent US 20120219508A1, August 9, 2012.
211. Dong, S.; Hirani, A.A.; Lee, Y.W.; Roman, M. Synthesis of FITC-labelled, folate-targeted cellulose nanocrystals. Technical Programming Archive. In *239th ACS National Meeting*, San Francisco, CA, March 21–25, 2010; p. CELL-061.

212. Roman, M.; Dong, S.; Hirani, A.; Lee, Y.W. Chapter 12: Cellulose nanocrystals for drug delivery. In *Polysaccharide Materials: Performance by Design*; Edgar, K.J., Heinze, T., Buchanan, C. Eds.; ACS Symposium Series 1017. American Chemical Society: Washington, DC, 2009.
213. Dong, S.; Roman, M. Synthesis of fluorescently-labelled, folate-targeted cellulose nanoconjugates. In *Proceedings of the 2009 MII Technical Conference and Review*, Macromolecules and Interfaces Institute: Blacksburg, VA. April 2009; p. 73.
214. Dong, S.; Roman, M. Fluorescently labelled cellulose nanocrystals for bioimaging applications. *J. Am. Chem. Soc.* 2007, 129 (45), 13810–13811.
215. Roman, M. Novel applications of cellulose nanocrystals: From drug delivery to micro-optics. In *Proceedings of the MII Technical Conference and Review*, Blacksburg, VA, October 22–24, 2007.
216. Dong, S.; Hirani, A.; Lee, Y.W.; Roman, M. Cellulose nanocrystals for the targeted delivery of therapeutic agents. In *Proceedings of the MII Technical Conference and Review*, Blacksburg, VA, October 22–24, 2007.
217. Dong, S.; Hirani, A.A.; Lee, Y.W.; Roman, M. Synthesis of FITC-labelled, folate-targeted cellulose nanocrystals. In *Proceedings of the MII Technical Conference and Review*, Blacksburg, VA, October 11–13, 2010.
218. Cho, H.J.; Lee, S.; Dong, S.; Roman, M.; Lee, Y.W. Cellulose nanocrystals as a novel nanocarrier for targeted drug delivery to brain tumor cells. *FASEB J.* 2011, 25, 762.2.
219. Colacino, K.R.; Dong, S.; Roman, M.; Lee, Y.W. Cellulose nanocrystals: A novel biomaterial for targeted drug delivery applications. *FASEB J.* 2011, 25, 762.3.
220. Aswathy, R.G.; Sivakumar, B.; Brahatheeswaran, D.; Raveendran, S.; Ukai, T.; Fukuda, T.; Yoshida, Y.; Maekawa, T.; Sakthikumar, D.N. Multifunctional biocompatible fluorescent carboxymethyl cellulose nanoparticles. *J. Biomater. Nanobiotechnol.* 2012, 3 (2), 254–261.
221. Yallapu, M.M.; Dobberpuhl, M.R.; Maher, D.M.; Jaggi, M.; Chauhan, S.C. Design of curcumin loaded cellulose nanoparticles for prostate cancer. *Curr. Drug Metab.* 2012, 13 (1), 120–128.
222. Neha, D. Novel cellulose nanoparticles for potential cosmetic and pharmaceutical applications. A Masters' Thesis, Department of Chemical Engineering, University of Waterloo, Waterloo, Ontario, Canada, 2010.
223. Baumgartner, S.; Kristl, J.; Peppas, N.A. Network structure of cellulose ethers used in pharmaceutical applications during swelling and at equilibrium. *Pharm. Res.* 2002, 19 (8), 1084–1090.
224. Sachiko, K.N.; Keiji, N. Biopolymer-based nanoparticles for drug/gene delivery and tissue engineering. *Int. J. Mol. Sci.* 2013, 14 (1), 1629–1654.
225. Cui, Z.; Mumper, R.J. Chitosan-based nanoparticles for topical genetic immunization. *J. Control. Release* 2001, 75 (3), 409–419.
226. Arhewoh, I.M.; Ahonkhai, E.I.; Okhamafe A.O. Optimising oral systems for the delivery of therapeutic proteins and peptides. *Afr. J. Biotech.* 2005, 4 (13), 1591–1597.
227. Okhamafe, A.O.; Amsden, B.; Chu, W.; Goosen, M.F.A. Modulation of protein release from chitosan-alginate microcapsules modified with the pH sensitive polymer-hydroxylpropyl methylcellulose acetate succinate (HPMCAS). *J. Microencapsul.* 1996, 13 (5), 497–508.
228. Amorim, M.J; Ferreira, J.P. Microparticles for delivering therapeutic peptides and proteins to the lumen of the small intestine. *Eur. J. Pharm. Biopharm.* 2001, 52 (1), 39–44.
229. Moghimi, S.M.; Hunter, A.C.; Murray, J.C. Long-circulating and target-specific nanoparticles: Theory to practice. *Pharm. Rev.* 2001, 53 (2), 283–318.
230. Adiseshaiah, P.P.; Hall, J.B.; McNeil, S.E. Nanomaterial standards for efficacy and toxicity assessment. *Wiley Interdiscip. Rev.* 2010, 2 (1), 99–112.
231. Mansouri, S.; Lavigne, P.; Corsi, K.; Benderdour, M.; Beaumont, E.; Fernandes, J.C. Chitosan-DNA nanoparticles as non-viral vectors in gene therapy: Strategies to improve transfection efficacy. *Eur. J. Pharm. Biopharm.* 2004, 57 (1), 1–8.
232. Thomas, M.; Klibanov, A.M. Non-viral gene therapy: Polycation-mediated DNA delivery. *Appl. Microbiol. Biotechnol.* 2003, 62 (1), 27–34.
233. Li, S.D.; Huang, L. Gene therapy progress and prospects: Non-viral gene therapy by systemic delivery. *Gene Ther.* 2006, 13 (18), 1313–1319.
234. Dang, J.M.; Leong, K.W. Natural polymers for gene delivery and tissue engineering. *Adv. Drug Deliv. Rev.* 2006, 58 (4), 487–499.
235. Xu, F.J.; Ping, Y.; Ma, J.; Tang, G.P.; Yang, W.T.; Li, J.; Kang, E.T.; Neoh, K.G. Comb-shaped copolymers composed of hydroxypropyl cellulose backbones and cationic poly((2-dimethyl amino) ethyl methacrylate) side chains for gene delivery. *Bioconjug. Chem.* 2009, 20 (8), 1449–1458.
236. Czaja, W.K.; Young, D.J.; Kawecki, M.; Malcolm-Brown, R. Jr. The future prospects of microbial cellulose in biomedical applications. *Biomacromolecules* 2007, 8 (1), 1–12.

29 Starch-Based Polymeric Biomaterial
Drug Delivery

Akhilesh Vikram Singh and Ashok M. Raichur

CONTENTS

29.1 Introduction 575
29.2 Native and Modified Starches 575
29.3 Pharmaceutical Application of Native and Modified Starches 576
    29.3.1 Native Starches in Drug Development and Drug Delivery 576
    29.3.2 Modified Starches in Solid Dosage Form 577
    29.3.3 Modified Starches in Sustained/Controlled Delivery 578
    29.3.4 Modified Starches in Nanoparticulate Delivery 579
    29.3.5 Modified Starches an Vaccine and Gene Delivery 579
    29.3.6 Modified Starch as Plasma Volume Expander 580
29.4 Conclusions 580
References 580

29.1 INTRODUCTION

Biomaterials are generally natural substances apart from food or drugs present in therapeutic or diagnostic systems that are in contact with tissue or biological fluids.[1] Polysaccharide-derived polymers display perfect biocompatibility and biodegradability, which are the basic requisites for polymers used as biomaterials. In carbohydrate-based polymeric systems, the drug is embedded in a polymer matrix and surgically implanted. The matrix may be biodegradable or nonbiodegradable. Hydrophilic polysaccharide carriers remain a highly popular design of modified release dosage form. Drug delivery systems containing polysaccharide excipient system was found to be capable of assuming different formulations and use in process methods to provide a variety of different release modalities independent of matrix-dimension.

29.2 NATIVE AND MODIFIED STARCHES

Starches are polysaccharides, composed of a number of monosaccharide or sugar (glucose) molecules linked together with α-D-(1–4) and/or α-D-(1–6) linkages. The starch consists of two main structural components as shown in Figure 29.1. The two components are amylose, which is essentially a linear polymer in which glucose residues are α-D-(1–4) linked typically constituting 15%–20% of starch, and amylopectin, which is a larger branched molecule with α-D-(1–4) and α-D-(1–6) linkages and is a major component of starch. Some starches contain high percentage of amylopectin and are known as "waxy starch." Industrially, starch is either used as normal extracted form called "native starch" or with certain modifications to achieve specific properties referred to as "modified starch." Worldwide, the main sources of starch are maize (82%), wheat (8%), potatoes (5%), and cassava (5%). The total utilization of dried starch in the world in 2008 was around 66.5 million tons with annualized global growth of 2%–3%.

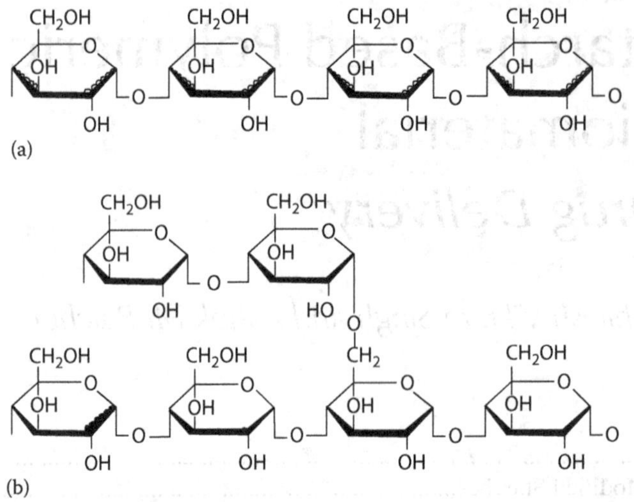

FIGURE 29.1 The molecular structure of starch containing (a) amylose and (b) amylopectin.

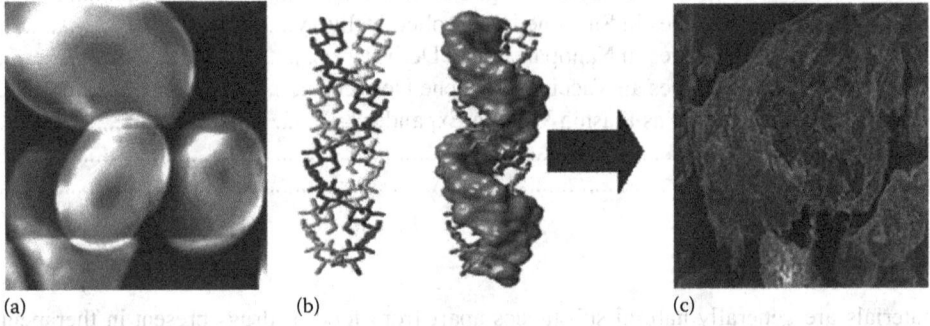

FIGURE 29.2 Change in the structure of (a, b) native starch and (c) after modification.

Modified starches consist of starch with low to very low level of substituent groups (depending upon the industrial demand). Chemical modification has been the main approach to produce modified starch in the last century. Many developments of chemical modification of starches have been introduced in food, pharmaceutical, and textile industries. Physical modification involves pregelatinization, and heat treatment of starch. Pregelatinized starches are precooked starches that can be used as a thickener in cold water. While the heat-treatment processes include heat–moisture and annealing treatments, both of which cause a physical modification of starch without any gelatinization, damage to granular integrity or loss of birefringence takes place (shown in Figure 29.2). Enzymatic modification of starch involves hydrolysis of some part of starch into a low molecular weight of starch called maltodextrin, or dextrin using amylolytic enzymes. They are widely used in food and pharmaceutical industries. Table 29.1[2,3] contains a brief description of modified starches and their subsequent application in pharmaceutical industry.

29.3 PHARMACEUTICAL APPLICATION OF NATIVE AND MODIFIED STARCHES

29.3.1 Native Starches in Drug Development and Drug Delivery

Native starches due to their swelling and gelling behavior at body temperature were explored as a binder and a disintegrant in pharmaceutical formulation development. Out of all the starches, maize, potato, and now pea finds official status in many pharmacopoeias worldwide for their use

TABLE 29.1
Application of Various Starches in Pharmaceutical Formulation Development

Starch Type	Pharmaceutical Application	Active Drug	References
Cassava, cocoyam, and maize starch	Binder	Paracetamol	[1]
Cyperus esculentus Starch	Binder	Metronidazole	[2]
Trapa bispinosa Roxb. Starch	Binder	Diclofenac sodium	[3]
Ginger starch	Binder	Acetaminophen	[4]
Sorghum starch	Binder	Sodium carbonate	[5]
Musa paradisiaca starch	Binder	Paracetamol and chloroquine	[6]
Ensete ventricosum starch	Binder and disintegrant	Paracetamol and chloroquine	[7]
Breadfruit starch	Disintegrant	Paracetamol	[8]
Yam starch	Disintegrant	Hydrochlorthiazide	[9]
Corn starch	Binder and disintegrant	Hydrochlorthiazide	[10]
Abelmoschus starch	Disintegrant	Paracetamol	[11]

as a binder and a disintegrant. Binders are used in wet granulation step of tablet manufacturing and their nature should be polymeric. The starches are either dispersed or dissolved in water or hydroalcoholic medium to form a mucilaginous gel. The binder increases the mechanical strength and disintegration time of solid dosage form. The native starches were first time introduced as disintegrants in tablet manufacturing worldwide and is still the most widely used material. Starches act as disintegrants by its swelling property and capillary uptake action. It has been found and established that starch swells at body temperature in biological medium, which exerts a hydrostatic pressure in tablet that helps in dispersion. Apart from official starches various other starch sources were evaluated recently for binder and disintegrant property. In Table 29.2, a brief detail about these new starches is tabulated.[4–14]

29.3.2 Modified Starches in Solid Dosage Form

Starch is used as a versatile excipient in the development of pharmaceutical solid dosage form. In pharmaceutical industry, starch obtained from maize and potato was exclusively used in the formulation development and regulatory approval was obtained for it to be used as a tablet binder and disintegrant.[15,16] The native starches form a gel mass after heating at their gelatinization temperature, which is used in the granulation step of solid dosage form development. The starch gel holds the entire solid drug particle by forming a suitable chemical and physical bonding and acts as a binder. In the same way the native starches also act as a disintegrant due to their swelling characteristics.[17]

In recent times, modified starches especially sodium carboxymethylated derivative, commercially known as sodium starch glycolate (SSG), and pregelatinized starches have been evaluated as super disintegrant (i.e., agent for dispersing the solid drug within a few minutes or making a dispersible tablet). Further, many researchers have explored various other starch sources for sodium carboxymethyl derivative of starch, which is officially known as SSG.[18–21]

The pregelatinized form of modified starch is prepared by heating the native starch at their gelatinization temperature using techniques like solvent-based processing, oxidation, hydrolysis, and cross-linking. Pregelatinization of natural starches has been used to produce cold-water swellable forms with improved flowability. Recently, many researchers focused their research work on pregelatinized modification and understanding the mechanism by which the disintegration occurs. They found that during the change from natural to pregelatinized starch, there is an improvement in disintegrating property, which could be as a result of subsequent generation of a higher swelling force due to the absorption of large quantities of water into the tablet mass.[22–24]

TABLE 29.2
Types of Modified Starches and Their Relevance to Pharmaceutical Application

Types of Modification	Method and Change in Starch Structure	Change in Physicochemical Property	Drug Delivery Application
1. Chemical modification			
(i) Oxidation	It causes breaking of glucosidic linkage and oxidation of –OH group into –COOH and –COCl groups	Decreased swelling and paste viscosity	Binder
(ii) Cross-linking	The free –OH group is cross-linked with Sodium tri-meta phosphate, citric acid, $POCl_3$, EPI	Decreased swelling and gelatinization	Controlled/sustained delivery
(iii) Esterification	The free –OH group is replaced with ester groups like acetate, suucinate, phosphate	Reduced granule swelling and viscosity	Controlled/sustained delivery
(iv) Etherification	Substitution of hydroxypropyl group on starch backbone	Higher viscosity and decreased enzyme digestibility	Plasma volume expander and gene delivery application
(v) Grafting	Grafting of vinyl-acrylic derivative	High gelling and mucoadhesive value	Mucoadhesive/controlled drug delivery
2. Physical modification			
(i) Pregelatinization	Produced by cooking, drum drying, and solvent processing	Increased hydration and swelling	Excipient and controlled delivery
(ii) Extrusion	Produced by cooking extruder	High water solubility and absorption	Sustained release delivery
(iii) Spray drying	Produced by spray drying technique	Low viscosity and cold water solubility	Excipient and controlled delivery
3. Enzymatic modification			
(i) Enzyme hydrolysis	Produced by acid/enzymatic hydrolysis of starch	Increased crystallization	In syrup preparation

Some modified starches were also investigated for their direct compressible property, i.e., binders to be used without wet granulation step. The native starch has one disadvantage that the flowing property either falls in poor or very poor category. To overcome this problem, some modifications like pregelatinization, acetylation, oxidation, acid treatment, or spray drying is required. These modifications lead to development of highly crystallizable, flowable, and compressible derivative of starch. Highly compressible starches are in high demand for making direct compressible tablet as it reduces the overall cost and manufacturing life cycle.[25–27]

29.3.3 MODIFIED STARCHES IN SUSTAINED/CONTROLLED DELIVERY

In sustained release system, researchers try to achieve a slow release of the drug over an extended period of time, while in controlled drug delivery, control and constant drug level is maintained in the target tissue or cell by using a group of polymers. Various types of modified starches like pregelatinized, acetylated derivatives, cross-linked, or grafted were evaluated for sustained and controlled property.

The pregelatinized starch obtained from spray-drying technique changed the insoluble form into a high-viscosity gel, which controls the drug release behavior and mechanism was dominated by polymer erosion control. The only restriction with these starches is that the starch having higher amylopectin proportion (waxy starch type) would be a better option as compared to high amylose containing starch. This is because the amylopectin seems to be responsible for the cohesive character of the gel while amylose provides hardness to the dosage form.[28–30]

The basic idea behind cross-linking is the toughening of starch granules by treatment with di- or polyfunctional reagents capable of reacting with the free hydroxyl groups present in the starch molecules. Sodium trimetaphosphate (STMP), monosodium phosphate, sodium tripolyphosphate), epichlorohydrin (EPI), phosphoryl chloride ($POCl_3$), a mixture of adipic acid and acetic anhydride, and vinyl chloride are the important food-grade cross-linking agents used for making cross-linked starch. EPI cross-linked high amylose has been used for the controlled release of contramid.[31] Cross-linked starches modified with STMP and $POCl_3$ were extensively evaluated for sustained delivery, and the reason for this modification is because of a change in amylopectin chain entanglement and porosity.[32–36]

Starch modified with acetic anhydride in the presence of aqueous/organic solvent produces acetylated starch. The highly substituted starch acetate, which is chemically hydrophobic in nature, forms small cracks after dissolution, which is responsible for its sustained release activity.[37–40]

Grafting is a very common and established technique to alter the physicochemical properties of the polysaccharides. Acrylic acid derivative grafting on starch backbone has been well studied in recent times for bioadhesive and controlled drug delivery applications. It was found that the grafted starch forms a rigid gel when it came in contact with biological fluids; simultaneously the release behavior of the graft copolymers was found to be non-Fickian, suggesting that the release was controlled by a combination of tablet erosion and the diffusion of the drug from the swollen matrix.[41–45]

29.3.4 Modified Starches in Nanoparticulate Delivery

Nanoparticles are solid, colloidal particles consisting of macromolecular substances that vary in size from 10 to 100 nm. The drug is dissolved, entrapped, adsorbed, attached, or encapsulated into the nanoparticle matrix. The nanoparticle matrix can be made up of biodegradable materials such as polymers or proteins. Depending on the method of preparation, nanoparticles can be obtained with different properties and release characteristics for the encapsulated therapeutic agents.[46] Starch modified with acids produces starch nanocrystals with good solubility property. Some researchers also reported the formulation of starch nanoparticles by precipitation with ethanol and microfluidization techniques. Starch nanoparticle research is still in a nascent stage and more focus is needed from application point of view.[47] Recently some researchers reported synthesis of starch nanoparticle with spray drying and freeze-drying methods.[48] Starch nanoparticles are also used for targeted delivery of anticancer drugs. In this context, a new magnetically controlled starch nanoparticle was synthesized for the delivery of cisplatin.[49] They concluded that the release profiles were greatly influenced by pH and temperature of the release medium as well as external magnetic field. Simi and Abraham[50] have synthesized hydrophobic-grafted starch nanoparticle and successfully delivered indomethacin. They found that a significant amount of drug can be loaded with starch-based nanoparticle and proposed that it can be used as a good carrier for controlled delivery. Some researchers prepared propyl starch-based nanoparticles and loaded some drugs for their transdermal delivery. The encapsulation efficacy was found to be much better for hydrophobic drugs with a null initial burst effect.[51,52]

29.3.5 Modified Starches an Vaccine and Gene Delivery

Gene therapy, an approach for treatment or prevention of diseases associated with defective gene expression, involves the insertion of a therapeutic gene into cells, followed by expression and production of the required proteins. In gene therapy for malignant tumors, highly efficient gene transfer to the tumor and minimization of transfer to normal tissues are extremely important. Degradable starch microspheres are evaluated as embolic agents used in trans-arterial chemotherapy for hepatocellular carcinoma and metastatic liver tumors. Hyperbranched cationic amylopectin derivatives conjugated with 1,2-ethylenediamine showed better blood compatibility and lower cytotoxicity when compared to synthetic polymers for gene delivery. The amylopectin derivatives bind and condense with plasmid DNA showing higher transfection efficiency.[53] Immunization is highly effective in protecting the body against systemic infection. Immunization is generally administered through either parenteral or oral route in the form of vaccines. It is found that mucosal vaccination (oral) has several advantages over parenteral vaccination because it can provoke both mucosal and circulating antibody. Powder formulations based on spray-dried mixtures of starch and poly(acrylic acid) has been successfully used for intranasal delivery of heat-inactivated influenza virus. Human serum albumin entrapped in modified starch has also been evaluated for successful delivery of this virus.[54–58]

29.3.6 MODIFIED STARCH AS PLASMA VOLUME EXPANDER

The plasma volume expanders are the substances that are transfused to maintain fluid volume of the blood in event of great necessity, supplemental to the use of whole blood and plasma. Some starch derivatives like hydroxyethyl starch[59] and acetyl starch,[60] which are a group of colloids, are used to provide sustained intravascular volume expansion. Hydroxyethyl starches are high-polymeric compounds obtained via hydrolysis and subsequent hydroxyl ethylation of glucose units substituted at carbon number 2, 3, and 6 of starch. Recently some waxy starches[61] were also evaluated for plasma volume expander, but more research is needed to establish them as a good substitute for the synthetic polymers.

29.4 CONCLUSIONS

In spite of recent developments and research in synthetic polymers applied in pharmaceutical field, biomaterials, particularly polysaccharides like starch, cellulose, gums and mucilages, have a greater role to play in pharmaceutical formulation development. It is concluded from the discussion in this entry that because of some limitations at the molecular level, there is a need to modify the parent structure of starch to make it available for pharmaceutical and other industrial uses. Physically and chemically modified starches showed very promising results in the delivery of therapeutic agents. The modified starches have enormous potential to be used in drug delivery because of its easy modification and biodegradability.

REFERENCES

1. Langer, R.; Peppas, N.A. Advances in biomaterials, drug delivery, and bionanotechnology. *AIChE J.* **2003**, *49* (12), 2990–3006.
2. Singh, A.V.; Nath, L.K.; Singh, A. Pharmaceutical, food and non-food applications of modified starches: A critical review. *EJEAFChe* **2010**, *9* (7), 1214–1221.
3. Abbas, K.A.; Khalil, S.K.; Hussin, A.S.M. Modified starches and their usages in selected food products: A review study. *J. Agric. Sci.* **2010**, *2* (2), 90–100.
4. Uhumwangho, M.U.; Okor, R.S.; Eichie, F.E.; Abbah, C.M. Influence of some starch binders on the brittle fracture tendency of paracetamol tablets. *Afr. J. Biotechnol.* **2006**, *5* (20), 1950–1953.
5. Manek, R.V.; Builders, P.F.; Kolling, W.M.; Emeje, M.; Kunle, O.O. Physicochemical and binder properties of starch obtained from *Cyperus esculentus*. *AAPS PharmSciTech.* **2012**, *13* (2), 379–388.
6. Singh, A.V.; Singh, A.; Nath, L.K.; Pani, N.R. Evaluation of *Trapa bispinosa* Roxb. Starch as pharmaceutical binder in solid dosage form. *Asian Pacific J. Trop. Biomed.* **2011**, *1* (1), S86–S89.
7. Ibezim, E.C.; Ofoefule, S.I.; Omeje, E.O.; Onyishi, V.I.; Odoh, U.E. The role of ginger starch as a binder in acetaminophen tablets. *Sci. Res. Essays* **2008**, *3* (2), 46–50.
8. Garr, J.S.M.; Bangudu, A.B. Evaluation of sorghum starch as tablet excipients. *Drug Dev. Ind. Pharm.* **1991**, *17* (1), 1–6.
9. Esezobo, S.; Ambujam, V. An evaluation of starch obtained from plantain *Musa paradisiaca* as a binder and disintegrant for compressed tablets. *J. Pharm. Pharmacol.* **1982**, *34* (12), 761–765.
10. Gebre-Mariam, T.; Nikolayev, A.S. Evaluation of starch obtained from *Ensete ventricosum* as a binder and disintegrant for compressed tablets. *J. Pharm. Pharmacol.* **1993**, *45* (4), 317–320.

11. Adebayo, S.A.; Brown-Myrie, E.; Itiola, A.O. Comparative disintegrant activities of breadfruit starch and official corn starch. *Powder Technol.* **2008**, *181* (2), 98–103.
12. Nattapulwat, N.; Purkkao, N.; Suwithayapanth, O. Evaluation of native and carboxymethyl yam (*Dioscorea esculenta*) starches as tablet disintegrants. *Silpakorn U. Sci. Tech. J.* **2008**, *2* (2), 18–25.
13. Kottke, M.K.; Chueh, H.R.; Rhodes, C.T. Comparison of disintegrant and binder activity of three corn starch products. *Drug Dev. Ind. Pharm.* **1992**, *18* (20), 2207–2223.
14. Ramu, G.; Mohan, G.K.; Jayaveera, K.N.; Suresh, N.; Prakash, K.C.; Ramesh, B. Evaluation of abelmoschus starch as disintegrant. *Indian J. Nat. Prod. Res.* **2010**, *1* (3), 342–347.
15. Ochubiojo, E.M.; Rodrigues, A. Starch: From food to medicine. In *Scientific, Health and Social Aspects of the Food Industry*; Valdez, B., Ed.; InTech Publishers: Croatia, Europe, 2012; 355–380.
16. Kaur, L.; Singh, J.; Liu, Q. Starch—A potential biomaterial for biomedical applications. In *Nanomaterials and Nanosystem for Biomedical Applications*; Springer Publications, Dordrecht, the Netherlands, 2012; 83–98.
17. Michaud, J. Starch based excipients for pharmaceutical tablets. *Pharm. Chem.* **2002**, *1*, 42–44.
18. Shah, U.; Augsburger, L. Multiple sources of sodium starch glycolate, NF: Evaluation of functional equivalence and development of standard performance tests. *Pharm. Dev. Technol.* **2002**, *7* (3), 345–359.
19. Singh, A.V.; Singh, A.; Nath, L.K. Microwave assisted synthesis and evaluation of modified pea starch as tablet superdisintegrant. *Curr. Drug Deliv.* **2011**, *8* (2), 203–207.
20. Singh, A.V.; Nath, L.K.; Guha, M.; Kumar, R. Microwave assisted synthesis and evaluation of cross-linked carboxymethylated sago starch as superdisintegrant. *Pharmacol. Pharm.* **2011**, *2* (1), 42–46.
21. Singh, A.V.; Nath, L.K.; Pandey, R.D.; Das, A. Syntheisis and application of carboxymethyl moth bean starch as superdisintegrant in pharmaceutical formulation. *J. Polym. Mater.* **2010**, *27* (2), 173–178.
22. Alebiowu, G.; Itiola, O.A. The influence of pregelatinized starch disintegrants on interacting variables that act on disintegrant properties. *Pharm. Tech.* **2003**, *27* (8) 28–32.
23. Adedokun, O.M.; Itiola, O.A. Material properties and compaction characteristics of natural and pregelatinized forms of four starches. *Carbohydr. Polym.* **2010**, *79* (4), 818–824.
24. Zgoda, M.M.; Kolodziejczyk, M.K.; Nachajski, M.J. Starch and its derivatives as excipients in oral and parenteral drug form technology. *Polim. Med.* **2009**, *39* (1), 31–45.
25. Bos, C.E.; Bolhuis, G.K.; Lerk, C.F.; Duineveld, C.A.A. Evaluation of modified rice starch, a new excipient for direct compression. *Drug Dev. Ind. Pharm.* **1992**, *18* (1), 93–106.
26. Afolabi, A.T.; Olu-Owolabi, B.; Kayode, O.; Adebowale, O.K.; Lawal, S.O.; Akintayo, O.C. Functional and tableting properties of acetylated and oxidised finger millet (*Eleusine coracana*) starch. *Starch/Stärke* **2012**, *64* (4), 326–337.
27. Raatikainen, P.; Korhonen, O.; Peltonen, S.; Paronen, P. Acetylation enhances the tabletting properties of starch. *Drug Dev. Ind. Pharm.* **2002**, *28* (2), 165–175.
28. Hermen, J.; Remon, J.P.; Vilder, D.J. Modified starches as hydrophilic matrices for controlled oral delivery-I. Production and characterisation of thermally modified starches. *Int. J. Pharm.* **1989**, *56* (1), 51–63.
29. Hermen, J.; Remon, J.P. Modified starches as hydrophilic matrices for controlled oral delivery. II. In vitro drug release evaluation of thermally modified starches. *Int. J. Pharm.* **1989**, *56*, 65–70.
30. Peerapattana, J.; Phuvarit, P.; Srijesdaruk, V.; Preechagoon, D.; Tattawasart, A. Pregelatinized glutinous rice starch as a sustained release agent for tablet preparations. *Carbohydr. Polym.* **2010**, *80* (2), 453–459.
31. Lenaerts, V.; Moussa, I.; Dumoulin, Y.; Mebsout, F.; Chouinard, F.; Szabo, P.; Mateescu., Cartilier, L.; Marchessault, R. Cross-linked high amylase starch for controlled release of drugs: Recent advances. *J. Control. Release* **1998**, *53* (1–3), 225–234.
32. Onofre, F.O.; Mendez-Montealvo, G.; Wang, Y.I. Sustained release properties of cross linked corn starches with varying amylose contents in monolithic tablets. *Starch/Stärke* **2010**, *62* (3–4), 165–172.
33. O'Brien, S.; Wang, Y.J.; Vervaet, C.; Remon, J.P. Starch phosphates prepared by reactive extrusion as a sustained release agent. *Carbohydr. Polym.* **2009**, *76* (4), 557–566.
34. O'Brien, S.; Wang, Y.J. Effects of shear and pH on starch phosphates prepared by reactive extrusion as a sustained release agent. *Carbohydr. Polym.* **2009**, *77* (3), 464–471.
35. Anwar, E.; Khotimah, H.; Yanur, A. An approach on pregelatinized cassava starch phosphate esters as hydrophilic polymer excipient for controlled release tablet. *J. Med. Sci.* **2006**, *6* (6), 925–929.
36. Mulhbacher, J.; Ispas-Szabo, P.; Lenaerts, V.; Mateescu, A.M.; Cross-linked high amylose starch derivatives as matrices for controlled release of high drug loadings. *J. Control. Release* **2001**, *76* (1–2), 51–58.
37. Korhonen, O.; Kanerva, H.; Vidgen, M.; Urtti, A.; Ketolainen, J. Evaluation of novel starch-acetate diltiazem controlled release tablets in healthy human volunteer. *J. Control. Release* **2004**, *95* (3), 515–520.
38. Pohja, S.; Suihko, E.; Vidgen, M.; Paronen, P.; Ketolainen, J. Starch acetate as a tablet matrix for sustained drug release. *J. Control Release* **2004**, *94* (2–3), 293–302.

39. Singh, A.V.; Nath, L.K. Evaluation of acetylated moth bean starch as a carrier for controlled drug delivery. *Int. J. Biol. Macromol.* **2012**, *50* (2), 362–368.
40. Singh, A.V.; Nath, L.K. Evaluation of chemically modified hydrophobic sago starch as a carrier for controlled drug delivery. *Saudi. Pharm. J.* **2012**, *21* (2), 193–200.
41. Singh, A.V.; Nath, L.K. Evaluation of acrylamide grafted moth bean starch as controlled release excipient. *Carbohydr. Polym.* **2012**, *87* (4), 2677–2682.
42. Geresh, S.; Gilboa, Y.; Peisahov-Korol, J.; Gdalevsky, G.; Voorspoels, J.; Remon, J.P.; Kost, G. Preparation and characterization of bioadhesive grafted starch copolymers as platforms for controlled drug delivery. *J. Appl. Polym. Sci.* **2002**, *86* (5), 1157–1162.
43. Echeverria, I.; Silva, I.; Goni, I.; Gurruchaga, M. Ethyl methacrylate grafted on two starches as polymeric matrices for drug delivery. *J. Appl. Polym. Sci.* **2005**, *96* (2), 523–536.
44. Silva, I.; Gurruchaga, M.; Goni, I. Drug release from microstructured grafted starch monolithic tablets. *Starch/Starke* **2011**, *63* (12), 808–819.
45. Shaikh, M.M.; Lonikar, S.V. Starch–acrylics graft copolymers and blends: Synthesis, characterization, and applications as matrix for drug delivery. *J. Appl. Polym. Sci.* **2009**, *114* (5), 2893–2900.
46. Sahoo, S.K.; Labhasetwar, V. Nanotech approaches to drug delivery and imaging. *Drug Discov. Today* **2003**, *8* (24), 1112–1120.
47. Rodrigues, A.; Emeje, M. Recent applications of starch derivatives in nanodrug delivery. *Carbohydr. Polym.* **2012**, *87* (2), 987–994.
48. Corre, D.L.; Bras, J.; Dufresne, A. Starch nanoparticles: A review. *Biomacromolecules* **2010**, *11* (5), 1139–1153.
49. Shi, A.; Li, D.; Wang, L.; Adhikari, B. Rheological properties of suspensions containing cross-linked starch nanoparticles prepared by spray and vacuum freeze drying methods. *Carbohydr. Polym.* **2012**, *90* (4), 1732–1738.
50. Likhitkar, S.; Bajpai, A.K. Magnetically controlled release of cisplatin from superparamagnetic starch nanoparticles. *Carbohydr. Polym.* **2012**, *87* (1), 300–308.
51. Cimi, C.K.; Abraham, T.E. Hydrophobic grafted and cross-linked starch nanoparticles for drug delivery. *Bioprocess Biosyst. Eng.* **2007**, *30* (3), 173–180.
52. Santander-Ortega, M.J.; Stauner, T.; Loretz, B.; Ortega-Vinuesa, J.L.; Bastos-González, D.; Wenz, C.; Schaefer, U.F.; Lehr, C.M. Nanoparticles made from novel starch derivatives for transdermal drug delivery. *J. Control. Release* **2010**, *141* (1), 85–92.
53. Shiba, H.; Okamato, T.; Fugawa, Y.; Misawa, T.; Yanaga, K.; Ohashi, T.; Eto, Y. Adenovirus vector-mediated gene transfer using degradable starch microspheres for hepatocellular carcinoma in rats. *J. Surg. Res.* **2006**, *133* (2), 193–196.
54. Henry, S.D.; Wegen, P.; Metselar, H.J.; Scholte, B.J.; Tilanus, H.W.; Laan, L.J.W. Hydroxyethyl starch–based preservation solutions enhance gene therapy vector delivery under hypothermic conditions. *Liver Transpl.* **2008**, *14* (12), 1708–1717.
55. Zhou, Y.; Yang, B.; Ren, X.; Liu, Z.; Deng, Z.; Chen, L.; Deng, Y.; Zhang, L.M.; Yang, L. Hyperbranched cationic amylopectin derivatives for gene delivery. *Biomaterials* **2012**, *33* (18), 4731–4740.
56. Heritage, P.L.; Loomes, L.M.; Jianxiong, J.; Brook, M.A.; Underdown, B.J.; Mcdermott, M.R. Novel polymer-grafted starch microparticles for mucosal delivery of vaccines. *Immunology* **1996**, *88* (1), 162–168.
57. Mcdermott, M.; Heritage, P.L.; Bartzoka, V.; Brook, M.A. Polymer grafted starch microparticles for oral and nasal immunization. *Immunol. Cell Biol.* **1998**, *76* (3), 256–262.
58. Coucke, D.; Schotsaert, M.; Libert, C.; Pringels, E.; Vervaet, C.; Foreman, P.; Saelens, X.; Remon, J.P. Spray-dried powders of starch and crosslinked poly(acrylic acid) as carriers for nasal delivery of inactivated influenza vaccine. *Vaccine* **2009**, *27* (8), 1279–1286.
59. Gallandat Huet, R.C.; Siemons, A.W.; Baus, D.; Rooyen-Butijn, W.T.; Haagenaars, J.A.; Oeveren, W.; Bepperling, F. A novel hydroxyethyl starch (Voluven) for effective perioperative plasma volume substitution in cardiac surgery. *Can. J. Anaesth.* **2000**, *47* (12), 1207–1215.
60. Behne, M.; Thomas, H.; Bremerich, D.H.; Lischke, V.; Asskali, F.; Förster, H. The pharmacokinetics of acetyl starch as a plasma volume expander in patients undergoing elective surgery. *Anesth. Analg.* **1998**, *86* (4), 856–860.
61. Bhattacharya, A.; Akhter, S.; Shahnawaz, S.; Siddiqui, A.W.; Ahmad, M.Z. Evaluation of Assam Bora rice starch as plasma volume expander by polymer analysis. *Curr. Drug Deliv.* **2010**, *7* (5), 436–441.

30 Biodegradable Polymers
Drug Delivery Applications

Satish Shilpi and Sanjay K. Jain

CONTENTS

30.1	Introduction	584
30.2	Degradation of Polymers	584
30.3	Selection Criteria of Polymers for Drug Delivery Systems	585
30.4	Classification of Biodegradable Polymers	586
	30.4.1 Natural Polymers	588
	30.4.1.1 Collagen	588
	30.4.1.2 Gelatin	588
	30.4.1.3 Albumin	588
	30.4.1.4 Alginate	588
	30.4.1.5 Fibrin	590
	30.4.1.6 Dextran	590
	30.4.1.7 Chitosan	590
	30.4.1.8 Starch	591
	30.4.1.9 Hyaluronic Acid	591
	30.4.1.10 Cyclodextrins	591
	30.4.2 Synthetic Polymers	592
	30.4.2.1 Polylactic Acid	592
	30.4.2.2 Polyglycolic Acid	593
	30.4.2.3 Poly(lactide-co-glycolide)	593
	30.4.2.4 Polyhydroxybutyrate	593
	30.4.2.5 Poly(propylene fumarate)	594
	30.4.2.6 Polycaprolactone	594
	30.4.2.7 Polydioxanone	594
	30.4.2.8 Polyanhydrides	595
	30.4.2.9 Polyamides	595
	30.4.2.10 Polyacetals	595
	30.4.2.11 Polyorthoesters	596
	30.4.2.12 Polyurethanes	596
	30.4.2.13 Polycarbonates	596
	30.4.2.14 Polyphosphazenes	596
	30.4.2.15 Polyphosphoesters	597
30.5	Biodegradable Polymers in Drug Delivery	597
	30.5.1 Implants and Depots	597
	30.5.2 Microspheres and Nanospheres	598
	30.5.3 Dendrimers	601
	30.5.4 Polymeric Hydrogels for Intravesical Drug Delivery	603
	30.5.5 Floating Drug Delivery Systems	606
	30.5.6 Polymersomes	607
	30.5.7 Micelles/Reverse Micelles	608

30.6 Conclusion 608
Acknowledgments..609
References..609

30.1 INTRODUCTION

Biodegradable polymers can be either natural or synthetic. However, synthetic polymers offer greater advantages and possibilities of use because they can be molded into anything you need them to be. The main advantages of synthetic biodegradable polymers are their strength, degradability, adhesiveness, non-immunogenicity, non-inflammatory, and non-toxicity. They can be easily sterilized and have good shelf life. Biodegradable polymers are designed to degrade upon disposal by the action of living organisms. They are the most versatile class of biomaterials, being extensively applied in medicine and biotechnology, as well as in the food and cosmetics industries, surgical devices, implants, artificial organs, prostheses and sutures, drug-delivery systems, tissue engineering, dental applications, cardiovascular applications, bone replacement, enzymes and cells immobilization, biosensors, bioadhesives, ocular devices, and materials for orthopedic applications.[1,2] Biodegradation of polymers takes place through the action of enzymes and/or chemical deterioration associated with living organisms. This event occurs in two steps. The first one is the fragmentation of the polymers into lower molecular mass species by means of either abiotic reactions, i.e., oxidation, photodegradation, or hydrolysis, or biotic reactions, i.e., degradation by microorganisms. This is followed by bioassimilation of the polymer fragments by microorganisms and their mineralization.[1–3]

Natural polymers and some of the synthetic polymers have biodegradable properties and are utilized in designing of drug delivery systems for the purpose of controlled release of drugs or other active ingredients. The purpose of polymers in the system is to immobilize active drug molecules and protect them from the environmental physiological conditions, then deliver drugs to the pathological cell or other specific body organ in order to increase the effectiveness of drugs and thus reducing undesirable side effects resulting in an increase in bioavailability of active molecules in the body.[2,4] Polymers that are used to formulate drug delivery systems should have some specific properties such as biocompatibility, non-toxicity, and have a well-defined structure. Taking this point in mind, many polymers have been invented with the desired physiological properties, which are utilized in the development of various drug delivery systems such as biodegradable drug delivery systems, diffusion controlled, and responsive drug delivery systems. Biodegradability of polymers can easily be modified by incorporating a variety of labile groups such as ester, orthoester, anhydride, carbonate, amide, urea, and urethane in their backbone.[5–12]

30.2 DEGRADATION OF POLYMERS

Biodegradation is a process in which the enzymes produced by microorganisms breakdown or assimilate the organic substances. Such living organisms use these organic substances as a carbon and food/energy source. Biological degradation of polymers occurs either in the presence of oxygen (aerobic biodegradation) or in the absence of oxygen (anaerobic biodegradation) in which carbon dioxide, water, and biomass are the ultimate end products in both the processes while methane, an additional byproduct, is also formed in the anaerobic biodegradation.

Biodegradation normally refers to an attack by microorganisms on water insoluble polymers. This implies that the biodegradation of plastics is usually a heterogeneous process. Due to the lack of water solubility and the size of the polymer molecules, microorganisms are unable to transport the polymeric material directly into the cells where most biochemical processes take place; rather, they must first excrete extracellular enzymes, which de-polymerize the polymers outside the cells. As a consequence, if the molar mass of the polymers can be sufficiently reduced to generate water-soluble intermediates, they can be transported into the microorganisms and fed into the

Biodegradable Polymers

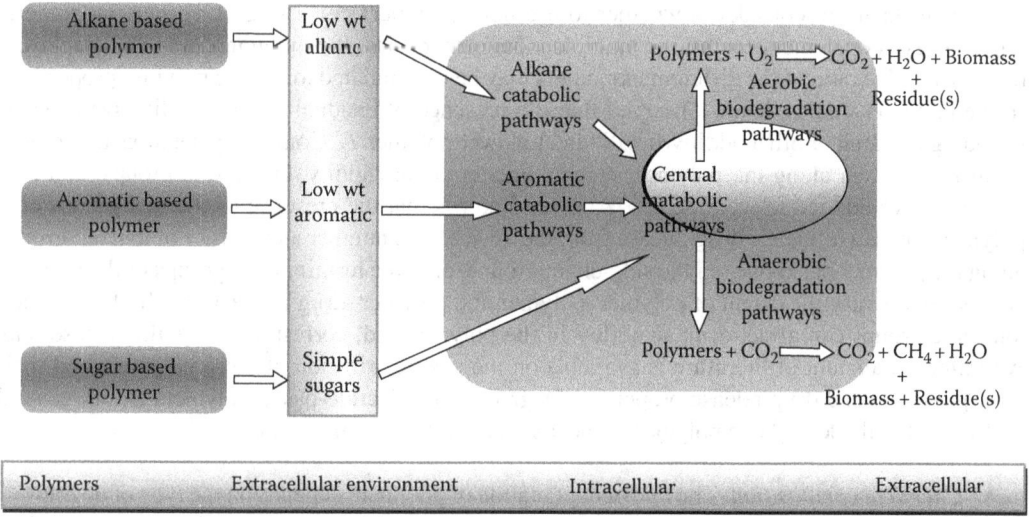

FIGURE 30.1 Polymer biodegradation pathways.

appropriate metabolic pathway. As a result, the end products of these metabolic processes include water, carbon dioxide, and methane (in the case of anaerobic degradation), together with a new biomass (Figure 30.1). The extracellular enzymes are too large and unable to penetrate deeply into the polymer material, and consequently act on the polymer surface; thus, the biodegradation of plastics is usually a surface erosion process.[1,4] In many cases, the above process may occur but in many cases the primary process of biodegradation may be non-biotic, i.e., chemical and physical processes also act on the polymer, either in parallel or as a first stage solely on the polymer. These non-biotic effects include chemical hydrolysis, thermal polymer degradation, and oxidation or scission of the polymer chains by irradiation (photodegradation). For some materials, these effects are used directly to induce the biodegradation process, for example, poly(lactic acid), but they must also be taken into account when biodegradation is caused predominantly by extracellular enzymes. Because of the coexistence of biotic and non-biotic processes, the entire mechanism of polymer degradation could also be referred to as environmental degradation. Environmental factors not only influence the polymer to be degraded, they also have a crucial influence on the microbial population and on the activity of the different microorganisms themselves. Parameters such as humidity, temperature, pH, salinity, the presence or absence of oxygen, and the supply of different nutrients have important effects on the microbial degradation of polymers, and so these conditions must be considered when the biodegradability of plastics is tested.[1,4,13,14]

30.3 SELECTION CRITERIA OF POLYMERS FOR DRUG DELIVERY SYSTEMS

For the biomedical applications and development of drug delivery systems, polymers that have already received regulatory approval are mostly used, apart from polymer physicochemical properties such as bulk hydrophilicity, morphology, and structure, solute solubility in polymer and degradation rate of both the polymer and the device are to be kept in mind for the selection of proper polymers.

The glass transition temperature (T_g) and melting temperature (T_m) can affect the mass transport rates through the polymer as well as the polymer processing characteristics and the stability of the dosage form. Below the glass transition temperature, the polymer will exist in an amorphous, glassy state. When exposed to temperatures above T_g, the polymer will experience an increase in free volume that permits greater local segmental chain mobility along the polymer backbone. Consequently, mass transport through the polymer is faster at temperatures above T_g.[4,5,15]

Another factor to consider is whether to use homopolymers consisting of a single monomeric repeat unit or copolymers containing multiple monomer species. If copolymers are to be employed, then the relative ratio of the different monomers may be manipulated to change polymer properties.[5] In the previous studies, it was observed that the presence of residual solvent or dissolved solutes including the drug or other additives will tend to lower polymer T_g. Conversely, features that hinder segmental motion along the polymer, such as greater chain rigidity, bulky side groups, and ring structures, would tend to increase T_g.[15] Another factor is that the presence of charged groups on a polymer can also influence drug release from the device. The number and density of ionized groups along the polymer backbone, on the side-chain groups, or at the terminal end groups of the polymer chains can all vary the extent of polymer–polymer and polymer–drug interactions. In this manner, ionizable groups can affect drug solubility in the polymer and, correspondingly, the release rate. An immense amount of literature is available on the characterization of these polymers and their biodegradation and drug release properties. Degradation of lactide-based polymers and in general all hydrolytically degradable polymers depends on the following properties:

- *Chemical composition*: The rate of degradation of polymers depends on the type of degradable bonds present on the polymer. In general, the rate of degradation of different chemical bonds follows as anhydride > esters > amides.
- *Crystallinity*: Polymers having higher crystallinity will have slower rate of degradation.
- *Hydrophilicity*: Polymers with a lot of hydrophobic groups will likely degrade at a slower rate than a polymer which is hydrophilic in nature.[6,15]

30.4 CLASSIFICATION OF BIODEGRADABLE POLYMERS

Polymers are categorized on the basis of their structural design applicable to drug delivery applications, for example, linear polymers, block copolymers, branched polymers, and cross-linked polymers (Figure 30.2). Linear polymers are generally hydrophilic in nature and known as water-soluble polymers and more applicable to make conjugates with drugs and other active micromolecules.[11] Block copolymers are the class of linear polymers but they have different architectures, which are used in building supramolecular structures such as polymeric micelles. Branched copolymers are characterized by the presence of branch points and more than two end groups, which comprise a class of polymers residing between linear polymers and polymer networks.

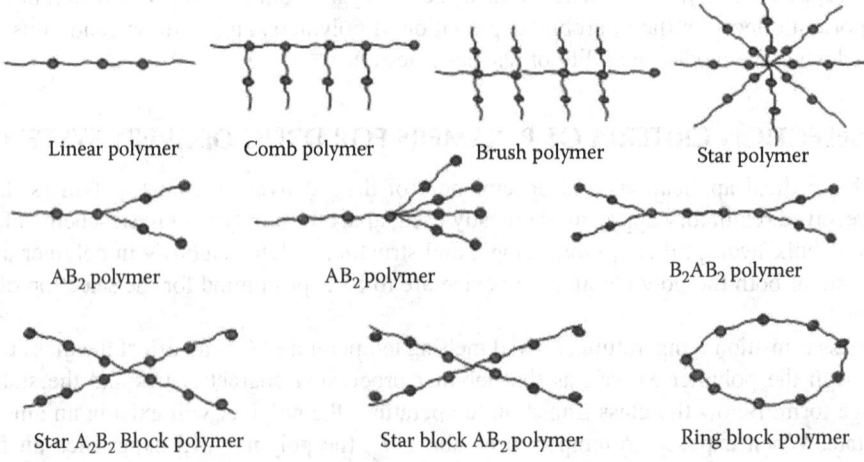

FIGURE 30.2 Polymer type based on structural configuration.

Generally, branched polymers have significantly different physical properties from linear polymers and polymer networks, such as solubility, melting point, and mechanical behavior.[8-11] Branched polymers only refer to those polymers prepared via controlled polymerization techniques and may produce undesirable branching. Hyperbranched, graft, star-shaped polymers, comb polymers, brush polymers, and dendrimers are some examples of branched polymers. Because the majority of cross-linked polymers, including interpenetrating polymer networks or semi-interpenetrating polymer networks, involve chemical cross-linking techniques to form drug delivery systems beyond nanoscale in size.[11,12]

Polymers are mainly classified into two categories, natural and synthetic polymers (Table 30.1). Natural (water soluble) polymers are mostly obtained from natural sources. Naturally derived polymers with special focus on polysaccharides and proteins have become attractive in the biological applications of controlled release systems. Polysaccharides are a class of biopolymers constituted by either one or two alternating monosaccharides, which differ in their monosaccharide units in the length of a chain, in the types of the linking units, and in the degree of branching.[16]

Drug delivery systems based on biodegradable aliphatic polyesters have advanced remarkably over the past few decades. Commonly used polymers such as poly(ε-caprolactone) (PCL), poly (lactide acid) (PLA), and poly (lactic-co-glycolic acid) (PLGA) are FDA approved and well known for their biodegradability, biocompatibility, and non-toxic properties, which make them suitable as matrices for controlled release drug delivery systems.

To enhance the desirable properties of a polymer as a matrix for a controlled drug delivery system, efforts have been made to improve its hydrophilicity, biodegradation rate, and drug stability. Hydrophilic blocks play vital role in polymeric drug delivery systems, for example, poly(ethylene oxide)

TABLE 30.1
Classification of Biodegradable Polymers

Polymer	Examples	Decomposition Product
Naturally occurring polymers		
Polysaccharides	Hyaluronic acid	Oligosaccharides
	Carrageenan, agarose, and cyclodextrin	—
	Chitosan	Glucosamine and N-acetyl-glucosamine
	Dextran and starch	Glucose
	Alginate	d-Mannuronic and l-Gulonic acids
Polypeptide and protein	Collagen, gelatin, and albumin	α-Amino acids
Synthetic polymers		
Polyamide	Polyamino acids and polypeptides	α-Amino acids
Polyesters	Poly(glycolide), poly(d,l-lactide), and poly(d,l-lactide-co-glycolide)	Lactic acid and glycolic acid
	Poly(ε-caprolactone), poly(dioxanone), and poly(hydroxybutyrate)	5 Hydroxyhexanoic acid
Polyanhydride	Poly(sebacic acid)	Sebacic acid
	Poly(adipic acid)	Adipic acid
	Poly(terephthalic acid)	Phthalic acid
Polyorthoester	2,2-Diethoxytetrahydrofuran-co-1,6-hexanediol	Alcohol
Polyphosphazenes	Polyphosphazene	Phosphoric acid and ammonia
Polyphosphoester	Polyphosphate, polyphosphonate, and polyphosphite	
Polyurethanes		
Polycarbonate	Poly(1,3-dioxan-2-one)	1,3-Propanediol and carbonic acid
Polydihydropyrans		
Polyacetals		

(PEO) and poly(ethylene glycol) (PEG) are the most commonly used hydrophilic blocks. This FDA-approved PEO has low toxicity and biocompatibility.[10] One of the primary advantages of attachment of the PEO moiety is its effectiveness against protein adsorption to hydrophobic surfaces and this PEO block's length extends circulation time and reduces uptake by phagocytes.[17] The character in which the PEO block system achieves longer circulation in the blood compartment is called "stealth character."[18] PEG grafted to surfaces of nanospheres proved to reduce thrombogenicity and to increase their dispersion stability in aqueous medium, due to steric repulsion effects of tethered PEG strands.[19]

30.4.1 Natural Polymers

30.4.1.1 Collagen

Collagen is the most abundant polypeptide in the human body and is a major component of ligament, cartilage, skin, bone, and blood vessels that is fabricated from glycine–proline–(hydroxy)-proline repeat units to form a triple helix molecular structure.[12] There are 28 types of collagen molecules, which have been isolated and characterized. However, types I, II, III, and IV are the most heavily investigated. Among various types of collagen, type-I has been widely used in both medical and pharmaceutical applications.[11,12,15] Because of their pharmaceutical applications due to the fulfillment of many requirements of a drug delivery system such as good biocompatibility, low antigenicity, and degradability by enzymes collagenase and metalloproteinase, they are good for the development of drug delivery systems apart from their use in the fabrication of drug delivery carriers, they are also used as a promising matrix for gene therapy and tissue engineering. They are also used in combination with PLA, PLGA, PCL, and chitosan for the preparation of nano- or microparticles for drug delivery.[12,20,21]

30.4.1.2 Gelatin

Gelatin is a common natural water-soluble polymer and it is created by denaturing collagen. It has been used in pharmaceutical and medical applications due to its admirable properties such as biodegradability, biocompatibility, and low antigenicity. In addition, gelatin can be easily modified due to its isoelectric point that allows it to change from negative to positive charge in an appropriate physiological environment. Molecular structure of gelatin is shown in Figure 30.3a. This property of gelatin is utilized as a potential tool to generate new concepts in the drug delivery and other pharmaceutical and biomedical applications.[15] Gelatin is one of the natural polymers used as support material for tissue culture, gene delivery, and in tissue engineering. Gelatin-based systems such as films, nanoparticles, and microspheres have the ability to control release of bioactive agents such as drugs, protein, and therapeutic peptides.[15,13]

30.4.1.3 Albumin

Albumin is an abundant water-soluble blood protein, comprising about 50% of total plasma mass in the body. Albumin carries hydrophobic fatty acids in the blood stream as well as help in maintaining blood pH. It binds water, cations (such as Ca^{2+}, Na^+, and K^+), fatty acids, hormones, bilirubin, thyroxin, and drugs. The main function of albumin is to regulate the colloidal osmotic pressure of blood. As albumin is essentially ever-present in the body, nearly all tissues have enzymes that can degrade it and make it a promising polymer for biomedical applications. Due to its high drug-binding properties it may lead to sustained release of drugs, and this property of albumin is utilized in drug delivery application. It can be used as drug delivery vehicles, i.e., gels, microspheres, nanoparticles, etc.

30.4.1.4 Alginate

Alginate is also a naturally occurring linear polysaccharide extracted from seaweed, algae, and bacteria. The fundamental chemical structure of alginate is composed of (1–4)-b-ᴅ-mannuronic acid

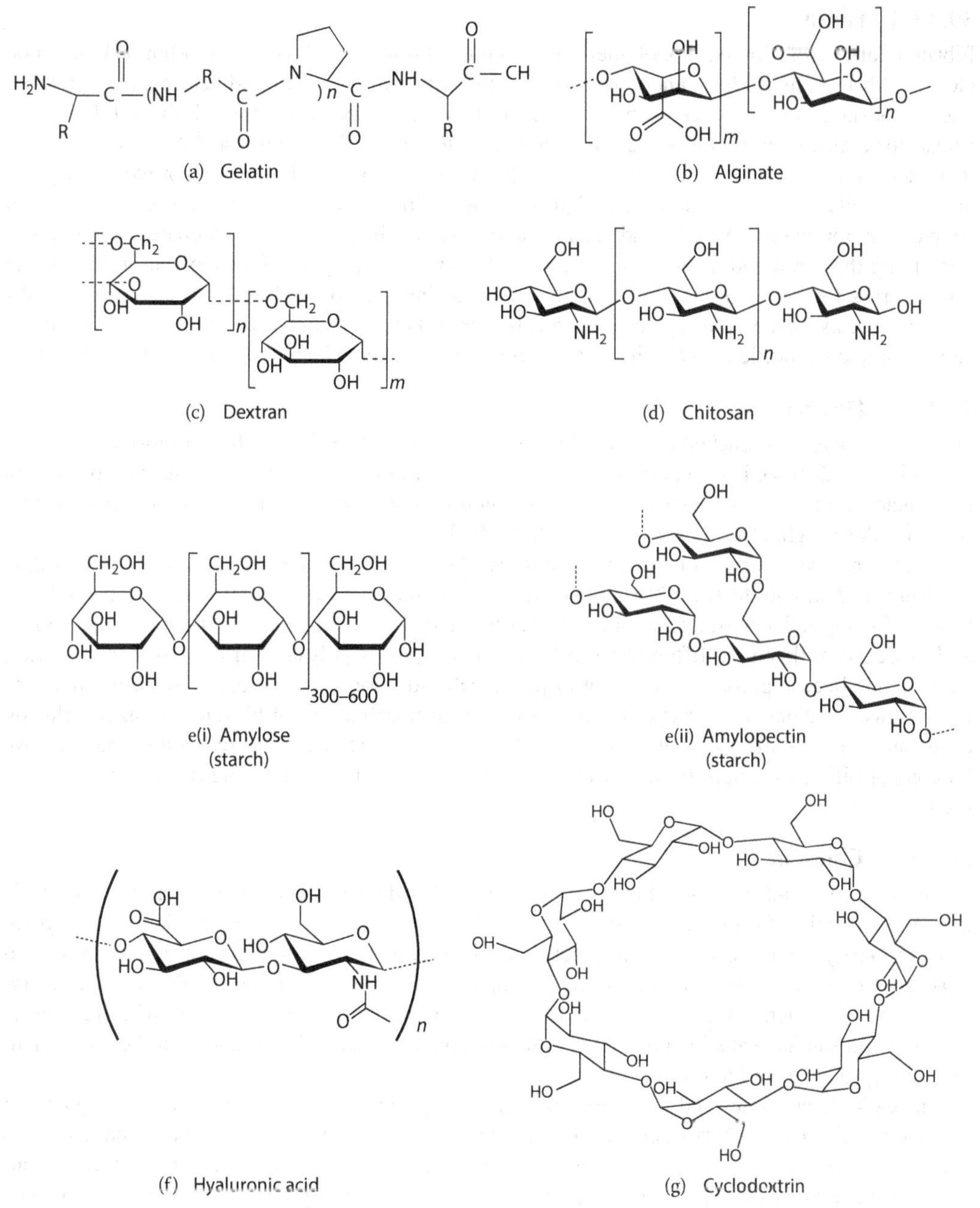

(a) Gelatin
(b) Alginate
(c) Dextran
(d) Chitosan
e(i) Amylose (starch)
e(ii) Amylopectin (starch)
(f) Hyaluronic acid
(g) Cyclodextrin

FIGURE 30.3 Molecular structure of natural polymers.

(M) and (1–4)-a-L-guluronic acid (G) units in the form of homopolymeric (MM or GG blocks) and heteropolymeric sequences (MG or GM blocks), which is shown in Figure 30.3b.[20]

Alginate and its derivatives are widely used by many pharmaceutical scientists for the drug delivery and tissue engineering applications due to their many unique properties such as biocompatibility, biodegradability, low toxicity, non-immunogenicity, water solubility, relatively low cost, gelling ability, stabilizing properties, and high viscosity in aqueous solutions. Alginate-based systems have also been successfully used as a matrix for the encapsulation of stem cells and for controlled release of proteins, genes, and drugs.[5–12,15,20]

30.4.1.5 Fibrin

Fibrin, a large cross-linked biopolymer composed of fibronectin, plays an important role in blood clotting, fibrinolysis, cellular and matrix interactions, inflammation, wound healing, angiogenesis, and neoplasia. In the presence of the enzyme thrombin, cleavage of an internal fibrin linker yields linear fibrils that laterally associate into nanofibers (10–200 nm) that form a clot. This clot can be degraded by a complex cascade of enzymes. It has been shown to be biocompatible, biodegradable, injectable, and able to enhance cell proliferation. Fibrin glues have been studied as a surgical supplement for tissue sealants and haemostatic agents. Fibrin glues are prepared as solutions containing thrombin and fibronectin separately that are mixed right before application. Thrombin rapidly cross-links the fibronectin into a fibrin clot closing the wound. Fibrin has also been investigated for use as a drug delivery device and cell carrier. Due to its potential of cross-linking, fibrin can be uniquely modified, so that its material properties can be adapted for desired applications.[5]

30.4.1.6 Dextran

Dextran is a natural, branched polymer of glucose linked by a 1–6-linked glucoyranoside, and some branching of 1,3-linked side chains (Figure 30.3c). Dextran is synthesized from sucrose by certain lactic acid bacteria, the best known being *Leuconostoc mesenteroides* and *Streptococcus mutans*. Its molecular weight ranges from 3000 to 2,000,000 Da.

There are two commercial preparations available, namely dextran 40 kilodaltons (kDa) (Rheomacrodex) and dextran 70 kDa (Macrodex). In pharmaceutics, dextran has been used as a model of drug delivery due to its unique characteristics, i.e., water solubility, biocompatibility, and biodegradability that differentiate it from other types of polysaccharides. In recent studies, dextran has been regarded as a potential polysaccharide polymer that can sustain the delivery of proteins, vaccines, and drugs. It can be used in immobilization of biosensors, size-exclusion chromatography matrices, coating to protect metal nanoparticles from oxidation and improve biocompatibility, as a nutrient and volume expanders and maintaining the osmotic pressure in the body.[15,20,22]

30.4.1.7 Chitosan

Chitosan (2-amino-2-deoxy-D-glucose) is a polysaccharide and polycationic polymer, which is derived from chitin (β-linked *N*-acetyl-D-glucosamine) after its partial *N*-deacetylation and hydrolysis, primarily from crustacean and insect shells. It consists of repeating units of glucosamine and *N*-acetyl-glucosamine (Figure 30.3d), the proportions of which determine the degree of deacetylation of the polymer. Chitosan has primary and secondary hydroxyl groups along with free amino groups. Chitosan has valuable properties as biomaterials because it is considered to be biocompatible, biodegradable, and non-toxic.[3,15]

The cationic character and the potential functional group make it an attractive biopolymer for many biomedical and pharmaceutical applications. As a pharmaceutical excipient, chitosan has been used in many formulations like powders, tablets, emulsions, and gels. Furthermore, a controlled release of incorporated drugs can be assured by making nano- or microsphere, chitosan film, hydrogel, and transdermal patches. Chitosan also shows mucoadhesive, antimicrobial, anticancer, antifungal, hemostatic, antioxidant, anti-inflammatory, and hypocholesteromic properties.[23] During the encapsulation process using synthetic polymers, the protein is generally exposed to the conditions which might cause its denaturation or deactivation. Therefore, a biocompatible alternative such as chitosan is desirable for such applications. It is a promising bioadhesive material at physiological pH. This polymer possesses OH and NH_2 groups that can give rise to hydrogen bonding. These properties are considered essential for mucoadhesion.[24] In one study, chitosan with different ratios of anionic polymers is used to prepare mucoadhesive tablets for vaginal delivery of metronidazole.[24,25]

30.4.1.8 Starch

Starch is a heterogeneous biodegradable polymer consisting of tens to hundreds to several thousand monosaccharide units (α-D-glucose units). All of the common polysaccharides contain glucose as the monosaccharide unit. The anhydrous glucose units are mainly linked by α-(1,4) bonds and to some extent by α-(1,6) linkages. The biopolymer consists of two distinguished structural forms: amylose and amylopectin (Figure 30.3e(i) and e(ii)).[26,27]

Amylose is mainly found as a long linear polymer containing about several hundred α-(1,4)-linked glucose units (up to 6000 AGUs), with a molecular weight of 105–106 g mol^{-1}. In the solid state, the chains very easily form single or double helices. In contrast, amylopectin is a highly branched molecule with a molecular weight of 107–109 g mol^{-1}. The branched polymer contains α-(1,4)-linked glucose units but has additional α-(1,6)-glucosidic branching points, which are believed to occur every 10–60 glucose units, i.e., 5% of the glucose moieties are branched.[26,28–30] Modified starches such as dextrin, acid-treated starch, alkaline-treated starch, oxidized starch, acetylated distarch phosphate, starch acetate, hydroxypropyl starch, hydroxypropyl distarch phosphate, etc., were tested for general applicability in the preparation of pharmaceutical dosage form, for example, directly compressible controlled release matrix systems and other drug delivery systems. The advantages of the material include ease of tablet preparation, the potential of a constant release rate (zero order) for an extended period of time, and its ability to incorporate high percentages of drugs with different physicochemical properties.[26,30,31]

30.4.1.9 Hyaluronic Acid

Hyaluronic acid is a naturally occurring biodegradable and biocompatible linear polysaccharide with a wide molecular weight range of 1000–10,000,000 Da, which serves important biological functions in bacteria and higher animals including humans. Naturally occurring hyaluronic acid may be found particularly as an intercellular space filler, greatest concentrations found in the vitreous humor of the eye, epithelial, and neural tissues and in the synovial fluid of articular joints. It comprises polyanionic disaccharide units consisting of glucuronic acid and N-acetyl glucosamine joined alternately by β-1-3 and β-1-4 glycosidic bonds (Figure 30.3f). It has antiaging, wound healing, anti-inflammatory, and tumor metastatic properties.[5,15] Due to these properties, it is now widely used in the target drug delivery to tumor cell using novel nanocarriers. The viscoelastic property of hyaluronic acid solution is important for its use as a biomaterial and is controlled by the concentration and molecular weight of the hyaluronic acid chains. Hyaluronan is gaining popularity as a biomaterial scaffold in tissue engineering.[15,32,33]

30.4.1.10 Cyclodextrins

They are cyclic oligosaccharides consisting of six to eight glucose units joined through α-1,4 glucosidic bonds (Figure 30.3g). Cyclodextrins are produced from starch by means of enzymatic conversion. They are used in food, pharmaceutical, drug delivery, and chemical industries, as well as agriculture and environmental engineering. The most common pharmaceutical application of cyclodextrin is to enhance the solubility, stability, safety, and bioavailability of drug molecules. Cyclodextrins remain intact during their passage throughout the stomach and small intestine of the gastrointestinal tract. However, in colon, they undergo fermentation in the presence of vast colonic microfloras into small monosaccharides and thus absorbed from these regions.[34,35] β-Cyclodextrins are degraded to a very small extent in the small intestine but are completely digested in the large intestine. Most bacterial strains found abundantly in human beings are capable of degrading cyclodextrin polysaccharides. Cyclodextrins are of wide interest for the formulation of a broad range of delivery devices from the most classical dosage forms to the newest drug delivery carriers and the ability to architecture makes it possible to use cyclodextrins not only as drug solubilizers but also a true delivery system.[34,36,37]

30.4.2 Synthetic Polymers

30.4.2.1 Polylactic Acid

Polylactide (PLA) possesses chiral molecules, polylactides exist in four forms: poly(L-lactic acid) (PLLA), poly(D-lactic acid) (PDLA), poly(D,L-lactic acid) (PDLLA)—a racemic mixture of PLLA and PDLA, and mesopoly(lactic acid).[5,15] Figure 30.4a illustrates the molecular structure of PLA.

PLA is a thermoplastic biodegradable polymer produced synthetically by polymerization of lactic acid monomers or cyclic lactide dimers. This material has lower tensile strength, higher elongation, and a much more rapid degradation time, making it more attractive as a drug delivery system. PLA is degraded by hydrolysis, which is also called bulk degradation (the breaking of a chemical bond by adding water to it) of the backbone esters of the polymer. Poly(L-lactide) is about 37% crystalline, with a melting point of 175°C–178°C and a glass transition temperature of 60°C–65°C.

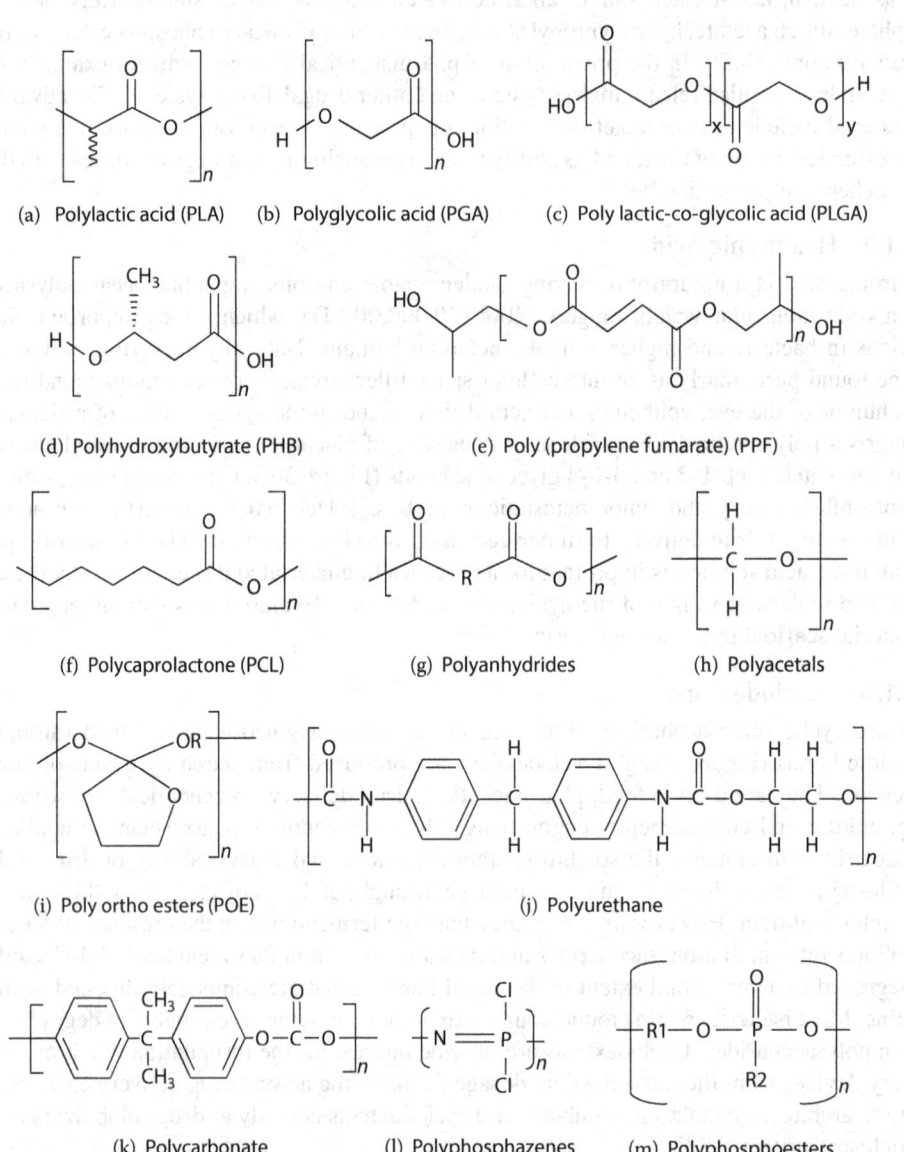

FIGURE 30.4 Molecular structure of synthetic biodegradable polymers.

Biodegradable Polymers

The degradation time of PLLA is much slower than that of PDLLA, requiring more than 2 years to be completely absorbed. Copolymers of L-lactide and DL-lactide have been prepared to disrupt the crystallinity of L-lactide and accelerate the degradation process. The final crystallinity and mechanical properties of the polymer depend on the stereochemistry of the polymer backbone. PLA has a number of biomedical applications, such as sutures, stents, dialysis media, and drug delivery devices.[15,38–40]

30.4.2.2 Polyglycolic Acid

Polyglycolide is the simplest linear aliphatic polyester. Glycolide monomer is synthesized by the dimerization of glycolic acid. Ring-opening polymerization yields high molecular weight materials. Figure 30.4b illustrates the molecular structure of polyglycolic acid (PGA). PGA is hard, tough, and highly crystalline (45%–55%), with a high melting point of 220°C–225°C and a glass transition temperature of 35°C–40°C.

Because of its high degree of crystallization, unlike closely related polyesters such as PLA, PGA is insoluble in most common polymer solvents except highly fluorinated organics such as hexafluoroisopropanol. The low solubility and high melting point of PGA limit its use for drug delivery applications, since it cannot be made into films, rods, capsules, or microspheres using solvent or melt techniques. Most research has been focused on short-term tissue engineering scaffolds and the utilization of PGA as a filler material coupled with other degradable polymer networks. PGA is often fabricated into a mesh network and has been used as a scaffold for bone, cartilage, tendon, tooth, vaginal, intestinal, lymphatic, and spinal regeneration.[5,8,15]

30.4.2.3 Poly(lactide-co-glycolide)

In the past two decades, poly lactic-*co*-glycolic acid (PLGA) has been among the most attractive polymeric candidates used to fabricate devices for drug delivery and tissue engineering applications. It is a copolymer of both L- and DL-lactides (Figure 30.4c), which has been developed for both device and drug delivery applications.[15] The ratio of glycolide to lactide at different compositions allows control of the degree of crystallinity of the polymers.[3,20,16] When the crystalline PGA is copolymerized with PLA, the degree of crystallinity is reduced and as a result this leads to increase in rates of hydration and hydrolysis. It can therefore be concluded that the degradation time of the copolymer is related to the ratio of monomers used in synthesis. In general, the higher the content of glycolide, the quicker the rate of degradation. A copolymer of 50:50 ratio of glycolide and DL-lactide degrades faster than either homopolymer. Copolymers of L-lactide with 25%–70% glycolide are amorphous due to the disruption of the regularity of the polymer chain by the other monomer. A copolymer of 90:10 glycolide–lactide was developed by Ethicon as an absorbable suture material under the trade name Vicryl®. It absorbs within 3–4 months but has a slightly longer strength-retention time.[3,5]

The possibility to tailor the polymer degradation time by altering the ratio of the monomers used during synthesis has made PLGA a common choice in the production of a variety of biomedical devices such as grafts, sutures, implants, prosthetic devices, micro-, and nanoparticle devices for controlled delivery of small molecule drugs, proteins, vaccines, genes, antigens, and other macromolecules in commercial use and in research.

30.4.2.4 Polyhydroxybutyrate

Polyhydroxybutyrate (PHB) is a biopolymer, which is present in all living organisms. Many bacteria produce PHB in large quantities as storage material. The PHB homopolymer is crystalline and brittle, with a melting point of 175°C, whereas the copolymers of PHB with polyhydroxyvalerate are less crystalline. It is non-toxic and is totally biodegradable. The polymer is primarily a product of carbon assimilation (from glucose or starch) and is employed by microorganisms as a form of energy storage molecule to be metabolized when other common energy sources are not available.[5,16,41]

PHB and its copolymers have attracted much attention because they are produced biosynthetically from renewable resources. Figure 30.4d shows the molecular structure of PHB. Microcapsules from PHB have been prepared by various techniques and investigated for the release of drugs in a sustained manner. PHB has also been suggested as a suitable matrix for drug delivery in veterinary medicine, for instance, in the rumen of cattle.[42] These polymers typically require the presence of enzymes for biodegradation but can degrade in a range of environments and are under consideration for several biomedical applications.[14,43]

30.4.2.5 Poly(propylene fumarate)

Poly(propylene fumarate) (PPF) is an unsaturated linear biocompatible, biodegradable, osteoconductive polyester with fumarate double bonds that can be cross-linked in situ. PPF scaffolds can be used to fill irregularly shaped defects with minimal surgical intervention. It is a high-strength polymeric biomaterial that possesses the unique ability to be cross-linked through the unsaturated bonds in its backbone (Figure 30.4e).[5,15] As PPF can be cross-linked, polymer degradation is dependent on molecular weight, cross-linker, and cross-linking density. PPF is a liquid, which becomes solid during cross-linking; therefore, it is found to be favorable in biomedical applications such as filling bone defects and the depot, long-term delivery of ocular drugs. For osteogenic tissue engineering, PPF is often mixed with ceramics such as hydroxyapatite or alumoxane to create stronger, more bioactive scaffolds.

Recent research has focused on the use of PPF to fill irregular-shaped bone defects such as ear ossicle or mandibular defects. In both circumstances, PPF-based scaffolds allow the design of structures that may not be attainable from non-cross-linkable degradable polymers.[8,42]

30.4.2.6 Polycaprolactone

Polycaprolactone (PCL) is obtained by ring-opening polymerization of the six-membered lactone, ε-caprolactone (Figure 30.4f), which yields a semicrystalline polymer with a melting point of 59°C–64°C and a glass transition temperature of 60°C with great organic solvent solubility. Anionic, cationic, coordination, or radical polymerization routes are all applicable for synthesis.[14,42,43]

PCL exhibits high permeability to low molecular species at body temperature. These properties, combined with recognized biocompatibility, make PCL a promising candidate for biodegradable controlled drug delivery systems. PCL degradation proceeds through hydrolysis of backbone ester bonds as well as by enzymatic attack. Hence, PCL degrades under a range of conditions, biotically in soil, lake waters, in vivo, and in phosphate buffer solutions. Hydrolysis of PCL yields 6-hydroxycaproic acid, an intermediate of the oxidation, which enters the citric acid cycle and is completely metabolized. Hydrolysis, however, proceeds by homogeneous erosion at a much slower rate than PLA and PLGA. Hydrolysis of PCL is faster at basic pH and higher temperatures. It is most suitable for long-term implant delivery device or other drug delivery.[3,15,42] Capronor®, a contraceptive, represents such a system able to deliver levonorgestrel in vivo for over a year. It is used in the development of micronized and nano-sized drug delivery vehicles, but its degradation time is more, i.e., 2–3 years, which is a major issue for PCL products to be FDA approved for this use. It is used as a blend or copolymerized with other polymers such as PLLA, PDLLA, PLGA, and polyethers, which is an alternative way to accelerate overall polymer erosion. PCL has low tensile strength (23 MPa) but very high elongation at breakage (4700%), making it a very good elastic biomaterial. PCL and its composites have been used as tissue engineering scaffolds for the regeneration of bone, ligament, cartilage, skin, nerve, and vascular tissues. A recent advancement using PCL hybrid scaffolds has been used in interfacial tissue engineering.[14,44]

30.4.2.7 Polydioxanone

Polydioxanone (PDS) is obtained by a ring-opening polymerization of the p-dioxanone monomer. It is characterized by a glass transition temperature in the range of −10°C to 0°C and a degree of crystallinity of about 55%. Materials prepared with PDS show enhanced flexibility due to the presence of an ether oxygen within the backbone of the polymer chain. When used in vivo, it degrades into monomers with low toxicity and also has a lower modulus than PLA or PGA. PDS has demonstrated no acute or toxic effects on implantation.[3–5,42]

30.4.2.8 Polyanhydrides

Polyanhydrides have been synthesized via the dehydration of diacid molecules by melt polycondensation. The glass transition point (T_g) of this polymer is several degrees below body temperature, it becomes a soft, sticky material when placed in the body. Polyanhydrides are surface-eroding polymers, which contain two carbonyl groups bound together by an ether bond (Figure 30.4g), and have been almost exclusively studied for biomedical applications. There are three main classes of polyanhydrides: aliphatic, unsaturated, and aromatic. These classes are determined by examining their R groups (the chemistry of the molecule between the anhydride bonds).

The degradation of their anhydride bond is highly dependent on polymer backbone chemistry. In fact, the degradation rate can vary by over six orders of magnitude based on monomer chemistry.[3,5,8,15,42] Degradation times can be adjusted from days to years according to the degree of hydrophobicity of the monomer selected. They degrade by surface erosion rather than by bulk hydrolytic degradation. This allows for precision tuning of payload release rate by which polyanhydrides have found excellent in vivo compatibility and are significantly favorable in drug delivery applications. Polyanhydrides have been used for the delivery of chemotherapeutics, antibiotics, vaccines, and proteins. Polyanhydrides are often fabricated into microparticles or nanoparticles to allow for injectable, oral, or aerosol delivery.[1,4,8]

30.4.2.9 Polyamides

The synthetic aliphatic polyamides are polymeric compounds frequently referred to as Nylons, which form an important group of polycondensation polymers. They are linear molecules that are semicrystalline and thermoplastic in nature. A typical polyamide chain consists of amide groups separated by alkane segments and the number of carbon atoms separating the nitrogen atoms, which defines the particular polyamide type.[3,23,42] It has excellent physicochemical and physicomechanical properties such as abrasion resistance, biodegradability, ease of processing, higher melting points, and heat resistance than many other semicrystalline polymers such as polyethylene. Their biocompatibility and non-toxicity make them attractive for use in the design and development of drug delivery systems.[4,7,8]

30.4.2.10 Polyacetals

Polyacetals also known as polyoxymethylene (POM) are degradable polymers in which two ether bonds are connected to the same carbon molecule (geminal, Figure 30.4h). It is a semicrystalline polymer (75%–85% crystalline) with a melting point of 175°C. The molecular closeness of the normally stable ether bonds conveys hydrolytic instability close to that seen for polyanhydrides and gives polyacetals surface-eroding properties. Polyacetals are normally subdivided into two subgroups: polyacetals and polyketals.[1,5,42]

Both polyacetals and polyketals have gained attraction in biomedical research because their degradation products possess no carboxylic acids yielding significantly milder pH microenvironments, and their degradation is acid catalyzed. Milder pH microenvironments allow for the delivery of acid and hydrolytically sensitive biomolecules. Acid-catalyzed degradation allows for intracellular active biomolecules delivery, because particle-based delivery vehicles are stable under normal physiological pH 7.4, but rapidly degrade when they reach lysosomal pH.[4] Polyacetals are used in making drug delivery systems, which are further used in drug delivery, gene delivery, and tissue engineering. To produce tissue engineering scaffolds from polyacetals, cyclic polyacetal monomers with two ester acrylate end groups have been synthesized that can then be cross-linked.[3,5,11,23]

30.4.2.11 Polyorthoesters

Polyorthoesters (POEs) are hydrophobic, surface-eroding polymers that have three geminal ether bonds that are acid-sensitive but stable to base. Like polyacetals, control of POE backbone chemistry allows for the synthesis of polymers with varied acid-catalyzed degradation rates and material properties.[1,3] Figure 30.4i demonstrates the molecular structure of POE.

These polymers degrade by surface erosion, and degradation rates can be controlled by incorporation of acidic or basic excipients. These polymers were specifically developed for drug delivery. Although four classes of POEs have been developed (POE I–IV), POE IV incorporates short segments of lactic or glycolic acid into the polymer backbone to expedite degradation because POE I–III possess too slow erosion rates to be clinically relevant as drug delivery vehicles. POE IV polymers have been used for the delivery of drugs, proteins and peptides, DNA, vaccines, antiproliferative drugs, and in tissue engineering.[4-6,11]

30.4.2.12 Polyurethanes

Polyurethanes are composed of a chain of organic units joined by carbamate (urethane) links (Figure 30.4j). Polyurethane polymers are formed by combining two bi- or higher functional monomers. One contains two or more isocyanate functional groups and the other contains two or more hydroxyl groups. It has been used extensively in prostheses such as cardiac assist devices, small vascular shunts, and tracheal tubes. The self-setting system is an injectable liquid that not only polymerizes at physiological temperatures creating a biomaterial that has been shown to be mechanically similar to bone cements but also promotes favorable cell adhesion and proliferation. Under most conditions, pure polyurethanes are degradation resistant making them poor candidates for drug delivery and many tissue engineering applications.[1]

30.4.2.13 Polycarbonates

Polycarbonates are linear polymers that have two geminal ether bonds and a carbonyl bond. Although this bond is extremely hydrolytically stable, research has shown in vivo degradation to be much more rapid presumably due to enzymatic degradation, which causes these polymers to be surface eroding. Polycarbonate has a glass transition temperature of about 147°C. They are also known by their market names such as Lexan®, Makrolon®, and Makroclear®.[1,4,5,11] Figure 30.4k shows the molecular structure of polycarbonates.

A modification in polycarbonate yields poly(trimethylene carbonate), which has a glass transition temperature of 17°C. It is an elastomeric aliphatic polymer with great flexibility and has slow degradation profile. Its metabolite into biocompatible substances, i.e., 1,3-propanediol and carbonic acid, makes it an ideal polymer for fabricating microparticles, discs, and gels for the systemic delivery of angiogenic agents, antibiotics and other active molecules.[1,5] Poly(trimethylene carbonate) can be copolymerized with PLA, PCL, polyether, and poly(L-glutamic acid) to allow for the fabrication of sutures, depot, micelles, and polymersomes with superior mechanical and degradation properties for delivery of active biomolecules. Polycarbonates have been widely used for tissue engineering applications. Attaching a cyclohexane or propylene group instead of trimethylene in the monomer backbone produces polymer with strong mechanical properties while maintaining the biocompatibility of their degradation products.[5,15,20,42]

30.4.2.14 Polyphosphazenes

Polyphosphazenes are biodegradable polymers, their backbone is completely inorganic consisting of phosphorous and nitrogen bonded linearly through alternating single and double bonds (Figure 30.4l). With the help of two phosphorous side groups, more than 500 derivatives of polyphosphazene have been synthesized via esterification, etherification, or amidification. Certain side groups, such as amino acid esters, glucosyl, glyceryl, glycolate, lactate, and imidazole, have been found to sensitize hydrolysis of the backbone to allow for the design of clinically relevant biomaterials. By changing disubstituted polyphosphazene side groups for one particular system we can change their physicochemical properties.[1,5,15,20]

Another unique feature is that polyphosphazenes degrade into neutral products that have been found to have a pH buffering effect when combined with polymers, such as polyesters, that have highly acidic degradation products. Polyphosphazenes have allowed for them to be fabricated into particles, micelles, microneedle coatings, and gels. They have been used in the delivery of different drugs, chemotherapeutics, growth factors, DNA proteins, and vaccines. Prior research has found that certain polyphosphazenes are showed strong immunoactivating properties and hold great potential as adjuvant as well as non-specific immune boosting substances. Although many rapidly degrading polyphosphazenes have shown promise in drug delivery applications, more hydrophobic side group substitutions have allowed for the use of polyphosphazenes in tissue engineering applications. Polyphosphazene scaffolds have been composed of films, fibers, and sintered microspheres as drug delivery systems.[11,15,20,42]

30.4.2.15 Polyphosphoesters

Polyphosphoesters form another interesting class of biomaterials that is composed of phosphorous-incorporated monomers (Figure 30.4m). These polymers consist of phosphates with two R groups (one in the backbone and one side group) and can be synthesized by a number of routes including ring-opening polymerization, polycondensation, and polyaddition.[7,13,44]

Polyphosphoesters have great biocompatibility and similarity to biomacromolecules such as RNA and DNA. Relatively rapid hydrolytic cleavage of the phosphate bonds in the backbone leads to the production of bioresorbable or excretable phosphates, alcohols, and diols. Polyphosphoesters are divided into two different classes: polyphosphonates (alkyl/aryl R groups) and polyphosphates (alkoxy/aryloxy R groups). Because of the flexibility in choosing R groups, polymers of significantly varying physical properties and degradation rates can be synthesized. To enhance physical properties, polyphosphoesters are commonly copolymerized with polyethers and polyesters. Polyphosphoesters and polyphosphoester composites have shown significant promise in chemotherapy, tissue engineering, DNA delivery.[23,45–48] They are also used in the formation of drug delivery devices, i.e., nanoparticles, micelles, films, and gels for these applications. Their chemical flexibility and similarity to biomacromolecules give them great potential for future applications.[15,23,44,45]

30.5 BIODEGRADABLE POLYMERS IN DRUG DELIVERY

Most of the drug delivery systems are designed using various biodegradable and/or non-biodegradable polymers. These polymers especially biodegradable polymers play important roles in the modification of drug release, drug absorption, distribution, and elimination for improving the drug stability, efficacy, safety, and patient compliance. Various drug delivery systems such as implants, depots, microspheres, nanospheres, dendrimers, polymersomes, micelle systems, etc., are developed for parenteral and oral administration.

30.5.1 Implants and Depots

Implants are the drug delivery systems, which act as a depot or a reservoir of drug and deliver it over a prolonged period of time in a controlled manner upon insertion into the body, intramuscular, subdermal, intracranial, or other organ-specific routes. The subcutaneously implantable drug delivery devices offer one unique advantage of recovery opportunity. This feature enables a readily reversible termination of drug delivery whenever so required. Clinically, implant systems are recommended in situations where chronic therapy is indicated, such as hormone replacement therapy and chemical castration in the treatment of prostate cancer. Implants may be in the form of tiny rods impregnated with drug substances or a liquid, which gel following administration and release drug in a sustained or controlled release manner.[11,49] Implants intended for the parenteral route are prepared from a variety of biodegradable polymeric materials including polysaccharides, polylactic acid-*co*-glycolic acid, and the non-biodegradable methacrylates. Biodegradable materials, such as polylactic acid-*co*-glycolic acid, are preferred as they exclude the need for surgical removal of the implant after treatment ends. PLGA in a 75:25 molar ratio (the ratio of lactide to glycolide) is used

TABLE 30.2
Biodegradable Polymers Used in the Preparation of Depots/Implants and Their Application

Depot/Implant	Polymer Used	Drug/Therapeutic Active Molecules	Purpose	References
Implant	Beta-tricalcium phosphate	Gentamicin		[51]
Implant	PLGA	BCA	Ex-vivo release profile	[49]
Depot	Gelatin–chondroitin	Granulocyte-macrophage colony-stimulating factor	Systemic anti-tumor immune responses	[53]
Depot	Biodegradable polymer	Clonidine	Anti-inflammatory and analgesic	[52]
Implant	Dialkylaminoalkyl-amine–poly(vinyl alcohol)-γ-PLGA	Insulin	Antidiabetic	[54]
Scaffold depot	PLGA	Recombinant human growth hormone (rhGH)	Protein preservation	[55]
Thermally triggered in situ depot	Elastin-like polypeptides	Cefazolin and vancomycin	Provide sustained release of antibiotic	[17]
Scaffold depots	Poly(ester urethane)urea	Insulin-like growth factor-1 (IGF-1) and hepatocyte growth factor (HGF)	To provide long-term growth-factor delivery	[18]
Scaffolds implant	Polylactoglycolide	Multipotent mesenchymal stromal cells from the pulp of human deciduous teeth (SHED cells)	Bone tissue regeneration for bone defect replacement	[19]

as a drug carrier. Profact® depot implant has been designed for 2- and 3-month buserelin release. Various drugs can be incorporated into PLGA or PLA implants.[1,49–51] Drugs with low molecular weight, antibiotics, antiviral drugs, anticancer drugs, analgesics, and steroids have been reported to be incorporated into PLGA and PLA polymers for preparation of implants.[49,50,52] Table 30.2 shows different types of biodegradable polymers used in the depot or implant drug delivery system. They can also deliver peptides and proteins, such as LHRH analogue.[17,54]

Alginate-based systems have also been used as depots for bioactive agent-loaded liposomes, for slow drug release, showed greater increase in efficacy in comparison to polymeric-based systems or liposome-based systems alone. PLGA is used in Lupron Depot® for the treatment of advanced prostate cancer. A study by Stenekes and coworkers demonstrated the successive encapsulation of a drug-loaded liposome depot into a dextran polymer-based material, which releases liposome content in a sustained manner over a period of 100 days.[50,52,54] Biodegradable collagen-based systems have also been used in gene therapy. In this case, collagen scaffolds were fabricated in which bioactive agents encapsulated liposomes embedded inside them. The combination of these two technologies (i.e., liposomes and collagen-based system) has improved storage stability, prolonged the drug release rate, and increased the therapeutic efficacy.[12] Collagen has also been used as a depot payload delivery device in the local extended release of antibiotics, DNA, siRNA, and proteins. It has been reported that a liposome-loaded bioactive compound embedded into PEG-gelatin gel functions as a porous scaffold gelatin-based temporary depot with controlled drug release over prolonged periods of time.[17,18,54,55]

30.5.2 Microspheres and Nanospheres

Microspheres are small spherical particles, with diameters in the micrometer range (typically 1–1000 μm). They are also referred to as microparticles, microcapsules, microbeads, or beads (Figure 30.5). Microspheres bearing a drug dispersed or dissolved throughout particle

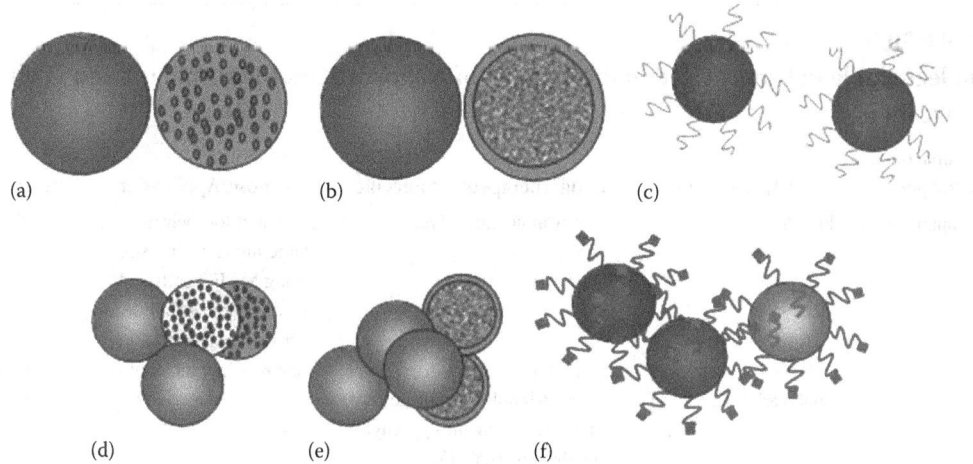

FIGURE 30.5 Nano- and microparticulate drug carrier systems. (a) Microsphere, (b) microcapsule, (c) surface modified nanoparticles, (d) nanosphere, (e) nanocapsule, and (f) surface modified targeting molecules conjugated nanoparticles.

matrix have the potential for controlled release of drug and they also protect them. These carriers received much attention as carriers not only in prolonged and controlled release formulations but also for their carrier potential in drug targeting and increase in therapeutic efficiency particularly of cytotoxic drugs. Microspheres can be manufactured from various natural and synthetic materials (Table 30.3). Both synthetic and natural polymers are employed as carrier materials. Synthetic polymers such as PLGA, methyl methacrylate, lactide, glycolide and their copolymers, polyanhydrides, polycaprolactone, etc., are widely used in preparation of microspheres. The natural polymers used for the purpose include chitosan, albumin, gelatin, starch, collagen, carrageenan, guar gum, etc. The microsphere can be prepared by using any of the appropriately selected methods including in situ polymerization, solvent evaporation, coacervation phase separation, spray drying, spray congealing, etc., but the choice of the technique depends on the nature of the polymer used, the drug, the intended use, and the duration of therapy. The choice of method is based on the particle size requirement, drug or the protein process stability, and reproducibility of the released nonsteroidal anti-inflammatory drugs, diflunisal and diclofenac sodium, which have been incorporated into PLGA microspheres and investigated for the treatment of rheumatoid arthritis, osteoarthritis, and related diseases. PLGA microspheres also reported for delivery of protein and peptides and DNA materials.[19,56]

Proteins and polysaccharides have been extensively investigated for targeted drug delivery. Albumin in the form of microspheres consisting of a large number of functional groups offers sites for attachment of the drug as well as high payloads can be obtained by chemically binding the drug to the matrix. Drug release from albumin microspheres can be controlled by changing their cross-linking density and the drug albumin ratio. To avoid the risk of viral contamination with albumin, the milk protein casein is an alternate and used as a carrier for drugs. The biodegradation rate of casein microspheres is found to be very slow when compared with albumin. The anionic group of carboxylated or sulfonated dextran interacts with the basic group of drug such as Adriamycin® to form the ionic salt complex, which exchanges the free drug for other cations. Aliphatic polyesters have been extensively investigated as drug carriers. All polyesters are degrade by homogeneous erosion following random hydrolytic degradation process.

Nanospheres also known as nanospheres, nanocapsules, nanocrystals, or nanoparticulates may be defined as solid core spherical particulates, which are nanometric in size. They contain

TABLE 30.3
Biodegradable Polymers Used in Preparation of Nanoparticles/Microspheres and Their Application

Nanoparticle/ Microspheres	Polymer Used	Drug/Therapeutic Molecule	Purpose/Application	References
Nanoparticles	PLGA	Indocyanine green dye	Evaluated for their in vitro targeting characteristics using MCF-7 cells and in-vivo biodistribution performance on tumors	[56]
	Polyester/ polycarbonate	5-Indolyl derivative, (2-(1H-Indol-5-yl) thiazol-4-yl) 3,4,5trimethoxyphenyl methanone (LY293)	Treat resistant melanoma	[57]
	PCL	2-Ethylhexyl-p-methoxycinnamate, octocrylene, and benzophenone-3	Increases the retention of UV absorbers in the skin	[58]
	Polyalkylcyanoacrylate	Cisplatin	Anticancer activity	[59]
	PLGA	Ricin (immunotoxin)	Antitumor	[55]
	PLGA	Anti-proliferative agent with heparin, didodecyl methyl-ammonium bromide (DMAB), fibrinogen, or combinations	Intraluminal therapy of restenosis	[60]
	PLGA	Rolipram	Treatment of inflammatory bowel disease	[61]
	PLGA and PLA	Paclitaxel	Anticancer	[62]
	Chitosan	Doxorubicin	Anticancer	[63]
	Hematoporphyrin modified albumin	Doxorubicin	Anticancer	[64]
	Chitosan	Tacrine	Anti-Alzheimer	[65]
	Chitosan	Doxorubicin	Anticancer	[66]
	Gelatin	NF-kappaB decoy oligonucleotide	Drug targeting approach	[67]
Microparticles/ microsphere	Gelatin	Protein	Modulation of release profile	[68]
	Albumin-dextran	Doxorubicin	Anticancer	[69]
	PLGA	Protective effect of recombinant staphylococcal enterotoxin A	Vaccination against *Staphylococcus aureus* infection	[56]
	PLGA	Model protein, bovine serum albumin	Improved drug targeting	[70]
	Hyaluronan polymers (hyaluronic acid)	Metronidazole and prednisolone hemisuccinate sodium salt	Evaluation of drug release kinetic	[71]
	Sulfopropyl dextran	Mitomycin C and doxorubicin	To improve the therapeutic index for cancer chemotherapy of selected solid tumors under special conditions	[72]
	Albumin	Doxorubicin	Anticancer	[73]

(*Continued*)

TABLE 30.3 (Continued)
Biodegradable Polymers Used in Preparation of Nanoparticles/Microspheres and Their Application

Nanoparticle/ Microspheres	Polymer Used	Drug/Therapeutic Molecule	Purpose/Application	References
	Gelatin	Transforming growth factor-beta1 (TGF-beta1)	Promoted bone regeneration	[74]
	Biodegradable polymer	Tetanus and diphtheria	Immunogenicity of single-dose tetanus and diphtheria vaccines based on controlled release	[75]
	Poly(3-hydroxybutyrate-co-3-hydroxyvalerate)	Ibuprofen	Prolong the drug release with reduced initial burst release	[76]
	Polyphosphazene	Hydroxyapatite (nHAp) scaffold	Bone tissue regeneration	[77]
	Polyanhydride	Model proteins from bovine serum (albumin (BSA), immunoglobulin G (IgG), and fibrinogen (Fg)	In-vivo performance of surface adsorbed protein on microparticles	[78]

drug embedded into the matrix or adsorbed onto the surface. In the case of nanocapsules, the drug is essentially encapsulated within the central void surrounded by an embryonic continuous polymeric sheath (Figure 30.5). They consist of drug bearing natural and synthetic or semisynthetic polymers and can be targeted systematically to the specific cells or tissues (cancerous cells) of specific body organs. They are easily uptaken by cells due to their nanometric particle size (1–200 nm). Various synthetic biodegradable polymers such as polylactic acid, PLA, PLGA, polycyanoacrylate (PCA), PCL, etc., have been studied as carrier materials (Table 30.3). When nanoparticles are administered intravenously they may be easily recognized and opsonized by immune cellular component(s). To reduce opsonization and clearance, PEGylated nanoparticles have been investigated in which PEG was used to modify the surface of nanoparticles.[19,70] Paclitaxel loaded in PLGA-PEG nanoparticles exhibited similar apoptotic cell death as Taxol®, in HeLa cancer cell lines.[70] Another nanoparticle-based FDA approved formulation is Abraxane®.[73] This is based on albumin nanoparticles of paclitaxel, which evades the hypersensitivity reaction associated with Cremophor® EL, a solvent used in conventional paclitaxel formulation. Polyacetal (polyketal) nanoparticles have been used to directionally deliver siRNA, DNA, proteins, and vaccines in the treatment of different types of diseases like ischemic heart disease and cancer.[23,50,57] Polymers of natural origin such as gelatin and polysaccharides including chitosan, cyclodextrins, and dextran are also extensively used for fabrication of nanoparticle-based drug delivery systems.

30.5.3 Dendrimers

Dendrimers are core–shell macromolecules having a unique three-dimensional structure made up of highly organized polymeric branches surrounding a core. Successive layers of the monomers making up the branches (called generations) can be added to increase the molecular size and the number of available surface groups. Dendrimers have distinct advantages due to their highly ordered structure, narrow size distribution, and availability of a large number of functional groups for attachment of drug molecules or targeting ligands and a high degree of control on the drug release properties. Dendrimers can allow both stable loading of a large number of drug molecules as well as grafting of specific ligands on the surface to enable active targeting of specific cells and tissues.

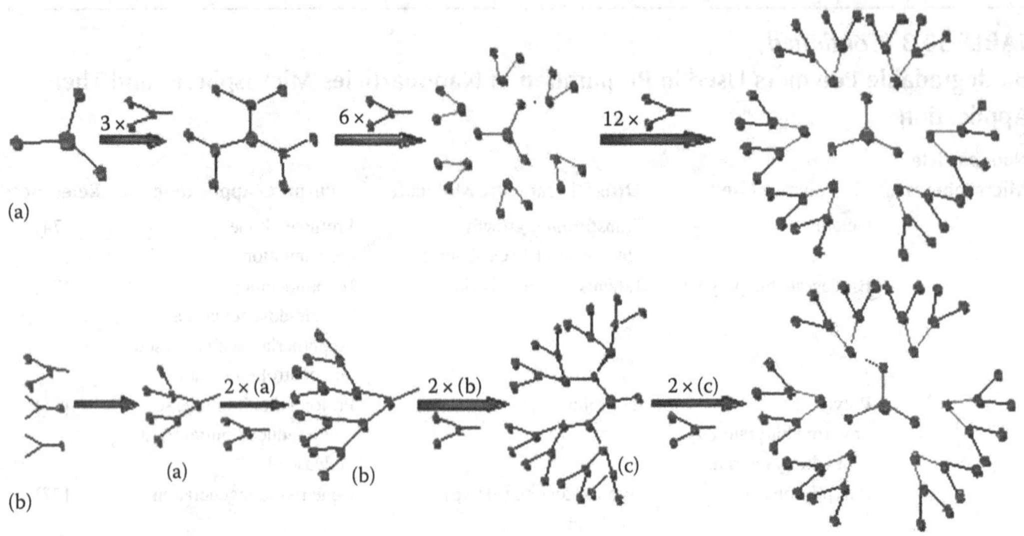

FIGURE 30.6 (a) The divergent growth method of dendrimers. (b) The convergent growth method of dendrimers.

Divergent approach initiated by Tomalia et al.[79] and the convergent approach given by Hawker and Frechet[80] have been applied for the synthesis of dendrimers (Figure 30.6). Nowadays, dendrimers have various potential applications ranging from tissue engineering to anticancer, antiviral, antibacterial drugs, and vaccines to gene delivery. The precise synthesis technique facilitates the emergence of several kinds of dendrimer backbones with good water solubility and biocompatibility. PAMAM dendrimers are biocompatible and non-toxic in nature especially when their surface is modified with anionic or neutral groups such as carboxylic or hydroxylic moieties, which are synthesized by the divergent approach and are the first complete dendrimer family to be commercialized. Polypropyleneimine (PPI) dendrimer, which has multiple cationic amine groups, is another commercialized material. But these polyesters, i.e., PPI and PAMAM dendrimers, are not biodegradable in nature (Table 30.4).[85]

Biodegradable polyester dendrimers incorporating monomers, such as glycerol, succinic acid, phenylalanine, and lactic acid, have been prepared by Grinstaff et al.[87] Compared to linear polymers, dendrimers exhibit unique architecture, which can provide several advantages for drug delivery applications, the internal cavity of the den-drimer provides a location for hydrophobic drug to achieve non-covalent binding. The multivalency of dendrimers can be used to attach combination of drug molecules, targeting groups, and solubilizing groups to the periphery of the dendrimers and the more globular shape of dendrimers could affect their biological properties and polydispersity of dendrimers should provide reproducible pharmacokinetic behavior. Dendrimer drug delivery offers a uniform and promising protocol for drug entrapment, conjugation, and controlled release and also covers factors such as the chemical property and size of branched units (generation number "G," Figure 30.7), which always exerts the most critical influence on the quality of the resultant system, such as size, drug loading capability, efficiency, and safety.[79,80] Paclitaxel-loaded polyglycerol dendrimers also show high encapsulation efficiency. Poly(lysine) dendrimers modified with sulfonated naphthyl groups have been found to be useful as antiviral drugs against the herpes simplex virus.[81–84] In another study, poly(lysine) dendrimers with mannosyl surface groups are effective inhibitors of the adhesion of *E. coli* to horse blood cells in a hemagglutination assay, making these structures promising antibacterial agents.[81]

In another study, dendritic amidoamine side chains of different generations were covalently attached to the polysaccharide chitosan in an attempt to combine the biological activity of chitosan in gene delivery,

TABLE 30.4
Biodegradable Polymers Used in Preparation of Dendrimers and Their Application

Dendrimer Type	Polymer Used	Drug/Therapeutic Molecule	Purpose/Application	References
PEGylated polylysine dendrimers	Polylysine	Doxorubicin	HSBA linker based targeting of DOX loaded dendrimer to solid tumor	[81]
Poly(ethylene glycol)-block-poly(L-lysine) dendrimer	Poly-L-lysine	Plasmid DNA	To provide stability and delivery of pDNA	[82]
Cellobiose–polylysine dendrimer	Poly-L-lysine	Cyclic oligopeptide of human AIDS virus	AIDS vaccine preparation and delivery	[83]
Poly-L-lysine dendrimer	Poly-L-lysine	Antiangiogenesis	Antitumor activity	[84]
CMCht/PAMAM dendrimer	6-Azido-6-deoxy-chitosan	Plasmid DNA	Gene delivery	[85]
β-Cyclodextrin-based biodegradable dendrimers	β-Cyclodextrin and 2,3-di-o-methacrylated-6-methacrylated	Methotrexate	Anticancer	[86]

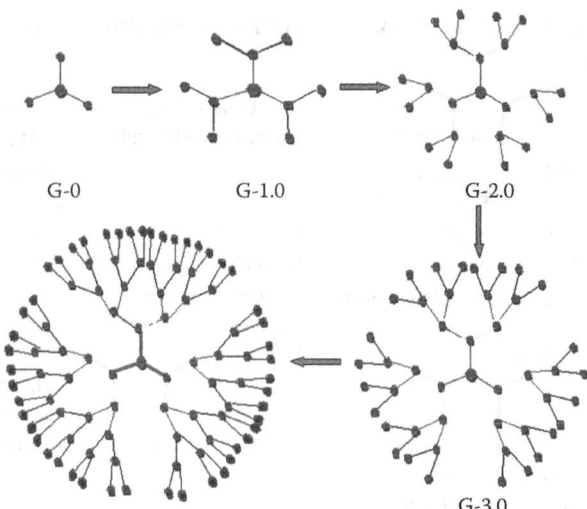

FIGURE 30.7 Various generations of dendrimers.

antibacterial activity, and wound healing activity with the delivery benefits found for dendrimers. While several chitosan–dendrimer hybrids have been synthesized and characterized and found to be useful as antibacterial agents, carriers in drug delivery systems, and in other biomedical applications.[85]

30.5.4 Polymeric Hydrogels for Intravesical Drug Delivery

Hydrogels are three-dimensional macromolecular, covalently cross-linked networks of hydrophilic polymers that can hold a large fraction of an aqueous solvent within their structures. They are particularly suitable for biomedical applications, including controlled drug delivery; because of their ability to simulate biological tissues. Depending on the polymer used these systems can imbibe water content ranging from 30% to 90%. Because of their highly swollen nature, hydrogel membranes are usually quite permeable to hydrophilic drugs and high molecular weight proteinaceous agents, such as insulin, aprotinin, tumor antigenesis factor, and luteinizing hormone. Since alginate is anionic, fabrication of alginate hydrogels has successively been achieved through a reaction with cross-linking agents such as divalent or trivalent cations mainly calcium ions, water-soluble carbodiimide, and/or glutaraldehyde. The cross-linking methodology was conducted at room temperature and physiological pH. Interleukin-2, which is a highly effective anticancer drug, is among the success obtained in delivering a combination of drug-loaded liposome and injectable dextran hydrogel.[88] Serum albumin was conjugated to poly-(ethylene glycol) (PEG) and cross-linked to form mono-PEGylated albumin hydrogels. These hydrogels were used as a basis for drug carrying tissue engineering scaffold materials, based on the natural affinity of various drugs and compounds for the tethered albumin in the polymer network.[89,90]

Some polymers undergo abrupt changes in solubility in response to increases in environmental temperature. This property is utilized in the preparation of temperature-sensitive drug delivery systems. Temperature-sensitive hydrogels made of polymers, which swell upon changing the temperature, and the release as well as mechanical characteristics of drug and hydrogels are altered with the change in the temperature of the external environment. Thermosensitive polymers can be used for the preparation of hydrogel, which is injected in its liquid form in the body cavity and then forms a gel in situ inside the cavity at elevated body temperature. Table 30.5 shows different types of biodegradable polymers, which are used in the above mentioned hydrogels for drug delivery. The triblock copolymer

TABLE 30.5
Biodegradable Polymers Used in Preparation of Hydrogel Drug Delivery Systems and Their Application

Hydrogel Type	Polymer Used	Drug/Active Therapeutic Molecules	Purpose/Application	References
Temperature-sensitive	PLGA-PEG-PLGA	Insulin	Antidiabetes	[91]
pH/temperature-sensitive	PAE-PCL-PEG-PCL-PAE	Insulin	Antidiabetes	[92]
Temperature-sensitive	PEG-PLGA-PEG	Plasmid DNA	Gene delivery	[93]
Temperature-sensitive	Polyphosphazenes	Doxorubicin, paclitaxel	Anticancer	[94]
Swelling trigger release hydrogel	Poly-N-isopropylacrylamide	Sustained transdermal drug release		[95]
Hydrogel	Gelatin	Rifampicin	Antitubercular drug delivery and tissue regeneration	[89]
Hydrogel	Methoxy-poly(ethylene glycol)-b-poly(ε-caprolactone)-b-poly[2-(dimethylamino)ethyl methacrylate]	pDNA	Gene therapy	[96]
Hydrogel	Gelatin	siRNA		[97]
Hydrogel	Polyethyleneimine dextran	siRNA		[98]
	Alginate	Polyproline-rich synthetic peptides	Bone tissue regeneration	[99]
	Hyaluronic acid	Morphogenetic protein-2	Osteogenic activity	[88]
	Hyaluronic acid	Paclitaxel	Antitumors	[100]
Chitosan scaffolds	Chitosan	Angiogenesis growth factor	Angiogenesis in the case of myocardial infarction treatment	[90]
Triple cross-linking network hydrogel	Thiolated chitosan (CS-TGA), β-glycerophosphate (β-GP), and poly(ethylene glycol) diacrylate	—	Evaluation of non-cytotoxic properties	[101]

{poly (ethylene glycol)–poly [lactic acid-*co*-glycolic acid]–poly (ethylene glycol)} (PEG-PLGA-PEG) was used to make them more hydrophobic in nature to improve delivery of both hydrophilic as well as hydrophobic drugs. PLGA-PEG-PLGA, PAE-PCL-PEG-PCL-PAE, and PEG-PLGA-PEG block copolymers are other examples, which have been used to prepare pH-sensitive hydrogels for the delivery of insulin and pDNA.[91–95] This hydrogel composition was highly advantageous due to its ease of formulation, biocompatibility, and biodegradability. A further aspect of this gel is its reversible gelling property, i.e., it reverts back into liquid form at low temperature. Biodegradable in situ gelling dextran hydrogels have been used for sustained delivery of interleukin (IL-2) to tumors in mice. Hydrogels can be used to carry even hydrophobic drugs, such as injectable PLGA-PEG-PLGA gel carrying PEGylated camptothecin.[91] Another biodegradable in situ gelling PLGA-PEG-PLGA-based hydrogel (OncoGel) was used for local delivery of paclitaxel to solid tumors in animal models and results showed targeted delivery of the drug to tumor tissues and sustained release.[91,93] pH change occurs at many specific or pathological body sites, it is one of the important environmental parameters for drug delivery systems. In this system, the swelling depends not only on the chemical composition but also on the pH of the surrounding medium. The pH-sensitive polymers show dramatic changes in the pH and in the composition of the external solutions. Poly(methacrylic acid), poly(L-glutamic acid), alginate-*N,O* carboxymethyl chitosan, and PAE-PCL-PEG-PCL-PAE are some examples of pH-sensitive polymers, which have been used in the preparation of hydrogels.[90,96] The mechanism of drug release from the different types of hydrogels is clearly explained in Figure 30.8.

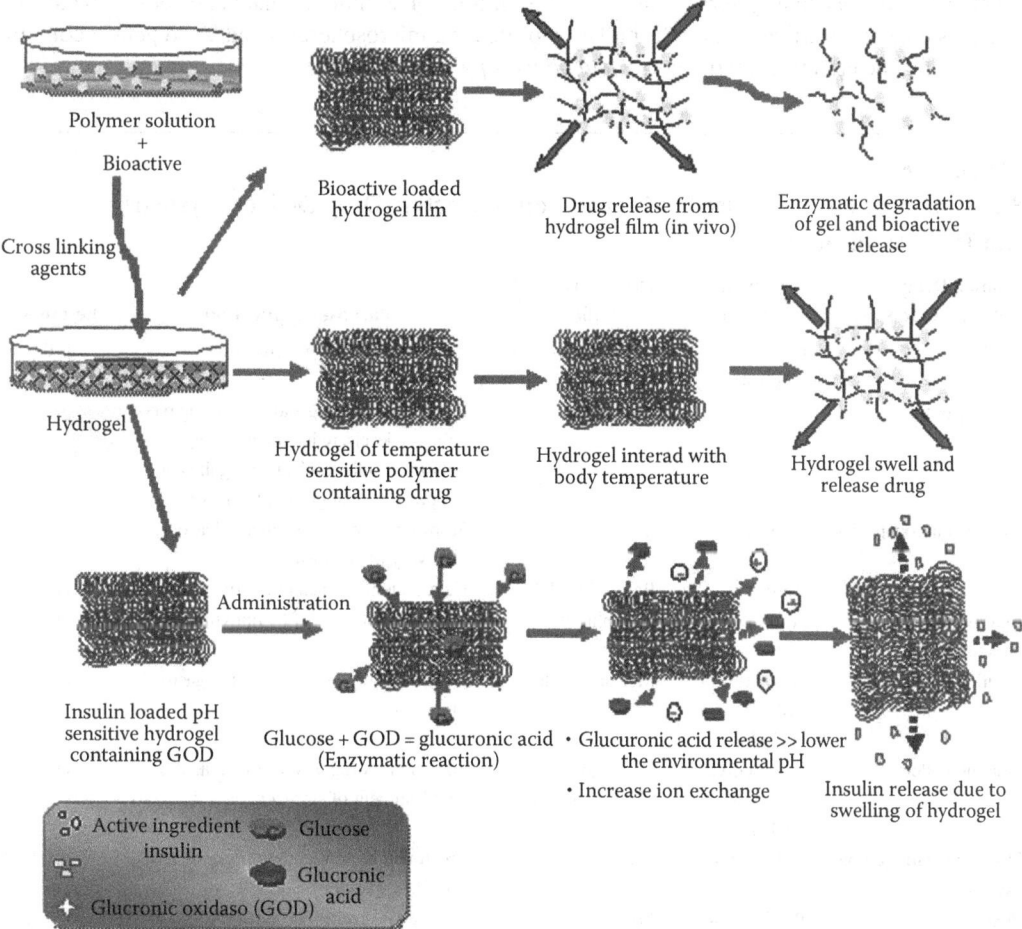

FIGURE 30.8 Drug release pattern of different hydrogels.

30.5.5 Floating Drug Delivery Systems

Several approaches are currently being utilized to enhance gastric retention time (GRT). They include floating drug delivery systems (FDDS), also known as hydrodynamically balanced systems (HBS), swelling and expanding systems, and polymeric bioadhesive systems. Floating drug delivery systems either float due to their low density than stomach contents or due to the gaseous phase formed inside the system after they come in contact with the gastric environment and release their loaded content continuously in a sustained and controlled manner. Non-effervescent floating systems in the class of floating drug delivery systems are usually prepared from gel-forming or highly swellable biodegradable polymers, polysaccharides, or other matrix-forming polymers (Table 30.6). When these dosage forms come in contact with an aqueous medium, the hydrocolloids imbibe water and start to hydrate thereby forming a gel at the surface, which controls the trafficking of drug out and passage of solvent into the dosage form. The drug in dosage form dissolves in and diffuses out with the diffusing solvent forming a "receding boundary" within the gel structure.

Effervescent floating delivery systems employ matrices from swelling polymers like Methocel® or chitosan and effervescent components such as sodium bicarbonate and tartaric or citric acid that gasify at body temperature. The matrices are prepared in such a manner that when they come in contact with gastric fluid, carbon dioxide is generated, and retained or entrapped within the hydrocolloid gel. This leads to an upward drift of the dosage form and maintains it in a floating condition. It can be formulated in single and bilayer tablets. The concept has been judiciously utilized to develop floating capsule systems consisting of a mixture of sodium alginate and sodium bicarbonate. This concept is utilized in the formulation of floating microspheres, beads, and pellets containing drug for the effective treatment of *Helicobacter pylori* infection.

TABLE 30.6
Biodegradable Polymers Used in Preparation of Floating Drug Delivery Systems and Their Application

Floating Drug Delivery System Type	Polymer Used	Drug/Therapeutic Molecule	Purpose/Application	References
Beads	Calcium alginate	Antihypertensive drug	Chronotherapy of hypertension	[102]
Floating beads	Pectin	Gliclazide	Reduction in fasting and non-fasting blood glucose levels, reduction in fasting plasma insulin level, and a significant improvement in glucose tolerance	[103]
Floating microcapsules	Alginate	Simvastatin	Stomach-specific sustained release delivery of simvastatin	[104]
Floating microspheres	Chitosan	Ranitidine HCl	Gastroretentive sustained drug release	[105]
Agar gel network floating tablet	Agar	Theophylline	Controlled release of theophylline	[106]
Floating gel beads	Alginate and ethyl cellulose	Metronidazole	Controlled drug delivery to the gastric mucosa	[107]
Floating pellets	Polymethyl methacrylate, gelatin	Verapamil hydrochloride	To increase drug absorption in acidic environment of stomach	[108]
Floating in situ gelling system	Gellan gum	Amoxicillin	Treatment of peptic ulcer disease caused by *Helicobacter pylori*	[109]
Floating gel beads	Pectin/sodium alginate	Loratadine	To evaluate controlled release of loratadine from polymeric gel beads	[110]

30.5.6 POLYMERSOMES

Polymersomes, self-assembled polymer shells composed of block copolymer amphiphiles. These synthetic amphiphiles with amphiphilicity similar to lipids constitute a new class of drug carriers. They are spontaneously formed in aqueous media, as unilamellar vesicles up to tens of microns in diameter. Amphiphilic block copolymers form a range of self-assembled aggregates including spherical, rod-like, tubular micelles, lamellae, or vesicles, depending on polymer architecture and preparation conditions. Polymers having low hydrophobicity (less than 50%) favor the formation of micelles, however, intermediate level of hydrophobicity (50%–80%) favors the formation of vesicles. Polymeric vesicles, which have a liposome-like structure with a hydrophobic polymer membrane and hydrophilic inner cavity, are called polymersomes.

The polymersomes offer some advantages over liposomes, not only in vesicle stability but also in the regulation of membrane thickness. Their particle size can be controlled by changing the chain length of the polymers. Both hydrophilic and hydrophobic drugs can be incorporated. Systemic circulation time is relatively longer than lipid vesicles. They offer enhanced mechanical stability (due to intra and intermolecular hydrophobic interwinding). Polymersomes are applicable in diverse fields such as drug delivery systems, transfection vectors, and protective shells for sensitive enzymes. Hydrophobic blocks are poly(ethyl ethylene) (PEE) and polybutadiene (PBD), which can be cross-linked subsequently to enhance stability.[111,112] Biodegradable PLA and PCL have been utilized considering the need for disposal in vivo and controlled drug release (Table 30.7). It has been proved that triblock copolymers can also form polymersomes.[113]

TABLE 30.7
Biodegradable Polymers Used in Preparation of Polymersomes and Micelles and Their Applications

Drug Delivery Carriers System	Polymer Used	Therapeutic Molecule	Purpose/Application	References
Flexible polymersomes	Poly(caprolactone)–poly(ethylene glycol)–poly(caprolactone) copolymer	Anticancer	Targeting melanomas and basal cell carcinomas	[111]
Polymersomes	Biodegradable polymer	Antisense oligonucleotides	Found efficient nuclear uptake	[112]
Polymersomes	PCL and PEG	Insulin	Enhanced therapeutic efficacy of insulin	[113]
Amphiphilic block copolymer micelles	Poly(ethylene glycol)–poly(caprolactone)	Norcantharidin	Anticancer	[114]
Amphiphilic block copolymer micelles	PLGA-PEG	Doxorubicin	Treatment of hepatocellular carcinoma	[115]
Thermosensitive polymeric micelles	Poly(N-isopropylacrylamide-co-acrylamide)-b-poly(DL-lactide)	Docetaxel	Thermal targeted antitumor drug delivery	[116]
Polymeric micelles	Poly(ethylene glycol)–poly(lactide)	Cyclosporin A	Solubility enhancement of poorly soluble cyclosporin A and further improving oral absorption of the drug	[117]
pH-responsive polymeric micelles	Biodegradable polymer	Protoporphyrin	Antitumor therapy	[118]
Polymeric micelles	PEG-b-poly(epsilon-caprolactone)	Rapamycin	Controlled release and bioavailability enhancement	[119]

30.5.7 Micelles/Reverse Micelles

Polymeric micelles are nanosized, supramolecular core–shell structures that are made up of polymer chains and are usually spontaneously formed by self-assembly of amphiphiles in an aqueous environment at the concentration of amphiphiles above critical micelle concentration (CMC), generally as a result of hydrophobic or ion pair interactions between polymer segments. The core of the micelles is either the hydrophobic part or the ionic part. The main driving force for micelle formation in aqueous solution is the effective interaction between the hydrophobic parts of the surfactant molecules, whereas interaction opposing micellization may include electrostatic repulsive interaction between charged head groups of ionic surfactants, repulsive osmotic interactions between chain-like polar head groups, such as oligoethylene oxide chains, or steric interaction between bulky head groups. In aqueous solution, surfactants aggregate in different forms, such as spherical micelles, worm-like micelle, bilayer fragments, vesicles, or inverted structures.[114–116]

Micelles have attracted the attention of researchers as drug delivery systems, because of their nanometric size, ability to solubilize hydrophobic drugs, and site-specific delivery by passive and active targeting. Biodegradable amphiphilic block copolymers such as poly(caprolactone)–poly(ethylene glycol)–poly(caprolactone), poly(ethylene glycol)–poly(caprolactone-*co*-polymer),[114] and poly (ethylene glycol)–poly(lactide)[117] have been used to incorporate some anticancer drugs and target them to their site of action upon parenteral administration (Table 30.7). Although much research has been carried out on micelles for parenteral delivery, their potential for oral drug administration remains largely untapped.

The micelles are excellent drug carriers, because they can hold drugs firmly in their inner cores (apolar) and control the drug release by changing the molecular structure of the inner cores. Micelles are also used as a non-viral carrier system for gene delivery. Polyethylene glycol–poly-aspartic acid as an anion block copolymer mixed with polyethylene glycol poly L-lysine (PEG-PLL) as a cation block copolymer forms micelles with a diameter of dozens of nanometers. Micelles of 80 nm in diameter were formed with plasmid DNA and PEG-PLL. Higher expression of luciferase in cultured cells was reported in the case of micelles when compared to conventional solution of the drug.[118,119]

30.6 CONCLUSION

This entry has focused on some of the widely studied natural and synthetic biodegradable polymers used for designing novel drug delivery systems for the purpose of targeted and controlled release applications. With the help of advances in polymer synthesis chemistry and technology, more defined biocompatible polymers are becoming available, and such polymers will contribute to new generations of biomimetic nanostructures and vehicles for carrying biomolecules or therapeutic drugs.

Biodegradable polymers are becoming the backbone of the field of drug delivery. However, many of the future challenges we face, such as gene therapy or delivery of protein and peptide drug, must require degradable polymer systems, which are prepared with unique requirements for specific applications. In this case, biodegradable polymers play important role, they not only deliver drug safely or effectively but also are able to be broken down and removed after they have served their function in the body. Many biodegradable polymers such as PLGA, PCL, PLA, chitosan, albumin, gelatin, alginate, etc., are widely used in fabrication of novel drug delivery systems like nanoparticles, microspheres, dendrimers, micelles, hydrogels, and polymersomes, which can securely deliver drugs, genes, proteins, vaccines, enzymes, and other biomolecules for the treatment of different diseases. Thus, we can state that biodegradable polymers have a remarkable impact on the science of drug delivery and promise to have an even greater impact on human health care.

ACKNOWLEDGMENTS

The author Satish Shilpi acknowledges the Indian Council of Medical Research, New Delhi, for the award of ICMRSRF (Grant: 45/08/2008-/Nan/BMS Dated 18/02/2008).

REFERENCES

1. Li, Y.Q.; You, H.B. Expert review on polymer architecture and drug delivery. *Pharm. Res.* 2006, *23* (1), 1–30.
2. Bret, D.U.; Nair, L.S.; Cato, T. Laurencin biomedical applications of biodegradable polymers. *Polym. Phys.* 2011, *49* (12), 832–864.
3. Mishra, N.; Goyal, A.K; Khatri, K.; Vaidya, B.; Paliwal, R.; Rai, S.; Mehta, A.; Tiwari, S.; Vyas, S.; Vyas, S.P. Biodegradable polymer based particulate carrier(s) for the delivery of proteins and peptides. *Antiinflamm. Antiallergy Agents Med. Chem.* 2008, *7* (4), 240–251.
4. Vroman, I.; Tighzert, L. Biodegradable polymers. *Materials* 2009, *2* (2), 307–344.
5. Chandra, R.; Rustgi, R. Biodegradable polymers. *Prog. Polym. Sci.* 1998, *23* (7), 1273–1335.
6. Nela, A.; David, H. Rationalizing the design of polymeric biomaterials. *Tibtech* 1999, *17* (10), 409–422.
7. Shanmugam, S.; Manavalan, R.; Venkappayya, D.; Sundramoorthy, K.; Mounnissamy, V.M.; Hemalatha, S.; Ayyappan, T. Natural Polymers and their applications. *Nat. Prod. Radiance* 2005, *4* (6), 478–481.
8. Bhowmik, D.C.; Chandira, M.R.; Jayakar, B. Role of nanotechnology in novel drug delivery system. *J. Pharm. Sci. Technol.* 2009, *1* (1), 20–35.
9. Kathryn, E.U.; Scott, M.C.; Robert, S.L.; Kevin, M.S. Polymeric systems for controlled drug release. *Chem. Rev.* 1999, *99* (11), 3181–3198.
10. Majeti, N.K.; Ravikumar, N.V., Domb, A.J. Biodegradable block copolymers. *Adv. Drug Del. Rev.* 2001, *53* (1), 23–44.
11. Wood, D.A. Biodegradable drug delivery system. *Int. J. Pharm.* 1980 *7*(1), 1–18.
12. Berisio, R.; Vitagliano, L.; Mazzarella, L.; Zagari, A. Crystal structure of the collagen triple helix model [(Pro-Pro-Gly)10]3. *Protein Sci.* 2002, *11* (2), 262–270.
13. Isabelle, V.; Lan, T. Biodegradable polymers. *Materials* 2009, *2* (2), 307–344.
14. Leja, K.; Lewandowicz, G. Polymer biodegradation and biodegradable polymers—A review. *Polish J. Environ. Stud.* 2010, *19* (2), 255–266.
15. Hayash, T. Biodegradable polymers for biomedical uses. *Prog. Polym. Sci.* 1994, *19* (1), 663–702.
16. Pouton, C.W.; Akhtar, S. Biosynthetic polyhydroxyalkanoates and their potential in drug delivery. *Adv. Drug Deliv. Rev.* 1996, *18* (2), 133.
17. Vakhrushev, I.V.; Antonov, E.N.; Popova, A.V.; Konstantinova, E.V.; Karalkin, P.A.; Kholodenko, I.V.; Lupatov, A.Y.; Popov, V.K.; Bagratashvili, V.N.; Yarygin, K.N. Design of tissue engineering implants for bone tissue regeneration of the basis of new generation polylactoglycolide scaffolds and multipotent mesenchymal stem cells from human exfoliated deciduous teeth (SHED cells). *Bull. Exp. Biol. Med.* 2012, *153* (1), 143–147.
18. Nelson, D.M.; Baraniak, P.R.; Ma, Z.; Guan, J.; Mason, N.S.; Wagner, W.R.; Controlled release of IGF-1 and HGF from a biodegradable polyurethane scaffold. *Pharm. Res.* 2011, *28* (6), 1282–1293.
19. Ferdous, A.J.; Stembridge, N.Y.; Singh, M. Role of monensin PLGA polymer nanoparticles and liposomes as potentiator of ricin A immunotoxins in vitro. *J. Control. Release* 1998, *50* (1–3), 71–78.
20. Huayu, T.; Zhaohui, T.; Xiuli, Z.; Xuesi, C.; Xiabin, J. Biodegradable synthetic polymers: Preparation, functionalization and biomedical application. *Prog. Polym. Sci.* 2012, *37* (2), 237–280.
21. Kelly, L.S.; Martin, E.S.; Karen, E.T. Bioerodible polymers for delivery of macromolecules. *Adv. Drug Deliv. Rev.* 1990, *4* (3), 343–357.
22. Crepon, B.; Jozeeonvicz, J.; Chytry, V.; Rihova, B.; Kopecek, J. Enzymatic degradation and immunogenic properties of derivatized dextrans. *Biomaterials* 1991, *12* (6), 550–554.
23. Lee, K.Y.; Ha, W.S.; Park, W.H. Blood compatibility and biodegradability of partially N-acylated chitosan derivatives. *Biomaterials* 1995, *16* (16), 1211–1216.
24. Peppas, N.A.; Huang, Y. Nanoscale technology of mucoadhesive interactions. *Adv. Drug Deliv. Rev.* 2004, *56* (11), 1675–1687.
25. Geresh, S.; Gdalevsky, G.Y.; Gilboa, I.; Voorspoels, J.; Remon, J.P.; Kost, J. Bioadhesive grafted starch copolymers as platforms for peroral drug delivery: A study of theophylline release. *J. Control. Release* 2004, *94* (2–3), 391–399.
26. Dubief, D.; Samain, E.; Dufresne, A. Polysaccharide microcrystals reinforced amorphous poly(β-hydroxyoctanoate) nanocomposite material. *Macromolecules* 1999, *32* (18), 5765–5771.

27. Bjork, E.; Edman, P. Characterization of degradable starch microspheres as a nasal delivery system for drugs. *Int. J. Pharm.* 1990, *62* (2–3), 187–192.
28. Fredriksson, H.; Silverio, J.; Andersson, R.; Eliasson, A.C.; Aman, P. The influence of amylase and amylopectine characteristics on gelatinization and retrogradation properties of different starches. *Carbohydr. Polym.* 1998, *35* (3–4), 119–134.
29. Imam, S.H.; Gordon, S.H.; Shogren, R.L.; Greene, R.V. Biodegradation of starch-poly(β-hydroxybutyrate-co-valerate) composites in municipal activated sludge. *J. Environ. Polym. Degr.* 1995, *3* (4), 205–213.
30. Van Soest, J.J.G.; Hulleman, S.H.D.; de Wit, D.; Vliegenthart, J.F.G. Crystallinity in starch bioplastics. *Ind. Crops Prod.* 1996, *5* (1), 11–22.
31. Myllarinen, P.; Buleon, A.; Lahtinen, R.; Forssell, P. The crystallinity of amylose and amylopectin films. *Carbohydr. Polym.* 2002, *48* (1), 41–48.
32. Grainger, D.A.; Meyer, W.R.; Decherney, A.H.; Diamond, M.P.; The use of hyaluronic acid polymers to reduce postoperative adhesions. *J. Gynecol. Surg.* 1991, *7* (2), 97–101.
33. Hunt, J.A.; Joshi, H.N.; Stella, V.J.; Topp, E.M. Diffusion and drug release of polymer films prepared from ester derivatives of hyaluronic acid. *J. Control. Release* 1990, *12* (2), 159–169.
34. Loftsson, T. Cyclodextrins and the biopharmaceutics classification system. *J. Incl. Phenom. Macrocycl. Chem.* 2002, *44* (1–4), 63–67.
35. Loftsson, T.; Duchene, D. Cyclodextrins and their pharmaceutical applications. *Int. J. Pharm.* 2007, 329 (1–2), 1–11.
36. Uekama, K. Design and evaluation of cyclodextrin based drug formulation. *Chem Pharm. Bull.* 2004, *52* (8), 900–915.
37. Loftsson, T.; Brewster, M.E.; Masson, M. Role of cyclodextrins in improving oral drug delivery. *Am. J. Drug Deliv.* 2004, *2* (4), 261–275.
38. Sodergard, A.; Stolt, M. Properties of lactic acid based polymers and their correlation with composition. *Progr. Polym. Sci.* 2002, *27* (6), 1123–1163.
39. Auras, R.; Harte, B.; Selke, S. An overview of polylactides as packaging materials. *Macromol. Biosci.* 2004, *4* (9), 835–864.
40. Mochizuki, M.; Hirami, M. Structural effects on biodegradation of aliphatic polyesters. *Polym. Adv. Technol.* 1997, *8* (4), 203.
41. Miller, N.D.; Williams, D.F. On the biodegradation of poly-beta-hydroxybutyrate (PHB) Homopolymer and poly-beta-hydroxybutyrate-hydroxyvalerate coploymers. *Biomaterials* 1987, *8* (2), 129–137.
42. Nishida, H.; Tokiwa, Y. Distribution of poly(β-hydroxy-butyrate) and poly(ε-caprolactone) aerobic degrading microorganisms in different environments. *J. Environ. Polym. Degrad.* 1993, *1* (3), 227–233.
43. Huang, C.; Shetty, A.S.; Wang, M.S. Biodegradable plastics: A review. *Polym. Technol.* 1990, *10* (1), 23–30.
44. Tokiwa, Y.; Suzuki, T. Hydrolysis of polyesters by lipases. *Nature* 1977, *270* (5632), 76–78.
45. Joseph, J.G. Biomedical application of functional polymers. *React. Funct. Polym.* 1999, *39*, 99–138.
46. Kim, S.; Kim J-H.; Jeon, O.; Kwon, I.C.; Park, K. Engineered polymers for advanced drug delivery. *Eur. J. Pharm. Biopharm.* 2009, *71* (3), 420–430.
47. Qiu, Y.; Park, K. Environment-sensitive hydrogels for drug delivery. *Adv. Drug Deliv. Rev.* 2001, *53* (3), 321–339.
48. Lu, Y.; Chen, S.C. Micro and nano-fabrication of biodegradable polymers for drug delivery. *Adv. Drug Deliv. Rev.* 2004, *56* (11), 1621–1633.
49. Ghalanbor, Z.; Korber, M.; Bodmeier, R. Protein release from poly(lactide-co-glycolide) implants prepared by hot-melt extrusion: Thioester formation as a reason for incomplete release. *Int. J. Pharm.* 2012, *438* (1–2), 302–306.
50. Golumbek, P.T.; Azhari, R.; Jaffee, E.M.; Levitsky, H.I.; Lazenby, A.; Leong, K.; Pardoll, D.M. Controlled release, biodegradable cytokine depots: A new approach in cancer vaccine design. *Canc. Res.* 1993, *53* (24), 5841–5844.
51. Thoma, K.; Alex, R.; Randzio, J. Biodegradable gentamicin-depot implants made of beta-tricalcium phosphate ceramics.3. In vivo studies on drug release, tissue tolerance, and biodegradation. *Pharmazie* 1991, *46* (4), 266–270.
52. Packhaeuser, C.B.; Kissel, T. On the design of in situ forming biodegradable parenteral depot systems based on insulin loaded dialkylaminoalkyl-amine-poly(vinylalcohol)-g-poly (lactide-co-glycolide) nanoparticles. *J. Control. Release* 2007, *123* (2), 131–140.
53. Beall, D.P.; Deer, T.R.; Wilsey, J.T.; Walsh, A.J.; Block, J.H.; McKay, W.F.; Zanella, J.M. Tissue distribution of clonidine following intraforaminal implantation of biodegradable pellets: Potential alternative to epidural steroid for radiculopathy. *Pain Phys.* 2012, *15* (5), 701–710.

54. Carrasquillo, K.G.; Costantino, H.R.; Cordero, R.A.; Hsu, C.C.; Griebenow, K. On the structural preservation of recombinant human growth hormone in a dried film of a syntheticbiodegradable polymer. *J. Pharm. Sci.* 1999, *88* (2), 166–173.
55. Adams, S.B.J.; Shamji, M.F.; Nettles, D.L.; Hwang, P.; Setton, L.A. Sustained release of antibiotics from injectable and thermally responsive polypeptide depots. *J. Biomed. Mater. Res. B Appl. Biomater.* 2009, *90* (1), 67–74.
56. Chen, L.; Li, S.; Wang, Z.; Chang, R.; Su, J.; Han, B. Protective effect of recombinant staphylococcal enterotoxin A entrapped in polylactic-co-glycolic acid microspheres against Staphylococcus aureus infection. *Vet. Res.* 2012, *43* (1), 1–20.
57. Mundra, V.; Lu, Y.; Danquah, M.; Li, W.; Miller, D.D.; Mahato, R.I. Formulation and characterization of polyester/polycarbonate nanoparticles for delivery of a novel microtubule destabilizing agent. *Pharm. Res.* November 2012, *29* (11), 3064–3074.
58. do Nascimento, D.F.; Silva, A.C.; Mansur, C.R.; Presgrave, R.F.; Alves, E.N.; Silva, R.S.; Ricci-Júnior, E.; de Freitas, Z.M.; dos Santos, E.P. Characterization and evaluation of poly(epsilon-caprolactone) nanoparticles containing 2-ethylhexyl-p-methoxycinnamate, octocrylene, and benzophenone-3 in antisolar preparations. *J. Nanosci. Nanotechnol.* 2012, *12* (9), 7155–7166.
59. Egea, M.A.; Gamisans, F.; Valero, J.; Garcia, M.E.; Garcia, M.L. Entrapment of cisplatin into biodegradable polyalkylcyanoacrylate nanoparticles. *Farmaco* 1994, *49* (3), 211–217. http://www.ncbi.nlm.nih.gov/pubmed/7519014.
60. Song, C.; Labhasetwar, V.; Cui, X.; Underwood, T.; Levy, R.J. Arterial uptake of biodegradable nanoparticles for intravascular local drug delivery: Results with an acute dog model. *J. Control. Release* 1998, *54* (2), 201–211.
61. Lamprecht, A.; Ubrich, N.; Yamamoto, H.; Schafer, U.; Takeuchi, H.; Maincent, P.; Kawashima, Y.; Lehr, C.M. Biodegradable nanoparticles for targeted drug delivery in treatment of inflammatory bowel disease. *Pharmacol. Exp. Ther.* 2001, *299* (2), 775–781. http://www.ncbi.nlm.nih.gov/pubmed/11602694J.
62. Xie, J.; Wang, C.H. Self-assembled biodegradable nanoparticles developed by direct dialysis for the delivery of paclitaxel. *Pharm. Res.* 2005, *22* (12), 2079–2090.
63. Yuan, H.; Bao, X.; Du, Y.Z.; You, J.; Hu, F.Q. Preparation and evaluation of SiO(2)-deposited stearic acid-g-chitosan nanoparticles for doxorubicin delivery. *Int. J. Nanomed.* 2012, *7*, 5119–5128.
64. Chang, J.E.; Shim, W.S.; Yang, S.G.; Kwak, E.Y.; Chong, S.; Kim, D.D.; Chung, S.J.; Shim, C.K. Liver cancer targeting of doxorubicin with reduced distribution to the heart using hematoporphyrin-modified albumin nanoparticles in rats. *Pharm. Res.* 2012, *29* (3), 795–805.
65. Wilson, B.; Samanta, M.K.; Santhi, K.; Kumar, K.P.; Ramasamy, M.; Suresh, B. Chitosan nanoparticles as a new delivery system for the anti-Alzheimer drug tacrine. *Nomedicine* 2010, *6* (1), 144–152.
66. Kimura, Y.; Sawai, N.; Okuda, H. Antitumour activity and adverse reactions of combined treatment with chitosan and doxorubicinin tumour-bearing mice. *J. Pharm. Pharmacol.* 2001, *53* (10), 1373–1378.
67. Zillies, J.C.; Zwiorek, K.; Hoffmann, F.; Vollmar, A.; Anchordoquy, T.J.; Winter, G.; Coester, C. Formulation development of freeze-dried oligonucleotide loaded gelatin nanoparticles. *Eur. J. Pharm. Biopharm.* 2008, *70* (2), 514–521.
68. Weiner, A.A.; Moore, M.C.; Walker, A.H.; Shastri, V.P. Modulation of protein release from photocrosslinked networks by gelatin microparticles. *Int. J. Pharm.* 2008, *360* (1–2), 107–114.
69. Sinha, V.R.; Trehan, A. Biodegradable microspheres for protein delivery. *J. Control. Release* 2003, *90* (3), 261–280.
70. Fahmy, T.M.; Samstein, R.M.; Harness, C.C.; Mark, S.W. Surface modification of biodegradable polyesters with fatty acid conjugates for improved drug targeting. *Biomaterials* 2005, *26* (28), 5727–5736.
71. Esposito, E.; Menegatti, E.; Cortesi, R. Hyaluronan-based microspheres as tools for drug delivery: A comparative study. *Int. J. Pharm.* 2005, *288* (1), 35–49.
72. Cheung, R.Y.; Rauth, A.M.; Yu, W.X. In vivo efficacy and toxicity of intratumorally delivered mitomycin C and its combination with doxorubicin using microsphere formulations. *Anticancer Drugs* 2005, *16* (4), 423–433.
73. Jones, C.; Burton, M.A.; Gray, B.N. Albumin microspheres as vehicles for the sustained and controlled release of doxorubicin. *J. Pharm. Pharmacol.* 1989, *41* (12), 813–816.
74. Hong, L.; Tabata, Y.; Miyamoto, S.; Yamada, K.; Aoyama, I.; Tamura, M.; Hashimoto, N.; Ikada, Y. Promoted bone healing at a rabbit skull gap between autologous bone fragment and the surrounding intact bone with biodegradable microspheres containing transforming growth factor-beta1. *Tissue Eng.* 2000, *6* (4), 331–340.

75. Gupta, R.K.; Griffin, P.J.; Rivera, R.; Siber, G.R. Development of an animal model to assess the immunogenicity of single-dose tetanus and diphtheria vaccines based on controlled release from biodegradable polymer microspheres. *Dev. Biol. Stand.* 1998, *92*, 277–287.
76. Wang, C.; Ye, W.; Zheng, Y.; Liu, X.; Tong, Z.; Fabrication of drug-loaded biodegradable microcapsules for controlled release by combination of solvent evaporation and layer-by-layer self-assembly. *Int. J. Pharm.* 2007, *338* (1–2), 165–173.
77. Nukavarapu, S.P.; Kumbar, S.G.; Brown, J.L.; Krogman, N.R.; Weikel, A.L.; Hindenlang, M.D.; Nair, L.S.; Allcock, H.R.; Laurencin, C.T. Polyphosphazene/nano-hydroxyapatite composite microsphere scaffolds for bone tissue engineering. *Biomacromolecules* 2008, *9* (7), 1818–1825.
78. Carrillo-Conde, B.; Garza, A.; Anderegg, J.; Narasimhan, B. Protein adsorption on biodegradable polyanhydride microparticles. *J. Biomed. Mater. Res. A* 2010, *95* (1), 40–48.
79. Tomalia, D.A.; Baker, H.; Dewald, J.; Hall, M.; Kallos, G.; Martin, S.; Roeck, J.; Ryder, J.; Smith, P. Dendritic macromolecules: Synthesis of starburst dendrimers. *Macromolecules* 1986, *19* (9), 2466–2468.
80. Hawker, C.J.; Frechet, J.M.J. Preparation of polymers with controlled molecular architecture. A new convergent approach to dendritic macromolecules. *J. Am. Chem. Soc.* 1990, *112* (2), 7638–7647.
81. Kaminskas, L.M.; Kelly, B.D.; McLeod, V.M.; Sberna, G.; Owen, D.J.; Boyd, B.J.; Porter, C.J. Characterisation and tumour targeting of PEGylated polylysine dendrimers bearing doxorubicin via a pH labile linker. *J. Control. Release* 2011, *152* (2), 241–248.
82. Choi, Y.H.; Liu, F.; Kim, J.S.; Choi, Y.K.; Park, J.S.; Kim, S.W. Poly(ethylene glycol)-grafted poly-L-lysine as polymeric gene carrier. *J. Control. Release* 1998, *54* (1), 39–48.
83. Huricha, B.; Kaname, K.; Shinichi, T.; Naohiko, F.; Masaya, S.; Tomohiro, M.; Kohsaku, O.; Gereltu, B.; Toshiyuki, U. Synthesis of an oligosaccharide–polylysine dendrimer with reducing sugar terminals leading to acquired immunodeficiency syndrome vaccine preparation. *J. Polym. Sci. Part A Polym. Chem.* 2005, *43* (11), 2195–2206.
84. Al-Jamal, K.T.; Al-Jamal, W.T.; Akerman, S.; Podesta, J.E.; Yilmazer, A.; Turton, J.A.; Bianco, A. et al. Systemic antiangiogenic activity of cationic poly-L-lysine dendrimer delays tumor growth. *Proc. Natl. Acad. Sci. USA* 2010, *107* (9), 3966–3971.
85. Deng, J.; Zhou, Y.; Xu, B.; Mai, K.; Deng, Y.; Zhang, L.M. Dendronized chitosan derivative as a biocompatible gene delivery carrier. *Biomacromolecules* 2011, *12* (3), 642–649.
86. Jianbin, T.; Xingping, W.; Xinping, W.; Meihua, S.; Weiwei, M.; Youqing, S. β-Cyclodextrin-based biodegradable dendrimers for drug delivery. *J. Control. Release* 2011, *152* (Suppl. 1), e1–e132.
87. Grinstaff, M.W. Biodendrimers: New polymeric biomaterials for tissue engineering. *Chemistry* 2002, *8* (13), 2839–2846.
88. Bhakta, G.; Rai, B.; Lim, Z.X.; Hui, J.H.; Stein, G.S.; van Wijnen, A.J.; Nurcombe, V.; Prestwich, G.D.; Cool, S.M. Hyaluronic acid-based hydrogels functionalized with heparin that support controlled release of bioactive BMP-2. *Biomaterials* 2012, *33* (26), 6113–6122.
89. Mintao, X.; Hongtao, H.; Yuanquan, J.; Jichun, L.; Hailong, H.; Xiaojian, Y. Biodegradable polymer-coated, gelatin hydrogel/bioceramics ternary composites for antitubercular drug delivery and tissue regeneration. *J. Nanomater.* 2012, Article ID 530978, 8 pages. http://dx.doi.org/10.1155/2012/530978.
90. Zhou, D.; Xiong, L.; Wu, Q.; Guo, R.; Zhou, Z.; Zhu, Q.; Jiang, Y.; Huang, J. Effects of transmyocardial jet revascularization with chitosan hydrogel on channel patency and angiogenesis in canine infarcted hearts. *J. Biomed. Mater. Res. A* February 2013, *101* (2), 567–574.
91. Kim, J.J.; Park, K. Modulated insulin delivery from glucosesensitive hydrogel dosage forms. *J. Control. Release* 2001, *77* (1–2), 39–47.
92. Huynh, D.P.; Nguyen, M.K.; Pi, B.S. Functionalized injectable hydrogels for controlled insulin delivery. *Biomaterials* 2008, *29* (16), 2527–2534.
93. Li, Z.; Yin, H.; Zhang, Z.; Liu, K.L.; Li, J. Supramolecular anchoring of DNA polyplexes in cyclodextrin-based polypseudorotaxane hydrogels for sustained gene delivery. *Biomacromolecules* 2012, *13* (10), 3162–3172.
94. Kang, G.D.; Cheon, S.H.; Song, S.C. Controlled release of doxorubicin from thermosensitive poly(organophosphazene) hydrogels. *Int. J. Pharm.* 2006, *319* (1–2), 29–36.
95. Kim, M.; Jung, B.; Park, J.H.; Hydrogel swelling as a trigger to release biodegradable polymer microneedles in skin. *Biomaterials* 2012, *33* (2), 668–678.
96. Li, Z.; Ning, W.; Wang, J.; Choi, A.; Lee, P.; Tyagi, P.; Huang, L. Controlled gene delivery system based on thermosensitive biodegradable hydrogel. *Pharm. Res.* 2003, *20* (6), 882–884.
97. Saito, T.; Tabata, Y.; Preparation of gelatin hydrogels incorporating small interfering RNA for the controlled release. *J. Drug Target.* December 2012, *20* (10), 864–872.
98. Nguyen, K.; Dang, P.N.; Alsberg, E. Functionalized, biodegradable hydrogels for control over sustained and localized siRNA delivery to incorporated and surrounding cells. *Acta. Biomater.* 2012, *9* (1), 4487–4495.

99. Rubert, M.; Monjo, M.; Lyngstadaas, S.P.; Ramis, J.M. Effect of alginate hydrogel containing polyproline-rich peptides on osteoblast differentiation. *Biomed. Mater.* October 2012, *7* (5), 055003. http://www.ncbi.nlm.nih.gov/pubmed/22782012.
100. Bajaj, G.; Kim, M.R.; Mohammed, S.I.; Yeo, Y. Hyaluronic acid-based hydrogel for regional delivery of paclitaxel to intraperitoneal tumors. *J. Control. Release* 2012 *158* (3), 386–392.
101. Chen, C.; Wang, L.; Deng, L.; Hu, R.; Dong, A. Performance optimization of injectable chitosan hydrogel by combining physical and chemical triple crosslinking structure. *J. Biomed. Mater. Res. A* 2013, *101*, 684–693.
102. Kshirsagar, S.J.; Patil, S.V.; Bhalekar, M.R. Statistical optimization of floating pulsatile drug delivery system for chronotherapy of hypertension. *Int. J. Pharm. Investig.* 2011, *1* (4), 207–213.
103. Awasthi, R.; Kulkarni, G.T. Development of novel gastroretentive floating particulate drug delivery system of gliclazide. *Curr. Drug Deliv.* 2012, *9* (5), 437–451.
104. Premchandani, T.A.; Barik, B.B. Preparation and statistical optimization of alginate based stomach specific floating microcapsules of simvastatin. *Acta Pol. Pharm.* 2012, *69* (4), 751–761.
105. Hooda, A.; Nanda, A.; Jain, M.; Kumar, V.; Rathee, P. Optimization and evaluation of gastroretentive ranitidine HCl microspheres by using design expert software. *Int. J. Biol. Macromol.* 2012, *51* (5), 691–700.
106. Desai, S.; Bolton, S. A floating controlled-release drug delivery system: In vitro-in vivo evaluation. *Pharm. Res.* 1993, *10* (9), 1321–1325.
107. Murata, Y.; Kofuji, K.; Kawashima, S.; Preparation of floating alginate gel beads for drug delivery to the gastric mucosa. *J. Biomater. Sci. Polym. Ed.* 2003, *14* (6), 581–588.
108. Sawicki, W.; Głód, J. Preparation of floating pellets with verapamil hydrochloride. *Acta Pol. Pharm.* 2004, *61* (3), 185–190.
109. Rajinikanth, P.S., Balasubramaniam, J., Mishra, B. Development and evaluation of a novel floating in situ gelling system of amoxicillin for eradication of *Helicobacter pylori*. *Int. J. Pharm.* 2007, *335* (1–2), 114–122.
110. Mishra, S.K.; Pathak, K. Formulation and evaluation of oil entrapped astroretentive floating gel beads of loratadine. *Acta Pharm.* 2008, *58* (2), 187–197.
111. Rastogi, R.; Anand, S.; Koul, V. Flexible polymerosomes an alternative vehicle for topical delivery. *Colloids Surf. B Biointerfaces* 2009, *72* (1), 161–166.
112. Kim, Y.; Tewari, M.; Pajeroski, D.J.; Sen, S.; Jason, W.; Sirsi, S.; Lutz, G.; Discher, D.E. Efficient nuclear delivery and nuclear body localization of antisense oligo-nucleotides using degradable polymersomes. *Conf. Proc. IEEE Eng. Med. Biol. Soc.* 2006, *1*, 4350–4353.
113. Rastogi, R.; Anand, S.; Koul, V. Polymerosomes of PCL and PEG demonstrate enhanced therapeutic efficacy of insulin. *Curr. Nanosci.* 2009, *5* (4), 409–416.
114. Chen, S.F.; Lu, W.F.; Wen, Z.Y.; Li, Q.; Chen, J.H. Preparation, characterization and anticancer activity of norcantharidin-loaded poly(ethylene glycol)-poly (caprolactone) amphiphilic block copolymer micelles. *Pharmazie* 2012, *67* (9), 781–788.
115. Jin, C.; Yang, W.; Bai, L.; Wang, J.; Dou, K. Preparation and characterization of targeted DOX-PLGA-PEG micelles decorated with bivalent fragment HAb18 F(ab')2 for treatment of hepatocellular carcinoma. *J. Control. Release* 2011, *152* (Suppl. 1), 14–15.
116. Zhang, Y.; Li, X.; Zhou, Y.; Wang, X.; Fan, Y.; Huang, Y.; Liu, Y. Preparation and evaluation of poly (ethylene glycol)-poly(lactide) micelles as nanocarriers for oral delivery of cyclosporine A. *Nanoscale Res. Lett.* 2010, *5* (6), 917–925.
117. Koo, H.; Lee, H.; Lee, S.; Min, K.H.; Kim, M.S.; Lee, D.S.; Choi, Y.; Kwon, I.C.; Kim, K.; Jeong, S.Y. In vivo tumor diagnosis and photodynamic therapy via tumoral pH-responsive polymeric micelles. *Chem. Commun. (Camb).* 2010, *46* (31), 5668–5670.
118. Mi, Y.; Yitao, D.; Leyang, Z.; Xiaoping, Q.; Xiqun, J.; Baorui, L. Novel thermosensitive polymeric micelles for docetaxel delivery. *J. Biomed. Mater. Res. Part A* 2007, *81* (4), 847–857.
119. Jaime, A.; Yanez, M.; Laird, F.; Yusuke, O.; Glen, S.K.; Davies, N.M. Pharmacometrics and delivery of novel nanoformulated PEG-b-poly(ε-caprolactone) micelles of rapamycin. *Canc. Chemother. Pharmacol.* 2008, *61* (1), 133–144.

Section IV

Characterization

Section IV

Characterization

31 Encapsulation Field Polymers
Fourier Transform Infrared Spectroscopy (FTIR)

Oana Lelia Pop, Dan Cristian Vodnar, and Carmen Socaciu

CONTENTS

31.1 Introduction 617
31.2 Bioencapsulation: General Data 618
31.3 FTIR Spectroscopy 618
31.4 Polymeric Biomaterials Used for Microencapsulation: Chemical Structure and Properties 619
    31.4.1 Classification 619
    31.4.2 Chemical Composition and Properties 620
        31.4.2.1 Homogeneous Carbohydrates (Glycans) 620
        31.4.2.2 Heterogeneous Carbohydrates 623
        31.4.2.3 Proteins 626
31.5 FT–MIR–ATR Fingerprint of the Biopolymers 626
    31.5.1 Materials and Methods 626
        31.5.1.1 FT–MIR–ATR Spectra of Individual Polymers 628
        31.5.1.2 Heterogeneous Carbohydrates 629
31.6 Identification of Polymer Structure by FT–MIR–ATR Analysis 635
31.7 Conclusion 635
Acknowledgments 635
References 636

31.1 INTRODUCTION

Bioencapsulation is an emerging technology applied to bioactive molecules to be protected and released under controlled conditions. It has many potential uses in agriculture, food industry, pharmacy, and biomedicine, actually not fully exploited. The most used matrices to build microcapsules or microspheres are polymers, either synthetic or natural, the last ones, known as "biopolymers," being preferred for their biocompatibility and good acceptance in food and cosmetics.

A proper, nondestructive, and fast method is required to fingerprint the specific functional groups of the matrix and to identify the modifications induced by encapsulation or the behavior during processing. The higher performance of Fourier Transform Infrared Spectroscopy (FTIR) recommends this method, especially with high reproducibility. FTIR spectroscopy is based on interferometry and makes use of a beamsplitter to divide the infrared radiation into two beams, with one beam being directed to a fixed mirror and the other to a moving mirror. When these two beams are reflected back to the beamsplitter and recombine, they undergo constructive and destructive interference due to the path difference between the two mirrors, yielding an interferogram. No guidelines or appropriate documentation of the characteristics of the biomaterials applied is available.

In this context, we used Fourier Transform Infrared–Mid Infrared–Attenuated Total Reflectance (FT–MIR–ATR) spectroscopy to characterize nine of the most important hydrophilic polymers used for microencapsulation, easy to form gels, namely, alginate, κ-carrageenan, chitosan, pullulan, dextrin, microcrystalline cellulose, hydroxypropyl methylcellulose, guar gum, and gelatin. For the attenuated total reflection–infrared (ATR–IR) spectroscopy, the infrared radiation is passed through an infrared transmitting crystal with a high refractive index, allowing the radiation to reflect within the attenuated total reflectance (ATR) element several times. The IRPrestige-21 FTIR Spectrometer, from Schimatzu, used in all the determinations is equipped with, the MIRacle single reflection horizontal ATR accessory, with a single reflection crystal plate and a high-pressure clamp.

In this entry, are presented, in the this order, general data about bioencapsulation and FTIR spectroscopy, followed by a presentation of the most important biopolymers used in this technology, their chemical structure. Then the integral FT–MIR–ATR spectra and the specific. In this entry, are presented, in the this order, general data about bioencapsulation and FTIR spectroscopy, followed by a presentation of the most important biopolymers used in this technology, their chemical structure. Then the integral FT–MIR–ATR spectra and the specific "fingerprint region" specific to each one is included. Finally, the identification of specific peaks that allow the identification of each biopolymer is presented.

These data can be useful to evaluate the identity and authenticity of each biopolymer, their stability, and modifications in the encapsulated form, during storage and specific degradations during targeted action.

31.2 BIOENCAPSULATION: GENERAL DATA

Bioencapsulation is a technology, being used since three decades, which uses bioactive molecules to be inserted and immobilized on specific supports (matrices). Encapsulation technology is now well developed and accepted within the pharmaceutical, chemical, cosmetic, foods, and printing industries.[1–4] The encapsulation of active components has become very attractive being adequate for food ingredients as well as for chemicals, drugs, or cosmetics to be released in a controlled way.

The bioactive substance that is encapsulated is called the core material, "active ingredient" or "agent," or internal phase. The material encapsulating the core is referred to be a matrix, as support material, coating, shell, or wall material.

Microcapsules can be classified in three basic categories according to their morphology as mono-cored (mononuclear), poly-cored (poly-nuclear), and matrix types. Mono-cored (mononuclear) microcapsules contain the shell around the core.

For controlled release, bioencapsulation is not just an added value technique, but is also a source of new ingredients with unexpected properties. The growing interest by food technologists in the enormous potential of bioencapsulation is demonstrated by the exponential increase in the number of publications (nonscientific and scientific articles and patents).[5–8]

Microencapsulation can be used to prevent excessive degradation of a sensitive ingredient or to reduce flashing off of volatile flavors during processing, in order to save on an expensive ingredient; in this case the cost-in-use must be lower than the nonencapsulated ingredient.[9–13]

However, the most important aspect of bioencapsulation, from the first laboratory tests, is an understanding of the industrial constraints and requirements to make a bioencapsulation process viable, from the transition to full-scale production to the marketing of the final product.[14]

31.3 FTIR SPECTROSCOPY

There is a diversity of both natural and synthetic polymeric systems that have been analyzed and studied for the entrapment of bioactive molecules and their controlled release. The use of FT-MIR spectroscopy to characterize their specific fingerprint is a pre-requisite for an appropriate characterization of microcapsules.

Infrared spectroscopy is a valuable, nondestructive technique, based on absorption, emission, or scattering of infrared radiation, which gives structural and quantitative information about functional groups in biomolecules, extensively used nowadays in biomaterials and life sciences.[12,15] The MIR region is mostly used to register spectra, from wavenumbers of 700–4000 cm^{-1}.

With the advent of FTIR, major drawbacks of "classical" dispersive Infrared spectroscopy could be put off. Briefly, FTIR no longer measures one wavelength after the other and is not like dispersive spectrometers, it applies interferometric modulation of radiation. FTIR spectrometers have several advantages over conventional dispersive IR instruments, including a dramatic improvement in the signal-to-noise ratio obtained by multiplexing (simultaneous detection of all frequencies), reduction in scan time, and higher energy throughput.[16] Another important advantage is the excellent wavelength reproducibility due to an internal reference laser, which allows spectral data manipulations with a very high degree of accuracy.

In conclusion, the most important advantages of FT–MIR spectroscopy to be used in biopolymer characterizations are the spectra can be obtained instantly in a solid form, liquid form, and in different solutions, the sample aliquots are small (less than 1 g), the operation of the equipment is simple, cheap, and the interpretation is done due to data processing software attached to instruments. No light scattering or fluorescent effects may interfere.[17–20]

31.4 POLYMERIC BIOMATERIALS USED FOR MICROENCAPSULATION: CHEMICAL STRUCTURE AND PROPERTIES

31.4.1 Classification

The material (matrix) that ensures the protection and the controlled release of the entrapped bioactive compound in a microcapsule is called shell material, wall material, or coating material.

A major problem in the microencapsulation field is the absence of guidelines for documentation of the characteristics of the materials applied. It is claimed to be mandatory because it is now widely accepted that the characteristics of the polymer is a dominant factor in determining the capsule properties.[21]

A large range of different materials can be used for encapsulation, either hydrophilic or hydrophobic, from polyelectrolytes, synthetic or natural,[22,23] to inorganic nanoparticles,[24] lipids,[25] dyes,[26] multivalent ions, and biopolymers.[27] A classification of the most used biopolymers for bioencapsulation is presented in Table 31.1.

TABLE 31.1
Classification of the Most Used Biopolymers for Bioencapsulation

		Homogeneous Carbohydrates Natural or Modified	Heterogenic Carbohydrates
Hydrophilic materials	Carbohydrates	Dextrins	Gums: xanthan, guar
		Starch	Alginates
		Cellulose and modified derivatives	Carrageenans
		Modified starch and cyclodextrins	Chondroitin sulfate, dextran sulfate
			Chitosan
		Pullulan	Agarose
	Proteins	Animal proteins	Vegetable proteins
		Collagen	Gluten
		Gelatin	Zein
		Casein	Polylysine
		Fibrin	
Hydrophobic materials	Waxes	Lipids	Polymers
	Bee wax	Phospholipids (Lecithin)	Shellac

Synthetic polymers (polyvinyl alcohol, polylactic acid,) are very popular due to their higher stability and lower price, but the biocompatible natural polymers are more and more applied, as well as cellular tissue matrices (e.g., bladder submucosa and small intestinal submucosa).

Biopolymers are polymers generated from renewable natural sources, often biodegradable and nontoxic. They can be produced by biological systems (i.e., microorganisms, plants, and animals), or chemically modified from biological starting materials (e.g., cellulose, starch, natural fats, or oils).

31.4.2 CHEMICAL COMPOSITION AND PROPERTIES

31.4.2.1 Homogeneous Carbohydrates (Glycans)

31.4.2.1.1 Dextrin

Dextrin is derived from starch by partial thermal degradation under acidic conditions. The only structural difference between dextrin and starch is that dextrin is a smaller highly branched molecule.

Dextrin consists of both linear (amylose) and branched (amylopectin) components. A schematic chemical structure of the dextrin molecule is shown in Figure 31.1. Amylose consists of glucose units joined mainly by α-glycosidic links through glucose residue carbons 1 and 4 (α-(1–4) linkage).[28] Amylopectin is a highly branched polymer with one of the highest molecular weights. Amylopectin is composed of linear chains of α-(1–4)-D-glucose residues (as in amylose, but shorter) connected together through α-(1–6)-linkages.

The molecular weight of dextrin may range from 800 to 70,000, whereas starch molecules are larger with a molecular weight of up to several millions. Dextrin has been found to adsorb on naturally hydrophobic minerals such as coal, talc, molybdenite, and others, but not at the hydrophilic surface of quartz or pyrite.[29] It was observed that when hydrophilic minerals locked hydrophobic grains at their surfaces, an intermediate level of dextrin adsorption occurred. Several hypotheses have been made based on the adsorption mechanism of dextrin adsorption at different mineral surfaces. It has been proposed that some adsorption phenomena are involved in dextrin adsorption such as (1) hydrophobic bonding between nonpolar sites of the dextrin molecule and the hydrophobic surface, (2) hydrogen bonding with surface sites, and (3) reactions with metal hydroxy species at

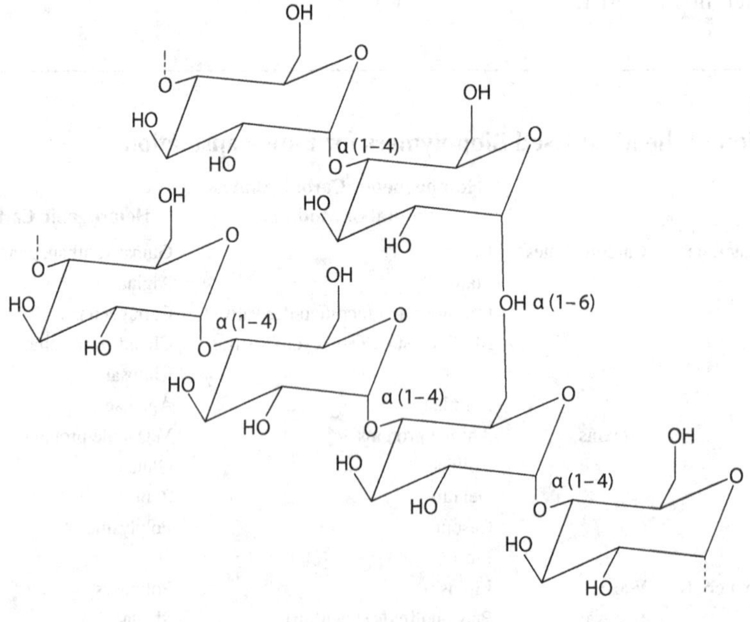

FIGURE 31.1 Dextrin with α (1→4) and α (1→6) glycosidic bond.

Encapsulation Field Polymers

mineral surfaces.[29] Dextrin has been used in the encapsulation of different food components, such as oils[30] and other active molecules, in order to protect them from the unfriendly environment.

It is evident that dextrin consists of numerous hydrophilic OH groups, which are projected away from hydrophobic hydrocarbon rings containing an oxygen atom as part of the ring structure. This is why dextrin has the tendency to hydrogen bond with other hydrophilic molecules.

31.4.2.1.2 Microcrystalline Cellulose

Microcrystalline cellulose is a naturally occurring polymer obtained from cellulose by a partial purification, and is a depolymerized nonfibrous form, inodorous and tasteless crystalline powder. Microcrystalline cellulose is composed of units connected by a 1–4 beta-glycosidic bonds as can be observed in Figure 31.2.

The cellulose hydrolysis in order to obtain the microcrystalline cellulose can be made using mineral acid, enzymes, and microorganisms. Microcrystalline cellulose is used, among other cellulosic materials in almost all areas of industries[31] but has a greater importance in foods, cosmetics, and pharmaceutics. In the pharmaceutical industry, it is used as a filling and binding agent[32] and in the encapsulation area it is used mostly as a filler.

31.4.2.1.3 Hydroxypropyl Methylcellulose

Hydroxypropyl methylcellulose is a synthetic modification of the natural polymer, cellulose, fibrous or granular, free-flowing powder, odorless, and tasteless. The representation of the molecular structure of hydroxypropyl methylcellulose can be seen in Figure 31.3.

When hydroxypropyl methylcellulose is dissolved in water, colloids are formed. The compound is nontoxic but with oxidizing agents it is combustible. Unlike methylcellulose, hydroxypropyl methylcellulose in an aqueous solution, displays a thermal gelation property.[33] This happens when the solution is heated up to a critical temperature; the solution congeals into a nonflowable but semi-flexible mass. Typically, this critical temperature point is inversely related to (1) the solution concentration and (2) the concentration of the methoxy group within the hydroxypropyl methylcellulose molecule.[34]

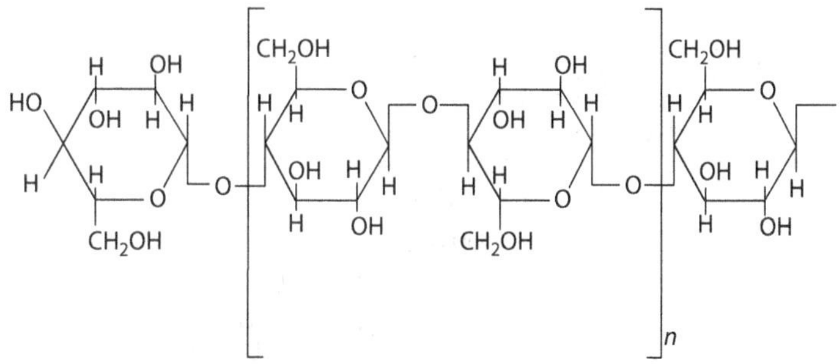

FIGURE 31.2 Structural formula of microcrystalline cellulose.

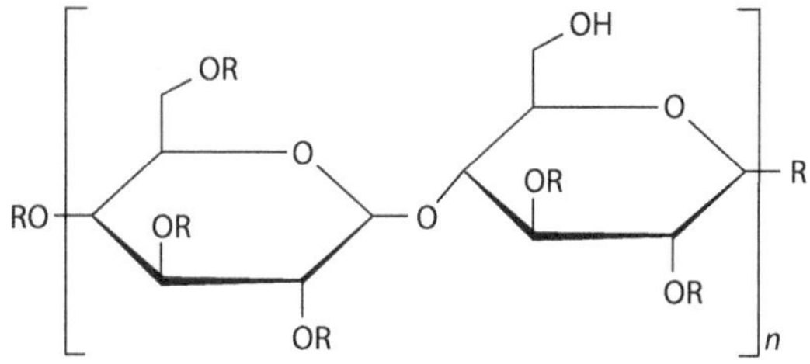

FIGURE 31.3 Molecular structure of hydroxypropyl methylcellulose, R = H or CH_3.

Hydroxypropyl methylcellulose is mostly used in pharmaceutics[35] as a drug tablet formulation and release tablet matrix. In foods it is used as an additive filler, gelling agent, film former, and stabilizer. Hydroxypropyl methylcellulose carries code E464 in the European Union, making it a known and listed food additive. It is an important source of soluble dietary fiber, and recognized as a nontoxic material.[34] The polymer is soluble in cold water and some organic solvents, over the entire biological pH range. From aqueous solutions hydroxypropyl methylcellulose forms transparent, tough, and flexible films. The literature[36] reports the first paper about hydroxypropyl methylcellulose in 1960, in a patent.

31.4.2.1.4 Pullulan

Pullulan is a neutral glucan (like amylose, dextran, cellulose), with a chemical structure more or less depending on carbon source, producing microorganisms (different strains of *Aureobasidium pullulans*) and the fermentation conditions.[37]

The basic structure (Figure 31.4) is a linear α-glucan one, made from three glucose units linked to α-1,4 in maltotriose units, which are linked in a α-1,6 way.

It is water soluble, insoluble in organic solvents, and not hygroscopic; its aqueous solutions are stable and show a relatively low viscosity compared with other polysaccharides. It has the specific characteristic to form thermo-stable, transparent, elastic, antistatic films easily, with extremely low oxygen permeability.[38] It is nontoxic, edible, biodegradable, and biocompatible polymer. Pullulan has many applications in food packaging due to the property of forming oxygen-impermeable follies and films, and also is used as composition filming agent in pharmaceutics (retard tablets, capsules, and microcapsules, and disaggregating and local administration agents) and in cosmetics (hydrating creams and gels).[38] The textiles industry uses pullulan as an antistatic agent, the paper industry for

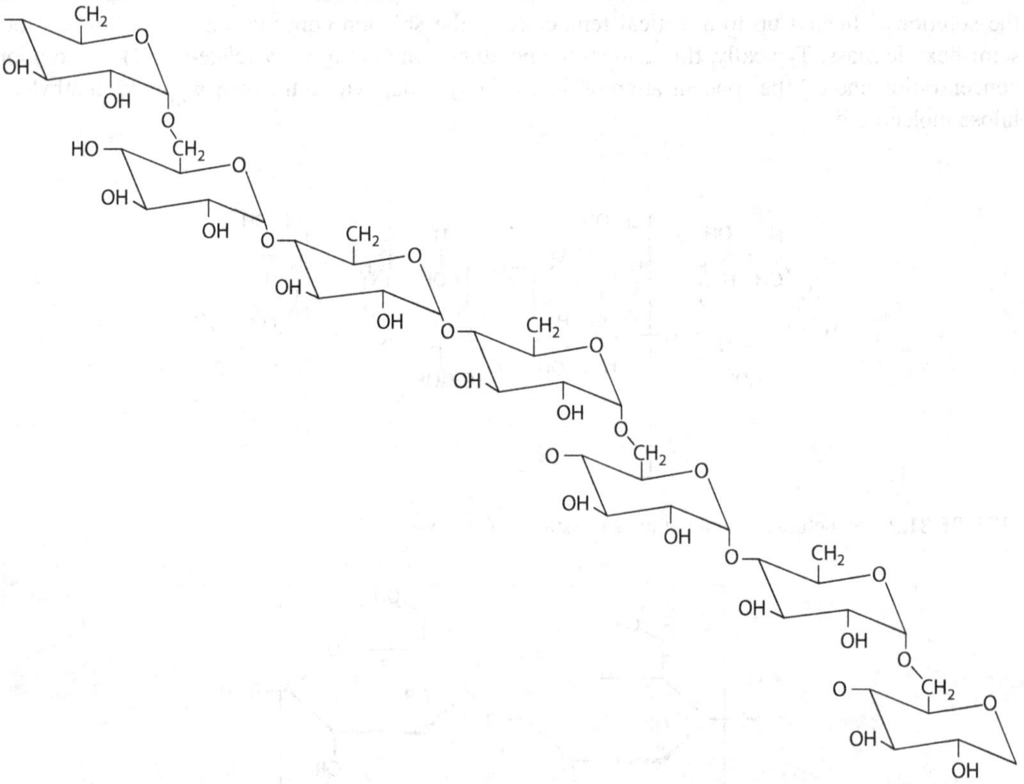

FIGURE 31.4 Chemical structure of a representative portion of pullulan, made of α-1,4 maltotriose units linked α-1,6.

coating and in composition, and in agriculture as a slow release fertilizer. The cross-linked pullulan microparticles are successfully utilized as intermediates in obtaining ionic chromatographic supports, in chromatographic separations through size exclusion of substances with various molecular weights, and in healing of infected wounds.[37]

31.4.2.2 Heterogeneous Carbohydrates

Carbohydrates are the most used hydrophilic materials for bioencapsulation. There are several reasons that make the utilization of these types of biomaterials very common, such as easy to find on the market, relatively low cost, very convenient functional properties that make these biomaterials accessible to many bioencapsulation techniques, and make the bioencapsulation of a large range of active materials suitable.

31.4.2.2.1 Alginates

Alginates are natural gums and biopolymers widely used in food and beverage as complexing and forming gel agents, in pharmaceutical industry as additive, and other applications.[39] Alginates (E400-E404) are produced by brown seaweeds.

Alginate belongs to linear unbanked polysaccharides. It contains two uronic acids, α-L-guluronic (G) and β-D-mannuronic (M), in varying amounts and alternating sequence, as can be seen in the delineate structure (Figure 31.5).[40] Alginates are not random copolymers but, according to the source, algae consist of blocks of similar and strictly alternating residues (i.e., MMMMMM, GGGGGG, and GMGMGMGM), each of which have different conformational preferences and behavior; as examples, the M:G ratio of alginate from different sources can have considerable difference ratio numbers.

The amount and sequences of α-L-guluronic and β-D-mannuronic acids influence the molecular weight and default the physical properties of the alginates. Alginate can be cross-linked into divalent solutions such as Ca^{2+}, Sr^{2+}, or Ba^{2+}.[41] The monovalent ions and Mg^{2+} do not induce gelation,[42] but a stronger gelation is induced by Ca^{+2} as is represented in Figure 31.6a, also by Sr^{2+} and Ba^{2+}.

Alginate's solubility and water-holding capacity depend on pH (precipitating below about pH 3.5), molecular weight (lower molecular weight calcium alginate chains with less than 500 residues showing increasing water binding with increasing size), ionic strength (low ionic strength increasing the extended nature of the chains), and the nature of the ions present. When alginate is coordinated to sodium, it is a very flexible chain. When sodium is replaced by calcium, each calcium ion (gray dots in Figure 31.6b) coordinates to two alginate chains, linking them together. The flexible chains (Figure 31.6b) become less flexible and form a huge network—a gel. This happens within seconds after the alginate mixture is dripped into the water bath with the calcium ions.

The literature[42] reports some of the properties that make the alginate so useful in so many areas, namely, mild encapsulation process, high gel porosity, suitability for diverse coatings applications,

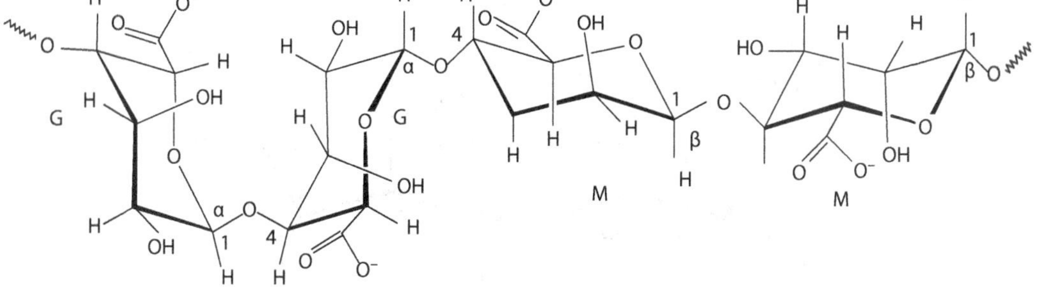

FIGURE 31.5 Chemical structure of alginate showing the α-L-guluronic and β-D-mannuronic acids blocks.

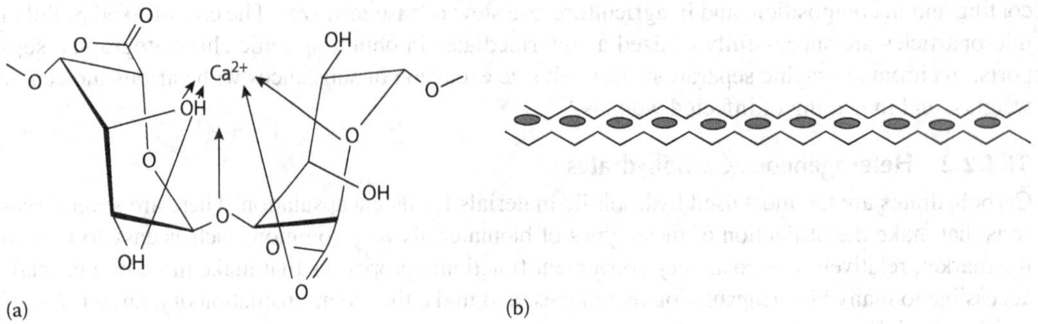

FIGURE 31.6 (a) The representation of alginate cross-linked with Ca^{2+} and (b) Ca^{2+} bound in alginate chain.

biodegradation, and dissolution under physiological conditions for a controlled and targeted release. All of these properties, in addition to the nonimmunogenicity of alginate, have led to an increased use of this polymer as a protein delivery system.

31.4.2.2.2 Carrageenan

Carrageenan (E407) is a generic term for different sulfated polysaccharides prepared by alkaline extraction (and modification) from red seaweed (*Rhodophyceae*), water-soluble polymers, typically form highly viscous aqueous solutions. Carrageenans are now defined in terms of chemical structure. Carrageenan consists of alternating 3-linked-β-D-galactopyranose and 4-linked-α-D-galactopyranose units. κ-Carrageenan (kappa-carrageenan) consists of an alternating linear chain of –(1→3)-β-D-galactopyranose-4-sulfate–(1→4)-3,6-anhydro-α-D-galactopyranose–(1→3) (Figure 31.7). It is soluble in hot water (>75°C) and even low concentrations (0.1–0.5%) yield high viscosity solutions[43]; viscosity increases with molecular weight. Different carrageenans are known, such as λ-, μ-, ν-, κ-, and ι-carrageenans.[44]

Gelation in the κ- and ι-carrageenan solutions is the result of a coil-to-helix molecular transition followed by aggregations that occur upon cooling. However, the exact gelation mechanism has not been established.[45] What is well known is that gel strength and gelation temperature both depend on the ionic strength, that is, κ-carrageenan is sensitive to K^+ and ι-carrageenan to Ca^{2+}.

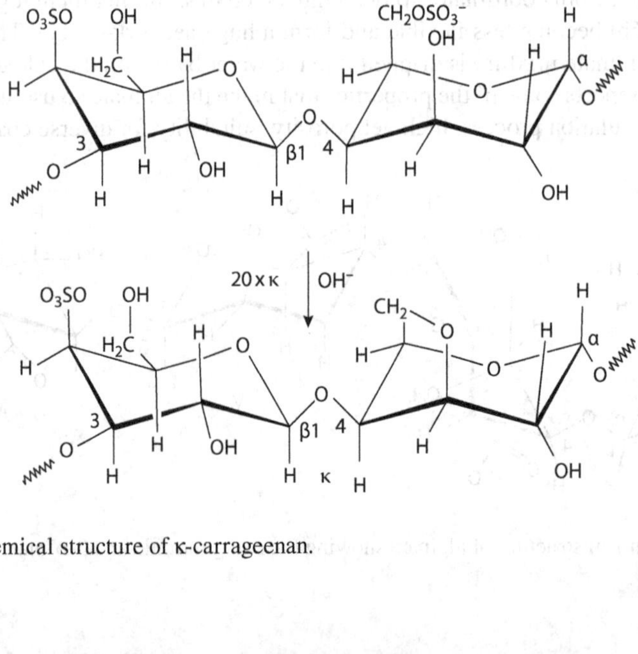

FIGURE 31.7 Chemical structure of κ-carrageenan.

The strongest gels of κ-carrageenan are formed with K⁺ rather than Li⁺, Na⁺, Mg²⁺, Ca²⁺, or Sr²⁺.[46] Formation of a gel using this polymer occurs because of temperature changes. In the bioencapsulation process the active material is mixed with the heat-sterilized polymer solution at 40°C–50°C. κ-Carrageenan produces fragile gels that are not able to stand stresses of internal bacterial growth.[47]

Carrageenans are frequently used in combination with starch in gelled dairy desserts, due to the namely milk reactivity, which has been attributed to an electrostatic interaction between a positively charged region of κ-casein and the negatively charged sulfate groups of κ-carrageenan.[48] It was reported that[49] κ-carrageenan increases the charge density of casein micelles below the coil–helix transition temperature, leading to the absorption of κ-carrageenan on casein micelles.

Carrageenans specifically tailored for water-thickening applications are usually lambda types or the sodium salt of mixed lambda and kappa. They dissolve in either cold or hot water to form viscous solutions. Their high water viscosities are desirable, and the high molecular weight and hydrophilicity of lambda contribute to this.[50] For gelling applications, a low viscosity in hot solution is usually desirable for ease in handling, and, fortunately, high gel strength carrageenans (mixed calcium and potassium salts of kappa or iota) fulfill this requirement because of their small hydrophilicity and the effect of the calcium ions.

31.4.2.2.3 Chitosan

Chitosan is a nontoxic and biocompatible linear polysaccharide derivate from chitin. Chitin is found in the exoskeleton of crustaceans, in fungal cell wall, and in some biomaterials.[51,52] The partial deacetylation of chitin, in the presence of hot alkali, leads to the formation of chitosan. Chitosan (Figure 31.8) is formed by 80% β-(1,4)-2-amino-2-deoxy D-glucose and 20% β-(1,4)-2-acet-amido-2-deoxy-D-glucose repeating units. Unlike insoluble chitin, chitosan is soluble in acid solution, conditions that lead to the protonation of amino groups that are responsible for its polyelectrolyte character.[53] Both deacetylation degree (DD) and molar mass have been shown to influence properties of chitosan in solution.[54]

Chitosan is characterized by unique properties, including polyoxysalt formation, ability to form films, gels, chelate metal ions, and optical structural characteristics.[55] The ability of chitosan to form gels in N-methylmorpholine N-oxide and the application in controlled release drug are reported.[56] Cross-linking reagents for chitosan are dialdehyde compounds, glutaraldehyde, and glyoxal. The literature[57,58] reports also the usage of sodium tripolyphosphate or sodium citrate solutions as cross-linking agents for chitosan.

Chitosan has been evaluated for a wide area of applications such as pharmaceutical, agricultural, food industry, and as a waste water treatment agent.[55] The literature[59] refers to chitosan as a biocompatible material used in implantation or injections in ophthalmology, in cosmetics, in photography, as artificial skin,[55] in molecular imagistic, and in bioencapsulation field.[60] The use of chitosan in bioencapsulation field is related to the ability of some human enzymes to metabolize this polymer[61] and due to its biocompatibility and bioadhesivity favored by the positive charges at physiological pH.[62,63] From industrial point of view, the usage of this polymer is profitable because it is abundant, has a low cost production, and there is no conflict with the environmental regulations.[64]

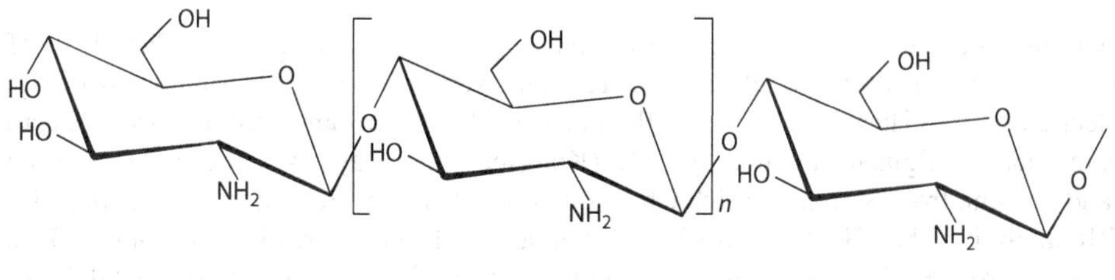

FIGURE 31.8 Chemical structure of chitosan.

31.4.2.2.4   Guar Gum

Guar gum is the endosperm polysaccharide of the seed of *Cyamopsis tetragonoloba*, belonging to the family *Luguminosae*, known as Cluster Bean Guar seed was industrially processed into guar gum in 1940s and 1950s in United States.[65] In the food, pharmaceuticals, cosmetics, and textiles industry, guar gum is used as stabilizer and thickening agent. It is a galactomannan consisting of backbone of α-1,4-D-mannopyranosyl units, with every second unit bearing a β-1,6-D-galactopyranosyl unit (Figure 31.9). It is easily available and cheap. It is a high molecular weight hydrocolloidal heteropolysaccharide composed of galactan and mannan units.

31.4.2.3   Proteins

Proteins are a class of biomaterials, more used in bioencapsulation, in association with their functional and nutritional properties.[9] Proteins may act not only as matrices but also as emulsifiers, favoring the simple or complex coacervation as bioencapsulation technique.

31.4.2.3.1   Gelatin

Gelatin is a natural, protein polymer with good biodegradability and film-forming property. It is an amphoteric, with an isoelectric pH ranging from 5 to 9.

Gelatin is the product of denaturation or disintegration of collagen I, lacking cystine in acidic environment. The representation of the chemical structure of gelatin can be seen in Figure 31.10.

The amino acid analysis of gelatin is variable, particularly for the minor constituents, depending on raw material and process used, but proximate values by weight are glycine 21%, proline 12%, hydroxyproline 12%, glutamic acid 10%, alanine 9%, arginine 8%, aspartic acid 6%, lysine 4%, serine 4%, leucine 3%, valine 2%, phenylalanine 2%, threonine 2%, isoleucine 1%, hydroxylysine 1%, and methionine and histidine <1% with tyrosine <0.5%.

Although gelatin is by far the major hydrocolloid used for gelling, current concerns about the need generated by vegetarians and certain religions, has recently encouraged the search for alternatives. The combination of the melt in the mouth, elastic and emulsification characteristics of gelatin gels is, however, difficult to reproduce.

The cross-linking of gelatin with aldehydes is being used in the encapsulation field. In particular, treatment of gelatin films with glutaraldehyde is receiving considerable study in order to improve their thermal resistance, decrease their solubility in water, and to improve the mechanical properties. An occasional phenomenon is the loss of gelatin solubility in time at room temperature. Furthermore, the preservation of foods using smokes rich in aldehydes can have unwanted reactions with gelatin. When heated, gelatin undergoes not only structural and mechanical but also physicochemical transformations such as partial or complete loss of solubility in water caused by cross-linking and lowering of molecular weight caused by thermal and thermo-oxidative destruction. These changes are observed in heating gelatin above 140°C. Above 170°C, gelatin is not easily soluble in hot water.

31.5   FT–MIR–ATR FINGERPRINT OF THE BIOPOLYMERS

31.5.1   MATERIALS AND METHODS

Materials investigated: Nine different polymers, namely, dextrin from Dow Chemicals (Berlin, Germany); microcrystalline cellulose from Rettenmayer (Rosenberg, Germany); hydroxypropyl methylcellulose from Harke Pharma (Muelheim an der Ruhr, Germany); pullulan from Hayashibara (Okayama, Japan); alginate supplied by FMC (Kongsberg, Norway); κ-carrageenan, chitosan, guar gum, and gelatine from Sigma–Aldrich Chemie GmbH (Munich, Germany) were analyzed.

FTIR analysis: The FTIR spectra were obtained with an IRPrestige 21 Fourier Transform Spectrometer from Schimatzu, equipped with the MIRacle single reflection horizontal ATR accessory, with the crystal plate and purge tubes attached to the base, and a MIRacle pressure clamp.

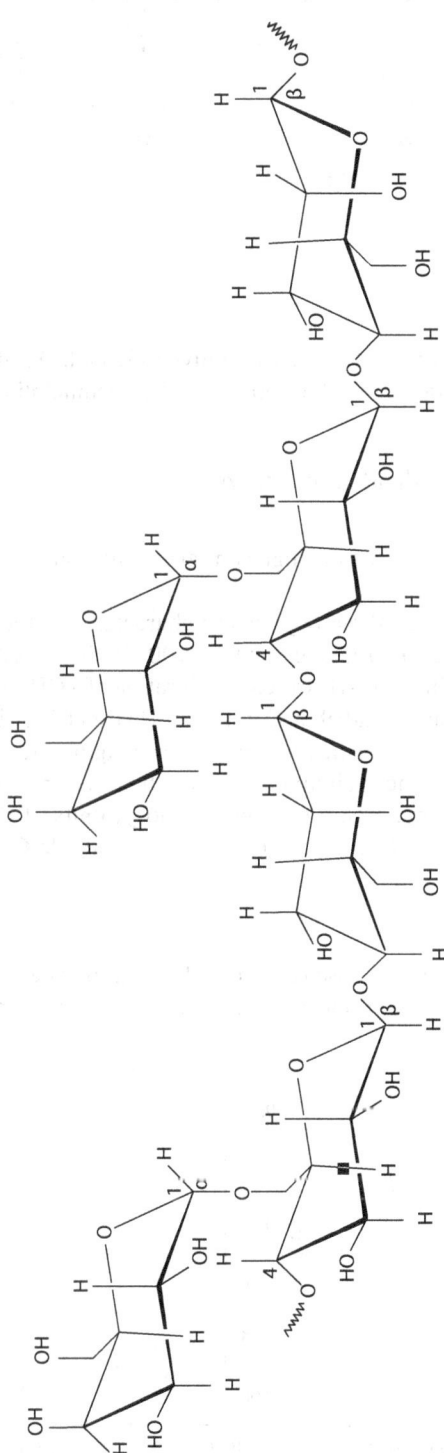

FIGURE 31.9 Chemical structure of guar gum.

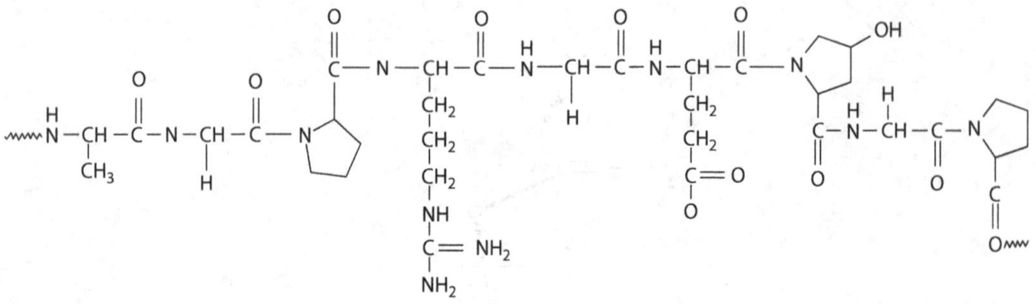

FIGURE 31.10 Chemical structure of gelatin.

Between measurements the crystal was cleaned with ethylic alcohol. All data were recorded at room temperature, in the spectral range of 4000–600 cm^{-1}, by accumulating 128 scans with a resolution of 4 cm^{-1}.

31.5.1.1 FT–MIR–ATR Spectra of Individual Polymers

31.5.1.1.1 Dextrin

The spectrum (Figure 31.11b) of dextrin shows the high similarity with starch FTIR profile (Figure 31.11a).

In the general spectra (Figure 31.12a), three regions are characteristic: the fingerprint region (1200–900 cm^{-1}), region II (1410–1330 cm^{-1}), and region III (2900–3500 cm^{-1}). At 2900–2923 cm^{-1} is located the absorbance corresponding to C–H stretching vibration of –CH$_2$ groups, whereas the large band between 3100 and 3500 cm^{-1}, is attributed to the O–H stretching vibrations.

The most characteristic bands from the fingerprint region (Figure 31.12b) are located at 1148, 1074, 990, 925, and 858 cm^{-1}. The assignment of each band is presented in Table 31.2. The peak located at 1148 cm^{-1} is characteristic to glucopyranosyl units of the polysaccharide, whereas the dominant peak is seen at 990 cm^{-1}, corresponding to simple C–O bonds from the α-D-glucopyranoside.[66]

31.5.1.1.2 Microcrystalline Cellulose

The FTIR spectrum of microcrystalline cellulose (Figure 31.13) highlights a slightly different fingerprint region, whereas the regions II and III have similar shapes and wavenumbers.

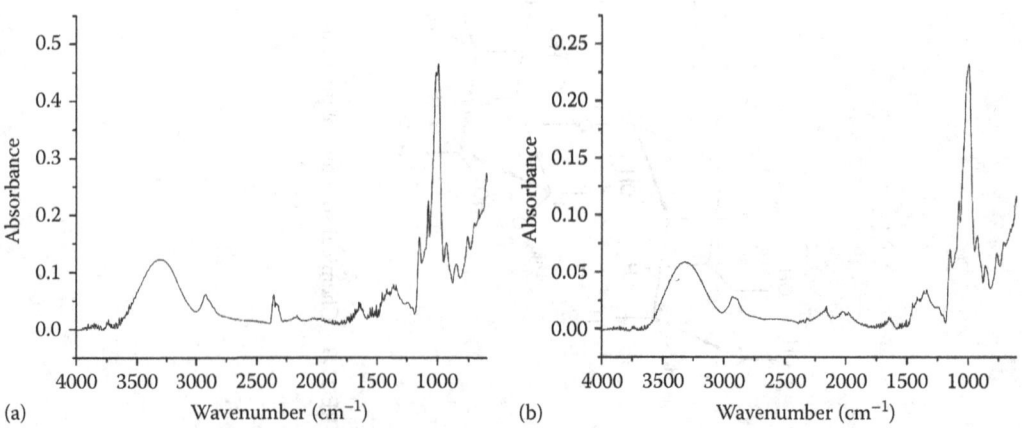

FIGURE 31.11 FTIR absorbance spectra of (a) starch and (b) dextrin.

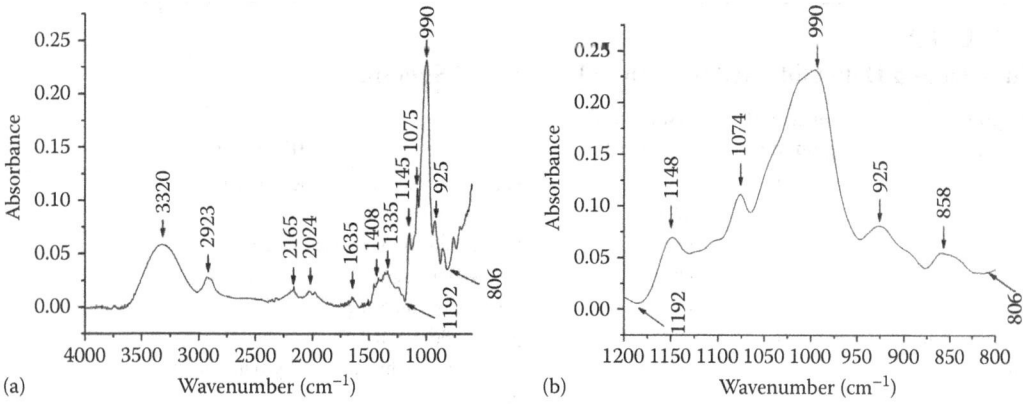

FIGURE 31.12 FTIR absorbance spectra of dextrin (arbitrary values). (a) From 900 to 4000 cm^{-1} and (b) the fingerprint region from 900 to 1200 cm^{-1}.

In the fingerprint region (1500–800 cm^{-1}), dominant peaks were observed at 1128, 1109, and 1054 cm^{-1}, corresponding to n(CO), n(CC), ring. The presence of C–O–C asymmetric valence vibration, in the ring n(C–O–C) and β-(1→4) linkages is assigned to the band at 1159 cm^{-1}.[67]

The literature data[31,67] report that the peaks located at 1375 and 1312 cm^{-1} are associated with the intermolecular hydrogen bonding at the C$_6$ group and OH. The band at 1426 cm^{-1} correspond to ring vibrations δ(CH$_2$), δ(CH), the band at about 900 cm^{-1} is attributed to the asymmetric ring stretching in cellulose due to the C–O bonds between β-glucose.

31.5.1.1.3 Hydroxypropyl Methylcellulose

Similarly, the three regions are well defined; the most differentiated part of the spectra is the fingerprint region. The main absorbance band of hydroxypropyl methylcellulose (Figure 31.14) is located at 1050 cm^{-1} and is attributed to n(CO), ring vibration.

Beside this dominant peak, we can notice at 1450, 1372–1312, band corresponding to ring deformations δ(CH$_2$), δ(CH). At 1200 cm^{-1}, the evidence of metoxy group vibrations, at 1106 cm^{-1}, combinations of vibrations for n$_{asym}$(C–O–C) n(CO), n(CC), δ(OCH).

The band at about 945 cm^{-1} can be attributed to the asymmetric ring stretching in cellulose due to the C–O bonds between β-glucose.

31.5.1.1.4 Pullulan

The FTIR absorbance spectra of pullulan can be seen in Figure 31.15, being similar, mostly, to dextrin. Information on the glucopyranosyl units' conformation in the polysaccharide can be acquired in the fingerprint region located between 1250 and 800 cm^{-1}.

The most intense peak at 994 cm^{-1} indicates the α-(1→4) linkages in the chain together with the peak from 930 cm^{-1}, indicating the α-(1→6) linkage type of inter-unit bonds and angles.[68] The existence of α-(1→4) and α-(1→6) glycosidic linkages in the pullulan spectra can be set also by the band at 1150 and 1079 cm^{-1}.[69,70] The peak at 1640 cm^{-1} is the result of H–O–H stretching.

31.5.1.2 Heterogeneous Carbohydrates

Comparing with the glucan-type of carbohydrates, the heterogeneous polysaccharides have a different fingerprint region (1700–900 cm^{-1}), characterized by three specific subregions, located between 900–100, 1300–1400, and around 1600 cm^{-1}.

TABLE 31.2
Summarized Data Obtained from the FT–MIR–ATR Spectra

Biopolymer Name	Fingerprint Region (cm^{-1})	Wavenumber (cm^{-1})	Attribution
Dextrin	1200–800	1148	Glucopyranosyl units in the polymer
		1074	Glucopyranosyl units
		990	C–O bonds from the α-D-glucopyranoside
		925	α-(1→4) linkages
		858	α-D-glucopyranoside
Microcrystalline cellulose	1500–800	1426	δ(CH$_2$), δ(CH), RINF
		1375–1312	Hydrogen bonding at the C$_6$ group and OH
		1159	n(C–O–C), ring
			β-(1→4) linkages
		1128, 1109	n(CO), n(CC), ring
		1054	C–O–C cyclic ether stretching vibration
		900	C–O bond, β-(1→4) linkages
Hydroxypropyl methylcellulose	1500–800	1450, 1372–1312	δ(CH$_2$), δ(CH), ring
		1200	Methoxy groups
		1106	n$_{asym}$(C–O–C)
			n(CO), n(CC), δ(OCH)
		1050	C–O–C cyclic ether stretching vibration
		945	Stretching of C–O bonds
Pullulan	1250–800	1150	Stretching of α-(1–6) glycosidic linkages
		1079	Glucopyranosyl units
		994	Stretching of α-(1→4) linkages
		930	Stretching of α-(1→6) linkages
		848	Stretching of α-D-glucopyranoside, α-(1–4)-D-glicosidic
Alginate	1700–900	1604	C=O stretch
		1404	C–O single bond stretch from phenol group
		1317	Stretching of C–O bond
		1124	also C–O single bond band
		1087	C–O–C cyclic ether stretching vibration
		1027	Stretching of C–O bonds, indicators of guluronic and mannuronic acids
		945	Stretching of C–O bonds
Carrageenan	1350–800	1216	SO$_3^-$ group stretching
		1153	C–O stretch of cyclic ethers
		1063	C–O–C cyclic ether stretching vibration
		1021	C–O–C (cyclic ether) stretching vibration
		965	trans-disubstituted alkenes
		924, 900	C–O stretch of polyhydroxy groups attached to carbons
		842	D-galactose-4-sulfate, 3,6-anhydro-D-galactose, glycosidic linkage
Chitosan	1650–600	1630–1560	N–H bends
		1373–1317	Stretching of C–O bond
		1148	C–O stretch of cyclic ethers
		1059	C–O–C cyclic ether stretching vibration
		1027	C–O groups of the sugar structure
		895	N–H bending

(Continued)

Encapsulation Field Polymers

TABLE 31.2 (Continued)
Summarized Data Obtained from the FT–MIR–ATR Spectra

Biopolymer Name	Fingerprint Region (cm⁻¹)	Wavenumber (cm⁻¹)	Attribution
Guar gum	1200–800	1143	Stretching of primary C–OH
		1059	C–O–C cyclic ether stretching vibration
		1017	–CH$_2$ twisting vibration
		860	Glycosidic linkage (in galactose and mannose)
Gelatin	1650–800	1635	C=O stretch
		1558	Amide I band
		1540–1510	Amide I band
		1456–1339	Carboxyl bands
		1238	Stretching of C–O
		1079	Symmetrical deformation of CH$_2$

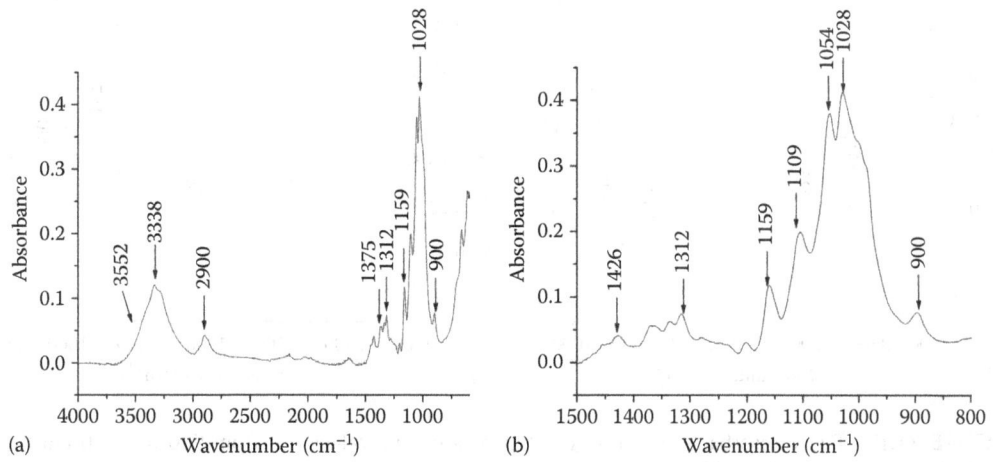

FIGURE 31.13 FTIR absorbance spectra of microcrystalline cellulose (arbitrary values) (a) from 900 to 4000 cm⁻¹ and (b) the fingerprint region from 800 to 1500 cm⁻¹.

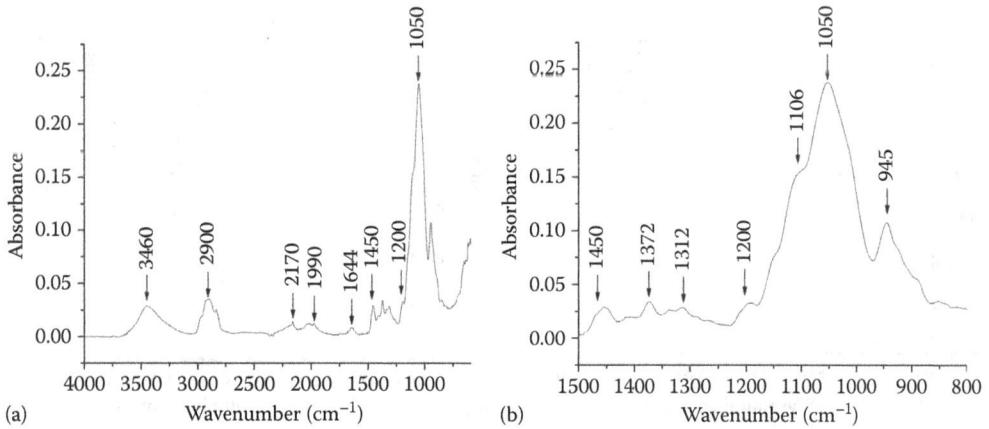

FIGURE 31.14 FTIR absorbance spectra of hydroxypropyl methylcellulose (arbitrary values) (a) from 900 to 4000 cm⁻¹ and (b) the fingerprint region from 800 to 1500 cm⁻¹.

31.5.1.2.1 Alginate

Beside the region III bands, at 3244 cm^{-1} representing hydroxyl groups (OH– stretching) and at 2926 cm^{-1} assigned to C–H and C–O stretching of the carboxyl group, the alginate spectrum (Figure 31.16) has its specific fingerprint region (I + II) from 945 to 1700 cm^{-1}.

At 1604 and 1404 cm^{-1}, two intense peaks corresponding to carboxyl group (COO– asymmetric and symmetric stretching) were identified[71]; Other peaks were identified at 1317, 1124, and 1087 cm^{-1}, characteristic to oligo-sugars whereas the intense peak at 1027 and 945 cm^{-1} is specific to the C–O bonds of the saccharide structure[72] and indicate the presence of guluronic and mannuronic acids, respectively.[73]

31.5.1.2.2 κ-Carrageenan

The FTIR spectrum of κ-carrageenan (Figure 31.17) is specific in the fingerprint region from 1350 to 800 cm^{-1}. Apart from alginate, the peak at 1604 cm^{-1} in less intense, specifically, in

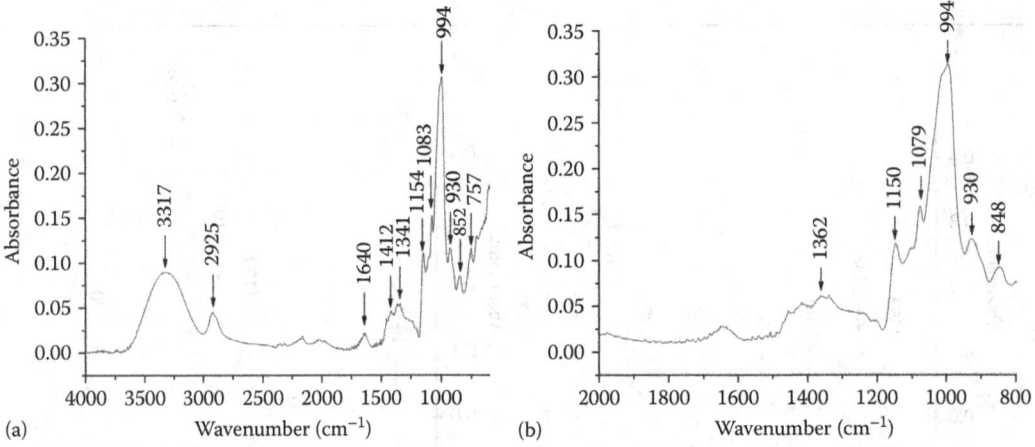

FIGURE 31.15 FTIR absorbance spectra of pullulan (arbitrary values) (a) from 900 to 4000 cm^{-1} and (b) the fingerprint region from 800 to 2000 cm^{-1}.

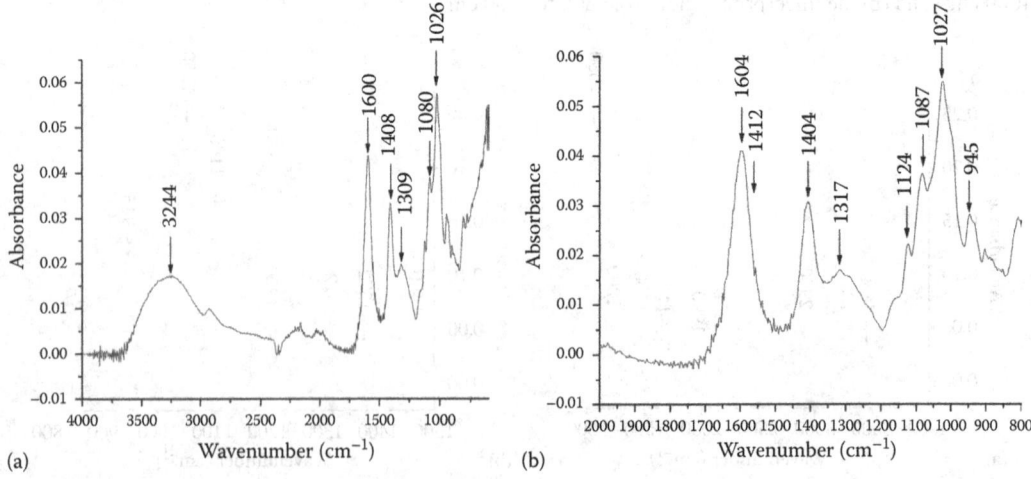

FIGURE 31.16 FTIR absorbance spectra of sodium alginate (arbitrary values) (a) from 900 to 4000 cm^{-1} and (b) the fingerprint region from 900 to 1700 cm^{-1}.

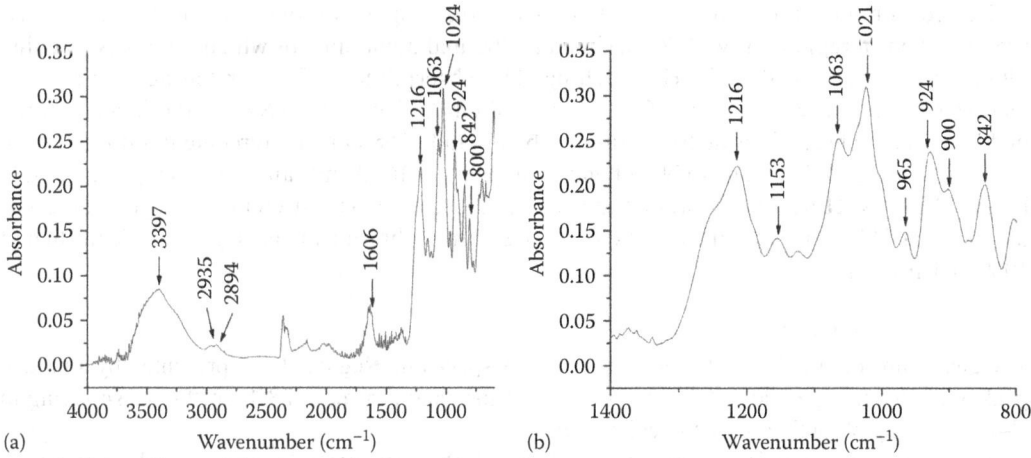

FIGURE 31.17 FTIR absorbance spectra of κ-carrageenan (arbitrary values) (a) from 900 to 4000 cm^{-1} and (b) the fingerprint region from 800 to 1350 cm^{-1}.

this case a new peak at 1216 cm^{-1} is observed, as well at 1153 cm^{-1}, corresponding to sulfate stretching and the cyclic ether stretching, respectively.[74–76] The most dominant peaks were noticed at 1063 and 1021 cm^{-1}, corresponding to C–O–C cyclic ether stretching vibration. The peaks at 924 and 900 cm^{-1} was assigned to the stretch of polyhydroxy groups attached to carbons. The band from 842 cm^{-1} can be attributed to glycosidic linkage from D-galactose-4-sulfate, 3,6-anhydro-D-galactose.[77,78]

A broad band due to the hydrogen bound OH group appears in region III, between 3200 and 3400 cm^{-1} attributed to the complex vibration stretching, associated with free, inter- and intramolecular-bounded hydroxyl groups.

31.5.1.2.3 Chitosan

The chitosan FT–MIR–ATR spectra (Figure 31.18) have also a specific fingerprint, easy to identify comparatively with alginate or carrageenan.

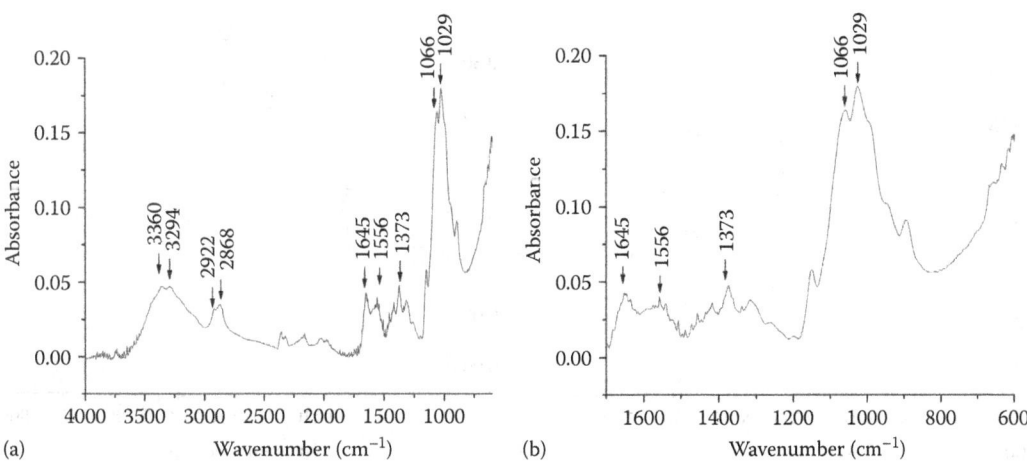

FIGURE 31.18 FTIR absorbance spectra of chitosan (arbitrary values) (a) from 900 to 4000 cm^{-1} and (b) the fingerprint region from 800 to 1650 cm^{-1}.

The concomitant presence of OH and N–H stretching vibrations specific to amine NH_2 are identified in region III, with 2 maxima at 3360 and 3294 cm^{-1} in which the OH stretching vibration are overlapped by N–H stretching. The absorption of C–H stretching of methyl or methylene group of chitosan is at 2922 and 2868 cm^{-1}. The amine (NH_2) band is also identified by specific absorptions at 1650 (amide I band) and 1558 cm^{-1} in agreement with literature data.[79] The amide II band due to N–H bending appears at 1640 cm^{-1} and is overlapped by amide I band.[80] The N–H bending vibration appears at 895 cm^{-1}.[81] The stretching of cyclic ethers is identified at 1148 cm^{-1} and the specific C–O stretching vibrations from sugars are identified at 1059 and 1027 cm^{-1}.

31.5.1.2.4 Guar Gum

The guar gum (Figure 31.19) shows a less complex spectrum. Region II is represented by a unique peak at 3326 cm^{-1} specific to OH stretching vibrations and a peak at 2898 cm^{-1} corresponding to C–H stretching of methyl or methylene groups.

The fingerprint region is less spread, from 800 to 1200 cm^{-1} spectral region typical for anomeric carbohydrate vibrations, and includes maxima at 1143, 1059, and 1017 cm^{-1}, the last ones being dominant, specific to CH_2OH starching mode and CH_2 twisting vibration. Finally the peaks at 860 and 780 cm^{-1} correspond to α- and β-forms of galactose and mannose. Other important peaks are observed at 1154, 1093, and 1021 cm^{-1}, which are due to C–O–C stretching from glycosidic linkages and O–H bending from alcohols.[82,83]

31.5.1.2.5 Gelatin

The only protein characterized by FT-MIR-ATR was gelatin (Figure 31.20). The spectrum was completely different, especially in the fingerprint region, which is located between 1650 and 1000 cm^{-1}. The concomitant presence of OH and N–H stretching vibrations specific to amine NH_2 are identified in region III, with a large band having a maximum at 3281 cm^{-1} in which the OH stretching vibration are overlapped by N–H stretching. The absorption of C–H asymmetric and symmetric stretching of methyl or methylene group of chitosan is at 2925 and 2863 cm^{-1}. As in guar gum, a peak at 2356 cm^{-1} was also identified, as an indicator of carbonyl group. The amine (NH_2) band is also identified by specific absorptions at 1635 (amide I band) and 1540–1510 cm^{-1} in agreement with literature data.[79] Multiple peaks are seen also at 1456–1339 cm^{-1}, 1238 and 1079 cm^{-1}, and specific to carboxyl bands, C–O single bond and symmetrical deformations of CH_2 groups.

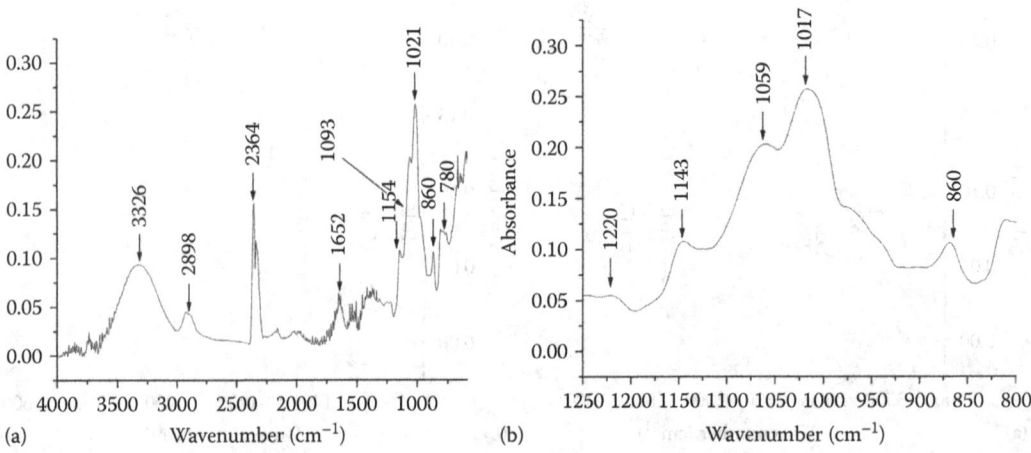

FIGURE 31.19 FTIR absorbance spectra of guar gum (arbitrary values) (a) from 900 to 4000 cm^{-1} and (b) the fingerprint region from 800 to 1200 cm^{-1}.

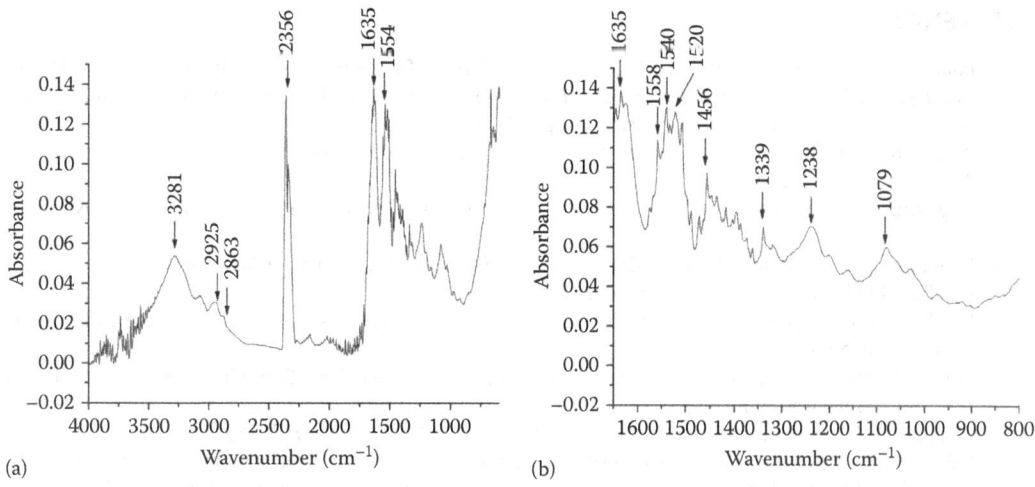

FIGURE 31.20 FTIR absorbance spectra of gelatin (arbitrary values) (a) from 900 to 4000 cm^{-1} and (b) the fingerprint region from 1000 to 1650 cm^{-1}.

31.6 IDENTIFICATION OF POLYMER STRUCTURE BY FT–MIR–ATR ANALYSIS

In Table 31.2 is summarizes the data obtained from the FT–MIR–ATR spectra, pointing out the main wavenumbers corresponding to specific peaks and bands. According to this table, the bolded numbers are useful indicators for the recognition of the biopolymer.

Generally, one can see a very good discrimination between the protein-based polymer (gelatin) and the carbohydrate-based polymers. Among homogeneous and heterogeneous carbohydrates, one can also discriminate by specific differences, for example, between dextrin and pullulan versus celluloses, among alginate–guar gum and carrageenan versus chitosan.

31.7 CONCLUSION

The FTIR characterization of dextrin, microcrystalline cellulose, hydroxypropyl methylcellulose, pullulan, alginate, carrageenan, chitosan, guar gum, and gelatin biomaterials was successfully performed. Also specific markers for each tested biomaterials were identified.

We used the FT–MIR–ATR spectroscopy to characterize nine of the most important hydrophilic carbohydrates or protein biopolymers, which can easily form gels, homogeneous glycans (dextrin, microcrystalline, and hydroxypropyl methylcellulose, pullulan) or heterogeneous carbohydrates (alginate, κ-carrageenan, chitosan, guar gum). These were characterized by their specific fingerprints comparatively with the protein-matrix of gelatin. Based on the specific MIR bands and peaks identified in the whole spectra or in the fingerprint region, the authentication of each individual polymer is possible.

These data can be useful not only to recognize the matrix used for building microcapsules but also to study their stability and modifications during storage or specific conversions by chemical treatments or degradations during their targeted action.

Further research is needed to identify concomitantly the biopolymer matrix and the encapsulated active molecules by this nondestructive FTIR spectroscopy.

ACKNOWLEDGMENTS

We express our gratitude to BRACE GmbH for the donation of some analyzed biopolymers (alginate, chitosan, microcrystalline cellulose, hydroxypropyl methylcellulose, dextrin, and pullulan) used in this entry.

REFERENCES

1. Poncelet, D.; Blitz, J.; Gun'ko, V. Microencapsulation: Fundamentals, methods and applications. In *Surface Chemistry in Biomedical and Environmental Science*; J.P. Blitz and V.M. Gun'ko, Eds.; Springer: Dordrecht, the Netherlands, 2006; pp. 23–34.
2. Augustin, M.A.; Sanguansri, L.; Margetts, C.; Young, B. Microencapsulation of food ingredients. *Food Australia* 2001, *53*, 220–223.
3. Augustin, M.A.; Sanguansri, L. Encapsulation of food ingredients, US Patent 20030185960, April 24, 2008.
4. Vandamme, T.; Poncelet, D.; Subra-Paternault, P. *Microencapsulation: des Sciences aux Technologies* Ed. Tec&Doc., Lavoisier: Paris, France, 2007.
5. Lacik, I. Polymer chemistry in diabetes treatment by encapsulated islets of langerhans. *Aus. J. Chem.* 2006, *59* (8), 508–524.
6. Benita, S. *Microencapsulation—Methods and Industrial Applications*, 2nd edn.; C. Press, Ed.; Taylor & Francis: Boca Raton, FL, 2006.
7. Finch, C.A. Polymers for microcapsule walls. *Chem. Ind.* 1985, *22* (0), 752–756.
8. Lakkis, J.M. Encapsulation and controlled release in bakery applications. In *Encapsulation and Controlled Release Technologies in Food Systems*; Jamileh M. Lakkis, Ed.; Blackwell Publishing: Ames, IA, 2007; pp. 113–133.
9. Nedovic, V.; Kalusevic, A.; Manojlovic, V.; Levic, S.; Bugarski, B. An overview of encapsulation technologies for food applications. *Procedia Food Sci.* 2011, *1*, 1806–1815.
10. Gibbs, B.F.; Kermasha, S.; Alli, I.; Mulligan, C.N. Encapsulation in the food industry: A review. *Int. J. Food Sci. Nutr.* 1999, *50* (3), 213–224.
11. Malafaya, P.B.; Silva, G.A.; Reis, R.L. Natural-origin polymers as carriers and scaffolds for biomolecules and cell delivery in tissue engineering applications. *Adv. Drug Deliv. Rev.* 2007, *59* (4–5), 207–233.
12. Shahidi, F.; Han, X.Q. Encapsulation of food ingredients. *Crit. Rev. Food Sci. Nutr.* 1993, *33* (6), 501–547.
13. Jackson, L.S.; Lee, K. Microencapsulation in food industry. *Lebensmittel Wissenschaft und Technologie* 1991, *24* (4), 289–297.
14. Gouin, S. Microencapsulation: Industrial appraisal of existing technologies and trends. *Trends Food Sci. Amp. Technol.* 2004, *15* (7–8), 330–347.
15. Trif, M.; Schumacher, M.A.; Socaciu, C.; Diehl, H.A. Determination of encapsulated seabuckthorn oil oxidation using FTIR-ATR spectroscopy. *Bull. USAMV-CN* 2007, *64* (1–2), 51–56.
16. Sedman, J.F.; Van De Voort, R.; Ismail, A.A. Attenuated total reflectance spectroscopy: Principles and applications in infrared analysis of food. In *Spectral Methods in Food Analysis*; Mossoba, M.M., Ed.; Marcel Dekker: New York, 1999.
17. Jackson, M.; Mantsch, H. The use and misuse of FTIR spectroscopy in the determination of protein structure. *Crit. Rev. Biochem. Mol. Biol.* 1995, *30* (2), 95–120.
18. Salzer, R.; Steiner, G.; Mantsch, H.H.; Mansfield, J.; Lewis, E.N. Infrared and Raman imaging of biological and biomimetic samples. *Fresenius' J. Anal. Chem.* 2000, *366* (6), 712–726.
19. Mansfield, C.D.; Mantsch, H.H.; Rutt, H.N. Application of infrared spectroscopy in the measurement of breath trace compounds: A review. *Can. J. Anal. Sci. Spectrosc.* 2002, *41* (1), 14–28.
20. Prado, B.M.; Kim, S.; Ozen, B.F.; Mauer, L.J. Differentiation of carbohydrate gums and mixtures using Fourier transform infrared spectroscopy and chemometrics. *J. Agric. Food Chem.* 2005, *53* (8), 2823–2829.
21. de Vos, P.; Bucko, M.; Gemeiner, P.; Navratil, M.; Svitel, J.; Faas, M.; Strand, B.L. et al. Multiscale requirements for bioencapsulation in medicine and biotechnology. *Biomaterials* 2009, *30* (13), 2559–2570.
22. Donath, E.; Sukhorukov, G.B.; Caruso, F.; Davis, S.A.; Mohwald, H. Novel hollow polymer shells by colloid-templated assembly of polyelectrolytes. *Angew. Chem. Int. Ed.* 1998, *37* (16), 2202–2205.
23. Shenoy, D.B.; Antipov, A.A.; Sukhorukov, G.B.; Möhwald, H. Layer-by-layer engineering of biocompatible, decomposable, core-shall structure. *Biomacromolecules* 2003, *4* (2), 265–272.
24. Caruso, F. Nanoengineering of particle surfaces. *Adv. Mat.* 2001, *13* (1), 11–22.
25. Moya, S.; Donath, E.; Sukhorukov, G.B.; Auch, M.; Baumler, H. Lipid coating on polyelectrolyte surface modified colloidal particles and polyelectrolyte capsules. *Macromolecules* 2000, *33* (12), 4538–4544.
26. Dai, Z.; Voigt, A.; Donath, E.; Mohwald, H. Novel encapsulated functional dye particles based on alternately adsorbed multilayers of active oppositely charged macromolecular species. *Macromol. Rapid Commun.* 2001, *22* (10), 756–762.

27. Draget, K.I. Alginates. *Handbook of Hydrocolloids*; Philips, G.O., Williams, P.A., Eds.; Woodhead: Cambridge, U.K., 2009, pp. 807–828, EC (2004).
28. Tester, R.F.; Qi, X. β-limit dextrin: Properties and applications. *Food Hydrocolloids* 2011, *25* (8), 1899–1903.
29. Brossard, S.K.; Du, H.; Miller, J.D. Characteristics of dextrin adsorption by elemental sulfur. *J. Colloid Interface Sci.* 2008, *317* (1), 18–25.
30. Kagami, Y.; Sugimura, S.; Fujishima, N.; Matsuda, K.; Kometani, T.; Matsumura, Y. Oxidative stability, structure, and physical characteristics of microcapsules formed by spray drying of fish oil with protein and dextrin wall materials. *J. Food Sci.* 2003, *68* (7), 2248–2255.
31. Keshk, S.M.A.S.; Haija, M.A. A new method for producing microcrystalline cellulose from *Gluconacetobacter xylinus* and kenaf. *Carbohydr. Polym.* 2011, *84* (4), 1301–1305.
32. Abushammala, H.; Hashaikeh, R.; Cooney, C. Microcrystalline cellulose powder tableting via networked cellulose-based gel material. *Powder Technol.* 2012, *217* (0), 16–20.
33. Sarkar, N. Thermal gelation properties of methyl and hydroxypropyl methylcellulose. *J. Appl. Poly. Sci.* 1979, *24* (4), 1073–1087.
34. Fahs, A.; Brogly, M.; Bistac, S.; Schmitt, M. Hydroxypropyl methylcellulose (HPMC) formulated films: Relevance to adhesion and friction surface properties. *Carbohydr. Polym.* 2010, *80* (1), 105–114.
35. Klayraung, S.; Viernstein, H.; Okonogi, S. Development of tablets containing probiotics: Effects of formulation and processing parameters on bacterial viability. *Int. J. Pharm.* 2009, *370* (1–2), 54–60.
36. Burdock, G.A. Safety assessment of hydroxypropyl methylcellulose as a food ingredient. *Food Chem. Toxicol.* 2007, *45* (12), 2341–2351.
37. Leathers, T.D. Biotechnological production and applications of pullulan. *Appl. Microbiol. Biotechnol.* 2003, *62* (5–6), 468–473.
38. Singh, R.S.; Saini, G.K.; Kennedy, J.F. Pullulan: Microbial sources, production and applications. *Carbohydr. Polym.* 2008, *73* (4), 515–531.
39. Ciofani, G.; Raffa, V.; Pizzorusso, T.; Menciassi, A.; Dario, P. Characterization of an alginate-based drug delivery system for neurological applications. *Med. Eng. Amp. Phys.* 2008, *30* (7), 848–855.
40. Pongjanyakul, T.; Sungthongjeen, S.; Puttipipatkhachorn, S. Modulation of drug release from glyceryl palmitostearate-alginate beads via heat treatment. *Int. J. Pharm.* 2006, *319* (1–2), 20–28.
41. Pongjanyakul, T.; Priprem, A.; Puttipipatkhachorn, S. Investigation of novel alginate-magnesium aluminum silicate microcomposite films for modified-release tablets. *J. Control. Release* 2005, *107* (2), 343–356.
42. Gombotz, W.R.; Wee, S.F. Protein release from alginate matrices. *Adv. Drug Deliv. Rev.* 1998, *31* (3), 267–285.
43. Nunez-Santiago, M.C.; Tecante, A.; Garnier, C.; Doublier, J.L. Rheology and microstructure of κ-carrageenan under different conformations induced by several concentrations of potassium ion. *Food Hydrocolloids* 2011, *25* (1), 32–41.
44. Stortz, C.A. MM3 Potential energy surfaces of trisaccharide models of λ-, μ-, and ν-carrageenans. *Carbohydr. Res.* 2006, *341* (15), 2531–2542.
45. Medina-Torres, L.; Brito-De La Fuente, E.; Gomez-Aldapa, C.A.; Aragon-Pita, A.; Toro-Vazquez, J.F. Structural characteristics of gels formed by mixtures of carrageenan and mucilage gum from *Opuntia ficus indica*. *Carbohydr. Polym.* 2006, *63* (3), 299–309.
46. Haug, I.J.; Draget, K.I.; Smidsrod, O. Physical behaviour of fish gelatin- κ-carrageenan mixtures. *Carbohydr. Polym.* 2004, *56* (1), 11–19.
47. Audet, P.; Paquin, C.; Lacroix, C. Immobilized growing lactic acid bacteria with κ-carrageenan—Locust bean gum gel. *Appl. Microbiol. Biotechnol.* 1988, *29* (1), 11–18.
48. Verbeken, D.; Thas, O.; Dewettinck, K. Textural properties of gelled dairy desserts containing κ-carrageenan and starch. *Food Hydrocolloids* 2004, *18* (5), 817–823.
49. Spagnuolo, P.A.; Dalgleish, D.G.; Goff, H.D.; Morris, E.R. Kappa-carrageenan interactions in systems containing casein micelles and polysaccharide stabilizers. *Food Hydrocolloids* 2005, *19* (3), 371–377.
50. Thrimawithana, T.R.; Young, S.; Dunstan, D.E.; Alany, R.G. Texture and rheological characterization of kappa and iota carrageenan in the presence of counter ions. *Carbohydr. Polym.* 2010, *82* (1), 69–77.
51. Basavaraju, K.C.; Damappa, T.; Rai, S.K. Preparation of chitosan and its miscibility studies with gelatin using viscosity, ultrasonic and refractive index. *Carbohydr. Polym.* 2006, *66* (3), 357–362.
52. Rinaudo, M. Chitin and chitosan: Properties and applications. *Prog. Polym. Sci.* 2006, *31* (7), 603–632.
53. Lamarque, G.; Lucas, J.-M.; Viton, C.; Domard, A. Physicochemical behavior of homogeneous series of acetylated chitosans in aqueous solution: Role of various structural parameters. *Biomacromolecules* 2004, *6* (1), 131–142.

54. Bastos, D.S.; Barreto, B.N.; Souza, H.K.S.; Bastos, M.; Rocha-Leao, M.H.M.; Andrade, C.T.; Gonçalves, M.P. Characterization of a chitosan sample extracted from Brazilian shrimps and its application to obtain insoluble complexes with a commercial whey protein isolate. *Food Hydrocolloids* 2010, *24* (8), 709–718.
55. Kumar, M.N.V.R. A review of chitin and chitosan applications. *Reactive Funct. Polym.* 2000, *46* (1), 1–27.
56. Jayakumar, R.; Prabaharan, M.; Nair, S.V.; Tokura, S.; Tamura, H.; Selvamurugan, N. Novel carboxymethyl derivatives of chitin and chitosan materials and their biomedical applications. *Prog. Mat. Sci.* 2010, *55* (7), 675–709.
57. Tiwary, A.; Rana, V. Cross-linked chitosan films: Effect of cross-linking density on swelling parameters. *Pak. J. Pharm. Sci.* 2010, *23* (4), 443–448.
58. Sharma, V.; Marwaha, R.K.; Dureja, H. Permeability evaluation through chitosan membranes using Taguchi design. *Sci. Pharm.* 2010, *78* (42), 977–983.
59. Berger, J.; Reist, M.; Mayer, J.M.; Felt, O.; Peppas, N.A.; Gurny, R. Structure and interactions in covalently and ionically crosslinked chitosan hydrogels for biomedical applications. *Eur. J. Pharm. Biopharm.* 2004, *57* (1), 19–34.
60. Agrawal, P.; Strijkers, G.J.; Nicolay, K. Chitosan-based systems for molecular imaging. *Adv. Drug Deliv. Rev.* 2010, *62* (1), 42–58.
61. Muzzarelli, R.A.A. Human enzymatic activities related to the therapeutic administration of chitin derivatives. *Cell. Mol. Life Sci.* 1997, *53* (2), 131–140.
62. Berger, J.; Reist, M.; Mayer, J.M.; Felt, O.; Gurny, R. Structure and interactions in chitosan hydrogels formed by complexation or aggregation for biomedical applications. *Eur. J. Pharm. Biopharm.* 2004, *57* (1), 35–52.
63. Dutta, P.K.; Tripathi, S.; Mehrotra, G.K.; Dutta, J. Perspectives for chitosan based antimicrobial films in food applications. *Food Chem.* 2009, *114* (4), 1173–1182.
64. Peter, M.G. Applications and environmental aspects of chitin and chitosan. *J. Macromol. Sci.* 1995, *32* (4), 629–640.
65. Mudgil, D.; Barak, S.; Khatkar, B.S. Effect of enzymatic depolymerization on physicochemical and rheological properties of guar gum. *Carbohydr. Polym.* September 1, 2012, *90* (1), 224–228.
66. Garcia, H.; Barros, A.S.; Gonzalves, C.; Gama, F.M.; Gil, A.M. Characterization of dextrin hydrogels by FTIR spectroscopy and solid state NMR spectroscopy. *Eur. Polym. J.* 2008, *44* (7), 2318–2329.
67. Adel, A.M.; Abd El-Wahab, Z.H.; Ibrahim, A.A.; Al-Shemy, M.T. Characterization of microcrystalline cellulose prepared from lignocellulosic materials. Part II: Physicochemical properties. *Carbohydr. Polym.* 2011, *83* (2), 676–687.
68. Mitic, Z.; Nikolic, G.S.; Cakic, M.; Premovic, P.; Ilic, L. FTIR spectroscopic characterization of Cu(II) coordination compounds with exopolysaccharide pullulan and its derivatives. *J. Mol. Struct.* 2009, *924–926* (0), 264–273.
69. Mitic, Z.; Cakic, M.; Nikolic, G.M.; Nikolic, R.; Nikolic, G.S.; Pavlovic, R.; Santaniello, E. Synthesis, physicochemical and spectroscopic characterization of copper(II)-polysaccharide pullulan complexes by UV-VIS, ATR-FTIR, and EPR. *Carbohydr. Res.* 2011, *346* (3), 434–441.
70. Wu, J.; Zhong, F.; Li, Y.; Shoemaker, C.F.; Xia, W. Preparation and characterization of pullulan-chitosan and pullulan-carboxymethyl chitosan blended films. *Food Hydrocolloids* 2013, *30* (1), 82–91.
71. Dai, Y.N.; Li, P.; Zhang, J.P.; Wang, A.Q.; Wei, Q. A novel pH sensitive N-succinyl chitosan/alginate hydrogel bead for nifedipine delivery. *Biopharm. Drug Dispos.* 2008, *29* (3), 173–184.
72. Puttipipatkhachorn, S.; Pongjanyakul, T.; Priprem, A. Molecular interaction in alginate beads reinforced with sodium starch glycolate or magnesium aluminum silicate, and their physical characteristics. *Int. J. Pharm.* 2005, *293* (1–2), 51–62.
73. Draget, K.I. Alginates. In *Handbook of Hydrocolloids*; Philips, G.O., Williams, P.A., Eds.; Woodhead Publishing: Cambridge, U.K., 2000; pp. 379–395.
74. Arman, M.; Qader, S.A.U. Structural analysis of kappa-carrageenan isolated from *Hypnea musciformis* (red algae) and evaluation as an elicitor of plant defense mechanism. *Carbohydr. Polym.* 2012, *88* (4), 1264–1271.
75. Hadi, H.; Muhamad, I.I. The effect of nanoparticles on gastrointestinal release from modified κ-carrageenan nanocomposite hydrogels. *Carbohydr. Polym.* 2012, *89* (1), 138–145.
76. Sankalia, M.G.; Mashru, R.C.; Sankalia, J.M.; Sutariya, V.B. Stability improvement of alpha-amylase entrapped in kappa-carrageenan beads: Physicochemical characterization and optimization using composite index. *Int. J. Pharm.* 2006, *312* (1–2), 1–14.
77. Pranoto, Y.; Lee, C.M.; Park, H.J. Characterizations of fish gelatin films added with gellan and κ-carrageenan. *LWT—Food Sci. Technol.* 2007, *40* (5), 766–774.

78. Pascalau, V.; Popescu, V.; Popescu, G.L.; Dudescu, M.C.; Borodi, G.; Dinescu, A.; Perhaita, I.; Paul, M. The alginate/κ-carrageenan ratio's influence on the properties of the cross linked composite films. *J. Alloys Compounds*, 2012, *536*, S418–S434.
79. Kadir, M.F.Z.; Aspanut, Z.; Majid, S.R.; Arof, A.K. FTIR studies of plasticized poly(vinyl alcohol)/chitosan blend doped with NH_4NO_3 polymer electrolyte membrane. *Spectrochim. Acta Part A Mol. Biomol. Spectrosc.* 2011, *78* (3), 1068–1074.
80. Osman, Z.; Arof, A.K. FTIR studies of chitosan acetate based polymer electrolytes. *Electrochim. Acta* 2003, *48* (8), 993–999.
81. Fernandez, C.; Ausar, S.F.; Badini, R.G.; Castagna, L.F.; Bianco, I.D.; Beltramo, D.M. An FTIR spectroscopy study of the interaction between alpha s-casein-bound phosphoryl groups and chitosan. *Int. Dairy J.* 2003, *13* (11), 897–901.
82. Sharma, R.K.; Lalita. Synthesis and characterization of graft copolymers of N-Vinyl-2-Pyrrolidone onto guar gum for sorption of Fe^{2+} and Cr^{6+} ions. *Carbohydr. Polym.* 2011, *83* (4), 1929–1936.
83. Das, D.; Ara, T.; Dutta, S.; Mukherjee, A. New water resistant biomaterial biocide film based on guar gum. *Bioresour. Technol.* 2011, *102* (10), 5878–5883.

Section V

Applications

Section V

Applications

32 Encapsulation Technologies for Modifying Food Performance

Maria Inês Ré, Maria Helena Andrade Santana, and Marcos Akira d'Ávila

CONTENTS

32.1 Introduction 643
32.2 Encapsulation Technologies... 644
32.3 Encapsulating Systems 649
    32.3.1 Spray-Dried Microparticles 649
        32.3.1.1 Concept, Structure, and Properties 649
    32.3.2 Gel Microparticles 653
        32.3.2.1 Concept, Structure, and Properties 653
    32.3.3 Liposomal Systems 657
        32.3.3.1 Basic Principles 657
        32.3.3.2 Applications in Foods 660
    32.3.4 Emulsified Systems 663
        32.3.4.1 Concept, Structure, and Properties 663
        32.3.4.2 Preparation and Characterization of Microemulsions 666
        32.3.4.3 Preparation and Characterization of Nanoemulsions 668
32.4 Applications of Encapsulated Active Ingredients in Foods.. 669
    32.4.1 Flavors 669
    32.4.2 Antioxidants.. 670
    32.4.3 Omega-3 Fatty Acids 671
    32.4.4 Vitamins and Minerals 672
    32.4.5 Probiotics 673
32.5 Encapsulation Challenges... 676
References... 677

32.1 INTRODUCTION

The food industry faces serious challenges in the twenty-first century. Consumers are demanding more from the foods they eat. Now foods must not only taste good and aid immediate nutrition, but also assist in mitigating disease, provide clear health benefits, and help to reduce health care costs. To meet these demands, food manufacturers must prepare safe, healthy, and convenient foods that are of good value and of great taste. The global functional foods market is a dynamic and growing segment of the food industry, being expected to represent 5% of the total global food market in 2010 (Market Research, 2004).

To date, a number of national authorities, academic bodies, and the industry have proposed definitions for functional foods. Although the term "functional food" has already been defined several times, there is no universally accepted definition for this emerging food category. Definition ranges from the very simple to the more complex (Siró et al., 2008): "foods that, by virtue of physiologically active components, provide health benefits beyond basic nutrition (International Life Sciences Institute, 1999)"; "food similar in appearance to a conventional food, consumed as part of the usual diet, with demonstrated physiological benefits, and/or to reduce the risk of chronic disease beyond basic nutritional functions" (Health Canada, 1998).

Examples of functional foods that have potential benefits for health and whose market has grown tremendously include baby food, ready meals, snacks, soft drinks such as energy and sport drinks, meat products, and spreads.

A functional benefit is usually obtained by fortification with an active (functional) ingredient. Examples of functional ingredients are flavors, vitamins, minerals, enzymes, peptides, bioactive lipids, antioxidants, and probiotic microorganisms. Functional ingredients come in a variety of molecular and physical forms, with different polarities (polar, nonpolar, amphiphilic), molecular weights (low to high), and physical states (solid, liquid). They are rarely used directly in their pure form. Instead, they are often incorporated into some form of delivery system generated by encapsulation technologies.

A delivery system must perform a number of different roles. First, it serves as a vehicle for carrying the functional ingredient to the desired site of action. Second, it may have to protect the functional ingredient from chemical or biological degradation (e.g., oxidation) during processing, storage, and utilization; this maintains the functional ingredient in its active state. Third, it may have to be capable of controlling the release of the functional ingredient, either responding to specific environmental conditions that trigger release (e.g., pH, ionic strength, or temperature) or varying the release rate. Fourth, the delivery system has to be food-grade and also compatible with the physicochemical and qualitative attributes (i.e., appearance, texture, taste, and shelf life) of the final product (Weiss et al., 2006).

The characteristics of the delivery systems are one of the most important factors influencing the efficacy of functional ingredients in many industrial products. The delivery systems can be used to deliver a host of ingredients in a range of food formulations. The demand for encapsulation technologies is growing at around 10% annually, driven by both the increasing fortification with healthy ingredients and the consumer demand for novel products (Brownlie, 2007). Encapsulation technologies are attracting growing interest because they can decrease costs for food makers, particularly those using sensitive ingredients like probiotics, by reducing the need for preservatives. Encapsulation also allows manufacturers of food and beverages, as well as other consumer products, to add into their formulations ingredients that would be normally used in traditional processing.

32.2 ENCAPSULATION TECHNOLOGIES

Encapsulation is a topic of interest in a wide range of scientific and industrial areas, varying from pharmaceutics to agriculture, and from pesticides to enzymes. The development of these technologies is characterized by strong fundamental research and several industrial applications, demonstrated by the growing number of scientific papers and patent applications. The European network of patent databases (esp@cenet) has approximately 600 patent documents worldwide containing "microencapsulation" in the title or in the abstract, whereas the U.S. Patent and Trademark Office has approximately 160 patents and 140 applications with at least one claim containing the term "microencapsulation."* In the past 5 years, there have been 200–300 peer-reviewed articles per year on microencapsulation in the *Web of Science*®† database (Figure 32.1), in contrast to 1600 papers in the previous 20 years.

Encapsulation involves the coating or entrapment of a desired component (active or core material) within a secondary material (encapsulating, carrier, coating, shell, or wall material) to prevent or delay the release of the active or core ingredient until a certain time or a set of conditions is

---

* Data related to a search performed on the U.S. Patent and Trademark Office: 1976–Jan. 2009 (patents) and 2001–Jan. 2009 (patent applications); and on the espac@cenet: 1970–Jan. 2009.
† Registered trademark of Thomson Reuters, New York.

Encapsulation Technologies for Modifying Food Performance

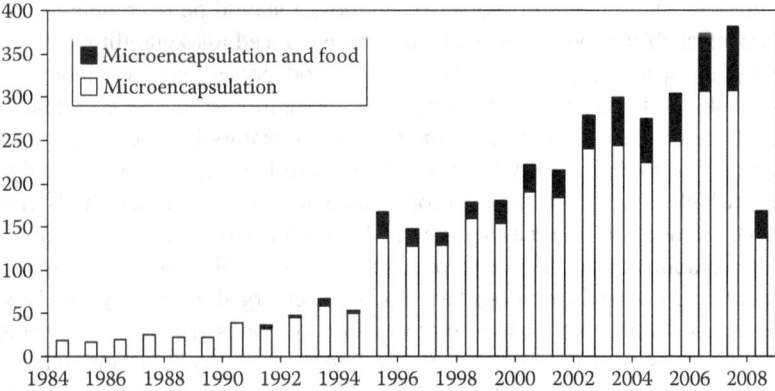

FIGURE 32.1 Evolution of the number of scientific publications in the Web of Science database (up to July 2008) on "microencapsulation" and "microencapsulation of food ingredients." Microencapsulation applied to food represents approximately 20% of the papers published in the microencapsulation area.

achieved. Encapsulation can potentially offer numerous benefits to the materials being encapsulated. Various properties of active ingredients may be changed by encapsulation. For example, handling and flow properties can be changed by converting a liquid to a solid encapsulated form, and hygroscopic materials can be protected from moisture.

Encapsulation systems in food applications are typically used with at least one of the following purposes:

- To solve formulation problems arising from a limited chemical or physical stability of the active ingredient
- To overcome incompatibility between active ingredient and food matrix
- To control the release of a sensorial active compound
- To assist or enhance the absorption of a nutrient

Examples of food additives that may benefit from encapsulation and controlled release are flavors, minerals, and lipids, among others.

There are several items to consider when choosing or developing an encapsulated food ingredient. Each of these is very important to the success of the product. The molecular characteristics of the ingredient, its desired function, the type of coating, the required protection rate, the particle size, and the processing conditions must all be clearly defined. In this regard, there are some important questions that must be answered:

- What are the molecular structure, size, and charge of the ingredient?
- What is it that you are trying to achieve with the encapsulation?
- When or how do you want the core ingredient to be released?
- Do you want the coating to melt and release the core ingredient at a certain temperature? Do you want the coating to break away by mechanical action or dissolving in water?
- Is a change in pH such as in the gastrointestinal (GI) tract going to release the core? Or is there another way your core ingredient will be released?

A critical step in developing encapsulated food products is to determine the encapsulant formulation that meets the desired stability and release criteria. The GRAS (generally recognized as safe) encapsulating material must stabilize the core material, must not react with or deteriorate the active ingredient, and should release it under the specific conditions based on product application. In addition, delivery systems should be developed from inexpensive ingredients, since the additional costs associated with the encapsulation of the active ingredient should be overcome by its benefits, for example, improved shelf life, better bioavailability, or enhanced marketability.

The variety of encapsulating materials allows food producers to select compounds that work for water or fat-soluble food ingredients; dissolve, melt, or rupture to release core material; and provide textural characteristics to satisfy consumer palates. Commonly used encapsulating materials are carbohydrates due to their ability to absorb and retain flavors, cellulose (based on its permeability), gums (which offer good gelling properties and heat resistance), lipids (based on their hydrophobicity), and usually gelatin as a protein, which is nontoxic, inexpensive, and commercially available. Some combinations of these encapsulating agents are also commonly used.

New encapsulating materials for foods have not really emerged in recent years. However, a great effort has been made with respect to food proteins. Food proteins have been engineered as a range of new GRAS matrices with the potential to incorporate nutraceutical compounds and provide controlled release via the oral route. The advantages of food protein matrices include their high nutritional value, abundant renewable sources, and acceptability as naturally occurring food components degradable by digestive enzymes. In addition, food proteins can be used to prepare a wide range of matrices and multicomponent matrices in the form of hydrogels, micro- or nanoparticles, all of which can be tailored for specific applications in the development of innovative functional food products, as recently reviewed (Chen et al., 2006).

Encapsulation and microencapsulation are often used interchangeably when discussing the process technology. Microencapsulation is encapsulation at the microscale, producing delivery devices ranging from 1 to 1000 μm in size, generally less than 200 μm.

Delivery devices can have many morphologies, depending upon the materials and methods used in their preparation. In general, one can distinguish between two main groups of device architecture, depending on the way the core (solid or liquid) is distributed within the system (Figure 32.2):

1. A reservoir system, in which the core is largely concentrated near the center and enveloped by a continuous film (wall) of the encapsulating material
2. A matrix system, in which the core is finely dispersed throughout a continuous matrix of the encapsulating material

The active constituent/encapsulating material ratio is usually high in reservoir systems (between 0.70 and 0.95), whereas for matrix systems, this ratio is generally lower than 1.5 (more commonly between 0.2 and 0.35). The delivery devices defined in 1 are often referred to as *microcapsules* and those described in 2 are called *microspheres*.

Delivery devices do not necessarily have a spherical shape, as illustrated in Figure 32.2e. A great variety of shapes can be obtained when a solid core material is encapsulated by a shell. Particle size is an important characteristic of these structures because it is one of the many parameters that can be tailored to control release rates of encapsulated ingredients. However, the production of microcapsules often gives a certain particle size polydispersity. The active ingredient release kinetics depends on the particle size distribution. It is thus necessary to determine both the mean particle size and the size distribution for the targeted delivery.

The release of the core material from a delivery device may be programmed to be immediate, delayed, pulsatile, or prolonged over an extended period of time. In general, the release depends upon the architecture and the physical structure of the device, as well as upon the barrier properties of the encapsulating material used to form the system.

Delivery devices can be formulated to release the core material for food applications through a variety of mechanisms to meet product performance requirements, which may be based on temperature or solvent effects, diffusion, degradation, or particle fractures. For example, the encapsulating material can be fractured by external or internal forces such as chewing (Figure 32.3). Melting of the core or the encapsulating material by means of an appropriate solvent or thermally is another way of controlling the release of the active ingredient. Solvent release is based upon the solubilization of

Encapsulation Technologies for Modifying Food Performance 647

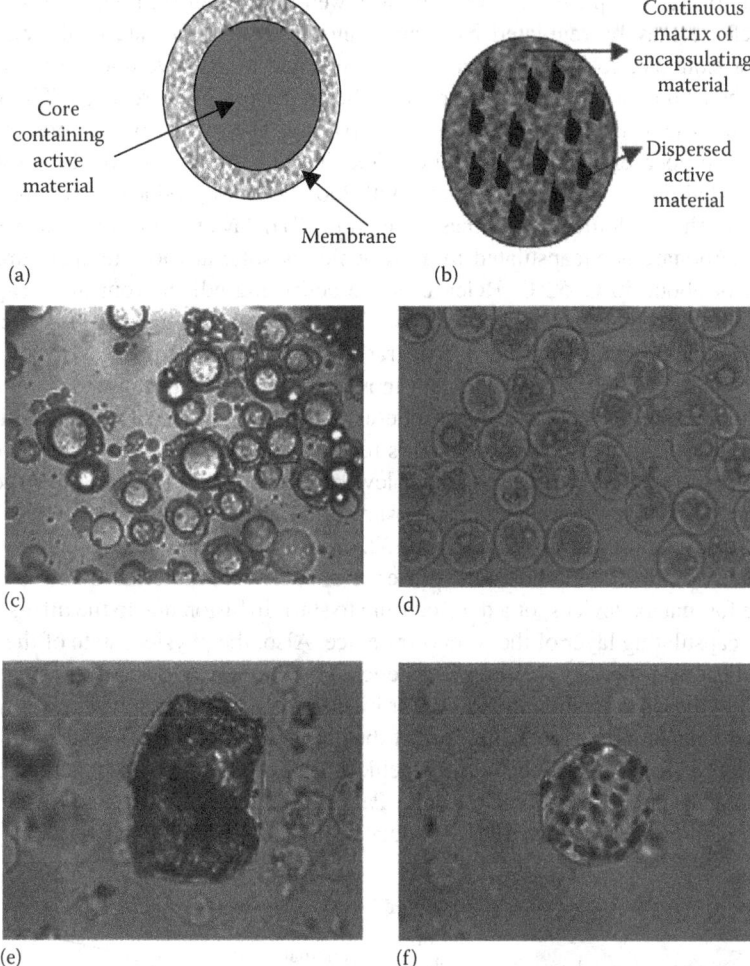

FIGURE 32.2 Schematic diagram of microcapsules morphology: (a) reservoir system (simple wall); (b) matrix system; (c) simple wall (liquid core); (d) multicore; (e) simple wall (solid and irregular core); and (f) matrix (solid core dispersed into the polymeric matrix).

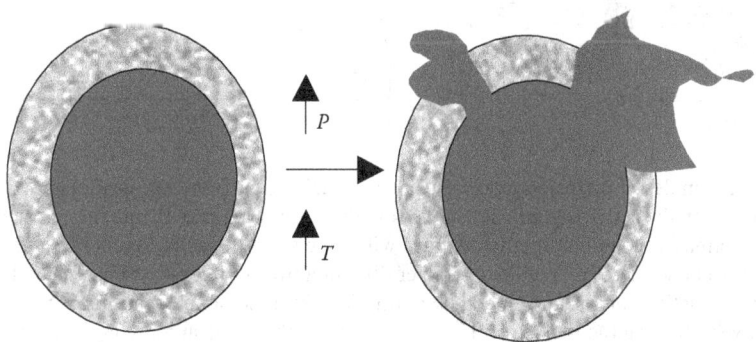

FIGURE 32.3 Release from reservoir device (microcapsule) fractured by mechanical forces.

the encapsulant (water is typically the solvent), followed by subsequent release of the encapsulated ingredient. Release may be regulated by controlling the dissolution rate of the encapsulant and pH effects. For example, coating materials can be selected to dissolve upon consumption, slowly or quickly in the acidic gastric medium, or only when a certain pH is reached. Thermal release is commonly used for fat capsules and occurs during baking. The release of the core material may be delayed until the proper temperature is reached, delaying a chemical reaction. For example, sodium bicarbonate is a baking ingredient that reacts with food acids to produce leavening agents, which give baked goods their volume and lightness of texture. To delay and control the leavening process, the sodium bicarbonate is encapsulated in a fat, which is solid at room temperature but melts at a temperature of about 50°C–52°C. Release of the active ingredient from microcapsules can be accomplished through biodegradation processes, if the encapsulating agent is sensitive to enzymatic actions. For example, lipid coatings may be degraded by the action of lipases.

Diffusion is another important mechanism in release into foods because it is dominant in controlled release from matrix systems (microspheres). Diffusion occurs when the active ingredient passes through the encapsulating material. This mechanism can occur on a macroscopic scale (as through pores in the matrix) or on a molecular level, by permeation through the structuring material. Examples of diffusion-release systems are shown in Figure 32.4.

The typical release profiles shown in Figure 32.4 for reservoir and matrix delivery systems may present variations: a burst effect due to the presence of some core material too close to the external device surface for matrix devices, or a delayed time to start diffusion due to the diffusion of the core through the encapsulating layer of the reservoir device. Also, the physical state of the core material (dissolved or dispersed) defines the release kinetics. For example, a reservoir system in which the active core is not dissolved results in zero-order kinetics (constant flow), whereas it results in first-order kinetics (exponentially decreasing flow) if the core is dissolved in the encapsulated material.

Encapsulation of food ingredients can be achieved by physical, physicochemical, or chemical techniques (Shahidi and Han, 1993; Desai and Park, 2005a; Champagne and Fustier, 2007). The various encapsulation technologies allow product formulators to make delivery devices from less

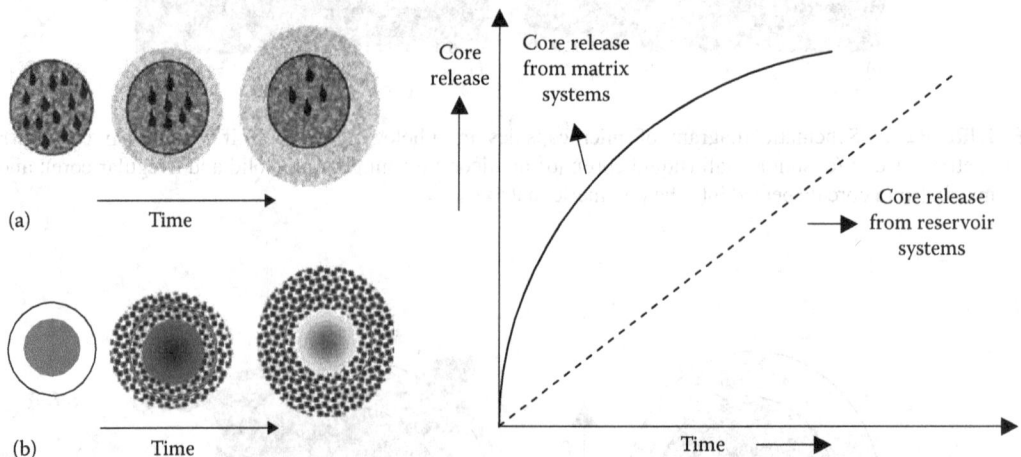

FIGURE 32.4 Examples of diffusion-release systems. (a) Diffusion from a typical matrix delivery system: Diffusion occurs when the active ingredient passes from the structuring matrix into the external environment. As the release continues, the rate normally declines with time since the active agent has a progressively longer distance to travel and therefore requires a longer diffusion time to release. (b) Diffusion from a reservoir system, whether the active core is surrounded by a film or membrane of an encapsulating material. The only structure effectively limiting the release of the core material is the encapsulating layer surrounding the core. Since this layer is uniform and has a constant thickness, the diffusion rate of the core can be kept fairly stable throughout the lifetime of the delivery system.

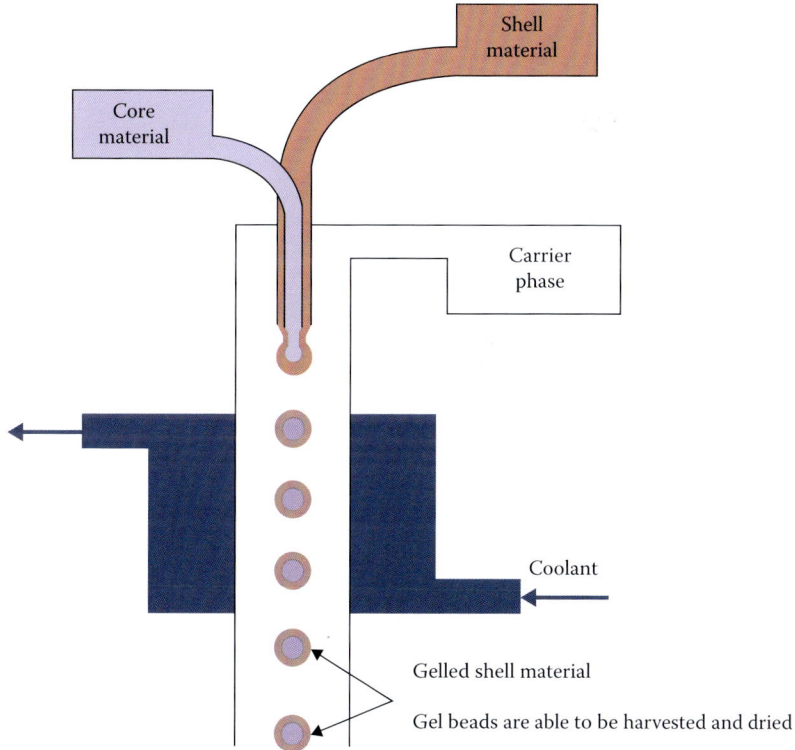

FIGURE 10.7 Schematics of a submerged nozzle process. (From Hávarri, M., Marañón, I., and Villarán, M.C., Published in DOI: 10.5772/50046 under CC BY 3.0 license, Available from: http://dx.doi.org/10.5772/50046, © 2012.)

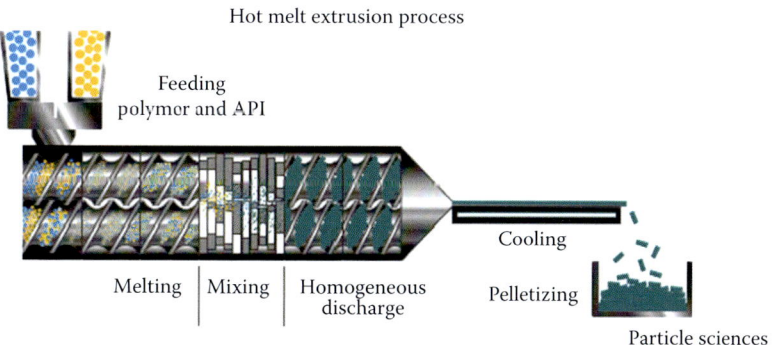

FIGURE 11.4 Schematic presentation of HME process. (From Particle Sciences, *Technical Brief*, 3, 2011.)

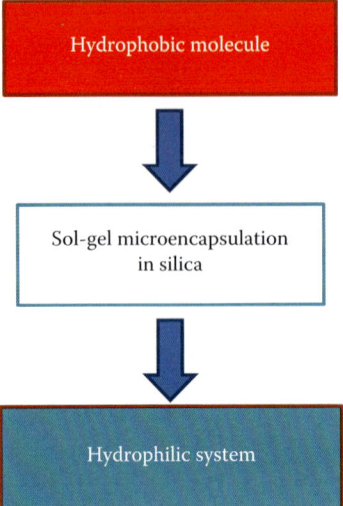

SCHEME 18.1 Once entrapped in hydrophilic silica microparticles, hydrophobic organic molecules can be formulated and incorporated into the aqueous phase.

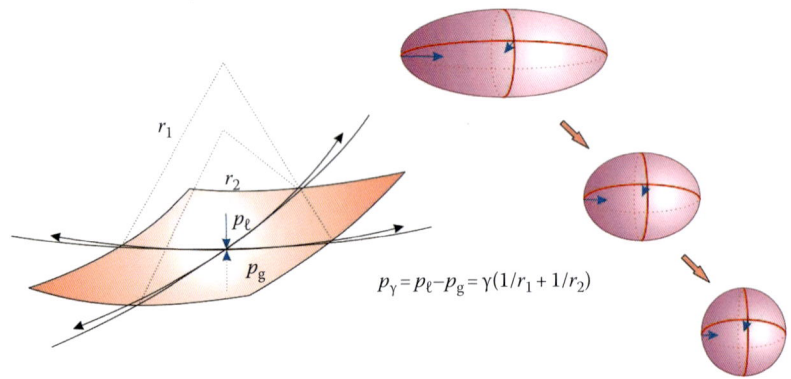

FIGURE 20.1 The Laplace pressure exerted by curved interfaces is proportional to the sum of the principal curvatures. The blue arrows signify the direction and magnitude of the interfacial forces, explaining why a deformed droplet restores its spherical shape.

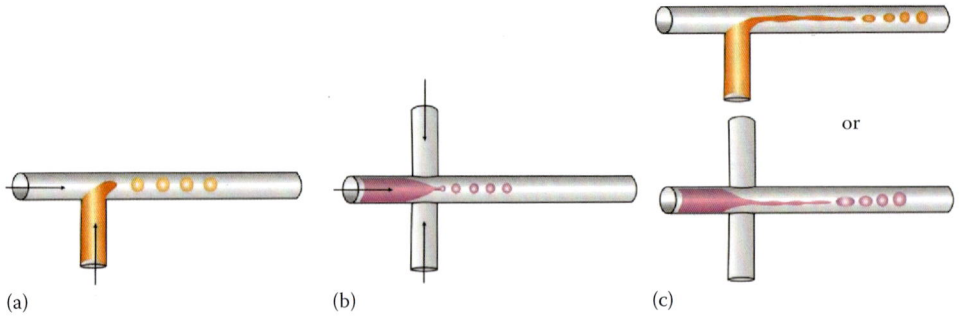

FIGURE 20.2 Droplet breakup (a) in cross flowing streams, (b) in coflowing streams, and (c) in elongational flows.

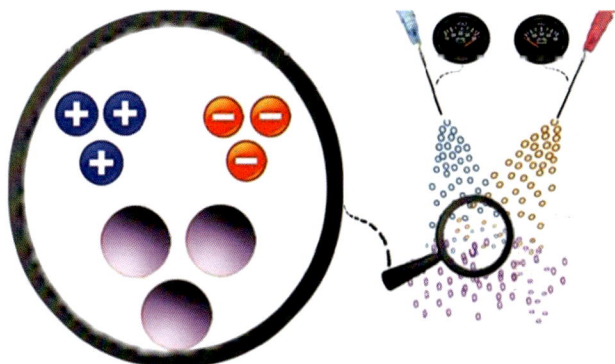

FIGURE 22.13 Bipolar electrospray setup (reactive electrospray).

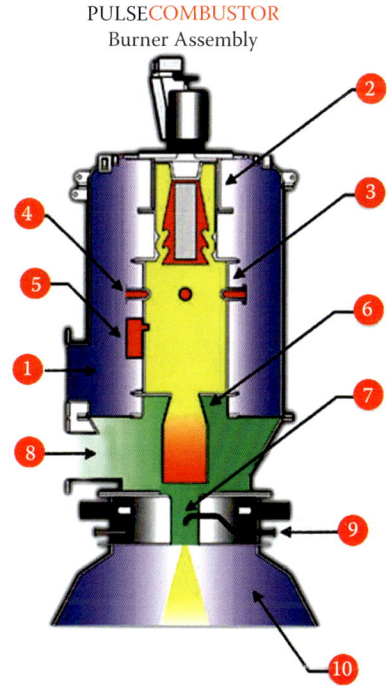

FIGURE 23.1 Pulse spray drying basic process operation diagram. Here's how pulse combustion spray driers work: Air is (1) pumped into the pulse combustor's outer shell at low pressure, where it flows through the patented unidirectional air valve. (2) The air enters a tuned combustion chamber (a *Helmholtz resonator*) (3) where fuel is added (4) to the spray dryer. The air valve closes. The fuel–air mixture is ignited by a pilot (5) and explodes, creating hot air pressurized to about 3 psi above combustion fan pressure. The hot gases rush down the spray dryer's tailpipe (6) toward the atomizer (7). The air valve reopens (2) and allows the next air charge to enter. The fuel valve admits fuel, and the mixture explodes in the hot chamber. This spray drying cycle is controllable from about 80 to 110 Hz. Just above the spray dryer's atomizer, quench air (8) is blended in to achieve the desired product contact temperature. The exclusive PCS atomizer releases the liquid (9) into a carefully balanced gas flow, which dynamically controls atomization, drying, and particle trajectory. The atomized liquid enters a conventional tall-form drying chamber (10). Downstream of the spray-drying process, the suspended powder is retrieved using standard collection equipment, such as cyclones and baghouses.

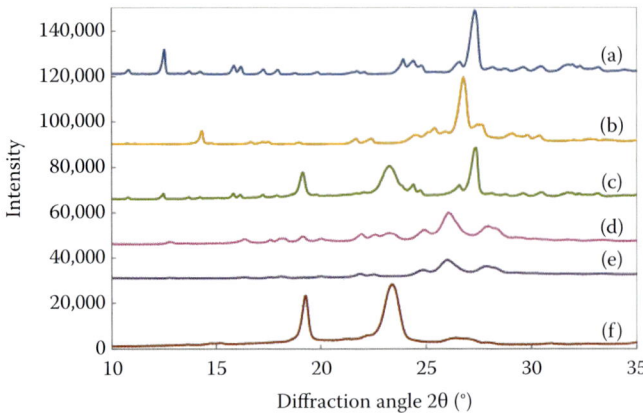

FIGURE 24.13 X-ray diffractograms of (a) unprocessed quercetin, (b) SAS-processed quercetin, (c) physical mixture of quercetin-Pluronic® F127 (mass ratio 1 g:1 g), (d) SAS-processed mixture of quercetin-Pluronic® F127 (mass ratio 2 g:1 g), (e) SAS-processed mixture of quercetin-Pluronic® F127 (mass ratio 1 g:1 g), and (f) unprocessed Pluronic® F127. For clarity, all diffractograms were vertically displaced by arbitrary amounts.

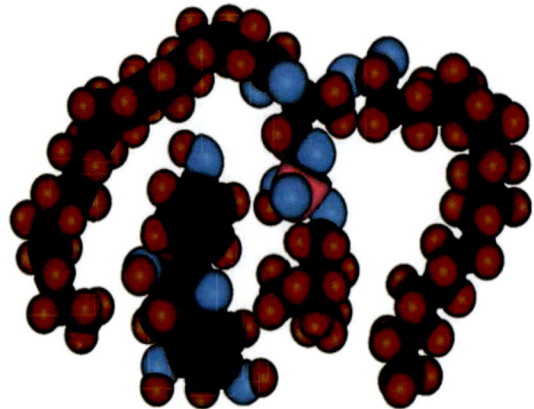

FIGURE 36.12 Molecular modeling of the interaction between a polyphenolic molecule and a phospholipid. (Reprinted from *Fitoterapia*, 81(5), Semalty, A., Semalty, M., Rawat, M.S.M., and Franceschi, F., Supramolecular phospholipids–polyphenolics interactions: The PHYTOSOME® strategy to improve the bioavailability of phytochemicals, 306–314. Copyright 2010, with permission from Elsevier.)

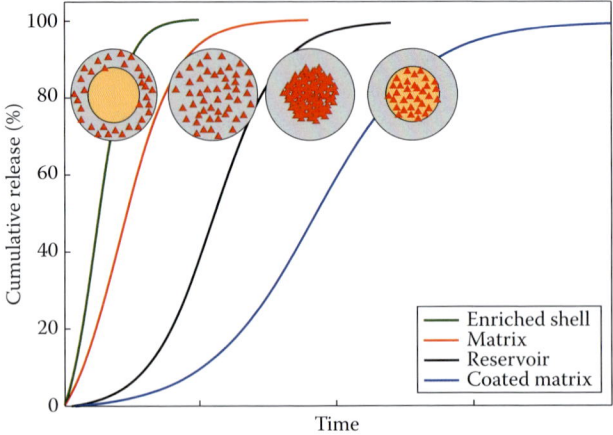

FIGURE 37.3 Schematic representation of cumulative release profiles of the main encapsulation architectures (reservoir type, matrix type, coated matrix type).

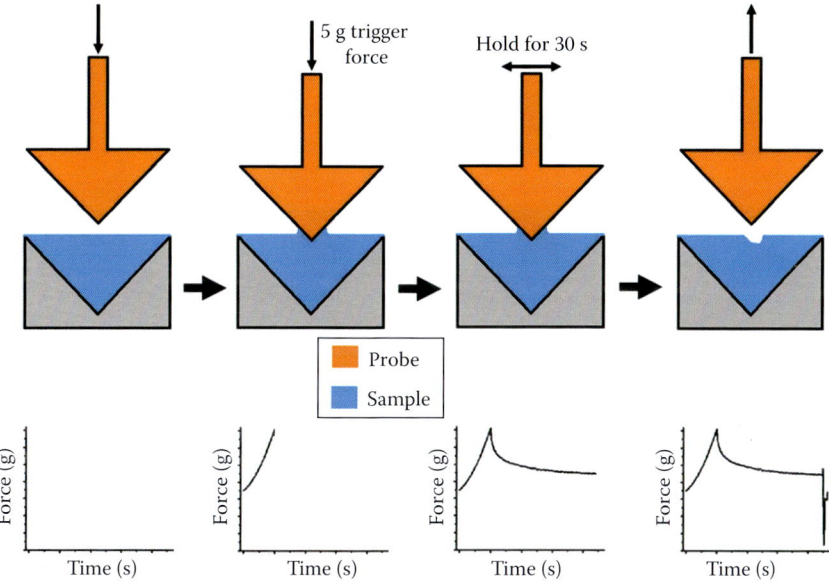

FIGURE 46.13 Schematic diagram of stress relaxation study.

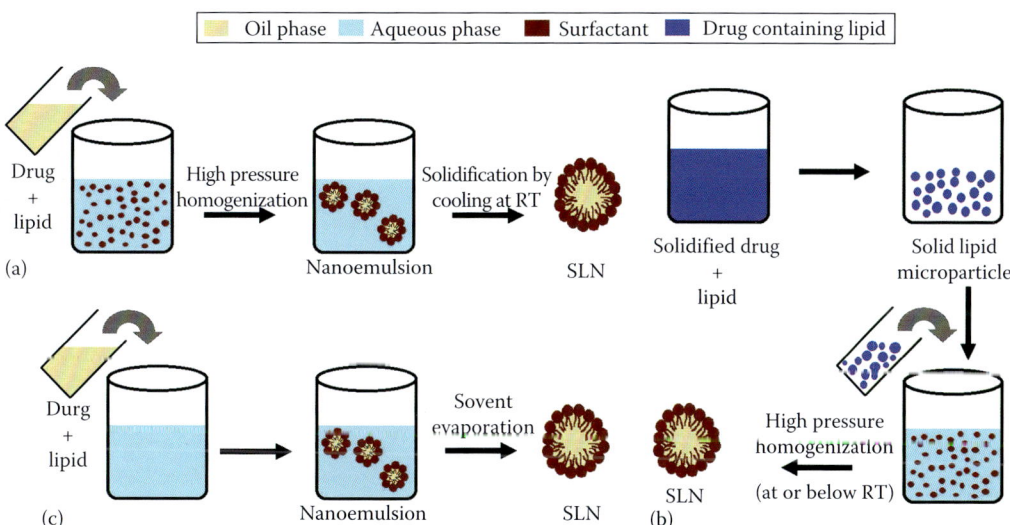

FIGURE 58.5 Schematic representation of the methods of preparation of solid lipid nanoparticles. (a) Hot homogenization method. (From Brugarolas, T. et al., *Soft Matter*, 9(38), 904664, 2013.) (b) Cold homogenization method. (From He, C.X. et al., *Exp. Opin. Drug Deliv.*, 7(4), 44565, 2010.) (c) Solvent evaporation method. (From Choi, C.H. et al., *Adv. Mater.*, 25(18), 2536, 2013.)

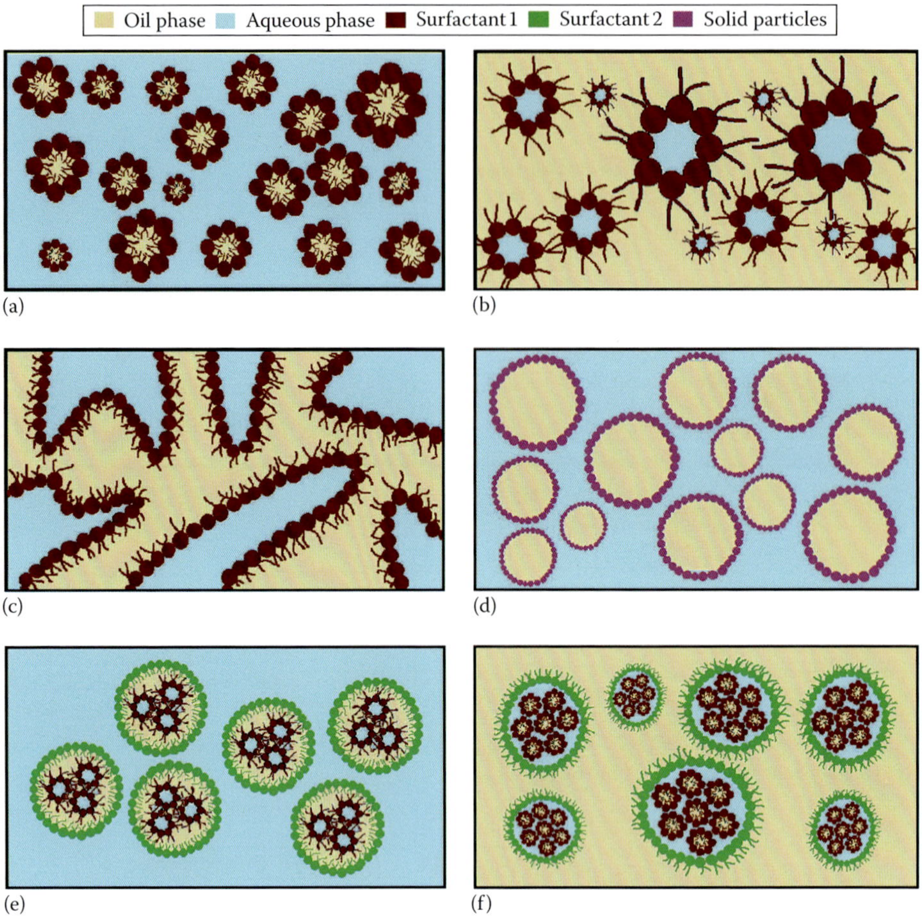

FIGURE 58.6 Schematic representation of different types of emulsions: (a) oil-in-water, (b) water-in-oil, (c) bicontinuous, (d) pickering emulsion, and double emulsions: (e) w/o/w, and (f) o/w/o. (From Markovic, N. et al., *Hydrocarbon Gels: Rheological Investigation of Structure,* ACS Publications, 2002; Zoumpanioti, M. et al., *Biotechnol. Adv.,* 28(3), 39576, 2010; Rogovina, L.Z. and Slonimskii, G.L., *Russ. Chem. Rev.,* 43(6), 50377, 1974.)

FIGURE 60.8 Laundry microcapsules use in textiles.

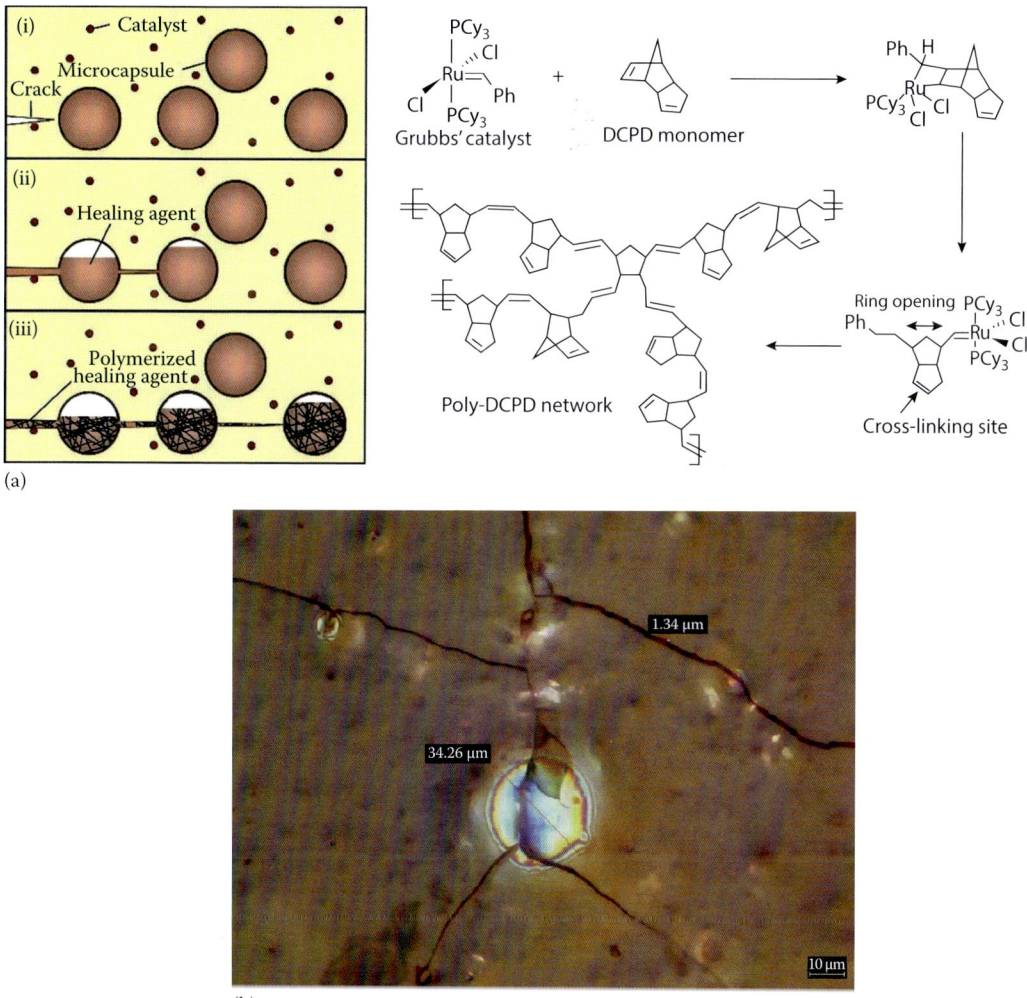

FIGURE 60.18 (a) Principle of self-healing system basically solution. (b) After provoking stress cracks, microcapsules will break and the core content refills the capillaries. (From White, S.R. and Sottos, N.R., *Nature*, 409, 794, 2001.)

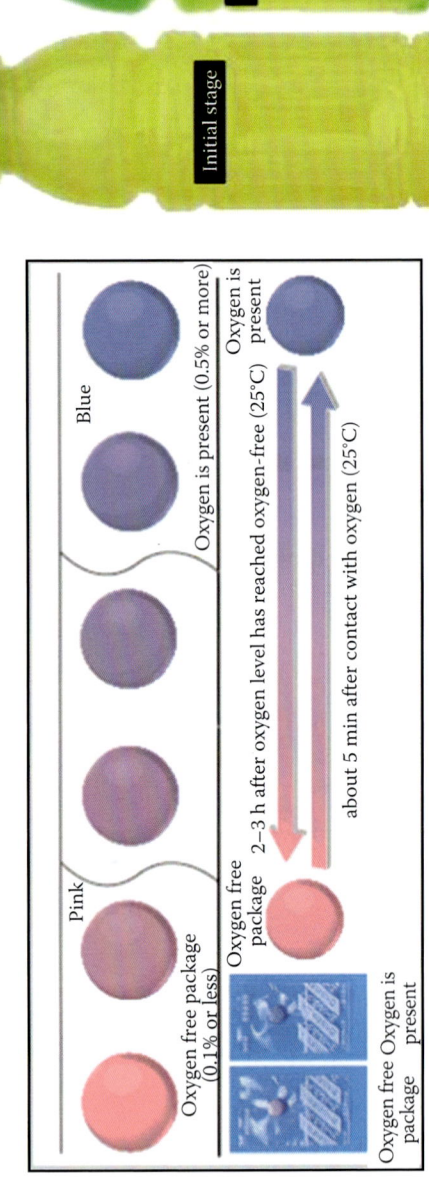

FIGURE 61.4 Optical indicator of the presence or absence of oxygen. (From http://www.mgc.co.jp/eng/products/abc/ageless/eye.html, http://machinedesign.com/archive/oxygen-living-color.)

than one micrometer to several thousand micrometers in size. Each technology offers specific attributes, such as high production rates, large production volume, high product yield, and different capital and operating costs. Other process variables include degree of flexibility in the selection of the encapsulating material and differences in the device size and morphology. The selection of an encapsulation technique is governed by the properties (physical and chemical) of core and encapsulating materials and the intended application of food ingredients.

The material to be encapsulated can be solid or liquid. The principle of most encapsulation technologies is quite simple, combining three consecutive steps (Poncelet and Dreffier, 2007):

1. The active ingredient is mixed within an encapsulating material, in most cases a polymer solution.
2. The resultant formulation, in a liquid form, is dispersed into fine droplets by dripping (drop-by-drop), spraying, or emulsification. This step increases the liquid surface available for further transformations (evaporation, cooling, gelation, chemical reactions) and favors the generation of dispersed (liquid or solid) delivery structures.
3. The oil or aqueous droplets are stabilized in a third step. Oil droplets can be made of a molten material and can be transformed into solid particles by cooling. Depending on their formulation, aqueous droplets can be subjected to a number of solidification processes such as gelation, polymerization, or crystallization.

When the active ingredient is in a solid form (solid particles), encapsulation can also be achieved by spraying a coating solution onto the particles surface, by many different processes that will keep the particles in motion (e.g., agitated particles bed in a fluid bed or pan rotating bed), followed by a consecutive step of stabilization of the coating by solidification or membrane formation.

32.3 ENCAPSULATING SYSTEMS

Delivery systems can be solid or liquid, depending on the food matrix where they are introduced. Some examples of *solid systems* are spray-dried and gel microparticles, whereas *liquid systems* include liposomal and emulsified systems. Each of these delivery systems has its own specific advantages and disadvantages for encapsulation, protection, and delivery of food ingredients. These aspects are briefly discussed in the following, together with a description of the basic principles of each technique, its physicochemical characteristics, and the current challenges for its application in foods.

32.3.1 SPRAY-DRIED MICROPARTICLES

32.3.1.1 Concept, Structure, and Properties

Spray drying (SD) is a preservation technique commonly used in the food industry, mainly for dairy products. By decreasing water content and water activity when converting liquids into powders, this technique increases the storage stability of products, minimizes the risk of chemical or biological degradations, and also reduces the storage and transport costs.

SD is a unique drying process since it involves both particle formation and drying. From a microstructural viewpoint, the formation of spray-dried powders involves the droplet formation from a liquid state followed by a solidification operation driven by solvent evaporation, as schematically represented in Figure 32.5.

Liquid atomization is a decisive stage in SD, defining the evaporation surface. It covers the process of liquid bulk breakup into millions of individual droplets forming a spray. To illustrate, consider the division of one liquid droplet with an initial diameter of 1 cm into N droplets of an equal final diameter of 100 μm. For the same liquid volume, this disintegration mechanism generates 10^6 droplets of 100 μm. The superficial area of the liquid, for example, the available surface to heat and mass transfer between the liquid and the drying air is thereby increased 100 times.

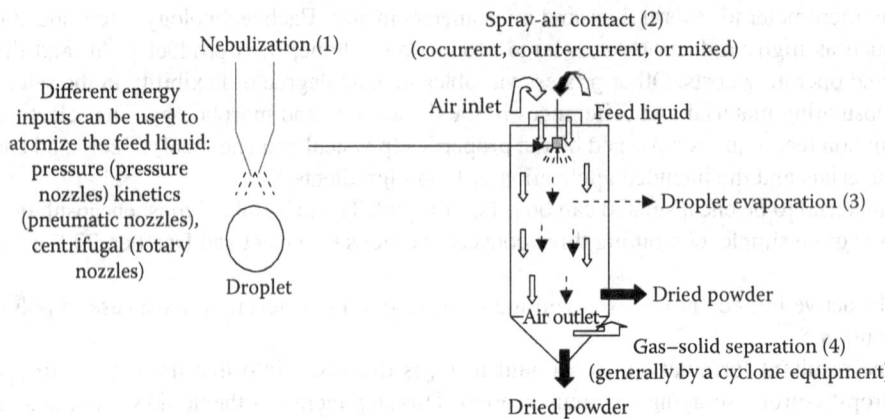

FIGURE 32.5 The main stages involved in the spray-drying process: 1, nebulization of feed liquid into small droplets (spray formation); 2, spray-air contact (mixing and flow), which can be made by several modes: cocurrent (the liquid is nebulized in the same direction as the airflow, as illustrated), countercurrent (liquid droplets and hot air flowing in opposite directions), or mixed flow (the liquid is sprayed upward and only remains in the hot zone for a short time); 3, droplet evaporation; and 4, separation of dried product from the air.

Liquid nebulization into small droplets can be carried out by kinetic pressure or centrifugal energy. Conventional atomizer or nebulizer devices include centrifugal or pneumatic nozzles and rotary discs. Selection of the atomizer configuration is one of most important choices in the SD design. It depends on the nature and viscosity of the liquid to be sprayed and has a significant effect on the size distribution of the final dry particles. As a general rule, an increase in the energy available for nebulization (i.e., rotary nozzle speed, nozzle pressure, or air-to-liquid ratio in a pneumatic nozzle) reduces the particles size (Masters, 1985). However, for pneumatic nozzles, there is an optimum air-to-liquid ratio, above which an increase in energy input does not increase the efficiency of the nozzle to disintegrate the liquid and represents a waste of energy (Ré et al., 2004). By modifying the atomization and the drying conditions, as well as the liquid formulation, it is possible to alter and control properties of spray-dried powders such as particle size, solubility, dispersibility, moisture content, hygroscopicity, flowability, and bulk density.

Spray-dried particles are produced as a very fine powder. Mean particle sizes range from few microns to several tens of microns, with a relatively narrow size distribution, which actually depends on the spray–air contact mode. Particles in the range of 1–50 µm in diameter are typically produced in the cocurrent mode, the most usual contact mode to dry thermosensitive products. Larger particles, ranging from 50 to 200 µm can be obtained in a countercurrent mode due to their agglomeration inside the drying chamber. Spray-dried particles may be spherical or present superficial indentations formed by shrinkage of the droplets during the early stages of the drying process due to the effect of a surface-tension-driven viscous flow (Masters, 1985). The interested reader is referred to Chapter 9 for further details on the principles of SD and its use as a preservation method in the food industry.

SD is used as a microencapsulation method because structured microparticles can be created starting from complex liquid mixtures comprising an active ingredient dissolved, dispersed, or emulsified with an encapsulating material in an aqueous or an organic solvent (water is used in most cases). An appropriate combination of the active ingredient with a range of encapsulating materials and other formulating ingredients, such as surfactants, can give the in-use properties of the encapsulated product. In a single operation, after nebulization of the liquid feed into the spray dryer, the active ingredient can be entrapped into the main component of the structure, for example, the encapsulating material. This process permits the isolation of microspheres or microcapsules depending on the initial formulation (Figure 32.6), as discussed by Ré (2006).

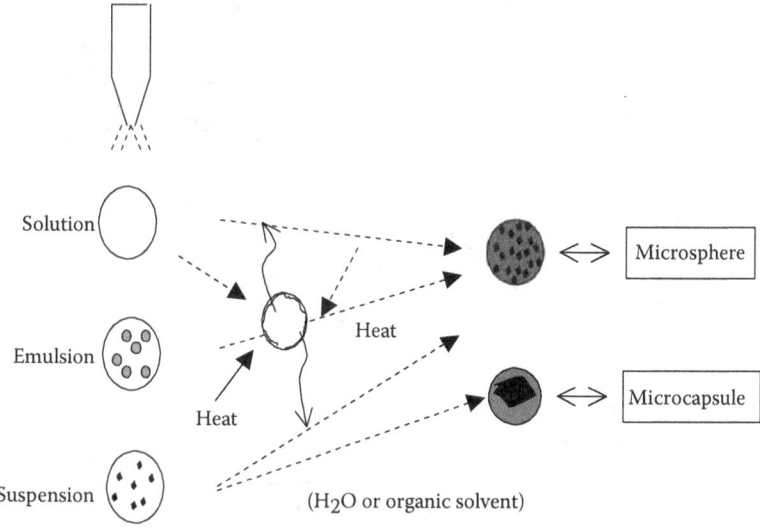

FIGURE 32.6 Architecture of spray-dried particles, depending on the initial formulation (solution, suspension, or emulsion). (From Ré, M.I., *Drying Technol.*, 24, 433, 2006. With permission.)

An important step in developing spray-dried microcapsules for food application is the choice of the encapsulating material. This greatly affects the encapsulation efficiency, appropriate thermal or dissolution release, mechanical strength, microcapsule stability, and compatibility with the food product. The criteria for this choice are mainly based on the solubility in water at an acceptable level, the film-forming and emulsifying properties of the limited number of food-grade materials available. Carbohydrates such as starches, maltodextrins, corn syrup solids, and gum arabic are used extensively in spray-dried encapsulations of food ingredients as the encapsulating material (Ré, 1998). Proteins such as whey proteins, soy proteins, and sodium caseinate have also been used (Gharsallaoui et al., 2007). However, the selection of the encapsulating material for spray-dried microcapsules production has traditionally involved trial-and-error procedures. Firstly, a material is chosen and microcapsules are produced using different mass proportions of encapsulating material to active ingredient and various drying conditions. The spray-dried powders are then evaluated for encapsulation efficiency, stability under different storage conditions, and degree of protection provided to the core material, among other physical characterizations (particle size, powder density, powder flowability, etc.). This procedure is costly and time consuming. Some efforts have been made to develop quantitative methods for selecting the most suitable wall materials for spray-dried microcapsules, mainly for lipid encapsulation.

A method proposed by Matsuno and Adachi (1993) for screening the most suitable wall materials for lipid encapsulation is based on measurements of the drying rate of an emulsion as a function of its moisture content. A suitable material for this application should possess a high emulsifying activity, a high stability, a tendency to form a fine and dense network during drying, and, at the same time, should not permit lipid separation from the emulsion during dehydration. Because the isothermal drying rate is governed by the water diffusion rate, the drying rate may reflect the sample matrix characteristics, i.e., the finer and denser the matrix, the lower the drying rate. According to this method, a characteristic drying curve as a function of moisture content for a suitable group of encapsulating materials is presented in Figure 32.7 (type 1 curve). This curve has been interpreted in terms of the ability of the wall material to form a dense network. The drying rate decreases rapidly as the water content decreases, reflecting a rapid formation of a dense skin and a good protection of the core ingredient against oxygen transfer and possible deterioration. Some materials presenting this type of drying curves are maltodextrin and gum arabic, which are considered as the most suitable for microencapsulation by SD. According to this method, materials that do not form

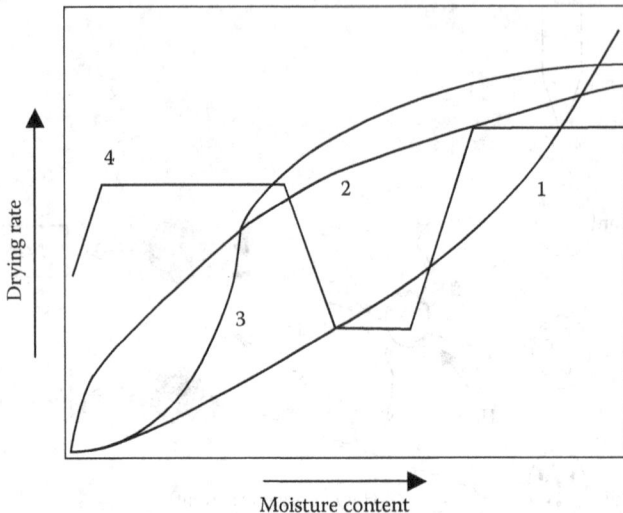

FIGURE 32.7 Schematic representation of isothermal drying curve for the selection of encapsulating materials for spray-dried microcapsules (Matsuno and Adachi method). Type 1 curve corresponds to materials that form fine, dense, two-dimensional skins immediate upon drying. Types 2, 3, and 4 curves correspond to materials that do not form dense skins at an early stage of drying. (From Pérez-Alonso, C. et al., *Carbohydr. Polym.*, 53, 197, 2003. With permission.)

dense skins at early stages of drying are unsuitable for efficient lipid encapsulation (types 2, 3, and 4 curves). Characteristic type 2 materials are caseinate and albumin, type 3 materials are low molecular weight saccharides that do not crystallize readily like glucose, and type 4 materials are those that easily crystallize upon dehydration such as mannitol (Matsuno and Adachi, 1993). However, not all the materials showing an early decreasing rate, in which water evaporation is controlled by diffusion mechanisms, are suitable for lipid encapsulation when used alone. Therefore, it is desirable to determine an optimal combination of materials that will provide excellent emulsifying capacity and very low oxygen diffusion. In this respect, as analyzed by Pérez-Alonso et al. (2003), the Adachi and Masuno method does not allow for an effective discrimination between materials showing similarly shaped drying curves.

Another method, proposed by Pérez-Alonso et al. (2003), uses the activation energy of carbohydrate polymer blends dried isothermally, as a discriminating parameter for selecting the most suitable mixture as wall material for spray-dried microcapsules. The activation energy provides a measure of the necessary energy required for evaporating a mass of water from the material to be dried. This method requires the knowledge of the drop volume shrinkage of every conceivable blend, which can be achieved as follows: A drop of a blend constituent aqueous solution is put on a glass slide and micrographs of the X–Y, X–Z, and Y–Z planes of the drop are taken. The area of each plane is calculated and approximated to that of a sphere. These steps are repeated as the drops are dried isothermally at intervals of approximately 10% moisture content decrease (determined by drop mass loss). The experimental points are then fitted with a polynomial reported for each blend constituent and assumed additive volumes of the blend constituents to determine the drop shrinkage of studied blends.

Despite these efforts, screening for new wall materials, such as milled citrus fruit fibers as a potential replacement for maltodextrin-type carriers (Chiou and Langrish, 2007), is still mainly done by trial and error.

When selecting SD to produce an encapsulated food ingredient, one is generally looking for high production in a short time and for a product in a powder form. Spray-dried microparticles are commonly used to encapsulate flavors or lipids, and the release mechanism is generally linked to the dissolution of the encapsulating agents. The encapsulated ingredients are, in general, rapidly released due to the dissolution of the spray dried structures at the time of consumption by dispersing the powder in a wet formulation (e.g., instant beverages). The challenge is how to encapsulate thermosensitive and volatile compounds by a drying process operation, avoiding thermal degradation and volatile losses during drying, and generating an encapsulated product in a powder form with good flowability, stability, and acceptable shelf life after drying.

32.3.2 GEL MICROPARTICLES

32.3.2.1 Concept, Structure, and Properties

Gel microparticles are formed from the concept of gelation as an encapsulation technology. Gelation is based on the formation of a solution, dispersion, or the emulsification of the core material in an aqueous solution containing a hydrophilic polymer (hydrocolloid) capable of forming a gel under an external action, either physical or chemical.

There are many techniques for physical gelation, whose use depends on whether the hydrocolloid can gel in water without additives (thermal gelation) or ions are required to aid gelation (ionotropic gelation). In thermal gelation, typically, a solution is made by dissolving a hydrocolloid in powder form in water at high temperature and then cooling to room temperature. As the solution cools, enthalpically stabilized chain helices may form from segments of individual chains, leading to a three-dimensional network (Burey et al., 2008). Examples of such systems are gelatin and agar.

Ionotropic gelation occurs via cross-linking of hydrocolloid chains with ions, generally cation-mediated gelation of negatively charged polysaccharides. Examples of such systems are alginate, carrageenan, and pectin. A typical example is the formation of alginate beads by dropping an alginate solution into a bath containing calcium chloride to form the insoluble calcium-alginate. There are two main methods by which ionotropic gelation can be done, namely external and internal gelation. External gelation involves the introduction of a hydrocolloid solution into an ionic solution, with gelation occurring via diffusion of ions from outside into the hydrocolloid solution. This method is the easiest and most often used one for encapsulation by ionotropic gelation. However, it can often cause inhomogeneous gelation of gel particles due to its diffusion-based mechanism, which constitutes a drawback. Surface gelation often occurs prior to core gelation, and the former can inhibit the latter, leading to gel particles with firm outer surfaces and soft cores (Chan et al., 2006). Internal gelation overcomes the main disadvantage of the external gelation method, as it requires the dispersion of ions prior to their activation to cause gelation of hydrocolloid particles. This usually involves the addition of an inactive form of the ion that will cause cross-linking of the hydrocolloid, which is then activated, for example, by a change in pH, after the ion dispersion is sufficiently complete. Internal gelation is particularly useful in alginate systems, which can gel rapidly and may become inhomogeneous if gelation occurs before adequate ion dispersion has occurred (Poncelet et al., 1992; Chan et al., 2002). For example, in the production of alginate particles by external gelation, the alginate solution is extruded as droplets into a solution of a calcium salt. For internal gelation, an insoluble calcium salt is added to the alginate–drug solution and the mixture extruded into oil (Liu et al., 2002). The latter is acidified to bring about the release of Ca^{2+} from the insoluble salt for cross-linking with the alginate. Despite their homogeneity, internal gelated matrices may be more permeable, resulting in lower encapsulation efficiencies and faster release rates (Vanderberg and De La Noüe, 2001), which may also be overcome by manipulating the pH of the medium and the amount of calcium salt used.

Whatever the technique used (thermal or ionotropic gelation), gel particles are generally formulated in a two-step procedure involving a droplet formation and hardening. The droplet formation step determines the mean size and the size distribution of the resulting gel particles. In the following, the main procedures used for droplet formation—droplet extrusion, nebulization (spray), and emulsification—are described.

32.3.2.1.1 Droplet Extrusion

Extrusion denotes feeding the hydrogel solution, typically containing the active material to be encapsulated, through a single or plurality of pathways directly into the continuous gelation bath. Henceforth, for simplicity, the hydrogel solution containing the active material to be encapsulated (generally an aqueous dispersion) is referred to only as "hydrogel" solution.

In the droplet extrusion technique, also referred to as the drop method, hydrogel solutions are extruded through a small tube or needle (Figure 32.8a), permitting the formed droplets to freely fall into a gelation bath. The droplets may be cross-linked by addition of an appropriate cross-linker to the receiving solution. The size of the droplets, and thus the size of the subsequent gel particles, depends upon the diameter of the needle, the flow rate of the solution, its viscosity, and the concentration of the ionic solution. Typical gel particle sizes obtained using the conventional syringe-drop method are 0.5–6 mm, and on the scale of hundreds of microns, if modified techniques suitable for large-scale processes are used to disperse the hydrocolloid solution into droplets (Burey et al., 2008). The main difference among the reported techniques for a mass production of small narrowly dispersed or monosized hydrocolloid gel particles lies in the way the drops are formed, i.e., electrostatic dripping (Figure 32.8b), jet breakup through mechanical vibrations (Figure 32.8c), jet-cutting (Figure 32.8d), and jet and rotating disc atomizers (Figure 32.8e and f, respectively).

The electrostatic technique uses an electric potential difference to pull the droplets from a needle tip (Figure 32.8b). The electrostatic potential difference is established between the needle feeding the solution and the gelling bath. In the absence of an electric field, a droplet forming on a needle tip will grow until its mass is large enough to escape the surface tension at the needle–droplet interface. With the introduction of an electric field, a charge is induced on the droplet surface. Mutual charge repulsion results in an outwardly directed force acting downward on the forming droplet. The additional electrostatic force pulls the droplet from the needle tip at a much lower mass and hence size. Capsule size may be controlled by adjusting the magnitude of the voltage. The higher the voltage, the higher the electrical force pulling the droplet, and therefore, the smaller the obtained droplet and, consequently, the capsule. The literature reports different capsule ranges that can be achieved using this method, for example, 40–2500 μm according to Burey et al. (2008) or 50–800 μm according to Poncelet and Dreffier (2007).

In the laminar jet breakup technique, a laminar flow of the hydrocolloid solution is converted into a succession of identical droplets by the action of an ultrasound vibrating nozzle (Figure 32.8c). Jet-cutting is another method, in which the fluid is pressed through a nozzle in the form of a liquid jet. This jet is cut into uniform cylindrical segments by a means of a rotating cutting tool (Figure 32.8d). Due to surface tension effects, these segments form spherical beads while falling down. The diameter of the resulting bead is determined by the number of cutting wires, the number of rotations of

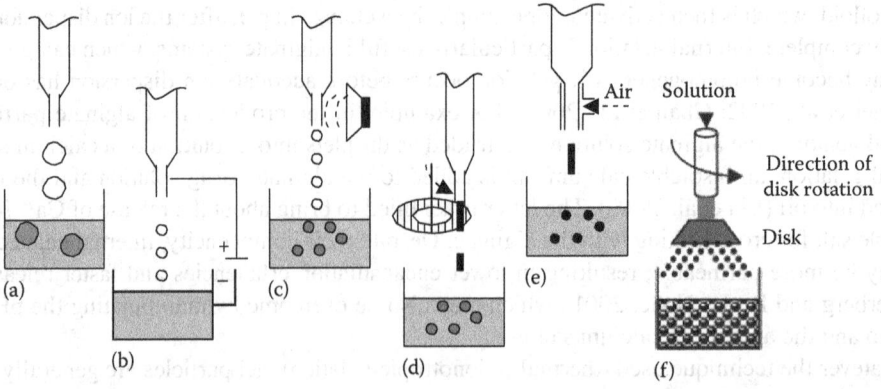

FIGURE 32.8 Schematic representation of the different techniques for drop formation employed in gelation: (a) droplet extrusion; (b) electrostatic dripping; (c) laminar jet breakup; (d) jet-cutting; (e) jet nebulizer; and (f) disk nebulizer.

the cutting tool, and the mass flow through the nozzle, which, in turn, depends on both the nozzle diameter and the fluid velocity.

32.3.2.1.2 Nebulization

The dispersion of the hydrocolloid solution may also be achieved by jet nebulizers using, for example, a coaxial air stream that pulls droplets from a needle tip into a gelling bath (Figure 32.8e). Small quantities of gel particles ranging in size down to around 400 µm are achieved by this method (Herrero et al., 2006). Droplet formation is aided by the shear energy of gas flow, used to overcome the viscous and surface tension forces of the fluid. The viscosity and surface tension of a liquid being nebulized can thus alter the properties of the aerosol generated (Figure 32.8f).

32.3.2.1.3 Emulsification

In the emulsion technique, solutions are mixed and dispersed into a nonmiscible phase. For food applications, vegetable oils are used as the continuous phase. In some cases, emulsifiers are added to form a better emulsion, since such chemicals lower the surface tension, resulting in smaller droplets. After emulsion formation, gelating and/or membrane formation is initiated by cooling and/or addition of a gelling agent to the emulsion, or by introducing a cross-linker. In a last step, the gel particles formed are washed to remove oil (Chan et al., 2002).

Stirring is the most straightforward method to generate droplets of a dispersed phase in a continuous phase to produce an emulsion. In the simplest approach, the continuous phase is poured into a vessel and stirred by an impeller (Figure 32.9a). The dispersed phase is then added, dropwise or all at once, under agitation at a sufficient speed to reach the desired droplet size. The final droplet size of the liquid–liquid dispersion in stirred vessels depends on parameters such as the physicochemical characteristics of the two phases (e.g., viscosity, interfacial tension, and stabilizer concentration), the preparation conditions of the emulsion (e.g., temperature, addition order of the components), and the stirring system (e.g., shear rate, design of the stirrer, and containing vessel). Hydrocolloid solution droplet sizes normally range from 0.2 to 80 µm, although they can be as large as 5000 µm, and gel particles can range from 10 to 3000 µm, as summarized in a recent review (Burey et al., 2008).

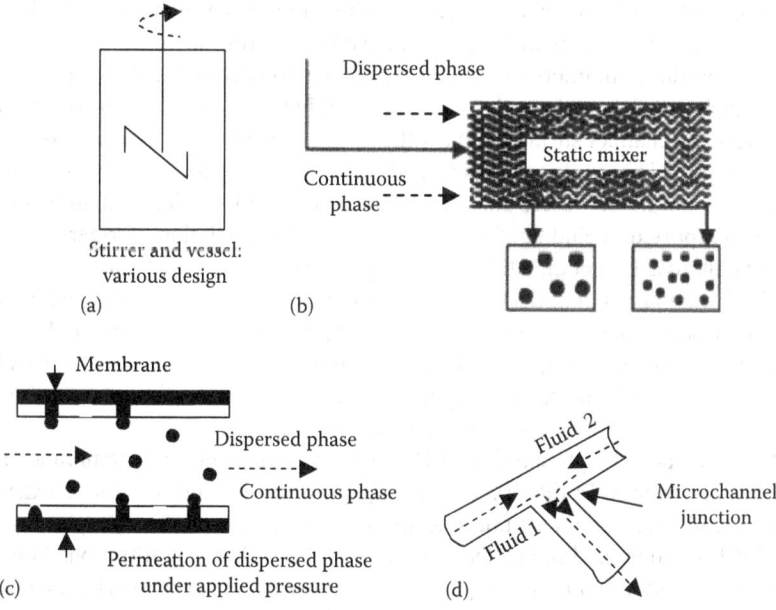

FIGURE 32.9 Schematic diagram of different emulsification processes to produce gel particles: (a) mechanical stirring; (b) static mixing; (c) membrane emulsification; and (d) microchannel emulsification.

A number of approaches have been proposed for a continuous emulsification and improved control of product size distribution: static mixers, membrane emulsification, and microchannel emulsification. For the last two, emulsions are produced by extruding a liquid through many individual pores or microchannels.

Static mixers consist of a series of geometric mixing elements fixed within a pipe. The particular arrangement of these mixing elements repeatedly splits and recombines the stream of fluid passing through the tube (Figure 32.9b). Recombination occurs through impingement of the substreams, creating turbulence and inducing back mixing. Static mixers installed in tubes allow continuous production, and have already been used to produce gel particles (Belyaeva et al., 2004).

Membrane emulsification is a relatively new method for the preparation of spherical particles with a highly uniform size distribution, and it can be used to extrude a hydrocolloid solution into a gelation bath (Oh et al., 2008). The method involves the use of a porous membrane with a highly uniform pore size. The dispersed phase is pressed through the membrane pores, whereas the continuous phase flows along the membrane surface. Droplets grow at pore outlets until they detach upon reaching a certain size (Figure 32.9c). A low-pressure drop is applied to force the dispersed phase to permeate through the microporous membrane into the continuous phase. Details on such methods using membrane emulsification are summarized in several review papers (Joscelyne and Trägardh, 2000; Vladisavljevic and Williams, 2005). The distinguishing feature is the fact that the resulting droplet size is primarily controlled by the choice of the membrane and not by the generation of turbulent droplet breakup. The technique is highly attractive given its simplicity, potentially lower energy demands, need for less surfactant, and the resulting narrow droplet-size distributions. It is applicable to both oil-in-water (O/W) and water-in-oil (W/O) emulsions, and it has been recently applied to prepare microgel particles with a uniform size distribution (Wang et al., 2005; Zhou et al., 2007).

Microfluidic methods have recently been mentioned as an emerging technology for the production of monosized gel particles (Amici et al., 2008). Typically, the term microfluidics refers to the manipulation of fluids in systems or devices having a network of chambers and reservoirs connected by channels, whose typical cross-sectional dimensions range from 1.0 to 500 μm, the so-called "microchannels." Materials such as silicon, cofired ceramic, glass, quartz, and polymers have been explored for the fabrication of microfluidic systems (Nguyen and Zhigang, 2005). Due to the large surface-to-volume ratio and low inertial forces encountered at the microscale, highly precise and specific flow manipulation and control can be achieved by appropriate microfluidic design.

Emulsions are produced in microfluidic devices when two immiscible fluids, flowing in two separate microchannels, are forced through a microchannel junction, and the flow of one fluid (usually the fluid that wets the channel surface) breaks the flow of the other to form microdroplets (Figure 32.9d). Drop formation is reproducible, and the drop size can be regulated by operating on factors such as flow rates, fluid viscosities, and surfactant concentration. Gel formation in microfluidic systems has been reported, including, for instance, the thermal gelation of κ-carrageenan (Walther et al., 2005) and agarose (Xu et al., 2005). Ionotropic gelation can also be achieved in microchannels through both the internal and the external approaches (Huang et al., 2007; Amici et al., 2008).

Alginate microspheres have been extensively used as delivery system because they are very easy to prepare, the process is very mild, and virtually any ingredient can be encapsulated (Vladisavljevic and Williams, 2005). That is the reason why they are chosen to exemplify the use of the membrane and channel emulsification processes to produce gel particles.

Figure 32.10 illustrates the preparation of Ca-alginate gel particles by membrane emulsification by two different approaches. In the first one, an aqueous Na-alginate solution is extruded through a hydrophobic membrane into an oil phase to form a W/O emulsion. The droplets are then gelled by adding a $CaCl_2$ solution, and the gelled particles can be colleted by filtration. This process was used by Weiss et al. (2004), enabling the formation of uniform gel particles with an adjustable mean diameter (20–300 μm). In the second approach, Ca-alginate gel particles can be produced by internal gelation; in brief, a dispersion of an aqueous Na-alginate solution containing $CaCO_3$ is extruded through a microporous membrane into an oil phase. The gelation is initiated by the addition of

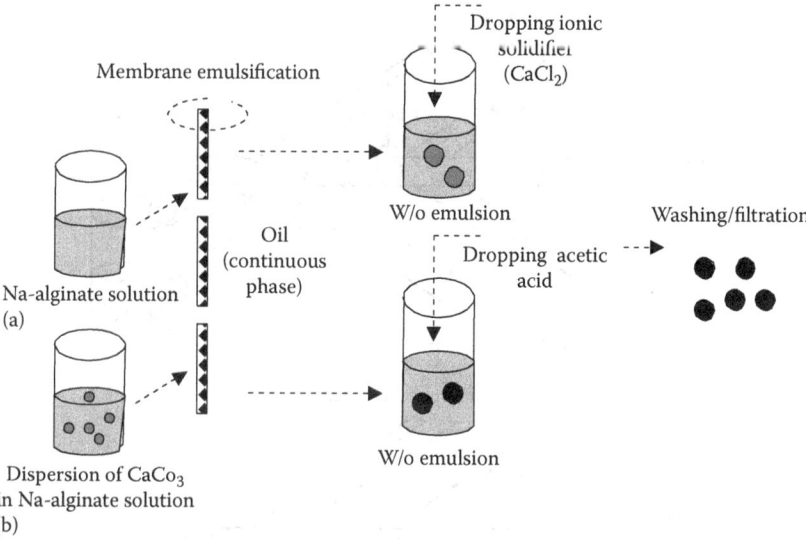

FIGURE 32.10 Illustration of the production of alginate microgels by membrane emulsification: (a) external gelation and (b) internal gelation.

acetic acid into the emulsion, dissolving $CaCO_3$ to release Ca^{2+} and form Ca-alginate. Using the second approach, Liu et al. (2003) obtained Ca-alginate gel particles with a mean size of 55 μm (with a coefficient of variation of 27%), using a nickel membrane with a pore size of 2.9 μm.

Figure 32.11 illustrates the preparation of Ca-alginate gel particles by microchannel emulsification. To produce alginate gel particles by external gelation, an alginate aqueous solution flows through a flow-focusing channel, and an alginate droplet is formed from the balance of interfacial and viscous drag forces resulting from the continuous (oil) phase flowing past the alginate solution. It immediately reacts with an adjacent $CaCl_2$ drop that is extruded into the main flow channel by another flow-focusing channel located downstream in relation to the site of the alginate drop creation. This procedure has been used in the literature, generating monosized alginate beads within a range of 50–200 μm, depending on flow conditions (Hong et al., 2007). To produce alginate drops in a microfluidic system by internal gelation (Amici et al., 2008), two aqueous streams, one acidic and one containing alginate and calcium carbonate, can merge immediately prior to entering a channel where a continuous flow of an oil phase breaks the flow of the aqueous phase to form microdroplets (Figure 32.11b). Several variations of both approaches have been developed in the literature by using new microfluidic systems (Liu et al., 2006; Choi et al., 2007).

32.3.3 Liposomal Systems

32.3.3.1 Basic Principles

Liposomes are conceptually biomimetic model systems. They allow studies of the lipid matrix of biomembranes, as well as the investigation of membrane embedded proteins and certain fundamental aspects of organelles in a biomimicking environment, outside the living cell. Structurally, liposomes are self-assembled colloidal particles in which an aqueous nucleus is enclosed by one or several concentric phospholipid bilayers. Phospholipids are amphiphilic molecules, which in aqueous solution form energy-favorable structures as a result of hydrophilic and hydrophobic interactions. Depending on the size and the number of bilayers, liposomes are classified as multilamellar vesicles (MLVs) or large and small unilamellar vesicles (LUVs and SUVs). Other classifications for larger liposomes include the plurilamellar vesicles (PLVs), when nonconcentric bilayers enclose various aqueous compartments, and oligolamellar vesicles (OLVs), in which small vesicles are included in the structure of

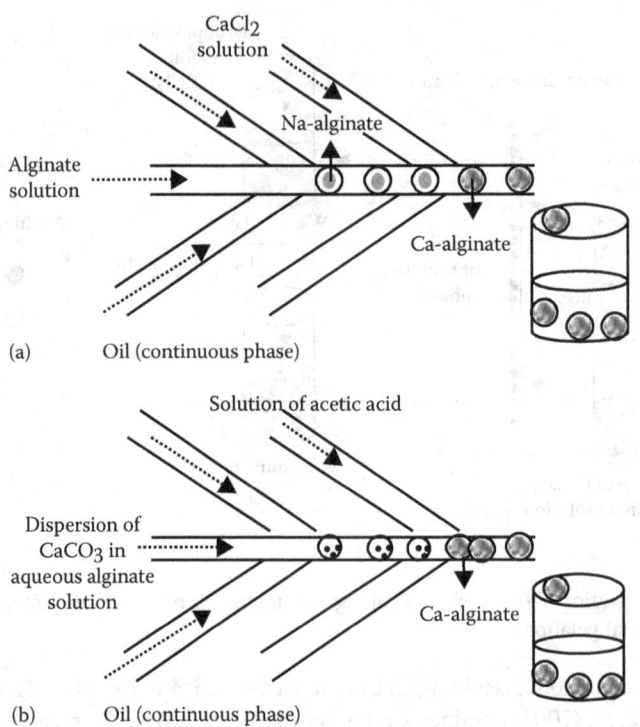

FIGURE 32.11 Illustration of the microfluidic production of alginate microgels by (a) internal gelation and (b) external gelation.

a large vesicle. The size of unilamellar liposomes may vary from 20 to 500 nm approximately, and the thickness of one lipid bilayer is about 4 nm. Due to the presence of hydrophilic and hydrophobic domains in the structure, liposomes are able to encapsulate water-soluble molecules in the aqueous nucleus, water-insoluble within the lipid membrane, or amphiphiles between the two domains. Therefore, liposomes are a powerful solubilizing system for a wide range of compounds.

Figure 32.12 shows schematically the phospholipid aggregation in bilayer, and the vesiculation, which occurs in excess of water, forming the liposomal structure. Due to their structure, chemical composition, and size, all of which can be well controlled by preparation processes, liposomes exhibit several properties that are useful in a large range of applications. The most important properties are colloidal size, bilayer phase behavior, mechanical properties, permeability, and charge density. Liposomes also allow surface modifications through the attachment of ligands and bounded or grafted polymers. Figure 32.13 shows the various modifications of liposome surfaces.

For historical reasons, when no physical or chemical surface modification is introduced, liposomes are called conventional to be distinguished from the surface-modified liposomes.

Liposomes can be made entirely from naturally occurring substances, and are therefore biodegradable and nonimmunogenic. In some cases, the introduction of synthetic lipids is useful for specific characteristics such as stability and charge. The lipid composition, colloidal characteristics, and surface modifications also allow liposomes to have functional properties, such as stability, controlled release of the encapsulated molecules, specific targeting, and controlled pH and temperature sensitivity.

From the thermodynamic point of view, liposomes are not stable structures, and so they cannot form themselves spontaneously. To produce liposomes, some energy from extrusion, homogenization, or sonication must be dissipated into the system. Subsequently, after formation, liposomes can aggregate, fuse, and form larger structures that eventually settle out of the liquid. Thermodynamically stable systems, such as micelles, stay at the same phase forever. The precursor phase of liposomes is

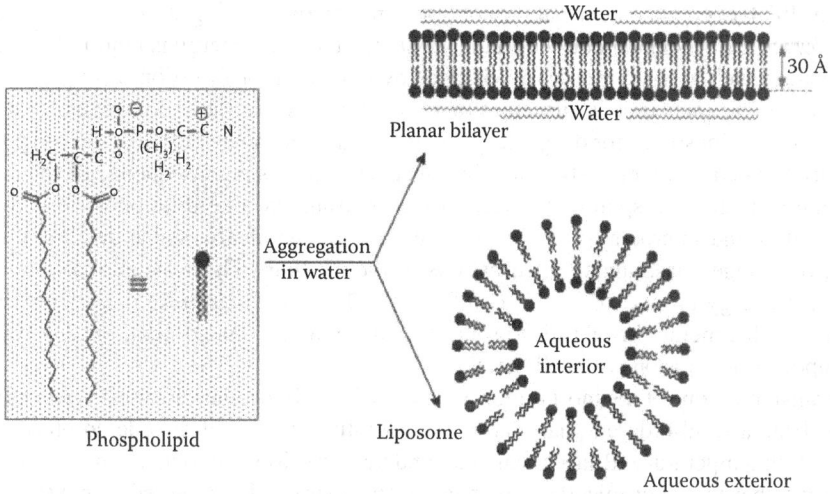

FIGURE 32.12 Phospholipid aggregation in planar bilayer and vesiculation in liposomal structure.

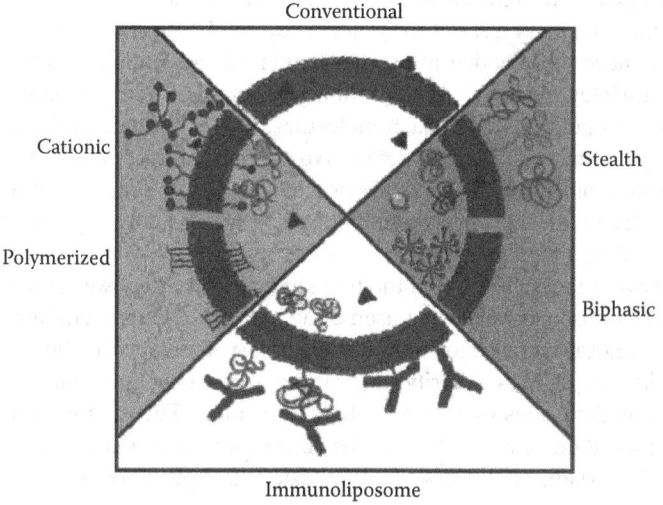

FIGURE 32.13 Surface modifications in liposomes.

the symmetric bilayer composed of self-assembled phospholipid molecules with comparable areas of the polar and nonpolar portions. Liposomes are formed when, at high concentration, the self-assembled molecules form a long-range ordered liquid-crystalline lamellar phase, which diluted in excess water is dispersed in stable colloidal particles. These particles retain the short-range symmetry of their original parent phase.

An extension of the Tanford's (1980) treatment of the shapes of micelles, based on the molecular-shape analysis and the concept of the shape parameter (ratio between the nonpolar and polar cross-section areas), was introduced by Israelachvili (1991) for a qualitative understanding of the topology of lipid aggregates with different lipid compositions. However, the liposome models are approximate, and a rigorous thermodynamic analysis fails because liposomes are not at the thermodynamic equilibrium. The rigorous analysis would yield a very narrow size distribution, which is never observed in practice, as well as the spontaneous formation, while a high-energy process is typically needed to produce liposomes.

Helfrich (1973) explained the instability of liposomes by the bending elasticity. Symmetric membranes prefer to be flat (spontaneous curvature equal to zero), and energy is required to curve them. Various factors, such as asymmetrically changes by ionization or insertion of molecules like surfactants, change the spontaneous curvature to nonzero values, but the spontaneous formation does not produce stable liposomes for drug encapsulation. Stable liposomes for drug-delivery must be in a kinetically trapped and thermodynamically unstable state. As a consequence of that, liposomes maintain their integrity in spite of changes in the environment like dilution or in vivo administration. Micelles and microemulsions, which are thermodynamically stable systems disintegrate, aggregate, or change phase under perturbations of the medium. The vesiculation accumulates an excess of free energy around 10–50 kT (1 kT = $4.11 \cdot 10^{-21}$ J at T = 298 K) in the curvature of the liposomes, which generate instabilities promoting fusion and disintegration, or may become bioavailable upon vesicle fusion with cell membranes.

Phase transition is one of the most important properties of liposomes. It happens when lipid bilayers change from a solid-ordered phase, at low-temperature, to a fluid-disordered phase, above the phase transition temperature. Bilayers can also undergo transitions into different liquid-crystalline phases, such as hexagonal or micellar. The phase transitions can be triggered by physical or chemical factors, resulting in different effects on the liposomes. Thus, transitions from gel to liquid-crystalline phase caused by temperature or ionic strength changes cause liposome leakage.

Physicochemical stability determines the shelf life stability of liposomes. By optimizing the size distribution, pH, and ionic strength, as well as by the addition of antioxidants and chelant agents, the stability of liposomes can be preserved for years. They can be stored in a frozen or in dried form, but cryoprotectants have to be added to prevent fusion. Electrostatic and steric stabilization also reduce fusion and disintegration by freezing. Biological in vivo stabilization is obtained by reducing the interactions of liposomes with macromolecules, blood protein, and disintegrating enzymes, as well as adverse pH conditions. Therefore, in vivo stability of liposomes depends on the route of administration. Steric stabilization is generally provided by the protective coating from the grafting of the liposome surface with inert hydrophilic polymers such as polyethylene glycol (Lasic, 1995) or hyaluronic acid (Eliaz and Szoka, 2001).

Since their discovery in the 1960s and the first studies where they were used as a drug delivery system in the 1970s, liposomes have been used as an encapsulation system in a myriad of applications ranging from material science to analytical chemistry, food, and medicine.

Numerous books and reviews describe the various physicochemical and biological aspects of liposomes, as well as their construction and characterization. This literature has been written by scientists who composed the scientific base of the design and construction of liposomes for the various applications (Gregoriadis, 1988; Lasic, 1993; Lasic and Papahadjopoulus, 1998).

32.3.3.2 Applications in Foods

Gomez-Hens and Fernandez-Romero (2006) depict the number of articles and patents published in the 1990–2004 period regarding the use of the liposomal systems in five areas: drug and gene delivery systems, biochemical and biotechnological applications, cosmetics, nutrition, and foods. Figure 32.14 shows the evolution of articles and patents from 2004 to 2008. In both cases, the tendency is the same: there have been significant advances in the applications of liposomes in the biomedical and pharmaceutical industries related to therapeutic drugs, opposed to their applications in foods, which are presently at an early stage of development.

The applications of liposomes in foods fulfill similar requirements to their applications in pharmaceuticals, and have been focused in the following categories: formulation aid, processing, preservation, stabilizer, nutritional supplement, and nutraceutical carrier.

Liposomes aid in formulation because they are powerful in solubilizing nonwater-soluble compounds, enhancing their bioavailability. Furthermore, liposomes entrap hydrophilic molecules into their interior, and hydrophobic molecules into their lipophilic membrane, encapsulating various nutritional molecules. The first example of liposomes in foods is human milk, which has been

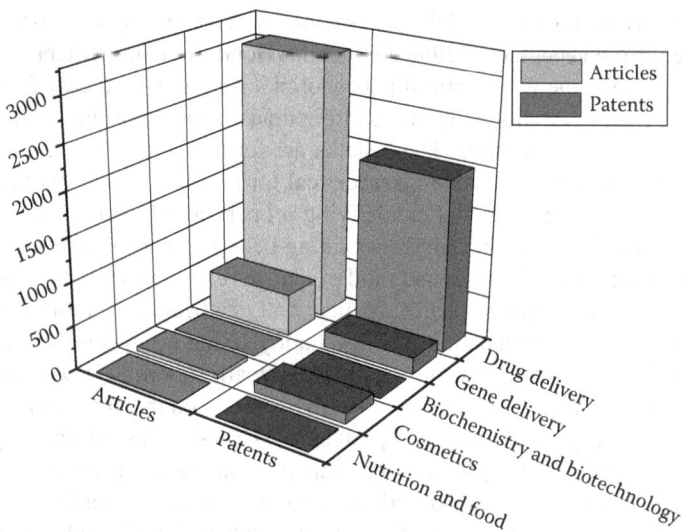

FIGURE 32.14 Articles and patents published on liposomal delivery systems in the 2004–2008 period (last year incomplete). (Courtesy of SciFinder Scholar®, American Chemical Society, Columbus, OH.)

studied for years. Electron microscopic studies show the presence of liposomes along with emulsion droplets and casein micelles (Roger and Anderson, 1998). Liposomes, as a microstructural component of breast milk, may play an important role in enhanced nutrient absorption, colloidal stability, and immunogenicity (Keller, 2001).

In the processing of foods, liposomes accelerate cheese ripening and increase the yield in bioconversion through uniform distribution of hydrophilic enzymes in hydrophobic medium.

Applications of liposomes in cheese ripening were developed by the 1980s (El Soda, 1986). Enhancement of proteolysis by encapsulated cyprosins was evident 24 h after manufacture of Manchego cheese. Addition of encapsulated cyprosins to milk perceptibly accelerated the development of flavor intensity in experimental cheese through 15 days of age without enhancing bitterness (Picon et al., 1996). The capability of neutral and charged liposomes to entrap the proteolytic enzyme neutrase, and the stability of the preparation, were evaluated in the ripening of Saint-Paulin cheese milk (Alkhalaf et al., 1989).

Liposomes also promote sustained release of antimicrobial peptides assuring protection of the formulation. Liposome-entrapped nisin retained higher activity against *Listeria innocua* and improved stability in cheese production, proving to be a powerful inhibitor in the growth of *L. innocua* in cheese, while not preventing the detrimental effect of nisin on the actual cheese-ripening process. Coencapsulation of calcein and nisin, and calcein and lysozyme demonstrated that production and optimization of stable nanoparticulate aqueous dispersions of polypeptide antimicrobials for microbiological stabilization of food products depend on selection of suitable lipid–antimicrobial combinations (Benech et al., 2002; Were et al., 2003).

Besides solubilization, the encapsulation capability of liposomes protects labile compounds from chemical degradation, light oxidation during storage, and harmful compounds from the environment. The stabilizing capability of liposomes also preserves the taste and flavors of foods during processing and storage. Liposomes improve the nutritional effects of foodstuff by entrapping nutritionally important compounds, such as vitamins, polyunsaturated fatty acids, minerals, and antioxidants.

Health benefits of nutrients by changing their kinetics of release and enhancing their bioavailability are obtained by liposome encapsulation. Recently, immunoliposomes, which are liposomes containing antibodies for site-specific targeting, have been studied for nutrient targeting regulation. A useful model involving the leptin protein and immunoliposomes was used to illustrate the nutrient regulation of the endocrine system (Xianghua and Zirong, 2006).

Liposome as a carrier matrix in foods has become an attractive system, because they can be constructed entirely from acceptable edible compounds (food-grade ingredients), like proteins and carbohydrates. Lecithin is the main natural phospholipid, routinely extracted from nutrients, such as egg yolks and soybeans. Additionally, the phospholipids in the liposome matrix are also versatile nutraceuticals for functional foods. The benefits are for the brain, liver, and blood circulation. Phosphatidylcholine is a highly effective nutraceutical for recovery of the liver following toxic or chronic viral damage. It has exceptional emulsifying properties, which the liver draws to produce the digestive bile fluid. The lung and intestinal lining cells use phosphatidylcholine to make the surfactant coating essential for their gas and fluid exchange functions. Phosphatidylcholine exhibits potentially lifesaving benefit against pharmaceutical and deathcap mushroom poisoning, alcohol-damaged liver, and the chronic hepatitis B. Phosphatidylserine establishes its benefits for higher brain functions such as memory, learning and words recall, mood elevation, and action against stress. Phosphatidylserine also has a salutary revitalizing effect on the aging brain, and may also be helpful to children with cognitive and mood problems. The fast access of glycerophosphocholines to the human brain and its capacity to sharpen mental performance also makes it well suited for drink formulations. The nutraceutical properties of phospholipids are described extensively by Kidd (2001). Therefore, the product value comes from the health benefits of the phospholipids associated with the benefits of the selected nutrient. This combined phospholipid–nutrient approach is suited to produce chewable tablets, confectionery products, cookies, granulates, spreads, bars, and emulsified or purely aqueous phase beverages.

Although liposomes carrying nutrients are ingested via the GI system, the oral route also offers a way through the sublingual mucosal membranes. In the first case, the adverse conditions of the environment (low pH of the stomach, surfactant action of bile salts, and the presence of lipases) destabilize the conventional liposome formulation. The sublingual route avoids the first pass liver clearance and metabolism offering a direct uptake of nutrients into the bloodstream through the mucosal membranes. Additionally, the sublingual administration avoids swallowing difficulties from the ingestion of tablets or large capsules by old people or children, in addition to being an alternative for personal preference.

The performance of the sublingual administration of nutrients has been demonstrated using the CoEnzyme Q10 in spray formulation, compared to the powder formulation in hard gelatin capsules. The results showed increased bioavailability of 100% for CoQ10 over endogenous levels with the sublingual spray compared to 50% increase over baseline levels with a two-piece gelatin capsule as measured by area under the curve in 24 h (Gibaldi, 1991). Additionally, the time of onset of the spray formulation administration was shorter than the capsule one, adding benefits to the treatments which require immediate onset, like the cardiogenic supplement CoQ10, diet aids, pain treatment, fever, or insomnia (Rowland and Towzer, 1995). Keller (2001) listed some products that have been formulated using this novel oral liposomal delivery system.

The main issue in liposome encapsulation for food industry is the scaling up of the processes at an acceptable cost. Methods of liposome formation now exist that do not make use of sonication (Batzri and Korn, 1973; Kirby and Gregoriadis, 1984; Zhang et al., 1997) or of any organic solvents (Frederiksen et al., 1997; Zheng et al., 1999), and allow the continuous production of microcapsules on a large scale. Nowadays, liposome encapsulation can also become a routine process in the food industry (Gregoriadis, 1987; Kirby and Law, 1987; Kim and Baianu, 1991; Gouin, 2004).

The great advantage of liposomes over other encapsulation technologies is the stability imparted by liposomes to water-soluble materials in high water activity application: spray-dried, extruded, and fluidized beds impart great stability to food ingredients in the dry state but release their content readily in high water activity application, giving up all protective properties.

Microfluidization techniques have been shown to be an effective and solvent-free continuous method for the production of liposomes with high encapsulation efficiency. The method can process a few hundred liters per hour of aqueous liposomes on a continuous basis. The process has been reported in the literature (Vuillemard, 1991; Maa and Hsu, 1999; Zheng et al., 1999).

Multitubular systems represent a scalable version of the Bangham method. It is adequate to prepare liposomes for food application, due to its simplicity and easy scaling up (Tournier et al., 1999; Carneiro and Santana, 2004; Latorre et al., 2007).

Dry liposomes circumvent the drawback of liposome stability in large-scale production, storage, and shipping of encapsulated food ingredients. Freeze drying the liposome suspension can only be carried out by high price encapsulated ingredients in a niche market, due to the considerable cost of large-scale freeze-drying processes. Moreover, not all liposome formulations can be freeze dried, and the reconstitution of the wet formulation is not always straightforward and usually requires complex steps and processes (Lasic, 1993).

These problems are reduced when the bioactive ingredient is incorporated in a lipid matrix by spray drying and, subsequently, liposomes are made through mechanical stirring (Alves and Santana, 2004). The operational conditions modulate the crystallinity of the lipid matrix and the efficiency of incorporation of the bioactive compound. Conventional or special mechanical stirrers can be used to adjust the size and distribution of liposomes.

Supercritical fluid offers another attempt to avoid the use of organic solvent in the production of liposomes. Basically, the process involves the solubilization of the phospholipids under supercritical condition, followed by the release of the supercritical mixture into an aqueous phase containing the dissolved active ingredient, which results in the formation of liposomes containing the active ingredient in their aqueous cores. Although this method is scientifically interesting, the encapsulation efficiency reported so far is limited to 15% (Frederiksen et al., 1997), which would have to be dramatically increased for this technique to become interesting from an industrial point of view.

Liposomes can also be associated with other technologies. Many hydrophilic and hydrophobic compounds, including various vitamins and antioxidants, can be dispersed in other matrices by liposomes encapsulation.

32.3.4 Emulsified Systems

32.3.4.1 Concept, Structure, and Properties

Emulsions are defined as a mixture of two immiscible liquids wherein one phase is dispersed in the other in the form of droplets. Food emulsions usually contain an aqueous phase (polar), an oil phase (apolar), and surfactants that are added to stabilize the system by reducing the interfacial tension between dispersed and continuous phases. When oil droplets are present in the emulsion they are called O/W emulsions. On the other hand, when the aqueous phase is dispersed, the emulsion is called W/O, and sometimes it is referred to as inverse emulsion. Figure 32.15 shows a schematic picture of O/W and W/O emulsions stabilized by surfactant molecules. These molecules are adsorbed on the interface and oriented according to the type of emulsion. In the case of O/W emulsions, the hydrophilic head is "dissolved" in the continuous phase, whereas in W/O emulsions it is in the dispersed phase. Formation of O/W or W/O emulsions depends on the migration of surfactant molecules to the interface, which stabilize the droplets, and coalescence, leading to the destruction of the droplets. Thus, this is a competing process and the phase that presents higher coalescence rates will become the continuous phase (Evans and Wennerstrom, 1994). The rates of migration and coalescence depend on the components chemical structure, leading to specific surfactant conformations when adsorbed on the oil–water interface. One practical parameter is the hydrophilic–lipophilic balance (HLB) of the surfactant, which is a number that relates the number of hydrophobic and hydrophilic groups in a surfactant molecule. In general, surfactants with HLB from 3 to 9 tend to stabilize W/O emulsions, whereas molecules with HLB higher than 9 tend to stabilize O/W emulsions. Therefore, it is possible to prepare emulsions with droplet volume fractions up to 90% provided that an adequate surfactant or a mixture of surfactants is used. Sometimes, emulsions are also stabilized by adding high molecular weight components, such as long chain polymers and proteins, which adsorb on the interface acting as a surfactant. Small solid particles also tend to adsorb on the oil– water interface and can act as surfactant in an emulsion.

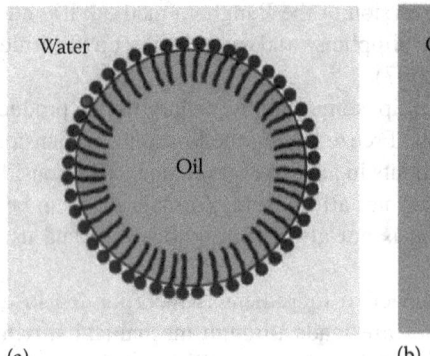
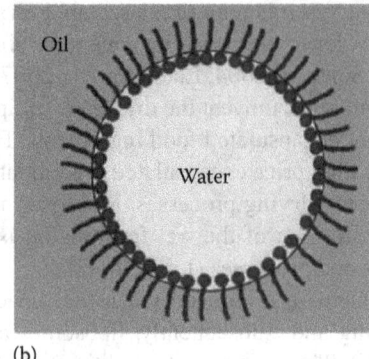

(a)   (b)

FIGURE 32.15 Schematic picture of (a) O/W and (b) W/O emulsions stabilized by surfactant molecules.

Another type of emulsion that has gained interest in food applications is the so-called multiple emulsion, which is basically an emulsion contained in a droplet. For example, a water-in-oil-in-water (W/O/W) emulsion means a multiple emulsion of water droplets inside an oil droplet that is dispersed in a continuous water phase. Recently, potential industrial applications in encapsulating active food components were recognized. One of the main difficulties in applying multiple emulsions is their low stability, which limits the applicability when prolonged stability and release are necessary (Muschiolik, 2007). Figure 32.16 shows a schematic representation of a W/O/W emulsion.

The fact that the dispersed phase can be composed of either oil or water shows that emulsions can be used to encapsulate both lipophilic and hydrophilic bioactives (Flanagan and Singh, 2006). Food systems based on emulsions are recognized to have great potential in delivering functional components such as omega-3, ®-carotene, fatty acids, phytosterols, and antioxidants, among others (McClements et al., 2007). The main preparation methods are based on addition and stirring components, mechanical mixing, homogenization, and heating, which make these systems suited for industrial scale-up.

Emulsified systems can be classified according to their thermodynamic stability and their droplets size. Macroemulsions (or simply emulsions) are metastable systems, i.e., the system is not in thermodynamic equilibrium, and it will breakdown into two distinct phases if sufficient time is allowed. However, emulsions that keep their kinetic stability for periods of months or years can be prepared by using appropriate components and amounts (McClements et al., 2007). This is the most common type of emulsion, and it is found in many food systems such as milk and salad dressing. Macroemulsions are usually polydisperse, with droplet sizes in the range of 1–100 μm. The main destabilization mechanisms in macroemulsions are droplets creaming, flocculation, and coalescence.

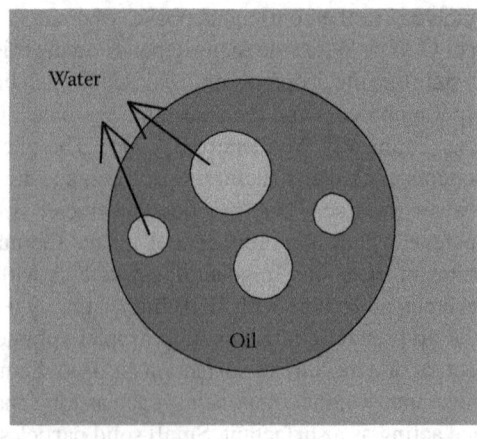

FIGURE 32.16 Schematic representation of a W/O/W emulsion.

Microemulsions, on the other hand, are thermodynamically stable emulsions. Therefore, at a given temperature, pressure, and composition, these systems keep their morphological characteristics and are not affected by the destabilization mechanisms cited above. Microemulsions are usually monodisperse, with droplets in a nanoscale range (10–100 nm) (Flanagan and Singh, 2006). Thermodynamic stability is achieved by the proper choice of the components, as well as their proportions, leading to a negative overall free energy of mixing (Evans and Wennerstrom, 1994). Usually, large amounts of surfactants are required, and different surfactants and cosurfactants are generally used. In microemulsions, surfactants are important because they not only decrease the interfacial tension between the oil and water phases, but also affect the energy balance of the system through the formation of self-assembled structures in the continuous phase, such as micelles. Furthermore, the chemical structure of the surfactant affects the interface spontaneous curvature, which is an important factor in determining the droplets size, as well as the type of emulsion.

Microemulsions exhibit different phase behavior under equilibrium, which is classified using the Winsor classification system. A Winsor I system means that the microemulsion coexists with an oil-rich region. When there is a water-rich region present in the system, the microemulsion is said to be a Winsor II system. In case there are both oil-rich and water-rich regions coexisting with the microemulsion, one speaks of a Winsor III system. Finally, a Winsor IV system means that there is no phase coexistence and only the microemulsified phase is observed. This phase behavior is desired for food delivery systems and, as said earlier, depends on the proper choice of system components, as well as temperature conditions (Flanagan and Singh, 2006). Figure 32.17 shows a schematic representation of the different microemulsion regimes.

Depending on its composition, a single-phase microemulsified system can also exhibit different morphologies. The three main structures are O/W, W/O, and bicontinuous. The latter is a structure in which both oil and water exist as a continuous phase, but all three structures have a surfactant monolayer in the interface separating both phases. These structures are shown in Figure 32.18. Usually, an O/W microemulsion is formed when oil concentration is low, and a W/O microemulsion is formed when water concentration is low. Bicontinuous systems are formed when the amounts of water and oil are similar (Lawrence and Rees, 2000).

Microemulsions as food systems have great potential, which can be attested in patented products (Bauer et al., 2002; Allgaier et al., 2004; Chanamai, 2007). Incorporation of proteins in microemulsions might also have impact in food applications in the future (Rohloff et al., 2003). Studies on microemulsions applied to the pharmaceutical field might also be of interest when seeking food applications, since biocompatible components are used in this field. Studies in this area have been summarized in review articles (Lawrence and Rees, 2000; Rane and Anderson, 2008).

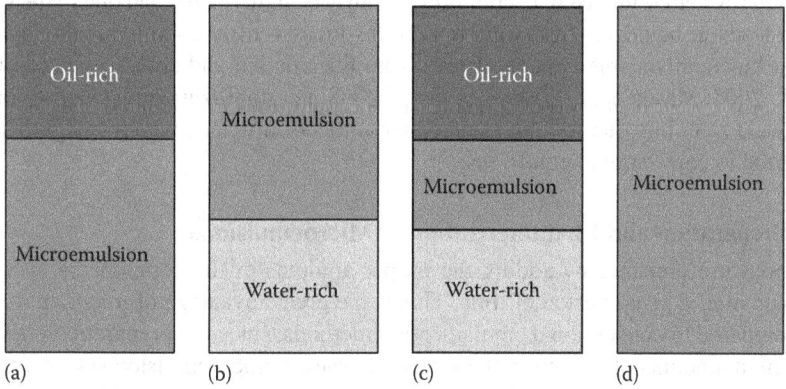

FIGURE 32.17 Schematic representation of the microemulsion regimes: (a) Winsor I, (b) Winsor II, (c) Winsor III, and (d) Winsor IV.

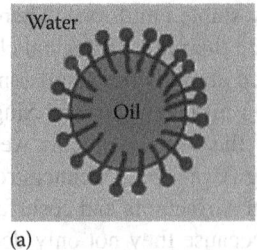 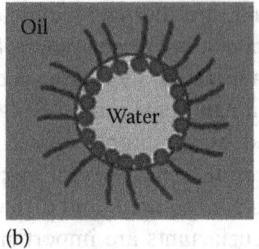 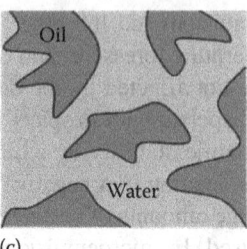

FIGURE 32.18 Schematic representation of the main single-phase microemulsion systems: (a) O/W, (b) W/O, and (c) bicontinuous.

Recently, metastable emulsions with nanosized droplets have started to receive attention due to their technological potential in pharmaceutical and food industries (Solans et al., 2005), and their fundamental properties have been studied (Mason et al., 2006b). Such systems are called nanoemulsions or miniemulsions. Basically, it is an emulsion with droplet sizes in the range of 50–200 nm. Nanoemulsions have the same physical appearance as a microemulsion, i.e., they have droplets in the nanoscale range and usually exhibit transparency and low viscosity. Although they can lose stability through coalescence and flocculation, the main destabilization mechanism is Ostwald ripening, due to the high Laplace pressure of the droplets (Porras et al., 2004).

Nanoscale emulsions have gained technological interest because the transport efficiency of functional components in emulsion food systems is increased when droplets are in the nanoscale (Spernath and Aserin, 2006). In addition, these emulsions are transparent, and they have lower viscosity when compared to conventional emulsions, which make them suitable for use in beverages, for example. In recent years, considerable research effort was made to understand the physical properties, preparation, phase behavior, and stability of micro- and nanoemulsions.

Nanoemulsions are metastable structures. This fact confers to nanoemulsions advantages and disadvantages for functional food applications in comparison with microemulsions. Main advantages are that nanoemulsions do not require the use of large amounts of surfactants and there is a wider range of possibilities of combination of different components for a given system (Solans et al., 2005). Furthermore, concentrated systems can be prepared, and their rheological properties can be explored for different food applications (Mason et al., 2006b). The main disadvantages are the limited kinetic stability of the system, which has to be monitored to keep the desired properties for a sufficient period of time for a given application. This is an important factor to determine the shelf life of food products based on nanoemulsions.

Although the system is metastable, creaming stability is highly enhanced due to the small droplet sizes, which leads to homogeneous systems even for low-viscosity continuous phases. It has been reported that kinetically stable nanoemulsions from flocculation and coalescence can be prepared (Solans et al., 2005; Mason et al., 2006a). In fact, the main destabilization mechanism in nanoemulsions is Ostwald ripening, due to high Laplace pressures of droplets, which is significantly higher when compared to conventional emulsions.

32.3.4.2 Preparation and Characterization of Microemulsions

Microemulsions are prepared by adding the proper amounts of the components, which form the microemulsion after a given period of time. This is the great advantage of microemulsion preparation when compared to conventional emulsification methods, since the preparation does not require the input of high amounts of energy in the system. However, microemulsion systems can lose their characteristic morphology with variations in temperature and composition. Therefore, the range of the parameters that maintain the microemulsion characteristics has to be determined to define the applicability range of a given system.

Preparation methods for microemulsions consist essentially in adding and mixing the components to the system in different ways and conditions to form a microemulsion. Usually, a single surfactant is not sufficient to decrease the interfacial tension up to the point at which spontaneous emulsification occurs. Thus, one or more cosurfactants are used, which are usually amphiphilic molecules with different HLB of the main surfactant, or alcohols, but their applicability can be limited as food systems due to toxicity issues. Although the system to be formed is thermodynamically favorable, i.e., the microemulsion is formed spontaneously, it is usually necessary to overcome kinetic energy barriers. The main preparation methods are the low energy emulsification and the phase inversion temperature (PIT) method. The first method can be achieved by (1) adding water in a mixture of oil and surfactant; (2) adding oil in a mixture of water and surfactant; and (3) by mixing all components together at once. It has been reported in the literature that the order of ingredient addition can play a significant role in the formation of the microemulsion (Flanagan and Singh, 2006).

In the PIT method, an initial emulsion, for example, a W/O emulsion, is heated up to a temperature, called PIT, for which the interfacial tension between the oil and water phases reaches a minimum. At this point there is an inversion, and an O/W emulsion is formed. The system is then cooled while stirring and a stable microemulsion is formed. Sometimes high-pressure homogenization is used to prepare microemulsions, but this method is limited due to the high heat dissipation involved.

Characterization of microemulsions requires the construction of phase diagrams, which can be done by titration methods (Lawrence and Rees, 2000). Ternary or pseudoternary phase diagrams are usually built by varying the amount of the components and observing the phase behavior of the system to identify the region where a clear and isotropic emulsion is formed, i.e., a Winsor IV system. In general, pseudoternary diagrams are found since microemulsion systems usually contain cosurfactants. Figure 32.19 shows a schematic pseudoternary phase diagram of a microemulsion system, indicating the phase behavior for a given composition. Each axis corresponds to the volume or mass fraction of each component or group of components, which are usually water, oil, and surfactant/cosurfactant.

The microstructure (or nanostructure) characterization of microemulsions can be performed by using dynamic light scattering (DLS) to determine droplet sizes. Scanning electron microscopy (SEM), small-angle x-ray scattering (SAXS), small-angle neutron scattering, and nuclear magnetic resonance (NMR) can be used to determine other structural features such as the presence of wormlike reverse micelles and other liquid-crystalline phases.

Microstructure and dynamical behavior of microemulsions strongly affect the macroscopic properties of microemulsions. Therefore, it is important to characterize the macroscopic properties of a determined system. In food applications, it is very important to characterize the rheological

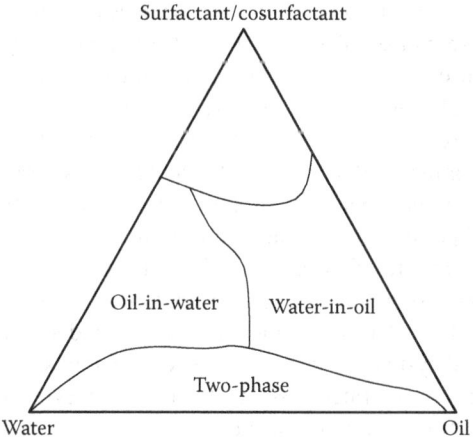

FIGURE 32.19 Schematic pseudoternary diagram showing regions of two-phase and microemulsion regimes.

properties of microemulsions, which can be performed using conventional rheometers. Techniques such as conductive and dielectric properties measurements can be used to determine the type of microemulsion formed (W/O or O/W) and monitor percolation, phase inversion, and other structural and dynamical features. Optical features of microemulsions are important for food applications since it is necessary for the system to be visually appealing when considering a commercial functional food system (Flanagan and Singh, 2006).

32.3.4.3 Preparation and Characterization of Nanoemulsions

Nanoemulsions can be prepared by low-energy methods based on PIT, which is similar to the one described in Section 32.3.4.2, but in this case the system, when cooled, keeps its morphology in a nonequilibrium state (Förster et al., 1995; Morales et al., 2003; Izquierdo et al., 2004, 2005). Other low-energy methods resulting in nanoemulsions are the phase inversion composition and autoemulsification methods. The former is similar to the PIT method, but the phase inversion occurs by modifying the composition of the system, leading to a kinetically stable nanoemulsion (Forgiarini et al., 2001; Porras et al., 2004). The latter is based on the dilution of an initial stable microemulsion, usually at the bicontinuous phase, resulting in a nanoemulsion (Pons et al., 2003; Wang et al., 2007, 2008).

Preparation using low-energy methods results in nanoemulsions lying in a region close to thermodynamic equilibrium. Therefore, their properties are very similar to a Winsor IV microemulsion system, and a distinction between this type of nanoemulsion and a thermodynamic stable microemulsion has been contested by some research groups (Mason et al., 2006b). However, stability of nanoemulsions prepared by low-energy methods, showing their nonequilibrium character, has been studied and summarized in review articles, and such emulsions have gained wide acceptance due to their potential in applications in food systems and pharmaceuticals (Tadros et al., 2004; Solans et al., 2005; Gutiérrez et al., 2008).

Another method that has gained importance is the preparation under high shear. In the past, this method was limited due to the lack of high-pressure homogenizers that were able to generate shear rates high enough to break droplets in to nanometer sizes. Therefore, most nanoemulsions prepared using this method were limited to laboratory research using homebuilt homogenizers. Recently, affordable high-pressure homogenizers with the capacity to generate high shear, such as those based on microchannel flows, appeared in the market. Consequently, the industrial interest in nanoemulsions is expected to increase. This method is particularly interesting for food applications, since most of the conventional food emulsions are prepared in this way, and high-pressure homogenization operations are suited for scale-up when industrial production rates are desired.

Nanoemulsions prepared by high shear methods form metastable systems far from equilibrium, and despite the comparable droplet sizes, their characteristics are not similar to microemulsions. The main advantage of these nanoemulsions is the fact that kinetically stable systems can be prepared using considerably lower amounts of surfactants, while exhibiting physical properties similar to the counterpart microemulsions, such as low viscosity and transparency (Mason et al., 2006b). Therefore, nanoemulsions prepared under high shear can be advantageous if surfactant cost is an important issue when developing food delivery systems. Also, it is possible to develop concentrated systems, which can exhibit unique rheological properties that could be explored in food applications.

The main morphological parameters in nanoemulsions are their average droplet size, droplet size distribution, and droplet volume fraction. Usually, droplet sizes are measured using DLS, and the stability of the system can be monitored by size measurements in a time interval. NMR can be used to measure droplet sizes and droplet interactions in concentrated systems, and it has been used in stability studies of emulsified systems. It can also be used when in situ measurements are required, and it is able to characterize concentrated systems, the emulsion type (W/O or O/W), and flow and mixing properties, although this technique has more widespread use in characterizing conventional emulsions, but it is feasible for applications in nanoemulsions. Morphological aspects of concentrated nanoemulsions have been studied using small-angle neutron scattering (Mason et al., 2006a).

Rheological and optical properties are important when seeking food applications. Therefore, the dependence of viscosity and other viscoelastic properties on droplet sizes and droplet concentrations are important in nanoemulsion characterization.

Reports of nanoemulsions applied as drug delivery systems can be found in the literature (Solans et al., 2005). However, few studies are found concerning applications in food systems, but this number can be expected to grow due to the increase in the availability of homogenization systems of ultrahigh shear based on microchannel flow and other recently developed preparation methods (Kentish et al., 2008; Yuan et al., 2008). Furthermore, fundamental research on nanoemulsion formation through high shear has recently appeared (Meleson et al., 2004). Reviews on the fundamentals and potential applications of nanoemulsions can be found in the literature (Solans et al., 2005; Mason et al., 2006b; Gutiérrez et al., 2008). Recently, a patent was filed claiming production of W/O nanoemulsion for food applications (Del Gaudio et al., 2007).

32.4 APPLICATIONS OF ENCAPSULATED ACTIVE INGREDIENTS IN FOODS

Encapsulated ingredients are used in many food applications. They can be incorporated in beverages, dairy products, baked products, and manufactured meats, including infant and other specialized formulations. Examples of food products in which encapsulated ingredients can be incorporated are UHT milk, cheese, ice cream, margarines, muesli bars, yoghurt, infant foods, dietetic food supplements, spreads, health drinks, mayonnaise, baked products, and breakfast cereals.

Encapsulation may be used to deliver traditional active ingredients, such as flavors, vitamins, minerals, sweeteners, and antioxidants, or relatively novel ones, such as probiotic microorganisms. SD and gel microparticles, liposome and emulsified systems, which are under focus in this chapter, have been used for some of these applications. Their functionality as delivery systems is discussed in Sections 32.4.1 through 32.4.5.

32.4.1 Flavors

One of the largest food applications is the encapsulation of flavors. Flavors can be among the most valuable ingredients in any food formula. Even small amounts of some aromatic substances can be expensive, and because they are usually sensitive and volatile, preserving them is often a top concern of food manufacturers. Encapsulating these high-cost materials can result in cost savings, as the loss through storage and processing is limited.

Encapsulation of flavors has been attempted and commercialized using many different methods such as SD, spray chilling or spray cooling, extrusion, coacervation, and molecular inclusion. Among them, SD is the widely used commercial process in large-scale production of encapsulated flavors and volatiles. One of the reasons is the large-scale capacity of production in a continuous mode. Microencapsulation of flavors by SD presents the challenge of removing water by evaporation, while retaining substances that are much more volatile than water, which is the case for most organic compounds. However, SD microcapsules with a high retention of aromatic compounds can be obtained, due to the phenomenon known as selective diffusion (Thijssen, 1971). This concept is based on the fact that the diffusion of water in concentrated solutions behaves differently from the diffusion of other substances. According to this concept, favorable conditions for obtaining high volatiles retention can be created in SD due to the rapid decrease of the water concentration at the drying droplet surface in contact with the hot drying air. Once the droplet surfaces have dried sufficiently, selective diffusion comes into effect because the diffusion coefficients of organic compounds in the surface region become much lower than that of water. When volatile substances are encapsulated, successful microencapsulation relies on achieving high retention of the core material during processing and storage. Many studies have been carried out on the influence of encapsulating material compositions and the operating conditions on the encapsulation efficiency and on the controlled release of encapsulated flavors. Most frequently used carriers include carbohydrates, gums, and food proteins. Each group of materials has certain advantages and disadvantages. For this reason, many coatings are actually composite formulations. Thorough reviews of the technology have been published (Madene et al., 2006), some of them, with special emphasis (Ré, 1998), are dedicated to the encapsulation of flavors by SD (Gharsallaoui et al., 2007).

A subject of increasing interest in this area concerns the development of alternative and inexpensive polymers that may be considered as natural, like gum arabic (good emulsifying properties), and could encapsulate flavors with a good efficiency. For example, sugar-beet pectin (Drusch, 2007) has been regarded as an alternative emulsifying encapsulating material for flavors.

Another area of interest to optimize the encapsulation efficiency of food flavors and oils by SD is the submicronization of the droplets oil of the emulsion. It has been well documented that emulsion droplet size has a pronounced effect on the encapsulation efficiency of different core materials by SD (Jafari et al., 2008). The findings clearly show that reducing emulsion size can result in encapsulated powders with higher retention of volatiles and lower content of unencapsulated oil at the surface of powder particles. The presence of oil on the surface of the powder particles is the most undesirable property of encapsulated powders, and it has been pointed out as a frequent problem with the quality of spray-dried products. This surface oil not only deteriorates the wettability and dispersability of the powder, but it is also readily susceptible to oxidation and to the development of rancidity.

Much of the work in this area has been done in emulsions having a droplet size of more than 1 μm, and the application of submicron (nano) emulsions in encapsulation of oils and flavors is relatively new in the literature. Some works have been carried out to determine the influence of submicron emulsions produced by different emulsification methods on encapsulation efficiency and to investigate the encapsulated powder properties after SD for different emulsion droplet sizes and surfactants. The process has been referred to as "nanoparticle encapsulation" since a core material in nanosize range is encapsulated into a matrix of micron-sized powder particles (Jafari et al., 2008). This area of research is developing. Some patents were filed in the past describing microemulsion formulations applied to flavor protection (Chung et al., 1994; Chmiel et al., 1997) and applications in flavored carbonated beverages (Wolf and Havekotte, 1989). However, there is no clear evidence on how submicron or nanoemulsions can improve the encapsulation efficiency and stability of food flavors and oils into spray-dried powders.

SD of emulsions has been traditionally used for the encapsulation of flavors; however, novel encapsulation and delivery properties can be achieved by encapsulating flavors into liposomes. Because liposomes have the ability to carry hydrophilic and fat-based flavors, they protect them from degradation during processing and storage and also increase the longevity of the flavor in the system where they are being used. Therefore, their use in the beverage industry has been widespread (Reineccius, 1995).

The bioflavor compounds of blue cheese, obtained from fermentation of *Aspergillus* spp., were encapsulated in soy lecithin liposomes and spray-dried to obtain the powder form by Santana et al. (2005). A sensory evaluation was performed, by adding the liposome-bioflavor powder in a base of light cream cheese, which was spread on toasts. Flavor intensity, acceptance by the consumers, and purchasing intention were the tests done in the sensory evaluation. The results showed that the encapsulation maintained the characteristic flavor of blue cheese and the product was classified by the consumers as acceptable. The dried liposome-stabilized flavor was useful to add in foods and to be kept in storage.

32.4.2 Antioxidants

There is a growing demand for delivery of antioxidants through functional foods with the concomitant challenge of protecting their bioactivity during food processing and subsequent passage through the GI tract. Antioxidants such as lycopene, and ®-carotene can be encapsulated by SD. Blends of sucrose and gelatin have been successfully used in encapsulation of lycopene

(Shu et al., 2006) and blends of maltodextrin and starches can be used to encapsulate ®-carotene (Loksuwan, 2007).

The product quality has been analyzed with respect to the retention of the antioxidant activity of the spray-dried powder. Polyphenolic compounds present in several extracts (grape seed, apple polyphenolic extract, or olive-leaf) were also encapsulated by SD in protein–lipid emulsions (Kosaraju et al., 2008) and chitosan (Kosaraju et al., 2006).

The common carotenoids in fruits and vegetables (licopen, lutein, zeaxantin, astaxantin, and ®-criptoxantina) are used as ingredients in foods. They substitute artificial colorants and are also functional ingredients due to their pro-vitamin A activity, apart from the fact that they act as antioxidants (Fernandez-Garcia et al., 2007). The potential market of water-dispersible carotenoids is broad, including ice creams, soups, disserts, meat products, and animal foods (Delgado-Vargas et al., 2000). Nevertheless, carotenoids lose color under oxidation, as they suffer isomerization easily under heat, acidic pH, or light exposition. The necessity of protection as well as the lipophilic or amphiphilic nature of carotenoids makes them attractive to liposome encapsulation. However, the approach of the studies of carotenoids in liposomes only focuses upon their oxidant activity and interactions into the lipid bilayer (Socaciu et al., 2000, 2002; Kostecka-Gugala et al., 2003; Gruszecki and Strzalka, 2005; Jemiola-Rzeminka et al., 2005; McNulty et al., 2007; Sujak et al., 2007). Applications of carotenoids in liposomes are intended to increase their longevity in the foods, as well as to increase their oral bioavailability due to the presence of lipids as coadjuvant in the formulation (Fernandez-Garcia et al., 2007; Parada and Aguilera, 2007).

Reports of conventional O/W emulsions used to encapsulate lycopene and ®-carotene are found in the literature (Ribeiro et al., 2006; Santipanichwong and Suphantharika, 2007). Stability of emulsified lycopene was evaluated after its incorporation in liquid food matrices (Ribeiro et al., 2003). Microemulsions have been applied to increase the efficiency of antioxidants such as ascorbic acid (Moberger et al., 1987). Furthermore, microemulsions have successfully been used for lycopene solubilization (Spernath et al., 2002). Applications of microemulsions to increase stability of antioxidants have been reported (Moberger et al., 1987; Yi et al., 1991).

32.4.3 OMEGA-3 FATTY ACIDS

Omega-3 acids are considered essential to human health, but cannot be manufactured by the human body and must therefore be obtained from food. These acids are naturally present in most fishes and certain plant oils such as soybean and canola, which are foods that people rarely consume in large quantities. Moreover, the direct addition of omega-3 fatty acids to many foods is prevented due to some characteristics (fishy flavors, readily oxidized), which together reduce the sensory acceptability of foods containing fatty acids, limit shelf life, and potentially reduce the bioavailability of the acids. Encapsulation responds to the challenges of omega-3 fatty acid delivery and extends the reach of its health benefits.

These acids can be encapsulated by SD. The success of the encapsulation is based on the ability of the spray-dried powder to provide, firstly, a good retention of these compounds within the structures and, secondly, a good oxidative stability. The goal is to find the appropriate carriers to create good barrier properties as shown in the literature. For example, glucose syrup was used in combination with proteins such as whey protein isolate or soy protein isolate (Rusli et al., 2006) or with sugar beet pectin (Drusch et al., 2007) to encapsulate fish oil, leading to a product that was more stable to oxidation than bulk oils. Other formulation compositions such as a blend of modified celluloses (methylcellulose and hydroxypropylmethylcellulose) with good emulsifying properties and maltodextrin (Kolanowski et al., 2004) have also been tested to encapsulate fish oil. Another strategy recently developed was based on the production of multilayer membranes of lecithin and chitosan around the oil droplets of the O/W emulsion before SD (Klinkesorn et al., 2006). Despite research being in progress, the influence of various process variables on oil oxidation during the emulsifying and drying stages is still not well known.

Encapsulation of omega-3 fatty acids using O/W emulsions was recently reported in the literature (Lee et al., 2006). The potential for emulsion-based delivery systems of omega-3 molecules in different types of foods such as yogurts, milk, and ice cream has been recognized (McClements et al., 2007). However, emulsion technology to encapsulate these molecules is difficult, requiring the development of antioxidant technologies to stabilize the system due to the complex oxidation reaction of omega-3 molecules (McClements and Decker, 2000).

Omega-3 fatty acids can also be protected against oxidation when encapsulated within liposomes (Haynes et al., 1991; Wallach and Mathur 1992).

32.4.4 Vitamins and Minerals

Encapsulating vitamins and minerals offers several benefits. It increases their stability when exposed to air, heat, or moisture. Many of these micronutrients are often destroyed in the baking process. Loss through processing and storage is prevented, resulting in the ability to use less of these products and thus save cost. Finally, many of these micronutrients have undesirable flavors or odors, which can be masked, keeping the micronutrients available to be absorbed in the GI tract.

Fortifying foods with minerals and vitamins is becoming more and more common. Mineral deficiency is one of the most important nutritional problems in the world. The best method to overcome this problem is to make use of an external supply, which may be nutritional or supplementary, like the fortification of foods with highly bioavailable mineral sources. Major interests of mineral encapsulation are linked to the fact that this technique enables to reduce mineral reactions with other ingredients, when they are added to dry mixes to fortify a variety of foods, and it can also incorporate time-release mechanisms of the minerals into the formulations. For example, iron is the most difficult mineral to add to foods and ensure adequate absorption, and iron bioavailability is severely affected by interactions with food ingredients (e.g., tannins, phytates, and polyphenols). Additionally, iron catalyses the oxidative degradation of fatty acids and vitamins (Schrooyen et al., 2001).

Liposomes have been used to encapsulate bioactive vitamins and minerals. Milk enriched with ferrous sulfate encapsulated in liposomes enabled an increase in the iron concentration compared to free iron. The encapsulated ferrous sulfate was stable to heat sterilization (100°C, 30 min) and storage at 4°C for 1 week.

Furthermore, liposomes provided the same bioavailability as the free sulfate, adding the advantage of being coated with a phospholipid membrane, which kept the iron from contacting with the other components of food, thus preventing undesirable interactions (Boccio et al., 1997; Uicich et al., 1999; Lysionek et al., 2000, 2002; Shuqin and Shiying, 2005).

Orange juice, cereals, and even candies are fortified with vitamins and minerals such as vitamin C and calcium. Vitamin C, also known as ascorbic acid or ascorbate, is added extensively to many types of foods for two quite different purposes: as a vitamin supplement to reinforce dietary intake of vitamin C, and as an antioxidant, to protect the sensory and nutritive quality of the food itself. Encapsulation of vitamin C improves and broadens its applications in the food industry. Spray-dried structures encapsulating vitamin C can be produced by using several carriers as encapsulating materials, among them, methacrylate copolymers named Eudragit®.* The resulting delivery systems were able to offer a controlled release at different pH values due to the Eudragit characteristics (Esposito et al., 2002). Chitosan, a hydrophilic polysaccharide also used as a dietary food additive, was used to encapsulate vitamin C by SD (Desai and Park, 2005b). Chitosan was cross-linked with a nontoxic cross-linking agent, tripolyphosphate.

SD was also used to formulate calcium microparticles using cellulose derivatives and polymethacrylic acid as encapsulating materials (Oneda and Ré, 2003), to modify the dissolution rate of calcium from calcium citrate and calcium lactate. Microparticulate systems with incorporated

---

* Registered trademark of Rohm GmbH & Co. KG, Darmstadt, Germany.

time-release mechanism were obtained to modify the calcium release from these commercial salts used in fortification of diet.

Liposomes composed of phosphatidylcholine, cholesterol, and dl-ζ-tocopherol improved shelf life of vitamin C from a few days up to 2 months, especially in the presence of common food components which normally speed up decomposition, such as copper ions, ascorbate oxidase, and lysine (Kirby et al., 1991). Calcium lactate was also encapsulated in lecithin liposomes, in this case to prevent undesirable calcium–protein interactions (Champagne and Fustier, 2007). The liposomal calcium levels of fortified soymilk were equivalent to those found in cow's milk. A synergistic effect of coencapsulation of vitamins A and D in liposomes promoted calcium absorption in the GI tract (Champagne and Fustier, 2007).

W/O emulsions based on olive oils were also used to encapsulate vitamin C (Mosca et al., 2008). Solubilization of vitamin E in microemulsions based on polyoxyethylene (POE) surfactants was reported in the past (Chiu and Yang, 1992). Phase behavior studies were conducted on microemulsions based on different oil phases such as limonene, medium-chain triglycerides (MCT), short-chain alcohols, polyols, and different surfactants (Garti et al., 2001; Papadimitriou et al., 2008; Zhang et al., 2008). In addition, the phase behavior of microemulsions prepared with food grade components based on lecithin has been investigated (Patel et al., 2006), showing potential for applications in encapsulating vitamins and minerals.

32.4.5 Probiotics

According to the Food and Agriculture Organization (FAO) of the United States and the World Health Organization (WHO), a probiotic is a live microorganism which, when administered in adequate amounts, confers a health benefit to the host. The FAO/WHO Expert Consultation lists benefits that had substantial support from peerreviewed publication of human studies (FAO/WHO, 2001).

Probiotics may consist of a single strain or a mixture of several strains. Most common are lactic acid bacteria from the *Lactobacillus* and *Bifidobacterium* genera. Species of bacteria and yeasts used as probiotics include *Bifidobacterium bifidum*, *B. breve*, *Lactobacillus casei*, *L. acidophilus*, *Saccharomyces boulardii*, and *Bacillus coagulan*, among others (Champagne et al., 2005).

Dietary supplements containing viable probiotic microorganisms (referred herein as probiotics) are increasing in popularity in the marketplace as their health benefits become recognized. Probiotics are sensitive to various environmental conditions such as pH, moisture, temperature, and light. When these conditions are not properly controlled, the product viability (measured in colony forming units or CFU), and therefore its efficacy, can be substantially reduced. The viability and stability of probiotics have both been a marketing and technological challenge for industrial producers.

Losses of microorganisms occur during manufacture and during the product shelf life. In addition, probiotics with good characteristics for effectiveness against disease and other conditions may not have good survival characteristics during transit through the GI tract. The probiotic cultures encounter gastric juices in the stomach ranging from pH 1.2 (on an empty stomach) through pH 5.0. These cultures stay in the stomach from around 40 min up to 5 h. In the stomach and the small intestine, these probiotics also encounter bile salts, and hydrolytic and proteolytic enzymes, which are able to kill them. These cultures are able to grow or survive only when they reach higher pH regions of the gastrointestine. During this transit, probiotics also have to compete with resident bacteria for space and nutrients. In addition, they have to avoid being flushed out of the tract by normal peristaltic action and being killed by antimicrobials produced by other microorganisms. Ability to adhere to surfaces, such as intestinal mucosal layer and the epithelial cell walls of the gut, is an important characteristic of a probiotic. The term "colonization" is used, and it means that the microorganism has mechanisms that enable it to survive in a region of the gastrointestine on an ongoing basis.

Intensive research efforts have been focused on protecting the viability of probiotic cultures both during product manufacture and storage, and through the gastric transit until the target site is reached. Protection may be achieved by several ways, among them, encapsulation.

At least five encapsulation methods have been investigated to protect probiotics: spray coating, SD, extrusion droplets, emulsion, and gel-particle technologies, including spray chilling. Several reviews in the literature are dedicated to probiotic (Mattila-Sandholm et al., 2002; Champagne et al., 2005) or, more specifically, to probiotic encapsulation (Kailasapathy, 2002; Krasaekoopt et al., 2003; Anal and Singh, 2007). In the specific case of fermented dairy products, a through discussion of both probiotics and prebiotics is given in Chapter 20.

SD is rarely considered for cell immobilization because of the high mortality resulting from simultaneous dehydration and thermal inactivation of microorganisms. Despite this limitation, several works have been evaluated SD as a process for encapsulating probiotics. Technical alternatives have been proposed to increase thermal resistance of the microorganisms during the dehydration process, such as the proper adjustment and control of the inlet and outlet drying temperatures (O'Riordan et al., 2001), the use of complex thermoprotector (prebiotic) carbohydrates as encapsulating materials (Rodriguez-Huezo et al., 2007), or even a previous encapsulation of the microorganisms by another technique, as proposed by Oliveira et al. (2007). These authors encapsulated probiotics (*B. lactis* and *L. acidophilus*) in a casein–pectin complex formed by complex coacervation, and the wet encapsulated microorganisms were dried by SD.

It has been demonstrated that a variety of probiotic cultures can be protected via encapsulation by SD in a variety of carriers, including whey protein (Picot and Lacroix, 2004), a matrix of gelatin, soluble starch, skim milk, or gum arabic (Lian et al., 2003), and cellulose phthalate, which is an enteric release pharmaceutical compound (Favaro-Trindade and Grosso, 2002), and a matrix of protective colloids (whey protein isolate, mesquite gum, and maltodextrin in a 17:17:66 ratio) associated to aguamiel as a prebiotic thermoprotector (Rodriguez-Huezo et al., 2007). In fact, the protective effect exerted by SD encapsulation against stressful conditions (the environment in the food product, during storage, and during the passage through the stomach or intestinal tract) may vary with the carriers or encapsulating materials and the microorganisms but, in all cases, the thermal resistance of strains is a critical parameter that should always be taken into consideration if SD is the intended method for encapsulation (Picot and Lacroix, 2004; Su et al., 2007).

Most of the literature reported on the encapsulation of probiotics has investigated the use of gel particles for improving their viability in food products and intestinal tract. The bacterial cells are dispersed into the hydrocolloid solution before gelation.

Entrapping probiotic bacteria in gels with ionic cross-linking is typically achieved with polysaccharides (alginate, pectin, and carrageenan). By far, the most commonly used material for this purpose is alginate, and the most commonly reported encapsulation procedure is based on the calcium-alginate gel formation. The droplets form gel spheres instantaneously (sodium-alginate in calcium chloride) entrapping the cells in a three-dimensional lattice of ionically cross-linked alginate. The success of this method is due to the gentle environment it provides for the entrapped material, cheapness, simplicity, and its biocompatibility (Anal and Singh, 2007). In the past 10 years, there have been 93 peer-reviewed articles on "probiotics encapsulation" (Web of Science database), out of which 47 especially on "encapsulation using alginate."

Various researchers have studied factors affecting the gel particles characteristics and their influence on the encapsulation of probiotics, such as concentrations of alginate and $CaCl_2$, timing of hardening of the gel particles, and cell concentrations (Chandramouli et al., 2004). Most of them have shown that probiotics can be protected in calcium-alginate beads, what is generally demonstrated by an increase in the survival of bacteria under different harsh conditions, compared with free microorganisms. Alginates also demonstrate easy release of the encapsulated bacteria when suspended in an alkaline buffer. However, the degree of protection might depend on the gel particle size, suggesting that these microorganisms should be encapsulated within a specific gel particle size range. For example, very large calcium-alginate beads (>1 mm) can negatively affect the textural and sensorial properties of food products in which they are added (Hansen et al., 2002), whereas reduction of the sphere size to less than 100 µm would be advantageous for texture considerations, allowing direct addition of encapsulated probiotics to a large number of foods. However, it has been demonstrated that particles smaller than 100 µm do not significantly protect the probiotics in simulated gastric fluid, compared with free cells (Hansen et al., 2002). One limitation for cell loading in small particles is also the large size of microbial cells, typically 1–4 µm, or particles freeze-dried culture (more than 100 µm). On the other hand, there are evidences in the literature that calcium-alginate gel particles with mean diameters of 450 µm (Chandramouli et al., 2004) and 640 µm (Shah and Ravula, 2000) could protect probiotics from adverse gastric conditions. In the latter case, the particles, after being freeze dried, also protected the viability of the microorganisms in fermented frozen dairy desserts. In fact, Chandramouli et al. (2004) found an optimal particle size of 450 µm for the calcium-alginate gel particles to protect the cells (*L. acidophilus*), when testing gel particles of different sizes (200, 450, and 1000 µm).

The composition of the alginate also influences bead size (Martinsen et al., 1989). Alginates are heterogeneous groups of polymers, with a wide range of functional properties. Alginates with a high content of guluronic acid blocks (G blocks) are preferable for capsules formation because of their higher mechanical stability and better tolerance to salts and chelating agents.

In addition to the reports of benefits of encapsulation in protecting probiotics against the stressful conditions of the GI tract, there is increasing evidence that the procedure is helpful in protecting the probiotic cultures destined to be added to foods. For example, encapsulation technologies have been used satisfactorily to increase the survival of probiotics in high acid fermented products such as yoghurts (Krasaekoopt et al., 2003), including Ca-alginate gel particles. Other reported food vehicles for delivery of encapsulated probiotic bacteria are cheese, ice cream, and mayonnaise (Kailasapathy, 2002).

Despite the suitability of alginate as entrapment matrix material, this system has some limitation due to its low stability in the presence of chelating agents such as phosphate, lactate, and citrate. The chelating agents share affinity for calcium and destabilize the gel (Kailasapathy, 2002). Special treatments, such as coating the alginate particles, can be applied to improve the properties of encapsulated gel particles. Coated beads not only prevent cell release but also increase mechanical and chemical stability. It has been reported that cross-linking with cationic polymers, coating with other polymers, mixing with starch, and incorporating additives can improve stability of beads (Krasaekoopt et al., 2003). For example, alginate can be coated with chitosan, a positively charged polyamine. Chitosan forms a semipermeable membrane around a negatively charged polymer such as alginate. This membrane, like alginate, does not dissolve in the presence of Ca^{2+} chelators or antigelling agents, and thus enhances the stability of the gel and provides a barrier to cell release (Krasaekoopt et al., 2004, 2006; Urbanska et al., 2007).

Various other polymer systems have been used to encapsulate probiotic microorganisms. |-Carrageenan (Adhikari et al., 2003), gellan gum, gelatin, starch, and whey proteins (Reid et al., 2007) have also been used as gel encapsulating systems for probiotics. An increasing interest in developing new compositions of gel particles to improve the viability of the probiotic microorganisms to harsh conditions (thermotolerance, acid-tolerance, etc.) is marked by the more recent researches reported in the literature. Some of these systems include alginate plus starch (Sultana et al., 2000), alginate plus methylcellulose (Kim et al., 2006), alginate plus gellan (Chen et al., 2007), alginate–chitosan–enteric polymers (Liserre et al., 2007), alginate-coated gelatin (Annan et al., 2008), gellan plus xanthan (McMaster et al., 2005), |-carrageenan with locust bean gum (Muthukumarasamy et al., 2006), and alginate plus pectin plus whey proteins (Guerin et al., 2003). In some cases, systems have been developed not only to provide better probiotic viability but also to deliver a prebiotic synergy (Iyer and Kailasapathy, 2005; Crittenden et al., 2006).

Improving the number of possibilities to encapsulate probiotics is a important tool even because, in recent years, the consumer demand for non-dairy-based probiotic products has increased (Prado et al., 2008), and the application of probiotic cultures in nondairy products represents a great challenge, because they may represent new hostile environment for probiotics (heat-processed foods, storage at room temperature, more acid foods like fruit juices, etc.).

Emulsified systems have also been investigated to protect probiotics. Incorporation of *L. acidophilus* in a W/O/W emulsion was recently reported and the protective effect of the probiotic in a low pH environment was evaluated (Shima et al., 2006). Lactic acid bacteria were encapsulated in sesame oil emulsions and, when subjected to simulated high gastric or bile salt conditions, a significant increase in survival rate was observed (Hou et al., 2003).

32.5 ENCAPSULATION CHALLENGES

The challenges in developing an encapsulated food ingredient commercially viable depend on selecting appropriate and food grade (GRAS) encapsulating materials, selecting the most appropriate process to provide the desired size, morphology, stability, and release mechanism, and economic feasibility of large-scale production, including capital, operating, and other miscellaneous expenses, such as transportation and regulatory costs.

However, the development of any encapsulation technique must not be treated as an isolated operation but as part of an overall process starting with ingredient production followed by processes, including encapsulation, right through to liberation and utilization of the ingredient. Furthermore, a selection has to be made between batch, semicontinuous, and continuous encapsulation processes, resulting in a difficult choice for process designers. Cost is often the main barrier of the implementation of encapsulation, and multiple benefits are generally required to justify the cost of encapsulation. Indeed, in the food industry, regulations with respect to ingredients, processing methods, and storage conditions are tight, and the price margin is much lower than in, for example, the pharmaceutical industry.

This procedure is something of an art, as Asajo Kondo asserts in *Microcapsule Processing and Technology* (Kondo, 1979):

> Microencapsulation is like the work of a clothing designer. He selects the pattern, cuts the cloth, and sews the garment in due consideration of the desires and age of his customer, plus the locale and climate where the garment is to be worn. By analogy, in microencapsulation, capsules are designed and prepared to meet all the requirements in due consideration of the properties of the core material, intended use of the product, and the environment of storage.

Encapsulation technology remains something of an art, although firmly grounded in science. Combining the right encapsulating materials with the most efficient production process for any given core material and its intended use requires extensive scientific knowledge of all the materials and processes involved and a good feel for how materials behave under various conditions.

Continuing research is clearly necessary to improve and extend the technology to the encapsulation to a wide variety of beneficial ingredients. Researchers are investigating the next generation of encapsulation technologies, including

- The development of *new, natural food materials and encapsulated products* that can be used by food manufacturers, among them, nonproteinaceous materials to eliminate allergens, that protect the encapsulated ingredients while they travel through the body to a targeted site in GI tract
- The increase in the range of *processing techniques*, with special interest for processes producing in continuous mode with high productivity
- The potential use of *coencapsulation methodologies*, where two or more bioactive ingredients can be combined to have a synergistic effect
- The *targeted delivery* of bioactives to various parts of the GI tract
- The trial of new ways of *incorporating bioactives into foods* with minimal loss of bioactivity and without compromising the quality of the food that is used as a delivery vehicle
- The understanding of the self-assembly and stabilization of *nanoemulsions* during food processing

These developments will give food manufacturers new opportunities to produce a greater variety of innovative functional foods that promote the health and well being of consumers.

REFERENCES

Adhikari, K., Mustapha, A., and Grun, I.U. 2003. Survival and metabolic activity of microencapsulated *Bifidobacterium longum* in stirred yogurt. *J. Food Sci.* 68:275–280.

Alkhalaf, W., El Soda, M., Gripon, J.-C., and Vassal, L. 1989. Acceleration of cheese ripening with liposomes-entrapped proteinase: Influence of liposomes net charge. *J. Dairy Sci.* 72:2233–2238.

Allgaier, J., Willner, L., Richter, D., Jakobs, B., Sottmann, T., and Strey, R. 2004. Method for increasing the efficiency of surfactants with simultaneous suppression of lamellar mesophases and surfactants with an additive added thereto. U.S. Patent 2004054064-A1, filed Aug. 19, 2003, and issued Mar. 18, 2004.

Alves, G.P. and Santana, M.H.A. 2004. Phospholipid dry powders produced by spray drying processing: Structural, thermodynamic and physical properties. *Powder Technol.* 145:141–150.

Amici, E., Tetradis-Meris, G., Pulido de Torres, C., and Jousse, F. 2008. Alginate gelation in microfluidic channels. *Food Hydrocolloids* 22:97–104.

Anal, A.K. and Singh, H. 2007. Recent advances in microencapsulation of probiotics for industrial applications and targeted delivery. *Trends Food Sci. Technol.* 18:240–251.

Annan, N.T., Borza, A.D., and Hansen, L.T. 2008. Encapsulation in alginate-coated gelatin microspheres improves survival of the probiotic *Bifidobacterium adolescentis* 15703T during exposure to simulated gastro-intestinal conditions. *Food Res. Int.* 41:184–193.

Batzri, S. and Korn, E.D. 1973. Single bilayer liposomes prepared without sonication. *Biochim. Biophys. Acta* 298:1015–1019.

Bauer, K., Neuber, C., Schmid, A., and Voelker, K.M. 2002. Oil in water microemulsion. U.S. Patent 6426078-B1, filed Feb. 26, 1998, and issued July 30, 2002.

Belyaeva, E., Della Valle, D., Neufeld, R.J., and Poncelet, D. 2004. New approach to the formulation of hydrogel beads by emulsification/thermal gelation using a static mixer. *Chem. Eng. Sci.* 59:2913–2920.

Benech, R.-O., Kheadr, E.E., Lacroix, C., and Fliss, I. 2002. Antibacterial activities of nisin Z encapsulated in liposomes or produced in situ by mixed culture during cheddar cheese ripening. *Appl. Environ. Microbiol.* 68:5607–5619.

Boccio, J.R., Zubillaga, M.B., Caro, R.A., Gotelli, C.A., and Weill, R. 1997. A new procedure to fortify fluid milk and dairy products with high bioavailable ferrous sulfate. *Nutr. Rev.* 55:240–246.

Brownlie, K. 2007. Marketing perspective of encapsulation technologies in food applications. In *Encapsulation and Controlled Release Technologies in Food Systems*, ed. J.M. Lakkis. Ames, IA: Blackwell Publishing Professional, pp. 213–233.

Burey, P., Bhandari, B.R., Howes, T., and Gidley, M.J. 2008. Hydrocolloid gel particles: Formation, characterization, and application. *Crit. Rev. Food Sci. Nutr.* 48:361–377.

Carneiro, A.L. and Santana, M.H.A. 2004. Production of liposomes in a multitubular system useful for scaling-up of processes. *Prog. Colloid Polym. Sci.* 128:273–277.

Champagne, C.P. and Fustier, P. 2007. Microencapsulation for the improved delivery of bioactive compounds into foods. *Curr. Opin. Biotechnol.* 18:184–190.

Champagne, C.P., Gardner, N.J., and Roy, D. 2005. Challenges in the addition of probiotic cultures to foods. *Crit. Rev. Food Sci. Nutr.* 45:61–84.

Chan, L., Lee, H., and Heng, P. 2002. Production of alginate microspheres by internal gelation using an emulsification method. *Int. J. Pharm.* 241:259–262.

Chan, L.W., Lee, H.Y., and Heng, P.W.S. 2006. Mechanisms of external and internal gelation and their impact on the functions of alginate as a coat and delivery system. *Carbohydr. Polym.* 63:176–187.

Chanamai, R. 2007. Microemulsions for use in food and beverage products. U.S. Patent 087104-A1, filed Oct. 6, 2006, and issued Apr. 19, 2007.

Chandramouli, V., Kailasapathy, K., Peiris, P., and Jones, M. 2004. An improved method of microencapsulation and its evaluation to protect *Lactobacillus* spp. in simulated gastric conditions. *J. Microbiol. Methods* 56:27–35.

Chen, L., Remondetto, G.E., and Subirade, M. 2006. Food protein-based materials as nutraceutical delivery systems. *Trends Food Sci. Technol.* 17:262–283.

Chen, M.J., Chen, K.N., and Kuo, Y.T. 2007. Optimal thermotolerance of *Bifidobacterium bifidum* in gellan-alginate microparticles. *Biotechnol. Bioeng.* 98:411–419.

Chiou, D. and Langrish, T.A.G. 2007. Development and characterisation of novel nutraceuticals with spray drying technology. *J. Food Eng.* 82:84–91.

Chiu, Y.C. and Yang, W.L. 1992. Preparation of vitamin E microemulsion possessing high resistance to oxidation. *Colloids. Surf.* 63:311–322.

Chmiel, O., Traitler, H., and Vopel, K. 1997. Food microemulsion formulations. WO Patent 96/23425, filed Jan. 24, 1996, and issued Aug. 8, 1996.

Choi, C.H., Jung, J.H., Rhee, Y.W., Kim, D.P., Shim, S.E., and Lee, C.S. 2007. Generation of monodisperse alginate microbeads and in situ encapsulation of cell in microfluidic device. *Biomed. Microdevices* 6:855–862.

Chung, S.L., Tan, C.-T., Tuhill, I.M., and Scharpf, L.G. 1994. Transparent oil-in-water microemulsion flavor or fragrance concentrate, process for preparing same, mouthwash or perfume composition containing said transparent microemulsion concentrate, and process for preparing same. U.S. Patent 5283056, filed July 1, 1993, and issued Feb. 1, 1994.

Crittenden, R., Weerakkody, R., Sanguansri, L., and Augustin, M. 2006. Symbiotic microcapsules that enhance microbial viability during nonrefrigerated storage and gastrointestinal transit. *Appl. Environ. Microbiol.* 72:2280–2282.

Del Gaudio, L., Lockhart, T.P., Belloni, A., Bortolo, R., and Tassinari, R. 2007. Process for the preparation of water-in-oil and oil-in-water nanoemulsions. WO Patent 2007/112967-A1, filed Mar. 28, 2007, and issued Oct. 11, 2007.

Delgado-Vargas, F., Jimenez, A.R., and Paredes-Lopez, O. 2000. Natural pigments: Carotenoids, anthocyanins and betalains—Characteristics, biosynthesis, processing and stability. *Crit. Rev. Food Sci. Nutr.* 40:173–189.

Desai, K.G. and Park, H.J. 2005a. Recent developments in microencapsulation of food ingredients. *Drying Technol.* 23:1361–1394.

Desai, K.G. and Park, H.J. 2005b. Encapsulation of vitamin C in tripolyphosphate cross-linked chitosan microspheres by spray drying. *J. Microencapsul.* 22:179–192.

Drusch, S. 2007. Sugar beet pectin: A novel emulsifying wall component for microencapsulation of lipophilic food ingredients by spray-drying. *Food Hydrocolloids* 21:1223–1228.

Drusch, S., Serfert, Y., Scampicchio, M., Schmidt-Hansberg, B., and Schwarz, K. 2007. Impact of physicochemical characteristics on the oxidative stability of fish oil microencapsulated by spray-drying. *J. Agric. Food Chem.* 55:11044–11051.

El Soda, M. 1986. Acceleration of cheese ripening: Recent advances. *J. Food Prot.* 49:395–399.

Eliaz, R.E. and Szoka, F.C. Jr., 2001. Liposome-encapsulated doxorubicin targeted to CD44: A strategy to kill CD44-overexpressing tumor cells. *Cancer. Res.* 61:2592–2601.

Esposito, E., Cervellati, F., Menegatti, E., Nastruzzi, C., and Cortesi, R. 2002. Spray dried Eudragit microparticles as encapsulation devices for vitamin C. *Int. J. Pharm.* 242:329–334.

Evans, D.F. and Wennerstrom, H. 1994. *The Colloidal Domain—Where Physics, Chemistry, Biology and Technology Meet*. New York: Wiley-VCH.

FAO/WHO. 2001. Evaluation of health and nutritional properties of powder milk and live lactic acid bacteria. Food and Agriculture Organization of the United Nations and World Health Organization Expert Consultation Report. Cordoba, Argentina. Available at http://www.who.int/foodsafety/publications/fs_management/en/probiotics.pdf, accessed Feb. 16, 2009.

Favaro-Trindade, C.S. and Grosso, C.R.F. 2002. Microencapsulation of L-acidophilus (La-05) and B-lactis (Bb-12) and evaluation of their survival at the pH values of the stomach and in bile. *J. Microencapsul.* 19:485–494.

Fernandez-Garcia, E., Minguez-Mosquera, M.I., and Perez-Galvez, A. 2007. Changes in composition of the lipid matrix produce a differential incorporation of carotenoids in micelles. Interaction effect of cholesterol and oil. *Innov. Food Sci. Emerg. Technol.* 8:379–384.

Flanagan, J. and Singh, H. 2006. Microemulsions: A potential delivery system for bioactives in food. *Crit. Rev. Food Sci. Nutr.* 46:221–237.

Forgiarini, A., Esquena, J., Gozales, C., and Solans, C. 2001. Formation of nanoemulsions by low-energy emulsification methods at constant temperature. *Langmuir* 17:2076–2083.

Förster, T., Rybinski, W.V., and Wadle, A. 1995. Influence of microemulsion phases on the preparation of fine-disperse emulsions. *Adv. Coll. Int. Sci.* 58:119–149.

Frederiksen, L., Anton, K., van Hoogevest, P., Keller, H.R., and Leuenberger, H. 1997. Preparation of liposomes encapsulating water-soluble compounds using supercritical CO_2. *J. Pharm. Sci.* 86:921–928.

Garti, N., Yaghnur, A., Leser, M.E., Clement, V., and Watzke, H.J. 2001. Improved oil solubilization in oil/water food grade microemulsions in the presence of polyols and ethanol. *J. Agric. Food Chem.* 49:2552–2562.

Gharsallaoui, A., Roudaut, G., Chambin, O., Voilley, A., and Saurel, R. 2007. Applications of spray-drying in microencapsulation of food ingredients: An overview. *Food Res. Int.* 40:1107–1121.
Gibaldi, M. 1991. *Biopharmaceutics Clinical Pharmacokinetics*, 4th edn. Philadelphia, PA: Lea & Febiger.
Gomez-Hens, A. and Fernandez-Romero, J.M. 2006. Analytical methods for the control of liposomal delivery systems. *Trends Anal. Chem.* 25:167–177.
Gouin, S. 2004. Microencapsulation: Industrial appraisal of existing technologies and trends. *Trends Food Sci. Technol.* 15:330–347.
Gregoriadis, G. 1987. Encapsulation of enzymes and other agents liposomes. In *Chemical Aspects in Food Enzymes*, ed. A.J. Andrews. London, U.K.: Royal Society of Chemistry.
Gregoriadis, G., ed. 1988. *Liposomes as Drug Carriers*. Chichester, U.K.: John Wiley & Sons Ltd.
Gruszecki, W.I. and Strzalka, K. 2005. Carotenoids as modulators of lipid membrane physical properties. *Biochim. Biophys. Acta* 1740:108–115.
Guerin, D., Vuillemard, J.C., and Subirade, M. 2003. Protection of bifidobacteria encapsulated in polysaccharide-protein gel beads against gastric juice and bile. *J. Food Prot.* 66:2076–2084.
Gutiérrez, J.M., González, C., Maestro, A., Solè, I., Pey, C.M., and Nolla, J. 2008. Nanoemulsions: New applications and optimization of their preparation. *Curr. Opin. Colloid Interface Sci.* 13:245–251.
Hansen, L.T., Allan-Wojtas, P.M., Jin, Y.L., and Paulson, A.T. 2002. Survival of Ca-alginate microencapsulated *Bifidobacterium* spp. in milk and simulated gastrointestinal conditions. *Food Microbiol.* 19:35–45.
Haynes, L.C., Levine, H., and Finley, J.W. 1991. Liposome composition for stabilization of oxidizable substances. U.S. Patent 5015483, filed Sep. 02, 1989, and issued May 14, 1991.
Health Canada. 1998. Policy Paper—Nutraceuticals/functional foods and health claims on foods. Available at http://www.hc-sc.gc.ca/fn-an/label-etiquet/claims-reclam/nutrafunct_foods-nutra-fonct_aliment-eng.php, accessed Feb. 16, 2009.
Helfrich, W. 1973. Elastic properties of lipid bilayers: Theory and possible experiments. *Z. Naturforsch.* 28C:693–703.
Herrero, E.P., Martin Del Valle, E.M., and Galan, M.A. 2006. Development of a new technology for the production of microcapsules based in atomization processes. *Chem. Eng. J.* 117:137–142.
Hong, J.S., Shin, S.J., Lee, S.H., Wong, E., and Cooper-White, J. 2007. Spherical and cylindrical microencapsulation of living cells using microfluidic devices. *Korea-Aust. Rheol. J.* 19:157–164.
Hou, R.C.W., Lin, M.Y., Wang, M.M.C., and Tzen, J.T.C. 2003. Increase of viability of entrapped cells of *Lactobacillus delbrueckii* ssp bulgaricus in artificial sesame oil emulsions. *J. Dairy Sci.* 86:424–428.
Huang, K.S., Lai, T.H., and Lin, Y.C. 2007. Using a microfluidic chip and internal gelation reaction for monodisperse calcium alginate microparticles generation. *Front. Biosci.* 12:3061–3067.
International Life Sciences Institute. 1999. Safety assessment and potential health benefits of food components based on selected scientific criteria. ILSI North America Technical Committee on Food Components for Health Promotion. *Crit. Rev. Food Sci. Nutr.* 39:203–316.
Israelachvili, J.N. 1991. *Intramolecular and Surface Forces*. New York: Academic Press.
Iyer, C. and Kailasapathy, K. 2005. Effect of co-encapsulation of probiotics with prebiotics on increasing the viability of encapsulated bacteria under in vitro acidic and bile salt conditions and in yogurt. *J. Food Sci.* 70:18–23.
Izquierdo, P., Esquena, J., Tadros, T.F. et al. 2004. Phase behavior and nano-emulsion formation by the phase inversion temperature method. *Langmuir* 20:6594–6598.
Izquierdo, P., Feng, J., Esquena, J. et al. 2005. The influence of surfactant mixing ratio on nano emulsion formation by the pit method. *J. Colloid Interface Sci.* 285:388–394.
Jafari, S.M., Assadpoor, E., Bhandari, B., and He, Y. 2008. Nano-particle encapsulation of fish oil by spray drying. *Food Res. Int.* 41:172–183.
Jemiola-Rzeminka, M., Pasenkiewicz-Gierula, M., and Strzalka, K. 2005. The behaviour of ®-carotene in the phosphatidylcholine bilayer as revealed by a molecular simulation study. *Chem. Phys. Lipids* 135:27–37.
Joscelyne, S.M. and Trägardh, G. 2000. Membrane emulsification—A literature review. *J. Membr. Sci.* 169:107–117.
Kailasapathy, K. 2002. Microencapsulation of probiotic bacteria: Technology and potential applications. *Curr. Issues Intest. Microbiol.* 3:39–48.
Keller, B.C. 2001. Liposomes in nutrition. *Trends Food Sci. Technol.* 12:25–31.
Kentish, S., Wooster, T.J., Ashokkumar, M., Balachandran, S., Mawson, R., and Simons, L. 2008. The use of ultrasonics for nanoemulsion preparation. *Innovat. Food Sci. Emerg. Technol.* 9:170–175.
Kidd, P.M. 2001. Phospholipids, nutrients for life. *Total Health* 23:Sept.–Oct. issue.
Kim, C.J., Jun, S.A., and Lee, N.K. 2006. Encapsulation of *Bacillus polyfermenticus* SCD with alginate-methylcellulose and evaluation of survival in artificial conditions of large intestine. *J. Microbiol. Biotechnol.* 16:443–449.

Kim, H.H.Y. and Baianu, I.C. 1991. Novel liposome microencapsulation techniques for food applications. *Trends Food Sci. Technol.* 2:55–61.

Kirby, C.J. and Gregoriadis, G. 1984. Dehydration-rehydration vesicles: A simple method for high yield drug entrapment in liposomes. *Biotechnology* 2:979–984.

Kirby, C.J. and Law, B. 1987. Development in the microencapsulation of enzymes in food technology. In *Chemical Aspect of Food Enzymes*, ed. A.T. Andrews. London, U.K.: Royal Society of Chemistry.

Kirby, C.J., Whittle, C.J., Rigby, N., Coxon, D.T., and Law, B.A. 1991. Stabilization of ascorbic acid by microencapsulation in liposomes. *Int. J. Food Sci. Technol.* 26:437–449.

Klinkesorn, U., Sophanodora, P., Chinachoti, P., Decker, E.A., and McClements, J. 2006. Characterization of spray-dried tuna oil emulsified in two-layered interfacial membranes prepared using electrostatic layer-by-layer deposition. *Food Res. Int.* 39:449–457.

Kolanowski, W., Laufenberg, G., and Kunz, B. 2004. Fish oil stabilisation by microencapsulation with modified cellulose. *Int. J. Food Sci. Nutr.* 55:333–343.

Kondo, A. 1979. *Microcapsule Processing and Technology*. New York: Marcel Dekker, Inc.

Kosaraju, S.L., D'ath, L., and Lawrence, A. 2006. Preparation and characterisation of chitosan microspheres for antioxidant delivery. *Carbohydr. Polym.* 64:163–167.

Kosaraju, S.L., Labbett, D., Emin, M., Konczak, L., and Lundin, L. 2008. Delivering polyphenols for healthy ageing. *Nutr. Diet.* 65:48–52.

Kostecka-Gugala, A., Latowski, D., and Strzalka, K. 2003. Thermotropic phase behaviour of alpha-dipalmitoylphosphatidylcholine multibilayers is influenced to various extents by carotenoids containing different structural features—Evidence from differential scanning calorimetry. *Biochim. Biophys. Acta* 1609:193–202.

Krasaekoopt, W., Bhandari, B., and Deeth, H. 2003. Evaluation of encapsulation techniques of probiotics for yoghurt. *Int. Dairy J.* 13:3–13.

Krasaekoopt, W., Bhandari, B., and Deeth, H. 2004. The influence of coating materials on some properties of alginate beads and survivability of microencapsulated probiotic bacteria. *Int. Dairy J.* 14:737–743.

Krasaekoopt, W., Bhandari, B., and Deeth, H.C. 2006. Survival of probiotics encapsulated in chitosan-coated alginate beads in yoghurt from UHT and conventionally treated milk during storage. *LWT—Food Sci. Technol.* 39:177–183.

Lasic, D.D. 1993. *Liposomes: From Physics to Applications*. New York: Elsevier.

Lasic, D.D. 1995. Pharmacokinetics and antitumor activity of anthracyclines precipitated in sterically (stealth) liposomes. In *Stealth Liposomes*, eds. D.D. Lasic and F.J. Martin. Boca Raton, FL: CRC Press.

Lasic, D.D. and Papahadjopoulos, D., eds. 1998. *Medical Applications of Liposomes*. New York: Elsevier.

Latorre, L.G., Carneiro, A.L., Rosada, R.S., Silva, C.L., and Santana, M.H.A. 2007. A mathematical model describing the kinetic of cationic liposome production from dried lipid films adsorbed in a multitubular system. *Braz. J. Chem. Eng.* 24:1–10.

Lawrence, M.J. and Rees, G. 2000. Microemulsion-based media as novel drug delivery systems. *Adv. Drug Deliv. Rev.* 45:89–121.

Lee, S., Hernandez, P., Djordjevic, D. et al. 2006. Effect of antioxidants and cooking on stability of n-3 fatty acids in fortified meat products. *J. Food Sci.* 71:C233–C238.

Lian, W.C., Hsiao, H.C., and Chou, C.C. 2003. Viability of microencapsulated bifidobacteria in simulated gastric juice and bile solution. *Int. J. Food Microbiol.* 86:293–301.

Liserre, A.M., Ré, M.I., and Franco, B.D.G.M. 2007. Microencapsulation of *Bifidobacterium animalis* subsp. *lactis* in modified alginate-chitosan beads and evaluation of survival in simulated gastrointestinal conditions. *Food Biotechnol.* 21:1–16.

Liu, K., Ding, H.J., Chen, Y., and Zhao, X.Z. 2006. Shape-controlled production of biodegradable calcium alginate gel microparticles using a novel microfluidic device. *Langmuir* 22:9453–9457.

Liu, X.D., Bao, D.C., Xue, W. et al. 2003. Preparation of uniform calcium alginate gel beads by membrane emulsification coupled with internal gelation. *J. Appl. Polym. Sci.* 87:848–852.

Liu, X.D., Yu, X.W., Zhang, Y., Xue, W.M. et al. 2002. Characterization of structure and diffusion behaviour of Ca-alginate beads prepared with external or internal calcium sources. *J. Microencapsul.* 10:775–782.

Loksuwan, J. 2007. Characteristics of microencapsulated beta-carotene formed by spray drying with modified tapioca starch, native tapioca starch and maltodextrin. *Food Hydrocolloids* 21:928–935.

Lysionek, A.E., Zubillaga, M.B., Sarabia, M.I. et al. 2000. Study of industrial microencapsulated ferrous sulfate by means of the prophylactic-preventive method to determine its bioavailability. *J. Nutr. Sci. Vitaminol.* 6:125–129.

Lysionek, A.E., Zubillaga, M.B., Salgueiro, M.J., Pineiro, A., Caro, R.A., Weill, R., and Boccio, J.R. 2002. Bioavailability of microencapsulated ferrous sulfate in powdered milk produced from fortified fluid milk: A prophylactic study in rats. *Nutrition* 18:279–281.

Maa, Y.F. and Hsu, C. 1999. Performance of sonication and microfluidization for liquid-liquid emulsification. *Pharm. Dev. Technol.* 4:233–240.

Madene, A., Jacquot, M., Scher, J., and Desobry, S. 2006. Flavour encapsulation and controlled release—A review. *Int. J. Food Sci. Technol.* 41:1–21.

Market Research. 2004. Global market review of functional foods—Forecasts to 2010. Available at http://www.marketresearch.com, accessed July 20, 2008.

Martinsen, A., Skjak-Braek, C., and Smidsrod, O. 1989. Alginate as immobilization material. I. Correlation between chemical and physical properties of alginate gel beads. *Biotechnol. Bioeng.* 33:79–89.

Mason, T.G., Graves, S.M., Wilking, J.N., and Lin, M.Y. 2006a. Extreme emulsification: Formation and structure of nanoemulsions. *Condens. Matter Phys.* 9:193–199.

Mason, T.G., Wilking, J.N., Meleson, K., Chang, C.B., and Graves, S.M. 2006b. Nanoemulsions: Formation, structure, and physical properties. *J. Phys. Condens. Matter* 18:635–666.

Masters, K. 1985. *Spray Drying Handbook*. New York: Halsted Press.

Matsuno, R. and Adachi, S. 1993. Lipid encapsulation technology—Techniques and applications to food. *Trends Food Sci. Technol.* 4:256–261.

Mattila-Sandholm, T., Myllarinen, P., Crittenden, R., Mogensen, G., Fondén, R., and Saarela, M. 2002. Technological challenges for future probiotic foods. *Int. Dairy J.* 12:173–182.

McClements, D.J. and Decker, E.A. 2000. Lipid oxidation in oil-in-water emulsions: Impact of molecular environment on chemical reactions in heterogeneous food systems. *J. Food Sci.* 65:1270–1282.

McClements, D.J., Decker, E.A., and Weiss, J. 2007. Emulsion-based delivery systems for lipophylic bioactive components. *J. Food Sci.* 72:109–124.

McMaster, L.D., Kokott, S.A., Reid, S.J., and Abratt, V. 2005. Use of traditional African fermented beverages as delivery vehicles for *Bifidobacterium lactis* DSM 10140. *Int. J. Food Microbiol.* 102:231–237.

McNulty, H.P., Byun, J., Lockwood, S.F., Jacob, R.F., and Mason, R.P. 2007. Differential effects of carotenoids on lipid peroxidation due to membrane interactions: X-ray diffraction analysis. *Biochim. Biophys. Acta* 1768:167–174.

Meleson, K., Graves, S., and Mason, T.G. 2004. Formation of concentrated nanoemulsions by extreme shear. *Soft Matter* 2:109–123.

Moberger, L., Larsson, K., Buchheim, W., and Timmen, H. 1987. A study on fat oxidation in a microemulsion system. *J. Disper. Sci. Technol.* 8:207–215.

Morales, D., Gutierrez, J.M., Garcia-Celma, M.J., and Solans, Y.C. 2003. A study of the relation between bicontinuous micro emulsions and oil/water nano-emulsion formation. *Langmuir* 19: 7196–7200.

Mosca, M., Ceglie, A., and Ambrosone, L. 2008. Antioxidant dispersions in emulsified olive oils. *Food Res. Int.* 41:201–207.

Muschiolik, G. 2007. Multiple emulsions for food use. *Curr. Opin. Colloid Interface Sci.* 12:213–220.

Muthukumarasamy, P., Allan-Wojtas, P., and Holley, R.A. 2006. Stability of *Lactobacillus reuteri* in different types of microcapsules. *J. Food Sci.* 71:20–24.

Nguyen, N.T. and Zhigang, W. 2005. Micromixers—A review. *J. Micromech. Microeng.* 15:R1–R6.

Oh, J.K., Drumright, R., Siegwart, D.J., and Matyjaszewski, K. 2008. The development of microgels/nanogels for drug delivery applications. *Prog. Polym. Sci.* 33:448–477.

Oliveira, A.C., Moretti, T.S., Boschini, C., Baliero, J.C.C., Freitas, O., and Favaro-Trindade, C.S. 2007. Stability of microencapsulated B lactis (Bl 01) and L acidophilus (LAC 4) by complex coacervation followed by spray drying. *J. Microencapsul.* 24:685–693.

Oneda, F. and Ré, M.I. 2003. The effect of formulation variables on the dissolution and physical properties of spray dried microspheres containing organic salts. *Powder Technol.* 130:377–384.

O'Riordan, K., Andrews, D., Buckle, K., and Conway, P. 2001. Evaluation of microencapsulation of a *Bifidobacterium* strain with starch as an approach to prolonging viability during storage. *J. Appl. Microbiol.* 91:1059–1066.

Papadimitriou, V., Pispas, V., Syriou, S. et al. 2008. Biocompatible microemulsions based on limonene: Formulation, structure, and applications. *Langmuir* 24:3380–3386.

Parada, J. and Aguilera, J.M. 2007. Food microstructure affects the bioavailability of several nutrients. *J. Food Sci R Conc. Rev. Hypoth. Food Sci.* 72:21–32.

Patel, N., Schmid, U., and Lawrence, M.J. 2006. Phospholipid-based microemulsions suitable for use in foods. *J. Agric. Food Chem.* 54:7817–7824.

Pérez-Alonso, C., Báez-González, J.G., Beristain, C.I., Vernon-Carter, E.J., and Vizcarra-Mendonza, M.G. 2003. Estimation of the activation energy of carbohydrate polymers blends as selection criteria for their use as wall material for spray-dried microcapsules. *Carbohydr. Polym.* 53:197–203.

Picon, A., Serrano, C., Gaya, P., Medina, M., and Nunhez, M. 1996. The effect of liposome-encapsulated cyprosins on manchego cheese ripening. *J. Dairy Sci.* 79:1694–1705.

Picot, A. and Lacroix, C. 2004. Encapsulation of bifidobacteria in whey protein-based microcapsules and survival in simulated gastrointestinal conditions and in yoghurt. *Int. Dairy J.* 14:505–515.

Poncelet, D. and Dreffier, C. 2007. Les methods de microencapsulation de A à Z (ou presque). In *Microencapsulation: Des sciences aux technologies*, eds. T. Vandamme, D. Poncelet, and P. Subra-Paternault. Paris, France: Tec&Doc (Editions), pp. 23–33.

Poncelet, D., Lencki, R., Beaulieu, C., Halle, J.P., Neufeld, R.J., and Fournier, A. 1992. Production of alginate beads by emulsification internal gelation. I. Methodology. *Appl. Microbiol. Biotechnol.* 38:39–45.

Pons, R., Carrera, I., Caelles, J., Rouch, J., and Panizza, P. 2003. Formation and properties of miniemulsions formed by microemulsions dilution. *Adv. Colloid Interface Sci.* 106:129–146.

Porras, M., Solans, C., Gonzalez, C., Martinez, A., Guinart, A., and Gutierrez, J.M. 2004. Studies of formation of W/O nano-emulsions. *Colloids Surf. A* 249:115–118.

Prado, F.C., Parada, J.L., Pandey, A., and Soccol, C.R. 2008. Trends in non-dairy probiotic beverages. *Food Res. Int.* 41:111–123.

Rane, S.S. and Anderson, B.D. 2008. What determines drug solubility in lipid vehicles: Is it predictable? *Adv. Drug Deliv. Rev.* 60:638–656.

Ré, M.I. 1998. Microencapsulation by spray drying. *Drying Technol.* 16:1195–1236.

Ré, M.I. 2006. Formulating drug delivery systems by spray drying. *Drying Technol.* 24:433–446.

Ré, M.I., Messias, L.S., and Schettini, H. 2004. The influence of the liquid properties and the atomizing conditions on the physical characteristics of the spray-dried ferrous sulfate microparticles. Paper presented at the *Annual International Drying Symposium*, Campinas, Brazil, Aug. 2004, 1174–1181.

Reid, A.A., Champagne, C.P., Gardner, N., Fustier, P., and Vuillemard, J.C. 2007. Survival in food systems of *Lactobacillus rhamnosus* R011 microentrapped in whey protein gel particles. *J. Food Sci.* 72:31–37.

Reineccius, G.A. 1995. Liposomes for controlled release in the food industry. In *Encapsulation and Controlled Release of Food Ingredients*, American Chemical Society Symposium Series 590, eds. S.J. Risch and G.A. Reineccius. Washington, DC: American Chemical Society, pp. 113–131.

Ribeiro, H.S., Ax, K., and Schubert, H. 2003. Stability of lycopene emulsions in food systems. *J. Food Sci.* 68:2730–2734.

Ribeiro, H.S., Guerrero, J.M.M., Briviba, K., Rechkemmer, G., Schuchmann, H.P., and Schubert, H. 2006. Cellular uptake of carotenoid-loaded oil-in-water emulsion in colon carcinoma cell in vitro. *J. Agric. Food Chem.* 54:9366–9369.

Rodriguez-Huezo, M.E., Duran-Lugo, R., Prado-Barragan, L.A. et al. 2007. Pre-selection of protective colloids for enhanced viability of *Bifidobacterium bifidum* following spray-drying and storage, and evaluation of aguamiel as thermoprotective prebiotic. *Food Res. Int.* 40:1299–1306.

Roger, J.A. and Anderson, K.E. 1998. The potential of liposomes in oral drug delivery. *Crit. Rev. Ther. Drug. Carrier Syst.* 15:421–481.

Rohloff, C.M., Shimek, J.W., and Dungan, S.R. 2003. Effect of added alpha-lactalbumin protein on the phase behavior of AOT-brine-isooctane systems. *J. Colloid Interface Sci.* 261:514–523.

Rowland, M. and Towzer, T.N. 1995. *Clinical Pharmacokinetics: Concepts and Applications*, 3rd edn. Baltimore, MD: Williams & Williams.

Rusli, J.K., Sanguansri, L., and Augustin, M.A. 2006. Stabilization of oils by microencapsulation with heated protein-glucose syrup mixtures. *J. Am. Oil Chem. Soc.* 83:965–997.

Santana, M.H.A., Martins, F., and Pastore, G.M. 2005. Processes for stabilization of bioflavors through encapsulation in cyclodextrins and liposomes. Brazilian Patent Application, PI 0403279-9 A, filed Dec. 2005 (in Portuguese).

Santipanichwong, R. and Suphantharika, M. 2007. Carotenoids as colorants in reduced-fat mayonnaise containing spent brewer's yeast beta-glucan as fat replacer. *Food Hydrocolloids* 21:565–574.

Schrooyen, P.M.M., van der Meer, R., and De Krif, C.G. 2001. Microencapsulation: Its application in nutrition. *Proc. Nutr. Soc.* 60:475–479.

Shah, N.P. and Ravula, R. 2000. Microencapsulation of probiotic bacteria and their survival in frozen fermented dairy desserts. *Aust. J. Dairy Technol.* 55:139–144.

Shahidi, F. and Han, X.Q. 1993. Encapsulation of food ingredients. *Crit. Rev. Food Sci. Nutr.* 33:501–547.

Shima, M., Morita, Y., Yamashita, M., and Adachi, S. 2006. Protection of *Lactobacillus acidophilus* from the low pH of a model gastric juice by incorporation in a W/O/W emulsion. *Food Hydrocolloids* 20:1164–1169.

Shu, B., Yu, W., Zhao, Y., and Liu, X. 2006. Study on microencapsulation of lycopene by spray-drying. *J. Food Eng.* 76:664–669.

Shuqin, X. and Shiying, X. 2005. Ferrous sulfate liposomes: Preparation, stability and application in fluid milk. *Food Res. Int.* 38:289–296.

Siró, I., Kálpona, E., Kálpona, B., and Lugasi, A. 2008. Functional food. Product development, marketing and consumer acceptance—A Review. *Appetite* 51:456–467.

Socaciu, C., Bojarski, P., Aberle, L., and Diehl, H.A. 2002. Different ways to insert carotenoids into liposomes affect structure and dynamics of bilayer differently. *Biophys. Chem.* 99:1–15.

Socaciu, C., Jessel, R., and Diehl, H.A. 2000. Competitive carotenoid and cholesterol incorporation into liposomes: Effects on membrane phase transition, fluidity, polarity and anisotropy. *Chem. Phys. Lipids* 106:79–88.

Solans, C., Izquierdo, P., Nolla, J., Azemar, N., and Garcia-Celma, M.J. 2005. Nano-emulsions. *Curr. Opin. Colloid Interface Sci.* 10:102–110.

Spernath, A. and Aserin, A. 2006. Microemulsions as carriers for drugs and nutraceuticals. *Adv. Colloid Interface Sci.* 128–130:47–64.

Spernath, A., Yaghmur, A., Aserin, A., Hoffman, R.E., and Garti, N. 2002. Food-grade microemulsions based on nonionic emulsifiers: Media to enhance lycopene solubilization. *J. Agric. Food Chem.* 50:6917–6922.

Su, L.C., Lin, C.W., and Chen, M.J. 2007. Development of an oriental-style dairy product coagulated by microcapsules containing probiotics and filtrates from fermented rice. *Int. J. Dairy Technol.* 60:49–54.

Sujak, A., Strzalka, K., and Gruszecki, W.I. 2007. Thermotropic phase behaviour of lipid bilayers containing carotenoid pigment canthaxanthin: A differential scanning calorimetry study. *Chem. Phys. Lipids* 145:1–12.

Sultana, K., Godward, G., Reynolds, N., Arumugaswamy, R., Peiris, P., and Kailasapathy, K. 2000. Encapsulation of probiotic bacteria with alginate-starch and evaluation of survival in simulated gastrointestinal conditions and in yoghurt. *Int. J. Food Microbiol.* 62:47–55.

Tadros, T., Izquierdo, R., Esquena, J., and Solans, C. 2004. Formation and stability of nanoemulsions. *Adv. Colloid Interface Sci.* 108:303–318.

Tanford, C. 1980. *The Hydrophobic Effect: Formation of Micelles and Biological Membranes*. New York: Wiley-Interscience.

Thijssen, H.A.C. 1971. Flavour retention in drying preconcentrated food liquids, *J. Appl. Chem. Biotechnol.* 21:372–376.

Tournier, H., Schneider, M., and Guillot, C. 1999. Liposomes with enhanced entrapment capacity and their use in imaging. U.S. Patent 5,980,937, filed Aug. 12, 1997 and issued Nov. 9, 1999.

Uicich, R., Pizarro, F., Almeida, C. et al. 1999. Bioavailability of microencapsulated ferrous sulfate in fluid cow milk: Studies in human beings. *Nutr. Rev.* 19:893–897.

Urbanska, A.M., Bhathena, J., and Prakash, S. 2007. Live encapsulated *Lactobacillus acidophilus* cells in yogurt for therapeutic oral delivery: Preparation and in vitro analysis of alginate-chitosan microcapsules. *Can. J. Physiol. Pharmacol.* 85:884–893.

Vanderberg, G.W. and De La Noüe, J. 2001. Evaluation of protein release from chitosan-alginate microcapsules produced using external or internal gelation. *J. Microencapsul.* 18:433–441.

Vladisavljevic, G.T. and Williams, R.A. 2005. Recent developments in manufacturing emulsions and particulate products using membranes. *Adv. Colloid Interface Sci.* 113:1–20.

Vuillemard, J.-C. 1991. Recent advances in the large-scale production of lipid vesicles for use in food products: Microfluidization. *J. Microencapsul.* 8:547–562.

Wallach, D.F.H. and Mathur, R. 1992. Method of making oil filled paucilamellar lipid vesicles. U.S. Patent 5160669, filed Oct. 16, 1990, and issued Nov. 03, 1992.

Walther, B., Cramer, C., Tiemeyer, A. et al. 2005. Drop deformation dynamics and gel kinetics in a co-flowing water-in-oil system. *J. Colloid Interface Sci.* 286:378–386.

Wang, L., Li, X., Zhang, G., Dong, J., and Eastoe, J. 2007. Oil-in-water nanoemulsions for pesticide formulations. *J. Colloid Interface Sci.* 314:230–235.

Wang, L., Mutch, K.J., Eastoe, J., Heenan, R.K., and Dong, J. 2008. Nanoemulsions prepared by a two-step low-energy process. *Langmuir* 24:6092–6099.

Wang, L.Y., Ma, G.H., and Su, Z.G. 2005. Preparation of uniform sized chitosan microspheres by membrane emulsification technique and application as a carrier of protein drug. *J. Control. Release* 106:62–75.

Weiss, J., Kobow, K., and Muschiolik, G. 2004. Preparation of microgel particles using membrane emulsification. Abstracts of Food Colloids, Harrogate, U.K., B-34.

Weiss, J., Takhistov, P., and McClements, D.J. 2006. Functional materials in food nanotechnology. *J. Food Sci.* 71:107–116.

Were, L.M., Bruce, B.D., Davidson, P.M., and Weiss, J. 2003. Size, stability, and entrapment efficiency of phospholipid nanocapsules containing polypeptide antimicrobials. *J. Agric. Food Chem.* 51:8073–8079.

Wolf, P.A. and Havekotte, M.J. 1989. Microemulsions of oil in water and alcohol. U.S. Patent 4,835,002, filed July 10, 1987, and issued May 30, 1989.

Xianghua, Y. and Zirong, X. 2006. The use of immunoliposome for nutrient target regulation (a review). *Crit. Rev. Food Sci. Nutr.* 46:629–638.

Xu, S., Nie, Z., Seo, M. et al. 2005. Generation of monodisperse particles by using microfluidics: Control over size, shape, and composition. *Angew. Chem. Int. Ed.* 44:724–728.

Yi, O.S., Han, D., and Shin, H.K. 1991. Synergistic antioxidative effects of tocopherol and ascorbic-acid in fish oil lecithin water-system. *J. Am. Oil Chem. Soc.* 68:881–883.

Yuan, Y., Gao, Y., Zhao, J., and Mao, L. 2008. Characterization and stability evaluation of beta-carotene nanoemulsions prepared by high pressure homogenization under various emulsifying conditions. *Food. Res. Int.* 41:61–68.

Zhang, H., Feng, F., Li, J. et al. 2008. Formulation of food-grade microemulsions with glycerol monolaurate: Effects of short-chain alcohols, polyols, salts and nonionic surfactants. *Eur. Food Res. Technol.* 226:613–619.

Zhang, L., Liu, J., Lu, Z., and Hu, J. 1997. Procedure for preparation of vesicles with no leakage from water-in-oil emulsion. *Chem. Lett.* 8:691–692.

Zheng, S., Alkan-Onyuksel, H., Beissinger, R.L., and Wasan, D.T. 1999. Liposome microencapsulations without using any organic solvent. *J. Disper. Sci. Technol.* 20:1189–1203.

Zhou, Q.Z., Wang, L.Y., Ma, G.H., and Su, Z.G. 2007. Preparation of uniform-sized agarose beads by microporous membrane emulsification technique. *J. Colloid Interface Sci.* 311:118–127.

33 Microencapsulation
Probiotics

Dan Cristian Vodnar, Oana Lelia Pop, and Carmen Socaciu

CONTENTS

33.1 Introduction 685
33.2 Probiotic Bacteria 686
33.3 Microencapsulation of Probiotic Bacteria 686
33.4 Gastrointestinal Passage of Microcapsules 687
33.5 Encapsulation in Alginate...687
    33.5.1 Introduction 687
    33.5.2 Survivability of Microencapsulated *L. plantarum* ATCC 8014 in Alginate
        during Exposure to Simulated Gastro-Intestinal Juice..687
33.6 Chitosan-Coated Alginate Microcapsules..690
    33.6.1 Introduction 690
    33.6.2 Survivability of Microencapsulated *L. plantarum* ATCC 8014
        in Chitosan-Coated Alginate Microcapsules during Exposure
        to Simulated Gastro-Intestinal Juice 690
33.7 Other Coat Materials for Alginate Microcapsules 692
    33.7.1 Polylysine-Coated Alginate Microcapsules..692
    33.7.2 Alginate–Polylysine–Alginate Microcapsules 692
    33.7.3 Alginate–Chitosan–Alginate Microcapsules 693
    33.7.4 Whey Protein 693
    33.7.5 Palm Oil..693
33.8 Other Applied Matrices 693
33.9 Conclusion 693
References..694

33.1 INTRODUCTION

Probiotics are a group of bacteria recognized as good or friendly bacteria and are used to reduce potentially harmful bacteria in the intestine. In order to ensure the minimum therapeutic level, which provides the health benefits of probiotic consumption, the use of encapsulation techniques is needed. For this reason, microencapsulation of probiotic bacteria is currently used as a method to improve the stability of probiotic microorganisms in functional food products. There is a need for microencapsulation not only to help the probiotic bacteria to survive in the food product but also during the passage through the human gastrointestinal tract. In the microencapsulation process are used different biopolymers that ensure good protection of the stability and viability of probiotic bacteria. These biopolymers are non-cytotoxic, non-antimicrobial, biodegradable, and do not require the use of organic solvents for microcapsules preparation.

The survivability evaluation of microencapsulated *Lactobacillus plantarum* ATCC 8014 in 1%, 1.5%, and 2% of alginate and chitosan-coated alginate microcapsules along the gastro-intestinal tract will be a step forward in understanding the microencapsulation efficiency, offering an effective way for increasing the life span.

33.2 PROBIOTIC BACTERIA

Probiotics are live microorganisms recognized as good or friendly bacteria, which settle in the intestine medium and render healthful effects on the host.[1] Thus, these live microorganisms can improve microbial balances in intestine and exert positive effects on the host.[2] Various health benefits have been attributed to probiotics such as antimutagenic and anticarcinogenic properties, anti-infection properties, immune system stimulation, and serum cholesterol reduction.[3,4] Lactobacillus and bifidobacteria are the most commonly used probiotics and are extensively investigated for their beneficial importance.[5]

In order to provide health benefits, probiotic bacteria are expected to be at a level of 10^7 cfu/g of food product at the point of delivery.[6] *L. plantarum* is one of the most widely used lactic bacteria, showing a homofermentative metabolism, moderate acid tolerance and is considered as a GRAS (Generally Regarded as Safe) organism[7] and is marked as probiotic.[8] Probiotics have to be stable in the product, and must survive the passage of the digestive tract in large numbers, showing the ability to adhere and colonize in the intestine system.[9]

33.3 MICROENCAPSULATION OF PROBIOTIC BACTERIA

Microencapsulation is defined as a packaging technology for solids, liquids, and gaseous materials in miniature. The resultant products of encapsulation are capsules that can control the release of their contents under the influence of specific conditions.[10] Microencapsulation of probiotic bacteria has received remarkable attention because of its growing and promising potential against many diseases.[11,12] This technique has the potential to be a useful tool in improving the survival of probiotics over the course of the gastrointestinal tract protecting probiotic bacteria cells, or other therapeutic live cells, for oral delivery from the harsh acid conditions and delivering them with improved survival rate.[13]

The size of a microcapsule varies from a few microns to 1 mm, and it consists of a semipermeable, spherical, thin, and strong membrane surrounding a solid/liquid core.[14] The microencapsulation technique can be used for many applications in the food industry, including the core material stabilization, control of oxidation reaction, masking undesirable flavors, colors, and odors, extending the shelf-life, and protecting the components against nutritional losses.[14]

The presence of varying conditions in the human digestive tract not only makes the designing of probiotic bacteria release system challenging but also gives the opportunity to produce a highly targeted system. Due to the micron size of encapsulated bacteria, nanoencapsulation is not an option. The current size of the microcapsules typically ranges from tens of microns[15] to millimeters,[16] and survival of probiotics is dependent on microcapsules size.

The production of microcapsules containing probiotic bacteria could be done by extrusion, emulsion, and spray drying.[17] Generally, the emulsion and extrusion methods refer to the cross-linking of the polymer solution after suspension in oil or dropping into cross-linker.[17] Extrusion can be done not only using a syringe and a needle but also could be done utilizing systems with vibrating nozzles,[18] air-atomizing nozzles,[19] and spinning disk atomization.[20] The microencapsulation emulsion method has the advantage of producing very small microcapsules (less than 100 µm).

After preparation of the capsules, usually it is desirable to store, isolate as a dry powder. For this reason, a number of drying procedures have been used, such as spray drying, air drying, freeze drying, and fluid bed drying.[17] The spray drying of bacteria produced microcapsules with good protection during transit of gastric juice, with a loss of 2 log cfu/mL.[21] The drying could be a promising method for long-time storage (further studies), having the advantage of reducing the size of microcapsules and improving the storage properties. Freeze drying consists of sublimation of ice under vacuum. The advantage of freeze drying is that bacteria can survive well with the addition of cryoprotectants.[22] Fluid bed drying attains drying by the suspension of the load in air with extra heat to induce the evaporation of the solvent.[23] This process is a quick one but is not recommended for strictly anaerobic bacteria because the process is carried out under aerobic conditions and could damage the viability.

33.4 GASTROINTESTINAL PASSAGE OF MICROCAPSULES

Microencapsulation of probiotic bacteria has received remarkable attention because of its growing and promising potential against many diseases.[11,12] This technique has the potential to be a useful tool in improving the survival of probiotics over the course of the gastrointestinal tract protecting probiotic bacteria cells, or other therapeutic live cells, for oral delivery from the harsh acid conditions and delivering them with improved survival rate.[13]

In order to design delivery systems for microencapsulated probiotics, with the aim of providing a controlled release, it is necessary to consider the complexity of the gastrointestinal tract.[17] Normally, the release from microencapsulated formula will be triggered by degradation, disintegration, or dissolution.[24]

Once ingested, microencapsulated probiotic bacteria will pass through the esophagus[25] and reach the stomach, where the greatest viability loss of bacteria is registered due to the high acid level. The acidity level of the gastric juice varies between subjects, being dependent on many factors such as age, nutrition, metabolism, and stress. The literature studies report not only a different range of stomach pH of 1–2.5[25] but also a pH of 5 for patients with gastric diseases.[26] Because of the wide range of pH, the controlled release of bacteria from microencapsulated formulas may be ineffective in some patients, since specific pH requirements are needed.

After stomach passage, the microcapsules will dip into the small intestine for a reported period of 0.5–9.5 h.[27] The estimated pH of the small intestine has been reported to be in the range of 6.15–7.88.[28] After passage through the small intestine, the microcapsules will reach the large intestine in a medium with a pH of 5.26–7.02.[28] The passage time of pharmaceuticals in the large intestine has been reported to be 6–32 h,[27] but longer times have also been noted.[29]

33.5 ENCAPSULATION IN ALGINATE

33.5.1 Introduction

Alginate is commonly obtained from brown seaweed as a natural polysaccharide, linear heteropolysaccharide of D-mannuronic and L-guluronic acid, which in the presence of divalent cations forms a physical hydrogel.[30] The most widely used encapsulating material is alginate, because of its biocompatibility, non-toxicity, cheapness, simplicity, gelatin conditions, and low immunogenicity.[31–33] The alginate solution in contact with calcium bath instantaneously polymerized with precipitation of calcium alginate followed by a more gradual gelation of the interior of microcapsules as calcium ions permeate through the alginate systems.[14] Because of the cooperative nature of cross-linking by the polyguluronic units and ions, gelation of alginate occurs rapidly.[30]

Bacteria (1–3 µm size) are well retained in the alginate gel matrix, which has a pore size less than 17 nm.[34] The size of the microcapsules depends on the polymer solution viscosity, the nozzle diameter, and the distance between the outlet and hardening bath solution.[35]

The literature studies[18,36] reported different factors affecting microcapsules preparation, such as concentration of alginate, hardening bath, gelation time, as well as concentration of probiotic bacteria.

33.5.2 Survivability of Microencapsulated *L. plantarum* ATCC 8014 in Alginate during Exposure to Simulated Gastro-Intestinal Juice

Figure 33.1 represents the dynamics of bacterial cell density (log × cfu/mL) at different time intervals (30, 60, 90, and 120 min) after incubation in simulated gastric juice (pH 2) of alginate

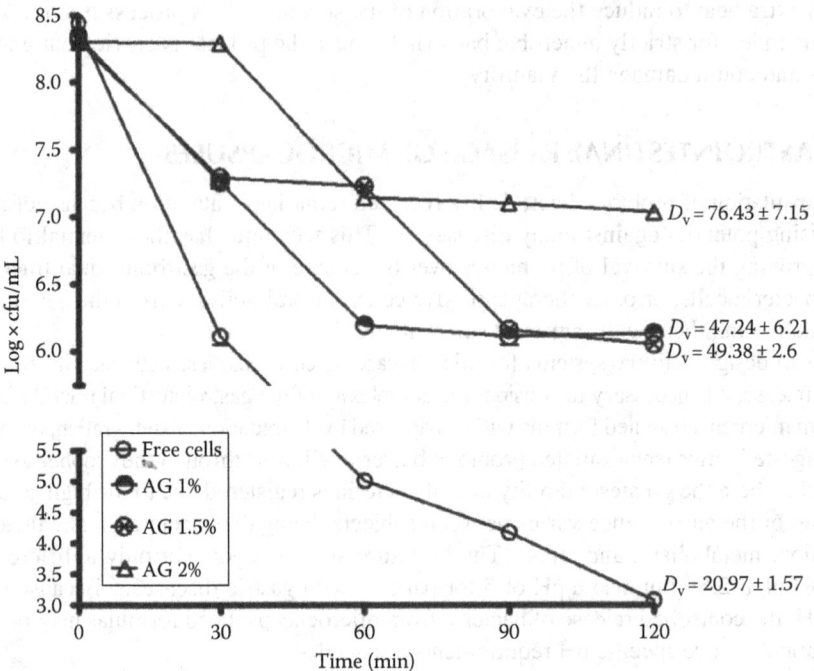

FIGURE 33.1 Dynamics of bacterial cell density (log × cfu/mL) at different periods (after 30, 60, 90, and 120 min) after incubation of microcapsules containing bacteria cells in alginate (AG) matrix (1%, 1.5%, and 2%) in simulated gastric juice (pH 2). The decimal reduction time values (D_v) (min) are represented for each variant.

beads (1%, 1.5%, and 2%) containing bacterial cells of *L. plantarum* ATTC 8014. To correlate the different types of microcapsules with the viability, the decimal reduction time value (D_v) was calculated for each trial, representing the time (min) required to destroy 90% or one log cycle of the bacteria.

The initial bacterial cell population, before microencapsulation, was in the range of 8.3–8.44 log × cfu/mL, in all variants of microcapsules, demonstrating a very good incorporation yield, and no significant loss of viability for *L. plantarum* ATCC 8014, 99.6% of cells being successfully entrapped (Figure 33.1). Different dynamics for the decrease in cell viability can be observed, after 30, 60, 90, and 120 min of incubation in acidic environment, related to matrix concentration. The microcapsules built from alginate 2% proved to protect the cell viability better than alginate 1.5% and 1%. The D_v time was 76.43 for bacteria encapsulated in alginate 2%, while free cells had only a D_v value of 20.97 min.

The results are in agreement with those of Kim et al.,[37] who reported that the pH 1.2, for the non-encapsulated strain of *Lactobacillus acidophilus*, leads to complete death after 1 h of incubation, while encapsulated strains maintained above 10^6 cfu/mL after 2 h. Additionally, Mokarram et al.[38] reported that at pH 1.5 after 2 h of incubation, the strains of *L. acidophilus* and *Lactobacillus rhamnosus* built with alginate maintained above 10^7–10^8 cfu/mL. Chandramouli et al.[18] reported a higher survival of *Lactobacillus* ssp. immobilized in alginate microcapsules in low pH environments.

The non-encapsulated *L. plantarum* ATCC 8014 (Figure 33.1) was sensitive to the acidic environment (pH 1.5) and the ingestion of unprotected strains might result in reduced viability (5 log reduction after 2 h). According to this study, in agreement with the previous findings of Murata et al.,[39,40] the 2% alginate capsules provided the best protection of bacteria viability in simulated gastric juice.

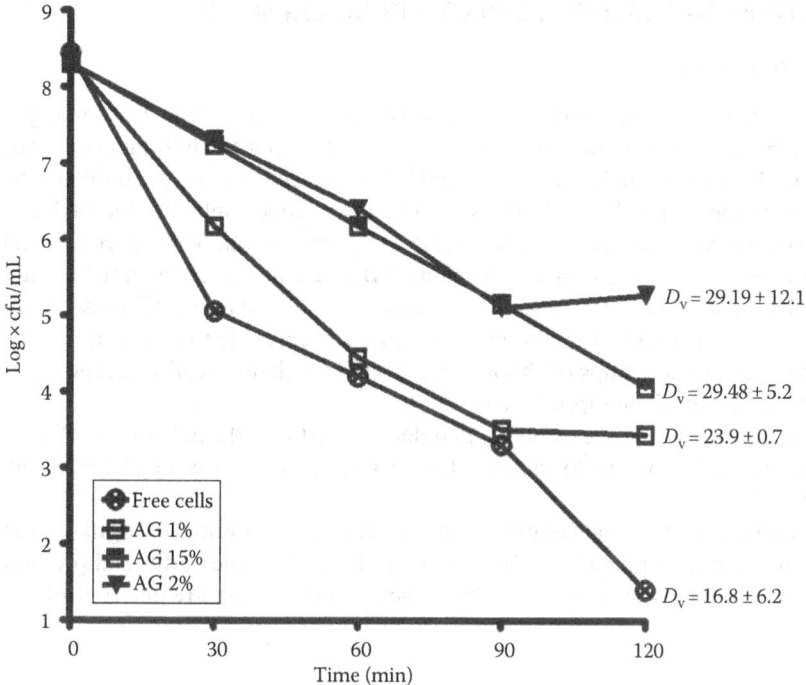

FIGURE 33.2 Dynamics of bacterial cell density (log × cfu/mL) at different periods (30, 60, 90, and 120 min) after successive incubation of alginate (AG) microcapsules containing *L. plantarum* ATCC 8014 (AG1%, AG 1.5%, and AG 2%) in simulated gastric juice (pH 1.5) for 60 min and in simulated intestinal juice at 37°C for 2 h. The decimal reduction time values (D_v) (min) are represented for each variant.

In order to exert positive health effects and to be considered probiotic bacteria, *L. plantarum* ATCC 8014 should resist the stressful conditions of the stomach and upper intestine[41] to the level of 10^7 cfu/g product when ingested.

Figure 33.2 represents the dynamics of bacterial cell density and viability (expressed as log × cfu/mL) at different time intervals (30, 60, 90, and 120 min) after successive incubation of alginate microcapsules containing *L. plantarum* ATCC 8014 in simulated gastric juice (pH 1.5) for 60 min and in simulated intestinal juice at 37°C for 2 h.

We noticed a gradual swelling and disintegration of beads in the intestinal environment, after 30, 60, 90, and 120 min, from the successive incubations in acidic and neutral environments. Good similarities of density curves were observed in intestinal media compare to gastric media (Figure 33.1).

In general, the D_v values of bacteria incubated in simulated gastrointestinal juice were lower than values registered in simulated gastric juice. This may be attributable to the difference in resistance of bacteria to environmental acidity and to their cytoplasmic and membrane composition.[42] The cells microencapsulated in the alginate matrix survived better than free cells (Figure 33.2). The microcapsules containing alginate 2% provide better protection for the cell viability (D_v: 29.19 min), implicitly increasing their survivability.

The results are in agreement with those of Kim et al.,[37] who demonstrated that microencapsulation using alginate 2% may be an effective way to increase the survival of bacteria in simulated intestinal juice. The strongest decrease in viability was noticed after 60 min of incubation in simulated gastric juice, which resulted in an initial reduction in viable cells during the first hour of exposure to simulated intestinal juice. Anal and Singh[14] and Vodnar et al.[39] showed that the formation of a hydrogel barrier by sodium alginate retards the permeation of the gastric fluid into the cells. Thus, microencapsulation matrix containing alginate 2% offered more protection to the cells during sequential incubation in simulated gastric juice followed by simulated intestinal juice.

33.6 CHITOSAN-COATED ALGINATE MICROCAPSULES

33.6.1 INTRODUCTION

Chitosan is a linear polysaccharide with negative charge arising from its amine groups, which are obtained by diacetylation of chitin with a cationic character.[43] Chitosan can be obtained from crustacean shells, insect cuticles, and from fungi membranes. The names chitin and chitosan refer to two types of copolymers, containing two monomer residues anhydro-N-acetyl-D-glucosamine and anhydro-D-glucosamine, respectively.[14] Chitosan polymers can polymerize by means of cross-linking in the presence of anions and polyanions.[34] Because its efficiency in increasing viability of probiotic cells is not satisfactory, it is mostly used as a coating material. Chitosan coating provides stability to alginate microparticles for effective microencapsulation of therapeutic live cells. The positively charged amino groups of chitosan and negatively charged carboxylic acid groups of alginate reduce the leakage of entrapped materials.[44]

Many research studies were carried out in order to investigate the potentiality of chitosan-coated alginate microcapsules system to increase the survival and stability of entrapped live probiotic cells.[36,45–49]

In the probiotic microencapsulation research, the incorporation in alginate microcapsules and different coatings application is frequently used. Besides the additional protection that the coatings can offer to the entrapped bacteria, they could provide greater control over bacterial release.

33.6.2 SURVIVABILITY OF MICROENCAPSULATED *L. PLANTARUM* ATCC 8014 IN CHITOSAN-COATED ALGINATE MICROCAPSULES DURING EXPOSURE TO SIMULATED GASTRO-INTESTINAL JUICE

Several studies report the effect of chitosan coating on the viability of probiotics in simulated gastric juice, describing different results. The viability of free cells and immobilized *L. plantarum* ATCC 8014 in chitosan-coated alginate microcapsules during exposure to simulated gastric juice was evaluated and the results are shown in Figure 33.3.

Encapsulation in chitosan-coated alginate microcapsules protected the survival of *L. plantarum* ATCC 8014 during exposure to gastric juice (Figure 33.3). Microencapsulated *L. plantarum* ATCC 8014 was resistant to simulated gastric conditions in trials that used chitosan-coated alginate microcapsules. In microcapsules built with alginate 2%, the survival rate of bacteria was higher compared with the capsules containing 1.5% and 1% alginate. Thus, the survivability rate increased proportionally with the concentration of alginate. Our results suggested that microcapsules with chitosan (0.4%)-coated alginate (2%) increased the number of survival cells after 120 min of exposure to simulated gastric juice in comparison with microcapsules with alginate 1.5% and 1%. Thus, chitosan-coated alginate microcapsules protect the viability of *L. plantarum* ATCC 8014 in all trials; after 120 min of exposure to simulated gastric juice, survivability was higher than 10^7 cfu/mL, which represents the minimum time needed to provide health benefits to the consumer.[50] However, there was a rapid loss of viability for the free probiotic bacteria in simulated gastric juice, the initial number of 8.3 log cfu/mL for free strains decreased to less than 3.1 log cfu/mL after exposure for 2 h. The D_v time was 92.32 min for bacteria encapsulated in chitosan-coated alginate 2% microcapsules, while free cells had only a D_v value of 22.35 min. The D_v value for chitosan-coated alginate (1.5% and 1%) microcapsules was almost the same around 88 min. It can be noted that alginate-coated chitosan microcapsules (Figure 33.3) exert more protection on the viability of *L. plantarum* ATCC 8014 than uncoated alginate microcapsules (Figure 33.1). The decimal reduction time

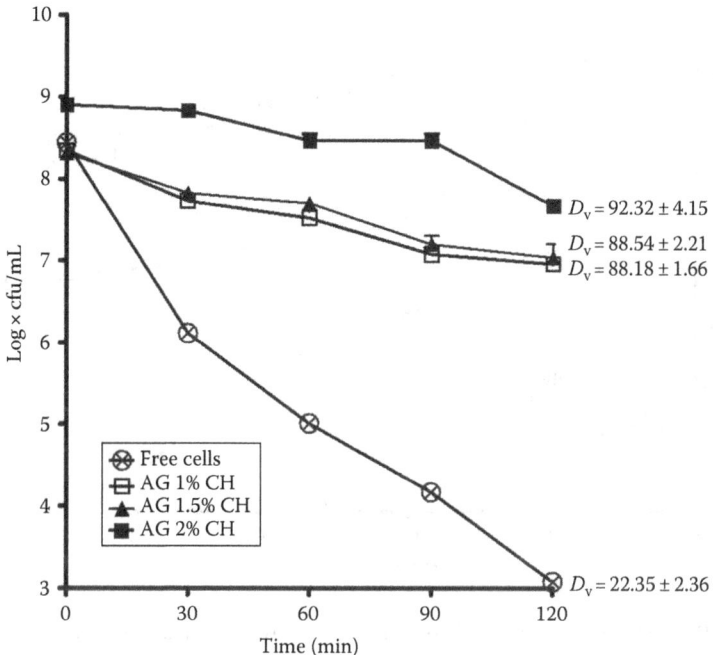

FIGURE 33.3 Dynamics of bacterial cell density (log × cfu/mL) at different periods (after 30, 60, 90, and 120 min) after incubation of microcapsules containing bacteria cells in chitosan-coated alginate matrix (AG 1% CH, AG 1.5% CH, and AG 2% CH) in simulated gastric juice (pH 2). The decimal reduction time values (D_v) (min) are represented for each variant.

decreased to 16 min for the chitosan-coated alginate microcapsules compared with uncoated alginate microcapsules.

Some reports have indicated differences among strains of probiotic bacteria with respect to their survival in acid environment.[51] Krasaekoopt et al.[52] found that encapsulation in alginate coated with chitosan was the best treatment to protect *L. plantarum* ATCC 8014 bacteria under all conditions tested.

Chitosan-coated alginate microcapsules were the most effective in protecting probiotic bacteria from bile salt.[53] The chitosan coating provides protection in bile salt solution because an ion exchange reaction takes place when the beads absorb bile salt.[40] Krasaekoopt et al.[52] found that microencapsulation with alginate coated with chitosan was the best treatment to protect *L. plantarum* ATCC 8014.

The viability of immobilized or free cells of *L. plantarum* ATTC 8014 in simulated intestinal juice was evaluated and the results are plotted in Figure 33.4. In the case of free *L. plantarum* ATTC 8014, the initial average viable count of 8.3 log × cfu/mL was reduced to 3.3 log × cfu/mL after 90 min and the average viable number was further reduced to 1.41 log × cfu/mL after 120 min. The survival percentage of microencapsulated bacteria after exposure to simulated intestinal juice for 120 min was the highest in trials with chitosan-coated alginate (2%) microcapsules, being 83.73%, while in chitosan-coated alginate (1%) microcapsules it was 58.67%. Coated alginate 2% microcapsules exert a protective effect on the survivability of probiotic bacteria after 120 min of simulated intestinal juice exposure, enhancing the number of *L. plantarum* ATCC 8014 by 20% when coated with chitosan (Figure 33.4), while the utilization of uncoated microcapsules of alginate 2% registered a diminished survival percentage by 20% (Figure 33.2). Lee et al.[36] showed that uncoated

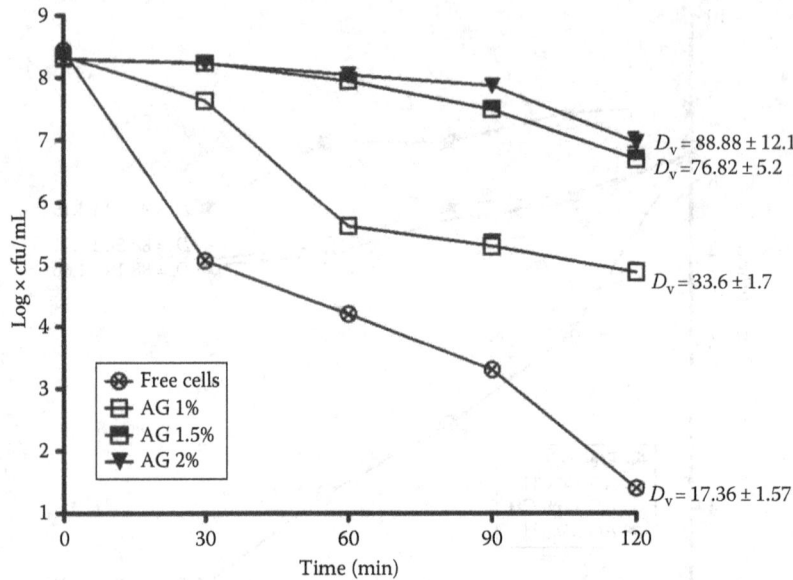

FIGURE 33.4 Dynamics of bacterial cell density (log × cfu/mL) at different periods (30, 60, 90, and 120 min) after successive incubation of chitosan-coated alginate (AG) microcapsules containing *L. plantarum* ATCC 8014 (AG 1% CH, AG 1.5% CH, and AG 2% CH) in simulated gastric juice (pH 1.5) for 60 min and in simulated intestinal juice at 37°C for 2 h. The decimal reduction time values (D_v) (min) are represented for each variant.

alginate microcapsules presented a lower survival rate (25%) of *Lactobacillus bulgaricus* KFRI 673 after exposure to simulated gastric juice.

On the other hand, literature studies report that the survivability and stability of probiotic bacteria loaded into chitosan-coated alginate microcapsules are largely dependent on the molecular weight of chitosan.[30] Thus, the microcapsules prepared with high molecular weight chitosan provided a higher survival rate (46%) compared with the microcapsules made with low molecular weight chitosan (36%).[36]

33.7 OTHER COAT MATERIALS FOR ALGINATE MICROCAPSULES

33.7.1 POLYLYSINE-COATED ALGINATE MICROCAPSULES

The literature reports show an improvement in survival and stability of microencapsulated probiotic bacteria during exposure to harsh acid conditions when polylysine has been used as a coating material. Several studies reported on alginate–polylysine microencapsulation systems coupled with bifidobacteria to investigate their survival in vitro and in vivo after oral administration. After 16 weeks of storage at 4°C, the stability of polylysine-coated alginate microcapsules successfully maintained the survival of bifidobacteria at over 10^7 cfu/g.[19,54]

The study reported by Lin et al.[55] used alginate microcapsules coated with poly-L-lysine followed by a further coating with an alginate layer. It was found that this kind of capsules does not increase the number of *Escherichia coli* strain during exposure to gastric juice. The strain used had high tolerance to acid and decreases over 120 min at pH 2 to 1 log/mL.

33.7.2 ALGINATE–POLYLYSINE–ALGINATE MICROCAPSULES

One of the most promising formulations for an efficient microencapsulation of probiotic bacteria system is alginate–polylysine–alginate microcapsules. This kind of microcapsules has been successfully used for live cell therapy and other biomedical applications.[36,56-58] This microcapsule showed a very good mechanical stability and has been used to protect *L. acidophilus*. Oral administration of alginate–polylysine–alginate microcapsules incorporating *L. acidophilus* to mice resulted in a significant suppression of colon tumor incidence and tumor multiplicity, reducing the tumor size.[59]

33.7.3 Alginate–Chitosan–Alginate Microcapsules

Alginate–chitosan–alginate microcapsules were used to protect the *E. coli* strain, engineered genetically to contain the gene encoding urease for oral therapy of uremia. After 24 h of exposure to simulated gastric juice, the microcapsule remains intact with spherical shape. After 2 h of exposure to simulated gastric juice, the survival rate was much higher in trial with alginate–chitosan–alginate microcapsules 55% than for free cells (8.4%).[60]

33.7.4 Whey Protein

Besides this, other coating materials have also been used. The whey protein has been used to coat wet alginate microcapsules containing *L. plantarum*. After 90 min of exposure to simulated gastric juice, an improved survival rate of 6 log × cfu/mL was registered. The suggested rationale for this is that whey protein provides a layer around the microcapsules, which cannot be degraded by stomach.[17]

33.7.5 Palm Oil

Palm oil has been used as a coating material in order to improve the bacterial viability during gastric passage. Melted palm oil is cold sprayed around microcapsules, providing a layer on the surface of the microcapsules, reducing the rate of water uptake. The study reported by Ding and Shah[15] showed that this protective effect is not as great as that with chitosan.

33.8 OTHER APPLIED MATRICES

The cellulose acetate phthalate, xanthan gum,[47] starch,[57] and gelatin[48] were incorporated into alginate microcapsules, showing potential for improvement of probiotic survival during their passage through the gastrointestinal tract.

33.9 CONCLUSION

Microencapsulation has the potential to be a useful tool in improving the survival of probiotics during passage through gastrointestinal tract. In this study, chitosan-coated and uncoated alginate microcapsule systems have been developed to improve the stability and survivability of *L. plantarum* ATCC 8014 during passage through gastrointestinal tract.

It appears that the chitosan-coated alginate microcapsules have more effective applications for oral delivery of *L. plantarum* ATCC 8014, because the chitosan coat showed good results in terms of the survival and stability of encapsulated probiotic bacteria.

The potential of other coat materials for alginate microcapsules was also discussed. In order to advance the science of probiotic microencapsulation, it is important to use new polymers as encapsulation matrices or coating materials and to add excipients into the matrices with a view to increase the protection of bacteria and to reduce the production costs.

REFERENCES

1. Gilliland, S.E. Health and nutritional benefits from lactic acid bacteria. *FEMS Microbiol. Rev.* 1990, 7 (1–2), 175–188.
2. Fuller, R. Probiotics in man and animals. *J. Appl. Bacteriol.* 1989, 66, 365–378.
3. Daly, C.; Davis, R. The biotechnology of lactic acid bacteria with emphasis on application in food safety and human health/agricultural and food science in Finland. *Agric. Food Sci.* 1998, 7 (2), 251–265.
4. Mombelli, B.; Gismondo, M.R. The use of probiotics in medical practice. *Int. J. Antimicrob. Agents* 2000, 16 (4), 531–536.
5. Rokka, S.; Rantamäki, P. Protecting probiotic bacteria by microencapsulation: Challenges for industrial applications. *Eur. Food Res. Technol.* 2010, 231 (1), 1–12.
6. Lee, K.Y.; Heo, T.R. Survival of *Bifidobacterium longum* immobilized in calcium alginate beads in simulated gastric juices and bile salt solution. *Appl. Environ. Microbiol.* 2000, 66 (2), 869–873.
7. Brinques, G.B.; Ayub, M.A.Z. Effect of microencapsulation on survival of *Lactobacillus plantarum* in simulated gastrointestinal conditions, refrigeration, and yogurt. *J. Food Eng.* 2011, 103 (2), 123–128.
8. de Vries, M.C.; Vaughan, E.E.; Kleerebezem, M.; de Vos, W.M. *Lactobacillus plantarum*-survival, functional and potential probiotic properties in the human intestinal tract. *Int. Dairy J.* 2006, 16 (9), 1018–1028.
9. Champagne, C.P.; Gardner, N.J.; Roy, D. Challenges in the addition of probiotic cultures to foods. *Crit. Rev. Food Sci. Nutr.* 2005, 45 (1), 61–84.
10. Anal, A.K.; Stevens, W.F.; Remunan-Lopez, C. Ionotropic cross-linked chitosan microspheres for controlled release of ampicillin. *Int. J. Pharm.* 2006, 312 (1–2), 166–173.
11. McIntosh, G.H.; Royle, P.J.; Playne, M.J. A probiotic strain of *L. acidophilus* DMH-induced large intestinal tumors in male Sprague-Dawley rats. *Nutr. Cancer* 1999, 35, 153–159.
12. Chang, T.M.; Prakash, S. Therapeutic uses of microencapsulated genetically engineered cells. *Mol. Med. Today* 1998, 4 (5), 221–227.
13. Chang, T.M. Therapeutic applications of polymeric artificial cells. *Nat. Rev. Drug Discov.* 2005, 4 (3), 221–235.
14. Anal, A.K.; Singh, H. Recent advances in microencapsulation of probiotics for industrial applications and targeted delivery. *Trends Food Sci. Technol.* 2007, 18 (5), 240–251.
15. Ding, W.K.; Shah, N.P. An improved method of microencapsulation of probiotic bacteria for their stability in acidic and bile conditions during storage. *J. Food Sci.* 2009, 74 (2), M53–M61.
16. Sun, W.; Griffiths, M.W. Survival of bifidobacteria in yogurt and simulated gastric juice following immobilization in gellan-xanthan beads. *Int. J. Food Microbiol.* 2000, 61 (1), 17–25.
17. Cook, M.T.; Tzortzis, G.; Charalampopoulos, D.; Khutoryanskiy, V.V. Microencapsulation of probiotics for gastrointestinal delivery. *J. Control. Release* 2012, 162 (1), 56–67.
18. Chandramouli, V.; Kailasapathy, K.; Peiris, P.; Jones, M. An improved method of microencapsulation and its evaluation to protect *Lactobacillus* spp. in simulated gastric conditions. *J. Microbiol. Methods* 2004, 56 (1), 27–35.
19. Cui, J.H.; Goh, J.S.; Kim, P.H.; Choi, S.H.; Lee, B.J. Survival and stability of bifidobacteria loaded in alginate poly-L-lysine microparticles. *Int. J. Pharm.* 2000, 210 (1–2), 51–59.
20. Senuma, Y.; Lowe, C.; Zweifel, Y.; Hilborn, J.G.; Marison, I. Alginate hydrogel microspheres prepared by spinning disk atomization. *Biotechnol. Bioeng.* 2000, 67 (5), 616–622.
21. Gharsallaoui, A.; Roudaut, G.L.; Chambin, O.; Voilley, A.E.; Saurel, R.M. Applications of spray-drying in microencapsulation of food ingredients: An overview. *Food Res. Int.* 2007, 40 (9), 1107–1121.
22. Saarela, M.; Virkajärvi, I.; Alakomi, H.L.; Mattila-Sandholm, T.; Vaari, A.; Suomalainen, T.; Mättö, J. Influence of fermentation time, cryoprotectant and neutralization of cell concentrate on freeze-drying survival, storage stability, and acid and bile exposure of *Bifidobacterium animalis* ssp. lactis cells produced without milk-based ingredients. *J. Appl. Microbiol.* 2005, 99 (6), 1330–1339.
23. Faure, A.; York, P.; Rowe, R.C. Process control and scale-up of pharmaceutical wet granulation processes: A review. *Eur. J. Pharm. Biopharm.* 2001, 52 (3), 269–277.
24. Gombotz, W.R.; Wee, S. Protein release from alginate matrices. *Adv. Drug Deliv. Rev.* 1998, 31 (3), 267–285.
25. Evans, D.F.; Pye, G.; Bramley, R.; Clark, A.G.; Dyson, T.J.; Hardcastle, J.D. Measurement of gastrointestinal pH profiles in normal ambulant human subjects. *Gut* 1988, 29, 1035–1041.
26. Fordtran, J.S.; Walsh, J.H. Gastric acid secretion rate and buffer content of the stomach after eating. Results in normal subjects and in patients with duodenal ulcer. *J. Clin. Invest.* 1973, 52 (3), 645–657.
27. Coupe, A.; Davis, S.; Wilding, I. Variation in gastrointestinal transit of pharmaceutical dosage forms in healthy subjects. *Pharm. Res.* 1991, 8 (3), 360–364.

28. Press, A.G.; Hauptmann, I.A.; Hauptmann, L.; Fuchs, B.; Fuchs, M.; Ewe, K.; Ramadori, G. Gastrointestinal pH profiles in patients with inflammatory bowel disease. *Aliment. Pharmacol. Ther.* 1998, *12* (7), 673–678.
29. Sathyan, G.; Hwang, S.; Gupta, S.K. Effect of dosing time on the total intestinal transit time of non-disintegrating systems. *Int. J. Pharm.* 2000, *204* (1–2), 47–51.
30. Islam, M.A.; Yun, C.H.; Choi, Y.J.; Cho, C.S. Microencapsulation of live probiotic bacteria. *J. Microbiol. Biotechnol.* 2010, *20* (10), 1367–1377.
31. Thomas, S. Alginate dressings in surgery and wound management. *J. Wound Care* 2000, *9*, 56–60.
32. Joki, T.; Machluf, M.; Atala, A.; Zhu, J.; Seyfried, N.T.; Dunn, I.F.; Abe, T.; Carroll, R.S.; Black, P.M. Continous release of endostatin from microencapsulated engineered cells for tumor therapy. *Nat. Biotechnol.* 2001, *19* (1), 35–39.
33. Cirone, P.; Bourgeois, J.M.; Austin, R.C.; Chang, P.L. A novel approach to tumor suppression with microencapsulated recombinant cells. *Hum. Gene Ther.* 2002, *13* (10), 1157–1166.
34. Klein, J.; Stock, J.; Vorlop, K.D. Pore size and properties of spherical Ca-alginate biocatalysts. *Appl. Microbiol. Biotechnol.* 1983, *18* (2), 86–91.
35. Anal, A.K.; Stevens, W.F. Chitosan-alginate multilayer beads for controlled release of ampicillin. *Int. J. Pharm.* 2005, *290* (1–2), 45–54.
36. Lee, J.S.; Cha, D.S.; Park, H.J. Survival of freeze-dried *Lactobacillus bulgaricus* KFRI 673 in chitosan-coated calcium alginate microparticles. *J. Agric. Food Chem.* 2004, *52* (24), 7300–7305.
37. Kim, S.J.; Cho, S.Y.; Kim, S.H.; Song, O.J.; Shin, I.I.S.; Cha, D.S.; Park, H.J. Effect of microencapsulation on viability and other characteristics in *Lactobacillus acidophilus* ATCC 43121. *LWT—Food Sci. Technol.* 2008, *41* (3), 493–500.
38. Mokarram, R.R.; Mortazavi, S.A.; Najafi, M.B.H.; Shahidi, F. The influence of multi stage alginate coating on survivability of potential probiotic bacteria in simulated gastric and intestinal juice. *Food Res. Int.* 2009, *42* (8), 1040–1045.
39. Vodnar, D.C.; Socaciu, C.; Rotar, A.M.; Stănilă, A. Morphology, FTIR fingerprint and survivability of encapsulated lactic bacteria (*Streptococcus thermophilus* and *Lactobacillus delbrueckii* subsp. *bulgaricus*) in simulated gastric juice and intestinal juice. *Int. J. Food Sci. Technol.* 2010, *45* (11), 2345–2351.
40. Murata, Y.; Toniwa, S.; Miyamoto, E.; Kawashima, S. Preparation of alginate gel beads containing chitosan nicotinic acid salt and the functions. *Eur. J. Pharm. Biopharm.* 1999, *48* (1), 49–52.
41. Mater, D.D.; Bretigny, L.; Firmesse, O.; Flores, M.J.; Mogenet, A.; Bresson, J.L.; Corthier, G. *Streptococcus thermophilus* and *Lactobacillus delbrueckii* subsp. *bulgaricus* survive gastrointestinal transit of healthy volunteers consuming yogurt. *FEMS Microbiol. Lett.* 2005, *250* (2), 185–187.
42. Begley, M.I.; Gahan, C.G.M.; Hill, C. The interaction between bacteria and bile. *FEMS Microbiol. Rev.* 2005, *29* (4), 625–651.
43. Mortazavian, A.; Razavi, S.H.; Ehsani, M.R.; Sohrabvandi, S. Principles and methods of microencapsulation of probiotic microorganisms. *Iran. J. Biotechnol.* 2007, *5* (1), 1–18.
44. Huguet, M.L.; Neufeld, R.J.; Dellacherie, E. Calcium-alginate beads coated with polycationic polymers: Comparison of chitosan and DEAE-dextran. *Process Biochem.* 1996, *31* (4), 347–353.
45. Chen, H.; Ouyang, W.; Jones, M.; Metz, T.; Martoni, C.; Haque, T.; Cohen, R.; Lawuyi, B.; Prakash, S. Preparation and characterization of novel polymeric microcapsules for live cell encapsulation and therapy. *Cell Biochem. Biophys.* 2007, *47* (1), 159–167.
46. Iyer, C.; Phillips, M.; Kailasapathy, K. Release studies of *Lactobacillus casei* strain Shirota from chitosan-coated alginate-starch microcapsules in ex vivo porcine gastrointestinal contents. *Lett. Appl. Microbiol.* 2005, *41* (6), 493–497.
47. Sultana, K.; Godward, G.; Reynolds, N.; Arumugaswamy, R.; Peiris, P.; Kailasapathy, K. Encapsulation of probiotic bacteria with alginate-starch and evaluation of survival in simulated gastrointestinal conditions and in yoghurt. *Int. J. Food Microbiol.* 2000, *62* (1–2), 47–55.
48. Urbanska, A.M.; Bhathena, J.; Prakash, S. Live encapsulated *Lactobacillus acidophilus* cells in yogurt for therapeutic oral delivery: Preparation and in vitro analysis of alginate–chitosan microcapsules. *Can. J. Physiol. Pharmacol.* 2007, *85* (9), 884–893.
49. Vodnar, D.C.; Socaciu, C. Green tea increases the survival yield of Bifidobacteria in simulated gastrointestinal environment and during refrigerated conditions. *Chem. Cent. J.* 2012, *6* (1), 61. doi:10.1186/1752-153X-6-61.
50. Ouwehand, A.C.; Salminen, S.J. The health effects of cultured milk products with viable and non-viable bacteria. *Int. Dairy J.* 1998, *8* (9), 749–758.

51. Kailasapathy, K. Survival of free and encapsulated probiotic bacteria and their effect on the sensory properties of yoghurt. *LWT—Food Sci. Technol.* 2006, *39* (10), 1221–1227.
52. Krasaekoopt, W.; Bhandari, B.; Deeth, H. The influence of coating materials on some properties of alginate beads and survivability of microencapsulated probiotic bacteria. *Int. Dairy J.* 2004, *14* (8), 737–743.
53. Chavarri, M.; Maranon, I.; Ares, R.; Ibanez, F.C.; Marzo, F.; Villaran, M.D.C. Microencapsulation of a probiotic and prebiotic in alginate-chitosan capsules improves survival in simulated gastro-intestinal conditions. *Int. J. Food Microbiol.* 2010, *142* (1–2), 185–189.
54. Cui, J.H.; Cao, Q.R.; Lee, B.J. Enhanced delivery of bifidobacteria and fecal changes after multiple oral administrations of bifidobacteria-loaded alginate poly-L-lysine microparticles in human volunteers. *Drug Deliv.* 2007, *14* (5), 265–271.
55. Lin, J.Z.; Yu, W.; Liu, X.; Xie, H.; Wang, W.; Ma, X. in vitro and in vivo characterization of alginate-chitosan-alginate artificial microcapsules for therapeutic oral delivery of live bacterial cells. *J. Biosci. Bioeng.* 2008, *105* (6), 660–665.
56. Lanza, R.P.; Kühtreiber, W.M.; Ecker, D.; Staruk, J.E.; Chick, W.L. Xenotransplantation of porcine and bovine islets without immunosuppression using uncoated alginate microspheres. *Transplantation* 1995, *59* (10), 1377–1384.
57. Sun, A.; O'Shea, G.; Goosen, M. Injectable microencapsulated islet cells as a bioartificial pancreas. *Appl. Biochem. Biotechnol.* 1984, *10* (1), 87–99.
58. Sun, A.M.; Goosen, M.F.; O'Shea, G. Microencapsulated cells as hormone delivery systems. *Crit. Rev. Ther. Drug Carrier Syst.* 1987, *4* (1), 1–12.
59. Xie, Z.P.; Huang, Y.; Chen, Y.L.; Jia, Y. A new gel casting of ceramics by reaction of sodium alginate and calcium iodate at increased temperatures. *J. Mater. Sci. Lett.* 2001, *20* (13), 1255–1257.
60. Kitajimaa, H.; Sumidaa, Y.; Tanakab, R.; Yukib, N.; Takayamab, H.; Fujimuraa, M. Early administration of *Bifidobacteriumbreve* to preterm infants: Randomised controlled trial. *Arch. Dis. Child Fetal Neonatal Ed.* 1997, *76* (2), F101–F107.

34 Organogels as Food Delivery Systems

Tarun Garg, Goutam Rath, and Amit K. Goyal

CONTENTS

34.1	Introduction	698
34.2	Advantages of Organogels	700
34.3	Limitations/Challenges of Organogels	701
34.4	Mechanism of Organogelation	702
	34.4.1 Kinetic Approach	703
	34.4.2 Statistical Approach	703
34.5	Physicochemical Properties of the Organogels	703
	34.5.1 Biocompatibility	703
	34.5.2 Biodegradability	704
	34.5.3 Thermostability	704
	34.5.4 Thermoreversibility	704
	34.5.5 Viscoelasticity	704
	34.5.6 Nonbirefringence	705
	34.5.7 Chirality Effects	705
	34.5.8 Optical Clarity	705
	34.5.9 Mechanical Properties	705
	34.5.10 Processability	705
	34.5.11 Loading Capacity Release Kinetics	705
	34.5.12 Stability and Safety	706
34.6	Factors Affecting Organogels	706
34.7	Fabrication of Organogels	706
	34.7.1 Fluid-Filled Fiber Method	706
	34.7.2 Solid Fiber Method	708
	34.7.3 Hydration Method	708
34.8	Types of Organogelators	709
34.9	Types of Organogels	709
	34.9.1 Sorbitan Monostearate Organogels	709
	34.9.2 L-Alanine Derivative Organogels	711
	34.9.3 Eudragit Organogel	711
	34.9.4 Microemulsion-Based Organogel	711
	34.9.5 Lecithin Organogel	712
	34.9.6 Pluronic Lecithin Organogels	712
34.10	Characterization of Organogels	713
	34.10.1 Morphological Study	713
	34.10.2 Determination of Phase Transition Temperatures	713
	34.10.3 Differential Scanning Calorimetry	713

34.10.4	Scattering Techniques	713
34.10.5	Swelling Studies	714
34.10.6	Structural Features	714
34.10.7	Homogeneity Test	714
34.10.8	Determination of the Solvent Absorbency	714
34.10.9	Stability Testing	714
34.10.10	Spreadability	715
34.10.11	Ball Indentation	715
34.10.12	Uniaxial Tensile Testing	715
34.10.13	Viscoelasticity	715
34.10.14	pH Determination	715
34.10.15	Active Constituent Content	715
34.10.16	In Vitro Release Studies	716
34.10.17	In Vitro Permeation Study	716
34.10.18	Antimicrobial Test	716
34.11	Potential Food Applications of Organogels	716
34.11.1	Novel Organogel Compositions in Foods, Beverages, Nutraceutical, Pharmaceuticals, Pet Food, or Animal Feed	716
34.11.2	Replacement of Saturated Animal Fat	717
34.11.3	Organogels as an Oil Structuring Alternatives to the Fat Crystal Network	718
34.11.4	Self-Assembly Network of Organogels Has Been Proposed as Oil Migration Inhibitor for Chocolate Products	719
34.11.5	Organogel for Capturing Heavy Metals	722
34.12	Conclusions	722
References		722

34.1 INTRODUCTION

Food ingredients and nutraceutical components cannot simply be incorporated into foods in their regular forms due to chances of physical, chemical, or biological degradation. So, it is necessary to encapsulate them first in a suitable delivery system before they can be successfully introduced into a food matrix.[1] A wide variety of improved technologies are available in the food industry for the encapsulation, protection, release, and enhanced bioavailability of food ingredients and nutraceutical components that are vital to the development of future foods.[2] A variety of different materials such as carbohydrates, proteins, lipids, minerals, surfactants, and water are used for the fabrication of delivery systems for active food components by using a range of different processing operations.[3] A huge number of delivery systems with different compositions and structure have been developed, but only few possible systems are used in the food industry. In the food industry, ingredients of delivery systems must be of food grade, inexpensive, and fabrication technique must be economical, reproducible, and robust.[4]

Why food components are encapsulated?

- To protect against chemical, physical, and biological degradation
- To improve storage, handling, and utilization
- To extend their shelf-life
- To successfully incorporate them into the food matrix
- To remove flavors
- To deliver them to a particular site of action
- To control their adverse effects on quality attributes
- To exhibit their maximum therapeutic activity

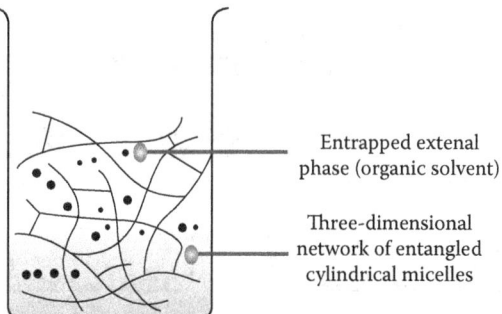

FIGURE 34.1 Structure of organogels.

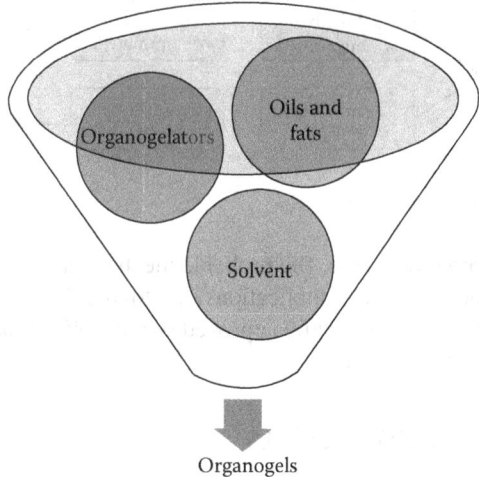

FIGURE 34.2 Essential components of organogels.

Organogel (oils and fats), with potential food applications, has only very recently been developed as a delivery system. It can be defined as an organic liquid entrapped within an anhydrous thermoreversible, three-dimensional supramolecular gel networks[5] (Figure 34.1). Organogel is a bicontinuous, semisolid formulation having an external apolar solvent phase immobilized within the spaces available in a three-dimensional networked structure consisting of the solid component, known as gelator. Several organogelators (responsible for the gelation) are able to gel oils at concentrations below 10% or as low as 0.5%–2% in some cases.[6] General categories of network-forming edible oils include fatty acid, fatty alcohol, lecithin, phytosterols, monoacylglycerol, diacylglycerol, triglycerol, sorbitan monosterate, waxes, wax esters, oryzanol, etc. Organogelators such as ethylcellulose (EC), ceramides, stearic acid, stearyl alcohol, CLW, 12-hydroxy stearic acid, and ricinelaidic acid are well known for their ability to structure edible oils (Figure 34.2). Among these, EC, β-sitosterol, γ-oryzanol, or mixed ceramides have been successfully used in the food systems.[7] Organogel consists of macromolecules in the form of twisted matted strands in which each unit is bound together by strong molecular van der Waal forces, hydrogen bond, and chemical bond, leading to the formation of crystalline amorphous regions throughout the solid phase. Due to this property, the members of the organogel do not form semisolid phase upon standing condition.[8] Figure 34.3 presents a flow chart compiling various accepted classification of organogels based on the nature of gelator, oils, solvents, and intermolecular interactions.

The motivation to compile this chapter mainly arose from the observation of excellent publications regarding the application of organogels for drug delivery systems. There is a huge knowledge

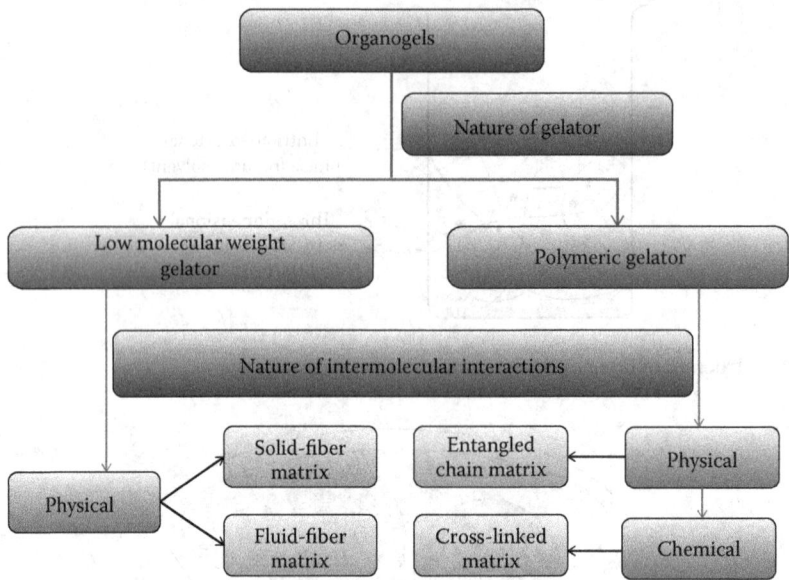

FIGURE 34.3 Classification of organogels.

on the theory of different organogelators, the probable mechanisms of their formation, their characterization techniques, and numerous applications in food science. The development of food grade and GARS (generally regarded as safe) approved organogelators has opened the potential for organogel use in a variety of food products.

34.2 ADVANTAGES OF ORGANOGELS

There are several advantages associated with the organogels, which has drawn the interest of researchers toward this wonderful carrier system with respect to drug, food, cosmetics, and other applications.

- The structural integrity of organogels is maintained for longer time period, so it does not revert back to semisolids on standing.
- The apolar phase being organic, the molds do not grow even if the gel becomes contaminated.[9]
- Organogels offer opportunities for absorption of extensive range of bioactive molecules, specially food ingredients with different physiochemical characters such as chemical nature, solubility, molecular weight, size, etc.
- Organogels are thermodynamically stable, so they can be kept in a closed vessel for a very long time.[10]
- Structural integrity of organogels is preserved for longer duration of time period.
- Organogel materials are safe for longer duration of period due to its biocompatible, biodegradable, and nonimmunogenic nature.
- Naturalness of organogels formation by quality of self-assembled supermolecular organization of surfactant molecule.
- Both hydrophobic and hydrophilic food substances can be successfully incorporated into the organogels system.[11]
- They are less slippery and can be easily detached or removed from the skin or target site.
- Enhancement of skin penetration and effective transport of the molecules into the skin.
- Existence of well-balanced hydrophilic and lipophilic characters.

Organogels as Food Delivery Systems

- Various types of substances are successfully incorporated into these systems.
- Poorly water-soluble food ingredients can be easily formulated by using this system.[12]
- Resistance from moisture and microbial contamination.
- Efficient partition within the skin.
- Easier quality monitoring.
- Ease of administration.
- Avoidance of first-pass effects.
- They show site-specific food or other substances to overcome systemic adverse effects.
- They are safer to use and can obtain regulatory approval easily.[13]

34.3 LIMITATIONS/CHALLENGES OF ORGANOGELS

Despite the advantages of organogels, it suffers from some limitations or challenges, which mainly affects the stability and therapeutic performance of the organogels. So, many efforts are undertaken in this field to control the same. Some important limitations or drawbacks along with their possible solution of these systems are discussed in the text following:

- Gelling will not be proper if any contaminant is present in the substances of organogels.
- Difficult to predict the molecular structure of a potential gelator and its preferential gelling solvent.[14]
- They are expensive and require special storage conditions.
- Some organogels are greasy by nature.
- They show instability in higher temperature conditions.
- They are difficult to obtain in large quantities.
- In some cases, they often shrink naturally if the gel stands for longer durations.
- Swelling (when the gel takes up increasing volumes of liquid) and syneresis (some of its liquid is pressed out) problems are also associated with these systems.[15]
- The possibility of local irritation and skin's low permeability.
- Biobased porous organogels and organogel-based materials are quickly saturated on their surface and change their density, their porosity as well as their polarity.
- They are hard to recycle, which leads to additional costs.
- Properties of organogels mainly depend upon the quality of organogelators, oils and fats, solvents, and various processing parameters such as temperature, pH, charge, etc.[16]

Today, the discovery of gelator remains serendipitous and is usually followed by screening of different solvent systems potentially compatible with gelation. Prognosis of gelation potential of a given molecule is due to its propensity toward chemical or physical intermolecular interactions; however, no generalizations have been possible so far. Many factors such as steric effects, rigidity, and polarity can counter the molecule's aggregating tendency. Control over the gelation process as well as the conception of new gelling molecules remain important challenges while making new organogelators.[17] Another challenge of organogels is the phenomenon of syneresis (when a gel stands for some time, it often shrinks naturally, and some of the liquid within is pressed out); this at times limits the use of these gels in food applications. Thus, it is important to better understand the structural factors that influence syneresis in an attempt to minimize it. Syneresis may be due to the two-stage nature of the organogelators (e.g., 12-hydroxystearic acid [12HSA]) gelation process. Initially, there is a rapid increase in the opacity, followed by a slight decrease over a period of time. This decrease in opacity has been attributed to the fiber agglomeration into bundles resulting in larger aggregates, as well as larger pores.[18] Fiber agglomeration may affect both gel strength and solvent-holding capacity. Many studies on hydrogels have supported the fact that the gel strength and water-holding capacity are directly related to the microstructure of the gel. In agar gels, for example, lower gel strength is correlated with more severe syneresis. Therefore, it is important to keep in mind the oil-binding capacity and crystallinity of the gel matrix, which may change with regard to organogelators concentration, storage temperature, and time.[19] The gel properties of organogels can be tuned to an extent, that is, the introduction of new functionality requires covalent modification. The surface of currently used adsorbent, biobased porous organogels-based materials gets quickly saturated, thus limiting the trapping of fluids in high quantity.[20] Their density, their porosity as well as their polarity cannot be controlled; this sometimes raises concerns for optimization. Currently available adsorbent porous materials are either mineral-based (activated carbon, zeolite, clay) or synthetic polymers, which makes it hard to recycle, leading to additional costs. Moreover, the trapped products are known to be hard to extract from said materials. It is also noteworthy to mention that these organogels have greasy properties and are less stable to temperature and henceforth need proper storage conditions.[21] Presently, however, there are no known industrial applications of organogels in the food industry due to many issues. These include the lack of low-cost and effective self-assembled fibrillar networks, which are food grade, as well as a lack of apprehension of the mechanistic aspects of these systems, which is critical to the control of the gelation process. It is the gelation process that defines the gel structure and thus the appearance, structure, and stability of the material.[22]

34.4 MECHANISM OF ORGANOGELATION

The fundamental mechanism of organogels occurs due to the addition of a polar solvent to a non-polar solvent containing organogelators. There are two possible mechanisms that are proposed for the formation of organogels depending upon their physical intermolecular interactions, namely, the fluid-filled fiber and the solid fiber mechanism. The main difference is in the starting materials, that is, surfactant in apolar solvent versus solid organogelators in apolar solvent.[23] Surfactant or surfactant mixture forms reverse micelles when mixed with an apolar solvent. When a polar solvent (e.g., water) is added to the reverse micelles, it forms tubular reverse micelle structures. As more polar solvent is added, the reverse micelles elongate and entangle to form organogels. On the other hand, gel formation via solid-fiber matrix forms when the mixture of organogelators in apolar solvent is heated to give apolar solution of organogelators and then cooled down below the solubility limit of the organogelators. The organogelators precipitate out as fibers, forming a three-dimensional network, which then immobilizes the apolar solvent to produce organogels[24] (Figure 34.4).

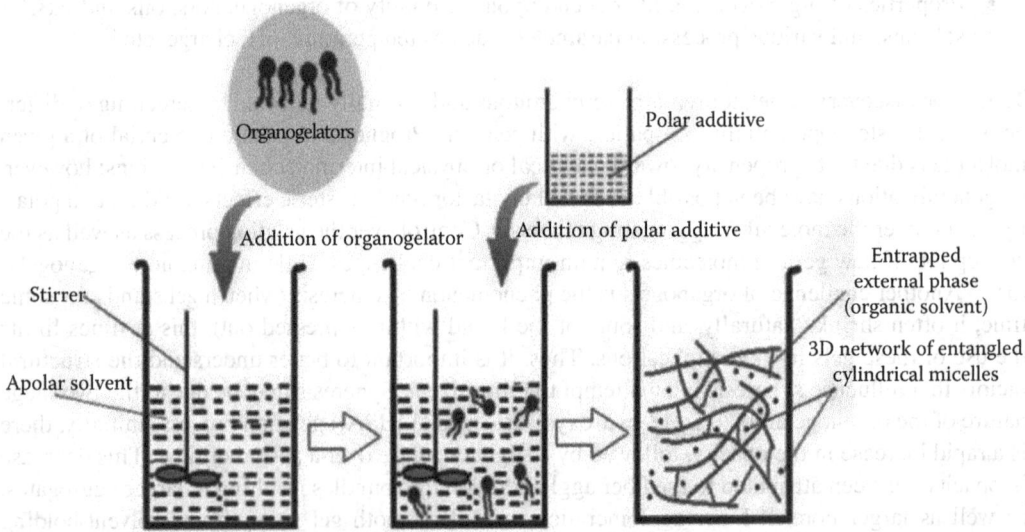

FIGURE 34.4 Mechanism of organogelation.

The gel formation depends upon the accurate prediction of gelation parameters (such as time, rate, and structure of organogelators, its nature, and type of organic solvent used) of a broad range of materials, which is highly sought after for both monetary and intellectual reasons. Utilizing mathematical approaches, there are two approaches that entail the gelation mechanism.

34.4.1 Kinetic Approach

The kinetic is also known as coagulation approach, which preserves the integrity of all structures formed during network creation. In general, the application of kinetic treatment for gelation although results in large, unwieldy, and dense sets of equations, but the answers given by the are discernibly better than those given by the statistical approach.[25] A major drawback associated with the kinetic approach is the assumption of the gel as essentially one giant, rigid molecule, and cannot actively simulate characteristic structures of gels such as elastic and dangling chains.[26]

34.4.2 Statistical Approach

This is currently the most widely used approach. The statistical approach views the phase change from liquid to gel as a uniform process throughout the fluid. There is promotion of equal probability of polymerization reactions to occur throughout the solution. Statistical theories try to determine the fraction of the total possible bonds that need to be made before an infinite polymer network can appear.[27] The classic statistical theory first developed by Flory was based on two critical assumptions. No intermolecular reactions occur. That is, no cyclic molecules form during polymerization leading up to gelation. Every reactive unit has the same reactivity regardless of other factors. For example, a reactive group –B– on a 50-mer (a polymer with 50 monomer units) has the same reactivity as another group –B– on a 5000-mer.[28]

Based on these two approaches, the gelation process of organogels is assumed to take place upon the addition of trace amount of water into the apolar phase containing the organogelators. Initially, the organogelator molecules are randomly dispersed in the apolar phase. The addition of water causes the organogelators molecules to assemble in a spherical reverse micellar form. Further addition of water causes the aggregation of these short tubular or cylindrical micelles.[29] The water molecules bind stoichiometrically to the hydrophilic head of the organogelators molecules acting as a bridge between two adjacent organogelators molecules to form the linear networks with the hydrogen bonds between polar molecules and the negative group of an organogelators molecule. A further increase in a small amount of water results in the formation of long, flexible, and worm-like tubular micellar structure. Thousands of such tubular microstructures overlap and entangle with each other to form a three-dimensional gel network, which possesses viscoelasticity and thermoreversibility. The apolar phase gets entrapped in the spaces between the entangled reverse micelles.[30]

34.5 PHYSICOCHEMICAL PROPERTIES OF THE ORGANOGELS

Organogels play a critical role in drug as well as food delivery. It protects the active component, extends its stability, and offers opportunities for absorption of a wide range of bioactive molecules, specially food ingredients with different physiochemical characters such as chemical nature, solubility, molecular weight, size, etc., due to its various physicochemical properties[31] (Figure 34.5). This section attempts to discuss about the various physicochemical properties of the organogels.

34.5.1 Biocompatibility

Biocompatibility is the capability of the delivery system to perform in a specific application without provoking a harmful immune or inflammatory reaction. The organogels possess acceptable biocompatibility and toxicity profile. Initially organogel is nonbiocompatible, because it was developed by noncompatible constituents but now research on organogels using various biocompatible constituents has opened up new magnitudes for the use of the same in various biomedical as well as food applications.[32]

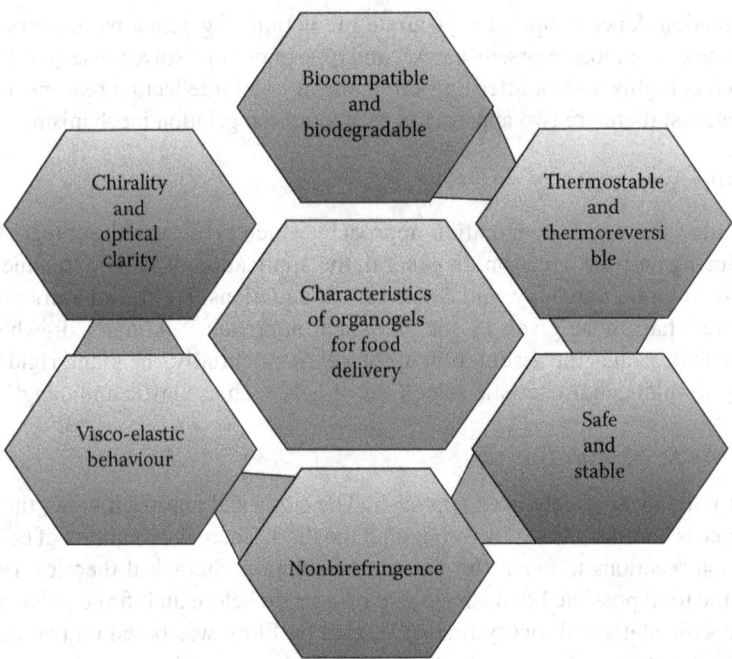

FIGURE 34.5 Schematic showing characters of organogels required for effective food delivery.

34.5.2 BIODEGRADABILITY

The organogel material is biodegradable in nature and the degradation product is free of any toxic substance and is easily eliminated from the body. The degradation rate can be controlled by using suitable various biocompatible constituents and advanced technologies.[33]

34.5.3 THERMOSTABILITY

The organogels are intrinsically thermostable in nature. The gelators have the ability to undergo self-assembly under suitable conditions, so there is a decrease in the total free energy of the system and this renders the organogels a low-energy thermostable system. Due to this property of the organogels, they have been anticipated as a delivery vehicle for bioactive agents and for food applications where a longer shelf-life is desirable.[34]

34.5.4 THERMOREVERSIBILITY

As the organogels are heated up above a critical temperature, there is a disruption in the physical interactions among the gelator molecules due to the increase in the thermal energy within the organogels, so they lose the solid matrix-like structure and start flowing. But as the heated organogels systems are consequently cooled down, the physical interaction between the organogelators is overcome and the organogels return to the more stable alignment.[5]

34.5.5 VISCOELASTICITY

The organogel materials having both viscous and elastic properties specially follow Maxwell model of viscoelasticity. At low shear rate, organogels behave like a solid, but at high shear rate, the physical interacting points among the fiber structures start getting weakened and organogels start flowing. This behavior may be best explained by the plastic flow behavior.[35]

34.5.6 NONBIREFRINGENCE

When the polarized light does not pass through the matrix, it is regarded as nonbirefringent. Organogels do not allow the polarized light to pass through the matrix and appears as a dark matrix that is regarded as nonbirefringent. This can be accounted to the isotropic nature of the organogels, which does not allow the polarized light to pass through the matrix.[36]

34.5.7 CHIRALITY EFFECTS

The presence of chirality in the gelators mainly affects the growth and the stability of the solid-fiber networks as well as thermoreversibility of the gels. It also helps in the formation of a compact molecular packing, which provides a thermodynamic and kinetic stability to the organogels system. It has been found that a good solid-fiber gelator has a chiral center, whereas chirality does not have any effect on fluid-fiber gelators.[37]

34.5.8 OPTICAL CLARITY

Depending on the composition of the organogels, the organogels may be transparent or opaque in nature. The lecithin organogels are transparent in nature, while the sorbitan monostearate organogels are opaque in nature.[38]

34.5.9 MECHANICAL PROPERTIES

Mechanical properties of the organogels match that of the tissue at the implantation site, or are sufficient to shield encapsulated active constituents from damaging compressive or tensile forces without inhibiting appropriate biomechanical cues and to survive under physiological conditions. Immediately after administration, the organogels provide a minimal level of biomechanical function that should improve progressively until normal tissue function has been restored, at which point the construct should have fully integrated with the surrounding host tissue.[39]

34.5.10 PROCESSABILITY

The organogels processability relatively easy and malleability into the desired shape as according to the need. They are capable of being produced into a sterile product. Spontaneity of organogels formation by virtue of self-assembled supermolecular arrangement of surfactant molecule makes the process very simple and easy to handle.[40]

34.5.11 LOADING CAPACITY RELEASE KINETICS

Loading capacity release kinetics is defined as the amount of active substance that can be mixed into the delivery system. The organogels have a maximum loading capacity, so the active constituents are released continuously for longer duration after insertion into the body. The active constituents are dispersed homogenously throughout the delivery system or in discrete areas and avoid an initial burst effect. They have well-balanced hydrophilic and lipophilic character. Their micellar structure can contain both water-soluble and oil-soluble ingredients. Both hydrophobic and hydrophilic substances can be delivered. It provides opportunities for incorporation of wide range of substances with different physiochemical characters. The release of substances from the organogels show controlled release behavior to allow the appropriate dose of substance to reach the cells over a given period of time.[41]

34.5.12 Stability and Safety

The stability of the incorporated active substances at physiological temperature with respect to physical, chemical, and biological activity is to be considered. They possess dimensional stability, chemical stability, and biological activity over a prolonged period of time. Use of biocompatible, biodegradable, and nonimmunogenic material makes them safe for longer term applications. Organic solvents could be of natural origin: for example, sunflower oil, mustard oil, etc. They are safer to use and subject to chances of easy regulatory approval.[42]

34.6 FACTORS AFFECTING ORGANOGELS

There are various physicochemical parameters (Figure 34.6) that can be varied over a large range to produce organogels, which have the desired property depending upon the active substances to be delivered or the site at which it is to be given and other factors. The factors affecting organogels have to be monitored carefully in order to optimize the formulation. Table 34.1 describes the list of such factors and their effects.

34.7 FABRICATION OF ORGANOGELS

Preparation of an organogel involves the simple step of initially heating the apolar phase with a gelator to form an organic solution/dispersion to allow complete dissolution, followed by cooling. This decreases the solubility of the gelator in the apolar phase due to the reduction in the gelator–solvent interactions, resulting in the gelator molecule coming out of the solution forming a gel.[56] Gelator–gelator interactions lead to gelator self-assembly into well-defined aggregates such as tubules, rods, and fibers. The physical organogels are held together by noncovalent forces. The application of heat from external source causes the gel to transform its phase to sol state due to the dissolution of the gelator aggregate in the organic liquid. Whereas, cooling the hot sol phase results in gelation.[57] The methods used for preparation of an organogels can be broadly classified as:

34.7.1 Fluid-Filled Fiber Method

This method (Figure 34.7) involves the dissolution of surfactant and cosurfactant mixtures in apolar solvent, resulting in the formation of reverse micelles. Thereafter, addition of water forms tubular

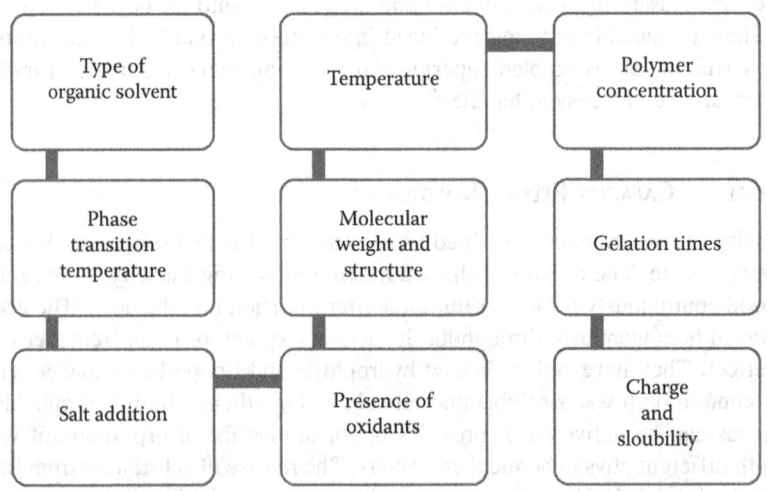

FIGURE 34.6 Factors affecting organogels.

TABLE 34.1
Various Parameters Affecting the Property of Organogels

S. No.	Factors	Effects on Organogel's Property
1.	Type of organic solvent	• Determines the strength and possibility of organogelation. • Polar solvent increases the cross-sectional area of the spherical micelles upon incorporation. • Nonaqueous solvent (polyethylene glycol) replaces water of the bacterial cellulose hydrogel without destroying it's shape.[43]
2.	Phase transition temperature (PTT)	• It gives an insight into nature of microstructures that form the gelling cross-linked network. For example, a narrow PTT range is indicative of homogenous microstructures within the gel. • For determination of PTT, hot-stage microscopy and high-sensitivity differential scanning calorimetry are the accurate and sensitive techniques.[44]
3.	Salt addition	• Salt may attract part of water of hydration of the polymer, allowing more formation of intermolecular secondary bond, this is known as salting-out.[45]
4.	Temperature	• The effect of temperature depends on the chemistry of the polymer and its mechanism of interaction with the medium. • If the temperature is reduced once the gel is in the solution, degree of hydration is reduced and gelation occurs.[46]
5.	Molecular weight	• Molecular weight of solvent affects gel formation. • Low molecular weight polymer requires a high concentration to build up viscosity and set to gel possibly.[47]
6.	Presence of oxidants	• Strong oxidants such as cerium (IV) ammonium nitrate and weak oxidants like nitric oxide, NO, can induce gelation. • An oxidation-induced planarization to trigger gelator self-assembly and gelation through donor–acceptor π-stacking interaction.[48]
7.	Molecular structure	• The self-assembly capability could be customized by modifying the chemical structure of the polymer backbone.[49]
8.	Polymer concentration	• Polymeric organogelators can induce gelation even at very low concentrations (less than 20 g/L).[50]
9.	Gelation times	• Varies depending on the organogelators and medium. • One can promote or delay gelation by influencing the molecular self-assembly of organogelators in a system.[51]
10.	Surfactants	• Gel characteristics can be varied by adjusting the proportion and concentration of the ingredients. • Poloxamer 407 is a polyoxyethylene that functions as a surfactant.[52]
11.	Charge	• The presence of charged groups on a polymer favors mucoadhesion. • Polyanions, particularly polycarboxylates, are preferred to polycations.[33]
12.	Solubility	• Mucoadhesive swells on contact with moisture, increasing the mobility of polymer molecules at the interface and exposing more sites for bond formation.[54]
13.	Spatial configuration	• It favors change in entanglement and interaction after the polymer and mucin has interpenetrated.[55]

reverse micelles. The elongated tubular reverse micelle gets entangled to form a three-dimensional network, which immobilizes apolar solvent after the addition of water into tubular reverse micelles. The main advantages of this technique are easy processing, cheap, time saving, and no need of expert for fabrication of organogels through the same. However, its quality is mainly dependent upon the nature of surfactant and the cosurfactant. Maintaining the formed reverse micelles for longer

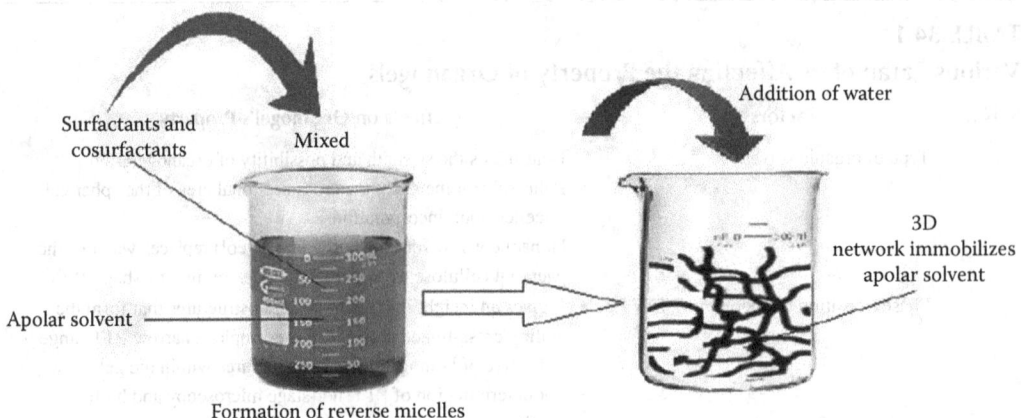

FIGURE 34.7 Organogel fabrication by fluid-filled fiber technique.

duration is also a challenging task.[58] Bhattacharya et al. developed span 80–tween 80 mixture-based organogels for the first time by fluid-filled fiber mechanism. In this technique, span 80 and tween 80 were used as surfactant and cosurfactant, respectively. The surfactant mixtures were dissolved in oil followed by the addition of water, which led to the formation of organogels at specific compositions. Microscopic results indicated that the gels contained clusters of water-filled spherical structures. XRD study indicated the amorphous nature of the organogels and these were found to be stable enough to be used in food as well as pharmaceutical formulation.[59]

34.7.2 Solid Fiber Method

Apolar solvent and solid organogelators are mixed together and heated to effect dissolution of gelators in the liquid. After cooling to room temperature, organogelators precipitate out as fibers that undergo physical interactions (Figure 34.8) among one other, thereby forming a three-dimensional network structure, which immobilizes apolar solvent. The physical organogels, held together by noncovalent forces, are thermoreversible.[1] This technique is simple, comparatively inexpensive, and produces stable organogels as compared to fluid-filled fiber technology. But organogelators may influence the property of organogels, and heating as well as cooling parameters also affects the strength of organogel network.[60] Hailong et al. developed a food-grade organogel loaded with curcuminoids compound, this is a well-known nutraceutical manufactured by using solid fiber technology. In this technique, monostearin functioned as an organogelator to form organogel and stabilize the metastable curcuminoids. This organogel has a stand-alone potential for curcuminoids delivery as well as to be used as the oil phase of emulsions for downstream functional foods application.[61]

34.7.3 Hydration Method

This method involves the direct hydration of the inorganic chemical, which produces dispersed phase of the dispersion. In addition to water as a vehicle, other agents such as propylene glycol, propyl gallate, and hydroxyl propyl cellulose may be used to enhance gel formation. The major advantages of this technique are easy, inexpensive, reproducible, accurate, stable, and structure network maintenance for longer duration of period but chances of contamination is more due to use of inorganic chemicals and water solvent.[62] Richards et al. prepared pluronic lecithin organogels (PLOs) as a base by hydration technique for the delivery of bioactive polyunsaturated fatty acids from fish oil, eicosapentaenoic acid (EPA), and docosahexaenoic acid (DHA). The result showed that a PLO-based gel is capable of delivering EPA and DHA via a repeat finite dosing regimen.[63]

FIGURE 34.8 Organogel fabrication by solid fiber technique.

34.8 TYPES OF ORGANOGELATORS

The role of organogelators in the design and stability of organogels is proved from the earlier discussion. The organogelators may be categorized into two groups based on their capability of hydrogen bounding. Some organogelators form hydrogen bounding (including amino acids, amide, urea moieties, and carbohydrates) and some do not (including anthracene, anthraquinone, and steroid based molecules).[64] Several organogelators (responsible for the gelation) are able to gel oils at concentrations below 10% or as low as 0.5%–2% in some cases. The following organogelators such as EC, ceramides, stearic acid, stearyl alcohol, candelilla wax, 12-hydroxy stearic acid, and ricinelaidic acid are well known for their ability to structure organogels. Among of these, EC, β-sitosterol, γ-oryzanol, or mixed ceramides are successfully used in the food.[65] Table 34.2 represents the various properties of organogelators and their effects on characteristics of prepared organogels.

34.9 TYPES OF ORGANOGELS

The last decade has seen an increase in researchers devoted to the discovery and synthesis of plethora of diverse molecules of physical organogels, which can gel organic solvents at low concentrations. The gelator molecules immobilize large volumes of liquid following their self-assembly into a variety of aggregates such as rods, tubules, fibers, and platelets. It is due to the various interesting properties of these gels that have found applications in industrial uses.[73] Still, only a few organogels are now being studied as delivery vehicles. Some of these are discussed in the following.

34.9.1 Sorbitan Monostearate Organogels

A sorbitan monostearate (hydrophobic nonionic surfactant) gel involves a number of organic solvents such as hexadecane, isopropyl myristate, and vegetable oils. Firstly, organogelators are dissolved or dispersed in hot apolar solvent to form an organic solution/dispersion and gelation occurs after subsequent cooling. Heating causes complete dissolution of organogelators, whereas cooling decreases the solvent–gelator affinities, such that at the gelation temperature, the surfactant molecules self-assemble into toroidal inverse vesicles. Further cooling also results in the conversion of the toroid's into rod-shaped tubules.[76] These tubules then aggregate to form a three-dimensional network, which immobilizes the solvent. Sorbitan monostearate gels are opaque,

TABLE 34.2
Properties of Organogelators and Their Effects on Prepared Organogels

Types of Organogelators	Properties of Organogelators	Properties of Synthesized Organogel
4-tertbutyl-1-aryl Cyclohexanol derivatives (arylcyclohexanol derivatives)[66]	Solid at room temperature, low solubility in apolar solvent	Either transparent or turbid mainly depends on the type of the apolar solvent
Polymeric gelators (e.g., poly(ethylene glycol), polycarbonate, polyesters, poly(alkylene))[67]	Induce organogelation even at very low concentrations, Low sol-gel processing temperature	Good gel strength, excellent mechanical properties
Gemini gelators (e.g., N-lauroyl-L-lysine ethyl ester)[68]	High ability of immobilizing apolar solvents	Good mechanical properties, good organogelation property
Boc-Ala(1)-Aib(2)-β-Ala(3)-OMe (synthetic tripeptide)[69]	Capability to undergo self-association	Produce thermoreversible transparent gel
Low molecular weight gelators (e.g., fatty acids and n-alkanes)[70]	High ability of immobilizing apolar solvents at small concentration (<2%), produce either solid-fiber matrix or fluid fiber matrix	Good mechanical properties, excellent gel strength
(R)-12-Hydroxystearic acid (HAS)[71]	Low molecular weight, undergoes self-assembly when crystallized in oil	Form supramolecular structure, good mechanical properties, excellent gel strength
Candelilla wax[72]	As a food additive, candelilla wax has the Enumber E 902 and is used as a glazing agent	Produce thermo reversible transparent gel, Increase the gel strength
β-Sitosterol[60]	Solid at room temperature, low solubility in apolar solvent	Produce thermoreversible transparent gel, good organogelation property
γ-Oryzanol[60]	Produce either solid-fiber matrix or fluid fiber matrix, low molecular weight	Produce thermoreversible transparent gel, increased gelation rate, excellent mechanical properties
Ceramides[73]	Family of waxy lipid molecule, Capability to undergo self-association	Good mechanical properties, excellent gel strength
EC[74]	Soluble in a nonpolar substance such as a vegetable oil	Increase the gel strength, replace saturated fat in a variety of food products

thermoreversible semisolids, and they are stable at room temperature for weeks. Murdan et al. prepared organogels by dissolving/dispersing the sorbitan monostearate (10%w:v) as gelator in the organic solvent at 60°C and allowing the resulting solution/suspension to cool. Cooling the sol phase resulted in reduction in the solubility of the gelator with consequent lowered solvent–surfactant affinities.[77] As a result, the surfactant molecules self-assembled into aggregates, which interact forming a three-dimensional network that immobilizes the solvent and thus caused gelation. The resulting gels were found to be opaque, thermoreversible semisolids, with a smooth *silky* feel, while their microstructures (as revealed by light microscopy) consist of tubular aggregates dispersed in the fluid phase.[78] It is noteworthy to mention that the organogels so formed are not genuine due to the existence of additives such as the hydrophilic surfactant, polysorbate 20, which increases gel's stability and modifies the gel microstructure.[79] Correspondingly, sorbitan monopalmitate also gels organic solvents to give opaque thermoreversible semisolids. Like sorbitan monostearate gels, the microstructure of the palmitate gels comprises of an interconnected network of rod-like tubules. Unlike the stearate gels, however, the addition of small amounts of a polysorbate monoester causes a large increase in tubular length instead of the "clustering effect" seen in stearate gels.[78] Hailong et al. developed food-grade organogels by using Span 20 (sorbitan monolaurate), Span 40 (sorbitan monopalmitate), Span 60 (sorbitan monostearate), and Span 80 (sorbitan mono-oleate) with high concentration of curcuminoids. The result showed that the bioaccessibility of curcuminoids was not affected by the formation of organogel, and the fasted state generated higher bioaccessibility than the fed state. This organogel has a stand-alone potential for curcuminoids delivery as well as to be used as the oil phase of emulsions for downstream functional foods application.[61]

34.9.2 L-Alanine Derivative Organogels

In situ L-alanine derivative organogel is prepared using N-lauroyl-L-alanine methyl ester (LAM), which predominantly gels in organic solvents such as soybean oil and medium-chain triglycerides. Normally, these systems exist in the gel state at room temperature, but upon addition of ethanol to a gelator/solvent solution, they cause inhibition of gelation. This is because the ethanol disrupts the formation of hydrogen bonds (essential for gelator self-assembly into aggregates) between the gelator molecules.[33] This means that a solution of LAM in an organic solvent can remain in the sol phase at room temperature when some ethanol is added to the mixture. But, when this sol phase (20% LAM plus 14% ethanol in soybean oil) was placed in phosphate-buffered saline at 37°C, it turned into an opaque gel within 2 min owing to the diffusion of the hydrophilic ethanol into the aqueous buffer and due to the formation of gelator–gelator hydrogen bonds.[32] Thus, theoretically, such a LAM/ethanol/soybean oil solution could form gels in situ following its subcutaneous injection, due to ethanol diffusion away from the formulation, into the surrounding tissues. The main advantage of in situ forming gels is their injectability at room temperature. Once a drug-containing gel is formed in situ, it could act as a sustained-release implant.[36]

34.9.3 Eudragit Organogel

Eudragit organogels are generally the mixtures of Eudragit (L or S) and polyhydric alcohols, such as glycerol, propylene glycol, and liquid polyethylene glycol, containing high concentrations (30% or 40% w/w) of Eudragit.[80] Active ingredient–containing gels were prepared by dissolving it in propylene glycol, pouring the resulting solution into Eudragit powder (contained in a mortar), and immediately mixing with a pestle for 1 min. Gel viscosity was found to increase with increasing concentration of Eudragit and to decrease with increasing content of loading active ingredient. The loading of Eudragit organogels should be kept low to maintain gel rigidity and stability.[81]

34.9.4 Microemulsion-Based Organogel

Microemulsion-based gels are prepared by initially dissolving solid gelatin in a hot w/o microemulsion (which was composed of water and isooctane) followed by cooling. The gelatin would dissolve in the water droplets of the w/o microemulsion and that cooling of the system would result in gelation of the water droplets, which would lead to clouding of the system and possibly phase separation. Thus, microemulsion gelled to a transparent semisolid with a high viscosity and a high electroconductivity.[82] Yu and Huang developed novel organogel-based nanoemulsions for oral delivery of curcumin and improvement of its bioavailability. In this study, curcumin organogel was used as the oil phase in the curcumin nanoemulsions formulation. Tween 20 was selected as the emulsifier on the basis of maximum in vitro bio accessibility of curcumin in the nanoemulsions. The in vivo pharmacokinetics analysis on mice confirmed that the oral bioavailability of curcumin in the nanoemulsions was increased by nine-fold compared with unformulated curcumin. This novel formulation approach may also be used for oral delivery of other poorly soluble nutraceutical with high loading capacity, which has significant impact on functional foods, dietary supplements, and pharmaceutical industries.[41]

34.9.5 LECITHIN ORGANOGEL

Lecithin is an important component of all living cells and is recognized by Food and Drug Administration as GRAS. Its unique lipid molecular structure performs versatile functions. It has a wide variety of roles in pharmaceuticals, cosmetics, and food industries as an emulsifier, viscosity modifier, stabilizer, solubilizer, and penetration enhancer. Lecithin involves phospholipid molecules, which are self-assembled to form the microstructure of organogel.[83] The unsaturated nonpolar part of lecithin molecules distress the critical packing parameter and assist in the enlargement of reverse micellar structures followed to three-dimensional long tubular networks. Lecithin organogels (LOs) are thermodynamically stable, clear, viscoelastic, biocompatible, and isotropic gels, which immobilizes the continuous or macroscopic external organic phase, thus turning a liquid into a gel.[84] A LO is formed when small amounts of water or other polar substances, such as glycerol, ethylene glycol, or formamide, are added to a nonaqueous solution of lecithin. The organogel formation of lecithin takes place with the initial step of gelation, which occurs with the addition of a trace amount of water to the lecithin solution of organic solvent. Initially, the lecithin molecules are randomly dispersed in the organic medium. With the addition of small amount of water, the lecithin molecules orient in spherical reverse micellar form. Additional water makes the short tubular or cylindrical micellar aggregate.[85] The water molecules bind stoichiometrically to the hydrophilic head of the lecithin molecules and act as a bridge between two adjacent lecithin molecules to form the linear networks with the hydrogen bonds between polar molecules and phosphate groups of lecithin molecules. A further increase in a small amount of water results in the formation of long, flexible, and worm-like tubular micellar structures. Thousands of such tubular microstructures overlap and entangle with each other to form a three-dimensional gel network, which possesses viscoelasticity and thermoreversibility.[86] The organic liquids get entrapped in the spaces between the entangled reverse micelles. Despite the advantages of using LOs, there are some potent disadvantages. LOs require lecithin of highest purity, or else no gelling will occur. This in turn increases the cost of manufacturing. It is also difficult to obtain in large quantities inclusion of pluronics as cosurfactant, which can make organogelling feasible with lecithin of relatively less purity.[87] Esposito et al. developed of mono-olein aqueous dispersions (MADs) and lecithin organogels (ORGs) as percutaneous delivery systems for curcumin (CUR). The results indicated that ORG induces a rapid and intense initial penetration of CUR probably due to a strong interaction between the peculiar supramolecular aggregation structure of phospholipids in the vehicle and the lipids present in the stratum corneum.[88]

34.9.6 PLURONIC LECITHIN ORGANOGELS

PLO is an opaque, yellow gel, composed of isopropyl palmitate, soy lecithin, water, and the hydrophilic polymer, Pluronic F127. The difference between PLO and its precursor, lecithin gels, is the presence of Pluronic F127 (a hydrophilic polymer that gels water) and the greater amount of water compared with the oil. PLO gel looks and feels like a cream but is actually a gel. When the aqueous phase (Pluronic gel) is combined with the lecithin oil base, it creates an emulsion that forms due to the pluronic gel and retains viscosity of that gel at room temperature.[89] Accudel™ is a cream from imprims pharmaceuticals that "carries" drugs through the skin, penetrating to the problem site. It delivers ketoprofen using PLO as the drug carrier with the efficiency to accommodate different size of molecules and large quantities of active drugs. Low toxicity and biodegradable components are nonimmunogenic which are "Generally Regarded as Safe" (GRAS) by the US FDA.[90] Richards et al. used PLO as a base for the delivery of bioactive polyunsaturated fatty acids from fish oil, EPA, DHA, and ketoprofen. The result showed that a PLO-based gel is capable of delivering EPA and DHA via a repeat finite dosing regimen, although there is evidence for the retention of these very lipophilic molecules within the gel matrix. Although to a lesser extent than EPA and DHA, ketoprofen was also substantially retained.[63]

34.10 CHARACTERIZATION OF ORGANOGELS

The characterization of organogels is relatively complicated on account of their interior structural design build-up on the self-associated supramolecules when compared to its preparation. This is also a daunting task owing to the varied polar–nonpolar interactions and high sensitivity. However, different characterization studies have been attempted by various researchers, which have resulted in an expanding pool of experience and expertise. We have therefore realized the need for condensed background information on physical, chemical, and biological characterization of organogels.

34.10.1 Morphological Study

There are numerous microscopy techniques for defining gel structures, which include scanning electron microscope (SEM) and transmission electron microscope. Microscopic techniques can directly determine the physical parameters of the gel matrix. These include measurements of pore diameter, wall thickness, and shape of the gel network.[91] Use of SEM can distinguish between gels that have a fibrous network as opposed to those that have a three-dimensional cross-linked structure. It must be noted that microscopy techniques may not yield quantitatively accurate results.[92] If a high vacuum is used during imaging, the liquid solvent can be removed from the gel matrix-inducing strain to the gel, which leads to physical deformation. Use of an environmental SEM, which operates at higher pressures, can yield higher quality imaging.[93]

34.10.2 Determination of Phase Transition Temperatures

The PTT of an organogel is measured using melting point technique.[94] The gel was introduced into the capillary tube (approx. 100 mm in length) by dipping the tube into the gel formulation. The sample was then drawn from the tube to approximately 0.3 cm from the lower end of the tube with the help of a syringe and then placed in the melting point apparatus with programmed temperature rise of 1°C min^{-1}. The temperature at which the gel melts into an isotropic liquid and flows down the capillary tube is assumed as its PTT.[95]

34.10.3 Differential Scanning Calorimetry

This is a reliable technique for measuring the strength of the intermolecular interactions in gels. Gel network strength is proportional to the magnitude of enthalpy change (ΔH). A higher ΔH means a more tightly bonded network, while a smaller enthalpy value means a network made of weaker bond.[96] Rocha et al. evaluated the ability of sugarcane wax to form organogels at different concentrations and temperatures, and also compare its rheological and thermal properties with the already studied organogel of candelilla wax (CLW). The results proposed that the increase of temperature leads to a breakage of the network and the dissolution of the solid material as an exothermic might suppress the visualization of the endothermic melting event.[97]

34.10.4 Scattering Techniques

Two scattering techniques for indirectly measuring gel parameters are small angle x-ray scattering (SARS/SAXS) and small angle neutron scattering (SANS). SARS works exactly like x-ray scattering (XRD) except that small angles (0.1°–10.0°) are used. The challenge with small angles is in separating the scattering pattern from the main beam. In SANS, the procedure is the same as SARS except that a neutron beam is used instead of an x-ray beam.[98] One advantage of using a neutron beam as opposed to an x-ray beam is an increased signal-to-noise ratio. It also provides the ability for isotope labeling, because the neutrons interact with the nuclei instead of the electrons. By analyzing the scattering pattern, direct information about the size of the material can be obtained. Both

SARS and SANS provide useful data on the atomic scale at 50–250 Å and 10–1000 Å, respectively. These distances are perfectly suited for studying the physical parameters of gels.[99]

34.10.5 Swelling Studies

It is also an important property of organogel characterization. It can be observed that the swelling degree increases when long polymer chains form intermolecular space within the polymeric network, while it decreases when there is a chain break and collapse in the organogels network.[90] However, sometimes the introduction of strong hydrophilic groups also increases the swelling degree. When temperature increases as result of decrease in the size of organogel, it is called low critical swelling temperature (LCST).[100] Pandey et al. found that volume change in thermosensible hydrogels is characterized by LCST of polymer chains, and established that polymer LCST can be higher or lower in copolymerization with hydrophilic monomers. In essence, some polymers with appropriate composition and cross-linked density can reach high swelling values in water and room temperature and collapse at LCST.[90]

34.10.6 Structural Features

Molecular architecture of organogels has been evaluated using NMR spectroscopy; hydrogen bonding has been established by FTIR spectroscopy. The knowledge of molecular packing within organogel network has been obtained using scanning and transmission electron microscopy.[101]

34.10.7 Homogeneity Test

Homogeneity test is used to determine the consistency of the gel with body organs or application site. Homogeneity of an organogel is tested by pressing hundred milligrams of an organogel between the thumb and the index finger in order to notice the consistency of gel that any coarse particles being attached or detached on finger.[102]

34.10.8 Determination of the Solvent Absorbency

Solvent absorbency is holding capacity of solvent by the organogels at a particular temperature. In this technique, completely dried organogels samples are cut into pieces and weighed till 0.4 g, after that each sample is placed inside a glass vial with top and was left to swell in a solvent at a constant temperature. The solvent absorbed in the organogels after swelling was calculated from the relation:% swelling (WC) = [(WS − WD)/WD)] × 100, Where, WC is the hold solvent in percentage absorbed by the organogels, and WS and WD are organogel weights after and before swelling, respectively.[103]

34.10.9 Stability Testing

Studies are carried out at three different storage conditions, that is, in refrigerated condition (RF; 5°C ± 3°C), room temperature (RT; 25°C ± 2°C/60% ± 5% RH), and elevated temperature (HT; 40°C ± 2°C/75% ± 5% RH) for a period of 6 months. Active content assay, gel strength, PTT, viscosity, extrusion force, pH value, and organoleptic characters are measured at different predetermined time points (i.e., twice in first month, and every month afterward) during the stability period of 6 months.[104] Physical stability is noted by microscopy or by visual inspection at periodic time intervals (i.e., twice in first month, and every month afterwards) for any perceptible change in physical appearance, that is, odor, color, consistency, phase separation, etc.

Organogels as Food Delivery Systems

34.10.10 Spreadability

Spreadability of organogels are determined using an apparatus consisting of a wooden block, with a fixed glass slide and movable glass slide with one end tied to weight pan rolled on the pulley, which was at a horizontal level with fixed slide. An excess of gel sample is placed between two glass slides and an X g weight is placed on slides for t minutes to compress the sample to a uniform thickness. The time (seconds) required to separate the two slides was taken as a measure of spreadability. It was calculated using the formula, $S = m \cdot l/t$, where, S is the spreadability in g cm/s, m is the weight tied to upper slide, l is the length of glass slide, and t is the time in seconds.[105]

34.10.11 Ball Indentation

Indentation hardness tests are used to determine the hardness of a material to deformation. Hardness or stiffness of the gel is measured by placing a metal ball on top of the material and the hardness of the material depends on the amount of indentation caused by the ball.[106]

34.10.12 Uniaxial Tensile Testing

In this technique, the tensile strength of the gel is measured in one direction. The two important measurements to make include the force applied per unit area and the amount of elongation under a known applied force. This test provides information on how a gel will respond when an external force is applied.[107]

34.10.13 Viscoelasticity

Different gels display different viscoelastic properties due to varying degrees of cross-linkage in a gel network. A material containing viscoelastic properties undergoes time-dependent structural changes in response to a physical deformation. Viscosity can be thought of as a time-dependent process of a material deforming to a more relaxed state while elasticity is an instantaneous process.[108] The two techniques for measuring viscoelasticity are broadband viscoelastic spectroscopy and resonant ultrasound spectroscopy. In both techniques, a damping mechanism is resolved with both differing frequency and time in order to determine the viscoelastic properties of the organogel. When the organogels are sheared at low shear rates, they behave like a solid and show an elastic property. With a subsequent increase in shear rate, the tubule structures are weakened. The gelator molecules break the physical interactions from the tubule until the shear stress is high enough to destroy the organogel structure, termed as plastic flow behavior.[109]

34.10.14 pH Determination

A value characteristic of an aqueous solution is its pH value, which represents conventionally its acidity or alkalinity. The pH of prepared organogel formulation is determined employing digital bench pH meter.[110]

34.10.15 Active Constituent Content

In this technique, accurately weighed organogels were transferred into a volumetric flask, dissolved in suitable solvent. The solution was filtered through Whatman filter paper No. 42, 1 ml of the above filtrate was pipetted out and diluted to 10 ml with solvent. The content of the encapsulated substance was estimated spectrophotometrically by using standard curve. The active substance content is measured by disrupting the gel to extract out the active constituent and then measured spectroscopically.[78]

34.10.16　IN VITRO RELEASE STUDIES

There is at present no pharmacopeia method described for the in vitro analysis of an organogel. Literature data is suggestive of using a dialysis tubing (molecular weight cutoff: 50 kDa) and accurately placing a weighed amount of organogel and placing it in a beaker containing the dissolution medium for a stipulated period of time for 1 h. Upon completion of 1 h, the sample tubings are transferred into another beaker containing fresh dissolution medium. Such changeovers are conducted at every 30-min interval and the samples collected are subjected to spectroscopic analysis using UV-visible spectrophotometer.[111]

34.10.17　IN VITRO PERMEATION STUDY

Permeation studies are accomplished by insertion the fabricated organogels with suitable skin or synthetic membrane in between receptor and donor compartment in a vertical diffusion cell such as Franz diffusion cell or keshary-chien diffusion cell. The system is applied to the hydrophilic side of the membrane and then mounted in the diffusion cell with lipophilic side in contact with receptor fluid. The receiver compartment is maintained at specific temperature (usually 32°C ± 5°C for skin) and is continuously stirred at a constant rate. The samples are withdrawn at different time intervals, and equal amount of buffer is replaced each time. The samples are diluted appropriately and absorbance is determined spectrophotometrically. Then the amount of drug permeated per centimeter square at each time interval is calculated.[112]

34.10.18　ANTIMICROBIAL TEST

The organogels are subjected to antimicrobial testing as suggested by various pharmacopoeias based on requirement using Petri plate method.[113]

34.11　POTENTIAL FOOD APPLICATIONS OF ORGANOGELS

The exceptional physical, functional, and nutritional properties of organogels have attracted the attention of the food and pharmaceutical industries. Organogels are used for delivery of drugs, foods, nutraceutical and many other pharmaceutical ingredients through various routes such as oral, parenteral, transdermal, topical, etc.

34.11.1　NOVEL ORGANOGEL COMPOSITIONS IN FOODS, BEVERAGES, NUTRACEUTICAL, PHARMACEUTICALS, PET FOOD, OR ANIMAL FEED

For regulatory acceptance, it is liable that the structured organogel must comprise of an edible emulsifier, an edible organic phase, an edible water-soluble polymer, and an edible polar phase. The GRAS FDA status has been given to few organogels like PLO, but their applicability extension for foods, beverages, nutraceutical, pharmaceuticals, pet food, or animal feed has not yet received approval.[114] Some of the causes for delayed regulatory approvals can be the impurities and contaminants that adsorb to their surfaces; these impurities can also contribute to ambiguous analytical results. Difficulty of interpreting data generated by a variety of unfamiliar techniques without a substantial history of acceptance in scientific literature may also lead to delays in getting regulatory approvals. Due to these difficulties, regulatory agencies have been less straightforward in the development of standards for characterization and the subsequent nonclinical application of organogels.[115] The edible agents (phospholipid organogel) used for the purpose of food and nutraceutical delivery which could be almond oil, ascorbic acid, an amphiphilic guest molecules, bromelain, citric acid, caffeine, capsaicin, green tea extract, grape-seed extract, a nutritional compound, potassium sorbate, natural polar antioxidants, tocopherols, tocotrienols, sterols, phytosterols, sterol esters, stanols, stanol esters, isoflavones, saw palmetto, sea weed extract, rosemary extract, lavender oil, peppermint oil, a vitamin, a mineral, a pharmaceutical, retinol, emulsifiers, a polar guest molecule, a nonpolar guest molecule, and an enzyme.[116] The edible organic phase can be soybean oil, canola oil, rapeseed oil, sunflower oil, corn oil, cottonseed oil, linseed oil, safflower oil, palm oil, cocoa butter, coconut oil, peanut oil, or any clinically approved vegetable oil. To extend such wide range of food and nutraceutical, attempts are being made. These organogels can hold active ingredients, yet restrict the diffusion of the active ingredients to facilitate their controlled release. However, the incorporation of these cubic crystalline phases can at times be difficult.[60] For instance, monoglycerides have some associated undesirable physical characteristics such as a high melting point, which transforms them into pastes or waxy solids at room temperature. Furthermore, the equilibration time needed to transform the monoglycerides into structured organogels may require several hours or days, since the diffusion of water through the solid monoglycerides is delayed. Another problem is that the processes used to form the cubic, liquid crystalline phases are cumbersome, since such processes require long holding times, high manufacturing temperatures, and high shear processes that are not economically or commercially viable.[117] To overcome such issues, lecithin organogels are mostly preferred due to their thermodynamically stable and viscoelastic properties. A limitation of such organogel is the requirement of highly pure lecithin that is expensive and not easily obtained. Another limitation in the formation of lecithin organogels is the polymer that is typically used. For instance, the synthetic polymer, such as pluronics, has been used to formulate lecithin organogels at an amount of about 30%–40%. However, pluronics are nonionic triblock copolymers, which may be characterized as a skin irritant, are not biobased, not allowed in food systems, and are not inexpensive compounds. In such cases, phospholipid organogels have been the most promising alternative.[118]

34.11.2 Replacement of Saturated Animal Fat

The structuring of edible oils using the polymer EC has been shown to have important applications in food systems as a substitute for fats which can be rich in trans- and saturated fatty acids. EC-based organogels in frankfurters investigated by Zetzl et al. were shown to alter the textural properties from that of a control formulated with liquid soybean oil. Vegetable oils have the capacity to produce stable meat emulsions, but the very small size of the fat globules in the system leads to products with a firmer texture.[74] Organogels can have differing mechanical properties depending on the formulation. Their incorporation into meat emulsions was researched so as to gain a broader knowledge of how their physical properties affect the qualities of the food product. The World Health Organization (WHO) recommends limiting the amount of dietary fat to 15%–30% of total daily energy intake, with saturated fats comprising only 10% of that total, and increasing the amount of healthier, polyunsaturated fats (WHO 2013). Challenging for the processed meat industry, these saturated fats play a functional role in the products they produce. Saturated fats contribute to the texture, mouth feel, juiciness, and sensory acceptability of products such as frankfurters.[119] In order to attain a healthier lipid profile in such a product, the use of unsaturated vegetable oils has been suggested. Canola oil was chosen as it contains the lowest saturated fatty acid levels when compared to other vegetable oils. This would ultimately lead to the highest reduction in saturated fat levels in the frankfurters.[120] Choi et al. confirm that vegetable oils can successfully replace fats of animal origin in controls, but this type of replacement is seldom used in the industry. They do, however, cite that textural and yield properties change significantly.[114] Currently, very little information exists on the replacement of saturated fats in emulsion style meat products with structured oils. Simply replacing beef fat with canola oil in all-beef frankfurters produced a product that was significantly harder. The gelling of the canola oil (organogel) before introduction into the meat led to lowering of the hardness values that resembled a beef fat control. Cohesiveness, chewiness, and gumminess values could also be altered to be similar to that of a beef fat control. Shear force values of organogel frankfurters with high EC concentrations (12%–14%) were also tailored to more closely match a beef fat control.[121]

34.11.3 Organogels as an Oil Structuring Alternatives to the Fat Crystal Network

The repeated consumption of high levels of saturated and trans-fats in the diet pose a chronic threat to human health by increasing, for instance, the risk of cardiovascular diseases and type II diabetes. The recent nutritional guidelines indicate reduction in the saturated fat intake to be less than 7% of total calories in order to decrease the risk of a cardiovascular disease in the population of developed countries, where these levels are significantly higher.[122] With increasing awareness, Canada has recently implemented strict labeling regulations whereby, in order to claim "trans-fat free," one serving must not only contain less than 0.2 g trans-fats per serving, but also less than 2 g of saturated fats per serving.[123] Pernetti et al. divided the structuring techniques for organic solvents (including edible oils) into two categories: suspensions and networks. The structuring networks are formed by specific molecular mechanisms, which hold these structures via a combination of covalent, electrostatic, steric, van der Waals, or hydrogen-bonding interactions. Apart from the conventional solid fat crystal network, there are a plethora of other compounds such as biopolymers and amphiphilic molecules that undergo self-assembly, self-assembled low molecular weight organogelators, etc., that are capable of forming networks to structure organic solvents.[124] Among them, the self-assembled low molecular weight organogelators are currently considered to be the most exciting option for application in food systems. Recently, the mono- and diacylglycerols, fatty acids, fatty alcohols, waxes, wax esters, sorbitan esters, and phytosterols have emerged as the food-grade alternatives to trans-fats. Research has shown that sometimes, the combination of these systems exhibits synergism and is more effective when paired together. Some of such combinations currently being studied are fatty acids and fatty alcohols, lecithin and sorbitan tri-stearate, and phytosterols and oryzanol.[125] Sitosterol and oryzanol (<2% w/w) has been developed as healthier alternatives to the trans-fats. These have been used to solidify edible oil and were patented in 2005. Further findings suggested it to be an appetite suppressant. This is due to its extended survival in the human digestive tract; reaching the ileum unabsorbed, it triggers the ileal break and leads to feelings of satiety. Despite the fact that this type of product has already been designed, more work will be required in order to exploit the full potential of structuring edible oils via organogelation.[126] Extensive research needs to be done in order to determine nutritional benefits, processing considerations, as well as the potential for application in more complex composite food products (i.e., multiphasic foods). The replacement of unhealthy saturated and trans-fatty acids with organogels may decrease the risk of coronary heart disease, type II diabetes, and metabolic syndrome. This has been supported by the clinical trials. It has been investigated that the phase separation of water in oil emulsions can be delayed by the formation of a gel network in the continuous oil phase.[127] This postulation has been of great interest to those working in the field of lipid technology developing alternative methods to structure edible oils. The ability of the longer chain trehalose diesters to gel olive oil attests to its potential use as a food or cosmetic additive that can be prepared using food-approved enzymatic synthesis approaches. On similar platform, Pernetti et al. have investigated organogels with food systems in order to structure edible oils without the need for high levels of saturated or trans fatty acids. But, unfortunately, mere replacement of the saturated lipid fraction with an unsaturated one does not help in attaining the food quality characteristics as expected by consumers.[124] It is interesting to mention here that the final structure of the product, as well as its sensory properties, is highly related to the fat crystallization degree (solid/liquid ratio). It is actually the solid lipids, normally existing in a three-dimensional architecture made of fat crystal network, which establishes the physical properties of the product. Lipid molecules upon crystallization aggregate to form fat crystals, which interact in the fashion similar to gels, to form clusters. The clusters aggregate into flocks and finally weak links between the flocks form in the final macroscopic network. The latter could then interact with other food constituents in multicomponent systems.[128] Therefore, the most worthy strategy would be to design a healthier fatty food that would act as the suitable replacement to the plastic structure made of saturated fats with unsaturated oils solidified through the addition of molecules forming self-assembly networks. This has been possible due to the successful application of organogels, which has been efficiently quoted as "the fat of the future." The attempt for using organogels in food formulations has ever since been increasing. Studies have been carried out to study the effect of organogels on the stability of the oils.[129] The major disadvantage of lipid oxidation in these oils causes not only the formation of undesirable flavors and/or color changes, making foods unacceptable by consumers, but also the loss of nutritional components as well as the formation of toxic compounds. It has been suggested that the occurrence of crystallization events induces a cascade of temperature-dependent events, such as solute concentration and changes in physicochemical properties in the liquid phases surrounding crystals affecting the lipid stability.[130] The role of the cod liver oil structuration on its oxidative stability has been studied by Pieve et al. with the intent of using organogels in foods. The purpose of studying oxidative stability of organogels made of increasing amounts of saturated monoglycerides and cod liver oil was its richness in PUFA, cod liver oil is particularly prone to oxidation as well as it is a source of essential polyunsaturated fatty acid. Results obtained from their study confirmed the ability of saturated monoglycerides to structure liquid oils, thereby opening interesting perspectives for the production of novel health value-added foods.[131] Although, the organogel network seemed to have a double effect on the oxidative stability of cod liver oil yet, despite structuring, it was quite ineffective in slowing down the initial steps of the oxidation reaction, but it did appear as a hurdle against the development of secondary oxidation products. In the light of these, it can be inferred that organogels are quite ineffective in protecting oxidizing molecules from oxygen, which easily diffuses into the gel network.[132] However, when oxidation reactions require reactant mobility, organogels can effectively hinder the development of oxidative reactions. Thus, oil structuring by the addition of monoglycerides can be apparently a promising strategy to extend the shelf-life of the product by slowing down the formation of secondary oxidation products. However, with time, the accumulation of primary oxidizing compounds may adversely affect human health without affecting product acceptability. The presence of "silent" and reactive species in foods represents a hot topic, which poses some important ethical issues especially in the research area targeted at the development of new formulation strategies.[133]

34.11.4 Self-Assembly Network of Organogels Has Been Proposed as Oil Migration Inhibitor for Chocolate Products

Chocolates, the word, itself waters the mouth. But imagine a situation where in you gallop a chocolate and with unpalatable taste you need to spit it out. One of the major causes of chocolate confectionary spoilage is the migration of oil in filled chocolate confections. These filled portions are basically unsaturated liquid triacylglycerols from the soft filling (typically containing large proportions of highly mobile oils), which diffuses into the cocoa butter and migrate through the coating layer toward the outer surface of the product, making it unpalatable.[134] Not only this, as the oil migrates through the chocolate, the oil acts as a solvent and high melting cocoa butter triacylglycerols are dissolved and travel with the oil toward the product surface, leading to the dilution and softening of the coating. These recrystallizing triacylglycerol molecules (formed due to oil migration) have a higher degree of mobility and are more likely to arrange themselves in the most stable and densely packed β-VI polymorphic form. These newly formed and relatively large fat crystals grow and protrude outward from the chocolate surface known as oil migration–induced surface fat bloom. They are characterized by a dull white or gray film on the surface of the chocolate.[135] Among the different polymeric forms of cocoa butter that have been identified, β-V is the stable form. Actually this is not the most stable polymorph (β-VI is most sable), but due to its long stability of 6–12 months, it is suitable commercially. These fat blooms are one of the primary causes of consumer dissatisfaction and rejection of confectionary products. In order to maintain the quality of this type of chocolate confections during storage, it is crucial to restrict the migration of oil between its components.[136] Efforts have been growing at large pace to develop methods to slow the oil migration (and fat bloom) in confectionary products; yet, no efficient or practical solution has been identified. Henceforth, a combined effort is being put forth by lipid scientists and confectionary technologists to search for ways to control and/or prevent oil migration.[137] A winning strategy for limiting oil migration through confections would be to gel any of the free oil, which would be prone to freeze-thaw deterioration. Theoretically, by immobilizing liquid oils within a gel network, the movement and migration of triacylglycerols out of or through the material will be greatly reduced and oil migration–induced fat bloom will be slowed. The crystallinity and oil-binding capacity of 12-HSA for vegetable oil organogels has been continuously monitored by many researchers.[138] Rogers et al. observed longer fibers and a less branched network when the gels were stored at 30°C. They also found the enhancement of 12-HAS network crystallinity at 30°C with fewer inclusions of liquid oil. Hydroxylated fatty acids, including 12-HSA, and ricinelaidic acid, are capable of forming oil gels at concentrations as low as 0.3%. Long, thin molecular fibers are capable of spanning hundreds of microns in length. Fiber formation occurs, because the 12-HSA stacks via a zigzag pattern of hydrogen bonds. The physical appearance of these systems may be completely translucent or opaque, depending on the microscopic arrangement of the fibers.[139] 12-HSA is not considered a food-grade additive as it is derived from castor oil, and is considered an irritant due to its laxative effects. It is notable here to mention that 12-HSA has not yet been approved as a food-grade additive despite the fact that it is derived from a natural product (castor seed oil), but it is quite often used as a prototype for the study of the network formation process as well as the material properties of the resulting organogel.[71] Rogers et al. also investigated the mobility of canola oil gelled with varying concentrations of 12-HSA at different isothermal crystallization temperatures. They conducted experiments by placing a disk-shaped organogel sample onto a filter paper and measuring the time-dependent mass of oil drawn into the paper via capillary forces; this showed that the mobility of canola oil triacylglycerols and their migration rate varied significantly with the gel setting temperature.[139] Similarly, Dibildox-Alvarado et al. monitored and studied the oil migration effect from a 60:40 (w/w) mixture of peanut oil and chemically interesterified and hydrogenated palm oil. They proposed that the relative oil loss as determined gravimetrically was inversely proportional to the cooling rate.[140] A study reported by Hughes et al. showed an unexpected result upon the incorporation of 12-HSA to a stimulated fat-based cream filling containing canola oil and interesterified hydrogenated palm oil (IHPO). The model chocolate confection showed an increasing rate of oil migration through the nontempered cocoa butter rather than decreasing it.[141] The authors suggest that probably the 12-HSA interfered with the crystallization of the IHPO in the early stages of migration, and allowed a more rapid migration of the oil. Based on these hypotheses, 12-HAS was applied as a low molecular mass edible organogelators to restrict the mobility of edible oils for the prevention of oil migration in chocolate confectionary products by Marty et al. In order to test their hypothesis, they monitored the effects of the incorporation of 12-HSA into a model of a filled chocolate confection using a scanner imaging technique.[142] They created a model of a cream-filled chocolate confection using nontempered cocoa butter to represent the chocolate coating and a (60:40 canola oil: IHPO) to represent the filling. The model filling was stained with 0.015% w/w Nile red so that the migration of oil was visually observable. 0.0%, 0.5%, or 2.0% 12-HSA was fully dissolved into hot cocoa butter and/or the hot filling, and the mixtures were held for 30 min at 85°C to ensure that the crystal memory was erased. Using a preheated pipette, the melted fats were deposited into custom-made glass-bottomed molds.[143] The cocoa butter (containing 0.0%, 0.5%, or 2.0% 12-HSA) was crystallized at 5°C for 24 h before being transferred to a storage incubator at 20°C in order to form a 12-HSA network with optimal oil-binding capacity. Cocoa butter samples were crystallized for 6 days at 20°C until a strong signal corresponding to the βV crystal polymorph (stable polymorph) was identified by x-ray diffraction. At this point, the dyed model filling (mixed with 0.0%, 0.5%, or 2.0% 12-HSA and held at 85°C for 30 min) was deposited into the mold using a preheated pipette. Samples were scanned immediately after solidification (crystallization at 5°C).[144] The migration of the dyed model filling into and through the model coating was tracked by scanning the glass-bottomed sample molds daily using a flatbed scanner and the position of the moving front of oil migration was determined quantitatively using graph pad prism. This experiment proposed that the incorporation of 12-HSA into the filling of the model chocolate confection significantly increased the rate of oil migration through nontempered cocoa butter.[145] This was due to the influence of 12-HSA on the crystallization of IHPO and vice versa. The increased oil migration in the presence of 12-HSA may be attributed to its effects on the formation, packing, and morphology of the IHPO crystals such that the fat crystal network has a decreased ability to structure and entrap oils.[146]

TABLE 34.3
Encapsulation of Food Ingredients by Organogels

Name (Examples)	Use of Food Ingredients	Potential Advantages of Encapsulation
Flavors (salt, sugar, vinegar, soy sauce, citrus oil)	• Used to control the taste or aroma of food products	• Inhibit chemical degradation • Control the release profile • Ease of handling and use • Allow incorporation in aqueous media
Antimicrobials (essential oils)	• Used to kill or inhibit the growth of micro-organisms such as bacteria, molds, and yeast in food industry	• Increase their efficacy • Mask off flavors • Increase their ease of storage, transport, and utilization • Increase their compatibility with food matrix
Antioxidants (carotenoids)	• Provide prevention from oxidation by altering its proximity to the chemical reactants	• Improve their ease, handling, and use • Improve their compatibility with food matrix • Avoid chemical degradation
Bioactive lipids (phytosterols, fatty acid, carotenoids)	• Useful in growth, development, and other normal body functions	• Increase bioavailability • Controlled delivery in GIT • Avoid chemical degradation • Improve ease of utilization
Bioactive peptides (cholecystokinin)	• Provide energy and essential nutrients to the diet and improve biological functions	• Control release profile and bioactivity • Reduce bitterness and astringency • Retard degradation in stomach
Bioactive carbohydrates (dietary fibers)	• Used in cholesterol reduction, modulation of blood glucose level, and prevention from prebiotic effects	• Controlled delivery in GIT • Improved product texture • Avoid adverse interaction of ingredients
Essential minerals (iron, calcium)	• Maintenance of proper human health	• Enhance bioavailability • Prevent precipitation • Prevent undesirable oxidative reaction • Reduce off-flavor
Vitamins (Vitamin D)	• Useful in growth, development and other normal body functions	• Increase bioavailability • Controlled delivery in GIT • Avoid chemical degradation • Improve ease of utilization
Probiotics (lactic acid bacteria)	• Control blood pressure, cholesterol and improve immune functions	• Avoid degradation in stomach • Improve cell viability in product

34.11.5 ORGANOGEL FOR CAPTURING HEAVY METALS

Nii et al. attempted to formulate an organogel of dodecyl acrylate loaded with a metal extractant di-2-ethylhexyl phosphoric acid or ethylhexylphosphonic acid for the sorption of Zn (II) and Cu (II) from their hydrochloric acid solutions. The mechanism of metal sorption for each metal–extractant system was found to be similar to that of metal separation in liquid–liquid extraction, while the metal uptake process was film-diffusion controlled.[147] During the sorption–desorption cycle (repeated five times), more than 98% of the extractant was retained. Substantially high percentage of metal ion sorption and desorption was achieved in the cyclic use of the organogel. The results suggested the feasibility of extractant-impregnated organogels for separating metal ions from their aqueous solutions. Table 34.3 shows the selected examples of active ingredients that are encapsulated by organogels for use in the food industry.[148]

34.12 CONCLUSIONS

Based on the evidence described in this book chapter, it is clear that organogels exhibit significant promise and potential for a wide variety of applications in both the food and pharmaceuticals industries. Organogel networks have confirmed the ability to protect part of their contents through the initial phases of the human digestive route. Yet again, the capability of organogel networks to delay the release of lipophilic bioactive compounds must also be discovered in detail in order to evaluate their full potential in this area. The time-dependent response of organogel networks releases compounds at a specific and predetermined point during the digestive process. The use of nutraceutical-loaded food-grade organogels could be an integral part of our diets in the near future; therefore, much more work is required to develop this technology into marketable consumer products and help to improve our overall health.

REFERENCES

1. Gibbs BF, Kermasha S, Alli I, Mulligan CN. Encapsulation in the food industry: A review. *Int J Food Sci Nutr.* 50(3) (1999):213–224.
2. Shahidi F, Han XQ. Encapsulation of food ingredients. *Crit Rev Food Sci Nutr.* 33(6) (1993):501–547.
3. Augustin MA, Hemar Y. Nano- and micro-structured assemblies for encapsulation of food ingredients. *Chem Soc Rev.* 38(4) (2009):902–912.
4. Wei X, Luo M, Xie Y, Yang L, Li H, Xu L et al. Strain screening, fermentation, separation, and encapsulation for production of nattokinase functional food. *Appl Biochem Biotechnol.* 168(7) (2012):1753–1764.
5. Murdan S. Organogels in drug delivery. *Expert Opin Drug Deliv.* 2(3) (2005):489–505.
6. Ajayaghosh A, Praveen VK, Vijayakumar C. Organogels as scaffolds for excitation energy transfer and light harvesting. *Chem Soc Rev.* 37(1) (2008):109–122.
7. Ajayaghosh A, Praveen VK. Pi-organogels of self-assembled p-phenylenevinylenes: Soft materials with distinct size, shape, and functions. *Acc Chem Res.* 40(8) (2007):644–656.
8. Jang K, Ranasinghe AD, Heske C, Lee DC. Organogels from functionalized T-shaped pi-conjugated bisphenazines. *Langmuir.* 26(16) (2010):13630–13636.
9. George M, Weiss RG. Chemically reversible organogels: Aliphatic amines as "latent" gelators with carbon dioxide. *J Am Chem Soc.* 123(42) (2001):10393–10394.
10. Angelico R, Ceglie A, Colafemmina G, Lopez F, Murgia S, Olsson U et al. Biocompatible lecithin organogels: Structure and phase equilibria. *Langmuir.* 21(1) (2005):140–148.
11. Luo X, Xiao W, Li Z, Wang Q, Zhong J. Supramolecular organogels formed by monochain derivatives of succinic acid. *J Colloid Interface Sci.* 329(2) (2009):372–375.
12. Luo X, Liu B, Liang Y. Self-assembled organogels formed by mono-chain L-alanine derivatives. *Chem Commun (Camb).* 17 (2001):1556–1557.
13. Luo X, Chen Z, Xiao W, Li Z, Wang Q, Zhong J. Two-component supramolecular organogels formed by maleic N-monoalkylamides and aliphatic amines. *J Colloid Interface Sci.* 362(1) (2011):113–117.
14. Wu H, Xue L, Shi Y, Chen Y, Li X. Organogels based on J- and H-type aggregates of amphiphilic perylenetetracarboxylic diimides. *Langmuir.* 27(6) (2011):3074–3082.

15. Baddeley C, Yan Z, King G, Woodward PM, Badjic JD. Structure-function studies of modular aromatics that form molecular organogels. *J Org Chem.* 72(19) (2007):7270–7278.
16. Jiao T, Huang Q, Zhang Q, Xiao D, Zhou J, Gao F. Self-assembly of organogels via new luminol imide derivatives: Diverse nanostructures and substituent chain effect. *Nanoscale Res Lett.* 8(1) (2013):278.
17. Garcia Velazquez D, Gonzalez Orive A, Hernandez Creus A, Luque R, Gutierrez Ravelo A. Novel organogelators based on amine-derived hexaazatrinaphthylene. *Org Biomol Chem.* 9(19) (2011):6524–6527.
18. Mallia VA, George M, Blair DL, Weiss RG. Robust organogels from nitrogen-containing derivatives of (R)-12-hydroxystearic acid as gelators: Comparisons with gels from stearic acid derivatives. *Langmuir.* 25(15) (2009):8615–8625.
19. Spandagos C, Goudoulas TB, Luckham PF, Matar OK. Surface tension-induced gel fracture. Part 1. Fracture of agar gels. *Langmuir.* 28(18) (2012):7197–7211.
20. Wang XJ, Xing LB, Cao WN, Li XB, Chen B, Tung CH et al. Organogelators based on TTF supramolecular assemblies: Synthesis, characterization, and conductive property. *Langmuir.* 27(2) (2011):774–781.
21. Funkhouser GP, Tonmukayakul N, Liang F. Rheological comparison of organogelators based on iron and aluminum complexes of dodecylmethylphosphinic acid and methyl dodecanephosphonic acid. *Langmuir.* 25(15) (2009):8672–8677.
22. Ayabe M, Kishida T, Fujita N, Sada K, Shinkai S. Binary organogelators which show light and temperature responsiveness. *Org Biomol Chem.* 1(15) (2003):2744–2747.
23. Yemloul M, Steiner E, Robert A, Bouguet-Bonnet S, Allix F, Jamart-Gregoire B et al. Solvent dynamical behavior in an organogel phase as studied by NMR relaxation and diffusion experiments. *J Phys Chem B.* 115(11) (2011):2511–2517.
24. Wang D, Hao J. Self-assembly fibrillar network gels of simple surfactants in organic solvents. *Langmuir.* 27(5) (2011):1713–1717.
25. Annamalai PK, Dagnon KL, Monemian S, Foster EJ, Rowan SJ, Weder C. Water-responsive mechanically adaptive nanocomposites based on styrene-butadiene rubber and cellulose nanocrystals—Processing matters. *ACS Appl Mater Interfaces.* 6(2) (2014):967–976.
26. Li JL, Liu XY. Microengineering of soft functional materials by controlling the fiber network formation. *J Phys Chem B.* 113(47) (2009):15467–15472.
27. Kwon SK, Song JJ, Cho CG, Park SW, Choi SJ, Oh SH et al. Polycaprolactone spheres and theromosensitive Pluronic F127 hydrogel for vocal fold augmentation: In vivo animal study for the treatment of unilateral vocal fold palsy. *Laryngoscope.* 123(7) (2013):1694–1703.
28. Flory J, Emanuel E. Interventions to improve research participants' understanding in informed consent for research: A systematic review. *JAMA.* 292(13) (2004):1593–1601.
29. Iwata Y, Furuichi K, Hashimoto S, Yokota K, Yasuda H, Sakai N et al. Pro-inflammatory/Th1 gene expression shift in high glucose stimulated mesangial cells and tubular epithelial cells. *Biochem Biophys Res Commun.* 443(3) (2014):969–974.
30. Singh M, Tan G, Agarwal V, Fritz G, Maskos K, Bose A et al. Structural evolution of a two-component organogel. *Langmuir.* 20(18) (2004):7392–7398.
31. Tu T, Fang W, Sun Z. Visual-size molecular recognition based on gels. *Adv Mater.* 25(37) (2013):5304–5313.
32. Motulsky A, Lafleur M, Couffin-Hoarau AC, Hoarau D, Boury F, Benoit JP et al. Characterization and biocompatibility of organogels based on L-alanine for parenteral drug delivery implants. *Biomaterials.* 26(31) (2005):6242–6253.
33. Wang K, Jia Q, Han F, Liu H, Li S. Self-assembled L-alanine derivative organogel as in situ drug delivery implant: Characterization, biodegradability, and biocompatibility. *Drug Dev Ind Pharm.* 36(12) (2010):1511–1521.
34. Escuder B, Marti S, Miravet JF. Organogel formation by coaggregation of adaptable amidocarbamates and their tetraamide analogues. *Langmuir.* 21(15) (2005):6776–6787.
35. Jones DS, Muldoon BC, Woolfson AD, Andrews GP, Sanderson FD. Physicochemical characterization of bioactive polyacrylic acid organogels as potential antimicrobial implants for the buccal cavity. *Biomacromolecules.* 9(2) (2008):624–633.
36. Patra T, Pal A, Dey J. Birefringent physical gels of N-(4-n-alkyloxybenzoyl)-L-alanine amphiphiles in organic solvents: The role of hydrogen-bonding. *J Colloid Interface Sci.* 344(1) (2010):10–20.
37. Samanta SK, Bhattacharya S. Excellent chirality transcription in two-component photochromic organogels assembled through J-aggregation. *Chem Commun (Camb).* 49(14) (2013):1425–1427.
38. Lai WC, Tseng SC. Novel polymeric nanocomposites and porous materials prepared using organogels. *Nanotechnology.* 20(47) (2009):475606.
39. Mark JE. Some novel polymeric nanocomposites. *Acc Chem Res.* 39(12) (2006):881–888.

40. Coradin T, Livage J. Aqueous silicates in biological sol-gel applications: New perspectives for old precursors. *Acc Chem Res.* 40(9) (2007):819–826.
41. Yu H, Huang Q. Improving the oral bioavailability of curcumin using novel organogel-based nanoemulsions. *J Agric Food Chem.* 60(21) (2012):5373–5379.
42. Kim YS, Cho YG, Odkhuu D, Park N, Song HK. A physical organogel electrolyte: Characterized by in situ thermo-irreversible gelation and single-ion-predominent conduction. *Sci Rep.* 3 (2013):1917.
43. Yebra-Pimentel I, Martinez-Carballo E, Regueiro J, Simal-Gandara J. The potential of solvent-minimized extraction methods in the determination of polycyclic aromatic hydrocarbons in fish oils. *Food Chem.* 139(1–4) (2013):1036–1043.
44. El Khoury ED, Patra D. Ionic liquid expedites partition of curcumin into solid gel phase but discourages partition into liquid crystalline phase of 1,2-dimyristoyl-sn-glycero-3-phosphocholine liposomes. *J Phys Chem B.* 117(33) (2013):9699–9708.
45. Saber Tehrani M, Aberoomand Azar P, Mohammadiazar S. A single step technique for preparation of porous solid phase microextraction fibers by electrochemically co-deposited silica based sol-gel/Cu nanocomposite. *J Chromatogr A.* 1278 (2013):1–7.
46. Chiang PR, Lin TY, Tsai HC, Chen HL, Liu SY, Chen FR et al. Thermosensitive hydrogel from oligopeptide-containing amphiphilic block copolymer: Effect of Peptide functional group on self-assembly and gelation behavior. *Langmuir.* 29(51) (2013):15981–15991.
47. Pragatheeswaran AM, Chen SB. Effect of chain length of PEO on the gelation and micellization of the pluronic F127 copolymer aqueous system. *Langmuir.* 29(31) (2013):9694–9701.
48. Xiong YL, Blanchard SP, Ooizumi T, Ma Y. Hydroxyl radical and ferryl-generating systems promote gel network formation of myofibrillar protein. *J Food Sci.* 75(2) (2010):C215–C221.
49. Wilson SL, Guilbert M, Sule-Suso J, Torbet J, Jeannesson P, Sockalingum GD et al. A microscopic and macroscopic study of aging collagen on its molecular structure, mechanical properties, and cellular response. *FASEB J.* 28(1) (2014):14–25.
50. Akash MS, Rehman K, Sun H, Chen S. Assessment of release kinetics, stability and polymer interaction of poloxamer 407-based thermosensitive gel of interleukin-1 receptor antagonist. *Pharm Dev Technol.* 19(3) (2014):278–284.
51. Gramlich WM, Holloway JL, Rai R, Burdick JA. Transdermal gelation of methacrylated macromers with near-infrared light and gold nanorods. *Nanotechnology.* 25(1) (2013):014004.
52. O'Sullivan S, Arrigan DW. Impact of a surfactant on the electroactivity of proteins at an aqueous-organogel microinterface array. *Anal Chem.* 85(3) (2013):1389–1394.
53. Mas A, Lopez ML, Alvarez-Serrano I, Pico C, Veiga ML. Electrochemical performance of Li((4−x)/3)Mn((5−2x)/3)Fe(x)O$_4$ (x = 0.5 and x = 0.7) spinels: Effect of microstructure and composition. *Dalton Trans.* 42(27) (2013):9990–9999.
54. Souza AL, Andreani T, de Oliveira RN, Kiill CP, Santos FK, Allegretti SM et al. in vitro evaluation of permeation, toxicity and effect of praziquantel-loaded solid lipid nanoparticles against Schistosoma mansoni as a strategy to improve efficacy of the schistosomiasis treatment. *Int J Pharm.* 463(1) (2014):31–37.
55. Costa A, Madeira M, Lima Santos J, Plieninger T, Seixas J. Fragmentation patterns of evergreen oak woodlands in Southwestern Iberia: Identifying key spatial indicators. *J Environ Manage.* 133C (2013):18–26.
56. Rosman BM, Barbosa JA, Passerotti CP, Cendron M, Nguyen HT. Evaluation of a novel gel-based ureteral stent with biofilm-resistant characteristics. *Int Urol Nephrol.* 46(6) (2014):1053–1058.
57. Rest C, Mayoral MJ, Fucke K, Schellheimer J, Stepanenko V, Fernandez G. Self-assembly and (hydro) gelation triggered by cooperative pi-pi and unconventional CHX hydrogen bonding interactions. *Angew Chem Int Ed Engl.* 53(3) (2014):700–705.
58. Gallyas F. A cytoplasmic gel network capable of mediating the conversion of chemical energy to mechanical work in diverse cell processes: A speculation. *Acta Biol Hung.* 61(4) (2010):367–379.
59. Bhattacharya C, Kumar N, Sagiri SS, Pal K, Ray SS. Development of span 80-tween 80 based fluid-filled organogels as a matrix for drug delivery. *J Pharm Bioallied Sci.* 4(2) (2012):155–163.
60. Sawalha H, den Adel R, Venema P, Bot A, Floter E, van der Linden E. Organogel-emulsions with mixtures of beta-sitosterol and gamma-oryzanol: Influence of water activity and type of oil phase on gelling capability. *J Agric Food Chem.* 60(13) (2012):3462–3470.
61. Yu H, Shi K, Liu D, Huang Q. Development of a food-grade organogel with high bioaccessibility and loading of curcuminoids. *Food Chem.* 131 (2012):44–54.
62. Bhatia A, Singh B, Wadhwa S, Raza K, Katare OP. Novel phospholipid-based topical formulations of tamoxifen: Evaluation for antipsoriatic activity using mouse-tail model. *Pharm Dev Technol.* 19(2) (2014):160–163.

63. Richards H, Thomas CP, Bowen JL, Heard CM. In-vitro transcutaneous delivery of ketoprofen and polyunsaturated fatty acids from a pluronic lecithin organogel vehicle containing fish oil. *J Pharm Pharmacol.* 58(7) (2006):903–908.
64. Haino T, Hirai Y, Ikeda T, Saito H. Photoresponsive two-component organogelators based on trisphenylisoxazolylbenzene. *Org Biomol Chem.* 11(25) (2013):4164–4170.
65. Afrasiabi R, Kraatz HB. Small-peptide-based organogel kit: Towards the development of multicomponent self-sorting organogels. *Chemistry.* 19(47) (2013):15862–15871.
66. Nishimata T, Sato Y, Mori M. Palladium-catalyzed asymmetric allylic substitution of 2-arylcyclohexenol derivatives: Asymmetric total syntheses of (+)-crinamine, (-)-haemanthidine, and (+)-pretazettine. *J Org Chem.* 69(6) (2004):1837–1843.
67. Suzuki M, Hanabusa K. Polymer organogelators that make supramolecular organogels through physical cross-linking and self-assembly. *Chem Soc Rev.* 39(2) (2010):455–463.
68. Nakahara H, Hasegawa A, Uehara S, Akisada H, Shibata O. Solution properties of gemini surfactant of decanediyl-1-10-bis (dimethyltetradecylammonium bromide) in aqueous medium. *J Oleo Sci.* 62(11) (2013):905–912.
69. Adams DJ. Dipeptide and tripeptide conjugates as low-molecular-weight hydrogelators. *Macromol Biosci.* 11(2) (2011):160–173.
70. Das UK, Banerjee S, Dastidar P. Primary ammonium monocarboxylate synthon in designing supramolecular gels: A new series of chiral low-molecular-weight gelators derived from simple organic salts that are capable of generating and stabilizing gold nanoparticles. *Chem Asian J.* 8(12) (2013): 3022–3031.
71. Fameau AL, Houinsou-Houssou B, Novales B, Navailles L, Nallet F, Douliez JP. 12-Hydroxystearic acid lipid tubes under various experimental conditions. *J Colloid Interface Sci.* 341(1) (2010):38–47.
72. Asumadu-Mensah A, Smith KW, Ribeiro HS. Solid lipid dispersions: Potential delivery system for functional ingredients in foods. *J Food Sci.* 78(7) (2013):E1000–E1008.
73. Maula T, Artetxe I, Grandell PM, Slotte JP. Importance of the sphingoid base length for the membrane properties of ceramides. *Biophys J.* 103(9) (2012):1870–1879.
74. Zetzl AK, Marangoni AG, Barbut S. Mechanical properties of ethylcellulose oleogels and their potential for saturated fat reduction in frankfurters. *Food Funct.* 3(3) (2012):327–337.
75. Willemen HM, Marcelis AT, Sudholter EJ, Bouwman WG, Deme B, Terech P. A small-angle neutron scattering study of cholic acid-based organogel systems. *Langmuir.* 20(6) (2004):2075–2080.
76. Murdan S, Gregoriadis G, Florence AT. Sorbitan monostearate/polysorbate 20 organogels containing niosomes: A delivery vehicle for antigens? *Eur J Pharm Sci.* 8(3) (1999):177–186.
77. Murdan S, Gregoriadis G, Florence AT. Inverse toroidal vesicles: Precursors of tubules in sorbitan monostearate organogels. *Int J Pharm.* 183(1) (1999):47–49.
78. Murdan S, van den Bergh B, Gregoriadis G, Florence AT. Water-in-sorbitan monostearate organogels (water-in-oil gels). *J Pharm Sci.* 88(6) (1999):615–619.
79. Murdan S, Gregoriadis G, Florence AT. Novel sorbitan monostearate organogels. *J Pharm Sci.* 88(6) (1999):608–614.
80. Alasino RV, Leonhard V, Bianco ID, Beltramo DM. Eudragit E100 surface activity and lipid interactions. *Colloids Surf B Biointerfaces.* 91 (2012):84–89.
81. Dave VS, Fahmy RM, Bensley D, Hoag SW. Eudragit((R)) RS PO/RL PO as rate-controlling matrix-formers via roller compaction: Influence of formulation and process variables on functional attributes of granules and tablets. *Drug Dev Ind Pharm.* 38(10) (2012):1240–1253.
82. Liu H, Wang Y, Han F, Yao H, Li S. Gelatin-stabilised microemulsion based organogels facilitates percutaneous penetration of cyclosporin A in vitro and dermal pharmacokinetics in vivo. *J Pharm Sci.* 96(11) (2007):3000–3009.
83. Peacock GF, Sauvageot J. Evaluation of the stability of acetaminophen in pluronic lecithin organogel and the determination of an appropriate beyond-use date. *Int J Pharm Compd.* 16(5) (2012):428–430.
84. Jadhav KR, Kadam VJ, Pisal SS. Formulation and evaluation of lecithin organogel for topical delivery of fluconazole. *Curr Drug Deliv.* 6(2) (2009):174–183.
85. Lehman PA, Raney SG. in vitro percutaneous absorption of ketoprofen and testosterone: Comparison of pluronic lecithin organogel vs. pentravan cream. *Int J Pharm Compd.* 16(3) (2012):248–252.
86. Franckum JP, Ramsay DP, Das NGPMR, Das SKPM. Pluronic lecithin organogel for local delivery of anti-inflammatory drugs. *Int J Pharm Compd.* 8(2) (2004):101–105.
87. Glisson JKMP, Wood RLP, Kyle PB, Cleary JDPF. Bioavailability of promethazine in a topical pluronic lecithin organogel: A pilot study. *Int J Pharm Compd.* 9(3) (2005):242–246.

88. Esposito E, Ravani L, Mariani P, Huang N, Boldrini P, Drechsler M et al. Effect of nanostructured lipid vehicles on percutaneous absorption of curcumin. *Eur J Pharm Biopharm.* 86(2) (2014):121–132.
89. Bramwell BL, Williams LA. The use of Pluronic lecithin organogels in the transdermal delivery of drugs. *Int J Pharm Compd.* 16(1) (2012):62–63.
90. Pandey M, Belgamwar V, Gattani S, Surana S, Tekade A. Pluronic lecithin organogel as a topical drug delivery system. *Drug Deliv.* 17(1) (2010):38–47.
91. Maiti DK, Banerjee A. A synthetic amino acid residue containing a new oligopeptide-based photosensitive fluorescent organogel. *Chem Asian J.* 8(1) (2013):113–120.
92. Huang X, Terech P, Raghavan SR, Weiss RG. Kinetics of 5alpha-cholestan-3beta-yl N-(2-naphthyl) carbamate/n-alkane organogel formation and its influence on the fibrillar networks. *J Am Chem Soc.* 127(12) (2005):4336–4344.
93. Woerly S, Doan VD, Sosa N, de Vellis J, Espinosa A. Reconstruction of the transected cat spinal cord following NeuroGel implantation: Axonal tracing, immunohistochemical and ultrastructural studies. *Int J Dev Neurosci.* 19(1) (2001):63–83.
94. Yao X, Ju J, Yang S, Wang J, Jiang L. Temperature-driven switching of water adhesion on organogel surface. *Adv Mater.* 26(12) (2014):1895–1900.
95. Bhatia A, Singh B, Raza K, Wadhwa S, Katare OP. Tamoxifen-loaded lecithin organogel (LO) for topical application: Development, optimization and characterization. *Int J Pharm.* 444(1–2) (2013):47–59.
96. Ahmed J, Varshney SK, Auras R. Rheological and thermal properties of polylactide/silicate nanocomposites films. *J Food Sci.* 75(2) (2010):N17–N24.
97. Rocha JCB, JD L. Thermal and rheological properties of organogels formed by sugarcane or candelilla wax in soybean oil. *Food Res Int.* 50 (2013):318–323.
98. Shaikh IM, Jadhav SL, Jadhav KR, Kadam VJ, Pisal SS. Aceclofenac organogels: In vitro and in vivo characterization. *Curr Drug Deliv.* 6(1) (2009):1–7.
99. Huang X, Raghavan SR, Terech P, Weiss RG. Distinct kinetic pathways generate organogel networks with contrasting fractality and thixotropic properties. *J Am Chem Soc.* 128(47) (2006):15341–15352.
100. Pandey MS, Belgamwar VS, Surana SJ. Topical delivery of flurbiprofen from pluronic lecithin organogel. *Indian J Pharm Sci.* 71(1) (2009):87–90.
101. Simmons BA, Taylor CE, Landis FA, John VT, McPherson GL, Schwartz DK et al. Microstructure determination of AOT + phenol organogels utilizing small-angle X-ray scattering and atomic force microscopy. *J Am Chem Soc.* 123(10) (2001):2414–2421.
102. Agrawal V, Gupta V, Ramteke S, Trivedi P. Preparation and evaluation of tubular micelles of pluronic lecithin organogel for transdermal delivery of sumatriptan. *AAPS PharmSciTech.* 11(4) (2010):1718–1725.
103. Lo Nostro P, Fratoni L, Ninham BW, Baglioni P. Water absorbency by wool fibers: Hofmeister effect. *Biomacromolecules.* 3(6) (2002):1217–1224.
104. Garg T, Singh O, Arora S, Murthy R. Scaffold: A novel carrier for cell and drug delivery. *Crit Rev Ther Drug Carrier Syst.* 29(1) (2012):1–63.
105. Madan JR, Sagar B, Chellappan DK, Dua K. Development and evaluation of transdermal organogels containing nicorandil. *Antiinflamm Antiallergy Agents Med Chem.* 12(3) (2013):246–252.
106. Baig MS, Dowling AH, Fleming GJ. Hertzian indentation testing of glass-ionomer restoratives: A reliable and clinically relevant testing approach. *J Dent.* 41(11) (2013):968–973.
107. Jayaramudu J, Reddy GS, Varaprasad K, Sadiku ER, Ray SS, Rajulu AV. Structure and properties of poly (lactic acid)/Sterculia urens uniaxial fabric biocomposites. *Carbohydr Polym.* 94(2) (2013):822–828.
108. Hashizaki K, Sakanishi Y, Yako S, Tsusaka H, Imai M, Taguchi H et al. New lecithin organogels from lecithin/polyglycerol/oil systems. *J Oleo Sci.* 61(5) (2012):267–275.
109. Pal A, Srivastava A, Bhattacharya S. Role of capping ligands on the nanoparticles in the modulation of properties of a hybrid matrix of nanoparticles in a 2D film and in a supramolecular organogel. *Chemistry.* 15(36) (2009):9169–9182.
110. Woodall R, Arnold JJ, McKay D, Asbill CS. Effect of formulation pH on transdermal penetration of antiemetics formulated in poloxamer lecithin organogel. *Int J Pharm Compd.* 17(3) (2013):247–253.
111. Sahu A, Choi WI, Tae G. A stimuli-sensitive injectable graphene oxide composite hydrogel. *Chem Commun (Camb).* 48(47) (2012):5820–5822.
112. Goyal G, Garg T, Malik B, Chauhan G, Rath G, Goyal AK. Development and characterization of niosomal gel for topical delivery of benzoyl peroxide. *Drug Deliv*, 2013.
113. Garg T, Singh S, Goyal AK. Stimuli-sensitive hydrogels: An excellent carrier for drug and cell delivery. *Crit Rev Ther Drug Carrier Syst.* 30(5) (2013):369–409.
114. Choi BD, Choi YJ. Nutraceutical functionalities of polysaccharides from marine invertebrates. *Adv Food Nutr Res.* 652012):11–30.

115. Dong Y, Guha S, Sun X, Cao M, Wang X, Zou S. Nutraceutical interventions for promoting healthy aging in invertebrate models. *Oxid Med Cell Longev.* 2012 (2012):718491.
116. Lupi FR, Gabriele D, Baldino N, Mijovic P, Parisi OI, Puoci F. Olive oil/policosanol organogels for nutraceutical and drug delivery purposes. *Food Funct.* 4(10) (2013):1512–1520.
117. Iwanaga K, Sumizawa T, Miyazaki M, Kakemi M. Characterization of organogel as a novel oral controlled release formulation for lipophilic compounds. *Int J Pharm.* 388(1–2) (2010):123–128.
118. Kumar R, Katare OP. Lecithin organogels as a potential phospholipid-structured system for topical drug delivery: A review. *AAPS PharmSciTech.* 6(2) (2005):E298–E310.
119. Yin A, Alfadhli E, Htun M, Dudley R, Faulkner S, Hull L et al. Dietary fat modulates the testosterone pharmacokinetics of a new self-emulsifying formulation of oral testosterone undecanoate in hypogonadal men. *J Androl.* 33(6) (2012):1282–1290.
120. Pietrasik Z, Wang H, Janz JA. Effect of canola oil emulsion injection on processing characteristics and consumer acceptability of three muscles from mature beef. *Meat Sci.* 93(2) (2013):322–328.
121. Azadbakht L, Haghighatdoost F. Canola oil consumption and bone health. *J Res Med Sci.* 17(12) (2012):1094–1095.
122. German JB, Dillard CJ. Saturated fats: A perspective from lactation and milk composition. *Lipids.* 45(10) (2010):915–923.
123. German JB, Dillard CJ. Saturated fats: What dietary intake? *Am J Clin Nutr.* 80(3) (2004):550–559.
124. Pernetti M, van Malssen KF, Floter E, Bot A. Structuring of edible oils by alternatives to crystalline fat. *Curr Opin Coll Interface Sci.* 12 (2007):221–231.
125. Bot A, Gilbert EP, Bouwman WG, Sawalha H, den Adel R, Garamus VM et al. Elucidation of density profile of self-assembled sitosterol + oryzanol tubules with small-angle neutron scattering. *Faraday Discuss.* 158 (2012):223–238; discussion 39–66.
126. Gulaboski R, Mirceski V, Mitrev S. Development of a rapid and simple voltammetric method to determine total antioxidative capacity of edible oils. *Food Chem.* 138(1) (2013):116–121.
127. Ribeiro HM, Morais JA, Eccleston GM. Structure and rheology of semisolid o/w creams containing cetyl alcohol/non-ionic surfactant mixed emulsifier and different polymers. *Int J Cosmet Sci.* 26(2) (2004):47–59.
128. Braun DM, Wang L, Ruan YL. Understanding and manipulating sucrose phloem loading, unloading, metabolism, and signalling to enhance crop yield and food security. *J Exp Bot.* 65(7) (2014):1713–1735.
129. Hammer KD, Birt DF. Evidence for contributions of interactions of constituents to the anti-inflammatory activity of hypericum perforatum. *Crit Rev Food Sci Nutr.* 54(6) (2014):781–789.
130. Veeck AP, Klein B, Ferreira LF, Becker AG, Heldwein CG, Heinzmann BM et al. Lipid stability during the frozen storage of fillets from silver catfish exposed in vivo to the essential oil of Lippia alba (Mill.) NE Brown. *J Sci Food Agric.* 93(4) (2013):955–960.
131. Pieve SD, Calligaris S, Panozzo A, Arrighetti G, Nicoli MC. Effect of monoglyceride organogel structure on cod liver oil stability. *Food Res Int.* 44 (2011):2978–2983.
132. Shu C, Liu MQ, Wang SS, Li L, Ye LW. Gold-catalyzed oxidative cyclization of chiral homopropargyl amides: Synthesis of enantioenriched gamma-lactams. *J Org Chem.* 78(7) (2013):3292–3299.
133. Wegner HA, Ahles S, Neuburger M. A new gold-catalyzed domino cyclization and oxidative coupling reaction. *Chemistry.* 14(36) (2008):11310–11313.
134. Verna R. The history and science of chocolate. *Malays J Pathol.* 35(2) (2013):111–121.
135. van Mechelen JB, Goubitz K, Pop M, Peschar R, Schenk H. Structures of mono-unsaturated triacylglycerols. V. The beta'(1)-2, beta'-3 and beta(2)-3 polymorphs of 1,3-dilauroyl-2-oleoylglycerol (LaOLa) from synchrotron and laboratory powder diffraction data. *Acta Crystallogr B.* 64(Pt 6) (2008):771–779.
136. van Mechelen JB, Peschar R, Schenk H. Structures of mono-unsaturated triacylglycerols. II. The beta2 polymorph. *Acta Crystallogr B.* 62(Pt 6) (2006):1131–1138.
137. Parnami N, Garg T, Rath G, Goyal AK. Development and characterization of nanocarriers for topical treatment of psoriasis by using combination therapy. *Artif Cells Nanomed Biotechnol.* 42(6) (2014):406–412.
138. Toro-Vazquez JF, Morales-Rueda J, Torres-Martinez A, Charo-Alonso MA, Mallia VA, Weiss RG. Cooling rate effects on the microstructure, solid content, and rheological properties of organogels of amides derived from stearic and (R)-12-hydroxystearic acid in vegetable oil. *Langmuir.* 29(25) (2013):7642–7654.
139. Rogers MA, Marangoni AG. Solvent-modulated nucleation and crystallization kinetics of 12-hydroxystearic acid: A nonisothermal approach. *Langmuir.* 25(15) (2009):8556–8566.
140. Elena DA, Rodrigues JN, AG L. Effects of crystalline microstructure on oil migration in a semisolid fat matrix. *Crystal Growth and Design.* 4(4) (2004):731–736.

141. Hughes BH, Muzzy HM, Laliberte LC, Grenier HS, Perkins LB, Skonberg DI. Oxidative stability and consumer acceptance of fish oil fortified nutrition bars. *J Food Sci.* 77(9) (2012):S329–S334.
142. Marty S, Baker K, Dibildox-Alvarado E, JN R. Monitoring and quantifying of oil migration in cocoa butter using a flatbed scanner and fluorescence light microscopy. *Food Res Int.* 38 (2005):1189–1197.
143. Osman H, Usta IM, Rubeiz N, Abu-Rustum R, Charara I, Nassar AH. Cocoa butter lotion for prevention of striae gravidarum: A double-blind, randomised and placebo-controlled trial. *BJOG.* 115(9) (2008):1138–1142.
144. Terech P, Aymonier C, Loppinet-Serani A, Bhat S, Banerjee S, Das R et al. Structural relationships in 2,3-bis-n-decyloxyanthracene and 12-hydroxystearic acid molecular gels and aerogels processed in supercritical CO(2). *J Phys Chem B.* 114(35) (2010):11409–11419.
145. Zhang H, Smith P, Adler-Nissen J. Effects of degree of enzymatic interesterification on the physical properties of margarine fats: Solid fat content, crystallization behavior, crystal morphology, and crystal network. *J Agric Food Chem.* 52(14) (2004):4423–4431.
146. Adam-Berret M, Boulard M, Riaublanc A, Mariette F. Evolution of fat crystal network microstructure followed by NMR. *J Agric Food Chem.* 59(5) (2011):1767–1773.
147. Nii S, Okumura S, Kinoshita T, Ishigaki Y, Nakano K, Yamaguchi K, Akita S. Extractant-impregnated organogel for capturing heavy metals from aqueous solutions. *Sep Purif Technol.* 73(2) (2010):250–255.
148. Ray S, Das AK, Banerjee A. Smart oligopeptide gels: In situ formation and stabilization of gold and silver nanoparticles within supramolecular organogel networks. *Chem Commun (Camb).* (26) (2006):2816–2818.

35 β-Lactoglobulin
Bioactive Nutrients Delivery

Li Liang and Muriel Subirade

CONTENTS

35.1 Introduction 729
35.2 Background of the Structure of β-LG and Its Response to Physicochemical Properties 730
35.3 Ligand-Binding Sites of Native β-LG 731
35.4 Hydrophobic Ligand-Binding Properties of β-LG 731
    35.4.1 Fatty Acids 732
    35.4.2 Vitamins 733
    35.4.3 Other Ligands 733
35.5 Amphiphilic and Hydrophilic Ligand-Binding Properties of β-LG 734
35.6 Questions Regarding the Potential Use of β-LG as a Carrier of Bioactive Molecules 734
Acknowledgments 736
References 736

35.1 INTRODUCTION

Increasing consumption of functional foods has been proposed as an alternative to classical pharmacology for improving public health. Such foods contain bioactive nutrients, that is, vitamins, polyphenols, probiotics, and peptides that offer benefits beyond basic nutritional functions, including the possibility for delaying and preventing chronic diseases. However, many of these nutrients are poorly soluble in water and sensitive to environmental factors associated with food processing or the gastrointestinal tract. They must therefore be encapsulated using effective carriers in order to make them compatible with food product systems and to provide protection against destructive factors.[1] Except in the case of bioactive molecules dissolved in the inner phase of an emulsion-based carrier, entrapped bioactive nutrients generally interact with the carrier material. Clarification of the underlying interaction mechanisms is therefore important for the development of effective carriers as delivery systems for bioactive nutrients.

Food proteins have been widely used as carrier materials for the preparation of nano/microparticles because of their nutritional value and their ability to form gels and emulsions and to interact with polysaccharides.[2,3] Beta-lactoglobulin (β-LG) is the major whey protein in milk, exhibiting an affinity to a wide range of compounds, including fatty acids, vitamins, phospholipids, polyphenols, indole derivatives, peptides, and other proteins. Since this protein is classified as a member of the lipocalin superfamily, of which one of the distinguishing properties is the ability to bind small hydrophobic molecules, most reviews have focused on its binding of hydrophobic molecules, such as retinol, fatty acid, cholesterol, and vitamin D.[4,5] In this entry, we emphasize bioactive molecules of small molecular mass and with different physiochemical properties, and to discuss their binding properties on β-LG and the potential of β-LG-ligand complexes for use as effective delivery systems for bioactive molecules.

35.2 BACKGROUND OF THE STRUCTURE OF β-LG AND ITS RESPONSE TO PHYSICOCHEMICAL PROPERTIES

Bovine β-LG is a small, soluble globular protein of well-known structure containing 162 amino acid residues for a molecular mass of about 18 kDa. Under physiological condition, it exists predominantly as a dimer. The monomer is folded into a calyx called the β-barrel, which consists of eight antiparallel β-strands (A–H) starting from the N-terminal end of the peptide (Figure 35.1).[6] The bend in strand A allows it to form a connection with strand H. Strands A–D form one surface of the β-barrel and strands E–H form the other.[7] The quite short BC, DE, and FG loops connect the β-strands at the closed end of the calyx, whereas the AB, CD, EF, and GH loops are significantly longer and more flexible and located at the open end. The EF loop acts as a lip at the entrance of the cavity. An α-helix, which follows strand H, and a ninth β-strand (I) located on the outer surface of the β-barrel near the C terminus complete the structure. Strand I on the opposite side of strand A from strand H forms an antiparallel interaction with the same strand of the second monomer in dimer formation.[5]

The structure of β-LG is dependent on the pH of the solutions. At pH 7.0, β-LG undergoes the so-called Tanford transition, which involves a conformational change, an opening–closing shift of the EF loop (residues 85–90) triggered by protonation of the Glu89 (glutamate) residue.[8,9] At neutral pH, β-LG exists as a mixture of monomers and oligomers, of which the oligomeric state depends on concentration and pH. As the pH nears the isoelectric point of nearly 5.2, β-LG undergoes a dimer-to-octamer transition, which dissociates at pH <3 to yield primarily monomers.[10] It has been reported that β-LG native structure is stable and remains mostly intact at acidic pH,[11] but the molecule undergoes irreversible base-induced unfolding above pH 9.[12]

β-LG is a heat-sensitive protein, showing a multistep mechanism of structural transitions.[13–15] When heated at physiological concentrations (<5 wt%) from room temperature to above 60°C, the protein dimer first dissociates to the native monomer, of which the structure is then disrupted to form a molten globule, that is, a partly unfolded state with less tight packing of the side chains while remaining a secondary structure similar to that of the native state.[13,15,16] Beyond 80°C, the structure

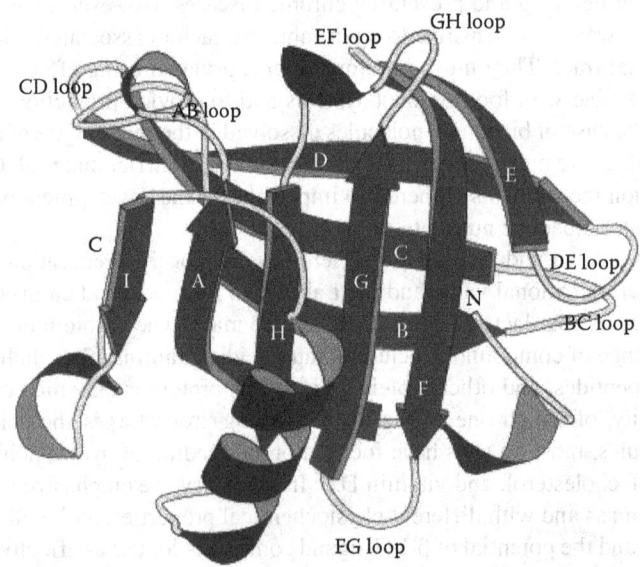

FIGURE 35.1 Structure of β-LG with β strands and joining loops labeled. (Reprinted from *Structure*, 5(1), Brownlow, S., Morais, C.J.H., Cooper, R., Flower, D.R., Yewdall, S.J., Polikarpov, I., North, A.C.T., and Sawyer, L., Bovine β-lactoglobulin at 1.8 Å resolution-still an enigmatic lipocalin, 481–495. Copyright 1997, with permission from Elsevier.)

of β-LG undergoes irreversible modifications, exposing both the inner hydrophobic residues and the sulfhydryl groups to intermolecular interactions. Thermally induced structural transitions are dependent on concentration and pH. Irreversible aggregation of β-LG is increased at higher concentration and pH closer to pI.[17] The resistance of native β-LG to thermal unfolding is pH-dependent, being greater at acidic pH.[18]

In addition to pH and temperature, the native structure of β-LG is also sensitive to ionic strength, high-pressure, and the presence of surfactant, and other physicochemical factors. The protein structural features and transitions have a major impact on its functional properties, including emulsification, gelling properties, and ligand binding.

35.3 LIGAND-BINDING SITES OF NATIVE β-LG

β-LG has a primary ligand-binding site at the internal cavity inside the β-barrel. Tryptophan at position 19 (Trp19) is hidden in an apolar environment at the bottom of the calyx and appears to be involved in the interaction of the protein with hydrophobic ligands. Its fluorescence contributes to about 80% of the total fluorescence of β-LG[19] and has proven useful for studying β-LG-ligand binding properties. Access of ligands to the internal cavity is under the control of the Tanford transition. At neutral pH, EF loop is in the opening conformation, giving hydrophobic ligands access to the internal cavity. As the solution pH decreases, protonation of Glu89 shifts the EF loop to the closed conformation.[19,20] In addition, electrostatic interaction between Lys83 (strand E) and Glu74 (stand D) is disrupted, which combined with electrostatic repulsion among positive charges softens the structure of β-LG near the cavity entrance.[19] These factors might combine to expel ligands from the cavity, decreasing ligand-induced Trp19 fluorescence quenching at acidic pH.[21]

Among the external portions of β-LG that have been postulated as potential sites for ligand binding are the outer surface near Trp19–Arg124, the surface hydrophobic pocket in the groove between the α-helix and the β-barrel,[22] a site near the aperture of the β-barrel and a site at the monomer–monomer interface of the protein dimer.[23] The site near the aperture of the β-barrel is sensitive to Tanford transition, whereas the monomer–monomer interface site is dependent on the monomer/dimer equilibrium.

Due to the existence of multiple binding sites, β-LG can bind various types of ligands. Each site might bind ligands of similar structure and physicochemical properties. Local conformational changes in β-LG are dependent on environmental factors, which imply the possibility of modulating the protein ligand-binding properties.[23,24] Structural modification and unfolding may destroy ligand-binding sites of native β-LG while creating new binding sites by exposing internal hydrophobic residues to the solvent phase.[25,26] Investigation of the response of β-LG–ligand interactions to various physicochemical stimuli or stress has been used to gain insight into binding mechanisms.[21]

35.4 HYDROPHOBIC LIGAND-BINDING PROPERTIES OF β-LG

On the basis of structural similarity, β-LG and retinol-binding protein are classified as members of the lipocalin family, which are characterized by a structure possessing a conserved folding pattern: the eight-stranded antiparallel β-barrel with a repeated + 1 topology.[27] Together with the fatty-acid-binding proteins and the avidins, the lipocalins make up part of the calycin superfamily, which is marked by extensive structural similarities. The avidin barrel, although eight-stranded, is less elliptical in cross section than the lipocalin barrel structure, whereas the fatty-acid-binding protein structure comprises a ten-stranded antiparallel and discontinuous β-barrel.[28] One of the distinguishing molecular properties of the calycin superfamily is the ability to bind small and principally hydrophobic molecules within the internal cavity. β-LG is known to bind several hydrophobic ligands such as fatty acids, vitamins, cholesterol, and porphyrin species.[4,29,30] It has been proposed that β-LG could be used as a versatile carrier of small hydrophobic molecules in controlled delivery applications, although some controversy still surrounds the reported β-LG-ligand-binding properties.

35.4.1 Fatty Acids

When isolated from milk using nondenaturing method, β-LG contains endogenously bound fatty acids, mainly palmitic, oleic, myristic, and butyric.[31,32] Its interactions with fatty acids are mainly hydrophobic and appear to require a hydrocarbon terminus. The affinity of fatty acids for β-LG is dependent on the length in the aliphatic chain and is maximal for palmitic acid (Figure 35.2). Structural constraints imposed by C=C double bonds may slightly affect fatty acid binding to β-LG.[33] The internal cavity of the molecule, which appears to be the primary binding site for fatty acids, allows the hydrocarbon chain stretch out into the center of the protein.[34] Binding of fatty acids longer than 16 carbons is limited due to steric hindrance. Electrostatic interaction is also involved by binding the fatty acid carboxyl group to both Lys-60 and Lys-69 at the entrance to the β-LG cavity.[34]

However, several other sites on β-LG have been reported to bind fatty acids. At pH 7, increasing dimer formation as a result of increased concentration has been found to eliminate palmitate binding sites on the monomer, while forming a higher affinity pocket.[35] Fatty acids have also been found to bind to the external site in the groove between the α-helix and the β-barrel.[36]

The ability of β-LG to bind fatty acids has been exploited as an emulsifying agent in food technology or as a fatty acid carrier in cell culture.[4] Formation of complexes with β-LG provides partial protection of the w-3 polyunsaturated fatty acid docosahexaenoic acid (DHA) against oxidation degradation.[37] Complexation with fatty acids increases the resistance of β-LG to thermal and urea denaturation[38,39] and to proteolytic degradation,[40] suggesting that ligand binding may be an important factor in the stabilization of the protein structure,[4] and hence in the modification of its functional properties.

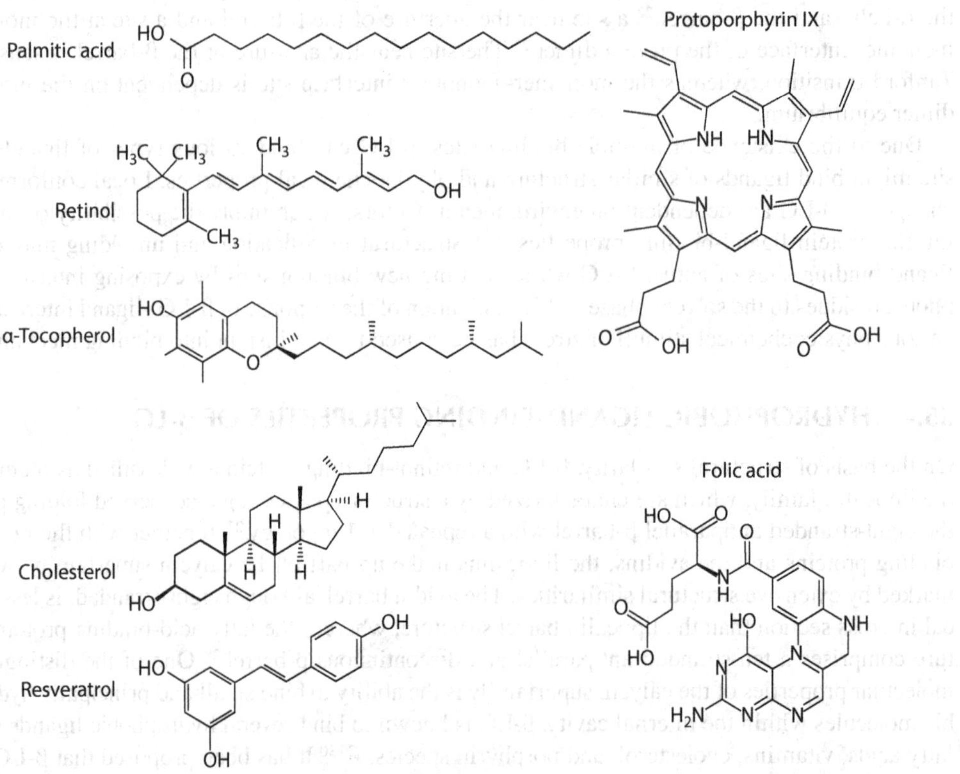

FIGURE 35.2 Chemical structure of fatty acids, vitamins, cholesterol, protoporphyrin IX, and resveratrol.

35.4.2 Vitamins

Futterman and Heller first reported the formation of β-LG complexes with retinol (vitamin A, Figure 35.2) in 1972.[41] The β-LG dimer binds two retinol molecules, whereas the monomer binds one.[42] This protein and retinol-binding protein are known as kernel lipocalins, which share three characteristic conserved sequence motifs.[43] By inference from the structural similarities of these proteins, it has been proposed that retinol and its derivatives bind within the internal cavity of β-LG, with the retinol β-ionone ring located close to Trp19 at the bottom of the calyx. The retinol hydroxyl group can interact with Lys70,[44,45] which contributes a minor additional force to β-LG-retinol binding. The carboxyl group of retinoic acid can of course interact with the protonated amino group of the lysine residue by electrostatic attraction.[19] However, an external site has been reported for the binding of retinol and retinoic acid, at which Phe136 and Lys141, respectively, replace the roles of Trp19 and Lys70 within the internal cavity.[46,47]

Interactions of β-LG with other lipophilic vitamins have also been reported. Vitamin D binds to β-LG at a stoichiometric ratio of 2:1. One site is within the internal cavity, which receives the ligand aliphatic tail, whereas the ligand 3-OH group forms a hydrogen bond with the carbonyl group of Lys60. The other site is located at the surface pocket between α-helix and β-strand I, which provides a strong hydrophobic force.[48] The existence of the external site is substantiated by switching off the lip of calyx at low pH and thermally disrupted the calyx structure, which results in a stoichiometric binding ratio of 1:1.[49] The affinity of this site for vitamin D is similar to that of the calyx.[29] Alpha-tocopherol (Figure 35.2), the most abundant and active form of vitamin E, also has two binding sites on β-LG, one inside the internal cavity and the other with lower affinity on the outer surface,[21,50] contrary to previous report of binding only to the internal cavity.[51]

The formation of complexes with β-LG improves the solubility of α-tocopherol and its stability against oxidative degradation[50] and protects retinol from degradation induced by heat, oxidation, and irradiation.[52] Furthermore, binding of retinol or retinoic acid increases the stability of β-LG during heating and increases its resistance to tryptic hydrolysis.[53] Intestinal uptake of retinol is enhanced specifically in the presence of β-LG[54] and a receptor and binding sites for retinol-complexed β-LG in germ cell plasma membranes have been identified.[55] Milk with vitamin D effectively enhances uptake of this vitamin, and little doubt remains that β-LG is largely responsible.[56] These findings support the hypothesis that β-LG could be used as a carrier to protect hydrophobic vitamins against destructive factors and deliver them to different cells and tissues.

35.4.3 Other Ligands

Cholesterol (Figure 35.2), an essential structural component of mammalian cell membranes and a precursor for the biosynthesis of steroid hormones, bile acids, and vitamin D, can also bind to β-LG with a stoichiometry of 2:1, albeit with an affinity less than that of vitamin D.[29] The protein internal cavity is a proved binding site of cholesterol, of which the compound 3-OH forms a hydrogen bond with the carbonyl oxygen of Pro38 at the mouth of the calyx,[5] while the aromatic ring chromophore positions itself in close proximity to Trp19. An external site in the surface hydrophobic pocket formed by the α-helix and the β-barrel is also likely.[29] However, protoporphyrin IX (Figure 35.2), an important precursor of biologically essential prosthetic groups such as heme and chlorophylls, is reported to bind on β-LG at a different site,[30] possibly near the aperture of the interior β-barrel, as suggested by the pH dependence of this interaction.[23]

In summary, as a lipocalin, β-LG contains a primary site in the internal cavity of the β-barrel and a second site in the groove between the α-helix and β-barrel for external binding of small hydrophobic molecules, although several surface pockets have also been suggested as ligand-binding sites. The ligand-binding capability of the internal cavity of the β-barrel may be limited to structures that have a hydrophobic "tail."[57] It seems to accommodate linear hydrophobic molecules such as fatty acids[5] and single-ring aromatic compounds such as retinol.[58,59] Large molecular size may be the reason why protoporphyrin IX binds to a site different from that of retinol. Furthermore, both ligand and β-LG structures are dependent on environmental factors such as concentration, pH, and temperature. These factors, together with differences in protein purification or complex preparation,[60] may have resulted in conflicting reports regarding β-LG/ligand interactions.

35.5 AMPHIPHILIC AND HYDROPHILIC LIGAND-BINDING PROPERTIES OF β-LG

In 1987, Farrell et al. reported binding of β-LG with *p*-nitrophenyl phosphate and its derivatives and pyridoxal phosphate at a stoichiometry of 1 molecule per protein monomer.[59] A variety of polar compounds including 1-anilinonaphthalene-8-sulfonate (ANS), polyphenols, and vitamin B were subsequently found to bind to β-LG.[24,61,62] Our studies suggest that the amphiphilic compound resveratrol (Figure 35.2) and the water-soluble compound folic acid (Figure 35.2) both form a 1:1 complex with β-LG,[62,63] with the former bound to the outer surface of β-LG near Trp19–Arg124 and the latter bound in a hydrophobic pocket in the groove between the α-helix and the β-barrel.[21] A greater stoichiometric ratio has been found for phenolic compounds: 1.70 ± 0.13 for daidzein and 1.74 ± 0.17 for myricetin.[24] Insensitivity to acidic pH suggests their binding sites are the same as those of resveratrol or folic acid. The amphiphilic compound ANS is an exception, binding to both the internal and surface sites.[61,64] Ligands that bind to the outer surface of β-LG tend to bear multiple hydrophilic groups at different extremities.

Affinities of different bioactive molecules for β-LG vary widely (Table 35.1). It is generally accepted that affinities of amphiphilic or hydrophilic molecules for external sites are weaker than those of hydrophobic molecules for the internal cavity (~10^5 M^{-1} compared to 10^6 M^{-1}). However, Riihimaki et al. reported constants of ~10^6 M^{-1} for the binding of phenolic compounds to external sites.[24] It has been reported that protein–ligand interactions are reversible, with equilibriums existing between monomeric and associated ligands, between monomeric ligands and protein-ligand complexes, and possibly between associated ligands and protein–ligand complexes.[65] It is thus possible for different ligands to compete for the same site of β-LG molecule. Preferential binding of one ligand to β-LG depends not only on their relative binding constants but also upon their relative solubility or dissolved protein concentration.[7] A ligand that does not bind as strongly might be present at sufficient concentration for preferential binding to the site. This should be carefully considered when investigating the site of ligand binding on β-LG using competition experiments.

The formation of complexes with β-LG has an impact on physicochemical properties of amphiphilic or hydrophilic compounds. It provides a significant increase in the hydrosolubility of *trans*-resveratrol, a slight increase in its photostability,[63] and considerable protection of folic acid against photodecomposition.[62] Although it decreases the antioxidant activity of tea polyphenol epigallocatechin-3-gallate (EGCG), the protein complexation could delay the bioactive molecule degradation.[26,66] With an affinity for EGCG about 3.5 fold higher than that of native β-LG,[26] heat-induced β-LG nanoparticles could suppress the bitterness and astringency of this compound and sustain its release during digestion.[67]

35.6 QUESTIONS REGARDING THE POTENTIAL USE OF β-LG AS A CARRIER OF BIOACTIVE MOLECULES

It has been suggested that ligand-binding proteins have potential applications as means of encapsulation in systems designed for the protection and delivery of bioactive molecules.[68] As a member of the proteins, β-LG possesses multiple sites for binding a variety of ligands with different structures and physicochemical properties. β-LG is a water-soluble protein and can form soluble complexes with both lipophilic and amphiphilic nutrients. By forming complexes with β-LG, bioactive molecules can be protected under otherwise destructive conditions. Its protection of folic acid has been

TABLE 35.1
Binding Constants of Various Ligands on β-LG Obtained Using Ligand-Induced Trp Fluorescence Quenching

Ligands	Solubility	Binding Site	Binding Constant (M^{-1})	Reference
Palmitate	Hydrophobic	Internal cavity	1×10^7	[33]
Oleic acid	Hydrophobic	Internal cavity	8×10^6	[33]
Retinol	Hydrophobic	Internal cavity	2×10^7	[19]
Protoporphyrin IX	Hydrophobic	Possibly near the aperture of the interior β-barrel	2×10^6	[23]
Resveratrol	Amphiphilic	The outer surface near Trp19–Arg124	2×10^5	[63]
Folic acid	Hydrophilic	A hydrophobic pocket between the α-helix and the β-barrel	4×10^5	[62]
EGCG	Hydrophilic	Uncertain exosite	1×10^4	[66]
Myricetin	Hydrophilic	Uncertain exosite	3×10^6	[24]

found superior to that provided by bovine serum albumin (BSA) and α-lactalbumin (Figure 35.3).[69] β-LG-bound retinol and retinoic acid have been found less sensitive to light-induced oxidation than the same compounds bound to BSA.[53] This protein therefore might be useful as a natural vehicle for fortifying dietary products with a variety of bioactive molecules, particularly in the case of aqueous systems such as low-fat foods.[70]

The ligand-binding sites of β-LG, including all the external sites as well as the internal cavity, are solvent-accessible. Resveratrol bound to the outer surface near Trp19–Arg124 is in a less hydrophobic environment compared to 75% ethanol.[63] With the tip of the isoprene tail protruding from the

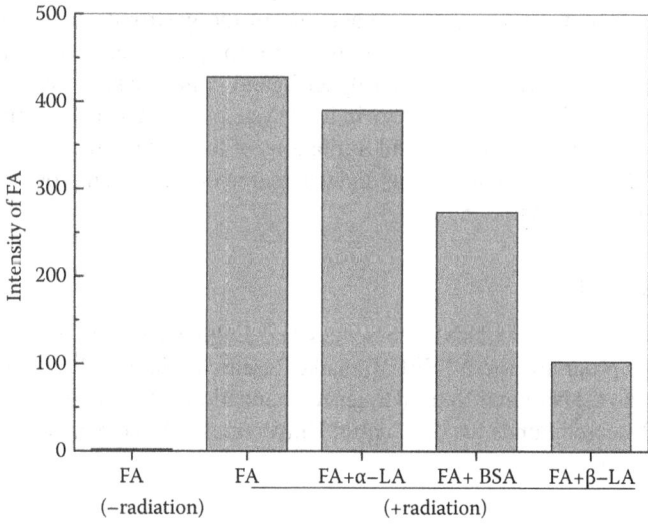

FIGURE 35.3 Fluorescence intensity (λ_{max} = 450 nm) of 10 μM folic acid (FA) in the absence or presence of β-LG, bovine serum albumin (BSA), or α-lactalbumin (α-LA) before and after 110 min of irradiation at an excitation wavelength of 348 nm (protein concentration is 20 μM). (Reprinted from *Food Chemistry*, 141(2), Liang, L., Zhou, P., and Subirade, M., Protective effect of bovine serum albumin, β-lactoglobulin and α-lactalbumin on photodegradation of folic acid, 655–1572. Copyright 2013, with permission from Elsevier.)

central calyx, retinol bound to β-LG is more exposed to the solvent than it is when bound to retinol-binding protein.[19] This partial exposure should be similar for other hydrophobic ligands bound to the internal cavity of β-LG. This might be the reason why β-LG only delays rather than preventing degradation of some bound bioactive molecules. It has been reported that the formation of β-LG-pectin nanocomplexes could improve the protection of DHA and vitamin D against oxidation degradation.[37,71] Therefore, it might be necessary to develop more complicated particulate carrier systems based on β-LG–ligand complexes for protecting bioactive molecules.

Binding of ligands to the outer surface near Trp19–Arg124 withstands changes in pH and can be improved by heat. Binding to the hydrophobic pocket in the groove between the α-helix and the β-barrel withstands changes in pH and temperature.[21] However, the opposite is observed for binding of ligands (e.g., fatty acids, vitamins, and cholesterol) to the internal cavity.[21,24,33,49] The involvement of hydrophilic interactions (i.e., hydrogen bonding and electrostatic interaction) makes binding to this site somewhat sensitive to changes in ionic strength. Binding of retinol and cis-parinaric acid to β-LG has been found to decrease under high-pressure treatment exceeding 150 and 200 MPa, respectively.[72] β-LG-retinol binding properties were not affected in the presence of anionic surfactants such as sodium dodecyl sulfate or nonionic surfactants such as Triton X-100[73] but were enhanced substantially in the presence of cationic surfactants.[74] The environmental responses must be considered when designing effective carriers based on β-LG for encapsulation and controlled release of bioactive molecules.

Ligand functional groups essential for biological activity (e.g., the hydroxyl groups that underlie the antioxidant activity of EGCG) may be shielded in complexes with β-LG. When binding to β-LG occurs via these groups, the physiological activity of the compound is temporarily suspended.[66] Although β-LG in native state is resistant to breakdown by acid and pepsin in stomach, it is digested by proteases in the small intestine. Additional study is needed in order to determine whether or not enzymatic hydrolysis releases bound ligands from β-LG or its hydrolyzed products.

A carrier used for the encapsulation of bioactive molecules added to functional foods should satisfy the following requirements: generally recognized as safe, high-loading capacity and efficiency, compatible with food product systems, physical stability under conditions encountered during food processing and storage, ensuring the chemical stability of the bioactive molecule under conditions encountered during food processing and storage and in the gastrointestinal tract, bringing about controlled release of the bioactive molecule in response to specific environmental conditions.[75] In order to develop effective carriers based on β-LG-ligand complexes, the release of encapsulated bioactive molecules from the protein complex in food systems and during digestion must be studied, in order to further define the advantages and limitations of the system, based on previous studies of the mechanisms of β-LG-ligand interactions and the degree of protection provided against chemical degradation of the payload compound.

ACKNOWLEDGMENTS

This work was supported by the Natural Sciences and Engineering Research Council of Canada (NSERC) discovery program, the NSERC Canada Research Chair in Proteins, Bio-systems and Functional Foods, and the National Natural Science Foundation of China (NSFC Projects 31201291), the Fundamental Research Funds for the Central Universities (China, No. JUSRP211A31).

REFERENCES

1. McClements, D.J.; Decher, E.A.; Park, Y.; Weiss, J. Structural design principles for delivery of bioactive components in nutraceuticals and functional foods. *Crit. Rev. Food Sci. Nutr.* 2009, *49* (6), 577–606.
2. Chen, L.Y.; Remondetto, G.E.; Subirade, M. Food protein-based materials as nutraceutical delivery systems. *Trends Food Sci. Technol.* 2006, *17* (5), 272–283.
3. Kanakis, C.D.; Hasni, I.; Bourassa, P.; Tarantilis, P.A.; Polissiou, M.G.; Tajmir-Riahi, H.A. Milk β-lactoglobulin complexes with tea polyphenols. *Food Chem.* 2011, *127* (3), 1046–1055.

4. Perez, M.D.; Calvo, M. Interaction of β-lactoglobulin with retinol and fatty acids and its role as a possible biological function for this protein: A review. *J. Dairy Sci.* 1995, *78* (5), 978–988.
5. Kontopidis, G.; Holt, C.; Sawyer, L. Invited review: Beta-lactoglobulin: Binding properties, structure, and function. *J. Dairy Sci.* 2004, *87* (4), 785–796.
6. Brownlow, S.; Morais, C.J.H.; Cooper, R.; Flower, D.R.; Yewdall, S.J.; Polikarpov, I.; North, A.C.T.; Sawyer L. Bovine β-lactoglobulin at 1.8 Å resolution-still an enigmatic lipocalin. *Structure* 1997, *5* (4), 481–495.
7. Kontopidis, G.; Holt, C.; Sawyer, L. The ligand-binding site of bovine β-lactoglobulin: Evidence for a function? *J. Mol. Biol.* 2002, *318* (4), 1043–1055.
8. Qin, B.Y.; Bewley, M.C.; Creamer, L.K.; Baker, H.M.; Baker, E.N.; Jameson, G.B. Structural basis of the tanford transition of bovine β-lactoglobulin. *Biochemistry* 1998, *37* (40), 14014–14023.
9. Sakurai, K.; Goto, Y. Dynamics and mechanism of the tanford transition of bovine β-lactoglobulin studied using heteronuclear NMR spectroscopy. *J. Mol. Biol.* 2006, *356* (2), 483–496.
10. Sakurai, K.; Goto, Y. Principal component analysis of the pH-dependent conformational transitions of bovine β-lactoglobulin monitored by heteronuclear NMR. *PNAS* 2007, *104* (39), 15346–15351.
11. Fogolari, F.; Ragona, L.; Zetta, L.; Romagnoli, S.; De Kruif, K.G.; Molinari, H. Monomeric bovine β-lactoglobulin adopts a β-barrel fold at pH 2. *FEBS Lett.* 1998, *436* (2), 149–154.
12. Taulier, N.; Chalikian, T.V. Characterization of pH-induced transitions of β-lactoglobulin: Ultrasonic, densimetric, and spectroscopic studies. *J. Mol. Biol.* 2001, *314* (4), 873–889.
13. Qi, X.L.; Holt, C.; Mcnulty, D.; Clarke, D.T.; Brownlow, S.; Jones, G.R. Effect of temperature on the secondary structure of β-lactoglobulin at pH 6.7, as determined by CD and IR spectroscopy: A test of the molten globule hypothesis. *Biochem. J.* 1997, *324* (Pt. 1), 341–346.
14. Seo, J.A.; Hedoux, A.; Guinet, Y.; Paccou, L.; Affouard, F.; Lerbret, A.; Descamps, M. Thermal denaturation of β-lactoglobulin and stabilization mechanism by trehalose analyzed from raman spectroscopy investigations. *J. Phys. Chem. B* 2010, *114* (19), 6675–6684.
15. De wit, J.N. Thermal behavior of bovine β-lactoglobulin at temperatures up to 150°C. *Trends Food Sci. Technol.* 2009, *20* (1), 27–34.
16. Carrotta, R.; Bauer, R.; Waninge, R.; Rischel, C. Conformational characterization of oligomeric intermediates and aggregates in β-lactoglobulin heat aggregation. *Protein Sci.* 2001, *10* (7), 1312–1318.
17. Nicolai, T.; Britten, M.; Schmitt, C. β-Lactoglobulin and WPI aggregates: Formation, structure and applications. *Food Hydrocolloids* 2011, *25* (8), 1945–1962.
18. Kella, N.K.D.; Kinsella, J.E. Enhanced thermodynamic stability of β-lactoglobulin at low pH. A possible mechanism. *Biochem. J.* 1988, *255* (1), 113–118.
19. Cho, Y.; Batt, C.A.; Sawyer, L. Probing the retinol-binding site of bovine β-lactoglobulin. *J. Biol. Chem.* 1994, *269* (15), 11102–11107.
20. Ragona, L.; Fogolari, F.; Catalano, M.; Ugolini, R.; Zetta, L.; Molinari, H. EF loop conformational change triggers ligand binding in β-lactoglobulins. *J. Biol. Chem.* 2003, *278* (40), 38840–38846.
21. Liang, L.; Subirade, M. Study of the acid and thermal stability of β-lactoglobulin–ligand complexes using fluorescence quenching. *Food Chem.* 2012, *132* (4), 2023–2029.
22. Sawyer, L.; Brownlow, S.; Polikarpov, I.; Wu, S.Y. β-Lactoglobulin: Structural studies, biological Clues. *Int. Dairy J.* 1998, *8* (2), 65–72.
23. Tian, F.; Johnson, K.; Lesar, A.E.; Moseley, H.; Ferguson, J.; Samuel, I.D.W.; Mazzini, A.; Brancaleon, L. The pH-dependent conformational transition of β-lactoglobulin modulates the binding of protoporphyrin IX. *Biochim. Biophys. Acta* 2006, *1760* (1), 38–46.
24. Riihimaki, L.H.; Vainio, M.J.; Heikura, J.M.S.; Valkonen, K.H.; Virtanen, V.T.; Vuorela, P.M. Binding of phenolic compounds and their derivatives to bovine and reindeer β lactoglobulin. *J. Agric. Food Chem.* 2008, *56* (17), 7721–7729.
25. Mousavi, S.H.; Bordbar, A.K.; Haertlé, T. Changes in structure and in interactions of heat-treated bovine β-lactoglobulin. *Protein Pept. Lett.* 2008, *15* (8), 818–825.
26. Shpigelman, A.; Israeli, G.; Livney, Y.D. Thermally-induced protein–polyphenol co-assemblies: Beta lactoglobulin-based nanocomplexes as protective nanovehicles for EGCG. *Food Hydrocolloids* 2010, *24* (8), 735–743.
27. Flower, D.R.; North, A.C.T.; Attwood, T.K. Structure and sequence relationships the lipocalins and related proteins. *Protein Sci.* 1993, *2* (5), 753–761.
28. Flower, D.R. The lipocalin protein family: Structure and function. *Biochem. J.* 1996, *318* (Pt. 1), 1–14.
29. Wang, Q.W.; Allen, J.C.; Swaisgood, H.E. Binding of vitamin D and cholesterol to β-lactoglobulin. *J. Dairy Sci.* 1997, *80* (6), 1054–1059.
30. Dufour, E.; Marden, M.C.; Haertlé, T. β-Lactoglobulin binds retinol and protoporphyrin IX at two different binding sites. *FEBS Lett.* 1990, *277* (1–2), 223–226.

31. Perez, D.M.; Diaz de Villegras, C.; Sanchez, L.; Aranda, P.; Ena, J.M.; Calvo, M. Interaction of fatty acids with β-lactoglobulin and albumin from ruminant milk. *J. Biochem.* (Tokyo) 1989, *106* (6), 1094–1097.
32. Taheri-Kafrani, A.; Asgari-Mobarakeh, E.; Bordbar, A.K.; Haertlé, T. Structure–function relationship of β-lactoglobulin in the presence of dodecyltrimethyl ammonium bromide. *Colloid Surface B* 2010, *75* (1), 268–274.
33. Frapin, D.; Dufour, E.; Haertle, T. Probing the fatty acid binding site of β-lactoglobulins. *J. Protein Chem.* 1993, *12* (4), 443–449.
34. Wu, S.Y.; Perez, M.D.; Puyol, P.; Sawyer, L. β-Lactoglobulin binds palmitate within its central cavity. *J. Biol. Chem.* 1999, *274* (1), 170–174.
35. Wang, Q.W.; Allen, J.C.; Swaisgood, H.E. Protein concentration dependence of palmitate binding to β-lactoglobulin. *J. Dairy Sci.* 1998, *81* (1), 76–81.
36. Narayan, M.; Berliner, L.J. Mapping fatty acid binding to β-lactoglobulin:Ligand binding is restricted by modification of Cys 121. *Protein Sci.* 1998, *7* (1), 150–157.
37. Zimet, P.; Livney, Y.D. Beta-lactoglobulin and its nanocomplexes with pectin as vehicles for ω-3 polyunsaturated fatty acids. *Food Hydrocolloids* 2009, *23* (4), 1120–1126.
38. Puyol, P.; Ptrez, M.D.; Peiro, J.M.; Calvo, M. Effect of retinol and fatty acid binding to bovine β-lactoglobulin on its resistance to thermal denaturation. *J. Dairy Sci.* 1994, *77* (6), 1494.
39. Creamer, L.K. Effect of sodium dodecyl sulfate and palmitic acid on the equilibrium unfolding of bovine β–lactoglobulin. *Biochemistry* 1995, *34* (21), 7170–7176.
40. Puyol, P.; Ptrez, M.D.; Mata, L.; Ena, J.M.; Calvo, M. Effect of retinol and fatty acid binding by bovine β-lactoglobulin on its resistance to trypsin digestion. *Int. Dairy J.* 1993, *3* (7), 589.
41. Futterman, S.; Heller, J. The enhancement of fluorescence and the decreased susceptibility to enzymatic oxidation of retinol complexed with bovine serum albumin, β-lactoglobulin and the retinol-binding protein of human plasma. *J. Biol. Chem.* 1972, *247* (16), 5168–5172.
42. Fugate, R.D.; Song, P.S. Spectroscopic characterization of β-lactoglobulin-retinol complex. *Biochim. Biophys. Acta* 1980, *625* (1), 28–42.
43. Flower, D.R.; North, A.C.T.; Sansom, C.E. The lipocalin protein family: Structural and sequence overview. *Biochim. Biophys. Acta* 2000, *1482* (1–2), 9–24.
44. Horwitz, J.; Heller, J. Properties of the chromophore binding site of retinol-binding protein from human plasma. *J. Biol. Chem.* 1974, *249* (15), 4712–4719.
45. Dufour, E.; Haertle, T. Alcohol-induced changes of beta-lactoglobulin-retinol-binding stoichiometry. *Protein Eng.* 1990, *4* (2), 185–190.
46. Monaco, H.L.; Zanotti, G.; Spadon, P.; Bolognesi, M.; Sawyer, L.; Eliopoulos, E.E. Crystal structure of the trigonal form of bovine β-lactoglobulin and of its complex with retinol at 2–5 A resolution. *J. Mol. Biol.* 1987, *197* (4), 695–706.
47. Lange, D.C.; Kothari, R.; Patel, R.C.; Patel, S.C. Retinol and retinoic acid bind to a surface cleft in bovine β-lactoglobulin: A method of binding site determination using fluorescence resonance energy transfer. *Biophys. Chem.* 1998, *74* (1), 45–51.
48. Yang, M.C.; Guan, H.H.; Liu, M.Y.; Lin, Y.H.; Yang, J.M.; Chen, W.L.; Chen, C.J.; Mao, S.J.T. Crystal structure of a secondary vitamin D_3 binding site of milk β-lactoglobulin. *Proteins* 2008, *71* (3), 1197–1210.
49. Yang, M.C.; Guan, H.H.; Yang, J.M.; Ko, C.N.; Liu, M.Y.; Lin, Y.H.; Huang, Y.C.; Chen, C.J.; Mao, S.J.T. Rational design for crystallization of β-Lactoglobulin and Vitamin D3 complex: Revealing a secondary binding site. *Cryst. Growth Des.* 2008, *8* (12), 4268–4276.
50. Liang, L.; Tremblay-Hébert, V.; Subirade, M. Characterisation of the β-lactoglobulin/α-tocopherol complex and its impact on α-tocopherol stability. *Food Chem.* 2011, *126* (3), 821–826.
51. Allen, J.C.; Wang, Q.; Swaisgood, H.E. Binding of vitamin K and vitamin E to bovine beta-lactoglobulin. In *IFT Annual Meeting*, Chicago, IL, July 24–28, 1999.
52. Hattori, M.; Watabe, A.; Takahashi, K. β-Lactoglobulin protects β-ionone related compounds from degradation by heating, oxidation, and irradiation. *Biosci. Biotechnol. Biochem.* 1995, *59* (12), 2295–2297.
53. Shimoyamada, M.; Yoshimura, H.; Tomida, K.; Watanabe, K. Stabilities of bovine β-lactoglobulin/retinol or retinoic acid complexes against tryptic hydrolysis, heating and light-induced oxidation. *Lebensm. Wiss. u. Technol.* 1996, *29* (8), 763–766.
54. Said, H.M.; Ong, D.E.; Shingleton, J.L. Intestinal uptake of retinol: Enhancement by bovine milk β-lactoglobulin. *Am. J. Clin. Nutr.* 1989, *49* (4), 690–694.
55. Mansouri, A.; Haertle, T.; Gerard, A.; Gerard, H.; Gueant, J.L. Retinol free and retinol complexed β-lactoglobulin binding sites in bovine germ cells. *Biochim. Biophys. Acta* 1997, *1357* (1), 107–114.

56. Yang, M.C.; Chen, N.C.; Chen, C.J.; Wu, C.Y.; Mao, S.J.T. Evidence for β-lactoglobulin involvement in vitamin D transport in vivo—Role of the c turn (Leu Pro Met) of β-lactoglobulin in vitamin D binding. *FEBS J.* 2009, *276* (8), 2251–2265.
57. Barbiroli, A.; Bonomi, F.; Ferranti, P.; Fessas, D.; Nasi, A.; Rasmussen, P.; Lametti, S. Bound fatty acids modulate the sensitivity of bovine β-Lactoglobulin to chemical and physical denaturation. *J. Agric. Food Chem.* 2011, *59* (10), 5729–5737.
58. Robillard, K.A., Jr.; Wishnia, A. Aromatic hydrophobes and β-lactoglobulin A. Thermodynamics of binding. *Biochemistry* 1972, *11* (21), 3835–3840.
59. Farrell, H.M.; Behe, M.J.JR.; Enyeart, J.A. Binding of p-nitrophenyl phosphate and other aromatic compounds by β-lactoglobulin. *J. Dairy Sci.* 1987, *70* (2), 252–258.
60. Beringhelli, T.; Eberini, I.; Galliano, M.; Pedoto, A.; Perduca, M.; Sportiello, A.; Fontana, E.; Monaco, H.L.; Gianazza, E. pH and ionic strength dependence of protein (un)folding and ligand binding to bovine β-lactoglobulins A and B. *Biochemistry* 2002, *41* (51), 15415–15422.
61. Collini, M.; D'Alfonso, L.; Baldini, G. New insight on beta-lactoglobulin binding sites by 1-anilinonaphthalene-8-sulfonate fluorescence decay. *Protein Sci.* 2000, *9* (10), 1968–1974.
62. Liang, L.; Subirade, M. β-Lactoglobulin/folic acid complexes: Formation, characterization, and biological implication. *J. Phys. Chem. B* 2010, *114* (19), 6707–6712.
63. Liang, L.; Tajmir-Riahi, H.A.; Subirade, M. Interaction of β-lactoglobulin with resveratrol and its biological implications. *Biomacromolecules* 2008, *9* (1), 50–56.
64. Collini, M.; D'Alfonso, L.; Molinari, H.; Ragona, L.; Catalano, M.; Baldini, G. Competitive binding of fatty acids and the fluorescent probe 1–8-anilinonaphthalene sulfonate to bovine β-lactoglobulin. *Protein Sci.* 2003, *12* (8), 1596–1603.
65. Cogan, U.; Kopelman, M.; Mokady, S.; Shinitzky, M. Binding affinities of retinol and related compounds to retinol binding proteins. *Eur. J. Biochem.* 1976, *65* (1), 71–78.
66. Zorilla, R.; Liang, L.; Remondetto, G.; Subirade, M. Interaction of epigallocatechin–3-gallate with β-lactoglobulin: Molecular characterization and biological implication. *Dairy Sci. Technol.* 2011, *91* (5), 629–644.
67. Shpigelman, A.; Cohen, Y.; Livney Y.D. Thermally-induced β-lactoglobulin-EGCG nanovehicles: Loading, stability, sensory and digestive-release study. *Food Hydrocolloids* 2012, *29* (1), 57–67.
68. De Wolf, F.A.; Brett, G.M. Ligand-binding proteins: Their potential for application in systems for controlled delivery and uptake of ligands. *Pharmacol. Rev.* 2000, *52* (2), 207–236.
69. Liang, L.; Zhou, P.; Subirade M. Protective effect of bovine serum albumin, β-lactoglobulin and α-lactalbumin on photodegradation of folic acid. *Food Chem.* 2013, *141* (2), 655–1572.
70. Swaisgood, H.E.; Wang, Q.; Allen, J.C. Protein ingredient for carrying lipophilic nutrients. US Patent 6,290,974 B1, 2001.
71. Ron, N.; Zimet, P.; Bargarum, J.; Livney Y.D. Beta-lactoglobulin/polysaccharide complexes as nanovehicles for hydrophobic nutraceuticals in non-fat foods and clear beverages. *Int. Dairy J.* 2010, *20* (10), 686–693.
72. Dufour, E.; Hoa, G.H.; Haertlé, T. High-pressure effects on beta-lactoglobulin interactions with ligands studied by fluorescence. *Biochim. Biophys. Acta* 1994, *1206* (2), 166–172.
73. Taheri-Kafrani, A.; Bordbar, A.K.,; Mousavi, S.H.A.; Haertle, T. β-Lactoglobulin structure and retinol binding changes in presence of anionic and neutral detergents. *J. Agric. Food Chem.* 2008, *56* (16), 7528–7534.
74. Sahihi, M.; Bordbar, A.K.; Ghayeb, Y. Thermodynamic stability and retinol binding property of β-lactoglobulin in the presence of cationic surfactants. *J. Chem. Thermodynamics* 2011, *43* (8), 1185–1191.
75. McClements, D.J.; E.A.; Park, Y.; Weiss J. Structural design principles for delivery of bioactive components in nutraceuticals and functional foods. *Crit. Rev. Food Sci. Nutr.* 2009, *49* (6), 577–606.

36 Encapsulation of Polyphenolics

Florence Edwards-Lévy and Aude Munin-César

CONTENTS

Abbreviations 741
36.1 Introduction 742
36.2 Physical Methods 746
    36.2.1 Spray Drying 746
    36.2.2 Atomization by Supercritical Fluids 747
36.3 Physicochemical Methods 748
    36.3.1 Cooling of Emulsions 748
    36.3.2 Emulsification-Solvent Removal Methods 748
    36.3.3 Acidic Precipitation 750
    36.3.4 Methods Based on Ionic Interactions 750
        36.3.4.1 Ionic Gelation 750
        36.3.4.2 Layer-by-Layer Process 751
        36.3.4.3 Complex Coacervation 751
    36.3.5 Methods Based on Hydrophobic Interactions 751
        36.3.5.1 Micelles 751
        36.3.5.2 Liposomes 753
36.4 Chemical Methods 755
    36.4.1 *In Situ* Polymerization 755
    36.4.2 Interfacial Cross-Linking Using Acid Dichlorides 755
36.5 Other Encapsulation Methods 755
    36.5.1 Molecular Encapsulation 755
    36.5.2 Cocrystallization 757
    36.5.3 Encapsulation in Yeasts 757
36.6 Conclusion 757
References 757

ABBREVIATIONS

CD	cyclodextrin
CMC	carboxymethylcellulose
DNA	deoxyribonucleic acid
EC	ethylcellulose
EGCG	epigallocatechin gallate
LbL	layer by layer
NLC	nanostructured lipid carrier
PCL	poly(ε-caprolactone)
PHBV	poly(3-hydroxybutyrate-co-3-hydroxyvalerate
PLGA	poly(lactic-co-glycolic acid)
PLA	polylactic acid
PLL	poly-L-lysine

PVA   polyvinylic acid
ROS   reactive oxygen species
sc-CO_2 supercritical carbon dioxide
SDS   sodium dodecylsulfate
SOD   superoxide dismutase
TPP   tripolyphosphate
UV   ultraviolet

36.1 INTRODUCTION

Polyphenolics are aromatic compounds bearing several phenol groups. These secondary metabolites are produced in plants by the shikimate pathway and/or the polyacetate pathway (Figure 36.1).[1,2] This large family of molecules (several thousands) has been grouped into various classes, and their structures vary around a few chemical skeletons in terms of degree of oxidation, hydroxylation, methylation, glycosylation, and possible connection to other molecules (carbohydrates, lipids, proteins, other phenolic compounds, etc.). Among the different classes, flavonoids are the most abundant, and have been divided into subclasses (flavones, flavanols, flavanones, anthocyanins, proanthocyanidols, etc.).

Natural polyphenols like flavonoids or tannins possess a well-known protective effect towards oxidation, which can be used against pathologies in which free radicals are incriminated, like cancer.[3,4] In cosmetic formulations, these antioxidant molecules are useful to protect the constitutive elements of skin, like collagen or elastin, against degradations or cross-linking reactions responsible for a decrease in elasticity and the appearance of wrinkles with ageing.[5]

The physiological protective system of the body against oxidative damage is constituted of enzymes like superoxide-dismutase (SOD). The function of this enzyme is to reduce the superoxide radical, which is among the most powerful oxidative reactive oxygen species (ROS), into hydrogen peroxide, which in turn can easily generate other toxic ROSs, but whose concentration is regulated by other enzymes like catalase and glutathione peroxidases. When the regulating system works well, it leads to the elimination of ROSs and the production of water and molecular oxygen.

When the natural enzymatic antioxidant system is not efficient enough, the production of ROSs exceeds the antioxidant capacity of the body, and the oxidative free radicals start attacking biological molecules, forming secondary radicals and initiating damages in cell membranes, in DNA molecules, in circulating lipoproteins; thus these are implied not only in global ageing and neurodegenerative pathologies,[6] but also in mutagenesis, carcinogenesis, cell mortality, atheroma formation, etc.[7–9]

The mechanism by which polyphenols are active against oxidation caused by ROS is triple.[10] First, these molecules act as free radical scavengers, and this property has been extensively studied.[10–16] The strong reducing character of polyphenols is due to their ability to give electrons or hydrogen atoms to free radicals. The aryloxy radical ArO• formed when the polyphenol loses a hydrogen radical that is stabilized by mesomeric resonance (Figure 36.2), and is much less deleterious for biological molecules *in vivo*.

Another activity of polyphenols against oxidation is related to their chelating property toward metal ions,[10] which masks the catalytic effect of metals on the reactions responsible for the formation of free radicals. Polyphenols are known to inhibit, by chelating iron ions, the Fenton reaction by which ROSs are produced *in vivo* from hydrogen peroxide.[17] This metal-chelating ability of polyphenols is related to the presence of hydroxyl and carboxyl groups on the molecule, associated to the presence of aromatic rings of strong nucleophilic character.

Many polyphenols have the property of interacting with proteins,[18–21] mainly through hydrophobic interactions and formation of hydrogen bonds.[22–24] This property accounts for a third antioxidant mechanism. Some polyphenols have been shown to inhibit ROS-generating enzymes like xanthine oxidase, cyclo-oxygenase, or lipo-oxygenase, by complexing the enzyme.

Encapsulation of Polyphenolics

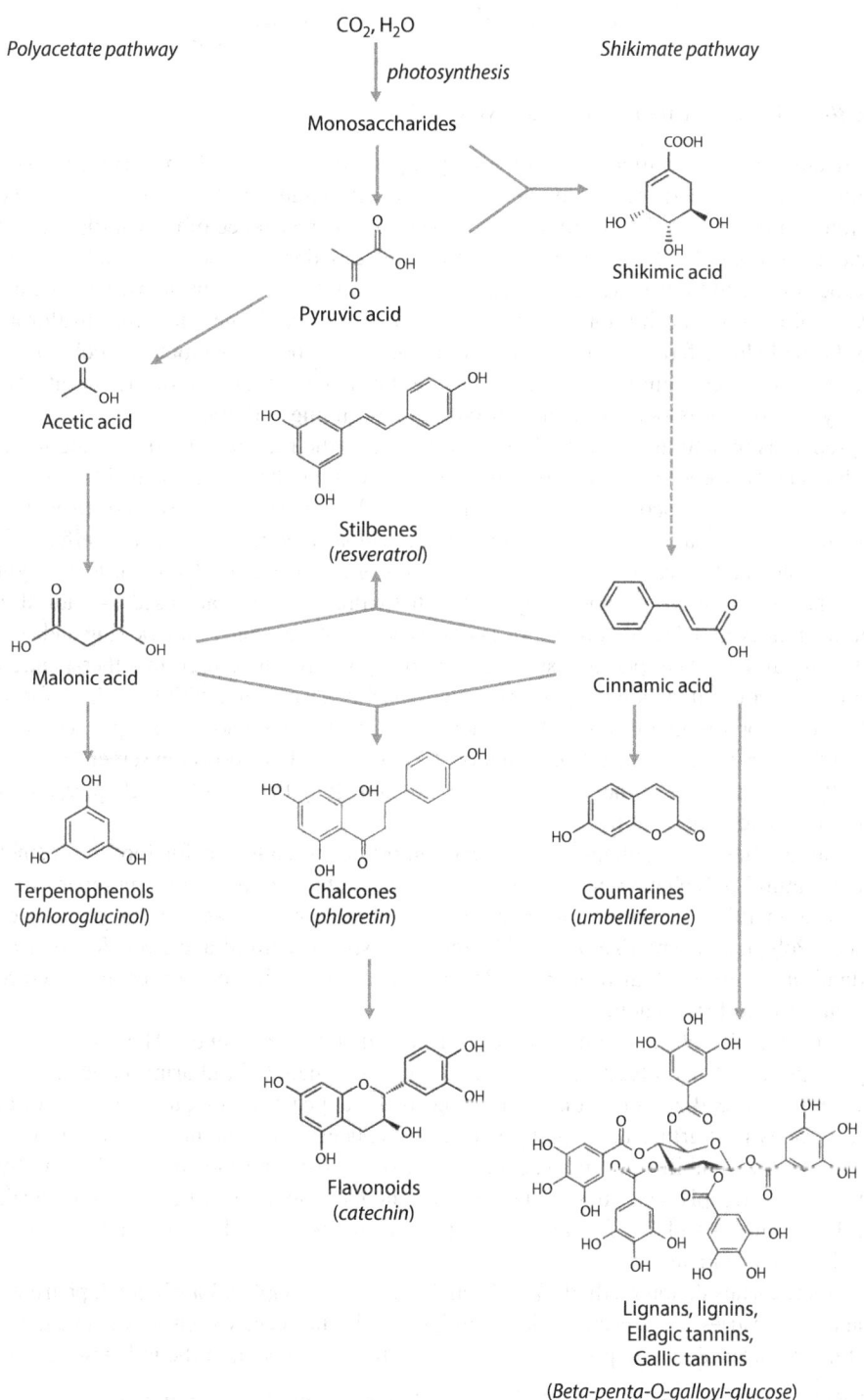

FIGURE 36.1 Biosynthesis of plant polyphenols.

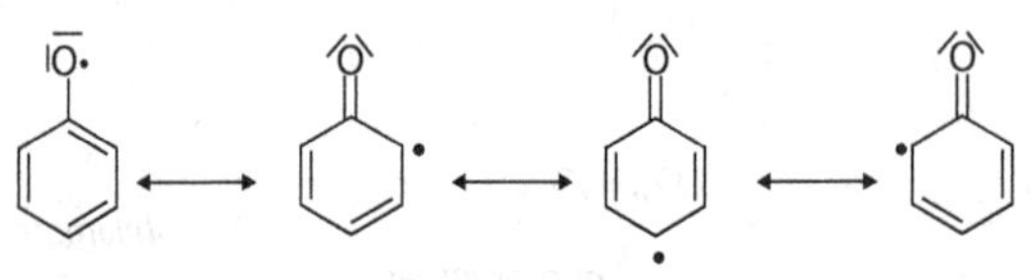

FIGURE 36.2 Mesomeric resonance of the aryloxy radical.

Also related to this "tanning" effect on proteins, polyphenolics are often astringent, and present a vasculoprotective effect; these polyphenols are named "vitaminic P factors" and can be found incorporated in pharmaceutical formulations to increase the resistance of the capillaries (esculetol, rutin, and diosmin). Other have anti-inflammatory and antiedema properties,[25] and these activities can also be related to their general tendency to form complexes with proteins, which can lead to enzymatic inhibition. The inhibitory effect of flavonoids toward enzymes, like hyaluronidase,[26] catechol-O-methyltransferase,[27] and cyclo-oxygenase,[28] have been documented. Polyphenols have also been shown to act against Alzheimer's disease through a modulation of the activity of the sirtuin family of proteins associated with their radical-scavenging activity.[29]

The properties of natural polyphenols are not limited to their antioxidant power and to their protein-binding ability. Depending on their structure, some of them have been named "phytoestrogens" because of hormone-like activities linked to a structural similarity with estradiol (isoflavones like genistein, flavanones, chalcones, etc.).[30,31] Some of these molecules present a sedative effect (vitexin),[32] others have antiplatelet activity (dicoumarol, resveratrol), or hepatoprotective effect (silymarin).[33] Some polyphenolics are very toxic compounds (aflatoxin), and some are used as natural coloring substances (anthocyanidols and curcumin). The presence of at least one aromatic ring and conjugated double bonds can give them photosensitizing properties, which can be used in a therapeutic method named PUVA therapy, in which the polyphenol (coumarin) is irradiated with UVA light for the treatment of psoriasis or vitiligo by "photochemotherapy"[34] Other polyphenolics like quercetin are able to protect lipids against UVC-induced peroxidation, by acting as a UV-absorbing screen.[35]

Table 36.1 groups some of the polyphenols mentioned along the text of this chapter, and presents their structures and main activities.

The potentialities of polyphenolics are very important in health, or for food or cosmetics.[36,37] The use of natural polyphenols for human health is the subject of a growing number of papers in the literature, mainly concerning the treatment of degenerative diseases, inflammatory diseases, and cancer. Polyphenols are also used in the food industry as natural additives, for their coloring, antioxidant, and antimicrobial properties. Their radical-scavenging properties and UV-protective activity can be used for cosmetics.

However, several problems limit the use of these valuable compounds. The problem of lack of stability should first be resolved before using natural polyphenols in pharmaceutics or cosmetics. These compounds oxidize very quickly, leading to the appearance of unwanted colors and to a decrease in activity. Furthermore, many natural polyphenolic compounds present limited water solubility. This low solubility, often associated with low intestinal permeability and instability in the gastrointestinal tract (pH, enzymes, and other nutrients), results in insufficient oral bioavailability. Finally, the astringent and bitter taste of many polyphenolics should be masked before use in food or for oral pharmaceutical forms.

Many microencapsulation methods have been developed, for applications in food, pharmaceutical, and cosmetic industries.[38–41] Encapsulation of polyphenolics has been extensively studied. The application of microencapsulation to polyphenolics can reach one of several of the following goals:

- Stabilizing the encapsulated molecule by isolating it from the environment
- Protecting the user from the unpleasant taste of the encapsulated compound
- Hydrophilizing a poorly water-soluble compound by formulating in an hydrophilic particle
- Controlling the release of the encapsulated compound
- Increasing oral bioavailability of the encapsulated compound

TABLE 36.1
Some of the Polyphenols of Interest Mentioned in This Chapter, Their Structures, and Main Biological Properties

Polyphenol (Class)	Structure	Biological Properties
Catechin (flavanol)		Antioxidant, enzyme inhibitor, protective against oxidative stress
Chlorogenic acid (hydroxycinnamic acid)		Antihypertensive
Curcumin		Yellow dye, antihypertensive, antitumoral
Ellagic acid (ellagitannin)		Antitumoral, protective against oxidative stress
Epigallocatechin gallate (catechin, gallic ester)		Antioxidant, anticancer, anti-HIV, enzyme inhibitor
Genistein (isoflavone)		Phytoestrogen, antioxidant
Luteolin (flavone)		Anti-inflammatory, anticancer, antioxidant

(*Continued*)

TABLE 36.1 (*Continued*)
Some of the Polyphenols of Interest Mentioned in This Chapter, Their Structures, and Main Biological Properties

Polyphenol (Class)	Structure	Biological Properties
Naringenin (flavanone)		Anti-inflammatory, protective effect against oxidative damage
Proanthocyanidin (condensed tannin)		Radical scavenger, vasculoprotective, antimutagenic
Quercetin (flavonol)		UV protectant, anti-inflammatory, antiviral, antitumoral, hepatoprotectant
Resveratrol (stilbene)		Anticarcinogenic, anti-inflammatory, antiplatelet, protective effect against oxidative stress

Numerous encapsulation methods have been developed, based on physical, physicochemical, or chemical principles. This chapter will present examples of polyphenol encapsulation, classified as a function of the method used for encapsulation.

36.2 PHYSICAL METHODS

36.2.1 Spray Drying

Spray-drying is probably one of the oldest encapsulation methods. The process involves the dispersion of the compound to be encapsulated in a solution of coating agent, followed by atomization of the liquid into droplets and subsequent drying of the droplets by evaporating the solvent in a warmed gas current, in a spray-drying apparatus. The coating agent initially present in each droplet is deprived of its solvent and precipitates in the form of microparticles, containing the encapsulated compound.[42,43] This process, inexpensive and versatile, is widely used in the food industry.

Encapsulation of Polyphenolics

FIGURE 36.3 Scanning electron micrograph of procyanidin microcapsules prepared by a spray-drying process with gum arabic. (Zhang, L., Mou, D., and Du, Y.: Procyanidins: Extraction and micro-encapsulation. *J. Sci. Food Agric.* 2007. 87(12). 2192–2197. Copyright Wiley-VCH Verlag GmbH & Co. KGaA. Reproduced with permission.)

The coating agent can be of various nature, and encapsulation of polyphenolics by spray-drying has be performed using proteins like caseinate for *Quercus resinosa* infusions,[44] gum arabic for epigallocatechin gallate (EGCG)[45] or procyanidins (Figure 36.3),[46] maltodextrins for anthocyanins[47,48] or procyanidins from grape seeds,[46] chitosan for olive tree leaves extract[49] or a *Paeonia rockii* extract,[50] carrageenan for various plant extracts,[51,52] hypromellose for *Quercus resinosa* infusions,[53] mixtures of protein and phospholipid for polyphenol-rich extracts.[54] A recent review on encapsulation of fruit extracts by spray drying demonstrates the effect of additives on the antioxidant activity of the resulting particles.[55] For each formulation, the optimal relative concentrations are the ones leading to high encapsulation efficiency and masking of the astringency, associated with a protection of the polyphenolic activity, for nutraceutical or pharmaceutical applications.

Spray drying is also a dehydrating method that can be used to stabilize plant extracts, such as a Yerba mate (*Ilex paraguaiensis*) extract[56] or a soybean extract,[57] without any coating agent.

36.2.2 Atomization by Supercritical Fluids

The specific properties of supercritical fluids can be exploited to develop alternative encapsulation methods or drying processes avoiding the use of classical solvents.[58] For carbon dioxide, the supercritical region can be reached at moderate temperatures and pressures. Supercritical carbon dioxide (sc-CO_2) is a polar, nontoxic, and inert fluid, and its elimination from the product is done easily by simple depressurization. Sc-CO_2 is mixed with the solution containing the substance, and the gas-saturated solution is atomized through a nozzle. Fine droplets are formed, and the heated gas evaporates the solvent, giving a free-flowing powder of unaltered dry substance. Green tea extracts have been treated by supercritical-assisted atomization, and this process involving low temperature in oxygen-free atmosphere gave promising results for the gentle drying of fragile compounds.[59]

36.3 PHYSICOCHEMICAL METHODS

36.3.1 Cooling of Emulsions

The first step of the processes based on cooling of emulsions is to disperse or dissolve the substance to be encapsulated in a hot melted lipid phase (waxes, stearates, and glycerides), or in a hot aqueous solution of hydrophilic polymers, which can form a gel upon cooling (gelatin, agarose, and glucans). Then the warm liquid is emulsified in a heated continuous phase. Finally, the system is cooled, and solid particles are formed. For an optimal encapsulation rate, the polarity of the continuous phase and that of the dispersed phase have to be chosen adequately as a function of the solubility of the encapsulated substance. Several anthocyanins from black currant have been stabilized by this method, using β-glucan to form hydrophilic gel beads. A further freeze-drying step produced a dry matter suitable for utilization as food ingredient.[60] EGCG could be immobilized into lipid-coated nanoparticles produced by cooling of very fine emulsions produced by ultrasonication.[61] The resulting particles improved the oral bioavailability of EGCG for the treatment of Alzheimer's disease, and appear as good candidates for clinical trials (Figure 36.4). Quercetin is a natural flavonoid that can be used in therapeutics for its anti-inflammatory, antiviral, antitumor, or hepatoprotectant effects. Many papers deal with encapsulation of quercetin to provide a delivery system suitable for this poorly water-soluble compound. For this purpose, nanostructured lipid carriers (NLCs) containing quercetin have been developed. For the preparation of quercetin-loaded NLCs, a melted lipid phase containing quercetin is emulsified in an aqueous phase by sonication in the presence of lecithin as a surfactant. An aqueous suspension of NLCs is formed upon cooling of the dispersion. These particles have been shown to enhance quercetin's bioavailability by the oral route[62,63] or by the topical route.[64,65]

36.3.2 Emulsification-Solvent Removal Methods

In the emulsification-solvent removal methods, the coating agent is dissolved in a solvent, and the organic solution is added to a continuous phase. The solvent is then removed from the internal phase, either by evaporation or by diffusion in the external phase.

In the solvent evaporation method, the organic solution of coating agent is emulsified in an aqueous external phase in which the solvent is not miscible; then evaporation of the solvent, by heating under vacuum, produces micro- or nanoparticles as a function of the stirring speed used for

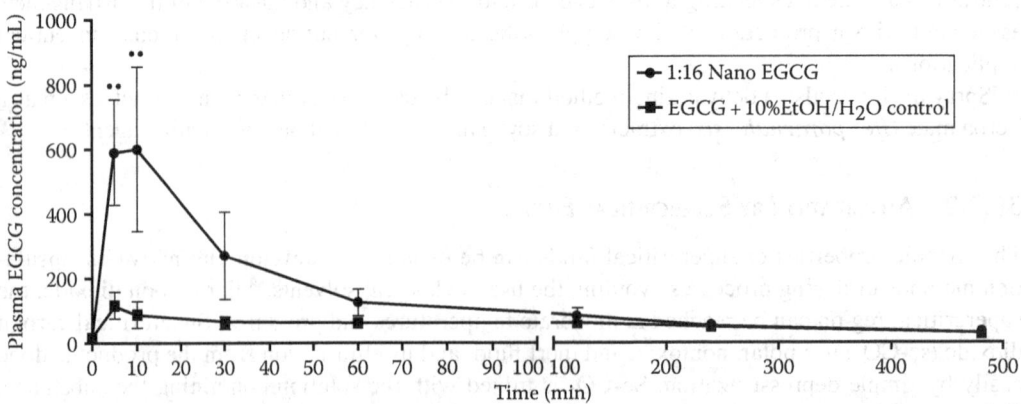

FIGURE 36.4 EGCG pharmacokinetic curve (mean plasma concentration ± SEM versus time) for nanoparticulate EGCG formulation (Nano EGCG, $n = 3$) and free EGCG in 10% ethanol solution ($n = 3$). (Reprinted from *Int. J. Pharm.*, 389(1–2), Smith, A., Giunta, B., Bickford, P.C., Fountain, M., Tan, J., and Shytle, R.D., Nanolipidic particles improve the bioavailability and α-secretase inducing ability of epigallocatechin-3-gallate (EGCG) for the treatment of Alzheimer's disease, 207–212. Copyright 2010, with permission from Elsevier.)

the emulsification step. This method has been applied to the development of various polyphenolic microparticles for a sustained release after oral administration. For example, resveratrol has been encapsulated in poly(3-hydroxybutyrate-co-3-hydroxyvalerate) and poly(ε-caprolactone) (PCL) microparticles,[66] ethylcellulose (EC) microparticles could be used to stabilize tea polyphenols[67] or bayberry polyphenols,[68] and porous kafirin microspheres have been applied to the encapsulation of catechin and condensed tannins from sorghum.[69]

The solvent evaporation method can also be used to produce nanoparticles from polylactic polymers, mainly to improve the oral bioavailability of polyphenolics by the oral route. For example, curcumin,[70,71] EGCG,[72] or ellagic acid[73] has been encapsulated in PLGA nanoparticles, and quercitrin[74] or quercetin[75] has been immobilized in polylactic acid (PLA) nanoparticles, to improve the aqueous solubility, to increase the intestinal absorption, and to increase globally the oral efficacy of these compounds. Ellagic acid in PLGA-PCL nanoparticles was shown to be more active against nephrotoxicity than the free molecule.[76]

Polylactide-co-glycolide (PLGA) microparticles releasing curcumin for 4 weeks after subcutaneous injection can be used as slow-release systems for cancer prevention (Figure 36.5).[77]

Nanoparticles prepared from the solvent evaporation method can also improve the transdermal penetration of polyphenolics. PLGA nanoparticles containing quercetin and the hypoglycemic voglibose were incorporated in a polymeric film, for transdermal delivery in complications associated with diabetes.[78] Recently, nanoparticles prepared from PLGA associated with hyaluronic acid were used to encapsulate quercetin.[79] The resulting nanosystem allowed an enhanced transdermal drug permeation and effective scavenging of free radicals.

In the emulsification-solvent diffusion method, also called nanoprecipitation, the polymer solution is prepared in a hydromiscible solvent, and injected in an aqueous phase in which the polymer is insoluble; when the solvent diffuses in the aqueous phase, the polymer becomes insoluble in the mixture and precipitates in the form of nanoparticles. This method has been applied to the encapsulation of resveratrol in PEG-PCL nanoparticles,[80] or EGCG in PEG-PLA nanoparticles,[81] for chemoprevention or chemotherapy. The oral bioavailability and the hepatoprotectant effect of naringenin have been increased by encapsulation in Eudragit-polyvinyl alcohol (PVA) nanoparticles.[82] pH-responsive polymeric nanoparticles containing quercetin and targeted against cancer cells by a folate moiety have been recently developed (Figure 36.6).[83] The same Eudragit-PVA nanoparticles have been successfully tested to enhance the bioavailability of quercetin.[84] Cationic

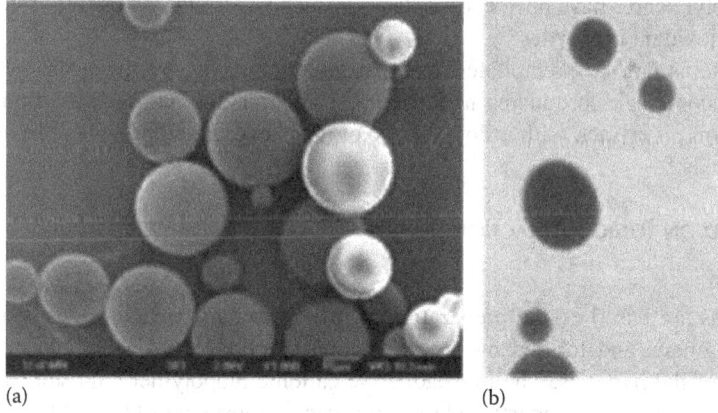

(a) (b)

FIGURE 36.5 (a) Scanning electron microscopy image of curcumin-loaded microparticles prepared by rapid solvent removal process. 1000× magnification and (b) optical microscopy image of curcumin-loaded microparticles prepared by rapid solvent removal process. 400× magnification. Bar is 10 μm. (Shahani, K. and Panyam, J.: Highly loaded, sustained-release microparticles of curcumin for chemoprevention. *J. Pharm. Sci.* 2011. 100(7). 2599–2609. Copyright Wiley-VCH Verlag GmbH & Co. KGaA. Reproduced with permission.)

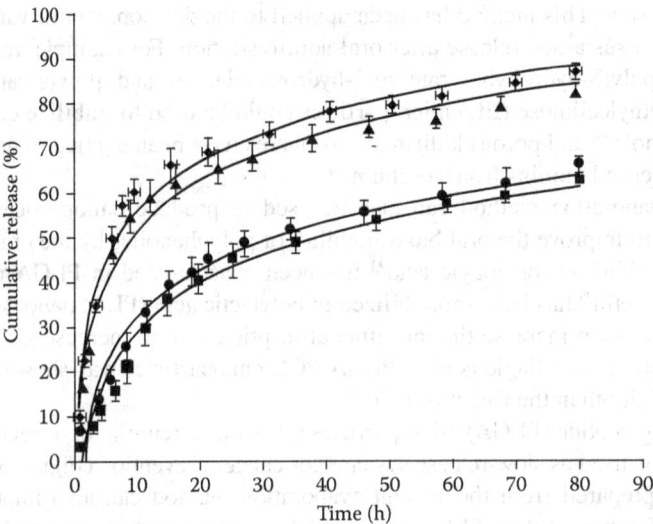

FIGURE 36.6 *In vitro* release profiles of quercetin from folated nanoparticles (at pH 5.8 (♦) and 7.4 (●)) and nonfolated nanoparticles (at pH 5.8 (▲) and 7.4 (■)) at 37°C (*n* = 3). (Reprinted from *Eur. J. Med. Chem.*, 50, Khoee, S. and Rahmatolahzadeh, R., Synthesis and characterization of pH-responsive and folated nanoparticles based on self-assembled brush-like PLGA/PEG/AEMA copolymer with targeted cancer therapy properties: A comprehensive kinetic study, 416–427. Copyright 2012, with permission from Elsevier Masson SAS.)

chitosan- and anionic alginate-coated PLGA nanoparticles prepared through a nanoprecipitation process turned out to be effective in protecting resveratrol against light-exposure degradation, while releasing this bioactive polyphenol in a controlled manner.[85]

A supercritical fluid can be used as the external phase of a nanoprecipitation process. In this case, the method is named *"supercritical antisolvent processing"*; green tea polyphenols have been encapsulated in a biodegradable polylactide-PCL copolymer by a supercritical antisolvent process.[86]

36.3.3 Acidic Precipitation

Sodium caseinate or calcium caseinate can be used for the preparation of casein beads by acidic precipitation. Green tea polyphenols have been encapsulated in such beads with a positive effect on the preservation of the antioxidant properties.[87]

Heat-treated whey proteins can be precipitated inside a microemulsion, by the hydrolysis of glucono-δ-lactone into gluconic acid, producing nanoparticles. This method has been used for the encapsulation of a date palm pit extract, with a slow release of the encapsulated polyphenols, and a potentiality for nutraceuticals.[88]

36.3.4 Methods Based on Ionic Interactions

36.3.4.1 Ionic Gelation

The ionic gelation methods are based on the preparation of micro- or nanoparticles consisting of ionically cross-linked polymers, and formed by the interaction of oppositely charged ions with the polymers constitutive of the particles. The mucoadhesive cationic biopolymer chitosan can be ionically cross-linked in the form of nanoparticles, by using tripolyphosphate (TPP) as the negatively charged ion. This system has been applied to the encapsulation of a flavonoid-rich extract from *Elsholtzia splendens*,[89] from *Ilex paraguaiensis* (Figure 36.7)[90] and from green tea,[91] but also to catechin and epigallocatechin,[92] for stabilization purposes. The cellular uptake of EGCG was shown to increase when encapsulated in nanoparticles of chitosan ionically cross-linked with casein phosphopeptide.[93]

FIGURE 36.7 Scanning electron micrographs of *Ilex paraguaiensis* microspheres prepared by ionic gelation of chitosan with TPP. (Reprinted from *Carbohydr. Polym.*, 84(2), Harris, R., Lecumberri, E., Mateos-Aparicio, I., Mengíbar, M., and Heras, A., Chitosan nanoparticles and microspheres for the encapsulation of natural antioxidants extracted from Ilex paraguariensis, 803–806. Copyright 2011, with permission from Elsevier.)

Calcium is a divalent cation that can be used to cross-link alginate or pectinate in the form of gel beads. These beads can be a suitable immobilization system for plant extracts[94,95] or for pure polyphenolics like resveratrol,[96] for controlled release after oral administration.

Other ionic interactions can be exploited to create microgels or nanogels. Chitosan and carboxymethyl chitosan have been associated to formulate nanoparticles containing tea polyphenols for sustained release of these antitumor compounds,[97] Gelatin–dextran conjugates have been cross-linked with TPP for tea polyphenol delivery.[98]

36.3.4.2 Layer-by-Layer Process

The layer-by-layer process (LbL) is based on the self-assembly of successive layers onto a core of mineral or organic nature.[99] The number of layers can be set to obtain the desired thickness or porosity for the membrane. The alternate layers are constituted of compounds with an affinity one to another, like polyelectrolytes of opposite charges, or proteins and polyphenols. EGCG and gelatin have been successively deposited onto manganese carbonate cores[100] or onto gelatin nanoparticles (Figure 36.8),[101] on the basis of the affinity of EGCG for proline-rich domains of proteins. The resulting assemblies prolonged the life-time and enhance the activity of the encapsulated flavonoid.

36.3.4.3 Complex Coacervation

Complex coacervation is a phase-separation process, in which two water-soluble polymers are brought to physicochemical conditions leading to their desolvation. The desolvated polymers separate from the equilibrium phase, in the form of a coacervate, that tend to deposit onto dispersed core materials. The polymers forming the coacervate can be two polyelectrolytes susceptible of bearing opposite net charges, and the dispersed cores can be oily droplets of an L/H emulsion. Pectin has been associated to soy proteins to encapsulate propolis, a polyphenol-rich compound produced by bees, for stabilization, water dispersibility, and controlled release in food (Figure 36.9).[102]

36.3.5 Methods Based on Hydrophobic Interactions

36.3.5.1 Micelles

Micelles are nanometric molecular aggregates of amphiphilic molecules, presenting a hydrophobic core and a polar surface.[103] They spontaneously appear in an aqueous phase if the concentration of the amphiphilic molecule is above the critical micellar concentration (CMC), and they have the property of solubilizing hydrophobic molecules in aqueous environments.

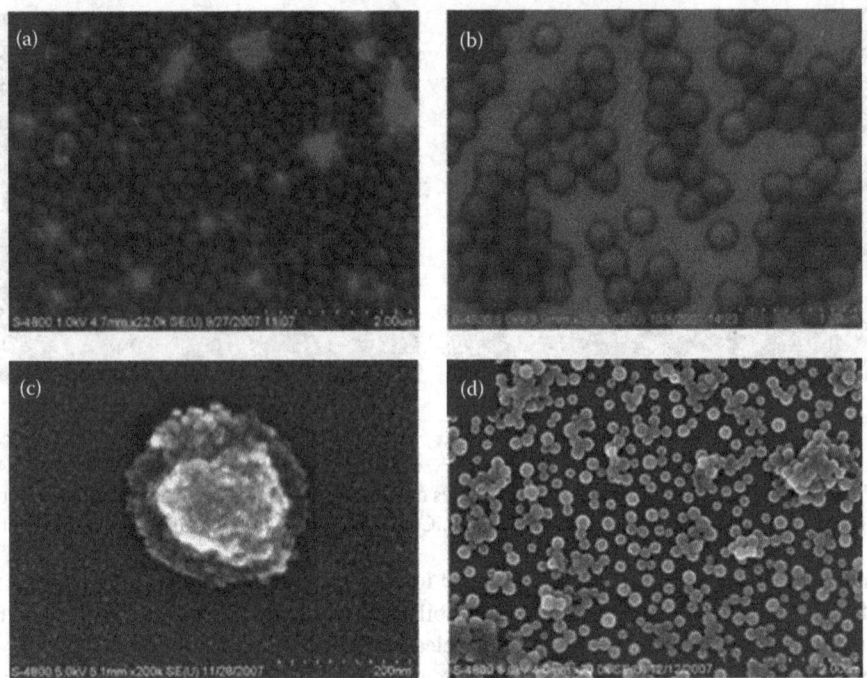

FIGURE 36.8 SEM images of (a) 200 nm and (b–d) 300 nm gelatin A nanoparticles: (a) uncoated, (b) coated with (polyglutamic acid/poly-l-lysine)$_2$ shells, (c) one nanoparticle coated with (dextran sulfate/protamine sulfate)$_2$SiO$_2$ shell, and (d) uncoated nanoparticles after loading EGCG. (Reprinted with permission from Shutava, T.G., Balkundi, S.S., Vangala, P., Steffan, J.J., Bigelow, R.L., Cardelli, J.A. et al., Layer-by-layer-coated gelatin nanoparticles as a vehicle for delivery of natural polyphenols, *ACS Nano.*, 3(7), 1877–1885. Copyright 2009 American Chemical Society.)

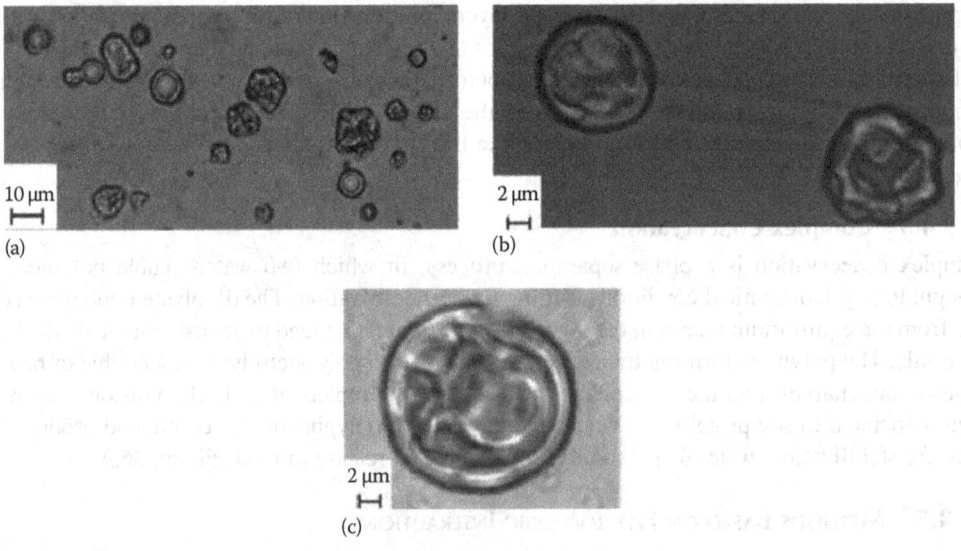

FIGURE 36.9 Pectin-soy protein microcapsules containing propolis and obtained at pH 4 produced with: (a) concentration of 2.5 g/100 mL (40×), (b) concentration of 5 g/100 mL (100×), (c) concentration of 5 g/100 mL (100×). (Reprinted from *LWT—Food Sci. Technol.*, 44(2), Nori, M.P., Favaro-Trindade, C.S., Matias de Alencar, S., Thomazini, M., de Camargo Balieiro, J.C., and Contreras Castillo, C.J., Microencapsulation of propolis extract by complex coacervation, 429–435. Copyright 2011, with permission from Elsevier.)

Encapsulation of Polyphenolics

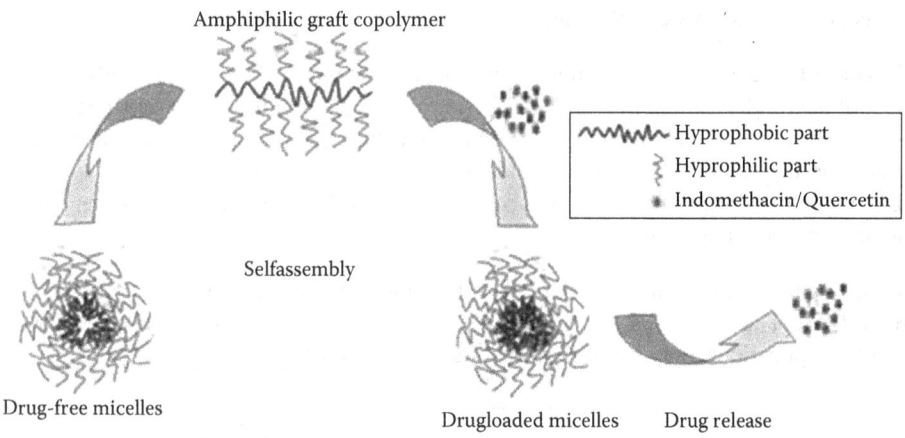

FIGURE 36.10 Schematic illustration of self-assembly amphiphilic graft copolymers including loading and release of drugs. (Reprinted from *Int. J. Pharm.*, 460(1–2), Bury, K. and Neugebauer, D., Novel self-assembly graft copolymers as carriers for anti-inflammatory drug delivery, 150–157. Copyright 2014, with permission from Elsevier.)

The amphiphilic molecules constituting the micelles can be surfactants, like sodium dodecyl sulfate (SDS) for the solubilization of artemisinin and curcumin[104] or for the delivery of polyphenolic fractions from *Hamamelis virginiana*.[105] The micelles can also be formed of synthetic block copolymers. For example, copolymers of caprolactone 2-(methacryloyloxy)ethyl ester units with poly(meth)acrylic acid side chains have been self-assembled in micelles of quercetin associated with indomethacin for the treatment of inflammatory diseases (Figure 36.10).[106] Resveratrol has been loaded into PCL-PEG micelles for its protective effect against oxidative stress.[107] An aqueous formulation of luteolin has been developed with the same PCL-PEG micelles, with potential anticancer effect.[108]

Curcumin shows very interesting therapeutic properties in the fields of cancer, AIDS, or inflammation, but its use is limited by a low aqueous solubility and poor bioavailability. Nanocurc® is a dispersion of curcumin, in micelles of synthetic block copolymer of N-isopropylacrylamide, N-vinylpyrrolidone, and poly(ethylene glycol)acrylate. Nanocurc is a water-soluble version of curcumin, with increased bioavailability and activity against tumor growth and metastases.[109] The *in vivo* studies suggest that Nanocurc allows curcumin to cross the blood–brain barrier and exert an activity against inflammatory damages in the brain associated with Alzheimer's disease.[110]

36.3.5.2 Liposomes

Liposomes are spherical vesicles composed of one or several phospholipid bilayers separated by aqueous compartments. Their structure makes them theoretically able to encapsulate water-soluble compounds in the aqueous compartment, or lipophilic substances associated with the lipid bilayer. These biocompatible vesicles are used as drug-delivery systems for drugs by different routes.[103]

The localization of the polyphenolic compounds inside liposomes depends on their lipophilic character.[111,112] Some of these molecules are located in the hydrophobic bilayer of the liposomal membrane and are associated with the apolar moieties of phospholipids through hydrophobic interactions; more hydrophilic polyphenols associate with the polar heads of phospholipids through hydrogen bondings.

Liposomes have been proposed for the delivery of several polyphenolic compounds, such as resveratrol,[113] curcumin,[114] quercetin,[115] catechin and EGCG,[116] algal phlorotannins,[117] etc.

Various positive results were obtained:

- Prolonged efficacy, due to the improvement of drug stability,[118] associated with a prolongation of the residence time *in vivo*[119] and a controlled release of the active substance[119]
- Crossing of biological barriers: improvement of cellular uptake,[120] efficient use of the nasal route for delivery to the brain,[121,122] and increase of skin penetration[123]
- Bioavailability improvement by oral route (Figure 36.11)[124] or parenteral route,[125] leading to an increased therapeutic effect

Phytosomes® are molecular associations of phospholipids complexed with polyphenols (Figure 36.12).[126] These systems, facilitating the crossing of biomembranes, were first investigated for cosmetic applications, but also possess a potential use for drug delivery.

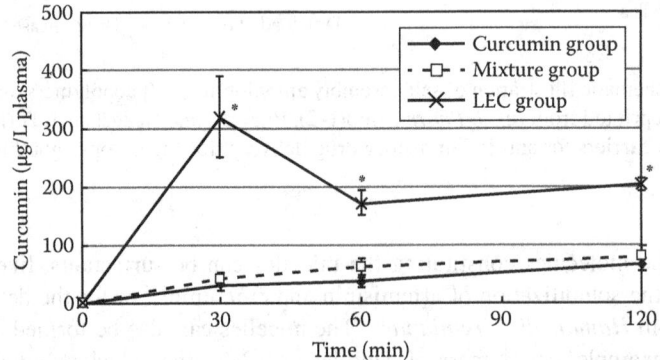

FIGURE 36.11 Concentration of curcumin in rat plasma after a single oral administration of: curcumin, a mixture of curcumin and lecithin, and curcumin liposomes(LEC) (100 mg curcumin/kg body weight). Curcumin group (●), mixture group (□), and curcumin liposome group (×). The asterisks represent a significant difference at $P < 0.01$ (vs curcumin group). Values are represented as means ± SEMs ($n = 7$). (Reprinted with permission from Takahashi, M., Uechi, S., Takara, K., Asikin, Y., and Wada, K., Evaluation of an oral carrier system in rats: Bioavailability and antioxidant properties of liposome-encapsulated curcumin, *J. Agric. Food Chem.*, 57(19), 9141–9146. Copyright 2009 American Chemical Society.)

FIGURE 36.12 (See color insert.) Molecular modeling of the interaction between a polyphenolic molecule and a phospholipid. (Reprinted from *Fitoterapia*, 81(5), Semalty, A., Semalty, M., Rawat, M.S.M., and Franceschi, F., Supramolecular phospholipids–polyphenolics interactions: The PHYTOSOME® strategy to improve the bioavailability of phytochemicals, 306–314. Copyright 2010, with permission from Elsevier.)

When the surface of liposomes is functionalized with antibodies, the resulting immunoliposomes convey most of the encapsulated molecules to specific targeted cells, like cancer cells. The side effects of the anticancer drugs are thus reduced, and the action of the drug is potentialized. Recently, HER2-targeted immunoliposomes improved the bioavailability and selectivity of curcumin and resveratrol, with a dramatic effect on the proliferation of human breast cancer cell lines.[127]

36.4 CHEMICAL METHODS

36.4.1 IN SITU POLYMERIZATION

In situ polymerization reactions are performed in an emulsion system where an organic solution of the monomer (bearing, e.g., vinylic or acrylic moieties) is the dispersed phase and an aqueous solution containing a surfactant is the continuous phase. Methyl methacrylate has been used for the preparation of quercetin-loaded nanoparticles in a miniemulsion system.[128] Quercetin was shown to slower the polymerization reaction, but this effect could be attenuated by the addition of ascorbic acid.

36.4.2 INTERFACIAL CROSS-LINKING USING ACID DICHLORIDES

Microcapsules of chemically cross-linked polymers can be prepared in an emulsion system, using bifunctional cross-linkers, such as acid dichlorides, to form bridges between chemical groups of the polymer molecules around each droplet, at the interface of the emulsion. This method has been applied to proteins[129] and polysaccharides or oligosaccharides,[130] but it was also applied to the stabilization of proanthocyanidins from grape seeds by cross-linking of some of the phenolic groups with terephthaloyl chloride (Figure 36.13).[131] The immobilized proanthocyanidins exhibited a high stability (more than 5 months at 45°C), and the cross-linking reaction preserved enough phenolic groups to provide good radical-scavenging properties to the proanthocyanidin microcapsules. Moreover, the particles slowly degraded in plasma by the action of esterases. This enzymatic hydrolysis unmasked previously cross-linked phenolic groups, leading to an increase of the scavenging activity *in vivo*.

36.5 OTHER ENCAPSULATION METHODS

36.5.1 MOLECULAR ENCAPSULATION

Cyclodextrins (CDs) are cyclic oligomers composed of a few glucopyranose units linked by α(1–4) bonds.[132] These molecules are produced from starch by the action of bacteria (*Bacillus macerans*). CDs of less than 6 glucopyranose units cannot occur naturally due to steric issues. The most used natural CDs are α-CD (6 glucopyranose units), β-CD (7 units), and γ-CD (8 units). Their conical

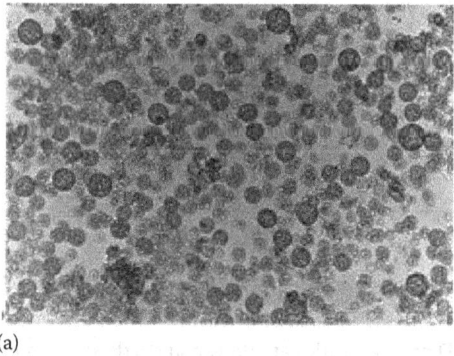
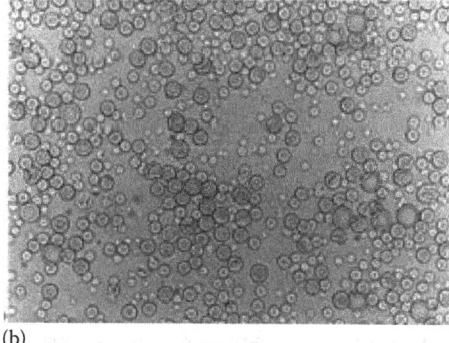

(a) (b)

FIGURE 36.13 Optical photomicrographs of proanthocyanidin microcapsules: (a) prepared at pH 9.8, (b) prepared at pH 11 (same magnification). (Reprinted from *Int. J. Pharm.*, 171(2), Andry, M.-C., Vezin, H., Dumistracel, I., Bernier, J.L., and Lévy, M.-C., Proanthocyanidin microcapsules: Preparation, properties and free radical scavenging activity, 217–226. Copyright 1998, with permission from Elsevier.)

lipophilic inner cavities enable them to host guest molecules, when the guest and the cavity are sterically compatible. The complexes often possess an increased solubility in water, and the resorption of the guest molecules occurs after dissociation of the complexes. Chemically modified CDs have been designed, with the aim of modifying the physicochemical properties, increasing the inclusion capacity, and reducing the toxicity of the CDs.

Many papers describe the use of natural or modified CDs to encapsulate polyphenolic compounds. A review of the literature concerning the inclusion of polyphenolics in CDs was very recently published.[133] In the numerous papers available on the subject, molecular inclusion has been shown to improve the chemical stability, the antioxidant activity, the solubility, and/or the bioavailability of various natural polyphenolics (Figure 36.14).

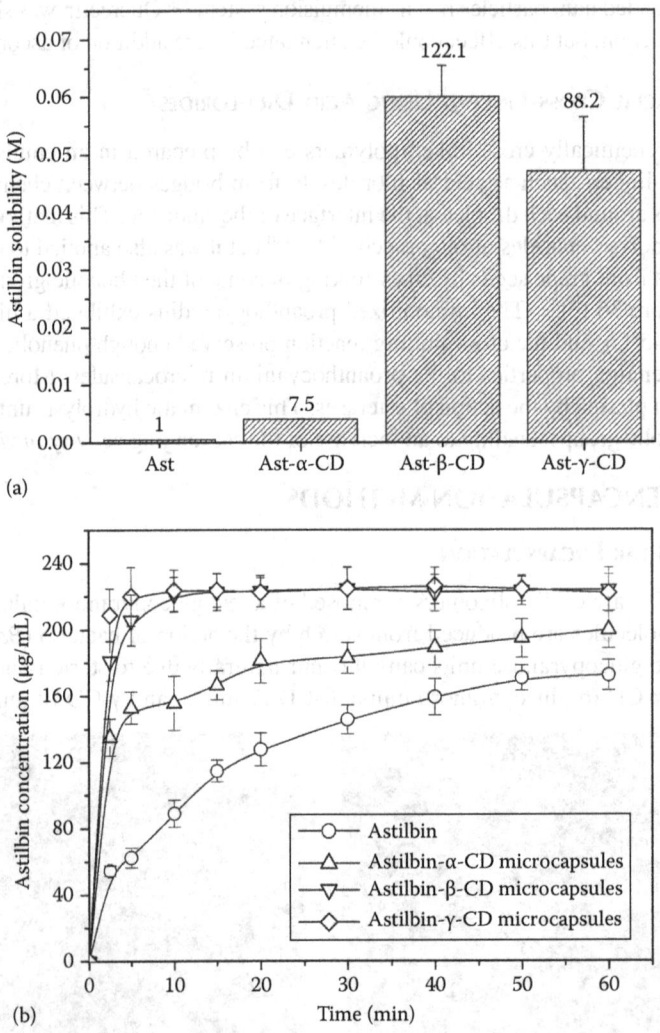

FIGURE 36.14 (a) Solubility of astilbin in different CD microcapsules at 298 K and (b) dissolution profiles of astilbin and its CD microcapsules in water at 310 K. The microcapsules of astilbin and CDs were prepared by a freeze-drying method at a molar ratio of 1:1. (Reprinted with permission from Zhang, Q.-F., Nie, H.-C., Shangguang, X.-C., Yin, Z.-P., Zheng, G.-D., and Chen, J.-G., Aqueous solubility and stability enhancement of astilbin through complexation with cyclodextrins, *J. Agric. Food Chem.*, 61(1), 151–156. Copyright 2013 American Chemical Society.)

Molecular complexation is theoretically not limited to inclusion in CDs,[135] but it seems that no other complexing systems have been applied to polyphenolics yet.

36.5.2 Cocrystallization

The cocrystallization method consists in the incorporation of the molecule of interest in a saturated sucrose solution, followed by the formation of sucrose microcrystals, containing the molecule in a crystallized or amorphous state. After a drying step and a grinding operation, the encapsulated molecule contained in the resulting granular solid presents improved solubility and stability.

This method has been applied to the encapsulation of a Yerba mate extract, reducing its hygroscopicity without affecting the high solubility of this polyphenol-rich extract,[136] and improving its stability.[137]

36.5.3 Encapsulation in Yeasts

Yeast cells (*Saccharomyces cerevisiae*) are surrounded by a membrane protecting the inside from evaporation or oxidation. This membrane can be reversibly permeabilized by immersion in a hypotonic NaCl solution. The holes created in the membrane allow the loading of a molecule in the cell, according to the method of Bishop.[138] Chlorogenic acid[139] or curcumin[140] have been encapsulated in yeast cells, leading to a better stability of the encapsulated polyphenol toward thermal stress. This simple and cheap process is promising for applications in the food industry, because of the compatibility of the encapsulant with food products.

36.6 CONCLUSION

The encapsulation of polyphenolics is generating a growing number of scientific papers, some of which are discussed in this chapter. The published studies show that the major obstacles to the use of these powerful natural compounds, that is, their limited stability, and their poor aqueous solubility limiting their bioavailability, can be overcome by encapsulation. Many innovative solutions are available, and have been shown to be effective in potentiating and protecting the action of polyphenolics.

REFERENCES

1. Bruneton J. *Pharmacognosie: Phytochimie, Plantes Médicinales*. Paris, France: Tec & Doc Lavoisier; 2009.
2. Macheix J-J, Fleuriet A, Jay-Allemand C. *Les composés phénoliques des végétaux: Un exemple de métabolites secondaires d'importance économique*. Lausanne, Switzerland: PPUR Presses Polytechniques; 2005. 212p.
3. Ding Y, Yao H, Yao Y, Fai LY, Zhang Z. Protection of dietary polyphenols against oral cancer. *Nutrients* juin 14, 2013;5(6):2173–2191.
4. He S, Sun C, Pan Y. Red wine polyphenols for cancer prevention. *Int J Mol Sci*. mai 20, 2008;9(5):842–853.
5. Mukherjee PK, Maity N, Nema NK, Sarkar BK. Bioactive compounds from natural resources against skin aging. *Phytomedicine*. déc 15, 2011;19(1):64–73.
6. Barnham KJ, Masters CL, Bush AI. Neurodegenerative diseases and oxidatives stress. *Nat Rev Drug Discov*. 2004;3(3):205–214.
7. Cadenas E, Davies KJA. Mitochondrial free radical generation, oxidative stress, and aging. *Free Radic Biol Med*. 2000;29(3–4):222–230.
8. Harman D. Aging: A theory based on free radical and radiation chemistry. *J Gerontol*. 1956;11(3):298–300.
9. Sies H. Oxidative stress: Oxidants and antioxidants. *Exp Physiol*. 1997;82(2):291–295.
10. Leopoldini M, Russo N, Toscano M. The molecular basis of working mechanism of natural polyphenolic antioxidants. *Food Chem*. 2011;125(2):288–306.

11. Hanasaki Y, Ogawa S, Fukui S. The correlation between active oxygens scavenging and antioxidative effects of flavonoids. *Free Radic Biol Med.* 1994;16(6):845–850.
12. Rice-Evans CA, Miller NJ, Paganga G. Structure-antioxidant activity relationships of flavonoids and phenolic acids. *Free Radic Biol Med.* 1996;20(7):933–956.
13. Van Acker SABE, Van Den Berg D-J, Tromp MNJL, Griffioen DH, Van Bennekom WP, Van Der Vijgh WJF et al. Structural aspects of antioxidant activity of flavonoids. *Free Radic Biol Med.* 1996;20(3):331–342.
14. Burda S, Oleszek W. Antioxidant and antiradical activities of flavonoids. *J Agric Food Chem.* 2001;49(6):2774–2779.
15. Silva MM, Santos MR, Caroço G, Rocha R, Justino G, Mira L. Structure-antioxidant activity relationships of flavonoids: A re-examination. *Free Radic Res.* 2002;36(11):1219–1227.
16. Dugas Jr. AJ, Castañeda-Acosta J, Bonin GC, Price KL, Fischer NH, Winston GW. Evaluation of the total peroxyl radical-scavenging capacity of flavonoids: Structure-activity relationships. *J Nat Prod.* 2000;63(3):327–331.
17. Buettner GR, Jurkiewicz BA. Catalytic metals, ascorbate and free radicals: Combinations to avoid. *Radiat Res.* 1996;145(5):532–541.
18. Hagerman AE, Butler LG. The specificity of proanthocyanidin-protein interactions. *J Biol Chem.* 1981;256(9):4494–4497.
19. McManus JP, Davis KG, Beart JE, Gaffney SH, Lilley TH, Haslam E. Polyphenol interactions. Part 1. Introduction; some observations on the reversible complexation of polyphenols with proteins and polysaccharides. *J Chem Soc Perkin 2.* 1985;(9):1429–1438.
20. Richard T, Lefeuvre D, Descendit A, Quideau S, Monti JP. Recognition characters in peptide-polyphenol complex formation. *Biochim Biophys Acta Gen Subj.* 2006;1760(6):951–958.
21. Siebert KJ, Troukhanova NV, Lynn PY. Nature of polyphenol–protein interactions. *J Agric Food Chem.* 1996;44(1):80–85.
22. Cos P, Ying L, Calomme M, Hu JP, Cimanga K, Van Poel B et al. Structure-activity relationship and classification of flavonoids as inhibitors of xanthine oxidase and superoxide scavengers. *J Nat Prod.* 1998;61(1):71–76.
23. Kim HP, Mani I, Iversen L, Ziboh VA. Effects of naturally-occurring flavonoids and biflavonoids on epidermal cyclooxygenase and lipoxygenase from guinea-pigs. *Prostaglandins Leukot Essent Fat Acids.* 1998;58(1):17–24.
24. Haslam E. *Practical Polyphenolics: From Structure to Molecular Recognition and Physiological Action.* Cambridge, U.K.: Cambridge University Press; 1998. 446p.
25. Li J, Gang D, Yu X, Hu Y, Yue Y, Cheng W et al. Genistein: The potential for efficacy in rheumatoid arthritis. *Clin Rheumatol.* 2013;32(5):535–540.
26. Silva JC, Rodrigues S, Feás X, Estevinho LM. Antimicrobial activity, phenolic profile and role in the inflammation of propolis. *Food Chem Toxicol.* 2012;50(5):1790–1795.
27. Kang KS, Yamabe N, Wen Y, Fukui M, Zhu BT. Beneficial effects of natural phenolics on levodopa methylation and oxidative neurodegeneration. *Brain Res.* 2013;1497:1–14.
28. Mushtaq M, Wani SM. Polyphenols and human health—A review. *Int J Pharma Bio Sci.* 2013;4(2):B338–B360.
29. Jayasena T, Poljak A, Smythe G, Braidy N, Münch G, Sachdev P. The role of polyphenols in the modulation of sirtuins and other pathways involved in Alzheimer's disease. *Ageing Res Rev.* Sept 2013;12(4):867–883.
30. Leclercq G, Jacquot Y. Interactions of isoflavones and other plant derived estrogens with estrogen receptors for prevention and treatment of breast cancer—Considerations concerning related efficacy and safety. *J Steroid Biochem Mol Biol.* 2014;139:237–244.
31. Louw A, Joubert E, Visser K. Phytoestrogenic potential of cyclopia extracts and polyphenols. *Planta Med.* 2013;79(7):580–590.
32. Da Silva Morrone M, De Assis AM, Da Rocha RF, Gasparotto J, Gazola AC, Costa GM et al. Passiflora manicata (Juss.) aqueous leaf extract protects against reactive oxygen species and protein glycation in vitro and ex vivo models. *Food Chem Toxicol.* 2013;60:45–51.
33. Zhang A, Sun H, Wang X. Recent advances in natural products from plants for treatment of liver diseases. *Eur J Med Chem.* 2013;63:570–577.
34. Conforti F, Marrelli M, Menichini F, Bonesi M, Statti G, Provenzano E et al. Natural and synthetic furanocoumarins as treatment for vitiligo and psoriasis. *Curr Drug Ther.* 2009;4(1):38–58.

35. Saija A, Tomaino A, Trombetta D, Pellegrino ML, Tita B, Messina C et al. "In vitro" antioxidant and photoprotective properties and interaction with model membranes of three new quercetin esters. *Eur J Pharm Biopharm.* 2003;56(2):167–174.
36. El Gharras H. Polyphenols: Food sources, properties and applications—A review. *Int J Food Sci Technol.* 2009;44(12):2512–2518.
37. Fang Z, Bhandari B. Encapsulation of polyphenols—A review. *Trends Food Sci Technol.* 2010;21(10):510–523.
38. Arshady R. *Microspheres Microcapsules and Liposomes: Preparation and Chemical Applications.* London, U.K.: Citus Books; 1999. 616p.
39. Benita S. *Microencapsulation: Methods and Industrial Applications.* Hoboken, NJ: Taylor & Francis; 2006. 783p.
40. Shanthi CN, Gupta R, Mahato AK. Traditional and emerging applications of microspheres: A review. *Int J PharmTech Res.* 2010;2(1):675–681.
41. Vandamme TF, Poncelet D, Subra-Paternault P. *Microencapsulation: Des sciences aux technologies.* Paris, France: Lavoisier Tec & Doc; 2007. 355p.
42. Patel RP, Patel MP, Suthar AM. Spray drying technology: An overview. *Ind J Sci Technol.* 1999;2(10):44–47.
43. Vehring R. Pharmaceutical particle engineering via spray drying. *Pharm Res.* 2008;25(5):999–1022.
44. Rocha-Guzmán NE, Gallegos-Infante JA, González-Laredo RF, Harte F, Medina-Torres L, Ochoa-Martínez LA et al. Effect of high-pressure homogenization on the physical and antioxidant properties of Quercus resinosa infusions encapsulated by spray-drying. *J Food Sci.* 2010;75(5):N57–N61.
45. Rocha S, Generalov R, Pereira MDC, Peres I, Juzenas P, Coelho MAN. Epigallocatechin gallate-loaded polysaccharide nanoparticles for prostate cancer chemoprevention. *Nanomedicine.* 2011;6(1):79–87.
46. Zhang L, Mou D, Du Y. Procyanidins: Extraction and micro-encapsulation. *J Sci Food Agric.* 2007;87(12):2192–2197.
47. Ersus S, Yurdagel U. Microencapsulation of anthocyanin pigments of black carrot (Daucus carota L.) by spray drier. *J Food Eng.* 2007;80(3):805–812.
48. Robert P, Gorena T, Romero N, Sepulveda E, Chavez J, Saenz C. Encapsulation of polyphenols and anthocyanins from pomegranate (Punica granatum) by spray drying. *Int J Food Sc Technol.* 2010;45(7):1386–1394.
49. Kosaraju SL, D'ath L, Lawrence A. Preparation and characterisation of chitosan microspheres for antioxidant delivery. *Carbohydr Polym.* 2006;64(2):163–167.
50. Sansone F, Picerno P, Mencherini T, Porta A, Lauro MR, Russo P et al. Technological properties and enhancement of antifungal activity of a Paeonia rockii extract encapsulated in a chitosan-based matrix. *J Food Eng.* janv 2014;120:260–267.
51. Krishnaiah D, Sarbatly R, Hafiz AMM, Hafeza AB, Rao SRM. Study on retention of bioactive components of Morinda citrifolia L. using spray-drying. *J Appl Sci.* 2009;9(17):3092–3097.
52. Krishnaiah D, Sarbatly R, Mohan Rao SR, Nithyanand RR. Optimal operating conditions of spray dried noni fruit extract using κ-carrageenan as adjuvant. *J Appl Sci.* 2009;9(17):3062–3067.
53. Gavini E, Alamanni MC, Cossu M, Giunchedi P. Tabletted microspheres containing Cynara scolymus (var. Spinoso sardo) extract for the preparation of controlled release nutraceutical matrices. *J Microencapsul.* 2005;22(5):487–499.
54. Kosaraju SL, Labbett D, Emin M, Konczak I, Lundin L. Delivering polyphenols for healthy ageing. *Nutr Diet.* 2008;65(Suppl. 3):S48–S52.
55. Krishnaiah D, Nithyanandam R, Sarbatly R. A critical review on the spray drying of fruit extract: Effect of additives on physicochemical properties. *Crit Rev Food Sci Nutr.* 2014;54(4):449–473.
56. Yatsu FKJ, Borghetti GS, Bassani VL. Technological characterization and stability of Ilex paraguariensis St. Hil. Aquifoliaceae (maté) spray-dried powder. *J Med Food.* 2011;14(4):413–419.
57. Georgetti SR, Casagrande R, Souza CRF, Oliveira WP, Fonseca MJV. Spray drying of the soybean extract: Effects on chemical properties and antioxidant activity. *LWT—Food Sci Technol.* 2008;41(8):1521–1527.
58. Cocero MJ, Martín A, Mattea F, Varona S. Encapsulation and co-precipitation processes with supercritical fluids: Fundamentals and applications. *J Supercrit Fluids.* 2009;47(3):546–555.
59. Meterc D, Petermann M, Weidner E. Drying of aqueous green tea extracts using a supercritical fluid spray process. *J Supercrit Fluids.* 2008;45(2):253–259.
60. Xiong S, Melton LD, Easteal AJ, Siew D. Stability and antioxidant activity of black currant anthocyanins in solution and encapsulated in glucan gel. *J Agric Food Chem.* 2006;54(17):6201–6208.

61. Smith A, Giunta B, Bickford PC, Fountain M, Tan J, Shytle RD. Nanolipidic particles improve the bioavailability and α-secretase inducing ability of epigallocatechin-3-gallate (EGCG) for the treatment of Alzheimer's disease. *Int J Pharm.* 2010;389(1–2):207–212.
62. Li H, Zhao X, Ma Y, Zhai G, Li L, Lou H. Enhancement of gastrointestinal absorption of quercetin by solid lipid nanoparticles. *J Control Release.* 2009;133(3):238–244.
63. Liu L, Tang Y, Gao C, Li Y, Chen S, Xiong T et al. Characterization and biodistribution in vivo of quercetin-loaded cationic nanostructured lipid carriers. *Colloids Surf B Biointerfaces.* 2014;115:125–131.
64. Bose S, Michniak-Kohn B. Preparation and characterization of lipid based nanosystems for topical delivery of quercetin. *Eur J Pharm* Sci. 2013;48(3):442–452.
65. Guo C-Y, Yang C-F, Li Q-L, Tan Q, Xi Y-W, Liu W-N et al. Development of a Quercetin-loaded nanostructured lipid carrier formulation for topical delivery. *Int J Pharm.* 2012;430(1–2):292–298.
66. Mendes JBE, Riekes MK, de Oliveira VM, Michel MD, Stulzer HK et al. PHBV/PCL microparticles for controlled release of resveratrol: Physicochemical characterization, antioxidant potential, and effect on hemolysis of human erythrocytes. *Sci World J* [Internet]. mai 1, 2012 [cité 21 févr 2014];2012. Disponible sur: http://www.hindawi.com/journals/tswj/2012/542937/abs/.
67. Li Y, Huang C, Cen Y, Xu S, Xu S. Preparation of tea polyphenols sustained-release microcapsule. *J Chin Med Mater.* 2000;23(5):281–284.
68. Zheng L, Ding Z, Zhang M, Sun J. Microencapsulation of bayberry polyphenols by ethyl cellulose: Preparation and characterization. *J Food Eng.* 2011;104(1):89–95.
69. Janet T, Taylor JRN, Belton PS, Amanda M. Kafirin microparticle encapsulation of catechin and sorghum condensed tannins. *J Agric Food Chem.* 2009;57(16):7523–7528.
70. Tsai Y-M, Jan W-C, Chien C-F, Lee W-C, Lin L-C, Tsai T-H. Optimised nano-formulation on the bioavailability of hydrophobic polyphenol, curcumin, in freely-moving rats. *Food Chem.* 2011;127(3):918–925.
71. Mukerjee A, Vishwanatha JK. Formulation, characterization and evaluation of curcumin-loaded PLGA nanospheres for cancer therapy. *Anticancer Res.* 2009;29(10):3867–3875.
72. Italia JL, Datta P, Ankola DD, Kumar MNVR. Nanoparticles enhance per oral bioavailability of poorly available molecules: Epigallocatechin gallate nanoparticles ameliorates cyclosporine induced nephrotoxicity in rats at three times lower dose than oral solution. *J Biomed Nanotechnol.* 2008;4(3):304–312.
73. Bala I, Bhardwaj V, Hariharan S, Kharade SV, Roy N, Kumar MNVR. Sustained release nanoparticulate formulation containing antioxidant-ellagic acid as potential prophylaxis system for oral administration. *J Drug Target.* 2006;14(1):27–34.
74. Kumari A, Yadav SK, Pakade YB, Kumar V, Singh B, Chaudhary A et al. Nanoencapsulation and characterization of Albizia chinensis isolated antioxidant quercitrin on PLA nanoparticles. *Colloids Surf B Biointerfaces.* 2011;82(1):224–232.
75. Kumari A, Yadav SK, Pakade YB, Singh B, Yadav SC. Development of biodegradable nanoparticles for delivery of quercetin. *Colloids Surf B Biointerfaces.* 2010;80(2):184–192.
76. Sonaje K, Italia JL, Sharma G, Bhardwaj V, Tikoo K, Kumar MNVR. Development of biodegradable nanoparticles for oral delivery of ellagic acid and evaluation of their antioxidant efficacy against cyclosporine A-induced nephrotoxicity in rats. *Pharm Res.* 2007;24(5):899–908.
77. Shahani K, Panyam J. Highly loaded, sustained-release microparticles of curcumin for chemoprevention. *J Pharm Sci.* juill 2011;100(7):2599–2609.
78. Bennet D, Marimuthu M, Kim S, An J. Dual drug-loaded nanoparticles on self-integrated scaffold for controlled delivery. *Int J Nanomedicine.* 2012;7:3399–3419.
79. Bennet D, Kim S. A transdermal delivery system to enhance quercetin nanoparticle permeability. *J Biomater Sci Polym Ed.* 2013;24(2):185–209.
80. Shao J, Li X, Lu X, Jiang C, Hu Y, Li Q et al. Enhanced growth inhibition effect of resveratrol incorporated into biodegradable nanoparticles against glioma cells is mediated by the induction of intracellular reactive oxygen species levels. *Colloids Surf B Biointerfaces.* 2009;72(1):40–47.
81. Siddiqui IA, Adhami VM, Bharali DJ, Hafeez BB, Asim M, Khwaja SI et al. Introducing nanochemoprevention as a novel approach for cancer control: Proof of principle with green tea polyphenol epigallocatechin-3-gallate. *Cancer Res.* 2009;69(5):1712–1716.
82. Yen F-L, Wu T-H, Lin L-T, Cham T-M, Lin C-C. Naringenin-loaded nanoparticles improve the physicochemical properties and the hepatoprotective effects of naringenin in orally-administered rats with CCl4-induced acute liver failure. *Pharm Res.* 2009;26(4):893–902.
83. Khoee S, Rahmatolahzadeh R. Synthesis and characterization of pH-responsive and folated nanoparticles based on self-assembled brush-like PLGA/PEG/AEMA copolymer with targeted cancer therapy properties: A comprehensive kinetic study. *Eur J Med Chem.* 2012;50:416–427.

84. Wu T-H, Yen F-L, Lin L-T, Tsai T-R, Lin C-C, Cham T-M. Preparation, physicochemical characterization, and antioxidant effects of quercetin nanoparticles. *Int J Pharm.* 2008;346(1–2):160–168.
85. Sanna V, Roggio AM, Siliani S, Piccinini M, Marceddu S, Mariani A et al. Development of novel cationic chitosan-and anionic alginate-coated poly(D,L-lactide-co-glycolide) nanoparticles for controlled release and light protection of resveratrol. *Int J Nanomedicine.* 2012;7:5501–5516.
86. Sosa MV, Rodríguez-Rojo S, Mattea F, Cismondi M, Cocero MJ. Green tea encapsulation by means of high pressure antisolvent coprecipitation. *J Supercrit Fluids.* 2011;56(3):304–311.
87. Dehkharghanian M, Lacroix M, Vijayalakshmi MA. Antioxidant properties of green tea polyphenols encapsulated in caseinate beads. *Dairy Sci Technol.* 2009;89(5):485–499.
88. Sadeghi S, Madadlou A, Yarmand M. Microemulsification-cold gelation of whey proteins for nanoencapsulation of date palm pit extract. *Food Hydrocolloids* 2014;35:590–596.
89. Lee J-S, Kim GH, Lee HG. Characteristics and antioxidant activity of elsholtzia splendens extract-loaded nanoparticles. *J Agric Food Chem.* 2010;58(6):3316–3321.
90. Harris R, Lecumberri E, Mateos-Aparicio I, Mengíbar M, Heras A. Chitosan nanoparticles and microspheres for the encapsulation of natural antioxidants extracted from Ilex paraguariensis. *Carbohydr Polym.* 2011;84(2):803–806.
91. Wisuitiprot W, Somsiri A, Ingkaninan K, Waranuch N. A novel technique for chitosan microparticle preparation using a water/silicone emulsion: Green tea model. *Int J Cosmet Sci.* 2011;33(4):351–358.
92. Dube A, Ng K, Nicolazzo JA, Larson I. Effective use of reducing agents and nanoparticle encapsulation in stabilizing catechins in alkaline solution. *Food Chem.* 2010;122(3):662–667.
93. Hu B, Ting Y, Zeng X, Huang Q. Bioactive peptides/chitosan nanoparticles enhance cellular antioxidant activity of (–)-epigallocatechin-3-gallate. *J Agric Food Chem.* janv 30, 2013;61(4):875–881.
94. Deladino L, Anbinder PS, Navarro AS, Martino MN. Encapsulation of natural antioxidants extracted from Ilex paraguariensis. *Carbohydr Polym.* 2008;71(1):126–134.
95. López Córdoba A, Deladino L, Martino M. Effect of starch filler on calcium-alginate hydrogels loaded with yerba mate antioxidants. *Carbohydr Polym.* 2013;95(1):315–323.
96. Das S, Ng K-Y. Resveratrol-loaded calcium-pectinate beads: Effects of formulation parameters on drug release and bead characteristics. *J Pharm Sci.* 2010;99(2):840–860.
97. Liang J, Li F, Fang Y, Yang W, An X, Zhao L et al. Synthesis, characterization and cytotoxicity studies of chitosan-coated tea polyphenols nanoparticles. *Colloids Surf B Biointerfaces.* 2011;82(2):297–301.
98. Zhou H, Sun X, Zhang L, Zhang P, Li J, Liu Y-N. Fabrication of biopolymeric complex coacervation core micelles for efficient tea polyphenol delivery via a green process. *Langmuir.* 2012;28(41):14553–14561.
99. De Cock LJ, De Koker S, De Geest BG, Grooten J, Vervaet C, Remon JP et al. Polymeric multilayer capsules in drug delivery. *Angew Chem Int Ed.* 2010;49(39):6954–6973.
100. Shutava TG, Balkundi SS, Lvov YM. (-)-Epigallocatechin gallate/gelatin layer-by-layer assembled films and microcapsules. *J Colloid Interf Sci.* 2009;330(2):276–283.
101. Shutava TG, Balkundi SS, Vangala P, Steffan JJ, Bigelow RL, Cardelli JA et al. Layer-by-layer-coated gelatin nanoparticles as a vehicle for delivery of natural polyphenols. *ACS Nano.* 2009;3(7):1877–1885.
102. Nori MP, Favaro-Trindade CS, Matias de Alencar S, Thomazini M, de Camargo Balieiro JC, Contreras Castillo CJ. Microencapsulation of propolis extract by complex coacervation. *LWT—Food Sci Technol.* 2011;44(2):429–435.
103. Kraft JC, Freeling JP, Wang Z, Ho RJY. Emerging research and clinical development trends of liposome and lipid nanoparticle drug delivery systems. *J Pharm Sci.* 2014;103(1):29–52.
104. Lapenna S, Bilia AR, Morris GA, Nilsson M. Novel artemisinin and curcumin micellar formulations: Drug solubility studies by NMR spectroscopy. *J Pharm Sci.* 2009;98(10):3666–3675.
105. Pazos M, Torres JL, Andersen ML, Skibsted LH, Medina I. Galloylated polyphenols efficiently reduce α-tocopherol radicals in a phospholipid model system composed of sodium dodecyl sulfate (sds) micelles. *J Agric Food Chem.* 2009;57(11):5042–5048.
106. Bury K, Neugebauer D. Novel self-assembly graft copolymers as carriers for anti-inflammatory drug delivery. *Int J Pharm.* 2014;460(1–2):150–157.
107. Lu X, Ji C, Xu H, Li X, Ding H, Ye M et al. Resveratrol-loaded polymeric micelles protect cells from Aβ-induced oxidative stress. *Int J Pharm.* 2009;375(1–2):89–96.
108. Qiu J-F, Gao X, Wang B-L, Wei X-W, Gou M-L, Men K et al. Preparation and characterization of monomethoxy poly(ethylene glycol)-poly(ε-caprolactone) micelles for the solubilization and in vivo delivery of luteolin. *Int J Nanomedicine.* 2013;8:3061–3069.
109. Bisht S, Mizuma M, Feldmann G, Ottenhof NA, Hong S-M, Pramanik D et al. Systemic administration of polymeric nanoparticle-encapsulated curcumin (NanoCurc) blocks tumor growth and metastases in preclinical models of pancreatic cancer. *Mol Cancer Ther.* 2010;9(8):2255–2264.

110. Ray B, Bisht S, Maitra A, Maitra A, Lahiri DK. Neuroprotective and neurorescue effects of a novel polymeric nanoparticle formulation of curcumin (NanoCurc™) in the neuronal cell culture and animal model: Implications for Alzheimer's disease. *J Alzheimers Dis*. 2011;23(1):61–77.
111. Nakayama T, Kajiya K, Kumazawa S. Chapter 4: Interaction of plant polyphenols with liposomes. *Advances in Planar Lipid Bilayers and Liposomes*. San Diego, CA: Academic Press; 2006;4:107–133.
112. Oteiza PI, Erlejman AG, Verstraeten SV, Keen CL, Fraga CG. Flavonoid-membrane interactions: A protective role of flavonoids at the membrane surface? *Clin Dev Immunol*. 2005;12(1):19–25.
113. Hung C-F, Chen J-K, Liao M-H, Lo H-M, Fang J-Y. Development and evaluation of emulsion-liposome blends for resveratrol delivery. *J Nanosci Nanotechnol*. 2006;6(9–10):2950–2958.
114. Agrawal R, Kaur IP. Inhibitory effect of encapsulated curcumin on ultraviolet-induced photoaging in mice. *Rejuvenation Res*. 2010;13(4):397–410.
115. Mandal AK, Das N. Sugar coated liposomal flavonoid: A unique formulation in combating carbontetrachloride induced hepatic oxidative damage. *J Drug Target*. 2005;13(5):305–315.
116. Fang J-Y, Hung C-F, Hwang T-L, Huang Y-L. Physicochemical characteristics and in vivo deposition of liposome-encapsulated tea catechins by topical and intratumor administrations. *J Drug Target*. 2005;13(1):19–27.
117. Shibata T, Ishimaru K, Kawaguchi S, Yoshikawa H, Hama Y. Antioxidant activities of phlorotannins isolated from Japanese Laminariaceae. *J Appl Phycol*. 2008;20(5):705–711.
118. Zou L-Q, Liu W, Liu W-L, Liang R-H, Li T, Liu C-M et al. Characterization and bioavailability of tea polyphenol nanoliposome prepared by combining an ethanol injection method with dynamic high-pressure microfluidization. *J Agric Food Chem*. 2014;62(4):934–941.
119. Zhong H, Deng Y, Wang X, Yang B. Multivesicular liposome formulation for the sustained delivery of breviscapine. *Int J Pharm*. 2005;301(1–2):15–24.
120. Kristl J, Teskač K, Caddeo C, Abramović Z, Šentjurc M. Improvements of cellular stress response on resveratrol in liposomes. *Eur J Pharm Biopharm*. 2009;73(2):253–259.
121. Priprem A, Watanatorn J, Sutthiparinyanont S, Phachonpai W, Muchimapura S. Anxiety and cognitive effects of quercetin liposomes in rats. *Nanomedicine*. 2008;4(1):70–78.
122. Tong-Un T, Wannanon P, Wattanathorn J, Phachonpai W. Quercetin liposomes via nasal administration reduce anxiety and depression-like behaviors and enhance cognitive performances in rats. *Am J Pharmacol Toxicol*. 2010;5(2):80–88.
123. Fang J-Y, Hwang T-L, Huang Y-L, Fang C-L. Enhancement of the transdermal delivery of catechins by liposomes incorporating anionic surfactants and ethanol. *Int J Pharm*. 2006;310(1–2):131–138.
124. Takahashi M, Uechi S, Takara K, Asikin Y, Wada K. Evaluation of an oral carrier system in rats: Bioavailability and antioxidant properties of liposome-encapsulated curcumin. *J Agric Food Chem*. 2009;57(19):9141–9146.
125. Seguin J, Brullé L, Boyer R, Lu YM, Ramos Romano M, Touil YS et al. Liposomal encapsulation of the natural flavonoid fisetin improves bioavailability and antitumor efficacy. *Int J Pharm*. 2013;444(1–2):146–154.
126. Semalty A, Semalty M, Rawat MSM, Franceschi F. Supramolecular phospholipids–polyphenolics interactions: The PHYTOSOME® strategy to improve the bioavailability of phytochemicals. *Fitoterapia*. juill 2010;81(5):306–314.
127. Catania A, Barrajón-Catalán E, Nicolosi S, Cicirata F, Micol V. Immunoliposome encapsulation increases cytotoxic activity and selectivity of curcumin and resveratrol against HER2 overexpressing human breast cancer cells. *Breast Cancer Res Treat*. 2013;141(1):55–65.
128. Bernardy N, Romio AP, Barcelos EI, Dal Pizzol C, Dora CL, Lemos-Senna E et al. Nanoencapsulation of quercetin via miniemulsion polymerization. *J Biomed Nanotechnol*. 2010;6(2):181–186.
129. Andry M-C, Edwards-Lévy F, Lévy M-C. Free amino group content of serum albumin microcapsules. III. A study at low pH values. *Int J Pharm*. 1996;128(1–2):197–202.
130. Pariot N, Edwards-Lévy F, Andry M-C, Lévy M-C. Cross-linked β-cyclodextrin microcapsules. II. Retarding effect on drug release through semi-permeable membranes. *Int J Pharm*. 2002;232(1–2):175–181.
131. Andry M-C, Vezin H, Dumistracel I, Bernier JL, Lévy M-C. Proanthocyanidin microcapsules: Preparation, properties and free radical scavenging activity. *Int J Pharm*. 1998;171(2):217–226.
132. Challa R, Ahuja A, Ali J, Khar RK. Cyclodextrins in drug delivery: An updated review. *AAPS PharmSciTech*. 2005;6(2):329–57.
133. Pinho E, Grootveld M, Soares G, Henriques M. Cyclodextrins as encapsulation agents for plant bioactive compounds. *Carbohydr Polym*. janv 30, 2014;101:121–135.

134. Zhang Q-F, Nie H-C, Shangguang X-C, Yin Z-P, Zheng G-D, Chen J-G. Aqueous solubility and stability enhancement of astilbin through complexation with cyclodextrins. *J Agric Food Chem.* janv 9, 2013;61(1):151–156.
135. Schneider H-J, Agrawal P, Yatsimirsky AK. Supramolecular complexations of natural products. *Chem Soc Rev.* 2013;42(16):6777–6800.
136. Deladino L, Anbinder PS, Navarro AS, Martino MN. Co-crystallization of yerba mate extract (Ilex paraguariensis) and mineral salts within a sucrose matrix. *J Food Eng.* 2007;80(2):573–580.
137. López-Córdoba A, Deladino L, Agudelo-Mesa L, Martino M. Yerba mate antioxidant powders obtained by co-crystallization: Stability during storage. *J Food Eng.* mars 2014;124:158–165.
138. Bishop JRP, Nelson G, Lamb J. Microencapsulation in yeast cells. *J Microencapsul.* 1998;15(6):761–773.
139. Shi G, Rao L, Yu H, Xiang H, Pen G, Long S et al. Yeast-cell-based microencapsulation of chlorogenic acid as a water-soluble antioxidant. *J Food Eng.* 2007;80(4):1060–1067.
140. Paramera EI, Konteles SJ, Karathanos VT. Stability and release properties of curcumin encapsulated in Saccharomyces cerevisiae, β-cyclodextrin and modified starch. *Food Chem.* 2011;125(3):913–922.

37 Encapsulation of Bioactive Compounds

Francesco Donsì, Mariarenata Sessa, and Giovanna Ferrari

CONTENTS

37.1 Introduction 765
37.2 Bioactive Compounds in Foods 766
    37.2.1 Phytochemicals 767
    37.2.2 Micronutrients 771
    37.2.3 Dietary Fibers 771
    37.2.4 Prebiotics and Probiotics 772
        37.2.4.1 Prebiotics 772
        37.2.4.2 Probiotics 773
37.3 Encapsulation Systems 775
37.4 Encapsulation Techniques 782
    37.4.1 Mechanical Processes 782
    37.4.2 Physicochemical Processes 784
37.5 Encapsulation of Probiotics 786
37.6 Conclusions and Perspectives 787
References 789

37.1 INTRODUCTION

The scientific and industrial interest for functional foods, where bioactive compounds are incorporated, has been significantly reinforced in recent years by the increasing demand from consumers for health promotion and disease prevention through diet and nutrition.

The main bioactive compounds for food functionalization include several classes of compounds, which were discovered to positively affect health. They can be classified as phytochemicals, micronutrients, dietary fibers, prebiotics, and probiotics. In particular, prebiotics compounds and probiotic microorganisms, owing to their contribution to regulate intestinal flora, have recently gained significant attention, because the human biota is considered to affect directly and indirectly several body functions. Despite not being single chemical molecules, but living organisms, probiotics are generally classified among the bioactive compounds, and therefore their encapsulation is also treated in this chapter.

The incorporation of bioactive compounds in foods is challenged by significant technological hurdles, which are related to the desired *in product* as well as *in body* behavior.[1,2]

The *in product* behavior is affected by the following:

- The efficient dispersion in the food matrix and the compatibility with it
- The reactivity of bioactive compounds, which in turn is responsible for high degradation rate and interaction with other food components
- The complex environment of food matrices, with different interfaces between aqueous and lipid phases, where bioactive compounds may adsorb
- The intense treatment conditions, to which food is subjected because of processing, preservation, or preparation (high or low temperatures, pH extremes, intense shearing, etc.)

The *in body* behavior is affected by the following:

- The release from the food matrix, preferably triggered by environmental changes, such as pH (chewing, gastrointestinal tract), temperature (body temperature, cooking), mechanical shear (chewing, mastication), enzymes (gastrointestinal tract) as well as addition of moisture (dissolution, chewing)
- The fate of the bioactive compounds during gastric and intestinal digestion
- The bioavailability of the active compounds, taking into account epithelial cells uptake, absorption in the blood stream and reaching the target sites

In order to control and regulate the *in product* as well as *in body* behavior, the bioactive compounds need a suitable encapsulation system, which is able not only to promote their efficient dispersion, their protection during most intense phases of processing, preservation, or preparation, but also to control their release during mastication (taste masking or enhancing) and gastrointestinal digestion, enhancing their bioaccessibility and bioavailabilty.[1]

In particular, controlled release of the bioactive compounds plays a double role, of preserving the bioactive compounds *in product*, and of maximizing their release in the intestinal tract to promote their uptake.

Encapsulation involves the immobilization of the bioactive compounds within a capsule, which completely embeds them or where they are dispersed in. The particles obtained upon encapsulation may have a size ranging from a few nanometers to a few millimeters, with smaller sizes preferred in order to increase the specific surface for the release of the encapsulated compounds (also known as payload) as well as the interaction with biological tissues.[2]

Taking into account the requirements in terms of physicochemical stability, mean droplet size, controlled or triggered release, as well as food compatibility and ease of production at industrial scale, the main technological and scientific efforts toward the development of efficient encapsulation systems for the food industry in recent years were addressed to

- Develop novel capsule architectures, with a compartmentalized structure able to host bioactive compounds with different interfacial properties[3]
- Exploit the functionality of existing food-grade ingredients, to replace artificial molecules with limited food acceptability[4]
- Design novel production processes, with low cost and high productivity, to produce at an industrial scale compatible with the food industry novel delivery systems[1]

It must be remarked that, because of the health beneficial properties of bioactive compounds, the interest for their encapsulation extends not only to the food and beverage industry, but also to the pharmaceutical and nutraceutical, as well as to the agricultural industry.

37.2 BIOACTIVE COMPOUNDS IN FOODS

In the recent years, a strong correlation between nutrition and chronic diseases, such as diabetes, obesity, cardiovascular diseases, hypertension, some types of cancer, osteoporosis, and dental diseases, became progressively more evident. The increased incidence of these chronic illnesses as well as their appearance in life earlier than before, which are considered to be the main consequence of increasing consumption of high-fat and energy-dense foods, regular intake of so-called junk foods and low fruit and vegetable consumption, sedentary lifestyles, and insufficient physical activity, impose a growing burden to health care systems, as recorded by the World Health Organization.[5]

Good nutrition is important for maintaining good health and promoting socio-economic development. Epidemiological studies have shown a positive relationship between dietary intake of whole grains, fruits, vegetables, fish, and fermented milk products and health status,[6-9] because of their content in bioactive compounds such as dietary fibers, phytosterols, carotenoids, peptides, bioactive lipids, and probiotics.

Bioactive compounds, as defined by Kitts,[10] are "extranutritional" constituents that are naturally present in small quantities in foodstuff of both plant and animal origin. In fact, these compounds cannot be considered nutrients in the classical sense of the term, which are capable to develop, grow, and maintain an organism alive but can be defined as substances that can modulate several important biological activities and functions of the human body. The biological activities modulated by the intake of bioactive compounds are different, including antioxidant and anti-inflammatory activity, modulation of detoxification enzymes, stimulation of the immune system, antibacterial and antiviral activity, antiproliferative and proapoptotic activity, etc.[11] However, the studies that allow to highlight the beneficial effects on human health are not yet conclusive and are generally focused on specific compounds. Still unclear are in fact many aspects related to their bioavailability, metabolism, and interaction with the food matrix.[12] In addition, some of these substances, which are reported to exhibit either a positive or negative effect on health, depending on the amount taken, are contained in foods, whose consumption should not be promoted, such as alcoholic beverages.

Nevertheless, consumers' interest toward the relationship between diet and health has increased the demand for functional foods, fostering a continuous advancement in the related science, and technology. Many bioactive compounds are highly lipophilic resulting in poor absorption and limited bioavailability; others once extracted from the animal or plant source tissue are chemically unstable. It has become apparent that current difficulties associated with the inclusion of bioactives in food matrices are one of the major problems that manufacturers struggle with when developing functional foods. There is a pressing need for edible delivery system to encapsulate, protect, and release bioactive compounds within the food industry.[13]

The most important types of bioactive compounds can be classified in phytochemicals, micronutrients, dietary fibers, prebiotics, and probiotics, which are briefly discussed in the following sections.

37.2.1 Phytochemicals

Phytochemicals are chemical compounds that naturally occur in plants. They are responsible for some organoleptic properties, such as the deep purple of blueberries and the smell of garlic. They are nonessential nutrients, meaning that they are not required by the human body for sustaining life, but have protective or disease preventive properties.[14] It is well known that the production of these chemicals by plants has a protective purpose; recent researches demonstrated that phytochemicals can also protect humans against several diseases.[15–17]

Without specific knowledge of their cellular actions or mechanisms, phytochemicals have been considered as drugs for millennia. For example, Hippocrates considered willow tree leaves able to abate fever. During the nineteenth and twentieth centuries, the main strategy of scientists for the cure of illnesses was to discover the active ingredients, which had medicinal or pesticidal properties. Examples of these discoveries include salicylic acid, morphine, and pyrethroids (pesticides). During the 1980s, many laboratories started to identify phytochemicals in plants that might be used as medicines. At the same time other scientists conducted epidemiological studies to determine the relationship between the consumption of certain phytochemicals and human health.[18] Although scientific evidence supports the health-promoting functions of phytochemicals, their beneficial effects are often lost owing to their poor solubility in aqueous and lipid phases, physicochemical instability under food processing conditions (temperature, light, oxygen, and interaction with food matrix ingredients) and in the gastrointestinal tract (pH, enzymes, and presence of other nutrients), as well as insufficient gastric residence time and low permeability within the gut, limiting their activity and potential benefits.[19] Today, the scientific research is focused on the study of protective mechanisms able to maintain the active molecular form until the time of consumption and to deliver it to the target sites within the organism,[20] through suitable delivery systems.

Table 37.1 presents a list of phytochemicals, divided into different classes, with their beneficial effects, the food sources and the technological hurdles related to their incorporation in food matrices.

TABLE 37.1
Food Sources, Beneficial Effects, and the Technological Hurdles Related to Food Incorporation of Phytochemicals

Class	Examples	Food Sources	Beneficial Effects	Technological Hurdle for Food Incorporation	References
Phenolic compounds					
Polyphenols					
Flavonoids	Quercetin	Capers, red onion, lovage, green tea, apple, broccoli	Antioxidant activity, antiinflammatory activity, bronchodilator, reduces the release of histamine	Poor solubility in aqueous phase and low chemical stability	[21,22]
	Chatechins	Cocoa/chocolate, tea, red wine, berries, peach, apple, pear, apricot, vinegar	Antiatherosclerotic effect, inhibit the oxidation of LDL, cancer chemopreventive activity	Sensitive to oxidation, light and pH, astringent and bitter taste, slightly soluble in water	[23,24]
Stilbenoids	Resveratrol	Red wine, red grapes, peanuts, cocoa/chocolate	Antiinflammatory activity, beneficial cardiovascular effects, anti atherosclerotic effects, antioxidant activity, chemoprotective advantages	Low solubility in aqueous and lipid phases, easily degraded by sunlight and susceptible to react with dissolved O_2, bitter taste	[25–27]
Curcuminoids	Curcumin	Indian spice turmeric	Antioxidant activity, chemopreventive and anticarcinogenic effects, antiinflammatory activity	Scarse solubility in water and oil phases, unstable at neutral-basic pH values	[28–30]
Aromatic acid					
Phenolic acid	Gallic acid	Blackberry, mango, chocolate, raspberry, vinegar, wine	Antifungal and antiviral properties, antioxidant activity, used as remote astringent in cases of internal haemorrhage	Sensitive to temperature, oxidation, light, and pH	[31,32]

(Continued)

TABLE 37.1 (Continued)
Food Sources, Beneficial Effects, and the Technological Hurdles Related to Food Incorporation of Phytochemicals

Class	Examples	Food Sources	Beneficial Effects	Technological Hurdle for Food Incorporation	References
Hydroxycinnamic acids	Caffeic acid	Coffee, barley grain, argan oil	Antioxidant activity, immunomodulatory and anti-inflammatory activity	Low stability in UV irradiation and O_2 presence, low aqueous solubility, and bitter taste	[33,34]
Terpenes					
Carotenoids	β-carotene	Carrots, pumpkins, sweet potatoes, cantaloupe, mango, papayas	Protection against photooxidative stress and prevention against skin damage	Susceptible to light, oxygen, and autooxidation	[35,36]
	Lycopene	Tomatoes, watermelon, papaya	Potential preventive and/or therapeutic effect in prostate cancer		[37–40]
	Lutein	Collardgreens, kale, spinach, eggs, asparagus, broccoli, carrots	Supports maintenance of eye health		[41]
Monoterpenes	Limonene	Essential oils of citrus fruit	Antimicrobial activity, potential chemopreventive agent	Chemically reactive species	[42,43]
Lipids	Omega-3-, -6 and -9 fatty acid	Vegetable oils, nuts, fish, eggs	Reduce blood triglyceride levels, decrease risks of cardiovascular diseases	Vulnerable to oxidation and rancidity	[44–46]
Organosulfides	Allicin, diallylsulfide Allylmethyltrisulfide	Garlic, leek, onion	Enhance detoxification of undesirable compounds; support maintenance of heart, immune, and digestive health	Chemically unstable and offensive odor	[47,48]

Polyphenols are a heterogeneous group of natural compounds, particularly known for their beneficial effects on human health. In nature, polyphenols are produced by the secondary metabolism of plants, where, in relation to their various chemical characteristics have different roles:

- Defense against herbivores, imparting unpleasant taste, and pathogens (phytoalexins)
- Mechanical support (lignins)
- Barrier against microbial invasion
- Attraction to pollinators improving pigmentation (anthocyanins)

The polyphenols are characterized by possessing at least one aromatic ring, where one or more hydroxyl groups are bound. More than 8000 different structures of polyphenols have been identified, including simple, low molecular weight compounds with a single aromatic ring, up to highly polymerized compounds with very high molecular weight. Polyphenols can be classified into two groups: flavonoids and nonflavonoids (phenolic acids, stilbenes, and lignans) in function of the number of phenolic rings and of the structural elements that bind these rings.[49–51]

The intake of polyphenols in the human diet varies in relation to the type, quantity, and quality of the plant foods consumed. Not only fruits and vegetables are particularly rich in polyphenols, but also tea, red wine, cocoa, and derivatives. However, the cooking process reduces considerably the polyphenol content of food.[52,53]

Polyphenols, such as curcumin, quercetin and resveratrol, are some examples of naturally occurring phytochemicals with proven antioxidant and anti-inflammatory activity,[54,55] beneficial cardiovascular effects,[49,51,56] antiatherosclerotic effects,[57,58] chemopreventive, and anticarcinogenic effects.[59–61] An abundance of mechanistic information has become available on how polyphenols derived from dietary sources, which have putative chemopreventive properties, interfere with tumor promotion and progression.[62] The effect of these bioactive compounds on the organism is influenced by their bioavailability, namely, their ability to be effectively absorbed by the human body. However, the polyphenols have a poor bioavailability and are rapidly metabolized by human body, losing their potential beneficial effects.[50,63]

To solve these problems and ensure that the polyphenols retain their beneficial properties even after ingestion, it is necessary to develop encapsulation systems that protect these compounds once extracted from plants by interaction with atmospheric agents (light, oxygen, and temperature), during the production, transformation, storage, and cooking of foods in which they are incorporated, and finally during the digestion process, in order to reach the target sites in a chemically stable form, which is able to exert beneficial effects.[26,27]

Carotenoids contribute to the yellow and red colors of many foods. Carotenoids containing oxygen are known as xanthophylls (e.g., lutein and zeaxanthin), while those without oxygen are known as carotenes (e.g., lycopene and β-carotene). The carotenoids have been reported to exhibit several potential health benefits: for example, lutein and zeaxanthin contribute to decrease age-related macular degeneration and cataracts[41] and lycopene contributes to decrease the risk of prostate cancer.[64] In their endogenous form in foods, carotenoids are generally stable. However, when incorporated as food additives, carotenoids are relatively unstable because they are susceptible to light, oxygen, and autooxidation.[65] Consequently, dispersion of carotenoids into food matrix can result in their rapid degradation.[38] An additional challenge to using carotenoids as ingredients in functional foods is their high melting point, making them crystalline at food storage.

Bioactive lipids, such as omega-3, -6, and -9 fatty acids, are unsaturated fatty acids. Their important physiological role has been attributed to their ability to decrease the risks of cardiovascular disease,[44] of diseases induced by immune response disorders (e.g., type 2 diabetes, inflammatory bowel diseases, and rheumatoid arthritis),[45,66] of mental disorders,[67,68] as well as to benefit infant development.[69–71] The growing list of disorders positively affected by bioactive lipids strongly suggests that large portions of the population would benefit from increased consumption of omega-3, -6 and -9 fatty acids, making them an excellent candidate for incorporation into functional foods.

However, numerous challenges exist in the production, transportation, and storage of fatty acid fortified functional foods, since these lipids are extremely susceptible to oxidative deterioration. Encapsulation of bioactive lipids has been found to be an excellent method for their stabilization.[46]

37.2.2 Micronutrients

Micronutrients are different from macronutrients (like carbohydrates, protein, and fat) because they are necessary only in very small quantities. Nevertheless, micronutrients are essential for good health and their deficiency can cause serious health problems. Micronutrients include dietary minerals, which are important for the smooth operation of all biological functions and cellular activity. Therefore, their consumption is associated with the reduction of the risk of certain types of diseases. The most important micronutrients are the following:

- *Calcium*, which reduces the risk of osteoporosis and of other disorders related to its deficiency in the diet, has a protective role against hypertension and certain types of cancer[72,73]
- *Potassium*, which reduces the risk of high blood pressure and stroke,[74] in combination with a low sodium diet
- *Magnesium*, which supports the maintenance of normal muscle and nerve functions, immune, and bone health[75]
- *Iron*, which helps human body to produce red blood cells and lymphocytes[76]
- *Chromium*, which reduces hyperglycemia and maintains the correct level of cholesterol and glucose in the blood[77,78]

The minerals are used to fortify various types of products, especially bread and baked goods, breakfast cereals, dairy products, and vegetables. Moreover, they are often mixed with some vitamins and homogeneously distributed in the products. However, in some formulations, it is necessary to adopt measures to protect the micronutrients by some factors that may cause their loss or decrease of bioavailability, such as the use of microencapsulation technique.[79–81]

37.2.3 Dietary Fibers

Dietary fibers represent a group of carbohydrates, isolable from different plant sources, that are resistant to hydrolysis by enzymes of the gastrointestinal tract. They are divided into two groups: water soluble fibers, among which the main are the β-glucans and arabinoxylans, and water insoluble fibers, which include lignins, celluloses, and hemicelluloses.

Dietary fibers exert both functional and metabolic effects, which make them important components of the diet. In addition to the increase of satiety and improvement of bowel function and disorders associated with it (constipation and diverticulosis), the intake of fibers with food has been related to the reduction in the risk of major chronic diseases, such as cardiovascular diseases,[7] thanks to the reduction of cholesterol,[82] of glucose in the blood and of insulin.[83]

Among the water soluble fibers, the most important group is represented by β-glucans, which have important positive effects on heart diseases, reducing the level of cholesterol and glucose in the blood, and which are able to promote the growth of lactobacilli and bifidobacteria present in the gastrointestinal tract. The β-glucans are present in all the cereals, but barley and oats are the ones that contain the highest amounts.

The fibers are used as ingredients in many processed foods, such as bread, pasta, cereal for breakfast, yogurt, and meat products. They may be mixed either in the form of concentrated isolated products, such as concentrates of oat fiber, or may be introduced in the formulations through the addition of flour or other ingredients that contain them.

The main difficulty related to their incorporation in food matrices derives from the consideration that dietary fibers have a high capacity to bind water, thus determining a reduction of the final volume of the product (in the case of bakery products) and, in general, an alteration of the organoleptic properties of the food. For this reason, encapsulation seems a promising technique to improve the inclusion of dietary fibers in functional foods, without modifying the sensory characteristics, such as color, appearance, flavor, and taste of the food itself.

37.2.4 Prebiotics and Probiotics

Prebiotics and probiotics are an area of growing scientific interest for the development and production of functional foods.[84] The terms prebiotics and probiotics themselves reveal important implications for human health that may result from their use and that are related to the intestinal microbiota. In fact, the target organ of the action of these "ingredients" is the intestines, but indirectly the whole body is the real beneficiary of their effects. The function of prebiotics and probiotics is to promote the proliferation and the balance of the bacterial composition that constitutes the intestinal ecosystem. The intestinal microbiota is made of hundreds of different bacterial species, whose multiple metabolic activities affect the state of health of the host.

Under psycho-physical, dietary, and environmental stress conditions, or after medication intake, there is an imbalance of microflora that makes the body susceptible to attack by pathogens. A proper diet is one of the main factors that influence the qualitative and quantitative composition of intestinal microflora. The most common approach involves the consumption of traditional foods, such as yogurt and fermented milk, which essentially contain probiotics, defined as live microorganisms, which can positively affect the host by improving its intestinal microbial balance.

The intestinal microflora, consisting of populations of lactobacilli and bifidobacteria, beneficial microorganisms, and of a high number of bacteria, such as clostridia and enteropathogenic, capable of causing health risks, has the primary role of recovering the energy through the fermentation processes that occur against specific substrates, especially carbohydrates, peptides, proteins, and some lipids, not digested in the upper part of the intestine.

Therefore, health benefits from the consumption of prebiotics and probiotics are observed when

- The composition of the intestinal flora is influenced, with an increase in the amount of probiotic bacteria in the gut
- The development of a limited number of probiotic bacteria is selectively stimulated in the colon, through the fermentation of nondigestible substrates, owing to the intake of prebiotic ingredients

37.2.4.1 Prebiotics

The prebiotics are not live microorganisms, but non-digestible food ingredients, which, when administered in adequate amounts, bring benefit to the consumer due to their ability to selectively promote the growth and/or the activity of one or more bacteria already present in the gastrointestinal tract or ingested together with the prebiotics.[85,86] In order to be defined as prebiotic, a substance must be resistant to attack by hydrochloric acid in the stomach and to the hydrolytic and enzymatic processes that occur in the duodenum, acting as a substrate for fermentation by intestinal bacteria, selectively stimulating their growth and/or activity.

Therefore, the effect of the prebiotic is essentially indirect because it acts selectively as a substrate for one or for a limited number of microorganisms causing a modification of the intestinal microflora; it is not the prebiotic itself but rather the changes that it promotes in the composition of the intestinal microflora, which will bring the expected benefits.[87]

As defined by Wang,[88] an ingredient can be classified as prebiotic if (1) it is resistant to the digestion in the upper gut tract, (2) it can be fermented by intestinal microbiota, (3) it brings beneficial effects to the host health, (4) it selectively stimulates the activity of probiotics, and (5) it is stable to food processing treatments.

TABLE 37.2
Food Sources and Beneficial Effects of the Most Important Prebiotics

Prebiotics	Food Sources	Beneficial Effects	References
Inulin	Chicory, banana, onion, garlic	Promotes colonic health, decreases amount of cholesterol and triglycerides	[90,91]
Lactulose	Milk	Beneficial effects on digestive health	[92]
Isomaltooligosaccharides (IMO)	Wheat, barley, corn, pulses, oats, tapioca, rice, potato and other starch sources	Reduce flatulence (i.e., generating least gas), low glycemic index, and anticaries activities	[93,94]
Xylooligosaccharides (XOS)	Bamboo shoots, milk, and honey	Antiallergy, anti-infection and anti-inflammatory properties, immunomodulatory, and antimicrobial activity	[95]
Soybean oligosaccharides (SOS)	Soybeans	Prevent constipation due to the production of short-chain fatty acids, improve absorption of calcium and other minerals, reduce risk of colon cancer	[96]
Fructooligosaccharides (FOS)	Banana, onion, chicory root, garlic, asparagus, wheat, and leeks	Promote calcium absorption, provide some energy to the body	[97]
Galactooligosaccharides (GOS)	Bovine milk	Excellent source for health-promoting bacteria, support the intestinal immune system, improve mineral absorption	[98,99]

Moreover, the prebiotics can be incorporated into food matrices if they are chemically stable during food processing, under conditions of low pH and high temperatures and to Maillard reactions.

Most of prebiotics are nondigestible oligosaccharides; they are obtained by extraction from plants (e.g., inulin from chicory), possibly followed by enzymatic hydrolysis (e.g., oligofructose from inulin), or by synthesis from mono or disaccharides.[89] Among all the prebiotics, inulin, and oligosaccharides are certainly the most studied and have been recognized as dietary fibers in the world. The most important prebiotics are presented in Table 37.2.

Prebiotics are increasingly used to achieve a double action: an improvement of the organoleptic properties of the food and a better balance of the nutritional benefits.[100]

The use of inulin and nondigestible oligosaccharides improves the taste and texture of food in which are added. In addition, these fibers are readily fermentable by specific bacteria such as lactobacilli and bifidus bacteria, resulting in an increase in their population with the simultaneous production of short chain fatty acids.[101] These fatty acids, especially butyrate, acetate, and propionate provide metabolic energy for the host.

37.2.4.2 Probiotics

Probiotics are defined by the World Health Organization as "live microorganisms that, when administered in adequate amounts, confer a health benefit on the host."[102] This definition refers to non-pathogenic microorganisms present in foods or added to them, and "excludes references to biotherapeutic agents and beneficial microorganisms not used in the food industry."

The main probiotic preparations available on the market are known as Lactic Acid Bacteria (LAB), for the most part represented by lactobacilli and bifidobacteria, which are important constituents, normally present, of the gastrointestinal microflora, and that produce lactic acid as a major metabolite. Moreover, also certain yeasts and bacilli are counted among probiotics. Table 37.3 reports the microorganisms recognized as probiotics and used in the production of functional foods.

TABLE 37.3
Microorganisms Used as Probiotics for the Production of Functional Foods

Lactobacilli	Bifidobacteria	Other Microorganisms
L. acidophilus	B. animalis	E. faecium
L. casei	B. breve	B. subtilis
L. johnsonii	B. infantis	E. coli
L. reuteri	B. longum	S. boulardii
L. rhamnosus	B. adolescentis	Cl. butyricum
L. salivarius	B. lactis	
L. plantarum	B. bifidum	
L. crispatus		

The bacteria normally used in the production of yogurt, in particular *Streptococcus thermophilus* and *Lactobacillus bulgaricus*, are not expected to survive and overcome the intestinal tract and therefore are not considered probiotics.[103]

There are many scientific evidences, supported by clinical studies, on the efficacy of probiotics in the prevention and treatment of gastrointestinal disorders, respiratory, and urogenital diseases.[104] Many microbial strains with probiotic properties are able not only to restore the intestinal microbial balance, but also to impart other beneficial effects on health, associated with the production of acids, bacteriocins and with the competition with pathogenic microorganisms. Among these, the main effects are the reduction of the level of cholesterol in the blood, the reduction of fecal enzymes, with potentially mutagenic activity that can induce the onset of tumors, the reduction of lactose intolerance, the increase of the response of the immune system, the increase of calcium absorption, and synthesis of vitamins.[105]

As evidenced in Figure 37.1, probiotics have been shown to work by the following mechanisms[106]:

- *Competition for nutrients*: Within the gut, beneficial and pathogenic microorganisms use the same types of nutrients, creating a general competition between bacteria for these nutrients. When a probiotic is administered, there is an overall reduction in nutrients available for pathogenic bacteria and consequently this minimizes the levels of pathogenic microorganisms.
- *Competition for adhesion sites*: Beneficial bacteria can attach to the gut wall and form colonies at various sites throughout the gut. This prevents pathogenic bacteria from gaining a foothold, resulting in their expulsion from the body.
- *Lactic acid production*: Probiotics produce lactic acid, which acts to reduce the gut pH, inhibiting the growth of pathogenic bacteria, which prefer a more alkaline environment.
- *Effect on immunity*: Probiotics have been shown to increase the levels of cell-signaling chemicals and the effectiveness of infection-fighting cells (white blood cells).

The probiotics used for the production of functional foods are incorporated as a supplement or added in advance to facilitate fermentation processes at the base of the preparation of the food itself. It appears evident that the physicochemical properties of a food used as carrier to deliver the probiotics within the gastrointestinal tract are an important factor that determines the survival and thus the potential beneficial effect of the probiotic.[107] Fat content, concentration and type of proteins and sugars, oxygen level, pH, and storage temperature of the product are some factors that affect the growth and survival of the probiotic. Formulating the food in order to achieve an appropriate pH value and a high buffer capacity, it is possible to increase the pH of the gastric tract and in this way improve the stability of the probiotic.[108]

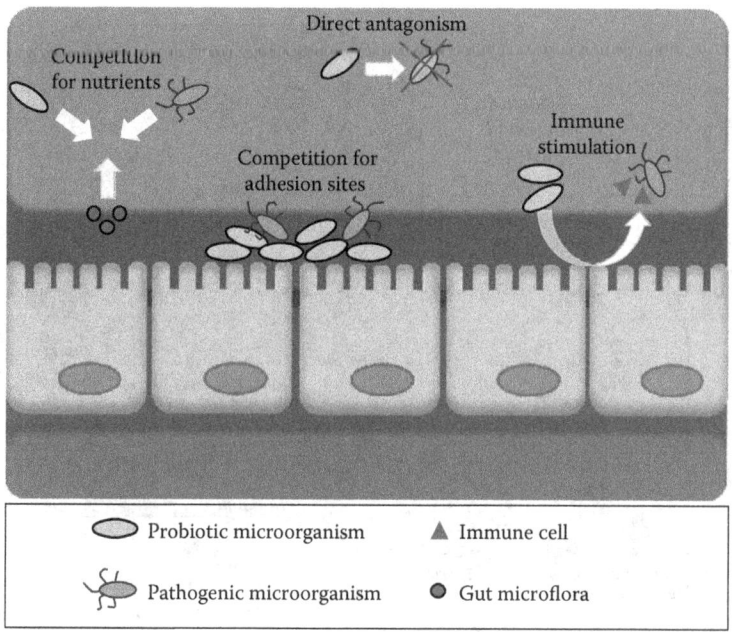

FIGURE 37.1 Mechanisms of action of probiotic microorganisms into the gut.

Therefore, it is necessary not only to ensure a high level of viable microorganisms in the food (at least 10^7 cfu/g) in order to have beneficial effects, but also ensure the protection of the probiotics during production, storage, and consumption of the food product. Moreover, it is necessary to protect them from the action of gastrointestinal acids and enzymes, from the adhesion to the intestinal epithelium and from the attacks of the antibiotics.[109]

Although probiotic cultures do not tend to markedly change the sensory properties of the products in which they are added, in many cases consumers have found the products, fermented with *L. delbrueckii* subsp. *bulgaricus*, too acidic and with a strong flavor of acetaldehyde (typical flavor of yogurt). For the production of probiotic products, it is necessary the development of probiotic cultures not able to alter the organoleptic properties, but to enhance the flavors of the foods in which they are added.[110]

Probiotics are mainly added in dairy products, but in the recent years, thanks to the use of different technologies, the food industry is trying to add probiotics in different type of foods, such as beverages,[111] baking products,[112,113] ice cream,[114–116] and chocolate.[117]

To ensure the viability and stability of probiotics and do not alter the organoleptic properties of food after their inclusion in food matrices, different strategies have been developed, ranging from the use of particular technologies, such as microencapsulation, which allows to protect bacteria, enclosing them in special coatings, from unfavorable conditions that may occur during preparation, production and storage of the product, to the use of special substrates that can be used selectively by probiotic microorganisms for their growth. Such substrates are represented by prebiotic ingredients and a growing interest is developing in a synbiotic use of probiotics and prebiotics for the functionalization of foods.[118]

37.3 ENCAPSULATION SYSTEMS

Encapsulation systems for bioactive compounds may take different configurations and architectures, depending on materials, fabrication process, dispersion medium, as well as interaction with the payload. However, despite an ultimate classification of encapsulation systems being difficult, Figure 37.2 tries to provide approximate categories,[2,119] identifying two broad classes of systems, one pertaining to matrix type and the other to core and shell type.

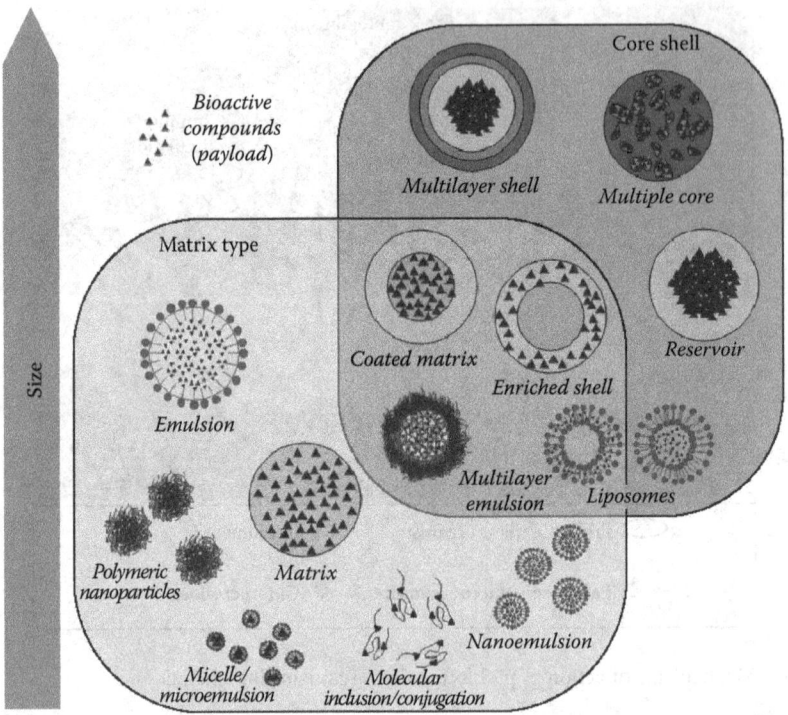

FIGURE 37.2 Classification of the main encapsulation systems for bioactive compounds.

In the *matrix type*, the payload is well dispersed within the encapsulation material, more or less homogeneously distributed, and, in general, it is present also on the surface of the matrix. In contrast, *core-shell type* systems are characterized by an external shell completely covering a core containing the bioactive compounds, which therefore are not in direct contact with the external environment.

As shown in Figure 37.2, core-shell systems encompass different architectures, the simplest of which is single core/single shell (also known as reservoir type), and the most complex ones comprise several cores and/or multiple-layers shell. Ideally, the shell material should be able to limit the diffusion of the payload when *in product*, while, once *in body*, should facilitate its release when needed. Therefore, waxes, fats, or proteins are primarily used as shell materials.[120,121] Because of their structural organization, with a higher degree of complexity than matrix type, core-shell systems are generally characterized by larger mean particle sizes and a more complex fabrication process, whose key step is the formation of the shell layer, which implies higher processing costs.[122,123]

In contrast, matrix-type systems are less expensive to fabricate but are not suitable for taste masking and are characterized by a lower loading capability.

Molecular inclusion or conjugation complexes, micelles, microemulsions, polymeric particles, and emulsions and nanoemulsions can all be classified as matrix type encapsulation systems.

For example, at molecular levels, micellization and molecular inclusion complexation represent some of the simplest approaches to fabricate matrix-type systems.

Micelles result from the spontaneous aggregation of amphiphilic molecules (emulsifier or surfactant) in a solvent, at a concentration level above the critical micelle concentration (CMC). If the surfactant concentration remains above the CMC, micelles are thermodynamically stabilized against disassembly.[124]

The use of micelles in foods is limited by the scarce loading capacity of bioactive compounds and the usage level of surfactants, which may significantly impact on the organoleptic properties of the end product, as well as by their potential metastability with respect to the solubility of bioactive compounds (dynamic formation and rupture of micelle structures), which may cause precipitation upon dilution or crystallization over long storage times.[175]

Inclusion complexes are instead based on the steric and hydrophobic interaction of the hydrophobic polysaccharide chain with hydrophobic molecules. For example, amylase was reported to be able to form a complex with conjugated linoleic acid owing to its helical molecular structure, whose moieties can form steric and hydrophobic interactions with compounds such as free fatty acids.[126] However, inclusion complexation is more efficient if the polysaccharide structure is a ring of well-defined size, such as for cyclodextrins, which are naturally occurring cyclic oligosaccharides derived from starch, with six, seven, or eight glucose residues assembled in a ring, characterized by external hydrophilic moieties and internal hydrophobic moieties.[127] Thanks to their architecture, cyclodextrins can accommodate in the internal cavity non-polar molecules that can fit their specific cavity. The inclusion of bioactive molecules in cyclodextrins may significantly increase their water solubility,[127] as shown for flavonols, such as quercetin and myricetin, which were inclusion-complexed with β-cyclodextrins,[128] as well as for lycopene in α-, β-, γ-cyclodexrins.[129] The inclusion complexation in cyclodextrins find application in deodoration processes or in taste masking of bitter molecule. The encapsulated compounds, which are efficiently protected by the presence of oxygen and radiation, are released under high moisture and low temperature conditions.[130] β-cyclodextrins have been approved for food use.[131] However, it must be highlighted that the use of cyclodextrins is limited by the high cost of the material as well as by the high ratio between encapsulant and encapsulated material.

At a scale larger than molecular, but still in the nanoscale range (<100 nm), *microemulsions* are self-assembling encapsulation systems, consisting of inner droplets stabilized by amphiphilic molecules, generating hydrophilic and hydrophobic regions of large interfacial area, which enable to host different guest molecules such as food additives, nutraceuticals, aromas, cosmetic compounds, active ingredients, and drugs.[132] They differ from emulsions and nanoemulsion because they are thermodynamically stable systems, with characteristic dimensions ranging from 5 to 50 nm and typical emulsifier to oil ratio larger than 1.[133]

Depending on the properties of the amphiphilic molecules used, of the temperature and of the oil and water fractions, O/W or W/O microemulsions can be formed, but O/W are those with highest potential application to the delivery of bioactive compounds in food products.[3,132] In comparison to emulsions, microemulsions require the use of relatively large amounts of surfactant,[134] with the consequence of their loading capacity being significantly lower.[125]

With a typical size ranging from nanometric (<100 nm) to submicrometric (<1 μm), *biopolymeric particles and nanoparticles*, made of proteins or polysaccharides, thanks to their excellent compatibility with foods, are able to efficiently encapsulate, protect and deliver bioactive compounds, forming different structures, such as random coils, sheets, or rods around the bioactive molecules.[135,136] The most suitable biopolymers for the incorporation into foods include (1) proteins, such as whey proteins, casein, gelatin, soy protein, zein, and (2) polysaccharides, such as starch, cellulose, and other hydrocolloids,[137] with the particle formulation depending on the desired particle functionality (size, morphology, charge, permeability, environmental stability), on end product compatibility and in general *in product* behavior, as well as on release properties and *in body* behavior.

Remarkably, steric and electrical characteristics of the biopolymeric particles, which are controlled by biopolymer properties and assembly conditions, enable the control of physicochemical stability, but also the interaction with other species or food ingredients with opposite charge, as well as the interaction with biological surfaces (i.e., mucoadhesiveness).[137]

In general, fabrication of biopolymer particles is based on the induced aggregation of homogeneous or heterogeneous molecules, controlling their intrinsic properties as well as the environmental conditions (i.e., temperature, pH, and ionic strength). In particular, self-assembly of proteins is induced by physical interactions due to van der Waals electrostatic, hydrophobic forces, and hydrogen bonding,[138] while self-assembly of polysaccharides by van der Waal and electrostatic forces as well as hydrogen bonding.[139] Subsequently to spontaneous association, a consolidation step of the particle structures, such as covalent bonding, is desired in order to prevent formation of larger structures or of a gelling network.

Emulsions and nanoemulsions are heterogeneous systems consisting of two immiscible liquids, with one liquid phase being dispersed as droplets into another continuous liquid phase and stabilized through an appropriate emulsifier. In particular, nanoemulsions are characterized by a nanometric size (<100 nm), while emulsions are in the submicrometric and micrometric range.

Oil in water (O/W) emulsions, which are of prevalent interest for encapsulating bioactive compounds and delivering them into food systems, are composed of oil droplets dispersed in an aqueous medium and are stabilized by a food-grade surfactant or biopolymeric layer, whose properties control the interfacial behavior (charge, thickness and droplet size, and rheology), as well as the response to environmental stresses (pH, ionic strength, temperature, and enzyme activity) of the encapsulation system.[136]

Differently from microemulsions, emulsions are kinetically stable, which require energy to be formed, and are subjected to several instability phenomena, including coalescence and gravitational separation. The most common approach to reduce their instability is to reduce their mean droplet size to such values that Brownian motion effects dominate over gravitational forces.[133] Therefore, nanoemulsions are highly stable to gravitational separation and show a lower tendency to droplet aggregation than conventional emulsions, because the strength of the net attractive forces acting between droplets usually decreases with decreasing droplet diameters.[133]

The encapsulation of bioactive compounds in O/W emulsions enables their efficient incorporation in foods. Depending on the properties of the bioactive compounds, and in particular on their lipo- or hydrophilicity, the payload localization within an O/W emulsion may change from a prevalent entrapment within the inner oil phase (bioactive-enriched core) or a prevalent concentration into the outer stabilizer film (bioactive-enriched shell). Remarkably, the localization of the bioactive compounds influences their stability, release, and bioavailability.

Similar to emulsions, *solid lipid particles* consist of emulsifier-coated lipid droplets dispersed within an aqueous phase, with the main difference that the lipid phase is at the solid or semi-solid state.

In comparison to emulsions, solid lipid particles exhibit an increased chemical protection against degradation, a higher encapsulation efficiency (>90%) and a better controlled release, due to the immobilization of the encapsulated bioactive compound in the solid lipid matrix.[140]

In addition, when the lipid phase consists of fats of different types and with different properties as well as crystallization kinetics, the encapsulation system is characterized by voids and defects of the solid fat crystalline structure, with high loading capability. In contrast, when the lipid phase consists of a more ordered crystalline structure, made of a single type of fats, less space will be available for the bioactive molecules, reducing the loading capability.[141] Therefore, typically two or more lipids with different melting points are used, such as mixtures of purified triglycerides, waxes, or fatty acids.[142] The main limitation to the use of solid lipid particles is in their fabrication, which is based on high-energy emulsification processes at temperatures above the melting point of the lipids, and requires a fine control of the lipid crystallization, which is significantly dependent on the temperature history of the system, on the presence of impurities in the lipid phase, as well as on droplet size.[143]

Nanostructure lipid carriers indicate lipid particles with a disperse phase made of a mixture of solid and liquid lipids. Owing to the decreased melting point of the lipid phase, such systems can be produced at lower temperatures, reducing the extent of degradation of the thermolabile compounds.[141]

Liposomes are spherical bilayers made of amphiphilic molecules, such as phospholipids, which are characterized by an inner hydrophilic water domain physically separated from the bulk of water. Owing to their structure, liposomes may enclose bioactive molecules of hydrophilic nature in the inner hydrophilic domain, as well as of lipophilic nature in the bilayer. Liposome fabrication is based on the spontaneous association of amphiphilic molecules in a lamellar phase, induced by proper selection of solvent, concentration, and temperature conditions, followed by its dispersion by intense mechanical disruption or the use of solvent evaporation techniques. Liposome can be structured in (1) single bilayers, forming unilamellar vesicles, (2) several concentric bilayers forming multilamellar vesicles as well as (3) non-concentric bilayers forming multi vesicular vesicles.[143] The use of liposomes in food industry is limited by the high costs of pure lecithins, which are the most suitable food-grade amphiphilic molecules, the low encapsulation efficiency, and the complicated and costly fabrication process.

Matrix-type systems, where an additional coating layer is applied, are defined as coated-matrix type.[2] They are usually fabricated in two-step processes, consisting in the formation of the matrix-type system (i.e., spray drying or emulsification), followed by the formation of a coating layer (i.e., fluidized-bed coating or layer-by-layer deposition).

In the case of emulsions, the deposition of a *multiple layer* of emulsifiers and/or polyelectrolytes may positively affect the absorption properties in the gastrointestinal tract,[144] and significantly improve emulsion stability by increasing the packing density at the interface.[145] For example, adsorption of layers of biopolymers, such as chitosan, on lipid-stabilized emulsion droplets was reported not only to improve the physical stability against thermal processing, freeze–thaw cycling and drying, as well as the chemical stability against lipid oxidation,[146] but also to enable controlled or triggered release.[147] Moreover, alternating layers of proteins and hydrocolloids are used, based on electrostatic interactions, in order to realize multiple layer systems, with improved stability to extremes of temperature, pH, and ionic strength.[148]

It must be remarked that, in practice, many of these encapsulation systems are not spherical in shape, and therefore, the terms microcapsules and nanocapsules are used in a general way to denote either irregular or spherical shapes.[149]

The type of encapsulation system, together with encapsulating material properties, significantly affect the release properties of the payload. In Figure 37.3, a schematic representation of the release profiles of the main classes of encapsulation types is reported.

In general, faster release kinetics are expected from those systems where the bioactive compounds are in close contact with the release medium, which occurs in particular for enriched shell and matrix-type systems. In contrast, when a shell layer is present, the release may be significantly slowed down, depending on the mechanism of permeabilization of the external shell. Therefore, it is possible to trigger the payload release by opportunely triggering the degradation of the external shell under specific environmental conditions.

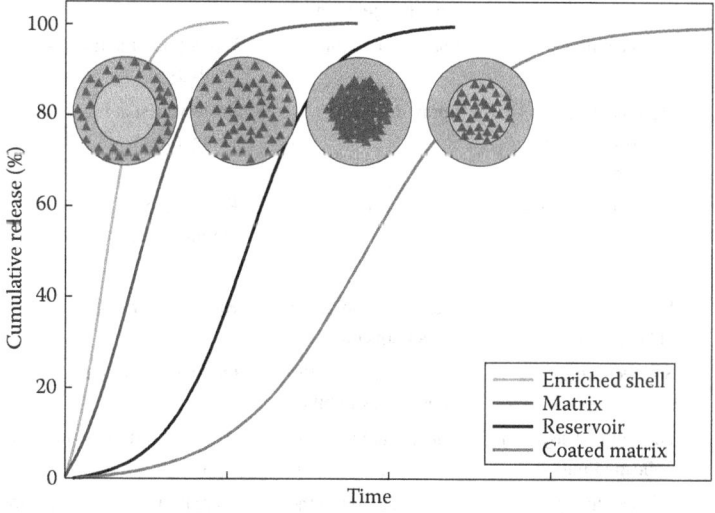

FIGURE 37.3 (See color insert.) Schematic representation of cumulative release profiles of the main encapsulation architectures (reservoir type, matrix type, coated matrix type).

In addition to the architecture of the encapsulation system, the other factors affecting the release kinetics of the encapsulated bioactive compounds can be classified in chemical and morphological factors. The chemical factors comprise the size of the bioactive molecules and the properties of the encapsulant material, such as its thermal stability, amorphous or crystalline structure, glass transition properties, melting point, phase transition, and ionic charge. The morphological factors instead include size distribution, shape, coating uniformity, moisture content, physicochemical stability, and hygroscopicity.[2]

Tables 37.4 and 37.5 provide a survey of recent application of micrometric and submicrometric encapsulation systems for bioactive compounds.

TABLE 37.4
Encapsulation of Bioactive Compounds at the Microscale

Bioactives	Formulation	Fabrication	Application	References
Core-shell systems				
Jasmine essential oil	Gelatin and gum arabic	Complex coacervation	Preservation at high temperature	[150]
Vanilla oil	Chitosan	Complex coacervation	Controlled release and thermo stability in spice industry	[151]
Sucralose	Gelatin and gum arabic	Double emulsion and coacervation	Preservation	[152]
Ascorbic acid	Gelatin and gum arabic	Double emulsion and coacervation	Preservation	[153]
Riboflavin	Whey protein/alginate gel	Cold gelation	Controlled release in beverages	[154]
Lycopene	Gelatin and gum arabic	Complex coacervation	Preservation in cake making	[155]
Matrix systems				
Natural vitamins	Gum arabic	Spray drying	Preservation	[156]
Black currant polyphenols	Maltodextrins and inulin	Spray drying	Protection of antioxidant activity	[157]
Bilberry extract	Whey proteins	Emulsion templating and hot gelation	Preservation	[158]
Bayberry extracts	Ethyl cellulose	Phase separation	Protection of antioxidant activity	[159]
Pomegranate polyphenols	Maltodextrin or soybean protein isolates	Spray drying	Preservation at high temperature	[160]
Cactus pear extracts	Maltodextrins and inulin	Spray drying	Preservation at high temperature	[161]
Liposomes				
Peptides, bacteriocines	Different phospholipids	Encapsulation in the inner aqueous core	Biopreservatives	[162]
Curcumin	Soy lecithin	Encapsulation in the phospholipid bilayer	Oral carrier (in rats)	[163]
Enzymes	Unsaturated soybean phospholipid	Encapsulation in the inner aqueous core	Alleviate bitterness in protein hydrolysates	[164]
Vitamin C	Soy lecithin	Encapsulation in the inner aqueous core	Increased vitamin protection and controlled release	[165]
Gallic acid	Nopal mucilage	Spray drying	Preservation	[166]

TABLE 37.5
Encapsulation of Bioactive Compounds at the Submicrometric Scale

Bioactives	Formulation	Fabrication	Application	References
Nanoemulsions				
β-carotene	Medium chain triglyceride oil and Tween 20–80	Emulsification via high pressure homogenization	Food functionalization	[167]
Curcumin	Medium chain triglycerols and Tween 20	Emulsification via high pressure homogenization	Food functionalization	[168]
D-limonene	Palm oil and soy lecithin	Emulsification via hot high pressure homogenization	Antimicrobial activity in foods	[42]
Molecular inclusion/conjugation				
Iron	Casein and whey protein isolates	Electrostatic bonds	Food functionalization	[169]
Resveratrol	β-lactoglobulin	Complexation	Increase of dispersibility in water and photostability	[170]
Conjugated Linoleic Acid	Amylase	Inclusion complex via hydrophobic interactions	Protection against oxidation and dispersion in aqueous phase	[126]
Quercetin	Chitosan	Complexation followed by ionic gelation	Increased of dispersibility in water and bioavailability	[171]
Quercetin and Myricetin	HP-β-cyclodextrins	Inclusion complexation	Increase of solubility and inhibition of oxidation	[128]
Kaempferol, quercetin and myricetin	HP-β-cyclodextrins	Inclusion complexation	Increase of solubility and inhibition of oxidation	[172]
Curcumin	HP-β-, M-β- and HP-γ-cyclodextrins	Inclusion complexation	Preservation of antioxidant activity	[173]
Curcumin	Hydrophobically modified starch	Hydrophobic interaction	Increased of dispersibility in water and anticarcinogenic activity	[174]
Micelles				
Curcumin	Casein micelles	Hydrophobic interactions	Food functionalization	[175]
Vitamins D2, D3	Casein micelles	Hydrophobic interactions	Food functionalization	[176,177]
Naringenin	Polyvinylpyrrolidone	Solvent evaporation method	Improved dissolution rate High physicochemical stability	[178]
Quercetin	Pluronic F68 + Lecithin	Solvent evaporation method High pressure homogenization	Improved dissolution rate	[179]
Citral	Tween 80	Micellization	Preservation of flavor in carbonated beverages	[180]
Phytosterols	Water/propylene glycol/R(+)-limonene/ ethanol/Tween 60	Microemulsion	Solubilization and dispersion in food systems	[181]
Lycopene	Water/propylene glycol/R(+)-limonene/ ethanol/Tween 60	Microemulsion	Solubilization and dispersion in food systems	[182]

(Continued)

TABLE 37.5 (Continued)
Encapsulation of Bioactive Compounds at the Submicrometric Scale

Bioactives	Formulation	Fabrication	Application	References
Curcumin	Tween 20/Glycerol monooleate/Medium chain triglycerides/water	Microemulsion	Solubilization and dispersion in food systems	[183]
Multiple layer nanoemulsions				
β-Carotene	Lactoferrin and beta-lactoglobulin	Multilayer emulsions	Preservation	[184]
Strawberry flavors	Pea protein isolate and pectin	Spray dried multilayer emulsions	Flavor preservation	[185]
Citral	Chitosan and ε-polylysine	Layer-by-layer emulsions	Flavor preservation	[186]
β-Carotene	Chitosan and soybean soluble polysaccharides	Two-stage deposition on emulsion by homogenization	Preservation	[187]
ω-3 Fatty acids	Lecithin and chitosan	Spray-dried multilayer emulsions	Increasing oxidative stability	[188]

In particular, Table 37.4 provides examples of core shell and of matrix-type systems, with details on the type of bioactive compounds, the formulation of the shell or the matrix, the fabrication technique, and the desired application. In Table 37.4 also liposomes are included, whose size, especially for food application is in the micrometric range, but they can be fabricated also at smaller size.

Table 37.5 gives example of encapsulation systems of sub-micrometric scale, providing the same details of Table 37.4.

37.4 ENCAPSULATION TECHNIQUES

The fabrication of encapsulation systems may be carried through mechanical processes, in general based on a top-down approach to the disruption of larger systems into homogeneously sized particles or droplets with desired properties, or by physicochemical processes, based on a bottom-up approach to the assembling of molecular building blocks into structured systems (i.e., micelles, microemulsions, and some biopolymeric nanoparticles), as well as by a mixed approach, where molecular assembly and comminution processes are combined together (i.e., liposomes, multilayer emulsions, and some biopolymeric nanoparticles).

Different techniques are used at industrial level, including spray drying and spray chilling, fluid bed coating, melt injection and melt infusion, extrusion, coacervation, crystallization, as well as emulsions and particles and liposomes production.[189,190]

Figure 37.4 shows a classification of the main encapsulation techniques as a function of the type of capsules that can be produced, and of their typical size. Interestingly, smaller size systems are usually matrix type, such as nanoemulsions, micelles, microemulsions, and molecular complexes. Only in the case of biopolymeric particles, it is possible to attain a certain degree of structural complexity at submicrometric size. In contrast, for larger size systems, different architectures are possible, through mechanical technologies, such as atomization and extrusion, together with mechanical or physicochemical coating approaches.

37.4.1 Mechanical Processes

Mechanical processes to the production of encapsulation systems are in general based on the disruption of a dispersion of the bioactive compounds in the encapsulant material into small particles or droplets, and their stabilization by drying, phase transition, or deposition of molecular layers on their surface.

Encapsulation of Bioactive Compounds

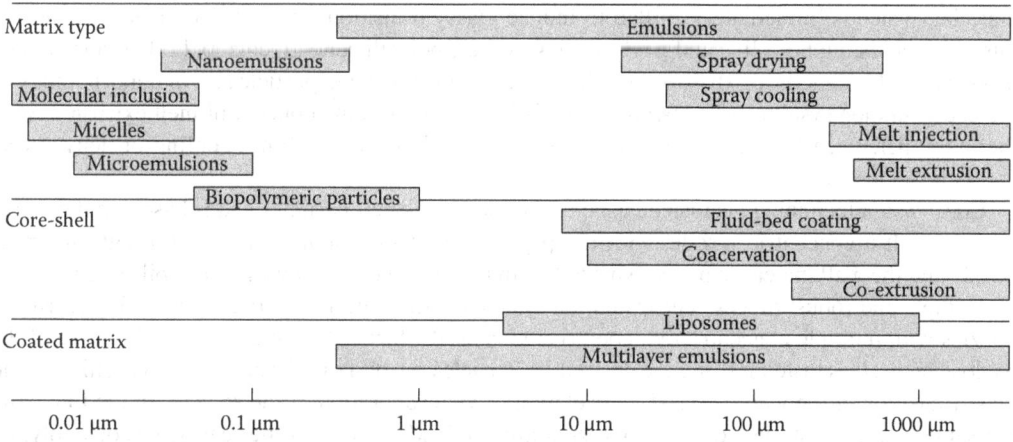

FIGURE 37.4 Main encapsulation techniques classified as a function of the prevalent architecture and mean particle size of the encapsulation system.

Spray drying is a relatively simple and inexpensive drying process that can be advantageously used to encapsulate bioactive and aroma compounds in foods.[191,192] In the spray-drying process, the payload is preliminarily finely dispersed or homogenized in a highly concentrated (up to 30%wt) biopolymeric aqueous solution, containing for example, starches, succinylated starches, cellulosics, gelatin, gums and proteins, or in organic solvent solutions, containing for example PLGA, ethyl cellulose, or acrylates, with the main requirement being ability of forming a glassy material upon drying. Eventually, another emulsifier might be added to the solution to improve payload dispersion. The dispersion is then sprayed in a drying chamber, forming fine droplets, which are rapidly dried upon contact with a cocurrent or countercurrent flow of hot gas, ultimately forming small micrometric droplets, which are collected in a cyclone or in a filter cloth. The payload of the produced microcapsules is dispersed in the matrix of the encapsulant material. The encapsulation systems obtained by spray drying are in general capsules of matrix type, whose properties are mainly affected by the properties of the emulsion and by the process conditions. However, especially for vitamins, aroma compounds, and probiotics, the high temperatures required for the complete solvent evaporation may induce significant thermal damages or volatilization.[193,194]

Spray chilling represents an alternative to spray drying, used to prevent the volatilization or degradation of thermolabile food additives. Spray chilling is also based on the preliminary dispersion of the payload in a solution, that can be made of low molecular weight polymers, resins, hydrogenated vegetable oils or waxes, and its atomization, with the capsule consolidation not being based on dehydration, but on glass transition or crystallization of the encapsulant material upon rapid chilling in a cooled gas flow.[195,196]

Fluid-bed coating is based on the deposition of a shell layer on preformed particles (e.g., from spray drying), therefore consisting in the final step of production of a core-shell or a coated matrix type architecture. The particles containing the payload are fluidized, and the coating material is sprayed over them at high pressure,[197] forming a shell layer that is then dried by solvent evaporation or crystallization. The coating material usually consists of starches, dextrins, protein derivatives, molasses, lipids, and waxes. Prior to spraying, the coating material is either melted or dispersed in a suitable solvent that can be easily evaporated, forming a viscous system with enhanced tendency to deposition and adhesion on the payload particles. The air or gas flow through the fluidized bed serves to chill the molten material, or to evaporate the solvent, causing its consolidation in a shell layer.[198] Fluid-bed coating is an inexpensive process, which is often associated with spray drying.[199]

Melt extrusion is a process used to encapsulate the bioactive compounds in a glassy, impermeable, dense matrix through the extrusion of a suitable dispersion of the payload in a molten material, which is immediately chilled to induce glassy transition. Typical materials used include fats, fatty acids, mono-, di- , and triglycerides, waxes, polyethylene glycols and other commercial products, such as Shellac. Extrusion conditions determine the mean particle size and morphology of the encapsulation system.[196,200] *Melt injection* is based on the same concept of melt extrusion, with the main difference that extrusion is carried out through a filter within a cooling or dehydrating medium.[196,200]

Coextrusion technology is instead used to produce core-shell particles and is based on the extrusion through a concentric nozzle, with the payload dispersion being extruded through the inner nozzle and the wall materials being extruded from the outer nozzle. Owing to its ability to fabricate in a simple and robust process encapsulation systems with multiple coating layers (using extrusion nozzles with multiple concentricity), coextrusion is used when a slow and controlled release of the payload, as well as taste masking are desired.[201] Particle consolidation occurs either via chilling and glass transition of the wall material, via gelation or via evaporation of the solvent. In the first case, the same coating materials of melt extrusion are used, while in the other cases, viscous polymer solutions are used, based on proteins and polysaccharides, gums, and other commercial polymers. The main disadvantages of melt extrusion and coextrusion processes are the high temperatures and the high shear rates attained in the extruder.[202]

The top-down approaches to the production of sub-micrometric particulate systems are mainly based on comminution by mechanical size reduction techniques, which require minimal use of chemical additives.[4]

Media milling is a process based on the action of grinding media, which are usually balls or beads of hard materials, which are rotated at a very high speed to generate strong shear forces to disintegrate larger particles into nanoparticles.[203] The milling chamber is loaded not only with the powder material to be comminuted and with the grinding media, but also with a dispersion medium (e.g., water) and a stabilizer, in order to prevent aggregation or coalescence phenomena. The efficiency of the milling process mainly depends on number and type (i.e., hardness) of the grinding media, the ratio of bioactive compound and stabilizer, the milling time, the milling speed, and the process temperature.[204]

Colloid milling is instead based on a high speed rotor/stator system, in which the processed material is grinded, dispersed, or emulsified as a consequence of its exposure to intense shear stresses, friction, and high frequency vibrations. Colloid mills are constituted by smooth or toothed rotors and stators, gear-rim dispersion machines, and intensive mixers.[205]

The main disadvantages of media and colloid milling are the difficulty in the removal of residual grinding media from the final product and the loss of bioactive compounds owing to adhesion to the inner surface of the milling chamber. However, they represent low cost and easily scalable methods of comminution of solid particles, with wide industrial application.[204]

High pressure homogenization is another mechanical disruption process for the production of sub-micrometric emulsions, suspensions, and dispersions, with significant advantages in terms of ease of operation, industrial scalability, reproducibility, and high throughput.[205,206] The process consists in the mechanical disruption of the disperse phase by application of high-intensity fluid-mechanical stresses, as a consequence of the flow of the continuous phase under high pressures (50–400 MPa) through a specifically designed homogenization chamber. The homogenization chamber can be realized in different geometries, ranging from a simple orifice plate to colliding jets and radial diffuser assemblies.[1,207–211]

37.4.2 Physicochemical Processes

Sub-micrometric systems can also be produced through physicochemical processes, in general based on bottom-up approaches to the spontaneous association of molecules or larger building blocks around the bioactive compounds, driven by the balance of attraction and repulsion forces tending to thermodynamic equilibrium. The entity of the involved forces can be controlled by environmental factors, such as temperature, concentration, pH, and ionic strength of the system,[212] with the required mechanical energy being limited to system agitation.[4] Since the encapsulation systems formed through bottom-up approaches exist at the thermodynamic equilibrium, any environmental modification would bring to their disassembly, therefore limiting their use and incorporation in real food systems, unless a stabilization step is added through physical or chemical changes, such as rapid quench cooling, sudden solvent dilution or evaporation, as well as chemical reactions.

In general, the fabrication of encapsulation systems via bottom-up approaches is based on the dissolution of the bioactive compound in a suitable organic solvent, followed by its precipitation through a non-solvent addition in the presence of stabilizers, such as surfactants or hydrocolloids.[4,213] Eventually, a polymer soluble in the internal solvent but insoluble in the external one may be added to the system.[4]

Phase inversion methods are based on spontaneous O/W nanoemulsion formation induced by controlling the interfacial behavior, from predominantly lipophilic to predominantly hydrophilic, of the surfactants at the O/W interface, in response to changes in system compositions or environmental conditions.[214,215]

For non-ionic surfactants, the change in surfactant behavior can be achieved by changing the temperature of the system[215]: at low temperatures, the volume of the hydrated surfactant head group is larger than that of the hydrophobic tail group, thus favoring the formation of O/W emulsions, while at high temperatures, the volume of the dehydrated surfactant head group is lower than the hydrophobic tail group, thus favoring the formation of W/O emulsions.[214] During cooling, the system crosses a point of minimal surface tension, which is referred to as phase inversion temperature,[215] around which the formation of nanometric droplets is promoted.[216] The change of surfactant behavior can also be achieved through environmental changes, such as salt concentration or pH value,[215] as well as surfactant concentration.[214]

Owing to the formulation requirements of phase inversion methods, the nanoemulsions are highly unstable to coalescence when conditions closer to the phase inversion are attained, such as upon incorporation in real foods, or during high temperatures treatments.[214]

The *phase separation method* is based on the spontaneous solvent-in-water emulsification induced by the addition of an aqueous solution containing a surfactant to a polar organic solvent, such as acetone or methylene chloride, containing the bioactive compound and a lipophilic polymer, owing to the affinity between the water and the polar solvent will induce. Further water addition is subsequently used to cause the diffusion of the solvent out of the emulsion droplets, inducing the precipitation of the lipophilic polymer and the encapsulation of the bioactive compound.[4,217]

Coacervation is a relatively simple approach to the fabrication of core-shell materials, exploiting the phase separation method. It is based on the electrostatic interaction of the components of an emulsion, to create microcapsules that are resistant to moisture and heating.[121] Basically, coacervation is a phase separation process in aqueous phase, which begins with the dispersion or emulsification of the payload in an aqueous solution of a primary biopolymer, such as gum arabic or gelatin.[119] Subsequently, the system is mixed with another aqueous solution of a secondary encapsulant biopolymer with opposite charge to the primary one, in order to promote the formation of complexes, which are not soluble in the original aqueous phase.[218] Mean droplet size of the capsules can be controlled through pH and temperature of the original solution, but depends also on the properties of the encapsulated compounds as well as of the encapsulant agent.[121] The capsules formed by coacervation method can be further consolidated by inducing polymer gelation through the addition of minerals and changing the temperature of the system,[219] or inducing polymer precipitation by changes of pH, temperature, or electrolyte composition.[4,136] Because of the structural complexity of coacervates system, their typical sizes are in the micrometric range.

The *layer-by-layer method* enables the fabrication of multiple-layer polyelectrolyte capsules and emulsions.[220] It consists of the spontaneous deposition of biopolymer stabilizing layers at the interface of particles or emulsion droplets, driven by the electrostatic attraction between the templating interface and the forming monolayers. In particular, the spontaneous deposition of a first layer of charged polyelectrolyte occurs around primary systems with opposite charge, producing a secondary particulate system coated with a two-layer interface. In addition, each deposited layer not only fully compensates the charge of the previous templating layer but imparts also an uncompensated counter-charge, generating a charge reversal. Once a monolayer is formed, the surface is saturated with polyelectrolytes, thus preventing further adsorption. As a consequence, the deposition of multiple biopolymer layers is possible,[146] improving the stability to environmental stresses in comparison to conventional single-layer systems.[221]

37.5 ENCAPSULATION OF PROBIOTICS

The main techniques of encapsulation of probiotic cells are based on spray drying, emulsification, and extrusion, which lead to matrix-type systems, as well as co-extrusion, and fluid-bed coating, which instead lead to core-shell systems. In both cases, the typical size of the encapsulation system is comprised between 1 and 5 µm in diameter, being dictated by the size of the microbial cells to be encapsulated.[222]

The design of encapsulation systems for probiotics is primarily aimed to preserve their viability as high as possible, protecting them from adverse environments and ensuring the release in the specific tract of the intestine, to which the probiotics are targeted.[223] In particular, the physicochemical properties of the capsules are the controlling factors of cell viability and targeted release. For example, capsules with shells that are insoluble in water are required to ensure physical stability in most of the food matrices and in the initial regions of the gastrointestinal tract.

Different materials are used to encapsulate probiotic cells, such as vegetable and animal proteins, starch and its derivatives, as well as maltodextrins. However, also other biopolymers are used.[222]

For example, *calcium or sodium alginates* are polysaccharides derived from algae, which are frequently used to encapsulate probiotics, because of their lack of toxicity, biocompatibility, low cost, and ease of use.[224] Upon drying, they form a porous structure, which is not resistant to acidic environments, and therefore is not suitable to ensure adequate stability in the gastric tract. However, enhanced stability can be achieved by blending calcium alginate with other biopolymers, or further coating alginate capsules with a layer of insoluble polymers.

Gellan or xanthan gums are polysaccharides of microbial origin, obtained from *Pseudomonas elodea* and *Xanthomonas campestris*, respectively. Their mixture was reported to be extremely suitable to encapsulate probiotics in systems with high resistance in acidic environments.[222]

K-carrageenan is a natural polymer, widely used in the food industry, which is extremely compatible with microbial cells, ensuring high viability after the encapsulation process. However, the resulting gel structures have limited physical stability, upon the stress conditions experimented in food transformation,[225] requiring its blending with other polymers.

Chitosan is a linear polysaccharide, which is typically used to form coating layers on preformed systems. For example, probiotics encapsulated in an alginate matrix, and further coated with a layer of chitosan, are characterized by an improved protection in the gastrointestinal tract, which enable them to reach with high viability the colon.[226] The main disadvantage of chitosan is an observed inhibitory effect on LAB.

Cellulose acetate phthalate is a polymer typically used as enteric coating,[227] because it is not soluble in water at acidic pH, but only at pH > 6.

Matrix-type systems to encapsulate probiotics use spray drying or spray freeze drying. Spray drying ensures a rapid and low cost process, which suits different industrial applications. However, its wide use is limited by the high temperatures required in the drying chamber, which significantly affect cell viability. The use of protective materials, such as granular starches, soluble fibers or trealose, which are thermal protector agents, is recommended to limit thermal damages on probiotic cells.

Moreover, an additional coating layer deposited in a fluid-bed coating systems integrated with the spray dryer can improve the capsule resistance in the gastrointestinal tract.[228]

Spray freeze drying instead combines the characteristics of the freeze-drying process with those of spray drying. The probiotic dispersion is sprayed in a chilled vapor of a cryogenic liquid, causing the formation of a dispersion of frozen droplets, which are subsequently spray dried,[228,229] with the advantage of a narrow particle size distribution, especially if compared to freeze-dried of a microbial dispersion followed by mechanical comminution, short freeze-drying times, and high cell viability. However, the main drawback of this technique is represented by the high process costs, which are between 30 and 50 times higher than spray drying.[230]

Emulsification of probiotics enables to encapsulate microbial cells in small-diameter capsules, maintaining a high viability. However, the encapsulated cells remain in aqueous phase, with consequent limitation of use as additive to liquid products. Typical coating materials are alginates, carrageenan, gellan, and xanthan gums, but also whey proteins have been used, to improve the compatibility with milk products as well as to overcome the regulations that restrict the use of excipients of microbial origin.[231]

Extrusion is a simple and low-cost process of encapsulation with core-shell architecture, which is able to preserve probiotic cell viability, owing to the limited use of harmful solvents and the small stresses exerted on microbial cells. However, its use on large scale is limited by the slow process of capsule fabrication. In contrast, owing to its easier scalability, fluid-bed coating is more widely used in the encapsulation of probiotic cells.

Table 37.6 reports some details, such as type of microbial cells, encapsulant materials, and encapsulation process, about the encapsulation systems for probiotics described by recent literature, in order to provide some indication on the main research trends.

From the analysis of the data reported in Table 37.6, it is evident that the consolidation of the outer layer of the encapsulation system, carried out through gelation or cross-linking processes, is a fundamental step not only toward the improvement of viability preservation but also toward the controlled release of the probiotics in the gastrointestinal tract.

Therefore, the adequate selection of the encapsulant materials, of the process of encapsulation, as well as of the consolidation method enables to enhance the protection of the payload also during intense transformation processes, including extremes of temperature and pH, as well as high shear stresses.

37.6 CONCLUSIONS AND PERSPECTIVES

The encapsulation of bioactive compounds represents an efficient approach of wide use in the food and nutraceutical industry to promote homogeneous dispersion of bioactive compounds in the end product, to protect them from interaction with other ingredients and from degradation during end product transformation, storage, and preparation, as well as to control their release where needed.

To date, different architectures of encapsulation systems are used, in order to tailor their properties on the requirements of the end product. For example, matrix-type systems, in general fabricated through simple and inexpensive processes, are indicated for sustained release applications. In addition, due to their simplicity, matrix systems can also be fabricated with smaller particle sizes, in the sub-micrometric range. In contrast, core-shell systems, with a higher structural complexity that is reflected in multiple-stage fabrication processes, are especially suitable for the enhanced protection of the bioactive compounds as well as for a triggered release in the intestinal tract. However, their complexity is also limiting their typical size, which usually is in the micrometric range.

Future trends are toward the utilization of the different encapsulation systems to prepare food additives and compare their stability and performance when introduced in real foods.

This could answer to two different needs, one related to the extension of the use of nutraceutical compounds in foodstuff to increase the content of nutrients with beneficial effects on human health, the other related to the design of novel foods tailored to fulfil peculiar dietetic requirements, namely, able to overcome specific problems related to the lack of essential nutrients in the diet of the population of certain countries and/or able to answer the demand for personalized nutrition of groups of the population as young or elder people.

TABLE 37.6
Review of Recently Investigated Encapsulation Systems for Probiotics, with Details on the Encapsulant Materials, Type of Probiotic Cells, Encapsulation Process, and Viability

Encapsulant Material	Microorganism	Technique	Viability	References
Microcapsules of calcium alginate with an external coating of whey proteins	• Lactobacillus plantarum 299v (1) • Lactobacillus plantarum 800 (2) • Lactobacillus plantarum CIP A159 (3)	Freeze drying	Reduction after 60 min at 37°C (Simulated Gastric Fluid) for coated capsules: 2.3 log CFU/g for (1), 4.2 log CFU/g for (2) and 3.3 log CFU/g for (3). Reduction after 60 min at 37°C (Simulated Gastric Fluid) for uncoated capsules: 7.8 log CFU/g for (1), 8.4 log CFU/g for (2) and 8.2 log CFU/g for (3). After 180 min at 37°C (Simulated Intestinal Fluid): only bacteria in coated capsules survive.	[232]
Casein-based microcapsules produced by enzymatic gelation with transglutaminase	• Lactobacillus F19 • Bifidobacterium BB12	Emulsification + freeze drying	*Lactobacillus* F19 survived in number significantly higher in the encapsulated form compared to unencapsulated cells; encapsulation improves survival of *Bifidobacterium* BB12 during storage to a maximum of 90 days in all conditions tested; for *Lactobacillus* F19 was not found a positive effect of the encapsulation during storage.	[231]
Microcapsules of calcium alginate with a coating of sodium alginate	• Lactobacillus acidophilus (1) • Lactobacillus rhamnosus (2)	Emulsification + gelation	After incubation in simulated gastric fluid (60 min) and simulated intestinal fluid (pH 7.25, 2 h), the number of surviving cells encapsulated into double layer coated alginate microspheres corresponds to 6.5 log CFU/mL for (1) and 7.6 log CFU/mL for (2), while corresponds to 2.3 (1) and 2.0 (2) log CFU/mL for unencapsulated cells.	[233]
Capsules of sodium alginate or amidated pectin and their combination	*Lactobacillus casei*	Extrusion	Pectin and alginate induce a sort of synergic effect in the consolidation of the trapping matrix. Capsules produced with 2%–3% pectin combined with 0.5% alginate tend to display optimal counts of *L. casei*, which were above the therapeutic requirement of 10^7 cfu/g in yoghurt, after 20 days of storage at 4°C.	[234]
Microcapsules of sodium alginate and corn starch	*Lactobacillus acidophilus* LA1	Emulsification	The microorganisms survived better in the encapsulated form at high temperatures and at high salt concentrations. The unencapsulated cells were completely destroyed at 90°C whereas the microencapsulated cells reduced by 4.14 log cycles. After 3 h incubation in simulated intestinal fluid, the unencapsulated and encapsulated cells registered 5.47 and 2.16 log cycle reduction, respectively.	[235]
Capsules of maltodextrin	• Lactobacillus acidophilus (1) • Lactobacillus rhamnosus (2)	Spray drying	Maximum survival at 100°C: 81.17%. Maximum survival at 130°C: 55%.	[236]
Microcapsules of whey proteins gelled with transglutaminase	*Bifidobacterium bifidum* F-35	Gelation vs. Spray drying	The gelation technique allows to produce capsules bigger and denser, which degrade more slowly in simulated gastric fluids and provide better protection for the cells, compared to the capsules produced by spray drying.	[237]
Capsules of trehalose consolidated with monosodium glutamate	*Lactobacillus rhamnosus* GG	Spray drying	Spray drying of *L. rhamnosus* in trehalose resulted in a survival rate of 69%, but the addition of glutamate has significantly increased the survival rate to 80.8%.	[238]

The approach described is particularly valuable considering that the strong interaction between nutrition and health is receiving nowadays a renewed attention due to the awareness of researchers, medical doctors, and public administrators responsible of the health care, that there is a pressing need for preventing illness, reducing the number of hospitalization, dealing with the problem of the ageing of the population of developed countries in order to reduce the costs and, at the same time, ensuring the survival of the public health care management system.

Moreover, the introduction on the market of novel food products whose consumption is beneficial to prevent health risks available for a wide range of consumers and not only limited to niche markets due to the elevated costs of industrial transformation could also sustain the agro-food industry of developed countries in the very aggressive globalized scenario.

REFERENCES

1. Donsì, F.; Sessa, M.; Ferrari, G., Nanometric-size delivery systems for bioactive compounds for the nutraceutical and food industries. In *Bio-Nanotechnology: A Revolution in Food, Biomedical and Health Sciences*, Bagchi, D., Bagchi, M, Moriyama, H., Shahidi, F., (Eds.) John Wiley & Sons, Oxford, U.K., 2013, p. 619.
2. Lakkis, J. M., Introduction. In *Encapsulation and Controlled Release Technologies in Food Systems*, Lakkis, J. M., (Ed.) Blackwell Publishing Ltd, Oxford, U.K., 2007, pp. 1–12.
3. Garti, N.; Yuli-Amar, I., Micro- and nano-emulsions for delivery of functional food ingredients. In *Delivery and Controlled Release of Bioactives in Foods and Nutraceuticals*, Garti, N., (Ed.) Woodhead Publishing Limited, Cambridge, U.K., 2008, pp. 149–183.
4. Acosta, E., Bioavailability of nanoparticles in nutrient and nutraceutical delivery. *Current Opinion in Colloid and Interface Science* (2009): *14*, 3–15.
5. WHO; FAO *Diet, Nutrition and the Prevention of Chronic Diseases: Report of a Joint WHO/FAO Expert Consultation*; World Health Organisation, Geneva, Switzerland, 2003.
6. Kris-Etherton, P. M.; Hecker, K. D.; Bonanome, A.; Coval, S. M.; Binkoski, A. E.; Hilpert, K. F.; Griel, A. E.; Etherton, T. D., Bioactive compounds in foods: Their role in the prevention of cardiovascular disease and cancer. *American Journal of Medicine* (2002): *113*, 71–88.
7. Satija, A.; Hu, F. B., Cardiovascular benefits of dietary fiber. *Current Atherosclerosis Reports* (2012): *14*, 505–514.
8. Zamora-Ros, R.; Fedirko, V.; Trichopoulou, A.; Gonzalez, C. A.; Bamia, C.; Trepo, E.; Noethlings, U. et al., Dietary flavonoid, lignan and antioxidant capacity and risk of hepatocellular carcinoma in the European prospective investigation into cancer and nutrition study. *International Journal of Cancer* (2013): *133*, 2429–2443.
9. Ford, D. W.; Jensen, G. L.; Hartman, T. J.; Wray, L.; Smiciklas-Wright, H., Association between dietary quality and mortality in older adults: A review of the epidemiological evidence. *Journal of Nutrition in Gerontology and Geriatrics* (2013): *32*, 85–105.
10. Kitts, D. D., Bioactive substances in food—Identification and potential uses. *Canadian Journal of Physiology and Pharmacology* (1994): *72*, 423–434.
11. Kris-Etherton, P. M.; Lefevre, M.; Beecher, G. R.; Gross, M. D.; Keen, C. L.; Etherton, T. D., Bioactive compounds in nutrition and health research methodologies for establishing biological function: The antioxidant and anti-inflammatory effects of flavonoids on atherosclerosis. *Annual Review of Nutrition* (2004): *24*, 511–538.
12. Carrato, B.; Sanzini, E., Biologically-active phytochemicals in vegetable food. *Annali dell'Istituto superiore di sanita* (2005): *41*, 7–16.
13. McClements, D. J.; Decker, E. A.; Weiss, J., Emulsion-based delivery systems for lipophilioc bioactive components. *Journal of Food Science* (2007): *72*, R109–R124.
14. Kanazawa, K., Bioavailability of non-nutrients for preventing lifestyle-related diseases. *Trends in Food Science and Technology* (2011): *22*, 655–659.
15. Gillespie, S.; Gavins, F. N. E., Phytochemicals: Countering risk factors and pathological responses associated with ischaemia reperfusion injury. *Pharmacology and Therapeutics* (2013): *138*, 38–45.
16. Su, Z.-Y.; Shu, L.; Khor, T. O.; Lee, J. H.; Fuentes, F.; Kong, A.-N. T., A Perspective on dietary phytochemicals and cancer chemoprevention: Oxidative stress, Nrf2, and epigenomics. *Natural Products in Cancer Prevention and Therapy* (2013): *329*, 133–162.

17. Dao, C. A.; Patel, K. D.; Neto, C. C., Phytochemicals from the fruit and foliage of Cranberry (Vaccinium macrocarpon)—Potential benefits for human health. *Emerging Trends in Dietary Components for Preventing and Combating Disease* (2012): *1093*, 79–94.
18. Wildman, R. E. C., *Handbook of Nutraceuticals and Functional Foods*. CRC Press, New York, 2001.
19. Bell, L. N., Stability testing of nutraceuticals and functional foods. In *Handbook of Nutraceuticals and Functional Foods*, Wildman, R. E. C., (Ed.) CRC Press, New York, 2001, pp. 501–516.
20. Wang, X.; Jiang, Y.; Huang, Q., Encapsulation technologies for preserving and controlling the release of enzymes and phytochemicals. In *Encapsulation and Controlled Release Technologies in Food Systems*, Lakkis, J. M., (Ed.) Wiley-Blackwell, Hoboken, NJ, 2007.
21. Wang, H. K., The therapeutic potential of flavonoids. *Expert Opinion on Investigational Drugs* (2000): *9*, 2103–2119.
22. Lamson, D. W.; Brignall, M. S., Antioxidants and cancer, Part 3: Quercetin. *Alternative Medicine Review: A Journal of Clinical Therapeutic* (2000): *5*, 196–208.
23. Kielhorn, S.; Thorngate, J. H., Oral sensations associated with the flavan-3-ols (+)-catechin and (-)-epicatechin. *Food Quality and Preference* (1999): *10*, 109–116.
24. Siess, M. H.; Le Bon, A. M.; Canivenc-Lavier, M. C.; Suschetet, M., Mechanisms involved in the chemoprevention of flavonoids. *Biofactors* (2000): *12*, 193–199.
25. Jang, M. S.; Cai, E. N.; Udeani, G. O.; Slowing, K. V.; Thomas, C. F.; Beecher, C. W. W.; Fong, H. H. S. et al., Cancer chemopreventive activity of resveratrol, a natural product derived from grapes. *Science* (1997): *275*, 218–220.
26. Sessa, M.; Tsao, R.; Liu, R.; Ferrari, G.; Donsi, F., Evaluation of the stability and antioxidant activity of nanoencapsulated resveratrol during in vitro digestion. *Journal of Agricultural and Food Chemistry* (2011): *59*, 12352–12360.
27. Sessa, M.; Balestrieri, M. L.; Ferrari, G.; Servillo, L.; Castaldo, D.; D'Onofrio, N.; Donsi, F.; Tsao, R., Bioavailability of encapsulated resveratrol into nanoemulsion-based delivery systems. *Food Chemistry* (2014): *147*, 42–50.
28. Chauhan, D. P., Chemotherapeutic potential of curcumin for colorectal cancer. *Current Pharmaceutical Design* (2002): *8*, 1695–1706.
29. Donsi, F.; Wang, Y.; Li, J.; Huang, Q., Preparation of curcumin sub-micrometer dispersions by high-pressure homogenization. *Journal of Agricultural and Food Chemistry* (2010): *58*, 2848–2853.
30. Sharma, D.; Sukumar, S., Big punches come in nanosizes for chemoprevention. *Cancer Prevention Research* (2013): *6*, 1007–1010.
31. Umadevi, S.; Gopi, V.; Vellaichamy, E., Inhibitory effect of gallic acid on advanced glycation end products induced up-regulation of inflammatory cytokines and matrix proteins in H9C2 (2–1) cells. *Cardiovascular Toxicology* (2013): *13*, 396–405.
32. Kratz, J. M.; Andrighetti-Frohner, C. R.; Leal, P. C.; Nunes, R. J.; Yunes, R. A.; Trybala, E.; Bergstrom, T.; Monte Barardi, C. R.; Oliveira Simoes, C. M., Evaluation of anti-HSV-2 activity of gallic acid and pentyl gallate. *Biological and Pharmaceutical Bulletin* (2008): *31*, 903–907.
33. Fathi, M.; Mirlohi, M.; Varshosaz, J.; Madani, G., Novel caffeic acid nanocarrier: Production, characterization, and release modeling. *Journal of Nanomaterials* (2013).
34. Variyar, P. S.; Ahmad, R.; Bhat, R.; Niyas, Z.; Sharma, A., Flavoring components of raw monsooned arabica coffee and their changes during radiation processing. *Journal of Agricultural and Food Chemistry* (2003): *51*, 7945–7950.
35. Cao-Hoang, L.; Fougere, R.; Wache, Y., Increase in stability and change in supramolecular structure of beta-carotene through encapsulation into polylactic acid nanoparticles. *Food Chemistry* (2011): *124*, 42–49.
36. Biesalski, H. K.; Obermueller-Jevic, U. C., UV light, beta-carotene and human skin—Beneficial and potentially harmful effects. *Archives of Biochemistry and Biophysics* (2001): *389*, 1–6.
37. Arab, L.; Steck, S., Lycopene and cardiovascular disease. *American Journal of Clinical Nutrition* (2000): *71*, 1691S–1695S.
38. Ribeiro, H. S.; Ax, K.; Schubert, H., Stability of lycopene emulsions in food systems. *Journal of Food Science* (2003): *68*, 2730–2734.
39. Kucuk, O.; Sarkar, F. H.; Djuric, Z.; Sakr, W.; Pollak, M. N.; Khachik, F.; Banerjee, M.; Bertram, J. S.; Wood, D. P., Effects of lycopene supplementation in patients with localized prostate cancer. *Experimental Biology and Medicine* (2002): *227*, 881–885.
40. Krinsky, N. I., Overview of lycopene, carotenoids, and disease prevention. *Proceedings of the Society for Experimental Biology and Medicine* (1998): *218*, 95–97.

41. Stringham, J. M.; Bovier, E. R.; Wong, J. C.; Hammond, B. R., Jr., The Influence of dietary lutein and zeaxanthin on visual performance. *Journal of Food Science* (2010): *75*, R24–R29.
42. Donsi, F.; Sessa, M.; Ferrari, G., Nanoencapsulation of essential oils to enhance their antimicrobial activity in foods. *Journal of Biotechnology* (2010): *150*, S67–S67.
43. Crowell, P. L., Prevention and therapy of cancer by dietary monoterpenes. *Journal of Nutrition* (1999): *129*, 775S–778S.
44. Lluis, L.; Taltavull, N.; Munoz-Cortes, M.; Sanchez-Martos, V.; Romeu, M.; Giralt, M.; Molinar-Toribio, E. et al., Protective effect of the omega-3 polyunsaturated fatty acids: Eicosapentaenoic acid/docosahexaenoic acid 1:1 ratio on cardiovascular disease risk markers in rats. *Lipids in Health and Disease* (2013): *12*.
45. Galgani, J. E.; Uauy, R. D.; Aguirre, C. A.; Diaz, E. O., Effect of the dietary fat quality on insulin sensitivity. *British Journal of Nutrition* (2008): *100*, 471–479.
46. Garg, M.; Mishra, D.; Agashe, H.; Jain, N. K., Ethinylestradiol-loaded ultraflexible liposomes: Pharmacokinetics and pharmacodynamics. *Journal of Pharmacy and Pharmacology* (2006): *58*, 459–468.
47. Fukushima, S.; Takada, N.; Hori, T.; Wanibuchi, H., Cancer prevention by organosulfur compounds from garlic and onion. *Journal of Cellular Biochemistry* (1997): 100–105.
48. Hirsch, K.; Danilenko, M.; Giat, J.; Miron, T.; Rabinkov, A.; Wilchek, M.; Mirelman, D.; Levy, J.; Sharoni, Y., Effect of purified allicin, the major ingredient of freshly crushed garlic, on cancer cell proliferation. *Nutrition and Cancer-an International Journal* (2000): *38*, 245–254.
49. Curin, Y.; Andriantsitohaina, R., Polyphenols as potential therapeutical agents against cardiovascular diseases. *Pharmacological Reports* (2005): *57*, 97–107.
50. D'Archivio, M.; Filesi, C.; Vari, R.; Scazzocchio, B.; Masella, R., Bioavailability of the polyphenols: Status and controversies. *International Journal of Molecular Sciences* (2010): *11*, 1321–1342.
51. Habauzit, V.; Morand, C., Evidence for a protective effect of polyphenols-containing foods on cardiovascular health: An update for clinicians. *Therapeutic Advances in Chronic Disease* (2012): *3*, 87–106.
52. Kita, A.; Bakowska-Barczak, A.; Hamouz, K.; Kulakowska, K.; Lisinska, G., The effect of frying on anthocyanin stability and antioxidant activity of crisps from red- and purple-fleshed potatoes (Solanum tuberosum L.). *Journal of Food Composition and Analysis* (2013): *32*, 169–175.
53. Perla, V.; Holm, D. G.; Jayanty, S. S., Effects of cooking methods on polyphenols, pigments and antioxidant activity in potato tubers. *Lwt-Food Science and Technology* (2012): *45*, 161–171.
54. Tipoe, G. L.; Leung, T.-M.; Hung, M.-W.; Fung, M.-L., Green tea polyphenols as an anti-oxidant and anti-inflammatory agent for cardiovascular protection. *Cardiovascular and Hematological Disorders Drug Targets* (2007): *7*, 135–144.
55. Bognar, E.; Sarszegi, Z.; Szabo, A.; Debreceni, B.; Kalman, N.; Tucsek, Z.; Sumegi, B.; Gallyas, F., Jr., Antioxidant and anti-inflammatory effects in RAW264.7 macrophages of malvidin, a major red wine polyphenol. *Plos One* (2013): *8*.
56. Quinones, M.; Miguel, M.; Aleixandre, A., Beneficial effects of polyphenols on cardiovascular disease. *Pharmacological Research* (2013): *68*, 125–131.
57. Widmer, R. J.; Freund, M. A.; Flammer, A. J.; Sexton, J.; Lennon, R.; Romani, A.; Mulinacci, N.; Vinceri, F. F.; Lerman, L. O.; Lerman, A., Beneficial effects of polyphenol-rich olive oil in patients with early atherosclerosis. *European Journal of Nutrition* (2013): *52*, 1223–1231.
58. Kurosawa, T.; Itoh, F.; Nozaki, A.; Nakano, Y.; Katsuda, S. I.; Osakabe, N.; Tsubone, H.; Kondo, K.; Itakura, H., Suppressive effect of cocoa powder on atherosclerosis in Kurosawa and Kusanagi-hypercholesterolemic rabbits. *Journal of Atherosclerosis and Thrombosis* (2005): *12*, 20–28.
59. Stoner, G. D.; Mukhtar, H., Polyphenols as cancer chemopreventive agents. *Journal of Cellular Biochemistry. Supplement* (1995): *22*, 169–180.
60. Araujo, J. R.; Goncalves, P.; Martel, F., Chemopreventive effect of dietary polyphenols in colorectal cancer cell lines. *Nutrition Research* (2011): *31*, 77–87.
61. Giftson, J. S.; Jayanthi, S.; Nalini, N., Chemopreventive efficacy of gallic acid, an antioxidant and anticarcinogenic polyphenol, against 1,2-dimethyl hydrazine induced rat colon carcinogenesis. *Investigational New Drugs* (2010): *28*, 251–259.
62. D'Incalci, M.; Steward, W. P.; Gescher, A. J., Use of cancer chemopreventive phytochemicals as antineoplastic agents. *Lancet Oncology* (2005): *6*, 899–904.
63. Hu, M., Commentary: Bioavailability of flavonoids and polyphenols: Call to arms. *Molecular Pharmaceutics* (2007): *4*, 803–806.
64. Basu, A.; Imrhan, V., Tomatoes versus lycopene in oxidative stress and carcinogenesis: Conclusions from clinical trials. *European Journal of Clinical Nutrition* (2007): *61*, 295–303.

65. Xianquan, S.; Shi, J.; Kakuda, Y.; Yueming, J., Stability of lycopene during food processing and storage. *Journal of Medicinal Food* (2005): *8*, 413–422.
66. Calder, P. C., Omega-3 fatty acids and inflammatory processes. *Nutrients* (2010): *2*, 355–374.
67. Mozurkewich, E.; Chilimigras, J.; Klemens, C.; Keeton, K.; Allbaugh, L.; Hamilton, S.; Berman, D.; Vazquez, D.; Marcus, S.; Djuric, Z.; Vahratian, A., The mothers, Omega-3 and mental health study. *BMC Pregnancy and Childbirth* (2011): *11*.
68. Goren, J. L.; Tewksbury, A. T., The use of omega-3 fatty acids in mental illness. *Journal of Pharmacy Practice* (2011): *24*, 452–471.
69. Seida, J. C.; Mager, D. R.; Hartling, L.; Vandermeer, B.; Turner, J. M., Parenteral omega-3 fatty acid lipid emulsions for children with intestinal failure and other conditions: A systematic review. *Journal of Parenteral and Enteral Nutrition* (2013): *37*, 44–55.
70. Shek, L. P.; Chong, M. F.-F.; Lim, J. Y.; Soh, S.-E.; Chong, Y.-S., Role of dietary long-chain polyunsaturated fatty acids in infant allergies and respiratory diseases. *Clinical and Developmental Immunology* (2012): *2012*, 730568–730568.
71. Huffman, S. L.; Harika, R. K.; Eilander, A.; Osendarp, S. J. M., Essential fats: How do they affect growth and development of infants and young children in developing countries? A literature review. *Maternal and Child Nutrition* (2011): *7*, 44–65.
72. Major, G. C.; Alarie, F.; Dore, J.; Phouttama, S.; Tremblay, A., Supplementation with calcium plus vitamin D enhances the beneficial effect of weight loss on plasma lipid and lipoprotein concentrations. *American Journal of Clinical Nutrition* (2007): *85*, 54–59.
73. Abbott, R. D.; Curb, J. D.; Rodriguez, B. L.; Sharp, D. S.; Burchfiel, C. M.; Yano, K., Effect of dietary calcium and milk consumption on risk of thromboembolic stroke in older middle-aged men—The Honolulu Heart Program. *Stroke* (1996): *27*, 813–818.
74. Aburto, N. J.; Hanson, S.; Gutierrez, H.; Hooper, L.; Elliott, P.; Cappuccio, F. P., Effect of increased potassium intake on cardiovascular risk factors and disease: Systematic review and meta-analyses. *British Medical Journal* (2013): *346*.
75. Cherbuin, N.; Kumar, R.; Sachdev, P. S.; Anstey, K. J., Dietary mineral intake and risk of mild cognitive impairment: The PATH through life project. *Frontiers in Aging Neuroscience* (2014): *6*.
76. Domellof, M.; Thorsdottir, I.; Thorstensen, K., Health effects of different dietary iron intakes: A systematic literature review for the 5th Nordic Nutrition Recommendations. *Food and Nutrition Research* (2013): *57*.
77. Cekic, V.; Vasovic, V.; Jakovljevic, V.; Lalosevic, D.; Cabo, I.; Mikov, M.; Sabo, A., Effect of chromium enriched fermentation product of barley and brewer's yeast and its combination with rosiglitazone on experimentally induced hyperglycaemia in mice. *Srpski Arhiv Za Celokupno Lekarstvo* (2011): *139*, 610–618.
78. Liu, J.; Bao, W.; Jiang, M.; Zhang, Y.; Zhang, X.; Liu, L., Chromium, selenium, and zinc multimineral enriched yeast supplementation ameliorates diabetes symptom in streptozocin-induced mice. *Biological Trace Element Research* (2012): *146*, 236–245.
79. Nicolae, A.; Gabaldon Hernandez, J. A.; Martinez San Martin, A., Evaluation of the calcium yield in the microencapsulation process. *Romanian Biotechnological Letters* (2013): *18*, 8685–8688.
80. Li, Y. O.; Diosady, L. L., Microencapsulation and its application in micronutrient fortification through "engineered" staple foods. *Agro Food Industry Hi-Tech* (2012): *23*, 18–21.
81. Wegmuller, R.; Zimmermann, M. B.; Buhr, V. G.; Windhab, E. J.; Hurrell, R. E., Development, stability, and sensory testing of microcapsules containing iron, iodine, and vitamin a for use in food fortification. *Journal of Food Science* (2006): *71*, S181–S187.
82. Brown, L.; Rosner, B.; Willett, W. W.; Sacks, F. M., Cholesterol-lowering effects of dietary fiber: A meta-analysis. *American Journal of Clinical Nutrition* (1999): *69*, 30–42.
83. Salmeron, J.; Ascherio, A.; Rimm, E. B.; Colditz, G. A.; Spiegelman, D.; Jenkins, D. J.; Stampfer, M. J.; Wing, A. L.; Willett, W. C., Dietary fiber, glycemic load, and risk of NIDDM in men. *Diabetes Care* (1997): *20*, 545–550.
84. Figueroa-Gonzalez, I.; Quijano, G.; Ramirez, G.; Cruz-Guerrero, A., Probiotics and prebiotics—Perspectives and challenges. *Journal of the Science of Food and Agriculture* (2011): *91*, 1341–1348.
85. Van Loo, J. A. E., Prebiotics promote good health: The basis, the potential, and the emerging evidence. *Journal of Clinical Gastroenterology* (2004): *38*, S70–S75.
86. Roberfroid, M., Prebiotics: The concept revisited. *Journal of Nutrition* (2007): *137*, 830S–837S.
87. Teitelbaum, J. E.; Walker, W. A., Nutritional impact of pre- and probiotics as protective gastrointestinal organisms. *Annual Review of Nutrition* (2002): *22*, 107–138.
88. Wang, Y., Prebiotics: Present and future in food science and technology. *Food Research International* (2009): *42*, 8–12.

89. Crittenden, R. G.; Playne, M. J., Production, properties and applications of food-grade oligosaccharides. *Trends in Food Science and Technology* (1996); *7*, 353–361.
90. Komninou, D.; Ayonote, A.; Richie, J. P.; Rigas, B., Insulin resistance and its contribution to colon carcinogenesis. *Experimental Biology and Medicine* (2003): *228*, 396–405.
91. Kaur, N.; Gupta, A. K., Applications of inulin and oligofructose in health and nutrition. *Journal of Biosciences* (2002): *27*, 703–714.
92. Ballongue, J.; Schumann, C.; Quignon, P., Effects of lactulose and lactitol on colonic microflora and enzymatic activity. *Scandinavian Journal of Gastroenterology. Supplement* (1997): *222*, 41–44.
93. Minami, T.; Miki, T.; Fujiwara, T.; Kawabata, S.; Izumitani, A.; Ooshima, T.; Sobue, S.; Hamada, S., Caries-inducing activity of isomaltooligosugar (IMOS) in in vitro and rat experiments. *Shoni shikagaku zasshi. The Japanese Journal of Pedodontics* (1989): *27*, 1010–1017.
94. Mussatto, S. I.; Mancilha, I. M., Non-digestible oligosaccharides: A review. *Carbohydrate Polymers* (2007): *68*, 587–597.
95. Tateyama, I.; Hashii, K.; Johno, I.; Iino, T.; Hirai, K.; Suwa, Y.; Kiso, Y., Effect of xylooligosaccharide intake on severe constipation in pregnant women. *Journal of Nutritional Science and Vitaminology* (2005): *51*, 445–448.
96. Espinosa-Martos, I.; Ruperez, P., Soybean oligosaccharides. Potential as new ingredients in functional food. *Nutricion Hospitalaria* (2006): *21*, 92–96.
97. Dominguez, A. L.; Rodrigues, L. R.; Lima, N. M.; Teixeira, J. A., An overview of the recent developments on fructooligosaccharide production and applications. *Food and Bioprocess Technology* (2014): *7*, 324–337.
98. Tomomatsu, H., Health-effects of oligosaccharides. *Food Technology* (1994): *48*, 53–53.
99. Bodera, P., Influence of prebiotics on the human immune system (GALT). *Recent Patents on Inflammation and Allergy Drug Discovery* (2008): *2*, 149–153.
100. Franck, A., Technological functionality of inulin and oligofructose. *British Journal of Nutrition* (2002): *87*, S287–S291.
101. Nelson, A. L., Properties of high-fiber ingredients. *Cereal Foods World* (2001): *46*, 93–97.
102. Sanders, M. E., Probiotics: Definition, sources, selection, and uses. *Clinical Infectious Diseases* (2008): *46*, S58–S61.
103. Senok, A. C.; Ismaeel, A. Y.; Botta, G. A., Probiotics: Facts and myths. *Clinical Microbiology and Infection* (2005): *11*, 958–966.
104. Gardiner, G. E.; Bouchier, P.; O'Sullivan, E.; Kelly, J.; Collins, J. K.; Fitzgerald, G.; Ross, R. P.; Stanton, C., A spray-dried culture for probiotic Cheddar cheese manufacture. *International Dairy Journal* (2002): *12*, 749–756.
105. Scholz-Ahrens, K. E.; Ade, P.; Marten, B.; Weber, P.; Timm, W.; Asil, Y.; Glueer, C.-C.; Schrezenmeir, J., Prebiotics, probiotics, and synbiotics affect mineral absorption, bone mineral content, and bone structure. *Journal of Nutrition* (2007): *137*, 838S–846S.
106. Hemarajata, P.; Versalovic, J., Effects of probiotics on gut microbiota: Mechanisms of intestinal immunomodulation and neuromodulation. *Therapeutic Advances in Gastroenterology* (2013): *6*, 39–51.
107. Shah, N. P., Functional cultures and health benefits. *International Dairy Journal* (2007): *17*, 1262–1277.
108. Kailasapathy, K.; Harmstorf, I.; Phillips, M., Survival of Lactobacillus acidophilus and Bifidobacterium animalis ssp lactis in stirred fruit yogurts. *LWT-Food Science and Technology* (2008): *41*, 1317–1322.
109. Ranadheera, R. D. C. S.; Baines, S. K.; Adams, M. C., Importance of food in probiotic efficacy. *Food Research International* (2010); *43*, 1–7.
110. Granato, D.; Branco, G. F.; Cruz, A. G.; Fonseca Faria, J. d. A.; Shah, N. P., Probiotic dairy products as functional foods. *Comprehensive Reviews in Food Science and Food Safety* (2010): *9*, 455–470.
111. Shah, N. P.; Ding, W. K.; Fallourd, M. J.; Leyer, G., Improving the stability of probiotic bacteria in model fruit juices using vitamins and antioxidants. *Journal of Food Science* (2010): *75*, M278–M282.
112. Zhang, L.; Huang, S.; Ananingsih, V. K.; Zhou, W.; Chen, X. D., A study on Bifidobacterium lactis Bb12 viability in bread during baking. *Journal of Food Engineering* (2014): *122*, 33–37.
113. Cote, J.; Dion, J.; Burguiere, P.; Casavant, L.; Van Eijk, J., Probiotics in bread and baked products: A new product category. *Cereal Foods World* (2013): *58*, 293–296.
114. Mohammadi, R.; Mortazavian, A. M.; Khosrokhavar, R.; da Cruz, A. G., Probiotic ice cream: Viability of probiotic bacteria and sensory properties. *Annals of Microbiology* (2011): *61*, 411–424.
115. Cruz, A. G.; Antunes, A. E. C.; Sousa, A. L. O. P.; Faria, J. A. F.; Saad, S. M. I., Ice-cream as a probiotic food carrier. *Food Research International* (2009): *42*, 1233–1239.

116. Kailasapathy, K.; Sultana, K., Survival and beta-D-galactosidase activity of encapsulated and free Lactobacillus acidophilus and Bifidobacterium lactis in ice-cream. *Australian Journal of Dairy Technology* (2003): *58*, 223–227.
117. Possemiers, S.; Marzorati, M.; Verstraete, W.; Van de Wiele, T., Bacteria and chocolate: A successful combination for probiotic delivery. *International Journal of Food Microbiology* (2010): *141*, 97–103.
118. Ziemer, C. J.; Gibson, G. R., An overview of probiotics, prebiotics and synbiotics in the functional food concept: Perspectives and future strategies. *International Dairy Journal* (1998): *8*, 473–479.
119. Jyothi, N. V. N.; Prasanna, P. M.; Sakarkar, S. N.; Prabha, K. S.; Ramaiah, P. S.; Srawan, G. Y., Microencapsulation techniques, factors influencing encapsulation efficiency. *Journal of Microencapsulation* (2010): *27*, 187–197.
120. Ezhilarasi, P. N.; Karthik, P.; Chhanwal, N.; Anandharamakrishnan, C., Nanoencapsulation techniques for food bioactive components: A review. *Food and Bioprocess Technology* (2013): *6*, 628–647.
121. Dong, Q.-Y.; Chen, M.-Y.; Xin, Y.; Qin, X.-Y.; Cheng, Z.; Shi, L.-E.; Tang, Z.-X., Alginate-based and protein-based materials for probiotics encapsulation: A review. *International Journal of Food Science and Technology* (2013): *48*, 1339–1351.
122. Shewan, H. M.; Stokes, J. R., Review of techniques to manufacture micro-hydrogel particles for the food industry and their applications. *Journal of Food Engineering* (2013): *119*, 781–792.
123. Oxley, J. D., Coextrusion for food ingredients and nutraceutical encapsulation: Principles and technology. *Encapsulation Technologies and Delivery Systems for Food Ingredients and Nutraceuticals* (2012): 131–150.
124. Letchford, K.; Burt, H., A review of the formation and classification of amphiphilic block copolymer nanoparticulate structures: Micelles, nanospheres, nanocapsules and polymersomes. *European Journal of Pharmaceutics and Biopharmaceutics* (2007): *65*, 259–269.
125. Narang, A. S.; Delmarre, D.; Gao, D., Stable drug encapsulation in micelles and microemulsions. *International Journal of Pharmaceutics* (2007): *345*, 9–25.
126. Yang, Y.; Gu, Z. B.; Zhang, G. Y., Delivery of bioactive conjugated linoleic acid with self-assembled amylose-CLA complex. *Journal of Agricultural and Food Chemistry* (2009): *57*, 7125–7130.
127. Fang, Z. X.; Bhandari, B., Encapsulation of polyphenols—A review. *Trends in Food Science and Technology* (2010): *21*, 510–523.
128. Lucas-Abellan, C.; Fortea, I.; Gabaldon, J. A.; Nunez-Delicado, E., Encapsulation of quercetin and myricetin in cyclodextrins at acidic pH. *Journal of Agricultural and Food Chemistry* (2008): *56*, 255–259.
129. Patricia Blanch, G.; Luisa Ruiz Del Castillo, M.; Del Mar Caja, M.; Perez-Mendez, M.; Sanchez-Cortes, S., Stabilization of all-trans-lycopene from tomato by encapsulation using cyclodextrins. *Food Chemistry* (2007): *105*, 1335–1341.
130. Zeller, B. L.; Saleeb, F. Z.; Ludescher, R. D., Trends in development of porous carbohydrate food ingredients for use in flavor encapsulation. *Trends in Food Science and Technology* (1998): *9*, 389–394.
131. Astray, G.; Gonzalez-Barreiro, C.; Mejuto, J. C.; Rial-Otero, R.; Simal-Gandara, J., A review on the use of cyclodextrins in foods. *Food Hydrocolloids* (2009): *23*, 1631–1640.
132. Spernath, A.; Aserin, A., Microemulsions as carriers for drugs and nutraceuticals. *Advances in Colloid and Interface Science* (2006): *128*, 47–64.
133. McClements, J. D., *Food Emulsions: Principles, Practices and Techniques*. 2nd edn.; CRC Press, Boca Raton, FL, 1999.
134. de Campo, L.; Yaghmur, A.; Garti, N.; Leser, M. E.; Folmer, B.; Glatter, O., Five-component food-grade microemulsions: Structural characterization by SANS. *Journal of Colloid and Interface Science* (2004): *274*, 251–267.
135. Mezzenga, R.; Schurtenberger, P.; Burbidge, A.; Michel, M., Understanding foods as soft materials. *Nature Materials* (2005): *4*, 729–740.
136. Chen, L. Y.; Remondetto, G. E.; Subirade, M., Food protein-based materials as nutraceutical delivery systems. *Trends in Food Science and Technology* (2006): *17*, 272–283.
137. Jones, O. G.; McClements, D. J., Functional biopolymer particles: Design, fabrication, and applications. *Comprehensive Reviews in Food Science and Food Safety* (2010): *9*, 374–397.
138. Damodaran, S., Protein stabilization of emulsions and foams. *Journal of Food Science* (2005): *70*, R54–R66.
139. Rinaudo, M., Main properties and current applications of some polysaccharides as biomaterials. *Polymer International* (2008): *57*, 397–430.
140. Fathi, M.; Mozafari, M. R.; Mohebbi, M., Nanoencapsulation of food ingredients using lipid based delivery systems. *Trends in Food Science and Technology* (2012): *23*, 13–27.

141. Iqbal, M. A.; Md, S.; Sahni, J. K.; Baboota, S.; Dang, S.; Ali, J., Nanostructured lipid carriers system: Recent advances in drug delivery. *Journal of Drug Targeting* (2012): *20*, 813–830.
142. Schubert, M. A.; Muller-Goymann, C. C., Characterisation of surface-modified solid lipid nanoparticles (SLN): Influence of lecithin and nonionic emulsifier. *European Journal of Pharmaceutics and Biopharmaceutics* (2005): *61*, 77–86.
143. Sagalowicz, L.; Leser, M. E., Delivery systems for liquid food products. *Current Opinion in Colloid and Interface Science* (2010): *15*, 61–72.
144. McClements, D. J.; Li, Y., Structured emulsion-based delivery systems: Controlling the digestion and release of lipophilic food components. *Advances in Colloid and Interface Science* (2010): *159*, 213–228.
145. Wackerbarth, H.; Schon, P.; Bindrich, U., Preparation and characterization of multilayer coated microdroplets: Droplet deformation simultaneously probed by atomic force spectroscopy and optical detection. *Langmuir* (2009): *25*, 2636–2640.
146. Grigoriev, D. O.; Miller, R., Mono- and multilayer covered drops as carriers. *Current Opinion in Colloid and Interface Science* (2009): *14*, 48–59.
147. Ogawa, S.; Decker, E. A.; McClements, D. J., Production and characterization of O/W emulsions containing droplets stabilized by lecithin-chitosan-pectin mutilayered membranes. *Journal of Agricultural and Food Chemistry* (2004): *52*, 3595–3600.
148. Gu, Y. S.; Decker, E. A.; McClements, D. J., Application of multi-component biopolymer layers to improve the freeze-thaw stability of oil-in-water emulsions: beta-Lactoglobulin-iota-carrageenan-gelatin. *Journal of Food Engineering* (2007): *80*, 1246–1254.
149. Arshady, R., In the name of particle formation. *Colloids and Surfaces a-Physicochemical and Engineering Aspects* (1999): *153*, 325–333.
150. Lv, Y.; Yang, F.; Li, X. Y.; Zhang, X. M.; Abbas, S., Formation of heat-resistant nanocapsules of jasmine essential oil via gelatin/gum arabic based complex coacervation. *Food Hydrocolloids* (2014): *35*, 305–314.
151. Yang, Z. M.; Peng, Z.; Li, J. H.; Li, S. D.; Kong, L. X.; Li, P. W.; Wang, Q. H., Development and evaluation of novel flavour microcapsules containing vanilla oil using complex coacervation approach. *Food Chemistry* (2014): *145*, 272–277.
152. Rocha-Selmi, G. A.; Theodoro, A. C.; Thomazini, M.; Bolini, H. M. A.; Favaro-Trindade, C. S., Double emulsion stage prior to complex coacervation process for microencapsulation of sweetener sucralose. *Journal of Food Engineering* (2013): *119*, 28–32.
153. Comunian, T. A.; Thomazini, M.; Alves, A. J. G.; de Matos, F. E.; Balieiro, J. C. D.; Favaro-Trindade, C. S., Microencapsulation of ascorbic acid by complex coacervation: Protection and controlled release. *Food Research International* (2013): *52*, 373–379.
154. Wichchukit, S.; Oztop, M. H.; McCarthy, M. J.; McCarthy, K. L., Whey protein/alginate beads as carriers of a bioactive component. *Food Hydrocolloids* (2013): *33*, 66–73.
155. Rocha-Selmi, G. A.; Favaro-Trindade, C. S.; Grosso, C. R. F., Morphology, stability, and application of lycopene microcapsules produced by complex coacervation. *Journal of Chemistry* (2013).
156. Romo-Hualde, A.; Yetano-Cunchillos, A. I.; Gonzalez-Ferrero, C.; Saiz-Abajo, M. J.; Gonzalez-Navarro, C. J., Supercritical fluid extraction and microencapsulation of bioactive compounds from red pepper (*Capsicum annum* L.) by-products. *Food Chemistry* (2012): *133*, 1045–1049.
157. Bakowska-Barczak, A. M.; Kolodziejczyk, P. P., Black currant polyphenols: Their storage stability and microencapsulation. *Industrial Crops and Products* (2011): *34*, 1301–1309.
158. Betz, M.; Kulozik, U., Microencapsulation of bioactive bilberry anthocyanins by means of whey protein gels. *11th International Congress on Engineering and Food (Icef11)* (2011): *1*, 2047–2056.
159. Zheng, L. Q.; Ding, Z. S.; Zhang, M.; Sun, J. C., Microencapsulation of bayberry polyphenols by ethyl cellulose: Preparation and characterization. *Journal of Food Engineering* (2011): *104*, 89–95.
160. Robert, P.; Gorena, T.; Romero, N.; Sepulveda, E.; Chavez, J.; Saenz, C., Encapsulation of polyphenols and anthocyanins from pomegranate (Punica granatum) by spray drying. *International Journal of Food Science and Technology* (2010): *45*, 1386–1394.
161. Saenz, C.; Tapia, S.; Chavez, J.; Robert, P., Microencapsulation by spray drying of bioactive compounds from cactus pear (Opuntia ficus-indica). *Food Chemistry* (2009): *114*, 616–622.
162. Malheiros, P. D.; Daroit, D. J.; Brandelli, A., Food applications of liposome-encapsulated antimicrobial peptides. *Trends in Food Science and Technology* (2010): *21*, 284–292.
163. Takahashi, M.; Uechi, S.; Takara, K.; Asikin, Y.; Wada, K., Evaluation of an oral carrier system in rats: Bioavailability and antioxidant properties of liposome-encapsulated curcumin. *Journal of Agricultural and Food Chemistry* (2009): *57*, 9141–9146.

164. Nongonierma, A. B.; Abrlova, M.; Fenelon, M. A.; Kilcawley, K. N., Evaluation of two food grade proliposomes to encapsulate an extract of a commercial enzyme preparation by microfluidization. *Journal of Agricultural and Food Chemistry* (2009): *57*, 3291–3297.
165. Kirby, C. J.; Whittle, C. J.; Rigby, N.; Coxon, D. T.; Law, B. A., Stabilization of ascorbic-acid by microencapsulation in liposomes. *International Journal of Food Science and Technology* (1991): *26*, 437–449.
166. Medina-Torres, L.; Garcia-Cruz, E. E.; Calderas, F.; Laredo, R. F. G.; Sanchez-Olivares, G.; Gallegos-Infante, J. A.; Rocha-Guzman, N. E.; Rodriguez-Ramirez, J., Microencapsulation by spray drying of gallic acid with nopal mucilage (Opuntia ficus indica). *LWT-Food Science and Technology* (2013): *50*, 642–650.
167. Yuan, Y.; Gao, Y.; Zhao, J.; Mao, L., Characterization and stability evaluation of b-carotene nanoemulsions prepared by high pressure homogenization under various emulsifying conditions. *Food Research International* (2008): *41*, 61–68.
168. Wang, X.; Jiang, Y.; Wang, Y. W.; Huang, M. T.; Ho, C. T.; Huang, Q., Enhancing anti-inflammation activity of curcumin through o/w nanoemulsions. *Food Chemistry* (2008): *108*, 419–424.
169. Sugiarto, M.; Ye, A.; Singh, H., Characterisation of binding of iron to sodium caseinate and whey protein isolate. *Food Chemistry* (2009): *114*, 1007–1013.
170. Liang, L.; Tajmir-Riahi, H. A.; Subirade, M., Interaction of beta-Lactoglobulin with resveratrol and its biological implications. *Biomacromolecules* (2008): *9*, 50–56.
171. Zhang, Y. Y.; Yang, Y.; Tang, K.; Hu, X.; Zou, G. L., Physicochemical characterization and antioxidant activity of quercetin-loaded chitosan nanoparticles. *Journal of Applied Polymer Science* (2008): *107*, 891–897.
172. Mercader-Ros, M. T.; Lucas-Abellan, C.; Fortea, M. I.; Gabaldon, J. A.; Nunez-Delicado, E., Effect of HP-beta-cyclodextrins complexation on the antioxidant activity of flavonols. *Food Chemistry* (2010): *118*, 769–773.
173. Tomren, M. A.; Masson, M.; Loftsson, T.; Tonnesen, H. H., Studies on curcumin and curcuminoids XXXI. Symmetric and asymmetric curcuminoids: Stability, activity and complexation with cyclodextrin. *International Journal of Pharmaceutics* (2007): *338*, 27–34.
174. Yu, H. L.; Huang, Q. R., Enhanced in vitro anti-cancer activity of curcumin encapsulated in hydrophobically modified starch. *Food Chemistry* (2010): *119*, 669–674.
175. Sahu, A.; Kasoju, N.; Bora, U., Fluorescence study of the curcumin-casein micelle complexation and its application as a drug nanocarrier to cancer cells. *Biomacromolecules* (2008): *9*, 2905–2912.
176. Forrest, S. A.; Yada, R. Y.; Rousseau, D., Interactions of vitamin D-3 with bovine beta-lactoglobulin A and beta-casein. *Journal of Agricultural and Food Chemistry* (2005): *53*, 8003–8009.
177. Semo, E.; Kesselman, E.; Danino, D.; Livney, Y. D., Casein micelle as a natural nano-capsular vehicle for nutraceuticals. *Food Hydrocolloids* (2007): *21*, 936–942.
178. Kanaze, F. I.; Kokkalou, E.; Niopas, I.; Barmpalexis, P.; Georgarakis, E.; Bikiaris, D., Dissolution rate and stability study of flavanone aglycones, naringenin and hesperetin, by drug delivery systems based on polyvinylpyrrolidone (PVP) nanodispersions. *Drug Development and Industrial Pharmacy* (2010): *36*, 292–301.
179. Gao, L.; Liu, G. Y.; Wang, X. Q.; Liu, F.; Xu, Y. F.; Ma, J., Preparation of a chemically stable quercetin formulation using nanosuspension technology. *International Journal of Pharmaceutics* (2011): *404*, 231–237.
180. Choi, S. J.; Decker, E. A.; Henson, L.; Popplewell, L. M.; McClements, D. J., Inhibition of citral degradation in model beverage emulsions using micelles and reverse micelles. *Food Chemistry* (2010): *122*, 111–116.
181. Garti, N.; Yaghmur, A.; Leser, M. E.; Clement, V.; Watzke, H. J., Improved oil solubilization in oil/water food grade microemulsions in the presence of polyols and ethanol. *Journal of Agricultural and Food Chemistry* (2001): *49*, 2552–2562.
182. Spernath, A.; Yaghmur, A.; Aserin, A.; Hoffman, R. E.; Garti, N., Food-grade microemulsions based on nonionic emulsifiers: Media to enhance lycopene solubilization. *Journal of Agricultural and Food Chemistry* (2002): *50*, 6917–6922.
183. Wang, Y.; Donsì, F.; Huang, Q., Self-assembling amphiphilic system for curcumin delivery. *in preparation* (2011).
184. Mao, Y. Y.; Dubot, M.; Xiao, H.; McClements, D. J., Interfacial engineering using mixed protein systems: Emulsion-based delivery systems for encapsulation and stabilization of beta-carotene. *Journal of Agricultural and Food Chemistry* (2013): *61*, 5163–5169.
185. Gharsallaoui, A.; Roudaut, G.; Beney, L.; Chambin, O.; Voilley, A.; Saurel, R., Properties of spray-dried food flavours microencapsulated with two-layered membranes: Roles of interfacial interactions and water. *Food Chemistry* (2012): *132*, 1713–1720.

186. Yang, X. Q.; Tian, H. X.; Ho, C. T.; Huang, Q. R., Stability of citral in emulsions coated with cationic biopolymer layers. *Journal of Agricultural and Food Chemistry* (2012): *60*, 402–409.
187. Hou, Z. Q.; Gao, Y. X.; Yuan, F.; Liu, Y. W.; Li, C. L.; Xu, D. X., Investigation into the physicochemical stability and rheological properties of beta-carotene emulsion stabilized by soybean soluble polysaccharides and chitosan. *Journal of Agricultural and Food Chemistry* (2010): *58*, 8604–8611.
188. Shaw, L. A.; McClements, D. J.; Decker, E. A., Spray-dried multilayered emulsions as a delivery method for omega-3 fatty acids into food systems. *Journal of Agricultural and Food Chemistry* (2007): *55*, 3112–3119.
189. Parada, J.; Aguilera, J. M., Food microstructure affects the bioavailability of several nutrients. *Journal of Food Science* (2007): *72*, R21–R32.
190. de Vos, P.; Faas, M. M.; Spasojevic, M.; Sikkema, J., Encapsulation for preservation of functionality and targeted delivery of bioactive food components. *International Dairy Journal* (2010): *20*, 292–302.
191. Ubbink, J.; Kruger, J., Physical approaches for the delivery of active ingredients in foods. *Trends in Food Science and Technology* (2006): *17*, 244–254.
192. Champagne, C. P.; Fustier, P., Microencapsulation for the improved delivery of bioactive compounds into foods. *Current Opinion in Biotechnology* (2007): *18*, 184–190.
193. Fuchs, M.; Turchiuli, C.; Bohin, M.; Cuvelier, M. E.; Ordonnaud, C.; Peyrat-Maillard, M. N.; Dumoulin, E., Encapsulation of oil in powder using spray drying and fluidised bed agglomeration. *Journal of Food Engineering* (2006): *75*, 27–35.
194. Gharsallaoui, A.; Roudaut, G.; Chambin, O.; Voilley, A.; Saurel, R., Applications of spray-drying in microencapsulation of food ingredients: An overview. *Food Research International* (2007): *40*, 1107–1121.
195. Okuro, P. K.; de Matos Junior, F. E.; Favaro-Trindade, C. S., Technological challenges for spray chilling encapsulation of functional food ingredients. *Food Technology and Biotechnology* (2013): *51*, 171–182.
196. Nedovic, V.; Kalusevic, A.; Manojlovic, V.; Levic, S.; Bugarski, B., An overview of encapsulation technologies for food applications. *11th International Congress on Engineering and Food (Icef11)* (2011): *1*, 1806–1815.
197. Takei, N.; Unosawa, K.; Matsumoto, S., Effect of the spray-drying process on the properties of coated films in fluidized bed granular coaters. *Advanced Powder Technology* (2002): *13*, 333–342.
198. Lopez-Rubio, A.; Gavara, R.; Lagaron, J. A., Bioactive packaging: Turning foods into healthier foods through biomaterials. *Trends in Food Science and Technology* (2006): *17*, 567–575.
199. Barbosa-Canovas, G. V.; Juliano, P., Physical and chemical properties of food powders. *Encapsulated and Powdered Foods* (2005): *146*, 39–71.
200. Vaz, C. M.; van Doeveren, P.; Reis, R. L.; Cunha, A. M., Soy matrix drug delivery systems obtained by melt-processing techniques. *Biomacromolecules* (2003): *4*, 1520–1529.
201. Henrist, D.; Van Bortel, L.; Lefebvre, R. A.; Remon, J. P., in vitro and in vivo evaluation of starch-based hot stage extruded double matrix systems. *Journal of Controlled Release* (2001): *75*, 391–400.
202. Breitenbach, J., Melt extrusion: From process to drug delivery technology. *European Journal of Pharmaceutics and Biopharmaceutics* (2002): *54*, 107–117.
203. Merisko-Liversidge, E.; Liversidge, G. G.; Cooper, E. R., Nanosizing: A formulation approach for poorly-water-soluble compounds. *European Journal of Pharmaceutical Sciences* (2003): *18*, 113–120.
204. Singh, S. K.; Srinivasan, K. K.; Gowthamarajan, K.; Singare, D. S.; Prakash, D.; Gaikwad, N. B., Investigation of preparation parameters of nanosuspension by top-down media milling to improve the dissolution of poorly water-soluble glyburide. *European Journal of Pharmaceutics and Biopharmaceutics* (2011): *78*, 441–446.
205. Schultz, S.; Wagner, G.; Urban, K.; Ulrich, J., High-pressure homogenization as a process for emulsion formation. *Chemical Engineering and Technology* (2004): *27*, 361–368.
206. Liedtke, S.; Wissing, S.; Muller, R. H.; Mader, K., Influence of high pressure homogenisation equipment on nanodispersions characteristics. *International Journal of Pharmaceutics* (2000): *196*, 183–185.
207. Stang, M.; Schuchmann, H.; Schubert, H., Emulsification in high-pressure homogenizers. *Engineering in Life Sciences* (2001): *1*, 151–157.
208. Donsi, F.; Annunziata, M.; Vincensi, M.; Ferrari, G., Design of nanoemulsion-based delivery systems of natural antimicrobials: Effect of the emulsifier. *Journal of Biotechnology* (2012): *159*, 342–350.
209. Donsi, F.; Sessa, M.; Ferrari, G., Effect of emulsifier type and disruption chamber geometry on the fabrication of food nanoemulsions by high pressure homogenization. *Industrial and Engineering Chemistry Research* (2012): *51*, 7606–7618.
210. Donsi, F.; Sessa, M.; Mediouni, H.; Mgaidi, A.; Ferrari, G., Encapsulation of bioactive compounds in nanoemulsion-based delivery systems. In *11th International Congress on Engineering and Food*, Saravacos, G.; Taoukis, P.; Krokida, M.; Karathanos, V.; Lazarides, H.; Stoforos, N.; Tzia, C.; Yanniotis, S., (Eds.) 2011; Vol. 1, pp 1666–1671.

211. Donsì, F.; Ferrari, G.; Maresca, P., High-pressure homogenization for food sanitization. In *Global Issues in Food Science and Technology*, Barbosa-Cánovas, G. V.; Mortimer, A.; Lineback, D.; Spiess, W.; Buckle, K., (Eds.) Academic Press, Burlington, MA, 2009.
212. Sanguansri, P.; Augustin, M. A., Nanoscale materials development—A food industry perspective. *Trends in Food Science and Technology* (2006): *17*, 547–556.
213. Gao, Y.; Li, Z. G.; Sun, M.; Guo, C. Y.; Yu, A. H.; Xi, Y. W.; Cui, J.; Lou, H. X.; Zhai, G. X., Preparation and characterization of intravenously injectable curcumin nanosuspension. *Drug Delivery* (2011): *18*, 131–142.
214. Rao, J. J.; McClements, D. J., Stabilization of phase inversion temperature nanoemulsions by surfactant displacement. *Journal of Agricultural and Food Chemistry* (2010): *58*, 7059–7066.
215. Fernandez, P.; Andre, V.; Rieger, J.; Kuhnle, A., Nano-emulsion formation by emulsion phase inversion. *Colloids and Surfaces A-Physicochemical and Engineering Aspects* (2004): *251*, 53–58.
216. Roger, K.; Cabane, B.; Olsson, U., Formation of 10–100 nm size-controlled emulsions through a sub-PIT cycle. *Langmuir* (2010): *26*, 3860–3867.
217. Horn, D.; Rieger, J., Organic nanoparticles in the aqueous phase—Theory, experiment, and use. *Angewandte Chemie-International Edition* (2001): *40*, 4331–4361.
218. Singh, S. S.; Siddhanta, A. K.; Meena, R.; Prasad, K.; Bandyopadhyay, S.; Bohidar, H. B., Intermolecular complexation and phase separation in aqueous solutions of oppositely charged biopolymers. *International Journal of Biological Macromolecules* (2007): *41*, 185–192.
219. Li, B. Z.; Wang, L. J.; Li, D.; Bhandari, B.; Li, S. J.; Lan, Y. B.; Chen, X. D.; Mao, Z. H., Fabrication of starch-based microparticles by an emulsification-crosslinking method. *Journal of Food Engineering* (2009): *92*, 250–254.
220. Yang, W. J.; Trau, D.; Renneberg, R.; Yu, N. T.; Caruso, F., Layer-by-layer construction of novel biofunctional fluorescent microparticles for immunoassay applications. *Journal of Colloid and Interface Science* (2001): *234*, 356–362.
221. Guzey, D.; McClements, D. J., Formation, stability and properties of multilayer emulsions for application in the food industry. *Advances in Colloid and Interface Science* (2006): *128*, 227–248.
222. Burgain, J.; Gaiani, C.; Linder, M.; Scher, J., Encapsulation of probiotic living cells: From laboratory scale to industrial applications. *Journal of Food Engineering* (2011) *104*, 467–483.
223. Picot, A.; Lacroix, C., Encapsulation of bifidobacteria in whey protein-based microcapsules and survival in simulated gastrointestinal conditions and in yoghurt. *International Dairy Journal* (2004): *14*, 505–515.
224. Krasaekoopt, W.; Bhandari, B.; Deeth, H., Evaluation of encapsulation techniques of probiotics for yoghurt. *International Dairy Journal* (2003): *13*, 3–13.
225. Chen, M. J.; Chen, K. N., Application of probiotic encapsulation in dairy products. In *Encapsulation and Controlled Release Technologies in Food System*, Lakkis, J. M., (Ed.) Blackwell Publishing Ltd, Oxford, U.K., 2007, pp. 83–107.
226. Chavarri, M.; Maranon, I.; Ares, R.; Ibanez, F. C.; Marzo, F.; del Carmen Villaran, M., Microencapsulation of a probiotic and prebiotic in alginate-chitosan capsules improves survival in simulated gastro-intestinal conditions. *International Journal of Food Microbiology* (2010): *142*, 185–189.
227. Mortazavian, A. M.; Aziz, A.; Ehsani, M. R.; Razavi, S. H.; Mousavi, S. M.; Sohrabvandi, S.; Reinheimer, J. A., Survival of encapsulated probiotic bacteria in Iranian yogurt drink (Doogh) after the product exposure to simulated gastrointestinal conditions. *Milchwissenschaft-Milk Science International* (2008): *63*, 427–429.
228. Semyonov, D.; Ramon, O.; Kaplun, Z.; Levin-Brener, L.; Gurevich, N.; Shimoni, E., Microencapsulation of Lactobacillus paracasei by spray freeze drying. *Food Research International* (2010): *43*, 193–202.
229. Wang, Z. L.; Finlay, W. H.; Peppler, M. S.; Sweeney, L. G., Powder formation by atmospheric spray-freeze-drying. *Powder Technology* (2006): *170*, 45–52.
230. Zuidam, N. J.; Shimoni, E., Overview of Microencapsulates for use in food products or processes and methods to make them. *Encapsulation Technologies for Active Food Ingredients and Food Processing* (2010): 3–29.
231. Heidebach, T.; Foerst, P.; Kulozik, U., Influence of casein-based microencapsulation on freeze-drying and storage of probiotic cells. *Journal of Food Engineering* (2010): *98*, 309–316.
232. Gbassi, G. K.; Vandamme, T.; Ennahar, S.; Marchioni, E., Microencapsulation of Lactobacillus plantarum spp in an alginate matrix coated with whey proteins. *International Journal of Food Microbiology* (2009): *129*, 103–105.
233. Mokarram, R. R.; Mortazavi, S. A.; Najafi, M. B. H.; Shahidi, F., The influence of multi stage alginate coating on survivability of potential probiotic bacteria in simulated gastric and intestinal juice. *Food Research International* (2009): *42*, 1040–1045.

234. Sandoval-Castilla, O.; Lobato-Calleros, C.; Garcia-Galindo, H. S.; Alvarez-Ramirez, J.; Vernon-Carter, E. J., Textural properties of alginate-pectin beads and survivability of entrapped Lb. casei in simulated gastrointestinal conditions and in yoghurt. *Food Research International* (2010): *43*, 111–117.
235. Sabikhi, L.; Babu, R.; Thompkinson, D. K.; Kapila, S., Resistance of microencapsulated Lactobacillus acidophilus LA1 to processing treatments and simulated gut conditions. *Food and Bioprocess Technology* (2010): *3*, 586–593.
236. Anekella, K.; Orsat, V., Optimization of microencapsulation of probiotics in raspberry juice by spray drying. *LWT-Food Science and Technology* (2013): *50*, 17–24.
237. Zou, Q.; Liu, X.; Zhao, J.; Tian, F.; Zhang, H.-p.; Zhang, H.; Chen, W., Microencapsulation of Bifidobacterium bifidum F-35 in whey protein-based microcapsules by transglutaminase-induced gelation. *Journal of Food Science* (2012): *77*, M270–M277.
238. Sunny-Roberts, E. O.; Knorr, D., Cellular injuries on spray-dried Lactobacillus rhamnosus GG and its stability during food storage. *Nutrition and Food Science* (2011): *41*, 191–200.

38 Encapsulation of Flavors, Nutraceuticals, and Antibacterials

Stéphane Desobry and Frédéric Debeaufort

CONTENTS

38.1 Introduction 802
38.2 Capsules Matrices 803
    38.2.1 Carbohydrates 804
    38.2.2 Proteins 804
38.3 Encapsulation Methods 804
    38.3.1 Spray-Drying 804
    38.3.2 Freeze-Drying 805
    38.3.3 Spray-Cooling and Spray-Chilling 805
    38.3.4 Extrusion 805
    38.3.5 Coacervation 805
38.4 Controlled Release 806
    38.4.1 Release Controlled by Diffusion 806
    38.4.2 Release Controlled by Matrix Degradation 806
    38.4.3 Release Controlled by Swelling 806
    38.4.4 Release Controlled by Melting 806
38.5 Edible Films for Volatile Molecules (Flavors, Essential Oils) 806
    38.5.1 Flavor Compounds and Matrices Involved in Edible Films and Coatings 807
    38.5.2 Retention and Release of Volatile Compounds in Edible Films and Coatings 809
    38.5.3 Barrier Performances of Edible Films and Coatings to Aroma Compounds 814
    38.5.4 Protection of Flavor Compounds against Chemical Degradation by Edible Films and Coatings 815
    38.5.5 Volatile Compound Effect on Structural and Physical–Chemical Properties of Edible Films and Coatings 818
        38.5.5.1 Impact on Film Appearance and Transparency 818
        38.5.5.2 Changes in Film Microstructure Induced by Volatile Compound Encapsulation 818
        38.5.5.3 Influence of Volatile Compound Encapsulation on Film Mechanical Properties 820
        38.5.5.4 Impact of Flavor Compound Encapsulation on the Barrier Properties of Film 821
38.6 Edible Films and Coatings for Nonvolatile Molecule Encapsulation (Peptides, Polyunsaturated Fatty Acids, Antioxidants, and Antibacterials) 823
    38.6.1 Nutraceuticals 823
    38.6.2 Antibacterials 824

38.6.3 Edible Materials for Nutraceuticals or Antibacterial Molecule Encapsulation........ 825
    38.6.3.1 HPMC 825
    38.6.3.2 Polyacetic Acid (PLA) 825
38.6.4 Plasticizers 826
38.6.5 Molecular Diffusion from Film and Coating to Food Surface 826
38.6.6 Chemical Structure of Migrant and Matrix State 827
38.7 Conclusion 827
References 828

38.1 INTRODUCTION

Edible coatings provide a physical barrier against mass transport from the environment to food and from food to the environment, as shown in the previous chapters, and these barrier properties are important for food passive protection. Consumers ask, nowadays, for better food safety and for higher nutritional and flavor properties. In recent years, active packaging has been developed to extend food shelf life by increasing coatings' positive effect. For example, more activity can be provided to edible coatings by adding active compounds, such as flavors, antibacterials, or nutraceuticals. Active compounds can be incorporated directly in the edible polymer matrix or can be encapsulated to better protect their activity and properties. The Eastern countries are the most advanced in this particular domain, North America follows, and Europe is now developing more and more active material solutions.

Active compound stability in different foods has been of increasing interest due to its relationship with food quality and acceptability. Manufacturing and storage processes, packaging materials, and ingredients in foods often cause modifications in overall flavor and nutritional value by reducing an active compound's activity or intensity or by producing newly formed components. Many factors such as physicochemical properties, concentration, and interaction of active molecules with food components affect the resulting quality. As a result, it is beneficial to encapsulate active ingredients prior to use.

Encapsulation is the technique by which one material or mixture of materials is coated with or entrapped within another material or system. The coated material is called the active or core material, and the coating material is called the shell, wall material, carrier, matrix, or encapsulant. The development of microencapsulation products started in the 1950s in the research of pressure-sensitive coatings for the manufacture of carbonless copying paper (Green and Scheicher, 1955). Encapsulation technology is now well developed and accepted within the pharmaceutical, chemical, cosmetic, foods, and printing industries. In food products, fats and oils, aroma compounds and oleoresins, vitamins, minerals, colorants, and enzymes have been encapsulated, while in films of coatings oils, aroma compounds, antimicrobials, and enzymes have been encapsulated.

Encapsulation processes are receiving more and more interest in food application, and recently, evolution from micro- to nanoencapsulation has been observed (Shimoni, 2009). The main interest in reducing the size of the capsules was to limit the impact of capsule addition on food physical properties. Laboratories and industrial development teams have shown the high efficiency of this size reduction (Imran et al., 2010a,b). This evolution is nevertheless limited by suspicion of possible toxicity of some nanoparticles for the human body after inhalation, skin permeation, and ingestion (Bouwmeester et al., 2009). Nanoparticle absorption through cell membranes may oxidize cellular compounds or develop inflammation. This risk exists in both cases of free nanocapsules or nanocapsules incorporated in edible films or coatings. A lot of studies are required now to better understand the real effect of nanocapsules incorporated in foods. In particular, the effect of solid or "soft" nanoparticles should be compared more precisely, because some studies have reported a negligible effect of soft and organic compounds easily degradable by the cell compared to solid mineral particles.

These studies will make possible the development of clear legislation in the countries that use nanoparticles. Waiting for that, the industry is not very confident in this particular technology, and only a few applications of nanoparticle incorporation in an edible matrix exist.

In this chapter, after describing the encapsulation matrix and techniques, two parts will be presented to cover the main aspects of encapsulation in edible films and coating. The first concerns volatile compound encapsulation, and the second concerns nonvolatile molecules such as nutraceuticals and antimicrobials.

38.2 CAPSULES MATRICES

Encapsulation of biomolecules can be achieved using two main methods. The first consists in making capsules in which the compound is included as a core or entrapped in a polymeric matrix (Figure 38.1a and b).

These encapsulation methods are used for large spectra applications. The protective efficiency of the two systems is highly different, with much more efficient barrier action in the core system, compared to easier release, and lower cost in the matrix system.

The second method consists of developing films or coatings in which the biomolecules are directly included and entrapped just as a matrix but on a larger scale (Figure 38.1c). In this matrix film system, the film or coating contains aroma, nutraceuticals, and active compounds. The release is completely controlled by the molecular diffusivity.

To increase the potentiality of the films, more and more applications combine the two previous methods (Figure 38.1a through c) to obtain film/capsules systems as presented in Figure 38.1d. In this structured film or coating, the amount of volatile compound can be reduced thanks to the optimized preservation, low release from the capsules, and low transfer through the film. These systems allow negligible oxidation of the active compounds due to reduced oxygen diffusivity.

Matrix material has to be chosen to face the high number of performance requirements placed on microcapsules when a limited number of encapsulating materials and methods exist. Since the late 1980s, they have been prepared from a large range of materials including proteins, carbohydrates, lipids, or gums (Brazel, 1999). Each group of materials has certain advantages and disadvantages, and these materials are combined to produce high-quality systems. Functional characteristics of an effective wall material used to encapsulate active components must satisfy the following conditions:

Be inert regarding the encapsulated molecules.
Allow a complete elimination of solvent used for matrix or capsule formation.
Bring maximum protection of the active ingredient against external factors.
Produce a stable system before solidification of capsules within a film.
Release the active compounds at the time and the place desired.

Characteristics of major wall materials used for encapsulation are reported in the following sections.

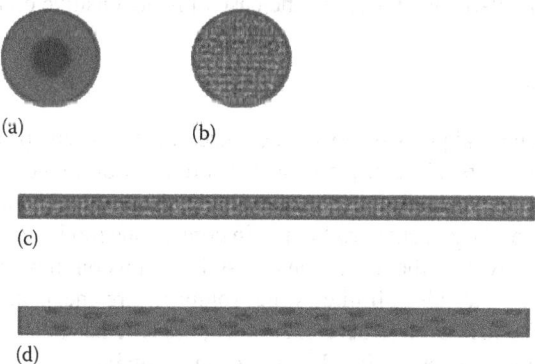

FIGURE 38.1 (a) Core, (b) matrix in nano-/microcapsules systems, (c) matrix film, and (d) film or coating with nano-/microcapsules entrapped.

38.2.1 Carbohydrates

Carbohydrates are used extensively in spray-dried encapsulations of food ingredients as the encapsulating support—that is, the wall material or carrier. The ability of carbohydrates to form a protective matrix, complemented by their diversity and low cost, makes them the first choice for encapsulation. The main limit for their application is their high water sensitivity and fast hydration when in contact with water.

The preferred carbohydrate matrices are starch and starch-based ingredients (modified starches, maltodextrins, β-cyclodextrins). Maltodextrins possess matrix-forming properties important in a wall system. In selecting the wall materials for encapsulation, maltodextrin is a good compromise between cost and effectiveness, as it is bland in flavor, has low viscosity at high solid ratios, and is available in different average molecular weights. Gums and thickeners are generally bland or tasteless, but they can have a pronounced effect on the taste and flavor of foods. In general, hydrocolloids decrease sweetness, with much of the effect being attributed to viscosity and hindered diffusion. Gum arabic is also an excellent encapsulating material. Its solubility, low viscosity, emulsification characteristics, and good retention of active compounds make it very versatile for most encapsulation methods such as spray-drying. Its application is nevertheless limited due to its high cost compared to other carbohydrates, and its availability and cost are subject to high variation.

38.2.2 Proteins

Although food hydrocolloids are widely used as microencapsulants, food proteins (i.e., sodium caseinate, whey protein isolates (WPIs), and soy protein isolates) have good ability to produce efficient encapsulation matrices. WPIs provide a good barrier against oxidation and an effective basis for microencapsulation by spray-drying. In combination with maltodextrins and corn syrup solids, whey proteins have been reported to be one of the most effective encapsulation materials during spray-drying. In such a system, whey proteins served as emulsifying and film-forming agents, while the carbohydrates acted as a matrix-forming material.

Protein-based materials such as polypeptone, soy protein, milk proteins, and gelatin derivatives are able to form stable emulsions with hydrophobic compounds. However, their solubilities in cold water, the potential to react with carbonyls, and their high cost limit potential applications.

38.3 ENCAPSULATION METHODS

Encapsulation is accomplished by a large variety of methods (Madene et al., 2006). The two major industrial processes are spray-drying and extrusion, freeze-drying, however, coacervation, and adsorption techniques are also widely used in the case of heat-sensitive compounds.

38.3.1 Spray-Drying

Spray-drying is the commercial process most widely used in large-scale production of encapsulated molecules. However, the merits of the process have ensured its dominance, including availability of equipment, low process cost, wide choice of carrier solids, good retention of volatiles, good stability of the finished product, and large-scale production in continuous mode. Production of encapsulated powders by spray-drying involves the formation of a stable emulsion in which the wall material acts as a stabilizer. When core materials of limited water solubility are encapsulated by spray-drying, the resulting capsules are of matrix-type structure. As such, the core has been shown to be organized in small wall-material-coated droplets embedded in the wall matrix.

The main disadvantage of spray-drying is that some low-boiling point aromatics or heat-sensible molecules can be lost, and some core material may also be on the surface of the capsule, where it is subject to oxidation. Another problem is that the product is a very fine powder,

Encapsulation of Flavors, Nutraceuticals, and Antibacterials

typically in the range of 10–100 μm in diameter, which needs further processing, to make agglomeras that are more readily soluble for liquid consumption.

38.3.2 FREEZE-DRYING

The freeze-drying technique is one of the most useful processes for drying thermosensitive substances. This technique is based on low-temperature dehydration under vacuum, avoiding water phase transition and oxidation. The dried mixture has a porous structure and must be ground, resulting in heterogeneous particles. This drying technique is less attractive than others, because its cost is up to 50 times higher than spray-drying. Freeze-drying, nevertheless, gives excellent preservation results and is adapted to high-value encapsulated molecules.

38.3.3 SPRAY-COOLING AND SPRAY-CHILLING

The loss of heat-sensitive material during the spray-drying process has led to a number of alternative methods for dehydration of sprayed microcapsules. Spray-cooling and spray-chilling are similar to spray-drying, where the core material is dispersed in a liquefied coating or wall material and atomized at low temperature.

38.3.4 EXTRUSION

Encapsulation via extrusion is slightly more expensive compared to spray-drying. The principal advantage of the extrusion method is the high density of the produced matrix allowing a great stability of encapsulated materials against oxidation. In extrusion, there are two steps. First, the feed is extruded without cooking. Second, the raw ingredients are cooked by the combined action of heat, mechanical shearing, and pressure.

38.3.5 COACERVATION

Coacervation consists in separating from the solution the colloidal particles that agglomerate into a separate liquid phase called coacervate. Coacervation can be simple or complex. Simple coacervation involves only one type of polymer with an addition of strongly hydrophilic agents to the colloidal solution. For complex coacervation, two or more types of polymers are used. Active molecules are entrapped in the matrix during coacervate formation by adjusting precisely the ratio between the matrix polymer and the entrapped molecule (Figure 38.2).

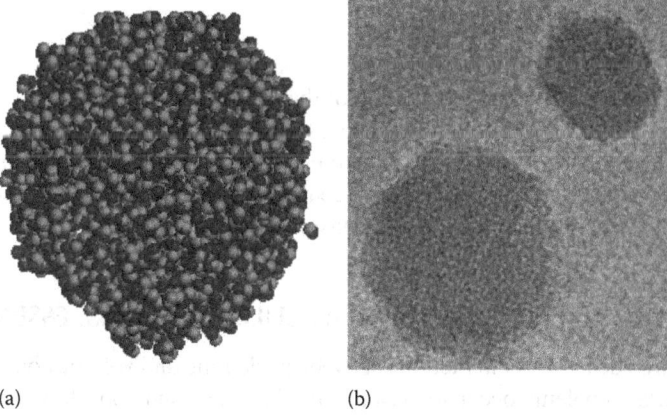

(a)   (b)

FIGURE 38.2 Gum arabic/ß-lactoglobulin coacervates: (a) molecular model and (b) photograpgh of scanning electron microscopy (SEM).

38.4 CONTROLLED RELEASE

Controlled release is a method by which one or more active agents or ingredients are made available at a desired site and time and at a specific rate. For matrix systems encapsulating an active compound, release depends on diffusion process, particle type and geometry, and controlled degradation or dissolution of matrix material.

Compared to molecules added in food, advantages of controlled release are that the active ingredients are released at controlled rates over a prolonged period of time; loss of ingredients during processing, cooking, and even digestive molecular destruction can be strongly reduced; molecule bioavailability can be increased for nanocapsules.

38.4.1 Release Controlled by Diffusion

Diffusion is the main regulation phenomena for controlled release in stable matrix. The diffusion process is well known, thanks to thousands of experimental studies related to mass transfer through films, coatings, and capsules. The macromolecular network density of the matrix and the molecular weight of the encapsulated compounds are the key factors for release controlled by diffusion. The chemical potential difference at each side of the matrix is the major driving force influencing diffusion. The principal steps in the release of a compound from a matrix system are diffusion of the active agent to matrix surface; component partition between matrix and food; and finally, transport away from the matrix surface.

38.4.2 Release Controlled by Matrix Degradation

The release of an active compound from a matrix-type delivery system may be controlled by a combination of diffusion and erosion. Heterogeneous erosion occurs when degradation is confined to a thin layer at the delivery system surface, whereas homogenous erosion is a result of degradation occurring at a uniform rate throughout the polymer matrix.

38.4.3 Release Controlled by Swelling

In swelling-controlled systems, the molecule dispersed in a polymeric matrix is unable to diffuse to any significant extent within the matrix. When the matrix polymer is placed in a thermodynamically compatible medium, the polymer swells owing to absorption of fluid from the medium. In the swollen matrix, the encapsulated molecule is able to diffuse due to larger intermolecular spaces.

38.4.4 Release Controlled by Melting

This mechanism of release involves the melting of the capsule wall to release the active material. It is readily accomplished in the food industry because there are numerous materials of low melting point that are approved for food use (lipids, modified lipids, or waxes). In such applications, coated particles are stored at temperatures well below coating melting point, and then heated above this temperature during preparation, cooking, and consumption.

38.5 EDIBLE FILMS FOR VOLATILE MOLECULES (FLAVORS, ESSENTIAL OILS)

Even if most research deals with interactions between packaging and volatile compounds in the case of beverages, similar problems occur for viscous liquids, gels, and solid foods. One possible way suggested to lower interactions between aroma compounds and plastics is to retain flavor molecules inside the food product by using an additional barrier. This could be an edible film or coating or a thin layer having a high selectivity against aroma transfer, and that can be eaten along with the protected food (Miller and Krochta, 1997; Debeaufort et al., 2002). This entails some kind of macroencapsulation of the product. The main application is using only barrier properties of the coating to retain aroma within the food; however, the film or coating could be used as a carrier or support for flavors at the surface of the product. So, edible films serve many purposes, including permitting the production of a dry free-flowing flavor (most flavors are liquids), protection of the flavoring from interaction with the food, or deleterious reactions such as oxidation, confinement during storage, and, finally, controlled release (Reineccius, 2009). The degree to which the edible film meets these requirements depends upon the process used to form the film around the flavoring and the film composition.

These volatile compounds will be released rapidly when the consumer tastes the product. Several products are already on the market using this technology for flavoring. For instance, there is a roasted peanut with a curry-flavored coating that is instantaneously dissolved in the mouth and gives immediately the perception of the Indian spice. Another example designed for children is a multi-sugar-coated sweet in which each layer of the coating contains different tastes and flavors separated by arabic gum or hydrocolloid layers to prevent migrations of aroma compounds from one layer to another. For this application, diffusivity of volatile compounds should be very low and with a high affinity for the coating, which should be highly soluble in the mouth. As previously outlined, edible films and coatings can deliver and maintain desirable concentrations of color, flavor, spiciness, sweetness, saltiness, and so forth. Several commercial films, especially Japanese pullulan-based films, are available in a variety of colors, with spices and seasonings included (Guilbert and Gontard, 2005). For instance, Laohakunjit and Kerdchoechuen (2007) coated milled rice with sorbitol–rice starch coatings, containing 25% natural pandan leaf extract (*Pandanus amaryllifolius* Roxb.). The rice starch coating containing pandan extract allowed production of jasmine-flavored rice after cooking. Recently, Origami Foods (Pleasantown, California) commercialized vegetable and fruit edible films as alternatives to the seaweed sheets (nori) traditionally used for sushi and other Asian cuisine. They are made from broccoli, tomato, carrot, mango, apple, peach, pear, as well as a variety of other fruit and vegetable products, and they can contain spices, seasonings, colorants, flavors, vitamins, and other beneficial plant-derived compounds (Martin-Belloso et al., 2009).

38.5.1 FLAVOR COMPOUNDS AND MATRICES INVOLVED IN EDIBLE FILMS AND COATINGS

The edible packaging materials are mainly composed of a film-forming substance that provides cohesiveness to the matrix (continuous network) or a barrier substance that lowers impermeability. These are usually polysaccharides, proteins, or lipids used alone or as mixtures. Only few substances have simultaneously good film-forming and barrier properties as, for example, wheat gluten–based films, which have satisfactory mechanical resistance and very low oxygen permeability. Moreover, the permeability of D-limonene (one of the main compounds of citrus flavor) in whey protein–based films is lower than in ethylene vinyl alcohol (EVOH) or polyvinylidene chloride (PVDC) films (Fayoux et al., 1997a,b; Miller et al., 1998). Because edible packaging containing volatile compounds is consumed along with the foods, their edibility and safety are essential. For this reason, flavor compounds can be used to obtain active edible packaging because they can act as antimicrobials, antioxidants, and flavoring agents. In particular, essential oils can be added to edible films and coatings to modify flavor, aroma, and odor, as well as to introduce antimicrobial properties (Sánchez-González et al., 2010c).

Very little published data exist on the incorporation of plant essential oils into edible films and coatings. Essential oils are regarded as alternatives to chemical preservatives, and their use in foods meets the demands of consumers for minimally processed natural products, as reviewed by Burt (2004). Vanillin has been used recently as a bacteriostatic rather than a bactericidal agent in fresh-cut apples (Rupasinghe et al., 2006). Essential oils have also been evaluated for their ability to protect food against pathogenic bacteria in contaminated apple juice (Friedman et al., 2004; Raybaudi-Massilia et al., 2006) and other foods, and they are used as flavoring agents in baked goods, sweets, ice cream, beverages, and chewing gum (Fenaroli, 1995; Burt, 2004). Sánchez-González et al. (2009, 2010a,b) introduced tea tree and bergamot essential oils in chitosan or hydroxypropylmethylcellulose edible films at a range of 0% to 3% (w/w) in the film-forming suspension for antimicrobial properties. They displayed the antimicrobial efficiency at the higher concentration of bergamot on *Penicillium italicum* and at the lowest rate of tea tree oil on *Listeria monocytogenes*. However, no information was given on the concentration in the film after drying or on the sensory impact.

McHugh et al. (1996) developed the first edible films made from fruit purees, which were shown to be a promising tool for improving quality and extending shelf life of minimally processed fruit. Rojas-Grau et al. (2006) recently investigated the effect of plant essential oils on antimicrobial and physical properties of apple puree edible films. Alginate-apple puree films, containing plant essential oils, were further explored as edible coatings by Rojas-Grau et al. (2007) with the aim of studying the effect of lemongrass, oregano oil, and vanillin on native psychrophilic aerobic bacteria, yeasts, molds, and inoculated *Listeria innocua* in fresh-cut "Fuji" apples. Coatings with essential oils seemed to effectively inhibit the growth of *L. innocua* inoculated on apple pieces as well as psychrophilic aerobic bacteria, yeasts, and molds. In some cases, essential oils like thymol and carvacrol were added to a bio-based coating as antimicrobial agents and not as flavoring compounds (Ben Arfa et al., 2007; Del Nobile et al., 2008). Ponce et al. (2008) used oleoresins containing both volatile and nonvolatile compounds extracted from oregano, rosemary, olive, capsicum, garlic, onion, and cranberries in edible films based on sodium caseinate and carboxymethylcellulose and chitosan. These authors displayed that the use of chitosan enriched with rosemary and olive did not introduce deleterious effects on the sensorial acceptability of squash. Chitosan enriched with rosemary and olive improved the antioxidant protection of the minimally processed squash, offering a great advantage in the prevention of browning reactions that typically result in quality loss in fruits and vegetables. These coatings provide both antioxidant and antimicrobial properties of coatings at 1% oleoresin content without too much sensory disturbance.

Gelatin- and chitosan-based edible films incorporated with clove essential oil were tested for antimicrobial activity against six selected microorganisms: *Pseudomonas fluorescens*, *Shewanella putrefaciens*, *Photobacterium phosphoreum*, *Listeria innocua*, *Escherichia coli*, and *Lactobacillus acidophilus*. The clove-containing film inhibited all these microorganisms irrespective of the film matrix (Gómez-Estaca et al., 2010). The effect on the microorganisms during this period was in accordance with biochemical indexes of quality, indicating the viability of these films for fish preservation.

These authors also tested on 18 bacterial strains other essential oils: fennel (*Foeniculum vulgare* Miller), cypress (*Cupressus sempervirens* L.), lavender (*Lavandula angustifolia*), thyme (*Thymus vulgaris* L.), herb-of-the-cross (*Verbena officinalis* L.), pine (*Pinus sylvestris*), and rosemary (*Rosmarinus officinalis*). Antioxidant properties as well as light barrier properties of gelatin-based edible films containing oregano or rosemary aqueous extracts have been assessed by Gómez-Estaca et al. (2009). The essential oil polyphenols–protein interaction was found to be more extensive when tuna-skin gelatin was employed. However, this did not clearly affect the antioxidant properties of the films, although it could affect diffusion of phenolic compounds in the essential oil from film to food. The light barrier properties were improved by the addition of oregano or rosemary extracts, irrespective of the type of gelatin employed. The shelf life of cold-smoked sardine (*Sardina pilchardus*) was improved also by gelatin-based films using, singly or in combination, high pressure (300 MPa/20°C/15 min), and films enriched by adding an extract of oregano (*Origanum vulgare*) or rosemary, (*Rosmarinus officinalis*) or by adding chitosan (Gomez-Estaca et al., 2007). Gelatin seems to be a good way for encapsulating both antimicrobial or antioxidant volatile essential oils. Seaweeds extracts such as alginates and carrageenans have been extensively studied by Hambleton et al. (2008, 2009a,b, 2010, 2012) and Fabra et al. (2008, 2009). These authors displayed that both carrageenans and alginates are able to be used as films or coatings having encapsulated volatile compounds such as aroma or essential oils. The adding of lipids such as acetylated monoglycerides, beeswax, and oleic acid used alone or as mixture allows reducing moisture transfer but affects their ability to retain and release volatile compounds. Lipid could have positive or negative influences, depending on the polarity of the volatile compounds. Zein-based monolayer and multilayer films were loaded with spelt bran and thymol (35% w/w) to obtain edible polymeric materials. Various composite systems were developed to control thymol release (Mastromatteo et al., 2009). Madene et al. (2006) described the process for encapsulation of sensitive volatile compounds. Encapsulation can be employed to retain aroma in a food product during storage, protect the flavor from undesirable interactions with food, minimize flavor/flavor interactions, guard against light-induced reactions and/or oxidation, to increase flavors shelf-life, and to allow a controlled release. Incorporation of small amounts of flavors into foods can greatly influence finished product quality, cost, and consumer satisfaction. The food industry is continuously developing ingredients, processing methods, and packaging materials to improve flavor preservation and delivery. The stability of the matrices is an important condition to preserve the properties of the flavor materials. Many factors such as the kind of wall material, the ratio of the core material to wall material, the encapsulation method, and storage conditions affect the antioxidative stability of the encapsulated flavor. According to the technological process used, the matrices of encapsulation will present various shapes (films, spheres, particles irregular), various structures (porous or compact), and various physical structures (vitreous or crystalline dehydrated solid, rubbery matrix) that will influence the diffusion of flavors properties or external substances (oxygen, solvent) and the stability of the product during its storage.

38.5.2 Retention and Release of Volatile Compounds in Edible Films and Coatings

Food matrix ingredients, among them food proteins, have little flavor of their own but are known to bind and trap aroma compounds. In function, the nature and strength of the binding, and release of aroma compounds in the gas phase will be more or less decreased. The mechanism of flavor binding was dependent on the role of the protein structure as well as the type of flavor compound (aldehyde, alcohol, ketone, and ester) involved in the binding process (Heng et al., 2004). The most extensively studied proteins are milk proteins, known for their emulsifying properties. For instance, the affinity for β-lactoglobulin increases with hydrophobic chain length or overall hydrophobicity of flavor compounds, except for terpenes, main constituents of essential oils. However, it was not possible to find a simple explanation for the binding strength for aroma compounds from different chemical classes. Among the other food proteins, β-lactalbumin, caseins, bovine serum albumin, and soy proteins are studied to a lesser extent for their binding properties toward flavor compounds. β-Lactalbumin was found to bind ketones and aldehydes but with a poor flavor binding capacity compared to other whey proteins. That explains why whey protein films are often desired for flavor or essential oil encapsulation. Moreover, whey proteins have both film-forming properties and emulsifying capacity.

Carbohydrates can have a measurable influence on the release and perception of flavors. Carbohydrates change the volatility of compounds relative to water, but the effect depends on the interaction between the particular volatile molecule and the particular carbohydrate. As a general rule, carbohydrates, especially polysaccharides, decrease the volatility of compounds relative to water by a small to moderate amount, as a result of molecular interactions. However, some carbohydrates, especially the monosaccharides and disaccharides, exhibit a salting-out effect, causing an increase in volatility relative to water (Godshall, 1997).

Lipids are rather homogeneous, hydrophobic, and nonpolar materials, existing in aqueous mediums in the form of distinct regions. In these systems, flavor compounds are distributed between the lipid and the aqueous phases, following the physical laws of partition (Solms et al., 1973). Lipids are carriers and release modulators of aroma, but they are also flavor precursors. Lipid oxidation as well as lipolysis generates numerous short-chain compounds to which we are extremely sensitive because of their low levels of olfactory detection. The global effect of lipids on aroma compound release is to decrease their volatility. Generally, aroma compounds are hydrophobic and show greater affinity for the lipidic phase than for the aqueous or vapor phase. The influence of the physicochemical properties of both the flavor compound and the fat content has been demonstrated by the determination of air–oil partition coefficients. Solid fat content and crystallinity of fat tend to improve the controlled release by diffusion fall. However, temperature induces the fat melting or solid fat content that is unfavorable to flavor retention.

A food product with active molecules on the surface and a food product coated with an edible film with active volatile molecules are compared in Figure 38.3. In food products, surface contamination and oxidation are the most probable and need to be prevented (Han, 2002).

In both systems, with and without edible film, the food layers that do not contain active agents (flavoring, antimicrobial, and antioxidant) have initially very large volume compared to the volume of thin films, coatings, or food surfaces where active volatile molecule was dispersed. Because of almost an infinite volume of food layer without active molecules, migration of active molecules from surface into food will be favored. If the agent is sprayed onto the surface of food, the initial surface concentration will be very high and start to decrease due to dissolution and diffusion of the agent toward the center of the food (Marcuzzo et al., 2011). Therefore, solubility (or partition coefficient) and diffusion coefficient (diffusivity) of the agent in a food are very important characteristics to maintain the surface concentration needed to hinder microorganism growth, oxidation or flavor loss oxidation, or flavor loss during the expected shelf life. Otherwise, active molecule concentration in the surface layer will be reduced or depleted. The release rate must be controlled to prevent the early depletion of active molecules due to fast migration. Application of edible coatings or films with active molecules could allow a very low diffusion rate of active molecules into the food to maintain its efficiency on the surface. Moreover, the effectiveness of the surface function could allow a lower active molecule concentration in the food. If the active molecule dispersed in the edible film or coating is an aroma compound, the factors that influence flavor release have to be considered. The release rate of the volatile agent from the packaging system is highly dependent on the volatility, which relates to the chemical interaction between the volatile agent and packaging materials. The absorption rate of headspace volatiles into a food surface is related to the composition of the food, as the ingredients undergo chemical interactions with the gaseous agents. Because most volatile agents are generally lipophilic, the lipid content of the food is an important factor of headspace concentration. The use of volatile antimicrobial agents has many advantages. This system can be used effectively for highly porous foods, powdered, shredded, irregularly shaped, and particulate foods, such as ground beef, shredded cheese, small fruits, and mixed vegetables (Han, 2002).

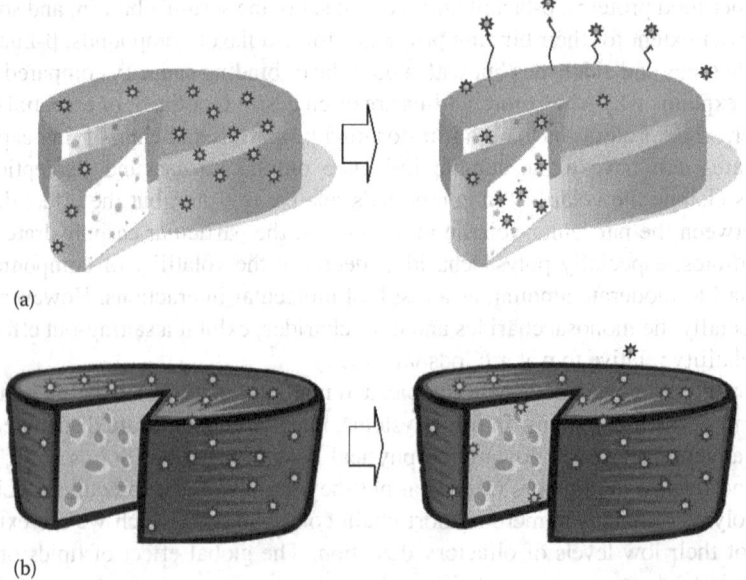

FIGURE 38.3 Retention and controlled release of aroma compounds added at the food surface or entrapped in edible coatings/films. (a) Deposition of volatile-active compound on food surface and (b) food coated with an edible film encapsulating the volatile-active compounds.

The ability of edible films to entrap or to retain flavor compounds during film storage and above all during film processing is of importance. If flavors are lost during drying of hydrocolloid-based film-forming liquid (solution, emulsion, or suspension), the resultant aroma will become unbalanced. The more volatile or the less interactive volatile compounds will be preferentially lost during the film/coating process, inducing a loss of the fresh aroma notes. Using emulsions or suspension with hydrocolloids interacting strongly with the aroma compounds tends to reduce the aroma loss during drying. When edible films are enriched with essential oils, the drying temperatures usually employed to form the edible coating are high enough to volatilize a high percentage of the aromatic components. Sánchez-González (2010a) showed that D-limonene (main component of the bergamot essential oil) loss during drying ranged from 39% to 99% when added from 0.5% to 3% (w/w) in chitosan films. Moreover, Monedero et al. (2010) found that losses during drying are tremendously greater than during film storage. Soy protein–beeswax–oleic acid emulsified films, dried at moderate temperature (20°C, 45% relative humidity, RH), lost up to 98% of encapsulated n-hexanal, but this could be reduced by using a higher amount of beeswax. Beeswax particles decreased by half the aroma diffusivity in the protein-based matrix. However, the greater the loss during drying, the faster is the release to the vapor phase. Most of the remaining n-hexanal in the film after drying was lost after 35 days storage in an open ventilated chamber (Figure 38.4). Similar trends were observed by

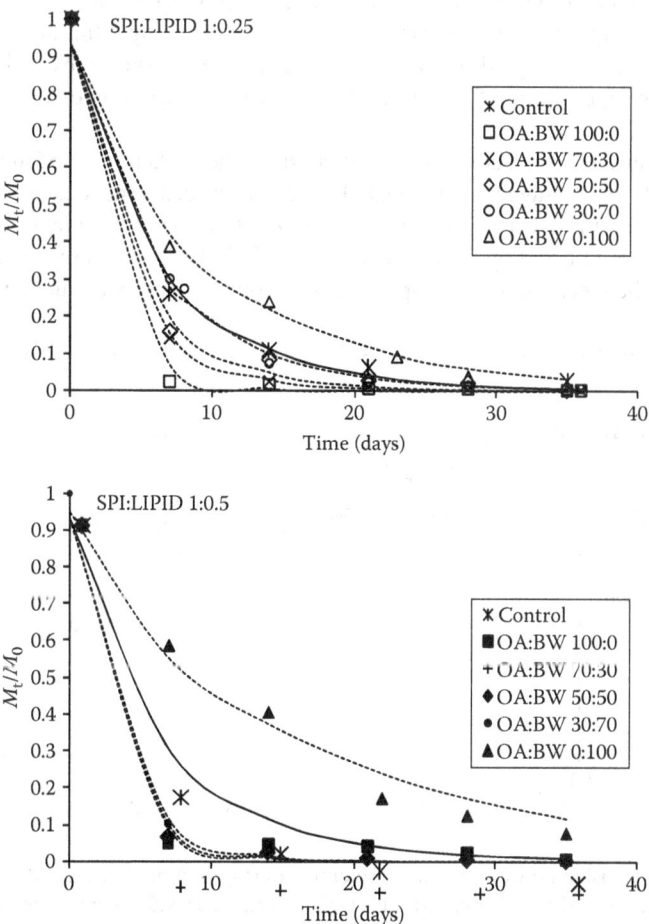

FIGURE 38.4 n-Hexanal release kinetic for the control film and films containing soy protein isolate (SPI) combined with oleic acid (OA) and beeswax (BW) in SPI:lipid ratio 1:0.25 and 1:0.5 (experimental data: symbols, and fitted model: lines). (From Monedero, M. et al., *J. Food Eng.*, 100, 128, 2010. With permission.)

Tunc and Duman (2011) when the montmorillonite (nanoclays) content in chitosan films increased. Solid particles such as beeswax or montmorillonite probably reduce the diffusivity of the volatile compound because of increasing "tortuosity" and then delay of the aroma loss by the films. A significant increase of thymol release rate with the increase of the bran concentration was observed by Mastromatteo et al. (2009). So, in this case, incorporation of nanoparticles (spelt bran) had the opposite effect and was unfavorable for volatile compound retention.

The advantage of substituting essential oils or aroma compounds for their corresponding food-grade oleoresins could lie in the introduction of other nonvolatile components positively affecting food quality (Ponce et al., 2008).

Flavor retention by the film matrix during processing depends as much on the temperature as on the film or coating composition. On one hand, temperature increases the aroma volatility according to Henry's laws (Buttery et al., 1971) and the diffusivity. On the other hand, many film constituents allow reduced volatility and diffusivity. Polyols used as plasticizers in film formulation are very good for aroma support and are able to significantly reduce flavor release. Acacia gum provides the same advantages for flavor retention and could also strengthen film mechanical properties. Recent studies showed that the addition of nanoparticles such as nanoclays is significant to improve the retention efficiency of chitosan-based edible films (Tunc and Duman, 2011). Usually, the mass partition coefficient K_{mass} (ratio between aroma concentration in air and aroma concentration in film) is inversely proportional to the retention capacity during film processing.

As displayed in Figure 38.5, the partition coefficient (during film storage) can be related to carvacrol retention during film processing (casting and drying) by chitosan-based films (more than 200 recipes in a 25°C–80°C drying temperature range) at various humidities (0%–98% RH).

The release of various aroma compounds (ethyl esters, methylketones, and alcohols), from either carrageenan-based, or carrageenan-acetylated monoglycerides emulsion-based, or acetylated-monoglycerides-based films strongly differs (Marcuzzo et al., 2010). In lipid films, the aroma compound release is more affected by factors related to diffusivity, whereas in carrageenan emulsified films, the affinity between volatile compounds and polymer preponderantly influences sorption

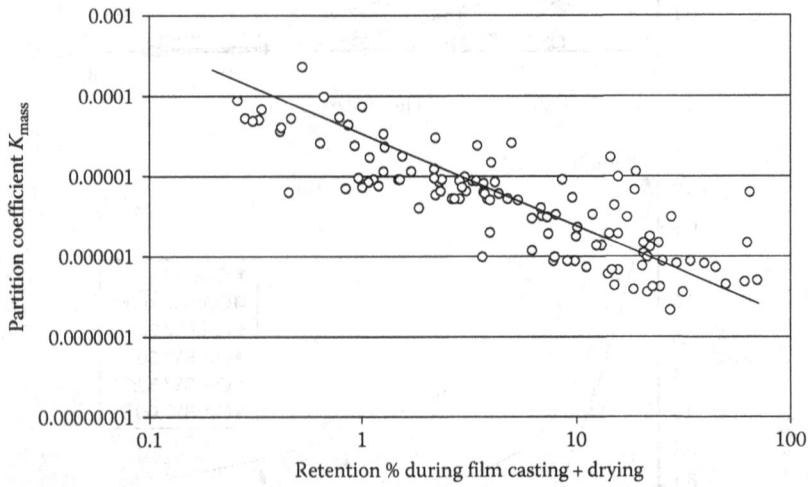

FIGURE 38.5 Relationship between carvacrol retention during film processing and air/film partition coefficient during film storage. (From Kurek, M. and Debeaufort, F., Development of an antimicrobial coating for packaging films: Physico-chemical and microbiological approaches, Doctoral school intermediate PhD report, University of Burgundy, Dijon, France, 2010.)

phenomena and thus the release. Carrageenan films resulted in possible encapsulating matrixes: they display better performances for retention of more polar aroma compounds than pure lipid or emulsified films. Carrageenan films were able to retain volatile compounds during film processing and released them gradually with time.

The surrounding medium of encapsulated aroma such as polysaccharides, proteins, lipids, and salts could play an important role in release in liquid media and then on aroma compound retention by the film matrix. Different behaviors have been observed in the presence of salt or sucrose molecules of aroma compound volatility in food products, observing that some of them presented a "salting out" effect (favored release by volatilization), others an opposite "salting in" effect, and for some others no modification (Lubbers et al., 1998; Van Ruth et al., 2002). Similar effects could be observed in the release of the aroma compounds encapsulated in films. While the "salting out" effect should accelerate the release, the "salting in" could decrease the rate of the release.

The effect of aqueous media (containing 0.9% of NaCl) and temperature (25°C and 37°C) on the release of encapsulated aroma compounds (n-hexanal and D-limonene) in ι-carrageenan-based films with and without lipid have been studied by Fabra et al. (2011). D-limonene was released quickly from water at higher temperatures. However, no effect of temperature was observed on n-hexanal release in water. Only between 40% and 56% of the hexanal was released, depending on film composition, while D-limonene was completely retained in the fat material (Table 38.1). Salt presence in the liquid release media favored significantly and tremendously the aroma retention by carrageenan-based films. Retention was 4 to 15 times higher according to aroma compound hydrophobicity and fat particle presence in the film matrix.

In a similar way, Sánchez-González et al. (2011) studied chitosan films enriched with different concentrations of bergamot oil and the migration of D-limonene (the major oil component) into five liquid simulated foods (water, 10% ethanol, 50% ethanol, 95% ethanol, and isooctane). D-Limonene migration (release) was significant in 95% ethanol, whereas in the other food simulants, its release remained less impressive. Composite films remained intact with isooctane CH-BO, and no release of limonene was observed. Polarity of simulant and migrant seems to be a key factor to explain these results. However, in the reports of both Sánchez-González et al. (2010c) and Fabra et al. (2011), chitosan- or carrageenan-based films immersed in liquid media probably swelled and some of their soluble components (plasticizers) probably migrated into the contacting liquid. Because film integrity was not maintained, the release phenomenon could not be attributed only to the partition (solubility) and diffusion coefficient of the aroma compounds. Other parameters must also be taken into account to better explain and describe the release phenomenon of aroma compounds in contact to liquid or gel foods, such as swelling, partial dissolution of film component, counterdiffusion of other solute from food to film, and osmotic gradients.

TABLE 38.1
Percentage of n-Hexanal and D-Limonene Retained in the Film at the Equilibrium of Release Kinetics in Liquid Media

Type of Film	Aqueous Medium	Retention of n-Hexanal (%)	Retention of D-Limonene (%)
Carrageenans without lipid	Water	9.3 (1.6)	6.2 (1.7)
	0.9% NaCl	57 (3)	79 (2)
Carrageenans with lipid	Water	11 (1)	7 (3)
	0.9% NaCl	44 (2)[e]	100

Note: Mean value (standard deviation).

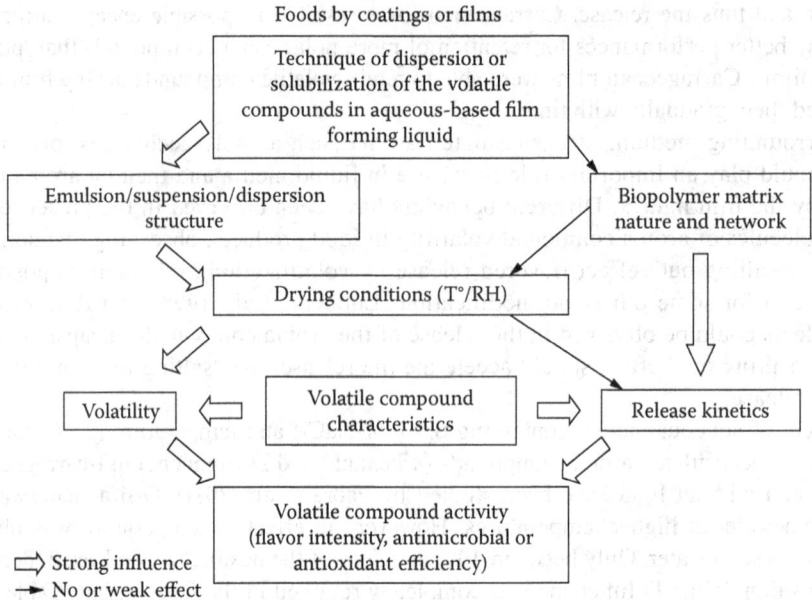

FIGURE 38.6 Overall mechanism of flavor release in air from edible films: main influencing factors.

Volatile compound retention by edible films is complex, and many factors could affect it and interact. Figure 38.6 is a simplified summary view of the main parameters influencing the flavor release from edible films.

38.5.3 BARRIER PERFORMANCES OF EDIBLE FILMS AND COATINGS TO AROMA COMPOUNDS

D-Limonene is the flavor compound most studied. It is considered as the typical aroma probe of citrus juice and soda beverage, and many data are available in the literature. Figure 38.7 gives a comparison between D-limonene permeabilities of edible and plastic films. Polysaccharide-based films are of the same order as heat sealable polymers such as polyolefins, while protein-based films and coatings have barrier properties as good as the best plastics. The permeability of wheat gluten and glycerol or whey protein and glycerol films is about 10^{-17} g m^{-1} s^{-1} Pa^{-1} and from 10^{-14} to 10^{-16} g m^{-1} s^{-1} Pa^{-1} in the case of polysaccharides and other protein-based films (Miller et al., 1998; Quezada-Gallo, 1999; Hambleton et al., 2010).

Although D-limonene is probably the flavor compound most studied in the field of polymer permeability, it is not often used in the flavoring of solid food products. So, Quezada-Gallo et al. (1999a) and Debeaufort and Voilley (1995) studied the mass transfers through edible films of several other molecules, commonly found in cheese, fruits, and dairy products. Table 38.2 displays the permeability of some flavor compounds through edible films. It seems that permeability is always much lower through protein-based than from carbohydrate films, from 100 to 100,000 times lower. These results allow considering their use as a protecting layer against scalping and permeation through classically used plastic films. Edible barriers are able to retain food flavors within the food matrix due to very low permeabilities. However, because of a lack of aroma permeability data, no trend or general rules could be given about the barrier efficiency of edible barriers. Interactions between film constituents and aroma compounds have a great impact on the permeability mechanism. Very few papers focused on the aroma transfers related to interactions, sorption, and diffusion phenomena. It seems that chemical affinities between flavor compounds and film components have more influence on the permeability value than the classical

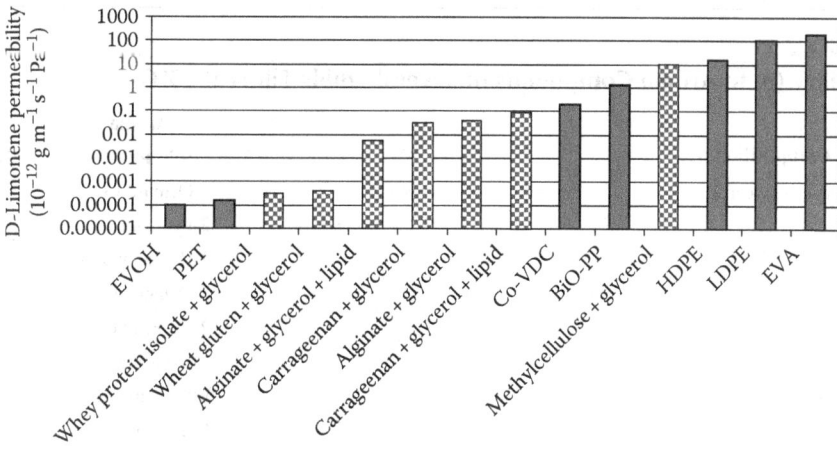

FIGURE 38.7 D-Limonene permeability (10–12 g m^{-1} s^{-1} Pa^{-1}) of some edible and plastic films at 25°C and 0% relative humidity (RH) (thickness ranging from 60 to 130 μm). (From Becker, K. et al., *Parfurmerie und Kosmetik*, 68, 268, 1987; Franz, R., *Packag. Tech. Sci.*, 6, 91, 1993; Debeaufort, F., Étude des Transferts de Matière au Travers de Films d'Emballages: Perméation de l'Eau et de Substances d'Arôme en Relation avec les Propriétés Physico-chimiques des Films Comestibles, PhD Dissertation, ENS.BANA, Université de Bourgogne, Dijon, France, 1994; Debeaufort, F. and Voilley, A., *J. Agr. Food Chem.*, 42, 2871, 1994; Debeaufort, F. and Voilley, A., *Cellulose*, 2, 1, 1995; Kobayashi, M. et al., *J. Food Sci.*, 60, 205, 1995; Paik, J.S. and Writer, M.S. *J. Agr. Food Chem.*, 43, 175, 1995; Miller, K.S. and Krochta, J.M., *Trends Food Sci. Technol.*, 8, 228, 1997; Miller, K.S. et al., *J. Food Sci.*, 63, 244, 1998; Quezada-Gallo, J.A. et al., *J. Agr. Food Chem.*, 47, 108, 1999; Quezada-Gallo, A. et al., *New Developments in the Chemistry of Packaging Materials*, ACS Books, Dallas, TX, 1999; Quezada-Gallo, J.A. et al., Influence de la Structure et de la Composition de Réseaux Macromoléculaires sur les Transferts de Molécules Volatiles (eau et Arômes). Application aux Emballages Comestibles et Plastiques, PhD Dissertation, Université de Dijon, Dijon, France, 1999; Hambleton, A. et al., *Biomacromolecules*, 9(3), 1058, 1999; Hambleton, A. et al., *Food Hydrocolloids*, 23(8), 2116, 2009; Hambleton, A. et al., *J. Food Eng.*, 93, 80, 2009. With permission.) (EVOH, ethylene vinyl alcohol; PET, polyethylene terephtalate; Co-VDC, polyvinylidene chloride copolymer; Bio-PP, bioriented polypropylene; HDPE and LDPE, high- and low-density polyethylene; EVA, ethylene vinyl acetate.)

sorption-diffusion models. Aroma induces significant changes in physical-chemical and structural properties of polysaccharide or protein-based edible films when entrapped or when transferred as described following.

38.5.4 Protection of Flavor Compounds against Chemical Degradation by Edible Films and Coatings

The most common problem occurring during flavor storage is deterioration due to oxidation. Any flavoring containing citrus oils or oils based on aldehydes is susceptible to oxidative reactions and development of off-flavors. Thus, an important function of an edible film is protection of the flavoring from oxygen.

Choosing an encapsulating agent with adequate barrier properties (to oxygen) will lead to more stable dry flavors. Because D-limonene is sensitive to oxidative degradation, several studies were focused on the possibility of encapsulating this aroma compound. Different ingredients were chosen as wall materials, such as starch, maltodextrin, and gum arabic (Wyler and Solms, 1982; Anandaraman and Reineccius, 1986; Bertolini et al., 2001). Generally, lemon oils are added to food in the form of water-in-oil emulsions. Djordevic et al. (2008) studied the possibility of

TABLE 38.2
Permeability (p) to Aroma Compounds of Several Edible Films at 25°C

Edible Film Composition	Aroma Compounds	P (10^{-12} g m^{-1} s^{-1} Pa^{-1})
Methylcellulose + glycerol	1-Octen-3-ol	122
	2-Pentanone	19
	2-Heptanone	39
	2-Octanone	338
	2-Nonanone	420
	Ethyl acetate	128
	Ethyl butyrate	119
	Ethyl isobutyrate	106
	Ethyl hexanoate	668
	D-Limonene	10
ι-Carrageenan + glycerol	Ethyl acetate	0.046
	Ethyl butyrate	1.25
	Ethyl hexanoate	<0.014
	2-Hexanone	0.196
	1-Hexanol	<0.0011
	Cis-3-hexenol	<0.012
	n-Hexanal	0.0189
	D-Limonene	0.0291
ι-Carrageenan + glycerol + acetylated monoglycerides + beeswax (emulsion)	Ethyl acetate	0.022
	Ethyl butyrate	0.33
	Ethyl hexanoate	<0.014
	2-Hexanone	0.111
	1-Hexanol	<0.0011
	Cis-3-hexenol	<0.017
	n-Hexanal	<0.00005
	D-Limonene	0.0803
ι-Carrageenan + glycerol + oleic acid + beeswax (emulsion)	Ethyl acetate	0.83
	Ethyl butyrate	0.75
	Ethyl hexanoate	194
	2-Hexanone	<0.0011
	1-Hexanol	1.12
	Cis-3-hexenol	<0.0012
Sodium alginate + glycerol	n-Hexanal	0.0134
	D-Limonene	0.00034
Sodium alginate + glycerol + acetylated monoglycerides + beeswax (emulsion)	Ethyl butyrate	0.033
	Ethyl hexanoate	134.2
	2-Hexanone	0.094
	n-Hexanal	<0.00005
	D-Limonene	0.000053

(Continued)

TABLE 38.2 (Continued)
Permeability (p) to Aroma Compounds of Several Edible Films at 25°C

Edible Film Composition	Aroma Compounds	P (10^{-12} g m^{-1} s^{-1} Pa^{-1})
Wheat gluten + glycerol	1-Octen-3-ol	4.6
	2-Pentanone	0.12
	2-Heptanone	0.50
	2-Octanone	<0.005
	Ethyl acetate	0.059
	Ethyl butyrate	0.670
	Ethyl isobutyrate	0.04
	Ethyl hexanoate	<0.005
	D-Limonene	0.00004
Sodium caseinate + glycerol	Ethyl acetate	0.006
	Ethyl butyrate	0.19
	Ethyl hexanoate	<0.0004
Sodium caseinate + glycerol + oleic acid (emulsion)	Ethyl acetate	36.1
	Ethyl butyrate	1732.7
	Ethyl hexanoate	1632.1
	2-Hexanone	1.85
	1-Hexanol	3116
	Cis-3-hexenol	404
Sodium caseinate + glycerol + oleic acid + beeswax (emulsion)	Ethyl acetate	0.73
	Ethyl butyrate	134.7
	Ethyl hexanoate	<0.0004
	2-Hexanone	18.7
	1-Hexanol	238.8
	Cis-3-hexenol	639.1
Sodium caseinate + glycerol + beeswax (emulsion)	Ethyl acetate	0.0061
	Ethyl butyrate	0.08
	Ethyl hexanoate	<0.00045
	2-Hexanone	<0.00003
	1-Hexanol	<0.005
	Cis-3-hexenol	<0.0004
Whey protein isolate + glycerol	D-Limonene	0.0003

Sources: Debeaufort, F., Étude des Transferts de Matière au Travers de Films d'Emballages: Perméation de l'Eau et de Substances d'Arôme en Relation avec les Propriétés Physico-chimiques des Films Comestibles, PhD Dissertation, ENS.BANA, Université de Bourgogne, Dijon, France, 1994; Debeaufort, F. and Voilley, A., *J. Agr. Food Chem.*, 42, 2871, 1994; Debeaufort, F. and Voilley, A., *Cellulose*, 2, 1, 1995; Kobayashi, M. et al., *J. Food Sci.*, 60, 205, 1995; Miller, K.S. and Krochta, J.M., *Trends Food Sci. Technol.*, 8, 228, 1997; Miller, K.S. et al., *J. Food Sci.*, 63, 244, 1998; Quezada-Gallo, A. et al., *New Developments in the Chemistry of Packaging Materials*, ACS Books, Dallas, TX, 1999; Quezada-Gallo, J.A. et al., *J. Agr. Food Chem.*, 47, 108, 1999; Quezada-Gallo, J.A., Influence de la Structure et de la Composition de Réseaux Macromoléculaires sur les Transferts de Molécules Volatiles (eau et Arômes). Application aux Emballages Comestibles et Plastiques, PhD Dissertation, Université de Dijon, Dijon, France, 1999; Hambleton, A. et al., *Biomacromolecules*, 9(3), 1058, 2008; Hambleton, A. et al., *Food Hydrocolloids*, 23(8), 2116, 2009; Hambleton, A. et al., *J. Food Eng.*, 93, 80, 2009; Fabra, M.J. et al., *Carbohydr. Polymer*, 76, 325, 2009; Fabra, M.J. et al., *Biomacromolecules*, 9(5), 9, 1406, 2008.

stabilizing oil in water emulsions with whey proteins instead of gum arabic, to inhibit D-limonene degradation. Formation of the limonene oxidation products limonene oxide and carvone were less in the WPI—than gum Arabic (GA)—stabilized emulsions. This was in agreement with Kim and Morr (1996), who found that limonene oxide, carvone, and carveol formation in microencapsulated orange oil was less in emulsions stabilized with WPI and soy protein isolate than emulsions stabilized with GA. Among wall materials that can be applied to preserve aroma compounds, κ-carrageenans and wheat gluten were selected for their useful gas barrier properties. Hambleton et al. (2008) showed that carrageenans-based edible film prevents the oxidation of n-hexanal to hexanoic acid in oxidative conditions. Marcuzzo et al. (2011) also showed that both carrageenan and wheat gluten–based edible films seem to prevent D-limonene oxidation. Wheat gluten–based film protected D-limonene from degradative reactions, and the increase in carvone release was probably due to oxidation in the headspace once D-limonene was released and not within the matrix as was confirmed by Fourier transform infrared (FTIR) analysis of edible film containing the aroma compounds.

The oxygen permeability is then a key factor for an improved efficiency of film and coating matrices to protect aroma compounds encapsulated or initially present in coated food products. Table 38.3 gives the oxygen permeability of the more studied edible films and coatings. Protein-based edible films are the more efficient barriers against oxygen transfer and are more suitable for flavor protection. Protein also interacts to a greater extent with flavor compounds. But the oxygen protection strongly varies with the moisture level. Increasing water activity or relative humidity tends to increase from 10 to 10^5 times the oxygen permeability. For example, oxygen permeability of collagen film rises from $6.6 \cdot 10^{-19}$ to $14.68 \cdot 10^{-15}$ g m^{-1} s^{-1} Pa^{-1} when water activity increased from 0 to 0.93, but this behavior is also observed for water-sensitive plastic films such as EVOH or nylon. Plasticization of food macromolecules by water increases molecular mobility that favors both oxygen diffusivity and aroma release and oxidation.

38.5.5 VOLATILE COMPOUND EFFECT ON STRUCTURAL AND PHYSICAL–CHEMICAL PROPERTIES OF EDIBLE FILMS AND COATINGS

Incorporating volatile active compounds may affect significantly the structural organization of the film-forming substance network and, consequently, its physical-chemical properties.

38.5.5.1 Impact on Film Appearance and Transparency

Incorporation of aroma compounds or essential oils always decreases the transparency and gloss of edible films made of carbohydrate or protein, because of emulsion structure formation during drying. Solubility of most aroma compounds and essential oils is much lower than the amount added to film forming suspensions. All the authors who measured optical properties of films containing essential oils or flavor compounds observed this behavior, regardless of the nature of the matrix (chitosan, HPMC, methylcellulose, soy protein isolate, whey protein isolate, carrageenans, sodium alginate) or the volatile compounds (tea tree oil, bergamot, cloves, ginger, onion, carvacrol, thymol, etc.) (Pranoto et al., 2005a; Fabra et al., 2008, 2009, 2011; Hambleton et al., 2009a,b, 2010; Atarés et al., 2010; Sanchez Gonzalez et al., 2009, 2010a,b; Tunc and Duman, 2011). Changes in appearance could generally be related to the microstructure of the film containing the volatile compound.

38.5.5.2 Changes in Film Microstructure Induced by Volatile Compound Encapsulation

Hambleton et al. (2009a) showed that incorporation of 3% of n-hexanal in alginate matrices induced a more homogeneous structure as observed by an environmental scanning electron microscope (ESEM). To the contrary, n-hexanal provoked a more heterogeneous distribution of the emulsified fat (beeswax + acetylated monoglycerides) in the film cross section. This was attributed to the competition between the emulsifier (glycerol monostearate) and n-hexanal,

TABLE 38.3
Oxygen Permeability of Edible Films and Coatings Compared to Common Plastic Films (10^{-15} g m^{-1} s^{-1} Pa^{-1})

Film	Oxygen Permeability	T (°C)	a_w
Low-density polyethylene	16.050	23	0
Polyesters	0.192	23	0
Ethylene vinyl alcohol	0.003	23	0
Polyvinylidene chloride	0.0066	23	0
Methylcellulose + glycerol	8.352	30	0
Carrageenan + glycerol	0.72	25	0
Carrageenan + glycerol + GBS	0.86	25	0
Alginate + glycerol	9.4	25	0
Alginate + glycerol + GBS	2.7	25	0
Chitosan + glycerol	0.009	25	0
Starch alginate	0.087	23	0
Beeswax	7.68	25	0
Carnauba wax	1.296	25	0
Collagen	0.00066	25	0
Zein + glycerol	0.56	38	0
Wheat gluten + glycerol	0.016	25	0
Soy protein isolate + glycerol	0.1	25	0
Casein + glycerol	0.0115	25	0
Casein + sorbitol	0.0132	25	0
Peanut protein + glycerol	0.034	23	0
Fish myofibrillar protein + glycerol	0.031	25	0
Low-density polyethylene	30.85	23	0.5
Polyesters	0.192	23	1
Ethylene vinyl alcohol	0.096	23	0.95
Polyvinylidene chloride	0.084	23	0.95
Pectin + glycerol	21.44	25	0.96
Starch + glycerol	17.36	25	1
Hydroxypropylmethylcellulose	4.88	25	0.50
Chitosan + glycerol	7.552	25	0.93
Collagen	0.48	25	0.63
Collagen	14.68	25	0.93
Wheat gluten + glycerol	20.64	25	0.95
Whey protein + glycerol	33	23	0.34
Whey protein + glycerol	435	23	0.56
Fish myofibrillar protein + glycerol	27.93	25	0.93

Sources: Gennadios, A., *Protein-Based Edible Films and Coatings*, CRC Press, Boca Raton, FL, 2002; With kind permission from Springer Science+Business Media: *Edible Films and Coatings for Food Applications*, 2009, Embuscado, M.E. and Huber, K.C.

which is amphipolar. The same behavior was also observed for carrageenan-based films and *n*-hexanal encapsulated (Hambleton et al., 2009b). Sánchez-González et al. (2009) observed a more heterogeneous structure with increasing tea tree oil content in hydroxypropyl methylcellulose (HPMC) films. The essential oil was at a concentration higher than the solubility limit of the volatile in the film-forming suspension, which resulted in an emulsion of tea tree oil droplets in the HPMC network. While a continuous structure was observed for the HPMC film, the presence of tea tree oil caused discontinuities associated with the formation of two phases in the matrix: lipid droplets embedded in a continuous polymer network. Lipid droplets, whose number increased with the tea tree oil concentration, were homogenously distributed across the film. This reveals that very little creaming occurred during the film drying, probably due to the highly viscous effect of HPMC.

38.5.5.3 Influence of Volatile Compound Encapsulation on Film Mechanical Properties

The oil droplets or discontinuities usually induce a loss of mechanical properties, such as a decrease of the tensile strength and Young (elastic) modulus as given in Table 38.4.

The addition of the tea tree oil in the 0.5%–3% concentration range caused a significant decrease in the elastic modulus and in tensile strength at break of HPMC films, although with no significant effect on deformation at break (Sánchez-González et al., 2009). As elongation did not increase, the loss of mechanical properties cannot be attributed to a plasticization of HPMC by tea tree oil. This coincides with the results reported by other authors when adding essential oil to a chitosan matrix (Pranoto et al., 2005a,b; Zivanovic et al., 2005) and is in agreement with the effect of the structural discontinuities provoked by the incorporation of the oil on the mechanical behavior. These discontinuities reduced the film's resistance to fracture. Bergamot oil (0.5%) added to chitosan films reduced the tensile strength and elongation two and three times, respectively (Sánchez-González et al., 2010). Cinnamon oil seemed to have some plasticizing effect on soy protein isolate–based film, making them more extensible as the oil content increased (Atarés et al., 2010). On the contrary, cinnamon oil may have caused some degree of rearrangement in the protein network, thus strengthening it and increasing the film resistance to elongation. This effect was not observed when ginger oil was added. Films with ginger oil were less resistant and

TABLE 38.4
Influence of Volatile Compound Incorporation on Mechanical Properties of Edible Films

Film Matrix	Volatile Compounds	Elongation (%)	Tensile Strength (MPa)	Elastic Modulus (MPa)
Soy protein isolate	/	4	11	412
	Cinnamon oil 1%	7.5	14.1	354
	Ginger oil 1%	3	8	340
Chitosan	/	22	113	2182
	Bergamot oil 3%	1.7	22	682
	Tea tree oil 2%	8	54	653
Alginate	/	4.1	66.1	/
	Garlic oil 0.4%	2.7	38.7	/
Sodium alginate	/	4.7	55	3280
	n-hexanal 1%	1.5	28	2247
Sodium alginate + GBS (emulsion)	/	2.1	31	2320
	n-hexanal 1%	1.9	20	1605
Carrageenan	/	1.2	8.8	95
	n-hexanal 1%	1.6	15	1259
Carrageenan + GBS (emulsion)	/	2.6	10	927
	n-hexanal 1%	2.4	10	751
HPMC	/	0.1	59	1697
	Tea tree oil 40%	0.11	42	956

Sources: Atarés, L. et al., *J. Food Eng.*, 99, 384, 2010; Sánchez-González, L. et al., J. Food Eng., 98, 443, 2010b; Sánchez-González, L. et al., *Food Hydrocolloids*, 23, 2102, 2009; Pranoto, Y. et al., LWT—Food Sci. Technol., 38(8), 859, 2005b; Hambleton, A. et al., Food *Chem.*, 2012.

less elastic than those with cinnamon oil ($p < 0.01$). The discontinuities in the protein matrix may imply a decrease in the deformability of the films with ginger oil, because these reach the break point at lower deformation. This tendency could be explained by the fact that lipids are unable to form a cohesive and continuous matrix. A very different behavior was observed by Hambleton et al. (2012) for carrageenan films incorporating n-hexanal. In fact, the addition of incorporated n-hexanal tends to increase the elastic modulus and tensile strength. This is probably due to the stabilizing effect of the aroma compound on the film matrix. n-Hexanal interacts with κ-carrageenan's lateral chains and plays a stabilizing role in the interface due to its amphipolar character leading to a much more homogeneous structure that increases the film's stiffness. Nevertheless, n-hexanal does not affect the film's capacity to stretch as the elongation percentages are not significantly different from the same film without n-hexanal. The addition of incorporated n-hexanal has the same effect as the presence of fat material; it reduces the elastic modulus and tensile strength, contrary to ι-carrageenan-based film. Incorporated n-hexanal weakly interacts with the sodium alginate, because this type of film has a well-organized structure in an egg-box model, stabilized by divalent ions to form stronger gels and thus stronger films. But n-hexanal interacts with the other components in the film like glycerol, which being a polyol has a great affinity for flavors of this type. These interactions lead to a reduction of the stiffness and to the film's resistance to elongation. The presence of both n-hexanal and fat material has a significant effect, reducing elastic modulus and tensile strength more than in other types of films, probably because the aroma compound interacts primarily with the fat material when added to the film. Film plasticization can occur during aroma transfer, and it depends on the aroma concentration as observed by Quezada-Gallo et al. (1999a,b) for permeation of 2-heptanone and 2-pentanone through methylcellulose-based films. The transfer rate of 2-heptanone strongly increased for an aroma gradient higher than 10 μg mL^{-1}. Mechanical film properties changed upon exposure to flavor concentrations higher than 10 μg mL^{-1}. 2-Heptanone and 2-pentanone increased film elongation, suggesting that polymer plasticization occurred. The ketone group of the flavor compound interacts with hydroxyl groups of methylcellulose. 2-Heptanone forms weak hydrogen bonds with methylcellulose. This likely widens the spaces among the polymer chains, resulting in swelling and plasticization of the film network that decreases the mechanical properties and enhances the transfer of volatiles and water vapor.

38.5.5.4 Impact of Flavor Compound Encapsulation on the Barrier Properties of Film

Both microstructure and mechanical properties are affected by flavor compound or essential oil encapsulation, including permeability of films to other volatile compounds such as water vapor, oxygen, or other aroma compounds. This suggests that the choice of matrix for encapsulation of a volatile compound on the basis of its oxygen barrier performance, for instance, could not be counted on because of permeability changes due to encapsulation. When 1% n-hexanal was encapsulated, oxygen permeability of carrageenan-glycerol and carrageenan-glycerol-lipid edible films increased 15% and 100%, respectively (Hambleton et al., 2008). Incorporation of 1% n-hexanal in a sodium alginate film induced a doubling in oxygen permeability and a 10-fold increase when the sodium alginate film contained lipid emulsions (Hambleton et al., 2009a). On the contrary, Rojas-Grau et al. (2007) did not show any change in the oxygen permeability of alginate-apple puree film when oregano, carvacrol, lemongrass oil, citral, cinnamon oil, or cinnamaldehyde were added in a range of 0.1%–0.5%. In these cases, the low content of encapsulated volatile compounds did not disturb the network structure.

The effect of RH on flavor transfer rates is probably as important as the effect of temperature. Moisture increases the mass transfer rates of gases and vapors through hydrophilic biopolymer films. Miller et al. (1998) observed substantial increases of D-limonene permeability through whey protein films when RH increased, increasing 2–20-fold when RH varied from 40% to 80%. Quezada-Gallo (1999) observed similar behavior for the permeability of 2-pentanone, 2-heptanone, and ethyl esters through both methylcellulose and wheat gluten films. This behavior was attributed to plasticization of the biopolymer network by water. But, if the effect of water on the permeability of volatile compounds is well known and seems obvious, the effect of aroma compounds or essential oils on the water vapor permeability (WVP) is quite a bit more complicated as displayed in Table 38.5.

WVP of soy protein isolate or alginate-based films without lipid emulsions tends to increase when essential oils or aroma compounds were encapsulated, whereas that of sodium alginate plus lipid, carrageenan-based, chitosan, or HPMC decreased. Atarés et al. (2010) showed that the addition of small proportions of ginger and cinnamon essential oils resulted in a reduction in the water vapor barrier properties of soy protein isolate films. This could be due to the interactions of oil components with some protein tails that could promote a decrease in the hydrophobic character of the protein matrix. Because the amount of oil incorporated is very low, the lipid discontinuities seem not to be relevant to increasing the tortuosity factor for transfer of water molecules, responsible for the reduction of WVP, although a further increase in cinnamon oil content resulted in WVP decrease. The effectiveness of cinnamon oil, as compared to ginger oil, in reducing WVP at a fixed protein-to-oil ratio, suggests that the former remains partially integrated in the protein network of the dry films. The difficulties in integrating the essential oils or aromas in a hydrophilic network may be due to matrix disruptions and creation of void spaces at the protein-essential oil interface. Therefore, it cannot be assumed that the WVP of edible films is reduced simply by adding a hydrophobic component such as an aroma or essential oil to the formulation, but the impact of lipid addition on the microstructure of emulsified film is a determining factor in water barrier efficiency. In the case of HPMC, chitosan, or carrageenan, the previous explanation for increasing the hydrophobicity does not fit, because WVP increased with essential oil content. The WVP values showed a significant decrease in line with the increase in tea tree oil concentration, following a linear trend reaching a maximum WVP reduction of about 40% with an incorporation of 2% of

TABLE 38.5
Influence of Volatile Compound Incorporation on Water Vapor Permeability of Edible Films

Film Matrix	Volatile Compounds	WVP (10^{-10} g/m s Pa)	T (°C)	ΔRH (%)
Soy protein isolate	/	1.38	25	33–53
	Cinnamon oil 1%	1.52	25	33–53
	Ginger oil 1%	1.89	25	33–53
Chitosan	/	12.4	25	100–54
	Bergamot oil 3%	6.5	25	100–54
	Tea tree oil 2%	7.4	25	100–54
Alginate	/	2.35	25	0–50
	Garlic oil 0.4%	72.64	25	0–50
Sodium alginate	/	2.13	25	30–84
	n-hexanal 1%	2.96	25	30–84
Sodium alginate + GBS (emulsion)	/	1.68	25	30–84
	n-hexanal 1%	1.21	25	30–84
Carrageenan	/	23.5	25	30–84
	n-hexanal 1%	22.2	25	30–84
Carrageenan + GBS (emulsion)	/	29	25	30–84
	n-hexanal 1%	25.3	25	30–84
HPMC	/	8.2	25	100–54
	Tea tree oil 40%	5.5	25	100–54

Sources: Atarés, L. et al., *J. Food Eng.*, 99, 384, 2010; Sánchez-González, L. et al., *Food Hydrocolloids*, 23, 2102, 2009; Sánchez-González, L. et al., *J. Food Eng.*, 98, 443, 2010b; Pranoto, Y. et al., Food Res. Int., 38, 267, 2005a; Hambleton, A. et al., Food Chem., 2012. With permission.

the essential oil (Sánchez-González et al., 2010a,b). This behavior is expected as an increase in the hydrophobic compound fraction usually leads to an improvement in the water barrier properties of films, as was previously reported for essential oil addition in chitosan films (Zivanovic et al., 2005).

Incorporation of *n*-hexanal in an ι-carrageenan film induced a twofold decrease in permeability of ethyl acetate, ethyl butyrate, and 2-hexanone and a sixfold decrease for the D-limonene. The encapsulated *n*-hexanal interacted with CH_2OH or sulfated groups of ι-carrageenan lateral chains inducing a lower permeability as displayed by Hambleton et al. (2008, 2010). When the film contained a lipid dispersed in the carrageenan matrix, the behavior of the aroma compound permeability varied, decreasing from 30% (2-hexanone), twofold (ethyl acetate), 100-fold (D-limonene), but decreasing by 15-fold for ethyl butyrate. The higher the log K, the greater the hydrophobicity of the aroma compound, thereby limiting the ethyl butyrate solubility in the hydrophilic ι-carrageenan matrix. On the contrary, it promotes interactions with the fat. Ethyl butyrate is then more retained in fat globules. The films with fat are also less uniform, fat globule particle size increases, and there are less open spaces in the film structure to facilitate the aroma diffusivity. Therefore, it limits the aroma compound transfer. Alginate-lipid emulsion films exhibit a significant increase in ethyl butyrate (×2000), ethyl hexanoate (×10,000), and D-limonene permeability (×300), whereas permeability drop fourfold for 2-hexanone. Permeability of aroma compounds is much more complex than that of gases or water vapor. It depends on the solubility of the volatile compound in the edible film, on their vapor pressure, volatility, hydrophobicity (often expressed as the partition coefficient between water and octanol), solubility (governed by the chemical nature of both film and volatile compound), molecular mobility of the matrix, volatile diffusion coefficient, and many external parameters such as temperature, pressure, moisture levels, and so forth. Therefore, previous observations when aroma or essential oils are added in edible films cannot easily predict what would happen in other situations or conditions because the physical-chemical explanations have not always been reported.

38.6 EDIBLE FILMS AND COATINGS FOR NONVOLATILE MOLECULE ENCAPSULATION (PEPTIDES, POLYUNSATURATED FATTY ACIDS, ANTIOXIDANTS, AND ANTIBACTERIALS)

Over the last few years, consumer demand for foods of natural origin, high quality, elevated safety, longer shelf life, fresh taste, and appearance has been strongly increasing. The minimally processed and easy-to-eat foods are also requested. Currently, there is an escalating tendency to employ environmentally friendly packaging materials with the intention of substituting nondegradable materials, thus reducing the environmental impact resulting from waste accumulation. To address the environmental issues, and concurrently extend the shelf life and food quality while reducing packaging waste has catalyzed the exploration of new bio-based packaging materials such as edible and biodegradable films. One of the approaches is to use renewable biopolymers such as polysaccharides, proteins, gums, lipids, and their derivatives, from animal and plant origin as described previously, not only to form capsules but also to form films and coatings able to encapsulate particular biomolecules. Such biodegradable/edible packaging/coatings not only ensure food safety but at the same time are good source of nutrition.

38.6.1 NUTRACEUTICALS

In parallel with natural foods and wrapping materials, consumer demand is more and more focused on healthy foods to provide more than the basic needs. Encapsulation of nutraceuticals is at present a very hot topic. A great number of research articles report on matrix and encapsulation processes. Milk proteins are largely used for encapsulation and controlled delivery using the micelle structures or any other nanostructure (Semo et al., 2007; Livney, 2010). These proteins are often conjugated with polysaccharides to improve the encapsulation and controlled release properties. Maltodextrins and other modified starches are also largely used as encapsulation matrices due to their low cost and ease of use. Cyclodextrins showed high potentiality for encapsulation but are limited in use by their industrial cost.

Several nutraceutical molecules can be incorporated in edible coatings such as vitamins, peptides, polyunsaturated fatty acids (PUFA), or antioxidants to increase the food nutritional value. The main problem in incorporating nutraceuticals in food is related to stability during storage. These reactive molecules rapidly lose their activity due to oxidation or other chemical reactions. Edible films and coatings are more and more used to protect these active biomolecules from contact with foods. While incorporated in coatings or encapsulated, their bioactive effect is preserved, and nanoencapsulation could even increase the molecule's bioavailability. All earlier cited techniques are commonly used for nutraceutical encapsulation with good efficiency. The main research topics are now the controlled release of the nutrient, and also, and this is the more complex problem, there are concerns about the bioavailability and the focused release of the active compound (Chen et al., 2006; Kosaraju et al., 2006; Gonnet et al., 2010; Nair et al., 2010). As an example, release of PUFA in the brain to limit Alzheimer's disease is a very hot research topic. Polyphenols are used as active compounds to reduce oxidative stress. Fang and Bandhari (2010) reviewed research on the application of polyphenols. The unpleasant taste of most phenolic compounds can be completely masked by encapsulation. The technologies of encapsulation of polyphenols are commonly spray-drying, coacervation, liposome entrapment, inclusion complexion, cocrystallization, nanoencapsulation, freeze-drying, yeast encapsulation, and use of emulsions. In parallel to the development of preservation techniques, advanced research is being developed on flavonoid functionalization to improve their nutraceutical use. Common research on simultaneous functionalization and encapsulation should be developed to accelerate commercialization.

38.6.2 ANTIBACTERIALS

Considering antibacterial molecules, postprocess contamination caused by product mishandling and faulty packaging is responsible for about two-thirds of all microbiologically related recalls in the United States, with most of these recalls originated from contamination of ready-to-eat food products. Antimicrobial agents are components that hinder growth of microorganisms, sometimes called food preservatives. According to the definition used by the Commission of the European Communities, preservatives are substances that extend the shelf life of foodstuffs by protecting them against deterioration caused by microorganisms (EU Directive 95/2/EC). Similar rules are applied in the United States, where the U.S. Food and Drug Administration (FDA) defines preservatives as any chemical that when added to food tends to prevent or retard deterioration. Antimicrobials are used in food to control natural spoilage and to prevent or control growth of microorganisms, including pathogenic microorganisms (da Silva Malheiros et al., 2010a,b; Drulis-Kawa and Dorotkiewicz, 2010).

Natural antimicrobials can be defined as substances produced by living organisms in their fight with other organisms for space and their competition for nutrients. The main sources of these compounds are plants (secondary metabolites in essential oils and phytoalexins), microorganisms (bacteriocins and organic acids), and animals (lysozyme from eggs, lactoferrins from milk). Across the various sources, the same types of active compounds can be encountered (e.g., enzymes, peptides, and organic acids). Reducing the need for antibiotics, controlling microbial contamination in food, improving shelf-life extension technologies to eliminate undesirable pathogens, delaying microbial spoilage, decreasing the development of antibiotic resistance by pathogenic microorganisms, or strengthening immune cells in humans are some of the benefits (Tajkarimi et al., 2010). Most approved food antimicrobials have limited application due to pH or food component interactions. They are amphiphilic and can solubilize or be bound by lipids or hydrophobic proteins in foods, making them less available to inhibit microorganisms in the food product.

The term *bacteriocin* is mostly used to describe the small, heat-stable cationic peptides synthesized by Gram-positive bacteria, namely lactic acid bacteria (LAB), which display a wider spectrum of inhibition (Cotter et al., 2005). The bacteriocins produced by LAB offer several desirable properties that make them suitable for food preservation. They are generally recognized as safe (GRAS) substances, not active and nontoxic on eukaryotic cells, become inactivated by digestive proteases, and have little influence on the gut microbiota. Bacteriocins are usually pH and heat-tolerant, and they have a relatively broad antimicrobial spectrum against many food-borne pathogenic and spoilage bacteria.

Nisin has been increasingly used as a bio-preservative for direct incorporation in food as well as in active/edible films. Nisin effectively inhibits Gram-positive bacteria and outgrowth spores of *Bacillus* and *Clostridium*. If nisin is efficient against particular microorganisms, its activity rapidly decreases as it hydrolyses in the food product and bacterial inactivation stops. All studies showed a restart of bacterial growth a few days after nisin incorporation. Encapsulation and controlled release is an efficient way to avoid this resumption of bacterial growth, because active nisin is slowly delivered into the food or onto the food surface. In a recent study for edible films (Sebti et al., 2007) using HPMC/chitosan and incorporating the pure nisin, the author evaluated the effect of nisin on the physical characteristics of films.

38.6.3 Edible Materials for Nutraceuticals or Antibacterial Molecule Encapsulation

The matrix used for nutraceuticals or antimicrobial molecules is of prime importance to allow good preservation or controlled release of these active compounds.

38.6.3.1 HPMC

Cellulose-based materials are being widely used as they offer advantages like edibility, biocompatibility, barrier properties, and aesthetic appearance as well as being nontoxic, nonpolluting, and having low cost (Vasconez et al., 2009). HPMC edible films are attractive for food applications because HPMC is a readily available nonionic edible plant derivative shown to form transparent, odorless, tasteless, oil-resistant, water-soluble films with very efficient oxygen, carbon dioxide, aroma, and lipid barriers, but with moderate resistance to water vapor transport. HPMC is used in the food industry as an emulsifier, film former, protective colloid, stabilizer, suspending agent, or thickener. HPMC is approved for food uses by the FDA (21 CFR 172.874) and the EU (EU, 1995); its safety in food use has been affirmed by the Joint Food and Agriculture Organization (FAO)/World Health Organization (WHO) Expert Committee on Food Additives (JECFA) (Burdock, 2007). The tensile strength of HPMC films is high, and flexibility is neither too high nor too fragile, which make them suitable for edible coating purposes (Imran et al., 2010b).

38.6.3.2 Polyacetic Acid (PLA)

As a GRAS and biodegradable material, and also because of its biosorbability and biocompatible properties in the human body, polyacetic acid (PLA) and its copolymers (especially polyglycolic acid) attracted the pharmaceutical and medical scientist researchers as a carrier for releasing various drugs and agents like bupivacaine and many others. In food domains, little research has been done on the suitability of PLA as an active packaging polymer. PLA is a new corn-derived polymer and needs time to be an accepted and effective active packaging material in the market.

Van Aardt et al. (2007) studied the release of antioxidants from loaded poly (lactide-co-glycolide) (PLGA) (50:50) films with 2% α-tocopherol, and a combination of 1% butylated hydroxytoluene (BHT) and 1% butylated hydroxyanisole (BHA) into water, oil (food stimulant: Miglyol 812), and milk products at 4°C and 25°C in the presence and absence of light. They concluded that in a water medium, PLGA (50:50) showed hydrolytic degradation of the polymer, and release of BHT into the water. In Miglyol 812, no degradation or antioxidant release took place, even after 8 weeks at 25°C. Milkfat was stabilized to some extent when light exposed dry whole milk and dry buttermilk was exposed to antioxidant loaded PLGA (50:50). They also suggested potential use of degradable polymers as a unique active packaging option for sustained delivery of antioxidants, which could be of benefit to the dairy industry by limiting the oxidation of high-fat dairy products, such as ice cream mixes.

38.6.4 PLASTICIZERS

Plasticizers impressively affect the physical properties of biopolymer films (Zhang and Han, 2008). The plasticizer helps to decrease inherent brittleness of films by reducing intermolecular forces, increasing the mobility of polymer chains, decreasing the glass transition temperature of these materials, and improving their flexibility (Zhang and Han, 2008; Galdeano et al., 2009). Thus, it is important to study the effect of commonly used polyol glycerol on the homogenous dispersion of Nisaplin® (nisin, salt, and milk solid) for the formation of composite active films of improved quality. However, plasticizers generally cause increased water permeability, so they must be added at a certain level to obtain a film with desired flexibility, thickness, and transparency without significant decrease of mechanical strength and barrier properties to mass transfer (Möller et al., 2004; Jongjareonrak et al., 2006; Brindle and Krochta, 2008).

38.6.5 MOLECULAR DIFFUSION FROM FILM AND COATING TO FOOD SURFACE

Migration from capsules included in films or coatings can be represented as shown in Figure 38.8.

Diffusion and erosion lead to slow release of the antibacterials, allowing food stabilization for a longer period than by adding the antibacterials directly into the food. Quantitative measurement of the rate at which a diffusion process occurs is usually expressed in terms of diffusivity (also called the *diffusion coefficient*), expressed in m² s⁻¹. The classical theory used to model the diffusion process is based on Fick's laws (Crank, 1975; Stannet, 1978). Diffusion in a homogeneous media is based upon the assumption that the rate of transfer, R, of a migrant passing perpendicularly through the unit area of a section is proportional to the concentration gradient between the two sides of the packaging:

$$R = \frac{dM}{Adt} = -D\frac{dC}{dx}$$

where
  D is the diffusion coefficient (m² s⁻¹)
  A is the film area (m²)

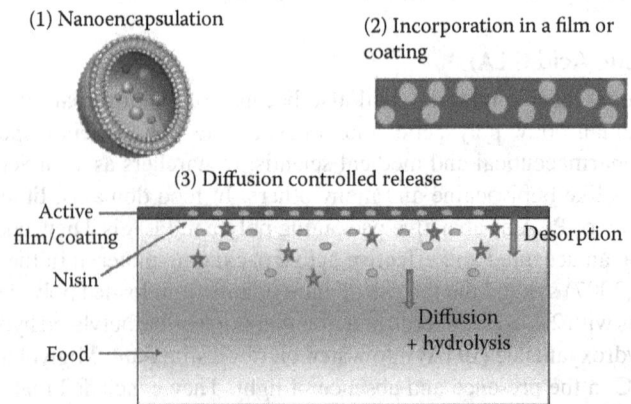

FIGURE 38.8 Nanoencapsulation, film/coating inclusion, and controlled release into food.

In general, D is a function of the local diffusant concentration, C (g m^{-3}), t is time (s), and x is the thickness of the film or coating (m). The amount of package components that may migrate from a packaging material into liquid or solid food depends on the chemical and physical properties of food and polymer. Various factors like migrant concentration, molecular weight, solubility, diffusivity, partition coefficient between polymer and food, time, temperature, polymer and food composition, and structures (density, crystallinity, chain branching) are the main controlling factors in migration.

Legally, polymers for packaging are regulated through global or specific migration levels. Global migration measures the total amount of all compounds migrating into food simulants independently of migrant composition. Specific migration concerns a given migrant. Several studies have measured global and specific migration from packaging materials to foods (Baner, 1991; Jamshidian et al., 2010). In the case of active films, the global migration limit does not apply, because active compound migration is required. Regulations are now more precise about the possible activity of the packaging and coating.

38.6.6 Chemical Structure of Migrant and Matrix State

The chemical structure of an encapsulated molecule is an important parameter that can influence the partition coefficient and then the controlled release into a food. Alcohols and short-chained esters had higher partition coefficients in the oil/polymer system, than in the water/polymer system. Several studies have attempted to model the relationship between the encapsulated molecule, the composition of the food, and the partition coefficient (Arab Tehrany and Desobry 2004). It is also known that matrix crystallinity and glass transition of the matrix are key factors for an efficient controlled release of an active compound. A controlled transition from glassy to rubbery state (temperature, water activity) leads to the best system for good food preservation. A lot of work still has to be done to allow perfect control of an active compound release.

38.7 CONCLUSION

A hot topic in the functional food and pharmaceutical industry is the efficient encapsulation of high value-added ingredients, such as PUFA, flavors, vitamins, and health-promoting ingredients, in relation to improved functionalities. In addition, because many of the most popular nutritional ingredients on the market today have unpleasant sensory characteristics, keeping the objectionable flavors out of products can sometimes be as important as keeping the enjoyable flavors in. Numerous developments have been made in the field of encapsulated food flavors. This is because of several favorable properties of the encapsulated form of flavors: ease in handling and mixing; stability against air, light, and evaporation; masking of undesirable tastes and aromas; and delivery of ingredients at the desired stage and at a specifically targeted release sites. Advances in the development of new wall materials and microencapsulation methods have paved the way for value-added ingredients of higher quality, consistency, and enhanced performance and improved prices. Each encapsulation process, generally developed to solve a particular problem encountered by product development, presents advantages and disadvantages. The relationships among problems, capabilities, and encapsulation methods were presented. Microencapsulation by spray-drying is the most economical and flexible way that the food industry can encapsulate ingredients. Thus, this technology is now becoming available to satisfy the increasingly specialized needs of the market. In addition, the fluid-bed process is a promising encapsulation technique for large-scale production of flavor powders to be applied in the food industry. The choice of an appropriate technique of encapsulation depends on the properties of the compounds, the degree of stability required during storage and processing, the properties of the food components, the specific release properties required, the maximum obtainable molecule load in the powder, and the production cost.

REFERENCES

Anandaraman, S., Reineccius, G.A. 1986. Stability of encapsulated orange peel oil. *Food Technology*, 11: 88–93.

Arab Tehrany, E., Desobry, S. 2004. Partition coefficient in food/packaging systems. *Food Additives and Contaminants*, 21(12): 1186–1202.

Atarés, L., De Jesús, C., Talens, P., Chiralt, A. 2010. Characterization of SPI-based edible films incorporated with cinnamon or ginger essential oils. *Journal of Food Engineering*, 99: 384–391.

Baner, A.L. 1991. Prediction of solute partition coefficients between polyolefins and alcohols using the regular solution theory and group contribution methods. *Industrial and Engineering Chemistry Research*, 30(7): 1506–1515.

Becker, K., Koszinowski, J., Piringer, O. 1987 Permeation von Riech- und Aromastoffen Durch Polyolefine, *Parfurmerie und Kosmetik*, 68: 268–278.

Ben Arfa, A., Chrakabandhu, Y., Preziosi-Belloy, L., Chalier, P., Gontard, N. 2007. Coating papers with soy protein isolates as inclusion matrix of carvacrol. *Food Research International*, 40: 22–32.

Bertolini, A.C., Siani, A.C., Grosso, C.R.F. 2001. Stability of monoterpenes encapsulated in gum arabic by spray-drying. *Journal of Agricultural and Food Chemistry*, 49: 780–785.

Bouwmeester, H., Dekkers, S., Noordam, M.Y., Hagens, W.I., Bulder, A.S., de Heer, C., ten Voorde, S., Wijnhoven, S., Marvin, H., Sips, A. 2009. Review of health safety aspects of nanotechnologies in food production. *Regulatory Toxicology and Pharmacology*, 53(1): 52–62.

Brazel, C.S. (1999). Microencapsulation: Offering solution for the food industry. *Cereal Foods World*, 44: 388–393.

Brindle, L.P., Krochta, J.M. 2008. Physical properties of whey protein hydroxypropyl methylcellulose blend edible films. *Journal of Food Science*, 73(9): 446–454.

Burdock, G.A. 2007. Safety assessment of hydroxypropyl methylcellulose as a food ingredient. *Food and Chemical Toxicology*, 45(12): 2341–2351.

Burt, S., 2004. Essential oils: Their antibacterial properties and potential applications in foods—A review. *International Journal of Food Microbiology*, 94: 223–253.

Buttery, R.G., Bomben, J.L., Guadagni, D.G., Ling, L.C. 1971. Volatilities of organic flavor compounds in foods. *Journal of Agricultural and Food Chemistry*, 19(6): 1045–1048.

Chen, L., Remondetto, G.E., Subirade, M. 2006. Food protein-based materials as nutraceutical delivery systems. *Trends in Food Science and Technology*, 17(5): 272–283.

Cotter, P.D., Hill, C., Ross, P.R. 2005. Bacteriocins: Developing innate immunity for food. *Nature Reviews Microbiology*, 3(10): 777–788.

Crank, J. 1975. *Mathematics of Diffusion*, 2nd edn. Clavedon Press, Oxford, England.

Da Silva Malheiros, P., Joner Daroit, D., Brandelli, A. 2010b. Food applications of liposome-encapsulated antimicrobial peptides. *Trends in Food Science and Technology*, 21(6): 284–292.

Da Silva Malheiros, P., Joner Daroit, D., Pesce da Silveira, N., Brandelli, A. 2010a. Effect of nanovesicle-encapsulated nisin on growth of *Listeria monocytogenes* in milk. *Food Microbiology*, 27(1): 175–178.

Debeaufort, F. 1994. Étude des Transferts de Matière au Travers de Films d'Emballages: Perméation de l'Eau et de Substances d'Arôme en Relation avec les Propriétés Physico-chimiques des Films Comestibles. PhD Dissertation, ENS.BANA, Université de Bourgogne, Dijon, France.

Debeaufort, F., Tesson, N., Voilley, A. 1995. Aroma compounds and water vapour permeability of edible films and polymeric packagings. In *Food and Packaging Materials—Chemical Interactions*, Ackermann, P., Jägerstad, M., and Ohlsson, T. (eds.). The Royal Society of Chemistry, Cambridge, U.K., pp. 169–175.

Debeaufort, F., Voilley, A. 1994. Aroma compound and water vapor permeability of edible films and polymeric packagings, *Journal of Agricultural and Food Chemistry*, 42: 2871–2875.

Debeaufort, F., Voilley, A. 1995. Methylcellulose-based edible films and coatings: 1. Effect of plasticizer content on water and 1-octen-3-ol sorption and transport. *Cellulose*, 2: 1–10.

Debeaufort, F., Quezada-Gallo, J.A., Voilley, A. 2002. Edible films and coatings as aroma barrier. In *Protein-Based Edible Films and Coatings*, Gennadios, A. (ed.). CRC Press, Boca Raton, FL, Chapter 24, pp. 579–600.

Del Nobile, M.A., Conte, A., Incoronato, A.L., Panza, O. 2008. Antimicrobial efficacy and release kinetics of thymol from zein films. *Journal of Food Engineering*, 89: 57–63.

Djordevic, D., Cercaci, L., Alamed, J., McClements, D.J., Decker, E.A. (2008). Chemical and physical stability of protein- and gum arabic- stabilized oil-in-water emulsions containing limonene. *Journal of Food Science*, 73: C167–C172.

Drulis-Kawa, Z., Dorotkiewicz-Jach, A. 2010. Liposomes as delivery systems for antibiotics. *International Journal of Pharmaceutics*, 387(1–2): 187–198.
Embuscado, M.E., Huber, K.C. 2009. *Edible Films and Coatings for Food Applications*. Springer Science, New York, pp. 403.
Fabra, M.J., Chambin, O., Assifaoui, A., Debeaufort, F. 2011. Influence of temperature and salt concentration on the release in liquid media of aroma compounds encapsulated in edible films. *Journal of Controlled Release*.
Fabra, M.J., Hambleton, A., Talens, P., Debeaufort, F., Chiralt, A., Voilley, A. 2008. Aroma barrier properties of sodium caseinate-based edible films. *Biomacromolecules*, 9(5): 9, 1406–1410.
Fabra, M.J., Hambleton, A., Talens, P., Debeaufort, F., Chiralt, A., Voilley, A. 2009. Influence of interactions on the water and aroma permeabilities of iota-carrageenan-oleic acid-beeswax edible films used for flavour encapsulation. *Carbohydrate Polymers*, 76: 325–332.
Fang, Z., Bhandari, B. 2010. Encapsulation of polyphenols—A review. *Trends in Food Science and Technology*, 21(10): 510–523.
Fayoux, S., Seuvre, A.M., Voilley, A. 1997a. Aroma transfers in and through plastic packagings: Orange juice and d-limonene. A review. Part 1: Orange juice aroma sorption. *Packaging Technology Science*, 10: 69–82.
Fayoux, S., Seuvre, A.M., Voilley, A. 1997b. Aroma transfers in and through plastic packagings: Orange juice and d-limonene. A review. Part 2: Overall sorption mechanism and parameter—A literature survey. *Packaging Technology Science*, 10: 145–160.
Fenaroli, G. (ed.). 1995. *Fenaroli's Handbook of Flavor Ingredients*. CRC Press, Boca Raton, FL.
Franz, R. 1993. Permeation of volatile organic compounds across polymer films. Part I: Development of a sensitive test method suitable for high barrier packaging films at very low permeant vapour pressures, *Packaging Technology Science*, 6: 91.
Friedman, M., Henika, P.R., Levin, C.E., Mandrell, R.E. 2004. Antibacterial activities of plant essential oils and their components against *Escherichia coli* O157:H7 and *Salmonella enterica* in apple juice. *Journal of Agricultural and Food Chemistry*, 52: 6042–6048.
Galdeano, M.C., Mali, S., Grossmann, M.V.E., Yamashita, F., Garcia, M.A. 2009. Effects of plasticizers on the properties of oat starch films. *Materials Science and Engineering C*, 29(2): 532–538.
Gennadios, A. 2002. *Protein-Based Edible Films and Coatings*. CRC Press, Boca Raton, FL, pp. 639.
Godshall, M.A. 1997. How carbohydrate influence food flavor. *Food Technology*, 51(1): 63–67.
Gómez-Estaca, J., López de Lacey, A., López-Caballero, M.E., Gómez-Guillén, M.C., Montero, P. 2010. Biodegradable gelatin–chitosan films incorporated with essential oils as antimicrobial agents for fish preservation. *Food Microbiology*, 27(7): 889–896.
Gómez-Estaca, J., Montero, P., Fernández-Martín, F., Alemán, A., Gómez-Guillén, M.C. 2009. Physical and chemical properties of tuna-skin and bovine-hide gelatin films with added aqueous oregano and rosemary extracts. *Food Hydrocolloids*, 23(5): 1334–1341.
Gómez-Estaca, J., Montero, P., Giménez, B., Gómez-Guillén, M.C. 2007. Effect of functional edible films and high pressure processing on microbial and oxidative spoilage in cold-smoked sardine (*Sardina pilchardus*). *Food Chemistry*, 105(2): 511–520.
Gonnet, M., Lethuaut, L., Boury, F. 2010. New trends in encapsulation of liposoluble vitamins. *Journal of Controlled Release*, 146(3): 276–290.
Green, B.K., Scheicher, L. 1955. Pressure sensitive record materials. US Patent no. 2, 217, 507, Ncr C.
Guilbert, S., Gontard, N. 2005. Agro-polymers for edible and biodegradable films: Review of agricultural polymeric materials, physical and mechanical characteristics. In *Innovations in Food Packaging*, Han, J.H. (ed.). Elsevier Academic Press, Oxford, U.K., pp. 263–276.
Hambleton, A., Debeaufort, F., Beney, L., Karbowiak, T., Voilley, A. 2008. Protection of active aroma compound against moisture and oxygen by encapsulation in biopolymeric emulsion-based edible films. *Biomacromolecules*, 9(3): 1058–1063.
Hambleton, A., Debeaufort, F., Bonnotte, A., Voilley, A. 2009b. Influence of alginate emulsion-based films structure on its barrier properties and on its protection of microencapsulated aroma compound. *Food Hydrocolloids*, 23(8): 2116–2124.
Hambleton, A., Fabra, M.J., Debeaufort, F., Brun-Dury, C., Voilley, A. 2009a. Interface and aroma barrier properties of iota-carrageenan emulsion-based films used for encapsulation of active food compounds. *Journal of Food Engineering*, 93: 80–88.
Hambleton, A., Perpiñan-Saiz, N., Fabra, M.J., Voilley, A., Debeaufort, F. 2012. The Schroeder paradox or how the state of water affects the moisture transfers through edible films. *Food Chemistry*, 132(4): 1629–2230.

Hambleton, A., Voilley, A., Debeaufort, F. 2010. Transport parameters for aroma compounds through i-carrageenan and sodium alginate-based edible films. *Food Hydrocolloids*, 10.1016/j.foodhyd.2010.10.010.

Han, J. 2002. Protein-based edible films and coatings carrying antimicrobial agents. In *Protein-Based Films and Coatings*, Gennadios, A. (ed.). CRC Press, Boca Raton, FL, pp. 485–500.

Heng, L., Van Koningsveld, G.A., Gruppen, H., Van Boekel, M., Vincken, J.P., Roozen, J.P., Voragen, A.G. 2004. Protein-flavour interactions in relation to development of novel protein foods. *Trends Food Science and Technology*, 15(3): 217–224.

Imran, M., El-Fahmy, S., Revol-Junelles, A.M., Desobry, S. 2010a. Cellulose derivative based active coatings: Effects of nisin and plasticizer on physico-chemical and antimicrobial properties of hydroxypropyl methylcellulose films. *Carbohydrate Polymers*, 81: 219–225.

Imran, M., Revol-Junelles, A.-M., Martyn, A., Tehrany, E.A., Jacquot, M., Linder, M., Desobry, S. 2010b. Active food packaging evolution: Transformation from micro- to nanotechnology. *Critical Reviews in Food Science and Nutrition*, 50(9): 799–821.

Jamshidian, M., Arab Tehrany, E., Imran, M., Jacquot, M., Desobry, S. 2010. PLA: Production, application and controlled release. *Comprehensive Reviews: Food Science and Food Safety*, 9(5): 552–571.

Jongjareonrak, A., Benjakul, S., Visessanguan, W., Tanaka, M. 2006. Effects of plasticizers on the properties of edible films from skin gelatin of bigeye snapper and brownstripe red snapper. *European Food Research and Technology*, 222(3–4): 229–235.

Kim, Y.D., Moor, C.V. 1996. Microencapsulation properties of gum arabic and several food proteins: Spray-dried orange oil emulsion particles. *Journal of Agricultural Food Chemistry*, 44(5): 1314–1320.

Kobayashi, M., Kanno, T., Hanada, K., Osanai, S.I. 1995. Permeability and diffusivity of d-limonene vapor in polymeric sealant films. *Journal of Food Science*, 60: 205–209.

Kosaraju, S.L., D'ath, L., Lawrence, A. 2006. Preparation and characterisation of chitosan microspheres for antioxidant delivery. *Carbohydrate Polymers*, 64(2): 163–167.

Kurek, M., Debeaufort, F. 2010, Development of an antimicrobial coating for packaging films: Physico-chemical and microbiological approaches. Doctoral school intermediate PhD report, University of Burgundy, Dijon, France.

Laohakunjit, N., Kerdchoechuen, O. 2007. Aroma enrichment and the change during storage of non-aromatic milled rice coated with extracted natural flavour. *Food Chemistry*, 101: 339–344.

Livney, Y.D. 2010. Milk proteins as vehicles for bioactives. *Current Opinion in Colloid and Interface Science*, 15(1–2): 73–83.

Lubbers, S., Landy, P., Voilley, A. 1998. Retention and release of aroma compounds in foods containing proteins. *Food Technology*, 52(5): 68–74, 208–214.

Madene, A., Jacquot, M., Scher, J., Desobry, S. 2006. Flavour encapsulation and controlled release—A review. *International Journal of Food Science and Technology*, 41(1): 1–21.

Marcuzzo, E., Debeaufort, F., Hambleton, A., Sensidoni, A., Tat, L., Beney, L., Voilley, A. 2011. Encapsulation of aroma compounds in biopolymeric emulsion emulsion-based edible films to prevent oxidation. *Food Research International*.

Marcuzzo, E., Sensidoni, A., Debeaufort, F., Voilley, A. 2010. Encapsulation of aroma compounds in biopolymeric emulsion based edible films to control flavour release. *Carbohydrate Polymers*, 80(3): 984–988.

Martin-Belloso, O., Rojas-Grau, M.A., Soliva-Fortuny, R. 2009. Delivery of flavour and active ingredients using edible films and coatings. In *Edible Films and Coatings for Food Applications*, Embuscado, M.E. and Huber, K.C. (eds.). Springer Science, New York, pp. 295–313.

Mastromatteo, M., Barbuzzi, G., Conte, A., Del Nobile, M.A. 2009. Controlled release of thymol from zein based film. *Innovative Food Science and Emerging Technologies*, 10(2): 222–227.

McHugh, T.H., Huxsoll, C.C., Krochta, J.M. 1996. Permeability properties of fruit puree edible films. *Journal of Food Science*, 61: 88–91.

Miller, K.S., Krochta, J.M. 1997. Oxygen and aroma barrier properties of edible films: A review. *Trends in Food Science and Technology*, 8: 228–237.

Miller, K.S., Upadhyaya, S.K., Krochta, J.M. 1998. Permeability of d-limonene in whey protein films. *Journal of Food Science*, 63: 244–247.

Möller, H., Grelier, S., Pardon, P., Coma, V. 2004. Antimicrobial and physicochemical properties of chitosan–HPMC-based films. *Journal of Agricultural and Food Chemistry*, 52(21): 6585–6591.

Monedero, M., Hambleton, A., Talens, P., Debeaufort, F., Chiralt, A., Voilley, A. 2010. Study of the retention and release of n-hexanal from soy protein isolate-lipid composite films. *Journal of Food Engineering*, 100: 128–133.

Nair, H.B., Sung, B., Yadav, V.R., Kannappan, R., Chaturvedi, M.M., Aggarwal, B.B. 2010. Delivery of anti-inflammatory nutraceuticals by nanoparticles for the prevention and treatment of cancer. *Biochemical Pharmacology*, 80(12): 1833–1843.

Paik, J.S., Writer, M.S. 1995. Prediction of flavor sorption using the Flory-Huggins equation. *Journal of Agriculture and Food Chemistry*, 43: 175–178.

Ponce, A.G., Roura, S.I., del Valle, C.E., Moreira, M.R. 2008. Antimicrobial and antioxidant activities of edible coatings enriched with natural plant extracts: In vitro and in vivo studies. *Postharvest Biology and Technology*, 49: 294–300.

Pranoto, Y., Rakshit, S.K., Salokhe, V.M. 2005b. Enhancing antimicrobial activity of chitosan films by incorporating garlic oil, potassium sorbate and nisin. *LWT—Food Science and Technology*, 38(8): 859–865.

Pranoto, Y., Salokhe, V.M., Rakshit, S.K. 2005a. Physical and antibacterial properties of alginate-based edible film incorporated with garlic oil. *Food Research International*, 38: 267–272.

Quezada-Gallo, A., Debeaufort, F., Voilley, A. 1999b. Mechanism of aroma transport through edible and plastic packagings. In *New Developments in the Chemistry of Packaging Materials*, Rish, S. (ed.). ACS Books, Dallas, TX, pp. 125–140.

Quezada-Gallo, J.A. 1999. Influence de la Structure et de la Composition de Réseaux Macromoléculaires sur les Transferts de Molécules Volatiles (eau et Arômes). Application aux Emballages Comestibles et Plastiques. PhD Dissertation, Université de Dijon, Dijon, France.

Quezada-Gallo, J.A., Debeaufort, F., Voilley, A. 1999a. Interactions between aroma and edible films. 1. Permeability of methylcellulose and polyethylene films to methyl ketones. *Journal of Agriculture and Food Chemistry*, 47: 108–113.

Raybaudi-Massilia, R., Mosqueda-Melgar, J., Martín-Belloso, O. 2006. Antimicrobial activity of essential oils on *Salmonella Enteritidis*, *Escherichia coli*, and *Listeria innocua* in fruit juices. *Journal of Food Protection*, 69: 1579–1586.

Reineccius, G. 2009. Edible films and coatings for flavour encapsulation. In *Edible Films and Coatings for Food Applications*, Embuscado, M.E. and Huber, K.C. (eds.). Springer Science, New York, pp. 269–294.

Rojas-Grau, M., Raybaudi-Massilia, R.M., Soliva-Fortuny, R.S., Avena-Bustillos, R.J., McHugh, T.H., Martın-Belloso, O. 2007. Apple puree-alginate edible coating as carrier of antimicrobial agents to prolong shelf-life of fresh-cut apples. *Postharvest Biology and Technology*, 45: 254–264.

Rojas-Grau, M.A., Avena-Bustillos, R., Friedman, M., Henika, P., Martın-Belloso, O., McHugh, T., 2006. Mechanical, barrier and antimicrobial properties of apple puree edible films containing plant essential oils. *Journal of Agricultural and Food Chemistry*, 54: 9262–9267.

Rupasinghe, H.P., Boulter-Bitzer, J., Ahn, T., Odumeru, J. 2006. Vanillin inhibits pathogenic and spoilage microorganisms in vitro and aerobic microbial growth in fresh-cut apples. *Food Research International*, 39: 575–580.

Sánchez-González, L. 2010c. Caracterizacion y aplicacion de recubrimientos antimicrobianos a base de polisacaridos y aceites esenciales. PhD Dissertation, Universidad Politecnica de Valencia, Valencia, Spain, pp. 309.

Sánchez-González, L., Chafer, M., Chiralt, A., Gonzalez-Martinez, C. 2010a. Physical properties of edible chitosan films containing bergamot essential oil and their inhibitory action on *Penicillium italicum*. *Carbohydrate Polymers*, 82: 277–283.

Sánchez-González, L., Cháfer, M., González-Martínez, C., Chiralt, A., Desobry, S. 2011. Study of the release of limonene present in chitosan films enriched with bergamot oil in food simulants. *Journal of Food Engineering*, 105(1), 138–143.

Sánchez-González, L., Gonzalez-Martinez, C., Chiralt, A., Chafer, M. 2010b. Physical and antimicrobial properties of chitosan–tea tree essential oil composite films. *Journal of Food Engineering*, 98: 443–452.

Sánchez-González, L., Vargas, M., Gonzalez-Martinez, C., Chiralt, A., Chafer, M. 2009. Characterization of edible films based on hydroxypropylmethylcellulose and tea tree essential oil. *Food Hydrocolloids*, 23: 2102–2109.

Sebti, I., Chollet, E., Degraeve, P., Noel, C., Peyrol, E. 2007. Water sensitivity, antimicrobial, and physicochemical analyses of edible films based on HPMC and/or chitosan. *Journal of Agriculture and Food Chemistry*, 55(3): 693–699.

Semo, E., Kesselman, E., Danino, D., Livney, Y.D. 2007. Casein micelle as a natural nano-capsular vehicle for nutraceuticals. *Food Hydrocolloids*, 21(5–6): 936–942.

Shimoni, E. 2009. Nanotechnology for foods: Delivery systems. In *Global Issues in Food Science and Technology*, Gustavo, B.-C., Alan, M., David, L., Walter, S., Ken, B. and Paul, C. Eds., 411–424.

Solms, J., Osman-Ismail, F., Beyler, M. 1973. The interaction of volatiles with food components. *Canadian Institute of Food Science and Technology Journal*, 6: A10–A16.

Stannet. 1978. The transport of gases in synthetic polymeric membranes—An historic perspective. *Journal of Membrane Science*, 3: 97–115.

Tajkarimi, M.M., Ibrahim, S.A., Cliver, D.O. 2010. Antimicrobial herb and spice compounds in food. *Food Control*, 21(9): 1199–1218.

Tunc, S., Duman, O. 2011. Preparation of active antimicrobial methyl cellulose/carvacrol/montmorillonite nanocomposite film and investigation of carvacrol release. *LWT—Food Science and Technology*, 44: 465–472.

Van Aardt, M., Duncan, S.E., Marcy, J.E., Long, T.E., O'Keefe, S.F., Sims, S.R. 2007. Release of antioxidants from poly(lactide-co-glycolide) films into dry milk products and food simulating liquids. *International Journal of Food Science and Technology*, 42(11): 1327–1337.

van Ruth, S.M., King, C., Giannouli, P. 2002. Influence of lipid fraction, emulsifier fraction, and mean particle diameter of oil-in-water emulsions on the release of 20 aroma compounds. *Journal of Agricultural and Food Chemistry*, 50(8): 2365–2371.

Vasconez, M.B., Flores, S.K., Campos, C.A., Alvarado, J., Gerschenson, L.N. 2009. Antimicrobial activity and physical properties of chitosan-tapioca starch based edible films and coatings. *Food Research International*, 42(7): 762–769.

Wyler, L., Solms, J. 1982. Starch flavour complexes III. Stability of dried starch-flavor complexes and other dried flavour preparations. *Lebensmittel-Wissenschaft und Technologie*, 15(2): 93–97.

Zhang, Y., Han, J.H. 2008. Sorption isotherm and plasticization effect of moisture and plasticizers in pea starch film. *Journal of Food Science*, 73(7): 313–324.

Zivanovic, S., Chi, S., Draughon, A.F. 2005. Antimicrobial activity of chitosan films enriched with essential oils. *Journal of Food Science*, 70(1): 1145–1151.

39 Encapsulation of Aroma

Christelle Turchiuli and Elisabeth Dumoulin

CONTENTS

39.1 Introduction 834
39.2 Properties of Microcapsules and Choice of Support Material 835
    39.2.1 Properties of Microcapsules 835
    39.2.2 Choice of Support Material 837
        39.2.2.1 Searched Properties for Supports 837
        39.2.2.2 Examples of Supports 837
    39.2.3 In Summary 838
39.3 Some Encapsulation Processes 839
    39.3.1 Introductive Remarks 839
    39.3.2 Emulsification 840
        39.3.2.1 Preparation/Composition 840
        39.3.2.2 Equipments 841
        39.3.2.3 In Conclusion 841
    39.3.3 Spray-Drying 841
        39.3.3.1 Principle of Spray-Drying 842
    39.3.4 Influence of Operating Conditions 843
        39.3.4.1 Characteristics of the Resulting Aroma Powder 844
    39.3.5 In Summary 845
        39.3.5.1 Spray-Drying of Aroma Emulsions 847
    39.3.6 Spray-Cooling 849
        39.3.6.1 Principle 849
    39.3.7 Freeze-Drying 852
        39.3.7.1 Principle 852
        39.3.7.2 Main Characteristics 852
    39.3.8 Cocrystallization 853
        39.3.8.1 Principle 853
        39.3.8.2 Main Characteristics 853
    39.3.9 Coacervation 853
        39.3.9.1 Principle 853
        39.3.9.2 Methods of Coacervation 854
    39.3.10 Extrusion 854
        39.3.10.1 Principle 854
    39.3.11 Molecular Inclusion 856
        39.3.11.1 Principle 856
        39.3.11.2 Production 856
    39.3.12 Coating: Impregnation 856
        39.3.12.1 Principle of Impregnation 856
        39.3.12.2 Principle of Coating in Turbine/Drum/Screw (Endless) or in Fluid Bed 856

39.4 Examples of Studies Using Complementary Processes 858
    39.4.1 Vegetable Oil Encapsulation Using Spray-Drying and Fluid Bed Agglomeration 858
    39.4.2 Aroma Encapsulation into Powder for Chewing Gum 859
    39.4.3 Delayed Release of Flavor Material.. 861
    39.4.4 Production of Soluble Beverage Powder... 861
39.5 Conclusion 861
References.. 863

39.1 INTRODUCTION

Aromas are used in food, pharmaceutical, chemical, and cosmetic industries to create or to modify the sensory perception (taste, odor) of food products, objects, or atmosphere, to finally reach human beings or animals. They are used with different concentrations and quantities, either for direct consumption (a dish, a drink, and a medicine) or as a component at an industrial level.

Aroma compositions (called aroma) are complex mixtures of different organic molecules of different size, molecular weight (MW), structure, and polarity (i.e., hydrophilic, hydrophobic groups), with various physicochemical properties. Depending on these properties, they are more or less sensitive to heat, light, and oxygen, and more or less soluble in solvents (water, alcohol), and are characterized by their diffusivity in various media, their relative volatility, and polarity. They give a global complex characteristic, the *aroma profile*, which may be modified because of volatility (balance between volatile/nonvolatile components), oxidation (off-flavors), and diffusivity of some of the molecules (Richard and Multon, 1992).

Very often, aroma molecules are hydrophobic oils. Their characteristic taste or smell may be pleasant, strong (if pure), or not, and may need to be modified, masked, and diluted. The aroma will be used in association with other substances, in an environment which may be a liquid, a paste, a solid, a gas, submitted to different usage conditions (temperature, pH), according to the final objectives.

Some examples of products containing aroma are given to show the variety of environmental conditions to deliver its characteristics:

- In food products: liquid yogurts (a_w, pH), beverages (a_w, pasteurization), aromatized tea (hot water), spiced pizzas (cooking), drinks in vending machine (dosage, rapid dissolution), chewing-gum with menthol (flavor and gum), coated products (meat/fish, frying), enriched bars, animal feed, etc.
- In chemicals, cosmetics, medicine, agriculture: washing products (solid, liquid), coated seeds, fertilizers (protection against insects), controlled-release medicines to mask a bad taste (saliva), perfumes (liquid, solid), perfumed objects (sponges, scarf), scratch, aerosols (liquid, powder, gas), antimicrobial agents (essential oils), cleaning agents (i.e., dissolution of oil in limonene), etc.

Due to the aroma composition and the use in various media, the stability, the intensity, the risks of losses, and the conservation of profile will represent important parameters during preparation (dosage, formulation, pH) and processing operations such as heating, cooking, freezing, scratching, dissolution, etc., including storage and delivery conditions. It means that an efficient protection is necessary to maintain aroma properties till utilization; to prevent or reduce the action of atmosphere, oxygen, water, or heat; the interactions with other components; the modifications of structure; and dosage. Then the products with protected aroma will have to deliver the right balance of flavor/smell (no off-flavor) under specific stress conditions of temperature, humidity, mechanical stress, often with the desired odor on a long-term basis.

There is not one unique problem and solution, but many, all specific.

Encapsulation of Aroma

In the nature, we may observe many products (flowers, fruits, vegetables) existing with a « protection » (Richard and Multon, 1992):

- Vegetable cells are good encapsulation tools to release active molecules at a given time.
- Membranes, skin, shell of fruits, nuts, and flowers control the transfer of water and gas, and give protection against oxygen/air, heat, and offer a physically resistant structure.

The microcapsules of aroma will imitate nature with a film, a skin, to be a *barrier* for the active aroma composition (core). The aroma core material will be finally dispersed homogenously (multicore) in the liquid or solid support material (~20%–30% of total weight): if soluble in water as a solution; if not soluble in water as an emulsion, or in a solid/past/powder. In some cases, a high core loading is obtained (90%) with specific techniques (coacervation, coating, and coextrusion). Or multicore capsules are prepared initially by emulsion, then transformed into powder by spray-drying, extrusion, etc.

The aroma support material (or wall) chosen as the barrier (skin) will be often a film-forming and surface-active polymer, with several roles (in liquid or solids); the main benefits for the aroma are protection, dosage, stability, and possible controlled release in different media and end use conditions.

The different steps to consider must be well defined: the preparation of microcapsules (composition, process), their storage, and their final use.

The total encapsulation process will be chosen and achieved with the close collaboration of several partners:

- The aroma specialist for knowledge on aroma properties and behavior
- The process engineer to propose different processes according to final use and shape and aroma behavior, with the necessary technical help of supports providers
- The final user, industry, and/or consumer, to define the needs in terms of utilization, concentration, medium, etc.

In this chapter, different aspects of aroma encapsulation will be considered: the choice of support material depending on the searched properties of microcapsules and the production of microcapsules, and examples of aroma encapsulation to produce complex structured products.

39.2 PROPERTIES OF MICROCAPSULES AND CHOICE OF SUPPORT MATERIAL

The general definitions of encapsulation (micro-nano encapsulation) may be applied to any active component X to be encapsulated: it is placed in a medium where its mobility and reactivity are reduced, controlled. X is protected inside the microcapsule with a minimal fraction of "unprotected X" on the surface.

39.2.1 Properties of Microcapsules

The core or the active component X is usually dispersed/placed in the encapsulating agent called support, wall, carrier, shell, matrix, skin, or film. In the case of aroma, the protection is applied to a mixture of organic components, some of them possibly volatile at normal pressure and temperature, with chemical functions more or less sensitive to oxidation, hydrolysis; often concentrated; oil or water soluble. The final capsules with different sizes will exist in products as

- A liquid as an emulsion or a gel (nm, µm)
- A solid as powder particles (µm, mm), with shape of spheres, sticks, crystals
- A gas as particles to reach lungs

Microcapsules must allow, according to Shahidi and Han (1993),

1. Reducing and controlling the reactivity of the aroma with its outside environment: light, oxygen, water, and other components
2. Decreasing the evaporation or transfer rate of the aroma components to the environment (modifying composition, profile, intensity) during all processing operations and storage, depending on temperature/heat, water content (a_w, atmosphere), and duration
3. Promoting defined handling of the aroma:
   a. Dilution in a support (if necessary)
   b. Achievement of uniform dispersion in the host/support material, at the right concentration
   c. Change into a shape, with size and outside surface, allowing mixing (in liquid, paste, solid) with other ingredients, sometimes incompatible if no encapsulation
   d. Conversion of the liquid aroma into a solid form easy to associate with other constituents (i.e., in a support)
4. Controlling the final release of the aroma: under specific operating conditions, with uniform distribution, masking other tastes, smell, with lasting effect (proper delay)

The wanted final release of aroma from microcapsule structure may be rapid or progressive (in water, solvent, saliva, washing, atmosphere), by mechanical breaking (scratch), physical or chemical action, diffusion, dissolution, or melting, according to the final use (Figure 39.1).

The parameters of release will be related to

- The structure and support composition (including coating) and properties (solubility, permeability, hygroscopicity, glass structure Tg)
- The water activity a_w, water content, relative humidity (RH) of atmosphere/medium in contact
- The temperature, pH, agitation (or not)
- The time in relation with reactions kinetics
- The preserved storage conditions (t, T, RH, O_2, etc.), between preparation of encapsulated aroma and final use

For example, during the storage of aroma powders, the unwanted release of encapsulated molecules (especially components with different volatilities) will depend on atmosphere, RH, and composition

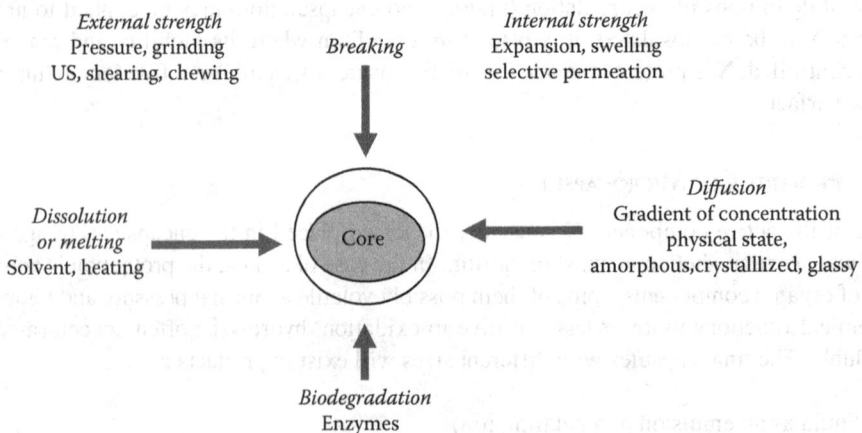

FIGURE 39.1 Modes of core release from capsule (including time, pH, temperature).

Encapsulation of Aroma

of supports/emulsifiers: this may be related to the water adsorption by wall materials, the possible modification of structure (collapse, glass transition) and appearance (or not) of cracks.

39.2.2 Choice of Support Material

39.2.2.1 Searched Properties for Supports

The choice of the aroma encapsulating supports (matrix) is the base for the success of encapsulation, along the different steps: from the formulation through the encapsulation process followed by storage and finally use/release. For example, a good support for encapsulation process may show bad properties during storage!

The properties of supports to consider for the composition depend on the final use:

- No reaction with aroma during process and storage; natural components (increasing demand), food/pharmaceutical grade
- Capacity to retain active component during process and storage, to limit diffusion, evaporation of aroma, along the wanted shelf-life
- Good protection against environment (heat, light, humidity, oxygen) to avoid physical changes (collapse, crystallization, reactions, etc.)
- Capacity to release active component in the end (operating conditions)
- Agreement for integration into the final formula

Then depending on the *processes* of encapsulation and on the wanted final physical structure, other properties to consider for supports are

- *Solubility* in solvents used in the concerned industry, food, pharmacy, cosmetics, chemistry
- Providing easy *dispersion* or emulsion of aroma, with the desired concentration
- Capacity to *eliminate solvent* (or other constituents) used in process (i.e., water, alcohol) if necessary
- *Rheological properties* in concentrated solutions, for easy use (i.e., pumping)

And finally but important from an industrial point of view, the final decision will take care of costs (i.e., raw materials, process, and storage conditions), *security*, and *legislation*.

The main idea is to build a protective network for the aroma, made of long chains, with possible polar, hydrophobic, and hydrophilic properties, with defined physical state (solid, liquid, crystallized, amorphous) in the working conditions (i.e., temperature, a_w). The necessary good knowledge of the aroma constituent properties provided by the aroma suppliers will complete the data given by the supports suppliers.

39.2.2.2 Examples of Supports

The following examples are commonly applied especially for food products (Risch and Reineccius, 1988, 1995; Karbowiak et al., 2007; Augustin and Hemar, 2009; De Vos et al., 2010).

39.2.2.2.1 Polysaccharides

Maltodextrins (DE < 20) are cornstarch derivates, with high solubility in water. They are neutral (white, tasteless), source of reducing environment, and they have poor emulsifying properties. If the DE (equivalent dextrose) increases, the viscosity decreases and the hygroscopicity increases.

They are used as support for retention of volatile substances, mainly in spray-drying and extrusion.

Modified starch is a good emulsifier because of the presence of both hydrophilic and hydrophobic groups, and is able to retain oil and hydrophilic components.

It is used in spray-drying, for retention of volatile substances as aroma.

Cyclodextrins α, β, γ have a molecular structure with hydrophobic cavity. Their solubility in water is low. They lead to a very stable inclusion complex (process, storage, and use).

Dextrose is used as texture and storage agent, with high water solubility (sweet taste), good heat stability, and low hygroscopicity. It is used in extrusion, cocrystallization processes.

Lactose is commonly used in pharmaceutical components. In powders, its ability to crystallize in high RH must be considered for storage.

Celluloses from plants, (hydroxymethyl, methyl, hydroxypropyl, ethyl, etc.) may be added to fatty acids.

Chitosan (from shellfish, mushroom) and inulin (oligofructose, chicoree, agave) are natural supports soluble in water.

Soybean soluble polysaccharide (SSPS) is used as emulsifier in spray-drying.

39.2.2.2.2 Gums

They are polymers with long chains, able to form gels, films, and to control crystallization. They are
Algae extracts (alginate, agar, and carrageenan), and plant exudates (acacia, mesquite gums).

Acacia gum (polysaccharide, proteins [5%], and fibers) brings emulsifying, film forming, and antioxidant properties, with low viscosity in water solution.

Their respective viscosity in water (1%) is: gum guar 3.5 Pa·s; locust bean 3; tragacanthe 0.7; carrageenan 0.5; acacia 5×10^{-3}.

They are used in spray-drying and are associated, for example, to maltodextrins.

39.2.2.2.3 Lipids

They are used as a barrier to water transfer (film, coating).
Wax is used for coating of components soluble in water.
Acetoglycerides have a water permeability that is a function of the degree of acetylation.
Lecithin (lipid mix) is a good emulsifier, providing low encapsulation temperature.
Liposomes (in soya, egg, and milk fat) are also used in cosmetics. They are single or multilayered vesicles with an aqueous phase enclosed within a phospholipid-based membrane.

39.2.2.2.4 Proteins

Milk proteins, caseinates, egg albumin, gluten, zein, and soy are the examples.

Proteins have the ability to assemble at interfaces. Whey proteins serve as emulsifying and film-forming agents against oxygen, aroma diffusion. Sodium caseinate is reported to be superior to whey proteins with regard to emulsifying properties and resistance to heat-induced denaturation. It is associated to lactose in spray-drying (Vega and Roos, 2006).

Gelatin is soluble in water, and gives thermally reversible gels, not easy to dry.

They are used in spray-drying, in coacervation (with gum); for coating (with sugar).

39.2.3 In Summary

The encapsulation support will be chosen according to the final use conditions (temperature, pH, delayed release, controlled water content, etc.), the mode of preparation, and the solubility properties (water, solvent). It may need to be natural, edible (i.e., food, medicine), neutral or with color, taste, odor; with hydrophilic properties (–OH groups) and/or hydrophobic, more or less polar (Figure 39.2).

The active aroma may be entrapped in

- An amorphous matrix (glassy state) (fast cooling or drying), the glass being less permeable to organic components and gases (i.e., O_2) for $T < T_g$
- In a medium with fat, crystals (sugar)
- Into linked polymers (coacervation), into liposomes

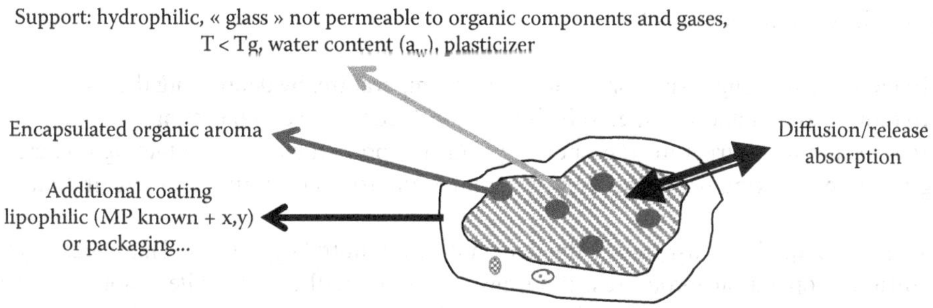

Encapsulation of Aroma

FIGURE 39.2 Schematic aroma encapsulation in powder particle.

Usually, several supports will be mixed in specific proportions, to use their complementary properties (and cost!) with the help—if necessary—of other necessary agents for emulsification, reticulation, and antioxidant action. For example, acacia gum can be associated to a less expensive carbohydrate, provided an oil/gum ratio of 1 is maintained to keep good encapsulation efficiency (McNamee et al., 2001). And it has also to be taken in consideration that the whole composition (supports + aroma) will be integrated as a part of the final product composition, including regulations limits.

39.3 SOME ENCAPSULATION PROCESSES

39.3.1 INTRODUCTIVE REMARKS

For aroma, the decision to invest in a delivery technology is driven by the balance between the cost and the added value as compared to a standard liquid flavor.

Several works may be consulted according to the objectives: Dziezak, 1988; Marion and Audrin, 1988; Risch and Reineccius, 1988; Richard and Multon, 1992; Risch and Reineccius, 1995; De Roos, 2003; Gouin, 2004; Onwulata, 2005; Madene et al., 2006; Lakkis, 2007; Bouquerand et al., 2012.

The research of the best aroma encapsulation process very often needs studies at different scales, from laboratory via pilot scale to industry scale, with the usual but not easy engineering problem of scaling-up. The choice of the process depends on the needed aroma protection and the possible supports for the wanted final use and release.

Again choice and study will be done in close cooperation between the aroma and supports suppliers, the process engineer, and the final producer/user. Very often preliminary trials will be done using aroma model molecules (i.e., one or two, or oils), at a pilot scale, with different supports to study. Then the proper scale and conditions will be studied and chosen.

The choice will depend also on the wanted production scale: either a limited production in batches for high value product (reproducibility, traceability), or a continuous controlled process with high capacity. That means also to find equipments and conditions adapted to the different scales, and often the need of specific lines (cleaning, contamination).

The process may lead to a liquid (emulsion, suspension, slurry) or a solid (paste, powder) or a gas where the aroma will be protected until release in the wanted conditions. Solid forms represent 95% of the delivery systems, of which 78% are obtained by spray-drying (Bouquerand et al., 2012).

The first encapsulation process to examine will be the liquid emulsion, used as it is or as a first step for other encapsulation form. Then several processes are using an initial formulated liquid to prepare, in the end, a dry product like powders. Starting with a liquid state provides a good initial homogeneous aroma dispersion, which will be more or less maintained in the powdered form.

Why to produce aroma powders?

- To facilitate the long-term storage at ambient temperature, by decreasing the water activity
- To reduce weight and volume, to facilitate worldwide trade and handling
- To give physical structure (i.e., porosity, surface properties) to facilitate agglomeration, granulation, mixing with other powders, and to improve flowability and rehydration

The final product quality must satisfy the demands of the initial objectives. That means that along all the different operations/processes, the transformations will be controlled: not only to define, measure, optimize the operating conditions but to be able to analyze the constituents, to describe the history of evolution of physical and chemical structures. As usual, adapted analyses are a very important complementary part of processing.

39.3.2 Emulsification

Emulsions represent a *liquid encapsulation form*, a macroscopic dispersion of two nonmiscible liquids. For aroma encapsulation, the *continuous phase* is usually the solution of supports and the *dispersed phase* is the core aroma material (or aroma solubilized/dispersed in an oil base). They are often « *oil in water* » *emulsions*:
→ polar hydrophilic liquid phase + lipophilic nonpolar liquid phase ←
Aroma emulsions have specific characteristics and needs:

- Size and size distribution (narrow) of dispersed oil drops (microcapsules/nanocapsules) (100–0.1 µm)
- Stability/protection of microcapsules during emulsion preparation and storage

Microcapsules containing fish oil with proteins as well were prepared using oil-in-water-in oil (O/W/O) double emulsification followed by gelation method (heat or enzyme) (Cho et al., 2003). The oil phase may be the base to include oil-soluble ingredients.

When the emulsion is the first step before further processing (i.e., spray-drying for powders with pumping, pulverization, and drying), stability properties of the emulsion will have to be adapted.

39.3.2.1 Preparation/Composition

Several components are associated to obtain the dispersion and stability of small regular oil drops into the continuous phase, by acting more or less rapidly and permanently at the interface:

- Emulsifiers: to facilitate the formation/dispersion of oil drops (glycerides, proteins, lecithin, etc.). They are adsorbed on the periphery of oil drops (interface oil/aqueous phase), to decrease the surface tension of the drops and to form a barrier to prevent their coalescence. They are amphiphilic molecules including both hydrophilic and lipophilic groups. They may have a role of protection against oxidation.
- Stabilizers: to maintain the dispersion by different phenomena such as steric repulsion, electrostatic interactions, viscosity, gelling effects (polyolosides, pectins, modified starches, gums, etc.).

The rate v of separation (rising or falling) of drops (mean radius r) is predicted by the Stokes law:

$$v = 2 r^2 g (d_d - d_c)/9\mu$$

Encapsulation of Aroma

where

g is the acceleration due to gravity
d_d, d_c are the densities of the two phases, dispersed and continuous
μ is the viscosity of the continuous phase

Decreasing the drop radius represents one way for stabilization of emulsion.

- Thickeners: weighing agents to improve the stability, by decreasing the difference of density between dispersed and continuous phases. They are used to increase the density of oil in which they are soluble (usually $d_d < d_c$).

These different components may sometimes interact together in a *positive or negative* way on the emulsion size, because of their structure or concentration.

39.3.2.2 Equipments

They must bring mechanical energy, shear forces, to break the oil aroma phase into small regular drops (initial coarse emulsion), then to decrease more or less the dispersed drop size (fine emulsion) to improve the stability of emulsion, directly linked to the diameter of dispersed drops. Different techniques such as ultrasound treatment, mixers (agitator, Ultra Turrax), homogenizers (with pressure), and membrane (Microfluidizer®) are used in relation with the desired final emulsion size, the composition of the emulsion, the volumes to produce (100 mL or 10 L), and with an energy consumption linked to energy density concept (Schubert et al., 2009).

During the emulsion preparation, some air may be incorporated, able to participate in aroma oxidation along time (or air inclusion in further spray-dried powder). So some sort of deaeration (i.e., under vacuum) and preservatives may be necessary.

And a part of the energy used to disrupt oil into small drops is transformed into heat that is transferred to the emulsion, increasing the temperature, which must be controlled (i.e., refrigeration) to prevent losses of volatile molecules.

The preparation of high volumes of emulsions for industrial capacity production (different from laboratory scale) needs special study and adapted equipment (Turchiuli et al., 2013b).

39.3.2.3 In Conclusion

The emulsion represents a liquid aroma encapsulation form, inexpensive, suitable for stable concentrated liquid with low volume (drinks, sauces, cosmetics) (Figure 39.3). But it may be also the preliminary step before other transformation to give a dried emulsion.

The nanoemulsion (<100 nm) is used for « clear, transparent » drinks/liquids, however, with possibility of coalescence and loss of clarity with time.

39.3.3 Spray-Drying

Spray-drying is a convective drying technique used to transform a feed in liquid or slurry form into a dry free-flowing powder. The aroma encapsulated and dispersed in a liquid form will be transformed into an aroma encapsulated in solid powder particles, with minimal modification of the aroma properties and dosage. The dry powder well dosed in aroma (~20%–25% total solid [TS]), will be easy to handle, to incorporate into a dry system, or to dissolve again.

Safety considerations will appear with flavor formulations containing high levels of low boiling components with explosion risks.

Many research works are published concerning aroma encapsulation by spray-drying. Some of the works worthy of mention are the following: Kerkhof and Thijssen, 1974; King et al., 1984;

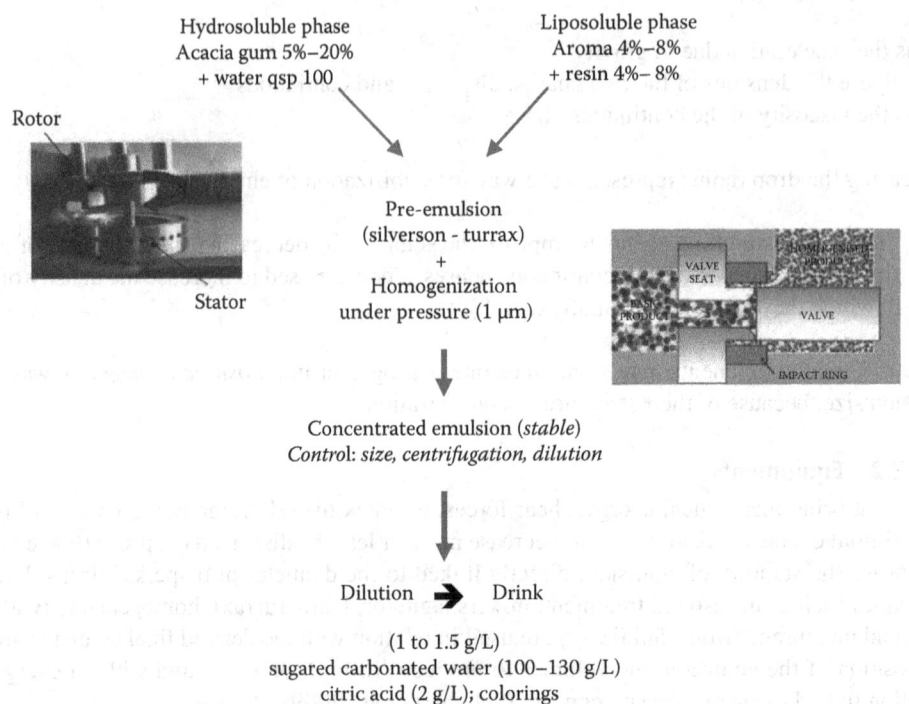

FIGURE 39.3 Preparation of concentrated emulsion for drinks.

Masters, 1985; Risch and Reineccius, 1988; Rosenberg et al., 1990; King, 1995; Risch and Reineccius, 1995; Dumoulin and Bimbenet; 1998; Re, 1998; Gibbs et al., 1999; Finney et al., 2002; Gouin, 2004; Reineccius, 2004; Vega and Roos, 2006; Vandamme et al., 2007; Jafari et al., 2008; Jafari, 2009; Drusch et al., 2012).

39.3.3.1 Principle of Spray-Drying

The formulated liquid feed containing the aroma (solution, emulsion, slurry) is dispersed with an atomizer (bifluid or high-pressure nozzle; centrifugal wheel) into spherical drops into a hot air flow (or inert nitrogen if absence of O_2 is required for oxidation/safety reasons). The contact configuration between hot air and sprayed drops is usually cocurrent. The hot air brings the necessary energy to evaporate water (solvent) from liquid drops through their surface. The heat and water mass transfers between drops and air are favored by the large contact surface, and the air flow is used to transport the resulting dry drops/particles of aroma powder along trajectories, with time-temperature histories. Spray-drying is a *continuous process,* with possible air treatment (dehydration, fines) and partial outlet air recycling.

When aqueous solution/emulsion drops are formed with atomizers, followed by solvent (water) evaporation in air, the risk of modification of the structure and composition of the solution/emulsion in the drop is limited for different reasons.

Spray-drying permits removal of water by evaporation through the drop surface while retaining volatile aroma (low concentration). This is based on the relative diffusivity of water and aroma in liquid droplets, which varies with the dry matter content (varying along drying) and temperature (Rulkens and Thijssen, 1972, Coumans et al., 1994). When water concentration in drops decreases (upon drying), both diffusion coefficients of water and aroma decrease. From a certain TS concentration, the aroma diffusivity, however, decreases faster than the water diffusivity (Figure 39.4). So, by fast drying, the drop surface layer becomes selective, which is favorable to volatile aroma retention.

Encapsulation of Aroma

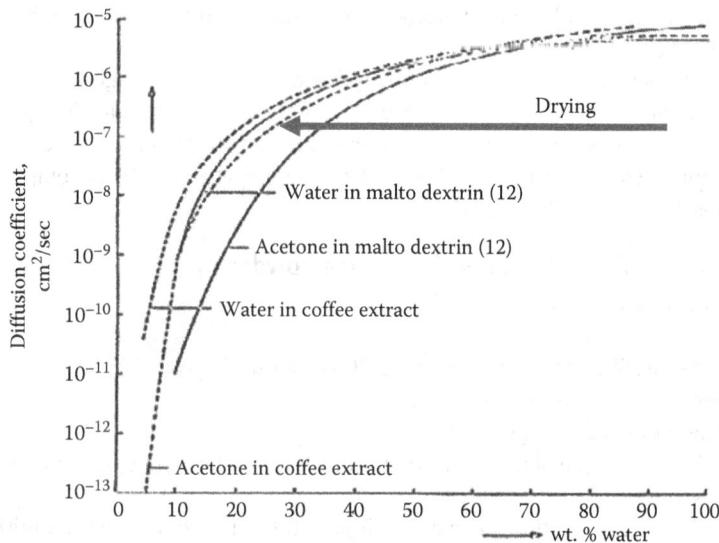

FIGURE 39.4 Thijssen theory of selective diffusion. Diffusion coefficients of water and acetone in coffee extracts and maltodextrin solutions (0.1%w/w acetone, 25°C) (*acetone: volatile molecule*).

During drying, the temperature of the drops remains below the wet bulb temperature of the drying gas. However, even if the drying time is short (some seconds) and the drop temperature low because of high water evaporation rate, some molecules of volatile aroma may migrate together with water molecules to the drop surface, where they may either disappear and/or stay unprotected at the surface.

The fast drying of drop surface layer depends on good conditions of air/drops mixing, in close connection with the viscosity, surface tension and size of liquid drops, and optimized operating conditions, without stopping further fast drying. And it is necessary to have a narrow size distribution of drops (spraying system/feed properties) to have a homogeneous drop drying behavior. The formation of spherical drops with a wanted size and narrow size distribution will depend on the physical properties of the sprayed liquid (viscosity, surface tension) associated to the chosen spraying system (sheets, ligaments broken in drops), and controlling liquid movement/mixing inside the drops before solid particle formation.

39.3.4 Influence of Operating Conditions

1. **A fine emulsion** (1 μm) (aroma in an aqueous solution of supports) leads to a better aroma retention and a lower fraction of aroma on surface of powder particles, compared to bigger ones.
2. The rapid formation of **a dry layer at the drop surface** (decreased a_w) must limit the diffusion of volatile components (selective diffusion compared to water). That will happen with:
    a. High temperature of drying air (high evaporation rate) limited by the temperature reached by drying drops (volatile evaporation)
    b. Optimal high TS concentration/viscosity of feed liquid to dry (limited by pumping and spraying need!) fixing rapidly the drop structure, with optimal drop size
3. A low controlled temperature for particles along drying process (i.e., no sticking) and an optimal short drying time will limit the losses of volatile components.

For example, during drop drying, main losses concern the more volatile constituents of aroma. To compensate these losses, the initial mix may be enriched in these constituents. This is also linked to the size of aroma molecule and aroma concentration in the liquid to dry, and to the possible migration of aroma into the drop.

Operating conditions depend on the chosen equipments (spray-dryer geometry, atomizer type), airflow rate, and temperature levels (difference inlet/outlet, feed flow rate) fixing the drying rate. For a given inlet air temperature, a higher outlet temperature corresponds to faster drying but higher powder temperature. And according to chosen supports, drying may be accompanied by surface cracking and modified aroma retention.

39.3.4.1 Characteristics of the Resulting Aroma Powder

The powder characteristics will be defined as

- Size and size distribution of particles (i.e., 20–100 μm).
- Water content, water activity (i.e., <0.2).
- Aroma fraction in 100 g of powder.
- Fraction of aroma on particle surface (nonprotected), surface state, smooth or with asperities (microscopy).
- Aroma well dispersed in the particles (with possibility to reconstitute the initial emulsion by dissolving the powder).
- Physicochemical and end-use properties of the bulk powder such as rehydration, density (bulk and packed), flowability, hygroscopicity, mechanical resistance, porosity, etc. Depending on the emulsion composition, the gas content of the drop, the drying conditions, and the dry particles may be compact or hollow, with a different distribution of aroma.

Figure 39.5 gives the spray-drying typical conditions, and the different operations with corresponding variables are summarized in Table 39.1. Some schemes for atomizers (nozzles, wheel),

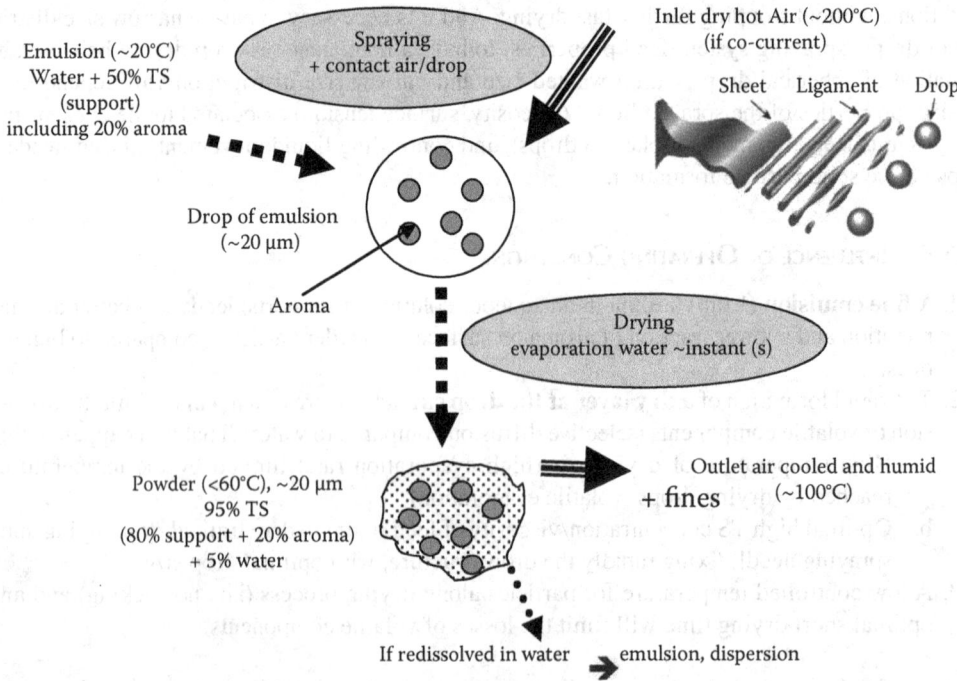

FIGURE 39.5 Schematic spray-drying process for dried emulsion.

TABLE 39.1
Spray-Drying Process and Main Parameters

Operations	Variables
	Total solid (TS) content
Dispersion, support + water	Composition
+ emulsifier	Viscosity, pH, temperature
+ stabilizer	% support/aroma (<20%)
+ aroma	
	Equipment (agitator, pressure, membrane)
Homogenization **emulsion** O/W	Size emulsion, size distribution
	Viscosity, surface tension
	Equipment (chamber, geometry, etc.)
Spraying into **drops**	Atomizers: nozzle (air, pressure; position, number); rotary, etc.
+	Shape/size, size distribution of drops
	Trajectory of drops
Drying **hot air**	Flow rate/pressure drying air
	Temperature air inlet/outlet
	Flow rate/temperature product
	Powder yield (TS), sticking
>> *Aroma encapsulated in powder*	Water content, aw; composition, retention, degradation,% surface; structure (amorphous, crystallized)
(+ agglomeration in fluid bed)	Size, density, wettability, flowability, surface state
	Temperature, Tg, MP

spray-dryers (one, two steps) are given in Figures 39.6 through 39.8. The choice of spray system, spray-dryer, and operating conditions will impact the resulting powder properties.

The spray-dried powders have a small particle size (10–100 μm) with poor handling properties. Very often in the food industry, at the outlet of the spray-drying chamber, the spray-dried particles are modified by additional treatment such as agglomeration, allowing the modification/increase of powder size/porosity and improvement of solubility/dispersibility properties (i.e., instant powder, decrease of fine particles proportion) (Buffo et al., 2002).

Computational fluid dynamic simulations have been used to model the drop/particle trajectories in spray-drying (Gianfrancesco et al., 2010), but optimization is still complex.

39.3.5 In Summary

With the spray-drying process, the main objectives are to obtain an aroma powder with a « wanted quality »:

- By easy elimination of solvent from a liquid, an emulsion, a slurry (drying properties)
- In a short time (i.e., a few seconds)
- By building solid particles with homogeneous size (i.e., 100 μm) and shape
- Maintaining the integrity of constituents (keeping aroma profile), with minimal aroma loss
- Giving to the final product, properties adapted to the use.

That will determine the choice of the spraying mode and drying parameters, including the liquid feed composition.

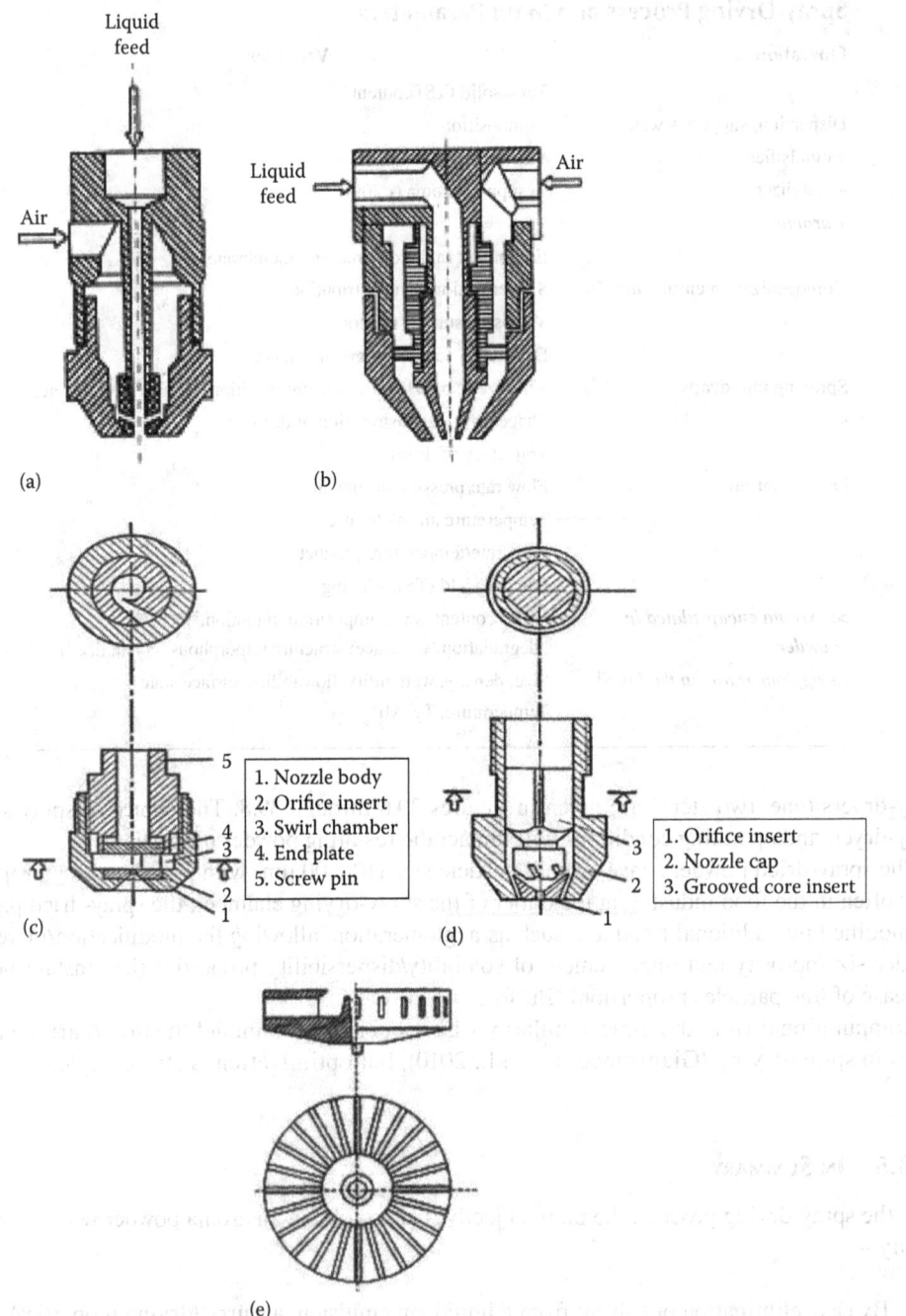

FIGURE 39.6 Types of atomizers in spray-drying for structured liquid drops. Two-fluid nozzle with (a) internal air/liquid mixing and (b) external mixing; pressure nozzle with (c) internal, (d) external air/liquid mixing, (e) rotary/centrifugal atomizer. (Adapted from Pisecky, J., *Handbook of Milk Powder Manufacture*, Niro A/S/, Copenhagen, Denmark, 1997, 261p.)

Encapsulation of Aroma

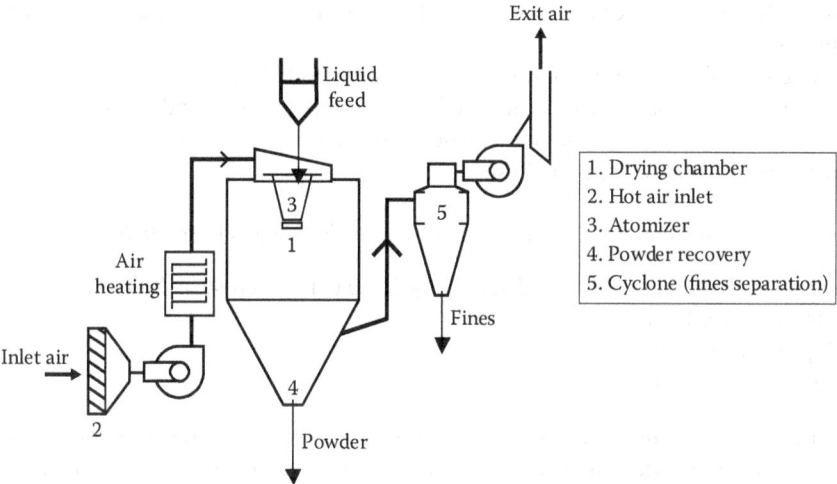

FIGURE 39.7 Spray-drying process scheme (one stage). Hot air drying/drops *cocurrent* = no high T for dry product. Residence time ~10 s, continuous process). (Adapted from Pisecky, J., *Handbook of Milk Powder Manufacture,* Niro A/S/, Copenhagen, Denmark, 1997, 261p.)

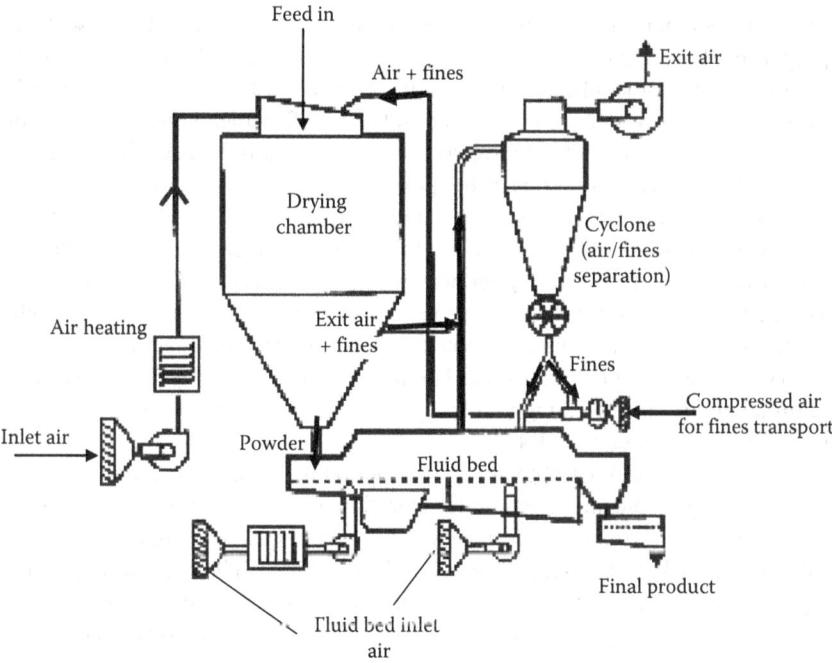

FIGURE 39.8 Spray-drying process (two stages) with external fluid bed. (Adapted from Pisecky, J., *Handbook of Milk Powder Manufacture,* Niro A/S/, Copenhagen, Denmark, 1997, 261p.)

39.3.5.1 Spray-Drying of Aroma Emulsions

The aromas in emulsion are dried with a « support » with the following properties:

- Edible, nontoxic and soluble in water (usual solvent, or ethanol) leading to a viscosity (optimal concentration) suitable for pumping and spraying
- Without action on aroma; neutral (i.e., odor, color, taste)

- Good ability to form a crust/layer at the surface of drops (~easy to dry), avoiding sticky problems (Tg)
- Low hygroscopicity for further storage (maintain capsule intact)
- Able to release aromas when further used in controlled release conditions
- Particle size and size distribution of particles linked to spraying in drops (~20 μm)
- Studied cost in relation with final use

In food, the supports (alone or mixed, more or less soluble in water) are the following:

- Glucides, maltodextrins, modified starch, cyclodextrins, sucrose, celluloses
- Lipids, wax, lecithins
- Milk proteins, gelatine
- Gums, algae extracts

The load in extract or aroma in the support (in powder) is usually 10%–30% w/w of TS. Lipids, proteins, gums, and modified starch act as emulsifying and film-forming agents. The carbohydrates act as matrix material, which may be protective against oxidation during storage (high DE) (Sheu and Rosenberg, 1995; Re, 1998; Baranauskiene et al., 2006, 2007).

Some examples of encapsulation studies are given, using simple *model aroma molecules* that are more or less volatile and soluble in water.

- Different food supports were tested (maltodextrin, inulin, and acacia gum with emulsifying properties) for the encapsulation by spray-drying (air T 180°C–90°C) of model lipophilic component (α-tocopherol dispersed in olive oil, 1/4) to prepare dry emulsions in powders. The emulsions (2 μm) contained 40%w/w TS, which contain 8%w/w olive oil, 2%w/w α-tocopherol, and 30%w/w support. No significant influence of the nature of these carriers on the physical properties of the powders and on the spray-drying efficiency was observed: size, size distribution, densities, flowability; and 73% of the initial oil was recovered in powder, with 5% of the oil phase at the particle surface (nonencapsulated). But due to its emulsifying and film-forming properties, the use of acacia gum, in combination with maltodextrin and/or inulin, helped to get more stable initial emulsions with monodispersed size distribution (~2 μm), and higher powder yield for spray-drying. When redissolved in water, dry emulsions led to reconstituted emulsions with a size distribution similar to that of the initial emulsion, indicating that spraying of the emulsion into drops did not modify its structure (Turchiuli et al., 2014).
- The presence of low MW carbohydrates in supports may contribute, via their amorphous state (T < Tg), to possible modification/deterioration of quality and functionality of powders linked with temperature and RH, especially during the spray-drying process and storage: crystallization, stickiness, caking (Vega and Roos, 2006). The determination of Tg will be important in predicting the onset of deteriorative reactions. And the choice of supports (proteins adsorption) may help to modify the powder surface (Jayasundera et al., 2009).

One example of study used aroma molecules as models with different MW and volatility (diacetyl (DA) and vanillin (VA), citral, linalyl acetate); and different supports such as maltodextrin, sucrose, skim, and whole milk (Senoussi et al., 1994, 1995).

The replacement (partly) of maltodextrin by sucrose led to more regular particles and better powder wettability. Flowability and regular shape of skim milk powder were better than that of whole milk due to absence/presence of fat.

It was observed higher retention of VA compared to DA (MW: VA > DA; volatility VA < DA) and better retention with whole milk (due to fat phase) compared to skim milk. The partial replacement of maltodextrin by sucrose (smaller MW) is favorable to departure of aroma molecules with water during spray-drying.

Encapsulation of Aroma

The study of volatile diacetyl retention (BP 88°C) in various constituents of milk (lactose, fat, proteins) during spray drying showed the importance of amorphous/crystalline ratio in lactose for aroma retention that is in turn linked to glass transition phenomena, temperature, and water/RH. The presence of microcrystals in drops (concentration, fast drying) may be a barrier to diffusion of volatile molecules, therefore increasing the retention in the final spray-dried powder.

During spray-drying and storage, the milk proteins showed a high retention capacity. During storage of powders in different humid atmospheres, volatile diacetyl losses followed lactose crystallization, higher with high RH. Changes are related to the difference T − Tg. That shows the importance of controlled atmosphere humidity (i.e., 0.22°C ≫ 0.53°C, 25°C) during powder storage to avoid crystallization (i.e., lactose/skim milk) and aroma loss.

Several studies have shown (see Table 39.2) the importance of retention efficiency in spray-drying encapsulation:

- The *properties of* aroma molecules such as solubility in water, size and shape of molecules, volatility, MW, and polarity. And the importance of matrix choice, especially in relation with glass transition temperature/crystallization, which can be the source of possible modification of powder structure. Also comparing volatiles and nonvolatiles (D-limonene, fish oil), the surface oil content of nonvolatile encapsulated powders was much higher: volatile compounds can be evaporated and removed during spray-drying (Jafari et al., 2007b).
- The size of feed emulsion transformed in emulsion drops with spraying systems. The applied forces to change a film/sheet of emulsion in drops are able to break or to associate the aroma oil drops of the emulsions inside the drops, and consequently to modify the aroma dispersion. Some research works (Table 39.2) (Risch and Reineccius 1988; Soottitantawat et al., 2005a,b) show that an emulsion of ~1 µm (compared to 4 µm) is more stable, adapted to good aroma retention in spray-dried powder, with a low fraction on dried particle surface The use of smaller emulsions needing more energy is not required for high retention in these studies. And the presence of smaller drops in capsules into particles may provide greater surface for oxidation in case of oxygen penetration.
- The emulsifiers and emulsifying techniques to produce nanoparticle encapsulated powders (Jafari et al., 2007a). In the case of fish oil encapsulated in maltodextrin combined with modified starch or whey protein concentrate, spray-dried powders were obtained from nanoemulsions (210–280 nm) prepared by microfluidization, with a good efficiency (Jafari, 2009). And the saturation of the carrier solution of wall materials influenced the flavor retention and surface oil content (Penbundiktul et al., 2012).
- Spray-drying of multiple emulsions results in a double-layered microcapsule, providing better protection. For example, orange oil in water emulsion is further encapsulated in another oil phase to form a double emulsion (O-W-O). Spray-drying of this double emulsion mixed with aqueous solution of lactose and caseinate provides a secondary efficient coating (Edris and Bergnstähl, 2001).

39.3.6 Spray-Cooling

39.3.6.1 Principle

In that process, aroma encapsulated (uniform dispersion) in molten wax or fat with surfactant is sprayed in drops at an airflow rate at ambient temperature or cold. Objective is to rapidly solidify the fat, avoiding crystallization.

The spray-dryer is equipped with a heated nozzle (or centrifugal atomizer) to spray the liquid.

The main supports for food products are vegetable fat and stearine (MP 45°C–122°C); monoglycerides and diglycerides (MP 45°C–67°C); and vegetable hydrogenated oils (MP 32°C–42°C). The final products can be used in bakery, in soups. Other supports, solid at working temperature, may be chosen as a function of domain of utilization, viscosity, and crystallization properties.

TABLE 39.2
Examples of Studies on Aroma Encapsulation by Spray-Drying, with Some Important Parameters

Aroma		Carriers Emulsifiers		Feed/Emulsion		Spray-Drying Air Temperature In/Out, Flow Rate Feed Flow Rate Powder Size Range	References Main Objectives
Molecule	Properties MW, BP, S/W	Product	Properties, Role	Composition % TS, Viscosity (mPa·s) % Aroma	Flavor Size in Emulsion → Powder		
D-limonene Linalool	Low solub. 1589 mg/L water	Modified starch (HiCap100)	Carriers	~10% to 40% TS (viscosity × 60)	Polytron + homogenizer, 8000 rpm, High-pressure Homog. 1 (GA) to 5 μm = >Nonsaturated or saturated overnight	3–9 mL/min feed 180°C–120°C Two fluid nozzle atomizer	Penbundiktul et al. (2012) Effect of saturation of carrier solution on high retention and low surface oil
Linalyl acetate (= Bergamot oil)	30 mg/L	Acacia gum		~10%w/w TS Flavor/solid 1:4 w/w			
Cardamon spice oleoresin (cooking, acid beverages)	Balance volatile (terpenoids)/ nonvolatile. Chemical changes for T > 149°C;pH <4	Acacia gum Modified starch (HiCap) Maltodextrin + Tween 80	Wall + emulsifier + for good emulsion	Oleoresin: 5% of carrier	Shear homogenizer 5 min, 3000 rpm, + 2 drops Tween	Buchi 190 178°C/120°C Nozzle air 5 bar 300 g/h liq. feed	Krishnan et al. (2005) Content and stability (6 w) (volatile/ nonvolatile). = >GA good protection
D-limonene Ethyl butyrate Ethyl propionate	Terpene, MW 136; 176°C Low sol. Ester MW 116; 116°C; 6.7 10⁻³ v/v Ester MW 102; 102°C 1.7 10⁻² v/v	MD 15–20 Acacia gum Mod starch (HI-CAP 100) SSPS (soybean soluble polysaccharide) SAIB sucrose acetoisobutyrate)	Carrier Emulsifier Emulsifier Emulsifier Weighing agent for esters (density)	40% w/w wet basis (10% E, 30% MD) GA-MD ~180 mPa·s SSPS-MD ~2600 HI-CAP-MD ~60–90	25% TS (mass) 0.8–4.09 μm →1 to 3 μm	200/110 ± 10°C Air 110 kg/h Centrifugal 30,000 rev/min Feed 45 mL/min Powder 42 to 55 μm	Soottitantawat et al. (2003) Role of size of aroma feed emulsion, with different carriers/ aromas, for powders % aroma retention % aroma on surface

(Continued)

TABLE 39.2 (Continued)
Examples of Studies on Aroma Encapsulation by Spray-Drying, with Some Important Parameters

Aroma		Carriers Emulsifiers		Feed/Emulsion		Spray-Drying Air Temperature In/Out, Flow Rate Feed Flow Rate Powder Size Range	References Main Objectives
Molecule	Properties MW, BP, S/W	Product	Properties, Role	Composition % TS, Viscosity (mPa·s) % Aroma	Flavor Size in Emulsion → Powder		
d-limonene	Low solubility MW 136 BP 176°C	Maltodextrin + Acacia gum (emulsion) Or +Soybean polys. SSPS (emuls.)	Carrier Emulsifier 10%w/w in liquid feed	10% to 30% TS (w/w) Flavor/emulsifier 0.25–1	Polytron homogenizer, 3 min + Microfluidizer (82.8 MPa)	45 mL/min feed 150/74°C–100°C Centrifugal atomizer	Liu et al. (2001) Very good retention for D-limonene (GA) For Et. But. good with SSPS or adjusting density (stable emulsion)
Ethylbutyrate	6 g/L, 20°C MW 116 BP 121°C	SAIB sucrose acetoisobutyrate +Triglycerides MCT	To regulate density of et. but. (stable emulsion)				
Citral	BP 228°C	Acacia gum	Emulsifier	30%–40%–50% TS	Shear homogenizer 20 min	Leaflash 350/100°C Nozzle 70 kg/h liq. feed Powder 50–150 μm	Bhandari et al. (1992) Optimal ratio ME/GA (3/2) for 82% encapsulation
Linalyl acetate	1.34 g/L, 37°C BP 220°C	Maltodextrin 17	Carrier MD/GA 4:1; 3:2; 2:3; 0:1	Flavor/support 1:4	Or + pressure homog. 1–5 μm		
Orange oil	0.9 g/mL, 25°C Proportion 80/20 0.83 mg/L	Acacia gum Modified starch Glyceral abebate, glycerol tribenzoate, brominated vegetable oil	Carriers Possible weighing agents to increase oil density (solubility)	Carrier:flavor ratio 4:1 37.5% TS	Coarse (whisk) 4 μm Medium coarse (high shear mixer) 2.5 μm Medium fine (+ high speed) 1.8 μm Fine, homogenizer 1.7 μm Microfluidizer 0.9 μm	200°C/100°C Niro	Risch and Reineccius (1988) Emulsion size decrease related to better flavor retention and less surface oil for powder

Note: MW, molecular weight (g); BP, boiling point (°C); S/W, solubility in water; MD, maltodextrin; GA, acacia gum; E, emulsifier.

Usually capsules have a large size (0.5–2 mm) and they are nonporous, not soluble in water. Active component may be partly found at the particle surface. The main advantages are the absence of solvent, a low consumption of energy, and no degradation of heat-sensitive components. But possible sticking on walls may lead to losses.

This process may be used for a double encapsulation of spray-dried particles.

Another possibility could be the atomization of emulsions (simple or double) with a twin-fluid nozzle, in cold atmosphere such as −30°C, to produce solid particles. This process is used to study the atomization process (drops size distribution, shape) through the microstructure of the solid particles.

39.3.7 FREEZE-DRYING

39.3.7.1 Principle

The aqueous aroma composition (solid, liquid, paste) includes components/supports able to form by freezing an amorphous matrix or a gel to immobilize aroma compounds. It will be transformed into an aroma powder, without modification of aroma, for easy storage. After freezing of the aroma composition, there is drying by sublimation of ice, usually under vacuum. The driving force for sublimation is the difference in pressure between the water vapor pressure at the ice interface and the partial pressure of water vapor in the chamber. Because of the low processing temperatures, thermal degradations reactions are excluded.

Atmospheric pressure may be used if the water vapor pressure of gas is inferior to vapor pressure of ice.

39.3.7.2 Main Characteristics

Usually the structure of the matrix is preserved during the whole process, which is a long one. Temperature and pressure are low (Coumans et al., 1994; Buffo and Reineccius, 2001).

The main parameters are

- High initial concentration in dissolved solids, to reduce the volume fraction of dispersed aroma, but giving a very dense product
- Freezing rate not too high (pure ice, aroma components concentrated in matrix phase)
- Sample size not too big (low thickness)
- Nature of volatile component: the retention increases with MW

Volatile retention occurs through entrapment in the amorphous matrix of hydrogen-bonded carbohydrate molecules and volatile retention regions are quite small.

The equipment is made of a chamber with plates, which can be either refrigerated or heated, under vacuum, with cold trap for water vapor (sublimation).

Freeze-drying consists of several steps (Figure 39.9):

- Freezing (AB) where most of water is frozen (~ −20°C, ice crystals) (internal or external) leaving a concentrated matrix phase (with aroma)
- Vacuum (BC), then heating for ice sublimation (CD, leaving open pores) and desorption (last% water…) (DE)
- Ambient pressure (in dry inert gas) (EF)

The final residual moisture content (low) is determined by secondary drying phase.

It is also possible to prepare a spray freeze-dried powder. Atomized drops are immediately frozen in liquid nitrogen and the frozen solvent removed by sublimation.

This technique is long and costly, requiringing specific storage and transport.

Encapsulation of Aroma

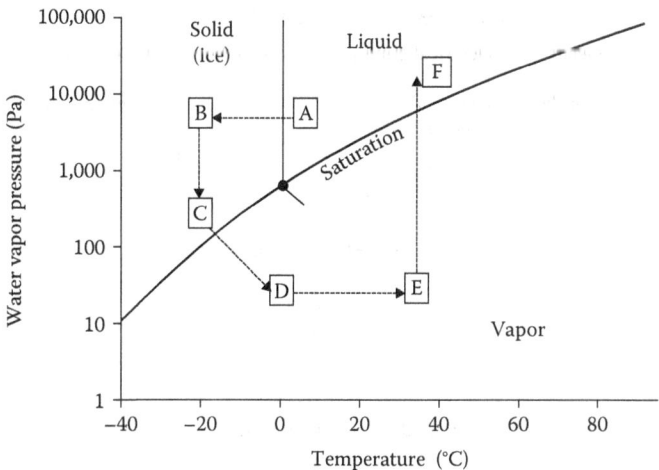

FIGURE 39.9 Freeze-drying principle (AB freezing; BC vacuum; CD heating/ice sublimation; DE water desorption; EF ambient pressure).

39.3.8 Cocrystallization

39.3.8.1 Principle

The solution of support able to crystallize (i.e., sucrose/lactose in water) is concentrated till it reaches supersaturation. Then the aroma is added into the supersaturated sugar syrup under vigorous agitation. After nucleation, the crystallization is operated with emission of a substantial amount of heat. Then *solid crystals* (3–30 μm) *with entrapped flavor and agglomerates* are formed and separated, dried, and sieved (Chen et al., 1988; Beristain et al., 1996; Kim et al., 2001).

39.3.8.2 Main Characteristics

The aroma is trapped in between microcrystals.

In the case of sucrose and water, the aroma must be soluble or dispersible in water. The encapsulated aroma is easy to dissolve, but the solution has a sugar taste! Addition of an antioxidant may be recommended (porous structure).

The principle may be adapted to other supports able to crystallize.

39.3.9 Coacervation

39.3.9.1 Principle

The process is based on the phase separation of one or several hydrocolloids from an initial solution and the subsequent deposition of this newly formed coacervation phase around the active ingredient (aroma) suspended or emulsified in the same reaction media. Then the hydrocolloid shell may be cross-linked with appropriate cross-linkers. The phase separation will be provoked by reaction between two different colloids by varying temperature, pH, and medium composition (solvent, salts). This process was tested in the 1950s for carbonless copy paper (dye in particles ~20 μm) (Risch and Reineccius, 1988; Gouin, 2004; Leclercq et al., 2009).

This is used for encapsulation of oils or lipohydro-soluble products. The capsules are very rich in internal phase (oil/aroma; 50%–90%), with very thin wall ("core/shell" system). If a final dry product is needed, the drying of capsules (fluid bed, freeze-drying) is delicate with possible loss of molecules with low MW.

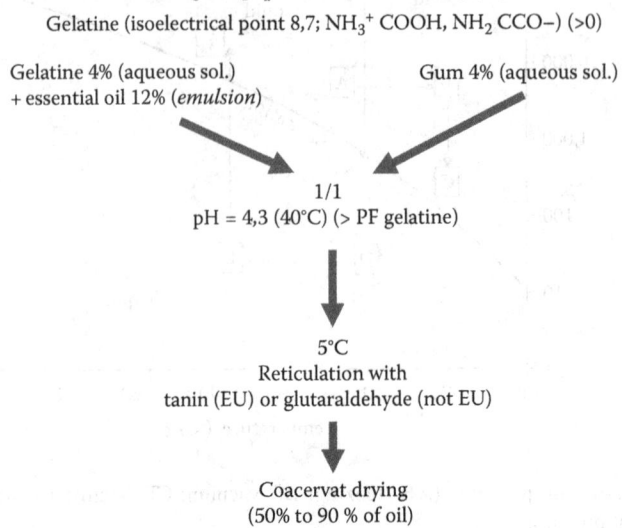

FIGURE 39.10 Example of complex coacervation.

39.3.9.2 Methods of Coacervation

Several methods exist, not all of them are applicable to food products, which use the evaporation of solvents, or noncompatible polymers, or interfacial polymerization.

In simple coacervation or gelification (500 µm to 2 mm), an emulsion of oil in an aqueous solution of a polymer/substance able to form a gel, is prepared. By changing pH, temperature, or adding salts, the substance will precipitate around the drops (alginate/$CaCl_2$; gelatine hot/cooled oil). Then particles are separated and dried. Essential oils in zein (proteins) nanospherical particles (100 nm) were prepared by phase separation, and then lyophilized (Parris et al., 2005).

In complex coacervation (20 µm to 1 mm), for example, aqueous solutions of active component (AC), polyanion (–) and polycation (+) are mixed. The two polymers with opposite charges (electrostatic interactions) will interact to form a deposit of coacervate at the surface of AC (i.e., acacia gum, alginate; CMC with gelatine; proteins and anionic polysaccharides) (De Kruif et al., 2004). Reticulation may be provoked by dilution, and modification of pH, temperature (Figure 39.10). Gelatine and acacia gum (opposite charge at low pH) were used to encapsulate flavor lipophilic oil to be used in frozen foods and released upon heating (Yeo et al., 2005), with liquid or solid core (Leclercq et al., 2009).

39.3.10 EXTRUSION

39.3.10.1 Principle

The aroma to encapsulate is dispersed in the melted or dissolved support (i.e., melted sucrose 110°C–130°C), in premix, in screw or injected at the end. Then the dispersion is extruded (forced to pass through a hole with studied size and shape, a needle tip) into a medium where the matrix is rapidly transformed in a gel, a precipitate, solid, with surface cleaned/dried by the medium (i.e., isopropylic alcohol for sugar). The main parameters are good mixing and controlled operating conditions such as mechanical forces, temperature (relatively low), pressure, and residence time, usually with no expansion at the outlet (Figure 39.11).

The shape of extruded solid products may be monodisperse microspheres, or small sticks (~300 µm to 1 mm).

If the mix contains water (matrix, water, aromas), extractive drying was achieved by removing water from the formed rigid drops by means of contact with water-absorbing second phase as PEG400 (Kerkhof and Thijssen, 1974). In another example, the aqueous emulsion of core (volatile esters,

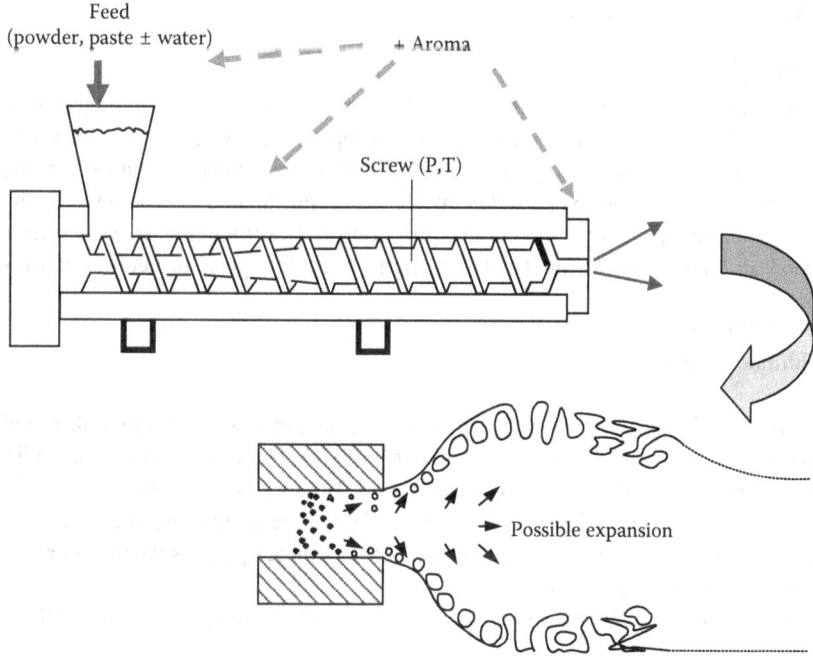

FIGURE 39.11 Extrusion process scheme.

paprika oleoresin)/wall (acacia gum, CMC) materials was injected/sprayed in a cold dehydrating liquid as absolute ethanol. The resulting slurry was filtered and then dried under vacuum (50°C). Retention was improved with high solid concentration, low core-to-shell ratio (Zilberboim et al., 1986).

The main supports in food products are glassy carbohydrates alone or mixed: dextrose, corn syrup, glycerin, maltodextrins, sucrose added with emulsifiers, monoglycerides, gum, pectin, lecithin. Cyclodextrin complexed flavors may be prepared prior to extrusion (Bhandari et al., 2001).

The extrusion products typically contain 8%–10% of flavor oil, but the final washed surface makes the product very stable (Rish and Reineccius, 1988, Qi and Xu, 1999). Carbohydrate matrices in the glassy state have very good barrier properties (i.e., atmosphere gases). Dripping and jet breakup are the methods to form microcapsules (Whelehan and Marison, 2011). Parameters are screw temperatures and speed, and residence time distribution (Yuliani et al., 2006).

One example of active component encapsulation (embedding) is described as a continuous process (several screws sections) with a matrix composition made of a hydrophobic component for controlling the release (i.e., fat, wax, paraffin, etc.), a material plasticizable at low shear (i.e., starch, cyclodextrin). The active component is added (5%–20% w) to the melt matrix at low temperature with a reduced postextrusion drying and expansion. Particles are extruded through a die with multiple apertures (i.e., 0.5–7 mm). They may be covered with additional film forming substance (i.e., wax, fat, etc.) (Van Lengerich, 2003).

In co-extrusion, two concentric jets are used (i.e., double fluid nozzle) (Schlameus, 1995; Gouin, 2004):

- An internal jet with product to encapsulate
- An external jet made of the wall product, liquid, melted (wax, fat), or in solution in water (hydrocolloid)

At the edge of the nozzle, Rayleigh instabilities lead to formation of round beads. The wall solidifies by cooling (wax), in a cooled organic solvent (gelatine); in $CaCl_2$ solution if alginate. The obtained capsules are spherical, with uniform size (>500 μm) and the internal phase may be important as 90% of the capsule mass.

39.3.11 MOLECULAR INCLUSION

39.3.11.1 Principle

The molecule of cyclodextrin (CD) (usually β) is a cut cone made of seven glucopyranose units; the external part is hydrophilic and the internal part is hydrophobic (MW 1145; solubility 1.85 g/100 mL water (25°C)(α-CD 12.7 g/100 mL; γ-CD 25.6 g/100 mL). The internal cavity (diameter 5–8 Å) is nonpolar, and may accept a nonpolar « host » molecule of similar size as aroma molecules (6%–15% w/w). This is a selection by the host molecule configuration. If formed, *the aroma/cyclodextrin complexes are very stable* to evaporation, oxidation, light, heat, with good protection of aroma profile.

39.3.11.2 Production

Several techniques may be used:

- In liquid state, the aroma dissolved in a solvent soluble in water (i.e., ethanol) is added drop by drop in an aqueous solution (i.e., + ethanol) of CD. Then the final complex CD/oil will precipitate (kind of cocrystallization) (lemon oil, Bhandari et al., 1998).
- In solid state, CD in paste (powder, water) is mixed (kneading) with the product to encapsulate, with just the quantity of water, some of the complex paste is vacuum- or spray-dried (lemon oil, Bhandari et al., 1999).
- Aroma vapor may flow through an aqueous solution of CD, with precipitation of the complex.

In all cases, the product may be filtrated and dried to get a final solid. The stable final product is interesting for strong flavors or tastes.

CD may be used in combination with carbohydrates (spray-drying, extrusion) to ensure both encapsulation and further protection (Qi and Xu, 1999).

39.3.12 COATING: IMPREGNATION

39.3.12.1 Principle of Impregnation

The aroma is dispersed at the surface of a support in powder (sugar, salt, modified starch, and silica gel) in a mixer (i.e., sucrose with vanilla).

The process is simple, economical but with bad protection of aroma. The product needs an additional protection as packaging against temperature and light.

Microcapsules for textile finishing or for incorporation (fragrance, essential oil) into the textile fibers have to be resistant to mechanical and thermal stress, with impermeable and pressure-sensitive walls. In situ polymerization of aminoaldehyde polymers may be chosen with good results (Boh et al., 2011).

39.3.12.2 Principle of Coating in Turbine/Drum/Screw (Endless) or in Fluid Bed

Coating can be used with solid particles containing aroma as an additional protection (i.e., smooth or hard structure). An additional possibility may be to include aroma (same, different) in the coating layer (well protected).

This is realized by agitation/fluidization of solid particles (individual) in an airflow rate—hot or cold. Then the coating product is sprayed (nozzle) on the moving particles, giving progressively a homogeneous layer.

For air fluid bed particle coating, spraying may be done from above (top) or under (bottom) the fluidized bed of particles (100 μm to mm). The added layer is usually thin (~no size change) and may need final drying (Figure 39.12). Several layers may be formed successively. Usually, particle size is superior to 100 μm for a good fluidization, which implies spraying of small coating drops in relation with this size to achieve uniform coating, avoiding agglomeration.

The coating products are melted (no solvent; hot-melt coating) or dissolved in volatile solvent (usually water): derivatives from cellulose, dextrins, emulsifiers, lipids; derivatives of proteins or starch.

Encapsulation of Aroma

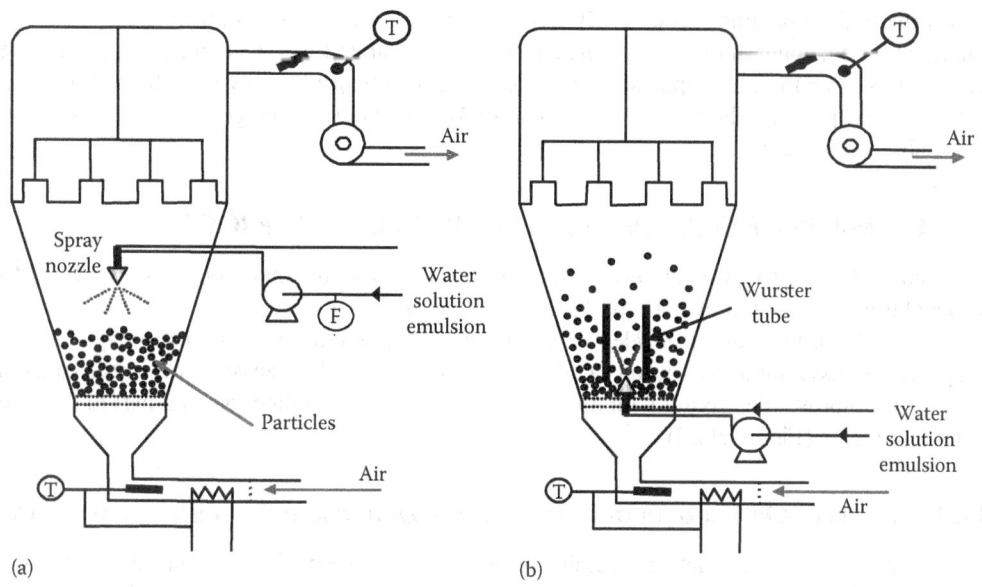

FIGURE 39.12 Coating of solid particles in fluid bed (top spray (a) or Wurster bottom mode (b)).

Among the polysaccharides, cellulose derivatives are often used as good barrier properties against gas as oxygen, with good mechanical properties but often permeable to water. A wide variety of possible coatings offer many possibilities for controlled release (Clark and Shen, 2004).

In a fluid bed, many parameters will be considered (Guignon et al., 2002, 2003; Karbowiak et al., 2007) (Table 39.3):

- The particles to coat (composition, size, density, surface state, load) and the coating (composition, adhesion to particle surface, spraying and drying properties, flow rate)
- The fluid bed (geometry, spraying system, airflow rate and temperature, continuous/batch)

TABLE 39.3
Coating Operation: Optimization Parameters for Solid Products

Product	Coating Agent	Chosen Process
Physical state	Solution or powder	Chosen equipment
Composition	Composition	Continuous, noncontinuous
Water content, fat content	Fat content	
Shape	Water or solvent content	Static, dynamic (agitation)
Dimensions	(modified drying time)	Volume
Density	Solubility in water of coating constituents	Temperature range
Texture (resistance, stickiness)		
Fines presence	Melting point or gel temperature	Flow rate
Surface state	Density	Residence time
	Viscosity, flowability limit	
Hygroscopicity	Size (if powder particles)	Energy flow rate
Surface tension	Necessary pretreatment	Air temperature and RH
Temperature	(mix, emulsion)	Coating recycling
(for easy dispersion, absorption)	Surface tension	
Production flow rate	Flow rate	

Another method to coat fine particles (~20 µm; with aroma) may consist of forming a suspension of particles in a liquid containing the coating agent (lipid), saturated with CO_2 at a supercritical pressure. Then by spraying the suspension and expansion, microcapsules may be collected in the gas after this decompression (Perrut, 2003; Gouin, 2004). The main advantage is the low temperature and the absence of water.

39.4 EXAMPLES OF STUDIES USING COMPLEMENTARY PROCESSES

The encapsulated aroma has been protected by one process, but the final product may need further additional treatment to improve or modify the final end use properties.

For example, in the case of powders, the controlled agglomeration of particles may contribute to improve the dissolution rate (porous agglomerate), to increase the size of particles, to change the density or to improve the mixing with other particles, or to add a protection (coating) to delay and control the release (Figure 39.13).

39.4.1 VEGETABLE OIL ENCAPSULATION USING SPRAY-DRYING AND FLUID BED AGGLOMERATION

The objective was to encapsulate a vegetable oil (representing a model for aroma compound) into a powder. The oil (ISIO4) (VO) will be at a concentration of 5% TS into a matrix made of maltodextrin MD (DE 12) and acacia gum (AG) (3/2 MD/AG). The composition of matrix was defined in a previous research work on encapsulation of aroma by spray-drying (Bhandari et al., 1992; Turchiuli et al., 2005; Fuchs et al., 2006).

The searched end-use properties (brief) for the powder with 5% aroma were the following:

- A low water content to avoid sticking during process and storage, and to get stability
- A high solubility in water for complete dissolution
- Controlled size, size distribution, and density facilitating good mixing with other powders, without fine particles (source for dust)
- A good flowability for easy handling, dosage, and transport
- A protection regarding oxidation during storage

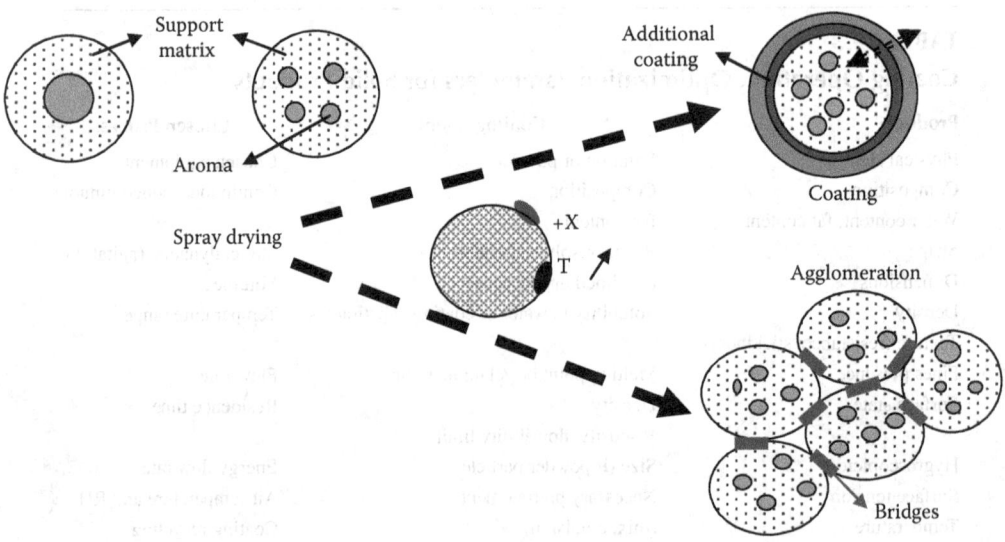

FIGURE 39.13 Complementary processes for aroma encapsulated powders: spray-drying, fluid bed coating/agglomeration.

Encapsulation of Aroma

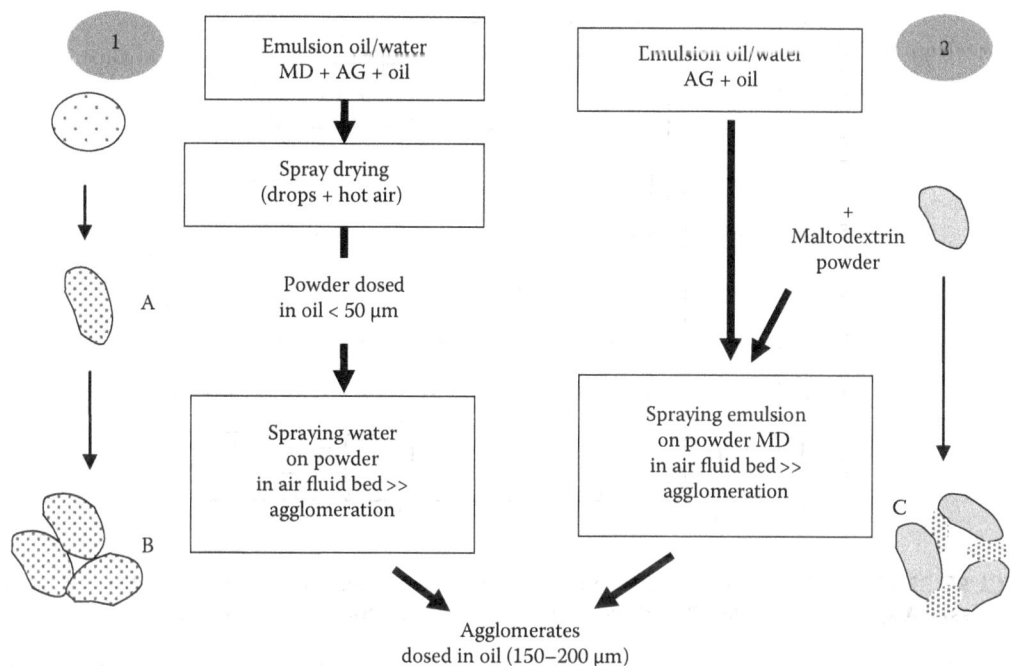

FIGURE 39.14 Two processing ways for dry emulsion preparation: three powders A,B,C (MD maltodextrin; AG acacia gum).

Three processes were tested at a pilot scale (Figure 39.14):

- *A) Spray-drying of a formulated emulsion MD/AG/VO: continuous process, with a powder yield of 65%, and a final formulated powder with small mean size (30 μm)
- *B) Agglomeration of spray-dried fine powder in a fluid bed: batch process, with a yield of 89%, giving porous agglomerates (200 μm), with low mechanical resistance but good wettability
- *C) Direct agglomeration in fluid bed of MD powder sprayed with formulated emulsion AG/VO: batch process, with a yield of 89%, giving porous agglomerates (240 μm), with both good mechanical resistance and good flowability

The three powders A, B, C, were well dosed in oil (>4.4 g/100 g TS), with low losses of oil during the different processes (>88% of initial oil in the powder), and showing a good protection of oil against oxidation (Figure 39.15). The oil on the particles surface was inferior to 2% of total oil for sprayed and agglomerated powder, and 6% for the maltodextrin agglomerated with emulsion.

In summary, the tested encapsulation method is applicable for other components and different concentrations (aromas, antioxidants, or other oil substances). Other possible supports may be proposed to replace maltodextrin and acacia gum as modified starch (Buffo et al., 2002), proteins, gums, or protective agents.

39.4.2 Aroma Encapsulation into Powder for Chewing Gum

The main industrial objective was to replace liquid concentrated aroma by a powder with same sensorial characteristics as liquid (aroma profile, intensity, etc.) to be produced at industrial scale. That meant

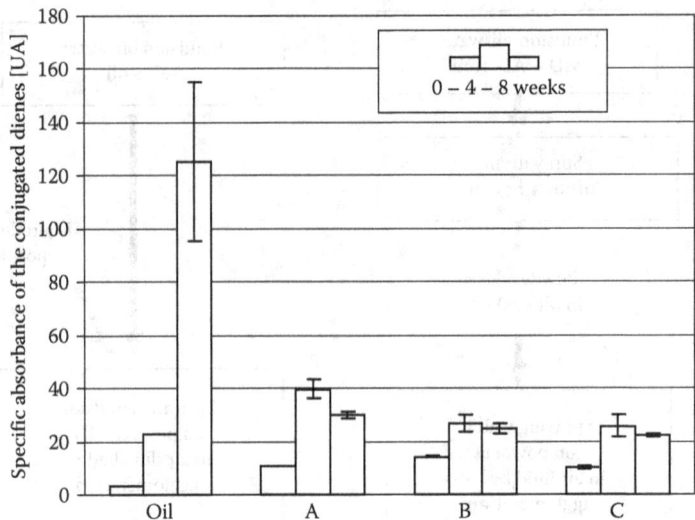

FIGURE 39.15 Oxidation tests (accelerated 60°C; conjugated dienes in extracted oil (234 nm); 0, 4, 8 weeks)—**A** spray-dried powder; **B** agglomerated A; **C** MD agglomerated with emulsion.

protection and stability of aroma (processing, storage, use) with a final controlled release during chewing (i.e., temperature, humidity, mechanical stress) (Turchiuli and Dumoulin, 2013a) (Figure 39.16).

It was proposed to prepare aroma powders with ~20% aroma in TS, with different particle size, structure, and distribution of aroma within the solid matrix.

The liquid aroma to consider was a complex mixture (M2). For simplification, a representative model mixture (M1) with three main molecules T, C, F, with different volatilities T > C > F (M1 representing 30% M2), was first used:

$$T \text{ (ester) } 96.57\% w/w + C \text{ (aldehyde) } 0.93\% + F \text{ (lactone) } 2.5\%$$

The encapsulation of M1 was studied in a matrix made of maltodextrin DE12 and acacia gum (ratio MD/GA = 3/2).

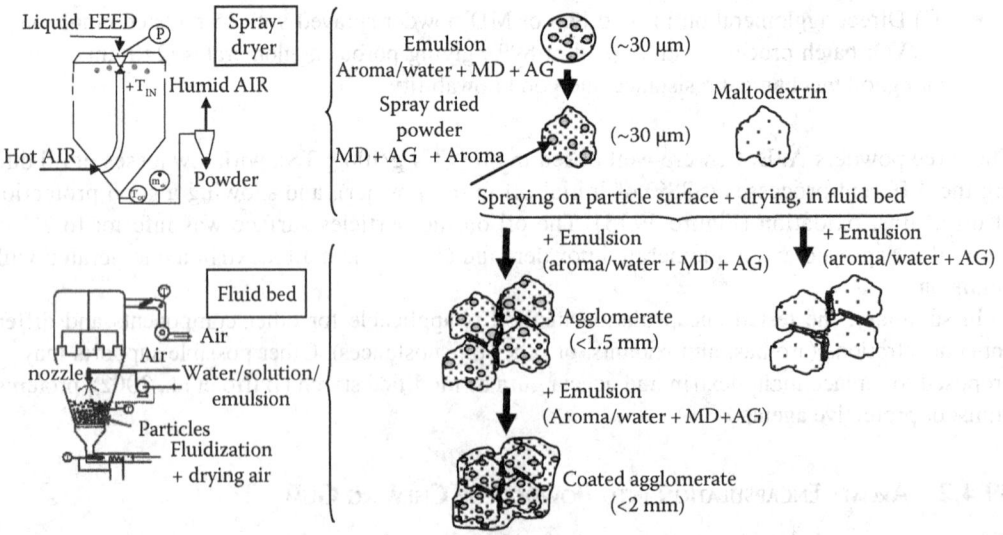

FIGURE 39.16 Aroma in chewing gum powder (MD maltodextrin; AG acacia gum).

First an aqueous emulsion (<5 μm, 40%TS) was spray-dried to obtain a dry emulsion (SD) (20% aroma, 80% MD/GA). Then some trials were done to agglomerate the spray-dried particles, in a fluid bed by spraying water on the surface of the fluidized particles (SDA). Finally, in the fluid bed, the agglomerates were coated by spraying the emulsion on them (SDAC). In the three cases, the total composition of the final powder was preserved.

At a pilot scale, the powder yield for the three studied processes was superior to 75%, with good physical properties for the powders.

The aroma concentration in dry matter was 15% in the spray-dried powder and agglomerates (compared to 20% in the initial emulsion). It was less than 9%–15% for the coated agglomerates, showing some aroma loss.

Tests for aroma perception in a gum paste (0.6%w/w aroma) to compare with liquid aroma were realized by the aroma producer. The perception was good for SP and SPA powders. For the coated agglomerates, there were changes in aroma perception. Probably due to the long coating process, it means that in that case, more aroma protection is needed in the coating, modifying the final global composition.

A spray-dried powder was prepared at a larger scale (×10) using the real aroma M2, and the results for sensorial tests were encouraging.

39.4.3 DELAYED RELEASE OF FLAVOR MATERIAL

First a base powder was prepared containing the flavor ingredient encapsulated in a matrix (water soluble or partially hydrophilic). The initial aqueous emulsion may be formed with gelatine, acacia gum, a plasticizer (glycerol), and an emulsifier. Then the drying was as a thin layer by drum drying. Secondly, the powder was ground to get a high-density powder (no holes), which will be further coated (fluid bed) with one or two layers of a water-insoluble material (i.e., polyvinylacetate, shellac, zein). The main objective was to prevent the flavor ingredient to migrate/diffuse into the product (food, toothpaste). The release will be at a given temperature by breaking, chewing, and brushing, delivering a high perception of flavors. Several different ways are described in the work by Merrit et al. (1985).

A double-encapsulated flavor powder (spray-dried) is prepared by secondary fat coating process, with better resistance to moisture and oxygen than single encapsulated powder (Cho and Park, 2002).

39.4.4 PRODUCTION OF SOLUBLE BEVERAGE POWDER

One example is given for encapsulating an aroma/flavor for a beverage (Liu and Rushmore, 1996).

An oil-in-water emulsion was prepared from coffee aroma incorporated in coffee/vegetable oil (5%–20% w), and water-soluble coffee solids forming the aqueous continuous phase (50%–75% TS). Individual drops of emulsion (nozzle/N_2, 0.4–1 mm; core of coffee oil) were sprayed on soluble coffee powder (fluid bed, pan coater), with little water content modification of powder (<4%w). The capsules were attached to the coffee powder surface (0.1%–1% w aroma), that were able to release the aroma only by dissolution in the hot water cup. The film-forming agent may be other supports such as maltodextrin, acacia gum, carbohydrates, tea, or cocoa solids, vegetables. And other flavors may be used, for example, for instant soups.

39.5 CONCLUSION

For the aroma encapsulation, a wide variety of possible supports and techniques exist to prepare a stable product with controlled aroma composition, taking into account the different steps of production, storage, and final use.

The choice for an aroma encapsulation project will be decided according to the initial objectives, with a multidisciplinary cooperation between the aroma and supports suppliers, the process

TABLE 39.4
Summary of Techniques for Aroma Encapsulation in Food Products with Some Examples of Use

Operation L liquid, S solid, P paste	Characteristics % = Active Component in Support	Size and Use (μm)
Emulsion L	Oil/water	(0.1–1)
		Drinks, dressings before drying
Spray-drying		(5–200)
L → S	Carbohydrates, proteins, gums	Citrus fruits
Hot air	10%–50%	Cinnamon, butter aroma
Cold air	Fats with low melting point	Pastry
Coacervation	Emulsion + polymerization	(5–5000)
L → L/S	(+ drying)	Citrus fruits, raspberry
	80%–90%	Vanilla, mint, coffee, onion
Inclusion	In cyclodextrins	(10–100)
L → L/S	6%–15%	Spices, onion, garlic
		Remove bitterness
Extrusion	Dispersion in matrix + extrusion in medium	(1–1000)
L/P → S	(for surface washing)	Drinks
	8%–20%	Cakes, desserts
Coextrusion	Two concentric streams (heart + wall) and	(>500)
L → S	hardening	
	90%	
Coating fluid bed	Solid particles + coating	(>100)
S + L → S	Lipid, cellulose, proteins, starch	
Coating mix	Mixing sugar + AC +…	Sugar + vanilla
S → S		
Cocrystallization	Supersaturated sugar solution	(5–25)
S + L → S	10%–25%	Drinks

Notes: L, liquid; P, paste; S, solid; AC, active component.

and chemical engineers, according to the objectives of properties for the final product, the release conditions (Table 39.4). They have to exchange with their different complementary knowledge and competencies, and find a compromise between the sciences, the choice, and limits of both formulation and process, to reach the objectives of both industry and consumers.

From the beginning of the project, the following general questions must be considered in the following order—with some possible comeback at any moment:

- What will be the functions of the encapsulated ingredients (aroma, oil) provided to the final product? What is the final physical state for use in liquid/solid/paste, and in which environment?
- What are the particle size, density, and stability requirements for the encapsulated ingredient (i.e., if mixed, mouthful factor)?
- What is the optimal concentration of the active material in the microcapsule/final product?
- By which mechanism will the aroma be released from the microcapsule?

From the answers to the previous questions, it is possible to define some possible materials, which can be used as support related to the searched protection and final use and release, with the constraints for cost and legislation (security, etc.).

Considering all these answers, one or more processes will be proposed with the following questions for each:

- To which processing conditions must the encapsulated ingredient resist before releasing its content? What is the best time to introduce the aroma in the process? What are the risks of losses of some aroma constituents, modifying the final aroma profile?
- What will be the resulting possible lifetime, with which packaging, for which storage conditions?

The final choice for the whole process will be done after research with pilot experiments, then possible scale-up to industrial trials (batch, continuous), with the necessary multiple analyses, according to the type of products to consider, for food, chemistry, and medicine. The questions of cost and security must be included all along the chosen process. Then a new tailored product will be born!

REFERENCES

Augustin M.A. and Hemar Y., 2009. Nano- and micro-structured assemblies for encapsulation of food ingredients. *Chemical Society Reviews*, 38, 902–912.

Baranauskiene R., Bylaite E., Zukauskaite J., and Venskutonis R.P., 2007. Flavor retention of peppermint essential oil spray-dried in modified starches during encapsulation and storage. *Journal of Agricultural and Food Chemistry*, 55, 3027–3036.

Baranauskiene R., Venkustonis P.R., Dewettinck K., and Verhé R., 2006. Properties of oregano, citronella and marjoram flavors encapsulated into milk protein-based matrices. *Food Research International*, 39, 413–425.

Beristain C.I., Vasquez A., Garcia H.S., and Vernon-Carter E.J., 1996. Encapsulation of orange peel oil by co-crystallization. *LWT*, 29, 645–647.

Bhandari B., D'Arcy B., and Young G., 2001. Flavour retention during high temperature short time extrusion cooking process: A review. *International Journal of Food Science and Technology*, 36, 453–461.

Bhandari B.R., D'Arcy B.R., and Le Thi Bich L., 1998. Lemon oil to β-cyclodextrin ratio effect on the inclusion efficiency of β-cyclodextrin and the retention of oil volatiles in the complex. *Journal of Agricultural and Food Chemistry* 46, 1494–1499.

Bhandari B.R., D'Arcy B.R., and Padukka I., 1999. Encapsulation of lemon oil by paste method using beta-cyclodextrin: Encapsulation efficiency and profile of oil volatiles. *Journal of Agricultural and Food Chemistry*, 47, 5194–5197.

Bhandari B.R., Dumoulin E.D., Richard H.M.J., Noleau I., and Lebert A., 1992. Flavor encapsulation by spray drying: Application to citral and linatyl acetate. *Journal of Food Science*, 57, 1, 217–221.

Boh B., Staresinic M., and Sumiga B., October 5–8, 2011. Synthesis and applications of scented microcapsules in textile products. *XIX International Conference on Bioencapsulation,* Amboise, France.

Bouquerand P.E., Dardelle G., and Erni P., 2012. Chapter 18: An industry perspective on the advantages and disadvantages of different flavour delivery systems. In: *Encapsulation Technologies and Delivery Systems for Food Ingredients and Nutraceuticals.* Eds. N. Garti and D.J. Clements, Woodhead Publishing, Oxford, pp. 453–487.

Buffo R., Probst K., Zehentbauer G., Luo Z., and Reineccius G.A., 2002. Effects of agglomeration on the properties of spray-dried encapsulated flavours. *Flavour and Fragrance Journal*, 17, 292–299.

Buffo R. and Reineccius G.A., 2001. Comparison among assorted drying processes for the encapsulation of flavors. *Perfumer and Flavorist*, 26, 58–67.

Chen A.C., Veiga M.F., and Rizzuto A.B., November 1988. Co-crystallization: An encapsulation process. *Food Technology*, 42(11), 87–90.

Cho Y.H. and Park J., 2002. Characteristics of double-encapsulated Fflavor powder prepared by secondary fat coating process. *Food Chemistry and Toxicology*, 67, 968–972.

Cho Y.H., Shim H.K., and Park J., 2003. Encapsulation of fish oil by an enzymatic gelation process using transglutaminase cross-linked proteins. *Journal of Food Science*, 68(9), 2717–2723.

Clark J. and Shen C., 2004. Fast flavor release coating for confectionery. Patent WO/2004/077956.

Coumans W.J., Kerkhof P.J., and Bruin S., 1994. Theoretical and practical aspects of aroma retention in spray drying and freeze drying. *Drying Technology*, 12(1&2), 99–149.

De Kruif C.G., Weinbreck F., and De Vries R., 2004. Complex coacervation of proteins and anionic polysaccharides. *Current Opinion in Colloid and Interface Science*, 9, 340–349.

De Roos K.B., 2003. Effect of texture and microstructure on flavour retention and release. *International Dairy Journal*, 13, 593–605.

De Vos P., Faas M.M., Spasojevic M., and Sikkema J., 2010. Encapsulation for preservation of functionality and targeted delivery of bioactive food components. *International Dairy Journal*, 20, 292–302.

Drusch S., Regier M., and Bruhn M., 2012. Chapter 7: Recent advances in the microencapsulation of oils high in polyunsaturated fatty acids. In: *Novel Technologies in Food Science*, Eds. A. McElhatton, and P.J. do Amaral Sobral. Springer, New York, pp. 159–181.

Dumoulin E. and Bimbenet J.J., 1998. Chapter 10: Spray drying and quality changes. In: *The Properties of Water in Foods*. ISOPOW 6. Ed. D.S. Ried. Blackie Academic and Professional, London, U.K. pp. 209–232.

Dziezak J.D., April 1988. Microencapsulation and encapsulated ingredients. *Food Technology*, 42(4), 136–151.

Edris A. and Bergnstähl B., 2001. Encapsulation of orange oil in a spray dried double emulsion. *Nahrung/Food*, 45(2), 133–137.

Fiess M., 1992. Les cyclodextrines ou le piégeage moléculaire. *RIA*, 475, 48–50.

Finney J., Buffo R., and Reineccius G.A., 2002. Effects of type atomization and processing temperatures on the physical properties and stability of spray-dried flavors. *Journal of Food Science: Food Engineering and Physical Properties*, 67, 1108–1114.

Fuchs M., Turchiuli C., Bohin M., Cuvelier M.E., Ordonnaud C., Peyrat-Maillard M.N., and Dumoulin E., 2006. Encapsulation of oil in powder using spray drying and fluidised bed agglomeration. *Journal of Food Engineering*, 75, 27–35.

Gianfrancesco A., Turchiuli C., Flick D., and Dumoulin E., 2010. CFD modeling and simulation of maltodextrin solutions spray drying to control stickiness. *Food and Bioprocess Technology*, 3, 946–955.

Gibbs B.F, Kermasha S., Alli I., and Mulligan C.N., 1999. Encapsulation in the food industry: A review. *International Journal of Food Sciences and Nutrition*, 50, 213–224.

Gouin S., 2004. Microencapsulation: Industrial appraisal of existing technologies and trends. *Trends in Food Science and Technology*, 15, 330–347.

Guignon B., Duquenoy A., and Dumoulin E., 2002. Fluid bed encapsulation of particles and practice. *Drying Technology*, 20(2), 419–447.

Guignon B., Regalado E., Duquenoy A., and Dumoulin E. 2003. Helping to choose operating parameters for a coating fluid bed process. *Powder Technology*, 130, 193–198.

Jafari S.M., 2009. Encapsulation of nano-emulsions by spray drying. PhD, Pub LAP Lambert, Saarbrücken, Germany.

Jafari S.M., Assadpoor E., He Y., and Bhandari B., 2008. Encapsulation efficiency of food flavours and oils during spray drying. *Drying Technology*, 26 (7), 816–835.

Jafari S.M., He Y., and Bhandari B., 2007a. Encapsulation of nanoparticles of d-limonene by spray drying: Role of emulsifiers and emulsifying techniques. *Drying Technology*, 25, 1069–1079.

Jafari S.M., He Y., and Bhandari B., 2007b. Role of particle size on the encapsulation efficiency of oils during spray drying. *Drying Technology*, 25, 1081–1089.

Jayasundera M., Adhikari B., Aldred P., and Ghandi A., 2009. Surface modification of spray dried food and emulsion powders with surface-active proteins: A review. *Journal of Food Engineering*, 93, 266–277.

Karbowiak T., Debeaufort F., and Voilley A., Avril-Mai 9–17, 2007. Les emballages comestibles: Nature, fonctionnalité et utilisations. *Industries Alimentaires Agricoles*, 124 (4/5), 9–17.

Kerkhof P.J.A.M. and Thijssen H.A.C., 1974. Retention of aroma components in extractive drying of aqueous carbohydrate solutions. *Journal of Food Technology*, 9, 415–423.

Kim S.S., Han Y.J., Hwang T.J., Roh H.J., Hahm T.S., Chung M.S., and Shin S.G., 2001. Microencapsulation of ascorbic acid in sucrose and lactose by cocrystallization. *Food Science and Biotechnology*, 10 (2), 101–107.

King C.J., 1995. Spray drying: Retention of volatile compounds revisited. *Drying Technology*, 13(5–7), 1221–1240.

King C.J., Kieckbusch T.G., and Greenwald C.G., 1984. Food-quality factors in spray drying. In: *Advances in Drying*, Ed. A.S Mujumdar, Hemisphere Publisher Corporation, New York, Vol. 3, pp. 70–120.

Krishnan S., Kshirsagar A.C., and Singhal R.S., 2005. The use of gum arabic and modified starch in the microencapsulation of a food flavoring agent. *Carbohydrate Polymers*, 62, 309–315.

Lakkis J.M., 2007. Encapsulation and controlled release technologies in food systems. Ed. J.M. Lakkis, Blackwell Publisher, Ames, Iowa, 240pp.

Leclercq S., Harlander K.R., and Reineccius G., 2009. Formation and characterization of microcapsules by complex coacervation with liquid or solid aroma cores. *Flavour Fragrance Journal*, 24, 17–24.

Liu R.T. and Rushmore D.F., 1996. Process for making encapsulated sensory agents. US Patent 5,580,593.

Liu X.D., Atarashi T., Furuta T., Yoshii H., Aishima S., Ohkawara M., and Linko P., 2001. Microencapsulation of emulsified hydrophobic flavors by spray drying. *Drying Technology*, 19(7), 1361–1374.

Madene A., Jacquot M., Scher J., and Desobry S., 2006. Flavour encapsulation and controlled release: A review. *International Journal of Food Science and Technology*, 41, 1–21.

Marion J.P. and Audrin A., October 10–24, 1988. L'encapsulation d'arômes en images. *RIA*, 411, 41–46.

Masters K., 1985. *Spray Drying*, 4th edn., Longman Scientific and Technical, John Wiley & Sons, New York, 696p.

McNamee B.F., O'Riordan E.D., and O'Sulivan M., 2001. Effect of partial replacement of gum arabic with carbohydrates on its microencapsulation properties. *Journal of Agriculture and Food Chemistry*, 49, 3385–3388.

Merrit C.G., Wingerd W.H., and Keller D.J., 1985. Encapsulated flavorant material, method for its preparation, and food and other composition incorporating same. US Patent 4,515,769.

Onwulata C. (Ed.), 2005. *Encapsulated and Powdered Foods*. CRC, Taylor & Francis, Boca Raton, FL, 514p.

Parris N., Cooke P.H., and Hicks K.B., 2005. Encapsulation of essential oils in zein nanospherical particles. *Journal of Agriculture and Food Chemistry*, 53, 4788–4792.

Penbunditkul P., Yoshii H., Ruktanonchai U., Charinpanitkul T., and Soottitantawat A., 2012. The loss of OSA-modified starch emulsifier property during the high-pressure homogeniser and encapsulation of multi-flavour bergamot oil by spray drying. *International Journal of Food Science and Technology*, 47, 2325–2333.

Perrut M., 2003. Method for encapsulating fine solid particles in the form of microcapsules. Brevet FR2811913, US2003157183.

Pisecki J., 1997. *Handbook of Milk Powder Manufacture*. Niro A/S/, Copenhagen, Denmark, 261p.

Qi Z.H. and Xu A., 1999. Starch base ingredients for flavour encapsulation. *Cereal Foods World*, 44(7), 460–465.

Re M.I., 1998. Mcroencapsulation by spray drying. *Drying Technology*, 16(6), 1195–1236.

Reineccius G.A., 2004. The spray drying of food flavors. *Drying Technology*, 22(6), 1289–1324.

Richard H. and Multon J.J., coord. 1992. *Les arômes alimentaires*, Tech. & Doc. Lavoisier, Paris, France, 438p.

Risch S.J. and Reineccius G.A. (Eds.), 1995. *Encapsulation and Controlled Release of Food Ingredients*, ACS Symposium Series 590, Washington, DC, 214p.

Risch S.J. and Reineccius G.A. (Eds.), 1988. *Flavor Encapsulation*, ACS Symposium Series 370, Washington, DC, 202p.

Rosenberg M., Kopelman I.J., and Talmon Y., 1990. Factors affecting retention in spray drying microencapsulation of volatile materials. *Journal of Agricultural and Food Chemistry*, 38, 1288–1294.

Rulkens W.H. and Thijssen H.A.C., 1972. The retention of organic volatiles in spray drying aqueous carbohydrate solutions. *Journal of Food Technology*, 7, 95–105.

Schlameus W., 1995. Centrifugal extrusion encapsulation. In: *Encapsulation and Controlled Release of Food Ingredients*. Eds. S.J. Risch and G.A. Reineccius, ACS Symposium Series 590, Washington, DC, pp. 96–103.

Schubert H., Engel R., and Kempa L., 2009. Chapter 1: Principles of structured food emulsion: Novel formulations and trends. In: *Global Issues and Food Science and Technology*, Eds. G. Barbosa Canovas et al., Elsevier Inc., Amsterdam, pp. 3–20.

Senoussi A., Bhandari B., Dumoulin E., and Berk Z., 1994. Flavour retention in different methods of spray drying. In: *Developments in Food Engineering*, Ed. T. Yano, R. Matsuno, and K. Nakamura, Blackie Academic and Professional, London, pp. 433–435.

Senoussi A., Dumoulin E., and Berk Z. 1995. Retention of diacetyl in milk during spray-drying and storage. *Journal of Food Science*, 60(5), 894–897, 905.

Shahidi F. and Han X., 1993. Encapsulation of food ingredients. *Critical Reviews in Food Science and Nutrition*, 33(6), 501–547.

Sheu T.Y. and Rosenberg M., 1995. Microencapsulation by spray drying ethyl caprylate in whey protein and carbohydrate wall systems. *Journal of Food Science*, 60(1), 98–103.

Soottitantawat A., Bigeard H., Yoshii H., Furuta T., Ohgawara M., and Linko P., 2005a. Influence of emulsion and powder size on the stability of encapsulated D-limonene by spray drying. *Innovative Food Science and Emerging Technologies*, 6, 107–114.

Soottitantawat A., Takayama K., Okamura K., Muranaka D., Yoshii H., and Furuta T., 2005b. Microencapsulation of L-menthol by spray drying and its release characteristics. *Innovative Food Science and Emerging Technologies*, 6, 163–170.

Soottitantawat A., Yoshii H., Furuta T., Ohkawara M., and Linko P., 2003. Microencapsulation by spray drying: Influence of emulsion size on the retention of volatile compounds. *Journal of Food Science*, 68, 2256–2262.

Turchiuli C. and Dumoulin E., 2013a. Chapter 14: Aroma encapsulation in powder by spray drying, and fluid bed agglomeration and coating. In: *Advances in Food Process Engineering Research and Applications*. Eds. S. Yanniotis, P. Taoukis, N.G. Stoforos, and T. Karathanos, Springer, New York, pp. 255–265.

Turchiuli C., Fuchs M., Bohin M., Cuvelier M.E., Ordonnaud C., Peyrat-Maillard M.N., and Dumoulin E., 2005. Oil encapsulation by spray drying and fluidised bed agglomeration. *Innovative Food Science and Emerging Technologies*, 6, 29–35.

Turchiuli C., Jimenez Munguia M., Hernandez Sanchez M., Cortes Ferre H., and Dumoulin E. 2014. Use of different supports for oil encapsulation in powder by spray drying. *Powder Technology*, 255, 103–108.

Turchiuli C., Lemarie N., Cuvelier M.E. and Dumoulin E., 2013b. Production of fine emulsions at pilot scale for oil compounds encapsulation. *Journal of Food Engineering*, 115, 452–458.

Van Lengerich B.H., 2003. Embedding and encapsulation of controlled release particles. Patent EP 1342548.

Vandamme T., Poncelet D., and Subra-Paternault P., 2007. Microencapsulation. Des sciences aux technologies. Ed. Tec & Doc, Lavoisier, Paris, 355p.

Vega C. and Roos Y.H., 2006. Invited review: Spray-dried dairy and dairy-like emulsions. Compositional considerations. *Journal of Dairy Science*, 89, 383–401.

Whelehan M. and Marison I.W. October 5–8, 2011. Microencapsulation by dripping and jet break-up. *XIX International Conference on Bioencapsulation*, Amboise, France.

Yeo Y., Bellas E., Firestone W., Langer R., and Kohane D.S., 2005. Complex coacervates for thermally sensitive controlled release of flavor compounds. *Journal of Agriculture and Food Chemistry*, 53, 7518–7525.

Yuliani S., Torley P.J., D'Arcy B., Nicholson T., and Bhandari B., 2006. Extrusion of mixtures of starch and d-limonene encapsulated with β-cyclodextrine. *Food Research International*, 39, 318–331.

Zilberboim R., Kopelman I.J., and Talmon Y., 1986. Microencapsulation by a dehydrating liquid: Retention of paprika oleoresin and aromatic esters. *Journal of Food Science*, 51(5), 1301–1306; 1307–1310.

40 Molecular (Cyclodextrin) Encapsulation of Volatiles and Essential Oils

Paulo José Salústio, Maria Graça Miguel, and Helena Cabral-Marques

CONTENTS

40.1	Introduction	867
40.2	Extrusion	869
40.3	Fluid Bed Coating	869
40.4	Spray-Chilling/Spray-Cooling	870
40.5	Spray-Drying	870
40.6	Emulsification and Multiple Emulsions Encapsulation	871
40.7	Coacervation	871
40.8	Liposomes	873
40.9	Cocrystallization	874
40.10	Interfacial Polymerization	874
40.11	Molecular Encapsulation with Cyclodextrins	874
40.12	Flavors/Fragrances and Essentials Oils	877
40.13	Conclusion	899
References		899

40.1 INTRODUCTION

An essential oil (EO) is internationally defined as the product obtained by hydro-, steam-, or dry-distillation of a plant or of some of its parts, or by a suitable mechanical process without heating, as in the case of *Citrus* fruits (AFNOR, 1998; Council of Europe, 2010). Vacuum distillation; solvent extraction combined offline with distillation; simultaneous distillation extraction; supercritical fluid extraction; microwave-assisted extraction and hydro-distillation; and static, dynamic, and high concentration capacity headspace sampling are other techniques used for extracting the volatile fraction from aromatic plants, although the products of these processes cannot be termed EOs (Faleiro and Miguel, 2013).

For food and beverage consumption, there is a list of EO products generally recognized as safe (GRAS) approved by Food and Drug Administration (FDA) (Viuda-Martos et al., 2011; Prakash et al., 2012). National and international pharmacopoeias possess monographs of EOs for medical uses. The maximum quantities and uses of some EOs as well as their single components are regulated by the International Fragrance Association (IFRA), the Bundesinstitut für Risikobewertung (BfR), the Research Institute for Fragrance Materials (RIFM), and the Scientific Committee Consumer's Safety (SCCS). Physical standards of EOs are also specified by the Association Française de Normalisation (AFNOR) as well as the International Organization for Standardization (ISO) (Turek and Stintzing, 2013). This requirement in the control and standardization of EOs is due to the great variability in the chemical composition of EOs, which depends on the plant health, growth stage, edaphic and climate factors, harvesting time, part of plant used, and agronomic conditions, among other factors (Miguel et al., 2003; Figueiredo et al., 2008).

The biological properties of EOs (antimicrobial, antiviral, nematicidal, antifungal, insecticidal, antioxidant, and anti-inflammatory) are responsible for their uses in the pharmaceutical, agricultural, and nutritional fields (Prakash et al., 2012). In addition, EOs have also been extensively used in spa, cosmetics, toiletries, and in many household products, due to their unique odors. However, the volatilization of EOs, generally with intense aroma, may be undesirable. In addition, they can undergo alterations, namely, auto-oxidation, giving rise to changes in their sensory characteristics and production of allergenic products after air exposure (Matura et al., 2005; Sansukcharearnpon et al., 2010). Therefore, encapsulation techniques have been developed in order to mask the unpleasant and intense tastes and odors of EOs, to control the release of the components of the EOs, to protect them from the light, air, and aggressive body fluids, such as gastric acid (São Pedro et al., 2009; Sansukcharearnpon et al., 2010).

Encapsulation is the technique by which one material or mixture of materials is coated or entrapped within another material or system. The coated or entrapped material is called active or core material, and the coating material is known as shell, wall material, carrier, or encapsulant (Madene et al., 2006).

In encapsulation process, microspheres and microcapsules can be obtained. Microspheres are microbeads composed of a biopolymer gel network entrapping an active, whereas microcapsules are constituted by an active ingredient (small droplets of liquid or particles) that is inside a hollow and involved by thin walls(s) (Madene et al., 2006; Umer et al., 2011). The simplest of the microcapsules may have a core surrounded by a wall of uniform or nonuniform thickness. The core material may have one or several different types of ingredients. The wall may be single or multilayered (Madene et al., 2006).

According to the encapsulation processes used, the matrices of encapsulation can show diverse shapes (films, spheres, irregular particles), structures (porous or compact), physical structures (amorphous or crystalline dehydrated solid, rubbery or glassy matrix). This diversity is responsible for the different diffusion of flavors (Madene et al., 2006).

EOs may be encapsulated, allowing their isolation from their environment and a control of their release. Chemical and mechanical processes are two main processes of encapsulation of flavor compounds. Mechanical processes include extrusion, fluid bed coating, spray-chilling/spray-cooling, and spray-drying. Chemical processes include coacervation, molecular inclusion, emulsification, liposome, cocrystallization, and interfacial polymerization (Edrits and Bergnstahl, 2001; Gouin, 2004; Madene et al., 2006; van Soest, 2007; Zuidan and Shimoni, 2010).

For encapsulating EOs, a range of different materials have been used: proteins (sodium caseinate, whey proteins, soy protein, gelatine, pea protein, silk fibroin); carbohydrates (starches and their derivatives such as maltodextrines and CDs; corn syrup solids; acacia gums; chitosan; agar; pectin; alginate; carrageenan; sodium carboxymethyl celluluse); lipids (fatty acids, fatty alcohols, waxes, including beeswax, carnauba wax, candellia wax, glycerides, phospholipids such as lecithin, cholesterol among others); inorganics (silicates, clays, calcium sulphate, etc.); synthetics (acrylic polymers, poly(vinylpyrrolidone), among others) (Madene et al., 2006; Zuidan and Shimoni, 2010; Nedovic et al., 2011; Xiao et al., 2014).

An ideal coating material may have the following characteristics: good rheological characteristics at high concentration and easy workability during encapsulation; the ability to disperse or emulsify the active and stabilize the emulsion produced; nonreactivity with core both during processing and storage; ability to completely release the solvent or other materials used during the encapsulation under drying or other desolventization conditions; solubility in solvents acceptable in the food, and pharmaceutical industry (e.g., ethanol, water); cheap and at least food- or pharmaceutical-grade status, if the application is in food or pharmaceutical stuffs (Goud and Park, 2005; Poshdri and Kuna, 2010).

Food, textile, agricultural, and pharmaceutical industries are the main areas of application of flavor and EO microcapsules (Xiao et al., 2014).

40.2 EXTRUSION

In extrusion, carbohydrates, such as starch, maltodextrins, and CDs, are melted at elevated temperature (generally above 100°C) and low water contents and are intensively mixed with the active agent. Basically, two processes to encapsulate active agent in a carbohydrate melt can be distinguished. One is melt injection, in which the melt is pressed through one or more orifices (filter) and then quenched by isopropanol, and also liquid nitrogen, that is, a cold and dehydrating solvent. This is a vertical, screwless extrusion process. The coating material hardens on contact with the dehydrating solvent, thereby encapsulating the active agent. The other utilizes screws in a horizontal position. Extruders are thermomechanical mixers that consist of one or more screws in a barrel. The transport of material within the extruder occurs by rotational and sometimes oscillatory movement of the screws. In the beginning, the pressure is lower and a gradual increase in pressure is achieved via the screw design to melt, further homogenize, and compress the extrudate. In the final part of the barrel, there is a continuous high pressure to ensure a uniform delivery rate of molten material out of the extruder (van Soest, 2007; Zuidan and Shimoni, 2010).

Extrusion has been used for volatile and unstable flavors. The shelf-life of flavor oils could be extended from several months to 5 years. The particles formed have relatively high dimensions (500–1000 μm) (van Soest, 2007). However, diffusion of flavors from extruded carbohydrates can be enhanced in the presence of structural defects such as crakes, thin wall, or pores formed during or after processing (Madene et al., 2006).

The use of flexible films incorporated with antimicrobials, such as EOs, has attracted great interest from manufacturers and demanding consumers, because it is a way of avoiding microbial food spoilage. One procedure to obtain this kind of biofilms is by extrusion as reported by several authors with successful results (Pelissari et al., 2009, 2011; Solano and Gante, 2012; Woranuch and Yoksan, 2013).

40.3 FLUID BED COATING

In fluid bed coating, there is an application of an uniform layer of shell material (polysaccharides, proteins, emulsifiers, fats, complex formulations, enteric coating, powder coatings, yeast cell extract, etc.) onto solid particles. Aqueous solutions of hydrocolloids (gums and proteins), ethanolic solutions of synthetic polymers, and melted fats/waxes have been used as coating formulations in fluidized bed microencapsulation. The powder particles are suspended by an air stream at a specific temperature and sprayed with an atomized, coating material. Over time, each particle will be gradually covered. The coating material must have an adequate viscosity, must also be thermally stable, and be able to form a film over a particle surface (Zuidam and Shimoni, 2010).

The solvent of the coating material must be evaporated, which is controlled by several factors such as spray rate, the solvent content of the coating solution, the air flow, the humidity of the air inlet in the chamber, and the temperature of the coating solution, atomized air, and the material in the chamber (Gouin, 2004; Poshadri and Kuna, 2010; Zuidam and Shimoni, 2010). Fluidized bed processes for coating particles with fat and waxes using supercritical carbon dioxide as the solvent have been used with the advantage to reduce the energy to the minimal needed to "evaporate" the solvent (Gouin, 2004).

The atomized coating droplets must be significantly smaller than the particle to be coated in order to obtain a uniform and complete coating and avoiding agglomeration (Gouin, 2004).

Spray dried particles, such as spray-dried flavor microcapsules, can also be further coated by fluidized bed, generally with a fat layer to impart better protection and shelf-life (Gouin, 2004; Poshadri and Kuna, 2010).

40.4 SPRAY-CHILLING/SPRAY-COOLING

In spray-cooling or spray-chilling, the active or core is mixed with the carrier and atomized using cool air (van Soest, 2007). The matrix is usually a regular, hydrogenated, or fractionated vegetable oil. This methodology is not a true microencapsulation process, because a significant amount of active is located at the surface of the microcapsules, having a direct access to the environment (Gouin, 2004). This is a matrix-type encapsulate, because the active is much more dispersed over the carrier material, either in the form of relatively small droplets or more homogenously distributed over the encapsulate (Zuidam and Shimoni, 2010).

Although spray-cooling or spray-chilling is one of the least expensive methods, it is not the most adequate for encapsulating volatile perfumes, due to the matrix-type encapsulate (van Soest, 2007). Nevertheless, two products were patented, in which oleoresin cinnamon and lemon oil was submitted to spray-cooling using molten hardened vegetable fat or a molten nonself-emulsifying glyceryl monostereate after being spray-dried in the presence of gum acacia and modified gelatine, respectively (Smith and Lambrou, 1974).

40.5 SPRAY-DRYING

Spray-drying is an operation by which a liquid product is atomized in a hot gas current to instantaneously obtain a powder with particle sizes range from 10–50 µm to 2–3 mm, depending on the starting feed material and operating condition. The gas can be air or more rarely nitrogen. The liquid may be a solution, an emulsion, or a suspension (Gharsallaoui et al., 2007).

The microencapsulation by spray-drying involves four steps: preparation of the dispersion or emulsion; homogenization of the dispersion; atomization of the mass into the drying chamber; and dehydration of the atomized particles (Shahidi and Han, 1993; Gharsallaoui et al., 2007).

Spray-drying is a method largely used for protecting aroma chemicals against oxidation or degradation and at the same time to convert liquids into free-flowing solids. This technology has been particularly used in the food industry (van Soest, 2007).

The criteria for selecting a wall material are based on the physicochemical properties such as solubility, molecular weight, glass/melting transition, crystallinity, diffusibility, film-forming, and emulsifying properties (Gharsallaoui et al., 2007). The selection of wall materials generally involves trial-and error procedures in which the microcapsules are formed. Encapsulation efficiency, stability under different storage conditions, degree of protection of the core material, and surface observation by scanning microscopy of microcapsules, among other assays, are evaluated (Gharsallaoui et al., 2007).

Typical shell materials include gum acacia, maltodextrins, CDs, hydrophobically modified starch, and mixtures thereof. Other polysaccharides (alginate, carboxymethylcellulose, guar gum) and proteins (whey proteins, soy proteins, and sodium caseinate) may also be used nevertheless with lower application because of their low solubilities in water, although the addition of a small amount of these low solubility hydrocolloids has shown some beneficial effects on the stability of encapsulated ingredients (Gouin, 2004).

Only very few examples of encapsulation of flavors are reported: oregano EO using gum arabic, maltodextrin and modified starch as wall materials (Botrel et al., 2012), inulin (Beirão-da-Costa et al., 2013); rosemary EO of rosemary using gum arabic/starch/maltodextrin/inulin (Fernandes et al., 2014), gum arabic (Fernandes et al., 2013a,b); orange EOs using starch derivatives of taro (*Colocasia esculenta* L. Schott) and rice (Verdalet-Guzmán et al., 2013); and *Pterodon emarginatus* EOs using gum arabic and maltodextrin (Alves et al., 2014).

The main drawback of spray-drying is the high temperatures needed. New technologies using compressed carbon dioxide as solvent do not need the utilization of relative high temperatures with the advantages to be nontoxic, environmentally friendly, and eliminated completely from the final product by depressurization (Varona et al., 2013). Particles from gas-saturated solutions

(PGSS)-drying and spray-drying were used for the encapsulation of lavandin (*Lavandula hybrida*) oil used as carrier materials for the encapsulation soybean lecithin, *n*-octenyl succinic anhydride modified starch, and poly-caprolactone. According to the authors, the encapsulation of this oil by PGSS-drying showed higher antibacterial activity than particles formed by spray-drying with a similar EO load (Varona et al., 2013).

40.6 EMULSIFICATION AND MULTIPLE EMULSIONS ENCAPSULATION

There are two main combinations of emulsions: water-in-oil (W/O) or oil-in-water (O/W) simple emulsions and water-in-oil-in-water (W/O/W) and oil-in-water-in-oil (O/W/O) double emulsions also called multiple emulsions. Double emulsions are complex soft colloidal systems in which droplets of the dispersed phase have still smaller droplets. Each dispersed globule is separated from the aqueous phase by a layer of oil-phase compartments (Garti, 1997; Nisisako et al., 2005). In both cases, emulsions need to be stabilized through a set of lipophilic and hydrophilic surfactants dissolved in each phase according to their affinity (Nisisako et al., 2005).

Multiple emulsions may be interesting ways for releasing bioactive compounds in a controlled rate, useful in cosmetic, pharmacy, agricultural, and industrial chemicals; nevertheless, their commercial applications have been limited due to their thermodynamic instability and unexpected fast release of encapsulated bioactive molecules (Yoshida et al., 1999; Beer et al., 2013).

Examples of multiple W/O/W emulsions used for EO encapsulation include those of orange oil (Edrits and Bergnstahl, 2001) and *Zanthoxylum limonella* (Banerjee et al., 2013). The O/W/O emulsions are also checked for encapsulation of volatile EOs, such as limonene (Cho and Park, 2003) and *Satureja hortensis* oil (Hosseini et al., 2013).

40.7 COACERVATION

Coacervation is defined as the separation of colloid systems into two liquid phases, according to the IUPAC (International Union of Pure and Applied Chemistry) definition (Martins, 2012).

The coacervation method has widely been employed for the preparation of microcapsules. This process comprises five basic steps:

1. Dissolution: the creation of an aqueous solution containing two different polymers, generally a protein and a polysaccharide, usually at a temperature above the gelling point and pH that is above the isoelectric point of protein.
2. Emulsification/dispersion: emulsification of hydrophobic material (e.g., EO) in the solution reported earlier, the emulsion is stabilized by the two polymers.
3. Coacervation is the separation into two liquid phases (an insoluble polymer-rich phase and an aqueous phase that is depleted in both polymers) due to the attractive electrostatic interactions between oppositely charged polymers caused by lowering the solution pH below the gelling temperature.
4. Gelation consists of the wall formation due to deposition of the polymer-rich phase around the droplets of the hydrophobic material, induced by controlled cooling below the gelling temperature.
5. Hardening and rinsing/filtering/drying is a process that needs a stabilizing agent for hardening the microcapsule walls; the stabilizing agent and the oil that were not encapsulated or adsorbed on the surface are washed out; the capsules obtained are dried to obtain a powder sample (Xiao et al., 2014); to form self-sustaining microcapsules in the coating hardening, thermal, cross-linking, or desolvation techniques can be used (Martins, 2012).

Two main coacervation techniques can be considered: aqueous, which can only be used to encapsulate water-insoluble materials (hydrophobic core materials presented in solid or liquid state), and organic in which the organic phase permits the encapsulation of hydro-soluble material, requiring the utilization of organic solvents (Martins, 2012).

Coacervation in aqueous phase can be classified into simple and complex. In simple coacervation, the polymer is salted out by the action of electrolytes (sodium sulfate) or desolvated by the addition of an organic miscible water solvent, such as ethanol, or by increasing/decreasing temperature. In these cases, the macromolecule–macromolecule interactions are promoted, instead of the macromolecule–solvent interaction (Martins, 2012). Complex coacervation is defined as a liquid–liquid phase separation promoted by electrostatic interactions, hydrogen bonding, hydrophobic interactions, and polarization-induced attractive interactions occurring between two oppositely charged polymers in aqueous solution (Xiao et al., 2014). This technique is based on the ability of cationic and anionic water-soluble polymers to interact in water to form a liquid polymer-rich phase called complex coacervate (Martins, 2012).

Simple coacervation shows some advantages when compared to the complex coacervation: it is cheaper than complex coacervation, because in the induction of the phase separation, simple coacervation utilizes inexpensive inorganic salts, whereas complex coacervation uses more expensive hydrocolloids, and complex coacervation is more sensitive to even small pH changes than simple coacervation (Sutaphanit and Chitprasert, 2014).

Microcapsules produced by complex coacervation may be classified into mononuclear microcapsules (one spot core droplets included in a single polymer shell), and multinuclear microcapsules (aggregates of mononucleated capsules, which are made of multiple small entities core droplets included in a single polymer shell) (Xiao et al., 2013).

The wall material used in microencapsulation of EOs is subdivided into protein (gelatin, whey protein, fibroin, and soybean proteins isolate), polysaccharide (gum arabic, chitosan, pectin, agar, alginate, carrageenan, and sodium carboxymethylcellulose), lipids (waxes, paraffin, oils), inorganics (silicates, clays), and synthetic polymers [acrylic polymers, poly(vinylpyrrolidone)] (Martins, 2012; Xiao et al., 2014). The polysaccharides have the advantages: they possess good solubility in water and low viscosity at high concentrations (Xiao et al., 2014). The proteins used are abundant, cheap, possess good amphiphilic and functional properties such as water solubility, and emulsifying and foaming capacity. At the same time, they present nontoxic biocompatibility and biodegradability (Xiao et al., 2014).

Gelatin used as wall material in complex coacervation needs hardening with glutaraldehyde or formaldehyde. Both aldehydes are considered toxic to humans and must be carefully used. For this reason, several studies have been carried out to find safer cross-linking agents such as transglutaminase, tannic acid, and glycerol (Xiao et al., 2014).

Xiao et al. (2014) have reviewed the application of microcapsules produced by coacervation of EOs in food, agriculture, textile, and pharmaceutics. In food and pharmaceutical industries, the utilization of microcapsules of EOs obtained by coacervation has practically the same goals: protect the core material of degradation, improve the stability, mask a bad taste or unpleasant smell, control flavor release at the right place and right time, avoid undesirable incompatibilities and increase bioavailability, and reduce toxicity and irritation of the gastrointestinal tract. However, in textile industry, the EOs encapsulated by coacervation have shown to be a good way to meet important psychological and emotional needs, to have more ecofriendly antibacterial agents by contact of textile with skin with the advantage to offer well-being, and as mosquito repellents (Xiao et al., 2014).

EOs of lemongrass (Leimann et al., 2009), vetiver (Prata et al., 2008), *Zanthoxylum limonella* (Maji et al., 2007), citronella (Specos et al., 2010), thyme (Martins et al., 2011), jasmine (Lv et al., 2014), basil (Sutaphanit and Chitprasert, 2014) and EO components, such as thymol and carvacrol (Martins et al., 2012), and linalool (Lopez et al., 2012) are some examples of microencapsulated volatiles obtained by coacervation.

40.8 LIPOSOMES

Liposomes generally made from phospholipids may form membrane-like vesicles with selective permeability for small molecules. The diameters of vesicles can range from 25 nm to 10 μm (van Soest, 2007). They can contain one bilayer forming unilamellar vesicles (ULVs), several concentric bilayers forming multilamellar vesicles or nonconcentric bilayers forming multivesicular vesicles (MVVs). Both hydrophobic and hydrophilic ingredients can be entrapped (Bilia et al., 2014). Beyond the protection of bioactive compounds in liposomes against degradation, for lipophilic compounds such as EOs, this type of encapsulation also increases solubilization (Bilia et al., 2014).

Eugenia uniflora L oil from Brazil was incorporated in multilamellar liposomes in the presence of cryoprotectors, such as sucrose and trehalose, in order to preserve the liposomal structure during the dehydration process. The authors concluded that the oil was effectively incorporated in the liposomes (Yoshida et al., 2010).

Valenti et al. (2001) prepared a formulation of liposomal *Santolina insularis* EO. The authors evaluated the influence of the vesicular inclusion on the stability and in vitro antiviral activity of the EO. However, they concluded that free EO was more effective than liposomal oil, even though the vesicular inclusion had greatly improved oil stability and permitted its delivery in an efficacious composition even after 1 year of storage.

The antiviral activity of the oil obtained from *Artemisia arborescens* L was also studied by Sinico et al. (2005). They studied the influence of vesicle structure and composition on the antiviral activity, against herpes simplex virus type 1 (HSV-1), of the vesicle-incorporated oil, multilamellar vesicles (MLV), and small unilamellar vesicles (SUV) positively charged liposomes prepared by the film method and sonication. The liposomal incorporation of *A. arborescens* EO enhanced the in vitro antiherpetic activity mainly when vesicles were made with hydrogenated soy phosphatidylcholine.

The antimicrobial activity of *Origanum dictamnus* L oil and their main components or mixtures increased when previously encapsulated in liposomes constituted by egg L-α-phosphatidylcholine and cholesterol prepared by thin-film method when compared to those not encapsulated (Liolios et al., 2009).

Zanthoxylum tingoassuiba oil, with significant antimicrobial activity against gram-positive bacteria and dermatophyte fungi, when encapsulated into multilamellar liposomes prepared with dipalmitoyl phosphatidylcholine by thin-film hydration, revealed that the oil showed an incomplete release profile from liposomes: according to the authors, such EO-loaded liposomes will be useful in pharmaceutical applications to improve EO targeting to cells. The same authors also reported the importance of the liposome size for the antimicrobial activity of the EO (Detoni et al., 2009). These authors suggested a reduction in the liposome diameter as a promising strategy for improving the in vitro interaction between the EO and the targeted cell. Later on, those authors assayed the EO of the same plant encapsulated into multilamellar and unilamellar liposomes for characterizing and evaluating their oxidative stability as well as their activity against glioma cells, due to the richness of the oil in α-bisabolol (Detoni et al., 2012). Liposomes presented significant apoptotic-inducing activity for glioma cells.

Studies in vivo showed that *Ligusticum chuanxiong* Hort prepared as a liposomal formulation possessed therapeutic effects on formed hypertrophic scars in the rabbit ear model. According to these results, the authors suggested that this formulation may be an effective cure for human hypertrophic scars (Zhang et al., 2012a).

The influence of size, the liposome composition, and lamellarity on the entrapment efficiency of EO of *Anethum graveolens* was studied by Ortan et al. (2009), using multilamellar and unilamellar liposomes prepared by the thin hydration method. They observed a good incorporation of the oil remaining stable for 1 year, and their size distribution showed only slight modifications during the same period.

Bergamot EO has anticancer activity against neuroblastoma cells, and this activity was enhanced when it was encapsulated in extruded multilamellar liposomes (Celia et al., 2013).

It has been shown that liposomes formed by the supercritical process exhibited higher encapsulation efficiency and smaller particle size with a unimodal size distribution than those prepared by thin-film dispersion. In this process, liposomal materials and EOs (rose and *Atractylodes macrocephala*) were dissolved in a mixture of supercritical carbon dioxide/ethanol and the solution was sprayed into aqueous solution containing surfactants to stabilize small particles by minimizing flocculation and agglomeration (Wen et al., 2010, 2011).

Nanoliposomes compared to liposomes provide more surface area and have the potential to increase solubility, enhance bioavailability, and improve controlled release. The principal constituents of nanoliposomes are phospholipids: for example, soya, rapeseed, and marine lecithin used by Zhang et al. (2012b). Jiménez et al. (2014) incorporated antimicrobial volatile compounds (orange EO and limonene) into soy and rapeseed nanoliposomes. These were then added to starch sodium caseinate film, forming dispersions. The antimicrobial activity of these films was not observed probably due to the encapsulation, which made difficult their release from the matrix (Jiménez et al., 2014).

40.9 COCRYSTALLIZATION

Cocrystallization consists of introducing the aromatic compounds into a saturated solution of sucrose. The spontaneous crystallization of this syrup is made at temperatures higher than 120°C and with a low degree of humidity. The crystal structure is modified and small crystal aggregates (lower than 30 µm) form, trapping the aromatic compounds. The sugars form an oxygen barrier, extending the shelf-life of aroma chemicals. Such processes are cheap, offering a good economic and flexible technique owing to its simplicity (van Soest, 2007; Munin and Edwards-Lévy, 2011).

40.10 INTERFACIAL POLYMERIZATION

In the interfacial polymerization technique, a wall is formed from monomers that are dissolved in the two separated phases (oil and water phase) and they polymerize at the interface of emulsion droplets. The use of these methods is limited, since the preferred matrix or coating materials are nonrenewable or nonfood grade, such as polyesters, polyamides, polyurethanes, polyacrylates, or polyureas, almost always accomplished by traces of toxic monomers (van Soest, 2007).

Some examples of microencapsulated EO are those reported by Scarfato et al. (2007), in which the polyurea microcapsules containing the EO of lemon balm, lavender, sage, and thyme with antigerminative activity were prepared by interfacial polymerization in O/W emulsion.

Encapsulation of galangal EO with antimicrobial activity against *Staphylococcus aureus* in another example of polyurethane-urea microcapsules was carried out by interfacial polymerization at oil–water interface in O/W emulsion (Podshivalov et al., 2013).

40.11 MOLECULAR ENCAPSULATION WITH CYCLODEXTRINS

CD were first isolated in 1891 by Villiers as degradation products of starch from a medium of *Bacillus amylobacter*. It was Schardinger, however, between 1903 and 1911, who did much of the characterization of Villiers cellulosines and determined that they were cyclic oligosaccharides. French (1957) published a review on these substances, being followed by monographs in 1965 and 1968 by Thoma and Stewart and by Caesar, respectively (Thoma and Stewart, 1965; Caesar, 1986). Parent/natural CDs are obtained through the enzymatic reaction on the starch. They are cyclic oligosaccharides formed from glucopyranose units linked by α-1,4 bonds, forming torus-like macrorings. The natural CDs are αCD (Schardinger's α-dextrin, cyclomaltohexaose, cyclohexaglucan, cyclohexaamylose, αCD, ACD, C6A) comprising six glucopyranose units, βCD (Schardinger's β-dextrin, cyclomaltoheptaose, cycloheptaglucan, cycloheptaamylose, βCD, BCD, C7A) comprising seven of such units, and γCD (Schardinger's γ-dextrin, cyclomaltooctaose, cyclooctaglucan, cyclooctaamylose, γCD, GCD, C8A) that comprises eight of such units (Figure 40.1). These substances are crystalline, homogeneous, and nonhygroscopic (Szejtli, 1998).

Modified CDs have also been synthesized for enhancing the binding abilities of EOs or other natural products to CDs. 2-Hydroxypropyl-βCD, 2-O-methyl-βCD, quinoline- and naphthol-modified-βCD,

Molecular (Cyclodextrin) Encapsulation of Volatiles and Essential Oils

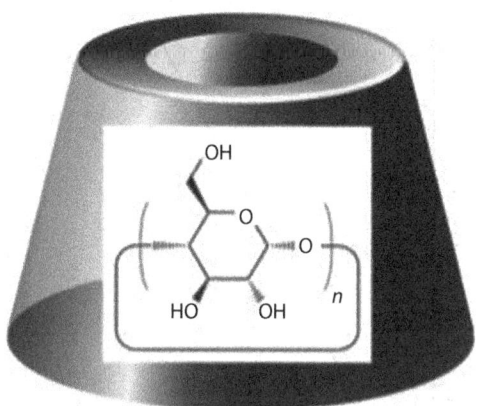

$n = 6(\alpha CD); 7(\beta CD); 8(\gamma CD)$

FIGURE 40.1 Schematic CDs structure. (From Salústio, P.J. et al., *AAPS PharmSciTech*, 12(4), 1276, 2011; Szejtli, J., *Chem. Rev.*, 98, 1743, 1998.)

triazole-modified-βCD, and azido-βCD are only some examples (Khan et al., 1998; Schönbeck et al., 2010; Holm et al., 2011; Faugeras et al., 2012; Li et al., 2014).

The CDs as pharmaceutical products have been used for the enhancement of the stability, solubility, dissolution rate, bioavailability, and oral absorption or to modulate biological activity of drugs (Frömming and Szejtli, 1994; Al-Omar et al., 1999; García-Rodriguez et al., 2001; LeBlanc et al., 2008; Arana-Sánchez et al., 2010). They have also been used to reduce or prevent GI and ocular irritation, to reduce or eliminate unpleasant smells or tastes or volatility, to prevent drug–drug or drug–additive interactions, or to convert oily and liquid drugs into microcrystalline or amorphous powders (Uekama, 2004; Carrier et al., 2007). Due to the characteristics of the CDs structure, these substances allow inclusion complexes formation with a large variety of apolar and hydrophobic molecules by taking up the whole molecule or rather some nonpolar parts into its hydrophobic cavity (it is noteworthy that hydrophobicity and geometry of guest structures affect the inclusion) (Szejtli, 1998). During the complexation phenomena, the apolar CD cavity occupied by high enthalpy water molecules (polar–apolar interaction) is readily substituted by appropriate "guest molecules," which are less polar than them (Figure 40.2).

Higuchi and Connors (1965) have established a classification of the complexes from the phase solubility profiles (Figure 40.3) obtained from the interaction between the guest and the host when in solution (Higuchi and Connors, 1965).

Thus, in a simplified and summarized view, A-type curves indicate the formation of soluble inclusion complexes. B type suggests the formation of inclusion complexes with poor solubility. B_S-type response denotes complexes of limited solubility and a B_I-type curve indicates the formation of insoluble complexes. A-type curves are subdivided into A_L-type (linear increase of drug solubility as a function of CD concentration), A_P-type (positively deviating isotherms), and A_N-type (negatively deviating isotherms) subtypes (Higuchi and Connors, 1965; Cabral-Marques, 2010). The complexation process is most frequently a 1:1 host/guest ratio; however, 2:1, 1:2, 2:2, or even

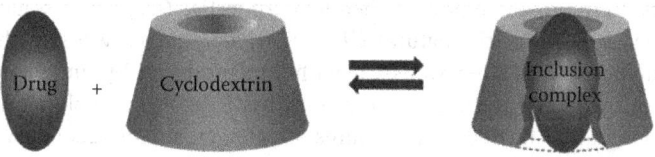

FIGURE 40.2 Interaction of a drug with a CD to form an inclusion complex. (From Salústio, P.J. et al., *AAPS PharmSciTech*, 12(4), 1276, 2011.)

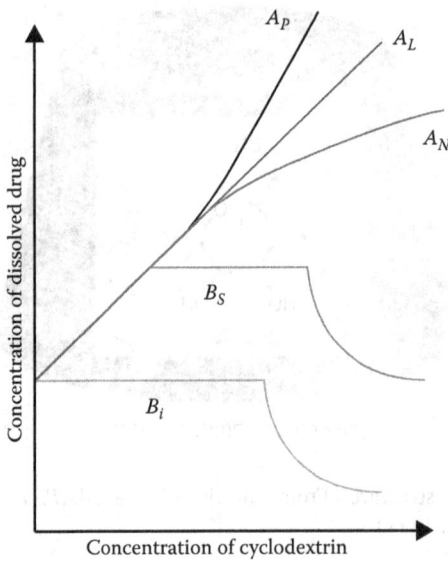

FIGURE 40.3 Phase solubility profiles and classification of complexes according to Higuchi and Connors. S_0 is the intrinsic solubility of the substrate (the dissolved drug or guest) in the aqueous complexation medium when no ligand (CD) is present. (From Cabral-Marques, H.M., *Flavour Frag. J.*, 25, 313, 2010.)

more complicated associations, and higher-order equilibrium exists, almost always simultaneously. Equilibrium established between dissociated and associated species can be expressed by the complex formation constant (Szejtli, 1998) also called stability (association or equilibrium) constant and noted as K_s or K_a or $K_{1:1}$. These constants for the inclusion of EOs or fragrance materials with CD may be determined by several methods, for example: phase solubility studies (Waleczek et al., 2003; Daletos et al., 2008), NMR, UV-VIS spectrophotometry, HPLC (Cabral Marques, 1994a), TLC (Lederer and Leipzig-Pagani, 1996), and static head-space method (Saito et al., 1999). The CD association with the drug or guest molecules (D), and the dissociation of the formed inclusion complexes, is governed by a thermodynamic equilibrium (Szejtli, 1998):

$$D + CD \leftrightarrow D_{CD} \tag{40.1}$$

$$K_{1:1} = \frac{[DC_D]}{[CD] \cdot [D]} \quad \text{or} \quad K_{1:1} = \frac{\text{slope}}{S_0(1-\text{slope})} \tag{40.2}$$

The different interactions between different chemical species (molecules, ions, and radicals) do not involve covalent bonds (Charoenchaitrakool et al., 2002). The driving forces between hosts and guests, which have been proposed to justify the complex formation are hydrogen bonds, van der Waals forces, hydrophobic interactions, and the release of "high energy water" molecules from the cavity (Cabral Marques, 1994b; Sun et al., 2006).

In the last years, to improve the physicochemical properties (i.e., better solubility, stability, and inclusion formation abilities) of the natural CDs, many derivatives were synthesized by various molecular manipulations. These derivatives usually are produced by aminations, esterifications, or etherifications of their primary and secondary hydroxyl groups, involving also polymeric and amphiphilic CD (long alkyl or fluoroalkyl chains at primary and/or secondary sides of the CDs) (Szejtli, 1992; Szente and Szejtli, 1999).

The complexation processes can be achieved by several techniques, such as physical mixing (Menezes et al., 2012), coprecipitation (Sapkal et al., 2007; Nieddu et al., 2014), complexation in a slurry (Fages et al., 2007; Menezes et al., 2012, 2014), complexation in paste (Gil et al., 2004; Menezes et al., 2012, 2014), extrusion (Yano and Kleinebudde, 2010), dry mixing (Higashi et al., 2009), solution/suspension with remotion of the solvents by drying methods (spray-drying, freeze-drying) (Salústio et al., 2009), sealed-heating method (Nieddu et al., 2014) and supercritical fluids (Junco et al., 2002). In order to enhance the complex efficiency of the CDs, different methods can be used such as: ionization of the drug, salt formation, formation of metal complexes, and addition of organic cosolvents to the aqueous complexation media and the use of supercritical fluids, as all promote the enhancement of the intrinsic solubility of the drug-favoring complexation (Loftsson and Duchêne, 2007; Loftsson et al., 2007; Junco et al., 2002). There are also references to the inclusion of EO-βCD into a sol-gel process, which consists on the introduction of the suspension containing the EO and βCD into a colloidal silica, under continuous stirring, at room temperature. This procedure was used for encapsulating mint and lavender oils, which according to the authors permitted the EOs to be much more protected against the humidity, temperature, and solar light, among other factors (Răileanu et al., 2013).

Several works have been published involving CDs and EOs/volatiles. The association of those compounds (either as inclusion complexes or not) often advantageously modifies various physicochemical properties of the encapsulated molecules such as aqueous solubility and stability. Cabral-Marques (2010) has recently reviewed some issues on flavors/fragrances and/or EO and volatiles using CDs encapsulation approach.

More recently, cross-linked βCD polymers have been used for encapsulating *Lavandula angustifolia* and *Mentha piperita* EOs, using *epi*chlorohydrin as polymerization agent (Ciobanu et al., 2012, 2013).

40.12 FLAVORS/FRAGRANCES AND ESSENTIALS OILS

The flavor/aromatizing characteristics of different substances are dependent on the presence of their several components. To maintain the intensity of these flavor/aromatizing components, it is appropriate to include all components or some of them into the CD cavity without altering its characteristics. Hydrophobic molecules can be fully or partially complexed with those host agents. A list of flavor–βCD complexes loading is shown in Table 40.1 (Astray et al., 2009; Cabral-Marques, 2010).

The fragrance molecules due to their volatile nature can be included into βCD cavity in order to retain fragrances for longer time. The physicochemical properties of the compounds after inclusion change: e.g., the vapor pressure of the volatile substance is reduced and the stabilities against light and air are enhanced (Wang and Chen, 2010). These observed properties, improvement proved that complexation has occurred (Wang and Chen, 2010). When inclusion complexes are produced, the stability (decreased volatility) and water solubility of the complexed substances increases. The linalool and benzyl acetate (terpenic fragrances and aromatic compounds) were complexed with βCD and 2-hydroxypropyl-β-cyclodextrin(2-HPβCD), providing its controlled release, and allowing its conversion from liquid to powder form (improving handling properties for solid and semisolid formulations). The complexes (especially 2-HPβCD) obtained can be used also in cosmetic delivery systems (Numanoğlu et al., 2007). In the case of thyme EO (good antimicrobial agent to preserve food), complexation with βCD was carried out in order to prevent its high volatility and reactivity. This molecular complex with antifungal activity allows the EO release to the atmosphere of interest, passively and controlled when exposed to high relative humidity. The optimization of the use of natural antimicrobials as food preservatives will facilitate the transition through a more clean and sustainable environment (Del Toro-Sánchez et al., 2010). A wide range of antimicrobial compounds (e.g., o-methoxycinnamaldehyde, *trans,trans*-2,4-decadienal, cinnamic acid, and citronellol) have limited solubility in water and in aqueous medians (e.g., juices) that may be improved using the ability of CDs, allowing their use as viable natural alternatives to common preservatives (e.g., sodium benzoate). These complexes showed similar storage stability either in glass or in polyethylene terephthalate (PET) containers, allowing more options on beverages' packaging (Samperio et al., 2010).

TABLE 40.1
Flavor Content in βCD Complexes

Flavor Material	Flavor Load of Complex (%)
Anise oil	9.00
Basil oil	10.72
Laurel leaf oil	10.80
Benzaldehyde	8.70
Caraway oil	10.50
Carrot oil	8.82
Celery oil	10.00
Cinnamon oil	8.76
Coriander oil	7.72
Dill oil	6.92
Smoke oil	12.20
Garlic oil	10.20
Lemon oil	8.75
Marjoram oil	8.00
Mustard oil	10.92
Onion oil	10.20
Orange oil	9.20
Mint oil	9.70
Raspberry oil	8.66
Sage oil	8.20
Sweet cumin oil	10.00
Tarragon oil	10.23
Thyme oil	9.60
Vanilla	6.20

Sources: Astray, G. et al., *Food Hydrocolloids*, 23,1631, 2009; Cabral-Marques, H.M., *Flavour Frag. J.*, 25, 313, 2010.

An interesting application of lavender oils with βCD at a textile level was achieved by anchoring this inclusion compound. The inclusion compounds are fixed onto fabric by the traditional pad method to obtain the medical textile with aromatherapy effect. The sedative effects for emotion and the pharmaceutical effects of EOs were shown. The results of sensorial evaluations have shown that the odor of the fabrics can last for more than 30 days (Wang and Chen, 2010). Other methods for loading fragrances into fabrics such as immersion in their hydro-alcoholic solutions can also be used. Cotton, wool, and polyester fabrics finished with CDs (by the intermediate of polycarboxylic acids) were impregnated with six different fragrant molecules (β-citronellol, camphor, menthol, *cis*-jasmone, benzyl acetate, geraniol) and citronella oil. The odor retention capacity of most of the treated samples lasted for 1 year compared to untreated textiles that lasted only for 1 or 2 weeks. The final product textile containing fragrance-CDs complexes are able to keep the smell even after washings with water. The efficiency of the different CDs varied in the order γCD > βCD > αCD, and the durability of the fragrant effect was directly dependent on the amount of CD grafted onto the fabrics. The polyester fabrics showed better performance (Martel et al., 2002). Sometimes the fragrance retention phenomenon was not only due to guest-host complexation, but was also dependant on nonspecific interactions of the substrates with the surface modified fiber. By choosing the most appropriated CD, and by adjusting the amount of CD to graft onto the fabrics, it will be possible to tailor the desired effect, depending on its ability to capture or to release the odor molecules (Martel et al., 2002). The storage conditions for hydrophobic flavors under inclusion complex with βCD can be predicted, allowing the determination of their shelf-life. Having this in mind, two hydrophobic components (thymol and cinnamaldehyde) were complexed with βCD (1:1 molar ratio) and the influence of the water or absorption by the host and their complexes on the release of compounds was analyzed. The GAB (Guggenheim–Anderson–de Boer) model was also used to accurately describe the sorption isotherms of the βCD and its complexes. When adsorption equilibrium was achieved, the water adsorbed for both complexes was less than the correspondent to the βCD (Figure 40.4) shows the % of release of thymol or cinnamaldehyde inclusion complexes at 84% and 97% RH, respectively, as a function of time. The thymol and cinnamaldehyde released (%) were calculated by the following equation:

$$\% \text{ Released}(\%R) = \frac{\Delta H_S}{\Delta H_0} \quad (40.3)$$

where
ΔH_0 is the pure molecule heat of melting
ΔH_S is the heat of melting of the molecule as complex

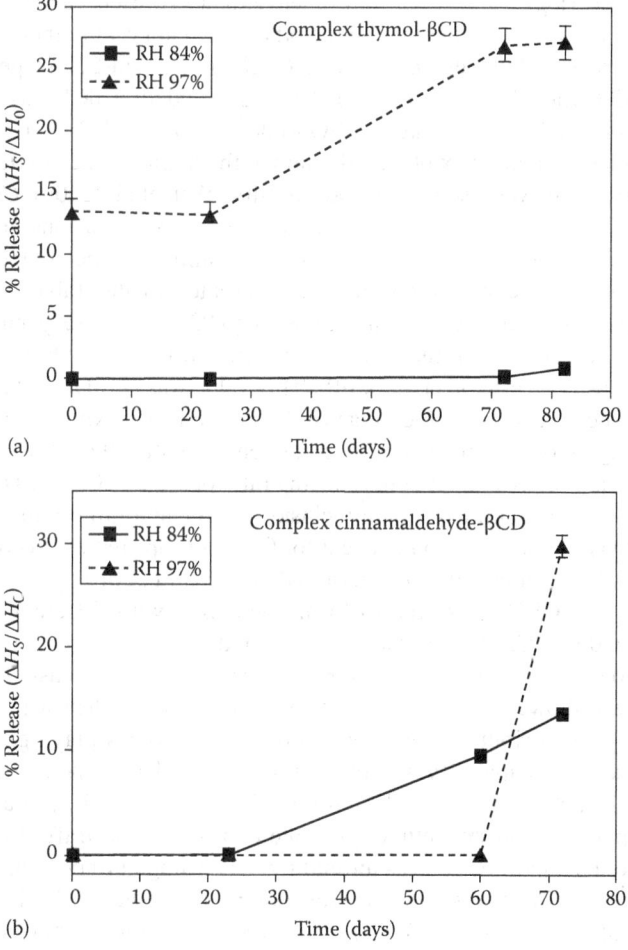

FIGURE 40.4 Release of (a) thymol and (b) cinnamaldehyde from the complexes as a function of time. (From Cevallos, P.A.P. et al., *J. Food Eng.*, 99, 70, 2010.)

At 84% RH, the release of thymol (Figure 40.4a) was very low after 70 days of storage and then it increased slightly. But at RH = 97%, the release of thymol increased almost linearly with time from 20 up to 70 days of storage. Cinnamaldehyde release was detected after 23 storage days at 84% RH and then it increased continuously, while at 97% RH started only after 60 days of storage and then increased abruptly (Figure 40.4b). The different behaviors between both molecules could be attributed to the different spatial structure of those components, which affects the kinetics of their release and to the physical structure of the material upon hydration, which could retard the release of the ligand. The results show the relevance of selecting appropriated storage conditions for hydrophobic flavors encapsulated in βCD or for predicting the shelf-life of functional products formulated with nanoencapsulated compounds (Cevallos et al., 2010).

Some volatile compounds are also released from the inclusion complexes in the same range temperature as the water of hydration, and therefore, their concentrations under inclusion complexes are higher. This is the case observed with the EOs from *Lauraceae* and *Rosaceae* botanical families (*Cinnamomum cassia* L. and *Rosa damascena* L.) that were complexed with βCD (2.4%–3.4% of EO as inclusion complexes) by crystallization from alcohol-water solution method. The higher hydrophobicity of cinnamaldehyde (log$P$ 1.8) compared with other compounds (e.g., benzaldehyde and o-methoxy cinnamaldehyde with log$P$ of 1.7 and 1.5, respectively) is responsible for its higher concentration in the recovered *C. cassia* EO. By contrary, the main volatile compound from *R. damascena* EO (log$P$ value of 1.7 for β-phenylethanol) presents a lower relative concentration in the recovered EO when compared with the other compounds of higher hydrophobicity, such as citronellol and α,α-dimethylphenethyl acetate (log$P$ of 2.8 and 2.7, respectively) (Hădărugă et al., 2006). The EO from *R. damascena*, Miller was included in βCD cavity by the saturated water solution-stirring method (mass ratio of EO to βCD 1:10; 60°C; 2 h). The inclusion rate was up to 80% and the inclusion complex obtained retained the main characteristic of the components of the EO and improved its water solubility and stability (Wen et al., 2009). *Cinnamomum verum* EO (CvEO) and βCD were prepared by coprecipitation process (in various ratios). The composition of the oil extracted from the inclusion complex was similar to the composition of the initial oil. Recoveries of inclusion complex powders are presented in the Table 40.2. The maximum CvEO-βCD interaction was achieved at a ratio 20:80 with 93.77% recovery, but the data presented in Figure 40.5 show the proportion that achieved maximum load of CvEO with βCD was 15:85 (117.2 mg of oil/g of βCD) (Petrović et al., 2010). This value was similar to that observed for ratio 20:80. This inclusion efficiency is in the range of the theoretical maximum loading for βCD with EO of 8%–12% (Pagington, 1986) which is in agreement with other flavors (Table 40.1). The results obtained in this work reveal that high starting ratio of the CvEO to βCD produces the maximum recovery of the oil powder, maximum inclusion of EO, and minimum noncomplexed βCD, suggesting that βCD is a good complexing agent for CvEO encapsulation, keeping its organoleptic properties as well as its pharmacological activity (Petrović et al., 2010).

Litsea cubeba EO (LCEO) was able to form complexes with different CDs such as βCD, DMβCD, HPβCD, and HEβCD by the suspension method. The LCEO–CDs inclusion complexes' molar ratio stoichiometry was 1:1 and their $K_{1:1}$ decreased with increasing temperatures. The complexation process showed to be an exothermic and enthalpy-driven process accompanied by a negative entropic contribution with the van der Waals forces playingan important role in this process. The inclusion capability was in the following order: HPβCD > HEβCD > DMβCD > βCD (Wang et al., 2009). A new inclusion complex containing L. grata leaf EO and βCD (LglEO-βCD), prepared by slurry method, presented an important draft of new compounds for the treatment of orofacial pain in mice as animal models (using antinociceptive activity). A possible CNS action of EO-βCD was evaluated on Fos protein labelled by immunofluorescence, showing a significant ($p < 0.05$ or $p < 0.001$) activation of the motor cortex, locus ceruleus (LC), nucleus raphe magnus (NRP), and periaqueductal gray (PAG), when compared to the control group (vehicle) (Figure 40.6). Despite the fact that EO-βCD increases significantly the Fos protein marking in the motor cortex CNS area, EO-βCD treated mice did not indicate any significant

TABLE 40.2
Recovery of the Powder (Complex) at Various *C. verum* EO to βCD Ratios

C. verum Oil-βCD	Starting Material[a] (g, db*)	Recovered Powder (g, db*)	Recovery (%)
5:95	4.836 ± 0.04	4.047 ± 0.24	83.68
10:90	5.102 ± 0.01	4.309 ± 0.35	84.46
15:85	5.408 ± 0.03	5.047 ± 0.17	93.32
20:80	5.743 ± 0.02	5.385 ± 0.14	93.77

Source: Petrović, G.M. et al., *J. Med. Plant Res.*, 4(14), 1382, 2010.
Note: db*, dry weight basis.
[a] Total amount of dry βCD (4.591 g) plus *C. verum* oil used.

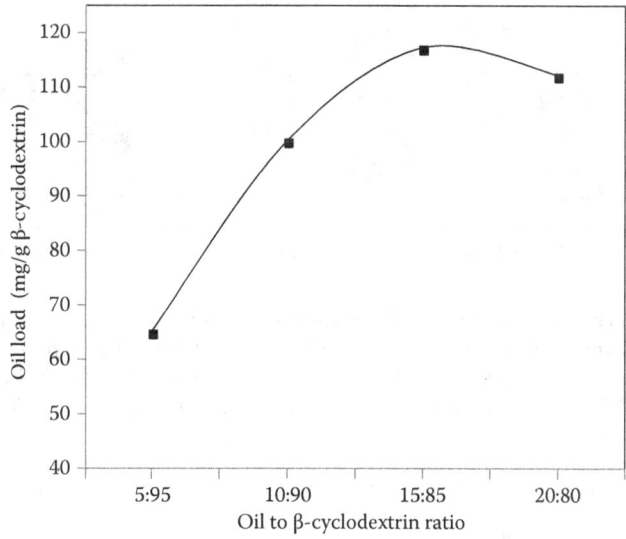

FIGURE 40.5 Flavor oil load of βCD as a function of the initial EO to βCD ratio. (From Petrović, G.M. et al., *J. Med. Plant Res.*, 4(14), 1382, 2010.)

motor performance alterations in the rota-rod apparatus. Antinociceptive profile might be linked to the presence of some terpenoids, such as camphor, borneol, and β-caryophyllene, and to the activation of the motor cortex, NRP, and PAG (cerebral areas involved in pain modulation). The incorporation of natural products like EOs or related compounds in drug-delivery systems (such as CDs) improves some of their intrinsic characteristics, leading to an improvement on their plasma half-life, effectiveness, and decrease in side effects and/or toxicity (Siqueira-Lima et al., 2014).

Pure eugenol (EG), representing 90% to 95% of the total essential clove oil (ECO) amount (Schmid, 1972; Briozzo et al., 1989), was complexed with either βCD or HPβCD. The inclusion complexes obtained showed differences between them. The total ECO (concentration refers to EG) dissolved increased with βCD concentrations up to 2 mM (Figure 40.7, filled circles). At this point, the maximum level of ECO in solution was reached (12 mM). Above 2 mM, the ECO concentration decreased until reaching a constant level of 6 mM, leading to a B_S-type phase-solubility profile. Over 2 mM βCD concentration, a gradual increase in ECO-βCD insoluble complexes was noted, until reaching a maximum ECO level of 550 mg/g precipitate (Figure 40.7, open circles) (Hernández-Sánchez et al., 2012).

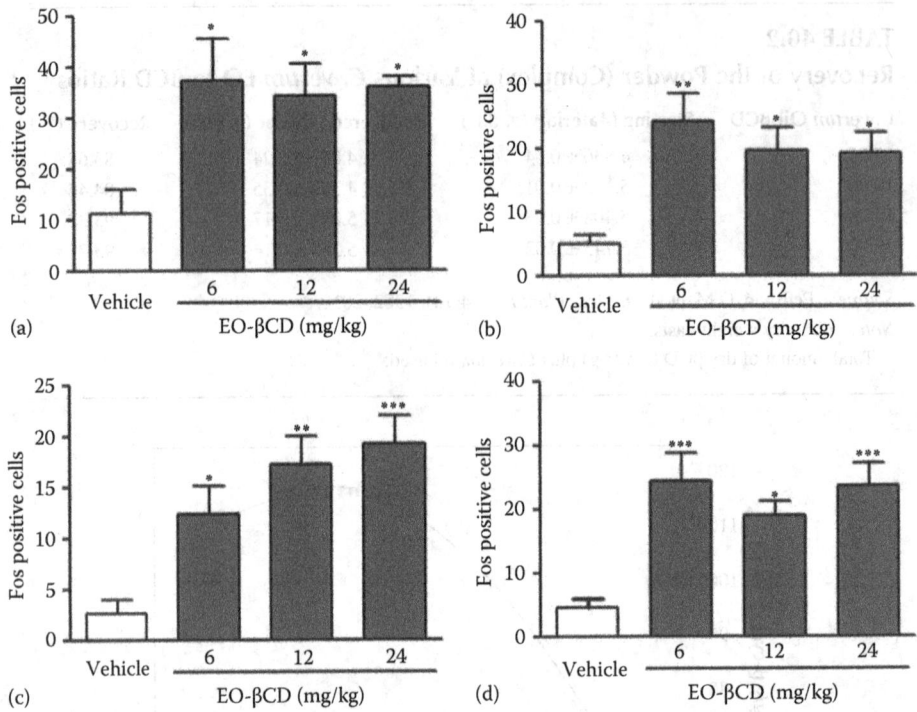

FIGURE 40.6 Neurons Fos positive in the: (a) motor cortex, (b) locus ceruleus, (c) nucleus raphe magnus, and (d) periaqueductal grey. Vehicle (control) or EO-βCD (6, 12 and 24 mg/kg) were administered orally 1.5 h before perfusion. Values represent mean ± SEM ($n = 6$ per group). $*p < 0.05$, $**p < 0.01$ or $***p < 0.001$ versus control (one-way ANOVA followed by Tukey's test). (From Siqueira-Lima, P.S. et al., *Basic Clin. Pharmacol. Toxicol.*, 114(2), 188, 2014.)

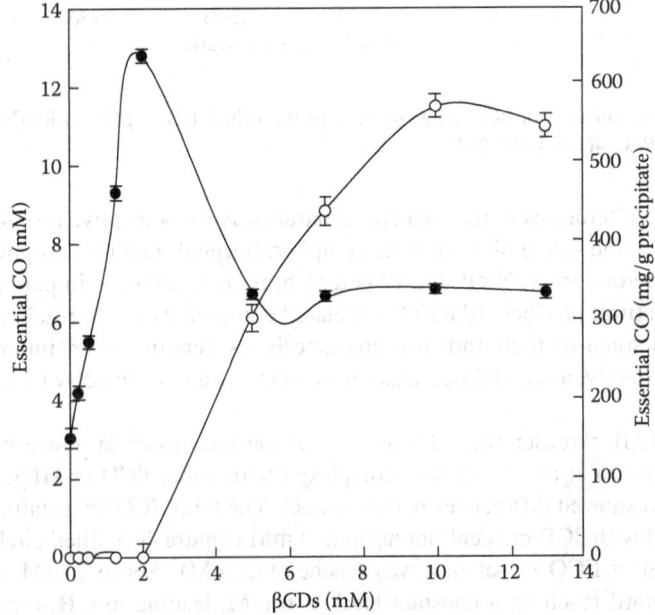

FIGURE 40.7 Phase solubility diagram for ECO in the presence of increasing βCD concentrations: soluble (●) and insoluble (○) ECO-βCD complexes. (From Hernández-Sánchez, P. et al., *Food Nutr. Sci.*, 3, 716, 2012.)

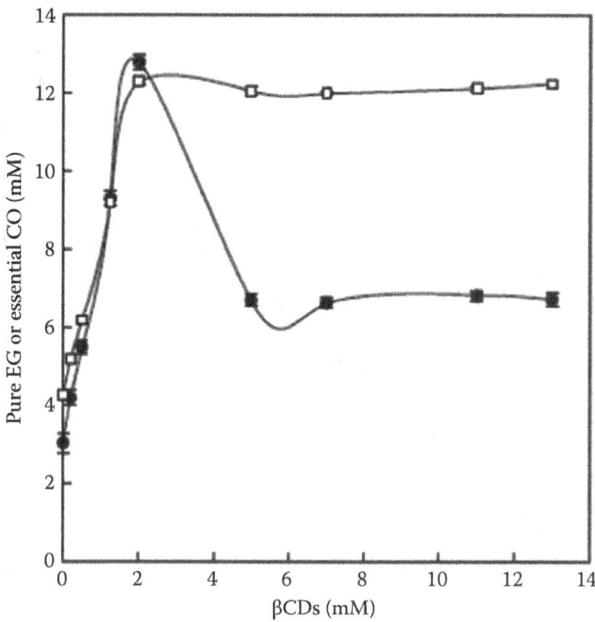

FIGURE 40.8 Phase solubility diagram for pure EG and ECO in the presence of increasing βCD concentrations: pure EG-βCD complexes (□) and soluble ECO-βCD complexes (●). (From Hernández-Sánchez, P. et al., *Food Nutr. Sci.*, 3, 716, 2012.)

The profile of pure EG using βCD (Figure 40.8, open squares) was not exactly the same as that obtained for ECO (Figure 40.8, filled circles). In the case of pure EG, the saturation level of dissolved compound was reached with 2 mM βCD concentration, maintaining a plateau up to 12 mM. The ECO profile also reached its maximum solubility level at 2 mM but decreased at about 6 mM maintaining a plateau up to 12 mM (Hernández-Sánchez et al., 2012).

The ECO in the presence of HPβCD showed a linear response with CD concentration increasing (Figure 40.9, filled circles) corresponding to an A_L-type profile, according to Higuchi and Connors (1965). The Kc value for the complexes formation between HPβCDs and ECO and pure EG was 2005 ± 199 M^{-1} (Figure 40.9, filled circles) and 4555 ± 225 M^{-1} (Figure 40.9, open squares), respectively. This linear relationship with a slope value lower than 1 suggests a 1:1 stoichiometry for ECO-HPβCDs complexes (Hernández-Sánchez et al., 2012).

The results suggest that both βCD and its derivative HPβCD could be an alternative to solve the practical application of the ECO and at the same time facilitate the controlled release of its constituents in food and medical industries (Hernández-Sánchez et al., 2012).

Stable nonhygroscopic microcrystalline substance was obtained when the content of the Diapulmon (camphor, l-menthol, eucalyptus oil and quinine dissolved in sunflower oil) was complexed with βCD. This powder, when sprinkled on hot water, gradually releases the included volatile compounds, achieving desired pharmacological effects (Gál-Füzy et al., 1984). Flavor retention promoted by α-, β-, and γCDs in order to provide a better basis for selecting which CD to use in a given flavor application was analyzed. Employing a broad range of flavor molecules (Table 40.3), γCD generally achieved the highest initial flavor retention, while α- and βCD yielded lower but similar results to one another. However, losses of volatiles were greatest for γCD and least in the case of αCD, during storage. The results suggest that CD encapsulation through spray-drying involves matrix entrapment as well as molecular inclusion. For example, the less stable flavor compounds when complexed with α- and βCD show the best overall retention during spray-drying and storage. Thus, CDs prove to be useful in stabilizing flavor compounds that cannot be effectively stabilized using other techniques (Reineccius et al., 2002). The same authors, in other studies, analyzed the thermal influence in the processed foods components to check for the damages.

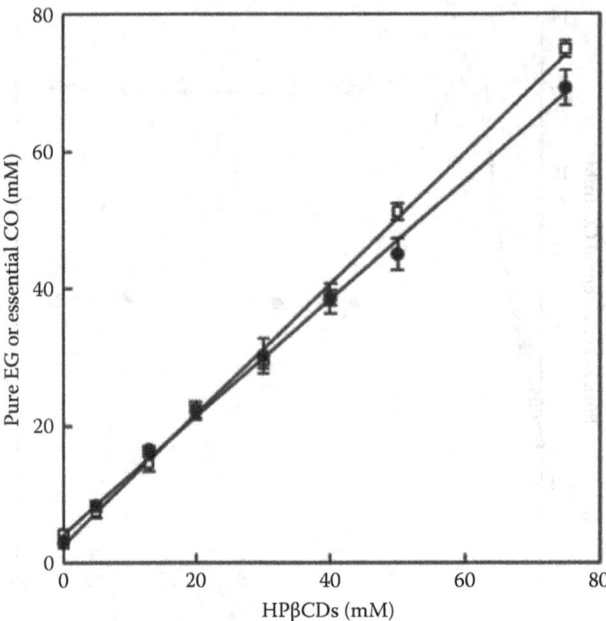

FIGURE 40.9 Phase solubility diagram for ECO and pure EG with HPβCD: ECO-HPβCD (●) and pure EG-HPβCD (□). (From Hernández-Sánchez, P. et al., *Food Nutr. Sci.*, 3, 716, 2012.)

To reduce these bad effects on several flavors—benzaldehyde (characteristic of cherry flavor), citral (lemon flavor), l-menthol (mint flavor), and vanillin (vanilla flavor)—the CD efficacy was studied. These flavor compounds were added to selected food applications (hard candy, fruit leathers, angel food cake) in the form of both flavor-βCD complex and liquid formulation. The flavor compounds retained after processing (cooking of the hard candy, fabrication of the fruit leather, or baking of the angel food cake) are presented in Table 40.4. Benzaldehyde, citral, and menthol retention flavor values after processing were about 20-, 26-, and 86-fold more efficiently retained, respectively, under inclusion complexes solid form when compared with the same free compounds in the liquid formulations (Figure 40.10). Only the angel food cake application did not benefit significantly from including the flavor in βCD. It has been shown that dry CDs will not give up their guest molecules until they thermally decompose. However, adding water to the system results in the release of the guest molecule, even at room temperature, albeit in the form of a food matrix/CD equilibrium (Reineccius et al., 2004).

Considering flavor retention across the applications, it appears that the more severe the heat treatment given to the product, the more it benefits from including the flavoring into the CD. The hard candy is the most severely heated, followed by the fruit leather, and finally, the angel food cake.

However, when products were subjected to sensory evaluation, results indicate that the benefits of enhanced flavor retention achieved by the flavor/CD complexation could be mitigated by poor release from molecular-inclusion complexes during consumption (Reineccius et al., 2004).

The greasy, oily or liquid, coffee aroma concentrates (natural and synthetic) under microcrystalline stable inclusion complexes form are very important for using as additives (maintain sensory properties). By this reason, inclusion complexes between coffee flavors (natural and synthetic) and βCD were produced in this work. These inclusion complexes release its flavor immediately when upon contact with water (Szente and Szejtli, 1986).

Applying conventional (thermogravimetry: TG; Differential thermal analysis: DTA; evolved gas detection: EGD) and combined (thermogravimetry-mass spectrometry: TG-MS) techniques, the βCD inclusion complexes with thymol and *Lippia sidoides Cham* EO (LsCEO) extract produced by kneading method were characterized. Its formations were detected through released gas

TABLE 40.3
Composition of Model Flavor

Compound	FEMA[a] nr	Nominal Molecular Weight	% Composition (Weight Basis)	% Composition (Molar Basis)
System I comprising broad range of flavor molecules				
Acetaldehyde	2003	44	1.30	3.58
Acetal	2002	118	3.48	3.58
2,3-butanedione	2370	86	2.53	3.58
2,3-pentanedione	2841	100	2.95	3.58
Dimethyl sulfide	2746	62	1.83	3.58
Dimethyl disulfide	3536	94	2.77	3.58
Ethyl acetate	2414	88	2.59	3.58
Ethyl butyrate	2427	116	3.42	3.58
Ethyl valerate	2462	130	3.83	3.58
Ethyl hexanoate	2439	144	4.25	3.58
Ethyl heptanoate	2437	158	4.66	3.58
2-hexenal	2560	98	2.89	3.58
2-heptenal	3165	112	3.14	3.40
2-octenal	3215	126	3.72	3.58
2-nonenal	3213	140	4.13	3.58
2-decenal	2366	154	4.54	3.58
2-methylpyrazine	3309	94	2.77	3.58
2,5-dimethylpyrazine	3272	108	3.18	3.58
2-methyl-3-ethylpyrazine	3155	122	3.60	3.58
3,5(6)-dimethyl-2-ethylpyrazine Isomers	3149	136	4.01	3.58
Benzaldehyde	2127	106	3.12	3.58
Limonene	2633	136	4.01	3.58
Methyl salicylate	2745	152	4.48	3.58
4-hydroxy-2,5-dimethyl-3(2H)-furanone	3174	128	3.77	3.58
Eugenol	2467	164	4.83	3.58
Isoeugenol	2468	164	4.83	3.58
Vanillin	3107	152	4.48	3.58
Ethylvanillin	2464	166	4.89	3.58
Total			100	100
System II comprising labile flavor molecules				
Ammonium isovalerate	2054	140	17.15	16.22
2-furfuryl mercaptan	2493	114	13.57	15.76
Dimethyl sulfide	2746	62	5.82[b]	12.42[b]
Methional	2747	104	6.52[c]	8.30[c]
3-methylthio-1-hexanol	3438	148	17.63	15.77
2-methyl-4-propyl-1,3-oxathiane	3578	160	19.04	15.75
Isomer A (81%)	—	—	15.42	12.76
Isomer B (19%)	—	—	3.62	2.99
Limonene/sulphur reaction product[d]	—	(170)	20.27	15.78
2,8-epithio-p-menth-1-ene (28%)	—	168	5.68	4.42
2,8-epithio-p-menthane (38%)	—	170	7.70	6.00

Source: Reineccius, T.A. et al., *J. Food Sci.*, 67, 3271, 2002.

[a] Flavor and Extract Manufacturers Association.
[b] Loss occurred after formulation but before spray-drying.
[c] Methional used in model flavor system contained 48% dimer.
[d] Commercial product comprised 34% monoterpenes and 66% terpene/sulfur adducts.

TABLE 40.4
Flavor Retention in Food Applications

Flavor Compound/ Food Application	Liquid Flavor (g Total)			CD Flavor (g Total)			Average Retention in Application	
	Initial Load	Final Load		Initial Load	Final Load		Liquid Flavor	Flavor/CD Complex
		Rep. 1	Rep. 2		Rep. 1	Rep. 2		
Benzaldehyde/ fruit leather	0.15	3.70×10^{-04}	4.30×10^{-04}	0.23	1.20×10^{-02}	1.30×10^{-02}	0.30%	5.4%
Citral/fruit leather	0.15	4.40×10^{-03}	3.80×10^{-03}	0.16	0.12	0.11	2.70%	72%
l-Menthol/ hard candy	0.30	3.00×10^{-03}	2.50×10^{-03}	0.28	0.20	0.24	0.90%	79%
Vanillin/angel food cake	0.30	0.19	0.19	0.31	0.21	0.23	63%	71%

Source: Reineccius, T.A. et al., *J. Food Sci.*, 69, FCT58, 2004.

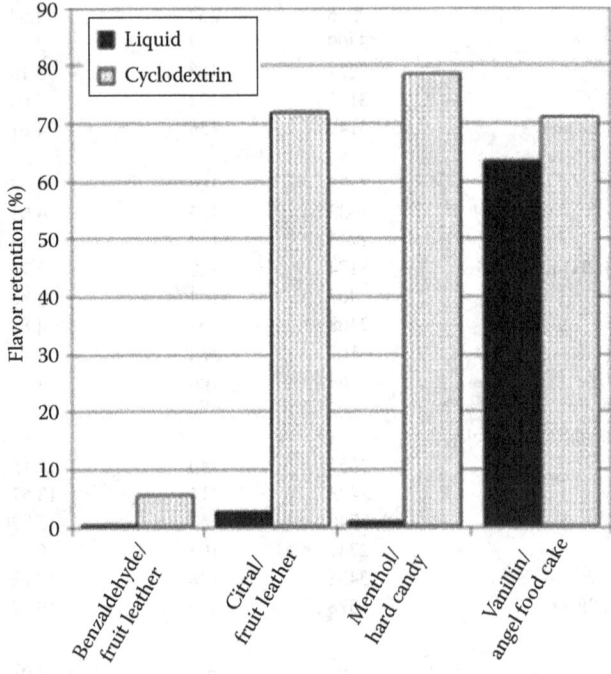

FIGURE 40.10 Retention of flavor constituents through processing in various applications when added either as CD complex or as liquid formulation. (From Reineccius, T.A. et al., *J. Food Sci.*, 69, FCT58, 2004.)

analysis. The combined methods also allowed to prove the complex formation and to selectively follow the liberation of the entrapped guest. The thermoanalytical results were completed with powder X-ray diffraction (XRD) experiments (Figure 40.11). The most intensive diffraction lines for βCD appeared in the x-ray pattern of the physical mixture (curve 3), denying any interaction between the components in solid phase after their gentle mixing. The XRD patterns of both the thymol-βCD and EO-βCD are remarkably different from any of the previous ones giving evidence on the

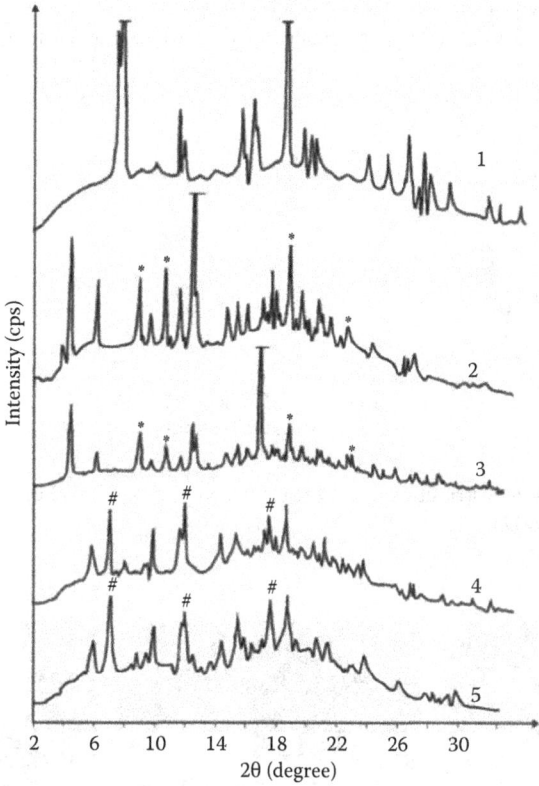

FIGURE 40.11 X-ray diffraction patterns of thymol-βCD and EO-βCD samples. 1—thymol, 2—βCD, 3—thymol-βCD physical mixture, 4—thymol-βCD complex, 5—EO-βCD complex (*—βCD and #—inclusion complex). (From Fernandes, L.P. et al., *J. Therm. Anal. Calorim.*, 78, 557, 2004.)

inclusion complex formation, but they are similar to each other because of the high thymol content of the EO sample (curves 4 and 5). All results obtained with TG-MS measurements were similar to those obtained with the conventional TG, DTA, and EGD runs. Also the x-ray results supported the existence of these new entities (Fernandes et al., 2004).

EO-βCD ratios, 1:10, 1.33:10, 2:10 m/m (IC-A, IC-B, and IC-C) inclusion complexes between LsCEO and βCD, were prepared using the slurry method. The solvent was removed by spray-drying technique. Correlation found between the EO-βCD ratio and the total oil retention in the powders, resulted in an optimal EO-βCD starting ratio (1:10 m/m). The evidence of the inclusion complex formation was made by thermal analysis, and morphological changes upon complexation and during storage were followed by scanning electron microscopy (SEM) and XRD. The particle size distribution of the powder samples varied from 0.375 to 52.63 mm; however, the representative particle size was about 10 to 12 mm (Figure 40.12). The SEM images (Figure 40.13) showed amorphous powder samples for either native CD itself or *Lippia sidoides*–βCD inclusion complex. After 6 weeks of storage, the morphology of the inclusion complex did not show any changes (Fernandes et al., 2009).

According to Figure 40.14, the amount of released oil from the IC-A was less compared to the released amount from IC-B and IC-C samples. These results are in agreement with the TG curves, where crescent mass variations in the same temperature interval were observed for the studied samples. Despite of the very similar thermal behavior among the samples, the different EO-βCD compositions resulted acrescentaram complexes with slightly different retardation properties. Thus, the oil was found to be more strongly bonded to βCD in the IC-A sample than in the IC-B and IC-C samples (Fernandes et al., 2009).

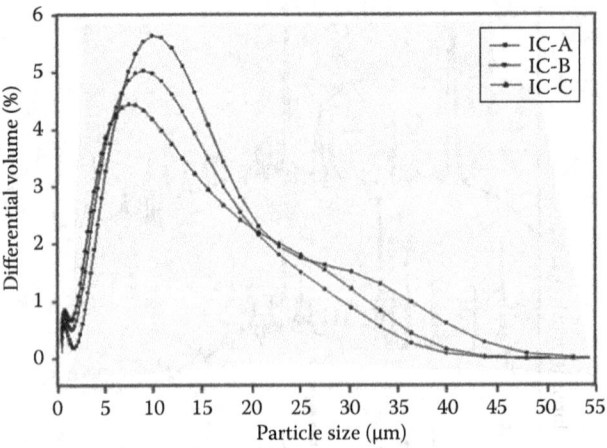

FIGURE 40.12 Particle size distribution of the inclusion complexes. (From Fernandes, L.P. et al., *J. Therm. Anal. Calorim.*, 95, 855, 2009.)

FIGURE 40.13 SEM images of (a) native, (b) spray-dried CDs, and (c) spray-dried inclusion complex at zero time ($N = 200$). (From Fernandes, L.P. et al., *J. Therm. Anal. Calorim.*, 95, 855, 2009.)

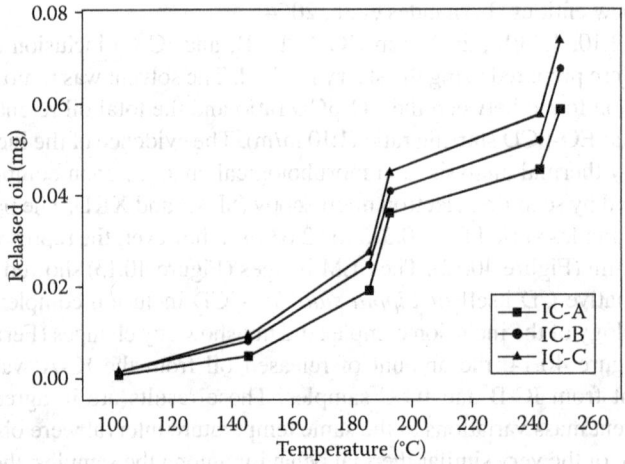

FIGURE 40.14 Release profiles of *Lippia sidoides* EO-βCD complexes. (From Fernandes, L.P. et al., *J. Therm. Anal. Calorim.*, 95, 855, 2009.)

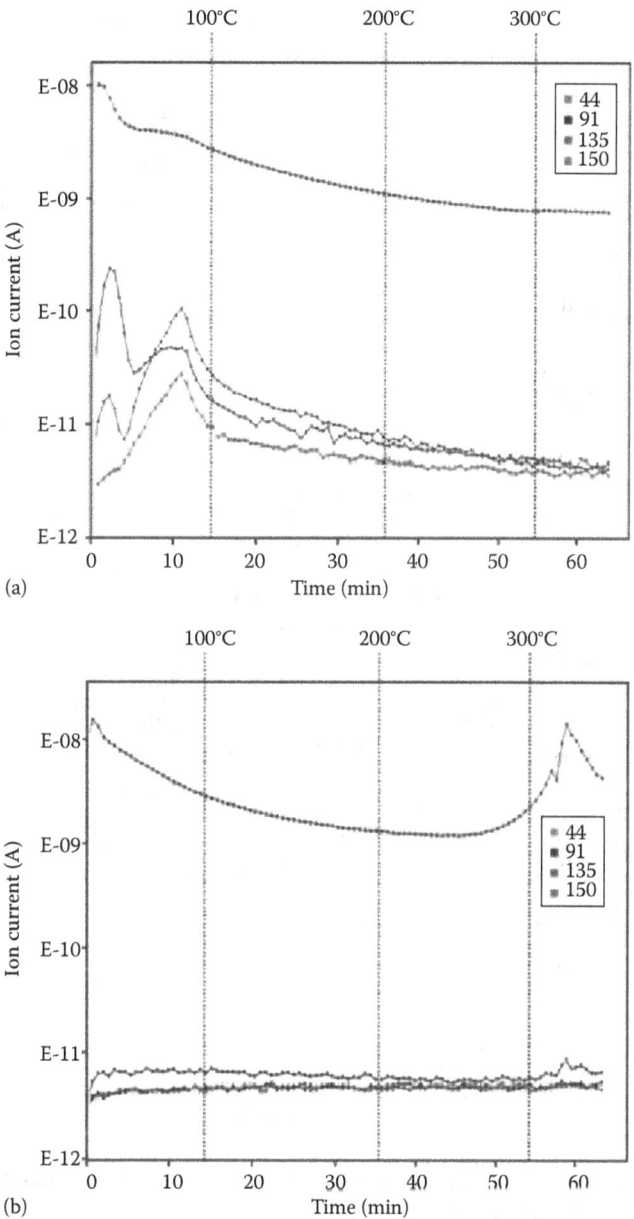

FIGURE 40.15 MID curves: (a) *Lippia sidoides* EO and (b) βCD. (From Fernandes, L.P. et al., *J. Therm. Anal. Calorim.*, 95, 855, 2009.)

In the multiple ion detection (MID) curves of the EO (Figure 40.15a), the presence of EO is confirmed by the lines showing double peak at room temperature; when temperature increases above 115°C, evaporation of the EO occurs and the peak disappears. This phenomenon is observed when the EO is not included. In the MID curve of βCD (Figure 40.15b), $m/z = 91$, 135, and 150 are running practically on the baseline. Consequently, they may be considered as selective ones for the presence of any constituents of the EO; however, they are not representative for the fragmentation of βCD. On the contrary, curve at $m/z = 44$ shows a higher signal intensity around 310°C, which is representative for the thermal decomposition of the glucopyranose unit (Fernandes et al., 2009).

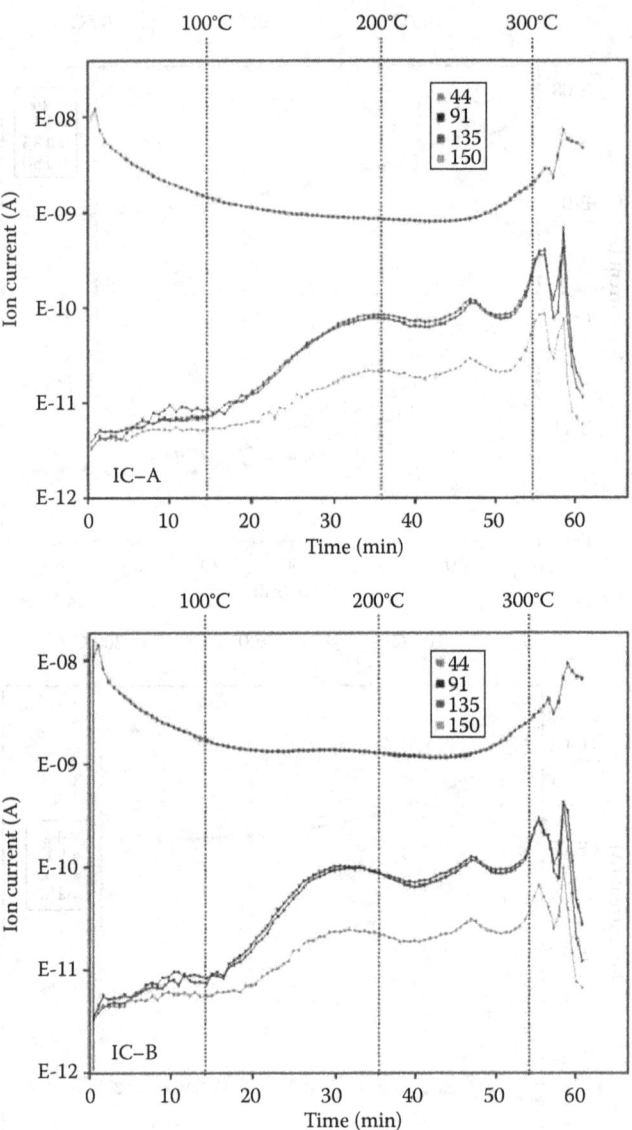

FIGURE 40.16 MID curves of inclusion complexes: IC-A, IC-B, and IC-C. (From Fernandes, L.P. et al., *J. Therm. Anal. Calorim.*, 95, 855, 2009.)

The shapes of the MID lines of the inclusion complexes (Figure 40.16) showed the release of uncomplexed oil and continuous evaporation of the EO below 105°C. One can see that the shape of the samples (IC-A, IC-B, and IC-C) MID profiles is very similar to each other. According to Figure 40.16, curves at m/z = 91, 135, and 150 showed a first maximum around 195°C to 200°C and a second one at 265°C, indicating the degradation of the inclusion complex. Nevertheless, the third maximum around 305°C to 310°C and a forth one at 320°C suggests that a part of the EO escapes from the βCD cavity only when the sugar derivative decomposes. This suggests the high thermal stability of the EO–inclusion complex. The thermal profiles of the complexes with different EO-βCD proportions were slightly different to each other. The IC-A sample was the most stable regarding the release of the EO amount and its storage time influenced the release of the oil amount. The thermal stabilities of the inclusion complexes were influenced not only by the different EO-βCD ratio, but also by the storage conditions (Fernandes et al., 2009). The $K_{1:1}$ obtained by static and dynamic head-space methods were in reasonable agreement with the corresponding values in the literature. In addition, the release profiles of fragrance materials from 2-HPβCD aqueous solution were investigated using the method previously cited. It was found that the suppression of the fragrance materials release was dependent on their K_c (Saito et al., 1999).

Mentha x villosa oil and βCD were used to produce inclusion complexes (1:9 m/m oil-βCD ratio) by coprecipitation and kneading methods. The CG/MS chromatographic profile of the total oil is shown in Figure 40.17.

The pure *Mentha x villosa* oil, physical mixture, and its inclusion complexes are shown through EGD curves (Figure 40.18). The shape of the evaporation peak of the pure *Mentha x villosa* oil showed the presence of fractions with different volatility and tension (Figure 40.18, curve A). The curve of the physical mixture presents the thermoanalytical characteristics of the individual components

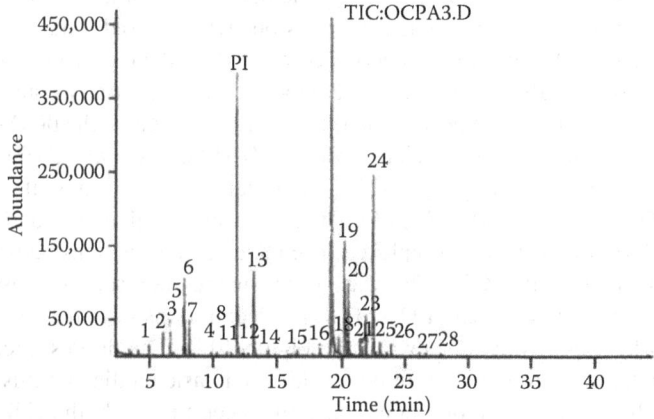

FIGURE 40.17 CG/MS chromatogram of the total oil extracted from the complex powder. (From Martins, A.P. et al., *J. Therm. Anal. Calorim.*, 88, 363, 2007.)

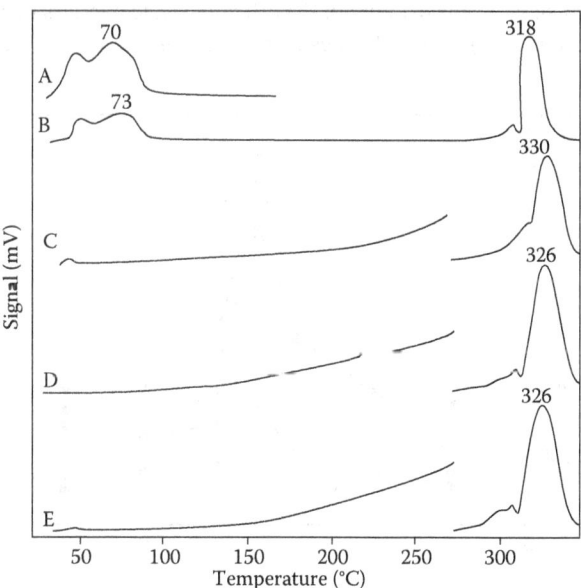

FIGURE 40.18 EGD curves of *Mentha x villosa*–βCD samples; A—pure oil, B—freshly prepared physical mixture, C—1-month stored physical mixture, D—coprecipitated and E—kneaded complexes. (From Martins, A.P. et al., *J. Therm. Anal. Calorim.*, 88, 363, 2007.)

such as the broad overlapping peaks below 100°C correspond to the evaporation of the oil extract and the peak over 300°C indicates the decomposition of the βCD ring (Figure 40.18, curve B). The thermal features of the 1-month stored physical mixture are similar to that of the complexes, proving that complexation occurs on storage (Figure 40.18, curves C–E). These results indicate the formation of the inclusion complex between *Mentha x villosa* and βCD. The slight deviation observed from the baseline corresponds to the evaporation of the oil components from their inclusion complex (Martins et al., 2007).

In other study, it was proved that the same oil previously cited can be successfully complexed with βCD by the coprecipitation method using hydroethanolic medium with a 96% powder recovery. The product obtained was similar to the original *Mentha x villosa* oil with respect to the major volatile components. The retention of volatiles was approximately 78%, while the total volatile compounds in the original oil was 99%. From the detected compounds in the original oil, 13 are monoterpenes, 10 are sesquiterpenes, 1 is ester, 1 is phenylpropanoid, and 3 were not identified. The major component of the oil is piperitenone oxide and its degree of complexation was 72%. Twelve compounds were totally complexed, 11 compounds were partially complexed, and 5 compounds were not complexed. It is supposed that 3 were only adsorbed on the βCD surface and 2 were absent. The calculated complexation efficiency was 13.6% (Martins et al., 2007).

S. sclarea L. EO (SEO) was investigated by fluorescence spectroscopy after its complexation with βCD, DMβCD, 2-HEβCD, and 2-HPβCD. The results showed that βCD and its derivatives can react with SEO to form the inclusion complexes. The inclusion interactions between βCD and SEO were far stronger than three other CDs. The effect of pH on the fluorescence emission of SEO in the absence or presence of CDs, in which SEO and CDs concentrations were held constant at 1.0×10^{-8} and 4.0×10^{-8} mol/L, respectively, is shown in the Figure 40.19. The fluorescence intensity of SEO itself in neutral media was stronger than in both acidic and basic media and this parameter in four CDs in neutral media was always more remarkably enhanced than in both acidic and basic media. Because the major components of SEO contain ester group, which is easy to react with an acid and a base, the SEO is most stable in neutral media. However, in any medium, it was noted that the fluorescence enhancement in four CDs followed the order: βCD > DMβCD > 2-HEβCD > 2-HPβCD (Tian et al., 2008). This sequence is explained: the effect of steric barriers becomes larger with βCD derivatives, which prevents the guest molecules from entering into the CDs cavities.

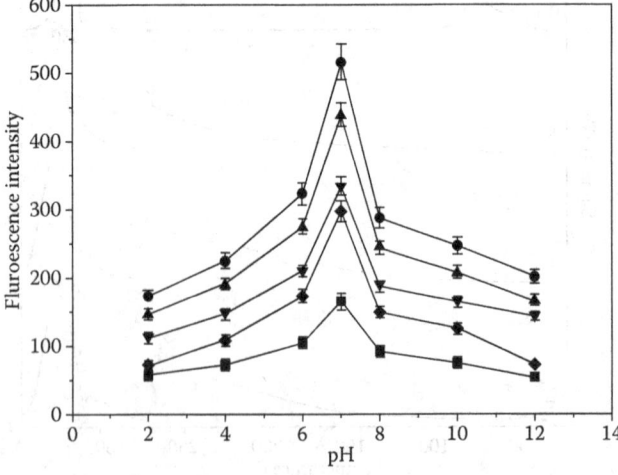

FIGURE 40.19 Influence of the pH on the fluorescence intensity of SEO (filled square); SEO-βCD complex (closed circle); SEO-DMβCD complex (filled triangle); SEO-2-HEβCD complex (filled inverted triangle); SEO-2-HPβCD complex (open square); [SEO] = 1.0×10^{-8} mol/L, and [CDs] = 4.0×10^{-3} mol/L. Each value represents the mean ± SD ($n = 3$). (From Tian, X.N. et al., *Eur. Food Res. Technol.*, 227, 1001, 2008.)

TABLE 40.5
Formation Constants (K_c) Value of SEO-CDs under Different Temperatures (mol/L)$^{-1}$

Temperature (°C)	25	35	45
βCD	164.5 ± 7.2	138.4 ± 5.5	114.6 ± 3.4
DMβCD	116.6 ± 3.9	89.7 ± 2.7	75.6 ± 1.9
2-HEβCD	85.2 ± 2.4	68.1 ± 1.8	57.6 ± 1.7
2-HPβCD	64.6 ± 2.6	52.8 ± 2.1	41.4 ± 1.4

Source: Tian, X.N. et al., *Eur. Food Res. Technol.*, 227, 1001, 2008.

The different inclusion capacity promoted by four CDs on SEO is shown in the Table 40.5. The values obtained for different powders showed that interaction between SEO and βCD was far stronger than that with the other three CDs (Tian et al., 2008). In addition, the formation constant K_c decreases with increasing temperature, indicating that the affinity of CDs for SEO decreased, as expected for an exothermic process.

The thermodynamic parameters (ΔH, ΔS, and ΔG) values obtained for the formation of inclusion complex (Table 40.6) were determined from temperature dependence of apparent formation constants, by using classical van't Hoff equation and plotting ln K versus $1/T$ (Tong, 2001):

$$\ln K = -\frac{\Delta H}{RT} + \frac{\Delta S}{R} \qquad (40.4)$$

The corresponding enthalpy and entropy can be obtained from the slop and intercept, respectively. The results indicate the marked tendency of SEO to complex with CDs in water. The negative values of enthalpy changes indicate that the interaction processes of SEO with CDs are exothermic. The enthalpy of the system was decreased largely, which implied that the main driving forces of inclusion reaction were van der Waals forces and the extrusion of "high-energy water" molecules from the CDs cavities. The Gibbs free energy change for the interactions that take place during the inclusion process may be found by the following equation:

$$\Delta G = -RT \ln K \qquad (40.5)$$

The Gibbs free energy at temperature 298 K is calculated as a negative value (Table 40.6), indicating that the inclusion process is a spontaneous one. The results also showed that SEO-βCDs inclusion complexes are very stable and can be used as an additive in foods and as a new formulation to optimize its pharmacological profile (Tian et al., 2008).

TABLE 40.6
Thermodynamic Parameters of SEO-CDs Complexes

Host	$-\Delta G_{25}$ (KJ mol^{-1})	$-\Delta H$ (KJ mol^{-1})	$-\Delta S$ (JK^{-1} mol^{-1})
βCD	12.63 ± 0.12	14.29 ± 0.5	5.57 ± 1.1
DMβCD	11.75 ± 0.1	17.18 ± 0.2	18.26 ± 0.42
2-HEβCD	10.99 ± 0.08	15.73 ± 0.37	15.77 ± 1.3
2-HPβCD	10.29 ± 0.14	17.49 ± 0.22	24.07 ± 0.01

Source: Tian, X.N. et al., *Eur. Food Res. Technol.*, 227, 1001, 2008.

The αCD, βCD, and HPβCD were used to increase the solubility of a range of plant-derived EO compounds (carvacrol = Carv, eugenol = Eug (EG), linalool = Lin and 2-pentanoylfuran = Pentfuran) and its influence on the antimicrobial activity of EO components on a range of microorganisms was tested.

The optimal fractional inhibitory concentration indices (FICI) for a number of combinations of individual EO compounds with regard to different microorganisms are shown in Table 40.7. EG and Lin combination displayed a synergistic activity (FICI = 0.375) for *S. aureus*. Carv and EG, Carv and Lin, Carv and Pentfuran exhibited a useful addictive effect with a FICI of 0.75 against *E. coli*, *B. subtilis*, and *S. cerevisiae*, respectively. However, no indifferent or antagonistic effects were observed for any of these combinations (Liang et al., 2012).

CDs such as HPβCD was also analyzed if they had a further enhancing activity with regard to changes to the membrane fatty acid profile of the microorganisms studied with regard to Carv. Addition of HPβCD alone did not significantly affect the content of the major fatty acids in the various microorganisms (Table 40.8). Addition of Carv revealed a marked change in the profile of the major fatty acids, which was enhanced by the addition of HPβCD (Table 40.8). The lipid profile of *E. coli* was principally composed of palmitic, (C16:0), palmitoleic, (C16:1), cis-10-heptadecenoic, (C17:1cis), oleic, (C18:1cis), and α-linolenic (C18:3cis) acids. When *E. coli* was treated with Carv alone (no CD), palmitic (C16:0), and palmitoleic (C16:1) acids increased. However, the relative levels of cis-10-heptadecenoic, (C17:1cis), oleic, (C18:1cis), and α-linolenic (C18:3cis) acids decreased; a trend further enhanced by the addition of HPβCD (Table 40.8). This shows that *E. coli* responds to hydrophobic stress, by synthesizing longer chained fatty acids at the expense of the shorter chained fatty acids. The exposure of the yeast *S. cerevisiae* to Carv resulted in a clear increase in the concentration of the saturated fatty acids: myristic (C14:0), palmitic (C16:0), and stearic (C18:0) acids. Again, these changes were further enhanced by the combined exposure to both Carv and HPβCD. It was also noted that all the microorganisms exhibited a reduction of unsaturated fatty acids, at a final concentration of 2 MIC and in most cases, the highest reduction was observed when the cells were treated with Carv containing HPβCD (Table 40.8).

In order to visually observe the effects of Carv on the integrity of the microbial cultures, a SEM of cells treated with HPβCD, Carv, and Carv in combination with HPβCD was carried out.

TABLE 40.7
Effects of CD Solubilizers on the Antimicrobial Activity of Combined EO Compounds

	MIC (lg/mL) and FICI of the Complex EOs and the Trend of Antimicrobial Activity											
Solubilizers	Escherichia Coli			Staphylococcus Aureus			Bacillus Subtilis[a]			Saccharomyces Cerevisiae		
Control	Carv	Eug	FICI	Eug	Lin	FICI	Carv	Lin	FICI	Carv	Pentfuran	FICI
	0.25	1.25	0.75	0.625	0.313	0.375	0.25	0.625	0.75	0.25	0.313	0.75
αCD	0.25	0.625	0.625	0.625	0.313	0.375	0.25	0.625	0.75	0.125	0.313	0.5
			↑			←			←			↑
						→			→			
βCD	0.25	0.625	0.625	0.625	0.313	0.375	0.25	0.625	0.75	0.125	0.313	0.5
			↑			←			←			↑
						→			→			
HPβCD	0.25	0.625	0.625	0.625	0.313	0.375	0.25	0.313	0.625	0.125	0.313	0.5
			↑			←			↑			↑
						→						

Source: Liang, H. et al., *Food Chem.*, 135, 1020, 2012.

[a] The trend of antimicrobial activity of selected complex EOs after adding solubilizers. ↑, increase of antimicrobial activity; ↓, decrease of antimicrobial activity; ←→, basically unchanged.

TABLE 40.8
Changes in the Percentage of Principal Fatty Acids of Cells[a]

Microorganisms	Fatty Acids	Untreated	Containing HP βCD	Treated with Carvacrol	Treated with Complex Carvacrol—HP βCD
Escherichia coli	C16:0	8.49 ± 1.1	8.17 ± 0.98	12.58 ± 2.13	15.69 ± 1.67
	C16:1	15.45 ± 2.11	15.02 ± 2.35	18.56 ± 0.89	19.97 ± 2.33
	C17:1cis	17.70 ± 1.32	17.11 ± 1.45	12.36 ± 2.21	10.33 ± 1.24
	C18:1cis	11.02 ± 0.98	11.34 ± 0.78	9.87 ± 1.09	9.01 ± 2.20
	C18:3cis	12.12 ± 0.80	12.08 ± 1.01	9.62 ± 0.87	8.73 ± 0.53
Staphylococcus aureus	C14:1cis	2.45 ± 0.22	2.13 ± 0.32	1.96 ± 0.68	1.77 ± 0.89
	C15:0	37.24 ± 2.20	39.28 ± 1.98	49.99 ± 2.10	57.66 ± 2.09
	C17:0	15.46 ± 1.23	16.61 ± 1.20	21.76 ± 3.02	23.78 ± 2.23
	C18:1cis	13.82 ± 1.75	13.34 ± 0.78	10.15 ± 2.01	ND[b]
	C18:3cis	16.32 ± 1.77	15.78 ± 1.34	9.92 ± 2.09	ND
Bacillus subtilis	C14:1cis	39.88 ± 3.02	41.98 ± 2.09	49.78 ± 2.02	53.26 ± 1.99
	C15:0	8.13 ± 0.99	8.40 ± 0.56	10.16 ± 1.19	11.65 ± 2.05
	C15:1cis	13.41 ± 1.20	14.92 ± 2.09	5.04 ± 1.30	2.24 ± 0.92
	C16:1	17.22 ± 2.14	16.79 ± 2.87	9.65 ± 0.99	5.09 ± 1.01
	C18:1cis	9.21 ± 1.22	8.96 ± 1.03	5.86 ± 0.82	3.28 ± 0.98
Saccharomyces cerevisiae	C14:0	0.19 ± 0.11	0.11 ± 0.09	6.99 ± 1.21	8.16 ± 2.01
	C16:0	3.12 ± 1.21	2.65 ± 0.99	17.37 ± 1.67	20.29 ± 1.89
	C16:1	0.66 ± 0.18	0.53 ± 0.10	5.60 ± 0.99	6.91 ± 1.24
	C18:0	27.28 ± 2.11	28.48 ± 2.40	48.23 ± 3.01	52.76 ± 2.78
	C18:1cis	3.60 ± 0.79	3.16 ± 1.03	0.52 ± 0.09	0.40 ± 0.03

Source: Liang, H. et al., *Food Chem.*, 135, 1020, 2012.

[a] Means of triplicate determinations ± SD.
[b] ND, not detectable.

Addition of HPβCD to any of the microbial cultures did not change the cellular morphology (comparison of Figures 40.20a and b; 40.20e and 40.21f; 40.21a and b; and 40.21e and f). The exposure to Carv at 2 × MIC revealed a significant detrimental effect on the morphology of cell membranes (Figures 40.20c and g and 40.21c and g). An incomplete cell structure and deformed shape was observed in treated *E. coli* and *S. aureus* (Figure 40.20c and g). A different degree of deformation was observed in treated *B. subtilis* and *S. cerevisiae*, showing a collapse of the cell structure (Figure 40.21c and g). The exposure to Carv in the presence of HPβCD caused an even greater deterioration of the cell structure of all cultures studied here (Figures 40.20d and h and 40.21d and h). The application of twice the MIC of Carv in the presence of HPβCD caused an almost complete collapse of the cell structure accompanied by cell lysis in *E. coli*, *B. subtilis*, and *S. cerevisiae* (Figures 40.20d and 40.21d and h) while *S. aureus* did not appear to lyse under these conditions (Figure 40.20h). These results, in combination with the fatty acid profiles, strongly support the notion that the EO-derived compounds studied here affect the membrane lipid composition and ultimately cause the complete deterioration of the cellular membranes at significantly elevated concentrations. Thus, all results of this study showed that CDs increase the solubility of EO compounds in an aqueous environment and enhanced their antimicrobial activity (Liang et al., 2012).

Thymol and its isomer Carv, main constituents of thyme and oregano oils, have strong and unpleasant fragrance and they can cause, if inhaled, respiratory irritation. Knowing that thymol may be used as antiseptic in healing ointments, syrups for the treatment of the respiratory system and preparations of inhalation, these drugs were successfully tested for decreasing their odor by creating complexes with βCD (prepared by kneading method) as proved by two-way ANOVA

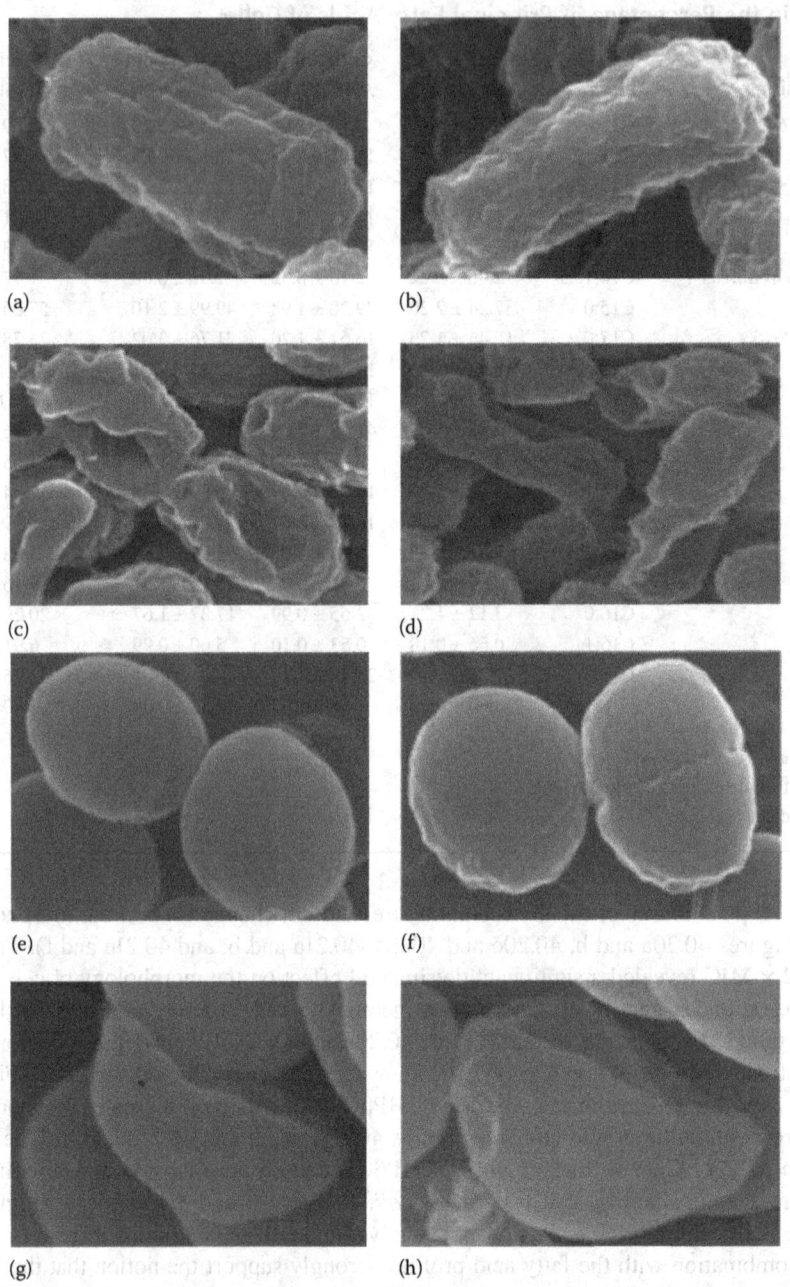

FIGURE 40.20 SEM images of *E. coli* cells. (a) Untreated cells (100,000×), (b) cells after treatment with 50 mmol/L HP βCD (100,000×), (c) 2 × MIC carvacrol (80,000×), and (d) 50 mmol/L HP βCD + 2 × MIC carvacrol (45,000×). SEM images of *S. aureus* cells. (e) Untreated cells (80,000×), (f) cells after treatment with 50 mmol/L HP βCD (80,000×), (g) 2 × MIC carvacrol (100,000×), and (h) 50 mmol/L HP βCD + 2 × MIC carvacrol (100,000×). All images shown are representative of replicate cultures. (From Liang, H. et al., *Food Chem.*, 135, 1020, 2012.)

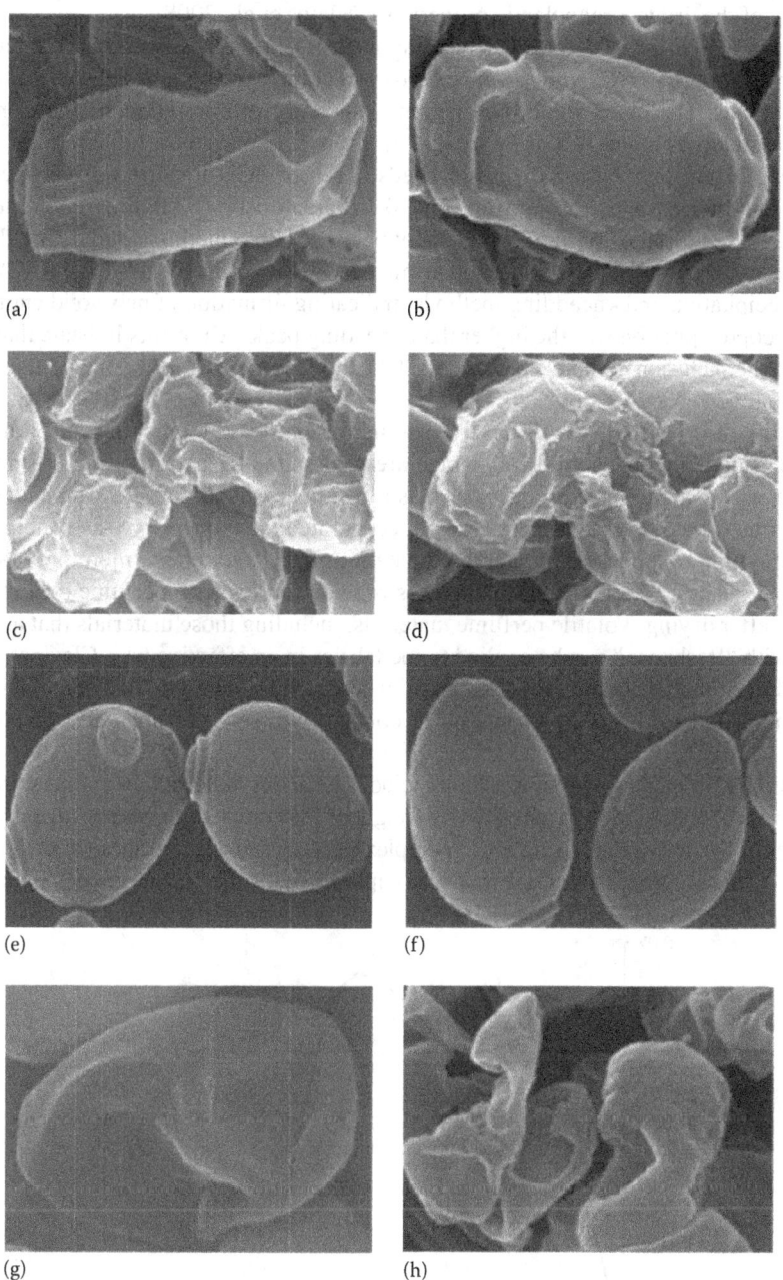

FIGURE 40.21 SEM images of *B. subtilis* cells. (a) Untreated cells (40,000×), (b) cells after treatment with 50 mmol/L HP βCD (40,000×), (c) 2 × MIC carvacrol (25,000×), and (d) 50 mmol/L HP βCD + 2 × MIC carvacrol (40,000×). SEM images of *S. cerevisiae* cells. (e) Untreated cells (13,000×), (f) cells after treatment with 50 mmol/L HP βCD (11,000×), (g) 2 × MIC carvacrol (20,000×), and (h) 50 mmol/L HP βCD + 2 × MIC carvacrol (11,000×). All images shown are representative of replicate cultures. (From Liang, H. et al., *Food Chem.*, 135, 1020, 2012.)

statistical analysis that took place after sensorial tests of thymol and Carv performed on 20 individuals. They felt less intense scent on these complexes than the pure samples, promising future improvement of the drugs' organoleptic properties (Daletos et al., 2008).

Lemongrass oleoresin (biological active component is citral or its isomers, geranial and neral) extracted from lemongrass (*Cymbopogon citratus*) was complexed with βCD by coprecipitation and kneading methods. The phase solubility diagram showed B_S profile (Higuchi and Connors, 1965) for lemongrass oleoresin and βCD (Figure 40.22). From phase solubility diagram, it was possible observed that 20 mg lemongrass oleoresin needs 7 mg of βCD to form soluble inclusion complexes. The band for carbonyl group (C=O) peaks characteristic (FTIR) of lemongrass oleoresin (1738 cm^{-1}) showed a reduction in their intensity peaks for both methods, indicating that carbonyl group was used for complexation. The exothermic peak (DSC) to βCD (130°C–140°C) can be also seen in coprecipitation and kneadding methods, indicating formation of new solid entity. The peak intensity for coprecipitation was the higher than kneading peak. All results indicate that both methods promote the inclusion complexes between βCD and lemongrass oleoresin and the best method was coprecipitation (Nur Ain et al., 2011).

An invention was based on the discovery that many volatile odiferous substances are capable of forming inclusion compounds with CD, which are well suited to incorporating in synthetic resins prior to heat processing to form products such as films, fibers, moldings, etc (Shibanai et al., 1982).

The perfume/CD complexes are preferably incorporated into solid, dryer-activated, fabric treatment (conditioning) composition, preferably containing fabric softeners, more preferably cationic and/or nonionic fabric softeners. The complexes provide fabrics with perfume benefits when they are rewetted after drying. Volatile perfume materials, including those materials that are commonly associated with "freshness," can be applied to the fabrics in an effective way. Clay provides protection for said perfume/CD complexes, especially when certain materials like some nonionic fabric softeners and/or fatty acids are present and in contact with said perfume/CD complexes (Banks et al., 1992).

An exemplary antiperspirant composition includes a carrier material; an antiperspirant active in an amount from about 5% to less than 19%, by weight of the composition and on an anhydrous basis; and plurality of particles comprising a CD complexing material and a fragrance material, wherein the percent of the fragrance material that is complexed with the CD is greater than about 75%,

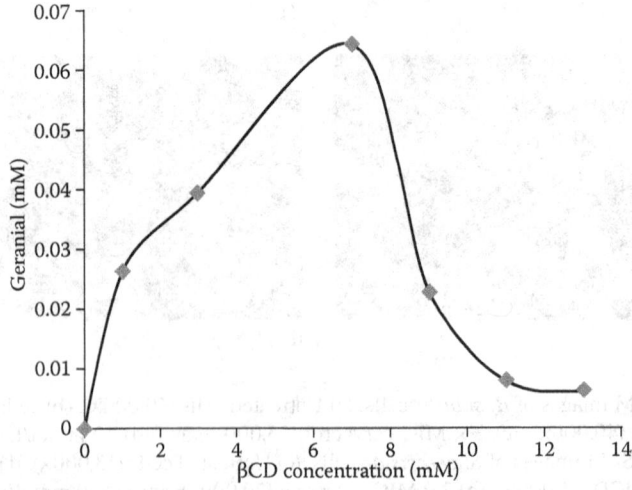

FIGURE 40.22 Solubility of geranial as a function of β CD in ethanol:water (25:75 v/v) solution at 30°C. Each data point is the mean of three measurements. (From Nur Ain, A.H. et al., Encapsulation of Lemongrass (Cymbopogon citratus) oleoresin with β-cyclodextrin: phase solubility study and its characterization, *Second International Conference on Biotechnology and Food Science IPCBEE,* vol. 7, Singapore, 2011.)

so that the perceptibility of the fragrance is minimized prior to its release, wherein the composition has a residue grade of less than about 35 (Scavone et al., 2008).

In this invention, the aqueous fragrance emulsions were prepared from a hydrophobic fragrance substance, a CD derivative, and a polysorbate. The emulsions were stable despite containing a high proportion of fragrance and could be used to prepare uniform coatings on substrates (Regiert et al., 2010).

An exemplary personal care product includes a composition that is applied to the body or clothing, or an article applied against the body; a plurality of particles associated with the composition or a component of the article, the plurality of particles, at least some of the plurality of particles comprising a CD complexing material and a first fragrance material, wherein the percent of the first fragrance material that is complexed with the CD is greater than about 90%, so that the perceptibility of the first fragrance is minimized prior to its release; and a second fragrance material that is not complexed with the CD and that is different from the first fragrance material, wherein the composition or article does not contain an antiperspirant active (Scavone et al., 2013).

40.13 CONCLUSION

The increasing demand for natural products for daily use in diverse fields such as pharmaceutical and cosmetics, textile, and agroindustries has led to the development of methods for increasing the lifetime of such natural products, protecting their properties from environmental factors, controlling their release and/or bioavailability and to mask the flavor or flavors that sometimes these natural products possess. Volatiles and EOs are some of these natural products largely used that need to be protected from the environmental factors (light, relative high temperature, moisture), but they also need to be protected due to their strong aroma and flavor. The molecular inclusion into the cavity of the CD, i.e., the molecular encapsulation of these EOs has the advantage of enhancing its solubility in water and in biofluids with the consequent of its bioavailability increase. Nevertheless, this process is dependent on several factors including the physicochemical properties of the EOs or their components. For this reason, the selection of the best CD for encapsulating them is needed in order to improve the stability of the complexes, but at the same time with the adequate capacity of release of the EO. Diverse types of CDs and their derivatives and methodologies of encapsulation have been used for efficiently encapsulating these EOs, which may constitute a promising pathway for the development of products friendlier to the people.

REFERENCES

AFNOR NF T 75-006 (February 1998) derived from a NF EN ISO 9235:2014. NF T 75-006 (février 1998) Matières premières aromatiquesd'originenaturelle—Vocabulaire. Aromatic natural raw materials. Vocabulary. (Indice sde classement: T75-006). http://www.boutique.afnor.org/resources/extraits/3177581CD.pdf.

Al-Omar, A., Abdou, S., De Robertis, L., Marsura, A., and Finance, C. (1999). Complexation study and anticellular activity enhancement by doxorubicin cyclodextrin complexes on a multidrug-resistant adenocarcinoma cell line. *Bioorganic* and *Medicinal Chemistry Letters*, 9, 1115–1120.

Alves, S. F., Borges, L. L., Santos, T. O., Paula, J. R., Conceição, E. C., and Bara, M. T. F. (2014). Microencapsulation of essential oil from fruits of Pterodon emarginatus using gum Arabic and maltodextrin as wall materials: Composition and stability. *Drying Technology*, 32, 96–105.

Arana-Sánchez, A., Estarrón-Espinosa, M., Obledo-Vázquez, E.N., Padilla-Camberos, E., Silva-Vázquez, R., and Lugo-Cervantes, E. (2010). Antimicrobial and antioxidant activities of Mexican oregano essentials oils (*Lippia graveolens* H. B. K.) with different composition ehen microencapsulated in β-cyclodextrins. *Letters in Applied Microbiology*, 50, 585–590.

Astray, G., Gonzalez-Barreiro, C., Mejuto, J. C., Rial-Otero, R., and Simal-Gandara, J. (2009). A review on the use of cyclodextrins in foods. *Food Hydrocolloids*, 23, 1631–1640.

Banerjee, S., Chattopadhyay, P., Ghosh, A., Goyary, D., Karmakar, S., and Veer, V. (2013). Influence of process variables on essential oil microcapsule properties by carbohydrate polymer-protein blends. *Carbohydrate Polymers*, 93, 691–697.

Banks, T. J., Benvegnu, F., Gardlik, J. M., and Trinh, T. (1992). Treatment of fabric with perfume/cyclodextrin complexes. U.S. Patent No. US5094761 A.

Beer, S., Dobler, D., Gross, A., Ost, M., Elseberg, C., Maeder, U., Schmidts, T. M., Keusgen, M., Fiebich, M., and Runker, F. (2013). In line monitoring of the preparation of water-in-oil-in-water/W/O/W) type multiple emulsions via dielectric spectroscopy. *International Journal of Pharmaceutics*, 441, 643–647.

Beirão-da-Costa, S., Duarte, C., Bourbon, A. I., Pinheiro, A. C., Januário, M. I. N., Vicente, A. A., Beirão-da-Costa, M. L., and Delgadillo, I. (2013). Inulin potential for encapsulation and controlled delivery of oregano essential oil. *Food Hydrocolloids*, 33, 199–206.

Bilia, A. R., Guccione, C., Isacchi, B., Righeshi, C., Firenzuoli, F., and Bergonzi, M. C. (2014). Essential oils loaded in nanosytems: A developing strategy for a successful therapeutics approach. *Evidence-Based Complementary and Alternative Medicine*, pp. 1–14.

Botrel, D. A., Borges, S. V., Fernandes, R. V. B., Viana, A. D., Costa, J. M. C., and Marques, G. R. (2012). Evaluation of spray drying conditions on properties of microencapsulated oregano essential oil. *International Journal of Food Science and Technology*, 47, 2289–2296.

Briozzo, J. L., Chirife, J., Herzage, L., and Dáquino, M. (1989). Antimicrobial activity of clove oil dispersed in a concentrated sugar solution. *Journal of Applied Bacteriology*, 66, 69–75.

Cabral Marques, H. M. (1994a). Structure and properties of cyclodextrins. Inclusion complex formation. *Revista Portuguesa de Farmácia*, XLIV, 77–84.

Cabral Marques, H. M. (1994b). Applications of cyclodextrins. Thermodynamic aspects of cyclodextrin complexes. *Revista. Portuguesa de Farmácia*, XLIV, 85–96.

Cabral-Marques, H. M. (2010). A review on cyclodextrin encapsulation of essential oils and volatiles. *Flavour and Fragrance Journal*, 25, 313–326.

Caesar, G. V. (1986). The Schardinger dextrins. In *Starch and its Derivatives*; Radley, J. A. Ed.; Chapman & Hall: London, U.K., Chapter 10, pp. 290.

Carrier, R. L., Miller, L. A., and Ahmed, I. (2007). The utility of cyclodextrins for enhancing oral bioavailability. *Journal of Controlled Release*, 123, 78–99.

Celia, C., Trapasso, E., Locatelli, M., Navarra, M., Ventura, C. A., Wolfram, J., Carafa, M., et al. (2013). Anticancer activity of liposomal bergamot essential oil (BEO) on human neuroblastoma cells. *Colloids and Surfaces B:* Biointerfaces, 112, 548–553.

Cevallos, P. A. P., Buera, M. P., and Elizalde, B. E. (2010). Encapsulation of cinnamon and thyme essential oils components (cinnamaldehyde and thymol) in β-cyclodextrin: Effect of interactions with water on complex stability. *Journal of Food Engineering*, 99, 70–75.

Charoenchaitrakool, M., Dehghani, F., and Foster, N. R. (2002). Utilization of supercritical carbon dioxide for complex formation of ibuprofen and methyl-β-cyclodextrin. *International Journal of Pharmaceutics*, 239, 103–112.

Cho, Y.-H. and Park, J. (2003). Evaluation of process parameters in the O/W/O multiple emulsion method for flavour encapsulation. *Journal of Food Science*, 68, 534–538.

Ciobanu, A., Mallard, I., Landy, D., Brabie, G., Nistor, D., and Fourmentin, S. (2012). Inclusion interactions of cyclodextrins and crosslinked cyclodextrin polymers with linalool and camphor in Lavandula angustifolia essential oil. *Carbohydrate Polymers*, 87, 1963–1970.

Ciobanu, A., Mallard, I., Landy, D., Brabie, G., Nistor, D., and Fourmentin, S. (2013). Retention of aroma compounds from Mentha piperita essential oil by cyclodextrins and crosslinked cyclodextrin polymers. *Food Chemistry*, 138, 291–297.

Council of Europe. *European Pharmacopoeia*, 7th edn.; European Directorate for the Quality of Medicines: Strasbourg, France, 2010.

Daletos, G., Papaioannou, G., Miguel, G., and Cabral Marques, H. (2008). Improvement of organoleptic properties of thymol and carvacrol using β-cyclodextrin. In *Proceedings of the 14th International Cyclodextrin Symposium, Kyoto, Japan*, Ueda, H. Ed.; The Society of Cyclodextrins: Tokyo, Japan, pp. 291–295.

Del Toro-Sánchez, C. L., Ayala-Zavala, J. F., Machi, L., Santacruz, H., Villegas-Ochoa, M. A., Alvarez-Parrilla, E., and González-Aguilar, G. A. (2010). Controlled release of antifungal volatiles of thyme essential oil from β-cyclodextrin capsules. *Journal of Inclusion Phenomena and Macrocyclic Chemistry*, 67, 431–441.

Detoni, C. B., Cabral-Albuquerque, E. C. M., Hohlemweger, S. V. A., Samapio, C., Barros, T. F., and Velozo, E. S. (2009). Essential oil from *Zanthoxylum tingoassuiba* loaded into multilamellar liposomes useful as antimicrobial agents. *Journal of Microencapsulation*, 26, 684–691.

Detoni, C. B., Oliveira, D. M., Santos, I. E., São Pedro, A., El-Bacha, R., Velozo, E. S., Ferreira, D., Sarmento, B., and Cabral-Albuquerque, E. C. M. (2012). Evaluation of thermal oxidative stability and antiglioma activity of *Zanthoxylum tingoassuiba* essential oil entrapped into multi- and unilamellar liposomes. *Journal of Liposome Research*, 22, 1–7.

Edris, A. and Bergnstahl, B. (2001). Encapsulation of orange oil in a spray dried double emulsion. *Nahrung/Food*, 45, 133–137.

Fages, J., Rodier, E., Chamayou, A., and Baron, M. (2007). Comparative study of two processes to improve the bioavailability of an active pharmaceutical ingredient: Kneading and supercritical technology. *Kona-Powder Part*, 25, 217–229.

Faleiro, M. L. and Miguel, M. G. (2013). Use of essential oils and their components against multidrug-resistant bacteria. In *Fighting Multidrug Resistance with Herbal Extracts, Essential Oils and their Components*; Rai, M. K. and Kon, K. V. Ed., Copyright © 2013 Elsevier Inc., New York, Chapter 6, pp. 65–94.

Faugeras, P.-A., Boëns, B., Elchinger, P.-H., Brouillette, F., Montplaisir, D., Zerrouki, R., and Romain Lucas, R. (2012). When cyclodextrins meet click chemistry. *European Journal of Organic Chemistry*, 22, 4087–4105.

Fernandes, L. P., Éhen, Zs., Moura, T. F., Novák, C. S., and Sztatisz, J. (2004). Characterization of Lippia sidoides oil extract-β-cyclodextrin complexes using combined thermoanalytical techniques. *Journal of Thermal Analysis and Calorimetry*, 78, 557–573.

Fernandes, L. P., Oliveira, W. P., Sztatisz, J., Szilágyi, I. M., and Novák, C. S. (2009). Solid state studies on molecular inclusions of Lippia sidoides essential oil obtained by spray drying. *Journal of Thermal Analysis and Calorimetry*, 95, 855–863.

Fernandes, R. V. B., Borges, S. V., and Botrel, D. A. (2013a). Influence of spray drying operating conditions on microencapsulated rosemary essential oil properties. *Ciência e Tecnologia de Alimentos*, 33(Suppl. 1), 171–178.

Fernandes, R. V. B., Borges, S. V., and Botrel, D. A. (2014). Gum Arabic/starch/maltodextrin/inulin as wall materials on the microencapsulation of rosemary essential oil. *Carbohydrate Polymers*, 101, 524–532.

Fernandes, R. V. B., Borges, S. V., Botrel, D. A., Silva, E. K., Costa, J. M. G., and Queiroz, F. (2013b). Microencapsulation of rosemary essential oil: Characterization of particles. *Drying Technology*, 31, 1245–1254.

Figueiredo, A C., Barroso, J. G., Pedro, L. G., and Scheffer, J. J. C. (2008). Factors affecting secondary metabolite production in plants: Volatile components and essential oils. *Flavour and Fragrance Journal*, 23, 213–226.

French, D. (1957). The Schardinger dextrins. *Advances in Carbohydrate Chemistry*, 12, 189–260.

Frömming, K.-H. and Szejtli, J. (1994). *Cyclodextrins in Pharmacy*; Kluwer Academic: Dordrecht, the Netherlands.

Gál-Füzy, M., Szente, L., Szejtli, J., and Harangi, J. (1984). Cyclodextrin-stabilized volatile substances for inhalation therapy. *Pharmazie*, 39(8), 558–559.

García-Rodriguez, J. J., Torrado, J., and Bolás, F. (2001). Improving bioavailability and anthelmintic activity of albendazole by preparing albendazole–cyclodextrin complexes. *Parasite*, 8, S188–S190.

Garti, N. (1997). Double emulsions—Scope, limitations and new achievements. *Colloids and Surfaces*, 123–124, 233–246.

Gharsallaoui, A., Roudaut, G., Chambin, O., Voilley, A., and Saurel, R. (2007). Applications of spray-drying in microencapsulation of food ingredients: An overview. *Food Research International*, 40, 1107–1121.

Gil, A., Chamayou, A., Leverd, E., Bougaret, J., Baron, M., and Couarraze, G. (2004). Evolution of the interaction of a new chemical entity, eflucimibe, with gamma-cyclodextrin during kneading process. *European Journal of Pharmaceutical Sciences*, 23, 123–129.

Goud, K. and Park, H. J. (2005). Recent developments in microencapsulation of food ingredients. *Drying Technology*, 23, 1361–1394.

Gouin, S. (2004). Microencapsulation: Industrial appraisal of existing Technologies and trends. *Trends in Food Science and Technology*, 15, 330–347.

Hădărugă, N. G., Hădărugă, D. I., Păunescu, V., Tatu, C., Ordodi, V. L., Bandur, G., and Lupea, A. X. (2006). Essential oil from *Lauraceae* and *Rosaceae*/β-cyclodextrin complexes. *Annals of the Faculty of Engineering Hunedoara*, Tome IV, Fascicole 3, 105–112.

Hernández-Sánchez, P., López-Miranda, S., Lucas-Abellán, C., and Núñez-Delicado, E. (2012). Complexation of Eugenol (EG), as main component of clove oil and as pure compound, with β- and HP-β-CDs. *Food and Nutrition Sciences*, 3, 716–723.

Higashi, T., Nishimura, K., Yoshimatsu, A., Ikeda, H., Arima, K., Motoyama, K., Hirayama, F., Uekama, K., and Arima, H. (2009). Preparation of four types coenzyme Q10/gamma-cyclodextrin supramolecular complexes and comparison of their pharmaceutical properties. *Chemical and Pharmaceutical Bulletin*, 57, 965–970.

Higuchi, T. and Connors, K. A. (1965). Phase-solubility techniques. *Advances in Analytical Chemistry and Instrumentation*, 4, 117–212.

Holm, R., Madsen, J. C., Shi, W., Larsen, K. L., Städe, L. W., and Westh, P. (2011). Thermodynamics of complexation of tauro- and glycol-conjugated bile salts with two modified β-cyclodextrins. *Journal of Inclusion Phenomena and Macrocyclic Chemistry*, 69, 201–211.

Hosseini, S. M., Hosseini, H., Mohammadifar, M.A., Mortazavian, A. M., Mohammadi, A., Khosravi-Darani, K., Shojaee-Aliabadi, S., Dehghan, S., and Khaksar, R. (2013). Incorporation of essential oil in alginate microparticles by multiple emulsion/ionic gelation process. *International Journal of Biological Macromolecules*, 62, 582–588.

Jiménez, A., Sánchez-González, L., Desobry, S., Chiralt, A., and Tehrany, E. A. (2014). Influence of nanoliposomes incorporation on properties of film forming dispersions and films based on corn starch and sodium caseinate. *Food Hydrocolloids*, 35, 159–169.

Junco, S., Casimiro, T., Ribeiro, N., Nunes da Ponte, M., and Cabral-Marques, H. M. (2002). Optimisation of supercritical carbon dioxide systems for complexation of naproxen: Beta-cyclodextrin. *Journal of Inclusion Phenomena and Macrocyclic Chemistry*, 44(5), 69–73.

Khan, A. R., Forgo, P., Stine, K. J., and D'Souza, V. T. (1998). Methods for selective modifications of cyclodextrins. *Chemical Reviews*, 98, 1977–1996.

LeBlanc, B. W., Boné, S., Hoffman, G., De, G., Deeby, T., McCready, H., and Loeffelmann, K. (2008). β-Cyclodextrins as carriers of monoterpenes into the hemolinph of the honey bee (*Apis mellifera*) for integrated pest management. *Journal of Agriculture and Food Chemistry*, 56, 8565–8573.

Lederer, M. and Leipzig-Pagani, E. (1996). A simple alternative determination for the formation constant for the inclusion complex between rutin and β-cyclodextrin. *Analytica Chimica Acta*, 329, 311–314.

Leimann, F. V., Gonçalves, O. H., Machado, R. A. F., and Bolzan, A. (2009). Antimicrobial activity of microencapsulated lemongrass essential oil and the effect of experimental parameters on microcapsules size and morphology. *Materials Science and Engineering C*, 29, 430–436.

Li, N., Chen, Y., Zhang, Y.-M., Wang, L.-H., Mao, W.-Z., and Liu, Y. (2014). Molecular binding thermodynamics of spherical guests by β-cyclodextrin bearing aromatic substituents. *Thermochimica Acta*, 576, 18–26.

Liang, H., Yuan, Q., Vriesekoop, F., and Lv, F. (2012). Effects of cyclodextrins on the antimicrobial activity of plant-derived essential oil compounds. *Food Chemistry*, 135, 1020–1027.

Liolios, C. C., Gortzi, O., Lalas, S., Tsaknis, J., and Chinou, I. (2009). Liposomal incorporation of carvacrol and thymol isolated from the essential oil of *Origanum dictamnus* L. and in vitro antimicrobial activity. *Food Chemistry*, 112, 77–83.

Loftsson, T. and Duchêne, D. (2007). Cyclodextrins and their pharmaceutical applications. *International Journal of Pharmaceutics*, 329, 1–11.

Loftsson, T., Hreinsdóttir, D., and Másson, M. (2007). The complexation efficiency. *Journal of Inclusion Phenomena and Macrocyclic Chemistry*, 57, 545–552.

Lopez, M. D., Maudhuit, A., Pascual-Villalobos, M. J., and Poncelet, D. (2012). Development of formulations to improve the controlled-release of linalool to be applied as an insecticide. *Journal of Agricultural and Food Chemistry*, 60, 1187–1192.

Lv, Y., Yang, F., Xueying, L., Zhang, X., and Abbas, S. (2014). Formulation of heat-resistant nanocapsules of jasmine essential oil via gelatine/gum Arabic based complex coacervation. *Food Hydrocolloids*, 35, 305–314.

Madene, A., Jacquot, M., Scher, J., and Desobry, S. (2006). Flavour encapsulation and controlled release a review. *International Journal of Food Science and Technology*, 41, 1–21.

Maji, T. K., Barruah, J., Dube, S., and Hussain, M. R. (2007). Microencapsulation of *Zanthoxylum limonella* oil (ZLO) in glutaraldehyde crosslinked gelatine from mosquito repellent application. *Bioresource Technology*, 98, 840–844.

Martel, B., Morcellet, M., Ruffin, D., Vinet, F., and Weltrowski, M. (2002). Capture and controlled release of fragrances by CD finished textiles. *Journal of Inclusion Phenomena and Macrocyclic Chemistry*, 44, 439–442.

Martins, A. P., Craveiro, A. A., Machado, M. I. L., Raffin, F. N., Moura, T. F., Novák, C. S., and Éhen, Z. (2007). Preparation and characterization of Mentha x villosa Hudson oil-β-Cyclodextrin complex. *Journal of Thermal Analysis and Calorimetry*, 88, 363–371.

Martins, I. M., Rodrigues, S. N., Barreiro, M. F., and Rodrigues, A. E. (2011). Release of thyme oil from polylactide microcapsules. *Industrial and Engineering Chemistry Research*, 50, 13752–13761.

Martins, I. M., Rodrigues, S. N., Barreiro, M. F., and Rodrigues, A. E. (2012). Release studies of thymol and *p*-cymene from polylactide microcapsules. *Industrial and Engineering Chemistry Research*, 51, 11565–11571.

Martins, I. M. D. (2012). Microencapsulation of thyme oil by coacervation: Production, characterization and release evaluation. PhD thesis. Faculdade de Engenharia da Universidade do Porto, Porto, Portugal.

Matura, M., Sköld, M., Börje, A., Andersen, K. E., Bruze, M., Frosch, P., Goossens, A. et al. (2005). Selected oxidized fragrance terpenes are common contact allergens. *Contact Dermatitis*, 52, 320–328.

Menezes, P. P., Serafini, M. R., Quintans-Júnior, L. J., Silva, G. F., Oliveira, J. F., Carvalho, F. M. S., Souza, J. C. C. et al. (2014). Inclusion complex of (-)-linalool and β-cyclodextrin. *Journal of Thermal Analysis and Calorimetry*, 115, 2429–2437.

Menezes, P. P., Serafini, M.R., Santana, B. V., Nunes, R. S., Quintans, Jr. L. J., Silva, G. F., Medeiros, I. A. et al. (2012). Solid-state β-cyclodextrins complexes containing geraniol. *Termochimica Acta*, 548, 45–50.

Miguel, M. G., Guerrero, C., Rodrigues, H., Brito, J., Duarte, F., Venâncio, F., and Tavares, R. (2003). Essential oils of Portuguese *Thymus mastichina* (L.) L. subsp. *mastichina* grown on different substrates and harvested on different dates. *Journal of Horticultural Science and Biotechnology*, 78, 355–358.

Munin, A. and Florence Edwards-Lévy, F. (2011). Encapsulation of natural polyphenolic compounds: A review. *Pharmaceutics*, 3(4), 793–829.

Nedovic, V., Kalusevic, A., Manojlovic, V., Levic, S., and Bugarski, B. (2011). Na overview of encapsulation Technologies for food applications. *Procedia Food Science*, 1, 1806–1815.

Nieddu, M., Rassu, G., Boatto, G., Bosi, P., Trevisi, P., Giunchedi, P., Carta, A., and Gavini, E. (2014) Improvement of thymol properties by complexation with cyclodextrins: In vitro and in vivo studies. *Carbohydrate Polymers*, 102, 393–399.

Nisisako, T., Okushima, S., and Torii, T. (2005). Controlled formulation of monodisperse double emulsions in a multiple-phase microfluidic system. *Soft Matter*, 1, 23–27.

Numanoğlu, U., Şen, T., Tarimci, N., Kartal, M., Koo, O. M.Y., and Onyuksel, H. (2007). Use of cyclodextrins as a cosmetic delivery system for fragrance materials: Linalool and benzyl acetate. *AAPS PharmSciTech*, 8(4), Article 85.

Nur Ain, A. H., Farah Diyana, M. H., and Zaibunnisa, A. H. (2011). Encapsulation of Lemongrass (Cymbopogon citratus) oleoresin with β-cyclodextrin: Phase solubility study and its characterization. *Second International Conference on Biotechnology and Food Science IPCBEE*, vol. 7, Singapore.

Ortan, A., Câmpeanu, G. H., Dinu-Pîrvu, C., and Popescu, L. (2009). Studies concerning the entrapment of *Anethum graveolens* essential oil in liposomes. *Roumanian Biotechnological Letters*, 14, 4411–4417.

Pagington, J. S. (1986). β-cyclodextrin and its uses in the flavour industry. In *Developments in Food Flavors*, Birch, G.G. and Lindley, M-G. Eds.; Elsevier Applied Science: Great Britain, U.K., pp. 131–150.

Pelissari, F. M., Grossmann, M. V. E., Yamashita, F., and Pineda, E. A. G. (2009). Antimicrobial, mechanical, and barrier properties of cassava starch-chitosan films incorporated with oregano essential oil. *Journal of Agricultural and Food Chemistry*, 57, 7499–7504.

Pelissari, F. M., Yamashita, F., and Grossmann, M. V. E. (2011). Extrusion parameters related to starch, chitosan active films prperties. *International Journal of Food Science and Technology*, 46, 702–710.

Petrović, G. M., Stojanović, G. S., and Radulović, N. S. (2010). Encapsulation of cinnamon oil in β-cyclodextrin. *Journal of Medicinal Plant Research*, 4(14), 1382–1390.

Podshivalov, A. V., Bronnikov, S., Zuev, V. V., Jiamrungraksa, T., and Charuchinda, S. (2013). Synthesis and characterization of polyurethane-urea microcapsules containing galangal essential oil: Statistical analysis of encapsulation. *Journal of Microencapsulation*, 30, 198–203.

Poshadri, A. and Kuna, A. (2010). Microencapsulation technology: A review. *Journal of Research ANGRAU*, 38, 86–102.

Prakash, B., Singh, P., Kedia, A., and Dubey, N. K. (2012). Assessment of some essential oils as food preservatives based on antifungal, antiaflatoxin, antioxidant activities and in vivo efficacy in food system. *Food Research International*, 49, 201–208.

Prata, A. S., Menut, C., Leydet, A., Trigo, J. R., and Grosso, C. R. F. (2008). Encapsulation and release of a fluorescent probe, khusimyl dansylate, obtained from vetiver oil by complex coacervation. *Flavour and Fragrance Journal*, 23, 7–15.

Răileanu, M., Todan, L., Voicescu, M., Ciuculescu, C., and Maganu, M. (2013). A way for improving the stability of the essential oils in an environmental friendly formulation. *Materials Science and Engineering C*, 33, 3281–3288.

Regiert, M.-E., Kupka, M., and Sigl, H. (2010). Aqueous emulsion containing a cyclodextrin derivative, a perfume, and a polysorbate. U.S. Patent No 20100280133.

Reineccius, T. A., Reineccius, G. A., and Peppard, T. L. (2002). Encapsulation of flavors using cyclodextrins: Comparison of flavor retention in alpha, beta, and gamma types. *Journal of Food Science*, 67, 3271–3279.

Reineccius, T. A., Reineccius, G. A., and Peppard, T. L. (2004). Utilization of β-cyclodextrin for improved flavor retention in thermally processed foods. *Journal of Food Science*, 69, FCT58–FCT62.

Saito, Y., Tanemura, I., Sato, T., and Ueda, H. (1999). Interaction of fragrance materials with 2-hydroxypropyl-beta-cyclodextrin by static and dynamic head-space methods. *International Journal of Cosmetic Science*, 21(3), 189–198.

Salústio, P. J., Feio, G., Figueirinhas, J. L., Pinto, J. F., and Cabral-Marques, H. M. (2009). The influence of the preparation methods on the inclusion of model drugs in a β-cyclodextrin cavity. *European Journal of Pharmaceutics and Biopharmaceutics*, 71, 377–386.

Salústio, P. J., Pontes, P., Conduto, C., Sanches, I., and Cabral-Marques, H. M. (2011). Advanced technologies for oral controlled release: Cyclodextrins for oral controlled release. *AAPS PharmSciTech*, 12(4), 1276–1292.

Samperio, C., Boyer, R., Eigel III, W. N., Holland, K. W., Mckinney, J. S., O'Keefe, S. F., Smith, R., and Marcy, J. E. (2010). Enhancement of plant essential oils' aqueous solubility and stability using alpha and beta cyclodextrin. *Journal of Agricultural and Food Chemistry*, 58, 12950–12956.

Sansukcharearnpon, A., Wanichwecharungruang, S., Leepipatpaiboon, N., Kerdcharoen, T., and Arayachukeat, S. (2010). High loading fragrance encapsulation bases on a polymer-blend: Preparation and release behaviour. *International Journal of Pharmaceutics*, 391, 267–273.

São Pedro, A., Cabral-Albuquerque, E., Ferreira, D., and Sarmento, B. (2009). Chitosan: An option for development of essential oil delivery systems for oral cavity care? *Carbohydrate Polymers*, 76, 501–508.

Sapkal, N.P., Kilor, V. A., Bhusari, K. P., and Daud, A. S. (2007). Evaluation of some methods for preparing glicazide beta-cyclodextrin inclusion complexes. *Tropical Journal of Pharmaceutical Research*, 6, 833–840.

Scarfato, P., Avallone, E., Iannelli, P., Feo, V., and Acierno, D. (2007). Synthesis and characterization of polyurea microcapsules containing essential oils with anti-germinative activity. *Journal of Applied Polymer Science*, 105, 3568–3577.

Scavone, T. A., Leblanc, M. J., and Sanker, L. A. (2008). Antiperspirant compositions comprising cyclodextrin complexing material. U.S. Patent No 20080213203.

Scavone, T. A., Leblanc, M. J., Sanker, L. A., and Switzer, A. G. (2013). Personal care product comprising cyclodextrin as fragrance-complexing material. U.S. Patent No CA2679504 C.

Schmid, R. (1972). A resolution of the *Eugenia-Syzygium* controversy (Myrtaceae). *American Journal of Botany*, 59, 423–436.

Schönbeck, C., Westh, P., Madsen, J. C., Larsen, K. L., Städe, L. W., and Holm, R. (2010). Hydroxylpropyl-substituted β-cyclodextrin: Influence of degree of substitution on the thermodynamics of complexation with tauroconjugated and glycoconjugated bile salts. *Langmuir*, 26, 17949–17957.

Shahidi, F. and Han, X. Q. (1993). Encapsulation of food ingredients. *Critical Review in Food Science and Nutrition*, 33, 501–547.

Shibanai, I., Horikoshi, K., and Nakamura, N. (1982) Inclusion compound from volatile odiferous substance with cyclodextrin. U.S. Patent No CA1124180 A1.

Sinico, C., Logu, A., Lai, F., Valenti, D., Manconi, M., Loy, G., Bonsignore, L., and Fadda, A. M. (2005). Liposomal incorporation of *Artemisia arborescens* L essential oil and in vitro antiviral activity. *European Journal of Pharmaceutics and Biopharmaceutics*, 59, 161–168.

Siqueira-Lima, P. S., Araújo, A. A. S., Lucchese, A. M., Quintans, J. S. S., Menezes, P. P., Barreto, P. A., deLuccaJúnior, W., Santos, M. R. V., Bonjardim, L. R., and Quintans-Júnior, L. J. (2014). β-cyclodextrin complex containing Lippia grata leaf essential oil reduces orofacial nociception in mice-evidence of possible involvement of descending inhibitory pain modulation pathway. *Basic and Clinical Pharmacology and Toxicology*, 114(2), 188–196.

Smith, R.A. and Lambrou, A. (1974). Encapsulated flavoring composition. US Pattern No US3819838.

Solano, A. C. V. and Gante, C. R. (2012). Two different processes to obtain antimicrobial packaging containing natural oils. *Food Bioprocess Technology*, 5, 2522–2528.

Specos, M. M. M., Garcia, J. J., Tornesello, J., Marino, P., Della Vecchia, M., Tesoriero, M.V. D., and Hermida, L. G. (2010). Microencapsulated citronella oil for mosquito repellent finishing of cotton textiles. *Transactions of the Royal Society of Tropical Medicine and Hygiene*, 104, 653–658.

Sun, D. Z., Li, L., Qiu, X. M., Liu, F., and Yin, B.-L. (2006). Isothermal titration calorimetry and 1H NMR studies on host–guest interaction of paeonol and two of its isomers with β-cyclodextrin. *International Journal of Pharmaceutics*, 316, 7–13.

Sutaphanit, P. and Chitprasert, P. (2014). Optimisation of microencapsulation of holy basil essential oil in gelatin by response surface methodology. *Food Chemistry*, 150, 313–320.

Szejtli, J. (1992). The properties and potential uses of cyclodextrin derivatives. *Journal of Inclusion Phenomena and Molecular Recognition in Chemistry*, 14, 25–36.

Szejtli, J. (1998). Introduction and general overview of cyclodextrin chemistry. *Chemical Reviews*, 98, 1743–1753.

Szente, L. and Szejtli, J. (1986). Molecular encapsulation of natural and synthetic coffee flavor with β-cyclodextrin. *Journal of Food Science*, 51, 1024–1027.

Szente, L. and Szejtli, J. (1999). Highly soluble cyclodextrin derivatives: Chemistry, properties, and trends in development. *Advanced Drug Delivery Reviews*, 36, 17–28.

Thoma, J. A. and Stewart, L. S. (1965). Cycloamyloses. In *Starch, Chemistry and Technology I*; Whistler, R. L. and Paschall, E. F., Eds.; Academic Press: New York, p. 209.

Tian, X.-N., Jiang, Z.-T., and Li, R. (2008). Inclusion interactions and molecular microcapsule of Salvia sclarea L. essential oil with β-cyclodextrin derivatives. *European Food Research and Technology*, 227, 1001–1007.

Tong, L. H. (2001). *Cyclodextrin Chemistry-Theory and Application*. Science Press: Beijing, China, pp. 163–176.

Turek, C. and Stintzing, F. C. (2013). Stability of essential oils: A review. *Comprehensive Reviews in Food Science and Food Safety*, 12, 40–53.

Uekama, K. (2004). Design and evaluation of cyclodextrin-based drug formulation. *Chemical and Pharmaceutical Bulletin*, 52, 900–915.

Umer, H., Nigam, H., Tamboli, A. M., and Nainar, M. S.M. (2011) Microencapsulation: Process, techniques and applications. *International Journal of Research in Pharmaceutical and Biomedical Sciences*, 2, 474–481.

Valenti, D., de Logu, A., Loy, G., Sinico, C., Bonsignore, L., Cottiglia, F., Garau, D., and Fadda, A. M. (2001). Liposome-incorporated *Santolina insularis* essential oils: Preparation, characterization and in vitro antiviral activity. *Journal of Liposome Research*, 11, 73–90.

van Soest, J. J. G. (2007). Encapsulation of fragrances and flavours: A way to control odour and aroma in consumer prducts. In *Flavours and Fragrances: Chemistry, Bioprocessing and Sustainability*; Berger, R. G. Ed.; Springer-Verlag: Berlin, Germany, pp. 439–455.

Varona, S., Rojo, S. R., Martín, A., Cocero, M. J., Serra, A. T., Crespo, T., and Duarte, C. M. M. (2013) Antimicrobial activity of lavandin essential oil formulations against three pathogenic food-borne bacteria. *Industrial Crops and Products*, 42, 243–250.

Verdalet-Guzmán, I., Martínez-Ortiz, L., and Martínez-Bustos, F. (2013). Characterization of mew sources of derivative starches as wall materials of essential oil by spray drying. *Food Science and Technology*, 33, 757–764.

Viuda-Martos, M., Mohamady, M. A., Fernández-López, J., Abd El-Razik, K. A., and Omer, E. A. (2011). In vitro antioxidant and antibacterial activities of essential oils obtained from Egyptian aromatic plants. *Food Control*, 22, 1715–1722.

Waleczek, K. J., Cabral Marques, H. M., Hempel, B., and Schmidt, P. C. (2003). Phase solubility studies of pure (2)-α-bisabolol and camomile essential oil with β-cyclodextrin. *European Journal of Pharmaceutics and Biopharmaceutics*, 55(2), 247–251.

Wang, C. X. and Chen, Sh. L. (2010). Fragrance-release property of β-cyclodextrin inclusion compounds and their application in aromatherapy. *Journal of Industrial Textiles*, 40, 13–32.

Wang, Y., Jiang, Z.-T., and Li, R. (2009). Complexation and molecular microcapsules of *Litsea cubeba* essential oil with β-cyclodextrin and its derivatives. *European Food Research and Technology*, 228, 865–873.

Wen, Z., Liu, B., Zheng, Z.-K., You, X.-K., Pu, Y.-T., and Li, Q. (2009). Preparation and characterization of β-cyclodextrin inclusion compound of essential oil from *Rosa damascena* miller. *Food Science*, 30, 29–32.

Wen, Z., Liu, B., Zheng, Z, You, X., Yitao, Pu, Y., and Li, Q. (2010). Preparation of liposomes entrapping essential oil from Atractylodes macrocephala Koidz by modified RESS technique. *Chemical Engineering Research and Design*, 88(8), 1102–1107.

Wen, Z., You, X., Jiang, L., Liu, B., Zheng, Z., Pu, Y., and Cheng, B. (2011). Liposomal incorporation of rose essential oil by a supercritical process. *Flavour and Fragrance Journal*, 26, 27–33.

Woranuch, S. and Yoksan, R. (2013). Eugenol-loaded chitosan nanoparticles: II. Application in bio-based plastics for active packaging. *Carbohydrate Polymers*, 96, 586–592.

Xiao, Z., Zhu, G., Zhou, R., and Niu, Y. (2013). A review of the preparation and application of flavour and essential oils microcapsules based on complex coacervation technology. *Journal of the Science Food and Agriculture*, 94(8), 1482–1494.

Yano, H. and Kleinebudde, P. (2010). Improvement of dissolution behavior for poorly water-soluble dug by application of cyclodextrin in extrusion process: Comparison between melt extrusion and wet extrusion. *AAPS PharmSciTech*, 11, 885–893.

Yoshida, K., Sekine, T., Matsuzaki, F., Yanaki, T., and Yamaguchi, M. (1999) Stability of vitamin A in oil-in-water-in-oil-type multiple emulsions. *Journal of the American Oil Chemists' Society*, 76, 1–6.

Yoshida, P. A., Yokota, D., Foglio, M. A., Rodrigues, R. A. F., and Pinho, S. C. (2010). Liposomes incorporating essential oil of Brazilian cherry (*Eugenia uniflora* L.): Characterization of aqueous dispersions and lyophilized formulations. *Journal of Microencapsulation*, 27, 416–425.

Zhang, H., Ran, X., Hu, C.-L., Qin, L.-P., Lu, Y., and Peng, C. (2012a). Therapeutic effects of liposome-enveloped Ligusticum chuanxiong essential oil on hypertropic scars in the rabbit ear model. *PlosOne*, 7, 1–8.

Zhang, H. Y., Tehrany, E. A., Kahn, C. J. F., Ponçot, M., Linder, M., and Cleymand, F. (2012b). Effects of nanoliposomes based on soya, rapeseed and fish lecithins on chitosan thin films designed for tissue engineering. *Carbohydrate Polymers*, 88, 618–627.

Zuidam, N. J. and Shimoni, E. (2010). Overview of microencapsulates for use in food products or processes and methods to make them. In *Encapsulation Technologies for Active Food Ingredients and Food Processing*; Zuidam, N. J. and Nedovic, V. A. Eds.; © Springer Science+Business Media LLC, New York, Chapter 2, pp. 1–29.

41 Microencapsulation
Artificial Cells

Thomas Ming Swi Chang

CONTENTS

41.1 Methods of Preparation 908
    41.1.1 Principles of Methods of Preparations 908
    41.1.2 Membranes Used for Encapsulation 909
    41.1.3 Variations in Contents 909
41.2 Microencapsulation of Bioactive Sorbents 909
    41.2.1 Clinical Uses 909
    41.2.2 Microencapsulation of Immunosorbents 910
41.3 Microencapsulation of Cells or Microorganisms 910
    41.3.1 Procedure 910
    41.3.2 High Concentrations of Smaller Cells 911
    41.3.3 Microencapsulated Islets for Diabetes Mellitus 911
    41.3.4 Microencapsulated Hepatocytes for Liver Failure 911
    41.3.5 Microencapsulation of Genetically Engineered Microorganisms 911
    41.3.6 Microencapsulation of Cholesterol-Removing Microorganisms 911
41.4 Microencapsulation of Enzymes and Multienzyme Systems 912
    41.4.1 Urea Removal 912
    41.4.2 Enzyme Therapy 912
    41.4.3 Multienzyme System 912
41.5 Red Blood Cell Substitutes 912
41.6 Biodegradable Artificial Cells 912
41.7 General Discussion 913
    41.7.1 Present Status 913
    41.7.2 Future Perspectives 913
References 913

Artificial cells prepared from microencapsulation of biologically active materials were first reported in 1957 and 1964.[1-3] They were first prepared as ultrathin polymer membranes of cellular dimensions microencapsulating the proteins and enzymes extracted from biological cells.[2,3] This was followed by the encapsulation of biological cells, adsorbents, magnetic materials, drugs, vaccines, hormones, and many other biologically active materials for applications in biotechnology and medicine (Figure 41.1).[3-7] Major interest came after its successful use for treating poisoning, kidney failure, and liver failure in patients.[4,8] The second impetus in artificial cell research came in the 1980s, a time of increasing international interest in all areas of biotechnology. Many groups, therefore, investigated microencap-sulation of cells, microorganisms, enzymes, and other biotechnological materials.[8-11] In the late 1980s AIDS caused by H.I.V. in transfused blood led to extensive research into artificial red blood cell substitutes.

Like biological cells, artificial cells contain biologically active materials. However, the content of artificial cells can be more varied than biological cells (Figure 41.1). The membranes of artificial cells also can be extensively varied using synthetic or biological materials. Permeability can be controlled over a wide range, and this allows the enclosed material to be retained and separated from

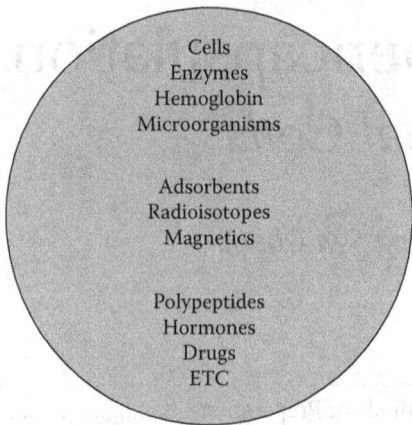

FIGURE 41.1 Variations in materials microencapsulated inside artificial cells. Biotechnological materials such as cell culture, enzymes, hemoglobin, microorganisms, and others are used. Materials include adsorbents, radioisotopes, magnetic materials, and others, and therapeutic agents include polypeptides, hormones, medications, and others. (From Chang, T.M.S., *Artif. Cells Blood Substit. Immobil. Biotechnol.*, 22(1), vii, 1994.)

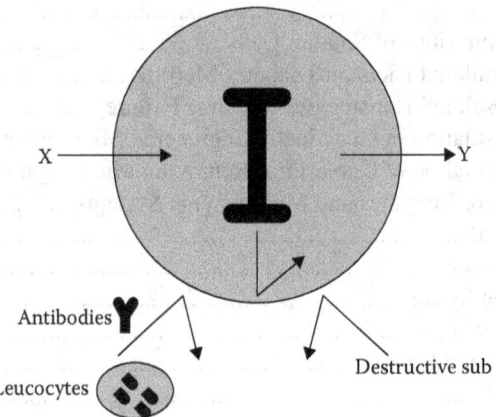

FIGURE 41.2 Basic principle of artificial cells: Artificial cells are prepared to have some of the properties of biological cells. Like biological cells, artificial cells contain biologically active materials (I). The enclosed material (I) can be retained and separated from undesirable external materials, such as antibodies, leukocytes, and destructive substances. The large surface area and the ultra-thin membrane allow selected substrates (X) and products (Y) to permeate rapidly. Mass transfer across 100 mL of artificial cells can be 100 times higher than that for a standard hemodialysis machine. The synthetic membranes are usually made of ultrathin synthetic polymer membranes for this type of artificial cell. (From Chang, T.M.S., *Artif. Cells Blood Substit. Immobil. Biotechnol.*, 22(1), vii, 1994.)

undesirable external materials (Figure 41.2). The large surface area and the ultrathin membrane allow selected substrates and products to permeate rapidly (Figure 41.2), and mass transfer across 100 mL of artificial cells can be 100 times higher than that for a standard hemodialysis machine.

41.1 METHODS OF PREPARATION

41.1.1 Principles of Methods of Preparations

Many methods are available to prepare artificial cells. The most commonly used approaches are based on the following principles. Small artificial cells in the micron dimensions are prepared by emulsification procedures that are usually modifications of the original basic procedures.[2-6] Here materials for microencapsulation are dissolved or suspended in an aqueous solution. An emulsion is formed, and membranes are formed on the surface of each microdroplet. The microcapsules are resuspended in an aqueous medium. Smaller artificial cells of nanometer dimensions are formed based on the same principles except the initial emulsion formed are much smaller. Larger artificial cells, especially those in the millimeter dimensions, are prepared based on modifications of the original drop method.[2,5-7]

These are generally used to microencapsulate cells or microorganisms in tissue engineering. A spray technique can also be used to encapsulate particulate matter.[2] This approach has now been developed into techniques for large-scale production. Artificial cells containing sorbents are usually prepared based on the original method of ultrathin membrane coating of sorbent granules.[8,12]

41.1.2 Membranes Used for Encapsulation

Different types of synthetic polymers are used, and variations in configuration are possible.[10] A single, ultrathin polymer membrane is the most common one. The unlimited variations in polymers available allow for possible variations in permeability, biocompatibility, and other characteristics.[2-11] Artificial cells can also be made to contain smaller "intracellular compartments."[5-7] Others can form solid polymer microspheres containing microdroplets of biologically active materials.[5] Liquid hydrocarbons form microdroplets containing biologically active materials that are useful in biotechnology and other applications.[13] Membranes formed from biodegradable or biological materials are useful for artificial cells that need to be degraded, and different materials are used, including protein membrane artificial cells and polyhemoglobin.[3,5-7]

The use of lipids, including lipid–protein membranes, concentric lipid membranes, and submicron ultrathin lipid membranes is another common approach.[8,14] Biodegradable synthetic polymers offer another approach. The first one used was polylactide.[15] Many types of polylactides and polyglycolic acids are now used for artificial cells.[16] Other synthetic biodegradable polymers, such as polyanhydride, also are used.[17] The use of biodegradable artificial cells has become an active field.

41.1.3 Variations in Contents

Varied contents are possible. The following provides examples of microencapsulation of biologically active materials for use in biotechnology and medicine.

41.2 MICROENCAPSULATION OF BIOACTIVE SORBENTS

Bioactive sorbents represent the simplest form of artificial cells already used in routine clinical applications for humans. Sorbents such as activated charcoal, resins, and immunosorbents could not be used in direct blood perfusion because particulate embolism and blood cells were removed. However, sorbents such as activated charcoal inside artificial cells no longer cause particulate embolism and blood cells removal.[4,8,9] This application was developed and used successfully in patients.[9,12,18] For example, the hemoperfusion device now used in patients contains 70 g of artificial cells. Each artificial cell is formed by applying an ultrathin coating of collodion membrane or other polymer membranes on each of the 100-μ-diameter activated-charcoal microspheres. The mass transfer for this small device is many times higher than that for a standard dialysis machine.

41.2.1 Clinical Uses

Microencapsulation has become a routine treatment for adult and pediatric patients who have been poisoned.[18,19] This applies to the many cases in which the toxin can be adsorbed by activated charcoal. In kidney failure, tests show this is more effective than hemodialysis in removing organic waste metabolites.[8,12,19] It is used two ways: In series with dialysis, it shortens dialysis time and improves dialysis-resistant symptoms;[8,12,19] In series with a small ultrafiltrator it replaces the dialysis machine.[12] Here oral adsorbents can control potassium and phosphates. A urea removal system is under development to complete the hemoperfusion–ultrafiltrator approach. The detoxifying functions of hemoperfusion resulted in temporary recoveries from coma in a grade IV hepatic coma patient.[20] Other groups also support this finding method.[19] Hemoperfusion also is very effective in detoxification. It is being studied as part of an artificial liver system. Hemoperfusion when used with a chelating agent, desferroxaimine, is effective in lowering high aluminum levels in patients.[19]

41.2.2 MICROENCAPSULATION OF IMMUNOSORBENTS

Immunosorbents, like other sorbents described above, also have problems when they contact blood directly. These problems include embolisms of particulates and adverse effects on blood cells. The same ultrathin coating used in sorbents has been applied to immunosorbents to prevent these problems.[21] This has been tested clinically in patients.[10]

41.3 MICROENCAPSULATION OF CELLS OR MICROORGANISMS

The first encapsulation of biological cells, which was based on a drop method, was reported in 1965.[5] It was proposed that "... protected from ... immunological process ... encapsulated endocrine cells might survive and maintain an effective supply of hormone. For organ deficiency ... cultures of liver cells ... in artificial cells."[5] This original drop method for cell encapsulation involves chemically crosslinking the surface of aqueous droplets that contain cells.[5,6] This was modified into the following drop technique that uses milder physical crosslinking, which resulted in alginate–polylysine–alginate (APA) microcapsules containing cells.[22,23] Alginates are heteropolymer carboxylic acids, coupled by $1 \rightarrow 4$ glycosidic bonds of β-D-mannuramic (M) and findxb.alpha;-L-gluronic acid unit (G). Alkali and magnesium alginate are soluble in water, whereas alginic acids and the salts of polyvalent metal cations are insoluble. Thus, gel spheres can be formed by drops of sodium alginate solution entering a calcium chloride solution.

41.3.1 PROCEDURE

The general method of encapsulation is described in the following procedure.[22–24] Cells or microorganisms are suspended in the sodium alginate solution. The suspension is pumped through a 23-G stainless steel needle. Sterile compressed air, forced through a 16-G coaxial stainless steel needle, shears the droplets coming out of the tip of the 23-G needle. Each droplet falls into the sterile ice cold solution of calcium chloride (1.40%, pH 7.20, heat sterilized). Upon contact with the calcium chloride buffer, alginate gelation is immediate. The droplets gel for 15 min in the calcium chloride solution. After gelation in the calcium chloride solution, alginate-gel beads were suspended for 10 min in a 0.05% polylysine solution. The positively charged polylysine forms a complex with the surface alginate to create a semi-permeable membrane. The beads are washed and placed in an alginate solution (0.10%) for 4 min. The alginate neutralizes any excess polylysine on the surface. The alginate–poly-L-lysine–alginate capsules are washed in a 3.00% citrate bath (3.00% in 1:1 HEPES:buffer saline, pH 7.20) to liquefy the gel in the microcapsules. The APA microcapsules formed are stored at 4°C for use in experiments.

This method results from extensive research to improve the original drop method.[5,6] This works well for larger cell aggregates such as islets. However, it is not as suitable for encapsulating high concentrations of smaller cells such as hepatocytes and microorganisms. Some cells or microorganisms remain on the surface of the microcapsule membrane and could result in the rejection of the whole artificial cell.[24]

41.3.2 HIGH CONCENTRATIONS OF SMALLER CELLS

We have devised a new method to allow for more complete encapsulation of higher concentrations of smaller cells in artificial cells.[25,26] The procedure follows:

- Small calcium alginate gel microspheres containing entrapped cells are formed. Like the general procedure, some cells protrude out of the surface of the smaller calcium alginate gel microspheres. These gel microspheres are resuspended in alginate solution to repeat the same droplet formation.
- This entraps the small microspheres within larger calcium alginate gel microspheres and eliminates cells protruding out of the larger alginate gel microspheres.
- The next two steps are the same as the previous process. In the last step, the entire content of the microcapsule was liquified by citrate. This also liquified the small calcium alginate gel microspheres inside the microcapsule. This way the hepatocytes in the smaller gel microsphere are released to float freely inside the microcapsule.

41.3.3 MICROENCAPSULATED ISLETS FOR DIABETES MELLITUS

This author persuaded Connaught Laboratory, Toronto, Canada, of insulin fame to develop this approach for diabetes. It was carried out at Connaught and later in other centers.[22,23,27] The research showed that islets inside artificial cells are prevented from immunorejection after implantation into animals. Islets can remain viable and continued to secrete insulin to control the glucose levels of diabetic rats. They improved the biocompatibility by the use of an APA membrane.[23] One group used a special alginate to further improve the biocompatibility.[27]

41.3.4 MICROENCAPSULATED HEPATOCYTES FOR LIVER FAILURE

We found that artificial cells containing hepatocytes increased the survival time of fulminant hepatic failure rats.[28] Xenografts of rat hepatocytes in artificial cells were not immunorejected in mice.[29] Instead, the viability of the enclosed liver cells increased after intraperitoneal implantation.[29] This was because the hepatotrophic factor secreted by the encapsulated hepatocytes accumulates in the artificial cells.[30] After implantation, hepatocytes in artificial cells can lower the high bilirubin level in the Gunn rats.[31,32] Reports by another group also support this finding.[33]

41.3.5 MICROENCAPSULATION OF GENETICALLY ENGINEERED MICROORGANISMS

We studied microencapsulated genetically engineered microorganisms by using *Escherichia coli* DH5 cells with the *Klebsiella aerogens* urease gene.[34,35] Overall, urea removal efficiency of microencapsulated genetically engineered bacteria is 10–30 times higher than the best available urea removal systems available. Urea and ammonia removal are needed in kidney failure and liver failure as well as environmental decontamination and the regeneration of water supply in space travel. Standard dialysis is usually complex and expensive, and several alternatives, including adsorbents or enzyme/adsorbents, have not been effective.

41.3.6 MICROENCAPSULATION OF CHOLESTEROL-REMOVING MICROORGANISMS

We selected *Pseudomonas pictorum* (ATCC #23328) as another model system because it can degrade cholesterol.[36,37] The standard encapsulation method does not result in a high-porosity membrane that would allow lipoprotein–cholesterol to cross. Therefore, we devised a modified method to prepare high-porosity agar microspheres. There was no evidence of leakage of the enclosed bacteria. Open pore agar beads were incubated in serum, and the bacterial action did not significantly differ between the encapsulated and free bacteria. Bacterial action was a limiting step in the overall reaction. For practical applications, a suitable bacteria with higher rates of cholesterol removal is needed. This should be available in the future with the help of genetic engineering. Many other researchers are now studying the microencapsulation of cells and microorganisms.[38,39]

41.4 MICROENCAPSULATION OF ENZYMES AND MULTIENZYME SYSTEMS

Artificial cells protect the enclosed enzyme from immunological rejection or tryptic enzymes.[3–8] However, substrates must be able to equilibrate rapidly into the artificial cells for conversion into products that can diffuse out.

41.4.1 UREA REMOVAL

We showed that artificial cells containing urease can convert urea to ammonia, which is then removed by ammonia adsorbent.[3–8] This approach has been developed further by us and other groups.[10] Ammonia adsorbents with better adsorbing capacity are required to improve this approach.

41.4.2 ENZYME THERAPY

Artificial cells have been used in hereditary enzyme defects, including our earliest use of a replacement for catalase in acatalasemic mice.[40] This also has been studied for asparagine removal in the treatment of leukemia in animals.[41] We used phenylalanine ammonia lyase artificial cells in phenylketonuria rats.[42] Later, we found an extensive enterore-circulation of amino acids in the intestine.[43] This allows enzyme artificial cells to be used orally to selectively remove specific amino acids from the body, as in phenylketonuria.[43] We also studied the oral administration of artificial cells containing xanthine oxidase.[44] This resulted in a decrease in systemic hypoxanthine in a pediatric patient with hypoxanthinuria (Lesch–Nyhan disease).

41.4.3 MULTIENZYME SYSTEM

Most enzymes in biological cells function as complex enzyme systems. We have prepared artificial cells that contain multienzyme systems with cofactor recycling.[45] This approach can convert metabolic wastes such as urea and ammonia into essential amino acids such as leucine, isoleucine, and valine, which are required by the body.[46] We have also prepared artificial cells containing hemoglobin with pseudoperoxidase activity and glucose oxidase to remove bilirubin.[47,48]

41.5 RED BLOOD CELL SUBSTITUTES

Two major approaches exist: modified hemoglobin and perfluorochemicals. Detailed reviews in the field are available.[49,50] This is a large area that will be discussed under a separate title.

41.6 BIODEGRADABLE ARTIFICIAL CELLS

Biodegradable artificial cells also are used for drug delivery. This was discussed under the section on preparation. We have used crosslinked protein and biodegradable polylactide artificial cells.[3,5–8,15] Many groups are extending these approaches for drug delivery use (e.g., medications, hormones, peptides, and proteins).[10,16,17]

We prepared lipid–protein and lipid–polymer artificial cells to encapsulate biologically active materials.[51] Later, Gregoriadis prepared concentric lipid membrane liposomes containing enzymes.[14] Liposomes feature multiple lipid layers and onion-skin-like microspheres originally used by Bangham for basic membrane research. Workers in liposomes turned to preparing small submicron artificial cells with a single bilayer lipid membrane.[14] These lipid membrane artificial cells are no longer concentric lipid membrane liposomes. Some still continue to call these "liposomes."[14] This has created some confusion in the field. The most extensive research is in its use for drug delivery.[14]

41.7 GENERAL DISCUSSION

41.7.1 Present Status

The present uses of artificial cells in biotechnology and medicine includes the following:

- Hemoperfusion for acute poisoning—routine treatment in patients
- Hemoperfusion for aluminium and iron overload—routine treatment in patients
- Supplement to hemodialysis in end-stage renal failure—routine treatment in patients
- Artificial liver support: hemoperfusion and hybrid systems—experimental
- Red blood cell substitutes for transfusion—Phase I and Phase II clinical trials
- Blood group antibodies removal (immunosorbents)—clinical trial
- Hereditary enzyme deficiency—clinical trial
- Clinical laboratory analysis—clinical application
- Production of monoclonal antibodies—development
- Diabetic mellitus and other endocrine diseases—animal experiment, clinical trial started
- Drug delivery systems—clinical application and experimental
- Conversion of cholesterol into carbon dioxide—experimental
- Bilirubin removal—experimental
- Production of fine biochemicals—development
- Food and aquatic culture—development
- Conversion of wastes into useful products—experimental
- Other biotechnological and medical applications

Artificial cells can contain an unlimited number of biologically active materials (Figure 41.1). Therefore, many other areas of applications and research exist. For example, the author has enclosed magnetic and biological materials together inside artificial cells.[4] This allows for localization with external magnetic fields.[4] Kato applied a magnetic field outside the body of animals.[11] This can direct magnetic artificial cells containing radioactive materials and chemotherapeutic agents to specific sites of bladder cancer. Magnetic artificial cells are also used in bioreactors. Others have used artificial cells in the laboratory analysis of free and protein-bound hormones in patients.[11] We have studied its use for one-shot vaccine and for removing large lipophilic molecules from small hydrophilic molecules.[10] Still others have used artificial cells for industrial aquatic culture for shrimps and lobsters. We have also studied artificial cells containing hepatic microsome and cytosol.[10]

41.7.2 Future Perspectives

The author wrote in his 1972 book on artificial cells: "'Artificial cell' is a concept; the examples described … are but physical examples for demonstrating this idea. In addition to extending and modifying the present physical examples, completely different systems could be made available to further demonstrate the clinical implications of the idea of 'artificial cells'."[8] An entirely new horizon is waiting impatiently to be explored. This future perspective is even more valid now.

REFERENCES

1. Chang, T.M.S. In *Encyclopedia of Human Biology*; Dulbecco, R., Ed.; Academic: San Diego, CA, 1991; *1*: 377–383.
2. Chang, T.M.S. Research Report for Honours Physiology; Medical Library: McGill University: Montreal, Quebec, Canada, 1957.

3. Chang, T.M.S. Semipermeable microcapsules. *Science* **1964**, *146* (3643), 524–525.
4. Chang, T.M.S. Semipermeable aqueous microcapsules ("artificial cells"): With emphasis on experiments in an extracorporeal shunt system. *Trans. Am. Soc. Artif. Intern. Organs* **1966**, *12* (1), 13–19.
5. Chang, T.M.S. PhD thesis, Semipermeable microcapsules. PhD thesis, McGill University: Montreal, Quebec, Canada, 1965.
6. Chang, T.M.S., MacIntosh, F.C., Mason, S.G. Semipermeable aqueous microcapsules. I. Preparation and properties. *Can. J. Physiol. Pharmacol.* **1966**, *44* (1), 115–128.
7. Chang, T.M.S., MacIntosh, F.C., Mason, S.G. Encapsulated hydrophilic compositions and methods of making them. Canadian Patent 873815, Jun 22, 1971.
8. Chang, T.M.S. *Artificial Cells*; C.C Thomas: Springfield, IL, 1972.
9. Chang, T.M.S. Editorial. *Artif. Cells Blood Substit. Immobil. Biotechnol.* **1994**, *22* (1), vii.
10. Chang, T.M.S. Recent advances in artificial cells with emphasis on biotechnological and medical approaches based on microencapsulation. In *Microcapsules and Nanoparticles in Medicine and Pharmacology*; Donbrow, M., Ed.; CRC: Boca Raton, FL, 1992; pp. 323–339.
11. Chang, T.M.S. Artificial cells with emphasis on bioencapsulation in biotechnology. *Biotechnol. Annu. Rev.* **1995**, *1*, 267–296, doi:10.1016/S1387-2656(08)70054-1.
12. Chang, T.M.S. Kidney. *Int.* **1975**, *7*, S5387–S5392.
13. May, S.W., Li, N.N. The immobilization of urease using liquid-surfactant membranes. *Biochem. Biophys. Res. Commun.* **1972**, *47* (5), 1179–1185.
14. Gregoriadis, F. *Liposomes as Drug Carriers: Recent Trends and Progress*; John Wiley & Sons: New York, 1989.
15. Chang, T.M.S. Biodegradable semipermeable microcapsules containing enzymes, hormones, vaccines, and other biologicals. *J. Bioeng.* **1976**, *1* (1), 25–32.
16. Jalil, R., Nixon, J.R. Biodegradable poly(lactic acid) and poly(lactide-co-glycolide) microcapsules: Problems associated with preparative techniques and release properties. *J. Microencapsul.* **1990**, *7* (3), 297–325.
17. Mathiowitz, E., Langer, R. Polyanhydride microspheres as drug delivery systems. In *Microcapsules and Nanoparticles in Medicine and Pharmacology*; Donbrow, M., Ed.; CRC: Boca Raton, FL, 1992; pp. 99–123.
18. Chang, T.M.S., Coffey, J.F., Barre, P., Gonda, A., Dirks, J.H., Levy, M., Lister, C. Microcapsule artificial kidney: Treatment of patients with acute drug intoxication. *Can. Med. Assoc. J.* **1973**, *108* (4), 429–433.
19. Winchester, J.F. Hemoperfusion. In *Replacement of Renal Function by Dialysis*; Maher, J.F., Ed.; Kluwer Academic: Boston, MA, 1988; pp. 439–459.
20. Chang, T.M.S. Haemoperfusions over microencapsulated adsorbent in a patient with hepatic coma. *Lancet* **1972**, *2* (7791), 1371–1372.
21. Chang, T.M.S. Blood compatible coating of synthetic immunoadsorbents. *Trans. Am. Soc. Artif. Intern. Organs* **1980**, *26* (1), 546–549.
22. Lim, F., Sun, A.M. Microencapsulated islets as bioartificial endocrine pancreas. *Science* **1980**, *210* (4472), 908–910.
23. Goosen, M.F.A., O'Shea, G.M., Gharapetian, H.M., Chou, S., Sun, A.M. Optimization of microencapsulation parameters: Semipermeable microcapsules as a bioartificial pancreas. *J. Biotechnol. Bioeng.* **1985**, *27* (2), 146–150.
24. Wong, H., Chang, T.M.S. The microencapsulation of cells within alginate poly-L-lysine microcapsules prepared with the standard single step drop technique: Histologically identified membrane imperfections and the associated graft rejection. *Biomater. Artif. Cells Immobil. Biotechnol.* **1991**, *19* (4), 675–686.
25. Wong, H., Chang, T.M.S. A novel two step procedure for immobilizing living cells in microcapsules for improving xenograft survival. *Biomater. Artif. Cells Immobil. Biotechnol.* **1991**, *19* (4), 687–697.
26. Chang, T.M.S., Wong, H. Method for encapsulating biologically active material including cells. U.S. Patent 5084350, Jan 28, 1992.
27. Soon-Shiong, P., Otterlie, M., Skjak-Braek, G., Smidsrod, O., Heintz, R., Lanza, R.P., Espevik, T. An immunologic basis for the fibrotic reaction to implanted microcapsules. *Transpl. Proc.* **1991**, *23* (1 Pt. 1), 758–759.
28. Wong, H., Chang, T.M.S. Bioartificial liver: Implanted artificial cells microencapsulated living hepatocytes increases survival of liver failure rats. *Int. J. Artif. Organs* **1986**, *9* (5), 335–336.
29. Wong, H., Chang, T.M.S. The viability and regeneration of artificial cell microencapsulated rat hepatocyte xenograft transplants in mice. *Biomater. Artif. Cells Artif. Organs* **1988**, *16* (4), 731–739.

30. Kashani, S., Chang, T.M.S. Physicochemical characteristics of hepatic stimulatory factor prepared from cell free supernatant of hepatocyte cultures. *Biomater. Artif. Cells Immo-bil. Biotechnol.* **1991**, *19* (3), 579–598.
31. Bruni, S., Chang, T.M.S. Hepatocytes immobilised by microencapsulation in artificial cells: Effects on hyperbilirubinemia in gunn rats. *J. Biomater. Artif. Cells Artif. Organs* **1989**, *17* (4), 403–411.
32. Bruni, S., Chang, T.M.S. Encapsulated hepatocytes for controlling hyperbilirubinemia in Gunn rats. *Int. J. Artif. Organs* **1991**, *14* (4), 239–241.
33. Dixit, V., Darvasi, R., Arthur, M., Brezina, M., Lewin, K., Gitnick, G. Restoration of liver function in Gunn rats without immunosuppression using transplanted microencapsulated hepatocytes. *Hepatology* **1990**, *12* (6), 1342–1349.
34. Prakash, S., Chang, T.M.S. Genetically engineered *E. coli* cells containing K. aerogenes gene, microencapsulated in artificial cells for urea and ammonia removal. *Biomater. Artif. Cells Immobil. Biotechnol.* **1993**, *21* (5), 629–636.
35. Praskan, S., Chang, T.M.S. Preparation and in vitro analysis of microencapsulated genetically engineered *E. coli* DH5 cells for urea and ammonia removal. *J. Biotechnol. Bioeng.* **1995**, *46* (6), 621–626.
36. Garofalo, F., Chang, T.M.S. Immobilization of P. pictorum in open pore agar, alginate and polylysine-alginate microcapsules for serum cholesterol depletion. *Biomater. Artif. Cells Artif. Organs* **1989**, *17* (3), 271–289.
37. Garofalo, F., Chang, T.M.S. Effects of mass transfer and reaction kinetics on serum cholesterol depletion rates of free and immobilized *Pseudomonas pictorum*. *Appl. Biochem. Biotechnol.* **1991**, *27* (1), 75–91.
38. Goosen, M.F.A., King, G.A., McKnight, C.A., Marcotte, N. Animal cell engineering using alginate polycation microcapsule of controlled membrane molecular weight cut-off. *J. Membr. Sci.* **1989**, *40*, 233.
39. Goosen, M., Ed. *Fundamentals of Animal Cell Encapsulation and Immobilization*; CRC: Boca Raton, FL, 1992.
40. Chang, T.M.S., Poznansky, M.J. Semipermeable microcapsules containing catalase for enzyme replacement in acatalasaemic mice. *Nature* **1968**, *218* (5138), 243–245.
41. Chang, T.M.S. The in vivo effects of semipermeable microcapsules containing l-asparaginase on 6C3HED lymphosarcoma. *Nature* **1971**, *229* (5280), 117–118.
42. Bourget, L., Chang, T.M.S. Phenylalanine ammonia-lyase immobilized in microcapsules for the depletion of phenyl-alanine in plasma in phenylketonuric rat model. *Biochim. Biophys. Acta.* **1986**, *883* (3), 432–438.
43. Chang, T.M.S., Bourget, L., Lister, C. New theory of enterore circulations of amino acids and its use for depleting unwanted amino acids using oral enzyme artificial cells as in removing phenylalanine in phenylketonuria. *Artif. Cells Blood Substit. Immobil. Biotechnol.* **1995**, *23* (1), 25.
44. Palmour, R.M., Goodyer, P., Reade, T., Chang, T.M.S. Microencapsulated xanthine oxidase as experimental therapy in Lesch-Nyhan disease. *Lancet* **1989**, *2* (8664), 687–688.
45. Chang, T. M.S. Recycling of NAD(P) by multienzyme systems immobilized by microencapsulation in artificial cells. *Methods Enzymol.* **1987**, *136*, 67–82, doi:10.1016/S0076-6879(87)36009-4.
46. Gu, K.F., Chang, T.M.S. Conversion of ammonia or urea into essential amino acids, L-leucine, L-valine, and L-isoleucine using artificial cells containing an immobilized multienzyme system and dextran-NAD+. *Appl. Biochem. Biotechnol.* **1990**, *26* (2), 263–269.
47. Daka, J.N., Chang, T.M.S. Bilirubin removal by the pseudoperoxidase activity of free and immobilized hemoglobin and hemoglobin co-immobilized with glucose oxidase. *Biomater. Artif. Cells Artif. Organs* **1989**, *17* (5), 553–562.
48. Chang, T.M.S., Daka, J.N, Removal of Bilirubin by the Pseudoperoxidase Activity of Immobilized Hemoglobin. U.S. Patent 4820416, April 11, 1989.
49. Chang, T.M.S., Ed. *Blood Substitutes and Oxygen Carriers*; Marcel Dekker: New York, 1992.
50. Chang, T.M.S. *Artif. Cells Blood Substit. Immboil. Biotechnol.* **1994**, *22*, 123, doi:10.3109/10731199 409117408.
51. Chang, T.M.S. *Fed. Proc. Fed. Am. Soc. Exp. Biol.* **1969**, *28*, 461.

42 Cell Encapsulation

James Blanchette

CONTENTS

42.1 Introduction 917
42.2 Key Material Properties 917
42.3 Common Encapsulation Materials 920
    42.3.1 Alginates 920
    42.3.2 Poly(Ethylene Glycol) 922
    42.3.3 Chitosan 923
    42.3.4 Poly(Lactic-co-Glycolic Acid) 925
References 926

42.1 INTRODUCTION

Cell encapsulation is a research area which has emerged as an exciting approach for numerous biomedical applications including delivery of therapeutics and regenerative medicine. The seminal study of Lim and Sun[1] published in 1980, using encapsulated islets of Langerhans to temporarily restore euglycemia in diabetic rats, offered a glimpse at the potential of combining living tissue with materials for transplantation. One of the primary motivations for cell encapsulation is isolation of cells from components of the immune system following transplantation. This could facilitate transplantation of allogeneic or xenogeneic tissue or reduce the need for immunosuppressive drugs. Materials can also be used to support cell function, and this is a primary requirement in tissue engineering. Materials can not only provide the desired shape for the tissue being formed but can also be designed to guide differentiation of encapsulated cells toward the desired phenotypes.

42.2 KEY MATERIAL PROPERTIES

There are a number of important properties to consider when selecting or designing a material for cell encapsulation. Their optimal values will vary with the intended application, but these critical properties include toxicity/immunogenicity, diffusivity of nutrients and waste products, degradation rate, presentation of cell-binding sequences, and mechanical properties. Figure 42.1 shows a schematic of encapsulated cells and these important material properties.

Obviously, the material selected should not cause loss of viability for either the encapsulated cells or surrounding tissue at the implantation site. This trait can be expanded to include the desire that the material not elicit a vigorous immune response as well. Many capsules are designed to prevent contact between immune cells and surface antigens on the transplanted cells. This only requires that the material have pores small enough to prevent cell motility or that another strategy (such as a conformal coating on cells) is used to physically interfere with the potential interaction.[2] If immune cells are activated by the surface of the material, secreted cytokines could pass through the material and damage encapsulated cells. Cytokine transport can also go in the other direction, with cytokines secreted by the encapsulated cells triggering an immune response if they are not autologous cells. What can and cannot freely diffuse through the material is represented by the material's molecular weight cutoff (MWCO). This represents a molecular mass, commonly expressed in kilodaltons, above which diffusion is impaired by the material. Materials with a low

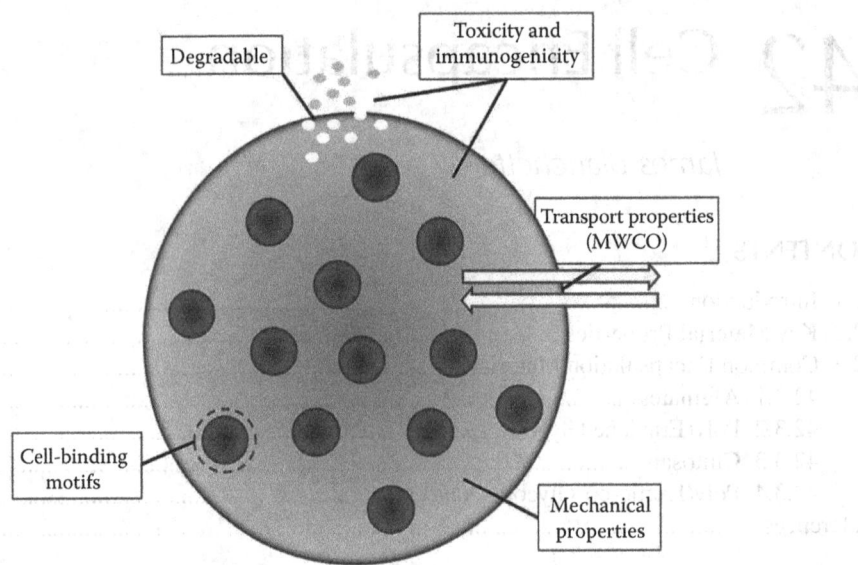

FIGURE 42.1 Schematic of encapsulated cells with five key material properties that must be considered to design a successful system.

MWCO may be able to prevent transport of antibodies, but it is not possible to block all components of the immune system while still allowing nutrient transport. Hydrophilic materials tend have a higher biocompatibility as they resist protein adsorption. Protein adsorption can initiate a cascade of events culminating in fibrotic encapsulation of the capsule as discussed in a recent review.[3] The use of natural materials for the capsule increases the risk of triggering an immune response due to antigens in the material or insufficient removal of immunogenic compounds from the material such as endotoxin.

Another potential consequence of activating the innate immune system is fibrous encapsulation of the material. This would limit the diffusion of nutrients and waste products to and from the encapsulated cells. The material itself must not present a barrier to this diffusion or insufficient access to nutrients, such as oxygen and glucose, combined with accumulation of metabolic waste would lead to loss of cell viability. Because many encapsulation materials do not allow cell migration, blood vessels and the lymphatic system cannot penetrate the capsule to deliver nutrients and eliminate waste. Highly porous materials or degradable materials may allow access of the circulatory and lymphatic systems to the cells eventually, but there will be a period when the encapsulated cells will be reliant on diffusion. Selecting a material which will allow maintenance of viability during this period is critical. In addition to allowing nutrient transport, the MWCO for a material must be selected to ensure passage of any molecules essential to function. For example, materials for encapsulation of islets must have a MWCO large enough to allow insulin (5.8 kDa) diffusion. The high water content of hydrogels aids in the diffusion of these compounds and is one reason for their popularity as a cell encapsulation material. MWCO can be modified by adjusting the pore size in the material. For hydrogels, this can be accomplished by modifying the polymer volume fraction in the gel or the cross-linking density, for example.

As mentioned above, some capsules should degrade to allow access of the encapsulated cells to the surrounding tissue. This is often the case in tissue engineering applications. It is ideal for the degradation rate of the capsule to match the rate at which the desired tissue replaces it. For applications like encapsulation of islets, the integrity of the capsule must not be compromised. Nondegradable materials are therefore desired to maintain the immunoprotective effect of the material. Materials which can be modified to control their degradation rate can be optimized for a wide range of cell encapsulation strategies. Degradation products also need to be evaluated for their potential toxicity and immunogenicity to avoid issues mentioned above.

The use of a synthetic material for the capsule presents a foreign microenvironment to encapsulated cells. Anchorage-dependent cells require engagement of membrane proteins like integrins to prevent apoptosis. This is the motivation behind functionalizing these materials to include ECM proteins or peptide sequences. Many natural materials like collagen are selected because such sequences are already present. By presenting the appropriate sequences in the appropriate ratios to the cells, the physiological environment can be recreated more accurately inside the capsule. This can impact not only cell viability but also differentiation of multipotent cells. These sequences can also be used to allow migration of cells into a material which is a component of some tissue engineering strategies. Apoptosis resulting from insufficient or improper cell–cell and cell–extracellular matrix (ECM) interactions is termed anoikis. Failure to provide these interactions to anchorage-dependent cells within the capsule will lead to cell death and failure of the implant. Encapsulating cells as multi-cell aggregates is another strategy to prevent anoikis, but as aggregates increase in dimensions, nutrient transport to central cells may be insufficient. The relationship between these potential causes of cell death is shown in Figure 42.2a and images of adipose-derived stem cells encapsulated in poly(ethylene glycol) as dispersed or aggregated cells are shown in Figure 42.2b.

The relationship between aggregate dimensions and cell stress can be seen when encapsulating islets. Islets are roughly 200 μm in diameter when isolated from the pancreas. The inability to revascularize these cell masses following encapsulation can lead to insufficient nutrient supply for central cells. If oxygen can only penetrate through a tissue layer with a small number of cells, any

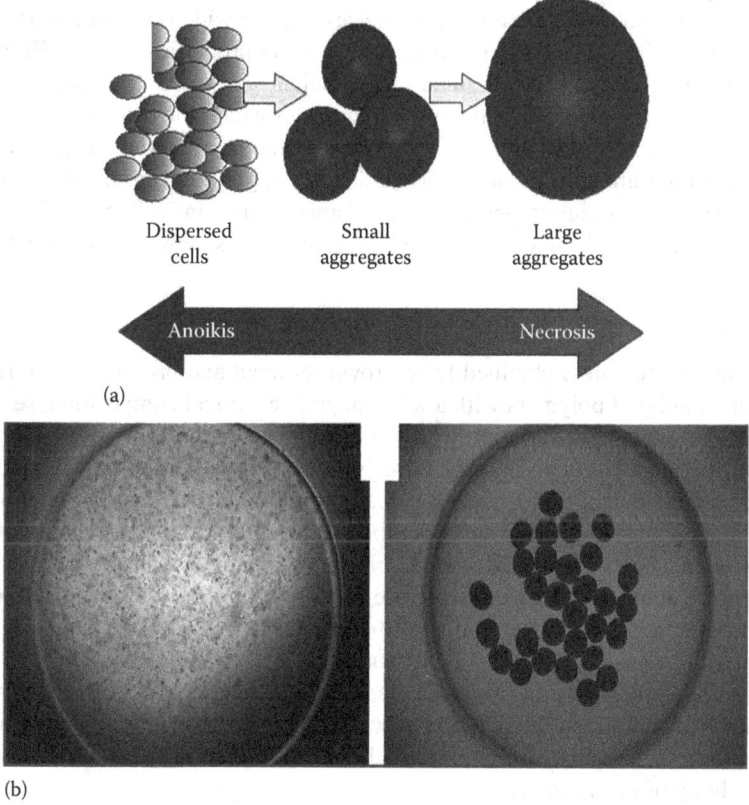

FIGURE 42.2 (a) Images for potential causes of cell death. (b) Images of adipose-derived stem cells encapsulated in poly(ethyleneglycol) as dispersed or aggregated cells.

tissue mass with a radius greater than this will see hypoxic or anoxic conditions in its core. Studies have shown that when native islets are transplanted they experience central necrosis, and that islet size plays an important role in performance.[4,5] A number of research groups have investigated the ability of islets to be dissociated and reaggregated into smaller tissue masses. By breaking down the native islet structure into single cells or small aggregates of cells, the hope is that no necrotic core will be formed due to insufficient oxygen reaching the center of large islets. Small aggregates or single cells are more likely to experience anoikis, which makes interaction with the encapsulation material perhaps more important. This is an example where both the material's transport properties and ability to interact with encapsulated cells are critical to success.

The mechanical properties of the capsule must be carefully considered. Not only does the capsule need sufficient strength to resist mechanical failure and rupture but the elasticity of cell substrates have been shown to influence cell behavior. The mechanism by which substrate elasticity influences differentiation of multipotent cells is not clear. A number of recent studies have analyzed this in both two- and three-dimensional (3D) culture systems,[6–9] Integrin engagement is thought to be required to transmit the elasticity of the capsule material to the cells inside. Matrix elasticity has been shown to influence the lineage commitment of naïve, mesenchymal stem cells.[10] Another important consideration is how the mechanical properties of a capsule impact the transmission of external mechanical forces to the cells. Many cells respond to mechanical loads and this can be necessary for maintenance of phenotype or to guide differentiation.[11]

42.3 COMMON ENCAPSULATION MATERIALS

Alginate showed promising early results and is the most widely used material for cell encapsulation. A number of synthetic polymers are attractive choices due the wide range of functionalities which can be introduced into the material to customize the material's properties. Specific ECM components or decellularized ECM isolated from human or nonhuman sources are gaining in popularity due to their close match to the physiological environment from which the cells were obtained. For tissue engineering applications, cells are sometimes seeded onto the material rather than encapsulated within it. These studies are still instructive to guide the design of encapsulation systems with that material as the desired material properties do not change significantly. Some of the most common encapsulation materials are discussed in detail below with a focus on ways to control key properties.

42.3.1 Alginates

Alginates are natural polymers obtained from brown seaweed and bacteria.[12,13] Alginate is a collective term for a family of polymers with a wide range in chemical composition, sequential structure and molecular size. This variation leads to a range of different properties; the origin and composition of the most common alginates are summarized by Thu et al.[14] Alginate is a linear polysaccharide copolymer of 1–4-linked β-D-mannuronic acid (M) and α-L-guluronic acid (G) monomers. The monomers are arranged in a block-wise pattern along the chain with homopolymeric regions of M and G termed M- and G-blocks, respectively. These M- and G-blocks are interspaced with regions of alternating structure (MG-blocks). Since alginates are derived naturally, there is a potential lot-to-lot variability of the material as it is isolated from different sources. Another issue with the use of alginates is activation of the immune response from samples that have not been sufficiently purified to remove compounds like endotoxins, polyphenols, and various proteins.[15] The purification and sterilization processes are time-consuming and typically accomplished by filtration techniques.[16,17] Removal of these compounds also leads to a modest increase in the hydrophilicity of the alginate.

Alginates are able to form gels by the binding of divalent cations to the G-units, and this interaction can influence the material's degradation rate. The affinity for various divalent cations for the G-unit is: $Pb^{2+} > Cu^{2+} > Cd^{2+} > Ba^{2+} > Sr^{2+} > Ca^{2+} > Co^{2+} = Ni^{2+} = Zn^{2+} > Mn^{2+}$. Alginate gelling conditions are mild, and the gelation process can be reversed by extracting calcium ions. This is done by adding citrate or by rinsing the gel with a sodium chloride solution.[18] Alginate hydrogels can be enzymatically degraded in a controlled and tunable fashion. Humans do not produce alginases, however, so these enzymes need to be introduced with a protein delivery system. A study[19] developed alginate scaffolds with tunable degradation rates by incorporating alginate lyase-loaded microspheres that released the enzymes over time to degrade the scaffold. These scaffolds were successfully used to culture neural progenitor cells and increased their proliferation rate compared to when such cells were cultured in alginate scaffolds without microspheres.

In 1980, Lim and Sun[1] originally described the treatment of diabetes in vivo with alginate-based capsules. The study exhibited a complete correction of the condition for up to 3 weeks, while the morphology and functionality of the cells was maintained for over 15 weeks. Furthermore, Sun et al.[20] showed induced normoglycemia without immunosuppression up to 800 days with the transplantation of encapsulated porcine islets into spontaneously diabetic monkeys. Both adipose-derived adult stem cells as well bone marrow-derived mesenchymal stem cells have been shown to survive and differentiate in these scaffolds.[15,21] Alginate scaffolds have also been used in combination with embryonic stem cells to generate hepatocytes and endothelial cells.[22,23] Alginate has been used extensively in the culturing of chondrocytes, hepatocytes, and Schwann cells for nerve regeneration.[24-28] These studies show that the transport properties of alginate are sufficient to maintain the encapsulated cells for days or perhaps weeks, but improved oxygen transport can lead to extension of cell survival and function. A common modification to alginate capsules to achieve this is incorporation of perfluorocarbon (PFC) into the capsule. A recent study by Johnson et al.[29] tracked viability of islets cultured in low oxygen conditions encapsulated in alginate or alginate containing a PFC emulsion and demonstrated less necrosis in central regions of the islets when PFC was present.

A potential limitation in using alginate gels in tissue engineering is the lack of cellular interaction. Alginates discourage protein adsorption due to their hydrophilic character and are unable to specifically interact with mammalian cells.[30] The cell adhesion ligand RGD can be covalently coupled into alginate gels to enhance cell adhesion and address this stress. These modified alginate gels have been demonstrated to provide for the adhesion and proliferation of encapsulated skeletal muscle cells.[31] Covalent binding of avidin to alginate capsules allows subsequent functionalization of the capsule with biotinylated molecules. This strategy was employed to modify an alginate capsule with lectin and is a flexible technique to introduce a range of cell-binding molecules.[32] Hydroxyapatite has also been blended with alginates to create an environment more closely matching native bone in bone regeneration strategies. Ca^{2+} ions present in the apatite cross-link with the alginate form a composite which supports viability and maintained the osteoblast phenotype for encapsulated cells.[33]

The cross-linking density of the ionic cross-linked gels can be readily manipulated by varying the M-to-G ratio and molecular weight of the polymer chain. This will impact both the pore size and the mechanical properties of the material. In general, alginates rich in G residues form strong, brittle gels, while M-rich alginates form softer, more elastic gels. The elastic modulus of an alginate gel depends on the number and the strength of the cross-links and the length and stiffness of chains between cross-links.[14,34-36] The modulus also depends on the cross-linking ions as the gel strength scales with the affinity between polymer and cross-linking ions. Alginate hydrogels lose more than 60% of their initial mechanical strength within 15 hr of exposure to physiological buffers due to the loss of divalent ions during ion exchange.[35] This problem can be avoided by adding calcium ions to the surrounding fluid. Typically, calcium or barium is used to form rigid, biocompatible beads from alginate droplets. An ionic cross-link is formed between the carboxylic acid group (i.e., guluronic acids), found on the polymer backbone, and the cation.[37] Barium, which forms strong cross-links with alginate, produces stronger gels when compared to calcium. The capsule strength can also be tuned by adjusting other factors including: polymer purity, polymer molecular weight, and gelation time.[38,39]

The stability of alginate capsules can be improved by adding a polycation layer to the outer surface. While successful transplants have been shown with barium alginate capsules in the absence of a polycation layer, most alginate gels, in the absence of additional polycation coatings, are not stable for long-term use due to the slow leakage of the cations out of the capsule.[40–42] Poly-L-lysine (PLL) is the most common cation used to form these surface coatings. While PLL was originally applied to provide an immunoprotective outer membrane as it has been shown to improve capsule stability, studies have shown that it can be toxic for certain cell types, can induce fibrotic encapsulation, can activate macrophages, and can also trigger the compliment system.[1,43–47]

42.3.2 Poly(Ethylene Glycol)

Poly(ethylene glycol) (PEG) is a polyether which is also referred to as poly(ethylene oxide) or polyoxyethylene depending on the chain's molecular weight. The terminal groups are initially hydroxyls which are replaced when the chains are cross-linked. To form microcapsules, the terminal groups are replaced with specific active groups that can cross-link chains under specific conditions such as the incidence of UV or visible light.[48] PEG is approved by the Food and Drug Administration (FDA) for use in a number of medical materials and is typically selected due to its biocompatibility and high degree of swelling in aqueous environments. This hydrophilic nature decreases the adsorption of proteins, which consequently reduces activation of an immune response following implantation and formation of a fibrous capsule around the material. Synthetic polymers, such as PEG, avoid the issue of source variability that accompanies natural polymers. Due to its properties, PEG has also been used to coat a number of materials and has also been attached to molecules to increase their circulation time in the bloodstream. PEG has been used to coat alginate/PLL capsules to shield the positive PLL charge from the in vivo environment to reduce toxicity. PEG has also been used to form conformal coatings, where a thin PEG barrier is directly attached to the outer membrane of cells.[49] Surface, shape, and pore size may vary between coated cell masses, thereby reducing reproducibility of results.

PEG-based hydrogels have been used for encapsulation of a number of cell types in addition to islets. The list includes marrow stromal cells,[50] chondrocytes,[51–53] osteoblasts,[54] and valvular interstitial cells.[55] The PEG hydrogel network structure is a flexible platform allowing physical properties, such as the swelling ratio and mesh size, to be easily manipulated through modification of the fabrication parameters. Two of these parameters are the macromer molecular weight and percentage of macromer in solution.[53,56–58] The network cross-linking density of a hydrogel controls many of its properties, such as diffusion coefficients, mechanical behavior, and rate of degradation, with significant effects on the behavior of entrapped cells.[59,60] Hydrogels with low cross-linking density have a larger mesh size, or the distance between cross-links, which allows faster diffusion of nutrients and waste to and from encapsulated cells.[60] The swelling ratio of a hydrogel, the ratio of its swollen weight to its dry weight, is related to the cross-linking density and is a measure of how much water is retained by the hydrogel.[52] Covalently cross-linked PEG networks are appealing candidates for an encapsulation material due to the "stealth" properties of PEG. In addition to the immunological concerns mentioned above, interaction with proteins could encourage cell binding to the surface of the capsule. This layer of cells on the surface would create additional diffusion limitations for the encapsulated tissue. For encapsulated islets, the response to glucose changes would be slowed and more of the encapsulated tissue may fail due to insufficient nutrient supply. Cells and biological molecules are readily encapsulated via solution photopolymerization at room temperature and physiological pH in PEG gels with multiple geometric configurations. Using previously developed models for diffusion in PEG gels,[61,62] the average diffusivity in the gel can be estimated relative to that in water as a function of the solute size and gel mesh size. The hydrophilic properties of PEG will facilitate transport of critical solutes to maintain cell viability.

In many applications, biodegradable or bioerodible photopolymerized hydrogels are required. Sawhney et al.[63] developed photopolymerizable, bioerodible hydrogels based on

PEG-*co*-poly(α-hydroxy acid) diacrylate macromers. As PEG does not degrade in the body naturally, proteolytically degradable, photopolymerized hydrogels have also been developed for tissue engineering applications. Proteolytically degradable peptides that are cleaved by enzymes involved in cell migration, such as collagenase and plasmin, can be copolymerized with PEG as a BAB block copolymer (with the degradable site as the A block), then terminated with acrylate groups. The resultant polymers were found to rapidly degrade in the presence of the targeted protease in a dose-dependent manner, but remained stable in the presence of other proteases.[64] Furthermore, cells were able to degrade these materials during migration.[65] This scheme may allow the rate of material degradation to match that of tissue formation. If the degradable peptide sequence in the hydrogel is a target sequence of matrix metalloproteinases (MMPs), the degradation of the capsule will occur by a mechanism similar to turnover of the ECM.

The major limitation of PEG hydrogels as scaffolds for tissue engineering is lack of cell-specific adhesion. The resistance to protein adsorption creates a blank environment surrounding the encapsulated cells which can trigger anoikis for anchorage-dependent cells. To overcome this limitation, PEG macromers have been modified with integrin-binding moieties found in the ECM surrounding the encapsulated cell to make the hydrogels more biomimetic. For example, hydrogels been modified with a variety of cell adhesion ligands such as RGD, a fibronectin-derived peptide KQAGDV, or a laminin-derived peptide YIGSR to enhance cell adhesion.[66–70] The pathways initiated by integrin binding may also be triggered by genetic modification to extend viability of encapsulated cells.[71]

In combination with stem cells, PEG scaffolds have been evaluated for their suitability as potential replacements for bone, cartilage, nerve, liver, and vascular tissue. A great deal of research has been published using PEG scaffolds seeded with stem cells to generate bone and cartilage.[50,72–78] For bone applications, studies explored using mesenchymal stem cells combined with a variety of cues, including RGD peptides, bone morphogenetic protein (BMP), and heparin, to promote osteogenic differentiation.[50,72–75] Similar approaches were used to produce cartilage by the addition of bioactive molecules such as chondroitin sulfate, transforming growth factor-β, and BMPs.[76–78] For the treatment of Parkinson's disease or spinal cord injury, nerve tissue was engineered using PEG scaffolds.[79–82] PEG scaffolds were used for coculture of neural progenitor cells and endothelial cells to engineer the nerve tissue.[79] The addition of endothelial cells allowed formation of microvasculature.[80] The addition of bFGF (basic fibroblast growth factor) and collagen was used to promote neuronal differentiation of precursor cells when cultured in PEG scaffolds.[81,82]

The permeability and mechanical properties of the hydrogels depend on the length of the PEG chains, the polymer volume fraction in the gel, and the cross-linking density. The cross-linking density of PEG hydrogels can be increased by increasing the concentration of macromer in PEG solutions, decreasing the molecular weight of the PEG macromers, or by using branched PEG structures instead of linear structures, with corresponding increases in compressive modulus.[52,60,83] The PEG macromer has also been modified with fumaric acid to form hydrogels made of oligo(poly(ethylene glycol) fumarate), which are photo-cross-linkable, injectable, and can be prepared with compressive moduli as high as cartilage.[60,84] One challenge is that the mechanical properties of PEG hydrogels formed through photopolymerization (a cytocompatible method commonly used for cell encapsulation) are inhomogeneous when tested at different locations in the material. This increases when cells are present and interact with the scaffold. These issues are discussed in detail in a recent article discussing the use of PEG hydrogels for 3D culture of cells.[85]

42.3.3 CHITOSAN

Chitosan is a natural biopolymer that has many desirable characteristics as a scaffold or encapsulation material. It is a biodegradable, semicrystalline polysaccharide obtained by *N*-deacetylation of chitin, which is harvested from the exoskeleton of marine crustaceans. Chitosan is composed of glucosamine and *N*-acetyl glucosamine which are linked by glycosidic bonds. Being structurally similar to ECM components, chitosan provides cell–ECM interactions which guide cell behavior.

The degree of deacetylation, indicating the free amine groups along the chitosan backbone, is a key parameter, which changes its physicochemical properties such as solubility, chain conformation, and electrostatic properties.[86] Chitosan possesses a number of characteristics which make it suitable for implantation, such as inducing a minimal foreign body reaction, an intrinsic antibacterial nature, and the ability to be molded into highly porous structures of varied geometry suitable for cell growth and osteoconduction.[87,88]

VandeVord et al.[89] examined the biocompatibility of chitosan in mice. Their data imply that chitosan has a chemotactic effect on immune cells, but that effect does not lead to a humoral immune response. Also, results suggest that the specific responses reported may have been caused by contaminating proteins/polysaccharides from the source organism. This type of contamination has been reported with other polysaccharides studied for implant use. As discussed earlier with alginates, the need to use highly purified grades of these biomaterials is critical to avoid toxicity and fibrotic encapsulation.[90,91]

Chitosan exhibits pH-sensitive behavior due to the large quantities of amino groups on its chains. It is a pH-dependent, cationic polymer, which is insoluble in aqueous solutions above pH 7. However, in dilute or weak acids (pH < 6), the protonated free amino groups of glucosamine increase the solubility of the molecule.[92] Above pH 6.2, chitosan solutions form a hydrated gel-like precipitate.[93,94] Due to its cationic nature and predictable degradation rate, chitosan-based materials bind growth factors and release them in a controlled manner.[95] Chitosan is degraded, depending on degree of deacetylation, by enzymes such as lysozyme, N-acetyl-D-glucosaminidase, and lipases.[96] *In vivo*, chitosan is degraded by enzymatic hydrolysis, primarily by lysozyme which appears to target acetylated residues.[97,98] Degradation kinetics seems to be inversely related to the degree of deacetylation.[97] Lysozyme breaks down the chitosan polymer chain, diminishing its molecular weight until it becomes short enough to be processed by cells. Glucosamines, the final degradation products of chitosan, are nontoxic, nonimmunogenic, and noncarcinogenic.[98] *In vivo*, the final degradation products undergo normal metabolism pathways and may be incorporated into glycoproteins or excreted as carbon dioxide gas during respiration.[99,100]

Due to its cationic amine groups, chitosan provides a suitable environment for cell adhesion through interaction with negatively charged ECM components like glycosaminoglycans and proteoglycans.[101] Chondrocytes encapsulated in injectable, chitosan hydrogels repaired non-weight-bearing defects in sheep, with good integration with the surrounding tissue.[102] This can be attributed to chitosan's structural similarity to major components of the ECM of bone and cartilage. To customize the adherent ability for seeding cells, chitosan allows for a wide range of molecules to be introduced within the material. Conjugation of chitosan with biologically active, RGD-containing peptides or laminin peptides allows further customization to develop desirable scaffold materials for tissue regeneration.[103] Kuo and Lin[104] hybridized chitin and chitosan through genipin cross-linking and subsequently freeze-dried the constructs to enhance chondrocytic attachment and growth. The resultant scaffold was then coated with hydroxyapatite to modify the surface chemistry that generated positive effects on the cell number, the content of glycosaminoglycans, and the collagen level for 28-day cultivation of bovine knee chondrocytes.

The cell-binding properties of chitosan have led to numerous studies where chitosan is used as a surface coating. Chitosan was employed as a surface modification of poly(ε-caprolactone) scaffolds and in vitro studies with fibroblasts showed significantly improved cell attachment and proliferation when chitosan was present.[105] In another study, polyurethane scaffolds were prepared and surface-modified with chitosan for the same reason.[106] This study showed that on the modified scaffold a monolayer of endothelial intima was formed. The incorporation of collagen with chitosan as a chitosan–collagen scaffold enhanced the resultant scaffold's ability to support cell attachment and is a strategy similar to the incorporation of peptide sequences.[107]

Chitosan can be incorporated into composite materials to create scaffolds with cellbinding and mechanical properties that vary from pure chitosan. Chitosan cross-linked with collagen has been investigated as a candidate for use as a matrix to support a bioartificial liver.[108] The study showed good compatibility of the scaffold with the hepatocytes as the presence of chitosan provided amino groups for cell adhesion. The composition of the cross-linked chitosan and collagen scaffold is closer to the native tissue and has been shown to exhibit higher mechanical strength than chitosan alone. The addition of collagen creates a scaffold with a tensile strength of 1.91 MPa and a Young's modulus of 7.11 MPa. Hydrated, porous chitosan membranes have a tensile modulus typically below 1 MPa. Zhang et al. reinforced chitosan scaffolds by addition β-tricalcium phosphate (β-TCP) to increase their mechanical strength. Pure chitosan scaffolds were soft, spongy, and very flexible with a maximum strain of ~20% before the loss of elasticity. With the addition of β-TCP, the compressive modulus was increased from 0.967 to 2.292 MPa and yield strength from 0.11 to 0.208 MPa.[109] Chitosan–alginate composites can also be formed due to the negative charge on alginate.[110] The strong ionic bonding between the amine groups of the chitosan and the carboxyl group of the alginate stabilize the scaffold despite a high porosity (~92%). This is achieved while reinforcing the scaffold; it displays a compressive modulus of 8.16 MPa and a yield strength of 0.46 MPa, which is about three times the value for pure chitosan scaffold. Also, cell–material interaction studies indicated that osteoblasts seeded on the chitosan–alginate scaffold attached well and promoted the deposition of minerals for bone formation relative to pure chitosan scaffolds.

42.3.4 Poly(Lactic-co-Glycolic Acid)

Poly(lactic-co-glycolic acid) (PLGA) is a copolymer that consists of monomers of glycolic acid and lactic acid connected by ester bonds. It is an FDA-approved polymer that is attractive for cell encapsulation and as a tissue engineering scaffold. Two key properties are its biocompatibility (as it degrades into natural compounds) and the ability to modulate its degradation rate. This degradation rate controls many of the key properties for use of PLGA with cells. The transport and mechanical properties will vary significantly once PLGA is placed in a physiological environment. This material would only be used for cell encapsulation when immunoisolation and/or cell support is only desired for a period of weeks.

PLGA is hydrolytically unstable, and although insoluble in water, they degrade by hydrolytic attack of their ester bonds resulting in the formation of lactic and glycolic acids, which can be removed from the body by normal metabolic pathways.[111–114] Other factors that affect degradation include hydrophobicity and molecular weight.[115–117] The degradation of PLGA is therefore affected by the ratio of hydrophilic poly(glycolic acid) (PGA) to hydrophobic poly(lactic acid) (PLA).[116] The biocompatibility of these polymers also has been demonstrated in biological applications.[118–121] These reports implied that the rate of degradation might affect cellular interaction including cell proliferation, tissue synthesis, and host response.[122,123] However, details of the potential effects of the acidic by-products on the 3D cell culture or upon in vivo host response have not been studied sufficiently. Conventionally, the rate of hydrolytic degradation for these biopolymers is controlled by altering their physical properties such as their molecular weights, degree of crystallinity, and glass transition temperature.[114,124]

Numerous studies have been conducted using PLGA scaffolds seeded with mesenchymal stem cells to differentiate into osteogenic, cartilage, neural, liver, or adipose cells.[125–133] Studies suggest that PLGA can possibly be used as a bioscaffold which can help in maintaining the viability of the cells as well as help in differentiation of stem cells when incorporated with ligands, peptides, or growth factors. PLGA scaffolds have been modified with bioactive ligands, such as galactose (a specific ligand for the asialoglycoprotein receptor on hepatocytes), to help binding and attachment of cells. Galactosylated PLGA has been processed to examine hepatocyte-specific cellular binding to the surface.[134] The results demonstrated that conjugation of galactose on PLGA supported cell adhesion as well as cell viability as compared to control PLGA. In another study, the RGD peptide was immobilized onto the surface of PLGA for enhancing cell adhesion and function for bone regeneration. The extent of cell adhesion was substantially enhanced when RGD was present and the level of alkaline phosphatase activity, a marker of osteoblast function, was also elevated.[135]

REFERENCES

1. Lim, F., Sun, A. M. Microencapsulated islets as bioartificial endocrine pancreas. *Science.* **1980**, *210*, 908–910.
2. Cruise, G. M., Hegre, O. D., Lamberti, F. V., Hager, S. R., Hill, R., Scharp, D. S., Hubbell, J. A. In vitro and in vivo performance of porcine islets encapsulated in interfacially photopolymerized poly(ethylene glycol) diacrylate membranes. *Cell Transplant.* **1999**, *8*, 293–306.
3. Hernández, R. M., Orive, G., Murua, A., Pedraz, J. L. Microcapsules and microcarriers for in situ cell delivery. *Advanced Drug Delivery Reviews.* **2010**, *62*, 711–730.
4. Giuliani, M., Moritz, W., Bodmer, E., Dindo, D., Kugelmeier, P., Lehmann, R., Gassmann, M., Groscurth, P., Weber, M. Central necrosis in isolated hypoxic human pancreatic islets: Evidence for postisolation ischemia. *Cell Transplant.* **2005**, *14*, 67–76.
5. Lehmann, R., Zuellig, R. A., Kugelmeier, P., Baenninger, P. B., Moritz, W., Perren, A., Clavien, P. A., Weber, M., Spinas, G. A. Superiority of small islets in human islet transplantation. *Diabetes.* **2007**, *56*, 594–603.
6. Du, J., Chen, X., Liang, X., Zhang, G., Xu, J., He, L., Zhan, Q., Feng, X. Q., Chien, S., Yang, C. Integrin activation and internalization on soft ECM as a mechanism of induction of stem cell differentiation by ECM elasticity. *Proceedings of the National Academy of Sciences of the United States of America.* **2011**, *108*, 9466–9471.
7. Parekh, S. H., Chatterjee, K., Lin-Gibson, S., Moore, N. M., Cicerone, M. T., Young, M. F., Simon, C. G. Jr. Modulus-driven differentiation of marrow stromal cells in 3D scaffolds that is independent of myosin-based cytoskeletal tension. *Biomaterials.* **2011**, *32*, 2256–2264.
8. Kumachev, A., Greener, J., Tumarkin, E., Eiser, E., Zandstra, P. W., Kumacheva, E. High-throughput generation of hydrogel microbeads with varying elasticity for cell encapsulation. *Biomaterials.* **2011**, *32*, 1477–1483.
9. Zemel, A., Rehfeldt, F., Brown, A. E., Discher, D. E., Safran, S. A. Cell shape, spreading symmetry and the polarization of stress-fibers in cells. *Journal of Physics: Condensed Matter.* **2010**, *22*, 194110.
10. Engler, A. J., Sen, S., Sweeney, H. L., Discher, D. E. Matrix elasticity directs stem cell lineage specification. *Cell.* **2006**, *126*, 677–689.
11. Shav, D., Einav, S. The effect of mechanical loads in the differentiation of precursor cells into mature cells. *Annals of the New York Academy of Sciences.* **2010**, *1188*, 25–31.
12. Smidsrod, O., Skjak-Baek, G. Alginate as immobilization matrix for cells. *Trends in Biotechnology.* **1990**, *8*, 71–78.
13. Johnson, F. A., Craig, D. Q. M., Mercer, A. D. Characterization of the block structure and molecular weight of sodium alginates. *Journal of Pharmacy and Pharmacology.* **1997**, *49*, 639–643.
14. Thu, B., Smidsrod, O., Skjak-Baek, G. Alginate gels—Some structure–function correlations relevant to their use as immobilization matrix for cells. *Progress in Biotechnology.* **1996**, *11*, 19–30.
15. de Vos, P., Andersson, A., Tam, S. K., Faas, M. M., Halle, J. P. Advances and barriers in mammalian cell encapsulation for treatment of diabetes. *Immunology, Endocrine and Metabolic Agents—Medical Chemistry.* **2006**, *6*, 139–153.
16. Vandenbossche, G. M. R., Remon, J.-P. Influence of the sterilization process on alginate dispersion. *Journal of Pharmacy and Pharmacology.* **1993**, *45*, 484–486.
17. Zimmermann, U., Klock, G., Federlin, K., Hannig, K., Kowalski, M., Bretzel, R. G., Horcher, A., Entenmann, H., Sieber, U., Zekorn, T. Production of mitogen-contamination free alginates with variable ratios of mannuronic acid to guluronic acid by free-flow electrophoresis. *Electrophoresis.* **1992**, *13*, 269–274.
18. LeRoux, M. A., Guilak, F., Setton, L. A. Compressive and shear properties of alginate gel: Effects of sodium ions and alginate concentration. *Journal of Biomedical Materials Research.* **1999**, *47*, 46–53.
19. Ashton, R. S., Banerjee, A., Punyani, S., Schaffer, D. V., Kane, R. S. Scaffolds based on degradable alginate hydrogels and poly(lactide-*co*-glycolide) microspheres for stem cell culture. *Biomaterials.* **2007**, *28*, 5518–5525.
20. Sun, Y., Ma, X., Zhou, D., Vacek, I., Sun, A. M. Normalization of diabetes in spontaneously diabetic cynomologus monkeys by xenografts of microencapsulated porcine islets without immunosuppression. *Journal of Clinical Investigation.* **1996**, *98*, 1417–1422.
21. Awad, H. A., Wickham, M. Q., Leddy, H. A., Gimble, J. M., Guilak, F. Chondrogenic differentiation of adipose-derived adult stem cells in agarose, alginate, and gelatin scaffolds. *Biomaterials.* **2004**, *25*, 3211–3222.
22. Maguire, T., Novik, E., Schloss, R., Yarmush, M. Alginate-PLL microencapsulation: Effect on the differentiation of embryonic stem cells into hepatocytes. *Biotechnology and Bioengineering.* **2006**, *93*, 581–591.

23. Gerecht-Nir, S., Cohen, S., Ziskind, A., Itskovitz-Eldor, J. Three-dimensional porous alginate scaffolds provide a conducive environment for generation of well-vascularized embryoid bodies from human embryonic stem cells. *Biotechnology and Bioengineering*. **2004**, *88*, 313–320.
24. Hannouche, D., Terai, H., Fuchs, J. R., Terada, S., Zand, S., Nasseri, B. A., Petite, H., Sedel, L., Vacanti, J. P. Engineering of implantable cartilaginous structures from bone marrow-derived mesenchymal stem cells. *Tissue Engineering*. **2007**, *13*, 87–99.
25. Dvir-Ginzberg, M., Gamlieli-Bonshtein, I., Agbaria, R., Cohen, S. Liver tissue engineering within alginate scaffolds: Effects of cell-seeding density on hepatocyte viability, morphology and function. *Tissue Engineering*. **2003**, *9*, 757–766.
26. Mosahebi, A., Wiberg, M., Terenghi, G. Addition of fibronectin to alginate matrix improves peripheral nerve regeneration in tissue-engineered conduits. *Tissue Engineering*. **2003**, *9*, 209–218.
27. Mosahebi, A., Fuller, P., Wiberg, M., Terenghi, G. Effect of allogeneic Schwann cell transplantation on peripheral nerve regeneration. *Experimental Neurology*. **2002**, *173*, 213–223.
28. Prang, P., Muller, R., Eljaouhari, A., Heckmann, K., Kunz, W., Weber, T., Faber, C., Vroemen, M., Bogdahn, U., Weidner, N. The promotion of oriented axonal regrowth in the injured spinal cord by alginate-based anisotropic capillary hydrogels. *Biomaterials*. **2006**, *27*, 3560–3569.
29. Johnson, A. S., O'Sullivan, E., D'Aoust, L. N., Omer, A., Bonner-Weir, S., Fisher, R. J., Weir, G. C., Colton, C. K. Quantitative assessment of islets of langerhans encapsulated in alginate. *Tissue Engineering Part C Methods*. **2011**, *17*, 435–449.
30. Smentana, K. Cell biology of hydrogels. *Biomaterials*. **1993**, *14*, 1046–1050.
31. Rowley, J. A., Madlambayan, G., Mooney, D. J. Alginate hydrogels as synthetic extracellular matrix materials. *Biomaterials*. **1999**, *20*, 45–53.
32. Sultzbaugh, K. J., Speaker, T. J. J. A method to attach lectins to the surface of spermine alginate microcapsules based on the avidin biotin interaction. *Microencapsulation*. **1996**, *13*, 363–376.
33. Tampieri, M. S. A., Landi, E., Celotti, G., Roveri, N., Mattioli-Belmonte, M., Virgili, L., Gabbanelli, F., Biagini, G. HA/alginate hybrid composites prepared through bio-inspired nucleation. *Acta Biomaterialia*. **2005**, *1*, 343–351.
34. Draget, K. I., Skjak-Baek, G., Smidsrod, O. Alginate based new materials. *International Journal of Biological Macromolecules*. **1997**, *21*, 47–55.
35. Lee, K. Y., Rowley, J. A., Eiselt, P., Moy, E. M., Bouhadir, K. H., Mooney, D. J. Controlling mechanical and swelling properties of alginate hydrogels independently by cross-linker type and cross-linking density. *Macromolecules*. **2000**, *33*, 4291–4294.
36. de Groot, M., Schuurs, T., van Schilfgaarde, R. Causes of limited survival of microencapsulated pancreatic islet grafts. *Journal of Surgical Research*. **2004**, *121*, 141–150.
37. Martinsen, A., Skjak-Braek, G., Smidsrod, O. Alginate as immobilization material: I. Correlation between chemical and physical properties of alginate gel beads. *Biotechnology and Bioengineering*. **1989**, *33*, 79–89.
38. Grant, G. T., Morris, E. R., Rees, D. A., Smith, P. J. C., Thom, D. Biological interactions between polysaccharides and divalent cations: The egg-box model. *FEBS Letters*. **1973**, *32*, 195–198.
39. Smidsrod, O. Molecular basis for some physical properties of alginates in the gel state. *Journal of Chemical Society, Faraday Transactions*. **1974**, *57*, 263–274.
40. Stabler, C., Wilks, K., Sambanis, A., Constantinidis, I. The effects of alginate composition on encapsulated betaTC3 cells. *Biomaterials*. **2001**, *22*, 1301–1310.
41. Simpson, N. E., Stabler, C. L., Simpson, C. P., Sambanis, A., Constantinidis, I. The role of the $CaCl_2$-guluronic acid interaction on alginate encapsulated betaTC3 cells. *Biomaterials*. **2004**, *25*, 2603–2610.
42. Morch, Y. A., Donati, I., Strand, B. L., Skjak-Baek, G. Effect of Ca^{2+}, Ba^{2+}, and Sr^{2+} on Alginate Microbeads. *Biomacromolecules*. **2006**, *7*, 1471–1480.
43. Benson, J. P., Papas, K. K., Constantinidis, I., Sambanis, A. Towards the development of a bioartificial pancreas: Effects of poly-L-lysine on alginate beads with BTC3 cells. *Cell Transplant*. **1997**, *6*, 395–402.
44. de Vos, P., De Haan, B., van Schilfgaarde, R. Effect of the alginate composition on the biocompatibility of alginate–polylysine microcapsules. *Biomaterials*. **1997**, *18*, 273–278.
45. Darquy, S., Pueyo, M. E., Capron, F., Reach, G. Complement activation by alginate–polylysine microcapsules used for islet transplantation. *Artificial Organs*. **1994**, *18*, 898–903.
46. Pueyo, M. E., Darquy, S., Capron, F., Reach, G. In vitro activation of human macrophages by alginate-polylysine microcapsules. *Journal of Biomaterials Science, Polymer Edition*. **1993**, *5*, 197–203.
47. Strand, B. L., Ryan, T. L., In't Veld, P., Kulseng, B., Rokstad, A. M., Skjak-Brek, G., Espevik, T. Poly-L-lysine induces fibrosis on alginate microcapsules via the induction of cytokines. *Cell Transplant*. **2001**, *10*, 263–275.

48. Lee, K., Mooney, D. Hydrogels for tissue engineering. *Chemical Reviews*. **2001**, *101*, 1869–1877.
49. Hill, R. S., Cruise, G. M., Hager, S. R., Lamberti, F. V., Yu, X., Garufis, C. L., Yu, Y. et al. Immunoisolation of adult porcine islets for the treatment of diabetes mellitus. The use of photopolymerizable polyethylene glycol in the conformal coating of mass-isolated porcine islets. *Annals of the New York Academy of Sciences*. **1997**, *831*, 332–343.
50. Nuttelman, C. R., Tripodi, M. C., Anseth, K. S. In vitro osteogenic differentiation of human mesenchymal stem cells photoencapsulated in PEG hydrogels. *Journal of Biomedical Materials Research*. **2004**, *68*, 773–782.
51. Bryant, S. J., Anseth, K. S. The effects of scaffold thickness on tissue engineered cartilage in photocrosslinked poly(ethylene oxide) hydrogels. *Biomaterials*. **2001**, *22*, 619–626.
52. Bryant, S., Anseth, K. Hydrogel properties influence ECM production by chondrocytes photoencapsulated in poly(ethylene glycol) hydrogels. *Journal of Biomedical Materials Research*. **2002**, *59*, 63–72.
53. Rice, M. A., Anseth, K. S. Encapsulating chondrocytes in copolymer gels: Bimodal degradation kinetics influence cell phenotype and extracellular matrix development. *Journal of Biomedical Materials Research A*. **2004**, *70*, 560–568.
54. Burdick, J. A., Anseth, K. S. Photoencapsulation of osteoblasts in injectable RGD-modified PEG hydrogels for bone tissue engineering. *Biomaterials*. **2002**, *23*, 4315–4323.
55. Masters, K. S., Shah, D. N., Leinwand, L. A., Anseth, K. S. Crosslinked hyaluronan scaffolds as a biologically active carrier for valvular interstitial cells. *Biomaterials*. **2005**, *26*, 2517–2525.
56. Metters, A. T., Anseth, K. S., Bowman, C. N. A statistical kinetic model for the bulk-degradation of PEG-b-PLA hydrogel networks. *Journal of Physical Chemistry B*. **2000**, 104, 7043–7049.
57. Martens, P. J., Bryant, S. J., Anseth, K. S. Tailoring the degradation of hydrogels formed from multivinyl poly(ethylene glycol) and poly(vinyl alcohol) macromers for cartilage tissue engineering. *Biomacromolecules*. **2003**, *4*, 283–292.
58. Temenoff, J. S., Park, H., Jabbari, E., Sheffield, T. L., LeBaron, R. G., Ambrose, C. G., Mikos, A. G. In-vitro osteogenic differentiation of marrow stromal cells encapsulated in biodegradable hydrogels. *Journal of Biomedical Materials Research A*. **2004**, *70*, 235–244.
59. Lowman, A. M. Peppas, N. A., Hydrogels, in *Encyclopedia of Controlled Drug Delivery*, Mathiowitz, E., Ed., Wiley, New York, **1999**, *1*, 397–418.
60. Bryant, S. J., Anseth, K. S., Lee, D. A., Bader, D. L. Crosslinking density influences the morphology of chondrocytes photoencapsulated in PEG hydrogels during the application of compressive strain. *Journal of Orthopedic Research*. **2004**, *22*, 1143–1149.
61. Mason, M. N., Metters, A. T., Bowman, C. N., Anseth, K. S. Predicting controlled-release behavior of degradable PLA-b-PEG-b-PLA hydrogels. *Macromolecules*. **2001**, *34*, 4630–4635.
62. Watkins, A. W. Anseth, K. S. Investigation of molecular transport and distributions in poly(ethylene glycol) hydrogels with confocal laser scanning microscopy. *Macromolecules*. **2005**, *38*, 1326–1334.
63. Sawhney, A. S., Pathak, C. P., Hubble, J. A. Bioerodible hydrogels based on photopolymerized poly(ethylene glycol)-co-poly(a-hydroxy acid) diacrylate macromers. *Macromolecules*. **1993**, *26*, 581–587.
64. West, J. L., Hubbell, J. A. Polymeric biomaterials with degradation sites for proteases involved in cell migration. *Macromolecules*. **1999**, *32*, 241–244.
65. Mann, B. K., Gobin, A. S., Tsai, A. T., Schmedlen, R. H., West, J. L. Smooth muscle cell growth in photopolymerized hydrogels with cell adhesive and proteolytically degradable domains: Synthetic ECM analogs for tissue engineering. *Biomaterials*. **2001**, *22*, 3045–3051.
66. Peyton, S. R., Raub, C. B., Keschrumrus, V. P., Putnam, A. J. The use of poly(ethylene glycol) hydrogels to investigate the impact of ECM chemistry and mechanics on smooth muscle cells. *Biomaterials*. **2009**, *27*, 4881–4893.
67. Hubbell, J. A., Massia, S. P., Desai, N. P., Drumheller, P. D. Endothelial cell-selective materials for tissue engineering in the vascular graft via a new receptor. *Biotechnology*. **1991**, *9*, 568–572.
68. Massia, S. P., Hubbell, J. A. Human endothelial cell interactions with surface-coupled adhesion peptides on a nonadhesive glass substrate and two polymeric biomaterials. *Journal of Biomedical Materials Research*. **1991**, *25*, 223–242.
69. Hern, D. L., Hubbell, J. A. Incorporation of adhesion peptides into nonadhesive hydrogels useful for tissue resurfacing. *Journal of Biomedical Materials Research*. **1998**, *39*, 266–276.
70. Mann, B. K., Tsai, A. T., Scott-Burden, T., West, J. L. Modification of surfaces with cell adhesion peptides alters extracellular matrix deposition. *Biomaterials*. **1999**, *20*, 2281–2286.
71. Blanchette, J. O., Langer, S. J., Sahai, S., Topiwala, P. S., Leinwand, L. L., Anseth, K. S. Use of integrin-linked kinase to extend function of encapsulated pancreatic tissue. *Biomedical Materials*. **2010**, *5*, 061001.

72. Benoit, D. S. W., Anseth, K. S. Heparin functionalized PEG gels that modulate protein adsorption for hMSC adhesion and differentiation. *Acta Biomaterialia.* **2005**, *1*, 461–470.
73. Benoit, D. S. W., Collins, S. D., Anseth, K. S. Multifunctional hydrogels that promote osteogenic human mesenchymal stem cell differentiation through stimulation and sequestering of bone morphogenic protein 2. *Advanced Functional Materials.* **2007**, *17*, 2085–2093.
74. Shin, H., Zygourakis, K., Farach-Carson, M. C., Yaszemski, M. J., Mikos, A. G. Modulation of differentiation and mineralization of marrow stromal cells cultured on biomimetic hydrogels modified with Arg–Gly–Asp containing peptides. *Journal of Biomedical Materials Research.* **2004**, *69*, 535–543.
75. Buxton, A. N., Zhu, J., Marchant, R., West, J. L., Yoo, J. U., Johnstone, B. Design and characterization of poly(ethylene glycol) photopolymerizable semi-interpenetrating networks for chondrogenesis of human mesenchymal stem cells. *Tissue Engineering.* **2007**, *13*, 2549–2560.
76. Varghese, S., Hwang, N. S., Canver, A. C., Theprungsirikul, P., Lin, D. W., Elisseeff, J. Chondroitin sulfate based niches for chondrogenic differentiation of mesenchymal stem cells. *Matrix Biology.* **2008**, *27*, 12–21.
77. Salinas, C. N., Cole, B. B., Kasko, A. M., Anseth, K. S. Chondrogenic differentiation potential of human mesenchymal stem cells photoencapsulated within poly(ethylene glycol)–arginine–glycine–aspartic acid–serine thiol–methacrylate mixed-mode networks. *Tissue Engineering.* **2007**, *13*, 1025–1034.
78. Hwang, N. S., Kim, M. S., Sampattavanich, S., Baek, J. H., Zhang, Z., Elisseeff, J. Effects of three dimensional culture and growth factors on the chondrogenic differentiation of murine embryonic stem cells. *Stem Cells.* **2006**, *24*, 284–291.
79. Royce Hynes, S., McGregor, L. M., Ford Rauch, M., Lavik, E. B. Photopolymerized poly(ethylene glycol)/poly(L-lysine) hydrogels for the delivery of neural progenitor cells. *Journal of Biomaterials Science.* **2007**, *18*, 1017–1030.
80. Ford, M. C., Bertram, J. P., Hynes, S. R., Michaud, M., Li, Q., Young, M., Segal, S. S., Madri, J. A., Lavik, E. B. A macroporous hydrogel for the coculture of neural progenitor and endothelial cells to form functional vascular networks in vivo. *Proceedings of the National Academy of Sciences of the United States of America.* **2006**, *103*, 2512–2517.
81. Mahoney, M. J., Anseth, K. S. Three-dimensional growth and function of neural tissue in degradable polyethylene glycol hydrogels. *Biomaterials.* **2006**, *27*, 2265–2274.
82. Mahoney, M. J., Anseth, K. S. Contrasting effects of collagen and bFGF-2 on neural cell function in degradable synthetic PEG hydrogels. *Journal of Biomedical Materials Research A.* **2007**, *81*, 269–278.
83. Sontjens, S. H., Nettles, D. L., Carnahan, M. A., Setton, L. A., Grinstaff, M. W. Biodendrimer-based hydrogel scaffolds for cartilage tissue repair. *Biomacromolecules.* **2006**, *7*, 310–316.
84. Suggs, L. J., Kao, E. Y., Palombo, L. L., Krishnan, R. S., Widmer, M. S., Mikos, A. G. Preparation and characterization of poly(propylene fumarate-*co*-ethylene glycol) hydrogels. *Journal of Biomaterials Science.* **1998**, *9*, 653–666.
85. Kloxin, A. M., Tibbitt, M. W., Anseth, K. S. Synthesis of photodegradable hydrogels as dynamically tunable cell culture platforms. *Nature Protocols.* **2010**, *5*, 1867–1887.
86. Kumar, M. N. A review of chitin and chitosan applications. *Reactive and Functional Polymers.* **2000**, *46*, 1–27.
87. Kim, I. Y., Seo, S. J., Moon, H. S., Yoo, M. K., Park, I. Y., Kim, B. C., Cho, C. S. Chitosan and its derivatives for tissue engineering applications. *Biotechnological Advances.* **2008**, *26*, 1–21.
88. Aimin, C., Chunlin, H., Juliang, B., Tinyin, Z., Zhichao, D. Antibiotic loaded chitosan bar: An in vitro, in vivo study of a possible treatment for osteomyelitis. *Clinical Orthopaedics and Related Research.* **1999**, *366*, 239–247.
89. VandeVord, P. J., Matthew, H. W., DeSilva, S. P., Mayton, L., Wu, B., Wooley, P. H. Evaluation of the biocompatibility of a chitosan scaffold in mice. *Journal of Biomedical Materials Research.* **2002**, *59*, 585–590.
90. de Vos, P., Wolters, G. H., Fritschy, W. M., van Schilfgaarde, R. Obstacles in the application of microencapsulation in islet transplantation. *International Journal Artificial Organs.* **1993**, *16*, 205–212.
91. de Vos, P., De Haan, B., van Schilfgaarde, R. Effect of the alginate composition on the biocompatibility of alginate–polylysine microcapsules. *Biomaterials.* **1997**, *18*, 273–278.
92. Madihally, S. V., Matthew, H. W. T. Porous chitosan scaffolds for tissue engineering. *Biomaterials.* **1999**, *20*, 1133–1142.
93. Ruel-Gariépy, E., Leroux, J. C. In situ-forming hydrogels—Review of temperature-sensitive systems. *European Journal of Pharmaceutics and Biopharmaceutics.* **2004**, *58*, 409–426.
94. Di Martino, A., Sittinger, M., Risbud, M. V. Chitosan: A versatile biopolymer for orthopaedic tissue engineering. *Biomaterials.* **2005**, *26*, 5983–5990.

95. Muzzarelli, R. A., Mattioli-Belmonte, M., Tietz, C., Biagini, R., Ferioli, G., Brunelli, M. A., Fini, M., Giardino, R., Ilari. P., Biagini, G. *Biomaterials.* **1994**, *15*, 1075–1081.
96. Muzzarelli, R. A. A. Chitins and chitosans for the repair of wounded skin, nerve, cartilage and bone. *Carbohydrate Polymers.* **2009**, *76*, 167–182.
97. Tomihata, K., Ikada, Y. In vitro and in vivo degradation of films of chitin and its deacetylated derivatives. *Biomaterials.* **1997**, *18*, 567–575.
98. Muzzarelli, R. A. Human enzymatic activities related to the therapeutic administration of chitin derivatives. *Cellular and Molecular Life Sciences.* **1997**, *53*, 131–140.
99. Abarrategi, A., Civantos, A., Ramos, V., Sanz Casado, J. V., López-Lacomba, J. L. Chitosan film as rhBMP-2 carrier: Delivery properties for bone tissue application. *Biomacromolecules.* **2008**, *9*, 711–718.
100. Ma, J., Wang, H., He, B., Chen, J. A preliminary in vitro study on the fabrication and tissue engineering applications of a novel chitosan bilayer material as a scaffold of human neofetal dermal fibroblasts. *Biomaterials.* **2001**, *22*, 331–336.
101. Zhang, Z., Wang, S., Tian, X., Zhao, Z., Zhang, J., Lv, D. A new effective scaffold to facilitate peripheral nerve regeneration: Chitosan tube coated with maggot homogenate product. *Medical Hypotheses.* **2010**, *74*, 12–14.
102. Hao, T., Wen, N., Cao, J. K., Wang, H. B., Lu, S. H., Liu, T., Lin, Q. X., Duan, C. M., Wang, C. Y. The support of matrix accumulation and the promotion of sheep articular cartilage defects repair in vivo by chitosan hydrogels. Osteoarthritis Cartilage/OARS, *Osteoarthritis Research Society.* **2010**, *18*, 257–265.
103. Mochizuki, M., Kadoya, Y., Wakabayashi, Y., Kato, K., Okazaki, I., Yamada, M., Sato, T., Sakairi, N., Nishi, N., Nomizu, M. Laminin-1 peptide-conjugated chitosan membranes as a novel approach for cell engineering. *FASEB Journal.* **2003**, *17*, 875–877.
104. Kuo, Y. C., Lin, C. Y. Effect of genipin-crosslinked chitin–chitosan scaffolds with hydroxyapatite modifications on the cultivation of bovine knee chondrocytes. *Biotechnology and Bioengineering.* **2006**, *95*, 132–144.
105. Mei, N., Chen, G., Zhou, P., Chen, X., Shao, Z. Z., Pan, L. F., Wu, C. G. Biocompatibility of poly(ε-caprolactone) scaffold modified by chitosan—The fibroblasts proliferation in vitro. *Journal of Biomaterials Applications.* **2005**, *19*, 323–339.
106. Zhu, Y., Gao, C., He, T., Shen, J. Endothelium regeneration on luminal surface of polyurethane vascular scaffold modified with diamine and covalently grafted with gelatin. *Biomaterials.* **2004**, *25*, 423–430.
107. Cuy, J. L., Beckstead, B. L., Brown, C. D., Hoffman, A. S., Giachelli, C. M. Adhesive protein interactions with chitosan: Consequences for valve endothelial cell growth on tissue-engineering materials. *Journal of Biomedical Materials Research A.* **2003**, *67*, 538–547.
108. Wang, X. H., Li, D. P., Wang, W. J., Feng, Q. L., Cui, F. Z., Xu, Y. X., Song, X. H., van der Werf, M. Crosslinked collagen/chitosan matrix for artificial livers. *Biomaterials.* **2003**, *24*, 3213–3220.
109. Zhang, Y., Zhang, M. Microstructural and mechanical characterization of chitosan scaffolds reinforced by calcium phosphates. *Journal of Non-Crystalline Solids.* **2001**, *282*, 159–164.
110. Li, Z., Ramay, H. R., Hauch, K. D., Xiao, D., Zhang, M. Chitosan–alginate hybrid scaffolds for bone tissue engineering. *Biomaterials.* **2005**, *26*, 3919–3928.
111. Griffith, L. G. Polymeric biomaterials. *Acta Materialia.* **2000**, *48*, 263–277.
112. Tice, T. R., Tabibi, E. S., Parenteral drug delivery: Injectables. *Treatise on Controlled Drug Delivery: Fundamentals Optimization, Applications*, Marcel Dekker, New York, **1991**, 315–339.
113. Wu, X. S., *Encyclopedic Hand Book of Biomaterials, Bioengineering*, Marcel Dekker, New York, **1995**, 1015–1054.
114. Lewis, D. H., *Biodegradable Polymers as Drug Delivery Systems*, Marcel Dekker, New York, **1990**, 1–41.
115. Lu, L., Garcia, C. A., Mikos, A. G. In vitro degradation of thin poly(DL-lactic-co-glycolic acid) films. *Journal of Biomedical Materials Research.* **1999**, *46*, 236–244.
116. Lu, L., Peter, S. J., Lyman, M. D., Lai, H.-L., Leite, S. M., Tamada, J. A., Uyama, S., Vacanti, J. P., Langer, R., Mikos, A. G. In vitro and in vivo degradation of porous poly (DL-lactic-co-glycolic acid) foams. *Biomaterials.* **2000**, *21*, 1837–1845.
117. Miller, R. A., Brady, J. M., Cutright, D. E. Degradation rates of oral resorbable implants (polylactates and polyglycolates): Rate modification with changes in PLA/PGA copolymer ratio. *Journal of Biomedical Materials Research.* **1977**, *11*, 711–719.
118. Zentner, G. M., Rathi, R., Shih, C., McRea, J. C., Seo, M. H., Oh, H., Rhee, B. G., Mestecky, J., Moldoveanu, Z., Morgan, M., Weitman, S. Biodegradable block copolymers for delivery of proteins and water-insoluble drugs. *Journal of Controlled Release.* **2001**, *72*, 203–215.

119. Hasirci, V., Lewandrowski, K., Gresser, J. D., Wise, D. L., Trantolo, D. J. Versatility of biodegradable biopolymers. Degradability and an in vivo application. *Journal of Biotechnology*. **2001**, *86*, 135–150.
120. Kweon, H., Yoo, M. K., Park, I. K., Kim, T. H., Lee, H. C., Lee, H. S., Oh, J. S., Akaike, T., Cho, C. S. A novel degradable polycaprolactone networks for tissue engineering. *Biomaterials*. **2003**, *24*, 801–808.
121. Rizzi, S. C., Heath, D. J., Coombes, A. G., Bock, N., Textor, M., Downes, S. Biodegradable polymer/hydroxyapatite composites: Surface analysis and initial attachment of human osteoblasts. *Journal of Biomedical Materials Research*. **2001**, *55*, 475–486.
122. Babensee, J. E., Anderson, J. M., McIntire, L. V., Mikos, A. G. Host response to tissue engineered devices. *Advanced Drug Delivery Reviews*. **1998**, *33*, 111–139.
123. Lewandrowski, K. U., Grosser, J. D., Wise, D. L., Trantolo, D. J., Hasirci, V. Tissue responses to molecularly reinforced polylactide–coglycolide implants. *Journal of Biomaterials Science, Polymer Edition*. **2000**, *11*, 401–414.
124. Cohen, S., Alonso, M. J., Langer, R. Novel approaches to controlled release antigen delivery. *International Journal Technology Assessment in Health Care*. **1994**, *10*, 121–130.
125. Chastain, S. R., Kundu, A. K., Dhar, S., Calvert, J. W., Putnam, A. J. Adhesion of mesenchymal stem cells to polymer scaffolds occurs via distinct ECM ligands and controls their osteogenic differentiation. *Journal of Biomedical Materials Research*. **2006**, *78*, 73–85.
126. Graziano, A., d'Aquino Cusella-De, R., Angelis, M. G., Laino, G., Piattelli, A., Pacifici, M., De Rosa, A., Papaccio, G. Concave pit-containing scaffold surfaces improve stem cell-derived osteoblast performance and lead to significant bone tissue formation. *PLoS ONE*. **2007**, *2*, e496.
127. Kim, H., Kim, H. W., Suh, H. Sustained release of ascorbate-2-phosphate and dexamethasone from porous PLGA scaffolds for bone tissue engineering using mesenchymal stem cells. *Biomaterials*. **2003**, *24*, 4671–4679.
128. Sun, H., Qu, Z., Guo, Y., Zang, G., Yang, B. In vitro and in vivo effects of rat kidney vascular endothelial cells on osteogenesis of rat bone marrow mesenchymal stem cells growing on polylactide–glycoli acid (PLGA) scaffolds. *Biomedical Engineering Online*. **2007**, *6*, 41.
129. Yoon, E., Dhar, S., Chun, D. E., Gharibjanian, N. A., Evans, G. R. In vivo osteogenic potential of human adipose-derived stem cells/poly lactide-*co*-glycolic acid constructs for bone regeneration in a rat critical-sized calvarial defect model. *Tissue Engineering*. **2007**, *13*, 619–627.
130. Levenberg, S., Huang, N. F., Lavik, E., Rogers, A. B., Itskovitz-Eldor, J., Langer, R. Differentiation of human embryonic stem cells on three-dimensional polymer scaffolds. *Proceedings of the National Academy of Sciences of the United States of America*. **2003**, *100*, 12741–12746.
131. Teng, Y. D., Lavik, E. B., Qu, X., Park, K. I., Ourednik, J., Zurakowski, D., Langer, R., Snyder, E. Y. Functional recovery following traumatic spinal cord injury mediated by a unique polymer scaffold seeded with neural stem cells. *Proceedings of the National Academy of Sciences of the United States of America*. **2002**, *99*, 3024–3029.
132. Neubauer, M., Hacker, M., Bauer-Kreisel, P., Weiser, B., Fischbach, C., Schulz, M. B., Goepferich, A., Blunk, T. Adipose tissue engineering based on mesenchymal stem cells and basic fibroblast growth factor in vitro. *Tissue Engineering*. *11*, 1840–1851.
133. Bhang, S. H., Lim, J. S., Choi, C. Y., Kwon, Y. K., Kim, B. S. The behavior of neural stem cells on biodegradable synthetic polymers. *Journal of Biomaterial Science, Polymer Edition*. **2007**, *18*, 223–239.
134. Yoon, J. J., Nam, Y. S., Kim, J. H., Park, T. G. Surface immobilization of galactose onto aliphatic biodegradable polymers for hepatocyte culture. *Biotechnology and Bioengineering*. **2002**, *78*, 1–10.
135. Yoon, J. J., Song, S. H., Lee, D. S., Park, T. G. Immobilization of cell adhesive RGD peptide onto the surface of highly porous biodegradable polymer scaffolds fabricated by a gas foaming/salt leaching method. *Biomaterials*. **2004**, *25*, 5613–5620.

43 Cell Immobilization Technologies for Applications in Alcoholic Beverages

Argyro Bekatorou, Stavros Plessas, and Athanasios Mallouchos

CONTENTS

43.1 Introduction	933
43.2 Cell Immobilization Techniques	934
43.2.1 Immobilization on a Solid Carrier Surface	934
43.2.2 Immobilization by Entrapment in a Porous Matrix	936
43.2.3 Carrier-Free Immobilization	938
43.2.4 Mechanical Containment behind a Barrier	938
43.3 Effect of Immobilization on Microbial Cells	939
43.3.1 Effect of Immobilization on Product Flavor	939
43.3.2 Low-Temperature Fermentation with Immobilized Cells	940
43.4 Wine Making by Immobilized Cells	941
43.4.1 Malolactic Fermentation of Wine by Immobilized Cells	942
43.5 Brewing by Immobilized Cells	943
43.6 Ethanol and Distillates Production by Immobilized Cells	946
43.7 Cider Making by Immobilized Cells	947
43.8 Production of Cheese Whey–Based Beverages by Immobilized Cells	948
43.9 Conclusions	949
References	949

43.1 INTRODUCTION

Whole cell immobilization is defined as "the physical confinement or localization of intact cells to a certain region of space with preservation of some desired catalytic activity."[1,2] Immobilization resembles the conditions in which microbial cells are found in nature, taking into account that they are usually found adhered and growing on different kinds of surfaces. The considerable research and industrial interest in the use of immobilized cells for alcoholic and other food- and fuel-related fermentation applications is due to the numerous advantages that such technologies offer compared to conventional free cell systems. Specifically, the advantages of immobilized cells for alcoholic beverages production include: (1) achievement of higher cell densities in the bioreactors, and therefore increased substrate uptake, higher productivities, and shorter process times, (2) protection against shear forces and stress (pH, temperature, substrate concentration and end-product inhibition, presence of heavy metals, etc.), leading to extended operational stability of the biocatalyst, (3) feasibility of continuous processing (Figure 43.1), easy product recovery, and reusability of the biocatalyst, (4) feasibility of low-temperature fermentation, which can lead to improved product quality, (5) reduction of secondary fermentation (maturation) times, (6) reduced contamination risk due to the higher cell densities and increased fermentation activity, (7) reduction of investment and energy costs due to the construction of smaller bioreactors, less separation and filtration requirements, and higher productivities.[2–6]

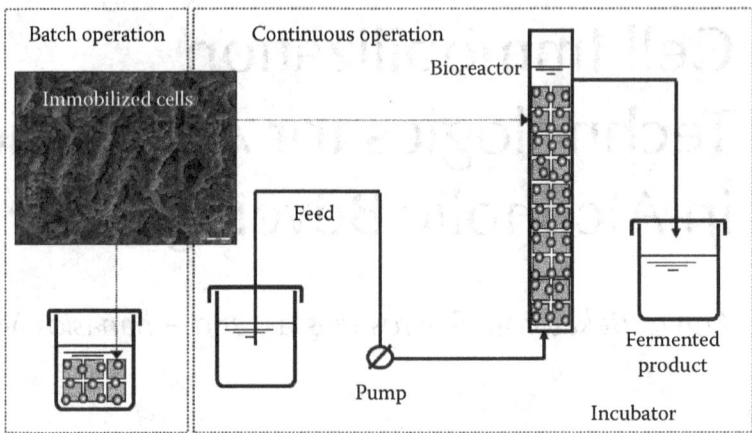

FIGURE 43.1 Batch and continuous fermentation by immobilized cells.

Various materials have been proposed as carriers for cell immobilization of wine, beer, cider, distillates, and ethanol production, in various process designs and bioreactor configurations. Carriers include organic and inorganic natural or synthesized products as well as waste materials. Organic materials can be synthetically made (e.g., synthetic polymers) or extracted from natural sources (e.g., cellulose, polymeric hydrogels, etc.). Natural carriers can be used with minimum or no pretreatment, and can be products such as wood, pieces, and parts of fruit, etc., or wastes and by-products of food-grade purity, such as sawdust, spent grains, crop residues, etc. A carrier is suitable for cell immobilization for use in alcoholic beverages production, when various prerequisites are satisfied such as: (1) good stability against enzymes, temperature, solvents, pressure, and shearing forces, (2) large surface area, (3) porous structure or functional groups for cells to adhere to, (4) ease of handling and regeneration, (5) ability to protect cells and extend their viability and activity, (6) allowing easy substrate and product transfer, and (7) the immobilization technique should be easy, cost-effective, and suitable for scale-up. Last but not least, a suitable cell immobilization carrier should not affect product quality by remaining residues and be readily accepted by consumers.[2,4]

Nevertheless, industrial use of immobilized cells is still limited and it will depend on the development of processes that can be readily scaled up.[2] Data available in the literature on materials and techniques used for viable cell immobilization for application in alcoholic beverages and food-grade ethanol production are highlighted and discussed.

43.2 CELL IMMOBILIZATION TECHNIQUES

Cell immobilization has been used in all types of alcoholic beverages production and various other biotechnological processes, and therefore, many such techniques have been developed, which can be grouped into the following four major categories:[1,2,7–9] (1) immobilization on a solid carrier surface, (2) immobilization by entrapment in a porous matrix, (3) carrier-free immobilization, and (4) containment behind barriers.

43.2.1 Immobilization on a Solid Carrier Surface

Cell immobilization on a solid carrier can be done by physical adsorption due to electrostatic forces or by covalent binding between the cell membrane and the carrier. Immobilization can also be done by growth of the cells into natural cavities on a surface, and therefore, containment of the cells due to entrapment and/or a combination with electrostatic and other weak forces (Figure 43.2). The strength with which the cells are bonded to the carrier as well as the depth of the biofilm varies

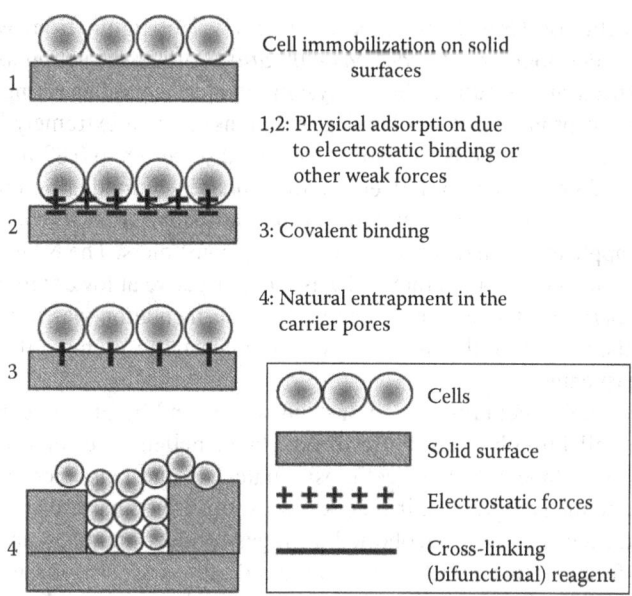

FIGURE 43.2 Basic types of cell immobilization on solid carrier surfaces. (Adapted from Kourkoutas, Y. et al., *Food Microbiol.*, 21(4), 377, 2004.)

depending on the cell strain and the nature of the carrier surface.[2] In this type of immobilization, the adhered viable cells may grow and escape, depending on the conditions applied, and therefore may be in equilibrium with free cells in suspension. Such immobilized cell systems have been extensively used mainly due to the ease of the immobilization technique, which in some cases can be achieved by simple contact of a cell suspension with the carrier for a small time period (Figure 43.3). Examples of solid carriers used in this type of immobilization are cellulosic materials such as DEAE-cellulose, wood, sawdust, delignified sawdust, cereal bran, etc., and inorganic materials such as polygorskite, montmorilonite, hydromica, porous porcelain, porous glass, pumice stone, etc. Solid materials like glass or cellulose can also be treated with various chemicals (polycations, chitosan, etc.) to improve their characteristics as cell binding carriers.[2]

Cellulosic materials are very popular as immobilization carriers for alcoholic beverage production, due to their food-grade purity, low cost, and availability all year round. Delignified cellulosic materials (DCMs) have been successfully used as carriers for the development of immobilized cell biocatalysts for use in various bioprocesses related to food and fuel industries such as alcoholic and lactic acid fermentations for alcoholic beverages and dairy products production.[2,10,11]

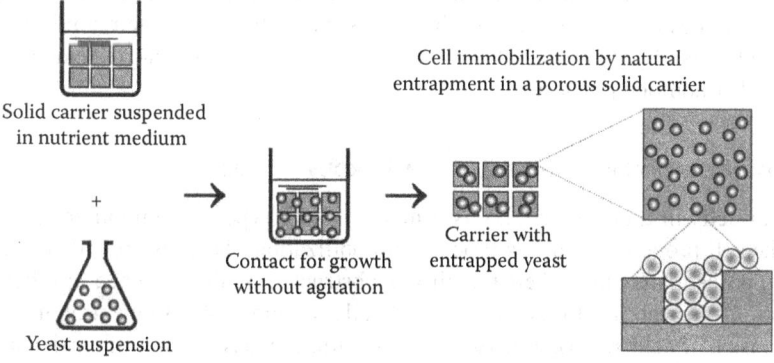

FIGURE 43.3 A simple technique for cell immobilization by natural entrapment on a porous solid carrier.

Nano/microporous cellulose (NMC) prepared after removal of lignin from wood cellulose was found suitable for the development of *"cold pasteurization"* processes acting as a biofilter for cell removal. It was also used successfully as biocatalyst in food fermentations acting as both cell immobilization carrier and as promoter of biochemical reactions, even at extremely low temperatures.[10] The cumulative surface area of the NMC pores was found to be 0.8 to 0.89 $m^2\ g^{-1}$ as indicated by porosimetry analysis. This surface is relatively small compared with other porous materials such as γ-alumina; however, using a natural organic material is attractive from the point of view that it is safer for bioprocess applications and is better accepted by consumers. The NMC/immobilized yeast biocatalyst increased the fermentation rate and was more effective at lower temperatures compared with free cells. Furthermore, the activation energy E_a of fermentation was found to be 28% lower than that of free cells, indicating that it is an excellent material to promote the catalytic action of cells for alcoholic fermentation.

Other natural materials, including agroindustrial wastes and by-products, used as carriers for this type of viable cell immobilization are dried gluten pellets, brewer's spent grains, whole cereal grains, composite biocatalysts of cellulosic materials and gels, pieces and parts of dried or fresh fruit, grape stems and grape skins, cork pieces, olive pits, natural sponges, etc. All these materials were used successfully in alcoholic beverages production at research level (wine making and malolactic fermentation (MLF), brewing, distillates, and whey-based alcoholic drinks), showing a good potential of larger-scale application due to advantages such as their low cost, easy industrial preparation, good operational stability, improved quality of the produced beverages compared to free cells, and easier acceptance by consumers. To facilitate commercialization for industrial or home-scale fermentations, research on the development of active dried ready-to-use immobilized biocatalysts was carried out, mainly aiming to optimize freeze-drying or simpler, mild, and cost-effective thermal drying.[2] For example, no protecting medium was needed for freezing and freeze-drying of *Saccharomyces cerevisiae* cells immobilized on DCM and gluten pellets. The immobilized biocatalysts retained their viability during storage and showed high productivity and stability for glucose, wine, and beer fermentations,[12–15] leading to products of similar quality to those produced by fresh immobilized cells and of improved quality compared to free cells.

Inorganic materials are to some extent advantageous for use as yeast immobilization carriers for alcoholic fermentation, because they are usually abundant and cheap materials that can improve productivity and in most cases product flavor. They can withstand shear forces; they can be easily recovered and reused; and therefore, they can facilitate the development of large-scale continuous operations. Porous inorganic materials, like γ-alumina pellets and the mineral kissiris (a cheap, porous volcanic rock found in Greece, which contains mainly 70% SiO_2), ceramic chamotte (clay), hydroxylapatite, etc., have been evaluated as carriers for yeast immobilization.[2,16] They were found suitable for ambient and low-temperature fermentation, increasing both ethanol productivity and biocatalytic stability of the yeast and leading to products of improved aroma. However, their composition (that liberates mineral residues in the processed media) limits their applications related to food. On the other hand, they are suitable for distillates, potable and fuel-grade alcohol production, biodiesel, or other nonfood purposes.[2]

43.2.2 Immobilization by Entrapment in a Porous Matrix

Entrapment of cells in a porous matrix is a more definite type of immobilization that does not depend on the cell properties.[9] In this type of immobilization, the cells are either allowed to penetrate into a porous matrix until their mobility is obstructed by the presence of other cells, or the porous material is formed in situ into a culture of cells (Figure 43.4). Both entrapment methods are based on inclusion of cells in a rigid network, which allows mass transfer of nutrients and metabolites. At research level, this approach for cell immobilization has been by far the most popular for various applications, including alcoholic beverages production, with main advantage that it is a

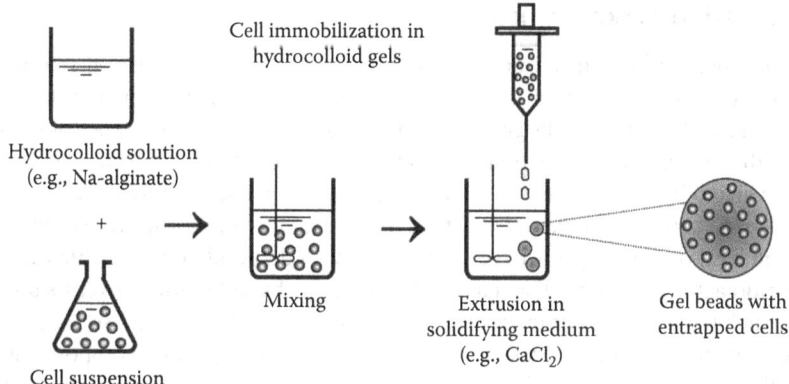

FIGURE 43.4 The common extrusion technique for immobilized cells/polysaccharide gel beads generation.

simple technique that proceeds under very mild conditions and is therefore compatible with most living cells. The most frequently used than any other materials for this application are hydrogels of natural polysaccharides found widely in nature as constituents of cell walls of plants, crustaceans or insects, such as alginate salts, cellulose, k-carrageenan, agar, pectic acid, and chitosan, or synthetic polymeric matrixes such as polyacrylamide. Other examples include gelatin, collagen, and polyvinyl alcohol (PVA).[2,9,17–19]

Entrapment of cells in Ca-alginate beads is the most widely used and studied technique for immobilizing living cells, although it has some limitations such as (1) low stability (e.g., disruption by substances, which have a high affinity for Ca^{2+} such as phosphate or citrate), (3) high porosity, which although allows easy mass transfer, it may lead to leakage of nutrients and limit their use to cells, and (3) wide pore size distribution, which makes controlled release difficult.[19] Cell growth in the gel depends initially on diffusion limitations (substrates, metabolites, and oxygen), affected by the porosity of the material, and later on the impact of the accumulated cell mass. Due to these limitations, cells near the outer surface may behave differently compared to cells inside the beads; they may multiply and be released from the beads.[4] In that case, the fermentation system would comprise of immobilized cells and free cells in suspension. To avoid this problem, double-layer beads have been developed where the external layer prevents the cells from escaping.[2,20] Composites of alginate gels with other organic or inorganic materials have also been developed for beverage fermentations with more than one entrapped microbial species. For example, alginate beads were used to entrap *S. cerevisiae* or *O. oeni* cells functioning as a protective environment for the deposition of silica gel membranes.[21] The biocatalyst was evaluated for both alcoholic fermentation of glucose and MLF. Bacterial cellulose-alginate (BCA) composite sponge was used as a yeast carrier for ethanol production.[22] The biocatalyst had an asymmetric structure, with a thin and dense outer layer covering a macroporous interior, which was effective for yeast immobilization. Kregiel et al.[16] immobilized yeasts in foamed alginate gels to study the induced growth, vitality, and metabolic activity alterations, as well as the potential for fermentation applications. Scale-up of ethanol production from molasses using yeast immobilized in alginate-based microporous and mesoporous zeolite composites was also attempted.[23]

The enumeration of biomass entrapped in a gel matrix is critical for application of biotechnological processes using viable immobilized cells. Such methods usually are gravimetric or include determination of proteins, DNA, NADH, and ATP, which are expressed as biomass concentration. For example, a reliable, accurate, and rapid luminometric method (ATP determination) was developed for estimating active biomass of brewing, wine-making and ethanol-producing yeast strains, immobilized in alginate, pectate, and k-carrageenan hydrogels.[24]

43.2.3 CARRIER-FREE IMMOBILIZATION

Carrier-free immobilization can be done by self-aggregation by flocculation (natural) or artificially induced using cross-linking agents. Cell flocculation has been defined by many authors as an aggregation of cells to form a larger unit or the property of cells in suspensions to adhere in clumps and sediment rapidly.[2,25] Flocculation can be considered as an immobilization technique, since the large size of the aggregates facilitates their potential use in bioreactors, such as packed-bed, fluidised-bed, and continuous stirred-tank reactors. It is the most simple and inexpensive immobilization method, although it is very difficult to predict and control, which is essential for maximizing bioreactor efficiency. The natural flocculation characteristics of yeasts are affected by many factors such the genetic characteristics of the strain, the structure and surface charges of the cell wall, the growth phase, the process temperature, the medium pH and composition, etc.[8] Weak flocculation ability may result in cell washout of the bioreactor, resulting in low cell concentration and therefore insufficient fermentation rates. Artificial flocculating agents or cross-linkers can be used to enhance aggregation in cell cultures that do not naturally flocculate. On the other hand, high flocculation activity may result in low concentrations of active cells due to nutrient diffusion limitations to the cells in the core of large aggregates.[8] Yeast flocculation is a property of major importance for the brewing industry as it affects fermentation productivity and beer quality in addition to yeast removal and recovery. The importance of flocculation properties of *S. cerevisiae* for alcoholic beverage production and mechanisms and factors affecting it has been widely reviewed.[2,8,25,26]

A new trend of carrier-free immobilization is the coimmobilization of cells in the form of biocapsules. In this immobilization technique, the immobilization matrix is provided by one of the microorganisms to be immobilized without need for an external support and the associated costs. For example, a special procedure to immobilize *S. cerevisiae* on the filamentous fungus *Penicillium chrysogenum* (GRAS) was developed via a simple and inexpensive, natural coimmobilization process, for use in ethanol production and wine making.[27–29] The resulting biocapsules were spherical, hollow, smooth, elastic, and strong, and were used with no loss of integrity during the fermentation processes. The fungus died during the fermentation process, remaining as an inert yeast carrier, facilitating reuse of the biocapsules.

43.2.4 MECHANICAL CONTAINMENT BEHIND A BARRIER

Containment of cells behind a barrier can be achieved either by the use of semipermeable membranes, to isolate the cells from the bulk liquid, or by entrapment of cells in a microcapsule, or by cell immobilization on the interaction surface of two immiscible liquids. The cells can be immobilized on the membrane (as in the case of biosensors), or they can be allowed to grow into a void enclosed by the membrane (as in the case of membrane reactor systems). However, growth must be controlled to prevent excessive build-up of biomass that could cause membrane rupture.[2,8,9] This type of immobilization is ideal when cell-free product and minimum transfer of compounds are required,[17] and it is widely used in cell recycling and continuous processes.[30,31] In wine making, constructions using yeasts confined by microfiltration membranes have been developed and are available in the market, such as the "Millispark" cartridge, which was developed for secondary fermentation of sparkling wine inside the bottle.[2] The major disadvantages of membrane immobilization techniques are mass transfer limitations (supply of oxygen and nutrients to the cells and the removal of carbon dioxide), and possible membrane plugging caused by cell growth.[9,32,33]

An important technique of cell entrapment behind a barrier is microencapsulation, which involves inclusion of cells in polymeric microspheres of 1 to 1000 μm size range. This technique has numerous advantages such as higher cell densities and increased cell survival and productivity.[34] Microencapsulation has found application in various biotechnological processes such as encapsulation of probiotics for food production, development of encapsulated biocatalysts for fermentation processes, and environmental bioremediation. The microspheres have larger specific surface area to allow good diffusion of nutrients and metabolites. Additionally, microencapsulation allows easy separation of cells during fermentation processes and minimizes cell wash-out, but the microspheres must be mechanically strong to withstand shear forces and other destructive conditions such as exposure to acids, gases, and solvents.[34]

43.3 EFFECT OF IMMOBILIZATION ON MICROBIAL CELLS

Alterations in cell growth, physiology, and metabolic activity may be induced by cell immobilization, such as effects on the energetic metabolism activation, targeted protein expression to support the altered metabolic behavior of immobilized cells, altered growth rates, increased substrate uptake and product yield, increase in storage polysaccharides, altered yield of fermentation by-products including flavor compounds, higher intracellular pH values, increased tolerance against toxic and inhibitory compounds, increased hydrolytic enzyme activities, modifications in the nucleic acid contents, etc. Parameters that have been considered responsible for these alterations include mass transfer limitations, disturbances in the growth pattern, surface tension and osmotic pressure effects, reduced water activity, cell-to-cell communication, changes in the cell morphology, altered membrane permeability, etc.[2,5,35–37] Generally, it is considered that it is the microenvironment inside the immobilization matrix that affects the physiology and metabolic behavior of yeasts and not the nature of the matrix.[38] For example, the insufficient space for growth in the immobilization matrix was considered responsible for differences in the morphology between free and immobilized yeast cells, while the increased viability and activity of the immobilized cells during storage has been attributed to the protective effect of the matrix.[36,37,39] Altered intracellular pH values observed between free and immobilized yeast were attributed to the increased permeability of the cell membranes to protons, higher ATP consumption, increased glucose uptake, and increased glycolytic activity.[2] The increased ethanol tolerance and lower substrate inhibition of immobilized yeasts was attributed to the protective effect of the matrix or to modified structural features that affect permeability, such as the fatty acid composition of cell membranes. Additionally, osmotic stress caused by the immobilization techniques was found to lead to intracellular production of pressure regulating compounds such as polyols, which lead to decreased water activity and consequently higher tolerance to toxic compounds.[2,36,37] Finally, due to these reasons, cells immobilized in various types of carriers showed enhanced viability and stability during freezing, freeze-drying, and thermal drying, which was exploited for the development of ready-to-use dried biocatalysts for commercial distribution regarding alcoholic fermentation applications.[12–15,40–42]

43.3.1 Effect of Immobilization on Product Flavor

The flavor of fermented foods depends highly on the metabolic activity of the culture used. Especially, the amino acid and the lipid metabolism in yeasts have a crucial contribution to flavor, because they are linked to the production of flavor-active compounds such as esters, alcohols, carbonyl compounds, fatty acids, and nonvolatile components.[2,43] The increased ester and decreased fusel alcohol (mainly amyl alcohols) formation that has been observed during fermentations using immobilized cells, or the improved ratios of esters to higher alcohols, especially at low temperatures, is considered to have a great impact on beverage quality and technology.[2,44,45]

The aroma of wine is the result of a complex combination of varietal, fermentation-, and maturation-derived compounds that give each wine its distinctive character. A considerable fraction of these aroma-related compounds is produced during the primary fermentation of grape must. These compounds are mainly acetate esters of higher alcohols, ethyl esters of fatty acids, higher alcohols, fatty acids, ketones and aldehydes, and sulfur compounds.[45,46] Among these, esters are considered particularly important for wine aroma due to their characteristic fruity and floral odors and because their concentrations usually exceed their odor threshold values in wines. Higher alcohols have higher threshold values and contribute mainly to the complexity of wine aroma at concentrations lower than 400 mg L^{-1}, above which they are considered off-flavors.[47] Among carbonyl compounds, acetaldehyde, smelling like green apple, freshly cut grass, or green leaves, has a great impact on wine quality. In most wines, above threshold values, it is considered an off-odor. Finally, among fatty acids, the ones that mainly affect wine aroma are acetic, propanoic, butanoic, 3-methylbutanoic, hexanoic, octanoic, and decanoic acids.[48] The concentrations of these compounds are affected by any factor that affects the progress of fermentation, such as the yeast strain used, immobilization technique, fermentation temperature, composition of grape must, dissolved oxygen levels, etc. Most of the research works investigating the effect of fermentation by immobilized cells on wine aroma focus on the analysis of the major volatile compounds in wines (acetaldehyde, ethyl acetate, propanol, isobutanol, and amyl alcohols) as well as minor volatile compounds with low threshold values, using various gas chromatography techniques. To evaluate the effect of cell immobilization and other process parameters on wine flavor, the analytical results are compared with sensory evaluations by expert panelists.[45]

Beer is produced by a number of complex biochemical (germination, enzyme hydrolysis of carbohydrates and proteins, etc.) and technological processes (malting, kilning, mashing, fermentation, maturation), which all affect the flavor of the final product. The fermentation stage only contributes more than 600 flavor-active compounds in beer.[49] The flavor of beer is affected by the same groups of compounds as in wine and other fermented products, along with carbon dioxide, ethanol, and glycerol that control the overall effect of the minor constituents. Amino acid metabolism is also a key to the formation of the mentioned compounds and since it is affected by immobilization technology as mentioned earlier, this technology has become interesting for controlling or altering flavor, leading to the production of beers with characteristic flavor profiles.[2,50,51] A few but very important changes also occur during the maturation (lagering) stage of the brewing process. One of the key compounds in beer maturation is diacetyl, a compound with an undesired butter flavor above threshold values. Diacetyl is produced from α-acetolactate by an oxidative nonenzymatic reaction and is slowly converted to the flavorless derivatives acetoin and 2,3-butanediol by yeast metabolism during the maturation stage. This process, known as "diacetyl rest," is time consuming and energy demanding, since it must take place at very low temperatures (~0°C) to avoid degradation of product quality.[52] The combination of yeast immobilization and low-temperature fermentation can reduce primary fermentation time, accelerate the removal of diacetyl (maturation), and improve overall product quality. Therefore, it is has been reported as a promising technological strategy for significant reduction of production costs.[2,18,53]

43.3.2 Low-Temperature Fermentation with Immobilized Cells

A large amount of studies report the selection or improvement of psychrophilic, psychrotolerant, or cold-adapted yeasts for low-temperature alcoholic beverages or ethanol production, and many more describe the development of low-temperature fermentation processes employing immobilized cell systems. Metabolic and physiological changes in yeast are not only induced in psychrophilic species evolved in cold environments, but are also common during growth or fermentation processes at low temperatures. *S. cerevisiae* is naturally found in environments, such as the surface of fruit, which can be subjected to low temperatures. In alcoholic fermentation processes, these yeasts can be exposed to temperatures around 10°C to 12°C, while industrial strains may be stored at very low temperatures (4°C) at which viability is maintained but growth is restricted.[54] In *S. cerevisiae*, low temperatures induce the expression of genes, which display a cold-sensitivity phenotype, including induction of fatty acid desaturases, proteins involved in pre-rRNA processing and ribosome biogenesis, specific amino-acid-rich cell-wall proteins, altered nitrogen metabolism, changes in the membrane fatty acids, alterations in aroma-related biochemical reactions, etc.[54]

Regarding alcoholic beverage production, low-temperature fermentation possesses a number of advantages such as production of beers of superior quality, ability to ferment low acidity musts to produce more malic and succinic acid, glycerol, and β-phenylethanol and less acetic acid.[54,55] However, these processes are not commonly used due to the increased risk of stuck and sluggish fermentations. The use of immobilized yeasts has facilitated the development of productive low-temperature fermentation processes as demonstrated by numerous published works.[2,18,54] The greatest impact of such technologies is considered to be the improved flavor of the products, which was mainly attributed to the better ratios of esters to higher alcohols on total volatiles. Specifically, wines produced at low temperatures have aromas with more fruity notes due to increased synthesis or reduced conversion of esters. The use of immobilized cells for very low-temperature fermentation (below 10°C) led to wines of improved aroma due to this effect.[2,43,45,56–60] The use of natural food-grade supports for cell immobilization such as DCM and gluten pellets proved to be effective for low-temperature wine making, with significantly improved fermentation productivity and product quality compared to free cells. To facilitate commercialization of such biocatalysts, freeze-drying techniques were also evaluated for the production of ready-to-use active dry formulations.[15]

Beer, on the other hand, is produced by more complex biochemical and technological processes, which all affect its flavor. Yeast amino acid metabolism, a key to the development of beer flavor as described earlier, is affected by process temperature and use of cell immobilization techniques. Therefore, technologies based on these features as well as other process conditions and strain selection have been developed to control beer flavor.[2,51] The combination of immobilized yeast and low-temperature primary fermentation was found to produce beers with low diacetyl amounts, therefore indicating potential of low-cost industrial application since maturation is a high-energy-consuming process.[2] Finally, Perpete and Collin[61] showed that during alcohol-free beer production, the enzymatic reduction of *worty* flavor (caused by Strecker aldehydes) by brewer's yeast was improved by cold contact fermentation.

43.4 WINE MAKING BY IMMOBILIZED CELLS

Although research on wine making by immobilized cells is extensive, industrial applications are still limited, mainly due to the strong traditional character of this product and consumer susceptibility. The objectives of using immobilized cells in wine making are to improve fermentation productivity, reduce maturation time, reduce production cost and installation size, improve flavor, and produce novel products of distinct characters.[2] A variety of materials and techniques have been proposed for wine yeast immobilization, involving organic, inorganic, natural, or synthesized immobilization carriers that may be used with or without modifications to optimize their characteristics.

Inorganic materials that have been proposed for yeast immobilization for batch and continuous primary fermentation in wine making, leading to significant improvement of process kinetics even at extremely low temperatures (0°C–10°C), include kissiris,[56,62,63] γ-alumina,[63,64] polygorskite, montmorilonite, hydromica, porous porcelain, porous glass, glass pellets covered with a layer of alginates, etc.[2,65] Despite their good attributes, the use of such biocatalysts is limited by the possibility to transfer undesirable residues in the final product (e.g., Al in the case of γ-alumina or kissiris). Efforts to remove such residues have also been reported, while wines produced by these techniques can be used as raw materials for distillates production, since mineral residues do not distil.[64,66] Apart from safety and quality, consumer acceptance is another main drawback for using inorganic materials in wine making.

To avoid the presence of undesirable residues in wine released by inorganic materials, food-grade organic carriers were also evaluated for wine making, mainly polysaccharide hydrogels. Ca-alginate gels were considered suitable for laboratory- and pilot-scale fermentation and under real vinification conditions, with interesting results regarding the formation of glycerol and secondary fermentation by-products.[2] Nevertheless, their use does not offer a good industrial choice because of their high cost and low stability that leads to cell wash-out and release of residues in wine.

Practical applications of alginates concern mainly secondary fermentation inside the bottle in sparkling wine production for easy clarification and removal of cells.[65,67,68] Other applications are the treatment of sluggish and stuck fermentations.[69]

Natural supports for yeast immobilization, such as DCM,[44] gluten pellets,[57] and cork pieces,[70] were found to be effective for both ambient and low-temperature wine making with significantly increased fermentation rates compared to free cells. The produced wines were of improved quality, which was demonstrated by both sensorial tests and chemical analysis of aroma volatiles.[45] These materials are of food-grade purity, cheap, abundant, and easy to prepare industrially. They can be easily accepted by consumers, and compared to other natural supports like fruit pieces, they have higher operational and mechanical stability. Freeze-dried biocatalysts consisting of yeast cells immobilized on gluten pellets or DCM were also evaluated for producing wines of similar quality to those made by fresh immobilized cells.[15]

The idea to use fruit parts and pieces as yeast immobilization supports for wine making was also attractive; because fruits are the natural habitats of yeasts, they can be easily accepted by consumers, and can affect flavor, leading to new types of wines. Cells immobilized on apple, pear and quince pieces,[2] fresh sugarcane pieces,[71] dried raisin berries,[72] and grape skins[73] were successfully used for batch and continuous wine making at ambient and extremely low temperatures. Raisin berries and grape skins are fully compatible with wine, and consumer acceptance is not an issue. Although the use of fruit in high volume bioreactors is problematic, such materials could still be employed in low-capacity processes due to the fine quality of the produced wines, which can add value to the product.

In search of natural supports for yeast immobilization suitable for wine making, starch-containing materials such as potatoes,[74] whole wheat,[42] corn,[75] barley grains,[76] and starch gels[77] were evaluated, leading to increased productivities even at extremely low temperatures compared to free cells. Finally, for economical and sustainable food-grade wine biocatalysts, various agro-industrial by-products and residues have been proposed as wine yeast carriers, including brewer's spent grains,[58] watermelon rind pieces,[78] grape pomace,[79] grape seeds, skins, stems, and corn cobs,[80] rendering all of the discussed advantages in wine making, with main drawback usually a negative effect on product color.

Other applications of immobilization techniques in wine making include the use of immobilized yeasts for color or acidity correction and immobilized enzymes for aroma release from bonded precursors. For example, chitin, chitosan, diethyl-amino-ethyl chitosan, and acrylic beads have been used for enzyme immobilization (e.g., β-glucosidase, α-arabinosidase, and α-rhamnosidase) to enhance the aroma of wines, musts, fruit juices, and beverages through the breakage of glycosidic linkages of rhamnose with aromatic compounds (e.g., monoterpenes and norisoprenoids) present in glycosidic form.[81–83] Yeasts immobilized in k-carragenate and alginate gels were used effectively for color correction treatments as well as to delay accelerated browning during storage of sherry pale white wines.[84,85] An *S. cerevisiae* stain immobilized in double-layer alginate-chitosan beads was used for bioreduction of volatile acidity of acidic wines with a volatile acidity higher than 1.44 g L^{-1} acetic acid, with no detrimental impact on wine aroma.[86] Finally, a few applications of membrane technology in wine making have been reported. Takaya et al.[87] studied the efficiency of two membrane bioreactor systems for continuous dry wine making: a single-vessel bioreactor in which cells where entrapped by a cross-flow type microfilter and a two-vessel system of a continuous stirred tank reactor and a membrane bioreactor. The double-vessel was found more suitable for dry wine making.

43.4.1 MALOLACTIC FERMENTATION OF WINE BY IMMOBILIZED CELLS

MLF is a secondary process that usually occurs in red wines, or wines with high acidities, during the maturation period. During MLF, L-malic acid is converted to L-lactic acid and carbon dioxide by bacteria such as *Leuconostoc*, *Lactobacillus*, and *Pediococcus spp.*

Most bacteria convert malic acid to lactic acid with an intermediate formation of pyruvic acid, while *Oenococcus oeni* (previously classified as *Leuconostoc oenos*) expresses the malolactic enzyme to directly convert malic acid in one-step reaction. *O. oeni* has been extensively studied for controlled MLF of wine due to its higher tolerance to ethanol, low pH, and SO_2.[2,88,89] Yeasts like *Schizosaccharomyces pombe* and *Saccharomyces* strains can also convert malic acid through a maloethanolic-type fermentation.[90] Lactic acid is less acidic than malic acid, and as a consequence, MLF leads to improvement of the sensory properties and biological stability of the wines. Additionally, the production of various other by-products of the MLF reaction may affect wine flavor positively.

MLF may occur in bottled wines that have not been adequately preserved with SO_2, by transformation of residual sugars by wild microflora, causing undesirable turbidity and development of off-flavors. The use of selected immobilized lactic acid bacteria can accelerate controlled MLF by higher cell densities and increased tolerance to inhibitory wine constituents, leading to flavor improvements. Immobilization can also facilitate reuse of the MLF biocatalyst and application of continuous processing.[2] Most of the initial efforts to conduct controlled MLF in wine involved the use of immobilized *O. oeni* in alginate gels.[65] Later works investigated the use of a variety of other carriers and bioreactor designs to optimize immobilized cell systems for practical MLF applications. Carriers, such as polyacrylamide, k-carrageenan, silica gel, pactate gels, chitosan, PVA (Lentikats) positively charged cellulose sponge, composites of alginates with organo-silica or charcoal, polyurethane foams, gel-like membranes, and even oak chips, were used as immobilization supports for various species with satisfactory results.[91-96] The use of immobilized bacteria in must inoculated with free yeast or coimmobilized bacteria and yeast biocatalyst can allow simultaneous alcoholic and MLF. An integrated wine-making process, including sequential alcoholic fermentations and MLF operated continuously, was developed by Genisheva et al.[97] *S. cerevisiae* cells immobilized either on grape stems or on grape skins, and *O. oeni* cells immobilized on grape skins only, were employed for a high yield production of dry white wine having a good physicochemical quality and 67% reduced malic acid concentration. Similar results were obtained by the use of three cultures (two *S. cerevisiae* spp. and *L. delbrueckii*) immobilized in alginate gel beads packed in near-horizontal acrylic columns.[98] The use of immobilized *O. oeni* on DCM also led to improvements of MLF in wine making,[99] and this effort was further enhanced by the development of a two-layer composite biocatalyst for simultaneous alcoholic and MLF.[100] The biocatalyst consisted of DCM with entrapped *O. oeni* cells, covered with starch gel containing an alcohol resistant and cryotolerant *S. cerevisiae* strain. The significance of such composite biocatalysts is the feasibility of two or three bioprocesses in the same bioreactor, thus reducing production cost in the food industry.[100]

Most of these efforts claimed possible industrial application of the immobilized biocatalysts due to their high operational stability, faster conversion of malic acid compared to free cells, and increased tolerance to SO_2 and ethanol inhibition. Understanding the nature and factors that affect MLF, which is a very complex process, as well as its effect on flavor formation, is crucial for optimizing wine technology and applying immobilization techniques at full scale.

43.5 BREWING BY IMMOBILIZED CELLS

In brewing, research has also focused on cell immobilization techniques to facilitate continuous processing, reduce maturation time, and produce alcohol-free beer. Brewing requires long fermentation times, large-scale fermentation, maturation at very low temperature, and big storage capacity and it is therefore a highly energy consuming process.[2,101,102] Rapid maturation of beer has been attempted by employing immobilization techniques in batch and continuous processes with claimed potential industrial application. Another advantage of immobilization is the reduced filtration needs, since the concentration of free cells in the produced is usually very low.[2] Nevertheless, continuous beer fermentation has not seen significant industrial application, because process characteristics, such as simplicity of design, low investment costs, flexible operation, effective process control, and good product quality, have not been yet achieved. The application of effective, cheap, and sustainable carrier materials for yeast immobilization could significantly lower the investment costs of continuous fermentation systems given that the correct sensory characteristics are achieved in the short time typical for such systems.[102,103] The research efforts to optimize brewing with immobilized cells are extensive, and several materials have been proposed as yeast immobilization carriers such as polysaccharide gels (mainly aginates), polyethylene film, polyethylene rings, PVA, PVC, DCM, gluten pellets, DEAE-cellulose, waste materials such as wood sawdust, spent grains and corncobs, and a few inorganic materials. Following, the most recent of these efforts are highlighted.

A complete continuous beer fermentation system consisting of a main fermentation reactor (gas-lift) and a maturation reactor (packed-bed) containing yeast immobilized on spent grains and corncobs, respectively, was evaluated.[103] It was found that by fine tuning of process parameters (residence time, aeration), it was possible to adjust the flavor profile of the final product, which was of a regular quality according to consumer evaluation and comparisons of analytical and sensorial profiles. Continuous, primary bottom beer fermentation maintaining a stable and high-level yeast activity during long-term operation was also evaluated in a bioreactor filled with liquid nonabsorbent carrier particles, resulting in improved fermentation performance.[104] Two innovative brewing processes, high gravity batch and complete continuous beer fermentation systems, were studied, showing a significant influence of the variables such as concentration and temperature on the ethanol yield and consequently on the productivity of the high gravity batch process. The technological feasibility of continuous production of beer based on yeast immobilization on cheap alternative carriers (delignified spent grains and corncob cylinders) was also demonstrated. The influence of process parameters on fermentation performance and quality of the obtained beers was studied by sensorial analysis. No significant difference in the degree of acceptance between the obtained products and some traditional market brands was observed.[105] Optimization of process parameters and monitoring of flavor formation was studied for continuous brewing with yeasts immobilized on spent grains at 7°C to 15°C in a bubble column reactor with high-gravity all-malt wort (15°Plato). As the fermentation temperature was increased, the degree of fermentation, rate of sugar consumption, ethanol volumetric productivity, consumption of free amino nitrogen, and the ratio of higher alcohols to esters increased.[106] The continuous fermentation of more concentrated worts (16.6 and 18.5°Plato) resulted in beers with unbalanced flavor profiles due to excessive ethyl acetate formation.[107] The potential of application of nonaggressive LentiKat (R) technique for brewer's yeast immobilization on PVA was assessed.[108] High cell loads achieved by this procedure and immobilization procedure had no adverse effect on cell viability. The immobilized cells exhibited high fermentation activity in both laboratory- and pilot-scale fermentations in three successive gas-lift bioreactors, indicating good potential of immobilized cells for development of continuous primary beer fermentation.

Several methods have also been developed in order to meet the increasing demand for alcohol-free beer over the last decade, due to health issues, safety in the workplace or during driving, and strict social regulations. These methods include alcohol removal from the product or limited fermentation of wort. In the case of limited fermentation, production is most efficient when immobilized cells are employed.[2,109] However, nonalcoholic beer suffers from flavor defects as well as improper body and foaming properties. Therefore, production of alcohol-free beer with satisfactory sensory characteristics has recently given rise to increased technological and economic interest.[109] Such systems have already been successfully applied. Van Iersel et al.[110] used a system for production of nonalcohol beer by limited fermentation, achieved by low temperatures and anaerobic conditions, using immobilized *S. cerevisiae* in a packed-bed reactor. Ethanol contents lower than 0.08% were obtained, while the production of esters and alcohols was stimulated. Flavor formation and cell physiology during alcohol-free beer production by limited fermentation in a controlled down-flow packed-bed using immobilized *S. cerevisiae* on polystyrene coated with DEAE-cellulose was also investigated.[110] The system was characterized as highly controllable, and optimal flavor was achieved by introduction of regular aerobic periods to stimulate yeast growth and temperature variations to control the growth rate and flavor formation.[111] To optimize lab-scale continuous alcohol-free beer production, experiments were carried out using real wort and a mimicking model medium, in a continuously operating gas-lift reactor with brewing yeast immobilized on spent grains. The results also suggested that the process parameters represent a powerful tool in controlling the degree of fermentation and flavor formation by the immobilized biocatalyst.[112] Finally, the influence of production strains (bottom-fermenting *S. pastorianus* and *S. cerevisiae* with disruption in the KGD2 gene), carrier materials (spent grains and corncobs), bioreactor designs (packed-bed and gas-lift), and mixing regimes (ideally mixed and plug flow) on the formation of flavor-active compounds was demonstrated during alcohol-free beer production.[113]

Among polysaccharide hydrogels, alginate gels are the most extensively studied supports for brewer's yeast immobilization.[2] A two-stage reactor system was proposed for continuous secondary fermentation of wort in laboratory scale using immobilized yeast:[114] an up-flow gas-lift bioreactor for main fermentation, and column packed-beds reactors with yeast entrapped in three different polysaccharide hydrogels (Ca-alginate, Ca-pectate, and k-carrageenan). All three carriers were found suitable for continuous secondary fermentation of green beers produced by continuous main fermentation. Patkova et al.[115] used Ca-alginate entrapped yeast to ferment high gravity wort in half the time needed for fermentation by free yeast. They also observed that when the original wort gravity was increased, the specific rate of ethanol production remained constant and the viability did not fall below 95% of living cells, confirming protection of the immobilized cells against osmotic stress. The influence of immobilized yeasts on fermentation parameters and beer quality using a continuous gas-lift bioreactor system with brewer's yeast entrapped in Ca-pectate or k-carrageenan was evaluated. The produced beers had suitable flavor, with low levels of diacetyl, an optimum ratio of higher alcohols to esters and maximum specific rate of sugar utilization.[50] The feasibility to explore and potential uses of immobilization on sensorial characteristics of stout beer (color, flavor, and headspace compounds) was evaluated in batch fermentation using yeast microencapsulated in alginate. Free and immobilized yeasts fermentation showed no significant difference for all process variables of interest. The profile of headspace compounds was different, perhaps because of changes in yeast's behavior and the presence of secondary metabolites.[116]

As in the case of wine making, biocatalysts prepared by immobilization of the alcohol resistant and cryotolerant strain *S. cerevisiae* AXAZ-1 on DCM and gluten pellets were found suitable for batch and continuous fermentation of wort at ambient and low temperatures.[117] The immobilized yeast showed important operational stability with no decrease of activity even at very low temperatures (below 5°C). Batch fermentations at various temperatures were faster than those of free cells and those usual in commercial brewing, while beer produced by the immobilized yeast contained lower amounts of diacetyl and polyphenols as well as lower bitterness and pH compared to beer produced by free cells. The fruity aroma of beers obtained at low temperatures was attributed to improved ratios of higher alcohols to esters. The use of these immobilized biocatalysts was considered advantageous for brewing, and their possible commercialization in active freeze-dried form was evaluated.[13,40] The freeze-dried biocatalysts retained their viability during long fermentation periods (13–14 months), and the produced beers were clear with low concentrations of suspended cells and lower diacetyl contents compared to beers produced by free cells. Bekatorou et al.[118] used the same *S. cerevisiae* strain immobilized on dried figs for batch beer fermentations at low and room temperatures (3°C–20°C) with similar results. The produced green beers had a fine clarity, and were sweet, smooth with a special fruity fig-like aroma and taste, clearly distinct from other commercial products. The safety, low cost, and consumer acceptance of such natural carriers are unquestionable.[2] Finally, a biocatalyst, prepared by immobilization of the strain on the delignified brewer's spent grains, was used for brewing at very low temperatures, also resulting in beers with fine clarity, excellent quality, and mature character after the end of primary fermentation.[119] Fermentation times were low (only 20 days at 0°C), with high ethanol and beer productivities and low vicinal diketone, DMS, and amyl alcohol concentrations. GC and GC–MS analysis showed significant quantitative differences in the composition of aroma volatiles, revealing an impact of fermentation temperature on sensory properties.

A few inorganic materials have been used as immobilization supports for brewing including porous ceramics, diatomaceous silica (*kieselghur*), porous brick pieces, and porous glass. Porous glass beads were used for the development of continuous brewing processes for rapid maturation of beer.[52,120] The fermentation times were reduced by half compared to the conventional batch processes, and the proposed technology was demonstrated as feasible for application in brewing if combined with a heat treatment system to reduce the relatively high diketone contents. Finally, a novel material, the foam ceramic with enormous surface area and good mechanical properties, was applied as yeast immobilization carrier for continuous secondary fermentation of beer.[121] The results indicated that dilution rate was the primary factor for reducing diacetyl, which could be promptly decreased to permission level after about 6 h. Therefore, the secondary fermentation cycle for beer maturation, and the associated costs, could be cut significantly.

Scale-up processes and industrial application of immobilized cell systems for brewing have been reported and reviewed by various authors.[2,52,101,122]

43.6 ETHANOL AND DISTILLATES PRODUCTION BY IMMOBILIZED CELLS

The requirement for ethanol as an additive in the beverage industries is high and so is the pursuit for development of efficient bioethanol production systems, including the use of immobilized cells. Research on food-grade alcohol production usually focuses on controlling the formation of volatile by-products, since their concentration in the fermented broths is critical for the production of good-quality distillates and alcoholic beverages. The nature of the immobilization carrier does not usually affect the distillate composition, since nonvolatile constituents do not distil. Therefore, the requirement for food-grade purity carriers is not essential due to the employment of the distillation step. Immobilized microbes such as conventional, psychrotolerant, cold-adapted, thermotolerant, alcohol-resistant, and genetically modified yeasts (e.g., *S. cerevisiae*, *S. diastaticus*, *Kluyveromyces marxianus*, *Candida* spp., etc.), and bacteria (e.g., *Zymomonas mobilis*) have been evaluated for food- and fuel-grade ethanol production. These works involved various types of immobilization carriers, as described earlier, synthetic substrates (e.g., glucose media) or waste effluents (e.g., molasses and cheese whey), in various process designs, to evaluate use of these biocatalysts for bioethanol production.[2,54]

Production of alcohol can be done by single, double, or fractional distillation of the fermented liquids. The quality of the distillates depends on the quality of the raw materials, which is generally considered to be good in the case of low-temperature fermentations by immobilized cells, due to the improved profiles of volatile compounds, as discussed earlier. In distillates production, inorganic materials such as γ-alumina pellets and kissiris can be very useful as immobilization carriers, since they are cheap, abundant, stable, and can be easily reused. Moreover, mineral residues in the fermented liquid are not an issue, since they do not distil. For that reason, batch and continuous low-temperature fermentation of grape must using the cryotolerant and alcohol-resistant strain *S. cerevisiae* AXAZ-1 immobilized on γ-alumina, and kissiris was successfully performed for improved wine-based distillates production.[64,66] Other natural materials proposed as immobilization carriers for the strain were DCM,[10] gluten pellets,[2] orange peel,[123] and brewer's spent grains,[124] used for alcoholic fermentation of glucose media or molasses, at ambient, low as well as high temperatures (40°C). Kopsahelis et al.[125] further proposed an integrated cost-effective system for continuous alcoholic fermentation of sterilized and nonsterilized molasses at 30°C to

40°C, with the same yeast strain immobilized on brewer's spent grains, in two types of bioreactors, a multistage fixed-bed tower (MFBT) and a packed-bed reactor (PB). The MFBT bioreactor gave better results regarding ethanol concentration, productivity, and conversion. Higher ethanol productivity was obtained in the case of nonsterilized molasses, with no contamination observed during 32 days of continuous operation, which was considered particularly interesting for industrial application.

In the same manner, ethanol production from various single and mixed sugar substrates and industrial effluents was reported, using yeasts immobilized on natural materials such as corncobs,[126] maize stem ground tissue,[127] carob pod,[128] pine wood chips,[129] woven cotton fiber,[130] as well as in the form of yeast biocapsules.[27] Other research groups studied the effect of composite materials or chemically modified immobilization carriers for efficient bioethanol production. For example, the effect of derivatization was evaluated for the development of delignified corncob grits derivatized with 2-(diethylamino)ethyl chloride hydrochloride as carrier to immobilize S. cerevisiae. The biocatalyst produced ethanol under optimized derivatization and adsorption conditions between yeast cells and the DEAE-corncobs.[131] Similarly, high bioethanol productivity was reported by immobilized S. bayanus onto cross-linked graft carboxymethylcellulose-g-poly(N-vinyl-2-pyrrolidone) copolymer beads,[132] which was increased when the percentage of N-vinyl-2-pyrrolidone in copolymer increased. Efficient bioethanol production from glucose–xylose mixtures was also obtained using Scheffersomyces stipitis free or immobilized in silica-hydrogel films, and cocultures with S. cerevisiae.[133] A BCA sponge fabricated by a freeze-drying process was successfully used as yeast cell carrier for ethanol fermentation,[22] exhibiting several advantageous properties, such as high porosity, appropriate pore size, strong hydrophilicity, and high mechanical, chemical, and thermal stabilities. Scale-up of ethanol production from molasses using yeast immobilized in alginate-mesoporous zeolite composite carriers exhibited much shorter fermentation times, higher ethanol productivity, and better operational durability, indicating potential for commercial applications.[23]

The impact of high temperature (30°C–45°C) on ethanol fermentation by immobilized K. marxianus was evaluated by Du Le et al.[134] and Eiadpum et al.[135] The immobilized yeast demonstrated faster sugar assimilation and higher ethanol level in the fermentation broth in comparison with the free yeast. The altered response of the free and immobilized yeast to thermal stress was attributed to the observed changes in the cell membrane fatty acid composition.

Finally, various attempts have been made to optimize alcoholic fermentation processes using immobilized Z. mobilis cells. Altuntas and Ozcelik[136] performed continuous ethanol production in a stirred bioreactor using coimmobilized amyloglucosidase and Z. mobilis cells, for simultaneous saccharification of starch and fermentation to ethanol. Bioethanol production from cane molasses was also studied using a Z. mobilis strain entrapped in Luffa cylindrica L. sponge discs and Ca-alginate gel beads.[137] The immobilization carriers were found to be equally good for ethanol production, but Luffa was considered advantageous due to its lower cost, nondestructive nature, and no environmental hazard imposed.

43.7 CIDER MAKING BY IMMOBILIZED CELLS

Cider production is a complex process consisting of the alcoholic fermentation step by yeasts, followed by MLF by lactic acid bacteria. Traditional cider is produced by natural fermentation of apple juice by the apple wild microflora, leading to an unstable product of variable quality. Recent studies have been focusing on the use of selected starter cultures and novel technologies, including the use of immobilization techniques, in order to increase productivity, accelerate maturation, and improve cider quality and stability. The immobilized cell systems used for cider production usually involve coimmobilized species in the same bioreactor for one-step process, or a series of bioreactors containing different species to conduct sequential alcoholic fermentation and MLF.[2] The proposed carrier materials, microbial species, and process designs are generally similar to those applied in

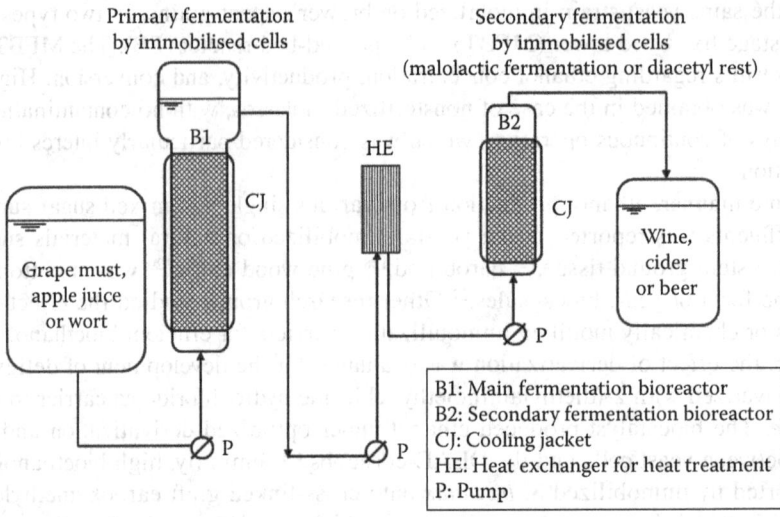

FIGURE 43.5 A schematic process design involving two immobilized cell bioreactors for sequential primary and secondary fermentation for wine, cider, or beer production. (Adapted from Kourkoutas, Y. et al., *Food Microbiol.*, 21(4), 377, 2004.)

wine making. Although most of the research efforts proved to be efficient for both main and secondary fermentation, further research is needed for application at industrial scale to produce ciders with improved and controlled flavor profiles (Figure 43.5).

43.8 PRODUCTION OF CHEESE WHEY–BASED BEVERAGES BY IMMOBILIZED CELLS

Whey is the main liquid waste of the dairy industries and is produced in large quantities worldwide. Its high organic load makes disposal in biological treatment plants impossible, while discarding imposes a serious environmental threat. Partial exploitation of whey includes its use as animal feed, as raw material for protein supplement formulations, and as industrial food additive. The cost-effective microbial conversion of whey into products of added value is very important from both economical and environmental points of view.[138,139] Recent advances include the use of selected cultures, single or mixed, including thermotolerant and genetically modified strains (e.g., to express β-galactosidase activity), in appropriate process designs, to produce or recover products of added value from whey, such as single-cell protein (SCP), organic acids, enzymes, lactose, and ethanol.

The bioconversion of whey into ethanol has been mainly attempted using immobilized *K. marxianus* strains in various types of bioreactor systems.[140–143] The proposed processes enabled high ethanol yields and productivities from whey. The natural mixed culture kefir, consisting of lactose fermenting yeasts and bacteria, has also been used to conduct alcoholic and lactic acid fermentation of whey, depending on the process conditions applied. Kefir immobilized on DCM was found suitable for continuous whey fermentation supplemented with 1% raisin extract or molasses.[144] Industrial scale-up process of whey alcoholic fermentation, promoted by raisin extracts, was successfully developed using suspended kefir cells.[138] The fermented whey could be exploited as raw material to produce kefir-like whey-based drinks, and food-grade and fuel alcohol. The development of this technology was supported by the easily precipitated biomass in the form of about 1 mm aggregates that led to the avoidance of centrifugal separators, which are high-cost equipment.[138] The recent advances in lactic acid production by microbial fermentation processes, including cell immobilization and cell recycling techniques, have been reviewed.[6,145]

43.9 CONCLUSIONS

A huge number of immobilization carriers and techniques have been proposed by various research groups for alcoholic beverages production due to the numerous advantages associated with such techniques (e.g., increased fermentation rates and yields, high cell densities in the bioreactors, reduced risk of contamination, feasibility of continuous processing, biocatalyst recycling, easier product separation, etc.). However, for full-scale application, features like the product quality, mechanical strength and operational stability of the biocatalyst, installation and production cost, flavor control, and consumer acceptance still need to be optimized.

REFERENCES

1. Karel S.F., Libicki S.B. and Robertson C.R. The immobilization of whole cells-engineering principles. *Chemical Engineering Science* 40 (8) (1985): 1321–1354.
2. Kourkoutas Y., Bekatorou A., Banat I.M., Marchant R. and Koutinas A.A. Immobilization technologies and support materials suitable in alcohol beverages production: A review. *Food Microbiology* 21 (4) (2004): 377–397.
3. Mussatto S.I., Dragone G., Guimaraes P.M.R., Silva J.P.A., Carneiro L.M., Roberto I.C., Vicente A., Domingues L. and Teixeira J.A. Technological trends, global market, and challenges of bio-ethanol production. *Biotechnology Advances* 28 (6) (2010): 817–830.
4. Freeman A. and Lilly M.D. Effect of processing parameters on the feasibility and operational stability of immobilized viable microbial cells. *Enzyme and Microbial Technology* 23 (5) (1998): 335–345.
5. Junter G.-A. and Jouenne T. 2.36—Immobilized viable cell biocatalysts: A paradoxical development. *Comprehensive Biotechnology* 2 (2011): 491–505.
6. Kosseva M.R. Immobilization of microbial cells in food fermentation processes. *Food and Bioprocess Technology* 4 (6) (2011): 1089–1118.
7. Verbelen P.J., De S.D.P., Delvaux F., Verstrepen K.J. and Delvaux F.R. Immobilized yeast cell systems for continuous fermentation applications. *Biotechnology Letters* 28 (19) (2006): 1515–1525.
8. Pilkington P.H., Margaritis A., Mensour N.A. and Russell I. Fundamentals of immobilized yeast cells for continuous beer fermentation: A review. *Journal of the Institute of Brewing* 104 (1) (1998): 19–31.
9. Scott C.D. Immobilized cells: A review of recent literature. *Enzyme and Microbial Technology* 9 (2) (1987): 66–72.
10. Koutinas A.A., Sypsas V., Kandylis P., Michelis A., Bekatorou A., Kourkoutas Y., Kordulis C. et al. Nano-tubular cellulose for bioprocess technology development. *Plos One* 7 (4) (2012): e34350.
11. Koutinas A.A., Papapostolou H., Dimitrellou D., Kopsahelis N., Katechaki E., Bekatorou A. and Bosnea L.A. Whey valorisation: A complete and novel technology development for dairy industry starter culture production. *Bioresource Technology* 100 (15) (2009): 3734–3739.
12. Bekatorou A., Koutinas A.A., Kaliafas A. and Kanellaki M. Freeze-dried *Saccharomyces cerevisiae* cells immobilized on gluten pellets for glucose fermentation. *Process Biochemistry* 36 (6) (2001): 549–557.
13. Bekatorou A., Koutinas A.A., Psarianos K. and Kanellaki M. Low-temperature brewing by freeze-dried immobilized cells on gluten pellets. *Journal of Agricultural and Food Chemistry* 49 (1) (2001): 373–377.
14. Iconomopoulou M., Kanellaki M., Psarianos K. and Koutinas A.A. Delignified cellulosic material supported biocatalyst as freeze-dried product in alcoholic fermentation. *Journal of Agricultural and Food Chemistry* 68 (3) (2000): 958–961.
15. Iconomopoulou M., Psarianos K., Kanellaki M. and Koutinas A.A. Low temperature and ambient temperature wine making using freeze-dried immobilized cells on gluten pellets. *Process Biochemistry* 37 (7) (2002): 707–717.
16. Kregiel D., Berlowska J. and Ambroziak W. Adhesion of yeast cells to different porous supports, stability of cell-carrier systems and formation of volatile by-products. *World Journal of Microbiology and Biotechnology* 28 (12) (2012): 3399–3408.
17. Park J.K. and Chang H.N. Microencapsulation of microbial cells. *Biotechnology Advances* 18 (4) (2000): 303–319.
18. Willaert R. and Nedovic V. Primary beer fermentation by immobilised yeast–A review on flavour formation and control strategies. *Journal of Chemical Technology and Biotechnology* 81 (8) (2006): 1353–1367.

19. Smidsrod O. and Skjak-Braek G. Alginate as immobilization matrix for cells. *Trends in Biotechnology* 8 (3) (1990): 71–78.
20. Tanaka H., Irie S. and Ochi H. A novel immobilization method for prevention of cell leakage from the gel matrix. *Journal of Fermentation and Bioengineering* 68 (3) (1989): 216–219.
21. Callone E., Campostrini R., Carturan G., Cavazza A. and Guzzon R. Immobilization of yeast and bacteria cells in alginate microbeads coated with silica membranes: Procedures, physico-chemical features and bioactivity. *Journal of Materials Chemistry* 18 (40) (2008): 4839–4848.
22. Kirdponpattara S. and Phisalaphong M. Bacterial cellulose-alginate composite sponge as a yeast cell carrier for ethanol production. *Biochemical Engineering Journal* 77 (2013): 103–109.
23. Zheng C., Sun X., Li L. and Guan N. Scaling up of ethanol production from sugar molasses using yeast immobilized with alginate-based MCM-41 mesoporous zeolite composite carrier. *Bioresource Technology* 115 (2012): 208–214.
24. Navratil M., Domeny Z., Hronsky V., Sturdic E., Smogrovicova D. and Gemeiner P. Use of bioluminometry for determination of active yeast biomass immobilized in ionotropic hydrogels. *Analytical Biochemistry* 284 (2) (2000): 394–400.
25. Zhao X.-Q. and Bai F.-W. Yeast flocculation: New story in fuel ethanol production. *Biotechnology Advances* 27 (6) (2009): 849–856.
26. Jin Y.-L. and Speers A.R. Flocculation of Saccharomyces cerevisiae. *Food Research International* 31 (6–7) (1998): 421–440.
27. Garcia-Martinez T., Puig-Pujol A., Peinado R.A., Moreno J. and Mauricio J.C. Potential use of wine yeasts immobilized on *Penicillium chrysogenum* for ethanol production. *Journal of Chemical Technology and Biotechnology* 87 (3) (2012): 351–359.
28. Garcia-Martinez T., Lopez de Lerma N., Moreno J., Peinado R.A., Carmen M.M. and Mauricio J.C. Sweet wine production by two osmotolerant Saccharomyces cerevisiae strains. *Journal of Food Science* 78 (6) (2013): M874–M879.
29. Lopez de Lerma N., Garcia-Martinez T., Moreno J., Mauricio J.C. and Peinado R.A. Volatile composition of partially fermented wines elaborated from sun dried Pedro Ximénez grapes. *Food Chemistry* 135 (4) (2012): 2445–2452.
30. Lebeau T., Jouenne T. and Junter G.-A. Simultaneous fermentation of glucose and xylose by pure and mixed cultures of *Saccharomyces cerevisiae* and Candida shehatae immobilized in a two-chambered bioreactor. *Enzyme and Microbial Technology* 21 (1997): 265–272.
31. Kargupta K., Siddhartha D. and Sanyal S.K. Analysis of the performance of a continuous membrane bioreactor with cell recycling during ethanol fermentation. *Biochemical Engineering Journal* 1 (1) (1998): 31–37.
32. Lebeau T., Jouenne T. and Junter G.-A. Diffusion of sugars and alcohols through composite membrane structures immobilising viable yeast cells. *Enzyme and Microbial Technology* 22 (1998): 434–438.
33. Gryta M. The assessment of microorganism growth in the membrane distillation system. *Desalination* 142 (1) (2002): 79–88.
34. Rathore S., Desai M.P., Liew C.V., Chan L.W. and Heng P.W.S. Microencapsulation of microbial cells. *Journal of Food Engineering* 116, (2) (2013): 369–381.
35. Westrin B.A. and Axelsson A. Diffusion in gels containing immobilized cells: A critical review. *Biotechnology and Bioengineering* 38 (5) (1991): 439–446.
36. Junter G.-A., Coquet L., Vilain S. and Jouenne T. Immobilized-cell physiology: Current data and the potentialities of proteomics. *Enzyme and Microbial Technology* 31 (2002): 201–212.
37. Junter G.-A. and Jouenne T. Immobilized viable microbial cells: From the process to the proteome… or the cart before the horse. *Biotechnology Advances* 22 (8) (2004): 633–658.
38. Jamai L., Sendide K., Ettayebi K., Errachidi F., Hamdouni-Alami O., Tahri-Jouti M.A., McDermott T. and Ettayebi M. Physiological difference during ethanol fermentation between calcium alginate-immobilized Candida tropicalis and Saccharomyces cerevisiae. *FEMS Microbiology Letters* 204 (2) (2001): 375–379.
39. Melzoch K., Rychtera M. and Habova V. Effect of immobilization upon the properties and behavior of Saccharomyces cerevisiae cells. *Journal of Biotechnology* 32 (1) (1994): 59–65.
40. Bekatorou A., Soupioni M.J., Koutinas A.A. and Kanellaki M. Low-temperature brewing by freeze-dried immobilized cells. *Applied Biochemistry and Biotechnology* 97 (2) (2002): 105–121.
41. Tsaousi K., Velli A., Akarepis F., Bosnea L, Drouza C., Koutinas A.A. and Bekatorou A. Low-temperature winemaking by thermally dried immobilized yeast on delignified brewer's spent grains. *Food Technology and Biotechnology* 49 (3) (2011): 379–384.
42. Kandylis P., Manousi M.E., Bekatorou A. and Koutinas A.A. Freeze-dried wheat supported biocatalyst for low temperature wine making. *LWT—Food Science and Technology* 43 (10) (2010): 1485–1493.

43. Mallouchos A., Komaitis M., Koutinas A.A. and Kanellaki M. Wine fermentations by immobilized and free cells at different temperatures. Effect of immobilization and temperature on volatile by-products. *Food Chemistry* 80 (1) (2003): 109–113.
44. Bardi E. and Koutinas A.A. Immobilization of yeast on delignified cellulosic material for room temperature and low-temperature wine making. *Journal of Agricultural and Food Chemistry* 42 (1) (1994) 221–226.
45. Mallouchos A. and Argyro B. Wine fermentations by immobilized cells. Effect on wine aroma. In *Microbial Implication for Safe and Qualitative Food Products*. Psarianos, C. and Kourkoutas, C. Eds. Kerala, India: Research Signpost, 2008.
46. Schreier P. and Jennings W.G. Flavor composition of wines: A review. *Critical Reviews in Food Science and Nutrition* 12 (1) (1979): 59–111.
47. Vidrih R. and Hribar J. Synthesis of higher alcohols during cider processing. *Food Chemistry* 67 (3) (1999): 287–294.
48. Etievant X.P. *Volatile Compounds in Foods and Beverages*. New York: Henk Maarse, 1991.
49. Russell I. and Stewart G.G. Contribution of yeast and immobilization technology to flavor development in fermenting beverages. *Food Technology* 46 (11) (1992): 146.
50. Smogrovicova D. Chapter 23—Formation of beer volatile compounds at different fermentation temperatures using immobilized yeasts. *Flavour Science* (2014): 129–131.
51. Branyik T., Vicente A.A., Dostalek P. and Teixeira J.A. A review of flavour formation in continuous beer fermentations. *Journal of the Institute of Brewing* 114 (1) (2008): 3–13.
52. Yamauchi Y., Okamoto T., Murayama H., Kajino K., Amikura T., Hiratsu H., Nagara A., Kamiya T. and Inoue T. Rapid maturation of beer using an immobilized yeast bioreactor. 1. Heat conversion of alpha-acetolactate. *Journal of Biotechnology* 38 (2) (1995a): 101–108.
53. Moll M. Fermentation and maturation of beer with immobilised yeasts. *Journal of the Institute of Brewing* 112 (4) (2006): 346–346.
54. Kanellaki M., Bekatorou A. and Koutinas A.A. Low-temperature production of wine, beer and distillates using cold-adapted yeasts. In *Cold-Adapted Yeasts: Biodiversity, Adaptation Strategies and Biotechnological Significance*, Buzzini, P. and Margesin, R. Eds. pp. 417–439. Heidelberg, Germany: Springer, 2014.
55. Kanellaki M. and Koutinas A.A. Low temperature fermentation by cold-adapted and immobilized yeast cells. In *Biotechnological Applications of Cold-Adapted Organisms*, Margesin, R. and Schinner, F. Eds. pp. 117–145. Berlin, Heidelberg: Springer-Verlag, 1999.
56. Bakoyianis V., Kana K., Kalliafas A. and Koutinas A.A. Low-temperature continuous wine-making by kissiris-supported biocatalyst: Volatile by-products. *Journal of Agricultural and Food Chemistry* 41 (3) (1993): 465–468.
57. Bardi E.P., Bakoyianis V., Koutinas A.A. and Kanellaki M. Room temperature and low temperature wine making using yeast immobilized on gluten pellets. *Process Biochemistry* 31 (5) (1996a): 425–430.
58. Mallouchos A., Loukatos P., Bekatorou A., Koutinas A.A. and Komaitis M. Ambient and low temperature winemaking by immobilized cells on brewer's spent grains: Effect on volatile composition. *Food Chemistry* 104 (3) (2007): 918–927.
59. Kourkoutas Y., Kanellaki M., Koutinas A.A. and Tzia C. Effect of fermentation conditions and immobilization supports on the wine-making. *Journal of Food Engineering* 69 (2005): 115–123.
60. Yajima M. and Yokotsuka K. Volatile compound formation in white wines fermented using immobilized and free yeast. *American Journal of Enology and Viticulture* 52 (3) (2001): 210–218.
61. Perpete P. and Collin S. How to improve the enzymatic worty flavour reduction in a cold contact fermentation. *Food Chemistry* 70 (4) (2000): 457–462.
62. Bakoyianis V., Kanellaki M., Kalliafas A. and Koutinas A.A. Low temperature wine-making by immobilized cells on mineral kissiris. *Journal of Agricultural and Food Chemistry* 40 (7) (1992): 1293–1296.
63. Bakoyianis V., Koutinas A.A., Aggelopoulos K. and Kanellaki M. Comparative study of kissiris, γ-alumina and Ca-alginates as supports of cells for batch and continuous wine making at low temperatures. *Journal of Agricultural and Food Chemistry* 45 (12) (1997): 4884–4888.
64. Loukatos P., Kiaris M., Ligas I., Bourgos G., Kanellaki M., Komaitis M. and Koutinas A.A. Continuous wine making by γ-alumina-supported biocatalyst. *Applied Biochemistry and Biotechnology* 89 (1) (2000): 1–13.
65. Colagrande O., Silva A. and Fumi M.D. Recent applications of biotechnology in wine production. *Biotechnology Progress* 10 (1) (1994): 2–18.

66. Loukatos P., Kanellaki M., Komaitis M., Athanasiadis I. and Koutinas A.A. A new technological approach proposed for distillate production using immobilized cells. *Journal of Bioscience and Bioengineering* 95 (1) (2003): 35–39.
67. Busova K., Magyar I. and Janky F. Effect of immobilized yeasts on the quality of bottle-fermented sparkling wine. *Acta Alimentaria* 23 (1) (1994): 9–23.
68. Torresi S., Frangipane M.T. and Anelli G. Biotechnologies in sparkling wine production. Interesting approaches for quality improvement: A review. *Food Chemistry* 129 (3) (2011): 1232–1241.
69. Silva S., Ramon P.F., Silva P., Maria de Fatima T. and Strehaiano P. Use of encapsulated yeast for the treatment of stuck and sluggish fermentations. *Journal International Des Sciences De La Vigne Et Du Vin* 36 (3) (2002): 161–168.
70. Tsakiris A., Kandylis P., Bekatorou A., Kourkoutas Y. and Koutinas A.A. Dry red wine-making using yeast immobilized on cork pieces. *Applied Biochemistry and Biotechnology* 162 (5) (2010): 1316–1326.
71. Reddy V.L., Reddy P.L., Wee Y.-J. and Reddy O.V.S. Production and characterization of wine with sugarcane piece immobilized yeast biocatalyst. *Food and Bioprocess Technology* 4 (1) (2011): 142–148.
72. Tsakiris A., Bekatorou A., Koutinas A.A., Marchant R. and B.I.M. Immobilization of yeast on dried raisin berries for use in dry white wine making. *Food Chemistry* 87 (2004): 11–15.
73. Mallouchos A., Reppa P., Aggelis G., Kanellaki M., Koutinas A.A. and Komaitis M. Grape skins as a natural support for yeast immobilization. *Biotechnology Letters* 24 (16) (2002): 1331–1335.
74. Kandylis P. and Koutinas A.A. Extremely low temperature fermentations of grape must by potatoes supported yeast-strain AXAZ-1. A contribution is performed to catalysis of alcoholic fermentation. *Journal of Agricultural and Food Chemistry* 56 (9) (2008): 3317–3327.
75. Kandylis P., Mantzari A., Koutinas A.A. and Kookos I.K. Modelling of low temperature wine-making, using immobilized cells. *Food Chemistry* 133 (4) (2012): 1341–1348.
76. Kandylis P., Dimitrellou D. and Koutinas A.A. Winemaking by barley supported yeast cells. *Food Chemistry* 130 (2) (2012): 425–431.
77. Kandylis P., Goula A. and Koutinas A.A. Corn starch gel for yeast cell entrapment a view for catalysis of wine fermentation. *Journal of Agricultural and Food Chemistry* 56 (24) (2008): 12037–12045.
78. Reddy V.L., Reddy H.K.Y., Reddy P.A.L. and Reddy V.S.O. Wine production by novel yeast biocatalyst prepared by immobilization on watermelon (Citrullus vulgaris) rind pieces and characterization of volatile compounds. *Process Biochemistry* 43 (7) (2008): 748–752.
79. Genisheva Z., Macedo S., Mussatto S.I., Teixeira J.A. and Oliveira J.M. Production of white wine by Saccharomyces cerevisiae immobilized on grape pomace. *Journal of the Institute of Brewing* 118 (2) (2012): 163–173.
80. Genisheva Z., Mussatto S.I., Oliveira J.M. and Teixeira J.A. Evaluating the potential of wine-making residues and corn cobs as support materials for cell immobilization for ethanol production. *Industrial Crops and Products* 34 (1) (2011): 979–985.
81. Spagna G., Barbagallo R.N., Casarini D. and Pifferi P.G. A novel chitosan derivative to immobilize a-L-rhamnopyranosidase from *Aspergillus niger* for application in beverage technologies. *Enzyme and Microbial Technology* 28 (2001): 427–438.
82. Gonzalez-Pombo P., Farina L., Carrau F., Batista-Viera F. and Brena B.M. Aroma enhancement in wines using co-immobilized. *Aspergillus niger* glycosidases. *Food Chemistry* 143 (2014): 185–191.
83. Gallifuoco A., D'Ercole L., Alfani F., Cantarella M., Spagna G. and Pifferi P.G. On the use of chitosan-immobilized β-glucosidase in wine-making: Kinetics and enzyme inhibition. *Process Biochemistry* 33 (2) (1998): 163–168.
84. Merida J., Lopez-Toledano A. and Medina M. Immobilized yeasts in kappa-carragenate to prevent browning in white wines. *European Food Research and Technology* 225 (2) (2007): 279–286.
85. Lopez-Toledano A., Merida J. and Medina M. Colour correction in white wines by use of immobilized yeasts on kappa-carragenate and alginate gels. *European Food Research and Technology* 225 (5–6) (2007): 879–885.
86. Vilela A., Schuller D., Mendes-Faia A. and Corte-Real M. Reduction of volatile acidity of acidic wines by immobilized Saccharomyces cerevisiae cells. *Applied Microbiology and Biotechnology* 97 (11) (2013): 4991–5000.
87. Takaya M., Matsumoto N. and Yanase H. Characterization of membrane bioreactor for dry wine production. *Journal of Bioscience and Bioengineering* 93 (2) (2002): 240–244.
88. Versari A., Parpinello G.P. and Cattaneo M. Leuconostoc oenos and malolactic fermentation in wine: A review. *Journal of Industrial Microbiology and Biotechnology* 23 (1999): 447–455.
89. Zhang D. and Lovitt R.W. Strategies for enhanced malolactic fermentation in wine and cider maturation. *Journal of Chemical Technology and Biotechnology* 81 (7) (2006): 1130–1140.

90. Redzepovic S., Orlic S., Madjak A., Kozina B., Volschenk H. and Viljoen-Bloom M. Differential malic acid degradation by selected strains of Saccharomyces during alcoholic fermentation. *International Journal of Food Microbiology* 83 (1) (2003): 49–61.
91. Kosseva M., Beschkov V., Kennedy J.F. and Lloyd L.L. Malolactic fermentation in Chardonnay wine by immobilized *Lactobacillus casei* cells. *Process Biochemistry* 33 (8) (1998): 793–797.
92. Rodriguez-Nogales M.J., Vila-Crespo J. and Fernandez-Fernandez E. Immobilization of *Oenococcus oeni* in lentikats (R) to develop malolactic fermentation in wines. *Biotechnology Progress* 29 (1) (2013): 60–65.
93. Maicas S. The use of alternative technologies to develop malolactic fermentation in wine. *Applied Microbiology and Biotechnology* 56 (2001): 35–39.
94. Iorio G., Catapano G., Drioli E., Rossi M. and Rella R. Malic enzyme immobilization in continuous capillary membrane reactors. *Journal of Membrane Science* 22 (1985): 317–324.
95. Hong S.-K., Lee H.-J., Park H.-J., Hong Y.-A., Rhee I.-K., Lee W.-H., Choi S.-H., Lee O.-S. and Park H.-D. Degradation of malic acid in wine by immobilized Issatchenkia orientalis cells with oriental oak charcoal and alginate. *Letters in Applied Microbiology* 50 (5) (2010): 522–529.
96. Guzzon R., Carturan G., Krieger-Weber S. and Cavazza A. Use of organo-silica immobilized bacteria produced in a pilot scale plant to induce malolactic fermentation in wines that contain lysozyme. *Annals of Microbiology* 62 (1) (2012): 381–390.
97. Genisheva Z., Mota A., Mussatto S.I., Oliveira J.M. and Teixeira J.A. Integrated continuous winemaking process involving sequential alcoholic and malolactic fermentations with immobilized cells. *Process Biochemistry* 49 (1) (2014): 1–9.
98. Aaron R.T., Davis R.C., Hamdy M.K. and Toledo R.T. Continuous alcohol/malolactic fermentation of grape must in a bioreactor system using immobilized cells. *Journal of Rapid Methods and Automation in Microbiology* 12 (2) (2004): 127–148.
99. Agouridis N., Kopsahelis N., Plessas S., Koutinas A.A. and Kanellaki M. *Oenococcus oeni* cells immobilized on delignified cellulosic material for malolactic fermentation of wine. *Bioresource Technology* 99 (18) (2008): 9017–9020.
100. Servetas I., Berbegal C., Camacho N., Bekatorou A., Ferrer S., Nigam P., Drouza C. and Koutinas A.A. Saccharomyces cerevisiae and *Oenococcus oeni* immobilized in different layers of a cellulose/starch gel composite for simultaneous alcoholic and malolactic wine fermentations. *Process Biochemistry* 48 (9) (2013): 1279–1284.
101. Masschelein C.A., Ryder D.S. and Simon J.-P. Immobilized cell technology in beer production. *Critical Reviews in Biotechnology* 14 (1994): 155–177.
102. Branyik T., Silva D.P., Baszczynski M., Lehnert R. and Almeida e Silva J.B. A review of methods of low alcohol and alcohol-free beer production. *Journal of Food Engineering* 108 (4) (2012): 493–506.
103. Branyik T., Silva D.P., Vicente A.A., Lehnert R., Silva J.B., Almeida E., Dostalek P. and Teixeira J.A. Continuous immobilized yeast reactor system for complete beer fermentation using spent grains and corncobs as carrier materials. *Journal of Industrial Microbiology and Biotechnology* 33 (12) (2006): 1010–1018.
104. Inoue T. and Mizuno A. Preliminary study for the development of a long-life, continuous, primary fermentation system for beer brewing. *Journal of the American Society of Brewing Chemists* 66 (2) (2008): 80–87.
105. Silva D.P., Branyik T., Dragone G., Vicente A.A., Teixeira J.A. and Almeida e Silva J.B. High gravity batch and continuous processes for beer production: Evaluation of fermentation performance and beer quality. *Chemical Papers* 62 (1) (2008): 34–41.
106. Dragone G., Mussatto S.I. and Almeida e Silva J.D. Influence of temperature on continuous high gravity brewing with yeasts immobilized on spent grains. *European Food Research and Technology* 228 (2) (2008): 257–264.
107. Dragone G., Mussatto S.I. and Almeida e Silva J.B. High gravity brewing by continuous process using immobilised yeast: Effect of wort original gravity on fermentation performance. *Journal of the Institute of Brewing* 113 (4) (2007): 391–398.
108. Bezbradica D., Obradovic B., Leskosek-Cukalovic I., Bugarski B. and Nedovic V. Immobilization of yeast cells in PVA particles for beer fermentation. *Process Biochemistry* 42 (9) (2007): 1348–1351.
109. Sohrabvandi S., Mousavi S.M., Razavi S.H., Mortazavian A.M. and Rezaei K. Alcohol-free beer: Methods of production, sensorial defects, and healthful effects. *Food Reviews International* 26 (4) (2010): 335–352.
110. Van Iersel M.F.M., Van Dieren B., Rombouts F.M. and Abee T. Flavor formation and cell physiology during the production of alcohol-free beer with immobilized Saccharomyces cerevisiae. *Enzyme and Microbial Technology* 24 (7) (1999): 407–411.

111. Van Iersel M.F.M., Brouwer P.E., Rombouts F.M. and Abee T. Influence of yeast immobilization on fermentation and aldehyde reduction during the production of alcohol free beer. *Enzyme and Microbial Technology* 26 (8) (2000): 602–607.
112. Lehnert R., Novak P., Macieira F., Kurec M., Teixeira J.A. and Tomas B. Optimisation of lab-scale continuous alcohol-free beer production. *Czech Journal of Food Sciences* 27 (4) (2009): 267–275.
113. Mota A., Novak P., Macieira F., Vicente A.A., Teixeira J.A., Smogrovicova D. and Branyik T. Formation of flavor-active compounds during continuous alcohol-free beer production: The influence of yeast strain, reactor configuration, and carrier type. *Journal of the American Society of Brewing Chemists* 69 (1) (2011): 1–7.
114. Domeny Z., Smogrovicova D., Gemeiner P., Sturdik E., Patkova J. and Malovikova A. Continuous secondary fermentation using immobilized yeast. *Biotechnology Letters* 20 (11) (1998): 1041–1045.
115. Patkova J., Smogrovicova D., Domeny Z. and Bafrncova P. Very high gravity wort fermentation by immobilized yeast. *Biotechnology Letters* 22 (14) (2000): 1173–1177.
116. Almonacid S.F., Najera A.L., Young M.E., Simpson R.J. and Acevedo C.A. A comparative study of stout beer batch fermentation using free and microencapsulated yeasts. *Food and Bioprocess Technology* 5 (2) (2012): 750–758.
117. Bardi E., Koutinas A.A., Soupioni M. and Kanellaki M. Immobilization of yeast on delignified cellulosic material for low temperature brewing. *Journal of Agricultural and Food Chemistry* 44 (2) (1996): 463–467.
118. Bekatorou A., Sarellas A., Ternan N.G., Mallouchos A., Komaitis M., Koutinas A.A. and Kanellaki M. Low-temperature brewing using yeast immobilized on dried figs. *Journal of Agricultural and Food Chemistry* 50 (25) (2002): 7249–7257.
119. Kopsahelis N., Kanellaki M. and Bekatorou A. Low temperature brewing using cells immobilized on brewer's spent grains. *Food Chemistry* 104 (2) (2007): 480–488.
120. Tata M., Bower P., Bromberg S., Duncombe D., Fehring J., Lau V., Ryder D. and Stassi P. Immobilized yeast bioreactor systems for continuous beer fermentation. *Biotechnology Progress* 15 (1) (1999): 105–113.
121. Cheng J., Liu J., Shao H. and Qiu Y. Continuous secondary fermentation of beer by yeast immobilized on the foam ceramic. *Research Journal of Biotechnology* 2 (3) (2007): 40–42.
122. Mensour N.A., Margaritis A., Briens C.L., Pilkington H. and Russell I. Developments in the brewing industry using immobilized yeast cell bioreactor systems. *Journal of the Institute of Brewing* 103 (6) (1997): 363–370.
123. Plessas S., Bekatorou A., Koutinas A.A., Soupioni M., Banat I.M. and Marchant R. Use of Saccharomyces cerevisiae cells immobilized on orange peel as biocatalyst for alcoholic fermentation. *Bioresource Technology* 98 (4) (2007): 860–865.
124. Kopsahelis N., Agouridis N., Bekatorou A. and Kanellaki M. Comparative study of spent grains and delignified spent grains as yeast supports for alcohol production from molasses. *Bioresource Technology* 98 (7) (2007): 1440–1447.
125. Kopsahelis N., Papachronopoulos A., Bosnea L., Bekatorou A., Tzia C. and Kanellaki M. Alcohol production from sterilized and non-sterilized molasses by Saccharomyces cerevisiae immobilized on brewer's spent grains in two types of continuous bioreactor systems. *Biomass and Bioenergy* 46 (2012): 809–809.
126. Laopaiboon L. and Laopaiboon P. Ethanol production from sweet sorghum juice in repeated-batch fermentation by Saccharomyces cerevisiae immobilized on corncob. *World Journal of Microbiology and Biotechnology* 28 (2) (2012): 559–566.
127. Razmovski R. and Vucurovic V. Bioethanol production from sugar beet molasses and thick juice using Saccharomyces cerevisiae immobilized on maize stem ground tissue. *Fuel* 92 (1) (2012): 1–8.
128. Yatmaz E., Turhan I. and Karhan M. Optimization of ethanol production from carob pod extract using immobilized *Saccharomyces cerevisiae* cells in a stirred tank bioreactor. *Bioresource Technology* 135 (2013): 365–371.
129. Dhabhai R., Chaurasia S.P. and Dalai A.K. Efficient bioethanol production from glucose-xylose mixtures using co-culture of Saccharomyces cerevisiae immobilized on canadian pine wood chips and free Pichia stipitis. *Journal of Biobased Materials and Bioenergy* 6 (5) (2012): 594–600.
130. Chen Y., Liu Q., Zhou T., Li B., Yao S., Wu J. and Ying H. Ethanol production by repeated batch and continuous fermentations by Saccharomyces cerevisiae immobilized in a fibrous bed bioreactor. *Journal of Microbiology and Biotechnology* 23 (4) (2013): 511–517.
131. Lee S.-E., Lee C.G., Kang D.H., Lee H.-Y. and Jung K.-H. Preparation of corncob grits as a carrier for immobilizing yeast cells for ethanol production *Journal of Microbiology and Biotechnology* 22 (12) (2012): 1673–1680.

132. Gokgoz M. and Yigitoglu M. High productivity bioethanol fermentation by immobilized Saccharomyces bayanus onto carboxymethylcellulose g poly(N-vinyl-2-pyrrolidone) beads. *Artificial Cells Nanomedicine and Biotechnology* 41 (2) (2013): 137–143.
133. De Bari I., De Canio P., Cuna D., Liuzzi F., Capece A. and Romano P. Bioethanol production from mixed sugars by Scheffersomyces stipitis free and immobilized cells, and co-cultures with *Saccharomyces cerevisiae*. *New Biotechnology* 30 (6) (2013): 591–597.
134. Du Le H., Pornthap T. and Van Viet M.L. Impact of high temperature on ethanol fermentation by *Kluyveromyces marxianus* immobilized on banana leaf sheath pieces. *Applied Biochemistry and Biotechnology* 171 (3) (2013): 806–816.
135. Eiadpum A., Limtong S. and Phisalaphong M. High-temperature ethanol fermentation by immobilized coculture of *Kluyveromyces marxianus* and *Saccharomyces cerevisiae*. *Journal of Bioscience and Bioengineering* 114 (3) (2012): 325–329.
136. Altuntas E.G. and Ozcelik F. Ethanol production from starch by co-immobilized amyloglucosidase— Zymomonas mobilis cells in a continuously-stirred bioreactor. *Biotechnology and Biotechnological Equipment* 27 (1) (2013): 3506–3512.
137. Behera S., Mohanty R.C. and Ray R.C. Ethanol fermentation of sugarcane molasses by *Zymomonas mobilis* MTCC 92 immobilized in *Luffa cylindrica* L. sponge discs and Ca-alginate matrices. *Brazilian Journal of Microbiology* 43 (4) (2012): 1499–1507.
138. Koutinas A.A., Athanasiadis I., Bekatorou A., Psarianos C., Kanellaki M., Agouridis N. and Blekas G. Kefir-yeast technology: Industrial scale-up of alcoholic fermentation of whey, promoted by raisin extracts, using kefir-yeast granular biomass. *Enzyme and Microbial Technology* 41 (5) (2007): 576–582.
139. Koutinas A.A., Bekatorou A., Nigam P., Banat I.M. and Marchant R. Whey utilization and SCP production. In *Advances in Cheese Whey Utilization*, Ma Esperanza Cerdan, Ma Isabel Gonzalez-Siso and Manuel Bacerra, Eds. p. 147–161. Kerala, India: Research Signpost, 2008.
140. Dale C.M., Eagger A. and Okos M.R. Osmotic inhibition of free and immobilized *K. marxianus* anaerobic growth and ethanol productivity in whey permeate concentrate. *Process Biochemistry* 29 (7) (1994): 535–544.
141. Gabardo S., Rech R., Zachia A. and Marco A. Performance of different immobilized-cell systems to efficiently produce ethanol from whey: Fluidized batch, packed-bed and fluidized continuous bioreactors. *Journal of Chemical Technology and Biotechnology* 87 (8) (2012): 1194–1201.
142. Kourkoutas Y., Dimitropoulou S., Kanellaki M., Marchant R., Nigam P., Banat I.M. and Koutinas A.A. High-temperature alcoholic fermentation of whey using *Kluyveromyces marxianus* IMB3 yeast immobilized on delignified cellulosic material. *Bioresource Technology* 82 (2002): 177–181.
143. Ozmihci S. and Kargi F. Fermentation of cheese whey powder solution to ethanol in a packed-column bioreactor: Effects of feed sugar concentration. *Journal of Chemical Technology and Biotechnology* 84 (1) (2009): 106–111.
144. Kourkoutas Y., Psarianos C., Koutinas A.A., Kanellaki M., Banat I.M. and Marchant R. Continuous whey fermentation using kefir yeast immobilized on delignified cellulosic material. *Journal of Agricultural and Food Chemistry* 50 (9) (2002): 2543–2547.
145. Abdel-Rahman M.A., Tashiro Y. and Sonomoto K. Recent advances in lactic acid production by microbial fermentation processes. *Biotechnology Advances* 31 (6) (2013): 877–902.

44 Enzyme Immobilization in Biodegradable Polymers for Biomedical Applications

S.A. Costa, Helena S. Azevedo, and Rui L. Reis

CONTENTS

44.1 Introduction 957
44.2 Enzymes in Medicine 958
    44.2.1 Clinical Diagnosis 958
    44.2.2 Enzyme Therapy...958
44.3 Enzyme Immobilization Technology 961
    44.3.1 Methods for Immobilizing Enzymes in Polymeric Carriers 961
        44.3.1.1 Adsorption..961
        44.3.1.2 Ionic Binding 962
        44.3.1.3 Covalent Binding by Chemical Coupling 962
        44.3.1.4 Crosslinking...964
        44.3.1.5 Entrapment and Encapsulation 965
        44.3.1.6 Protein Fusion to Affinity Ligands and Enzymatic Conjugation 965
    44.3.2 Enzyme Immobilization in Biomedical Applications 967
        44.3.2.1 Biologically Functional Surfaces...967
        44.3.2.2 Enzyme Delivery 969
        44.3.2.3 Bioreactors for Extracorporeal Enzyme Therapy............................971
        44.3.2.4 Diagnostic Assays and Biosensors...972
        44.3.2.5 Advantages and Disadvantages of Immobilized Enzymes..............973
44.4 Conclusions and Future Perspectives..974
Acknowledgments..974
References..974

44.1 INTRODUCTION

During the last three decades, enzymology and enzyme technology have progressed considerably and, as a result, there are many examples of industrial applications where enzymes, in the native or immobilized form, are being used. These include food industry, materials processing, textiles, detergents, biochemical and chemical industries, biotechnology, and pharmaceutical uses.[1] The overall impact of enzymes on industrial applications is, however, still quite limited due to their relative instability under operational conditions, which may involve high temperatures, organic solvents, and exposure to other denaturants. Various approaches, including, among others, addition of additives,[2,3] chemical modification,[4,5] protein engineering,[6] and enzyme immobilization,[1,7] have been assessed for their ability to increase the stability of enzymes toward heat or denaturants.[7,8] The use of enzymes in medical applications has been less extensive as those for other types of industrial applications. For example, pancreatic enzymes have been in use since the nineteenth century for the treatment of digestive disorders. At present, the most successful applications of enzymes in medicine are extracellular, such as topical uses, removal of toxic substances, and the treatment of life-threatening disorders within the blood circulation.[9] The production of therapeutic enzymes has progressed, but the costs of enzyme production, isolation, and purification are still too high to make them available for clinical applications. Furthermore, the ability to store unstable enzymes for long periods of time is also a limitation for their more widespread use. Most applications in the biomedical field are still in the state of basic studies rather than definite applications, owing to the absence of the necessary information on toxicology, hemolysis, allergenicity, immunological reactions, and chemical stability of the system *in vivo*.[9,10]

This chapter will focus on the importance of using enzymes in medical applications and, in particular, the use of immobilized enzymes. For that, some aspects of enzyme immobilization technology, including the traditional physical and chemical methods and new immobilization methodologies, based on biological and genetic engineering approaches, will be reviewed. Several examples will be given of immobilized enzymes in various support materials and using different immobilization strategies, according to the objective of their application in the biomedical field.

44.2 ENZYMES IN MEDICINE

Since the mid-1950s there has been a considerable increase in both measurement of enzyme activities and the use of purified enzymes in clinical practice. In recent years, many enzymes have been isolated and purified, and this made it possible to use enzymes to determine the concentration of substrates and products of clinical importance. A further development, arising from the increased availability of purified enzymes, has been targeted to enzyme therapy.

44.2.1 CLINICAL DIAGNOSIS

The measurement of enzyme activities in serum is of major importance as an aid in diagnosis, being used as means of monitoring progress after therapy, recovery after surgery, and detection of transplant rejection. Urine can be also analyzed for determination of enzyme activities since the detection of certain enzymes in urine may indicate kidney damage or failure.[11] On the other hand, the concentration of certain metabolites in serum or urine may be determined by enzymatic methods. The method consists of using an enzyme to transform a metabolite into its product and then estimate the amount of transformed substrate. The use of enzymatic methods presents several advantages, such as the high specificity of the enzyme to estimate the concentration of the metabolite in the presence of other substances, avoiding the need of purification steps prior to chemical analysis. In addition, enzymatic reactions are performed at mild conditions, allowing the analysis of labile compounds that would be degraded by harsher chemical methods. The cost of purified enzymes may be, however, too high to support routine analysis, but the use of immobilized enzymes allows for enzyme reuse and the application of immobilized enzymes for diagnostic assays and as biosensors will be further discussed in Section 44.3.2.4. The determination of serum metabolites in serum by enzymatic methods includes a wide range of substances such as glucose, uric acid, urea, cholesterol, cholesterol esters, triglycerides, and creatine, among others.[11]

44.2.2 ENZYME THERAPY

Many inborn metabolic disorders are associated with the absence of activity of one particular enzyme normally found in the body. Of the 1250 autosomal recessive human genetic diseases, over 200 involve errors in metabolism that result from specific known enzyme deficiencies.[12] Table 44.1 lists some examples of inborn errors or disorders in metabolism due to enzyme deficiencies.

The initial identification of the disease may be difficult to determine and normally requires a tissue biopsy. For some genetic diseases, DNA probes are now available that can be used on small amounts of blood, cells, or amniotic fluid. Albinism, for example, is often caused by the absence of tyrosinase, an enzyme essential for the production of cellular pigments. Tyrosinase is

TABLE 44.1
Some Examples of Inborn Errors and Disorders Associated with Enzyme Deficiencies

Inborn Error/Disorder	Enzyme Deficiency	Frequency (%)	References
Gaucher disease	Glucocerebrosidase	0.003–0.002	[11,13–15]
Acatalasemia	Catalase	0.004–0.004	[11]
Hypophosphatasia	Alkaline phosphatase	≈0.001	[11,15]
Glycogen storage disease type Ia	Glucose 6-phosphatase	0.001	[11,15]
Alkaptonuria	Homogentisate 1,2-dioxygenase	0.0001–0.001	[11]
Phenylketonuria	Phenylalanine 4-monooxygenase	0.005–0.01	[11,15,16]
Fructosuria	Fructokinase	0.0008	[11]
Pentosuria	L-Xylulose reductase	0.04 in Ashkenazi Jews	[11]
Tay-Sachs	β-N-acetyl-D-hexosaminidase	0.0003	[11,13–15]
Infantile neuronal ceroid lipofuscinosis (INCL)	Palmitoyl protein thioesterase 1 (PPT1)	≈0.013 live births	[17,18]
Cystic fibrosis	Cystic fibrosis transmembrane conductance regulator	0.03–0.05 Caucasians, rare in other ethnic groups	[11]
Albinism	Tyrosinase	0.02 schoolchildren in Zimbabwe	[16,19,20]
Glucose-6-phosphate dehydrogenase deficiency	Glucose-6-phosphate dehydrogenase	0.5–26	[11,21,22]
Neonatal jaundice	Biotinidase	0.0025	[23,24]
Prolidase deficiency (PD)	Prolidase	—	[25]
Pompe's disease	α-glucosidase	—	[13,14]
Severe combined immunodeficiency (SCID)	Adenosine deaminase	—	[12,15,26]
Xanthinuria	Xanthinine oxidase	—	[12]

a copper-containing enzyme that catalyzes the first two rate-limiting steps in the melanin biosynthetic pathway, the oxidation of tyrosine to dopa and the subsequent dehydrogenation of dopa to dopaquinone. The human tyrosinase gene, encoding 529 amino acids, consists of five exons spanning more than 50 Kb of DNA in chromosomes.[20] When homozygous mutations of the tyrosine gene result in complete absence of melanogenic activity, such a patient, categorized as tyrosinase-negative oculocutaneous albinism, will never develop any melanin pigment in the skin, hair, and eyes throughout his or her life.[20] Some inborn errors in metabolism are relatively harmless, e.g., albinism or alkaptonuria, but others must be detected early if the defect is to be circumvented. This is the case of phenylketonuria, where the enzyme phenylalanine 4-monooxygenase (enzyme that converts phenylalanine into tyrosine) is missing. Phenylketonuria results in the accumulation of phenylalanine, which may cause mental retardation. Patients with phenylalanine 4-monooxygenase deficiency must follow a phenylalanine-free diet in order to avoid the accumulation of deleterious effects. Phenylalanine is, however, an amino acid essential to maintain growth and protein turnover, and thus it must be supplied in a minimal amount required to maintain normal metabolism. Such a dietary scheme is normally carried out[11,16] by lowering the amount of protein consumed, but this may cause a deficiency in other essential amino acids. A possible alternative to this therapy is to replace the missing enzyme. It may be difficult to find, however, an enzyme with the same function from a human source since the direct administration of enzymes from other sources into the body would cause an adverse immunological response. A possible approach to circumvent this problem involves the isolation of the enzyme within a microcapsule, fiber, or gel, which will protect the enzyme from proteolysis and avoid the undesirable immunological response (see Section 44.3.2.2).

The pharmacological properties of enzymes have been employed to replace enzymes that are missing or defective as a consequence of an inherited disease or malfunction of an organ where they are normally synthesized or to accomplish a certain biological effect that is dependent on the catalytic activity of the enzyme. Therefore, depending on the treatment, the administration of enzymes as therapeutic agents can be subdivided into two categories[9,11]: (1) the topical application of an enzyme as an extracellular agent and (2) the intracellular applications of enzymes to treat metabolic deficiency and related disease. The main areas where enzyme therapy has been applied are the degradation of necrotic tissue by the use of proteolytic enzymes, removal of toxic compounds from the blood, treatment of genetic deficiency diseases and cancer, and treatment of pancreatic insufficiency.[11] In Table 44.2 are given some examples of areas where enzyme therapy may be used.

Enzymes may be administered either intra- or extracorporeally, depending on the objective. If the enzyme is to be used for the removal or transformation of a substance present in the blood (e.g., toxic metabolite or a blood clot), then it is only necessary for the enzyme to be present in the blood and not necessary for the enzyme to enter the intracellular compartments. This type of application may be either intra- or extracorporeal using a bypass as in kidney dialysis. These systems will be described in more detail in Section 44.3.2.3. For intracellular therapy, it is necessary for the enzyme to be taken up by the appropriate target cells.

Although the attempts made with enzyme therapy in clinical trials have so far had limited success, it is reasonable to assume that the delivery of enzymes (discussed in Section 44.3.2.2) would constitute a feasible approach for the treatment of certain diseases in the near future.

TABLE 44.2
Enzymes with Therapeutic Importance for Medical Applications

Enzyme	Typical Applications	References
Lysozyme	Recommended in treatment of certain ulcers, measles, multiple sclerosis, some skin diseases, and postoperative infections (antibacterial agent)	[27–29]
Urease	Biosensor and artificial kidneys	[30–32]
Catalase	Treatment of acatalasemia and removal of hydrogen peroxide in human cells	[33–35]
Glucose-6-phosphate dehydrogenase	Treatment of jaundice	[36–38]
Collagenase	Skin ulcers	[11]
Glucose oxidase	Glucose test in blood and urine	[39,40]
Asparaginase	Anticancer agent (leukemia)	[41–43]
α-amylase, protease, lipase	Digestive aids	[44]
Chymotrypsin and pepsin	Catalyzes the hydrolysis of peptide bonds of proteins in the small intestine	[45]
Trypsin	Anti-inflammatory agent, wound cleanser	[44,46,47]
Streptokinase	Anti-inflammatory agent, dissolution of blood clots in myocardial infarction	[45]
Hyaluronidase	Hydrolyses polyhyaluronic acid, a relatively impermeable polymer found between human cells; administered to increase diffusion of coinjected compounds, e.g., antibiotics, adrenaline, heparin, and local anesthetic in surgery and dentistry	[48–51]
Heparinase	Removal of heparin after surgery. Production of heparin oligosaccharides (wound healing and tumor netastasis properties)	[52,53]
Urokinase	Prevention and removal of blood clots	[54–56]
Streptodornase	Anti-inflammatory agent	[57]
Tissue plasminogen activator (TPA)	Dissolution of blood clots	[58]
Tyrosinase	Enzyme essential for the production of cellular pigments	[20,59,60]
Bilirubin oxidase	Treatment of neonatal jaundice	[61]
Penicillinase	Removal of allergenic form of penicillin from allergic individuals	[16]
Alkaline phosphatase	Treatment of hypophosphatasia	[15]

44.3 ENZYME IMMOBILIZATION TECHNOLOGY

The term *immobilized enzyme* was adopted in 1971[62] at the first Enzyme Engineering Conference. It describes enzymes physically confined at or localized in a certain region of space with retention of catalytic activity and which can be used repeatedly and continuously.[63] The immobilization of biocatalysts (not only enzymes but also other bioactive molecules such as growth factors and hormones, cellular organelles, microbial cells, and plant and animal cells) is attracting worldwide attention in biotechnology applications. In general, immobilized biocatalysts are more stable and easier to handle compared with their free counterparts.[64] At present, applications of immobilized biocatalysts include the production of useful compounds by stereospecific or regiospecific bioconversion, the production of energy by biological processes, the selective treatment of specific pollutants to solve environmental problems, continuous analyses of compounds with a high sensitivity and specificity, and medical uses such as new types of drugs for enzyme therapy or artificial organs.[14] Immobilized enzymes are already being used in medical applications for clinical diagnosis and also for intra- and extracorporeal enzyme therapy.[62] Applications in clinical analysis are mainly related to biosensors, which have been used to detect the presence of various organic compounds for many years. For example, glucose oxidase and catalase have been used to measure blood glucose concentration, and cholesterol oxidase and cholesterol esterase to determine cholesterol levels.[31] In addition, enzymes can be immobilized on different prosthetic devices or used extracorporeally (e.g., artificial heart, artificial lung, artificial kidney, equipment for hemodialysis and specific blood purification) as surface modifiers in order to increase the biocompatibility of these devices and to prevent blood clotting.[9]

44.3.1 METHODS FOR IMMOBILIZING ENZYMES IN POLYMERIC CARRIERS

Various methods have been developed[10,65] for the immobilization of biocatalysts, which are being used extensively today. A wide range of support materials has also been employed for enzyme immobilization. The support type can be classified according to their chemical composition, such as organic or inorganic supports, and the former can be further classified into natural or synthetic matrices.[66] Immobilization techniques can be divided into different categories: physical,[67] chemical,[68] enzymatic,[69] and genetic engineering methods.[53]

44.3.1.1 Adsorption

The adsorption of an enzyme onto a support or film material is the simplest method of obtaining an immobilized enzyme. Basically, the enzyme is attached to the support material by noncovalent linkages and does not require any preactivation step of the support. The interactions formed between the enzyme and the support material will be dependent on the existing surface chemistry of the support and on the type of amino acids exposed at the surface of the enzyme molecule. Enzyme immobilization by adsorption involves, normally, weak interactions between the support and the enzyme such as ionic or hydrophobic interactions, hydrogen bonding, and van der Waals forces (see Figure 44.1).[70,71]

Most of the support materials available have sufficient surface-charge properties suitable for immobilization by adsorption. They include inorganic carriers[62] (ceramic, alumina, activated carbon, kaolinite, bentonite, porous glass), organic synthetic carriers[72] (nylon, polystyrene), and natural organic carriers[1] (chitosan, dextran, gelatin, cellulose, starch). The method consists of simply mixing an aqueous solution of enzyme with the support material for a period of time, after which the excess enzyme is washed away from the immobilized enzyme on the support. The procedure requires strict control of the pH and ionic strength, because these can alter the charges of the enzyme and the support and therefore affect the level of adsorption. A simple shift in pH can cancel ionic interactions and promote the release of the enzyme from the support. The main advantages of adsorption are the method simplicity, the little effect on the conformation/activity of the biocatalyst, and the

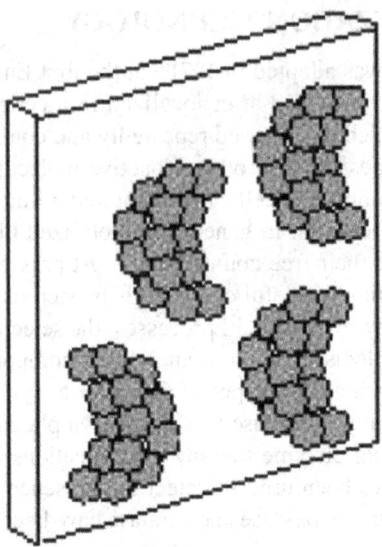

FIGURE 44.1 Biocatalysts bound to a carrier by adsorption.

possibility of regenerating inactive enzyme by addition of fresh enzyme. The main disadvantage is the desorption of the biocatalyst from the support due to the weak interactions established. The enzyme desorption can easily occur by changes in the environment medium such as pH, temperature, solvent, and ionic strength or in the case of extended reactions.[73]

44.3.1.2 Ionic Binding

Immobilization via ionic binding is based, mainly, on ionic binding of enzyme molecules or active molecule to solid supports containing ionic charges. In this method, the amount of enzyme bound to the carrier and the activity after immobilization depends on the nature of the carrier. Figure 44.2 shows how the enzyme is bound to the carrier. In some cases, physical adsorption may also take place. The main difference between ionic binding and physical adsorption is the strength of the interaction, which is much stronger for ionic binding, although less strong than covalent binding. The preparation of immobilized enzymes using ionic binding is based on the same procedure as described for physical adsorption.[74,75] The ionic nature of the binding forces between the enzyme and the support also depends on pH variations, support charge, enzyme concentrations, and temperature. The supports used for ionic binding may be based on polysaccharide derivatives[64] (e.g., diethylaminoethylcellulose, dextran, chitosan, carboxymethylcellulose), synthetic polymers[10] (e.g., polystyrene derivatives, polyethylene vinylalcohol), and inorganic materials[62] (e.g., ambertite, alumina, silicates, bentonite, sepiolite, silica gel, etc.). The immobilization by ionic binding has the advantage that changes in the enzyme conformation only occur in a small extent, resulting in immobilized enzymes with high enzymatic activities. The main disadvantage is the possible interference of other ions, and special attention should be paid in maintaining the correct ionic strength and pH conditions in order to prevent their easy detachment.[10]

44.3.1.3 Covalent Binding by Chemical Coupling

The covalent binding method is based on the binding of enzymes, or other active molecules, to a support or matrix by means of covalent bonds.[10] The bond is normally formed between a functional group present on the support surface and amino acid residues on the surface of the enzyme. Those which are most often involved in covalent binding are[65] the amino (NH_2) group of lysine or arginine, carboxyl (CO_2H) group of aspartic acid or glutamic acid, hydroxyl (OH) group of serine

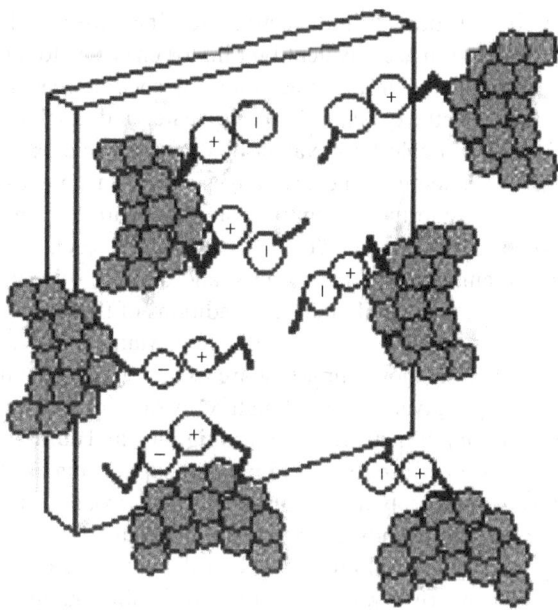

FIGURE 44.2 Biocatalysts bound to a carrier by ionic binding.

or threonine, and sulphydryl (SH) group of cysteine.[76,77] There are many reaction procedures for joining an enzyme to a material with a covalent bond (diazotation, amino bond, Schiff's base formation, amidation reactions, thiol-disulfide, peptide bond, and alkylation reactions). The connection between the support and the biocatalyst can be achieved either by direct linkage between the components or via an intercalated link of different length, the so-called spacer or harm. The advantage of using a spacer molecule is that it gives a greater degree of mobility to the coupled biocatalysts so that its activity can, under certain circumstances, be higher than if it is bound directly to the support (see Figure 44.3). It is important to choose a method that will not involve

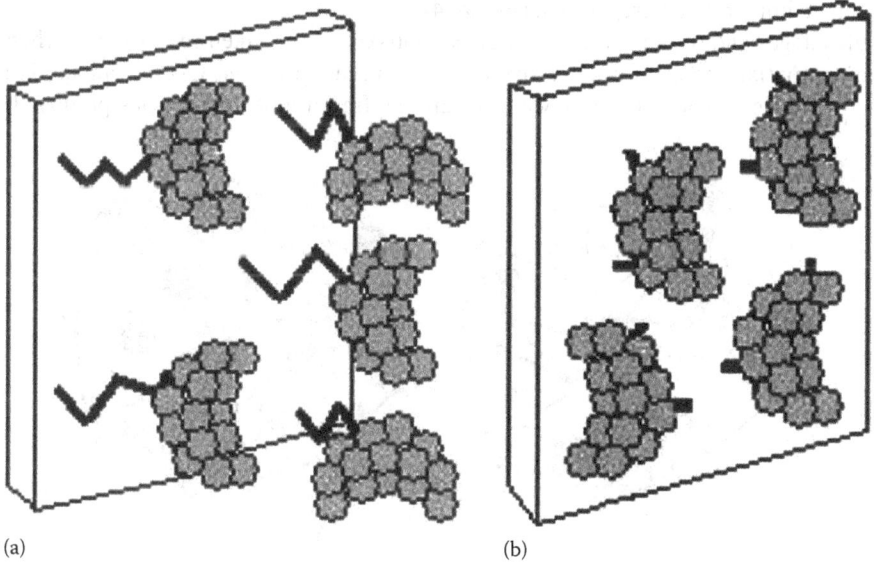

(a) (b)

FIGURE 44.3 Covalent bond between the biocatalysts and a carrier (a) with and (b) without spacer.

the reaction with the amino acids present in the active site, since this could inactivate the enzyme. Basically, two steps are involved in the covalent binding of enzymes to a support material. First, functional groups on the support material are activated by specific reagents (e.g., cyanogen bromide, carbodiimide, aminoalkylethoxysilane, isothiocyanate, and epichlorohydrin, etc.). A large range of support materials is available for covalent binding, and this extensive range reflects the fact that no ideal matrix exists. Therefore, the advantages and disadvantages of a given matrix must be taken into account when considering the appropriate procedure for a given enzyme immobilization. Immobilization of enzymes through covalent attachment has also been demonstrated[73] to induce higher resistance to temperature, denaturants, and organic solvents in several cases. The extent of these improvements may depend on other conditions of the system, e.g., the nature of the enzyme, type of support, and the method of immobilization. Many factors may influence the selection of the support, and some of the more important are its cost and availability, the binding capacity (amount of enzyme bound per given weight of matrix), hydrophilicity (the ability to incorporate water into the matrix and stability of matrix), structural rigidity, and durability during applications. Natural polymers, which are very hydrophilic, are popular support materials for enzyme immobilization since the residues in these polymers contain hydroxyl groups, which are ideal functional groups for participating in covalent bonds. A frequently encountered disadvantage of immobilization by covalent binding is that it places great stress on the enzyme. The necessary harshness of the immobilization procedure nearly always leads to considerable changes in conformation and a resultant loss of catalytic activity.[72,78,79]

44.3.1.4 Crosslinking

The crosslinking method is based on the formation of covalent bonds between the enzyme or active molecules, by means of bi- or multifunctional reagents.[80,81] The individual biocatalytic units (enzymes, organelles, whole cells) are joined to one another with the help of bi- or multifunctional reagents (e.g., glutardialdehyde, glutaraldehyde, glyoxal, diisocyanates, hexamethylene diisocyanate, toluene diisocyanate, etc.). Enzyme crosslinking involves normally the amino groups of the lysine but, in occasional cases, the sulfhydryl groups of cysteine, phenolic OH groups of tyrosine, or the imidazol group of histidine can also be used for binding. Figure 44.4 shows how the biocatalysts can be linked by a simple crosslinking process (Figure 44.4a) and also by co-crosslinking, in which inert molecules are incorporated in the high-polymer network in order to improve the mechanical and enzymatic immobilized preparation (Figure 44.4b).

The advantages and disadvantages of a given matrix must be taken into account when considering the appropriate procedure for a given enzyme immobilization. One advantage is the simplicity of the process. The main disadvantages are the fragility of the particles produced in some

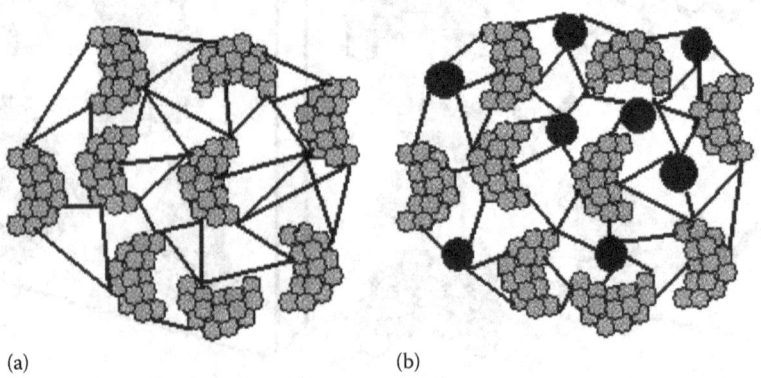

(a)   (b)

FIGURE 44.4 Biocatalysts immobilized by means of (a) crosslinking and (b) co-crosslinking with inert molecules incorporated.

cases and diffusion limitations. Since crosslinking and co-crosslinking usually involve covalent bonds, immobilized biocatalysts in this way frequently undergo changes in the conformation with a resultant loss of activity. The isomerization of glucose process is a very important example of the industrial application using biocatalysts crosslinked with glutaraldehyde. Some of the immobilized preparations used in these large-scale processes are produced simply by glutaraldehyde treatment of bacterial cell masses that have formed fine particles.[10]

44.3.1.5 Entrapment and Encapsulation

The entrapment method for immobilization consists of the physical trapping of the active components into a film, gel, fiber, coating, or microencapsulation[73] (see Figure 44.5). This method can be achieved by mixing an enzyme or active molecule with a polymer and then crosslinking the polymer to form a lattice structure that traps the enzyme. Microencapsulated enzymes are formed by enclosing enzymes solution within spherical semipermeable polymer membranes with controlled porosity.

While the encapsulation of dyes, drugs, and other chemicals has been known for some time, it was not until the mid-1960s that such a method was first applied to enzymes. Since that first report, a number of other enzymes have been successfully immobilized via microencapsulation, using a number of different materials and methods to prepare the microcapsules. The advantages of this immobilization method are the extremely large surface area between the substrate and the enzyme, within a relatively small volume, and the real possibility of simultaneous immobilization. The major disadvantages of this method include the occasional inactivation of enzyme during microencapsulation and the high enzyme concentration required. In addition, to retain the enzyme, the pore size needs to be very low and these systems tend to be very diffusion limited.[72]

44.3.1.6 Protein Fusion to Affinity Ligands and Enzymatic Conjugation

As described before, there are many methods for protein immobilization, but some of them require chemical modification of the matrix, which may result in material degradation, especially when biodegradable polymers are used. In addition, these modifications, necessary to attach the enzyme to the matrix, often result in the loss of enzyme activity as well as the inclusion of toxic organic

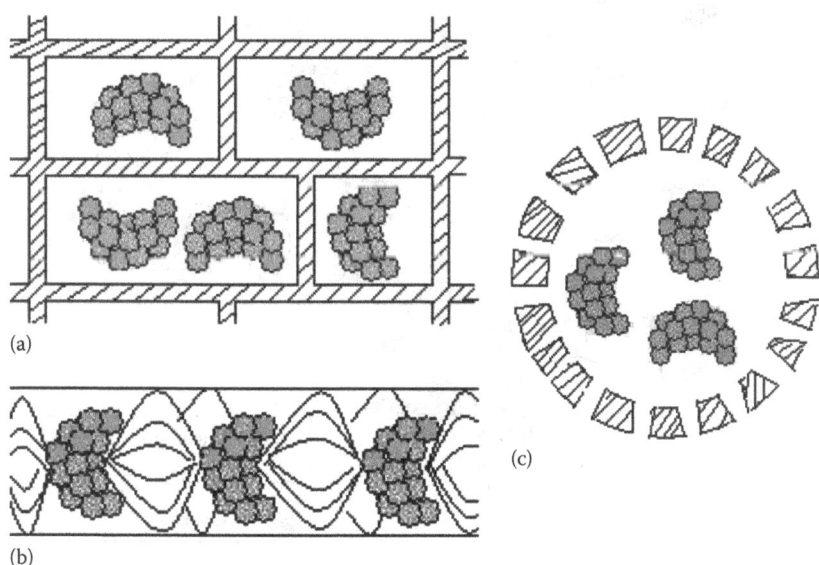

FIGURE 44.5 Enzyme encapsulation in a (a) matrix, (b) fiber, and (c) capsule.

compounds, which have to be removed before the system can be used in biomedical applications. In this type of application, the efficacy of immobilized biomolecules for stimulating specific cell responses (e.g., proliferation or differentiation) depends on the mode by which these modulators are presented to the target cells. In these cases, it is important to ensure the correct orientation and full bioactivity of the molecules when they are immobilized. Covalent binding by chemical coupling, again, might hinder ligand–receptor interaction or prevent receptor dimerization and capping on target cells.

In nature, there are certain protein molecules, such as lectins,[14,82] avidin,[83] immunoglobulin G (IgG) binding domains of protein G and protein A,[84,85] and carbohydrate binding modules[86,87] (present in many polysaccharide-degrading enzymes), that bind with high affinity and specifically to certain molecules or solid surfaces. These binding domains may be used as affinity tags for immobilizing proteins to affinity adsorbents. In this technique, DNA encoding a polypeptide affinity tag is fused to the gene of interest, and the expression of the gene results in a fusion protein. Such a fusion protein could be immobilized by the specific binding of the affinity tag to an affinity adsorbent (Figure 44.6a). With this method, the conformational changes in the protein upon immobilization are minimal, and the immobilized biomolecule could retain high activity. In addition, since fusion proteins are specifically immobilized on the support materials, these methods normally allow the immobilization of high densities of ligands and can also simplify the immobilization procedure.

The use of genetic engineering techniques to construct chimeric proteins, containing a functional domain displaying bioactivity together with an affinity domain, has proven to be a very

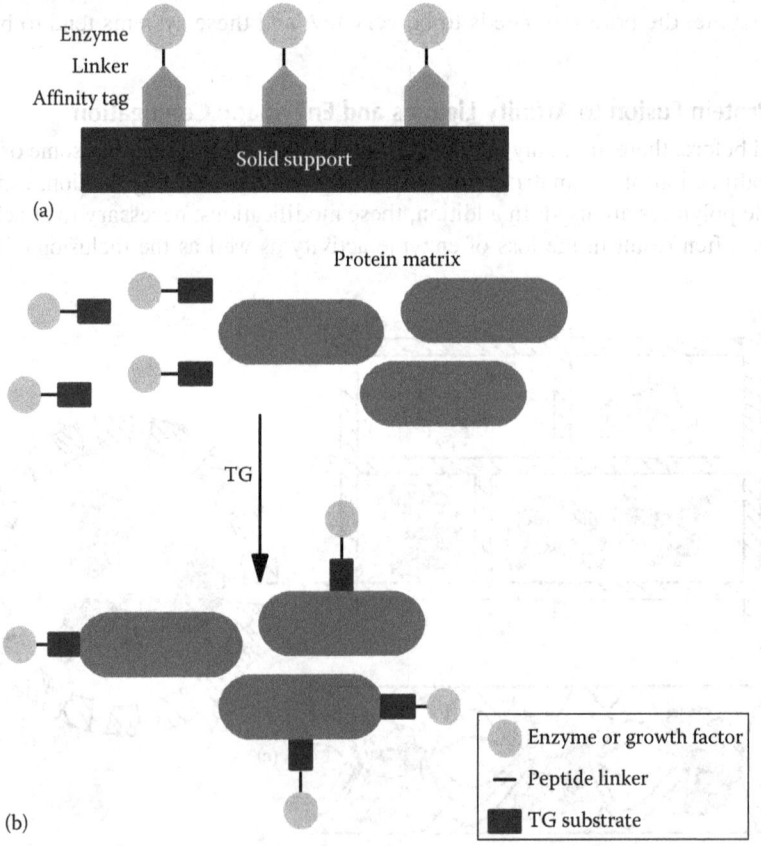

FIGURE 44.6 (a) Enzyme immobilization to solid matrices via protein fusion to an affinity ligand and (b) enzymatic conjugation catalyzed by transglutaminase.

useful approach for immobilizing biomolecules on solid materials. Several proteins, including enzymes,[53,88,89] antibodies,[90,91] cytokines,[92] and streptavidin,[23] have been immobilized on the surface of cellulosic matrices by fusing genetically these proteins to cellulose-binding domains.

By using protein fusion techniques, bifunctional proteins can be prepared without changing their activity and binding properties.[53,87,89] This approach has found various applications in biotechnology, diagnostics, and medicine for the purification and immobilization of biologically active proteins. It can be used, for instance, to promote the attachment of several mammalian cells to different surfaces by fusing a variety of peptides, growth factors, and cytokines to a specific binding domain, and this might be useful for the activation and growth of progenitor cells in culture.

The binding affinity between streptavidin and biotin is among the strongest noncovalent bonds known to exist ($K_D = 10^{-15}\ M$).[83] Therefore, the high affinity coupling of the biotin–avidin system has been used to immobilize different biomolecules on the surface of biomaterials and biosensors.[94,95] It consists of using avidin as a bridge between the biotinylated surface and biotinylated short ligand molecules.

The immobilization of biomolecules on protein matrices may also be achieved via an enzymatic reaction catalyzed by transglutaminase (TG) enzyme. TG catalyzes the acyl transfer reaction between the γ-carboximine group of a peptide-bound glutaminyl residue and a primary amino group of various protein substrates. The result of this reaction is the formation of an irreversible crosslinked, insoluble supramolecular structure.[96] In addition, TG can be used to bind glutamine-containing peptides or polypeptides to NH_2 surfaces,[69] and this methodology is schematically represented in Figure 44.6b. This approach was used by Sakiyama et al.[97,98] to incorporate heparin-binding peptides and to design a growth factor delivery system. In the latter work, they developed a growth factor fusion protein, containing a bi-domain peptide consisting of a β-nerve growth factor (β-NGF) and an exogenous factor XIIIa (substrate for enzymatic crosslinking). The fusion protein was covalently immobilized within a three-dimensional cell in-growth matrix based on fibrin using the transglutaminase activity of factor XIIIa.

44.3.2 Enzyme Immobilization in Biomedical Applications

Biomaterials can be combined with biomolecules, such as enzymes and growth factors, to yield biologically functional systems. There is a wide and diverse range of materials and methods available for enzyme immobilization on or within the biomaterial. The methods for immobilizing enzymes and other biomolecules are the same as described in Section 44.3.1, but the choice of the method depends largely on the final application. Furthermore, the criteria for selecting the immobilization methods should also take into account that the immobilized enzyme should retain an acceptable level of activity over a certain period of time in terms of economic or clinical aspects.

The methods used for the administration of immobilized enzymes may be divided into two principal groups[9]: immobilized enzymes that are intended for prolonged circulation and enzymes that must be necessarily present in different tissues and organs of the body. In the second case, the immobilized enzyme is intended for local deposition during the treatment of discrete lesions (e.g., thrombi, tumors, atherosclerotic injuries) or of the individual organs.[9]

Table 44.3 describes some examples of immobilized enzymes in various biomaterial supports for different biomedical applications.

44.3.2.1 Biologically Functional Surfaces

Biomaterials, especially when used in tissue engineering applications, must have the capacity to induce tissue regeneration/repair in order to achieve a more rapid recovery of the defect. At present, the existing scaffolds are not satisfactory in achieving rapid and full recovery of the defect. The attachment of cells to biomaterials, and their subsequent spreading, are mediated by extracellular

TABLE 44.3
Examples of Immobilized Enzymes in Various Biomaterials for Different Biomedical Applications

Enzyme	Biomaterial (Carrier)	Immobilization Method	Application	References
Alkaline phosphatase	pHEMA[a]	Entrapment	Induction of bone and cartilage mineralization	[99]
	Glassy carbon	Protein fusion to affinity tag	Biosensor/enzyme-linked immunoassays	[95]
α-amylase	pHEMA[a]	Covalent binding	Fixed-bed reactor	[100]
	Starch polymeric blend	Encapsulation	Tailor the degradation rate of starch-based biomaterials	[101]
β-galactosidase	Gelatin	Encapsulation	Not specified	[102]
Glucose oxidase	Proteins	Enzymatic conjugation	Biosensor	[69]
α-chymotrypsin	Cellophane	Plasma modification/ covalent binding	Not specified	[103]
Acetylcholinesterase Choline oxidase	pHEMA[a]	Entrapment Covalent binding Ionic interactions	Biosensor	[104]
Papain	Polyethylene Glass surfaces	Plasma modification Covalent binding	Not specified	[105]
Lysozyme	Gelatin and succinylated gelatin	Adsorption	Reduction of prosthetic valve endocarditis	[106]
	Poly(ethylene glycol)– poly(aspartic acid) copolymer	Ionic interactions Entrapment/ionic binding	Delivery of lysozyme as a lytic enzyme	[28]
Heparinase	Sepharose	Covalent binding	Bioreactor for extracorporeal elimination of heparin from blood	[52]
Phospholipase A$_2$	Agarose	Not mentioned	Bioreactor for treatment of hypercholesterolemia (reduce plasma cholesterol)	[107]
Bilirubin oxidase	Agarose	Covalent binding	Bioreactor for removal of bilirubin from blood	[61]
Prolidase	PLGA[b] microspheres	Encapsulation	Enzyme replacement therapy	[25]

[a] Poly(2-hydroxy ethyl methacrylate).
[b] Poly(D,L-lactide-co-glycolide).

matrix (ECM) glycoproteins such as fibronectin, collagen, etc. ECM glycoproteins contain short sequences with cell attachment properties, which interact with the integrin family of cell surface receptors. The peptide sequence Arg-Gly-Asp (RGD) has been identified as being capable of interacting with cell surface receptors. Thus, the immobilization of biologically active molecules on the surface of biomaterials for presenting effectors to target cells or to induce a particular effect is of great interest, since the immobilization of active agents presents the advantage of providing a continuous and localized stimulus for cell proliferation. Unlike nonimmobilized active agents, which are often consumed by cells, immobilized biomolecules remain bound to a substrate that is not consumed by cells and thus remain available to stimulate growth of additional cells. This is particularly useful in perfusion cultures in which growth medium is continuously added and removed to allow long-term cell proliferation.

Biodegradable polymers have been used as scaffolding materials for various tissue engineering applications because they can provide the support on which cells and tissues can adhere, but they can also guide and regulate the proliferation and activities of the adhered cells. However, the intrinsic hydrophobic property of some of these polymers[108] restricts their applications as cell colonizing materials. Many methods have been used to modify the properties of polymer surface, such as plasma treatment-induced grafting polymerization, ozone oxidation, and immobilization of enzymes, and special biologically active agents have been used to introduce reactive groups onto polymeric surfaces.[109]

Various strategies have been developed to incorporate bioactive agents on the surface of biomaterials for controlling cell and tissue responses. The immobilization process can be involved, enriching surfaces for enhancing the cellular adhesion. Biomolecules such as enzymes, antibodies, antigens, peptides, or drugs have been immobilized on or within polymeric systems. An example of an adhesive protein is fibronectin, which is able to promote cellular adhesion through binding to integrin receptors, and this interaction has also been shown to play a role in cell growth, differentiation, and overall regulation of cell function.[77,110] Hern and Hubbell[111] showed that the incorporation of the adhesion peptide RGD into a nonadhesive hydrogel proved to be useful for tissue resurfacing. There are numerous other adhesion peptides for targeting particularly desirable cell types and to modulate biological responses.

Urokinase has been widely used for the clinical treatment of thrombogenetic disease and hemorrhoidal disease. Artificial organ materials, on which urokinase was immobilized for its fibrinolytic activity, have been developed for blood-compatible materials. For example, Liu et al.[55] immobilized urokinase by encapsulation in poly(2-hydroxyethyl methacrylate) and König et al.[56] introduced urokinase on the surface of the polytetrafluoroethylene using plasma modification technique by covalent bond. Another example of immobilized urokinase application was reported by Kato and coworkers,[54] who had used urokinase immobilized in a Teflon catheter for treatment of thrombosis.

Most of the studies found in the literature, regarding the surface functionalization of biomaterials with biological molecules, include the incorporation of adhesion and differentiation factors. The same approach may be used to incorporate specific enzymes able to regulate a number of cell functions. For instance, it is known that mitogen activated protein (MAP) kinase, upon activation by dual phosphorylation at threonine and tyrosine residues, is able to activate downstream targets that have been implicated in controlling gene expression, cell differentiation, and proliferation.[112] This enzyme may be immobilized on the surface of biomaterials to control cell response, but other possibilities using different enzymes remain unexplored.

44.3.2.2 Enzyme Delivery

Although purified enzymes are now available for some enzyme deficiency diseases (see Table 44.1), there are many problems in delivering the enzyme to the required site under such conditions that it will remain stable and active for a reasonable time. Normally, quite large amounts of enzyme are necessary with high level of purity and in a nonimmunogenic form.[9,14,113,114] In addition, many enzymes when administered are inactivated or degraded fairly rapidly. The delivery of therapeutic molecules requires, therefore, efficient strategies to have a precise control on their release profile according to specific locations. It might be possible to control the release of such molecules by creating delivery systems sensitive to changes in pH, temperature, or salt concentration or to the feedback provided by cells. The concept of enzyme-activated drugs in therapy is scientifically, as well as clinically, attractive, as it allows the chemist and enzymologist full intellectual rein in designing interlinked systems.[114] As therapeutic drugs, enzymes possess several attributes such as high specificity toward substrate, high solubility for preparing liquid formulations, and optimum activity under physiological conditions.[115] The administration of enzymes, in cases of enzyme deficiency and inborn errors of metabolism and in the treatment of certain types of cancer, appears to offer a successful form of therapy. Cancer therapy based on the delivery of enzymes to tumor sites has advanced in several directions since antibody-directed enzyme/prodrug therapy was first described.[116] Nanospheres, nanocapsules, liposomes, micelles, and other nanoparticulates are frequently referred to as carriers for delivery of therapeutic and diagnostics agents.[113]

Asparagine is an essential amino acid for certain types of leukemias that lack asparagines synthetase activity. The activity of L-asparaginase is to degrade asparagine into aspartate and ammonia. Therefore, asparaginase has been of interest to biochemists and clinicians as a possible cancer therapeutic agent. Some success has been achieved in administering asparaginase in capsules made of nylon and polyurea to mice and rats. Although asparaginase has been found to be effective in the treatment of some patients,[117] it may have several serious side effects.[41] Relatively high concentrations of the enzyme are needed for it to be clinically effective. These levels cause a wide range of toxic effects on several organs including the liver, pancreas, kidneys, and brain. The enzyme may also be recognized as foreign by the body and potentially severe immunogenic responses will be stimulated, resulting in hypersensitivity reactions. To overcome these problems, immobilized enzyme derivatives have been prepared on various supports for extracorporeal treatment (see Section 44.3.2.3).

Enzymes may be also used in cancer therapy as prodrug activators. This therapy consists basically in using a drug that has been chemically modified so that it remains inactive until specifically activated by an enzyme at the target site.[11]

Another example of an anticancer enzyme is hyaluronidase.[50] Hyaluronidase is a globular enzyme of endoglycosidase action, which can depolymerize hyaluronic acid in the organism, decreasing its viscosity and increasing tissue permeability. Hyaluronidase has been utilized extensively as an adjunct in anticancer chemotherapy regimens, suggesting that hyaluronidase has intrinsic anticancer properties against tumor growth.[51] *In vitro* studies in tissue culture with tumor spheroids and *in vivo* tests using animal models demonstrated the beneficial effect of hyaluronidase for the penetration of drugs into tumor tissue. Later, in a prospective clinical trial, hyaluronidase significantly improved the outcomes of patients with bladder carcinoma if the enzyme was administrated topically together with mitomycin C.[49,50] Hyaluronidase is also used for local application (subcutaneous injections) during treatment of joint disease, in dermatology, and in ophthalmology.[48] Hyaluronidase had been used in ophthalmology with the aim of formation of a thinner scar and to prevent necrosis after paravataes with zytostatics.

Trypsin has also been used to remove dead tissue from wounds, burns, and ulcers to speed the growth of new tissue and skin grafts, as well as to inhibit the growth of some contaminant organisms. The inappropriate activation of trypsinogen within the pancreas leads to development of pancreatitis.[47] Once trypsin is activated, it is capable of activating many other digestive proenzymes. These activated pancreatic enzymes further enhance the auto-digestion of the pancreas. Many materials, such as nylon, polysulfone, glycidyl methacrylate, chitosan, cellulose, and cellulose derivatives, have been used for trypsin immobilization.[44,46,47]

Enzyme therapy has also been tested for pancreatic insufficiency and cystic fibrosis. Pancreatic insufficiency can be alleviated by administrating orally enteric-coated microspheres containing lipase, amylase, and proteases. A special polymer coating protects the enzymes at low pH, such as in the stomach, and then releases them in the intestine at physiological pH.[11,14]

Lysozyme is a good example of an enzyme that catalyzes chemical reactions in the cell. Lysozyme acts to kill bacteria by cleaving the covalent bond between the alternating polysaccharides that compose peptidoglycan in bacterial cell walls.[27] The human salivary defense proteins and lysozyme are known to exert a wide antimicrobial activity against a number of bacterial, viral, and fungal pathogens *in vitro*. Therefore, these proteins, alone or in combinations, have been incorporated as preservatives in foods and pharmaceuticals as well as in oral health care products to restore saliva's own antimicrobial capacity in patients with dry mouth. These antimicrobials used in oral health care products, such as dentifrices, mouth rinses, moisturizing gels, and chewing gums, have been purified from bovine colostrum. Other studies had been reported with lysozyme bound to chitosan, silica gel by means of physical adsorption, crosslinking to a polystyrene divinylbenzene matrix by the formation of ionic bindings, and by covalent attachment to nonporous glass beads.[29] Harada and Kataoka[28] described lysozyme immobilized into poly(ethylene glycol)–poly(aspartic acid) micelle. Lysozyme was selected as a model protein to incorporate into the micelle because it has a high isoelectric point (pI = 11), is positively charged over a wide range of pH, and has practical usage in drug delivery application as a lytic enzyme. Chen and Chen[118] prepared immobilized lysozyme by carbodiimide method to form amide bonds with an enteric coating polymer (hydroxypropyl methyl-cellulose acetate succinate [AS-L]) as the carrier, which shows reversibly soluble-insoluble characteristics with pH changes.

The glucose-6-phosphate dehydrogenase enzyme catalyzes the oxidation of glucose-6-phosphate to 6-phosphogluconate while concomitantly reducing the oxidized form of nicotinamide adenine dinucleotide phosphate ($NADP^+$) to nicotinamide adenine dinucleotide phosphate (NADPH). NADPH, a required cofactor in many biosynthetic reactions, maintains glutathione in its reduced form. Reduced glutathione acts as a scavenger for dangerous oxidative metabolites in the cell. With the help of the enzyme glutathione peroxidase, reduced glutathione also converts harmful hydrogen peroxide to water. Red blood cells rely heavily on glucose-6-phosphate dehydrogenase activity because it is the only source of NADPH that protects the cells against oxidative stresses. People deficient in glucose-6-phosphate dehydrogenase are not prescribed, therefore, with oxidative drugs because their red blood cells undergo rapid hemolysis under this stress.[36] In Greece, glucose-6-phosphate dehydrogenase deficiency is the main cause of severe neonatal jaundice. The deficiency of this enzyme affects all races; the highest prevalence is among persons of African, Asian, or Mediterranean descent.[37] Study of immobilized glucose-6-phosphate dehydrogenase has been reported. Kotorman et al.[38] immobilized glucose-6-phosphate dehydrogenase from yeast on polyacrylamide beads possessing carboxylic functional groups activated by a water-soluble carbodiimide. They verified highest operational stability of immobilized glucose-6-phosphate dehydrogenase.

In relation to the use of immobilized enzymes with therapeutic purposes, catalase is one of the most interesting because it is employed to accelerate healing as well as to correct hereditary deficiencies and, in combination with hydrogen peroxide, as an antiseptic against anaerobes.[33,34] The catalase enzyme has the ability to decompose hydrogen peroxide into oxygen and water, playing a central role in controlling the hydrogen peroxide concentration in human cells. More than 98% of blood catalase is localized in erythrocytes.[119] These cells, with their high catalase level, provide a general protection against the toxic concentration of this small hydrogen peroxide molecule. The deficiency of catalase could cause acatalasemia.[35,119] Several methods have been developed for the immobilization of catalase.[1,74,120] Immobilization is often accompanied by changes in the enzymatic activity, optimum pH, affinity to the substrate, and stability. The extent of these changes depends on the enzyme, carrier support, and the immobilization conditions.[121] The shift in the optimum pH, from acidic or alkaline to neutral pHs, may be useful for biomedical applications since it will allow the use of some enzymes (more active at low or high pHs) under more physiological conditions.

The approach developed by Sakiyama-Elbert et al.,[98] consisting in a cell-triggered growth factor delivery system, may also be used for the release of other important therapeutic molecules.

44.3.2.3 Bioreactors for Extracorporeal Enzyme Therapy

Extracorporeal shunts have been proposed[72,122] for the treatment of several clinical conditions. The most likely applications for enzymatic treatment are the removal of urea during kidney failure, removal of toxins (e.g., paracetamol) during liver failure, or the reduction of key metabolites from the circulation to treat cancer.

Urease is one of the most important enzymes in biomedical applications. Urease is an enzyme that catalyzes the hydrolysis of urea to form ammonia and carbon dioxide. Urea is one of the main metabolic end products, and the removal of its excess has been a major problem for patients suffering from renal failure. Hence, its immobilization by entrapment has been investigated by many workers for applications in biosensors and as artificial kidneys. The most attention has been given to the development of enzyme reactors, where the urea would be removed and the dialysis fluid prepared for further use.[30,31] The use of this enzyme is often limited due to its high cost, availability in small amounts, instability, and the limited possibility of feasible recovery of these biocatalysts from a reaction mixture. Numerous synthetic and natural polymeric supports have been used for urease immobilization, and their uses in medical and technical fields are well reported. The covalent bond of urease in different supports has been reported in many studies. Some commonly used supports are chitosan-poly(glycidil methacrylate), carboxymethylcellulose, polyurethane, sepharose-2B, polyacrylamide, ion exchange resins, copolymers of polyglycidylmethacrylate, calcium alginate beads, poly(vinyl alcohol) (PVA), hydroxyapatite, 2-dimethylaminoethylmethacrylate, poly(ethylene glycol dimethacrylate/2-hydroxy ethylene methacrylate) microbeads, poly(caprolactone)/starch, and poly(orthoesters).[32,109,123–125]

As mentioned before in Section 44.3.2.2, L-asparaginase has been used for treating leukemias and disseminating cancers that require asparagines for growth, but this treatment presents several serious side effects. To overcome these problems, immobilized enzyme derivatives have been prepared on various supports for extracorporeal treatment. Blood can be passed over the immobilized enzyme, thus depleting the asparagine supply needed by the cancer cells. The enzyme does not come into direct contact with the organs to which it is toxic, and hypersensitivity reactions do not occur. With this type of treatment, however, the blood plasma must be first separated from the cells to minimize cell damage and then passed through a separate column containing the immobilized enzyme. This process requires that the blood remain outside the body for relatively long periods of time, resulting in the denaturation and depletion of many plasma proteins.[41,42] Some techniques have been developed to minimize this problem, such as the use of a porous hollow-fiber plasmapheresis device. With this system, the plasma can be separated from the whole blood and contact with immobilized enzyme in one passage, thus minimizing its time outside the body and then reducing the damage to the plasma proteins. Adsorption techniques have been used to immobilize asparaginase onto hollow fibers after first coating the fibers with albumin and then crosslinking the enzyme with glutaraldehyde.[41,126] Maciel and Minim[43] also reported that the use of L-asparaginase covalently attached to nylon tubing may constitute a useful system to be used in clinical applications.

Bilirubin oxidase is also an example of enzyme used in extracorporeal applications.[61] All human newborns accumulate bilirubin to levels greater that those in adults, and 20% accumulate enough to stain their skin, resulting in jaundice. Bilirubin binds to cellular and mitochondrial membranes, causing cell death in a variety of tissues. Clinically, bilirubin toxicity may lead to mental retardation, cerebral palsy, deafness, seizures, or death. The most common treatments for jaundiced infants are phototherapy and exchange transfusion. This technique presents serious problems such as hypoglycemia, hypocalcemia, acidosis, transmission of infectious, etc. Lavin et al.[61] reported the use of a highly specific enzyme to remove bilirubin from the bloodstream using a small reactor (extracorporeal circuit) containing bilirubin oxidase covalently immobilized in agarose beads. These researchers obtained good results for the removal of the bilirubin in humans and in genetically jaundiced rats.

Heparinase, an enzyme that degrades heparin into small polysaccharides, has also been immobilized into an extracorporeal device (artificial kidney bioreactor) to eliminate the anticoagulant properties of heparin (used to prevent clotting in the device) before the blood returns to the patient.[52]

44.3.2.4 Diagnostic Assays and Biosensors

Isolated or combined enzymes are being used in medicine as useful tools for clinical analysis. About 50 different enzymes are used in different aspects of clinical diagnoses, and for most of these, much higher levels of purity are required than for most industrial enzymes. Two of the major enzymes used are peroxidase from horseradish and alkaline phosphatase from beef intestinal mucosa, both being required for immunoassays. The enzymes may be used in test strips, ELISA, biosensors, and autoanalyzers.[11]

The serum uric acid concentration is an important index for clinical diagnosis of gout, leukemia, toxemia of pregnancy, and severe renal impairment.[127] A number of enzymes are assayed in serum and urine for diagnostic purposes; the more frequently used ones are discussed below.

Alkaline phosphatases (ALP) are a group of enzymes found primarily in the liver (isoenzyme ALP-1) and bone (isoenzyme ALP-2). There are also small amounts produced by cells lining the intestines (isoenzyme ALP-3), the placenta, and the kidney (in the proximal convoluted tubules). What is measured in the blood is the total amount of alkaline phosphatase released from these tissues into the blood. As the name implies, this enzyme works best at an alkaline pH (pH 10), and thus the enzyme itself is inactive in the blood. Alkaline phosphatase acts by splitting off phosphorus (an acidic mineral), creating an alkaline pH. The primary importance of measuring alkaline phosphatase is to check the possibility of bone or liver diseases.[128]

Another application of immobilized enzymes is the development of improved sensing devices.[9] Because of their high specificity for given substances, enzymes and monoclonal antibodies are particularly suitable for use as sensors.[129] The membrane-covered electrode described by Clark in 1959 is the dominating sensor for the measurement of dissolved oxygen.[130] Numerous modifications of the original concept have been developed. For instance, biosensors using enzymes have been used to detect the presence of various organic compounds, and recent developments have proven to be both rapid and highly selective. They have been used in important applications such as in clinical laboratories, fermentation processes, and pollution monitoring.[131] Most of them have used a free or immobilized enzyme and an ion-sensitive electrode that measures indirectly (e.g., by temperature or color changes produced by an enzymatic reaction) the presence of a product whose formation is catalyzed by the enzyme. The biosensors usually have immobilized biological molecules attached to the surface of a transducer that allows an electronic or optical signal to be converted into an appropriate signal. This type of biosensor could be used to measure glucose, sucrose, lactose, L-lactate, galactose, L-glutamate, L-glutamine, choline, ethanol, methanol, hydrogen peroxide, starch, uric acid, etc., by using specific enzymes.[31,127]

Glucose oxidase is normally used to assay glucose concentration. Glucose sensors are the biosensors that have attracted much interest in both research and applications fields. One particularly important medical application of improved biosensors could be in the treatment of diabetic patients for whom proper levels of insulin and glucose must be maintained. For instance, small implantable devices for sampling blood to determine the levels of glucose and regulate the delivery of insulin could be developed using this enzyme.[39] A great variety of immobilization methods (e.g., encapsulation, entrapment) and transducers have been developed to construct the glucose sensors with better performance and practicability since the work of Clark and Lyons in 1962.[40] Several materials have been used, such as polyethylene terephthalate (PET), polyacrylamide, N-isopolyacrylamide, sol-gel, poly(2-hydroxyethyl methacrylate), alginate, artificial resins, glass, etc.[31,78,132]

For example, Zhang and Cass[133] have also immobilized alkaline phosphatase on a nanoporous nickel-titanium film for sensor applications.

A number of other enzymes have been described with great potential for medical applications, including carboxypeptidases, collagenase, fibrinolysin, pepsin,[44] streptokinase,[45] subtilisin, thrombin, tissue plasminogen activator, α-amylase, α-galactosidase, glucoamylase, lactase (-galactosidase), pectinase, pancreatin, phospholipases, cholesterol esterase and other DNases, RNases, phosphatases, esterases, sulfatases, isomerases, glucose isomerase, superoxide dismutase, cholesterol esterease, creatine kinase, and penicillin acylase.[45,58,114,134–136]

44.3.2.5 Advantages and Disadvantages of Immobilized Enzymes

The use of immobilized enzymes normally offers several advantages over free enzymes, such as increased stability, localization, and retention of the molecules at the material surface, which enables easier handling, repeated use, and decreased cost. Other important advantages of using therapeutic immobilized enzymes are the prolonged blood circulation lifetime without the loss of specific activity[9] and the lower immunogenicity.[11,137] This advantage is particularly important for delivering enzymes or other biomolecules and may constitute an alternative and suitable method for the enzyme replacement therapy. However, some limitations have been attributed to the use of immobilized enzymes in biomedical applications, such as mass transfer resistances (substrate in and product out), adverse biological responses of enzyme support surfaces (*in vivo* or *ex vivo*), fouling by other biomolecules, greater potential for product inhibition, and sterilization difficulties.[126,137] Although the preparation of sterile immobilized enzyme systems may be complex, sterilization may be achieved by filtrating all the reagents and protein solutions through 0.2-μm filters and working under aseptic conditions.

A very important issue regarding the use of enzymes or other products derived from biological or biotechnological processes in medical applications is to ensure that these therapeutic products do not contain any pyrogenic material, toxins, or infectious agents able to cause harmful effects.[138] For that, it is necessary to perform a complete examination of the products to test their safety in terms of local tolerance, toxicity, carcinogenicity, and immunogenicity, among other pharmacological safety tests.[138,139] Taking into account the diversity in the range of products and the uncertainty about the regulatory status of some of them, it is necessary to design safety evaluation programs to provide useful information to the responsible of clinical trials and to ensure patient safety.

44.4 CONCLUSIONS AND FUTURE PERSPECTIVES

Enzyme instability, combined with the high cost associated with their isolation and purification, had been restricting the general use of therapeutic enzymes on a clinical basis. With the advances made in recombinant DNA technology, it is possible, by means of using adequate expression systems, to put available in larger quantities many enzymes, both for the assay of metabolites and for enzyme replacement therapy. The immobilization of enzymes on support materials had contributed largely to the success of diagnosis and enzyme therapy approaches.

The recent progress in biological science had revealed many types of therapeutic proteins able to regulate various cell functions. On the other hand, the development of new immobilization strategies, such as selective immobilization of proteins to self-assembled monolayers presenting active site-directed capture ligands[140] or protein immobilization within specific locations (protein patterning),[82] may constitute the basis of future immobilization methods. This is particularly important for biomedical applications where it is necessary to control the densities of immobilized proteins, the binding strength, and, most important, their binding orientation. The success of therapeutic agent delivery strategies will be mainly dependent on the developments in these research fields.

ACKNOWLEDGMENTS

S. A. Costa and H. S. Azevedo thank the Portuguese Foundation for Science and Technology for providing them postdoctoral scholarships SFRH/BPD/8469/2002 and SFRH/BPD/5744/2001, respectively. This work was partially supported by FCT Foundation for Science and Technology, through funds from the POCTI and/or FEDER programs.

REFERENCES

1. Cetinus, S.A. and Oztop, H.N., Immobilization of catalase into chemically crosslinked chitosan beads, *Enzyme Microb. Technol.*, 32, 889, 2003.
2. Costa, S.A. et al., Studies of stabilization of native catalase using additives, *Enzyme Microb. Technol.*, 30, 387, 2002.
3. Matsumoto, M. et al., Effects of polyols and organic solvents on thermostability of lipase, *J. Chem. Technol. Biotechnol.*, 70, 188, 1997.

4. O'Fagain, C., Enzyme stabilization—Recent experimental progress, *Enzyme Microb. Technol.*, 33, 137, 2003.
5. Khajeh, K. et al., Chemical modification of bacterial alpha-amylases: Changes in tertiary structures and the effect of additional calcium, *Biochim. Biophys. Acta Protein Struct. Mol. Enzymol.*, 1548, 229, 2001.
6. Minshull, J. et al., Engineered protein function by selective amino acid diversification, *Methods*, 32, 416, 2004.
7. Bilkova, Z. et al., Oriented immobilization of chymotrypsin by use of suitable antibodies coupled to a nonporous solid support, *J. Chromatogr. A*, 852, 141, 1999.
8. Fagain, C.O., Understanding and increasing protein stability, *Biochim. Biophys. Acta*, 1252, 1, 1995.
9. Torchilin, V.P., Immobilised enzymes as drugs, *Adv. Drug Del. Rev.*, 1, 41, 1987.
10. Kennedy, J.F., Handbook of enzyme technology, in *Principles of Immobilization of Enzymes*, 3rd edn., Wiseman, A., Ed., Prentice Hall Ellis Harwood, New York, 1995, p. 235.
11. Price, N.C. and Stevens, L., *Fundamentals of Enzymology. The Cell and Molecular Biology of Catalytic Proteins*, 3rd edn., Oxford University Press Inc., New York, 1999.
12. Kellems, R.E. et al., Adenosine deaminase deficiency and severe combined immunodeficiencies, *TIG*, October, 278, 1985.
13. Tager, J.M., Biosynthesis and deficiency of lysosomal-enzymes, *Trends Biochem. Sci.*, 10, 324, 1985.
14. Poznansky, M.J., Enzyme-protein conjugates: New possibilities for enzyme therapy, *Pharmacol. Ther.*, 21, 53, 1983.
15. Goldberg, D.M., Enzymes as agents for the treatment of disease, *Clin. Chim. Acta*, 206, 45, 1992.
16. Bailey, J.E. and Ollis, D.F., *Biochemical Engineering Fundamentals*, 2nd edn., McGraw-Hill, Singapore, 1986.
17. Das, A.K. et al., Biochemical analysis of mutations in palmitoyl-protein thioesterase causing infantile and late-onset forms of neuronal ceroid lipofuscinosis, *Hum. Mol. Genet.*, 10, 1431, 2001.
18. de Grey, A.D., Bioremediation meets biomedicine: Therapeutic translation of microbial catabolism to the lysosome, *Trends Biotechnol.*, 20, 452, 2002.
19. Lund, P.M., Distribution of oculocutaneous albinism in Zimbabwe, *J. Med. Genet.*, 33, 641, 1996.
20. Nakamura, E. et al., A novel mutation of the tyrosinase gene causing oculocutaneous albinism type 1 (OCA1), *J. Dermatol. Sci.*, 28, 102, 2002.
21. Weng, Y.H. et al., Hyperbilirubinemia in healthy neonates with glucose-6-phosphate dehydrogenase deficiency, *Early Hum. Dev.*, 71, 129, 2003.
22. Vulliamy, T. et al., The molecular basis of glucose-6-phosphate dehydrogenase deficiency, *Trends Genet.*, 8, 138, 1992.
23. Schulpis, K.H. et al., The effect of neonatal jaundice on biotinidase activity, *Early Hum. Dev.*, 72, 15, 2003.
24. Nyhan, W.L., Multiple carboxylase deficiency, *Int. J. Biochem.*, 20, 363, 1988.
25. Genta, I. et al., Enzyme loaded biodegradable microspheres in vitro ex vivo evaluation, *J. Control. Rel.*, 77, 287, 2001.
26. Brewerton, L.J. et al., Polyethylene glycol-conjugated adenosine phosphorylase: Development of alternative enzyme therapy for adenosine deaminase deficiency, *Biochim. Biophys. Acta*, 1637, 171, 2003.
27. Brouwer, J. et al., Determination of lysozyme in serum, urine, cerebrospinal-fluid and feces by enzyme-immunoassay, *Clin. Chim. Acta*, 142, 21, 1984.
28. Harada, A. and Kataoka, K., Novel polyion complex micelles entrapping enzyme molecules in the core: Preparation of narrowly-distributed micelles from lysozyme and poly(ethylene glycol)-poly(aspartic acid) block copolymer in aqueous medium, *Macromolecules*, 31, 288, 1998.
29. Crapisi, A. et al., Enhanced microbial cell-lysis by the use of lysozyme immobilized on different carriers, *Process Biochem.*, 28, 17, 1993.
30. Higa, O.Z. and Kumakura, M., Preparation of polymeric urease discs by an electron beam irradiation technique, *Biomaterials*, 18, 697, 1997.
31. Karube, I. and Nomura, Y., Enzyme sensors for environmental analysis, *J. Mol. Catal. B Enzym.*, 10, 177, 2000.
32. Ayhan, F. et al., Optimization of urease immobilization onto non-porous HEMA incorporated poly(EGDMA) microbeads and estimation of kinetic parameters, *Biores. Technol.*, 81, 131, 2002.
33. Akertek, E. and Tarhan, L., Characterization of immobilized catalases and their application in pasteurization of milk with H_2O_2, *Appl. Biochem. Biotechnol.*, 50, 291, 1995.
34. Emerson, D. et al., A catalase microbiosensor for detecting hydrogen peroxide, *Biotechnol. Tech.*, 10, 673, 1996.

35. Goth, L., A novel catalase mutation (a G insertion in exon 2) causes the type B of the Hungarian acatalasemia, *Clin. Chim. Acta*, 311, 161, 2001.
36. Zaitseva, E.A. et al., Stabilization mechanism of glucose-6-phosphate dehydrogenase, *Biocatalysis*, 41, 127, 2000.
37. Reclos, G.J. et al., Evaluation of glucose-6-phosphate dehydrogenase activity in two different ethnic groups using a kit employing the haemoglobin normalization procedure, *Clin. Biochem.*, 36, 393, 2003.
38. Kotorman, M. et al., Coenzyme production using immobilized enzymes. III. Immobilization of glucose-6-phosphate dehydrogenase from bakers' yeast, *Enzyme Microb. Technol.*, 16, 974, 1994.
39. Traitel, T. et al., Characterization of glucose-sensitive insulin release systems in simulated in vivo conditions, *Biomaterials*, 21, 1679, 2000.
40. Eggins, B., *Biosensor: An Introduction*, John Wiley & Sons, New York, 1999.
41. Gombotz, W. et al., Immobilized enzymes in blood-plasma exchangers via radiation grafting, *Radiat. Phys. Chem.*, 25, 549, 1985.
42. Stecher, A.L. et al., Stability of L-asparaginase: An enzyme used in leukemia treatment, *Pharm. Acta Helv.*, 74, 1, 1999.
43. Maciel, R. and Minim, L.A., Adaptive control of an open tubular heterogeneous enzyme reactor for extracorporeal leukaemia treatment, *J. Process Control*, 6, 317, 1996.
44. Bolte, G. et al., Peptic-tryptic digests of gliadin: Contaminating trypsin but not pepsin interferes with gastrointestinal protein binding characteristics, *Clin. Chim. Acta*, 247, 59, 1996.
45. Koneracka, M. et al., Direct binding procedure of proteins and enzymes to fine magnetic particles, *J. Mol. Catal. B Enzym.*, 18, 13, 2002.
46. Guo, W. and Ruckenstein, E., Crosslinked mercerized cellulose membranes for the affinity chromatography of papain inhibitors, *J. Membr. Sci.*, 197, 53, 2002.
47. Hirota, M. et al., Significance of trypsin inhibitor gene mutation in the predisposition to pancreatitis, *Int. Congr. Ser.*, 1255, 41, 2003.
48. Maksimenko, A.V. et al., Chemical modification of hyaluronidase regulates its inhibition by heparin, *Eur. J. Pharm. Biopharm.*, 51, 33, 2001.
49. Pillwein, K. et al., Hyaluronidase additional to standard chemotherapy improves outcome for children with malignant brain tumors, *Canc. Lett.*, 131, 101, 1998.
50. St. Croix, B. et al., Reversal of intrinsic and acquired forms of drug resistance by hyaluronidase treatment of solid tumors, *Canc. Lett.*, 131, 35, 1998.
51. Lin, G. and Stern, R., Plasma hyaluronidase (Hyal-1) promotes tumor cell cycling, *Canc. Lett.*, 163, 95, 2001.
52. Langer, R. et al., An enzymatic system for removing heparin in extracorporeal therapy, *Science*, 217, 261, 1982.
53. Shpigel, E. et al., Immobilization of recombinant heparinase I fused to cellulose-binding domain, *Biotechnol. Bioeng.*, 65, 17, 1999.
54. Kato, H. et al., External venous shunt as a solution to venous thrombosis in microvascular surgery, *Br. J. Plast. Surg.*, 54, 164, 2001.
55. Liu, L.S. et al., Biological-activity of urokinase immobilized to cross-linked poly(2-hydroxyethyl methacrylate), *Biomaterials*, 12, 545, 1991.
56. König, U. et al., Plasma modification of polytetrafluoroethylene for immobilization of the fibrinolytic protein urokinase, *Surf. Coat. Technol.*, 119, 1011, 1999.
57. Miller, J.M. et al., Streptokinase and streptodornase in the treatment of surgical infections, *Lancet*, 261, 220, 1953.
58. Dempfle, C.E. et al., Plasminogen activation without changes in tPA and PAI-1 in response to subcutaneous administration of ancrod, *Thromb. Res.*, 104, 433, 2001.
59. Xu, Y. et al., Diverse roles of conserved asparagine-linked glycan sites on tyrosinase family glycoproteins, *Exp. Cell Res.*, 267, 115, 2001.
60. Chen, T.H. et al., Enzyme-catalyzed gel formation of gelatin and chitosan: Potential for in situ applications, *Biomaterials*, 24, 2831, 2003.
61. Lavin, A. et al., Enzymatic removal of bilirubin from blood: A potential treatment for neonatal jaundice, *Science*, 230, 543, 1985.
62. Hartmeier, W., *Immobilized Biocatalysts: An Introduction*, Springer-Verlag, Berlin, Germany, 1988.
63. Kragl, U., Immobilized enzymes and membrane reactor, in *Industrial Enzymology*, Godfrey, T. and Wet, S., Eds., Macmillan Press, London, U.K., 1996.
64. Swaisgood, H.E., Immobilized enzymes: Applications to bioprocessing of food, in *Food Enzymology*, Fox, P.F., Ed., Elsevier Science Publishers LTD, Essex, U.K., 1991.

65. Yang, Y. et al., Covalent bonding of collagen on poly(L-lactic acid) by gamma irradiation, *Nucl. Instrum. Meth. Phys. Res. Sect. B Beam Interact. Mater. Atoms*, 207, 165, 2003.
66. Oswald, P.R. et al., Properties of a thermostable beta-glucosidase immobilized using tris(hydroxymethyl) phosphine as a highly effective coupling agent, *Enzyme Microb. Technol.*, 23, 14, 1998.
67. Dybko, A. et al., Efficient reagent immobilization procedure for ion-sensitive optomembranes, *Sens. Actuat. B Chem.*, 39, 207, 1997.
68. Puleo, D.A. et al., A technique to immobilize bioactive proteins, including bone morphogenetic protein-4 (BMP-4), on titanium alloy, *Biomaterials*, 23, 2079, 2002.
69. Josten, A. et al., Enzyme immobilization via microbial transglutaminase: A method for the generation of stable sensing surfaces, *J. Mol. Catal. B Enzym.*, 7, 57, 1999.
70. Akgol, S. et al., Immobilization of catalase via adsorption onto L-histidine grafted functional pHEMA based membrane, *J. Mol. Catal. B Enzym.*, 15, 197, 2001.
71. de Oliveira, P.C. et al., Immobilisation studies and catalytic properties of microbial lipase onto styrene-divinylbenzene copolymer, *Biochem. Eng. J.*, 5, 63, 2000.
72. Rosevear, A. et al., *Immobilized Enzymes and Cells*, Adam Hilger, Philadelphia, PA, 1987.
73. Bickerstaff, G.F., *Enzymes in Industry and Medicine*, Cambridge University Press, Cambridge, U.K., 1991.
74. Solas, M.T. et al., Ionic adsorption of catalase on bioskin—Kinetic and ultrastructural studies, *J. Biotechnol.*, 33, 63, 1994.
75. Torres, R. et al., Reversible immobilization of invertase on Sepabeads coated with polyethyleneimine: Optimization of the biocatalyst's stability, *Biotechnol. Prog.*, 18, 1221, 2002.
76. Chae, H.J. et al., Optimization of protease immobilization by covalent binding using glutaraldehyde, *Appl. Biochem. Biotechnol.*, 73, 195, 1998.
77. Quirk, R.A. et al., Poly(L-lysine)-GRGDS as a biomimetic surface modifier for poly(lactic acid), *Biomaterials*, 22, 865, 2001.
78. Arica, Y. and Hasirci, V.N., Immobilization of glucose-oxidase in poly(2-hydroxyethyl methacrylate) membranes, *Biomaterials*, 8, 489, 1987.
79. Harold, E.S., Immobilized enzymes: Applications to bioprocessing of food enzymology, in *Food Enzymology*, Fox, P.F., Ed., Elsevier Applied Science, New York, 1991, p. 322.
80. Eldin, M.S.M. et al., Immobilization of penicillin G acylase onto chemically grafted nylon particles, *J. Mol. Catal. B Enzym.*, 10, 445, 2000.
81. Albayrak, N. and Yang, S.T., Immobilization of beta-galactosidase on fibrous matrix by polyethyleneimine for production of galacto-oligosaccharides from lactose, *Biotechnol. Prog.*, 18, 240, 2002.
82. Blawas, A.S. and Reichert, W.M., Protein patterning, *Biomaterials*, 19, 595, 1998.
83. Clare, D.A. et al., Molecular design, expression, and affinity immobilization of a trypsin-streptavidin fusion protein*(1), *Enzyme Microb. Technol.*, 28, 483, 2001.
84. Kondo, A. and Teshima, T., Preparation of immobilized enzyme with high-activity using affinity tag based on protein-a and protein-G, *Biotechnol. Bioeng.*, 46, 421, 1995.
85. Shpigel, E. et al., Expression, purification and applications of staphylococcal protein A fused to cellulose-binding domain, *Biotechnol. Appl. Biochem.*, 31, 197, 2000.
86. Boraston, A.B. et al., Carbohydrate-binding modules: Diversity of structure and function, in *Recent Advances in Carbohydrate Bioengineering*, Svenson, B., Ed., The Royal Society of Chemistry, Cambridge, U.K., 1999, p. 202.
87. Kobatake, E. et al., Production of the chimeric-binding protein, maltose-binding protein-protein A, by gene fusion, *J. Biotechnol.*, 38, 263, 1995.
88. Ong, E. et al., Enzyme immobilization using a cellulose-binding domain: Properties of a beta-glucosidase fusion protein, *Enzyme Microb. Technol.*, 13, 59, 1991.
89. Richins, R.D. et al., Expression, immobilization, and enzymatic characterization of cellulose-binding domain-organophosphorus hydrolase fusion enzymes, *Biotechnol. Bioeng.*, 69, 591, 2000.
90. Berdichevsky, Y. et al., Matrix-assisted refolding of single-chain Fv-cellulose binding domain fusion proteins, *Protein Expr. Purif.*, 17, 249, 1999.
91. Reinikainen, T. et al., Comparison of the adsorption properties of a single-chain antibody fragment fused to a fungal or bacterial cellulose-binding domain, *Enzyme Microb. Technol.*, 20, 143, 1997.
92. Doheny, J.G. et al., Cellulose as an inert matrix for presenting cytokines to target cells: Production and properties of a stem cell factor-cellulose-binding domain fusion protein, *Biochem. J.*, 339, 429, 1999.
93. Le, K.D. et al., A streptavidin-cellulose-binding domain fusion protein that binds biotinylated proteins to cellulose, *Enzyme Microb. Technol.*, 16, 496, 1994.

94. Cannizzaro, S.M. et al., A novel biotinylated degradable polymer for cell-interactive applications, *Biotechnol. Bioeng.*, 58, 529, 1998.
95. Zhang, J.K. and Cass, A.E.G., Electrochemical analysis of immobilised chemical and genetic biotinylated alkaline phosphatase, *Anal. Chim. Acta*, 408, 241, 2000.
96. Wilhelm, B. et al., Transglutaminases: Purification and activity assays, *J. Chromatogr. B Biomed. App.*, 684, 163, 1996.
97. Sakiyama, S.E. et al., Incorporation of heparin-binding peptides into fibrin gels enhances neurite extension: An example of designer matrices in tissue engineering, *FASEB J.*, 13, 2214, 1999.
98. Sakiyama-Elbert, S.E. et al., Development of growth factor fusion proteins for cell-triggered drug delivery, *FASEB J.*, 15, 1300, 2001.
99. Filmon, R. et al., Poly(2-hydroxy ethyl methacrylate)-alkaline phosphatase: A composite biomaterial allowing in vitro studies of bisphosphonates on the mineralization process, *J. Biomater. Sci. Polym. Ed.*, 11, 849, 2000.
100. Arica, M.Y. et al., Chapter 12: Covalent immobilization of alpha-amylase onto pHEMA microspheres: Preparation and application to fixed bed reactor, *Biomaterials*, 16, 761, 1995.
101. Azevedo, H.S. and Reis, R.L., Understanding the enzymatic degradation of biodegradable polymers and strategies to control their degradation rate, in *Biodegradable Systems in Medical Functions: Design, Processing, Testing and Applications*, Reis, R.L. and Roman, J.S., Eds., CRC Press, Boca Raton, FL, 2004.
102. Fuchsbauer, H.L. et al., Influence of gelatin matrices cross-linked with transglutaminase on the properties of an enclosed bioactive material using beta-galactosidase as model system, *Biomaterials*, 17, 1481, 1996.
103. Martinez, A.J. et al., Immobilized biomolecules on plasma functionalized cellophane. I. Covalently attached alpha-chymotrypsin, *J. Biomater. Sci. Polym. Ed.*, 11, 415, 2000.
104. Kok, F.N. et al., Immobilization of acetylcholinesterase and choline oxidase in/on pHEMA membrane for biosensor construction, *J. Biomater. Sci. Polym. Ed.*, 12, 1161, 2001.
105. Ganapathy, R. et al., Immobilization of papain on cold-plasma functionalized polyethylene and glass surfaces, *J. Biomater. Sci. Polym. Ed.*, 12, 1027, 2001.
106. Srinivas, S.S. and Rao, K.P., Controlled release of lysozyme from succinylated gelatin microspheres, *J. Biomater. Sci. Polym. Ed.*, 12, 137, 2001.
107. Labeque, R. et al., Enzymatic modification of plasma low density lipoproteins in rabbits: A potential treatment for hypercholesterolemia, *Proc. Nat. Acad. Sci. USA*, 90, 3476, 1993.
108. Zhu, Y.B. et al., Surface modification of polycaprolactone with poly(methacrylic acid) and gelatin covalent immobilization for promoting its cytocompatibility, *Biomaterials*, 23, 4889, 2002.
109. Ma, Z.W. et al., Protein immobilization on the surface of poly-L-lactic acid films for improvement of cellular interactions, *Eur. Polym. J.*, 38, 2279, 2002.
110. Ewert, S. et al., Biophysical properties of human antibody variable domains, *J. Mol. Biol.*, 325, 531, 2003.
111. Hern, D.L. and Hubbell, J.A., Incorporation of adhesion peptides into nonadhesive hydrogels useful for tissue resurfacing, *J. Biomed. Mater. Res.*, 39, 266, 1998.
112. Asthagiri, A.R. et al., A rapid and sensitive quantitative kinase activity assay using a convenient 96-well format, *Anal. Biochem.*, 269, 342, 1999.
113. Torchilin, V.P. and Trubetskoy, V.S., Which polymers can make nanoparticulate drug carriers long-circulating?, *Adv. Drug Del. Rev.*, 16, 141, 1995.
114. Sherwood, R.F., Advanced drug delivery reviews: Enzyme prodrug therapy, *Adv. Drug Del. Rev.*, 22, 269, 1996.
115. Liang, J.F. et al., ATTEMPTS: A heparin/protamine-based delivery system for enzyme drugs, *J. Contr. Rel.*, 78, 67, 2002.
116. Bagshawe, K.D. et al., Developments with targeted enzymes in cancer therapy, *Curr. Opin. Immunol.*, 11, 579, 1999.
117. Mori, T. et al., Enzymatic properties of microcapsules containing asparaginase, *Biochim. Biophys. Acta*, 321, 653, 1973.
118. Chen, J.P. and Chen, Y.C., Preparations of immobilized lysozyme with reversibly soluble polymer for hydrolysis of microbial cells, *Bioresour. Technol.*, 60, 231, 1997.
119. Goth, L., A new type of inherited catalase deficiencies: Its characterization and comparison to the Japanese and Swiss type of acatalasemia, *Blood Cells Mol. Dis.*, 27, 512, 2001.
120. Costa, S.A. et al., Immobilization of catalases from Bacillus SF on alumina for the treatment of textile bleaching effluents, *Enzyme Microb. Technol.*, 28, 815, 2001.

121. Petro, M. et al., Immobilization of trypsin onto "molded" macroporous poly(glycidyl methacrylate-co-ethylene dimethacrylate) rods and use of the conjugates as bioreactors and for affinity chromatography, *Biotechnol. Bioeng.*, 49, 355, 1996.
122. Mullerschulte, D. and Daschek, W., Application of radiation grafted media for lectin affinity separation and urease immobilization—A novel-approach to tumor-therapy and renal-disease diagnosis, *Radiat. Phys. Chem.*, 46, 1043, 1995.
123. Chellapandian, M. and Krishnan, M.R.V., Chitosan-poly (glycidyl methacrylate) copolymer for immobilization of urease, *Process Biochem.*, 33, 595, 1998.
124. Ibim, S.M. et al., Controlled macromolecule release from poly(phosphazene) matrices, *J. Contr. Rel.*, 40, 31, 1996.
125. Rejikumar, S. and Devi, S., Preparation and characterization of urease bound on crosslinked poly(vinyl alcohol), *J. Mol. Catal. B Enzym.*, 4, 61, 1998.
126. Hoffman, A.S. et al., Immobilization of enzymes and antibodies to radiation grafted polymers for therapeutic and diagnostic applications, *Radiat. Phys. Chem.*, 27, 265, 1986.
127. Liu, J.G. and Li, G.X., Application of biosensors for diagnostic analysis and bioprocess monitoring, *Sens. Actuat. B Chem.*, 65, 26, 2000.
128. Saheki, S. et al., Intestinal type alkaline-phosphatase hyperphosphatasemia associated with liver-cirrhosis, *Clin. Chim. Acta*, 210, 63, 1992.
129. Delvaux, M. and Demoustier-Champagne, S., Immobilisation of glucose oxidase within metallic nanotubes arrays for application to enzyme biosensors, *Biosens. Bioelectron.*, 18, 943, 2003.
130. Yang, X.R., Measurements of dissolved-oxygen in batch solution and with flow-injection analysis using an enzyme electrode, *Biosensors*, 4, 241, 1989.
131. Gould, B.J., Enzymes in clinical analysis: Principles updated, in *Handbook of Enzyme Biotechnology*, 3rd edn., Wiseman, A., Ed., Prentice Hall Ellis Harwood, New York, 1995, p. 311.
132. Qingwen, L. et al., Immobilization of glucose oxidase in sol-gel matrix and its application to fabricate chemiluminescent glucose sensor, *Mater. Sci. Eng. C Biomim. Supramol. Syst.*, 11, 67, 2000.
133. Zhang, J.K. and Cass, A.E.G., A study of his-tagged alkaline phosphatase immobilization on a nanoporous nickel-titanium dioxide film, *Anal. Biochem.*, 292, 307, 2001.
134. Murai, A. et al., Control of postprandial hyperglycaemia by galactosyl maltobionolactone and its novel anti-amylase effect in mice, *Life Sci.*, 71, 1405, 2002.
135. Posthaus, H. et al., Novel insights into cadherin processing by subtilisin-like convertases, *FEBS Lett.*, 536, 203, 2003.
136. Headon, D.R. and Walsh, G., The industrial-production of enzymes, *Biotechnol. Adv.*, 12, 635, 1994.
137. Hoffman, A.S., Biologically functional materials, in *Biomaterials Science: An Introduction to Materials in Medicine*, Ratner, B.D., Hoffman, A.S., Schoen, F.J., and Lemons, J.E., Eds., Academic Press, San Diego, CA, 1996, p. 124.
138. Dayan, A.D., Safety evaluation of biological and biotechnology-derived medicines, *Toxicology*, 105, 59, 1995.
139. Sims, J., Assessment of biotechnology products for therapeutic use, *Toxicology Lett.*, 120, 59, 2001.
140. Hodneland, C.D. et al., Selective immobilization of proteins to self-assembled monolayers presenting active site-directed capture ligands, *Proc. Nat. Acad. Sci. USA*, 99, 5048, 2002.

45 Emulsion-Solvent Removal System for Drug Delivery

Wasfy M. Obeidat

CONTENTS

45.1 Introduction 982
45.2 Production of Microparticles Using the Emulsion Solvent Removal Methods 982
    45.2.1 Rational for Microencapsulation by Solvent Evaporation/Extraction Method 982
    45.2.2 Emulsion Formation and Microencapsulation Process 986
    45.2.3 Microencapsulation Techniques 988
        45.2.3.1 O/W Emulsion Technique 988
        45.2.3.2 $W_I/O/W_{II}$ Emulsion Technique 990
    45.2.4 Modes of Solvent Removal 993
        45.2.4.1 Removal by Evaporation 993
        45.2.4.2 Removal by Extraction 994
        45.2.4.3 Removal by Combined Evaporation and Extraction 995
45.3 Formulation and Operational Factors Affecting Microspheres Properties Prepared by the Emulsion Solvent Removal Method 996
    45.3.1 Selection of Solvents 996
    45.3.2 Selection of Matrix Formers 997
        45.3.2.1 Biodegradable Polymers 997
        45.3.2.2 Nonbiodegradable 998
    45.3.3 Type and Role of the Emulsifier 999
    45.3.4 Agitation Intensity 1001
45.4 Microspheres Characterization 1001
    45.4.1 Microsphere Particle Size 1001
    45.4.2 Drug Loading Capacity and Encapsulation Efficiency 1003
    45.4.3 Drug Release from Microspheres 1004
        45.4.3.1 Factors Affecting Drug Release 1004
        45.4.3.2 Kinetics of Drug Release from Matrix Microspheres 1005
45.5 Potential Application of Medications Containing Microspheres 1009
    45.5.1 Oral Delivery 1009
    45.5.2 Parenteral Delivery 1012
45.6 Conclusions 1013
Acknowledgments 1013
References 1014

> Life affords no higher pleasure than that of surmounting difficulties, passing from one stage of success to another, forming new wishes and seeing them gratified.
>
> **Samuel Johnson**

45.1 INTRODUCTION

Several chemically and physically based techniques of microencapsulation have been developed over decades,[1–3] aiming to protect small drug molecules, proteins, and DNA from the environment,[4–7] to spatially or temporally control their release rates and patterns,[8] to mask their undesirable tastes and odours,[9] and to encapsulate cells.[10,11] Being a versatile and straightforward way, the solvent removal (evaporation/extraction) methods are the simplest and the most popular way for microspheres preparation to accomplish these goals. Microsphere/microcapsule formation by the solvent evaporation and extraction methods were patented early in 1970s by Vranken and Claeys,[12,13] and the first appearance of the technique in literature goes back to late seventies.[14] A wide variety of active pharmaceutical ingredients have been encapsulated; unionized and lipophilic small drug molecules are commonly encapsulated using simple emulsion technique, while hydrophilic or ionized compounds, such as proteins and peptides are often incorporated using double complex emulsion.[15–17] Despite the fact that controlled release can be achieved using common dosage forms such as tablets, rugged matrix microspheres manufactured by the emulsion solvent removal technique offer incentives in preparing liquid suspensions and injectable dosage forms with better control of drug release.[18,19] Efforts have been made to develop drug carriers capable of delivering the active molecules specifically to the intended target organ (drug targeting), while increasing the therapeutic efficacy and reducing the total amount of drug administered. Lupron Depot® (leuprolide acetate, TAP Pharmaceuticals Inc.), indicated for the treatment of prostate cancer, endometriosis, uterine fibroids, and central precocious puberty, is the first commercial product based on poly(lactic-co-glycolic) acid (PLGA) polymers[20] manufactured via the double emulsion solvent evaporation method. Later, more products were introduced, such as Vivitro® (naltrexone, Cephalon)[21,22] and Consta Risperdal® (Risperidone, Janssen Pharmaceuticals),[23–26] which were manufactured using the simple oil-in-water (O/W) emulsion solvent removal methods.

Microencapsulation processes based on the solvent removal method have been reviewed in few book chapters[1,3] and review articles[8,27–32]; however, the current chapter presents classical expertize and essential old results, in addition to an up-to-date findings pertaining to this method, and the potential use of produced microparticles in the delivery of drugs to human.

45.2 PRODUCTION OF MICROPARTICLES USING THE EMULSION SOLVENT REMOVAL METHODS

45.2.1 Rational for Microencapsulation by Solvent Evaporation/Extraction Method

Microencapsulation products (microparticles) can be defined as small entities (1–1000 μm in diameter) that contain an active core material surrounded by a shell or embedded into a matrix structure. Commercial microparticles have a diameter 3 and 800 μm, and contain 10%–90% w/w core.[2] From a morphological point of view, a microparticle can be in the form of either a microcapsule or a microsphere. The microcapsule looks very similar to a hen's egg, but in reduced size. Therefore, a reservoir-type system that contains a well-defined envelope surrounding a core is known as a microcapsule. The envelope can be named differently depending on the type of industry; it can be referred to as shell, membrane, wall, or coating. The internal core material also goes by a number of terms such as payload, encapsulant, fill, active ingredient, or internal phase. Alternatively, when the core is distributed more, or less uniformly throughout a monolithic continuous matrix structure with no well-defined envelope, then a microsphere is the preferred term. In this type, the core material particles are in direct contact and buried to varying depths inside matrix forming material. The core material may be molecularly or macroscopically dispersed within the matrix; however, sometimes the microsphere can contain mixture of coarse and molecular-sized cores. Most shells and matrices are organic polymers, but lipids, waxes, and proteins are also used. In pharmaceutical applications, cores are predominately solid drugs; however, liquid drug forms and gas have been encapsulated as well.[33,34] According to some authors,[3] the magnitude of the drug loading dictates the type of the microparticles obtained using the solvent evaporation technique. Hence, microspheres or microcapsules are obtained when drug loading is low or high, respectively. This can be apparent best when particles are viewed under the light microscope where a drug-free polymer layer similar to microcapsule envelope is observed, especially at high drug loading. Moreover, the type of encapsulation method yield, in most cases, microcapsules or a microspheres. For example, interfacial polymerization and coacervation methods almost always produce microcapsules, whereas solvent evaporation/extraction method may result in either type, depending on the formulation and processing factors. For example, a novel one-step method has been designed to use solvent evaporation to create double-walled microspheres, which are essentially microcapsules.[35] Drug encapsulation using solvent evaporation/extraction method, which depends on precipitating small polymer particles from an emulsion, undoubtedly attracted many researchers in the field of pharmaceuticals owing to its multiple advantages. The removal of the solvent via evaporation or extraction or via a combination of both methods is usually conducted at ambient temperatures or at slightly higher temperatures. Therefore, solvent evaporation/extraction technique is suitable for encapsulation of temperature-sensitive drug compounds that undergo degradation at elevated temperatures usually experienced by other encapsulation methods such as spray-drying.[36] Compared to polymerization reactions or complex coacervation, the emulsion solvent removal technique is deprived of use of chemical initiators, coacervating agent,[8,37] and crosslinking agents.[2] The emulsion solvent removal technique, through adjustment of certain crucial parameters, can provide a control of the average particle size and particle size distribution of the microsphere products,[38–42] an issue that is difficult to satisfy using spray-drying[36] or coacervation methods.[2] Microspheres produced using the emulsion solvent removal technique are rugged enough to withstand fracture compared to microcapsules produced by coacervation and polymerization reactions, which experience various degrees of shell discontinuities and fragilities.

Microspheres of particular compositions and properties that aid in controlling the drug absorption process to sustain adequate and effective plasma drug levels over an extended period of time are known as "delayed or sustained release microparticulate delivery systems." It is well known that the controlled release drug delivery systems offer numerous advantages over the conventional ones. Microspheres prepared using the emulsion solvent removal technique demonstrated successes in many aspects in the controlled delivery of medicaments administered orally, or parenterally. The optimization of drug concentration in plasma, and the reduction of side effects, particularly for drugs with narrow therapeutic indices, are known to improve patient's compliance and to reduce accompanying toxic effects.[43,44] Multiparticulate delivery systems are considered to be ideal systems in this regard since they are usually fabricated using biocompatible materials and able to encapsulate and control the release behavior of variety of drugs of different physicochemical properties. Compared to single unit conventional controlled release systems, orally administered multiparticulate systems are readily distributed over a large area, providing uniform drug release that is less dependent on gastric transit time.[45] Classical zero (or near zero) order kinetics of drug release that are often perceived as being ideal in controlled/sustained drug delivery were established through the homogeneous dispersion of insoluble particle within the microsphere's polymeric matrix.[34,46,47] Owing to their rapid dissolution, drug molecules or particles localized close to the surface of microspheres can be utilized as a loading dose to achieve immediate leap in drug concentration in the blood prior to the relatively slow prolonged release through the various depths within the microsphere matrix structure. This initial rapid release can be of particular utility for some antibiotics and vaccines that require bursts of the drug at specified intervals. Generally, two types of microspheres are believed to exist as far as the control of drug release is concerned—single and multiple purpose microspheres. Single purpose microspheres, full or hollowed, are single walled microspheres that have been extensively studied and are intended to perform single function—sustain, modify, or target drug release. Single purpose microspheres are usually comprised of single polymer, or mixtures of polymers of similar properties, and they have been extensively reported.[48–54] Multiple purpose microspheres are usually made of two (or more) materials in the form of homogeneous or heterogeneous matrices depending on polymers miscibilities, or they can be in the form of coated simple microsphere. Such microspheres are usually intended to control drug release in a sustained manner, and to achieve additional specific mission. Matrix microspheres comprised of a rate controlling component infiltrated with pH or temperature sensitive components, adhesive components, or adjuvants to protect the encapsulated drugs[55–60] are multipurpose microspheres. Camptothecine-loaded polymeric microspheres in thermosensitive hydrogel were prepared for treatment of colorectal peritoneal carcinomatosis. In addition to the sustained release pattern of the fabricated microspheres, Camptothecine-loaded microsphere in hydrogel induced a stronger antitumor effect compared to free Camptothecine by increasing apoptosis of tumor cells and inhibiting microvessel density of tumor tissue.[57] Price and Obeidat[61] employed a single step process for the preparation of cellulose acetate butyrate microspheres having enteric and controlled release characteristics using the nonaqueous solvent evaporation method comprising (O_I/O_{II} emulsion). Thereofore, in addition to the near zero order drug release kinetics for an appropriate time scale, microspheres also exhibited enteric propereties to minimize drug release in acidic pH. Scanning electron microscopy before and after dissolution studies, along with dissolution studies demonstrated that microspheres, resemble a porous reservoir system where the drug is concentrated in the central part of the microspheres. Figures 45.1 and 45.2[62] show blank enteric microspheres before and after extraction with simulated intestinal fluid (pH 7.5) where globules of cellulose acetate phthalate polymer apparent on the outer surface of microspheres created cavities and empty pockets after simulated intestinal fluid extraction. Similar approach was adopted later by other authors for purposes of extending drug release and targeting to specific location within the gastrointestinal tract.[63]

Multiple purpose systems also include multiple wall or layered microspheres. An interesting multilayer microcapsule (microsphere) was first prepared by Mathiowitz and Langer[64] in a single step process. The approach relies in on a modification of Harkin's equation for spreading equilibrium

FIGURE 45.1 Scanning electron micrograph of blank microsphere (without drug) containing of 10% CAP-90% CAB381-20, shows globules of CAP polymer on the microsphere surface. (Reproduced from Obeidat, W.M. and Price, J.C., *J. Microencapsul.*, 21(1), 47, 2004.)

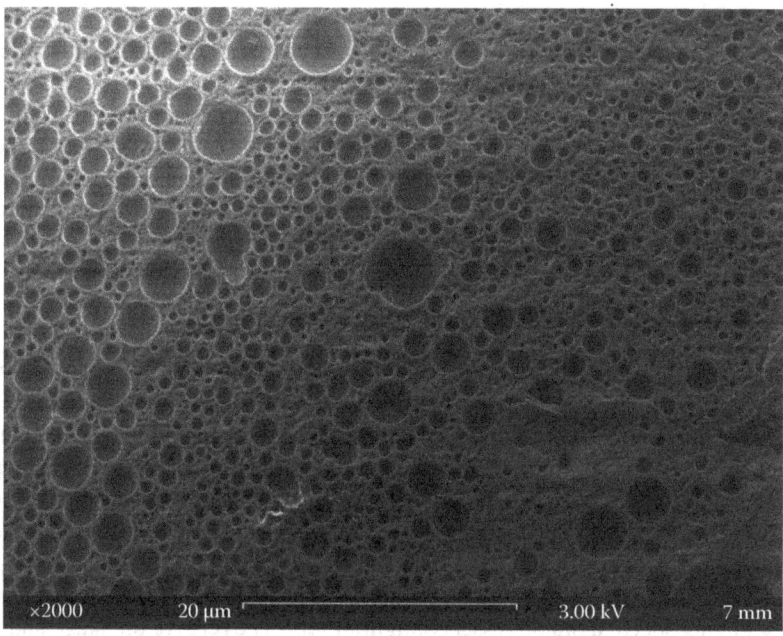

FIGURE 45.2 Scanning electron micrograph of blank microsphere (without drug) containing of 10% CAP-90% CAB381-20 shows cavities after CAP leaches out. (Reproduced from Obeidat, W.M. and Price, J.C., *J. Microencapsul.*, 21(1), 47, 2004.)

between two fluids emulsified in a certain solvent. Introducing two or three different degradable, or nondegradable, hydrophilic, or hydrophobic polymer solutions dissolved in a volatile organic solvent into a nonsolvent continuous phase creates a stable emulsion in which phase separation occurs within each droplet. The drug is dispersed or dissolved in the polymer solution. As the solvent evaporates, polymer solutions become more concentrated, and mutual miscibility reduces to a point of engulfment of one polymer by the other (s) creating multiwalled microspheres. Microspheres are distinguished by extremely uniform dimensioned layers of polymers.[35,65–69] Similarly, Wen and Anderson[70] prepared double wall microspheres using two biodegradable polymers by the O/W emulsification solvent extraction process. However, the systems described suffer take long solvent evaporation times (3 days at 90°C) when hydrophilic polymers dissolved in aqueous media are employed. Also, despite the advantages to be gained from multiwalled microsphere systems, they are limited to relatively low levels of drug loadings, since sufficient polymer must be present in order to form a sufficiently stable microsphere. For costly drugs such as synthetic peptides/proteins, or for highly potent water soluble drugs, high efficiency in drug loading is a required feature in clinically efficacious microencapsulation techniques. In addition, the degradation of proteins upon direct contact with certain polymers, such as PLGA, would affect the encapsulation efficiency of intact proteins and the stability of microsphere product.[71–74] These problems have been minimized by Mcginity and Iwata by the production of unique multiphase polymeric microspheres in which stable and unstable compounds are dispersed using microemulsions.[75] These multiphase microspheres include numerous, tiny microemulsions in the form of W/O_I fine droplets dispersed throughout a biodegradable polymer matrix. The process involves three steps; the preparation of the water-in-oil (W/O) microemulsion (aqueous solution of a water-soluble molecular compound in a fixed oil such as soybean oil or safflower oil). The W/O_I microemulsion is dispersed in polymer-organic solvent solution such as (PLGA-acetonitrile) to form W/O_I/O_{II} emulsion, which is then poured gently through a narrow nozzle into agitated light mineral oil containing stabilizer (Span 80) to form a W/O_I/O_{II}/O_{III} multiple emulsion. The multiple emulsion system is agitated by a stainless steel propeller for 24 h to evaporate the acetonitrile, and for further solvent removal, the microspheres were dried under reduced pressure for 48 h. Double wall microspheres can be also prepared where polymers are melted and combined with the substance to be incorporated, and then cooled to form layered microspheres.[76] Morris and Warburton[77] described the formation of a three-layer microsphere composed of acacia/ ethylcellulose/acacia prepared by using a $W_I/O/W_{II}$ multiple emulsion solvent evaporation technique. Results of interfacial rheological studies revealed the formation of a rigid bi-polymer film at the interface between an aqueous solution containing a water-soluble polymer and a nonaqueous solution of an oil-soluble polymer. Therefore, small spherical bodies are formed upon making a $W_I/O/W_{II}$ emulsion from these solutions.

45.2.2 Emulsion Formation and Microencapsulation Process

An understanding of emulsion properties is crucial since emulsion formation is the first step in all emulsion-based methods, including the emulsion solvent removal methods. Emulsions result from the deformation of the interface between the two immiscible phases, frequently called oil and water, to such an extent that droplets form. Further disruption of large droplets is a critical step in emulsification, and therefore in encapsulation. The rates of breakage and coalescence of dispersed liquid droplets determine the time evolution of the droplets size distribution. These rates are affected by the physicochemical properties of the two phases including interfacial tensions, viscosities and densities, the volume fraction of the phases, the type and concentration of the stabilizer, and the quality of agitation (e.g., stirring speed, agitator/vessel configuration).[78] The relative breakage and coalescence rates are determined by properly selected emulsifiers or stabilizers (e.g., surface active agents and small polymeric molecules) that lower the interfacial tension, and reduce the mechanical energy needed to achieve intended droplets sizes. Emulsifiers adsorb onto the interface between the two phases forming a protective film that would prevent drops coalescence. Therefore, immediate coalescence between colliding drops is reduced.[78] For large scale production of emulsions, colloidal mills and high pressure homogenizers are often used.[28] Stirring is the simplest and the most straightforward method exclusively used in bench preparation. Increasing the mixing or stirring speed results in a narrow particle size distribution and an exponential decrease in microsphere mean size.[79-81] Vigorous mixing results in lower microspheres polydispersity indices.

Static mixers or rotor-stator machines are commonly used in emulsion preparation. Such instruments are described under the name of homogenizers or "Ultra Turrax," and they consist of repeatedly split baffles, or other flow obstacles installed in a shaft. Baffles eliminate vortex formation, and therefore prevent aggregation and lumping. Emulsion droplets impaction of the baffles lead to their breakup, which consequently reduces the average particle size of microspheres.[82] In static mixing, turbulent flow constantly acts on the disperse phase and, therefore a continuous change in the size of droplets occurs, especially before an appreciable magnitude of solvent evaporation from emulsion droplets takes place. Because of the different turbulent levels within the dispersion within an agitated vessel, a distribution of droplet sizes is obtained.[83] However, when viscous forces predominate over the turbulent velocity and pressure forces, coalescence rates exceeds droplet break up rates.

Extrusion, with laminar flow pattern and a nearly constant initial droplet size, can also be used for emulsion preparation. The procedure involves feeding or injecting the internal phase through single or multichannel pathways directly into the continuous phase. When the internal phase leaves the pathways, discrete droplets are formed at the site of contact with the slowly flowing continuous phase. More advanced technology such as high pressure homogenizers and ultrasonication can also be used. In the high pressure homogenizers, the pre-emulsion is usually fed under high pressure (10–40 MPa) via a pump and is forced through a narrow (e.g., 0.1 mm) valve slit within very short time (0.1 ms). The potential energy is converted into kinetic energy as the pumped liquid obtains a high velocity. The kinetic energy is dissipated into heat during passage through the valve. In ultrasonication, high frequency vibrations are applied. Ultrasound is transmitted through a medium via pressure waves by inducing vibrational motion of the molecules, which alternately compress and stretch the molecular structure of the medium due to a time-varying pressure.

Liquids irradiated with ultrasound can produce bubbles. These bubbles oscillate and grow a little more during the expansion phase of the sound wave than they shrink during the compression phase. Under the proper conditions these bubbles can undergo cavitation, violent bubble collapse or implosion, leading to droplets particle size reduction. The use of ultrasonication in emulsification is more efficient than rotor-stator systems.

Although, the particle size of the prepared microcapsule/microsphere is primarily dependent on the initially formed emulsion droplets, especially its particle size and particle size distribution,[28] the initial size of the emulsion may differ from the final size of the microspheres or microcapsules. The change in droplet size distribution after emulsification can be due to coalescence or isothermal distillation, unless the dispersed (internal) phase is completely insoluble in the continuous one. Microsphere size may affect the rate of drug release, drug encapsulation efficiency, product syringeability, in vivo fate in terms of uptake by phagocytic cells and biodistribution of the particles after subcutaneous injection, or intranasal administration.[3] The extent of size reduction of emulsion droplets depends on the viscosities of the dispersed (internal) and continuous phases, the interfacial tension between the two phases and their volume ratio, the geometry, position, number of the impeller(s), and the size ratio of impeller and mixing vessel.[84] Surface active agents such as polysorbates or viscosity-enhancing stabilizers such as polyvinyl alcohol (PVA) that are generally added to the continuous phase prevent rapid coalescence of the emulsion droplets during early emulsification process. Increasing the stabilizer concentration frequently leads to decreased emulsion's droplet size and droplet size distribution.[85,86] With macromolecular stabilizers, the viscosity of the continuous phase even increases, amplifying—for a given stirring rate—the shear forces acting upon the emulsion droplets and thus minimizing their sizes.[84,87]

There are several ways to efficiently encapsulate drugs using the emulsion solvent removal methods, depending majorly on the physico-chemical properties of the drug and its interplay with the matrix former, and the solvents employed in the internal and continuous phases, as will be discussed in more details later. In the process of encapsulation by the emulsion solvent removal methods, polymers and oligomers are dissolved in a suitable volatile organic solvent or in solvents mixture to form polymeric solution phase that is then emulsified with an immiscible phase. One of these phases is called the internal or disperse phase, which is being engulfed by the other phase that is called a continuous phase, which contain the stabilizer. Here, it is worth mentioning that water immiscibility of the organic solvent is not an absolute prerequisite for making an emulsion.[88] The medicament is usually dissolved or dispersed as finely pulverised solid in the internal phase, and the resulting solution or dispersion is then emulsified with the continuous phase under agitation to form discrete droplets that contain the medicament.[70,89] When an aqueous internal phase is emulsified with an oily continuous phase, an emulsion of the W/O type is produced. An O/W emulsion is formed when an oily internal phase is emulsified with an aqueous continuous phase. The size of the internal phase droplet, which—to a great extent—determines the size of the microspheres produce, is dependent on the intensity the system is agitated upon emulsification. In particular, the initial agitation speed (during the first few minutes) is believed to be most influential in the final microspheres size distribution. For lipophilic compounds, an aqueous continuous phase may be comfortably chosen, while for hydrophilic compounds, a continuous phase made of hydrophobic organic liquids is a more delicate choice.[90–92] Once the emulsion is stabilized, agitation is maintained while the organic solvent diffuses from the embryonic particles into the aqueous continuous hardening bath. Diffused solvent is removed by extraction or evaporation at the water/air interface. Polymer phase separation at the microsphere surface, particle coalescence, and drug loss into the hardening bath take place during microspheres hardening.

The solubility of the disperse phase solvent in the continuous phase has an impact on the pattern of solvent removal. Solvent removal is usually performed by extraction, or evaporation, or by both procedures. In both procedures, the disperse phase solvent must be slightly soluble and therefore partitions in the continuous phase, leading to precipitation of the matrix material.

Solvent evaporation will only play a dominant role in microsphere hardening kinetics when the extraction capacity of the continuous phase has been saturated; and when the diffusion rate of organic solvents from the dispersed phase to the continuous phase is fast compared with the solvent evaporation kinetics.[93] In solvent extraction, the amount and composition of the continuous phase are chosen so that the entire volume of the solvent can be dissolved.[28]

45.2.3 Microencapsulation Techniques

45.2.3.1 O/W Emulsion Technique

One of the simplest methods to encapsulate drugs is by the O/W emulsion/solvent removal technique. In this technique, the matrix formers (usually polymers and oligomers) are dissolved in a suitable organic solvent (or mixture of solvents) to form polymeric solution phase. The drug is dissolved or dispersed as finely pulverised solid in the polymeric solution phase. The resulting solution or suspension is emulsified with the continuous aqueous phase, which contains the emulsifier (surface active agent) under agitation to form discrete droplets of O/W emulsion type. Polymer solution phase constitutes the oil internal phase and is immiscible with the continuous aqueous phase. Generally, the continuous aqueous phase should have a low dissolving power for the drug. Once the emulsion is stabilized, agitation is maintained and the solvent present in the internal phase dissolves and diffuses through the continuous phase and then almost completely evaporates,[94–99] or being extracted at the water/air interface aiding the hard microspheres formation.[100,101] Figure 45.3 depicts the microencapsulation by emulsion solvent evaporation method.

O/W emulsion technique is suitable for the encapsulation of hydrophobic (poorly water soluble) compounds since they partition favorably into the oil phase,[70,89] and the technique has been employed for the encapsulation of various drugs, such as steroids,[14,102–106] thioridazine,[107] dexamethasone,[108] chlorpromazine,[109] methadone,[110] and anticancer agents—aclarubicin,[111–113] lomustine,[79,114,115] and paclitaxel.[116,117]

The encapsulation of hydrophilic (water soluble) compounds using O/W emulsion technique is farfetched since they may not dissolve in the organic solvent and will diffuse into the aqueous continuous phase during emulsion, leading to a considerable drug loss. It has been reported that salicylic acid, theophylline, and caffeine could not be encapsulated in PLA polymers using O/W emulsion technique due to their high aqueous solubilities.[118] However, since the O/W emulsion technique is an attractive option for encapsulation of hydrophilic drug compounds, various approaches have been employed to tackle their low encapsulation efficiencies. Lowering of the solubility of the hydrophilic drug in the aqueous continuous phase by adjustment of the pH, the aqueous phase[118] has shown some success, especially with drugs with appreciable ionizable groups within their chemical structures. Otherwise, lowering the aqueous solubility of the drug by chemically modifying it into a lipophilic prodrug improved its encapsulation efficiency.[119] Minimizing drug loss can also be brought about by saturating the continuous phase with the drug,[82,120] by employing water-immiscible organic solvent with significant water solubility to rapidly precipitate the polymer,[121] or by increasing the concentration of the matrix material solution; thereby restricting drug migration from the solidifying microspheres to the external phase.[122–128] High encapsulation efficiency of a hydrophilic drug can also be achieved by formation of solid dispersion of the drug in a biodegradable polymer by dissolving them in an organic solvent. The produced solid dispersion is dissolved in an organic solvent that is water-immiscible and is nonsolvent for the drug. The resulting oil phase is then emulsified into an aqueous phase containing emulsifying agent to give O/W emulsion.[129] Elevated temperatures or reduced pressure to promote the evaporation of the solvent may lead to better drug entrapment by reducing the time for drug escape, and to enhance fast microsphere solidification.[108,130–132] However, elevated temperatures influence both the aqueous drug solubility and the viscosity of the emulsion.

Combination of rapid solvent extraction and evaporation was adopted to minimize drug loss and improve the encapsulation efficiency of neurotensin peptide.[133] Neurotensin was encapsulated in poly-dl-lactic acid at high concentrations by the O/W emulsion solvent removal methods using an internal phase consisting of mixed solvent system. One of the solvents is water immiscible

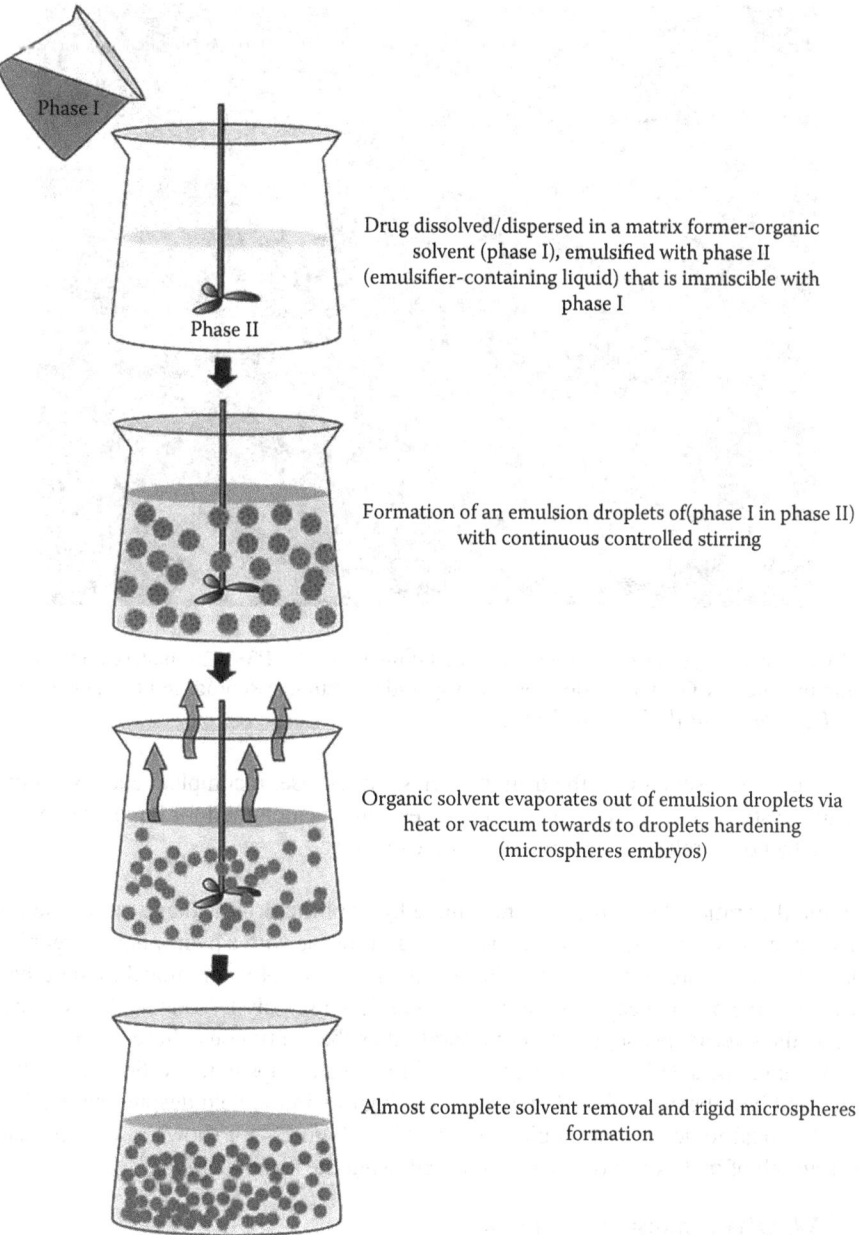

FIGURE 45.3 Schematic diagram of microsphere formation by the single emulsion solvent evaporation method.

(e.g., methylene chloride [MC]) and one is water miscible (e.g., ethanol). Microspheres are formed due to the quick phase separation within the oil droplets after the water miscible solvent being partitioned into the outer aqueous phase and the fast evaporation of the water immiscible solvent.

Another approach used to avoid drug loss to the continuous aqueous medium is to employ an insoluble form of the drug dispersed within the organic solvent or mixtures of organic solvents to form solid-in-oil-in-water (S/O/W) emulsion technique. In fact, the technique was applied very early in the encapsulation of hydrophobic drug compounds,[104,118,134–140] but later, it was applied to hydrophilic drugs such as haloperidol,[141] camptothecin and its derivatives,[142–144] levonorgestrel,[145] and beta estradiol.[146,147] For best results, the drug compound must be in very low particle size (micronized), to

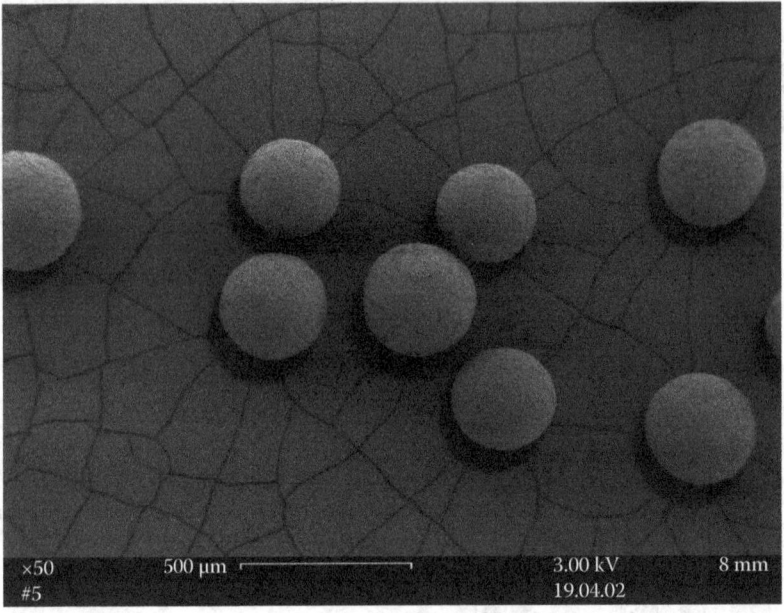

FIGURE 45.4 Scanning electron micrograph of cellulose esters (CAB381–20) matrix microspheres of size 250–355 μm prepared by O_I/O_2 emulsion solvent evaporation method. (Reproduced from Obeidat, W.M. and Price, J.C., *J. Microencapsul.*, 21(1), 47, 2004.)

obtain complete encapsulation of the drug particles. Otherwise, incomplete coats with the appearance of drug crystals on the surface of microspheres could be obtained for large drug particle sizes, which will affect drug release rates from the microspheres.

- Yet, another rationale approach to minimize hydrophilic drug leakage out of the forming microspheres is to replace the aqueous continuous phase with a hydrophobic organic liquid phase, that is, to prepare O_I/O_{II} emulsion. The modified solvent removal method has been utilized in the microencapsulation of phenobarbitone in poly(L-lactic acid), using acetonitrile as the internal phase, and light mineral oil as the continuous phase.[148] This technique has proven to be a delicate approach to maximize the encapsulation efficiency hydrophilic drugs and have been employed by many investigators in the field despite the need remove the additional toxic organic liquids.[34,61,89–92,149–157] Figure 45.4 shows a scanning electron micrograph of matrix microspheres prepared using the same method.[62]

45.2.3.2 $W_I/O/W_{II}$ Emulsion Technique

This is another innovative emulsion technique proposed for the efficient encapsulation of hydrophilic drug compounds involves the formation of double emulsions (multiple emulsions).[158] In this technique, an aqueous core solution (W_I) is emulsified in a polymer-organic solvent solution (O) to form the "primary W/O emulsion," which is further emulsified in an external aqueous solution (W_{II}), giving rise to the double emulsion of $W_I/O/W_{II}$ type. Evaporation or extraction of the organic solvent yields a solid microcapsule with an aqueous core. The organic phase in (O) acts as a barrier between the two aqueous compartments, W_I and W_{II}, to prevent the diffusion of hydrophilic drug compounds out of the core toward the external aqueous solution. Figure 45.5 depicts the microencapsulation by the $W_I/O/W_{II}$ emulsion technique.

The technique requires few steps and a rigid control of all parameters including the temperature and the viscosity of the primary W/O emulsion. Otherwise, a wide size distribution of microspheres is produced. The $W_I/O/W_{II}$ emulsions have many of the attributes of W/O emulsions; however,

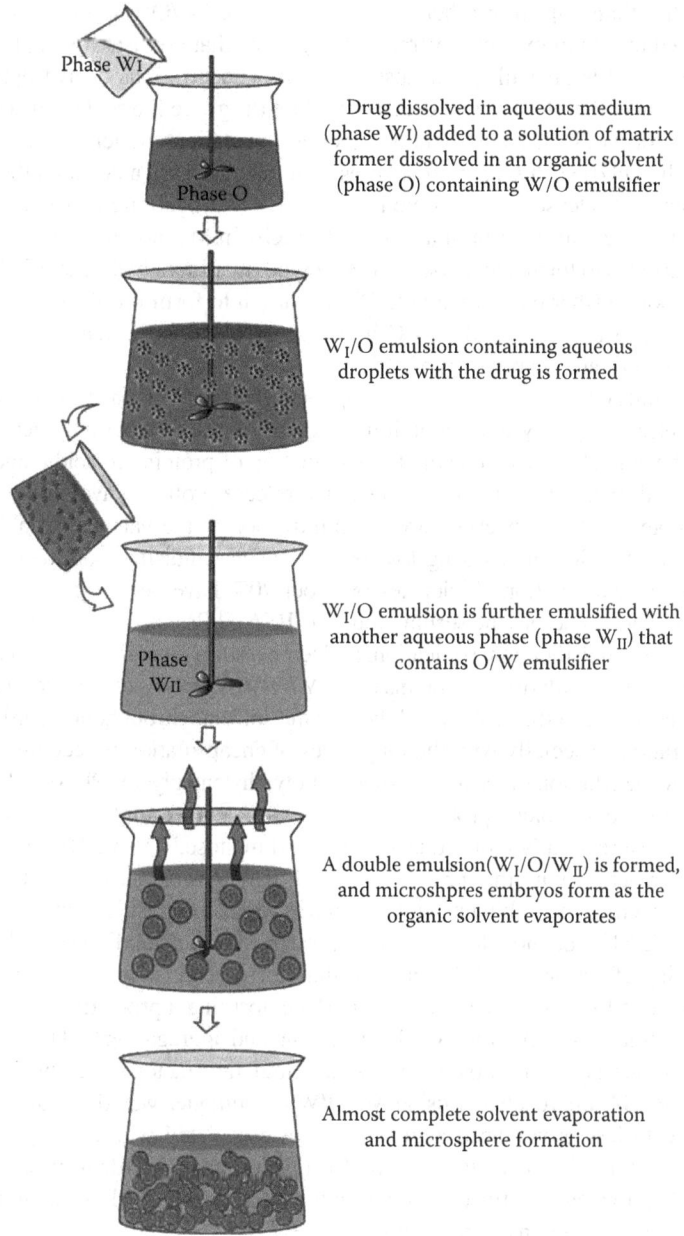

FIGURE 45.5 Schematic diagram of microsphere formation by the multiple emulsion solvent evaporation method.

their lower viscosity is derived from water being the external phase. A drug that reaches the oil middle phase often dissolves out into the external aqueous solution so that the incorporation efficiency of the active ingredient into microspheres becomes low. The technique is particularly suited for quite hydrophilic drug compounds, such as peptides and proteins, although hydrophobic compounds have been encapsulated as well.[159–161] The creation of a pH gradient between the inner and the outer aqueous phases[162,163] and/or the presence of salts in the multiple emulsion[162,164] lead to an increase in drug retention within the microspheres. For example, the enhancement of the entrapment of enoxaparin sodium, a low molecular weight heparin, within PLGA microspheres prepared by $W_I/O/W_{II}$ technique, was attributed mainly to the addition of phosphate buffer solution of 10 mM (pH 7.4) in the W_{II} phase rather than adjusting other parameters.[165] The $W_I/O/W_{II}$ emulsion technique overcomes issues related to oil removal by several washing steps that are usually employed in the O_I/O_{II} technique.[158] However, it is difficult to encapsulate high concentrations of hydrophilic drugs using the $W_I/O/W_{II}$ technique; therefore potent drugs (low dose drug) are the preferred cores. The internal phase can be in the form of hydrogel to obtain higher encapsulation efficiency. For example, acryloyl hydroxyethyl starch (AcHES)-PLGA microspheres were prepared via a novel method to encapsulate insulin and to sustain its release.[166] In this method, AcHES hydrogel microparticles were allowed to swell with vortex mixing with insulin in acetic acid. PLGA in MC polymer phase was added to the swollen AcHES particles to form a dispersion of W/O type (insulin in hydrogel)/(PLGA in MC). This primary emulsion was further emulsified with PVA solution to form the $W_I/O/W_{II}$ emulsion before solvent extraction that was performed at 4°C in an ice bath. Then the temperature was gradually elevated to 39°C to facilitate the removal of MC.

The $W_I/O/W_{II}$ emulsion technique has been employed frequently to encapsulate peptides and proteins. Controlled release depot systems containing proteins in the form of polymer microcapsules or microspheres are beneficial to avoid enzymatic degradation of proteins in acidic and alkaline media of the gastrointestinal tract, and to provide a sustained release profile to avoid repeated injections of the short acting proteins.[167-169] However, careful optimization of the various variables is required to obtain successful results. Significant drug loss occurs during emulsification, hardening, and washing,[170] although high encapsulation efficiencies of about 90% have been reported for certain model proteins such as lysozyme and bovine serum albumin (BSA).[73] Protein aggregation and denaturation may also occur during emulsification, sonication, and temperature increase.[171-173] Gombotz et al.[174] and Shah[175] worked on the encapsulation of proteins using W/O or $W_I/O/W_{II}$ techniques. However, in many cases, direct contacts between the protein and the organic solvents throughout the microsphere matrix may cause denaturation, especially with the long times of encapsulation procedures. Protein stability can be enhanced by the addition of excipients such as polyethylene glycol (PEG) and sugars to prevent aggregation and stabilize the folded protein structure.[167,176-179] For example, α-chymotrypsin aggregation and inactivation were greatly reduced using PEG and maltose,[178] while BSA aggregation in PLA microspheres was prevented with the incorporation of PEG.[180] Surfactants were also found to exhibit potentialities in protecting recombinant human (rhGH) against emulsification-induced aggregation. For example, pluronic F127, but not pluronic F68, experienced an outstanding as well as concentration-dependent stabilizing effect on rhGH. It could be that during emulsification, pluronics compete for adsorption and oriented location at the biphasic interface, forming a protective barrier that prevents a protein exposure to organic solvent, followed by unfolding and aggregating.[181] The amount of aggregation of recombinant human erythropoietin (EPO), a stimulating factor of red blood cell production, encapsulated into PLGA microspheres using $W_I/O/W_{II}$ technique, was dependent on the homogenization method, with less pronounced aggregation was associated with high speed homogenizers, compared with ultrasonication or vortex mixing. Excipients, such as hydroxypropyl, β-cyclodextrin, L-arginine, or BSA, provided a distinct reduction in the aggregation of EPO, and human serum albumin was reported to protect against EPO degradation.[182]

Although appears to be less efficiently encapsulated than protein, DNA and DNA vaccines have been encapsulated using the $W_I/O/W_{II}$ emulsion technique.[183-190] Alpar et al.[191] prepared vaccine containing microparticles using the double emulsion technique. The vaccine was suspended or dissolved in an aqueous solution containing PVA and a phospholipid (lecithin) and is emulsified with a biodegradable polymer dissolved in an organic phase. The resultant emulsion is then dropped into a second aqueous phase, also containing PVA and optionally also the phospholipid with vigorous stirring to form the $W_I/O/W_{II}$ emulsion with subsequent organic solvent evaporation. Hurtado et al.[192] produced antigen-PLGA microspheres having a native conformational epitope, particularly an antigen derived from the G protein of the human respiratory syncytial virus (RSV), using $W_I/O/W_{II}$ techniques. The encapsulated antigens showed controlled biphasic release, thus mimicking a multiple dose schedule vaccination. A second antigen could be adsorbed on blank PLGA microspheres and/or on antigen-loaded PLGA microspheres.

The $W_I/O/W_{II}$ emulsion technique has been also employed to encapsulate variety of drug compounds, including adriamycin and tobramycin,[193] leuprolide acetate (Lupron Depot®)15,20, nucleotides,[194] thyrotropin-releasing hormone,[195] growth hormone,[98,196] RNA,[197] and calcitonin.[198]

45.2.4 Modes of Solvent Removal

One of the important features of solvent removal methods of encapsulation is the rate at which the organic solvent is removed from the microdroplets of the emulsion. Generally the solvent can be removed either by evaporation or extraction, or by combined procedures.

45.2.4.1 Removal by Evaporation

The rate of evaporation of volatile solvent from the solidifying microspheres can be controlled by adjusting the temperature of the microsphere dispersion, the agitation intensity, and the diameter of the impellers employed in the process.[93] Higher temperatures will facilitate the evaporation of the solvent from the emulsion/air interface and thereby maintaining a high concentration gradient for the solvent between the embryonic particles and the continuous phase.[199,200] The solvent evaporation method was described very early,[94–99] where O/W emulsions were prepared using polymeric materials as disperse phases and an aqueous solution containing a hydrophilic colloid as a continuous phase. Aqueous solutions are frequently used as continuous phase, even for the microencapsulation of hydrophilic compounds since hydrophobic organic liquids are hard to be totally removed, thereby leaving undesired residues.The rate of evaporation of the internal organic phase solvent has a significant impact on the final microspheres properties.[201–204] Unlike slow solvent evaporation rate, rapid evaporation favors more porous microspheres morphologies,[28] or hollow microsphere with porous walls,[120,205,206] wider sizes distributions, and decreased particles densities,[199,200,207] all of which affect drugs release rates. Rapid solvent evaporation cause solidification of microsphere polymer matrix at rates exceeding polymer coalescence rates after solvent evaporation and pore formation. In addition, rapid microsphere solidification results in insufficient mixing time to reduce droplet size. Nevertheless, rapid solvent evaporation might be needed in situations where the drug partitions easily into the continuous phase. On the other hand, denser microspheres are usually obtained at slower evaporation rates all of which affects drug release rates.[28]

An ideal solvent should evaporate relatively fast, yet should aid the formation of spherical, intact, and uniform microspheres with high drug encapsulation efficiency. Methylene chloride, MC, is the most extensively employed solvent for encapsulation using the emulsion solvent evaporation. Ethyl formate, EF, has shown an interesting behavior as a solvent in manufacturing PLGA microspheres.[88] Ethyl formate has lower vapor pressure, higher boiling point, and evaporates in water faster than MC, which is attributed to its higher solubility in water compared to MC, rendering more molecules of the solvent exposed to the air–liquid interface. Ethyl acetate, EA, is potentially less toxic than MC, and has been considered frequently as alternative to MC in microencapsulation of several drugs. For example, fentanyl was encapsulated in PLGA microspheres using O/W emulsion technique followed by solvent evaporation. PLGA and fentanyl were dissolved in EA and then emulsified by drop in W phase containing 3% PVA as an emulsifier. EA was removed at 35°C by evaporation, and monolithic microspheres containing fentanyl were obtained.[205] However, due to its partial miscibility in water (EA is 4.5 times higher than that of MC); it seems difficult to obtain microspheres through solvent evaporation once introduced directly into the continuous aqueous phase. Hence, sudden extraction of a big quantity of EA from the dispersed phase would make the polymer precipitate into fibre-like agglomerates.[208] This problem can be avoided by pre-saturating the aqueous solution with EA.[209] Otherwise; the disperse phase containing EA can be emulsified with a little quantity of aqueous solution, which is then added to a larger quantities of aqueous solution.[208] Despite the success in preventing rapid solvent removal using these method, concerns over drug encapsulation efficiency and the morphology of the manufactured microspheres may still be raised.[89]

The unstirred gas phase that exists over the liquid/gas interface can affect the overall mass transport resistance of organic solvent in the internal phase at certain conditions.[93] When nitrogen gas was blown over the liquid/gas interface of dispersions containing methylene chloride (MC, b.p. 40°C), acetonitrile (ACN, b.p. 82°C) or ethyl acetate (EA, b.p. 77°C) at varying temperature, the overall permeability coefficient (cm/s) (sum of the permeabilities in liquid and gas phases), was increased for all solvents. However, for MC at 25°C and 35°C, the increase was not substantial (<15%), indicating that the gas phase experienced negligible resistance at those temperatures, thus favoring liquid side transport control. The highly water soluble solvent, ACN, with lower vapor pressure and lower Henry's law constant compared to MC, was found to be gas phase transport limited at 25°C. On the other hand, at 25°C, EA, with a higher boiling point and lower vapor pressure than MC, but medium Henry's law constant, the transport control was in both liquid and gas phases. Therefore, for industrial applications in closed vessels, flushing of the surface of the liquid phase is for replacing unstirred gas phase has been suggested for appropriate solvent evaporation rates.[210]

Solvent evaporation can be brought about at reduced pressures instead of elevated temperatures.[130] Reduced pressure of 460 or 160 mmHg promoted the evaporation of MC from lidocaine,[132] and BSA[211] encapsulated within PLA at 25°C. In both cases, microsphere mean size and encapsulation efficiency decreased at reduced pressure, whereas the drug release profile remained similar to those microspheres prepared at atmospheric pressure. However, when progesterone was encapsulated in PLA, drug release was slower for microspheres prepared at reduced pressure (200 mmHg) compared to those manufactured at atmospheric pressure.[212]

In certain processes, solvent evaporation may be performed via sublimation by freeze drying after the emulsification process. During lyophilization, simple or double emulsions are cooled to a temperature that freezes the drug-polymer-solvent (microspheres) and not the continuous phase. The suspension is then subjected to sublimation permitting the removal of the solvent from the continuous phase and that entrapped in the microspheres.[213] The process improves the yield of water soluble drugs and prevents the hydrolysis of the drug or the polymer. However, residual solvents pose technological problems, and the formation of ice crystals within the microspheres can contribute to cracking or complete fracture of the microspheres, especially occurring at high drying temperatures, resulting in large initial burst drug release.

45.2.4.2 Removal by Extraction

The solvent present in the internal phase (embryonic microspheres) can be extracted, usually, using a relatively large continuous phase volume.[96,100,101,214] The amount and the composition of the continuous phase are chosen so that the entire volume of the disperse phase solvent can be dissolved in, and therefore extracted from the internal phase. Therefore, the rate of solvent removal from the oil phase is controlled by the volume of the continuous phase and the solubility of the organic solvent in this phase. The process is frequently performed in two steps. The drug-matrix-organic solvent dispersion is first mixed with a small amount of continuous phase to yield an emulsion of desired droplet size and size distribution. Then, a further continuous phase and/or additional extraction agents are added at an amount sufficient to absorb the entire solvent leaching from the solidifying microspheres. The entire amount of the large volume of the continuous phase in the second step can be added at once, or it may be added continuously over an extended period of time. According to some workers, the use of either "all at once" or "continuous" solvent extraction procedures insignificantly influence microspheres characteristics.[199,206] However, the addition of an O/W emulsion to high amounts of the continuous extracting medium all at once was found to enhance detainment of highly water soluble drugs.[215]

As mentioned earlier, the magnitude of the solubility of the organic solvent in the continuous phase influences its initial extraction rate and microsphere solidification time, all being affected by the temperature. Fast precipitation of the matrix former due to the initial efflux of the organic solvent to the continuous phase is advantageous for achieving high encapsulation efficiencies.[121,216] Rapid initial extraction aids in rapid skin formation on microsphere's periphery, and therefore minimizing drug loss. However, very rapid solvent extraction can reduce the encapsulation efficiency due to the formation of either hollowed microcapsules with dense shell around the droplets,[147,217] or collapsed particles especially obtained at low matrix former concentrations. Otherwise, aggregation of embryonic microspheres may result since the solvent diffuses out of the droplets before a stable emulsion is developed.[218] Direct solvent extraction from W/O emulsion using ethanol or methanol precluded the formation of lysozyme-loaded microspheres of copolymers of hydrophilic poly(ethylene glycol) blocks and hydrophobic poly(butylene terephthalate) blocks due to massive polymer precipitation. However, when a $W_I/O/W_{II}$ emulsion is employed instead, with extracting solvents, such as ethanol, methanol, and mixtures of methanol and water were incorporated in the second aqueous phase (W_{II}), the rate of extraction of the inner phase water was controlled, and lysozyme-containing microspheres stabilized using PVA were obtained.[219] Overly rapid solidification can also be overcome by pre-saturation of the continuous phase with the organic solvent, allowing sufficient time for microspheres hardening and shrinkage.[220] For example, when two partially water miscible solvents, such as benzyl alcohol (BA) and ethyl acetate (EA), were used for microspheres preparation, EA was extracted faster in aqueous continuous phase than BA, which was not easily removed by evaporation owing to its high boiling point (BA b.p. 205.3°C). Therefore, to control the rate of EA diffusion and to enhance BA extraction, EA (about 20%–70% of saturation point of EA in the extraction medium) was added to the extraction medium prior to addition of the emulsion. Thus, when the emulsion was added to the aqueous extracting medium, extraction of the more rapidly extracted solvent is retarded and more of slowly extracted solvent is removed.[221,222]

Compared to the evaporation procedure, extraction of organic solvent occurs relatively more rapidly, therefore, microspheres produced by solvent extraction are more porous usually resulting in faster release of drugs. Hence, for sustained release purposes, solvent evaporation is preferred.[223] Solvent extraction could be more appropriate and effective compared to evaporation method for encapsulation of delicate and sensitive drugs such as proteins and peptides, enzymes, hormones, and antigens that are susceptible to thermal degradation at temperatures above room temperature (i.e., 20°C).[224–226]

45.2.4.3 Removal by Combined Evaporation and Extraction

Solvent extraction and evaporation procedures may be combined to improve the economic efficiency of the microencapsulation process. The two-step solvent removal can be performed in a way that extraction is the first step. A sufficient quantity of extracting liquid is first added to induce skin formation on microspheres surfaces by removing part of the solvent from the emulsion droplets, while the remaining solvent is removed by evaporation. Otherwise, solvent removal process starts with solvent evaporation step with or without reduced pressure, followed by a washing step where the residual organic solvent is extracted until hardening of the microcapsules.[99,103] The combined procedure, minimizes operation time, and reduces volumes of extracting fluids consumed compared to an extraction process alone. Extraction alone or when combined with evaporation, often result in microspheres that are better in shape (not hollowed) with reduced loss of water soluble active ingredient when compared to those prepared solely with evaporation procedure.[206] For example, since MC is more soluble in methanol than in water, extraction using methanol, and the combined procedure using a mixed solvent system (water and methanol), was found superior to evaporation procedure using water alone in PLGA microspheres containing ovalbumin.[227] In a similar way, levonorgestrel-loaded PLGA microspheres prepared using the modified solvent extraction/evaporation method using acetone and chloroform mixture were nonporous with encapsulation efficiencies close to 100%, while those prepared with the solvent evaporation process using MC were porous with poor encapsulation efficiencies.[145] Mixed solvents have been also utilized to achieve both solvent extraction and evaporation for the encapsulation of BSA or lysozyme into PLGA. For example, PLGA was dissolved in a solvent mixture of strongly polar, acetonitrile (ACN), and moderately polar, MC, and emulsified with paraffin oil, to form $W/O_I/O_{II}$ emulsion, with proteins incorporated in aqueous phase. MC was extracted by the paraffin oil (in-oil drying), whereas ACN, which is not soluble in paraffin, was evaporated during the purging with air and the thereafter stepwise pressure reduction (300/50 mm Hg).[228] However, the process may take long time to efficiently remove the solvent; which is undesirable, particularly if the process is to be operated continuously. Fluid bed drying under nitrogen flow applied after the extraction step to remove the aqueous solution was found to reduce the lengthy time required for more traditional methods of drying and reduces the amount of degradation of the microspheres.[229] The two-step solvent removal method was utilized to encapsulate pre-fabricated small spherical particles such as insulin particles,[230] to prepare sustained release nonsteroidal anti-inflammatory and lidocaine PLGA microspheres,[231] to prepare cyclosporin micro and nanospheres,[232] and to produce microcapsules containing several active ingredients.[233–238]

45.3 FORMULATION AND OPERATIONAL FACTORS AFFECTING MICROSPHERES PROPERTIES PREPARED BY THE EMULSION SOLVENT REMOVAL METHOD

The wide spread and popularity of the emulsion solvent removal method in microspheres manufacturing are attributed to its inherent simplicity, operated in most cases at room temperature and normal atmospheric conditions. However, the physicochemical phenomena governing this process are complex due to the existence of a considerable number of processing and formulation parameters that profoundly affect the properties of the product obtained.[29] The following formulation and processessing parameters greatly influence the microspheres products:

45.3.1 Selection of Solvents

Of prime importance in solvent removal method is the selection of organic solvents making out the oily phase. The selected organic solvent should dissolve the matrix former and should, ideally, dissolve the active ingredient when employed as an internal phase. The solubility of matrix formers, such as polymers, in the organic solvent determines the morphology of precipitated polymer during the solvent removal process. The solvent should be of high degree of immiscibility with the continuous phase (solvent evaporation) with lower boiling point, or should have certain magnitude of solubility in the continuous phase (solvent extraction) since solubility of the organic solvent in the continuous phase impacts its initial extraction during microparticle preparation. The safety concerns about the use of organic solvents in microparticles preparation has received considerable attention due to the potential toxic effects brought by such solvents and their residuals in the final product microparticles even after drying.[239,240] MC is the most commonly used solvent in the solvent extraction/evaporation method owing to its good solvation power and low boiling point. However, MC is a group 2 solvent according to the ICH guidelines, and its use of MC is not recommended in routine microsphere manufacturing from the environmental and human safety perspectives.[241] Nevertheless, MC is employed in commercial microsphere preprations where its level in the final product is kept very low to meet acceptable residual solvent levels dictated by the FDA (Food and Drug Administration). For example, a commercial microsphere-based product, the Lupron Depot® contains less than 50 ppm MC per dose,[242] which is below accepted USP levels of 600 ppm. Ethyl acetate (EA) and ethyl formate (EF) are class 3 solvents that are less toxic and of lower risk to human health than MC, and therefore are considered a more preferred solvents.[241] However, the use of EA leads to a decreased drug encapsulation efficiency.[243] Other suitable solvent choices could be N-methyl-2-pyrrolidone (NMP), dimethylsulfoxide (DMSO), 2-pyrrolidone and PEG 400, which are acceptable for human use at 50 mg or less per day without justification.[244]

Moreover, safety issues are of concern also for the type of oil that can be used. Most commonly used oil for O_I/O_{II} emulsions is paraffin/mineral oils that is known to cause severe lipoid pneumonia,[245,246] and therefore must be excluded from injectible formulations. Arachis oil (peanut oil) is vegitable oil, which has been suggested by different pharmacopoeia as standard oil for injections. However, purified arachis oil may contain peanut proteins that are allergenic.[247] In rare cases, anaphylactic reaction and hypotension may occur after parenteral administration of arachis oil; however, severe reactions may occur in persons with allergy to this oil. Since the European pharmacopeia (PhEur) monograph does not contain a test for residual peanut proteins, an analytical monitoring specification should be developed.[748] Allergies also present in other vegetable oils such as sesame[249] and almond oils,[250] and therefore, they may be used for potential patients only when devoid of allergic reactions. Oils are usually removed from the solidified microspheres using organic solvents such as hexane, which themselves may present problems in terms of completeness of removal.

45.3.2 Selection of Matrix Formers

The selection of a matrix former or a carrier material for the encapsulation of drugs is dependent on the intended route of administration, the intended rates and duration of drug release, the drug dose, and whether targeting to certain body sites is a potential goal. The selection of matrix former type and molecular weight are of importance from microsphere manufacturing point of view since the size of the emulsion droplets, and the final size of the microcapsules are dependent of matrix formers solutions properties.

Polymers are the most commonly employed matrix formers; however other materials like proteins,[251] lipids[252-254] polysaccharides such as chitosan[255-257] and alginates,[258] and have been also studied, although at a lower frequency.

Most commonly employed polymers can be categorized as follows:

45.3.2.1 Biodegradable Polymers

Owing to the excellent biocompatibility, biodegradability, and low immunogenicity and toxicity,[259] polyesters such as poly(lactic acid) (PLA), poly(lactic-co-glycolic acid) (PLGA) and to a lesser extent poly(glycolic acid) (PGA), are the most frequently employed polymers as matrix formers in experimental and commercial encapsulation of therapeutics and antigens.[259,260] The polymers are FDA approved for use in humans[261] and are of considerable interest for parenteral use since they degrade in the body to hydroxycarboxylic acid then to carbon dioxide and water.[262-264] Depending on their degradation characteristics, they have shown to be able to achieve prolonged release of actives. The degradation rate can be regulated by changing their molecular weights, chemical compositions, and crystallinites.[265] For hydrophobic drugs, PLA–PEG–PLA block polymers are employed instead of PLGA, owing to their higher porosities and a faster release.[266] Blends of high molecular weight (MWT) polymer with a small portion of a low (MWT) polymer, or the use of low (MWT) polymer altogether could be optimistically used to manipulate release rates.[267,268] Lupron Depot® (leuprolide acetate, TAP Pharmaceuticals Inc.); the first commercial microspheres product is based on PLGA polymer that is manufactured via $W_I/O/W_{II}$ emulsion solvent evaporation.[20]

Despite the successful use of PLGA polymers in microencapsulation industry, they have been found not universally suited for different applications. The bulk hydrolysis of PLGA polymers creates an acidic environment within the microparticles, which can be detrimental to proteins and nucleic acids.[17,269] The direct contact with PLGA monomer and dimer residues with organic solvents; such as acetonitrile or MC, promotes protein degradation.[73,74] Likewise, proteins in direct contact with the polymer will enhance polymeric degradation over time.[71,72] Other limitations for PLGA polymers are the apparent inability of controlling drug release rate,[270,271] and the existence of large initial "burst release" upon dissolution wherein about 60% to 80% of the protein load is released within 24 h.[7] The release rates of a drug from matrix microspheres made of degradable polymers is inversely dependent on polymer molecular weight,[265,272] the aqueous solubility of the drug, and porosity of the matrix. For lactic and glycolic acid based microspheres experiencing a lag time, drug release does not occur until sufficient reduction of their molecular weights by hydrolysis to a critical level is achieved.[110] The incorporation of amine-containing drugs into PLA, PLG, and PLGA polymers[107,273] and the increase in the environmental pH and ionic strength[72,82] enhance the rate of polymers hydrolysis and molecular weights reduction. On the other hand, the enhanced degradation and the production of acidic metabolites has been utilized to improve the solubility of a drug compound that is slightly water soluble at pH 6 to 8, but readily soluble in an acidic medium (pH < 3).[274] Gamma irradiation dramatically reduces the molecular weight of PLGA polymers without affecting the drug release rates from the microspheres, which might suggest that the other factors—drug solubility and matrix porosity—are the determinant factors in controlling drug release rates.[275,276]

Other biodegradable polymers include polyanhydrides, which are suitable for the fabrication implantable controlled release microspheres. Since polyanhydrides undergo hydrolysis in aqueous medium; often their microspheres are prepared using O_I/O_{II} emulsion technique.[277,278]

45.3.2.2 Nonbiodegradable

45.3.2.2.1 Celluloses

Ethyl cellulose (EC) is the most widely used cellulose ester for retarding drug release orally and to protect the drug in the gastrointestinal tract.[279–283] It is degradable, biocompatible, and FDA approved. Ethyl cellulose is a cellulose derivative in which part of the hydroxyl sites are substituted with ethoxy groups[284] and is readily available in various grades. Other cellulose esters, such as cellulose acetate butyrate (CAB),[284,285] and cellulose acetate propionate (CAP)[284,285] have also been investigated in microsphere formulations with the potential to achieve controlled drug release.[150,156] The hydroxyl sites are substituted with acetate and butyryl ester groups in CAB, or with acetate and propionyl ester groups in CAP. CAB and CAP polymers are hydrophobic, nonswelling. Microspheres of cellulose esters have been made using O_I/O_{II} and O/W emulsion techniques.[80] Cellulose ethers, such as hydroxyl propylcellulose (HPC), have been employed, however, to lesser extents.[286]

Drug release rate from cellulose esters microspheres is dependent on the porosity of the matrix, and on the solubility of the drug in the matrix material or in the dissolution medium. Hydrophobic drugs exhibit prolonged release rates that can be modified using hydrophilic polymers such as PEG4000[80] and HPC.[286,287] Cellulose acetate phthalate (CAPh) is a pH dependent cellulose ester, frequently employed alone,[288,289] or in combination with other polymers to obtain microspheres with enteric properties.[61]

45.3.2.2.2 Acrylics

Methacrylic acids and their esters (Eudragits®)[290] have received much attention in the formulation of microspheres for oral drug delivery. Eudragit®RL100 and Eudragit®RS100 are ammonio methacrylate copolymer with a low content of quaternary ammonium groups as salts that render the polymers permeable. Eudragit® RL100 with higher percentage (8.8%–12%) of quaternary ammonium groups, has higher permeability, and therefore swells more at physiological pH values compared to Eudragit®RS100. Thus, microspheres made of Eudragit®RL100 exhibit faster drug release rates.[291] Due to the swelling properties of both polymers in the aqueous phase, it has been found that more acceptable microspheres can be produced using O_I/O_{II} rather than O/W emulsion technique.[34,149,152–154]

The unique solubility profiles of Eudragit®L, S, and E have suggested their use in microencapsulation for controlled release applications with enteric properties.[292,293] Eudragit®L100 and Eudragit®S100 are anionic copolymers based on methacrylic acid and methyl methacrylate. The ratio of the free carboxyl groups to the ester groups is approx. 1:1 in Eudragit®L100 and approx. 1:2 in Eudragit®S100. The free carboxylic acid groups make the polymers pH sensitive. Eudragit®S100 enjoys the same properties as in Eudragit®L100, except that the former is soluble at pH > 7 (Rowe, 2006), while Eudragit®L100 is soluble at pH 6 to 7. Microspheres made of these polymers are expected to prevent, or at least minimize drug release at the gastric fluid pH, and preferably control its release at intestinal fluid pH. They have been employed either individually or in combination with other rate controlling matrices. For example, Eudragit®S100 has been used to encapsulate ketoprofen,[294] indomethacin and theophylline,[155] and Spodoptera frugiperda nucleopolyehedrovirus.[295] Similarly, Eudragit®L100 has been employed for the encapsulation of β-galactosidase.[293] Polymers combinations of Eudragit®S100 or Eudragit®L100, with Eudragit®RL100 or ethyl cellulsoe have been used to control the release of theophylline and indomethacin, respectively, using the emulsion solvent removal techniques.[157,296] Another polymer, Eudragit®E, is a cationic polymer based on dimethylaminoethyl methacrylate and neutral methacrylic acid esters. It is soluble in gastric fluid as well as in weakly acidic buffer solutions up to pH 5.[297] Eudragit®E microspheres were prepared by solvent removal methods for the delivery of bacampicillin[298] and nitrendipine.[299]

45.3.2.2.3 Lipids

Lipids, oily materials and waxes, have been utilized similar to polymers to produce solid lipid microparticles.[252,254,300–304] In manufacturing a lipid component or a mixture of lipid components containing a drug compound are heated to melting or dissolved in an organic solvent.[305–307] Separately an aqueous solution containing surfactants and possibly a co-surfactant is prepared, and heated to a temperature equal at least to the melting temperature of the mixture of lipid components. The hot aqueous solution is then admixed under mild stirring with the mixture of lipid components, obtaining an emulsion, which is then poured under stirring in water of 2°C to 10°C, obtaining the formation of well dispersed lipid microspheres. Morgan and Blagdon[308] prepared edible microcapsules containing multi liquid cores. In the process, a W/O emulsion, with the drug compound dissolved in an inner aqueous phase, is spray cooled for the solidification of the fat phase and the entrapment of the aqueous phase as minute droplets within a microcapsule. The process, however, leads to very unstable microcapsules; from which the aqueous phase migrated to the outer part of the microcapsule. In addition, the release of the drug compound cannot be controlled in the microcapsules. On the other hand, Coyne et al.[309] provided a microcapsule comprised of a solidified hydrophobic shell matrix (fats, oils, waxes, or resins) containing encapsulated gelled, or cross-linked aqueous beads using hydrocolloids. Thus the drug compound is double encapsulated in the microcapsules. The advantage of such system is that the release rate of a water soluble drug compound in a conventionally spray cooled fat matrix microcapsule is usually not controlled by the melting of the fat matrix, but rather by the diffusion of water into the microcapsule and subsequent migration of the active ingredient outside the microcapsule.

45.3.2.2.4 Proteins

Proteins have been utilized to form microspheres for drug delivery.[251] Mostly, proteins are cross linked in solution using glutaraldehyde, or hardened at elevated temperatures. Mathiowitz et al.[310–312] produced microspheres made of Zein proteins that are hydrophobic, biodegradable, and can be modified proteolytically or chemically to endow them with desirable properties. Microspheres containing solid zinc insulin, soluble insulin, or vasopressin are formed by phase separation of simple W/O or double emulsion techniques followed by solvent removal. The process does not involve use of temperatures and agents that degrade most labile proteins. Suslick[313] produced surface modified microparticles with a novel protein shell and a surface coating via emulsification technique. The protein shell consists of cross-linked albumin or other proteins with functional moieties for cross linking. The surface coating consisting of polyethylene glycol, a protein, or an antibody is covalently or electrostatically adsorbed to the cross-linked protein shell.

45.3.2.2.5 Polysaccharides

Polysaccharides such as chitosan[256,257] and starch[314–316] have been also employed as matrix formers in the fabrication of microparticles by the solvent removal method, however, to lower extents compared to polymers.

45.3.3 Type and Role of the Emulsifier

A suitable emulsifier acceptable for therapeutic use is needed to produce a stable emulsion owing in major to the lowering of the emulsions' interfacial tension. High concentrations result in a significant reduction of the interfacial tension leading to the formation of smaller drops with narrower and more uniform size distribution. However, in emulsion solvent removal methods, the role of the emulsifier is only temporal; to provide a short term stabilization of droplets to prevent initial aggregation and coalescence before adequate volume of the organic solvent is being removed. Once the droplets surfaces (microspheres embryos) become hardened, they should no longer aggregate or coalesce, and the emulsifier role in emulsification process is over. However, the emulsifier molecules existing on the surface of the finished product and in the continuous phase apparently impact microspheres behavior. Most commonly used emulsifiers for O/W emulsions are PVA, methylcellulose, gelatin, polysorbate 80, and sodium dodecyl sulphate. Magnesium stearate and span83 are commonly used in W/O or in O_I/O_{II} emulsion techniques. Emulsifiers are known to enhance the solubility of drugs. The appearance of lomustine and progesterone crystals both on the surface of PLA microspheres and within the aqueous continuous phase was attributed to the enhanced solubilization of the drugs by PVA and methylcellulose.[79] Similarly, PVA promotes insulin growth over PLA microspheres, and therefore resulted in marked rapid instantaneous insulin up to 88%. This "burst" release was found to be lower when other emulsifier types (e.g., gelatin) were employed.[317]

Increasing the surfactant concentration reduces the average droplets sizes of emulsions, and therefore finer microspheres sizes. The observed reductions in droplets sizes of the emulsion can be explained by the lowering of the interfacial tensions according to Gibbs equation.[78,318,319] However, no further reduction of the average droplets sizes of emulsions occurs beyond a limiting or critical surfactant concentration.[320-322] Mixtures of surfactants at optimum ratios were found provide better control microspheres sizes compared to individual surfactants.[323] The size distribution of microspheres often narrows with increasing surfactant concentration.[78] However, some workers reported that the size distribution becomes significantly wider as the concentration of the emulsifier increased.[319] The increase in surfactant concentration was found to reduce microsphere drug content. This is due, in part, to the enhanced drug solubilization away from the microsphere matrix. Additionally, adsorbed surfactants would stabilize droplet interface of the emulsions, and therefore would reduce solvent loss and polymer solidification rate, and eventually allowing more time for drug loss from the forming microspheres. For example, it was reported that the encapsulation efficiency of all-trans retinoic acid (atRA) in PLGA polymer decreases with the increase in PVA,[322,324] or in sodium oleat (SO)[322,325] concentrations. Furthermore, the co-encapsulation of a surfactant would reduce drug encapsulation efficiency.[326] For example, the co-encapsulation of a surfactant (poloxamer 188 or 331) lead to reduced encapsulation of BSA within PLGA microspheres, accompanied with faster release rates.

Frequently, the use of mixed emulsifiers resulted in improved microsphere shape and drug entrapment efficiency. For example, Cavalier et al.[139] reported that at hydrocortisone/PLA ratio of less than 0.4, microcapsules could be prepared using only PVA. However, when this ratio was increased to 0.6, irregular and large aggregates resulted due to the formation of unstable emulsion, and therefore a combination of PVA and hydroxypropylmethylcellulose (HPMC) as polymeric emulsifiers was required. Conversely, higher encapsulation efficiency of all-trans retinoic acid (atRA) in PLGA was obtained in microspheres stabilized using PVA alone compared to those stabilized with a combination of PVA and SO, especially at higher percentage of SO.[322] Surfactants also play a role in modifying the release kinetics of drugs from microspheres formulations. According to some authors,[323,326,327] surfactants offer another mean of modulating microsphere particle size and dissolution profile. Polar surfactant regions often interact with aqueous media, and therefore enhance drug dissolution. Bouissou et al.[328] pointed out that the residual amounts of PVA and Triton X-100 raised the hydrophilicity of the microspheres. On the other hand, higher PVA concentrations were found to decrease drug release rate from PLGA microspheres. In addition, microspheres containing SO contributed to a relatively slower release compared to microspheres stabilized by PVA alone.[322] It was reported also that other emulsifier properties, such as the HLB value can influence the encapsulation efficiency, size, and morphology of microspheres.[329]

45.3.4 AGITATION INTENSITY

During the emulsification of oil (O) and water (W), a large amount of new interface is created (creation of smaller droplets). The surface free energy, ΔG, required for this increase in interfacial area ΔA can be estimated from the product of the interfacial tension, γ_{ow}, and the area finite change, expressed in terms of the Gibbs free energy (J/m²) as

$$\Delta G = \gamma_{ow}\Delta A \ (\text{constant } T, P, n)$$

In addition, a positive entropy change, ΔS, occurs as one phase is dispersed into another, e.g., (O/W):

$$\Delta G_{formation} = \gamma_{ow}\Delta A - T\Delta S$$

However, since the droplets sizes are not very small, the entropy change is small, and therefore, $\gamma_{ow}\Delta A \gg T\Delta S$, and an energy input is needed to break up the droplets into smaller ones. In part, this may be in terms of heating, but it is mainly in terms of mechanical energy using very high shear rate instruments, such as a high-speed blender, a homogenizer or an ultrasonic probe.[330]

Reported quite frequently, especially for low-coalescing system, higher emulsification agitation rates often result in droplets with smaller diameters and with narrow size distribution, which after solvent removal, end up with small microspheres produce that is narrowly distributed, most likely log-normally.[19,78,331–334] Higher agitation intensities are required to decrease microspheres diameters when polymer solution phase of higher viscosity is used.[121,151] For a high-coalescing system, high agitation rates increase the velocity of approach of colliding droplets, and therefore can increase the drop coalescence rate,[78] since only a small number of collisions result in immediate coalescence.[335] As mentioned earlier, surfactants can modulate droplet particle size of the emulsion, by decreasing it at any given agitation intensity. The decrease in the droplet size can be explained by Gibbs equation, where the lowering of the interfacial tension increases the newly created surface areas at constant energy input. Otherwise, the decrease in size can be explained by the lowering of the interfacial tension, resulting in an increase of the Weber number and, thus, a decrease of the Sauter mean diameter.[78,336]

45.4 MICROSPHERES CHARACTERIZATION

45.4.1 MICROSPHERE PARTICLE SIZE

The microspheres population produced by the emulsion solvent removal methods is often polydisperse. Microspheres can be viewed under the light microscope to examine the general shape, presence of the encapsulant, and to detect aggregations of microspheres. Nearly monosized fractions (>38 μm) can be safely separated from the bulk product by ordinary sieving methods. Smaller sizes become more difficult to separate due to electrostatic attraction. Microspheres particle sizes and size distribution are dependent on the agitation intensity employed during emulsification,[331] surfactant concentration,[78] core properties and its ratio to the matrix former,[337] and the volumes of the internal,[152,338] and continuous phases.[73,339] Larger microsphere sizes are generally produced when cores of large sizes are used compared to small and medium core sizes.[340] Also, as the core to polymer ratio increased, the geometric mean diameter of the microspheres increased as well.[150] At constant core to polymer ratio and at constant mixing conditions, microsphere sizes tend to get larger as polymer concentration,[341] molecular weight,[150,342] or polymer solution viscosity[151,319] increases. Actually, both the increase in polymer concentration and molecular weight would increase the polymer solution phase viscosity. Therefore, viscous forces and hence emulsion's droplets volumes increase, leading to an increase in microspheres sizes. An empirical equation that relates the viscosities of the dispersed and continuousphases, (η_d) and (η_c), respectively, to the average diameter (Sauter's diameter), d_{32}, is given by[343–345]

$$d_{32} = C\emptyset \left(\frac{\rho_c N^2 D^3}{\gamma} \right)^{-3/5} \left(\frac{\eta_d}{\eta_c} \right)^{0.25}$$

where
  D is the diameter of the agitator (m)
  ρ_c is the density of the continuous phase (kg/m³)
  N is the agitation rate (turns/s)
  γ is the interfacial tension between the dispersed and the continuous phase (N/m)
  $\emptyset$ is the phase volume ratio of the dispersed and the continuous phases
  C is a constant

Constant C is experimentally determined and its value is affected by factors related to agitation conditions. An increase in the phase volume ratio, $\emptyset$, will, generally, result in a decrease of the ratio of the viscosities of the two phases, (η_d/η_c). This can, in part, explain some of the controversies present in the literature regarding modification of $\emptyset$ and the resulting changes microspheres sizes. Depending on the experimental conditions and the formulation components, an increase in $\emptyset$ can result in a decrease in the size of microspheres[339,342,346] due to the decrease in the disperse phase viscosity, and hence the decrease in (η_d/η_c). However, an increase in $\emptyset$ has been reported to result in an increase,[347,348] or in no significant influence on size of microspheres.[81] This might be attributed to the increased viscosity of the internal phase in systems where a better solubility of the matrix former (polymer) in the "good solvent" occurs. Therefore, the ratio (η_d/η_c) increases, resulting in larger microspheres diameters.

The sizes of microspheres prepared under reduced pressure have been found in some studies to be smaller compared to those prepared under atmospheric pressure,[132,211] which can be explained according to Laplace equation, greater pressure difference across the droplet under reduced pressure.

Attempts have been made to investigate new techniques and equipments to yield microsphere products that are monosized with narrow size distributions using the emulsion solvent removal method.[38–40,42,349] Uniform size microspheres are obtained when a polymer solution phase is pumped at a given rate through a stainless steel needle seated at different angles within TEFLON™-coated polyethylene tubing protruding through the tubing wall. For microcapsule formation, PVP solution in reverse osmosis water solution is pumped at a constant rate through the polyethylene tubing and past the needle. According to this technique, controlling parameters such as the interfacial tension, needle gauge number, angle of needle orientation and the viscosity of PVA solution, would control microspheres sizes.[350] Uniform inorganic microspheres were obtained by injecting an aqueous solution containing a particle forming material into an organic solvent through a macromolecular membrane of a hydrophobic surface. The membrane has pores that are substantially uniform in size formed by either a corpuscular or laser beam, and extending in the direction of thickness, and substantially straight through the membrane.[351] Also, ultrasonication of an O/W or an O_I/O_{II} emulsion prepared using a propeller stirrer or turbine stirrer yielded fine microsphere particles having an average particle diameter of about 0.1 to 10 μm.[113] In addition, the simultaneous continuous introduction of oily and aqueous phases into the zone of high shear mixers; such as colloid mill, was found to result in an emulsion having a high volume of oil phase droplets of substantially uniform size.[352]

45.4.2 Drug Loading Capacity and Encapsulation Efficiency

Encapsulation efficiency is defined as the percentage of drug incorporated into the microspheres relative to the total drug added. Loading capacity refers to the percentage of drug incorporated into the microsphere relative to the total weight of all components. For a drug and a polymer composing a microsphere, the encapsulation efficiency and drug loading can be determined as follows:

$$\text{Encapsulation efficiency}(\%) = \frac{\text{actual drug loading}}{\text{theoretical drug loading}} \times 100$$

$$\text{Theoretical drug loading}(\%) = \frac{\text{initial drug ass}}{\text{initial drug mass} + \text{initial polymer mass}} \times 100$$

$$\text{Actual drug loading}(\%) = \frac{\text{encapsulated drug mass}}{\text{initial drug mass} + \text{initial polymer mass}} \times 100$$

Higher encapsulation efficiency of a drug is achieved when it partitions favorably into the internal phase.[3] Frequently, theoretical and actual drug loading capacities are dissimilar due to various parameters influencing drug encapsulation efficiency. Increasing the quantity of drug (theoretical loading) improves the encapsulation efficiency,[342] since the loss of drug into the continuous phase is constant at constant operating conditions. However, when the drug is overly loaded, its encapsulation efficiency will decrease, since it transferes to the continuous phase upon microspheres saturation. Higher drug loadings would generally increase the geometric mean diameter of microspheres at constant processing conditions,[340] with smaller microspheres having higher drug loadings.[175,353] However, insignificant difference in drug content in microspheres size fractions different was also reported.[337] Micronized drug core that is dispersed in the polymer solution phase, often yields higher drug loading or encapsulation efficiency compared to larger size drug core. However, larger drug crystals were found to yield larger microsphere sizes with greater drug loadings compared to medium and small drug crystals.[340] Formulation components, such as drug to polymer ratio, drug solubility in the continuous phase and in the extracted organic solvent would affect drug loadings capacities.[152] Increasing viscosity of the internal phase (via increasing polymer concentration, molecular weight, and/or viscosity grade) improves the drug encapsulation efficiency by restricting its diffusion. For example, higher concentrations of ethyl cellulose resulted in a higher aspirin[354] or ethyl benzoate[342] encapsulation efficiencies. Processing time and temperature influence drug detainment within the micro spheres since both potentially increase the solubility and diffusion rate of the drug toward the continuous phase.

The volume of the inner aqueous solution relative to that of the organic solvent in $W_I/O/W_{II}$ emulsion was found to be a critical parameter that determines the encapsulation efficiency of medicaments. For example, increasing the volume fraction of the internal aqueous phase was found to lower the encapsulation efficiency of ovalbumin in PLA.[355] This could be due to the increased probability of contact between the internal drug solution and the continuous extraction phase. Also, at low aqueous to organic phase volume ratios (≤ 0.1 mL/mL), the encapsulation efficiencies of BSA,[356] and an antigen[357] within PLGA microspheres were found to be about 90%, and 100%, respectively.

Drug loading and encapsulation efficiency can be determined by dissolving the microspheres containing the drug in an appropriate solvent, and the drug can be analyzed using the suitable method that guarantees its quantification solely without interferences with microsphere components. It could be more convenient, especially where interference existed, to leach out the drug using a nonsolvent for the polymer.

45.4.3 Drug Release from Microspheres

45.4.3.1 Factors Affecting Drug Release

Many factors are known to impact the release rates and profiles of medicaments from microspheres. These parameters may include the following: physicochemical properties of the drug and the matrix former, porosity of the individual microsphere, thickness of matrix former, particle size and size distribution of microspheres, core particle size, and medicament loading affects. Release rates for many microspheres (especially nonerodible ones) were found inversely proportional to the square of the particle size as a result of the increased surface area exposed to the dissolution medium. Often, a linear relationship exists between the mean microsphere size and the time for 50% of the encapsulated drug to be released.[109,341] Therefore, as microsphere size decreases, the surface area-to-volume ratio of the particle increases, thus, the rate of drug release out of the microsphere increases.

Dissolution rate for a given microspheres particle size fraction increases with increasing drug loading.[212,358] At low loading of a small particle size drug, drug release is a result of its dissolution into the matrix followed by diffusion to the dissolution medium.[359] At high drug loading, the relatively small drug particles are crowded together. Therefore, particles dissolution creates channels within the microsphere matrix resulting in an increased leaching of the drug.[150] At a given drug loading, drug release is more rapid from microspheres containing large and medium drug crystals, and slower and more predictable from microspheres containing micronized drug. Unlike small drug particles that are well detained, large drug particles may not be wholly encapsulated inside the microspheres. Therefore particles are in direct contact with dissolution medium, so they dissolve and increase the access to the dissolution medium.[150]

The ratio of drug to polymer significantly affects the rate of drug release from microspheres, where highest drug release rates are obtained for the microsphere with high drug to polymer ratio.[337,360–362] The drug release kinetics and mechanisms can also be affected by the actual drug loading. For example, betamethasone release profile from high drug loaded microspheres was fitted best to Higuchi equation with a release mechanism of diffusion and erosion. At middle drug loadings, release profile best agreed with Hixcon-Crowell equation and controlled by diffusion and erosion as well. However, at low drug loading release profile were fitted to logarithm normal distribution equation with mechanism of purely Fickian diffusion.[363]

The particle size of drug cores impacts their release rates. The release of the drug in simulated intestinal fluid was very rapid from microspheres containing large and medium drug crystals, while release was slower and more predictable from microspheres made from micronized drug.[340]

The release rate is inversely related to physicochemical properties of polymers, molecular weights,[150,361] concentrations,[341] viscosity grades,[364] composition and degradation rate.[242] An increase in the molecular weight, concentration, or viscosity grade of the polymer results in slower drug release rates from microspheres. This can be attributed in major to the increase of the viscosity of the polymer solution phase, which therefore yields denser and less porous microspheres with low drug release rates. Price and Obeidat investigated the influence of the initial polymer solution (internal) phase of microspheres properties and drug release. The increase in polymers solutions viscosities for CAB381-2 and CAB381-20 was accompanied by a decrease in drug release rates. A fairly linear relationship existed between both T30% and the initial polymer solution viscosities, as shown in Figure 45.6.[151] Results suggest that adjusting the viscosity of polymer solutions could predictably optimize the release rates of encapsulated drugs from matrix microspheres.

The volume of the inner aqueous solution relative to that of the organic solvent in $W_I/O/W_{II}$ emulsion was found affect the release characteristics of BSA,[356] or an antigen[357] from PLGA microspheres. At low aqueous to organic phase volume ratios (≤0.1 mL/mL), the microspheres exhibited a triphasic release: initial burst, lag phase with little or no release, and a second release phase,

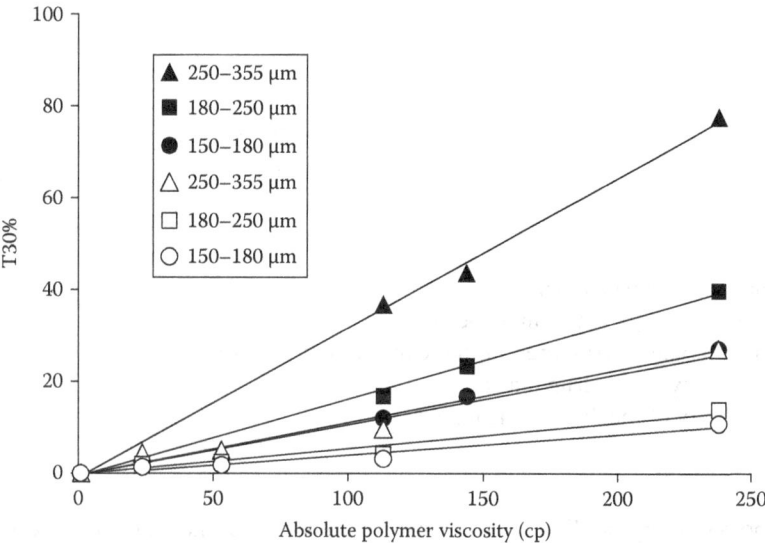

FIGURE 45.6 Effect of the apparent viscosity of polymer solutions on the T30% for different microsphere size fractions prepared using CAB381-2 (empty symbols) with polymer phase viscosities of 24, 55, 115cp and CAB381-20 (filled symbols) with viscosities 115, 155, 240 cp. (From Obeidat, WM. and Price, J.C., *J. Microencapsul.*, 20(1), 57, 2003.)

for both the protein and the antigen. Higher aqueous to organic phase volume ratios generally alter the release profile into a more continuous pattern in vivo.[356,365]

Drug release can be modified incorporation of additives on microspheres cores or surfaces that would either control polymers degradation,[274] or change the pH of the microsphere environment. For example, the release rate of a core material ketotifen hydrogen fumarate was modified by the use of an alkaline agent, for example, sodium hydroxide that was incorporated either in the water or in the oil phase.[366]

45.4.3.2 Kinetics of Drug Release from Matrix Microspheres

Drug release from a nonbiodegradable microsphere or matrix microparticle is a mass transport phenomenon involving diffusion of drug molecules from the region of high concentration within the microsphere to a region of low concentration in the surrounding environment. Mathematical models used here to describe the kinetics of drug release from microparticles are usually based for the case of a drug intimately mixed with the release rate controlling membrane in a monolithic device, that is, microspheres. The drug can exist as dissolved in or dispersed throughout the monolith where these particular cases lead to different release characteristics. Often, drug release rarely follows zero order, but it can be described by Higuchi models for both types of microspheres: microspheres with homogenous and heterogeneous (granular) matrix structure. For biodegradable microspheres, drug release is due to drug diffusion through water-filled networks of pores and channels coupled with the bulk erosion of the microspheres by hydrolysis of the polymer. However, the kinetics of drug release from biodegradable microspheres is beyond the scope of this chapter, where description of kinetics of drug release will be provided for nonbiodegradable microspheres only.

45.4.3.2.1 Release of Drug from a Spherical Pellet of a Homogeneous Matrix

Such system is obtained using the emulsion solvent removal method and other methods of encapsulation such as congealable disperse phase encapsulation procedure in molten wax. Higuchi[367] derived two equations for predicting the release rate from spherical matrices.

For $C_s \ll A$

$$a_0^3 - 3a'^2 a_0 + 2a'^3 = \frac{6a_0 D C_s t}{A} \qquad (45.1)$$

where
  a_0 is the radius of the whole pellet
  a' is the radius of the part still unextracted
  A is the total amount of drug present in the matrix per unit time
  D is the diffusivity of the drug matrix
  C_s is the solubility of the drug in the matrix
  t is the time

By dividing both sides by a_0^3, Equation 45.1 can be transformed into a dimensionless relationship,

$$1 - 3\left(\frac{a'}{a_0}\right)^2 + 2\left(\frac{a'}{a_0}\right)^3 = \frac{6 D C_s t}{A a_0^2} = Kt \qquad (45.2)$$

The left hand expression is dimensionless in terms of (a'/a_0) and independent of any unit of measure.

$$\left(\frac{a'}{a_0}\right)^3 = \text{Residual fraction of drug in pellet}$$

The equation for the release of the drug from spherical pellet of homogenous matrix can be written as follows:

$$1 + 2(F) - 3(F)^{2/3} = \frac{6 D C_s t}{A a_0^2} = Kt \qquad (45.3)$$

where
  $F = \left(\dfrac{a'}{a_0}\right)^3$ is the residual fraction of drug in pellet
  a_0, a', A, D, C_s, and t are the same as above

A plot of the dimensionless left hand expression values as a function of relative time yields a straight line, passing through the origin with a slope of K.

A similar equation for drug release from spherical matrix was derived by Baker and Lonsdale[368] and is given as follows:

$$\frac{3}{2}\left[1 - (1 - F')^{2/3}\right] - (F') = \frac{3 D C_s t}{C_0 r_0^2} = K't \qquad (45.4)$$

where

$F' = \dfrac{M_t}{M_\infty}$ is the amount of drug released in time t divided by the total amount of drug in the matrix

K' is the constant $= \dfrac{3DC_s}{C_0 r_0^2}$

D is the diffusivity of the drug in the dissolution fluid
C_s is the solubility of the drug in the dissolution fluid
C_0 is the total concentration of drug initially present in the matrix
r_0 is the initial radius of the spherical matrix
t is the time

Plotting the left-hand side of Equation 45.4 versus time yields a straight line through the origin with slope, K'.

45.4.3.2.2 Leaching of Drug from a Spherical Granular Pellet (Heterogeneous Matrix)

The rate of leaching of the drug into the external solvent (e.g., intestinal fluid) when the solid drug is dispersed within a granular spherical pellet is expressed by

$$1 + 2\left(\dfrac{a'}{a_0}\right)^3 - 3\left(\dfrac{a'}{a_0}\right)^2 = \dfrac{6DV_{sp}C_s t}{\tau A a_0^2} \qquad (45.5)$$

$$\left(\dfrac{a'}{a_0}\right)^3 = \text{Residual fraction}$$

The equation can be written as follows:

$$1 + 2(F) - 3(F)^{2/3} = \dfrac{6DV_{sp}C_s t}{A a_0^2} = kt \qquad (45.6)$$

where

$k = \dfrac{6D V_{sp} C_s t}{A a_0^2}$

$F = \left(\dfrac{a'}{a_0}\right)^3$ is the residual fraction of drug in pellet

a_0, a', A, D, C_s, and t are the same as mentioned earlier

τ is the tortuosity of the matrix pore structure

$V_{sp} = \dfrac{1}{\text{drug density}}$ is the specific volume

Typical dissolution plots for matrix microspheres with uniform properties are shown in Figures 45.7[61] and 45.8,[150] while Higuchi plots are shown in Figure 45.9.[151]

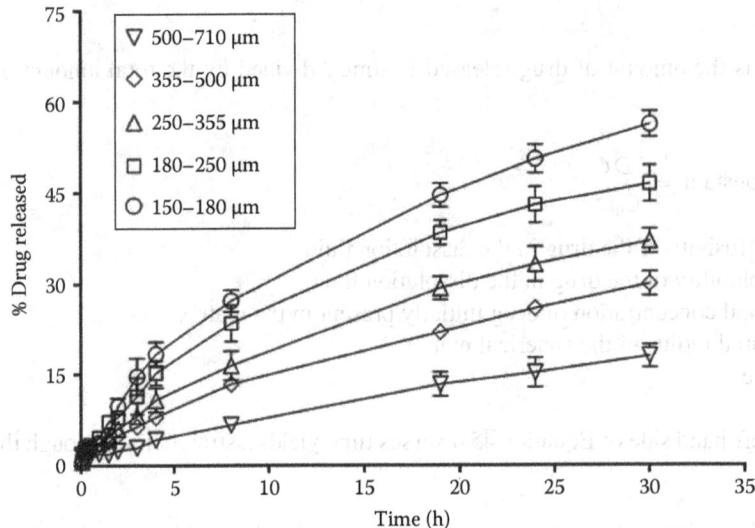

FIGURE 45.7 Release profiles of different size fractions of theophylline microspheres prepared using 15% CAB381-2 in acetone (240 cp). (From Price, J.C. and Obeidat, W.M., Microspheres and related processes and pharmaceutical compositions. In US patent 20060099256, 2006.)

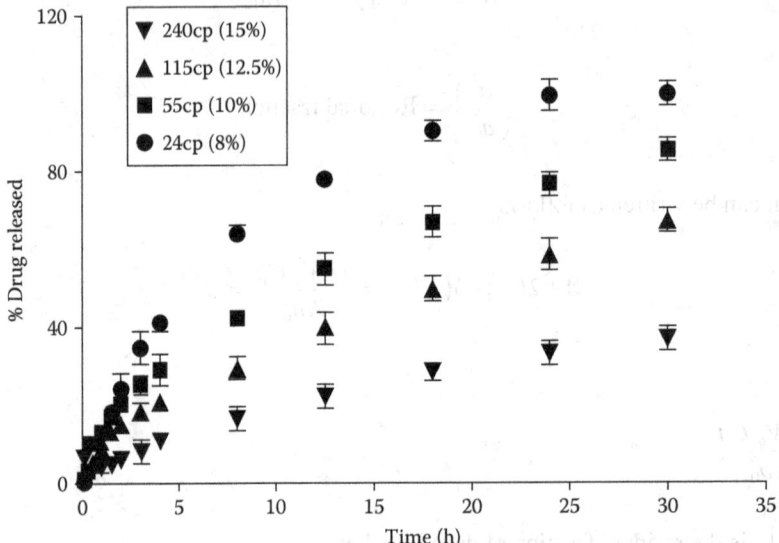

FIGURE 45.8 Release profile of theophylline from 250 μm microspheres in different CAB381-2 concentrations (8%, 10%, 12.5%, 15%) with respective viscosities (24, 55, 115, 240cps). (From Obeidat, W.M. and Price, J.C., *J. Microencapsul.*, 20(1), 57, 2003.)

It is worthy to mention that in a population of microparticles, individual release rate constants are often distributed log-normally, and drug release profiles frequently approximate first order kinetics although individual microcapsules may exhibit constant release pattern.[369–371] Therefore, the dissolution kinetics of a monodisperse sample of the microparticle popoulation will closely approximate that of individual microparticles. Monodisperse size can be approximated by taking narrow sieve cuts of the population.[18,19]

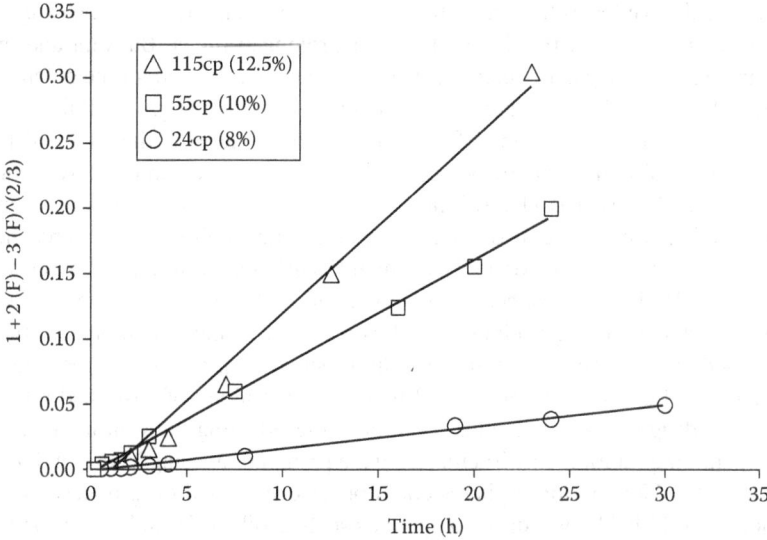

FIGURE 45.9 Higuchi plots for dissolution of 250 μm theophylline microspheres prepared using different CAB381-2 concentrations (10%, 12.5%, 15%), with respective viscosities (55, 115, 240) cp. (From Obeidat, W.M. and Price, J.C., *J. Microencapsul.*, 20(1), 57, 2003.)

45.5 POTENTIAL APPLICATION OF MEDICATIONS CONTAINING MICROSPHERES

Although encapsulation using the emulsion solvent removal methods has been investigated for applications in transdermal,[372–376] nasal,[377–381] pulmonary,[382–387] and ophthalmic drug delivery,[50,388–391] the vast majority of papers addressed the most common routes; the oral and the parenteral, for the delivery of medications.

45.5.1 ORAL DELIVERY

Nondisintegrating multiparticulate systems, such as microspheres and pellets, have many advantages that make them good candidates as carriers for oral drug delivery. For example, they tend to distribute randomly and uniformly throughout the gastrointestinal tract. The distribution behavior of multiparticulate systems affects their gastric emptying and their residence times in the gastrointestinal tract. Unlike nondisintegrating single unit dosage forms, which may exhibit an all-or-none emptying process, multiparticulate systems are emptied either with a slow phase followed by bolus pattern in fast state, or in a linear or bolus patterns in fed state.[392,393] Hence, multiparticulate systems have longer reproducible gastric residence; resulting in less variation between individuals with regards to gastric residence time and drug absorption than do single unit systems. Moreover, multiparticulate systems produce less mucosal irritation than single unit systems.[392,394] Furthermore, despite the requirement of few technical issues for their large scale production, multiparticulate systems can be tailored to achieve various release types at various sites in the gastrointestinal tract.

Most of the published papers and patents focused on the utilization of the emulsion solvent removal methods in the production of microspheres of the single or the multipurpose types. Majority of investigated microspheres are intended to achieve controlled release of medications by using matrix formers to sustain, or modify drug release by the incorporation of pH sensitive components, adhesive components, temperature sensitive components, or adjuvants to protect the encapsulated drugs.[35,51,54,55,57,63,80,283]

Gastric retentive dosage forms have recently gained interest in pharmaceutical drug delivery since they enhancing the residence of the dosage form for prolonged times. Buoyant and mucoadhesive microspheres are examples of gastric retentive that have been investigated for the potential continuous release of drugs. Buoyant or Hollow cores microspheres tend to float at gastric fluids, thus enhancing microspheres residence. The use of two different solvents that differed in the rate of diffusion usually leads to the formation of hollow microspheres. Hollow floating microspheres were found effective for controlled release of nonsteroidal anti-inflammatory drugs, such as piroxicam,[395] and diclofenac sodium.[396] Many other drugs have also been incorporated into hollow core microspheres; such as theophylline,[34,397,398] glipizide,[399] metoprolol succinate,[400] carbidopa/levodopa,[401] aspirin, griseofulvin and p-nitroaniline.[402] Hollow microspheres of rosiglitazone maleate were prepared by O/W emulsion-solvent diffusion technique using Eudragit®S100 as a matrix former. Entrapment efficiency of the drug was increased up to 89.71% as a result of salting out effect produced by the salts added in the continuous medium. High floating ability (>12 h) was shown along of sustained release drug profile.[403] Site-specific drug delivery to stomach can be achieved using buoyant microspheres. Hollow microspheres containing ranitidine hydrochloride were prepared by using Eudragit®RLPO with high loading (80 +/− 4.0%). Hollow microspheres could provide sustained drug release (about 24 h) and floating for more than 12 h.[404] Also, drug compounds such as ofloxacin and clarithromycin indicated for the eradication of Helicobacter pylori have been incorporated into hollow floating microsphere to obtain site-specific sustained actions of the medication in stomach.[405,406]

Mucoadhesive microspheres containing mucoadhesive polymers are proposed to stick to gastric mucous membranes for the delivery of various drugs.[407–411] Once microspheres stick firmly to the mucous membranes, their gastric residence times are prolonged until removed by turnover of mucins.[392] Mucoadhesive microspheres for targeting 5-fluorouracil to the colon without the release in the stomach or small intestine were prepared using Assam Bora rice starch, a natural mucoadhesive polymer, by a double emulsion solvent evaporation method. 5-FU release pattern exhibited slow and extended release over longer periods of time with reduced systemic side effects.[412] Illum et al.[413] prepared an adhesive delivery system consisting of cross linked albumin or PLGA microspheres prepared using W/O emulsification method followed by fibrial bacterial adhesion material obtained from E. coli that is attached to the microsphere surfaces by adsorption or by covalent linkage. Gemcitabine microparticles were prepared using chitosan, polyethylene oxide, or carbopol as the mucoadhesive polymer and Eudragit®L100–55 as the enteric polymer by a double emulsion method.[414] Amoxicillin mucoadhesive microspheres containing carbopol-934P as mucoadhesive polymer and ethyl cellulose as carrier polymer showed that amoxicillin mucoadhesive microspheres adhered in vitro more strongly to the gastric mucous layer than amoxicillin powder and can be retained in the gastrointestinal tract for an extended period of time (>12 h).[415]

Hollow-bioadhesive microspheres have also been proposed. The combination may be advantageous in the treatment of stomach diseases by lengthening drug retention time in the stomach. The combination seems, for a glance, to create conflicting results since floating works best in the presence of abundant gastric juice, which may render the microspheres nonmucoadhesive.[392] However, since mucoahdesive polymers may require a lag time to hydrate and swell to interact with gastric mucins, it might be possible that byoyanct microspheres will provide sufficient floating, and therefore longer residence of the microspheres to develop mucoadhesion.[416]

Microspheres of enteric properties have been proposed for site specific drug delivery to the colon,[292,417] or the small intestine[289,418,419] or the stomach.[298,420] Enteric microspheres may be entirely fabricated using one or more pH sensitive polymers,[155,288,289,293–295,297,418,421] or it may be combined with another release controlling non-pH sensitive polymer.[157,296] Eudragit®P-4135F is a pH sensitive polymer that becomes soluble in aqueous solution of pH > 7.2. The polymer was utilized for the design of colonic delivery of drugs that are sensitive to the gastrointestinal proteolytic degradation. Calcitonin was encapsulated into Eudragit®P-4135F using $W_I/O/W_{II}$ emulsion technique.[422] Calcitonin was incorporated in the internal aqueous phase and carboxyfluorescein was encapsulated similarly to follow microspheres dissolution behavior. Plasma levels of carboxyfluorescein after oral microspheres administration proved a sustained release in a rat model, where maximum concentration of carboxyfluorescein appeared around 4 h, compared to 60 min for carboxyfluorescein solution. Tacrolimus was encapsulated into Eudragit®P-4135F using an O/W emulsification technique, since it was found superior to the O_I/O_{II} technique in terms of encapsulation efficiency.[423] Generally, in both microspheres types, Eudragit®P-4135F was found to limit drug leakage at pH 6.8 to levels below 20% within 4 to 6 h.

Enteric properties may be also utilized to avoid contact of drug with gastric mucosa or with gastric fluids to prevent mucosal damages or drug breakdown. The acid-labile proton pump inhibitor, pantoprazole was encapsulated using two different types of enteric-coating polymers; Eudragit®S100 and hydroxypropyl methylcellulose phthalate.[418] Alao, enteric microspheres of papain were prepared by $W_I/O/W_{II}$ emulsion solvent evaporation using hydroxypropyl methylcellulose phthalate (HPMCP), Eudragit®L100 and Eudragit®S100, to avoid gastric inactivation of papain.[424]

Microspheres have also shown potential as carriers for oral vaccine delivery. The protective polymer coating of microspheres is believed to partially shield the antigens from destructive pH of the stomach, and the high levels of proteases and bile salts in the intestine. Furthermore, microspheres smaller than 10 μm in diameter are thought to be at least partially taken up from the intestine into Peyer's patches where they can induce both mucosal and systemic immune responses.[425,426]

Microsphere particulate systems prepared via the emulsion solvent removal methods can be further processed into compressed tablets that disintegrate into the gastrointestinal tract into discrete particles. In doing so, dual advantages of tablet dosing as well as controlled release particulate therapy can be gained. However, this is highly dependent on the ability of the microparticles to withstand compression into tablets and still maintain controlled release properties. Microcapsules and matrix microspheres have been tabletted to control the release of drugs, and to avoid gastric irritation.[427] The dissolution properties of cellulose acetate butyrate microspheres containing succinyl sulfathiazole were found essentially unchanged at compression pressures up to 351 MPa with $T_{50\%}$ values ranging from 121 to 132 min compared to 130 min for uncompressed microspheres.[428] However, when propranolol HCl was encapsulated in cellulose acetate butyrate and then compressed into tablets, the drug release from compressed tablets was always faster than from uncompressed microspheres; indicating that rupture of some of the microspheres had occurred, and larger changes in release rate and $T_{50\%}$ percent were observed with larger microspheres sizes.[429] In addition, theophylline ethylcellulose microcapsules were found to rupture and lose their sustained release properties.[430] Therefore, for successful tabletting of microspheres, the microspheres should be capable of resisting mechanical stresses developed during compression. Stresses can lead to microcapsule wall fracture,[429,430] or can yield nondisintegrated matrix due to microcapsule fusion.[431] To overcome problems commonly encountered during microparticles compression, novel microparticles-containing beads were suggested. The beads are formed by ionotropic gelation of the charged polysaccharide, chitosan, or sodium alginate, in solutions of the counterion, tripolyphosphate (TPP), or calcium chloride ($CaCl_2$), respectively. Chitosan beads disintegrate and release the microparticles in the acidic medium, while calcium alginate beads rapidly disintegrate in simulated intestinal fluids, thus allowing pH-dependent release. The method is simple and rapid, and maintains the physical properties of microparticles.[394]

Undesirable bitter taste is one of the important formulation problems encountered in many drugs, an issue that greatly restrict the further development of oral preparations and clinical applications of these drugs. Microspheres made by the emulsion methods have been also utilized in providing such a tool to minimize, or completely mask undesirable drug taste, and therefore to improve patient compliance. Trimebutine maleate was encapsulated in acid soluble, polyvinylacetal diethylaminoacetate through the $W_I/O/W_{II}$ emulsion technique. The pH of the second aqueous phase was the critical factor in achieving a high loading efficiency; nearly 90% (w/w) loading efficiency was obtained at pH > 10. In vitro, trimebutine was completely released within 10 min at pH 1.2, whereas at pH 6.8, the pH in the mouth, only small quantities of trimebutine were released in the initial 1 to 2 min. Taste masking of the microspheres was also confirmed through a gustatory sensation test in healthy volunteers.[163] Eudragit®S100 and Eudragit®L100 microspheres were utilized in masking the unpleasant taste of roxithromycin and clarithromycin, respectively. For both microspheres preparations, the ratio of the polymer to the drug influences microspheres properties, and generally higher ratios improved the bitter taste test results in vitro.[432,433]

45.5.2 Parenteral Delivery

Microspheres prepared by solvent removal have been extensively researched as carriers for the parenteral delivery of small drug molecules, and also for macromolecules such as polypeptides, proteins, and vaccines. Success has been made in parenteral sustained release depots. Lupron Depot®[20] and Consta Risperdal®[23] are few examples of successful parenteral microsphere formulations. Biodegradable microspheres may have utility as a parenteral formulation to provide a continuous delivery of therapeutic proteins or a pulsatile delivery of protein-based vaccines. This pattern of delivery of proteins and vaccines will provide improvements in patient care and, perhaps, increased efficacy, by eliminating the need for repeated administration usually encountered in vacccination protocols. In recent years, much work has focused on developing vaccines adjuvants to replace or complement existing aluminum salts for vaccine potentiation. The term adjuvant has been used to describe any molecule that improves the immune response to co-administered antigen. Aluminum salts are effective with many antigens, but repeat administration is necessary to achieve protection against infection. Their main mode of adjuvanticity is due to their ability to provide a short-term depot effect for absorbed proteins,[434] slowly "leaking" antigen to the body's immune system. In contrast to aluminum salts, polymeric controlled microsphere may be designed to release entrapped antigens for very long times following a single immunization, thereby eliminating the need for booster doses in many cases. Therefore, single administration of microspheres-based vaccine delivery vehicles have been studied,[435–437] with the majority employing PLG, PLA, PLGA polymers,[435,438–442] or polyanhydride polymers.[242] Both PLGA and polyanhydrides polymers are hydrophobic biodegradable polymers, approved for human use. However, unlike PLGA, polyanhydrides degrade by surface erosion, releasing the encapsulated antigen at the microsphere surface.[443]

Microspheres have been also investigated for passive and active targeted delivery of drug compounds. Passive delivery can be achieved using either large microspheres (>7 µm) that are trapped by the first capillary bed encountered (e.g., the lungs after IV injection), or using small microspheres (1–5 µm) that are trapped by the reticuloendothelial system consisting of phagocytic cells located mainly in liver, spleen, lymph nodes, and lungs.[444] Microspheres are of high utility in vaccination in particular by virtue of their depot and protection effect, in addition to their targetability. Microspheres, 1 to 10 µm in diameter, can be passively targeted to antigen-presenting cells such as macrophages and dendritic cells in a number of tissues leading to direct intracellular delivery of antigen for processing by the major histocompatibility complex (MHC) class II pathway. In addition, encapsulation of antigen within the particulates[445–447] can lead to antigen presentation by the MHC class I pathway as well. Rifampin loaded-PLGA microspheres made by O/W emulsion technique were found to be more effective at reducing *M. tuberculosis* intracellular growth than equivalent doses of rifampin given as a free drug when microspheres were targeted to the infected macrophages.[448] It has been shown that microspheres enhance antigen-specific T-helper lymphocyte responses[449] leading to an enhancement in antigen-specific antibody responses. Furthermore, it has been shown that microparticles can provide a novel way of inducing a cytotoxic T cell response using synthetic peptides. Peptides (e.g., influenza A virus nucleoprotein) entrapped in poly lactide-co-glycolide (PLGA) microparticles were able to induce cytotoxic T cell response. Peptide entrapped in microparticles of mean size <500 nm were better inducers than larger microparticles (mean >2 µm and above).[447] It was reported that T-cell activation in response to antigen-encapsulating microspheres has 100 to 1000 fold better than antigen alone.[450] A subcutaneous injection into mice of a PLGA microspheres containing staphylococcal enterotoxin B toxoid induced an immune response approximately 500 times compared with that seen with nonencapsulated toxoid, as measured by circulating immunoglobulin G (IgG) antitoxin titers. Staphylococcal enterotoxin B toxoid injected as a mixture with empty PLGA microspheres was no more effective as an immunogen than toxoid alone. Antigen-containing microspheres of 1 to 10 μm in diameter exhibited stronger adjuvant activity than those greater than 10 μm, owing to the delivery of the smaller size microspheres into the draining lymph nodes within macrophages.[451] Using $W/O_I/O_{II}$, ovalbumin antigen encapsulated in the poly epsilon-caprolactone microparticles was found to induce both antibody and cell-mediated immune responses.[452] Immunization with PLGA and PLA nanoparticles and microparticles encapsulating stabilized tetanus toxoid elicited early as well as high antibody titers in experimental animals, which persisted for more than 5 months and were higher than those obtained with saline tetanus toxoid.[453]

The stability of vaccines was enhanced in several ways against destabilizing solvents and polymers. By careful selection of the PLGA copolymer composition and molecular weight, an oil (mineral oil)-based cores of tetanus toxoid surrounded by outer polymer shells (PLGA) potentially protecting the bioactive material against water-mediated inactivation processes.[437] However, adjuvants such as trehalose,[59,454] threonyl muramyl dipeptide,[60] BSA,[17,182] and PEG 400[167] can also be used for this purpose.

Furthermore, the release of proteins and vaccines from microspheres can be optimized in terms of rates and patterns. An antigen-loaded PLGA microspheres prepared by the $W_I/O/W_{II}$ emulsion technique at low aqueous to organic phase volume ratios (≤0.1 mL/mL), exhibited a triphasic release. An in vivo continuous release pattern of proteins was observed at higher aqueous to organic phase volume ratios,[356] or at low polymer molecular weight and high loading for vaccines.[242] However, a pulsatile-release PLGA microsphere formulation was desirable for vaccine formulations since vaccines are usually administered over several injections separated in time by several months.[455,456]

45.6 CONCLUSIONS

Owing to its simplicity and robustness, the emulsion solvent removal methods have been successfully employed to encapsulate a wide variety of drugs, from small molecules to proteins, into rigid microspheres that can be tailored for specific applications. Many of these applications were covered briefly throughout this chapter, where they range from oral and parenteral controlled drug release and targeting. This, in large, is attributed the advances in polymer science and the continuously evolving new polymers as carriers that satisfy specific applications, and also to the novel techniques that control the sizes of the microspheres products. However, despite the huge numbers of papers and patents in this domain, only limited microspheres products have been commercialized in the last few years. Therefore, efforts should be made to find solutions of the identified limitations whenever possible. Such limitations such as the low encapsulation efficiency, the burst effect, the incomplete drug release, and stability of proteins need to be researched intensively to find suitable formulations (polymers, adjuvants, stabilizers, and solvents), and suitable handling and manufacturing procedures. The exposure to toxic organic solvents is an important limitation that should be minimized since toxic organic solvents may adversely effect the stability and/or the activity of the encapsulated drug, and residual amounts of these solvents in the final product must be tightly regulated and kept way below the accepted levels for safety issues.

The future of microspheres-based drug delivery systems is expected to find a fruitful era with the successful entrapement of numerous novel peptides and protein drugs produced via genetic engineering dedicated for various applications. This is especially true with the introduction of new instruments capable of manufacturing uniform monosized microspheres in a continuous production pattern. However, this highly dependent on resolving or lessen the few drawbacks of the emulsion solvent removal methods.

ACKNOWLEDGMENTS

The author would like to acknowledge the support of Jordan University of Science and Technology.

REFERENCES

1. Deasy PB. *Microencapsulation and Related Drug Processes*. New York: Marcel Dekkar; 1984.
2. Thies C. A survey of microencapsulation processes. In: Benita S, ed. *Microencapsulation: Methods and Industrial Applications*. New York: Marcel Dekker; 1996. pp. 1–21.
3. Kreitz M, Brannon-Peppas L, Mathiowitz E. Microencapsulation. In: Mathiowitz E, ed. In: *Encyclopedia of Controlled Drug Delivery*. New York: John Wiley & Sons, Inc.; 1999. pp. 493–546, vol. 2.
4. Arica B, Arica MY, Kas HS, Hincal AA, Hasirci V. In-vitro studies of enteric coated diclofenac sodium-carboxymethylcellulose microspheres. *Journal of Microencapsulation*. November–December 1996;13(6):689–699. PubMed PMID: 8933354. Epub 1996/11/01. eng.
5. Castellanos IJ, Crespo R, Griebenow K. Poly(ethylene glycol) as stabilizer and emulsifying agent: A novel stabilization approach preventing aggregation and inactivation of proteins upon encapsulation in bioerodible polyester microspheres. *Journal of Controlled Release*. February 14, 2003;88(1):135–145. PubMed PMID: 12586511. Epub 2003/02/15. eng.
6. Ando S, Putnam D, Pack DW, Langer R. PLGA microspheres containing plasmid DNA: Preservation of supercoiled DNA via cryopreparation and carbohydrate stabilization. *Journal of Pharmaceutical Sciences*. January 1999;88(1):126–130. PubMed PMID: 9874713.
7. Bouissou C, Van der Walle C. Poly(lactic-co-glycolic acid) Microspheres. In: Uchegbu IF, Schätzlein AG, eds. *Polymers in Drug Delivey*. Boca Raton, FL: CRC Press, Taylor & Francis Group, LLC; 2006. pp. 81–101.
8. Watts PJ, Davies MC, Melia CD. Microencapsulation using emulsification/solvent evaporation: An overview of techniques and applications. *Critical Reviews in Therapeutic Drug Carrier System*. 1990;7(3):235–259. PubMed PMID: 2073688.
9. Bakan JA, Powell TC, Szotak PS. Recent advances using microencapsulation for taste-masking of bitter drugs. In: Donbrow M, ed. *Microcapsules and Nanoparticles in Medicine and Pharmacy*. Boca Raton, FL: CRC Press; 1992. pp. 149–156.
10. Lim F, Sun AM. Microencapsulated islets as bioartificial endocrine pancreas. *Science*. November 21, 1980;210(4472):908–910. PubMed PMID: 6776628.
11. Walter E, Dreher D, Kok M, Thiele L, Kiama SG, Gehr P et al. Hydrophilic poly(DL-lactide-co-glycolide) microspheres for the delivery of DNA to human-derived macrophages and dendritic cells. *Journal of Controlled Release*. September 11, 2001;76(1–2):149–168. PubMed PMID: 11532321.
12. Vranken MN, Claeys DA. Inventors. US35239061970.
13. Vranken MN, Claeys DA. Inventors. US35239071970.
14. Beck LR, Cowsar DR, Lewis DH, Cosgrove RJ, Jr., Riddle CT, Lowry SL et al. A new long-acting injectable microcapsule system for the administration of progesterone. *Fertility and Sterility*. May 1979;31(5):545–551. PubMed PMID: 446779.
15. Ogawa Y, Yamamoto M, Okada H, Yashiki T, Shimamoto T. A new technique to efficiently entrap leuprolide acetate into microcapsules of polylactic acid or copoly(lactic/glycolic) acid. *Chemical and Pharmaceutical Bulletin*. March 1988;36(3):1095–1103. PubMed PMID: 3136939.
16. Mehta RC, Thanoo BC, Deluca PP. Peptide containing microspheres from low molecular weight and hydrophilic poly(d,l-lactide-co-glycolide). *Journal of Controlled Release*. 1996;41(3):249–257.
17. Zhu G, Mallery SR, Schwendeman SP. Stabilization of proteins encapsulated in injectable poly (lactide-co-glycolide). *Nature Biotechnology*. January 2000;18(1):52–57. PubMed PMID: 10625391.
18. Tomlinson E. Microsphere delivery systems for drug targeting and controlled release. *International Journal of Pharm Tech Production Manufacturing*. 1983;4:49–57.
19. Price JC. Diffusion controlled release systems: Polymeric microcapsules. In: Tarcha PJ, ed. *Polymers for Controlled Drug Delivey*. Boca Raton, FL: CRC Press, Taylor & Francis Group, LLC; 1990. pp. 1–14.
20. Okada H, Heya T, Ogawa Y, Shimamoto T. One-month release injectable microcapsules of a luteinizing hormone-releasing hormone agonist (leuprolide acetate) for treating experimental endometriosis in rats. *The Journal of Pharmacology and Experimental Therapeutics*. February 1988;244(2):744–750. PubMed PMID: 3126294.
21. Brittain HA, Dickason DA, Hotz J, Lyons SL, Ramstack MJ, Wright SG. Polymorphic forms of naltrexone. EU patent 754164, 2004.
22. Dean RL. The preclinical development of Medisorb Naltrexone, a once a month long acting injection, for the treatment of alcohol dependence. *Frontiers in Bioscience*: A *Journal and Virtual Library*. January 1, 2005;10:643–655. PubMed PMID: 15569605.
23. Rickey ME, Ramstack JM, Lewis DH, Mesens J. Preparation of extended shelf-life biodegradable, biocompatible Microparticles containing a biologically active agent. US patent 5792477, 1998.

24. Wright SG, Rickley ME, Ramstack JM, Lyons SL, Hotz JM. Method for preparing microparticles having a selected polymer molecular weight. US patent 6379704, 2001.
25. Ramstack JM, Riley MG, Zale SE, Hotz JM, Johnson OL. Preparation of injectable suspensions having improved injectability. US patent 6667061, 2003.
26. Mesens J, Rickey ME, Atkins TJ, Microencapsulated 3-Piperidinyl-substituted 1,2-benzisooxazoles and 1,2-benzisothiazoles. US patent 6544559, 2003.
27. Arshady R. Microspheres and microcapsules, a survey of manufacturing techniques. 3. Solvent evaporation. *Polymer Engineering Science*. 1990;30:915–924.
28. Freitas S, Merkle HP, Gander B. Microencapsulation by solvent extraction/evaporation: Reviewing the state of the art of microsphere preparation process technology. *Journal of Controlled Release*. February 2, 2005;102(2):313–332. PubMed PMID: 15653154.
29. Lassalle V, Ferreira ML. PLA nano- and microparticles for drug delivery: An overview of the methods of preparation. *Macromolecular Bioscience*. June 7, 2007;7(6):767–783. PubMed PMID: 17541922.
30. Li M, Rouaud O, Poncelet D. Microencapsulation by solvent evaporation: State of the art for process engineering approaches. *International Journal of Pharmaceutics*. November 3, 2008;363(1–2):26–39. PubMed PMID: 18706988.
31. Wischke C, Schwendeman SP. Principles of encapsulating hydrophobic drugs in PLA/PLGA microparticles. *International Journal of Pharmaceutics*. December 8, 2008;364(2):298–327. PubMed PMID: 18621492.
32. Obeidat WM. Recent patents review in microencapsulation of pharmaceuticals using the emulsion solvent removal methods. *Recent Patents on Drug Delivery and Formulation*. November 2009;3(3):178–192. PubMed PMID: 19925442.
33. Takenaka H, Kawashima Y, chikamatsu Y, Ando Y. Reactivity and stability of microencapsulated placental alkaline phosphatase. *Chemical and Pharmaceutical Bulletin*. February 1982;30(2):695–701. PubMed PMID: 7094153.
34. Stithit S, Chen W, Price JC. Development and characterization of buoyant theophylline microspheres with near zero order release kinetics. *Journal of Microencapsulation*. November–December 1998;15(6):725–737. PubMed PMID: 9818950.
35. Pekarek KJ, Jacob JS, Mathiowitz E. Double-walled polymer microspheres for controlled drug release. *Nature*. January 20, 1994;367(6460):258–260. PubMed PMID: 8121490.
36. Johansen P, Merkle HP, Gander B. Technological considerations related to the up-scaling of protein microencapsulation by spray-drying. *European Journal of Pharmaceutics and Biopharmaceutics: Official Journal of Arbeitsgemeinschaft fur Pharmazeutische Verfahrenstechnik eV*. November 2000;50(3):413–417. PubMed PMID: 11072199.
37. Thomasin C, Johansen P, Alder R, Bemsel R, Hottinger G, Altorfer H et al. A contribution to overcoming the problem of residual solvents in biodegradable microspheres prepared by coacervation. *European Journal of Pharmaceutics and Biopharmaceutics: Official Journal of Arbeitsgemeinschaft fur Pharmazeutische Verfahrenstechnik eV*. 1996;42:16–24.
38. Muramatsu N, Kondo T. An approach to prepare microparticles of uniform size. *Journal of Microencapsulation*. March–April 1995;12(2):129–136. PubMed PMID: 7629655.
39. Shiga K, Muramatsu N, Kondo T. Preparation of poly(D,L-lactide) and copoly(lactide-glycolide) microspheres of uniform size. *The Journal of Pharmacy and Pharmacology*. September 1996;48(9):891–895. PubMed PMID: 8910847.
40. Kawakatsu T, Komori H, Nakajima M, Kikuchi Y, Yonemoto T. Production of monodispersed oil-in-water emulsion using crossflow-type silicon microchannel plate. *Journal of Chemical Engineering of Japan*. 1999;32:241–244.
41. Berkland C, King M, Cox A, Kim K, Pack DW. Precise control of PLG microsphere size provides enhanced control of drug release rate. *Journal of Controlled Release*. July 18, 2002;82(1):137–147. PubMed PMID: 12106984.
42. Sugiura S, Nakajima M, Seki M. Preparation of monodispersed emulsion with large droplets using microchannel emulsification. *Journal of American Oil Chemical Society*. 2002;79:515–519.
43. Malahy B. The effect of instruction and labeling on the number of medication errors made by patients at home. *American Journal of Hospital Pharmacy*. June 1966;23(6):283–292. PubMed PMID: 5943307.
44. Li SP, Kowalski CR, Feld KM, Grim WM. Recent advances in microencapsulation technology and equipment. *Drug Development and Industrial Pharmacy*. 1988;14:353–376.
45. Vilivalam VD, Adeyeye CM. Development and evaluation of controlled-release diclofenac microspheres and tabletted microspheres. *Journal of Microencapsulation*. July–August 1994;11(4):455–470. PubMed PMID: 7931945.

46. Li Z, Li L, Liu Y, Zhang H, Li X, Luo F et al. Development of interferon alpha-2b microspheres with constant release. *International Journal of Pharmaceutics*. May 30, 2011;410(1–2):48–53. PubMed PMID: 21419205.
47. Su ZX, Shi YN, Teng LS, Li X, Wang LX, Meng QF et al. Biodegradable poly(D, L-lactide-co-glycolide) (PLGA) microspheres for sustained release of risperidone: Zero-order release formulation. *Pharmaceutical Development and Technology*. August 2011;16(4):377–384. PubMed PMID: 20370594.
48. Tinsley-Bown AM, Fretwell R, Dowsett AB, Davis SL, Farrar GH. Formulation of poly(D,L-lactic-co-glycolic acid) microparticles for rapid plasmid DNA delivery. *Journal of Controlled Release*. May 15, 2000;66(2–3):229–241. PubMed PMID: 10742583.
49. Thatcher JE, Welch T, Eberhart RC, Schelly ZA, DiMaio JM. Thymosin beta4 sustained release from poly(lactide-co-glycolide) microspheres: Synthesis and implications for treatment of myocardial ischemia. *Annals of the New York Academy of Sciences*. October 2012;1270:112–119. PubMed PMID: 23050826.
50. Yandrapu S, Kompella UB. Development of sustained-release microspheres for the delivery of SAR 1118, an LFA-1 antagonist intended for the treatment of vascular complications of the eye. *Journal of Ocular Pharmacology and Therapeutics: The Official Journal of the Association for Ocular Pharmacology and Therapeutics*. March 2013;29(2):236–248. PubMed PMID: 23256487. Pubmed Central PMCID: 3601676.
51. Shahzad MK, Ubaid M, Raza M, Murtaza G. The formulation of flurbiprofen loaded microspheres using hydroxypropylmethycellulose and ethylcellulose. *Advances in Clinical and Experimental Medicine: Official Organ Wroclaw Medical University*. March–April 2013;22(2):177–183. PubMed PMID: 23709373.
52. Gaignaux A, Reeff J, Siepmann F, Siepmann J, De Vriese C, Goole J et al. Development and evaluation of sustained-release clonidine-loaded PLGA microparticles. *International Journal of Pharmaceutics*. November 1, 2012;437(1–2):20–28. PubMed PMID: 22903047.
53. Hu L, Zhang H, Song W. An overview of preparation and evaluation sustained-release injectable microspheres. *Journal of Microencapsulation*. 2013;30(4):369–382. PubMed PMID: 23140260.
54. Gokhale KS, Jonnalagadda S. Preparation and evaluation of sustained release infliximab microspheres. *PDA Journal of Pharmaceutical Science and Technology/PDA*. May–June 2013;67(3):255–266. PubMed PMID: 23752752.
55. Joshi RV, Nelson CE, Poole KM, Skala MC, Duvall CL. Dual pH- and temperature-responsive microparticles for protein delivery to ischemic tissues. *Acta Biomaterialia*. May 2013;9(5):6526–6534. PubMed PMID: 23402764. Pubmed Central PMCID: 3702271.
56. Fundueanu G, Constantin M, Ascenzi P, Simionescu BC. An intelligent multicompartmental system based on thermo-sensitive starch microspheres for temperature-controlled release of drugs. *Biomedical Microdevices*. August 2010;12(4):693–704. PubMed PMID: 20414809.
57. Islam MA, Jiang HL, Quan JS, Arote RB, Kang ML, Yoo HS et al. Mucoadhesive and pH-sensitive thiolated Eudragit microspheres for oral delivery of Pasteurella multocida antigens containing dermonecrotoxin. *Journal of Nanoscience and Nanotechnology*. May 2011;11(5):4174–4181. PubMed PMID: 21780423.
58. Varde NK, Pack DW. Microspheres for controlled release drug delivery. *Expert Opinion on Biological Therapy*. January 2004;4(1):35–51. PubMed PMID: 14680467.
59. Moynihan JS, Blair J, Coombes A, D'Mello F, Howard CR. Enhanced immunogenicity of a hepatitis B virus peptide vaccine using oligosaccharide ester derivative microparticles. *Vaccine*. March 15, 2002;20(13–14):1870–1876. PubMed PMID: 11906777.
60. Puri N, Kou JH, Sinko PJ. Adjuvancy enhancement of muramyl dipeptide by modulating its release from a physicochemically modified matrix of ovalbumin microspheres I. in vitro characterization. *Journal of Controlled Release*. 2000;69(1):53–67.
61. Price JC, Obeidat WM. Microspheres and related processes and pharmaceutical compositions patent US 20060099256. 2006.
62. Obeidat WM, Price JC. Evaluation of enteric matrix microspheres prepared by emulsion-solvent evaporation using scanning electron microscopy. *Journal of Microencapsulation*. February 2004;21(1):47–57. PubMed PMID: 14718185.
63. El-Bary AA, Aboelwafa AA, Al Sharabi IM. Influence of some formulation variables on the optimization of pH-dependent, colon-targeted, sustained-release mesalamine microspheres. *AAPS PharmSciTech*. March 2012;13(1):75–84. PubMed PMID: 22130789. Pubmed Central PMCID: 3299443.
64. Mathiowitz E, Langer R. Preparation of multiwall polymeric microcapsules. US patent 4861627, 1989.
65. Mathiowitz E, Jacob J, Chickering IDE, Pekarek KJ. Preparation of multiwall polymeric microcapsules from hydrophilic polymers. US patent 5985354, 1999.

66. Leach K, Noh K, Mathiowitz E. Effect of manufacturing conditions on the formation of double-walled polymer microspheres. *Journal of Microencapsulation.* March–April 1999;16(2):153–167. PubMed PMID: 10080110.
67. Mathiowitz E, Jacob J, Chickering IDE, Pekarek KJ. Preparation of multiwall polymeric microcapsules from hydrophilic polymers. US patent 6511749, 2003.
68. Mathiowitz E, Jacob J, Chickering IDE, Leach KJ. Multiwall polymeric microcapsules from hydrophilic polymers. US patent 6528035, 2003.
69. Rahman NA, Mathiowitz E. Localization of bovine serum albumin in double-walled microspheres. *Journal of Controlled Release.* January 8, 2004;94(1):163–175. PubMed PMID: 14684280.
70. Wen J, Anderson AB. Microparticle containing matrices for drug delivery. US patent 20070275027, 2007.
71. Williams DF, Mort E. Enzyme-accelerated hydrolysis of polyglycolic acid. *Journal of Bioengineering.* August 1977;1(3):231–238. PubMed PMID: 210160.
72. Makino K, Ohshima H, Kondo T. Mechanism of hydrolytic degradation of poly(L-lactide) microcapsules: Effects of pH, ionic strength and buffer concentration. *Journal of Microencapsulation.* July–September 1986;3(3):203–212. PubMed PMID: 3508186.
73. Sah H. A new strategy to determine the actual protein content of poly(lactide-co-glycolide) microspheres. *Journal of Pharmaceutical Sciences.* November 1997;86(11):1315–1318. PubMed PMID: 9383747.
74. Crotts G, Park TG. Protein delivery from poly(lactic-co-glycolic acid) biodegradable microspheres: Release kinetics and stability issues. *Journal of Microencapsulation.* November–December 1998;15(6):699–713. PubMed PMID: 9818948.
75. Mcginity JW, Iwata M. Preparation and uses of multi-phase microspheres. US patent 5288502, 1994.
76. Mathiowitz E, Langer R. Multiwall polymeric microspheres. US patent 5912017, 1999.
77. Morris NJ, Warburton B. Three-ply walled w/o/w microcapsules formed by a multiple emulsion technique. *The Journal of Pharmacy and Pharmacology.* August 1982;34(8):475–479. PubMed PMID: 6126555.
78. Kentepozidou A, Kiparissides C. Production of water-containing polymer microcapsules by the complex emulsion/solvent evaporation technique. Effect of process variables on the microcapsule size distribution. *Journal of Microencapsulation.* November–December 1995;12(6):627–638. PubMed PMID: 8558385.
79. Benita S, Benoit JP, Puisieux F, Thies C. Characterization of drug-loaded poly(d,l-lactide) microspheres. *Journal of Pharmaceutical Sciences.* December 1984;73(12):1721–1724. PubMed PMID: 6527243.
80. Babay D, Hoffman A, Benita S. Design and release kinetic pattern evaluation of indomethacin microspheres intended for oral administration. *Biomaterials.* November 1988;9(6):482–488. PubMed PMID: 3224134.
81. Gabor F, Ertl B, Wirth M, Mallinger R. Ketoprofen-poly(D,L-lactic-co-glycolic acid) microspheres: Influence of manufacturing parameters and type of polymer on the release characteristics. *Journal of Microencapsulation.* January–February 1999;16(1):1–12. PubMed PMID: 9972498.
82. Bodmeier R, McGinity JW. Polylactic acid microspheres containing quinidine base and quinidine sulphate prepared by the solvent evaporation technique. I. Methods and morphology. *Journal of Microencapsulation.* October–December 1987;4(4):279–288. PubMed PMID: 3504509.
83. Chatzi EG, Gavrielides AD, Kiparissides C. Generalized model for prediction of the steady-state drop size distributions in batch stirred vessels. *Industrial and Engineering Chemistry Research.* 1989;28:1704–1711.
84. Sansdrap P, Moës AJ. Influence of manufacturing parameters on the size characteristics and the release profiles of nifedipine from poly(DL-lactide-co-glycolide) microspheres. *International Journal of Pharmaceutics.* 1993;98:157–164.
85. Chanana GD, Sheth BB. Particle size reduction of emulsions by formulation design. I: Effect of polyhydroxy alcohols. *Journal of Parenteral Science and Technology: A Publication of the Parenteral Drug Association.* May–June 1993;47(3):130–134. PubMed PMID: 8360805.
86. Chanana GD, Sheth BB. Particle size reduction of emulsions by formulation design-II: Effect of oil and surfactant concentration. *PDA Journal of Pharmaceutical Science and Technology/PDA.* March–April 1995;49(2):71–76. PubMed PMID: 7780748.
87. Jeffery H, Davis SS, O'Hagan DT. The preparation and characterization of poly(lactide-co-glycolide) microparticles. II. The entrapment of a model protein using a (water-in-oil)-in-water emulsion solvent evaporation technique. *Pharmaceutical Research.* March 1993;10(3):362–368. PubMed PMID: 8464808.
88. Sah H. Ethyl formate—Alternative dispersed solvent useful in preparing PLGA microspheres. *International Journal of Pharmaceutics.* February 15, 2000;195(1–2):103–113. PubMed PMID: 10675688.

89. Herrmann J, Bodmeier R. Biodegradable, somatostatin acetate containing microspheres prepared by various aqueous and non-aqueous solvent evaporation methods. *European Journal of Pharmaceutics and Biopharmaceutics: Official Journal of Arbeitsgemeinschaft fur Pharmazeutische Verfahrenstechnik eV.* January 1998;45(1):75–82. PubMed PMID: 9689538.
90. Mosier B. Method of preparing microspheres for intravascular delivery. US patent 4492720, 1985.
91. Wang HT, Schmitt E, Flanagan DR, Linhardt RJ. Influence of formulation methods on the in vitro controlled release of protein from poly(ester) microspheres. *Journal of Controlled Release.* 1991;17:23–31.
92. Sturesson C, Carlfors J, Edsman K, Andersson M. Preparation of biodegradable poly(lactic-co-glycolic) acid microspheres and their in vitro release of timolol maleate. *International Journal of Pharmaceutics.* 1993;89:235–244.
93. Wang J, Schwendeman SP. Mechanisms of solvent evaporation encapsulation processes: Prediction of solvent evaporation rate. *Journal of Pharmaceutical Sciences.* October 1999;88(10):1090–1099. PubMed PMID: 10514360.
94. Vrancken M, Claeys D. Process for encapsulating water and compounds in aqueous phase by evaporation. US patent 3523906, 1970.
95. Kitajima M, Yamaguchi T, Kondo A, Muroya N, Tagata G. Encapsulation method. US patent 3691090, 1972.
96. Pampus G, Schnoring H, Schon N, Witte J. Process for the production of microgranulates. US patent 3737337, 1973.
97. Fukushima M, Inaba Y, Kobari S, Morishita M. Process for preparing microcapsules. US patent 3891570, 1975.
98. Morishita M, Inaba Y, Fukushima M, Hattori Y, Kobari S, Matsuda T. Process for encapsulation of medicaments. US patent 3960757, 1976.
99. Tice TR, Lewis DH. Microencapsulation process. US patent 4389330, 1983.
100. Vrancken M, Claeys D. Method for encapsulating water and compounds in aqueous phase by extraction. US patent 3523907, 1970.
101. Abe J, Morishita M, Inaba Y, Fukushima M, Kobari S, Nagata A. Preparation of microcapsules. US patent 3943063, 1976.
102. Tice TR, Gilley RM. Preparation of injectable controlled release microcapsules by a solvent evaporation process. *Journal of Controlled Release.* 1985;2:343–352.
103. Beck LR, Cowsar DR, Lewis DH, Tice T. Injectable, long-acting microparticle formulation for the delivery of anti-inflammatory agents. US patent 4530840, 1985.
104. Cowsar DR, Tice TR, Gilley RM, English JP. Poly (lactide-co-glycolide) microspheres for controlled release of steroids. *Methods Enzymology.* 1985;112:101–116.
105. Bums PJ, Steiner JV, Sertich PL, Pozor MA, Tice TR, Mason DW et al. Evaluation of biodegradable microspheres for the controlled release of progesterone and estradiol in an ovulation control program for cycling mares. *Journal of Equine Veterinary Science.* 1993;13:521–524.
106. Hill VL, Paserini N, Craig DQM, Vickers M, Anwar J, Feely LC. Investigation of progesterone loaded poly(d,l-lactide) microspheres using TMDCS, SEM and PXRD. *Journal of Thermal Analysis.* 1998;54:673–685.
107. Maulding HV, Tice TR, Cowar DR, Fong JW, Pearson JE, Nazareno JP. Biodegradable microcapsules: Acceleration of polymeric excipient hydrolytic rate by incorporation of a basic medicament. *Journal of Controlled Release.* 1986;3:103–117.
108. Thote AJ, Chappell JT, Jr., Gupta RB, Kumar R. Reduction in the initial-burst release by surface crosslinking of PLGA microparticles containing hydrophilic or hydrophobic drugs. *Drug Development and Industrial Pharmacy.* January 2005;31(1):43–57. PubMed PMID: 15704857.
109. Suzuki K, Price JC. Microencapsulation and dissolution properties of a neuroleptic in a biodegradable polymer, poly(d,l-lactide). *Journal of Pharmaceutical Sciences.* January 1985;74(1):21–24. PubMed PMID: 2858575.
110. Cha Y, Pitt CG. A one-week subdermal delivery system for l-methadone based on biodegradable microcapsules. *Journal of Controlled Release.* 1988;7:69–78.
111. Wada R, Hyon S-H, Ikada Y, Nakao Y, Yoshikawa H, Muranishi S. Lactic acid oligomer microspheres containing an anticancer agent for selective lymphatic delivery: I. in vitro studies. *Journal of Bioactive and Compatible Polymers.* 1988;3:126–136.
112. Yoshikawa H, Nakao Y, Takada K, Muranishi S, Wada R, Tabata Y et al. Targeted and sustained delivery of aclarubicin to lymphatics by lactic acid-oligomer microsphere in rat. *Chemical and Pharmaceutical Bulletin.* March 1989;37(3):802–804. PubMed PMID: 2752497.
113. Gen S, Muranishi S, Ikada Y, Yoshikawa H. Polylactic acid microspheres and process for producing the same. US patent 4994281, 1991.

114. Benoit JP, Benita S, Puisieux F, Thies C. Stability and release kinetics of drugs incorporated within microspheres. In: Davis SS, Illum L, McVie JG, Tomlinson E, eds. *Microspheres and Drug Therapy Pharmaceutical, Immunological and Medical Aspects*. Amsterdam, the Netherlands: Elsevier; 1984.
115. Bissery M-C, Valeriote F, Thies C. In vito and in vivo evaluation of CCNU-loaded microspheres prepared from poly((±)-lactide) and poly(β-hydroxybutyrate). In: Davis SS, Illum L, McVie JG, Tomlinson E, eds. *Microspheres and Drug Therapy Pharmaceutical, Immunological and MedicalAspects*. Amsterdam, the Netherlands: Elsevier; 1984. pp. 217–27.
116. Gupte A, Ciftci K. Formulation and characterization of Paclitaxel, 5-FU and Paclitaxel + 5-FU microspheres. *International Journal of Pharmaceutics*. May 200419;276(1–2):93–106. PubMed PMID: 15113618.
117. Xie M, Zhou L, Hu T, Yao M. Intratumoral delivery of paclitaxel-loaded poly(lactic-co-glycolic acid) microspheres for Hep-2 laryngeal squamous cell carcinoma xenografts. *Anti-Cancer Drugs*. April 2007;18(4):459–466. PubMed PMID: 17351398.
118. Bodmeier R, McGinity JW. The preparation and evaluation of drug-containing poly(dl-lactide) microspheres formed by the solvent evaporation method. *Pharmaceutical Research*. December 1987;4(6):465–471. PubMed PMID: 3508558.
119. Seki T, Kawaguchi T, Endoh H, Ishikawa K, Juni K, Nakano M. Controlled release of 3′,5′-diester prodrugs of 5-fluoro-2′-deoxyuridine from poly-L-lactic acid microspheres. *Journal of Pharmaceutical Sciences*. November 1990;79(11):985–987. PubMed PMID: 2149864.
120. Bodmeier R, McGinity JW. Polylactic acid microspheres containing quinidine base and quinidine sulphate prepared by the solvent evaporation technique. II. Some process parameters influencing the preparation and properties of microspheres. *Journal of Microencapsulation*. October–December 1987;4(4):289–297. PubMed PMID: 3504510.
121. Bodmeier R, McGinity JW. Solvent selection in the preparation of poly(dl-lactide) microspheres prepared by the solvent evaporation method. *International Journal of Pharmaceutics*. 1988;43:179–186.
122. Okada H, Ogawa Y, Yashiki T. Prolonged release microcapsule and its production. US patent 4652441, 1987.
123. Ogawa Y, Okada H, Yashiki T. Prolonged release microcapsules. US patent 4917893, 1990.
124. Ogawa Y, Takada S, Yamamoto M. Method for producing microcapsule. US patent 4954298, 1990.
125. Metha RC, Thanoo BC, DeLuca PP. Peptide containing microspheres from low molecular weight and hydrophilic poly(DL-lactide-co-glycolide). *Journal of Controlled Release*. 1996;41:249–257.
126. Rafati H, Coombes AG, Adler J, Holland J, Davis SS. Protein-loaded poly(DL-lactide-co-glycolide) microparticles for oral administration: Formulation, structural and release characteristics. *Journal of Controlled Release*. 1997;43:89–102.
127. Nagai A, Ohtani S, Takechi N. Production of microspheres. US patent 5851451, 1998.
128. Igari Y, Takada S, Kosakai H. Sustained release microcapsules of a bioactive substance and a biodegradable polymer. US patent 6419961, 2002.
129. Kobayashi M, Matsukawa Y, Nishioka Y, Suzuki T. Method for producing sustained release microsphere preparation. US patent 5556642, 1996.
130. Masao K, Asaji K, Noriynki M, Tsutomu Y. Encapsulation method. US patent 3691090, 1972.
131. Beck LR, Pope VZ, Tice TR, Gilley RM. Long-acting injectable microsphere formulation for the parenteral administration of levonorgestrel. *Advances in contraception: The Official Journal of the Society for the Advancement of Contraception*. June 1985;1(2):119–129. PubMed PMID: 3939509.
132. Chung TW, Huang YY, Liu YZ. Effects of the rate of solvent evaporation on the characteristics of drug loaded PLLA and PDLLA microspheres. *International Journal of Pharmaceutics*. January 16, 2001;212(2):161–169. PubMed PMID: 11165073.
133. Yamakawa I, Machida R, Watanabe S. Production process of microspheres. EU patent 0461630, 1995.
134. Beck LR, Cowsar DR, Lewis DH, Gibson JW, Flowers CE, Jr. New long-acting injectable microcapsule contraceptive system. *American Journal of Obstetrics and Gynecology*. October 1, 1979;135(3):419–426. PubMed PMID: 114054.
135. Beck LR, Pope VZ, Cowsar FR, Lewis DH, Tice TR. Evaluation of a new three-month injectable contraceptive microsphere systeminprimates (baboon). *Contraceptive Delivery System*. 1980;1:79–86.
136. Beck LR, Ramos RA, Flowers CE, Jr., Lopez GZ, Lewis DH, Cowsar DR. Clinical evaluation of injectable biodegradable contraceptive system. *American Journal of Obstetrics and Gynecology*. August 1, 1981;140(7):799–806. PubMed PMID: 7020421.
137. Beck LR, Pope VZ, Flowers CE, Jr., Cowsar DR, Tice TR, Lewis DH et al. Poly(DL-lactide-co-glycolide)/norethisterone microcapsules: An injectable biodegradable contraceptive. *Biology of Reproduction*. February 1983;28(1):186–195. PubMed PMID: 6830939.

138. Fong JW, Nazareno JP, Pearson JE, Maulding HV. Evaluation of biodegradable microspheres prepared by a solvent evapoartion process using sodium oleate as emulsifier. *Journal of Controlled Release*. 1986;3:119–130.
139. Cavalier M, Benoit JP, Thies C. The formation and characterization of hydrocortisone-loaded poly((+/-)-lactide) microspheres. *The Journal of Pharmacy and Pharmacology*. April 1986;38(4):249–253. PubMed PMID: 2872287.
140. Cong H, Beck LR. Preparation and pharmacokinetic evaluation of a modified long-acting injectable norethisterone microsphere. *Advances in Contraception: The Official Journal of the Society for the Advancement of Contraception*. June–September 1991;7(2–3):251–256. PubMed PMID: 1950722.
141. Kino S, Mizuta H, Osajima T. Sustained release microsphere preparation containing antipsychotic drug and production process thereof. US patent 5656299, 1997.
142. Shenderova A, Burke TG, Schwendeman SP. Stabilization of 10-hydroxycamptothecin in poly(lactide-co-glycolide) microsphere delivery vehicles. *Pharmaceutical Research*. October 1997;14(10):1406–1414. PubMed PMID: 9358554.
143. Shenderova A, Burke TG, Schwendeman SP. The acidic microclimate in poly(lactide-co-glycolide) microspheres stabilizes camptothecins. *Pharmaceutical Research*. February 1999;16(2):241–248. PubMed PMID: 10100309.
144. Ertl B, Platzer P, Wirth M, Gabor F. Poly(D,L-lactic-co-glycolic acid) microspheres for sustained delivery and stabilization of camptothecin. *Journal of Controlled Release*. September 20, 1999;61(3):305–317. PubMed PMID: 10477803.
145. Wang SH, Zhang LC, Lin F, Sa XY, Zuo JB, Shao QX et al. Controlled release of levonorgestrel from biodegradable poly(D,L-lactide-co-glycolide) microspheres: In vitro and in vivo studies. *International Journal of Pharmaceutics*. September 14, 2005;301(1–2):217–225. PubMed PMID: 16040213.
146. Birnbaum DT, Kosmala JD, Henthorn DB, Brannon-Peppas L. Controlled release of beta-estradiol from PLAGA microparticles: The effect of organic phase solvent on encapsulation and release. *Journal of Controlled Release*. April 3, 2000;65(3):375–387. PubMed PMID: 10699296.
147. Mogi T, Ohtake N, Yoshida M, Chimura R, Kamaga Y, Ando S et al. Sustained release of 17-Beta-estradiol frompoly(lactide-co-glycolide) microspheres in vitro and in vivo. *Colloids Surface B: Biointerfaces*. 2000;17:153–165.
148. Jalil R, Nixon JR. Microencapsulation using poly(L-lactic acid). I: Microcapsule properties affected by the preparative technique. *Journal of Microencapsulation*. October–December 1989;6(4):473–484. PubMed PMID: 2585239.
149. Kawata M, Nakamura M, Goto S, Aoyama T. Preparation and dissolution pattern of Eudragit RS microcapsules containing ketoprofen. *Chemical and Pharmaceutical Bulletin*. June 1986;34(6):2618–2623. PubMed PMID: 3769077.
150. Shukla AJ, Price JC. Effect of drug loading and molecular weight of cellulose acetate propionate on the release characteristics of theophylline microspheres. *Pharmaceutical Research*. November 1991;8(11):1396–1400. PubMed PMID: 1798676.
151. Obeidat WM, Price JC. Viscosity of polymer solution phase and other factors controlling the dissolution of theophylline microspheres prepared by the emulsion solvent evaporation method. *Journal of Microencapsulation*. January–February 2003;20(1):57–65. PubMed PMID: 12519702. Epub 2003/01/10. eng.
152. Bolourtchian N, Karimi K, Aboofazeli R. Preparation and characterization of ibuprofen microspheres. *Journal of Microencapsulation*. August 2005;22(5):529–538. PubMed PMID: 16361196.
153. Obeidat WM, Price JC. Preparation and in vitro evaluation of propylthiouracil microspheres made of Eudragit RL 100 and cellulose acetate butyrate polymers using the emulsion-solvent evaporation method. *Journal of Microencapsulation*. May 2005;22(3):281–289. PubMed PMID: 16019914. Epub 2005/07/16. eng.
154. Mateovic-Rojnik T, Frlan R, Bogataj M, Bukovec P, Mrhar A. Effect of preparation temperature in solvent evaporation process on Eudragit RS microsphere properties. *Chemical and Pharmaceutical Bulletin*. January 2005;53(1):143–146. PubMed PMID: 15635253.
155. Obeidat WM, Price JC. Preparation and evaluation of Eudragit S 100 microspheres as pH-sensitive release preparations for piroxicam and theophylline using the emulsion-solvent evaporation method. *Journal of Microencapsulation*. March 2006;23(2):195–202. PubMed PMID: 16754375.
156. Obeidat WM, Obaidat IM. Effect of the dispersion of Eudragit S100 powder on the properties of cellulose acetate butyrate microspheres containing theophylline made by the emulsion-solvent evaporation method. *Journal of Microencapsulation*. May 2007;24(3):263–273. PubMed PMID: 17454437.

157. Obeidat WM, Obeidat SM, Alzoubi NM. Investigations on the physical structure and the mechanism of drug release from an enteric matrix microspheres with a near-zero-order release kinetics using SEM and quantitative FTIR. *AAPS PharmSciTech.* 2009;10(2):615–623. PubMed PMID: 19444619. Pubmed Central PMCID: 2690803.
158. Ogawa Y, Takada S, Yamamoto M. Sustained release microcapsule. US patent 5330767, 1994.
159. Giunchedi P, Alpar HO, Conte U. PDLLA microspheres containing steroids: Spray-drying, o/w and w/o/w emulsifications as preparation methods. *Journal of Microencapsulation.* March–April 1998;15(2):185–195. PubMed PMID: 9532524.
160. Dhanaraju MD, Jayakumar R, Vamsadhara C. Influence of manufacturing parameters on development of contraceptive steroid loaded injectable microspheres. *Chemical and Pharmaceutical Bulletin.* August 2004;52(8):976–979. PubMed PMID: 15304994.
161. Dhanaraju MD, Rajkannan R, Selvaraj D, Jayakumar R, Vamsadhara C. Biodegradation and biocompatibility of contraceptive-steroid-loaded poly (DL-lactide-co-glycolide) injectable microspheres: In vitro and in vivo study. *Contraception.* August 2006;74(2):148–156. PubMed PMID: 16860053.
162. Blanco MJ, Fattal E, Gulik A, Dedieu JC, Roques BP. Characterization and morphological analysis of cholecystokinin derivative peptide-loaded poly(lactide-co-glycolide) microspheres prepared by a water-in-oil-in-water emulsion solvent evaporation method. *Journal of Controlled Release.* 1997;43:81–87.
163. Hashimoto Y, Tanaka M, Kishimoto H, Shiozawa H, Hasegawa K, Matsuyama K et al. Preparation, characterization and taste-masking properties of polyvinylacetal diethylaminoacetate microspheres containing trimebutine. *The Journal of Pharmacy and Pharmacology.* October 2002;54(10):1323–1328. PubMed PMID: 12396292.
164. Deng XM, Li XH, Yuan ML, Xiong CD, Huang ZT, Jia WX et al. Optimization of preparative conditions for poly-DL-lactide- polyethylene glycol microspheres with entrapped Vibrio cholera antigens. *Journal of Controlled Release.* March 29, 1999;58(2):123–131. PubMed PMID: 10053185.
165. Oliveira SS, Oliveira FS, Gaitani CM, Marchetti JM. Microparticles as a strategy for low-molecular-weight heparin delivery. *Journal of Pharmaceutical Sciences.* May 2011;100(5):1783–1792. PubMed PMID: 21374614.
166. Deluca P, Jiang G, Woo B. (Poly(acryloyl-hydroxyethyl starch)-PLGA composition microspheres. US patent 20070122487, 2007.
167. Pean JM, Boury F, Venier-Julienne MC, Menei P, Proust JE, Benoit JP. Why does PEG 400 co-encapsulation improve NGF stability and release from PLGA biodegradable microspheres? *Pharmaceutical Research.* August 1999;16(8):1294–1299. PubMed PMID: 10468034.
168. Rosa GD, Iommelli R, La Rotonda MI, Miro A, Quaglia F. Influence of the co-encapsulation of different non-ionic surfactants on the properties of PLGA insulin-loaded microspheres. *Journal of Controlled Release.* November 3, 2000;69(2):283–295. PubMed PMID: 11064135.
169. Aubert-Pouessel A, Venier-Julienne MC, Clavreul A, Sergent M, Jollivet C, Montero-Menei CN et al. in vitro study of GDNF release from biodegradable PLGA microspheres. *Journal of Controlled Release.* March 24, 2004;95(3):463–475. PubMed PMID: 15023458.
170. Pays K, Giermanska-Kahn J, Pouligny B, Bibette J, Leal-Calderon F. Double emulsions: How does release occur? *Journal of Controlled Release.* February 19, 2002;79(1–3):193–205. PubMed PMID: 11853931.
171. Sah H. Stabilization of proteins against methylene chloride/water interface-induced denaturation and aggregation. *Journal of Controlled Release.* March 29, 1999;58(2):143–151. PubMed PMID: 10053187.
172. Sah H. Protein behavior at the water/methylene chloride interface. *Journal of Pharmaceutical Sciences.* December 1999;88(12):1320–1325. PubMed PMID: 10585229.
173. van de Weert M, Hoechstetter J, Hennink WE, Crommelin DJ. The effect of a water/organic solvent interface on the structural stability of lysozyme. *Journal of Controlled Release.* September 3, 2000;68(3):351–359. PubMed PMID: 10974389.
174. Gombotz W, Huang JW, Lawter JR, Pankey S, Pettit D. Prolonged release of GM-CSF. US patent 5942253, 1999.
175. Shah S. Biodegradable microparticles for the sustained delivery of therapeutic drugs. EU patent 0975334, 2003.
176. Jiang W, Schwendeman SP. Stabilization of a model formalinized protein antigen encapsulated in poly(lactide-co-glycolide)-based microspheres. *Journal of Pharmaceutical Sciences.* October 2001;90(10):1558–1569. PubMed PMID: 11745714.
177. Lam XM, Duenas ET, Cleland JL. Encapsulation and stabilization of nerve growth factor into poly(lactic-co-glycolic) acid microspheres. *Journal of Pharmaceutical Sciences.* September 2001;90(9):1356–1365. PubMed PMID: 11745788.

178. Sturesson C, Carlfors J. Incorporation of protein in PLG-microspheres with retention of bioactivity. *Journal of Controlled Release.* July 3, 2000;67(2–3):171–178. PubMed PMID: 10825551.
179. Wolf M, Wirth M, Pittner F, Gabor F. Stabilisation and determination of the biological activity of L-asparaginase in poly(D,L-lactide-co-glycolide) nanospheres. *International Journal of Pharmaceutics.* April 30, 2003;256(1–2):141–52. PubMed PMID: 12695020.
180. Higaki M, Azechi Y, Takase T, Igarashi R, Nagahara S, Sano A et al. Collagen minipellet as a controlled release delivery system for tetanus and diphtheria toxoid. *Vaccine.* April 30, 2001;19(23–24):3091–3096. PubMed PMID: 11312003.
181. Wei G, Lu LF, Lu WY. Stabilization of recombinant human growth hormone against emulsification-induced aggregation by Pluronic surfactants during microencapsulation. *International Journal of Pharmaceutics.* June 29, 2007;338(1–2):125–132. PubMed PMID: 17336005.
182. Morlock M, Koll H, Winter G, Kissel T. Microencapsulation of rh-erythropoietin, using biodegradable poly(D,L-lactide-co-glycolide). *European Journal of Pharmaceutics and Biopharmaceutics.* 1997;43:29–36.
183. Farrar GH, Jones DH, Clegg JCS. Microencapsulated DNA for gene therapy. EU patent 0965336, 1999.
184. Cohen H, Levy RJ, Gao J, Fishbein I, Kousaev V, Sosnowski S et al. Sustained delivery and expression of DNA encapsulated in polymeric nanoparticles. *Gene Therapy.* November 2000;7(22):1896–1905. PubMed PMID: 11127577.
185. Farrar GH, Tinsley BAM, Jones DH. Encapsulation of bioactive agents. US patent 6309569, 2001.
186. Jones DH, Farrar GH, Clegg JCS. Method of making microencapsulated DNA for vaccination and gene therapy. US patent 6270795, 2001.
187. Benoit MA, Ribet C, Distexhe J, Hermand D, Letesson JJ, Vandenhaute J et al. Studies on the potential of microparticles entrapping pDNA-poly(aminoacids) complexes as vaccine delivery systems. *Journal of Drug Targeting.* 2001;9(4):253–266. PubMed PMID: 11697029.
188. Farrar GH, Tinsley BAM, Jones DH. Encapsulation of bioactive agents. US patent 6565777, 2003.
189. Dunne M, Bibby DC, Jones JC, Cudmore S. Encapsulation of protamine sulphate compacted DNA in polylactide and polylactide-co-glycolide microparticles. *Journal of Controlled Release.* September 19, 2003;92(1–2):209–219. PubMed PMID: 14499198.
190. Jones DH, Farrar GH, Clegg JCS. Method of making microencapsulated DNA for vaccination and gene therapy. US patent 6743444, 2004.
191. Alpar HO, Baillie LWJ, Williamson ED. Composition vaccinalze a base de particules. EU patent 1162945, 2003.
192. Hurtado P, Ferret E, Perez A, Asin MA, Libon C, Nguyen NT. Novel vaccine composition for the treatment of respiratory infectious diseases. EU patent 1972348, 2008.
193. Hyon SH, Ikada Y. Polylactic acid type microspheres containing physiologically active substance and process for preparing the same. US patent 5100669, 1992.
194. Amos M, Sandra G. Long-acting treatment by slow-release delivery of antisense oligodeoxyribonucleotides from biodegradable microparticles. WO 1994023699, 1994.
195. Nagai A, Ohtani S, Takechi N. Sustained release microspheres and preparation thereof. US patent 6036976, 2000.
196. Auer H, Bernstein H, Ganmukhi MM, Johnson OL, Khan AM. Composition for sustained release of human growth hormone. EU patent 0831787, 2001.
197. Eyles J, Westwood A, Elvin SJ, Healey GD. Pharmaceutical composition. US patent 20080138431, 2008.
198. Woo BH, Dagar SH, Yang KY. Sustained-release microspheres and methods of making and using same. US patent 20080131513, 2008.
199. Yang YY, Chung TS, Bai XL, Chan WK. Effect of preparation conditions on morphology and release profiles of biodegradable polymeric microspheres containing protein fabricated by double-emulsion method. *Chemical Engineering Science.* 2000;55:2223–2236.
200. Yang YY, Chia HH, Chung TS. Effect of preparation temperature on the characteristics and release profiles of PLGA microspheres containing protein fabricated by double-emulsion solvent extraction/evaporation method. *Journal of Controlled Release.* October 3, 2000;69(1):81–96. PubMed PMID: 11018548.
201. Thies C. Formation of degradable drug-loaded microparticles by in-liquid drying processes. In: Donbrow M, ed. *Microcapsules and Nanoparticles in Medicine and Pharmacy.* Ann Arbor, MI: CRC; 1991.
202. Arshady R. Preparation of biodegradable microspheres and microcapsules: 2. Polyactides and related polyesters. *Journal of Controlled Release.* 1991;17:1–22.

203. Crotts G, Park TG. Preparation of porous and nonporous biodegradable polymeric hollow microspheres. *Journal of Controlled Release.* 1995;35:91 105.
204. Li WI, Anderson KM, Mehta RC, Deluca PP. Prediction of solvent removal profile and effect on properties for peptide-loaded PLGA microspheres prepared by solvent extraction/evaporation method. *Journal of Controlled Release.* 1995;37:199–214.
205. Choi HS, Seo SA, Khang G, Rhee JM, Lee HB. Preparation and characterization of fentanyl-loaded PLGA microspheres: In vitro release profiles. *International Journal of Pharmaceutics.* March 2, 2002;234(1–2):195–203. PubMed PMID: 11839450.
206. Jeyanthi R, Thanoo BC, Metha RC, DeLuca PP. Effect of solvent removal technique on the matrix characteristics of polylactide/glycolide microspheres for peptide delivery. *Journal of Controlled Release.* 1996;38:235–244.
207. Yang CY, Tsay SY, Tsiang RC. Encapsulating aspirin into a surfactant-free ethyl cellulose microsphere using non-toxic solvents by emulsion solvent-evaporation technique. *Journal of Microencapsulation.* March–April 2001;18(2):223–236. PubMed PMID: 11253939.
208. Freytag T, Dashevsky A, Tillman L, Hardee GE, Bodmeier R. Improvement of the encapsulation efficiency of oligonucleotide-containing biodegradable microspheres. *Journal of Controlled Release.* October 3, 2000;69(1):197–207. PubMed PMID: 11018557.
209. Sah H. Microencapsulation techniques using ethyl acetate as a dispersed solvent: Effect of its extraction rate on the characteristics of PLGA microspheres. *Journal of Controlled Release.* 1997;47:233–245.
210. Nagai A, Ohtani S, Takechi N. Production of microspheres. EU patent 1142567, 2002.
211. Chung TW, Huang YY, Tsai YL, Liu YZ. Effects of solvent evaporation rate on the properties of protein-loaded PLLA and PDLLA microspheres fabricated by emulsion-solvent evaporation process. *Journal of Microencapsulation.* July–August 2002;19(4):463–471. PubMed PMID: 12396383.
212. Izumikawa S, Yoshioka S, Aso Y, Takeda Y. Preparation of poly(l-lactide) microspheres of different crystalline morphology and effect of crystalline morphology on drug release rate. *Journal of Controlled Release.* 1991;15:133–140.
213. Sato T, Kanke M, Schroeder HG, DeLuca PP. Porous biodegradable microspheres for controlled drug delivery. I. Assessment of processing conditions and solvent removal techniques. *Pharmaceutical Research.* January 1988;5(1):21–30. PubMed PMID: 3244604.
214. Cleland J, Lim A, Powell MF. Methods and compositions for microencapsulation of adjuvants. US patent 5643605, 1997.
215. Gilley R, Tice T. Microencapsulation process and products therefrom. US patent 5407609, 1995.
216. Mao S, Shi Y, Li L, Xu J, Schaper A, Kissel T. Effects of process and formulation parameters on characteristics and internal morphology of poly(d,l-lactide-co-glycolide) microspheres formed by the solvent evaporation method. *European Journal of Pharmaceutics and Biopharmaceutics: Official Journal of Arbeitsgemeinschaft fur Pharmazeutische Verfahrenstechnik eV.* February 2008;68(2):214–223. PubMed PMID: 17651954.
217. Birnbaum DT, Brannon-Peppas L. Microparticle drug delivery systems. In: Braun AM, ed. *Drug Delivery Systems in Cancer Therapy.* Totowa, NJ: Humana Press; 2003. pp. 117–35.
218. Kawashima Y, Iwamoto T, Niwa T, Takeuchi H, Hino T. Role of the solvent-diffusion-rate modifier in a new emulsion solvent diffusion method for preparation of ketoprofen microspheres. *Journal of Microencapsulation.* July–September 1993;10(3):329–340. PubMed PMID: 8377091
219. Bezemer JM, Radersma R, Grijpma DW, Dijkstra PJ, van Blitterswijk CA, Feijen J. Microspheres for protein delivery prepared from amphiphilic multiblock copolymers. 1. Influence of preparation techniques on particle characteristics and protein delivery. *Journal of Controlled Release.* July 3, 2000;67(2–3):233–248. PubMed PMID: 10825557.
220. Soppimath KS, Aminabhavi TM. Ethyl acetate as a dispersing solvent in the production of poly(DL-lactide-co-glycolide) microspheres: Effect of process parameters and polymer type. *Journal of Microencapsulation.* May–June 2002;19(3):281–292. PubMed PMID: 12022494.
221. Atkins TJ, Herbert PF, Ramstack M, Strobel J. Preparation of biodegradable microparticles containing a biologically active agent. US patent 5650173, 1997.
222. Rickey ME, Ramstack M, Lewis D. Preparation of biodegradable, biocompatible microparticles containing a biologically active agent. US patent 6290983, 2001.
223. Yeo Y, Baek N, Park K. Microencapsulation methods for delivery of protein drugs. *Biotechnology and Bioprocess Engineering.* 2001;6:213–230.
224. Dawson GF, Koppenhagen F. Production of microparticles. US patent 20030180368, 2003.
225. Brown LR, Gombotz WR, Healy MS. Very low temperature casting of controlled release microspheres. US patent 5019400, 1991.

226. Ferrell TM, Markland P, Staas JK, Tice TR. Injectable buprenorphine microparticle compositions and their use. EU patent 1555023, 2005.
227. Yeh MK, Coombes AG, Jenkins PG, Davis SS. A novel emulsification–solvent extraction technique for production of protein loaded biodegradable microparticles for vaccine and drug delivery. *Journal of Controlled Release*. 1995;33:437–445.
228. Viswanathan NB, Thomas PA, Pandit JK, Kulkarni MG, Mashelkar RA. Preparation of non-porous microspheres with high entrapment efficiency of proteins by a (water-in-oil)-in-oil emulsion technique. *Journal of Controlled Release*. March 8, 1999;58(1):9–20. PubMed PMID: 10021485.
229. Cleland JL, Jones AJS, Powell MF. Method for drying microspheres. US patent 6080429, 2000.
230. Brown LR, Mcgeehan JK, Yuanxi Q, Rashba SJ, Scott TL. Pulmonary delivery of spherical insulin microparticles. WO 20080026068, 2008.
231. Vaugn WM, Van Hamont JE, Setterstrom JA. Sustained release non-steroidal, anti-inflammatory and lidocaine PLGA microspheres. US patent 6217911, 2001.
232. Ramtoola Z. Controlled release biodegradable micro- and nanospheres containing cyclosporin. US patent 5641745, 1997.
233. Geary R, Schlameus H. Microparticulate pharmaceutical delivery system. US patent 5382435, 1995.
234. Boedeker EC, Brown WR, Reid RH, Thies C, John EVH. Microparticle carriers of maximal uptake capacity by both M cells and non-M cells. US patent 5693343, 1997.
235. Tajima M, Watabe K, Yoshimoto T. Delayed drug-releasing microspheres. US patent 5993855, 1999.
236. Hural J, Johnson ME, Spies G. Microparticles and methods for delivery of recombinant viral vaccines. WO 2002092132, 2002.
237. Flashner B, Hinchcliffe M, Lerner IE, Parness H, Smith A, Tzafriri A. Microparticle pharmaceutical compositions for intratumoral delivery. US patent 20040092577, 2004.
238. Liggins R, Toleikis P, Guan D. Microparticles with high loadings of a bioactive agent. US patent 20080124400, 2008.
239. Benoit JP, Courteille F, Thies C. A physicochemical study of the morphology of progesterone-loaded poly (d,l-lactide) microspheres. *International Journal of Pharmacy*. 1986;29:95–102.
240. Brannon-Peppas L, Vert M. Polylactic and polyglycolic acids as drug delivery carriers. In: Wise DL, ed. *Handbook of Pharmaceutical Controlled Release Technology*. New York: Marcel Dekker; 2000. pp. 99–130.
241. ICH Q3C Guidance (International Conference on Harmonisation). Impurities: Guideline for Residual Solvents, 2011.
242. Langer R, Cleland JL, Hanes J. New advances in microsphere-based single-dose vaccines. *Advanced Drug Delivery Reviews*. October 13, 1997;28(1):97–119. PubMed PMID: 10837567.
243. Herrmann J, Bodmeier R. Somatostatin containing biodegradable microspheres prepared by a modified solvent evaporation method based on w/o/w-multiple emulsions. *International Journal of Pharmaceutics*. 1995;126:129–138.
244. Elkharraz K, Ahmed AR, Dashevsky A, Bodmeier R. Encapsulation of water-soluble drugs by an o/o/o-solvent extraction microencapsulation method. *International Journal of Pharmaceutics*. May 16, 2011;409(1–2):89–95. PubMed PMID: 21356287.
245. Perings SM, Hennersdorf M, Koch J-A, Perings C, Kelm M, Heintzen MP et al. Lipoid pneumonia following attempted suicide by intravenous injection of lamp oil. *Med Klin*. 2001;96:685–688.
246. Simmons A, Rouf E, Whittle J. Not your typical pneumonia: A case of exogenous lipoid pneumonia. *Journal of General Internal Medicine*. November 2007;22(11):1613–1616. PubMed PMID: 17846847. Pubmed Central PMCID: 2219803.
247. Bernard H, Mondoulet L, Drumare MF, Paty E, Scheinmann P, Thai R et al. Identification of a new natural Ara h 6 isoform and of its proteolytic product as major allergens in peanut. *Journal of Agricultural and Food Chemistry*. November 200714;55(23):9663–9669. PubMed PMID: 17949050.
248. EMEA (European Medicines Agency). Final position paper on the allergenic potency of herbal medicinal products containing soya or peanut proteins. EA/HMPWP/37/04, London, 11 June 2004.
249. Gangur V, Kelly C, Navuluri L. Sesame allergy: A growing food allergy of global proportions? *Annals of Allergy, Asthma and Immunology: Official Publication of the American College of Allergy, Asthma, and Immunology*. July 2005;95(1):4–11; quiz -3, 44. PubMed PMID: 16095135.
250. Roux KH, Teuber SS, Sathe SK. Tree nut allergens. *International Archives of Allergy and Immunology*. August 2003;131(4):234–244. PubMed PMID: 12915766.
251. Lu B, Zhang JQ, Yang H. Lung-targeting microspheres of carboplatin. *International Journal of Pharmaceutics*. October 20, 2003;265(1–2):1–11. PubMed PMID: 14522113.

252. Siekmann B, Westesen K. Solid lipid particles, particles of bioactive agents and methods for the manufacture and use thereof. US patent 5885486, 1999.
253. Reithmeier H, Herrmann J, Gopferich A. Lipid microparticles as a parenteral controlled release device for peptides. *Journal of Controlled Release*. June 15, 2001;73(2-3):339-350. PubMed PMID: 11516510.
254. Gasco MR. Solid lipid microspheres having a narrow size distribution and method for producing them. EU patent 0526666, 2002.
255. Illum L, Ping H. Inventors Gastroretentive controlled release microspheres for improved drug delivery. US patent 6207197, 2001.
256. Kato Y, Onishi H, Machida Y. Application of chitin and chitosan derivatives in the pharmaceutical field. *Current Pharmaceutical Biotechnology*. October 2003;4(5):303-309. PubMed PMID: 14529420.
257. Garces GJ, Viladot P-JI. Microcapsules. US patent 6818296, 2004.
258. Chan LW, Heng PW. Effects of poly(vinylpyrrolidone) and ethylcellulose on alginate microspheres prepared by emulsification. *Journal of Microencapsulation*. July-August 1998;15(4):409-420. PubMed PMID: 9651863.
259. Shive MS, Anderson JM. Biodegradation and biocompatibility of PLA and PLGA microspheres. *Advanced Drug Delivery Reviews*. October 13, 1997;28(1):5-24. PubMed PMID: 10837562.
260. Smith A, Hunneyball IM. Evaluation of poly(lactic acid) as a biodegradable drug delivery system for parenteral administration. *International Journal of Pharmaceutics*. 1986;30:215-220.
261. Chulia D, Deleuil M, Pourcelot Y. *Powder Technology and Pharmaceutical Processes*. Amsterdam, the Netherlands: Elsevier Science; 1994.
262. Brady JM, Cutright DE, Miller RA, Barristone GC. Resorption rate, route, route of elimination, and ultrastructure of the implant site of polylactic acid in the abdominal wall of the rat. *Journal of Biomedical Materials Research*. March 1973;7(2):155-166. PubMed PMID: 4267379.
263. Thanoo BC, Doll WJ, Mehta RC, Digenis GA, DeLuca PP. Biodegradable indium-111 labeled microspheres for in vivo evaluation of distribution and elimination. *Pharmaceutical Research*. December 1995;12(12):2060-2064. PubMed PMID: 8786990.
264. Tamber H, Johansen P, Merkle HP, Gander B. Formulation aspects of biodegradable polymeric microspheres for antigen delivery. *Advanced Drug Delivery Reviews*. January 10, 2005;57(3):357-376. PubMed PMID: 15560946.
265. Wada R, Tabata Y, Hyon S-H, Ikada U. Preparation of poly(lactic acid) microspheres containing anticancer drugs. *Bulletin of the Institute for Chemical Research, Kyoto University*. 1988;66(3):241-250.
266. Ruan G, Feng SS. Preparation and characterization of poly(lactic acid)-poly(ethylene glycol)-poly(lactic acid) (PLA-PEG-PLA) microspheres for controlled release of paclitaxel. *Biomaterials*. December 2003;24(27):5037-5044. PubMed PMID: 14559017.
267. Hutchinson FG. Continuous release pharmaceutical compositions. EU patent 0058481, 1986.
268. Bodmeier R, Oh KH, Chen H. The effect of the addition of low molecular weight poly(d,l-lactide) on drug release from biodegradable poly(d,l-lactide) drug delivery systems. *International Journal of Pharmaceutics*. 1989;51:1-8.
269. Spenlehauer G, Vert M, Benoit JP, Boddaert A. In vitro and in vivo degradation of poly(D,L lactide/glycolide) type microspheres made by solvent evaporation method. *Biomaterials*. October 1989;10(8):557-563. PubMed PMID: 2605288.
270. Wang C, Ge Q, Ting D, Nguyen D, Shen HR, Chen J et al. Molecularly engineered poly(ortho ester) microspheres for enhanced delivery of DNA vaccines. *Nature Materials*. March 2004;3(3):190-196. PubMed PMID: 14991022.
271. Little SR, Lynn DM, Ge Q, Anderson DG, Puram SV, Chen J et al. Poly-beta amino ester-containing microparticles enhance the activity of nonviral genetic vaccines. *Proceedings of the National Academy of Sciences of the United States of America*. June 29, 2004;101(26):9534-9539. PubMed PMID: 15210954. Pubmed Central PMCID: 470709.
272. Ogawa Y, Yamamoto M, Takada S, Okada H, Shimamoto T. Controlled-release of leuprolide acetate from polylactic acid or copoly(lactic/glycolic) acid microcapsules: Influence of molecular weight and copolymer ratio of polymer. *Chemical and Pharmaceutical Bulletin*. April 1988;36(4):1502-1507. PubMed PMID: 3138032.
273. Cha Y, Pitt CG. The acceleration of degradation-controlled drug delivery from polyester microspheres. *Journal of Controlled Release*. 1989;8(3):259-265.
274. Iwasa S, Nakagawa Y, Takada S. Substained release microcapsule of physiologically active compound which is slightly water soluble at pH 6 to 8. US patent 6113941, 2000.

275. Tsai DC, Howard SA, Hogan TF, Malanga CJ, Kandzari SJ, Ma JK. Preparation and in vitro evaluation of polylactic acid-mitomycin C microcapsules. *Journal of Microencapsulation.* July–September 1986;3(3):181–193. PubMed PMID: 3149671.
276. Spenlehauer G, Vert M, Benoit JP, Chabot F, Veillard M. Biodegradable cisplatin microspheres prepared by the solvent evaporation method: Morphology and release characteristics. *Journal of Controlled Release.* 1988;7:217–229.
277. Mathiowitz E, Saltzman WM, Domb A, Dor P, Langer R. Polyanhydride microspheres as drug carriers. II. Microencapsulation by solvent removal. *Journal of Applied Polymer Science.* 1988;35:755–774.
278. Howard MA, 3rd, Gross A, Grady MS, Langer RS, Mathiowitz E, Winn HR et al. Intracerebral drug delivery in rats with lesion-induced memory deficits. *Journal of Neurosurgery.* July 1989;71(1):105–112. PubMed PMID: 2567778.
279. Fatome M, Courteille F, Laval JD, Roman V. Radioprotective activity of ethylcellulose microspheres containing WR 2721, after oral administration. *International Journal of Radiation Biology and Related Studies in Physics, Chemistry, and Medicine.* July 1987;52(1):21–29. PubMed PMID: 3036725.
280. Uddin MS, Hawlader MN, Zhu HJ. Microencapsulation of ascorbic acid: Effect of process variables on product characteristics. *Journal of Microencapsulation.* March–April 2001;18(2):199–209. PubMed PMID: 11253937.
281. Cheu SJ, Chen RR, Chen PF, Lin WJ. In vitro modified release of acyclovir from ethyl cellulose microspheres. *Journal of Microencapsulation.* September–October 2001;18(5):559–565. PubMed PMID: 11508761.
282. Dinarvand R, Mirfattahi S, Atyabi F. Preparation, characterization and in vitro drug release of isosorbide dinitrate microspheres. *Journal of Microencapsulation.* January–February 2002;19(1):73–81. PubMed PMID: 11811761.
283. Sengel CT, Hascicek C, Gonul N. Development and in-vitro evaluation of modified release tablets including ethylcellulose microspheres loaded with diltiazem hydrochloride. *Journal of Microencapsulation.* March 2006;23(2):135–152. PubMed PMID: 16754371.
284. Acros Organics. http://www.acros.com, 2015.
285. Eastman Chemical Company. http://www.eastman.com, 2015.
286. Tsujiyama T, Suzuki N, Kawata M, Uchida T, Goto S. Preparation and pharmacokinetic and pharmacodynamic evaluation of hydroxy propyl cellulose-ethyl cellulose microcapsules containing piretanide. *Journal of Pharmacobio-Dynamics.* June 1989;12(6):311–323. PubMed PMID: 2778624.
287. Obeidat WM. Viscosity of the polymer solution phase and other rational approaches to control matrix-microsphere properties. PhD dissertation, UGA, GA, USA, 2002.
288. Silva JP, Ferreira JP. Effect of drug properties on the release from CAP microspheres prepared by a solvent evaporation method. *Journal of Microencapsulation.* January–February 1999;16(1):95–103. PubMed PMID: 9972506.
289. Dalmoro A, Lamberti G, Titomanlio G, Barba AA, d'Amore M. Enteric micro-particles for targeted oral drug delivery. *AAPS PharmSciTech.* December 2010;11(4):1500–1507. PubMed PMID: 20931307. Pubmed Central PMCID: 3011076.
290. Evonik Industries. http://www.evonik.com, 2015.
291. Goto S, Kawata M, Nakamura M, Maekawa K, Aoyama T. Eudragit RS and RL (acrylic resins) microcapsules as pH insensitive and sustained release preparations of ketoprofen. *Journal of Microencapsulation.* October–December 1986;3(4):293–304. PubMed PMID: 3508190.
292. Rodriguez M, Vila-Jato JL, Torres D. Design of a new multiparticulate system for potential site-specific and controlled drug delivery to the colonic region. *Journal of Controlled Release.* October 30, 1998;55(1):67–77. PubMed PMID: 9795017.
293. Squillante E, Morshed G, Bagchi S, Mehta KA. Microencapsulation of beta-galactosidase with Eudragit L-100. *Journal of Microencapsulation.* March–April 2003;20(2):153–167. PubMed PMID: 12554371.
294. Re MI. Biscans BPomokwapbaq-esdm. *Powder Technology.* 1999;101:120–133.
295. Villamizar L, Barrera G, Cotes AM, Martinez F. Eudragit S100 microparticles containing Spodoptera frugiperda nucleopolyehedrovirus: Physicochemical characterization, photostability and in vitro virus release. *Journal of Microencapsulation.* 2010;27(4):314–324. PubMed PMID: 19839785.
296. Chandran S, Sanjay KS, Ali Asghar LF. Microspheres with pH modulated release: Design and characterization of formulation variables for colonic delivery. *Journal of Microencapsulation.* August 2009;26(5):420–431. PubMed PMID: 18821120.
297. McGinity JW. *Aqueous Polymeric Coatings for Pharmaceutical Dosage Forms*, 2nd edn. New York: Marcel Dekker; 1997.

298. Bogataj M, Mrhar A, Kristl A, Kozjek F. Eudragit E microspheres containing bacampicillin: Preparation by solvent removal methods. *Journal of Microencapsulation*. July–September 1991;8(3):401–406. PubMed PMID: 1941447.
299. Yang M, Cui F, You B, You J, Wang L, Zhang L et al. A novel pH-dependent gradient-release delivery system for nitrendipine: I. Manufacturing, evaluation in vitro and bioavailability in healthy dogs. *Journal of Controlled Release*. August 11, 2004;98(2):219–229. PubMed PMID: 15262414.
300. Gasco MR. Method for producing solid lipid microspheres having a narrow size distribution. US patent 5250236, 1993.
301. Domb AJ. Lipospheres for controlled delivery of substances. US patent 5188837, 1993.
302. Siekmann B, Westesen K. Solid lipid particles, particles of bioactive agents and methods for the manufacture and use thereof. US patent 5785976, 1998.
303. Westesen K, Siekmann B. Solid lipid particles, particles of bioactive agents and methods for the manufacture and use thereof. US patent 6207178, 2001.
304. Gasco MR. Pharmaceutical composition in form of solid lipidic microparticles suitable to parenteral administration. EU patent 0988031, 2003.
305. Rozier A. Fluid ophthalmic composition based on lipid microparticles containing at least one active principle. EU patent 0437368, 1991.
306. Domb AJ. Liposphere carriers of vaccines. US patent 5340588, 1994.
307. Hedley M, Lunsford L, Putnam D. Microparticles for delivery of nucleic acid. US patent 20020182258, 2002.
308. Blagdon PA, Morgan R. Methods of encapsulating liquids in fatty matrices, and products thereof. US patent 5204029, 1993.
309. Coyne B, Faragher J, Gouin S, Hansen CB, Ingram R, Isak T et al. Microcapsules. US patent 20070042184, 2007.
310. Bernstein H, Mathiowitz E, Morrel E, Schwaller K. Method for producing protein microspheres. US patent 5271961, 1993.
311. Mathiowitz E, Bernstein H, Morrel E, Schwaller K. Method for producing protein microspheres. EU patent 0499619, 1996.
312. Beck T, Bernstein H, Mathiowitz E, Morrel E, Schwaller K. Protein microspheres and methods of using them. US patent 5679377, 1997.
313. Suslick KS, Toublan FJ, Boppart SA, Marks DL. Surface modified protein microparticles. US patent 7217410, 2007.
314. Brook M, Heritage P, Jiang J, Loomes LM, Mcdermott MR, Underdown BJ. Microparticle delivery system with a functionalized silicone bonded to the matrix. US patent 5571531, 1996.
315. Gustafsson NO, Jönsson M, Laakso T, Larsson K, Reslow M. Vaccine composition comprising an immunologically active substance embedded in microparticles consisting of starch with reduced molecular weight. WO 2002028371, 2002.
316. Reslow M, Björn S, Drustrup J, Gustafsson NO, Jönsson M, Laakso T. A controlled-release, parenterally administrable microparticle preparation. EU patent 1328258, 2008.
317. Kwong AK, Chou S, Sun AM, Sefton MV, Goosen MFA. In vitro and in vivo release of insulin from poly(lactic acid) microbeads and pellets. *Journal of Controlled Release*. 1986;4:47–62.
318. Jeong YI, Song JG, Kang SS, Ryu HH, Lee YH, Choi C et al. Preparation of poly(DL lactide-co-glycolide) microspheres encapsulating all-trans retinoic acid. *International Journal of Pharmaceutics*. June 18, 2003;259(1–2):79–91. PubMed PMID: 12787638.
319. Song M, Li N, Sun S, Tiedt LR, Liebenberg W, de Villiers MM. Effect of viscosity and concentration of wall former, emulsifier and pore-inducer on the properties of amoxicillin microcapsules prepared by emulsion solvent evaporation. *Farmaco*. March 2005;60(3):261–267. PubMed PMID: 15784247.
320. Jalil R, Nixon JR. Microencapsulation using poly (L-lactic acid) II: Preparative variables affecting microcapsule properties. *Journal of Microencapsulation*. January–March 1990;7(1):25–39. PubMed PMID: 2308052.
321. Lee SC, Oh JT, Jang MH, Chung SI. Quantitative analysis of polyvinyl alcohol on the surface of poly(D, L-lactide-co-glycolide) microparticles prepared by solvent evaporation method: Effect of particle size and PVA concentration. *Journal of Controlled Release*. May 20, 1999;59(2):123–132. PubMed PMID: 10332048.
322. Cirpanli Y, Unlu N, Calis S, Hincal AA. Formulation and in-vitro characterization of retinoic acid loaded poly (lactic-co-glycolic acid) microspheres. *Journal of Microencapsulation*. December 2005;22(8):877–889. PubMed PMID: 16423759.

323. Thakare M, Israel B, Garner ST, Ahmed H, Garner P, Elder D et al. Formulation parameters and release mechanism of theophylline loaded ethyl cellulose microspheres: Effect of different dual surfactant ratios. *Pharmaceutical Development and Technology*. September–October 2013;18(5):1213–1219. PubMed PMID: 21991996.
324. Conti B, Genta T, Modena F, Pavenotti F. Investigation on process parameters involved in PLGA microspheres preparation. *Drug Development and Industrial Pharmacy*. 1995;21:615–622.
325. Fong JW, Maulding HV, Visscher GE, Nazareno JP, Pearson JEWDACS. Enhancing drug release from poly(lactide) microspheres by using base in the microencapsulation process. In: Lee P, Good WR, eds. *Controlled Release Technology, Pharmaceutical Applications*. Washington, DC: ACS Symposium Series; 1987.
326. Blanco D, Alonso MJ. Protein encapsulation and release from poly(lactide-co-glycolide) microspheres: Effect of the protein and polymer properties and of the co-encapsulation of surfactants. *European Journal of Pharmaceutics and Biopharmaceutics: Official Journal of Arbeitsgemeinschaft fur Pharmazeutische Verfahrenstechnik eV*. May 1998;45(3):285–294. PubMed PMID: 9653633.
327. Ranjha NM, Khan H, Naseem S. Encapsulation and characterization of controlled release flurbiprofen loaded microspheres using beeswax as an encapsulating agent. *Journal of Materials Science Materials in Medicine*. May 2010;21(5):1621–1630. PubMed PMID: 20217193.
328. Bouissou C, Rouse JJ, Price R, van der Walle CF. The influence of surfactant on PLGA microsphere glass transition and water sorption: Remodeling the surface morphology to attenuate the burst release. *Pharmaceutical Research*. June 2006;23(6):1295–1305. PubMed PMID: 16715359.
329. Dinarvand R, Moghadam SH, Sheikhi A, Atyabi F. Effect of surfactant HLB and different formulation variables on the properties of poly-D,L-lactide microspheres of naltrexone prepared by double emulsion technique. *Journal of Microencapsulation*. March 2005;22(2):139–151. PubMed PMID: 16019900.
330. Goodwin JW. *Colloids and Interfaces with Surfactants and Polymers*, 2nd edn. Chichester, U.K,.: Wiley; 2009.
331. Kawashima Y, Niwa T, Handa T, Takeuchi H, Iwamoto T, Itoh K. Preparation of controlled-release microspheres of ibuprofen with acrylic polymers by a novel quasi-emulsion solvent diffusion method. *Journal of Pharmaceutical Sciences*. January 1989;78(1):68–72. PubMed PMID: 2709323.
332. Zhu KJ, Jiang HL, Du XY, Wang J, Xu WX, Liu SF. Preparation and characterization of hCG-loaded polylactide or poly(lactide-co-glycolide) microspheres using a modified water-in-oil-in-water (w/o/w) emulsion solvent evaporation technique. *Journal of Microencapsulation*. March–April 2001;18(2):247–260. PubMed PMID: 11253941.
333. Brime B, Ballesteros MP, Frutos P. Preparation and in vitro characterization of gelatin microspheres containing Levodopa for nasal administration. *Journal of Microencapsulation*. November–December 2000;17(6):777–784. PubMed PMID: 11063424.
334. Mateovic T, Kriznar B, Bogataj M, Mrhar A. The influence of stirring rate on biopharmaceutical properties of Eudragit RS microspheres. *Journal of Microencapsulation*. January–February 2002;19(1):29–36. PubMed PMID: 11811756.
335. Shinar R. On the behaviour of liquid dispersions in mixing vessels. *Journal of Fluid Mechanics*. 1961;10:259–275.
336. Shinar R, Church JM. Predicting particle size in agitated dispersions. *Industrial and Engineering Chemistry*. 1960;52(3):253–256.
337. Pongpaibul Y, Maruyama K, Iwatsuru M. Formation and in-vitro evaluation of theophylline-loaded poly(methyl methacrylate) microspheres. *The Journal of Pharmacy and Pharmacology*. August 1988;40(8):530–533. PubMed PMID: 2907004.
338. Schlicher EJAM, Postma NS, Zuidema J, Talsma H. Preparation and characterisation of Poly (d,l-lactic-co-glycolic acid) microspheres containing desferrioxamine. *International Journal of Pharmaceutics*. 1997;153:235–245.
339. Jeffery H, Davis SS, O'Hagan DT. The preparation and characterisation of poly(lactide-co-glycolide) microparticles: I. Oil-in-water emulsion solvent evaporation. *International Journal of Pharmaceutics*. 1991;77:169–175.
340. Shukla AJ, Price JC. Effect of drug (core) particle size on the dissolution of theophylline from microspheres made from low molecular weight cellulose acetate propionate. *Pharmaceutical Research*. May 1989;6(5):418–421. PubMed PMID: 2748534.
341. Pongpaibul Y, Price JC, Whitworth CW, 10, 10,. Preparation and evaluation of controlled release indomethacin microspheres. *Drug Development and Industrial Pharmacy*. 1984;10(10):1597–1616.
342. André-Abrant A, Taverdet J-L, Jay J. Microencapsulation par évaporation de solvant. *European Polymer Journal*. 2001;37:955–967.

343. Calderbank PH. Physical rate processes in industrial fermentation. Part I: The interfacial area in gas-liquid contacting with mechanical agitation. *Transactions of the Institution of Chemical Engineers.* 1958;36:443–463.
344. Hinze J. Fundamentals of the hydrodynamic mechanism of splitting in dispersion processes. *AIChE J.* 1955;1:289–295.
345. Davies GA. Mixing and coalescence phenomena in liquid–liquid systems. In: Thornton JD, ed. *In Science and Practice of Liquid-Liquid Extraction*, 1st edn. Oxford, U.K: Clarendon Press; 1992.
346. Jeyanthi R, Mehta RC, Thanoo BC, DeLuca PP. Effect of processing parameters on the properties of peptide-containing PLGA microspheres. *Journal of Microencapsulation.* March–April 1997;14(2):163–174. PubMed PMID: 9132468.
347. Lee JH, Park TG, Choi HK. Development of oral drug delivery system using floating microspheres. *Journal of Microencapsulation.* November–December 1999;16(6):715–729. PubMed PMID: 10575624.
348. Seo SA, Khang G, Rhee JM, Kim J, Lee HB. Study on in vitro release patterns of fentanyl-loaded PLGA microspheres. *Journal of Microencapsulation.* September–October 2003;20(5):569–579. PubMed PMID: 12909542.
349. Sugiura S, Nakajima M, Iwamoto S, Seki M. Interfacial tension driven monodispersed droplet formation from microfabricated channel array. *Langmuir.* 2001;17:5562–5566.
350. Amsden B, Liggins R. Methods for microsphere production. US patent 6224794, 2001.
351. Hirano A, Ipponmatsu M, Nishigaki M, Tsurutani T. Uniform inorganic micropheres and production thereof. US patent 5376347, 1994.
352. Rourke JK. Continuous production of Emulsions and microcapsules of uniform particle size. US patent 5643506, 1997.
353. Al-Azzam W, Pastrana EA, Griebenow K. Co-lyophilization of bovine serum albumin (BSA) with poly(ethylene glycol) improves efficiency of BSA encapsulation and stability in polyester microspheres by a solid-in-oil-in-oil technique. *Biotechnology Letters.* 2002;24:1367–1374.
354. Yang CY, Tsay SY, Tsiang RC. An enhanced process for encapsulating aspirin in ethyl cellulose microcapsules by solvent evaporation in an O/W emulsion. *Journal of Microencapsulation.* May–June 2000;17(3):269–277. PubMed PMID: 10819416.
355. Uchida T, Yoshida K, Ninomiya A, Goto S. Optimization of preparative conditions for polylactide (PLA) microspheres containing ovalbumin. *Chemical and Pharmaceutical Bulletin.* September 1995;43(9):1569–1573. PubMed PMID: 7586084.
356. Cohen S, Yoshioka T, Lucarelli M, Hwang LH, Langer R. Controlled delivery systems for proteins based on poly(lactic/glycolic acid) microspheres. *Pharmaceutical Research.* June 1991;8(6):713–720. PubMed PMID: 2062800.
357. Cleland JL, Lim A, Barron L, Duenas ET, Powell MF. Development of a single-shot subunit vaccine for HIV-1: Part 4. Optimizing microencapsulation and pulsatile release of MN rgp120 from biodegradable microspheres. *Journal of Controlled Release.* 1997;47:135–150.
358. Jalil R, Nixon JR. Microencapsulation using poly(DL-lactic acid). III: Effect of polymer molecular weight on the release kinetics. *Journal of Microencapsulation.* July–September 1990;7(3):357–374. PubMed PMID: 2384838.
359. Haleblian J, Runkel R, Mueller N, Christopherson J, Ng K. Steroid release from silicone elastomer containing excess drug in suspension. *Journal of Pharmaceutical Sciences.* April 1971;60(4):541–545. PubMed PMID: 4108479.
360. Chiao CS, Price JC. Formulation, preparation and dissolution characteristics of propranolol hydrochloride microspheres. *Journal of Microencapsulation.* March–April 1994;11(2):153–159. PubMed PMID: 8006762.
361. Elkheshen SA, Radwan MA. Sustained release microspheres of metoclopramide using poly(D,L-lactide-co-glycolide) copolymers. *Journal of Microencapsulation.* July–August 2000;17(4):425–435. PubMed PMID: 10898083.
362. Jelvehgari M, Atapour F, Nokhodchi A. Micromeritics and release behaviours of cellulose acetate butyrate microspheres containing theophylline prepared by emulsion solvent evaporation and emulsion non-solvent addition method. *Archives of Pharmacal Research.* July 2009;32(7):1019–1028. PubMed PMID: 19641883.
363. Song X, Song SK, Zhao P, Wei LM, Jiao HS. beta-methasone-containing biodegradable poly(lactide-co-glycolide) acid microspheres for intraarticular injection: Effect of formulation parameters on characteristics and in vitro release. *Pharmaceutical Development and Technology.* September–October 2013;18(5):1220–1229. PubMed PMID: 22295954.
364. Deasy PB, Brophy MR, Ecanow B, Joy MM. Effect of ethylcellulose grade and sealant treatments on the production and in vitro release of microencapsulated sodium salicylate. *The Journal of Pharmacy and Pharmacology.* January 1980;32(1):15–20. PubMed PMID: 6102120.

365. Cleland JL. Protein delivery from biodegradable microspheres. In: Sanders L, Hendren W, eds. *Protein Delivery: Physical Systems*. New York: Plenum Press; 1997.
366. Fong JW. Process for preparation of microspheres and modification of release rate of core material. US patent 4479911, 1984.
367. Higuchi T. Mechanism of sustained-action medication. Theoretical analysis of rate of release of solid drugs dispersed in solid matrices. *Journal of Pharmaceutical Sciences*. December 1963;52:1145–1149. PubMed PMID: 14088963.
368. Baker RW, Londsdale HK. Controlled release: Mechanisms and rates. In: Tanquary AC, Lacey RE, eds. *Controlled Release of Biologically Active Agents*. New York: Plenum Press; 1974.
369. Thies C, Dappert T. Statistical models for controlled release microcapsules: Rationale and theory. *Journal of Membrane Science*. 1978;4:99–113.
370. Hoffman A, Donbrow M, Benita S. Direct measurements on individual microcapsule dissolution as a tool for determination of release mechanism. *The Journal of Pharmacy and Pharmacology*. October 1986;38(10):764–766. PubMed PMID: 2878999.
371. Hoffman A, Donbrow M, Gross ST, Benita S, Bahat R. Fundamentals of release mechanism interpretation in multiparticulate systems: Determination of substrate release from single microcapsules and relation between individual and ensemble release kinetics. *International Journal of Pharmaceutics*. 1986;29:195–211.
372. Ozsoy Y, Caybasi P, Terzioglu N, Ozkirimli S, Tuncel T. In vitro studies on the transdermal formulation containing pentoxifylline microspheres with poly-L-lactide. *Bollettino Chimico Farmaceutico*. Jnuary–February 2002;141(1):29–32. PubMed PMID: 12064054.
373. Hameed A, Brothwood T, Bouloux P. Delivery of testosterone replacement therapy. *Current Opinion in Investigational Drugs*. October 2003;4(10):1213–1219. PubMed PMID: 14649214.
374. Peyre M, Fleck R, Hockley D, Gander B, Sesardic D. In vivo uptake of an experimental microencapsulated diphtheria vaccine following sub-cutaneous immunisation. *Vaccine*. June 23, 2004;22(19):2430–2437. PubMed PMID: 15193406.
375. Haddadi A, Farboud ES, Erfan M, Aboofazeli R. Preparation and characterization of biodegradable urea-loaded microparticles as an approach for transdermal delivery. *Journal of Microencapsulation*. September 2006;23(6):698–712. PubMed PMID: 17118885.
376. Jorizzo J, Grossman R, Nighland M. Tretinoin microsphere gel in younger acne patients. *Journal of Drugs in Dermatology: JDD*. August 2008;7(8 Suppl):s9–s13. PubMed PMID: 18724649.
377. Hunter SK, Andracki ME, Krieg AM. Biodegradable microspheres containing group B Streptococcus vaccine: Immune response in mice. *American Journal of Obstetrics and Gynecology*. November 2001;185(5):1174–1179. PubMed PMID: 11717653.
378. Yildiz A, Okyar A, Baktir G, Araman A, Ozsoy Y. Nasal administration of heparin-loaded microspheres based on poly(lactic acid). *Farmaco*. November–December 2005;60(11–12):919–924. PubMed PMID: 16243322.
379. Jaganathan KS, Vyas SP. Strong systemic and mucosal immune responses to surface-modified PLGA microspheres containing recombinant hepatitis B antigen administered intranasally. *Vaccine*. May 8, 2006;24(19):4201–4211. PubMed PMID: 16446012.
380. Brandhonneur N, Loizel C, Chevanne F, Wakeley P, Jestin A, Le Potier MF et al. Mucosal or systemic administration of rE2 glycoprotein antigen loaded PLGA microspheres. *International Journal of Pharmaceutics*. May 21, 2009;373(1–2):16–23. PubMed PMID: 19429284.
381. Gungor S, Okyar A, Erturk-Toker S, Baktir G, Ozsoy Y. Ondansetron-loaded biodegradable microspheres as a nasal sustained delivery system: In vitro/in vivo studies. *Pharmaceutical Development and Technology*. June 2010;15(3):258–265. PubMed PMID: 22716466.
382. Mohamed F, van der Walle CF. PLGA microcapsules with novel dimpled surfaces for pulmonary delivery of DNA. *International Journal of Pharmaceutics*. March 27, 2006;311(1–2):97–107. PubMed PMID: 16414217.
383. Doan TV, Olivier JC. Preparation of rifampicin-loaded PLGA microspheres for lung delivery as aerosol by premix membrane homogenization. *International Journal of Pharmaceutics*. December 1, 2009;382(1–2):61–66. PubMed PMID: 19682562.
384. Thomas C, Gupta V, Ahsan F. Particle size influences the immune response produced by hepatitis B vaccine formulated in inhalable particles. *Pharmaceutical Research*. May 2010 ;27(5):905–919. PubMed PMID: 20232117.
385. Doan TV, Couet W, Olivier JC. Formulation and in vitro characterization of inhalable rifampicin-loaded PLGA microspheres for sustained lung delivery. *International Journal of Pharmaceutics*. July 29, 2011;414(1–2):112–117. PubMed PMID: 21596123.

386. Gupta V, Ahsan F. Influence of PEI as a core modifying agent on PLGA microspheres of PGE(1), a pulmonary selective vasodilator. *International Journal of Pharmaceutics*. July 15, 2011;413(1–2):51–62. PubMed PMID: 21530623. Pubmed Central PMCID: 3123904.
387. Lee HY, Mohammed KA, Goldberg EP, Nasreen N. Arginine-conjugated albumin microspheres inhibits proliferation and migration in lung cancer cells. *American Journal of Cancer Research*. 2013;3(3):266–277. PubMed PMID: 23841026. Pubmed Central PMCID: 3696533.
388. Rong X, Mo X, Ren T, Yang S, Yuan W, Dong J et al. Neuroprotective effect of erythropoietin-loaded composite microspheres on retinal ganglion cells in rats. *European Journal of Pharmaceutical Sciences: Official Journal of the European Federation for Pharmaceutical Sciences*. July 17, 2011;43(4):334–342. PubMed PMID: 21621611.
389. Rong X, Yang S, Miao H, Guo T, Wang Z, Shi W et al. Effects of erythropoietin-dextran microparticle-based PLGA/PLA microspheres on RGCs. *Investigative Ophthalmology and Visual Science*. September 2012;53(10):6025–6034. PubMed PMID: 22871834.
390. Kadam RS, Tyagi P, Edelhauser HF, Kompella UB. Influence of choroidal neovascularization and biodegradable polymeric particle size on transscleral sustained delivery of triamcinolone acetonide. *International Journal of Pharmaceutics*. September 15, 2012;434(1–2):140–147. PubMed PMID: 22633904. Pubmed Central PMCID: 3573139.
391. Shinde UA, Shete JN, Nair HA, Singh KH. Eudragit RL100 based microspheres for ocular administration of azelastine hydrochloride. *Journal of Microencapsulation*. 2012;29(6):511–519. PubMed PMID: 22375685.
392. Hwang SJ, Park H, Park K. Gastric retentive drug-delivery systems. *Critical Reviews in Therapeutic Drug Carrier Systems*. 1998;15(3):243–284. PubMed PMID: 9699081.
393. Dressman JB, Lennernäs H. *Oral Drug Absorption: Prediction and Assessment*. New York: Marcel Dekker; 2000.
394. Bodmeier R, Chen HG, Paeratakul O. A novel approach to the oral delivery of micro- or nanoparticles. *Pharmaceutical Research*. May 1989;6(5):413–417. PubMed PMID: 2748533.
395. Joseph NJ, Lakshmi S, Jayakrishnan A. A floating-type oral dosage form for piroxicam based on hollow polycarbonate microspheres: In vitro and in vivo evaluation in rabbits. *Journal of Controlled Release*. February 19, 2002;79(1–3):71–79. PubMed PMID: 11853919.
396. Bv B, R D, S B, Abraham S, Furtado S, V M. Hollow microspheres of diclofenac sodium—A gastroretentive controlled delivery system. *Pakistan Journal of Pharmaceutical Sciences*. October 2008;21(4):451–454. PubMed PMID: 18930869.
397. Miyazaki Y, Yakou S, Takayama K. Comparison of gastroretentive microspheres and sustained-release preparations using theophylline pharmacokinetics. *The Journal of Pharmacy and Pharmacology*. June 2008;60(6):693–698. PubMed PMID: 18498704.
398. Zhao L, Wei YM, Yu Y, Zheng WW. Polymer blends used to prepare nifedipine loaded hollow microspheres for a floating-type oral drug delivery system: In vitro evaluation. *Archives of Pharmacal Research*. March 2010;33(3):443–450. PubMed PMID: 20361310.
399. Pandya N, Pandya M, Bhaskar VH. Preparation and in vitro characterization of porous carrier-based glipizide floating microspheres for gastric delivery. *Journal of Young Pharmacists: JYP*. April 2011;3(2):97–104. PubMed PMID: 21731353. Pubmed Central PMCID: 3122053.
400. Raut NS, Somvanshi S, Jumde AB, Khandelwal HM, Umekar MJ, Kotagale NR. Ethyl cellulose and hydroxypropyl methyl cellulose buoyant microspheres of metoprolol succinate: Influence of pH modifiers. *International Journal of Pharmaceutical Investigation*. July 2013;3(3):163–170. PubMed PMID: 24167789. Pubmed Central PMCID: 3807984.
401. Choudhary H, Agrawal AK, Malviya R, Yadav SK, Jaliwala YA, Patil UK. Evaluation and optimization of preparative variables for controlled-release floating microspheres of levodopa/carbidopa. *Die Pharmazie*. March 2010;65(3):194–198. PubMed PMID: 20383939.
402. Thanoo BC, Sunny MC, Jayakrishnan A. Oral sustained-release drug delivery systems using polycarbonate microspheres capable of floating on the gastric fluid. *The Journal of Pharmacy and Pharmacology*. January 1993;45(1):21–24. PubMed PMID: 8094440.
403. Rane BR, Gujarathi NA, Patel JK. Biodegradable anionic acrylic resin based hollow microspheres of moderately water soluble drug rosiglitazone maleate: Preparation and in vitro characterization. *Drug Development and Industrial Pharmacy*. December 2012;38(12):1460–1469. PubMed PMID: 22356275.
404. Singh V, Chaudhary AK. Preparation of Eudragit E100 microspheres by modified solvent evaporation method. *Acta Poloniae Pharmaceutica*. November–December 2011;68(6):975–980. PubMed PMID: 22125964.

405. Ali J, Hasan S, Ali M. Formulation and development of gastroretentive drug delivery system for ofloxacin. *Methods and Findings in Experimental and Clinical Pharmacology*. September 2006;28(7):433–439. PubMed PMID: 17003848.
406. Rajinikanth PS, Karunagaran LN, Balasubramaniam J, Mishra B. Formulation and evaluation of clarithromycin microspheres for eradication of Helicobacter pylori. *Chemical and Pharmaceutical Bulletin*. December 2008;56(12):1658–1664. PubMed PMID: 19043235.
407. Liu H, Pan W, Ke P, Dong Y, Ji L. Preparation and evaluation of a novel gastric mucoadhesive sustained-release acyclovir microsphere. *Drug Development and Industrial Pharmacy*. September 2010;36(9):1098–105. PubMed PMID: 20545521.
408. Kasliwal N, Negi JS, Jugran V, Jain R. Formulation, development, and performance evaluation of metoclopramide HCl oro-dispersible sustained release tablet. *Archives of Pharmacal Research*. October 2011;34(10):1691–1700. PubMed PMID: 22076769.
409. Garg Y, Pathak K. Design and in vitro performance evaluation of purified microparticles of pravastatin sodium for intestinal delivery. *AAPS PharmSciTech*. June 2011;12(2):673–682. PubMed PMID: 21594729. Pubmed Central PMCID: 3134671.
410. Shadab, Ahuja A, Khar RK, Baboota S, Chuttani K, Mishra AK et al. Gastroretentive drug delivery system of acyclovir-loaded alginate mucoadhesive microspheres: Formulation and evaluation. *Drug Delivery*. May 2011;18(4):255–264. PubMed PMID: 21110695.
411. Swain S, Meher D, Patra CN, Sruti J, Dinda SC, Rao ME. Design and characterization of sustained release mucoadhesive microspheres of tolterodine tartrate. *Current Drug Delivery*. August 2013;10(4):413–426. PubMed PMID: 23215776.
412. Ahmad MZ, Akhter S, Anwar M, Ahmad FJ. Assam Bora rice starch based biocompatible mucoadhesive microsphere for targeted delivery of 5-fluorouracil in colorectal cancer. *Molecular Pharmaceutics*. November 5, 2012;9(11):2986–2994. PubMed PMID: 22994847.
413. Illum L, Williams P, Caston AJ. Adhesive drug delivery composition. EU patent 0442949, 1993.
414. Lim JH, You SK, Baek JS, Hwang CJ, Na YG, Shin SC et al. Surface-modified gemcitabine with mucoadhesive polymer for oral delivery. *Journal of Microencapsulation*. 2012;29(5):487–496. PubMed PMID: 22783823.
415. Patel JK, Chavda JR. Formulation and evaluation of stomach-specific amoxicillin-loaded carbopol-934P mucoadhesive microspheres for anti-Helicobacter pylori therapy. . *Journal of Microencapsulation*. June 2009;26(4):365–376. PubMed PMID: 18720199.
416. Liu Y, Zhang J, Gao Y, Zhu J. Preparation and evaluation of glyceryl monooleate-coated hollow-bioadhesive microspheres for gastroretentive drug delivery. *International Journal of Pharmaceutics*. July 15, 2011;413(1–2):103–109. PubMed PMID: 21540088.
417. Omar S, Aldosari B, Refai H, Gohary OA. Colon-specific drug delivery for mebeverine hydrochloride. *Journal of Drug Targeting*. December 2007;15(10):691–700. PubMed PMID: 18041637.
418. Comoglu T, Gonul N, Dogan A, Basci N. Development and in vitro evaluation of pantoprazole-loaded microspheres. *Drug Delivery*. June 2008;15(5):295–302. PubMed PMID: 18763160.
419. Liu YH, Zhu X, Zhou D, Jin Y, Zhao CY, Zhang ZR et al. pH-sensitive and mucoadhesive microspheres for duodenum-specific drug delivery system. *Drug Development and Industrial Pharmacy*. July 2011;37(7):868–874. PubMed PMID: 21231900.
420. Raffin RP, Colome LM, Pohlmann AR, Guterres SS. Preparation, characterization, and in vivo antiulcer evaluation of pantoprazole-loaded microparticles. *European Journal of Pharmaceutics and Biopharmaceutics: Official Journal of Arbeitsgemeinschaft fur Pharmazeutische Verfahrenstechnik eV*. June 2006;63(2):198–204. PubMed PMID: 16531029.
421. Nilkumhang S, Basit AW. The robustness and flexibility of an emulsion solvent evaporation method to prepare pH-responsive microparticles. *International Journal of Pharmaceutics*. July 30, 2009;377(1–2):135–141. PubMed PMID: 19515519.
422. Lamprecht A, Yamamoto H, Takeuchi H, Kawashima Y. pH-sensitive microsphere delivery increases oral bioavailability of calcitonin. *Journal of Controlled Release*. July 23, 2004;98(1):1–9. PubMed PMID: 15245884.
423. Lamprecht A, Yamamoto H, Takeuchi H, Kawashima Y. Design of pH-sensitive microspheres for the colonic delivery of the immunosuppressive drug tacrolimus. *European Journal of Pharmaceutics and Biopharmaceutics: Official Journal of Arbeitsgemeinschaft fur Pharmazeutische Verfahrenstechnik eV*. July 2004;58(1):37–43. PubMed PMID: 15207535.
424. Sharma M, Sharma V, Panda AK, Majumdar DK. Enteric microsphere formulations of papain for oral delivery. *Yakugaku Zasshi: Journal of the Pharmaceutical Society of Japan*. 2011;131(5):697–709. PubMed PMID: 21532266.

425. Eldridge JH, Meulbroek JA, Staas JK, Tice TR, Gilley RM. Vaccine-containing biodegradable microspheres specifically enter the gut-associated lymphoid tissue following oral administration and induce a disseminated mucosal immune response. *Advances in Experimental Medicine and Biology*. 1989;251:191–202.
426. Eldridge JH, Hammond CJ, Meulbroek JA, Staas JK, Gilley RM, Tice TR. Controlled vaccine release in the gut-associated lymphoid tissues. I. Orally administered biodegradable microspheres target the Peyer's patches. *Journal of Controlled Release*. 1990;11:205–214.
427. Bakan JA. Microencapsulation. In: Lachman L, Lieberman HA, Kanig JL, eds. *The Theory and Practice of Industrial Pharmacy*, 3rd edn. Lea and Feibger, Philadelphia. 1990.
428. Sayed HA, Price JC. Tablet properties and dissolution characteristics of compressed cellulose acetate butyrate microcapsules containing succinyl sulfathiazole. *Drug Development and Industrial Pharmacy*. 1986;12:577–587.
429. Chiao CS, Price JC. Effect of compression pressure on physical properties and dissolution characteristics of disintegrating tablets of propranolol microspheres. *Journal of Microencapsulation*. March–April 1994;11(2):161–170. PubMed PMID: 8006763.
430. Lin SY. Effect of excipients on tablet properties and dissolution behavior of theophylline-tableted microcapsules under different compression forces. *Journal of Pharmaceutical Sciences*. March 1988;77(3):229–232. PubMed PMID: 3373426.
431. Fassihi AR. Consolidation behavior of polymeric substances in non-disintegrating solid matrices. *International Journal of Pharmaceutics*. 1988;44:249–256.
432. Gao Y, Cui FD, Guan Y, Yang L, Wang YS, Zhang LN. Preparation of roxithromycin-polymeric microspheres by the emulsion solvent diffusion method for taste masking. *International Journal of Pharmaceutics*. August 2, 2006;318(1–2):62–69. PubMed PMID: 16647230.
433. Harshada SA, Dharmendra MR, Shyamala B, Sohail A, Gopal SG. Dry suspension formulation of taste masked antibiotic drug for pediatric use. *Journal of Applied Pharmaceutical Science*. 2012;2:1–6.
434. Gupta RK, Relyveld EH, Lindblad EB, Bizzini B, Ben-Efraim S, Gupta CK. Adjuvants—A balance between toxicity and adjuvanticity. *Vaccine*. 1993;11(3):293–306. PubMed PMID: 8447157.
435. Cleland JL. Solvent evaporation processes for the production of controlled release biodegradable microsphere formulations for therapeutics and vaccines. *Biotechnology Progress*. January–February 1998;14(1):102–107. PubMed PMID: 9496674.
436. Cleland JL. Single-administration vaccines: Controlled-release technology to mimic repeated immunizations. *Trends in Biotechnology*. January 1999;17(1):25–29. PubMed PMID: 10098275.
437. Sanchez A, Gupta RK, Alonso MJ, Siber GR, Langer R. Pulsed controlled-released system for potential use in vaccine delivery. *Journal of Pharmaceutical Sciences*. June 1996;85(6):547–552. PubMed PMID: 8773947.
438. Cleland JL. Design and production of single-immunization vaccines using polylactide polyglycolide microsphere systems. In: Powell MF, Newman MJ, eds. *Vaccine Design: The Subunit and Adjuvant Approach*. New York: Plenum Press; 1995.
439. Hanes J, Chiba M, Langer R. Polymer microspheres for vaccine delivery. In: Powell MF, Newman MJ, eds. *Vaccine Design: The Subunit and Adjuvant Approach*. New York: Plenum Press; 1995.
440. Cleland JL, Lim A, Daugherty A, Barron L, Desjardin N, Duenas ET et al. Development of a single-shot subunit vaccine for HIV-1. 5. programmable in vivo autoboost and long lasting neutralizing response. *Journal of Pharmaceutical Sciences*. December 1998;87(12):1489–1495. PubMed PMID: 10189254.
441. Shi L, Caulfield MJ, Chern RT, Wilson RA, Sanyal G, Volkin DB. Pharmaceutical and immunological evaluation of a single-shot hepatitis B vaccine formulated with PLGA microspheres. *Journal of Pharmaceutical Sciences*. April 2002;91(4):1019–1035. PubMed PMID: 11948541.
442. Sturesson C, Artursson P, Ghaderi R, Johansen K, Mirazimi A, Uhnoo I et al. Encapsulation of rotavirus into poly(lactide-co-glycolide) microspheres. *Journal of Controlled Release*. June 2, 1999;59(3):377–389. PubMed PMID: 10332067.
443. Hanes J, Chiba M, Langer R. Degradation of porous poly(anhydride-co-imide) microspheres and implications for controlled macromolecule delivery. *Biomaterials*. January–February 1998;19(1–3):163–172. PubMed PMID: 9678864.
444. Guyton AC, John EH. *Textbook of Medical Physiology*, 12th edn. Philadelphia, PA: Saunders Elsevier; 2010.
445. Men Y, Thomasin C, Merkle HP, Gander B, Corradin G. A single administration of tetanus toxoid in biodegradable microspheres elicits T cell and antibody responses similar or superior to those obtained with aluminum hydroxide. *Vaccine*. May 1995;13(7):683–689. PubMed PMID: 7668038.

446. Moore A, McGuirk P, Adams S, Jones WC, McGee JP, O'Hagan DT et al. Immunization with a soluble recombinant HIV protein entrapped in biodegradable microparticles induces HIV-specific CD8+ cytotoxic T lymphocytes and CD4+ Th1 cells. *Vaccine*. December 1995;13(18):1741–1749. PubMed PMID: 8701587.
447. Nixon DF, Hioe C, Chen PD, Bian Z, Kuebler P, Li ML et al. Synthetic peptides entrapped in microparticles can elicit cytotoxic T cell activity. *Vaccine*. November 1996;14(16):1523–1530. PubMed PMID: 9014294.
448. Barrow EL, Winchester GA, Staas JK, Quenelle DC, Barrow WW. Use of microsphere technology for targeted delivery of rifampin to Mycobacterium tuberculosis-infected macrophages. *Antimicrobial Agents and Chemotherapy*. October 1998;42(10):2682–2689. PubMed PMID: 9756777. Pubmed Central PMCID: 105919.
449. Ramachandra L, Song R, Harding CV. Phagosomes are fully competent antigen-processing organelles that mediate the formation of peptide:class II MHC complexes. *Journal of Immunology*. March 15, 1999;162(6):3263–3272. PubMed PMID: 10092778.
450. Kim KK, Pack DW. Microspheres for Drug Delivery. In: Ferrari M, Lee AP, Lee LJ, eds. *BioMEMS and Biomedical Nanotechnology. Biological and Biomedical Nanotechnology*. Springer-Verlag, New York; 2006.
451. Eldridge JH, Staas JK, Meulbroek JA, Tice TR, Gilley RM. Biodegradable and biocompatible poly(DL-lactide-co-glycolide) microspheres as an adjuvant for staphylococcal enterotoxin B toxoid which enhances the level of toxin-neutralizing antibodies. *Infection and Immunity*. September 1991;59(9):2978–2986. PubMed PMID: 1879922. Pubmed Central PMCID: 258122.
452. Slobbe L, Medlicott N, Lockhart E, Davies N, Tucker I, Razzak M et al. A prolonged immune response to antigen delivered in poly (epsilon-caprolactone) microparticles. *Immunology and Cell Biology*. June 2003;81(3):185–191. PubMed PMID: 12752682.
453. Raghuvanshi RS, Katare YK, Lalwani K, Ali MM, Singh O, Panda AK. Improved immune response from biodegradable polymer particles entrapping tetanus toxoid by use of different immunization protocol and adjuvants. *International Journal of Pharmaceutics*. October 1, 2002;245(1–2):109–121. PubMed PMID: 12270248.
454. Cleland JL, Duenas ET, Park A, Daugherty A, Kahn J, Kowalski J et al. Development of poly-(D,L-lactide—coglycolide) microsphere formulations containing recombinant human vascular endothelial growth factor to promote local angiogenesis. *Journal of Controlled Release*. May 14, 2001;72(1–3):13–24. PubMed PMID: 11389981.
455. Cleland JL, Barron L, Daugherty A, Eastman D, Kensil C, Lim A et al. Development of a single-shot subunit vaccine for HIV-1. 3. Effect of adjuvant and immunization schedule on the duration of the humoral immune response to recombinant MN gp120. *Journal of Pharmaceutical Sciences*. December 1996;85(12):1350–1357. PubMed PMID: 8961152.
456. Kaslow DC, Shiloach J. Production, purification and immunogenicity of a malaria transmission-blocking vaccine candidate: TBV25H expressed in yeast and purified using nickel-NTA agarose. *Biotechnology*. May 1994;12(5):494–499. PubMed PMID: 7764708.

46 Organogels in Controlled Drug Delivery

*V.K. Singh, B. Behera, Sai S. Sagiri, Kunal Pal,
Arfat Anis, and Mrinal K. Bhattacharya*

CONTENTS

46.1 Introduction 1035
46.2 Mechanisms of Organogellation 1036
    46.2.1 Solid Fiber Mechanism 1036
    46.2.2 Fluid Filled Structure Mechanism 1038
    46.2.3 Polymeric Matrix Mechanism 1039
    46.2.4 Microemulsion-Based Mechanism 1040
46.3 Characterization Techniques 1040
    46.3.1 Accelerated Stability Study 1040
    46.3.2 Microscopic Studies 1041
    46.3.3 Molecular Properties 1046
    46.3.4 Mechanical Properties 1047
    46.3.5 Thermal Properties 1053
    46.3.6 Impedance Spectroscopy 1054
    46.3.7 Biocompatibility Studies 1055
46.4 Applications of Organogels 1056
    46.4.1 Topical Delivery 1056
    46.4.2 Dermal and Transdermal 1059
    46.4.3 Parenteral Delivery 1059
    46.4.4 Nasal Delivery 1059
    46.4.5 Oral Delivery 1060
46.5 Conclusion 1060
Acknowledgment 1060
References 1060

46.1 INTRODUCTION

Gels have been defined as the semisolid formulations. Gels usually exhibit viscoelastic properties. Structurally, gels contain two components: one being the solid component and the other being the liquid component.[1] The solid component is regarded as gelator. The gelator undergo self-assembly to subsequently form a three-dimensional (3D) network.[2] The liquid component is accommodated within this 3D fibrous network. The associative forces and the surface active forces (surface tension) amongst the liquid component and the gelator do not allow the leakage of the liquid component.[3] Depending on the polarity of the liquid component, the gels may be broadly classified into two groups, namely, hydrogels and organogels. Hydrogels contain polar liquid whereas organogels contain apolar liquid.[4] Organogels have often been sub-classified as oleogels and amphiphilogels. Oleogels are the formulations that contain vegetable oil (e.g., corn oil, mustard oil, olive oil, sesame oil, and castor oil) or mineral oil (e.g., liquid paraffin) as the liquid component.[5] Both the solid (e.g., span 40 and span 60) and the liquid (e.g., Tween 80 and Tween 20) components in amphiphilogels are surfactant molecules. In the recent past, there has been an exponential increase in the use of organogels as matrices for controlled drug delivery.[2] This may be attributed to the advantages of the organogels over the conventional delivery vehicles. The main advantages of the gels include easy preparation methodology, conversion of the liquid formulations into semisolid formulations, and its ability to accommodate both hydrophilic and hydrophobic drugs.[6] It is also possible to modulate the release behavior of the drugs, incorporated within the gels, by altering the chemical and physical properties of the gels.[7] Since the external phase of the organogels is apolar in nature, the chances of microbial contamination are reduced many folds.[8] Even the biphasic organogels has apolar solvents as the external phase, that is, the aqueous phase is totally protected from the external environment with the apolar phase.[9]

Categorically, the increase in the research on organogels is toward synthesizing and searching of new gelator molecules having the ability to form physical organogels.[10] This is due to the advantages associated with the physical organogels as compared to the organogels formed by chemical reactions. The first and foremost advantage is the easy method of preparation of the physical organogels. Secondly, there are chances of residual monomers in the organogel structure formed by chemical reactions.[11] Since the monomers are usually highly reactive in nature, the unreacted monomers may induce adverse reactions when in contact with the human body. The third advantage of the physical organogels is the thermoreversibility and the shear-thinning properties.[12] These two properties are being well explored in the pharmaceutical, nutraceutical, and cosmetic industries. Thermoreversibility helps in easy preparation of the organogels, while the shear-thinning property improves the shelf-life of the organogels.[13] The current review discusses about the mechanisms of organogel formation, methods of characterization, and applications of the organogels in controlled delivery.

46.2 MECHANISMS OF ORGANOGELLATION

Till date, the gelation mechanisms have been broadly categorized into four groups, *viz.,* solid fiber mechanism, fluid filled structure mechanism, polymer matrix mechanism, and microemulsion based mechanism.[2] The formation of organogels is dependent on the physicochemical properties of the gelator molecules. In general, low molecular weight organogelators (LMOGs) results in the formation of organogels either by solid fiber or fluid filled structure mechanisms.[14] On the other hand, if the gelator molecules are able to undergo polymeric reactions to form a 3D matrix, polymer matrix organogels are formed.[15] If the gelator molecules are dissolved above a critical concentration in the aqueous phase of the water-in-oil microemulsion for inducing gelation, the mechanism of gelation is regarded as microemulsion-based organogels. The current section has devoted to understand the mechanisms of gelation.[16]

46.2.1 SOLID FIBER MECHANISM

The solid fiber mechanism for the preparation of the organogels is the most widely studied organogelation mechanism. As per the mechanism, the solid LMOGs are dissolved in apolar solvents at higher temperatures.[17] As the temperature is lowered, there is a corresponding decrease in the solubility parameter of the gelator molecules due to the decrease in the affinity amongst the gelator molecules and the apolar solvent.[1] This results in the precipitation of the gelator molecules, which initially forms nuclei for the formation of the 3D mesh structure. As the temperature is further lowered, the precipitated gelator molecules start forming fiber like structure, due to aggregation of the precipitated gelator molecules by self-assembly.[18] The fibers radiate out from the nucleation point and form a 3D meshed structure of the gelator molecules. The 3D network structure, hence formed, prevent the free flow of the apolar solvent and thus prevent phase separation.[15] This may be associated with the affinity amongst the gelator molecules and the apolar solvent. The formation of the 3D network of the molecules is dependent on a variety of parameters, namely, physico-chemical properties of the gelator, concentration of the gelator used and the interactions amongst the gelators, and the apolar solvent. The schematic diagram of the mechanism has been shown in Figure 46.1.

Till date, many organogelators have been identified and reported for the development of organogels by solid fiber matrix mechanism. Some of the commonly studied gelators are amino acids,[19,20] dendrimers,[21,22] fatty acids,[23] nucleotides,[24] organometallic compounds,[25] steroids,[26,27] and amide or urea compounds.[28,29] Typically, the critical gelation concentration (CGC), that is, minimum amount of gelator concentration needed to induce gelation, for the preparation of organogels by solid fiber mechanism has been reported to be <15%. In some instances, the gelation may be induced even at concentration as low as 0.1% (e.g., sugar derived gelators). These type of organogelators are referred as super gelators.[30,31] The CGC is dependent on the starting materials of the formulations. The CGC for a gelator molecule may be different for different apolar solvents. Although a wide array of organogelators exists, the mechanism of gelation is simple and follows the same protocol for the formation of the organogels. The organogels formed by solid fiber matrix mechanism are robust and usually result in the formation of formulations with higher stability. The robustness and the higher stability of the organogels may be associated with the formation of junction points in the solid network of gelator molecules. These junction points impart crystallinity to the organogels. This explains the pseudocrystalline nature of the organogels.[32,33] The junction points of the solid network aggregates are held together by inter-molecular physical interactions. The precipitated gelator molecules self-assemble to form aggregates and subsequently form fibers of gelator molecules. As the length of the formed fibers increases, they ultimately form 3D network structure. The self-assembly of the gelator molecules are predominantly driven by hydrogen bonding, van der Waal forces, π-π stacking, and metal-coordination bonds. Hydrogen bonding is the predominant force involved in the formation of the fibers in nonaqueous solvents. Sometimes, van der Waal forces alone may induce gelation, for example, long chain alkanes gelates short chain alkanes and organic solvents.[34] Unfortunately, the gels formed by van der Waal forces are not stable for long periods. The shelf-life of the gels may be increased by increasing the chain length of the organogelator. π-π stacking plays an important role when aromatic ring is present in the backbone of the organogelators (e.g., cholesterol and 3,5-diaminobenzoate derivatives).[35,36] Cholesterol molecules self-assemble primarily via van der Waal's interaction. The presence of additional functional groups in the central backbone may alter the self-assembly mechanism. It has been observed that if hydroxyl group is present at the C3 position, the mechanism of self-assembly is via hydrogen bonding. The stability of

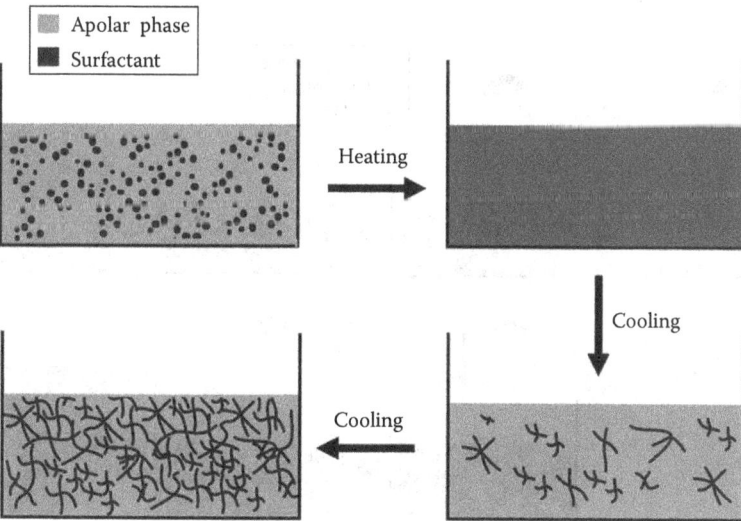

FIGURE 46.1 Mechanism of organogel formation by solid fiber mechanism.

the organogels prepared using C3 substituted hydroxyl derivative of cholesterol has been found to be more than the organogels derived from virgin cholesterol. This may be explained to the synergistic interactions exerted by the van der Waal's interaction and hydrogen bonding.

Chirality is another important factor that may affect the gelation efficiency of the organogelators. Chiral gelators have been observed to be good gelators as against the nonchiral gelators. Chiral gelators help forming helical or twisted fibers with improved gelation and stability of the organogels.[37] It is neither necessary nor sufficient for inducing organogelation.

The growth of the fibers during the formation of organogels is either unidimensional (1D) or bidimensional (2D). Unidimensional growth of the fibers results in the high aspect ratios of length to width ratio whereas the formation of fibers with low aspect ratios is associated with the 2D growth of the fibers. Hence, the formation of long fibers is usually reported in formulations, which promote unidimensional growth of the fibers and formation of microplatelet like arrangement is associated with bidimensional growth of the fibers.[16,38,39] Typical examples of the gelator molecules exhibiting unidimensional and bidimensional fiber growth are L-alanine fatty acid derivatives and hexatriacontane (C36 alkane), respectively.[39,40]

46.2.2 Fluid Filled Structure Mechanism

Fluid filled structure mechanism, like solid fiber mechanism, also involves nonspecific physical interactions for inducing gelation. The physical interactions involved in this type of mechanism are weaker than the solid fibrous gels. This may be attributed to the transient network of organogelators in fluid filled structure based matrices. The junction points in fluid filled structure based matrices are simple physical entanglements as against the microcrystalline domain interaction of solid fiber gels.[41] Due to this reason, the crystallinity of the solid fiber gels is higher than the fluid filled gels. This results in the robust physical properties of the solid fibrous gels as compared to transient and dynamic nature of the fluid filled gels. The transient and dynamic nature of fluid filled gels is due to the exchange of the gelator molecules amongst the fluid filled structures and the solvent.[15] The transient network of the fluid matrices hence allows or facilitates the incorporation of water or polar solvent within the matrix.[42] The schematic representation of the fluid filled structure mechanism has been shown in Figure 46.2.

Lecithin organogels (LOs) are the best examples of polar solvent containing organogels. As lecithin is mixed with apolar solvents, lecithin self-assembles to form isotropic reverse micelles. With the addition of water or polar solvent to the mixture, the reverse micelles start accommodating water

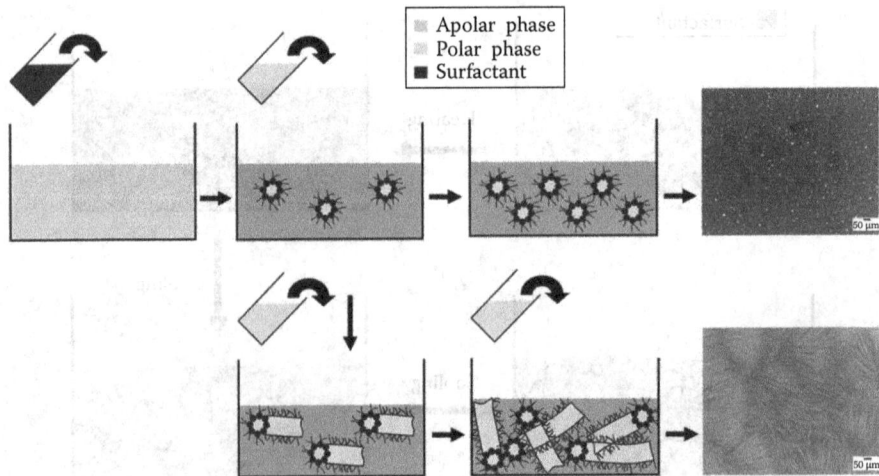

FIGURE 46.2 Mechanism of organogel formation by fluid filled structures.

within the core. Depending on the concentration of lecithin in the formulation, the reverse micelles either remain as spherical droplets or form reverse micellar tubules. Typically, the reverse micellar tubules are formed at higher concentrations of lecithin. The formation of water filled structures is associated with the formation of hydrogen bonds amongst the phosphate groups of lecithin and the polar solvent. The fluid filled structures entangle with each other and form a networked architecture.[43] The polar solvent may either be water, glycerol, formamide, and ethylene glycol.[44,45] Every polar solvents cannot induce gelation (e.g., ethyl alcohol and diethylene glycol).[43] In general, solvents possessing hydrogen atom donating ability readily form gels rather than the solvents having hydrogen atom accepting ability.[43,46]

Fatty acid derived organogeators (e.g., stearic acid, 12-hydroxy stearic acid [HSA], sorbitan monostearate (Span 60), sorbitan monopalmitate [Span 40]) can form both anhydrous and water containing organogels. In general, these organogelators are being termed as amphiphiles. These amphiphiles can form organogels with or without polar solvent. The organogelators are solubilized in hot apolar solvent at a temperature higher than their melting point and subsequently cooled to room-temperature. During cooling, the organogelators self-assemble to form network of fibers.[47] In case of polar solvent containing organogels, polar solvent is added drop-wise into the hot gelator-solvent mixture under homogenization.[3] Classification of these gels is tricky as they resemble the solid fibrous gels in the method of preparation and microstructure. But there is a dynamic movement of the gelator molecules between the bulk liquid and fibrous network of these organogels. This suggests the transient nature of the organogels, which is a primary feature of fluid fiber matrix organogels. In addition to this, the ability to incorporate water or polar solvents within gel architecture indicates the fluidity of the network. The water phase is accommodated within the opposing polar groups of the amphiphilic bilayer.[48] These features enable them to be grouped under fluid fiber matrix organogels.

Span 80-Tween 80 based organogels also come under this category. Span 80 (sorbitan monooleate) and Tween 80 (polyoxyethylene sorbitan monooleate) when mixed in 1:2 proportion, works best as an organogelator when dissolved in vegetable oils (e.g., sunflower oil and palm oil).[1,49] Drop-wise addition of water into the gelator/co-gelator-oil mixture results in the formation of spherical reverse-micellar droplets or reverse micellar fluid filled fibers. These droplets/fibers undergo self-assembly to induce gelation. Similarly, addition of water to the hot Span 60 and Span 40 leads to the formation of gels via fluid filled structure mechanism.[50]

46.2.3 POLYMERIC MATRIX MECHANISM

Chemically active organogelators are responsible for the formation of 3D network structure and gelation of apolar solvents. The mechanism of formation of organogel has been shown schematically in Figure 46.3. These organogels possess remarkable stability due to the polymerization of the organogelators. Polymeric gels are being formed from linear to hyperbranched and star shaped polymers. The organogelators undergo polymerization via chemical and/or physical interactions and entrap the organic solvent within the crosslinked matrices. Polymerization of the organogelators is being governed by stereoregularity of the organogelators and supramolecular self-assembly

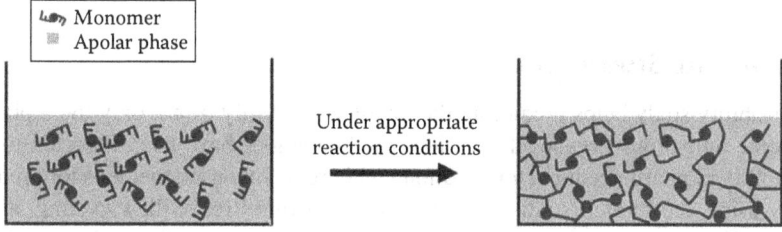

FIGURE 46.3 Mechanism of organogel formation by polymerization reaction.

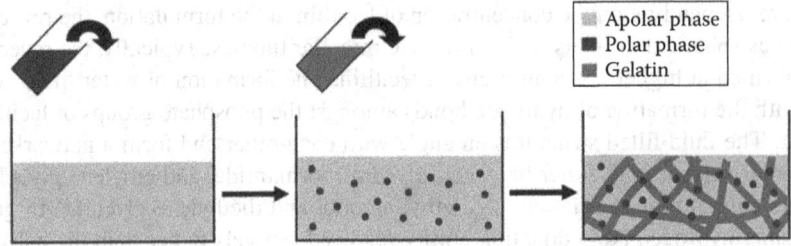

FIGURE 46.4 Mechanism of organogel formation by gelation of microemulsions.

in apolar solvents. Due to this reason, quite often, polymeric organogels are termed as supramolecular organogels. Polymeric organogelators involve physical and/or chemical interactions in the formation of organogels.

L-lysine derivatives undergo polymerization via hydrogen bonding and van der Waals forces to form supramolecular organogels. Poly (ethylene) organogels are formed by dissolving the polymer in apolar solvent at 130°C and subsequently "shock cooled." This leads to the partial precipitation of the polymer to form colorless organogel.[51] Similarly, cyclohexane derivatives,[52] diacetylene derivatives,[53] poly (acrylic acid), and methacrylate derivates[54] also undergo self-assembly during polymerization to form organogels. Interestingly, polymeric urea derivatives can induce gelation for both polar and apolar solvents and hence are termed as ambidextrous gelators.[55] Till date, polymeric organogelators have found limited applications in the field of drug delivery with exceptions such as poly (ethylene), poly (methacrylic acid), and methyl methacrylic acid co-polymers and poly (glyceryl methacrylic acid) polymers.[56]

46.2.4 MICROEMULSION-BASED MECHANISM

Microemulsion-based mechanism of organogelation is a relatively new methodology for developing organogels (Figure 46.4). Gelatin and lecithin are the only organogelators reported to develop organogels via this mechanism. Water-in-oil microemulsions are stabilized by gelatin and anionic surfactants (AOT, Triton-X-100, and Tween 85). Gelatin stabilizes the aqueous phase of the microemulsion thereby leading to the formation of organogels. Nonionic surfactants alone cannot form the organogels, but when used in combination with gelatin yields stable organogels. These organogels are electrically conductive and have been explored for ionotropic drug delivery.[57] Microemulsion-based lecithin organogels have also been reported. The applications of these organogels in controlled release formulations are yet to be explored.

46.3 CHARACTERIZATION TECHNIQUES

The developed organogels are thoroughly characterized to determine their suitability to be used as matrices for controlled delivery applications. This section describes the different methodologies employed for the characterizations of the organogels.

46.3.1 ACCELERATED STABILITY STUDY

Accelerated stability study helps predicting the long-term stability of not only the biphasic organogels but also the anhydrous organogels. Various methodologies have been reported to predict the long-term stability of pharmaceutical formulations. The frequently used methods for predicting long-term stability using accelerated studies include thermocycling, freeze/thaw cycling, and syneresis measurements. Thermocycling and freeze-thaw cycling tests involve alternative incubation of the samples at high and low temperatures. It is expected that the oxidative reactions will be prevalent

Organogels in Controlled Drug Delivery

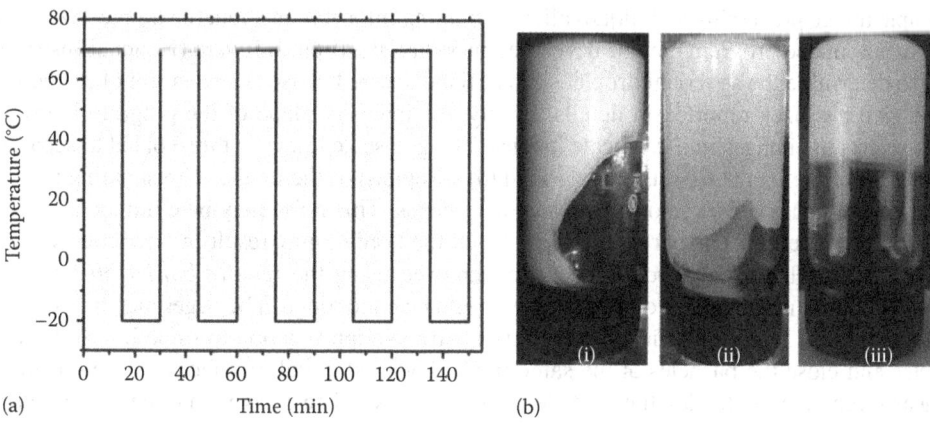

FIGURE 46.5 Accelerated stability study by freeze-thaw cycling. (a) Schematic representation of temperature variation and (b) Representative formulations: (i–ii) Destabilized gels and (iii) Stable gel.

at higher temperatures whereas at lower temperatures the physical damage to the interfacial layer is expected to be more prevalent. Many scientists have reported different protocols for conducting these tests. In general, thermocycling study is conducted by alternatively incubating the samples at 4°C and 45°C at an interval of 48 h. The test is conducted for 1 month.[58,59] The cyclic freeze-thaw test is conducted by alternatively incubating the samples at temperatures <−10°C and >10°C for predetermined periods. The samples are subjected to at least five cycles of freeze–thaw (Figure 46.5).[60] The syneresis of the liquid components is often also tested by centrifugation. In this test, the freeze-thawed samples are tested for syneresis after centrifugation at 100 × g for 15 min. The separated liquid, if any, is quantified by determining percentage syneresis, which provides information about the stability of the formulations.[61] Similar types of studies have also been extensively reported with slight modifications in the protocol.[62,63]

46.3.2 Microscopic Studies

Microscopy is a useful tool commonly used to visualize the microarchitecture of the formulations. The microscopic techniques have been broadly categorized into three groups, namely, optical (bright field microscopy, phase contrast microscopy, confocal laser scanning microscopy, fluorescent microscopy), electron (scanning electron microscopy), and scanning probe microscopy (atomic force microscopy). The optical and electron microscopic techniques involve the interaction of the electromagnetic radiation/electron beam with the specimen. The image is reconstructed by capturing and processing the signals associated with the scattering/transmitted/reflected radiations. On the other hand, the scanning probe microscopy is associated with the interaction of the sample interface with the scanning probe head.

Bright field microscopy is the simplest among all the microscopic techniques and is the most commonly used methodology to study the microstructures of the organogels. Transmitted white light illuminates the sample (present as a thin film) from below and the transmitted light is observed from the eye piece. Though the technique is simple, the bright field microscopes provide low contrast and low resolution images of the samples analyzed. The maximum magnification that can be achieved using this technique is only up to 1000 to 2000 times. The microstructure of the organogels (devoid of water) prepared with low-molecular weight gelator molecules shows the formation of 3D network structure made up of gelator fibers (Figure 46.7a). Many water containing organogel samples show the presence of spherical water droplets dispersed in an oil continuum phase (Figure 46.7b). Depending on the concentration and the nature of the gelator, fluid filled fibers may also be obtained.

Simple image processing techniques allow extracting information about the droplet size distribution. An automated program may be developed in Vision Assistant software (National Instruments, USA) to determine the spherical droplet sizes of the dispersed phase. The concept of extracting the droplet size has been reported in details. Briefly, the intensity plane of the program is extracted. This converts the images into grayscale images. The grayscale image is thresholded to separate the background of the image from the foreground (the droplets) of the image. Thresholding may introduce high frequency noises into the thresholded images. This noise may be eliminated by *remove small objects* function. The particles, which touch the border, may result in erroneous size distribution analysis. Hence the border objects are removed using the *remove border object* function. During thresholding, the edges of the particles might become rough. The edges may be smoothened using *convex hull* function, which uses a convex extrapolation function to smooth the edges of the particles and close the particles at the same time. Heywood circularity factor is used to filter out the near circular particles for the particle measurements.[18] The images may be calibrated using scaling information present either in the image itself or in a separate image. The same operations may be performed using the free image processing software ImageJ (NIH, USA). The only disadvantage of the ImageJ software is that all the steps have to be repeated for each and every image unlike in Vision Assistant software where a program is developed for analyzing multiple images (Figure 46.6).

Phase contrast microscopy allows improving the quality of the bright field microscopic images by increasing the contrast of the image (Figures 46.7c and 46.7d). In this technique, the analysis of the differences in the refractive indices of the different phases present in the samples is used to

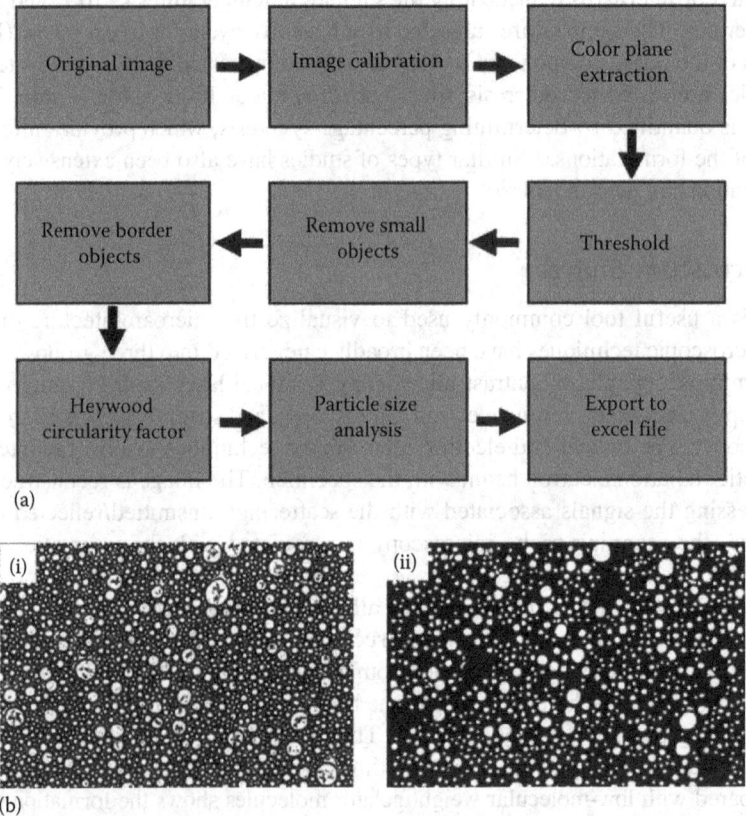

FIGURE 46.6 Particle size analysis using microscopic images. (a) Flowchart for particle size analysis and (b) Micrographs: (i) Original micrograph before analysis and (ii) Processed micrograph for size analysis.

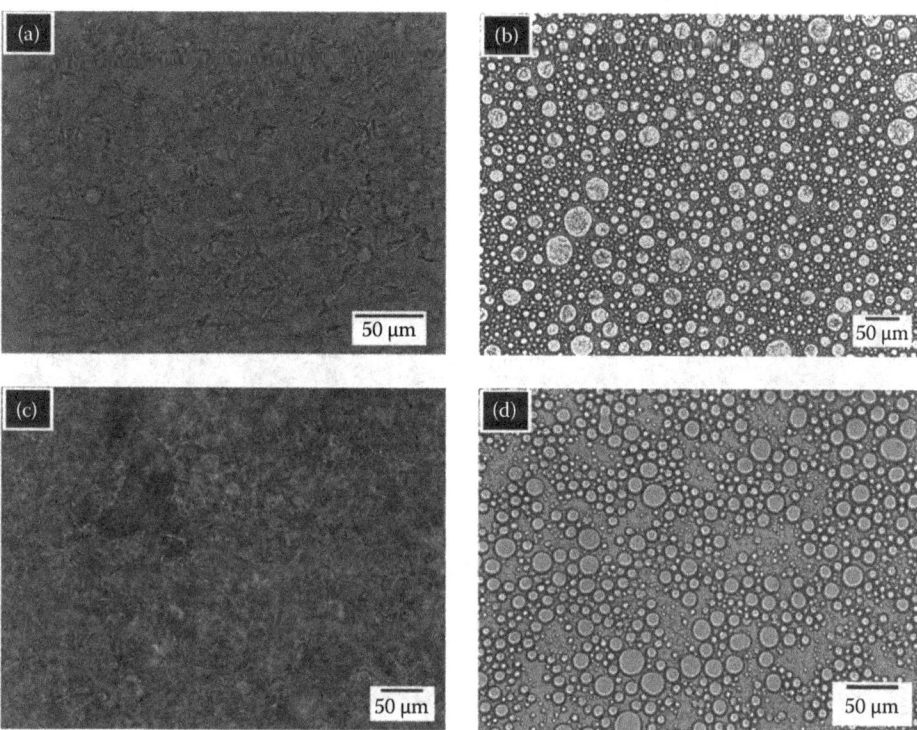

FIGURE 46.7 Bright field micrographs of (a) Organogel formed by solid fiber mechanism showing fiber structures and (b) Organogel formed by fluid filled mechanism showing globular structures; Phase contrast micrographs of (c) Organogel formed by solid fiber mechanism showing fiber structures and (d) Organogel formed by fluid filled mechanism showing globular structures.

manipulate the differences in terms of brightness in the image. This results in the increase in the contrast of the image.[64] Due to this reason, this microscopic technique allows visualizing finer structural details of the samples even from apparently looking transparent samples.[64,65]

Fluorescent microscopy (conventional and confocal) has been extensively studied to understand the distribution of different phases in the organogels. This technique is associated with the dissolution of a suitable fluorescent dye (s.a. fluoral yellow, rhodamine, FITC, and fluorescein) either in the apolar phase or in the aqueous phase. The fluorescent dyes absorb light waves of a particular wavelength and subsequently emits visible light of another wavelength. The emitted light waves are detected, which appear as a bright image against a dark background. Due to this reason, this microscopic technique helps indicating the distribution of the liquid phase in which the fluorescent dye has been dissolved. Figure 46.8a shows the fluorescence image of an organogel prepared by solid fiber mechanism. The solid fibers appear as black fibrous structure suggesting that the gelator fibers did not stain with the dye. The oil, containing fluorol yellow, fluorescence and appears as bright region. Figure 46.8b shows the fluorescent image of emulsion organogel. The aqueous phase did not contain any dye. Hence, the inner aqueous layer appeared as black droplets. The black droplets formed an interconnecting network. The confocal fluoresce microscopy works in a similar principal of fluorescence microscope but the resolution of the micrographs are many fold better. This may be associated with the use of laser source (which emits visible light of single wavelength) to excite the fluorophore. This eliminates the chances cross-fluorescence, which is a common phenomenon in fluorescence microscopy thereby improving the quality of the micrographs. The additional feature of the confocal fluorescence microscope that helps improving the quality of the image is the ability of the microscope to acquire the fluorescence signals from only the focal plane. This not only improves the quality of the image, but also allows reconstruction

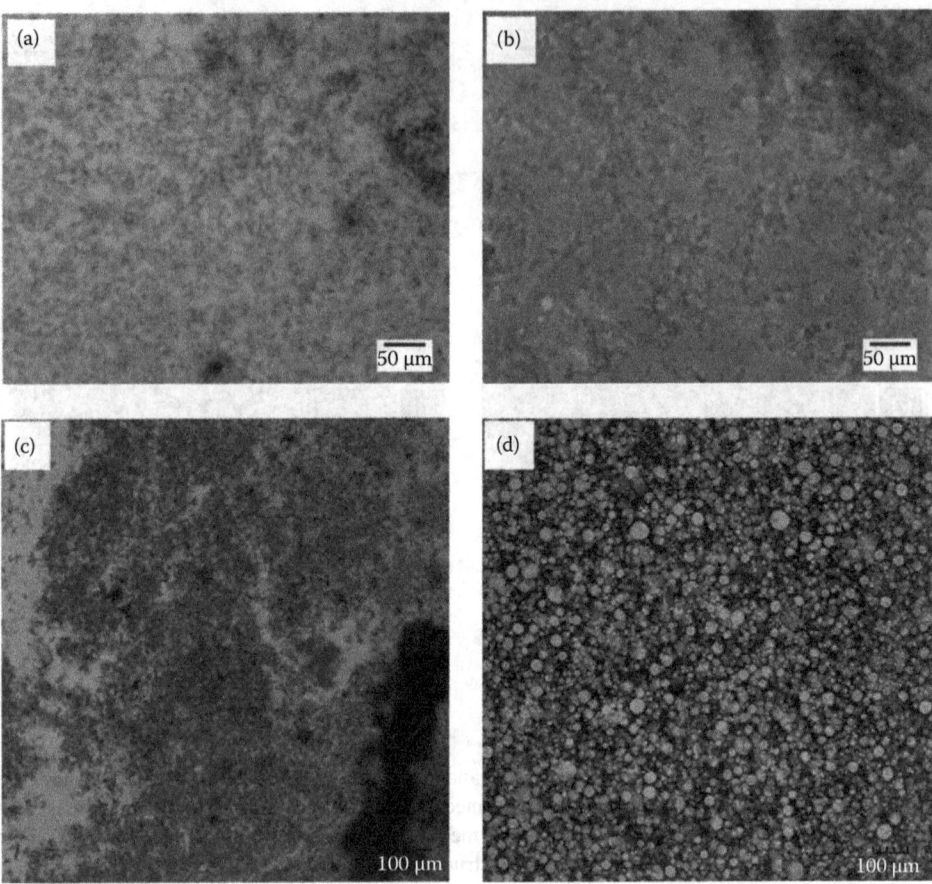

FIGURE 46.8 Fluorescence micrographs of (a) Organogel formed by solid fiber mechanism showing fiber structures and (b) Organogel formed by fluid filled mechanism showing globular structures. Confocal scanning laser micrographs of (c) Organogel formed by solid fiber mechanism showing fiber structures and (d) Organogel formed by fluid filled mechanism showing globular structures.

of 3D images from the stack of images taken from the various focal planes. Figures 46.8c and 46.8d show confocal microscopic images of organogels. Figure 46.8c is the confocal image of the organogel formed by the solid fiber mechanism. The oil phase has been stained with fluorol yellow. The micrograph shows the presence of black colored fibrillar structures that formed interconnecting network. The oil phase is immobilized within the 3D network formed. Figure 46.8d shows the confocal image of the fluid filled organogel whose aqueous phase has been stained with rhodamine dye (water soluble fluorophore). Since the aqueous phase has been stained, the aqueous phase appears as bright regions. The micrographs show that the aqueous phase appears as droplets and forms a 3D interconnecting structure. The oil phase (appearing as dark region) is immobilized in the 3D network.

The organogels may also be visualized under scanning electron microscope (SEM), environmental scanning electron microscope (ESEM), and atomic force microscope (AFM). The organogels are converted into xerogels before analysis under SEM and atomic force microscope (contact mode). Xerogels may be defined as the gel matrices, which is devoid of the solvent phase of the gels. The organogels are converted into xerogels by washing with organic liquids. Figure 46.9a shows the SEM micrograph of the organogel (xerogel) prepared by solid fiber mechanism. The solid gelator network can be clearly identified. Similar observations can also be made in the AFM micrograph (Figure 46.9c). Figure 46.9b shows the SEM micrograph of

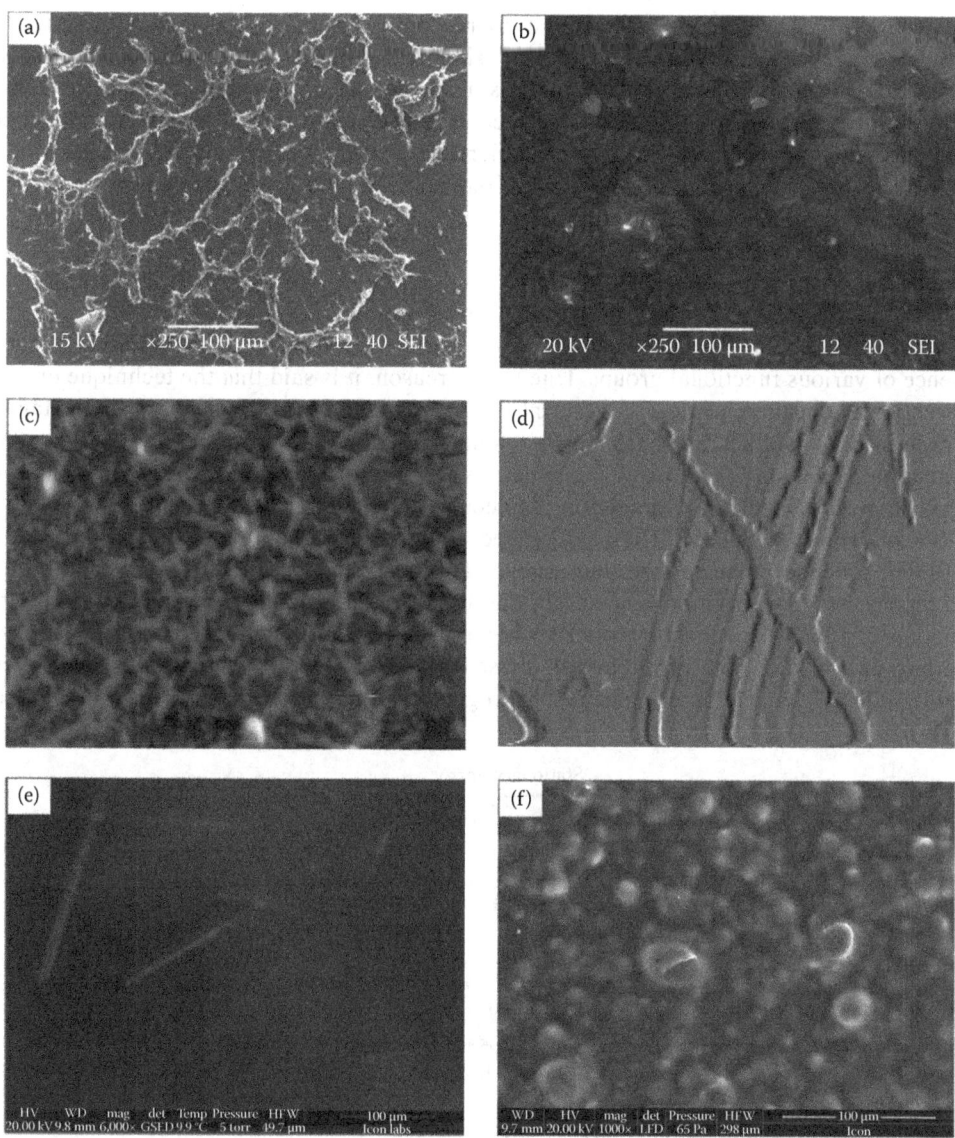

FIGURE 46.9 Scanning electron micrographs of (a) Organogel formed by solid fiber mechanism showing fiber structures and (b) Organogel formed by fluid filled mechanism showing globular structures. Atomic force micrographs of (c) Organogel formed by solid fiber mechanism showing fiber structures and (d) Organogel formed by fluid filled mechanism showing globular structures. Environmental scanning electron micrographs of (e) Organogel formed by solid fiber mechanism showing fiber structures and (f) Organogel formed by fluid filled mechanism showing globular structures.

organogel (xerogel) prepared by fluid filled mechanism using liquid gelator. The micrograph shows the presence of lamellar structure formed by the surfactant molecules. Figure 46.9d shows the AFM image of the fluid filled organogel (xerogel) prepared with liquid organogelator. The image shows the presence of long fiber-like interconnecting structures. The ESEM micrograph of the solid fiber organogel shows the presence of the fibrillar structure of the organogelators (Figure 46.9e). But the microstructure of the fluid filled organogels is completely different from what is seen from the SEM and AFM studies. The ESEM micrograph shows the presence of globular structures, which form a 3D architecture (Figure 46.9f). Similar observations are also made in the bright field and phase contrast microscopy. This may be associated with the fact that as the organogel is converted into xerogel using the harsh solvent treatment during analysis under SEM and AFM (contact mode), there is a drastic change in the molecular architecture due to the rearrangement of the gelator molecules. Hence, care should be taken while choosing the microscopic techniques for the characterization of the organogels. The analysis of the organogels is preferred using ESEM and AFM (noncontact mode) where there is no need for the conversion of the organogels into xerogels.

46.3.3 Molecular Properties

Infrared spectroscopic technique has been extensively used to have an understanding about the presence of various functional groups. Due to this reason, it is said that the technique provides information about the chemical fingerprint of the compound. It also helps in understanding the chemical interactions amongst the functional groups of the different components of the formulation. Based on the absorption characteristics of a formulation, the identification of the functional groups, molecular interactions, molecular compatibility, and chemical modifications can be confirmed.[66,67] The FTIR spectroscope allows analysis both in transmission (Figure 46.10) and reflection modes. Inter/intra-molecular hydrogen bonding and π–π stacking have been reported to play an important role in the formation of organogels.[68] Fan et al.[69] reported formation of tunable mechano-responsive organogels by ring-opening copolymerizations of N-carboxyanhydrides. The properties of these organogel systems were extensively investigated by ATR-FTIR. The intermolecular hydrogen bonding has been reported to be higher in

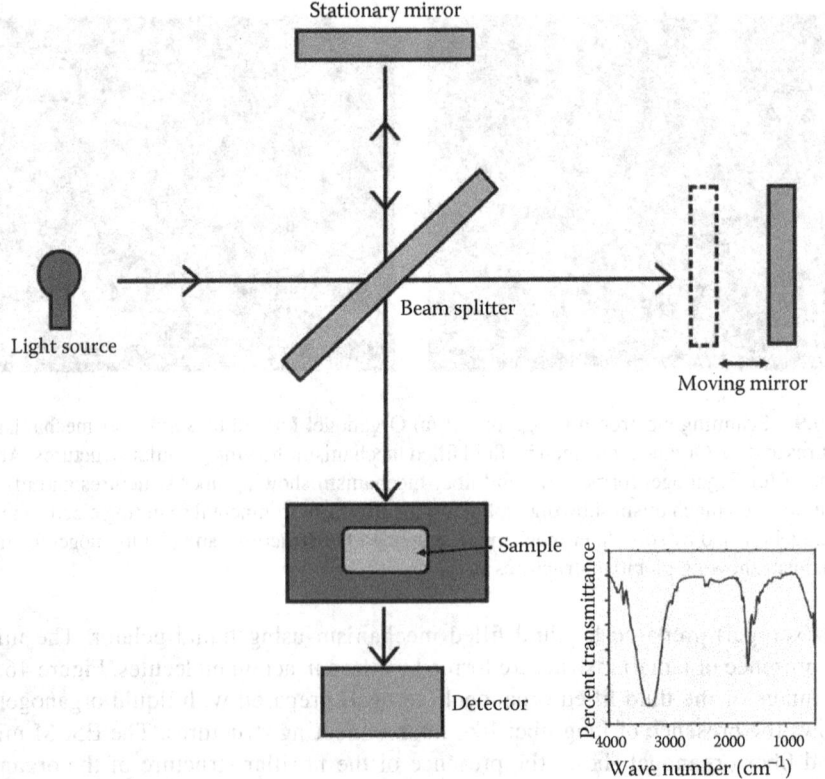

FIGURE 46.10 Schematic diagram for FTIR spectroscope and a representative FTIR spectrum of water containing organogel.

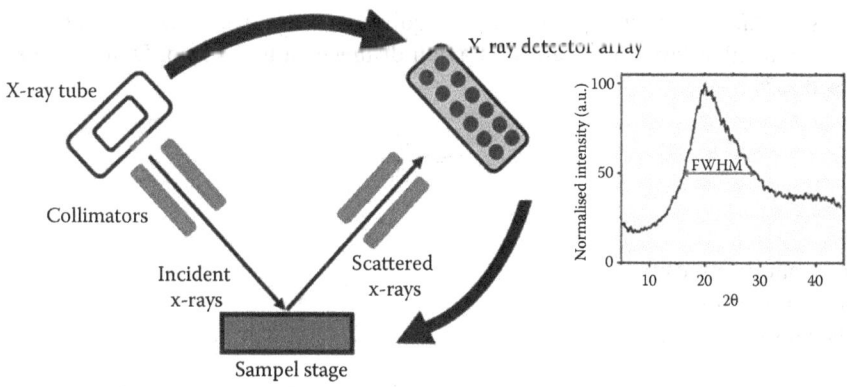

FIGURE 46.11 Schematic diagram for XRD and a representative XRD profile of an organogel.

organogels containing water. The peak due to hydrogen bonding appears at ~3200 cm^{-1} (shown as insert in Figure 46.10). In general, stronger hydrogen bonding results in the higher intensity of the peak.

X-ray diffraction (XRD) is one of the nondestructive tools to analyze the physico-chemical properties of the samples. The interaction of x-ray beam with the atoms present in the different planes is analyzed in this study (Figure 46.11). When an x-ray beam is diffracted from the sample, the distances between the planes of the atoms (d) can be theoretically calculated by applying Bragg's Law:

$$n\lambda = 2d \sin \theta \tag{46.1}$$

where
- n is the integer representing the order of the diffraction peak
- λ is the wavelength of the x-ray
- d is inter-plane distance of (i.e., atoms, ions, molecules) (the d-spacings)
- θ is the angle of incidence of the x-ray beam

In the above equation: λ is known and θ and n can be obtained from the XRD profile. Hence, the d-spacing can be easily calculated. The d-spacings are characteristic for an individual material and are regarded as the "molecular fingerprint" of the material, which helps in identification of the materials. Amorphous materials like organogels do not produce a distinct diffraction patterns, instead, they produce a broad peak. The broad peak associated with the organogels may be explained by the diffraction of the x-rays from the crystallite domains, which are randomly arranged and are far from each other. Change in the processing parameters of the organogels and/or addition of another component in the organogels might alter the crystallite properties and hence the crystallinity of the organogels. To have an understanding of the change in the crystallinity, the XRD profiles are normalized first. The width of the normalized profile at 50% intensity is regarded as full-width-at-half-maximum (FWHM) (shown as insert in Figure 46.11). FWHM is inversely related to the crystallinity, that is, lower the FWHM and higher is the crystallinity and vice-versa.

46.3.4 MECHANICAL PROPERTIES

In general, viscosity is defined as the fundamental property of liquids by virtue of which they show resistance to flow or shear. Viscosity has been broadly classified into two groups, namely, dynamic viscosity and kinematic viscosity. Dynamic viscosity (also known as absolute viscosity) is defined as the tangential force per unit area required to move one horizontal layer of the liquid with respect to another layer, maintained at a unit distance, at unit speed. Dynamic viscosity can be expressed as[70]

$$\tau = \eta \frac{\partial u}{\partial y} \qquad (46.2)$$

where
  η is the dynamic viscosity
  τ is the shearing stress
  $\partial u/\partial y$ is the shear velocity

Kinematic viscosity is defined as the ratio of dynamic viscosity of a fluid to its mass density. It is expressed as[71]

$$\nu = \frac{\eta}{\rho} \qquad (46.3)$$

where
  ν is the kinematic viscosity
  η is the dynamic viscosity
  ρ is the mass density

Based on viscosity of the samples, the flow of samples is broadly classified into three categories, namely, Newtonian, time independent non-Newtonian and time dependent non-Newtonian. Newtonian fluids show shear stress independent constant viscosity profile where as non-Newtonian fluids show a viscosity profile, which is dependent on the shear force and time. In time independent non-Newtonian fluids, the shear stress does not vary proportionally to the shear rate. The time independent non-Newtonian fluids show mainly three types of flow. A decreasing viscosity with an increase of shear rate is called shear thinning or pseudoplastic flow (Figure 46.12a). An increasing viscosity with an increase of shear rate is called shear thickening or dilatant flow. Some fluids need application of certain amount of force before any flow is induced that are known as Bingham plastics.

Time dependent non-Newtonian fluids show a change in viscosity with time under constant shear rate. They are further classified as thixotropic, showing decrease in viscosity with time, and as rheopectic, showing an increase in viscosity with time at constant shear rate.

The rheological profile obtained from the viscometer is fitted with Ostwald-de Waele Power law model. This mathematical model has been reported to have lowest standard error. The power law mathematical model is given by[72]

$$\tau = K \cdot \gamma^n \qquad (46.4)$$

where
  τ is the shear stress (Pa) at γ shear rate (s^{-1})
  K is the flow consistency index (Pa·s)
  n is the power law indices

The "n" value indicates the type of flow behavior. The formulations having "n" = 1 indicates a Newtonian flow behavior, $n < 1$ signifies pseudoplastic flow whereas $n > 1$ indicates a dilatant flow.[73] In general, physical organogels shows "n" values <1 suggesting pseudoplastic nature of the gels.[74–76]

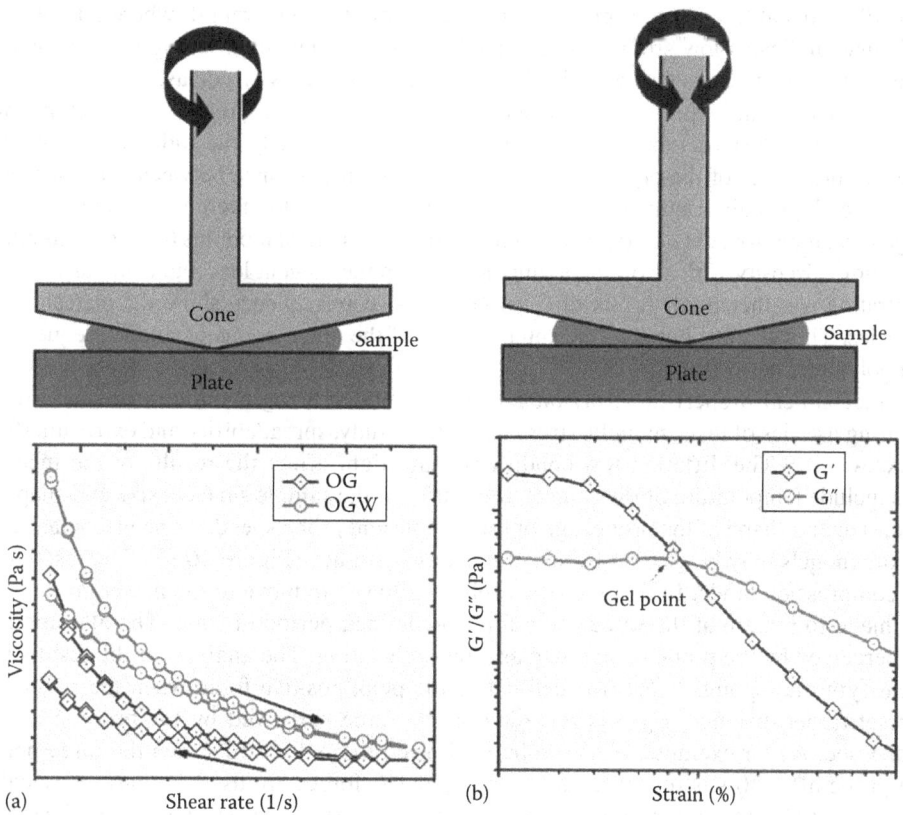

FIGURE 46.12 Mechanical behaviour of the organogels. (a) Viscosity and (b) Rheology.

Rheology is defined as the science of flow and small-scale deformation of materials under applied forces. The term is derived from the Greek words *"rheo"* meaning "to flow" and *"logos"* means "science."[12,77–79] Rheological measurements are performed by applying strain or stress at a constant frequency (e.g., 1 Hz). It is also possible to carry out rheological studies in the frequency sweep mode (e.g., between 0.1 and 100 Hz). The modulus of a material (G^*) is a measure of its stiffness and is defined as the ratio of stress to strain. G^* can be divided into two parts, *viz.*, storage modulus (G') and loss modulus (G''). Mathematically, G^* is expressed as

$$G^* = G' + iG'' \tag{46.5}$$

where
  G' describes the elastic properties and reflects the energy stored in the cycle
  G'' describes the viscous properties and reflects the energy lost in the cycle

As mentioned above, the rheological profile of a material may be studied either in the strain sweep mode or in the frequency sweep mode (Figure 46.12b). The strain sweep (generally performed from 0.01% to 500% at a frequency of 1 rad/s) gives information about the elastic modulus G', the viscous modulus G'' and the phase angle d. A larger value of G' in comparison to G'' indicates elastic properties of the sample. A phase angle of 0° means a perfectly elastic material and a phase angle of 90° means a perfectly viscous material. The frequency sweep (performed between

0.1 and 100 rad/s at fixed strain) gives information about the gel strength where a large slope of the G' curve indicates low strength and a small slope indicates high strength of the sample. A sample possessing both viscous and elastic character is termed as viscoelastic.

Toro-Vazquez et al.[80] reported the influence of the type of oil phase on the self-assembly process of γ-oryzanol+β-sitosterol tubules in organogel systems. In the study, the authors observed that the rheological properties of the organogels were dependent on a balance between hydrogen-bonding sites and the alkyl chain length. The molecular weight of the gelator altered the crystal growth during organogelation. Jones et al.[81] reported an increase in the storage modulus (G'), loss modulus (G'') and dynamic viscosity with a corresponding decrease in the tangent loss (tan delta) as the polymer concentration was increased. A plot of G' or G'' and sweep frequency showed a plateau at higher frequencies. This was attributed to the entanglement of the polymer chains and subsequent formation of polymeric network.

The mechanical properties of organogel can be studied using a static mechanical tester by performing a series of tests, namely, stress relaxation study, spreadability and extrusion (forward and backward).[82] The different test conditions that might affect the results of the mechanical studies include temperature of the sample, regularity of the sample surface, size and shape of the sample, size and shape of the probe, age of the sample and probe speed.[83] The viscoelastic nature of the organogels may be studied by stress relaxation studies (Figure 46.13). The test is carried out in compression mode. In this test, the probe is allowed to move to the preset target distance and is made to remain at the same place for a predefined period of time. The alteration in the stress perceived by the probe is regarded as stress relaxation. The analysis of the results is done by identifying few points.[84] $F_0(g)$ is defined as the peak positive force when the probe attains the preset target distance. $F_x(g)$ is regarded as the force perceived by the probe after a hold time of x sec. As for example, if the probe is allowed to hold for 30 s then the force perceived by the probe after 30 s is regarded as $F_{30}(g)$. F_0 and F_x forces are used as markers to calculate relaxation and % stress relaxation (% SR). The area under the curve (work done during relaxation) in between F_0 and F_x is calculated and recorded as the work done on relaxation (g.sec) of

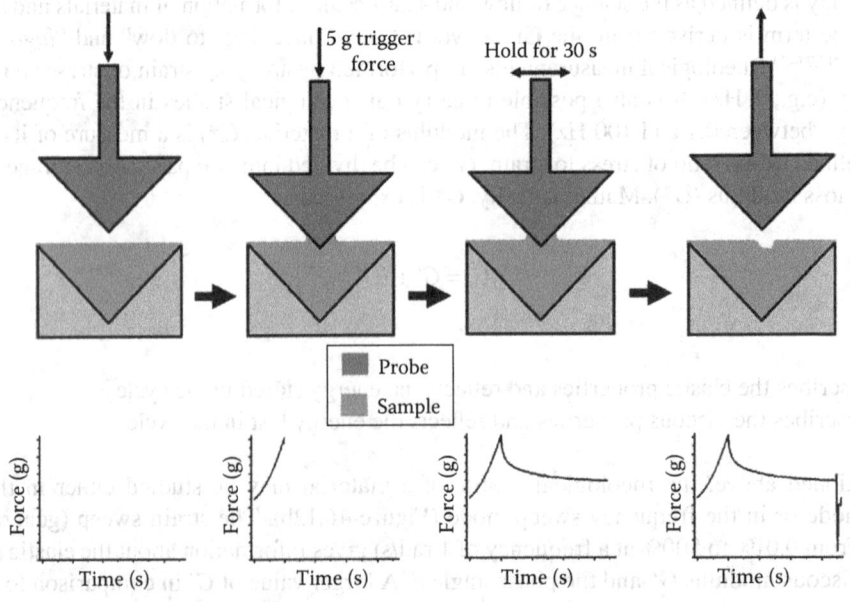

FIGURE 46.13   (See color insert.) Schematic diagram of stress relaxation study.

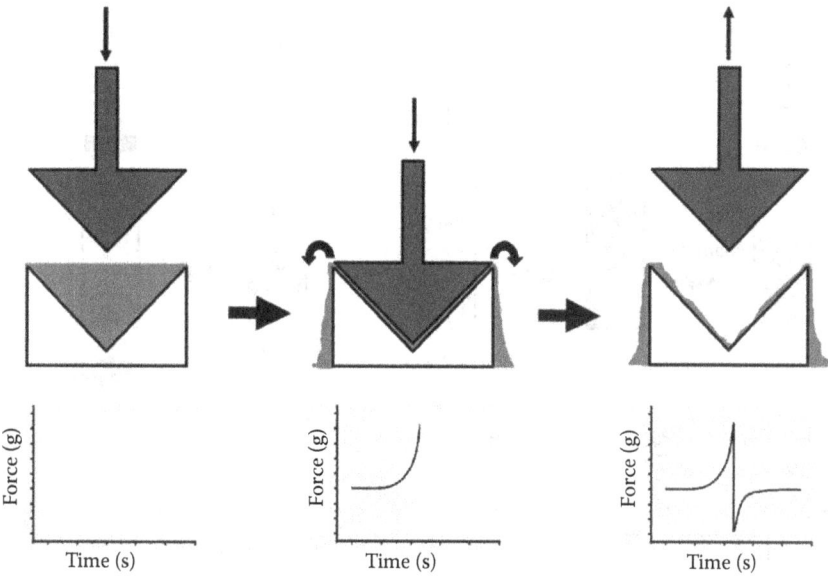

FIGURE 46.14 Schematic diagram of spreadability study.

the sample. The percent decrease in the force is regarded as %SR of the formulation.% SR is calculated using the following formula:[85]

$$\%SR = \frac{F_0 - F_x}{F_0} \times 100 \quad (46.6)$$

D_x (mm) is the distance travelled by the probe to attain x g of force. Usually, D_{25} is determined, i.e., the distance travelled by the probe to attain 25 g of force. D_x value provides information about the firmness of the gels. Lower D_x value indicates that for attaining x g of the force, the probe has to move a smaller distance. This can only happen when the firmness of the formulation is higher. The slope of the linear region of the stress-strain plot is defined as the modulus of deformability (g/s).[86]

Spreadability study of the formulations is performed by using the spreadability fixture (45° perpex conical probe). In this study, the male cone is allowed to move a preset distance into the female cone, filled with the formulations (Figure 46.14). This results in the formation of annular region between the male cone and the female cone. The formulations are forced to come out of the female cone through the formed annular region. The spreadability of the formulation is inversely related to the firmness. Higher the firmness of the formulation, lesser is the spreadability.[87] The positive peak force and the negative peak forces are defined as the firmness and the stickiness of the formulations, respectively. The positive and the negative area under the curve are defined as the cohesiveness and adhesiveness of the formulations, respectively.[88]

The extrusion test employs applying force to the sample until it starts flowing through an outlet. The maximum force required to extrude the sample is the measure of index of sample quality. Extrusion test may be conducted by two methods, one is forward extrusion study and the other is backward extrusion study. In forward extrusion study, the formulation is forced out of an annulus (a central hole) in the container as the formulation is pushed using a tightly fit plunger (Figure 46.15). The positive peak is the minimum force needed to start extrusion and is regarded as the firmness of the formulation. Thereafter, a plateau is generally obtained indicating the force needed to continue the extrusion remains same. The unevenness of the plateau may be associated with the heterogeneous nature of the formulations while passing through the annulus at any particular time.

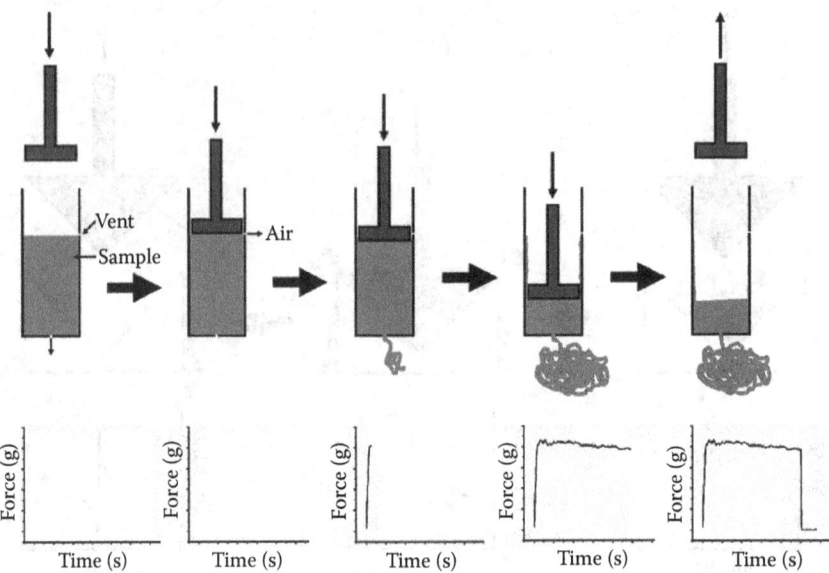

FIGURE 46.15 Schematic diagram of forward extrusion study.

Sometimes this unevenness may also happen due to the release of air pockets created during sample preparation. The total area under the extrusion curve is known as work of extrusion.

In backward extrusion study, the formulation is forced to move out of the annular ring formed in between the inner diameter of the sample container and the outer diameter of the plunger (Figure 46.16). As the plunger moves down into the formulation, there is an increase in the force perceived by the probe until the sample starts flowing through the annular ring. Thereafter, a plateau phase is seen. The average force achieved during the plateau phase in the force-time plot is regarded as index of firmness. The work done, that is, the area under the positive peak curve is regarded as consistency of the sample. The negative peak force gives the cohesiveness and the negative area under the curve gives the index of viscosity of the gels.

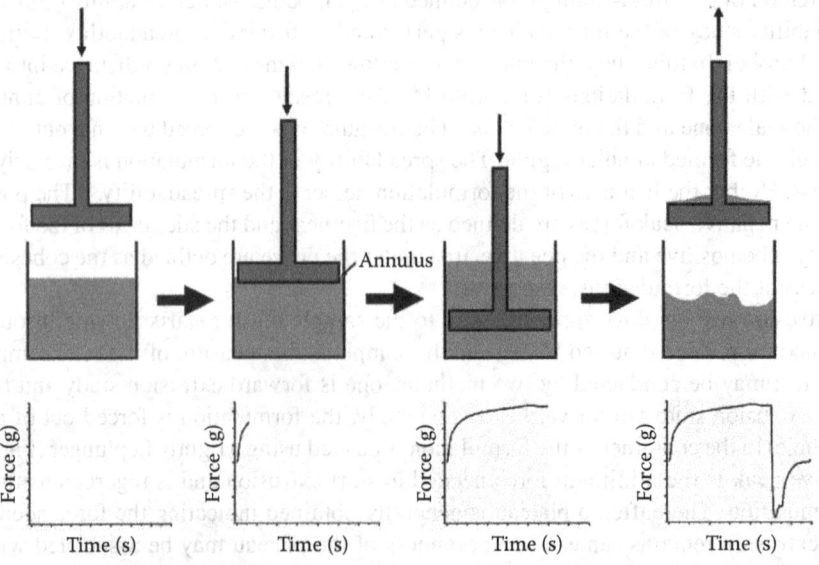

FIGURE 46.16 Schematic diagram for backward extrusion study.

46.3.5 THERMAL PROPERTIES

Melting point (T_m) of the organogels can be easily determined by drop-ball melting method (ASTM D127).[34] In the study, accurately weighed 2.0 g of the formulations are melted and poured in test tubes (volume capacity = 5 mL). The test tubes are subsequently incubated in a thermal cabinet maintained at 4°C for 10 min. A stainless steel ball having a diameter of 1 mm (0.11 g) is put on the surface of the gels. The gels are then heated at a heating rate of 1°C/min in a melting point determination apparatus. The temperature at which the ball reaches the bottom of the tube is taken as the T_m (Figure 46.17).

Differential scanning calorimetry (DSC) is an analytical thermal analysis technique, which measures the rate of heat flow and compares the differences between the heat flow rate of the test formulation and the reference is measured as a function of temperature. DSC allows measurement of the amount of energy absorbed or released under controlled heating and cooling rates. The plot between heat flow and temperature provides insight into the thermal characteristics of the organogels (Figure 46.18). The organogels (without water) shows a sharp endothermic peak,

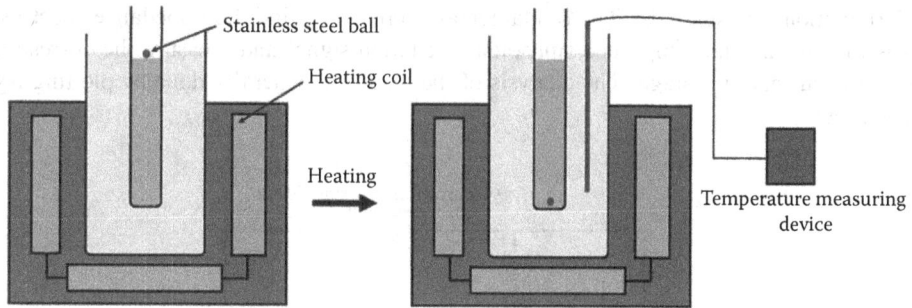

FIGURE 46.17 Schematic diagram of the melting point apparatus.

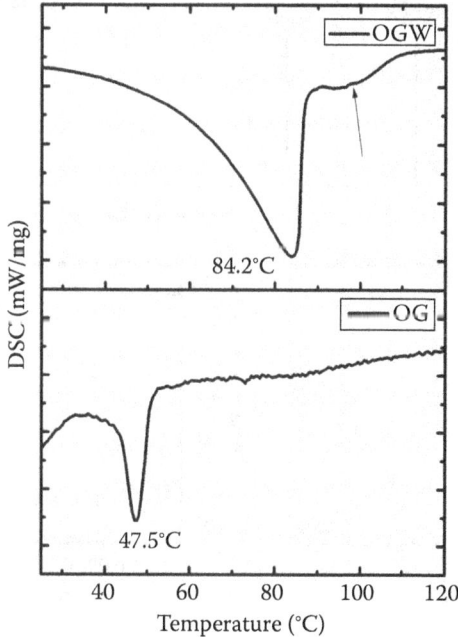

FIGURE 46.18 DSC of the anhydrous (OG) and water containing (OGW) organogels.

which corresponds to T_m. Organogels (containing water) usually show a broad endothermic peak having a peak maxima in the range of 80°C to 110°C. This broad endothermic peak is associated with the evaporation of the water molecules from the formulations. The melting endotherm is usually subsided within the broad endothermic peak due to evaporation of water. Durrschmidt et al.[89] reported the change in the phase, determination of melting point and fusion heat from the DSC profiles. They reported that an increase in the gelator or co-gelator results in the increase in the glass transition and melting point temperatures.

46.3.6 Impedance Spectroscopy

The electrical conductivity of a formulation is studied by impedance spectroscopy (Figure 46.19). In recent years, impedance spectroscopy has found widespread applications in the characterization of polymeric materials like organogels. The knowledge of electrical properties may assist in understanding the release of bioactive agents from the organogel matrices. The electrical conductivity can be measured in two forms (e.g., DC and AC).[90] The electrical properties s.a. AC conductivity, dielectric constant, tangential loss, and real and imaginary components of the dielectric property of the formulations are studied.[91] The fundamental approach behind the impedance spectroscopic studies is to apply a small amplitude sinusoidal excitation signal and measure the corresponding change in the current or voltage. The analysis of the results is generally done by plotting Nyquist and Bode Plots.

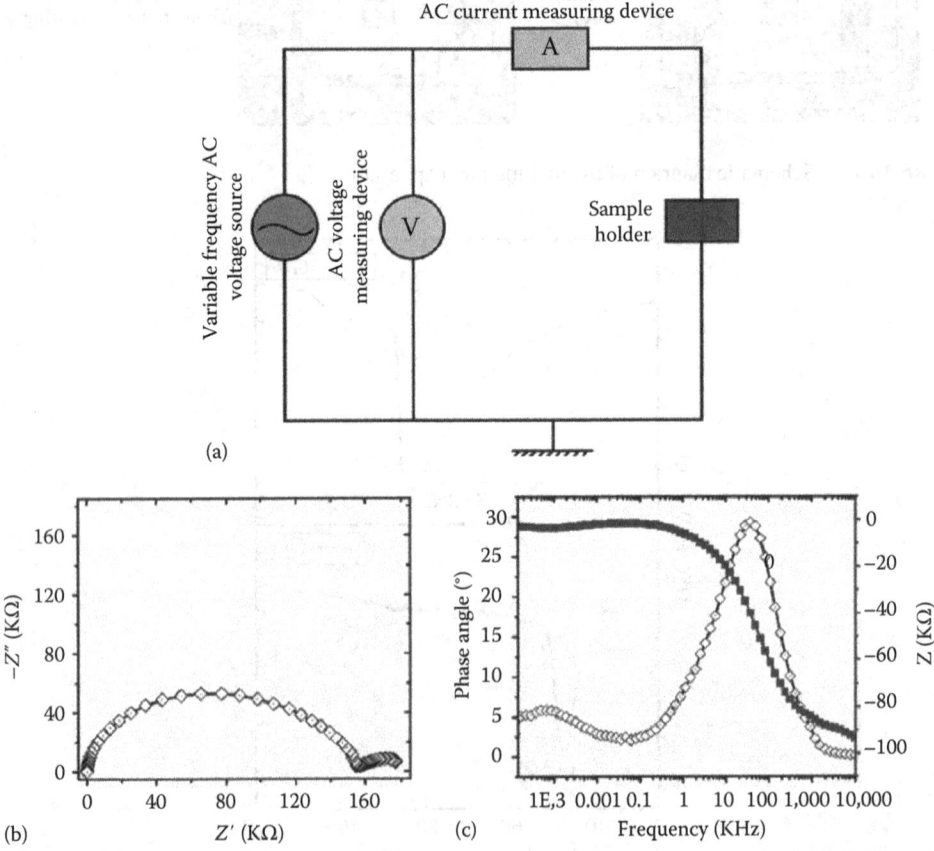

FIGURE 46.19 Impedance spectroscopy. (a) Schematic diagram of the impedance measurement device; analysis plots: (b) Nyquist plot and (c) Bode plot.

Nyquist plot, also known as Cole-Cole plot, is a plot between the real impedance component (Z') and the imaginary impedance component (Z''). Nyquist plot gives information about the bulk resistance (R_b) as well as the presence of grain boundary resistance (R_{gb}). The formation of a semicircular arc in the high frequency region is due to the bulk resistance of the material whereas the presence of an arc/spike in low frequency region is associated with the grain boundary effect.[92–94] The intercept of the arcs with the real axis determines the bulk and the grain boundary resistance, respectively. The grain boundary effect is observed due to the inhomogenity introduced by the fillers or crystalline phase of the polymers. Plasticizer reduces the inhomogeneity introduced by the crystalline phase and suppresses the grain boundary effects. This results in the increase in the bulk conductivity of the matrices.[95] The nonvertical spike in the complex impedance plane may be attributed to the roughness of the electrode–electrolyte interface.[90,96] Though, the Nyquist plot gives a quick overview of bulk resistance of the sample, it fails to divulge frequency dependent information. Also, though the bulk resistance and polarization resistance can be directly predicted from the Nyquist plot, the calculation of the electrode capacitance is possible only if the frequency is known. Bode plot, an alternate way to represent impedance results, eliminates the disadvantages of the Nyquist plot and provides better understanding about the electrical properties of the gels. The plot shows the frequency dependent impedance and phase shifts.

46.3.7 Biocompatibility Studies

The applications of organogels in controlled delivery are in its budding stage. Hence, not much information is available in the literature on the biocompatibility of the organogel-based formulations. The preliminary biocompatibility can be done by performing hemocompatibility studies (Figure 46.20).

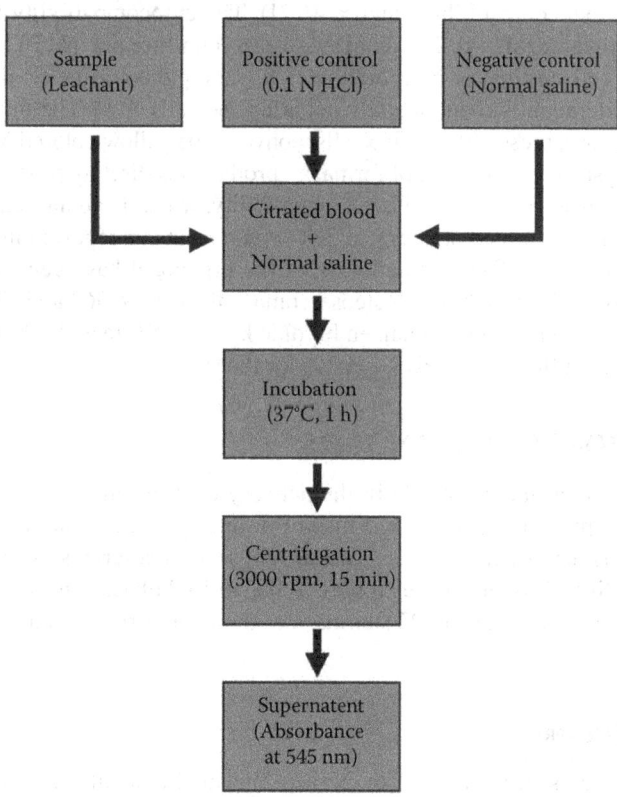

FIGURE 46.20 Schematic flowchart diagram of hemocompatibility test.

The test is usually performed to assess the extent of hemolysis of the citrated goat blood in the presence of the organogel extract.[3,8,97,98] In short, accurately weighed 1 g of the formulation is enclosed in dialysis tubing and is submerged in 50 mL of normal saline. This system is incubated at 37°C for 30 min in a shaker incubator so as to allow the leaching of the components from the organogels. 0.5 mL of the leachant is then diluted with 0.5 mL of diluted goat blood (prepared by diluting 8 mL of fresh goat blood with 10 mL of normal saline) followed by the addition of 9 mL of normal saline. The mixture is then incubated at 37°C for 1 h followed by centrifugation at 3000 rpm for 10 min. Positive and negative controls are also prepared in a similar way. 0.1 N hydrochloric acid and normal saline were used in place of the leachant for preparing positive and negative controls, respectively. The supernatant is analyzed at 545 nm using a UV-visible spectrophotometer. The test measures the extent of hemolysis in the presence of the organogel. Percent hemolysis is then calculated by the formula[97,99,100]

$$\%Hemolysis = \frac{OD_{test} - OD_{negative}}{OD_{positive} - OD_{negative}} \times 100 \tag{46.7}$$

where
  OD_{test} is the optical density of test sample
  $OD_{positive}$ is the optical density of positive control
  $OD_{negative}$ is the optical density of negative control

The organogel may be regarded as highly hemocompatible if the hemolysis is found to be less than 5. A value between 5 and 10 suggests the hemocompatible nature of the organogel whereas a value more than 20 indicates non-hemocompatible nature of the organogel.[101]

After getting some preliminary information from the hemocompatibility study, the extracts are used to estimate the cytocompatibility (Figure 46.21). The cytocompatibility test can be done by in vitro 3-[4,5-dimethylthiazol-2-yl]-2,5-diphenyl tetrazolium bromide (MTT) cell viability assay, which measures the cell survival or proliferation.[102] It is a rapid colorimetric method to determine the relative cell proliferation without actually counting the cells.[103–105] The mitochondrial enzyme succinate-dehydrogenase, present in the live cells, converts the yellow colored MTT dye into purple colored formazan crystals. The amount of formazan produced is directly proportional to number of viable cells and enables measurement of cytocompatibility, which is quantified by spectrophotometer.[106] MTT assay based cell mediated cytotoxicity study has been reported in variety of formulations including organogels.[107] Cytocompatibility of the organogel has been studied using human mesenchymal stem cells (MSCs), human osteosarcoma cells (e.g., MG63 and TE85), human histiocytic lymphoma U937 cell line, and human embryonic kidney cells (e.g., HEK293). Organogels are extracted in phosphate buffer saline (PBS, pH 7.2) for the study.

46.4 APPLICATIONS OF ORGANOGELS

Organogels has played an important role in the delivery of drugs and vaccines through different routes.[15] As discussed previously, organogels may be developed either containing only apolar phase or containing water as the dispersed phase. The current section describes the use of organogels for the delivery of bioactives through various routes. Table 46.1 tabulates some of the organogel-based controlled delivery systems. Figure 46.22 summarizes the various routes of administration of drugs using organogels.

46.4.1 Topical Delivery

Topical delivery systems are intended to deliver the medicaments to body surface such as the skin or the mucus membrane. These systems minimize the problems associated with the conventional drug delivery systems. Hence, the delivery route may be used as an alternative

Organogels in Controlled Drug Delivery

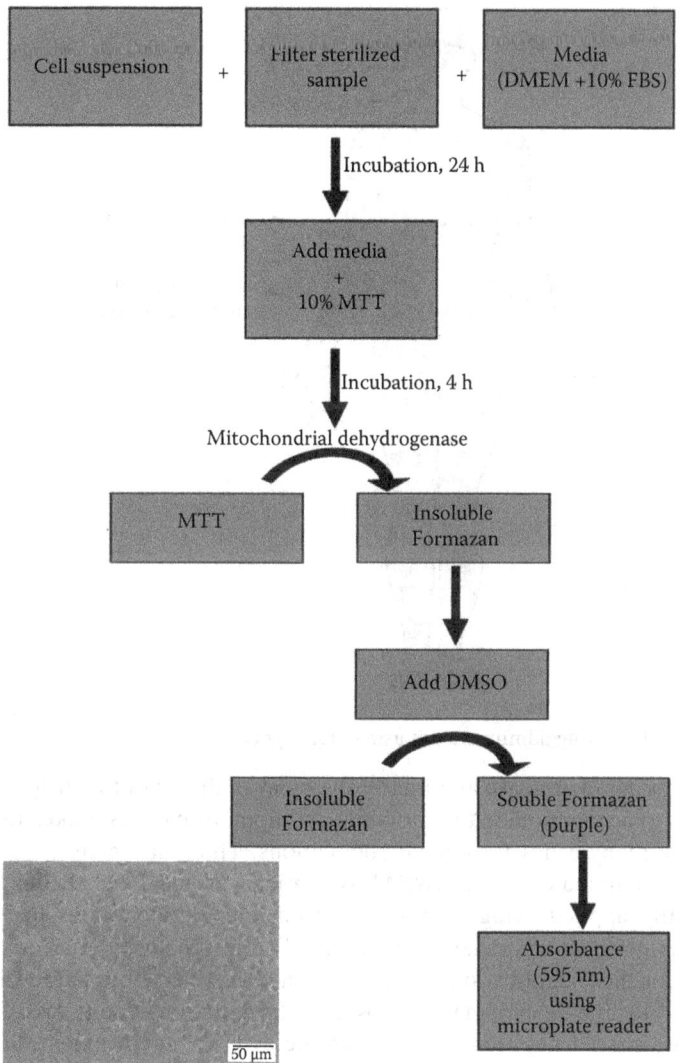

FIGURE 46.21 Schematic flowchart diagram for conducting MTT assay and micrograph (as insert) showing live cells after the cytotoxicity study.

TABLE 46.1
Some Organogel-Based Systems Used as Controlled Delivery Matrices

Gelator	Solvent	Bioactive Agent	Purpose	References
Stearyl acrylate	Oleyl alcohol	Indomethacin	Development of thermosensitive pulsatile delivery system	[108]
1,3:2,4-di-O-benzylidene-D-sorbitol	1,2 – Propylene glycol	5-fluorouracil	To study the host guest interaction	[109]
Plicosaaanol	Olive oil	Ferulic acid	Oral controlled delivery of ferulic acid	[110]
Gelatin stabilised microemulsion based organogels	Isopropyl myristate	Cyclosporin A	Safe and effective delivery through topical application	[111]
Sorbitan esters	Coconut, canola and corn oil	Curcuminoids	High bioaccessibility and high loading	[112]

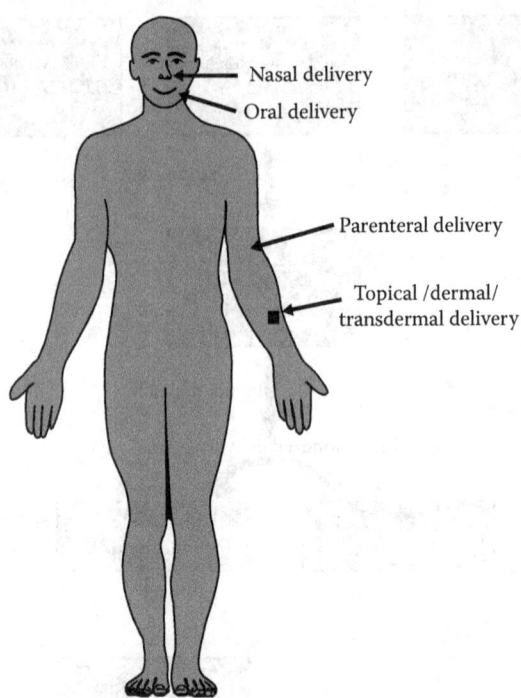

FIGURE 46.22 Route of drug administration using organogels.

pathway for improving the release and finally the bioavailability of the drugs.[113] The efficiency of the organogels to accommodate thermolabile and lipophilic drugs makes them ideal vehicle for the delivery of many drugs for topical applications. There are many classes of drugs (e.g., nonsteroidal anti-inflammatory drugs (NSAIDs), anticancer, local anesthetics, and antifungal), which are topically applied using organogels-based formulations so as to improve the bioavailability of the drugs. The oraganogels prepared from the sorbitan esters and lecithin are widely explored for these applications.[114] Lipogels, a class of organogels developed using oils, have been studied for the development of cosmetic formulations due to their lipidic composition. They act as a barrier between the skin and the external environment. Since oleogels are rich in fatty acid (e.g., oleic fatty acid) content, the organogel-based formulations help in the regeneration of stratum corneum.[115] The topical administration of NSAIDs eliminates the side effects of the gastrointestinal tract. Novel lecithin organogels loaded with aceclofenac have been prepared for topical application.[116] Lecithin-based microemulsion organogels has been successfully tested for the delivery of NSAIDs (e.g., ketoprofen, flurbiprofen, aceclofenac, and prioxicam).[117] Poloxamer lecithin organogels have shown sufficient properties as suitable carriers for controlled release applications.[118] Pluronic lecithin organogels have also been tested for the topical administration of flurbiprofen.[119] Liquid glyceryl fatty acid esters (e.g., solid glyceryl fatty acid esters) has been used as organogelators for formulating organogels for controlled delivery of piroxicam.[120]

Apart from NSAIDs, organogel-based formulations have been used for the topical delivery of antimicrobial agents. Fluconazole, commonly used antifungal drug, was successfully incorporated within microemulsion-based lecithin organogels.[121] Carbopol and PEG400 based formulations were prepared by high speed homogenization for the improved delivery of triclosan (antibacterial and antifungal agent) over the skin. The antimicrobial activities of the developed formulations were better than the commercially available products.[9]

Pluronic lecithin organogels have been tested for the topical delivery of ondansetron (5-HT_3 receptor antagonist). The formulations exhibited dose-dependent attenuation of nociceptive and inflammatory effects of intradermally injected capsaicin in humans.[122] Pluronic lecithin organogels have also been used for delivery of estrogen receptor antagonist.[123]

The lecithin organogels have also been explored for the delivery of synthetic retinoids for the treatment of topical tumors. Retinoids, being poorly soluble drug, was encapsulated in lecithin organogels. The results showed a slower diffusion of the drug from the organogels as compared to available conventional topical dosage systems.[124]

46.4.2 Dermal and Transdermal

A major research has been focused on the development of organogels for dermal delivery of drugs. Skin has been reported to be barriers for many drugs. Hence, development of formulations for transdermal drug delivery is of great interest in the pharmaceutical fraternity.[125] The delivery of drugs into the skin layers is regarded as cutaneous or dermal delivery, whereas the delivery of the drug across the skin layers is regarded as percutaneous or transdermal delivery.[126] Transdermal delivery system has been reported to increase the systemic circulation of drugs having short biological half-life. This is due to by-passing the first-pass metabolism of the drugs. Due to the increased bioavailability of the drugs, there is a reduction in the dosing frequency.[127] Most commonly used transdermal drug delivery systems (TDDS) include patches, ointments, and gels.[128] Lecithin organogels have been used for the transdermal application of Nicardipine hydrochloride, for the treatment of chronic stable angina and hypertension. The organogels showed enhanced skin permeating effect on the giunea pig skin and human stratum corneum[129] Pluronic lecithin organogels have been used for the transdermal delivery of methimazole, an antithyroid drug.[130] Lecithin organogels have been reported to improve the permeability of propranolol across the skin as compared to the conventional petroleum jelly-based formulations.[131]

Microemulsion-based organogels have been reported for transdermal delivery applications. Gelatin-based microemulsion organogels have been used as a delivery vehicle for cyclosporine, an immunosuppressant.[57] Butenafine hydrochloride has been successfully incorporated in the gelatin-based microemulsion organogels and used for controlled delivery applications.[132] Pluronic and lecithin organogels have been formulated for transdermal delivery of sumatriptan.[133] Sorbitan ester-based organogels has been reported as potential reservoir system for the effective transdermal drug delivery applications.[134]

46.4.3 Parenteral Delivery

Daily administration of oral drugs reduces patient compliance. This results in the noncompliance of the patients in drug consumption, which in turn, leads to increased risk of inconsistent drug intake. Parenteral injections offer solution to this problem. Recently, organogels have gained much attention due to the ease of preparation and high drug loading efficiency of hydrophobic drugs. The release of drugs through the organogels is usually controlled by diffusion.[135] Tyrosine based organogels loaded with rivastigmine were prepared for the treatment of Alzehimer's disease.[136] Another study proposed the development of safflower oil based oleogels using N-stearoyl L-alanine methyl ester as organogelator for the subcutaneous delivery of rivastigmine.[137] A novel L-alanine derivative based in situ-forming biodegradable oleogel implants have demonstrated great potential for sustained delivery of drugs.[138]

46.4.4 Nasal Delivery

Nasal drug delivery is gaining importance due to the large surface area, ease of accessibility and easy administration of the drug without any help of a medical professional.[139] Sorbitan ester-isopropyl myristate-water based organogels have been developed for nasal delivery of propanolol hydrochloride.[140] The presence of polysorbates as co-gelators in these organogels increased the viscosity of the gels.

46.4.5 ORAL DELIVERY

The absorption of drugs through lipid vehicles usually takes place in three possible ways. The presence of surfactants or low molecular weight organogelators enhances the solubilization of the drug in the intestinal environment. The interaction of the lipid vehicles with the enterocytes has also been reported to alter the drug transport mechanism. P-glycoproteins usually play an important role in this type of transport.[141] The lipid vehicles have the ability to bypass the hepatic circulation and gain entry directly to the systemic circulation via lymphatic circulation.

Organogels prepared from 12- Hydroxystearic acid as gelator and soyabean oil as solvent have been prepared for the oral delivery of ibuprofen, commonly used NSAIDs.[142] Another study reported that soybean oil and 12-hydroxy stearic acid based organogels are suitable for the delivery of both lipophilic and hydrophilic drugs.[143] Organogels have been encapsulated within hard gelatin capsules for the delivery of cyclosporine A, a potent immunosuppressant drug. The efficiency of the formulation was comparable to the marketed microemulsion based cyclosporine A formulations. But the ease of preparation of the organogel based product was the added advantage.[16] Poly(acrylic acid) organogels have been developed for local delivery to the oral cavity. The gels showed good mucoadhesive and rheological properties.[144] Pluronic lecithin oranogels have been used for both oral delivery of dexamethasone.[145]

46.5 CONCLUSION

The current review discusses about the various probable mechanisms of organogel formation. The use of organogels in food, pharmaceutical, and cosmetic industries is on the rise. This may be attributed to the inherent thermodynamic stability of the organogels. The properties (e.g., mechanical, chemical, and drug release) of the organogels may be easily tailored by slight alterations in the compositions of the organogels. Though there are few organogel-based commercial products available in the market, the research on organogels is still in its nascent stage. There is a need for an extensive research before the organogel-based products can be used for various applications. The unique properties presented by the organogels will tease the food, pharmaceutical, and cosmetic industries to develop organogel-based products, and it is only a matter of time when most of the industries will employ organogellation for product development.

ACKNOWLEDGMENT

The authors acknowledge the financial support received from Department of Biotechnology, New Delhi, India vide sanction order (BT/PR14282/PID/06/598/2010).

REFERENCES

1. Pradhan, S. et al., Palm oil-based organogels and microemulsions for delivery of antimicrobial drugs. *Journal of Applied Polymer Science*, 2013. Vol. **131**.
2. Sagiri, S. et al., Organogels as matrices for controlled drug delivery: A review on the current state. *Soft Materials*, 2013. **12**: 47–72.
3. Singh, V.K. et al., Castor oil and sorbitan monopalmitate based organogel as a probable matrix for controlled drug delivery. *Journal of Applied Polymer Science*, 2013. **130**: 1503–1515.
4. Calligaris, S. et al., Effect of palm oil replacement with monoglyceride organogel and hydrogel on sweet bread properties. *Food Research International*, 2013. **51**(2): 596–602.
5. Patel, A.R. et al., Preparation and rheological characterization of shellac oleogels and oleogel-based emulsions. *Journal of Colloid and Interface Science*, 2013. **411**: 114–121.
6. Satapathy, D. et al., Sunflower-oil-based lecithin organogels as matrices for controlled drug delivery. *Journal of Applied Polymer Science*, 2012. **129**: 585–594.
7. Pal, K. et al., Hydrogel-based controlled release formulations: Designing aconsiderations, characterization techniques and applications. *Polymer-Plastics Technology and Engineering*, 2013. **52**(14): 1391–1422.

8. Shah, D.K. et al., Development of olive oil based organogels using sorbitan monopalmitate and sorbitan monostearate: A comparative study. *Journal of Applied Polymer Science*, 2012. **129**: 793–805.
9. Gökçe, E.H. et al., A novel preparation method for organogels: High-speed Homogenization and micro-irradiation. *Aaps Pharmscitech*, 2013. **14**: 391–397.
10. Singh, V.K. et al., Olive oil based novel thermo-reversible emulsion hydrogels for controlled delivery applications. *Journal of Materials Science. Materials in Medicine*, 2013. **25**: 703–721.
11. Sun, T.L. et al., Physical hydrogels composed of polyampholytes demonstrate high toughness and viscoelasticity. *Nature Materials*, 2013. **12**(10): 932–937.
12. Han, L.-J. et al., Rheological properties of organogels developed by sitosterol and lecithin. *Food Research International*, 2013. **53**(1): 42–48.
13. Douaire, M. and I. Norton, Designer colloids in structured food for the future. *Journal of the Science of Food and Agriculture*, 2013. **93**: 3147–3154.
14. Sahoo, S. et al., Organogels: Properties and applications in drug delivery. *Designed Monomers and Polymers*, 2011. **14**(2): 95–108.
15. Vintiloiu, A. and J.-C. Leroux, Organogels and their use in drug delivery—A review. *Journal of Controlled Release*, 2008. **125**(3): 179–192.
16. Murdan, S., Organogels in drug delivery. *Expert Opinion on Drug Delivery*, 2005. **2**(3): 489–505.
17. George, M. and R.G. Weiss, Chemically reversible organogels via "latent" gelators. Aliphatic amines with carbon dioxide and their ammonium carbamates. *Langmuir*, 2002. **18**(19): 7124–7135.
18. Yan, N. et al., Pyrenyl-linker-glucono gelators. Correlations of gel properties with gelator structures and characterization of solvent effects. *Langmuir*, 2013. **29**(2): 793–805.
19. Brosse, N., D. Barth, and B. Jamart-Grégoire, A family of strong low-molecular-weight organogelators based on aminoacid derivatives. *Tetrahedron Letters*, 2004. **45**(52): 9521–9524.
20. Terech, P. and R.G. Weiss, Low molecular mass gelators of organic liquids and the properties of their gels. *Chemical Reviews*, 1997. **97**(8): 3133–3160.
21. Palui, G. et al., Organogelators from self-assembling peptide based dendrimers: Structural and morphological features. *Tetrahedron*, 2008. **64**(1): 175–185.
22. Ji, Y. et al., A dendron based on natural amino acids: Synthesis and behavior as an organogelator and lyotropic liquid crystal. *Angewandte Chemie*, 2005. **117**(37): 6179–6183.
23. Pal, A., Y.K. Ghosh, and S. Bhattacharya, Molecular mechanism of physical gelation of hydrocarbons by fatty acid amides of natural amino acids. *Tetrahedron*, 2007. **63**(31): 7334–7348.
24. Rosemeyer, H., E.-M. Stürenberg, and P. Herdewijn, Nucleolipids as potential organogelators. *Nucleosides, Nucleotides, and Nucleic Acids*, 2007. **26**(8–9): 995–999.
25. Tu, T. et al., An air-stable organometallic low-molecular-mass gelator: Synthesis, aggregation, and catalytic application of a palladium pincer complex. *Angewandte Chemie International Edition*, 2007. **46**(33): 6368–6371.
26. Edelsztein, V.C., G. Burton, and P.H. Di Chenna, Self-assembly of a silylated steroid-based organogelator and its use as template for the in situ sol–gel polymerization of tetraethyl orthosilicate. *Tetrahedron*, 2010. **66**(12): 2162–2167.
27. Terech, P. et al., Gels from small molecules in organic solvents: Structural features of a family of steroid and anthryl-based organogelators. *Faraday Discussions*, 1995. **101**: 345–358.
28. Fages, F., F. Vögtle, and M. Žinic, Systematic design of amide-and urea-type gelators with tailored properties, in *Low Molecular Mass Gelator*, Fages, F. ed., 2005, Berlin, Heidelberg: Springer. pp. 77–131.
29. Simic, V., L. Bouteiller, and M. Jalabert, Highly cooperative formation of bis-urea based supramolecular polymers. *Journal of the American Chemical Society*, 2003. **125**(43): 13148–13154.
30. Luboradzki, R. et al., An attempt to predict the gelation ability of hydrogen-bond-based gelators utilizing a glycoside library. *Tetrahedron*, 2000. **56**(49): 9595–9599.
31. Gronwald, O. and S. Shinkai, Sugar-integrated gelators of organic solvents. *Chemistry-A European Journal*, 2001. **7**(20): 4328–4334.
32. Takeno, H. et al., A structural study of an organogel investigated by small-angle neutron scattering and synchrotron small-angle X-ray scattering. *The Journal of Physical Chemistry B*, 2012. **116**(26): 7739–7745.
33. Tung, S.-H., Y.-E. Huang, and S.R. Raghavan, Self-assembled organogels obtained by adding minute concentrations of a bile salt to AOT reverse micelles. *Soft Matter*, 2008. **4**(5): 1086–1093.
34. Abdallah, D.J. and R.G. Weiss, n-Alkanes gel n-alkanes (and many other organic liquids). *Langmuir*, 2000. **16**(2): 352–355.
35. Žinic, M., F. Vögtle, and F. Fages, Cholesterol-based gelators, in *Low Molecular Mass Gelator*, Fages, F. ed., 2005, Berlin, Heidelberg: Springer. pp. 39–76.

36. Chow, H.-F. et al., Improving the gelation properties of 3, 5-diaminobenzoate-based organogelators in aromatic solvents with additional aromatic-containing pendants. *Tetrahedron*, 2007. **63**(2): 363–373.
37. Maitra, U. et al., Helical aggregates from a chiral organogelator. *Tetrahedron: Asymmetry*, 2001. **12**(3): 477–480.
38. Balasubramanian, R., A.A. Sughir, and G. Damodar, Oleogel: A promising base for transdermal formulations. *Asian Journal of Pharmaceutics*, 2012. **6**(1): 1.
39. Abdallah, D.J., S.A. Sirchio, and R.G. Weiss, Hexatriacontane organogels. The first determination of the conformation and molecular packing of a low-molecular-mass organogelator in its gelled state. *Langmuir*, 2000. **16**(20): 7558–7561.
40. Luo, X., B. Liu, and Y. Liang, Self-assembled organogels formed by mono-chain L-alanine derivatives. *Chemical Communications*, 2001. (17): 1556–1557. doi: 10.1039/B104428C.
41. Terech, P., Fibers and wires in organogels from low-mass compounds: Typical structural and rheological properties. *Berichte der Bunsengesellschaft für physikalische Chemie*, 1998. **102**(11): 630–1643.
42. Shchipunov, Y.A. and H. Hoffmann, Thinning and thickening effects induced by shearing in lecithin solutions of polymer-like micelles. *Rheologica Acta*, 2000. **39**(6): 542–553.
43. Shchipunov, Y.A. and E.V. Shumilina, Lecithin bridging by hydrogen bonds in the organogel. *Materials Science and Engineering: C*, 1995. **3**(1): 43–50.
44. Angelico, R. et al., Biocompatible lecithin organogels: Structure and phase equilibria. *Langmuir*, 2005. **21**(1): 140–148.
45. Shchipunov, Y.A., Lecithin organogel: A micellar system with unique properties. *Colloids and Surfaces A: Physicochemical and Engineering Aspects*, 2001. **183**: 541–554.
46. Scartazzini, R. and P.L. Luisi, Organogels from lecithins. *The Journal of Physical Chemistry*, 1988. **92**(3): 829–833.
47. Shah, D.K. et al., Development of olive oil based organogels using sorbitan monopalmitate and sorbitan monostearate: A comparative study. *Journal of Applied Polymer Science*, 2013. **129**(2): 793–805.
48. Hughes, N.E. et al., Potential food applications of edible oil organogels. *Trends in Food Science and Technology*, 2009. **20**(10): 470–480.
49. Sagiri, S.S. et al., Effect of composition on the properties of tween-80–span-80-based organogels. *Designed Monomers and Polymers*, 2012. **15**(3): 253–273.
50. Behera, B. et al., Modulating the physical properties of sunflower oil and sorbitan monopalmitate-based organogels. *Journal of Applied Polymer Science*, 2013. **127**(6): 4910–4917.
51. Bajaj, A., S. Gupta, and A. Chatterjee, Plastibase: A new base for patch testing of metal antigens. *International Journal of Dermatology*, 1990. **29**(1): 73–73.
52. De Loos, M. et al., Remarkable stabilization of self-assembled organogels by polymerization. *Journal of the American Chemical Society*, 1997. **119**(51): 12675–12676.
53. Tamaoki, N. et al., Polymerization of a diacetylene dicholesteryl ester having two urethanes in organic gel states. *Langmuir*, 2000. **16**(19): 7545–7547.
54. Beginn, U., S. Sheiko, and M. Möller, Self-organization of 3, 4, 5-tris (octyloxy) benzamide in solution and embedding of the aggregates into methacrylate resins. *Macromolecular Chemistry and Physics*, 2000. **201**(10): 1008–1015.
55. Dastidar, P., Supramolecular gelling agents: Can they be designed? *Chemical Society Reviews*, 2008. **37**(12): 2699–2715.
56. Jones, M.-C. et al., Self-assembled nanocages for hydrophilic guest molecules. *Journal of the American Chemical Society*, 2006. **128**(45): 14599–14605.
57. Kantaria, S., G.D. Rees, and M.J. Lawrence, Formulation of electrically conducting microemulsion-based organogels. *International Journal of Pharmaceutics*, 2003. **250**(1): 65–83.
58. El Araby, A.M. and Y.F. Talic, The effect of thermocycling on the adhesion of self-etching adhesives on dental enamel and dentin. *The Journal of Contemporary Dental Practice*, 2007. **8**(2): 1–11.
59. Lee, Y.K. et al., Changes of optical properties of dental nano-filled resin composites after curing and thermocycling. *Journal of Biomedical Materials Research Part B: Applied Biomaterials*, 2004. **71**(1): 16–21.
60. Arunyanart, T. and S. Charoenrein, Effect of sucrose on the freeze–thaw stability of rice starch gels: Correlation with microstructure and freezable water. *Carbohydrate Polymers*, 2008. **74**(3): 514–518.
61. Charoenrein, S. and N. Preechathammawong, Effect of waxy rice flour and cassava starch on freeze–thaw stability of rice starch gels. *Carbohydrate Polymers*, 2012. **90**(2): 1032–1037.
62. Wang, B. et al., Rheological properties of waxy maize starch and xanthan gum mixtures in the presence of sucrose. *Carbohydrate Polymers*, 2009. **77**(3): 472–481.

63. Arocas, A., T. Sanz, and S. Fiszman, Improving effect of xanthan and locust bean gums on the freeze-thaw stability of white sauces made with different native starches. *Food Hydrocolloids*, 2009. **23**(8): 2478–2484.
64. Johnson, S.A., Chapter 1—Phase contrast microscopy, in *Biomedical Optical Phase Microscopy and Nanoscopy*, Lisa L. Satterwhite, ed., 2013, Academic Press: Oxford, U.K. pp. 3–18.
65. Hirama, H. et al., Hyper alginate gel microbead formation by molecular diffusion at the hydrogel/droplet interface. *Langmuir*, 2013. **29**(2): 519–524.
66. Maatar, W., S. Alila, and S. Boufi, Cellulose based organogel as an adsorbent for dissolved organic compounds. *Industrial Crops and Products*, 2013. **49**(0): 33–42.
67. Zou, J. et al., Responsive organogels formed by supramolecular self assembly of PEG-block-allyl-functionalized racemic polypeptides into β-sheet-driven polymeric ribbons. *Soft Matter*, 2013. **9**: 5951–5958.
68. Zhang, L. et al., Pyrene-functionalized organogel and spacer effect: From emissive nanofiber to nanotube and inversion of supramolecular chirality. *Soft Matter*, 2013. **9**(33): 7966–7973.
69. Fan, J. et al., Tunable mechano-responsive organogels by ring-opening copolymerizations of N-carboxyanhydrides. *Chemical Science*, 2014. **5**: 141–150.
70. Viswanath, D.S., *Viscosity of Liquids: Theory, Estimation, Experiment, and Data*. 2007: Springer. pp. 6–7.
71. Torresi, R.M. et al., Convective mass transport in ionic liquids studied by electrochemical and electrohydrodynamic impedance spectroscopy. *Electrochimica Acta*, 2013. **93**: 32–43.
72. Neves, J. et al., Rheological properties of vaginal hydrophilic polymer gels. *Current Drug Delivery*, 2009. **6**(1): 83–92.
73. Jones, D.S. et al., Physicochemical characterization of bioactive polyacrylic acid organogels as potential antimicrobial implants for the buccal cavity. *Biomacromolecules*, 2008. **9**(2): 624–633.
74. Fang, P., R.M. Manglik, and M.A. Jog, Characteristics of laminar viscous shear-thinning fluid flows in eccentric annular channels. *Journal of Non-Newtonian Fluid Mechanics*, 1999. **84**(1): 1–17.
75. Sutar, P.B. et al., Development of pH sensitive polyacrylamide grafted pectin hydrogel for controlled drug delivery system. *Journal of Materials Science: Materials in Medicine*, 2008. **19**(6): 2247–2253.
76. Sae-kang, V. and M. Suphantharika, Influence of pH and xanthan gum addition on freeze-thaw stability of tapioca starch pastes. *Carbohydrate Polymers*, 2006. **65**(3): 371–380.
77. Lupi, F.R. et al., A rheological characterisation of an olive oil/fatty alcohols organogel. *Food Research International*, 2013. **51**(2): 510–517.
78. Sawalha, H. et al., The influence of the type of oil phase on the self-assembly process of γ-oryzanol+ β-sitosterol tubules in organogel systems. *European Journal of Lipid Science and Technology*, 2013. **115**: 295–300.
79. Ibrahim, M.M., S.A. Hafez, and M.M. Mahdy, Organogels, hydrogels and bigels as transdermal delivery systems for diltiazem hydrochloride. *Asian Journal of Pharmaceutical Sciences*, 2013. **8**(1): 48–57.
80. Toro-Vazquez, J.F. et al., Cooling rate effects on the microstructure, solid content, and rheological properties of organogels of amides derived from stearic and (R)-12-hydroxystearic acid in vegetable oil. *Langmuir*, 2013. **29**: 7642–7654.
81. Jones, D.S., A.F. Brown, and A.D. Woolfson, Rheological characterization of bioadhesive, antimicrobial, semisolids designed for the treatment of periodontal diseases: Transient and dynamic viscoelastic and continuous shear analysis. *Journal of Pharmaceutical Sciences*, 2001. **90**(12): 1978–1990.
82. Pinthus, E.J., P. Weinberg, and I.S. Saguy, Gel-strength in restructured potato products affects oil uptake during deep-fat frying. *Journal of Food Science*, 1992. **57**(6): 1359–1360.
83. Chockchaisawasdee, S., J.S. Mounsey, and C.E. Stathopoulos, Textural and rheological characteristics of sun-dried banana traditionally prepared in the North-East of Thailand. *Food Science and Technology Research*, 2010. **16**(4): 291–294.
84. Mandala, I., Physical properties of fresh and frozen stored, microwave-reheated breads, containing hydrocolloids. *Journal of Food Engineering*, 2005. **66**(3): 291–300.
85. Winter, H.H. and F. Chambon, Analysis of linear viscoelasticity of a crosslinking polymer at the gel point. *Journal of Rheology*, 1986. **30**(2): 367–382.
86. Lee, C.H., V. Moturi, and Y. Lee, Thixotropic property in pharmaceutical formulations. *Journal of Controlled Release*, 2009. **136**(2): 88–98.
87. Scholl, B.J., Attentive tracking of objects versus substances. *Psychological Science*, 2003. **14**(5): 498–504.
88. Dhawan, S., B. Medhi, and S. Chopra, Formulation and evaluation of diltiazem hydrochloride gels for the treatment of anal fissures. *Scientia Pharmaceutica*, 2009. **77**(2): 465.
89. Dürrschmidt, T. and H. Hoffmann, Organogels from ABA triblock copolymers. *Colloid and Polymer Science*, 2001. **279**(10): 1005–1012.

90. Pradhan, D.K., R. Choudhary, and B. Samantaray, Studies of structural, thermal and electrical behavior of polymer nanocomposite electrolytes. *Express Polymer Letters*, 2008. **2**: 630–638.
91. Pradhan, D.K., R. Choudhary, and B. Samantaray, Studies of dielectric and electrical properties of plasticized polymer nanocomposite electrolytes. *Materials Chemistry and Physics*, 2009. **115**(2): 557–561.
92. Leo, C., G. Subba Rao, and B. Chowdari, Studies on plasticized PEO–lithium triflate–ceramic filler composite electrolyte system. *Solid State Ionics*, 2002. **148**(1): 159–171.
93. Pradhan, D.K. et al., Effect of plasticizer on microstructure and electrical properties of a sodium ion conducting composite polymer electrolyte. *Ionics*, 2005. **11**(1): 95–102.
94. Wieczorek, W. et al., Modifications of crystalline structure of PEO polymer electrolytes with ceramic additives. *Solid State Ionics*, 1989. **36**(3): 255–257.
95. Ngai, K. and T. Ramakrishnan, *Non-Debye Relaxation in Condensed Matter*. 1987, World Scientific, Singapore.
96. Ramesh, S. and K. Wong, Conductivity, dielectric behaviour and thermal stability studies of lithium ion dissociation in poly (methyl methacrylate)-based gel polymer electrolytes. *Ionics*, 2009. **15**(2): 249–254.
97. Pal, K., A. Banthia, and D. Majumdar, Biomedical evaluation of polyvinyl alcohol–gelatin esterified hydrogel for wound dressing. *Journal of Materials Science: Materials in Medicine*, 2007. **18**(9): 1889–1894.
98. Pal, K., A.K. Banthia, and D.K. Majumdar, Preparation and characterization of polyvinyl alcohol-gelatin hydrogel membranes for biomedical applications. *AAPS PharmSciTech*, 2007. **8**(1): E142–E146.
99. Behera, B. et al., Span-60-based organogels as probable matrices for transdermal/topical delivery systems. *Journal of Applied Polymer Science*, 2012. **125**: 852–863.
100. Bhattacharya, C. et al., Development of span 80–tween 80 based fluid-filled organogels as a matrix for drug delivery. *Journal of Pharmacy and Bioallied Sciences*, 2012. **4**(2): 155.
101. Sagiri, S.S. et al., Lanolin-based organogels as a matrix for topical drug delivery. *Journal of Applied Polymer Science*, 2012. **128**: 3831–3839.
102. Mosmann, T., Rapid colorimetric assay for cellular growth and survival: Application to proliferation and cytotoxicity assays. *Journal of Immunological Methods*, 1983. **65**(1): 55–63.
103. Khor, K. et al., *Processing and Fabrication of Advanced Materials VIII*. 2001: World Scientific Publishing Co. Pte. Ltd., Singapore.
104. Borenfreund, E., H. Babich, and N. Martin-Alguacil, Comparisons of two in vitro cytotoxicity assays The neutral red (NR) and tetrazolium MTT tests. *Toxicology in Vitro*, 1988. **2**(1): 1–6.
105. van de Loosdrecht, A.A. et al., Cell mediated cytotoxicity against U 937 cells by human monocytes and macrophages in a modified colorimetric MTT assay: A methodological study. *Journal of Immunological Methods*, 1991. **141**(1): 15–22.
106. Vellonen, K.-S., P. Honkakoski, and A. Urtti, Substrates and inhibitors of efflux proteins interfere with the MTT assay in cells and may lead to underestimation of drug toxicity. *European Journal of Pharmaceutical Sciences*, 2004. **23**(2): 181–188.
107. Wang, K. et al., Self-assembled l-alanine derivative organogel as in situ drug delivery implant: Characterization, biodegradability, and biocompatibility. *Drug Development and Industrial Pharmacy*, 2010. **36**(12): 1511–1521.
108. Tokuyama, H. and Y. Kato, Preparation of thermosensitive polymeric organogels and their drug release behaviors. *European Polymer Journal*, 2010. **46**(2): 277–282.
109. Wang, H. et al., Host–guest interactions of 5-fluorouracil in supramolecular organogels. *European Journal of Pharmaceutics and Biopharmaceutics*, 2009. **73**(3): 357–360.
110. Lupi, F.R. et al., Olive oil/policosanol organogels for nutraceutical and drug delivery purposes. *Food and Function*, 2013. **4**(10): 1512–1520.
111. Liu, H. et al., Gelatin-stabilised microemulsion-based organogels facilitates percutaneous penetration of Cyclosporin A in vitro and dermal pharmacokinetics in vivo. *Journal of Pharmaceutical Sciences*, 2007. **96**(11): 3000–3009.
112. Yu, H. et al., Development of a food-grade organogel with high bioaccessibility and loading of curcuminoids. *Food Chemistry*, 2012. **131**(1): 48–54.
113. Lam, P.-L. et al., Development of hydrocortisone succinic acid/and 5-fluorouracil/chitosan microcapsules for oral and topical drug deliveries. *Bioorganic and Medicinal Chemistry Letters*, 2012. **22**(9): 3213–3218.
114. Rogers, M.A., A.J. Wright, and A.G. Marangoni, Nanostructuring fiber morphology and solvent inclusions in 12-hydroxystearic acid/canola oil organogels. *Current Opinion in Colloid and Interface Science*, 2009. **14**(1): 33–42.
115. Gallardo, V., M. Muñoz, and M. Ruiz, Formulations of hydrogels and lipogels with vitamin E. *Journal of Cosmetic Dermatology*, 2005. **4**(3): 187–192.

116. Shaikh, I.M. et al., Topical delivery of aceclofenac from lecithin organogels: Preformulation study. *Current Drug Delivery*, 2006. **3**(4): 417–427.
117. Nasseria, A.A. et al., Lecithin–stabilized microemulsion–based organogels for topical application of ketorolac tromethamine. II. In vitro release study. *Iranian Journal of Pharmaceutical Research*, 2003. **117**: 123.
118. Dowling, T.C. et al., Relative bioavailability of ketoprofen 20% in a poloxamer-lecithin organogel. *American Journal of Health-System Pharmacy*, 2004. **61**(23): 2541–2544.
119. Pandey, M., V. Belgamwar, and S. Surana, Topical delivery of flurbiprofen from Pluronic Lecithin organogel. *Indian Journal of Pharmaceutical Sciences*, 2009. **71**(1): 87.
120. Pénzes, T. et al., Topical absorption of piroxicam from organogels—In vitro and in vivo correlations. *International Journal of Pharmaceutics*, 2005. **298**(1): 47–54.
121. Jadhav, K.R., V.J. Kadam, and S.S. Pisal, Formulation and evaluation of lecithin organogel for topical delivery of fluconazole. *Current Drug Delivery*, 2009. **6**(2): 174–183.
122. Giordano, J., C. Daleo, and S.M. Sacks, Topical ondansetron attenuates nociceptive and inflammatory effects of intradermal capsaicin in humans. *European Journal of Pharmacology*, 1998. **354**(1): R13–R14.
123. Bhatia, A. et al., Tamoxifen-loaded lecithin organogel (LO) for topical application: Development, optimization and characterization. *International Journal of Pharmaceutics*, 2013. **444**(1–2): 47–59.
124. Esposito, E., E. Menegatti, and R. Cortesi, Design and characterization of fenretinide containing organogels. *Materials Science and Engineering: C*, 2013. **33**(1): 383–389.
125. Shaikh, I.M. et al., Aceclofenac organogels: In vitro and in vivo characterization. *Current Drug Delivery*, 2009. **6**(1): 1–7.
126. Upadhyay, K.K. et al., Sorbitan ester organogels for transdermal delivery of Sumatriptan. *Drug Development and Industrial Pharmacy*, 2007. **33**(6): 617–625.
127. Vaughan, D.F., *Pharmacokinetics of Albuterol and Butorphanol Administered Intravenously and via a Buccal Patch.* 2003, Texas A&M University, College Station, TX.
128. Krotscheck, U., D.M. Boothe, and H. Boothe, Evaluation of transdermal morphine and fentanyl pluronic lecithin organogel administration in dogs. *Veterinary Therapeutics*, 2004. **5**: 202–211.
129. Aboofazeli, R., H. Zia, and T.E. Needham, Transdermal delivery of nicardipine: An approach to in vitro permeation enhancement. *Drug Delivery*, 2002. **9**(4): 239–247.
130. Hoffman, S.B., A.R. Yoder, and L.A. Trepanier, Bioavailability of transdermal methimazole in a pluronic lecithin organogel (PLO) in healthy cats. *Journal of Veterinary Pharmacology and Therapeutics*, 2002. **25**(3): 189–193.
131. Bhatnagar, S. and S.P. Vyas, Organogel-based system for transdermal delivery of propranolol. *Journal of Microencapsulation*, 1994. **11**(4): 431–438.
132. Zhao, X.-Y. et al., Rheological properties and microstructures of gelatin-containing microemulsion-based organogels. *Colloids and Surfaces A: Physicochemical and Engineering Aspects*, 2006. **281**(1–3): 67–73.
133. Agrawal, V. et al., Preparation and evaluation of tubular micelles of pluronic lecithin organogel for transdermal delivery of sumatriptan. *Aaps Pharmscitech*, 2010. **11**(4): 1718–1725.
134. Rogers, M.A. et al., A novel cryo-SEM technique for imaging vegetable oil based organogels. *Journal of the American Oil Chemists' Society*, 2007. **84**(10): 899–906.
135. Bonacucina, G. et al., Colloidal soft matter as drug delivery system. *Journal of Pharmaceutical Sciences*, 2009. **98**(1): 1–42.
136. Bastiat, G. et al., Tyrosine-based rivastigmine-loaded organogels in the treatment of Alzheimer's disease. *Biomaterials*, 2010. **31**(23): 6031–6038.
137. Vintiloiu, A. et al., In situ-forming oleogel implant for rivastigmine delivery. *Pharmaceutical Research*, 2008. **25**(4): 845–852.
138. Motulsky, A. et al., Characterization and biocompatibility of organogels based on l-alanine for parenteral drug delivery implants. *Biomaterials*, 2005. **26**(31): 6242–6253.
139. Dondeti, P., H. Zia, and T.E. Needham, Bioadhesive and formulation parameters affecting nasal absorption. *International Journal of Pharmaceutics*, 1996. **127**(2): 115–133.
140. Pisal, S. et al., Effect of organogel components on in vitro nasal delivery of propranolol hydrochloride. *Aaps Pharmscitech*, 2004. **5**(4): 92–100.
141. Hall, S.D. et al., Molecular and physical mechanisms of first-pass extraction. *Drug Metabolism and Disposition*, 1999. **27**(2): 161–166.
142. Iwanaga, K. et al., Characterization of organogel as a novel oral controlled release formulation for lipophilic compounds. *International Journal of Pharmaceutics*, 2010. **388**(1–2): 123–128.

143. Iwanaga, K. et al., Application of organogels as oral controlled release formulations of hydrophilic drugs. *International Journal of Pharmaceutics*, 2012. **436**(1–2): 869–872.
144. Jones, D.S. et al., An examination of the rheological and mucoadhesive properties of poly(acrylic acid) organogels designed as platforms for local drug delivery to the oral cavity. *Journal of Pharmaceutical Sciences*, 2007. **96**(10): 2632–2646.
145. Willis-Goulet, H.S. et al., Comparison of serum dexamethasone concentrations in cats after oral or transdermal administration using Pluronic Lecithin Organogel (PLO): A pilot study. *Veterinary Dermatology*, 2003. **14**(2): 83–89.

47 Microparticulate Drug Delivery Systems

*Hemant Kumar Singh Yadav, M. Navya, Abhay Raizaday,
V. Naga Sravan Kumar Varma, and H.G. Shivakumar*

CONTENTS

47.1 Introduction 1069
47.2 Classification 1069
47.3 Advantages of Microencapsulation 1070
47.4 Materials for Microencapsulation 1071
    47.4.1 Core Material 1071
    47.4.2 Coating Materials 1071
47.5 Microencapsulation Techniques 1072
    47.5.1 Polymerization 1073
        47.5.1.1 Interfacial Polymer 1073
        47.5.1.2 In Situ Polymerization 1073
    47.5.2 Solvent Evaporation 1073
        47.5.2.1 Choice of Materials 1075
        47.5.2.2 Variables and Factors 1077
    47.5.3 Single Emulsion Method 1078
    47.5.4 Double Emulsion Method 1080
        47.5.4.1 W/O/W Double Emulsion 1080
        47.5.4.2 W/O/W Double Emulsion Solvent Evaporation Technique 1080
        47.5.4.3 S/O/W Double Emulsion Technique 1080
        47.5.4.4 S/O/O Double Emulsion Technique 1081
        47.5.4.5 W/O/O Double Emulsion Technique 1081
        47.5.4.6 O/O/O Extraction Method 1081
    47.5.5 Coacervation 1081
        47.5.5.1 Simple Coacervation 1082
        47.5.5.2 Complex Coacervation 1082
    47.5.6 Coacervation Phase Separation 1082
        47.5.6.1 Step 1: Formation of Three Immiscible Chemical Phases 1082
        47.5.6.2 Step 2: Depositing the Liquid Polymer Coating on the Core Material 1083
        47.5.6.3 Step 3: Rigidizing the Coating 1083
    47.5.7 Modified Melt-Dispersion Method 1084
    47.5.8 Internal Gelation 1084
    47.5.9 Supercritical Fluid Technology 1085
        47.5.9.1 Rapid Expansion of Supercritical Solution 1085
        47.5.9.2 Gas Antisolvent (GAS) Process 1085
        47.5.9.3 Particles from a Gas-Saturated Solution (PGSS) 1085

- 47.5.10 Fluidized-Bed Technology 1086
- 47.5.11 Centrifugal Extrusion 1086
- 47.5.12 Air Suspension Coating 1086
- 47.5.13 Jet Excitation 1088
- 47.5.14 Spray-Drying and Congealing 1088
  - 47.5.14.1 Airflow Patterns 1089
  - 47.5.14.2 Spray Nozzles 1089
- 47.5.15 Electrospraying 1089
- 47.5.16 Ultrasonic Atomization 1090
- 47.6 Evaluation and Characterization 1090
  - 47.6.1 Physicochemical Evaluation 1090
    - 47.6.1.1 Preformulation Studies 1090
  - 47.6.2 Characterization 1091
    - 47.6.2.1 Particle Size 1091
    - 47.6.2.2 Tapped Density and Compressibility Index 1091
    - 47.6.2.3 Surface Morphology 1091
    - 47.6.2.4 Powder X-Ray Diffraction (XRD) 1091
    - 47.6.2.5 Percentage Yield 1091
    - 47.6.2.6 Percentage of Drug Content and Drug Entrapment (%) 1092
    - 47.6.2.7 Swelling Measurements 1092
  - 47.6.3 Drug Release Studies 1092
    - 47.6.3.1 In Vitro Methods 1092
    - 47.6.3.2 In Vivo Methods 1093
    - 47.6.3.3 In Vitro Mucoadhesion Test 1093
    - 47.6.3.4 In Situ Bioadhesivity Studies 1094
    - 47.6.3.5 Floating Behavior 1094
    - 47.6.3.6 In Vitro–In Vivo Correlations 1095
- 47.7 Microparticulate Drug Delivery Systems 1095
  - 47.7.1 Double-Walled Microspheres 1095
  - 47.7.2 Floating Microspheres 1096
  - 47.7.3 Solid Lipid Microparticles (SLMs) 1097
    - 47.7.3.1 Applications 1097
  - 47.7.4 Targeted Microsheres 1098
  - 47.7.5 Magnetic Microspheres 1099
  - 47.7.6 Bioadhesive Microspheres 1099
  - 47.7.7 Radioactive Microspheres 1101
- 47.8 Applications of Microencapsulations 1101
  - 47.8.1 Cell Immobilization 1101
  - 47.8.2 Beverage Production 1101
  - 47.8.3 Protection of Molecules from Other Compounds 1101
  - 47.8.4 Drug Delivery 1101
  - 47.8.5 Quality and Safety in Food, Agricultural, and Environmental Sectors 1102
  - 47.8.6 Agriculture 1102
  - 47.8.7 Soil Inoculation 1102
  - 47.8.8 Applications of Microcapsules in Building Construction Materials 1102
  - 47.8.9 Defense 1102
- 47.9 Marketed Products 1103
- 47.10 Recent Work in the Field OF Microencapsulation 1103
- References 1108

Microparticulate Drug Delivery Systems

47.1 INTRODUCTION[1-5]

With advances in biotechnology, genomics, and combinatorial chemistry, a wide variety of new, more potent, and specific therapeutics are being created. Most of the therapeutic agents possess the problem of poor solubility or high potency or poor stability, etc. Hence, there is a need for a proper drug delivery system that can overcome these challenges and make the therapeutic agent successful in pharmaceutical, clinical, and marketing aspects. In spite of the large number of drug delivery systems being introduced, the search for a much better one still continues. To obtain maximum therapeutic efficacy, it becomes necessary to deliver the agent to the target tissue in the optimum amount in the right period of time, thereby causing little toxicity and minimal side effects.

Controlled-release systems can fulfill most of all these required measures and can master the troubles of conventional delivery systems. One such approach of controlled delivery, which is most leisurely and convenient, is the microparticulate drug delivery system.

Microparticulate drug delivery systems represent one of the frontier areas of science, which involves a multidisciplinary scientific approach, contributing to human health care. Microparticles, because of their unique properties, occupy special position in the field of drug delivery technology. They provide several advantages over conventional varieties. The microparticles can be molded into various novel drug delivery systems such as microbeads, floating microspheres, implants, microemulsions, targeted microspheres, etc. Development of new polymers has further widened the scope of microparticulate drug delivery systems. The constant increase in the number of research works on microencapsulation stands as an evidence to this statement. The global market for microspheres in all fields of its applications was estimated to total nearly $2.2 billion in 2012 and is projected to increase to $2.4 billion in 2013; the market should total $4.4 billion by 2018 and have a 5-year compound annual growth rate for of 12.6%.

Microencapsulation is described as a process of enclosing micron-sized particles of solids or droplets of liquids or gasses in an inert shell, which in turn isolates and protects them from the external environment. The products obtained by this process are called microparticles.

The lowest particle size of microparticles is 1 μm and the largest size is 1 mm. Commercial microparticles have a diameter range of 3 to 800 μm. Microparticles may be formulated as microcapsules or as microspheres that differ in morphology and internal structure. In addition other terms such as beads, microbes are also being used alternatively.

This chapter includes an overview of microencapsulation, microparticulate drug delivery systems: their components, preparation techniques, evaluation, and types of microparticulate systems.

47.2 CLASSIFICATION[3]

Microcapsules can be classified based on three basic categories according to their morphology as follows (Figure 47.1):

1. Mononuclear (core-shell) microcapsules contain the shell around the core.
2. Polynuclear capsules have many cores enclosed within the shell.
3. Matrix types in which the core material is distributed homogeneously in the shell material.

In addition to these three basic morphologies, microcapsules can also be mononuclear with multiple shells, or they may form clusters of microcapsules.

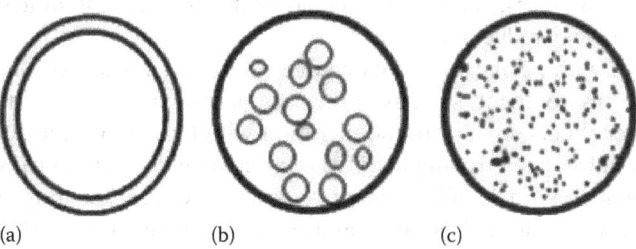

FIGURE 47.1 Morphology of microcapsules: (a) mononuclear, (b) polynuclear, and (c) matrix.

47.3 ADVANTAGES OF MICROENCAPSULATION[2,3,6]

The reasons for microencapsulation are countless. Overall, there are two main objectives of microencapsulation:

1. In some cases, the core must be isolated from its surroundings.
   Example: isolating vitamins from the deteriorating effects of oxygen.
2. In other cases, the objective is not to isolate the core completely, but to control the rate at which it leaves the microcapsule.
   a. Microencapsulation provides the means of altering colloidal and surface properties.
   b. To convert liquid active components into a dry solid system.
   c. To separate incompatible components for functional reasons.
   d. To protect the immediate environment of the microcapsules from the active components.
   e. To attain the sustained or prolonged release of the drug.
   f. For masking the organoleptic properties like taste and odor of many drugs and thus improve patient compliance.
   g. This technique can be used for converting liquid drugs in a free-flowing powder.
   h. Microencapsulation can be employed to change the site of absorption. This application has been useful for those drugs that have the toxicity at lower pH.
   i. The drugs, which are sensitive to oxygen, moisture, or light, can be stabilized by microencapsulation.
   j. Incompatibility among the drugs can be prevented by microencapsulation.
   k. Vaporization of many volatile drugs, for example, methyl salicylate and peppermint oil, can be prevented by microencapsulation.
   l. Many drugs have been microencapsulated to reduce toxicity and GI irritation.
   m. Alteration in site of absorption can also be achieved by microencapsulation.
   n. Toxic chemicals such as insecticides may be microencapsulated to reduce the possibility of sensitization of factorial person.
   o. In some cases, the core must be isolated from its surroundings.
   p. In the case of microencapsulated cells and enzymes in biotechnology, high-molecular-weight components can be retained in microcapsules, while low-molecular by-products and substrate residues are extracted through semipermeable microcapsule walls.

The purpose of microencapsulation can also be defined by the permeability:

1. Microcapsules with impermeable walls are used in products where isolation of active substances is needed, followed by a quick release under defined conditions. The effects achieved with impermeable microcapsules include: separation of reactive components, protection of sensitive substances against environmental effects, reduced volatility of highly volatile substances, conversion of liquid ingredients into a solid state, taste and odor masking, and toxicity reduction.
2. Microcapsules with permeable walls enable prolonged release of the active components into the environment, such as in the case of prolonged release drugs, perfumes, deodorants, repellents, etc., or immobilization with locally limited activity of microencapsulated substances. Examples of the latter include microencapsulated fertilizers and pesticides with locally limited release to reduce leaching into the ground water, or microencapsulated catalysts and enzymes for chemical and biotechnological processes.

47.4 MATERIALS FOR MICROENCAPSULATION[2-4]

Preparation of microspheres should satisfy certain criteria, like a basic understanding of the general properties of microcapsules, such as the nature of the core and coating materials, the stability and release characteristics of the coated materials, and the microencapsulating methods.

47.4.1 CORE MATERIAL

The core material defined as the specific material to be coated can be liquid or solid in nature. The composition of the core material is varied, as the liquid core can include dispersed and/or dissolved material. The solid core can be a mixture of active constituents, stabilizers, diluents, recipients, and release-rate retardants or accelerators. The ability to vary the core material composition provides definite flexibility, and utilization of this characteristic often allows effectual design and development of the desired microcapsule properties.

47.4.2 COATING MATERIALS

The coating material is an inert substance, which coats on core with desired thickness. Generally, hydrophilic polymers, hydrophobic polymers, (or) a combination of both are used for the microencapsulation process. The film thickness can be varied considerably, depending on the system. Surface area of the material to be coated and other physical characteristics of the coating materials used in microencapsulation methods are amenable, to some extent, to in situ modification. The list of a specific coating material is a lengthy. Table 47.1 provides the list of various polymers employed in microencapsulation. The typical coating properties such as cohesiveness, permeability, moisture

TABLE 47.1
List of Various Polymers Employed in Microencapsulation

Biodegradable Polymers	Based on Water Solubility	Enteric Coating Polymers
Synthetic	*Water soluble*	*Polymethacrylates*
Polyesters	Gelatin	Methacrylic acid/
Poly(ortho esters)	Gum Arabic	Ethyl acrylate
Polyanhydrides	Starch	
Polyphosphazenes	Polyvinyl pyrrolidone	*Cellulose esters*
	Carboxymethyl	Cellulose acetate phthalate (CAP)
Natural polymers	Cellulose	Cellulose acetate trimellitate (CAT)
Chitosan	Hydroxyethyl	HPMC acetate succinate
Hyaluronic acid	Cellulose	HPMC phthalate
Alginic acid	Methylcellulose	
	Arabino galactan	*Polyvinyl derivatives*
	Polyvinyl alcohol	Polyvinyl acetate phthalate
	Polyacrylic acid	
	Water insoluble	
	Ethylcellulose	
	Polyethylene	
	Polymethacrylate	
	Polyamide(nylon)	
	Poly (ethylene vinyl acetate)	
	Cellulose nitrate	
	Silicones	
	Poly lactide-*co*-glycolide	

sorption, solubility, stability, and clarity must be considered in the selection of the proper microcapsule coating material. The selection of the appropriate coating material dictates, to a major degree, the resultant physical and chemical properties of the microcapsules.

Properties of coating materials are

- Stabilization of core material
- Inert toward active ingredients
- Controlled release under specific conditions
- Film-forming, pliable, tasteless, and stable
- Nonhygroscopic, no high viscosity, and economical
- Soluble in an aqueous media or solvent, or melting
- Capable of forming a film that is cohesive with the core material
- Chemically compatible and nonreactive with the core material
- Provide the desired coating properties, such as strength, flexibility, impermeability, optical properties, and stability

The thickness that can be applied to small spherical particles must be of prime consideration just as the smallness of microcapsules allows unique properties and formulations to be accomplished. The thinness of the resultant coatings also can present unique problems. Examples of various coat materials are given:

1. Vegetable gums: gum arabic, agar, sodium alginate, carrageenan, and dextran sulfate
2. Celluloses: ethyl cellulose, methylcellulose, nitrocellulose, carboxy methyl cellulose, hydroxyethylcellulose, cellulose acetate phthalate, cellulose acetate butyrate phthalate
3. Proteins: collagen, gelatin, casein, fibrinogen, hemoglobin, and polyamino acids
4. Waxes: paraffin, rosin shellac, tristerium, monoglyceride, beeswax, oils, fats, and hardened oils
5. Condensation polymers: nylon, teflon, polymethane, polycarbonate, amino resins, alkyl resins, and silicone resins
6. Homopolymer: polyvinyl chloride, polyethylene, polystyrene, polyvinyl acetate, and poly vinyl alcohol
7. Copolymers: maleic anhydride copolymer with ethylene or vinyl methyl ether, acrylic acid copolymers, and methacrylic acid copolymers (Eudragit)
8. Curable polymers: epoxy resins, nitroparaffin, and nitrated polystyrene

47.5 MICROENCAPSULATION TECHNIQUES

Various techniques are available for the encapsulation of core materials (Table 47.2). There are many factors to consider when selecting the encapsulation process. Broadly, the methods are divided into two types, such as physical and chemical methods.

TABLE 47.2
Microencapsulation Techniques

Chemical Methods	Physicochemical Methods	Physical Methods
Interfacial polymerization	Coacervation and phase separation	Spray-drying and spray-congealing
In situ polymerization	Supercritical CO_2-assisted microencapsulation	
Solvent evaporation		Fluidized bed technology
		Air suspension coating
		Pan coating
		Centrifugal extrusion
		Jet excitation

Any technique selected for the preparation of microspheres should satisfy certain criteria:

- The ability to incorporate reasonably high concentrations of the drug
- Stability of the preparation after synthesis with a clinically acceptable shelf-life
- Controlled particle size and dispersibility in aqueous vehicles for injection
- Release of active reagent with a good control over a wide time scale
- Biocompatibility with a controllable biodegradability and susceptibility to chemical modification

47.5.1 Polymerization[7,8]

47.5.1.1 Interfacial Polymer

In interfacial polymerization, the two reactants in a polycondensation meet at an interface and react rapidly. The substances used are multifunctional monomers. Generally used monomers include multifunctional isocyanates and multifunctional acid chlorides. The basis of this method is the classical Schottenn–Baumann reaction between an acid chloride and a compound containing an active hydrogen atom, such as an amine or alcohol, polyesters, polyurea, and polyurethane. Under the right conditions, thin, flexible walls/shell will be formed at or on the surface of the droplet or particle by polymerization of the reactive monomers.

- The multifunctional monomer dissolves in the liquid core material.
- It will be dispersed in an aqueous phase containing dispersing agent.
- A coreactant multifunctional amine will be added to the mixture.
- This results in rapid polymerization at interface and generation of the capsule shell takes place. For example, A polyurea shell will be formed when isocyanate reacts with amine, polynylon.

Polyamide shell will be formed when acid chloride reacts with amine.

When isocyanate reacts with hydroxyl containing monomer, it produces a polyurethane shell.

47.5.1.2 In Situ Polymerization

In this process, no reactive agents are added to the core material; polymerization occurs exclusively in the continuous phase and on the continuous phase side of the interface formed by the dispersed core material and continuous phase. Initially, a low-molecular-weight prepolymer will be formed; as time goes on, the prepolymer grows in size, and it deposits on the surface of the dispersed core material, thereby generating a solid capsule shell.

Examples:

a. Encapsulation of various water-immiscible liquids with shells formed by the reaction at the acidic pH of urea with formaldehyde in aqueous media.
b. Cellulose fibers are encapsulated in polyethylene while immersed in dry toluene.

47.5.2 Solvent Evaporation[4,9–11]

Microencapsulation by solvent evaporation technique is widely used in pharmaceutical industries. It facilitates a controlled release of a drug, which has many clinical benefits. Different kinds of drugs have been successfully encapsulated.

For example: hydrophobic drugs such as cisplatin, lidocaine, naltrexone, and progesterone, and hydrophilic drugs such as insulin, proteins, peptide, and vaccine.

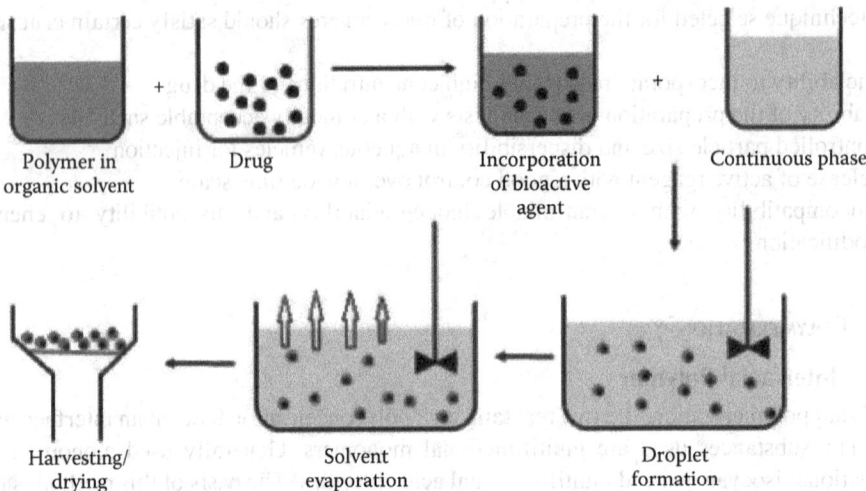

FIGURE 47.2 Schematic overview over the four principal process steps in microsphere preparation by solvent extraction/evaporation.

Microsphere preparation by solvent extraction/evaporation basically consists of four major steps as shown in Figure 47.2. They are the following:

(i) Incorporation of Bioactive Compounds: Bioactive compounds may be added to the solution of the matrix material by either codissolution in a common solvent, dispersion of finely pulverized solid material, or emulsification of an aqueous solution of the bioactive compound immiscible with the matrix material solution. Codissolution may require a cosolvent to fully dissolve the drug in the matrix containing solvent. Dispersion of the solid or dissolved bioactive material in the matrix-containing solution may be achieved by ultrasonication, impeller or static mixing, high-speed rotor–stator mixing, or microfluidization.

(ii) Droplet Formation: The droplet formation step determines the size and size distribution of the resulting microspheres. The main procedures used for droplet formation in microsphere production are the following:
Stirring: The extraction phase is filled into a vessel and agitated by an impeller. The drug/matrix dispersion is then added, dropwise or all at once, under agitation at a speed sufficient to reach the desired droplet size.
Static mixing: Static mixers consist of baffles or other flow obstacles installed in a tube. The baffle arrangement repeatedly splits and recombines the stream of fluid passing through the tube. Recombination occurs through impingement of the substreams, creating turbulence and inducing back-mixing.
Extrusion: Extrusion denotes feeding the drug/matrix dispersion through a single or a plurality of pathways directly into the continuous extraction phase. Upon leaving the pathway(s), discrete droplets of the drug/matrix dispersion are formed within the slowly flowing continuous phase, which also transports the droplets away from the site of their formation.

Extrusion is distinguished from static mixing by the droplet-forming mechanism and the prevailing flow regime. In extrusion, the flow is mainly laminar and the droplets are formed directly at the site of introduction of the dispersed phase into the continuous phase and do not change their dimension thereafter (given that coalescence is negligible). On the contrary, static mixing relies mainly on turbulent flow, which constantly acts on the disperse phase and thus causes the size of the droplets to change over the whole length of the mixer. Therefore, extrusion is considered to allow for more uniform and better controlled microsphere sizes than static mixing.

Jet excitation: A longitudinal oscillation imposed on a liquid stream causes periodic surface instabilities, which break up the liquid into a chain of uniform droplets.

(iii) Solvent Removal: In both solvent extraction and evaporation, the solvent of the disperse phase, that is, the drug/matrix dispersion must be slightly soluble in the continuous phase so that partitioning into the continuous phase can occur, leading to precipitation of the matrix material. In solvent evaporation, the capacity of the continuous phase is insufficient to dissolve the entire volume of the disperse phase solvent. Therefore, the solvent must evaporate from the surface of the dispersion to yield sufficiently hardened microspheres. In solvent extraction, the amount and composition of the continuous phase are chosen so that the entire volume of the disperse phase solvent can be dissolved.

Generally, a continuous phase that is a nonsolvent for the microencapsulated bioactive compound is favorable. For lipophilic compounds, aqueous solutions may be comfortably chosen while the use of hydrophobic, organic liquids is preferred as continuous phase for the encapsulation of hydrophilic compounds. The ideal rate of solvent removal depends on a variety of factors like the type of matrix material, drug and solvent as well as the desired release profile of the microspheres.

(vi) Harvest and Drying: Separation of the solidified microspheres from the continuous phase is usually done either by filtration or centrifugation. The particles may then be rinsed with appropriate liquids to remove adhering substances such as dispersion stabilizers or nonencapsulated drugs. Rinsing may involve elevated temperatures or the use of extraction agents to reduce the amount of residual solvent in the microspheres. Finally, the microspheres are dried either at ambient conditions or under reduced pressure, heat, or by lyophilization to yield a free-flowing powder.

47.5.2.1 Choice of Materials[12]

47.5.2.1.1 Dispersed Phase

47.5.2.1.1.1 Polymer The biodegradability or biocompatibility is an essential property for the polymer used in pharmaceutical applications. Nonbiodegradable polymers with good biocompatibility are also used as drug carriers. The choice of polymer used as drug carrier depends also on the desired drug release rate, which is essentially determined by the polymer's physical properties. If one polymer cannot offer a satisfying drug release, a single polymer, called copolymer, can be synthesized from two different polymers. The properties of the copolymer are improved, since it has two segments in the chain. For example, copolymer (PEG/PLA) synthesized from polyethylene glycol (PEG) and polylactic acid homopolymers (PLA) in order to increase the degradation rate. Table 47.3 lists outs various polymers employed in solvent evaporation and their properties.

47.5.2.1.1.2 Solvent The commonly used solvents for solvent evaporation method are given in Table 47.4. For the technique of microencapsulation by solvent evaporation, a suitable solvent should meet the following criteria:

- Being able to dissolve the chosen polymer
- Being poorly soluble in the continuous phase
- Have a high volatility and a low boiling point
- Have low toxicity

TABLE 47.3
Polymers Commonly Used for Microencapsulation Using Solvent Evaporation Technique

Polymer	Full Name	Properties
PLGA, PLG	Poly(lactic-*co*-glycolic acid) or poly(lactide-*co*-glycolide)	Good biodegradability and biocompatibility
PLA	Poly(lactic acid) or polylactide	Good biodegradability and biocompatibility, slow degradation rate compared to PLGA
PEG	Poly(ethylene glycol)	Often synthesized with PLGA or with PLA to form a copolymer with fast degradation rate
PHB	Poly-3-hydroxybutyrate	Bacterial storage polyester; slower degradation rate than polylactic polymers
PHB-HV	Poly-3-hydroxybutyrate with hydroxyvalerate	Bacterial storage polyester; slower degradation rate than polylactic polymers
EC	Ethyl cellulose	Degradable, biocompatible, approved by FDA for pharmaceutical application; low cost
PMMA	Polymethyl methacrylate	Nondegradable but biocompatible, approved by FDA; bone cement material; low cost
	Eudragit	Enteric coating polymer

Source: Ming, L. et al., *Int. J. Pharm.*, 363(1–2), 26, 2008.

TABLE 47.4
List of Solvents Commonly Used for Microencapsulation by Solvent Evaporation

Name	Vapor Pressure (mbar) at 20°C	Boiling Point (°C)	Solubility in Water (gm/L) at 20°C
Chloroform	212	61	8
Dichloromethane (methylene chloride)	453	39	20
Ethyl acetate	100	77	90
Ethyl formate	259	54	105

Source: Ming, L. et al., *Int. J. Pharm.*, 363(1–2), 26, 2008.

47.5.2.1.1.3 Alternative Components Cosolvent is used to dissolve the drug that is not totally soluble in the solvent in the dispersed phase. For example, organic solvents miscible with water, such as methanol and ethanol, are the common choices:

- Porosity generator, called also porosigen or porogen, is used to generate the pores inside the microspheres, which consequently increases the degradation rate of polymer and improves the drug release rate. For example, organic solvents such as hexane, which do not dissolve poly (lactic acid) and poly (lactic-co-glycol acid), can be incorporated into microspheres to form pores.

47.5.2.1.2 Continuous Phase

47.5.2.1.2.1 Surfactant The surfactant, also called tension-active agent, reduces the surface tension of continuous phase, avoids the coalescence and agglomeration of drops, and stabilizes the emulsion. A suitable surfactant should be able to give microspheres a regular size and a small size distribution, guaranteeing a more predictable and stable drug release. The increase of surfactant concentration reduces the size of microspheres. The addition of surfactant lowers the surface tension of the continuous phase and the diminution of the latter one decreases the particle size.

For example, nonionic: partially hydrolyzed polyvinyl alcohol (PVA), methylcellulose, tween, and span.

Anionic: sodium dodecyl sulfate (SDS).

Cationic: cetyltrimethyl ammonium bromide (CTAB).

Among these surfactants, partially hydrolyzed PVA is mostly used, because it gives the smallest microspheres.

47.5.2.1.2.2 Alternative Components Besides the surfactant, the antifoam is sometimes added into the aqueous phase in the case of strong agitation, because the foaming problem will disturb the formation of microspheres. When the stirring speed increases, more air is entrained and foam forms. So antifoams of silicon and nonsilicon constituents are used to increase the rate at which air bubbles are dissipated.

Recent studies show that it is possible to prepare microspheres without surfactant by replacing it with an amphiphilic biodegradable polymer. The advantage is to avoid the potential harm of surfactant residue on the surface or inside the final microspheres.

For example, PLA oligomers prepared by direct condensation of D,L-lactic acid have an amphiphilic surfactant similar structure, since the polymers are composed of a hydrophobic polyester chain ended with a carboxylic acid group, which can be ionized to form hydrophilic carboxylate polar heads at neutral pH in water.

47.5.2.2 Variables and Factors[11,13,14]

- *Polymer concentration*: A saturated solution of polymer produces smooth and high-yield microspheres. Increased viscosity of the drug/matrix dispersion yields larger microspheres, because higher shear forces are necessary for droplet disruption. The undissolved polymer produces irregular and rod-shaped particles. The amount of polymer directly affects swelling index.
- *Stabilizer concentration*: Increasing the stabilizer concentration frequently leads to decreased microsphere sizes. Higher stabilizer concentrations will yield a large excess of material that is adsorbed on the surface of newly formed droplets, thus preventing coalescence.
- *Temperature*: Preparation at lower temperatures provides porous microspheres having higher porosity, with a rough surface.
- *Extent of loading*: When the loading is high, the proportion of larger particles formed will also be high. The size of the microspheres formed may however be a function of many other factors.
- *Baffles*: Both the size distribution of the microspheres and the median particle size indicated will be significantly smaller when side baffles are present in the apparatus.
- *Effect of agitation rate*: As the agitation rate decreases, the microsphere size and size distribution increases. The tendency of the droplets to coalesce and aggregate at the slower agitation rates may result in larger mean microsphere diameters. These low agitation rates may decrease the uniformity of the mixing force throughout the emulsion mixture, hence resulting in a wider size distribution of the final microspheres. As the rotation speed/agitation rate increases, the average particle size decreases, while maintaining its morphology and width of the size distribution as it produces smaller emulsion droplets through stronger shear forces and increased turbulence. The stirring speed directly affects swelling index also. The optimum rotation speed for any experimental system has to be judged from the results of particle size and size distribution and the drug content.
- *Dispersed phase volume*: The microsphere yield declines as larger volumes of solvent or less viscous dispersed phase solutions are used. When very dilute dispersed phase is employed, the addition of two phases together produces rapid intermixing of the phases, resulting in immediate precipitation of drug and polymer before droplets could form.

TABLE 47.5
Impact of Parameters and Operating Conditions on the Properties of Microspheres

Increase of the Following Parameters	Size	Morphology	Encapsulation Efficiency
Viscosity of the dispersed phase	Bigger diameter	Smoother surface	Increases; slower drug release
Polymer concentration			Increases
Volume fraction of dispersed phase to continuous phase	Decreases or no influence		Increases
Quantity of drug in the dispersed phase		More porous and irregular shape	Decreases due to the formation of big pores
Concentration of porosigen	Smaller diameter	Coarser surface with larger pores	No impact
Increase of agitation rate	Smaller diameter		
Increase of temperature	Bigger diameter	Coarser surface	Decreases
Solvent removal rate			Increases due to faster solidification of particles
Low solubility of the polymer in organic solvent			High encapsulation efficiency
High solubility of organic solvent in water			
High solubility of the polymer in organic solvent			Low encapsulation efficiency
Low solubility of organic solvent in water			

Source: Ming, L. et al., *Int. J. Pharm.*, 363(1–2), 26, 2008.

Hence, no microspheres will be produced. When more viscous dispersed phases are used, some intermixing of the phases, as well as droplet formation occurs to yield microspheres.

- *Volume of aqueous phase*: With larger volumes, solidification of particles occurs faster as compared to that with small volumes of aqueous phase. Thus, particles can be separated after shorter stirring times.
- *Diffusion rate of solvent*: Few research works reported that when the diffusion rate of solvent out of emulsion droplet is too slow, microspheres coalesce together. Conversely, when the diffusion rate of solvent is too fast, the solvent may diffuse into the aqueous phase before stable emulsion droplets are developed, causing the aggregation of embryonic microsphere droplets.

Table 47.5 briefs about the impact of parameters and operating conditions on the properties of microspheres.

- Depending on the drug properties (e.g., solubility, stability), the drug-containing polymer phase is then emulsified into an external oil (o/o or w/o/o) or aqueous phase (o/w or w/o/w).

47.5.3 Single Emulsion Method[6,15,16]

For insoluble or poorly water-soluble drugs or hydrophobic drugs, the oil-in-water (o/w) method is frequently used. The process is as follows (Figure 47.3):

- The polymer is dissolved in a water-immiscible, volatile organic solvent such as dichloromethane.
- The drug is dissolved or suspended in the polymer solution.
- The resulting mixture is emulsified in a large volume of water in the presence of an emulsifier.

Microparticulate Drug Delivery Systems

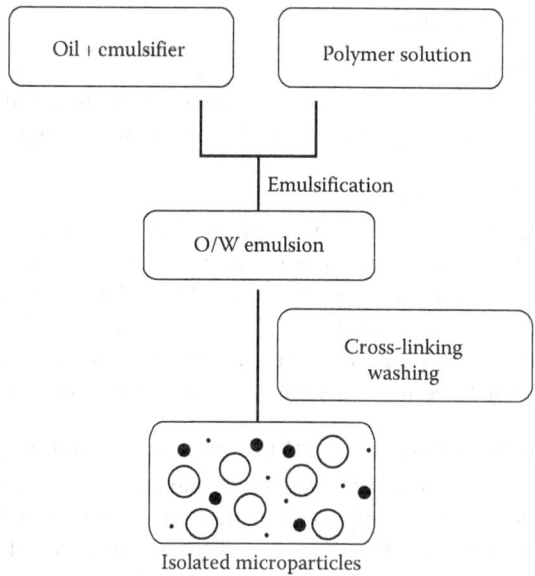

FIGURE 47.3 Schematic procedure of single emulsion technique.

- The solvent in the emulsion is removed by either evaporation at elevated temperatures or extraction in a large amount of water, resulting in formation of compact microparticles.

Disadvantage: This method is not suitable for the encapsulation of high hydrophilic drugs, because the hydrophilic drugs may diffuse out or partition from the dispersed oil phase into the aqueous phase, leading to poor encapsulation efficiencies. There are two main reasons:

- The hydrophilic drug may not be dissolved in the organic solvent.
- The drug will diffuse into the continuous phase during emulsion, leading to a great loss of drug.
- Accumulation of drug crystals on the surface of microparticles produces a burst release of the drug upon administration. The partitioning effect of the drug can be reduced by chemically modifying it to a lipophilic prodrug prior to its incorporation into the organic phase. Otherwise, modification of the continuous phase of the emulsion to reduce leakage of the drug from the oily droplets can be employed.

In an attempt to encapsulate hydrophilic drugs (e.g., peptides and proteins), the following alternative methods have been proposed:

1. *The o/w cosolvent method*: When the drug is not soluble in the main organic solvent, a second solvent called cosolvent is necessary to dissolve the drug.
2. *The o/w dispersion method*: The drug is dispersed in the form of solid powder in the solution of polymer and organic solvent.
3. *The o/o nonaqueous solvent evaporation method*: In this method, water-miscible organic solvents are employed to dissolve the drug and polymer, whereas hydrophobic oils are used as a continuous phase of the o/o emulsion.

47.5.4 DOUBLE EMULSION METHOD[6,15]

The double emulsion method has been considered better, since the method is relatively simple, convenient in controlling process parameters, and has the ability to produce with simple instrument. The process can efficiently encapsulate highly water-soluble compounds including proteins and peptides.

47.5.4.1 W/O/W Double Emulsion

In the water in oil in water (w/o/w) double emulsions, the internal and external aqueous phases are separated by an oil layer. For their formation and stability, at least two surfactants, one having a low HLB to form the primary water in oil (w/o) emulsion and the other having a higher HLB to achieve secondary emulsification, are required to emulsify water in oil emulsion into water. These emulsion (w/o/w) systems, being less viscous, are excellent candidates for controlled release of hydrophilic drugs due to the existence of a middle oil layer that acts as a liquid membrane.

47.5.4.2 W/O/W Double Emulsion Solvent Evaporation Technique

In water/oil/water double emulsion solvent evaporation method, an aqueous solution or suspension of the drug (internal aqueous phase, W_1) is emulsified in a solution of polymer in organic solvent. The resulting primary emulsion (W_1/O) is then dispersed in a second aqueous phase (external aqueous phase, W_2) containing suitable emulsifier(s) to form double emulsion (W_1/O/W_2). Removal of the volatile organic solvent leads to the formation of solid microparticles. The volatile organic solvent used in the preparation of the microparticles by the double emulsion solvent evaporation method should be of low boiling point to facilitate the removal of residual solvent. Various solvents that can be used for the preparation of microparticles are acetonitrile, ethyl acetate, chloroform, and benzene and methylene chloride. The procedure is briefly illustrated in Figure 47.4.

In w/o/w double emulsion solvent extraction technique, the solvent can be removed from the microparticles through extraction of the solvent present in the internal phase. This can be achieved by the addition of w/o/w emulsion to a third solution, which is a nonsolvent of the polymer but miscible with both water and the organic solvent. Under these conditions, the solvent contained in the polymer droplets is extracted into the aqueous medium.

In a novel method reported by Ereitas et al.,[17] a flow through ultrasonic cell was used to prepare primary emulsion and static mixer was used to prepare double emulsion.

47.5.4.3 S/O/W Double Emulsion Technique

Using solids, a double emulsion of solid in oil in water (s/o/w) can be prepared, in which the internal solid phase and external aqueous phase are separated by an oil layer. Initially, the solid pharmaceuticals or biopharmaceuticals are dispersed in the polymer solution to form a primary emulsion. Then the dispersion is introduced into a large volume of aqueous solution containing emulsifying agent, such as PVA or PEG.

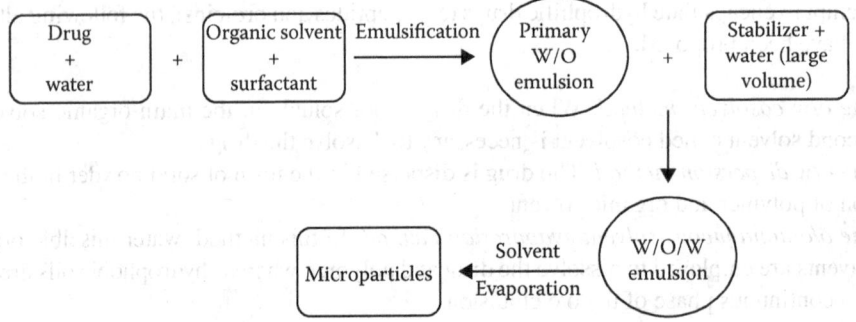

FIGURE 47.4 Schematic procedure of double emulsion solvent evaporation method.

Advantage: The s/o/w technique has gained recognition, since it improves protein stability during encapsulation, because the first w/o emulsion step is avoided.

47.5.4.4 S/O/O Double Emulsion Technique

Similarly, in the solid in oil in oil (s/o/o) double emulsions, the internal solid phase and external oil phase are separated by an oil phase. The s/o/o emulsion can be prepared by dispersing solid particles in an organic solvent, for example, methylene chloride, and then mixed with a polymer solution. Alternatively, the dispersion of solids in organic solvent can be coated with a suitable polymer using a coacervation technique.

47.5.4.5 W/O/O Double Emulsion Technique

In the water-in-oil-in-oil (w/o/o) double emulsions, the internal aqueous phase and external oil phase are separated by an oil phase. In the w/o/o method, water-soluble compounds are first dissolved in the aqueous phase, which is emulsified in an oil to form a stable emulsion. This primary emulsion is then dispersed in a solution of polymer in organic solvent to form a w/o/o emulsion.

Advantages:

- Water-soluble compounds cannot diffuse into the processing medium, since the oil is outer processing medium.
- The total protein encapsulation efficiency is high in the microparticles produced by w/o/o emulsification technique, since oil is the outer processing medium and protein cannot diffuse into the processing medium.

47.5.4.6 O/O/O Extraction Method[18]

Depending on the drug properties (e.g., solubility, stability), the drug-containing polymer phase is emulsified into an external oil (o/o or w/o/o) or aqueous phase (o/w or w/o/w). Major problems of both microencapsulation methods are the use of toxic solvents (e.g., methylene chloride, chloroform, acetonitrile, etc.) and residual solvents in the microparticles. Because of toxicity issues, the solvent selection for the microparticles preparation has received some attention and a number of nonchlorinated solvents (e.g., ethyl acetate or acetonitrile) have been evaluated. According to ICH guideline, 50 mg or less per day (corresponding to 5000 ppm or 0.5%) of class 3 solvents is acceptable for human use without justification. There are several examples of class 3 solvents like dimethylsulfoxide (DMSO), 2-pyrrolidone, or PEG 400 (poly (e-caprolactone).

Advantages:

- Among o/o/o, o/o/w, w/o/w method to prepare microparticles, the o/o/o-method gives the highest encapsulation efficiency.
- Less toxic and nonflammable solvents are used in the proposed o/o/o method.
- The higher polymer concentration in the solution results in a smaller preparation volume and lower solvent consumption.

47.5.5 COACERVATION[8]

For this method, the core material should not react or dissolve in water (maximum solubility 2%). The steps involved are the following:

Dispersion: The core material is dispersed in the solution. The particle size will be defined by dispersion parameters such as stirring speed, stirrer shape, surface tension, and viscosity. Size range that can be obtained by this simple technique is 2 to 1200 µm.

Coacervation: Coacervation starts with a change in the pH value of the dispersion, for example, by adding H_2SO_4, HCl, or organic acids. The result is a reduction of the solubility of the dispersed phases (shell material). The shell material (coacervate) starts to precipitate from the solution, forming a continuous coating around the core droplets.

Cooling and hardening phase: The shell material is cooled down to harden and forms the final capsule. Hardening agents like formaldehyde can be added to the process. The microcapsules are now stable in the suspension and ready to be dried.

Drying phase: The suspension is dried in a spray-dryer or in a fluidized bed drier. Spray-drying is a suitable method for heat-sensitive products. The atomized particles assume a spherical shape.

47.5.5.1 Simple Coacervation[3]
Simple coacervation can be effected either by mixing two colloidal dispersions, one having a high affinity for water, or it can be induced by adding a strongly hydrophilic substance such as alcohol or sodium sulfate.

47.5.5.2 Complex Coacervation[3]
Complex coacervation can be induced in systems having two dispersed hydrophilic colloids of opposite electric charges. Neutralization of the overall positive charges on one of the colloids by the negative charge on the other is used to bring about separation of the polymer-rich complex coacervate phase. A typical complex coacervation process using gelatin and gum arabic colloids is given in Figure 47.5.

47.5.6 Coacervation Phase Separation[3,7]

Bungenberg de Jong and Kruyt (1929) and Bungenberg de Jong (1949) defined this as partial desolvation of a homogeneous polymer solution into a polymer-rich phase (coacervate) and the poor polymer phase (coacervation medium). The general outline of the process consists of three steps carried under continuous agitation, as given in Figure 47.6.

47.5.6.1 Step 1: Formation of Three Immiscible Chemical Phases
The immiscible chemical phases are (1) a liquid manufacturing vehicle phase, (2) a core material phase, and (3) a coating material phase. To form the three phases, the core material is dispersed in a solution of the coating polymer, the solvent for the polymer being the liquid manufacturing vehicle phase. The coating material phase, an immiscible polymer in a liquid state, is formed by utilizing one of the methods of phase separation coacervation, that is,

- By changing the temperature of the polymer solution (e.g., ethyl cellulose in cyclohexane12 (*N*-acetyl P-amino phenol as core)
- By adding a salt (e.g., addition of sodium sulfate solution to gelatin solution in vitamin encapsulation)

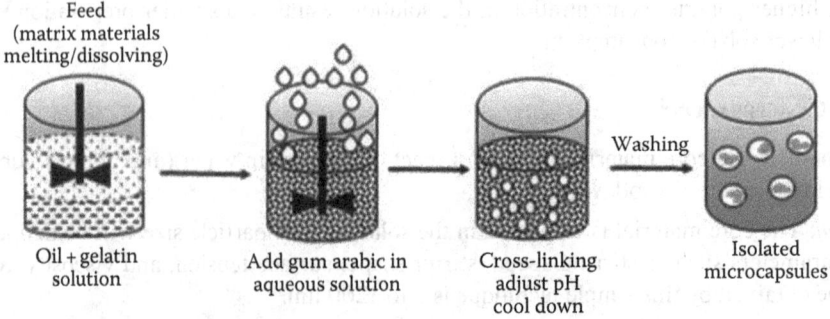

FIGURE 47.5 General process scheme of complex coacervation using gelatin/gum arabic.

Microparticulate Drug Delivery Systems

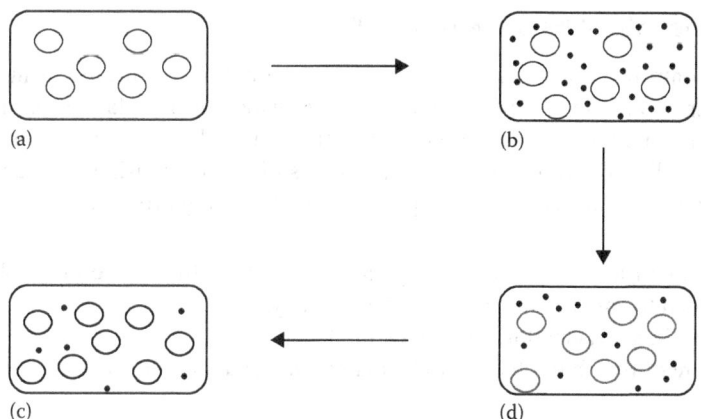

FIGURE 47.6 Steps involved in coacervation process: (a) core material dispersion in solution of shell polymer, (b) separation of coacervate from solution, (c) coating of core material by microdroplets of coacervate, and (d) coalescence of coacervate to form continuous shell around core particles.

- By adding a nonsolvent (e.g., addition of isopropyl ether to methyl ethyl ketone solution of cellulose acetate butyrate)
- By adding incompatible polymer to the polymer solution (e.g., addition of polybutadiene to the solution of ethylcellulose in toluene; methylene blue as core material)
- By inducing a polymer–polymer interaction (e.g., interaction of gum Arabic and gelatin at their isoelectric point)

47.5.6.2 Step 2: Depositing the Liquid Polymer Coating on the Core Material

This is accomplished by controlled, physical mixing of the coating material (while liquid) and the core material in the manufacturing vehicle. Deposition of the liquid polymer coating around the core material occurs if the polymer is adsorbed at the interface formed between the core material and the liquid vehicle phase, and this adsorption phenomenon is a prerequisite to effective coating. The continued deposition of the coating material is promoted by a reduction in the total free interfacial energy of the system, brought about by the decrease of the coating material surface area during coalescence of the liquid polymer droplets.

47.5.6.3 Step 3: Rigidizing the Coating

This is usually done by thermal, cross-linking, or desolvation techniques, to form a self-sustaining microcapsule. Few examples of cross-linking agents include

- Derivatives of ethylene glycol di(meth)acrylate like, ethylene glycol diacrylate, di(ethylene glycol) diacrylate, tetra (ethylene glycol) diacrylate, ethylene glycol dimethacrylate, di(ethylene glycol) dimethacrylate, and tri(ethylene glycol) dimethacrylate
- Derivatives of methylene bisacrylamide like N, N-methylenebis acrylamide, N,N-methylene bisacrylamide, N,N-(1,2-dihydroxyethylene) bisacrylamide, glutaraldehyde, and sodium tripolyphosphate
- Chitosan at pH 7.0
- Polyethylenimine (PEI) under basic conditions (pH 10.5)

Increasing the amount of cross-linking agent leads to an increase in the encapsulation ability. Few studies have reported that the particle size of the microcapsule was not affected by the difference of cross-linking agents.

47.5.7 MODIFIED MELT-DISPERSION METHOD[19]

Preparation of controlled-release lipid microspheres is commonly achieved using melt-dispersion (congealable disperse phase encapsulation) technique, which utilizes fats of animals and/or vegetable origin as the matrix. This technique is especially suitable for drugs, which are insoluble in water. Bhoyar et al.[19] have prepared lipid microspheres of naproxen using carnauba wax as a lipid carrier by modified melt-dispersion technique using the following procedure:

- The lipid is first melted on the heating mantle. Lipid modifiers like glycerol stearate can also be added to enhance the entrapment efficiency.
- The drug is then dispersed in the molten lipid.
- Temperature of this fatty phase is to be maintained at about 10°C above the melting point of lipid.
- Then, external phase containing a dispersant (polyvinyl alcohol) is maintained at about 5°C above the melting point of lipid.
- The fatty phase is then poured into this solution while continuously stirring to form an o/w emulsion.
- The resultant emulsion is then agitated for 3 min while maintaining the temperature.
- Hardening of the oily internal phase and formation of emulsion microspheres is accomplished by pouring, twice the volume of ice-cold acidified water (4°C) into the beaker and stirring continuously for further 15 min in an ice bath.
- Microspheres obtained are filtered under vacuum and washed with phosphate suitable buffer to remove the unentrapped drug and dried in room temperature for 24 h.

The release of drug from microspheres is influenced by number of factors such as nature of carrier, drug concentration, dispersant concentration, stirring speed, stirring time, external-phase temperature, and external phase.

47.5.8 INTERNAL GELATION[20]

Emulsification/internal gelation has been suggested as an alternative to extrusion/external gelation in the encapsulation of several compounds including sensitive biologicals such as protein drugs. This is a simple and economic method for microencapsulation proposed for producing small-diameter alginate microspheres in large quantity.

The difficulty in using dispersion/external gelation techniques with ionic polysaccharide is that the calcium source ($CaCl_2$) is insoluble in the oil phase. As an alternative, internal gelation of the dispersed alginate droplets may be initiated by releasing Ca^{2+} from an insoluble complex (calcium salt) through pH reduction. By controlling the conditions under which the water-in-oil dispersion is produced, the bead size can be controlled from a few microns to millimeters in diameter. Microencapsulation by the emulsification/gelation method involves two major steps:

- The formation of stable droplets of the polymer solution with drug incorporated in as an emulsified system
- The subsequent solidification of the droplets

These two steps have a significant effect on size and encapsulation efficiency of microparticles. Stirring speed is the most important parameter for controlling the drug/matrix dispersion's droplet size in the continuous phase. It was shown that increasing the stirring speed generally results in decreased microparticle size, as it produces smaller emulsion droplets through stronger shear forces and increased turbulence.

47.5.9 SUPERCRITICAL FLUID TECHNOLOGY[6,7]

Supercritical fluids are highly compressed gasses that possess several advantageous properties of both liquids and gases. A small change in temperature or pressure causes a large change in the density of supercritical fluids near the critical point. In the case of carbon dioxide, the supercritical region can be achieved at moderate pressures and temperatures (T_c = 304.2 K, P_c = 7.38 MPa). The most widely used methods for microencapsulation are as follows.

47.5.9.1 Rapid Expansion of Supercritical Solution

In this process, supercritical fluid containing the active ingredient and the shell material are maintained at high pressure and then released at atmospheric pressure through a small nozzle. The sudden drop in pressure causes desolvation of the shell material, which is then deposited around the active ingredient (core) and forms a coating layer. An overview of process involved is given in Figure 47.7.

The disadvantage of this process is that both the active ingredient and the shell material must be very soluble in supercritical fluids. In general, very few polymers are soluble in supercritical fluids such as CO_2. The solubility of polymers can be enhanced by using cosolvents or nonsolvents. This increases the solubility in supercritical fluids, but the shell materials do not dissolve at atmospheric pressure.

47.5.9.2 Gas Antisolvent (GAS) Process

This process is also called supercritical fluid antisolvent (SAS). Here, supercritical fluid is added to a solution of shell material and the active ingredients and maintained at high pressure. This leads to a volume expansion of the solution that causes supersaturation such that precipitation of the solute occurs. Thus, the solute must be soluble in the liquid solvent, but should not dissolve in the mixture of solvent and supercritical fluid. On the other hand, the liquid solvent must be miscible with the supercritical fluid.

This process is unsuitable for the encapsulation of water-soluble ingredients, as water has low solubility in supercritical fluids. It is also possible to produce submicron particles using this method.

47.5.9.3 Particles from a Gas-Saturated Solution (PGSS)

This process is carried out by mixing core and shell materials in supercritical fluid at high pressure. During this process, supercritical fluid penetrates the shell material, causing swelling. When the mixture is heated above the glass transition temperature (T_g), the polymer liquefies. Upon releasing the pressure, the shell material is allowed to deposit onto the active ingredient. In this process, the core and shell materials may not be soluble in the supercritical fluid. Preformed microparticles are often used for the entrapment of active materials using supercritical fluids under pressure. When the pressure is released, the microparticles shrink and return to their original shape and entrap the ingredients.

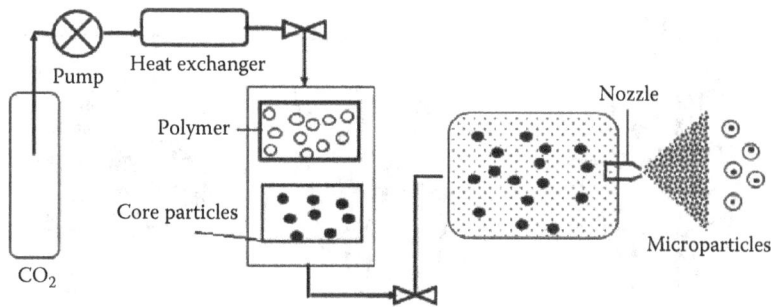

FIGURE 47.7 Microencapsulation by rapid expansion of supercritical solutions (RESS).

47.5.10 Fluidized-Bed Technology[7]

In this method, solid particles to be encapsulated are suspended on a jet of air and then covered by a spray of liquid coating material. The capsules are then moved to an area where their shells are solidified by cooling or solvent vaporization. The process of suspending, spraying, and cooling is repeated until microparticles of desired thickness are obtained. Different types of fluid-bed coaters include top spray, bottom spray, and tangential spray, as depicted in Figure 47.8:

1. *Top spray*: In the top spray system, the coating material is sprayed downward on to the fluid bed such that as the solid or porous particles move to the coating region, they become encapsulated. Top spray fluid-bed coaters produce higher yields of encapsulated particles than either bottom or tangential sprays.
2. *Bottom spray*: This process is known as the Wurster process/bottom spray when the spray nozzle is located at the bottom of the fluidized bed of particles. As the particles move upward through the perforated bottom plate and pass the nozzle area, they are encapsulated by the coating material. Although it is a time-consuming process, the multilayer coating procedure helps in reducing particle defects.
3. *Tangential spray*: It consists of a rotating disk at the bottom of the coating chamber, with the same diameter as the chamber. During the process, the disk is raised to create a gap between the edge of the chamber and the disk. The tangential nozzle is placed above the rotating disk through which the coating material is released. The particles move through the gap into the spraying zone and are encapsulated. As they travel a minimum distance, there is a higher yield of encapsulated particles.

47.5.11 Centrifugal Extrusion[10]

In this process, a jet of core liquid is surrounded by a sheath of coat solution or melt. The core and the shell materials should be immiscible with one another.

Initially, dual fluid stream of liquid core and shell materials is pumped through concentric tubes. As the jet moves through the air, due to vibration, it breaks into droplets of core, each coated with the wall solution, as given in Figure 47.9. While the droplets are in flight, a molten wall may be hardened or a solvent may be evaporated from the wall solution.

47.5.12 Air Suspension Coating[7,5]

Microencapsulation by air suspension technique involves dispersing of solid, particulate core materials in a supporting air stream, and the spray-coating on the air suspended particles:

- Within the coating chamber, particles are suspended on an upward moving air stream.
- Just sufficient air is permitted to rise through the outer annular space to fluidize the settling particles. Most of the rising air (usually heated) flows inside the cylinder, causing the

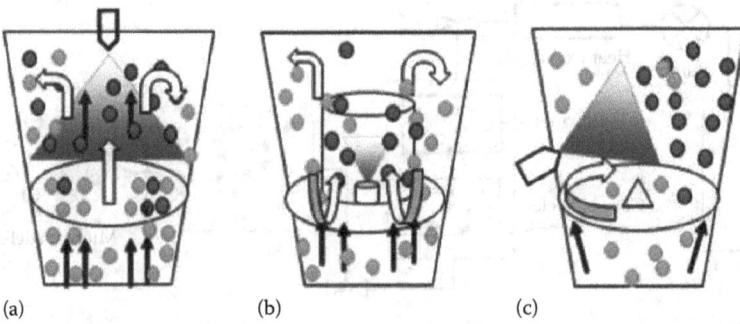

FIGURE 47.8 Schematics of a fluid-bed coater: (a) top spray, (b) bottom spray, and (c) tangential spray.

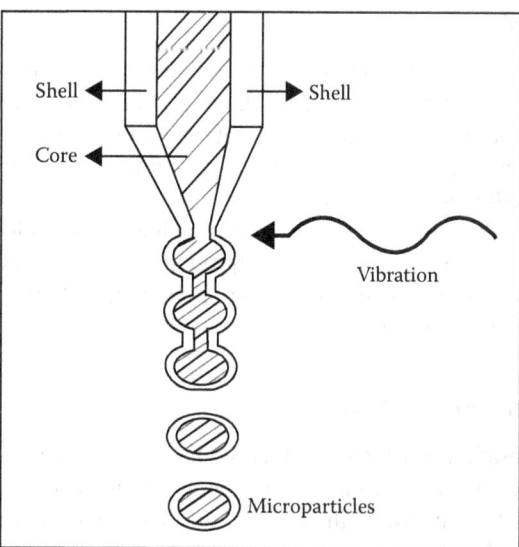

FIGURE 47.9 Representation of a typical centrifugal extrusion.

particles to rise rapidly. At the top, as the air stream diverges and slows, they settle back onto the outer bed and move downward to repeat the cycle. The particles pass through the inner cylinder many times in a few minutes methods. Thus, the design of the chamber and its operating parameters affect a recirculating flow of the particles.

- During recirculation, the particles pass through the coating zone portion of the chamber, where a coating material, usually a polymer solution, is spray-applied to the moving particles.
- During each pass through the coating zone, the core material receives an increment of coating material.
- The cyclic process is repeated, perhaps several hundred times during processing, depending on the purpose of microencapsulation, the coating thickness desired, or whether the core material particles are thoroughly encapsulated.
- The supporting air stream also serves to dry the product while it is being encapsulated. Drying rates are directly related to the volume temperature of the supporting air stream.

The process variables that can affect the process are

- Concentration of the coating material or, if in solid form, then its melting point
- Solubility, surface area, density, volatility, and melting point of the core material
- Application rate of coating material
- Temperature of air stream
- Amount of air required to fluidize the core material

Air-suspension coating of particles by solutions or melts gives better control and flexibility. The particles are coated while suspended in an upward-moving air stream. They are supported by a perforated plate having different patterns of holes inside and outside a cylindrical insert. The air suspension process offers a wide variety of coating material candidates for microencapsulation. The process has the capability of applying coatings in the form of solvent solutions, aqueous solution, emulsions, dispersions, or hot melts in equipment ranging in capacities from one pound to 990 pounds. Core materials comprised of micron or submicron particles can be effectively encapsulated by air suspension techniques, but agglomeration of the particles to some larger size is normally achieved.

47.5.13 Jet Excitation[6]

The main principle involved is that when a longitudinal oscillation is imposed on a liquid stream, it causes periodic surface instabilities, which break up the liquid into a chain of uniform droplets. Uniform droplets are produced from a range of excitation wavelengths corresponding to 7–36 times the liquid jet radius. This principle was recently used by Berkland et al.[21] to produce uniform PLGA microparticles. The procedure was as follows.

A 5% (w/v) solution of PLGA in DCM was fed through a nozzle to form a cylindrical jet. The nozzle was excited by an ultrasonic transducer of adjustable frequency. The particles were collected in 1% (w/v) PVA solution for solvent extraction/evaporation. Figure 47.10 gives the schematic representation of microencapsulation by jet excitation.

47.5.14 Spray-Drying and Congealing[9]

Spray-drying is a unit operation by which a liquid product is atomized in a hot gas current to instantaneously obtain a powder. Microencapsulation by spray-drying is conducted by dispersing a core material in a coating solution, in which the coating substance is dissolved and in which the core material is insoluble, and then by atomizing the mixture into air stream. The air, usually heated, supplies the latent heat of vaporization required to remove the solvent from the coating material, thus forming the microencapsulated product. The equipment components of a standard spray dryer include an air heater, atomizer, main spray chamber, blower or fan, cyclone, and product collector.

Spray-drying produces, depending on the starting feed material and operating conditions, a very fine powder (10–50 lm) or large size particles (2–3 mm). The main advantages are the ability to handle labile materials because of the short contact time in the dryer; in addition, the operation is economical. In modern spray-dryers, the viscosity of the solutions to be sprayed can be as high as 300 mPa·s.

Spray-drying and spray-congealing processes are similar: both involve dispersing the core material in a liquefied coating substance and spraying or introducing the core–coating mixture into some environmental condition, whereby relatively rapid solidification (and formation) of the coating is affected. The principal difference between the two methods is the means by which

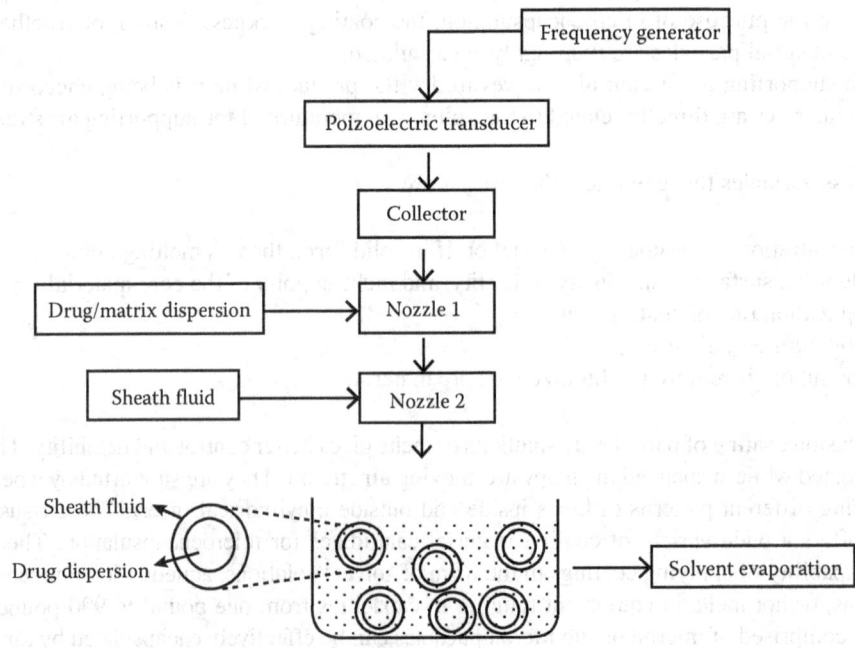

FIGURE 47.10 Microencapsulation by jet excitation.

coating solidification is accomplished. Coating solidification in the case of spray-drying is effected by rapid evaporation of a solvent in which the coating material is dissolved. Coating solidification in spray-congealing methods, however, is accomplished by thermally congealing a molten coating material or by solidifying a dissolved coating by introducing the coating–core material mixture into a nonsolvent.

47.5.14.1 Airflow Patterns

The initial contact between spray droplets and drying air controls evaporation rates and product temperatures in the dryer. There are three modes of contact:

Cocurrent: Drying air and particles move through the drying chamber in the same direction. Product temperatures on discharge from the dryer are lower than the exhaust air temperature, and hence, this is an ideal mode for drying heat-sensitive products.

Countercurrent: Drying air and particles move through the drying chamber in opposite directions. This mode is suitable for products that require a degree of heat treatment during drying. The temperature of the powder leaving the dryer is usually higher than the exhaust air temperature.

Mixed flow: Particle movement through the drying chamber experiences both cocurrent and countercurrent phases.

47.5.14.2 Spray Nozzles

Two fluid nozzles: In two fluid nozzles, atomization of a liquid is assisted by a compressed gas.

Centrifugal atomizer: In centrifugal atomizer, the droplet is obtained because of the fast rotation of the feed.

Pressure nozzle: The droplet in pressure nozzle is created by passing the pressurized feed into an orifice.

Three-fluid nozzle[22]: The three-fluid nozzle (3N) has a unique, three-layered concentric structure that consists of inner and outer liquid passages and an outermost gas passage, and can therefore spray two different solvents separately. Furthermore, the inner and outer liquid passages are individually connected with the centre and peripheral nozzles, which may allow a solvent sprayed via the outer peripheral nozzle to coat another solvent sprayed via the inner centre nozzle during a very brief atomizing process. Thus, the 3N spray-drying method is expected to be a novel microencapsulation technique. On the other hand, the four-fluid nozzle (4N), which has two liquid and two gas passages, has been utilized to prepare microparticles containing submicron-size polymers by antisolvent, to improve drug solubility and to design the matrix microparticles that modified the release profile.

47.5.15 ELECTROSPRAYING[23]

Electrospraying is another attractive technique for fabricating nanoscale to microscale particles that are suitable for drug delivery systems as well as many other applications. Electrospraying uses a strong electric field to break up the liquid containing the material of interest into a continuous stream of finely dispersed particles. Size distribution and surface morphology of the particles produced by electrospraying can be controlled by adjusting the electrospraying operation and formulation parameters.

This technique has many notable advantages over conventional preparation methods:

- It is a one-step process that does not require any templates or surfactants and can be performed at ambient temperature and pressure.
- Less drug is lost compared to other processes. Over 90% of drug was reported to be encapsulated by this method.
- The drug is more homogenously distributed in the microspheres. If drug is not homogenously distributed or if some part of the drug agglomerates in a certain region of the carrier, there will be ups and downs in release rate, which is not desirable.

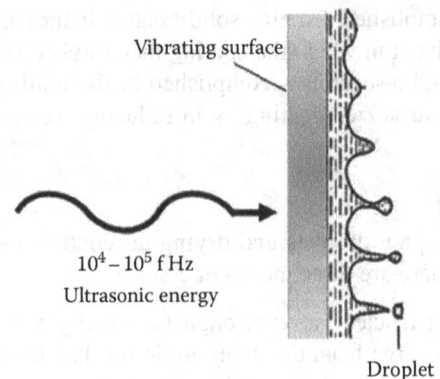

FIGURE 47.11 Ultrasonic atomization mechanism.

47.5.16 ULTRASONIC ATOMIZATION[24]

Unlike conventional atomizing nozzles that rely on pressure and high-velocity motion to produce droplets, ultrasonic atomizers use only low ultrasonic vibrational energy for atomization. This feature responds to the need to reduce the energy request in manufacturing processes. The mechanism of ultrasonic atomization is illustrated in Figure 47.11.

Ultrasonic atomization is accomplished by several means:

- By focusing high-frequency ultrasonic energy on the surface of a liquid in a bowl-shaped transducer (0.4–10.0 MHz)
- By ultrasonically vibrating a surface over which the liquid flows (18–100 kHz)
- By feeding the fluid into the active zone of a whistle (8–30 kHz)

Small droplets of a uniform size may be formed by feeding the fluid at a controlled rate through a small orifice in the tip of a horn vibrating ultrasonically in a longitudinal mode. The process parameters influencing ultrasonic liquid processing are energy and intensity, pressure, temperature, and viscosity.

Association of ultrasound to spray-drying is a powerful tool. Disadvantages of typical of a spray-dryer using a pneumatic nozzle to generate aerosol are: lack of control over the mean droplet size, broad droplet distributions, and risk of clogging in the case of suspensions. They can be overcome by employing ultrasonic energy to obtain the generation of droplets with a relatively uniform size distribution. Moreover, the spray-congealing method assisted by ultrasounds easily yields spherical microparticles.

47.6 EVALUATION AND CHARACTERIZATION[25]

47.6.1 PHYSICOCHEMICAL EVALUATION

47.6.1.1 Preformulation Studies

Preformulation studies can be carried out using FT-IR, DSC, IR spectroscopy, thin layer chromatography, UV-FTTR, etc. The preformulation studies help in studying the following parameters:

- The physical stability of drugs
- Physicochemical interactions between polymer and drug
- Effect of method of preparation on drug stability if any
- Degradation of drug

The study is generally carried on pure drug, physical mixture, formulations, and empty microspheres; the identical peaks in IR spectrum or FT-IR spectrum corresponding to the functional groups features confirm that neither the polymer nor the method of preparation has affected the

Microparticulate Drug Delivery Systems

drug stability. In thin layer chromatographic studies, the R_f values of the prepared microspheres can be compared with the R_f value of the pure drug. The values indicate the drug stability.

47.6.2 Characterization

47.6.2.1 Particle Size

Size can be measured using an optical microscope, and mean particle size is calculated by measuring 200 to 300 particles with the help of a calibrated ocular micrometer. Different sizes of microspheres and their distribution in each batch are measured by sieving in a mechanical shaker, using a nest of standard sieves American Society for Testing and Materials (ASTM) and the shaking period of 15 min. Particle size distribution is determined and the mean particle size of microspheres is calculated by using the following formula:

Mean particle size = Σ(mean particle size of the fraction × weight fraction)/Σ(weight fraction)

The particle size distribution of microcapsules can also be measured using laser diffraction particle sizing, Mastersizer and Zetasizer. For microcapsules prepared for inhalation, particle size measurement of the dry powders formulated can be measured more accurately by the dry dispersion method. The Scirocco accessory of the Malvern Mastersizer allows particle size measurement of dry powders, with particle flow controlled by a variable feed-rate vibrating tray.

47.6.2.2 Tapped Density and Compressibility Index

The tapping method is used to determine the tapped density and percentage compressibility index, as follows:

$$\text{Tapped density} = \frac{\text{Mass of microspheres}}{\text{Volume of microspheres after tapping}}$$

$$\% \text{ Compressibility index} = \left[1 - \frac{V}{V}\right] \times 100$$

47.6.2.3 Surface Morphology

The external and internal morphology of the microspheres is studied by scanning electron microscopy (SEM). Composite particles are fixed to a brass stage via double-face carbon tape and coated with platinum (current: 30 mA and time: 90 s) using sputtering equipment. Morphological studies can also be carried out using opticalstereo microscopy, transmission electron microscopy, and confocal laser scanning microscopy. Protein and additive distribution into microspheres was achieved by confocal microscopy.

47.6.2.4 Powder X-Ray Diffraction (XRD)

In order to further confirm the physical state, XRD studies can be performed on samples placed in a low background silicon holder, using x-ray diffractometer. The XRD patterns of crystalline compounds show high intensity peaks while that of amorphous compounds show broad, diffused maxima peak.

47.6.2.5 Percentage Yield

The yield of the microparticles is determined by dividing the weight of the prepared microparticles by the original amount of the polymer and drug used, and the results are expressed as a percentage according to the equation

$$\% \text{ Yield} = \frac{\text{Weight of hollow microspheres}}{\text{Weight of drug taken} + \text{Total polymer weight}} \times 100$$

The theoretical amount is the sum of weight of all the nonvolatile solid ingredients used in the process.

47.6.2.6 Percentage of Drug Content and Drug Entrapment (%)

A drug encapsulation efficacy experiment is performed to determine the actual amount of drug entrapped in the microspheres. A fixed amount of microspheres containing a drug is dissolved in a suitable solvent such as ethanol, methanol, etc., by ultrasonication or magnetic stirring. The solution is then filtered through a 5 µm membrane filter. Finally, drug concentration is determined by a suitable analytical method. Drug content is calculated according to following equation:

$$\% \text{ Drug content} = \frac{\text{Weight of drug in microspheres}}{\text{Weight of microspheres recovered}} \times 100$$

The percentage of drug entrapment can be calculated by the following equation:

$$\% \text{ Drug entrapment} = \frac{\text{Calculated drug concentration}}{\text{Theortical drug content}} \times 100$$

47.6.2.7 Swelling Measurements

The swelling study of the microparticles was conducted in suitable buffer solutions. Briefly, the diameters of the dried microparticles before and after incubation with buffers are measured. The percentage of swelling at different time intervals is determined by the difference between the diameter of microparticles at time t (D_t) and the initial time ($t = 0$ [D_0]), as given in the following equation:

$$\% \text{ Swelling} = \frac{D_t - D_0}{D_0} \times 100$$

47.6.3 DRUG RELEASE STUDIES

47.6.3.1 In Vitro Methods[26]

In vitro drug release studies have been employed as a quality control procedure in pharmaceutical production, in product development, etc. Sensitive and reproducible release data derived from physicochemically and hydrodynamically defined conditions are necessary; however, no standard in vitro method has yet been developed. Different workers have used apparatus of varying designs and under varying conditions, depending on the shape and application of the dosage form developed.

47.6.3.1.1 Beaker Method

The dosage form in this method is made to adhere to the bottom of the beaker containing the medium and is stirred uniformly using overhead stirrer. Volume of the medium used in the literature for the studies varies from 50 to 500 mL and the stirrer speed from 60 to 300 rpm.

47.6.3.1.2 Interface Diffusion System[27]

This method was developed by Dearden and Tomlinson. It consists of four compartments. One compartment represents the oral cavity, and initially contains an appropriate concentration of drug in a buffer. The next compartment represents the buccal membrane and the third compartment represents body fluid. The last compartment represents protein binding. Before use, the solutions in first, second, and fourth compartments are saturated with each other. Samples were withdrawn and returned to the first compartment with a syringe.

47.6.3.1.3 Modified Keshary–Chien Cell[28]

A specialized apparatus was designed in the laboratory. It comprised of a Keshary–Chien cell containing distilled water (50 mL) at 37°C as dissolution medium. TMDDS (Trans Membrane Drug Delivery System) was placed in a glass tube fitted with a 10# sieve at the bottom, which reciprocated in the medium at 30 strokes per min.

47.6.3.1.4 Dissolution Apparatus

Standard USP or BP dissolution apparatus has been used to study in vitro release profiles using rotating elements, paddle, and basket. Dissolution medium used for the study varied from 100 to 500 ml and speed of rotation from 50 to 100 rpm.

47.6.3.2 In Vivo Methods

Methods for studying the permeability of intact mucosa comprise of techniques that exploit the biological response of the organism locally or systemically and those that involve direct local measurement of uptake or accumulation of penetrate at the surface. The most widely used methods include in vivo studies using animal models, buccal absorption tests, and perfusion chambers for studying drug permeability.

47.6.3.2.1 Animal Models[29]

Animal models are used mainly for the screening of the series of compounds, investigating the mechanisms and usefulness of permeation enhancers or evaluating a set of formulations. Animal models such as the dog, rats, rabbits, cat, hamster, pigs, and sheep have been reported. In general, the procedure involves anesthetizing the animal followed by administration of the dosage form. In case of rats, the esophagus is ligated to prevent absorption pathways other than oral mucosa. At different time intervals, the blood is withdrawn and analyzed.

47.6.3.2.2 Buccal Absorption Test

The buccal absorption test was developed by Beckett and Triggs in 1967. It is a simple and reliable method for measuring the extent of drug loss of the human oral cavity for single and multicomponent mixtures of drugs. The test has been successfully used to investigate the relative importance of drug structure, contact time, initial drug concentration, and pH of the solution while the drug is held in the oral cavity.

47.6.3.3 In Vitro Mucoadhesion Test[30]

The mucoadhesive property of the optimized microspheres prepared by different methods is evaluated by an in vitro mucoadhesion testing method known as the wash-off method. A rat stomach mucosa is tied onto the glass slide using a thread. In this method, microspheres are spread onto wet rinsed tissue specimen and the prepared slide is hung onto one of the grooves of an USP tablet disintegrating test apparatus. The disintegrating test apparatus is switched on and the tissue specimen is given up and down movements for 2 h in the beaker of the disintegration test apparatus, which contained the stimulated gastric fluid (pH 1.2). The microspheres remaining at the surface of gastric mucosa are then collected, and the percentage of the remaining microspheres is calculated. The experiment is performed in triplicate. The percentage mucoadhesion is calculated by the following formula:

$$\% \text{ Mucoadhesion} = \frac{\text{Weight of adhered microspheres}}{\text{Weight of applied microsperes}} \times 100$$

47.6.3.4 In Situ Bioadhesivity Studies[30]

Bioadhesivity testing is done by a novel in situ method. A freshly cut 5 to 6 cm long piece of small intestine of rat is obtained and cleaned by washing with isotonic saline. The piece is cut open and the mucosal surface is exposed. Known weights of microspheres are added evenly on the mucosal surface. The intestinal piece is maintained at 80% relative humidity (RH) for 30 min in a desiccator. The piece is taken out and suitable buffer is allowed to flow over the intestinal piece for about 2 m at a rate of 20 mL/min. The perfusate is collected and dried to get the particles not adhered. The percent of bioadhesion is estimated by the ratio of amount applied to amount of microparticles adhered to mucosal surface.

47.6.3.5 Floating Behavior

In formulating floating microparticulate drug delivery system, the key point to be evaluated is the floating ability of microspheres.

47.6.3.5.1 In Vitro Evaluation

47.6.3.5.1.1 Visual Observation Method Each sample of microspheres is placed into 500 ml of simulated gastric fluid (SGF, pH 1.2, without enzymes) filled in basket-type dissolution apparatus. Paddle rotation speed is set at 100 rpm, temperature is to be maintained at 37°C ± 0.5°C. The number of floating microspheres has to be counted visually after 24 h. The percentage of floating microspheres was calculated according to the following equation:

$$\%F = \frac{N_f}{N_t} \times 100$$

where
  F is the floating percent
  N_f is the number of floating microspheres
  N_t is the total number of the microspheres

47.6.3.5.1.2 Resultant Weight Method The magnitude and the direction of total force F correspond to the vectorial sum of the buoyancy (F_{buoy}) and gravity (F_{grav}) forces acting on the object.

$$F = F_{buoy} - F_{grav}$$
$$= d_f v_g - d_s v_g$$
$$= (d_f - M/V) v_g$$

where
  F is the total vertical force (resultant weight force of the object)
  g is the acceleration of gravity
  d_f is the fluid density
  d_s is the object density
  M is the object mass
  V is the object volume

The total force F acting vertically on an immersed object may be used to quantify the object floating or nonfloating capabilities. A positive total force F signifies that the object is able to float, whereas a negative F means that the object sinks. Moreover, the larger the total force F value of the object is, the more excellent its floating capabilities are. The total force F determines resultant-weight of the object in immersed conditions.

47.6.3.5.2 In Vivo Evaluation of Floating Ability of Microspheres In vivo evaluation of the floating behavior is carried out by radiolabeling of microspheres. A radioactive label, which has a short half-life, is to be selected: for example, technetium.

47.6.3.6 In Vitro–In Vivo Correlations

Correlations between in vitro dissolution rates and the rate and extent of availability as determined by blood concentration and or urinary excretion of drug or metabolites are referred to as "in vitro–in vivo correlations." Such correlations allow one to develop product specifications with bioavailability.

47.7 MICROPARTICULATE DRUG DELIVERY SYSTEMS[31–43]

47.7.1 DOUBLE-WALLED MICROSPHERES

The use of conventional single-polymer microspheres is severely undermined by several limitations, including the initial burst release caused by rapid release of drug found on or near the external surface, difficulty in achieving zero-order drug release, and a lack of time-delayed or pulsatile release of drugs. Since an important goal of drug delivery systems is to attain well-controlled drug release rates, double-walled microspheres with a drug-encapsulating particle core surrounded by a drug-free shell layer are introduced. These double-walled microspheres often exhibit a reduction in the initial burst release as compared to single-polymer microspheres, and provide a sustained drug release that is tunable by adjusting the shell material or thickness. In addition, these microspheres enable the encapsulation of multiple drugs in the core and shell phases, and allow their release in various stages, thus achieving synergistic therapeutic effects. For example, the parallel or sequential release of multiple drugs would be useful for expediting a variety of growth factor–driven tissue regenerative processes in tissue engineering or formulating a successful tumor inhibition strategy in cancer therapy. Recently, the development of triple-walled microspheres has also gained significant interest, and such multilayered drug delivery systems could provide a versatile approach to deliver several drugs and control their respective drug release profiles.

Advantages: Double-walled microspheres offer several advantages as compared to more conventional monolithic microspheres. They are the following:

- Drug encapsulated in the core of double-walled microspheres may overcome the problem of high initial burst release commonly encountered in microspheres.
- Higher drug loads with improved drug stability may be achieved by using materials in the core phase that offer increased drug solubility.
- Drug release rates may be controlled by controlling the shell material or thickness.
- Drugs can be released either in a sequential or simultaneous manner by selectively loading them into the core or shell phase, thereby potentially enhancing drug efficacy.
- The encapsulation efficiency of double-walled microspheres will be higher than that in single polymer microspheres.

Fabrication: Precision particle fabrication (PPF) technique has been widely used in fabrication of double-walled microspheres with unprecedented control of the particle size and morphology. The two polymeric solutions are passed through a coaxial nozzle to produce a jet of core surrounded by an annular stream of outer polymer and disrupted into uniform double-walled droplets by an ultrasonic transducer controlled by a frequency generator.

An important advantage of the PPF technique is the capability of producing monodispersed double-walled microspheres with uniform shell thickness by varying the flow rates of core and shell polymer solutions.

Moreover, microspheres with two agents loaded in the respective core and shell phases can be fabricated in a single step, and the agents can be released either in a sequential or parallel manner.

The efficient production of fully formed double-walled microspheres depended on the mass ratio of both polymers, and it occurred only above a critical polymer mass ratio. The core concentration, on the other hand, does not seem to have a significant effect on the critical polymer mass ratio needed for forming double-walled microspheres. Partial encapsulation of core by shell polymer was found in microspheres when a shell flow rate below the critical polymer mass ratio was selected. Above the critical polymer mass ratio, fully formed double-walled microspheres were observed with no sign of partial encapsulation configuration. The molecular weight of the shell layer did not influence the subsequent drug release from the microspheres as evidenced by the release profiles obtained experimentally.

Many groups have produced double-walled microspheres from a variety of materials including bulk- and surface-eroding polymers, and investigated their degradation behavior:

- In one study, the degradation of double-walled microspheres with a core of poly(1,3-bis-(p-carboxyphenoxypropane)-co-(sebacic anhydride)) 20:80 (P(CPP:SA)20:80) and an external coat of poly(L-lactic acid) (PLLA) was monitored in vitro and in vivo for 6 months. The inner core of the more hydrolytically labile P(CPP:SA) 20:80 degraded first, while the shell layer remained relatively intact.
- In another study, the degradation of double-walled microspheres consisting of a poly(ortho ester) (POE) core surrounded by a poly(D,L-lactic-co-glycolic acid) 50:50 (PLGA) shell layer was examined. Similar to the previous study, preferential degradation of the POE core was observed, and formation of hollow microspheres became pronounced after the first week of incubation. In an attempt to limit water penetration into the inner core phase, a surface-eroding polymer, poly(1,6-bis-(p-carboxyphenoxyhexane)) (PCPH), was used to encapsulate a PLGA core. However, the slow eroding PCPH shell layer could not prevent water penetration, and the PLGA core was completely eroded by 6 weeks of incubation.
- The dual-responsive hollow polymeric microspheres, which could be responsive simultaneously to two environmental stimuli, are expected to be more suitable for the controlled release of drugs. For example, the pH and temperature dual-responsive hollow polymeric microspheres could be responsive to the pH and temperature changes in their environment. Temperature-responsive outer shell and pH-responsive inner shell that encapsulate magnetic nanoparticles are widely used for the delivery of therapeutic compounds to the targeting specific sites.

47.7.2 Floating Microspheres

Drugs that are easily absorbed from the gastrointestinal tract (GIT) and have a short half-life are eliminated quickly from the blood circulation, so they require frequent dosing. To avoid this drawback, the oral sustained-/controlled-release formulations have been developed in an attempt to release the drug slowly into the GIT and maintain an effective drug concentration in the serum for longer period of time. However, such oral drug delivery devices have a physiological limitation of gastric retention time (GRT). Variable and short gastric emptying time can result in incomplete drug release from the drug delivery system in the absorption zone (stomach or upper part of small intestine), leading to diminished efficacy of the administered dose.

Advantages: Floating microspheres can overcome the aforementioned limitations. Floating drug delivery is of particular interest for drugs, which

1. Act locally in the stomach
2. Are absorbed primarily in the stomach
3. Exhibit poor solubility at an alkaline pH
4. Have a narrow window of absorption

However, many floating systems previously reported are single-unit systems such as hydrodynamically balanced systems (HBS), which are unreliable in prolonging the GRT owing to their "all-or-nothing" emptying process and, thus, may result in high variability in bioavailability and local irritation due to a large amount of drug delivered at a particular site of GIT. In contrast, multiple-unit microparticulate dosage forms (e.g., microspheres) have the advantages of passing through the GIT uniformly, which not only avoids the vagaries of gastric emptying but also provides an adjustable release, and reduced intersubject variability in absorption and risk of local irritation were achieved consequently.

Fabrication: Various multiple-unit floating systems have been developed in different forms and are based on various principles, such as air compartment multiple-unit system, microparticles based on porous carriers, hollow microspheres (microballoons), and oil-entrapped gel beads prepared by gelation method

47.7.3 Solid Lipid Microparticles (SLMs)

With regard to particulate matter, solid lipid-based microparticulate systems have attracted the attention of the researchers in the last decade owing to their biocompatibility, biodegradability, physical stability, low cost of ingredients, ease of preparation and scale-up, high dispersibility in an aqueous medium, high entrapment of lipophilic drugs, and extended drug release. SLM usually consist of particles (in the dimensional range of microns) composed of a solid hydrophobic fat matrix in which the active is dissolved or, more rarely, dispersed.

SLM combines the *advantages* of different drug delivery systems such as: polymeric microparticles, fat emulsions, and liposome. In this respect, they have been successfully proposed for the delivery of a variety of drugs including: antibiotics, anti-inflammatory compounds, vaccines, and adjuvant. Among the lipids, phospholipids, triacyl glycerols, waxes, fatty acids, or their mixtures can be used. SLM can be administered by subcutaneous, oral, intramuscular, topical, or pulmonary ways.

Fabrication: The preparative strategies for SLM and solid lipid nanoparticles (SLNs) can present substantial differences. For instance, the high-pressure homogenization that is largely employed for SLN, is scarcely applied to the production of SLM. SLMs are usually prepared by various processes including hot melt emulsification, solvent evaporation (also called in-liquid drying process), high-pressure homogenization, membrane emulsification, or spray cooling/congealing.

47.7.3.1 Applications

1. *Transdermal delivery*: With regard to SLNs, an enhanced penetration of active substances was generally found owing to the formation of an occlusive film due to their small dimensions. It is recognized that SLMs can protect the embedded compounds from photodegradation, retain the bioactive substance on the skin surface, act as film forming reducing the transepidermal water loss, or deliver gradually the active substance to the skin surface sustaining the drug effect.
2. *Probiotics*: Packaging probiotic microorganisms within SLMs may protect them and is an interesting alternative to other conveyance systems, such as polymers and polysaccharides, because the microparticles will deliver the bioactive compound at approximately

the melting point of the carrier material. SLMs produced with lipid materials are easily digested by the lipases in the intestines, releasing the probiotic in the vicinity of the intestines where they are required. The viability of microorganisms within lipid matrices was reported to be greater than that of free microorganisms.

3. *Pulmonary drug delivery*: The production of SLMs has been previously studied as a respiratory drug delivery vehicle for both poorly water-soluble drugs, such as budesonide and for more water-soluble drugs, such as salbutamol. There are many advantages in using SLMs, with the most significant being the ability to control release after deposition. Furthermore, SLMs should be well tolerated in vivo, since they are made of physiological compounds. Of course, the toxicity of the surfactants and other excipients, used for their manufacture, needs to be considered.

Despite the ease in production and the possibility of industrial scale manufacture, SLMs or nanoparticles (SLNs) can present some disadvantages like the following:

- Low encapsulation capacity.
- Expulsion of core material during storage due to the crystalline structure and polymorphic arrangements.
- They show some physical instability and early burst releases of encapsulated drugs as proteins or water-insoluble drugs as ibuprofen. This phenomenon could be due to the location of a part of the associated drug at the outer surface of the particles or by the presence of surfactants at the interface.

47.7.4 Targeted Microsheres

The fabrication of microspheres, based on bio- and synthetic polymers, has an increasing interest in scientific society. Especially, biopolymers with responsive stimuli have been widely used in the domain of biomaterials, such as proteins, polysaccharides, and nucleic acids as basic components in living organic systems. On the other hand, many research groups deal with the synthetic polymers that have been modified, aiming at improving their properties in comparison with biopolymers. These smart drug delivery systems are susceptible to a rapid response in the changes of the local environment, such as temperature, pH, magnetic field, and last ionic strength.

Among the controlled drug delivery systems, pH-sensitive system has a special interest in view of the fact that both extracellular tumor (pH 6.8) and endosomes (pH 5.5) are more acidic than normal tissues (pH 7.4). The physical stimuli, such as temperature, or magnetic field will affect the level of various energy sources and alter molecular interactions at critical points. Treating the synthesized microspheres under the above environments can affect their behavior via the interactions of the polymeric chains.

Applications: Drugs can be targeted in different ways using microparticles.

1. Targeting to tissues:
    a. Antitumor microparticles are administered intra-arterially and target an organ or body cavity, that is, peritoneum.
    b. Therapeutic drug delivery of anticancer drugs. For example, doxorubicin and 5-fluorouracil.
    c. As markers for analysis/detection. For example, detect tumors, infected cells.
    d. Antipathogen lymphocytes.
2. Intracellular delivery:
    a. Gene delivery. For example, delivery of plasmid DNA.
    b. Antisense therapy For example, closing production of certain proteins by delivery of anti-sense oligonucleotides to bind ribosomal mRNA.

c. Intracellular toxins for cancer therapy.
   d. Ribozyme delivery.
   e. Drug delivery to cell organelles. For example, mitochondria.
   f. Vaccine adjuvant, that is, biodegradable PLA and PLA coglycolic acid microspheres also act as immune adjuvant by providing a depot formulation of the antigen at the site of administration. The antigen is thus continuously released to antigen-presenting cells.

47.7.5 Magnetic Microspheres

Magnetic microspheres have been reported to successfully target the cancer cells. Magnetic microspheres are generally formed by coating superparamagnetic iron oxide nanoparticles with a shell of hydrophilic or hydrophobic polymer, which would minimize the toxicity of and immunological response to iron oxide nanoparticles and obtain a colloidal or stable composite system. Magnetic polymer microspheres have been widely used in the fields of biotechnology and biomedical engineering, such as cell isolation, protein immobilization, enzyme immobilization, immunoassay RNA and DNA purification, biochemical assays, drug targeting and delivery, hyperthermia for cancer therapy, and so on. These magnetic polymer microspheres have been increasingly exploited for biomedical applications, with the multifunctional therapeutic advantages of targeted and controlled drug delivery and thermal therapy.

Applications: By applying an external magnetic field, the drug-loaded magnetic polymer microspheres can be rapidly delivered and accumulated in a specific zone of the body, which can sharply increase the drug concentration at the desired target zone and consequently greatly reduce the adverse effects typical of nontarget chemotherapy systems for cancer therapy. Moreover, the polymer coating can also provide a sustained-release effect for drugs. Meanwhile, due to their magnetically induced excellent heating ability, thermal therapy for the tumor covered with these magnetic polymer microspheres can also be achieved under an ac magnetic field.

Fabrication: There are different routes in the preparation of magnetic polymer microspheres. The most frequently used methods can be divided into two categories:

1. Synthesis of polymer microspheres in the presence of magnetic nanoparticles, such as suspension polymerization or its modified versions, dispersion polymerization, surface-initiated radical polymerization, acid-catalyzed condensation polymerization, emulsion polymerization, mini-/microemulsion polymerization, in situ oxidative polymerization, inverse emulsion cross-linking, emulsion/double emulsion–solvent evaporation, and supercritical fluid extraction of o/w miniemulsion
2. Incorporation of magnetic nanoparticles into presynthesized polymer microspheres, such as the swelling and thermolysis technique and the swelling and penetration process

47.7.6 Bioadhesive Microspheres

An intimate contact of the microspheres with the absorbing membranes helps in prolonging the residence time. It can be achieved by coupling bioadhesion characteristics to microspheres and developing novel delivery systems referred to as "bioadhesive microspheres." Adhesion can be defined as sticking of drug to the membrane by using the sticking property of the water-soluble polymers. Adhesion of drug delivery device to the mucosal membrane such as buccal, ocular, rectal, nasal, etc. can be termed as bioadhesion. The term "bioadhesion" describes materials that bind to biological substrates, such as mucosal membranes.

Bioadhesive microspheres include microparticles and microcapsules (having a core of the drug) of 1 to 1000 μm in diameter and consisting either entirely of a bioadhesive polymer or having an outer coating of it, respectively. Microspheres, in general, have the potential to be used for targeted and controlled-release drug delivery; but coupling of bioadhesive properties to microspheres has additional advantages like

- Efficient absorption
- Enhanced bioavailability of the drugs due to a high surface-to-volume ratio
- A much more intimate contact with the mucus layer
- Specific targeting of drugs to the absorption site achieved by anchoring plant lectins, bacterial adhesions, and antibodies, etc., on the surface of the microspheres

Fabrication: Common preparation techniques discussed in earlier sections are used in formulating bioadhesive microspheres. An important criterion is the use of bioadhesive polymers like chitosan, hyaluronic acid, starch, lysophosphatidylcholine along with degradable starch microspheres.

Applications: Bioadhesive microspheres can be tailored to adhere to any mucosal tissue including those found in eye, nasal cavity, and urinary, and GIT, thus offering the possibilities of localized as well as systemic controlled release of drugs:

1. *Ocular*: To prolong the residence time of drugs in the preocular area, bioadhesive drug delivery system has been developed taking advantage of the presence of a mucin–glycocalyx domain in the external portion of the eye. Various bioadhesive drug delivery systems employed for ocular delivery of drugs include the semisolids, viscous liquids, solids/inserts, and the particulate drug delivery system, including bioadhesive microspheres and liposomes. The advantages of microspheres: increased residence time and decreased frequency of administration.
2. *Nasal*: The nasal cavity offers a large, highly vascularized subepithelial layer for efficient absorption. Also, blood is drained directly from nose into the systemic circulation, thereby avoiding first-pass effect. However, nasal delivery of drugs has certain limitations due to the mucociliary clearance of therapeutic agents from the site of deposition, resulting in a short residence time for absorption. Use of bioadhesive drug delivery system increases the residence time of formulations in nasal cavity, thereby improving absorption of drugs. The excellent absorption-enhancing properties of bioadhesive microspheres are now being used extensively for both low-molecular-weight as well as macromolecular drugs like proteins. Chitosan and starch are the two most widely employed bioadhesive polymers for nasal drug delivery. It has been reported that the clearance half-life was 25% greater for chitosan microspheres than for starch microspheres.
3. *Vaginal*: The vaginal route has been frequently used for delivery of therapeutic and contraceptive agents to exert a local effect (antifungal, spermicidal) and for the systemic delivery of drugs. It has been used for the delivery of drugs, which are susceptible to GI degradation or hepatic metabolism following per-oral delivery: for example, estrogens and progestogens for the treatment of postmenopausal symptoms and for contraception. This route has also been explored for the delivery of therapeutic peptides, for example, calcitonin and for microbicidal agents to help prevent the transmission of human immunodeficiency virus and other sexually transmitted diseases (STDs).
4. *Colon*: Colon drug delivery has been used for molecules aimed at local treatment of colonic diseases and for delivery of molecules susceptible to enzymatic degradation such as peptides. Bioadhesive microspheres can be used during the early stages of colonic cancer (when systemic prevention of possible metastasis in the blood is still not necessary), for enhancing the absorption of peptide drugs and vaccines, for the localized action of steroids, and drugs with a high hepatic clearance, for example, budesonide

and for the immunosuppressive agents such as cyclosporine. Colon-specific bioadhesive microspheres can be used for protection of peptide drugs from the enzyme-rich part of the GIT and to release the biologically active drug at the desired site for its maximum absorption.

5. *GI epithelium*: Microspheres prepared with bioadhesive and bioerodible polymers undergo selective uptake by the M cells of Peyer's patches in GI mucosa. This uptake mechanism has been used for the delivery of protein and peptide drugs, antigens for vaccination, and plasmid DNA for gene therapy. Moreover, by keeping the drugs in close proximity to their absorption window in the GI mucosa, the bioadhesive microspheres improve the absorption and oral bioavailability of drugs like furosemide and riboflavin.

47.7.7 RADIOACTIVE MICROSPHERES

Radiotherapy microspheres, sized 10 to 30 nm, are larger than capillaries and get trapped in first capillary bed when they come across. They are injected to the arteries that lead to tumor of interest. So, in all these conditions, radioactive microspheres deliver high-radiation dose to the targeted areas without damaging the normal surrounding tissues. It differs from drug delivery system, as radioactivity is not released from microspheres but acts within a radioisotope typical distance and the different kinds of radioactive microspheres are α emitters, β emitters, and γ emitters. It offers new solutions for patients who need drugs delivered directly to tumors, diabetic ulcers, and other disease sites.

47.8 APPLICATIONS OF MICROENCAPSULATIONS[10,44]

47.8.1 CELL IMMOBILIZATION

In plant cell cultures, microencapsulation, by mimicking cell natural environment, improves efficiency in production of different metabolites used for medical, pharmacological, and cosmetic purposes. Human tissues are turned into bioartificial organs by encapsulation in natural polymers and transplanted to control hormone-deficient diseases such as diabetes and severe cases of hepatic failure. In continuous fermentation processes, immobilization is used to increase cell density, productivity and to avoid washout of the biological catalysts from the reactor. This has already been applied in ethanol and solvent production, sugar conversion, or wastewater treatment.

47.8.2 BEVERAGE PRODUCTION

Today, beer, wine, vinegar, and other food drinks production are using immobilization technologies to boost yield, improve quality, change aromas, etc.

47.8.3 PROTECTION OF MOLECULES FROM OTHER COMPOUNDS

Microencapsulation is often a necessity to solve simple problem like the difficulty to handle chemicals (detergents dangerous if directly exposed to human skin) as well as many other molecule inactive or incompatible if mixed in any formulation. Moreover, microencapsulation also allows preparing many formulations with lower chemical loads, reducing significantly processes' cost.

47.8.4 DRUG DELIVERY

After designing the right biodegradable polymers, microencapsulation has permitted controlled-release delivery systems. These revolutionary systems allow controlling the rate, duration, and distribution of the active drug. With these systems, microparticles sensitive to the biological environment are designed to deliver an active drug in a site-specific way (stomach, colon, and specific organs). One of the main advantages of such systems is to protect sensitive drug from drastic environment (pH) and to reduce the number of drug administrations for patient.

47.8.5 QUALITY AND SAFETY IN FOOD, AGRICULTURAL, AND ENVIRONMENTAL SECTORS

The development of the "biosensors" has been enhanced by encapsulated biosystems used to control environmental pollution, food cold chain (abnormal temperature change).

47.8.6 AGRICULTURE

One of the most important applications of microencapsulated products is in the area of crop protection. Nowadays, insect pheromones are becoming viable as a biorational alternative to conventional hard pesticides. Specifically, sex-attractant pheromones can reduce insect populations by disrupting their mating process. Hence, small amounts of species-specific pheromone are dispersed during the mating season, raising the background level of pheromone to the point where it hides the pheromone plume released by its female mate. Polymer microcapsules, polyurea, gelatin, and gum arabic serve as efficient delivery vehicles to deliver the pheromone by spraying the capsule dispersion. Further, encapsulation protects the pheromone from oxidation and light during storage and release.

47.8.7 SOIL INOCULATION

For example, *Rhizobium* is a very interesting bacterium, which improves nitrate adsorption and conversion. But inoculation is often unsuccessful, because cells are washed out by rain. By cell encapsulation processes, it is possible to maintain continuous inoculation and higher cell concentration. This list is not exhaustive, the *nutraceuticals* world could be the last mentioned because of the growing interest and increasing demand we have to face in ingredients with health benefits, which often require improvement of their efficiency and stability (e.g., probiotics, vitamins, etc.) by protecting and offering targeting release of the active materials.

47.8.8 APPLICATIONS OF MICROCAPSULES IN BUILDING CONSTRUCTION MATERIALS

An analysis of scientific articles and patents shows numerous possibilities of adding microencapsulated active ingredients into construction materials, such as cement, lime, concrete, mortar, artificial marble, sealants, paints and other coatings, and functionalized textiles.

47.8.9 DEFENSE

One of the important defense applications of microencapsulation technology is in self-healing polymers and composites. They possess microencapsulated healing agents embedded within the matrix and offer tremendous potential for providing long-lived structural materials. The microcapsules in self-healing polymers not only store the healing agent during quiescent states, but provide a mechanical trigger for the self-healing process when damage occurs in the host material and the capsules rupture. The microcapsules posses sufficient strength to remain intact during processing of the host polymer, yet rupture when the polymer is damaged. High bond strength to the host polymer combined with a moderate strength microcapsule shell is required. To provide long shelf-life, the capsules must be impervious to leakage and diffusion of the encapsulated healing agent for considerable time. These combined characteristics are achieved with a system is based on the in situ polymerization of urea-formaldehyde microcapsules encapsulating dicyclopentadiene healing agent. The addition of these microcapsules to an epoxy matrix also provides a unique toughening mechanism for the composite system. Such microcapsules have tremendous application in aerospace area for making self-repairable spacecrafts. Such self-healing spacecrafts open up the possibility of longer duration missions by increasing the lifetime of a spacecraft.

47.9 MARKETED PRODUCTS[45–47]

Table 47.6 provides the list of commercially available products, and microencapsulation technique along with their clinical use and technique employed in preparation.

47.10 RECENT WORK IN THE FIELD OF MICROENCAPSULATION

Over the last few years, innumerable numbers of research works have been carried out in the field of microencapsulation. Some works focused on developing microcapsules with new polymers, while few research works lead to development of new encapsulation techniques or improving the already existing techniques. A brief list of few research works on microencapsulation carried out in the recent times is given in Table 47.7.

TABLE 47.6
List of Marketed Products

Drug	Company	Technology	Indication
Naltrexone	Janssen®/Alkermes, Inc.	Double emulsion (oil in water)	Schizophrenia; bipolar I disorder
Leuprolide	Alkermes	Double emulsion (oil in water)	Alcohol dependence
Octreotide	TAP Takeda	Double emulsion (water in oil in water)	Prostate cancer/endometriosis
Somatropin	Genentech/Alkermes	Cryogenic spray-drying	Growth deficiencies
Triptorelin	Pfizer Ferring	Phase separation	Prostate cancer
Buserelin	Sanofi-Aventis	N/A	Endometriosis
Lanreotide	Ipsen-Beafour	Phase separation	Acromegaly
Bromocriptine	Novartis	Spray dry	Parkinsonism
Minocycline	Orapharma	N/A	Periodontitis
Cephalexin	Lupin	N/A	Bacterial infections
Cefadroxil	Lupin	N/A	Bacterial infections
Ciprofloxacin	Bayer, Inc.	N/A	Microbial infection
Lactobacillus reuteri	Micropharma Ltd.	N/A	Hypercholesterolemia
Potassium chloride	KV Pharmaceutical Co	N/A	Hypokalemia
Paracetamol	Mayne Pharma	N/A	Pain and fever
Cyclosporin A	Novartis Int. AG	N/A	Rheumatoid arthritis
Saquinavir	Roche	N/A	HIV
Ritonavir	Abbott Laboratories	N/A	HIV
Vitamins A, E, and F	Tagra Biotechnologies Ltd.	N/A	Aging
Ginseng and hamamelis extracts, arnica	DS Laboratories	N/A	Propionibacterium acnes bacteria and fungi

Sources: Qingxing, X. et al., *J. Control. Release,* 163, 130, 2012; Pengcheng, D. et al., *Colloids Surf. B,* 102, 1–8, 2013; Ninan, M. et al., *Int. J. Pharm.,* 358, 82, 2008.

TABLE 47.7
Recent Work in the Field of Microencapsulation

Sl. no.	Drug	Polymer	Technique	Conclusion	References
1.	Protein	Poly(-caprolactone) (PCL)	Multiple emulsion (w/o/w) with solvent evaporation/extraction method	MC (methylene chloride) used as a solvent of PCL in a microencapsulation process for therapeutic protein by a nontoxic solvent belonging to Class 3.	[48]
2.	Simvastatin	Poly (D,L-lactic-co-glycolide) (PLGA),	Electrospraying method	Preparation of microsphere with electrospraying technique produces smaller size microsphere more easily when compared to the emulsification and solvent evaporation methods.	[48]
3.	Lysozyme	Poly (lactic-co-glycolic acid) (PLGA) and poly (L-lactide (PLLA)	Double-walled microspheres	Lysozyme released from DW microspheres between 40 to 100 days of microsphere incubation in vitro.	[49]
4.	Protein	Poly(lactic-co glycolic acid) (PLGA)	S/O/W (solid-in-oil-in water) and addition of an oily additive, vitamin E (Vit E)	The protein was released in vitro in its bioactive form for more than 3 months. Vitamin E (Vit E), useful from a technological point of view, by promoting additional protein protection and also from a pharmacological point of view, because of its antioxidant and antiproliferative properties.	[50]
5.	Vitamin c	Chitosan	Inversed w/o micro-emulsion	The study indicated that the vitamin C release rates decrease with increasing cross-link density of the capsules and increase with increasing concentration of vitamin C.	[51]
6.	Doxorubicin and Gene encoding the p53 tumor suppressor protein	Poly(D,L-lactic-co-glycolic acid) (PLGA) core surrounded by a poly(lactic acid) (PLA) shell	PPF technique	PPF method is capable of producing double-walled microspheres and encapsulating dual agents for combined modality treatment, such as gene therapy and chemotherapy.	[52]
7.	Bovine serum albumin	Poly(lactide-co-glycolide) (PLG) and poly(D,L-lactic acid) (PDLL)	PPF technology (doubled walled microsphere)	BSA in vitro release rates were retarded by the presence of the drug-free PDLL shell. Moreover, increasing PDLL shell thickness resulted in decreasing BSA release rate. With a 14 μm thick PDLL shell, an extended period of constant-rate release was achieved.	[53]
8.	Glibenclamide	Cellulose acetate	Emulsion solvent evaporation method	In vitro release studies revealed that the drug release was sustained up to 12 h.	[54]

(Continued)

TABLE 47.7 (Continued)
Recent Work in the Field of Microencapsulation

Sl. no.	Drug	Polymer	Technique	Conclusion	References
9.	Gentamicin sulfate	Alginate–pectin	Supercritical-assisted atomization (SAA)	All formulations showed an initial burst effect in the first 6 h of application (40%–65% of GS loaded), and microparticle produced with a GS/alginate/pectin ratio of 1:3:1 exhibited the ability to release GS continuously over 6 days.	[55]
10.	Alginate–pectin	Furosemide	Self-microemulsifying system (SMES) core using a vibrating nozzle technology	The best-shaped microcapsules with highest encapsulation efficiency for furosemide-loaded SMES were obtained from the shell-formation phase with an A/P ratio of 25:75, containing 10% lactose.	[56]
11.	Ovalbumin	Calcium-alginate and calcium-yam-alginate. methylated N-(4-N,N dimethylaminocinnamyl) chitosan (TM6-CM CS) was used to coat microparticles	Electrospraying technique	The particles demonstrated a greater swelling and mucoadhesive properties than did uncoated microparticles. The in vitro release from the microparticles indicated that the coated microparticles resulted in more sustained release than uncoated microparticles. The in vivo oral administration demonstrated that at the same amount of 250 µg OVA, coated microparticles exhibited the highest in vivo adjuvant activity in both IgG and IgA immunogenicity.	[57]
12.	Salmeterolxinafoate	Wheat germ agglutinin (WGA)-anchored salmeterolxinafoate(SalX)-loaded nanoparticles-in-micropaticles system (NiMS)	Ionotropic gelation technique combined with a spray drying method	Results suggest that it is possible to control drug release from a lectin-anchored drug delivery system using a specific sugar, and that the designed novel WGA-SalX-NiMS may be a suitable formulation for chronotherapy of asthma.	[58]
13.	Nifedipine	PVA release modifier	Spray-drying technique	The in vitro release profile showed a burst release followed by controlled release. A more prolonged release can be achieved by increasing the PVA:nifedipine ratio.	[59]
14.	Aspirin	Poly(l-lactic-co-glycolic acid) (PLGA) and poly(l-lactic acid) (PLLA)		The selective dissolution technique was used to improve the phase separation, and successfully fabricate core–shell microspheres for controlled delivery of drug with reduced initial burst release. These microspheres showed sustained release of aspirin for at least 456 h with a little burst release (3.49%).	[60]

(Continued)

TABLE 47.7 (Continued)
Recent Work in the Field of Microencapsulation

Sl. no.	Drug	Polymer	Technique	Conclusion	References
15.	Risperidone, methyl prednisolone acetate and Paclitaxel	Poly(lactide-co-glycolide) (PLGA)	Hydrogel template method	For all three drugs, release was sustained for weeks, and the in vitro release profile of risperidone was comparable to that of microparticles prepared using the conventional emulsion method. The hydrogel template method provides a new approach of manipulating microparticles.	[61]
16.	Bovine serum albumin	Alginate poloxamer	Spray-drying	Aqueous cross-linking led to a significant size increase, whereas ethanolic cross-linking did not. The substantial drug loss during aqueous $CaCl_2$ cross-linking could be avoided by using aqueous $ZnSO_4$ or ethanolic $CaCl_2$ solutions. Protein release from microparticles cross-linked with ethanolic $CaCl_2$ solutions was much faster than in the case of aqueous $CaCl_2$ solutions, probably due to the lower calcium content.	[62]
17.	Verapamil Hcl	Chitosan	Spray-drying and precipitation techniques	The spray-drying technique was superior over precipitation technique in providing higher VRP entrapment efficiency and smaller burst release followed by a more sustained one over 6 h.	[63]
18.	Ethenzamide (etz)	Ethylcellulose	Three-fluid nozzle (3N) spray drying technique.	It was found that 3N can obtain smaller particles than 4N. The results for contact angle and drug release corresponded, thus suggesting that 3N-PostMix particles are more effectively coated by ethylcellulose, and can achieve higher-level controlled release than 4N-postmix particles, while 3N-premix particles are not encapsulated with pure ethylcellulose, leading to rapid release.	[64]
19.	Dexamethasone	Acrylic acid	Microfluidics. Three main parts involved: a temperature control, a coflow dripping element, and a congealing element.	The obtained results demonstrated that the dexamethasone did not affect the general characteristics of SLM, confirming the robustness of the microfluidic procedure in view of the production of SLM for biopharmaceutical and biotech protocols.	[65]
20.	Simvastatin	Poly (D,L-lactic-co-glycolide) (PLGA),	Electrospraying method	The drug was continuously released from the microspheres for >3 weeks.	[66]

(*Continued*)

TABLE 47.7 (Continued)
Recent Work in the Field of Microencapsulation

Sl. no.	Drug	Polymer	Technique	Conclusion	References
21.	—	Superparamagnetic iron oxide nanoparticles, PEG and hydroxypropyl-b-cyclodextrin (HPbCD)	Spray drying	The addition of $(NH_4)_2CO_3$ and magnesium stearate (MgST) to the formulation improved the aerodynamic properties of the Trojan particles and resulted in a mass median aerodynamic diameter (MMAD) of 2.2 ± 0.8 lm.	[67]
22.	Antisense oligonucleotides	Poly(D,L-lactide) or poly (D,L-lactide-co-glycolide) (PLGA)	O/o/o-solvent extraction microencapsulation method	This new method has the advantages of less toxic solvents, much lower preparation volume, and solvent consumption and high encapsulation efficiencies when compared to the classical w/o/w method.	[18]
23.	Progesterone	PLGA	Ammonolysis-based microencapsulation process	The structural integrity of ester bonds in PLGA is greatly affected by the type and volume of a dispersed solvent, the molar ratio of ammonia to solvent, and the interaction between solvent volume and ammonia concentration.	[68]
24.	Ibuprofen	Cetyl alcohol	A hot melt dispersion method	Silica nanoparticles had no influence on thermic profile, crystalline state of ibuprofen and lipid, they had an influence on the kinetics of drug release related to the increase of size of the composite solid-lipid microparticles prepared.	[69]

REFERENCES

1. Microspheres: Technologies and Global Markets BCC. Res AVM073B (2013). (online) http://www.bccresearch.com/market -research/advanced -materials/microspheres -global-markets-avm073b.html, Date accessed: 03/06/2014.
2. Hammad U., Hemlata N., Asif M.T., and Nainar, MSM. Microencapsulation: Process, techniques and applications. *IJBRA* 2(2) (2011):472–481.
3. Nitika A., Ravinesh M., Chirag G., and Manu A. Microencapsulation—A novel approach in drug delivery: A review. *Indo-Global J. Pharm. Sci* 2(1) (2012):1–20.
4. Malakar J., Das T., and Ghatak S. Microencapsulation: An indispensable technology for drug delivery system. *IRJP* 2(3) (2012):8–13.
5. Van-Thanh T., Jean-Pierre B., and Marie-Claire V.-J. Why and how to prepare biodegradable, monodispersed, polymeric microparticles in the field of pharmacy? *Int. J. Pharm.* 407(1–2) (2011):1–11.
6. Deepak K.M., Ashish K.J., and Prateek K.J. A review on various techniques of microencapsulation. *Int. J. Pharm. Chem Sci.* 2(2) (2013):962–977.
7. Venkata Naga J.N., Muthu Prasanna P., Suhas Narayan S., Surya Prabha K., Seetha Ramaiah P., and Srawan G.Y. Microencapsulation techniques, factors influencing encapsulation efficiency. *J. Microencap.* 27(3) (2010):187–197.
8. Shekhar K., Naga Madhu M., Pradeep B., and David B. A review on microencapsulation. *Int. J. Pharm. Sci. Rev. Res.* 5(2) (2010):58–62.
9. Bansode S.S., Banarjee S.K., Gaikwad D.D., Jadhav S.L., and Thorat R.M. Microencapsulation: A review. *Int. J. Pharm. Sci. Rev. Res.* 1(2) (2010):38–43.
10. Jyothi Sri S., Seethadevi A., SuriaPrabha K., Muthuprasanna P., and Pavitra P. Microencapsulation: A review. *Int. J. Pharm. Bio. Sci.* 3(1) (2013):509–531.
11. Sergio F., Hans P.M., and Bruno G. Microencapsulation by solvent extraction/evaporation: Reviewing the state of the art of microsphere preparation process technology. *J. Control. Release* 102(2) (2005):313–332.
12. Ming L., Olivier R., and Denis P. Microencapsulation by solvent evaporation: State of the art for process engineering approaches. *Int. J. Pharm.* 363(1–2) (2008):26–39.
13. Pandya N., Pandya M., and Bhaskar V.H. Preparation and *in-vitro* characterization of porous carrier-based glipizide floating microspheres for gastric delivery. *J. Young Pharmacists* 3(2) (2011):97–104.
14. Perumal D. Microencapsulation of ibuprofen and Eudragit RS 100 by the emulsion solvent diffusion technique. *Int. J Pharm.* 218(1–2) (2001):1–11.
15. Tapan Kumar G., Chhatrapal C., Ajazuddin, Amit A., Hemant B., and Dulal Krishna T. Prospects of pharmaceuticals and biopharmaceuticals loaded microparticles prepared by double emulsion technique for controlled delivery. *Saudi Pharm. J.* 21(2) (2013):125–141.
16. Zohra M., Yann P., and Alf L. Oil-in-oil microencapsulation technique with an external perfluorohexane phase. *Int. J. Pharm.* 338(1–2) (2007):231–237.
17. Ereitas S., Rudolf B., Merkle H.P., and Gander B. Flow through ultrasonic emulsification combined with static micromixing for aseptic production of microspheres by solvent extraction. *Eur. J. Pharm. Biopharm.* 61(3) (2005):181–187.
18. Khaled E., Abid Riaz A., Andrei D., and Roland B. Encapsulation of water-soluble drugs by an o/o/o-solvent extraction microencapsulation method. *Int. J. Pharm.* 409(1–2) (2011):89–95.
19. Bhoyar P.K., Morani D.O., Biyani D.M., Umekar M.J., Mahure J.G., and Amgaonkar Y.M. Encapsulation of naproxen in lipid-based matrix microspheres: Characterization and release kinetics. *J Young Pharmacists* 3(2) (2011):105–111.
20. Mahmoud M.A., Saleh Abd E.R., Sayed H.A., and Mohamed A.I. Emulsification/internal gelation as a method for preparation of diclofenac sodium–sodium alginate microparticles. *Saudi Pharm. J.* 21(1) (2013):61–69.
21. Berkland C., Kim K., and Pack D.W. Fabrication of PLG microspheres with precisely controlled and monodisperse size distributions. *J. Control. Release* 73(1) (2001):59–74.
22. Keita K., Toshiyuki N., and Kazumi D. Preparation of sustained-release coated particles by novel microencapsulation method using three-fluid nozzle spray drying technique. *Eur. J. Pharm. Sci.* 51 (2014):11–19.
23. Subrata Deb N., Sora S., Alexandar S., Young Ki M., and Byong Taek L. Preparation and characterization of PLGA microspheres by the electrospraying method for delivering simvastatin for bone regeneration. *Int. J. Pharm.* 443(1–2) (2013):87–94.

24. Annalisa D., Anna Angela B., Gaetano L., and Matteod' A. Intensifying the microencapsulation process: Ultrasonic atomization as an innovative approach. *Eur. J. Pharm. Biopharm.* 80(3) (2012):471–477.
25. Mukund J.Y., Kantilal B.R., and Sudhakar R.N. Floating microspheres: A review. *Braz. J. Pharm.* 48(1) (2012):17–30.
26. Jain N.K. *Advances in Controlled and Novel Drug Delivery*, 3rd edn., p. 71. CBS Publication, Delhi, India, 2004.
27. Narasimha R.R., Sampath K.M., Dileep K.K., and Vineeth P. Formulation and evaluation of capcitabine microspheres. *IJPT* 3(2) (2011):2599–2632.
28. Tanuja S and Padma V. Buccoadhesive tablets of nifedipine: Standardization of a novel buccoadhesive-erodible carrier. *Drug Dev. Ind. Pharm.* 20(19) (1994):3005–3014.
29. Alagusundaram M., Madhu S.C.C., and Umashankari K. Microspheres as a novel drug delivery system-a review. *Int. J. Chem. Tech. Res.* 1(3) (2009):526–534.
30. Hemlata K., Hari K.S.L., and Amanpreet K. Mucoadhesive microspheres as carriers in drug delivery: A review. *Int. J. Drug Dev. Res.* 4(2) (2012):21–34.
31. Qingxing X., Shi En C., Chi-Hwa W., and Daniel W.P. Mechanism of drug release from double-walled PDLLA(PLGA) microspheres. *Biomaterials* 34 (2013):3902–3911.
32. Qingxing X., Yujie X., Chi-Hwa W., and Daniel W.P. Monodisperse double-walled microspheres loaded with chitosan-p53 nanoparticlesand doxorubicin for combined gene therapy and chemotherapy. *J. Control. Release* 163 (2012):130–135.
33. Pengcheng D., Tingmei W., and Peng L. Double-walled hollow polymeric microspheres with independent pH and temperature dual-responsive and magnetic-targeting function from onion-shaped core-shell structures. *Colloids Surf. B* 102 (2013):1–8.
34. Ninan M., Lu X., Qifang W., Xiangrong Z., Wenji Z., Yang L., Lingyu J., and Sanming L. Development and evaluation of new sustained-release floating microspheres. *Int. J. Pharm.* 358 (2008): 82–90.
35. Valentina I., Gilberto C., Marcello R., Santo S., and Eliana L. In vivo detection of lipid-based nano- and microparticles in the outermost human stratum corneum by EDX analysis. *Int. J. Pharm.* 447 (2013):204–212.
36. Paula K.O., Marcelo T., Júlio C.C.B., Roberta D.C.O.L., and Carmen S.F.-T. Co-encapsulation of Lactobacillus acidophilus with inulin orpolydextrose in solid lipid microparticles provides protection and improves stability. *Food Res. Int.* 53 (2013):96–103.
37. Chambi H.N.M., Alvim I.D., Barrera-Arellano D., and Grosso C.R.F. Solid lipid microparticles containing water-soluble compoundsof different molecular mass: Production, characterization and release profiles. *Food Res. Int.* 41 (2008):229–236.
38. Laurent P., Mike R., Coralie G., Jean-Marie D., Franc, O., and Philippe L. New solid lipid microparticles for controlled ibuprofen release: Formulation and characterization study. *Int. J. Pharm.* 422 (2012):59–67.
39. Lorenzo C., Stefania M., and Claudio N. Design, production and optimization of solid lipid microparticles (SLM) by a coaxialmicrofluidic device. *J Control. Release* 160 (2012):409–417.
40. Ai-Zheng C., Li L., Shi-Bin W., Xiao-Fen L., Yuan-Gang L., Chen Z., Guang-Ya W., and Zheng Z. Study of Fe3O4–PLLA–PEG–PLLA magnetic microspheres based on supercriticalCO2: Preparation, physicochemical characterization, and drug loading investigation. *J. Supercrit. Fluids* 67 (2012):139–148.
41. Priya P., Sivabalan M., Balaji M., Rajashree S., and Muthu Dhanabakiam H. Microparticle: A novel drug delivery system. *Int. J. Pharm. Biol. Sci.* 3(2) (2013):310–326.
42. Jaspreet K.V., Kaustubh T., and Sanjay G. Bioadhesive microspheres as a controlled drug delivery system. *Int. J. Pharm.* 255 (2003):13–32.
43. Pavan Kumar B., Sarath Chandiran I., Bhavya B., and Sindhuri M. Microparticulate drug delivery system: A review. *Ind. J. Pharm. Sci. Res.* 1(1) (2011):19–37.
44. Rama D., Shami T.C., and Bhasker R.K.U. Microencapsulation technology and applications. *Defence Sci. J.* 59(1) (2009):82–95.
45. Kumar R. and Palmieri M.J. Jr. Points to consider when establishing drug product specifications for parenteral microspheres. *AAPS J.* 12(1) (2010):27–32.
46. Priya P., Sivabalan M., Balaji M., Rajashree S., and Muthu D.H. Microparticle: A novel drug delivery system. *IJPBS* 3(2) (2013):310–326.
47. Lam P.L. and Gambari R. Advanced progress of microencapsulation technologies: In vivo and in vitro models for studying oral and transdermal drug deliveries. *J Control. Release* 178(C) (2014):25–45.
48. Bordesa C., FrEvillea V., Ruffinb E., Marotea P., Gauvrita J.Y., Brianc S., and Lanteria P. Determination of poly(-caprolactone) solubility parameters: Application to solvent substitution in a microencapsulation process. *Int. J. Pharm.* 383(1–2) (2010):236–243.

49. Lauren E.K., Huaping T., Siddharth J., Steven R.L., Jason W.F., and Kacey G.M. Protein bioactivity and polymer orientation is affected by stabilizer incorporation for double-walled microspheres. *J. Control. Release* 141(2) (2010):168–176.
50. Casalengua P.C., Jiang C., Bravo-Osuna I., Budd A.T., Molina-Martínez I.T., Michael J.Y., and Herrero-Vanrell R. Retinal ganglion cells survival in a glaucoma model by GDNF/Vit E PLGA microspheres prepared according to a novel microencapsulation procedure. *J. Control. Release* 156(1) (2011):92–100.
51. Haixia W., Haifeng S., Agnes C.C., and John H.X. Microencapsulation of vitamin C by interfacial/emulsion reaction: Characterization of release properties of microcapsules. *J Control. Release* 152(1) (2011):e78–e79.
52. Xu Q., Xia Y., Chi-Hwa W., and Daniel W.P. Monodisperse double-walled microspheres loaded with chitosan-p53 nanoparticles and doxorubicin for combined gene therapy and chemotherapy. *J. Control. Release* 163(2) (2012):130–135.
53. Yujie X., Pedro F.R., and Daniel W.P. Controlled protein release from monodisperse biodegradable double-wall microspheres of controllable shell thickness. *J. Control. Release* 172(3) (2013):707–714.
54. Sarath C.I., Balagani P.K., and Korlakanti N.J. Characterization of Glibenclamide loaded cellulose acetate microparticles prepared by an emulsion solvent evaporation method. *J. Pharm. Res.* 7(8) (2013):766–773.
55. Rita P.A., Giulia A., Teresa M., Paola R., Renata A., Sara L., Giovanna D.P., Amalia P., Ernesto R., and Pasquale D.G. Design and production of gentamicin/dextransmicroparticles by supercritical Assistedatomisation for the treatment of wound bacterial infections. *Int. J. Pharm.* 440(2) (2013):188–194.
56. Zvonar A., Bolko K., and GaSperlin M. Microencapsulation of self-microemulsifying systems: Optimization of shell-formation phase and hardening process. *Int. J. Pharm.* 437 (2012):294–302.
57. Suksamran T., Ngawhirunpat T., Rojanarata T., Warayuth S., Pitaksuteepong T., and Praneet O. Methylated N-(4-N,N-dimethylaminocinnamyl) chitosan-coated electrospray OVA-loaded microparticlesfor oral vaccination. *Int. J. Pharm.* 448(1–2) (2013):19–27.
58. Li H., Wen-Feng D., Jian-Yuan Z., Xi-Ming X., and Feng-Qian L. Triggering effect of N-acetylglucosamine on retarded drug release from a lectin-anchored chitosan nanoparticles-in-microparticles system. *Int. J. Pharm.* 449(1–2) (2013):37–43.
59. Aparna S., Wai K.N., Reginald B.H.T., and Chan S.Y. Development of controlled release inhalable polymeric microspheres for treatment of pulmonary hypertension. *Int. J. Pharm.* 450(1–2) (2013):114–122.
60. Chao-Da X., Xiang-Chun S., and Ling T. Modified emulsion solvent evaporation method for fabricating core–shell microspheres. *Int. J. Pharm.* 452(1–2) (2013):227–232.
61. Lu Y., Sturek M., and Park K. Microparticles produced by the hydrogel template method for sustained drug delivery. *Int. J. Pharm.* 461(1–2) (2014):258–269.
62. Katrin M., Juergen S., and Roland B. Novel preparation techniques for alginate–poloxamermicroparticles controlling protein release on mucosal surfaces. *Eur. J. Pharm. Sci.* 45(3) (2012):358–366.
63. Mamdouh A.M., Noha M.Z., Samar M., and Ahmed S.G. Bioavailability enhancement of verapamil HCl via intranasal chitosan microspheres. *Eur. J. Pharm. Sci.* 51(2014):59–66.
64. Keita K., Toshiyuki N., and Kazumi D. Preparation of sustained-release coated particles by novel microencapsulation method using three-fluid nozzle spray drying technique. *Eur. J. Pharm. Sci.* 51 (2014):11–19.
65. Lorenzo C., Stefania M., and Claudio N. Design, production and optimization of solid lipid microparticles (SLM) by a coaxial microfluidic device. *J. Control. Release* 160(3) (2012):409–417.
66. Subrata D.N., Sora S., Alexandar S., Young K.M., and Byong T.L. Preparation and characterization of PLGA microspheres by the electrospraying method for delivering simvastatin for bone regeneration. *Int. J. Pharm.* 443(1–2) (2013):87–94.
67. Frederic T., Carsten E., and Anne M.H. Superparamagnetic iron oxide nanoparticles (SPIONs)-loaded Trojan microparticles for targeted aerosol delivery to the lung. *Eur. J. Pharm. Biopharm.* 86(1) (2014):98–104.
68. Sunju H., Minjung L., Sunhwa L., and Hongkee S. Investigation on structural integrity of PLGA during ammonolysis-based microencapsulation process. *Int. J. Pharm.* 419(1–2) (2011):60–70.
69. Laurent L.P., Mike M.R., Coralie C.G., Jean-Marie J.M.D., Françoise F.Q., and Philippe P.L. New solid lipid microparticles for controlled ibuprofen release: Formulation and characterization study. *Int. J. Pharm.* 422(1–2) (2012):59–67.

48 Colloid Drug Delivery Systems

Monzer Fanun

CONTENTS

48.1	Colloids	1111
48.2	Micelles	1112
48.3	Dendrimers	1114
48.4	Liquid Crystals	1114
48.5	Cubosomes and Hexosomes	1115
48.6	Gels	1116
48.7	Emulsions	1116
48.8	Multiple Emulsions	1117
48.9	Nanoemulsions	1117
48.10	Microemulsions	1117
48.11	Liposomes	1118
48.12	Niosomes	1119
48.13	Polymersomes	1120
48.14	Nanocapsules	1120
48.15	Microspheres	1121
48.16	Aerosols	1121
48.17	Foams	1122
References		1122

48.1 COLLOIDS

Colloids are a dispersion of one phase to another, and its size ranges from 1 nm to 1 μm, and in a system, discontinuities are found at distances of that order. These nanoscale dispersed systems have considerable application in drug delivery and formulation methods.[1,2] Understanding the properties of colloids can be used as a tool to access rational criteria for formulation development. Colloidal properties provide an idea about absorption, cell membrane interaction, and drug molecule behavior in the surrounding environment. Such properties play a major role in drug-loading capacity and transport mechanisms as charge and colloidal size ranges are set for this capacity. The colloidal particles greatly affect the rate of sedimentation, osmotic pressure as well as the stability and biocompatibility of colloidal drug carrier. The colloidal properties also provide a selection basis for excipients and manufacturing process criteria. These properties include the particle properties studied by the Tyndall effect, turbidity, and dynamic light scattering. Optical properties are widely used to observe the size, shape, and structure of colloidal particles. The kinetic properties, which include Brownian motion, diffusion, osmosis, sedimentation, and viscosity, deal with motion of particles with respect to the dispersion medium. These properties provide a detailed idea on the movement of colloidal carrier within the body as well as the transportation criteria of colloidal carrier across the cell membrane. The physicochemical properties such as physical state, lyophilicity, and lyophobicity are helpful in the selection of colloidal carrier for a particular drug delivery. Electrical properties, which include electrical conductivity and surface charges, are related with migration of particle and give an idea about the selection basis for excipients and the interaction of colloidal carrier with a cell membrane. The magnetic properties of colloids deal with specific delivery of drug inside the body, which helps to increase the efficacy of drug and reduces the toxicity.

The problems experienced by free drugs delivery are instability, poor solubility, toxicity, nonspecificity, and inability to cross blood–brain barrier. The problem experienced by free drug can be overcome by the use of drug delivery system. The development of delivery systems has had a huge impact on our ability to treat numerous diseases. To attain highest pharmacological effects with least side effects of drugs, it should be delivered to target sites without significant distribution to nontarget areas. Colloidal drug delivery carriers are known to improve the solubility of poorly soluble drugs; to provide a microenvironment for the protection of fragile drugs, such as proteins and peptides, that are very often of large size and need to be protected from hydrolysis or enzymatic degradation and absorption through membrane; to increase the drug efficacy and reduce their toxicity; to achieve the targeted delivery of drug by conjugating a specific vector to the carrier; and to have emerged as efficient vehicles for drug delivery, which allow sustained or controlled release for transdermal, topical, oral, nasal, intravenous, ocular, parenteral, and other administration routes of drugs.[1,2] The colloidal delivery systems have been aimed at improving therapeutics by virtue of their submicron size and targetability. The effectiveness of a drug therapy is often governed by the extent to which temporal and distribution control could be achieved. Temporal control is the ability to manipulate the period over which drug release is to take place and/or the possibility of triggering the release at a specified time during the treatment in response to some stimuli such as temperature, pH, and so on. Distribution control is to direct the delivery system to the desired site of action. Colloidal drug delivery systems include micelles, dendrimers, liquid crystals, cubosomes, hexosomes, emulsions, multiple emulsions and self-emulsifying drug delivery systems (SEDDS), nanoemulsions, microemulsions and self-microemulsifying drug delivery systems (SMEDDSs), liposomes, niosomes, polymesomes, nanocapsules, solid–lipid nanoparticles, microspheres, aerosols, and foams. Figure 48.1 presents a lipid-based drug nanocarrier and its nanoscale dimensional comparison with biological cell, DNA, lipid bilayers, and atoms. In this entry, we have attempted to present a broad view over the past years on these colloidal drug delivery systems as solubilization and dissolution enhancers of poorly soluble drugs, as a medium for generating new drug delivery systems, and as delivery systems themselves.[1,2]

48.2 MICELLES

Micelles are the self-assembling nanosized colloidal particles, stealth properties with a hydrophobic core and hydrophilic shell with control size and composition. Micelles are dynamic structures with a liquid core, so they cannot be assumed to be having a definitive rigid shape. For scientific consideration, they are usually regarded as having sphericity. Micelles are formed only when the concentration of the surfactant in the solution increases above critical micellar concentration. Micellar shape may be affected by factors such as concentration, temperature, and presence of added electrolyte. Thus, they may undergo a transition when any of the factors are changed. Micelles also exhibit polydispersity in size. Among the micelle-forming compounds besides surfactants, amphiphilic copolymers, that is, polymers consisting of hydrophobic and hydrophilic blocks, are gaining an increasing attention.[3] The interest has specifically been focused on the potential application of polymeric micelles in the major areas in drug delivery, drug solubilization, controlled drug release, drug targeting, and diagnostic purpose. Increased solubility of a less soluble organic substance in surfactant solution has been applied since long.[1] This increased solubility of a solubilizate in surfactant solution was due to some form of attachment of the solubilizate to the exterior of the micelle or solution in it. There is a marked difference in the behavior of polar and nonpolar solutes.[1] The solubilizate may be present at different sites in the micelles, dependent on its chemical nature. It is usually observed that the nonpolar solubilizates are accommodated in the hydrocarbon core and the semipolar and polar solubilizates are present in the palisade layer.

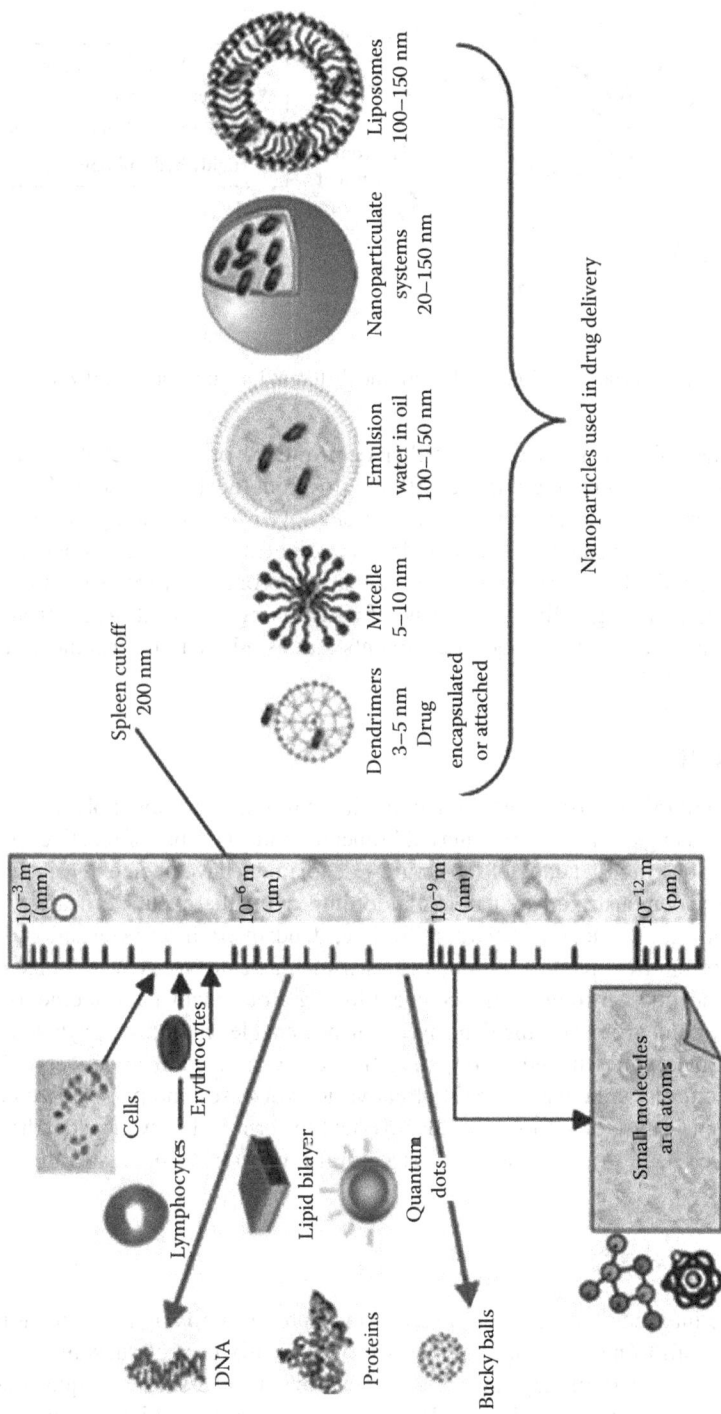

FIGURE 48.1 Lipid-based drug nanocarriers and their nanoscale dimensional comparison with biological cell, DNA, lipid bilayers, and atoms.

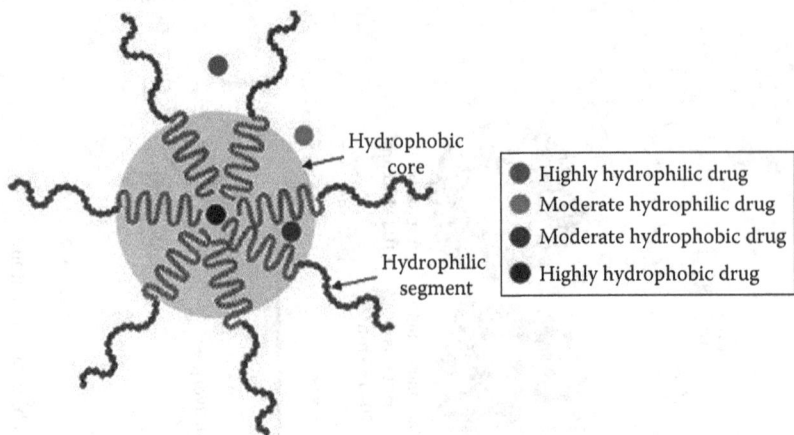

FIGURE 48.2 Simplified structure of a micelle with the different locations of solubilizate.

The location in the palisade layer can be either deep buried or short penetration. Figure 48.2 depicts the simplified structure of a micelle with the different locations of solubilizate. The application of micelles as drug delivery systems can be employed for enhancing the solubilization of potent but hydrophobic and sparingly soluble drug candidates. A low water concentration inside the micellar core retards the degradation of drugs that are prone to hydrolysis. Micelles facilitate controlled and sustained drug delivery by virtue of partitioning toward it. Apart from these major applications, they may also serve as good excipients, act as adjuncts in vaccines, facilitate taste masking, and so on.[4–6]

48.3 DENDRIMERS

Dendrimers have a highly branched, nanoscale architecture with very low polydispersity and high functionality, comprising a central core, internal branches, and a number of reactive surface groups. Because of their unique highly adaptable structures, dendrimers have been extensively investigated for drug delivery and demonstrated great potential for improving therapeutic efficacy.[7–11] Figure 48.3 presents a schematic of a dendritic structure. To date, dendrimers have been tailored to deliver a variety of drugs for the treatment of various diseases such as cancers. A wide range of drugs can be delivered by dendrimers through either covalent linkages or noncovalent interactions. The presence of a number of end groups on the dendrimer surface enables high drug payload and assembly of multiple functional entities including targeting ligand for targeted drug delivery. Highly adaptable dendritic structures can be engineered to treat various diseases and personalized medicine. In addition to the extensive use of off-the-shelf dendrimers for drug delivery, new dendritic structures are also being developed to obtain additional structural properties to meet specific needs in drug delivery.[7–11]

48.4 LIQUID CRYSTALS

Liquid crystals are intermediate states of matter or mesophases, halfway between an isotropic liquid and a solid crystal. In nature, some substances, or even mixtures of substances, present these mesomorphic states. These liquid crystalline phases exhibit a local disorder ("liquid-like" behavior) and are dynamic at a molecular level, but a long-range order exists, which endows it with unique rheological, mass transport, and optical properties. Liquid crystals offer a number of useful properties for the drug delivery. Solubilization of drug in the liquid crystalline is similar to the solubilization of drug in micelles. Simultaneously, increase in viscosity of the system helps to provide more

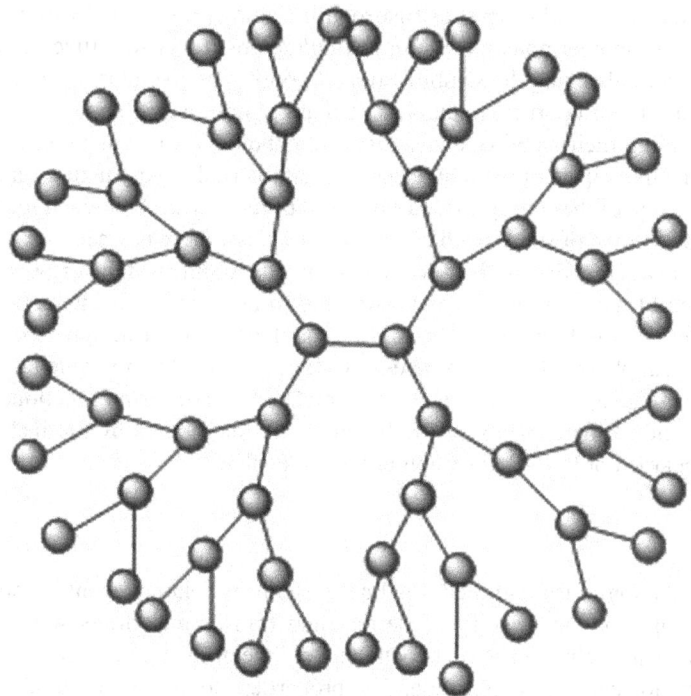

FIGURE 48.3 A schematic of a dendritic structure.

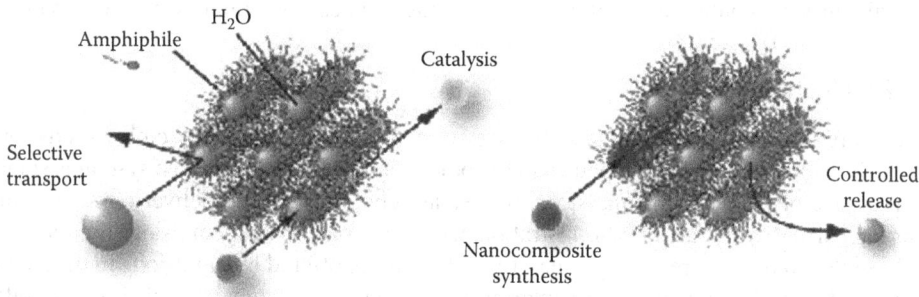

FIGURE 48.4 Functional liquid crystal assembly.

localized effects in parenteral (intramuscular), topical, or oral administration. The phase transitions of liquid crystals can be achieved either by temperature or by dilution. The systems can be tailored in such a way the transitions can be achieved at body temperature or in contact with the body fluids. Figure 48.4 presents a schematic of functional liquid crystal assembly.[12–14]

48.5 CUBOSOMES AND HEXOSOMES

Cubosomes, hexosomes, and micellar cubosomes are colloidal aqueous dispersions of surfactant-like lipids with confined inner nanostructures. The formation of these dispersed submicron-sized particles with embedded inverted-type mesophases that display nanostructures closely related to those observed in biological membranes is receiving much attention in pharmaceutical, food, and cosmetical applications. Owing to their unique physicochemical characteristics, they represent an interesting colloidal family that has excellent potential to solubilize bioactive molecules with different physicochemical properties (hydrophilic, amphiphilic, and hydrophobic molecules). In particular, there is an enormous interest in testing the possibility to utilize these dispersions as drug nanocarriers for enhancing the solubilization of poorly water-soluble drugs and for improving their bioavailability. It is important to emphasize that the optimal utilization of these dispersions requires understanding their stability under different conditions (such as presence of salt and pH variation), studying the impact of loading drugs on their internal nanostructures, and fully comprehending the interaction of these drug-loaded dispersed particles with biological interfaces to ensure the efficient transportation of solubilized drugs. Nevertheless, there is a need for further investigations in the future to address the challenges of enhancing the stability of the dispersed particles after administration and of modulating their nanostructure to optimize their interaction with different biological surfaces. It was found that the addition of short charged designer peptide surfactants mimicking biological phospholipids can be used to functionalize the nonlamellar mesophases and to enhance the loading capacity of charged active molecules. The term "functionalization" means to control the solubilization capacity of the liquid crystalline phases by the inclusion of specific anchors such as charged or long-chain amphiphilic molecules.[15–19]

48.6 GELS

The use and formulation of three-dimensional gel systems as gelating agents, matrices in patches, and wound dressings for the delivery of dermal and transdermal drugs were investigated.[20–25] Requirements for such a delivery system depend on both the drug characteristics and the conditions in the place of the topical drug application. The properties, composition, duration of drug effect, uses, and side effects of the products that are applicable for the skin are considered. The most applied gels belong to the classes of cellulose derivatives, chitosan, carageenan, polyacrylates, polyvinyl alcohol, polyvinylpyrrolidone, and silicones, which can be further modified to alter the drug delivery or three-dimensional configuration of the gel, including synthesis of stimuli-sensitive polymers.[20–25]

48.7 EMULSIONS

Emulsions are thermodynamically unstable system consisting of two immiscible liquids, oil and water, one is dispersed as minute globules in the other continuous medium. The system is stabilized by the presence of surfactants. Emulsions are extensively used as drug delivery systems particularly for topical and oral route of administration. Generally, oil-in-water emulsions are intended for internal use due to obvious reasons. Some of the advantages offered by emulsions in drug delivery are solubilization of hydrophobic drug, high drug release rate, and enhanced chemical stability or protection of drug from hydrolysis in aqueous environment. Pharmaceutical emulsions contain less amount of surfactants (phospholipids, lecithins), which are natural in origin and thus nontoxic in nature. Oil components are always selected from the oils of natural origin such as soya bean oil, cottonseed oil, coconut oil, corn oil, sesame oil, cod liver oil, olive oil, and linseed oil or mixture of different fatty acids in different proportions.[26] SEDDS are ideally isotropic mixtures of oils and surfactants, and sometimes including cosolvents. They emulsify under mild agitation conditions, just like those encountered in the gastrointestinal tract and produces fine emulsion/lipid droplets, ranging from 100 nm to less than 50 nm. Substantial interfacial disruption and/or ultra-low oil–water interfacial tension are the primary necessities to achieve the self-emulsification. SEDDS have been shown their ability to improve the oral bioavailability of less water-soluble and hydrophobic drugs. Generally, SEDDS are prepared by dissolving drugs in oils containing suitable solubilizing agents. SEDDS are generally formulated by using triglyceride oils and nonionic surfactants at surfactant concentrations more than 25%. As an improvement of conventional liquid-SEDDS, solid-SEDDS (S-SEDDS) may also be developed. S-SEDDS may reduce production cost, having simple manufacturing process, and they are stable as well as patient compliant. S-SEDDS may also be modified to include solid dosage forms for oral as well as for parenteral administration.[26–33]

48.8 MULTIPLE EMULSIONS

Multiple emulsions are complex emulsion systems where both oil-in-water (o/w) and water-in-oil (w/o) emulsion types exist simultaneously. The simplest multiple emulsions, popularly known as "double emulsions," are ternary systems consisting of either water-in-oil-in-water (w/o/w) or oil-in-water-in-oil (o/w/o) structures. Multiple emulsions are generally prepared either by a one-step emulsification or by a two-step emulsification process. Multiple emulsions-based formulations have been widely used in the pharmaceutical industry as vaccine adjuvants, sustained release, and parenteral drug delivery systems. However, o/w/o systems have found wide application in the cosmetic and food industries. Many particulate drug delivery systems (nanoparticles, microspheres, liposomes) have been prepared by using multiple emulsions during one of the developmental steps. The poor stability and polydispersibility of multiple emulsions (w/o/w) has been a major challenge for its pharmaceutical applications. Therefore, mechanism of its stability, different formulation methods used to enhance monodispersibility, and evaluation of the finished product ensuring quality were investigated.[34-38] Multiple emulsions are versatile drug carriers, which have a wide range of applications in controlled drug delivery, particulate-based drug delivery systems, targeted drug delivery, taste masking, bioavailability enhancement, enzyme immobilization, overdosage treatment/detoxification, red blood cell substitute, lymphatic delivery, and shear-induced drug release formulations for topical application. Surface-modified fine multiple emulsions containing biopolymers are used as parenteral and transarterial delivery vehicles for various anticancer drugs and antigens employed in cancer chemotherapy and immunotherapy. Some novel applications include encapsulation of a drug by the evaporation of the intermediate phase, leaving behind capsules formed of polymers (polymerosomes), solid particles (colloidosomes), and vesicles.[34-38]

48.9 NANOEMULSIONS

Much attention has been paid to potential pharmaceutical uses of nanoemulsions as novel drug delivery systems. Nanoemulsions are transparent or translucent systems that have a dispersed-phase droplet size range of typically 20–200 nm; although in earlier cases, these systems have also been called microemulsions. Nanoemulsions are thermodynamically stable, isotropically clear dispersions of oil and water stabilized with the help of surfactant and cosurfactant. These systems are attractive as pharmaceutical formulations, and they are drug carrier systems for oral, topical, and parenteral administration. Especially, these dosage forms have been suggested as carriers for peroral peptide-protein drugs. It was hypothesized that formulating a nanoemulsion of the drug would help to increase the bioavailability of the drug due to the high solubilization capacity as well as the potential for enhanced absorption. On the other hand, the small droplet size, high kinetic stability, and optical transparency of nanoemulsions compared with conventional emulsions give them advantages for their use in many technological applications. The majority of works on nanoemulsion applications deal with the preparation of polymeric nanoparticles using a monomer as the disperse phase (the so called miniemulsion polymerization method).[39-43]

48.10 MICROEMULSIONS

Microemulsions are transparent systems of two immiscible fluids, stabilized by an interfacial film of surfactant or a mixture of surfactants, frequently in combination with a cosurfactant. These systems could be classified as water-in-oil, bicontinuous, or oil-in-water type depending on their microstructure, which is influenced by their physicochemical properties and the extent of their ingredients.[1,2,44-49] SMEDDSs form transparent microemulsions with a droplet size of less than 50 nm. Oil is the most important excipient in SMEDDSs because it can facilitate self-emulsification and increase the fraction of lipophilic drug transported through the intestinal lymphatic system, thereby increasing absorption from the gastrointestinal tract. Long-chain and medium-chain

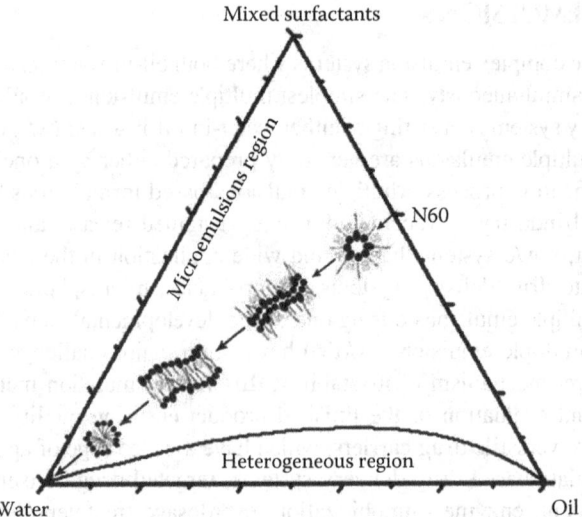

FIGURE 48.5 Schematic presentation (not to scale) of possible packings of drug along dilution line N60 at the different dilution regions: water-in-oil, bicontinuous, and oil-in-water microemulsions. N60 means that the weight ratio of oil/mixed surfactants equals 4/6.

triglyceride oils, modified or hydrolyzed vegetable oils, have been explored widely. Novel semisynthetic medium-chain triglyceride oils have surfactant properties and are widely replacing the regular medium-chain triglyceride. Microemulsions are characterized by ultra-low interfacial tension between the immiscible phases and offer the advantage of spontaneous formation; thermodynamic stability; simplicity of manufacture; solubilization capacity of lipophilic, hydrophilic, and amphiphilic solutes; improved solubilization and bioavailability of hydrophobic drugs; the large area per volume ratio for mass transfer; and the potential for permeation enhancement. It was demonstrated that microemulsions enhance the solubilization capacity and dissolution efficiency of poorly soluble drugs; the solubilization capacity and dissolution efficiency of drugs are reliant on the microstructure of the microemulsions; the solubilized drugs may influence the boundaries of structural regions and the transition point between different microemulsion's microstructures; drug extent and route of delivery could be influenced by the microstructure of the microemulsions; drug delivery systems generated in microemulsions improved drug release and compatibility; the extent and rate of drug delivery is dependent on the generated system preparation method in microemulsions; and the generated system could influence the selection of a delivery route.[44–49] Figure 48.5 presents schematically (not to scale) of possible packings of drug along dilution line N60 at the different dilution regions: water-in-oil, bicontinuous, and oil-in-water microemulsions. N60 means that the weight ratio of oil/mixed surfactants equals 4/6.

48.11 LIPOSOMES

Advances in liposome technologies for conventional and nonconventional drug deliveries have helped pharmacologists with a tool to increase the therapeutic index of several drugs by improving the ratio of the therapeutic effect over the drug's side effect. Thanks to liposome versatility, effective formulations able to deliver hydrophobic drugs, to prevent drug degradation, to alter biodistribution of their associated drugs, to modulate drug release, and, most of all, to selectively target the carrier to specific areas have been obtained. Moreover, by gradually increasing the complexity of the formulation, sophisticated membrane models have been developed, and new landscapes on characterization possibilities of membrane-associated biomacromolecules have been discovered. Basic conventional liposomes are multilamellar or unilamellar vesicles composed of phospholipids,

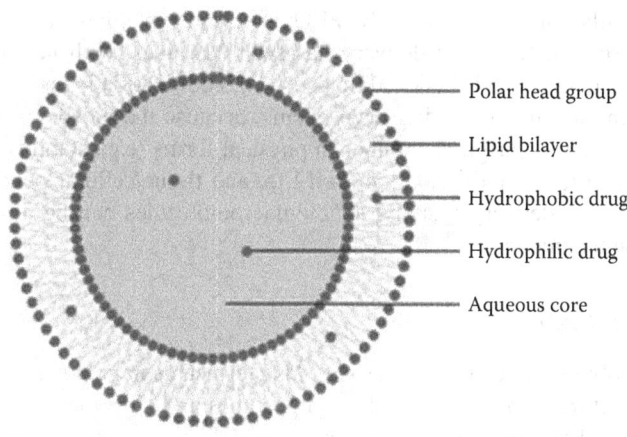

FIGURE 48.6 Unilamellar liposomal vesicle.

but several different constituents are often included to modulate the fluidity or the permeability of the bilayer. Figure 48.6 presents a unilamellar liposomal vesicle. With the aim of increasing bioavailability in blood, differently alkylated poly(ethylene glycol)-based polymers are usually added to liposomal formulations. Variations in formulation and in preparation methods may affect the efficacy of the active principle that has to be delivered or the functionality of the biomacromolecule that has to be studied, and this is why a universal liposome preparation protocol cannot be developed.[50–54]

48.12 NIOSOMES

Similar to liposomes, nonionic surfactant-based vesicles (niosomes) are vesicularly formed from the self-assembly of nonionic amphiphiles in aqueous media. This self-assembly into closed bilayers is rarely spontaneous and usually requires an energy input in the form of physical agitation or heat. The bilayer assembly ensures minimum contact of hydrophobic parts of the molecule from the aqueous solvent, and the hydrophilic head groups are completely in contact with the same. Like liposomes, niosomes represent a highly flexible platform due to their unique bilayer structure. First, the bilayer structure is composed of two layers of nonionic amphiphiles, with their hydrophobic tails facing one another forming the lipophilic core region while their hydrophilic heads are exposed to the interfacial aqueous environment. Therefore, such a kind of unique structure makes niosomes a promising drug carrier because the interior is able to encapsulate aqueous solutes, and hydrophobic molecules can be incorporated within the lipophilic region of the membrane. Second, unstable drugs (e.g., peptides, proteins, and genetic materials) can be isolated from the adverse environment by being encapsulated into the interior vehicles. Third, the bilayer provides great potentials for surface modification to achieve special functions. The low cost, greater stability, and resultant ease of storage of nonionic surfactants have led to the exploitation of these compounds as alternatives to phospholipids. The niosome formation is affected by surfactant, presence of additives (usually cholesterol), nature of the drug, and physical parameters such as hydration temperature, agitation, size reduction techniques used, and so on. Niosome formation is also based on reduction of interfacial tension as other vesicular assemblies. Typically, vesicles (closed lamellar structures) are formed in the water-rich phase of the binary phase diagram. Almost all the methods used in liposomal surface modification can be easily transferred to the application of niosomal system. For instance, PEGylation has been widely used in liposomal drug delivery systems for the past two decades, which is used to decrease the interaction between liposomes and serum proteins, with a consequent long-circulation effect. Similarly, PEGylated niosomes have also been developed for specific delivery purposes. Niosomal delivery has seen advances, with new applications in biomacromolecular drug delivery. It is believed that surface modification is one of the most important techniques applied in niosomal drug delivery systems, because the niosomal systems can thereby be tailored for some specific purposes—either in physical forms (e.g., stability, drug release) or in biological performance (e.g., biodistribution, half-life, and tissue/cellular penetration). Therefore, multifunctional niosomal delivery systems for biomacromolecules can be developed using some suitable modification techniques.[55–59]

48.13 POLYMERSOMES

Polymersomes are polymer-based bilayer vesicles, also termed as nanometer-sized "bags" by scientists. The bilayer structure displayed is similar to liposomes and niosomes as shown in Figure 48.7. They can be considered as liposomes but of nonbiological origin. Amphiphilic block copolymers can form various vesicular architectures in solution. They can have different morphologies such as uniform common vesicles, large polydisperse vesicles, entrapped vesicles, or hollow concentric vesicles.[60–67]

48.14 NANOCAPSULES

Nanocapsules are submicron-sized polymeric colloidal particles with a therapeutic agent of interest encapsulated within their polymeric matrix or adsorbed or conjugated on the surface. They are mainly based on polymeric materials either of natural or of synthetic origin. Natural substances such as proteins, albumin, gelatin, and legumin, and polysaccharides, starch, alginates, and agarose, have been widely studied. Synthetic hydrophobic polymers from the ester class (polylactic acid) and polyglycolic acid copolymers along with ε-caprolactone have been investigated.[68–76] Figure 48.8 presents a conventional versus surface-functionalized nanocapsules. The theoretical model for a polymeric nanocapsule is a vesicle, where an oily or an aqueous core is surrounded by a thin polymeric wall. Those devices are stabilized by surfactants, such as phospholipids, polysorbates, poloxamers, and cationic surfactants. Nanocapsules composed of different raw materials have been described, including polyesters and polyacrylates, as polymers, triglycerides, large size alcohols, and mineral oil, as oily cores. Several bioactive molecules have been loaded into these systems, such as antitumorals, antibiotics, antifungals, antiparasitics, anti-inflammatories, hormones, steroids, proteins, and peptides. As a general rule, the type of raw materials used to compound nanocapsules can influence their morphological and functional characteristics, which may influence the in vitro release and/or the in vivo response.[68–76]

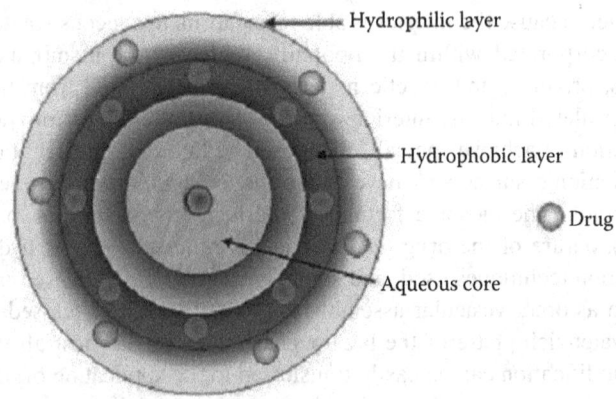

FIGURE 48.7 Drug-loaded bilayer polymerosome.

Colloid Drug Delivery Systems

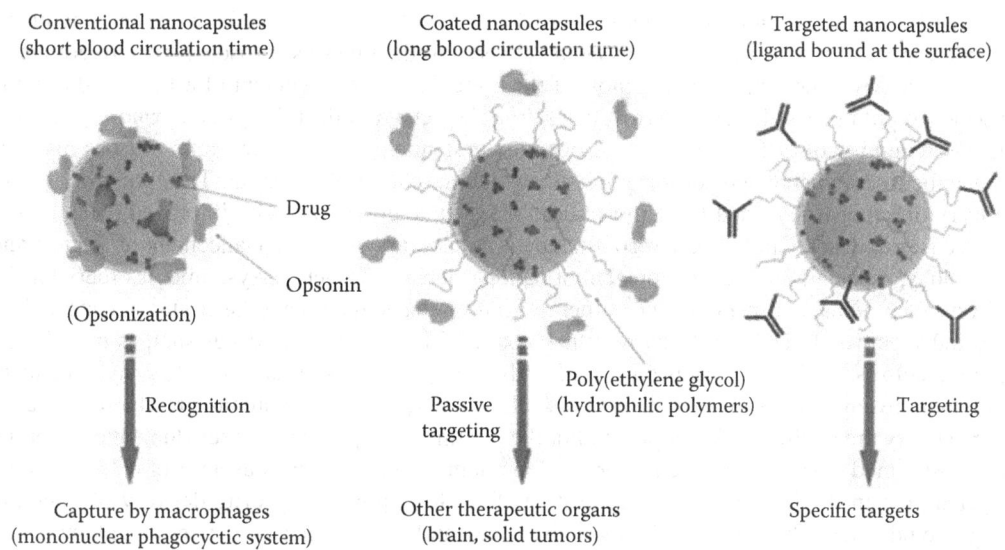

FIGURE 48.8 Conventional versus surface-functionalized nanocapsules.

48.15 MICROSPHERES

There has been considerable interest of late using protein- or polymer-based microsphere as drug carrier. Several methods of microsphere preparation such as single and double emulsion, phase separation, coacervation, spray drying and congealing, and solvent evaporation techniques are used nowadays. The inclusion of drugs in microparticulate carriers clearly holds significant promise for improvement in the therapy of several disease categories. The microspheres were characterized with respect to physical, chemical, and biological parameters. There are a lot of application of microsphere as drug delivery, which includes dermatology, vaccine adjuvant, ocular delivery, brain targeting, gene therapeutics, cancer targeting, magnetic targeting, and many more. This is confident that microparticulate technology will take its place, along with other drug delivery technologies, in enhancing the effectiveness, convenience, and general utility of new and existing drugs.[77–80]

48.16 AEROSOLS

The term "aerosol" describes systems in which liquids are dispersed in air, similar in principle to emulsions, stabilized by the presence of surfactant. Inhalation drug delivery is gaining increasing popularity in treatment of lung diseases due to its advantages over oral administration such as less side effect and quicker action onset. However, therapeutic results of inhaled medications are dependent upon effective deposition at the target site in the respiratory tract. Determining the regional and local deposition characteristics within the respiratory tract is a critical first step for making accurate predictions of the dose received and the resulting topical and systemic health effects. Furthermore, the diversifying areas of pharmaceutical research and the growing interaction between them make inhalation drug delivery a very much multidisciplinary effort, necessitating inputs from engineering and computer techniques from medicine and physiology. Aerosol drug delivery of pharmaceutical colloidal preparations is a novel mode of drug delivery, which has shown promise in the treatment of various local and systemic disorders. Colloidal preparations delivered through nebulizers, pressurized metered dose inhalers, and dry powder inhalers are the major aerosol delivery systems, and drugs are dispersed directly from the formulations (solutions, suspensions, and dry powders) to the lungs. Noninvasive administration of drugs directly to the lungs by various aerosol delivery techniques results in rapid absorption across bronchopulmonary mucosal membranes. Particle size and size distribution are an important factor in efficient aerosol delivery of medicament into the lungs. A wide variety of medicinal agents has been delivered to the lungs as aerosols for the treatment of diverse diseases; however, currently, most of them are for the management of asthma and chronic obstructive pulmonary diseases. Aerosolized drug delivery into deep lungs is expanding with the increased number of different diseases. Local and systemic delivery of drugs for various diseases is now focused on using aerosol formulations, which have a lot of potential. The future of aerosol delivery of nanoparticles and large molecules for systemic conditions with improved patient compliance is promising. Currently, aerosol therapy is expanding with the advancement of science and technology, especially in the development of nanoparticles, to target the systemic disorders for the delivery of proteins and peptides, gene therapy, pain management, nanotherapeutics, cancer therapy, and vaccines. In addition to current therapeutics for asthma, other drugs such as mucolytics, antituberculosis, antibiotics, drugs for sexual dysfunction, drugs for otitis media, fentanyl for cancer pain, tobramycin, opioids for pain, interferons, alpha-1 antitrypsin, and human growth hormone for lung delivery are in clinical development. For the treatment of specific diseases (lung and systemic) with costly medicines, it is desired to provide medicinal colloidal drugs as aerosol delivery to the targeted area in the human respiratory system. Therefore, pulmonary drug delivery of colloidal drugs would extend the new era of drug delivery research with increased patient compliance, and reduce the total cost of chronic human diseases. With advanced research, it is anticipated that the world will know more about colloidal drug delivery technology as well as the potential applications of deep lung delivery of those drugs. Pulmonary delivery of large molecules for chronic diseases is advancing rapidly and may become successful in the near future. Therefore, deep lung delivery of drugs needs to be focused not only on lung diseases but also for conditions in which rapid onset is desirable such as cancer pains, allergic reactions, brain disorders, and cardiovascular disorders. It is worthwhile to mention that researchers will develop more novel therapeutics, efficient delivery devices, and better formulation to deliver drugs in to the deep lungs for various types of diseases in the near future.[81–85]

48.17 FOAMS

A foam is a dispersion of a gas in a liquid or a solid. The formation of foam relies on the surface activity of the surfactants, polymers, proteins, and colloidal particles to stabilize the interface. Thus, the foamability increases with increasing surfactant concentration up to critical micelle concentration because above critical micelle concentration, the unimer concentration in the bulk remains nearly constant. The structure and molecular architecture of the foam is known to influence foamability and its stability. The packing properties at the interface are not excellent for very hydrophilic or very hydrophobic drug. The surfactant promoting a small spontaneous curvature at interface is ideal for foams. Nonionic surfactants are the most commonly used one. The main advantage with foams is its site-specific delivery and multiple dosing of the drug.[86–90]

REFERENCES

1. Fanun, M., Ed. *Colloids in Drug Delivery*; Taylor & Francis: New York, 2010; *Colloids in Biotechnology*; Taylor & Francis: New York, 2010.
2. Burgess, D.J. *Colloids and Colloid Drug Delivery System, Encyclopedia of Pharmaceutical Technology*; 3rd edn.; Informa: New York, 2007; pp. 636–647.
3. Rapoport, N. Physical stimuli-responsive polymeric micelles for anti-cancer drug delivery. *Prog. Polym. Sci.* 2007, *32* (8–9), 962–990.
4. Qiu, L.Y.; Bae, Y.H. Self-assembled polyethylenimine-graft-poly(ε-caprolactone) micelles as potential dual carriers of genes and anticancer drugs. *Biomaterials* 2007, *28* (28), 4132–4142.
5. Bae, Y.; Kataoka, K. Intelligent polymeric micelles from functional poly(ethylene glycol)-poly(amino acid) block copolymers. *Adv. Drug Deliv. Rev.* 2009, *61* (10), 768–784.

6. Wei, H.; Cheng, S.X.; Zhang, X.Z.; Zhuo, R.X. Thermo-sensitive polymeric micelles based on poly(N-isopropylacrylamide) as drug carriers. *Prog. Polym. Sci.* 2009, *34* (9), 893–910.
7. Svenson, S.; Chauhan, A.S. Dendrimers for enhanced drug solubilization. *Nanomedicine* 2009, *3* (5), 679–702.
8. Mintzer, M.A.; Grinstaff, M.W. Biomedical applications of dendrimers: A tutorial. *Chem. Soc. Rev.* 2011, *40* (1), 173–190.
9. Percec, V.; Wilson, D.A.; Leowanawat, P.; Wilson, C.J.; Hughes, A.D.; Kaucher, M.S.; Hammer, D.A. et al. Self-assembly of janus dendrimers into uniform dendrimersomes and other complex architectures. *Science* 2010, *328* (5981), 1009–1014.
10. Astruc, D.; Boisselier, E.; Ornelas, C. Dendrimers designed for functions: From physical, photophysical, and supramolecular properties to applications in sensing, catalysis, molecular electronics, photonics, and nanomedicine. *Chem. Rev.* 2010, *110* (4), 1857–1959.
11. Taratula, O.; Garbuzenko, O.B.; Kirkpatrick, P.; Pandya, I.; Savla, R.; Pozharov, V.P.; He, H.; Minko, T. Surface-engineered targeted PPI dendrimer for efficient intracellular and intratumoral siRNA delivery. *J. Contr. Release* 2009, *140* (3), 284–293.
12. Fong, W.K.; Hanley, T.; Boyd, B.J. Stimuli responsive liquid crystals provide 'on-demand' drug delivery in vitro and in vivo. *J. Contr. Release* 2009, *135* (3), 218–226.
13. Lee, K.W.Y.; Nguyen, T.H.; Hanley, T.; Boyd, B.J. Nanostructure of liquid crystalline matrix determines in vitro sustained release and in vivo oral absorption kinetics for hydrophilic model drugs. *Int. J. Pharm.* 2009, *365* (1–2), 190–199.
14. Boyd, B.J.; Khoo, S.M.; Whittaker, D.V.; Davey, G.; Porter, C.J.H. A lipid-based liquid crystalline matrix that provides sustained release and enhanced oral bioavailability for a model poorly water soluble drug in rats. *Int. J. Pharm.* 2007, *340* (1–2), 52–60.
15. Garg, G.; Saraf, S.; Saraf, S. Cubosomes: An overview. *Biol. Pharm. Bull.* 2008, *30* (2), 350–353.
16. Yaghmur, A.; Glatter, O. Characterization and potential applications of nanostructured aqueous dispersions. *Adv. Colloid Interface Sci.* 2009, *147–148*, 333–342.
17. Rizwan, S.B.; Boyd, B.J.; Rades, T.; Hook, S. Bicontinuous cubic liquid crystals as sustained delivery systems for peptides and proteins. *Expert Opin. Drug Deliv.* 2010, *7* (10), 1133–1144.
18. Libster, D.; Aserin, A.; Yariv, D.; Shoham, G.; Garti, N. Soft matter dispersions with ordered inner structures, stabilized by ethoxylated phytosterols. *Colloids Surf. B Biointerfaces* 2009, *74* (1), 202–215.
19. Swarnakar, N.K.; Jain, V.; Dubey, V.; Mishra, D.; Jain, N.K. Enhanced oromucosal delivery of progesterone via hexosomes. *Pharm. Res.* 2007, *24* (12), 2223–2230.
20. Dash, M.; Chiellini, F.; Ottenbrite, R.M.; Chiellini, E. Chitosan—A versatile semi-synthetic polymer in biomedical applications. *Prog. Polym. Sci.* 2011, *36* (8), 981–1014.
21. Ryu, J.H.; Chacko, R.T.; Jiwpanich, S.; Bickerton, S.; Babu, R.P.; Thayumanavan, S. Self-cross-linked polymer nanogels: A versatile nanoscopic drug delivery platform. *J. Am. Chem. Soc.* 2010, *132* (48), 17227–17235.
22. Malmsten, M.; Bysell, H.; Hansson, P. Biomacromolecules in microgels—Opportunities and challenges for drug delivery. *Curr. Opin. Colloid Interface Sci.* 2010, *15* (6), 435–444.
23. Jayakumar, R.; Menon, D.; Manzoor, K.; Nair, S.V.; Tamura, H. Biomedical applications of chitin and chitosan based nanomaterials—A short review. *Carbohydr. Polym.* 2010, *82* (2), 227–232.
24. Saunders, B.R.; Laajam, N.; Daly, E.; Teow, S.; Hu, X.; Stepto, R. Microgels: From responsive polymer colloids to biomaterials. *Adv. Colloid Interface Sci.* 2009, *147–148*, 251–262.
25. Hamidi, M.; Azadi, A.; Rafiei, P. Hydrogel nanoparticles in drug delivery. *Adv. Drug Deliv. Rev.* 2008, *60* (15), 1638–1649.
26. Malmsten, M. *Surfactants and Polymers in Drug Delivery*; Marcel Dekker, Inc.: New York, 2002.
27. Kohli, K.; Chopra, S.; Dhar, D.; Arora, S.; Khar, R.K. Self-emulsifying drug delivery systems: An approach to enhance oral bioavailability. *Drug Discov. Today* 2010, *15* (21–22), 958–965.
28. Buyukozturk, F.; Benneyan, J.C.; Carrier, R.L. Impact of emulsion-based drug delivery systems on intestinal permeability and drug release kinetics. *J. Contr. Release* 2010, *142* (1), 22–30.
29. El Maghraby, G.M. Self-microemulsifying and microemulsion systems for transdermal delivery of indomethacin: Effect of phase transition. *Colloids Surf. B Biointerfaces* 2010, *75* (2), 595–600.
30. Li, Y.; Le Maux, S.; Xiao, H.; McClements, D.J. Emulsion-based delivery systems for tributyrin, a potential colon cancer preventative agent. *J. Agric. Food Chem.* 2009, *57* (19), 9243–9249.
31. Balakrishnan, P.; Lee, B.J.; Oh, D.H.; Kim, J.O.; Hong, M.J.; Jee, J.P.; Kim, J.A.; Yoo, B.K.; Woo, J.S.; Yong, C.S.; Choi, H.G. Enhanced oral bioavailability of dexibuprofen by a novel solid self-emulsifying drug delivery system (SEDDS). *Eur. J. Pharm. Biopharm.* 2009, *72* (3), 539–545.

32. Frelichowska, J.; Bolzinger, M.A.; Valour, J.P.; Mouaziz, H.; Pelletier, J.; Chevalier, Y. Pickering w/o emulsions: Drug release and topical delivery. *Int. J. Pharm.* 2009, *368* (1–2), 7–15.
33. Tang, B.; Cheng, G.; Gu, J.C.; Xu, C.H. Development of solid self-emulsifying drug delivery systems: Preparation techniques and dosage forms. *Drug Discov. Today* 2008, *13* (13–14), 606–612.
34. Cohen-Sela, E.; Teitlboim, S.; Chorny, M.; Koroukhov, N.; Danenberg, H.D.; Gao, J.; Golomb, G. Single and double emulsion manufacturing techniques of an amphiphilic drug in PLGA nanoparticles: Formulations of mithramycin and bioactivity. *J. Pharm. Sci.* 2009, *98* (4), 1452–1462.
35. Cohen-Sela, E.; Chorny, M.; Koroukhov, N.; Danenberg, H.D.; Golomb, G. A new double emulsion solvent diffusion technique for encapsulating hydrophilic molecules in PLGA nanoparticles. *J. Contr. Release* 2009, *133* (2), 90–95.
36. Hanson, J.A.; Chang, C.B.; Graves, S.M.; Li, Z.; Mason, T.G.; Deming, T.J. Nanoscale double emulsions stabilized by single-component block copolypeptides. *Nature* 2008, *455*, 85–88.
37. Ho, M.L.; Fu, Y.C.; Wang, G.J.; Chen, H.T.; Chang, J.K.; Tsai, T.H.; Wang, C.K. Controlled release carrier of BSA made by W/O/W emulsion method containing PLGA and hydroxyapatite. *J. Contr. Release* 2008, *128* (2), 142–148.
38. Zhang, X.Q.; Intra, J.; Salem, A.K. Comparative study of poly (lactic-co-glycolic acid)-poly ethyleneimine-plasmid DNA microparticles prepared using double emulsion methods. *J. Microencapsul.* 2008, *25* (1), 1–12.
39. Kong, M.; Chen, X.G.; Kweon, D.K.; Park, H.J. Investigations on skin permeation of hyaluronic acid based nanoemulsion as transdermal carrier. *Carbohydr. Polym.* 2011, *86* (2), 837–843.
40. Shakeel, F.; Ramadan, W. Transdermal delivery of anticancer drug caffeine from water-in-oil nanoemulsions. *Colloids Surf. B Biointerfaces* 2010, *75* (1), 356–362.
41. Rapoport, N.Y.; Kennedy, A.M.; Shea, J.E.; Scaife, C.L.; Nam, K.H. Controlled and targeted tumor chemotherapy by ultrasound-activated nanoemulsions/microbubbles. *J. Contr. Release* 2009, *138* (2), 268–276.
42. Ahmed, M.; Ramadan, W.; Rambhu, D.; Shakeel, F. Potential of nanoemulsions for intravenous delivery of rifampicin. *Pharmazie* 2008, *63* (11), 806–811.
43. Kumar, M.; Misra, A.; Babbar, A.K.; Mishra, A.K.; Mishra, P.; Pathak, K. Intranasal nanoemulsion based brain targeting drug delivery system of risperidone. *Int. J. Pharm.* 2008, *358* (1–2), 285–291.
44. Fanun, M. Ed. *Microemulsions—Properties and Applications*; Taylor & Francis: New York, 2010.
45. Stubenrauch, C. Ed. *Microemulsions: Background, New Concepts, Applications, Perspectives*; Wiley: New York, 2009.
46. Kumar, P.; Mital, K.L.; Eds. *Handbook of Microemulsion Science and Technology*; Marcel Dekker: New York, 1999.
47. Kunieda, H.; Solans, C. How to prepare microemulsions: Temperature-insensitive microemulsions. In *Industrial Applications of Microemulsions*; Solans, C., Kunieda, H., Eds.; Marcel Dekker: New York, 1997; pp. 21–45.
48. Lawrence, M.J.; Rees, G.D. Microemulsion-based media as novel drug delivery systems. *Adv. Drug Deliv. Rev.* 2000, *45* (1), 89–121.
49. Kreilgaard, M. Influence of microemulsions on cutaneous drug delivery. *Adv. Drug Deliv. Rev.* 2002, *54* (Suppl. 1), S77–S98.
50. Maruyama, K. Intracellular targeting delivery of liposomal drugs to solid tumors based on EPR effects. *Adv. Drug Deliv. Rev.* 2011, *63* (3), 161–169.
51. Schäfer, J.; Höbel, S.; Bakowsky, U.; Aigner, A. Liposome-polyethylenimine complexes for enhanced DNA and siRNA delivery. *Biomaterials* 2010, *31* (26), 6892–6900.
52. Abu Lila, A.S.; Ishida, T.; Kiwada, H. Recent advances in tumor vasculature targeting using liposomal drug delivery systems. *Expert Opin. Drug Deliv.* 2009, *6* (12), 1297–1309.
53. Christensen, D.; Agger, E.M.; Andreasen, L.V. Liposome-based cationic adjuvant formulations (CAF): Past, present, and future. *J. Liposome Res.* 2009, *19* (1), 2–11.
54. Schroeder, A.; Kost, J.; Barenholz, Y. Ultrasound, liposomes, and drug delivery: Principles for using ultrasound to control the release of drugs from liposomes. *Chem. Phys. Lipids* 2009, *162* (1–2), 1–16.
55. Manosroi, J.; Lohcharoenkal, W.; Götz, F.; Werner, R.G.; Manosroi, W.; Manosroi, A. Transdermal absorption enhancement of n-terminal tat-GFP fusion protein (TG) loaded in novel low-toxic elastic anionic niosomes. *J. Pharm. Sci.* 2011, *100* (4), 1525–1534.
56. Karim, K.; Mandal, A.; Biswas, N.; Niosome: A future of targeted drug delivery systems. *J. Adv. Pharm. Technol. Res.* 2010, *1* (4), 374–380.
57. Azeem, A.; Anwer, M.K.; Talegaonkar, S. Niosomes in sustained and targeted drug delivery: Some recent advances. *J. Drug Target* 2009, *17* (9), 671–689.

58. Attia, I.A.; El-Gizawy, S.A.; Fouda, M.A.; Donia, A.M. Influence of a niosomal formulation on the oral bioavailability of acyclovir in rabbits. *AAPS PharmSciTech* 2007, *8* (4), 106.
59. Aggarwal, D.; Pal, D.; Mitra, A.K.; Kaur, I.P. Study of the extent of ocular absorption of acetazolamide from a developed niosomal formulation, by microdialysis sampling of aqueous humor. *Int. J. Pharm.* 2007, *338* (12), 21–26.
60. Lomas, H.; Johnston, A.P.; Such, G.K.; Zhu, Z.; Liang, K.; van Koeverden, M.P.; Alongkornchotikul, S.; Caruso, F. Polymersome-loaded capsules for controlled release of DNA. *Small* 2011, *7* (14), 2109–2019.
61. Brinkhuis, R.P.; Rutjes, F.P.J.T.; Van Hest, J.C.M. Polymeric vesicles in biomedical applications. *Polym. Chem.* 2011, *2* (7), 1449–1462.
62. Egli, S.; Nussbaumer, M.G.; Balasubramanian, V.; Chami, M.; Bruns, N.; Palivan, C.; Meier, W. Biocompatible functionalization of polymersome surfaces: A new approach to surface immobilization and cell targeting using polymersomes. *J. Am. Chem. Soc.* 2011, *133* (12), 4476–4483.
63. Liu, G.; Ma, S.; Li, S.; Chen, R.; Meng, F.; Jiu, H.; Zhong, Z. The highly efficient delivery of exogenous proteins into cells mediated by biodegradable chimaeric polymersomes. *Biomaterials* 2010, *31*, 7575–7585.
64. Upadhyay, K.K.; Bhatt, A.N.; Mishra, A.K.; Dwarakanath, B.S.; Jain, S.; Schatz, C.; Le Meins, J.F. et al. The intracellular drug delivery and anti tumor activity of doxorubicin loaded poly(γ-benzyl l-glutamate)-b-hyaluronan polymersomes. *Biomaterials* 2010, *31* (10), 2882–2892.
65. Lo Presti, C.; Lomas, H.; Massignani, M.; Smart, T.; Battaglia, G. Polymersomes: Nature inspired nanometer sized compartments. *J. Mater. Chem.* 2009, *19*, 3576–3590.
66. Christian, D.A.; Cai, S.; Bowen, D.M.; Kim, Y.; Pajerowski, J.D.; Discher, D.E. Polymersome carriers: From self-assembly to siRNA and protein therapeutics. *Eur. J. Pharm. Biopharm.* 2009, *71* (3), 463–474.
67. Meng, F.; Zhong, Z.; Feijen, J. Stimuli-responsive polymersomes for programmed drug delivery. *Biomacromolecules* 2009, *10* (2), 197–209.
68. Yang, X.C.; Samanta, B.; Agasti, S.S.; Jeong, Y.; Zhu, Z.J.; Rana, S.; Miranda, O.R.; Rotella, V.M. Drug delivery using nanoparticle-stabilized nanocapsules. *Angew Chem. Int. Ed. Engl.* 2011, *50* (2), 477–481.
69. Mora-Huertas, C.E.; Fessi, H.; Elaissari, A. Polymer-based nanocapsules for drug delivery. *Int. J. Pharm.* 2010, *385* (1–2), 113–142.
70. Kamphuis, M.M.J.; Johnston, A.P.R.; Such, G.K.; Dam, H.H.; Evans, R.A.; Scott, A.M.; Nice, E.C.; Heath, J.K.; Caruso, F. Targeting of cancer cells using click-functionalized polymer capsules. *J. Am. Chem. Soc.* 2010, *132* (45), 15881–15883.
71. Chen, Y.; Chen, H.; Zeng, D.; Tian, Y.; Chen, F.; Feng, J.; Shi, J. Core/shell structured hollow mesoporous nanocapsules: A potential platform for simultaneous cell imaging and anticancer drug delivery. *ACS Nano* 2010, *4* (10), 6001–6013.
72. Delcea, M.; Yashchenok, A.; Videnova, K.; Kreft, O.; Möhwald, H.; Skirtach, A.G. Multicompartmental micro- and nanocapsules: Hierarchy and applications in biosciences. *Macromol. Biosci.* 2010, *10* (5), 465–474.
73. Shen, Y.; Jin, E.; Zhang, B.; Murphy, C.J.; Sui, M.; Zhao, J.; Wang, J. et al. Prodrugs forming high drug loading multifunctional nanocapsules for intracellular cancer drug delivery. *J. Am. Chem. Soc.* 2010, *132* (12), 4259–4265.
74. Matsusaki, M.; Akashi, M. Functional multilayered capsules for targeting and local drug delivery. *Expert Opin. Drug Deliv.* 2009, *6* (11), 1207–1217.
75. Huynh, N.T.; Passirani, C.; Saulnier, P.; Benoit, J.P. Lipid nanocapsules: A new platform for nanomedicine. *Int. J. Pharm.* 2009, *379* (2), 201–209.
76. Wang, Y.; Bansal, V.; Zelikin, A.N.; Caruso, F. Templated synthesis of single-component polymer capsules and their application in drug delivery. *Nano Lett.* 2008, *8* (6), 1741–1745.
77. Mundargi, R.C.; Rangaswamy, V.; Aminabhavi, T.M. pH-Sensitive oral insulin delivery systems using Eudragit microspheres. *Drug Dev. Ind. Pharm.* 2011, *37* (8), 977–985.
78. Yang, X.; Chen, L.; Han, B.; Yang, X.; Duan, H. Preparation of magnetite and tumor dual-targeting hollow polymer microspheres with pH-sensitivity for anticancer drug-carriers. *Polymer (Guildf)* 2010, *51* (12), 2533–2539.
79. Alagusundaram, M.; Madhu Sudana Chetty, C.; Umashankari, K. Microspheres as a novel drug delivery sysystem—A review. *Int. J. Chem. Tech. Res.* 2009, *1*, 526–534.
80. Sun, L.; Zhou, S.; Wang, W.; Li, X.; Wang, J.; Weng, J. Preparation and characterization of porous biodegradable microspheres used for controlled protein delivery. *Colloids Surf. A Physicochem. Eng. Asp.* 2009, *345* (1–3), 173–181.

81. Dolovich, M.B.; Dhand, R. Aerosol drug delivery: Developments in device design and clinical use. *Lancet* 2011, *377* (9770), 1032–1045.
82. Pilcer, G.; Amighi, K. Formulation strategy and use of excipients in pulmonary drug delivery. *Int. J. Pharm.* 2010, *392* (1–2), 1–19.
83. Denyer, J.; Dyche, T. The Adaptive Aerosol Delivery (AAD) technology: Past, present, and future. *J. Aerosol. Med. Pulm. Drug Deliv.* 2010, 23 (Suppl. 1), S1–S10.
84. Dudley, M.N.; Loutit, J.; Griffith, D.C. Aerosol antibiotics: Considerations in pharmacological and clinical evaluation. *Curr. Opin. Biotechnol.* 2008, *19* (6), 637–643.
85. Kleinstreuer, C.; Zhang, Z.; Donohue, J.F. Targeted drug-aerosol delivery in the human respiratory system. *Annu. Rev. Biomed. Eng.* 2008, *10*, 195–220.
86. Zhang, Y.; Zhang, J.; Jiang, T.; Wang, S. Inclusion of the poorly water-soluble drug simvastatin in mesocellular foam nanoparticles: Drug loading and release properties. *Int. J. Pharm.* 2011, *410* (1–2), 118–124.
87. Wu, C.; Wang, Z.; Zhi, Z.; Jiang, T.; Zhang, J.; Wang, S. Development of biodegradable porous starch foam for improving oral delivery of poorly water soluble drugs. *Int. J. Pharm.* 2011, *403* (1–2), 162–169.
88. Hegge, A.B.; Andersen, T.; Melvik, J.E.; Bruzell, E.; Kristensen, S.; Tønnesen, H.H. Formulation and bacterial phototoxicity of curcumin loaded alginate foams for wound treatment applications: Studies on curcumin and curcuminoides XLII. *J. Pharm. Sci.* 2011, *100* (1), 174–185.
89. Sirsi, S.R.; Borden, M.A. Microbubble compositions, properties and biomedical applications. *Bubble Sci. Eng. Technol.* 2009, *1* (1–2), 3–17.
90. Klasse, P.J.; Shattock, R.; Moore, J.P. Antiretroviral drug-based microbicides to prevent HIV-1 sexual transmission. *Annu. Rev. Med.* 2008, *59*, 455–71.

49 Melt Extrusion
Pharmaceutical Applications

James DiNunzio, Seth Forster, and Chad Brown

CONTENTS

49.1 Introduction 1127
49.2 Formulation Design and Characterization 1129
49.3 Applications 1135
49.4 Dissolution-Enhanced Products 1135
    49.4.1 Applications of Dissolution-Enhanced Dispersions 1136
49.5 Controlled Release Products 1138
    49.5.1 Controlled Release Excipients and Applications 1140
49.6 Foamed Products 1141
    49.6.1 Application of Foam Extrusion to Modified Release Formulations 1143
49.7 Directly Shaped Products 1144
    49.7.1 Applications of Directly Shaped Products by HME 1145
49.8 Conclusions 1145
References 1146

49.1 INTRODUCTION

Drug development pipelines have changed dramatically over the last 40 years, yielding a number of candidate molecules with challenging properties that limit drugability.[1-4] These properties range from physicochemical stability to permeability, with the most significant limitation for oral molecules being poor aqueous solubility. Driven by the advent of high throughput screening and the need to enhance molecular structure to facilitate target binding efficiency, compounds have evolved lower solubility, making many of the lead compounds in modern pipelines more insoluble than marble.[5] This has led to the characterization of many modern active pharmaceutical ingredients as "brick dust." These compounds are often physically delineated by a higher melting temperature and larger melting enthalpy.[6] Another defining characteristic of many modern new chemical entities has been a transition in lipophilicity,[7] increasing to high log P values thereby reducing the ability to wet and subsequently dissolve. Additionally, the higher degree of molecular complexity challenges the ability to generate a stable crystalline drug substance. Other non-ideal properties such as chemical instability, the need for nontraditional release profiles, and/or alternative delivery routes makes solid dispersion technology an ideal approach for enabling next-generation therapies.

Extrusion technology traces its origins back nearly 2000 years to the first uses of screw conveyors to move water in ancient times.[8] Since then, the technology has changed significantly and found utility in a range of applications including food manufacturing, industrial ceramics, and electronic components.[9,10] Today, modern extruders are available in a variety of different forms based on the number of screws in the system and direction in which the screws rotate. Broadly speaking, extruders are classified as single screw, twin screw, and multiple screw extruders. For the pharmaceutical industry, twin screw extruders are the most prevalent system, with the co-rotating version the most commonly leveraged sub-type within the class. It is also important to note that to date there are no reported applications of multi-screw extruders in the pharmaceutical manufacturing of drug products. Twin screw extruders provide several key advantages when compared to the single screw class and the multi-screw class. First, because of the twin screw configuration it is possible to achieve a higher degree of mixing than that of a single screw extruder. Secondly, twin screw extruders are available in an inter-meshing screw design which allows for self-wiping.[11] This concept, initially introduced by Erdmenger,[12] allows one screw to remove material from the surface of the other during the rotation. This leads to more effective conveyance along the length of the processing section, which translates into shorter and more uniform residence time distributions. These attributes are key to the success of dispersion manufacture, particularly in the case of pharmaceutical systems where a high degree of uniformity and purity are required. There is one limitation associated with co-rotating twin screw extruders when compared to single screw or counter-rotating extruders that should be noted. When compared to these systems, co-rotating extruders provide less consistent pumping action, which can impact performance in the production of directly shaped products. To compensate for this gear pumps can be added onto the end of the co-rotating extruder to allow for consistent pumping and enhanced flow behavior. A summary of the key attributes of commonly available extruder classes and subclasses is shown in Table 49.1. Although a detailed summary of the equipment is outside of the scope of this work, a number of useful references are cited here for the interested reader.[13,14]

To date, melt extrusion has served as the backbone for 14 publicly disclosed marketed products, with numerous more in development. As shown in Table 49.2, these products represent a range of applications, many of which provide some form of novel delivery behavior not available with traditional technologies. In cases where extrusion technology was not applied for a drug delivery benefit, it was leveraged for continuous processing that can more easily support high volume products while still being operated in the batch mode paradigm required by regulatory agencies.

This section is structured to provide the reader an overview of pharmaceutical applications of melt extrusion to enhance the delivery of pharmaceutical products. The first section on formulation design covers aspects related to the composition of pharmaceutical dispersions. It highlights the important properties of drugs and carriers to achieve the desired critical quality attributes while

TABLE 49.1

Comparison of Equipment Attributes for Classes and Subclasses of Extruders for Pharmaceutical Applications

Class Subclass	Single Screw	Twin Screw		Multi-screw
		Co-rotating	Counter Rotating	
Schematic				
Advantages	High pumping efficiency	Narrow residence time distribution	Very narrow residence time distribution	Excellent surface renewal for devolatilization
	Simple equipment design	High dispersive and distributive mixing	High pumping efficiency	
		Very high feeding efficiency	High feeding efficiency	
Disadvantages	Broad residence time distribution	Low pumping efficiency	Lower output than co-rotating extruder	Higher cost and complexity
	No self-wiping	Reduced degassing efficiency	Not as effective for distributive mixing when compared to co-rotating extruder	Not commonly used in Pharma
	Lower dispersive mixing			
	Reduced feeding efficiency			

TABLE 49.2
Marketed Pharmaceutical Products Using Melt Extrusion Technology

Product	Indication	Application	Company
Eudragit E PO (methacrylic acid copolymer)	Excipient	Excipient manufacturing	Evonik
Gris-PEG (griseofulvin)[a]	Anti-fungal	Dissolution enhanced	Pedinol Pharmacal
Rezulin (troglitazone)	Diabetes	Dissolution enhanced	Wyeth
Onmel (itraconazole)	Anti-fungal	Dissolution enhanced	Merz Pharmaceuticals
Norvir (ritonavir)	HIV therapy	Dissolution enhanced	Abbott Laboratories
Kaletra (ritonavir/lopinavir)	HIV therapy	Dissolution enhanced	Abbott Laboratories
Isoptin SR (verapamil)	Angina	CR/direct shape tab.	Abbott Laboratories
Opana ER (oxymorphone HCl)	Pain	CR/direct shape tab.	Endo Pharmaceuticals
Palladone (hydromorphone)	Pain	CR/direct shape pellet	Purdue Pharma
Ozurdex (dexamethasone)	Macular edema	CR/shaped device	Allergan
Zoladex (goserelin acetate implant)	Prostate cancer	CR/shaped device	AstraZeneca
Nuvaring (etonogestrel/ethinyl estradiol)	Contraceptive	CR/shaped device	Merck & Co., Inc.
Implanon (etonogestrel)	Contraceptive	CR/shaped device	Merck & Co., Inc.
Lacrisert (HPC rod)	Dry eye syndrome	Shaped device	Merck & Co., Inc.
Eucreas (vildagliptin/metformin HCl)	Diabetes	Melt granulation	Novartis
Zithromax (azythromycin)[b]	Antibiotic	Taste masking	Pfizer

[a] Undisclosed melt quench process.
[b] Melt spray congeal technology.

also describing strategies for the structured design of extruded products. The second section covers applications of melt extrusion for controlled release and targeted release, highlighting the use of this technology for dissolution enhancement, controlled release, foaming, and direct shaping. Finally, a concluding section summarizes the current state of melt extrusion technology and the future outlook for pharmaceutical manufacturing.

49.2 FORMULATION DESIGN AND CHARACTERIZATION

Pharmaceutical melt extrusion technology is primarily used in the production of solid dispersions.[15–19] As summarized in Figure 49.1, depending on the nature of the product being produced, the material can exist as one of several different types of solid dispersions based on the nature of the drug substance and carrier. In the case of crystalline solid dispersions, the drug substance is dispersed in the carrier phase as discrete crystalline particulates. For amorphous solid dispersions, they encompass amorphous drug dispersed in the carrier phase, which can include drug molecularly dispersed, as is the case of an amorphous solid solution, and also systems containing discrete domains of amorphous drug in the larger carrier phase. To denote systems where the drug is molecularly dispersed in the carrier, the term "solid solution" was developed. In practice however,

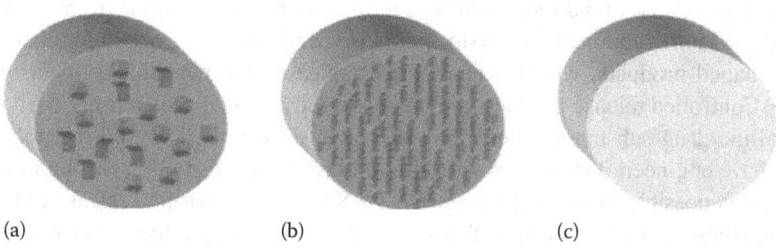

FIGURE 49.1 Types of solid dispersions: (a) crystalline solid dispersion, (b) amorphous solid dispersion, and (c) amorphous solid solution.

because of the drive to develop a homogenous amorphous system with the drug ideally molecularly dispersed, the terms "solid dispersion" and "solid solution" can be used interchangeably. Within this text, the term "solid dispersion" refers to a homogeneous system where ideally the drug is dispersed uniformly at a molecular level.

It is also important to note that each of these dispersions can be exploited to enhance the delivery of the drug substance. For example, the majority of dissolution enhanced products are prepared in the amorphous state to leverage the higher solubility of the material and increase release rates from the drug product. Furthermore, the homogeneous nature of the dispersion provides a kinetic barrier for recrystallization of the amorphous form that contributes to improved physical stability. However, in the case of controlled release and directly shaped products where the systems may be prepared to have specific structural and mechanical attributes, the use of solid dispersions wherein both the drug and polymer exist as crystalline materials within the system may be beneficial.

The resulting solid dispersion is a strong function of formulation and process, both of which are contoured to the drug product application. Composition considerations for each application can be broken down based on the primary function of each of the excipients, as shown in Table 49.3. For successful manufacturing using extrusion, it is necessary that one or more of the components be a thermoplastic material capable of facilitating molten flow within the process section. In the case of an amorphous dispersion for bioavailability enhancement applications, the polymeric carrier is amorphous and thermoplastic in nature. Materials such as copovidone,[20] polyvinyl caprolactam—polyvinyl acetate—polyethylene glycol graft copolymer,[21] and hypromellose acetate succinate[22] exhibit rheological properties necessary for extrusion within pharmaceutical formulations and can even be extruded independently. However, other materials do not exhibit ideal rheological characteristics. Amongst these are high molecular weight povidone and methacrylic acid copolymer,[23] which exhibit a narrow difference between glass transition temperature (T_g) and degradation temperature (T_{deg}). This ratio, $T_g:T_{deg}$, should ideally be less than 0.85 to provide a sufficient operational window. One strategy to address this is to include a plasticizer, which can be an additional component, such as triethyl citrate or dibutyl sebacate,[24,25] that provides molecular lubrication and lowers the T_g of the system. In other select cases, the drug substance may also function as a plasticizer; however, this requires specific properties of the active ingredient. In general, the addition of a plasticizer presents unique challenges, particularly for the development of amorphous dispersions where the reduced T_g can negatively impact physical stability. Additional risks for the use of small molecule plasticizers include concerns for uniformity, chemical stability, and volatility. A viable alternative to using plasticizers is the use of polymer blends to enhance the processability of formulations while maintaining higher T_gs.[26]

Many solid dispersions are prepared as multicomponent systems to engineer specific properties within the product.[27–30]

For amorphous systems where the drug substance exhibits a strong lipophilicity, wettability can be limited. To compensate for this, many formulations include surfactants to aid in surface wetting and dissolution of the formulation. Wettability can also be an important consideration in the preparation of controlled release and directly shaped products, contributing to the overall performance of the drug product. Multicomponent dispersions are also particularly important for controlled release and directly shaped products, where extended drug release and mechanical integrity are critical attributes.[27,28] Controlled release products and many directly shaped products are designed to incorporate crystalline drug substance where they may also utilize semi-crystalline polymers. These systems also have engineered structures with limited internal porosity.[31,32] Through the addition of pore formers, it is possible to facilitate drug release. Similarly, controlled release polymers can be engineered into these systems to design release rates from the drug product. Interplay with the shaping process may also influence the formulation. For injection molding and calendering operations, it is necessary to have a finished product that can be removed from the shaping cavity. Adhesion between the melt and surface during the shaping and cooling process can lead to surface defects and

TABLE 49.3
Functional Excipient Roles for Different Types of Dosage Forms Prepared by Melt Extrusion

Polymer	T_g or T_m (°C)	Grades	Notes
Hypromellose	170–180 (T_g)	Methocel® E5	Non-thermoplastic
			API must plasticize
			Excellent nucleation inhibition
			Difficult to mill
Vinylpyrrolidone	168 (T_g)	Povidone® K30	API must plasticize
			Potential for H-bonding
			Hygroscopic
			Residual peroxides
			Easily milled
Vinylpyrrolidone-vinylacetate copolymer	106 (T_g)	Kollidon® VA 64	Easily processed by melt extrusion
			No API plasticization required
			More hydrophobic than vinylpyrrolidone
			Processed around 130°C
Polyethylene glycol, vinyl acetate, vinyl caprolactam graft co-polymer	70 (T_g)	Soluplus®	Newest excipient for melt extruded dispersions
			Easily process by melt extrusion
			Low T_g can limit stability
			Not of compendial status
			Stable up to >180°C
Polymethacrylates	130 (T_g)	Eudragit® L100–55	Not easily extruded without plasticizer
		Eudragit® L100	Degradation onset is 155°C
			Ionic polymer soluble above pH 5.5
Hypromellose acetate succinate	120–135 (T_g)	AQOAT®-L	Easily extruded without plasticizer
		AQOAT®-M	Process temperatures >140°C
		AQOAT®-H	Ionic polymer soluble above pH 5.5 depending on grade
			Excellent concentration enhancing polymer
			Stable to 190°C depending on processing conditions
Amino methacrylate copolymer	56 (T_g)	Eudragit® E PO	Processing at ~100°C
			Degradation onset is >200°C
			Low T_g can limit stability
Methacrylic acid ester	65–70 (T_g)	Eudragit® RS	Extrudable at moderate temperatures, >100°C
		Eudragit® RL	Excellent CR polymer
Poly(ethylene vinylacetate)	35–205 (T_m)	Elvax® ATEVA	Extrudable at low temperatures, >60°C
			Excellent controlled release polymer but non-biodegradable
Poly(ethylene oxide)	<25–80 (T_m)	Polyox®	Mechanical properties ideal for abuse-deterrent applications and CR
			Process temperatures >70°C
			Excellent CR polymer
Poly(lactic-co-glycolic acid)	40–60 (T_m)	RESOMER®	Low melt viscosity for certain grades is challenging to process
			Biodegradation rate controlled by polymer chemistry
			Excellent for implantable systems

product failure. To prevent this, lubricants are often included in the formulation. These materials mitigate surface interactions and lead to more effective discharge from the die.

Formulation design is also influenced by the properties of the drug substance and the operating conditions during extrusion. Specifically, the mode of manufacturing is defined based on the melting temperature of the drug substance in relation to the processing temperature wherein two distinct regimes can be defined: the miscibility regime and the solubilization regime.[33]

In the miscibility regime, the processing temperature is greater than the melting temperature of the drug substance leading to the situation where lower viscosity molten drug substance must be mixed within the molten polymer. In this case, the ability to form a stable dispersion is a function of miscibility between the components and the ability of the extrusion process to homogeneously distribute the materials. When materials are selected with appropriate miscibility, the formation of an amorphous solid dispersion can be easily achieved. A major advantage for processing in this mode is the limited number of critical kinetic processes that occur during manufacturing, specifically the distributive mixing process during extrusion. In other processing variants in this regime, it is possible to produce bottom-up engineered particles by selecting immiscible drug: carrier combinations. In this application, the molten drug and carrier are intimately mixed in the extruder before the drug recrystallizes out during the subsequent cooling and mixing process.

Processing in the solubilization regime occurs when the operating temperatures are below the melting temperature of the drug substance. In this mode, the formation of an amorphous form is achieved by dissolution of the drug substance into the molten carrier. Careful control must be provided for the process and coupled with judicious selection of formulation additives. During operation in this regime, the molten polymer functions as a solvent, allowing the drug substance to dissolve within the carrier at a temperature below the melting temperature of the compound. The selection of the formulation becomes driven not only by the ability of the formulation to provide the necessary delivery characteristics but also the ability to solubilize the drug substance within the molten carrier. The diffusivity of drug in molten polymer can be related to key descriptors of the process and composition through the Stokes–Einstein equation. Changing formulation additives to enhance molten solubility or reduce viscosity leads to increased dissolution rates. These attributes can be screened in early development to aid in the selection of solubilizing material, with differential scanning calorimetry (DSC) and hot stage microscopy being well-utilized techniques.[20] The application of these methods are shown in Figures 49.2 and 49.3, respectively. Similarly, control of starting particle size governs specific surface area, which can be manipulated to control dissolution rates. With respect to process, temperature influences viscosity and solubility, where higher temperatures increase dissolution rates. Higher shear rates achieved by more aggressive screw design and higher screw speeds can also be shown to reduce the viscosity of most pharmaceutical systems and shorten characteristic transport path lengths leading to improved mass transfer. Noting that the preceding discussion focused on the application of solubilizing a drug within a carrier system, most crystalline systems are also produced in this processing mode; however, the formulation and process are designed to minimize dissolution rates. In this case, one seeks a formulation with limited molten solubility for drug within the carrier and is also operated at lower temperatures with minimal dispersive mechanical energy input where the free energy of drug dissolution in the carrier is greater than zero. This allows the drug substance to be distributed with the carrier while retaining the intrinsic crystalline structure of the feedstock.

Characterization of these dispersions is also a necessary step to establish manufacturing control of the product while also yielding valuable insight into the nature of the system. In general, the determination of the type of solid dispersion product is achieved through the use of several concurrent technologies aimed at identifying crystallinity, distribution, and chemical interactions of the components. For the assessment of crystallinity, the most commonly used technique is X-ray diffraction (XRD),[34–36] which elucidates the long range order of crystalline material. In the case of amorphous materials, a characteristic amorphous halo is observed. Similarly, different crystalline materials exhibit different diffraction patterns allowing for identification of the specific crystalline phase. Crystallinity confirmation and distribution within the dispersion can be assessed using DSC to determine crystalline material

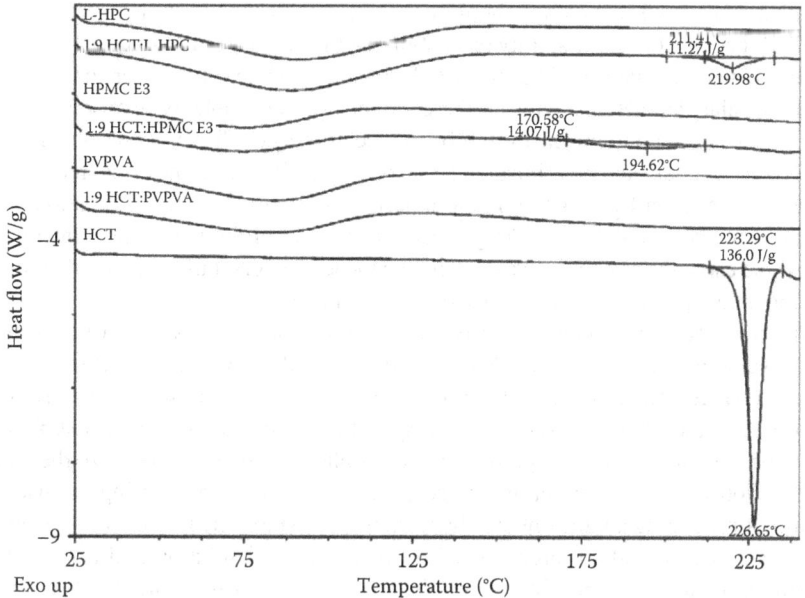

FIGURE 49.2 Application of DSC for assessment of melt solubilization. (From DiNunzio, J.C. et al., *Eur. J. Pharm. Biopharm.*, 74(2), 340, 2012.)

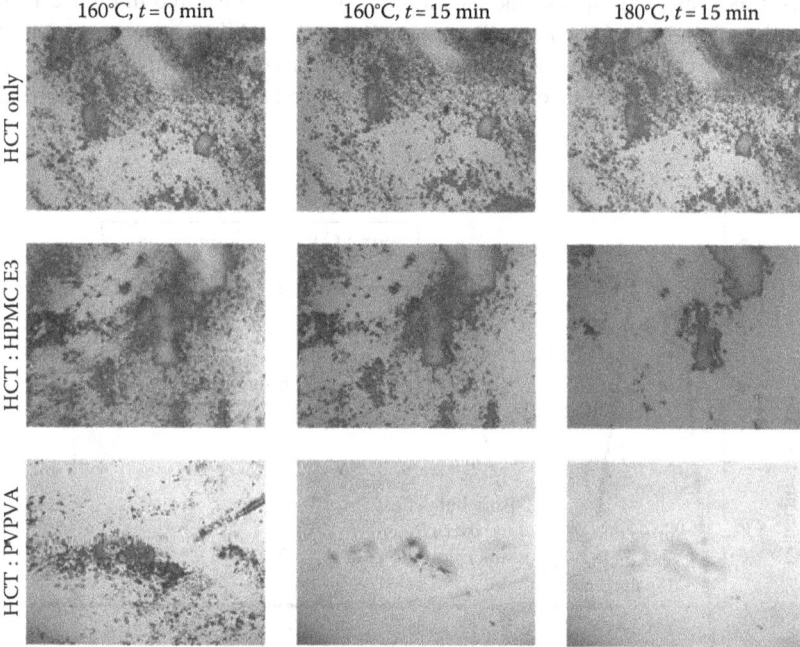

FIGURE 49.3 Application of hot stage microscopy for the assessment of melt solubilization. (From DiNunzio, J.C. et al., *Eur. J. Pharm. Biopharm.*, 74(2), 340, 2012.)

based on melting endotherm and modulated differential scanning calorimetry (mDSC) to identify glass transition temperature of the individual components or the resulting dispersion.[37–42] With this technique, it may be possible to identify the number of phases within the solid dispersion, where a homogeneous amorphous dispersion is indicated by a single glass transition temperature and a heterogeneous system is identifiable by multiple thermal events associated with glass transitions or melting.

The application of thermal techniques should be coupled with secondary isothermal techniques, in particular XRD, because of the possibility of solubilizing the drug substance within the polymer system during preparation and/or analysis. The ability to form and stabilize solid dispersions is also a function of the intermolecular interactions between the components, detectable by a variety of spectroscopic techniques[43–45] including Fourier transform infrared spectroscopy (FTIR), solid state nuclear magnetic resonance (NMR), Raman, and near infrared spectroscopy (NIR). With these techniques, it is possible to determine the transition from crystalline to amorphous and the resulting formation of drug–polymer interactions by the change of wavelengths associated with different chemical interactions. It also becomes possible using these techniques to quantitate the level of crystalline material within a formulation depending on the properties of the native crystalline material.

Discrimination between products as it relates to bioperformance is also a necessary consideration when developing a melt extruded dosage form, particularly when the technology is leveraged for solubility enhancement and controlled release. Specific factors must be considered both in the formulation design and characterization of the system. In the case of poorly soluble compounds, carrier selection is identified to both produce the amorphous form and maximize the apparent solubility.[38,40,46–53] Noting the considerations of polymer selection for processing as discussed previously, the aim of producing an amorphous dispersion is to exploit the free energy benefit associated with the absence of a crystal structure, which leads to rapid dissolution and is often termed "the spring." Without the aid of stabilizing materials, the excess drug in solution will often precipitate until reaching the solubility of the crystalline material. In order to reduce precipitation rates, the stabilizing carrier material used during extrusion should also interact molecularly with the drug substance to enhance its solubility. This behavior is commonly referred to as "the parachute" effect. Conceptually, this combined "spring and parachute" effect[54] is illustrated in Figure 49.4.

Conveniently, many of the polymers that most effectively stabilize supersaturation also provide suitable thermoplastic characteristics for extrusion. Two of the most commonly used materials are copovidone and hypromellose acetate succinate, which are non-ionic and enteric materials, respectively.

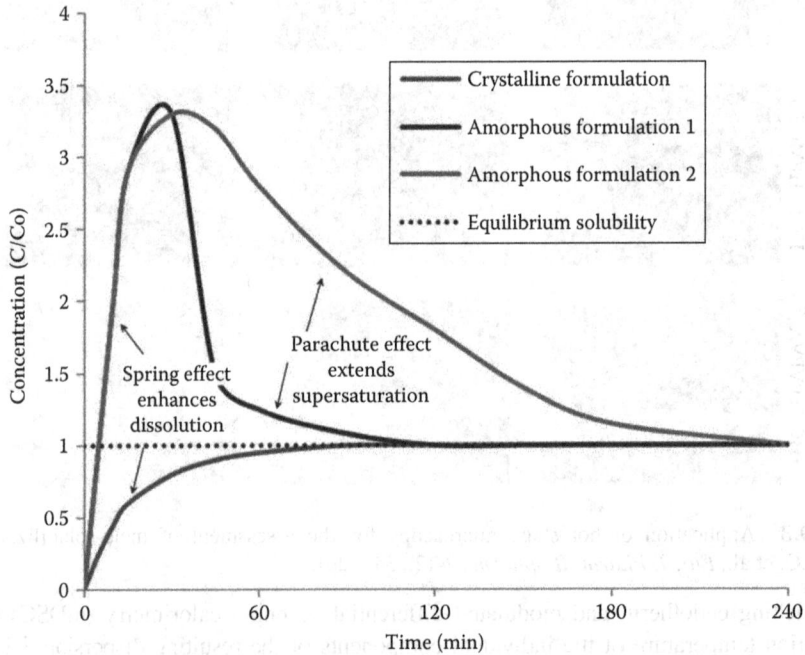

FIGURE 49.4 Spring and parachute performance of amorphous dispersions in comparison to crystalline material.

In order to optimally exploit the spring and parachute behavior, it is necessary to understand the origin of the phenomena. Numerous researchers have reported on the nature of these interactions, ranging from hydrogen bonding, hydrophobic interactions, ionic interaction of the carrier, and steric hindrance.[54–59] In reality, it is likely a combination of drug and polymer-specific interactions as well as contributions of drug substance properties such as glass transition temperature and recrystallization potential[60] that drive stabilization. Most importantly, a wealth of academic and industrial information has highlighted the use of these materials in solid dispersions for enhancing the oral bio-availability of poorly soluble compounds.

For controlled release products, the considerations are very different. For these systems, it is necessary that extrudable formulations lead to controlled release characteristics of the delivery system. In many of these cases, the drug substance is intended to retain its native form, i.e., crystalline structure, which requires processing of the formulation in a mode where the drug remains insoluble in the melt. Challenges further emerge in that many of the materials needed for controlled release lack desirable characteristics for extrusion. For example, hypromellose is a commonly used polymer for controlled oral delivery but is non-thermoplastic. This mandates the need for additional materials to aid in processability, with plasticizer and low melting point surfactants frequently meeting this need.[22,26,27,61,62] The presence of plasticizers may be leveraged to regulate the ability of the polymeric phase to control the release rate of the active due to changes in diffusivity, wettability, and water penetration into the system. In the case of other systems, which exhibit the necessary thermoplastic characteristics for processing, particularly polyethylene oxide (discussed in greater detail in the controlled release section), the semi-crystalline nature of the polymer can present additional challenges.[63] The level of crystallinity in a material impacts the microstructure of the system, leading to transport variations. Additionally, as changes occur on stability resulting in crystallization, a volume contraction occurs due to the ordering within the system that can impact the stability and distribution of other components in the formulations. These changes, both of formulation and physical stability, can also influence mechanical properties of the system, which is a key factor in the production of directly shaped products.[30] For these reasons, in the case of crystalline and semi-crystalline polymers it is necessary that they reach a similar endpoint to enable batch-to-batch consistency and achieve requisite stability. Another application for controlled release products prepared by extrusion is the case of amorphous controlled release products. Exploiting the increased solubility of the amorphous form, it becomes possible to tailor release properties from the dosage form. Careful design of these systems is necessary however, because of the complex interactions that occur within the dosage form and use environment.

Numerous publications have covered the design and implementation of extrusion operations, with several seminal references available for the interested reader.[11,64–67] The subsequent sections of this entry discuss the applications of melt extrusion for controlled delivery, focusing on the interplay between formulation design, characterization, and the desired drug product attributes for the delivery routes.

49.3 APPLICATIONS

Broadly speaking, melt extrusion is used to produce pharmaceutical products that can be broken down into four application classes: dissolution-enhanced products, controlled released products, foamed products, and shaped delivery systems. The subsequent sections describe the design of drug products to achieve the desired controlled release characteristics.

49.4 DISSOLUTION-ENHANCED PRODUCTS

In the absence of solubility, drug substances are poorly absorbed in vivo leading to ineffective therapies. One strategy to address this is the formation of an amorphous dispersion that leads to higher solubility and faster dissolution rates, whereby the drug substance becomes more bioavailable.

In other cases where crystalline interactions are favorable for dissolution, crystalline dispersions can be manufactured to provide enhanced wetting. This section describes examples of both types of dispersions, with a focus on the influence of formulation to achieve the desired attributes.

49.4.1 Applications of Dissolution-Enhanced Dispersions

Lopinavir and ritonavir are both antiviral compounds that exhibit low solubility, metabolic conversion, and efflux leading to low oral exposure, which are currently formulated as melt extruded solid dispersions.[68] Chronologically, ritonavir was developed prior to lopinavir utilizing a metastable polymorph, Form II. After initial agency approval and marketing, stability issues emerged and the product was pulled from the market and reformulated.[69] A soft gelatin capsule product was subsequently launched. As lopinavir was developed, liabilities associated with metabolic conversion and efflux were noted that limited oral bioavailability. Given the need for a fixed dose combination product with other antivirals to enhance treatment, lopinavir was combined with ritonavir, a powerful cytochrome P450 (CYP) enzyme and P-glycoprotein (PGP) efflux inhibitor. This product was also formulated as a soft-gelatin capsule; however, due to the storage restrictions and undesirable manufacturing process, researchers at Abbott looked to reformulate the product. Noting that both compounds exhibit molecular characteristics appropriate for use in melt extrusion, specifically low melting points that enable processing at moderate temperatures, a combination product of lopinavir and ritonavir was developed (Kaletra®) as well as a standalone drug product of ritonavir alone (Norvir®). In the design of the formulation researchers looked to design a system capable of forming a colloidal dispersion on exposure with an aqueous environment.[68] Additionally, the need for physical stability at room temperature storage conditions required the development of a formulation, where the drug loading did not exceed the solubility of API in the carrier matrix. Despite this constraint, the total number of dosage units administered compared with the softgel formulation was reduced from 6 units to 4. Another consideration in the design of the formulation was the utilization of formulation and processing conditions to minimize impurity formation in the solid dispersion. The incorporation of inviscid polymer, copovidone, and low levels of the surfactant, sorbitan monolaurate, enabled production at lower temperatures while also aiding in the release of intermediate from the calendering process. These changes also minimized impurities, which was further supplemented through the incorporation of customized screw elements,[70] shown in Figure 49.5, which minimized local excess energy build-up during extrusion that can lead to decomposition.

Crystalline solid dispersions are also effective in enhancing dissolution rates by creating solid dispersions with intimate contact between the drug and polymer. Historically, one of the earliest applications of thermally produced solid dispersions for dissolution enhancement was the development of Gris-PEG (griseofulvin ultramicrosize), a crystalline solid dispersion of griseofulvin prepared in polyethylene glycol.[71]

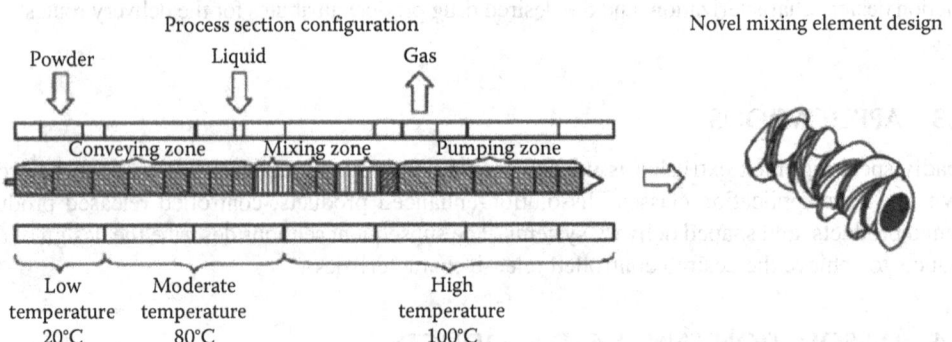

FIGURE 49.5 Screw elements and designs used for minimization of impurities during extrusion processing of ritonavir and lopinavir. (From Kessler, T. et al., *Process for Producing a Solid Dispersion of an Active Ingredient*, U.S.P.T. Office, Abbott GmbH & Co., KG, 2009.)

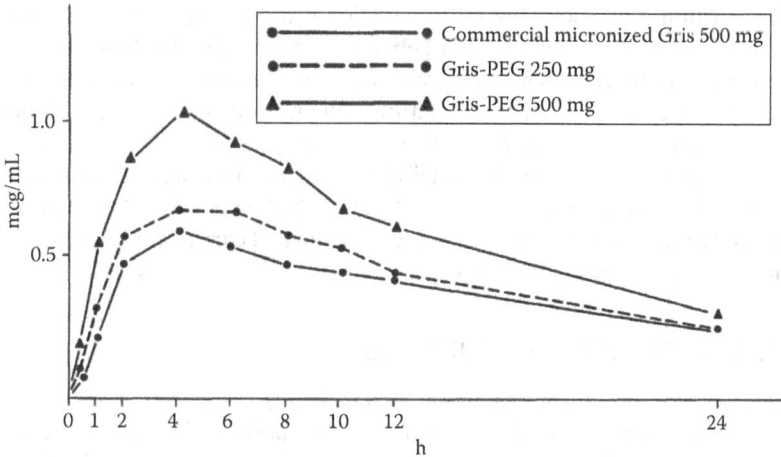

FIGURE 49.6 In vitro dissolution rates of griseofulvin: polyethylene glycol solid dispersions. (From Riegelman, S. and Chiou, W.L., *Increasing the Absorption Rate of Insoluble Drugs*, U.S.P.T. Organization, The Regents of the University of California, Oakland, CA, 1979, p. 9.)

Due to the rapid dissolution, illustrated in Figure 49.6, even greater exposures were possible when compared to conventionally formulated higher dose products. In this system, the particle size of the drug substance is set by the properties of the material prior to the extrusion operation. Achieving a high degree of dispersive mixing was key to uniform spatial distribution of the API in the carrier.

In another novel application of melt extrusion for the production of dissolution enhanced drug products, the extrusion process was used for the production of bottom-up nanoparticles as part of the development of dalcetrapib.[72] Outlined in a patent application, the inventors described the process of melting the drug within an immiscible carrier matrix. After intimate mixing, nucleation and growth of the drug substance were driven in the later stages of the process section via temperature reduction while the continued agitation from the screws aided in the control of the embedded crystalline drug particle size. As a result of the dispersion structure enhanced dissolution was observed while maintaining enhanced stability characteristics when compared to amorphous concepts. Similar process concepts have also been reported for the direct extrusion of engineered nanoparticles. Miller et al.[49] demonstrated the use of non-solubilizing carriers to densify particles by incorporation into melt-extruded matrices. In this particular process variant, carrier selection was critical to both dissolution enhancement and stabilization of the particles. As shown in Figure 49.7,[58]

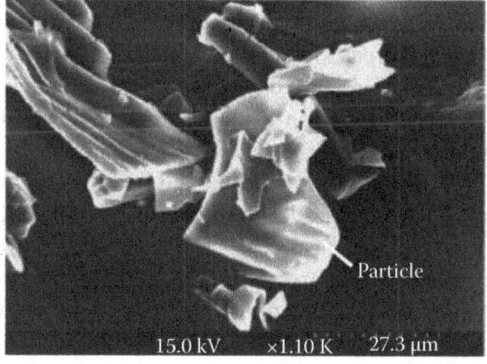 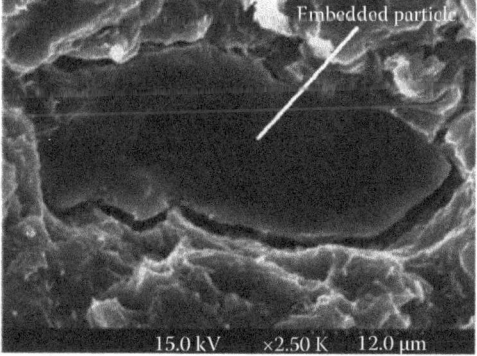

FIGURE 49.7 Embedded engineered particles in solid dispersions prepared by melt extrusion. (Adapted from Miller, D.A., Improved oral absorption of poorly water-soluble drugs by advanced solid dispersion systems, in *Division of Pharmaceutics*, The University of Texas at Austin, Austin, TX, 2007, p. 312.)

when processed within a non-solubilizing formulation using a single screw extruder to improve pumping, the particles can be maintained largely intact, preserving the dissolution benefits of the engineered materials. In other variants, researchers have prepared top-down nanoparticles in aqueous media and then used the combined devolatilization and compounding capabilities to generate nanoparticle-based solid dispersions. For such cases where it is necessary to preserve the particle properties, processing is concluded in the solubilization regime using a design intended to minimize dissolution within the molten carrier. Conversely, the production of bottom-up engineered particles relies on processing the miscibility regime where the drug and carrier are immiscible and processing conditions aid in phase separation and crystal growth.

49.5 CONTROLLED RELEASE PRODUCTS

Controlled release products, including those with delayed, sustained, or extended release, are designed for reduced frequency of dosing, reduced absorption variability, improved efficacy, or improved safety over immediate release formulations. In 2008, the controlled release market was estimated at US$21 billion globally and is anticipated to increase to US$29.5 billion in 2017. In the past, controlled release products have been developed to protect innovators from generic erosion of immediate release products while improving patient compliance or product performance. Innovators are now seeking earlier entry of controlled release products to increase patient benefit and sales.[73]

One area where melt extruded controlled release products have made a major impact is in the abuse deterrent space. By creating a system with intimate contact between the drug and polymer, as well as the use of highly deformable carrier materials, it is possible to engineer abuse deterrence. An epidemic of opioid dependency and abuse prompted the Food and Drug Administration (FDA) to issue a draft guidance describing the desired attributes for abuse-deterrent formulations, including limited ability to inhale or inject the active ingredient via crushing, grinding, melting, or dissolving of the drug product.[31] These recommendations apply to both immediate release and controlled release products; however, many new opioid-based products have been recently developed as controlled release systems to improve patient compliance. In April 2013, the FDA approved an abuse-deterrent formulation of OxyContin as the previous formulation was withdrawn from the market for safety concerns.[32] The FDA explicitly will not accept Abbreviated New Drug Applications (ANDAs) for the previous formulation.

Controlled release (CR) products can be broadly classified into two major categories: extended release and delayed release. While a striking range of innovative techniques exist, drug release from the dosage form is regulated by a number of methods based on diffusion, swelling, erosion, and osmotic pressure. Extended release is desired to maintain a therapeutic level of drug in the body over a longer time, reduce peak plasma concentrations, or minimize exposure variation. For these systems, drug products contain polymers that do not rapidly dissolve in the use environment, leading to a slower rate of erosion or swelling that creates a gel layer. Both erosion of the exterior layer and diffusion through the gel layer govern the release of the drug from the device. Other variants of the extended release technology can also rely on osmotic pressure to "push" the drug from the drug product in a controlled fashion. Similarly, a delayed release product has a drug release profile that does not begin immediately after administration and is desired if the drug substance irritates the upper gastrointestinal (GI) tract, degrades in the presence of low pH, or exhibits a large change in solubility with minor changes in gastric pH. Delayed release often exploits the pH-solubility of a polymer with a pK_a at or near the target physiological pH. On contact with an environment where the pH is greater than the pK_a the polymers ionize and dissolve to release the payload.

Compared to "conventional" CR formulation technologies, hot-melt extrusion (HME) offers several potential technical and business advantages. HME can be used to simultaneously generate an amorphous solid dispersion (ASD) and CR/ASD behavior, providing unique release profiles for low solubility molecules. It is also a convenient trend that a number of polymers used to generate amorphous solid dispersions with HME have inherent delayed or extended release functionality, allowing for a healthy selection of available materials. Crystalline solid dispersions produced with HME also tend to have high density, low porosity, and low specific surface area, which allows for a higher drug loading and lower polymer loading while maintaining comparable release characteristics from a smaller dosage form. Amorphous systems, by nature of the reduced surface area, may also benefit from improved physical stability. From a mechanical properties perspective, many polymeric excipients used in HME tend to deform plastically, frustrating attempts to grind or mill it for abuse. Further engineering of the drug product is also possible through the production of multi-layer systems and directly manufactured products via calendering and injection molding. From a business perspective, the technology can be adapted as a continuous process.

As discussed in the earlier sections, the desired solid state properties of the extrudate are critical to consider and monitor throughout the formulation and process development. This is particularly critical in the case of melt extruded dispersions for CR applications, where these systems can be intentionally designed as crystalline or amorphous dispersions. If a crystalline solid suspension is desired, the extrusion process must impart sufficient thermal and shear energy to melt the excipients, but not melt the API. To achieve this goal, the API may need to be added after melting and mixing the excipients, late in the extrusion process to limit time at high temperatures and exposure to high shear zones. The melting points and/or glass transition temperatures of excipients will be crucial to understand. For an amorphous solid dispersion, the HME process parameters must be appropriate to achieve a single, continuous phase by imparting sufficient thermal and shear energy, and then the system must be monitored for physical stability since the thermodynamically unstable amorphous state could result in API crystallization or phase separation could occur on stability.

It is often challenging to achieve immediate release with amorphous solid formulations due to the physical nature of HME particles and the polymers generally used for this purpose. Many of the commonly used excipients tend to be enteric polymers (e.g., HPMCAS) or binders (PVP, PVP/PVAc). Furthermore, the high density and low porosity nature of the dispersion results in lower specific surface area that drives slower release rates. Coupled with the lower disintegration rates for the monolithic HME products, release rates will tend to remain slower after introduction to the aqueous environment. However, for CR applications these attributes present unique advantages that can be exploited in the development process.

In design of CR products using melt extrusion, a number of commonly used oral controlled release excipients could be successfully processed as is or with the addition of plasticizers. Desired drug release profiles can be achieved by manipulating the polymer chemistry, grade and level of polymers, and geometric considerations of the drug product. Specific polymers used in CR HME applications include the following:

- Cellulose esters or ethers such as methylcellulose (MC), ethylcellulose (EC), hydroxyethyl cellulose (HEC), hydroxypropyl cellulose (HPC), hydroxypropyl methylcellulose/hypromellose (HPMC), hypromellose phthalate (HPMCP), hypromellose acetate phthalate (HPMCAP), hypromellose acetate succinate (HPMCAS), cellulose acetate (CA) and derivates
- Poly (methyl) methacrylates (Eudragit® L, S, E, RS/ RL)
- Polyethylene oxide (PEO)
- Polyvinyl pyrrolidone (PVP), polyvinyl acetate (PVAc), or combination as Kollidon SR®
- Cross-linked polyacrylic anhydride (PAA), for example, Carbopol®
- Starches or modified starches
- Long-chain fatty acids, mono- or di-glycerides
- Natural products such as gelatin, gums, chitosan, carnauba wax, and alginates

Some of these polymers have a narrow $T_g:T_{deg}$ ratios or have such high viscosities that require the addition of plasticizers such as phthalate or citrate esters, low molecular weight polyethylene glycol (PEG), and/or propylene glycol to process robustly or improve flow during extrusion.

Often a surfactant is required for bioavailability enhancement and can conveniently provide process enhancement as well due to a reduction of melt viscosity.[27,74] Similarly, drug substance in the formulation can also provide auto-plasticization. In more complex manufacturing design, fugitive plasticizers can be used to lower viscosity during extrusion. One unique advantage of fugitive plasticizers is that they are not retained in the drug product, which allows the formulation to maintain an elevated T_g. In the case of non-fugitive plasticizers, changes to the stability and release rate of a formulation are possible so it is important to understand the impact to performance across different levels.[75] Another strategy to improve performance is to develop formulations of multiple polymers, exploiting the beneficial aspects of the individual materials. This leads to improved processing as well as release and bioavailability characteristics.[76]

Beyond judicious polymer selection, the controlled release profile can be further modified by the addition of disintegrants or soluble components such as saccharides, salts, or pH-modifying excipients. For a selected formulation, the morphology and surface area of the HME particle or product (i.e., direct shaped pellet) can also be manipulated to change the drug release profile and likely will be important to control for consistent performance. Exemplary control of the pellet or tablet surface area and morphology can be achieved through die-face or strand pelletization. Pellets can then be spheronized to remove asperities and improve the consistency of release.[77]

49.5.1 Controlled Release Excipients and Applications

Hypromellose (HPMC) and polyethylene oxide (PEO) are commonly used as eroding hydrophilic matrices for sustained release. EC is commonly used as an insoluble matrix polymer. These polymers are often used in combination, e.g., HPMC/EC or PEO/EC, in order to tailor release profiles. All have been used to enhance solubility by producing solid dispersions.[31,32,38,63,78–80] Of these materials, PEO and EC are readily amenable to HME based on the physical properties, but the narrow processing window for HPMC and very high melt viscosity requires the use of high levels of plasticizer. In many cases the plasticizer level can be as much as 30% by weight.[81] While evaluating different formulations, HME process parameters should be considered as well. Of note is the study of HME process on guaifenesin release from EC matrices, which demonstrated a statistically significant effect due to changes in the matrix porosity induced by higher temperatures.[32] Soluble components in an insoluble matrix can be used to promote release by leaving voids upon dissolution that increase surface exposed for diffusion.[80]

Poly(methylmethacrylate) (PMMA) was originally developed and marketed in 1933 as a glass replacement called Plexiglas®. This polymer and derivatives were first used to coat oral solid dosage forms in the 1950s and were extruded for pharmaceutical applications in the late 1990s. A wide range of Eudragit polymers cover the physiological pH range and act as insoluble sustained release polymers, with the specific pH of release determined by the side group chemistry of the material grade. However, many grades have relatively low degradation temperatures and high melt viscosities, which complicates their use in extrusion applications. For example, Eudragit S100 and L100 functional groups degrade near the polymers' T_gs, so a plasticizer is required.[23] Several recommended plasticizers for Eudragit L100 and S100 are triethyl citrate (TEC), PEG 6000, or propylene glycol at concentrations ranging from 5% to 25% polymer weight. In specific cases the drug may also provide plasticizer functionality, concomitantly reducing viscosity, and providing therapeutic efficacy through improved release performance. Eudragit E and L 100-55 have precedence of use for solubility enhancement, and HME has been used to make crystalline suspensions for theophylline enteric pellets (L100, S100, L100-55), indomethacin controlled release (L100, S100), and sustained release of diltiazem hydrochloride and chlorpheniramine maleate (RSPO).[82–85] Although most CR HME systems are crystalline dispersions, CR amorphous solid solutions may also be achievable with the right stabilizing formulation.

Beyond the aforementioned polymers, many other excipients have been used for HME and CR applications. Notably, Kollidon SR is a physical blend of PVP and PVAc, which generates a matrix of insoluble (PVAc) and soluble (PVP) domains in the extrudate and has been used for combined CR and amorphous solid solution of ibuprofen[86] for sustained release. The addition of long fatty acids, glycerides, or waxes that slow drug release may be of interest since they act as polymer plasticizers and have demonstrated a reduction of in vitro food effect risk.

Abuse-deterrent products have additional requirements for demonstrating reduced ability to crush, cut, grate, or grind with common household implements like spoons, cutters, and coffee grinders, under hot or cold conditions. The particle size before and after these treatments should be compared to assess risk of inhaled use. Lab assessments of the extractability of the drug in common household solvents, especially ethanol, and syringeability/injectability are also required.[31] Several abuse-deterrent systems take advantage of PEO's significant plastic strain properties that make milling or crushing difficult. Further, the hydrogel nature makes dissolving or injecting the formulation challenging. These products usually provide controlled release characteristics to support once a day dosing and may also be formulated as such due to the matrix polymer viscosity. These PEO/drug formulations can be calendered using equipment such as Abbott/Soliqs Meltrex calendering technology (e.g., Isoptin SR)[87] or melt extruded/ compression molded as in Grünenthal's Intac® technology (e.g., Opana ER)[88,89] to produce a low porosity tablet that releases slowly. Other variants have been developed to provide immediate release characteristics through pelletization, illustrating that these systems can be used for immediate release or controlled release applications. Figure 49.8 highlights the Grünenthal Intac technology that provides improved abuse deterrence through greater crushing resistance.

49.6 FOAMED PRODUCTS

In addition to traditional extrusion compounding to produce solid dispersion or solid solutions for immediate and controlled release applications, foam extrusion has emerged as a means to provide additional product and process value in the pharmaceutical industry.[90–92] In general, foam extrusion can be utilized to produce extrudate formulations with improved milling properties,[93–95] increased dissolution rates,[93,96,97] and potentially improved in vivo performance. In addition, this technology can be employed to generate low density, buoyant dosage forms with or without entrapped gas that allow for targeted release of drug through gastric retention.[98–100] Finally, unique dosage forms, such as orally disintegrating tablets, can be generated to allow for the targeted delivery of actives.[101]

Foam extrusion, represented schematically in Figure 49.9, is achieved through (1) the uniform dissolution of gas into the polymer melt at high pressure, (2) the introduction of thermodynamic instability via a decrease in pressure or an increase in temperature at the die resulting in cell nucleation, (3) followed by cell growth and expansion, and finally (4) stabilization of the resulting foamed structure by cooling below the T_g of the material. Foaming of hot melt extrudates can be achieved using chemical agents such as sodium bicarbonate and citric acid that decompose with temperature to form carbon dioxide[9] or physical blowing agents such as carbon dioxide, nitrogen, or water.[90,92,100,102] Physical blowing agents have the advantage of being removed during processing and as such may have minimal impact on the physical and chemical stability or dissolution performance of the resulting foamed extrudate. In addition, because of the plasticizing property of physical blowing agents, they can be utilized as fugitive plasticizers allowing for processing at lower product temperatures in cases where the active ingredients or excipients are thermally labile, or to lower the viscosity of the extruded formulation potentially leading to increased productivity.[93–96,102] Unlike chemical blowing agents that are easily added to the pre-extruded feed at a desired ratio, physical blowing agents require additional equipment to process in a robust and safe manner. Typically, high pressure pumps capable of metering supercritical fluids are utilized to maintain pressure and control the delivery of known amounts of gas to the extruder. Importantly, a screw design must be considered to allow for the correct pressure profile and solubilization of gas into the polymer melt. A sufficient melt seal must be formed upstream of the injection site to prevent gas escaping through

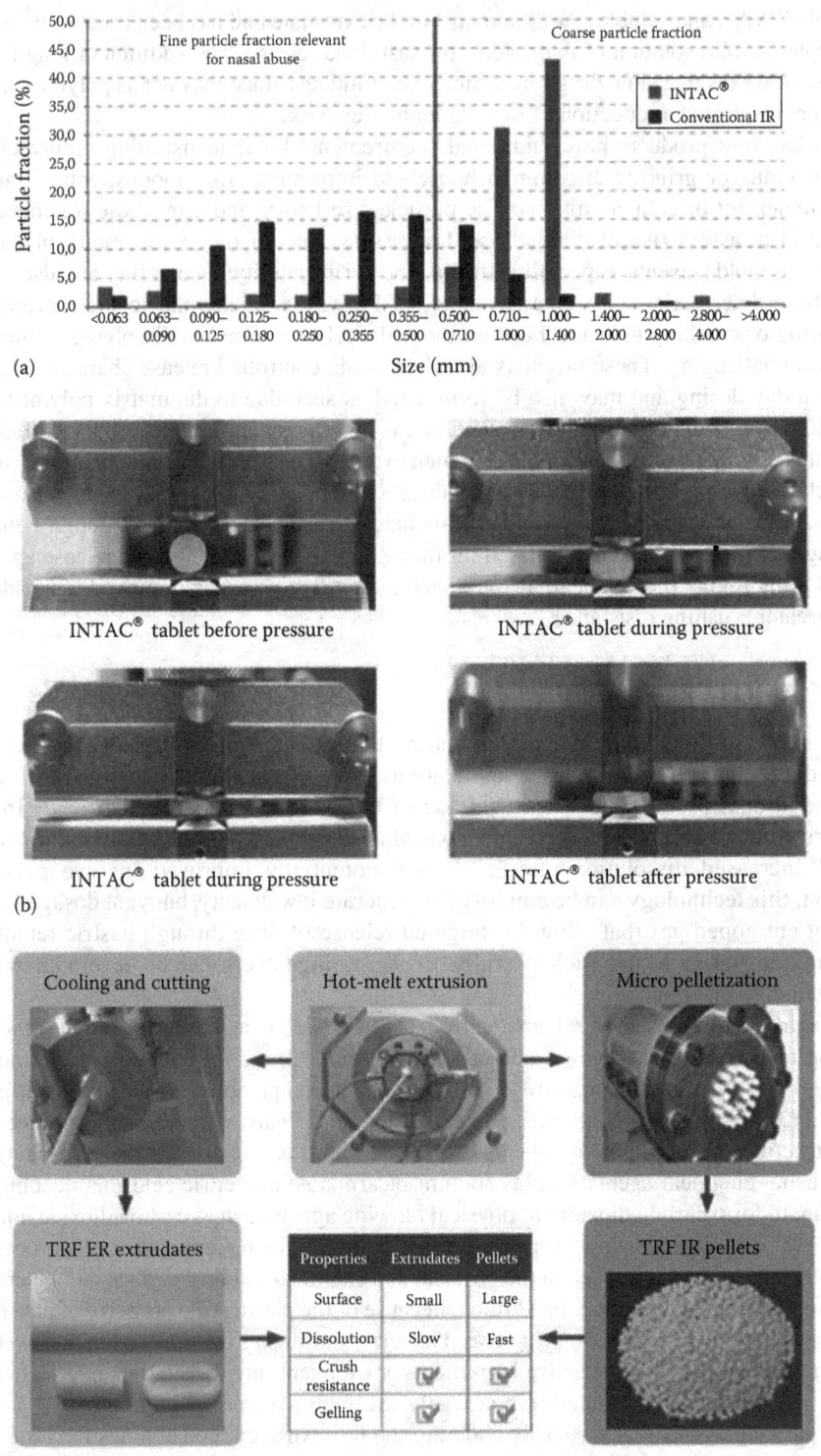

FIGURE 49.8 (a–c) Abuse-resistant formulation manufacturing and characterization. (Used with permission from Grünenthal GmbH, ©2013.)

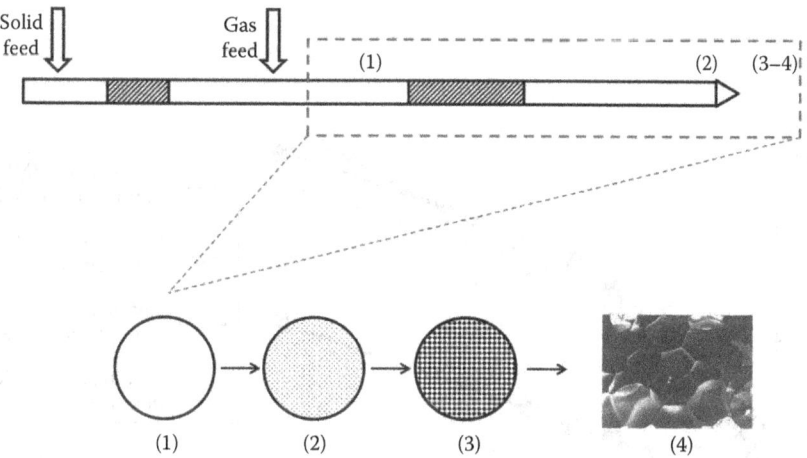

FIGURE 49.9 Schematic representation of foaming process: (1) dissolution of gas into polymer, (2) cell nucleation, (3) cell growth and expansion, and (4) SEM of hydroxypropyl methyl cellulose acetate succinate (HPMCAS) extrudate foamed with nitrogen gas.

the feed zone. In addition, a filled section must be present downstream of gas injection to allow adequate pressure and time for gas dissolution. Finally, one must utilize the extruder to cool the polymer melt to achieve a melt strength that is supportive of foaming.[103] If the melt strength is too low, the foam will collapse. If the melt viscosity is too high, there is the potential the gas will phase separate or the polymer will not achieve full expansion.

49.6.1 Application of Foam Extrusion to Modified Release Formulations

The most straightforward application of foam extrusion to control the release of active ingredients is the fabrication of low density floating dosage forms that can achieve gastric retention allowing for enhanced delivery of drugs that act locally (in stomach) or to improve the absorption of drugs with an absorption window limited to the stomach or upper gastrointestinal tract. Fukuda et al. prepared foam matrices of acetohydroxamic acid and chlorpheniramine maleate with Eudragit RS PO using 5–10 wt% sodium bicarbonate as a chemical foaming agent.[98] Sodium bicarbonate decomposes thermally into carbon dioxide and water as the foaming agents leaving behind sodium carbonate in the formulation matrix. Extrudate tablets were produced on a single screw Randcastle extruder and manually cut into tablets. Foamed tablets with apparent densities ranging from 0.6 to 1.0 g/cm^3 demonstrated the ability to remain buoyant in dissolution media (0.1 N HCl) for 24 h. In addition, both compounds in their work showed sustained release over this period of time, as shown in Figure 49.10. The release rate of chlorpheniramine maleate could be increased by adding Eudragit E PO (0%–65%) to the formulation or by decreasing the diameter of the extruder die (from 3 to 1 mm). The formulations were found to be physically stable at 40°C and 75% relative humidity with no changes in buoyancy or dissolution properties; however, unknown is the long-term impact of residual sodium carbonate in the formulation. Nakamichi et al. utilized water as a physical blowing agent to produce floating dosage forms of hydroxypropyl methylcellulose acetate succinate (HPMCAS, AQOAT MF grade) and nicardipine hydrochloride.[103] Formulations containing 10–30 wt% nicardipine hydrochloride were extruded in a co-rotating twin screw extruder and cut to desired lengths for testing. In addition, varying amounts of calcium phosphate dihydrate was added to the formulation as a nucleating agent. Nucleating agents are utilized to increase the number of sites for gas nucleation leading to an overall increase in cell number and a decrease in foam density. As the amount of calcium phosphate dihydrate increased from 0 to 12 wt.%, the porosity of the tablets increased from 5% to 72%. Foamed tablets retained their buoyancy for 4–6 h in pH

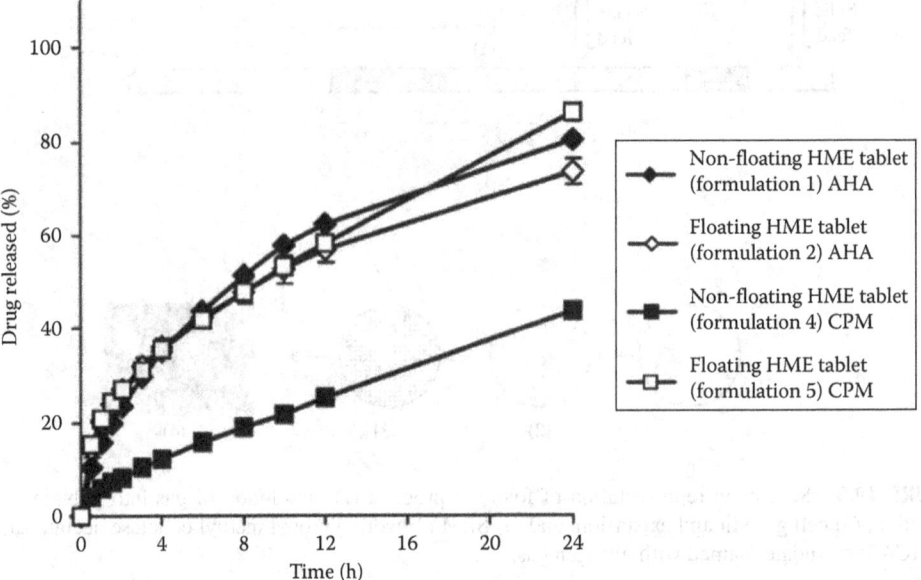

FIGURE 49.10 Controlled release profiles for conventional CR tablets produced by HME and porous floating tablets produced using HME. Acetohydroxamic acid (AHA) tested in 0.01 N HCl, chlorpheniramine maleate (CPM) tested in 0.1 N HCl. (Reprinted from Fukuda, M., Peppas, N.A., and McGinity, J.W., Floating hot-melt extruded tablets for gastroretentive controlled drug release system, *J. Control. Release*, 115, 121–129, Copyright 2006, with permission from Elsevier.)

1.2 media. The foamed tablets showed low rates of API release at pH 1.2 and rapid and complete release at pH 6.8 due to the enteric nature of the HPMCAS polymer. In addition to gastric retentive dosage forms, extrusion foaming can be utilized to create fast disintegrating oral tablets to target drug delivery in the buccal or sublingual space. Clarke describes the use of extrusion compounding coupled with injection molding and an inert physical blowing agent to form foamed dosage forms with rapid disintegration.[101] Clarke explored the ability to extrude, injection mold, and foam formulations of several polymers, sweeteners, flavorants, and disintegrants. The foaming process developed would allow for the continuous manufacture of orally disintegrating tablets as well as gastric retentive dosage forms.

49.7 DIRECTLY SHAPED PRODUCTS

Although the majority of pharmaceutical formulations are delivered orally, a large number of products have been developed for non-oral administration based on compound properties, delivery frequency, patient compliance, or specific site targeting. In order to prepare such systems, it is important that in addition to the desired release characteristics consideration must be given for geometry, mechanical behavior, and immunogenicity. Utilizing extrusion it is possible to prepare monolithic drug products for advanced non-oral applications, where the unique advantages of a continuous non-solvent process are used to more rapidly and effectively prepare products when compared to similar solvent-based approaches. To date products including subcutaneous filaments, intravitreal implants, buccal fibers, and vaginal rings have been developed using melt extrusion. Release mechanisms from directly shaped systems rely on similar principles to those outlined in the controlled release section, where delivery into the bulk is driven by diffusion, swelling, erosion, and/or pressure differentials coupled with the geometric design of the device.

Rod-shaped implants or vaginal rings can be made using HME, formulated as drug in polymer matrix or membrane-controlled reservoir systems. Several polymer options have been demonstrated

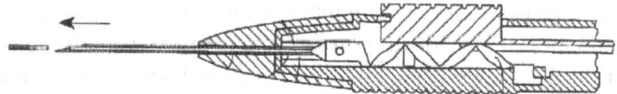

FIGURE 49.11 Ozurdex delivery system schematic. (From Weber, D. et al., *Methods and Apparatus for Delivery of Ocular Implants*, U.S.P.A.T. Organization, Allergan, Inc., p. 20, 2005.)

for commercial and development purposes. For flexible rings, low T_g polymers are preferred like ethylene vinyl acetate (EVA), poly-urethane elastomers, and silicones produced by reactive injection molding. Poly (lactide) (PLA),[104] poly (lactide-co-glycolide) (PLGA),[104] and polycaprolactone (PCL)[105] have been extruded and cut into strands or injection molded to produce bioabsorbable implants (e.g., with dexamethasone,[106] gentamicin sulfate,[107] and lysozyme[108]).

49.7.1 Applications of Directly Shaped Products by HME

One of the most successful implant systems developed to date is the Nuvaring®, which is a vaginal implant containing hormonal contraceptives that are delivered over a 21 day duration.[109] During the use of the product, insertion and removal are performed by the patient, which allows for the use of non-degradable polymers without compromising patient compliance. The circular ring is comprised of an inner core of EVA embedded with etonogestrel and ethinyl estradiol. Release rates of 120 and 15 µg/day, respectively, are achieved by diffusion through an exterior EVA layer with a thickness of approximately 110 µm.[109] During the manufacturing of this product, a dual extrusion setup into a coaxial die is used to prepare the drug-loaded core while a second layer of pure EVA is extruded at a product matching feed rate to achieve a rod intermediate. The rod is cut to length before the ends of the rod are welded together to form the final ring. The finished product exhibits the necessary mechanical strength to facilitate patient insertion and removal, while also yielding the desire hormone release profile, and is an excellent example of the complex product engineering that can be achieved with the melt extrusion platform.

Another recent example of a directly shaped drug product that illustrates the synergy between product, device, and delivery mode is Ozurdex®, an intravitreal implant that delivers dexamethasone over an extended period to treat macular degeneration.[110] Unlike vaginal implants that can be manually removed by patients, products that are implanted intravitreally cannot be removed by the patient and would potentially require significant invasive surgical procedures to remove. As a result, products such as Ozurdex must utilize biodegradable polymers to eliminate the need for post-delivery removal. To facilitate this, dexamethasone is processed with a biodegradable PLGA polymer to form a cylindrical extrudate with a diameter of 460 µm. The resulting cylinder is then loaded into a 22-guage hypodermic needle capable of delivering the drug product to the back of the eye without the need for post-implantation sutures. A representative schematic of the delivery device is shown in Figure 49.11. It is also important to minimize immune response of the drug product, which can be achieved through the implementation of terminal sterilization. Ozurdex is sterilized using gamma irradiation of the finished drug product, which does not negatively impact release characteristics of the product.

49.8 CONCLUSIONS

Melt extrusion has been established over the last 30 years as a viable platform for the production of solid dispersions to aid in the delivery of pharmaceutical compounds. Illustrated in this section, the technology can be used to support a range of delivery aspects for a number of new therapies. Being a low cost manufacturing technology, capable of achieving unique delivery attributes and supporting new production paradigms, it will remain a pivotal platform for the development of novel drug products. As the industry continues to evolve in many new areas, including personalized medicine and continuous manufacturing, melt extrusion is uniquely suited to support these paradigms.

By virtue of its well-established nature in a number of industries, scalability, and excellent process control, melt extrusion meets the needs of both biotechnology and small molecule discovery pipelines. Combining this with an ever-expanding compositional knowledge for the technology, melt extrusion will make a clear difference in the future of drug delivery.

REFERENCES

1. Ku, M.S. Use of the biopharmaceutical classification system in early drug development. *AAPS J.* 2008, *10*(1), 208–212.
2. Lipinski, C.A. Drug-like properties and the causes of poor solubility and poor permeability. *J. Pharmacol. Toxicol. Met.* 2000, 44(1), 235–249.
3. Lipinski, C.A. Lead- and drug-like compounds: The rule-of-five revolution. *Drug Discov. Today* 2004, *1*(4), 337–341.
4. Lipinski, C.A.; Lombardo, F.; Dominy, B.W.; Feeney, P.J. Experimental and computational approaches to estimate solubility and permeability in drug discovery and development settings. *Adv. Drug Deliv. Rev.* 2001, *46*(1–3), 3–26.
5. Alsenz, J.; Kansy, M. High throughput solubility measurement in drug discovery and development. *Adv. Drug Deliv. Rev.* 2007, *59*(7), 546–567.
6. Jain, P.; Yalkowsky, S.H. Prediction of aqueous solubility from SCRATCH. *Int. J. Pharm.* 2010, *385*(1–2), 1–5.
7. Di, L.; Fish, P.V.; Mano, T. Bridging solubility between drug discovery and development. *Drug Discov. Today* 2012, *17*(9–10), 486–495.
8. Ullrich, M. Historical development of co-rotating twin screw extruders. In *Co-rotating Twin-Screw Extruders*; Kohlgrüber, K., Ed.; Hanser Garnder Publications: Cincinnati, OH, 2008, pp. 1–7.
9. DiNunzio, J.C.; Martin, C.; Zhang, F. Melt extrusion: Shaping drug delivery in the 21st century. *Pharm. Technol.* 2010, *57*(SI), s30–s37.
10. Crowley, M.M.; Zhang, F.; Repka, M.A.; Thumma, S.; Upadhye, S.B.; Kumar Battu, S.; McGinity, J.W.; Martin, C. Pharmaceutical applications of hot-melt extrusion: Part I. *Drug Dev. Ind. Pharm.* 2007, *33*(9), 909–926.
11. Thiele, W. Twin-screw extrusion and screw design. In *Pharmaceutical Extrusion Technology*; Ghebre-Sellassie, I., Martin, C., Eds.; Informa Healthcare: New York, 2003, pp. 69–98.
12. Erdmenger, R. Mixing and kneading machine; US2670188. U.S. Pattern Organization, Farbenfabriken Bayer, 1954, p. 6.
13. Rauwendaal, C. *Polymer Extrusion*, Revised 4th ed.; Hanser: Cincinnati, OH, 2001, p. 781.
14. Tadmor, Z.; Gogos, C.G. *Principles of Polymer Processing*, 2nd edn.; Wiley-Interscience: Hoboken, NJ, 2006, p. 961.
15. Serajuddin, A.T.M. Solid dispersion of poorly water-soluble drugs: Early promises, subsequent problems, and recent breakthroughs. *J. Pharm. Sci.* 1999, *88*(10), 1058–1066.
16. Breitenbach, J.; Mägerlin, M. Melt-extruded solid dispersions. In *Pharmaceutical Extrusion Technology*; Ghebre-Sellassie, I., Martin, C., Eds.; Informa Healthcare: New York, 2003, pp. 245–260.
17. McGinity, J.W.; Repka, M.A.; Koleng, J.J.; Zhang, F. Hot-melt extrusion technology. In *Encyclopedia of Pharmaceutical Technology*; Swarbrick, J., Boylan, J.C., Eds.; Informa Healthcare: Hooboken, NJ, 2007, pp. 2004–2020.
18. McGinity, J.W.; Zhang, F. Melt-extruded controlled-release dosage forms. In *Pharmaceutical Extrusion Technology*; Ghebre-Sellassie, I., Martin, C., Eds.; Informa Healthcare: New York, 2003, pp. 183–208.
19. Leuner, C.; Dressman, J. Improving drug solubility for oral delivery using solid dispersions. *Eur. J. Pharm. Sci.* 2000, *50*(1), 47–60.
20. DiNunzio, J.C.; Brough, C.; Hughey, J.R.; Miller, D.A.; Williams, R.O., III; McGinity, J.W. Fusion production of solid dispersions containing a heat-sensitive active ingredient by hot melt extrusion and Kinetisol dispersing. *Eur. J. Pharm. Biopharm.* 2012, *74*(2), 340–351.
21. Djuris, J.; Nikolakakis, I.; Ibric, S.; Djuric, Z.; Kachrimanis, K. Preparation of carbamazepine-Soluplus solid dispersions by hot-melt extrusion, prediction of drug-polymer miscibility by thermodynamic model fitting. *Eur. J. Pharm. Biopharm.* 2013, *84*(1), 228–237.
22. Ghebremeskel, A.N.; Vemavarapu, C.; Lodaya, M. Use of surfactants as plasticizers in preparing solid dispersions of poorly soluble API: Selection of polymer–surfactant combinations using solubility parameters and testing the processability. *Int. J. Pharm.* 2007, *328*(2), 119–129.

23. Lin, S.-Y.; Yu, H.-L. Thermal stability of methacrylic acid copolymers of eudragits L, S, and L30D and the acrylic acid polymer of carbopol. *J. Polym. Sci. A* 1999, *37*(13), 2061–2067.
24. Repka, M.A.; Battu, S.K.; Upadhye, S.B.; Thumma, S.; Crowley, M.M.; Zhang, F.; Martin, C.; McGinity, J.W. Pharmaceutical applications of hot-melt extrusion: Part II. *Drug Dev. Ind. Pharm.* 2007, *33*(10), 1043–1057.
25. Repka, M.A.; Majumdar, S.; Kumar Battu, S.; Srirangam, R.; Upadhye, S.B. Applications of hot-melt extrusion for drug delivery. *Expert Opin. Drug Deliv.* 2008, *5*(12), 1357–1376.
26. Janssens, S.; de Armas, H.N.; Roberts, C.J.; Van den Mooter, G. Characterization of ternary solid dispersions of itraconazole, PEG 6000, and HPMC 2910 E5. *J. Pharm. Sci.* 2008, *97*(6), 2110–2120.
27. Repka, M.A.; McGinity, J.W. Influence of vitamin E TPGS on the properties of hydrophilic films produced by hot-melt extrusion. *Int. J. Pharm.* 2000, *202*(1–2), 63–70.
28. Repka, M.A.; McGinity, J.W. Physical-mechanical, moisture absorption and bioadhesive properties of hydroxypropyl cellulose hot-melt extruded films. *Biomaterials* 2000, *21*(14), 1509–1517.
29. Repka, M.A.; McGinity, J.W. Bioadhesive properties of hydroxypropyl cellulose topical films produced by hot-melt extrusion. *J. Control. Release* 2001, *70*(3), 341–351.
30. Repka, M.A.; Prodduturi, S.; Stodghill, S.P. Production and characterization of hot-melt extruded films containing clotrimazole. *Drug Dev. Ind. Pharm.* 2003, *29*(7), 757–765.
31. Crowley, M.M.; Fredersdorf, A.; Schroeder, B.; Kucera, S.; Prodduturi, S.; Repka, M.A.; McGinity, J.W. The influence of guaifenesin and ketoprofen on the properties of hot-melt extruded polyethylene oxide films. *Eur. J. Pharm. Sci.* 2004, *22*(5), 1409–1418.
32. Crowley, M.M.; Schroeder, B.; Fredersdorf, A.; Obara, S.; Talarico, M.; Kucera, S.; McGinity, J.W. Physicochemical properties and mechanism of drug release from ethyl cellulose matrix tablets prepared by direct compression and hot-melt extrusion. *Int. J. Pharm.* 2004, *269*(2), 509–522.
33. DiNunzio, J.C.; Zhang, F.; Martin, C.; McGinity, J.W. Melt extrusion. In *Formulating Poorly Water Soluble Drugs*; Miller, D.A., Watts, A.B., Williams, R.O., III, Eds. Springer: New York, 2012, pp. 311–362.
34. Bates, S.; Zografi, G.; Engers, D.; Morris, K.; Crowley, K.; Newman, A. Analysis of amorphous and nanocrystal-line solids from their X-ray diffraction patterns. *Pharm. Res.* 2006, *23*(10), 2333–2349.
35. Newman, A.; Engers, D.; Bates, S.; Ivanisevic, I.; Kelly, R.C.; Zografi, G. Characterization of amorphous API: polymer mixtures using x-ray powder diffraction. *J. Pharm. Sci.* 2008, *97*(11), 4840–4856.
36. Shah, B.; Kakumanu, V.K.; Bansal, A.K. Analytical techniques for quantification of amorphous/crystalline phases in pharmaceutical solids. *J. Pharm. Sci.* 2006, *95*(8), 1641–1665.
37. Craig, D.Q.M. A review of thermal methods used for the analysis of the crystal form, solution thermodynamics and glass transition behaviour of polyethylene glycols. *Thermochim. Acta* 1995, *248*(2), 189–203.
38. Six, K.; Berghmans, H.; Leuner, C.; Dressman, J.; Van Werde, K.; Mullens, J.; Benoist, L. et al. Characterization of solid dispersions of itraconazole and hydroxypropylmethylcellulose prepared by melt extrusion, part II. *Pharm. Res.* 2003, *20*(7), 1047–1054.
39. Six, K.; Leuner, C.; Dressman, J.; Verreck, G.; Peeters, J.; Blaton, N.; Augustijns, P.; Kinget, R.; Van den Mooter, G. Thermal properties of hot-stage extrudates of itraconazole and eudragit E100. Phase separation and polymorphism. *J. Ther. Anal. Calorimet.* 2002, *68*(2), 591–601.
40. Six, K.; Verreck, G.; Peeters, J.; Brewster, M.; Van Den Mooter, G. Increased physical stability and improved dissolution properties of itraconazole, a class II drug, by solid dispersions that combine fast- and slow-dissolving polymers. *J. Pharm. Sci.* 2004, *93*(1), 124–131.
41. van Drooge, D.J.; Hinrichs, W.L.; Visser, M.R.; Frijlink, H.W. Characterization of the molecular distribution of drugs in glassy solid dispersions at the nano-meter scale, using differential scanning calorimetry and gravimetric water vapour sorption techniques. *Int. J. Pharm.* 2006, *310*(1–2), 220–229.
42. Weuts, I.; Kempen, D.; Six, K.; Peeters, J.; Verreck, G.; Brewster, M.; Van den Mooter, G. Evaluation of different calorimetric methods to determine the glass transition temperature and molecular mobility below T_g for amorphous drugs. *Int. J. Pharm.* 2003, *259*(1–2), 17–25.
43. Gupta, P.; Thilagavathi, R.; Chakraborti, A.K.; Bansal, A.K. Role of molecular interaction in stability of celecoxib-PVP amorphous systems. *Mol. Pharm.* 2005, *2*(5), 384–391.
44. Lacoulonche, F.; Chauvet, A.; Masse, J. An investigation of flurbiprofen polymorphism by thermoanalytical and spectroscopic methods and a study of its interactions with poly-(ethylene glycol) 6000 by differential scanning calorimetry and modelling. *Int. J. Pharm.* 1997, *153*(2), 167–179.
45. Taylor, L.S.; Zografi, G. Spectroscopic characterization of interactions between PVP and indomethacin in amorphous molecular dispersions. *Pharm. Res.* 1997, *14*(12), 1651–1698.

46. Blagden, N.; de Matas, M.; Gavan, P.T.; York, P. Crystal engineering of active pharmaceutical ingredients to improve solubility and dissolution rates. *Adv. Drug Deliv. Rev.* 2007, *59*(7), 617–630.
47. Konno, H.; Handa, T.; Alonzo, D.E.; Taylor, L.S. Effect of polymer type on the dissolution profile of amorphous solid dispersions containing felodipine. *Eur. J. Pharm. Biopharm.* 2008, *70*(2), 493–499.
48. Matteucci, M.E.; Hotze, M.A.; Johnston, K.P.; Williams, R.O., III. Drug nanoparticles by antisolvent precipitation: Mixing energy versus surfactant stabilization. *Langmuir* 2006, *22*(21), 8951–8959.
49. Miller, D.A.; McConville, J.T.; Yang, W.; Williams, R.O., III; McGinity, J.W. Hot-melt extrusion for enhanced delivery of drug particles. *J. Pharm. Sci.* 2007, *96*(2), 361–376.
50. Vertzoni, M.; Dressman, J.; Butler, J.; Hempenstall, J.; Reppas, C. Simulation of fasting gastric conditions and its importance for the in vivo dissolution of lipophilic compounds. *Eur. J. Pharm. Biopharm.* 2005, *60*(3), 413–417.
51. Vertzoni, M.; Fotaki, N.; Kostewicz, E.; Stippler, E.; Leuner, C.; Nicolaides, E.; Dressman, J.; Reppas, C. Dissolution media simulating the intralumenal composition of the small intestine: Physiological issues and practical aspects. *J. Pharm. Pharmacol.* 2004, *56*(4), 453–462.
52. Vogt, M.; Kunath, K.; Dressman, J.B. Dissolution improvement of four poorly water soluble drugs by cogrinding with commonly used excipients. *Eur. J. Pharm. Biopharm.* 2008, *68*(2), 330–337.
53. Vogt, M.; Kunath, K.; Dressman, J.B. Dissolution enhancement of fenofibrate by micronization, cogrinding and spray-drying: Comparison with commercial preparations. *Eur. J. Pharm. Biopharm.* 2008, *68*(2), 283–288.
54. Guzmán, H.R.; Tawa, M.; Zhang, Z.; Ratanabanangkoon, P.; Shaw, P.; Gardner, C.R.; Chen, H.; Moreau, J.P.; Almarsson, O.; Remenar, J.F. Combined use of crystalline salt forms and precipitation inhibitors to improve oral absorption of celecoxib from solid oral formulations. *J. Pharm. Sci.* 2007, *96*(10), 2686–2702.
55. Alonzo, D.E.; Gao, Y.; Zhou, D.; Mo, H.; Zhang, G.G.; Taylor, L.S. Dissolution and precipitation behavior of amorphous solid dispersions. *J. Pharm. Sci.* 2011, *100*(8), 3316–3331.
56. Alonzo, D.E.; Zhang, G.G.; Zhou, D.; Gao, Y.; Taylor, L.S. Understanding the behavior of amorphous pharmaceutical systems during dissolution. *Pharm. Res.* 2010, *27*(4), 608–618.
57. DiNunzio, J.C.; Miller, D.A.; Yang, W.; McGinity, J.W.; Williams, R.O., III. Amorphous compositions using concentration enhancing polymers for improved bioavailability. *Mol. Pharm.* 2008, *5*(6), 968–980.
58. Miller, D.A. Improved oral absorption of poorly water-soluble drugs by advanced solid dispersion systems. In *Division of Pharmaceutics*; The University of Texas at Austin: Austin, TX, 2007, p. 312.
59. Miller, D.A.; DiNunzio, J.C.; Yang, W.; McGinity, J.W.; Williams, R.O., III. Targeted intestinal delivery of supersaturated itraconazole for improved oral absorption. *Pharm. Res.* 2008, *25*(6), 1450–1459.
60. Baird, J.A.; Van Eerdenbrugh, B.; Taylor, L.S. A classification system to assess the crystallization tendency of organic molecules from undercooled melts. *J. Pharm. Sci.* 2010, *99*(9), 3787–3805.
61. Miller, D.A.; DiNunzio, J.C.; Yang, W.; McGinity, J.W.; Williams, R.O., III. Enhanced in vivo absorption of itraconazole via stabilization of supersaturation following acidic-to-neutral pH transition. *Drug Dev. Ind. Pharm.* 2008, *34*(8), 890–902.
62. Wu, C.; McGinity, J.W. Influence of methylparaben as a solid-state plasticizer on the physicochemical properties of Eudragit® RS PO hot-melt extrudates. *Eur. J. Pharm. Biopharm.* 2003, *56*(1), 95–100.
63. Crowley, M.M.; Zhang, F.; Koleng, J.J.; McGinity, J.W. Stability of polyethylene oxide in matrix tablets prepared by hot-melt extrusion. *Biomaterials* 2002, *23*(21), 4241–4248.
64. Schenck, L.; Troup, G.M.; Lowinger, M.; Li, L.; McKelvey, C. Achieving a hot melt extrusion design space for the production of solid solutions. In *Chemical Engineering in the Pharmaceutical Industry: R&D to Manufacturing*; am Ende, D.J., Ed.; John Wiley & Sons, Inc.: New York, 2011.
65. Bessemer, B. Shape extrusion. In *Pharmaceutical Extrusion Technology*; Ghebre-Sellassie, I., Martin, C., Ed.; Marcel Dekker, Inc.: New York, 2003, pp. 209–224.
66. Doetsch, W. Material handling and feeder technology. In *Pharmaceutical Extrusion Technology*; Ghebre-Sellassie, I., Martin, C., Eds.; Marcel Dekker, Inc.: New York, 2003, pp. 11–134.
67. Dreiblatt, A. Process design. In *Pharmaceutical Extrusion Technology*; Ghebre-Sellassie, I., Martin, C. Ed.; Marcel Dekker, Inc.: New York, 2003, pp. 149–170.
68. Breitenbach, J. Melt extrusion can bring new benefits to HIV therapy: The example of Kaletra tablets. *Am. J. Drug Deliv.* 2006, *4*(2), 61–64.
69. Bauer, J.; Spanton, S.; Henry, R.; Quick, J.; Dziki, W.; Porter, W.; Morris, J. Ritonavir: An extraordinary example of conformational polymorphism. *Pharm. Res.* 2001, *18*(6), 859–866.
70. Kessler, T.; Breitenbach, J.; Schmidt, C.; Degenhardt, M.; Rosenberg, J.; Krull, H.; Berndl, G. *Process for Producing a Solid Dispersion of an Active Ingredient*; US20090302493. U.S. Patent Organization, Abbott GmbH & Co., KG, 2009.

71. Riegelman, S.; Chiou, W.L. *Increasing the Absorption Rate of Insoluble Drugs*; U.S. Patent Organization, The Regents of the University of California: Oakland, CA, 1979, p. 9.
72. Chatterji, A.; Desai,,D.; Miller, D.A.; Sandhu, H.K.; Shah, N.H. *A Process for Controlled Crystallization of an Active Pharmaceutical Ingredient from Supercooled Liquid State by Hot Melt Extrusion*; World Intellectual Property Organization, WO2012110469. 2012.
73. Crew, M.D.; Curatolo W.J.; Friesen, D.T.; Gumkowski, M.J.; Lorenz, D.A.; Nightingale, J.A.S.; Ruggeri, R.B.; Shanker, R.M. *Pharmaceutical Compositions of Cholesteryl Ester Transfer Protein Inhibitors*; U.S.P.T. Office, Pfizer, Inc., US20120277315. 2007, p. 74.
74. Bernard, B.L. *Cellulosic Films Incorporating a Pharmaceutically Acceptable Plasticizer with Enhanced Wettability*, World Intellectual Property Organization, Eastman Chemical Company, WO2006115712. 2006.
75. DiNunzio, J.C.; Brough, C.; Miller, D.A.; Williams, R.O.; McGinity, J.W. Applications of KinetiSol® dispersing for the production of plasticizer free amorphous solid dispersions. *Eur. J. Pharm. Sci.* 2010, *40*(3), 179–187.
76. Al bano, A.; Desai, D.; Dinunzio, J.; Go, Z.; Iyer, R.M.; Sandhu, H.K.; Shah, N.H. *Pharmaceutical Composition with Improved Bioavailability for High Melting Hydro-Phobic Compound*; World Intellectual Property Organization, World Intellectual Property Organization, F. Hoffmann-La Roche, WO2013087546. 2013.
77. Young, C.R.; Koleng, J.J.; McGinity, J.W. Production of spherical pellets by a hot-melt extrusion and spheronization process. *Int. J. Pharm.* 2002, *242*(1–2), 87–92.
78. Zhang, F.; McGinity, J.W. Properties of sustained-release tablets prepared by hot-melt extrusion. *Pharm. Dev. Technol.* 1999, *4*(2), 241–250.
79. Verreck, G.; Six, K.; Van den Mooter, G.; Baert, L.; Peeters, J.; Brewster, M.E. Characterization of solid dispersions of itraconazole and hydroxypropylmethylcellulose prepared by melt extrusion—Part I. *Int. J. Pharm.* 2003, *251*(1–2), 165–174.
80. Coppens, K.A.; Hall, M.J.; Koblinski, B.D.; Larsen, P.S.; Read, M.D.; Shrestha, M. Controlled release of poorly soluble drugs utilizing hot melt extrusion. In *Proceedings of the American Association of Pharmaceutical Scientists*, Los Angeles, CA, 2009.
81. Alderman, D.A.; Wolford, T.D. *Sustained Release Dosage Form Based on Highly Plasticized Cellulose Ether Gels*; U.S.P.T. Organization, The Dow Chemical Company, 1987, p. 5.
82. Zhu, Y.; Mehta, K.A.; McGinity, J.W. Influence of plasticizer level on the drug release from sustained release film coated and hot-melt extruded dosage forms. *Pharm. Dev. Technol.* 2006, *11*(3), 285–294.
83. Zhu, Y.; Shah, N.H.; Malick, A.W.; Infeld, M.H.; McGinity, J.W. Controlled release of a poorly water-soluble drug from hot-melt extrudates containing acrylic polymers. *Drug Dev. Ind. Pharm.* 2006, *32*(5), 569–583.
84. Schilling, S.U.; Lirola, H.L.; Shah, N.H.; Waseem Malick, A.; McGinity, J.W. Influence of plasticizer type and level on the properties of Eudragit® S100 matrix pellets prepared by hot-melt extrusion. *J. Microencap.* 2010, *27*(6), 521–532.
85. Schilling, S.U.; Shah, N.H.; Waseem Malick, A.; McGinity, J.W. Properties of melt extruded enteric matrix pellets. *Eur. J. Pharm. Biopharm.* 2010, *74*(2), 352–361.
86. Özgüney, I.; Shuwistkul, D.; Bodmeier, R. Development of kollidon SR mini-matrices prepared by hot-melt extrusion. *Eur. J. Pharm. Biopharm.* 2010, *73*(1), 140–145.
87. Roth, W.; Setnik, B.; Zietsch, M.; Burst, A.; Breitenbach, J.; Sellers, E.; Brennan, D. Ethanol effects on drug release from Verapamil Meltrex®, an innovative melt extruded formulation. *Int. J. Pharm.* 2009, *368*(1–2), 72–75.
88. Debenedetti, P.G.; Stillinger, F.H. Supercooled liquids and the glass transition. *Nature* 2001, *410*(259–267), 259.
89. Bartholomaeus, J.H.; Arkenau-Marić, E.; Gali, E. Opioid extended-release tablets with improved tamper-resistant properties. *Expert Opin. Drug Deliv.* 2012, *9*(8), 879–891.
90. Terife, G.; Faridi, N.; Wang, P.; Gogos, C.G. Polymeric foams for oral drug delivery—A review. *Plast. Eng.* November/December 2012, 32–39.
91. Listro, T. Foamed hot melt extrusion for solid molecular dispersions. *Innovation in Pharmaceutical Technology.* 2012, 43, 60–62.
92. Brown, C.; Mckelvey, C.; Faridi, N.; Gogos, C.; Suwardie, J.; Wang, P.; Young, M.; Zhu, L. Evaluation of foaming of pharmaceutical polymers by CO_2 and N_2 to enable drug products. *SPE ANTEC* 2011, 224–1228.
93. Verreck, G.; Decorte, A.; Heymans, K.; Adriaensen, J.; Cleeren, D.; Jacobs, A.; Liu, D. et al. The effect of pressurized carbon dioxide as a temporary plasticizer and foaming agent on the hot stage extrusion process and extrudate properties of solid dispersions of itraconazole with PVP-VA 64. *Eur. J. Pharm. Sci.* 2005, *26*(3–4), 349–358.

94. Verreck, G.; Decorte, A.; Heymans, K.; Adriaensen, J.; Liu, D.; Tomasko, D.L.; Arien, A. et al. The effect of supercritical CO_2 as a reversible plasticizer and foaming agent on the hot stage extrusion of itraconazole with EC 20 cps. *J. Supercrit. Fluids* 2007, *40*(1), 153–162.

95. Verreck, G.; Decortea, A.; Lib, H.; Tomaskob, D.; Ariena, A.; Peetersa, J.; Rombautc, P.; Van den Mooterc, G.; Brewstera, M.E. The effect of pressurized carbon dioxide as a plasticizer and foaming agent on the hot melt extrusion process and extrudate properties of pharmaceutical polymers. *J. Supercrit. Fluids* 2006, *38*(3), 383–391.

96. Lyons, J.G.; Hallinan, M.; Kennedy, J.E.; Devine, D.M.; Geever, L.M.; Blackie, P.; Higginbotham, C.L. Preparation of monolithic matrices for oral drug delivery using a supercritical fluid assisted hot melt extrusion process. *Int. J. Pharm.* 2007, *329*(1–2), 62–71.

97. Nagy, Z.K.; Sauceau, M.; Nyúl, K.; Rodier, E.; Vajna, B.; Marosi, G.; Fages, J. Use of supercritical CO_2-aided and conventional melt extrusion for enhancing the dissolution rate of an active pharmaceutical ingredient. *Polym. Adv. Technol.* 2012, *23*(5), 909–918.

98. Fukuda, M.; Peppas, N.A.; McGinity, J.W. Floating hot-melt extruded tablets for gastroretentive controlled drug release system. *J. Control. Release* 2006, *115*(2), 121–129.

99. Streubel, A.; Siepmann, J.; Bodmeier, R. Floating matrix tablets based on low density foam powder: Effects of formulation and processing parameters on drug release. *Eur. J. Pharm. Sci.* 2003, *18*(1), 37–45.

100. Nakamichi, K.; Yasuura, H.; Fukui, H.; Oka, M.; Izumi, S. Evaluation of a floating dosage form of nicardipine hydrochloride and hydroxypropylmethylcellulose acetate succinate prepared using a twin-screw extruder. *Int. J. Pharm.* 2001, *218*(1–2), 103–112.

101. Clarke, A.J. Novel pharmaceutical dosage forms and method for producing the same. US20050202090. U.S. Pattern Organization, 2005.

102. Verreck, G.; Brewster, M.E.; Van Assche, I. The use of supercritical fluid technology to broaden the applicability of hot melt extrusion for drug delivery applications. *Bull. Tech. Gattef.* 2012, *46*(105), 28–42.

103. Lee, C.H.; Lee, K.J.; Jeong, H.G.; Kim, S.W. Growth of gas bubbles in the foam extrusion process. *Adv. Polym. Technol.* 2000, *19*(2), 97–112.

104. Passerini, N.; Craig, D.Q.M. An investigation into the effects of residual water on the glass transition temperature of polylactide microspheres using modulated temperature DSC. *J. Control. Release* 2001, *73*(1), 111–115.

105. Wang, Y.; Rodriguez-Perez, M.A.; Reis, R.L.; Mano, J.F. Thermal and thermomechanical behaviour of polycaprolactone and starch/polycaprolactone blends for biomedical applications. *Macromol. Mater. Eng.* 2005, *290*(8), 792–801.

106. Li, D.; Guo, G.; Fan, R.; Liang, J.; Deng, X.; Luo, F.; Qian, Z. PLA/F68/Dexamethasone implants prepared by hot-melt extrusion for controlled release of anti-inflammatory drug to implantable medical devices: I. Preparation, characterization and hydrolytic degradation study. *Int. J. Pharm.* 2013, *441*(1–2), 365–372.

107. Gosau, M.; Müller, B. Release of gentamicin sulphate from biodegradable PLGA-implants produced by hot melt extrusion. *Pharmazie* 2010, *65*(7), 487–492.

108. Ghalanbor, Z.; Körber, M.; Bodmeier, R. Improved lysozyme stability and release properties of poly(lactide-co-glycolide) implants prepared by hot-melt extrusion. *Pharm. Res.* 2010, *27*(2), 371–379.

109. De Graaff, W.; Groen, J.S.; Kruft, M.A.B.; Van Laarhoven, J.A.H.; Vromans, H.; Zeeman, R. Drug delivery system based on polyethylene vinylacetate copolymers; US20140302115. U.S. Pattern Organization, N.V. Organon, 2013, p. 14.

110. Shiah, J.-G.; Bhagat, R.; Blanda, W.M.; Nivaggioli, T.; Peng, L.; Chou, D.; Weber, D.A. Ocular implant made by a double extrusion process; US8778381. U.S. Pattern Organization, Allergan, Inc., 2008, p. 39.

111. Weber, D.; Kane, I.; Rehal, M.; Lathrop, R.L., III; Aptek-arev, K. Methods and apparatus for delivery of ocular implants; US6899717. U.S. Pattern Organization, Allergan, Inc., 2005, p. 20.

50 Nanoparticles
Biomaterials for Drug Delivery

Abhijit Gokhale, Thomas Williams, and Jason M. Vaughn

CONTENTS

50.1 Introduction 1151
50.2 Drug–Polymer Conjugates 1152
    50.2.1 Description and Method of Manufacturing 1152
    50.2.2 Formulations 1153
50.3 Solid Lipid Nanoparticles and Nanostructured Lipid Carriers 1154
    50.3.1 Solid Lipid Nanoparticles 1154
        50.3.1.1 Description and Method of Manufacturing 1154
        50.3.1.2 Formulation 1155
    50.3.2 Nanostructured Lipid Carrier 1156
50.4 Liposomes, Polymerosomes, and Niosomes 1156
    50.4.1 Liposomes 1156
        50.4.1.1 Description and Method of Manufacturing 1156
        50.4.1.2 Formulations 1157
    50.4.2 Polymerosomes 1158
        50.4.2.1 Description and Method of Manufacturing 1158
        50.4.2.2 Formulations 1159
    50.4.3 Niosomes 1159
        50.4.3.1 Description and Method of Manufacturing 1159
        50.4.3.2 Formulations 1159
50.5 Lipid–Polymer Hybrids 1160
    50.5.1 Description and Method of Manufacturing 1160
    50.5.2 Formulations 1161
50.6 Dendrimers 1161
    50.6.1 Description and Method of Manufacturing 1161
    50.6.2 Formulations 1162
50.7 Conclusions 1164
References 1164

50.1 INTRODUCTION

Biodegradable nanoparticles have many pharmaceutical applications in different areas of human activities like biology and neurology such as targeted drug delivery into the brain for cancer treatment.[1] The developments in bionanoparticles and the understanding of their properties show an optimistic picture in the development of future medicines and related products.

Bionanoparticles are used as functional excipients in a variety of pharmaceutical applications such as drug delivery with enhanced bioavailability, biological labels, and tissue engineering to name a few.[2] Drugs of nanosize are often unstable and encapsulation or conjugation of nanodrugs with biopolymer particles of size less than 200 nm can provide the required stability and better control of drug release. For more sophisticated targeted delivery applications, the required particle size needs to be less than 80 nm. Because of the smaller size, nanoparticles can penetrate through smaller capillaries, sustaining the intracellular drug levels and deliver the drug into the tissues with stable release kinetics.[3]

This article reviews several cutting-edge advanced drug delivery systems. Biomaterials with a wide range of physico-chemical properties such as hydrophilic–lipophilic balance value, solubility parameter, polarity, hydrogen bonding capacity, molecular weight and biological properties such as the ability to inhibit transporters (e.g., p-glycoprotein) and genes (e.g., CYP3A) are used as excipients in these drug delivery systems.

50.2 DRUG–POLYMER CONJUGATES

50.2.1 Description and Method of Manufacturing

Polymers offer several advantages in drug delivery systems due to their versatile nature. Polymeric functional groups can be engineered to produce hydrophilic or hydrophobic components to alter the solubility and permeability of the polymers. Therefore, the solubility and permeability of drug molecules attached to the polymer can be increased. Also, the drug delivery systems (drug + polymer) can be formulated to achieve the delivery of two or more drugs from the same formulation, contain readily clearable polymer carriers, control the release of highly toxic drugs, and improve drug targeting to tissues or cells. Drug–polymer conjugates have been used to deliver several biologically active compounds like drug molecules, proteins, peptides, hormones, enzymes, etc.[4] Selecting the proper polymeric excipient for a given drug is very important to achieve the desired therapeutic effect.

Chemical conjugation of drugs to biopolymers can form stable bonds such as ester, amide, and disulfide. Hydrogen bonds (e.g., ester or amide) are stable and can deliver the drug to the targeted site without releasing it during its transport and prior to the cellular localization of the drug.

Polymer properties can be altered by the polymer-specific structural characteristics including molecular weight, coil structure, copolymer composition, variable polyelectrolyte charges, flexibility of polymer chains, and microstructure. The advantages of this type of drug delivery systems include diversity in composition (molecular structure and molecular weight), polymer water solubility and reactivity for conjugation with drug molecules, as these are typically used in hydrophilic systems with polar or ionic groups.

Ringsdorf[5] first proposed the idea of drug–polymer conjugates in 1975. His model was based on the covalent bond between the drug and a macromolecular backbone through a biodegradable linkage. The hydrophilic functional group makes the entire molecule soluble, the drug molecule is attached to the backbone through covalent bonds and then the third area is comprised of a transport system where a specific functional group helps the entire macromolecule to be transported to the target cells or the site of pharmacological action. The basic structure is shown in Figure 50.1.

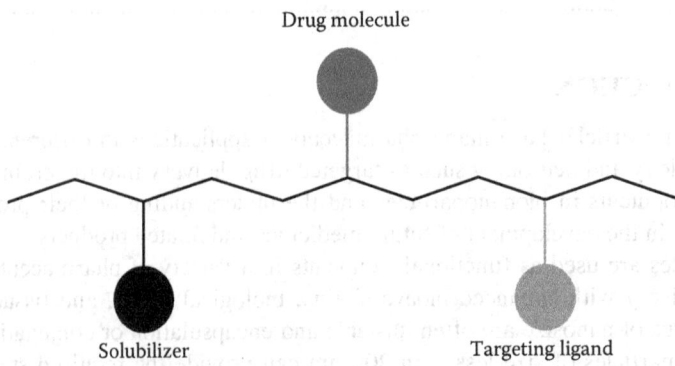

FIGURE 50.1 Ringsdorf model of synthetic polymer drugs.

The drug–polymer conjugates can be classified into two types based on the composition and the structure:

1. *Composition*: The conjugates can further be divided into three subtypes:[6]
   a. *Prodrugs*: This type of conjugate reacts with components inside the cells to form active substance(s). The selection of a suitable polymer and a targeting moiety is essential for the effectiveness of the prodrugs since the targeting ability vastly depends on the choice of antibody, ligand internalization and receptor expression.
   b. *Combination of substances*: This type of conjugate has two or more substances that react under specific intracellular conditions, forming an active drug.
   c. *Targeted drug delivery system*: This type of conjugate includes three components: a targeting moiety, a carrier, and one or more active component(s).
2. *Structure*: The drug–polymer conjugates can further be divided into six subtypes based on the structure of the polymer as well as the structure of the drug molecule used:[7]
   a. Linear polymer systems (LP)
   b. Linear polymer–linear spacer systems (LPLS)
   c. Linear polymer–branched spacer systems (LPBS)
   d. Branched polymer systems (BP)
   e. Branched polymer–linear spacer systems (BPLP)
   f. Branched polymer–branched spacer systems (BPBP)

Polymer–drug conjugates generally exhibit prolonged half-life, higher stability, water solubility, lower immunogenicity, and antigenicity. Other important aspects of the synthetic biopolymers used in this type of applications are (1) non-toxicity; (2) clearable from the body without any accumulation in tissues or organs; (3) controlled average molecular weight and molecular weight distribution, thereby avoiding any undesirable biological response due to low or high molecular weight polymer–drug conjugates; (5) high purity of polymers; and (6) possible sterilization of polymers with biodegradable characteristics.

50.2.2 Formulations

The commonly used functional groups for synthesizing these drug–polymer conjugates include carbonate, anhydride, urethane, orthoester, amide, and ester.[8] Since molecules with a low molecular weight (less than ~40,000 Da) are susceptible to excretion, the conjugates are synthesized with a molecular weight higher than 40,000 Da. At the same time high molecular weight conjugates cause problems with permeation and pinocytic capture by target cells other than phagocytes.

The first conjugate synthesized was in the field of cancer research in 1986. Anticancer drugs daunorubicin and doxorubicin were targeted to the tumor using this delivery system.[9] Since then several polymer–protein conjugates have been tested in clinical studies as part of the development phase.[10] The drug–polymer conjugate system is one of the most effective drug delivery systems for targeted drug delivery. Most of the research has been conducted to deliver anticancer or antitumor agents. High molecular weight prodrugs containing cytotoxic components have been developed to decrease peripheral side effects and to obtain a more specific administration of the drugs to the cancerous tissues. Polymer–drug conjugates have been tailored for activation by extra- or intracellular enzymes to release the parent drug in situ.

The polymeric drug delivery system can be designed for passive or active targeting. For cancer treatment, active targeting reduces side effects and enhances efficacy. Because the drug does not distribute throughout the body homogenously but gets delivered directly to the tumor cells, systemic side effects are reduced while enhancing the bioavailability, even for drugs with a relatively short half-life. Drug–polymer conjugates have been very successful in active targeting.[11]

Some of the most widely used polymers for these drug delivery systems are polysaccharides such as dextran and insulin. Cellulose-based polymers and polyarabogalactan are also seeing increased use. Other polymers like poly(D,L-lactic-co-glycolic acid), polylactic acids, N-(2-hydroxypropyl) methacrylamide, vinylic, acrylic polymers, and poly(α-amino acids) have been successfully used for drug conjugations. Poly(ethylene oxide)-β-poly(ε-caprolactone)-based drug conjugates for paclitaxel delivery were successfully demonstrated.[12] Star-shaped drug delivery systems with a poly[(p-iodomethyl) styrene] core and poly(tert-butyl acrylate) arms (PScorePtBuAarm) were used to conjugate cisplatin.[13]

Due to higher reactivity at physiological pH, the N-hydroxysuccinimide (NHS) ester makes an excellent choice for amine coupling reactions in bioconjugation synthesis. NHS ester compounds react with nucleophiles to release the NHS-leaving group and form an acylated product. Carboxyl groups activated with NHS esters are highly reactive with amine nucleophiles.[6]

Polymers containing hydroxyl groups (e.g., polyethylene glycol, PEG, etc.) can be modified to obtain anhydride compounds. PEG can be acetylated with anhydrides to form an ester terminating with free carboxylate groups. PEG and its succinimidyl succinate and succinimidyl glutarate derivatives can be further used for conjugation with drugs or proteins.[12]

In order to achieve good bioavailability, the conjugate needs to have balanced lipophilicity and hydrophilicity and to be of relatively low molecular weight to achieve the required absorption. The drug polymer–conjugate combines the advantages of both the hydrophilic and hydrophobic segments. Selecting polymeric structures like dendrimers, dendronized polymers, graft polymers, block copolymers, branched polymers, multivalent polymers, stars and hybrid glycol, and peptide derivatives increases the chance of success for producing adequate formulations.[14]

Although these conjugates have shown excellent results in various clinical trials, synthesizing such conjugates has been very difficult due to the complex nature of the drug molecules and the organic synthesis expert needed to produce these types of conjugates. It is always difficult to achieve good permeability with a minimal first-pass effect with macromolecules.[6]

50.3 SOLID LIPID NANOPARTICLES AND NANOSTRUCTURED LIPID CARRIERS

50.3.1 Solid Lipid Nanoparticles

50.3.1.1 Description and Method of Manufacturing

Nanoparticles made from solid lipids are gaining major attention in the field of oral drug delivery for small drug molecules as well as protein and peptides due to their lipophilic nature. Since most of the drugs coming out of drug discovery are poorly soluble, solid lipid nanoparticles (SLNs) provide an attractive alternative to deliver the drugs which can be readily absorbed along with SLNs through lipolysis in the small intestine.[15] There are several advantages of using SLNs for drug delivery including, but not limited to enhanced drug stability, improved bioavailability, controlled or targeted release, and aqueous solvent processing to name a few.

Excipients used in SLNs are mostly lipids with low melting points and naturally occurring in food or in the human body. Hence, they have better absorption characteristics. SLNs were initially manufactured in early 1990s using high shear mixing but this resulted in several processing issues including poor reproducibility and wide particle size distributions. To overcome these problems, techniques such as high pressure homogenization or microemulsion were used (with processes operating at elevated temperatures above the melting points of the lipids used in the formulation).[16] Typically much smaller particle sizes (less than 200 nm) are possible with the high pressure homogenization technique.

In the microemulsion method, the lipid component (fatty acids and/or glycerides) is melted, a mixture of water, cosurfactant(s) and a surfactant is heated to the same temperature as the lipid and added under mild stirring to the lipid melt. Once a thermodynamically stable microemulsion is produced, it is dispersed in cold aqueous media under constant mechanical mixing to avoid any

Nanoparticles

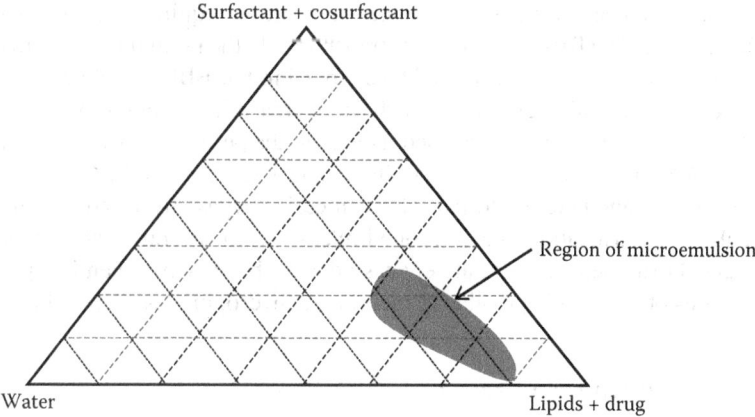

FIGURE 50.2 Phase diagrams to produce microemulsions of SLNs.

agglomeration of the solidified nanoparticles. The use of surfactants and cosurfactants in the cold aqueous media helps avoid agglomerate formation. However, in this approach, the formulation is diluted due to the presence of the cold aqueous media and hence not desirable when high solid contents are required.[17] Figure 50.2 shows the typical phase diagram to prepare the microemulsions at elevated temperature which in turn transform into nanoparticles upon cooling.

In high pressure homogenization, the formulation containing an emulsified lipid–drug mixture is cooled by the cavitation effect during pressure drop and it is not necessary to add another processing step to cool/solidify the lipid particles. A formulation with a high solid content is possible with this technique.[18]

50.3.1.2 Formulation

In order to formulate the SLN-based drug delivery system, the choice of lipids, surfactant, and cosurfactant is very important. The drug molecule should dissolve in molten lipid(s) to form the solid solution. The surfactant and cosurfactant should be able to emulsify the molten mixture of drug–lipid solution/dispersion in a nanometer particle size range (ideally less than 200 nm) to obtain a thermodynamically stable suspension. In general the loading capacity of drug in the lipid depends on the solubility of drug in the molten lipid.[17]

Commonly, lipids like glycerides (mono-, di-, or tri-), waxes (e.g., carnauba wax, beeswax, cetyl alcohol, emulsifying wax, cholesterol, cholesterol butyrate, etc.), and vegetable oils are used as carriers to dissolve/disperse the drug molecules. The surfactants (e.g., polysorbates, sodium dodecyl sulfate, castor oil, bile salts, etc.) and cosurfactants (e.g., alcohols, polyethylene glycol, cremophor, poloxamer, etc.) are used to emulsify the molten lipid mixture in aqueous media. The SLNs can then be lyophilized with excipients to obtain the dried powder. The list of excipients and formulations of the marketed products is found in Mehnert and Mader[19] and Pardeike et al.[20]

The oral delivery of proteins and peptides is very difficult due to the poor stability of large molecules in the gastrointestinal environment. Diffusion transport of such molecules through epithelial barriers is slow resulting in poor absorption in the blood stream. Further, these molecules are subject to the hepatogastrointestinal first-pass elimination. Hence, the physico-chemical properties of large molecules prohibit effective delivery through the oral route. SLNs show promising options for oral delivery of these molecules since the molecules can be completely "masked" in the gastrointestinal tract.[21] In the past two decades, several papers have been published detailing the successful formulation of proteins and peptides using SLNs.[22]

The use of SLNs is one of the potential drug delivery methods for targeted brain drug delivery.[23,24] Tightness of the endothelial vascular lining, referred to as the blood–brain barrier (BBB), provides a challenge to development of brain drug delivery. In the BBB, capillary endothelial cells protect the brain from foreign substances in the blood stream while letting required nutrients pass through for proper brain functioning. The BBB strictly limits transport into the brain through both physical (tight junctions) and metabolic (enzymes) barriers. Hence, passing the BBB to deliver the required drug molecules is an extremely challenging task. With the increase in lipophilicity exhibited by SLNs, permeability through the BBB can be increased. However, the permeability of drugs with molecular weight larger than 400 Da is low irrespective of its lipophilicity. Overall, SLNs have several advantages due to their size (nanometer-sized particles can easily bypass liver and spleen filtration) and lipophilicity (enhanced permeation and controlled release up to several weeks). Attaching ligands to SLNs for targeted drug delivery is relatively easy due to the versatile chemical properties of the lipids. With mixtures of lipids, it is possible to incorporate hydrophilic as well as lipophilic drugs.

50.3.2 Nanostructured Lipid Carrier

The second generation SLNs were developed to overcome the limitations of SLNs and they were termed as the nanostructured lipid carriers (NLCs). The primary difference between a SLN and a NLC is that the NLC also contains lipid in a liquid form. The NLC matrix is a solid at room temperature, but the formulation contains one or more liquid lipids adsorbed onto the solid lipids. Hence, to dissolve or disperse the drug molecule, NLCs offer more versatility in physico-chemical properties than SLNs. Since a larger pool of lipids can be used for formulation, drug loading capacity of NLCs is greater than SLNs and drug absorption and targeting can be greatly enhanced by the greater number of available lipids.

Several papers have been published presenting successful formulations of pharmaceutical drugs using the NLC approach. Itraconazole was formulated for pulmonary delivery using a hot high pressure homogenization technique.[20] The stability and physical properties of drug-loaded NLCs were found to be stable during nebulization. Hence, this drug delivery system was found to be promising for the pulmonary application of this poorly soluble drug.[20]

A comparison of the SLN and NLC drug delivery system was done by formulating clotrimazole using both systems. In both types of drug delivery systems, the formulation was developed by producing a single phase solid solution of drug and lipids. The drug release from NLCs was found to be faster than that of SLNs and also the stability of the NLC formulation was better than SLNs at room temperature.[25]

50.4 LIPOSOMES, POLYMEROSOMES, AND NIOSOMES

Liposomes, polymerosomes, and niosomes are part of the same structural family. Their properties, methods of manufacturing, and typical biopolymers used in formulations are discussed in this section.

50.4.1 Liposomes

50.4.1.1 Description and Method of Manufacturing

Liposomes are spherical vesicles consisting of a lipid bilayer that encapsulates water at the center. Water soluble drugs can be incorporated into the aqueous center of the liposome and lipophilic drugs can be incorporated into the lipid bilayer and hence liposomes can carry both hydrophilic as well as lipophilic drugs. The typical diameter of multilamellar vesicles (consisting of several concentric bilayers) ranges in size from 500 to 5000 nm, while the typical diameter of unilamellar (formed by a single bilayer) vesicles is around 100 nm in size. Figure 50.3 shows a typical liposome structure of multilamellar and unilamellar vesicles.

The predominant physical and chemical properties of a liposome are based on the net properties of the constituent phospholipids, including permeability, charge density, and steric hindrance. Physico-chemical properties of liposomes such as size, charge, and surface properties can be easily

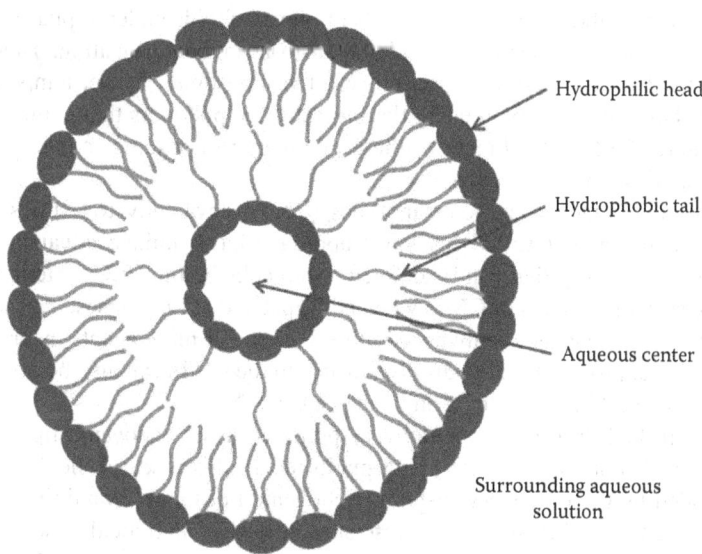

FIGURE 50.3 Liposome structure used for drug delivery systems.

manipulated by changing the composition of excipients in the lipid mixture before liposome preparation and/or by varying preparation methods.

Bangham published in 1967[26] that phospholipids in aqueous systems can form closed bilayered structures and since then numerous attempts have been made to use liposomes for pharmaceutical drug delivery. There are four commonly used methods to produce drug-loaded liposomes:

1. Liposomes are formed in an aqueous solution saturated with a soluble drug.
2. Use of organic solvents and solvent exchange mechanisms.
3. Use of lipophilic drugs.
4. pH gradient methods.

Due to interactions between water molecules and the hydrophobic phosphate groups of the phospholipids, the lipid bilayer closes in on itself. This process of liposome formation is spontaneous because the amphiphilic phospholipids self-associate into bilayers. Initially liposomes were made of phospholipids from the egg yolk but now with advances in materials science, a variety of synthetic materials are being used to produce liposomes.

Since liposomes are typically made from mostly naturally occurring lipids, they are biocompatible making them attractive for drug delivery. Liposome-incorporated pharmaceutical drugs are protected from the inactivating effect of external conditions; therefore drug stability and permeability are enhanced. One of the drawbacks of early liposomal systems in biological models was rapid excretion due to the small sizes. This challenge was overcome by attaching the polyethylene glycol molecule (PEGylation) to the liposome which makes it a so-called "stealth" molecule reducing opsonization (destruction by a phagocyte) and hence reduces clearance by the MPS, increasing the circulation half-life. Opsonization presents such a problem to the development of therapeutically useful liposomes that nearly all research reported in the literature involves PEG-coated or PEGylated liposomes.[27]

50.4.1.2 Formulations

Thousands of papers have been published presenting liposome-based drug delivery systems to overcome issues associated with pharmaceutical drugs like poor bioavailability, targeting of the drug to specific tissues, food effect, and poor stability in the gastrointestinal tract during oral drug delivery.

It is well known that membranes of various kinds of phospholipids undergo phase transitions, such as a gel-to-liquid crystalline transition and a lamellar-to-hexagonal transition. These phase transitions have been used for the design of stimuli-sensitive liposomes.[28] The, temperature sensitization of liposomes has been attempted using thermo-sensitive polymers that exhibit a lower critical solution temperature (CST). Attachment of these polymers to liposomes could give temperature-sensitive functionalities to the liposomes.[28]

Karchemski et al.[29] attached carbon nanotubes (CNTs) covalently to a liposome to produce CNTs–liposomes conjugates (CLC). This novel approach has a unique advantage of being able to load large amounts of drug that can be delivered into cells by the CLC system, thus preventing potential adverse systemic effects of CNTs when administered at high doses. This system provided a versatile and controlled means of enhanced delivery of one or more agents incorporated with the liposomes. With this approach, binding different drugs to the CNTs can also be possible by binding liposomes loaded with different contents on the same CNT.[29]

Ultrasound-controlled drug and gene delivery prepared using echogenic liposomes is gaining popularity. More and more articles are being published in this research field due to the unique advantages provided by these delivery systems. Echogenic liposomes containing entrapped therapeutic agents can be targeted to specific disease sites, allowing high local concentrations and low systemic toxicity. These types of liposomes carry an inert gas in the center of the liposome, which causes a burst due to ultrasonic pulses. Cavitation caused by ultrasound can destroy liposomes and cause local dose dumping locally at the targeted cell.[30]

Date et al.[31] presents a comprehensive review of attempts made to deliver drugs using liposomes and various biomaterials used to prepare complex liposomes in the treatment of parasitic diseases including malaria, leishmaniasis, and trypanosomiasis. Proteins and peptides are increasingly recognized as potential leads for the development of new therapeutics for a variety of human ailments. Proteins and peptides are typically delivered intravenously due to their poor stability in gastrointestinal fluids and poor permeability through gastrointestinal membranes. Liposomes provide an excellent alternative to deliver proteins and peptides orally since they can "hide" them during transport and limit the first-pass effect. Forssen and Willis[32] reviewed current efforts in using liposomes for targeted drug delivery by site-specific targeting using folate receptors, tailoring the properties of the liposomes to make them cell adhesion molecules, and targeting methods such as antibody targeting.

Liposomes are also used to deliver drugs to the lungs, as reviewed by Kellaway and Farr.[33] Lymphatic drug delivery using liposomes was reviewed by Forrest et al.[34] Drug delivery to macrophages using liposomes and the role of the physico-chemical properties such as surface charge, liposome size, concentration, composition, and inclusion of ligands on liposomes for uptake by macrophages are reviewed extensively by Ahsan et al.[35]

50.4.2 POLYMEROSOMES

50.4.2.1 Description and Method of Manufacturing

The basic structure of polymerosomes is similar to liposomes (Figure 50.3). Since polymers offer higher stability and structural variability compared to lipids, polymeric vesicles can eliminate the disadvantages associated with liposomes such as short half-life and fewer excipient choices. Polymers offer versatility of form and function due to the possibility of modification of tectons and hydrophilic block length, composition, and chemical structure. Since composition and molecular weight of the polymers can be varied, it allows not only the preparation of polymerosomes with different properties and responsiveness to stimuli but also with different membrane thicknesses and permeabilities. Polymerosomes typically have relatively thick and robust membranes (>50 nm) formed by amphiphilic block copolymers with a relatively high molecular weight. Relatively long blood circulation times of polymerosomes can be accomplished by the introduction of a hydrophilic surface layer [e.g., poly(ethylene glycol) (PEG) blocks].[36]

Polymerosomes are synthesized using various methods including self-assembly and polymer rehydration techniques. In the self-assembly technique, the block copolymers are first dissolved in organic solvents to form a single phase solution. The aqueous phase is then slowly added to precipitate the polymer in the form of a polymerosome. Since the polymer precipitation is uncontrolled, the particle size distribution varies depending on the rate of addition of water and the inherent properties of the polymer to precipitate out of the particular organic solvent. In the case of the polymer rehydration technique, the amphiphilic block copolymer film is hydrated to induce self-assembly. The block copolymers are dissolved in a suitable organic solvent similar to the self-assembly technique. The organic solvent is then evaporated by a technique such as rotovap to produce a thin dry polymeric film. The film is then hydrated by the addition of water.[37]

50.4.2.2 Formulations

The skin permeability of polymerosomes is excellent, and several articles have been published detailing drug delivery through the skin.[38] For a topical gel formulation, triblock copolymers of PEG and PCL (PCL–PEG–PCL) were synthesized by ring opening polymerization of ε-caprolactone with PEG. The particle size of these vesicles was found to be around 75–140 nm. The self-assembly technique using an acetone–water system was used. A skin permeation study of these polymer vesicles revealed the accumulation of these vesicles in the furrows of the stratum corneum. This research showed that drug delivery through the skin in the form of topical gel is possible.[39]

Rastogi et al.[40] published a study on enhancing the pharmacological efficacy of protein-loaded PCL–PEG-based polymerosomes. Insulin was selected as the model protein and was complexed with sodium deoxycholate, a naturally occurring bile salt. The prepared surfoplexes were efficiently encapsulated in PCL- and PEG-based polymer vesicles. It was found that with this drug delivery system, the initial burst effect was greatly reduced, and the therapeutic effect of insulin was increased by 2 h.

50.4.3 Niosomes

50.4.3.1 Description and Method of Manufacturing

Niosomes are similar to liposomes (Figure 50.3) and can be used to deliver both hydrophilic as well as lipophilic drug molecules. Niosomes are non-ionic surfactant vesicles. Niosome-based drug delivery is a relatively new field and few articles have been published. Similar to liposomes, niosomes can be formulated to produce prolonged systemic circulation of the drug and enhanced penetration into targeted tissues. Niosomes generally consist of a vesicle formed using non-ionic surfactants (e.g., Span-60) stabilized by the addition of cholesterol and a small amount of an anionic surfactant.[41]

50.4.3.2 Formulations

Kumar et al.[41] presented a study demonstrating the successful preparation of aceclofenac niosomes and their evaluation. Non-ionic surfactants (Span 60 and Span 20) along with cholesterol were used to produce niosomes. Excellent drug entrapment efficiency and controlled in vitro release were presented in this research article. Das and Palei[42] published the research on the encapsulation of rofecoxib in sorbitan ester niosomes. Niosomes prepared using Span 60, Span 20, and cholesterol were incorporated into a carbopol gel to produce a topical gel-based drug delivery system. The lower flux value of the niosomal gel as compared to the plain drug gel in pig skin indicated that the lipid bilayer of niosome was rate limiting in drug permeation and hence the controlled release was achieved.[42]

Sathali et al.[43] formulated a topical gel containing clobetasol propionate niosomes to prolong the duration of action and prevent side effects. The clobetasol propionate niosomes were prepared by altering the ratio of various non-ionic surfactants (Span 40, 60, and 80) to cholesterol by the thin film hydration method. The in vivo results showed that the niosomal gel had a sustained as well as a prolonged action.

Terzano et al.[44] developed Polysorbate 20 niosomes containing beclomethasone dipropionate for pulmonary delivery to patients with chronic obstructive pulmonary disease. Singh et al.[45] prepared niosomes containing the anti-inflammatory drug nimesulide and tested its physico-chemical properties including stability and in vitro release. It was shown that in vitro sustained drug release can be achieved with this approach. The research literature for niosomes in ocular drug delivery has been reviewed by Kaur et al.[46]

Niosome-based drug delivery for the oral delivery of peptides was investigated by Dufes et al.[47] Due to the BBB, vasoactive intestinal peptide (VIP), which is used for the treatment of Alzheimer's disease, has poor bioavailability. Also the VIP has a very short half-life which makes it necessary to increase the dose frequency. Dufes et al. developed glucose-bearing niosomes that encapsulate the VIP for delivery to specific brain areas. The authors concluded that glucose-bearing vesicles represent a novel tool for delivery of drugs across the BBB.[47]

50.5 LIPID–POLYMER HYBRIDS

50.5.1 Description and Method of Manufacturing

In lipid–polymer hybrid drug delivery systems, the drug is incorporated into a polymer in a matrix form or in a solubilized form and then the polymer particle is coated with a lipid film (Figure 50.4). Since drugs typically leak through polymeric matrix systems showing an initial burst in the release, such lipid film over a polymeric particle inhibits the drug diffusion outward. Polymeric nanoparticles provide robustness to the drug delivery system while the lipid coating provides the necessary lipophilicity to block the drug leakage.[48] Drug delivery of water soluble drugs to produce controlled or targeted release can be achieved using this system. Thus, in this drug delivery system, the polymeric core functions to encapsulate hydrophilic or lipophilic drug molecules and provides a robust structure, whereas the external lipid coating functions as a biocompatible shield, a template for surface modifications, and a barrier for preventing the fast leakage of water-soluble drugs.[49]

Several methods are used to produce lipid–polymer hybrids. Spray drying and spray freeze drying are two of the most attractive methods since they are cost-effective and readily scalable. In these methods, the drug and the polymer are dissolved in an organic solvent and then spray dried/spray freeze dried to produce matrix particles. The particles are then coated with a lipid film to produce

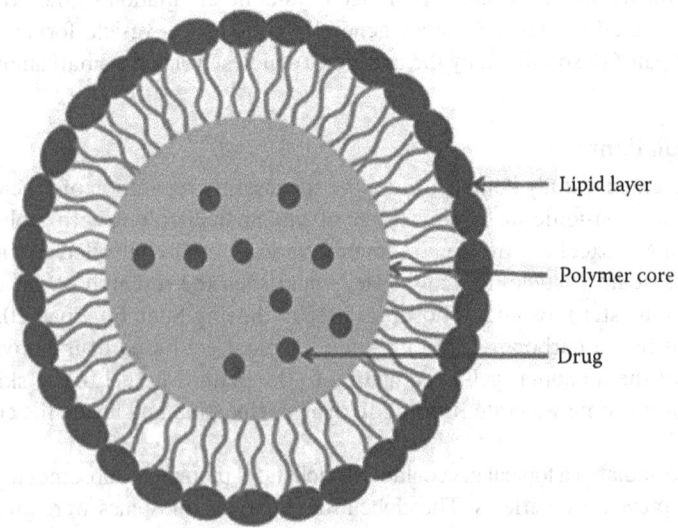

FIGURE 50.4 Lipid–polymer hybrid-based drug delivery system.

the lipid–polymer hybrids.[50] Spray freeze drying has been found to be more effective than spray drying since it produces porous nano-agglomerates which have superior properties (e.g., aqueous reconstitubility and aerosolization efficiency).[50]

50.5.2 Formulations

Lipid–polymer hybrid-based drug delivery systems have been studied largely to deliver water soluble drugs for targeted drug delivery. Cheow and Hadinoto[51] described the preparation of stable lipid–polymer hybrid nanoparticles encapsulating LEV, CIP, and OFX antibiotics. In this study, an extensive study of the effects of factors such as process parameters and excipients on drug loading and stability of the drug delivery system was carried out. The use of solvents and surfactants and the ionic interaction between the drug and the lipid were found to be most influential in preparing successful lipid–polymer hybrids with optimum drug loading.

Farokhzad et al.[52] reported the immunological characterization of lipid–polymer hybrid nanoparticles and proposed a method to control the levels of complement activation induced by these nanoparticles. In this method, the nanoparticle surface was modified by attaching methoxyl, carboxyl, and amine groups. It was found that the surface chemistry significantly affects human plasma and serum protein adsorption patterns.[52]

Wu et al.[53] demonstrates that a lipid–polymer hybrid drug delivery system loaded with doxorubicin is effective for tumor treatment in a well-established animal model. Tumor growth delay and tumor necrosis were observed in tumors treated with the lipid–polymer hybrid formulation of doxorubicin. It was found that these lipid–polymer hybrids carrying anticancer agents were useful for loco-regional treatment of breast cancer with an improved therapeutic index.

50.6 DENDRIMERS

50.6.1 Description and Method of Manufacturing

Dendrimers are globular in shape and have a three-dimensional highly controlled structure. These nanostructures have a narrow mass and size distribution.[54] Dendrimers are the fourth new class of polymer architectures after traditional linear, cross-linked, and branched types. Dendrimers can be used as well-defined scaffolding or as nanocontainers to conjugate, complex, or encapsulate therapeutic drugs or imaging moieties. Due to their structural and highly diversified physico-chemical properties, dendrimers are gaining popularity in pharmaceutical and biotech industries.

Dendrimers are synthesized using a bottom-up approach where polymeric layers are added stepwise around a central core. Each additional iteration leads to higher generation of dendrimers. Dendrimers can be constructed with a range of molecular weights and have a polyfunctional surface that facilitates the attachment of drugs and pharmacokinetic modifiers such as PEG or targeting moieties. Two distinct methods exist to synthesize dendrimers: The divergent approach developed by Tomalia[55] and Newkome[56] and the convergent approach developed by Hawker and Frechet:[57]

1. Divergent synthesis: In this type of synthesis the dendrimer growth starts from a polyfunctional core and proceeds radially outward with the stepwise addition of successive layers of building blocks. This approach involves assembling of monomeric modules into a radial, branch-upon-branch motif according to certain dendritic rules and principles.[55] Figure 50.5 shows a typical reaction and growth of the dendrimer.
2. Convergent synthesis: In this type of synthesis several dendrons (small functional groups) are reacted with a multifunctional core to obtain a final dendrimeric structure. Figure 50.6 shows a typical reaction and growth of the dendrimer. Hence, the growth begins at the surface of the dendrimer, and continues inward by gradually linking surface units together with more monomers.[55]

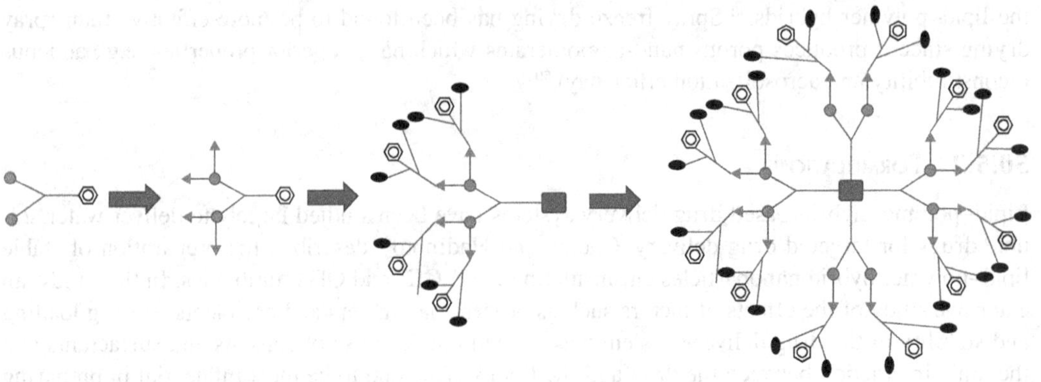

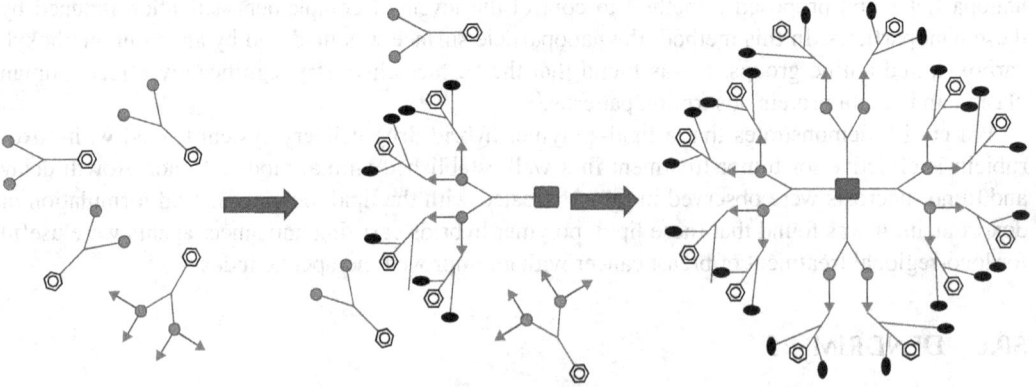

FIGURE 50.5 Divergent synthesis of dendrimers.

FIGURE 50.6 Convergent synthesis of dendrimers.

Since the conception of the dendrimers, a variety of dendrimeric structures have evolved as evidenced by the vast research in this field. Typical structures are liquid crystalline dendrimers, tecto-dendrimers, chiral dendrimers, PAMAM dendrimers, PAMAMOS dendrimers, hybrid dendrimers, peptide dendrimers, and glycodendrimers. Nanjwade et al.[58] reviewed these structures, biomaterials used, their properties, and applications. Various theoretical and computational approaches have been devised to predict the properties of dendrimeric structures so that the effective drug delivery system can be produced. These approaches include solubility parameters, Flory–Huggins theory, analytical predictions of partition coefficients, and molecular simulations. Huynh et al.[59] provided a review of the approaches for the optimization of drug–material pairs using important performance-related parameters including the size of the delivery particles, their surface properties, and the compatibility of the materials with the drug to be sequestered. The author[59] found that the analytical models are useful for fast prescreening during the development of delivery systems and drug derivatives. Molecular simulations were found to be more reliable.

50.6.2 Formulations

Dendrimer-based drug delivery systems have been successful in both improving bioavailability as well as targeting drugs. The main advantages for the use of dendrimers as drug carriers reside in their multivalent design and their well-defined structure that results from the stepwise, iterative approach used in their preparation. The anionic and neutral dendrimers have been shown to be less toxic than the cationic dendrimers but at the same time they suffer from lower permeability compared to cationic dendrimers. Typically the smaller size of dendrimers

(<8 nm) facilitates rapid renal clearance which in turn reduces toxicity. Also, surface coverage of the dendrimers with biocompatible groups such as PEG chains greatly reduced their cytotoxicity.[60]

Drugs are encapsulated in dendrimers primarily using two methods: by covalent dendrimer–drug conjugation or non-covalent encapsulation of drugs.

1. Covalent dendrimer–drug conjugation: In this method, a degradable covalent linkage between the drug and the dendrimer is formed to obtain a stronger relationship between the dendrimer and the drug. The drug release from a conjugate is governed largely by the nature of the linking bond or spacer between the drug and the scaffold and the targeted physiological domain for intended release. Ester and amide bonds might be cleavable by enzymes or under hydrolytic conditions; however, ester cleavage is generally more facile than amide cleavage for releasing drugs.[61]
2. Non-covalent drug encapsulation: In this method, the drug is physically encapsulated in a dendrimeric structure by virtue of a micellar structure. Although it is relatively easy to produce a drug delivery system with this approach, the drug release is often uncontrolled.[61]

Yang and Lopina[62] used Starburst® PAMAM dendrimers for delivery of Penicillin V. Drug delivery scaffolds were designed and built based on the polyethylene glycol polyamidoamine (PEG–PAMAM) star polymer. Penicillin V was coupled with the PAMAM dendrimer through a PEG spacer. The drug delivery scaffold based on the PEG–polyamidoamine star polymer proved to be feasible. The star polymers provided accessible ends which were used to chemically couple the desired small molecules. Ihre et al.[63] designed and synthesized dendritic polyester systems based on the 2,2-bis(hydroxymethyl) propanoic acid monomer unit as a possible versatile drug carrier. These systems were designed to meet the needs of their intended use as drug carriers, including water solubility, non-toxicity, and stability of the polymeric backbone. Several formulations were tried successfully that led to an improved solubility. Attachment of drugs to high molecular weight polymers can significantly improve both tumor targeting and therapeutic efficacy due to the enhanced permeability and the retention effect observed in the tumor tissue.

Kaminskas and Porter[64] reviewed the research work published on lymphatic drug delivery using dendrimers. It was concluded that after subcutaneous administration, larger dendrimers (>8 nm) were found to be drained from the injection site into the peripheral lymphatic capillaries and therefore have potential as lymphatic imaging agents for magnetic resonance and optical fluorescence lymphangiography and as vectors for drug-targeting to lymphatic sites of disease progression. Increasing hydrophilicity and reducing surface charge were found to enhance the drainage from subcutaneous injection sites. Specific access to the lymphatics from the interstitium is dictated by size where larger macromolecules or colloids are precluded from ready access to the vascular capillaries and instead enter the lymphatics via the large inter-endothelial gaps that open under the conditions of lymphatic filling.

Tremendous efforts have been taken to deliver proteins, peptides, genes, and antiretroviral drugs using dendrimers. Several authors[65-67] have reviewed various targeted drug delivery systems using dendrimers like folic acid–dendrimer conjugates, peptide–dendrimer conjugates, monoclonal antibody conjugated to PAMAM, glycosylation of dendrimers, boron neutron capture therapy, photothermal therapy, and gene therapy. Ojewole et al.[68] provided review articles on dendrimer-based antiretroviral drug delivery and Oliveir et al.[69] reviewed articles on dendrimer-based drug delivery for regenerative medicine.

Despite all the research efforts by the research community, it is still a challenge to prepare dendritic polymers that circulate in the blood long enough to accumulate at target sites, but that can also be eliminated from the body at a reasonable rate to avoid long-term buildup. In addition, the tissue localization of dendritic polymers is still difficult to predict in advance and more studies are required to determine the effect of peripheral dendritic groups on these properties.[70]

50.7 CONCLUSIONS

Each of the drug delivery system described in this review has unique requirements for biomaterials. For instance, drug–polymer conjugate systems offer unique advantages like ease of synthesis but the success of the formulation depends on the stability of hydrogen bonds between the drug and the polymer. SLNs and NLCs can mask the drug molecules completely and therefore are excellent in enhancing the drug permeability. However, an in-depth knowledge of lipids and surfactants is essential to form a solid solution of drug–lipid and nano-sized particles. Drug delivery systems like liposomes have been shown to be excellent carriers for targeted drug delivery as well as for enhancing bioavailability of drug molecules. PEGylation makes the drug delivery system "stealth" reducing the first-pass effect and increasing the permeability through the gastrointestinal lumen. Hence, PEGylation of drugs and carrier polymers is one of the most successful approaches to enhance bioavailability. Dendrimers provide the highest possible flexibility to the formulation scientist and can handle a variety of drug molecules, genes, imaging markers, etc.

REFERENCES

1. Hans, M.L., Lowman, A.M. Biodegradable nanoparticles for drug delivery and targeting. *Curr. Opin. Solid State Mat. Sci.* **2002**, *6* (4), 319–327.
2. Salata, O.V. Applications of nanoparticles in biology and medicine. *J. Nanobiotechnol.* **2003**, *2* (3), 1–6.
3. Schiavone, H., Palakodaty, S., Clark, A., York, P. Evaluation of SCF engineered particle-based lactose blends in passive dry powder inhalers. *Int. J. Pharm.* **2004**, *281* (1–2), 55–66.
4. Ringsdorf, H. Synthetic polymeric drugs. In *Polymeric Delivery Systems*; Gordon and Breach Science Publishers, Inc: New York, 1978, 197–225.
5. Ringsdorf, H. Structure and properties of pharmacologically active polymers. *J. Polym. Sci.* **1975**, *51* (1), 135–153.
6. Pasut, G., Veronese, F.M. Polymer–drug conjugation, recent achievements and general strategies. *Prog. Polym. Sci.* **2007**, *32* (8–9), 933–961.
7. Minko, T. Soluble polymer conjugates for drug delivery. *Drug Discov. Today Technol.* **2005**, *2* (1), 15–20.
8. Elvira, C., Gallardo, A., Cifuentes, A. Covalent polymer-drug conjugates. *Molecules* **2005**, *10* (1), 114–125.
9. Vasey, P., Kaye, S., Morrison, R., Twelves, C., Wilson, P., Duncan, R., Thomson, A. et al. Phase I clinical and pharmacokinetic study of PK1 [N-(2-hydroxypropyl)methacrylamide copolymer doxorubicin]: First member of a new class of chemotherapeutic agents-drug-polymer conjugates. *Clin. Cancer Res.* **1999**, *5* (1), 83–94.
10. Meerum Terwogt, J.M., ten Bokkel Huinink, W.W., Schellens, J.H., Schot, M., Mandjes, I.A., Zurlo, M.G., Rocchetti, M., Rosing, H., Koopman, J., Beijnen, J. Phase I clinical and pharmacokinetic study of PNU166945, a novel water soluble polymer conjugated prodrug of paclitaxel. *Anticancer Drugs* **2001**, *12* (4), 315–323.
11. Segal, E., Satchi-Fainaro, R. Design and development of polymer conjugates as anti-angiogenic agents. *Adv. Drug Deliv. Rev.* **2009**, *61* (13), 1159–1176.
12. Khandare, J., Minko, T. Polymer–drug conjugates: Progress in polymeric prodrugs. *Prog. Polym. Sci.* **2006**, *31* (4), 359–397.
13. Kowalczuk, A., Stoyanova, E., Mitova, V., Shestakova, P., Momekov, G., Momekova, D., Koseva. N. Star-shaped nano-conjugates of cisplatin with high drug payload. *Int. J. Pharm.* **2011**, *404* (1–2), 220–230.
14. Tomalia, D.A., Baker, H., Dewald, J., Hall, M., Kallos, G., Martin, S., Roeck, J., Ryder, J., Smith, P. A new class of polymers—starburst-dendritic macromolecules. *Polym. J.* **1985**, *17* (1), 117–132.
15. Souto, E.B., Muller, R.H. Lipid nanoparticles: Effect on bioavailability and pharmacokinetic changes. *Hand. Exp. Pharmacol.* **2010**, *197*, 115–141.
16. Speiser, P. Lipid nanopellets als Tragersystem fur Arzneimittel zur peroralen Anwendung. European Patent EP0167825, 1990.
17. Gasco, M. Method for producing solid lipid microspheres having a narrow size distribution. United States Patent US5188837, 1993.
18. Domb, A. Liposheres for controlled delivery of substances. United States Patent US5188837, 1993.

19. Mehnert, W., Mader, K. Solid lipid nanoparticles, production, characterization and applications. *Adv. Drug Deliv. Rev.* **2012**, *64* (Suppl.), 83–101.
20. Pardeike, J., Weber, S., Haber, T., Wagner, J., Zarfl, H., Plank, H., Zimmer, A. Development of an Itraconazole-loaded nanostructured lipid carrier (NLC) formulation for pulmonary application. *Int. J. Pharm.* **2011**, *419* (1–2), 329–338.
21. Ugwoke, M., Agu, R., Verbeke, N., Kinget, R. Nasal mucoadhesive drug delivery: Background, applications, trends and future perspectives. *Adv. Drug Deliv. Rev.* **2005**, *57* (11), 1640–1665.
22. Almeida, A., Souto, E. Solid lipid nanoparticles as a drug delivery system for peptides and proteins. *Adv. Drug Deliv. Rev.* **2007**, *59* (6), 478–490.
23. Blasi, P., Giovagnoli, S., Schoubben, A., Ricci, M., Rossi. C. Solid lipid nanoparticles for targeted brain drug delivery. *Adv. Drug Deliv. Rev.* **2007**, *59* (6), 454–477.
24. Fundaro, A., Cavalli, R., Bargoni, A., Vighetto, D., Zara, G., Gasco, M. Non-stealth and stealth solid lipid nanoparticles (SLN) carrying doxorubicin: Pharmacokinetics and tissue distribution after i.v. administration to rats. *Pharmacol. Res.* **2000**, *42* (4), 337–343.
25. Das, S., Ng, W.K., Tan, R.B. Are nanostructured lipid carriers (NLCs) better than solid lipid nanoparticles (SLNs): Development, characterizations and comparative evaluations of clotrimazole-loaded SLNs and NLCs? *Eur. J. Pharm. Sci.* **2012**, *47* (1), 139–151.
26. Bangham, A., Horne, R. Negative staining of phospholipids and their structural modification by surface-active agents as observed in the electron microscope. *J. Mol. Biol.* **1964**, *8* (5), 660–668.
27. Moghimia, S., Szebeni, J. Stealth liposomes and long circulating nanoparticles: Critical issues in pharmacokinetics, opsonization and protein-binding properties. *Prog. Lipid Res.* **2003**, *42* (6), 463–478.
28. Yatvin, M., Weinstein, J., Dennis, W., Blumenthal, R. Design of liposomes for enhanced local release of drugs by hyperthermia. *Science* **1978**, *202* (4374), 1290–1293.
29. Karchemski, F., Zucker, D., Barenholz, Y., Regev, O. Carbon nanotubes-liposomes conjugate as a platform for drug delivery into cells. *J. Control. Rel.* **2012**, *160* (2), 339–345.
30. Huang, S.L. Liposomes in ultrasonic drug and gene delivery. *Adv. Drug Deliv. Rev.* **2008**, *60* (10), 1167–1176.
31. Date, A., Joshi, M., Patravale, V. Parasitic diseases: Liposomes and polymeric nanoparticles versus lipid nanoparticles. *Adv. Drug Deliv. Rev.* **2007**, *59* (6), 505–521.
32. Forssen, E., Willis, M. Ligand-targeted liposomes. *Adv. Drug Deliv. Rev.* **1998**, *29* (3), 249–271.
33. Kellaway, I., Farr, S. Liposomes as drug delivery systems to the lung. *Adv. Drug Deliv. Rev.* **1990**, *5* (1–2), 149–161.
34. Cai, S., Yang, Q., Taryn, R., Bagby, M., Forrest, L. Lymphatic drug delivery using engineered liposomes and solid lipid nanoparticles. *Adv. Drug Deliv. Rev.* **2011**, *63* (10–11), 901–908.
35. Ahsan, F., Rivas, I.P., Khan, M.A., Torres Suarez, A.I. Targeting to macrophages: Role of physicochemical properties of particulate carriers-liposomes and microspheres-on the phagocytosis by macrophages. *J. Control. Rel.* **2002**, *79* (1–3), 29–40.
36. Romberg, B., Oussoren, C., Snel, C., Hennink, W., Storm, G. Effect of liposome characteristics and dose on the pharmacokinetics of liposomes coated with poly(amino acid)s. *Pharm. Res.* **2007**, *24* (12), 2394–2401.
37. Kita-Tokarczyk, K., Grumelard, J., Haefele, T., Meier, W. Block copolymer vesicles—Using concepts from polymer chemistry to mimic biomembranes. *Polymer* **2005**, *46* (11), 3540–3563.
38. Torchilin, V. Micellar nanocarriers: Pharmaceutical perspectives. *Pharm. Res.* **2007**, *24* (1), 1–16.
39. Rastogi, R., Ananda, S., Koul, V. Flexible polymerosomes—An alternative vehicle for topical delivery. *Coll. Surf. B Biointerf.* **2009**, *72* (1), 161–166.
40. Rastogi, R., Anand, S., Koul, V. Evaluation of pharmacological efficacy of 'insulin–surfoplex' encapsulated polymer vesicles. *Int. J. Pharm.* **2009**, *373* (1–2), 107–115.
41. Srinivas, S., Kumar, Y., Hemanth, A., Anitha, M. Preparation and evaluation of niosomes containing aceclofenac. *Dig. J. Nanomater. Biostruct.* **2010**, *5* (1), 249–254.
42. Das, M., Palei, N. Sorbitan ster niosomes for topical delivery of rofecoxib. *Indian J. Exp. Biol.* **2011**, *49* (6), 438–445.
43. Lingan, M., Sathali, A., Kumar, M., Gokila, A. Formulation and evaluation of topical drug delivery system containing clobetasol propionate niosomes. *Sci. Rev. Chem. Commun.* **2011**, *1* (1), 7–17.
44. Terzano, C., Allegra, L., Alhaique, F., Marianecci, C., Carafa, M. Non-phospholipid vesicles for pulmonary glucocorticoid delivery. *Eur. J. Pharm. Biopharm.* **2005**, *59* (1), 57–62.
45. Singh, C., Jain, C., Kumar, B. Formulation, characterization, stability And in vitro evaluation of nimesulide niosomes. *Pharmacophore* **2011**, *2* (3), 168–185.

46. Kaur, I., Garg, A., Singl, A., Aggarwal, D. Vesicular systems in ocular drug delivery: An overview. *Int. J. Pharm.* **2004**, *269* (1), 1–14.
47. Dufes, C., Gaillard, F., Uchegbu, I.F., Schatzlein, A.G., Olivier, J.C., Muller, J.M. Glucose-targeted niosomes deliver vasoactive intestinal peptide (VIP) to the brain. *Int. J. Pharm.* **2004**, *285* (1–2), 77–85.
48. Alphandary, P., Andremont, A., Couvreur, P. Targeted delivery of antibiotics using liposomes and nanoparticles: Research and applications. *Int. J. Antimicrob. Agents* **2000**, *13* (3), 155–168.
49. Cheow, W., Hadinoto, K. Enhancing encapsulation efficiency of highly watersoluble antibiotic in poly(lactic-*co*-glycolic acid) nanoparticles: modifications of standard nanoparticle preparation methods. *Coll. Surf. A Physicochem. Eng. Asp.* **2010**, *370* (1–3), 79–86.
50. Wang, Y., Kho, K., Cheow, W.S., Hadinoto, K. A comparison between spray drying and spray freeze drying for dry powder inhaler formulation of drug-loaded lipid–polymer hybrid nanoparticles. *Int. J. Pharm.* **2012**, *424* (1–2), 98–106.
51. Cheow, W.S., Hadinoto, K. Factors affecting drug encapsulation and stability of lipid–polymer hybrid nanoparticles. *Coll. Surf. B Biointerf.* **2011**, *85* (2), 214–220.
52. Salvador-Morales, C., Zhang, L., Langer, R., Farokhzad, O. Immunocompatibility properties of lipid–polymer hybrid nanoparticles with heterogeneous surface functional groups. *Biomaterials* **2009**, *30* (12), 2231–2240.
53. Wong, H., Rauth, A.M., Bendayan, R., Wu, X.Y. In vivo evaluation of a new polymer-lipid hybrid nanoparticle (PLN) formulation of doxorubicin in a murine solid tumor model. *Eur. J. Pharm. Biopharm.* **2007**, *65* (3), 300–308.
54. Kolhe, P., Khandare, J., Pillai, O., Kannan, S., Lieh-Lai, M., Kannan, R. Preparation, cellular transport, and activity of polyamidoamine-based dendritic nanodevices with a high drug payload. *Biomaterials* **2006**, *27* (4) 660–669.
55. Tomalia, D.A. A new class of polymers: Starburst-dendritic macromolecules. *Polym. J.* **1985**, *17* (1), 117–132.
56. Newkome, G. Cascade molecules: A new approach to micelles. *J. Org. Chem.* **1985**, *50* (11), 2003–2004.
57. Hawker, C., Frechet, J. Preparation of polymers with controlled molecular architecture. A new convergent approach to dendritic macromolecules. *J. Am. Chem. Soc.* **1990**, *112* (21), 7638–7647.
58. Nanjwade, B.K., Bechraa, H.M., Derkara, G.K., Manvi, F.V., Nanjwade, V.K. Dendrimers: Emerging polymers for drug-delivery systems. *Eur. J. Pharm. Sci.* **2009**, *38* (3), 185–196.
59. Huynh, L., Neale, C., Pomes, R., Allen, C. Computational approaches to the rational design of nanoemulsions, polymeric micelles, and dendrimers for drug delivery. *Nanomedicine* **2012**, *8* (1), 20–36.
60. Kurtoglu, Y., Mishra, M., Kannan, S., Kannan, R. Drug release characteristics of PAMAM dendrimer-drug conjugates with different linkers. *Int. J. Pharm.* **2010**, *384* (1–2), 189–194.
61. Liu, M., Frechet, J. Designing dendrimers for drug delivery. *Pharm. Sci. Technol. Today* **1999**, *2* (10), 393–401.
62. Yang, H., Lopina, S. Penicillin V-conjugated PEG-PAMAM star polymers. *J. Biomater. Sci. Polym. Ed.* **2003**, *14* (10), 1043–1056.
63. Ihre, H.R., Padilla De Jesus, O.L., Szoka, F.C. Jr., Frechet, J.M. Polyester dendritic systems for drug delivery applications: Design, synthesis, and characterization. *Bioconjug. Chem.* **2002**, *13* (3), 443–452.
64. Kaminskas, L., Porter, C. Targeting the lymphatics using dendritic polymers (dendrimers). *Adv. Drug Deliv. Rev.* **2011**, *63* (10–11), 890–900.
65. Wolinsky, J., Grinstaff, M. Therapeutic and diagnostic applications of dendrimers for cancer treatment. *Adv. Drug Deliv. Rev.* **2008**, *60* (9), 1037–1055.
66. Wong, A., DeWit, M.A., Gillies, E. Amplified release through the stimulus triggered degradation of self-immolative oligomers, dendrimers, and linear polymers. *Adv. Drug Deliv. Rev.* **2012**, *64* (11), 1031–1045.
67. Dufes, C., Uchegbu, I., Schatzlein, A. Dendrimers in gene delivery. *Adv. Drug Deliv. Rev.* **2005**, *57* (15), 2177–2202.
68. Ojewole, E., Mackraj, I., Naidoo, P., Govender, T. Exploring the use of novel drug delivery systems for antiretroviral drugs. *Eur. J. Pharm. Biopharm.* **2008**, *70* (3), 697–710.
69. Oliveir, J.M., Salgado, A.J., Sous, N., Mano, J.P., Reis, R.L. Dendrimers and derivatives as a potential therapeutic tool in regenerative medicine strategies—A review. *Prog. Polym. Sci.* **2010**, *35* (9), 1163–1194.
70. Gillies, E., Frechet, J.M. Dendrimers and dendritic polymers in drug delivery. *Drug Discov. Today* **2005**, *10* (1) 35–43.

51 Polymer Systems for Ophthalmic Drug Delivery

Sepideh Khoee and Frazaneh Hashemi Nasr

CONTENTS

51.1 Introduction 1168
51.2 Anatomy of the Eye 1169
    51.2.1 Cornea 1169
    51.2.2 Conjunctiva 1170
    51.2.3 Nasolacrimal Drainage System 1170
    51.2.4 Tear Film 1170
51.3 Challenges to Ocular Drug Delivery 1171
    51.3.1 Tear Film Barrier 1171
    51.3.2 Corneal Barrier 1172
    51.3.3 Blood Barriers 1173
        51.3.3.1 Blood-Aqueous Barrier 1173
        51.3.3.2 Blood-Retinal Barrier 1174
51.4 Routes of Ocular Drug Delivery 1174
    51.4.1 Topical Ocular Delivery 1174
        51.4.1.1 Aqueous Solutions 1174
        51.4.1.2 Suspensions 1175
        51.4.1.3 Emulsions 1175
        51.4.1.4 Ointments 1175
        51.4.1.5 Gels 1176
        51.4.1.6 Sol to Gel Systems 1176
    51.4.2 Ocular Inserts 1176
        51.4.2.1 Soluble Inserts 1177
        51.4.2.2 Insoluble Inserts 1177
        51.4.2.3 Nonerodible Inserts: Ocusert, Contact Lens 1177
        51.4.2.4 Erodible Ophthalmic Inserts 1178
    51.4.3 Subconjunctival Administration 1180
    51.4.4 Intravitreal Administration 1180
    51.4.5 Implants 1181
    51.4.6 Iontophoresis 1181
    51.4.7 Sprays 1182
    51.4.8 Contact Lens-Based Ophthalmic Drug Delivery 1182
        51.4.8.1 Drug-Eluting Conventional Contact Lenses to Absorb and Release Ophthalmic Drugs 1183
        51.4.8.2 Polymeric Hydrogels for Piggyback Contact Lens Combining with a Drug Plate or Drug Solution 1186
        51.4.8.3 Soft Contact Lenses Bearing Surface-Immobilized Liposomes and Pendant Cyclodextrins 1187
        51.4.8.4 Molecularly Imprinted Polymeric Hydrogels 1191

51.5   Nanotechnology in Drug Delivery 1193
51.6   Nanotechnology and Ocular Drug Delivery 1193
51.7   Therapeutic Significance in Ocular Delivery 1194
51.8   Various Nanoparticulate-Based Ophthalmic Drug Delivery Systems 1194
        51.8.1   Liposomes 1195
        51.8.2   Microemulsions 1197
        51.8.3   Nanosuspensions 1198
        51.8.4   Dendrimers 1201
        51.8.5   Niosomes 1204
        51.8.6   Discomes 1205
        51.8.7   Nanoparticle-Loaded Contact Lenses 1205
        51.8.8   Polymeric Nanoparticles 1208
51.9   Different Properties of Nanoparticulate-Based Polymeric Systems in Ocular Therapy 1211
        51.9.1   Mucoadhesive Polymers as the Penetration Enhancers in Ocular Drug Delivery ... 1212
        51.9.2   Biodegradable Polymers for Ocular Drug Delivery 1213
51.10 Conclusion 1214
References 1214

51.1   INTRODUCTION

Ocular drug delivery is one of the most fascinating and challenging tasks being faced by the pharmaceutical researchers for the past 20 years. The reason is that the eye is one of the most complicated and sophisticated organs of the body, so it is important to give special attention to eye diseases.

Most ocular diseases like dryness, conjunctiva, and eye flu are treated by topical drug application in the form of solutions, suspensions, and ointment. In the earlier period, drug delivery to the eye has been limited to topical application, redistribution into the eye following systemic administration, or direct intraocular/periocular injections. However, one of the major barriers of ocular medication is to obtain and maintain a therapeutic level at the site of action for a prolonged period of time.

Absorption and elimination of therapeutic active agents depend upon the physiochemical, microbiological, and pharmaceutical properties of dosage form and also depend upon the eye anatomy and physiology.

Drug absorption occurs through corneal and non-corneal pathways. Most non-corneal absorptions occur via the nasolacrimal duct and lead to non-productive systemic uptake, while most drugs transported through the cornea are taken up by the targeted intraocular tissue. Unfortunately, corneal absorption is limited by the drainage of the instilled solutions, lacrimation, tear turnover, metabolism, tear evaporation, non-productive absorption/adsorption, limited corneal area, poor corneal permeability, binding by the lacrimal proteins, enzymatic degradation, and the corneal epithelium itself. Therefore, topical delivery via eye drops that accounts for approximately 90% of all ophthalmic formulations is extremely inefficient, and in certain instances leads to serious side effects. Upon instillation, the drug mixes with the fluid present in the tear film and has a short residence time of approximately 2 min in the film. During this time, approximately 2%–5% of most drugs reach the target tissue, and the remaining drugs enter the systemic circulation through conjunctival or nasal uptake. The drug absorbed into the bloodstream escapes the first-pass metabolism and enters all major organs, with a potential for side effects.

To enhance the amount of active substance reaching the target tissue or exerting a local effect in the cul-de-sac, the residence time of the drug in the tear film should be lengthened. Moreover, once-a-day formulations should improve patient compliance. Numerous strategies have been developed to improve the bioavailibility, residence time and patient compliance, as well as, in order to minimize the toxic effect, the negative side effects and frequency of dosing.

Herein, we are focusing on various new drug delivery systems such as inserts, contact lenses (CLs), mucoadhesiveness, penetration enhancers, implants, particulate and vesicular systems like liposomes, niosomes, microemulsion, nanoparticles, iontophoresis, dendrimers, and so on.

51.2 ANATOMY OF THE EYE

The eye is a spherical structure with a wall made up of three layers: the outer part sclera, the middle parts choroid layer, ciliary body and iris, and the inner section nervous tissue layer retina. The sclera is a tough fibrous coating that protects the inner tissues of the eye, which is white except for the transparent area at the front, and the cornea allows light to enter into the eye.[1]

The eye structure is shown in Figure 51.1.[2]

51.2.1 Cornea

The cornea is the transparent, dome-shaped window covering the front of the eye. It is a powerful refracting surface, providing two-thirds of the eye's focusing power. Because there are no blood vessels in the cornea, it is normally clear and has a shiny surface. The adult cornea has a radius of approximately 7–8 mm that covers about one-sixth of the total surface area of the eye ball that is a vascular tissue to which provides nutrient and oxygen supplied via lacrimal fluid and aqueous humor as well as from blood vessels of the junction between the cornea and sclera.

The cornea is made of five layers—epithelium, bowman's layer, stroma, Descemet's membrane, and endothelium—which is the main pathway of drug permeation to the eye.[3,4] The epithelium is made up of 5–6 layers of cells. The corneal thickness is 0.5–0.7 mm in the central region. The main barrier of drug absorption into the eye is the corneal epithelium, in comparison to many other epithelial tissues (intestinal, nasal, bronchial, and tracheal) that are relatively impermeable.[3] The epithelium is squamous stratified (5–6 layer of cells) with a thickness of around 50–100 μm and turnover of about one cell layer every day. The basal cells are packed with a tight junction, in order to form not only an effective barrier to dust particles and most microorganisms, but also for drug absorption. The transcellular or paracellular pathway is the main pathway to penetrate drug across the corneal epithelium. The lipophilic drug chooses the transcellular route whereas the hydrophilic one chooses a paracellular pathway for penetration (passive or altered diffusion through intercellular spaces of the cells). The Bowman's membrane is an acellular homogeneous sheet 8–14 μm thick situated between the basement membrane of the epithelium and the stroma. Because this layer is very tough and difficult to penetrate, it protects the cornea from injury.

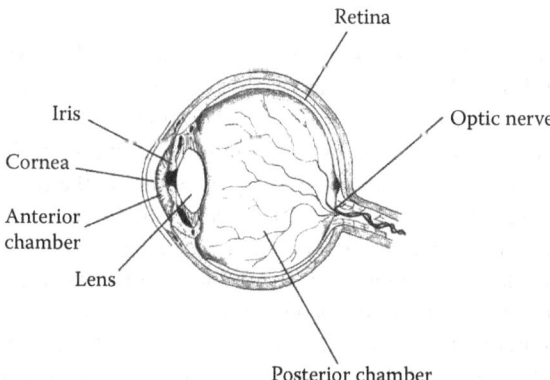

FIGURE 51.1 The structure of the eye showing the essential physiological features. (Reprinted from *Biomaterials*, 22(8), Lloyd, A.W., Faragher, R.G.A., and Denyer, S.P., Ocular biomaterials and implants, 769–785. Copyright 2001, with permission from Elsevier.)

The stroma, or substantia propria, is composed of around 90% of the corneal thickness that contains about 85% water and about 200–250 collagenous lamellae. The lamellae provide physical strength while permitting optical transparency of the membrane. The hydrophilic solutes diffuse through the stroma's open structure.

The Descemet's membrane is secreted by the endothelium and lies between the stroma and the endothelium.[5,6]

51.2.2 Conjunctiva

The conjunctiva protects the eye and is also involved in the formation and maintenance of the precorneal tear film. The conjunctiva is the thin, transparent tissue that covers the outer surface of the eye. It begins at the outer edge of the cornea, covering the visible part of the sclera, and lining the inside of the eyelids. It is nourished by tiny blood vessels that are nearly invisible to the naked eye. The conjunctiva also secretes oils and mucous that moistens and lubricates the eye.

The molecules up to 20,000 Da can cross the conjunctiva, while the cornea is restricted to molecules larger than 5,000 Da. The human conjunctiva absorbs about 2 and 30 times more drugs than the cornea and it has also been proposed that loss of drug by this route is a major path for drug clearance.

51.2.3 Nasolacrimal Drainage System

The nasolacrimal drainage system consists of three parts: the secretory system, the distributive system, and the excretory system. The secretory portion is composed of the lacrimal gland that secreted tears and is spread over the ocular surface by the eyelids during blinking. The secretory system is stimulated by blinking and temperature change due to tear evaporation and reflux secretors that have an efferent parasympathetic nerve supply and secrete in response to a physical and emotional state, for example, crying.

The distributive system consists of the eyelids and the tear meniscus around the lid edges of the open eye, which spreads tears over the ocular surface by blinking, thus preventing dry areas from developing.

The excretory part of the nasolacrimal drainage system consists of the lacrimal puncta; the superior, inferior, and common canaliculi; the lacrimal sac; and the nasolacrimal duct. In humans, the two puncta are the openings of the lacrimal canaliculi and are situated on an elevated area known as the lacrimal papilla. It is thought that tears are largely absorbed by the mucous membrane that lines the ducts and the lacrimal sac; only a small amount reaches the nasal passage.

51.2.4 Tear Film

The tear film is comprised of three layers: oil, water, and mucous. The lower mucous layer serves as an anchor for the tear film and helps it adhere to the eye. The middle layer is comprised of water. The upper oil layer seals the tear film and prevents evaporation.

The exposed part of the eye is covered by a thin fluid layer, the so-called precorneal tear film. The film thickness is reported to be about 3–10 μm depending on the measurement method used. The resident volume amounts to about 10 μL. According to the three layers theory, the precorneal tear film consists of a superficial lipid layer, a central aqueous layer, and an inner mucus layer. The lipids play an important role in reducing the evaporation rate in a way that normal tear osmolality can be maintained, even when the tear flow is quite low.[7,8]

The aqueous layer contains inorganic salts, glucose, and urea as well as retinol, ascorbic acid, various proteins, lipocalins (previously known as tear-specific prealbumins), immunoglobulins, lysozyme, lactoferrin, and glycoproteins.[9]

The osmolality of the tear film equals 310–350 mOsm/kg in normal eyes and is adjusted by the principal inorganic ions Na^+, K^+, Cl^-, HCO_3^-, and proteins. The mean pH value of normal tears is about 7.4. Depending on age and diseases, values between 5.2 and 9.3 have been measured.

The mucus layer is very sensitive to hydration and forms a gel-layer with viscoelastic rheological properties. It protects the epithelia from damage and facilitates the movements of the eyelids. Mucins improve the spreading of the tear film and enhance its stability and cohesion. Mucus is wiped over the surface of the eye by the upper eyelid during blinking. The mucus gel entraps bacteria, cell debris, and foreign bodies, forming mucous threads consisting of thick fibers arranged in bundles. These threads are transported during blinking to the inner canthus and expelled onto the skin.[3] The mucus layer can form a diffusion barrier to macromolecules depending on the degree of network entanglement. On the other hand, mucus can bind cationic substances because of the negative charges of mucins.[10]

51.3 CHALLENGES TO OCULAR DRUG DELIVERY

Ophthalmic drug delivery to the anterior segment of the eye would seem like an easy problem due to the easy accessibility of the eyes. Human eyes are, however, designed to minimize the penetration of harmful chemicals or microorganisms and the same protective mechanisms also create barriers to the delivery of drugs to the anterior chamber, thus making the problem of ophthalmic drug delivery to the anterior eye very challenging.[11]

Drug delivery in ocular therapeutics is a challenging problem and is a subject of interest to scientists working in the multidisciplinary areas pertaining to the eye.

Topical ocular drugs are generally administered in the form of eye drops. Ophthalmic formulations like solutions, suspensions, and ointments available in the market show drawbacks such as increased pre-corneal elimination, blurred vision, and high variability in efficiency. Actually, these conventional dosage forms suffer from the problems of poor ocular bioavailability, because of various anatomical and pathophysiological barriers prevailing in the eye.

The bioavailability of drugs administered as eye drops is severely limited by physiological constraints such as tear turnover and the blinking reflex. Further, drug loss due to nasolacrimal drainage, conjunctival absorption, and protein binding results again in poor bioavailability and systemic side-effects.[12]

Systemic absorption instead of ocular absorption may take place either directly from the conjunctival sac via local blood capillaries or after the solution flow to the nasal cavity. Therefore, most of the small molecular weight drug doses are absorbed into systemic circulation rapidly in a few minutes. This contrasts the low ocular bioavailability of less than 5%.[13]

Consequently, frequent instillation of eye drops is required, resulting in pulsed administration and patient noncompliance. Clearly, the main prerequisite for absorption of drugs into the eye is good corneal penetration and prolonged contact time with the corneal epithelium. Many intrinsic barriers, such as the cornea barrier, blood-aqueous barrier (BAB), and blood-retinal barrier (BRB), restrict ocular drug delivery (Figure 51.2).[14] The barriers related to topical drug administration route are as follows.

51.3.1 TEAR FILM BARRIER

After instillation, the flow of lacrimal fluid removes instilled compounds from the surface of the eye.[15] The tear film is only temporarily stable. The eyes cannot be kept open indefinitely. After 20–40 s, an unpleasant sensation compels humans to blink. The basal tear flow is approximately 1.2 µL/min (0.5–2.2 µL/min). This results in a tear turnover rate of 16%/min during waking hours. Reflex stimulation might increase lachrymation 100-fold, up to 300 µL/min.[16] Topical administration, mostly in the form of eye drops, is quickly washed away by the tear film after application. Gel-like mucus layer: Approximately 2–3 µL mucus is secreted daily.[17] Mucin present in the tear film has a protective role by forming a hydrophilic gel layer that moves over the glycocalyx of the ocular surface and clears cell debris, foreign bodies, and pathogens.[18] At the same time, it acts as a barrier to drug delivery systems (Figure 51.3).[15]

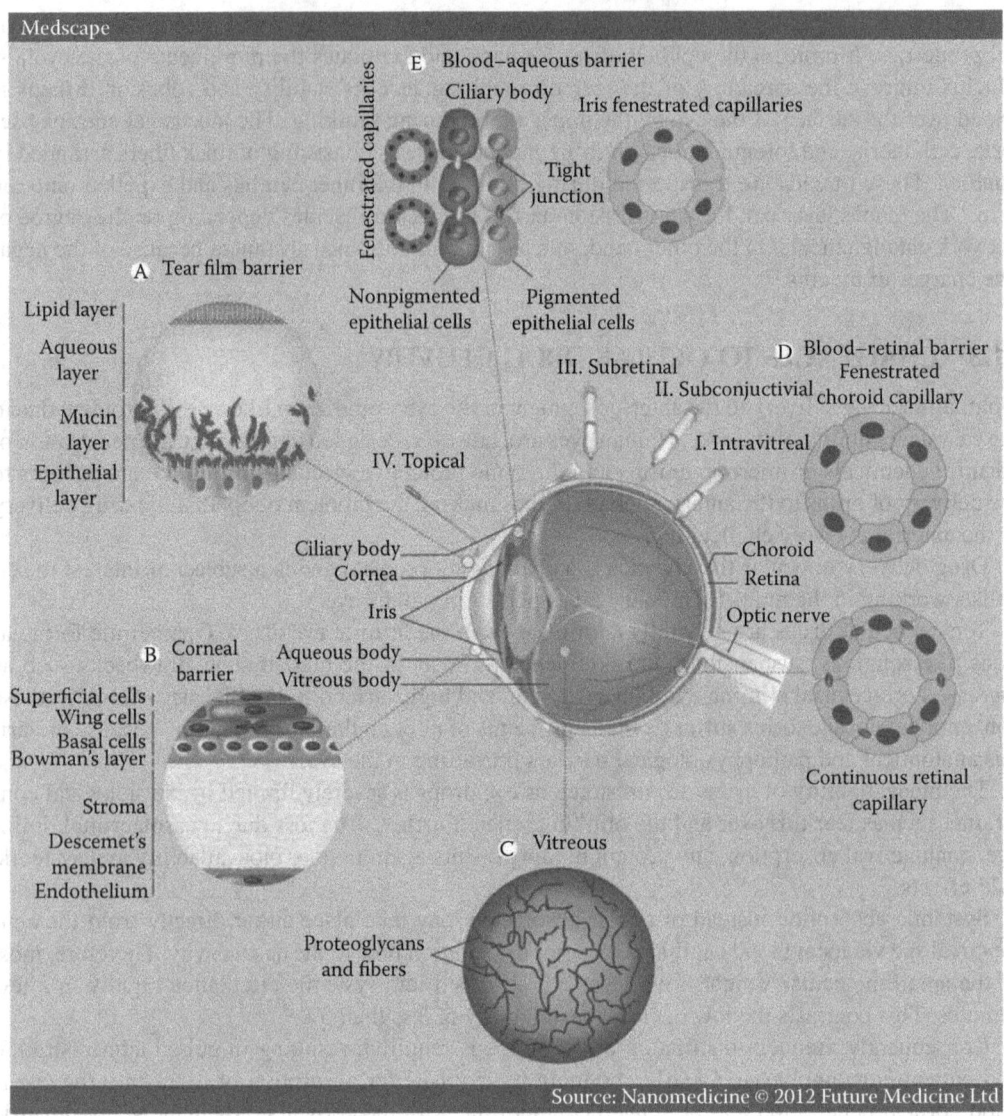

FIGURE 51.2 Major barriers to topical ocular drug delivery. (From Alqawlaq, S. et al., *Nanomedicine*, 7(7), 1067, 2012. With permission from Future Medicine Ltd.)

51.3.2 Corneal Barrier

The corneal epithelium (epithelium cornea anterior layer) is made up of epithelial tissue and covers the front of the cornea (Figure 51.4).[19] It acts as a barrier to protect the cornea, resisting the free flow of fluids from the tears, and prevents bacteria and also therapeutic drugs from entering the epithelium and corneal stroma. The epithelium of the cornea consists of five to six layers of cells packed closely and connected by tight junctions. The cornea is composed of five layers: epithelium, Bowman's membrane, stroma, Descemet's membrane, and endothelium, each of alternating polarity. This sandwich-like structure makes the cornea a crucial barrier to most lipophilic and hydrophilic drugs. To penetrate these layers, optimal lipophilicity for the permeant corresponds to log D values of 2–3.[20]

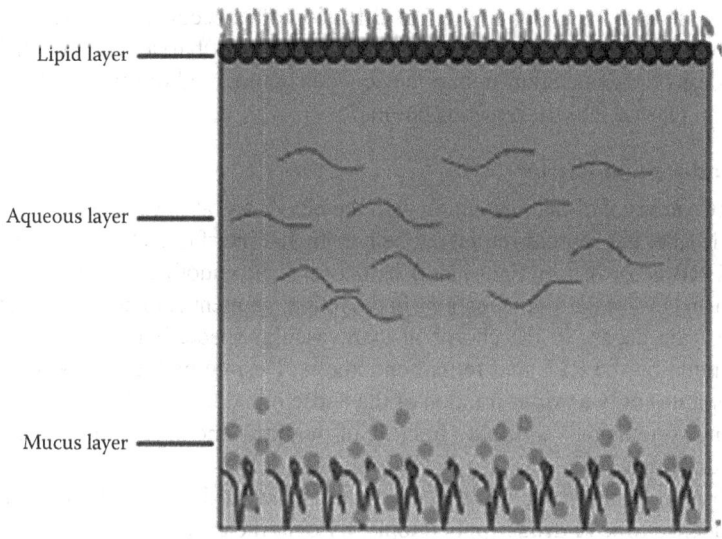

FIGURE 51.3 Tear film barrier: a high turnover rate and the gel-like mucus layer make tear film a barrier in topical ocular drug delivery. (Reprinted from *Drug Discov. Today*, 18(5–6), Gan, L., Wang, J., Jiang, M., Bartlett, H., Ouyang, D., Eperjesi, F., Liu, J., and Gan,Y., Recent advances in topical ophthalmic drug delivery with lipid-based nanocarriers, 290–297. Copyright 2013, with permission from Elsevier.)

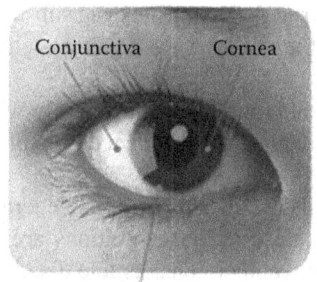
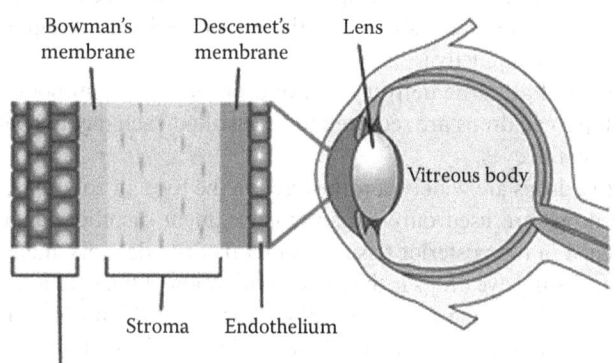

FIGURE 51.4 Cornea layers. (From Japan Tissue Engineering Co., Autologous cultured corneal epithelium, http://www.jpte.co.jp/english/business/Regenerative/cultured_corneal_epithelium.html [accessed December 2013]. With permission from Japan Tissue Engineering Co.)

51.3.3 Blood Barriers

51.3.3.1 Blood-Aqueous Barrier

There are two main structures involved in the formation of the BAB: the ciliary body and the iris (Figure 51.2).

The non-pigmented ciliary epithelium of the ciliary body produces the aqueous (the fluid in the anterior and posterior chambers). The components of the aqueous are quite different from those plasma exudates present in the ciliary body stroma. There are different concentrations of electrolytes, other small molecules, and a restricted set of proteins in low concentrations between the aqueous and the plasma. A barrier to the free diffusion of molecules is formed by tight junctions between the non-pigmented ciliary epithelial cells. This type of barrier is an epithelial barrier that prevents the access of plasma albumin into the aqueous humor, and limits also the access of hydrophilic drugs from plasma into the aqueous humor.

51.3.3.2 Blood-Retinal Barrier

The retina has two areas of direct interaction with the blood, one at the level of the retinal vessels and the other at the level of the choroid-retinal pigment epithelial interface. Thus, two structures comprise the BRB. The BRB is formed by tight junctions between the endothelial cells of the retinal vessels (the inner BRB) and by similar tight junctions in the retinal pigment epithelium (RPE, the outer BRB).

Drugs easily gain access to the choroidal extravascular space, but thereafter distribution into the retina is limited by the RPE and retinal endothelia. Despite its high blood flow, the choroidal blood flow constitutes only a minor fraction of the entire blood flow in the body. Therefore, without specific targeting systems only a minute fraction of the intravenous or oral drug dose gains access to the retina and choroid.[21]

Various efforts in ocular drug delivery have been made to improve the bioavailability and to prolong the residence time of drugs applied topically onto the eye.

51.4 ROUTES OF OCULAR DRUG DELIVERY

51.4.1 TOPICAL OCULAR DELIVERY

Typically topical ocular drug administration is accomplished by eye drops.[22,23] These are liquid preparations that contain drug substances that are used in ocular drug delivery. The drug substance must be active on the surface of the eye or the internal region of the eye after passage through the cornea or conjunctiva.[24]

First of all, drug delivery by eye drops suffers from poor patient compliance particularly when multiple eye drops are required to be instilled each day to overcome the short residence time of the drug in the eyes.

Eye drops are widely administered in the form of solutions, emulsion, and suspension. Generally, eye drops are used only for anterior segment disorders as adequate drug concentrations are not reached in the posterior tissues using this drug delivery method.[25] Another considerable disadvantage of using eye drops is the rapid elimination of the solution and their poor bioavailability in which less than 5% of the dose is absorbed after topical administration into the eye.[26] The dose is mostly absorbed to the systemic blood circulation via the conjunctival and nasal blood vessels.[27] The contact, and thereby duration of drug action, can be prolonged by formulation design.

The designing of experiments and parameters must be conducted to achieve the optimum formulation. Corneal absorption enhancement can be achieved best by increasing solution concentration and viscosity, increasing contact time of formulation in the cornea film, appropriate pK_a, and offering optimal lipid solubility of drug.[22] Commonly added viscosity enhancer agents to improve ocular bioavailability, these include various synthetic polymers such as carboxymethyl cellulose, hydroxyl methylcellulose, polyvinyl alcohol, hydroxypropyl methylcellulose, and carbomers. Recently, natural polymers have also been used to improve the bioavailability of drugs. Examples of these polymers are hyaluronic acid (HA), guar gum, xyloglucan gum, chitosan, gellan gum, pectin, etc. The rheological characteristics of a polymer should be implicated, such as no adverse effects, contact time of dosage formulation, and retention of dosage formulation ocular surface.[28,29]

51.4.1.1 Aqueous Solutions

A homogeneous solution dosage form offers many advantages including the simplicity of large-scale manufacture. The factors that must be taken into account while formulating an aqueous solution include the selection of appropriate salt of the drug substance, solubility, therapeutic concentration required, ocular toxicity, pK_a, and the effect of pH on stability and solubility, tonicity, buffer capacity, viscosity, compatibility with other formulation ingredients as well as packaging components, choice of preservative, ocular comfort, and ease of manufacturing. The stability of ophthalmic solutions and other dosage forms determine the shelf life and expiration date of the product. The drug product is analyzed for physical, chemical, and microbiological parameters throughout the shelf life. Typical physical parameters include pH, osmolality, viscosity, color, and appearance of the product. Chemical parameters include assays for the active and degradation product and preservative content. Microbiological parameters include sterility and antimicrobial preservative efficacy of the product and bioburden of all components.[30]

51.4.1.2 Suspensions

In most ophthalmic suspensions, the average particle size is less than 10 μm. The most efficient method of producing such particle size is by dry milling. However, milling may be desirable for potentially explosive ingredients. Other methods of particle size reduction include micro-pulverization, grinding, and controlled precipitation. An ophthalmic suspension contains many inactive ingredients such as dispersing and wetting agents, suspending agents, buffers, and preservatives. Wetting agents are surfactants that lower the contact angle between the solid surface and the wetting liquid. Suspending agents are used to prevent sedimentation by affecting the rheological behavior of a suspension. An ideal suspending agent should have certain attributes. It should produce a structured vehicle. It should be compatible with other formulation ingredients. It should be non-toxic. Generally, cellulosic derivatives such as methyl cellulose, carboxy methyl cellulose, and hydroxyl propyl methyl cellulose, synthetic polymers such as carbomers, poloxamers, and polyvinyl alcohol are used as suspending agents in ophthalmic suspensions. The selection of buffers and preservatives for suspension are similar to that of ophthalmic solutions.

51.4.1.3 Emulsions

Ophthalmic emulsions are generally dispersions of oily droplets in an aqueous phase. There should be no evidence of breaking or coalescence.

A pilocarpine emulsion in eye drop form (Piloplex) has prolonged therapeutic effect compared to pilocarpine hydrochloride eye drops such that it may be administered only twice, rather than four times, daily.[31-33]

In this formulation, pilocarpine is bound to a polymeric material and this complex makes up the internal, dispersed phase of the emulsion system. in vitro studies have indicated that the release time of 80% of the pilocarpine from this system is 6 h compared to 80% released in only 1 h from pilocarpine hydrochloride solution; thus, the prolonged therapeutic effect is apparently due to both an enhanced pulse entry of drug and to a prolongation of drug release from the vehicle.

51.4.1.4 Ointments

While ophthalmic solutions are by far the most preferred dosage forms, ophthalmic ointments are still being marketed for night time applications and where prolonged therapeutic actions are required.

Ophthalmic ointments are sterile, homogeneous, semi-solid preparations intended for application to the conjunctiva or the eyelids.

They are usually prepared from non-aqueous bases, for example, soft paraffin (Vaseline), liquid paraffin, and wool fat. They may contain suitable additives, such as antimicrobial agents, antioxidants, and stabilizing agents.

The major disadvantage of the ophthalmic ointments is that they cause blurred vision due to refractive index difference between the tears and the non-aqueous nature of the ointment and inaccurate dosing. Nevertheless, desirable attributes for ointment development should include factors as follows: They should not be irritating to the eye, should be uniform, should not cause excessive blurred vision, and should be easily manufactured. The typical manufacturing process for an ophthalmic ointment includes micronization and sterilization of the active agent by dry heat, ethylene oxide irradiation, or gamma irradiation. Antimicrobial preservatives (if required) such as chlorobutanol or parabens are dissolved in a mixture of molten petrolatum and mineral oil and cooled to about 40°C with continuous mixing to assure homogeneity. Sterilized and micronized active is then added aseptically to the warm sterilized petrolatum mineral oil mixture with continuous mixing until the ointment is homogeneous. The ointment is then filled into presterilized ophthalmic tubes.

51.4.1.5 Gels

The development of two soluble gels for the delivery of pilocarpine has been reported. One of these systems, described as a high viscosity acrylic vehicle, delivers a 24 h pilocarpine dose from a single, night-time placement in the cul-de-sac.[34,35]

As is usual with hydrophilic matrices, the effect is greatest immediately after instillation and diminishes with time.[34] The approach does allow for only a single administration per day, as opposed to a four times daily eye drop application, and a choice of night-time administration diminishes patient-perceived pilocarpine side effects, such as initial miosis and induced myopia.

A second ophthalmic gel, for the delivery of pilocarpine, is poloxamer 407.[36] This vehicle was chosen because of its low viscosity, optical clarity, and mucomimetic properties, and for its previous acceptability in ophthalmic preparations. This formulation enhanced pilocarpine activity, as indicated by miosis measurements in rabbits, compared to a pilocarpine aqueous solution of equal drug concentration.

51.4.1.6 Sol to Gel Systems

Frequent instillation of solution or higher drug concentration is needed to achieve the desired therapeutic response.[37,38] But this attempt is potentially dangerous if drug solution drained from the eye is systemically absorbed from the nasolacrimal duct.[39] To increase precorneal residence time and ocular bioavailability, different ophthalmic delivery systems such as viscous solutions, ointments, gels, suspensions, or polymeric inserts are used.[40] But because of blurred vision (e.g., ointments) or lack of patient compliance (e.g., inserts), these formulations have not been widely accepted.

The new concept of producing a gel in situ (e.g., in the cul-de-sac of the eye) was suggested for the first time in the early 1980s. Problems mentioned earlier can be overcome by using in situ gel forming ocular drug delivery system, prepared from polymer, exhibit sol-to-gel phase transition due to a change in a specific physicochemical parameter (pH, temperature, etc.) in their environment,[41] and as a result sustained drug release to the eye occurs.[42] Such formulations were used to deliver bioactive agents by instillation into the eye, which upon exposure to the eye temperature changes to the gel phase.[43] Thus, precorneal residence time of the delivery system is increased and ocular bioavailability is also enhanced. Methylcellulose (MC) solution is known to undergo thermoreversible sol-to-gel transition. Gel temperature of MC can be reduced by adding salts and other additives.[44]

Lin et al. prepared a series of alginate and Pluronic-based solutions as the in situ gelling vehicles for the ophthalmic delivery of pilocarpine.[45] The rheological properties, in vitro release, as well as in vivo pharmacological response of polymer solutions, including alginate, Pluronic solution, and alginate/Pluronic solution, were evaluated. The optimum concentration of alginate solution for the in situ gel-forming delivery systems was 2% (w/w) and that for Pluronic solution was 14% (w/w). The mixture of 0.1% alginate and 14% Pluronic solutions showed a significant increase in gel strength in the physiological condition; this gel mixture was also found to be free flowing at pH 4.0 and 25°C. Both in vitro release and in vivo pharmacological studies indicated that the alginate/Pluronic solution retained pilocarpine better than the alginate or Pluronic solutions alone.

The results demonstrated that the alginate/Pluronic mixture can be used as an in situ gelling vehicle to increase ocular bioavailability.

51.4.2 OCULAR INSERTS

Ocular inserts serve as an alternative approach to overcome the problems that were faced by other ocular drug delivery systems. They represent a significant advancement in the therapy of eye disease.

Ocular inserts are described as single, sterile, thin, and multilayered, drug impregnated, solid, or semisolid consistency devices, whose size and shape are especially designed for application in the eye. A polymeric support is a must for the ocular inserts that may or may not contain the drug. The drug is later entrapped or dispersed or the drug can be incorporated as a solution in the polymeric supports that have advantages as they increase the residence of the drug in the eye so a sustained release dosage form would be formulated. The body portion of the eye is sized in such a way so that it can position in lacrimal canaliculus of the eyelid. They can be classified on the basis of their solubility as insoluble, soluble, and bioerodible inserts. The drug release from the inserts would take place by following three procedures: (1) diffusion, (2) osmosis, and (3) bioerosion.[46,47]

51.4.2.1 Soluble Inserts

Soluble inserts consist of all monolytic polymeric devices that at the end of their release, the device dissolves or erodes. Soluble ophthalmic drug inserts is a soluble copolymer of acrylamide, N-vinyl pyrrolidone, and ethyl acrylate. It is a sterile thin film or wafer of oval shape. The system softens in 10–15 s after introduction into the upper conjunctival sac and gradually dissolves within 1 h, while releasing the drug. A soluble insert containing gentamycin sulfate and dexamethasone phosphate has been developed and Pilocarpine insert for glaucoma is also reported; but these systems have the drawback of blurred vision while the polymer is dissolving. A water soluble bioadhesive component in its formulation has been developed to decrease the risk of expulsion and ensure prolonged residence in the eye, combined with controlled drug release. They are bioadhesive ophthalmic drug inserts. A system based on gentamycin obtained by extrusion of a mixture of polymers, showing a release timer of about 72 h has been reported. Due to difficulty with self-insertion, foreign body sensation, only few insert products are listed and pharmaceutical manufacturers are not actively developing inserts for commercialization.

51.4.2.2 Insoluble Inserts

Insoluble inserts are polymeric systems into which the drug is incorporated as a solution or dispersion. Ophthalmic inserts (ocuserts) have been reported using alginate salts, poly(N-vinyl pyrrolidone), modified collagen, and hydroxyl propyl methyl cellulose. Ocufit is a silicone elastomer-based matrix that allows for the controlled release of an active ingredient over a period of at least 2 weeks. Osmotically controlled inserts have also been described, where release is by diffusion and is osmotically controlled.

51.4.2.3 Nonerodible Inserts: Ocusert, Contact Lens

51.4.2.3.1 Ocusert

Ocular inserts (ocuserts) are sterile, thin, and multilayered, drug impregnated, solid, or semisolid consistency devices (Figure 51.5)[48] that prolong residence time of drug with a controlled release manner and is negligible or less affected by nasolacrimal damage.[49] Inserts are available in different varieties depending upon their composition and applications.

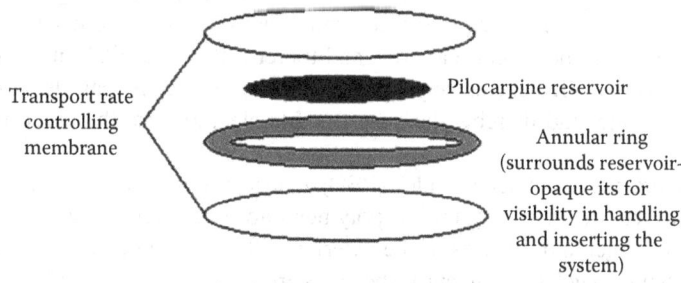

FIGURE 51.5 Ocusert structure. (From Kumari, A. et al., *J. Adv. Pharm. Technol. Res.*, 1(3), 291, 2010. With permission from Society of Pharmaceutical Education and Research.)

Ocusert® provides uniform controlled release (20 or 40 µg/h for 7 days) of pilocarpine as an ocular hypotensive drug and has been commercialized in 1974. Ocusert consists of two outer layers of ethylene-vinyl acetate copolymer (EVA) and an inner layer of pilocarpine in alginate gel within di-(ethylhexyl) phthalate for a release enhancer, sandwiched between EVA layers.[50]

The drug can be entrapped, dispersed, or can be incorporated as a solution in the polymeric supports that have advantages as they increase the residence of the drug in the eye; hence, a sustained release dosage form would be formulated. The drug release from the inserts would take place by following three procedures: diffusion, osmosis, and bioerosion.[51]

However, ocusert has not become widely used because of unsatisfactory intraocular pressure (IOP) control due to various causes, including difficulty of inserting the device, ejection of the device from eye, and irritation during insertion.[52]

51.4.2.3.2 Contact Lenses

The numerous and efficient barriers to local and systemic drug delivery to the eye make the development of sustained release systems a highly challenging issue.[53,54] Up to now, CLs can be considered as the devices that can stay on the eye for more time with good patient compliance. This fact together with their polymeric nature, which can be modulated up to a certain extent to fit specific purposes, has pointed out CLs as promising devices that accomplish the primary function of correcting vision deficiencies while playing a role as drug depots suitable for the management of ocular disorders. Neutral CLs could also be merely used as medicated bandages in patients with visual accuracy, performing only as drug delivery systems. From a regulatory perspective, CLs intended for both drug delivery and refractive correction would be considered drug-device combination products.[55–57]

CLs can absorb drugs when soaked in drug solutions. These drug saturated CLs are placed in the eye for releasing the drug for a long period of time. The hydrophilic CLs can be used to prolong the ocular residence time of the drugs. In humans, the Bionite lens that was made from hydrophilic polymer (2-hydroxy ethyl methacrylate) has been shown to produce a greater penetration of fluorescein.[28] Several kinds of polymers have been used for the preparation of these lenses. They are made up of hydrogels that absorb certain amounts of aqueous solution, and because of this property they have been found useful for drug delivery to the anterior of the eye.[58] For prolongation of ocular residence time of the drugs, hydrophilic CLs can be used.[29]

Various methods to prepare polymeric hydrogels for novel CL-based ophthalmic drug delivery systems are critically analyzed in Section 51.4.8. In addition, advantages and limitations of CLs as drug carriers and then a more extensive analysis of the approaches made to modify the polymeric networks at the nanoscale in order to improve drug loading and release performance will be discussed.

51.4.2.4 Erodible Ophthalmic Inserts

It is now common knowledge that the topical controlled delivery of ophthalmic drugs improves their ocular bioavailability with respect to traditional eye drops, by decreasing the rate of drug elimination from the precorneal area. When the controlled delivery is realized via an erodible insert, the drug residence time in the precorneal area, and thereby, the bioavailability will be maximized if the drug release is controlled exclusively by insert erosion, since any parallel release mechanism increases the release rate, and thereby, the dose fraction cleared from the precorneal area by tear fluid draining.

Drug release from poly (ethylene oxide) (PEO) matrices is elicited by water absorption into matrix, which converts the semi-crystalline polymer into a gel. The drug dissolves either completely or partly in the gel and diffuses to the exterior with a rate depending on its diffusivity and concentration gradients in gel. Concurrently, the gel is eroded at its surface with a rate depending on polymer molecular weight and hydrodynamics of dissolution medium. The release pattern depends on the relative rates of these processes. It is seen in Figure 51.6 that the release and erosion data virtually coincide, indicating that the release process soon became completely erosion-controlled.[59]

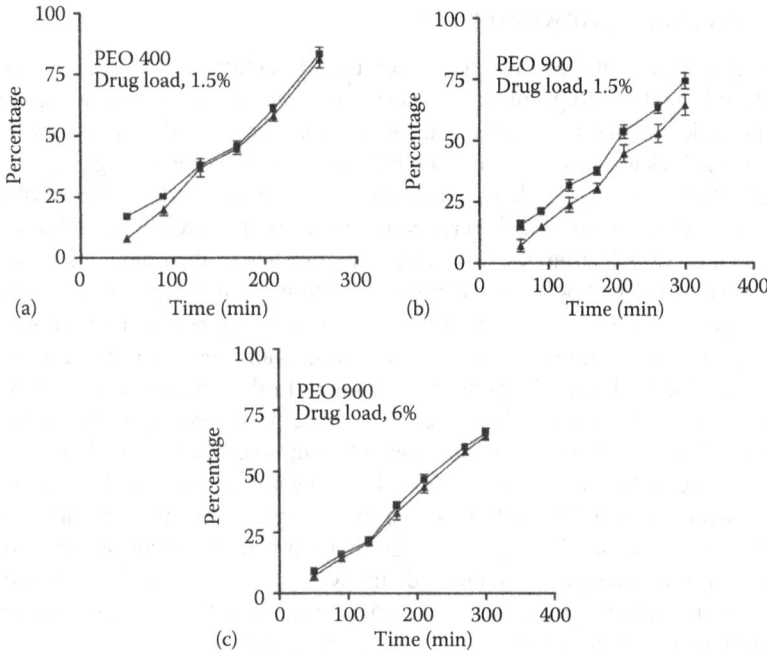

FIGURE 51.6 (a–c) Comparison of drug release and insert erosion kinetics for inserts based on PEO of different molecular weight, medicated with PDS. *Key*: ■, percent of released dose; ▲, percent of eroded insert. Each data point is the mean ± SD of at least three values. (Reprinted from *Eur. J. Pharm. Biopharm.*, 54(2), Di Colo, G. and Zambito, Y., A study of release mechanisms of different ophthalmic drugs from erodible ocular inserts based on poly(ethylene oxide), 193–199. Copyright 2002, with permission from Elsevier.)

The marketed devices of erodible drug inserts are Lacisert, soluble ocular drug insert (SODI), and Minidisc.

51.4.2.4.1 Lacisert

It is a sterile rod-shaped device made up of hydroxypropyl cellulose without any preservative used for the treatment of moderate to severe dry eye conditions, especially in patients who do not respond to other medicines. It is also used to treat other eye conditions, including corneal inflammation exposure, decreased corneal sensitivity, and repeated corneal exposures.[60] The Lacisert insert is an eye lubricant. It works by providing tear-like lubrication for the relief of dry eyes and eye irritation.

This device was introduced by Merck, Sharp, and Dohme in 1981. It weighs 5 mg and measures 12.7 mm in diameter with a length of 3.5 mm.[61]

51.4.2.4.2 Soluble Ocular Drug Insert

SODI is a small oval wafer developed for those could not use eye drops in weightless conditions. It is a sterile thin film of oval shape made from acrylamide, *N*-vinyl pyrrolidone, and ethyl acrylate called as ABE.[62] After introduction into cul-de-sacs where it is wetted by tear film, it softens in 10–15 s and assumes the curved configuration of the globe. During the following 10–15 min, the film turns into a viscous polymer mass thereafter in 30–60 min it becomes a polymer solution.[63]

51.4.2.4.3 Minidisc

The minidisc consists of a contoured disc with a convex front and concave back surface in contact with the eyeball. It is like a miniature CL with a diameter of 4–5 mm. The minidisc is made up of silicone-based prepolymer-α-ψ-bis(4-methacryloxy) butyl polydimethylsiloxane. Minidisc can be hydrophilic or hydrophobic to permit an extend release of both water soluble and insoluble drugs.[56]

51.4.3 SUBCONJUNCTIVAL ADMINISTRATION

Traditionally, subconjunctival injections have been used to deliver drugs at increased levels to the uvea. Currently, this mode of drug delivery has gained new momentum for various reasons. The progress in materials sciences and pharmaceutical formulation have provided new exciting possibilities to develop controlled release formulations to deliver drugs to the posterior segment and to guide the healing process after surgery (e.g., glaucoma surgery).[62] Secondly, the development of new therapies for macular degeneration (antibodies, oligonucleotides) must be delivered to the retina and choroid.[63,64]

After subconjunctival injection, the drug must penetrate across the sclera, which is more permeable than the cornea. Interestingly, the scleral permeability is not dependent on drug lipophilicity.[65] In this respect, it clearly differs from the cornea and conjunctiva. Even more interesting is the surprisingly high permeability of sclera to the large molecules of even protein size.[66] Thus, it would seem feasible to deliver drugs across the sclera to the choroid. However, delivery to the retina is more complicated, because in this case the drug must pass across the choroid and RPE. The role of blood flow is well characterized kinetically but based on existing information, there are good reasons to believe that drugs may be cleared significantly to the blood stream in the choroid. Pitkänen et al. showed that RPE is a tighter barrier than sclera for the permeation of hydrophilic compounds.[65] In the case of small lipophilic drugs, they have similar permeabilities. More complete understanding of the kinetics in sclera, choroid, and RPE should help to develop medications with optimal activity in the selected posterior target tissues. A combination of the kinetic knowledge and cell selective targeting moieties offer very interesting possibilities.

51.4.4 INTRAVITREAL ADMINISTRATION

Direct drug administration into the vitreous offers distinct advantage of more straightforward access to the vitreous and retina (Figure 51.7).[67] It should be noted, however, that delivery from the vitreous to the choroid is more complicated due to the hindrance by the RPE barrier. Small molecules

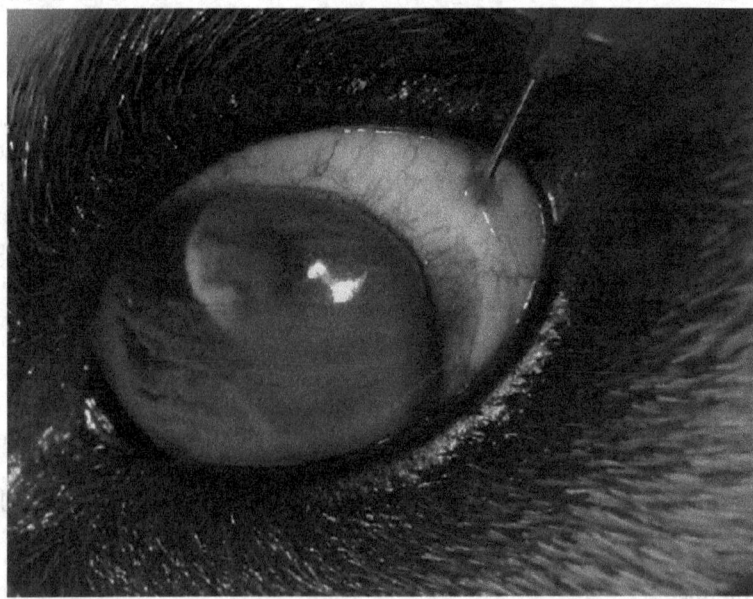

FIGURE 51.7 Intravitreal injection; distribution of triamcinolone on the lens surfaces and a large part of the vitreous. (Reprinted from *Vet. J.*, 176(3), Molleda, J.M., Tardón, R.H., Gallardo, J.M., and Martín-Suárez, E.M., The ocular effects of intravitreal triamcinolone acetonide in dogs, 326–332. Copyright 2008, with permission from Elsevier.)

are able to diffuse rapidly in the vitreous but the mobility of large molecules, particularly positively charged, is restricted.[68] Likewise, the mobility of the nanoparticles is highly dependent on the structure. In addition to the diffusive movement convection also plays a role.[69] The convection results from the eye movements. After intravitreal injection, the drug is eliminated by two main routes: anterior and posterior.[70] All compounds are able to use the anterior route. In this method, drug diffuses across the vitreous to the posterior chamber and, thereafter, eliminates via aqueous turnover and uveal blood flow. Posterior elimination takes place by permeation across the posterior blood–eye barrier. This requires adequate passive permeability (i.e., small molecular size, lipophilicity) or active transport across these barriers. For these reasons, large molecular weight and water-solubility tend to prolong the half-life in the vitreous.[71] Drugs can be administered to the vitreous also in controlled release formulations (liposomes, microspheres, implants) to prolong the drug activity.

51.4.5 IMPLANTS

Ocular implants have many advantages over more traditional methods of drug administration to the eye. The goal of the intraocular implant design is to provide prolonged activity with controlled drug release from the polymeric implant material. In addition, the implants have the benefit of bypassing the blood–ocular barriers to deliver constant therapeutic levels of drug directly to the site of action, avoidance of the side effects associated with frequent systemic and intravitreal injections, and smaller quantity of drug needed during the treatment. The ocular implants are classified as non-biodegradable and biodegradable devices.[71] Non-biodegradable implants can provide more accurate control of drug release and longer release periods than the biodegradable polymers do, but the non-biodegradable systems require surgical implant removal with the associated risks.

Intraocular administration of the implants always requires minor surgery. In general, they are placed intravitreally, at the pars plana of the eye (posterior to the lens and anterior to the retina).[72,73] Earlier, non-biodegradable polymers were used but they needed surgical procedures for insertion and removal. Biodegradable polymers such as poly(lactic acid) (PLA) are safe and effective to deliver drugs in the vitreous cavity and show no toxic signs.[74]

Release rates are typically well below toxic levels, and higher concentrations of the drug are therefore achieved without systemic side effects.[75] For chronic ocular diseases like cytomegalovirus (CMV) retinitis, implants are effective drug delivery systems.

Kato et al.[76] carried out rabbit studies with a scleral implant placed to the posterior pole. The device releases the drug, betamethasone, constantly for at least 3 months without detectable drug concentration in the aqueous humor. Interestingly, the implant showed more effective delivery to the macular region than the intravitreal implants. The episcleral system is a promising means for the treatment of the retinal and choroidal diseases.

Surodex1[77] and Posurdex1 (Allergan, USA)[78] are the biodegradable implants in clinical Phase III studies. They are nearly identical poly lactic-co-glycolic acid (PLGA) implants with different doses of dexamethasone (60 mg for Surodex1 and 700 mg for Posurdex1). Surodex1 was meant for the treatment of postoperative inflammation after filtering surgery on eyes with glaucoma placed underneath the scleral flap during the operation.

Semi-solid bioerodible implant materials would enable the delivery of soft implants with a needle and syringe. Heller[79] introduced such a material, poly(orthoester) IV, that shows long residence time after subconjunctival administration, an erosion-controlled drug release, and ocular biocompatibility. Depending on the ocular site of injection, the ocular lifetime of the drug ranges from 5 to 6 months.

51.4.6 IONTOPHORESIS

Ocular iontophoresis was first investigated in 1908 by the German investigator Wirtz, who passed an electric current through electrolyte-saturated cotton sponges placed over the globe for the treatment of corneal ulcers, keratitis, and episcleritis.[80]

Iontophoresis is a non-invasive technique in which a small electric current is applied to enhance ionized drug penetration into tissue.[81] The drug is applied with an electrode carrying the same charge as the drug, and the ground electrode, which is of the opposite charge, is placed elsewhere on the body to complete the circuit. The drug serves as a conductor of the current through the tissue.[82]

Pre-clinical and clinical studies have shown a 100-fold increase of drug bioavailability compared with topical drops.[83–85]

Therefore, ocular iontophoresis seemed to be the answer to the low bioavailability of drugs after topical administration and to the potential serious complications after intraocular injections used for the treatment of many eye disorders.

Despite its widespread use and study during the first 60 years of the twentieth century, iontophoresis was never fully adopted as standard procedure. The lack of carefully controlled trials and the paucity of toxicity data were among the reasons that precluded its acceptance as an alternative for drug delivery.

51.4.7 Sprays

In stark contrast with most other pharmacologic delivery methods (e.g., pills, intravenous), there is little control on what amount of drug is actually delivered to the target tissue (i.e., the ocular surface) when a physician prescribes a topical formulation. To overcome this problem, investigators have attempted to deliver medications by spraying the drug onto the eye, but initial efforts with such systems as atomizer sprays have failed due to the inability to control droplet size and flow dynamics for consistent and predictable administration. Major problems related to the physics of droplet ejection, such as dispersion, droplet evaporation, drag, and non-collimated flow turbulence, have held back such new approaches until recently.[28]

The micro-droplet piezoelectric ejection system may overcome these obstacles. The device provides collimated flow in a manner similar to ink-jet printers. This facilitates depositing of the drug on the ocular surface and the result differs markedly from eye dropper instillation or spray.

The delivery is based on the piezoelectric fluid ejection system, which monitors every application of a drug. Using sophisticated technology, the system is able to control the dosing and ejection of micro-droplets to enable direct application on the ocular surface. It can be used across a platform of topical medications. Glaucoma, antibiotic, and mydriatic drops have been successfully sprayed using this delivery system.

The technology has a light-emitting-diode optical targeting system that eliminates the need for the patient to tilt his or her head and delivers the spray within less than 30–40 ms, which is faster than the blink reflex.

Although not commonly used, some practitioners use mydriatics or cycloplegics alone or in combination in the form of eye spray. These sprays are used in the eye for dilating the pupil or for cycloplegic examination.

51.4.8 Contact Lens-Based Ophthalmic Drug Delivery

All of the approaches described earlier cannot eliminate or significantly reduce the loss of the drug to the systemic circulation through the nasal or the conjunctival route. Essentially, the drug released by the inserts mixes in the tear volume and can then flow out with the tear drainage or absorb into the conjunctiva, in addition to absorption into the cornea.

The exponential increase in the scientific publications and patents appearing in the last few years about medicated CLs helps to forecast that their use in the clinical arena is not far off.

Although CLs are mainly intended for the correction of ametropia problems, they have been also found suitable to be used as a type of sustained release drug delivery system.[86–89] CLs have the potential to overcome the low bioavailability of 5% or less and short residence time in the tear film of less than 5 min by providing extended drug delivery and an increased residence time in tears, which can increase the corneal bioavailability to higher than 50%.

TABLE 51.1
Classification of CLs according to Their Use

Classifications	Applications
Non-therapeutic use	Correction of refractive ametropia, aphakia, presbyopia
Specialized use	Treatment of keratoconus
Therapeutic use	Increased comfort and pain relief from exposed nerve endings
	Drug delivery for enhanced permeation and absorption
	Mechanical protection of the cornea
	Maintenance of corneal epithelial hydration
	Vision enhancement using plano or powered CLs to smooth an irregular corneal surface

Sources: Alvarez-Lorenzo, C. and Concheiro, A., Ocular drug delivery from nanostructured contact lenses, in *Nanostructured Biomaterials for Overcoming Biological Barriers*, Jose Alonso, M. and Csaba, N.S., Eds., The Royal Society of Chemistry, Cambridge, U.K., 2012; Shah, C. et al., *Ophtalmol. Clin. North Am.*, 16(1), 95, 2003; DeNaeyer, G.W., Exploring the therapeutic applications of contact lenses, New O.D. December 11–15, 2008.

The limited mixing in the post lens tear film between the lens and the cornea leads to a residence time of more than 30 min for drugs released from the lenses.[89,90]

On the other hand, the drug released from the posterior surface of the CL needs to diffuse radially over the long distance of the lens radius to reach tear-lake and then absorb into conjunctiva.

To reduce the frequency of dosing compared to eye drops, CLs must release drugs for an extended period of a few hours to possibly a few days.[91]

The drug released from the anterior surface of the CL is wasted due to absorption into the conjunctiva. There is, however, a possibility that the drying of the anterior surface leads to the formation of a thin glassy layer that could minimize the drug loss from the anterior surface. In any case, the drug released across the posterior surface will constitute at least 50% of the total amount released because the release conditions are likely "perfect sink" toward both the anterior and the posterior surfaces.[92] Perfect sink conditions imply that the time scale for the drug molecules to be absorbed by the cornea or the conjunctiva is much shorter than the time scale for diffusion across the CL, and thus the drug concentrations in both the pre- and post-lens tear film are very small.[93]

According to the FDA, CLs can be classified in three categories: non-therapeutic, specialized, and therapeutic use (Table 51.1).[86,94,95]

Since the first rigid and soft CLs (SCLs) based on methyl methacrylate (MMA) and 2-hydroxyethyl methacrylate (HEMA) that appeared in 1936 and 1954, respectively,[96–98] a plethora of monomers has been synthesized for preparing CLs searching for adequate optical and mechanical features, low density, enough oxygen permeability, and more recently anti-biofouling properties.[99–102]

Depending on the nature of the monomers and the proportion of water in the network, CLs are currently classified in two main groups: hydrophilic and hydrophobic. The oxygen permeability of hydrophilic SCLs increases as the content in water raises[103] and some with a degree of swelling above 35% have been approved for 7 days wearing. By contrast, silicone-based CLs (which first appeared in 1979) are more gas-permeable than water and thus an increase in water content diminishes oxygen permeability.[104]

There have been a number of attempts to use CLs for ophthalmic drug delivery.

51.4.8.1 Drug-Eluting Conventional Contact Lenses to Absorb and Release Ophthalmic Drugs

Conventional soft contact lenses (SCLs) have the ability to absorb a number of drugs. They can be loaded when the lenses are pre-soaked in the drug solution,[105–108] subsequently releasing them into the post-lens lacrimal fluid. As an alternative, one can also insert conventional SCLs into the eyes then apply eye drops or by dropping the ophthalmic solution in the concavity of the lens before insertion. By this means, drugs can be absorbed and released by the SCL, minimizing clearance and sorption through the conjunctiva.[86,109–112]

Loading by immersion in drug solutions requires a time that depends on the mesh size of the network (determined by the cross-linking density and the degree of swelling), the molecular size of the drug, and the concentration of the drug in the loading solution.[113] Diffusion until equilibrium can take less than 30 min but, most commonly, up to several days may be needed.[114] The uniform distribution of the drug throughout the network is necessary for reproducible performance during release. As the drug molecules diffuse out, the layers near to the surface can be reloaded from the deeper region of the lens, which acts as a reservoir making continuous release possible.[115,116]

The drug can be hosted in the SCLs by remaining free in the aqueous phase or establishing nonspecific interactions with the polymer network. Once on the eye, the drug is released to the post-lens tear film, between the cornea and the lens, where it can remain for a long time.

The drug release from such systems is very rapid in the beginning, and proceeds at a rate that declines exponentially with time.

Pre-soaked hydrophilic SCLs, as compared with an eye drop, extend the time that the dissolved drug will continue to be effective. For instance, SCLs pre-soaked in a 2% pilocarpine solution and placed on the cornea can maintain a significant reduction in IOP for almost 24 h.[117] However, 90% of the drug is desorbed within a half hour, and the increase in the pulse necessary to achieve a full day's hypotensive therapy produces significant pilocarpine side-effects, such as increased miosis and myopia.

In 2006, Li and Chauhan[118] investigated drug release from CLs into the pre- and post-lens tear films (PLTF) with the subsequent uptake of the drug by the cornea. Their results showed that the dispersion coefficient of the drug in the post-lens tear film was unaffected by the release of the drug from the lens. Furthermore, simulation results showed that drug delivery from a CL was more efficient than drug delivery by drops. In 2007, the same authors combined in vitro experiments with modeling to investigate the delivery of timolol maleate. in vitro experiments were conducted to create a transport model for releasing the drug from poly(2-hydroxyethyl methacrylate) (pHEMA) lenses. The transport model included drug adsorption on the polymer and drug diffusion in the bulk water. Results showed that at least 20% of the drug (timolol) that was entrapped in the lens entered the cornea, which is larger than the fractional uptake recorded from using eye drops.[106]

The research in this field is aimed at producing CLs with the ability to absorb high amounts of drugs and to deliver them in a controlled manner.

To enhance the potential of (pHEMA) hydrogels as an effective biomaterial for CL-based ophthalmic drug delivery system, Andrade-Vivero[119] incorporated 4-vinyl-pyridine (VP) and N-(3-aminopropyl) methacrylamide (APMA) in the CL structure to enhance the potential of (pHEMA) hydrogels as nonsteroidal anti-inflammatory drug (NSAID) delivery systems, such as diclofenac and ibuprofen. The incorporated monomers did not change the viscoelastic properties neither the state of water, but remarkably increased the amount of ibuprofen (up to 10-fold) and diclofenac (up to 20-fold) loaded.

In another study, Dracopoulos[120] analyzed the interactions of benzalkonium chloride (BAK) which is used as preservative in eye drops, with silicone containing (lotrafilcon A and galyfilcon A) and pHEMA-containing (etafilcon A and vifilcon A) hydrogel CLs. Four SCL types [Focus Monthly® (vifilcon A), Focus Night & Day® (lotrafilcon A), Acuvue® Advance with Hydraclear (galyfilcon A), and SUREVUE® (etafilcon A)] were soaked for 24 h in various concentrations of BAK (1%, 0.1%, 0.01%, and 0.001%) in 20 mL glass vials. Lens extracts showed increased levels of the back vertex distance variability of the cultured bovine lens, indicating that unknown chemical agents might be leached from CL polymers.

The drug loading yield and release rate determine the efficiency of SCLs as drug delivery systems.

51.4.8.1.1 Loading

In the absence of specific interaction mechanisms, the loading only takes place in the aqueous phase of the hydrogels by a simple equilibrium with the drug solution.

Most SCLs exhibit a limited loading. The amount load d can be estimated using the following equation[121]:

$$\text{loading}\left(\frac{V_s}{W_p}\right) \cdot C_0$$

where
V_s is the volume of water sorbed by the hydrogel
W_p the dried hydrogel weight
C_0 the concentration of drug in the loading solution

Therefore, the more water sorbed by CLs, the more the drug loading.

Nevertheless, the sustaining of the release is in general more efficient from low water CLs.[122]

As a result, the possibility of using drug-loaded SCLs depends on whether the drug and the hydrogel material can be matched so that the lens uptakes a sufficient quantity of drug and releases it in a controlled manner.

51.4.8.1.2 Release

Drug release rate from SCLs follows the diffusion laws that mainly depends on the thickness and degree of hydration of the lenses and the drug concentration in the network.

The following equation can be used to predict the release of a drug by diffusion through hydrophilic lenses[117]:

$$\frac{dM}{dT} = \frac{8DM_\infty}{l^2} \exp\left(-\frac{\pi^2 2Dt}{l^2}\right)$$

In this expression, M_∞ represents the total amount of drug released, l is the lens thickness, and D is the coefficient of drug diffusion, which is expected to remain constant if no changes in the swelling degree occur.

Results showed that the ability of conventional SCLs to be a drug reservoir strongly depends on the water content and thickness of the lens, the molecular weight of the drug, the concentration of the drug loading solution, the solubility of the drugs in the gel matrix, and the time the lens remains in it.[123]

For a given drug, the diffusion time decreases as the water content increases. For a given lens, the lower the molecular weight of the drug, the shorter the release time.[124] In this regard, the differences in the ability of silicon-containing (lotrafilcon and balafilcon) and (pHEMA) containing (etafilcon, alphafilcon, polymacon, vifilcon, and omafilcon) commercial CLs to absorb and release cromolyn sodium, ketorolac tromethamine, dexamethasone sodium, and ketotifen fumarate have been studied.[115]

However, CLs presoaked in medications provide a marginal means of delivery because therapeutics freely dispersed within the CL structure are rapidly released (i.e., burst-release), often leading to increased topical drug side effects and toxicity reactions.[125]

In addition, the maximum drug loading by pre-soaking the lenses in the drug solutions was finite, and they were only able to deliver most drugs for a period of a few hours. Thus, they cannot be used for long-term drug delivery.

51.4.8.2 Polymeric Hydrogels for Piggyback Contact Lens Combining with a Drug Plate or Drug Solution

Piggyback lens systems first emerged in the late 1960s. Westerhout,[126] a pioneer of the system, first published a paper on their use in 1973. The system consists of a rigid gas permeable (RGP) CL worn on top of an SCL liying directly on the cornea (Figure 51.8).[127]

Piggyback lenses are normally used where an RGP is indicated but cannot be tolerated by the patient or is inclined to decenter, for example, scarred corneas, high astigmatic ametropia, and ectatic disorders such as keratoconus and pellucid marginal degeneration. In the past, a combination of a soft hydrogel lens and a RGP was used but this inevitably led to hypoxia because of the low oxygen transmission through the combination of lenses. Lately with the advent of high Dk (a measure of their oxygen permeability) silicone hydrogel (SiHy) lenses and the discontinuation of the SoftPerm hybrid lens (after a product recall in 2010), piggyback lenses are enjoying something of a renaissance.

In this regard, this system can be used as another method to make ocular drugs deliverable by SCLs, and involves incorporating the drug solution or drug plate in a hollow cavity made by bonding two separate pieces of lens material.[128,129]

Bourlais et al.[22] created "drug plates" (8.0 mm diameter, 0.2 mm thickness, and 7.5 mm base curve) by freeze drying a mixture of levofloxacin (fluoroquinolon antibiotic) (20, 30, and 40 wt%) and polyvinyl alcohol (PVA), coated with a block styrene-(ethyl/butene)-styrene (SEBS) polymer solution (5.0, 7.5, and 10.0 wt%). The drug plate was placed between a 55% water content hydrophilic lens (poly-N-vinyl pyrrolidone) and a non-hydrophilic lens (copolymer of butyl acrylate/butyl methacrylate). in vivo assessment in albino rabbit eyes showed that the delivery of the drug to the aqueous humor with the piggyback drug delivery system was about 15 times more effective than frequently instilling eye drops.

The presence of a non-hydrophilic lens over the drug delivery system reduced evaporation from the drug plate. However, the oxygen and carbon dioxide permeability of the combined system was lower than that recommended for the safe daily wear of CLs, the CL was too thick to accept, and it proved ineffective at delivering medication for extended periods of time.

Another study on piggyback CL system is illustrated in Figures 51.9 and 51.10.[130]

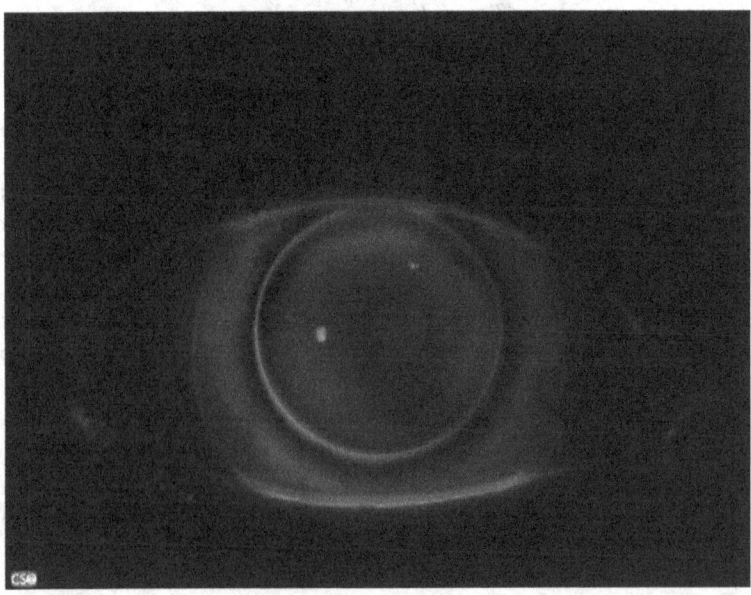

FIGURE 51.8 On-eye piggyback lens fitting. (Reprinted from *Cont. Lens Anterior Eye*, 36(1), Romero-Jiménez, M., Santodomingo-Rubido, J., Flores-Rodríguez, P., and González-Méijome, J.M., Which soft contact lens power is better for piggyback fitting in keratoconus? 45–48. Copyright 2013, with permission from Elsevier.)

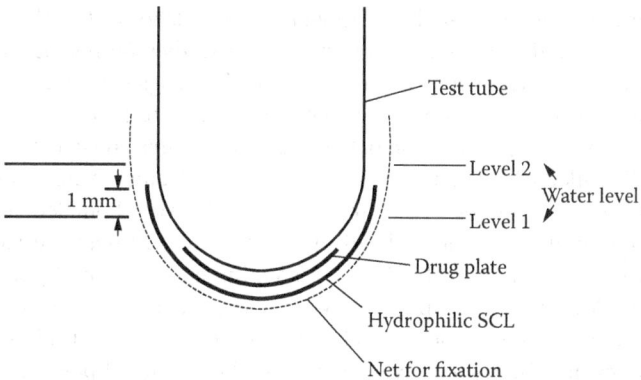

FIGURE 51.9 Drug base plates or drug plates containing levofloxacin were attached to the outside of the bottom of a test tube (diameter: 18 mm), and were covered with a hydrophilic SCL. This was fixed by placing a polypropylene net (mesh size: 2.5 × 2.5 mm) over it. (From *Acta Ophthalmol.*, 74(3), Sano, K., Tokoro, T., and Imai, Y., A new drug delivery system utilizing piggyback contact lenses, 243–248, 1996. Copyright Wiley-VCH Verlag GmbH & Co. KGaA. Reproduced with permission.)

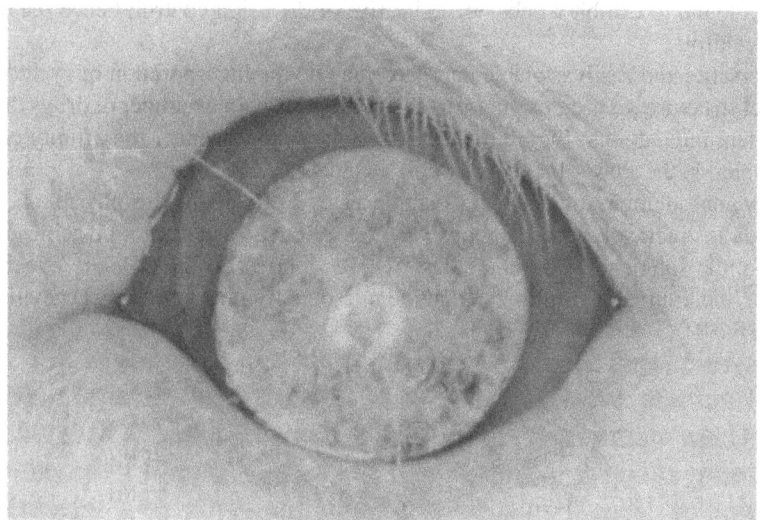

FIGURE 51.10 A hydrophilic SCL, a drug plate (containing 30%, levofloxacin, coated with 7.5% SEBS polymer), and a non-hydrophilic SCL were attached in this order to the left eye of rabbits. (From *Acta Ophthalmol.*, 74(3), Sano, K., Tokoro, T., and Imai, Y., A new drug delivery system utilizing piggyback contact lenses, 243–248, 1996. Copyright Wiley-VCH Verlag GmbH & Co. KGaA. Reproduced with permission.)

51.4.8.3 Soft Contact Lenses Bearing Surface-Immobilized Liposomes and Pendant Cyclodextrins

Drug delivery by injectable liposomes is well known in the pharmaceutical industry. Liposome suspensions of various compositions have been developed to enhance a sustained release of medications in the anterior segment of the eye,[131–133] but one of the major problems in such ocular applications is limited drug uptake, because liposome suspensions are quickly washed away by tearing action.

As many polymers cannot be loaded with diffusible drugs owing to insufficient solubility of the drug into the polymer or an inadequate diffusion rate of the drug through and out of the polymeric materials that constitute the biomedical device, it was proposed to deliver such drugs by loading them into liposomes and binding the intact liposomes onto the surface of devices.[134,135]

Chauhan et al. proposed to disperse dimyristoylphosphatidylcholine (DMPC) liposomes into the CL material.[136,137] However, the procedure suggested in this study requires the use of radicals for the polymerization of the CL matrix, which cannot be used with drugs sensitive to radicals.

Danion et al. immobilized intact liposomes onto SCLs (Figure 51.11).[138] In the first step, polyethylenimine was covalently bounded onto the hydroxyl groups available on the surface of a commercial CL (Hioxifilcon B). Then, NHS-PEG-biotin molecules were bounded onto the surface amine groups by carbodiimide chemistry. NeutrAvidin were bounded onto the PEG-biotin layer. Liposomes containing PEG-biotinylated lipids were docked onto the surface-immobilized NeutrAvidin. Consecutive addition of further NeutrAvidin and liposome layers enabled the fabrication of multilayers. Multilayers of liposomes were also produced by exposing CLs coated with NeutrAvidin to liposome aggregates produced by the addition of free biotin in solution.

The release kinetics of a fluorescent dye demonstrated that intact liposomes had been immobilized onto CL surfaces. The stability of surface-immobilized liposomes onto CL surfaces showed temperature dependence. Surface-bound liposomes can be stored up to 1 month at 4°C with little release of their content.

The problems of this method include the risk of the liposomes detaching from CL surface, and the multilayer scheme of the liposomes decreases the oxygen permeability, although the risk of the liposomes detaching from the surface can be decreased. The release rate of ophthalmic drugs from liposomes was found to exhibit a behavior indicative of diffusion control; hence the release profile is difficult to control.

Another versatile and easily scalable approach may be the incorporation of cyclodextrins (CDs) to the hydrogel structure. CDs can form inclusion complexes with a number of drugs through reversible non-covalent interactions (Figure 51.12).[139] In general, the stronger the affinity constant of the drug:CD complexes, the slower the dissociation kinetics.[140]

However, when solutions of drug:CD complexes are diluted in the physiological fluids, the decomplexation is practically instantaneous and controlled release cannot be achieved; such is the case of ophthalmic solutions containing CDs.[139,140] By contrast, if the CDs are attached to a polymeric network, the dilution is minimized and the microenvironment rich in CD cavities can release

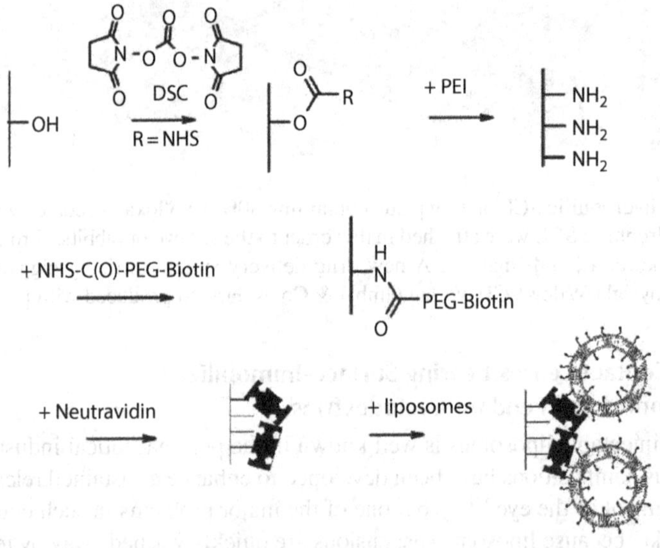

FIGURE 51.11 Chemical reactions leading to the attachment of liposomes onto the surfaces of soft contact lenses (not to scale). (From *J. Biomed. Mater. Res. A*, 82(1), Danion, A., Brochu, H., Martin, Y., and Vermette, P., Fabrication and characterization of contact lenses bearing surface-immobilized layers of intact liposomes, 41–51, 2007. Copyright Wiley-VCH Verlag GmbH & Co. KGaA. Reproduced with permission.)

$$D_f + CyD_f \xrightleftharpoons{K} DCyD$$

FIGURE 51.12 The interaction of drug, D, with a cyclodextrin; CyD, to form an inclusion complex; DCyD, with a binding constant of K; where $K = k_f/k_r$. (Reprinted from *Adv. Drug Deliv. Rev.*, 36(1), Stella, V.J., Rao, V.M., Zannou, E.A., and Zia, V., Mechanisms of drug release from cyclodextrin complexes, 3–16. Copyright 1999, with permission from Elsevier.)

the drug with a rate that is negatively correlated to the affinity constant.[141] In addition, the post-functionalization with CDs of already preformed lenses avoids changes in the mechanical and optical features of the starting networks.[142]

Dos Santos et al.[142] studied SCLs functionalized with pendant cyclodextrins for controlled drug delivery. The aim of this work was to develop acrylic hydrogels with high proportions of cyclodextrins maintaining the mechanical properties and the biocompatibility of the starting hydrogels, but notably improving their ability to load drugs and to control their release rate. Poly(hydroxyethylmethacrylate) hydrogels were prepared by copolymerization with glycidyl methacrylate (GMA) at various proportions and then β-cyclodextrin (βCD) was grafted to the network by reaction with the glycidyl groups under mild conditions. This led to networks in which the βCDs form no part of the structural chains but they are hanging on 2–3 ether bonds through the hydroxyl groups.

The values ranged from 3.2 nm, for the hydrogel prepared with 50 mM GMA, to 1.6 nm for the hydrogel prepared with 400 mM. These distances (a 14% greater when the hydrogel is fully swollen), together with the length of the glycidyl group and the flexibility of the low cross-linked network, makes the reaction of each βCD with two or even three glycidyl groups feasible.

The pendant βCDs did not modify the light transmittance, glass transition temperature, swelling degree, viscoelasticity, oxygen permeability, or surface contact angle of the hydrogels, but decreased their friction coefficient by 50% and improved diclofenac loading by 1300% and enhanced drug affinity 15-fold.

Diclofenac release from the hydrogels was evaluated in artificial lacrimal fluid using both freshly loaded discs and 1 month-stored discs. For each type of hydrogel, the drug release rate was similar in both cases. The hydrogels without pendant βCD rapidly delivered the complete dose by diffusion, the whole process finishing within 1 day. By contrast, the hydrogels with pendant βCDs were able to sustain the delivery for several days, even up to 2 weeks, particularly those containing high proportions of βCD (in which uncomplexation controls the release).

The hydrogels were able to prevent drug leakage to a common conservation liquid for SCLs and to sustain drug delivery in lacrimal fluid for 2 weeks. The hydrogels with pendant βCDs are particularly useful for the development of cytocompatible medicated implants or biomedical devices, such as drug-loaded SCLs.

Glisonia et al. developed two types of hydrophilic networks with conjugated beta-cyclodextrin (β-CD) with the aim of engineering useful platforms for the localized release of an antimicrobial 5,6-dimethoxy-1-indanone N4-allyl thiosemicarbazone (TSC) in the eye and its potential application in ophthalmic diseases.[143] Poly(2-hydroxyethyl methacrylate) SCLs displaying β-CD, namely, pHEMA-co-β-CD, and super-hydrophilic hydrogels (SHHs) of directly cross-linked hydroxypropyl-β-CD were synthesized and characterized regarding their structure (ATR/FT-IR), drug loading capacity, swelling, and in vitro release in artificial lacrimal fluid. The incorporation of TSC to the networks was carried out both during polymerization (DP method) and after synthesis (PP method). The first method led to similar drug loads in all the hydrogels, with minor drug loss during the washing steps to remove unreacted monomers, while the second method evidenced the influence of structural parameters on the loading efficiency (proportion of CD units, mesh size, swelling degree). Both systems provided a controlled TSC release for at least 2 weeks, TSC concentrations (up to 4000 µg/g dry hydrogel) being within an optimal therapeutic window for the antimicrobial ocular treatment (Figure 51.13).

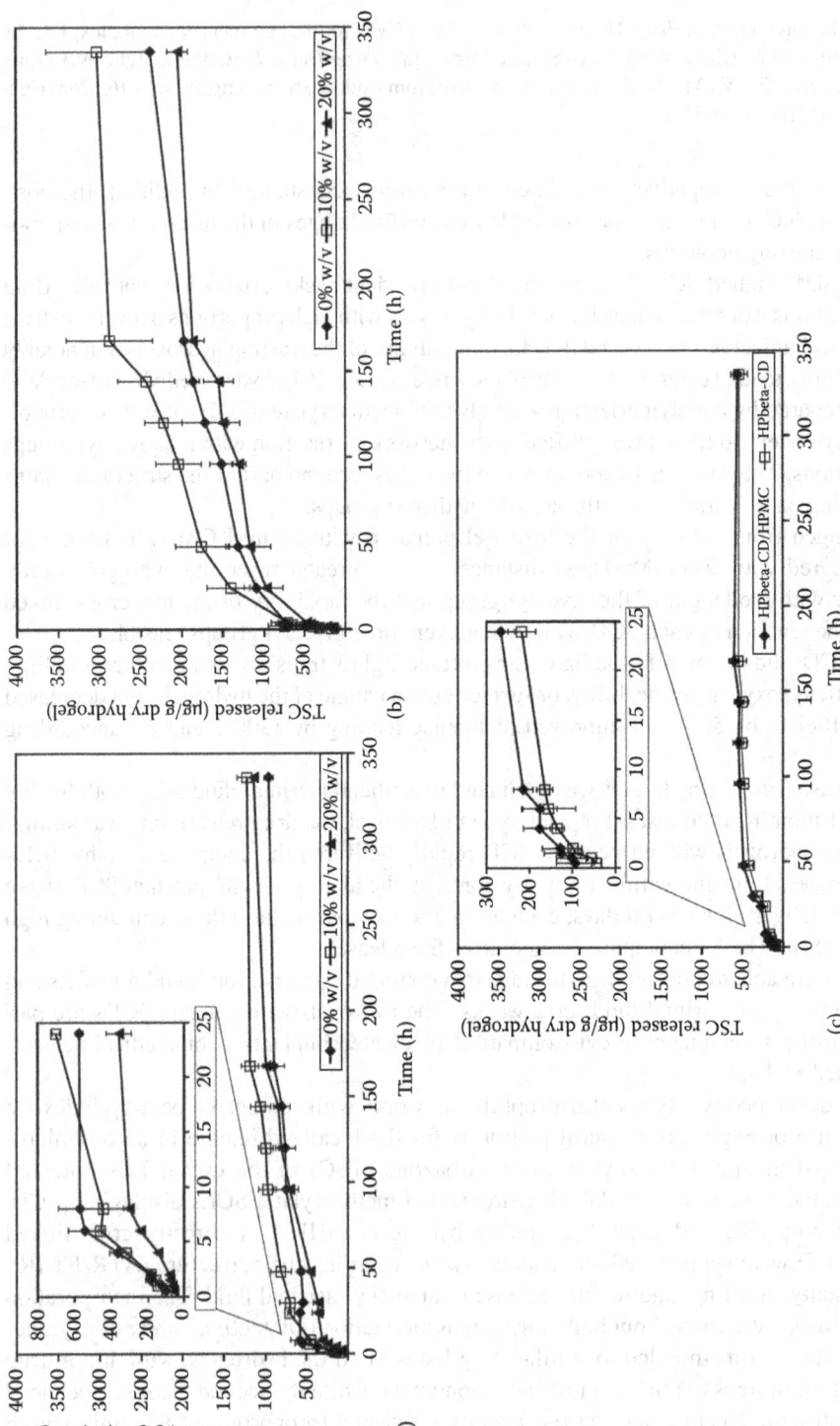

FIGURE 51.13 TSC release kinetics from different hydrogels over 2 weeks, at 25°C. (a) pHEMA-co-β-CD produced by the PP method, (b) pHEMA-co-β-CD produced by the DP method, and (c) HPβ-CD and HPβ-CD/HPMC SHHs. The artificial lacrimal fluid medium was replaced to maintain sink conditions every (a) 30 h, (b) 4 h, and (c) 8 h. Figure insets show the release during the first 24 h. All data were expressed as mean ± SD of at least three independent experiments. (Reprinted from *Carbohydr. Polym.*, 93(2), Glisonia, R.J., García-Fernández, M.J., Pinod, M., Gutkind, G., Moglioni, A.G., Alvarez-Lorenzo, C., Concheiro, A., and Sosnika, A., β-Cyclodextrin hydrogels for the ocular release of antibacterial thiosemicarbazones, 449–457. Copyright 2013, with permission from Elsevier.)

Regardless of the method of the drug loading (DP or PP) and in spite of their relatively fast swelling in artificial lacrimal fluid, pHEMA-co-β-CD SCLs sustained the release for several days (Figure 51.13a and b). CD-containing SCLs prepared applying the DP method released the whole amount of TSC that was initially solubilized in the reaction mixture (1000 μg/mL) after 1 week (Figure 51.13a). The slower release rate observed for networks containing 20% of mono-MA-β-CD would rely on their higher cross-linking density and their greater ability to form inclusion complexes. Pure pHEMA hydrogels loaded by the PP method (soaking in TSC suspension for 24 h) hosted 2380 μg TSC/g dry hydrogel (Figure 51.13b). The incorporation of 10% mono-MA-β-CD resulted in a pronounced increase of the loading capacity to 3025 μg TSC/g dry hydrogel. These data suggest that the formation of TSC/CD complexes favors the incorporation and retention of additional 700 μg of TSC within the network.

SHHs hosted smaller amounts of TSC, the loading capacity being approximately 530 μg TSC/g dry hydrogel (Figure 51.13c). This finding can be related to the greater hydrophilicity of the SHHs compared to the pHEMAco-β-CD networks, which determines that this hydrophobic drug can be hosted in the network only through the formation of an inclusion complex with the CD units, and not through unspecific hydrophobic interactions.

Microbiological tests against *Pseudomonas aeruginosa* and *Staphylococcus aureus* confirmed the ability of TSC-loaded pHEMA-co-β-CD network to inhibit bacterial growth.

51.4.8.4 Molecularly Imprinted Polymeric Hydrogels

Molecular imprinting, as first developed by Wulff and Mosbach and co-workers, is a method for the tailor-made preparation of polymeric systems with predetermined affinity and/or selectivity for specific molecules.[144–146]

A molecularly imprinted polymer is a polymer that has been processed using the molecular imprinting technique that controls the overall polymer morphology and macroporous structure (Figure 51.14).[147] The process usually involves initiating the polymerization of monomers in the presence of a template molecule that is extracted afterward, thus leaving complementary cavities behind. These polymers have affinity for the original molecule and have been used in new applications in the pharmaceutical field as carriers for drugs, peptides, and proteins that have gained increasing attention.[148–154]

Recently, some reviews presented a comprehensive introduction of molecularly imprinted therapeutic CLs.[155,156]

Novel approaches to enhance the capability of SCLs to load drugs and control release may make these optical devices behave as advanced drug delivery systems, fulfilling the requirements of both chronic and acute ocular diseases. Among those approaches, the application of the molecular imprinting technology during SCL manufacture enables the creation in the lens structure of imprinted pockets that memorize the spatial features and bonding preferences of the drug and provide the lens with a high affinity for a given drug. Imprinted SCLs could prolong the permanence of the drug in the precorneal area and provide sustained drug levels, increasing ocular bioavailability and avoiding systemic side-effects.

Alvarez-Lorenzo et al. demonstrated that molecularly imprinted hydrogel CLs, even slightly cross-linked (0.32–8.34 mol%), showed a 9- to 20-fold higher adsorption affinity for the template drug, timolol, than the corresponding non-imprinted lenses.[157,158]

Furthermore, they provided a 3.3-fold greater ocular bioavailability than those of non-imprinted ones in timolol release experiments in rabbit eyes.[159]

The higher affinity for the target drug, in addition to the higher drug loading capability, may play an important role in sustaining the release from imprinted polymers.[160]

Hiratani and Alvarez-Lorenzo prepared imprinted CLs made of HEMA or *N,N*-diethylacrylamide (DEAA), low cross-linker proportions, and a small proportion of functional monomer (methacrylic acid, MAA), which was able to interact with drug (timolol maleate) via ionic and hydrogen bonds. It was found that imprinted HEMA-based and DEAA-based CLs took up more timolol than the

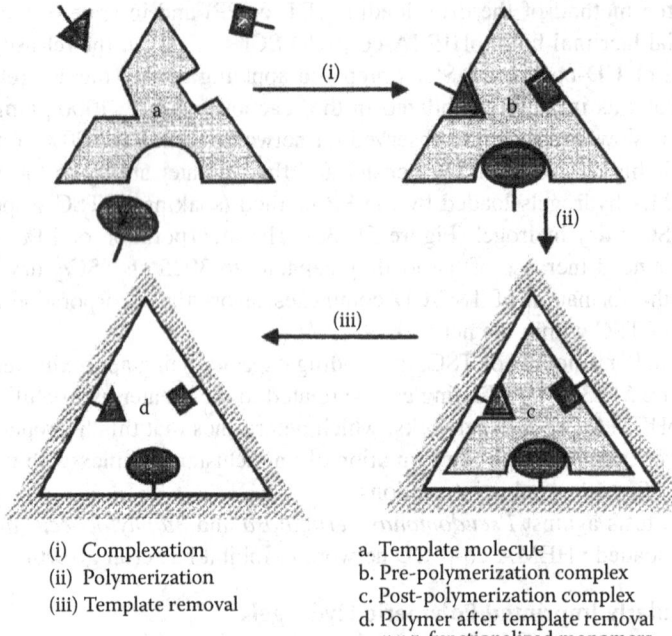

(i) Complexation
(ii) Polymerization
(iii) Template removal

a. Template molecule
b. Pre-polymerization complex
c. Post-polymerization complex
d. Polymer after template removal
x,y,z. functionalized monomers

FIGURE 51.14 Schematic representation of a molecular imprinting process. (Reprinted from *Int. J. Pharm.*, 195(1–2), Allender, C.J., Richardson, C., Woodhouse, B., Heard, C.M., and Brain, K.R., Pharmaceutical applications for molecularly imprinted polymers, 39–43. Copyright 2000, with permission from Elsevier.)

corresponding non-imprinted systems, the loaded lenses could sustain drug release in lacrimal fluid for more than 12 h, and the empty lenses could reload drug overnight for the next day use.[157,158]

Later, they designed imprinted HEMA-based hydrogel CL using acrylic acid (AA) as a functional monomer to load and to release norfloxacin for several hours or even days in a sustained way.[161] However, the duration of drug release in these imprinted hydrogel CLs using one functional monomer was limited to less than 1 day in vitro and in vivo experiments.

Ali and Byrne designed and synthesized hydrogel CLs that can release HA at a controlled rate for dry eye treatment.[162] Hydrogel films and CLs composed of nelfilcon A (a polymer synthesized by PVA modification), acrylamide (AM), *N*-vinyl pyrrolidone (NVP), and 2-(diethylamino) ethyl methacrylate (DEAEM) were biomimetically imprinted in the presence of HA, for the controlled release of HA over 24 h. The lenses were designed for the therapeutic delivery of HA to the eye surface, to improve the wettability of lenses, and to treat symptoms of dry eye. They demonstrated that by changing the mass content and relative proportions of the monomers AM, NVP, and DEAEM within the hydrogel, they can dramatically vary the diffusion coefficients and release profiles of the HA. The variations arise through the biomimetic imprinting process and not through structural changes in the hydrogel.

In another research, Malaekeh-Nikouei et al. studied the influence of methacrylic acid (MAA) as a comonomer and the application of a molecular imprinting technique on the loading and release properties of weakly cross-linked 2-hydroxyethyl methacrylate (HEMA) hydrogels, with a view toward their use as reloadable SCLs for the administration of prednisolone acetate (PA).[163] The hydrogels were prepared with HEMA (95.90–98.30 mol%) as a backbone monomer, ethylene glycol dimethacrylate (140 mM) as a cross-linker, and MAA (0, 50, 100, or 200 mM) as a functional monomer. Different PA/MAA molar ratios (0, 1:8, 1:6, and 1:4) in the feed composition of the hydrogels were also applied to study the influence of the molecular imprinting technique on their binding properties.

The hydrogels (0.4 mm thick) were synthesized by thermal polymerization at 60°C for 24 h in a polypropylene mold. The hydrogels were then characterized by the determination of their swelling and binding properties in water. Increasing the MAA content of the hydrogel and applying the molecular imprinting technique led to an increase in the loading capacity of the hydrogel. The optimized imprinted hydrogel showed the highest affinity for PA and the greatest ability to control the release process, sustaining it for 48 h. The results obtained clearly indicate that the incorporation of MAA as a comonomer increased the PA loading capacity of hydrogel. The data showed that the molecular imprinting technique also had a significant effect on the loading and release properties of the hydrogels.

Although all the results indicate drug-loading capacity of CLs was improved by the molecularly imprinting method, and the imprinted hydrogels showed a higher affinity for template molecules, but some drawbacks exist as well. The maximum drug loading was limited by the template molecules and functional monomers, and the CLs deform on the release of drug. In addition, it is difficult to adjust the release profile of drugs.

51.5 NANOTECHNOLOGY IN DRUG DELIVERY

Nanotechnology is the engineering and manufacturing of materials at the atomic and molecular scale. One important application of nanotechnology is for developing safer and more effective therapeutic modalities. In its strictest definition from the National Nanotechnology Initiative, nanotechnology refers to structures roughly in the 1–100 nm size in at least one dimension. To put this size range in perspective, a small molecule, a virus, a bacterium, and the cross-section of a human hair are around 1, 100, 1000, and 100,000 nm, respectively. Despite this size restriction, nanotechnology commonly refers to structures that are up to several hundred nanometers in size, and that are developed by top-down or bottom-up engineering of individual components.[164–167]

More than two dozen nanotechnology therapeutic products have been approved for clinical use.[168] Among these first-generation products, liposomes and polymer–drug conjugates are two dominant classes. The majority of these products were developed to improve the pharmaceutical properties or dosing of clinically approved drugs, which in some cases also provided life cycle extension of drugs after patent expiration.

The challenges faced in drug delivery are to develop means for administering drugs with release rates that vary according to therapeutic needs of a patient.

In this case, nanotechnology offers advantages that allow a more targeted drug delivery and controllable release of the therapeutic compound.[169,170] The aim of targeted drug delivery and controlled release is to manage better drug pharmacokinetics, pharmacodynamics, non-specific toxicity, immunogenicity, and biorecognition of systems in the quest for improved efficacy.[171]

51.6 NANOTECHNOLOGY AND OCULAR DRUG DELIVERY

For the effects of a drug to be maximized, the molecules of the drug need to reach specific locations within the target tissue. Because drug molecules typically cannot selectively reach their site of action, there is a need for carriers that can efficiently deliver the required amount of the drug to its target site. The eye, particularly the posterior segment, is composed of tissues that are difficult for drugs to penetrate because of structural peculiarities such as the barrier function. Thus, many research studies on nano-sized drug carriers have been conducted in the field of ophthalmology.[23,172,173]

The development of a drug delivery system that can be used for the posterior segment of the eye and that involves nanocarriers to overcome the issue of frequent intravitreal administration has received considerable attention.

Drug delivery systems with appropriate spatiotemporal characteristics are designed to allow drugs to affect the target tissue efficiently, and three classes of techniques are generally used: (1) absorption promotion—promotion of the passage of a drug through the tissue, which functions as a barrier; (2) controlled release—efficient time-controlled sustained release of a locally administered drug; and (3) drug targeting—to allow it to act efficiently and exclusively on the target tissue.[174]

51.7 THERAPEUTIC SIGNIFICANCE IN OCULAR DELIVERY

Even though the various drug delivery systems mentioned offer numerous advantages over conventional drug therapy, nonetheless, they are not devoid of pitfalls.

Recent progress in ocular drug delivery systems research has provided new insights into drug development, and the use of nanoparticles for drug delivery is thus a promising approach for the advanced therapy of ocular diseases.[171] Therefore, ocular drug targeting has the following major goals:

- To increase drug bioavailability of drug by increasing corneal contact time
- To enhance drug permeation to the eye
- To provide prolonged drug release
- To overcome the side effects produced by conventional system (e.g., eye drops)
- To provide controlled and sustained drug delivery
- To provide targeting within the ocular globe so as to prevent the loss of drug to other tissues
- To provide comfort
- To reduce frequency of administration
- Not to cause blurred vision

51.8 VARIOUS NANOPARTICULATE-BASED OPHTHALMIC DRUG DELIVERY SYSTEMS

Recently, an increasing number of scientists have turned to nanomaterial-based drug delivery systems to address the challenges faced by conventional methods.

Several dispersed systems have been developed, on different size scales, which can be listed as follows: nanoparticles, microemulsions, nanosuspensions, liposomes, dendrimers, niosomes, cyclodextrins, and so on, which can enhance the rate of ophthalmic drug delivery to a significant degree.

The incorporation of an ocular drug into smaller-sized dispersed systems offers several potential advantages, of which the most important ones are listed in Table 51.2.[175]

Meanwhile, the adverse effects, such as ocular toxicity, irritation, or vision interference of the delivery system should be taken into serious consideration.

TABLE 51.2
Potential Advantages of Nanoscaled Drug Delivery System in Ocular Drug Delivery

	Advantages
Administration	Self-administration by patients possible as eye drops
	No impairment of sight because of small dimensions of the delivery systems
Drug stability increase	Protection against metabolic enzymes, such as peptidases and nucleases
Drug clearance rate	Prolonged residence time at ocular surface or in intraocular tissues
	Possible uptake of particles and vesicles into corneal cells after topical administration
Drug release	Prolonged drug release, reducing the need for repeated instillation or injection
Drug targeting	Targeting possible toward affected tissues, reducing possible side effects and decreasing the required dose
	Possible penetration-enhanced effect

Source: Vandervoort, J. and Ludwig, A., *Nanomedicine*, 2(1), 11, 2007. With permission from Future Medicine Ltd.

51.8.1 LIPOSOMES

Liposomes are spherical self-closed structures, composed of curved lipid bilayers, which enclose part of the surrounding solvent into their interior (Figure 51.15).[176] The size of a liposome ranges from some 20 nm up to several micrometers and they may be composed of one or several concentric membranes, each with a thickness of about 4 nm. Liposomes possess unique properties owing to the amphiphilic character of the lipids, which make them suitable for drug delivery.[177]

Many potential mechanisms have been suggested for the formation of liposomes, and some of these are more complex than others.[178]

The use of colloidal drug delivery systems, such as nanoparticles and liposomes, is a suitable strategy to enhance the bioavailability of topically administered drugs[179,180] because they offer unique features while preserving the ease of delivery in liquid form.

Liposomal drug delivery systems not only enable the delivery of higher drug concentrations,[181] but also a possible targeting of specific cells or organs.[182–185] Harmful side effects can therefore be reduced owing to minimized distribution of the drug to non-targeted tissues. Like all other carrier systems, the use of liposomes in drug delivery has advantages and disadvantages. The amphiphilic character of the liposomes, with the hydrophobic bilayer and the hydrophilic inner core, enables solubilization or encapsulation of both hydrophobic and hydrophilic drugs. Along with their good solubilization power, a relatively easy preparation and a rich selection of physicochemical properties have made liposomes attractive drug carrier systems.

However, a complete saturation of the immune system[186] and interactions with lipoproteins[187,188] are some examples of potentially toxic and adverse effects.

Efficient drug delivery systems based on liposomes need to possess a large number of special qualities.[189] First, good colloidal, chemical, and biological stability is required. The fact that liposomes are non-equilibrium structures does not necessarily mean that they are unsuitable for drug delivery. On the contrary, a colloidally stable non-equilibrium structure is less sensitive to external changes than equilibrium structures, such as micelles. Hence, colloidally stable liposomes often work well in pharmaceutical applications.

These vesicular drug delivery systems can also carry a variety of molecules with therapeutic potential and offer different advantages for ocular drug delivery.[190] In fact, several experiments have shown that liposome entrapped drugs have improved ocular bioavailability.[191,192]

The behavior of liposomes as an ocular drug delivery system has been observed to be, in part, because of their surface charge. Positively charged liposomes seem to be preferentially captured at the negatively charged corneal surface, compared with neutral or negatively charged liposomes.[193]

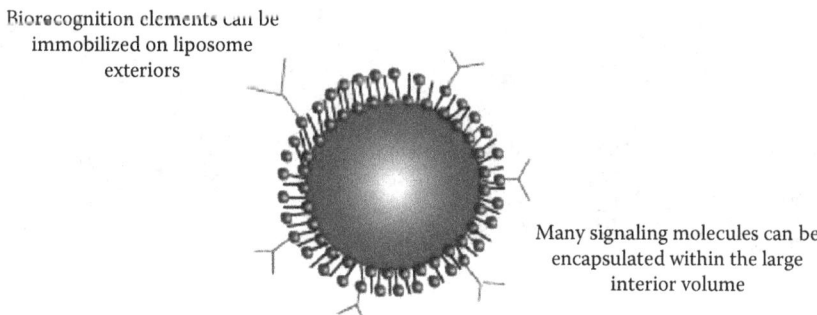

FIGURE 51.15 Diagram of a general liposome structure. Lipids form a bilayer entrapping an aqueous core. Biorecognition elements can be tagged to the outside membrane and highly water-soluble marker molecules can be entrapped in the inner volume. (Reprinted from *Talanta*, 68(5), Edwards, K.A. and Baeumner, A.J., Liposomes in analyses, 1421–1431. Copyright 2006, with permission from Elsevier.)

Li et al. modified the properties of liposome and brought a series of notable advantages for ocular drug delivery.[194] In their study, liposome coated with low molecular weight chitosan (LCH) was proposed and its in vitro and in vivo properties and potential use in ocular drug delivery were evaluated. LCH with a molecular weight of 8 kDa was prepared and coated on liposome loaded with diclofenac sodium. The LCH coating changed the liposome surface charge and slightly increased its particle size, while the drug encapsulation was not affected. After coating, the liposome displayed a prolonged in vitro drug release profile. LCH-coated liposome also demonstrated an improved physicochemical stability at 25°C in a 30-day storage period. The ocular bioadhesion property was evaluated by rabbit in vivo precorneal retention, and LCH-coated liposome achieved a significantly prolonged retention compared with non-coated liposome or drug solution. The LCH coating also displayed a potential penetration enhancing effect for the transcorneal delivery of the drug. In conclusion, the LCH coating significantly modified the properties of liposome and brought a series of notable advantages for ocular drug delivery.

The histopathology of the tested rabbit eyeballs after long-term irritation test to study the ocular tolerance is shown in Figure 51.16. Normal and healthy structures of ocular tissues were observed in all the tested eyes and there were no differences between the LCH-coated liposome (LCHL) treated group and the control group. Corneal and conjunctival epithelial cells maintained normal morphology and constructed integrated epithelium (Figure 51.16a through e). The basal cells of cornea remained abundant and were normally packed by junction complex (Figure 51.16a through c). Conjunctival lymphoid tissue was identified in all the conjunctivas without the abnormality of its size and location (Figure 51.16d through f). Normal levels of polymorphonuclear cells were observed in the conjunctival stroma (Figure 51.16d through f), indicating that there were no signs of inflammation. The histopathology confirmed that no ocular irritating effects were induced by LCHL compared with non-treated eyes. The combination of liposome and LCH, both of which are biocompatible, demonstrated a preferable ocular tolerance.

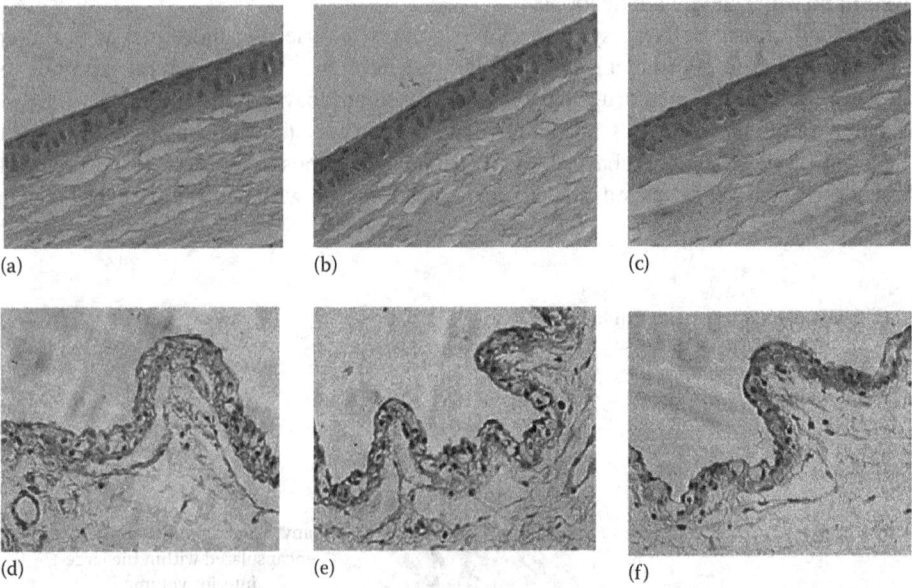

FIGURE 51.16 Histopathology microscopy of the ocular tissues after being treated with LCHL for 7 days: (a) cornea of non-treated, (b) treated with LCHL2, (c) LCHL3, (d) conjunctiva of non-treated, (e) treated with LCHL2, and (f) LCHL3. (Reprinted from *Int. J. Pharm.*, 379(1), Li, N., Zhuang, C., Wang, M., Sun, X., Nie, S., and Pan, W., Liposome coated with low molecular weight chitosan and its potential use in ocular drug delivery, 131–138. Copyright 2009, with permission from Elsevier.)

51.8.2 MICROEMULSIONS

Microemulsions are liquid dispersions of water and oil that are made homogenous, transparent (or translucent), and thermodynamically stable by the addition of relatively large amounts of a surfactant and a co-surfactant and having a diameter of the droplets in the range of 100–1000 Å (10–100 nm) (Figure 51.17).[195]

Microemulsions exhibit several advantages as a drug delivery system[196]:

1. Microemulsions are a thermodynamically stable system and the stability allows self-emulsification of the system whose properties are not dependent on the process followed.
2. They act as super solvents of drug. They can solubilize hydrophilic and lipophilic drugs including drugs that are relatively insoluble in both aqueous and hydrophobic solvents. This is due to the existence of microdomains of different polarity within the same single-phase solution.
3. The dispersed phase, lipophilic or hydrophilic (oil-in-water, O/W, or water-in-oil, W/O microemulsions) can behave as a potential reservoir of lipophilic or hydrophilic drugs, respectively. When the drug partitions between the dispersed and continuous phase, and when the system comes into contact with a semi-permeable membrane, the drug can be transported through the barrier. Drug release with pseudo-zero-order kinetics can be obtained, depending on the volume of the dispersed phase, the partition of the drug, and the transport rate of the drug.
4. The mean diameter of droplets in microemulsions is below 0.22 mm; they can be sterilized by filtration. The small size of droplet in microemulsions, for example, below 100 nm, yields a very large interfacial area, from which the drug can quickly be released into external phase when absorption (in vitro or in vivo) takes place, maintaining the concentration in the external phase close to initial levels.
5. Same microemulsions can carry both lipophilic and hydrophilic drugs.

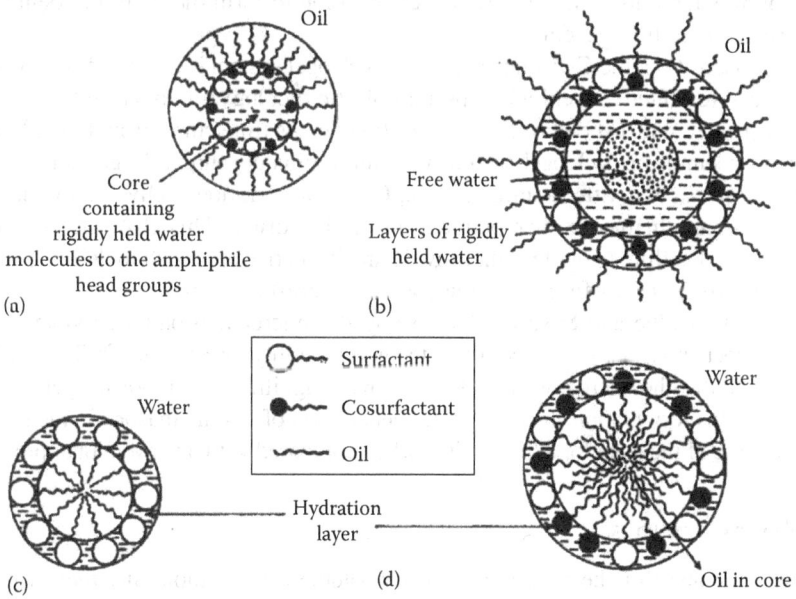

FIGURE 51.17 Pictorial representations of reverse micelles and microemulsions. (a) Reverse micelle, (b) water in oil microemulsion, (c) normal micelle, and (d) oil in water microemulsion. (Reprinted from *Adv. Colloid Interface Sci.*, 78(2), Moulik, S.P. and Paul, B.K., Structure, dynamics and transport properties of microemulsions, 99–195. Copyright 1998, with permission from Elsevier.)

6. Because of thermodynamic stability, microemulsions are easy to prepare and require no significant energy contribution during preparation. Microemulsions have low viscosity compared to other emulsions.
7. The use of microemulsion as delivery systems can improve the efficacy of a drug, allowing the total dose to be reduced and thus minimizing side effects.
8. The formation of microemulsion is reversible. They may become unstable at low or high temperature but when the temperature returns to the stability range, the microemulsion reforms.

Nevertheless, microemulsion-based systems have some disadvantages[197]:

1. Use of a large concentration of surfactant and co-surfactant necessary for stabilizing the nanodroplets.
2. Limited solubilizing capacity for high-melting substances.
3. The surfactant must be nontoxic for using pharmaceutical applications.
4. Microemulsion stability is influenced by environmental parameters such as temperature and pH. These parameters change upon microemulsion delivery to patients.

For the treatment of eye diseases, drugs are essentially delivered topically. O/W microemulsions have been investigated for ocular administration, to dissolve poorly soluble drugs, to increase absorption, and to attain prolonged release profile.[198]

The microemulsions containing pilocarpine were formulated using lecithin, propylene glycol, and PEG 200 as co-surfactant and isopropyl myristate as the oil phase. The formulations were of low viscosity with a refractive index lending to ophthalmologic applications.[199]

Zhu et al. investigated the potential of a microemulsion in situ electrolyte-triggered gelling system for the specific delivery of cyclosporine A (CsA) to external ocular tissue.[200] A CsA-loaded microemulsion was prepared using castor oil, Solutol HS 15 (surfactant), glycerol, and water. This microemulsion was then dispersed in a Kelcogel® solution to form the final microemulsion in situ electrolyte-triggered gelling system.

In vitro, the viscosity of the CsA microemulsion Kelcogel system increased dramatically on dilution with artificial tear fluid and exhibited pseudo-plastic rheology. in vivo results revealed that the $AUC0 \rightarrow 32$ h (the area under the plasma concentration versus time curve from 0 to 32 h) of corneal CsA for the microemulsion Kelcogel system was approximately threefold greater than for a CsA emulsion. Moreover, at 32 h after administration, CsA concentrations delivered by the microemulsion Kelcogel system remained at therapeutic levels in the cornea. This CsA microemulsion in situ electrolyte-triggered gelling system might provide an alternative approach to deliver prolonged precorneal residence time of CsA for preventing cornea allograft rejection.

In another study, Fialho and Silva-Cunha developed a microemulsion-based system that showed acceptable physicochemical behavior and presented good stability for 3 months.[201] The ocular irritation test used suggested that the microemulsion did not provide significant alteration to eyelids, conjunctiva, cornea, and iris. This formulation showed greater penetration of dexamethasone in the segment of the eye and also release of the drug for a longer time when compared with a conventional preparation.

51.8.3 Nanosuspensions

Nanosuspensions consist of the pure poorly water-soluble drug without any matrix material suspended in dispersion.[202] It is sub-micron colloidal dispersion of pure particles of drug stabilized by surfactants.[203] By formulating nanosuspensions, problems associated with the delivery of poorly water-soluble drugs and poorly water-soluble and lipid-soluble drugs can be solved. Nanosuspensions differ from nanoparticles,[204] which are polymeric colloidal carriers of drugs (nanospheres and nanocapsules), and from solid-lipid nanoparticles,[205] which are lipidic carriers of drug.

Conventionally, the drugs that are insoluble in water but soluble in the oil phase system are formulated in liposome, emulsion systems, but these lipidic formulation approaches are not applicable to all drugs. In these cases nanosuspensions are preferred.[206] In the case of drugs that are insoluble in both water and in organic media, instead of using lipidic systems nanosuspensions are used as a formulation approach. The nanosuspension formulation approach is most suitable for the compounds with high log p value, high melting point, and high dose.[207]

Therefore, an outstanding feature of the nanosuspension is the increase in saturation solubility and consequently an increase in the dissolution rate of the compound.[208]

Drug micro-particle suspensions can be milled by applying a high pressure homogenization process[209] leading to a product called nanosuspension.

Ophthalmic drug delivery, more than any other route of administration, may benefit to a full extent from the characteristics of nano-sized drug particles. The nano-size represents a state of matter characterized by higher solubility,[210] higher surface area available for dissolution,[211] higher dissolution rate,[212] higher bioadhesion,[213] and corneal penetration. It has been recommended that particles be less than 10 μm to minimize particle irritation to the eye, decrease tearing and drainage of instilled dose, and therefore increase the efficacy of an ocular treatment.

Kassema et al. used the high pressure homogenization method to prepare nanosuspensions of three practically insoluble glucocorticoid drugs: hydrocortisone, prednisolone, and dexamethasone.[214] The effect of particle size in the micron and nano-size ranges as well as the effect of viscosity of the nanosuspension on the ocular bioavailability was studied by measuring the IOP of normotensive Albino rabbits using shioetz tonometer. The results show that compared to solution and micro-crystalline suspensions, it is a common feature of the three drugs that the nanosuspensions always enhance the rate and extent of ophthalmic drug absorption as well as the intensity of drug action. In the majority of cases, nanosuspensions extend the duration of drug effect to a significant extent. The data presented confirms that nanosuspensions differ from micro-crystalline suspensions and solution as ophthalmic drug delivery systems and that the differences are statistically, highly, to very highly significant. The results confirmed also the importance of viscosity of nanosuspension especially in increasing the duration of drug action.

Eudragit RS100 (RS) and RL100 (RL) polymers are commonly used for the enteric coating of tablets and the preparation of controlled-release drug forms.[215] Their structures are shown in Figure 51.18. They are copolymers of poly(ethyl acrylate, methyl–methacrylate, and chlorotrimethyl–ammonioethyl methacrylate), containing an amount of quaternary ammonium groups between 4.5%–6.8% and 8.8%–12% for RS and RL, respectively. Both are insoluble at physiological pH values and capable of swelling,[216] thus representing good material for the dispersion of drugs.[217]

In this case, with the aim of improving the availability of sodium ibuprofen (IBU) at the intraocular level, Pignatello et al.[218] made IBU-loaded polymeric nanoparticle suspensions from inert polymer resins (Eudragit RS100®). The nanosuspensions were prepared by a modification of the quasi-emulsion solvent diffusion technique using variable formulation parameters (drug-to-polymer ratio, total drug and polymer amount, stirring speed). Nanosuspensions had mean sizes around 100 nm and a positive charge (ζ-potential of +40/+60 mV), which makes them suitable for ophthalmic applications. Stability tests (up to 24 months storage at 4°C or at room temperature) or freeze-drying were carried out to optimize a suitable pharmaceutical preparation. in vitro dissolution tests indicated a controlled release profile of IBU from nanoparticles. in vivo efficacy was assessed on the rabbit eye after the induction of an ocular trauma (paracentesis). An inhibition of the miotic response to the surgical trauma was achieved, comparable to a control aqueous eye-drop formulation, even though a lower concentration of free drug in the conjunctival sac was reached from the nanoparticle system. Drug levels in the aqueous humor were also higher after application of the nanosuspensions; moreover, IBU-loaded nanosuspensions did not show toxicity on ocular tissues.

Agnihotri and Vavia[219] prepared polymeric nanoparticle suspensions (NS) from poly(lactide-co-glycolide) and poly(lactide-*co*-glycolide-leucine) {poly[Lac(Glc-Leu)]} biodegradable polymers

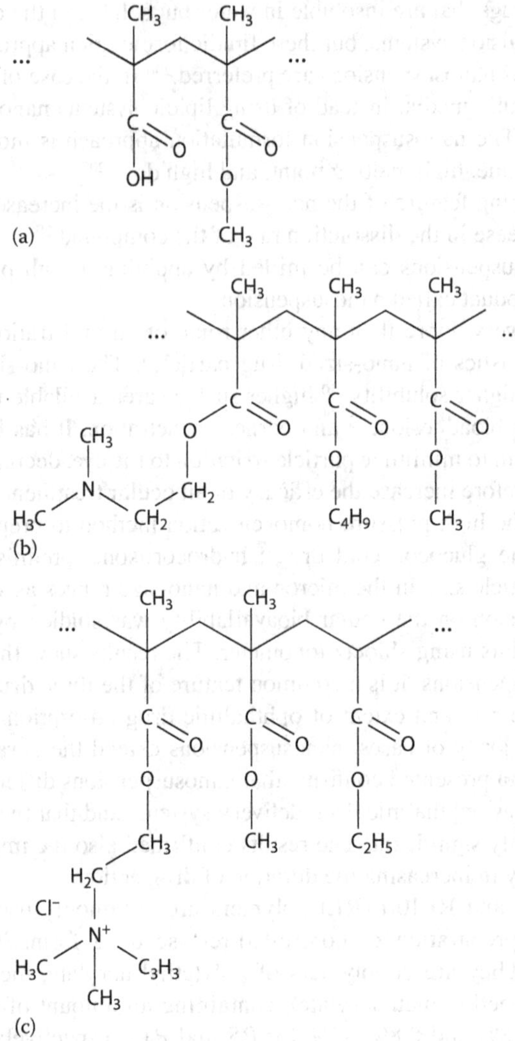

FIGURE 51.18 Structural monomer unit fragments of copolymers: (a) Eudragit L 100, (b) Eudragit E PO, and (c) Eudragit RL PO. (Reprinted with permission from *Mol. Pharm.*, 10(7), Moustafine, R.I., Bukhovets, A.V., Sitenkov, A.Y., Kemenova, V.A., Rombaut, P., and Van den Mooter, G., Eudragit E PO as a complementary material for designing oral drug delivery systems with controlled release properties: Comparative evaluation of new interpolyelectrolyte complexes with countercharged Eudragit L100 copolymers, 2630–2641. Copyright 2013 American Chemical Society.)

and loaded with diclofenac sodium (DS), with the aim of improving the ocular availability of the drug. NS were prepared by emulsion and solvent evaporation technique and characterized on the basis of physicochemical properties, stability, and drug release features. The nanoparticle system showed an interesting size distribution suitable for ophthalmic application. Stability tests (as long as 6 months' storage at 5°C or at 25°C/60% relative humidity) or freeze-drying were carried out to optimize a suitable pharmaceutical preparation. in vitro release tests showed an extended-release profile of DS from the nanoparticles. To verify the absence of irritation toward the ocular structures, blank NS were applied to rabbit eye and a modified Draize test performed. Polymer nanoparticles seemed to be devoid of any irritant effect on the cornea, iris, and conjunctiva for as long as 24 h after application, thus apparently a suitable inert carrier for ophthalmic drug delivery.

51.8.4 DENDRIMERS

Colloidal dosage forms, such as liposomes, nanoparticles, nanocapsules, microspheres, microcapsules,[22] and microemulsions,[220] can be used as drug reservoirs and can therefore prolong drug residence time on the cornea, consequently prolonging the drug's pharmacological activity.

However, liposomes or nanoparticles in contact with the ocular area tend to be removed by lacrimal drainage. Microparticulates are larger than nanoparticles, which may make them more appropriate for sustained or controlled drug release, but the larger size may make them less tolerable and therefore they may be eliminated from the eye by the flow of tears.[221]

To sustain drug release and prolong therapeutic activity, the colloidal particles must be retained in the ocular cul-de-sac after topical administration, and the entrapped drug released from the particles at an appropriate rate. To enhance particle retention in the ocular cul-de-sac, it is highly recommended to manufacture or coat the particles with bioadhesive materials.[222,223] These dosage forms can then bioadhere to the conjunctival epithelium and therefore increase contact time between these tissues and the drugs adsorbed or encapsulated into these carrier systems.

These bioadhesive polymers, however, are associated with problems like blurred vision and the formation of a veil in the corneal area, leading to loss of eyesight. To avoid these problems, dendrimers like poly(amidoamine) (PAMAM) (Figure 51.19)[224] are used, which are liquid or semi-solid polymers and have several amine, carboxylic, and hydroxyl surface groups, which increases with the generation number (G0, G1, G2, and so on). Because of this unique architecture, PAMAM

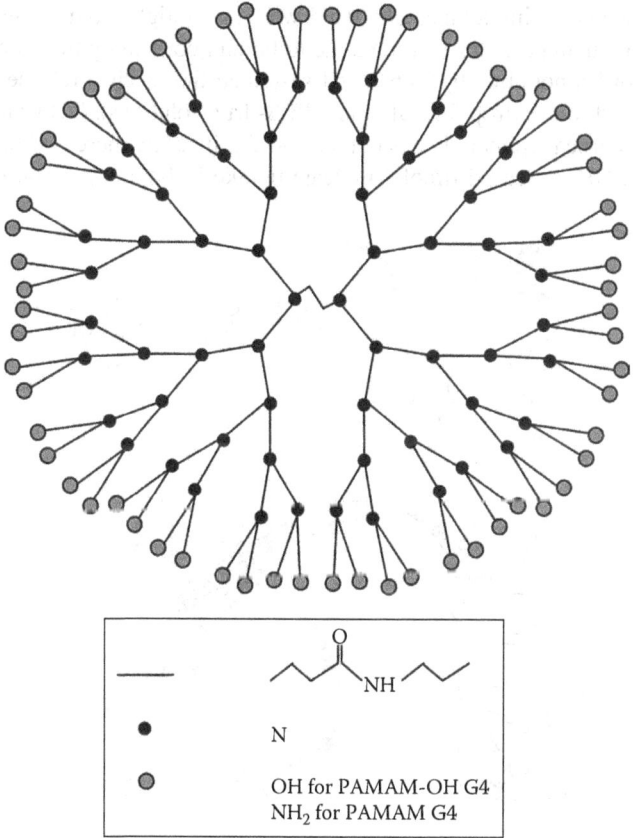

FIGURE 51.19 Schematic drawing of PAMAM dendrimers. (Reprinted from *Bioelectrochemistry*, 65(1), Klajnert, B., Sadowska, M., and Bryszewska, M., The effect of polyamidoamine dendrimers on human erythrocyte membrane acetylcholinesterase activity, 23–26. Copyright 2004, with permission from Elsevier.)

dendrimers are able to solubilize strongly and poorly water-soluble drugs into their inner zones containing cascading tiers of branch cells with radial connectivity to the initiator core and an exterior or surface region of terminal moieties.[225–227] Hence, greater possibilities can be explored by using dendrimers as ophthalmic drug delivery vehicles.

Vandamme and Brobeck determined the influence of a controlled incremental increase in size, molecular weight, and number of amine, carboxylate, and hydroxyl surface groups in several series of poly(amidoamine) dendrimers for controlled ocular drug delivery.[228] The duration of residence time was evaluated after solubilization of several series of PAMAM dendrimers (generations 1.5 and 2–3.5 and 4) in buffered phosphate solutions containing 2% (w/v) of fluorescein. The New Zealand albino rabbit was used as an in vivo model for qualitative and quantitative assessment of ocular tolerance and retention time after a single application of 25 μL of dendrimer solution to the eye (Figure 51.20).[229] The same model was also used to determine the prolonged miotic or mydriatic activities of dendrimer solutions, some containing pilocarpine nitrate and some tropicamide, respectively. Residence time was longer for the solutions containing dendrimers with carboxylic and hydroxyl surface groups. No prolongation of remanence time was observed when dendrimer concentration (0.25%–2%) increased. The remanence time of PAMAM dendrimer solutions on the cornea showed size and molecular weight dependency. This study allowed novel macromolecular carriers to be designed with prolonged drug residence time for the ophthalmic route.

Kadam et al.[230] prepared dendrimer hydrogel (DH), made from ultraviolet-cured polyamidoamine dendrimer G3.0 tethered with three polyethylene glycol (PEG, 12,000 Da)–acrylate chains (8.1% w/v) in pH 7.4 phosphate buffered saline (PBS) (Figure 51.21), and studied the delivery of brimonidine (0.1% w/v) and timolol maleate (0.5% w/v), two antiglaucoma drugs. DH was found to be mucoadhesive to mucin particles and nontoxic to human corneal epithelial cells. DH increased the PBS solubility of brimonidine by 77.6% and sustained the in vitro release of both drugs over 56–72 h. As compared to eye drop formulations (PBS-drug solutions), DH brought about substantially higher human corneal epithelial cells uptake and significantly increased bovine corneal transport for both drugs. DH increased timolol maleate uptake in bovine corneal epithelium, stroma,

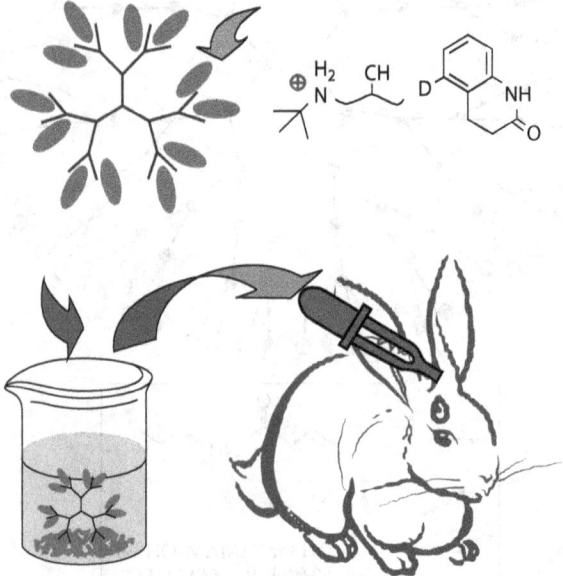

FIGURE 51.20 Ocular drug delivery by dendrimers. (Reprinted from *Eur. J. Med. Chem.*, 45(1), Spataro, G., Malecaze, F., Turrin, C.O., Soler, V., Duhayon, C., Elena, P.P., Majoral, J.P., and Caminade, A.M., Designing dendrimers for ocular drug delivery, 326–334, 2010. Copyright 2013, with permission from Elsevier Masson SAS. All rights reserved.)

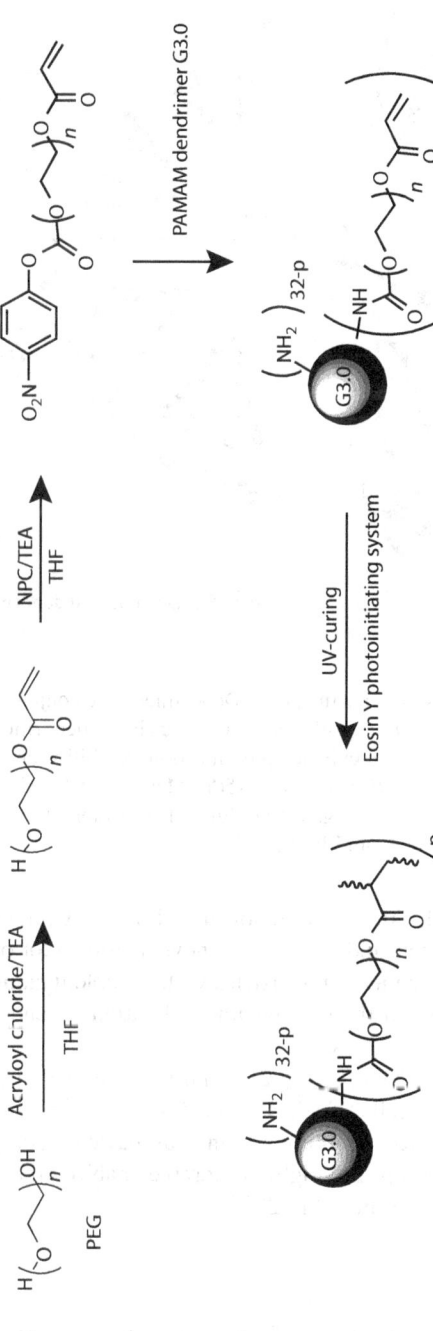

FIGURE 51.21 Synthesis route of photoreactive G3.0–PEG acrylate conjugates. *Abbreviations:* NPC, 4-nitrophenyl chloroformate; TEA, triethylamine; THF, tetrahydrofuran. (Reprinted from *Nanomed. Nanotechnol.*, 8(5), Kadam, R., Jadhav, G., Kompella, U.B., and Yang, H. Polyamidoamine dendrimer hydrogel for enhanced delivery of antiglaucoma drugs, 776–783. Copyright 2012, with permission from Elsevier.)

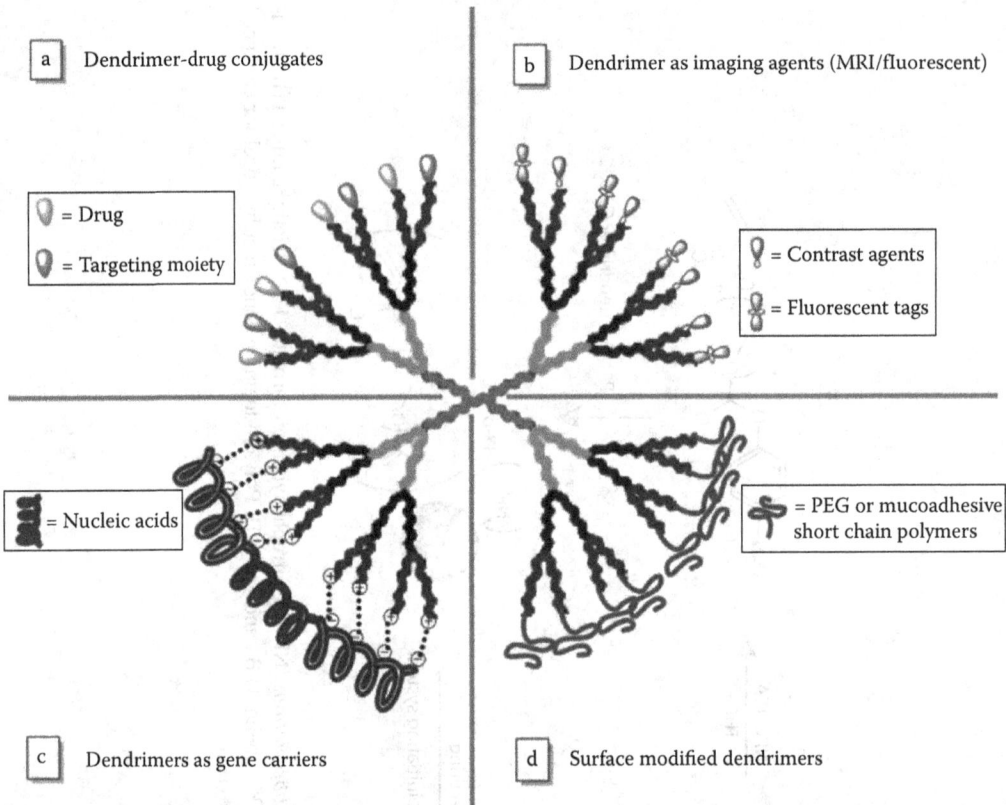

FIGURE 51.22 Potential applications of dendrimers. (a) Dendrimer drug conjugates, dendrimers linked to targeting moieties and imaging agents. (b) Encapsulation of the drugs in the dendritic interiors. (c) Dendrimers incorporated into various delivery systems for enhancing permeation, solubility, and so on. (d) Dendrimers as complexing agents. (Reprinted from *Drug Discov. Today*, 15(5), Menjoge, A.R., Kannan, R.M., and Tomalia, D.A., Dendrimer-based drug and imaging conjugates: Design considerations for nanomedical applications, 171–185. Copyright 2010, with permission from Elsevier.)

and endothelium by 0.4- to 4.6-fold. This work demonstrated that DH can enhance the delivery of antiglaucoma drugs in multiple aspects and represents a novel platform for ocular drug delivery.

Surface functionalization of dendrimer nanoparticles with suitable ligands, size, surface groups, and surface charge plays a key role on the cellular internalization by endocytosis, phagocytosis, micropinocytosis, or fluid-phase endocytosis.

Dendrimers have a well-defined shape, size (~2–20 nm), and narrow polydispersity. They are readily transported into and out of cells. Their surface functional groups enable multiple drug molecules to be attached to the surface, enabling a high drug payload, compared to typical linear polymers. The surface functional groups are highly tailorable enabling the attachment of multiple imaging moieties, drugs, and ligands (Figure 51.22).[231]

51.8.5 Niosomes

Non-ionic surfactant-based vesicles (niosomes) are formed from the self-assembly of non-ionic amphiphiles in aqueous media resulting in closed bilayer structures (Figure 51.23).[232] The assembly into closed bilayers is rarely spontaneous[233] and usually involves some input of energy such as physical agitation or heat. The result is an assembly in which the hydrophobic parts of the molecule are shielded from the aqueous solvent and the hydrophilic head groups enjoy maximum contact with the same.

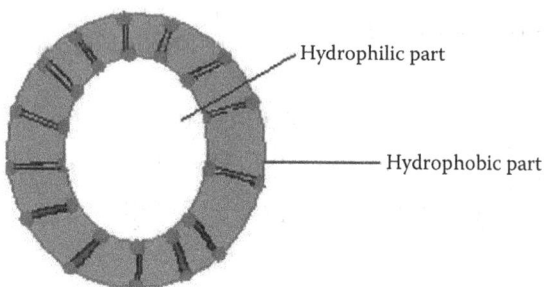

FIGURE 51.23 Niosome structure. (Reprinted from *Adv. Colloid Interface Sci.*, 183–184, Mahale, N.B., Thakkar, P.D., Mali, R.G., Walunj, D.R., and Chaudhari, S.R., Niosomes: Novel sustained release nonionic stable vesicular systems—An overview, 46–54. Copyright 2012, with permission from Elsevier.)

These structures are analogous to phospholipid vesicles (liposomes) and are able to encapsulate aqueous solutes and serve as drug carriers. The low cost, greater stability, and resultant ease of storage of non-ionic surfactants[234] has led to the exploitation of these compounds as alternatives to phospholipids.

Differences in characteristics exist between liposomes and niosomes, especially since niosomes are prepared from uncharged single-chain surfactant and cholesterol whereas liposomes are prepared from double-chain phospholipids (neutral or charged).

Different types of nonionic surfactants that are known to form vesicles are shown in Figure 51.24.[235]

Such amphiphiles by definition must possess a hydrophilic head group and a hydrophobic tail. The hydrophobic moiety may consist of one or two alkyl or perfluoroalkyl groups or in certain cases a single steroidal group.

Nonionic surfactant vesicles have been reported successfully, as ocular vehicles for cyclopentolate. In the in vivo study, niosomes, independent of their pH, significantly improved the ocular bioavailability of cyclopentolate, with respect to reference buffer solution, indicating that it can be used as an efficient vehicle for ocular drug delivery.[236]

51.8.6 Discomes

Modified forms of niosomes, the discomes, are also used in ophthalmic drug delivery systems. This phase consists of vesicles of 12–16 μm in diameter, which encapsulates aqueous solutes such as carboxyfluorescein (CF), and is derived from niosomes by the addition of non-ionic surfactant, which is Solulan C24. These large vesicles were found to be of two types: large vesicles that appear ellipsoid in shape and large vesicles that are truly discoid.

They have particular advantages for ocular drug delivery, since, as a result of their larger size, they can prevent their drainage into the systemic pool; also, their disc shape could provide a better fit in the cul-de-sac of the eye.[237] Discomes are prepared by the progressive incorporation of Solulan C24 into the vesicular dispersion that leads to the partitioning of this soluble surfactant into the lipid bilayer until a critical level is reached. This results in the formation of a large flattened disc-like structure (discomes) in place of the spherical structures.[236] The studies carried out by Vyas showed that the entrapment efficiency of timolol maleate (water soluble drug) is higher in discomes than niosomes.[238] Moreover, an increase in ocular bioavailability was observed when timolol maleate was entrapped in niosomes and discomes compared with maleate solution.

51.8.7 Nanoparticle-Loaded Contact Lenses

Conventional CLs have been used for years as a type of sustained release drug delivery system, usually used by soaking the lens in a drug solution and allowing retention of the drug in the 70% water component of the lenses. The drug was then allowed to release to the ocular surface once the

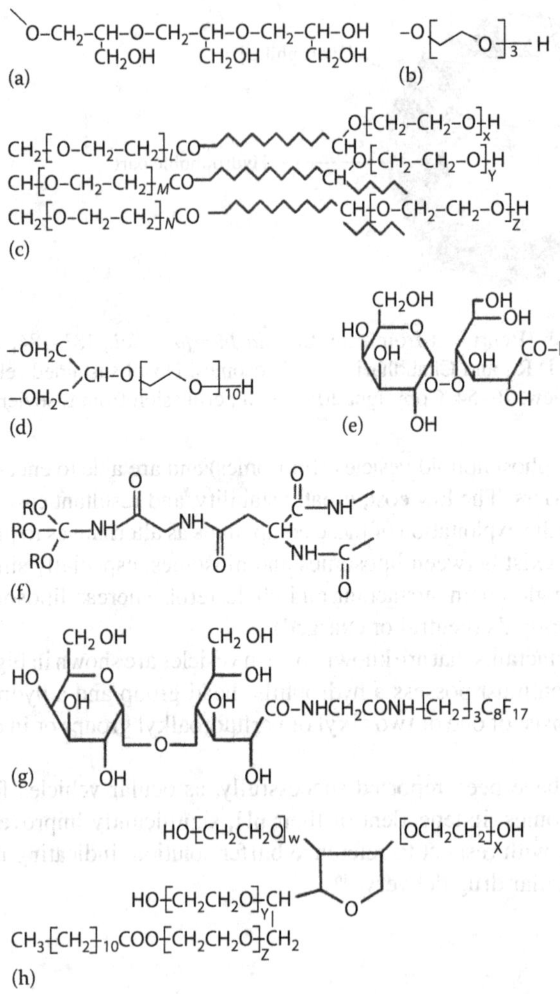

FIGURE 51.24 (a) Glycerol head group, (b–d) ethylene oxide head groups, (e, f) sugar head group + amino acid, (g) vesicle forming fluorinated surfactant, and (h) polysorbate 20. (Reprinted from *Int. J. Pharm.*, 172(1–2), Uchegbu, I.F. and Vyas, S.P., Non-ionic surfactant based vesicles (niosomes) in drug delivery, 33–70. Copyright 1998, with permission from Elsevier.)

contact had been placed on the eye. Although delivery of a drug via a CL does improve the contact time of the drug compared with topical eye drops, drugs released by a CL results in a burst of drug at therapeutic levels only for a few hours.[239,240]

Drug-eluting CLs composed of a polymer-drug film encapsulated within a poly-2-hydroxyethyl methacrylate (pHEMA) hydrogel have been described.[241,242] This CL has been shown to deliver ciprofloxacin with zero-order kinetics up to 4 weeks.[241] Also, poly-2-hydroxyethyl methacrylate hydrogels containing beta-cyclodextrin (pHEMA/beta-CD) have been investigated as a platform for the sustained release of ophthalmic drugs. In rabbit eyes, use of this CL allowed higher mean residence time and aqueous humor concentration of a drug compared to topical application. These specialized drug-eluting CLs may prove to be very beneficial in the treatment of ocular surface and anterior segment disease in veterinary patients.[242]

Graziacascone et al.[243] published a study on encapsulating lipophilic drugs inside nanoparticles and entrapping the particles in hydrogels. They used PVA hydrogels as hydrophilic matrices for the release of lipophilic drugs loaded in PLGA particles. They compared the drug release rates from hydrogels loaded with the particles with the delivery rates directly from the PLGA particles and found comparable results, which implies that the particles controlled the drug release rates. This current paper deals with the incorporation of drug-laden nanoparticles in a p-HEMA CL matrix in a manner such that the particle-laden gels are transparent and can release drugs at therapeutic rates.

As mentioned earlier, the essential idea is to encapsulate the ophthalmic drug formulations in nanoparticles and disperse these drug-laden particles in the lens material. Upon insertion into the eye, the lens will slowly release the drug into the pre-lens (the film between the air and the lens) and the post-lens (the film between the cornea and the lens) tear films, and thus provide drug delivery for extended periods of time. Gulsen and Chauhan[244] focus on dispersing stabilized microemulsion drops in (p-HEMA) hydrogels (Figure 51.25). The results of this study show that the p-HEMA gels loaded with a microemulsion that is stabilized with a silica shell are transparent and that these gels release drugs for a period of over 8 days. CLs made of microemulsion-laden gels are expected to deliver drugs at therapeutic levels for a few days. The delivery rates can be tailored by controlling the particle and the drug loading. It may be possible to use this system for both therapeutic drug delivery to the eyes and the provision of lubricants to alleviate eye problems prevalent in extended lens wear.

It has also been proposed to create colloid-laden hydrogel in situ in one step. Surfactant-laden hydrogels can be prepared by the addition of surfactants to the polymerizing mixture. A schematic of the microstructure of the surfactant-laden gels is shown in Figure 51.26.[245]

During the process of fabrication, the surfactants interact with polymer chains and form micelles creating hydrophobic cores, where the hydrophobic drugs will preferentially enter into. The drug transport is inhibited due to the presence of surfactant micelles. Kapoor et al. prepared Brij surfactant-laden p-HEMA hydrogels that can release Cyclosporine A (CyA) at a controlled rate for extended periods of time (20 days). Their results show that Brij surfactant-laden p-HEMA gels provide extended release of CyA and possess suitable mechanical and optical properties for CL applications. However, the hydrogels are not as effective for extended release of two other hydrophobic ophthalmic drugs, i.e., dexamethasone (DMS) and dexamethasone 21 acetate (DMSA), because of insufficient partitioning inside the surfactant aggregates.[246,247]

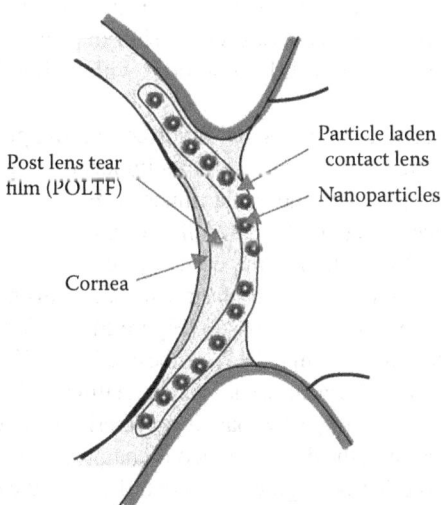

FIGURE 51.25 Schematic illustration of the particle-laden lens inserted in the eye. (Reprinted from *Int. J. Pharm.*, 292(1–2), Gulsen, D. and Chauhan, A., Dispersion of microemulsion drops in HEMA hydrogel: A potential ophthalmic drug delivery vehicle, 95–117. Copyright 2005, with permission from Elsevier.)

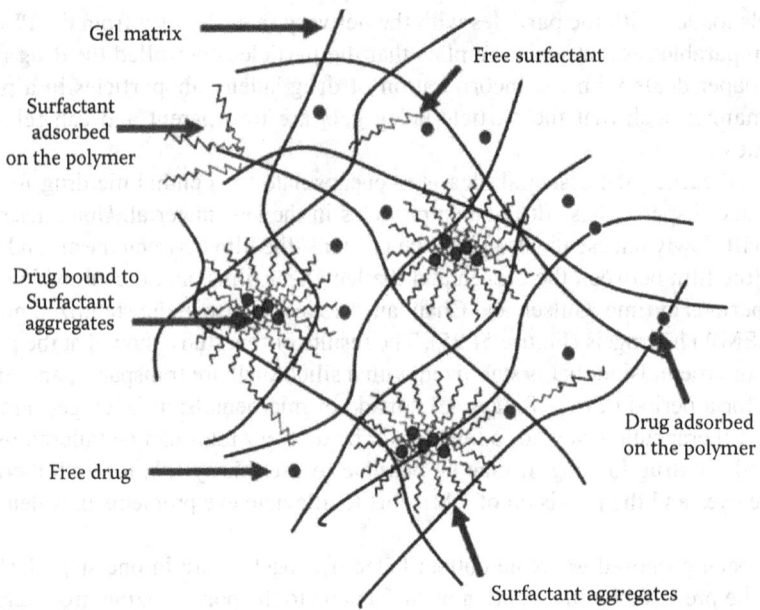

FIGURE 51.26 A schematic of the microstructure of the surfactant-laden gels. (From Hu, X. et al., *Int. J. Polym. Sci.*, 2011, 1, 2011. With permission from Hindawi.)

51.8.8 Polymeric Nanoparticles

For ophthalmic applications, properly formulated drug-loaded nanoparticles (DNPs) are reported to provide ease of application just like eye drop solutions, with the added advantage of being patient friendly, due to less frequent application and extended duration of retention in the extraocular portion. The drug may be attached to a nanoparticle matrix, or dissolved, encapsulated, and entrapped, giving rise to different terminologies as nanoparticles, nanospheres, or nanocapsules.[248]

The potential of polymeric nanoparticles as an ocular drug delivery system has been explored by a colloidal system consisting of an aqueous suspension of nanoparticles. These nanoparticles can be rapidly fabricated under extremely mild conditions with their ability to incorporate bioactive compounds.[249]

The stability of colloidal particles in biological fluids containing proteins and enzymes is a crucial issue, because the size of the nanoparticles plays an important role in its ability to interact with mucosal surfaces and, in particular, with the ocular mucosa (Figure 51.27).[248]

Drug-loaded polymeric micelles should be carefully designed in order to deliver the drug at the site of its action. If the site of drug action is the cornea or conjunctiva, polymeric micelles should be retained at the ocular surface long enough to ensure sustained drug release as well as more efficient drug permeation in the front eye surface tissue (Figure 51.28, Case 1).[250] At the same time, the adsorption of intact drug-loaded micelles may be hypothesized providing sustained drug release and prolonged therapeutic activity in the corneal tissue. When dealing with the drug whose site of action is aqueous humor associated tissues, the drug-loaded polymeric micelles must have potential to be adsorbed across the cornea, releasing the drug in aqueous chamber (Figure 51.28, Case 2). In the case of the diseases affecting the posterior eye segment, and in order to evade the invasive ocular injections thereby improving patient compliance, polymeric micelles should reach the posterior of the eye most probably by trans-scleral route around the conjunctiva, through the sclera, choroid, and finally retina (Figure 51.28, Case 3). However, after topical application, polymeric micelles will encounter different challenges that they have to overcome in order to successfully deliver the drug to the site of action.

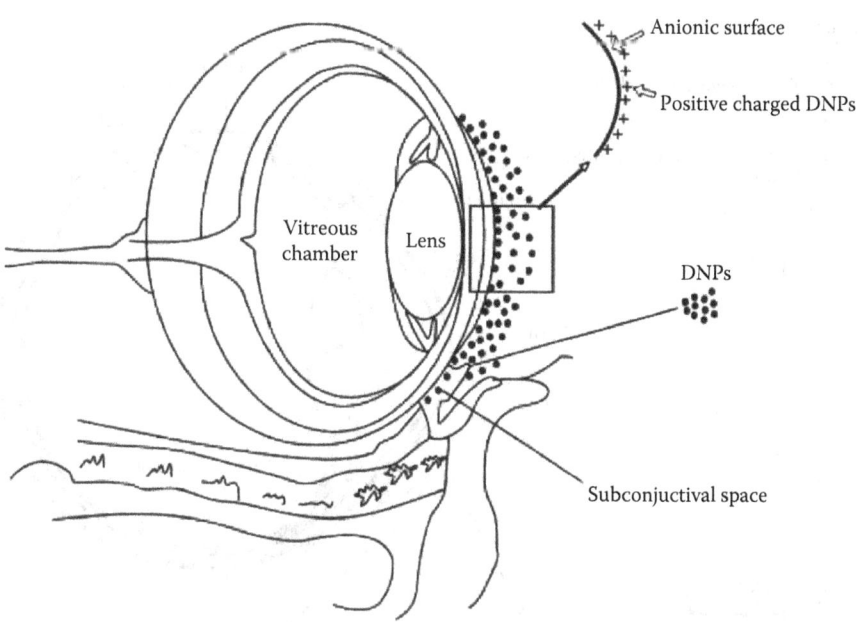

FIGURE 51.27 Interaction of DNPs to ocular surface and subconjunctival space of human eye. (Reprinted from *J. Contr. Release*, 136(1), Nagarwal, R.C., Kant, S., Singh, P.N., Maiti, P., and Pandit, J.K., Polymeric nanoparticulate system: A potential approach for ocular drug delivery, 2–13. Copyright 2009, with permission from Elsevier.)

The formulation of biodegradable polymers as a colloidal system holds significant promise for ophthalmic drug delivery. A colloidal system is suitable for poorly water soluble drugs, and would allow drop-wise administration while maintaining the drug activity at the site of action.[251] To achieve sustained drug release and prolonged therapeutic activity, the particles need to be retained in the ocular cul-de-sac after topical administration, and the consequent release of the drug from the particles at an appropriate rate. If the drug leaches out of the particles too fast or too slow, then there will be either little sustained drug release, or the concentration of the drug in the tears may be too low to allow adequate drug penetration into ocular tissues.[221] It is important that the particle size for ophthalmic applications be within the nano range because with larger sizes a scratching feeling of foreign body sensation might occur.[252] For effective retention in ocular cul-de-sac, it is essential to fabricate the particles with bioadhesive materials. Without bioadhesion, nanoparticles are eliminated from the precorneal site almost as quickly as aqueous solution.

DNPs of various sizes based on polymers and biomaterials such as poly(lactide-*co*-glycolide) (PLGA), poly(lactide) (PLA), poly ε-caprolactone, albumin, and chitosan have been developed and tested in various cell culture and animal models[253,254] in order to improve ocular drug delivery.

For instance, Yadav and Ahuja prepared nanoparticles using gum cordia as the polymer and to evaluate them for ophthalmic delivery of fluconazole.[255] A w/o/w emulsion containing fluconazole and gum cordia in aqueous phase, methylene chloride as the oily phase, and di-octyl sodium sulfosuccinate and polyvinyl alcohol as the primary and secondary emulsifiers, respectively, were cross-linked by the ionic gelation technique to produce a fluconazole-loaded nanoreservoir system. The formulation of nanoparticles was optimized using response surface methodology. Multiple response simultaneous optimizations using the desirability approach were used to find optimal experimental conditions. The optimal conditions were found to be concentrations of gum cordia (0.85%, w/v), di-octyl sodium sulfosuccinate (9.07%, w/v), and fluconazole (6.06%, w/v). On comparison of the optimized nanosuspension formulation with commercial formulation, it was found to provide comparable in vitro corneal permeability of

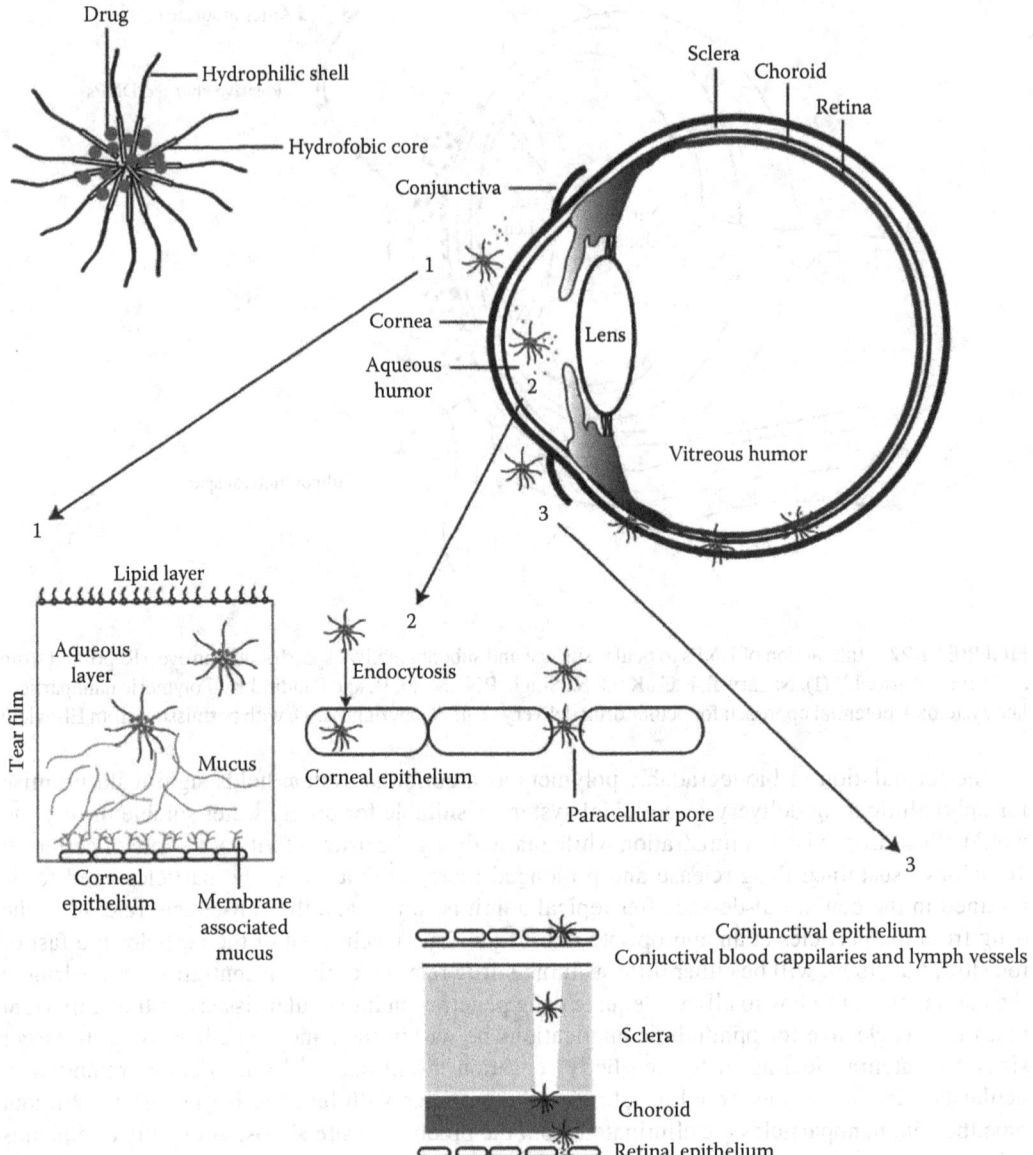

FIGURE 51.28 Fate of drug-loaded polymeric micelles after topical ocular application. (From Pepic, I. et al., *Chem. Biochem. Eng. Q.*, 26(4), 365, 2012. With permission from Croatian Society of Chemical Engineers.)

fluconazole across isolated goat cornea, indicating suitability of nanosuspension formulation in the ophthalmic delivery of fluconazole.

Bourges et al.[256] studied the kinetics of polylactide (PLA) nanoparticle (NP) localization within the intraocular tissues and to evaluate their potential to release encapsulated material. Environmental scanning electron microscopy (ESEM) showed the flow of the NPs from the site of injection into the vitreous cavity and their rapid settling on the internal limiting membrane. Histology demonstrated the anatomic integrity of the injected eyes and showed no toxic effects. A mild inflammatory cell infiltrate was observed in the ciliary body 6 h after the injection and in the posterior vitreous and retina at 18–24 h. The intensity of inflammation decreased markedly by 48 h. Confocal and fluorescence microscopy and immunohistochemistry showed that a transretinal movement of the NPs was gradually taking place with a later localization in the RPE cells. Rh (a fluorochrome) encapsulated within the injected NPs diffused and stained the retina and RPE cells. PLA NPs were still present within the RPE cells 4 months after a single intravitreous injection.

Gupta et al.[257] developed and evaluated a new colloidal system, i.e., poly(DL lactide-co-glycolide) (PLGA) nanoparticles for sparfloxacin ophthalmic delivery, to improve precorneal residence time and ocular penetration. Nanoparticles were prepared by nanoprecipitation technique and characterized for various properties such as particle size, zeta potential, in vitro drug release, statistical model fitting, stability, and so forth. Microbiological assay was carried out against *P. aeruginosa* using the cup-plate method. Precorneal residence time was studied in albino rabbits by gamma scintigraphy after radiolabeling of sparfloxacin by Tc-99m. Ocular tolerance of the developed nanosuspension was also studied by the Hen Egg Test-Chorioallantoic Membrane (HET-CAM) method. The developed nanosuspension showed a mean particle size in the range of 180–190 nm, suitable for ophthalmic application with zeta potential of –22 mV. in vitro release from the developed nanosuspension showed an extended release profile of sparfloxacin according to the Peppas model. Acquired gamma camera images showed good retention over the entire precorneal area for the developed nanosuspension compared with that of a marketed formulation.

The marketed drug formulation cleared very rapidly from the corneal region and reached the systemic circulation through the nasolacrimal drainage system, as significant radioactivity was recorded in the kidney and bladder after 6 h of ocular administration, whereas the developed nanosuspension cleared at a very slow rate and remained at the corneal surface for longer duration, as no radioactivity was observed in the systemic circulation. HET-CAM assay with 0 score in 8 h indicates the nonirritant property of the developed nanosuspension. The developed lyophilized nanosuspension was found to be stable for a longer duration of time than the conventional marketed formulation with a good shelf life.

Polymeric nanoparticles have many advantages such as the following:

- Easy to fabricate and characterize.
- Process is inexpensive so the final cost of formulation is reduced.
- Polymers are biocompatible, biodegradable, non-toxic, and nonimmunogenic, which is suitable for ocular application.
- Mostly polymers are water soluble such as chitosan and sodium alginate.
- Applicable to a broad category of small molecules (drugs), genes, proteins, and polynucleotides.
- Can be lyophilized and are stable after reconstitution with medium for application.
- Do not produce irritation to eye.
- Mucoadhesive property of polymer (e.g., chitosan) increases the retention time at the site, which is helpful for bioavailability improvement.

51.9 DIFFERENT PROPERTIES OF NANOPARTICULATE-BASED POLYMERIC SYSTEMS IN OCULAR THERAPY

Successful results have been obtained with polymeric colloidal nanoparticles. Treatment with these systems increases bioavailability, reduces the frequency of administration, and promotes targeting of drugs to specific sites.

Nanoparticles have been intensively investigated to deliver a variety of small and large chemical entities like drugs, polypeptides, proteins, vaccines, and genes for improved utilization and reduced toxic side effects. Since DNPs reside in the target tissues and in circulation for an extended period of time, the biocompatibility and biodegradability of the polymers used in their preparation are the two basic prerequisites. In order to improve ocular bioavailability and extended drug effect in targeted tissues, the implementation of mucoadhesive polymers is necessary. Due to interactions with the mucus layer or the eye tissues, an increase in the precorneal residence time of the preparation will be observed.

A variety of polymers fulfilling these requirements have been utilized, and their utility vis-à-vis the characteristics of the DNPs fabricated with them make an extremely interesting study in the development of an almost ideal therapeutics tool for the treatment of a variety of ocular disorders.

51.9.1 Mucoadhesive Polymers as the Penetration Enhancers in Ocular Drug Delivery

Bioadhesion can be defined as the process by which a natural or a synthetic polymer can adhere to a biological substrate. When the biological substrate is a mucosal layer, the phenomenon is known as mucoadhesion. The substrate possessing bioadhesive property can help in devising a delivery system capable of delivering a bioactive agent for a prolonged period of time at a specific delivery site.

Penetration enhancers can improve corneal barrier restrictions. From a safety point of view, it is important to maintain the viability of the living epithelial cells after application of the enhancer.[258] Among the penetration enhancers studied, cationic polymers such as chitosan, aminated gelatin, and poly-L-arginine are reported to increase the transepithelial absorption of peptide drugs by dissociation of tight junction assemblies that restrict the paracellular permeation in intestinal and nasal epithelia without producing significant epithelial damage. Thus, poly-cationic polymers may be useful penetration enhancers for ocular drug delivery.[254] The mechanism of mucoadhesion of chitosan is due to an ionic interaction between the positively charged amino groups of chitosan and the negatively charged sialic acid residues in mucus.

Various theories (electronic, adsorption, wetting, diffusion, or interpenetration) were proposed to explain bioadhesion or mucoadhesion. In order to be a good mucoadhesive adjuvant, the polymer of the drug delivery system must make intimate contact with the mucus layer. The polymer chains must be mobile and flexible enough to interdiffuse into the mucus and penetrate to a sufficient depth in order to create a (strong entangled) network. They should interact with mucins by hydrogen bonding, electrostatic and hydrophobic interactions.[259]

Polymer-related factors influencing mucoadhesion are hydration or degree of swelling, molecular weight, functional groups, molecular conformation or chain flexibility and mobility, and concentration.

Polymer hydration results in the relaxation of stretched, entangled, or twisted macromolecules, exposing the adhesive sites. Furthermore, chain interdiffusion is favored by polymer–water interactions dominating the corresponding polymer–polymer interactions.[260]

An overview of important polymers investigated by several research groups is given in Table 51.3.[258] The polymers are categorized according to mucoadhesive properties even if, due to various experimental approaches applied in the different studies, it is difficult to assign a mucoadhesive capacity to each polymer in order to allow comparison.

A general conclusion that can be drawn from Table 51.3 is that charged polymers both anionic and cationic demonstrate a better mucoadhesive capacity in comparison to non-ionic cellulose-ethers or polyvinyl alcohol (PVA).

Cationic polymers were probably superior mucoadhesives due to the ability to develop molecular attraction forces by electrostatic interactions with the negative charges of the mucus; the polycationic chitosan was investigated as an ophthalmic vehicle. The polymer is biodegradable, biocompatible, and non-toxic. It possesses antimicrobial and wound-healing properties. Moreover, chitosan exhibits a pseudoplastic and viscoelastic behavior.[261]

Khangtragool et al.[262] studied about chitosan as an ocular drug delivery vehicle for vancomycin. Their study demonstrated quite clearly that a 0.3% w/v chitosan solution in 1% aqueous L-lactic acid offers several advantages as a vehicle for the ophthalmic delivery of vancomycin. These advantages include controlled drug delivery for the eye, biocompatibility, storage stability, and cost

TABLE 51.3
Viscosifying Polymers Screened for Ocular Mucoadhesive Capacity

Polymer	Charge	Mucoadhesive Capacity
Poly(acrylic acid) (neutralized)	A	+++
Carbomer (neutralized)	A	+++
Hyaluronan	A	+++
Chitosan	C	++
Na carboxymethyl cellulose	A	++(+)
Poly(galacturonic acid)	A	++
Na alginate	A	++(+)
Pectin	A	++(+)
Xanthan gum	A	+
Xyloglucan gum	A	+
Scleroglucan	A	+
Poloxamer	NI	+(+)
Hydroxypropyl methylcellulose	NI	+
Methylcellulose	NI	+
Poly(vinyl alcohol)	NI	+
Poly(vinyl pyrrolidone)	NI	+

Source: Reprinted from *Adv. Drug Deliv. Rev.*, 57(11), Ludwing, A., The use of mucoadhesive polymers in ocular drug delivery, 1595–1639. Copyright 2005, with permission from Elsevier.

Notes: A, anionic; C, cationic; NI, non-ionic; +++, excellent; ++, good; +, poor/absent.

effectiveness. The results have shown that the 0.3% w/v chitosan solution enhances drug delivery due to a combination of viscosity and mucoadhesive effects. The sol-gel transition is not considered to exert a significant influence on the drug delivery. Additionally, the physical properties of the solution showed good compatibility when prepared extemporaneously as eye drops. Based on these findings, it is concluded that the 0.3% w/v chitosan solution shows considerable potential for the topical ocular administration of vancomycin. By enhancing the bioavailability of the drug, it holds the added attraction for health care teams of reducing the frequency of application of topical eye drops. Thus, it is both convenient to use and, due to its low cost, affordable by hospital administrations operating under strict budget limitations. In the wider context of this research, the conversion of chitosan, a relatively inexpensive derivative of a natural polymer (chitin) produced in Thailand, into a high value-added product for use in such a specialist application as this reflects the need for research to identify ways in which it can reduce the reliance on expensive imported products.

Besides chitosan, numerous polysaccharides were evaluated as mucoadhesive ophthalmic vehicles: polygalacturonic acid, xyloglucan, xanthan gum, gellan gum, pullulan, guar gum, scleroglucan, and carrageenan.

Due to interactions with the mucus layer or the eye tissues, an increase in the precorneal residence time of the preparation was observed. Some mucoadhesive polymers showed not only good potential to increase the bioavailability of the drug applied, but also protective and healing properties to epithelial cells.[263,264]

51.9.2 Biodegradable Polymers for Ocular Drug Delivery

Biodegradable ocular drug delivery devices are composed of biocompatible polymers, which degrade into nontoxic byproducts, or polymers that solubilize in vivo and can be eliminated safely by the human body.

Colloidal nanosystems based on biodegradable polymeric materials that combine the capabilities of stimulus response and molecular recognition promise significant improvements in the ocular delivery of therapeutic agents.

Biodegradable polymers can be combined with drugs in such a way that the drug is released into the eye in a very careful and controlled manner. The formulation of biodegradable or bioerodible polymers as water-based colloidal nanosystems holds significant promise for ophthalmic drug delivery. For retinal drug delivery, biodegradable polymers are preferable and in most cases required. Both lactic acid and glycolic acid are biodegradable and they are produced by the body and eliminated as carbon dioxide and water.

It has been found that biodegradable polymers can be combined with drugs in such a way that the drug is released into the eye in a very precise and controlled manner.

Various researches have concentrated on biodegradable polymers, many of natural origin.[251,265–269] Work on biodegradable cross-linkers that are commonly used for the construction of three-dimensional hydrogel networks has been reviewed.[270]

Materials based on poly(lactic) and poly(glycolic acid)s, or their copolymers and derivatives, have been formulated as implants[266] as micro/nanospheres and nanocapsules[267] and as films for glaucoma treatment.[265] The in vitro release of methotrexate from poly(D,L-lactide)-poly(ethylene glycol) diblock copolymer nanocapsules has been compared with that from poly(D,L-lactide) nanocapsules.[268] Biodegradable poly(lactide-*co*-glycolide) microparticles have shown sustained retinal delivery of celecoxib and have inhibited diabetes-induced retinal oxidative damage.[270] In these materials, the drug is released by bulk erosion of the matrix following the cleavage of the polymeric chains via autocatalytic acid/base and/or enzymatic hydrolysis; the products, lactic and glycolic acids, are metabolized to carbon dioxide and water. Comparative studies have shown a faster degradation of poly(lactic-*co*-glycolic acid) (PLGA) in vivo, suggesting a cell-mediated degradation process of the polymer, caused by giant cells and hydrolytic enzymes. Low molecular weight polymers tend to degrade rapidly; copolymers such as PLGA degrade faster than the corresponding homopolymers.

These materials are particularly suitable for implantable devices or injectable micro (nano) spheres.[72]

51.10 CONCLUSION

Improving the bioavailability of ocular drugs is a great challenge. It is also important to achieve an optimal drug concentration at the targeted site. Pre-corneal deficit factors have a crucial role in determining the bioavailability of a drug. These factors can include tear dynamics, impermeability of the corneal epithelium membranes, momentary residence in the fornix conjunctiva, and nonspecific absorption among other things. Therefore, myriad advances have been made to overcome these physiological barriers for the targeted ocular delivery of drugs.

REFERENCES

1. Jitendra, S.P.K.; Banik, A.; Dixit, S. A new trend: Ocular drug delivery system. *Pharm. Sci. Monitor* 2011, *2* (2), 1–25.
2. Lloyd, A.W.; Faragher, R.G.A.; Denyer, S.P. Ocular biomaterials and implants. *Biomaterials* 2001, *22* (8), 769–785.
3. Greaves, J.L.; Wilson, C.G. Treatment of diseases of the eye with mucoadhesive delivery systems. *Adv. Drug Deliv. Rev.* 1993, *11* (3), 349–383.
4. Klintworth, G.K. The cornea structure and macromolecules in health and disease: A review. *Am. J. Pathol.* 1977, *89* (3), 718–808.
5. Waugh, A.; Grant, A. *The Special Senses Ross and Wilson Anatomy and Physiology in Health and Illness*. Churchill Livingstone: London, U.K., 2012; pp. 197–207.

6. Chien, Y.W. Ocular drug delivery and delivery systems. In *Novel Drug Delivery Systems*, 2nd edn. Marcel Dekker: New York, 1996; pp. 269–270.
7. Robinson, J.C. Ocular anatomy and physiology relevant to ocular drug delivery. In *Ophthalmic Drug Delivery Systems*; Mitra, A.K., Ed. Marcel Dekker: New York, 1993; pp. 29–57.
8. King-Smith, P.E.; Fink, B.A.; Fogt, N.; Nichols, K.K.; Hill, R.M.; Wilson, G.S. The thickness of the human precorneal tear film: Evidence from reflection spectra. *Invest. Ophthalmol. Visual Sci.* 2000, *41* (11), 3348–3359.
9. Tiffany, J.M. Tears in health and disease. *Eye* 2003, *17* (8), 1–4.
10. Lee, V.H.L.; Robinson, J.R. Review: Topical ocular drug delivery: Recent developments and future challenges. *Ocul. Pharmacol.* 1986, *2* (1), 67–108.
11. Chauhan, A.; Bengani, L. Are contact lenses the solution for effective ophthalmic drug delivery? *Future Med. Chem.* 2012, *4* (17), 2141–2143.
12. Jain, D.; Carvalho, E.; Banerjee, R. Biodegradable hybrid polymeric membranes for ocular drug delivery. *Acta Biomater.* 2010, *6* (4), 1370–1379.
13. Urtti, A.; Salminen, L. Minimizing systemic absorption of topically administered ophthalmic drugs. *Surv. Ophthalmol.* 1993, *37* (6), 435–457.
14. Alqawlaq, S.; Huzil, J.T.; Ivanova, M.V.; Foldvari, M. Challenges in neuroprotective nanomedicine development: Progress towards noninvasive gene therapy of glaucoma. *Nanomedicine* 2012, *7* (7), 1067–1083.
15. Gan, L.; Wang, J.; Jiang, M.; Bartlett, H.; Ouyang, D.; Eperjesi, F.; Liu, J.; Gan,Y. Recent advances in topical ophthalmic drug delivery with lipid-based nanocarriers. *Drug Discov. Today* 2013, *18* (5–6), 290–297.
16. Mishima, S.; Gassef, A.; Klyce, S.D.; Baum, J.L. Determination of tear volume and tear flow. *Invest. Ophthalmol.* 1966, *5* (3), 264–276.
17. Ludwig, A. The use of mucoadhesive polymers in ocular drug delivery. *Adv. Drug Deliver. Rev.* 2005, *57* (11), 1595–1639.
18. Gipson, I.K.; Argueso, P. Role of mucins in the function of the corneal and conjunctival epithelia. *Int. Rev. Cytol.* 2003, *231*, 1–49.
19. Japan Tissue Engineering Co. Autologous cultured corneal epithelium. http://www.jpte.co.jp/english/business/Regenerative/cultured_corneal_epithelium.html (accessed December 2013).
20. Mannermaa, E.; Vellonen, K.S.; Urtti, A. Drug transport in corneal epithelium and blood-retina barrier: Emerging role of transporters in ocular pharmacokinetics. *Adv. Drug Deliv. Rev.* 2006, *58* (11), 1136–1163.
21. Pandey, H.; Kumar Sharma, U.; Pandey, A.C. Eudragit-based nanostructures: A potential approach for ocular drug delivery. *IJRDPL* 2012, *1* (2), 40–43.
22. Bourlais, C.; Acar, L.; Zia, H.; Sado, P.A.; Needham, T.; Leverge, R. Ophthalmic drug delivery systems—Recent advances. *Prog. Retin. Eye Res.* 1998, *17* (1), 33–58.
23. Gaudana, R.; Ananthula, H.K.; Parenky, A.; Mitra, A.K. Ocular drug delivery. *AAPS J.* 2010, *12* (3), 348–360.
24. Sikandar, M.K.; Sharma, P.K.; Visht, S. Ocural drug delivery system: An overview. *Int. J. Pharm. Sci. Res.* 2011, *2* (5), 1168–1175.
25. Patel, V.; Agrawal, Y.K. Current status and advanced approaches in ocular drug delivery system. *JGTPS* 2011, *2* (21), 131–148.
26. Tangri, P.; Khurana, S. Basics of ocular drug delivery systems. *Int. J. Res. Pharm. Biomed. Sci.* 2011, *2* (4), 1541–1552.
27. Sireesha, D.S.; Suriaprabha, K.; Prasanna, P.M. Advanced approaches and evaluation of ocular drug delivery system. *Am. J. Pharmatech. Res.* 2011, *1* (4), 72–92.
28. Bhargava, H.N.; Nicolai, D.W.; Oza, B.J. *Topical Suspensions Pharmaceutical Dosage Forms: Disperse Systems*. Marcel Dekker: New York, 1996; Vol. 2, pp. 183–241.
29. Ludwig, A.; Van Ootengm, M. Influence of viscolyzers on the residence of ophthalmic solution evaluated by slit lamp fluorometry. *Pharm. Sci.* 1992, *2*, 81–87.
30. Yusuf, A. Industrial perspective in ocular drug delivery. *Adv. Drug Deliv. Rev.* 2006, *58* (11), 1258–1268.
31. Blumenthal, J.; Ticho, U.; Zonis, S.; Gal, A.; Blank, I.; Mazor, Z. Further clinical trial with piloplex-A new long-acting pilocarpine salt. *Glaucoma* 1979, *63* (1), 145–148.
32. Mazor, A.; Ticho, U.; Rehany, U.; Rose, L. Piloplex-A new long-acting pilocarpine salt: B. Comparative study of the visual effects of pilocarpine and piloplex eyedrops. *Br. J. Ophthalmol.* 1979, *63* (1), 48–51.
33. Ticho, U.; Blumenthal, M.; Zonis, S.; Gal, A.; Blank, I.; Mazor, Z. Piloplex, a new long-acting pilocarpine polymer salt: A long-term study. *Br. J. Ophthalmol.* 1979, *63* (1), 45–47.

34. Goldberg, I.; Ashburn, F.S.; Kass, M.A.; Becker, B. Efficacy and patient acceptance of pilocarpine gel. *Am. J. Ophthalmol.* 1979, *88* (5), 843–846.
35. March, W.F.; Stewart, R.M.; Mandell, A.I.; Bruce, L.A. Duration of effect of pilocarpine gel. *Arch. Ophthalmol.* 1982, *100* (8), 1270–1271.
36. Miyazaki, S.; Ishii, K.; Takada, M. Use of fibrin film as a carrier for drug delivery: A long-acting delivery system for pilocarpine into the eye. *Chem. Pharm. Bull.* 1982, *30* (9), 3405–3407.
37. Bhowmik, M.; Bain, M.K.; Ghosh, L.K.; Chattopadhyay, D. Effect of salts on gelation and drug release profiles of methylcellulose based ophthalmic thermoreversible in-situ gels. *Pharm. Dev. Technol.* 2011, *16* (4), 385–391.
38. Schoenwald, R.D. Ocular drug delivery: Pharmacokinetic considerations. *Clin. Pharmacokinet.* 1990, *18* (4), 255–269.
39. Hongyi, Q.; Wenwen, C.; Chunyan, H.; Li, L.; Chuming, C.; Wenmin, L.; Chunjie, W. Development of a poloxamer analogs/carbopol-based in-situ gelling and mucoadhesive ophthalmic delivery system for puerarin. *Int. J. Pharm.* 2007, *337* (1–2), 178–187.
40. Ridell, A.; Evertsson, H.; Nilsson, S.; Sundelof, L.O. Ampiphillic association of ibuprofen and two non-ionic cellulose derivatives in aqueous solution. *J. Pharm. Sci.* 1999, *88* (11), 1175–1181.
41. Bain, M.K.; Bhowmik, M.; Maity, D.; Bera, N.K.; Ghosh, S.N.; Chattopadhaya, D. Control of thermo reversible gelation of methylcellulose using different molecular weight of PEG and NaCl for sustain delivery of ophthalmic drug. *J. Appl. Polym. Sci.* 2010, *118* (2), 631–637.
42. Hsiue, G.H.; Chang, R.W.; Wang, C.H.; Lee, S.H. Development of in situ thermosensitive drug vehicles for glaucoma therapy. *Biomaterials* 2003, *24* (13), 2423–2430.
43. Jeong, B.; Choi, Y.K.; Bae, Y.H.; Zentner, G.; Kim, S.W. New biodegradable polymers for injectable drug delivery systems. *J. Contr. Release* 1999, *62* (1–2), 109–114.
44. Qiu, Y.; Park, K. Environment-sensitive hydrogels for drug delivery. *Adv. Drug Deliv. Rev.* 2001, *53* (3), 321–339.
45. Lin, H.R.; Sung, K.C.; Vong, W.J. In situ gelling of alginate/pluronic solutions for ophthalmic delivery of pilocarpine. *Biomacromolecules* 2004, *5* (6), 2358–2365.
46. Kumari, A.; Sharma, P.K.; Garg, V.K. Ocular inserts: Advancement in therapy of eye diseases. *J. Adv. Pharm. Technol. Res.* 2010, *1* (3), 87–96.
47. Neefe, C.W. Contact lens for ocular drug delivery. US Patent 3786812, January 22, 1974.
48. Kumari, A.; Sharma, P.K.; Garg, V.K.; Garg, G. Ocular inserts—Advancement in therapy of eye diseases. *J. Adv. Pharm. Technol. Res.* 2010, *1* (3), 291–296.
49. Mishra, D.N.; Gilhotra, R.M. Design and characterization of bioadhesive in-situ gelling ocular insert of gatifloxacin sesquihydrate. *DARU* 2008, *16* (1), 1–8.
50. Saettone, M.F.; Salminen, L. Ocular inserts for topical delivery. *Adv. Drug Deliv. Rev.* 1995, *16* (1), 95–106.
51. Karthikeyan, D.; Bhowmick, M.; Pandey, V.P.; Nandhakumar, J.; Sengottuvelu, S.; Sonkar, S.; Sivakumar, T. The concept of ocular inserts as drug delivery systems: An overview. *Asian J. Pharm.* 2008, *2* (4), 192–200.
52. Sihvola, P.; Puustjarvi, T. Practical problems in the use of Ocusert®-pilocarpine delivery system. *Acta Ophthalmol.* 1980, *58* (6), 933–937.
53. Hughes, P.M.; Olejnik, O.; Chang-Lin, J.E.; Wilson, C.G. Topical and systemic drug delivery to the posterior segments. *Adv. Drug Deliv. Rev.* 2005, *57* (14), 2010–2032.
54. Kearns, V.R.; Williams, R.L. Drug delivery systems for the eye. *Expert Rev. Med. Dev.* 2009, *6* (3), 277–290.
55. Novack, G.D. Ophthalmic drug delivery: Development and regulatory considerations. *Clin. Pharm. Ther.* 2009, *85* (5), 539–543.
56. Leppard, S. Regulation of drug–device combination products in Europe. In *Drug-Device Combination Products. Delivery Technologies and Applications*; Lewis, A., Ed. Woodhead Publishing Limited and CRC Press LLC: Boca Raton, FL, 2010; pp. 464–495.
57. Greenbaum, J. Regulation of drug–device combination products in the USA. In *Drug-Device Combination Products. Delivery Technologies and Applications*; Lewis, A., Ed. Woodhead Publishing Limited and CRC Press LLC: Boca Raton, FL, 2010; pp. 496–529.
58. Rathore, K.S.; Nema, R.K.; Sisodia, S.S. An overview and advancement in ocular drug systems. *Int. J. Pharm. Sci. Res.* 2010, *1*, 11–23.
59. Di Colo, G.; Zambito, Y. A study of release mechanisms of different ophthalmic drugs from erodible ocular inserts based on poly(ethylene oxide). *Eur. J. Pharm. Biopharm.* 2002, *54* (2), 193–199.

60. Ratnam, V.G.; Madhavi, S.; Rajesh, P. Ocular drug delivery: An update review. *Int. J. Pharm. Bio. Sci.* 2011, *1* (4), 437–446.
61. Sultana, Y.; Jain, R.; Aqil, M.; Ali, A. Review of ocular drug delivery. *Curr. Drug Deliv.* 2006, *3* (2), 207–217.
62. Gomes dos Santos, A.L.; Bochot, A.; Doyle, A.; Tsapis, N.; Siepmann, J.; Siepmann, F.; Schmaler, J.; Besnard, M.; Behar-Cohen, F.; Fattal, E. Sustained release of nanosized complexes of polyethylenimine and anti-TGF-beta 2 oligonucleotide improves the outcome of glaucoma surgery. *J. Contr. Release* 2006, *112* (3), 369–381.
63. Bashshur, Z.F.; Bazarbachi, A.; Schakal, A.; Haddad, Z.A.; El Haibi, C.P.; Noureddin, B.N. Intravitreal bevacizumab for the management of choroidal neovascularization in age-related macular degeneration. *Am. J. Ophthalmol.* 2006, *142* (1), 1–9.
64. Zhou, B.; Wang, B. Pegaptanib for the treatment of age-related macular degeneration. *Exp. Eye Res.* 2006, *83* (3), 615–619.
65. Pitkänen, L.; Ranta, V.P.; Moilanen, H.; Urtti, A. Permeability of retinal pigment epithelium: Effect of permeant molecular weight and lipophilicity. *Investig. Ophthalmol. Vis. Sci.* 2005, *46* (2), 641–646.
66. Ambati, J.; Gragoudas, E.S.; Miller, J.W.; You, T.T.; Miyamoto, K.; Delori, F.C.; Adamis, A.P. Transscleral delivery of bioactive protein to the choroid and retina. *Invest. Ophthalmol. Vis. Sci.* 2000, *41* (5), 1186–1191.
67. Molleda, J.M.; Tardón, R.H.; Gallardo, J.M.; Martín-Suárez, E.M. The ocular effects of intravitreal triamcinolone acetonide in dogs. *Vet. J.* 2008, *176* (3), 326–332.
68. Pitkänen, L.; Ruponen, M.; Nieminen, J.; Urtti, A. Vitreous is a barrier in non-viral gene transfer by cationic lipids and polymers. *Pharm. Res.* 2003, *20* (4), 576–583.
69. Park, J.; Bungay, P.M.; Lutz, R.J.; Augsburger, J.J.; Millard, R.W.; Roy, A.S.; Banerjee, R.K. Evaluation of coupled convective–diffusive transport of drugs administered by intravitreal injection and controlled release implant. *J. Contr. Release* 2005, *105* (3), 279–295.
70. Maurice, D.M.; Mishima, S. Ocular pharmacokinetics. In *Handbook of Experimental Pharmacology*; Sears, M.L., Ed. Springer Verlag: Berlin, Germany, 1984; Vol. 69, pp. 16–119.
71. Bourges, J.L.; Bloquel, C.; Thomas, A.; Froussart, F.; Bochot, A.; Azan, F.; Gurny, R.; BenEzra, D.; Behar-Cohen, F. Intraocular implants for extended drug delivery: Therapeutic applications. *Adv. Drug Deliv. Rev.* 2006, *58* (11), 1182–1202.
72. Yasukawa, T.; Ogura, Y.; Sakurai, E.; Tabata, Y.; Kimura, H. Intraocular sustained drug delivery using implantable polymeric devices. *Adv. Drug Deliv. Rev.* 2005, *57* (14), 2033–2046.
73. Jaffe, G.J.; Martin, D.; Callanan, D. Fluocinolone acetonide implant (Retisert) for noninfectious posterior uveitis: Thirty-four-week results of a multicenter randomized clinical study. *Ophthalmology* 2006, *113* (6), 1020–1027.
74. Vishal, P.; Agrawal, Y.K. Current status and advanced approaches in ocular drug delivery system. *J. Global Trends Pharm. Sci.* 2011, *2* (2), 131–148.
75. Short, B.G. Safety evaluation of ocular drug delivery formulation: Techniques and practical considerations. *Toxicol. Pathol.* 2008, *36* (1), 49–62.
76. Kato, A.; Kimura, H.; Okabe, K.; Okabe, J.; Kunou, N.; Ogura, Y. Feasibility of drug delivery to the posterior pole of the rabbit eye with an episcleral implant. *Invest. Ophthalmol. Vis. Sci.* 2004, *45* (1), 238–244.
77. Seah, S.K.; Husain, R.; Gazzard, G. Use of surodex in phacotrabeculectomy surgery. *Am. J. Ophthalmol.* 2005, *139* (5), 927–928.
78. Kuppermann, B.D.; Blumenkranz, M.S.; Haller, J.A.; Williams, G.A.; Weinberg, D.V.; Chou, C.; Whitcup, S.M. Randomized controlled study of an intravitreous dexamethasone drug delivery system in patients with persistent macular edema. *Arch. Ophthalmol.* 2007, *125* (3), 309–317.
79. Heller, J. Ocular delivery using poly(ortho esters). *Adv. Drug Deliv. Rev.* 2005, *57* (14), 2053–2062.
80. Wirtz, R. Die ionentherapie in der augenheilkunde. *Klinische Monatsblatter fur Augenheilkunde* 1908, *46*, 543–579.
81. Wong, V.G. Biodegradable ocular implants. US Patent 4853224 A, August 1, 1989.
82. Baeyens, V.; Percicot, C.; Zignani, M. Ocular drug delivery in veterinary medicine. *Adv. Drug Deliv. Rev.* 1997, *28* (3), 335–361.
83. Eljarrat-Binstock, E.; Domb, A.J. Iontophoresis: A non-invasive ocular drug delivery. *J. Contr. Release* 2006, *110* (3), 479–489.
84. Bejjani, R.; Andrieu, C.; Bloquel, C. Electrically assisted ocular gene therapy. *Surv. Ophthalmol.* 2007, *52* (2), 196–208.

85. Eljarrat-Binstock, E.; Orucov, F.; Frucht-Pery, J. Methylprednisolone delivery to the back of the eye using hydrogel iontophoresis. *J. Ocul. Pharmacol. Ther.* 2008, *24* (3), 344–350.
86. Alvarez-Lorenzo, C.; Concheiro, A. Ocular drug delivery from nanostructured contact lenses. In *Nanostructured Biomaterials for Overcoming Biological Barriers*; Jose Alonso, M.; Csaba, N.S., Eds. The Royal Society of Chemistry: Cambridge, U.K., 2012.
87. Hiratania, H.; Alvarez-Lorenzo, C. The nature of backbone monomers determines the performance of imprinted soft contact lenses as timolol drug delivery systems. *Biomaterials* 2004, *25* (6), 1105–1113.
88. Heng, Z.; Chauhan, A. Effect of viscosity on tear drainage and ocular residence time. *Optom. Vis. Sci.* 2008, *85* (8), 715–725.
89. Creech, J.L.; Chauhan, A.; Radke, C.J. Dispersive mixing in the posterior tear film under a soft contact lens. *Ind. Eng. Chem. Res.* 2001, *40* (14), 3015–3026.
90. McNamara, N.A.; Polse, K.A.; Brand, R.J.; Graham, A.D.; Chan, J.S.; McKenney, C.D. Tear mixing under a soft contact lens: Effects of lens diameter. *Am. J. Ophthalmol.* 1999, *127* (6), 659–665.
91. Bengani, L.C.; Chauhan, A. Extended delivery of an anionic drug by contact lens loaded with a cationic surfactant. *Biomaterials* 2013, *34* (11), 2814–2821.
92. Gulsen, D.; Chauhan, A. Ophthalmic drug delivery through contact lenses. *Invest. Ophthalmol. Vis. Sci.* 2004, *45* (7), 2342–2347.
93. Jung, H.J.; Chauhan, A. Ophthalmic drug delivery by contact lenses. *Expert Rev. Ophthalmol.* 2012, *7* (7), 199–201.
94. Shah, C.; Raj, S.; Foulks, G.N. Evolution in therapeutic contact lenses. *Ophtalmol. Clin. North Am.* 2003, *16* (1), 95–101.
95. DeNaeyer, G.W. Exploring the therapeutic applications of contact lenses. New O.D. December 11–15, 2008.
96. Witcherle, O.; Lim, D. Hydrophilic gels for biological use. *Nature* 1960, *185* (4706), 117–118.
97. McMahon, T.T.; Zadnik, K. Twenty-five years of contact lenses: The impact on the cornea and ophthalmic practice. *Cornea* 2000, *19* (5), 730–740.
98. Munoa-Roiz, J.L.; Aramendia-Salvador, E. Historia y desarrollo de las lentes de contacto. http://www.oftalmo.com/publicaciones/lentes/cap2.htm (accessed February 2011).
99. Nicolson, P.C.; Vogt, J. Soft contact lens polymers: An evolution. *Biomaterials* 2001, *22* (24), 3273–3283.
100. Yamauchi, A. In *Gels Handbook*; Osada, Y.; Kajiwara, K.; Eds.; Academic Press: San Diego, CA, 2001; Vol. 3, pp. 166–179.
101. Goda, T.; Ishihara, K. Soft contact lens biomaterials from bioinspired phospholipid polymers. *Expert Rev. Med. Devices* 2006, *3* (2), 167–174.
102. Kopecek, J. Hydrogels: From soft contact lenses and implants to self-assembled nanomaterials. *J. Polym. Sci.* 2009, *47* (22), 5929–5946.
103. Kunzler, J.F.; McGee, J.A. Contact lens materials. *Chem. Ind.* 1995, *16*, 651–655.
104. Chekina, N.A.; Pavlyuchenko, V.N.; Danilichev, V.F.; Ushakov, N.A.; Novikov, S.A.; Ivanchev, S.S. A new polymeric silicone hydrogel for medical applications: Synthesis and properties. *Polym. Adv. Technol.* 2006, *17* (11–12), 872–877.
105. Duncan, R.; Izzo, L. Dendrimer biocompatibility and toxicity. *Adv. Drug Deliv. Rev.* 2005, *57* (15), 2215–2237.
106. Li, C.C.; Chauhan, A. Ocular transport model for ophthalmic delivery of timolol through p-HEMA contact lenses. *J. Drug Deliv. Sci. Technol.* 2007, *17* (1), 69–79.
107. Paula, A.V.; Elena, F.G.; Carmen, A.L.; Angel, C. Improving the loading and release of NSAIDs from pHEMA hydrogels by copolymerization with functionalized monomers. *J. Pharm. Sci.* 2007, *96* (4), 802–813.
108. Garcia, D.M.; Escobar, J.L.; Noa, Y.; Bada, N.; Hernaez, E.; Katime, I. Timolol maleate release from pH-sensible poly(2-hydroxyethyl methacrylate-co-methacrylic acid) hydrogels. *Eur. Polym. J.* 2004, *40* (8), 1683–1690.
109. Hehl, E.M.; Beck, R.; Luthard, K.; Guthoff, R. Improved penetration of aminoglycosides and fluoroquinolones into the aqueous humor of patients by means of acuvue contact lenses. *Eur. J. Clin. Pharmacol.* 1999, *55* (4), 317–323.
110. Sedlacek, J. Possibilities of application of eye drugs with the aid of gel contact lenses. *Cesk. Oftalmol.* 1965, *21*, 509–512.
111. Peterson, R.C.; Wolffsohn, J.S.; Nick, J.; Winterton, L.; Lally, J. Clinical performance of daily disposable soft contact lenses using sustained release technology. *Cont. Lens. Anterior Eye* 2006, *29* (3), 127–134.
112. Stefan, S.; Thilo, W.; Ralf, M.; Gerd, G. Combination of serum eye drops with hydrogels bandage contact lenses in the treatment of persistent epithelial defects. *Graefe's Arch Clin. Exp. Ophthalmol.* 2006, *244* (10), 1345–1349.

113. Refojo, M.F.; Leong, F.L.; Chan, I.M.; Tolentino, F.I. Absorption and release of antibiotics by a hydrophilic implant for scleral buckling. *Retina* 1983, *3* (1), 45–49.
114. Hsiue, G.H.; Gut, J.A.; Cheng, C.C. Poly(2-hydroxyethyl methacrylate) film as a drug delivery system for pilocarpine. *Biomaterials* 2001, *22* (13), 1763–1769.
115. Karlgard, C.C.S.; Wong, N.S.; Jones, L.W.; Moresoli, C. in vitro uptake and release studies of ocular pharmaceutical agents by silicon-containing and p-HEMA hydrogel contact lenses. *Int. J. Pharm.* 2003, *257* (1), 141–151.
116. Winterton, L.C.; Lally, J.M.; Sentell, K.B.; Chapoy, L.L. The elution of poly (vinyl alcohol) from a contact lens: The realization of a time release moisturizing agent/artificial tear. *J. Biomed. Mater. Res. B* 2007, *80* (2), 424–432.
117. Podos, S.M.; Becker, B.; Assef, C. Pilocarpine therapy with soft contact lenses. *Am. J. Ophthalmol.* 1972, *73* (3), 336–341.
118. Li, C.C.; Chauhan, A. Modeling ophthalmic drug delivery by soaked contact lenses. *Ind. Eng. Chem. Res.* 2006, *45* (10), 3718–3734.
119. Andrade-Vivero, P. Improving the loading and release of NSAIDs from pHEMA hydrogels by copolymerization with functionalized monomers. *J. Pharm. Sci.* 2007, *96* (4), 802–813.
120. Dracopoulos, A. In vitro assessment of medical device toxicity: Interactions of benzalkonium chloride with silicone-containing and p-HEMAcontaining hydrogel contact lens materials. *Eye Contact Lens* 2007, *33* (1), 26–37.
121. Kim, S.W.; Bae, Y.H.; Okano, T. Hydrogels, swelling, drug loading and release. *Pharm. Res.* 1992, *9* (3), 283–290.
122. Peng, C.C.; Chauhan, A. Extended cyclosporine delivery by silicone–hydrogel contact lenses. *J. Contr. Release* 2011, *154* (3), 267–274.
123. Xinming, L.; Yingde, C.; Lloyd, A.W.; Mikhalovsky, S.V.; Sandeman, S.R.; Howel, C.A.; Liewen, L. Polymeric hydrogels for novel contact lens-based ophthalmic drug delivery systems: A review. *Cont. Lens Anterior Eye* 2008, *31* (2), 57–64.
124. Wajs, G.; Meslard, J.C. Release of therapeutic agents from contact lenses. *Crit. Rev. Ther. Drug* 1986, *2* (3), 275–289.
125. Lesher, G.A.; Gunderson, G.G. Continuous drug delivery through the use of disposable contact lenses. *Optom. Vis. Sci.* 1993, *70* (12), 1012–1018.
126. Westerhout, D. The Combination lens and therapeutic uses of soft lenses. *Contact Lens J.* 1973, *4*, 3–10.
127. Romero-Jiménez, M.; Santodomingo-Rubido, J.; Flores-Rodríguez, P.; González-Méijome, J.M. Which soft contact lens power is better for piggyback fitting in keratoconus? *Cont. Lens Anterior Eye* 2013, *36* (1), 45–48.
128. Shirley, H.L.; Chang, M.D.; Gerald Lim, M.D. Secondary pigmentary glaucoma associated with piggyback intraocular lens implantation. Case report. *J. Cataract Refract. Surg.* 2004, *30* (10), 2219–2222.
129. Rootman, D.S.; Willoughby, R.P.N.; Bindlish, R.; Avaria, M.; Basu, P.K.; Krajden, M. Continuous-flow contact lens delivery of gentamicin to rabbit cornea and aqueous-humor. *J. Ocul. Pharmacol.* 1992, *8* (4), 317–323.
130. Sano, K.; Tokoro, T.; Imai, Y. A new drug delivery system utilizing piggyback contact lenses. *Acta Ophthalmol.* 1996, *74* (3), 243–248.
131. Guo, L.S.S.; Redemann, C.T.; Radhakrishnan, R.; Yau-Young, A. Liposomes with enhanced retention on mucosal tissue. US Patent No. 4839175, June 13, 1989.
132. Barber, R.F.; Shek, P.N. Tear-induced release of liposome entrapped agents. *Int. J. Pharm.* 1990, *60* (3), 219–227.
133. Popescu, M.C.; Weiner, A.L.; Carpenter-Green, S.S. Liposome gel compositions. International Patent No. WO85/03640, August 29, 1985.
134. Vermette, P.; Meagher, L.; Gagnon, E.; Griesser, H.J.; Doillon, C.J. Immobilized liposomes layers for drug delivery applications: Inhibition of angiogenesis. *J. Contr. Release* 2002, *80* (1–3), 179–185.
135. Vermette, P.; Griesser, H.J.; Kambouris, P.; Meagher, L. Characterization of surface-immobilized layers of intact liposomes. *Biomacromolecules* 2004, *5* (4), 1496–1502.
136. Gulsen, D.; Li, C.C.; Chauhan, A. Dispersion of DMPC liposomes in contact lenses for ophthalmic drug delivery. *Curr. Eye Res.* 2005, *30* (12), 1071–1080.
137. Chauhan, A.; Gulsen, D. Ophthalmic drug delivery system. US Patent No. 0241207, December 2, 2004.
138. Danion, A.; Brochu, H.; Martin, Y.; Vermette, P. Fabrication and characterization of contact lenses bearing surface-immobilized layers of intact liposomes. *J. Biomed. Mater. Res. A* 2007, *82* (1), 41–51.
139. Stella, V.J.; Rao, V.M.; Zannou, E.A.; Zia, V. Mechanisms of drug release from cyclodextrin complexes. *Adv. Drug Deliv. Rev.* 1999, *36* (1), 3–16.

140. Loftsson, T.; Duchene, D. Cyclodextrins and their pharmaceutical applications. *Int. J. Pharm.* 2006, *329* (1), 1–11.
141. Rodriguez-Tenreiro, C.; Alvarez-Lorenzo, C.; Rodriguez-Perez, A.; Concheiro, A.; Torres-Labandeira, J.J. Estradiol sustained release from high affinity cyclodextrin hydrogels. *Eur. J. Pharm. Biopharm.* 2007, *66* (1), 55–62.
142. Dos Santos, J.F.R.; Alvarez-Lorenzo, C.; Silva, M.; Balsa, L.; Couceiro, J.; Torres-Labandeira, J.J.; Concheiro, A. Soft contact lenses functionalized with pendant cyclodextrins for controlled drug delivery. *Biomaterials* 2009, *30* (7), 1348–1355.
143. Glisonia, R.J.; García-Fernández, M.J.; Pinod, M.; Gutkind, G.; Moglioni, A.G.; Alvarez-Lorenzo, C.; Concheiro, A.; Sosnika, A. β-Cyclodextrin hydrogels for the ocular release of antibacterial thiosemicarbazones. *Carbohydr. Polym.* 2013, *93* (2), 449–457.
144. Wulff, G. Molecular imprinting in cross-linked materials with the aid of molecular templates—A way towards artificial antibodies. *Angew. Chem. Int. Ed.* 1999, *34* (17), 1812–1832.
145. Mosbach, K.; Haupt, K.; Liu, X.C.; Cormack, P.A.G.; Ramstrom, O. Molecular imprinting: Status artis et quo vadere? *ACS Symp. Ser.* 1998, *703*, 29–48.
146. Sellergren, B.; Hall, A.J. Fundamental aspects on the synthesis and characterization of imprinted network polymers. In *Molecularly Imprinted Polymers*. Elsevier: Amsterdam, the Netherlands, 2001; p. 21.
147. Allender, C.J.; Richardson, C.; Woodhouse, B.; Heard, C.M.; Brain, K.R. Pharmaceutical applications for molecularly imprinted polymers. *Int. J. Pharm.* 2000, *195* (1–2), 39–43.
148. Byrne, M.E.; Park, K.N.; Peppas, A. Molecular imprinting within hydrogels. *Adv. Drug Deliv. Rev.* 2002, *54* (1), 149–161.
149. Bowman, M.; Allender, C.; Heard, C.; Brain, K. Molecularly imprinted polymers as selective sorbents for the preliminary screening of combinatorial libraries. *J. Mater. Sci. and Eng. C-Biomi Mater. Sens. Syst.* 1998, *25*, 37–43.
150. Hiratani, H.C.; Alvarez-Lorenzo, J.; Chuang, O.; Guney, A.; Grosberg, Y.; Tanaka, T. Effect of reversible cross-linker, N,N'-bis(acryloyl)cystamine, on calcium ion adsorption by imprinted gels. *Langmuir* 2001, *17* (14), 4431–4436.
151. Ito, K.; Chuang, J.; Alvarez-Lorenzo, C.; Watanabe, T.; Ando, N.; Grosberg, A.Y. Multiple point adsorption in a heteropolymer gel and the Tanaka approach to imprinting: Experiment and theory. *Prog. Polym. Sci.* 2003, *28* (10), 1489–1515.
152. Asanuma, H.; Hishiya, T.; Komiyama, M. Tailor-made receptors by molecular imprinting. *Adv. Mater.* 2000, *12* (14), 1019–1030.
153. Hilt, J.Z.; Byrne, M.E. Configurational biomimesis in drug delivery: Molecular imprinting of biologically significant molecules. *Adv. Drug Deliv. Rev.* 2004, *56* (11), 1599–1620.
154. Alvarez-Lorenzo, C.; Concheiro, A.J. Molecularly imprinted polymers for drug delivery. *Chromatogr. B Analyt. Technol. Biomed. Life Sci.* 2004, *804* (1), 231–245.
155. Alvarez-Lorenzo, C.; Yanez, F.; Concheiro, A. Ocular drug delivery from molecularly-imprinted contact lenses. *J. Drug Deliv. Sci. Technol.* 2010, *20* (4), 237–248.
156. White, C.J.; Byrne, M.E. Molecularly imprinted therapeutic contact lenses. *Expert Opin. Drug Deliv.* 2010, *7* (6), 765–780.
157. Alvarez-Lorenzo, C.; Hiratani, H.; Gomez-Amoza, J.L.; Martinez-Pacheco, R. Soft contact lenses capable of sustained delivery of timolol. *J. Pharm. Sci.* 2002, *91* (10), 2182–2192.
158. Hiratani, H.; Alvarez-Lorenzo, C. Timolol uptake and release by imprinted soft contact lenses made of N,N-diethylacrylamide and methacrylic acid. *J. Contr. Release* 2002, *83* (2), 223–230.
159. Hiratani, H.; Fujiwara, A.; Tamiya, Y.; Mizutani, Y.; Alvarez-Lorenzo, C. Ocular release of timolol from molecularly imprinted soft contact lenses. *Biomaterials* 2005, *26* (11), 1293–1298.
160. Hiratani, H.; Mizutani, Y.; Alvarez-Lorenzo, C. Controlling drug release from imprinted hydrogels by modifying the characteristics of the imprinted cavities. *Macromol. Biosci.* 2005, *5* (8), 728–733.
161. Alvarez-Lorenzo, C.; Yanez, F.; Barreiro-Iglesias, R.; Concheiro, A. Imprinted soft contact lenses as norfloxacin delivery systems. *J. Contr. Release* 2006, *113* (3), 236–244.
162. Ali, M.; Byrne, M.E. Controlled release of high molecular weight hyaluronic acid from molecularly imprinted hydrogel contact lenses. *Pharm. Res.* 2009, *26* (3), 714–726.
163. Malaekeh-Nikouei, B.; Abbasi Ghaeni, F.; Motamedshariaty, V.S.; Mohajeri, S.A. Controlled release of prednisolone acetate from molecularly imprinted hydrogel contact lenses. *J. Appl. Polym. Sci.* 2012, *126* (1), 387–394.
164. Farokhzad, O.C. Nanotechnology for drug delivery: The perfect partnership. *Expert Opin. Drug Deliv.* 2008, *5* (1), 927–929.
165. Whitesides, G.M. The 'right' size in nanobiotechnology. *Nat. Biotechnol.* 2003, *21* (10), 1161–1165.

166. Ferrari, M. Cancer nanotechnology: Opportunities and challenges. *Nat. Rev. Canc.* 2005, *5* (3), 161–171.
167. Farokhzad, O.C.; Karp, J.M.; Langer, R. Nanoparticle-aptamer bioconjugates for cancer targeting. *Expert Opin. Drug Deliv.* 2006, *3* (1), 311–324.
168. Wagner, V.; Dullaart, A.; Bock, A.K.; Zweck, A. The emerging nanomedicine landscape. *Nat. Biotech.* 2006, *24* (10), 1211–1217.
169. Vasir, J.K.; Labhasetwar, V. Targeted drug delivery in cancer therapy. *Technol. Canc. Res. Treat.* 2005, *4* (4), 363–374.
170. Yih, T.C.; Al-Fandi, M. Engineered nanoparticles as precise drug delivery systems. *J. Cell Biochem.* 2006, *97* (6), 1184–1190.
171. Sahoo, S.K.; Dilnawaz, F.; Krishnakumar, S. Nanotechnology in ocular drug delivery. *Drug Discov. Today* 2008, *13* (3), 144–151.
172. Liu, S.; Jones, L.; Gu, F.X. Nanomaterials for ocular drug delivery. *Macromol. Biosci.* 2012, *12* (5), 608–620.
173. Honda, M.; Asai, T.; Oku, N.; Araki, Y.; Tanaka, M.; Ebihara, N. Liposomes and nanotechnology in drug development: Focus on ocular targets. *Int. J. Nanomed.* 2013, *8* (1), 495–504.
174. Yasukawa, T.; Ogura, Y.; Tabata, Y.; Kimura, H.; Wiedemann, P.; Honda, Y. Drug delivery systems for vitreoretinal diseases. *Prog. Retin. Eye Res.* 2004, *23* (3), 253–281.
175. Vandervoort, J.; Ludwig, A. Ocular drug delivery: Nanomedicine applications. *Nanomedicine* 2007, *2* (1), 11–21.
176. Edwards, K.A.; Baeumner, A.J. Liposomes in analyses. *Talanta* 2006, *68* (5), 1421–1431.
177. Bergstrand, N. Liposomes for drug delivery from physicochemical studies to applications. Doctoral thesis, comprehensive summary, Acta Universitatis Upsaliensis, 2003.
178. Lasic, D.D. The mechanism of vesicle formation. *Biochem. J.* 1988, *256* (1), 1.
179. Calvo, P.; Alonso, M.J.; Vila-Jato, J.L.; Robinson, J.R. Improved ocular bioavailability of indomethacin by novel ocular drug carriers. *J. Pharm. Pharmacol.* 1996, *48* (11), 1147–1152.
180. Monem, A.S.; Ali, F.M.; Ismail, M.W. Prolonged effect of liposomes encapsulating pilocarpine HCl in normal and glaucomatous rabbits. *Int. J. Pharm.* 2000, *198* (1), 29–38.
181. Woodle, M.C.; Strom, G.; Eds. *Long Circulating Liposomes: Old Drugs, New Therapeutics*. Springer: New York, 1997.
182. Drummond, D.C.; Hong, K.L.; Park, J.W.; Benz, C.C.; Kiroptin, D.B. Liposomal targeting to tumors using vitamin and growth factor receptors. *Vitam. Horm.* 2001, *60*, 285–332.
183. Mastrobattista, E.; Koning, G.A.; Strom, G. Immunoliposomes for the targeted delivery of antitumor drugs. *Adv. Drug Deliv. Rev.* 1999, *40* (1), 103–127.
184. Park, Y.S. Tumor-directed targeting of liposomes. *Biosci. Rep.* 2002, *22* (2), 267–281.
185. Sudimack, J.; Lee, R.J. Targeted drug delivery via the folate receptor. *Adv. Drug Deliv. Rev.* 2000, *41* (2), 147–162.
186. Tardi, P.G.; Swatrz, E.N.; Harasum, T.O.; Cullis, P.R.; Bally, M.B. An immune response to ovalbumin covalently coupled to liposomes is prevented when the liposomes used contain doxorubicin. *J. Immunol. Methods* 1997, *210* (2), 137–148.
187. Du, H.; Chandaroy, P.; Hui, S.W. Grafted poly-(ethylene glycol) on lipid surfaces inhibits protein adsorption and cell adhesion. *Biochim. Biophys. Acta* 1997, *1326* (2), 236–248.
188. Malmsten, M.; Van Alstine, J.M. Adsorption of poly(ethylene glycol) amphiphiles to form coatings which inhibit protein adsorption. *J. Colloid Interface Sci.* 1996, *177* (2), 502–512.
189. Ulrich, A.S. Biophysical aspects of using liposomes as delivery vehicles. *Biosci. Rep.* 2002, *22* (2), 129–150.
190. Law, S.L.; Huang, K.J.; Chiang, C.H. Acyclovir-containing liposomes for potential ocular delivery. Corneal penetration and absorption. *J. Contr. Release* 2000, *63* (1), 135–140.
191. Chetoni, P.; Rossi, S.; Burgalassi, S.; Monti, D.; Mariotti, S.; Saettone, M.F. Comparison of liposome-encapsulated acyclovir with acyclovir ointment: Ocular pharmacokinetics in rabbits. *J. Ocul. Pharmacol. Ther.* 2004, *20* (2), 169–177.
192. Diebolda, Y.; Jarrina, M.; Saeza, V.; Carvalhob, E.L.S.; Oreaa, M.; Calongea, M.; Seijob, B.; Alonso, M.J. Ocular drug delivery by liposome–chitosan nanoparticle complexes (LCS-NP). *Biomaterials* 2007, *28* (8), 1553–1564.
193. Felt, O. Topical use of chitosan in ophthalmology: Tolerance assessment and evaluation of precorneal retention. *Int. J. Pharm.* 1999, *180* (2), 185–193.
194. Li, N.; Zhuang, C.; Wang, M.; Sun, X.; Nie, S.; Pan, W. Liposome coated with low molecular weight chitosan and its potential use in ocular drug delivery. *Int. J. Pharm.* 2009, *379* (1), 131–138.
195. Moulik, S.P.; Paul, B.K. Structure, dynamics and transport properties of microemulsions. *Adv. Colloid Interface Sci.* 1998, *78* (2), 99–195.

196. Vyas, S.P.; Khar, R.K. *Submicron Emulsions in Targeted and Controlled Drug Delivery.* Novel Carrier Systems; CBS Publishers and Distributors: New Delhi, India, 2002; pp. 282–302.
197. Shaji, J.; Reddy, M.S. Microemulsions as drug delivery systems. *Pharm. Times* 2004, *36* (3), 17–24.
198. Vandamme, T.F. Microemulsions as ocular drug delivery systems: Recent developments and future challenges. *Prog. Retin. Eye Res.* 2002, *21* (1), 15–34.
199. Hasse, A.; Keipert, S. Development and characterisation of microemulsions for ocular application. *Eur. J. Pharm. Biopharm.* 1997, *43* (2), 179–183.
200. Gana, L.; Gan,Y.; Zhu, C.; Zhang, X.; Zhu, J. Novel microemulsion in situ electrolyte-triggered gelling system for ophthalmic delivery of lipophilic cyclosporine A: In vitro and in vivo results. *Int. J. Pharm.* 2009, *365* (1–2), 143–149.
201. Fialho, S.L.; Silva-Cunha, A. New vehicle based on a microemulsion for topical ocular administration of dexamethasone. *Clin. Exp. Ophthalmol.* 2004, *32* (6), 626–632.
202. Muller, R.H.; Gohla, S.; Dingler, A.; Schneppe, T. Large-scale production of solid-lipid nanoparticles (SLN) and nanosuspension (Dissocubes). In *Handbook of Pharmaceutical Controlled Release Technology*; Wise, D. Ed. Marcel Dekker: New York, 2000; pp. 359–375.
203. Rabinow, B.E. Nanosuspensions in drug delivery. *Nat. Rev.* 2004, *3* (9), 785–796.
204. Rani, S.; Hiremath, R.; Hota, A. Nanoparticles as drug delivery systems. *Ind. J. Pharm. Sci.* 1999, *61* (2), 69–75.
205. Mehnertw, M.K. Solid lipid nanoparticles: Production, characterization and applications. *Adv. Drug Deliv. Rev.* 2000, *47* (2–3), 165–196.
206. Ali, H.S.; York, P.; Ali, A.M.; Blagden, N. Hydrocortisone nanosuspensions for ophthalmic delivery: A comparative study between microfluidic nanoprecipitation and wet milling. *J. Contr. Release* 2011, *149* (2), 175–181.
207. Patravale, V.B.; Date, A.A.; Kulkarni, R.M. Nanosuspensions: A promising drug delivery strategy. *J. Pharm. Pharcol.* 2004, *56* (7), 827–840.
208. Kocbek, P.; Baumgartner, S.; Kristl, J. Preparation and evaluation of nanosuspensions for enhancing the dissolution of poorly soluble drugs. *Int. J. Pharm.* 2006, *312* (1–2), 179–186.
209. Keck, C.M.; Muller, R.H. Drug nanocrystals of poorly soluble drugs produced by high pressure homogenization. *Eur. J. Pharm. Biopharm.* 2006, *62* (1), 3–16.
210. Muller, R.H.; Keck, C.M. Challenges and solutions for the delivery of biotech drugs. A review of drug nanocrystal technology and lipid nanoparticles. *J. Biotechnol.* 2004, *113* (1–3), 151–170.
211. Bisrat, M.; Nystrom, C. Physicochemical aspects of drug release. VIII. The relation between particle size and surface specific dissolution rate in agitated suspensions. *Int. J. Pharm.* 1988, *47* (1–3), 223–231.
212. Zhang, J.; Shen, Z.; Zhong, J.; Hu, T.; Chen, J.; Ma, Z.; Yun, J. Preparation of amorphous cefuroxime axetil nanoparticles by controlled nanoprecipitation method without surfactants. *Int. J. Pharm.* 2006, *323* (1–2), 153–163.
213. Yoncheva, K.; Lizarraga, E.; Irache, J.M. Pegylated nanoparticles based on poly (methyl vinyl ether-comaleic anhydride): Preparation and evaluation of their bioadhesive properties. *Eur. J. Pharm. Sci.* 2005, *24* (5), 411–419.
214. Kassema, M.A.; Abdel Rahman, A.A.; Ghorab, M.M.; Ahmeda, M.B.; Khalil, R.M. Nanosuspension as an ophthalmic delivery system for certain glucocorticoid drugs. *Int. J. Pharm.* 2007, *340* (1–2), 126–133.
215. Moustafine, R.I.; Bukhovets, A.V.; Sitenkov, A.Y.; Kemenova, V.A.; Rombaut, P.; Van den Mooter, G. Eudragit E PO as a complementary material for designing oral drug delivery systems with controlled release properties: Comparative evaluation of new interpolyelectrolyte complexes with countercharged Eudragit L100 copolymers. *Mol. Pharm.* 2013, *10* (7), 2630–2641.
216. Bodmeier, R.; Chen, H. Preparation and characterization of microspheres containing the anti-inflammatory agents, indomethacin, ibuprofen, and ketoprofen. *J. Contr. Release* 1989, *10* (2), 167–175.
217. Pignatello, R.; Amico, D.; Chiechio, S.; Giunchedi, P.; Spadaro, C.; Puglisi, G. Preparation and analgesic activity of RS100 microparticles containing diflunisal. *Drug Deliv.* 2001, *8* (1), 35–45.
218. Pignatello, R.; Bucolob, C.; Ferraraa, P.; Malteseb, A.; Puleoa, A.; Puglisi, G. Eudragit RS100® nanosuspensions for the ophthalmic controlled delivery of ibuprofen. *Eur. J. Pharm. Sci.* 2002, *16* (1), 53–61.
219. Agnihotri, S.M.; Vavia, P.R. Diclofenac-loaded biopolymeric nanosuspensions for ophthalmic application. *Nanomed. Nanotechnol.* 2009, *5* (1), 90–95.
220. Lv, F.F.; Zheng, L.Q.; Tung, C.H. Phase behavior of the microemulsions and the stability of the chloramphenicol in the microemulsion-based ocular drug delivery system. *Int. J. Pharm.* 2005, *301* (1), 237–246.
221. Ding, S. Recent developments in ophthalmic drug delivery. *Pharm. Sci. Technol.* 1998, *1* (8), 328–335.
222. Durrani, A.M.; Davies, N.M.; Thomas, M.; Kellaway, I.W. Pilocarpine bioavailability from a mucoadhesive liposomal ophthalmic drug delivery system. *Int. J. Pharm.* 1992, *88* (1), 409–415.

223. Davies, N.M.; Farr, S.J.; Hadgraft, J.; Kellaway, I.W. Evaluation of mucoadhesive polymers in ocular drug delivery: II. Polymer coated vesicles. *Pharm. Res.* 1992, *9* (9), 1137–1144.
224. Klajnert, B.; Sadowska, M.; Bryszewska, M. The effect of polyamidoamine dendrimers on human erythrocyte membrane acetylcholinesterase activity. *Bioelectrochemistry* 2004, *65* (1), 23–26.
225. Milhem, O.M. Polyamidoamine starburst dendrimers as solubility enhancers. *Int. J. Pharm.* 2000, *197* (1), 239–241.
226. Bhadra, D. A PEGylated dendritic nanoparticulate carrier of fluorouracil. *Int. J. Pharm.* 2003, *257* (1), 111–124.
227. Ooya, T.; Lee, J.; Park, K. Effects of ethylene glycol-based graft, star-shaped, and dendritic polymers on solubilization and controlled release of paclitaxel. *J. Contr. Release* 2003, *93* (2), 121–127.
228. Vandamme, T.F.; Brobeck, L. Poly(amidoamine) dendrimers as ophthalmic vehicles for ocular delivery of pilocarpine nitrate and tropicamide. *J. Contr. Release* 2005, *102* (1), 23–38.
229. Spataro, G.; Malecaze, F.; Turrin, C.O.; Soler, V.; Duhayon, C.; Elena, P.P.; Majoral, J.P.; Caminade, A.M. Designing dendrimers for ocular drug delivery. *Eur. J. Med. Chem.* 2010, *45* (1), 326–334.
230. Kadam, R.; Jadhav, G.; Kompella, U.B.; Yang, H. Polyamidoamine dendrimer hydrogel for enhanced delivery of antiglaucoma drugs. *Nanomed. Nanotechnol.* 2012, *8* (5), 776–783.
231. Menjoge, A.R.; Kannan, R.M.; Tomalia, D.A. Dendrimer-based drug and imaging conjugates: Design considerations for nanomedical applications. *Drug Discov. Today* 2010, *15* (5), 171–185.
232. Mahale, N.B.; Thakkar, P.D.; Mali, R.G.; Walunj, D.R.; Chaudhari, S.R. Niosomes: Novel sustained release nonionic stable vesicular systems—An overview. *Adv. Colloid Interface Sci.* 2012, *183–184*, 46–54.
233. Lasic, D.D. On the thermodynamic stability of liposomes. *J. Colloid Interface Sci.* 1990, *140* (1), 302–304.
234. Florence, A.T. New drug delivery systems. *Chem. Ind.* 1993, *24*, 1000–1004.
235. Uchegbu, I.F.; Vyas, S.P. Non-ionic surfactant based vesicles (niosomes) in drug delivery. *Int. J. Pharm.* 1998, *172* (1–2), 33–70.
236. Kaur, I.P. Vesicular systems in ocular drug delivery: An overview. *Int. J. Pharm.* 2004, *269* (1), 1–14.
237. Aggarwal, D. Development of a topical niosomal preparation of acetazolamide: Preparation and evaluation. *J. Pharm. Pharmacol.* 2004, *56* (12), 1509–1517.
238. Vyas, S.P. Discoidal niosome based controlled ocular delivery of timolol maleate. *Pharmazie* 1998, *53* (7), 466–469.
239. Ciolino, J.B.; Dohlman, C.H.; Kohane, D.S. Contact lenses for drug delivery. *Semin. Ophthalmol.* 2009, *24* (3), 156–160.
240. Weiner, A.L.; Gilger, B.C. Advancements in ocular drug delivery. *Vet. Ophthalmol.* 2010, *13* (6), 395–406.
241. Ciolino, J.B.; Hoare, T.R.; Iwata, N.G. A drug-eluting contact lens. *Invest. Ophthalmol. Vis. Sci.* 2009, *50* (7), 3346–3352.
242. Xu, J.; Li, X.; Sun, F. Cyclodextrin-containing hydrogels for contact lenses as a platform for drug incorporation and release. *Acta Biomater.* 2010, *6* (2), 486–489.
243. Graziacascone, M.; Zhu, Z.; Borselli, F.; Lazzeri, L. Poly(vinyl alcohol) hydrogels as hydrophilic matrices for the release of lipophilic drugs loaded in PLGA nanoparticles. *J. Mater. Sci.* 2002, *13* (1), 29–32.
244. Gulsen, D.; Chauhan, A. Dispersion of microemulsion drops in HEMA hydrogel: A potential ophthalmic drug delivery vehicle. *Int. J. Pharm.* 2005, *292* (1–2), 95–117.
245. Hu, X.; Hao, L.; Wang, H.; Yang, X.; Zhang, G.; Wang, G.; Zhang, X. Hydrogel contact lens for extended delivery of ophthalmic drugs. *Int. J. Polym. Sci.* 2011, *2011*, 1–9.
246. Kapoor, Y.; Thomas, J.C.; Tan, G.; John, V.T.; Chauhan, A. Surfactant-laden soft contact lenses for extended delivery of ophthalmic drugs. *Biomaterials* 2009, *30* (5), 867–878.
247. Kapoor, Y.; Chauhan, A. Drug and surfactant transport in Cyclosporine A and Brij 98 laden p-HEMA hydrogels. *J. Colloid Interface Sci.* 2008, *322* (2), 624–633.
248. Nagarwal, R.C.; Kant, S.; Singh, P.N.; Maiti, P.; Pandit, J.K. Polymeric nanoparticulate system: A potential approach for ocular drug delivery. *J. Contr. Release* 2009, *136* (1), 2–13.
249. Calvo, P.; Remuñán-López, C.; Vila-Jato, J.L.; Alonso, M.J. Novel hydrophilic chitosan–polyethylene oxide nanoparticles as protein carriers. *J. Appl. Polym. Sci.* 1997, *63* (1), 125–132.
250. Pepic, I.; Lovric, J.; Filipovic-Grcic, J. Polymeric micelles in ocular drug delivery: Rationale, strategies and challenges. *Chem. Biochem. Eng. Q.* 2012, *26* (4), 365–377.
251. Barbu, E.; Verestiuc, L.; Nevell, T.G.; Tsibouklis, J. Polymeric materials for ophthalmic drug delivery: Trends and perspectives. *J. Mater. Chem.* 2006, *16* (1), 3439–3443.
252. Zimmer, A.; Kreuter, J. Microspheres and nanoparticles used in ocular delivery systems. *Adv. Drug Deliv. Rev.* 1995, *16* (1), 61–73.

253. De Campos, A.M.; Sanchez, A.; Gref, R.; Calvo, P.; Alonso, M.J. The effect of PEG versus a chitosan coating on the interaction of drug colloidal carriers with the ocular mucosa. *Eur. J. Pharm. Sci.* 2003, *20* (1), 73–81.
254. Nemoto, E.; Ueda, H.; Akimoto, M.; Natsume, H.; Morimoto, Y. Ability of poly-L-arginine to enhance drug absorption into aqueous humor and vitreous body after instillation in rabbits. *Biol. Pharm. Bull.* 2007, *30* (9), 1768–1772.
255. Yadav, M.; Ahuja, M. Preparation and evaluation of nanoparticles of gum cordia, an anionic polysaccharide for ophthalmic delivery. *Carbohydr. Polym.* 2010, *81* (4), 871–877.
256. Bourges, J.L.; Gautier, S.E.; Delie, F.; Bejjani, R.A.; Jeanny, J.C.; Gurny, R.; BenEzra, D.; Behar-Cohen, F.F. Ocular drug delivery targeting the retina and retinal pigment epithelium using polylactide nanoparticles. *Invest. Ophthalmol. Vis. Sci.* 2003, *44* (8), 3562–3569.
257. Gupta, H.; Aqil, M.; Khar, R.K.; Ali, A.; Bhatnagar, A.; Mittal, G. Sparfloxacin-loaded PLGA nanoparticles for sustained ocular drug delivery. *Nanomed. Nanotechnol.* 2010, *6* (2), 324–333.
258. Ludwing, A. The use of mucoadhesive polymers in ocular drug delivery. *Adv. Drug Deliv. Rev.* 2005, *57* (11), 1595–1639.
259. Imam, M.E.; Hornof, M.; Valenta, C.; Reznicek, G.; Bernkop-Schnuerch, A. Evidence for the interpenetration of mucoadhesive polymers into the mucous gel layer. *S.T.P. Pharma Sci.* 2003, *13* (3), 171–176.
260. Mikos, A.G.; Peppas, N.A. Systems for controlled drug delivery of drugs: V. Bioadhesive systems. *S.T.P. Pharma Sci.* 1986, *2*, 705–716.
261. Alonso, M.J.; Sanchez, A. The potential of chitosan in ocular drug delivery. *J. Pharm. Pharmacol.* 2003, *55* (11), 1451–1463.
262. Khangtragool, A.; Ausayakhun, S.; Leesawat, P.; Laokul, C.; Molloy, R. Chitosan as an ocular drug delivery vehicle for vancomycin. *J. Appl. Polym. Sci.* 2011, *122* (5), 3160–3167.
263. Calonge, M. The treatment of dry eye. *Surv. Ophthalmol.* 2001, *45* (2), S227–S239.
264. Lee, J.W.; Park, J.H.; Robinson, J.R. Bioadhesive-based dosage forms: The next generation. *J. Pharm. Sci.* 2000, *89* (7), 850–866.
265. Huang, S.F.; Chen, J.L.; Yeh, M.K.; Chiang, C.H. Physicochemical properties and in vivo assessment of timolol-loaded poly(D,L-lactide-co-glycolide) films for long-term intraocular pressure lowering effects. *J. Ocul. Pharmacol. Ther.* 2005, *21* (6), 445–453.
266. Fialho, S.L.; Cunha, A.D. Manufacturing techniques of biodegradable implants intended for intraocular application. *Drug Deliv.* 2005, *12* (2), 109–116.
267. Andrieu-Soler, C.; Aubert-Pouessel, A.; Doat, M.; Picaud, S.; Halhal, M.; Simonutti, M.; Venier-Julienne, M.C.; Benoit, J.P.; Behar-Cohen, F. Intravitreous injection of PLGA microspheres encapsulating GDNF promotes the survival of photoreceptors in the rd1/rd1 mouse. *Mol. Vision* 2005, *11*, 1002–1011.
268. De Faria, T.J.; De Campos, A.M.; Senna, E.L. Preparation and characterization of poly(D,L-lactide) (PLA) and poly(D,L-lactide)-poly(ethylene glycol) (PLA-PEG) nanocapsules containing antitumoral agent methotrexate. *Macromol. Symp.* 2005, *229* (1), 228–233.
269. Ayalasomayajula, S.P.; Kompella, U.B. Subconjunctivally administered celecoxib-PLGA microparticles sustain retinal drug levels and alleviate diabetes-induced oxidative stress in a rat model. *Eur. J. Pharmacol.* 2005, *511* (2), 191–198.
270. Thomas, A.A.; Kim, I.T.; Kiser, P.F. Symmetrical biodegradable crosslinkers for use in polymeric devices. *Tetrahedron Lett.* 2005, *46*, 8921–8925.

52 Drug Delivery Systems
Oral Mucosal

Javier Octavio Morales

CONTENTS

52.1 Introduction	1226
52.2 The Oral Mucosa as a Site for Drug Delivery	1226
52.2.1 Physiology	1226
52.2.2 Buccal Mucosa	1226
52.2.3 Sublingual Mucosa	1228
52.3 Fundamentals of Mucoadhesion Relevant for Oral Mucosal Delivery	1229
52.3.1 Role of Mucins in Mucoadhesion	1229
52.3.2 Theories of Mucoadhesion and Desired Characteristics of Polymers	1230
52.3.2.1 Electronic Theory	1230
52.3.2.2 Adsorption Theory	1230
52.3.2.3 Wetting Theory	1230
52.3.2.4 Diffusion Theory: Interpenetration Mechanism	1231
52.3.2.5 Fracture Theory	1231
52.4 Drug Delivery Systems for Oral Mucosal Drug Delivery and the Use of Polymers for Their Manufacture	1232
52.4.1 Tablets	1232
52.4.2 Patches/Films	1234
52.4.3 Chewing Gums	1234
52.4.4 Lozenges	1235
52.4.5 Liquids, Gels, and Ointments	1235
52.4.6 Sprays	1235
52.5 Mucoadhesive Polymers for Oral Mucosal Delivery	1235
52.5.1 Conventional Mucoadhesive Polymers	1236
52.5.1.1 Polyacrylic Acid Derivatives	1236
52.5.1.2 Cellulose Derivatives	1243
52.5.1.3 Gums and Others	1243
52.5.2 Chitosan	1243
52.5.3 Polymers with Synergistic Characteristics	1244
52.5.3.1 Enzyme-Inhibiting Polymers	1244
52.5.3.2 Permeation-Enhancing Polymers	1244
52.5.4 New-Generation Mucoadhesive Polymers	1244
52.5.4.1 Thiolated Polymers	1244
52.5.4.2 Lectin-Mediated Polymers	1246
52.5.4.3 Bacterial Adhesion	1246
52.6 Conclusions	1247
References	1248

52.1 INTRODUCTION

There is a clear tendency in the field of drug delivery toward the increase in macromolecular therapeutic agents, demonstrated both in the scientific literature and in the number of products going through the approval process in regulatory agencies.[1] However, the delivery of these actives continues to be mostly through injectable formulations.[2,3] The issues associated with this delivery route known are,[4,5] and this has always driven the research for finding alternative and more convenient routes of delivery.

The oral route is undoubtedly the most widely investigated alternative administration route; however, it presents major concerns in the delivery of macromolecular actives. The gastrointestinal route can promote degradation in the stomach due to the acidic gastric pH. The intestine has issues arising from the presence of proteolytic enzymes and insufficient permeation toward these actives, all of which result in limited bioavailability.[6,7] Therefore, other routes of delivery have been investigated and the oral mucosal route presents a convenient alternative.

Advantages of the oral mucosal route of delivery include its capacity to bypass all the limitations associated with the oral route, ease of administration, relatively low content of enzymes, and adequate vascular drainage. As described in the following sections, most of the limitations of the oral mucosa epithelium arise from its stratified nature and its intercellular content characteristics. Nonetheless, due to its direct connection to systemic circulation, delivery systems could potentially be formulated to show either bolus-like or controlled release profiles for specific therapeutic needs.[8] Polymers used in the development of such delivery systems play a major role in the release profile, permeation enhancement, and the localization of the active in the vicinity of the absorbing mucosa. Among the various uses of polymers in delivery systems, their mucoadhesive nature is the most prominent application in the oral mucosal route and is the main focus of this entry. After describing the physiological considerations in the oral cavity mucosa, this entry will review the literature pertinent to the use of polymers in delivery systems for the oral mucosal route.

52.2 THE ORAL MUCOSA AS A SITE FOR DRUG DELIVERY

52.2.1 Physiology

The oral cavity presents different types of mucosa in various regions.[9] The masticatory mucosa covers those areas that are involved in mechanical processes, such as mastication or speech, and includes the gingival mucosa and hard palate. This masticatory region is stratified and has a keratinized layer on its surface. This most superficial layer of the mucosa is similar to the structure found in the epidermis and covers about 25% of the oral cavity.[10] The specialized mucosa covers about 15%, corresponding to the dorsum of the tongue, and is a stratified tissue with keratinized as well as non-keratinized domains.[11] Finally, the lining mucosa covers the remaining 60% of the oral cavity, consisting of the inner cheeks, floor of the mouth, and underside of the tongue. This lining epithelium is stratified and non-keratinized on its surface and contains both the buccal and sublingual mucosa.[12]

52.2.2 Buccal Mucosa

The buccal mucosa is located in the inner side of the cheeks and is a stratified epithelium having approximately 40–50 cell layers resulting in a thickness of 500–600 µm (Figure 52.1).[13] The epithelium is attached to the underlying structures by a connective tissue or lamina propria, separated by a basal lamina. The connective tissue and the lamina propria regions provide mostly mechanical support and no major barrier for penetration of actives.[14,15] The connective tissue also contains the blood vessels that drain into the lingual, facial, and retromandibular veins, which then open into the internal jugular vein.[14] By driving the active directly into systemic circulation via the jugular vein,

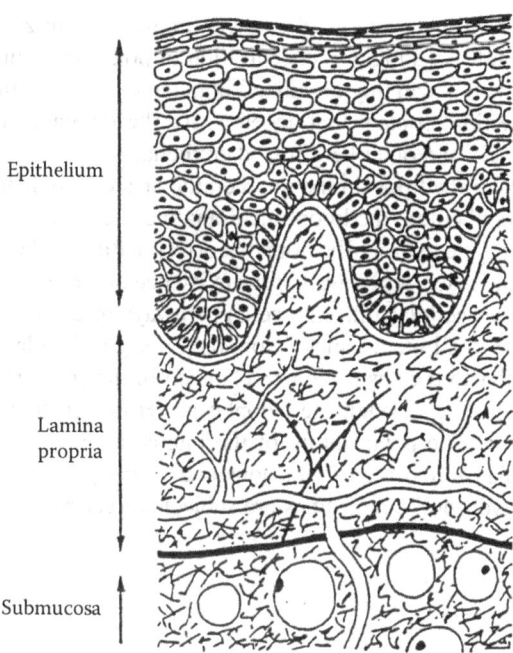

FIGURE 52.1 Diagram of a cross section of the buccal mucosa. (Adapted from Li, B. and Robinson, J.R.: *Drug Delivery to the Oral Cavity*, Ghosh, T.K. and Pfister, W.R., Eds., Marcel Dekker, Inc., New York, pp. 41–66, 2005. Copyright Wiley-VCH Verlag GmbH & Co. KGaA. Reproduced with permission.)

the buccal route is more attractive over the oral route due to the avoidance of the gastrointestinal tract conditions, such as gastric pH, enzyme content, and the first pass effect due to direct absorption into the portal vein. Once a given drug molecule reaches the connective tissue, it may be readily absorbed and distributed, thus the permeation barrier corresponds to the whole thickness of the stratified epithelium.[14]

The existence of membrane-coating granules (MCGs) in the epidermis has been well characterized and it is known to be the precursor of the keratin layer or stratum corneum.[16,17] Similarly, cytoplasmic MCGs of approximately 0.2 μm in diameter have been found in cells in the buccal epithelium; however, presence of a keratinized upper layer cannot be observed in cross sections of the epithelium. Nonetheless, the permeation barrier is believed to be related to the presence of these MCGs in the buccal mucosa.[18,19] Squier described these MCGs as organelles containing amorphous material that is extruded into the intercellular space after membrane fusion.[18] Additionally, contrary to the highly organized electron-dense lipid lamellae in epidermis, it has been reported that some MCGs in buccal mucosa contain roughly organized lipid lamellae domains.[20] Therefore, the intercellular space of the stratified non-keratinized buccal mucosa is filled with a combination of amorphous material presenting some domains where short stack of lipid lamellae can be observed. This major difference in composition of the intercellular space of buccal mucosa and epidermis is responsible for the difference in permeability toward exogenous molecules.[21]

By studying the penetration of horseradish peroxidase across buccal mucosa, it was demonstrated that the permeation barrier is located in the upper one third to one quarter of the epithelium.[22] After topical application, the horseradish peroxidase only permeated through the first 1–3 cell layers. However, when injected subepithelially it was found to permeate through as deep as the connective tissue and up as far as the MCGs zone.[22] This is well correlated with the presence of the lipid-rich domain of the epithelium in its upper region.[21,23,24] The lipid composition in the buccal epithelium has a higher content of phospholipids, cholesterol esters, and glycosylceramides, while the ceramides content is minimal, compared to the skin and keratinized regions of the oral cavity.[21] This composition results in a higher concentration of polar lipids in the intercellular space.[23] Therefore, it is not only due to the highly organized lipid lamellae found in the keratinized epithelia but also the nature of the lipid content that accounts for the increased permeation of the buccal mucosa compared to the skin and other keratinized epithelia.

Due to the polar nature of the lipids in the intercellular space, two different domains can be differentiated in the buccal epithelium: the lipophilic domain corresponding to the cell membranes of the stratified epithelium and the hydrophilic domain corresponding to the extruded content from the MCGs into the intercellular space. As a result of the existence of these two domains the paracellular (between cells) and the transcellular (through cells) pathways of absorption have been identified.[9] The lipophilic nature of the cell membranes favors the passage of molecules with high log P values across the cells. Similar to the absorption mechanism in the small intestine, it is believed that lipophilic molecules are carried through the cytoplasm.[16] However, there is still lack of evidence supporting this assumption. The polar nature of the intercellular space favors the penetration of more hydrophilic molecules across a more tortuous and longer path.[25,26] Additionally, it has been demonstrated that some hydrophilic molecules are subjected to carrier-mediated transport through the buccal mucosa.[27]

52.2.3 Sublingual Mucosa

The sublingual mucosa is also part of the lining mucosa in the oral cavity, and as such is a stratified non-keratinized epithelium. Similarly to the buccal epithelium, it presents a mitotically active basal cell layer, followed by an intermediate differentiating cell layer, and a superficial region where cells become flattened and increase in size.[28] However, the sublingual mucosa is much thinner than the buccal mucosa, having 8–12 cell layers followed by the basal lamina.[26] Drugs delivered through the sublingual mucosa need to pass the epithelium and basal lamina to reach the underlying connective tissue and be absorbed. This region is composed of collagen fibrils, nerve fibers, and blood vessels that allow for absorption of drugs into circulation. Blood vessels in the connective tissue drain into either the sublingual artery, supplying blood to the salivary glands and mucosae, or the lingual artery, which provides blood supply to not only the sublingual mucosa but also the tongue.

The biochemical composition of the sublingual mucosa, similar to that found in the buccal mucosa, has higher amounts of polar lipids, such as phospholipids, cholesterol esters, and glycosylceramides, in comparison to the composition found in the keratinized areas of the oral cavity. The polar lipids are found in the intercellular spaces and provide the epithelium with increased fluidity and a higher permeation to hydrophilic molecules.[23,29] Even though the sublingual mucosa is thinner than the buccal mucosa, its permeation barrier is also located in the upper region of the epithelium. MCGs are also responsible for the formation of the permeability barrier by fusing with the cellular membrane and extruding their content into the intercellular spaces of the superficial layers of the epithelium.[18] Therefore, due to the similarities in composition found for sublingual and buccal mucosa, the higher permeability shown by the sublingual mucosa is due to the difference in thickness.[28,30] Similar to the mechanisms described for the buccal mucosa, molecules permeating through the sublingual mucosa will cross either by the paracellular or transcellular route depending on the physicochemical properties of the molecule. In addition, the existence of carrier-mediated transport in the sublingual mucosa has only been reported for small molecules such as glucose, glutathione, some amino acids, and others.[28,31] The presence of a P-glycoprotein efflux transporter has been suggested in the sublingual mucosa[32,33]; however, most research in the literature indicates that passive diffusion is the most prominent permeation pathway.[34,35]

A distinct feature of the sublingual mucosa is its capacity to produce most of the saliva in the oral cavity. Saliva is produced mainly in the sublingual region by the submandibular, parotid, or sublingual salivary glands. The normal range of secretion of saliva is from 0.5 to 2 L/day.[36] Saliva concentrates in the posterior region of the tongue surface, the buccal mucosa, and in the anterior region of the floor of the mouth, providing sublingual delivery systems with a wet environment.[37]

52.3 FUNDAMENTALS OF MUCOADHESION RELEVANT FOR ORAL MUCOSAL DELIVERY

The term mucoadhesion is used to describe the specific interaction between a mucosa (soft tissue) and a material of interest. As seen in Figure 52.2, the mucoadhesive bond is composed of three regions, the surface of the mucoadhesive material, the first layer of the natural tissue, and the interfacial layer.[38] The mucoadhesive bond is constituted by interactions of the molecules in the delivery systems (typically polymers) and macromolecules present in the mucus layer known as mucins.

52.3.1 Role of Mucins in Mucoadhesion

The mucus is a viscous fluid covering the mucous membrane.[39] It is majorly composed of water and a high concentration of mucins, inorganic salts, proteins, lipids, and mucopolysaccharides. Mucins are the most important components in terms of mucoadhesion due to their interpenetration and entanglement with the polymer chains. Structurally, mucins are glycoproteins present in different mucosae with different structures specific to some regions in the body. These molecules are composed of a protein core and carbohydrate side chains, responsible for the non-covalent bonding that occurs when a mucoadhesive material is brought in contact.[40] The mucus may be visualized as a highly entangled system of macromolecular chains with physical junctions, probably corresponding to disulfide bonding and stabilized by interhydrogen bonding or other non-covalent bonds (Figure 52.3).

The types of bonds that can be elicited in the mucoadhesion phenomenon can be physical or mechanical, secondary chemical or ionic, and primary or covalent chemical bonds (like adhesion between two polymers). Physical bonding is believed to occur between the polymer chains and the tissue roughness; therefore, fluid materials behave better in this type of interaction because they can be successfully included in these anomalies. Secondary chemical bonds include hydrogen bonding and van der Waals forces and are believed to play a major role in the mucoadhesive bond between polymer chains and mucin molecules. Primary chemical bonds occur by chemical reaction of the mucin functional groups with the mucoadhesive material.[41] In order to better understand the process of mucoadhesion, different theories have been proposed to explain the phenomenon and are described in the following.

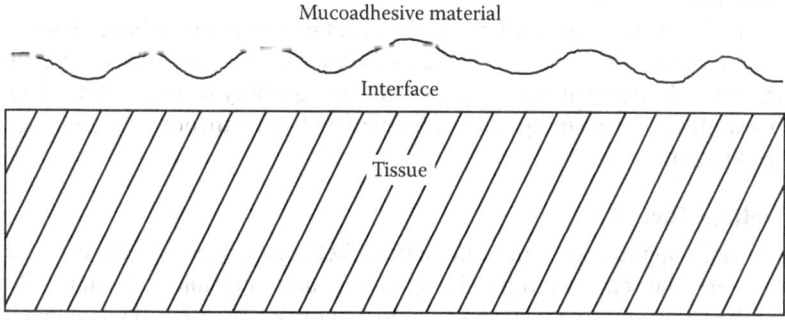

FIGURE 52.2 Basic scenario of mucoadhesion between a polymer and a soft tissue. (Reprinted from *J. Contr. Release*, 2, Peppas, N.A. and Buri, P.A., Surface, interfacial and molecular aspects of polymer bioadhesion on soft tissues, 257–275. Copyright 1985, with permission from Elsevier.)

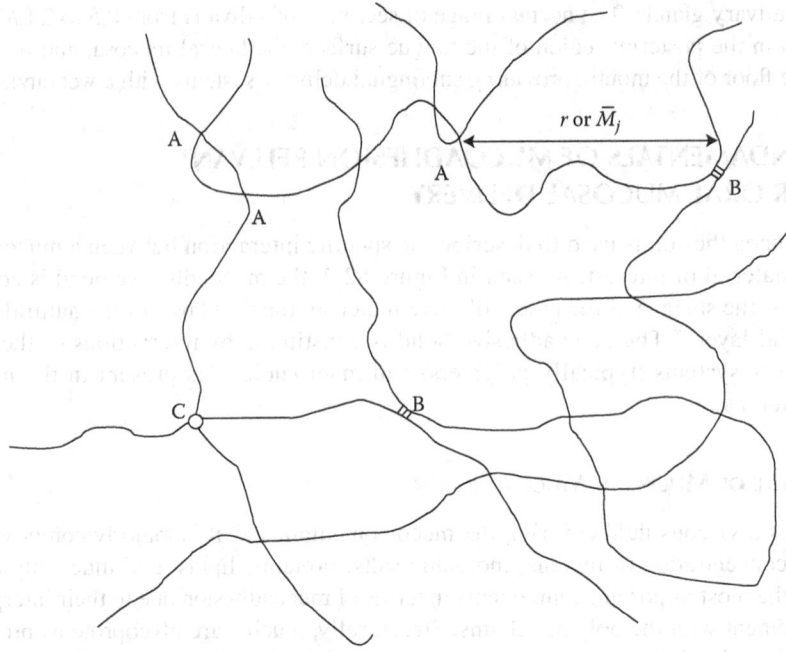

FIGURE 52.3 Representation of the cross-linked structure of the mucin network. (1) Possible cross-linking can occur through entanglements, (2) molecular associations, and (3) permanent covalent cross-links. On average, the end-to-end distance between two junctions, r, corresponds to a particular molecular weight, M_j. (Reprinted from *J. Contr. Release*, 2, Peppas, N.A. and Buri, P.A., Surface, interfacial and molecular aspects of polymer bioadhesion on soft tissues, 257–275. Copyright 1985, with permission from Elsevier.)

52.3.2 Theories of Mucoadhesion and Desired Characteristics of Polymers

52.3.2.1 Electronic Theory

The electronic theory is based on the assumption that the materials in contact (mucoadhesive and substrate) have different electronic structures.[42] In order to balance the Fermi levels, when the materials come in contact, electron transfer occurs creating a double layer of electrical charge. In this theory, the mucoadhesive bond is believed to be maintained by the attractive forces in this electrical double layer. However, it is still not fully understood whether the electrostatic forces are a cause or the result of the contact between the two materials.[43]

52.3.2.2 Adsorption Theory

According to this theory, the adhesive bond is the result of non-covalent bonds between the mucoadhesive and the substrate, such as van der Waals forces and hydrogen bonds.[41,44] Even though it is known that these bonds are weak individually, the large quantity of interactions between the two materials is what allows for a strong mucoadhesive bond. This theory is one of the most widely accepted in the literature.

52.3.2.3 Wetting Theory

This theory is more applicable to the interactions between a mucoadhesive liquid and a substrate. It uses interfacial tensions to predict spreading and thus mucoadhesion.[38,45,46] Spreading coefficients of the mucoadhesive material need to be positive in order to spontaneously displace the content in contact with the mucosa and create the mucoadhesive bond.[47] By determining surface and interfacial tensions, it is possible to calculate the work done in forming the mucoadhesive bond.

Drug Delivery Systems

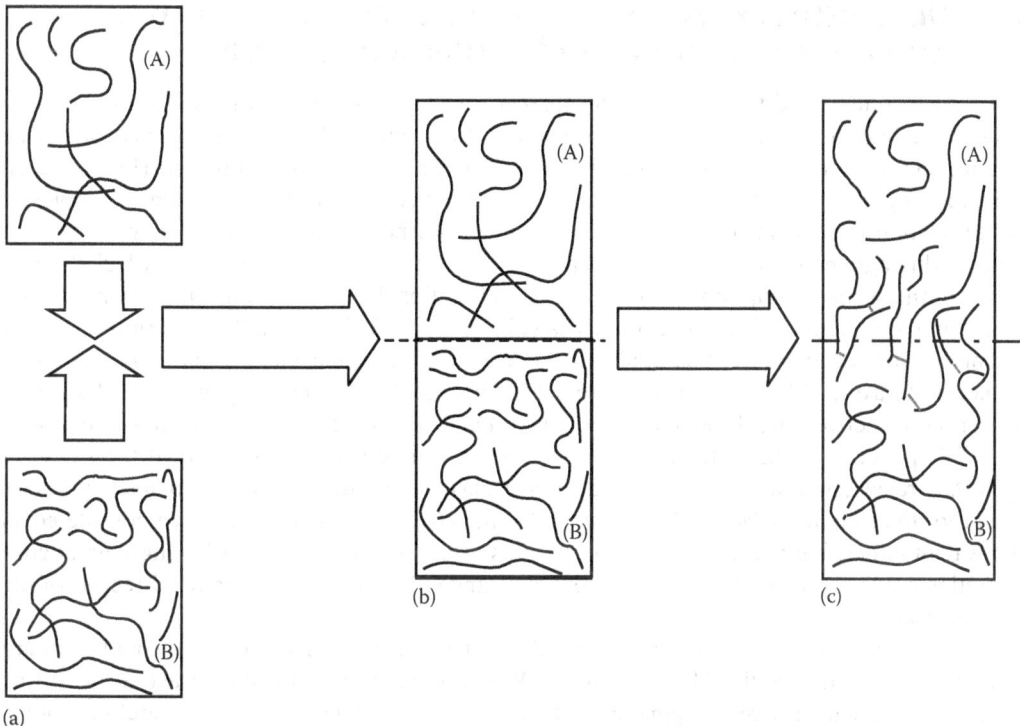

FIGURE 52.4 Molecular model of chain interpenetration during mucoadhesion of a polymer (A) with the mucin (B). Three stages are distinguished in the process: (a) contact, (b) interpenetration, and (c) interaction. (Reprinted from *J. Contr. Release*, 2, Peppas, N.A. and Buri, P.A., Surface, interfacial and molecular aspects of polymer bioadhesion on soft tissues, 257–275. Copyright 1985, with permission from Elsevier.)

52.3.2.4 Diffusion Theory: Interpenetration Mechanism

The interpenetration between the polymer chains and the mucin chains in the mucosa is believed to be the main physical mechanism for mucoadhesion, and together with the adsorption theory is the most accepted in the literature. This mechanism was first proposed in the polymer–polymer interface[48,49] but was later applied to mucoadhesion due to the polymeric nature of mucin.[38] Basically, during interpenetration the molecules of the mucoadhesive and the mucin molecules in the mucosa are brought into contact, and due to the concentration gradient the polymer chains penetrate into the mucin network with specific diffusion coefficients (Figure 52.4).

Cross-linking of either component will hinder interpenetration, but small chains and end chains that can freely move will still produce the entanglement. The depth of interpenetration needed to create an effective mucoadhesive bond has not been determined, but it is believed to be in the 0.2–0.5 μm range.[38]

52.3.2.5 Fracture Theory

This theory relates the force needed to separate the two surfaces to the mucoadhesive strength[50] and is widely applied in research, where force of detachment versus distance is normally measured. In general, the fracture stress (considered to be equivalent to the mucoadhesion stress) is calculated by dividing the fracture force by the area of contact in the mucoadhesive bond. In the same study, fracture energy (or work of mucoadhesion) can be obtained as the area under the curve in the force versus distance plot. Nowadays, instruments measure directly the force of mucoadhesion between two surfaces in contact as a function of distance and time.

52.4 DRUG DELIVERY SYSTEMS FOR ORAL MUCOSAL DRUG DELIVERY AND THE USE OF POLYMERS FOR THEIR MANUFACTURE

There are a number of different dosage forms that have been explored for the delivery of actives through the buccal and sublingual route; therefore, they can be classified in various ways. One practical classification system is by looking at the control over direction of release (Figure 52.5). First, the "multidirectional type" has no control over release in any direction. Some examples of this group of dosage forms include sprays, gels, mouthwashes, monolayered films, and uncoated tablets. Although these formulations could potentially be easier to manufacture, they lack the efficiency of the second group of platforms. The "unidirectional type" has a means to control drug release in the direction of the buccal mucosa (Figure 52.5). This is usually done in the form of an impermeable coating layer that hinders the release into the oral cavity. Examples of this group include multilayered films and coated tablets. Platforms in the second group are more efficient in delivering the active drug through the oral mucosa and constitute the trend for product development.[51] Depending on the style of coating the release can be partially directed (if the sides are open then release will also occur sideways) or exclusively unidirectional (if completely sealed except for the side in contact with the mucosa). The latter strategy has even been considered as one way of enhancing permeation[52]; however, this effect is only a result of having a great concentration gradient enclosed in the polymer matrix and only able to release in the direction of the oral mucosa.

Since transmucosal dosage forms will be administered and left inside the oral cavity they are desired to be small, thin, flexible, and tough.[53] While in the oral cavity, the dosage form will be subjected to mechanical stress originating from the tongue and mouth activities[54]; therefore, dosage forms need to be tough and elastic to withstand such stresses. Additionally, other desired characteristics of oral mucosal dosage forms include high drug loading capacity, absence of irritation, controlled release, and tasteless. Adequate mucoadhesive properties are needed, especially, for dosage forms that require prolonged contact with the mucosa for sustained release over long periods of time. Mucoadhesive polymers have been widely used for providing adhesion to the dosage form and will be further discussed in the following sections.

A number of products in later clinical trials and already commercialized are available and are listed in Table 52.1.[55,56] Tablets, films, and sprays are examples of products successfully commercialized and the reader is directed to References 55–57 for further information on such products.

52.4.1 Tablets

Tablets are the most common dosage form investigated in the literature as well as products under development.[58] Commonly manufactured by compression of powders to disks with a diameter of

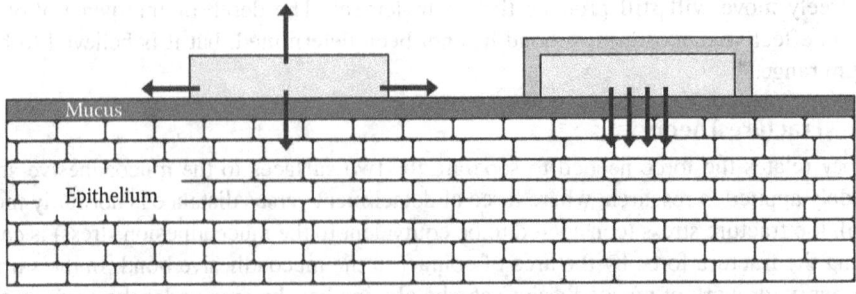

FIGURE 52.5 The two types of mucoadhesive platforms for oral mucosal release: multidirectional (on the left) and unidirectional (on the right) release.

TABLE 52.1
Examples of Drugs Administered through Oral Transmucosal Sites and Information Regarding Specific Site, Dosage Form, Product Name, and Manufacturer

Drug	Mucosal Site	Dosage Form	Product Name	Manufacturer
Asenapine	Sublingual	Tablet	Saphris®	Schering-Plough Corp.
Buprenorphine and naloxone	Sublingual	Tablet	Suboxone®	Reckitt Benckiser
Buprenorphine and naloxone	Sublingual	Film	Suboxone®	Reckitt Benckiser
Fentanyl citrate	Buccal	Tablet	Fentora®	Cephalon
Fentanyl citrate	Buccal	Film	Onsolis®	Meda Pharmaceuticals, Inc.
Fentanyl citrate	Buccal	Lozenge	Actiq®	Teva Pharmaceuticals
Fentanyl citrate	Sublingual	Spray	Subsys®	INSYS Therapeutics
Fentanyl citrate	Sublingual	Tablet	Abstral®	Orexo AB
Glyceryl trinitrate	Sublingual	Spray	Glytrin®	Sanofi-Aventis
Insulin	Buccal	Spray	Oral-lyn®	Generex Biotechnology Corp.
Miconazole nitrate	Buccal	Tablet	Oravig®	BioAlliance Pharma
Midazolam	Buccal	Solution	Buccolam®	ViroPharma
Nicotine	Buccal	Gum	Nicorette®	GlaxoSmithKline
Nitroglycerin	Buccal	Tablet	Nitroguard®	Forest Pharmaceuticals
Ondansetron	GIT	Film	Zuplenz®	Strativa Pharmaceuticals
Testosterone	Buccal	Tablet	Striant®	Columbia Laboratories
THC and cannabidiol	Buccal, sublingual	Solution	Sativex®	Almirall, S.A. (ALM) and GW Pharmaceuticals
Zolpidem tartrate	All	Spray	Zolpimist®	ECR Pharmaceuticals Co., Inc.
Zolpidem tartrate	Sublingual	Tablet	Edluar®	Orexo AB
Zolpidem tartrate	Sublingual	Tablet	Intermezzo®	Transcept Pharmaceuticals, Inc.

Sources: Derjaguin, B.V. and Smilga, V.P., *Adhesion: Fundamentals and Practice*, McLaren, London, U.K., 1969; Chickering, D.E. and Mathiowitz, E., Definitions, mechanisms, and theories of bioadhesion, in *Bioadhesive Drug Delivery Systems: Fundamentals, Novel Approaches, and Development*, Mathiowitz, E., Chickering, D.E., and Lehr, C.M., Eds., Marcel Dekker, Inc., New York, 1999.

approximately 5–8 mm,[59] tablets can be designed to release multidirectionally through the oral mucosa and into the oral cavity, as well as unidirectionally through either the buccal, gingival, or sublingual mucosa. Normally, mucoadhesive tablets are manufactured by direct compression but can sometimes be made by wet granulation. Tablets are formulated to remain in contact with the mucosa for a predetermined time that can range from minutes to a few hours. To control the release unidirectionally, tablets can be coated in all their faces except one with water insoluble and non-permeable polymers such as ethylcellulose[60] and hydrogenated castor oil.[61] Permeation enhancers in non-coated tablets are difficult to properly formulate due to the uncontrolled release into the oral cavity.[62] Tablets swell in contact with the water in saliva and soften to establish the mucoadhesive bond with the mucin molecules; however, one of their limitations is the potential for discomfort upon repeated administrations due to the relatively large size and thickness, and their poor flexibility.[58]

Alternative applications of tablets as oral mucosal devices include, for example, the addition of the active in microspheres prior to compression[63] and use of matrices that melt at body temperatures.[64,65] Giunchedi et al. have reported on the use of a spray drying technique to produce chitosan and drug-loaded microspheres to improve the flowability of chitosan and to further enhance the control over drug release.[63] Poloxamer 407 has been reported to improve the mucoadhesion of tablets for buccal delivery by softening at body temperature and allowing for a more intimate contact between the materials in the matrix and the mucins in saliva.[64] Furthermore, fast-melting tablets have allowed for the quick sublingual release of ketotifen.[65] By using a blend of polyethylene glycol 400 and 6000 the melting temperature of the matrix could be tailored and drug permeation enhanced.[65]

52.4.2 Patches/Films

Films, also known as patches, as dosage forms for oral mucosal delivery are usually more patient friendly and acceptable than tablets; however, their small size and reduced area of application result in a limited capacity to deliver large drug amounts.[66] They can be manufactured by solvent casting, direct milling, and hot-melt extrusion. Due to the versatility of the manufacturing processes and the film characteristics, permeation enhancers can be included in the formulation and their use is well described in the literature and has been reviewed previously.[16,67] Films can be designed differently in accordance with the type of polymers and the layering system utilized and are commonly classified into three groups.[56,62] First, films consisting of one soluble matrix layer, much like tablets, release drug multidirectionally and are more efficient in the treatment of local diseases, such as candidiasis and mucositis, rather than having a systemic target.[62] Due to their mucoadhesive nature and formulation, single-layer films can remain in the oral cavity and release into the oral cavity and the gastrointestinal tract for longer times than tablets. The water-soluble polymers used in the formulations of this type of films allow for the complete dissolution and erosion of the delivery system after complete release of the active. Second, films with a non-dissolvable backing layer that is impermeable to the passage of the active through its matrix. These films allow for drug release exclusively into the oral mucosa direction. As a consequence of their mucoadhesion and the presence of the backing layer, films with a non-dissolvable backing layer can be formulated to release the drug for extended periods of time between 8 and 24 h.[68] However, after the drug has been released and absorbed through the mucosa the backing layer remains and needs to be manually removed. Thus, the third type of films are comprised of a dissolvable backing layer as a step further in the optimization of films as oral mucosal delivery systems. These films share many characteristics of the second group except that films with the dissolvable backing layer do not need physical removal at the end due to complete erosion upon drug release.[62,69] It is important to note that films without a loaded drug can also be used for their occlusive properties as wound dressings.[56,70] In a study by Karlsmark et al., it was reported that the use of a cold sore patch, without a treating drug, was comparable to the effect of a conventional acyclovir cream 5% for the treatment of herpes simplex labialis.[71]

52.4.3 Chewing Gums

Chewing gums have been investigated for their use as drug delivery systems for the oral mucosal route.[72] Even though chewing gums are delivery systems that cannot control the direction of release, they are advantageous in terms of patient convenience and compliance due to the conventional use of non-medicated gums. Chewing gums have the possibility of controlling drug release for extended periods of time thus improving the variability in drug release and retention times. It has been suggested that due to the convenience of the system, patients could tailor the release rate by simply changing the rate and vigor of chewing, as well as halting therapy by expelling the gum altogether.[62] A novel antimicrobial decapeptide (KSL) has been incorporated into chewing gums for its local use as an antiplaque device.[73] In an in vitro/in vivo investigation, KSL-containing chewing gum demonstrated stability of the peptide and controlled release over 30 min.[74] Another novel application is the use of chewing gums for the delivery of chlorhexidine.[75] Chlorhexidine-containing chewing gum can be used for the treatment of gingivitis, periodontitis, oral, and pharyngeal infections, also as an inhibitor of plaque growth.[72] Nicotine-containing chewing gum has been shown to release 90% of its content within 30 min of chewing and the release extent and rate can be controlled by the rate and vigor of chewing.[76] Thus, despite the limitation of local delivery of chewing gums, they can be controlled by patients to meet their need for actives.[76]

52.4.4 LOZENGES

Lozenges share many characteristics of tablets, although they are designed to gradually dissolve on the back surface of the tongue. Similarly to tablets, lozenges can be manufactured by compression of sugar-based formulations and by molding.[77] From a formulation standpoint, lozenges pose the need for a sweetener or a taste-masking strategy due to the local action of the active in the tongue, throat, and oral mucosa. Lozenges dissolve usually in less than 10 min but can remain for as long as 30 min depending on the formulation. Since the residence time of the unit depends on the active saliva production by the patient, high inter- and intra-individual variation in absorption and bioavailability can be found.[62] The use of permeation enhancers is hindered by the more local effect as opposed to a localized transmucosal delivery.

52.4.5 LIQUIDS, GELS, AND OINTMENTS

Liquid dosage forms normally refer to solutions or suspensions containing a drug solubilized or suspended for a local action. Mouthwashes and mouth-fresheners are common commercially available products as liquid dosage forms. More than other delivery systems, liquid forms have a reduced retention time that normally depends on the patient exclusively. Thus, high variability can be observed in the amount of drug absorbed and delivered. A different approach consists of delivering drugs from liquid dosage forms by iontophoresis through the buccal mucosa. It has been described that after 5–10 min of administration, naltrexone appeared in pigs plasma in a system delivered buccally with the use of iontophoresis, and peaked at 90 min.[78]

Gels typically contain a polymer, an active substance, and any required excipient dissolved or suspended in an aqueous or non-aqueous base to yield a viscous liquid. Gels can be administered conveniently with the use of a finger or a syringe to a localized portion of the oral mucosa; however, this could result in a high variability in the amounts of drug administered and thus delivered.[79] Due to the inherent poor retention time of semisolid formulations, mucoadhesive polymers are usually utilized to increase retention time, providing a more sustained release of actives.[80] In an investigation of mucoadhesive polymer solutions, Vonarx et al. reported on the capacity of formulations to coat the esophagus under conditions mimicking the salivary flow and act as an esophageal bandage.[81] Polycarbophil and xanthan gum demonstrated excellent mucoadhesive potential, whereas croscarmellose sodium and poloxamer 407 had short retention times.

52.4.6 SPRAYS

Sprays are liquid dosage forms aerosolized into the oral cavity for transmucosal drug delivery. Due to the formation of small droplets upon aerosolization of the liquid, sprays can achieve shorter lag times and a faster onset of action than other liquid formulations. Oral-lyn™ is a micellar system utilizing a blend of surfactants as permeation enhancers and formulation stabilizers.[82,83] The product is one application of the RapidMist™ system patented by Generex Biotechnology Corporation that has also been used to develop products for vaccine delivery, pain management, and weight loss.

52.5 MUCOADHESIVE POLYMERS FOR ORAL MUCOSAL DELIVERY

As a consequence of the theories of mucoadhesion listed earlier and by experimental investigations, several parameters affecting mucoadhesion have been identified in the literature.[84,85] A first group of variables that determine the extent, force, and duration of the mucoadhesive bond is related to the polymer in the dosage form. The common characteristics usually listed for an ideal mucoadhesive polymer include the presence of strong hydrogen-bonding groups [–OH, –COOH]; strong anionic or cationic charges and high charge density; sufficient flexibility of the dosage form and polymer molecule to allow for interpenetration into the mucus layer; adequate surface tension to allow for the wetting in the interface polymer–mucus layer; pH characteristics rendering the polymer biocompatible with the mucosa; the polymer and its degradation products should be non-toxic, non-irritant, and free from leachable impurities; adequate mechanical properties; and an optimum molecular weight and cross-linking degree.[86,87] Many of these parameters will vary on a case by case basis depending on the nature of the polymer, the formulation content, and the type of dosage form.

Other factors controlling the mucoadhesive bond are environmentally (oral cavity) related factors. The formation and persistence in time of the mucoadhesive bond are highly dependent on the salivary content and characteristics. For example, depending on the saliva flow rate and method of determination, the pH of saliva has been estimated to range between 6.5 and 7.5.[59] The pH in the vicinity of the dosage form will determine the ionization state of the dosage form and thus directly affect the potential for mucoadhesion of the polymer. Recently, Kotagale et al. reported on the influence of pH modifiers in buccal tablet formulations containing Carbopol® 934, sodium alginate, and gelatin.[88] By varying the content of the three polymers as well as the addition of pH modifiers, the authors could vary the microenvironmental pH from 3.91 to 6.37 and observed an increase in mucoadhesion as the pH decreased, which was correlated with an increase in the concentration of Carbopol 934. Another factor to consider in the formulation of mucoadhesive dosage forms for oral mucosal delivery is the mucin turnover in the oral cavity. Mucin turnover determines the residence time of the mucoadhesive form in vivo and has been reported as being between 5 and 30 min in the oral cavity of humans.[89] Table 52.2 presents the most significant mucoadhesive polymers utilized in the literature for oral mucosal delivery. These are most of the so-called first and second generation (conventional) mucoadhesive polymers and further description follows.

52.5.1 Conventional Mucoadhesive Polymers

52.5.1.1 Polyacrylic Acid Derivatives

This group of anionic molecules constitutes most of the synthetic polymers used as mucoadhesives and includes carbopols, polycarbophil, polyacrylic acid, polyacrylate, poly(methylvinylether-co-methacrylic) acid, poly(2-hydroxyethyl methacrylate), poly(methacrylate), poly(alkyl cyanoacrylate), poly(isohexyl cyanoacrylate), and poly (isobutyl cyanoacrylate). Their mucoadhesive capacity is associated with the presence of the carboxylic groups responsible for the formation of hydrogen bonds. By investigating a series of weakly cross-linked hydrogels obtained by copolymerization of acrylic acid with acrylamide, Park and Robinson evaluated the effect of solution pH, copolymer composition, and cross-linking density on mucoadhesion measured in excised stomach tissue.[90] Mucoadhesion of polymers was the strongest under acidic pH and decreased sharply at pH above 4.0. This indicated the need of the carboxylic group in its acidic form for hydrogen bonding of the polymer with mucin molecules. Increase in the concentration of the cross-linker was also associated with a decrease in mucoadhesion. A hydrogel with a higher degree of cross-linking results in a more rigid polymeric network. In order for interpenetration of polymer and mucin chains to occur, a certain degree of flexibility of the polymer is required, and thus the detrimental effect of the cross-linker on mucoadhesion. An experimental device consisting of a cell containing a layer of polyacrylic acid and water as a penetrant was used to study the dependence of swelling of the polymer with respect to pH and the presence of mucin.[91] The lowest diffusion rates were observed at pH 4 and 5, correlating with the formation of an impermeable layer impeding the passage of water into the polymer matrix and highlighting the importance of the ionization state of the molecules.

TABLE 52.2
Properties and Characteristics of Some Representative Conventional Mucoadhesive Polymers

Mucoadhesive Polymer	Properties	Characteristics and Applications
Polycarbophil	Mw 2.2×10^5 η 2,000–22,500 cps (1% aqueous solution) κ 15–35 mL/g in acidic media (pH 1–3), 100 mL/g in neutral and basic media φ viscous colloid in cold water Insoluble in water, but swells to varying degrees in common organic solvents, strong mineral acids, and bases	Synthesized by lightly cross-linking 0.5%–1% w/w divinyl glycol Swellable depending on pH and ionic strength Swelling increases as pH increases At pH 1–3 absorbs 15–35 mL of water per gram but absorbs 100 mL per gram at neutral and alkaline pH Entangle the polymer with mucus on the surface of the tissue Hydrogen bonding between the non-ionized carboxylic acid and mucin
Carbopol/carbomer	Mw 1×10^6–4×10^6 η 29,400–39,400 cps at 25°C with 0.5% neutralized aqueous solution κ 5 g/cm³ in bulk, 1.4 g/cm³ tapped pH 2.5–3.0 φ water, alcohol, glycerin White, fluffy, acidic, hygroscopic powder with a slight characteristic odor	Synthesized by cross-linking allyl sucrose or allyl pentaerythritol Excellent thickening, emulsifying, suspending, and gelling agent Common component in bioadhesive dosage forms Gel loses viscosity on exposure to sunlight Unaffected by temperature variations, hydrolysis, oxidation, and resistant to bacterial growth It contributes no off-taste and may mask the undesirable taste of the formulation Incompatible with phenols, cationic polymers, high concentrations of electrolytes, and resorcinol
Sodium carboxymethyl cellulose (SCMC)	It is an anionic polymer made by swelling cellulose with NaOH and then reacting it with monochloroacetic acid Grades H, M, and L Mw 9×10^4–7×10^5 η 1200 cps with 1.0% solution ρ 0.75 g/cm³ in bulk pH 6.5–8.5 φ water	Emulsifying, gelling, and binding agent Sterilization in dry and solution form, irradiation of solution loses the viscosity Stable on storage Incompatible with strongly acidic solutions In general, stability with monovalent salts is very good; with divalent salts good to marginal; with trivalent and heavy metal salts poor, resulting in gelation or precipitation CMC solutions offer good tolerance of water miscible solvents, good viscosity stability over the pH range of 4–10, compatibility with most water-soluble non-ionic gums, and synergism with hydroxyethyl cellulose (HEC) and hydroxypropyl cellulose (HPC) Most CMC solutions are thixotropic; some are strictly pseudoplastic

(*Continued*)

TABLE 52.2 (Continued)
Properties and Characteristics of Some Representative Conventional Mucoadhesive Polymers

Mucoadhesive Polymer	Properties	Characteristics and Applications
	White to faint yellow, odorless, hygroscopic powder, or granular material having faint paper-like taste	All solutions show a reversible decrease in viscosity at elevated temperatures. CMC solutions lack yield value Solutions are susceptible to shear, heat, bacterial, enzyme, and UV degradation Good bioadhesive strength Cell immobilization via a combination of ionotropic gelation and polyelectrolyte complex formation (e.g., with chitosan) in drug delivery systems and dialysis membranes
Hydroxypropyl cellulose (HPC)	Mw 6×10^4–1×10^6 η 4–6500 cps with 2.0% aqueous solution pH 5.0–8.0 ρ 0.5 g/cm^3 in bulk Soluble in water below 38°C, ethanol, propylene glycol, dioxane, methanol, isopropyl alcohol, dimethyl sulfoxide, dimethyl formamide, etc. Insoluble in hot water White to slightly yellowish, odorless powder	Best pH is between 6.0 and 8.0 Solutions of HPC are susceptible to shear, heat, bacterial, enzymatic, and bacterial degradation It is inert and showed no evidence of skin irritation or sensitization Compatible with most water-soluble gums and resins Synergistic with CMC and sodium alginate Not metabolized in the body It may not tolerate high concentrations of dissolved materials and tends to be salting out It is also incompatible with the substituted phenolic derivatives such as methyl and propyl parahydroxy benzoate Granulating and film coating agent for tablet Thickening agent, emulsion Stabilizer, suspending agent in oral and topical solution or suspension
Hydroxypropylmethyl cellulose (HPMC)	Mw 8.6×10^4 η E15–15 cps, E4M–400 cps, and K4M–4000 cps (2% aqueous solution) φ Cold water, mixtures of methylene chloride and isopropyl alcohol Insoluble in alcohol, chloroform and ether Odorless, tasteless, white, or creamy white fibrous or granular powder	Mixed alkyl hydroxyalkyl cellulosic ether Suspending, viscosity-increasing and film-forming agent Tablet binder and adhesive ointment ingredient E grades (Methocel®) are generally suitable as film formers while the K grades are used as thickeners Stable when dry Solutions are stable at pH 3.0–11.0 Incompatible to extreme pH conditions and oxidizing materials

(Continued)

TABLE 52.2 (Continued)
Properties and Characteristics of Some Representative Conventional Mucoadhesive Polymers

Mucoadhesive Polymer	Properties	Characteristics and Applications
HEC	Available in grades ranging from 2 to 8,00,000 cps at 2%	Solutions are pseudoplastic and show a reversible decrease in viscosity at elevated temperatures
	Light tan or cream to white powder, odorless, and tasteless. It may contain suitable anticaking agents	HEC solutions lack yield value
	ρ 0.6 g/mL	Solutions show only a fair tolerance with water miscible solvents (10%–30% of solution weight)
	pH 6–8.5	Compatible with most water-soluble gums and resins
	φ in hot or cold water and gives a clear, colorless solution	Synergistic with CMC and sodium alginate
	Non-ionic	Susceptible for bacterial and enzymatic degradation
		Polyvalent inorganic salts will salt out HEC at lower concentrations than monovalent salts
		Shows good viscosity stability over the pH range of 2–12
		Used as suspending or viscosity builder
		Binder, film former
Xanthan gum	It is soluble in hot or cold water and gives visually hazy, neutral pH solutions	Xanthan gum is more tolerant of electrolytes, acids, and bases than most other organic gums
	It will dissolve in hot glycerin	It can, nevertheless, be gelled or precipitated with certain polyvalent metal cations under specific circumstances
	Solutions are typically in the 1500–2500 cps range at 1%; they are pseudoplastic and especially shear-thinning. In the presence of small amounts of salt, solutions shows good viscosity stability at elevated temperatures	Solutions show very good viscosity stability over the pH range of 2–12 and good tolerance of water-miscible solvents
	Solutions possess excellent yield value	It is more compatible with most non-ionic and anionic gums, featuring useful synergism with galactomannans
	Anionic	It is more resistant to shear, heat, bacterial, enzyme, and UV degradation than most gums
Guar gum	Obtained from the ground endosperms of the seeds of *Cyamopsis tetragonolobus* (family leguminosae)	Prolonged heating degrades viscosity. Bacteriological stability can be improved by the addition of mixture of 0.15% methyl paraben or 0.1% benzoic acid
	MW approx. 220 000	Stable in solution over a pH range of 1.0–10.5
	Forms viscous colloidal solution when hydrated in cold water. The optimum rate of hydration is between pH 7.5 and 9.0	Incompatible with acetone, tannins, strong acids, and the alkalis. Borate ions, if present in the dispersing water, will prevent hydration of guar
	η 2,000–22,500 cps (1% aqueous solution)	The FDA recognizes guar gum as a substance added directly to human food and has been affirmed as generally recognized as safe
	Non-ionic	Used as thickener for lotions and creams, as tablet binder, and as emulsion stabilizer

(*Continued*)

TABLE 52.2 (Continued)
Properties and Characteristics of Some Representative Conventional Mucoadhesive Polymers

Mucoadhesive Polymer	Properties	Characteristics and Applications
Hydroxypropyl guar	Φ in hot and cold water Gives high viscosity, pseudoplastic solutions that show reversible decrease in viscosity at elevated temperatures Lacks yield value Non-ionic	Compatible with high concentration of most salts Shows good tolerance of water miscible solvents Better compatibility with minerals than guar gum Good viscosity stability in the pH range of 2–13 More resistance to bacterial and enzymatic degradation
Chitosan	Prepared from chitin of crabs and lobsters by N-deacetylation with alkali Φ dilute acids to produce a linear polyelectrolyte with a high positive charge density and forms salts with inorganic and organic acids such as glutamic acid, hydrochloric acid, lactic acid, and acetic acid The amino group in chitosan has a pK_a value of ~6.5, thus, chitosan is positively charged and soluble in acidic to neutral solution with a charge density dependent on pH and the %DA value	Possesses cell-binding activity due to polymer cationic polyelectrolyte structure and to the negative charge of the cell surface Mucoadhesive agent due to either secondary chemical bonds such as hydrogen bonds or ionic interactions between the positively charged amino groups of chitosan and the negatively charged sialic acid residues of mucus glycoproteins or mucins Biocompatible and biodegradable Excellent gel-forming and film-forming ability Widely used in controlled delivery systems such as gels, membranes, and microspheres Widely used in controlled delivery systems such as gels, membranes, and microspheres Chitosan enhances the transport of polar drugs across epithelial surfaces. Purified qualities of chitosans are available for biomedical applications. Chitosan and its derivatives such as trimethyl chitosan (where the amino group has been trimethylated) have been used in non-viral gene delivery. Trimethyl chitosan, or quaternized chitosan, has been shown to transfect breast cancer cells. As the degree of trimethylation increases the cytotoxicity of the derivative increases. At approximately 50% trimethylation the derivative is the most efficient at gene delivery. Oligomeric derivatives (3–6 kDa) are relatively non-toxic and have good gene delivery properties

(Continued)

TABLE 52.2 (Continued)
Properties and Characteristics of Some Representative Conventional Mucoadhesive Polymers

Mucoadhesive Polymer	Properties	Characteristics and Applications
Carrageenan	Available in sodium, potassium, magnesium, calcium, and mixed cation forms Three structural types exist: Iota, kappa, and lambda, differing in solubility and rheology The sodium form of all three types is soluble in both cold and hot water Other cation forms of kappa and Iota are soluble only in hot water All forms of lambda are soluble in cold water	All solutions are pseudoplastic with some degree of yield value. Certain ca-iota solutions are thixotropic. Lambda is non-gelling; kappa can produce brittle gels; and Iota can produce elastic gels. All solutions show a reversible decrease in viscosity at elevated temperatures. Iota and lambda carrageenan have excellent electrolyte tolerance; kappa's being somewhat less. Electrolytes will, however, decrease solution viscosity. The best solution stability occurs in the pH range of 6–10. It is compatible with most non-ionic and anionic water-soluble thickeners. It is strongly synergistic with locust bean gum and strongly interactive with proteins. Solutions are susceptible to shear and heat degradation Excellent thermoreversible properties Used also for microencapsulation
Alginate	Purified carbohydrate product extracted from brown seaweed by the use of dilute alkali Insoluble in other organic solvents and acids where the pH of the resulting solution falls below 3.0 pH 7.2 η 20–400 cps (1% aqueous solution) φ Water forming a viscous, colloidal solution Occurs as a white or buff powder, which is odorless and tasteless	Stabilizer in emulsion, suspending agent, tablet disintegrant, and tablet binder Incompatible with acridine derivatives, crystal violet, phenyl mercuric nitrate and acetate, calcium salts, and alcohol in concentrations greater than 5%, and heavy metals Safe and non-allergenic It is also used as hemostatic agent in surgical dressings Excellent gel formation properties Microstructure and viscosity are dependent on the chemical composition Used as immobilization matrices for cells and enzymes, controlled release of bioactive substances, injectable microcapsules for treating neurodegenerative and hormone deficiency diseases Lacks yield value Solutions show fair to good tolerance of water miscible solvents (10%–30% of volatile solvents; 40%–70% of glycols) Compatible with most water-soluble thickeners and resins Its solutions are more resistant to bacterial and enzymatic degradation than many other organic thickeners Biocompatible

(Continued)

TABLE 52.2 (Continued)
Properties and Characteristics of Some Representative Conventional Mucoadhesive Polymers

Mucoadhesive Polymer	Properties	Characteristics and Applications
Poly(hydroxy butyrate), poly(ε-caprolactone) and copolymers	Biodegradable Properties can be changed by chemical modification, copolymerization, and blending	Used as a matrix for drug delivery systems, cell microencapsulation
Poly(orthoesters)	Surface-eroding polymers	Application in sustained drug delivery and ophthalmology
Poly (cyanoacrylates)	Biodegradable depending on the length of the alkyl chain	Used as surgical adhesives and glues Potentially used in drug delivery
Polyphosphazenes	Can be tailored with versatile side chain functionality	Can be made into films and hydrogels Applications in drug delivery
Poly(vinyl alcohol)	Biocompatible	Gels and blended membranes are used in drug delivery and cell immobilization
Poly(ethylene oxide)	Highly biocompatible	Its derivatives and copolymers are used in various biomedical applications
Poly(hydroxyethyl methacrylate)	Biocompatible	Hydrogels have been used as soft contact lenses, for drug delivery, as skin coating, and for immunoisolation membranes
Poly(ethylene oxide-b-propylene oxide)	Surfactants with amphiphilic properties	Used in protein delivery and skin treatments

Source: Reprinted from *J. Contr. Release*, 114(1), Sudhakar, Y., Kuotsu, K., and Bandyopadhyay, A.K., Buccal bioadhesive drug delivery—A promising option for orally less efficient drugs, 15–40. Copyright 2006, with permission from Elsevier.

Drug Delivery Systems

In further investigations, Nikonenko et al. used visible and IR spectroscopic ellipsometry to study the adsorption of mucin on poly(acrylic acid)-block-poly(methyl methacrylate) and poly(methyl methacrylate) surfaces at pH 3.0, 7.0, and 10.0.[92] IR ellipsometry experiments indicated that the highest density of hydrogen bonding-involved carboxylic groups is observed at pH 3.0.

Carbopols are other derivatives of polyacrylic acid and have a capacity to gel upon changes in pH.[93] For example, Carbopol 71G (granulated Carbopol 934P) 0.5% w/v forms a slightly viscous solution in water but gels upon increase in pH. This unique characteristic has been used in the development of formulations for ophthalmic drug delivery.[94] Formulations containing a combination of Carbopol 934P and Pluronic® F-127 resulted in solutions that had adequate rheological properties at pH 4.0 but gelled at higher pH as a suitable gelling system to be instilled in the eye.

Additionally, polymers such as polycarbophil and some carbopols have been identified as "multifunctional polymers" due to their capacity to inhibit certain proteolytic enzymes and are discussed in Section 52.5.4.[95,96]

52.5.1.2 Cellulose Derivatives

Usually found as non-ionic polymers, cellulose derivatives have been widely investigated in the literature as mucoadhesives, although most polymers present modest mucoadhesive force.[97,98] Examples from this group include methylcellulose, ethylcellulose (although non-adhesive), HPC, and HPMC. In comparison with chitosan and carbopol, HPMC of various grades showed very limited adhesion to mucin that increased with an increase in HPMC viscosity and was only superior to polyvinyl alcohol and polyvinylpyrrolidone.[98] One exception to this trend and the major mucoadhesive from the group is SCMC. SCMC is the sodium salt of a polycarboxymethyl ether of cellulose and the presence of the carboxylic group provides the polymer with great mucoadhesive properties comparable to those observed for polyacrylic acid.[99]

52.5.1.3 Gums and Others

The group of gums and semi-natural polymers include guar gum, xanthan gum, locust bean gum, gellan gum, tragacanth gum, carrageenan, pectin, and alginate. Comparatively with other gums, xanthan gum is more tolerant of electrolytes, acids, and bases than most other gums.[86] Most gums have adequate viscosity stability over a wide pH range and temperature. Particularly, carrageenan is an anionic polysaccharide compatible with most non-ionic and anionic water-soluble thickeners. Viscosity increases synergistically with locust bean gum and carrageenan is strongly interactive with proteins.[86]

52.5.2 CHITOSAN

Chitosan, and other synthetic polymethacrylates,[100,101] is a cationic polymer regularly regarded in the literature as an excellent mucoadhesive polymer.[102] Usually dissolved in dilute acids, chitosan has a linear polysaccharide structure consisting of β-(1→4)-linked 2-amino-2-deoxy-D-glucose residues, and is commonly produced by alkali N-deacetylation of chitin. Due to the cationic nature of chitosan, it aggregates mucins exhibiting great mucoadhesion.[102] Furthermore, chitosan presents a number of other advantages including its film forming ability, antimicrobial and wound healing properties, biocompatibility and biodegradability,[103] its ability to bind lipids and fatty acids, and its ability to enhance penetration by the aperture of tight junctions.[104]

Even though the mechanism by which chitosan is an excellent mucoadhesive is usually related to its ability to electrostatically interact with negatively charged mucins in the mucus, other authors have suggested the complexity of the interaction of chitosan and mucins and indicated that hydrogen bonding and hydrophobic effects may also play a role in the mucoadhesive bond.[105,106] In another study, by modifying the molecular structure of chitosan Sogias et al. have investigated the contribution of different domains to the mucoadhesive interaction.[102] Even though electrostatic attraction was found to be the most prominent interaction mechanism, the mucoadhesive bond is further reinforced by contributions from hydrogen bonding and hydrophobic effects. Additionally, the pH and the presence of other chemicals in solution had the ability to individually modify the contribution of each physical interaction to the chitosan–mucin bond.

In order to modify or enhance specific properties of chitosan, chemical derivatives have been synthesized by using the amines and hydroxyl groups in the molecular structure. Some examples include trimethyl chitosan,[107] glycol chitosan,[93] carboxymethyl chitosan,[108] half-acetylated chitosan,[109] and thiolated chitosan.[110]

52.5.3 Polymers with Synergistic Characteristics

52.5.3.1 Enzyme-Inhibiting Polymers

Some mucoadhesive polymers present an enzyme-inhibiting characteristic, which makes them a great choice for molecules that are particularly prone to enzymatic degradation, such as proteins and peptides. For example, polyacrylic acid operates through a competitive mechanism with proteolytic enzymes due to its strong affinity to divalent cations (Ca^{2+} and Zn^{2+}).[95,111] These cations are cofactors of metalloproteinases such as trypsin. Investigations by circular dichroism indicated that Ca^{2+} is depleted due to the presence of polyacrylate derivatives. This caused the secondary structure of trypsin to change and initiated further degradation of the enzyme.[95,111]

52.5.3.2 Permeation-Enhancing Polymers

Another synergistic effect featured by some polymers is their capacity to enhance permeation while acting as mucoadhesives. It has been described that specific polymers exhibit a permeation-enhancing effect by opening tight junctions.[112] Particularly, polycations such as chitosan can absorb water from the epithelial cells, which results in dehydration of the cells and subsequent shrinking.[113] This has been suggested as the mechanism by which tight junctions open and has been described as being reversible for most polycations studied, except for chitosan at the highest concentration tested (0.01%) resulting in an irreversible opening of tight junctions.[113] Even though tight junctions are not found in the oral mucosa, a similar effect over the first few cell layers can be expected.

52.5.4 New-Generation Mucoadhesive Polymers

If the conventional mucoadhesive polymers were characterized by adhering to the mucin non-specifically and lacking specificity, the new generation of mucoadhesive polymers is characterized by either binding covalently (thiomers) or simply by binding to the cell surface.

52.5.4.1 Thiolated Polymers

Thiolated polymers, also termed thiomers, are conventional mucoadhesive polymers chemically modified to contain a cysteine residue in the polymer chain and thus establish covalent disulfide bonds with mucin.[114] They can be manufactured to be either cationic (mostly thiolated chitosans) or anionic (carboxylic acid-containing polymers); however, their mucoadhesive extent will mostly be determined by their capacity to covalently bind to mucin. The polypeptide backbone of mucin can be divided into three major subunits: tandem repeat array, carboxyl-, and amino-terminal domains. While the amino-terminal domain contains some of the cysteine residues, the carboxyl-terminal domain contains more than 10% of the cysteine residues. These cysteine-rich regions are responsible for forming the large mucin oligomers and ultimately, the groups that allow for the covalent mucoadhesive bond formation with oral mucosal systems.[115]

Contrary to what has been reported for non-covalently binding mucoadhesive polymers, the disulfide bond is not influenced by ionic strength, electrostatic, or hydrophobic domain interactions.[116] The extent of the disulfide bond is strictly related to the concentration of thiolate anions, which is the group that will go through thiol/disulfide exchange reactions and oxidation processes. The concentration of the thiolate groups is mostly determined by the pK_a value of the thiol group, which is determined by the molecular structure of the thiolated polymer. It has been described that a chitosan–thiobutylamidine conjugate has a pK_a value of 9.9,[117] while a polyacrylate–cysteine conjugate has a pK_a of 8.35.[116] Other factors controlling the concentration of thiolate anions are the pH of the thiomer[118] and pH of the surrounding media, especially, in the surface of the oral mucosa. Another mechanism that has been reported to contribute to the enhanced mucoadhesion of thiolated polymers is the in situ cross-linking property they possess. Much like other strong adhesives are based on, the in situ cross-linking effect acts after the polymer chains have inserted into the mucin layer as a stabilizer of the mucoadhesive bond (Figure 52.6). It has been reported that this phenomenon also contributed to the sustained release of fluorescein isothiocyanate–dextran from the hyaluronic acid–cysteine ethyl ester matrix.[119] The in situ cross-linking effect has also been characterized by rheological studies where the sol–gel transition of thiolated chitosans was studied.[120] After 2 h at pH 5.5, the sol–gel transition of thiolated chitosans was completed due to the formation of highly cross-linked gels. In parallel, the content of the thiol group decreased significantly, indicating the formation of disulfide bonds within the polymer. Conversely, the rheological properties of unmodified chitosan remained constant throughout the observation period.[120]

Additionally to the high mucoadhesion exhibited by thiolated polymers, other advantages include their improved tensile strength, high cohesive properties, rapid swelling, and water uptake behavior.[116] A positive correlation between the adhesive properties and increasing amounts of the polymer in dry compacts of polycarbophil covalently modified with L-cysteine has been reported.[114] A study of the total work of adhesion revealed that the 16:1 and 2:1 polycarbophil–cysteine conjugates

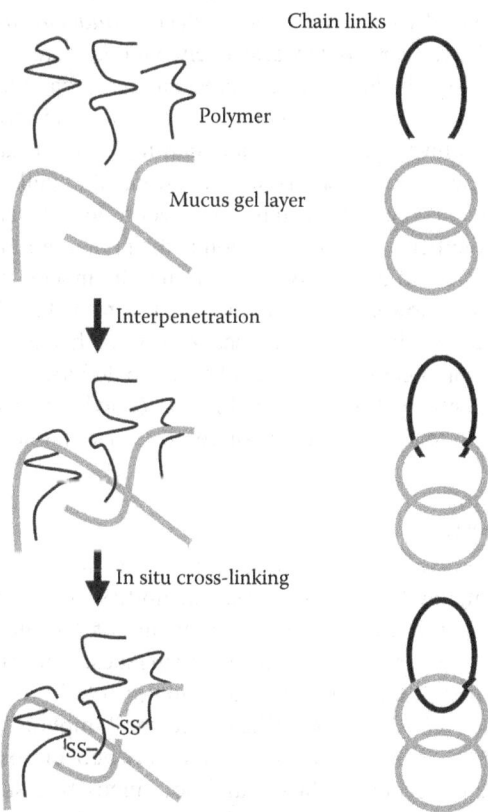

FIGURE 52.6 Representation of the in situ cross-linking effect of thiolated polymers and a comparison to chain links. (Reprinted from *Adv. Drug Deliv. Rev.*, 57(11), Bernkop-Schnürch, A., Thiomers: A new generation of mucoadhesive polymers, 1569–1582. Copyright 2005, with permission from Elsevier.)

achieved adhesions of 191 ± 47 and 280 ± 67 µJ, respectively, almost 2–3 times higher than that of the unmodified polymer (104 ± 21 µJ).

In another study, pituitary adenylate cyclase-activating polypeptide (PACAP) was successfully delivered via the buccal mucosa of pigs.[121] The formulation consisted of a polymeric matrix of thiolated chitosan, an enzyme inhibitor, and a permeation enhancer, and it was applied directly onto the buccal mucosa of healthy pigs. After the 6 h long experiments, formulations containing thiolated chitosan needed to be removed from the buccal mucosa, while formulations containing unmodified chitosan detached 4 h after application. This corroborated the high mucoadhesive characteristic reported previously for thiolated chitosan.[122]

52.5.4.2 Lectin-Mediated Polymers

Lectins are naturally proteins or glycoproteins with the potential to specifically bind to sugar residues.[123] This feature allows lectins to bind not only to mucin but also to glycoconjugates present on cell surfaces in the oral mucosa. This has been described as an anchoring system for drug delivery platforms to remain attached to the oral mucosa.[124,125] After initial mucosal cell binding, lectins can either remain on the cell surface or in the case of receptor-mediated adhesion they could become internalized by endocytosis.[126] Lectins from *Arachis hypogaea*, *Canavalia ensiformis*, and *Triticum vulgaris* have been found to bind to oral mucosal cells in vitro, with *T. vulgaris* exhibiting the highest extent of binding (approximately 7×10^9 molecules/cell).[127] Studies in rats again showed that *T. vulgaris* reached the highest level bound after 20 min (of about 30 µg) on buccal mucosa tissue that remained at similar levels after 2 h. Another possibility of lectins is their potential to selectively target damaged tissue. Although, most of the lectins investigated showed little difference in their binding capacity to specifically bind to damaged tissue over intact surfaces, *Maackia amurensis* lectin (specific to *N*-acetylneuraminic acid) showed potential by selectively binding more to damaged tissue.[128]

One concern that arises from the increase in interest and research in lectins as targeting moieties in mucoadhesive systems is their cytotoxicity potential.[129] An acute in vivo study, where lectins from potato and edible snail were investigated by intradermal injection in rabbits, showed no evidence of irritancy.[130] Nonetheless, when cytotoxicity was investigated in cells from oral mucosa (human tongue), a clear toxic effect could be evidenced for most lectins tested, particularly at higher concentrations and over 48 h.[131] Another potential limitation is the premature inactivation by the shedding off of mucus. However, this could be an advantage in that the mucus layer provides an initial yet fully reversible binding site followed by distribution of lectin-mediated drug delivery systems to the cell layer.[132] The potential for immunogenic reactions is another concern in the development of lectin-mediated mucoadhesion. Lectin-induced antibodies could block subsequent adhesive interactions between mucosal epithelial cell surfaces and lectin-loaded delivery systems.[133] Furthermore, the said antibodies could also render individuals susceptible to systemic anaphylaxis on subsequent exposure.

52.5.4.3 Bacterial Adhesion

The capacity of pathogenic bacteria to adhere to mucosal membranes has been exploited in the modification of new mucoadhesive polymers. The ability of bacteria to adhere to a specific target is rooted from particular cell–surface components or appendages, known as fimbriae, which promote adhesion to other cells or inanimate surfaces. Fimbriae are extracellular, long thread-like protein polymers of bacteria that play a major role in many diseases.[58] It has been reported that *Escherichia coli* adheres specifically to the lymphoid follicle epithelium of the ileal Peyer's patch in rabbits.[134] Similarly, different staphylococci possess the ability to adhere specifically to the surface of mucus gel layers and not mucus-free surfaces.[135] Thus, polymers have been modified by the attachment of these fimbriae to enhance mucoadhesion. An attachment protein derived from *E. coli*, K99-fimbriae, has been covalently attached to polyacrylic acid networks in an attempt to provide a novel polymer with enhanced adhesive properties (Figure 52.7).[136]

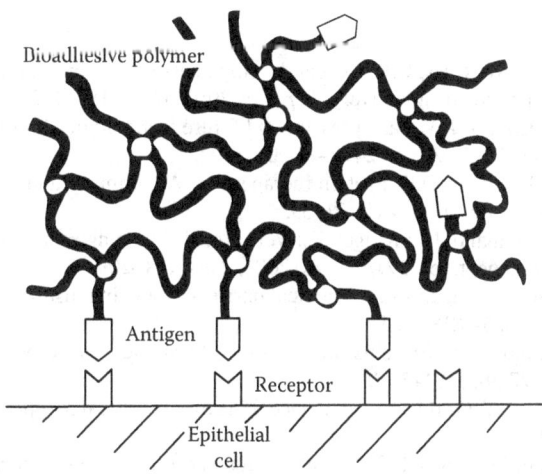

FIGURE 52.7 A diagram representing the mechanism of action of the specific adhesion of K99 fimbriae from *E. coli* to cell receptors. The carrier system consists of a polyacrylic acid matrix containing the fimbrial protein for adhesion. (Reprinted from *Eur. J. Pharm. Sci.*, 3(5), Bernkop-Schnürch, A., Gabor, F., Szostak, M.P., and Lubitz, W., An adhesive drug delivery system based on K99-fimbriae, 293–299. Copyright 1995, with permission from Elsevier.)

Some bacteria not only adhere to the epithelial cells but also invade host cells using a mechanism resembling phagocytosis.[137] Bioinvasive drug delivery systems have been developed based on this bacterial mechanism, where bacteria could be used as a vehicle to introduce drug compounds into host cells by means of multiple h1 chain integrin cell receptors, which are members of the cell adhesion molecule family.[138]

Not only specific adhesion but also controlled endo- and transcytosis of microorganisms into cells by bacterial adhesion is another potential development in bacterial adhesion. The mechanism follows signal transmission associated with the protein binding, which then triggers the intracellular translocation of the microorganism.[58] This feature could be applied for controlled cellular binding and internalization to deliver compounds specifically to the interior of cells.[139]

52.6 CONCLUSIONS

In the past few decades, polymers for oral mucosal delivery have thoroughly been used and investigated as mucoadhesive materials in the development of dosage forms. Due to their regulatory status, conventional mucoadhesives continue to be the most prominent type of polymers in the scientific body of literature as well as new patents and products. However, novel mucoadhesives continue to increase in number and potential applications and it is safe to say that in the next few years, development of novel products will continue due to their unique features. Selective interactions with thiol groups (Section 52.5.4.1), sugar residues (Section 52.5.4.2), and specific membrane receptors (Section 52.5.4.3) represent the targets of the new-generation mucoadhesives. Having proved their efficacy in selective binding, the main concern now in their development is demonstrating their safety for use in order to be used in products.

The oral mucosal route, although conventionally explored for small molecule delivery, has its greatest potential in the delivery of macromolecules. By avoiding the degradation concerns raised by oral delivery, and by eliminating the need of needles associated with injections, the oral transmucosal delivery presents as a patient friendly and effective route of administration. It is maybe not far-fetched to say that the smart use of conventional and the development of the new-generation mucoadhesive polymers will in the future drive the development of dosage forms for oral transmucosal delivery.

REFERENCES

1. Tan, M.L.; Choong, P.F.M.; Dass, C.R. Recent developments in liposomes, microparticles and nanoparticles for protein and peptide drug delivery. *Peptides* 2010, *31* (1), 184–193.
2. Singh, R.; Singh, S.; Lillard, J.W. Past, present, and future technologies for oral delivery of therapeutic proteins. *J. Pharm. Sci.* 2008, *97* (7), 2497–2523.
3. Leader, B.; Baca, Q.J.; Golan, D.E. Protein therapeutics: A summary and pharmacological classification. *Nat. Rev. Drug Discov.* 2008, *7* (1), 21–39.
4. Agu, U.; Ugwoke, I.; Armand, M.; Kinget, R.; Verbeke, N. The lung as a route for systemic delivery of therapeutic proteins and peptides. *Respir. Res.* 2001, *2* (4), 198–209.
5. Korytkowski, M. When oral agents fail: Practical barriers to starting insulin. *Int. J. Obes. Relat. Metab. Disord.* 2002, *26* (Suppl. 3), S18–S24.
6. Benson, H.A.E.; Namjoshi, S. Proteins and peptides: Strategies for delivery to and across the skin. *J. Pharm. Sci.* 2008, *97* (9), 3591–3610.
7. Heinemann, L.; Jacques, Y. Oral insulin and buccal insulin: A critical reappraisal. *J. Diabetes Sci. Tech.* 2009, *3* (3), 568–584.
8. Pather, S.I.; Rathbone, M.J.; Senel, S. Current status and the future of buccal drug delivery systems. *Expert Opin. Drug Deliv.* 2008, *5* (5), 531–542.
9. Wertz, P.W.; Squier, C.A. Cellular and molecular basis of barrier function in oral epithelium. *Crit. Rev. Ther. Drug Carrier. Syst.* 1991, *8* (3), 237–269.
10. Collins, L.M.C.; Dawes, C. The surface area of the adult human mouth and thickness of the salivary film covering the teeth and oral mucosa. *J. Dent. Res.* 1987, *66* (8), 1300–1302.
11. DeGrande, G.; Benes, L.; Horriere, F.; Karsenty, H.; Lacoste, C.; McQuinn, R.L.; Guo, J.; Scherrer, R. Specialized oral mucosal drug delivery systems: Patches. In *Oral Mucosal Drug Delivery*; Rathbone, M.J., Ed. Informa Healthcare: New York, 1996.
12. Squier, C.A.; Hill, M.W. Oral mucosa. In *Oral Histology, Development, Structure, and Function*; Ten Cate, A.R., Ed. Mosby, Inc.: St. Louis, MO, 1989; pp. 319–356.
13. Gandhi, R.B.; Robinson, J.R. Oral cavity as a site for bioadhesive drug delivery. *Adv. Deliv. Rev.* 1994, *3* (1–2), 43–74.
14. Squier, C.A.; Wertz, P.W. Structure and function of the oral mucosa and implications for drug delivery. In *Oral Mucosal Drug Delivery*; Rathbone, M.J., Ed. Informa Healthcare: New York, 1996.
15. Li, B.; Robinson, J.R. Preclinical assessment of oral mucosal drug delivery systems. In *Drug Delivery to the Oral Cavity*; Ghosh, T.K., Pfister, W.R., Eds. Marcel Dekker, Inc.: New York, 2005; pp. 41–66.
16. Nicolazzo, J.A.; Reed, B.L.; Finnin, B.C. Buccal penetration enhancers—How do they really work? *J. Contr. Rel.* 2005, *105* (1–2), 1–15.
17. Matoltsy, A.G.; Parakkal, P.F. Membrane-coating granules of keratinizing epithelia. *J. Cell Biol.* 1965, *24* (2), 297–307.
18. Squier, C.A. Membrane coating granules in nonkeratinizing oral epithelium. *J. Ultrastruct. Res.* 1977, *60* (2), 212–220.
19. Squier, C.A. Zinc iodide-osmium staining of membrane-coating granules in keratinized and non-keratinized mammalian oral epithelium. *Arch. Oral Biol.* 1982, *27* (5), 377–382.
20. Wertz, P.W.; Swartzendruber, D.C.; Squier, C.A. Regional variation in the structure and permeability of oral mucosa and skin. *Adv. Drug Deliv. Rev.* 1993, *12* (1–2), 1–12.
21. Law, S.; Wertz, P.W.; Swartzendruber, D.C.; Squier, C.A. Regional variation in content, composition and organization of porcine epithelial barrier lipids revealed by thin-layer chromatography and transmission electron microscopy. *Arch. Oral Biol.* 1995, *40* (12), 1085–1091.
22. Squier, C.A. The permeability of keratinized and nonkeratinized oral epithelium to horseradish peroxidase. *J. Ultrastruct. Res.* 1973, *43* (1–2), 160–177.
23. Squier, C.A.; Cox, P.S.; Wertz, P.W.; Downing, D.T. The lipid composition of porcine epidermis and oral epithelium. *Arch. Oral Biol.* 1986, *31* (11), 741–747.
24. Squier, C.A.; Wertz, P.W.; Cox, P.S. Thin-layer chromatographic analyses of lipids in different layers of porcine epidermis and oral epithelium. *Arch. Oral Biol.* 1991, *36* (9), 647–653.
25. Barnett, M.L.; Szabo, G. Gap junctions in human gingival keratinized epithelium. *J. Periodontal Res.* 1973, *8* (3), 117–126.
26. Harris, D.; Robinson, J.R. Drug delivery via the mucous membranes of the oral cavity. *J. Pharm. Sci.* 1992, *81* (1), 1–10.

27. Utoguchi, N.; Watanabe, Y.; Suzuki, T.; Maehara, J.; Matsumoto, Y.; Matsumoto, M. Carrier-mediated transport of monocarboxylic acids in primary cultured epithelial cells from rabbit oral mucosa. *Pharm. Res.* 1997, *14* (3), 320–324.
28. Goswami, T.; Jasti, B.; Li, X. Sublingual drug delivery. *Crit. Rev. Ther. Drug Carrier Syst.* 2008, *25* (5), 449–484.
29. Squier, C.A.; Cox, P.S.; Wertz, P.W. Lipid content and water permeability of skin and oral mucosa. *J. Investig. Dermatol.* 1991, *96* (1), 123–126.
30. Squier, C.A.; Hall, B.K. The permeability of skin and oral mucosa to water and horseradish peroxidase as related to the thickness of the permeability barrier. *J. Invest. Dermatol.* 1985, *84* (3), 176–179.
31. Tamai, I.; Tsuji, A. Carrier-mediated approaches for oral drug delivery. *Adv. Drug Deliv. Rev.* 1996, *20* (1), 5–32.
32. Wang, Y.; Zuo, Z.; Lee, K.K.H.; Chow, M.S.S. Evaluation of HO-1-u-1 cell line as an in vitro model for sublingual drug delivery involving passive diffusion—Initial validation studies. *Int. J. Pharm.* 2007, *334* (1–2), 27–34.
33. Jain, V.; Das, S.N.; Luthra, K.; Shukla, N.K.; Ralhan, R. Differential expression of multidrug resistance gene product, P-glycoprotein, in normal, dysplastic and malignant oral mucosa in India. *Int. J. Canc.* 1997, *74* (1), 128–133.
34. Chen, L.L.H.; Chetty, D.J.; Chien, Y.W. A mechanistic analysis to characterize oramucosal permeation properties. *Int. J. Pharm.* 1999, *184* (1), 63–72.
35. Chetty, D.J.; Chen, L.L.H.; Chien, Y.W. Characterization of captopril sublingual permeation: Determination of preferred routes and mechanisms. *J. Pharm. Sci.* 2001, *90* (11), 1868–1877.
36. Shojaei, A.H. Buccal mucosa as a route for systemic drug delivery: A review. *J. Pharm. Pharmaceut. Sci.* 1998, *1* (1), 15–30.
37. DiSabato-Mordarski, T.; Kleinberg, I. Measurement and comparison of the residual saliva on various oral mucosal and dentition surfaces in humans. *Arch. Oral Biol.* 1996, *41* (7), 655–665.
38. Peppas, N.A.; Buri, P.A. Surface, interfacial and molecular aspects of polymer bioadhesion on soft tissues. *J. Contr. Release* 1985, *2*, 257–275.
39. Forstner, J.; Taichman, N.; Kalnins, V.; Forstner, G. Intestinal goblet cell mucus: Isolation and identification by immunofluorescence of a goblet cell glycoprotein. *J. Cell Sci.* 1973, *12* (2), 585.
40. Horowitz, M.I. Gastrointestinal glycoproteins. In *The Glycoconjugates*; Horowitz, M.I., Pigman, W., Eds. Academic Press: New York, 1977; pp. 189–213.
41. Kinloch, A.J. The science of adhesion I: Surface and interfacial aspects. *J. Mater. Sci.* 1980, *15* (9), 2141–2166.
42. Derjaguin, B.V.; Smilga, V.P. *Adhesion: Fundamentals and Practice*. McLaren: London, U.K., 1969.
43. Chickering, D.E.; Mathiowitz, E. Definitions, mechanisms, and theories of bioadhesion. In *Bioadhesive Drug Delivery Systems: Fundamentals, Novel Approaches, and Development*; Mathiowitz, E., Chickering, D.E., Lehr, C.M., Eds. Marcel Dekker, Inc.: New York, 1999.
44. Hench, L.L.; Ethridge, E.C. *Biomaterials: An Interfacial Approach*. Academic Press: New York, 1982.
45. Baszkin, A.; Proust, J.E.; Monsenego, P.; Boissonnade, M.M. Wettability of polymers by mucin aqueous solutions. *Biorheology* 1990, *27* (3–4), 503–514.
46. Mikos, A.G.; Peppas, N.A. Measurement of the surface tension of mucin solutions. *Int. J. Pharm.* 1989, *53* (1), 1–5.
47. Kaelble, D.H.; Moacanin, J. A surface energy analysis of bioadhesion. *Polymer* 1977, *18* (5), 475–482.
48. Voyutskii, S.S. *Autohesion and Adhesion of High Polymers*. Wiley: New York, 1963.
49. Prager, S. The healing process at polymer–polymer interfaces. *J. Chem. Phys.* 1981, *75* (10), 5194–5198.
50. Kammer, H.W. Adhesion between polymers. *Rev. Acta Polymerica* 1983, *34* (2), 112–118.
51. Mizrahi, B.; Domb, A.J. Mucoadhesive polymers for delivery of drugs to the oral cavity. *Recent Pat. Drug Deliv. Formul.* 2008, *2* (2), 108–119.
52. Senel, S.; Kremer, M.J.; Kas, S.; Wertz, P.W.; Hıncal, A.A.; Squier, C.A. Enhancing effect of chitosan on peptide drug delivery across buccal mucosa. *Biomaterials* 2000, *21* (20), 2067–2071.
53. Dixit, R.P.; Puthli, S.P. Oral strip technology: Overview and future potential. *J. Contr. Release* 2009, *139* (2), 94–107.
54. Perumal, V.A.; Lutchman, D.; Mackraj, I.; Govender, T. Formulation of monolayered films with drug and polymers of opposing solubilities. *Int. J. Pharm.* 2008, *358* (1–2), 184–191.
55. Senel, S.; Rathbone, M.J.; Cansız, M.; Pather, I. Recent developments in buccal and sublingual delivery systems. *Expert Opin. Drug Deliv.* 2012, *9* (6), 615–628.

56. Hearnden, V.; Sankar, V.; Hull, K.; Juras, D.V.; Greenberg, M.; Kerr, A.R.; Lockhart, P.B.; Patton, L.L.; Porter, S.; Thornhill, M.H. New developments and opportunities in oral mucosal drug delivery for local and systemic disease. *Adv. Drug Deliv. Rev.* 2012, *64* (1), 16–28.
57. Patel, V.F.; Liu, F.; Brown, M.B. Advances in oral transmucosal drug delivery. *J. Contr. Release* 2011, *153* (2), 106–116.
58. Salamat-Miller, N.; Chittchang, M.; Johnston, T.P. The use of mucoadhesive polymers in buccal drug delivery. *Adv. Drug Deliv. Rev.* 2005, *57* (11), 1666–1691.
59. Rathbone, M.J.; Drummond, B.K.; Tucker, I.G. The oral cavity as a site for systemic drug delivery. *Adv. Drug Deliv. Rev.* 1994, *13* (1–2), 1–22.
60. Patel, V.M.; Prajapati, B.G.; Patel, M.M. Formulation, evaluation, and comparison of bilayered and multilayered mucoadhesive buccal devices of propranolol hydrochloride. *AAPS Pharm. Sci. Tech.* 2007, *8* (1), E147–E154.
61. Alur, H.H.; Beal, J.D.; Pather, S.I.; Mitra, A.K.; Johnston, T.P. Evaluation of a novel, natural oligosaccharide gum as a sustained-release and mucoadhesive component of calcitonin buccal tablets. *J. Pharm. Sci.* 1999, *88* (12), 1313–1319.
62. Madhav, N.V.S.; Shakya, A.K.; Shakya, P.; Singh, K. Orotransmucosal drug delivery systems: A review. *J. Contr. Release* 2009, *140* (1), 2–11.
63. Giunchedi, P.; Juliano, C.; Gavini, E.; Cossu, M.; Sorrenti, M. Formulation and in vivo evaluation of chlorhexidine buccal tablets prepared using drug-loaded chitosan microspheres. *Eur. J. Pharm. Biopharm.* 2002, *53* (2), 233–239.
64. Albertini, B.; Passerini, N.; Di Sabatino, M.; Monti, D.; Burgalassi, S.; Chetoni, P.; Rodriguez, L. Poloxamer 407 microspheres for orotransmucosal drug delivery. Part I: Formulation, manufacturing and characterization. *Int. J. Pharm.* 2010, *399* (1–2), 71–79.
65. Tayel, S.A.; Soliman, I.I.; Louis, D. Formulation of ketotifen fumarate fast-melt granulation sublingual tablet. *AAPS Pharm. Sci. Tech.* 2010, *11* (2), 679–685.
66. Hariharan, M.; Bogue, A. Orally dissolving film strips: The final evolution of orally dissolving dosage forms. *Drug Deliv. Tech.* 2009, *9* (2), 24–29.
67. Morales, J.O.; McConville, J.T. Manufacture and characterization of mucoadhesive buccal films. *Eur. J. Pharm. Biopharm.* 2011, *77* (2), 187–199.
68. Rodrigues, L.B.; Leite, H.F.; Yoshida, M.I.; Saliba, J.B.; Cunha, A.S. Jr.; Faraco, A.A.G. in vitro release and characterization of chitosan films as dexamethasone carrier. *Int. J. Pharm.* 2009, *368* (1–2), 1–6.
69. Zhang, H. Dissolvable backing layer for use with a transmucosal delivery device. US 7276246, October 2, 2007.
70. Sankar, V., Hearnden, V.; Hull, K.; Juras, D.V.; Greenberg, M.; Kerr, A.; Lockhart, P.; Patton, L.; Porter, S.; Thornhill, M. Local drug delivery for oral mucosal diseases: Challenges and opportunities. *Oral Dis.* 2011, *17* (Suppl. 1), 73–84.
71. Karlsmark, T.; Goodman, J.; Drouault, Y.; Lufrano, L.; Pledger, G.; the C.S.S. Group. Randomized clinical study comparing Compeed® cold sore patch to acyclovir cream 5% in the treatment of herpes simplex labialis. *J. Eur. Acad. Dermatol. Venereol.* 2008, *22* (10), 1184–1192.
72. Madhav, N.V.S.; Semwal, R.; Semwal, D.K.; Semwal, R.B. Recent trends in oral transmucosal drug delivery systems: An emphasis on the soft palatal route. *Expert Opin. Drug Deliv.* 2012, *9* (6), 629–647.
73. Na, D.H.; Faraj, J.; Capan, Y.; Leung, K.P.; DeLuca, P.P. Chewing gum of antimicrobial decapeptide KSL as a sustained antiplaque agent: Preformulation study. *J. Contr. Release* 2005, *107* (1), 122–130.
74. Faraj, J.A.; Dorati, R.; Schoubben, A.; Worthen, D.; Selmin, F.; Capan, Y.; Leung, K.; DeLuca, P.P. Development of a peptide-containing chewing gum as a sustained release antiplaque antimicrobial delivery system. *AAPS Pharm. Sci. Tech.* 2007, *8* (1), 177–185.
75. Imfeld, T. Chlorhexidine-containing chewing gum. *Schweizer Monatsschr. Zahnmed.* 2006, *116* (5), 476–483.
76. Semwal, R.; Semwal, D.K.; Badoni, R. Chewing gum: A novel approach for drug delivery. *J. Appl. Res.* 2010, *10* (3), 115–123.
77. Mendes, R.; Bhargava, H. Lozenges. In *Encyclopedia of Pharmaceutical Technology*; Swarbrick, J., Ed., 3rd edn. Informa Healthcare: New York, 2006; pp. 2231–2236.
78. Campisi, G.; Giannola, L.I.; Florena, A.M.; De Caro, V.; Schumacher, A.; Göttsche, T.; Paderni, C.; Wolff, A. Bioavailability in vivo of naltrexone following transbuccal administration by an electronically-controlled intraoral device: A trial on pigs. *J. Contr. Release* 2010, *145* (3), 214–220.
79. Squier, C.A.; Kremer, M.J. Biology of oral mucosa and esophagus. *J. Natl. Canc. Inst. Monogr.* 2001, (29), 7–15.

80. Jones, D.S.; Woolfson, A.D.; Djokic, J.; Coulter, W.A. Development and mechanical characterization of bioadhesive semi-solid, polymeric systems containing tetracycline for the treatment of periodontal diseases. *Pharm. Res.* 1996, *13* (11), 1734–1738.
81. Vonarx, V.; Eleouet, S.; Carre, J.; Ioss, P.; Gouyette, A.; Leray, A.M.; Merle, C.; Lajat, Y.; Patrice, T. Potential efficacy of a delta 5-aminolevulinic acid bioadhesive gel formulation for the photodynamic treatment of lesions of the gastrointestinal tract in mice. *J. Pharm. Pharm.* 1997, *49* (7), 652–656.
82. Bernstein, G. Delivery of insulin to the buccal mucosa utilizing the RapidMist system. *Expert Opin. Drug Deliv.* 2008, *5* (9), 1047–1055.
83. Modi, P.; Mihic, M.; Lewin, A. The evolving role of oral insulin in the treatment of diabetes using a novel RapidMist System. *Diabetes Metab. Res. Rev.* 2002, *18* (Suppl. 1), S38–S42.
84. Jiménez-castellanos, M.R.; Zia, H.; Rhodes, C.T. Mucoadhesive drug delivery systems. *Drug Dev. Ind. Pharm.* 1993, *19* (1–2), 143–194.
85. Junginger, H.E.; Verhoef, J.C.; Thanou, M. Drug delivery: Mucoadhesive hydrogels. In *Encyclopedia of Pharmaceutical Technology*; Swarbrick, J., Ed., 3rd edn. Informa Healthcare: New York, 2006, pp 1169–1182.
86. Sudhakar, Y.; Kuotsu, K.; Bandyopadhyay, A.K. Buccal bioadhesive drug delivery—A promising option for orally less efficient drugs. *J. Contr. Release* 2006, *114* (1), 15–40.
87. Ahuja, A.; Khar, R.K.; Ali, J. Mucoadhesive drug delivery systems. *Drug Dev. Ind. Pharm.* 1997, *23* (5), 489–515.
88. Kotagale, N.; Patel, C.; Parkhe, A.; Khandelwal, H.; Taksande, J.; Umekar, M. Carbopol 934-sodium alginate-gelatin mucoadhesive ondansetron tablets for buccal delivery: Effect of PH modifiers. *Indian J. Pharm. Sci.* 2010, *72* (4), 471–479.
89. Lai, S.K.; Wang, Y.Y.; Hanes, J. Mucus-penetrating nanoparticles for drug and gene delivery to mucosal tissues. *Adv. Drug Deliv. Rev.* 2009, *61* (2), 158–171.
90. Park, H.; Robinson, J.R. Mechanisms of mucoadhesion of poly acrylic acid hydrogels. *Pharm. Res.* 1987, *4* (6), 457–464.
91. Degim, Z.; Kellaway, I.W. An investigation of the interfacial interaction between polyacrylic acid and glycoprotein. *Int. J. Pharm.* 1998, *175* (1), 9–16.
92. Nikonenko, N.A.; Bushnak, I.A.; Keddie, J.L. Spectroscopic ellipsometry of mucin layers on an amphiphilic diblock copolymer surface. *Appl. Spectrosc.* 2009, *63* (8), 889–898.
93. Khutoryanskiy, V.V. Advances in mucoadhesion and mucoadhesive polymers. *Macromol. Biosci.* 2011, *11* (6), 748–764.
94. Lin, H.R.; Sung, K. Carbopol/pluronic phase change solutions for ophthalmic drug delivery. *J. Contr. Release* 2000, *69* (3), 379–388.
95. Lueßen, H.L.; Verhoef, J.C.; Borchard, G.; Lehr, C.M.; de Boer, A.B.G.; Junginger, H.E. Mucoadhesive polymers in peroral peptide drug delivery. II. Carbomer and polycarbophil are potent inhibitors of the intestinal proteolytic enzyme trypsin. *Pharm. Res.* 1995, *12* (9), 1293–1298.
96. Lueßen, H.L.; de Leeuw, B.J.; Pérard, D.; Lehr, C.M.; Bert, A.; de Boer, G.; Verhoef, J.C.; Junginger, H.E. Mucoadhesive polymers in peroral peptide drug delivery. I. Influence of mucoadhesive excipients on the proteolytic activity of intestinal enzymes. *Eur. J. Pharm. Sci.* 1996, *4* (2), 117–128.
97. Thongborisute, J.; Takeuchi, H. Evaluation of mucoadhesiveness of polymers by BIACORE method and mucin-particle method. *Int. J. Pharm.* 2008, *354* (1–2), 204–209.
98. Takeuchi, H.; Thongborisute, J.; Matsui, Y.; Sugihara, H.; Yamamoto, H.; Kawashima, Y. Novel mucoadhesion tests for polymers and polymer-coated particles to design optimal mucoadhesive drug delivery systems. *Adv. Drug Deliv. Rev.* 2005, *57* (1–2), 1583–1594.
99. Ludwig, A. The use of mucoadhesive polymers in ocular drug delivery. *Adv. Drug Deliv. Rev.* 2005, *57* (11), 1595–1639.
100. Fefelova, N.A.; Nurkeeva, Z.S.; Mun, G.A.; Khutoryanskiy, V.V. Mucoadhesive interactions of amphiphilic cationic copolymers based on [2-methacryloyloxyethyl]trimethylammonium chloride. *Int. J. Pharm.* 2007, *339* (1–2), 25–32.
101. Keely, S.; Rullay, A.; Wilson, C.; Carmichael, A.; Carrington, S.; Corfield, A.; Haddleton, D.M.; Brayden, D.J. In vitro and ex vivo intestinal tissue models to measure mucoadhesion of poly methacrylate and N-trimethylated chitosan polymers. *Pharm. Res.* 2005, *22* (1), 38–49.
102. Sogias, I.A.; Williams, A.C.; Khutoryanskiy, V.V. Why is chitosan mucoadhesive? *Biomacromolecules* 2008, *9* (7), 1837–1842.
103. Sandri, G.; Rossi, S.; Ferrari, F.; Bonferoni, M.C.; Muzzarelli, C.; Caramella, C. Assessment of chitosan derivatives as buccal and vaginal penetration enhancers. *Eur. J. Pharm. Sci.* 2004, *21* (2–3), 351–359.

104. Schipper, N.G.M.; Olsson, S.; Hoogstraate, J.A.; deBoer, A.G.; Vaarum, K.M.; Artursson, P. Chitosans as absorption enhancers for poorly absorbable drugs 2: Mechanism of absorption enhancement. *Pharm. Res.* 1997, *14* (7), 923–929.
105. Qaqish, R.; Amiji, M. Synthesis of a fluorescent chitosan derivative and its application for the study of chitosan–mucin interactions. *Carbohydr. Polym.* 1999, *38* (2), 99–107.
106. Deacon, M.P.; Davis, S.S.; White, R.J.; Nordman, H.; Carlstedt, I.; Errington, N.; Rowe, A.J.; Harding, S.E. Are chitosan–mucin interactions specific to different regions of the stomach? Velocity ultracentrifugation offers a clue. *Carbohydr. Polym.* 1999, *38* (3), 235–238.
107. Mourya, V.K.; Inamdar, N.N. Trimethyl chitosan and its applications in drug delivery. *J. Mater. Sci. Mat. Med.* 2009, *20* (5), 1057–1079.
108. Gujarathi, N.A.; Rane, B.R.; Patel, J.K. pH sensitive polyelectrolyte complex of O-carboxymethyl chitosan and poly acrylic acid cross-linked with calcium for sustained delivery of acid susceptible drugs. *Int. J. Pharm.* 2012, *436* (1–2), 418–425.
109. Sogias, I.A.; Williams, A.C.; Khutoryanskiy, V.V. Chitosan-based mucoadhesive tablets for oral delivery of ibuprofen. *Int. J. Pharm.* 2012, *436* (1–2), 602–610.
110. Anitha, A.; Deepa, N.; Chennazhi, K.P.; Nair, S.V.; Tamura, H.; Jayakumar, R. Development of mucoadhesive thiolated chitosan nanoparticles for biomedical applications. *Carbohydr. Polym.* 2011, *83* (1), 66–73.
111. Lueßen, H.L.; Lehr, C.M.; Rentel, C.O.; Noach, A.B.J.; de Boer, A.G.; Verhoef, J.C.; Junginger, H.E. Bioadhesive polymers for the peroral delivery of peptide drugs. *J. Contr. Release* 1994, *29* (3), 329–338.
112. Lehr, C.M.; Haas, J. Developments in the area of bioadhesive drug delivery systems. *Expert Opin. Biol. Ther.* 2002, *2* (3), 287–298.
113. Ranaldi, G.; Marigliano, I.; Vespignani, I.; Perozzi, G.; Sambuy, Y. The effect of chitosan and other polycations on tight junction permeability in the human intestinal Caco-2 cell line. *J. Nutr. Biochem.* 2002, *13* (3), 157–167.
114. Bernkop-Schnürch, A.; Schwarz, V.; Steininger, S. Polymers with thiol groups: A new generation of mucoadhesive polymers? *Pharm. Res.* 1999, *16* (6), 876–881.
115. Gum, J.R., Jr; Hicks, J.W.; Toribara, N.W.; Rothe, E.M.; Lagace, R.E.; Kim, Y.S. The human MUC2 intestinal mucin has cysteine-rich subdomains located both upstream and downstream of its central repetitive region. *J. Biol. Chem.* 1992, *267* (30), 21375–21383.
116. Bernkop-Schnürch, A. Thiomers: A new generation of mucoadhesive polymers. *Adv. Drug Deliv. Rev.* 2005, *57* (11), 1569–1582.
117. Bernkop-Schnürch, A.; Hornof, M.; Zoidl, T. Thiolated polymers—Thiomers: Synthesis and in vitro evaluation of chitosan–2-iminothiolane conjugates. *Int. J. Pharm.* 2003, *260* (2), 229–237.
118. Bernkop-Schnürch, A.; Gilge, B. Anionic mucoadhesive polymers as auxiliary agents for the peroral administration of poly peptide drugs: Influence of the gastric juice. *Drug Dev. Ind. Pharm.* 2000, *26* (2), 107–113.
119. Kafedjiiski, K.; Jetti, R.K.R.; Föger, F.; Hoyer, H.; Werle, M.; Hoffer, M.; Bernkop-Schnürch, A. Synthesis and in vitro evaluation of thiolated hyaluronic acid for mucoadhesive drug delivery. *Int. J. Pharm.* 2007, *343* (1–2), 48–58.
120. Hornof, M.D.; Kast, C.E.; Bernkop-Schnürch, A. In vitro evaluation of the viscoelastic properties of chitosan–thioglycolic acid conjugates. *Eur. J. Pharm. Biopharm.* 2003, *55* (2), 185–190.
121. Langoth, N.; Kahlbacher, H.; Schöffmann, G.; Schmerold, I.; Schuh, M.; Franz, S.; Kurka, P.; Bernkop-Schnürch, A. Thiolated chitosans: Design and in vivo evaluation of a mucoadhesive buccal peptide drug delivery system. *Pharm. Res.* 2006, *23* (3), 573–579.
122. Roldo, M.; Hornof, M.; Caliceti, P.; Bernkop-Schnürch, A. Mucoadhesive thiolated chitosans as platforms for oral controlled drug delivery: Synthesis and in vitro evaluation. *Eur. J. Pharm. Biopharm.* 2004, *57* (1), 115–121.
123. Sharon, N. Lectin-carbohydrate complexes of plants and animals: An atomic view. *Trends Biochem. Sci.* 1993, *18* (6), 221–226.
124. Smart, J.D. Lectin-mediated drug delivery in the oral cavity. *Adv. Drug Deliv. Rev.* 2004, *56* (4), 481–489.
125. Smart, J.D. Buccal drug delivery. *Expert Opin. Drug Deliv.* 2005, *2* (3), 507–517.
126. Lehr, C.M. Lectin-mediated drug delivery: The second generation of bioadhesives. *J. Contr. Release* 2000, *65* (1–2), 19–29.
127. Smart, J.D.; Nantwi, P.K.K.; Rogers, D.J.; Green, K.L. A quantitative evaluation of radiolabelled lectin retention on oral mucosa in vitro and in vivo. *Eur. J. Pharm. Biopharm.* 2002, *53* (3), 289–292.

128. Banchonglikitkul, C.; Smart, J.D.; Gibbs, R.V.; Cook, D.J. Lectins as targeting agents—The in vitro binding of lectins to lesions in the eye and mouth. *Br. J. Biomed. Sci.* 2002, *59* (2), 115–118.
129. Smart, J.D. Recent developments in the use of bioadhesive systems for delivery of drugs to the oral cavity. *Crit. Rev. Ther. Drug Carrier Syst.* 2004, *21* (4), 319–344.
130. Smart, J.D.; Nicholls, T.J.; Green, K.L.; Rogers, D.J.; Cook, J.D. Lectins in drug delivery: A study of the acute local irritancy of the lectins from *Solanum tuberosum* and *Helix pomatia*. *Eur. J. Pharm. Sci.* 1999, *9* (1), 93–98.
131. Smart, J.D.; Banchonglikitkul, C.; Gibbs, R.V.; Donovan, S.J.; Cook, D.J. Lectins in drug delivery to the oral cavity, in vitro toxicity studies. *STP Pharma Sci.* 2003, *13* (1), 37–40.
132. Wirth, M.; Gerhardt, K.; Wurm, C.; Gabor, F. Lectin-mediated drug delivery: Influence of mucin on cytoadhesion of plant lectins in vitro. *J. Contr. Release* 2002, *79* (1–3), 183–191.
133. Clark, M.A.; Hirst, B.H.; Jepson, M.A. Lectin-mediated mucosal delivery of drugs and microparticles. *Adv. Drug Deliv. Rev.* 2000, *43* (2–3), 207–223.
134. Inman, L.R.; Cantey, J.R. Specific adherence of *Escherichia coli* strain RDEC-1 to membranous M cells of the Peyer's patch in *Escherichia coli* diarrhea in the rabbit. *J. Clin. Invest.* 1983, *71* (1), 1–8.
135. Sanford, B.A.; Thomas, V.L.; Ramsay, M.A. Binding of staphylococci to mucus in vivo and in vitro. *Infect. Immun.* 1989, *57* (12), 3735–3742.
136. Bernkop-Schnürch, A.; Gabor, F.; Szostak, M.P.; Lubitz, W. An adhesive drug delivery system based on K99-fimbriae. *Eur. J. Pharm. Sci.* 1995, *3* (5), 293–299.
137. Haltner, E.; Easson, J.H.; Lehr, C.M. Lectins and bacterial invasion factors for controlling endo- and transcytosis of bioadhesive drug carrier systems. *Eur. J. Pharm. Biopharm.* 1997, *44* (1), 3–13.
138. Isberg, R.R.; Miller, V.; Falkow, S. Yersinia INV Nucleic Acids. US5338842, August 16, 1994.
139. Lehr, C.M. From sticky stuff to sweet receptors—Achievements, limits and novel approaches to bioadhesion. *Eur. J. Drug Metab. Pharm.* 1996, *21* (2), 139–148.

53 Polymeric Biomaterials for Controlled Drug Delivery

Sutapa Mondal Roy and Suban K. Sahoo

CONTENTS

53.1 Introduction 1255
53.2 Introduction to Biodegradable Polymers and Their Application 1256
     53.2.1 Biodegradable Polymers and Drug Delivery 1257
     53.2.2 Various Drug Delivery Systems 1257
53.3 Usefulness of Various Biodegradable Polymers in Drug Delivery 1258
     53.3.1 Polysaccharides 1258
          53.3.1.1 Alginic Acid or Alginates 1258
          53.3.1.2 Starch 1259
          53.3.1.3 Dextran 1259
          53.3.1.4 Pullulan 1260
          53.3.1.5 Hyaluronic Acid 1261
          53.3.1.6 Chitin and Chitosan 1261
     53.3.2 Polyurethane 1262
     53.3.3 Other Polymeric Systems 1262
53.4 Various Drug Delivery Systems: Application of Biodegradable Polymers 1263
     53.4.1 Liposomal Drug Delivery System 1263
     53.4.2 Nanoparticle Drug Delivery System 1265
     53.4.3 Other Drug Delivery Systems 1266
53.5 Concluding Remarks 1267
Acknowledgment 1267
References 1268

53.1 INTRODUCTION

Polymers, as already known, are large molecules created by linking a series of simple monomers. Apart from synthetically generated polymers, there are biopolymers that are the most important components of living organisms such as proteins, nucleic acids, and sugars. These biopolymers are the key factor in controlling and regulating several biochemical and biophysical functions of living cells, thus, in turn, participating in all major natural processes leading to be an integral part of cell machinery. These biopolymers can undergo highly cooperative interactions, which give rise to their property of nonlinear response to external stimuli. This high degree of cooperativity is thought to be the cause behind the strongly coherent biochemical and biophysical functions in living cell machinery.

The mechanism of the cooperative interaction of these biopolymers and their interaction with the external stimuli are a major field of study for generating synthetic polymers that can mimic the cooperative behavior of biopolymers. These polymers can then be utilized as biomaterials and can be employed to interface with biological systems for various functions of a living cell.

Biocompatibility is an essential requirement for a polymer to be characterized as a biomaterial. Biocompatibility of a material is its ability to execute a response in a specific application with respect to a particular host. The transient existence of the polymeric biomaterials is suitable for in vivo applications since it can overcome biostability and biocompatibility issues in the "long term." Among several applications, one of the major applications of the biodegradable polymers is in the field of drug delivery. In recent years, controlled and targeted delivery of pharmaceutics (or drugs) has become an extremely important field of research considering the fact that the delivery of the pharmaceutics in a very controlled manner and targeted only at the diseased region (tissues or cells) will increase the efficacy of the pharmaceutics by several folds and will also help to get rid of undesired side effects of the drugs.

In this entry, we aim to discuss the use of several biodegradable polymeric biomaterials and elaborate some of the biodegradable polymeric materials that can act as successful pharmaceutical drug delivery systems. We will also highlight the recent developments and ongoing research about the finding and synthesizing of new polymeric biomaterials to be used for drug delivery successfully.

53.2 INTRODUCTION TO BIODEGRADABLE POLYMERS AND THEIR APPLICATION

Polymeric, biodegradable materials with transient existence are mostly useful in different biomedical applications, because these polymers degrade into products that are either normal metabolites of the body or can be completely eliminated from the body with or without further metabolic transformation.[1,2] Polymeric biomaterials have been extensively monitored for their biotechnological, biochemical, pharmaceutical, and medical applications, viz. treatment, augmentation, or replacement of tissue, organ, etc.[3] These polymeric biomaterials are developed in such a way that the physical and chemical properties along with the inherent flexible nature of the biomaterials are retained. Since 1940, these synthetic polymeric biomaterials have found use in artificial corneal substitute, blood contacting devices, hip joint replacements, formation of intraocular lenses, etc.[4,5] Later, engineering of novel polymeric biomaterials and modification of existing polymers were mostly executed considering their bioactivity, biocompatibility, and transient existence.[6]

Biodegradable polymers can be broadly classified into two subclasses: natural and synthetic, considering their origin. Natural polymers, obtained directly from natural resources, are likely to be the obvious choice for all kinds of biomedical and pharmaceutical applications for their high degree of biocompatibility, biomimicking environments, unique mechanical properties, and most importantly biodegradability by any natural degradable mechanism. However, natural polymers have some inherent disadvantages associated with them such as risk of viral infection, antigenicity, unstable material supply, etc., which prevent them from being used for biomedical purposes.[7] In contrast, synthetic polymers offer extraordinary advantages over natural polymers with respect to their mode of synthesis. Owing to the flexibility in synthesis procedure, it is possible for the polymers to develop a wide range of properties with excellent reproducibility. Furthermore, fine tuning of the degradation rate of these polymers is highly feasible by varying their structure.

The most important factor to be considered for the biodegradable polymers is that they must be degraded to the products that are strictly nontoxic in nature. Use of biodegradable polymers is nowadays overpowering the use of polymeric biomaterials, which are mostly biostable, for medical purposes. Apart from the elimination of the long-term biocompatibility concern, these biodegradable polymers are found useful for surgical and short-term medical application procedures. Since biodegradable polymers that are once implanted in the system do not need to get eliminated, these polymers are found to be useful in orthopedic applications, where mechanically incompatible, metallic implants can lead to stress shielding compared with biodegradable implants, which may slowly transfer the load as it degrades.[8]

Drug delivery is one of the major applications of these biodegradable polymers, which are briefly elaborated in the following.

53.2.1 Biodegradable Polymers and Drug Delivery

Drug delivery and related pharmaceutical development offers the potential to enhance the therapeutic index of various drugs, mostly anticancer drugs, by increasing the drug concentration at the diseased site as well as by decreasing the exposure of the drugs toward normal host tissues. Drugs are generally administered into the body via an oral or intravenous (i.v.) route. In these modes of administration, the therapeutic concentration of a drug is maintained in the body by repeated administration. This mode of drug administration leads to a problem related to the concentration of the drug in the body. During each repeated administration, the concentration of drug in the body reaches an extreme level and then declines rapidly, particularly when the rate of elimination of the drug from the body is very high. This can lead to an undesirable situation, where at certain times the drug concentration in the body will be too low to provide therapeutic benefit, and at other times the concentration will be too high resulting in adverse side effects. This is particularly of great concern, when using highly active biotechnological drugs with narrow therapeutic windows.[9] Thus, the optimum concentrations of the therapeutic drugs do not reach the diseased site for the required amount of time, which in turn leads to a not-so-successful treatment.

A controlled and targeted drug delivery technology has thus been developed to resolve the problems that are significantly important in terms of reducing the undesired side effects of the body as well as rapid recovery. Controlled and targeted drug delivery is aimed to regulate the rate and the spatial localization of the therapeutic agents or drugs in the body to increase its concentration at the diseased site for optimum amount of time to increase its efficacy.[10–12]

In controlled drug delivery, generally, the therapeutic and bioactive agents are entrapped or encapsulated in an insoluble matrix of small dimensions (sub-nano, nano, micro) from where the therapeutic agents are released in a controlled manner. Biodegradable polymers are generally used as these insoluble matrixes. These biodegradable polymers are used, taking into consideration the following factors[13,14]:

- More specific drug targeting and delivery
- Better drug incorporation for controlled release
- Greater stability and shelf life
- Biocompatibility to the highest degree and safety
- Maximum biodistribution
- Reduction of toxicity of the drugs, while maintaining their therapeutic effect

53.2.2 Various Drug Delivery Systems

Apart from various biodegradable polymers that can act as drug delivery systems at various physical forms, the most important and widely used is the liposomal drug delivery system.[15,16] Liposomes are self-closed spherical particles with an aqueous core surrounded by one or more outer layers consisting of lipids arranged in a bilayer constellation.[17–19]

Owing to their versatile nature, liposomes can be used for diverse functions. They are formed spontaneously in aqueous solutions. Their similarities to cell membranes make them a very useful model in biophysical, biochemical studies, pharmaceutical chemistry, biomedical engineering, etc. High degree of biocompatibility, biodegradability, low toxicity, and immunogenicity of the liposomes make them the most popular drug delivery agent, especially as an i.v. drug carrier.[20] More than two decades ago, the potential use of liposomes as drug delivery agent was recognized.[21] Liposomes are easy to prepare in large quantities and they are versatile in nature. Their surface characteristics can be widely modified by changing the lipid composition and their size can be adjusted within a wide range staring from about 10 μm to as low as 20 nm. Classical liposomes consists of phospholipids, i.e., fatty acid esters and fat alcohol ethers of glycerol phosphatides.

They are generally negatively charged at physiological pH due to their phosphate groups. Cationic liposomes can also be prepared using lipid molecules bearing quaternary ammonium head group. Since the cellular membranes are negatively charged, these cationic liposomes interact strongly with cellular membranes.[22]

Though, liposomes seem to be the perfect drug delivery agent, their use as potent drug delivery agent was not that successful due to their pronounced instability in biological environment, especially in blood flow. To make the liposomes stable in the biological environment over a certain period of time, it requires to provide them more stability, or more precisely, steric stability. Steric stabilization can increase the stability of the liposomes significantly and can prolong their blood circulation time by several folds after their administration.[23] Most of the biodegradable polymers are generally used to increase the steric stability of the liposomes. Depending on the size of the polymer (R) and the distance between attachment or grafting points (D), the polymer molecules can adopt various conformations. These conformations over the membrane surface generally lead to steric stability of various degrees.

Natural biodegradable polymers, viz. proteins (collagen, gelatin, albumin, etc.) and polysaccharides (starch, dextran, chitosan, etc.), have also been used as drug delivery agent in their microparticle and nanoparticle forms, much due to their ability for controlled release, cost-effectiveness, ease of preparation, and chemical modification.[24] Moreover, these natural biodegradable polymers are safest because degradation of these polymers yield nutrients that can be assimilated by living cells.

This entry aims to discuss the use of several such biodegradable polymeric biomaterials, which can act as successful pharmaceutical drug delivery systems. It also highlights the recent developments and ongoing research about the finding and synthesizing of new polymeric biomaterials to be used for successful drug delivery.

53.3 USEFULNESS OF VARIOUS BIODEGRADABLE POLYMERS IN DRUG DELIVERY

There are several natural and synthetic biodegradable polymers that are being investigated and used for controlled drug delivery. In this section, these biodegradable polymers, which are, in some way, important in drug delivery will be mentioned, along with their physical and chemical properties.

53.3.1 Polysaccharides

Polysaccharides are composed of several monosaccharide repeating units and have a very high molecular weight. Their major advantages of use are availability, cost-effectiveness, and a wide range of properties and structures that can be easily modified, owing to the presence of reactive functional groups along the polymer chain. Their biodegradability, biocompatibility, and water solubility make them good candidates for drug delivery. There are several different types of polysaccharides having different functional groups, which will be discussed here.

53.3.1.1 Alginic Acid or Alginates

Alginic acid is a linear hetero polysaccharide, which is nonbranched, high-molecular-weight binary copolymer of (1–4) glycosidically linked β-D-mannuronic acid and α-L-guluronic acid monomers, given in Figure 53.1.[25,26] Natural alginic acid is generally obtained from the cell walls of brown algae. Its highly acidic nature allows it to undergo spontaneous formation of salts to form gels in the presence of divalent cations such as calcium ions. Gelation occurs by interaction of divalent cations with blocks of guluronic acid from different polysaccharide chains. This mild gelling property allows the encapsulation of various molecules that can act as drugs within alginate gels with minimal negative impact. Immobilization of cells in alginate is a well-established technology in a broad range of biotechnology and biomedical

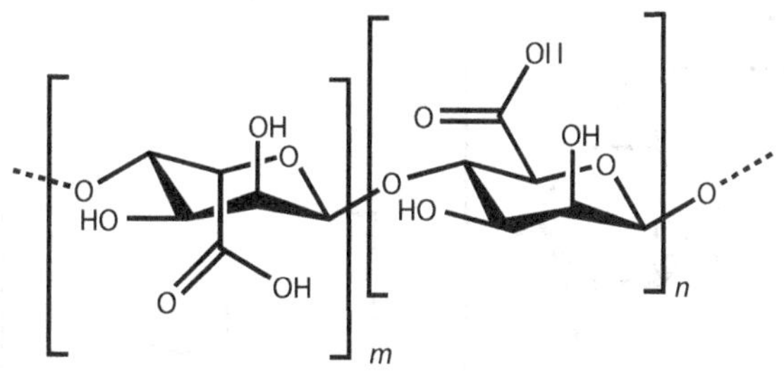

FIGURE 53.1 Chemical structure of alginates, where m and n stand for whole numbers signifying the number of monomers present in the polymeric form.

fields. The drug-delivery property of alginates generally depends on two factors: drug polymer interaction and chemical immobilization of the drug on the polymer backbone using reactive carboxylate groups.[27–29] Alginates can also be modified to increase its hydrophobicity in order to make them useful for delivery of proteins, DNA, etc., to increase the efficiency and targetability of these bioactive factors.

53.3.1.2 Starch

Starch is another widely available natural polymeric biomaterial, which is generally isolated from corn, wheat, potato, tapioca, rice, etc. The major carbohydrate reserve in the plants is in the form of starch. It mainly consists of two glucosidic macromolecules: 20%–30% of linear molecule "amylase" and 70%–80% of branched molecule "amylopectin." The chemical structure of starch is given in Figure 53.2.

Starch can be easily processed with proper modifications to form a variety of structures such as thin films, fibers, and porous matrices. Thus it has become a potentially useful polymer for thermoplastic biodegradable materials because of its low cost, availability, biocompatibility, biodegradability, and production from renewable resources.[30] Degradation of starch yields a variety of sugar molecules of low molecular weight, such as fructose and maltose, making it a successful drug delivery vehicle.[31] Starch microspheres have been investigated as a bioadhesive drug delivery system for the nasal delivery of proteins.[32]

53.3.1.3 Dextran

Dextran is a natural polysaccharide consisting of many glucose molecules coupled into long, branched chains of varying lengths from 3 to 2000 kD. The branching in dextran occurs mainly through the 1,6- and partly through the 1,3-glucosidic linkages. The chemical structure of dextran is

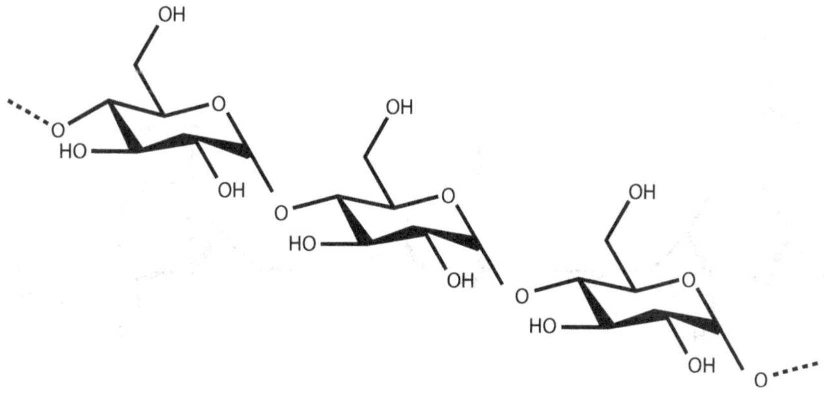

FIGURE 53.2 Chemical structure of starch.

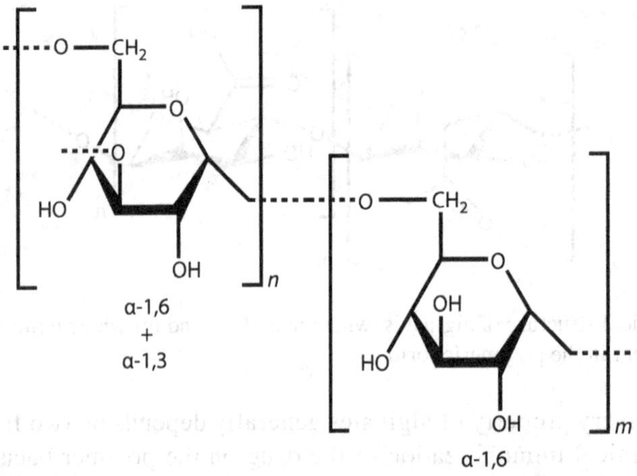

FIGURE 53.3 Chemical structure of dextran, where *m* and *n* stand for whole numbers signifying the number of monomers present in the polymeric form.

given in Figure 53.3. Dextran is synthesized from sucrose using some lactic-acid bacteria, such as *Leuconostoc mesenteroides* and *Streptococcus mutans*. Dextran is also synthesized, in the form of crystals, from the lactic acid bacterium *Lactobacillus brevis*. Dextrans are colloidal and hydrophilic in nature. They are inert to the in vivo environment and do not affect cell viability.[33]

Dextran is generally used for medical purposes as an antithrombotic (antiplatelet), to reduce blood viscosity, and as a volume expander in anemia.[34] It has been found out that dextran can be degraded by the enzyme dextranase in the colon, and so a polymeric prodrug for colonic drug delivery based on dextran can be possibly designed.

53.3.1.4 Pullulan

Pullulan is a naturally occurring, linear homopolysaccharide polymer consisting of maltotriose units (consisting of three glucose or D-glucopyranose units). The three glucose units or D-glucopyranose units in maltotriose are connected by α-(1 → 4) glycosidic linkages, whereas consecutive maltotriose units are connected to each other by α-(1 → 6) glycosidic bond (Figure 53.4). The backbone structure of pullulan resembles dextran, with both of them lending themselves as plasma expanders. Pullulan is an edible, bland, and tasteless polymer and that makes it useful in food and beverages as fillers, in pharmaceuticals as a coating agent, in breath fresheners or oral hygiene products.[36]

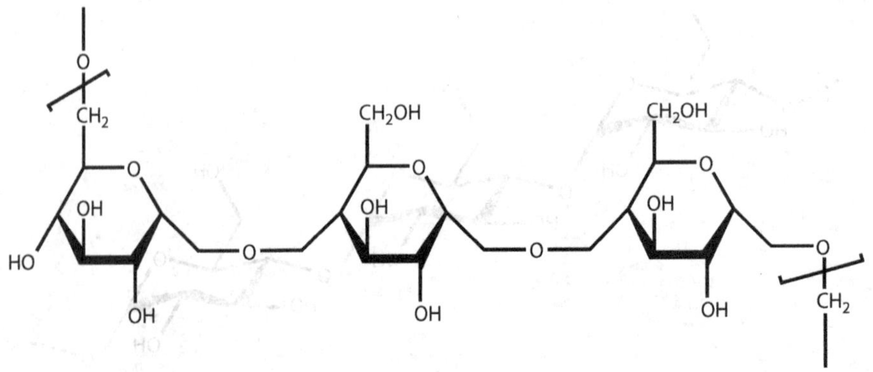

FIGURE 53.4 Chemical structure of pullulan.

Commercially, pullulan is produced by a fermentation process. A fungus *Aureobasidium pullulans* grows on a carbohydrate substrate. Then *A. pullulans* is harvested, followed by the rupture of cell with either enzyme or physical force. Pullulan then is extracted using simple water extraction process.[36] The process is completely eco-friendly and therefore pullulan can be used for drug delivery. Pullulan hydrogel microparticles or nanoparticles are used for oral administration of gastrosensitive drugs.

53.3.1.5 Hyaluronic Acid

Hyaluronic acid is another naturally occurring, linear polysaccharide, which is negatively charged. Chemically, hyaluronic acid consists of repeating disaccharide units, i.e., D-glucuronic acid and 2-acetamido-2-deoxy-D-glucose monosaccharide units and the chemical structure is given in Figure 53.5. Hyaluronic acid is mostly found in articular cartilage, connective tissues, and in synovial fluids of mammals. It is also abundantly present in the mesenchyme of developing embryos.

The polymer is water soluble and forms very viscous solutions. Hyaluronic acid is an ideal candidate for wound dressing applications because it can act as a scavenger for free radicals in wound sites, thereby modulating inflammation.[37] It can interact with a variety of biomolecules; it is a bacteriostat; and it can be recognized by receptors on a variety of cells associated with tissue repair. These characteristics of hyaluronic acid make it useful in several clinical applications—e.g., to protect delicate tissue in the eye during cataract extraction, corneal transplantation, and glaucoma surgery. It can also act as a vitreous substitute during retina reattachment surgery. Moreover, it can be used to relieve pain and improve joint mobility in patients suffering from osteoarthritis of the knee and to accelerate bone fracture healing.

53.3.1.6 Chitin and Chitosan

Chitin is one of the most abundant, naturally occurring polysaccharide. It is a (1 → 4) β-linked glycan composed of 2-acetamido-2-deoxy-D-glucose. The deacetylated derivative of chitin is known as chitosan, which is the most extensively studied polymer for biomedical applications. Chitosan is a semicrystalline linear copolymer polysaccharide, consisting of (1 → 4) β-linked D-glucosamine residues with some randomly distributed *N*-acetyl glucosamine groups. The chemical structure of chitin and chitosan are given in Figure 53.6.

Chitin is naturally occurring and is the principal component of the outer shells of crustaceans, cell walls of fungi, etc. The term chitosan is used to describe a series of polymers of different molecular weights (M_W) and degree of deacetylation (DD), defined in terms of the percentage of primary amino groups present in the polymer backbone. The DD of typical commercial chitosan is usually between 70% and 90% and the M_W is in between 10 and 1000 K.[38]

Although chitin is insoluble in common solvents, chitosan is completely soluble in aqueous solutions with a pH lower than 5.0.[39] It undergoes biodegradation in vivo enzymatically by lysozyme

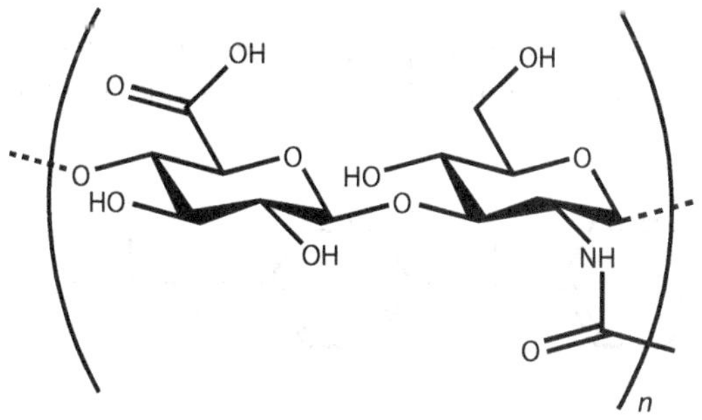

FIGURE 53.5 Chemical structure of hyaluronic acid, where *n* is the number of monomers present in the polymer.

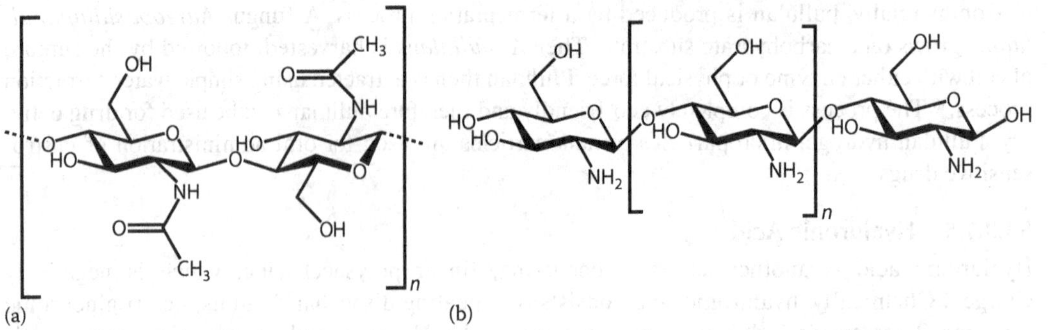

FIGURE 53.6 Chemical structure of (a) chitin and (b) chitosan, where n stands for the whole numbers signifying the number of monomers present in the polymer.

to nontoxic products.[39] The easy processability of chitosan and its versatile properties makes it an attractive material for various medical applications. Chitosan has been extensively used as a wound and burn dressing material due to its easy applicability, oxygen permeability, water absorptivity, hemostatic property, and ability to induce interleukin-8 from fibroblasts, which is involved in the migration of fibroblasts and endothelial cells.[40] Incorporation of antibacterial agents into these wound dressings significantly improves the performance of chitosan-based dressings.[41] Moreover, chitosan is most extensively used for drug delivery and controlled drug release. Chitosan alone or in presence of other polymeric biomaterials is widely used as useful tool for drug delivery.

53.3.2 Polyurethane

Polyurethane is a polymer composed of a chain of organic units joined by carbamate (urethane) linkage. Polyurethane polymers are formed by combining two or several bi- or higher-functional monomers. One contains two or more isocyanate functional groups (with formula –N=C=O) and the other contains two or more hydroxyl groups (with formula –OH).[42] The chemical structure of polyurethane is given in Figure 53.7.

Polyurethane is a unique material that offers the elasticity of rubber combined with the toughness and durability of metal. Because urethane is available in a very broad range of hardness (soft as an eraser to hard as a bowling ball), it allows the engineer to replace rubber, plastic, and metal with the ultimate in abrasion resistance and physical properties.

Urethane polymer, when fully reacted, is completely chemically inert. Polyurethane micelles are thus used for some drug delivery processes.

53.3.3 Other Polymeric Systems

There are some other polymers such as polyethylene glycol (PEG) and polyethylene oxide (Figure 53.8) which is synthetically prepared and biodegradable. PEG is the oligomer or polymer

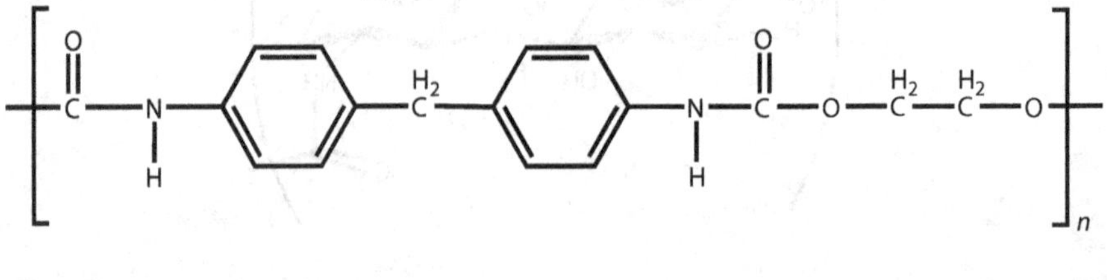

FIGURE 53.7 Chemical structure of polyurethane, where n stand for whole numbers signifying the number of monomers present in the polymeric form.

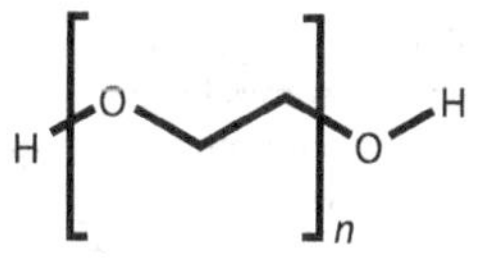

$$\text{H} \left[\text{O} - \text{CH}_2\text{CH}_2 \right]_n \text{O} - \text{H}$$

FIGURE 53.8 Chemical structure of PEG, where *n* stand for whole numbers signifying the number of monomers present in the polymeric form.

of ethylene oxide. The molecular mass of PEG is generally kept below 20000 g/mol. These polymers are liquid or low-melting solids, depending on their molecular weights. PEGs are prepared by polymerization of ethylene oxide. Depending on their molecular weight, these polymers find use in different applications, including drug delivery.

Elastin-like polypeptides (ELPs) are another class of synthetic biopolymers that consist of a repeating pentapeptide sequence that is represented in native elastin.[43] The peptide sequence in ELPs is VPGXG, where X is any amino acid, except proline. ELPs are water soluble and can form micron- or sub-micron-sized aggregates; moreover, they are biocompatible and nonimmunogenic, which in turn make it useful as a potent drug delivery system.[44,45]

53.4 VARIOUS DRUG DELIVERY SYSTEMS: APPLICATION OF BIODEGRADABLE POLYMERS

Although biopolymers as well as polymeric biomaterials are used for successful drug delivery, these polymers mostly cannot work alone. There are some conventional drug delivery vehicles that are used to deliver the therapeutics to the diseased site with considerable efficiency. Mostly, these biopolymers and polymeric biomaterials are used along with these drug delivery agents to increase the efficacy of the drugs. There are some other examples, too, where these polymers can independently act as effective and efficiently useful drug delivery agents. Liposomes or vesicles, as mentioned earlier, are one of the most important drug delivery system, which will be discussed in detail. Apart from liposomes, effectiveness of various nanoparticles, microparticles, and micelles as drug delivery vehicles will also be discussed. Nanoparticles and microparticles of the biopolymers are indeed an emerging field of study, considering their high degree of biodegradability, biocompatibility, specific targeting, and functionality.

53.4.1 Liposomal Drug Delivery System

Over the past two decades, liposomes (Figure 53.9) are used as drug delivery agents in vivo.[46,47] A very simplistic model of liposomal drug delivery has been provided here for better

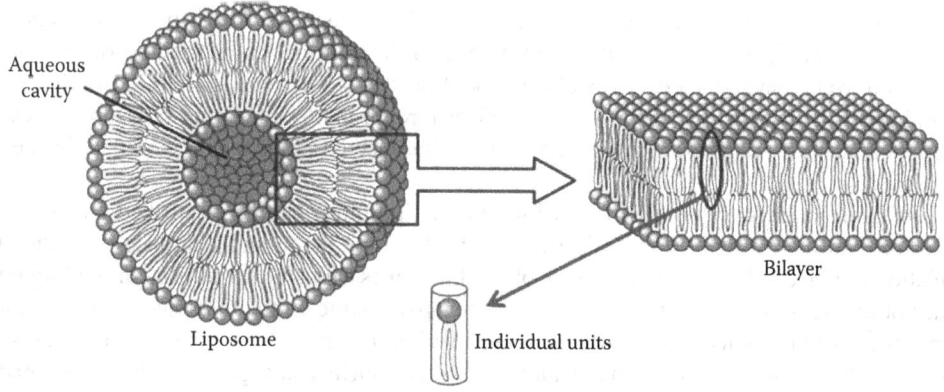

FIGURE 53.9 Cartoon diagram of liposome, the individual units are the lipid molecules.

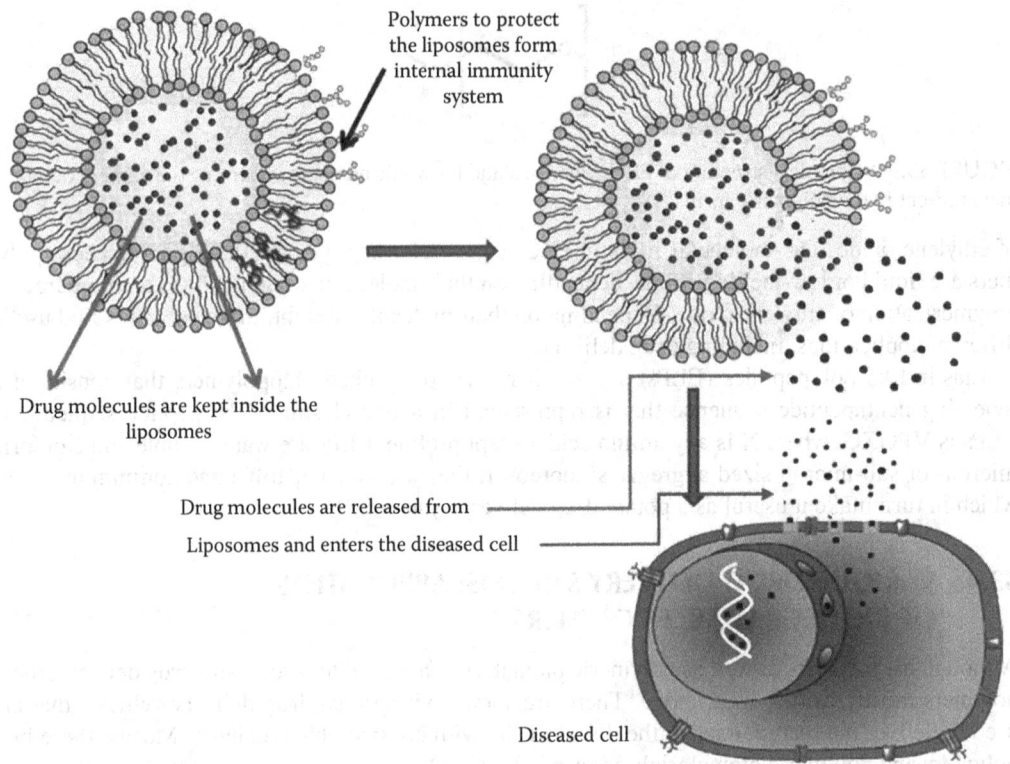

FIGURE 53.10 Cartoon diagram showing the path of liposomal drug delivery to cell.

understanding of the process (Figure 53.10). Being a drug delivery agent, the circulation lifetime, biodistribution, and clearance of liposomes from the body is the subject of extensive study. All of these parameters depend on the physico-chemical properties of the liposomes, such as liposome size, surface charge, liposomal membrane bilayer packing density, and lipid composition in the liposomes.[48,49]

To deliver the therapeutic drugs at the desired diseased site, the liposomes must be circulated by blood for a considerable amount of time. After distributing the drug, the liposomes have to be eliminated from the blood circulation. But, in fact, immediately after i.v. injection, liposomes become coated by proteins circulated in the blood. Some of these proteins change the integrity of the lipid bilayer causing rapid leakage of liposomes' inner contents. Other proteins promote the recognition of the liposomes and subsequently lead to the elimination of the liposomes from the blood. Moreover, phagocytic cells may also uptake liposomes. These cells are responsible for their removal from the circulation. Since the removal of foreign matter including liposomes is carried out by mononuclear phagocyte system, the bulk of the injected liposomes accumulate in the liver and spleen.

It has been found that the circulation lifetime of liposomes can be controlled by manipulating its size and lipid composition.[49] Liposome stability in the blood and clearance from the blood circulation are inversely related to the size of the liposomes, the amount of protein binding on the surface of the liposomes, presence of positive or negative charge on the liposome surface, presence of unsaturated lipid molecules in the liposomal membrane, etc.[50,51] In other words, small-sized vesicles (<200 nm) with saturated lipids and without any surface charge have a lesser tendency of protein binding. Moreover, charge neutral lipids in combination with cholesterol (closely packed membranes) as well as steric stabilization reduce the level of protein binding and thus increase the circulation lifetime of the liposomes. For achieving this steric stabilization, several suitable biopolymers as well as polymeric biomaterials are used. Here are some examples:

- The major steric stabilizing agent for liposomes is poly(ethylene glycol) or PEG. PEG-ylated liposomes of size ~100 nm containing various anti cancer drugs, e.g., doxorubicin, anthracyclines, and taxanes, have been used for the treatment of breast cancer. Polymeric coating of PEG over these small, rigid liposomes increases the terminal half life up to ~55 h in humans. Moreover, these PEG-ylated liposomes experience an inhibited interaction with plasma proteins and mono-nuclear phagocytes and consequently the liposomes show a dramatically prolonged circulation time hence become commercially extremely useful.[52,53]
- Low-molecular-weight chitosan-coated liposomes (LCHLs) designed by the drug cyclosporine A (CsA) can be encapsulated as a model drug inside LCHL.[54]
- It has been concluded from rigorous studies that LCHL can be used as a potential ocular drug carrier due to its property of prolonged drug retention, enhanced drug permeation, and biocompatibility.
- Small unilamellar vesicles (SUVs) of different lipids are coated with chitosan (CHT) by dropwise addition.[55] These CHT-coated liposomes are found to show increased drug encapsulation efficiency, nebulization efficiency, mucoadhesive property, and decrease in toxicity level of the drugs toward epithelial cells. Therefore, CHT-coated liposomes are considered to be advantageous for the drug delivery to the lungs by the method of nebulization.
- The negatively charged alginate can form excellent microcapsules with considerably enhanced stability, when it is added up with positively charged chitosan.[56]
- Chitosan, in combination with sodium alginate, is found to show an increase in the encapsulation efficiency. At low pH (gastric environment) alginate is found to shrink, so encapsulated drugs are not released. For the insulin delivery, it is initially encapsulated in liposomes and then the lipoinsulin is entrapped in chitosan–alginate microcapsule system. The aqueous interior of the liposome will preserve the structure and function of the insulin, whereas the lipid exteriors are protected by the biopolymer coating. Oral administration of these lipoinsulin-coated alginate–chitosan microcapsules is found to be indeed effective.[57]
- For the delivery of anticancer drug, ellagic acid, soybean-lecithin liposomes coated with chitosan–dextran sulfate can be used. Studies have proved that layer-by-layer electrostatic deposition of biopolymers, that is, positively charged chitosan and negatively charged dextran sulfate over negatively charged soybean lecithin liposomes, gives rise to a highly stable drug delivery unit, which can be used as potent delivery vehicle for anticancer drugs such as ellagic acid.[58]

53.4.2 Nanoparticle Drug Delivery System

The use of nanoparticles (NPs) in medicine and, more specifically, drug delivery is currently gaining momentum. Use of NPs, in pharmaceutical sciences, mainly focuses on the reduction of toxicity and undesired side effects arising from the drugs. To get rid of these toxic effects, use of drug delivery systems such as NPs made of biopolymers as well as polymeric biomaterials have become exceedingly popular.[59]

For drug targeting and delivery at the intended diseased site in the body, NPs of biodegradable polymers are needed as they are more specific to their action compared to other drug delivery systems.[60] A simplistic line diagram has been incorporated here for a better understanding of NPs-mediated drug delivery systems (Figure 53.11). Moreover, due to their biocompatibility and limited lifespan, these NPs of biopolymers and polymeric biomaterials are therapeutically demanding. Here are some examples:

- NPs formed by complex formation (1:1) of negatively charged dextran sulfate and positively charged chitosan have been developed for the delivery of insulin.[61]
- Vitamin B_{12} or $VitB_{12}$-coated dextran NP conjugates (150–300 nm) can be used for the delivery of insulin with significantly prolonged (~54 h) effect.[62]

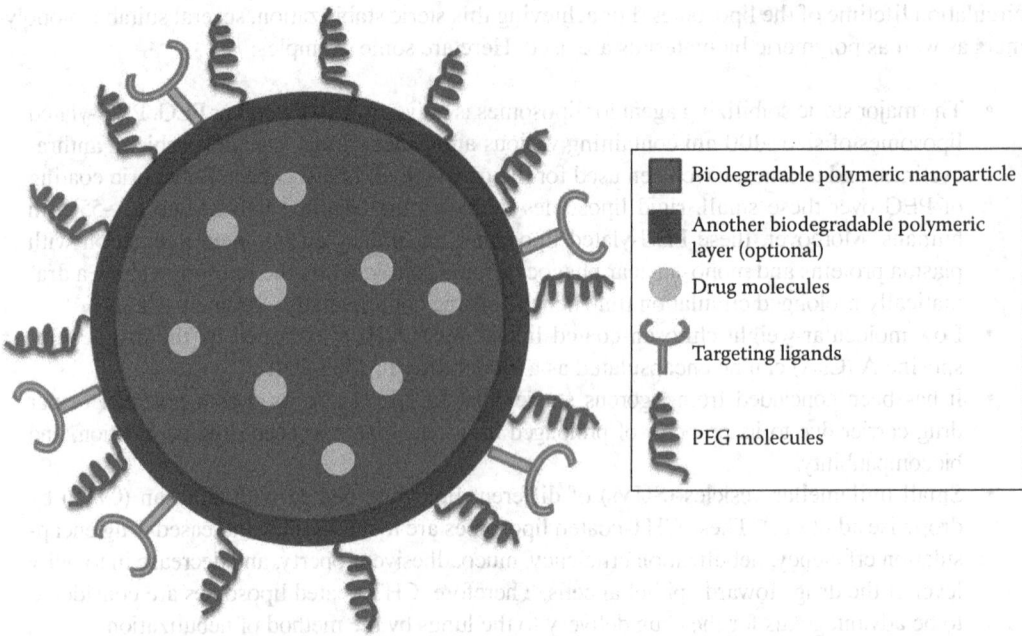

FIGURE 53.11 Cartoon diagram showing biodegradable polymeric nanoparticle drug delivery system. The nanoparticle drug delivery system has a dimension of about 50–100 nm.

- Cholesterol-bearing pullulan nanogels are capable of binding various hydrophobic substances and have been used for the delivery of some hydrophobic, anticancer therapeutics such as adriamycin and doxorubicin.[63,64]
- Drugs such as verapamil, which have an absorption window in the gastrointestinal (GI) tract, are very less soluble in the alkaline medium and which are intended for local action on the gastro-duodenal wall, are delivered in the gastrointestinal cavity with chitosan nanobeads.[65] Since chitosan has a high degree of bioadhesive and floating capabilities, chitosan nanobeads help a continuous and gradual release of the drugs in the upper part of the small intestine making a more uniform concentration of the drugs in the blood flow thereby increasing drug efficacy.[66]
- Trimethyl substituent of chitosan (TMC) and alginate-coated chitosan nanoparticles are used for oral vaccination diphtheria toxoid.

53.4.3 Other Drug Delivery Systems

There are some other methods of drug delivery such as modified chitosan cross-linked starch polymers, which are used for oral delivery of insulin.[68] Few important drug delivery systems are discussed here.

- *Buccal drug delivery*: Administration of drugs through buccal mucosa in the mouth provides a unique advantage of reduced chances of increased acidity, proteolytic activity, and toxicity of the GI tract. Chitosan, due to its significantly high mucoadhesive nature, can prolong the residence time of the therapeutics in the oral cavity and release them in a unidirectional way toward the mucosa.[70] Chitosan microspheres are used for the buccal delivery of ketoprofen and chlorhexidinediacetate for the treatment of periodontal disease, stomatitis, fungal/viral infections, etc.

- *Colon specific drug delivery*: Since chitosan gets eliminated in colon, enteric coating materials are used for colon specific drug delivery. Drugs, such as sodium diclofenac, an anti-inflammatory drug, are entrapped inside the core of the chitosan microsphere. These microspheres are then coated with enteric coating,[71] which is then used for drug delivery. In the colon, sodium diclofenac is found to be released over the period of time of ~12 h.
- *Mucosal vaccination*: Chitosan microparticles are able to associate with large amount of ovalbumin (model vaccine for diphtheria toxoid). Chitosan microparticles are not disintegrated in an acidic environment and protect the antigen against degradation by entrapping it into their porous structure.[72] The chitosan microparticles transport the associated ovalbumin into the Peyer's patches.[73] Since the uptake of antigen by Peyer's patches is an essential step in oral vaccination, these porous chitosan microparticles are a very promising vaccine delivery system.[74]
- *Intra-articular drug delivery*: For a sustained intra-articular drug release to minimize the frequency of intra-articular injection for treating osteoarthritis, a typical thermo-gelling polymer named elastin-like polypeptides (ELPs) are used. ELPs have the capability to entrap drugs at a temperature above their transition temperature. These ELPs can spontaneously aggregate into the joints (bone joints) upon injection forming a drug depot. From there the ELPs disaggregate over a considerable amount of time to prolong the drug release. This aggregate formation of ELPs increases the drug half-life at the bone joints by ~25 folds.[75]
- *Transdermal drug delivery*: A transdermal drug delivery system is a device known as the transdermal patch, which is an adhesive patch made of one or more types of polymers embedded with pharmaceutics (drugs). This transdermal patch is placed on the skin to deliver the specific dose of the embedded drug into the bloodstream over a controlled period of time. Polyurethanes, because of their unique elasticity, are among the most useful biopolymers for transdermal drug delivery.[76,77] Sodium alginate–xanthum gum films are also used for transdermal drug delivery, mostly for the delivery of anti-inflammatory drugs such as ketoprofen.[78] Since these anti-inflammatory drugs show an increased rate of GI toxicity, a transdermal delivery of the drugs is a better approach to minimize the GI toxicity.

53.5 CONCLUDING REMARKS

Thus, this entry summarizes the use of various biodegradable polymers as important drug delivery agents. Mostly, chitosan-based biopolymers are used for drug delivery purposes in pharmaceutical industry. But there are other biopolymers as well as polymeric biomaterials that can meet the need of the growing demand for more improved, biodegradable, and extremely targeted drug delivery. Since targeted drug delivery is the most important goal for these studies to minimize the undesired side effects that are exerted from these pharmaceutics, more and more studies are required to improve the process of drug delivery. Earlier it was only liposomal drug delivery, then came the PEG-coated liposomal drug delivery and now various biopolymer-coated liposomes and nanomaterials are being used as more stable drug delivery agents capable of targeting the diseased site and delivering the pharmaceutics at the desired site more accurately. We hope that further studies on using polymeric biomaterials as potent drug delivery agents will open up a vista of using these biopolymers for the betterment of medical and pharmaceutical science.

ACKNOWLEDGMENT

Dr. Sutapa Mondal Roy sincerely thanks and acknowledges CSIR, Government of India, for the fellowship and the financial support.

REFERENCES

1. Vroman, I.; Tighzert, L. Biodegradable polymers. *Materials* 2009, *2* (2), 307–344.
2. Leja, K.; Lewandowicz, G. Polymer biodegradation and biodegradable polymers—A review. *Polish J. Environ. Stud.* 2010, *19* (2), 255–266.
3. Williams, D.F. *The Williams Dictionary of Biomaterials*; Liverpool University Press: Liverpool, U.K., 1999.
4. Castner, D.G.; Ratner, B.D. Biomedical surface science: Foundations to frontiers. *Surf. Sci.* 2002, *500* (1–3), 28–60.
5. Tathe, A.; Ghodke, M.; Nikalje, A.P. A brief review: Biomaterials and their application. *Int. J. Pharm. Pharm. Sci.* 2010, *2* (4), 19–23.
6. Bezwada Rao, S. From biostable to biodegradable polymers for biomedical applications. *Polym. Mater. Sci. Eng.* 2009, *101*, 1044–1045.
7. Barbucci, R. (Ed.). *Integrated Biomaterial Science*; Kluwer/Plenum: New York, 2002.
8. Middleton, J.C.; Tipton, A.J. Synthetic biodegradable polymers as orthopedic devices. *Biomaterials* 2000, *21* (23), 2335–2346.
9. Leong, K.W.; Langer, R. Polymeric controlled drug delivery. *Adv. Drug Deliv. Rev.* 1988, *1* (3), 199–233.
10. Gupta, M.; Sharma, V. Targeted drug delivery system: A review. *Res. J. Chem. Sci.* 2011, *1* (2), 135–138.
11. Bertrand, N.; Leroux, J.C. The journey of a drug carrier in the body: An anatomo-physiological perspective. *J. Contr. Release* 2012, *161* (2), 152–163.
12. Hans, M.L.; Lowman, A.M. Biodegradable nanoparticles for drug delivery and targeting. *Curr. Opin. Solid State Mater. Sci.* 2002, *6* (4), 319–327.
13. Vasir, J.K.; Reddy, M.K.; Labhasetwar, V. Nanosystems in drug targeting: Opportunities and challenges. *Curr. Nanosci.* 2005, *1* (1), 47–64.
14. Nair, L.S.; Laurencin, C.T. Biodegradable polymers as biomaterials. *Prog. Polym. Sci.* 2007, *32* (8–9), 762–798.
15. Allen, T.M. Liposomal drug delivery. *Curr. Opin. Colloid Interface Sci.* 1996, *1* (5), 645–651.
16. Chonn, A.; Cullis, P. Recent advances in liposomal drug delivery systems. *Curr. Opin. Biotech.* 1995, *6* (6), 698–708.
17. Bangham, A.D.; Hill, M.W.; Miller, N.G.A. Preparation and use of liposomes as models of biological membranes. *Methods Membr. Biol.* 1974, *1*, 1–68.
18. Papahadjopoulos, D. (Ed.). Liposomes and their use in biology. *Ann. NY Acad. Sci.* 1973, *378*, 1–412.
19. Lasic, D.D. *Liposomes: From Physics to Applications*; Elsevier: Amsterdam, the Netherlands, 1993; pp. 1–575.
20. Szoka, F., Jr.; Papahadjopoulos, D. Comparative properties and methods of preparation of lipid vesicles (liposomes). *Annu. Rev. Biophys. Bioeng.* 1980, *9* (5), 467–508.
21. Sessa, G.; Weissmann, G. Phospholipid spherules (liposomes) as a model for biological membranes. *J. Lipid Res.* 1968, *9* (3), 310–318.
22. Finkelstein, E.I.; Chao, P.G.; Hung, C.T.; Bulinski, J.C. Electric field-induced polarization of charged cell surface proteins does not determine the direction of galvanotaxis. *Cell Motil. Cytoskel.* 2007, *64* (11), 833–846.
23. Papahadjopoulos, D.; Allen, T.M.; Gabizon, A.; Mayhew, E.; Matthay, K.; Huang, S.K.; Lee, K.D. et al. Sterically stabilized liposomes: Improvements in pharmacokinetics and antitumor therapeutic efficacy. *Proc. Natl. Acad. Sci. USA* 1991, *88* (24), 11460–11464.
24. Liu, Z.; Jiao, Y.; Wang, Y.; Zhou, C.; Zhang, Z. Polysaccharides-based nanoparticles as drug delivery systems. *Adv. Drug Deliv. Rev.* 2008, *60* (15), 1650–1662.
25. August, A.D.; Kong, H.J.; Mooney, D.J. Alginate hydrogels as biomaterials. *Macromol. Biosci.* 2006, *6* (8), 623–633.
26. Tønnesen, H.H.; Karlsen, J. Alginate in drug delivery system. *Drug Dev. Indus. Pharm.* 2008, *28* (6), 621–630.
27. Matricardi, P.; Di Meo, C.; Coviello, T.; Alhaique, F. Recent advances and perspectives on coated alginate microspheres for modified drug delivery. *Expert Opin. Drug Deliv.* 2008, *5* (4), 417–425.
28. Leonard, M.; Boisseson, R.D.; Hubert, M.; Dalenc, P.; Dellacherie, F.E. Hydrophobically modified alginate hydrogels as protein carriers with specific controlled release properties. *J. Contr. Release* 2004, *98* (3), 395–405.
29. Martins, S.; Sarmento, B.; Souto, E.B.; Ferreira, D.C. Insulin-loaded alginate microspheres for oral delivery—Effect of polysachharide reinforcement on physico-chemical properties and release profile. *Carbohydr. Polym.* 2007, *69* (4), 725–731.

30. Morrison, W.R.; Karkalas, J. Starch. *Methods in Plant Biochemistry: Carbohydrates*; Academic Press: London, U.K., 1990; p. 323.
31. Marques, A.P.; Reis, R.L.; Hunt, J.A. The biocompatibility of novel starch-based polymers and composites: In vitro studies. *Biomaterials* 2002, *23* (6), 1471–1478.
32. Illum, L.; Fisher, A.N.; Jabbal-Gill, I.; Davis, S.S. Bioadhesive starch microspheres and absorption enhancing agents act synergistically to enhance the nasal absorption of polypeptides. *Int. J. Pharm.* 2001, *222* (1), 109–119.
33. Hennink, W.E.; Franssen, O.; Van Dijk-Wolthuis, W.N.E.; Talsma, H. Dextran hydrogels for the controlled release of proteins. *J. Contr. Release* 1997, *48* (2–3), 107–114.
34. Dhaneshwar, S.S.; Kandpal, M.; Gairola, N.; Kadam, S.S. Dextran: A promising macromolecular drug carrier. *Ind. J. Pharm. Sci.* 2006, *68* (6), 705–714.
35. Rekha, M.R.; Sharma, C.P. Pullulan as a promising biomaterial for biomedical applications: A perspective. *Trends Biomater. Artif. Organs* 2007, *20* (2), 116–121.
36. Alemzadeh, I. The study on microbial polymers: Pullulan and PHB. *Iran. J. Chem. Chem. Eng.* 2009, *28* (1), 13–21.
37. Lloyd, L.L.; Kennedy, J.F.; Methacanon, P.; Paterson, M.; Knill, C.J. Carbohydrate polymers as wound management aids. *Carbohydr. Polym.* 1998, *37* (7), 315–322.
38. Rinaudo, M. Chitin and chitosan: Properties and applications. *Prog. Polym. Sci.* 2006, *31* (7), 603–632.
39. Khor, E.; Lim, L.Y. Implantable applications of chitin and chitosan. *Biomaterials* 2003, *24* (13), 2339–2349.
40. Ishihara, M.; Nakanishi, K.; Ono, K.; Sato, M.; Kikuchi, M.; Saito, Y.; Yura, H. et al. Photocrosslinkable chitosan as a dressing for wound occlusion and accelerator in healing process. *Biomaterials* 2002, *23* (3), 833–840.
41. Mi, F.W.; Wu, Y.B.; Shyu, S.S.; Schoung, J.Y.; Huang, Y.B.; Tsai, Y.H.; Hao, J.Y. Control of wound infections using a bilayer chitosan wound dressing with sustainable antibiotic delivery. *J. Biomed. Mater. Res.* 2001, *59* (3), 438–449.
42. Zdrahala, R.J.; Zdrahala, I.J. Biomedical applications of polyurethanes: A review of past promises, present realities, and a vibrant future. *J. Biomater. Appl.* 1999, *14* (1), 67–90.
43. Urry, D.W. Physical chemistry of biological free energy transduction as demonstrated by elastic protein-based polymers. *J. Phys. Chem. B* 1997, *101* (51), 11007–11028.
44. Herrero-Venrell, R.; Rincon, A.C.; Alonso, M.; Reboto, V.; Molina-Martinez, I.T.; Rodriguez-Cabello, J.C. Self-assembled particles of an elastin-like polymer as vehicles for controlled drug release. *J. Contr. Release* 2005, *102* (1), 113–122.
45. Wang, N.Z.; Urry, D.W.; Swaim, S.F.; Gillette, R.L.; Hoffman, C.E.; Hinkle, S.H.; Coolman, S.L.; Luan, C.X.; Xu, J.; Kemppainen, B.W. Skin concentrations of thromboxane synthetase inhibitor after topical application with bioelastic membrane. *J. Vet. Pharmacol. Ther.* 2004, *27* (1), 37–43.
46. Senior, J.H. Fate and behavior of liposomes in vivo: A review of controlling factors. *Crit. Rev. Ther. Drug Carrier Syst.* 1987, *3* (2), 123–193.
47. Moghimi, M.S.; Hunter, C.A.; Murray, C.J. Long-circulating and target-specific nanoparticles: Theory to practice. *Pharmacol. Rev.* 2001, *53* (2), 283–318.
48. Semple, S.C.; Chonn, A.; Cullis, P.R. Interactions of liposomes and lipid-based carrier systems with blood proteins: Relation to clearance behaviour in vivo. *Adv. Drug Deliv. Rev.* 1998, *32* (1–2), 3–18.
49. Drummond, D.C.; Meyer, O.; Hong, K.; Kirpotin, D.B.; Papahadjopoulos, D. Optimizing liposomes for delivery of chemotherapeutic agents to solid tumors. *Pharmacol. Rev.* 1999, *51* (4), 691–744.
50. Allen, T.M.; Start, D.D. Liposome pharmacokinetics. *Liposomes: Rational Design*; Janoff, A.S., Ed.; Marcel Dekker, Inc.: New York, 1999; pp. 63–87.
51. Gabizon, A.; Papahadjopoulos, D. The role of surface charge and hydrophilic groups on liposome clearance in vivo. *Biochim. Biophys. Acta* 1992, *1103* (1), 94–100.
52. Lasic, D.D.; Papahadjopoulos, D. Liposomes and biopolymer in drug and gene delivery. *Curr. Opin. Solid State Mater. Sci.* 1996, *1* (3), 392–400.
53. Park, J.W. Liposome based drug delivery in breast cancer treatment. *Breast Canc. Res.* 2002, *4*, 95–99.
54. Li, N.; Zhuang, C.Y.; Wang, M.; Sui, C.G.; Pan, W.S. Low molecular weight chitosan-coated liposomes for ocular drug delivery: In vitro and in vivo studies. *Drug Deliv.* 2012, *19* (1), 28–35.
55. Zaru, M.; Manca, M.L.; Fadda, A.M.; Antimisiaris, S.G. Chitosan-coated liposomes for delivery to lungs by nebulization. *Colloids Surf. B Biointerfaces* 2009, *71* (1), 88–95.
56. Gåserød, O.; Smidsrød, O.; Skjåk-Braek, G. Microcapsules of alginate-chitosan-I—A quantitative study of the interaction between alginate and chitosan. *Biomaterials* 1998, *19* (20), 1815–1825.

57. Ramadas, M.; Paul, W.; Dileep, K.J.; Anitha, Y.; Sharma, C.P. Lipoinsulin encapsulated alginate-chitosan capsules: Intestinal delivery in diabetic rats. *J. Microencapsul.* 2000, *17* (4), 405–411.
58. Madrigal-Carballo, S.; Lim, S.; Rodriguez, G.; Vila, A.O.; Krueger, C.G.; Gunasekaran, S.; Reed, J.D. Biopolymer coating of soyabean lecithin liposomes via layer-by-layer self-assembly as novel delivery system for ellagic acid. *J. Funct. Foods* 2010, *2* (2), 99–106.
59. Duncan, R. The dawning area of polymer therapeutics. *Nat. Rev. Drug Discov.* 2003, *2* (5), 347–360.
60. Ferrari, M. Cancer nanotechnology: Opportunities and challenges. *Nat. Rev. Canc.* 2005, *5* (3), 161–171.
61. Sarmento, B.; Ribeiro, A.; Veiga, F.; Ferreira, D. Development and characterization of new insulin containing polysachharide nanoparticles. *Colloids Surf. B Biointerfaces* 2006, *53* (2), 193–202.
62. Chalasani, K.B.; Russell-Jones, G.J.; Jain, A.K.; Diwan, P.V.; Jain, S.K. Effective oral delivery of insulin in animal models using vitamin B12-coated dextran nanoparticles. *J. Contr. Release* 2007, *122* (2), 141–150.
63. Akiyoshi, A.; Kobayashi, S.; Shichibe, S.; Mix, D.; Baudys, M.; Kim, S.W.; Sunamoto, J. Self-assembled hydrogel nanoparticle of cholesterol-bearing pullulan as a carrier of protein drugs: Complexation and stabilization of insulin. *J. Contr. Release* 1998, *54* (1–3), 313–320.
64. Na, K.; Seong-Lee, E.; Bae, Y.H. Adriamycin loaded pullulan acetate/sulfonamide conjugate nanoparticles responding to tumor pH: pH dependent cell interaction. *J. Contr. Release* 2003, *87*, 3–13.
65. Moes, A. Gastro retentive dosage forms. *Crit. Rev. Ther. Drug Carrier Syst.* 1993, *10* (2), 143–195.
66. Yang, L.; Eshraghi, J.; Fassihi, R. A new intragastric delivery system for the treatment of *Helicobacter pylori* associated gastric ulcer: In vitro evaluation. *J. Contr. Release* 1999, *57* (3), 215–222.
67. Slütter, B.; Jiskoot, W. Dual role of CpG as immune modulator and physical cross linker in ovalbumin loaded N-trimethyl chitosan (TMC) nanoparticles for nasal vaccination. *J. Contr. Release* 2010, *148* (1), 117–121.
68. Mahkam, M. Starch based polymeric carriers for oral-insulin delivery. *J. Biomed. Mater. Res.* 2010, *92* (4), 1392–1397.
69. Puratchikody, A.; Prasanth, V.V.; Mathew, S.T.; Kumar, A.B. Buccal drug delivery: Past, present and future—A review. *Int. J. Drug Deliv.* 2011, *3* (2), 171–184.
70. Senel, S.; Kremer, M.J.; Ka, S.H.; Wertz, P.W.; Hıncal, A.A.; Squier, C.A. Enhancing effect of chitosan on peptide drug delivery across buccal mucosa. *Biomaterials* 2000, *21* (20), 2067–2071.
71. Thakral, N.K.; Ray, A.R.; Majumdar, D.K. Eudragit S-100 entrapped chitosan microspheres of valdecoxib for colon cancer. *J. Mater. Sci. Mater. Med.* 2010, *21* (9), 2691–1699.
72. Van der Lubben, I.M.; Kersten, G.; Fretz, M.M.; Beuvery, C.; Coos Verhoef, J.; Junginger, H.E. Chitosan microparticles for mucosal vaccination against diphtheria: Oral and nasal efficacy studies in mice. *Vaccine* 2003, *21* (13), 1400–1408.
73. Bacon, A.; Makin, J.; Sizer, P.J.; Jabbal-Gill, I.; Hinchcliffe, M.; Illum, L. Carbohydrate biopolymers enhance antibody responses to mucosally delivered vaccine antigens. *Infect. Immun.* 2000, *68* (10), 5764–5770.
74. Van der Lubben, I.M.; Verhoef, J.C.; Van Aelst, A.; Borchard, G.; Junginger, H.E. Chitosan microparticles for oral vaccination. *Biomaterials* 2001, *22* (7), 687–694.
75. Betre, H.; Liu, W.; Zalutsky, M.R.; Chilkoti, A.; Kraus, V.B.; Setton, L.A. A thermally responsive biopolymer for intra-articular drug delivery. *J. Contr. Release* 2006, *115* (2), 175–182.
76. Prausnitz, M.R.; Langer, R. Transdermal drug delivery. *Nat. Biotechnol.* 2008, *26* (11), 1261–1268.
77. Guy, R.H. Drug delivery. *Handbook of Experimental Pharmacology*; Schäfer-Korting, M., Ed.; Springer: Heidelberg, Germany, 2010; pp. 197, 399–410.
78. Rajesh, N.; Siddaramaia, H.; Gowda, D.V.; Somashekar, C.K. Formulation and evaluation of biopolymers based transdermal drug delivery. *Int. J. Pharm. Pharm. Sci.* 2010, *2* (2), 142–147.

54 Nanogels
Chemical Approaches to Preparation

Sepideh Khoee and Hamed Asadi

CONTENTS

54.1 Introduction 1271
    54.1.1 Brief Overview of Nanogels 1271
54.2 Preparation of Nanogels 1272
    54.2.1 Preparation of Nanogels from Polymer Precursors 1272
        54.2.1.1 Disulfide-Based Cross-Linking 1272
        54.2.1.2 Amine-Based Cross-Linking 1275
        54.2.1.3 Click Chemistry-Based Cross-Linking 1279
        54.2.1.4 Imine Bonds-Induced Cross-Linking 1283
        54.2.1.5 Photo-Induced Cross-Linking 1284
        54.2.1.6 Physical Cross-Linking 1287
    54.2.2 Preparation of Nanogels via Monomer Polymerization 1287
        54.2.2.1 Heterogeneous Free Radical Polymerization 1288
        54.2.2.2 Precipitation Polymerization 1288
        54.2.2.3 Inverse (Mini) Emulsion Polymerization 1288
        54.2.2.4 Inverse Microemulsion Polymerization 1289
        54.2.2.5 Heterogeneous Controlled/Living Radical Polymerization 1290
        54.2.2.6 Atom-Transfer Radical Polymerization 1290
        54.2.2.7 Reversible Addition–Fragmentation Chain Transfer 1292
        54.2.2.8 Nanogels via Direct RAFT Polymerization 1292
        54.2.2.9 Nanogel Synthesis by RAFT Polymerization in Water 1293
        54.2.2.10 Nanogels Synthesis by Inverse RAFT Miniemulsion 1295
54.3 Applications of Nanogels 1295
54.4 Release Mechanisms from Nanogels 1297
54.5 Conclusion 1298
Acknowledgments 1299
References 1299

54.1 INTRODUCTION

54.1.1 BRIEF OVERVIEW OF NANOGELS

Polymer-based drug delivery systems (DDS) have attracted significant attention in biomedicine, pharmaceutics, and bio-nanotechnology. In particular, polymer-based DDS with controllable release of therapeutics and cell targeting have the potential to treat numerous diseases, including cancers, with a reduction in the side effects of the drugs. Several types of polymer-based DDS have been explored and nanogels are among the most promising DDS.[1–4] Nanogels are hydrogel networks that are confined to submicron-size. In other words, nanogels are defined as nanosized networks of chemically or physically cross-linked polymers composed of hydrophilic or amphiphilic chains.[5] Like hydrogels, nanogels are three-dimensional biocompatible materials with high water content. For drug delivery applications, key features including high water content/swellability, biocompatibility, and adjustable chemical/mechanical properties are particularly attractive. The large surface area provides space for functionalization and bioconjugation. In addition, they have a tunable size from submicrons to tens of nanometers, and their size[3,6] can be tuned to an optimal diameter for increased blood circulation time in vivo after IV administration. A smaller diameter (<200 nm) enables better cellular uptake and reduced NP uptake by mononuclear phagocyte system.[7,8] Finally, the interior network allows for the encapsulation of therapeutics.

These unique properties made nanogels as a topic of increasing interest across multi-disciplinary fields over the past decade as evidenced by the appearance of a number of excellent reviews on the preparation, properties, and applications of nanogels.[1,9–23] For example, the physical entrapment of bioactive molecules (including drugs, proteins, carbohydrates, and nucleic acids in the polymeric network) and their in vitro release kinetics have been extensively investigated. In addition, the incorporation of inorganic materials has been reported. Examples include quantum dots[24,25] and magnetic NPs[26,27] for optical and magnetic imaging, and gold nanorods for photodynamic therapy.[28]

54.2 PREPARATION OF NANOGELS

Given the immense application potentials of nanogels in various applications, significant efforts have been devoted to the design and synthesis of nanogels. Nanogels are prepared by various methods of copolymerization of hydrophilic or water-soluble monomers in the presence of difunctional or multifunctional cross-linkers.[13] They include photolithographic, micromolding, microfluidic, reverse micellar, membrane emulsification, and homogeneous gelation methods.[13] They are also prepared by heterogeneous free-radical polymerization in various media including precipitation,[29–31] inverse (mini)-emulsion, and inverse microemulsion.[13] Controlled/living radical polymerization (CRP) techniques[32–35] have been explored in various media including water, (inverse) miniemulsion, and dispersion for the preparation of cross-linked particles and gels with well-controlled polymer segments.[36–44] Here we only focus on chemical approaches that can synthesize nanogels in large quantities. With this restriction, we divide existing synthetic strategies into two main categories. One strategy involves the formation of nanogels using preformed polymers whereas the other entails the formation of nanogels via the direct polymerization of monomers. These two approaches are illustrated in Figure 54.1.[45]

54.2.1 Preparation of Nanogels from Polymer Precursors

Amphiphilic copolymers are known to self-assemble in solution to form various nanoscopic structures, thus providing a versatile platform to synthesize nanogels by simply locking the assembly. In this section, several examples of nanogel preparation through this methodology will be given and details discussed based on cross-linking reaction types employed to fix the self-assembled polymer.

54.2.1.1 Disulfide-Based Cross-Linking

Disulfide bonds are found in natural peptides and proteins and play an important role in keeping the structural stability and rigidity.[46,47] Due to its stability under certain conditions and its reversibility under others, disulfide-thiol chemistry is becoming increasingly popular in conventional polymer syntheses and provides a facile route to the preparation of recyclable cross-linking of micelles.[48,49] Moreover, disulfide can reversibly undergo reduction to thiols depending on the environmental thiol concentration. For example, the thiol concentration varies in our body, depending on the location as well as pathological conditions. It is about 10 μM in the plasma but about 10 mM in the cytosol. It is seven times higher around some tumors.[50,51] In addition, there

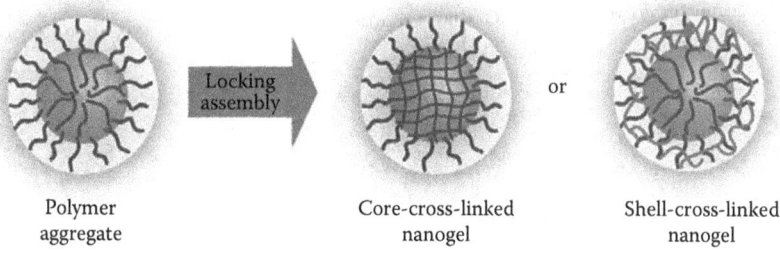

(a)   Nanogel prepared from assembled polymer precusor

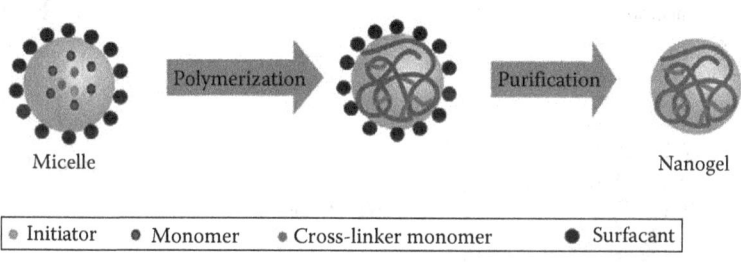

(b)   Nanogel synthesized from microemulsion/inverse microemulsion

FIGURE 54.1 Methods of nanogel synthesis: (a) the polymer precursor method and (b) the emulsion method. (Reprinted from *Adv. Drug Deliv. Rev.*, 64(9), Chacko, R.T., Ventura, J., Zhuang, J., and Thayumanavan, S., Polymer nanogels: A versatile nanoscopic drug delivery platform, 836–851. Copyright 2012, with permission from Elsevier.)

exist other methods of reducing disulfide bonds to thiols, including dithiothreitol, zinc dust, and UV light.[52] The thiol groups can be subsequently reacted with reactive groups such as activated disulfides, maleimides, iodoacetyl groups, and some thiol-containing biomolecules (e.g., antisense oligonucleotides).[53,54]

These unique properties suggest that the well-defined functional nanogels hold great potential as carriers for controlled drug delivery scaffolds to target specific cells. Drugs can be loaded into the nanogels. In a reducing environment, the nanogels will degrade to individual polymeric chains with a molecular weight below the renal threshold, thus it leads to controllably releasing the encapsulated drugs over a desired period of time.

Thayumanavan and coworkers prepared the nanogels system based on RAFT-synthesized copolymers of oligo(ethylene glycol) methacrylate (OEGMA) and pyridyl disulfide-derived methacrylate (PDS-MA) of different compositions and molecular weights.[55–57] The addition of deficient amounts of dithiothreitol (DTT) reduces a controlled percentage of PDS groups to thiols, which subsequently reacted with an equivalent amount of the remaining PDS groups to generate disulfide cross-links, and as a result nanogels were formed (Figure 54.2).[55] The authors demonstrated nanogels with different sizes can be easily obtained by varying polymer concentration and utilizing the lower critical solution temperature (LCST) behavior of polymers. The authors encapsulated a hydrophobic dye, Nile red, into the nanogels to test the encapsulation capability and the responsive release of hydrophobic molecules and also demonstrated the possibility of further functionalizing the nanogels using the remaining PDS groups. For this purpose, the authors functionalized the nanogels with thiol-modified fluorescein isothiocyanate (FITC) and a modified cell-penetrating peptide,

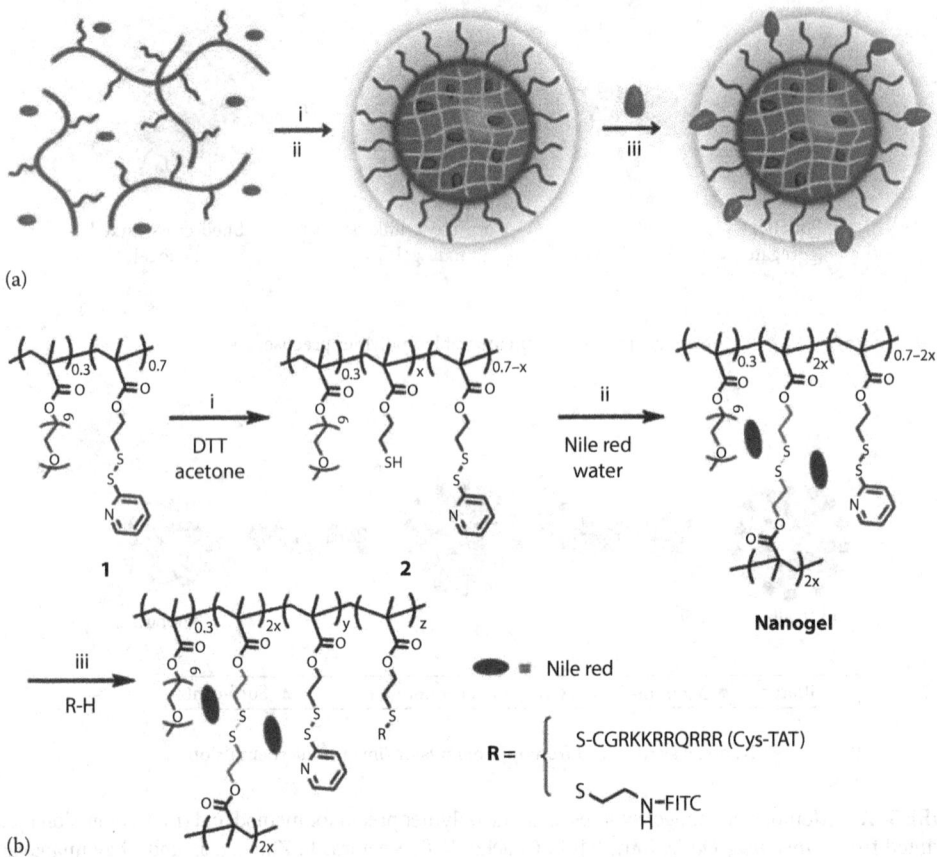

FIGURE 54.2 Design and synthesis of the polymer nanoparticles. (a) Schematic representation of the preparation of biodegradable nanogels with surface modification and (b) structure of the polymer and nanogel. (i) Cleavage of specific amount of PDS group by DTT. (ii) Nanogel formation by inter/intrachain cross-linking. (iii) Surface modification of nanogels with thiol-modified Tat peptide or FITC. (Reprinted with permission from *J. Am. Chem. Soc.*, 132(24), Ryu, J.H., Jiwpanich, S., Chacko, R., Bickerton, S., and Thayumanavan, S., Surface-functionalizable polymer nanogels with facile hydrophobic guest encapsulation capabilities, 8246. Copyright 2010 American Chemical Society.)

Tat-SH, containing a C-terminal cysteine. They also showed that the multifunctional nanoparticles are indeed interesting materials for drug delivery applications.

Control of the stability of the polymer associates as DNA delivery systems has been a longstanding challenge. The associate must maintain its structure during the circulation and efficiently dissociate to release DNA in order to exert biological effects inside the cell. The core–shell-type polyion complex (PIC) micelle with a disulfide cross-linked core was recently proposed, which has the ability to dissociate in response to chemical stimuli given at the site of the drug action. Kataoka et al.[58] reported the preparation of PIC composed of poly(ethylene glycol)-*block*-poly(L-lysine) (PEG-*b*-PLL) copolymers modified with thiol groups using *N*-succinimidyl 3-(2-pyridyldithio) propionate (SPDP) to construct PIC micelles with a disulfide cross-linked core for the delivery of plasmid DNA and antisense oligo-DNA (Figure 54.3).

In an another study, they reported a PIC micelle siRNA delivery system prepared from the block copolymer poly(ethylene glycol)-*block*-poly(L-lysine) (PEG-*b*-PLL) modified with the cross-linking reagent 2-iminothiolane (2-IT, Traut's reagent) (Figure 54.4).[59]

The resulting block copolymer, termed PEG-*b*-PLL(IM), was designed to contain cationic amidine groups for PIC formation with anionic siRNAs and also free sulfhydryls to allow disulfide

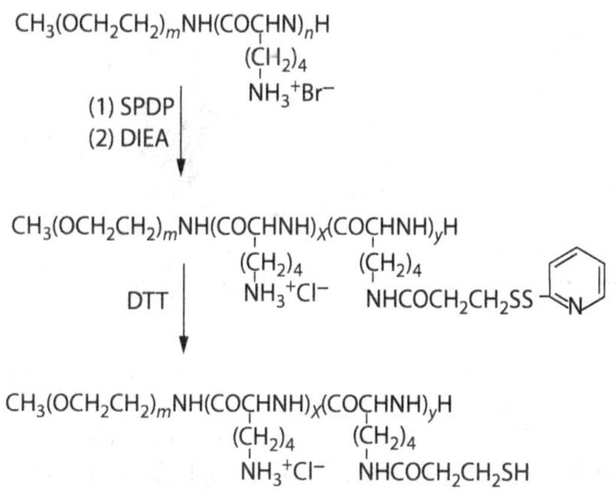

FIGURE 54.3 Introduction of thiol groups into the lysine residues of PEG-PLL. (Reprinted with permission from *Biomacromolecules*, 2(2), Kakizawa, Y., Harada, A., and Kataoka, K., Glutathione-sensitive stabilization of block copolymer micelles composed of antisense DNA and thiolated poly(ethylene glycol)-block-poly(l-lysine): A potential carrier for systemic delivery of antisense DNA, 491. Copyright 2001 American Chemical Society.)

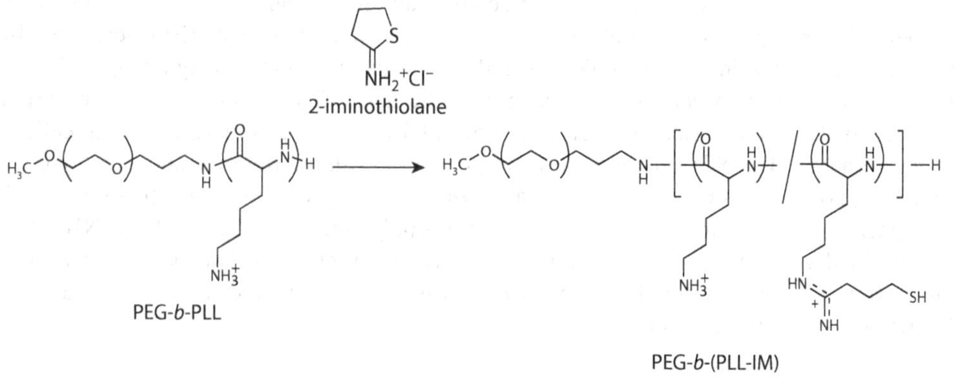

FIGURE 54.4 Preparation of iminothiolane-modified poly(ethylene glycol)-*block*-poly(L-lysine) [PEG-*b*-(PLL-IM)]. (Reprinted with permission *Biomacromolecules*, 10(1), Matsumoto, S., Christie, R., Nishiyama, N., Miyata, K., Ishii, A., Oba, M., Koyama, H., Yamasaki, Y., and Kataoka, K., Environment-responsive block copolymer micelles with a disulfide cross-linked core for enhanced siRNA delivery, 119. Copyright 2009 American Chemical Society.)

cross-linking in the micelle core for improved stability. Covalent disulfide cross-links are particularly attractive for micelle core stabilization because they are reversible and more susceptible to cleavage (reduction) at the subcellular site of activity where the levels of natural disulfide-reducing agents are higher than in the bloodstream. Disulfide cross-linked PIC micelle formation between siRNA and thiol-modified cationic block copolymer is shown in Figure 54.5.[60]

54.2.1.2 Amine-Based Cross-Linking

One of the most commonly used groups in the preparation of nanogels is amine groups due to their reactivity toward carboxylic acids, activated esters, isocyanates, iodides, and others. The Wooley group utilized amine cross-linkers to develop a methodology for the preparation of shell-cross-linked knedel-like structures (SCKs).[61] SCKs are essentially unimolecular polymer micelles, which are prepared by stabilizing the basic structure of the spherical micellar assembly through linking together of the hydrophilic portions of the chains within the micelle shell. They synthesized a variety of amphiphilic block copolymers in which poly(acrylic acid) were employed as the hydrophilic and

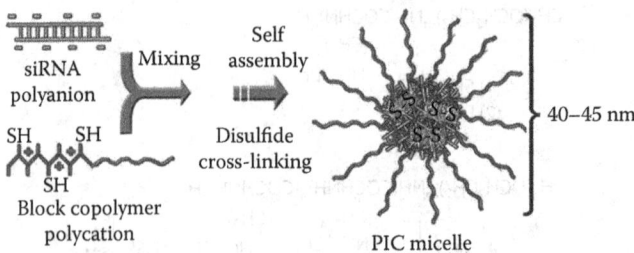

FIGURE 54.5 Preparation of disulfide cross-linked PIC micelles containing siRNA. (Reprinted with permission from *Biomacromolecules*, 12(9), Christie, R.J., Miyata, K., Matsumoto, Y., Nomoto, T., Menasco, D., Lai, T.ch., Pennisi, M. et al., Effect of polymer structure on micelles formed between siRNA and cationic block copolymer comprising thiols and amidines, 3174. Copyright 2011 American Chemical Society.)

cross-linkable block. Following the self-assembly of the block copolymers, the amidation of carboxylic acid with diamine cross-linkers resulted in cross-linking of the micellar assemblies and formation of nanogel networks. They also demonstrated that the remaining carboxylic groups on the shell can be converted to other functionalities for orthogonal surface modification.[62] In another study, they utilized diamine cross-linker containing acetal group for the preparation of pH-responsive SCKs. As detailed, after being allowed to assemble supramolecularly into micelles in water, the amphiphilic block copolymers of poly(acrylic acid) (PAA) and polystyrene (PS) were cross-linked through amidation reactions with a unique acetal-containing cross-linker (Figure 54.6).[63]

By using reaction between amines and carboxylic acid, Zhang and coworkers reported the preparation of ferrocene-based shell cross-linked (SCL) thermoresponsive hybrid micelles with antitumor efficacy. The SCL micelle consisting of a cross-linked thermoresponsive hybrid shell and a hydrophobic core domain was fabricated via a two-step process: micellization of poly(*N*-isopropylacrylamide-*co*-aminoethyl methacrylate)-*b*-polymethyl methacrylate P(NIPAAm-*co*-AMA)-*b*-PMMA in aqueous solution followed by cross-linking of the hydrophilic shell layer via the amidation reaction between the amine groups of AMA units and the carboxylic acid functions of 1,1′-ferrocenedicarboxylic acid.[64]

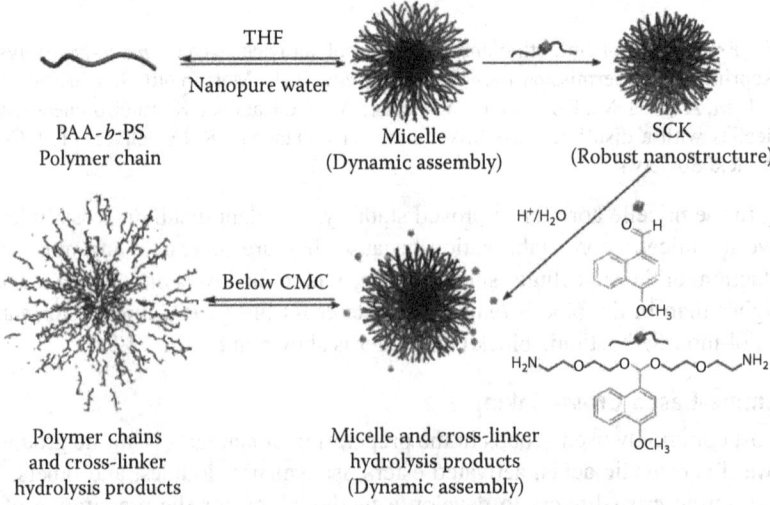

FIGURE 54.6 Illustration of SCKs that contain hydrolytically labile cross-links to allow for pH-triggered hydrolysis and nanostructure disassembly. (Reprinted with permission from *Macromolecules*, 41(18), Li, Y., Du, W., Sun, G., and Wooley, K.L., pH-responsive shell cross-linked nanoparticles with hydrolytically labile cross-links, 6605. Copyright 2008 American Chemical Society.)

In addition to carboxylic acids, activated esters also can be used for cross-linking with amines. In this regard, McCormick et al.[65] reported the synthesis of reversible SCL micelles by cross-linking of reactive N,N-acryloxysuccinimide (NAS) units incorporated within an ABC tri-block copolymer, poly(ethylene oxide)-block-[(N,N-dimethylacrylamide)-stat-(N-acryloxysuccinimide)]-block (N-isopropylacrylamide), [PEO-b-P(DMA-stat-NAS)-b-NIPAM], with cystamine, a reversible cross-linking agent. The McCormick group also reported the synthesis of pH-responsive shell cross-linked micelles consisting of α-methoxypoly (ethylene oxide)-b-poly[N-(3-aminopropyl) methacrylamide]-b-poly[2-(diisopropylamino)ethyl methacrylate] (mPEO-PAPMA PDPAEMA) by reaction with an amine-reactive polymeric cross-linking agent, NHS-PNIPAM-NHS (Figure 54.7).[66]

Pentafluorophenyl acrylate (PFPA) is another activated ester used for the preparation of nanogels. In this regard, Thayumanavan et al.[67] proposed a facile methodology for the preparation of water-dispersible nanogels based on pentafluorophenyl acrylate and polyethylene glycol methacrylate random copolymer and diamine cross-linkers. They envisaged that the addition of a calculated amount of diamine to a solution of the PPFPA-r-PPEGMA random copolymer will cause inter- and intrachain cross-linking amidation reactions to afford the nanogel. They also demonstrated the possibility of further functionalizing the nanogels using the remaining reactive PFP functionalities, as illustrated in Figure 54.8.

Furthermore, Davis, Boyer, and coworkers described the synthesis of pH-sensitive core–shell nanoparticles that contain both methanethiosulfonate and activated ester PFPA pendant

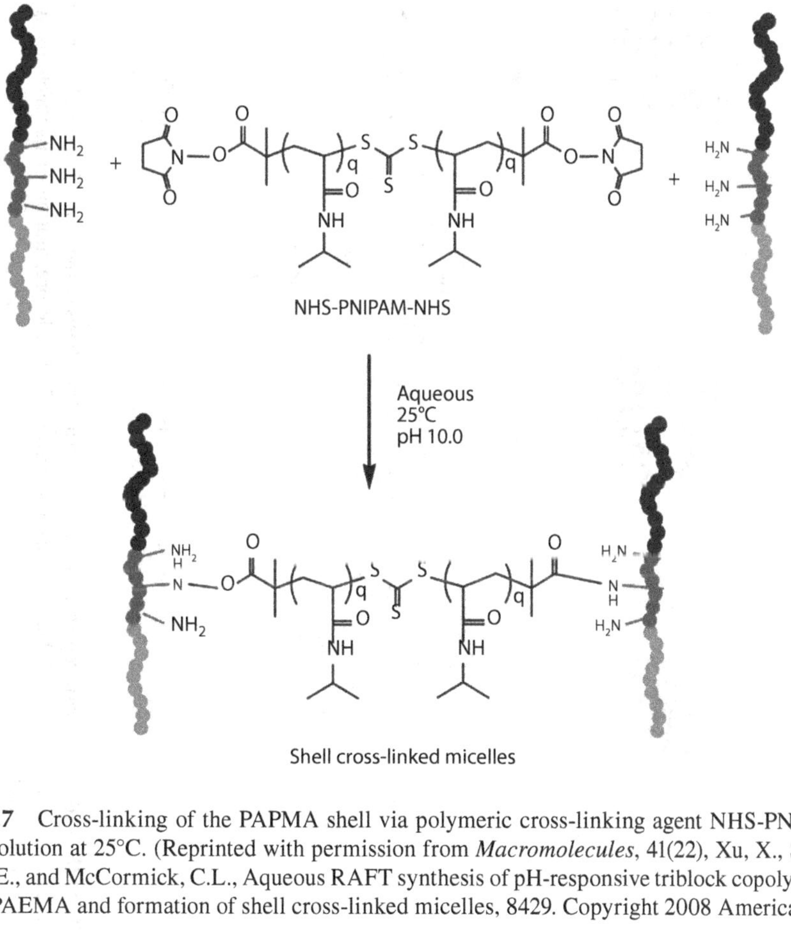

FIGURE 54.7 Cross-linking of the PAPMA shell via polymeric cross-linking agent NHS-PNIPAM-NHS in aqueous solution at 25°C. (Reprinted with permission from *Macromolecules*, 41(22), Xu, X., Smith, A.E., Kirkland, S.E., and McCormick, C.L., Aqueous RAFT synthesis of pH-responsive triblock copolymer mPEO-PAPMAPDPAEMA and formation of shell cross-linked micelles, 8429. Copyright 2008 American Chemical Society.)

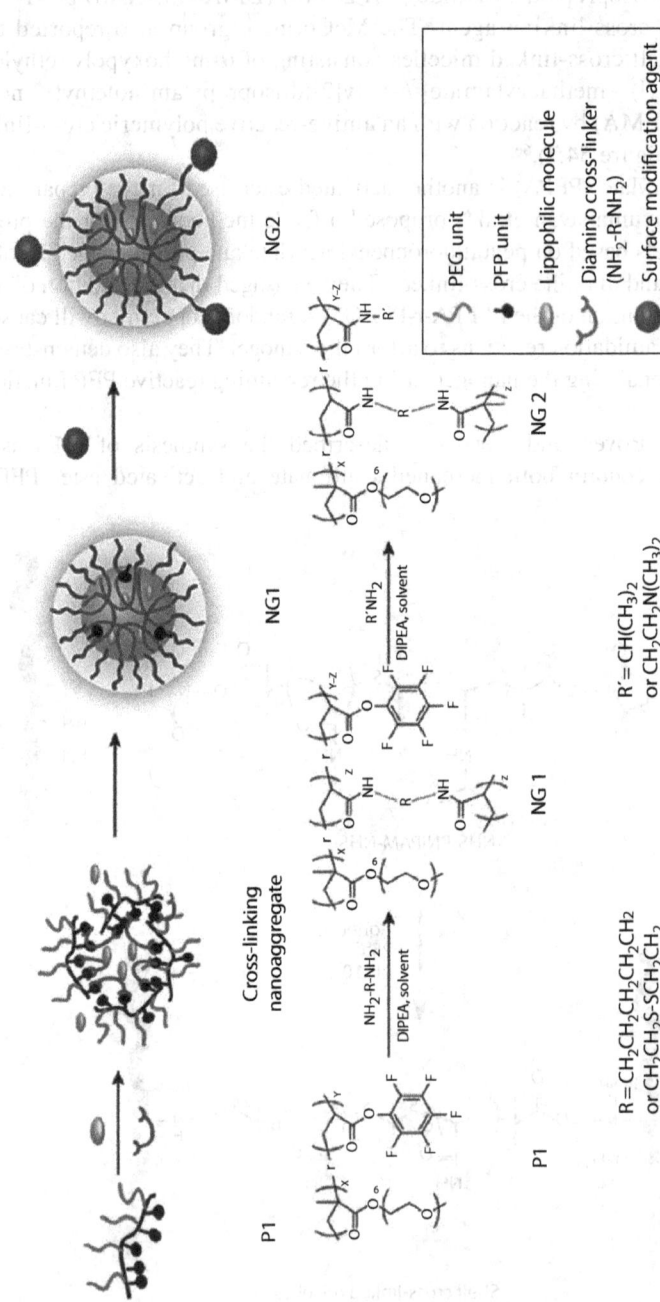

FIGURE 54.8 Schematic representation of design and synthesis of the cross-linked polymer nanogels. (Reprinted with permission from *ACS Macro Lett.*, 1(1), Zhuang, J., Jiwpanich, S., Deepak, V.D., and Thayumanavan, S., Facile preparation of nanogels using activated ester containing polymers, 175. Copyright 2012 American Chemical Society.)

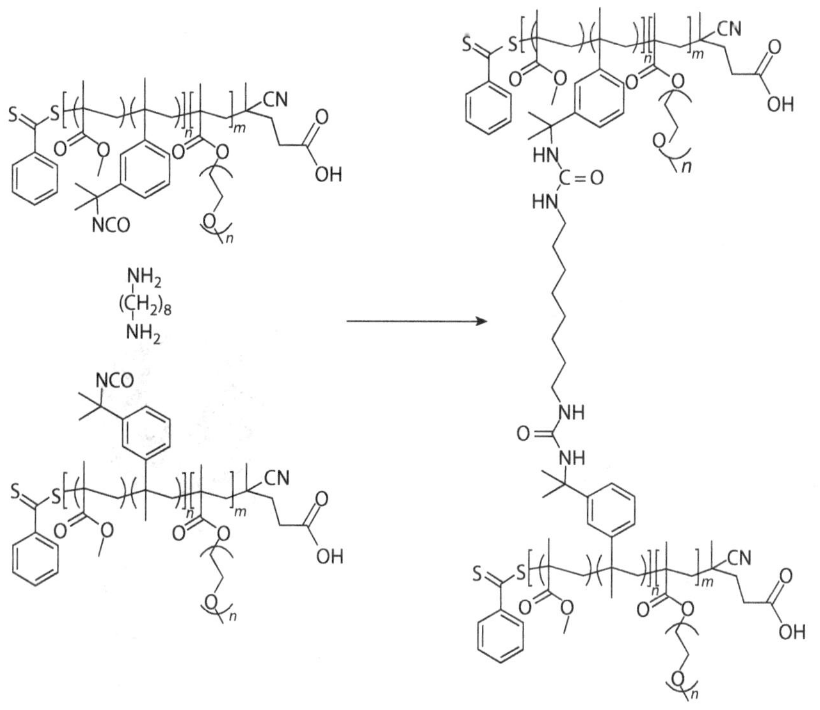

FIGURE 54.9 Synthesis of core cross-linked block copolymer micelles. (From *J. Mater. Chem.*, 21(34), Kim, Y., Pourgholami, M.H., Morris, D.L., and Stenzel, M.H., Triggering the fast release of drugs from cross-linked micelles in an acidic environment, 12777–12783. Copyright 2011. Reproduced by permission of The Royal Society of Chemistry.)

functionalities within core and poly oligo ethylene glycol (POEG) in shell. They used activated esters in the copolymer to cross-link the nanoparticles with difunctional amino cross-linkers bearing an acid cleavable bond (ketal) to generate pH-sensitive nanogels.[68]

Additionally, reactions of isocyanate with amines provide another cross-linking approach to make nanogels.[69] Cross-linked micelles with pH-responsive features were obtained by the addition of excess amounts of 1,8-diaminooctane to micellar aggregates consisting of poly (poly(ethylene glycol) methyl ether methacrylate)-*block*-poly(methyl methacrylate-*co*-poly (3-isopropenyl-a,a-dimethylbenzyl isocyanate)) [PPEGMEMA-*b*-P(MMA-*co*-TMI)] (Figure 54.9).

54.2.1.3 Click Chemistry-Based Cross-Linking

In the past several years, "click chemistry,"[70–76] as termed by Sharpless et al., has gained a great deal of attention due to its high specificity, quantitative yield, tolerance to a broad variety of functional groups, and applicability under mild reaction conditions. The Wooley and Hawker groups have reported a nanogel fabrication method utilizing click chemistry.[77,78] Alkynyl shell functionalized block copolymer micelles consist of diblock poly(acrylic acid)$_{80}$-*b*-poly(styrene)$_{90}$ were utilized as click-readied nanoscaffolds for the formation of nanogels. The following click reaction between click-readied micelles and azido dendrimers resulted in nanogel networks (Figure 54.10).[77]

They also found that only the first generation of azido terminated dendrimers successfully cross-linked the hydrophilic shell of the alkynyl functionalized micelles and that is because of the hydrophobic nature of the dendrimer of generations greater than one that is proved to be incompatible with the hydrophilic nature of the micelle corona within the aqueous reaction conditions, and as a result did not behave as effective cross-linkers. They exploited this incompatibility by changing

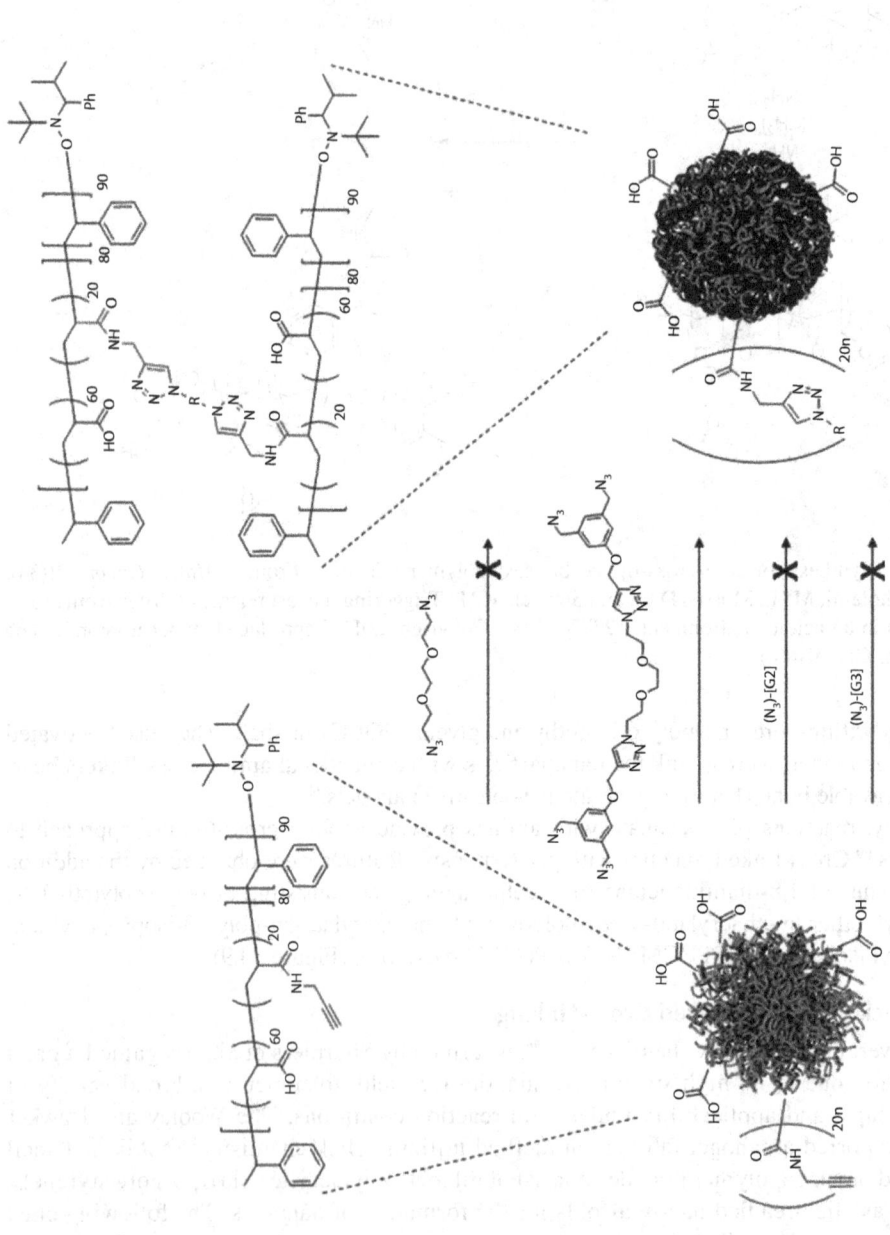

FIGURE 54.10 Synthesis of shell click cross-linked nanoparticles from click-readied micelles and dendrimers, where R represents the dendritic cross-linking unit, having the possibility of multiple cross-linkages and remaining azido functionalities. (Reprinted with permission from *J. Am. Chem. Soc.*, 127(48), Joralemon, M.J., O'Reilly, R.K., Hawker, C.J., and Wooley, K.L., Shell click-crosslinked (SCC) nanoparticles: A new methodology for synthesis and orthogonal functionalization, 16892. Copyright 2005 American Chemical Society.)

the cross-linking site from the hydrophilic shell to the hydrophobic core of the polymer micelle and synthesized the core click cross-liked micelles of amphiphilic diblock copolymers of poly(acrylic acid)-b-poly(styrene) (PAA-b-PS) that contained alkynyl functionality throughout the hydrophobic PS block.[79] The alkynyl core-functionalized block copolymer micelles and azido-terminated dendrimers were employed to construct nanogels with click reaction (Figure 54.11).

Liu and coworkers utilized click chemistry to prepare core cross-linked PIC micelles with thermoresponsive coronas.[80] As detailed, azide-containing monomer was incorporated into two oppositely charged backbones of two graft ionomers, P(MAA-co-AzPMA)-g-PNIPAM and P(QDMA-co-AzPMA)-g-PNIPAM, containing thermosensitive PNIPAM graft chains. The self-assembled PIC micelles in aqueous solution were subsequently core-stabilized via "click" reactions upon the addition of a difunctional reagent, propargyl ether (Figure 54.12).

Stenzel et al.[81] reported the preparation of click-cross-linked micelle structure as a carrier to deliver cobalt pharmaceuticals. They synthesized block copolymers of poly (propargyl methacrylate)-block-poly (poly (ethylene glycol) methyl ether methacrylate) [P(PAMA)-b-P(PEGMA)] with pendant alkyne groups (Figure 54.13), which self-assembled in aqueous solution into micelles that alkyne groups in the core took on two functions, acting as a ligand for $Co_2(CO)_8$ to generate a derivative of the antitumor agents based on (alkyne) $Co_2(CO)_6$ as well as an anchor point for the cross-linking of micelles via click chemistry.

They also synthesized core-cross-linked micelles by clicking bi-functional Pt(IV) anticancer drugs to isocyanate groups of micelle cores.[82] In a one-pot reaction, the incorporation of anticancer drug and core cross-linking was simultaneously carried out by using the highly effective reaction of isocyanate groups in the core of the polymeric micelles poly(oligo(ethylene glycol) methyl ether methacrylate)-block-poly(styrene-co-3-isopropenyl-R,R-dimethylbenzyl isocyanate) [POEGMA-block-P(STY-co-TMI)] with amine groups in the prepared platinum(IV) drug.

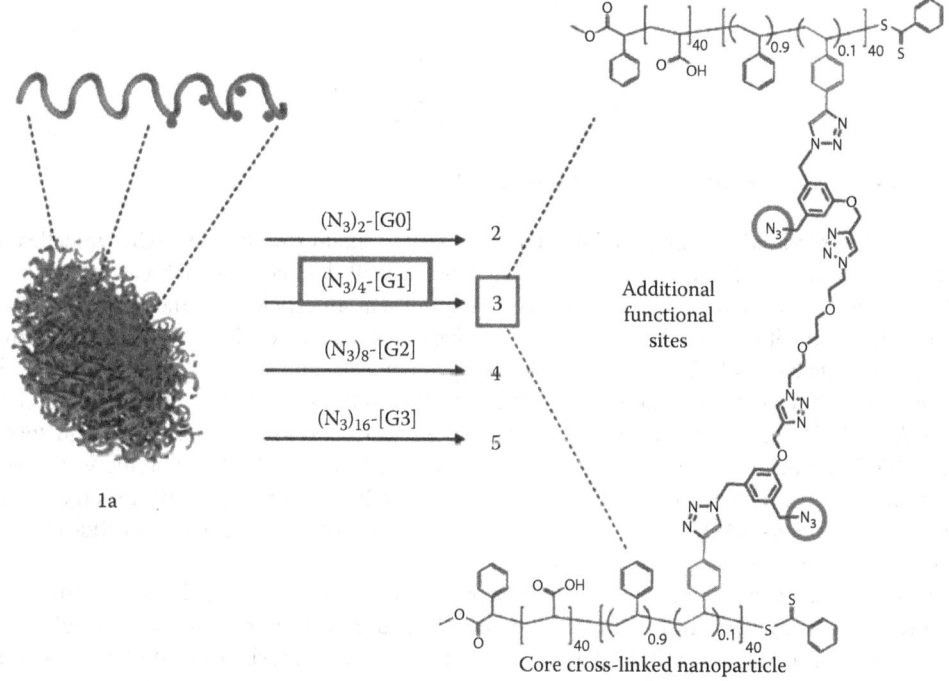

FIGURE 54.11 Synthesis of polymer nanoparticles from alkynyl-functionalized micelles and azido-terminated dendrimers. (From *New J. Chem.*, 31(5), O'Reilly, R.K., Joralemon M.J., Hawker C.J., and Wooley K.L., Preparation of orthogonally-functionalized core click cross-linked nanoparticles, 718–724. Copyright 2007. Reproduced by permission of The Royal Society of Chemistry.)

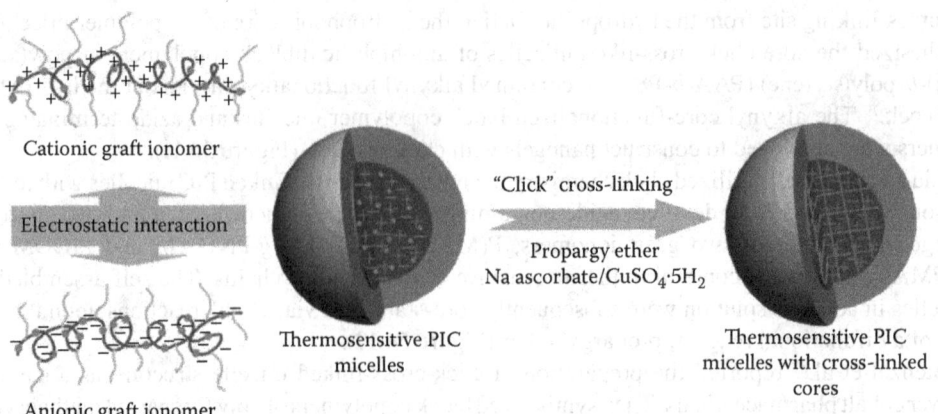

FIGURE 54.12 Schematic illustration of formation of thermosensitive polyion complex (PIC) micelles and their core cross-linking via click chemistry. (Reprinted with permission from *Macromolecules*, 41(4), Zhang, J., Zhou Y., Zhu, Z.H., Ge, Z.H., and Liu, S.H., Polyion complex micelles possessing thermoresponsive coronas and their covalent core stabilization via "click" chemistry, 1444. Copyright 2008 American Chemical Society.)

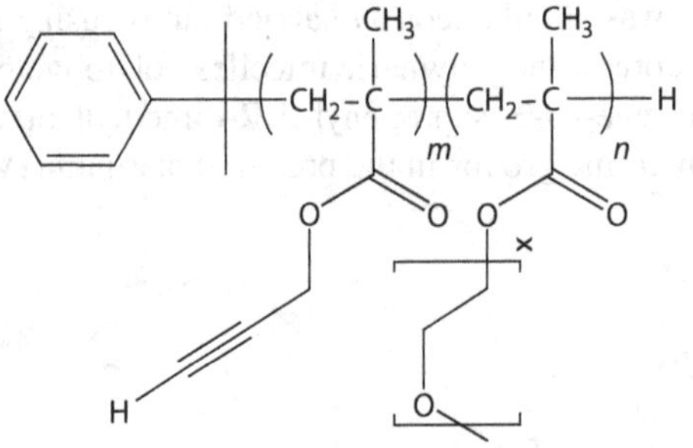

FIGURE 54.13 The structure of P(PAMA)-*b*-P(PEGMA).

Liu et al.[83] reported the preparation of two types of shell-cross-linked (SCL) micelles with inverted structures via click chemistry starting from a well-defined schizophrenic water-soluble triblock copolymer in purely aqueous solution. They present an efficient synthesis of two types of SCL micelles with either pH- or temperature-sensitive cores from a novel poly(2-(2-methoxyethoxy) ethyl methacrylate)-*b*-poly(2-(dimethylamino) ethyl methacrylate)-*b*-poly(2-(diethylamino) ethyl methacrylate) [PMEO$_2$MA-*b*-P(DMA-*co*-QDMA)-*b*-PDEA] triblock copolymer. First, PMEO$_2$MA-*b*-PDMA-*b*-PDEA was prepared and the DMA blocks were partially converted to a quaternized DMA (QDMA) block with click-cross-linkable moieties to form novel schizophrenic water-soluble triblock copolymers. The pH- or temperature-induced micellization and subsequent shell cross-linking of the P(DMA-*co*-QDMA) inner shell with the tetra-(ethylene glycol) diazide via click chemistry resulted in nanogel networks (Figure 54.14).

Recently, Liu's group also reported the fabrication of thermoresponsive cross-linked hollow poly(*N*-isopropylacrylamide) (PNIPAM) nanocapsules with controlled shell thickness via the combination of surface-initiated atom transfer radical polymerization (ATRP) and "click" cross-linking.[84] Cross-linked PNIPAM nanocapsules were fabricated by the "click" cross-linking of PNIPAM shell layer with a tri-functional molecule, 1,1,1-*tris*(4-(2-propynyloxy) phenyl)ethane. Due to the thermo-responsiveness of PNIPAM, cross-linked PNIPAM nanocapsules exhibit thermo-induced collapse/swelling transitions that make it possible to classify them as nanogels.

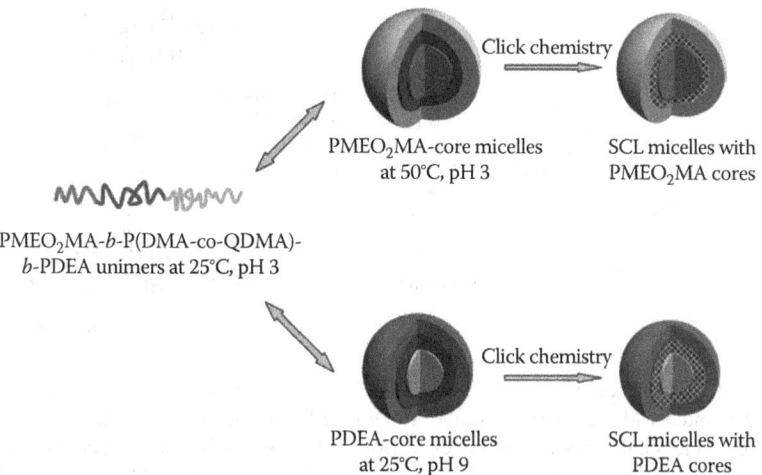

FIGURE 54.14 Schematic illustration of the schizophrenic micellization behavior of PMEO$_2$MA45-*b*-P(DMA0.65-*co*-QDMA0.35) 47-*b*-PDEA36 and the preparation of two types of SCL micelles with inverted structures in aqueous solution. (Reprinted with permission from *Langmuir*, 25(4), Jiang, X., Zhang, G., Narain, R., and Liu, S.H., Fabrication of two types of shell-cross-linked micelles with "Inverted" structures in aqueous solution from schizophrenic water-soluble ABC triblock copolymer via click chemistry, 2046. Copyright 2009 American Chemical Society.)

54.2.1.4 Imine Bonds-Induced Cross-Linking

Dynamic covalent bonds have been used to endow polymeric systems with the abilities to adapt their structures or compositions in response to external stimuli.[85] In this regards, Fulton et al. showed that dynamic covalent imine bonds can be used to cross-link linear polymer chains into core cross-linked star (CCS) polymer and nanogel nanoparticles.[86] Imine condensation reactions are a particularly appealing dynamic covalent reaction as it is possible to tune the stability of the imine bond by altering stereo electronic characteristics of the reaction partners, in particular the carbonyl-derived part.[87] They utilized RAFT polymerization to prepare novel aldehyde and amine functional styrenic- and methyl methacrylate-based copolymers and have demonstrated that these copolymers can cross-link through imine bond formation to prepare core cross-linked star polymers and spherical cross-linked nanogels. As a continuing effort, Fulton and coworkers reported the preparation of nanogels whose disassembly into their component polymer chains is triggered by the simultaneous application of two different stimuli by combination of imine and disulfide bonds.[88] They used acrylamide-based linear copolymers displaying pyridyl disulfide appendages and either aldehyde or amine functional groups for nanogel preparation (Figure 54.15).

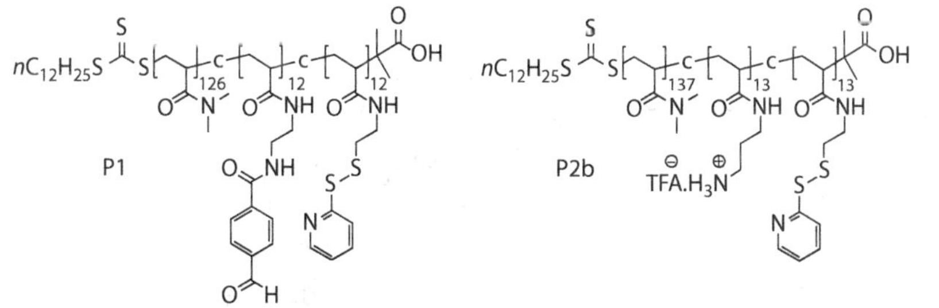

FIGURE 54.15 Polymer chains P1 and P2b are used for nanogel preparation through imine bond. (Reprinted with permission from *Macromolecules*, 45(6), Jackson, A.W. and Fulton, D.A., Triggering polymeric nanoparticle disassembly through the simultaneous application of two different stimuli, 2699. Copyright 2012 American Chemical Society.)

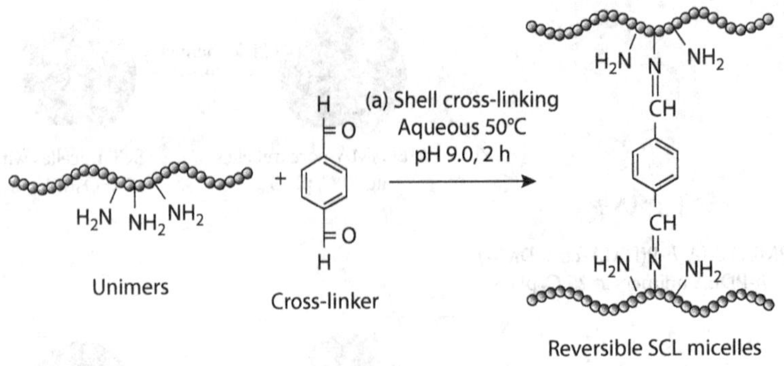

FIGURE 54.16 Cross-linking of mPEO113-PAPMA12-PNIPAM136 triblock copolymer micelles in aqueous solution. (Reprinted with permission from *Macromolecules*, 44(6), Xu, X., Flores, J.D., and McCormick, C.L., Reversible imine shell cross-linked micelles from aqueous RAFT synthesized thermoresponsive triblock copolymers as potential nanocarriers for "pH-triggered" drug release, 1327. Copyright 2011 American Chemical Society.)

Additionally, McCormick et al. utilized imine bonds for the preparation of reversible imine SCL micelles.[89] They synthesized a temperature-responsive triblock copolymer, α-methoxypoly(ethylene oxide)-*b*-poly(*N*-(3-aminopropyl) methacrylamide)-*b*-poly(*N*-isopropylacrylamide) (mPEO-PAPMA-PNIPAM), via aqueous RAFT polymerization. By increasing the solution temperature above the LCST of the PNIPAM block, the polymers self-assembled into micelles. Subsequently, the PAPMA shell was cross-linked with terephthaldicarboxaldehyde to generate SCL micelles with cleavable imine linkages (Figure 54.16).

54.2.1.5 Photo-Induced Cross-Linking

All of the cross-linking methods mentioned earlier need a cross-linking agent and/or catalyst, and the cross-linked micelles need to be purified to remove unreacted cross-linking agents and byproducts. In general, photo-cross-linking is a clean method, compared to the chemical methods, because no cross-linking agents are needed, and no byproducts are formed. As an alternative to the preceding cross-linking techniques, photo-induced cross-linking has been utilized to stabilize polymer assemblies that are functionalized with polymerizable or dimerizable units (Figure 54.17).[90]

For instance, He et al.[91] reported the preparation of photoresponsive nanogels based on photocontrollable cross-links. In this regards, they used double hydrophilic block copolymers containing coumarin, which is known to dimerize when treated with UV light >310 nm. The design of such photoresponsive nanogels is schematically illustrated in Figure 54.18. They combined the use of block copolymer self-assembly and a reversible photo-cross-linking reaction. Basically, with a water-soluble diblock copolymer, one block of which displays a LCST and bears photochromic

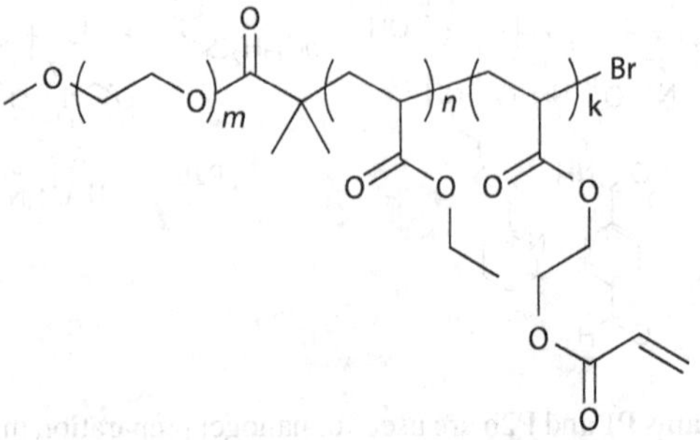

FIGURE 54.17 Polymer precursor with photopolymerizable functionality.

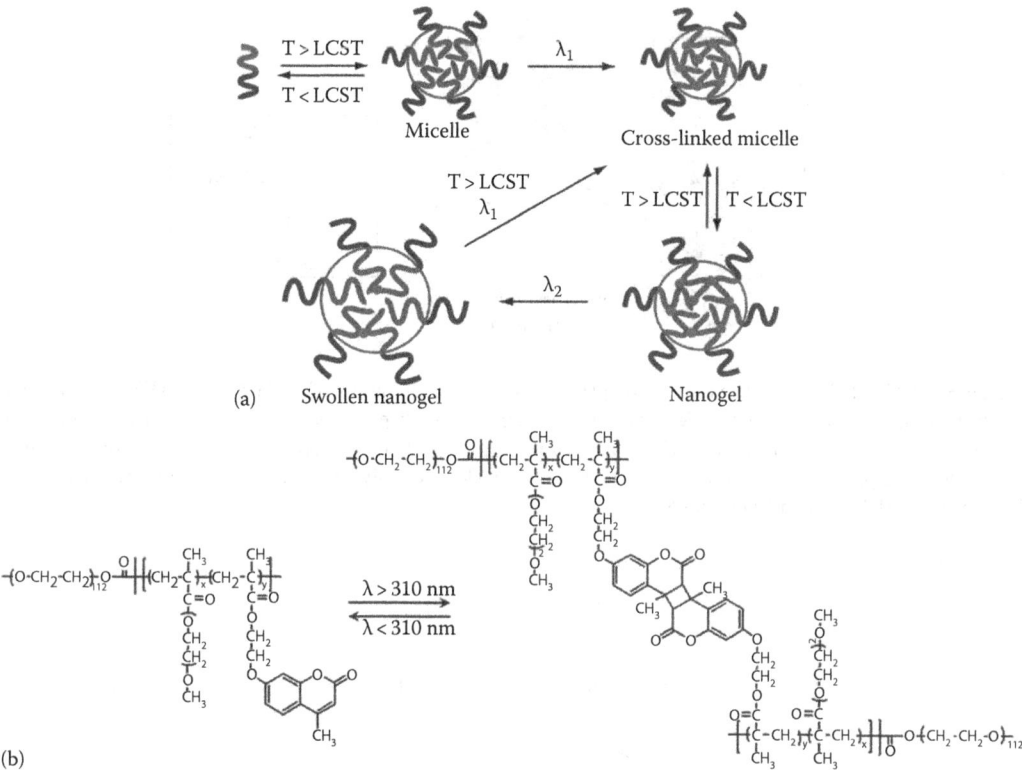

FIGURE 54.18 (a) Schematic illustration of the preparation and photocontrolled volume change of nanogel and (b) designed diblock block copolymer bearing coumarin side groups for the reversible photo-cross-linking reaction. (Reprinted with permission from *Macromolecules*, 42(13), He, J., Tong, X., and Zhao, Y., Photoresponsive nanogels based on photocontrollable cross-links, 4845. Copyright 2009 American Chemical Society.)

side groups, micelles can be obtained by heating the solution to T > LCST and cross-linked by the photoreaction of the chromophore upon illumination at λ_1. By cooling the solution to T < LCST, the preparation of nanogel is completed with cross-linked water-soluble nanoparticles. Because of the reversibility of the photoreaction, the cross-linking degree of the nanogel can be reduced in a controlled fashion by illumination at λ_2, which leads to the swelling of nanogel particles.

In another study, the Zhao group as a continuing effort reported the synthesis of both core and shell-cross-linked nanogels with the incorporation of photo cross-linkable moieties to either the core or shell of nanogels.[92]

Additionally, Yusa et al.[93] prepared a stimuli-responsive nanogel in water by a CCL technique using AB diblock copolymer micelles by photo-cross-linking of the micelle core (Figure 54.19). They utilized poly (ethylene glycol)-b-poly (2-(diethylamino) ethyl methacrylate-co-2-cinnamoyloxyethyl acrylate) [PEG-b-P(DEAEMA/CEA)], a pH-responsive block copolymer, for nanogel synthesis. While the solubility of DEAEMA depends on the pH, increasing the pH results in the formation of micelles that subsequently cross-link via UV irradiation made nanogels.

The photo-based cross-linking was also used for preparation of thermosensitive nanogels prepared from photocross-linkable copolymers of *N*-isopropylacrylamide (NIPAAm) and 2-dimethyl-maleinimido ethylacrylamide (DMIAAm) (Figure 54.20).[94]

As mentioned earlier, the principle of this method is based on the phase transition phenomena of temperature-sensitive polymers in combination with photochemistry. At elevated temperatures, the phase-separated structure of aqueous PNIPAAm copolymer solutions was fixed by UV irradiation.

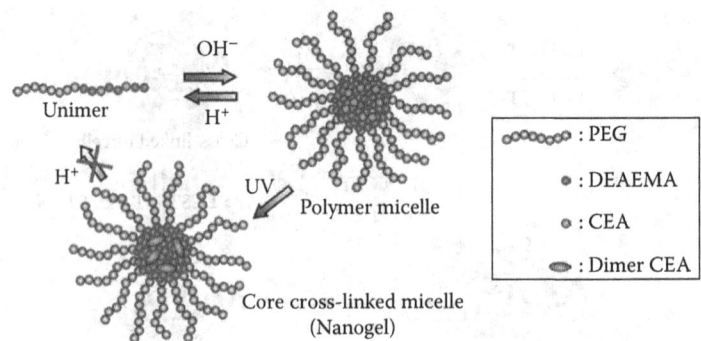

FIGURE 54.19 Schematic illustration of pH-dependent micellization of PEG-*b*-P(DEAEMA/CEA) and the formation of a nanogel with UV-irradiation. (Reprinted with permission from *Langmuir*, 25(9), Yusa, S.I., Sugahara, M., Endo, T., and Morishima, Y., Preparation and characterization of a pH-responsive nanogel based on a photo-crosslinked micelle formed from block copolymers with controlled structure, 5258. Copyright 2009 American Chemical Society.)

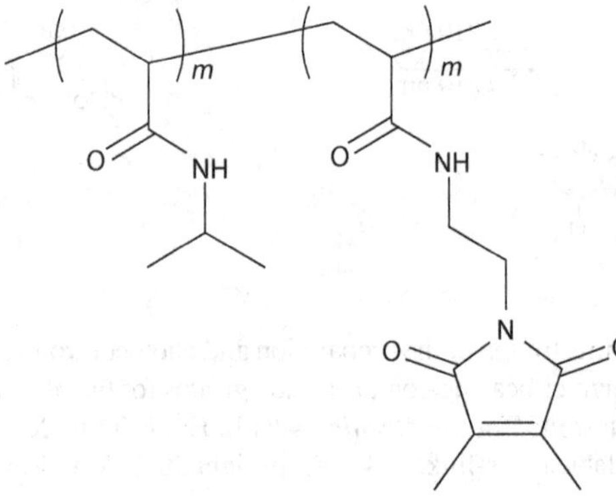

FIGURE 54.20 The structure of NIPAAm-DMIAAm copolymer.

Sugihara et al.[95] introduced an approach for the synthesis of thermoresponsive SCL micelles via living cationic polymerization and UV irradiation. The block copolymers are based on poly(2-ethoxyethyl vinyl ether)-*block*-poly(2-hydroxyethyl vinyl ether) (PEOVE-*b*-PHOVE) with methacryloyl groups derived from VEM in the PHOVE segment (shown in Figure 54.21). They showed that the resulting SCL micellar core was reversibly hydrated or dehydrated, depending on the solution temperature, because an aqueous PEOVE solution undergoes LCST-type phase separation around 20°C.

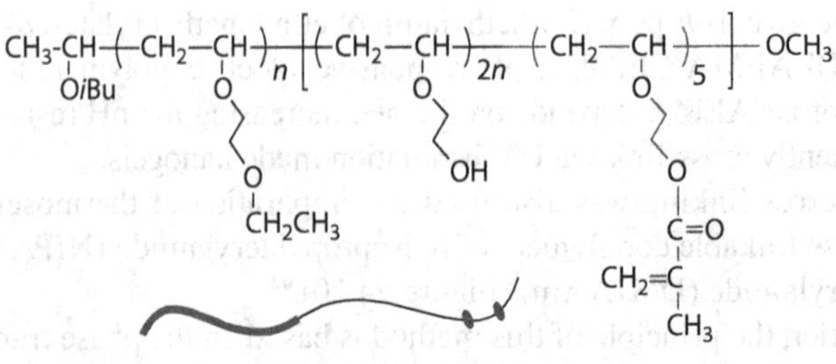

FIGURE 54.21 The structure of PEOVE-*b*-P(HOVE/VEM) synthesized by living cationic polymerization.

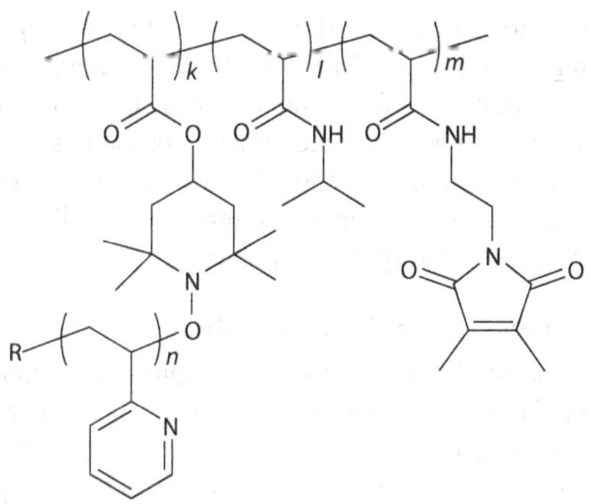

FIGURE 54.22 The structure of PNIPAAm graft terpolymer used for nanogel preparation by photo-induced cross-linking.

By using of this technique, Kuckling et al.[96] reported the preparation of temperature-responsive colloidal nanogels with a more complex structure. They prepared temperature-responsive colloidal nanogels with a pH-responsive shell by photo-cross-linking of poly(N-isopropylacrylamide) (PNIPAAm) graft terpolymers (Figure 54.22). They also demonstrated that it is possible to obtain a response to one stimulus without interfering with the other stimulus.

54.2.1.6 Physical Cross-Linking

The physical self-assembly of polymers was used by several research groups to produce various nanogels. This method usually involves a controlled aggregation of hydrophilic polymers capable of hydrophobic or electrostatic interactions and/or hydrogen bonding with each other. The preparation of nanogels is conducted in mild conditions and in aqueous media. A review regarding nanogel preparation from these associating polymers has recently been published.[19]

54.2.2 Preparation of Nanogels via Monomer Polymerization

While cross-linking preformed polymers, physically or chemically, represent an important strategy for the preparation of nanogels, especially for naturally occurring water-soluble polymers, the preparation of nanogels via direct monomer polymerization combines the two processes of polymerization and the formation of nanogels in a one-pot protocol. Thus, in a sense, synthesizing nanogels via monomer polymerization highlights an accelerated strategy, which in some cases involves polymerization-induced self-assembly of the in situ formed polymers. In some other cases, the formation of nanogels via polymerization requires the application of templating methods to realize the desired size and colloidal stability of the nanogels. Given the current intensive pursuit for highly efficient chemical approaches to the preparation of polymers and polymeric nanostructures, such an accelerated strategy of preparing nanogels plays an increasingly important role in modern polymer chemistry and materials.[97]

The predominantly used polymerization technique to fabricate nanogels is free-radical polymerization, because of its ease of manipulation, efficiency, tolerance of functionality, and adaptability to water-based heterogeneous systems. Recently, other polymerization techniques such as ring opening polymerization have been used for the synthesis of polyether-based nanogels.[23] By considering the recent development of polymerization techniques, it is not unreasonable to expect that the scope of polymerization techniques that can be used to fabricate nanogels will be significantly expanded.

Among the mentioned techniques, nanogel synthesis in water via direct polymerization of monomer is obviously advantageous because such a process uses water as the dispersant. However, this process requires the polymer to have a transition in solubility, a property that not every hydrophilic polymer has. A more general process for the synthesis of nanogels of hydrophilic polymers is inverse phase polymerizations, in which the continuous phase is oil and the dispersed phase is water containing hydrophilic monomers. The dispersed aqueous phase is stabilized by surfactant, which also serves as templates for the formation of nanogels.[98] But for some monomers, nanogel synthesis by direct polymerization in an aqueous dispersion system is impossible due to the good water solubility of both the monomer and the corresponding polymer.

54.2.2.1 Heterogeneous Free Radical Polymerization

Various heterogeneous polymerization reactions of hydrophilic or water-soluble monomers in the presence of either difunctional or multifunctional cross-linkers have been mostly utilized to prepare well-defined synthetic nanogels. They include precipitation, inverse (mini)emulsion, and inverse microemulsion polymerization utilizing an uncontrolled free radical polymerization process.

54.2.2.2 Precipitation Polymerization

Precipitation polymerization involves the formation of a homogeneous mixture at its initial stage and the occurrence of initiation and polymerization in the homogeneous solution. As the formed polymers are not swellable but soluble in the medium, the use of a cross-linker is necessary to cross-link polymer chains for the isolation of particles. As a consequence, the resulting cross-linked particles often have an irregular shape with high polydispersity (PDI).[99–102]

The preparation of microgels and nanogels based on PNIPAM and its derivatives by precipitation polymerization in water has been extensively explored for biomedical applications.[103–109] This is due to the thermosensitive properties of PNIPAM-nanogels that undergo volume change at LCST in water, at around 32°C. Above the LCST, PNIPAM hydrogels and nanogels are hydrophobic and expel water; below LCST, they are hydrophilic and swollen in water. These unique properties facilitate the loading of drugs into nanogels as well as enhance the controllable release of drugs encapsulated in nanogels.[110] However, the use of PNIPAM-based nanogels shows a certain limitation to biomedical applications due to a relatively narrow range of physical and chemical properties of PNIPAM.

54.2.2.3 Inverse (Mini) Emulsion Polymerization

Inverse miniemulsion polymerization is a water-in-oil heterogeneous polymerization process that forms kinetically stable macro-emulsions at, below, or around the critical micellar concentration (CMC). This process contains aqueous droplets (including water-soluble monomers) stably dispersed, with the aid of oil-soluble surfactants, in a continuous organic medium. Stable inverse miniemulsions are formed under high shear by either a homogenizer or a high-speed mechanical stirrer. Oil-soluble nonionic surfactants with hydrophilic–lipophilic balance value around four are used to implement colloidal stability of the resulting inverse emulsion. Upon addition of radical initiators, polymerization occurs within the aqueous droplets producing colloidal particles (Figure 54.23).[111]

Several reports have demonstrated the synthesis of hydrophilic or water-soluble particles of PHEMA,[112] PAA,[113] and PAAm,[112] temperature-sensitive hollow microspheres of PNIPAM,[114] core–shell nanocapsules with hydrophobic shell and hydrophilic interior,[115] and polyaniline nanoparticles.[116] This method has been explored to prepare cross-linked nanogels in the presence of difunctional cross-linkers for effective drug delivery. This is due to a facile confinement of water-soluble drugs in aqueous droplets dispersed in continuous organic solvents. Temperature- and pH-sensitive P(NIPAM-co-AA) minigels[117] and PAAm/PAA interpenetrating polymer network nanogels[118] were prepared in the presence of N,N'-methylenebisacrylamide. Their swelling properties were studied in water by measuring particle diameter upon change in temperature and pH. Stable, cross-linked, amphiphilic nanoparticles based on acrylated triblock copolymer of poly(ethylene glycol)-b-poly(propylene glycol)-b-poly(ethylene glycol) (PEO-b-PPO-b-PEO) were prepared by inverse emulsion photopolymerization

Nanogels

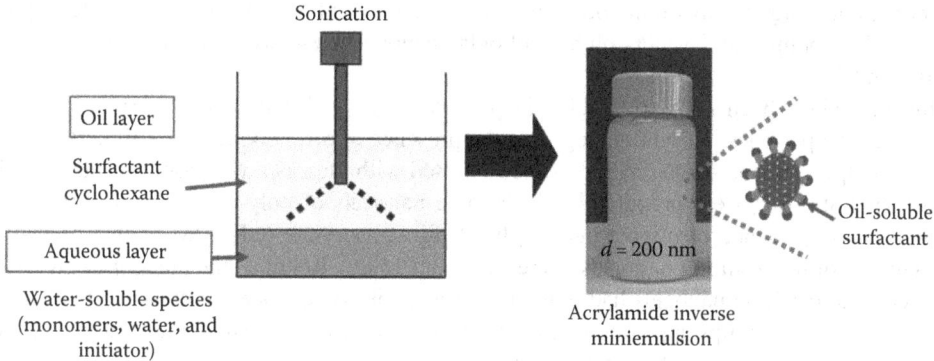

FIGURE 54.23 A schematic representation of inverse miniemulsion or microemulsion polymerization for the preparation of nanometer-sized particles of water-soluble and water-swellable polymers as well as cross-linked particles in the presence of cross-linkers. (Reprinted from *Polymer*, 50(19), Oh, J.K., Bencherif, S.A., and Matyjaszewski, K., Atom transfer radical polymerization in inverse miniemulsion: A versatile route toward preparation and functionalization of microgels/nanogels for targeted drug delivery applications, 4407–4423. Copyright 2009, with permission from Elsevier.)

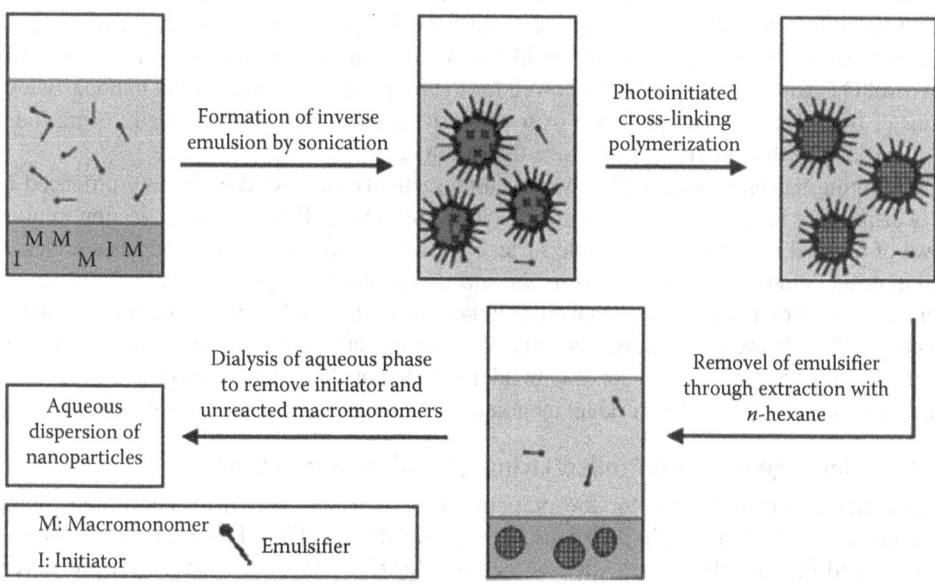

FIGURE 54.24 Illustration of preparation for microgels of PEO-*b*-PPO-*b*-PEO via inverse emulsion polymerization. (Reprinted with permission from *Langmuir*, 21(6), Missirlis, D., Tirelli, N., and Hubbell, J.A., Amphiphilic hydrogel nanoparticles. Preparation, characterization, and preliminary assessment as new colloidal drug carriers, 2605. Copyright 2005 American Chemical Society.)

(Figure 54.24). The hydrophobic PPO-rich domains enhanced the incorporation of doxorubicin (Dox), an amphiphilic anticancer drug, up to 9.8 wt%. The resulting microgels had a diameter of 50 and 500 nm.[119]

54.2.2.4 Inverse Microemulsion Polymerization

While inverse (mini)emulsion polymerization forms kinetically stable macro-emulsions at, below, or around the CMC, inverse microemulsion polymerization produces thermodynamically stable microemulsions upon further addition of emulsifier above the critical threshold. This process also involves aqueous droplets, stably dispersed with the aid of a large amount of oil-soluble surfactants in a continuous organic medium; polymerization occurs within the aqueous droplets, producing stable hydrophilic and water-soluble colloidal nanoparticles having a diameter of less than 50–100 nm.[120,121]

This method has been also explored for the preparation of well-defined nanoparticles,[122–127] magnetic polymeric particles,[27,128] and nanogels in the presence of difunctional cross-linkers.[129–134]

Poly(vinylpyrrolidone)-based nanogels incorporated with Dex as a water-soluble macromolecular carbohydrate drug were prepared.[129,130] Cationic nanogels of poly (HEA-co-AETMAC) were prepared in the presence of oligo(ethylene glycol) dimethacrylate (OEGDMA) as a cross-linker. The diameter of the resulting nanogels decreased from 150 to 40 nm as the amount of cross-linker increased. The resulting nanogels had potential for gene delivery, since the presence of quaternary ammonium ion side groups appeared to enhance the incorporation of DNA into nanogels via electrostatic association with the phosphate groups.[131]

Poly(amino acid)-based nanoparticles with different surface PEGylation were prepared. a,b-Poly(N-2-hydroxyethyl)-D,L-aspartamide (PHEAS) and PEG-modified PHEA (PHEAS-PEG) were functionalized with a methacrylate group and then polymerized by UV irradiation in inverse microemulsion. The resulting nanoparticles had a size of around 250 nm in diameter by TEM. The fluorescein-loaded PHEA-based nanoparticles were prepared in the presence of fluorescein sodium salts, and examined for cellular uptake using macrophage cells.[132]

Ultrafine hydrophilic PAAm-based nanogels incorporated with meta-tetra (hydroxyphenyl) chlorine (mTHPC) as a photosensitizer were prepared for photodynamic therapy. In the process, both AAm and N,N-methylene(bis acrylamide) as a cross-linker were directly emulsified into hexane/aerosol OT, without water, which allowed for the preparation of tiny PAAm nanoparticles with a diameter of 2–3 nm.[133] The presence of water in hexane/aerosol OT produced 20 nm diameter PAAm particles with a relatively broad size distribution.[134]

The resulting nanogels with both polymerization methods mentioned earlier are produced in the form of dispersions in organic solvents with oil-soluble surfactants. Biomedical applications require the removal of residual monomers, oil-soluble surfactants, and organic solvents. Nanogels are then redispersed in water before use. For comparison, the aqueous precipitation polymerization of water-soluble monomers in the presence of difunctional cross-linkers is also allowed for the preparation of microgels/nanogels.[13,44,135–138] However, in order to avoid microscopic gelation, the concentration of monomers in water (reaction media) should be kept low. In addition, this method also needs the purification of the resulting polymers by removal of residual monomers and initiators from reaction mixtures.

54.2.2.5 Heterogeneous Controlled/Living Radical Polymerization

CRP provides a versatile route for the preparation of (co) polymers with controlled molecular weight, narrow molecular weight distribution (i.e., Mw/Mn, or PDI < 1.5), designed architectures, and useful end-functionalities.[32–34] Various methods for CRP have been developed; however, the most successful techniques include ATRP,[35,139] stable free radical polymerization,[140] and reversible addition fragmentation chain transfer (RAFT) polymerization.[141,142] CRP techniques have been explored for the synthesis of gels[38,44] and cross-linked nanoparticles of well-controlled polymers in the presence of cross-linkers.

54.2.2.6 Atom-Transfer Radical Polymerization

ATRP is one of the most successful CRP techniques, enabling the preparation of a wide spectrum of polymers with predetermined molecular weight and relatively narrow molecular weight distribution (Mw/Mn < 1.5).[35,139] ATRP also allows for the preparation of copolymers with different chain architectures, such as block, random, gradient, comb-shaped, brush, and multi-armed star copolymers.[143–145] In addition, ATRP has been utilized to prepare polymer–protein/peptide bioconjugates,[146–148] polymer modified polysaccharides,[149,150] micellar nanoparticulates,[62,151] hydrogels and nanogel,[152] ligands stabilizing metal nanocrystals for cellular imaging,[153,154] surface-initiated brushes,[155,156] and emulsion particles.[157,158]

The mechanism of ATRP is based on a rapid dynamic equilibration between a minute amount of growing radicals and a large majority of dormant species.[159] In a normal ATRP process (Figure 54.25), transition metal complexes in a lower oxidation state (Cu(I)/Lm) are added directly to the reaction as an activator, which reacts reversibly with the dormant species (RX) to generate a deactivator [X–Cu(II)/Lm] and an active radical (R). The radical can propagate with the addition of monomers (M) and is rapidly deactivated by reacting with X–Cu(II)/Lm, regenerating Cu(I)/Lm and a halogen-terminated polymeric chain.

Recently, a versatile method for preparing and functionalizing well-defined biodegradable nanogels for targeted drug delivery applications has been developed.[13,154,160,161] The method utilizes ATRP, inverse miniemulsion polymerization,[111] and disulfide–thiol exchange.[162,163] The novel approach allows the preparation of biomaterials with many useful predeterminable site-specific features. As illustrated in Figure 54.26, the features include uniform network, high loading efficiency, novel distributed functionality including bromine end groups, and degradation by hydrolysis or through a disulfide–thiol exchange.[1]

Hydrogel NPs of PNIPAAm were prepared by precipitation polymerization via ATRP in water.[164] OEOMA, an analog of PEG has been polymerized by AGET ATRP in homogenous aqueous solution[165] and in heterogeneous conditions.[166] In this context, biodegradable cross-linked nanogels of well-controlled hydrophilic polymers were synthesized using ATRP in inverse miniemulsion in

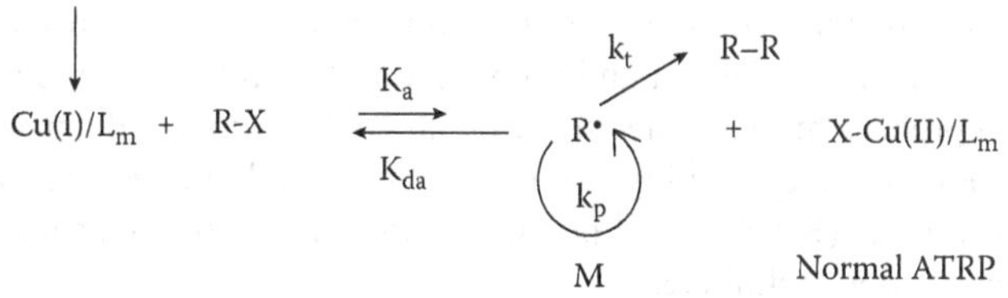

$$Cu(I)/L_m + R\text{-}X \underset{K_{da}}{\overset{K_a}{\rightleftharpoons}} R^\bullet \xrightarrow{k_t} R\text{-}R + X\text{-}Cu(II)/L_m$$

M, k_p

Normal ATRP

FIGURE 54.25 Mechanism of normal ATRP.

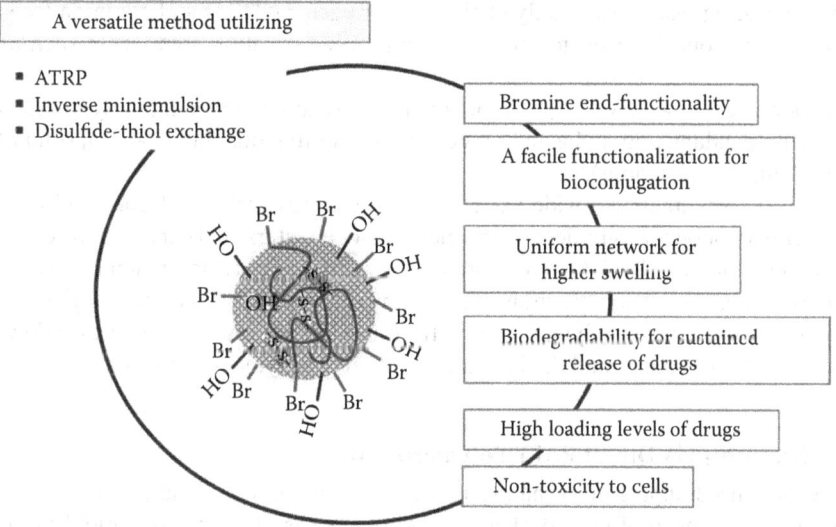

FIGURE 54.26 Illustration of unique features of biodegradable nanogels prepared by atom transfer radical polymerization in inverse miniemulsion of oligo(ethylene glycol) monomethyl methacrylate in the presence of a disulfide-functionalized dimethacrylate. (Reprinted from *Polymer*, 50(19), Oh, J.K., Bencherif, S.A., and Matyjaszewski, K., Atom transfer radical polymerization in inverse miniemulsion: A versatile route toward preparation and functionalization of microgels/nanogels for targeted drug delivery applications, 4407–4423. Copyright 2009, with permission from Elsevier.)

the presence of a disulfide-functionalized dimethacrylate (DMA) cross-linker.[166] The nanogels preserved a high degree of halide end-functionality that enabled further functionalization, including chain extension to form functional block copolymers. The nanogels were nontoxic to cells and degraded in a reducing environment to individual polymeric chains with a relatively narrow molecular weight distribution (Mw/Mn < 1.5), indicating the formation of a uniformly cross-linked network within the NPs. This uniform structure is expected to improve the controlled release of encapsulated species. The measured swelling ratio, degradation behavior, and colloidal stability of nanogels prepared by ATRP were superior to those prepared by conventional free radical inverse miniemulsion polymerization. In another report, these nanogels were loaded with Dox, an anticancer drug.[154] The nanogels released Dox in vitro upon exposure to glutathione, which degraded the nanogels.

Since ATRP results in polymers with a high degree of halide end-functionality, facile functionalization with various molecules is possible. Also, the preparing of functional nanogels using ATRP in inverse miniemulsion using functional ATRP initiators and copolymerization with functional monomers offer two approaches toward bioconjugation.[167]

54.2.2.7 Reversible Addition–Fragmentation Chain Transfer

The development of emerging technologies has imposed ever-increasing demand on the synthesis of well-defined polymer architectures with predictable properties and functions. Fortunately living/controlled radical polymerizations are becoming powerful tools to meet these challenges and have been widely used in the design and preparation of architecturally defined and functional macromolecules with predetermined molecular weight, composition, and narrow molecular weight distribution.[166–170] Among the major living/controlled radical polymerizations, reversible addition–fragmentation chain transfer (RAFT)[171–174] polymerization has been advocated by many to be the most versatile polymerization technique.

The major strengths of the RAFT approach include the following:

1. An ability to control the polymerization of a wide range of monomers in varying solvents, including water, using only chain transfer agents (CTAs) and common free radical initiators (without the need for any additional polymerization component such as metal catalysts).[171,175,176]
2. The tolerance to a wide variety of functional groups, allowing the facile synthesis of polymers with pendant, and alpha and omega end-group functionalities (an important feature for biological applications).[177–183]
3. The ability to synthesize a wide variety of architectures such as telechelic, block copolymers, graft copolymers, gradient copolymers, nanogels, stars, and dendritic structures.[184–187]
4. The compatibility of RAFT with a variety of established polymerization methods such as bulk, solution, suspension,[188] emulsion,[189,190] and dispersion[191,192] polymerizations.
5. The ability to perform polymerizations from a wide variety of substrates, allowing the modification of surfaces and the in situ generation of polymer conjugates.[177,193–196]

54.2.2.8 Nanogels via Direct RAFT Polymerization

The most widely used approach for nanogel preparation is direct radical polymerization of vinyl monomers in the presence of cross-linkers, because it is easy to carry out, rapid in polymerization rate, amenable to various reaction conditions, compatible with a variety of functionalities, and flexible in initiation methods. As an avenue to nanogels, the RAFT polymerization largely inherits the majority of the advantages of nanogel synthesis by traditional free-radical polymerization. More importantly, it offers additional merits that are unmatchable by the traditional free-radical approach. Firstly, the RAFT approach offers excellent control over molecular weight and molecular weight distribution of the constitutional polymers of the nanogels, which is an important structural parameter that determines the mechanical and responsive properties of the nanogels. Secondly, the RAFT approach allows uniform incorporation of cross-linkers into the nanogels, a feature that is in stark contrast to the nanogels synthesized by the free-radical process. Thirdly, the RAFT approach allows precise localization of functional groups into the nanogels and the direct formation of well-defined nanogel architectures, providing a versatile technique in the design of multi-functional and exquisite nanogel structures for value-added applications.[97]

54.2.2.9 Nanogel Synthesis by RAFT Polymerization in Water

An interesting approach that combines polymerization and self-assembly of the in situ produced polymers is nanogel synthesis by RAFT polymerization in water in the form of precipitation/dispersion polymerization. More importantly, it has three distinct features that distinguish it from other approaches: (1) In aqueous dispersion/precipitation polymerization systems, water is used as the sole dispersant without the addition of any organic solvents. This is a particularly important feature considering the current intensive pursuit for green polymerization processes; (2) The polymerization can be carried out at high solid content, much higher than many other nanogel synthetic approaches; (3) No surfactant is added to provide colloidal stability of the nanogels. Such a surfactant-free process is not only economic but also environmentally benign.

The first example of nanogel synthesis by direct RAFT polymerization under precipitation/dispersion polymerization condition was reported by An et al. in 2007 (Figure 54.27).[44] Two types of poly(N,N'-dimethylacrylamide)s (PDMAs) bearing a trithiocarbonate group were first synthesized by RAFT solution polymerization and were subsequently used as both stabilizers and RAFT agents for nanogel synthesis by RAFT precipitation/dispersion polymerization. These two types of

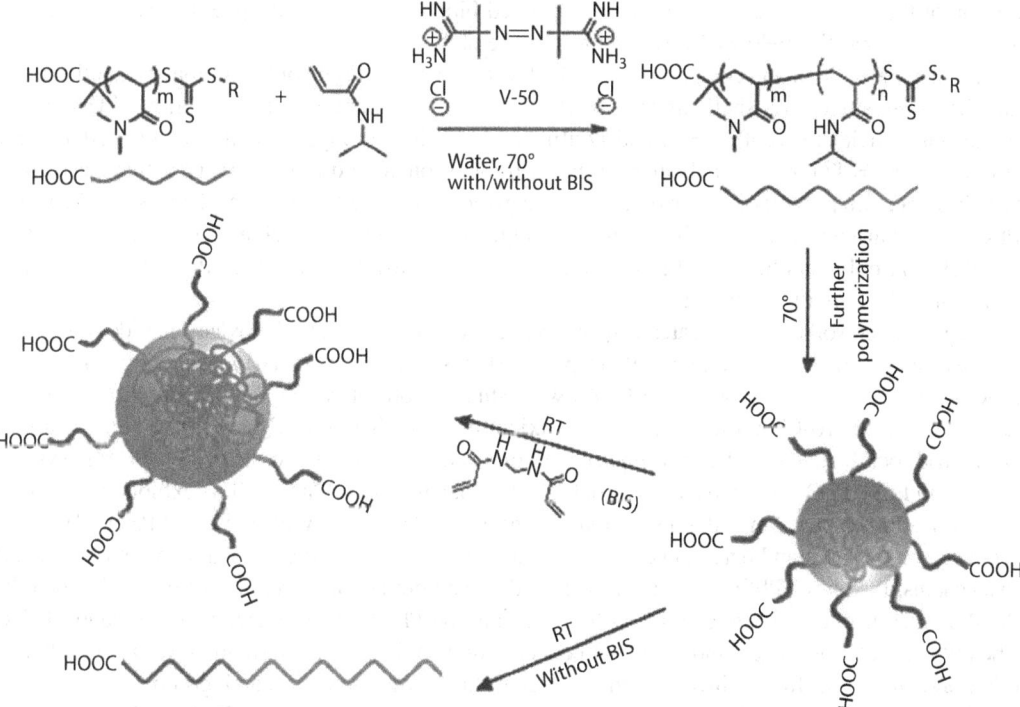

FIGURE 54.27 Macro-CTAs and schematic representation of the RAFT precipitation polymerization process. (Reprinted with permission from *J. Am. Chem. Soc.*, 129(46), An, Z., Shi, Q., Tang, W., Tsung, C.K., Hawker, C.J., and Stucky, G.D., Facile RAFT precipitation polymerization for the microwave-assisted synthesis of well-defined, double hydrophilic block copolymers and nanostructured hydrogels, 14493. Copyright 2007 American Chemical Society.)

PDMAs were either hydrophilic or amphiphilic due to the different R groups of the RAFT agents. When the cross-linker N,N'-methylenebisacrylamide was used, thermosensitive nanogels were produced. When no cross-linker was used, after cooling to room temperature, the nanoparticles dissociated into double hydrophilic block copolymers, which allowed for convenient characterization of the polymers. The use of both types of PDMAs permitted the production of well-defined block copolymers with low values of Mw/Mn, indicating a controlled nature of the RAFT process under precipitation/dispersion conditions.

In addition, the RAFT precipitation/dispersion polymerization process transferred the functional groups of the RAFT agents to those of the produced nanogels, allowing for precise localization of the functionalities into either the core or the shell. By using the carboxylic acid groups at the nanogel surface, bioconjugation of the nanogels with albumin was produced in this study.

Taking advantage of the combination of the RAFT process and click chemistry, heterofunctional nanogels was demonstrated by An et al.[197] By using an azide-functionalized CTA, telechelic PDMAs and PNIPAMs bearing the azide group at their α-end and the trithiocarbonate group at their ω-end were produced by RAFT synthesis. The ω-end trithiocarbonate group was subjected to a one-pot aminolysis and thiol-ene addition, and subsequently, the α-end azide group was reacted with functional alkyne by click chemistry. This approach to obtain heterofunctionalized nanogels indeed demonstrated that multifunctional nanogels with precise location can be produced by RAFT precipitation/dispersion polymerization and that further functionalization is possible to the functional groups installed by the tailored CTAs.

PNIPAM is perhaps the mostly studied thermosensitive polymers and as such the majority of nanogels studied are based on PNIPAM. Recently, there has been much interest in developing PNIPAM alternatives.[189–203] Polymers based on OEGMAs have come to the forefront as a new generation of thermosensitive polymers with improved biocompatibility, sharpness of responsiveness, antifouling properties below LCST, and bio-inertness.[204]

In this regard, Shen and coworkers recently developed a novel type of biocompatible, antifouling, and thermosensitive core–shell nanogels by RAFT aqueous dispersion polymerization.[205] Using both linear poly(ethylene glycol) (PEG) and nonlinear poly(ethylene glycol) methyl ether methacrylate (PEGMA, Mn = 475) polymers bearing either trithiocarbonate or dithioester CTAs, they performed RAFT dispersion polymerizations of di(ethylene glycol) methyl ether methacrylate (MEO_2MA) and PEGMA of varied ratios to obtain nanogels of adjustable thermosensitivities. The hydrophilic PEG or PPEGMA polymers became the shell, and the thermosensitive MEO_2MA-PEGMA copolymers constituted the core of the nanogels.

Rieger and coworkers investigated dispersion polymerization of N,N'-diethylacrylamide (DEAAm) for the formation of PEGylated thermally responsive block copolymers and nanogels using amphiphilic Macro-CTAs.[135] Two types of Macro-CTAs were studied both of which had a hydrophobic dodecyl tail. The first Macro-CTA was poly(ethylene oxide) monomethyl ether (Mn = 2000) functionalized with a trithiocarbonate (PEO-TTC). The second one is a block copolymer of PEO-b-PDMA synthesized from PEO-TTC with increased hydrophilic chain length. Both Macro-CTAs exhibited controlled polymerization of DEAAm under dispersion polymerization in water. While the use of PEO-TTC alone produced rather large and heterogeneous gel particles due to the insufficient stabilization of the short PEO chains, the use of PEO-b-PDMA-TTC afforded well-defined nanogels with PEO shell when the PDMA block had a sufficient chain length. Increasing the PDMA length decreased the nanogel size. The influence of cross-linker and monomer concentration on the nanogel formation was also studied in this work, which was further investigated in a recent study by the same research group.[206]

The thermosensitivity of poly($N,N0$-dimethylaminoethyl methacrylate) (PDMAEMA) was also used for nanogel synthesis. By using an amphiphilic Macro-CTA similar to the one used by Rieger and coworkers but with a much higher PEO molecular weight (Mn = 24,000), Yan and Tao reported the synthesis of PDMAEMA nanogels via dispersion polymerization in water.[207] Since PDMAEMA is also a pH-sensitive polymer, the prepared PEGylated cationic nanogels may find application in gene delivery.

Nanogels

The preparation of nanogels by direct RAFT polymerization is still a relatively young area though successful examples have been demonstrated and clear advantages have been established. Surprisingly, although RAFT dispersion polymerization has been well studied in organic media[208–214] and supercritical CO_2,[215–217] their study in water has been limited to a few monomers including those aforementioned and 2-hydroxypropyl methacrylate that was recently studied by Li and Armes.[218] Considering the fact that a variety of thermosensitive and other stimuli-responsive polymers have been extensively studied, significant expansion of nanogel synthesis by direct RAFT polymerization can be envisaged.

54.2.2.10 Nanogels Synthesis by Inverse RAFT Miniemulsion

Qi and coworkers reported the first study combining RAFT and inverse miniemulsion.[219] Their inverse miniemulsion system comprised cyclohexane as the continuous phase, B246SF as the surfactant, and aqueous solution containing a CTA, acrylamide, and costabilizer $MgSO_4$. They found that using a water-soluble initiator, 4,4′-azobis(4-cyanovaleric acid), afforded better control of the polymerization of acrylamide than using a lipophilic one, 2,2′-azobis(2-methylpropionitrile) (AIBN). RAFT control was realized up to 50% monomer conversion and after that significant deviation from RAFT control was observed. More recently, they have extended their RAFT inverse miniemulsion polymerization approach to other hydrophilic (co)polymers.[220–223]

Nanocapsules with hydrophilic polymer walls can be considered as hollow nanogels, which are useful for encapsulation of hydrophilic materials in the aqueous interior. Using RAFT-confined interfacial inverse miniemulsion polymerization, Lu and coworkers developed a novel approach to the synthesis of nanocapsules (Figure 54.28).[224] In their approach, an amphiphilic Macro-CTA was added to a conventional inverse miniemulsion such that both the surfactant and the amphiphilic Macro-CTA together stabilized the formed inverse miniemulsion. The location of the amphiphilic Macro-CTA at the oil–water interface was the initiating site for polymerization. After initiation of polymerization with the use of a water-soluble initiator, the formed oligomeric radicals would migrate to the interface for RAFT-controlled polymerization, and thus an interfacial layer of RAFT polymers was formed, which functioned as the wall of the nanocapsule. It was expected that the functionality and the thickness of the nanocapsules could be controlled through the RAFT polymerization process. In the control experiment where no Macro-CTA was added, polymerization led to the formation of solid particles instead of nanocapsules.

In a recent report, Wang and coworkers nicely extended Lu's approach by synthesizing SCL nanocapsules to improve the stability.[225] In this case, they used an amphiphilic PDMAEMA as the Macro-CTA to interfacially polymerize methylacrylic acid (MAA) in an inverse miniemulsion. The cross-linker they used was *bis*(acryloyloxyethyl) disulfide, which was dissolved in the continuous oil phase. The disulfide cross-linker could be degraded by reducing agents such as DTT as already demonstrated in numerous examples for the destruction of polymer nanoparticles.

54.3 APPLICATIONS OF NANOGELS

Nanogels show a lot of promise as delivery systems in particular due to their encapsulation stability, in addition to water solubility and biocompatibility. These nanocarriers have been utilized in a variety of fields including cancer drug delivery. The following examples from the literature demonstrate the diversity in applications and the versatility of these delivery systems. In 2010, Du et al. designed a pH-responsive charge-conversional nanogel for promoted tumoral-cell uptake and Dox delivery.[226] These nanogels were prepared from poly(2-aminoethyl methacrylate hydrochloride) (PAMA) and were subsequently treated with 2,3-dimethylmaleic anhydride (DMMA) to produce a negatively charged nanogel.

In another application, the delivery of therapeutic molecules to treat inflammatory disorders such as rheumatoid arthritis was investigated. Macrophage cells, of the immune system, are targeted for photodynamic therapy as a treatment for this disorder. Schmitt et al. developed chitosan-based

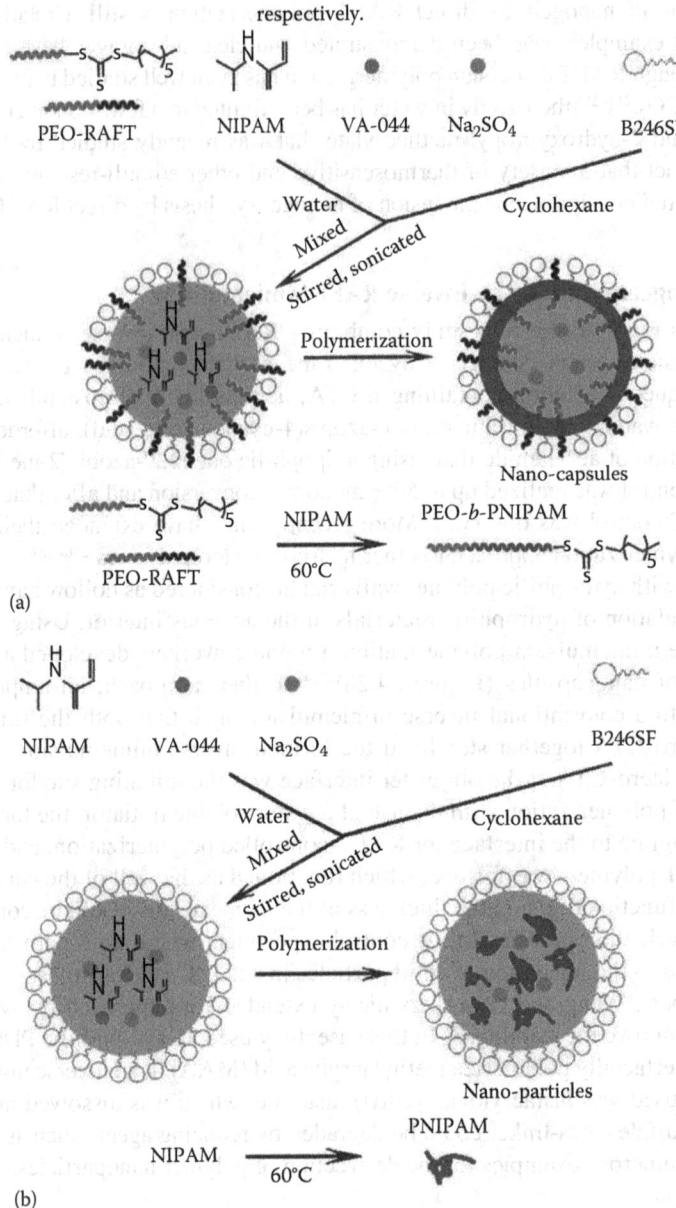

FIGURE 54.28 Suggested mechanisms of RAFT interfacial inverse miniemulsion polymerization and free-radical inverse miniemulsion polymerization with NIPAM as the monomer as illustrated in Part (a) and (b), respectively. (Reprinted with permission from *Macromolecules*, 43(1), Lu, F.J., Luo, Y.W., Li, B.G., Zhao, Q., and Schork, F.J., Synthesis of thermo-sensitive nanocapsules via inverse miniemulsion polymerization using a PEO–RAFT agent, 568. Copyright 2010 American Chemical Society.)

nanogels decorated with hyaluronate to target macrophages, which were then loaded with one of three different photosensitizers.[227] Their studies show that photodynamic therapy using these hyaluronate-chitosan nanogels is effective in the treatment of inflamed articular joints.

Nanogels have also been utilized for the delivery of local anesthetic drugs. Yin et al. designed a biodegradable delivery system where lidocaine was encapsulated into poly(ε-caprolactone)-poly(ethylene glycol)-poly(ε-caprolactone) (PCL-PEG-PCL or PCEC) nanoparticles.[228] Carriers were coated with hydrophilic thermosensitive Pluronic F-127 hydrogels to form the composite carrier nanogel system. The studies showed that nanogels effectively infiltrate into the wound for prolonging the effects of these anesthetics, which is useful for treatment during post-operative periods.

Physically cross-linked nanogels composed of hexadecyl groups bearing cationic cycloamylose have been developed for the delivery of pDNA.[229] Cycloamylose, a large cationic cyclodextrin ring, was employed for the complexation of pDNA. Phospholipase A2 was co-delivered with pDNA to help in endosome disruption by catalyzing the hydrolysis of membrane phospholipids triggering the release of the pDNA in the cytoplasm. It was found that the levels of protein expression due to the delivery of pDNA were enhanced with the co-delivery of PLA2 using the C16-catCA nanogels/PLA2 system. Hemolysis studies using sheep red blood cells showed that the activity for the native PLA2 was maintained when co-delivered with pDNA into the nanogels.

Tamura et al. studied the delivery of siRNA using a PIC based on PEGylated polyamine nanogels containing a chemically cross-linked core.[230]

In nucleic acid delivery complexes between amine-based carriers and nucleic acid, drugs are defined in terms of the ratio of amines on the carrier to the number of phosphates on the nucleic acid backbone. This ratio is referred to as the N/P ratio with a higher N/P ratio implying a greater number of amine-based complexing agents per nucleic acid. Gene silencing activity against firefly luciferase in HuH-7 cells showed that these nanogel/siRNA complexes possess high transfection efficiencies at low nitrogen to phosphate (N/P) ratios when compared with the uncrosslinked complexes. However, cellular uptake of the nanogels/siRNA complexes was found to be lower than that of oligofectamine/siRNA complexes. This was suggested to be due to steric hindrance between PEG chains and the cell membrane. The chemically cross-linked nanogel was composed of poly[2-(N,N-diethylaminoethyl) methacrylate] (PDEAMA), which served as a siRNA complexing functionality and an aid for endosomal escape, bifunctional ethylene glycol dimethacrylate as the cross-linker, and heterobifunctional PEG macromonomer. Due to their size, most DDSs are taken up into cells by endocytosis. However, once inside the endosomes, the drugs are either ineffective because their sites of action are at the cytosol or are eventually degraded in the lysosomes. Thus, in most cases in order to achieve efficacious drug delivery a vehicle should not only internalize into the cell but also be able to escape the endosome effectively. To achieve endosomal escape, proton sponge functionality was introduced into the nanogel so that during internalization and subsequent acidification of the endosome, these functionalities buffer the endosomal pH eventually causing disruption of the endosome. This is presumably due to the relative increase in osmotic pressure inside the endosome. Thus, it was reasoned that the polyamine core of this nanogel was protonated as a result of the decrease in pH in the endosomal compartment facilitating the escape of the nanocarriers into the cytoplasm through the proton sponge effect. Thus, PEGylated polyamine siRNA nanogels have potential applications in cancer therapy and for the treatment of genetic disorders.

54.4 RELEASE MECHANISMS FROM NANOGELS

Biological agents can be incorporated in nanogels by (1) physical entrapment, (2) covalent conjugation, or (3) controlled self-assembly and released from nanogels through (1) simple diffusion, (2) degradation of the nanogel, (3) a shift in the pH value, (4) displacement by counterions present in the environment, or (5) transitions induced by an external energy source.[5]

Examples include diffusion-controlled release of Dox from pluronic-based hydrogels.[231] There is also increased interest in developing nanogels that can release biological agents in response to environmental signals at the disease site. A change in pH value or the presence of a reducing environment can serve as chemical signals that trigger the release. For example, an acrylamide-based nanogel with acetal cross-links is stable at an extracellular pH value of 7.4, but degrades and releases entrapped protein at pH 5.0.[232] Similarly, a poly-[oligo(ethylene oxide)-methyl methacrylate] nanogel with disulfide cross-links degraded in the presence of a glutathione tripeptide commonly found in cells.[154] The degradation of these nanogels was shown to trigger the release of the encapsulated low-molecular-mass solutes rhodamine 6G and Dox. In another study, the dissolution of disulfide cross-linked HA nanogels and the release of siRNA was induced by adding glutathione.[233] Clearly, the release kinetics can be fine-tuned in each case by altering the number of cross-links.

Polyelectrolyte hydrogels that incorporate biological agents through electrostatic bonds can also release biological agents in response to environmental changes. For example, pH-sensitive nanogels based on PAA can release an oppositely charged protein in tumor sites or endosomal compartments upon acidification.[234–237] A different mechanism was proposed for the release of nucleotide drugs from cationic PEG-cl-PEI nanogels.[238] In this case, negatively charged biomacromolecules bound to nanogels can be displaced by negatively charged cellular components. For example, the interaction of cationic nanogels with cellular membranes can trigger the release of anionic 5′-triphosphates of nucleoside analogues.[239]

In summary, the combination of the release approaches can provide a very useful means for the control of the drug-release characteristics of the nanogel carriers. For example, in the nanogels, drug release can be decreased by cross-linking the polymer chains, and it can be adjusted and made responsive to environmental changes by introducing cleavable cross-links. Furthermore, this technology offers the possibility to control the drug-release profiles. In contrast to liposomes and insoluble nanoparticles, the hydrophilic nanogels swell as the drug is released, which should sustain the release of the drug from the inner layers of the nanogels as the amount released increases. This can be used to modify or eliminate batch release or even to achieve zero-order release kinetics of the drug from nanogels delivered to the disease site.[230,239]

54.5 CONCLUSION

Recent developments of nanogels as drug delivery carriers for biological and biomedical applications were reviewed in this entry. Among major synthetic strategies for the preparation of nanogels, here we only focus on chemical approaches. With this restriction, we divide existing synthetic strategies into two main categories. One strategy involves the formation of nanogels using preformed polymers whereas the other entails the formation of nanogels via direct polymerization of monomers.

The preparation of nanogels from polymer precursor is based on the simply locking of self-assembled amphiphilic copolymers. This versatile methodology is discussed in detail based on cross-linking reaction types employed to fix the self-assembled polymer in this entry.

Other strategies used for nanogel preparation via the direct polymerization of monomers are heterogeneous free radical polymerization and heterogeneous controlled/living radical polymerization. Heterogeneous polymerization of hydrophilic or water-soluble monomers in the presence of either difunctional or multifunctional cross-linkers is the most general method used for the preparation of well-defined synthetic nanogels. This method as well as free radical precipitation, inverse (mini)emulsion, and inverse microemulsion polymerization have been explored.

CRP techniques including ATRP and RAFT, which are used to prepare stable nanogels of well-controlled polymers, are discussed, and among the ATRP methods, a new method utilizing ATRP, inverse miniemulsion polymerization, and thiol-disulfide exchange reaction allowed for the development and functionalization of stable biodegradable nanogels of well-controlled water-soluble polymers in the presence of a disulfide-functionalized DMA is comprehensively explored. This approach provides the opportunity for the preparation of biomaterials with many useful features, such as the preservation of a high degree of halide end functionality for facile functional block copolymerization and bioconjugation; biodegradation in the presence of a biocompatible glutathione tripeptide; formation of uniform networks; larger and better swelling ratio, degradation behavior, and colloidal stability compared to conventional counterparts; high loading level of anticancer drugs; and enhanced circulation time in the blood.

Nanogel synthesis by direct RAFT polymerization is a promising route to industrial setup because of high monomer conversion and high solids content of such processes. The control of nanogel architectures can be realized through the rational selection of polymerization systems such as dispersion polymerization or inverse miniemulsion. Nanogel synthesis by dispersion polymerization in water is particularly attractive for obvious reasons and more efforts should be directed to expand this approach to a wider range of monomers and to the control of other nanogel architectures and morphologies. Although the use of organic solvents as the continuous phase in RAFT inverse miniemulsion polymerization is an inherent limitation, this method has provided a nice set of tools that can be used by researchers from multidisciplinary fields in the design, synthesis, and application of tailored nanogels for bioapplications.

In the last section, some examples in literature are mentioned that demonstrate diverse nanogels and versatile applications as DDSs. In addition, the drug release mechanisms from nanogels are discussed.

About the future of nanogels, it can be said that one future goal of nanogel research should be the improved the design of nanogels with specific targeting residues to enable highly selective uptake into specific cells, particularly cancer cells. Polymer chemists and biologists can learn from each other to elucidate the specific interactions between biomolecules and cellular integrin receptors, which in turn can be carefully attached to advanced delivery systems. Through collaboration, advanced nanogels with careful control over stability, size, biodegradability, and functionality for bioconjugation can be realized.

ACKNOWLEDGMENTS

The authors would like to thank the University of Tehran for partial support during preparation of this entry.

REFERENCES

1. Oh, J.K.; Bencherif, S.A.; Matyjaszewski, K. Atom transfer radical polymerization in inverse miniemulsion: A versatile route toward preparation and functionalization of microgels/nanogels for targeted drug delivery applications. *Polymer* **2009**, *50* (19), 4407–4423.
2. Zhang, H.; Mardyani, S.; Chan, W.C.W.; Kumacheva, E. Design of biocompatible chitosan microgels for targeted pH-mediated intracellular release of cancer therapeutics. *Biomacromolecules* **2006**, *7* (5), 1568–1572.
3. Jung, T.; Kamm, W.; Breitenbach, A.; Kaiserling, E.; Xiao, J.X.; Kissel, T. Biodegradable nanoparticles for oral delivery of peptides: Is there a role for polymers to affect mucosal uptake? *Eur. J. Pharm. Biopharm.* **2000**, *50* (1), 147–160.
4. Raemdonck, K.; Demeester, J.; De Smedt, S. Advanced nanogel engineering for drug delivery. *Soft Matter* **2009**, *5* (4), 707–715.
5. Kabanov, A.V.; Vinogradov, S.V. Nanogels as pharmaceutical carriers: Finite networks of infinite capabilities. *Angew. Chem. Int. Ed.* **2009**, *48* (30), 5418–5429.
6. Rao, J.P.; Geckeler, K.E. Polymer nanoparticles: Preparation techniques and size-control parameters. *Prog. Polym. Sci.* **2011**, *36* (7), 887–913.
7. Seymour, L.W.; Duncan, R.; Strohalm, J.; Kopecek, J. Effect of molecular weight (Mw) of N-(2-hydroxypropyl) methacrylamide copolymers on body distribution and rate of excretion after subcutaneous, intraperitoneal, and intravenous administration to rats. *J. Biomed. Mater. Res. A* **1987**, *21* (11), 1341–1358.
8. Davis, M.E.; Chen, Z.; Shin, D.M. Nanoparticle therapeutics: An emerging treatment modality for cancer. *Nat. Rev. Drug Discov.* **2008**, *7* (9), 771–782.
9. Lemieux, P.; Vinogradov, S.V.; Gebhart, C.L.; Guerin, N.; Paradis, G.; Nguyen, H.K.; Ochietti, B. et al. Block and graft copolymers and nanogel copolymer networks for DNA delivery into cell. *J. Drug Target.* **2000**, *8* (2), 91–105.
10. Vinogradov, S.V.; Bronich, T.K.; Kabanov, A.V. Nanosized cationic hydrogels for drug delivery: Preparation, properties and interactions with cells. *Adv. Drug Deliv. Rev.* **2002**, *54* (1), 135–147.
11. Nayak, S.; Lyon, L.A. Soft nanotechnology with soft nanoparticles. *Angew. Chem. Int. Ed.* **2005**, *44* (47), 7686–7708.
12. Vinogradov, S.V. Polymeric nanogel formulations of nucleoside analogs. *Expert Opin. Drug Deliv.* **2007**, *4* (1), 5–17.

13. Oh, J.K.; Drumright, R.; Siegwart, D.J.; Matyjaszewski, K. The development of microgels/nanogels for drug delivery applications. *Prog. Polym. Sci.* **2008**, *33* (4), 448–477.
14. Oh, J.K.; Lee, D.I.; Park, J.M. Biopolymer-based microgels/nanogels for drug delivery applications. *Prog. Polym. Sci.* **2009**, *34* (12), 1261–1282.
15. Motornov, M.; Roiter, Y.; Tokarev, I.; Minko, S. Stimuli-responsive nanoparticles, nanogels and capsules for integrated multifunctional intelligent systems. *Prog. Polym. Sci.* **2010**, *35* (1–2), 174–211.
16. Oh, J.K. Engineering of nanometer-sized cross-linked hydrogels for biomedical applications. *Can. J. Chem.* **2010**, *88* (3), 173–184.
17. Oishi, M.; Nagasaki, Y. Stimuli-responsive smart nanogels for cancer diagnostics and therapy. *Nanomedicine* **2010**, *5* (3), 451–468.
18. Sanson, N.; Rieger, J. Synthesis of nanogels/microgels by conventional and controlled radical crosslinking copolymerization. *Polym. Chem.* **2010**, *1* (7), 965–977.
19. Sasaki, Y.; Akiyoshi, K. Nanogel engineering for new nanobiomaterials: From chaperoning engineering to biomedical applications. *Chem. Rec.* **2010**, *10* (6), 366–376.
20. Tamura, A.; Nagasaki, Y. Smart siRNA delivery systems based on polymeric nanoassemblies and nanoparticles. *Nanomedicine* **2010**, *5* (7), 1089–1102.
21. Zha, L.; Banik, B.; Alexis, F. Stimulus responsive nanogels for drug delivery. *Soft Matter* **2011**, *7* (13), 5908–5916.
22. Hamidi, M.; Azadi, A.; Rafiei, P. Hydrogel nanoparticles in drug delivery. *Adv. Drug Deliv. Rev.* **2008**, *60* (15), 1638–1649.
23. Sisson, A.L.; Haag, R. Polyglycerol nanogels: Highly functional scaffolds for biomedical applications. *Soft Matter* **2010**, *6* (20), 4968–4975.
24. Hasegawa, U.; Nomura S.I.M.; Kaul, S.C.; Hirano, T.; Akiyoshi, K. Nanogel quantum dot hybrid nanoparticles for live cell imaging. *Biochem. Biophys. Res. Commun.* **2005**, *331* (4), 917–921.
25. Fukui, T.; Kobayashi, H.; Hasegawa, U.; Nagasawa, T.; Akiyoshi, K.; Ishikawa, I. Intracellular delivery of nanogel-quantum dot hybrid nanoparticles into human periodontal ligament cells. *Drug Metab. Lett.* **2007**, *1* (2), 131–135.
26. Chatterjee, J.; Haik, Y.; Chen C.J. Biodegradable magnetic gel: Synthesis and characterization. *Colloid Polym. Sci.* **2003**, *281* (9), 892–896.
27. Gupta Ajay, K.; Wells, S. Surface-modified superparamagnetic nanoparticles for drug delivery: Preparation, characterization, and cytotoxicity studies. *IEEE Trans. Nanobiosci.* **2004**, *3* (1), 66–73.
28. Das, M.; Sanson, N.; Fava, D.; Kumacheva, E. Microgels loaded with gold nanorods: Photothermally triggered volume transitions under physiological conditions. *Langmuir* **2007**, *23* (1), 196–201.
29. Kim, J.; Nayak, S.; Lyon, L.A. Bioresponsive hydrogel microlenses. *J. Am. Chem. Soc.* **2005**, *127* (26), 9588–9592.
30. Thornton, P.D.; McConnell, G.; Ulijn, R.V. Enzyme responsive polymer hydrogel beads. *Chem. Commun.* **2005**, *47* (47), 5913–5915.
31. Ulijn, R.V.; Bibi, N.; Jayawarna, V.; Thornton, P.D.; Todd, S.J.; Mart R.J.; Smith, A.M.; Gough, J.E. Bioresponsive hydrogels. *Mater. Today* **2007**, *10* (4), 40–48.
32. Matyjaszewski, K.; Davis, T.P. *Handbook of Radical Polymerization*; John Wiley & Sons Ltd: New York, 2002.
33. Davis, K.A.; Matyjaszewski, K. Statistical, gradient, block, and graft copolymers by controlled/living radical polymerizations. *Adv. Polym. Sci.* **2002**, *159*, 1–169.
34. Coessens, V.; Pintauer, T.; Matyjaszewski, K. Functional polymers by atom transfer radical polymerization. *Prog. Polym. Sci.* **2001**, *26* (3), 337–377.
35. Braunecker, W.A.; Matyjaszewski, K. Controlled/living radical polymerization: Features, developments, and perspectives. *Prog. Polym. Sci.* **2007**, *32* (1), 93–146.
36. Ide, N.; Fukuda, T. Nitroxide-controlled free-radical copolymerization of vinyl and divinyl monomers. Evaluation of pendant-vinyl reactivity. *Macromolecules* **1997**, *30* (15), 4268–4271.
37. Shim, S.E.; Oh, S.; Chang, Y.H.; Jin, M.J.; Choe, S. Solvent effect on TEMPO-mediated living free radical dispersion polymerization of styrene. *Polymer* **2004**, *45* (14), 4731–4739.
38. Tsarevsky, N.V.; Matyjaszewski, K. Combining atom transfer radical polymerization and disulfide/thiol redox chemistry: A route to well-defined (bio)degradable polymeric materials. *Macromolecules* **2005**, *38* (8), 3087–3092.
39. Li, W.; Gao, H.; Matyjaszewski, K. Influence of initiation efficiency and polydispersity of primary chains on gelation during atom transfer radical copolymerization of monomer and cross-linker. *Macromolecules* **2009**, *42* (4), 927–932.

40. Min, K.; Gao, H.; Yoon; J.A; Wu, W.; Kowalewski, T.; Matyjaszewski, K. One-pot synthesis of hairy nanoparticles by emulsion ATRP. *Macromolecules* **2009**, *42* (5), 1597–1603.
41. Gao, H.; Matyjaszewski, K. Synthesis of functional polymers with controlled architecture by CRP of monomers in the presence of cross-linkers: From stars to gels. *Prog. Polym. Sci.* **2009**, *34* (4), 317–350.
42. Gao, H.; Miasnikova, A.; Matyjaszewski, K. Effect of cross-linker reactivity on experimental gel points during ATRP of monomer and cross-linker. *Macromolecules* **2008**, *41* (21), 7843–7849.
43. Huang, J.; Cusick, B.; Pietrasik, J.; Wang, L.; Kowalewski, T.; Lin, Q.; Matyjaszewski, K. Synthesis and in situ atomic force microscopy characterization of temperature-responsive hydrogels based on poly(2-(dimethylamino)ethyl methacrylate) prepared by atom transfer radical polymerization. *Langmuir* **2007**, *23* (1), 241–249.
44. An, Z.; Shi, Q.; Tang, W.; Tsung, C.K.; Hawker, C.J.; Stucky, G.D. Facile RAFT precipitation polymerization for the microwave-assisted synthesis of well-defined, double hydrophilic block copolymers and nanostructured hydrogels. *J. Am. Chem. Soc.* **2007**, *129* (46), 14493–14499.
45. Chacko, R.T.; Ventura, J.; Zhuang, J.; Thayumanavan, S. Polymer nanogels: A versatile nanoscopic drug delivery platform. *Adv. Drug Deliv. Rev.* **2012**, *64* (9), 836–851.
46. Sun, K.H.; Sohn, Y.S.; Jeong, B. Thermogelling poly(ethylene oxide-b-propylene oxide-b-ethylene oxide) disulfide multiblock copolymer as a thiol-sensitive degradable polymer. *Biomacromolecules* **2006**, *7* (10), 2871–2877.
47. Castellani, O.F.; Martinez, E.N.; Anon, M.C. Role of disulfide bonds upon the structural stability of an amaranth globulin. *J. Agric. Food Chem.* **1999**, *47* (8), 3001–3008.
48. Christensen, L.V.; Chang, C.W.; Kim, W.J.; Kim, S.W.; Zhong, Z.Y.; Lin, C.; Engbersen, J.F.J.; Feijen, J. Reducible poly(amido ethylenimine)s designed for triggered intracellular gene delivery. *Bioconjug. Chem.* **2006**, *17* (5), 1233–1240.
49. Kakizawa, Y.; Harada, A.; Kataoka, K. Environment-sensitive stabilization of core–shell structured polyion complex micelle by reversible cross-linking of the core through disulfide bond. *J. Am. Chem. Soc.* **1999**, *121* (48), 11247–11249.
50. Saito, G.; Swanson, J.A.; Lee, K.D. Drug delivery strategy utilizing conjugation via reversible disulfide linkages: Role and site of cellular reducing activities. *Adv. Drug. Deliv. Rev.* **2003**, *55* (2), 199–205.
51. Lee, Y.; Koo, H.; Jin, G.W.; Mo, H.; Cho, M.Y.; Park, J.-Y.; Choi, J.S; Park, J.S. Poly(ethylene oxide sulfide): New poly(ethylene glycol) derivatives degradable in reductive conditions. *Biomacromolecules* **2005**, *6* (1), 24–26.
52. Kolano, C.; Helbing, J.; Bucher, G.; Sander, W.; Hamm, P. Intramolecular disulfide bridges as a phototrigger to monitor the dynamics of small cyclic peptides. *J. Phys. Chem. B* **2007**, *111* (38), 11297–11302.
53. Torchilin, V.P. Recent advances with liposomes as pharmaceutical carriers. *Nat. Rev. Drug Discov.* **2005**, *4* (2), 145–160.
54. Wang, L.; Kristensen, J.; Ruffner, D.E. Delivery of antisense oligonucleotides using HPMA polymer: Synthesis of a thiol polymer and its conjugation to water-soluble molecules. *Bioconjug. Chem.* **1998**, *9* (6), 749–757.
55. Ryu, J.H.; Jiwpanich, S.; Chacko, R.; Bickerton, S.; Thayumanavan, S. Surface-functionalizable polymer nanogels with facile hydrophobic guest encapsulation capabilities. *J. Am. Chem. Soc.* **2010**, *132* (24), 8246–8247.
56. Ryu, J.H.; Chacko, R.T.; Jiwpanich, S.; Bickerton, S.; Babu, R.P.; Thayumanavan, S. Self-cross-linked polymer nanogels. A versatile nanoscopic drug delivery platform. *J. Am. Chem. Soc.* **2010**, *132* (48), 17227–17235.
57. Jiwpanich, S.; Ryu, J.H.; Bickerton, S.; Thayumanavan, S. Noncovalent encapsulation stabilities in supramolecular nanoassemblies. *J. Am. Chem. Soc.* **2010**, *132* (31), 10683–10685.
58. Kakizawa, Y.; Harada, A.; Kataoka, K. Glutathione-sensitive stabilization of block copolymer micelles composed of antisense DNA and thiolated poly(ethylene glycol)-block-poly(L-lysine): A potential carrier for systemic delivery of antisense DNA. *Biomacromolecules* **2001**, *2* (2), 491–497.
59. Matsumoto, S.; Christie, R.; Nishiyama, N.; Miyata, K.; Ishii, A.; Oba, M.; Koyama, H.; Yamasaki, Y.; Kataoka, K. Environment-responsive block copolymer micelles with a disulfide cross-linked core for enhanced siRNA delivery. *Biomacromolecules* **2009**, *10* (1), 119–127.
60. Christie, R.J.; Miyata, K.; Matsumoto, Y.; Nomoto, T.; Menasco, D.; Lai, T.ch.; Pennisi, M. et al. Effect of polymer structure on micelles formed between siRNA and cationic block copolymer comprising thiols and amidines. *Biomacromolecules* **2011**, *12* (9), 3174–3185.
61. Huang, H.; Remsen, E.E.; Wooley, K.L. Amphiphilic core–shell nanospheres obtained by intramicellar shell crosslinking of polymer micelles with poly(ethylene oxide) linkers. *Chem. Commun.* **1998**, *13* (13), 1415–1416.

62. Joralemon, M.J.; Smith, N.L.; Holowka, D.; Baird, B.; Wooley, K.L. Antigen-decorated shell cross-linked nanoparticles: Synthesis, characterization, and antibody interactions. *Bioconjug. Chem.* **2005**, *16* (5), 1246–1256.
63. Li, Y.; Du, W.; Sun, G.; Wooley, K.L. pH-responsive shell cross-linked nanoparticles with hydrolytically labile cross-links. *Macromolecules* **2008**, *41* (18), 6605–6607.
64. Wei, H.; Quan, C.Y.; Chang, C.; Zhang, X.Z.; Zhuo, R.X. Preparation of novel ferrocene-based shell cross-linked thermoresponsive hybrid micelles with antitumor efficacy. *J. Phys. Chem. B* **2010**, *114* (16), 5309–5314.
65. Li, Y.; Lokitz, B.S.; Armes, S.P.; McCormick, C.L. Synthesis of reversible shell cross-linked micelles for controlled release of bioactive agents. *Macromolecules* **2006**, *39* (8), 2726–2728.
66. Xu, X.; Smith, A.E.; Kirkland, S.E.; McCormick, C.L. Aqueous RAFT synthesis of pH-responsive triblock copolymer mPEO-PAPMA-PDPAEMA and formation of shell cross-linked micelles. *Macromolecules* **2008**, *41* (22), 8429–8435.
67. Zhuang, J.; Jiwpanich, S.; Deepak, V.D.; Thayumanavan, S. Facile preparation of nanogels using activated ester containing polymers. *ACS Macro Lett.* **2012**, *1* (1), 175–179.
68. Duong, H.T.T.; Marquis, C.P.; Whittaker, M.; Davis, T.P.; Boyer, C. Acid degradable and biocompatible polymeric nanoparticles for the potential codelivery of therapeutic agents. *Macromolecules* **2011**, *44* (20), 8008–8019.
69. Kim, Y.; Pourgholami, M.H.; Morris, D.L.; Stenzel, M.H. Triggering the fast release of drugs from crosslinked micelles in an acidic environment. *J. Mater. Chem.* **2011**, *21* (34), 12777–12783.
70. Kolb, H.C.; Finn, M.G.; Sharpless, K.B. Click chemistry: Diverse chemical function from a few good reactions. *Angew. Chem. Int. Ed.* **2001**, *40* (11), 2004–2021.
71. Tsarevsky, N.V.; Sumerlin, B.S.; Matyjaszewski, K. Step-growth "click" coupling of telechelic polymers prepared by atom transfer radical polymerization. *Macromolecules* **2005**, *38* (9), 3558–3561.
72. Sumerlin, B.S.; Tsarevsky, N.V.; Louche, G.; Lee, R.Y.; Matyjaszewski, K. Highly efficient "click" functionalization of poly(3-azidopropyl methacrylate) prepared by ATRP. *Macromolecules* **2005**, *38* (18), 7540–7545.
73. Wu, P.; Feldman, A.K.; Nugent, A.K.; Hawker, C.J.; Scheel, A.; Voit, B.; Pyun, J.; Frehet, J.M.J.; Sharpless, K.B.; Fokin, V.V. Efficiency and fidelity in a click-chemistry route to triazole dendrimers by the copper(I)-catalyzed ligation of azides and alkynes. *Angew. Chem. Int. Ed.* **2004**, *43* (30), 3928–3932.
74. Hawker, C.J.; Wooley, K.L. The convergence of synthetic organic and polymer chemistries. *Science* **2005**, *309* (5738), 1200–1205.
75. Demko, Z.P.; Sharpless, K.B. A click chemistry approach to tetrazoles by huisgen 1,3-dipolar cycloaddition: Synthesis of 5-sulfonyl tetrazoles from azides and sulfonyl cyanides. *Angew. Chem. Int. Ed.* **2002**, *41* (12), 2110–2113.
76. Demko, Z.P.; Sharpless, K.B. A click chemistry approach to tetrazoles by huisgen 1,3-dipolar cycloaddition: Synthesis of 5-acyltetrazoles from azides and acyl cyanides. *Angew. Chem. Int. Ed.* **2002**, *41* (12), 2113–2116.
77. Joralemon, M.J.; O'Reilly, R.K.; Hawker, C.J.; Wooley, K.L. Shell click-crosslinked (SCC) nanoparticles: A new methodology for synthesis and orthogonal functionalization. *J. Am. Chem. Soc.* **2005**, *127* (48), 16892–16899.
78. O'Reilly, R.K.; Joralemon, M.J.; Wooley, K.L.; Hawker, C.J. Functionalization of micelles and shell cross-linked nanoparticles using click chemistry. *Chem. Mater.* **2005**, *17* (24), 5976–5988.
79. O'Reilly, R.K.; Joralemon M.J.; Hawker C.J.; Wooley K.L. Preparation of orthogonally-functionalized core click cross-linked nanoparticles. *New J. Chem.* **2007**, *31* (5), 718–724.
80. Zhang, J.; Zhou Y.; Zhu, Z.H.; Ge, Z.H.; Liu, S.H. Polyion complex micelles possessing thermoresponsive coronas and their covalent core stabilization via "click" chemistry. *Macromolecules* **2008**, *41* (4), 1444–1454.
81. Withey, A.B.J.; Chen, G.; Nguyen, T.L.U.; Stenzel, M.H. Macromolecular cobalt carbonyl complexes encapsulated in a click-cross-linked micelle structure as a nanoparticle to deliver cobalt pharmaceuticals. *Biomacromolecules* **2009**, *10* (12), 3215–3226.
82. Duong, H.T.T.; Huynh, V.T.; de Souza, P.; Stenzel, M.H. Core-cross-linked micelles synthesized by clicking bifunctional Pt(IV) anticancer drugs to isocyanates. *Biomacromolecules* **2010**, *11* (9), 2290–2299.
83. Jiang, X.; Zhang, G.; Narain, R.; Liu, S.H. Fabrication of two types of shell-cross-linked micelles with "Inverted" structures in aqueous solution from schizophrenic water-soluble ABC triblock copolymer via click chemistry. *Langmuir* **2009**, *25* (4), 2046–2054.

84. Wu, T.; Ge, Z.H.; Liu, S.H. Fabrication of thermoresponsive cross-linked poly(N-isopropylacrylamide) nanocapsules and silver nanoparticle-embedded hybrid capsules with controlled shell thickness. *Chem. Mater.* **2011**, *23* (9), 2370–2380.
85. Lehn, J.M. From supramolecular chemistry towards constitutional dynamic chemistry and adaptive chemistry. *Chem. Soc. Rev.* **2007**, *36* (2), 151–160.
86. Jackson, A.W.; Stakes, C.H.; Fulton, D.A. The formation of core cross-linked star polymer and nanogel assemblies facilitated by the formation of dynamic covalent imine bonds. *Polym. Chem.* **2011**, *2* (11), 2500–2511.
87. Corbett, P.T.; Leclaire, J.; Vial, L.; West, K.R.; Wietor, J.-L.; Sanders, J.K.M.; Otto, S. Dynamic combinatorial chemistry. *Chem. Rev.* **2006**, *106* (9), 3652–3711.
88. Jackson, A.W.; Fulton, D.A. Triggering polymeric nanoparticle disassembly through the simultaneous application of two different stimuli. *Macromolecules* **2012**, *45* (6), 2699–2708.
89. Xu, X.; Flores, J.D.; McCormick, C.L. Reversible imine shell cross-linked micelles from aqueous RAFT synthesized thermoresponsive triblock copolymers as potential nanocarriers for "pH-triggered" drug release. *Macromolecules* **2011**, *44* (6), 1327–1334.
90. Pioge, S.; Nesterenko, A.; Brotons, G.; Pascual, S.; Fontaine, L.; Gaillard, C.; Nicol, E. Core cross-linking of dynamic diblock copolymer micelles: Quantitative study of photopolymerization efficiency and micelle structure. *Macromolecules* **2011**, *44* (3) 594–603.
91. He, J.; Tong, X.; Zhao, Y. Photoresponsive nanogels based on photocontrollable cross-links. *Macromolecules* **2009**, *42* (13), 4845–4852.
92. He, J.; Yan, B.; Tremblay, L.; Zhao, Y. Both core- and shell-cross-linked nanogels: Photoinduced size change, intraparticle LCST, and interparticle UCST thermal behaviors. *Langmuir* **2011**, *27* (1), 436–444.
93. Yusa, S.I.; Sugahara, M.; Endo, T.; Morishima, Y. Preparation and characterization of a pH-responsive nanogel based on a photo-cross-linked micelle formed from block copolymers with controlled structure. *Langmuir* **2009**, *25* (9), 5258–5265.
94. Kuckling, D.; Vo, C.D.; Adler, H.-J.P.; Vollkel, A.; Collfen, H. Preparation and characterization of photocross-linked thermosensitive PNIPAAm nanogels. *Macromolecules* **2006**, *39* (4), 1585–1591.
95. Sugihara, S.H.; Ito, S.; Irie, S.; Ikeda, I. Synthesis of thermoresponsive shell cross-linked micelles via living cationic polymerization and UV irradiation. *Macromolecules* **2010**, *43* (4), 1753–1760.
96. Kuckling, D.; Vo, C.V.; Wohlrab, S.E. Preparation of nanogels with temperature-responsive core and pH-responsive arms by photo-cross-linking. *Langmuir* **2002**, *18* (11), 4263–4269.
97. An, Z.H.; Qiu, Q.; Liu, G. Synthesis of architecturally well-defined nanogels via RAFT polymerization for potential bioapplications. *Chem. Commun.* **2011**, *47* (46), 12424–12440.
98. Cheng, G.; Mi, L.; Cao, Z.Q.; Xue, H.; Yu, Q.M.; Carr, L.; Jiang, S.Y. Functionalizable and ultrastable zwitterionic nanogels. *Langmuir* **2010**, *26* (10), 6883–6886.
99. Guha, S.; Ray, B.; Mandal, B.M. Anomalous solubility of polyacrylamide prepared by dispersion (precipitation) polymerization in aqueous tert-butyl alcohol. *J. Polym. Sci. A* **2001**, *39* (19), 3434–3442.
100. Liu, T.; Desimone, J.M.; Roberts, G.W. Continuous precipitation polymerization of acrylic acid in supercritical carbon dioxide: The polymerization rate and the polymer molecular weight. *J. Polym. Sci. A* **2005**, *43* (12), 2546–2555.
101. Bai, F.; Yang, X.; Zhao, Y.; Huang, W. Synthesis of core–shell microspheres with active hydroxyl groups by two-stage precipitation polymerization. *Polym. Int.* **2005**, *54* (1), 168–174.
102. Li, W. H.; Stover, H.D.H. Mono- or narrow disperse poly(methacrylate-co-divinylbenzene) microspheres by precipitation polymerization. *J. Polym. Sci. A* **1999**, *37* (15), 2899–2907.
103. Duracher, D.; Elaissari, A.; Pichot, C. Preparation of poly(N-isopropylmethacrylamide) latexes kinetic studies and characterization. *J. Polym. Sci. A* **1999**, *37* (12), 1823–1837.
104. Hazot, P.; Chapel, J.P.; Pichot, C,; Elaissari, A.; Delair, T. Preparation of poly(N-ethyl methacrylamide) particles via an emulsion/precipitation process: The role of the crosslinker. *J. Polym. Sci. A* **2002**, *40* (11), 1808–1817.
105. Huang, G.; Gao, J.; Hu, Z.; St John, J.V.; Ponder, B.C.; Moro, D. Controlled drug release from hydrogel nanoparticle networks. *J. Control. Release* **2004**, *94* (2–3), 303–311.
106. William, H.; Blackburn, L.; Lyon, A. Size-controlled synthesis of monodisperse core/shell nanogels. *Colloid Polym. Sci.* **2008**, *286* (5), 563–569.
107. Gan, D.; Lyon, L.A. Tunable swelling kinetics in core–shell hydrogel nanoparticles. *J. Am. Chem. Soc.* **2001**, *123* (31), 7511–7517.
108. Jones, C.D.; Lyon, L.A. Synthesis and characterization of multiresponsive core–shell microgels. *Macromolecules* **2000**, *33* (22), 8301–8306.

109. Jones, C.D.; Lyon, L.A. Shell-restricted swelling and core compression in poly(N-isopropylacrylamide) core–shell microgels. *Macromolecules* **2003**, *36* (6), 1988–1993.
110. Huang, X.; Lowe, T.L. Biodegradable thermoresponsive hydrogels for aqueous encapsulation and controlled release of hydrophilic model drugs. *Biomacromolecules* **2005**, *6* (4), 2131–2139.
111. Antonietti, M.; Landfester, K. Polyreactions in miniemulsions. *Prog. Polym. Sci.* **2002**, *27* (4), 689–757.
112. Landfester, K.; Willert, M.; Antonietti, M. Preparation of polymer particles in nonaqueous direct and inverse miniemulsions. *Macromolecules* **2000**, *33* (7), 2370–2376.
113. Kriwet, B.; Walter, E.; Kissel, T. Synthesis of bioadhesive poly(acrylic acid) nano- and microparticles using an inverse emulsion polymerization method for the entrapment of hydrophilic drug candidates. *J. Control. Release* **1998**, *56* (1–3), 149–158.
114. Sun, Q.; Deng, Y. In situ synthesis of temperature-sensitive hollow microspheres via interfacial polymerization. *J. Am. Chem. Soc.* **2005**, *127* (23), 8274–8275.
115. Sankar, C.; Rani, M.; Srivastava, A.K.; Mishra, B. Chitosan based pentazocine microspheres for intranasal systemic delivery: Development and biopharmaceutical evaluation. *Pharmazie* **2001**, *56* (3), 223–226.
116. Marie, E.; Rothe, R.; Antonietti, M.; Landfester, K. Synthesis of polyaniline particles via inverse and direct miniemulsion. *Macromolecules* **2003**, *36* (11), 3967–3673.
117. Dowding, P.J.; Vincent, B.; Williams, E. Preparation and swelling properties of poly(NIPAM) "Minigel" particles prepared by inverse suspension polymerization. *J. Colloid Interface Sci.* **2000**, *221* (2), 268–272.
118. Owens III, D.E.; Jian, Y.; Fang, J.E.; Slaughter, B.V.; Chen, Y.-H.; Peppas, N.A. Thermally responsive swelling properties of polyacrylamide/poly(acrylic acid) interpenetrating polymer network nanoparticles. *Macromolecules* **2007**, *40* (20), 7306–7310.
119. Missirlis, D.; Tirelli, N.; Hubbell, J.A. Amphiphilic hydrogel nanoparticles. Preparation, characterization, and preliminary assessment as new colloidal drug carriers. *Langmuir* **2005**, *21* (6), 2605–2613.
120. Lovell, P.; El-Aasser, M.S. *Emulsion Polymerization and Emulsion Polymers*; John Wiley & Sons Ltd; West Sussex, U.K., 1997, p. 723.
121. Nagarajan, R.; Wang, C.-C. Theory of surfactant aggregation in water/ethylene glycol mixed solvents. *Langmuir* **2000**, *16* (12), 5242–5251.
122. Braun, O.; Selb, J.; Candau, F. Synthesis in microemulsion and characterization of stimuli-responsive polyelectrolytes and polyampholytes based on N-isopropylacrylamide. *Polymer* **2001**, *42* (21), 8499–8510.
123. Fernandez, V.V.A.; Tepale, N.; Sanchez-Diaz, J.C.; Mendizabal, E.; Puig, J.E.; Soltero, J.F.A. Thermoresponsive nanostructured poly(N-isopropylacrylamide) hydrogels made via inverse microemulsion polymerization. *Colloid Polym. Sci.* **2006**, *284* (4), 387–395.
124. Juranicova, V.; Kawamoto, S.; Fujimoto, K.; Kawaguchi, H.; Barton, J. Inverse microemulsion polymerization of acrylamide in the presence of N,N-dimethylacrylamide. *Angew. Makromol. Chem.* **1998**, *258* (1), 27–31.
125. Barton, J. Inverse microemulsion polymerization of oil-soluble monomers in the presence of hydrophilic polyacrylamide nanoparticles. *Macromol. Symp.* **2002**, *179* (1), 189–208.
126. Renteria, M.; Munoz, M.; Ochoa, J.R.; Cesteros, L.C.; Katime, I. Acrylamide inverse microemulsion polymerization in a paraffinic solvent: Rolling-M-245. *J. Polym. Sci. A Polym. Chem.* **2005**, *43* (12), 2495–2503.
127. Kaneda, I.; Sogabe, A.; Nakajima, H. Water-swellable polyelectrolyte microgels polymerized in an inverse microemulsion using a nonionic surfactant. *J. Colloid Interface Sci.* **2004**, *275* (2), 450–457.
128. Deng, Y.; Wang, L.; Yang, W.; Fu, S.; Elaissari, A. Preparation of magnetic polymeric particles via inverse microemulsion polymerization process. *J. Magn. Magn. Mater.* **2003**, *257* (1), 69–78.
129. Gaur, U.; Sahoo, S.K.; De, T.K.; Ghosh, P.C.; Maitra, A.; Ghosh, P.K. Biodistribution of fluoresceinated dextran using novel nanoparticles evading reticuloendothelial system. *Int. J. Pharm.* **2000**, *202* (1–2), 1–10.
130. Bharali, D.J.; Sahoo, S.K.; Mozumdar, S.; Maitra, A. Cross-linked polyvinylpyrrolidone nanoparticles: A potential carrier for hydrophilic drugs. *J. Colloid Interface Sci.* **2003**, *258* (2), 415–423.
131. McAllister, K.; Sazani, P.; Adam, M.; Cho, M.J.; Rubinstein, M.; Samulski, R.J.; DeSimone, J.M. Polymeric nanogels produced via inverse microemulsion polymerization as potential gene and antisense delivery agents. *J. Am. Chem. Soc.* **2002**, *124* (51), 15198–15207.
132. Craparo, E.F.; Cavallaro, G.; Bondi, M.L.; Mandracchia, D.; Giammona, G. PEGylated nanoparticles based on a polyaspartamide. Preparation, physico-chemical characterization, and intracellular uptake. *Biomacromolecules* **2006**, *7* (11), 3083–3092.

133. Gao, D.; Xu, H.; Philbert, M.A.; Kopelman, R. Ultrafine hydrogel nanoparticles: Synthetic approach and therapeutic application in living cells. *Angew. Chem. Int. Ed.* **2007**, *46* (13), 2224–2227.
134. Gao, D.; Agayan, R.R.; Xu, H.; Philbert, M.A.; Kopelman, R. Nanoparticles for two-photon photodynamic therapy in living cells. *Nano Lett.* **2006**, *6* (11), 2383–2386.
135. Rieger, J.; Grazon, C.; Charleux, B.; Alaimo, D.; Jerome, C. Pegylated thermally responsive block copolymer micelles and nanogels via in situ RAFT aqueous dispersion polymerization. *J. Polym. Sci. A Polym. Chem.* **2009**, *47* (9), 2373–2390.
136. Das, M.; Mardyani, S.; Chan, W.C.W.; Kumacheva, E. Biofunctionalized pH-responsive microgels for cancer cell targeting: Rational design. *Adv. Mater.* **2006**, *18* (1), 80–83.
137. Thronton, P.D.; Mart, R.J.; Ulijn, R.V. Enzyme-responsive polymer hydrogel particles for controlled release. *Adv. Mater.* **2007**, *19* (9), 1252–1256.
138. Kim, J.; Serpe, M.J.; Lyon, L.A. Photoswitchable microlens arrays. *Angew. Chem. Int. Ed.* **2005**, *44* (9), 1333–1336.
139. Matyjaszewski, K.; Xia, J. Atom transfer radical polymerization. *Chem. Rev.* **2001**, *101* (9), 2921–2990.
140. Hawker, C.J.; Bosman, A.W.; Harth, E. New polymer synthesis by nitroxide mediated living radical polymerizations. *Chem. Rev.* **2001**, *101* (12), 3661–3688.
141. Chiefari, J.; Chong, Y.K.; Ercole, F.; Krstina, J.; Jeffery, J.; Le, T.P.T.; Mayadunne, R.T.A. et al. Living free-radical polymerization by reversible addition-fragmentation chain transfer: The RAFT process. *Macromolecules* **1998**, *31* (16), 5559–5562.
142. Moad, G.; Rizzardo, E.; Thang, S.H. Living radical polymerization by the RAFT process—A third update. *Aust. J. Chem.* **2012**, *65* (8), 985–1076.
143. Sheiko, S.S.; Sumerlin, B.S.; Matyjaszewski, K. Cylindrical molecular brushes: Synthesis, characterization, and properties. *Prog. Polym. Sci.* **2008**, *33* (7), 759–785.
144. Yagci, Y.; Tasdelen, M.A. Mechanistic transformations involving living and controlled/living polymerization methods. *Prog. Polym. Sci.* **2006**, *31* (12), 1133–1170.
145. Hadjichristidis, N.; Iatrou, H.; Pitsikalis, M.; Mays, J. Macromolecular architectures by living and controlled/living polymerization. *Prog. Polym. Sci.* **2006**, *31* (12), 1068–1132.
146. Broyer, R.M.; Quaker, G.M.; Maynard, H.D. Designed amino acid ATRP initiators for the synthesis of biohybrid materials. *J. Am. Chem. Soc.* **2008**, *130* (3), 1041–1047.
147. Le Droumaguet, B.; Velonia, K. In situ ATRP-mediated hierarchical formation of giant amphiphile bionanoreactors. *Angew. Chem. Int. Ed.* **2008**, *47* (33), 6263–6266.
148. Loschonsky, S.; Couet, J.; Biesalski, M. Synthesis of peptide/polymer conjugates by solution ATRP of butylacrylate using an initiator-modified cyclic D-alt-L-peptide. *Macromol. Rapid Commun.* **2008**, *29* (4), 309–315.
149. Ostmark, E.; Harrisson, S.; Wooley K.L.; Malmstrom E.E. Comb polymers prepared by ATRP from hydroxypropyl cellulose. *Biomacromolecules* **2007**, *8* (4), 1138–1148.
150. Ifuku, S.; Kadla, J.F. Preparation of a thermosensitive highly regioselective cellulose/N-isopropylacrylamide copolymer through atom transfer radical polymerization. *Biomacromolecules* **2008**, *9* (11), 3308–3313.
151. Xu, P.; Van Kirk, E.A.; Murdoch, W.J.; Zhan, Y.; Isaak, D.D.; Radosz, M.; Shen, Y. Anticancer efficacies of cisplastin-releasing pH-responsive nanoparticles. *Biomacromolecules* **2006**, *7* (3), 829–835.
152. Oh, J.K.; Siegwart, D.J.; Lee, H.I.; Sherwood, G.; Peteanu, L.; Hollinger, J.O.; Kataoka, K.; Matyjaszewski, K. Biodegradable nanogels prepared by atom transfer radical polymerization as potential drug delivery carriers: Synthesis, biodegradation, in vitro release, and bioconjugation. *J. Am. Chem. Soc.* **2007**, *129* (18), 5939–5945.
153. Wang, M.; Dykstra, T.E.; Lou, X.; Salvador, M.R.; Scholes, G.D.; Winnik, M.A. Colloidal CdSe nanocrystals passivated by a dye-labeled multidentate polymer: Quantitative analysis by size-exclusion chromatography. *Angew. Chem. Int. Ed.* **2006**, *45* (14), 2221–2224.
154. Wang, M.; Felorzabihi, N.; Guerin, G.; Haley, J.C.; Scholes, G.D.; Winnik, M.A. Water-soluble CdSe quantum dots passivated by a multidentate diblock copolymer. *Macromolecules* **2007**, *40* (17), 6377–6384.
155. Huang, J.; Koepsel, R.R.; Murata, H.; Wu, W.; Lee, S.B.; Kowalewski, T.; Russell, A.J.; Matyjaszewski, K. Nonleaching antibacterial glass surfaces via "grafting onto": The effect of the number of quaternary ammonium groups on biocidal activity. *Langmuir* **2008**, *24* (13), 6785–6795.
156. Huang, J.; Murata, H.; Koepsel, R.R.; Russell, A.J.; Matyjaszewski, K. Antibacterial polypropylene via surface-initiated atom transfer radical polymerization. *Biomacromolecules* **2007**, *8* (5), 1396–1399.
157. Oh, J.K. Recent advances in controlled/living radical polymerization in emulsion and dispersion. *J. Polym. Sci. A Polym. Chem.* **2008**, *46* (21), 6983–7001.

158. Cunningham, M.F. Controlled/living radical polymerization in aqueous dispersed systems. *Prog. Polym. Sci.* **2008**, *33* (4), 365–398.
159. Matyjaszewski, K.; Tsarevsky, N.V.; Braunecker W.A.; Dong, H.; Huang, J.; Jakubowski, W.; Kwak, Y.; Nicolay, R.; Tang, W.; Yoon, J.A. Role of Cu^0 in controlled/"living" radical polymerization. *Macromolecules* **2007**, *40* (22), 7795–7806.
160. Oh, J.K.; Dong, H.; Zhang, R.; Matyjaszewski, K.; Schlaad, H. Preparation of nanoparticles of double-hydrophilic PEO-PHEMA block copolymers by AGET ATRP in inverse miniemulsion. *J. Polym. Sci. A Polym. Chem.* **2007**, *45* (21), 4764–4772.
161. Oh, J.K.; Siegwart, D.J.; Matyjaszewski, K. Synthesis and biodegradation of nanogels as delivery carriers for carbohydrate drugs. *Biomacromolecules* **2007**, *8* (11), 3326–3331.
162. Tsarevsky, N.V.; Matyjaszewski, K. Reversible redox cleavage/coupling of polystyrene with disulfide or thiol groups prepared by atom transfer radical polymerization. *Macromolecules* **2002**, *35* (24), 9009–9014.
163. Tsarevsky, N.V.; Bernaerts, K.V.; Dufour, B.; Du Prez, F.E.; Matyjaszewski, K. Well-defined (co)polymers with 5-vinyltetrazole units via combination of atom transfer radical (co)polymerization of acrylonitrile and "click chemistry"-type postpolymerization modification. *Macromolecules* **2004**, *37* (25), 9308–9313.
164. Kim, K.H.; Kim, J.; Jo, W.H. Preparation of hydrogel nanoparticles by atom transfer radical polymerization of N-isopropylacrylamide in aqueous media using PEG macro-initiator. *Polymer* **2005**, *46* (9), 2836–2840.
165. Oh, J.K.; Min, K.; Matyjaszewski, K. Preparation of poly(oligo(ethylene glycol) monomethyl ether methacrylate) by homogeneous aqueous AGET ATRP. *Macromolecules* **2006**, *39* (9), 3161–3167.
166. Oh, J.K.; Tang, C.B.; Gao, H.F.; Tsarevsky, N.V.; Matyjaszewski, K. Inverse miniemulsion ATRP: A new method for synthesis and functionalization of well-defined water-soluble/cross-linked polymeric particles. *J. Am. Chem. Soc.* **2006**, *128* (16), 5578–5584.
167. Siegwart, D.J.; Oh, J.K.; Gao, H.; Bencherif, S.A.; Perineau, F.; Bohaty, A.K.; Hollinger, J.O.; Matyjaszewski, K. Biotin-, pyrene-, and GRGDS-functionalized polymers and nanogels via ATRP and end group modification. *Macromol. Chem. Phys.* **2008**, *209* (21), 2180–2193.
168. Moad, G.; Rizzardo, E.E.; Thang, S.H. Radical addition–fragmentation chemistry in polymer synthesis. *Polymer* **2008**, *49* (5), 1079–1131.
169. Kakwere, H.; Perrier, S. Design of complex polymeric architectures and nanostructured materials/hybrids by living radical polymerization of hydroxylated monomers. *Polym. Chem.* **2011**, *2* (2), 270–288.
170. Boyer, C.; Stenzel, M.H.; Davis, T.P. Building nanostructures using RAFT polymerization. *J. Polym. Sci. A* **2011**, *49* (3), 551–595.
171. Moad, G.; Rizzardo, E.E.; Thang, S.H. Living radical polymerization by the RAFT process. *Aust. J. Chem.* **2005**, *58* (6), 379–410.
172. Moad, G.; Rizzardo, E.E.; Thang, S.H. Toward living radical polymerization. *Acc. Chem. Res.* **2008**, *41* (9), 1133–1142.
173. Moad, G.; Rizzardo, E.E.; Thang, S.H. Living radical polymerization by the RAFT process—A first update. *Aust. J. Chem.* **2006**, *59* (10), 669–692.
174. Moad, G.; Rizzardo, E.E.; Thang, S.H. Living radical polymerization by the RAFT process—A second update. *Aust. J. Chem.* **2009**, *62* (11), 1402–1472.
175. Convertine, A.J.; Lokitz, B.S.; Vasileva, Y.; Myrick, L.J.; Scales, C.W.; Lowe, A.B.; McCormick, C.L. Direct synthesis of thermally responsive DMA/NIPAM diblock and DMA/NIPAM/DMA triblock copolymers via aqueous, room temperature RAFT polymerization. *Macromolecules* **2006**, *39* (5), 1724–1730.
176. McCormick, C.L.; Lowe, A.B. Aqueous RAFT polymerization: Recent developments in synthesis of functional water-soluble (co)polymers with controlled structures. *Acc. Chem. Res.* **2004**, *37* (5), 312–325.
177. Boyer, C.; Bulmus, V.; Priyanto, P.; Teoh, W.Y.; Amal, R.; Davis, T.P. The stabilization and bio-functionalization of iron oxide nanoparticles using heterotelechelic polymers. *J. Mater. Chem.* **2009**, *19* (1), 111–123.
178. Boyer, C.; Liu, J.; Bulmus, V.; Davis, T.P.; Barner-Kowollik, C.; Stenzel, M.H. Direct synthesis of well-defined heterotelechelic polymers for bioconjugations. *Macromolecules* **2008**, *41* (15), 5641–5650.
179. Liu, J.; Bulmus, V.; Barner-Kowollik, C.; Stenzel, M.H.; Davis, T.P. Direct synthesis of pyridyl disulfide-terminated polymers by RAFT polymerization. *Macromol. Rapid Commun.* **2007**, *28* (3), 305–314.
180. Postma, A.; Davis, T.P.; Evans, R.A.; Li, G.; Moad, G.; O'Shea, M.S. Synthesis of well-defined polystyrene with primary amine end groups through the use of phthalimido-functional RAFT agents. *Macromolecules* **2006**, *39* (16), 5293–5306.

181. Scales, C.W.; Convertine, A.J.; McCormick, C.L. Fluorescent labeling of RAFT-generated poly(N-isopropylacrylamide) via a facile maleimide–thiol coupling reaction. *Biomacromolecules* **2006**, *7* (5), 1389–1392.
182. Wong, L.; Boyer, C.; Jia, Z.; Zareie, H.M.; Davis, T.P.; Bulmus, V. Synthesis of versatile thiol-reactive polymer scaffolds via RAFT polymerization. *Biomacromolecules* **2008**, *9* (7), 1934–1944.
183. York, A.W.; Scales, C.W.; Huang, F.; McCormick, C.L. Facile synthetic procedure for ω, primary amine functionalization directly in water for subsequent fluorescent labeling and potential bioconjugation of RAFT-synthesized (co)polymers. *Biomacromolecules* **2007**, *8* (8), 2337–2341.
184. Stenzel, M.H.; Davis, T.P. Star polymer synthesis using trithiocarbonate functional β-cyclodextrin cores (reversible addition–fragmentation chain-transfer polymerization). *J. Polym. Sci. A* **2002**, *40* (24), 4498–4512.
185. Quinn, J.F.; Chaplin, R.P.; Davis, T.P. Facile synthesis of comb, star, and graft polymers via reversible addition–fragmentation chain transfer (RAFT) polymerization. *J. Polym. Sci. A* **2002**, *40* (17), 2956–2966.
186. Liu, J.Q.; Liu, H.Y.; Jia, Z.F.; Bulmus, V.; Davis, T.P. An approach to biodegradable star polymeric architectures using disulfide coupling. *Chem. Commun.* **2008**, *48* (48), 6582–6584.
187. Jia, Z.F.; Wong, L.J.; Davis, T.P.; Bulmus, V. One-pot conversion of RAFT-generated multifunctional block copolymers of HPMA to doxorubicin conjugated acid- and reductant-sensitive crosslinked micelles. *Biomacromolecules* **2008**, *9* (11), 3106–3113.
188. Biasutti, J.D.; Davis, T.P.; Lucien, F.P.; Heuts, J.P.A. Reversible addition–fragmentation chain transfer polymerization of methyl methacrylate in suspension. *J. Polym. Sci. A* **2005**, *43* (10), 2001–2012.
189. Simms, R.W.; Davis, T.P.; Cunningham, M.F. Xanthate-mediated living radical polymerization of vinyl acetate in miniemulsion. *Macromol. Rapid Commun.* **2005**, *26* (8), 592–596.
190. Lansalot, M.; Davis, T.P.; Heuts, J.P.A. RAFT miniemulsion polymerization: Influence of the structure of the RAFT agent. *Macromolecules* **2002**, *35* (20), 7582–7591.
191. Barner, L.; Li, C.E.; Hao, X.; Stenzel, M.H.; Barner-Kowollik, C.; Davis, T.P. Synthesis of core-shell poly(divinylbenzene) microspheres via reversible addition fragmentation chain transfer graft polymerization of styrene. *J. Polym. Sci. A* **2004**, *42* (20), 5067–5076.
192. Chan, Y.; Bulmus, V.; Zareie, M.H.; Byrne, F.L.; Barner, L.; Kavallaris, M. Acid-cleavable polymeric core–shell particles for delivery of hydrophobic drugs. *J. Control. Release* **2006**, *115* (2), 197–207.
193. Barsbay, M.; Gueven, O.; Davis, T.P.; Barner-Kowollik, C.; Barner, L. RAFT-mediated polymerization and grafting of sodium 4-styrenesulfonate from cellulose initiated via γ-radiation. *Polymer* **2009**, *50* (4), 973–982.
194. Barsbay, M.; Gueven, O.; Stenzel, M.H.; Davis, T.P.; Barner-Kowollik, C.; Barner, L. Verification of controlled grafting of styrene from cellulose via radiation-induced RAFT polymerization. *Macromolecules* **2007**, *40* (20), 7140–7147.
195. Boyer, C.; Bulmus, V.; Liu, J.; Davis, T.P.; Stenzel, M.H.; Barner-Kowollik, C. Well-defined protein–polymer conjugates via in situ RAFT polymerization. *J. Am. Chem. Soc.* **2007**, *129* (22), 7145–7154.
196. Liu, J.; Bulmus, V.; Herlambang, D.L.; Barner-Kowollik, C.; Stenzel, M.H.; Davis, T.P. In situ formation of protein–polymer conjugates through reversible addition fragmentation chain transfer polymerization. *Angew. Chem. Int. Ed.* **2007**, *46* (17), 3099–3103.
197. An, Z.S.; Tang, W.; Wu, M.H.; Jiao, Z.; Stucky, G.D. Heterofunctional polymers and core shell nanoparticles via cascade aminolysis/Michael addition and alkyne–azide click reaction of RAFT polymers. *Chem. Commun.* **2008**, *48* (48), 6501–6503.
198. Lutz, J.F. Polymerization of oligo(ethylene glycol) (meth)acrylates: Toward new generations of smart biocompatible materials. *J. Polym. Sci. A* **2008**, *46* (11), 3459–3470.
199. Hoogenboom, R. Poly(2-oxazoline)s: A polymer class with numerous potential applications. *Angew. Chem. Int. Ed.* **2009**, *48* (43), 7978–7994.
200. Schlaad, H.; Diehl, C.; Gress, A.; Meyer, M.; Demirel, A.L.; Nur, Y.; Bertin, A. Poly(2-oxazoline)s as smart bioinspired polymers. *Macromol. Rapid Commun.* **2010**, *31* (6), 511–525.
201. Li, W.; Zhang, A.; Chen, Y.; Feldman, K.; Wu, H.; Schluter, A.D. Low toxic, thermoresponsive dendrimers based on oligoethylene glycols with sharp and fully reversible phase transitions. *Chem. Commun.* **2008**, *45* (45), 5948–5950.
202. Li, W.; Zhang, A.; Feldman, K.; Walde, P.; Schluter, A.D. Thermoresponsive dendronized polymers. *Macromolecules* **2008**, *41* (10), 3659–3667.
203. Fernandez-Trillo, F.; van Hest, J.C.M.; Thies, J.C.; Michon, T.; Weberskirch, R.; Cameron, N.R. Fine-tuning the transition temperature of a stimuli-responsive polymer by a simple blending procedure. *Chem. Commun.* **2008**, *19* (19), 2230–2232.

204. Lutz, J.F.; Akdemir, O.; Hoth, A. Point by point comparison of two thermosensitive polymers exhibiting a similar LCST: Is the age of poly(NIPAM) over? *J. Am. Chem. Soc.* **2006**, *128* (40), 13046–13047.
205. Shen, W.Q.; Chang, Y.L.; Liu, G.Y.; Wang, H.F.; Cao, A.N.; An, Z.S. Biocompatible, antifouling, and thermosensitive core–shell nanogels synthesized by RAFT aqueous dispersion polymerization. *Macromolecules* **2011**, *44* (8), 2524–2530.
206. Grazon, C.; Rieger, J.; Sanson, N.; Charleux, B. Study of poly(N,N-diethylacrylamide) nanogel formation by aqueous dispersion polymerization of N,N-diethylacrylamide in the presence of poly(ethylene oxide)-b-poly(N,N-dimethylacrylamide) amphiphilic macromolecular RAFT agents. *Soft Matter* **2011**, *7* (7), 3482–3490.
207. Yan, L.F.; Tao, W. One-step synthesis of pegylated cationic nanogels of poly(N,N'-dimethylaminoethyl methacrylate) in aqueous solution via self-stabilizing micelles using an amphiphilic macroRAFT agent. *Polymer* **2010**, *51* (10), 2161–2167.
208. Bathfield, M.; D'Agosto, F.; Spitz, R.; Charreyre, M.T.; Pichot, C.; Delair, T. Sub-micrometer sized hairy latex particles synthesized by dispersion polymerization using hydrophilic macromolecular RAFT agents. *Macromol. Rapid Commun.* **2007**, *28* (15), 1540–1545.
209. Houillot, L.; Bui, C.; Save, M.; Charleux, B.; Farcet, C.; Moire, C.; Raust, J.A.; Rodriguez, I. Synthesis of well-defined polyacrylate particle dispersions in organic medium using simultaneous RAFT polymerization and self-assembly of block copolymers. A strong influence of the selected thiocarbonylthio chain transfer agent. *Macromolecules* **2007**, *40* (18), 6500–6509.
210. Saikia, P.J.; Lee, J.M.; Lee, K.; Choe, S. Reaction parameters in the RAFT mediated dispersion polymerization of styrene. *J. Polym. Sci. A* **2008**, *46* (3), 872–885.
211. Chen, Z.; Wang, X.L.; Su, J.S.; Zhuo, D.; Ran, R. Branched methyl methacrylate copolymer particles prepared by RAFT dispersion polymerization. *Polym. Bull.* **2010**, *64* (4), 327–339.
212. Huang, C.Q.; Pan, C.Y. Direct preparation of vesicles from one-pot RAFT dispersion polymerization. *Polymer* **2010**, *51* (22), 5115–5121.
213. Raust, J.A.; Houillot, L.; Save, M.; Charleux, B.; Moire, C.; Farcet, C.; Pasch, H. Two dimensional chromatographic characterization of block copolymers of 2-ethylhexyl acrylate and methyl acrylate, P2EHA-b-PMA, produced via RAFT-mediated polymerization in organic dispersion. *Macromolecules* **2010**, *43* (21), 8755–8765.
214. Wan, W.M.; Pan, C.Y. One-pot synthesis of polymeric nanomaterials via RAFT dispersion polymerization induced self-assembly and re-organization. *Polym. Chem.* **2010**, *1* (9), 1475–1484.
215. Gregory, A.M.; Thurecht, K.J.; Howdle, S.M. Controlled dispersion polymerization of methyl methacrylate in supercritical carbon dioxide via RAFT. *Macromolecules* **2008**, *41* (4), 1215–1222.
216. Lee, H.; Terry, E.; Zong, M.; Arrowsmith, N.; Perrier, S.; Thurecht, K.J.; Howdle, S.M. Successful dispersion polymerization in supercritical CO_2 using polyvinylalkylate hydrocarbon surfactants synthesized and anchored via RAFT. *J. Am. Chem. Soc.* **2008**, *130* (37), 12242–12243.
217. Zong, M.M.; Thurecht, K.J.; Howdle, S.M. Dispersion polymerisation in supercritical CO_2 using macroRAFT agents. *Chem. Commun.* **2008**, *45* (45), 5942–5944.
218. Li, Y.T.; Armes, S.P. RAFT synthesis of sterically stabilized methacrylic nanolatexes and vesicles by aqueous dispersion polymerization. *Angew. Chem. Int. Ed.* **2010**, *49* (24), 4042–4046.
219. Qi, G.G.; Jones, C.W.; Schork, F.J. RAFT inverse miniemulsion polymerization of acrylamide. *Macromol. Rapid Commun.* **2007**, *28* (9), 1010–1016.
220. Qi, G.; Eleazer, B.; Jones, C.W.; Schork, F.J. Mechanistic aspects of sterically stabilized controlled radical inverse miniemulsion polymerization. *Macromolecules* **2009**, *42* (12), 3906–3916.
221. Ouyang, L.; Wang, L.S.; Schork, F.J. RAFT inverse miniemulsion polymerization of acrylic acid and sodium acrylate. *Macromol. React. Eng.* **2011**, *5* (3–4), 163–169.
222. Ouyang, L.; Wang, L.S.; Schork, F.J. Synthesis and nucleation mechanism of inverse emulsion polymerization of acrylamide by RAFT polymerization: A comparative study. *Polymer* **2011**, *52* (1), 63–67.
223. Liu, O.Y.; Wang, L.S.; Schork, F.J. Synthesis of well-defined statistical and diblock copolymers of acrylamide and acrylic acid by inverse miniemulsion raft polymerization. *Macromol. Chem. Phys.* **2010**, *211* (18), 1977–1983.
224. Lu, F.J.; Luo, Y.W.; Li, B.G.; Zhao, Q.; Schork, F.J. Synthesis of thermo-sensitive nanocapsules via inverse miniemulsion polymerization using a PEO–RAFT agent. *Macromolecules* **2010**, *43* (1), 568–571.
225. Wang, Y.; Jiang, G.H.; Zhang, M.; Wang, L.; Wang, R.J.; Sun, X.K. Facile one-pot preparation of novel shell cross-linked nanocapsules: Inverse miniemulsion RAFT polymerization as an alternative approach. *Soft Matter* **2011**, *7* (11), 5348–5352.

226. Du, J.Z.; Sun, T.M.; Song, W.J.; Wu, J.; Wang, J. A tumor-acidity-activated chargeconversional nanogel as an intelligent vehicle for promoted tumoral cell uptake and drug delivery. *Angew. Chem. Int. Ed.* **2010**, *49* (21), 3621–3626.

227. Schmitt, F.; Lagopoulos, L.; Kauper, P.; Rossi, N.; Busso, N.; Barge, J.; Wagnieres, G.; Laue, C.; Wandrey, C.; Juillerat-Jeanneret, L. Chitosan-based nanogels for selective delivery of photosensitizers to macrophages and improved retention in and therapy of articular joints. *J. Control. Release* **2010**, *144* (2), 242–250.

228. Yin, Q.Q.; Wu, L.; Gou, M.L.; Qian, Z.Y.; Zhang, W.S.; Liu, J. Long-lasting infiltration anaesthesia by lidocaine-loaded biodegradable nanoparticles in hydrogel in rats. *Acta Anaesthesiol. Scand.* **2009**, *53* (9), 1207–1213.

229. Toita, S.; Sawada, S.; Akiyoshi, K. Polysaccharide nanogel gene delivery system with endosome-escaping function: Co-delivery of plasmid DNA and phospholipase A2. *J. Control. Release* **2011**, *155* (1) 54–59.

230. Tamura, A.; Oishi, M.; Nagasaki, Y. Enhanced cytoplasmic delivery of siRNA using a stabilized polyion complex based on PEGylated nanogels with a cross-linked polyamine structure. *Biomacromolecules* **2009**, *10* (7), 1818–1827.

231. Missirlis, D.; Kawamura, R.; Tirelli, N.; Hubbell, J.A. Doxorubicin encapsulation and diffusional release from stable, polymeric, hydrogel nanoparticles. *Eur. J. Pharm. Sci.* **2006**, *29* (2), 120–129.

232. Murthy, N.; Xu, M.; Schuck, S.; Kunisawa, J.; Shastri, N.; Frechet, J.M. A macromolecular delivery vehicle for protein-based vaccines: Acid-degradable protein-loaded microgels. *Proc. Natl. Acad. Sci. U.S.A.* **2003**, *100* (9), 4995–5000.

233. Lee, H.; Mok, H.; Lee, S.; Oh, Y.K.; Park, T.G.; Target-specific intracellular delivery of siRNA using degradable hyaluronic acid nanogels. *J. Control. Release* **2007**, *119* (2), 245–252.

234. Yu, S.; Hu, J.; Pan, X.; Yao, P.; Jiang, M. Stable and pH-sensitive nanogels prepared by self-assembly of chitosan and ovalbumin. *Langmuir* **2006**, *22* (6), 2754–2759.

235. Varga, I.; Szalai, I.; Meszaros, R.; Gilanyi, T. Pulsating pH-responsive nanogels. *J. Phys. Chem. B* **2006**, *110* (41), 20297–20301.

236. Chang, C.; Wang, Z.C.; Quan, C.Y.; Cheng, H.; Cheng, S.X.; Zhang, X.Z.; Zhuo, R.X. Fabrication of a novel pH-sensitive glutaraldehyde cross-linked pectin nanogel for drug delivery. *J. Biomater. Sci. Polym. Ed.* **2007**, *18* (12), 1591–1599.

237. Oishi, M.; Sumitani, S.; Nagasaki, Y. On-off regulation of 19F magnetic resonance signals based on pH-sensitive PEGylated nanogels for potential tumor-specific smart 19F MRI probes. *Bioconjug. Chem.* **2007**, *18* (5), 1379–1382.

238. Vinogradov, S.V.; Kohli, E.; Zeman, A.D. Cross-linked polymeric nanogel formulations of 5′-triphosphates of nucleoside analogues: Role of the cellular membrane in drug release. *Mol. Pharm.* **2005**, *2* (6), 449–461.

239. Ng, E.Y.; Ng, W.K.; Chiam, S.S. Optimization of nanoparticle drug microcarrier on the pharmacokinetics of drug release: A preliminary study. *J. Med. Syst.* **2008**, *32* (2), 85–92.

55 Electrospinning Technology
Polymeric Nanofiber Drug Delivery

*Narendra Pal Singh Chauhan, Kiran Meghwal,
Priya Juneja, and Pinki B. Punjabi*

CONTENTS

55.1 Introduction 1312
55.2 Types of Electrospinning Process.. 1312
    55.2.1 Coaxial Electrospinning 1312
    55.2.2 Emulsion Electrospinning 1312
55.3 Surface Modification Techniques of Electrospun Nanofibers 1314
    55.3.1 Plasma Treatment 1314
    55.3.2 Wet Chemical Method 1315
    55.3.3 Surface Graft Polymerization 1316
    55.3.4 Co-Electrospinning of Surface-Active Agents and Polymers 1316
55.4 Drug Loading Method on the Nanofiber Surface for Biomedical Applications 1317
    55.4.1 By Physical Adsorption 1317
        55.4.1.1 Simple Physical Adsorption 1317
        55.4.1.2 Nanoparticles Assembly on the Surface 1318
        55.4.1.3 Layer-by-Layer Multilayer Assembly 1318
    55.4.2 Chemical Immobilization 1318
55.5 Principle of Drug Delivery 1319
55.6 Application of Drug Delivery 1319
    55.6.1 PLA/Captopril 1320
    55.6.2 Amoxicillin-Loaded Electrospun Nano-HAp/PLA 1320
    55.6.3 Wound Dressing.. 1321
        55.6.3.1 Characterization of Fiber Scaffolds 1322
        55.6.3.2 Cell Viability, Attachment, and Proliferation 1323
    55.6.4 Electrospun PU–Dextran Nanofiber 1323
        55.6.4.1 Characterization 1323
        55.6.4.2 Antimicrobial Test of the Composite Nanofibers 1323
    55.6.5 Rifampin PLLA Electrospun Fibers 1324
    55.6.6 Chitosan Nanofiber for Antibacterial Applications 1325
        55.6.6.1 Immobilization of Lysozyme-CLEA onto Electrospun CS Nanofibers ... 1326
        55.6.6.2 Antibacterial Measurement... 1327
55.7 Conclusion 1327
Acknowledgment 1328
References... 1328

55.1 INTRODUCTION

Electrospinning is a technique of producing ultrafine fibers with diameters starting from 10 nm to several microns either in the form of solution or molten liquid by the application of high voltage (30–50 kV). This process is conventional drawing of the fibers where external pressure is used. The features of electrospun nanofibers are high specific surface area, high porosity, and three-dimensional (3D) reticulate structures, which is quite similar to a natural extracellular matrix. Due to these features, electrospun nanofibers have attracted attention for their potential applications in different fields.[1–5] These features allow them to have a wide range of biomedical applications such as tissue engineering,[6] wound dressing,[7] biosensors,[8] and particularly for drug delivery applications.[5,9–12] Electrospinning technique has been used to fabricate nanofibers for drug encapsulation and release. In the late 1500s, Gilbert observed the elctrospraying process. He observed that when a suitably charged piece of amber was brought near to a droplet of water it formed a cone shape and a small droplet ejected from the tip of the cone. Later in the 1950s Cooley and many researchers patented electrospinning process. The various advantages associated with the electrospinning process include that long continuous fibers can be produced and process can be used on laboratory as well as industrial scale. The process also offers cost-effective considerations.

55.2 TYPES OF ELECTROSPINNING PROCESS

Chiefly, the electrospinning process may be of two types:

1. Coaxial[12]
2. Emulsion[13,14]

55.2.1 Coaxial Electrospinning

A coaxial setup allows for the injection of one solution into another at the tip of the spinneret by using a multiple solution feed system. The sheath fluid acts as a carrier, which draws in the inner fluid at the Taylor Cone of the electrospinning jet.[15] If the solutions are immiscible, then a core shell structure is usually observed. Miscible solutions, however, can result in porosity or a fiber with distinct phases due to phase separation during solidification of the fibers. Taylor cones may be stationary, but they are never static features. Their apices are always the source of emission of charged particles under a rich range of regimens. In this regimen, a steady jet issues continuously from the cone apex, eventually breaking into a spray of charged drops or electrospray (Figure 55.1). This steady regimen is not only the best known but also the simplest to analyze, but the limit of high electrical conductivity of the liquid, where the jet radius is typically smaller than 1 μm, enables a division of the problem into two regions. Outer domains (the cone), which are effectively hydrostatic, are slow and have little influence on jet formation, in which the liquid behaves as infinitely conducting, and an inner region, which is dynamic and where a very fine jet carrying a finite current and flow rate forms. Taylor cones of highly conducting liquids offer the only known scheme to produce submicrometer and nanometer jets (down to a diameter of about 10 nm).

55.2.2 Emulsion Electrospinning

Emulsions can be used to create a core shell or composite fibers without modification of the spinneret. However, these fibers are usually more difficult to produce as compared to coaxial spinning due to the greater number of variables, which must be accounted for in creating the emulsion. A water phase and an immiscible solvent phase are mixed in the presence of an emulsifying agent to form the emulsion. Any agent that stabilizes the interface between the immiscible phases can be used for this like surfactants such as sodium dodecyl sulfate, Triton, and nanoparticles. During the electrospinning process, the emulsion droplets within the fluid are stretched and gradually confined that leads to their coalescence. If the volume fraction of inner fluid is sufficiently high, then a continuous inner core can be formed.[16]

Electrospinning Technology

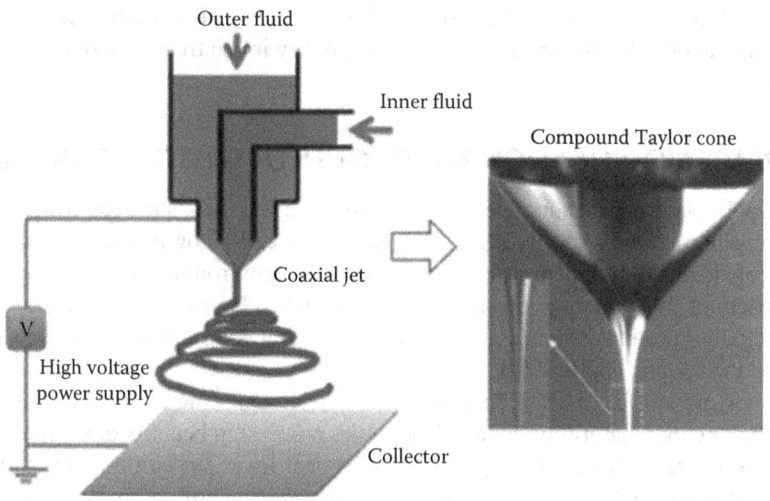

FIGURE 55.1 Diagram showing fiber formation by coaxial electrospinning.

Further, a modified electrospinning method was introduced for the creation of an elastomeric, fibrous sheet [fibrous composite sheet with two distinct submicrometer fiber populations: biodegradable poly(ester urethane) urea (PEUU) and poly(lactide-co-glycolide) (PLGA)], where the PLGA was loaded with the antibiotic tetracycline hydrochloride (PLGA–tet) capable of sustained antibacterial activity in vitro. Composite sheets were flexible with breaking strains exceeding 200%, tensile strengths of 5–7 MPa, and high suture-retention capacity. The blending of PEUU fibers markedly reduced the shrinkage ratio observed for PLGA–tet sheets in buffer from 50% to 15%, while imparting elastomeric properties to the composites. In the development of this material, a new approach to two-stream electrospinning (Figure 55.2) was used in which one component stream

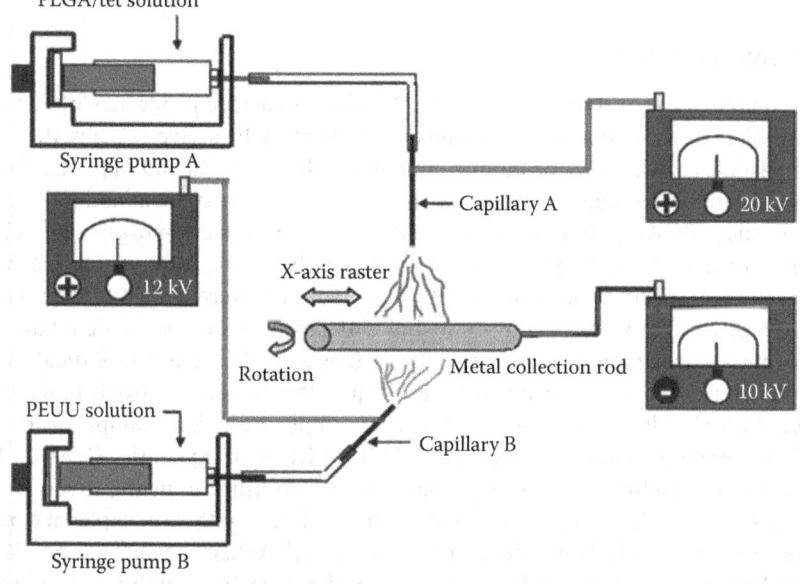

FIGURE 55.2 Two-stream electrospinning setup. (Reprinted from *Polymer*, 49(26), Agarwal, S., Wendorff, J.H., and Greiner, A., Use of electrospinning technique for biomedical applications, 5603–5621. Copyright 2008, with permission from Elsevier.)

provided for antibiotic release while the other provided mechanical properties deemed essential for the desired application. This material may find applicability in the treatment of temporary abdominal wall closure.

55.3 SURFACE MODIFICATION TECHNIQUES OF ELECTROSPUN NANOFIBERS

In order to apply electrospun nanofibers in biomedical uses, their surfaces have been modified chemically and physically with bioactive molecules and cell-recognizable ligands. This subsequently has provided bio-modulating or biomimetic microenvironments for contacting cells and tissues. A variety of functionalization strategies of synthetic electrospun nanofibers with bioactive molecules including proteins, nucleic acids, and carbohydrates have been employed for advanced biological and therapeutic applications.[18]

Synthetic polymers offer easier processability for electrospinning and more controllable nanofibrous morphology than natural polymers. Due to the processing benefits of synthetic polymers, a wide variety of natural polymers having unique biological functions can be immobilized onto the nanofibrous surface of synthetic polymers without compromising bulk properties. Such biologically functionalized synthetic nanofibers can direct enhanced, cell-specific phenotype and organization, because tissue regeneration process is strongly involved with diverse biochemical cues on the cell-contacting surface.[19] For continuous drug delivery applications, the electrospinning process enables a variety of hydrophobic therapeutic agents, to be directly incorporated within the bulk phase of nanoscale fibers for controlled release. For example, a biodegradable polymer solution containing hydrophobic anti-cancer drugs such as paclitaxel was directly electrospunned to produce drug-releasing nanofibrous mesh.[20] Alternatively, hydrophilic and charged macromolecular drugs such as proteins and nucleic acids were immobilized covalently and physically onto the modified surface of nanofibrous mesh for modulating the cellular functions. The electrospun nanofiber mesh so produced possesses a highly interconnected open nanoporous structure with a high specific surface area that offers an ideal condition for sustained and local drug delivery.[21] Various surface modification techniques for applying synthetic polymer nanofibers to tissue engineering and drug delivery are as follows:

55.3.1 PLASMA TREATMENT

Plasma is a complex energy source for the modification of surface properties of various materials including biomaterials. Plasma is composed of chemically active species that are excited and ionized particles, which may be atomic, and molecular, photons and radicals. These species are highly energetic to induce the chemical reactions.[22] Electrospun nanofibers composed of poly(glycolic acid) (PGA), poly(L-lactic acid) (PLLA), or poly(lactic-co-glycolic acid) (PLGA) have been modified with carboxylic acid groups through plasma glow discharge with oxygen and gas-phased acrylic acid.[23] Such hydrophilized nanofibers were shown to increase fibroblast adhesion and proliferation without compromising with physical and mechanical bulk properties. Air or argon plasma treatment has been widely used as a facile surface modification technique for many biomaterials, because using this technique, the surface hydrophilicity can be easily increased with simultaneous removal of surface impurities. For example, various electrospun nanofibers made of poly(ε-caprolactone) (PCL), PCL/hydroxyapatite (HAp), PS, and silk fibroin were surface modified by air or argon plasma, resulting in an improved cell adhesion and proliferation.[24–27] When PCL nanofibers were modified with argon plasma, rich-quality carboxylic acid groups could be produced on the surface.[28] When the surface-activated nanofibers were soaked in a simulated body fluid solution, the bone-like calcium phosphate mineralization occurred on the surface of nanofibers. This mineralized nanofibrous scaffold exhibited improved wettability with a bio-mimicking bone structure which indicated potential application for bone grafting (Figure 55.3).

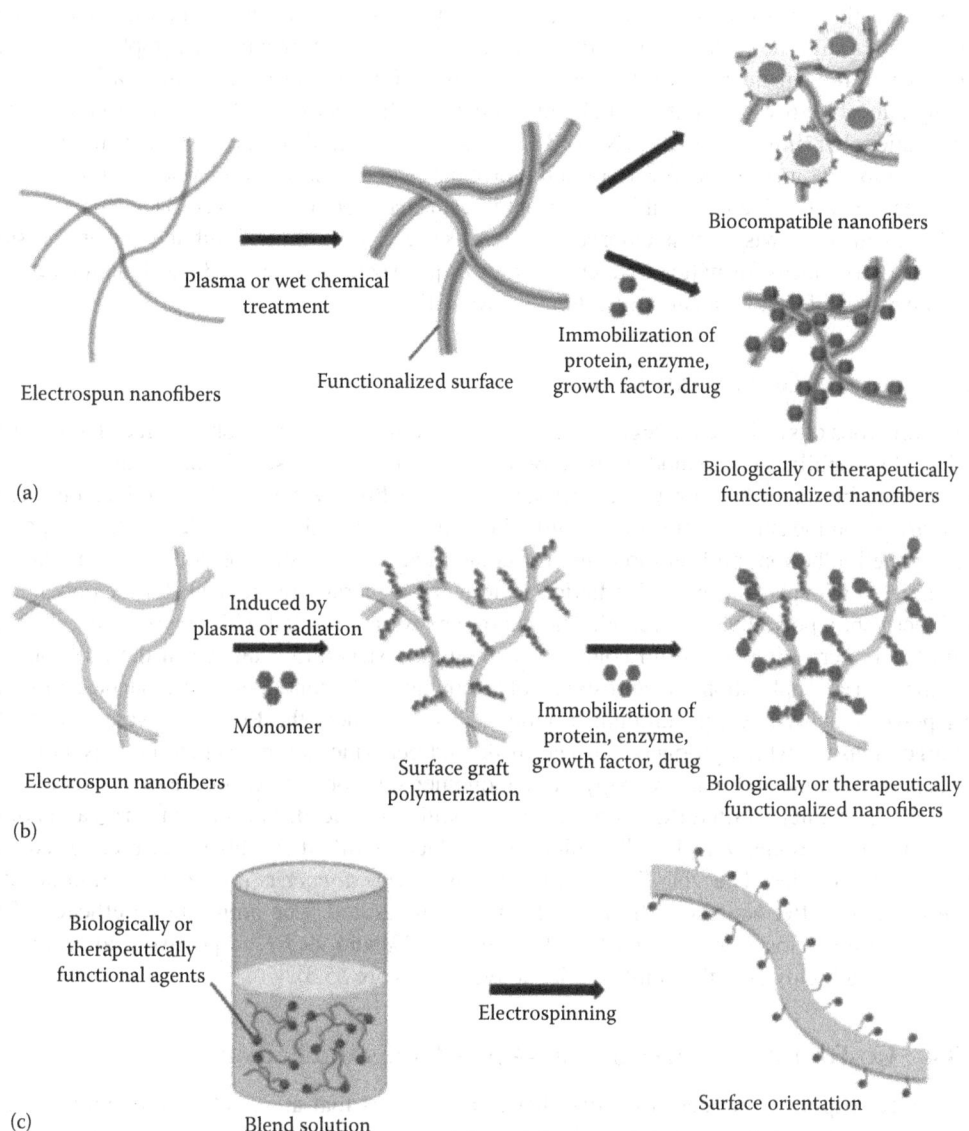

FIGURE 55.3 Surface modification techniques of electrospun nanofibers. (a) Plasma treatment or wet chemical method, (b) surface graft polymerization, and (c) co-electrospinning. (Reprinted from *Adv. Drug Deliv. Rev.*, 61(12), Hyuk, S.Y., Taek, G.K., and Tae, G.P., Surface-functionalized electrospun nanofibers for tissue engineering and drug delivery, 1033–1042. Copyright 2009, with permission from Elsevier.)

55.3.2 Wet Chemical Method

Since the plasma treatment for nanofibrous mesh cannot effectively modify the surface of buried nanofibers deeply located in the mesh due to the limited penetration depth of plasma in the nanopores, wet chemical etching methods can offer the flexibility for surface modification of thick nanofibrous meshes. When biodegradable polymeric nanofibrous meshes are surface modified using the partial hydrolysis method, as shown in Figure 55.3, special care must be taken. The duration of the hydrolysis and the concentration of hydrolyzing agents are important to optimally produce surface functional groups only by minimum change in the bulk properties.[29] Since carboxylic acids can chelate calcium ions, surface-induced nucleation and growth of minerals were shown to be enhanced on the surface-modified PLLA electrospun scaffold. PCL electrospun nanofibers were also used to modify the surface of thin PCL membrane for generating nanotopographical surface.[30] When the modified membrane was treated with 5 M NaOH, wettability was dramatically enhanced, showing almost zero water contact angle due to the capillary action on the highly rough surface. When National Institute of Health (NIH) 3T3 cells were cultured on the surface of the modified nanotopographical membrane, favorable cell morphology and adhesion was observed on the modified surface, possibly due to the unique hydrophilic surface topography.[19] NIH 3T3 cells are established from an NIH Swiss mouse embryo. These cells are highly contact inhibited and are sensitive to sarcoma virus focus formation and leukemia virus propagation. The "3T3" designation refers to the abbreviation of 3-day transfer, inoculum 3×10^5 cells.

55.3.3 Surface Graft Polymerization

Almost all types of synthetic biodegradable polymers retain their hydrophobic surface nature, often requiring hydrophilic surface modification for desired cellular responses. Surface graft polymerization has been introduced not only to confer surface hydrophilicity but also to introduce multifunctional groups on the surface for covalent immobilization of bioactive molecules for the purpose of enhanced cell adhesion, proliferation, and differentiation.[31–34] The surface graft polymerization is often initiated with plasma and UV radiation treatment to generate free radicals for the polymerization. Electrospun polyethylene terephthalate nanofibers were modified with poly(methacrylic acid) by graft polymerization in a mild condition without any structural damage in the bulk phase.[35] For antibacterial applications, electrospun polyurethane (PU) nanofibers were surface modified using poly(4-vinyl-N-hexyl pyridinium bromide).[36] In this study, the PU fibers were first treated with argon plasma, which produced surface oxide and peroxide groups. When the plasma-treated PU fibers were immersed in a 4-vinylpyridine monomer solution with exposure of UV irradiation, poly(4-vinylpyridine)–grafted PU fibers were successfully produced. Through quaternization of the grafted pyridine groups with hexylbromide, the surface-modified PU fibers were endowed with antibacterial activities. The viability of gram-positive *Staphylococcus aureus* and gram-negative *Escherichia coli* after contact with the PU fibers was measured. The antibacterial efficacy of the modified PU fibers for *S. aureus* and *E. coli* were 99.999% and 99.9%, respectively, after 4 h contact, indicating highly effective antibacterial activities (Figure 55.3).

55.3.4 Co-Electrospinning of Surface-Active Agents and Polymers

The co-electrospinning process uses two different solutions that are electrospun simultaneously using a spinneret with two coaxial capillaries to produce core/shell nanofibers. The core is then selectively removed and hollow fibers are formed. A well-stabilized co-electrospinning process can be achieved when both the solutions are sufficiently viscous and even spinnable and the solutions are immiscible. When PLLA solution blended with Hap nanocrystals was co-electrospun, HAp was exposed on the surface of the resultant fibers, giving rise to high surface free energy and low water contact angle.[37] These composite fibers exhibited a retarded degradation rate as compared to pure PLLA fibers due to the internal ionic bonding between ester groups in PLLA and calcium ions in HAp. In addition, a novel in situ peptide bio-functionalization method driven by an electric field was developed.[38] Firstly, an antimicrobial peptide, with three repeating units of three anionic amino acids, serine, glutamic acid, and another glutamic acid (SEE)$_3$, was terminally conjugated to polyethylene oxide (PEO). The addition of (SEE)$_3$-PEO conjugate to PEO solution decreased viscosity but increased the solution conductivity. During co-electrospinning of the PEO/(SEE)$_3$-PEO blend solution, electrically polarizable SEE segment had significant influence on fiber morphology. When the collector was connected as an anode, thick and inter-welded fiber morphology could be observed due to the high flow rate of the blend solution under an electric field (Figure 55.3).

55.4 DRUG LOADING METHOD ON THE NANOFIBER SURFACE FOR BIOMEDICAL APPLICATIONS

55.4.1 BY PHYSICAL ADSORPTION

Drug-friendly physical immobilization on the surface can be achieved by using surface-modified prefabricated nanofibrous meshes that have an extremely high surface area to volume ratio which results in higher drug loading amount per unit mass than any other devices. The immediate release of drugs from the nanofiber surface can enable facile dosage control of some therapeutic agents which suits some specific applications such as prevention of bacterial infection occurring within few hours after surgery.[39] In addition to the high drug-loading capacity, hierarchically organized structure such as drug-loaded nanoparticles adsorbed on the nanofibers can allow unique drug-releasing profiles, which nanofiber itself cannot achieve.[40,41] Figure 55.4 describes three different modes of physical drug loading methods on the surface of electrospun nanofibers for drug delivery application. These are simple physical adsorption, nanoparticles adsorption, and adsorption of multilayer assembly.

55.4.1.1 Simple Physical Adsorption

Physical surface adsorption is the simplest approach for loading drug on the nanofibrous mesh. The use of specific interaction between heparin and growth factor is a very typical example. Heparin, a highly sulfated glycosaminoglycan, has strong binding affinity with various growth factors such as fibroblast growth factor, vascular endothelial growth factor, heparin-binding epidermal growth factor, and transforming growth factor-β. This approach offers preservation of biological activity by preventing early degradation of growth factors. Heparin immobilization of biomaterial surface and subsequent attachment of growth factor can be the most efficient way for local delivery of growth factors and consequent mitogenic induction.[42]

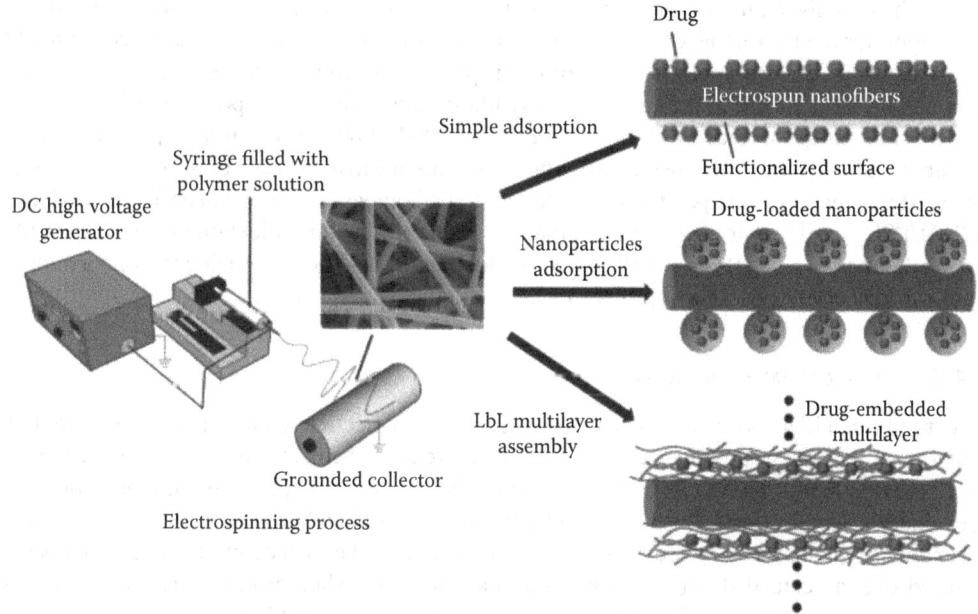

FIGURE 55.4 Three modes of physical drug loading on the surface of electrospun nanofibers. (Reprinted from *Adv. Drug Deliv. Rev.*, 61(12), Hyuk, S.Y., Taek, G.K., and Tae, G.P., Surface-functionalized electrospun nanofibers for tissue engineering and drug delivery, 1033–1042. Copyright 2009, with permission from Elsevier.)

The surface-modified nanofibrous meshes have been proven to be an efficient barrier for preventing postsurgical adhesion. As an adhesion barrier, a commercial antibiotic drug, Biteral®, was adsorbed to the electrospun PCL nonwoven sheet.[43] To load the drug, the drug solution was simply dropped on the PCL nonwoven sheet and left to be completely adsorbed. Due to the unique morphological feature of electrospun fibers, such drug loading can be a very straightforward and effective method. A rapid drug release profile is highly desirable for preventing infections at an early stage.

55.4.1.2 Nanoparticles Assembly on the Surface

The large interfacial areas of the nanofibers make possible the fabrication of high-performance devices. Any combination of both drug-encapsulated polymeric nanoparticles and nanofibers can be possible to have multifunctional performance. Electrospinning technique is a good method for the formation of such multifunctional devices in which functional nanoparticles were readily embedded within or adsorbed onto nanofibrous mesh.[44,45] An interesting assembly design using nanoparticles and nanofibers was reported for drug delivery applications[40] using electrospinning and electrospraying techniques. The assembly is made up of nanoparticles on nanofibers hierarchical structure having attractive forces between opposite charges. It was reported that poly(methyl methacrylate) (PMMA) solution was electrospun and PS solution was simultaneously electrosprayed from two separate counter-charged nozzles in a side-by-side fashion. Two oppositely charged electrohydrodynamic jets were encountered with neutralization, resulting in a composite structure consisting of electrospun PMMA nano- or microfibers uniformly combined with PS nano- or microparticles. When lipoic acid (antioxidant drug) or gold nanoparticles suspension were electrosprayed, electrospun PMMA fibers were successfully modified with the corresponding composition in an in situ manner.

55.4.1.3 Layer-by-Layer Multilayer Assembly

The method comprises an alternative layer-by-layer deposition of polyanions and polycations principally driven by an electrostatic force on charged substrates, resulting in self-assembled multilayer coating or free-standing film. This technique has attracted considerable attention due to the ease of its synthesis, universality for any complex structure of substrate, and the possibility of using any composition for the coating layer. Many bioactive agents and chemical drugs have been assembled for topical drug delivery applications mainly on planar substrates such as silicon wafer, quartz slides, and metal oxide due to the ease of their synthesis and analysis. The polyelectrolyte multilayer was deposited on the surface of electrospun PS fibers.[46] PS fibers were initially prepared by electrospinning and then the fiber surface was endowed with negative charges by sulfonation of phenyl groups. Moreover, gold nanoparticles were shown to be homogenously and densely assembled into the PAH/PSS multilayer. It was suggested that this facile surface modification of electrospun fibers with synthetic polymers (PAH and PSS) and biopolymer (DNA) could provide the opportunity for creating a variety of drug-releasing surfaces for biomedical applications.

55.4.2 CHEMICAL IMMOBILIZATION

In order to immobilize bioactive molecules on the surface of electrospun nanofibers, chemical modification must be done to produce reactive functional groups. Chemical immobilization of bioactive molecules on to the surface of electrospun nanofibers is preferred over physical immobilization in tissue engineering applications. This is so because the immobilized molecules are covalently attached to the nanofibers, and hence, they are not easily leached out from the surface-modified nanofibers when incubated over an extended period. Primary amine and carboxylate groups were most extensively employed to immobilize bioactive molecules onto the surface of nanofibers.[47] Carboxylic groups on the surface of polymeric nanofibers containing different amounts of polyacrylic acid were employed for conjugation with collagen.[23] In other studies, acrylic acid–immobilized nanofibers were conjugated to amine groups to prepare aminated nanofibers.[48] These nanofibers were further employed for in vitro cultivation of umbilical cord blood cells.[48,49] Hydroxyl groups were also employed for the chemical immobilization of bioactive molecules. Modified PS with a hydroxyl containing initiator was electrospun and alpha chymotrypsin was covalently attached on the surface.[50] Upon the advancement of electrospinning technology, nanofibrous polymer scaffolds have been diversified in their polymer sources, fiber size, porosity, and texture. Biodegradable synthetic polymers, that is, PLLA, PGA, PLGA, and PCL, as well as natural polymers, such as collagen, silk proteins, and fibrinogen, are fabricated into nanofibers through electrospinning.[51] These nanofiber-based polymer scaffolds have been extensively used for tissue-engineered cartilage or bone, skin, and nerves.[52–55] Transformations of hydrophobic surface either physically or chemically into hydrophilic one have been used to improve poor cell adhesive characteristics. Some bioactive proteins, that is, fibronectin, collagen, and laminin or Arg–Gly–Asp (RGD) peptide can be adsorbed on the polymer surface to increase cellular attachment.[56] Chemicals were also used to change surface characteristics. When PGA mesh scaffolds were treated with sodium hydroxide, the cell seeding density is increased, mainly due to the chemically modified surface with the hydrophilic functional groups.[57] Pre-treatment of electrospun PGA nanofibers in concentrated hydrochloric acid considerably improved cell attachment.[58]

55.5 PRINCIPLE OF DRUG DELIVERY

Drug delivery with polymer nanofibers is based on the principle that dissolution rate of a drug particulate increases with increased surface area of both the drug and the corresponding carrier, if necessary. For controlled drug delivery, in addition to their large surface area to volume ratio, polymer nanofibers also have other additional advantages. For example, unlike common encapsulation which involves controlled delivery systems, the polymer nanofibers have been used to improve the therapeutic efficacy and safety of drugs by delivering them to the site of action at a rate dictated by the need of the physiological environment. A wide variety of polymeric materials have been used as delivery matrices and the choice of the delivery vehicle polymer is determined by the requirements of the specific application.[59]

55.6 APPLICATION OF DRUG DELIVERY

The applications of some of the polymer nanofibers as drug delivery systems have been described (Table 55.1).

TABLE 55.1
Some of the Representative Electrospun Systems Used for Drug-Delivery Applications

Electrospun Mat	Drug
Poly(caprolactone), PCL	Diclofenac sodium
	Tetracycline hydrochloride
	Resveratrol
	Gentamycin sulfate
	Biteral
Poly(lactic acid), PLA	Tetracycline hydrochloride and mefoxin
Poly(caprolactone-D,L-lactide)	Diclofenac sodium
Poly(vinyl alcohol), PVA	Diclofenac sodium, tetracycline hydrochloride
	Sodium salicylate, naproxen, indomethacin
Poly(maleic anhydride-alt-2-methoxyethylvinylether)	Diclofenac sodium
Poly(lactide–glycolide), PLGA	Paclitaxel (anticancer), tetracycline hydrochloride
Poly(ethylene-*co*-vinylacetate)	Tetracycline hydrochloride
Gelatin	*Centella asiatica*-herbal
	Extract
Cellulose acetate	Vitamin A and E

Source: Reprinted from *Polymer*, 49(26), Agarwal, S., Wendorff, J.H., and Greiner, A., Use of electrospinning technique for biomedical applications, 5603–5621. Copyright 2008, with permission from Elsevier.

55.6.1 PLA/Captopril

The nanofiber membranes of PLLA/Captopril were prepared by the electrospinning technique.[60] Captopril is used widely for the treatment of hypertension and congestive heart failure.[61] Clinically, the general captopril tablets often need to be used three times a day for a long period. The side effects associated are vertigo and headache.[62] To avoid side effects, a novel formulation of captopril using electrospinning technique has been made for promising clinical applications. When the drug is dissolved in the polymer solution and electrospunned into the composite nanofibers, then it can make the drug to be loaded in the carrier of polymer nanofibers. It can not only achieve a relatively high bioavailability of the loaded drug but also minimize their severe side-effects. Further, more than one drug can be encapsulated directly into the electrospun fibers.[60]

55.6.2 Amoxicillin-Loaded Electrospun Nano-HAp/PLA

Drug-loaded halloysite nanotubes (HNTs), a naturally occurring clay material, can be incorporated within PLGA nanofibers by simply electrospinning the mixture solution of PLGA and HNTs drug particles.[5] The hybrid nanofibers formed as a consequence afford the drug with a significantly decreased burst release profile, and simultaneous incorporation of HNTs has greatly improved the mechanical durability of the nanofibers.[5,63–65] The incorporated HNTs themselves are a kind of drug carrier, which allows drug molecules to be encapsulated within the lumen of the HNTs.

Nano-HAp (n-HA) has been considered as an ideal inorganic drug carrier due to its high surface area to volume ratio, high surface activity, good biocompatibility, porosity, surface hydrophilicity, and strong ability to absorb a variety of chemical species.[66]

A model drug, amoxicillin (AMX) was first loaded onto the n-HA surface via physical adsorption. Then the AMX-loaded n-HA particles were mixed with PLGA solution for subsequent formation of electrospun AMX/n-HA/PLGA composite nanofibers (Scheme 55.1a).

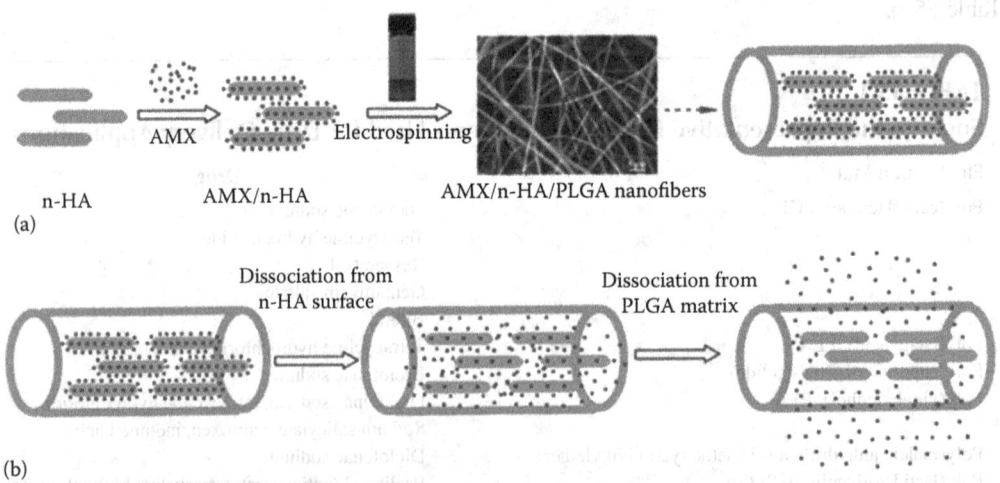

SCHEME 55.1 (a) Schematic illustration of the encapsulation and (b) release pathways of AMX within n-HA–doped PLGA nanofibers. (Reprinted from *Biomaterials*, 34(4), Zheng, F., Wang, S., Wen, S., Shen, M., Zhu, M., and Shi, X., Characterization and antibacterial activity of amoxicillin-loaded electrospun nano-hydroxyapatite/poly(lactic-co-glycolic acid) composite nanofibers, 1402–1412, 2013. Copyright 2008, with permission from Elsevier.)

TABLE 55.2
Apparent Density, Porosity, and Water Contact Angle of PLGA, n-HA/PLGA, and AMX/n-HA/PLGA Nanofibers

Sample	Apparent Density (g/cm³)	Porosity (%)	Water Contact Angle (°)
PLGA	0.357 ± 0.087	71.5 ± 6.9	139.2 ± 2.1
n-HA/PLGA	0.357 ± 0.067	71.4 ± 5.4	136.3 ± 2.2
AMX/n-HA/PLGA	0.315 ± 0.02	74.8 ± 1.6	137.2 ± 2.9

Source: Reprinted from *Biomaterials*, 34(4), Zheng, F., Wang, S., Wen, S., Shen, M., Zhu, M., and Shi, X., Characterization and antibacterial activity of amoxicillin-loaded electrospunnano-hydroxyapatite/poly(lactic-co-glycolic acid) composite nanofibers, 1402–1412, 2013. Copyright 2008, with permission from Elsevier.

Note: Data are representative of independent experiments and all data are given as mean ± SD, n = 5.

The loading of AMX onto n-HA (n-HA/AMX) and the formation of AMX/n-HA/PLGA composite nanofibers were characterized by using different techniques. The antimicrobial activity of the AMX/n-HA/PLGA nanofibers was investigated by using *S. aureus* as a model bacterium both in liquid and in solid medium.

The structural formula of n-HA is $Ca_{10}(PO_4)_6(OH)_2$. It is a principal inorganic ingredient of bone and teeth of the mammal.[68] The loading percentage of AMX onto n-HA was optimized by varying the respective concentration of AMX and n-HA. The apparent density, porosity, and water contact angle data of PLGA, n-HA/PLGA, and AMX/n-HA/PLGA nanofibers (data are representatives of independent experiments and all data are given as mean ± SD, n = 5) are listed in Table 55.2.

The incorporation of drug-loaded n-HA not only significantly improve the mechanical durability of the nanofibers but also appreciably weaken the initial burst release of the drug. The combination of the two pathways for the AMX dissociation, that is, first from n-HA surface to PLGA fiber matrix and second from PLGA fiber matrix to the release medium, is proven to be an efficient strategy to slow down the release rate of AMX. This is important for biomedical applications requiring the drug to maintain a long-term antibacterial efficacy.

55.6.3 Wound Dressing

The ideal wound dressing should minimize infection and pain, prevent excessive fluid loss, maintain a moist healing environment, promote epithelial restoration, and be biocompatible. However, dressings available in the market do not meet all the requirements necessary for an ideal dressing. In addition to the application of dressing, wound treatment includes irrigation of the affected area with an anesthetic solution followed by application of prophylactic antibiotics to prevent wound infection.

Based on these ideas, the concept of a dual drug scaffold wound dressing was put forward. Such a scaffold would offer a unique combination of the inherent properties of electrospun scaffolds such as promoting cell proliferation and simultaneously providing anesthetic and antibiotic activity for pain relief and healing.[69]

In a study, the drugs lidocaine hydrochloride (LH) and mupirocin have been selected to include in drug scaffold.[70] Lidocaine is a routinely used anesthetic in wound-related pain management. It has some antibacterial actions against *E. coli*, *S. aureus*, *Pseudomonas aeruginosa*, and *Candida albicans*. Mupirocin is a commonly used antibiotic in wound care for prophylaxis against cutaneous infection and elimination of carriage of *S. aureus*. It is effective against aerobic gram-positive and some gram-negative flora. The low incidence of adverse effects and rare cases of resistance

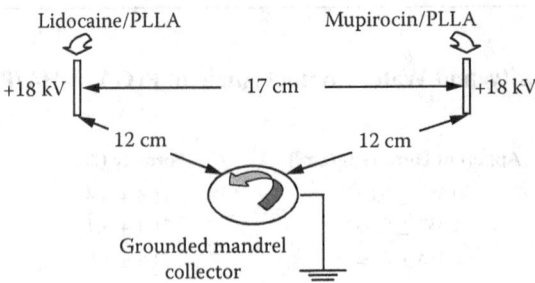

FIGURE 55.5 Schematic of the dual spinneret electrospinning apparatus. (Reprinted from *Int. J. Pharm.*, 364(1), Thakur, R.A., Florek, C.A., Kohn, J., and Michniak, B.B., Electrospun nanofibrous polymeric scaffold with targeted drug release profiles for potential application as wound dressing, 87–93. Copyright 2008, with permission from Elsevier.)

add to its attractiveness as an antibiotic of choice. It is formed as a 2% cream or ointment that is applied three times daily from 3 to 10 days as required. The wound is covered with gauze after its application.

To achieve the proposed dual release, a novel electrospinning technique with simultaneous electrospinning from dual spinnerets has been investigated and depicted in Figure 55.5.

55.6.3.1 Characterization of Fiber Scaffolds

Fiber scaffolds containing fibers of two unique compositions were obtained using the DS electrospinning apparatus. Macroscopically, the scaffold was a conformable and resilient structure having ultrafine cloth appearance. Fluorescence microscopy of the scaffold, which contained one fiber doped with Texas Red and another fiber without Texas Red, showed homogenous distribution of the two fibers (Figure 55.6). One can observe two larger diameter fibers: One of which is clearly fluorescent (Fiber 1) and another which is not fluorescent (Fiber 2).

The DS electrospinning apparatus could be used to electrospin a hybrid mesh of materials of varying degradation rates, mechanical properties, or chemical functionality. Here, the technique was used to create a mesh where one fiber was loaded with an antibiotic and another fiber was loaded with an anesthetic.

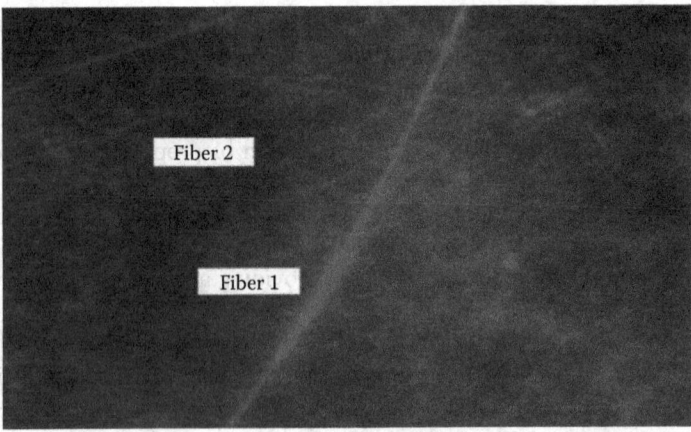

FIGURE 55.6 (Fiber 2) Light microscopy image of electrospun fiber scaffolds. (Fiber 1) Fluorescence image of electrospun fiber scaffolds. (Reprinted from *Int. J. Pharm.*, 364(1), Thakur, R.A., Florek, C.A., Kohn, J., and Michniak, B.B., Electrospun nanofibrous polymeric scaffold with targeted drug release profiles for potential application as wound dressing, 87–93. Copyright 2008, with permission from Elsevier.)

Above images are fabricated by DS technique. Fiber 1 is clearly visible by both imaging techniques while Fiber 2 is not visible by fluorescence microscopy.

55.6.3.2 Cell Viability, Attachment, and Proliferation

Wound-healing scaffolds should be able to support cell proliferation and viability for fast wound healing. Electrospun PLLA has been reported to support the growth of cells such as neural stem cells[71] and cardiac myocytes.[72] It is possible that inclusion of drugs may alter the cell proliferation in vivo.[73] However, it was found that the histopathologic appearance of wounded tissues infiltrated with lidocaine did not vary consistently in relation to collagenization, edema, or acute and chronic inflammatory processes. Lidocaine alone cannot alter wound healing or the breaking strength of the wounds.[74] The dual spinneret electrospinning technique facilitated the fabrication of a polymeric dressing with dual drug release kinetics that could have potential application for wound therapy. An anesthetic and LH were crystallized in the PLLA matrix and eluted through a burst release mechanism. This action would be useful if this mat was used as a wound dressing for immediate relief of pain. Mupirocin, an antibiotic, was released simultaneously with LH through a diffusion-mediated mechanism for extended antibiotic activity. The dual spinneret electrospinning technique achieved the required dual release profiles by allowing LH to crystallize in other PLLA fibers and also by maintaining mupirocin in the non-crystallized form within the PLLA matrix.

55.6.4 ELECTROSPUN PU–DEXTRAN NANOFIBER

Dextran is a versatile biomacromolecule for preparing electrospun nanofibrous membranes by blending with either water-soluble bioactive agents or hydrophobic biodegradable polymers for biomedical applications. Dextran is a bacterial polysaccharide made up of R-1, 6 linked D-glucopyranose residues with some R-1, 2-, R-1, 3-, or R-1, 4 linked side chains. Due to its biodegradability and biocompatibility, it has been used for various biomedical applications.[75] In comparison to other biodegradable polymers, dextran is inexpensive and readily available. Most importantly, dextran is soluble in both water and some organic solvents. This unique solubility characteristic of dextran put forth the possibility of directly blending it with biodegradable hydrophobic polymers such as PU to prepare composite nanofibrous membranes by using electrospinning method.

PU is a commonly used candidate for wound dressing applications because of its good barrier properties and oxygen permeability.[76] Ciprofloxacin HCl (CipHCl), a fluoroquinolone antibiotic, is one of the most widely used antibiotics in wound healing because of its low minimal inhibitory concentration for both gram-positive and gram-negative bacteria that cause wound infections.[77] The frequency of spontaneous resistance to ciprofloxacin is very low.[78] An electrospun nanofiber membrane containing antibiotic agents has been used as a barrier to prevent the postwound infections. The combination of both these properties can result in a perfect wound dressing material.

55.6.4.1 Characterization

The morphology of the electrospun PU, PU–dextran, and PU–dextran drug-loaded composite nanofibers was observed by using scanning electron microscopy shown in Figure 55.7.

55.6.4.2 Antimicrobial Test of the Composite Nanofibers

The composite drug-loaded nanofibers have been screened for their antimicrobial activity after an overnight incubation of the agar plate at 37°C. The bactericidal activity was reported by the measurement of zone of inhibition within and around the drug-loaded nanofiber mat for gram-positive and for gram-negative bacteria. As shown in Figure 55.8a (1, 2) for gram-positive bacteria *S. aureus* and *Bacillus subtilis* the mean diameter of inhibition ring of drug-loaded PU–dextran composite nanofibers was around 15 and 20 mm, respectively. In the case of gram-negative bacteria, the diameter of the inhibition zone reached 20 mm as shown in Figure 55.8b (1, 2, and 3, respectively). No bactericidal activity was detected for pristine PU and PU–dextran nanofibrous mats.

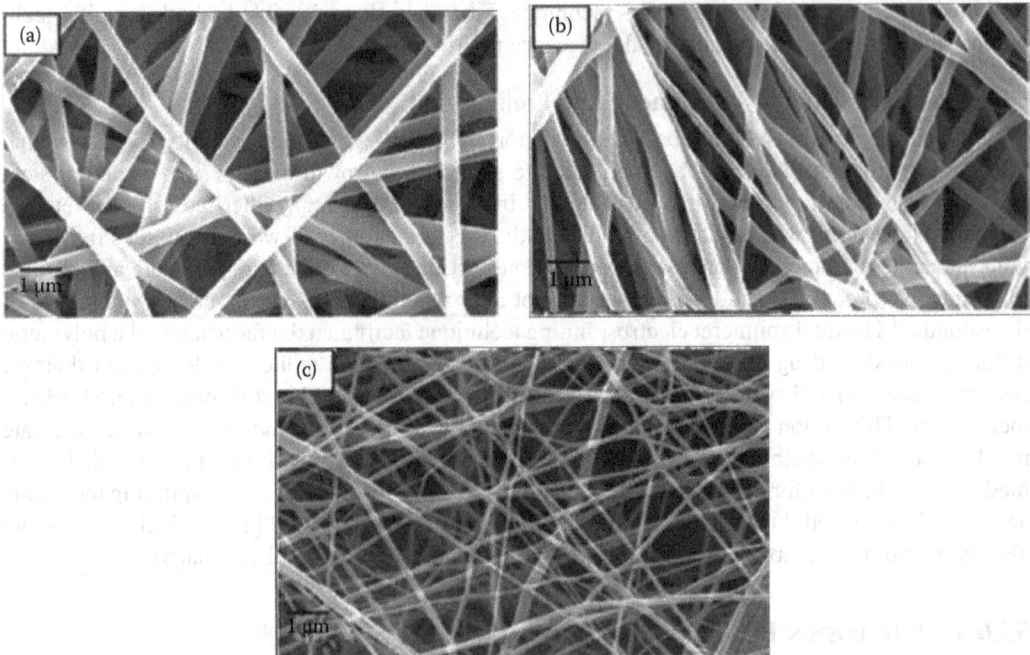

FIGURE 55.7 SEM images of electrospun (a) PU, (b) PU dextran, and (c) PU–dextran–drug nanofibrous mat. (Reprinted from *Carbohydr. Polym.*, 90(4), Unnithan, A.R., Barakat, N.A.M., Pichiah, P.B.T., Gnanasekaran, G., Nirmala, R., Cha, Y.S., Jung, C.H., El-Newehy, M., and Kim, H.Y., Wound-dressing materials with antibacterial activity from electrospun polyurethane–dextran nanofiber mats containing ciprofloxacin HCl, 1786–1793. Copyright 2012, with permission from Elsevier.)

The drug-loaded PU–dextran composite nanofibers showed excellent bactericidal activity against a wide range of bacteria, thereby avoiding exogenous infections effectively.

It is a known factor that the decontamination of exogenous organisms is a critical factor for a wound-healing material. The antibacterial property plays a crucial role for the electrospun-based wound dressing membranes. As the interconnected nanofibers create perfect blocks and pores in a nanofiber membrane, the nanofiber membrane should be able to prevent any bacteria from penetrating and therefore avoiding exogenous infections effectively. The results showed that a composite nanofiber mat is a good antibacterial membrane and it can be applied as a perfect wound dressing material.

55.6.5 Rifampin PLLA Electrospun Fibers

Rifampin, an antituberculosis drug is lipophilic and highly soluble in PLLA/chloroform/acetone solution. Electrospinning of PLLA/chloroform/acetone solutions usually resulted in fibers with a diameter range of 0.3–4.2 μm. The diameter size was greatly decreased by adding rifampin of small molecular size in the solution. Uniform fibers were obtained as shown in Figure 55.9. Red color of the fibers indicated uniform distribution of drug in the fibers. With increasing amount of rifampin, the color of the fibers became deeper. The average diameter was about 700 nm. No rifampin crystals were detected by optical or electronic microscopy, either on the surface of the fibers or outside the fibers, as seen in Figure 55.9, indicating that rifampin was perfectly included in the fibers.

When the solution jet was rapidly elongated and the solvent evaporated quickly, the drug remained compatible with PLLA. During the rapid evaporation of the solvent, phase separation took place quickly between the drug and PLLA. Therefore, a reasonable amount of the drug exists on the fiber surface.[80]

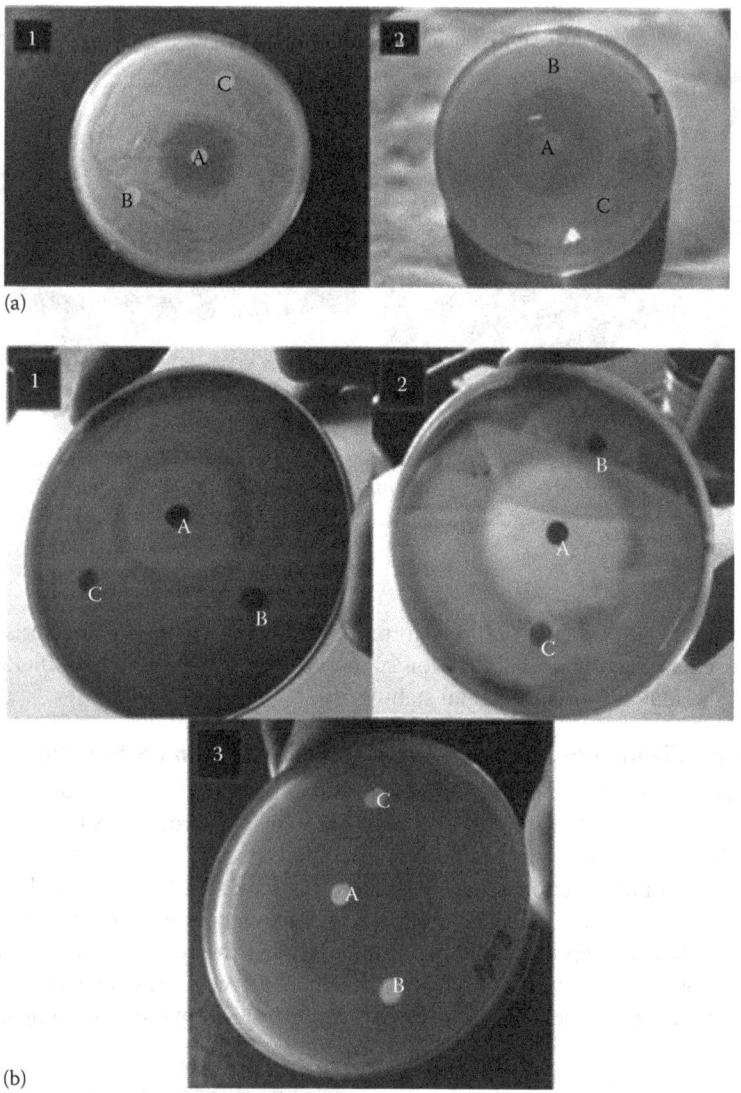

FIGURE 55.8 (a) Bactericidal activity of PU–dextran–drug nanofibrous mat with gram-positive *S. aureus* (1) and *B. subtilis* (2), respectively. PU–dextran–drug, PU–dextran, and pristine PU discs were denoted as A, B, and C, respectively, in the Petri plates. (b) Bactericidal activity of PU–dextran–drug nanofibrous mat with gram-negative *E. coli* (1), *Salmonella typhimurium* (2), and *Vibrio vulnificus* (3), respectively PU–dextran drug, PU–dextran, and pristine PU discs were denoted as A, B, and C, respectively, in the Petri plates. (Reprinted from *Carbohydr. Polym.*, 90(4), Unnithan, A.R., Barakat, N.A.M., Pichiah, P.B.T., Gnanasekaran, G., Nirmala, R., Cha, Y.S., Jung, C.H., El-Newehy, M., and Kim, H.Y., Wound-dressing materials with antibacterial activity from electrospun polyurethane–dextran nanofiber mats containing ciprofloxacin HCl, 1786–1793. Copyright 2012, with permission from Elsevier.)

55.6.6 Chitosan Nanofiber for Antibacterial Applications

Chitosan (CS) nanofibers with a diameter of 150–200 nm were fabricated from a mixed chitosan/poly (vinyl alcohol) solution by the electrospinning method. Hen egg-white lysozyme was immobilized on electrospun CS nanofibers via cross-linked enzyme aggregates (CLEAs) and used for effective and continuous antibacterial applications. CS has excellent biological properties, such as biodegradability, biocompatibility, antibacterial properties, non-toxicity, hydrophilicity, high

FIGURE 55.9 SEM photographs of PLLA electrospun fibers containing different amounts of rifampin: (a) 30 wt.% and (b) 50 wt.%. (Reprinted from *J. Contr. Release*, 92(3), Zeng, J., Xu, X., Chen, X., Liang, Q., Bian, X., Yang, L., and Jing, X., Biodegradable electrospun fibers for drug delivery, 227–231. Copyright 2003, with permission from Elsevier.)

mechanical strength, and excellent affinity to proteins. Owing to these characteristics, CS-based materials are used as enzyme immobilization supports. CLEA can be highly efficient biocatalysts with enhanced thermal and environmental stability.

55.6.6.1 Immobilization of Lysozyme-CLEA onto Electrospun CS Nanofibers

The prepared electrospun CS nanofibers contain a functional group with surface-reactive primary amino groups to allow for the immobilization of proteins. In order to effectively facilitate the covalent coupling and prevent enzyme deactivation, the selection of the optimum coupling time is important. The immobilization efficiency of the lysozyme-CLEA increased with increasing coupling time. The lysozyme-CLEA–immobilized CS nanofibers were confirmed with FE-SEM micrographs. As shown in Figure 55.10, after immobilizing the lysozyme, the cross-linked lysozyme aggregates on the CS nanofibers showed a rougher surface morphology (Figure 55.10b) than that of the control CS nanofibers (Figure 55.10a), confirming formation of strong covalent bonds

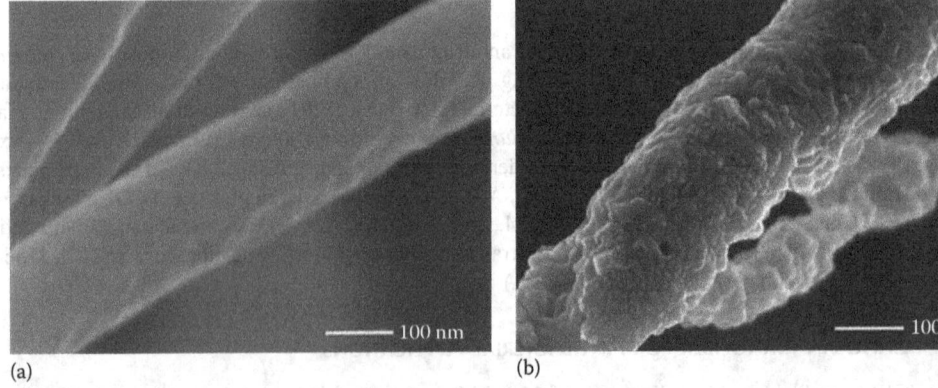

FIGURE 55.10 FE-SEM of the (a) control and (b) lysozyme-CLEA–immobilized CS nanofibers. (Reprinted from *Int. J. Biol. Macromol.*, 54, Park, J.M., Kim, M., Park, H.S., Jang, A., Mind, J., and Kima, Y.H., Immobilization of lysozyme-CLEA onto electrospun chitosan nanofiber for effective antibacterial applications, 37–43. Copyright 2013, with permission from Elsevier.)

TABLE 55.3
The Influence of Reusing the Lysozyme-CLEA–Immobilized CS Nanofibers on the Bacteriostasis Ratio of Four Different Pathogenic Bacteria

Number of Reuses	R (%)			
	S. aureus	B. subtilis	S. flexneri	P. aeruginosa
0	100	100	100	100
1	98.3	98.1	97.6	99.2
2	97.6	96.4	98.1	98.1
3	97.5	95.7	96.4	97.8
4	96.2	94.2	94.3	96.1
5	95.5	93.1	92.8	94.9
6	93.2	89.5	91.2	92.1
7	90.5	86.2	89.7	89.6
8	88.4	83.7	88.7	87.5
9	84.7	81.2	86.4	85.2
10	82.4	79.8	83.4	84.1

Source: Reprinted from *Int. J. Biol. Macromol.*, 54, Park, J.M., Kim, M., Park, H.S., Jang, A., Mind, J., and Kima, Y.H., Immobilization of lysozyme-CLEA onto electrospun chitosan nanofiber for effective antibacterial applications, 37–43. Copyright 2013, with permission from Elsevier.

between the amino groups of the lysozyme molecules and the CS nanofibers. The morphology comparison was useful in confirming the covalent attachments between the CS nanofibers and lysozyme-CLEA molecules.

55.6.6.2 Antibacterial Measurement

The antibacterial activities of the free and immobilized lysozymes were determined by using four pathogenic bacteria, viz. *S. aureus*, *B. subtilis*, *Shigella flexneri*, and *P. aeruginosa*. The stability of the antibacterial lysozyme-CLEA–immobilized CS nanofiber was evaluated by multiple reuses. Table 55.3 shows the effect of reusing the immobilized lysozyme-CLEA on the ratio of bacteriostasis for the four pathogenic bacteria. It can be seen that the antibacterial ratio of the immobilized lysozyme-CLEA decreased with increasing number of reuses, which might be associated with the leakage of the lysozyme from the CS nanofibers upon use. The lysozyme-CLEA–immobilized CS nanofibers retained 82.4%, 79.8%, 83.4%, and 84.1% of their antibacterial ratio after 10 cycles for *S. aureus*, *B. subtilis*, *S. flexneri*, and *P. aeruginosa*, respectively. These results mean that lysozyme-CLEA–immobilized CS nanofibers could be used as a promising material for enhanced and continuous antibacterial applications.[81]

55.7 CONCLUSION

Electrospun polymeric nanofibers are being extensively used for drug delivery applications because of their ability to produce thin fibers with large surface areas, with high porosity and 3D reticulate structure, simple functionalities, and rich mechanical properties and process ease in applying them to various purposes like tissue engineering, wound dressing, and biosensors. Employing electrospun nanofibers as drug delivery vehicles has found its base in the form of their unique functionality and inherent nanoscale morphological characteristics which makes them compatible to be used as such. Last but not the least, coaxial electrospinning techniques also have succeeded in achieving high drug loading, encapsulation and release and also facilitation of the solubilization of some insoluble and intractable drugs with the aid of hollow nanofibrous tubes.

ACKNOWLEDGMENT

The authors are thankful to Elsevier for copyright permission. Authors are also thankful Prof. Suresh C. Ameta, Department of Chemistry, Pacific College of Basic and Applied Sciences, PAHER University, Udaipur (Rajasthan) for suggestions.

REFERENCES

1. Doshi, J.; Reneker, D.H. Electrospinning process and applications of electrospun fibers. *J. Electrost.* 1995, *35* (2–3), 151–160.
2. Dzenis, Y. Spinning continuous fibers for nanotechnology. *Science* 2004, *304* (5679), 1917–1919.
3. Huang, Z.M.; Zhang, Y.Z.; Kotaki, M.; Ramakrishna, S. A review on polymer nanofibers by electrospinning and their applications in nanocomposites. *Compos. Sci. Technol.* 2003, *63*, 2223–2253.
4. Reneker, D.H.; Chun, I. Nanometre diameter fibres of polymer, produced by electrospinning. *Nanotechnology* 1996, *7* (3), 216–223.
5. Qi, R.; Guo, R.; Shen, M.; Cao, X.; Zhang, L.; Xu, J. Electrospun poly (lactic-co-glycolic acid)/halloysite nanotube composite nanofibers for drug encapsulation and sustained release. *J. Mater. Chem.* 2010, *20* (47), 10622–10629.
6. Li, W.J.; Laurencin, C.T.; Caterson, E.J.; Tuan, R.S.; Ko, F.K. Electrospun nanofibrous structure: A novel scaffold for tissue engineering. *J. Biomed. Mater. Res.* 2002, *60* (4), 613–621.
7. Khil, M.S.; Cha, D.I.; Kim, H.Y.; Kim, I.S.; Bhattarai, N. Electrospun nanofibrous polyurethane membrane as wound dressing. *J. Biomed. Mater. Res.* 2003, *67* (3), 675–679.
8. Ding, B.; Wang, M.; Wang, X.; Yu, J.; Sun, G. Electrospun nanomaterials for ultrasensitive sensors. *Mater. Today* 2010, *13* (11), 16–27.
9. Zeng, J.; Xu, X.; Chen, X.; Liang, Q.; Bian, X.; Yang, L. Biodegradable electrospun fibers for drug delivery. *J. Contr. Release* 2003, *92* (3), 227–231.
10. Luu, Y.; Kim, K.; Hsiao, B.; Chu, B.; Hadjiargyrou, M. Development of a nanostructured DNA delivery scaffold via electrospinning of PLGA and PLA-PEG block copolymers. *J. Contr. Release* 2003, *89* (2), 341–353.
11. Kenawy, E.R.; Bowlin, G.L.; Mansfield, K.; Layman, J.; Simpson, D.G.; Sanders, E.H. Release of tetracycline hydrochloride from electrospun poly (ethylene-covinylacetate), poly (lactic acid), and a blend. *J. Contr. Release* 2002, *81* (1–2), 57–64.
12. Jiang, H.; Hu, Y.; Li, Y.; Zhao, P.; Zhu, K.; Chen, W. A facile technique to prepare biodegradable coaxial electrospun nanofibers for controlled release of bioactive agents. *J. Contr. Release* 2005, *108* (2–3), 237–243.
13. Xu, X.; Yang, L.; Wang, X.; Chen, X.; Liang, Q.; Zeng, J. Ultrafine medicated fibers electrospun from W/O emulsions. *J. Contr. Release* 2005, *108* (1), 33–42.
14. Qi, H.; Hu, P.; Xu, J.; Wang, A. Encapsulation of drug reservoirs in fibers by emulsion electrospinning: Morphology characterization and preliminary release assessment. *Biomacromolecules* 2006, *7* (8), 2327–2330.
15. Alexander, V.B.; Alexander, L.Y.; Constantine, M.M. Co-electrospinning of core-shell fibers using a single-nozzle technique. *Langmuir* 2007, *23* (5), 2311–2314.
16. Xu, X.; Zhuang, X.; Chen, X.; Wang, X.; Yang, L.; Jing, X. Preparation of core-sheath composite nanofibers by emulsion electrospinning. *Macromol. Rapid Commun.* 2006, *27* (19), 1637–1642.
17. Agarwal, S.; Wendorff, J.H.; Greiner, A. Use of electrospinning technique for biomedical applications. *Polymer* 2008, *49* (26), 5603–5621.
18. Pham, Q.P.; Sharma, U.; Mikos, A.G. Electrospinning of polymeric nanofibers for tissue engineering applications. *Tissue Eng.* 2006, *12* (5), 1197–1211.
19. Hyuk, S.Y.; Taek, G.K.; Tae, G.P. Surface-functionalized electrospun nanofibers for tissue engineering and drug delivery. *Adv. Drug Deliv. Rev.* 2009, *61* (12), 1033–1042.
20. Xie, J.W.; Wang, C.H. Electrospun micro- and nanofibers for sustained delivery of paclitaxel to treat C6 glioma in vitro. *Pharm. Res.* 2006, *23* (8), 1817–1826.
21. Kim, T.G.; Lee, D.S.; Park, T.G. Controlled protein release from electrospun biodegradable fiber mesh composed of poly(epsilon-caprolactone) and poly (ethylene oxide). *Int. J. Pharm.* 2007, *338* (1–2), 276–283.
22. Borcia, C.; Borcia, G.; Dumitrascu, N. Surface treatment of polymers by plasma and uv radiation. *Rom. J. Phys.* 2011, *56* (1–2), 224–232.

23. Park, K.; Ju, Y.M.; Son, J.S.; Ahn, K.D.; Han, D.K. Surface modification of biodegradable electrospun nanofiber scaffolds and their interaction with fibroblasts. *J. Biomater. Sci. Polym. Ed.* 2007, *18* (4), 369–382.
24. Zhu, X.; Chian, K.S.; Park, M.B.E.C.; Lee, S.T. Effect of argon-plasma treatment on proliferation of human-skin-derived fibroblast on chitosan membrane in vitro. *J. Biomed. Mater. Res. A* 2005, *73* (3), 264–274.
25. Venugopal, J.; Low, S.; Choon, A.T.; Kumar, A.B.; Ramakrishna, S. Electrospun modified nanofibrous scaffolds for the mineralization of osteoblast cells. *J. Biomed. Mater. Res.* 2008, *85* (2), 408–417.
26. Jia, J.; Duan, Y.Y.; Yu, J.; Lu, J.W. Preparation and immobilization of soluble eggshell membrane protein on the electrospun nanofibers to enhance cell adhesion and growth. *J. Biomed. Mater. Res.* 2008, *86* (2), 364–373.
27. Prabhakaran, M.P.; Venugopal, J.; Chan, C.K.; Ramakrishna, S. Surface modified electrospun nanofibrous scaffolds for nerve tissue engineering. *Nanotechnology* 2008, *19* (45), 455102.
28. Yang, F.; Wolke, J.G.C.; Jansen, J.A. Biomimetic calcium phosphate coating on electrospun poly (epsilon-caprolactone) scaffolds for bone tissue engineering. *Chem. Eng. J.* 2008, *137* (1), 154–161.
29. Croll, T.I.; O'Connor, A.J.; Stevens, G.W.; Cooper-White, J.J. Controllable surface modification of poly(lactic-co-glycolic acid) (PLGA) by hydrolysis or aminolysis I. Physical, chemical, and theoretical aspects. *Biomacromolecules* 2004, *5* (2), 463–473.
30. Chen, F.; Lee, C.N.; Teoh, S.H. Nanofibrous modification on ultra-thin poly(epsiloncaprolactone) membrane via electrospinning. *Mater. Sci. Eng.* 2007, *27* (2), 325–332.
31. Turmanova, S.; Minchev, M.; Vassilev, K.; Danev, G. Surface grafting polymerization of vinyl monomers on poly(tetrafluoroethylene) films by plasma treatment. *J. Polym. Res.* 2008, *15* (4), 309–318.
32. Mori, M.; Uyama, Y.; Ikada, Y. Surface modification of polyethylene fiber by graft polymerization. *J. Polym. Sci. Polym. Chem.* 1994, *32* (9), 1683–1690.
33. Kou, R.Q.; Xu, Z.K.; Deng, H.T.; Liu, Z.M.; Seta, P.; Xu, Y.Y. Surface modification of microporous polypropylene membranes by plasma-induced graft polymerization of alpha-allylglucoside. *Langmuir* 2003, *19*, 6869–6875.
34. Liu, Z.M.; Xu, Z.K.; Wang, J.Q.; Wu, J.; Fu, J.J. Surface modification of polypropylene microfiltration membranes by graft polymerization of N-vinyl-2-pyrrolidone. *Eur. Polym. J.* 2004, *40* (9), 2077–2087.
35. Ma, Z.W.; Kotaki, M.; Yong, T.; He, W.; Ramakrishna, S. Surface engineering of electrospun polyethylene terephthalate (PET) nanofibers towards development of a new material for blood vessel engineering. *Biomaterials* 2005, *26* (15), 2527–2536.
36. Yao, C.; Li, X.S.; Neoh, K.G.; Shi, Z.L.; Kang, E.T. Surface modification and antibacterial activity of electrospun polyurethane fibrous membranes with quaternary ammonium moieties. *J. Membr. Sci.* 2008, *320* (1–2), 259–267.
37. Luong, N.D.; Moon, S.; Lee, D.S.; Lee, Y.K.; Nam, J.D. Surface modification of poly (L-lactide) eletrospun fibers with nanocrystal hydroxyapatite for engineered scaffold applications. *Mater. Sci. Eng. C* 2008, *28* (8), 1242–1249.
38. Sun, X.Y.; Shankar, R.; Borner, H.G.; Ghosh, T.K.; Spontak, R.J. Field-driven biofunctionalization of polymer fiber surfaces during electrospinning. *Adv. Mater.* 2007, *19* (1), 87–91.
39. Chen, C.; Lv, G.; Pan, C.; Song, M.; Wu, C.H.; Guo, D.D.; Wang, X.M.; Chen, B.A.; Gu, Z.Z. Poly(lactic acid) (PLA) based nanocomposites—A novel way of drug-releasing *Biomed. Mater.* 2007, *2*, L1–L4.
40. Park, C.H.; Kim, K.H.; Lee, J.C.; Lee, J. In-situ nanofabrication via electrohydrodynamic jetting of counter charged nozzles. *Polym. Bull.* 2008, *61* (4), 521–528.
41. Ma, Z.W.; He, W.; Yong, T.; Ramakrishna, S. Grafting of gelatin on electrospun poly (caprolactone) nanofibers to improve endothelial cell spreading and proliferation and to control cell orientation. *Tissue Eng.* 2005, *11* (7–8), 1149–1158.
42. Joung, Y.K.; Bae, J.W.; Park, K.D. Controlled release of heparin-binding growth factors using heparin-containing particulate systems for tissue regeneration. *Expert Opin. Drug Deliv.* 2008, *5* (11), 1173–1184.
43. Bolgen, N.; Vargel, I.; Korkusuz, P.; Menceloglu, Y.Z.; Piskin, E. In vivo performance of antibiotic embedded electrospun PCL membranes for prevention of abdominal adhesions. *J. Biomed. Mater. Res. B. Appl. Biomater.* 2007, *81* (2), 530–543.
44. Kim, H.W.; Song, J.H.; Kim, H.E. Nanofiber generation of gelatin-hydroxyapatite biomimetics for guided tissue regeneration. *Adv. Funct. Mater.* 2005, *15* (12), 1988–1994.
45. Rujitanaroj, P.O.; Pimpha, N.; Supaphol, P. Wound-dressing materials with antibacterial activity from electrospun gelatin fiber mats containing silver nanoparticles. *Polymer* 2008, *49* (21), 4723–4732.
46. Muller, K.; Quinn, J.F.; Johnston, A.P.R.; Becker, M.; Greiner, A.; Caruso, F. Polyelectrolyte functionalization of electrospun fibers. *Chem. Mater.* 2006, *18* (9), 2397–2403.

47. Chua, K.N.; Chai, C.; Lee, P.C.; Tang, Y.N.; Ramakrishna, S.; Leong, K.W.; Mao, H.Q. Surface-aminated electrospun nanofibers enhance adhesion and expansion of human umbilical cord blood hematopoietic stem/progenitor cells. *Biomaterials* 2006, *27* (36), 6043–6051.
48. Chua, K.N.; Chai, C.; Lee, P.C.; Ramakrishna, S.; Leong, K.W.; Mao, H.Q. Functional nanofiber scaffolds with different spacers modulate adhesion and expansion of cryo preserved umbilical cord blood hematopoietic stem/progenitor cells. *Exp. Hematol.* 2007, *35* (5), 771–781.
49. Ye, P.; Xu, Z.K.; Wu, J.; Innocent, C.; Seta, P. Nanofibrous membranes containing reactive groups: Electrospinning from poly(acrylonitrile-co-maleic acid) for lipase immobilization. *Macromolecules* 2006, *39* (3), 1041–1045.
50. Ma, Z.; Kotaki, M.; Inai, R.; Ramakrishna, S. Potential of nanofiber matrix as tissue-engineering scaffolds. *Tissue Eng.* 2005, *11* (1–2), 101–109.
51. Yoshimoto, H.; Shin, Y.M.; Terai, H.; Vacanti, J.P. A biodegradable nanofiber scaffold by electrospinning and its potential for bone tissue engineering. *Biomaterials* 2003, *24* (12), 2077–2082.
52. Li, W.; Tuli, R.; Okafor, C.; Derfoul, A.; Danielson, K.G.; Hall, D.J.; Tuan, R.S. A three-dimensional nanofibrous scaffold for cartilage tissue engineering using human mesenchymal stem cells. *Biomaterials* 2005, *26* (6), 599–609.
53. Rho, K.S.; Jeong, L.; Lee, G.; Seo, B.M.; Park, Y.J.; Hong, S.D.; Roh, S.; Cho, J.J.; Park, W.H.; Min, B.M. Electrospinning of collagen nanofibers: Effects on the behavior of normal human keratinocytes and early-stage wound healing. *Biomaterials* 2006, *27* (8), 1452–1461.
54. Yang, F.; Murugan, R.; Ramakrishna, S.; Wang, X.; Ma, Y.X.; Wang, S. Fabrication of nano-structured porous PLLA scaffold intended for nerve tissue engineering. *Biomaterials* 2004, *25* (10), 1891–1900.
55. Ho, M.H.; Wang, D.M.; Hsieh, H.J.; Liu, H.C.; Hsien, T.Y.; Lai, J.Y.; Hou, L.T. Preparation and characterization of RGD-immobilized chitosan scaffolds. *Biomaterials* 2005, *26* (16), 3197–3206.
56. Boland, E.D.; Telemeco, T.A.; Simpson, D.G.; Wnek, G.E.; Bowlin, G.L. Biomimetic nanofibrous scaffolds for bone tissue engineering. *J. Biomed. Mater. Res.* 2004, *71* (1), 144–152.
57. Gao, J.; Niklason, L.; Langer, R. Surface hydrolysis of poly(glycolic acid) meshes increases the seeding density of vascular smooth muscle cells. *J. Biomed. Mater. Res.* 1998, *42* (3), 417–424.
58. Mwale, F.; Wang, H.T.; Nelea, V.; Luo, L.; Antoniou, J.; Wertheimer, M.R. The effect of novel nitrogen-rich plasma polymer coatings on the phenotypic profile of notochordal cells. *Biomaterials* 2006, *27*, 2258–2264.
59. Rathinamoorthy, R. Nanofiber for drug delivery system—Principle and application. *P. T. J.* 2012, *61*, 45–48.
60. Wei, A.; Wang, J.; Wang, X.; Hou, D.; Wei, Q. Morphology and surface properties of poly(L-lactic acid)/captopril composite nanofiber membranes. *J. Eng. Fiber Fabr.* 2012, *7* (1), 129–135.
61. Kalia, K.; Narula, G.D.; Kannan, G.M.; Flora, S.J.S. Effects of combined administration of captopril and DMSA on arsenite induced oxidative stress and blood and tissue arsenic concentration in rats. *Comp. Biochem. Physiol. C. Toxicol. Pharmacol.* 2007, *144* (4), 372–379.
62. Abubakr, O.N.; Jun, S.Z. Recent progress in sustained: Controlled oral delivery of captopril: An overview. *Int. J. Pharm.* 2000, *194* (2), 139–146.
63. Qi, R.; Cao, X.; Shen, M; Guo, R.; Yu, J.; Shi, X. Biocompatibility of electrospun halloysite nanotube-doped poly (lactic-co-glycolic acid) composite nanofibers. *J. Biomater. Sci. Polym. Ed.* 2012, *23* (1–4), 299–313.
64. Qi, R.; Shen, M.; Cao, X.; Guo, R.; Tian, X.; Yu, J. Exploring the dark side of MTT viability assay of cells cultured onto electrospun PLGA-based composite nanofibrous scaffolding materials. *Analyst* 2011, *136* (14), 2897–2903.
65. Zhao, Y.; Wang, S.; Guo, Q.; Shen, M.; Shi, X. Hemocompatibility of electrospun halloysite nanotube and carbon nanotube-doped composite poly (lactic-co-glycolic acid) nanofibers. *J. Appl. Polym. Sci.* 2012, *127* (1–4), 4825–4832.
66. Zhang, J.; Wang, Q.; Wang, A. In situ generation of sodium alginate/hydroxyapatite nanocomposite beads as drug-controlled release matrices. *Acta. Biomater.* 2010, *6* (2), 445–454.
67. Zheng, F.; Wang, S.; Wen, S.; Shen, M.; Zhu, M.; Shi, X. Characterization and antibacterial activity of amoxicillin-loaded electrospun nano-hydroxyapatite/poly(lactic-co-glycolic acid) composite nanofibers. *Biomaterials* 2013, *34* (4), 1402–1412.
68. Wang, S.; Wen, S.; Shen, M.; Guo, R.; Cao, X.; Wang, J. Aminopropyltriethoxysilane-mediated surface functionalization of hydroxyapatite nanoparticles: Synthesis, characterization and in vitro toxicity assay. *Int. J. Nanomed.* 2011, *6*, 3449–3459.
69. Thakur, R.A.; Florek, C.A.; Kohn, J.; Michniak, B.B. Electrospun nanofibrous polymeric scaffold with targeted drug release profiles for potential application as wound dressing. *Int. J. Pharm.* 2008, *364* (1), 87–93.

70. Aydin, O.N.; Eyigor, M.; Aydin, N. Antimicrobial activity of ropivacaine and other local anaesthetics. *Eur. J. Anaesthesiol.* 2001, *18* (10), 687–694.
71. Yang, F.; Murugan, R.; Wang, S.; Ramakrishna, S. Electrospinning of nano/micro scale poly (L-lactic acid) aligned fibers and their potential in neural tissue engineering. *Biomaterials* 2005, *26* (15), 2603–2610.
72. Zong, X.; Bien, H.; Chung, C.Y.; Yin, L.; Fang, D.; Hsiao, B.S.; Chu, B.; Entcheva, E. Electrospun fine-textured scaffolds for heart tissue constructs. *Biomaterials* 2005, *26* (26), 5330–5338.
73. Martinsson, T.; Haegerstrand, A.; Dalsgaard, C.J. Ropivacaine and lidocaine inhibit proliferation of non-transformed cultured adult human fibroblasts, endothelial cells and keratinocytes. *Agents Actions* 1993, *40* (1–2), 78–85.
74. Drucker, M.; Cardenas, E.; Arizti, P.; Valenzuela, A.; Gamboa, A. Experimental studies on the effect of lidocaine on wound healing. *World J. Surg.* 1998, *22* (4), 394–397.
75. Hennink, W.E.; Van Nostrum, C.F. Novel cross-linking methods to design hydrogels. *Adv. Drug Delivery Rev.* 2002, *54* (1), 13–36.
76. Lakshmi, R.L.; Shalumon, K.T.; Sreeja, V.; Jayakumar, R.; Nair, S.V. Preparation of silver nanoparticles incorporated electrospun polyurethane nano-fibrous mat for wound dressing. *J. Macromol. Sci. Pure Appl. Chem.* 2010, *47* (10), 1012–1018.
77. Tsou, T.L.; Tang, S.T.; Huang, Y.C.; Wu, J.R.; Young, J.J.; Wang, H.J. Poly(2-hydroxyethyl methacrylate) wound dressing containing ciprofloxacin and its drug release studies. *J. Mater. Sci.* 2005, *16* (2), 95–100.
78. Dillen, K.; Vandervoort, J.; Van den Mooter, G.; Verheyden, L.; Ludwig, A. Factorial design, physicochemical characterization and activity of ciprofloxacin-PLGA nanoparticles. *Int. J. Pharm.* 2004, *275* (1–2), 171–187.
79. Unnithan, A.R.; Barakat, N.A.M.; Pichiah, P.B.T.; Gnanasekaran, G.; Nirmala, R.; Cha, Y.S.; Jung, C.H.; El-Newehy, M.; Kim, H.Y. Wound-dressing materials with antibacterial activity from electrospun polyurethane–dextran nanofiber mats containing ciprofloxacin HCl. *Carbohydr. Polym.* 2012, *90* (4), 1786–1793.
80. Zeng, J.; Xu, X.; Chen, X.; Liang, Q.; Bian, X.; Yang, L.; Jing, X. Biodegradable electrospun fibers for drug delivery. *J. Contr. Release* 2003, *92* (3), 227–231.
81. Park, J.M.; Kim, M.; Park, H.S.; Jang, A.; Mind, J.; Kima, Y.H. Immobilization of lysozyme-CLEA onto electrospun chitosan nanofiber for effective antibacterial applications. *Int. J. Biol. Macromol.* 2013, *54*, 37–43.

56 Polyelectrolyte Complexes
Drug Delivery Technology

Lankalapalli Srinivas

CONTENTS

- 56.1 Introduction 1334
- 56.2 Polyelectrolytes 1335
- 56.3 Classification of Polyelectrolytes 1335
- 56.4 Characterization of Polyelectrolytes 1335
- 56.5 Polyelectrolyte Complexes 1335
  - 56.5.1 Theoretical Aspects of PECs 1336
  - 56.5.2 Formation of PECs 1336
  - 56.5.3 Primary Complex Formation 1337
  - 56.5.4 Formation Process within Intracomplexes 1337
  - 56.5.5 Intercomplex Aggregation Process 1337
  - 56.5.6 Structures and Models of PECs 1337
  - 56.5.7 Factors Affecting the Formation of PECs 1338
    - 56.5.7.1 Effect of the Molecular Structure of Polyelectrolytes 1338
    - 56.5.7.2 Effect of Polyelectrolyte Concentration 1338
    - 56.5.7.3 Effect of Ionic Strength 1339
    - 56.5.7.4 Effect of Stoichiometric Ratio 1339
    - 56.5.7.5 Effect of Nature of the Ionic Group 1339
    - 56.5.7.6 Effect of Hydrophobicity 1340
    - 56.5.7.7 Effect of Charge Density 1340
    - 56.5.7.8 Effect of the Degree of Neutralization of the Polyelectrolyte 1340
    - 56.5.7.9 Influence of the Molecular Weight 1340
  - 56.5.8 Characterization of PECs 1340
    - 56.5.8.1 Potentiometry 1340
    - 56.5.8.2 Conductivity 1341
    - 56.5.8.3 Turbidimetry 1341
    - 56.5.8.4 Viscosity 1342
    - 56.5.8.5 Differential Scanning Calorimetry 1343
    - 56.5.8.6 Infrared Spectroscopy 1344
    - 56.5.8.7 X-Ray Diffraction 1345
    - 56.5.8.8 NMR Spectroscopy 1345
    - 56.5.8.9 Light Scattering 1345
    - 56.5.8.10 Particle Charge 1345
    - 56.5.8.11 Particle Size 1346
    - 56.5.8.12 Electron Microscopy 1347
    - 56.5.8.13 Drug Release 1347
  - 56.5.9 Applications of PECs in Drug Delivery 1347
    - 56.5.9.1 pH Stimuli 1348
    - 56.5.9.2 Ionic Strength Stimuli 1349

56.5.9.3	Solvent Stimuli	1349
56.5.9.4	Electrochemical Stimuli	1349
56.5.9.5	Temperature Stimuli	1350
56.5.9.6	Laser Light Stimuli	1350
56.5.9.7	Ultrasound Stimuli	1350
56.5.9.8	Magnetic Stimuli	1351
56.5.9.9	Mechanical Stimuli	1351
56.5.9.10	Biological Stimuli	1351

56.6 Conclusions ... 1351
Acknowledgments ... 1351
References ... 1351

56.1 INTRODUCTION

Competent drug delivery systems are of vital importance for efficient therapy. Great advances have been made over the past several years for development of novel drug delivery systems. Current state-of-art developments are witnessing a revolution in new techniques for drug delivery that are capable of modifying the therapeutic activity and/or targeting the drug to specific tissue, controlling the rate of drug delivery, and sustaining the duration. These advancements led to the development of various novel drug delivery systems, revolutionizing medication with several advantages.

Recent decades have experienced promising results on the appearance of polyelectrolyte polymers that respond in some desired way to changes in temperature, pH, and electric or magnetic field. The driving force behind these transitions include stimuli-like neutralization of charged groups by either a pH shift or the addition of an oppositely charged polymer, and changes in the efficiency of hydrogen bonding with an increase in temperature or ionic strength. These stimuli may lead to collapse of hydrogels and interpenetration of polymer network. These types of polymers not only convert the active substances into a nondeleterious form, which can be administered, but also have specific effect on the biodistribution, bioavailability, or absorption of the active substances and hence are increasingly gaining importance in modern pharmaceutical technology. Interaction of oppositely charged polyions and formation of polyelectrolyte complexes (PECs) have been the focus of intensive fundamental and applied research. Inter-PECs combine unique physicochemical properties with high biocompatibility. Many studies have been carried out on different polyion blends and types. These combinations may possess unique properties that are different from those of individual components.

PECs[1] are the association complexes mainly formed between oppositely charged particles (e.g., polymer–polymer, polymer–drug, polymer–surfactant, polymer–protein). These complexes are formed due to electrostatic interaction between oppositely charged polyions. This interaction avoids the use of chemical cross-linking agents that can reduce the possible toxicity and other undesirable effects of the reagents. The PECs formed between a poly acid and poly base are less affected by the pH variation of the dissolution medium. The complexation, between DNA and chitosan,[2] has been extensively studied in the development of delivery vehicle for gene therapy and oral vaccination.

The appearance of charge–charge interactions between different ionic polymers, polymers, and drugs was considered to be a negative event when these ionic polymers are used as excipients in pharmaceutical formulations. In these systems, release of drugs may be affected strongly by the appearance of charge–charge interactions. However, in recent research, these negative events of interaction between polymer–drug and polymer–polymer have proved promising for the development of efficient drug delivery systems.[3,4]

56.2 POLYELECTROLYTES

The polymers that contain a net negative or positive charge at near-neutral pH are called polyelectrolytes (PEs).[5] They are generally water soluble. Their solubility is affected by the electrostatic interactions between water and the charged monomer. Examples of such polymers include DNA, protein, certain derivatives of cellulose polymers, carrageenan, etc.

56.3 CLASSIFICATION OF POLYELECTROLYTES

The polyelectrolytes are classified as follows:[6]

1. Based on the origin
   a. Natural polyelectrolytes: for example, nucleic acids, poly (L-lysine), carrageenans
   b. Synthetic polyelectrolytes: for example, poly(vinylbenzyl trialkyl ammonium), poly(methacrylic acid)
   c. Chemically modified biopolymers: for example, chitosan, cellulose sulfate
2. Based on composition
   a. Homopolymers
   b. Copolymers
3. Based on molecular architecture
   a. Linear
      i. Integral type
      ii. Pendent type
   b. Branched
   c. Cross-linked
4. Based on electrochemistry
   a. Polyacids/polyanions: for example, poly(vinylsulfonic acid), poly(styrenesulfonic acid)
   b. Polybases/polycations: for example, poly(4-vinyl-N-alkyl-pyridimiun)
   c. Polyampholytes: for example, maleic acid/diallyamine copolymer

56.4 CHARACTERIZATION OF POLYELECTROLYTES

Polyelectrolytes are a group of macromolecules having many interesting properties with practically useful applications. PEs possess both the macromolecular and electrochemical properties.[7] The macromolecular characteristics are molar mass, molar mass distribution, chain architecture, electrochemical characteristics are kind of charges, charge density, charge distribution/charge-density distribution and the specific characters are like counterion binding. They are characterized[7,8] by size exclusion chromatography, viscosity, electrophoresis, conductivity, solubility and phase behavior, acid/base titration, osmotic pressure, diffusion, electric birefringence, adsorption, and attraction between like-charged chains.

56.5 POLYELECTROLYTE COMPLEXES

The interaction between oppositely charged polyelectrolytes in aqueous environment leads to the formation of PECs. These PECs meet the profile of requirements of biocompatible polymer systems and can be adapted to meet the various requirements like carrier substances and components for active substances. The initial studies on macromolecular complexes were reported by Kossel.[9] He described the formation of nucleoprotein complex due to interaction of proteins with nucleic acid and form heavy precipitates when mixed even at very high dilutions. In 1949, Fuoss and Sadek[10] reported mutual interactions of synthetic polyelectrolytes. Later, the layered structures of PECs reported by Decher et al.[11] attracted the attention of researchers due to easier preparation and higher stability, making it more popular and dominative for particulate drug delivery. These PECs showed functional application in various fields as potential carrier systems for drugs, enzymes, and genes.

56.5.1 Theoretical Aspects of PECs

There are extensive studies and reports on the properties of the polyelectrolytes[12] and on the formation of PECs.[13–15] The theories proposed by Overbeek and Voorn[16] and Debye–Huckel[17] for the mechanism of formation of PECs were based on the electrostatic forces and Flory–Huggins mixing free energies of the polyelectrolytes. In general, the backbones of any two polymers are not compatible and repel each other. However, the charge fraction of the polymers determines the type of interaction that occurs between these two polymers. When the charge fraction is low, the polymer backbone repulsion (Flory interaction parameter) occurs and the solution separates into two phases, each containing mostly one of the polymers. At high charge fraction, the attractive electrostatic interactions between the polymers dominate and they precipitate to form a complex. In an intermediate range of charge fraction, the equilibrium state can be a meso phase where the two polymers only separate microscopically.

Depending on the stoichiometry of the mixture (the relative concentrations, the relative chain lengths, and charge densities), one observes mainly two types of complex formations: a macroscopic phase separation between the solvent and the polymers or a partial aggregation of the polymer chains.[8] The complex formation observed from our earlier studies between chitosan and gum karya was shown in Figure 56.1.

56.5.2 Formation of PECs

This process involves mainly three steps as shown in Figure 56.2.[18]

FIGURE 56.1 PEC formation between gum karaya and chitosan. (From Srinivas, L. and Ramana Murthy, K., *Drug Develop. Ind. Pharm.*, 38(7), 815. Copyright 2012, with permission from Informa Healthcare.)

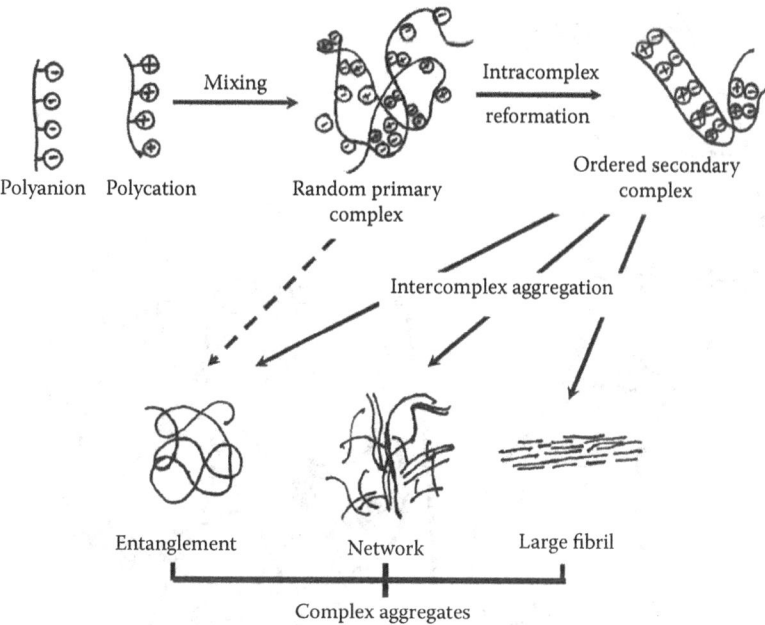

FIGURE 56.2 Schematic representation of the formation and aggregation of PECs. (From Tsuchida, E., *Pure Appl. Chem.*, 31(1), 1. Copyright 1994, with permission from Taylor & Francis.)

56.5.3 Primary Complex Formation

The forces responsible are Coulomb forces.

56.5.4 Formation Process within Intracomplexes

It involves the formation of new bonds and/or the correction of the distortions of the polymer chains.

56.5.5 Intercomplex Aggregation Process

It involves the aggregation of secondary complexes, mainly through hydrophobic interactions.

56.5.6 Structures and Models of PECs

PECs can adopt different structures, which can be categorized into three types: soluble, colloidally stable, and coacervate complexes. Combinations of polyions with significantly different molecular weights and weak ionic groups in a mixture of nonstoichiometric proportions result in water-soluble PECs.[19] Colloidally stable PEC formation between strong polyelectrolytes results in highly aggregated or macroscopic flocculated systems.[20] A coacervate is formed when the mutual binding of opposite polyelectrolytes is of moderate strength as a result of low charge density.

Two structural models[20–22] were discussed in the literature for PECs and are shown in Figure 56.3. They are ladder-like structure and show a scrambled egg model. In the ladder-like structure, the complex formation takes place on a molecular level via conformational adaptation. This structure consists of hydrophilic single-stranded and hydrophobic double-stranded segments. In the scrambled egg model, large numbers of chains are incorporated into particles architecture. This model refers to complexes that are a product of combination of polyions with strong ionic groups and comparable molar masses, yielding insoluble and highly aggregated complexes.

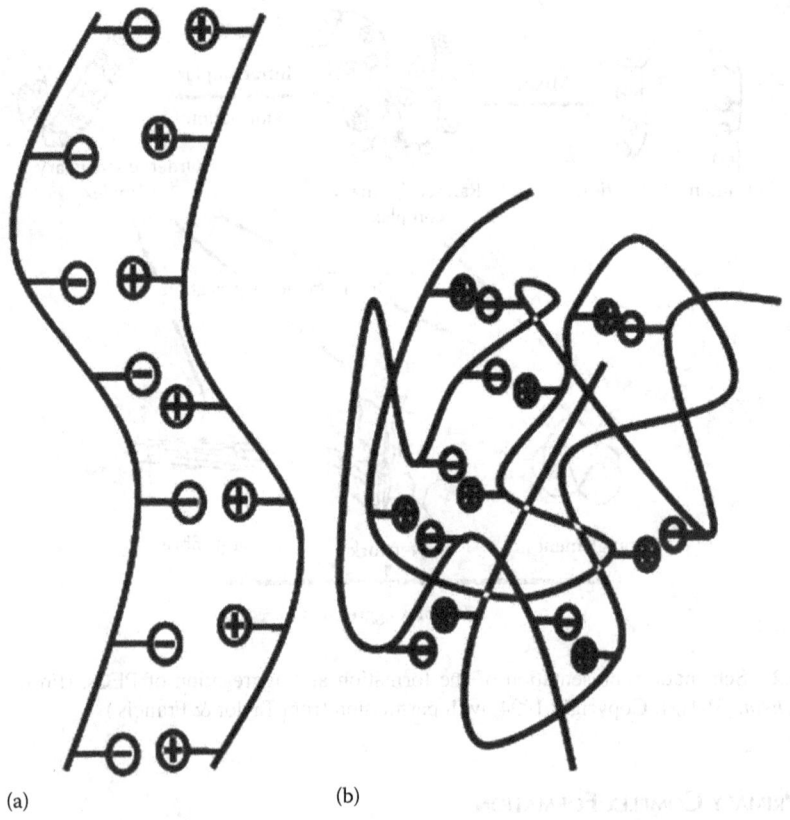

FIGURE 56.3 PEC models: (a) ladder model and (b) scrambled egg model. (With kind permission from Springer Science+Business Media: *Polyelectrolytes in Polyelectrolytes and Nanoparticles*, 2007, pp. 5–45, Koetz, J., Kosmella, S., Howard, G., Barth, D., and Harald, P., Springer, Berlin, Germany.)

56.5.7 Factors Affecting the Formation of PECs

A number of parameters are known to influence the formation of PECs.[23] The molecular structure of the polyelectrolytes, stoichiometric ratio between polyanion and polycation, nature of the ionic groups, hydrophobicity, charge density, degree of neutralization, concentration in solution, ionic strength, and the molecular weight of the PEs are the main parameters that influence the structure and properties of the resultant PEC.

56.5.7.1 Effect of the Molecular Structure of Polyelectrolytes

Tsuchida[8] showed that the composition of PECs formed between two strong PEs depends on the type of polyion used. The polyions are of two types. They are pendant type, where the charges are situated in the side groups of the polymer and integral type, where the charges are situated on the backbone of the polymer. The position of the ionic site on the PE chain has an effect on the complex formation. Insoluble complexes will form in the case of pendant-type polycation's interaction with polyanion, as the ionic group is slightly hindered by steric effects. In the integral type, the counterions of the polyanion may weaken the electrostatic interaction between PEs, leading to enhancement of the solubility of the complex.

56.5.7.2 Effect of Polyelectrolyte Concentration

The phase diagrams (Figure 56.4) made by Tsuchida[8] for a weak polyacid poly(methyl acrylic acid) (PMAA) and a strong polycation shows the effect of the concentration of the PEC. At low pH, the PEC formation is more favorable in dilute solution, but not in concentrated solution. This is assumed

Polyelectrolyte Complexes

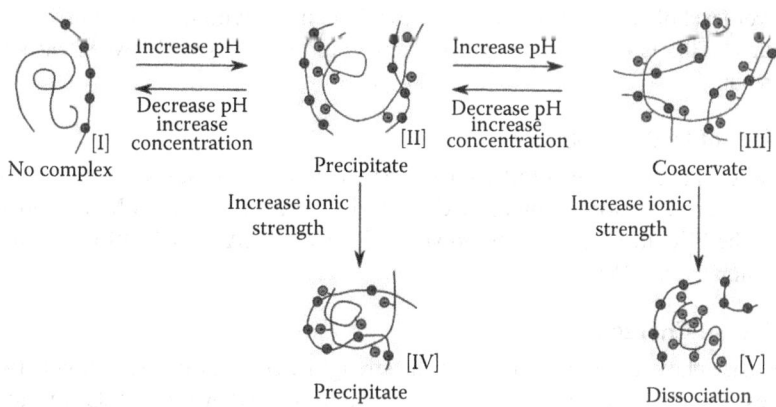

FIGURE 56.4 Schematic representation of the equilibrium for the phase changes of the PMAA-integral type polycation complex. (With kind permission from Springer Science+Business Media: *Adv. Polym. Sci.*, Interactions between macromolecules in solution and intermacromolecular complexes, 45, 1982, 1, Tsuchida, E. and Abe, K.)

to be due to the charge screening effect of concentrated PE solutions, that is, the range and strength of the electrostatic interaction is generally reduced. However, other researches have shown that in some cases, the concentration of polyelectrolyte can also have an effect on the level of aggregation of the PEC.[25,26] The level of aggregation results from two effects: the conformational changes and the incorporation of additional polyions. This second effect is favored by high polyelectrolyte concentrations. However, reports are showing that the increase of concentration of the PE solution increases the final degree of aggregation of the PEC.[27]

56.5.7.3 Effect of Ionic Strength

The ionic strength of PE solutions is of great importance as the presence of salt will screen charges and decrease electrostatic interactions. The influence of salt on PEC formation is demonstrated by the work of Dautzenberg.[25,26] The increase of the ionic strength reduces the electrostatic repulsion between the charges and leads to a more compact structure of the PEC. When the PEC is formed from the completely dissociated polyanion, a coacervate is formed. Further addition of salt will then break the coacervate through dissociation of the constituent chains. When salt is added to a PE solution, it affects the formation of the PEC. The presence of salt in the PE solution will therefore clearly have an effect on the PEC formation and on the characteristics of the final PEC. Addition of salt to a preformed PEC can also influence its properties.[8,28]

56.5.7.4 Effect of Stoichiometric Ratio

The yield of PEC is proportional to the amount of each polymer irrespective of the degrees of polymerization of the two PEs.[8,29] When the polyanion chains are partially covered by the polycation, they are then more reactive than the free polycation chains, due to changes in conformations, degree of dissociation, or the microenvironment. For strong polyelectrolytes in most cases, a 1:1 endpoint stoichiometry was found as reported by Srinivas et al. on the PEC formation, studied by calculating weight ratios of PEC formed between polyacrylic acid and polyvinyl pyrrolidone (PVP).[30] But the nonstoichiometric complex formation was observed by them in the case of gum karaya and chitosan.[31] This can be explained on the basis of presence of excess quantities of one polymer, which results in soluble complex due to screening of charges.

56.5.7.5 Effect of Nature of the Ionic Group

Different ionic groups on the polyelectrolytes show different binding capacities, thereby the PEC formation. PEC formation between a polycation- and a polyanion-containing carboxylic groups will have a lower level of aggregation compared to a PEC made with the same polyanion containing sulfonate groups.[8] This is due to the weaker electrostatic interactions between carboxylic groups and cationic groups.

56.5.7.6 Effect of Hydrophobicity
Tsuchida[8] showed that in the case of PECs formed between weak polyacids and polybases, the hydrophobicity can influence the formation of PEC. The PEC formed between PMAA and 10,10-ionene was more than the PEC formed between poly(acrylic acid) (PAA) and 10,10-ionene as the PMAA is more hydrophobic than PAA.

56.5.7.7 Effect of Charge Density
The charge density plays an important role in the strength electrostatic interaction between the two polyelectrolytes. The charge density will affect the stability and density of PEC. Vishalakshi et al.[32] showed a concentration-dependent effect of the charge density of the polyelectrolytes, by varying the charge density of the polycation.

56.5.7.8 Effect of the Degree of Neutralization of the Polyelectrolyte
The dissociation state of a weak PE directly affects the complexation ability and the composition of the resultant PEC.[8,29] It was observed that a low pH may lead to suppression of polyelectrolyte dissociation, thereby leading to no complex formation. Increasing the pH increased the ionic strength, dissociation, and complex formation. Higher pH and ionic strength will lead to complete dissociation of polyion and an equimolar PEC coacervate will be obtained.

56.5.7.9 Influence of the Molecular Weight
In the presence of low molecular weight polyions, often no stable complex formation occurs between two oppositely charged PEs. This suggests that the chain length of the PEs is one of the most important factors for the control of complexation.[8] The viscosity of PEC solutions increases with both the molecular weight and the concentration of the polymers.[29] The sequences of polymer involved in the complex formation are shorter, which explains the decrease of density of the PEC. However, the macromolecules are then freer and could complex with several other oppositely charged macromolecules and give a highly branched structure.

56.5.8 Characterization of PECs
Various methods have been used to investigate interactions between polymers. Measurements of turbidity, pH, pK_a, and ionic strength[33,34] as a function of weight ratio of polymer in the media,[35] viscosity,[36] light scattering,[37,38] infrared spectroscopy, nuclear magnetic resonance (NMR), thermal analysis, and powder x-ray diffraction[39] were employed to evaluate interpolymer complexation.

56.5.8.1 Potentiometry
Potentiometric methods are predominantly used for the experimental determination of ion activities by using ion-selective electrodes in combination with reference electrodes. The most important measurements in that field are based on the determination of the pH (H^+-ion activity) by using a single-rod glass electrode.[40] pH Measurements are mainly used to determine the degree of functionalization (called as the degree of substitution) and the acidity constants (pK_a values). Polyelectrolytes carry a net negative or positive charge on their monomer. They exhibit specific pH because of the presence of the charge. During the formation of PEC, there is a possibility of neutralization of charge or release of counter ions. Hence, the polyelectrolyte solutions exhibit variations in the pH profile when they interact with each other as shown in Figure 56.5. The potentiometric titrations can be useful for the determination of the degree of complexation.

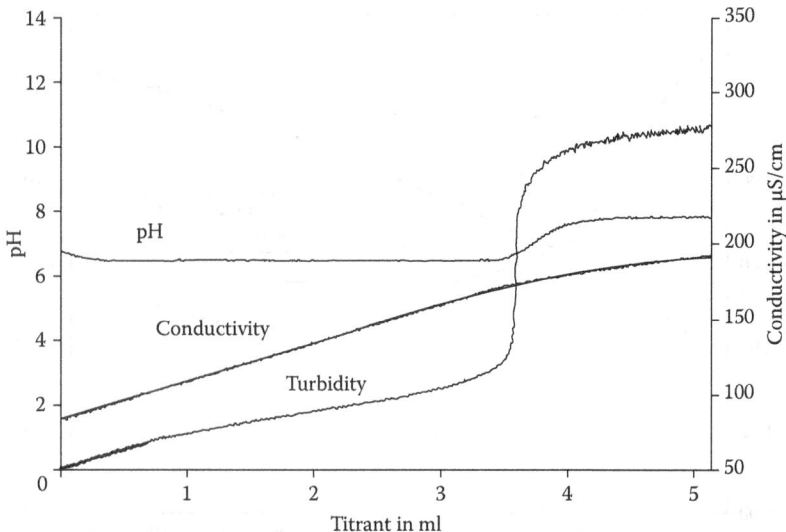

FIGURE 56.5 Potentiometric, conductometric, and turbidimetric titration curve of a polyelectrolyte titration. (With kind permission from Springer Science+Business Media: *Polyelectrolytes in Polyelectrolytes and Nanoparticles*, 2007, pp. 5–45, Koetz, J., Kosmella, S., Howard, G., Barth, D., and Harald, P., Springer, Berlin, Germany.)

56.5.8.2 Conductivity

Conductometry is applied to study properties of polyelectrolyte solutions and complexation reactions of polyelectrolyte systems. A number of characteristics of the conductivity for polyelectrolyte solutions are well established.[41,42] Polyion conductivity is the measure of degree of polymerization of the polyelectrolyte. Conductivity experiments are performed using conventional types of conductometers, with platinum or platinized electrodes and with cell constants in the order of magnitude of 0.1 to 10 cm^{-1}. Conductometric acid/base titrations can be applied to determine the total number of chargeable groups of polyacids or polybases. When the oppositely charged polyelectrolytes are mixed for the formation of PEC, the polyions exhibit variable conductivity due to neutralization of charge or release of counterions. Conductometric acid/base titrations have been reported for humic acid (HA),[43] and polyvinyl ether/maleic anhydride copolymers.[44] It our studies during the titration of gum karaya (GK) solution with chitosan (CH) solution, there was a gradual increase in the conductivity of the GK solution from 2.07 to 5.26 millisiemens (ms) with increasing additions of CH solution (Figure 56.6a). The pH decreased from 4.25 to 3.93. The CH solution was prepared in 2% v/v acetic acid in water and during the interaction of CH with GK, CH precipitates out by releasing the counterions, which could be the possible reason for increase in the conductivity with decrease in pH. In the case of polyacrylic acid (Carbopol) and PVP, conductivity decreased gradually (from 46.6 to 16.5 µs) and a sharp fall in conductivity (29.6–24.8 µs) was observed during complex formation (Figure 56.6b). Though there was a change in conductivity, very little change was observed in the case of pH (3.96–4.20).

56.5.8.3 Turbidimetry

In turbidimetric titration, a polyelectrolyte is added incrementally to another highly dilute polyelectrolyte solution and the intensity of the light scattered by or the turbidity due to complex or precipitate formation is measured. In this colloid titration, turbidity is recorded as a function of titrant volume and the end point indicated by maximum turbidity is used to ascertain the stoichiometry of complex formation as shown in Figure 56.5. Jiang and Zhu[45] used colloid titration to investigate PAA–gelatin complexation and Tsuboi et al.[46] used it to study complexation of papain with potassium poly(vinylsulfonate) (KPVS).

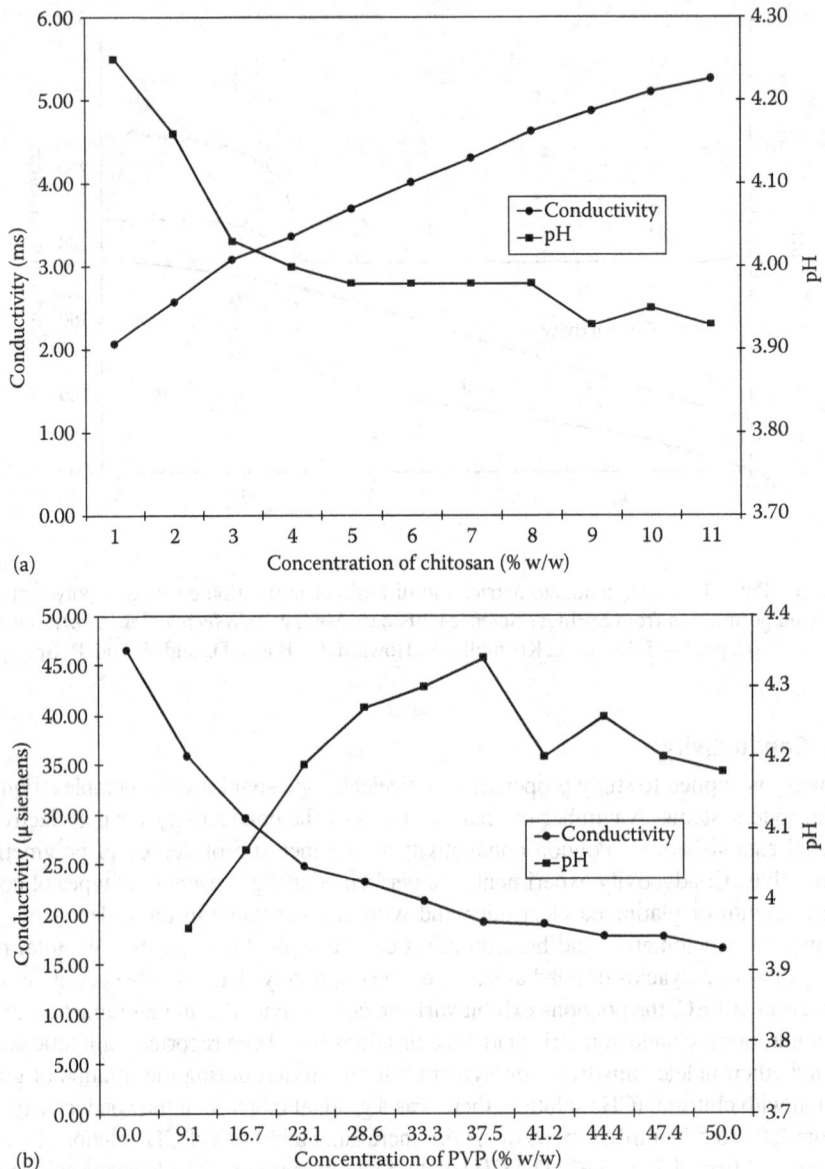

FIGURE 56.6 Conductivity and pH profiles of polyelectrolyte complexation between (a) gum karaya/chitosan and (b) polyacrylic acid/PVP. (From Srinivas, L. and Ramana Murthy, K., *Drug Develop. Ind. Pharm.*, 38(7), 815. Copyright 2012, with permission from Informa Healthcare.)

56.5.8.4 Viscosity

Rheology is the study of deformation and flow of materials, which is mainly used to investigate dynamic behavior and mechanical properties of complex coacervates. In contrast to neutral polymer solutions, polyelectrolyte solutions show reduced viscosity with increased weight concentration. The typical viscosity behavior of a polyelectrolyte solution made by quaternization of poly(styrene-*co*-4-vinylpyridine) with *n*-butyl bromide was reported by Fuoss and Cathers as in Figure 56.7.[47] They reported specific viscosities decreased strongly up to the 1:1 mixing ratio. The decrease of the viscosities during PEC formation results from three effects: (1) due to change of the conformation of the bound polyelectrolyte chains; (2) due to the release of the counterions, ionic strength of the free polyelectrolyte increases

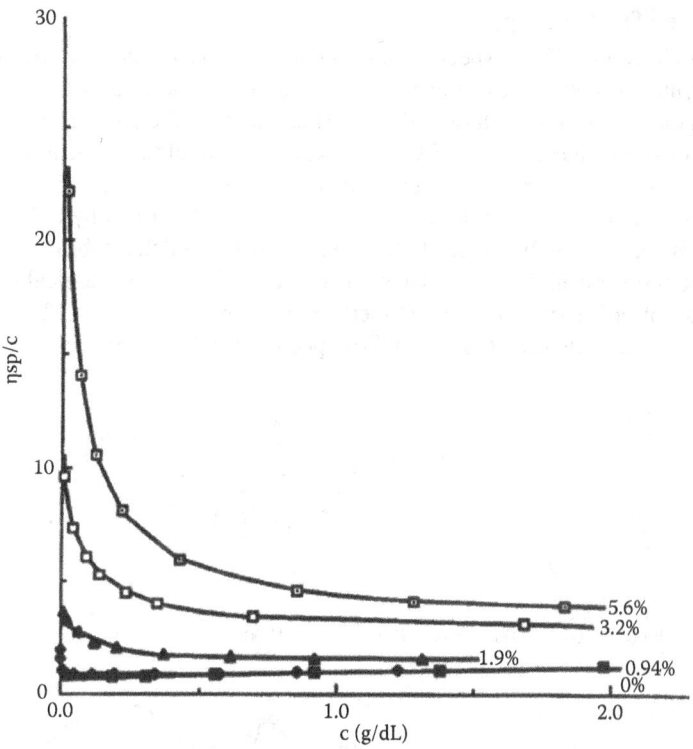

FIGURE 56.7 Viscosity curves for polyelectrolyte, 4-vinylpyridine/styrene copolymers quaternized with n-butyl bromide, in nitromethane/dioxane mixture (numbers indicate weight ratio of nitromethane/dioxane). (Reprinted with permission from *J. Polym. Sci.*, 4(2), Fuoss, R.M. and Cathers, G.I., Polyelectrolytes. III. Viscosities of n-butyl bromide addition compounds of 4-vinylpyridine-styrene copolymers in nitromethane-dioxane mixtures, 97. Copyright 1949 American Chemical Society.)

and viscosity decreases; and (3) dilution of the starting solutions according to the added amount of the cationic solutions. Masanori et al.[48] reported that intramolecular and intermolecular interactions on sulfonated polystyrene ionomers played a major role in the viscosity behavior of salt-free polyelectrolytes even in dilute solution. Viscosity curves for sulfonated polystyrene showed polyelectrolyte behavior (Figure 56.7[48]) in a low polarity and a polar solvent by viscosity measurements. It is suggested that the intermolecular interaction played a major role in the viscosity behavior of salt-free polyelectrolytes even in dilute solution. Weinbreck et al.[49] studied the viscoelastic properties of whey protein/gum acacia coacervates and verified that pH plays a major role in the microstructure of coacervates.

56.5.8.5 Differential Scanning Calorimetry

Differential scanning calorimetry (DSC) measures heat changes that occur during controlled changes in temperature, and has been used to study thermodynamic parameters associated with these changes between the sample versus an appropriate reference material in a furnace. A number of important physical changes in a polymer as well as drug substances may be characterized by DSC. These include glass transition temperature, crystallization temperature, melt temperature, and degradation or decomposition temperature. Chemical changes due to polymerization reactions, degradation reactions, complexation, and other reactions affecting the sample can be determined. Van de Weert et al.[50] applied calorimetry to study the complexation of heparin and lysozyme and observed a reduction in protein thermal stability. Ivinova et al.[51] used DSC to study the thermal denaturation of lysozyme and chymotrypsinogen without changing their enzymatic activity and reported that polyanions significantly reduce the initiation of this thermal denaturation.

56.5.8.6 Infrared Spectroscopy

Fourier transform infrared (FT-IR) spectroscopy can be used to characterize drug substances, polymer blends, polymer complexes, dynamics, surfaces, and interfaces, as well as chromatographic effluents and degradation products. It provides information about the complexation and interactions between the various constituents in the PECs.[52] It is capable of qualitative identification of the structure of unknown materials as well as the quantitative measurement of the components in a complex mixture. FT-IR spectra of physical mixture and PEC can be determined by FT-IR spectrophotometer using KBr disc method in the range of 4000 to 250 cm^{-1}. Since the stability and drug substance is very important in several applications, determination of their physicochemical stability is crucial. The FTIR spectra of polyacrylic acid, PVP, metformin hydrochloride, and PEC microparticles of metformin were shown in Figure 56.8. The FTIR spectra of polyacrylic acid and PVP have shown

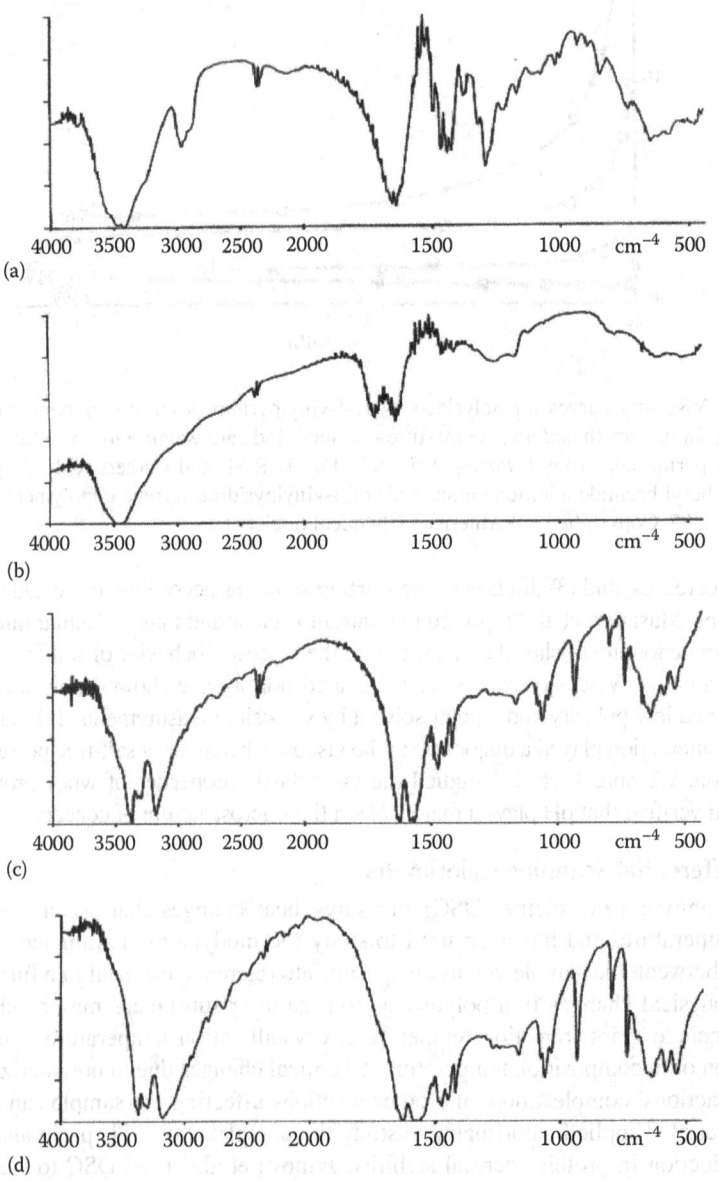

FIGURE 56.8 FTIR spectra of (a) PVP, (b) polyacrylic acid, (c) metformin, and (d) metformin PECs.

PEC formation due the hydrogen bond formation between the carboxyl group of polyacrylic acid and carbonyl group of PVP. The FTIR spectra of metformin showed characteristic –NH stretch at 3400 cm^{-1}, –C=N stretch at 1635 cm^{-1}, –N–CH$_3$ stretch at 2812 cm^{-1}, C–N–H bend at 1568 cm^{-1}, and –NH bend at 736 cm^{-1}. The metformin PEC spectra also showed all the characteristic peaks of metformin along with the characteristic peaks of carboxylic acid (COOH) stretch of carbopol at 1665 cm^{-1} and amide stretch of PVP at 1620 cm^{-1} with minor shift. The study indicated no alteration in chemical nature of metformin in PECs. Similar results were reported in the early works with diclofenac sodium PECs.[53]

56.5.8.7 X-Ray Diffraction

X-ray diffraction has become a powerful tool in the study of drug and polymer structures. X-ray diffraction involves placing the sample in the path of a monochromatized x-ray beam of low divergence. The scattered x-rays from the regularly placed atoms interfere with each other, giving strong diffraction signals in particular directions. The directions of the diffracted beams are related to the slope and dimensions of the unit cell of the crystalline lattice, and the diffraction intensity depends on the disposition of the atoms within the unit cell. The powder x-ray diffraction patterns of both physical mixture and complexes can be recorded using automated x-ray diffractometers. The samples are irradiated with monochromatized Cu kα radiation between two q angles. The time, voltage, and current are adjusted as per sample requirement. The x-ray diffractograms of crystalline drugs show sharp and intense peaks, and the amorphous drugs and polymers show smooth and low intense peaks. The x-ray diffractograms shown in Figure 56.9 for metformin HCl showed sharp peaks at 12.2°, 22.0°, 22.3°, 24.5°, 28.3°, 36.3°, and 37.1° angle (°2θ), indicating the crystallinity of the drug. The spectra of polyacrylic acid and PVP did not have any sharp peaks, indicating their amorphous nature. The spectra of metformin PECs showed peaks with low intensity and minor shift at 12.1°, 22.1°, 22.8°, 24.4°, 28.2°, 35.3°, and 37.2° angle (°2θ) may be due to fine dispersion and entrapment of the drug in the PEC.

56.5.8.8 NMR Spectroscopy

NMR spectroscopy is a most effective and significant method for observing the structure and dynamics of polymer complexes both in solution and in the solid state. The widest application of NMR spectroscopy is in the field of structure determination. The identification of certain atoms or groups in a molecule as well as their position relative to each other can be obtained by one-, two-, and three-dimensional NMR.[54] Information about the local dynamics, molecular and supramolecular organization, and the counterion condensation can be obtained. The counterion condensation and polyelectrolyte complexation between TEAC in poly(acrylic acid) was studied by Ni[55] using ^{13}C NMR.

56.5.8.9 Light Scattering

Light scattering[56,57] from solution of PECs allows the determination of the molecular parameters (molecular weight, dimensions, and shapes) of the scattering particles and thermodynamic quantities (virial coefficients, chemical potential, preferential adsorption coefficients, and excess free energies of mixing). Mihai et al.[58] reported the dependency of particle sizes and colloidal stability of PEC dispersions on polyanion structure and preparation mode was investigated by dynamic light scattering and atomic force microscopy.

56.5.8.10 Particle Charge

Particle charge plays a major role on the stabilization of colloidal systems. Especially when nanoparticles are stabilized by an adsorption layer of polyelectrolytes, zeta potential measurements are very useful.[22,59] The stabilization of the nanoparticles results from a combination of ionic and steric contributions. The zeta potential can be detected by means of electro-osmosis, electrophoresis, streaming potential, and sedimentation potential measurements. The potential drop across the mobile part of electric double layer can be determined experimentally, whenever one phase is made

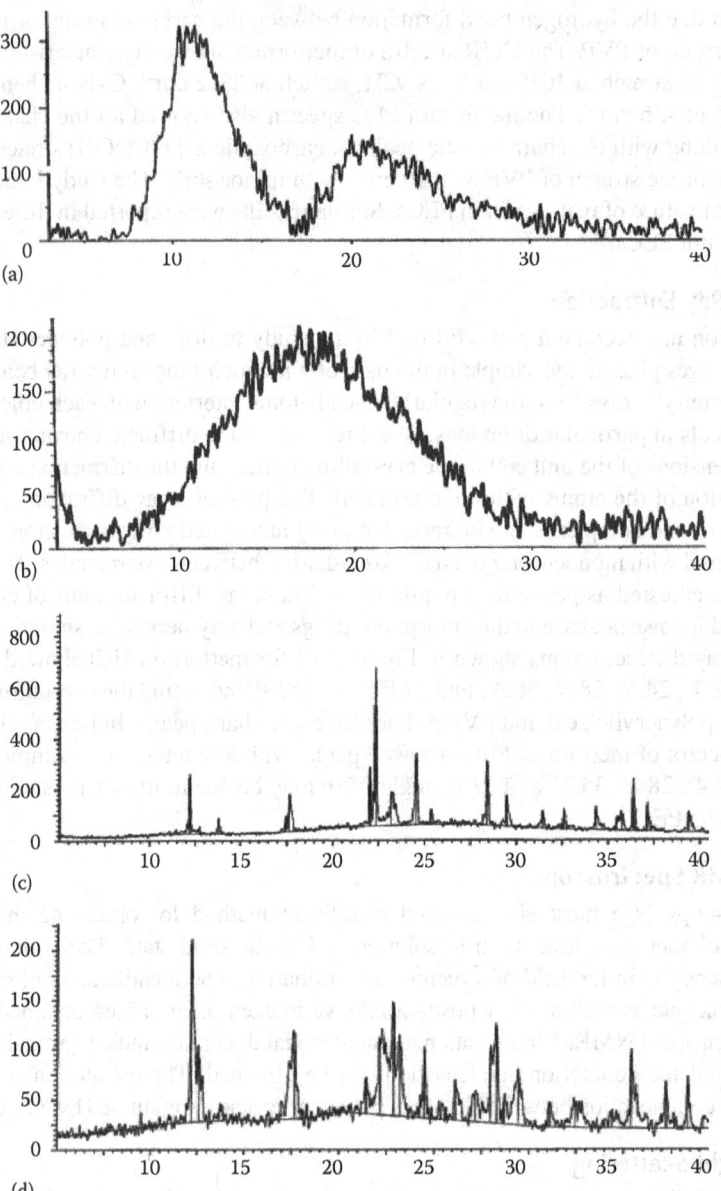

FIGURE 56.9 X-ray diffractograms of (a) PVP, (b) polyacrylic acid, (c) metformin, and (d) metformin PECs.

to move with respect to the other. This can occur when the particles settle under gravitation or in a centrifuge (sedimentation potential), or by putting the colloidal dispersion in an electrical field and measuring the speed of the particles (electrophoresis). Otherwise, the counterions can be moved by an electric field (electro-osmosis) or by forcing a fluid through it (streaming potential). Kötz and Komella. studied influence of charge by using zeta potential measurement for the PEC formation between polyanion and cation for encapsulation of kaolin.[60]

56.5.8.11 Particle Size

In general, polyelectrolyte complexation leads to formation of insoluble dense precipitate with molecular entrapment of drugs. These complexes are separated, dried, and size reduced to obtain the microparticles. The particle size of these microparticles plays an important role in drug release. The particle size and size distribution is determined by sieve analysis using standard sieves.[61] The particle size can also be determined by microscopy.[61]

56.5.8.12 Electron Microscopy

Electron microscopy is widely used for imaging objects by illuminating the sample with an electron beam, thus providing microscopic resolution at dimensions >10 nm. The size and shape of the colloidal PECs can be determined by using scanning electron microscopy (SEM) and transmission electron microscopy (TEM).[62] SEM images (Figure 56.10) for diclofenac sodium PEC microparticles prepared by using gum karaya and chitosan[31] showed discreteness of the particles with porous structures. Burgess has reported the applicability of SEM in determining the shape and stability of albumin/acacia and gelatin/acacia coacervate droplets. Pihlajamaa et al.[63] studied the complexation of heparin with collagen IX by EM and used this information to approximate binding sites for heparin. Burgess et al.[64] determined the shape and stability of albumin/acacia and gelatin/acacia coacervate droplets in relationship to the coalescence of coacervate droplets measured by SEM. For multilayer systems, whether assembled on colloids or on flat surfaces, EM is a powerful tool for monitoring layer formation and film thickness. Lvov et al.[65] used EM to verify the calculated film thickness of multicomponent protein multilayers PEI and PSS. Caruso and Mohwald[66] utilized EM to demonstrate uniform coating of polyelectrolyte-surface-modified polystyrene latex particles with FITC-labeled bovine serum albumin or immunoglobulin G multilayers.

56.5.8.13 Drug Release

The drug release character of the microparticles can be studied by in vitro diffusion or dissolution studies.[30,31,53] The diffusion studies[67] are performed by dialysis technique using diffusion cell apparatus with suitable medium and dialysis membrane. The diffusion coefficient and mean dissolution time can be calculated from the data to represent the drug release pattern.[53] Dissolution studies[67] are carried out by using United States Pharmacopoeia (USP) dissolution test apparatus in suitable medium. The drug release is characterized by calculating percentage drug dissolved in specific time. The kinetics and mechanism of drug release is the characteristic for the PEC microparticles that is generally studied by fitting the drug release data to suitable mathematical model.[68,69]

56.5.9 Applications of PECs in Drug Delivery

PECs have gained much attention in the past few years because of their potential applications in scientific and industrial interest.[70,71] These can be used as membranes,[72–74] for coating on films and fibers,[75] for isolation and fractionation of proteins,[76,77] for isolation of nucleic acid,[78–80] for binding pharmaceutical products,[81] as supports for catalyst[25], and for preparation of microcapsules for drug delivery.[82,83] Many of the applications are based on the functional properties of the polyelectrolytes[84]

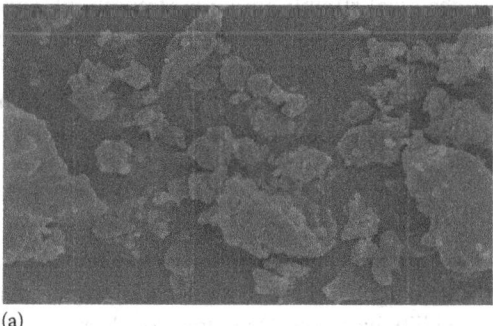 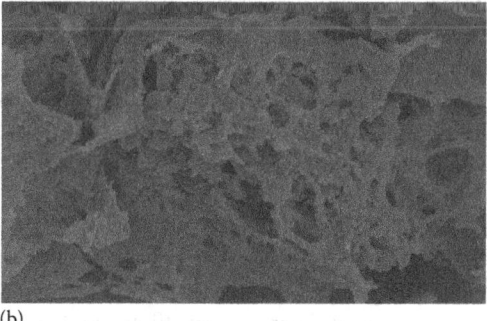

(a) (b)

FIGURE 56.10 SEM photographs of metformin (a) PEC microparticles and (b) porous surface structure.

TABLE 56.1
Practical Application of PECs

Polyelectrolyte Interactions	Description	Functional Application
Polyelectrolyte	Macromolecular interactions between oppositely charged natural, synthetic, or natural and synthetic polyelectrolytes lead to phase separation as a complex coacervate or a solid precipitate.	Surface modification[136] Flocculation[137] Encapsulation[138]
Drug	Polyelectrolytes react with drug molecules having basic or acid groups to form reversible complexes due to counterionic condensation.	Soluble complexes[139] Insoluble complexes[140]
Surfactant	Formation of highly ordered structures. This type of complex offers similarities with biological assemblies. Complexes can be of water-soluble and water-insoluble nature.	Surface modification[141] Micelle formation[142]
Protein	Interaction between polyelectrolytes and proteins results in formation of amorphous precipitates, complex coacervates, gels, fibers, or of soluble complexes.	Protein recovery[143] or separation[144] and release[145] Immobilization of enzymes[146]
Nucleic acid	Polyelectrolyte interactions play an important role in the inner organization of the cell and the binding of cells in the organism, structure of many cell membranes, and the intracellular reticulum, many of the properties of connective tissue.	DNA complexes[147] Targeted delivery[148]

and the interaction between protein–polyelectrolyte,[85] nucleic acid–polyelectrolyte,[86] surfactant–polyelectrolyte,[87] polyelectrolyte–polyelectrolyte,[88] and drug–polyelectrolyte.[89] The functional applications of PECs are summarized in Table 56.1.

The concept of PECs in the design of drug delivery systems may be useful due to the advancements made during the last two decades.[90–93] The active components will be encapsulated in the polymer matrix at the molecular level. They offer greater advantages for the drug substances in improving or/and altering physicochemical characters like stability, dissolution, etc. The active substance can be incorporated into the PECs by four ways.[94] In the first case, the active substance will be entrapped from the solution during precipitation of the complex. The active substance will be absorbed from the solution and gets incorporated into the already formed complex on contact in the second case. In the third case, the active substance may be chemically bound to at least one complex partner and precipitates during complexation. In the last case, the active compound itself may act as polyion and form PEC. The active substance from these PECs will be released either by passive diffusion due to solution equilibrium or by ion exchange mechanism or by charge interaction and slow decomplexation as well as breakdown and dissolution of the complex.

PECs have also emerged as new drug delivery systems that respond to physical, chemical, and biological stimuli[95] as shown in Figure 56.11.[96] The chemical stimuli are ionic strength, pH, and solvent; physical stimuli are light, magnetic field, ultrasound, mechanical action, and temperature; and biological stimuli are enzymatic reactions. These PECs are triggered by a variety of stimuli that have altered polyelectrolyte properties, which, in turn, modify the encapsulation and release of drugs. This stimuli-responsive PEC drug delivery is able to provide the dose of drugs on demand with reduced toxicity and increased efficacy.[97–100]

56.5.9.1 pH Stimuli

PECs prepared by using weak polyelectrolytes show response to pH due to protonation/deprotonation of charged groups.[101] Protonation, that is, accumulation of additional charges, results in stronger repulsion, which leads to capsule swelling and thereby permeability increases. Deprotonation decreases the interaction of polymers, causes shrinking, and thus lowers the permeability. The major advantage of this approach is its reversibility. Phyu et al.[102] prepared oral controlled release ibuprofen gel beads by

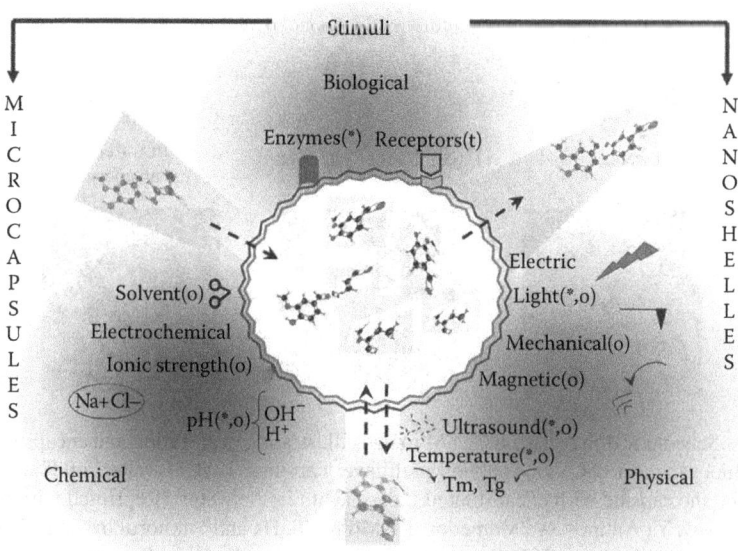

FIGURE 56.11 Schematic illustration of stimuli for microcapsule loading and release. (Reprinted from *Macromolecules*, 42(22), Lichter, J.A., VanVliet, K.J., and Rubner, M.F., Design of antibacterial surfaces and interfaces. Polyelectrolyte multilayers as a multifunctional platform, 8573–8586. Copyright 2009, with permission from Elsevier.)

PEC technique using soluble phosphorylated chitosan (PCS) and tripolyphosphate (TPP) at pH 4. The ibuprofen release from PCS gel beads was found to be increased as the pH of the dissolution medium increased. The release rate of ibuprofen at pH 7.4 was higher than the release rate at pH 1.4 due to the ionization of phosphate group and higher solubility of ibuprofen at pH 7.4 medium. The ability of the prepared copolymer to be used as drug carrier for colon-specific drug delivery system was estimated using ketoprofen as model drug. Macleod et al.[4] prepared PECs using pectin and chitosan. The amount of pectin relative to chitosan, required for optimal PEC formation, increased as the pH of the solution was reduced. From the reported results, it is applied as a film coat to tablets, and could be used to achieve bimodal drug release with colonic condition acting as a trigger for an increased rate of release.

56.5.9.2 Ionic Strength Stimuli

PECs are responsive to ionic strength because of screening of electrostatic interactions between oppositely charged polyelectrolytes. The encapsulation and release of active moieties in the PECs can be modified according to the requirements by changing the salt concentration.[103–108] The influence of ionic strength on polyelectrolyte multilayer capsules has been studied by measuring dye molecule diffusion through polyelectrolyte multilayers as a function of varying salt concentration.[108]

56.5.9.3 Solvent Stimuli

Lvov et al.[109] studied the influence of solvent ethanol by encapsulating urease using poly(styrenesulfonate, sodium salt) and poly(allylamine hydrochloride sodium salt) in 1:1 ethanol water mixture. In the presence of ethanol, urease activity was lower. Ethanol causes segregation of polyion network due to partial removal of the hydration water between the polyelectrolytes (Figure 56.12).[109]

56.5.9.4 Electrochemical Stimuli

Electrochemical stimulated systems have potential biomedical and micromechanical application. Electrochemical stimuli controlled loading/adsorption onto oppositely charged polyelectrolyte materials on porous microspheres that was previously reported by Malinova and Wandrey.[110] Electrochemical stimuli cause an increase in osmotic pressure of the redox-active polyelectrolyte

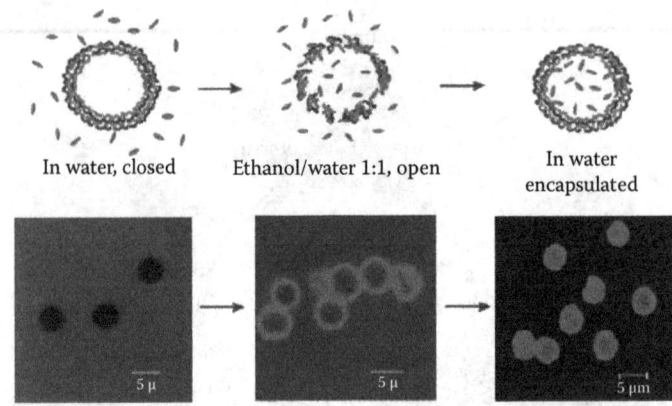

FIGURE 56.12 Schematic (top row) and CLSM images illustrating permeation and encapsulation of urease–fluorescein isothiocyanate (FITC) into polyion multilayer capsules. Left, in water; middle, in water/ethanol mixture 1:1; right, the capsule with encapsulated urease again in the water. (Reprinted with permission from *Nano Lett.*, 1(3), Lvov, Y., Antipov, A., Mamedov, A., Möhwald, H., and Sukhorukov, G.B., Urease encapsulation in nanoorganized microshells, 125. Copyright 2001 American Chemical Society.)

multilayers and swelling. The volumetric expansion is due to an influx of counterions and solvent molecules, which increases the osmotic pressure in the film. Changes in external parameters (pH, ionic strength, and ion species) result in changes in the ion flux and cause variations in swelling, which can be used for protecting ferrous metals.[111]

56.5.9.5 Temperature Stimuli

Temperature is important, because it allows producing mechanically strengthened PECs. Temperature is an external posttreatment parameter affecting material properties and polyelectrolyte multilayer formation.[105] Enhancement of mechanical properties and encapsulation efficiency due to shrinking of layers in response to thermal treatment has been reported by Köhler et al.[112] In the case of PECs, a significant reduction of the permeability for low-molecular-weight compounds accompanies heat treatment[113] due to the interplay of electrostatic and hydrophobic forces. PECs can be compressed or expanded in response to temperature stimuli, which can be enabled for encapsulation and release of active moieties.

56.5.9.6 Laser Light Stimuli

Light-responsive PECs have potential application for in vivo drug delivery. Light-sensitive capsules were reported by several groups.[114–118] One of the mechanisms is based on localized heating of nanoparticles upon exposure to light, because noble metal nanoparticles efficiently absorb laser energy and convert it into heat. Light response is not only useful for release but also for encapsulation in PECs. Bedard and coworkers reported on the light-induced shrinking and encapsulation of a fluorescently labeled polymer in a (PAH/poly[1-[4-(carboxy-4-hydroxyphenylazo)benzenesulfoamido]-1,2-ethanediyl, sodium salt] (PAzo))3/PAH/poly(vinylsulfonate) (PVS) microcapsule.[119] The encapsulation is attributed to rearrangements of polymers and changes in the permeability of the shell following exposure to light. The effectiveness of optical[120] encapsulation was shown to increase with increasing irradiation time.

56.5.9.7 Ultrasound Stimuli

Application ultrasound has been used for the synthesis of materials, coating carbon nanotubes and noble metals, destruction and fragmentation of contrast agents, gas release, polymer destruction, and in drug delivery.[121] Ultrasound application has effect on permeability of PEC[122] microcapsules.

Kolesnikova et al.[123] reported polyelectrolyte microcapsules with ZnO that proved to have potential as drug delivery systems with the possibility of releasing under the action of ultrasound.

56.5.9.8 Magnetic Stimuli

Lu et al. reported that magnetic field affects the permeability of PECs by acting on aggregates of nanoparticles.[124] Ferromagnetic cobalt nanoparticles containing a layer of gold (Co/Au) were incorporated into the assembly of PSS and PAH polyelectrolyte multilayer shells of microcapsules. Further, application of alternating magnetic fields with frequencies of 100 to 300 Hz and 1200 Oe strength resulted in increased permeability and release of the encapsulated content. Targeting by magnetic field is one of the main functions of magnetic nanoparticle incorporated into PECs. The therapeutic performance of capsules can be enhanced by magnetic field–triggered release.[125,126]

56.5.9.9 Mechanical Stimuli

Successful delivery of encapsulated drug is one of the main goals of drug delivery. The nanoparticle systems undergo deformation upon intracellular uptake,[127] leading to loss of encapsulated drug. Studies have been reported that mechanically stable carrier systems prepared with gold nanoparticle[128] and carbon nanotubes[129] can release the encapsulated substances upon mechanical rupture.[130]

56.5.9.10 Biological Stimuli

Biological stimuli induce a response to a biologically relevant molecule or phenomenon. Microcapsules containing phenylboronic acid in coating shows glucose-sensitive response,[131] which can be used for the delivery of insulin. Biodegradable polyelectrolyte multilayer capsules could be attractive for delivery of drugs with an intracellular target such as nucleic acids and proteins.[132] Enzymatically degradable PECs based on oppositely charged polypeptides and/or polysaccharides are having practical importance. De Geest et al.[133] in 2006 demonstrated that PECs consisting of dextran sulfate and poly-L-arginine could be degraded intracellularly by proteases upon phagocytosis by in vitro cultured cells. Similar results were reported when hyaluronidase[134] and chitosanase[135] was used to decompose capsules containing hyaluronic acid or chitosan as layer components.

56.6 CONCLUSIONS

An extensive research is going on in polyelectrolytes and PECs. There is a great potential in utilizing these PECs in ecology, biotechnology, medicine, and pharmaceutical technology. PECs are very interesting materials for different applications, because some of their properties, like swelling or permeability, can be easily modified by external stimuli. They may efficiently modify the release, and improve the stability and character of the drug substances due to their capacity to entrap the drug at molecular level. Hence, the PECs have great potential in the design of novel drug delivery systems for effective drug delivery.

ACKNOWLEDGMENTS

The author is thankful to the GITAM Institute of Pharmacy, GITAM University, for providing necessary support. The author expresses his gratitude to Prof. K.V. Ramana Murthy for his valuable guidance and support.

REFERENCES

1. Philip, B., Dautzenberg, H., Linow, K., Kotz, J., Dawydoff, W. Polyelectrolyte complexes-recent developments and open problems. *Prog. Polym. Sci.* **1989**, *14* (1), 91–172.
2. Roy, K., Mao, H.-Q., Huang, S.-K., Leong, K.W. Oral gene delivery with chitosan–DNA nanoparticles generates immunologic protection in a murine model of peanut allergy. *Nat. Med.* **1999**, *5* (4), 387–391.

3. Genta, I., Perugini, P., Modena, T., Pavanetto, F., Castelli, F., Muzzarelli, R.A.A., Muzzarelli, C., Conti, B. Miconazole-loaded 6-oxychitin–chitosan microcapsules. *Carbohydr. Polym.* **2003**, *52* (1), 11–18.
4. Macleod, G.S., Collett, J.H., Fell, J.T. The potential use of mixed films of pectin, chitosan and HPMC for bimodal drug release. *J. Control. Release* **1999**, *58* (3), 303–310.
5. Kubisa, P. Terminology of polymers containing ionizable or ionic groups and of polymers containing ions. *IUPAC Recommendations 2004* (DRAFT December 23, 2004).
6. Lankalapalli, S., Kolapalli, V.R.M. Polyelectrolyte complexes: A review on the applicability in drug delivery technology. *Ind. J. Pharm. Sci.* **2009**, *71* (5), 481–487.
7. Koetz, J., Kosmella, S., Howard, G., Barth, D., Harald, P. *Polyelectrolytes in Polyelectrolytes and Nanoparticles*, Springer: Berlin, Germany, 2007, pp. 5–45.
8. Tsuchida, E. Formation of polyelectrolyte complexes and their structures. *J. Macromol. Sci. Pure Appl. Chem.* **1994**, *31* (1), 1–15.
9. Kossel, A. Über einen peptoartigen Bestandteil des zellkerns. Hoppe-Seyler's. *Z. Physiol. Chem.* **1884**, *8*, 511–515.
10. Fuoss, R.M., Sadek, H. Mutual interaction of polyelectrolytes. *Science* **1949**, *110* (2865), 552–554.
11. Decher, G., Eckle, M., Schmitt, J., Struth, B. Layer-by-layer assembled multicomposite films. *Curr. Opin. Coll. Interface Sci.* **1998**, *3* (1), 32–39.
12. Petrak, K., Hara, M., Ed. *Polyelectrolytes Science and Technology*, Entry 5, Marcel Dekker: New York, 1993, pp. 265–298.
13. Webster, L., Huglin, M.B. Complex formation between polyelectrolytes in dilute aqueous solution. *Polymer* **1997**, *38* (6), 1373–1380.
14. Webster, L., Huglin, M.B. Observations on complex formation between polyelectrolytes in dilute aqueous solution. *Eur. Polym. J.* **1997**, *33* (7), 1173–1177.
15. Burgess, D.J. Practical analysis of complex coacervate systems. *J. Coll. Interfaces Sci.* **1990**, *140* (1), 227–238.
16. Overbeek, J.T.G., Voorn, M.J. Phase separation in polyelectrolyte solutions. Theory of complex coacervation. *J. Cell Comp. Physiol.* **1957**, *49* (1), 7–26.
17. Xavier, C., Jean-Francois, J. Adsorption of polyelectrolyte solutions on surface. A Debye-Huckel theory. *J. Phys. II France* **1996**, *6* (12), 1669–1686.
18. Kokufuta, E. Colloid titration behavior of poly (ethyleneimine). *Macromolecules* **1979**, *12* (2), 350–353.
19. Kabanov, V.A. Basic properties of soluble interpolyelectrolyte complexes applied to bioengineering and cell transformation. In *Macromolecular Complexes in Chemistry and Biology*, Dubin, P., Bock, J., Davies, R.M., Schulz, D.N., Thies, C., Eds., Springer-Verlag: Berlin, Germany, 1994, pp. 151–174.
20. Thünemann, A.F., Müller, M., Dautzenberg, H., Joanny, J.H., Löwen, H. Polyelectrolyte complexes. *Adv. Polym. Sci.* **2004**, *166* (10), 113–171.
21. Philipp, B., Dawydoff, W., Linow, K.J. Polyelektrolytkomplexe—Bildungsweise, Struktur und Anwendungsmö glichkeiten. *Zeitschrift für Chemie* **1982**, *22* (1), 1–13.
22. Radeva, T. Ed. *Physical Chemistry of Polyelectrolytes*, Surfactant Science Series 99, Marcel, 2001.
23. Schönhoff, M. Layered polyelectrolyte complexes. Physics of formation and molecular properties. *J. Phys. Condens. Matter.* **2003**, *15* (49), R1781–R1808.
24. Tsuchida, E., Abe, K. Interactions between macromolecules in solution and intermacromolecular complexes. *Adv. Polym. Sci.* **1982**, *45*, 1–130.
25. Dautzenberg, H., Koetz, J., Linow, B., Philipp, B., Rother, G. Eds. Static Light scattering of polyelectrolyte complex solution. In *Macromolecular Complexes in Chemistry and Biology*, Springer Berlin Heidelberg: Berlin, Germany, 1994, pp. 119–134.
26. Dautzenberg, H. Polyelectrolyte complex formation in highly aggregating systems. 1. Effect of salt: Polyelectrolyte complex formation in the presence of NaCl. *Macromolecules* **1997**, *30* (25), 7810–7815.
27. Dautzenberg, H., Rother, G., Hartmann, J. Light-scattering studies of polyelectrolyte complex formation. Effect of polymer concentration. *Macro-Ion Charac.* **1994**, *548*, 210–224.
28. Frugier, D., Audebert, R. Interaction between oppositely charged low ionic density polyelectrolytes. Complex formation or simple mixture. In *Macromolecular Complexes in Chemistry and Biology*, Springer: Berlin, Germany, 1994, pp. 135–149.
29. Iliopoulos, I., Audebert, R. Influence of concentration, molecular-weight and degree of neutralization of polyacrylic-acid on interpolymer complexes with polyoxyethylene. *Polym. Bull.* **1985**, *13* (2), 171–178.
30. Srinivas, L., Ramana Murthy, K.V. Preparation and evaluation of polyelectrolyte complexes for oral controlled drug delivery. *Asian J. Pharm.* **2010**, *4* (1), 69–78.

31. Srinivas, L., Ramana Murthy, K. Biopharmaceutical evaluation of diclofenac sodium controlled release tablets prepared from gum karaya–chitosan polyelectrolyte complexes. *Drug Develop. Ind. Pharm.* **2012**, *38* (7), 815–824.
32. Vishalakshi, B., Ghosh, S., Kalpagam, V. The effects of charge-density and concentration on the composition of polyelectrolyte complexes. *Polymer* **1993**, *34* (15), 3270–3275.
33. Argüelles-Monala, W., Cabrerab, G., Penichec, C., Rinaudod, M. Conductimetric study of the interpolyelectrolyte reaction between chitosan and polygalacturonic acid. *Polymer* **2000**, *41* (7), 2373–2378.
34. Van Leeuwen, H.P., Cleven, R.F.M.J., Valenta, P. Conductometric analysis of polyelectrolytes in solution. *Pure Appl. Chem.* **1991**, *63* (9), 1251–1268.
35. Takayama, K., Nagai, T. Application of interpolymer complexation of polyvinylpyrrolidone/carboxyvinyl polymer to control of drug release. *Chem. Pharm. Bull.* **1987**, *35* (12), 4921–4927.
36. Zhang, L.M. Synergistic blends from aqueous solutions of two cellulose derivatives. *Coll. Polym. Sci.* **1999**, *277* (9), 886–890.
37. Dautzenberg, H. Light scattering studies on polyelectrolyte complexes. *Macromol. Symp.* **2001**, *162* (1), 1–22.
38. Natalia, V.P., Nickolay, V.T. Structure and dynamics of the polyelectrolyte complex formation. *Macromolecules* **1997**, *30* (17), 4897–4904.
39. Anal, S., Capan, Y., Guven, O., Gogus, A., Dlakara, T., Hencal, A.A. Formulation and in-vitro-in-vivo evaluation of buccoadhesive morphine sulfate tablets. *Pharm. Res.* **1994**, *11* (2), 231–236.
40. Henze, G., Neeb, R. *Elektrochemische Analytik*, Springer: Berlin, Germany, 1986.
41. Kurucsev, T., Steel, B.J. The use of electrical transport measurements for the determination of counterion association in salt-free polyelectrolyte solutions. *Rev. Pure Appl. Chem.* **1961**, *17*, 149–157.
42. Rice, S.A., Nagasawa, M. *Polyelectrolyte Solutions*, Academic Press: New York, 1961.
43. Arai, S., Kumada, K. An interpretation of the conductometric titration curve of humic acid. *Geoderma* **1977a**, *19* (1), 21–35.
44. Varoqui, R., Strauss, U.P. Comparison of electrical transport properties of anionic polyelectrolytes and polysoaps. *J. Phys. Chem.* **1968**, *72* (7), 2507–2511.
45. Jiang, H.L., Zhu, K.J. Polyanion/gelatin complexes as pH-sensitive gels for controlled protein. *J. Appl. Polym. Sci.* **2001**, *80* (9), 1416–1425.
46. Tsuboi, A., Izumi, T., Hirata, M., Xia, J.L., Dubin, P.L., Kokufuta, E. Complexation of proteins with a strong polyanion in an aqueous salt free system. *Langmuir* **1996**, *12* (26), 6295–6303.
47. Fuoss, R.M., Cathers, G.I. Polyelectrolytes. III. Viscosities of n-butyl bromide addition compounds of 4-vinylpyridine-styrene copolymers in nitromethanedioxane mixtures. *J. Polym. Sci.* **1949**, *4* (2), 97–120.
48. Masanori, H., Jhi-Li, W., Lee, A.H. Effect of intra- and intermolecular interactions on solution properties of sulfonated polystyrene ionomers. *Macromolecules* **1988**, *21* (7), 2214–2218.
49. Weinbreck, F., Wientjes, R.H.W., Rheol, J. Rheological properties of whey protein/gum arabic coacervates. *J. Rheol.* **2004**, *48* (1), 1215–1228.
50. Van de Weert, M., Andersen, M.B., Frokjaer, S. Complex coacervation of lysozyme and heparin complex characterization and protein stability. *Pharm. Res.* **2004**, *21* (12), 2354–2359.
51. Ivinova, O.N., Izumrudov, V.A., Muronetz, V.I., Galaev, I.Y., Mattiasson, B. Influence of complexing polyanions on the thermostability of basic proteins. *Macromol. Biosci.* **2003**, *3* (3–4), 210–215.
52. Polu, A.R., Kumar, R. Impedence spectroscopy and FTIR studies of PEG based polymer electrolytes. *E- J. Chem.* **2011**, *8* (1), 347–353.
53. Srinivas, L. Preparation and evaluation of polyelectrolyte microparticles for controlled drug delivery. *Asian J. Chem.* **2012**, *24* (3), 1282–1288.
54. Everett, J.R. Nuclear magnetic resonance spectroscopy applications/pharmaceutical. In *Encyclopedia of Analytical Science*, Townshend, A. and Colin F. Poole, Eds., 2nd edn., Elsevier, U.K. 2005, pp. 321–332.
55. Ni, J.X. ^{13}C Nuclear magnetic relaxation of tetraethylammonium cation in poly(acrylic acid) solution. *New J. Chem.* **1999**, *23* (11), 1071–1073.
56. Reed, W.F. *Light Scattering Results on Polyelectrolyte Conformations, Diffusion and Interparticle Interactions and Correlations*, invited entry for ACS Symposium Series 548, *Macroion Characterization*. Schmitz, K. Ed., ACS, 1994, pp. 297–314.
57. Reed, C.E., Reed, W.F. Monte Carlo study of light scattering by linear polyelectrolytes. *J. Chem. Phys.* **1992**, *97* (10), 7766–7776.
58. Mihai, M., Dragan, E.S., Schwarz, S., Janke, A. Dependency of particle sizes and colloidal stability of polyelectrolyte complex dispersions on polyanion structure and preparation mode investigated by dynamic light scattering and atomic force microscopy. *J. Phys. Chem. B* **2007**, *111* (29), 8668–8675.

59. Kejian, D., Weimin, S., Zhang, H., Xianglei, P., Honggang, H. Dependence of zeta potential on polyelectrolyte moving through a solid-state nanopore. *Appl. Phys. Lett.* **2009**, *94* (1), 014101.
60. Kötz, J., Kosmella, S. Interactions between polyanion–polycation systems and kaolin. *J. Coll. Interface Sci.* **1994**, *168* (2), 505–513.
61. Lachman, L., Lieberman, H.A. *In The Theory and Practice of Industrial Pharmacy*, 3rd edn., Special Indian Edition, CBS Publishers and Distributers Pvt. Ltd: New Delhi, India, 2009, pp. 21–47.
62. McMullan, D. Scanning electron microscopy 1928–1965. *Scanning* **1995**, *17* (3), 175–185.
63. Pihlajamaa, T., Lankinen, H., Ylostalo, J., Valmu, L., Jaalinoja, J., Zaucke, F. Characterization of recombinant amino-terminal NC4 domain of human collagen IX—Interaction with glycosaminoglycans and cartilage oligomeric matrix protein. *J. Biol. Chem.* **2004**, *279* (23), 24265–24273.
64. Burgess, D.J., Dubin, P.L., Bock, J., Davis. Schulz. *Complex Coacervation Microcapsule Formation in Macromolecular Complexes in Chemistry and Biology*, Springer-Verlag: Berlin, Germany, 1994, pp. 285–300.
65. Lvov, Y., Ariga, K., Ichinose, I., Kunitake, T. Assembly of multicomponent protein films by means of electrostatic layer-by-layer adsorption. *J. Am. Chem. Soc.* **1995**, *117* (22), 6117–6123.
66. Caruso, F., Mohwald, H. Protein multilayer formation on colloids through a stepwise self-assembly technique. *J. Am. Chem. Soc.* **1999**, *121* (25), 6039–6046.
67. Patrick, J.S. Ed. *Martin's Physical Pharmacy and Pharmaceutical Sciences*, 5th edn., Indian Reprint, Lippincott Williams & Wilkins, 2006, pp. 301–354.
68. Costa, P., Lobo, J.M.S. Modeling and comparison of dissolution profiles. *Eur. J. Pharm. Sci.* **2001**, *13* (2), 123–133.
69. Yao, F., Weiyuan, J.K. Drug release kinetics and transport mechanisms of non-degradable and degradable polymeric delivery systems. *Expert Opin. Drug Deliv.* **2010**, *7* (4), 429–444.
70. Vergaro, V., Scarlino, F., Bellomo, C., Rinaldi, R., Vergara, D., Maffia, M., Baldassarre, F. et al. Drug-loaded polyelectrolyte microcapsules for sustained targeting of cancer cells. *Adv. Drug Deliv. Rev.* **2011**, *63* (9), 63847–63864.
71. Tripathy, S.K., Kumar, J., Nalwa, H.S. *Handbook of Polyelectrolytes and Their Applications*, American Scientific Publishers: Los Angeles, CA, 2002, p. 1200.
72. Senuma, M., Kuwabara, S., Kaeriyama, K., Hase, F., Shimura, Y. Polymer complex from copolymers of acrylonitrile and ionic vinyl benzyl compounds. *J. Appl. Polym. Sci.* **1986**, *31* (6), 1687–1697.
73. Sato, H., Maeda, M., Nakajima, A. Mechanochemistry and permeability of polyelectrolyte complex membranes composed of poly (vinyl alcohol) derivatives. *J. Appl. Polym. Sci.* **1979**, *23* (6), 1759–1767.
74. Harris, L.V., Harris, E.L.V., Angal, S. *Protein Purification Methods*, Oxford University Press: New York, 1993.
75. Yamamoto, H., Horita, C., Senoo, Y., Nishida, A., Ohkawa, K. Polyion complex fiber and capsule formed by self-assembly of chitosan and gellan at solution interfaces. *Macromol. Chem. Phys.* **2000**, *201* (1), 84–92.
76. Hirouki, Y., Takeshi, K. Adsorption of BSA on cross-linked chitosan. The equilibrium isotherm. *Chem. Eng. Jpn.* **1989**, *41* (1), B11–B15.
77. Dubin, P.L., Gao, J., Mattison, K. Protein purification by selective phase separation with polyelectrolytes. *Sep. Purif. Methods* **1994**, *23* (1), 1–16.
78. Cordes, R.M., Sima, W.B., Glatz, C.E. Precipitation of nucleic acids with poly (ethyleneimine). *Biotechnol. Prog.* **1990**, *6* (4), 283–285.
79. Atkinson, A., Jack, G.W. Precipitation of nucleic acids with polyethyleneimine and the chromatography of nucleic acids on immobilized polyethyleneimine. *Biochem. Biophys. Acta* **1973**, *308* (1), 41–52.
80. Jendrisak, J., Burgerss, R. The use of polyethyleneimine in protein purification. In *Protein Purification: Micro to Macro*, Burgess, R., Ed., Liss, Inc.: New York, 1987, pp. 75–97.
81. Chen, J., Jo, S., Park, K. Polysaccharide hydrogels for protein drug delivery. *Carbohydr. Polym.* **1995**, *28* (1), 69–76.
82. Artur, B., David, H. Carrageenan–oligochitosan microcapsules. Optimization of the formation process. *Coll. Surf. B Biointerfaces* **2001**, *21* (4), 285–298.
83. Murakami, R., Takashima, R. Mechanical properties of the capsules of chitosan–soy globulin polyelectrolyte complex. *Food Hydrocoll.* **2003**, *17* (6), 885–888.
84. Manning, G.S. Polyelectrolytes. *Ann. Rev. Phys. Chem.* **1972**, *23* (1), 117–140.
85. Cooper, C.L., Dubin, P.L., Kayitmazer, A.B., Turksen, S. Polyelectrolyte–protein complexes. *Curr. Opin. Coll. Interfaces Sci.* **2005**, *10* (1), 52–78.
86. Soliman, M., Allen, S., Davies, M.C., Alexander, C. Responsive polyelectrolyte complexes for triggered release of nucleic acid therapeutics. *Chem. Commun.* **2010**, *46* (30), 5421–5433.

87. Babak, V.G., Merkovich, E.A., Desbrières, J., Rinaudo, M. Formation of an ordered nanostructure in surfactant polyelectrolyte complexes formed by interfacial diffusion. *Polym. Bull.* **2000**, *45* (1), 77–81.
88. Tsuchida, E. Formation of interpolymer complexes. *J. Macromol. Sci. B* **1980**, *17* (4), 683–714.
89. Vert, M. Polyvalent polymeric drug carriers. *Crit. Rev. Ther. Drug Carrier Syst.* **1986**, *2* (3), 291–327.
90. Kabanov, V.A., Dubin, P. In *Macromolecular Complexes in Chemistry and Biology*, Springer: New York, 1994.
91. Leclercq, L., Boustta, M., Vert, M. A physico-chemical approach of polyanion-polycation interactions aimed at better understanding the in vivo behaviour of polyelectrolyte-based drug delivery and gene transfection. *J. Drug Target.* **2003**, *11* (3), 129–138.
92. Kekkonen, J., Lattu, H., Stenius, P. Adsorption kinetics of complexes formed by oppositely charged polyelectrolytes. *J. Coll. Interfaces Sci.* **2001**, *234* (2), 384–392.
93. Shojaei, A.H. Buccal mucosa as a route for systemic drug delivery. A review. *J. Pharm. Pharm. Sci.* **1998**, *1* (1), 15–30.
94. Krone, V., Magerstadt, M., Walch, A., Groner, A., Hoffmann, D. Pharmacological composition containing polyelectrolyte complexes in micro particulate form and at least on active agent. U.S. Patent 5,700,459, December 23, 1997.
95. Peyratout, C.S., Daehne, L. Tailor-made polyelectrolyte microcapsules from multilayers to smart containers. *Angew. Chem. Int. Ed.* **2004**, *43* (29), 3762–3783.
96. Delcea, M., Möhwald, H., Skirtach, A.G. Stimuli-responsive LbL capsules and nanoshells for drug delivery. *Adv. Drug Deliv. Rev.* **2011**, *63* (9), 730–747.
97. Lichter, J.A., VanVliet, K.J., Rubner, M.F. Design of antibacterial surfaces and interfaces. Polyelectrolyte multilayers as a multifunctional platform. *Macromolecules* **2009**, *42* (22), 8573–8586.
98. Sukhishvili, S.A. Responsive polymer films and capsules via layer-by-layer assembly. *Curr. Opin. Coll. Interfaces Sci.* **2005**, *10* (1), 37–44.
99. Tang, Z., Wang, Y., Podsiadlo, P., Kotov, N.A. Biomedical applications of layer-by layer assembly from biomimetics to tissue engineering. *Adv. Mater.* **2006**, *18* (24), 3203–3224.
100. Lynn, D.M. Peeling back the layers. Controlled erosion and triggered disassembly of multilayered polyelectrolyte thin films. *Adv. Mater.* **2007**, *19* (23), 4118–4130.
101. Sui, Z.J., Schlenoff, J.B. Phase separations in pH-responsive polyelectrolyte multilayers: Charge extrusion versus charge expulsion. *Langmuir* **2004**, *20* (14), 6026–6031.
102. Win, P.P., Shin-ya, Y., Hong, K.-J., Kajiuchi, T. Formulation and characterization of pH sensitive drug carrier based on phosphorylated chitosan (PCS). *Carbohydr. Polym.* **2003**, *53* (3), 305–310.
103. Ibarz, G., Dähne, L., Donath. E., Möhwald, H. Smart micro- and nanocontainers for storage, transport and release. *Adv. Mater* **2001**, *13* (17), 1324–1327.
104. Gao, C., Möhwald, H., Shen, J. Enhanced biomacromolecule encapsulation by swelling and shrinking procedures. *ChemPhysChem* **2004**, *5* (1), 116–120.
105. Buscher, K., Graf, K., Ahrens, H., Helm, C.A. Influence of adsorption conditions on the structure of polyelectrolyte multilayers. *Langmuir* **2002**, *18* (9), 3585–3591.
106. Fery, A., Scholer, B., Cassagneau, T., Caruso, F. Nanoporous thin films formed by salt-induced structural changes in multilayers of poly(acrylic acid) and poly (allylamine). *Langmuir* **2001**, *17* (13), 3779–3783.
107. McAloney, R.A., Sinyor, M., Dudnik. V., Goh, M.C. Atomic force microscopy studies of salt effects on polyelectrolyte multilayer film morphology. *Langmuir* **2001**, *17* (21), 6655–6663.
108. Antipov, A.A., Sukhorukov, G.B., Möhwald, H. Influence of the ionic strength on the polyelectrolyte multilayers' permeability. *Langmuir* **2003**, *19* (6), 2444–2448.
109. Lvov, Y., Antipov, A., Mamedov, A., Möhwald, H., Sukhorukov, G.B. Urease encapsulation in nanoorganized microshells. *Nano Lett.* **2001**, *1* (3), 125–128.
110. Malinova, V., Wandrey, C. Loading polyelectrolytes onto porous microspheres: Impact of molecular and electrochemical parameters. *J. Phys. Chem. B* **2007**, *111* (29), 8494–9501.
111. Zahn, R., Voros, J., Zambelli, T. Swelling of electrochemically active polyelectrolyte multilayers. *Curr. Opin. Coll. Interfaces Sci.* **2010**, *15* (6), 427–434.
112. Köhler, K., Shchukin, D.G., Möhwald, H., Sukhorukov, G.B. Thermal behavior of polyelectrolyte multilayer microcapsules: 1. The effect of odd and even layer number. *J. Phys. Chem. B* **2005**, *109* (39), 18250–18259.
113. Ibarz, G., Dähne, L., Donath, E., Möhwald, H. Controlled permeability of polyelectrolyte capsules via defined annealing. *Chem. Mater.* **2002**, *14* (10), 4059–4062.
114. Tao, X., Li, J., Möhwald, H. Self-assembly, optical behavior, and permeability of a novel capsule based on an azo dye and polyelectrolytes. *Chem. Eur. J.* **2004**, *10* (14), 3397–3403.

115. Ning, L., Kommireddy, D.S., Lvov, Y., Liebenberg, W., Tiedt, L.R., De Villiers, M.M. Nanoparticle multilayers: Surface modification of photosensitive drug microparticles for increased stability and in vitro bioavailability. *J. Nanosci. Nanotechol.* **2006**, *6* (9–10), 3252–3260.
116. Skirtach, A.G., Antipov, A.A., Shchukin, D.G., Sukhorukov, G.B. Remote activation of capsules containing Ag nanoparticles and IR dye by laser light. *Langmuir* **2004**, *20* (17), 6988–6992.
117. Radt, B., Smith, T.A., Caruso, F. Optically addressable nanostructured capsules. *Adv. Mater.* **2004**, *16* (23–24), 2184–2189.
118. De Geest, B.G., Skirtach, A.G., De Beer, T.R.M., Sukhorukov, G.B., Bracke, L., Baeyens, W.R.G., Demeester, J., De Smedt, S.C. Stimuli-responsive multilayered hybrid nanoparticle/polyelectrolyte capsules. *Macromol. Rapid Commun.* **2007**, *28* (1), 88–95.
119. Bedard, M., Skirtach, A.G., Sukhorukov, G.B. Optically driven encapsulation using novel polymeric hollow shells containing an azobenzene polymer. *Macromol. Rapid Commun.* **2007**, *28* (15), 1517–1521.
120. Koo, H.Y., Lee, H.J., Kim, J.K., Choi, W.S. UV-triggered encapsulation and release from polyelectrolyte microcapsules decorated with photoacid generators. *J. Mater. Chem.* **2010**, *20* (19), 3932–3937.
121. Shchukin, D.G., Gorin, D.A., Mohwald, H. Ultrasonically induced opening of polyelectrolyte microcontainers. *Langmuir* **2006**, *22* (17), 7400–7404.
122. Antipov, A.A., Sukhorukov, G.B., Leporatti, S., Radtchenko, I.L., Donath, E., Mohwald, H. Polyelectrolyte multilayer capsule permeability control. *Coll. Surf. Physicochem. Eng. Asp.* **2002**, *198–200*, 535–541.
123. Kolesnikova, T.A., Gorin, D.A., Fernandes, P., Kessel, S., Khomutov, G.B., Fery, A., Shchukin, D.C., Mohwald, H. Nanocomposite microcontainers with high ultrasound sensitivity. *Adv. Funct. Mater.* **2010**, *20* (7), 1189–1195.
124. Lu, Z., Prouty, M.D., Guo, Z., Golub, V.O., Kumar, C.S.S.R., Lvov, Y.M. Magnetic switch of permeability for polyelectrolyte microcapsules embedded with Co@Au nanoparticles. *Langmuir* **2005**, *21* (5), 2042–2050.
125. Wang, W., Liu, L., Ju, X.J., Zerrouki, D., Xie, R., Yang, L.H., Chu, L.Y. A novel thermo-induced self-bursting microcapsule with magnetic-targeting property. *ChemPhysChem* **2009**, *10* (4), 2405–2409.
126. Hu, S.H., Tsai, C.H., Liao, C.F., Liu, D.M., Chen, S.Y. Controlled rupture of magnetic polyelectrolyte microcapsules for drug delivery. *Langmuir* **2008**, *24* (20), 11811–11818.
127. Munoz-Javier, A., Kreft, O., Semmling, M., Kempter, S., Skirtach, A.G., Bruns, O.T., del Pino, P. et al. Uptake of colloidal polyelectrolyte-coated particles and polyelectrolyte multilayer capsules by living cells. *Adv. Mater.* **2008**, *20* (22), 4281–4287.
128. Bedard, M.F., Munoz-Javier, A., Mueller, R., del Pino, P., Fery, A., Parak, W.J., Skirtach, A.G., Sukhorukov, G.B. On the mechanical stability of polymeric microcontainers functionalized with nanoparticles. *Soft Matter* **2009**, *5* (1), 148–155.
129. Yashchenok, A.M., Bratashov, D.N., Gorin, D.A., Lomova, M.V., Pavlov, A.M., Sapelkin, A.V., Shim, B.S. et al. Carbon nanotubes on polymeric microcapsules free-standing structures and point-wise laser openings. *Adv. Funct. Mater.* **2010**, *20* (18), 3136–3142.
130. Caruso, M.M., Schelkopf, S.R., Jackson, A.C., Landry, A.M., Braun, P.V., Moore, J.S. Microcapsules containing suspensions of carbon nanotubes. *J. Mater. Chem.* **2009**, *19* (1), 6093–6096.
131. De Geest, B.G., Jonas, A.M., Demeester, J., De Smedt, S.C. Glucose-responsive polyelectrolyte capsules. *Langmuir* **2003**, *22* (11), 5070–5074.
132. She, S.J., Zhang, X.G., Wu, Z.M., Wang, Z., Li, C.X. Gradient cross-linked biodegradable polyelectrolyte nanocapsules for intracellular protein drug delivery. *Biomaterials* **2010**, *31* (23), 6039–6049.
133. De Geest, B.G., Vandenbroucke, R.E., Guenther, A.M., Sukhorukov, G.B., Hennink, W.E., Sanders, N.N., Demeester, J., De Smedt, S.C. Intracellularly degradable polyelectrolyte microcapsules. *Adv. Mater.* **2006**, *18* (8), 1005–1009.
134. Dimitrova, M., Affolter, C., Meyer, F., Nguyen, I., Richard, D.G., Schuster, C., Bartenschlager, R., Voegel, J.-C., Ogier, J., Baumert T.F. Sustained delivery of siRNAs targeting viral infection by cell-degradable multilayered polyelectrolyte films. *Proc. Natl. Acad. Sci. USA* **2008**, *205* (42), 16320–26325.
135. Itoh, Y., Matsusaki, M., Kida, T., Akashi, M. Enzyme-responsive release of encapsulated proteins from biodegradable hollow capsules. *Biomacromolecules* **2006**, *7* (10), 2715–2718.
136. Zhao, Q., Qian, J., An, Q., Gao, C., Gui, Z., Jin, H. Synthesis and characterization of soluble chitosan/sodium carboxymethyl cellulose polyelectrolyte complexes and the pervaporation dehydration of their homogeneous membranes. *J. Membr. Sci.* **2009**, *333* (1–2), 68–78.
137. Anders, L., Charlotte, W., Staffan, W. Flocculation of cationic polymers and nanosized particles. *Coll. Surf. A Physicochem. Eng. Asp.* **1999**, *159* (1), 65–76.

138. Akbuga, J., Bergisadi, N. 5-Fluorouracil-loaded chitosan microspheres: Preparation and release characteristics. *J. Microencapsul.* **1996**, *13* (2), 161–168.
139. Meerum Terwogt, J.M., ten Bokkel Huinink, W.W., Schellens, J.H., Schot, M., Mandjes, I.A., Zurlo, M.G., Rocchetti, M., Rosing, H., Koopman, F.J., Beijnen, J.H. Phase I clinical and pharmacokinetic study of PNU166945, a novel water-soluble polymer conjugated prodrug of paclitaxel. *Anticancer Drugs* **2001**, *12* (4), 315–323.
140. Vilches, A.P., Jimenez-Kairuz, A.F., Alovero, F., Olivera, M.E., Allemandi, D.A., Manzo, R.H. Release kinetics and up-take studies of model fluoroquinolones from carbomer hydrogels. *Int. J. Pharm.* **2002**, *246* (1–2), 17–24.
141. Pirogov, A.V., Shpak, A.V., Shpigun, O.A. Application of polyelectrolyte complexes as novel pseudostationary phases in MEKC. *Anal. Bioanal. Chem.* **2003**, *375* (8), 1199–1203.
142. Rafael, B.I., Carmen, A.L., Angel, C. Controlled release of estradiol solubilized in carbopol/surfactant aggregates. *J. Control. Release* **2003**, *93* (3), 319–330.
143. Shieh, J.Y., Glatz, C.E. Precipitation of proteins with polyelectrolytes. Roles of polymer molecular weight and modification of protein charge. *Polym. Preprints Am. Chem. Soc. Div. Polym. Chem.* **1991**, *32*, 606–607.
144. Dissing, U., Mattiasson, B. Partition of proteins in polyelectrolyte-neutral polymer aqueous two-phase systems. *Bioseparation* **1994**, *4* (1), 335–342.
145. Kim, T.H., Park, Y.H., Kim, K.J., Cho, C.S. Release of albumin from chitosan-coated pectin beads in vitro. *Int. J. Pharm.* **2003**, *250* (2), 371–383.
146. Dumitriu, S., Chornet, E. Immobilization of xylanase in chitosan-xanthan hydrogels. *Biotechnol. Prog.* **1997**, *13* (5), 539–545.
147. Richardson, S.C.W., Kolbe, H.J.V., Duncan, R. Potential of low molecular mass chitosan as a DNA delivery system: Biocompatibility, body distribution and ability to complex and protect DNA. *Int. J. Pharm.* **1999**, *178* (2), 231–243.
148. Jiang, G., Park, K., Kim, J., Kim, K.S., Oh, E.J., Kang, H., Han, S.E., Oh, Y.K., Park, T.G., Kwang Hahn, S. Hyaluronic acid-polyethyleneimine conjugate for target specific intracellular delivery of siRNA. *Biopolymers* **2008**, *89* (7), 635–642.

57 Polymeric Nano/Microparticles for Oral Delivery of Proteins and Peptides

S. Sajeesh and Chandra P. Sharma

CONTENTS

57.1 Introduction 1359
57.2 Barriers to Oral Delivery of Proteins/Peptides 1360
57.3 Strategy for Improved Oral Protein Delivery 1361
57.4 Polymeric Nano/Microparticles as a Possible Oral Peptide-Delivery System ... 1361
    57.4.1 Synthetic Biodegradable Polymeric Nano/Microparticles 1363
    57.4.2 Nonbiodegradable Synthetic Polymers 1367
    57.4.3 Natural and Protein-Based Polymers for Oral Peptide Delivery 1370
        57.4.3.1 Protein-Based Polymers for Oral Protein Delivery 1371
    57.4.4 Preparation of Nano/Microparticles 1372
        57.4.4.1 Nano/Microparticles Obtained by Polymerization of Monomers ... 1372
        57.4.4.2 Particles from Preformed Polymers 1374
57.5 Concluding Remarks 1375
References 1375

57.1 INTRODUCTION

Recent advancement in the field of pharmaceutical biotechnology and introduction of recombinant DNA technology have led to the production of a number of therapeutic peptides and proteins for the treatment of several life-threatening diseases (Table 57.1). A number of peptide-based therapeutics such as recombinant hormones, cytokines, vaccines, monoclonal antibodies, therapeutic enzymes, and the like have been recently approved for clinical use.[1] However, most of these peptides are administered by parenteral route. Inherent short half-lives of peptides and chronic therapy requirements in a majority of cases make their repetitive dosing necessary.[2] Frequent injections, oscillating blood drug concentrations, and low patient acceptability make even the simple parenteral administration of these drugs problematic.[3,4] In spite of significant advancement in the field of pharmaceutical research, development of a proper noninvasive delivery system for peptides remains a distant reality. Although there have been reports of successful delivery of various peptide therapeutics across nonoral mucosal routes (such as nasal and buccal), the oral route continues to be the most preferred route for drug administration.[5–7] The oral route, despite enormous barriers that exist in the gastrointestinal tract (GIT), has obvious advantages such as ease of administration, patient compliance, and cost effectiveness.[8,9]

TABLE 57.1
Commonly Used Biotech-Derived Pharmaceutical Products

Products	Application
Human insulin	Treatment of diabetes mellitus
Interferon-α	Leukemia, AIDS, and renal cell carcinoma
Interferon-β	AIDS, multiple sclerosis, and cancer
Erythropoietin	Anemia and chronic renal failure
Interleukin-2	Cancer treatment
Streptokinase	Heart attack
Monoclonal antibodies	Cancer treatment and septic shock
Tissue plasminogen activator	Acute myocardial infarction
Human growth hormone	Growth deficiency
Hepatitis B	Hepatitis B vaccine
Calcitonin	Osteoporosis

57.2 BARRIERS TO ORAL DELIVERY OF PROTEINS/PEPTIDES

Peptide-based biotechnology products are subject to the same hostile environment faced by all peptides in the GIT. The major problems associated with oral peptide delivery are the susceptibility to degradation by the hostile gastric environment; metabolism by luminal, brush border, and cytosolic peptidases; and poor permeability across the intestinal epithelium because of size, charge, and hydrophilicity.[10,11] Intestinal epithelium serves as a major barrier for the absorption of orally administered drugs and peptides into the systemic circulation. High-resistance epithelial cell barriers restrict the passage of various hydrophilic compounds from the small intestine into the human body. The high resistance is due to the formation of well-organized tight junctions that connect the cell plasma membranes by a network of apical localized seams. As the name implies, tight junctions exclude the paracellular passage of ions, peptides, and proteins. The paracellular route is the dominant pathway for passive transepithelial solute flow in the small intestine, and its permeability depends on the regulation of intercellular tight junctions. The utility of the paracellular route for oral drug delivery has remained unexplored because of a limited understanding of tight junction physiology and the lack of substances capable of increasing the tight junction permeability without irreversibly compromising intestinal integrity and function. The attempts made so far to find ways to increase paracellular transport by loosening intestinal tight junctions have been hampered by unacceptable side effects induced by the potential absorption enhancers.[12] Physiological considerations, such as gastric transit time, dilution, and interaction with intestinal debris, also influence peptide absorption across the intestinal epithelium. Furthermore, peptides absorbed through the hepatic portal vein have to negotiate with the first-pass metabolism in the liver.[9]

The nature of these barriers has now been expanded to include intracellular metabolism by cytochrome P450-3A4 as well as apically polarized efflux mediated by ATP-dependent P-glycoproteins.[13] Although, P-glycoprotein-mediated efflux systems are most commonly observed in tumor cells, they are also present in normal intestinal cells and act to reduce the intracellular accumulation or the transcellular flux of a wide variety of drugs, including peptides. Furthermore, peptides are associated with potential physical and chemical instability, and formulating peptide drugs are much more complex and demanding. Hence formulating a proper delivery system for proteins and peptides with optimal therapeutic effect and shelf life is an elusive goal for pharmaceutical scientists.

57.3 STRATEGY FOR IMPROVED ORAL PROTEIN DELIVERY

Successful oral delivery of protein involves overcoming the barriers of enzymatic degradation, achieving epithelial permeability, and taking steps to conserve bioactivity during formulation processing. The coadministration of enzyme inhibitors and permeation enhancers is an approach used to enhance the bioavailability of oral protein formulations.[12,14–16] Chemical modification of peptides and use of polymeric systems as carriers have also been attempted to overcome the inherent barriers.

Coadministration of protease inhibitors may retard the rate of degradation of peptides.[17] The slow rate of degradation may enhance the amount of peptides available for absorption. Enzyme inhibitors have been associated with systemic intoxication if they are absorbed and may affect the normal digestion of nutritive proteins. Use of absorption enhancers such as fatty acids, surfactants, and bile salts has been proposed to improve drug transport across intestinal epithelium.[18–23] An increase in paracellular transport is mediated by modulating tight junctions of the cells, and an increase in transcellular transport is associated with an increase in the fluidity of the cell membrane. Permeation enhancers that fall into the former category include calcium chelators, and those that fall into the latter category include surfactants and fatty acids. Calcium chelators act by inducing calcium depletion, thereby creating global changes in the cells, including disruption of actin filaments, disruption of adherent junctions, and diminished cell adhesion.[18] Surfactants act by causing exfoliation of the intestinal epithelium, thus compromising its barrier functions.[23] Use of these permeation enhancers has demonstrated that their enhancement is dose- and time-dependent. The utility of this approach for oral drug delivery has remained unexplored because of the limited understanding of tight junction physiology and the lack of substances capable of increasing the tight junction permeability without irreversibly compromising their integrity and function.

Chemical modification using poly(ethylene glycol) (PEG) and fatty acids has been proposed as a promising approach. Site-specific attachment of PEG to proteins such as insulin can significantly enhance the physical and pharmacological properties without negatively affecting its biological potency.[24] Moreover PEGylation may enhance the *in vivo* half-life of proteins, by protecting them from receptor-mediated uptake by the reticuloendothelial system (RES) and preventing recognition and degradation by proteolytic enzymes. However, use of polymer matrixes as carriers of proteins and peptides remains the most promising strategy in oral protein delivery.[25] In this chapter, we discuss the problems and prospects associated with polymeric oral protein-delivery systems.

57.4 POLYMERIC NANO/MICROPARTICLES AS A POSSIBLE ORAL PEPTIDE-DELIVERY SYSTEM

Polymeric drug carriers are particularly useful for formulating new drugs developed using biotechnology, because they can provide protection from degradation in the body and promote their penetration across biological barriers. Numerous attempts were made in the last few decades for formulating a suitable delivery system for peptides and proteins using polymeric carriers.[26,27] Polymeric carriers offer numerous advantages over conventional delivery systems, which include low cost, nonimmunogencity, and versatility. The traditional delivery systems or the so-called first generation polymeric systems are capable of delivering the active ingredient at a specific site in the body. However, application of such systems toward oral peptide delivery has been limited, since the enzymatic and absorption barriers may hinder the uptake of these bioactive agents.

Development of advanced drug-delivery systems, based on polymeric particulate systems, is an emerging area of research. These systems are designed to overcome the enzymatic and absorption barriers in the GIT. Recently much interest has been directed toward the development of small particles (10 μm or less) by oral route.[28] Depending on the nature of the polymer involved, these systems are reported to exhibit interesting properties, which can be exploited for developing oral delivery systems. For instance, polymeric nano/microparticles consisting of hydrophobic biodegradable polymers have been shown to translocate across the intestinal mucosa and hence can facilitate the absorption of peptides and proteins from the gut lumen (Scheme 57.1). M-cells, which are located on the surface of the Peyer's patches, are a possible pathway for transporting the nanoparticles through the epithelium of the gut.[29] On the other hand, particles composed of hydrophilic polymers have restricted movement across the intestinal epithelium under normal circumstances. Some of these materials are effective in enhancing the intestinal permeability and thereby improving the peptide absorption.[30] Net surface charge of the system is also regarded as a factor in deciding the fate of an oral delivery system. Using natural polymers as carriers is another promising prospect in oral peptide delivery. Chitosan, starch, alginate, dextran, and the like, are widely investigated for developing as drug-delivery systems.

Despite of the encouraging potential of polymeric nano/microparticles, formulating a marketable peptide-delivery system still remains a major challenge. In this chapter, we have attempted to review the prospects and problems associated with polymeric nano/microparticles toward oral peptide delivery. Polymers are classified under three different categories: (1) synthetic biodegradable polymers, (2) synthetic nonbiodegradable polymers, and (3) natural- and protein-based polymers (Table 57.2).

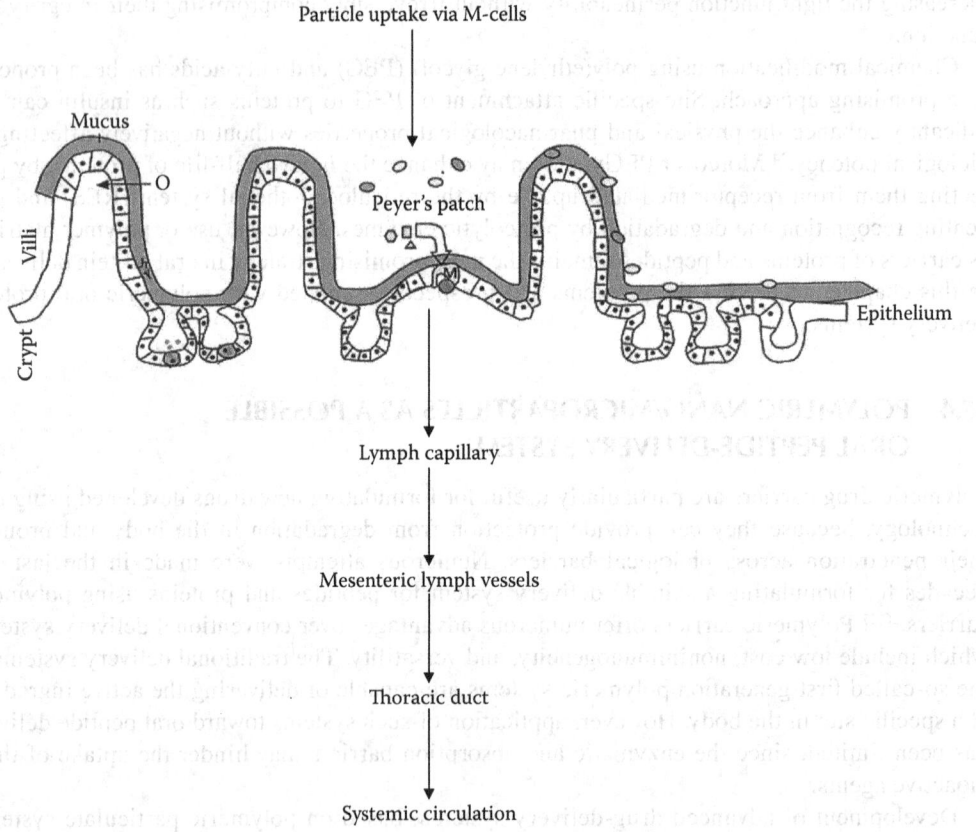

SCHEME 57.1 Schematic of transportation of nanoparticles through the epithelium of the gut.

TABLE 57.2
Polymers Investigated for Oral Peptide Delivery

Category	Polymers
Synthetic biodegradable polymers	Poly(glycolic acid), poly(lactic acid), and copolymers
	Poly(hydroxy butyrate)
	Poly(anhydrides)
	Poly(cyanoacrylates)
	Poly(ortho esters)
	Poly(phosphazenes)
Nonbiodegradable synthetic polymers	Poly(acrylic acid) and derivatives
	Poly(ethylene glycol)
	Poly(hydroxymethyl methacrylate)
	Poly(vinyl alcohol)/poly(vinyl acetate)
	Poly(vinyl pyrrolidone)
Natural polymers	Chitosan and derivatives
	Alginate
	Cellulose and starch
	Gelatin, casein, and collagen
	Pullulan
	Xanthan

57.4.1 SYNTHETIC BIODEGRADABLE POLYMERIC NANO/MICROPARTICLES

A large number of biodegradable polymers have been investigated as carriers in the design of controlled drug-delivery systems. Biodegradable nanoparticles are in much demand nowadays because of their selective uptake by Peyer's patches, which gives them great potential as carriers for oral peptide- or oral vaccine-delivery systems. Biodegradable polymers such as polylactide or glycolide, polyanhydrides, and the like, are capable of moving across intestinal epithelium.[31,32] It was proposed that the use of these nanoparticles would be beneficial for oral delivery because of their control release properties and their ability to protect drugs without exposing them to the gastric and intestinal fluids. However, a major issue with this approach is the insufficient absorption of nanoparticulate drug carriers to achieve an acceptable degree of bioavailability.[33]

Drugs administered through the gastrointestinal tract are normally transported into systemic circulation by the portal vein. As a consequence, compounds can sometimes undergo extensive metabolism during the first pass through the liver. First-pass metabolism may result in reduced bioavailability for many drugs compared with the parenteral administration, where drugs are directly released into the systemic circulation. If a drug is absorbed through the lymphatic system rather than by the portal circulation, it will find its way into the blood through the thoracic duct and will, therefore, avoid the first-pass effect.[34] A drug absorbed lymphatically is incorporated into chylomicrons (and other lipoproteins) produced by the fat digestion process. Numerous attempts were made to utilize this route for developing oral peptide-delivery system.

The idea of using biodegradable hydrophobic nanoparticles was proposed as a possible system to deliver proteins through the lymphatic circulation to achieve a higher degree of success. In the 1960s, Volkheimer described intestinal uptake of particulate matter on the absorption of intact starch particles.[35] This description awakened the hopes of achieving peptide and protein delivery using particles as carriers. Thereafter numerous studies were undertaken to establish the phenomenon of particle uptake, and now it is most accepted that the uptake happens mainly through phagocytosis.[36] Further studies carried out using model particles (polystyrene latex, polymethylmethacrylate) in the range of 50 nm to 3 μm revealed that the maximal absorption occurred with particles ranging from

50 to 100 nm in diameter, with particles above 1 µm being trapped in the Peyer's patches. Attempts were further made to study the mechanism responsible for the process and to explore their possible application in drug delivery.[37] It is proposed that M-cells of Peyer's patches are largely responsible for the uptake of these particles.[38]

Gut-associated lymphoid tissue (GALT) consists of lymphoid follicles arranged in single or in clusters to form distinct structures called Peyer's patches. The epithelium overlying the follicle is called the follicle-associated epithelium (FAE). This FAE contains specialized antigen sampling cells known as M-cells, which have unique structural features with sparse irregular microvilli on the apical side. In addition, they possess a basolateral cytoplasmic invagination that creates a pocket containing one or more lymphocytes and possibly macrophages. M-cells are thought to sample and transport antigens from the gut lumen to the underlying lymphoid cells and may elicit an immune response. While acting in this immunological surveillance role, it has been suggested that the M-cells absorb particles and consequently may be exploited for the delivery of therapeutic peptides and proteins or vaccines.[39–41]

Nanoparticles administered orally are absorbed, not only by way of the membranous epithelial cells (M-cells) of the Peyer's patches, but also by the much more numerous gut enterocytes. The absorption of the Particles is described as crossing either at the level of Peyer's patches or through the enterocyte layer. Usually particles absorbed through the Peyer's patches end up in the lymph and are carried through the mesenteric lymph vessels into the thoracic duct, which are further cleared into the bloodstream. Lymph from the intestinal lymphatic system drains through the thoracic lymph duct into the left internal jugular vein and then to the systemic circulation. Thus, the transport of drug through the intestinal lymphatic system may increase the percentage of drug that can gain access to the systemic circulation.[42] In addition, the process of intestinal lymphatic drug transport often continues over time periods longer than typically observed for drug absorption through the portal vein. Consequently, drug transport through the lymph may be utilized to prolong the time course of drug delivery to the systemic circulation.

The uptake is preceded by the interaction of the particles with the cell surface. Therefore, the nature of the polymer, mainly the hydrophobic or hydrophilic balance and the surface charge, will affect the uptake process to a large extent.[28] Physicochemical properties of particles govern their rate of uptake from the intestinal tract. The two main deciding factors are the size and the nature of the polymer used to make the particles.

Many studies regarding the size effects of nanoparticles absorption by intestinal epithelia have been performed using poly(styrene) standard particle suspensions. Particles with mean diameters of 50 and 100 nm showed a higher uptake in the rat intestine than larger particles.[43] The uptake of the nanoparticles was followed by their appearance in the systemic circulation and distribution to different tissues. After administration of equivalent doses of 33% of the 50 nm and 26% of the 100 nm, nanoparticles were detected in the intestinal mucosa and GALT.[44] In the case of 500 nm nanoparticles, only 10% were localized in intestinal tissues. Nanoparticles with size greater than 1 µm in diameter yielded only little uptake and had exclusive localization in Peyer's patches. Similarly poly(D,L-lactic-co-glycolide) (PLGA) nanoparticles with size 100 nm were uptaken from intestine, and efficiency was higher when compared with larger particles. It was clear from these experiments that the intestinal uptake of particles largely depends on the size.[44–46]

Another major factor affecting the particle uptake is the nature of the material used to prepare the particles. Uptake of nanoparticles prepared from hydrophobic polymers seems to be higher than that from particles with more hydrophilic surfaces.[32] Microspheres composed of polystyrene, poly(methylmethacrylate), poly(hydroxybutyrate), poly(D,L-lactide), poly(L-lactide), and poly(D,L-lactide-co-glycolide) were absorbed into the Peyer's patches of the small intestine, whereas those composed of ethyl cellulose, cellulose acetate hydrogen phthalate, and cellulose triacetate were not absorbed. Residual poly(vinyl alcohol) in the surface of PLGA nanoparticles significantly reduced the intercellular uptake, in spite of the smaller particle size.[47] Similarly, poloxamer coating of poly(styrene) nanoparticles caused a decrease of gastrointestinal uptake *in vivo*. Moreover, hydrophobic poly(styrene) nanoparticles seem to have a higher affinity for M-cells than for absorptive epithelia.[48] Another factor that influences the particle uptake is the surface charge. Generally, positively charged particles have higher uptake as compared with the negatively or neutrally charged species. Carboxylated poly(styrene) nanoparticles show significantly decreased affinity to intestinal epithelia, especially to M-cells, compared with positively charged and uncharged poly(styrene) nanoparticles.[49]

Specific strategies are being proposed to enhance the intestinal uptake of nanoparticles. These strategies include the surface modification of nanoparticles with some targeting ligands such as lectins. Lectins can be defined as proteins of nonimmune origin that bind to carbohydrates specifically and noncovalently. Lectins can increase the adherence of nanoparticles to the intestinal epithelium.[50] After binding to the cells, the lectins undergo cellular uptake and subsequently can also exhibit strong binding to nuclear pore membranes. Polystyrene microparticles coated with tomato lectin were shown to be specifically adhesive to enterocytes.[51]

Major questions to be asked here are as follows:

1. Is particle uptake sufficient enough to achieve therapeutic effectiveness?
2. Are biodegradable polymer nanoparticles capable of preserving the delicate structure of sensitive peptide drugs?
3. Are these particle systems safe for long-term therapeutic use?

The first and most important question to be asked here is whether the absorption of drug carriers via normal enterocytes and M-cells is sufficient to allow therapeutic effectiveness. This uptake may be correlated with the low percentage of M-cells present in the gut (0.1% of the epithelial cells), which may be insufficient to get an appreciable amount of bioavailability. Another major issue associated with solid-delivery systems is adsorption of proteins onto these solid-delivery devices.[52] The incorporation of peptides-based pharmaceuticals into solid-delivery matrices exposes them to a high surface to volume environment, creating ample opportunity for adsorption into the delivery devices. Adsorption of proteins may severely limit the amount of free unbound proteins that is available for release. Another major consequence of adsorption may be the surface-induced changes in the three-dimensional (3-D) structure of proteins that could result in loss of biological activity of proteins.[53] This loss of activity may also evoke some immune response in some cases, which in turn can be a major drawback. Several studies have indicated the loss of bioactivity of insulin and salmon calcitonin following the encapsulation onto the biodegradable matrix.[52,54]

Another major concern is the *in vivo* fate of these nanoparticles following their uptake by the Peyer's patches. A major aspect is that the exact mechanism of particle uptake is still unclear. A probable mechanism is that the particles absorbed through the Peyer's patches reaches mesenteric duct first and through cysterna chyli and thoracic duct reaches bloodstream. A major obstacle is the "RES clearance" of polymeric systems in the body.[55] Particulate carriers in the blood streams may be identified by a group of scavenger cells known as RES, which are located largely in organs such as liver, spleen, lung, and the like. Most colloidal carriers are rapidly removed from the circulation by phagocytic cells in liver and spleen. The recognition of particles by RES is mediated by interactions of blood components with the artificial surface of the carriers (opsonization) leading to activation of the complement system. Moreover, the increase in hydrophobicity or introduction of cationic charges significantly enhances the clearance of these particles from the blood streams.[56] However, these features are some of the prerequisites for the uptake of particles from the GIT. Hence optimization of surface properties or charge without compromising the properties remains a major challenge in this area. A recent area of concern is the long-term effects of these particles on the human body, and more intense investigations are required in this direction.

Linear polyesters of lactides and glycolides, poly(alkyl cyanoacrylates), polyanhydrides, and polyphosphazenes are some of the polymers commonly used in the development of polymeric particles.

Polylactic acid (PLA) and PLGA have been used for more than three decades for a variety of medical applications. Extensive research has been devoted to the use of these polymers as carriers for controlled drug delivery of a wide variety of bioactive agents. Mathiowitz and group introduced the concept of bioadhesive oral delivery system based on bioerodible polyanhydrides.[57] Bioerodible materials, which provide a continuously renewable cluster of carboxyl groups, might demonstrate a long duration of bioadhesiveness even though they do not possess flexible polymer chains as found in the case of hydrogels. Poly(fumaric-co-sebacic) [P(FA:SA)] was reported as the most bioadhesive polymer system from a series of thermoplastic materials evaluated.[58] Polyanhydride microspheres displayed strong interactions with mucus lining and cellular lining. P(FA:SA) microspheres transverse both mucosal absorptive epithelium and FAE covering the lymphoid tissues of Peyer's patches. This material was proposed as novel oral delivery system for insulin and other peptide drugs. P(FA:SA) microspheres increase the absorption of three model substances of widely varying molecular size: dicumarol, insulin, and plasmid DNA.

Poly(alkyl cyanoacrylate) nanocapsules were successfully used for oral administration of insulin in diabetic rats.[59] Insulin-loaded nanospheres (100 IU/kg of body weight) that were administered perorally in streptozotocin-induced diabetic rats provoked a 50% decrease of fasted-glycemia from the second hour up to 10–13 days. When ^{14}C-labeled nanospheres loaded with ^{125}I insulin were used, it was found that nanospheres increased the uptake of insulin or its metabolites in the GIT, blood, and liver, while the excretion was delayed when compared with ^{125}I insulin nonassociated to nanospheres; in addition, ^{14}C and ^{125}I radioactivities disappeared progressively as a function of time, parallel to the biological effect.

Poly(isobutyl cyanoacrylate) (PIBCA) nanocapsules were dispersed in a biocompatible microemulsion to facilitate the absorption of insulin following intragastric administration to diabetic rats.[60] Insulin-loaded PIBCA nanocapsules were prepared in situ in a biocompatible water-in-oil microemulsion by interfacial polymerization. Subcutaneous administration of insulin-loaded nanocapsules to diabetic rats demonstrated that the bioactivity of insulin was retained, and intragastric administration of insulin-loaded nanocapsules resulted in a significantly greater reduction in blood glucose levels of diabetic rats. *In vivo* performance of insulin-loaded PIBCA nanospheres with or without sodium cholate and Pluronic F68 surfactants were studied on alloxan-induced diabetic rats.[61] Administered orally, insulin-loaded (75 IU/kg) nanospheres, in the presence of surfactants, significantly reduced the mean blood glucose level for more than 8 h. These findings suggest the possible application of surfactant-incorporated polymeric nanoparticles for improving gastrointestinal absorption of insulin.

Nanoparticles prepared with a blend of a biodegradable polyester poly(ε-caprolactone) and a polycationic nonbiodegradable acrylic polymer [Eudragit (R) RS] were used for oral administration of insulin.[62] When administered orally by force-feeding to diabetic rats, insulin nanoparticles decreased fasted-glycemia in a dose-dependant manner with a maximal effect observed with 100 IU/kg. These insulin nanoparticles also increased serum insulin levels and improved the glycemic response to an oral glucose challenge for a prolonged period of time. Insulin-loaded poly(lactic-co-glycolic acid) nanoparticles were prepared by a double-emulsion solvent evaporation method.[63] After oral administration of 10 IU/kg nanoparticles, the plasma glucose level decreased significantly after 4 h ($p < 0.05$); 10 h later, the glucose level decreased to the lowest (52.4% ± 10.2%, $p < 0.01$), and the relative pharmacological bioavailability was 10.3% ± 0.8%. Biodegradable nanoparticles loaded with insulin–phospholipid complex were prepared by a novel reverse micelle-solvent evaporation method in which soybean phosphatidylcholine (SPC) was employed to improve the liposolubility of insulin and biodegradable polymers as carrier materials to control drug release.[64] Intragastric administration of the 20 IU/kg nanoparticles reduced fasting plasma glucose levels to 57.4% within the first 8 h of administration, and this reduction was continued for 12 h. Pharmacokinetic/pharmacodynamics (PK/PD) analysis indicated that 7.7% of oral bioavailability was relative to subcutaneous injection.

57.4.2 Nonbiodegradable Synthetic Polymers

Nano/microparticles composed of nonbiodegradable hydrophilic polymers are also of much interest in oral peptide delivery. Polymers such as polyvinyl alcohol (PVA), polyacrylic acid (PAA) and their derivatives, PEG, polyvinyl pyrrolidone (PVP), etc., are widely investigated for drug-delivery applications. Unlike biodegradable particles, these systems have restricted cell permeability under normal circumstances. Most of these systems are designed to release the encapsulated bioactive agent in the favorable regions of the GIT from where they can be absorbed. A very interesting and promising approach in oral peptide delivery is the use of bioadhesive or mucoadhesive polymers. The term bioadhesion refers to the adhesion between two biological materials or between any material and a biological material. Mucoadhesion refers to the interaction of a material with the mucosal surface.[65] It was proposed that if the intestinal transit of a delivery system could be delayed by mucoadhesion, followed by intimate contact at the brush border, might assist the uptake of a peptide or protein even without the intervention of a direct absorption enhancer. The intimate contact between a delivery system and the absorbing cell layer will improve both efficiency and effectiveness of the system. Mucoadhesive nano/microparticles are potential systems for drug-delivery applications, as they offer numerous advantages over conventional delivery systems. They can provide an intimate contact with the mucus layer and thereby enhance bioavailability of the drugs because of a high surface to volume ratio. Mucoadhesive nano/microparticles were proposed as a novel oral delivery system for poorly absorbable drugs such as peptides or proteins.[66–68]

GIT is coated with a continuous layer of protective secretions known as mucus. Intestinal mucus layer is composed of water (up to 95% by weight), glycoproteins (0.5%–5%), inorganic salts (about 1% by weight), carbohydrates, and lipids. Mucins represent more than 80% of the organic components of mucus, and they are O-linked glycoproteins.[69,70] Mucoadhesive delivery systems can penetrate the mucus layer and bind to the underlying epithelium. Several theories are proposed to explain the process of mucoadhesion; however, the mechanism involved is not fully elucidated yet. There are four main theories that describe the possible mechanisms of mucoadhesion: the electronic, the adsorption, the wetting, and the diffusion theory.[71] The electronic theory assumes that transfer of electrons occurs between the mucus and the mucoadhesive due to the differences in their electronic structures. The electron transfer between the mucus and the mucoadhesive leads to the formation of a double layer of electrical charges at the interface of the mucus and the mucoadhesive. This interaction results in attraction forces inside the double layer. The adsorption theory concerns the attraction between the mucus and the mucoadhesive achieved through molecular bonding caused by secondary forces such as hydrogen and Van der Waals bonds. The resulting attractive forces are considerably larger than the forces described in the electronic theory. The wetting theory correlates the surface tension of the mucus and the mucoadhesive with the ability of the mucoadhesive to swell and spread on the mucus layer and indicates that interfacial energy plays an important role in mucoadhesion. The wetting theory is significant, since the spreading of the mucoadhesive over the mucus is a prerequisite for the validity of all the other theories. The diffusion theory concerns the interpenetration to a sufficient depth and physical entanglement of the protein and polymer chains of the mucus and the mucoadhesive, depending on their molecular weight, degree of cross-linking, chain length, flexibility, and spatial conformation.[72–74]

There is no unified theory to explain the process of mucoadhesion. The total phenomenon of mucoadhesion is a combined result of all these theories. First, the polymer gets wet and swells (wetting theory) followed by the noncovalent (physical) bonds created within the mucus–polymer interface (electronic and adsorption theory). Then, the polymer and protein chains interpenetrate (diffusion theory) and entangle together to form further noncovalent (physical) and covalent (chemical) bonds (electronic and adsorption theory). Hydrophilic polymers usually display the property of mucoadhesion because of their large molecular weight and ability to interpenetrate and entangle through mucus gel layer. The presence of hydrophilic functional groups is an important criterion for mucoadhesion. Polymers containing hydrophilic functional groups such as carboxyl, amine, hydroxyl, and sulfate can interact with mucus layer to form noncovalent interactions.[75] Polyanionic polymers such as PAA interacts with mucus by hydrogen and van der Waals bonds created between the carboxylic groups and the sialic acid residues of mucin glycoproteins.[76,77] On the other hand, polycationic polymer such as chitosan exhibits strong mucoadhesive properties due to the formation of hydrogen and ionic bonds between the positively charged amino groups and the negatively charged sialic acid residues of mucin glycoproteins.[78] Polyacrylic acid–based systems (including their commercial forms such as Carbopol and Polycarbophil) are widely used in mucosal drug-delivery applications. Hydrogel microparticles of PAA or polymethacrylic acid (PMAA) prepared by the free radical polymerization of corresponding monomers with bifunctional cross-linking agents, such as ethylene glycol dimethacrylate, or similar bifunctional reagents were investigated for developing mucoadhesive delivery systems.[79,80] It has been proved that use of PEG-based adhesion promoters improves the chain flexibility and mobility of such delivery systems. Peppas and group used the strategy of PEG-grafted PMAA for the development of mucoadhesive drug-delivery systems.[81,82]

Recently a novel class of polymers, so-called thiomers, which in fact are capable of forming covalent interactions with mucus layer, was introduced in this area.[83] Thiolated polymers display thiol-bearing side chains and are based on thiol or disulfide exchange reactions, and through a simple oxidation process, disulfide bonds are formed between polymers and cysteine-rich subdomains of mucus glycoproteins. Mucoadhesive polymers such as PAA, alginate, and chitosan were modified with cysteine using water-soluble carbodimide to yield polymer–cysteine conjugate. It is reported that adhesive properties of these polymers were significantly enhanced with the immobilization of thiol groups. Thiolated nano- and microparticles were introduced recently, and these systems seem to be an attractive excipient for oral protein delivery.[84,85]

However, a major problem is the absorption of peptides released from the polymer particles. It is reported that some mucoadhesive polymers can enhance the permeability of epithelial tissues by loosening the tight intercellular junctions. Basically there are two types of passive diffusion processes through which a drug can be absorbed from the mucosal site into the bloodstream. The first mechanism is transcellular (intracellular) transport and is used by small molecules. The molecule diffuses from one side of the barrier, through the cell, to the other side. The second mechanism of transport is paracellular transport (intercellular). Paracellular transport is the passage of the molecules through adjacent cells in the layer. This movement is governed by the available space and environment between the cells.[78] Increasing the area available between the cells allows the molecules to move more easily across the layer. Paracellular transport is the primary route used by hydrophilic and charged molecules. The paracellular pathway found along the intestinal wall is an alternative pathway for peptide absorption.[86] This pathway normally does not allow the entry of peptides and nutrients by a specialized regions called "tight junction." In a current model of a tight junction, two major integral membrane proteins are found—occludin and claudin—each with four membrane spanning alpha-helices.[87] The junction depends upon extracellular calcium to maintain integrity. The permeability properties of tight junctions vary considerably in different epithelia, and epithelial cells can transiently alter their tight junctions in order to allow increased flow of solutes and water through breaches in the junction barriers. The tight junctions usually prevent the transport of protein through the paracellular pathway.[86]

Among hydrophilic polymers, PAA-based systems turned out to be of particular interest showing the capability of opening epithelial tight junctions, which are mainly responsible for limited paracellular uptake of hydrophilic macromolecules.[88,89] Calcium-binding ability of these polymers seems to be a major reason for such effects. Calcium chelators might disturb cell–cell adhesion by depleting extracellular calcium required for the interaction of components of adherent junctions. Further chelation of calcium may activate intracellular protein kinases, which ultimately can lead to the disruption of junctional integrity. Divalent ion-binding ability of these polymers may also help in reducing the proteolytic degradation of proteins in GIT.[90]

Ability of these materials to enhance the intestinal permeability was demonstrated using monolayers of caco-2 cells. Caco-2 cell models are widely used *in vitro* cell culture model for intestinal epithelium. Having originated from human colon adenocarcinoma cells, these cells can differentiate into polarized cells with distinct mucosal (apical) and serosal (basolateral) cell membrane domain. Although they have originated from colon cells, caco-2 cells have many properties of small intestine absorptive cells including microvilli, intercellular junctions, and many of the enzymes nutrient transporters and efflux transporters present in the small intestine. The tightness of the intercellular junctional complex can be characterized by measuring the transepithelial electrical resistance (TER).[91,92]

Numerous investigations using caco-2 cells demonstrated the ability of PAA-based systems to enhance the intestinal permeability. Luessen et al. demonstrated that carbopol and polycarbophil can reduce the TER of caco-2 cells at neutral pH and can improve the transport of hydrophilic markers across the epithelial barrier.[89] Kriwet et al. showed that PAA microparticles could widen the intercellular spaces in the monolayers of the caco-2 cells.[93] Microparticles of poly(methacrylic-g-ethylene glycol) hydrogels loaded with insulin and salmon calcitonin demonstrated that the polymer microparticles cause a significant increase in the permeability of these proteins across the cell monolayers.[94,95] This material was also found noncytotoxic and capable of opening the tight junctions in a reversible manner.[96]

Another advantage of using PAA-based systems is their pH-dependent release profile because of the protonation or deprotonation of carboxylic acid groups.[80] In acidic media, acid groups remain unionized and remain collapsed, protecting encapsulated drugs from the hostile gastric environment. In the small intestine region, where pH is above the pK_a of PAA, the acid groups get deprotonated and exist in the form of ionized acid groups. Repulsion caused by the adjacent ionized acid group causes the hydrogel network to swell, which in turn leads to the release of the encapsulated material from the system. So protein encapsulated in the particles is protected from the gastric environment and is released in the favorable regions of the intestinal tract.[97]

Peppas and his group have made numerous attempts to utilize poly(methacrylic-g-ethylene glycol) nano- and microspheres as oral protein-delivery system. Insulin-loaded pH-responsive poly(methacrylic-g-ethylene glycol) microspheres were administered orally to both healthy and diabetic Wistar rats. Within 2 h of administration of the insulin-containing polymers, strong dose-dependent hypoglycemic effects were observed in both healthy and diabetic rats up to 8 h.[98,99] Further studies carried out on particles composed of a 1:1 molar ratio of methacrylic acid or ethylene glycol units showed the pronounced hypoglycemic effects following oral administration to healthy rats and achieving a 9.5% pharmacological availability compared with subcutaneous insulin injection.[100]

Enteric polymers based on PMAA esters such Eudragit were also studied for developing oral protein-delivery systems. The effectiveness of Eudragit L100 microspheres containing a protease inhibitor was evaluated in normal and diabetic rats.[101] The dosage form based on insulin with protease inhibitor was administered orally with a 20 IU/kg insulin dose by force-feeding. Microspheres without protease inhibitor and with trypsin inhibitor (TI) or chymostatin (CS) produced no marked hypoglycemic response in both groups of rats. A significant continuous hypoglycemic effect was found after oral administration of microspheres containing aprotinin (AP) or Bowman-Birk inhibitor (BBI) in both normal and diabetic rats when compared with controls. The hypoglycemic effect of Eudragit S100 enteric-coated capsules containing sodium salicylate as an absorption promoter was studied in hyperglycemic beagle dogs.[102] This system demonstrated

25%–30% reduction in plasma glucose levels and about 12.5% relative to subcutaneous injection of regular soluble insulin.

Eudragit S100 microspheres were prepared using water-in-oil-in-water emulsion solvent-evaporation technique, and their application toward oral insulin delivery was evaluated.[103] Oral administration of PVA-stabilized microspheres in normal albino rabbits (equivalent to 6.6 IU insulin/kg of animal weight) demonstrated a 24% reduction in blood glucose level, with maximum plasma glucose reduction of 76% ± 3.0% in 2 h, and the effect continued up to 6 h.

57.4.3 Natural and Protein-Based Polymers for Oral Peptide Delivery

Polysaccharides are among the most versatile polymers because of their vast structural diversity and nontoxicity. Among polysaccharides, chitosan, alginate, pectin, hylauronic acid, and dextran have received much attention. Protein-based polymers such as albumin, casein, and gelatin have also been investigated for oral peptide delivery.

Among such materials, chitosan-based systems are of utmost importance. Chitosan [(1-4)-2-amino-2-deoxy-β-D-glucan], which is the deacetylated form of chitin, is of great interest as a functional material of high potential in various fields such as the biomedical field.[104,105] Chitosans have found application as biomaterials in tissue engineering and in controlled drug-release systems for various routes of delivery. Being cationic in nature chitosan possesses mucoadhesive properties. It has unique ability to control the release of active agents, avoids use of organic solvents for the fabrication, and has free amino groups available for cross-linking.[106] Excellent review articles outlining the major findings on the pharmaceutical applications of chitosan-based nano/microparticles have been published in the last few years.[107,108] However, it was Illum who first reported that chitosan can promote the transmucosal absorption of small polar molecules as well as peptide and protein drugs system.[109] Chitosan attracted the attention of pharmaceutical scientists as a mucoadhesive polymer that could be useful for peptide-drug delivery.

When protonated (pH 6.5), chitosan is able to increase the paracellular permeability of peptide drugs across mucosal epithelia. Chitosan in its protonated form is able to interact with the epithelial tight junctions and to provoke their opening allowing for paracellular permeation of hydrophilic macromolecular drugs.[110] Chitosan can bind tightly to the intestinal epithelium, inducing redistribution of F-actin and the tight junction protein ZO-1.[111] Chitosan nanoparticles were able to demonstrate reduction in TER value in caco-2 experiments.[112,113] However, the major drawback with chitosan is their limited solubility at pH above 6.5. At neutral pH, chitosan exist in nonprotonated form and was found ineffective in improving the permeability of intestinal epithelium.[111] Thanou et al. proposed that the problem of chitosan's ineffectiveness at neutral pH values could be tackled by derivatization at the amine group. Trimethyl chitosan chloride (TMC) was proved to increase substantially the intestinal absorption and bioavailability of peptide drugs. The mechanism by which TMC enhances intestinal permeability is similar to that of protonated chitosan. It reversibly interacts with the components of the tight junctions, leading to the widening of the paracellular routes.[114,115]

Chitosan nanocapsules were prepared by the solvent displacement technique using high (450 kDa) and medium (160 kDa) molecular weight chitosan glutamate as well as high-molecular weight chitosan hydrochloride (270 kDa). The nanocapsules were used for the oral delivery of salmon calcitonin.[116] The results of the *in vivo* studies, following oral administration to rats, indicated that chitosan nanocapsules were able to reduce the serum calcium levels significantly and to prolong this reduction for at least 24 h. Bioadhesive polysaccharide chitosan nanoparticles (CS-NP) were prepared by ionotropic gelation of chitosan with tripolyphosphate anions.[117] The ability of CS-NP to enhance intestinal absorption of insulin and increase the relative pharmacological bioavailability of insulin was investigated by monitoring the plasma glucose level of alloxan-induced diabetic rats after oral administration of various doses of insulin-loaded CS-NPs. CS-NP enhanced the intestinal absorption of insulin to a greater extent than the aqueous solution of chitosan *in vivo*. Above all, after administration of 21 IU/kg insulin in the CS-NP, the hypoglycemia was prolonged over 15 h, and the average pharmacological bioavailability relative to subcutaneous (SC) injection of insulin solution was up to 14.9%. Chitosan capsules were also exploited for colon-specific delivery of insulin.[118] Hydroxypropyl methyl cellulose phthalate was used to coat chitosan capsules to achieve colon-specific delivery. Capsules were administered orally to Wistar rats (20 IU). Hypoglycemic effect started after 6 h and lasted for 24 h. The bioavailability of insulin from the chitosan capsules was 5.3% compared with the intravenous one. Recently application of chitosan-based particles toward developing oral vaccine-delivery systems was demonstrated.[119] *In vivo* uptake of chitosan microparticles prepared by precipitation or coacervation method was studied on murine Peyer's patches using confocal laser scanning microscopy. Microparticles smaller than 10 μm were taken up by Peyer's patches, and chitosan microparticles seem to be the candidate for vaccination systems.

Alginate is another natural polymer of interest in this area.[120] Alginate is a copolymer of D-mannuronic acid and L-guluronic acid, and it can undergo sol–gel transition in the presence of divalent ions such as calcium, zinc, etc. This property has been explored for the fabrication of alginate microparticles. Calcium cross-linked alginate particles can protect peptides from gastric degradation via pH-dependent release mechanism.[121] Alginate-based systems were also found to enhance the paracellular absorption of mannitol across caco-2 cells by about three times.[122] Another important aspect is the formation of polyelectrolyte complex of alginate with polycationic polymers such as chitosan. Hari et al. utilized chitosan–calcium alginate microparticles for oral insulin delivery.[123] Chitosan coating can significantly modulate the release of bioactive agents from the matrix.[124]

Alginate microspheres prepared by an emulsion-based process was used for oral insulin delivery. Cyclodextrin-complexed insulin was encapsulated onto these microparticles, and *in vivo* studies were conducted on diabetic albino rats.[125] Radioimmunoassay (RIA) for serum insulin indicated absorption of insulin from the gastrointestinal region following oral administration of insulin formulation. Chitosan-coated alginate nanoparticles containing cyclodextrin–insulin complex demonstrated good glucose-lowering effect in alloxan-induced diabetic rats (unpublished data). Biodegradable microparticles were prepared with alginate by the piezoelectric ejection process, and lectin [wheat germ agglutinin (WGA)] was conjugated to alginate microparticles.[126] The hypoglycemic effects of alginate and WGA-conjugated alginate microparticles were examined after oral administration in streptozotocin-induced diabetic rats. Alginate–WGA microparticles enhance the intestinal absorption of insulin sufficient to drop the glucose level of blood. Ramdas et al. developed an oral formulation for insulin delivery based on liposome encapsulated alginate chitosan gel capsules. Following oral administration, this formulation delivered insulin in the neutral environment of the intestine, bypassing the acidic media in the stomach, and reduced blood glucose levels in diabetic rats.[127]

57.4.3.1 Protein-Based Polymers for Oral Protein Delivery

Some attempts were made to utilize protein-based polymers such as casein, gelatin, and albumin for developing oral delivery systems. However, the major problem associated with these polymers is their degradation caused by the proteolytic enzymes in the GIT. Calcium phosphate–PEG–insulin–casein (CAPIC)-based oral insulin-delivery system was developed by BioSante Pharmaceuticals Inc., and functional activity was tested in a nonobese diabetic (NOD) mice model.[128] Single doses of CAPIC formulation were tested in NOD mice under fasting or fed conditions to evaluate the glycemic activity. Microparticles displayed a prolonged hypoglycemic effect after oral administration to diabetic mice.

57.4.4 Preparation of Nano/Microparticles

With the recent advancement in the pharmaceutical technology, new and innovative techniques are being employed in the fabrication of polymeric nano/microparticles (Table 57.3). A number of review articles and chapters describing various aspects of particle preparation and characterization have been published in the last few decades.[129–133] Hence we are outlining some of the techniques used in the preparation of particles, which are also used for oral peptide delivery. Particles are prepared mainly by two processes. The first process is the in situ polymerization of monomer through suitable polymerization process to yield polymeric particulates. The second process is based on the dispersion of well-characterized preformed polymers of synthetic or natural origin using a suitable technique (Scheme 57.2).

57.4.4.1 Nano/Microparticles Obtained by Polymerization of Monomers

57.4.4.1.1 Emulsion Polymerization

Emulsion polymerization is a widely used method for nanoparticle preparation. This method is classified into two categories based on the nature of the continuous phase in the emulsion. In the first case, the continuous phase is aqueous (o/w emulsion), whereas in the other case continuous phase is organic (w/o emulsion). In either case, the monomer is emulsified in the nonsolvent phase

TABLE 57.3
Summary of Particle Manufacturing Techniques and Employed Polymers

Process	Polymers
Emulsion polymerization	Poly(alkyl methacrylate)
	Poly(alkyl cyanoacrylate)
	Poly(styrene)
	Poly(vinyl pyridine)
	Poly(acrolein)
Interfacial polymerization	Poly(alkyl cyanoacrylate)
	Poly(lysine) derivatives
Emulsification evaporation	Poly(lactic acid)
	Poly(lactide-*co*-glycolide)
	Poly(β-hydroxybutyrate)
	Ethyl cellulose
Solvent displacement	Poly(alkyl methacrylate)
	Poly(lactic acid)
	Poly(lactide-*co*-glycolide)
	Poly(1-caprolactone)
Salting out	Cellulose acetate phthalate
	Poly(alkyl methacrylate)
	Ethyl cellulose
	Poly(lactic acid)
	Poly(lactide-*co*-glycolide)
Desolvation and denaturation	Albumin
	Casein
	Gelatin
	Ethyl cellulose
Ionic gelation	Alginate
	Chitosan
	Carboxymethyl cellulose

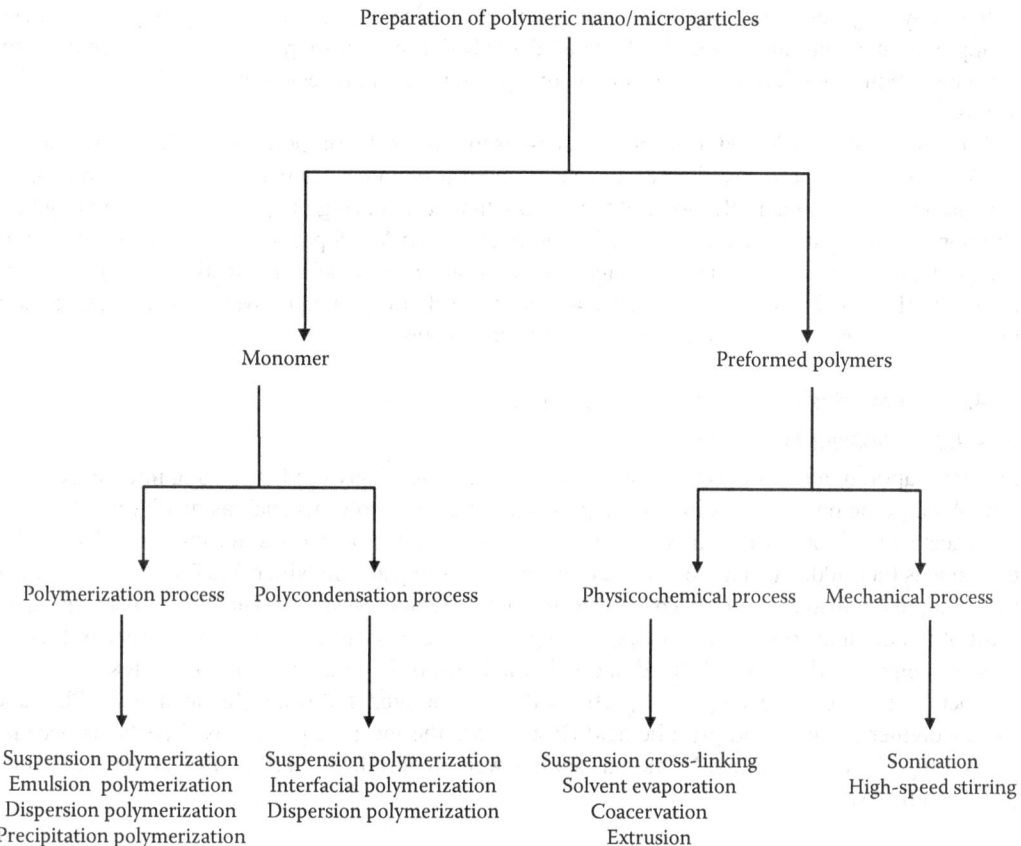

SCHEME 57.2 Preparation of polymeric nano/microparticles.

with surfactant molecules. The polymerization takes place in the presence of a chemical or physical initiator. The drug to be encapsulated may be incorporated in the reaction medium during the polymerization or can be subsequently added to the preformed particles. The advantage of this technique is that nanoparticles with smaller size (50–200 nm) can be obtained by this technique. Nanospheres of polymethylmethacrylate, poly(alkyl cyanoacrylate), polyacrylamide, etc., can be prepared by this technique.

57.4.4.1.2 Precipitation and Dispersion Polymerization

In precipitation polymerization technique, the monomer is completely miscible in the polymerization medium, but the medium is a precipitate for the resultant polymer. The polymerization medium will be a homogeneous reaction medium, but the polymerization will lead to the formation of a visible precipitate. Dispersion is similar to the precipitation polymerization, but the addition of one or more stabilizers to the polymerization medium leads to the formation of monodispersed particles.

Donini et al. reported the preparation of poly(methacrylic acid-g-polyethylene glycol) nanospheres by solution or precipitation polymerization.[134] The free radical polymerization of methacrylic acid with PEG macromer was carried out using a photoinitiator in water. Pluronics-based polymers were used to prevent the aggregation of these nanospheres and to render them redispersion ability.

57.4.4.1.3 Suspension Polymerization

Water-insoluble monomer involves the use of stabilizers and monomer-soluble initiators. This feature leads to the formation of polydisperse monomer droplets in water in the size of about 20–1000 μm followed by polymerization and direct conversion of droplets into corresponding polymer particles of approximately the same size. Skovby et al. described a suspension polymerization technique for the preparation of methacrylic acid using hydroxyapatite and magnesium hydroxide as suspending agents.[135,136]

We have recently adopted a novel chitosan-based ionic gelation process for the preparation of PMAA-based microparticles.[137–139] Methacrylic acid was polymerized in the presence of chitosan in aqueous medium, and particles were obtained spontaneously during the polymerization without the addition of any organic solvents and steric stabilizers. PMAA–CS particles displayed good protein encapsulation efficiency and demonstrated pH responsive release behavior at stimulated gastric and intestinal pH. Application of these microparticles toward oral protein delivery was evaluated using insulin and bovine serum albumin (BSA) as model proteins.

57.4.4.2 Particles from Preformed Polymers

57.4.4.2.1 Solvent Evaporation

Solvent evaporation is perhaps the easiest and the most extensively used method of microencapsulation. An organic phase consisting of the polymer solution in solvents such as dichloromethane (or ethyl acetate or chloroform) is mixed with the drug to form the primary water-in-oil emulsion. This emulsion is then added to a large volume of water containing an emulsifier like PVA or PVP to form the multiple emulsions (w/o/w). The so-formed double emulsion is then subjected to stirring until most of the organic solvent evaporates, leaving solid microspheres. The microspheres can then be washed, centrifuged, and lyophilized to obtain the free-flowing and dried microspheres.

There are numerous examples of particles that can be prepared using this technique. PLA and the copolymer of lactic and glycolic acid (PLGA) are the most frequently used particles because of their high biocompatibility. Numerous studies were undertaken to utilize PLGA matrix for oral protein delivery.[141]

57.4.4.2.2 Spray Drying

In this process, the drug may be dissolved or dispersed in the polymer solution and spray dried. The quality of spray-dried microspheres can be improved by the addition of plasticizers, for example, citric acid, which promote polymer coalescence on the drug particles and hence promote the formation of spherical and smooth-surfaced microspheres. The size of microspheres can be controlled by the rate of spraying, the feed rate of polymer drug solution, nozzle size, and the drying temperature. This method of microencapsulation is particularly less dependent on the solubility characteristics of the drug and polymer and is simple, reproducible, and easy to scale up.

57.4.4.2.3 Hot Melt Microencapsulation

This method was adopted to prepare polyanhydride copolymer of poly(bis(*p*-carboxy phenoxy) propane anhydride) with sebacic acid.[132] In this method, the polymer is first melted and then mixed with solid particles of the drug. The mixture is suspended in a nonmiscible solvent (e.g., silicone oil), continuously stirred, and heated to 5°C above the melting point of the polymer. Once the emulsion is stabilized, it is cooled until the polymer particles solidify. The resulting microspheres are washed by decantation with petroleum ether. The primary objective for developing this method is to develop a microencapsulation process suitable for the water-labile polymers, for example, polyanhydrides. Microsphere with diameter of 1–1000 μm can be obtained, and the size distribution can be easily controlled by altering the stirring rate.

57.4.4.2.4 Complex Coacervation

When oppositely charged polyelectrolytes with a relatively low-charge density are mixed at an appropriate temperature and pH concentration, a liquid polyelectrolyte complex called complex coacervate is formed. This technique is widely used in the preparation of polymeric particles.

Sodium alginate is a water-soluble polymer that gels in the presence of multivalent cations such as calcium, zinc, and barium. Alginate particles are usually produced by dropwise extrusion of sodium alginate solution into calcium chloride solution. The preparation of alginate nanoparticles was first achieved in a diluted aqueous sodium alginate solution in which gelation was induced by the addition of a low concentration of calcium. Alginate particles can be prepared by using a modified emulsification or internal gelation. The preparation of alginate nanoparticles by this method does not require specialized equipment and can be performed at an ambient temperature.

Numerous processes are reported for the preparation of chitosan-based particulate system, and excellent review articles have appeared for the same recently. Emulsion cross-linking, coacervation precipitation, spray drying, emulsion droplet coalescence, reverse micellar method, sieving method, ionic gelation process, etc., are widely used in the fabrication of chitosan-based nano- and microparticles. Ionic gelation process has generated much attention since the process involved is simple and mild. Moreover reversible physical cross-linking by ionic interaction between anionic and cationic groups can avoid the use of toxic cross-linkers and organic solvents. Ionotropic gelation method using small molecular weight counterions such as sodium tripolyphosphate and sodium citrate can be employed for the preparation of chitosan particles.

57.4.4.2.5 Thermal Denaturation

A number of water-soluble proteins denature when heated since they are heat-sensitive. The denaturation process causes protein chains to unfold and become chemically cross-linked. This property can be properly modulated to form protein microsphere. BSA microspheres are usually prepared by this technique.

57.5 CONCLUDING REMARKS

Several promising approaches are being developed to improve the efficacy of oral peptide-delivery systems. Polymeric particulate systems undoubtedly have enormous potential toward developing oral drug-delivery systems. Some of these carriers have been shown efficiency in improving the bioavailability of peptides and proteins either by transporting directly through the intestinal epithelium or by enhancing the permeability of the mucosal barriers. In spite of their advantages and shortcomings, polymeric systems continue to be the most promising systems for developing oral formulations for therapeutic peptides. However, most of these studies are still in preclinical phase, and more intense research is required in this direction to make a marketable oral protein formulation.

Another major aspect is that current encapsulation technologies are far from being optimal for a good manufacturing practice (GMP) environment and for making marketable products at industrial scale. Therefore, efforts are currently devoted to developing novel technologies of microencapsulation that can be readily scaled up.

More intense and innovative research is required in the area of polymer drug-delivery systems to render them as a possible solution to the problems associated with protein pharmaceuticals. A better understanding of the mechanisms of action of these novel vehicles will provide a basis for their further optimization, thus opening more exciting opportunities for improving the administration of macromolecules. A positive outcome in this direction will have a direct impact in the society and may help in alleviating human sufferings to a large extent.

REFERENCES

1. Walsh, G. Pharmaceutical biotechnology products approved within the European union. *Eur. J. Pharm. Biopharm.*, 55, 3, 2003.
2. Mcmartin, C. Pharmacokinetics of peptides and proteins: Opportunity and challenges. *Adv. Drug Deliv. Rev.*, 22, 39, 1992.
3. Humphrey, M.J., Ringrose, P.S. Peptides and related drugs: A review of their absorption, metabolism and excretion. *Drug Metabol. Rev.*, 17, 283, 1986.

4. Meibohm, B. Pharmacokinetics of protein- and nucleotide-based drugs. In *Biomaterials for Delivery and Targeting of Proteins and Nucleic Acid*. Mahato, R.I. (Ed.) CRC Press, Boca Raton, FL, 2005, p. 275.
5. Ingemann, I. et al. Peptide and protein drug delivery systems for non-parenteral routes of administration. In *Pharmaceutical Formulation Development of Peptides and Proteins*. Frokjaer, S. and Hovgaard, L. (Eds.) Taylor & Francis, London, U.K., 2000, p. 189.
6. Edman, P., Bjork, E. Nasal delivery of peptide drugs. *Adv. Drug Deliv. Rev.*, 8, 165, 1992.
7. Sood, A., Panchagnula, P. Peroral route: An opportunity for protein and peptide drug delivery. *Chem. Rev.*, 101, 3275, 2001.
8. Fix, J.A. Oral controlled release technology for peptides: Status and future prospects. *Pharm. Res.*, 13, 1760, 1996.
9. Schen, W.C. Oral peptide and protein delivery: Unfulfilled promises? *Drug Dis. Today*, 8, 607, 2003.
10. Lee, V.H.L., Yamamoto, Y. Penetration and enzymatic barriers to peptide and protein absorption. *Adv. Drug Deliv. Rev.*, 4, 171, 1990.
11. Woodley, J.F. Enzymatic barriers for GI peptides and protein delivery. *Cric. Rev. Ther. Drugs*, 11, 61, 1994.
12. Zhou, X.H. Overcoming enzymatic and absorption barriers to non-parentally administered protein and peptide drugs. *J. Contr. Release*, 29, 239, 1994.
13. Tandon V.R. P-glycoprotein: Pharmacological relevance. *Ind. J. Pharmacol.*, 38, 13, 2006.
14. Fanaso, A. Novel approaches for oral delivery of macromolecules. *J. Pharm. Sci.*, 87, 1351, 1998.
15. Saffran, M. et al. A new approach to the oral administration of insulin and other peptide drugs. *Science*, 233, 1081, 1986.
16. Gomez-Orellana, I. Strategies to improve oral drug bioavailability. *Expert Opin.*, 2, 119, 2005.
17. Bernkop-Schnurch, A. The use of inhibitory agents to overcome the enzymatic barrier to perorally administrated therapeutic peptides and proteins. *J. Contr. Release*, 52, 1, 1998.
18. Ward, P.D., Tippin, T.K., Thakker, D.R. Enhancing paracellular permeability by modulating epithelial tight junctions. *Pharm. Sci. Tech. Today*, 3, 346, 2000.
19. Daugherty, A.L., Mrsny, R.J. Regulation of the intestinal epithelial paracellular barrier. *Pharm. Sci. Tech. Today*, 2, 281, 1999.
20. Aungst, B.J. Intestinal permeation enhancers. *J. Pharm. Sci.*, 89, 429, 2000.
21. Mesiha, M.S., Ponnapula, S., Plakogiannis, F. Oral absorption of insulin encapsulated in artificial chyles of bile salts, palmitic acid and α-tocopherol dispersions. *Int. J. Pharm.*, 249, 1, 2002.
22. Aungst, B.J. et al. Enhancement of the intestinal absorption of peptides and non-peptides. *J. Contr. Release*, 41, 19, 1996.
23. Anderberg, E.K., Nystrom, C., Artusson, P. Epithelial transport of drugs in cell culture VII: Effects of pharmaceutical surfactants excipients and bile acids on transepithelial permeability in monolayers of human intestinal epithelial (caco-2) cells. *J. Pharm. Sci.*, 81, 879, 1992.
24. Katre, N.V. The conjugation of proteins with poly(ethylene glycol) and other polymers altering properties of proteins to enhance their therapeutic potential. *Adv. Drug Deliv. Rev.*, 10, 91, 1993.
25. Allémann, E., Leroux, J.C., Gurny, R. Polymeric nano- and microparticles for the oral delivery of peptides and peptidomimetics. *Adv. Drug Deliv. Rev.*, 34, 171, 1998.
26. Couvreur, P., Dubernet, C., Puisieux, F. Controlled drug delivery with nanoparticles: Current possibilities and future trends. *Eur. J. Pharm. Biopharm.*, 41, 2, 1995.
27. Soppimath, K.S., Aminabhavi, T.M., Kulkarni, A.R., Rudzinski, W.E. Biodegradable polymeric nanoparticles as drug delivery devices. *J. Contr. Release*, 70, 1, 2001.
28. Junga, T. et al. Biodegradable nanoparticles for oral delivery of peptides: Is there a role for polymers to affect mucosal uptake? *Eur. J. Pharm. Biopharm.*, 50, 147, 2000.
29. Florence, A.T. The oral absorption of micro- and nanoparticulates: Neither exceptional nor unusual. *Pharm. Res.*, 14, 259, 1997.
30. Junginger, H.E., Verhoef, J.C. Macromolecules as safe penetration enhancers for hydrophilic drugs—A fiction? *Pharm. Sci. Tech. Today*, 1, 370, 1998.
31. Florence, A.T. Nanoparticle uptake by the oral route: Fulfilling its potential? *Drug Dis. Today Tech.*, 2, 75, 2005.
32. Delie, F., Blanco-Príeto, M.J. Polymeric particulates to improve oral bioavailability of peptide drugs. *Molecules*, 10, 65, 2005.
33. Florence, A.T. Issues in oral nanoparticle drug carrier uptake and targeting. *J. Drug Target.*, 12, 65, 2004.
34. Rubas, W., Grass, G.M. Gastrointestinal lymphatic absorption of peptides and proteins. *Adv. Drug Deliv. Rev.*, 7, 15, 1991.

35. Volkheimer, G., Persorption von mikropartikeln. *Pathologe*, 14, 247, 1993.
36. Shakweh, M. et al. Particle uptake by Peyer's patches: A pathway for drug and vaccine delivery. *Expert Opin. Drug Deliv.*, 1, 141, 2004.
37. Hussain, N. et al. Recent advances in the understanding of uptake of micro particulates across the gastrointestinal lymphatics. *Adv. Drug Deliv. Rev.*, 50, 107, 2001.
38. Clark, M.A., Jepson, M.A., Hirst, B.H. Exploiting M cells for drug and vaccine delivery. *Adv. Drug Deliv. Rev.*, 50, 81, 2001.
39. Ermak, T.H., Giannasca, P.J. Microparticle targeting to M cells. *Adv. Drug Deliv. Rev.*, 34, 261, 1998.
40. Florence, A.T., Hussain, N. Transcytosis of nanoparticle and dendrimer delivery systems: Evolving vistas. *Adv. Drug Deliv. Rev.*, 50, S69, 2001.
41. Brayden, D. Oral vaccination in man using antigens in particles: Current status. *Eur. J. Pharm. Sci.*, 14, 183, 2001.
42. Wasan, K.M. The role of lymphatic transport in enhancing oral protein and peptide. *Drug Deliv. Ind. Pharm.*, 28, 1047, 2002.
43. Jani, P. et al. Nanoparticle uptake by the rat gastro-intestinal mucosa: Quantitation and particle size dependency. *J. Pharm. Pharmacol.*, 42, 821, 1990.
44. Desai, M.P., Labhasetwar, V., Amidon, G.L., Levy, R.J. Gastrointestinal uptake of biodegradable microparticles: Effect of particle size. *Pharm. Res.*, 13, 1838, 1996.
45. Shakweh, M., Besnard, M., Nicolas, V., Fattal, E. Poly(lactide-*co*-glycolide) particles of different physicochemical properties and their uptake by Peyer's patches in mice. *Eur. J. Pharm. Biopharm.*, 61, 1, 2005.
46. Ermak, T.H. et al. Uptake and transport of copolymer biodegradable microspheres by rabbit Peyer's patch M cells. *Cell Tissue Res.*, 279, 2, 1995.
47. Sahoo, S.K., Panyam, J., Prabha, S., Labhasetwar, V. Residual polyvinyl alcohol associated with poly(D,L-lactide-*co*-glycolide) nanoparticles affects their physical properties and cellular uptake. *J. Contr. Release*, 82, 105, 2002.
48. Jepson, M.A., Simmons, N.L., Hagan, D.T.O., Hirst, B.H. Comparison of poly(DL-lactide-*co*-glycolide) and polystyrene microsphere targeting to intestinal M cells. *J. Drug Target.*, 1, 245, 1993.
49. Jani, P., Halbert, G.W., Langridge, J., Florence, A.T. The uptake and translocation of latex nanospheres and microspheres after oral administer to rats. *J. Pharm. Pharmacol.*, 41, 809, 1989.
50. Jepson, M.A., Clark, M.A., Hirst, B.H.M. Cell targeting by lectins: A strategy for mucosal vaccination and drug delivery. *Adv. Drug Deliv. Rev.*, 56, 511, 2004.
51. Hussain, N., Jani, P., Florence, A.T. Enhanced oral uptake of tomato lectin-conjugated nanoparticles in the rat. *Pharm. Res.*, 14, 613, 1997.
52. Sinha, V.R., Tehran, A. Biodegradable microsphere for protein delivery, *J. Contr. Release*, 90, 261, 2003.
53. Weert, M.V., Hennink, W.E., Jiskoot, W. Protein instability in poly(lactic-*co*-glycolic acid) nanoparticles. *Pharm. Res.*, 17, 1159, 2000.
54. Ibrahim, M.A., Ismail, A., Fetouh, M.I., Gopferich, A. Stability of insulin during the erosion of poly(lactic acid) and poly(lactic-*co*-glycolic acid) microspheres. *J. Contr. Release*, 106, 241, 2005.
55. Owens, D.E., Peppas, N.A. Opsonization, biodistribution, and pharmacokinetics of polymeric nanoparticles. *Int. J. Pharm.*, 307, 93, 2006.
56. Carstensen, H., Muller, R.H., Muller, B.W. Particle-size, surface hydrophobicity and interaction with serum of parenteral fat emulsions and model-drug carriers as parameters related to RES uptake. *Clin. Nutr.*, 11, 289, 1992.
57. Mathiowitz, E. et al. Biologically erodable microspheres potential oral drug delivery systems. *Nature*, 386, 410, 1997.
58. Chickering, D., Jacob, J., Mathiowitz, E. Poly(fumaric-*co*-sebacic) microspheres as oral drug delivery systems. *Biotechnol. Bioeng.*, 52, 96, 1996.
59. Damge, C., Vranckx, H., Balschmidt, P., Couvreur, P. Poly(alkyl cyanoacrylate) nanospheres for oral administration of insulin. *J. Pharm. Sci.*, 86, 1997, 1403.
60. Watnasirichaikul, S., Rades, T., Tucker, I.G., Davies, N.M. *In-vitro* release and oral bioactivity of insulin in diabetic rats using nanocapsules dispersed in biocompatible microemulsion. *J. Pharm. Pharmacol.*, 54, 473, 2002.
61. Radwan, M.A. Enhancement of absorption of insulin-loaded polyisobutylcyanoacrylate nanospheres by sodium cholate after oral and subcutaneous administration in diabetic rats. *Drug Deliv. Int. Pharm.*, 9, 981, 2001.
62. Damge C., Maincent P., Ubrich N. Oral delivery of insulin associated to polymeric nanoparticles in diabetic rats. *J. Contr. Release*, 117(2):163–170. February 12, 2007 published online October 25.

63. Pan, Y. Study on preparation and oral efficacy of insulin-loaded poly(lactic-*co*-glycolic acid) nanoparticles. *Yao Xue Xue Bao.*, 37, 374, 2002.
64. Cui, F. et al. Biodegradable nanoparticles loaded with insulin-phospholipid complex for oral delivery: Preparation, in vitro characterization and in vivo evaluation. *J. Contr. Release*, 114, 242, 2006.
65. Akiyama, Y., Nagahara, N. Novel formulation approaches to oral mucoadhesive drug delivery systems. In *Bioadhesive Drug Delivery Systems—Fundamentals, Novel Approaches and Development.* Mathiowitz, E., Chickering, D.E., Lehr, C.M. (Eds.) Marcel Dekker, New York, 1998, p. 477.
66. Vasir, J.K., Tambwekar, K., Garg, S. Bioadhesive microspheres as a controlled drug delivery system. *Int. J. Pharm.*, 255, 13, 2003.
67. Lehr, C.M. Bioadhesion technologies for the delivery of peptide and protein drugs to the gastrointestinal tract. *Cri. Rev. Ther. Drug Carrier Syst.*, 11, 119, 1994.
68. Chowdary, K.P., Rao, Y.S. Mucoadhesive microspheres for controlled drug delivery. *Bio. Pharm. Bull.*, 27, 1717, 2004.
69. Huang, Y., Leobandung, W., Foss, A., Peppas, N.A. Molecular aspects of muco and bioadhesion: Tethered structures and site-specific surfaces. *J. Contr. Release*, 65, 63, 2000.
70. Ahuja, A., Khar, R.K., Ali, J. Mucoadhesive drug delivery systems. *Drug Dev. Ind. Pharm.*, 23, 489, 1997.
71. Dodou, D., Breedveld, P., Wieringa, P.A. Mucoadhesives in the gastrointestinal tract: Revisiting the literature for novel applications. *Eur. J. Pharm. Biopharm.*, 60, 1, 2005.
72. Peppas, N.A., Sahlin, J.J. Hydrogels as mucoadhesive and bioadhesive materials: A review. *Biomaterials*, 17, 1553, 1996.
73. Lowman, A.M., Peppas, N.A. Hydrogels. In *Encyclopedia of Controlled Drug Delivery*, Vol. 1. Mathiowitz, E. (Ed.) John Wiley & Sons, Inc., New York, 1999, p. 397.
74. Haas, J., Lehr, C.M. Developments in the area of bioadhesive drug delivery systems. *Expert Opin. Bio. Ther.*, 2, 287, 2002.
75. Leung, S.-H.S., Robinson, J.R. The contribution of anionic polymer structural features to mucoadhesion. *J. Contr. Release*, 5, 223, 1988.
76. Gu, J.M., Robinson, J.R., Leung, S.H. Binding of acrylic polymers to mucin/epithelial surfaces: Structure-property relationships. *Crit. Rev. Ther. Drug Carrier Syst.*, 5, 21, 1988.
77. Park, H., Robinson, J.R. Mechanism of mucoadhesion of PAA hydrogels. *Pharm. Res.*, 4, 457, 1987.
78. Bernkop-Schnurch, A. Chitosan and its derivatives: Potential excipients for peroral peptide delivery systems. *Int. J. Pharm.*, 194, 1, 2000.
79. Blanchette, J., Kavimandan, N., Peppas, N.A. Principles of transmucosal delivery of therapeutic agents. *Biomed. Pharmacother.*, 58, 142, 2004.
80. Peppas, N.A., Klier, J. Controlled release by using poly(methacrylic acid-*g*-ethylene glycol) hydrogels. *J. Contr. Release*, 16, 203, 1991.
81. Ascentiis, A.D., Degrazia, J.L., Bowman, C.N., Colombo, P., Peppas, N.A. Mucoadhesion of P (2-HEMA) is improved when linear PEO chains are added to polymer networks. *J. Contr. Release*, 33, 197, 1995.
82. Peppas, N.A., Kuys, K.B., Torres-Lugo, M., Lowman, A.M. PEG containing hydrogels in drug delivery. *J. Contr. Release*, 62, 81, 1999.
83. Leitner, V.M., Walker, G.F., Bernkop-Schnurch A. Thiolated polymers: Evidence for the formation of disulphide bonds with mucus glycoproteins. *Eur. J. Pharm. Biopharm.*, 56, 207, 2003.
84. Greindl, M., Bernkop-Schnurch, A. Development of a novel method for the preparation of thiolated polyacrylic acid nanoparticles. *Pharm. Res.*, 23, 2183, 2006.
85. Bernkop-Schnurch, A., Weithaler, A., Albrecht, K., Greimel, A. Thiomers: Preparation and in vitro evaluation of a mucoadhesive nanoparticulate drug delivery system. *Int. J. Pharm.*, 317, 76, 2006.
86. David, J., Brayden, D.J., O'Mahony, D.J. Novel oral drug delivery gateways for biotechnology products: Polypeptides and vaccines. *Pharm. Sci. Tech. Today*, 2, 67, 1998.
87. Anderson, J.M., Balda, M.S., Fanning, A.S. The structure and regulation of tight junctions. *Curr. Opin. Cell Bio. Suppl.*, 5, 772, 1993.
88. Borcharel, G. et al. The potential of mucoadhesive polymers in enhancing intestinal peptide drug abs. III: Effects of chitosan–glutamate and carbomer on epithelial tight junctions *in vitro*. *J. Contr. Release*, 39, 131, 1996.
89. Luessen, H.L. et al. Mucoadhesive polymers in peroral peptide drug delivery. IV. Polycarbophil and chitosan are potent enhancers of peptide transport across intestinal mucosa *in vitro*. *J. Contr. Release*, 45, 15, 1997.
90. Luessen, H.L. et al. Mucoadhesive polymers in peroral peptide drug delivery. II. Carbomer and poly carbophil are potent inhibitors of intestinal proteolytic enzyme trypsin. *Pharm. Res.*, 12, 129, 1995.

91. Artursson, P., Ungell, A.L., Löfroth, J.E. Selective paracellular permeability in two models of intestinal absorption: Cultured monolayers of human intestinal epithelial cells and rat intestinal segments. *Pharm. Res.*, 10, 1123, 1993.
92. Rubas, W. et al. Flux measurements across caco-2 monolayers may predict transport in human large intestinal tissue. *J. Pharm. Sci.*, 85, 165, 1996.
93. Kriwet, B., Kissel, T. Poly(acrylic acid) microparticles widen the intercellular spaces of caco-2 cell monolayers: Examination by confocal laser scanning microscopy. *Eur. J. Pharm. Biopharm.*, 42, 233, 1996.
94. Ichikawa, H., Peppas, N.A. Novel complexation hydrogels for oral peptide delivery: In vitro evaluation of their cytocompatibility and insulin-transport enhancing effects using caco-2 cell monolayers. *J. Biomed. Mater. Res.*, 67, 609, 2003.
95. Torres-Lugo, M., Garcia, M., Record, R., Peppas, N.A. pH sensitive hydrogels as GI tract absorption enhancers: Transport mechanism of salmon calcitonin and other model molecules using caco-2 cell model. *Biotech. Prog.*, 18, 612, 2002.
96. Torres-Lugo, M., García, M., Record, R., Peppas, N.A. Physicochemical behavior and cytotoxic effects of p(methacrylic acid-g-ethylene glycol) nanospheres for oral delivery of proteins. *J. Contr. Release*, 80, 197, 2002.
97. Ranjha, N.M., Doelkar, M. pH sensitive hydrogels for site specific delivery. I. Swelling behavior of crosslinked copolymers of acrylic acid and methacrylic acid. *S.T.P Pharm. Sci.*, 9, 335, 1999.
98. Lowman, A.M. et al. Oral delivery of insulin using pH responsive complexation gels. *J. Pharm. Sci.*, 88, 933, 1999.
99. Foss, A.C., Goto, T., Morishita, M., Peppas, N.A. Development of acrylic-based copolymers for oral insulin delivery. *Eur. J. Pharm. Biopharm.*, 57, 163, 2004.
100. Morishita, M. et al. Novel oral insulin delivery systems based on complexation polymer hydrogels: Single and multiple administration studies in type 1 and 2 diabetic rats. *J. Contr. Release*, 110, 587, 2006.
101. Morishita, I. et al. Hypoglycemic effect of novel oral microspheres of insulin with protease inhibitor in normal and diabetic rats. *Int. J. Pharm.*, 78, 9, 1992.
102. Hosny, E.A., Al-Shora, H.I., Elmazar, M.M.A. Oral delivery of insulin from enteric-coated capsules containing sodium salicylate: Effect on relative hypoglycemia of diabetic beagle dogs. *Int. J. Pharm.*, 237, 71, 2002.
103. Jain, D., Panda, A.K., Majumdar, D.K. Eudragit S100 entrapped insulin microspheres for oral delivery. *AAPS PharmSciTech*, 6, E100, 2005.
104. Paul, W., Sharma, C.P. Chitosan, a drug carrier for 21st century: A review. *STP Pharm. Sci.*, 10, 5, 2000.
105. Rao, S.B., Sharma, C.P. Use of chitosan as a biomaterial: Studies on its safety and haemostatic potential. *J. Biomat. Mater. Res.*, 34, 21, 1997.
106. Berger, J. et al. Structure and interactions in chitosan hydrogels formed by complexation or aggregation for biomedical applications. *Eur. J. Pharm. Biopharm.*, 57, 35, 2004.
107. Kas, H.S. Chitosan: Properties, preparations and application to microparticulate systems. *J. Microencapsul.*, 14, 689, 1997.
108. Agnihotri, S., Mallikarjuna, N.N., Aminabhavi, T.M. Recent advances on chitosan-based micro-and nanoparticles in drug delivery. *J. Contr. Release*, 100, 5, 2004.
109. Illum, L. Chitosan and its use as a pharmaceutical excipient. *Pharm. Res.*, 15, 1326, 1998.
110. Thanou, M., Verhoef, J.C., Junginger, H.E. Chitosan and its derivatives as intestinal absorption enhancers. *Adv. Drug Deliv. Rev.*, 50, S91, 2001.
111. Schipper, N.G.M. et al. Chitosan as absorption enhancers for poorly absorbable drugs 2: Mechanism of absorption enhancement. *Pharm. Res.*, 14, 923, 1997.
112. Artursson, P., Lindmark, T., Davis, S.S., Illum, L. Effect of chitosan on the permeability of monolayer of intestinal epithelial cells (caco-2). *Pharm. Res.*, 11, 1358, 1994.
113. Ma, Z., Lim, L.Y. Uptake of chitosan and associated insulin in caco 2 cell monolayers: A comparison between chitosan molecules and chitosan nanoparticles. *Pharm. Res.*, 20, 1812, 2003.
114. Kotze, A.F. et al. *N*-trimethyl chitosan chloride as a potential absorption enhancer across mucosal surfaces: In vitro evaluation in intestinal epithelial cells (caco-2). *Pharm. Res.*, 14, 1197, 1997.
115. Thanou, M. et al. *N*-trimethylated chitosan chloride (TMC) improves the intestinal permeation of the peptide drug buserelin in vitro (caco-2 cells) and in vivo (rats). *Pharm. Res.*, 17, 27, 2000.
116. Prego, C., Torres, D., Alonso, M.J. Chitosan nanocapsules as carriers for oral peptide delivery: Effect of chitosan molecular weight and type of salt on the in vitro behavior and in vivo effectiveness. *Nanosci. Nanotechnol.*, 6, 2921, 2006.

117. Pan, Y. et al. Bioadhesive polysaccharide in protein delivery system: Chitosan nanoparticles improve the intestinal absorption of insulin *in vivo. Int. J. Pharm.*, 249, 139, 2002.
118. Tozaki, H. et al. Chitosan capsules for colon drug delivery: Improvement of insulin absorption from rat colon. *J. Pharm. Sci.*, 86, 1016, 1997.
119. Lubben, I.M.V. et al. Chitosan microparticles for oral vaccination: Preparation characterization and preliminary in vivo uptake studies in murine payer's patches. *Biomaterials*, 22, 687, 2001.
120. Tønnesen, H.H., Karlsen, J. Alginate in drug delivery systems. *Drug Dev. Ind. Pharm.*, 28, 621, 2002.
121. Gombotzb, W.R., Wee, S.W. Protein release from alginate matrices. *Adv. Drug Del Rev.*, 31, 267, 1998.
122. Liu, P., Krishnan, T.R. Alginate-pectin-poly-L-lysine particulate as a potential controlled release formulation. *J. Pharm. Pharmcol.*, 51, 141, 1999.
123. Hari, P.R., Chandy, T., Sharma, C.P. Chitosan/calcium alginate beads for oral delivery of insulin. *J. Appl. Polymer Sci.*, 59, 1795, 1996.
124. Dileep, K.J., Roswen M.L., Sharma, C.P. Modulation of insulin from chitosan/alginate microspheres. *Trends Biomater. Art. Organs*, 12, 42, 1998.
125. Jerry, N., Anitha, Y., Sharma, C.P., Sony, P. In vivo absorption studies from an oral delivery system. *Drug Deliv.*, 8, 19, 2001.
126. Kim, B.Y., Jeong, J.H., Park, K., Kim, J.D. Bioadhesive interaction and hypoglycemic effect of insulin-loaded lectin-microparticle conjugates in oral insulin delivery system. *J. Contr. Release*, 102, 525, 2005.
127. Ramdas, M. et al. Lipoinsulin encapsulated alginate-chitosan capsules: Intestinal delivery in diabetic rats. *J. Microencapsul.*, 17, 405, 2000.
128. Morcol, T. et al. Calcium phosphate-PEG-insulin-casein (CAPIC) particles as oral delivery systems for insulin. *Int. J. Pharm.*, 277, 91, 2004.
129. Arshady, R. Manufacturing methodology of microspheres. In *Microspheres, Microcapsules and Liposomes, Vol 1: Preparation and Chemical Applications*. Arshady, R. (Ed.) Citus Book, London, U.K., 1999, p. 85.
130. De Jacghere, F., Doelkar, E., Gurny, R. Nanoparticles. In *Encyclopedia of Controlled Drug Delivery*, Vol II. Mathowitz, E. (Ed.) John Wiley & Sons, Inc., New York, 1999, p. 641.
131. Barratt, G. et al. Polymeric micro- and nanoparticles as drug carriers. In *Polymeric Biomaterials*, 2nd edn. Dumitriu, S. (Ed.) Marcel Dekker, New York, 2002, p. 753.
132. Mathiwotiz, E., Kreitz, M.R., Peppas, L.-B. Microencapsulation. In *Encyclopedia of Controlled Drug Delivery*, Vol II. Mathowitz, E. (Ed.) John Wiley & Sons, Inc., New York, 1999, p. 493.
133. Allémann, E., Gurny, R., Doelker, E. Drug loaded nanoparticles—Preparation methods and drug targeting issues. *Eur. J. Pharm. Biopharm.*, 39, 173, 1993.
134. Donini, C., Robinson, D.N., Colombo, P., Giordano, F., Peppas, N.A. Preparation of poly(methacrylic acid-g-poly(ethylene glycol))nanospheres from methacrylic monomers for pharmaceutical applications. *Int. J. Pharm.*, 245, 83, 2002.
135. Skovby, M.H.B., Kops, J. Preparation by suspension polymerization of porous beads for enzyme immobilization. *J. Appl. Polymer Sci.*, 39, 169, 1990.
136. Keiwet, B., Walker, E., Kissel, T. Synthesis of bioadhesive of poly(acrylic acid) nano and microparticles using an inverse emulsion polymerization method for entrapment of hydrophilic drug candidates. *J. Contr. Release*, 56, 149, 1998.
137. Sajeesh, S., Sharma, C.P. Cyclodextrin-insulin complex encapsulated polymethacrylic acid based nanoparticles for oral insulin delivery. *Int. J. Pharm.*, 325, 147, 2006.
138. Sajeesh, S., Sharma, C.P. Novel pH responsive polymethacrylic acid-chitosan-polyethylene glycol nanoparticles for oral peptide delivery. *J. Biomed. Mater. Res. Appl. Biomater.*, 76B, 298, 2006.
139. Sajeesh, S., Sharma C.P. Novel polyelectrolyte complex microparticles based on polymethacrylic acid-polyethylene glycol for oral peptide delivery. *J. Biomater. Sci. Polym. Ed.*, 18, 1125, 2007.
140. Sajeesh, S., Sharma, C.P. Interpolymer complex microparticles based on polymethacrylic acid—Chitosan for oral insulin delivery. *J. Appl. Polymer Sci.*, 99, 506, 2006.
141. Ravi Kumar, M.N.V. Nano and microparticles as controlled drug delivery devices. *J. Pharm. Pharmaceut. Sci.*, 3, 234, 2000.

58 Vegetable Oil–Based Formulations for Controlled Drug Delivery

*V.K. Singh, Sai S. Sagiri, K. Pramanik, Arfat Anis,
S.S. Ray, I. Banerjee, and Kunal Pal*

CONTENTS

58.1 Introduction 1381
58.2 Vegetable Oil–Based Formulations 1382
58.3 Different Vegetable Oil–Based Formulations 1384
    58.3.1 Micelles 1384
    58.3.2 Liposomes 1385
    58.3.3 Solid Lipid Nanoparticles 1387
    58.3.4 Emulsion 1388
        58.3.4.1 Microemulsions 1388
        58.3.4.2 Macroemulsions 1390
        58.3.4.3 Multiple Emulsions 1390
    58.3.5 Gels 1391
        58.3.5.1 Organogels 1391
        58.3.5.2 Emulgels 1393
        58.3.5.3 Bigels 1395
        58.3.5.4 Lyotropic Gels 1395
58.4 Applications of Vegetable Oil-Based Formulations 1397
58.5 Scope for New Formulations 1400
Acknowledgment 1400
References 1401

58.1 INTRODUCTION

Controlled drug delivery has emerged as a field of study that concentrates to treat ailments with improved safety, efficacy, convenience, and patient compliance. Controlled release formulations release drugs at a constant rate for longer durations. The main principle behind the release of the drug is diffusion-mediated. Other mechanisms like swelling and erosion may also govern the release process. Controlled delivery formulations require lower quantity of the drug for achieving therapeutic activity for prolonged time. This reduces the toxicity associated with the frequent and higher dosing. These systems help improving the pharmacokinetic and pharmacodynamic profiles of the drugs without compromising the therapeutic activity of the drugs. The controlled delivery formulations eliminate the potential of underdosing and overdosing and maintain the drug levels at the target sites within the therapeutic window. The less frequent dose requirement increases patient compliance.[1] Along with the potential advantages, these dosage forms suffer from some drawbacks as compared to the immediate-release formulations. The major disadvantages include dose dumping and lack of precision in the controlled release mechanism leading to the variation in dosing.

It is difficult to formulate controlled release formulations for the drugs with short half-life (<1 h) due to the requirement of high maintenance dose. The development of these formulations is a time-consuming process and difficult to manufacture on a large scale, which makes them more expensive than the conventional dosage forms. These are posing new challenges to the researchers to develop controlled delivery formulations with minimal drawbacks.

Vegetable oils and their analogues are being tried extensively for formulating oil-based formulations.[2] In recent years, the oil-based dosage forms have drawn considerable attention and interest of the formulation scientists.[3] The oil-based drug delivery system has experienced a renaissance since 1974, partly due to the fact that, approximately 40% to 70% of the new chemical entities selected for formulation development suffer from major drawbacks like poor solubility and dissolution rate limited absorption. This results in poor bioavailability of the drugs.[4–8] In general, these drugs show poor and variable absorption with unexpected toxicity along with costly and difficult development protocol.[9] Oil-based formulations improve the absorption of the poorly soluble drugs by increasing the solubilization capacity and preventing drug precipitation.[8,10] They improve the bioavailability of these drugs by keeping them in the dissolved state until they get absorbed.[11] Though the exact mechanisms for the improved absorption of the poorly soluble drugs are not known completely, it is assumed that the oil-based drug delivery systems solubilize the lipophilic drugs within lipid vesicles and excipients. This may result in improving the absorption, membrane permeability, and solubilization of the drugs.[12,13] Also, the oil-based formulations can be effective for delivering lipophilic drugs undergoing first pass metabolism (including anticancer and HIV protease inhibitors). Oils may provide a coating on the drug molecule that can mask the taste, impart protection from oxidation/light, and control the release behavior of the drug.[2] The saturated and unsaturated fatty acids present in the oils act as natural penetration enhancer, for example, oleic acid.[14–16] They may provide moisture retention property to the topical formulations. Most of the vegetable oils also have nutraceutical value, which improves the nutritional value of these formulations. The antioxidant properties of the vegetable oils impart additional benefit of improved stability of the formulations.

58.2 VEGETABLE OIL–BASED FORMULATIONS

Generally, vegetable oil-based delivery systems are composed of vegetable oil, surfactant, cosurfactant or water-soluble organic solvents, and an antioxidant (if required) (Figure 58.1).

Strickley,[17] Hauss,[5] and other researchers provided a detailed summary of the commercially available excipients in their recent reviews on lipid-based drug delivery. The most commonly used excipients in the oil-based formulations are listed in Table 58.1.

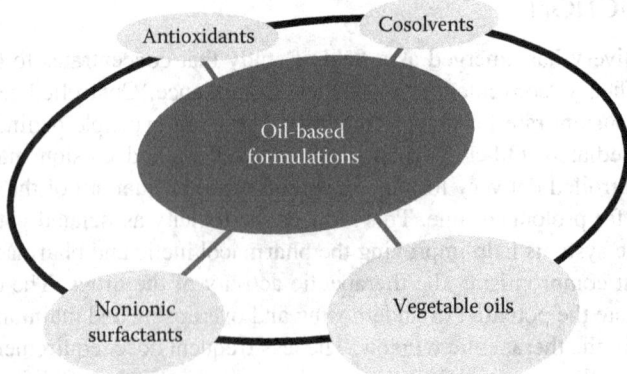

FIGURE 58.1 General composition of vegetable oil–based formulations. (From Hauss, D.J., *Adv. Drug Deliv. Rev.*, 59(7), 667, 2007.)

TABLE 58.1
Commonly Used Raw Materials for Preparing Vegetable Oil–Based Formulations

Compositions	Examples	References
Vegetable oils	Olive oil, soybean oil, sesame oil, peanut oil, sunflower oil, coconut oil, castor oil, cotton seed oil, canola oil, and corn oil.	[18,19]
Nonionic surfactants	Spans (Span 20, Span 40, Span 60, Span 80), Tweens (Tween 20, Tween 80), cyclodextrins, lecithin, sodium dodecyl sulfate (sodium lauryl sulfate), polyethoxylated alcohols, cetyltrimethylammonium bromide (CTAB), Triton X100, Brij 721, bile salts (sodium deoxycholate, sodium cholate), methylbenzethonium chloride (Hyamine), and poloxamer 407.	[20,21]
Cosurfactants/cosolvents	PEG 400, propylene glycol, glycerine, ethanol, propanol, isopropanol, butanol, isobutanol, pentanol, hexanol, heptanol, octanol sorbitol, and Cremophor RH40 (polyoxyl 40 hydrogenated castrol oil).	[22]
Antioxidants	Water soluble: ascorbic acid (vitamin C), glutathione, lipoic acid and uric acid. Oil soluble: α-tocopherol, β-carotene, butylated hydroxytoluene (BHT), butylated hydroxyanisole (BHA), propyl gallate carotenes, and ubiquinol (coenzyme Q).	[23,24]

Vegetable oils are primarily extracted from plant seeds. They are being investigated as a potential source of delivery vehicles due to their easy availability and improved stability during storage. The majority of the vegetable oils are primarily mixtures of triacylglycerides (also called as triglycerides) (90%–95% w/w) and small quantities of free fatty acids and phospholipids. These also contain some natural antioxidants to protect the oil from rancidity, which includes nonsaponifiable products (pigments and sterols), fat-soluble vitamins (tocopherols), and carotenoids.[2] Chemically, triacylglycerides are glycerol moieties in which three long-chain fatty acids are attached at the terminal hydroxy groups via ester linkages. The fatty acids in the vegetable oil are 14 and 22 carbons long with varying levels of unsaturation. Triglycerides can be classified as short-chain triglycerides (up to 5 carbons), medium-chain triglycerides (6–12 carbons), and long-chain triglycerides (>12 carbons). Vegetable oils are glyceride esters of medium-chain triglycerides or long-chain triglycerides. They contain different proportions of each type of fatty acid esters (Table 58.2).[2] Medium-chain triglycerides show higher solvent capacity compared to long-chain triglycerides.[18] Also, they are more stable to oxidation as compared to long-chain triglycerides, so are a popular choice for developing lipid-based products. Recent studies suggested that using a mixture of oils and/or mixture of surfactants showed better physico-chemical properties compared to the pure single oil and/or surfactant. Addition of salts in topical microemulsions is reported to have beneficial effect.[11,19] Partially hydrolyzed glycerides of high melting point (e.g., glyceryl monocaprylocaprate, glyceryl monostearate, glyceryl monooleate, glyceryl monolinoleate, etc.) containing saturated long-chain fatty acids are investigated for formulating controlled delivery systems.[2,20–22] Surfactants may cause irritation and impart toxicity by some possible interactions with the biological systems.[23] Ionic surfactants may produce more toxic effects as compared to the nonionic surfactants. Hence, nonionic surfactants are preferably used in oil-based delivery systems.[24] Several water soluble cosolvents are also added to improve the solvent capacity and dissolution of the drugs in the formulations. The saturated and unsaturated fatty acid esters (e.g., isopropyl myristate, isopropyl palmitate, and medium chain triglycerides) enhance the penetration of poorly soluble drugs through skin.[14–16] Oleic acid is one of the most popular penetration enhancer used in transdermal drug delivery applications. Few oil-soluble antioxidants are added in the formulations to protect the unsaturated fatty acids or the drug from oxidation.[25] Cao et al.[18] showed improved solubility of the drug molecules by providing hydrogen bonding sites upon addition of the aqueous solvents.

TABLE 58.2
Fatty Acid Composition of the Commonly Used Vegetable Oils

Vegetable Oils	SFA[a]	MUFA Oleic Acid	PUFA Linoleic Acid	Linolenic Acid	Ricinoleic Acid	References
Sesame oil	15	38.2	45	0.6	—	[33]
Olive oil	12.6	78.1	7.3	0.6	—	[34]
Castor oil	2	3	4	—	90	[35]
Coconut oil	91.6	6.2	1.6	—	—	[36]
Corn oil	15.8	30.5	52	1	—	[37]
Cotton seed oil	27.4	19	52.5	—	—	[37]
Grape seed oil	10	22	67	0.9	—	[38]
Palm oil	48	40	10	—	—	[34]
Safflower oil	10	14	75	—	—	[36]
Apricot kernel oil	5	69.2	25.4	—	—	
Sunflower oil	9	20	69	—	—	[34]
Soybean oil (unhydrogeneted)	14	23	51	7	—	[37]
Soybean oil (partially hydrogeneted)	15	43	35	3	—	[37]
Peanut oil	16	38	41	—	—	[34]
Canola oil	6	56	26	10	—	[34]

[a] SFA-Saturated fatty acids (capric acid, caprylic acid, myristic acid, lauric acid, palmitic acid, stearic acid).
Note: MUFA, monounsaturated fatty acids; PUFA, poly unsaturated fatty acids.

58.3 DIFFERENT VEGETABLE OIL-BASED FORMULATIONS

There are several oil-based formulations for the delivery of hydrophilic and lipophilic drugs, which majorly include emulsions and gels (Figure 58.2). Apart from these, various other novel formulations may also be named under this category such as micelles, reverse micelles, solid lipid nanoparticles, and liposomes.[26–28] Based on the physicochemical properties of the drug and patient requirements, the vegetable oil-based formulations may be delivered by various routes like oral, parenteral, topical, transdermal, pulmonary, or ocular, etc.,[5,17,29]

58.3.1 Micelles

Micelles (normal and reverse) are aggregates of surfactants, which are formed spontaneously in a liquid phase when the surfactant concentration is increased than the critical micelle concentration (CMC). Micellar solutions are formed in aqueous continuous systems, whereas, the reverse micelles are formed in oily continuous systems. In micelles (oil-in-water micelle), the polar heads of the surfactant lie outside in the aqueous phase, whereas the lipophilic hydrocarbon chains lie inside (Figure 58.3).[32] When the surfactant concentration is increased, the free energy of the system increases due to the inauspicious interactions between the water molecules and the lipophilic portions of the surfactant. The water molecules around the oil droplets structures themselves, thereby, resulting in the decrease in the entropy.[33] The reverse micelles (water-in-oil micelle) have opposite structure, that is, the polar heads lie at the centre, while the lipophilic tails are present outside in the oil phase (Figure 58.3). The surfactant concentration need not necessarily be higher than the CMC for the formation of reverse micelle.[33] Many scientists reported formation of reverse micelles by lecithin in different oil phases.[34–35]

The normal micelles and reverse micelles solubilize both hydrophilic and lipophilic drugs. The normal micellar solubilization increases the bioavailability of the poorly soluble drugs, whereas, the reverse micellar solubilization allows sustained release of the medicament.

Vegetable Oil–Based Formulations for Controlled Drug Delivery

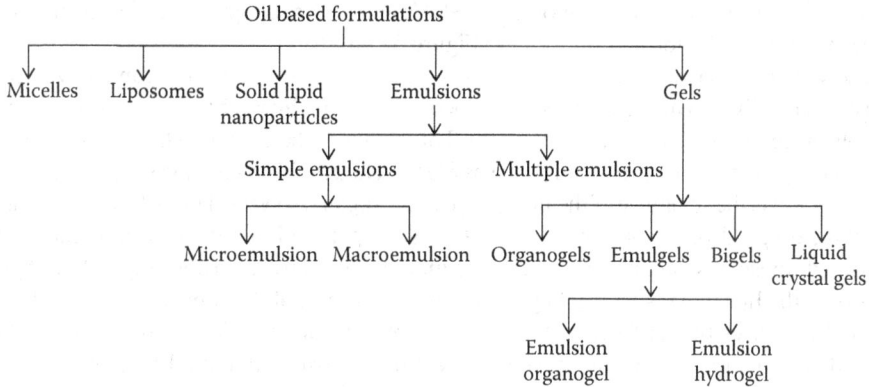

FIGURE 58.2 Classification of the vegetable oil–based formulations. (From dos Santos Guimarães, I. et al., *Conventional Cancer Treat.*, 2013; Schubert, M. and Müller-Goymann, C., *Eur. J. Pharm. Biopharm.*, 55(1), 125–131, 2003.)

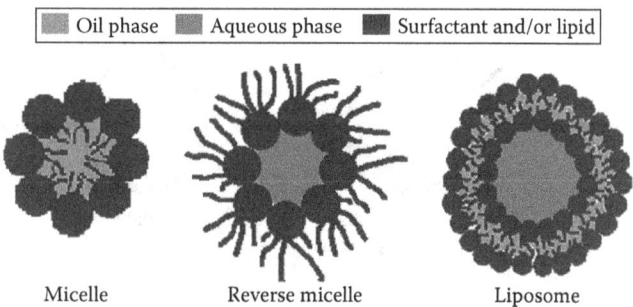

FIGURE 58.3 Schematic representation of the structures of micelles, reverse micelles, and liposomes. (From Tsuji, S. and Kawaguchi, H., *Langmuir*, 24(7), 3300, 2008; Frelichowska, J. et al., *Int. J. Pharm.*, 371(1–2), 5658, 2009.)

The reverse micelles get transformed into a liquid crystalline phase or vesicle dispersion, when it comes in contact with the aqueous body fluids. This reduces the rate of release of the solubilized drugs.[36]

58.3.2 Liposomes

Liposomes are bilayered phosphatidylcholine containing phospholipids.[39] They are formed spontaneously by dispersing phospholipids in water. They contain hydrophilic heads radiating outside and lipophilic tails lying inside (Figure 58.3). Based on the structural organization, they can be unilamellar (ULVs) and multilamellar vesicles (MLVs).[40] The unilamellar vesicles can be further divided into small (SUVs) and large unilamellar vesicles (LUVs).[41] They structurally differ from micelles, which are monolayered structures. They can be used to deliver both hydrophilic and lipophilic drugs. The hydrophilic drug may be incorporated in the central polar portion, while the amphiphilic and the lipophilic drugs may lie within the phospholipid bilayer.[41,42] Also, they provide protection to the encapsulated substances from degradation. The common methods for the preparation of liposomes are film-forming and organic solvent injection methods. The film-forming method employs dissolution of the lipophilic drug and lipid in an organic solvent. The solution is subsequently transferred in a round bottom flask. A thin film of lipid is formed by evaporating the organic solvent in a rotary evaporator. The film is then hydrated in an aqueous phase, which might contain hydrophilic drug. The hydration process results in the formation of primarily MLVs along with lower proportions of

SUVs and LUVs. The homogenous liposomes (SUVs and LUVs) are produced by further sonication or homogenization of the formed liposomes (Figure 58.4a).[43]

In the solvent injection method, the lipophilic drug and lipid dissolved in an organic solvent (e.g., ethanol) is injected rapidly into the aqueous phase, which may contain hydrophilic drug, using a syringe. The organic solvent gets diluted in the aqueous phase. This results in the change in the solubility of the lipids and subsequent precipitation of the lipid phase as SUVs. Though this method is simple and economical, the major concern is the solubility of the lipid phase in the organic solvent. This affects the overall encapsulation of the drug. Also, complete removal of the organic solvent is a difficult task (Figure 58.4b).[44]

PEGylated liposomes (stealth liposomes) are manufactured by incorporating poly-(ethylene glycol) (PEG) on the liposomal surface. Pegylation improves the stability and solubility and lowers the toxicity and increases the half-life of the drug. There are few PEGylated liposomal products like DOXIL® (doxorubicin), which is approved for treating Kaposi's sarcoma. Lipoplatin® (cisplatin) is in clinical trial stage for the treatment of several cancers.[45]

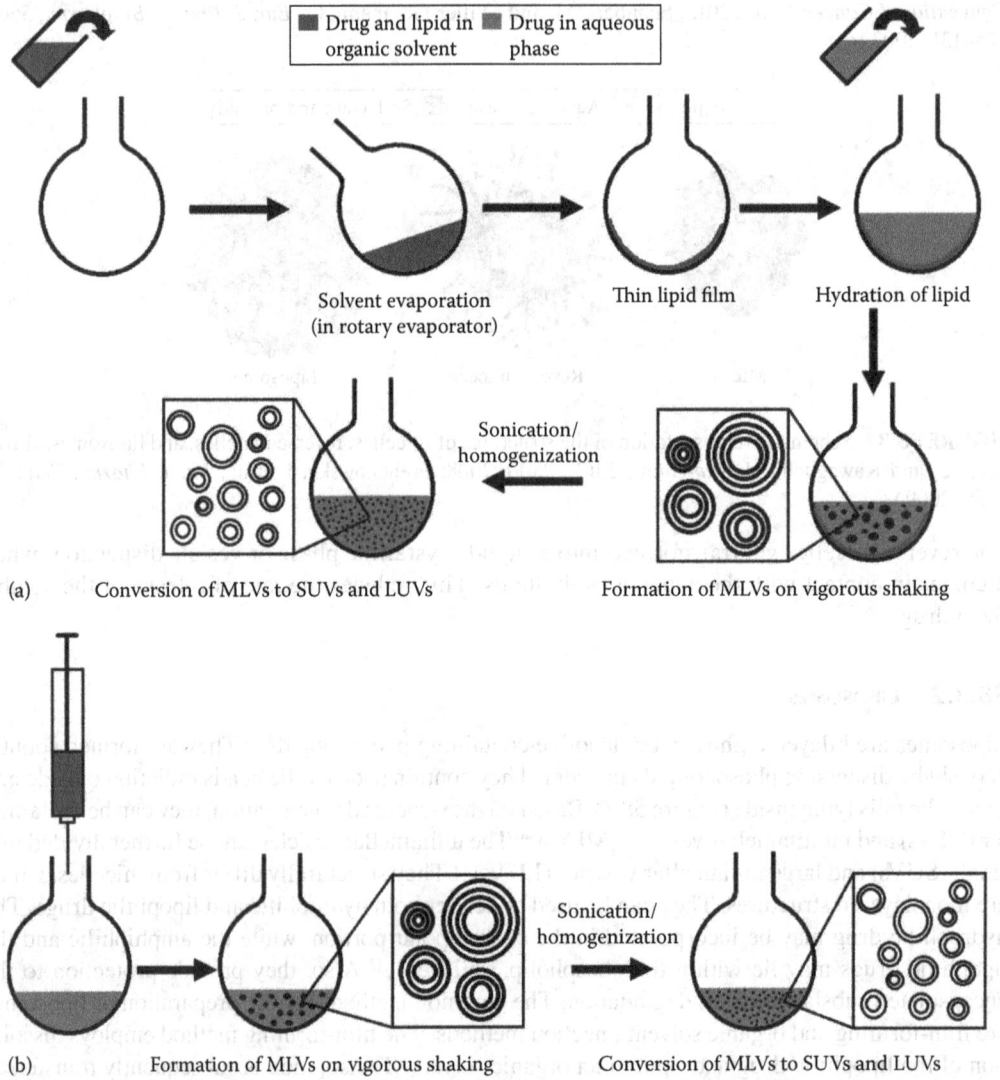

FIGURE 58.4 Schematic representation of the mechanisms of liposome formation: (a) film-forming method and (b) organic solvent injection method. (From Zhang, K. et al., *Powder Technol.*, 190(3), 39355, 2009.)

58.3.3 Solid Lipid Nanoparticles

Solid lipid nanoparticles are colloidal drug carriers. Nanoparticles (size ranging from 50 to 1000 nm) are made from solid lipid core that can solubilize lipophilic molecules and are dispersed in water or in aqueous surfactant solution.[46] Solid lipid nanoparticles possess many advantages like small size, large surface area, and high drug loading, which can be utilized for delivering bioactive agents.[47] As the particles are in the solid state at room temperature, the mobility of the incorporated drugs is reduced. This results in the release of the drug in a controlled manner.[48] The particles are stabilized using nontoxic surfactants (e.g., poloxamer and lecithin).[49] Solid lipid nanoparticles suffer from the drawback of lower loading capacity (only a few percent up to 50%).[42] The lipids that are well tolerated by the body are selected for preparing solid lipid nanoparticles (e.g., glycerides composed of fatty acids).

On large scale, the solid lipid nanoparticles are produced by high pressure homogenization, solvent evaporation, and by diluting microemulsion. High-pressure homogenization method involves two procedures. In both the methods, the drug is dissolved or dispersed in the bulk lipid.[50] The first method involves hot homogenization of the molten drug loaded lipid melt and the aqueous emulsifier phase maintained at the same temperature.[51] This results in formation of nanoemulsions, which forms solid particles when cooled to room temperature or to temperatures below the melting point of the lipid (Figure 58.5a).

In cold homogenization process, the drug containing bulk lipid is rapidly cooled using dry ice or liquid nitrogen. The solidified lipid is milled to obtain solid lipid microparticles, which are then dispersed in aqueous emulsifier solution (2°C–3°C) under stirring (Figure 58.5b). A pre-suspension is formed, which is homogenized under high pressure (at or below room temperature) to form solid lipid nanoparticles. In general, cold homogenization method produces larger size particles compared to hot homogenization method.[52]

In solvent evaporation method, the lipid is dissolved in a volatile organic solvent (e.g., cyclohexane or chloroform), which is dispersed in an aqueous phase to form an emulsion (Figure 58.5c). The lipid precipitates with the evaporation of the solvent and forms nanoparticle dispersion. The solid lipid nanoparticles are harvested after complete evaporation of the solvent.[53]

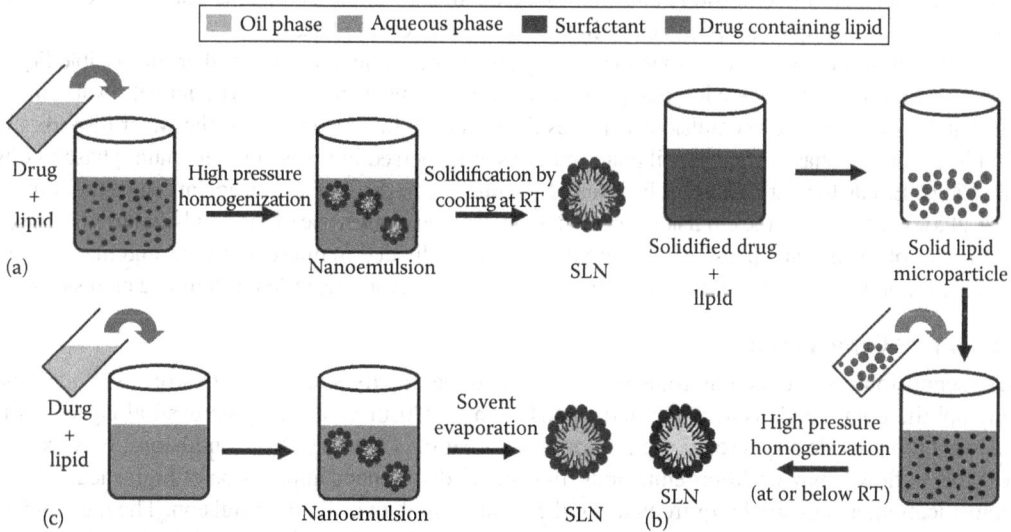

FIGURE 58.5 (See color insert.) Schematic representation of the methods of preparation of solid lipid nanoparticles. (a) Hot homogenization method. (From Brugarolas, T. et al., *Soft Matter*, 9(38), 904664, 2013.) (b) Cold homogenization method. (From He, C.X. et al., *Exp. Opin. Drug Deliv.*, 7(4), 44565, 2010.) (c) Solvent evaporation method. (From Choi, C.H. et al., *Adv. Mater.*, 25(18), 2536, 2013.)

Solid lipid nanoparticles are also prepared by the dilution of the lipid containing microemulsion. The lipid containing microemulsion is produced at a temperature above the melting point of the lipid. The hot microemulsion (65°C–70°C) is dispersed in chilled water maintained under mild mechanical stirring. The lipid phase precipitates and forms fine particles. In general, the volume ratio of hot microemulsion to chilled water is in the range of 1:25–1:50. Recently, Vectorpharma (Trieste, Italy) has developed solid lipid nanoparticles on a large scale using microemulsion technique.[47]

58.3.4 Emulsion

An emulsion is a mixture of two or more immiscible liquids, whose interface is stabilized by an emulsifier. Essentially, emulsion contains oil and water. Emulsions are inherently thermodynamically unstable system, often recognizable by their cloudy or white appearance since the substances do not mix homogenously. The mixed substances often separate due to reduction in the interfacial energy after a period of time. The possible mechanism of destabilization can be creaming, flocculation, or coalescence. The emulsions possess kinetic stability, which is lost after a period of time. The kinetic stability is associated with the presence of interfacial film around the dispersed droplets at the oil/water interface by the emulsifier. The interfacial film increases the interfacial viscosity, which retards the coalescence of the droplets.[54]

The stability of the emulsions may also be enhanced by adding solid particles (e.g., colloidal silica).[55] The solid particles remain dispersed in the continuous phase and prevent the coalescence of the dispersed droplets by binding to the surface of the interface. These types of emulsions are surfactant free and are called pickering emulsions (named after the scientist S.U. Pickering) (Figure 58.6d).[56] Tsuji and Kawaguchi[57] reported poly (N-isopropylacrylamide) stabilized thermosensitive pickering emulsions. Frelichowska et al.[58] compared a surfactant-based emulsion, an o/w pickering emulsion stabilized by silica particles and a solution in triglyceride oil for the topical delivery of lipophilic drug (e.g., retinol). Many emulsion-based products are used for topical applications (body lotions, creams, balms, pastes, and ointments). An emulsion can allow masking the unpleasant taste and odor of bitter and obnoxious drugs.

Broadly, emulsions are classified into two major groups depending on the number of internal phases, namely, simple emulsions and multiple emulsions. Simple emulsions contain one internal phase, whereas multiple emulsions contain two or more internal phases.[59]

Simple emulsions consist of dispersion of droplets of one liquid phase in another immiscible liquid phase. Based on the type of the internal phase, two types of simple emulsions exist, namely, water-in-oil (w/o) and oil-in-water (o/w) emulsions (Figures 58.6a and b). In w/o emulsions, the water disperses in the oil continuum phase, whereas, oil phase remains as dispersed in the water continuum phase in o/w emulsions. Emulsifiers are added to facilitate the reduction in the surface tension amongst the immiscible phases. The type of the emulsions formed is determined by the type of the emulsifier used.[60] Based on the size of the internal phase, simple emulsions can be divided in microemulsions and macroemulsions. If the emulsions contain more than two phases, then they are regarded as multiple emulsions.

58.3.4.1 Microemulsions

Microemulsions are defined as transparent, thermodynamically stable mixtures of oil, water, and amphiphilic compounds (surfactant and cosurfactant).[11] Cosurfactant may be used along with the surfactant to provide a synergistic action for emulsification. Contrary to the emulsions, they exhibit Newtonian flow behavior. Microemulsions, in general, do not need application of high shear during emulsification, an essential requirement for the formation of an ordinary emulsion. The mechanism behind the formation of microemulsion may be described by the reduction of the interfacial tension to a very low level at the oil-water interface. This depends on the critical proportions of the surfactant and cosurfactant, which brings flexibility at the oil-water interface. In microemulsions, the sizes of the dispersed droplets are in the range of 0.75 to 20 μm. As the wave length of the incidental light waves is smaller than the dispersed droplets, the light waves do not get refracted from

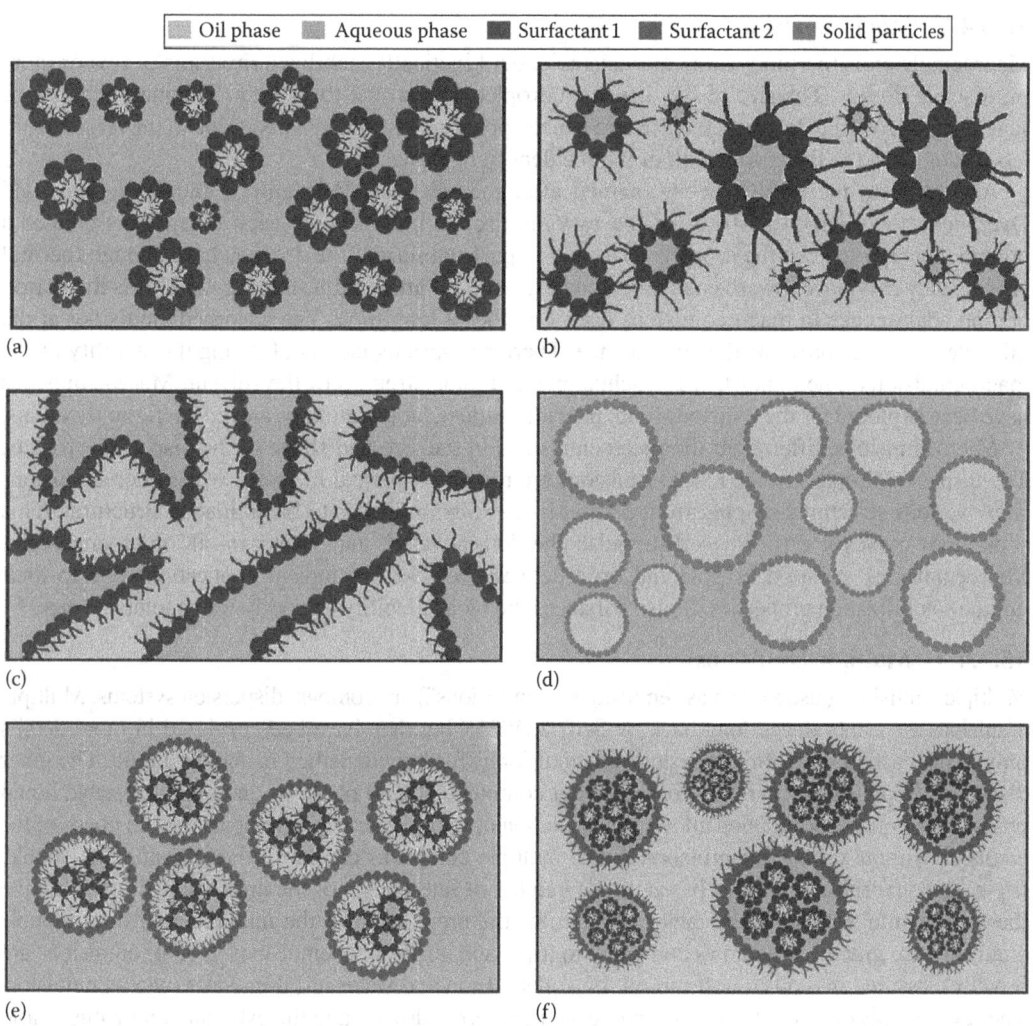

FIGURE 58.6 (**See color insert.**) Schematic representation of different types of emulsions: (a) oil-in-water, (b) water-in-oil, (c) bicontinuous, (d) pickering emulsion, and double emulsions: (e) w/o/w and (f) o/w/o. (From Markovic, N. et al., *Hydrocarbon Gels: Rheological Investigation of Structure*, ACS Publications, 2002; Zoumpanioti, M. et al., *Biotechnol. Adv.*, 28(3), 39576, 2010; Rogovina, L.Z. and Slonimskii, G.L., *Russ. Chem. Rev.*, 43(6), 50377, 1974.)

the droplet interface.[61] This explains clear appearance of the microemulsions. Due to this reason, microemulsions cannot be observed under an optical microscope.

The microemulsions can be broadly categorized into three types, namely,

1. Oil-in-water microemulsions
2. Water-in-oil microemulsions
3. Bicontinuous microemulsions

As the name suggests, oil phase remains dispersed in the water continuum phase to form an oil-in-water microemulsions and vice versa in the case of water-in-oil microemulsions. In the bicontinuous microemulsions, both the phases remain as the continuous phase (Figure 58.6c). The amphiphilic surfactant forms a monolayer at the interface of the two immiscible liquids by dissolving the hydrophobic tail in the oil phase and the hydrophilic head groups in the aqueous phase.

58.3.4.2 Macroemulsions

Macroemulsion is thermodynamically unstable, but kinetically stabilized mixtures of two or more immiscible liquids. The size of the dispersed droplet is usually 2 to 100 μm. The macroemulsions scatter light from the droplet interface and hence appear milky white. This is due to the larger droplets as compared to the wavelength of the incident light waves.

According to the DLVO theory (named after the scientists Derjaguin, Landau, Verwey, and Overbeek), the charged surfaces of the two immiscible phases repel each other. This creates a potential energy to keep the droplets apart and impart stability. If the droplets have enough thermal energy, they can move toward each other, coalesce and try attaining a stable state. This is the reason behind coalescence in macroemulsions over a period of long time. The adsorbed emulsifier at the oil-water interface prevents the coalescence. There are various factors affecting the stability of the macroemulsions. The major factors include pH and ionic strength of the solvent. Macroemulsions have been explored for their various food, pharmaceutical, biotechnology, and cosmetic applications.

Microemulsions differ from the macroemulsions in the size and shape of the dispersed droplets. The dispersed droplets in the macroemulsions are roughly spherical in shape, while microemulsions show various structures ranging from droplet-like swollen micelles to bicontinuous structures. This sometimes makes it difficult to distinguish the "oil-in-water" and "water-in-oil" microemulsions. Microemulsions are used as potential drug delivery vehicles for topical applications due to their unique solubilization properties. It also enhances the bioavailability of poorly water soluble drugs.[65]

58.3.4.3 Multiple Emulsions

Multiple emulsions (also known as "emulsions of emulsions") are complex dispersion systems. Multiple emulsions were introduced long back by Seifriz (1925), but they have been explored in more details only in the past 20 years. In this system, an emulsion is further emulsified in another liquid. The inner dispersed droplets are separated from the outer continuous liquid phase by another immiscible liquid phase. This type of formulation allows encapsulation of multiple active ingredients in each phase of the emulsion without cross-contamination.[66] The multiple emulsions can be further classified as double, triple, and quadruple emulsions based on the number of interphases of the immiscible drops.[66] As the thermodynamic stability is the major concern for the preparation of the multiple emulsions, double emulsions are greatly explored as compared to triple and quadrupole emulsions. Double emulsions are broadly classified into oil-in-water-in-oil (o/w/o) emulsion and water-in-oil-in-water (w/o/w) emulsion. The o/w/o emulsion consists of an oil phase as the internal droplet and the external continuous phase, which are separated by an aqueous phase. Likewise, the w/o/w emulsion consists of internal aqueous droplets and the external water continuous phase, which are separated by an oil phase. It is not a compulsion that the internal droplets and the external continuous phase should remain the same. A schematic representation of a w/o/w and o/w/o double emulsions is shown in Figures 58.6e and f.

In general, double emulsions are prepared by two-step emulsification method. First, a primary emulsion is prepared using two sets of emulsifiers: a hydrophobic emulsifier of low HLB (hydrophilic–lipophilic balance) value to stabilize the w/o interface and a hydrophilic surfactant of a high HLB value for stabilizing oil-in-water (o/w) interface. The primary emulsification is done by applying high shear force (ultrasonification, homogenization). In the secondary emulsification step, the primary emulsion is re-emulsified in either an aqueous phase or an oil phase. The w/o/w double emulsion is formed by emulsifying the w/o emulsions in aqueous solution of a high HLB value surfactant. Similarly, o/w/o double emulsion is formed by emulsifying o/w emulsion in an oil solution of low HLB value surfactant. The secondary emulsification is done by applying a low shear force to prevent breakage of the internal droplets in the external continuous phase. Else, it may result in the formation of a simple emulsion.[67] Hong et al.[68] reported one-step method of preparing w/o/w double emulsions.

The double emulsions have been explored to control the release of the bioactive components.[69] The oil/water phase behaves like a barrier and allows controlling the release pattern. Apart from this, they can provide a protective barrier from the environment (antioxidation)[70] and also can mask the obnoxious taste and flavor of the bioactive components.[71]

58.3.5 Gels

The word "gel" has been derived from the Greek word, gelatus, which means "to immobilize." In early days (1926), Jordon Lloyd defined gel as "colloidal substances that is easier to recognize than to define."[72] Gels are semisolid viscoelastic systems having both solid and liquid components linked together by physical or chemical bonds.[73] The solid components are said to be gelators.[74] The gel is defined as substantially dilute nonglassy thermoreversible system, which results from self-assembly of compounds capable of forming three-dimensional network. The confined bonds amongst the gelator molecules form a supramolecular three dimesional network configuration.[75] The liquid component is immobilized within the self-assembled networked structure.[73] Some gelators (e.g., Lecithin, Spans, or Tweens) form fluid-filled structures, which physically interacts with each other, to form a networked structure.[76] The dispersed phase is a three-dimensional network of solid matrix and holds the liquid continuous medium or dispersion medium.[77] Hence, gels do not have any fluidity.

Immobilization of the liquid constituent inside the networked arrangement of the solid constituent has been credited to the interfacial strain amongst the solid and liquid constituent.[78] The liquid phase may either be polar or apolar solvent. Based on the polarity, the gels can be broadly categorized into two types. If the liquid phase is polar in nature, then the gels can be called as hydrogels, otherwise, organogels.[79]

58.3.5.1 Organogels

Organogels are noncrystalline, semisolid viscoelastic systems, in which the organic liquid phase gets immobilized within the three-dimensional network developed from the self-assembly of the gelator molecules.[80-83] In organogels, the continuous phase is an apolar solvent, and the dispersed phase may either be none or any polar solvent. (Biphasic organogels may be called as emulsion organogels, discussed under separate heading "emulgels."). Organogels are thermodynamically stable and can accommodate higher amount of lipophilic drugs. The organogelators are generally organic molecules. Some of the commonly used organogelators include cholesteryl anthraquinone derivatives,[84] lecithin,[85] sorbitan monostearate,[86,87] and sorbitan monopalmitate,[86] etc. The liquid phase may either be organic solvent (e.g., cyclohexane, butyl acetate, benzene, tetralin, 1,2-dichloroethane or hexane, etc.)[88] or vegetable oils (e.g., groundnut oil, sunflower oil, mustard oil, castor oil, or sesame oil, etc.).[89] Organogels are often called oleogels, if the liquid phase is vegetable oil.[90] Apart from the conventional solid fat crystal network, the molecular self-assembly may be induced by the modified biopolymers and amphiphilic molecules.[91] In the recent years, the use of organogels as matrices for controlled drug delivery is on the rise. This is due to the beneficial effects of organogel, over the conventional drug delivery systems.[92] The major advantages include easy and economical method of preparation and its ability to hold both hydrophilic and hydrophobic drugs.[93] The release of the drug can be tailored by altering the physicochemical properties of the gels.[94] They possess several advantages over hydrogels, one of them being that the organogel-based formulations do not require preservatives. Absence of water makes them resistant to microbial attack.[95] Though various researches have undergone till date, very few commercial organogel formations are available.

Based on the type of linkage, organogels are classified into two categories, viz., physical and chemical organogels. Various physical forces like electrostatic, steric, van der Waals, or hydrogen-bonding interactions are responsible for the architecture of the physical organogels.[96] The chemical organogels are formed by chemical reactions like covalent bonding. Physical organogels is gaining more attention of the scientists compared to the chemical organogels because of the easy method of preparation, thermo-reversible nature, and shear-thinning properties.[97] Due to the thermo-reversible nature, organogels can regain their structure even after thermal shock. The shear-thinning property improves the shelf-life of the organogels and assures easy application.[98] A lower dose of the medicament is required to obtain a specific drug concentration at the site of action. These two properties make organogel a suitable candidate for pharmaceutical, nutraceutical, and cosmetic industries.

The organogels are generally prepared by solid fiber gelation, fluid-filled structure gelation, and apolar phase polymerization mechanisms.[92] Generally, low-molecular weight organogelators form organogels either by solid fiber or fluid-filled structure gelation mechanisms.[99] Polymer matrix–based organogels are formed if the organogelator undergo polymeric reactions.[73] Monophasic organogels are formed either by solid fiber mechanism or by polymer matrix gelation. The detail of the fluid-filled structure gellation is being discussed in detail under emulgel section.

The solid fiber mechanism is based on the dissolution of the organogelator in the apolar solvent at temperature higher than the melting point of the gelator.[100] As the temperature is lowered, the solubility of the gelator decreases and starts precipitating.[101] Initially, it forms nuclei for the formation of the radiating fibers, which on further cooling, forms an interconnecting network. This is due to the molecular self-assembly of the precipitated organogelator, which results in the formation of the three-dimensional network.[102] Figure 58.7a represents a schematic diagram of the mechanism of gel formation by solid fiber growth. Singh et al.[80] reported formation of sorbitan monopalmitate based acicular crystals that showed formation of long fibers. These long fibers self-assembled to form a three-dimensional network structure to induce gelation of castor oil. The confocal micrograph (Figure 58.7a) showed aggregated sorbitan monopalmitate fibers radiating out from the nucleation point, which formed a meshed structure to immobilize castor oil. The interactions amongst the organogelator and the apolar solvent govern the minimum organogelator concentration to induce gelation of the apolar solvents.[80]

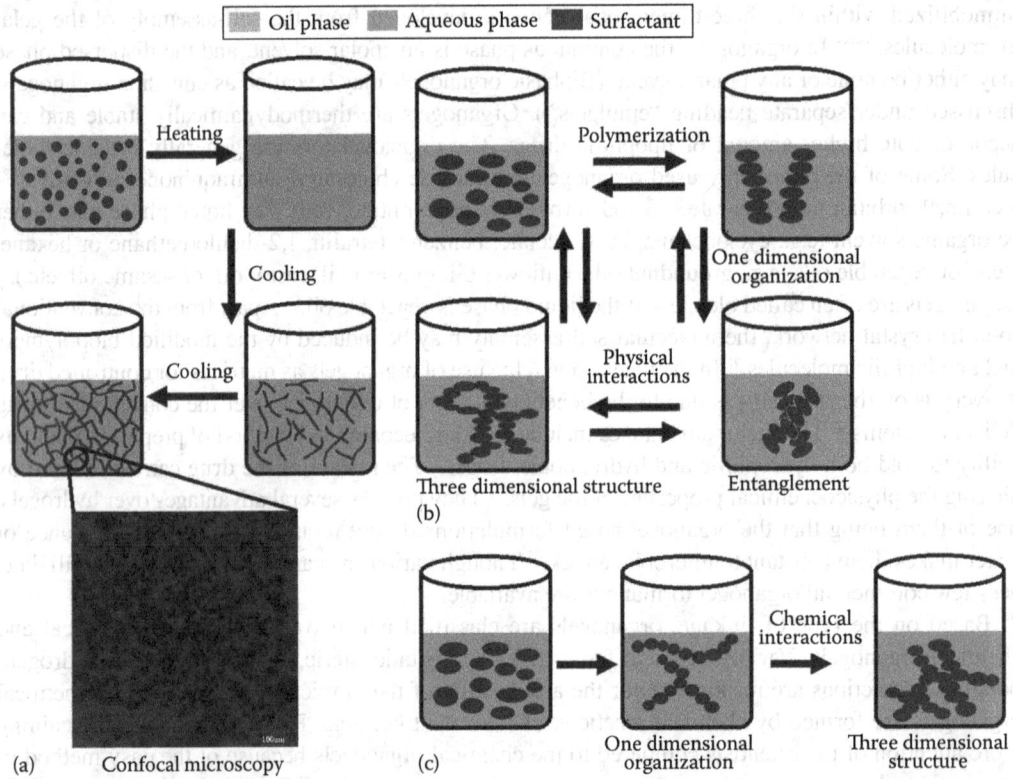

FIGURE 58.7 Schematic representation of the mechanisms of organogel formation. (a) Solid fiber mechanism. (From De Loos, M. et al., *J. Am. Chem. Soc.*, 119, 12675, 1997.) (b) Polymeric organogel formation by supramolecular polymer. (From Tamaoki, N. et al., *Langmuir*, 16(19), 7545, 2000.) (c) By conventional polymer. (From Tamaoki, N. et al., *Langmuir*, 16(19), 7545, 2000.)

Numerous organogels are reported till date using solid fiber-based mechanism. The commonly studied organogelators include sorbitan monopalmitate,[105,106] sorbitan monostearate,[106] amino acids,[107,108] dendrimers,[109,110] fatty acids,[111] and nucleotides.[112] The organogels formed by solid fiber mechanism are robust and stable due to the formation of junction points in the solid network of the organogelator molecules. This impart crystallinity to the organogels, which explains the pseudocrystalline nature of the organogels.[113,114] The molecular self-assembly of the organogelator occurs due to the weak interactions like hydrogen bonding, van der Waals forces, pi–pi stacking, and metal-coordination bonds. Hydrogen bonding is the predominant force involved in the formation of the fibers in nonaqueous solvents.

Many polymers (linear, hyperbranched, or star shaped) form organogels by physical entanglement or chemical crosslinking. The polymer, when dispersed in the apolar phase, undergo supramolecular self-assembly to form three-dimensional network structure. This causes gelation of the apolar phase. Due to this reason, the polymeric organogels are often termed as supramolecular organogels. The organogelator may also undergo polymerization governed by the stereoregulation of the organogelators and subsequent self-assembly. The mechanism of formation of organogel has been shown schematically in Figures 58.7b and c. The organogelation may be a result of physical or chemical interactions. The supramolecular cross-linking agents provide additional strength to the supramolecular organogels. The polymeric organogels add many advantages over other conventional gelation techniques like better gelation ability of the polar liquid, longer stability, and formation of more rigid/stable organogels.

L-Lysine derivatives form supramolecular organogels by hydrogen bonding and van der Waals forces. Poly (ethylene) organogels are prepared by dissolving poly (ethylene) in apolar solvent at 130°C and subsequently cooling to room temperature, which results in the partial precipitation of the polymer leading to the organogel formation.[115] Likewise, self-assembled polymeric organogels are reported for cyclohexane derivatives,[88] diacetylene derivatives,[116] poly (acrylic acid), and methacrylate derivatives.[117] Poly (ethylene), poly (methacrylic acid)-co-methyl methacrylic acid, and poly (glyceryl methacrylic acid) polymers are reported in the field of drug delivery.[118]

58.3.5.2 Emulgels

Emulgels are emulsions (water-in-oil or oil-in-water) that lose their flow due to gelation and formation of gelled structure. They can be named either as emulsion organogel and emulsion hydrogel, based on the type of the emulsion formed. The gels formed from the gelation of water-in-oil emulsions are named as emulsion organogel, and the gels formed from the oil-in-water emulsions are named as emulsion hydrogel. The formation of emulgel may either be due to physical interactions or chemical cross-linking. Since there are no chemical reactions involved in the preparation of the physical emulgels, gel-to-sol transition temperature is observed. Because of this thermoreversible nature, the physical emulgels possess self-healing capacity and are sensitive to various reaction conditions like pH, salt, temperature, rotation speed, and time of rotation. The physical emulgels also show pseudoplastic flow behavior, that is, they are present as solid-like structure at lower shear rates while they behave as fluid at higher shear rates.[119] These properties are of interest to use these emulgels as viscosity modifiers in cosmetic products (e.g., creams and lotions) and in controlled delivery of pharmaceuticals and nutraceuticals. These emulgels are made either from polymers (biopolymers and synthetic) or surfactants.[120] The emulgels are formed mainly by fluid-filled structure mechanism.

Fluid-filled structure mechanism is based on the gelation of the matrix by physical interactions. The physical interactions in the fluid-filled structures are weaker than the solid fibrous structures. This is due to the transient network of gelators in fluid-filled structure matrices.[121] These show momentary and vibrant nature due to the exchange of the gelator molecules amongst the fluid-filled structures and the solvent.[73] Due to the momentary network of the fluid-filled structure, water or polar solvent enters within the matrix.[122] The momentary nature of the gels is due to the dynamic movement of the gelator molecule between the liquid and fibrous network of the gels. A schematic

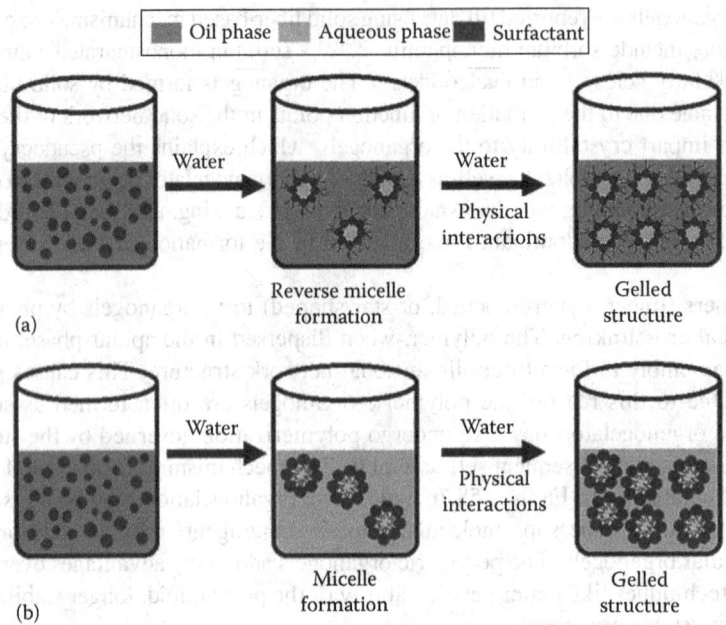

FIGURE 58.8 Schematic representation of the mechanisms of emulgel formation: (a) emulsion organogel by fluid-filled structures and (b) emulsion hydrogel by fluid-filled structures. (From De Loos, M. et al., *J. Am. Chem. Soc.*, 119, 12675, 1997.)

representation of the fluid-filled structure mechanism of emulsion organogel and emulsion hydrogel has been shown in Figures 58.8a and b, respectively.

Solid gelators (e.g., sorbitan monopalmitate and sorbitan monostearate) form both anhydrous and water containing organogels. These gelators are amphiphilic in nature and can form organogels with or without polar solvent. The monophasic organogel formation has been discussed in detail in organogel section. Organogel containing polar solvent is formed by adding the polar solvent dropwise into the hot gelator-solvent mixture under homogenization.[80] Singh et al.[80] reported the formation of emulsion organogel by adding water into the hot mixture of sorbitan monopalmitate and castor oil. Emulsion hydrogels are oil-in-water based hydrogels. They are also formed by similar mechanism. Here, the addition of polar solvent in the hot mixture of gelator and apolar solvent results into the formation of oil droplets dispersed into the water continuum phase as shown in the schematic diagram (Figure 58.8b). Singh et al. (2014) reported the formation of oil-in-water based emulsion hydrogels when water was added in the hot mixture of sorbitan monopalmitate and olive oil (Figure 58.9). Incorporation of water enables to deliver both hydrophilic and lipophilic drugs from the same formulation. When metronidazole was incorporated into sorbitan monopalmitate and castor oil emulsion organogels, the release of the drug improved significantly from 28% (monophasic organogel) to 84% (biphasic organogel) in a time span of 12 h.[80] Similarly, the release of metronidazole showed a significant increase in the release up to 71%–81% from the biphasic emulsion hydrogel.[123] The incorporation of water phase also improves the patient compliance by imparting easy washability of the emulsion gels.

The liquid gelators, Span 80-Tween 80 also forms emulsion organogels and emulsion hydrogels by fluid-filled fiber mechanism. It has been reported that Span 80 (sorbitan monooleate) and Tween 80 (polyoxyethylene sorbitan monooleate) mixed in the ratio of 1:2 w/w forms organogel with better firmness and architecture as compared to the other surfactant mixture ratios.[69,101] When water is added dropwise into the homogenous surfactant mixture and oil, it forms spherical reverse-micellar droplets. These droplets/fibers self-assemble to form three-dimensional architecture to immobilize apolar solvent.[124] Similarly, in case of hydrogels micellar structures are formed, which entraps the external liquid phase to flow and form hydrogel.

Vegetable Oil–Based Formulations for Controlled Drug Delivery

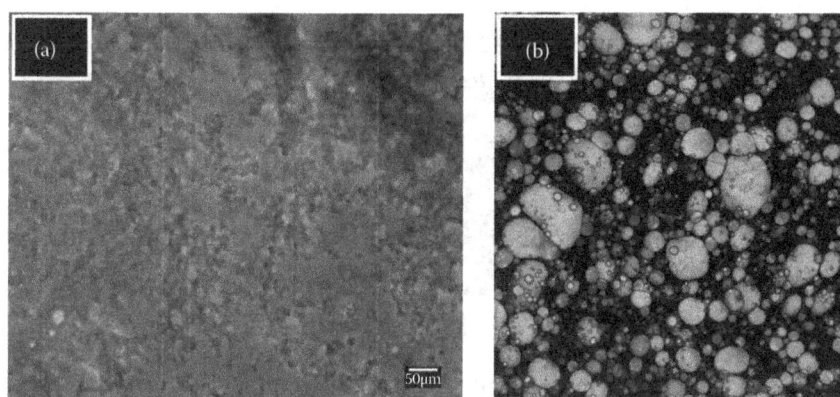

FIGURE 58.9 Fluorescent micrographs showing arrangement of oil in (a) emulsion organogel as the continuous phase and (b) emulsion hydrogel as the droplets.

58.3.5.3 Bigels

The oleogels are preferred formulations for drugs that are sparingly or poorly soluble in water, but the greasy feeling discourages the users for dermatological and cosmetic applications. Scientists are trying to eliminate this shortcoming by converting them into biphasic formulations. These formulations have good acceptability and easy washablilty. At the same time, the beneficial effects of the oil are not compromised. The emulsion gels are stabilized by adding an emulsifier. But on a time scale, the emulgels get destabilized. Phase separation and creaming is the most common phenomena often seen in emulsions after a certain time span. An approach to stabilize the emulsion gels is to immobilize the internal phase also, which came into existence recently.

Bigels are novel semisolid gel formulations in which two gels are mixed together to form a single formulation.[125] The two gels can either be identical or different types of colloidal gels[126,127] or a mixture of a hydrogel and an oleogel.[128,129] Bigels are uniform dispersion of the two gels that appear as a single gel, when inspected visually. A bigel differ from the emulsion (water-in-oil and oil-in-water), creams and emulgels structurally.[130] They possess significantly improved properties, which makes them a potent candidate to be utilized in pharmaceutical or cosmetic products. They do not show demixing of the two components and show good storage stability when kept at room temperature.[126] The stable structural organization is a result of entrapment of the internal liquid phase of emulsion gels within the three-dimensional network. The bigels are electrically conductive in nature due to the presence of pockets of aqueous phase.[128]

The microarchitecture of the bigel is studied by incorporation of fluorescent dyes in the oil and/or aqueous phases. The fluorescent dyes can be identified by their unique colors. Therefore, exact arrangement of the two phases can be easily studied. Schematic representation of the bicontinuous, aqueous phase continuous, and apolar continuous bigels has been shown in Figure 58.10.

As bigels are relatively newer formulations, there is no extensive study that explains some possible drawbacks associated to the system. Bigels may or may not be thermoreversible depending on the composition of the formulations.[125]

58.3.5.4 Lyotropic Gels

Liquid crystals are chemical systems that are prepared by self-assembly of solids and liquids. They are mesophases between the two phases, hence, they possess intermediate properties between those of a liquid and a solid.[133,135] Liquid crystals are majorly divided into two types:[134]

1. Thermotropic liquid crystals
2. Lyotropic liquid crystals

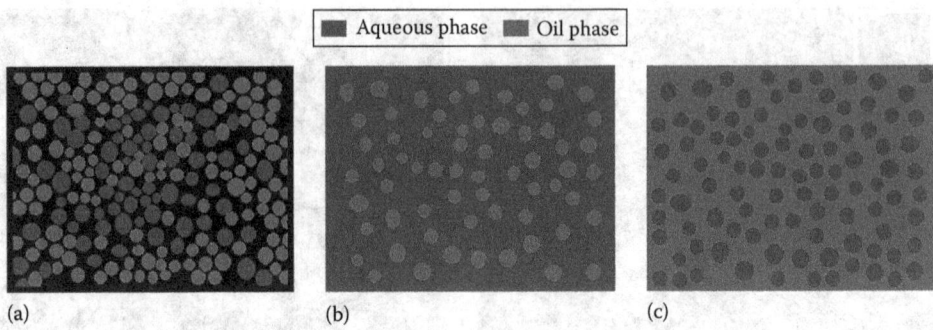

FIGURE 58.10 Schematic diagram representing different types of bigels: (a) both phases are dispersed, (b) oleogel dispersed in hydrogel, and (c) hydrogel dispersed in oleogel. (From From Garti, N. et al., *Food Funct.*, 3(7), 700, 2012; Hamley, I.W., *Soft Matter*, 6(9), 1863, 2010; Negrini, R. and Mezzenga, R., *Langmuir*, 27(9), 5296, 2011; Berni, M. et al., *Adv. Colloid Interface Sci.*, 98(2), 217, 2002.)

Thermotropic liquid crystals show ordering of the crystals based on the temperature of the system, whereas, the lyotropic liquid crystals show the ordering of the crystals based on the addition of the solvents.[136,137] Lyotropic liquid crystals are explored as potential drug delivery vehicles due to their biocompatible, biodegradable, bioadhesive, and nontoxic properties.

Lyotropic liquid crystals consist of at least two components:

1. An amphiphilic molecule
2. A hydrophilic solvent

The polar groups of the amphiphilic molecule form hydrogen bonding with the hydrophilic solvent. The nonpolar groups of the amphiphilic molecule interacts with the aliphatic chain by van der Waals interactions.[138] The amphiphilic molecule generally used in the formation of lyotropic liquid crystals can be long chain alcohols, deoxyribonucleic acid (DNA), gelatin, surfactants, and membrane phospholipid, whereas, the hydrophilic solvent is usually water.[139–143] Due to the presence of amphiphilic molecule, lyotropic liquid crystals are able to deliver hydrophilic, lipophilic, and amphiphilic drugs. The hydrophilic drugs can be accommodated in the aqueous phase, lipophilic drugs can be dissolved in the lipidic content, and the amphiphilic drugs may lie at the interface.[144] The mesophases exist as lamellar, cubic, and hexagonal structures,[136,145,146] The schematic diagrams of the lyotropic crystals are shown in Figure 58.11. The mesophases can be either oil-in-water (normal, type I) or water-in-oil (inverted, type II) type of formulations, which contain oil and water phases dispersed in the water and oil continuous phases, respectively.[147] There are numerous literatures reporting the application of lyotropic liquid crystals

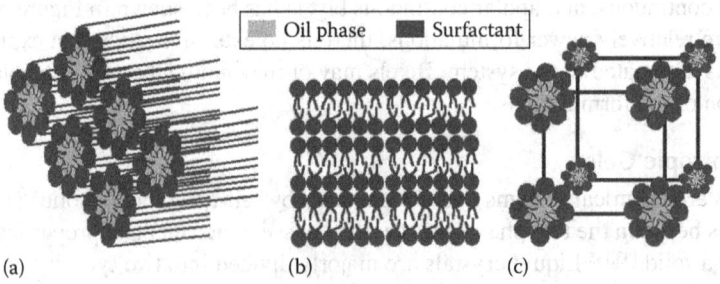

FIGURE 58.11 Schematic diagram of commonly formed mesophasic structures of lyotropic crystals: (a) hexagonal phase, (b) lamellar phase, and (c) cubic phase. (From Chen, W. et al., *J. Control. Release*, 142(1), 40, 2010.)

in drug delivery,[148–150] gene therapy,[151] protein,[152] and vaccine delivery. The hydrophilic drugs are released by diffusion-controlled process and are regulated by the symmetry of the aqueous phase and the mesophase.[153,154] Oral delivery of insulin has been reported from the self-assembled nanocubicles.[155] The lyotropic liquid crystals suffer from the major drawback of very high viscosity, which makes it very hard and renders difficulty during handling.[136,156] Overall, these systems may be considered as effective drug delivery systems with high solubilization capacity and good stability.

58.4 APPLICATIONS OF VEGETABLE OIL-BASED FORMULATIONS

In recent years, vegetable oil-based formulations are explored for their potential application in the delivery of active components. Micelles have been studied extensively for controlled and targeted drug delivery for the treatment of different types of cancer.[157–159] Nederberg et al.[160] reported delivery of paclitaxel from miktoarm copolymers and terpolymers and stereocomplexes micelles. Vrignaud et al.[161] developed reverse micelle-loaded lipid nanocarriers for the sustained release of doxorubicin hydrochloride. Koyamatsu et al.[162] showed pH-responsive release of proteins from reverse polymeric micelles. Lukyanov et al.[163] explored the use of micellar systems as delivery vehicles for poorly soluble drugs. Gaucher et al.[164] investigated block copolymer micelles for drug delivery applications.

Liposomes are reported to improve the solubility and chemical stability of the drugs,[165] active and passive drug targeting,[166] topical drug delivery,[167] cancer treatment,[168] HIV infections treatment,[169] and enhancing antimicrobial efficiency.[170] Park[171] investigated high drug encapsulation in liposomal anthracyclines and higher efficacy in PEGylated liposomal doxorubicin for breast cancer treatment. Fusogenic liposomes were reported for vaccine development.[172] Felnerova et al.[172] developed liposomes for antigens, nucleic acids, and drugs delivery. Pinto-Alphandary et al.[173] investigated liposomes for targeted delivery of antibiotics. Pavelic and Schubert[174] investigated polyacrylate gels as vehicles for liposomes (proliposomes and polyol dilution liposomes) and confirmed their applicability in vaginal drug delivery. Cao et al.[175] reported aptamer-functionalized liposomes for reversible cell-specific drug delivery. Sihorkar and Vyas[176] discussed about application of polysaccharide-based liposomes in drug delivery, targeting, and immunization.

Kakkar et al.[177] enhanced the oral bioavailability of curcumin by loading into solid lipid nanoparticles. Many scientists have reported the use of solid lipid nanoparticles as vehicles for oral,[178] pulmonary,[179] lymphatic,[180] and dermal[181] delivery of bioactive agents. Blasi et al.[182] developed solid lipid nanoparticles for targeted brain drug delivery. Muller et al.[183] and Wissing and Müller[184] reported the use of solid lipid nanoparticles in cosmetic and dermatological preparations. Jenning et al.[185] prepared vitamin A loaded solid lipid nanoparticles for topical drug delivery. Maia et al.[186] investigated solid lipid nanoparticles as drug carriers for topical glucocorticoids. Souto et al.[187] reported the development of a controlled release formulation based on solid lipid nanoparticles for topical delivery of clotrimazole. Shah et al. (2008) developed solid lipid nanoparticles as vehicles for topical delivery of tretinoin.[188] Cavalli et al.[189] used solid lipid nanoparticles for ocular delivery of tobramycin. Wissing and Müller[190] explored solid lipid nanoparticles as carrier for sunscreens.

Microemulsion-based formulations are reported to have beneficial effects in controlled delivery of drugs for oral,[29,191] parenteral,[192,193] pulmonary,[194] and topical[195,196] delivery. Nguyen et al.[197] developed lecithin-based microemulsion using rhamnolipid and sophorolipid biosurfactants. Ustundag Okur et al.[198] reported naproxen loaded microemulsion for transdermal drug delivery.[198] Shah et al.[199] developed rivastigmine-loaded microemulsions for delivery of the drug to brain via nasal delivery. Raza et al.[200] reported enhanced skin permeation and retention using dithranol-containing phospholipid microemulsion. Silicone-based macroemulsions are used as sun screens because of the nonirritating and lubricating properties of silicone.[201] In cell cultures, different combinations of macroemulsions and surfactants are used.[202] Liu et al.[203] prepared uniform-sized multiple emulsions by membrane emulsification for drug delivery applications. Sapei et al.[204] reported w/o/w double emulsions for controlled release in food applications.[204] ElShafei et al.[205] prepared w/o/w double emulsion using essential oils like eucalyptus, linalool, and marjoram for antifungal applications.

Gels are having tremendous and wide range of applications in food industry,[206–209] cosmetics,[208,210] lubricants, bioengineering,[211–216] bioseparations,[217–223] biosensors,[222] agriculture,[216] medicine,[224] nutraceuticals,[225] proteins,[226] and pharmaceutical industry.[227,228] Recently, gels are finding their applications in the preparation of immobilized enzymes,[229] suppositories,[230] probiotics,[231] medical devices like dermal patches[232] and wound dressings.[233] Miyazaki et al.[230] prepared thermally reversible xyloglucan gels as vehicles for rectal drug delivery.

12-Hydroxy stearic acid and soybean oil based organogels were explored as vehicles for oral controlled release of lipophilic as well as hydrophilic drugs.[234] Lupi et al.[235] explored olive oil and policosanol-based organogels for oral delivery of drug and nutraceutical. Sagiri et al.[236] prepared organogels using a mixture of surfactant Span 80-Tween 80 for iontophoretic drug delivery. Bhatia et al.[237] investigated lecithin and poloxamer (phospholipid)-based organogel for topical delivery of tamoxifen. Soy lecithin and palm oil–based organogels were investigated for controlled delivery of metronidazole.[238] Aceclofenac loaded organogels were reported for topical drug delivery for treating osteoarthritis.[239] Shah at al.[106] reported a comparative study on the effect of two different organogelators (sorbitan monopalmitate and sorbitan monostearate) on the physicochemical properties of organogels. Similarly, a comparative study was reported by Satapathy et al.[105] where they studied the effect of change in oil phase (mustard oil and Peanut oil) on physicochemical properties of Span 40 based organogels. Główka et al.[240] reported polymeric nanoparticles-embedded pluronic lecithin organogel for roxithromycin delivery to hair follicles. Singh et al.[80] investigated castor oil and Span 40 based organogel for controlled delivery of metronidazole for topical application.

Akram et al.[241] developed insulin emulgels for transdermal drug delivery. Khullar et al.[242] developed mefenamic acid based emulgel for topical delivery. Rahmani-Neishaboor et al.[243] investigated film-forming topical emulgel dressing for preventing hypertrophic scarring. Shen et al.[244] reported Cyclosporin A emulgel for ocular delivery. Badilli et al.[245] prepared topical emulgel for the treatment of psoriasis formulation containing inclusion complex of calcipotriol with cyclodextrin.

Almeida et al.[128] showed better moisturizing property of bigels compared to oleogel or hydrogel. Ibrahim et al.[129] reported a comparative study of diltiazem hydrochloride loaded organogels, hydrogels, and bigels as matrices for transdermal delivery systems. Singh et al.[133] reported guar gum based bigels for controlled drug delivery. Rehman et al.[246] developed fish oil and natural (sodium alginate) and synthetic (hydroxypropyl methylcellulose) polymer based bigel for transdermal drug delivery. Singh et al.[132] reported carbopol based bigels for topical delivery of metronidazole for the treatment of bacterial vaginosis.

pH-Responsive food-grade lyotropic liquid crystals were developed for controlled delivery of hydrophilic drug (e.g., phloroglucinol).[144] Estracanholli et al.[247] studied the release and skin permeation of celecoxib loaded liquid crystalline cubic phase systems for transdermal delivery. Ki et al.[248] reported injectable liquid crystal system that released euprolide for 1 month. Cohen-Avrahami et al.[249] reduced the problems of poor bioavailability and severe side effects associated with orally administered anti-inflammatory drugs (sodium diclofenac and celecoxib) by loading the drugs in cubic and lamellar liquid crystals mesophases, which can be used as transdermal delivery vehicles. Rizwan et al.[250] developed peptides and proteins containing bicontinuous cubic liquid crystals for sustained drug delivery. There are various commercially available vegetable oil–based products. Table 58.3 summarizes such vegetable oil–based formulations.

TABLE 58.3
Commercially Available Oil-Based Formulations

S.No	Brand Name	Active Substances	Oil-Based Ingredients	Uses	References
1	NEORAL® soft gelatin capsules	Cyclosporine; 25 mg, 100 mg	Corn oil-mono-di-triglycerides, polyoxyl 40 hydrogenated castor oil NF	Preventing the rejection of transplanted organs like kidney, liver, and heart	[42]
2	NEORAL® oral solution	Cyclosporine; 100 mg/mL	Corn oil-mono-di-triglycerides, polyoxyl 40 hydrogenated castor oil NF	Preventing the rejection of transplanted organs like kidney, liver, and heart	[42]
3	Oraqix® periodontal gel	Lidocaine 25 mg/mL and prilocaine 25 mg/mL	Eutectic mixture of lidocaine and prilocaine	Local anesthetic applied to prevent numbing of the gums during dental procedures	[251]
4	Intralipid® 10% and 20%	10% and 20% I.V. fat emulsion (10% soybean oil)	Egg yolk phospholipids	Energy source	[252]
5	Clinolipid®	Lipid injectable emulsion (olive oil NF and soybean oil USP)	Egg phospholipids NF	Source of calories and essential fatty acids for parenteral nutrition	[253]
6	KABIVEN®	Amino acids, electrolytes, dextrose and lipid Injectable emulsion	Intralipid® 20% (10% soybean oil)	Parenteral nutrition	[254]
7	DUREZOL®	Difluprednate ophthalmic emulsion	Castor oil	Treating pain and swelling following eye surgery	[255]
8	Estrasorb®	Estradiol topical emulsion	Soybean oil	Treating moderate to severe vasomotor symptoms associated with the menopause	[256]
9	BAL® in Oil ampules	Dimercarprol injection, USP	Peanut oil	Chelating agent used to treat poisoning with arsenic, gold, or mercury	[257]
10	ETHIODOL®	Ethiodized oil injection	37% Iodine in ethyl esters of the fatty acids of poppyseed oil 1%.	Radio-opaque contrast agent used to outline structures in radiological investigations	[258]
11	DermOtic® oil	Fluocinolone acetonide oil ear drops	Isopropyl myristate, light mineral oil, oleth-2, refined peanut oil NF	Treatment of chronic eczematous external otitis	[259]
12	Derma-Smoothe/FS®	Fluocinolone acetonide topical Oil, 0.01%	Isopropyl myristate, light mineral oil, oleth-2, refined peanut oil NF	Treatment of chronic eczematous external otitis	[259]
13	DEPOCYT®	Cytarabine liposome injection	Cholesterol, triolein, dioleoylphosphatidylcholine and dipalmitoylphosphatidylglycerol	Treating lymphomatous meningitis	[260]

(*Continued*)

TABLE 58.3 (Continued)
Commercially Available Oil-Based Formulations

S.No	Brand Name	Active Substances	Oil-Based Ingredients	Uses	References
14	DepoDur®	Morphine sulfate extended-release liposome injection	Cholesterol, dioleoylphosphatidylcholine, dipalmitoylphosphatidylglycerol, tricaprylin and triolein	Control of pain after major surgery	[260]
15	DOXIL®	Doxorubicin HCl liposome injection (intravenous infusion)	MPEG-DSPE, fully hydrogenated soy phosphatidylcholine, cholesterol	Treatment of AIDS-related Kaposi's sarcoma	[260,261]
16	EXPAREL®	Bupivacaine liposome injectable suspension	Cholesterol, dipalmitoylphosphatidylglycerol, tricaprylin, dierucoylphosphatidylcholine	Management of postsurgical pain	[262]
17	Marqibo®	Vincristine sulfate liposome Injection	Cholesterol	Treating Philadelphia chromosome negative acute lymphoblastic leukemia	[263]
18	AmBisome®	Amphotericin B liposome for injection	Hydrogenated soy phosphatidylcholine, cholesterol, distearoylphosphatidylglycerol,	Treatment of visceral leishmaniasis	[264]
19	ERAXIS®	Anidulafungin for injection	Semisynthetic lipopeptide synthesized from a fermentation product of *Aspergillus nidulans*	Treating Candida infections in the blood, stomach or esophagus	[265]
20	CANCIDAS®	Caspofungin acetate for injection	Semisynthetic lipopeptide (echinocandin) compound synthesized from a fermentation product of Glarea lozoyensis.	Antifungal medicine used to treat serious fungal infections in adults and children	[266]

58.5 SCOPE FOR NEW FORMULATIONS

Apart from the already established formulations, researchers are trying to develop novel oil-based formulations to combat the poor solubility and bioavailablity of NCE. Shevachman et al.[251] developed novel U-type microemulsions to improve the percutaneous permeability of diclofenac.[251] Shah et al.[252] used microwave heating for the preparation of solid lipid nanoparticles by microemulsion techniques, which resulted in improved particle characteristics. Ki et al.[248] reported sustained-release liquid crystal of injectable leuprolide using sorbitan monooleate. Recently, various novel oil-based drug delivery technologies are reported, which includes tocol emulsions,[253] solid lipid nanoparticles,[254] nanosuspensions,[255] lipid microbubbles,[256] sterically stabilized phospholipid micelles,[257] and environmentally responsive drug delivery systems for parenteral administration.[258,259]

ACKNOWLEDGMENT

The authors would like to acknowledge the financial support received from Science and Engineering Research Board (SERB) (SR/FT/LS-171/2009), sanctioned by the government of India.

REFERENCES

1. Kreilgaard M. Influence of microemulsions on cutaneous drug delivery. *Advanced Drug Delivery Reviews*. 2002;54(Supplement 1):S77–S98.
2. Jannin V, Musakhanian J, Marchaud D. Approaches for the development of solid and semi-solid lipid-based formulations. *Advanced Drug Delivery Reviews*. 2008;60(6):734–746.
3. Chen M-L. Lipid excipients and delivery systems for pharmaceutical development: A regulatory perspective. *Advanced Drug Delivery Reviews*. 2008;60(6):768–777.
4. Constantinides PP, Chaubal MV, Shorr R. Advances in lipid nanodispersions for parenteral drug delivery and targeting. *Advanced Drug Delivery Reviews*. 2008;60(6):757–767.
5. Hauss DJ. Oral lipid-based formulations. *Advanced Drug Delivery Reviews*. 2007;59(7):667–676.
6. Wasan KM. Formulation and physiological and biopharmaceutical issues in the development of oral lipid-based drug delivery systems. *Drug Development and Industrial Pharmacy*. 2001;27(4):267–276.
7. Attwood D, Currie L, Elworthy P. Studies of solubilized micellar solutions. I. Phase studies and particle size analysis of solutions formed with nonionic surfactants. *Journal of Colloid and Interface Science*. 1974;46(2):249–254.
8. Neslihan Gursoy R, Benita S. Self-emulsifying drug delivery systems (SEDDS) for improved oral delivery of lipophilic drugs. *Biomedicine and Pharmacotherapy*. 2004;58(3):173–182.
9. Charman WN, Rogge MC, Boddy AW, Berger BM. Effect of food and a monoglyceride emulsion formulation on danazol bioavailability. *The Journal of Clinical Pharmacology*. 1993;33(4):381–386.
10. Hörter D, Dressman JB. Influence of physicochemical properties on dissolution of drugs in the gastrointestinal tract. *Advanced Drug Delivery Reviews*. 2001;46(1–3):75–87.
11. Rane SS, Anderson BD. What determines drug solubility in lipid vehicles: Is it predictable? *Advanced Drug Delivery Reviews*. 2008;60(6):638–656.
12. O'Driscoll CM. Lipid-based formulations for intestinal lymphatic delivery. *European Journal of Pharmaceutical Sciences*. 2002;15(5):405–415.
13. Porter CJ, Trevaskis NL, Charman WN. Lipids and lipid-based formulations: optimizing the oral delivery of lipophilic drugs. *Nature Reviews Drug Discovery*. 2007;6(3):231–248.
14. Kogan A, Garti N. Microemulsions as transdermal drug delivery vehicles. *Advances in Colloid and Interface Science*. 2006;123:369–385.
15. Hua L, Weisan P, Jiayu L, Ying Z. Preparation, evaluation, and NMR characterization of vinpocetine microemulsion for transdermal delivery. *Drug Development and Industrial Pharmacy*. 2004;30(6):657–666.
16. Desai KGH. Enhanced skin permeation of rofecoxib using topical microemulsion gel. *Drug Development Research*. 2004;63(1):33–40.
17. Strickley RG. Solubilizing excipients in oral and injectable formulations. *Pharmaceutical Research*. 2004;21(2):201–230.
18. Cao Y, Marra M, Anderson BD. Predictive relationships for the effects of triglyceride ester concentration and water uptake on solubility and partitioning of small molecules into lipid vehicles. *Journal of Pharmaceutical Sciences*. 2004;93(11):2768–2779.
19. Djordjevic L, Primorac M, Stupar M, Krajisnik D. Characterization of caprylocaproyl macrogolglycerides based microemulsion drug delivery vehicles for an amphiphilic drug. *International Journal of Pharmaceutics*. 2004;271(1):11–19.
20. Reitz C, Strachan C, Kleinebudde P. Solid lipid extrudates as sustained-release matrices: The effect of surface structure on drug release properties. *European Journal of Pharmaceutical Sciences*. 2008;35(4):335–343.
21. Zhang Y-E, Schwartz JB. Effect of diluents on tablet integrity and controlled drug release. *Drug Development and Industrial Pharmacy*. 2000;26(7):761–765.
22. Jannin V, Pochard E, Chambin O. Influence of poloxamers on the dissolution performance and stability of controlled-release formulations containing Precirol® ATO 5. *International Journal of Pharmaceutics*. 2006;309(1):6–15.
23. Attwood D, Florence A. *Surfactant Systems: Their Chemistry, Pharmacy and Biology*. London, U.K.: Chapman & Hall, 1983.
24. Pouton CW, Porter CJ. Formulation of lipid-based delivery systems for oral administration: Materials, methods and strategies. *Advanced Drug Delivery Reviews*. 2008;60(6):625–637.
25. Pouton CW. Self-emulsifying drug delivery systems: Assessment of the efficiency of emulsification. *International Journal of Pharmaceutics*. 1985;27(2):335–348.

26. Gershanik T, Benita S. Self-dispersing lipid formulations for improving oral absorption of lipophilic drugs. *European Journal of Pharmaceutics and Biopharmaceutics*. 2000;50(1):179–188.
27. Bummer PM. Physical chemical considerations of lipid-based oral drug delivery—Solid lipid nanoparticles. *Critical Reviews™ in Therapeutic Drug Carrier Systems*. 2004;21(1):10–20.
28. Nielsen FS, Gibault E, Ljusberg-Wahren H, Arleth L, Pedersen JS, Müllertz A. Characterization of prototype self-nanoemulsifying formulations of lipophilic compounds. *Journal of Pharmaceutical Sciences*. 2007;96(4):876–892.
29. Constantinides PP. Lipid microemulsions for improving drug dissolution and oral absorption: Physical and biopharmaceutical aspects. *Pharmaceutical Research*. 1995;12(11):1561–1572.
30. Alexander A, Khichariya A, Gupta S, Patel RJ, Giri TK, Tripathi DK. Recent expansions in an emergent novel drug delivery technology: Emulgel. *Journal of Controlled Release*. 2013;171(2):122–132.
31. Skanji R, Andrieux K, Lalanne M, Caron J, Bourgaux C, Degrouard J et al. A new nanomedicine based on didanosine glycerolipidic prodrug enhances the long term accumulation of drug in a HIV sanctuary. *International Journal of Pharmaceutics*. 2011;414(1):285–297.
32. Rösler A, Vandermeulen GW, Klok H-A. Advanced drug delivery devices via self-assembly of amphiphilic block copolymers. *Advanced Drug Delivery Reviews*. 2012;64:270–279.
33. Letchford K, Burt H. A review of the formation and classification of amphiphilic block copolymer nanoparticulate structures: Micelles, nanospheres, nanocapsules and polymersomes. *European Journal of Pharmaceutics and Biopharmaceutics*. 2007;65(3):259–269.
34. Kang L, Cheong HH, Chan SY, Lim PFC. Applications of Ssmall-molecule gels—Drug delivery. In *Soft Fibrillar Materials: Fabrication and Applications*, eds. Liu XY, Jing-Liang Li. 2013; Wiley, ISBN: 9783527331628.
35. Martiel I, Sagalowicz L, Mezzenga R. Viscoelasticity and interface bending properties of lecithin reverse wormlike micelles studied by diffusive wave spectroscopy in hydrophobic environment. *Langmuir*. 2014;30(35):10751–10759.
36. Chime SA, Attama AA, Builders PF, Onunkwo GC. Sustained-release diclofenac potassium-loaded solid lipid microparticle based on solidified reverse micellar solution: In vitro and in vivo evaluation. *Journal of Microencapsulation*. 2012;30(4):335–345.
37. Lemons BG, Richens DT, Anderson A, Sedgwick M, Crans DC, Johnson MD. Stabilization of a vanadium (v)–catechol complex by compartmentalization and reduced solvation inside reverse micelles. *New Journal of Chemistry*. 2013;37(1):75–81.
38. Nucci NV, Pometun MS, Wand AJ. Site-resolved measurement of water-protein interactions by solution NMR. *Nature Structural and Molecular Biology*. 2011;18(2):245–249.
39. Ansel HC, Popovich NG, Allen LV. *Pharmaceutical Dosage Forms and Drug delivery systems*. Lea & Febiger, Philadelphia, PaA, 1990.
40. Samad A, Sultana Y, Aqil M. Liposomal drug delivery systems: An update review. *Current Drug Delivery*. 2007;4(4):297–305.
41. Allen TM, Cullis PR. Liposomal drug delivery systems: From concept to clinical applications. *Advanced Drug Delivery Reviews*. 2013;65(1):36–48.
42. Müller-Goymann C. Physicochemical characterization of colloidal drug delivery systems such as reverse micelles, vesicles, liquid crystals and nanoparticles for topical administration. *European Journal of Pharmaceutics and Biopharmaceutics*. 2004;58(2):343–356.
43. dos Santos Guimarães I, Daltoé RD, Herlinger AL, Madeira KP, Ladislau T, Valadão IC et al. Cancer treatment: Conventional and innovative approaches. *Conventional Cancer Treatment*. 2013; ISBN 978-953-51-1098-9.
44. Schubert M, Müller-Goymann C. Solvent injection as a new approach for manufacturing lipid nanoparticles–evaluation of the method and process parameters. *European Journal of Pharmaceutics and Biopharmaceutics*. 2003;55(1):125–131.
45. Immordino ML, Dosio F, Cattel L. Stealth liposomes: Review of the basic science, rationale, and clinical applications, existing and potential. *International Journal of Nanomedicine*. 2006;1(3):297.
46. Wissing S, Kayser O, Müller R. Solid lipid nanoparticles for parenteral drug delivery. *Advanced Drug Delivery Reviews*. 2004;56(9):1257–1272.
47. MuÈller RH, MaÈder K, Gohla S. Solid lipid nanoparticles (SLN) for controlled drug delivery—A review of the state of the art. *European Journal of Pharmaceutics and Biopharmaceutics*. 2000;50(1):161–177.
48. zur Mühlen A, Schwarz C, Mehnert W. Solid lipid nanoparticles (SLN) for controlled drug delivery–drug release and release mechanism. *European Journal of Pharmaceutics and Biopharmaceutics*. 1998;45(2):149–155.

49. Cavalli R, Morel S, Gasco M, Chetoni P. Preparation and evaluation in vitro of colloidal liposheres containing pilocarpine as ion pair. *International Journal of Pharmaceutics*. 1995;117(2):243–246.
50. Mehnert W, Mäder K. Solid lipid nanoparticles: Production, characterization and applications. *Advanced Drug Delivery Reviews*. 2012;64(Supplement):83–101.
51. Silva A, González-Mira E, García M, Egea M, Fonseca J, Silva R et al. Preparation, characterization and biocompatibility studies on risperidone-loaded solid lipid nanoparticles (SLN): High pressure homogenization versus ultrasound. *Colloids and Surfaces B: Biointerfaces*. 2011;86(1):158–165.
52. Bondì ML, Craparo EF. Solid lipid nanoparticles for applications in gene therapy: A review of the state of the art. *Expert Opinion on Drug Delivery*. 2010;7(1):7–18.
53. Liu D, Jiang S, Shen H, Qin S, Liu J, Zhang Q et al. Diclofenac sodium-loaded solid lipid nanoparticles prepared by emulsion/solvent evaporation method. *Journal of Nanoparticle Research*. 2011;13(6):2375–2386.
54. Williams JM. High internal phase water-in-oil emulsions: Influence of surfactants and cosurfactants on emulsion stability and foam quality. *Langmuir*. 1991;7(7):1370–1377.
55. Zhang K, Wu W, Meng H, Guo K, Chen J-F. Pickering emulsion polymerization: Preparation of polystyrene/nano-SiO_2 composite microspheres with core-shell structure. *Powder Technology*. 2009;190(3):393–400.
56. Chevalier Y, Bolzinger M-A. Emulsions stabilized with solid nanoparticles: Pickering emulsions. *Colloids and Surfaces A: Physicochemical and Engineering Aspects*. 2013;439(0):23–34.
57. Tsuji S, Kawaguchi H. Thermosensitive Pickering emulsion stabilized by poly (N-isopropylacrylamide)-carrying particles. *Langmuir*. 2008;24(7):3300–3305.
58. Frelichowska J, Bolzinger M-A, Pelletier J, Valour J-P, Chevalier Y. Topical delivery of lipophilic drugs from o/w Pickering emulsions. *International Journal of Pharmaceutics*. 2009;371(1–2):56–63.
59. Lorenceau E, Utada AS, Link DR, Cristobal G, Joanicot M, Weitz DA. Generation of polymerosomes from double-emulsions. *Langmuir*. 2005;21(20):9183–9186.
60. Valenta C, Auner BG. The use of polymers for dermal and transdermal delivery. *European Journal of Pharmaceutics and Biopharmaceutics*. 2004;58(2):279–289.
61. Damaschke N, Wedd M, Whybrew A, Blondel D. Particle sizing. *Handbook of laserTechnology and Applications*, ed. W. Jones, *Institute of Physics*. 2004;3:1931–1950.
62. McClements DJ. Nanoemulsions versus microemulsions: Terminology, differences, and similarities. *Soft Matter*. 2012;8(6):1719–1729.
63. Helgeson ME, Moran SE, An HZ, Doyle PS. Mesoporous organohydrogels from thermogelling photo-crosslinkable nanoemulsions. *Nature Materials*. 2012;11(4):344–352.
64. Brugarolas T, Tu F, Lee D. Directed assembly of particles using microfluidic droplets and bubbles. *Soft Matter*. 2013;9(38):9046–9058.
65. He C-X, He Z-G, Gao J-Q. Microemulsions as drug delivery systems to improve the solubility and the bioavailability of poorly water-soluble drugs. *Expert Opinion on Drug Delivery*. 2010;7(4):445–460.
66. Choi CH, Weitz DA, Lee CS. One step formation of controllable complex emulsions: From functional particles to simultaneous encapsulation of hydrophilic and hydrophobic agents into desired position. *Advanced Materials*. 2013;25(18):2536–2541.
67. Garti N, Bisperink C. Double emulsions: Progress and applications. *Current Opinion in Colloid and Interface Science*. 1998;3(6):657–667.
68. Hong L, Sun G, Cai J, Ngai T. One-step formation of w/o/w multiple emulsions stabilized by single amphiphilic block copolymers. *Langmuir*. 2012;28(5):2332–2336.
69. Sagiri SS, Behera B, Sudheep T, Pal K. Effect of composition on the properties of tween-80–span-80-based organogels. *Designed Monomers and Polymers*. 2012;15(3):253–273.
70. Poyato C, Navarro-Blasco I, Calvo MI, Cavero RY, Astiasarán I, Ansorena D. Oxidative stability of O/W and W/O/W emulsions: Effect of lipid composition and antioxidant polarity. *Food Research International*. 2013;51(1):132–140.
71. Patil V, Tambe V, Pathare B, Dhole S. Modern taste concealing techniques in pharmaceuticals: A Review. *World Journal of Pharmacy and Pharmaceutical Sciences*, 2014;3(8):293–316.
72. Lloyd DJ. The problem of gel structure. *Colloid Chemistry*. 1926;1:767–782.
73. Vintiloiu A, Leroux J-C. Organogels and their use in drug delivery—A review. *Journal of Controlled Release*. 2008;125(3):179–192.
74. George M, Weiss RG. Low molecular-mass organic gelators. *Molecular Gels*. 2006:449–551.
75. Markovic N, Dutta NK, Williams DRG, Matisons J, eds. *Hydrocarbon Gels*: Rheological Investigation of Structure. Polymer Gels: Fundamentals and Applications, ACS Publications, 2003;833:190–204.

76. Zoumpanioti M, Stamatis H, Xenakis A. Microemulsion-based organogels as matrices for lipase immobilization. *Biotechnology Advances*. 2010;28(3):395–406.
77. Rogovina LZ, Slonimskii GL. Formation, structure, and properties of polymer gels. *Russian Chemical Reviews*. 1974;43(6):503–523.
78. Landry CJT, Coltrain BK, Brady BK. In situ polymerization of tetraethoxysilane in poly (methyl methacrylate): Morphology and dynamic mechanical properties. *Polymer*. 1992;33(7):1486–1495.
79. Placin F, Colomès M, Desvergne JP. A new example of small molecular non-hydrogen bonding gelators for organic solvents. *Tetrahedron Letters*. 1997;38(15):2665–2668.
80. Singh VK, Pal K, Pradhan DK, Pramanik K. Castor oil and sorbitan monopalmitate based organogel as a probable matrix for controlled drug delivery. *Journal of Applied Polymer Science*. 2013;130(3):1503–1515.
81. Sahoo S, Kumar N, Bhattacharya C, Sagiri S, Jain K, Pal K et al. Organogels: Properties and applications in drug delivery. *Designed Monomers and Polymers*. 2010;14(2):95–108.
82. Schurtenberger P, Scartazzini R, Luisi P. Viscoelastic properties of polymerlike reverse micelles. *Rheologica Acta*. 1989;28(5):372–381.
83. Bastiat G, Plourde F, Motulsky A, Furtos A, Dumont Y, Quirion R et al. Tyrosine-based rivastigmine-loaded organogels in the treatment of Alzheimer's disease. *Biomaterials*. 2010;31(23):6031–6038.
84. Ostuni E, Kamaras P, Weiss RG. Novel X-ray Method for in situ determination of gelator strand structure: Polymorphism of cholesteryl anthraquinone-2-carboxylate. *Angewandte Chemie International Edition in English*. 1996;35(12):1324–1326.
85. Kumar R, Katare OP. Lecithin organogels as a potential phospholipid-structured system for topical drug delivery: A review. *AAPS PharmSciTech*. 2005;6(2):298–310.
86. Pisal S, Shelke V, Mahadik K, Kadam S. Effect of organogel components on in vitro nasal delivery of propranolol hydrochloride. *AAPS PharmSciTech*. 2004;5(4):92–100.
87. Murdan S, Andrýsek T, Son D. Novel gels and their dispersions—Oral drug delivery systems for ciclosporin. *International Journal of Pharmaceutics*. 2005;300(1):113–124.
88. De Loos M, van Esch J, Stokroos I, Kellogg RM, Feringa BL. Remarkable stabilization of self-assembled organogels by polymerization. *Journal of American Chemical Society*. 1997;119:12675–12676.
89. Bot A, den Adel R, Roijers EC. Fibrils of γ-oryzanol+ β-sitosterol in edible oil organogels. *Journal of the American Oil Chemists' Society*. 2008;85(12):1127–1134.
90. Patel AR, Schatteman D, De Vos WH, Lesaffer A, Dewettinck K. Preparation and rheological characterization of shellac oleogels and oleogel-based emulsions. *Journal of Colloid and Interface Science*. 2013;411:114–121.
91. Pernetti M, van Malssen KF, Flöter E, Bot A. Structuring of edible oils by alternatives to crystalline fat. *Current Opinion in Colloid and Interface Science*. 2007;12(4–5):221–231.
92. Sagiri S, Behera B, Rafanan R, Bhattacharya C, Pal K, Banerjee I et al. Organogels as matrices for controlled drug delivery: A review on the current state. *Soft Materials*. 2013;12(1):47–72.
93. Satapathy D, Biswas D, Behera B, Sagiri S, Pal K, Pramanik K. Sunflower-oil-based lecithin organogels as matrices for controlled drug delivery. *Journal of Applied Polymer Science*. 2012;129(2):585–594.
94. Pal K, Singh VK, Anis A, Thakur G, Bhattacharya MK. Hydrogel-based controlled release formulations: Designing considerations, characterization techniques and applications. *Polymer-Plastics Technology and Engineering*. 2013;52(14):1391–1422.
95. Shah DK, Sagiri SS, Behera B, Pal K, Pramanik K. Development of olive oil based organogels using sorbitan monopalmitate and sorbitan monostearate: A comparative study. *Journal of Applied Polymer Science*. 2012;129(2):793–805.
96. Singh VK, Ramesh S, Pal K, Anis A, Pradhan DK, Pramanik K. Olive oil based novel thermo-reversible emulsion hydrogels for controlled delivery applications. *Journal of Materials Science Materials in Medicine*. 2013.
97. Han L-J, Li L, Zhao L, Li B, Liu G-Q, Liu X-Q et al. Rheological properties of organogels developed by sitosterol and lecithin. *Food Research International*. 2013;53(1):42–48.
98. Douaire M, Norton I. Designer colloids in structured food for the future. *Journal of the Science of food and Agriculture*. 2013;93(13):3147–3154.
99. Sahoo S, Kumar N, Bhattacharya C, Sagiri S, Jain K, Pal K et al. Organogels: Properties and applications in drug delivery. *Designed Monomers and Polymers*. 2011;14(2):95–108.
100. George M, Weiss RG. Chemically reversible organogels via "latent" gelators. Aliphatic amines with carbon dioxide and their ammonium carbamates. *Langmuir*. 2002;18(19):7124–7135.
101. Pradhan S, Sagiri SS, Singh VK, Pal K, Ray SS, Pradhan DK. Palm oil-based organogels and microemulsions for delivery of antimicrobial drugs. *Journal of Applied Polymer Science*. 2013;131(6):39979.

102. Yan N, Xu Z, Diehn KK, Raghavan SR, Fang Y, Weiss RG. Pyrenyl-linker-glucono gelators. Correlations of gel properties with gelator structures and characterization of solvent effects. *Langmuir.* 2013;29(2):793–805.
103. Sagiri S, Behera B, Rafanan R, Bhattacharya C, Pal K, Banerjee I et al. Organogels as matrices for controlled drug delivery: A review on the current state. *Soft Materials.* 2014;12(1):47–72.
104. Suzuki M, Hanabusa K. Polymer organogelators that make supramolecular organogels through physical cross-linking and self-assembly. *Chemical Society Reviews.* 2010;39(2):455–463.
105. Satapathy D, Sagiri S, Pal K, Pramanik K. Development of mustard oil-and groundnut oil-based span 40 organogels as matrices for controlled drug delivery. *Designed Monomers and Polymers.* 2014;17(6):545–556.
106. Shah DK, Sagiri SS, Behera B, Pal K, Pramanik K. Development of olive oil based organogels using sorbitan monopalmitate and sorbitan monostearate: A comparative study. *Journal of Applied Polymer Science.* 2013;129(2):793–805.
107. Brosse N, Barth D, Jamart-Grégoire B. A family of strong low-molecular-weight organogelators based on aminoacid derivatives. *Tetrahedron Letters.* 2004;45(52):9521–9524.
108. Terech P, Weiss RG. Low molecular mass gelators of organic liquids and the properties of their gels. *Chemical Reviews.* 1997;97(8):3133–3160.
109. Palui G, Simon F-X, Schmutz M, Mesini PJ, Banerjee A. Organogelators from self-assembling peptide based dendrimers: Structural and morphological features. *Tetrahedron.* 2008;64(1):175–185.
110. Ji Y, Luo YF, Jia XR, Chen EQ, Huang Y, Ye C et al. A dendron based on natural amino acids: Synthesis and behavior as an organogelator and lyotropic liquid crystal. *Angewandte Chemie.* 2005;117(37):6179–6183.
111. Pal A, Ghosh YK, Bhattacharya S. Molecular mechanism of physical gelation of hydrocarbons by fatty acid amides of natural amino acids. *Tetrahedron.* 2007;63(31):7334–7348.
112. Rosemeyer H, Stürenberg E-M, Herdewijn P. Nucleolipids as potential organogelators. *Nucleosides, Nucleotides, and Nucleic Acids.* 2007;26(8–9):995–999.
113. Takeno H, Maehara A, Yamaguchi D, Koizumi S. A structural study of an organogel investigated by small-angle neutron scattering and synchrotron small-angle X-ray scattering. *The Journal of Physical Chemistry B.* 2012;116(26):7739–7745.
114. Tung S-H, Huang Y-E, Raghavan SR. Self-assembled organogels obtained by adding minute concentrations of a bile salt to AOT reverse micelles. *Soft Matter.* 2008;4(5):1086–1093.
115. Bajaj A, Gupta S, Chatterjee A. Plastibase: A new base for patch testing of metal antigens. *International Journal of Dermatology.* 1990;29(1):73.
116. Tamaoki N, Shimada S, Okada Y, Belaissaoui A, Kruk G, Yase K et al. Polymerization of a diacetylene dicholesteryl ester having two urethanes in organic gel states. *Langmuir.* 2000;16(19):7545–7547.
117. Beginn U, Sheiko S, Möller M. Self-organization of 3, 4, 5-tris (octyloxy) benzamide in solution and embedding of the aggregates into methacrylate resins. *Macromolecular Chemistry and Physics.* 2000;201(10):1008–1015.
118. Jones M-C, Tewari P, Blei C, Hales K, Pochan DJ, Leroux J-C. Self-assembled nanocages for hydrophilic guest molecules. *Journal of the American Chemical Society.* 2006;128(45):14599–14605.
119. Hanabusa K, Hirata T, Inoue D, Kimura M, Shirai H. Formation of physical hydrogels with terpyridine-containing carboxylic acids. *Colloids and Surfaces A: Physicochemical and Engineering Aspects.* 2000;169(1):307–316.
120. Peppas N, Bures P, Leobandung W, Ichikawa H. Hydrogels in pharmaceutical formulations. *European Journal of Pharmaceutics and Biopharmaceutics.* 2000;50(1):27–46.
121. Terech P. Fibers and wires in organogels from low-mass compounds: Typical structural and rheological properties. *Berichte der Bunsengesellschaft für physikalische Chemie.* 1998;102(11):1630–1643.
122. Shchipunov YA, Hoffmann H. Thinning and thickening effects induced by shearing in lecithin solutions of polymer-like micelles. *Rheologica Acta.* 2000;39(6):542–553.
123. Singh VK, Anis A, Al-Zahrani S, Pradhan DK, Pal K. FTIR, electrochemical impedance and iontophoretic delivery analysis of guar gum and sesame oil based bigels. *International Journal of Electrochemical Science.* 2014;9:5640–5650.
124. Behera B, Sagiri SS, Pal K, Srivastava A. Modulating the physical properties of sunflower oil and sorbitan monopalmitate-based organogels. *Journal of Applied Polymer Science.* 2013;127(6):4910–4917.
125. Rehman K, Zulfakar MH. Recent advances in gel technologies for topical and transdermal drug delivery. *Drug Development and Industrial Pharmacy.* 2013;40(4):433–440.
126. Di Michele L, Varrato F, Fiocco D, Sastry S, Eiser E, Foffi G. Aggregation dynamics, structure, and mechanical properties of bigels. *Soft Matter.* 2014;10(20):3633–3648.

127. Varrato F, Di Michele L, Belushkin M, Dorsaz N, Nathan SH, Eiser E et al. Arrested demixing opens route to bigels. *Proceedings of the National Academy of Sciences.* 2012;109(47):19155–19160.
128. Almeida I, Fernandes A, Fernandes L, Pena Ferreira M, Costa P, Bahia M. Moisturizing effect of oleogel/hydrogel mixtures. *Pharmaceutical Development and Technology.* 2008;13(6):487–494.
129. Ibrahim MM, Hafez SA, Mahdy MM. Organogels, hydrogels and bigels as transdermal delivery systems for diltiazem hydrochloride. *Asian Journal of Pharmaceutical Sciences.* 2013;8(1):48–57.
130. Rhee GJ, Woo JS, Hwang S-J, Lee YW, Lee CH. Topical oleo-hydrogel preparation of ketoprofen with enhanced skin permeability. *Drug Development and Industrial Pharmacy.* 1999;25(6):717–726.
131. Di Michele L, Varrato F, Kotar J, Nathan SH, Foffi G, Eiser E. Multistep kinetic self-assembly of DNA-coated colloids. *Nature Communications.* 2013;4:1–7.
132. Pradhan S, Sagiri SS, Singh VK, Pal K, Ray SS, Pradhan DK. Palm oil-based organogels and microemulsions for delivery of antimicrobial drugs. *Journal of Applied Polymer Science.* 2014;131(6):39979.
133. Singh VK, Banerjee I, Agarwal T, Pramanik K, Bhattacharya MK, Pal K. Guar gum and sesame oil based novel bigels for controlled drug delivery. *Colloids and Surfaces B: Biointerfaces.* 2014;123(2014):582–592.
134. Bara J, Funahashi M, Gin D, Goodby J, Kato T, Kerr R et al. Liquid crystalline functional assemblies and their supramolecular structures. *Recent Developments in Mercury Sience.* 2006;121, Springer.
135. Larsson K. Cubic lipid-water phases: Structures and biomembrane aspects. *The Journal of Physical Chemistry.* 1989;93(21):7304–7314.
136. Guo C, Wang J, Cao F, Lee RJ, Zhai G. Lyotropic liquid crystal systems in drug delivery. *Drug Discovery Today.* 2010;15(23):1032–1040.
137. Shah JC, Sadhale Y, Chilukuri DM. Cubic phase gels as drug delivery systems. *Advanced Drug Delivery Reviews.* 2001;47(2):229–250.
138. Garti N, Libster D, Aserin A. Lipid polymorphism in lyotropic liquid crystals for triggered release of bioactives. *Food and Function.* 2012;3(7):700–713.
139. Gin DL, Gu W, Pindzola BA, Zhou W-J. Polymerized lyotropic liquid crystal assemblies for materials applications. *Accounts of Chemical Research.* 2001;34(12):973–980.
140. Reppy MA, Gray DH, Pindzola BA, Smithers JL, Gin DL. A new family of polymerizable lyotropic liquid crystals: Control of feature size in cross-linked inverted hexagonal assemblies via monomer structure. *Journal of the American Chemical Society.* 2001;123(3):363–371.
141. Fernandes P, Mukai H, Laczkowski I. Magneto-optical effect in lyotropic liquid crystal doped with ferrofluid. *Journal of Magnetism and Magnetic Materials.* 2005;289:115–117.
142. Black CF, Wilson RJ, Nylander T, Dymond MK, Attard GS. Linear ds DNA partitions spontaneously into the inverse hexagonal lyotropic liquid crystalline phases of phospholipids. *Journal of the American Chemical Society.* 2010;132(28):9728–9732.
143. Hamley IW. Liquid crystal phase formation by biopolymers. *Soft Matter.* 2010;6(9):1863–1871.
144. Negrini R, Mezzenga R. pH-Responsive lyotropic liquid crystals for controlled drug delivery. *Langmuir.* 2011;27(9):5296–5303.
145. Berni M, Lawrence C, Machin D. A review of the rheology of the lamellar phase in surfactant systems. *Advances in Colloid and Interface Science.* 2002;98(2):217–243.
146. Blunk D, Biergans P, Bongartz N, Tessendorf R, Stubenrauch C. New speciality surfactants with natural structural motifs. *New Journal of Chemistry.* 2006;30(12):1705–1717.
147. Seddon JM. Structure of the inverted hexagonal (H_{II}) phase, and non-lamellar phase transitions of lipids. *Biochimica et Biophysica Acta (BBA)-Reviews on Biomembranes.* 1990;1031(1):1–69.
148. Mulet X, Kennedy DF, Conn CE, Hawley A, Drummond CJ. High throughput preparation and characterisation of amphiphilic nanostructured nanoparticulate drug delivery vehicles. *International Journal of Pharmaceutics.* 2010;395(1):290–297.
149. Makai M, Csanyi E, Erös I, Dekany I. Preparation and structural determination of lyotropic lamellar liquid crystalline systems of pharmaceutical importance. *Acta Pharmaceutica Hungarica.* 2002;73(2):71–76.
150. Chang C-M, Bodmeier R. Effect of dissolution media and additives on the drug release from cubic phase delivery systems. *Journal of Controlled Release.* 1997;46(3):215–222.
151. Rädler JO, Koltover I, Jamieson A, Salditt T, Safinya CR. Structure and interfacial aspects of self-assembled cationic lipid-DNA gene carrier complexes. *Langmuir.* 1998;14(15):4272–4283.
152. Sadhale Y, Shah JC. Biological activity of insulin in GMO gels and the effect of agitation. *International Journal of Pharmaceutics.* 1999;191(1):65–74.
153. Clogston J, Caffrey M. Controlling release from the lipidic cubic phase. Amino acids, peptides, proteins and nucleic acids. *Journal of Controlled Release.* 2005;107(1):97–111.

154. Boyd BJ, Whittaker DV, Khoo S-M, Davey G. Lyotropic liquid crystalline phases formed from glycerate surfactants as sustained release drug delivery systems. *International Journal of Pharmaceutics*. 2006;309(1–2):218–226.
155. Chung H, Kim J-s, Um J, Kwon I, Jeong S. Self-assembled "nanocubicle" as a carrier for peroral insulin delivery. *Diabetologia*. 2002;45(3):448–451.
156. Rosen M. *Delivery System Handbook for Personal Care and Cosmetic Products: Technology, Applications and Formulations*. William Andrew Inc., 2005;ISBN:0815515049.
157. Wang W, Cheng D, Gong F, Miao X, Shuai X. Design of multifunctional micelle for tumor-targeted intracellular drug release and fluorescent imaging. *Advanced Materials*. 2012;24(1):115–120.
158. Chen W, Meng F, Cheng R, Zhong Z. pH-Sensitive degradable polymersomes for triggered release of anticancer drugs: A comparative study with micelles. *Journal of Controlled Release*. 2010;142(1):40–46.
159. Xiao Y, Hong H, Javadi A, Engle JW, Xu W, Yang Y et al. Multifunctional unimolecular micelles for cancer-targeted drug delivery and positron emission tomography imaging. *Biomaterials*. 2012;33(11):3071–3082.
160. Nederberg F, Appel E, Tan JP, Kim SH, Fukushima K, Sly J et al. Simple approach to stabilized micelles employing miktoarm terpolymers and stereocomplexes with application in paclitaxel delivery. *Biomacromolecules*. 2009;10(6):1460–1468.
161. Vrignaud S, Anton N, Gayet P, Benoit J-P, Saulnier P. Reverse micelle-loaded lipid nanocarriers: A novel drug delivery system for the sustained release of doxorubicin hydrochloride. *European Journal of Pharmaceutics and Biopharmaceutics*. 2011;79(1):197–204.
162. Koyamatsu Y, Hirano T, Kakizawa Y, Okano F, Takarada T, Maeda M. pH-Responsive release of proteins from biocompatible and biodegradable reverse polymer micelles. *Journal of Controlled Release*. 2014;173(0):89–95.
163. Lukyanov AN, Torchilin VP. Micelles from lipid derivatives of water-soluble polymers as delivery systems for poorly soluble drugs. *Advanced Drug Delivery Reviews*. 2004;56(9):1273–1289.
164. Gaucher G, Dufresne M-H, Sant VP, Kang N, Maysinger D, Leroux J-C. Block copolymer micelles: preparation, characterization and application in drug delivery. *Journal of Controlled Release*. 2005;109(1):169–188.
165. Coimbra M, Isacchi B, van Bloois L, Torano JS, Ket A, Wu X et al. Improving solubility and chemical stability of natural compounds for medicinal use by incorporation into liposomes. *International Journal of Pharmaceutics*. 2011;416(2):433–442.
166. Torchilin VP. *Passive and Active Drug Targeting: Drug Delivery to Tumors as an Example*. Drug Delivery. Springer, 2010, pp. 3–53.
167. Pierre MBR, Costa IdSM. Liposomal systems as drug delivery vehicles for dermal and transdermal applications. *Archives of Dermatological Research*. 2011;303(9):607–621.
168. Jiang T, Zhang Z, Zhang Y, Lv H, Zhou J, Li C et al. Dual-functional liposomes based on pH-responsive cell-penetrating peptide and hyaluronic acid for tumor-targeted anticancer drug delivery. *Biomaterials*. 2012;33(36):9246–9258.
169. Sinha PK, van Griensven J, Pandey K, Kumar N, Verma N, Mahajan R et al. Liposomal amphotericin B for visceral leishmaniasis in human immunodeficiency virus-coinfected patients: 2-year treatment outcomes in Bihar, India. *Clinical Infectious Diseases*. 2011;53(7):e91–e98.
170. Alhajlan M, Alhariri M, Omri A. Efficacy and safety of liposomal clarithromycin and its effect on *Pseudomonas aeruginosa* virulence factors. *Antimicrobial Agents and Chemotherapy*. 2013;57(6):2694–2704.
171. Park JW. Liposome-based drug delivery in breast cancer treatment. *Breast Cancer Research*. 2002;4(3):95.
172. Felnerova D, Viret J-F, Glück R, Moser C. Liposomes and virosomes as delivery systems for antigens, nucleic acids and drugs. *Current Opinion in Biotechnology*. 2004;15(6):518–529.
173. Pinto-Alphandary H, Andremont A, Couvreur P. Targeted delivery of antibiotics using liposomes and nanoparticles: Research and applications. *International Journal of Antimicrobial Agents*. 2000;13(3):155–168.
174. Pavelić Ž, Škalko-Basnet N, Schubert R. Liposomal gels for vaginal drug delivery. *International Journal of Pharmaceutics*. 2001;219(1):139–149.
175. Cao Z, Tong R, Mishra A, Xu W, Wong GC, Cheng J et al. Reversible cell-specific drug delivery with aptamer-functionalized liposomes. *Angewandte Chemie International Edition*. 2009;48(35):6494–6498.
176. Sihorkar V, Vyas S. Potential of polysaccharide anchored liposomes in drug delivery, targeting and immunization. *Journal of Pharmacy and Pharmaceutical Science*. 2001;4(2):138–158.

177. Kakkar V, Singh S, Singla D, Kaur IP. Exploring solid lipid nanoparticles to enhance the oral bioavailability of curcumin. *Molecular Nutrition and Food Research*. 2011;55(3):495–503.
178. Das S, Chaudhury A. Recent advances in lipid nanoparticle formulations with solid matrix for oral drug delivery. *Aaps Pharmscitech*. 2011;12(1):62–76.
179. Weber S, Zimmer A, Pardeike J. Solid lipid nanoparticles (SLN) and nanostructured lipid carriers (NLC) for pulmonary application: A review of the state of the art. *European Journal of Pharmaceutics and Biopharmaceutics*. 2014;86(1):7–22.
180. Cai S, Yang Q, Bagby TR, Forrest ML. Lymphatic drug delivery using engineered liposomes and solid lipid nanoparticles. *Advanced Drug Delivery Reviews*. 2011;63(10):901–908.
181. Gokce EH, Korkmaz E, Dellera E, Sandri G, Bonferoni MC, Ozer O. Resveratrol-loaded solid lipid nanoparticles versus nanostructured lipid carriers: evaluation of antioxidant potential for dermal applications. *International Journal of Nanomedicine*. 2012;7:1841.
182. Blasi P, Giovagnoli S, Schoubben A, Ricci M, Rossi C. Solid lipid nanoparticles for targeted brain drug delivery. *Advanced Drug Delivery Reviews*. 2007;59(6):454–477.
183. Müller R, Radtke M, Wissing S. Solid lipid nanoparticles (SLN) and nanostructured lipid carriers (NLC) in cosmetic and dermatological preparations. *Advanced Drug Delivery Reviews*. 2002;54:S131–S155.
184. Wissing SA, Müller RH. Cosmetic applications for solid lipid nanoparticles (SLN). *International Journal of Pharmaceutics*. 2003;254(1):65–68.
185. Jenning V, Gysler A, Schäfer-Korting M, Gohla SH. Vitamin A loaded solid lipid nanoparticles for topical use: occlusive properties and drug targeting to the upper skin. *European Journal of Pharmaceutics and Biopharmaceutics*. 2000;49(3):211–218.
186. Maia CS, Mehnert W, Schäfer-Korting M. Solid lipid nanoparticles as drug carriers for topical glucocorticoids. *International Journal of Pharmaceutics*. 2000;196(2):165–167.
187. Souto E, Wissing S, Barbosa C, Müller R. Development of a controlled release formulation based on SLN and NLC for topical clotrimazole delivery. *International Journal of Pharmaceutics*. 2004;278(1):71–77.
188. Shah KA, Date AA, Joshi MD, Patravale VB. Solid lipid nanoparticles (SLN) of tretinoin: Potential in topical delivery. *International Journal of Pharmaceutics*. 2007;345(1):163–171.
189. Cavalli R, Gasco MR, Chetoni P, Burgalassi S, Saettone MF. Solid lipid nanoparticles (SLN) as ocular delivery system for tobramycin. *International Journal of Pharmaceutics*. 2002;238(1):241–245.
190. Wissing S, Müller R. Solid lipid nanoparticles as carrier for sunscreens: In vitro release and in vivo skin penetration. *Journal of Controlled Release*. 2002;81(3):225–233.
191. Sharma G, Wilson K, Van der Walle C, Sattar N, Petrie J, Ravi Kumar M. Microemulsions for oral delivery of insulin: design, development and evaluation in streptozotocin induced diabetic rats. *European Journal of Pharmaceutics and Biopharmaceutics*. 2010;76(2):159–169.
192. Park K-M, Kim C-K. Preparation and evaluation of flurbiprofen-loaded microemulsion for parenteral delivery. *International Journal of Pharmaceutics*. 1999;181(2):173–179.
193. Moreno MA, Ballesteros MP, Frutos P. Lecithin-based oil-in-water microemulsions for parenteral use: Pseudoternary phase diagrams, characterization and toxicity studies. *Journal of Pharmaceutical Sciences*. 2003;92(7):1428–1437.
194. Sommerville ML, Cain JB, Johnson Jr CS, Hickey AJ. Lecithin inverse microemulsions for the pulmonary delivery of polar compounds utilizing dimethylether and propane as propellants. *Pharmaceutical Development and Technology*. 2000;5(2):219–230.
195. Peltola S, Saarinen-Savolainen P, Kiesvaara J, Suhonen T, Urtti A. Microemulsions for topical delivery of estradiol. *International Journal of Pharmaceutics*. 2003;254(2):99–107.
196. Špiclin P, Homar M, Zupančič-Valant A, Gašperlin M. Sodium ascorbyl phosphate in topical microemulsions. *International Journal of Pharmaceutics*. 2003;256(1):65–73.
197. Nguyen TTL, Edelen A, Neighbors B, Sabatini DA. Biocompatible lecithin-based microemulsions with rhamnolipid and sophorolipid biosurfactants: Formulation and potential applications. *Journal of Colloid and Interface Science*. 2010;348(2):498–504.
198. Üstündağ Okur N, Apaydın Ş, Karabay Yavaşoğlu NÜ, Yavaşoğlu A, Karasulu HY. Evaluation of skin permeation and anti-inflammatory and analgesic effects of new naproxen microemulsion formulations. *International Journal of Pharmaceutics*. 2011;416(1):136–144.
199. Shah BM, Misra M, Shishoo CJ, Padh H. Nose to brain microemulsion-based drug delivery system of rivastigmine: Formulation and ex-vivo characterization. *Drug Delivery*. 2014:1–13.
200. Raza K, Negi P, Takyar S, Shukla A, Amarji B, Katare OP. Novel dithranol phospholipid microemulsion for topical application: Development, characterization and percutaneous absorption studies. *Journal of Microencapsulation*. 2011;28(3):190–199.

201. Leatherman MD, Policello GA, Peng WN, Zheng L, Wagner R, Rajaraman SK et al. Hydrolysis resistant organomodified disiloxane ionic surfactants. US 8367740 B2, 2013.
202. Zanatta C, Mitjans M, Urgatondo V, Rocha-Filho P, Vinardell M. Photoprotective potential of emulsions formulated with Buriti oil (*Mauritia flexuosa*) against UV irradiation on keratinocytes and fibroblasts cell lines. *Food and Chemical Toxicology*. 2010;48(1):70–75.
203. Liu W, Yang XL, Winston Ho W. Preparation of uniform-sized multiple emulsions and micro/nano particulates for drug delivery by membrane emulsification. *Journal of Pharmaceutical Sciences*. 2011;100(1):75–93.
204. Sapei L, Naqvi MA, Rousseau D. Stability and release properties of double emulsions for food applications. *Food Hydrocolloids*. 2012;27(2):316–323.
205. ElShafei GMS, El-Said MM, Attia HAE, Mohammed TGM. Environmentally friendly pesticides: Essential oil-based w/o/w multiple emulsions for anti-fungal formulations. *Industrial Crops and Products*. 2010;31(1):99–106.
206. Stefano Farris, Karen M, Schaich, LinShu Liu, and LP, Yam KL. Development of polyion-complex hydrogels as an alternative approach for the production of bio-based polymers for food packaging applications: A review. *Trends in Food Science and Technology*. 2009;20:316–332.
207. Hughes NE, Marangoni AG, Wright AJ, Rogers MA, Rush JWE. Potential food applications of edible oil organogels. *Trends in Food Science and Technology*. 2009;20(10):470–480.
208. Cavalieri F, Chiessi E, Finelli I, Natali F, Paradossi G, Telling MF. Water, solute, and segmental dynamics in polysaccharide hydrogels. *Macromolecular Bioscience*. 2006;6(8):579–589.
209. Bot A, Agterof WGM. Structuring of edible oils by mixtures of γ-oryzanol with β-sitosterol or related phytosterols. *JAOCS, Journal of the American Oil Chemists' Society*. 2006;83(6):513–521.
210. Morales ME, Gallardo V, Clarés B, García MB, Ruiz MA. Study and description of hydrogels and organogels as vehicles for cosmetic active ingredients. *Journal of Cosmetic Sciences*. 2009;60(6):627–636.
211. Ramanujan R, Ang K, Venkatraman S. Magnet–PNIPA hydrogels for bioengineering applications. *Journal of Materials Science*. 2009;44(5):1381–1387.
212. Berry J. Aqueous hydrogel lubricant. Google Patents, 1987.
213. Yang D, Tang L, Lang D. Fluorescent dyed lubricant for medical devices. US Patent 7,052,512, 2006, 2001.
214. Hawkins M, Pletcher D, Thomas B, Zhang K, Brinkerhuff H. Methods of preparing hydrogel coatings. Google Patents, 2007.
215. Wang W. Methods for making injectable polymer hydrogels. US20050281880, 2005.
216. Quong D. Encapsulated active material immobilized in hydrogel microbeads. US Patent 6,375,968, 2002.
217. Chacon D, Hsieh YL, Kurth MJ, Krochta JM. Swelling and protein absorption/desorption of thermosensitive lactitol-based polyether polyol hydrogels. *Polymer*. 2000;41(23):8257–8262.
218. Kim J, Park K. Smart hydrogels for bioseparation. *Bioseparation*. 1998;7(4):177–184.
219. Liang Y-Y, Zhang L-M, Jiang W, Li W. Embedding magnetic nanoparticles into polysaccharide-based hydrogels for magnetically assisted bioseparation. *ChemPhysChem*. 2007;8(16):2367–2372.
220. O'Connor CJ, Buisson YSL, Li S, Banerjee S, Premchandran R, Baumgartner T et al., editors. Ferrite synthesis in microstructured media: Template effects and magnetic properties. AIP, Atlanta, GA, 1997.
221. Huang R, Kostanski LK, Filipe CDM, Ghosh R. Environment-responsive hydrogel-based ultrafiltration membranes for protein bioseparation. *Journal of Membrane Science*. 2009;336(1–2):42–49.
222. Kumar A, Srivastava A, Galaev IY, Mattiasson B. Smart polymers: Physical forms and bioengineering applications. *Progress in Polymer Science*. 2007;32(10):1205–1237.
223. Kim JJ, K Park. Smart hydrogels for bioseparation. *Bioseparation*. 1998;7(4–5):177–184.
224. Hunt NC, Grover LM. Cell encapsulation using biopolymer gels for regenerative medicine. *Biotechnology Letters*. 2010;32(6):733–742.
225. Maltais A, Remondetto GE, Subirade M. Soy protein cold-set hydrogels as controlled delivery devices for nutraceutical compounds. *Food Hydrocolloids*. 2009;23(7):1647–1653.
226. Bromberg LE, Ron ES. Temperature-responsive gels and thermogelling polymer matrices for protein and peptide delivery. *Advanced Drug Delivery Reviews*. 1998;31(3):197–221.
227. Ferreira L, Vidal M, Gil M. Evaluation of poly (2-hydroxyethyl methacrylate) gels as drug delivery systems at different pH values. *International Journal of Pharmaceutics*. 2000;194(2):169–180.
228. Colombo P, Bettini R, Santi P, Peppas NA. Swellable matrices for controlled drug delivery: Gel-layer behaviour, mechanisms and optimal performance. *Pharmaceutical Science and Technology Today*. 2000;3(6):198–204.
229. Pollak A, Blumenfeld H, Wax M, Baughn RL, Whitesides GM. Enzyme immobilization by condensation copolymerization into crosslinked polyacrylamide gels. *Journal of the American Chemical Society*. 1980;102(20):6324–6336.

230. Miyazaki S, Suisha F, Kawasaki N, Shirakawa M, Yamatoya K, Attwood D. Thermally reversible xyloglucan gels as vehicles for rectal drug delivery. *Journal of Controlled Release*. 1998;56(1):75–83.
231. Anukam KC, Osazuwa E, Osemene GI, Ehigiagbe F, Bruce AW, Reid G. Clinical study comparing probiotic *Lactobacillus* GR-1 and RC-14 with metronidazole vaginal gel to treat symptomatic bacterial vaginosis. *Microbes and Infection*. 2006;8(12):2772–2776.
232. Wang C, Swerdloff RS, Iranmanesh A, Dobs A, Snyder PJ, Cunningham G et al. Transdermal testosterone gel improves sexual function, mood, muscle strength, and body composition parameters in hypogonadal men 1. *The Journal of Clinical Endocrinology and Metabolism*. 2000;85(8):2839–2853.
233. Eginton MT, Brown KR, Seabrook GR, Towne JB, Cambria RA. A prospective randomized evaluation of negative-pressure wound dressings for diabetic foot wounds. *Annals of Vascular Surgery*. 2003;17(6):645–649.
234. Iwanaga K, Kawai M, Miyazaki M, Kakemi M. Application of organogels as oral controlled release formulations of hydrophilic drugs. *International Journal of Pharmaceutics*. 2012;436(1–2):869–872.
235. Lupi FR, Gabriele D, Baldino N, Mijovic P, Parisi OI, Puoci F. Olive oil/policosanol organogels for nutraceutical and drug delivery purposes. *Food and Function*. 2013;4(10):1512–1520.
236. Sagiri SS, Kumar U, Champaty B, Singh VK, Pal K. Thermal, electrical, and mechanical properties of tween 80/span 80–based organogels and its application in iontophoretic drug delivery. *Journal of Applied Polymer Science*. 2014;132(6):41419.
237. Bhatia A, Singh B, Raza K, Wadhwa S, Katare OP. Tamoxifen-loaded lecithin organogel (LO) for topical application: Development, optimization and characterization. *International Journal of Pharmaceutics*. 2013;444(1–2):47–59.
238. Baran N, Singh VK, Pal K, Anis A, Pradhan DK, Pramanik K. Development and characterization of soy lecithin and palm oil based organogels. *Polymer Plastic Technology and Engineering*. 2014;53(9):865–879.
239. Raza K, Kumar M, Kumar P, Malik R, Sharma G, Kaur M et al. Topical delivery of aceclofenac: Challenges and promises of novel drug delivery systems. *BioMed Research International*. 2014;2014:406731.
240. Główka E, Wosicka-Frąckowiak H, Hyla K, Stefanowska J, Jastrzębska K, Klapiszewski Ł et al. Polymeric nanoparticles-embedded organogel for roxithromycin delivery to hair follicles. *European Journal of Pharmaceutics and Biopharmaceutics*. 2014;88(1):75–84.
241. Akram M, Naqvi SBS, Khan A. Design and development of insulin emulgel formulation for transdermal drug delivery and its evaluation. *Pakistan Journal of Pharmaceutical Sciences*. 2013;26(2):323–332.
242. Khullar R, Kumar D, Seth N, Saini S. Formulation and evaluation of mefenamic acid emulgel for topical delivery. *Saudi Pharmaceutical Journal*. 2012;20(1):63–67.
243. Rahmani-Neishaboor E, Jallili R, Hartwell R, Leung V, Carr N, Ghahary A. Topical application of a film-forming emulgel dressing that controls the release of stratifin and acetylsalicylic acid and improves/prevents hypertrophic scarring. *Wound Repair and Regeneration*. 2013;21(1):55–65.
244. Shen Y, Ling X, Jiang W, Du S, Lu Y, Tu J. Formulation and evaluation of cyclosporin A emulgel for ocular delivery. *Drug Delivery*. 2014(0):1–7.
245. Badilli U, Amasya G, Şen T, Tarimci N. Topical emulgel formulation containing inclusion complex of calcipotriol with cyclodextrin. *Journal of Inclusion Phenomena and Macrocyclic Chemistry*. 2014;78(1–4):249–255.
246. Rehman K, Mohd Amin MCI, Zulfakar MH. Development and physical characterization of polymer-fish oil bigel (hydrogel/oleogel) system as a transdermal drug delivery vehicle. *Journal of Oleo Science*. 2014;63(10):961–970.
247. Estracanholli ÉA, Praça FSG, Cintra AB, Pierre MBR, Lara MG. Liquid crystalline systems for transdermal delivery of Celecoxib: In vitro drug release and skin permeation studies. *AAPS PharmSciTech*. 2014:1–8.
248. Ki M-H, Lim J-L, Ko J-Y, Park S-H, Kim J-E, Cho H-J et al. A new injectable liquid crystal system for one month delivery of leuprolide. *Journal of Controlled Release*. 2014;185(0):62–70.
249. Cohen-Avrahami M, Shames AI, Ottaviani MF, Aserin A, Garti N. HIV-TAT Enhances the transdermal delivery of NSAID drugs from liquid crystalline mesophases. *The Journal of Physical Chemistry B*. 2014;118(23):6277–6287.
250. Rizwan SB, Boyd BJ, Rades T, Hook S. Bicontinuous cubic liquid crystals as sustained delivery systems for peptides and proteins. *Expert Opinion on Drug Delivery*. 2010;7(10):1133–1144.
251. Shevachman M, Garti N, Shani A, Sintov AC. Enhanced percutaneous permeability of diclofenac using a new U-type dilutable microemulsion. *Drug Development and Industrial Pharmacy*. 2008;34(4):403–412.

252. Shah RM, Malherbe F, Eldridge D, Palombo EA, Harding IH. Physicochemical characterization of solid lipid nanoparticles (SLNs) prepared by a novel microemulsion technique. *Journal of Colloid and Interface Science*. 2014;428(0):286–294.
253. Vatsraj S, Chauhan K, Pathak H. Formulation of a novel nanoemulsion system for enhanced solubility of a sparingly water soluble antibiotic, clarithromycin. *Journal of Nanoscience*. 2014;2014:268293.
254. de Jesus MB, Radaic A, Hinrichs WL, Ferreira CV, De Paula E, Hoekstra D et al. Inclusion of the helper lipid dioleoyl-phosphatidylethanolamine in solid lipid nanoparticles inhibits their transfection efficiency. *Journal of Biomedical Nanotechnology*. 2014;10(2):355–365.
255. Yang Z, Liu M, Chen J, Fang W, Zhang Y, Yuan M et al. Development and characterization of amphotericin B nanosuspensions for oral administration through a simple top-down method. *Current Pharmaceutical Biotechnology*. 2014;15(6):569–576.
256. Ren S-T, Kang X-N, Liao Y-R, Wang W, Ai H, Chen L-N et al. The ultrasound contrast imaging properties of lipid microbubbles loaded with urokinase in dog livers and their thrombolytic effects when combined with low-frequency ultrasound in vitro. *Journal of Thrombosis and Thrombolysis*. 2014;37(3):303–309.
257. Král P, Vuković L. *Computational Studies of Highly PEG-ylated Sterically Stabilized Micelles: Self-Assembly and Drug Solubilization. Intracellular Delivery II*. Springer, 2014;313–326.
258. Scherlund M, Malmsten M, Holmqvist P, Brodin A. Thermosetting microemulsions and mixed micellar solutions as drug delivery systems for periodontal anesthesia. *International Journal of Pharmaceutics*. 2000;194(1):103–116.
259. Chung J, Yokoyama M, Yamato M, Aoyagi T, Sakurai Y, Okano T. Thermo-responsive drug delivery from polymeric micelles constructed using block copolymers of poly (N-isopropylacrylamide) and poly (butylmethacrylate). *Journal of Controlled Release*. 1999;62(1):115–127.
260. Phuphanich S et al. A pharmacokinetic study of intra-CSF administered encapsulated cytarabine (DepoCyt®) for the treatment of neoplastic meningitis in patients with leukemia, lymphoma, or solid tumors as part of a phase III study. *Journal of Neuro-Oncology*. 2007;81(2):201–208.
261. Coukell AJ and Spencer CM. Polyethylene glycol-liposomal doxorubicin. *Drugs*. 1997;53(3):520–538.
262. Dasta J et al. Bupivacaine liposome injectable suspension compared with bupivacaine HCl for the reduction of opioid burden in the postsurgical setting. *Current Medical Research and Opinion*. 2012;28(10):1609–1615.
263. O'Brien S et al. High-dose vincristine sulfate liposome injection for advanced, relapsed, and refractory adult Philadelphia chromosome–negative acute lymphoblastic leukemia. *Journal of Clinical Oncology*. 2013;31(6):676–683.
264. Sundar S et al. Single-dose liposomal amphotericin B for visceral leishmaniasis in India. *New England Journal of Medicine*. 2010;362(6):504–512.
265. Petraitis V et al. Combination therapy in treatment of experimental pulmonary aspergillosis: *in vitro* and *in vivo* correlations of the concentration and dose-dependent interactions between anidulafungin and voriconazole by Bliss independence drug interaction analysis. *Antimicrobial Agents and Chemotherapy*. 2009;53(6):2382–2391.
266. Groll AH and Walsh TJ. Caspofungin: Pharmacology, safety and therapeutic potential in superficial and invasive fungal infections. *Expert Opinion on Investigational Drugs*. 2001;10(8):1545–1558.

59 Introduction to Commercial Microencapsulation

George A. Stahler

CONTENTS

59.1 Comparison Chart for Commercial Microcapsule 1416
    59.1.1 Capsule Technology..1416

Commercial scale encapsulation is an elusive term and can largely be defined based on the target market. Commercial success could be defined in terms of financial performance, pounds produced, or even manufacturing capability. Today's microencapsulation applications are present in a wide array of everyday items such as pharmaceutical, food, agricultural, and scents. In each case, the commercial scale is widely different. This chapter will be devoted to a product that has a 60-year history of commercial success. This product also has the distinction of being the first commercial product to use microcapsules. The product is the NCR PAPER* brand of carbonless paper. In this chapter, the aspects of developing a product based on innovation, applied technology, emphasis on quality, customer satisfaction, and manufacturing capabilities will be examined from the perceptive of microencapsulation. The history of carbonless paper also provides a prelude to new commercial applications for the microencapsulation process (Figure 59.1).

What began as a search for a better cash register receipt paper resulted in the invention of an office productivity phenomenon that continues to be an indispensable part of the business world, even in today's digital age.

In 1954, the National Cash Register Company (now NCR Corporation) and the Appleton Coated Paper Company (today's Appvion) collaborated to introduce the first carbonless paper. It was known then, as it is today, as the NCR PAPER* brand of carbonless paper. The product was simply promoted as a new paper product that "eliminates carbon paper." Some dubbed it the "no carbon required" paper. The first commercial sale took place on March 26, 1954, and marked the first use of the NCR PAPER* trademark.

The NCR PAPER* brand has been continuously manufactured by Appvion since 1954. The product revolutionized the forms industry by eliminating the mess and bother of carbon interleaves. Paper transactions became easier, faster, and cleaner for businesses, institutions, government, and service organizations of all kinds. Carbonless paper is used to make multipart business forms, invoices, and credit card receipts.

The late Barry Green was a scientist and inventor who developed the microencapsulation process used to create carbonless paper. While pursuing graduate studies in chemistry at Cornell University in the 1930s, Green began brainstorming about what could be accomplished if a system were produced composed of a liquid dispersed within a solid. He quickly learned that such a system was very rare outside living organisms. Years later while at the National Cash Register Company in Dayton, OH, Green applied his ideas to produce the first manmade, commercial example of a microencapsulated system. It was that system that led to the invention of carbonless paper.

---

* NCR PAPER is a registered trademark licensed to Appvion, Inc.

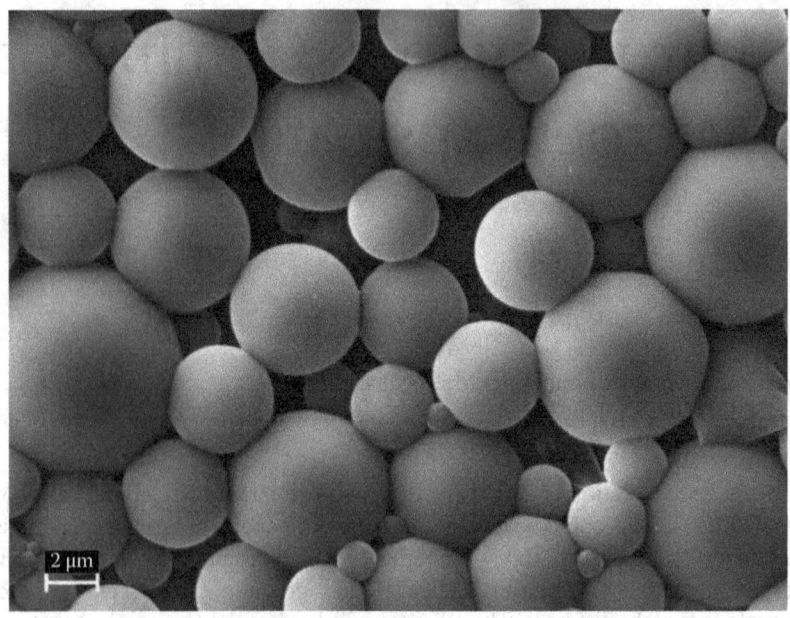

FIGURE 59.1 Polyacrylic capsule widely used today.

During 1952 and 1953, Barry Green worked with the late Lowell Schleicher, another NCR Corporation scientist, to develop and refine the microencapsulation system. They coinvented the system that is used to produce much of today's carbonless paper and filed the patent for the system on June 30, 1953.

During the development of microencapsulation, Green, Schleicher, and others at NCR went looking for paper companies capable of coating the pressure-sensitive microcapsules onto paper. Appleton's coating capabilities drew their attention, because the company's air knife coater did not require the application of pressure for coating, which would damage the capsules.

The carbonless system consists of colorformers and oil-filled microcapsules dispersed within a solid coating. In a typical three-part business form, three kinds of carbonless paper work together as a system to transfer images cleanly and clearly from one sheet to the next (Figure 59.2).

The top sheet is a coated back (CB) sheet, the back of which is covered with a coating made of millions of microscopic capsules containing colorless dyes. The last sheet is a coated front (CF) sheet coated on its front side with a coreactant or receiver material. The middle sheet is a coated front and back (CFB) sheet, front coated with receiver materials and back coated with dye capsules. As pressure from a pen or printer is applied to the top sheet, the dye capsules on the CB surface break, and the dye is released, reacting with the coating on the CF sheet and producing a bold image transfer (Figure 59.3).

The carbonless capsules were first produced at an NCR plant in Dayton, OH. Growing demand for the carbonless paper prompted the company to consider adding production capacity by building a new capsule plant. They chose a site in Portage, WI. Portage is nestled between the south-flowing Wisconsin River and north-flowing Fox River. The city has a history as a major trade route dating back well before European explorers came to the state. The short distance to Portage, a canoe between the two rivers gave early travelers access to the Mississippi River delta and to the Great Lakes and beyond into Canada. Although not a consideration in the site selection in 1967, the proximity to paper-coating operations in Appleton and a good interstate system probably played into the decision (Figure 59.4).

In 1970, NCR purchased the Appleton Coated Paper Company and merged it with the Combined Paper Mills to create Appleton Papers, Inc. Several owner changes through the years cumulated in

Introduction to Commercial Microencapsulation

FIGURE 59.2 Early advertisement for carbonless paper.

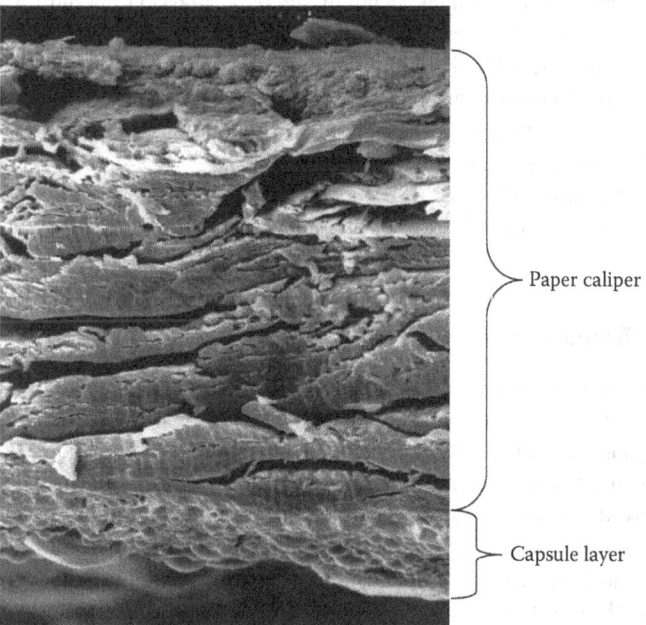

FIGURE 59.3 Cross-section of a carbonless paper with PAC capsules.

FIGURE 59.4 Original Portage microencapsulation plant, circa 1972.

2001 with the company becoming employee owned. During this period, the Portage plant completed four expansions to increase capacity and introduce new technology and products. Appleton built a second microencapsulation plant 1980 at Camp Hill, PA, to meet the peaking demand for the carbonless business forms industry. During this time, the company was operating eight production reactors and three pilot-scale reactors with a combined capacity to produce in excess of 130,000,000 lb of capsules per year. The Camp Hill facility operated for 20 years and supplied microcapsules for production of carbonless paper on the company's off-machine coaters and paper machines.

59.1 COMPARISON CHART FOR COMMERCIAL MICROCAPSULE

But just as carbonless paper utilized revolutionary technology to improve the processing of business information, new and competing technologies, such as laser, inkjet, thermal, voice, and data that do not use impact printing to create images, began to decrease market demand for carbonless paper in the 1990s. By 2000, Appleton Papers, Inc. decommissioned the capsule plant in Camp Hill and the Portage plant became the focus for the company's microcapsule production.

In recent years, Appvion has continued to leverage its core competency in microencapsulation and commercial scale production to bring new-to-world solutions and benefits to a variety of agricultural, industrial, and consumer products.

Today, the growing partnership with Procter & Gamble Company is well documented. Portage is preparing for another four decades as a worldclass, commercial-scale producer in the microencapsulation field.

59.1.1 Capsule Technology

The unique chemistry of microencapsulation is well documented. The aspects of how different capsule designs contributed to the commercial success of Appvion will be the focus for the following section. The gelatin capsule system developed in the 1950s for carbonless paper applications was produced on a simplified production line. Production on the gelatin line was manual and labor-intensive and involved process steps such as temperature control, agitation, and rate of additions, which were performed manually. Before computer automation, the operators would track recorders with circular charts to control agitator speeds and temperatures. Operators were trained to control addition rates by making turns on a valve at multiple times during the addition of a component. Quantities were added based on a simple volume measurement using a "stock pot" and "stand pipe." The plant operated three recirculation reactor lines and two coacervation tanks. The high water and gelatin content of the design made it susceptible to microbiological activity. The use of biocides and tight pH control were important considerations with the gelatin capsule. Crude by any manufacturing standards used today, the resulting capsules were a leading edge technology and met customer requirements longer than any of other systems used today (Figure 59.5).

The gelatin system consisted of 16 separate processing stages. Stages 1 through 2 took place in a "premix tank" and involved the preparation of the gelatin. A 10% solution of gelatin was heated to 60°F to swell the crystals to make them easier to melt. The solution temperature was raised to 115° to melt and dissolve the gelatin. Stages 3 through 7 again adjusted pH and added in the core solution. Milling or the emulsification of the oil-based core occurs next. At Stage 5, the gelatin and core material were pumped from the premix tank into a small 15-gallon "reactor" where impellers emulsified the oil. The material was returned into premix tank and recirculated for 1.5–2 h until a single oil droplet size of 3 µm was achieved. The solution was then transferred into the "coacervation tank" and the pH increased to the 5–8 range to prevent aggregation. The first addition is a copolymer of methyl vinyl ether and maleic anhydride (PVM/MA) that is put into the batch at this point and will work as a binder to hold aggregates together as they form during later stages. The pH is raised once again to balance charges in the solution. At this point, all raw materials either have no charge or a negative charge essentially repelling each other. Carboxyl methylcellulose (CMC) is added at Stage 9. The process is now at the critical coacervation phase. In Stage 10, the pH is lowered using acetic acid and aggregates begin to form. The size of the aggregate is controlled by the amount of first add PVM/MA, agitation speed of the impellers in the coacervation tank, and the solids of the solution. The final aggregate particles will have a size around 10 to 15 µm. The material in the tank is cooled to drop out the remaining gelatin and deposit it around the aggregate. Stage 12 is referred to as the hardening stage. Glutaraldehyde will be added to cross-link the gelatin and prevent it from breaking down. The final stages include a neutralization of residual charges using a second addition of PVM/MA, a buffering stage using soda ash, a final pH adjustment, and finally, bringing the batch back to room temperature. The system was complex and at any stage of the process, there was the real potential to cause a process failure.

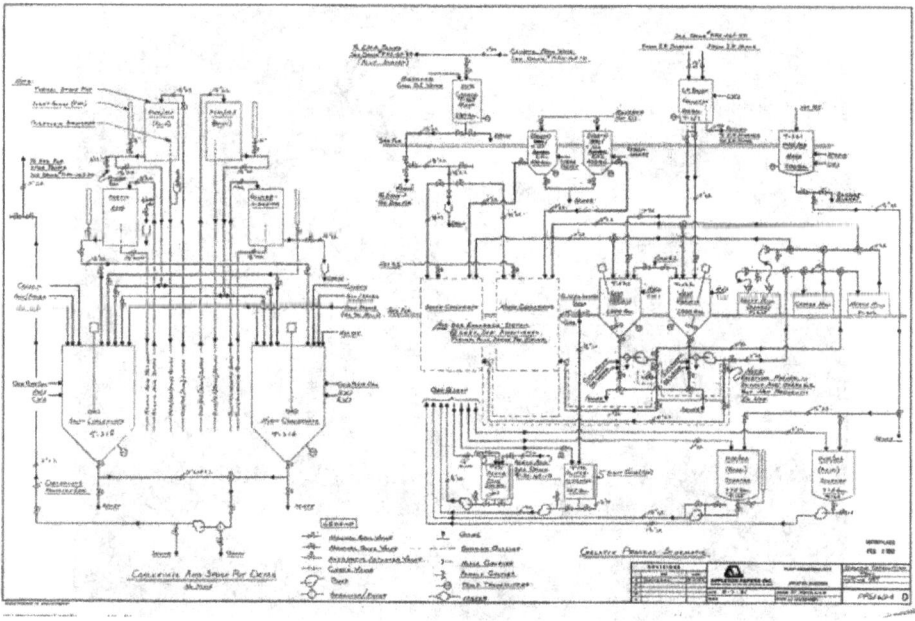

FIGURE 59.5 Original gelatin process schematic.

However, the longevity of the gelatin system illustrates the high degree of customer satisfaction with this product. The wall chemistry of the gelatin capsules translated into good coating properties for producing carbonless forms. The release mechanism for carbonless products is pressure.

The pressure generated from a typewriter, impact printer, or pen ruptures the capsule wall, which transfers the core (colorless ink) to a receptor through multiple layers of paper. The capsule is exposed to harsh conditions during processing including drying steps and numerous nip/pressure points during product manufacturing. The elasticity of the gelatin wall prevented the premature rupture of the capsule. The gelatin capsule was an aggregate of single droplets and as produced had a size in the 13 µm range. The size of the capsule allowed it to ride in the coating film and not get "lost" in the fibers of a paper sheet.

Low manufacturing solids of 23% finally led to the replacement of the gelatin capsule. Low solids equated to low production yields and inflated shipping cost. Customer requirements for high solid capsules due to changing technology on the paper machines finally caused the replacement of this gelatin capsule. The gelatin capsule was in production at Portage until 1996 and was commercially viable for over 40 years (Figure 59.6).

The demand for high solids coatings required a new innovative capsule. A single droplet capsule using an aminoplast wall was developed and introduced in 1975. The acrylic copolymer urea formaldehyde (ACUF) capsule went into production in 1975 at the Portage facility. The ACUF capsule was the first capsule produced at the newly built Camp Hill facility in 1981. The Portage facility saw its first building expansion and two new reactors at this time. A production milestone occurred—the process was automated and controlled by an Allen Bradley "Advisor" system. The new synthetic wall was composed of ethylene maleic anhydride, urea, and formaldehyde. Formula and physical properties were adjusted to provide a broader range of finished products. The ACUF production span lasted less than 10 years due to the cost, manufacturing complexities, and other shortcomings.

During this time period, AG Bayer developed Isocyanate wall chemistry for microcapsules. The system has very good properties but, due to cost, was never incorporated into Appleton's product line.

The polyacrylic copolymer (PAC) process incorporated new wall materials including proprietary emulsifiers and a melamine formaldehyde cross-linker (Figure 59.7). The polymers used

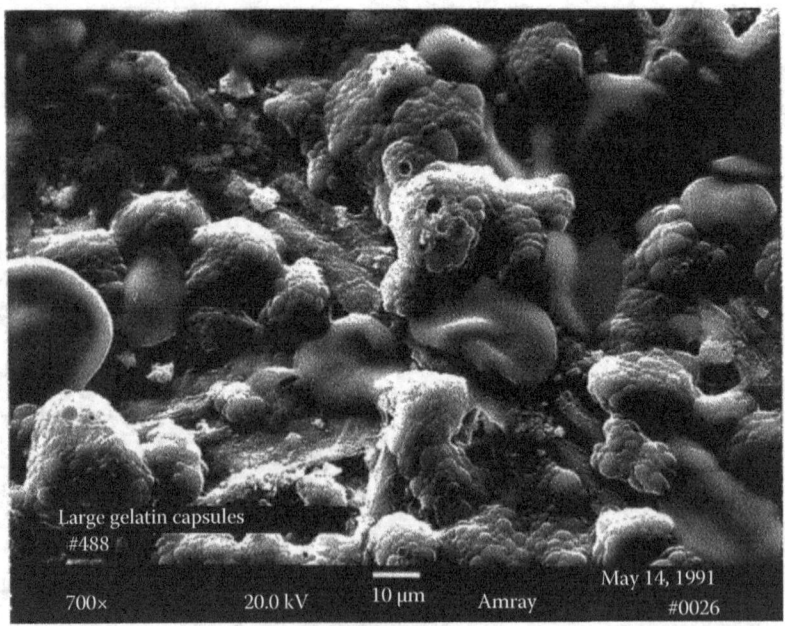

FIGURE 59.6 Gelatin aggregates on coated paper.

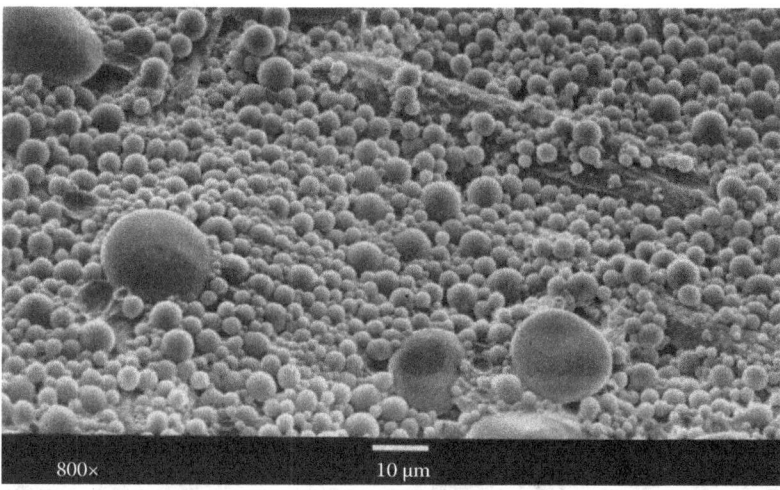

FIGURE 59.7 PAC capsules on carbonless business form.

in the wall formation of PAC capsules are more process-friendly than the gelatin. However, the key process criteria of rate of additions, pH, temperature, material quantity, and milling are just as unforgiving as in the gelatin system. The process can be explained in five phases. In the first phase, a solution is made up of equal parts of emulsifier and water. The solution is pH-adjusted and transferred into the reactor. In the reactor phase with the water phase, the core is added. The mixture is emulsified in situ using high horsepower motors and a series of impellers. The emulsion based on application is produced as a single droplet from 1 μm to over 20 μm. At Phase 3, a second water phase consisting of equal parts of water, emulsifier, and cross-linker is charged into the reactor. The charges in the system align the emulsifiers around the oil droplet and attract the cross-linker to begin forming the capsule wall. The cross-linker begins to self-polymerize and is driven by temperature. In Phase 4, the newly formed capsules are transferred into an "intermediate" one where the wall reaction is driven to completion by controlling time and temperature. The final phase is quenching the reaction (Figures 59.8 and 59.9).

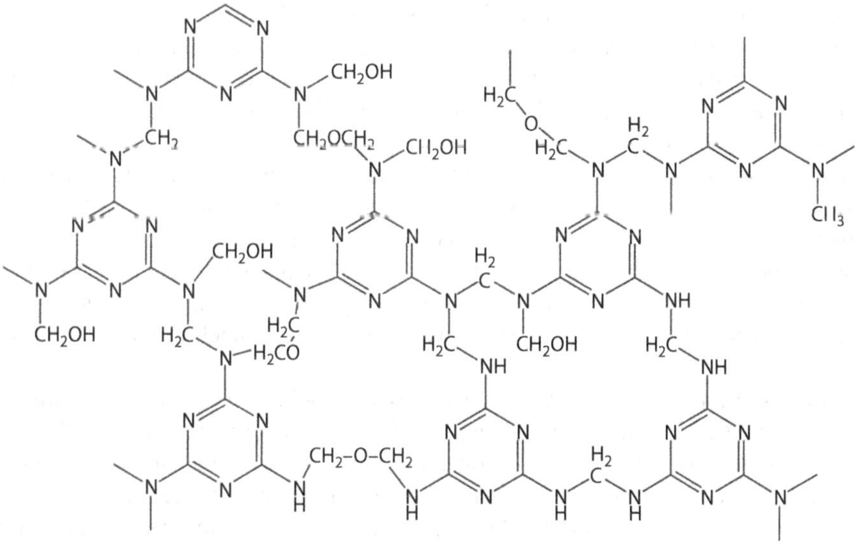

FIGURE 59.8 Melamine cross-linked structure. Main building block of the PAC wall.

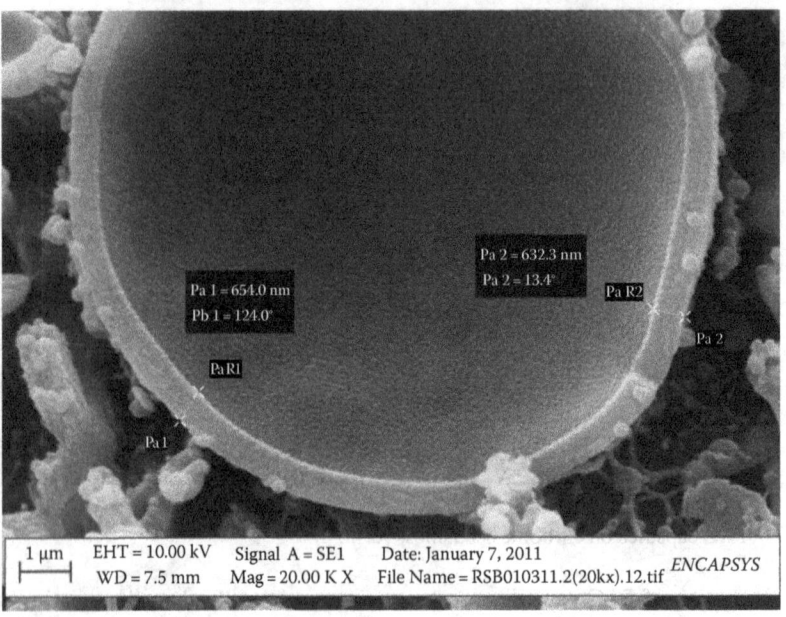

FIGURE 59.9 PAC capsule cross-section.

The wall material was received in bulk liquid form and streamlined the handling of raw materials. The system has proven to be easily adapted to new cores and new applications. During the 1990s, production optimization and cost reduction were major undertakings.

Process cycle times were dramatically reduced and output on existing equipment nearly doubled. Reformulation and the introduction of new materials significantly reduced material cost. Process thermal enhancements also reduced our environmental footprint by lowering volatile organic compounds (VOC) emissions. The capsule system replaced petroleum-based cores with soybean-based solvents. The plant successfully increased output, lowered operating cost, and became more environmentally responsible. Process automation and control has improved dramatically over the last decade, allowing continuous improvement in every aspect of the process. Process parameters can be tracked done to the second. Variable speed pumps and automated values control addition rates. Quantities are metered and weighed, offering a check and balance system. The accuracy of the additions ensures formula accuracy and tight inventory control. Today the PAC is the primary system in use at the Portage facility (Figure 59.10).

However, carbonless is not the entire story of Appvion microencapsulation. Leveraging of encapsulation knowledge, experience, and production capability are paramount in developing new products. Appvion has been able to bring new products to market using a solid core and modifying the PAC system to meet the demands of new markets. The Portage plant has added four production lines in the last 7 years. One reactor line was "supersized" to optimize capacity, logistics, and packaging. Other lines have been added to provide the capacity to quickly launch products into test markets (Figure 59.11).

During the 1990s, new microcapsule applications were being developed at Appvion. It seemed we would take one step forward and two steps backward, but we were learning. When the right opportunity presented itself in 2006, the momentum and teams were poised to succeed. For over 20 years, quality has been defined within Appleton as "meeting or exceeding customer expectation in all aspects of every transaction." This dedication to the customer was a key component in building a strong partnership and the commercial success of the perfume microcapsule or PMC product line.

Introduction to Commercial Microencapsulation

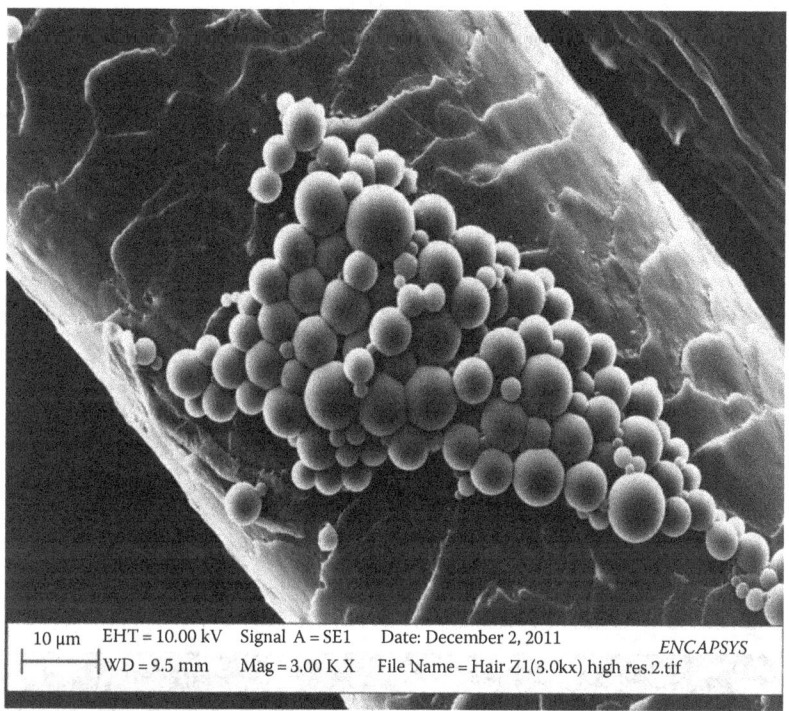

FIGURE 59.10 PAC capsules on a human hair.

FIGURE 59.11 Major capacity expansion at the Portage plant in 2010.

The capsule was designed to time release fragrance over an extended period, enhancing the desired effect of freshness associated with the original products. The PAC wall was used in the capsule production and was a straightforward proposition. The fragrance was originally encapsulated using existing equipment. The magic during the product development was the collaboration between the capsule manufacture and the customer. A new set of performance standards was

FIGURE 59.12 PMC microcapsules on fabric.

FIGURE 59.13 Encapsys encapsulation plant in Portage, 2013.

required and jointly developed. The format was so successful that "collaborative innovation" is now a cornerstone for developing new business within Encapsys (Figure 59.12).

Today the microencapsulation team incorporates marketing, financial, research and technology, logistics, and manufacturing into the Encapsys division of Appvion. In 2014, product lines began expanding beyond the PAC system with new wall chemistries under development. The advancements under development will provide for future commercial success and broadening platforms for micro-encapsulation (Figure 59.13).

60 Stable Core-Shell Microcapsules for Industrial Applications

Klaus Last

CONTENTS

60.1 Introduction 1423
60.2 Manufacture of Microcapsules 1426
60.3 New Formaldehyde-Free Systems 1429
60.4 Current Applications... 1434
    60.4.1 Self-Healing 1434
        60.4.1.1 Application of a Protective Coating 1435
        60.4.1.2 Anti-Friction Coatings 1435
        60.4.1.3 PCM Technology 1435
References... 1436

60.1 INTRODUCTION

There are countless physical and chemical methods of manufacturing microcapsules. Each method produces specific capsule systems with a highly individual profile of properties.

Microcapsules, which in our case mostly consist of a liquid core and a stable plastic shell, are defined as being on a scale of 0.5 to 1000 µm in diameter. Diameters of less than 60 µm are indistinguishable to the naked human eye (Figure 60.1).

- Nanocapsules   <0.5 µm
- Microcapsules   >0.5 µm to 1 mm
- Macrocapsules   >1 mm

Precise knowledge of the available systems and a great deal of experience is required to access the feasibility of a new system and its suitability for meeting the new requirements (Figure 60.2).[1]

One should really have all the available tools at one's disposal and know the important parameters for changing specific properties.

Typically, firms everywhere utilize only a few encapsulation techniques that suit their products and requirements.

Follmann manufactured and enhanced stable core-shell capsule systems for 25 years now. There are four main systems that have dominated the field of stable core-shell microcapsules for several decades (Figure 60.3).

Acrylate capsules produced by radical-induced polymerization. Gelatin capsules manufactured by coacervation. Polyurethane capsules produced by polyaddition of amines with diisocyanates. Aminoplast microcapsules made by polycondensation.

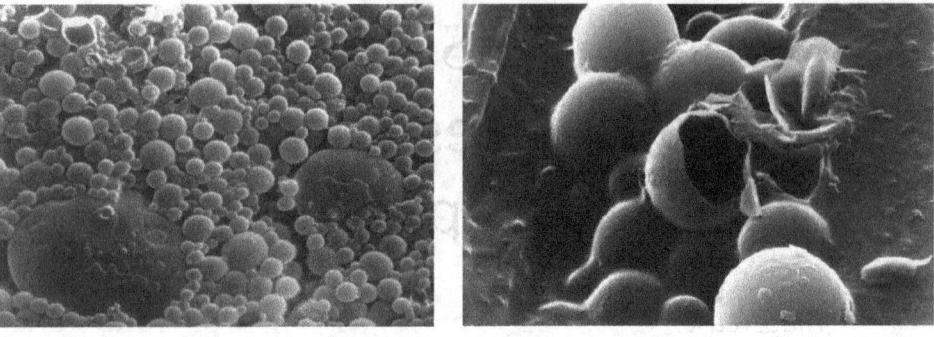

FIGURE 60.1 Aminoplastic microcapsules, definition of capsule size.

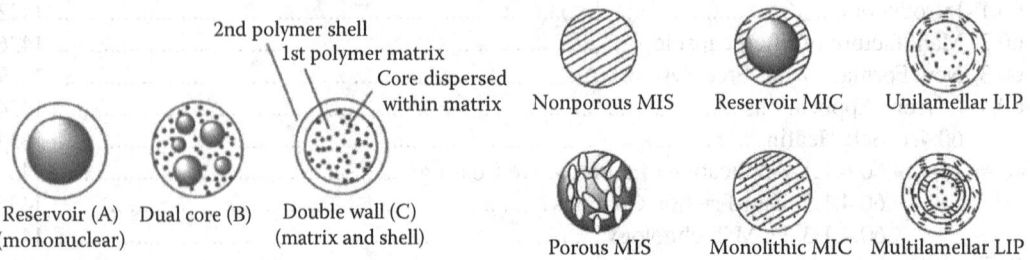

FIGURE 60.2 Examples of the morphology of various capsule systems. (From R. Arshady, *Microspheres, Microcapsules*, 1, 13, 1999.)

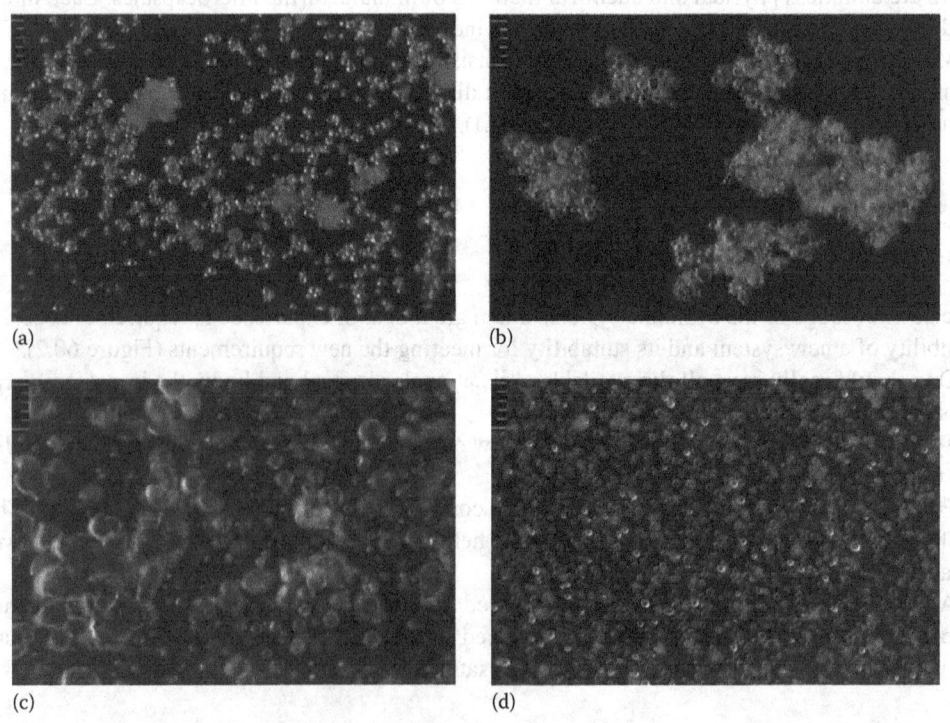

FIGURE 60.3 Shell materials: (a) acrylat, (b) gelatine, (c) polyurethan, and (d) aminoplast.

FIGURE 60.4 NCR laboratory building and a chemist at work.

History of the industrial application of microcapsule systems: In 1954, the Dayton, OH-based company National Cash Register (NCR) began the industrial manufacture of gelatin-based microcapsules for carbonless copy systems within a new laboratory building (Figure 60.4). By encapsulating dyes such as leucocrystal violet (LCV)—which is colorless at neutral pH and reacts in acidic solution to give a violet color—the dye precursor is released by mechanical means (e.g., pressure applied with a ball-point pen) and colors the acidic clay coated receiver paper in the ensuing reaction (Figure 60.5).

Figure 60.6 shows the historical changes in the market shares of this industrial application and the differences in quality that emerged.

Of the technologies used in the industry, the gelatin capsule had the largest market share in 1986. Over the last 20 years, the trend has been toward melamine–formaldehyde systems for the same application for reasons of cost, ease of manufacture, and quality.

To this day, big volumes of capsule slurry suspensions are manufactured using MF-technology, from which carbonless copy paper is produced. This application reached its peak with 2.6 million tons of carbonless copy paper produced worldwide.

In 2007, 1.9 million tonnes was produced, for which approximately 1600 tonnes of 50% microcapsule slurry was required.

Here in Europe, only small quantities are sold for special applications now, while in Asia, India, and Brazil this technology still sells in high volumes.

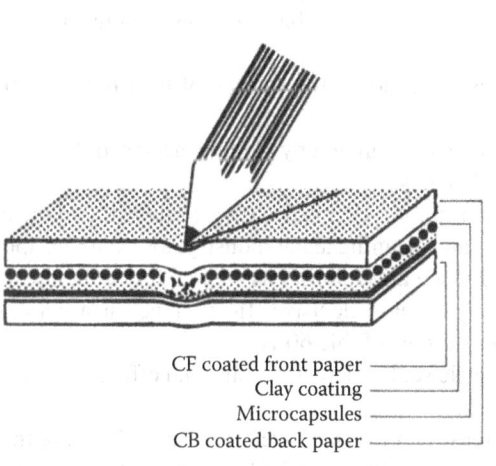 

FIGURE 60.5 Principle of carbonless copy paper. Coating weight 3 to 4 g/m^2; application: for example, credit transfer forms.

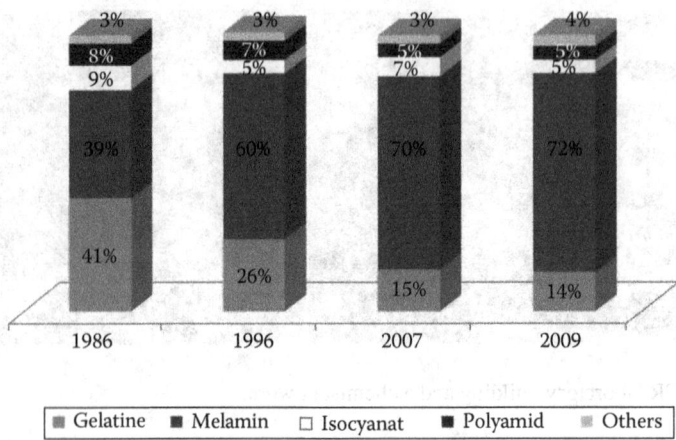

FIGURE 60.6 Technology progress for aminoplastic system.

The detergent and cleaning agent industry has emerged as a new area of application for this type of capsule. Increasingly, over roughly the last 8 years, encapsulated scented oils are being industrially used in fabric softeners. Nowadays they are also found in powder and liquid detergents.

60.2 MANUFACTURE OF MICROCAPSULES

The enzymes and scented oils added to detergents and cleaning agents are expansive raw materials. However, since only a small proportion of the fragrance adhered to the laundry, most of it is washed down the drain unused. Microencapsulation has proven to significantly increase effectiveness in this respect.

All major manufacturers of microcapsules with scented oils as the core material now use different variations of melamine-formaldehyde systems.

In Figure 60.7, a simplified description of how these capsule types are manufactured is shown.

A two-phase system is required for efficient oil-in-water encapsulation. The oil to be encapsulated should ideally be water-insoluble and preferably not react with the reactive components of the formulation.

In the first step, oil-in-water encapsulation by means of MF precondensates produces droplets in the form of an emulsion, which must be stabilized and must have a strong affinity for the MF precondensates used.

In an acidic curing phase, the precondensates enriched on the surface of the drops are further reacted to form a dense network.

The sizes of the particles and droplets are primarily regulated by the mixing speed, the viscosity of the emulsion, and the surface energies of the droplets.

The optimal capsule size in relation to the stability of the capsules varies according to the application. For use in fabric softeners in the detergent and cleaning agent industry, for example, approximately 30 μm is optimal to ensure chemical resistance and stability as well as sufficient boost.

The particle size distribution of the capsule systems has a decisive effect on the surface areas, the encapsulated volumes, and the resulting surface energies (Table 60.1).

If one reduces the system down to the nanometric scale, there are additional effects that can give the particles complete new properties.

It is a universal rule that as particle size decreases and surface area increases, the more the particles tend to agglomerate. This is a well known and widely researched problem with nanoparticles: unless these systems are stabilized, there are no individual particles or capsules. It is therefore crucial to offer the appropriate size distribution for the specific application. A size increases at a

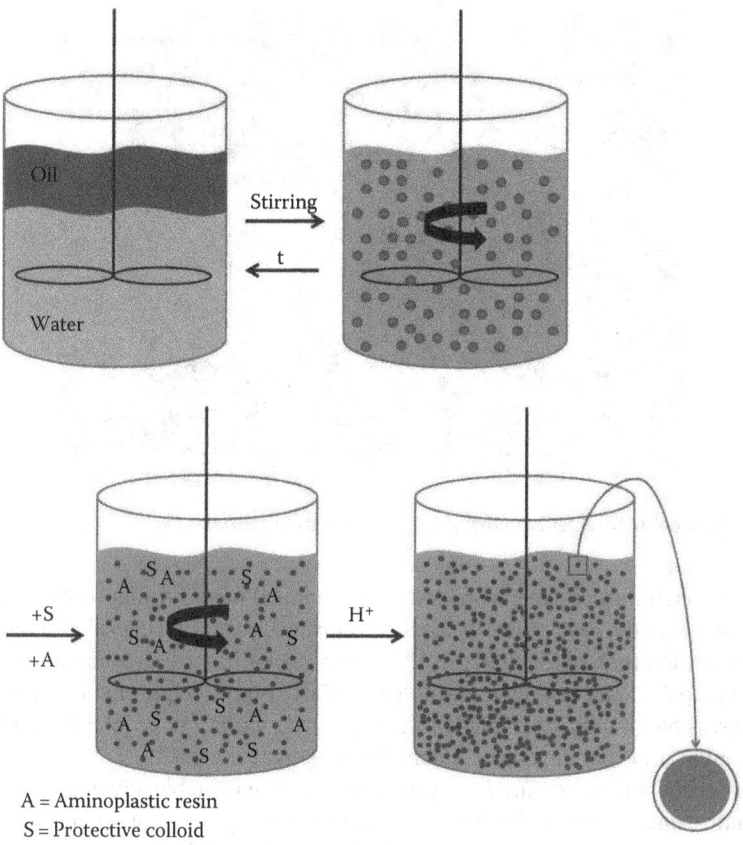

FIGURE 60.7 Manufacture of aminoplastic resin microcapsules.

TABLE 60.1
Effect of Particle Size on the Ratio of Surface Area to Volume

Diameter (µm)	Number of Particles × 1,000,000 Einh.	Spez. Surface (cm²)
1,000	0.002 = 2000	60
500	0.015	120
200	0.24	300
100	2	628
25	122	2,400
5	15,000	12,000
2	240,000	30,000
1	1,920,000	60,000
0.5	15,000,000	120,000
0.2	240,100,000	300,000
0.1	2,000,000,000	628,000

Note: Quantity and specific surface of 1 g capsule powder.

constant wall thickness and cross-linking density, the more mechanically unstable the capsules become. On the other hand, they have a considerably larger core volume. The outstanding quality of MF capsules is that they have excellent chemical resistance and density. In addition, these wall materials are odorless and the stability of the resulting capsules can, to a great extent, be controlled physically, chiefly through their size.

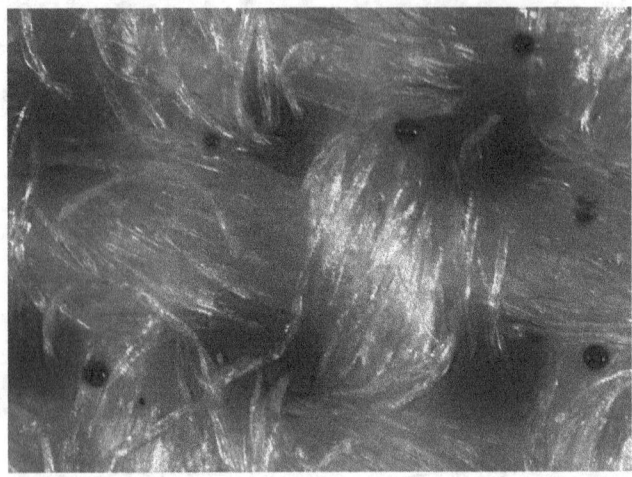

FIGURE 60.8 (See color insert.) Laundry microcapsules use in textiles.

The protective shell thus created is opened by mechanical means. Light rubbing or touching can suffice; for instance, in the case of textile surfaces finished with microcapsules (Figure 60.8).

The amount of literature on the field of MF microcapsules has greatly increased in recent years, with journals and patent disclosures focusing particularly on the requirement to reduce free formaldehyde. Progress had indeed been made in using a wide variety of co-condensation systems; for example, melamine-urea and melamine–latent formaldehyde donors.[3]

But there is no doubt that the pure MF system has some unbeatable advantages on account of its great functionality and reactivity. It's not for nothing that the pure cross-linked melamine-formaldehyde system is the hardest commercially available plastic of all.[4,8]

The investigations have shown that any sort of manipulation using co-condensates has drawbacks; for instance, lower chemical resistance.

The advantage of pure MF systems is linked with equilibrium conditions as a function of degree of curing, which in certain conditions can release more or less formaldehyde. An acidic environment, in particular, always brings about the hydrophilic splitting of methylene ether bridges from precondensates, a process in which formaldehyde is released. This effect correlates with the proportion of aminoplast resin in the wall, and lessens as the degree of cross-linking increases. In microencapsulation, the free formaldehyde value is higher than with MF-based surface coatings, since highly methylolated reagents have to be used and the curing reaction cannot be pushed too far. On parameter for controlling the cross-linking density can be analyzed via C-NMR spectroscopy, using the ratio of methylene bridges within an aminoplastic network. Figure 60.9 shows the reactions, relevant in MF-technology.

Unlike melamine, the precondensates are water-soluble, and in the acidic pH range they form extremely tight networks by further cross-linking under configurable optimized polycondensation conditions.

Special etherified melamine derivates meet the requirements for microcapsule manufacture. They are used to maintain the size distribution of the created droplets until the resin system fully coats them.

It is also advantageous to use special protective colloids, which influence the main two effects of droplet enlargement: Oswald ripening and coalescence. The main reactions of the splitting off of formaldehyde, which give rise to residual formaldehyde in the capsule dispersions and MF microcapsule systems used, are shown in Figure 60.10.

This fact limits the use of such systems in terms of field of application,[4,6,7] and many working groups around the world are looking into viable alternatives.

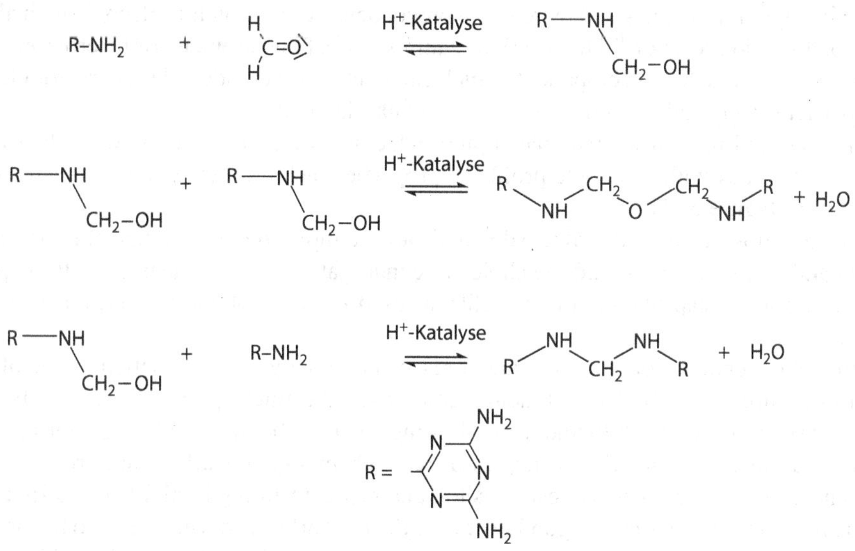

FIGURE 60.9 Precondensate manufacture.

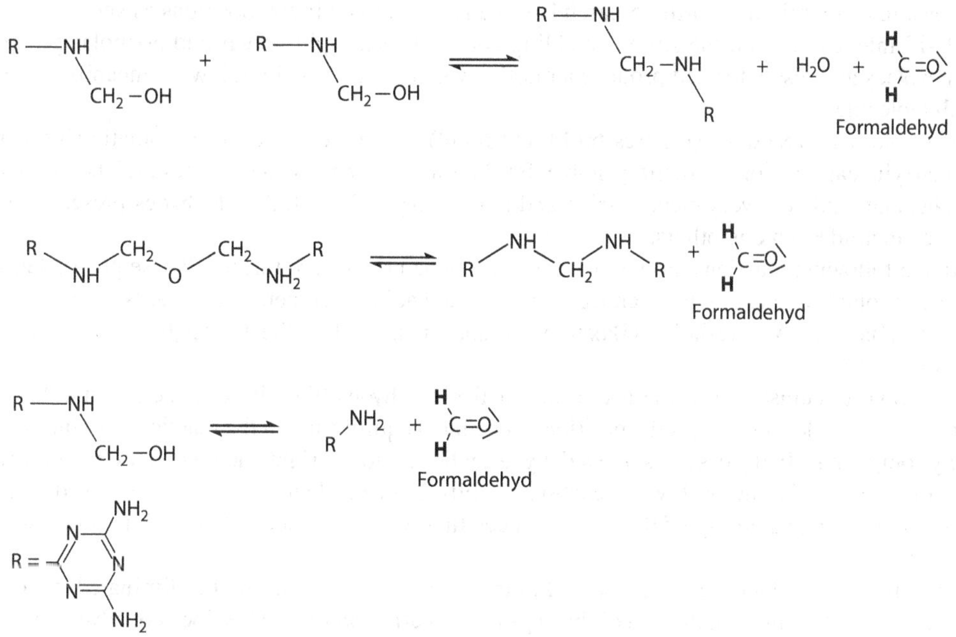

FIGURE 60.10 Formaldehyde equilibrium.

60.3 NEW FORMALDEHYDE-FREE SYSTEMS

Stable core-shell microcapsules, which have a similar profile of properties as MF capsules but do not use formaldehyde, are hitherto unknown in the free market. Follmann & Co. has succeeded in manufacturing stable microcapsules without the most reactive aldehyde-known formaldehyde, which are comparable with MF technology in terms of chemical and physical resistance.

The wall material is synthesized by reacting precondensates of aromatic alcohols with aldehydes in a polycondensation reaction.[9] The reaction requires at least one aromatic or heteroaromatic alcohol and at least one aldehydic component, which can contain other aldehydes apart from formaldehyde, in particular, glyoxal, suiccinaldehyde, and glutaraldehyde.

The simplest and interesting example is phloroglucinol, which by reaction with aldehydes produces precondensates with a suitable profile of properties owing to its symmetrical structure and high reactivity (Figure 60.11).

The molar ratios between the aldehydes and, for example, phloroglucinol, the pH value, the choice of catalysts, surfactants, and the choice of condensation process mean that all the possibilities for oil-in-water encapsulation are available, as used in the resol and novolac chemistry of our ancestors.

Example of dependence of gelation time of resorcinol-aldehyde condensation on the pH value[10] is shown in Figure 60.12. Under high acidic conditions, the linear precondensate with oligomer structure is preferred, in which further cross-linking can only be induced by an external aldehyde donor. In alkaline conditions, the principles of resol chemistry prevail, which proceeds through an A, B, and C stage in the form of resol, resitol, and resite, resulting in highly cross-linked structures. One reason for the extremely good results of the networks produced seems to be the strongly resonance-stabilized transition states of the intermediate benzylic cations in the acidic catalyzed reaction.

The secondary reactions proceed via stabilized carbocations and carboanions (depending on the pH value) and react by substitution, elimination, rearrangement, and disproportionation to form a complex reaction mixture. Little research has been done into their compositions as yet.

Pi–Pi interactions and the strong shielding zone between highly polar and nonpolar groups in the novolacs and resols formed produce a plastic, which seems to exhibit low permeability, even to small molecules.

The following reactions (Figures 60.13 and 60.14) appear relevant.[9] The aromatic ring makes the benzylic carbocation particulary stable. Predominantly linear structures result. The intermediate phenolate anion is very electron-rich and reacts very well with the aldehydes present to form quionen methides, among others.

In the following reactions, both ether and methylene bridges are formed. These precondensates are water-soluble and can only be entirely cross-linked at high temperatures. The Baekeland stages "resole" (Bakelite A), "resitole" (Bakelite B), and "resite" (Balelite C) apply analogously here (Figure 60.15).

A special circumstance in the use of alternative aldehydes like glyoxalic acid glutaraldehyde, glyoxal, succinaldehyde, etc., is the position of chemical equilibrium in the reaction with the polyhydroxy compounds. In the case of some aldehydes, it lies far to the right and gives a quantitative reaction. In other words, the backward reaction—splitting-off of aldehyde—is not observed because the thermodynamic stability of the intermediates than with formaldehyde-based aminoplast resin systems.

In addition, any aldehyde not consumed in the reaction can very easily be eliminated, for example, in alkaline treatment at the end of the capsule formation process. This does not adversely affect

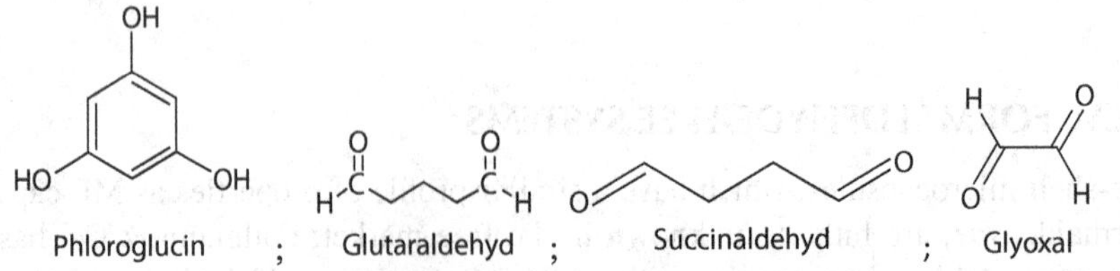

FIGURE 60.11 Structural formula selection of raw material examples.

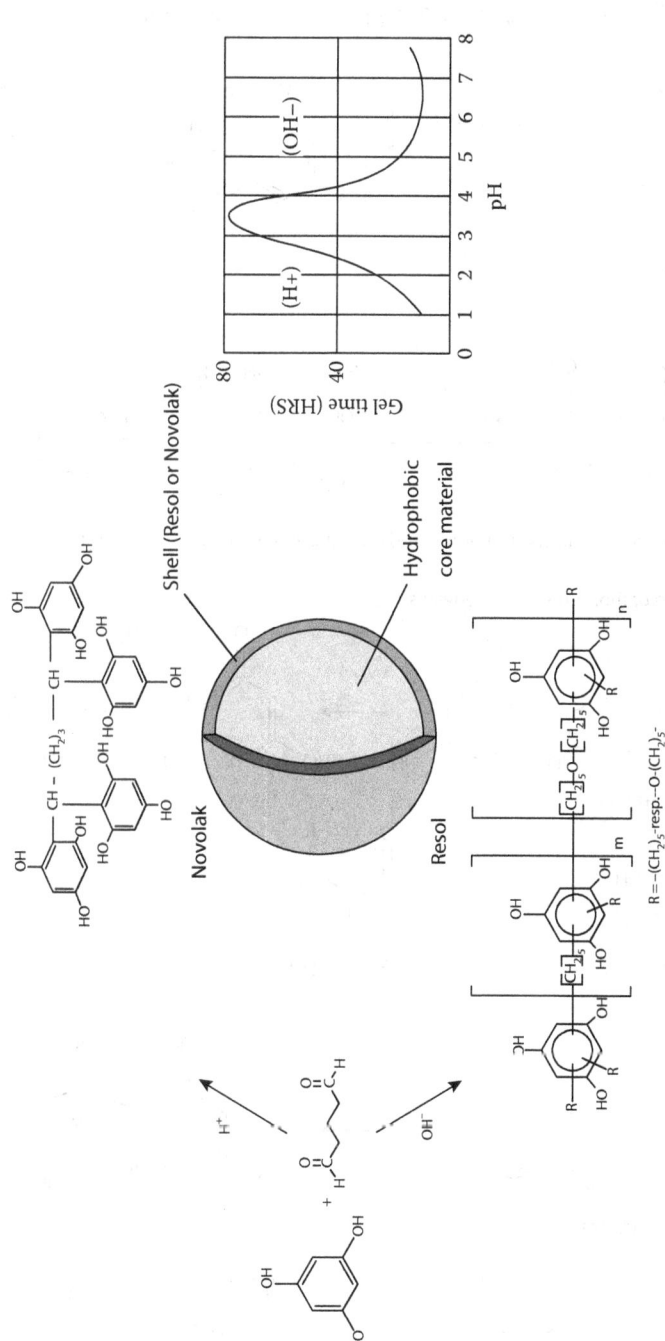

FIGURE 60.12 Novolac- and resole-formation. (With kind permission from Springer Science+Business Media: Resinol Chemistry, 2005, p. 183 ff, Durairaj, R.B.)

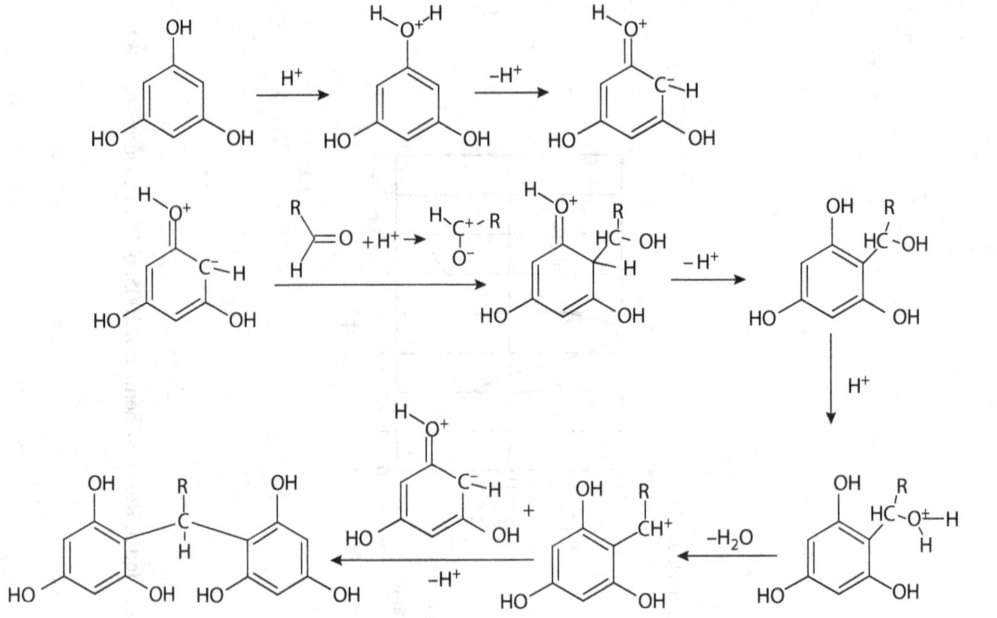

FIGURE 60.13 Proposed mechanism of acid catalyzed methylene bridge formation.

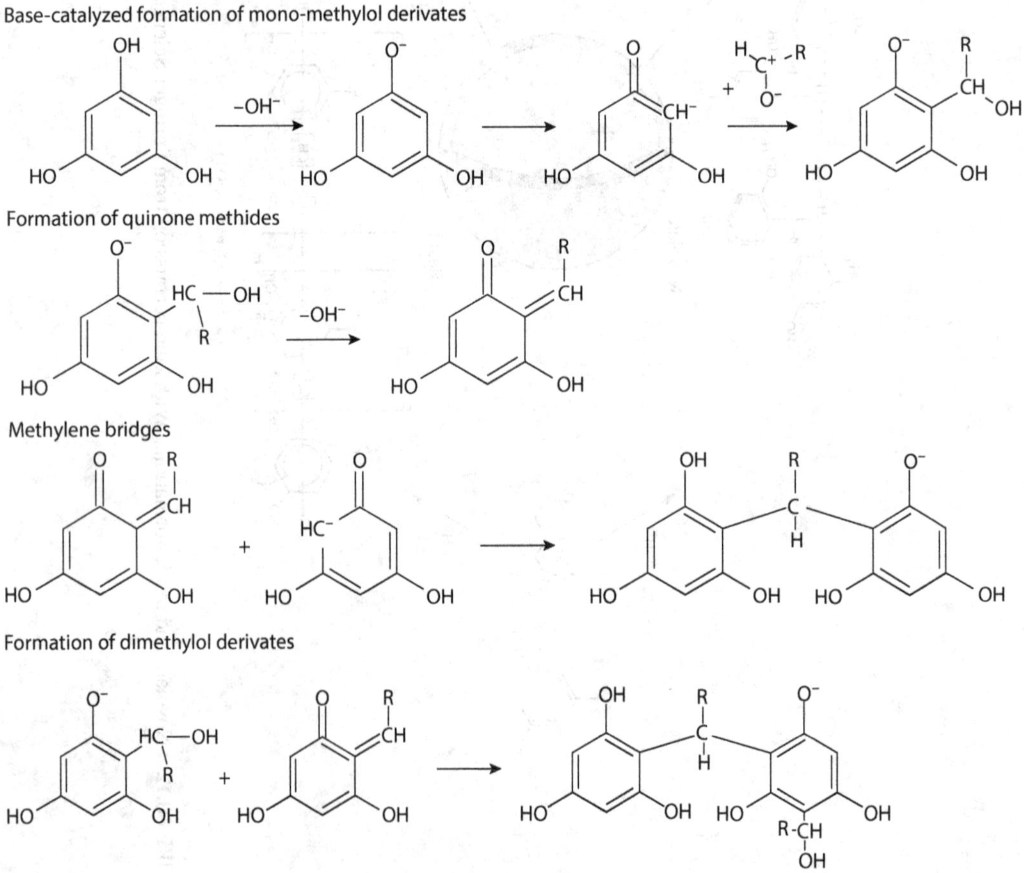

Base-catalyzed formation of mono-methylol derivates

Formation of quinone methides

Methylene bridges

Formation of dimethylol derivates

FIGURE 60.14 Acidic precondensate formation.

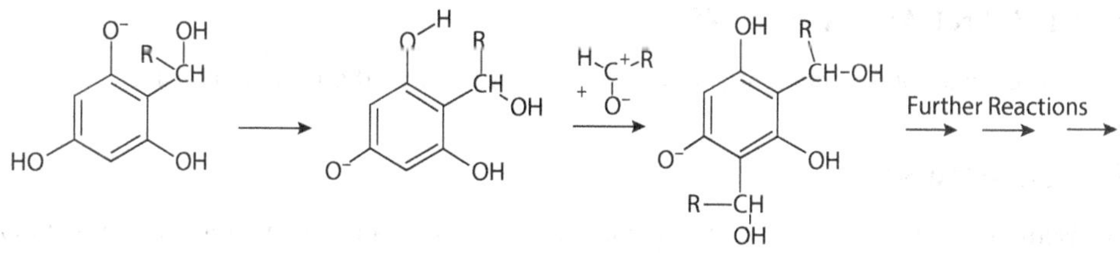

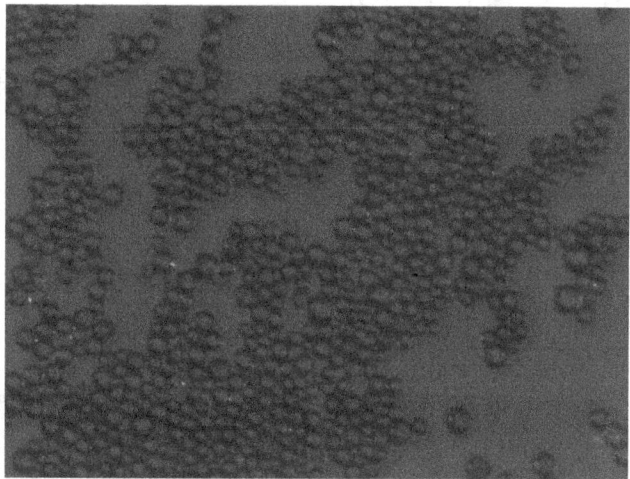

FIGURE 60.15 Alkaline precondensate formation.

the properties of the overall system. This new capsule wall concept opens a new chapter for different applications, which is just beginning.

Because of its very different toxicological evaluation, the microencapsulation technology can revitalize the fields of dentistry, medical technology, pharmaceuticals, cosmetics, detergents and cleaning agents, agriculture, and construction chemistry amongst others (Figure 60.16).

To conclude, it will be shown a way of manufacturing almost monodisperse particles[12] using such systems. This is successful on an industrial scale, and here displays particles with a diameter of approximately 3 μm with a very narrow size distribution (Figure 60.17).

FIGURE 60.16 Stable, formaldehyde-free microcapsules in an aqueous dispersion.

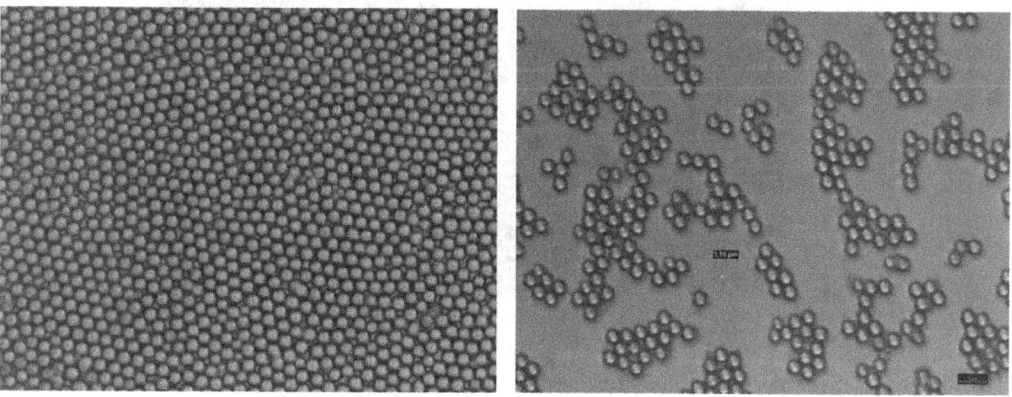

FIGURE 60.17 Almost monodisperse particles made from formaldehyde-free precusors by Follmann & Co.

60.4 CURRENT APPLICATIONS

It follows three examples of current applications in industry based on microcapsules.

60.4.1 Self-Healing

A low-viscosity reactive oil (e.g., dicyclopentadiene), which reacts when catalyzed to form cross-linked structures at even low temperature, was encapsulated (Figure 60.18). The catalyst was homogenously distributed throughout the coating matrix. If a stress crack arose, it inevitably ruptured a capsule and the low-viscosity reactive monomer filled the crack by capillary forces, come into contact with the catalyst, and healed the crack. Many variations and different generations of catalyst have been tried and tested since then. The paint industry, in particular, is trying to develop anti-corrosive self-healing paints on this basis (Figure 60.19).

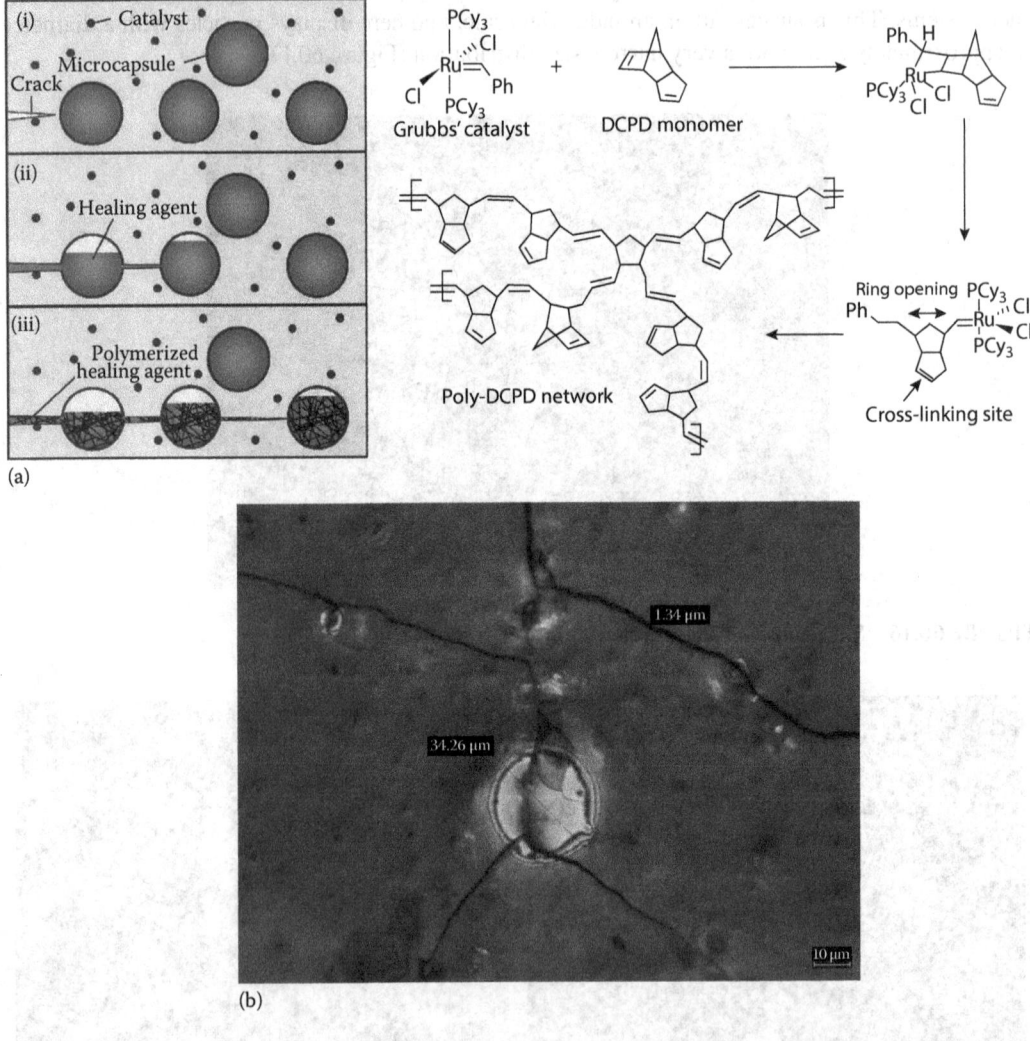

FIGURE 60.18 (See color insert.) (a) Principle of self-healing system basically solution. (b) After provoking stress cracks, microcapsules will break and the core content refills the capillaries. (From White, S.R. and Sottos, N.R., *Nature*, 409, 794, 2001.)

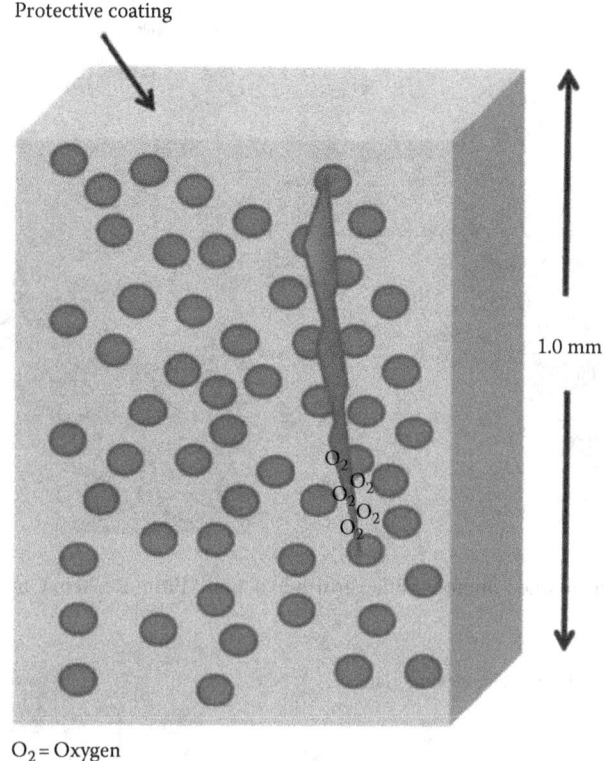

FIGURE 60.19 Self-healing through oxidation of the core material.

60.4.1.1 Application of a Protective Coating

The coating carries microcapsules containing substance reacting with oxygen. If a micro-crack occurs in the coating, also the microcapsule will break. The low viscosity liquid healing agent is released and fill the cracks by capillary forces. The healing agent solidifies by coming in contact with oxygen. The crack is closed.

60.4.1.2 Anti-Friction Coatings

Simplicity was the key to success. Anti-friction coatings are touch-dry lubricants with a formulation similar to that of conventional industrial paints or varnishes. They usually contain hydrophobic lubricants.

Microcapsules filled with lubricant are embedded in the coating. Under strong friction, the embedded microcapsules break and release the lubricant. A lubricating film is created between the contact surfaces. The appeal of the new system is its impressive longevity. The coefficient of friction of water-based microcapsule anti-friction coatings, as determined in fretting tests, was considerably lower than for Mo-disulfide coatings. The service life is significantly longer. Current fields of application are the contact surfaces in plastic/plastic office tracks, steel/steel applications and, above all, automotive applications; for example, on the tracks for the seat adjustment mechanism (to replace greases) window lifts, etc. (Figure 60.20).

The technology is still in its infancy and must undergo long-term testing in the automotive industry.

60.4.1.3 PCM Technology

Phase-change material technology utilizes encapsulated latent heat storage units, the core of which can absorb or emit heat when it changes from the solid to the liquid state and vice versa (Figure 60.21).

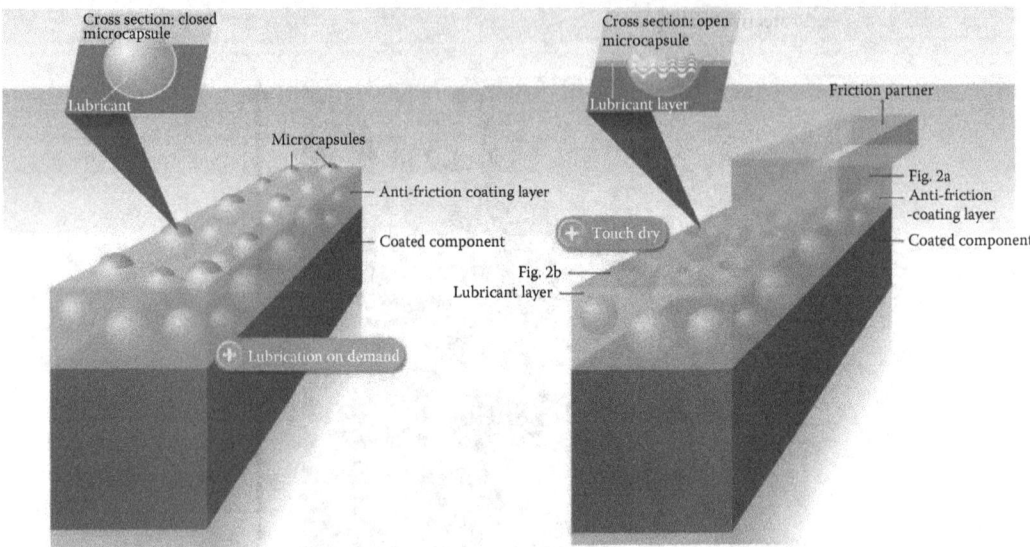

FIGURE 60.20 Lubricant filled microcapsule varnish in action. (From Bechem GmbH, Brochure, 2010.)

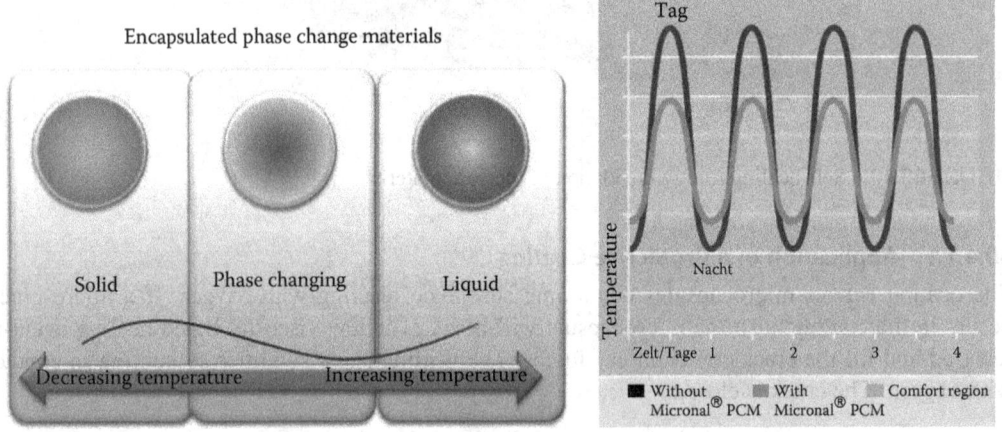

FIGURE 60.21 PCM principle. (From BASF, Brochure, 2008.)

The capsules are used, for example, in high concentrations for interior wall-boards to smooth out the temperature peaks and troughs of the rooms (day/night), thereby saving energy. The core material can be varied for a wide range of temperatures by means of the chain length of the material used, such as paraffin.

For several years, BASF has been working with acrylate-based microcapsules and offers various special products commercially.

These examples illustrate the upsurge that microcapsules have experienced in recent years, and not just in the detergent and cleaning agent industry. The new raw material concept favoring polycondensation based formaldehyde-free microcapsules is industrially feasible and waiting its market breakthrough.

REFERENCES

1. R. Arshady, *Microspheres, Microcapsules*, 1, 13, 77, 1999.
2. NCR-brochure, Encapsulation, Research and Development Facilities, p. 1, 2, 1966.

3. D.F. Müller, *Verfahrenstechnik*, 6, 409, 1972.
3. IFF, European Patent number EP 1719554, 2006.
4. Ciba, US 4100103, 1976 BASF, US 44 06816, 1983, BASF EP 0218887 B2, 1986.
5. C. Feldmann, *Angewandte Chemie*, 122, 1402–1437, 2010.
6. BASF, US 62 61483, 2001; BASF US 2003/0004226, 2003.
7. C. Geffroy, Givaudan, Chimia, 2011, 65, No. 3, p. 177 ff.; Ch. Quillet, Givaudan, WO 2008 09838, 2008.
8. I.H. Updegraff, Aminoresins, pp. 752–789; *Encyclopedia of Polymer Science and Engineering*, 2nd edn., Vol. 1, H.F. Mark et al., eds., John Wiley & Sons, New York, 1985.
9. Follmann & Co. WO 2010/102830 A2.
10. R.B. Durairaj, *Resinol Chemistry*, Springer-Verlag, Germany, p. 183 ff., 2005.
11. E. Campas, *Journal of Microencapsulation*, 25 (3), 154–169, May 2008.
12. A. Hui Lu, *Angewandte Chemie*, 123, 9187–9189, 2011.
13. Bechem GmbH, Brochure, 2010.
14. BASF, Brochure, 2008.
15. White, S.R. and Sottos, N.R., Autonomic healing of polymer composites, *Nature*, 409, 794–797, 2001.

61 Microencapsulation Applications in Food Packaging

*Artur Bartkowiak, Agnieszka Bednarczyk-Drag,
Wioletta Krawczynska, Agnieszka Krudos,
and Katarzyna Sobecka*

CONTENTS

61.1 Introduction 1439
61.2 Flavor Releasing Systems in Food Packaging 1440
61.3 Microencapsulated Coloring Agents and Pigments 1442
61.4 Antimicrobial Packaging 1446
61.5 Insect and/or Rodent-Resistant Food Packaging 1448
61.6 Conclusions/Future Perspectives 1450
References 1450

61.1 INTRODUCTION

Over the last decade, food packaging technology has gone through a fast and significant development especially in case of innovative materials, where some of them have been already commercially applied. The main role of food packaging is to secure proper preservation of food by protecting it from the influence of the various external and internal environmental conditions. Busy lifestyles and rising disposable income are resulting in increased consumption of packaged food. The days when food packaging was used simply to provide information about packed product, protect and provide convenient method of transportation are long gone. Current innovations of food packaging technology are focused on following functions: better protection, more efficient and long-lasting quality preservation, and enhanced safety in general for both consumer and environment. In addition, nowadays active and smart packaging facilitates increased functionality of packed product. The most recent packaging innovations on the market include unique features that capture the imagination of consumers providing an interactive communication experience that lasts long after a consumer has discarded the packaging. Most of such innovations lead to extending shelf life, and even further providing useful information that updates during the product's lifetime. Nowadays innovations in food packaging materials are perceived as combination with factors that will affect its further development and potential industrial application.

To fulfill all these requirements and challenges, the new tailored-made packaging materials using various encapsulation technologies have been recently developed. Encapsulation in most cases refers to an active component delivery system, designed for sustained release of specific products and additives, such as antimicrobials, colors, and aromas. With active and intelligent packaging growing in popularity among novel functional packaging solutions, encapsulation techniques are emerging as an efficient and safe mode of protecting and delivering specific products, which in most cases refer to additional functionalities of packaging.

This chapter presents selected solutions in food packaging where microencapsulation technologies have been applied such as (1) flavor releasing, (2) coloring agents and pigments, (3) antimicrobial, and (4) insect or rodent resistance.

61.2 FLAVOR RELEASING SYSTEMS IN FOOD PACKAGING

Flavors can be among the most valuable ingredients, and many food products on the market have their own distinct aromas. The aroma of a product can be a main driver in a consumer's purchase decision. For example, most people who purchase personal care products, for example, shower gels or shampoos, will firstly open the bottle in order to sample their aromas. In this context, packaging manufacturers and brand owners are seeking ways to communicate their offer while keeping their products properly packaged and safe. Aroma/flavor of a food product can also be a key factor in the consumer's experience with the product prior to consumption.

Many manufacturers have experimented for years in adding aroma into the headspace of a container to improve consumer perception. However, the degradation of unprotected flavors will often occur inside a package or a bottle, also due to their interactions within the container. To solve these problems, some manufacturers have considered adding encapsulated scented material to product packaging.

Encapsulation of flavors has been attempted and commercialized using many different methods such as spray-drying, spray chilling or spray-cooling, extrusion, freeze drying, coacervation, and molecular inclusion. The choice of appropriate microencapsulation technique depends upon the end use of the product and the processing conditions involved in the manufacturing product.[1] Microencapsulation process of flavorings as food ingredients is very similar to that of flavorings for food packaging applications. The microencapsulate coating contains flavor or fragrance scented compounds, which are mixed with a binding agent to create an emulsion in order to enable the adhesion to plastics and other common packaging materials.

Scented microencapsulated coatings help to keep flavors and fragrances fresh from oxidation and degradation. At the point of activation, when the microencapsulate material is physically altered to break open, the microencapsulate releases the intended scent. One example of microencapsulated material is what is typically referred to as "scratch and sniff." While this is an effective way to convey scent, the cost of printing a sticker with microencapsulates and then applying it to a product can be very expensive. On the other hand, during scratching of a paper product, the paper can actually wear away leaving the package looking as though it is damaged.

Various types of application methods may be used to apply microcapsules on food packaging. For example, microcapsules may be dispersed in a lacquer or ink that is coated on a can or container during the manufacturing processes. Alternatively, microcapsules may be sprayed on to a can or container such that the microcapsules form a film that dries or is cured on the substrate of the container. Due to the nature of the scented microencapsulate coating, the processes of air drying with or without light heat or curing using UV radiation are preferred to prevent the premature release of scented compounds. However, one of the best methods to apply microencapsulated flavor on packaging is the introduction of microcapsules into the structure of plastics package components, where the flavor controlled release will be activated during handling of the product.

A number of technologies and patents have been developed to help deliver aroma on a package. For example by the U.S.-based company ScentSational Technologies, LLC, which is the world's leader in developing, patenting, and licensing brand enhancing Scented Packaging technologies for food, beverage, pharmaceutical, personal care, and other consumer products. Using their CompelAroma® technologies encapsulated food grade aromas/flavors can form an integral part of the package, during manufacturing methods, including blow molding, injection molding, thermoforming, and extrusion. EncapScent™ Coatings (MEC's) technology involves applying a scented microencapsulate coating onto a packaging during filling or production of the finished product. The scented microcapsules are ruptured when the product is handled, releasing the product aroma during use. ScentSational Technologies has partnered with Sun Chemical, the world's largest manufacturer of inks and coatings to create a line of scented coatings for print applications. SunScent coatings are applied during printing and release brand and product aroma in store and during use. It can be printed directly onto film, cartons, paperboard, or other packaging materials on commercial print presses as the package is being printed.

As reported in the U.S. Patent Application No. 13/287122,[2] entitled "System and Method for Applying Aroma Releasing Material to Product Packaging," which discloses a system for creating a microencapsulate coating composition and the method of adding the scented microencapsulate coating to product packaging, the encapsulated particles are mixed with a binding agent to form an emulsion, to create a scented microencapsulate coating, which is then applied to packaging in areas that will be touched as the packaging is manipulated.

The physical contact ruptures the microencapsulate coating and releases the scented compounds contained therein. The aroma of the scented material can be any aroma that compliments the flavor of the product being consumed from the flexible bag or plastic bottles. For example, if the bag container holds chips, the scent can be of lime, jalapeno, barbeque, or anything else that complements the true aroma and/or flavor. When the beverage is a sports drink, the scented material can have the aroma of lime, vanilla, cherry, cola, or anything else that matches the anticipated aroma. This positive scent experience supersedes any negative scent perception caused by stale gases trapped in the headspace of the flexible bag and plastic bottle.

In some cases, the scented microencapsulate coating can be activated just during opening, by applying microencapsulate and then using a shrink band seal, which protects the scented microencapsulate coating from inadvertent contact prior to purchase.

This patent also disclosed the method of application the scented microencapsulate coating to a microwave tray surface. The scented microencapsulate coating breaks down and releases flavor during heating. In this manner, when a person opens the microwave door, they are presented with a strong pleasant scent that need not originate from the actual food. Also Yeo et al.[3] describes that microencapsulated flavors and aromas have been adapted for microwave and frozen foods packaging films to improve their appeal and release the flavor oil of interests during heating.

Microencapsulation of flavor/fragrance may help to control their release under specific conditions. Since the scented microencapsulate coating will be contacted when the cup container is being used, some of the binding agent and microencapsulated flavor will be released by the moisture, heat, and physical contact, when the cup is brought close to the face. In this case, if the cup container holds coffee, the microcapsules flavor can be of cinnamon, coco, vanilla, and anything else that compliments the true coffee flavor. It may also be preferable to add a sweet, sour, minty, or other sensation to the brim, which will deliver a mouth feel and enhance the drinker's experience. This positive scent experience displaces any negative scent perception caused by the synthetic material of the cup container.

A widely popular container or carton used to store liquid or solid food is coated/laminated with various barrier and sealant materials, for example, a high-density polyethylene (HDPE), a heat sealable polymer material such as low-density polyethylene (LDPE), or a heat-sealable barrier material such as an ethylene vinyl alcohol copolymer (EVOH), which can emit unwanted odors created during extrusion coating or package converting/sealing.

The invention published in U.S. Patent Application No. 09/828807,[4] entitled "Aromatized Carton with Delayed Release Fragrances," describes novel carton structures, which contain flavor/fragrance releasing microcapsules, which effectively address the problem of unwanted odor emissions from polymers, and enhance the aroma of the packaged liquid or solid, food, or non-food product by enhancing the aroma of its container. This patented solution includes a package for containing the liquid food product and delayed release fragrance microcapsules throughout the life of the product, incorporated into a cap, liner, or spout of the carton.

Also U.S. Patent Application No. 13/861936,[5] entitled "Structures and Methods for Controlling Fragrance Release Using Encapsulated Fragrance on Container Bodies," which is assigned to Crown Packaging Technology, Inc., discloses a system in which container or part of a container may comprise fragrance encapsulated in microcapsules configured to release fragrance emitting substance at least one predetermined period of time.

In addition to modifying perceived tastes, the microcapsules may be used to otherwise affect the consumer experience. Some food packaging systems may include different microcapsules that have two or more fragrances. In one embodiment, a can of regular flavor cola may have microcapsules with cherry fragrance and microcapsules with vanilla fragrance. The end user of such a product may perceive the regular flavor cola as having a cherry vanilla cola flavor.

Various kinds of active substances can be added to a packaging material to improve its functionality and give it new functions. Due to the environmental impact of packaging, several works have also been focused on development of edible films loaded with essential oils.[6]

Edible coatings cannot be typically considered packages but rather physical food protecting barriers, which additionally act as carriers of active and/or bioactive substances and controlled release of flavor molecules, giving them an added-value. Edible film can carry active components, such as flavors and other food additives, in the form of hard capsules, soft gel capsules, microcapsules, soluble strips, flexible pouches, coatings on hard particles, and others.[7]

Encapsulation with edible films allows the control of flavor loss, which strongly affects food quality during the processing or storage of food. Edible films could be applied to food as active packaging, with the aim of gradually releasing aroma compounds with time and thus of maintaining the characteristic flavor of food product. Microcapsules prepared from ι-carrageenans are interesting for good mechanical and gas barrier properties. To investigate the possibility of using edible films as flavor carriers, different aroma compounds were added to carrageenans, mainly typical flavors present in some food products, for example, ethyl acetate—pineapple; ethyl butyrate—fruits, pineapple; ethyl iso-butyrate—apple-like odor; ethyl hexanoate—fruity, banana, pineapple; ethyl octanoate—fruity, floral odor (wine apricot note); 2-pentanone—acetone-like; 2-heptanone—banana, slightly spicy odor; 2-octanone—floral and bitter, green fruity odor; and an n-hexanol—fruity and aromatic flavor).[8] According to Hambleton et al.,[9,10] the use of microcapsules with aroma compounds in hydrocolloid-based edible films confers to them the status of active films and the study on the relative encapsulation of aroma compound by iota-carrageenan matrix leads to promising potential applications for flavor encapsulation as well as flavoring by using food surface coating technologies.

61.3 MICROENCAPSULATED COLORING AGENTS AND PIGMENTS

Microcapsules with pigments were used for the first time on an industrial scale in 1954 by NCR Corporation™ (Appleton Papers). Two chemists, Lowell Schleicher and Barry Green, invented a system that had a liquid core, solid coat and when coated on a sheet of paper formed carbonless paper. Subsequent years have shown the importance of that invention. Nowadays encapsulation is used by every branch of the industry: packaging, food, pharmacy, medicine, or textile, to protect environment, etc. Microcapsules are the type of packaging that can be used to protect different substances, from active substances, coloring agents, and microorganisms to pesticides. In food industry, microcapsules are the most popular solution to protect active compounds against adverse conditions like temperature, light, humidity, radiation, to mask organoleptic properties like odor, taste, or color substances, and to obtain controlled release and avoid migration of substances to food or from food.[11]

Packaging pigments and color agents are crucial in food industry because color agents are very sensitive for storage and process conditions. Natural color agents like carotenoids, anthocyanin, and chlorophylls are responsible for the yellow, orange, and red color of foods.[12] Natural colorants are generally unstable, which restricts their commercial application.[13] Encapsulation is a method that can protect these unstable colorants from oxidation, light, temperature, pH, and redox agents and can also influence their particle size[14] (Figure 61.1). Microencapsulation efficiency and stability are dependent on the wall material composition. Wall material could be a natural or synthetic polymer, such as natural gum, proteins, maltodextrin, and starch, and, at the same time should be permitted for contact with food. A perfect coating material should have both film-forming and emulsifying properties, be biodegradable and possess specific solubility depending on conditions.[15]

Microencapsulation Applications in Food Packaging

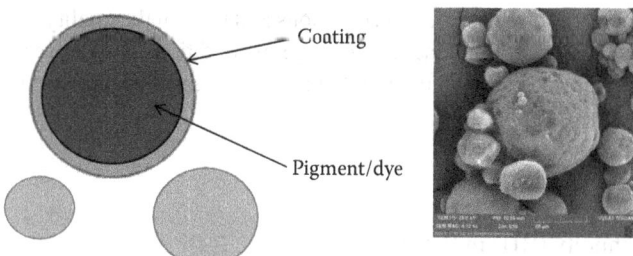

Capsul (CC BY 3.0) by *Wioletta Krawczyńska*

FIGURE 61.1 Schematic representation of a microcapsule containing pigment substance and SEM image of spray-dried microcapsules loaded with dye.

The most popular method of pigments encapsulation is spray-drying. Many papers present information about microencapsulation by spray-drying, and this method is most frequently used to protect pigments obtained from vegetables, fruits, flowers, and plants, for example, black carrot, curcumin,[16,17] beetroot,[18] or roselle.[16] Application of natural polymers as shell to protect beetroot color increased the half-life of spray dried dye about three times, whilst pink color was stabile for about 8 weeks.[19] In case of red roselle pigments stabilization, after microencapsulation, it was acceptable to extend the storage time under different water activities up to 4 months.[16]

One well-known problem with dyes in food industry and food packaging is their migration. Color migration is the largest problem in confectionery products where often different colors are used on decorated cakes, but is also important during food storage. Microcapsules ensure that the color does not change place from food to packaging and from packaging to food. The use of gelatin/alginate matrix as a layer could reduce the color migration from one layer to another. Encapsulation is also practical for foods that lose color. Created matrix could be used as a wall material that reduces discoloration. That solution is adapted to fish industry, especially surimi and other meet. The polymer could be applied on the surface to assure high quality and stability of product. Components used for the preparation of edible polymers are hydrocolloids, polypeptides, and lipids, their blends and composites. A polymer used for the surface of a packaging can modify its organoleptic properties. It does depend on various components, which are used to create the matrix and on applied additives. These compounds could be flavorings, colorings, and sweeteners.[20] Microencapsulation of pigments/color agents may help to control color release in function of various external conditions. If composition is applied as coating on cellulosic surface (paper, paperboard), then it also protects packaging from marks. Thus, it protects natural dye from discoloration as well as prevents its migration; dyes are more stable and more resistant to external conditions (Figure 61.2).

Another possibility to use microencapsulation for food packaging modification is to incorporate microcapsules loaded with dyes as indicators. Higher consumer awareness and the demand for quality and fresh food contributed to the search for new applications and finally to the creation of color control indicators. The main objective of those indicators is to give information to the costumers on

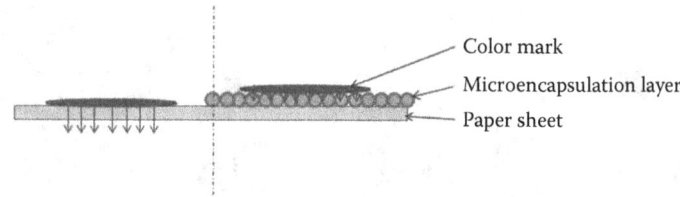

Microcapsuel layer (CC BY 3.0) by *Wioletta Krawczyńska*

FIGURE 61.2 Modified paper by microcapsule layer.

the packaging, storage, and transportation conditions and thus on the quality of food. Indicators are inks encapsulated in polymer coat that react to a change in conditions and deliver information by a change in color. Packaging indicators, depending on the type of external trigger, can be classified as follows:

- Time–temperature indicator
- Gas, oxygen indicator
- Fresh indicator
- Relative humidity (RH) indicator

Time–temperature indicators are used for products where the storage and transport in appropriate conditions is very important. This indicator informs on the difference between the optimal temperature and the actual temperature and how long a product was exposed to inappropriate temperature conditions. An indicator must be able to monitor one or more conditions: chill temperature, frozen temperature, temperature abuses, partial, or full history. The principle operation is based on the polymerization reaction or enzymatic fat hydrolysis or diffusion of the chemically modified dye from encapsulation layer.[21] Color change must be correlated to food shelf-life. The time–temperature indicator could be placed as a label or as a small packaging in shipping containers (Figure 61.3).

Another application of microcapsules is in both gas and oxygen indicators. Gas indicators are monitoring atmosphere parameter inside a package and tight seal packaging. It must have a contact with environment of the food in a package at all times.[22] Oxygen indicator changes the color when the oxygen is present. It indicates the potential leak, an increase in oxygen concentration or verifies the efficiency of oxygen absorber (Figure 61.4). Oxygen indicator could be used in a packaging made of polymers such as polyethylene terephthalate (PET), polypropylene (PP), and polyethylene (PE). The oxygen indicator gel (OIG) is used to determine the conditions in a critical area where the oxygen enters a bottle. The change in gel color in the area where the oxygen enters the product can be a visible indicator[23] (Figure 61.4).

Different researchers, Vu and Won,[24] investigated the problem of leakage of the dye from oxygen indicators, which are contained in an aqueous medium. They presented the results of the study on oxygen indicator films with added alginate. Introduced alginate diminished the dye leakage from 80.80% ± 0.45% to only 5.80% ± 0.06%. Reducing the leak of the dye from the oxygen indicator is due to ion binding ability of alginate. The solution presented by those authors is very simple and profitable.

Another example of the use of encapsulated dye as indicator is the measurement of relative humidity RH. The RH indicator is used in the packaging and transportation of sensitive components, such as food packaging, to regulate and control the humidity to protect books and paints and also to prevent the growth of bacteria and mould. Mills et al.[25] proposed encapsulation of

FIGURE 61.3 Schematic of time–temperature indicator created by The OnVu™. (From http://www.onvu.de/?page_id=25.)

Microencapsulation Applications in Food Packaging

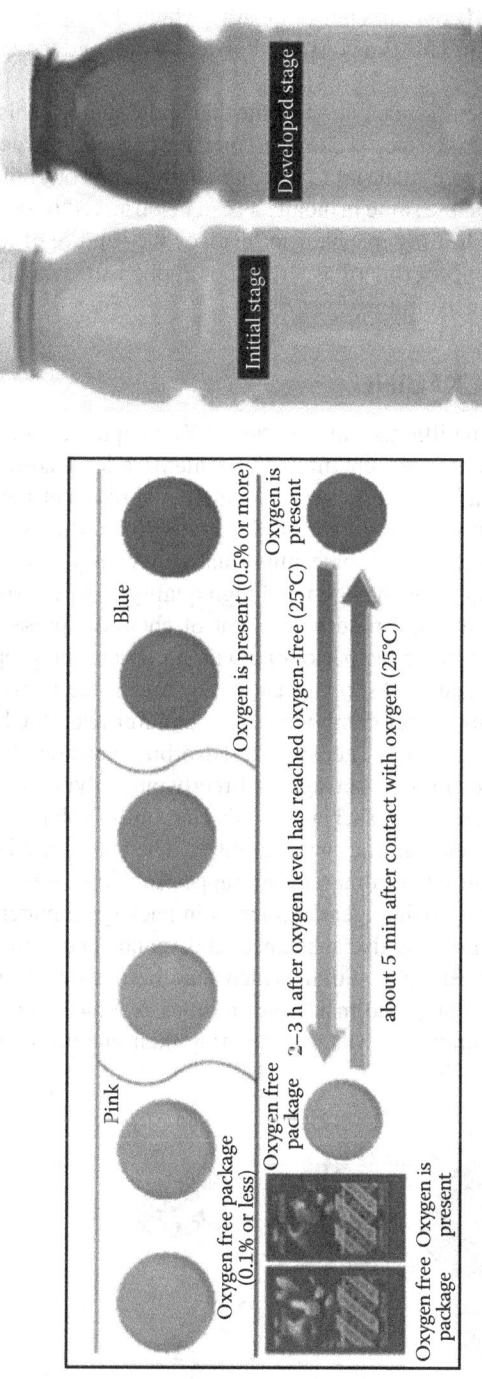

FIGURE 61.4 (See color insert.) Optical indicator of the presence or absence of oxygen. (From http://www.mgc.co.jp/eng/products/abc/ageless/eye.html, http://machinedesign.com/archive/oxygen-liv ng-color.)

compounds such as thiazine dye or methylene blue (MB) in a polymer such as hydroxyethylcellulose with urea. Obtained ink is blue when the RH reaches the level above 85%, and the color changes when the humidity is under 60% or in dry conditions. Under ambient conditions with RH below 85% MB is encapsulated in urea crystals as the pink trimer. When exposed to high RH, the urea crystals dissolve, releasing the MB into an environment in which the more stable form is the blue colored MB dimer and monomer.[25] This type of indicators could be used to control conditions during storage and ripening of fruits.

Odom et al.[26] presented a system for detection of mechanical damage. This system is based on core-shell microcapsules. When the surface is damaged then the capsules are destroyed and the core materials are released. Then the core material reacts and changes its color, thereby indicating the damage. The mechanism works on the same principle as can be observed in self-healing packaging. It was demonstrated that the self-healing mechanism involves the rupture of microcapsules occurring in thin coatings on surface of paperboard as a result of applied stress, with subsequent release of the core material with sealing capabilities.[27]

61.4 ANTIMICROBIAL PACKAGING

In developed countries, food borne illnesses affect about 30% people per year.[28,29] Therefore, it is important to preserve food well and to prevent microbial contamination leading to food spoilage. In spite of a number of commonly used food preservation methods, like thermal processing, drying, refrigeration, freezing, fermentation, salting, or modified atmosphere, this issue is still a challenge for food producers. Especially this has become important in recent years, when the consumers awareness has increased significantly as they demand high quality, safe, convenient, and nutritious food products with a long shelf-life and reduced content of chemical preservatives.[30] Hence the increasing interest in development of active packaging with antimicrobial properties. This type of packaging is able to inhibit pathogens and spoilage caused by microorganisms through the gradual release of contained therein antibacterial compounds.[31] Appendini and Hotchkiss[32] distinguished several types of antimicrobial packaging: (1) coating or adsorbing antimicrobial agents onto polymer surfaces, (2) incorporation of antimicrobial agents directly into polymers, (3) immobilization of antimicrobials to polymers, (4) addition of sachets or pads containing volatile antimicrobial agents, and (5) use of polymers with inherent antimicrobial activities. The antimicrobial activity of some of them is achieved by introducing an active compound entrapped in microcapsules. Figure 61.5 shows one concept of the released antimicrobial agent entrapped in packaging materials, which is called "BioSwitch." The idea is the release of active substance on demand. The antimicrobials are encapsulated in natural polymers, for example, starch, which may be destroyed by bacterial enzymes activity. So the active agents stay entrapped inside the capsules as long as the packed food product is free of microorganisms. Appearance of bacteria secreting their enzymes will cause progressive

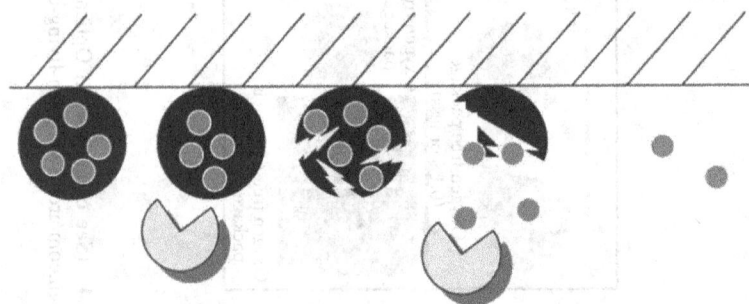

FIGURE 61.5 Schematic representation of antimicrobial active-packaging. Micro-organisms hydrolyze starch-based particles, causing release of the antimicrobial lysozyme, resulting in inhibition of microbial growth. (From De Jong, A.R. et al., *Food Addit. Contam.*, 22(10), 975, 2005.)

hydrolyzation of polysaccharides (matrix). In consequence, the antimicrobials will be released and inhibit the microbial growth.[22]

One of the technologies used in production of antimicrobial packaging is coating the surface of materials with active layer/coatings. Guarda and co-workers[33] encapsulated two natural antimicrobial agents, thymol and carvacrol, components of oregano and thyme essential oils. Microcapsules were prepared in the emulsion oil in water, where the aqueous phase was gum arabic solution and the oily phase was a mixture of soybean oil with antimicrobials. The coated packaging material was bi-axially oriented polypropylene (BOPP) film. The results showed that both encapsulated carvacrol and thymol can be successfully used as the active agents in antimicrobial food packaging. Tested films with active layers (10% of oils) were effective against *Staphylococcus aureus*, *Listeria innocua*, *Escherichia coli*, and *Saccharomyces cerevisiae*. Next group of researchers[29] determined antimicrobial activity of glass slides coated with cinnamaldehyde enclosed in polydiacetylene-N-hydroxysuccinimide (PDA-NHS) nanoliposomes. Using PDA, the scientists were guided by its natural ability to self-assemble into liposomes.[34] This antimicrobial activated on glass surfaces caused a significant inhibition of growth *E. coli* and *Bacillus cereus*.

Another example of antimicrobial packaging material was presented by.[35] They prepared the trilayer films using polycaprolactone (PCL) and methylcellulose (MC). Antibacterial agents were encapsulated in the MC, which was the internal layer between two layers of PCL films. Added mixtures of antimicrobials contained organic acids and essential oils. The trilayer films were manufactured using compression molding process. Next, these bioactive films were tested as packaging for broccoli inoculated with pathogenic bacteria and stored during 12 days at 4°C. There was observed antibacterial activity against *E. coli* and *Salmonella typhimurium*, and heir growth was totally inhibited after 12 and 7 days of storage, respectively. What is more, the growth of total aerobic microbiota was controlled in broccoli for 10 days.

Different examples of active food packaging with antimicrobial properties are edible coatings and films. Recently, more and more researchers have focused their attention on the development and improvement of this form of food protection. The increased interest is in response to dominating sales trends in the food market, where the growing popularity of the fresh-cut fruit, vegetables, and salads is visible.[36] Basically, edible coatings or films are packaging made entirely from edible components like: polysaccharides, proteins, and lipids.[37] The main objective of the edible coatings/films is to extend the shelf-life of fresh fruits or vegetables. Direct contact with food surface contributes to the gradual release of antimicrobial compound because moisture from food affects microcapsules membrane. In consequence, there is high concentration of released agents achieved between the food surface and capsules membrane and thereby, the inhibitory effect on pathogens and spoilage microorganisms growth may be more efficient.[38,39]

Mekkerdchoo et al.[39] worked on an edible film based on pectin with activity against food spoilage microorganisms. For this purpose, they incorporated into pectin film the antimicrobials encapsulated in liposome formed from lecithin. An advantage of using liposome encapsulation is their multilayer membrane where both polar compound (water or ethanol phase) and nonpolar compound (oil phase such as essential oil) can be encapsulated. This property was useful in that case because the active agents were garlic oil, clove oil, and pomegranate extracts.

Another groups of researchers worked on the multilayered antibacterial edible coatings to prolong the shelf-life of fresh-cut fruits: papaya,[40] pineapple,[41] and cantaloupe.[36] All of them used trans-cinnamaldehyde as an antimicrobial compound, which was encapsulated using β-cyclodextrin. Coatings were prepared by layer-by-layer procedure where fruits pieces were successively dipped into variety of solutions containing oppositely charged polyelectrolytes.[40] The base matrixes of those coatings were alginate in case of papaya and pectin with chitosan in case of pineaple and cantaloupe. All tested active multilayered edible coatings have given positive results and extended the shelf-life (antimicrobial protection against aerobes, psychrophilic microorganisms, yeast, and mold growth) as well as the physicochemical quality (maintaining original texture attributes) of fresh-cut fruits up to 15 days at 4°C.

Different kind of antimicrobial packaging system is presented in case of using volatile antimicrobial agents. The active compound is added into the package in separate sachets[42] or pads[43,44] and slowly released into the atmosphere leading to inhibition of microorganisms growth on food surfaces.[44] The volatile agent that is recently often tested as an active compound of antimicrobial packaging is allyl isothiocyanate (AITC), substance obtained from mustard seed.[43–45] AITC is characterized by a broad antimicrobial activity but on the other hand, it is highly volatile and pungent and accordingly its application in food production is limited.[44] The solution of this problem may be the encapsulation of AITC, which was successfully already used by some researchers. This agent was entrapped, for example, in cyclodextrins (CDs),[43] alginate,[44] and gum acacia[45] microcapsules. Generally encapsulation allowed to achieve odor reduction, decreased undesirable reactivity, and controlled release of the active component.[44] Another volatile antimicrobial compound used similarly was garlic oil, which mainly contains active allyl disulfide.[42] Scientists confirmed that encapsulation has positively influenced both the antimicrobial efficacy of the compounds and the quality of packed food: chopped beef,[45] chopped onion,[43] fresh-cut tomato,[42] and spinach.[44] Piercey and co-workers[43] noticed more efficient growth inhibition of *Penicillium expansum*, *E. coli*, *Listeria monocytogenes*, and fungi when AITC was entrapped in β-cyclodextrins than in case of free AITC or AITC entrapped in alpha-CD. Moreover, the group observed that release of encapsulated volatile antimicrobial compound was proportionally dependent on RH, which was connected with the impact of water on matrix destabilization.[42,44]

61.5 INSECT AND/OR RODENT-RESISTANT FOOD PACKAGING

Food packaging is the last line of defense available for producers against insect and/or rodent infestation of finished food products. Damage due to pests is equally important in stored raw materials and in semi-processed and final food products. Repelling activity of food packaging due to incorporation of microencapsulated active compounds has been more widely discussed in context of insects than rodents, nevertheless we also mention the second application here.

By definition, repellents are substances deterring an insect from flying to or coming into contact with any surface. They can act locally or at a distance and can be used to prevent the insects to enter packages.[46]

Insect species can be classified as penetrators, which damage the packaging materials by penetrating them, and invaders, which enter packaging through existing openings caused by sewing, folding, or damage.[47–51] To the former group, we can include adult forms of *Sitophilus* spp., *Rhyzopertha dominica* (F.), *Lasioderma serricorne* (F.), or *Stegobium paniceum* (L.), and larvae of *Plodia interpunctella* (Hübner) and *Ephestia cautella* (Walker). Some invader species infesting food products are *Tribolium* spp., *Cryptolestes ferrugineus* (Stephens), and *Oryzaephilus* spp.

Insects can access packaged products during transportation and storage, even though finished food products can be shipped from production uninfested. In recent times, more traditional packaging materials, that is, paper and cardboard, have been replaced by plastic films, which have more passive protection function against external damage. However those packaging materials also vary in resistance to penetration by insects, and multilayer films, 7-μm–thick aluminum foil and various plastic films, including 50-μm–thick LLDPE, are not a physical barrier for many stored-product insect species.[51,52]

Various active substances were tested as possible insect repellents, both synthetic and natural. The former ones include picaridin, pyrethrins synergized with piperonylbutoxide, methyl salicylate, and N,N-diethyl-m-toluamide (DEET), commonly used mosquito repellent.[46,49,53] Generally, insect infestation has been controlled by fumigation with methyl bromide, which is a toxic substance and exhibit adverse effects to humans if present at high concentrations. Some synthetic chemicals are not degradable in the environment and thus may cause damage to the ecosystem, especially methyl bromide that contributes to the depletion of ozone layer. Being natural, biodegradable, and environmentally friendly substances, essential oils have been studied as possible replacements for chemical pesticides. They are, however, volatile and susceptible to many physical factors (light, oxygen, temperature), and this decreases the times of protection, and their use in food packaging has many limitations. Thus, currently, the application of microencapsulation techniques of pest control agents has been extensively studied and developed to obtain prolonged activity of active substances, which reduce their toxicity as well as environmental impact.

Microencapsulation of essential oils stabilizes them from environmental damage and enables to control their release rate. By applying microencapsulation technique, it is possible to achieve constant concentration of the active substance included in microencapsulated essential oil in the target.[1,54,55] For example, in case of microencapsulated thyme oil used against *P. interpunctella*, the repellency percentage was maintained at high level of about 90% for at least 4 weeks.[56] It is worth noticing that the repelling effect of the active substance (as well as its release rate) was determined by both the type of coating medium and the combination of coating medium and wall materials.[57]

Several essential oils proved to be active as insect repellents, including thyme oil extracted from *Thymus vulgaris*,[56,58,59] citronella,[60] lemon,[61] ginger[62] and eucalyptus oils,[63] cinnamon oil[57] or cinnamon extract,[64] essential oils from *Mentha piperita*,[65] *Rosmarinus officinalis*,[66] *Ruta graveolens*,[67] or *Nigella sativa*[68] (for more, please see[69]). However, not all of the above mentioned essential oils have been applied directly to food packaging and tested against common insects of stored food products, so their applicability to food packaging remains to be assessed. As shown by Mohan et al.[70] a protein-enriched pea flour solution coating on polyethylene sheets active repelled *Sitophilus oryzae* (L.) and *R. dominica* (F.). The Indian meal moth, *P. interpunctella* (Hübner) was repelled by microencapsulated cinnamon oil applied on LDPE coated with an ink or a polyprophylene (PP) solution containing the microcapsules.[57]

Various techniques have been used for encapsulation of essential oils; however, here only those applied to insect repellent applications will be mentioned. The most extensively used microencapsulation technique for insect repellents microencapsulation is coacervation (however, other microencapsulation techniques may as well be used here, e.g., spray-drying) and the common encapsulants include gelatin, starch, gum arabic, whey protein isolate, soybean protein isolate, maltodextrin, poly(vinyl alcohol), or methyl cellulose.[57,61,62,71-76] For example Maji et al.[61] successfully microencapsulated *Zanthoxylum limonella* oil (ZLO) in glutaraldehyde cross-linked gelatin. In another application, Hussain and Maji[77] used genipin cross-linked chitosan-gelatin microcapsules and salting-out method for ZLO encapsulation. Solomon et al.[60] used formaldehyde cross-linked gelatin and a simple coacervation microencapsulation technique for microencapsulation of citronella oil.

Most microcapsules formed from natural polymers (gelatin, chitosan, poly (L-lactide), or alginate) have low mechanical resistance, low thermal stability, and short-term release properties compared with synthetic polymers (melamine, urea, etc.). Synthetic polymers, on the other hand, are not suitable for direct application to food system due to the toxicity of cross-linking agent and can be applied to the food packaging material in the way of printing or coating. This application was studied by Chung et al.[56] by in situ polymerization microencapsulation of thyme oil in melamine-formaldehyde microcapsules with various emulsifiers. The results showed long-lasting insect repellent effect of the thyme oil contained in the microcapsules.

On the other hand, the study of Kim et al.[57] proved that the use of gum arabic, whey protein isolate/maltodextrin and poly(vilyl alcohol) as encapsulants to obtain microencapsulated cinnamon oil by spray drying resulted in prolonged repellent effect against *P. interpunctella* (Hübner) larvae. In this case, LDPE films were coated with either inks containing the microcapsule emulsions or with PP incorporating microcapsule powders. Interestingly, the tensile properties of LDPE were not affected regardless of the method used.[57]

Rodent repellents are chemicals that keep certain animals away from target areas, preventing them from feeding or gnawing. The substances modify animal behavior by their taste or odor (or both) and may be used in protecting the area and packaged food from rodent infestation. Up to date, there have been limited reports on application of microencapsulated rodent repellent substances of natural origin to food packaging, although there have been some promising studies on this subject.

Commercial pesticides exhibit negative effect on human health by being toxic and environmentally unsafe. As with insect repellents, this led to resurgence in interest in more environmentally friendly and safe natural compounds of plant origin. One of such repellent is terpenes which, when incorporated into various thermoplastic and thermosetting polymers, mask the strong odor of products in a package made from the polymer and acts as animal repellents. Atkinson[78] used microencapsulation process for incorporation of terpenes into LLDPE melt via extrusion to obtain animal repellent plastic film, however not for food packaging application. Singla et al.[79] have shown that eucalyptus oil has a repellent activity against house rat (*Rattus rattus*) in the concentration of 5% as applied by spray. However, the spray had to be applied regularly in short intervals, as the concentration of eucalyptus oil decreased with time. Microencapsulation technique could be used to increase the persistence of the repellent effect for longer period of time. Interestingly some substances, as eucalyptus oil, may be used in food packaging for dual purpose, as they serve as both insect and rodent repellents.

61.6 CONCLUSIONS/FUTURE PERSPECTIVES

Expanding applications of innovative packaging materials and technologies in a wide spectrum of foods, both processed and fresh, is poised to drive growth in the market. Key benefits accelerating adoption of microencapsulation technologies to existing requirements and needs include improved delivery of immobilized components, increase in shelf life, and "tailored made" mechanical and morphological properties. Besides microencapsulation, recently nanoencapsulation is also making waves in the packaging functional additives for its ability to offer improved and enhanced physicochemical properties, including, for example, homogenous dispersability of functional additives. Recently nano-encapsulation has gained a big attention being often used in the development of new delivery systems for various applications, especially to modify the surface of various packaging materials. Furthermore, it is expected that nanotechnology will become increasingly important in the near future. However, existing health and environmental concerns may hinder public acceptance of new nano-technologies including the packaging sector.

Despite many successful implementations, the application of encapsulation technologies in food packaging continues to face challenges, such as issues related to stability of encapsulated additives during both storage and processing of packaging materials. Nowadays microencapsulation technologies still suffer from issues such as difficulty in controlling their mechanical resistance, inefficiencies in controlled release of active ingredients and conforming to standard size of final microcapsules. Furthermore, one of the main challenges in the future could be a lack in development of new materials due to very strict regulations related to food contact materials. Therefore, one could conclude there are still many challenges, which should be overcome in the future by the development of new solution based on microencapsulation techniques in this very interesting and dynamically growing market of functional/intelligent food packaging materials.

REFERENCES

1. Madene, A., Jacquot, M., Scher, J., and Desobry, S. 2006. Flavour encapsulation and controlled release—A review. *Int. J. Food Sci. Technol.* 41:1–21.
2. Landau, S.M. 2013. System and method for applying aroma releasing material to product packaging. U.S. Patent Application No. 13/287122.
3. Yeo, Y., Bellas, E., Firestone, W., Langer, R., and Kohane, D.S. 2005. Complex coacervates for thermally sensitive controller release of flavor compounds. *J. Agric. Food Chem.* 53 (19):7518–7525.
4. Thoman, B.J., Fairchild, T.M., Mabee, M.S., and Cleveland, C.S. 2002. Aromatized carton with delayed release fragrances. U.S. Patent Application No. 09/828807.
5. Ramsey, C.P., Young, P.A., and Abramowicz, D.A. 2013. Structures and methods for controlling fragrance release using encapsulated fragrance on container bodie's. US Patent Application No. 13/861936.
6. Mastromatteo, M., Mastromatteo, M., Conte, A., and Del Nobile, M.A. 2010. Advances in controlled release devices for food packaging applications. *Trends Food Sci. Technol.* 21:591–598.

7. Matalanis, A., Jones, O.G., and McClements, D.J. 2011. Structured biopolymerbased delivery systems for encapsulation, protection, and release of lipophilic compounds. *Food Hydrocolloids* 25(8):1865–1880.
8. Marcuzzo, E., Sensidoni, A., Debeaufort, F., and Voilley, A. 2010. Encapsulation of aroma compounds in biopolymeric emulsion based edible films to control flavour release. *Carbohydr. Polym.* 80:984–988.
9. Hambleton, A., Debeaufort, F., Beney, L., Karbowiak, T., and Voilley, A. 2008. Protection of active aroma compound against moisture and oxygen by encapsulation in biopolymeric emulsion-based edible films. *Biomacromolecules* 9(3):1058–1063.
10. Hambleton, A., Fabra, M-J., Debeaufort, F., Dury-Brun, C., and Voilley, A. 2009. Interface and aroma barrier properties of iota-carrageenan emulsion–based films used for encapsulation of active food compounds. *J. Food Eng.* 93:80–88.
11. Jyoti, V.N., Prasanna, M.P., Prabha, S.N.S.S.K., Seetha, R.P., and Sraawan, G.Y. 2010. Microencapsulation techniques, factors influencing encapsulation efficiency. *J. Microencapsul.* 27(3):187–197.
12. Özkan, E. and Bilek, S.E. 2014. Microencapsulation of natural food colourants. *Int. J. Nutr. Food Sci.* 3(3):145–156.
13. Shahidi, F. and Han, X.Q. 1993. Encapsulation of food ingridients. *Crit. Rev. Food Technol.* 33(6):501–504.
14. Santos, T., Meireles, M., and Angela, A. 2012. Carotenoid pigments encapsulation: Fundamentals, techniques and recent trend. *Open Chem. Eng. J.* 4:42–50.
15. Wang, X., Lu Zhaoxin, L., and Fengxia, L. 2009. Study on microencapsulation of curcumin pigments by spray drying. *Eur. Food Res. Technol.* 229:391–396.
16. Selim, K.A., Khalil, K.E., Abdel-Bary, M.S., and Abdel-Azeim, N.A. 2008. Extraction, encapsulation and utilization of red pigments from Roselle (*Hibiscus sabdariffa* L.) as natural food colourants. In: *Fifth Alex. Conference of Food Dairy Science and Technology, J. Food Sci. Technol.*, Alexandria, Egypt, pp. 7–20.
17. Aziz, H.A., Peh, K.K., and Tan, Y.T.F. 2007. Solubility of core materials in aqueous polymeric solution effect on microencapsulation of curcumin. *Drug Dev. Ind. Pharm.* 33(11):1263–1272.
18. Janiszewska, E. and Włodarczyk, J. 2013. Influence of spray drying condition on beetroot pigments retention after microencapsulation process. *Acta Agrophys.* 20(2):343–356.
19. Ersus, S. and Yurdagel, U. 2007. Microencapsulation of anthocyanin of black carrot (*Daucuscarota* L.) by spray drying. *J. Food Eng.* 80:805–812.
20. Subhas, C.S. and Pathik, M.S. 2014. Edible polymers: Challenges and opportunities. *J. Polym.* 2014:13, ID 427259.
21. Sykut, B., Kowalik, K., and Droździel, P. 2013. Modern packaging for food industry. *Eng. Sci. Technol.* 3(10):121–138.
22. De Jong, A.R., Boumans, H., Slaghek, T., Van Veen, J., Rijk, R., and Van Zandvoort, M. Active and intelligent packaging for food: Is it the future? *Food Addit. Contam.* 22(10) 2005:975–979.
23. Hoffman, J. September 15, 2005. Oxygen in living color. Machine design. Online (http://machinedesign.com/archive/oxygen-living-color).
24. Vu, C.H.T. and Won, K. 2013. Novel water-resistant UV-activated oxygen indicator for intelligent food packaging. *Food Chem.* 140:52–56.
25. Mills, A., Grosshans, P., and Hazafy, D. 2010. A novel reversible relative-humidity indicator ink based on methylene blue and urea. *Analyst* 135:33–35.
26. Odom, S.A., Jackson, A.C., Prokup, A.M., Chayanupatkul, S., Sottos, N.R., White, S.R., and Moore, S.J. 2007. Visual indication of mechanical damage using core–shell microcapsules. *Appl. Mater. Interface* 3:4547–4551.
27. Andersson, C., Järnström, L., Fogden, A., Mira, I., Voit, W., Żywicki, S., and Bartkowiak, A. 2009. Preparation and incorporation of microcapsules in functional coatings for self-healing of packaging board. *Packag. Technol. Sci.* 22:275–291.
28. Stringer, M. 2005. Summary report: Food safety objectives—Role in microbiological food safety management. *Food Control* 16:775–794.
29. Makwana, S., Choudhary, R., Dogra, N., Kohli, P., and Haddock, J. 2014. Nanoencapsulation and immobilization of cinnamaldehyde for developing antimicrobial food packaging material. *LWT—Food Sci. Technol.* 57(2):470–476.
30. Sung, S.-Y., Sin, L.T., Tee, T.-T., Bee, S.-T., Rahmat, A.R., W.A.W.A. Rahman, Tan, A.-C., and Vikhraman, M. 2013. Antimicrobial agents for food packaging applications. *Trends Food Sci. Technol.* 33:110–123.
31. Han, J.H. 2000. Antimicrobial food packaging. *Food Technol.* 50:56–65.

32. Appendini, P. and Hotchkiss, J. 2002. Review of antimicrobial food packaging. *Innovat. Food Sci. Emerg. Techn.* 3:113–126.
33. Guarda, A., Rubilar, F.J., Miltz, J., and Galotto, M.J. 2011. The antimicrobial activity of microencapsulated thymol and carvacrol. *Int. J. Food Microbiol.* 146:144–150.
34. Jung, K., Kim, T.W., Park, H.G., and Soh, H.T. 2010. Specific clorimetric detection of proteins using bidentate aptamer-conjugated polydiacetylene (PDA) liposomes. *Adv. Funct. Mater.* 20:3092–3097.
35. Takala, P.N., Salmieri, S., Boumail, A., Khan, R.A., Dang, V.K., Chauve, G., Bouchard, J., and Lacroix, M. 2013. Antimicrobial effect and physicochemical properties of bioactive trilayer polycaprolactone/methylcellulose-based films on the growth of foodborne pathogens and total microbiota in fresh broccoli. *J. Food Eng.* 116:648–655.
36. Martiñon, M.E., Moreira, R.G., and Castell-Perez, M.E., and Gomes, C. 2014. Development of a multilayered antimicrobial edible coating for shelf-life extension of fresh-cut cantaloupe (*Cucumis melo* L.) stored at 4°C. *LWT—Food Sci. Technol.* 56:341–350.
37. Pascall, M.A. and Lin, S.-J. 2013. The application of edible polymeric films and coatings in the food industry. *J. Food Proc. Technol.* 4:e116.
38. Quintavalla, S. and Vicini, L. 2002. Antimicrobial food packaging in meat industry. *Meat Sci.* 62:373–380.
39. Mekkerdchoo, O., Patipasena, P., and Borompichaichartkul, C. 2009. Liposome encapsulation of antimicrobial extracts in pectin film for inhibition of food spoilage microorganisms. *Asian J. Food Agro-Ind.* 2:817–838.
40. Brasil, I.M., Gomes, C., Puerta-Gomez, A., Castell-Perez, M.E., and Moreira, R.G. 2012. Polysaccharide-based multilayered antimicrobial edible coating enhances quality of fresh-cut papaya. *LWT—Food Sci. Technol.* 47:39–45.
41. Mantilla, N., Castell-Perez, M.E., Gomes, C., and Moreira, R.G. 2013. Multilayered antimicrobial edible coating and its effect on quality and shelf life of fresh-cut pineapple (*Ananas comosus*). *LWT—Food Sci. Technol.* 51:37–43.
42. Ayala-Zavala, J.F. and Gonzalez-Aguilar, G.A. 2010. Optimizing the use of garlic oil as antimicrobial agent on fresh-cut tomato through a controlled release system. *J. Food Sci.* 75:M398–M405.
43. Piercey, M.J., Mazzanti, G., Budge, S.M., Delaquis, P.J., Paulson, A.T., and Truelstrup Hansen, L. 2012. Antimicrobial activity of cyclodextrin entrapped allyl isothiocyanate in a model system and packaged fresh-cut onions. *Food Microbiol.* 30:213–221.
44. Seo, H.-S., Bang, J., Kim, H., Beuchat, L.R., Cho, S.Y., and Ryu, J.-H. 2012. Development of an antimicrobial sachet containing encapsulated allyl isothiocyanate to inactivate *Escherichia coli* O157:H7 on spinach leaves. *Int. J. Food Microbiol.* 159:136–143.
45. Chacon, P.A., Buffo, R.A., and Holley, R.A. 2006. Inhibitory effects of microencapsulated allyl isothiocyanate (AIT) against *Escherichia coli* O157:H7 in refrigerated, nitrogen packed, finely chopped beef. *Int. J. Food Microbiol.* 107:231–237.
46. Mullen, M.A. and Mowery, S.V. 2000. Insect-resistant packaging. *Int. Food Hyg.* 11:13–14.
47. Browditch, T.G. 1997. Penetration of polyvinyl chloride and polypropylene packaging films by *Ephestia cautella* (Lepidoptera: Pyralidae) and *Plodia interpunctella* (Lepidopters: Pyralidae) larvae, and *Tribolium confusum* (Coleoptera: Tenebrionidae) adults. *J. Econ. Entomol.* 90:1028–1031.
48. Highland, H.A. 1984. Insect infestation of packages. In: Baur, F.J. (Ed.), *Insect Management for Food Storage and Processing*. American Association of Cereal Chemists International, St. Paul, MN, pp. 309–320.
49. Highland, H.A. 1991. Protecting packages against insects. In: Gorham, J.R. (Ed.), *Ecology and Management of Food-Industry Pests*, FDA Technical Bulletin, 4. Association of Official Analytical Chemists, Arlington, VA, pp. 345–350.
50. Newton, J. 1988. Insects and packaging—A review. *Int. Biodeterior.* 24:175–187.
51. Riudavets, J., Salas, I., and Pons, M.J. 2007. Damage characteristics produced by insect pests in packaging film. *J. Stored Prod. Res.* 43:564–570.
52. Chung, S.K., Seo, J.Y., Lim, J.H., Park, H.H., Kim, Y.T., Song, K.H., Park, S.J., Han, S.S., Park, Y.S., and Park, H.J. 2011. Barrier property and penetration traces in packaging films against *Plodia interpunctella* (Hübner) larvae and *Tribolium castaneum* (Herbst) adults. *J. Stored Prod. Res.* 47:101–105.
53. Watson, E. and Barson, G. 1996. A laboratory assessment of the behavioural response of *Oryzaephilus surinamensis* (L.) (Coleoptera: Silvanidae) to three insecticides and the insect repellent N,N-diethyl-m-toluamide. *J. Stored Prod. Res.* 32:59–67.
54. Lakkis, J.M. 2007. *Encapsulation and Controlled Release Technologies in Food Systems*. Blackwell Publishing, Ames, IA, pp. 1–11.

55. Chang, C.P., Leung, T.K., Lin, S.M., and Hsu, C.C. 2006. Release properties of gelatin-gum Arabic microcapsules containing camphor oil with added polystyrene. *Colloid Surf. B* 50:136–140.
56. Chung, S.K., Seo, J.Y., Lim, J.H., Park, H.H., Yea, M.J., and Park, H.J. 2013. Microencapsulation of essential oil for insects repellent in food packaging system. *J. Food Sci.* 78(5):E709–713.
57. Kim, I.-H., Han, J., Na, J.H., Chang, P-S., Chung, M.S., Park, K.H., and Min, S.C. 2013. Insect-resistant food packaging film development using cinnamon oil and microencapsulation technologies. *J. Food Sci.* 78(2):E229–237.
58. Cloyd, R.A., Galle, C.L., Keith, S.R., Kalscheur, N.A., and Kemp, K.E. 2009. Effect of commercially available plant-derived essential oil products on arthropod pests. *J. Econ. Entomol.* 102:1567–79.
59. Aslan, İ., Özbek, H., Çalmaşur, Ö., and Şahìn, F. 2004. Toxicity of essential oil vapours to two greenhouse pests, *Tetranychu surticae* Koch and *Bemisia tabaci* Genn. *Ind. Crop. Prod.* 19:167–73.
60. Solomon, B., Sahle, F.F., Gebre-Mariam, T., Asres, K., and Neubert, R.H.H. 2012. Microencapsulation of citronella oil for mosquito-repellent application: Formulation and in vitro permeation studies. *Eur. J. Pharm. Biopharm.* 80:61–66.
61. Maji, T., Baruah, I., Dube, S., and Hussain, M.R. 2007. Microencapsulation of *Zanthoxylum limonella* oil (ZLO) in glutaraldehyde crosslinked gelatin for mosquito repellent application. *Bioresour. Technol.* 98:840–844.
62. Toure, A., Xiaoming, Z., Jia, C-S., and Zhijian, D. 2007. Microencapsulation and oxidative stability of ginger essential oil in maltodextrin/whey protein isolate (MD/WPI). *Int. J. Dairy Sci.* 2(4):387–392.
63. Erler, F., Ulug, I., and Yalcinkaya, B. 2006. Repellent activity of five essential oils against *Culexpipiens*. *Fitoterapia* 77:491–494.
64. Na, J.H, Hong, E.I., and Ryoo, M.I. 2008. Protection of chocolate products from Indian meal moth by adding cinnamon extract to the adhesive on the wrapping. *Korean J. Appl. Entomol.* 47(4):491–495.
65. Ansari, M.A., Vasudevan, P., Tandon, M., and Razdan, R.K. 2000. Larvicidal and mosquito repellent action of peppermint (*Mentha piperita*) oil. *Bioresour. Technol.* 71:267–271.
66. Prajapati, V., Tripathi, A.K., Aggarwal, K.K., and Khanuja, S.P.S. 2005. Insecticidal, repellent and oviposition-deterrent activity of selected essential oils against the filarial mosquito *Culexquinque fasciatus* (Say) (Diptera: Cullicidae). *Trop. Biomed.* 23:208–212.
67. Landolt, P.J., Hofstetter, R.W., and Biddick, L.L. 1999. Plant essential oils as arrestants and repellents for neonate larvae of the codling moth (Lepidoptera: Tortricidae). *Environ. Entomol.* 28:954–960.
68. Chaubey, M.K. 2007. Insecticidal activity of *Trachyspermum ammi* (Umbelliferae), *Anethum graveolens* (Umbelliferae) and *Nigella sativa* (Ranunculaceae) essential oils against stored-product beetle *Tribolium castaneum* Herbst (Coleoptera: Tenebrionidae). *Afr. J. Agric. Res.* 2:596–600.
69. Neroi, L.S., Olivero-Verbel, J., and Stashenko, E. 2010. Repellent activity of essential oils: A review. *Bioresour. Technol.* 101:372–378.
70. Mohan, S., Sivakumar, S.S., Venkatesh, S.R., and Raghavan, G.S.V. 2007. Penetration of polyethylene sheets coated with protein-enriched pea flour solution by two stored-products insects. *J. Stored Prod. Res.* 43:202–204.
71. Bachtsi, R.A. and Kipparissides, C. 1996. Synthesis and release studies of oil containing poly(vinyl alcohol) microcapsule prepared by coacervation. *J. Contr. Release* 38:49–58.
72. Jun-xia, X., Hai-Yan, Y., and Jian, Y. 2011. Microencapsulation of sweet orange oil by complex coacervation with soybean protein isolate/gum Arabic. *Food Chem.* 125:1267–1272.
73. Krishnan, S., Kshirsagar, A.C., and Singhal, R.S. 2005.The use of gum Arabic and modified starch in the microencapsulation of a food flavouring agent. *Carbohyd. Polym.* 62:309–315.
74. Salib, N.N., El-Menshaw, M.E., and Ismail, A.A. 1986. Preparation and evaluation of the release characteristics of methyl cellulose micropellets. *Pharm. Ind.* 38:577–580.
75. Sullad, A.G., Manjeshwar, L.S., and Aminabhavi, T.M. 2010. Controlled release of theophylline from interpenetrating blend microspheres of poly(vinyl alcohol) and methyl cellulose. *J. Appl. Polym. Sci.* 116:1226–1235.
76. Rosenblat, J., Magdassi, S., and Garti, N. 1989. Effect of electrolytes, stirring and surfactants in the coacervation and microencapsulation processes in the presence of gelatin. *J. Microencapsul.* 6:515–526.
77. Hussain, M.R. and Maji, T.K. 2008. Preparation of genipin cross-linked chitosan-gelatin microcapsules for encapsulation of *Zanthoxylum limonella* oil (ZLO) using salting-out method. *J. Microencapsul.* 25:414–420.
78. Atkinson, C.E., 1991. Animal repellant LLDPE. U.S. Patent number 5013551.
79. Singla, N., Thind, R.K., and Mahal, A.K. 2014. Potential of eucalyptus oil as repellent against house rat. *Rattus rattus*. *Sci. World J.* vol. 2014, pp. 1–7, Article ID 249284.

62 Microencapsulation of Phase Change Materials

*Jessica Giro-Paloma, Mònica Martínez,
A. Inés Fernández, and Luisa F. Cabeza*

CONTENTS

62.1 Introduction	1456
62.2 Types of Phase Change Materials (PCM)	1458
62.2.1 Organic PCM	1458
62.2.1.1 Paraffin Compounds	1458
62.2.1.2 Fatty Acids	1459
62.2.2 Inorganic PCM	1459
62.2.2.1 Salt Hydrates	1460
62.2.2.2 Salts	1460
62.2.2.3 Metals	1460
62.2.3 Eutectics	1461
62.3 Materials Used to Microencapsulate PCM	1461
62.4 Types of Microencapsulated Phase Change Materials (MPCM)	1462
62.5 Technologies Used to Encapsulate Organic PCM	1463
62.5.1 Chemical Methodology	1464
62.5.1.1 In situ Polymerization	1464
62.5.1.2 Emulsion Polymerization	1465
62.5.1.3 Suspension Polymerization	1465
62.5.1.4 Interfacial Polycondensation	1466
62.5.1.5 Dispersion Polymerization	1466
62.5.2 Physico-Chemical Methodology	1466
62.5.2.1 Sol-Gel Encapsulation or Core Templating	1466
62.5.2.2 Coacervation	1466
62.5.2.3 Supercritical CO_2-Assisted	1467
62.5.3 Physico-Mechanical Methodology	1467
62.5.3.1 One-Step Method	1467
62.5.3.2 Spray-Drying	1467
62.6 MPCM Characterization Techniques	1467
62.7 MPCM Applications	1470
62.8 Data	1475
62.9 Conclusions	1475
Acknowledgments	1475
References	1476

62.1 INTRODUCTION

Energy storage is an important area of research as is one of the key issues to improve the efficiency and the economic feasibility of energy systems. There are different types of energy storage[1,2]:

1. Mechanical energy storage that includes the gravitational energy storage, the compressed air energy storage, and the flywheels. The first two can be used for large-scale utility energy storage, and the flywheels are appropriate for intermediate storage.
2. Electrical storage that can be used through batteries when it is charged by connecting it to a source of direct electric current. When it is discharged the stored chemical energy converts into electrical energy.
3. Thermal energy storage[3-5] that is stored as a change in internal energy of a material as sensible or latent heat.
   a. Sensible heat storage (SHS): The thermal energy (Q) is stored by raising the temperature of a solid or liquid, using the heat capacity (C_p at constant pressure) and the change in temperature of the material during the charging/discharging process, T_f is the final temperature after heating and T_i is the initial temperature (before heating), m is the mass.

$$Q = \int_{T_i}^{T_f} mC_p dT = mC_p(T_f - T_i) \tag{62.1}$$

   b. Latent heat storage (LHS)[6,7]: Latent heat storage is based on the heat absorption or release when a storage material undergoes a phase change from solid to liquid, or liquid to gas (or vice versa); Δh_m is the enthalpy of fusion, dT is the temperature difference, a_m is the fraction melted, m is the mass of material, C_{sp} is the average specific heat between T_i and T_m (J kg^{-1} K^{-1}), and C_{lp} is the average specific heat between T_m and T_f (J kg^{-1} K^{-1}).

$$Q = \int_{T_i}^{T_m} mC_p dT + ma_m \Delta h_m + \int_{T_m}^{T_f} mc_p dT = mC_p(T_f - T_i)$$

$$= m\left[C_{sp}(T_m - T_i) + a_m \Delta h_m + C_{lp}(T_f - T_m)\right] \tag{62.2}$$

4. Thermochemical energy storage relies on the energy absorbed and released in breaking and reforming molecular bonds in a completely reversible chemical reaction, where a_r is the fractioned reacted, m is the mass, and Δh_r is the endothermic heat of reaction.

$$Q = a_r m \Delta h_r \tag{62.3}$$

Materials used in latent heat storage are known as phase change materials (PCM). The requirements for a PCM to be used as latent heat storage material are

- *Thermophysical properties*: Suitable phase-transition temperature, complete reversible freeze/melt cycle; low vapor pressure (P_v); high density (ρ), large change in enthalpy (ΔH) that is high latent heat of transition; large specific heat capacity (C_p); good heat transfer that means large thermal conductivity (k); no subcooling; small volume change (ΔV)
- *Kinetic properties*: High crystallization rate

- *Chemical properties*: Small volume pressure, low vapor pressure, compatibility with other materials, long-term chemical stability, nontoxic, no fire hazard
- *Economical*: Low price, recyclable, abundant

These materials have been studied for the last 40 years, and the most employed are hydrated salts, paraffin waxes, fatty acids, and eutectic mixtures. There are several applications for each PCM described in literature, as in textiles,[8] packaging, food transport, medical therapies, buildings,[9,10] industry, etc. The possible incorporation of PCM in building materials has attracted the attention of researchers in order to investigate the ability to reduce the energy consumption,[11] because improving the energy efficiency in buildings improves the total energy efficiency of a society, and it has an important benefit for the economy[12] and the environment. Nevertheless, it is crucial to select the proper PCM depending on the final application, although it is studied that in catering sectors, biomedical products, and construction areas it is possible to modify the phase change temperature of the PCM by modifying the weight ratio of it, creating a binary core material as Ma et al.[13,14] propose in their studies.

One of the applications for PCM is on the field of *food conservation* as the Lu and Tassou[15] and Oró et al.[16] studies. Following in this area, Oró et al.[17] studied the effect of placing PCM outside a freezer, and the results showed a good ice cream quality. Moreover, the container drinks also can be mixed with these types of materials to store energy, and release it when the systems needed. As an example of this, Oró et al.[18] studied chilly bins to enhance the thermal performance. Another application is the *refrigeration*,[19] where PCM can be applied on refrigeration systems by solar cooling as the study of Gil et al.[20] As it is possible to use PCM in cold storage applications, as Oró et al.[21] had evaluated the compatibility of PCM with some metallic materials. Besides the studies in metallic/PCM compatibility, Castellón et al.[22] evaluated the plastics/PCM ones, concluding that LDPE and PP show worst behavior compared with HDPE to encapsulate. For this reason, it was considered motivating the evaluation of Hardness and Young's Modulus with two different experimental procedures for different type of polymers applying nanoindentation, as Giro-Paloma et al.[23] studied. Subcooling is an ordinary problem in PCM cooling applications. It happens when a material is cooled and the crystallization does not start at the freezing temperature.[24] Huang et al.[25] and Günther et al.[26] evaluated the subcooling in PCM emulsions, remarking that in phase change slurries (PCS) applications, the changed nucleation and solidification behavior is critical. Hence, PCS are another important application widely employed in active pumping systems, and can be used for refrigeration,[27] air conditioning,[28-30] and cold storage[31] applications. Taking into account the *domestic heating applications* and the *domestic hot water*,[32-35] there are some investigations studying the way to evaluate the domestic demand and improve the possible losses by using a tank. Another important application in PCM area is the *solar thermal energy storage,* because it is used for space heating, power generation, and other applications.[36] The attention that had received this application is due to their large storage capacity and isothermal nature of the storage process.[37] The main disadvantage is that it is accessible only about 2000 h/year in abundant places. Hence, it is essential to study alternative methodologies to store solar thermal energy for the off hours. For this reason, the PCM usage is a key point to store thermal energy storage with the associated reversible heat transfer, as well as Farid et al.[38] exposed in their paper, where PCM have to be encapsulated in these systems to prevent the large drop in heat transfer rates during its melting and solidification process.

An important drawback that researchers had found when PCM is mixed with or incorporated to other materials is the leakage. Because of this, most times, it is necessary to use them microencapsulated to avoid leakage. Microencapsulation consists on enclosing PCM in a microscopic polymer container. There are many advantages on microencapsulating PCM, such as increasing the heat transfer area, reducing the PCM reactivity toward the outside environment, and controlling the changes in the storage material volume as phase change occurs. This methodology is solving several problems, but at the same time it has to be improved, investigating more in this field. Some of the aspects to be improved are related with the mechanical properties of the microcapsules because when they are used in active systems, they break when the slurries are pumped and the microcapsules collide among them. In the case of passive systems,[39] the principal problem is the compression force against the microcapsules when they are mixed with building materials such as mortar or concrete and integrated in a building wall.[40] To avoid the fracture of the microcapsules, it is important to evaluate the whole system shell/core.

Besides, due to the leakage in the liquid state when mixing the PCM with building construction materials,[7] and because of the thermal reliability, it was been investigated the possibility to make a polymeric container for the paraffin waxes, using the encapsulation technology and producing capsules named encapsulated phase change material.[41] When the sizes of these capsules are micrometers, these are called microencapsulated phase change material (MPCM), which consisted on a shell and a core made of the PCM. At this point, Kuznik et al.[42] exposed three relevant parameters for evaluating the MPCM quality, which are the mean diameter, the thickness of the shell, and the PCM mass percentage compared to the total mass of the capsule. The microcapsule skin has to be strong enough to hold the force generated in the system due to the volumetric changes during the phase change process of the PCM and due to the forces in the whole system. Moreover, the thickness of the coating material has to assure the efficiency of the encapsulated PCM. Therefore, after checking that the PCM chosen is a good material to store energy, it is essential the modeling of the system.[43]

62.2 TYPES OF PHASE CHANGE MATERIALS (PCM)

The three types of PCM are organic, inorganic, and eutectic PCM, as is described in Table 62.1.

Cabeza et al.[44] listed in their review numerous PCM specifying their type, melting temperature, heat of fusion, thermal conductivity, density, and their source. In Figure 62.1 some used PCM are plotted representing the melting point (°C) in front of the latent heat of fusion (kJ/kg). They are classified by their application in buildings, including those that are commercial in black color.[45]

62.2.1 Organic PCM

This type of PCM is divided in paraffin compounds and nonparaffin compounds (fatty acids mostly). Principal advantages of organic PCM are the chemical and thermal stability, they are noncorrosive, they are recyclable, and they have no subcooling. On the other hand, the disadvantages are their flammability, low thermal conductivity (k), and low phase change enthalpy.

62.2.1.1 Paraffin Compounds

They are tasteless and odorless white translucent solid hydrocarbons. The source of paraffin wax is a by-product of petroleum refinery. Typically, waxes are produced as extracted residues during the dewaxing of lubricant oil. It consists of carbon (C) and hydrogen (H) atoms joined by single bonds with the general formula: C_nH_{2n+2}, where n is the number of carbons (C). Hydrocarbons with more than 17C atoms per molecule, they are waxy solids at room temperature. It is considered a paraffin

TABLE 62.1
Classification of Phase Change Materials (PCM)

Organic	Inorganic	Eutectic
Paraffin compounds	Salt hydrates	Organic–organic
Fatty acids	Metals	Inorganic–inorganic
	Salts	Organic–inorganic

Microencapsulation of Phase Change Materials

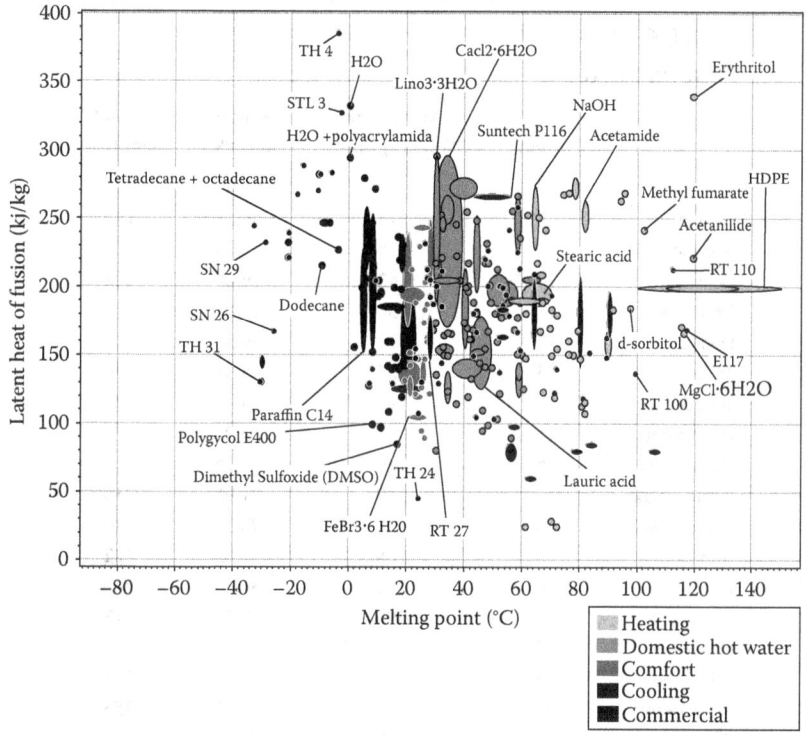

FIGURE 62.1 Plot of PCM latent heat of fusion (kJ/kg) versus melting point (°C). PCM are classified by their application in buildings.

wax when the atoms are in range of 20°C–40°C. Paraffin wax is used in the paper coating, candles manufacture, protective sealant for food products and beverages, glass-cleaning preparations, floor polishing, and stoppers for acid bottles. In Table 62.2 some paraffin waxes used in TES systems are listed detailing the number of C, the molecular weight, the melting point, and the latent heat of each one.[46] Moreover, in Figure 62.2 the melting point (°C) and the latent heat of fusion (kJ/kg) of some paraffin waxes as PCMs used for buildings applications is represented.

62.2.1.2 Fatty Acids

They are a carboxylic acid with a long hydrocarbon chain with a general formula $CH_3(CH_2)_{2n}COOH$, where n is between 4 and 14C. Fatty acids are nontoxic, with low corrosion activity, and color, chemically and thermally stable. Moreover, they can be obtained from natural products, not being a fossil fuel derivative. Fatty acids can be saturated (single bonds) or unsaturated (with one or more double bounds). The unsaturated fatty acids have lower melting points due to the molecular structure that allows closer molecular interactions. Moreover, unsaturated fatty acids have *cis* and *trans* chain configurations, in which the intermolecular interactions are weaker than in saturated ones. Besides, different fatty acids can be mixed to design PCMs with different melting temperatures.

62.2.2 Inorganic PCM

The two main subgroups for inorganic PCMs are the salt hydrates, salts, and the metals. Inorganic PCMs have fewer advantages than organic ones. The main advantage is that this type of PCM have higher phase change enthalpy, but the corrosion, subcooling, phase segregation, lack of thermal stability, and phase separation are the most important disadvantages.

TABLE 62.2
Paraffin Waxes

	Number of Carbons	Molecular Weight (g/mol[1])	Melting Point (°C)	Latent Heat (J/g[1])
Heptane	7	100	−90.55	141
Octane	8	114	−56.75	181
Nonane	9	128	−53.45	170
Decane	10	142	−29.65	202
Undecane	11	156	−25.55	177
Dodecane	12	170	−9.55	216
Tridecane	13	184	−5.35	196
Tetradecane	14	198	5.75	227
Pentadecane	15	212	9.95	207
Hexadecane	16	226	18.15	236
Heptadecane	17	240	21.95	214
Octadecane	18	254	28.15	244
Nonadecane	19	268	32.05	222
Eicosane	20	282	36.65	248
Heneicosane	21	296	40.25	213
Docosane	22	310	44.05	252
Tricosane	23	324	47.55	234
Tetracosane	24	338	64.85	255
Pentacosane	25	352	50.65	238
Hexacosane	26	366	53.55	250
Heptacosane	27	380	56.35	235
Octacosane	28	394	58.75	254
Nonacosane	29	408	63.25	239
Triacontane	30	422	65.45	252

62.2.2.1 Salt Hydrates

They have water molecules combined in a definite ratio as part of the crystal, as M·nH$_2$O formula, where M is the salt. They are solid at room temperature, and when the melting point is reached, the salt starts to dissolve in the crystal water. There is a situation that will decrease the storage capacity of the salt, known as incongruent melting, where sometimes the own water is insufficient to dissolve all the salt. As a consequence, the salt can precipitate at the bottom and only the salt dissolved in the solution would be involved in the phase change. A good solution to this problem is to build a container to prevent loss of water.

62.2.2.2 Salts

These type of inorganic PCM are those which the working temperature is ranged between 120°C and 600°C, as nitrates (NaNO$_3$, KNO$_3$), hydroxides (KOH), chlorides (NaCl, MgCl$_2$), carbonates (Na$_2$CO$_3$, K$_2$CO$_3$), and fluorides (KF).[9,44]

62.2.2.3 Metals

This subcategory of inorganic PCM includes the low melting metals and metal eutectics. They are good candidates to consider when the volume is a key point in a system (because of their high heat of fusion per unit volume) and when the required temperature of phase change has to be quite high. This type of PCM has high thermal conductivity, low specific heat, and low vapor pressure.

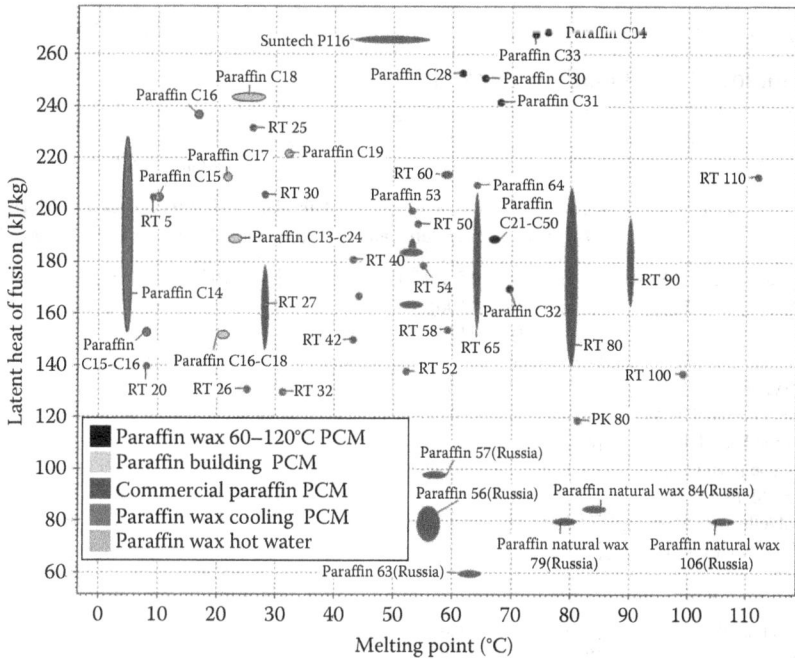

FIGURE 62.2 Paraffinic PCM classified by application in buildings by their melting point (°C) and their latent heat of fusion (kJ/kg).

62.2.3 Eutectics

They are a combination of chemical compounds or elements that have a single chemical composition, and solidify at a lower temperature than any other composition fabricated of the same ingredients. The combinations can be organic–organic, inorganic–inorganic, or organic–inorganic.[1] There are numerous eutectic mixtures suitable to be used as PCM, and they are usually preferred in cooling applications.

From the different PCM described, paraffin waxes are the most used PCM in MPCM systems, because of their latent heat and TES capacity, their abundance, low cost and large number of applications. Furthermore, these materials are very stable after several charging/discharging cycles,[47] and for this reason, the thermal reliability using paraffin waxes as PCM is satisfactory for MPCM[48] systems. Moreover, is well known they do exhibit slow thermal response due to their relatively low thermal conductivity, and this low thermal conductivity can be increased by introducing some additives such as graphite powder and metal particles.[49] The most employed paraffin wax PCM is n-octadecane (very used in building applications[30–32]). Moreover, n-hexadecane, n-docosane, n-eicosane, n-nonadecane, and fatty acids are widely employed as PCM in MPCM systems.

62.3 MATERIALS USED TO MICROENCAPSULATE PCM

There are a lot of studies about the shell of microcapsules. The shell of a MPCM has to be strong enough to stand all the forces they have to support in an active or passive system. There are a lot of possibilities for a shell such as polymeric, glass,[50] or metallic,[53] but we will focus on the polymeric ones. The most employed one is the thermoplastic acrylic polymer[54–56] polymethylmethacrylate (PMMA),[57–61,70] because it has good mechanical properties and good protection against the environment, it is cheap and inert. Moreover, BASF® has a variety of products with this polymeric shell for MCPM. In fact, there are some studies taking into account Micronal® sample, as Castellón et al.[62] and Tzvetkov et al.[63] ones. PMMA has good compatibility with a wide variety of PCM, for example with

TABLE 62.3
The Most Employed Combinations PCM/Shell

| | | | PCM | | | | |
| | | | Fatty Acids | | | | Mixture |
Shell	Paraffin	n-Octadecane	Palmitic Acid	Stearic Acid	Lauric Acid	Myristic Acid	Fatty Acids
Polymethyl methacrylate (PMMA)	[60,62,69]	[51,118,135]	[64]	[64,91]	[64]	[64]	
Melamine-formaldehyde (MF) resin	[65,80]	[106]					[96]
SiO$_2$	[66–68,93]	[114]				[92]	
PMMA+SiO$_2$	[70]						
Methyl methacrylate (MMA)		[55]					
Polyethyl methacrylate (PEMA)		[51]					
Styrene-methyl methacrylate	[86]						
Urea-formaldehyde (UF) resin		[134]					[96]
β-naphthol-formaldehyde							[96]
High density polyethylene (HDPE)	[68,71]				[71]		
HDPE/wood flour (HDPE/WF)	[49]						
Polycarbonate (PC)				[72]			
Polyvinyl chloride (PVC)	[71]				[71]		
Polyethylene terephthalate (PET)	[134]	[134]					
Polystyrene (PS)	[8]	[105]					
Polyurethane (PUR)	[71]	[134]			[71]		
Polyvinyl acetate (PVA)	[71]				[71]		
Styrene-butadiene-styrene (SBS)	[71]				[71]		[96]
Gelatin/arabic gum (G/AG)							
Titania		[53]					
AlOOH (Boehmite)			[104]				

fatty acids,[71,72] such as stearic acid, palmitic acid, myristic acid, and lauric acid (see Table 62.3). Alkan and Sari[64] studied their usage for latent heat thermal energy storage (LHTES).

Another material very used as shell in an MPCM system is melamine-formaldehyde resin (MF). Su et al.[65] studied the influence of the temperature in the deformation, for high-density polyethylene (HDPE), polystyrene (PS), and polyurethane (PUR), concluding that the yield point of MPCM decreases with the increase of temperature. Moreover, SiO$_2$ is also a very common nonpolymeric shell used in MPCM systems.[66–68]

62.4 TYPES OF MICROENCAPSULATED PHASE CHANGE MATERIALS (MPCM)

Microcapsules with a good mechanical resistance are crucial to allow reversible liquid-solid-liquid phase transitions, and to protect the PCM during the whole product life. The microencapsulation is a process of enclosing particles in a micrometer size contained in an inert shell, in order to isolate and protect them from the external environment. The combination of shell/core is the main point in the fabrication of these microcapsules, being the shell's purpose to protect the core, and the core's role to contain the PCM. The shell can be permeable, semi-permeable, and impermeable and the core can be in gas, liquid, and solid state. The shape can be spherical or irregular. A suitable shell material compatible with the core is required. Furthermore, the MPCM description depends on the core material, and also on the formation of the shell. Figure 62.3 shows the different types of MPCM, which are as follows:

- Mononuclear microcapsules contain the shell around the core
- Polynuclear capsules have many cores enclosed within the shell
- Matrix encapsulation in which the core material is distributed homogeneously into the shell material

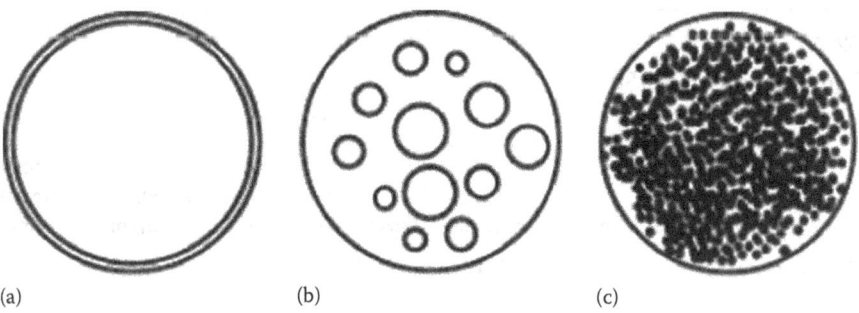

FIGURE 62.3 Types of microcapsule: (a) mononuclear, (b) polynuclear, and (c) matrix encapsulation.

Three main steps are required for microencapsulation preparation: to form the shell around the PCM, ensure there is no leakage, and guarantee that no desired materials are included in the system core/shell MPCM.

The inclusion of waxes into the microcapsules allowed to increase the heat transfer and to control the volume changes when the phase change is happening.

There are numerous studies about development of MPCM, with different combinations shell/core. Table 62.3 shows the most employed combinations between shell and PCM. Besides, other combinations shell/PCM can be possible as Table 62.4 describes.

Moreover, there are additional possibilities, with PMMA as a capsule with hexadecane,[58] heptadecane,[73] docosane,[54] MMA with *n*-pentadecane as PCM[61]; M-F resin with *n*-dodecanol[74]; polyurea as shell and butyl stearate as PCM,[75] and SiO_2, and nonadecane.[76] More possibilities are: PS with RT-31[77]; aminoplastics involving $C_{16}H_{33}Br$,[78] nickel with zinc[36]; stainless steel as shell and a eutectic as PCM[36]; and the combination with cryosol with RT-6; RT-10, and RT-20[30] as PCMs.

62.5 TECHNOLOGIES USED TO ENCAPSULATE ORGANIC PCM

The microencapsulation process includes two main steps: the emulsification step and the formation of the capsules. The emulsification step determines the size, and the size distribution of the microcapsules may be influenced at once by physical parameters such as the apparatus configuration, the stirring rate, the temperature, and the volume ratio of the two phases, and also, by the physicochemical properties such as the interfacial tension, the viscosities, the densities, and the chemical compositions of the two phases.

The formation of microcapsules is greatly affected by the surfactant, which influences not only the mean diameter but also the stability of the dispersion. The surfactants used in the system have two roles, the first one to reduce the interfacial tension between oil and aqueous phases allowing formation of smaller microcapsules and the other one to prevent coalescence by its adsorption on the oil–water interface and therefore by forming a layer around the oil droplets. The synthesis of a core/shell particle or other possible morphologies is mainly governed by the kinetic factors and thermodynamic factors.

The most employed methods to encapsulate the PCM are chemical, physico-chemical, and physico-mecanical methodologies. The emulsion and the in-situ polymerization are the two ones most used as chemical methodologies. Besides, in physico-chemical methodologies, the sol-gel and the coacervation are the most employed methods, and finally, for the physico-mechanical ones, the one-step is the most used. Some examples of these methodologies are shown in Table 62.5.

TABLE 62.4
Other Combinations PCM/Shell

Shell	PCM						
	n-Tetradecane	n-Octacosane	Tetradecane	Eicosane	Polethylene Glycol	Butyl Stearate + Paraffin	RT 27
Polymethyl methacrylate (PMMA)		[57,134]	[83]	[58]			
Polymethyl methacrylate-co-divinylbenzene [P(MMA-co-DVB)]						[14]	
Polyethyl methacrylate (PEMA)			[83]				
Acrylate						[56]	
Urea-formaldehyde (UF) resin	[85]						
High density polyethylene (HDPE)					[71]		
Polycarbonate (PC)	[84]						
Polyvinyl chloride (PVC)					[71]		
Polystyrene (PS)			[83]				[100]
Polyurethane (PUR)					[71]		
Polyurea + PUR						[13]	
Low density polyethylene ethyl + vinyl acetate (LDPE + EVA)							[100]
Polyvinyl acetate (PVA)			[83]		[71]		
Acrylonitrile–styrene–butadiene (ABS)	[84]						
Acrylonitrile-styrene copolymer (AS)	[84]						
Styrene-butadiene-styrene (SBS)					[71]		
Polysiloxane				[121]			
Gelatin/arabic gum (G/AG)							[103]
Sterilized gelatin/arabic gum (SG/AG)							[103]
Agar-agar/arabic gum (AA/AG)							[103]

62.5.1 Chemical Methodology

There are different techniques, such as in situ, emulsion, suspension, interfacial, and dispersion polymerization. The monomers polymerize around droplets of an emulsion and form a solid polymeric wall. In case of MPCM, this technique is extensively used as reported in the literature.

62.5.1.1 In Situ Polymerization

In this methodology the PCM emulsion has to be prepared and then the synthesis of the prepolymer solution has to be carried out mixing two monomers and water.[79] This pre-polymer

TABLE 62.5
List of Type of Polymerization with References

Type of Microencapsulation Technique	References
Coacervation technique	[65,103]
Emulsion polymerization	[51,54,57–59,73,83]
In situ cross-linking by coemulsification	[121]
In situ polymerization method	[13,74,80,85,104]
Interfacial polycondensation	[61,114]
One-step method	[99]
Sol-gel method	[66,68,92,93]
Spay drying	[100]
Suspension-like polymerization	[55,56,77,86]

has to be added to the emulsion in droplets form, while the emulsion is agitated during a fixed time. The microcapsules are obtained after cooling the emulsion and filtering. Finally, the MPCM has to be dried. This microencapsulation methodology can start either from monomers or from commercial pre-polymers. It is suggested to add modifying agents[80] to improve the mechanical properties. For instance, Boh et al.[81] used a modifier, with melamine–formaldehyde pre-polymers as wall materials, and styrene-maleic acid anhydride copolymers as modifying/emulsifying agents. Moreover, Boh and Šumiga[82] defined the in situ polymerization as the procedure where monomers or pre-condensates are added only to the aqueous phase of emulsion. Another study using this technique is the one of Yang et al.,[83] in which they evaluated the best shell to encapsulate tetradecane as PCM with polystyrene (PS), PMMA, polyethylmethacrylate (PEMA), and polyvinyl acetate (PVAc) as PCS. After this study, Yang et al.[84] had used the same PCM contained in different shell materials, acrylonitrile-styrene copolymer (AS), acrylonitrile-styrene-butadiene copolymer (ABS), polycarbonate (PC), concluding that all three shell materials could be used to microencapsulate *n*-tetradecane. Using this same PCM but in this case with urea and formaldehyde, Fang et al.[85] concluded that the *n*-tetradecane encapsulation is efficient enough with a good thermal stability and attractive for thermal energy storage and heat transfer applications. Besides, Chen et al.[66] synthesized by in situ polymerization paraffin/SiO_2 MPCMs.

62.5.1.2 Emulsion Polymerization

This method consists on adding the polymer in an oiled system with an emulsifier. To do so, it can be performed in a chemical, thermal, or in an enzymatic way. The last step is washing the emulsion, eliminating the oil to isolate the microcapsules. Sarı et al.[73] employed this methodology to prepare MPCM with PMMA as shell and *n*-heptadecane as PCM for TES.

62.5.1.3 Suspension Polymerization

This polymerization is governed by multiple simultaneous mechanisms such as particle coalescence and break-up, secondary nucleation and the diffusion of monomer to the interface, as Sánchez-Silva et al.[86] exposed in their study. The collective effect of these mechanisms confers the size, the structure, and the surface properties to the microcapsules. It was developed a method based on a free radical polymerization suspension process to fabricate nonpolar MPCM by Sánchez et al.[87] Also, these authors[88] studied the influence of the temperature in a reaction, stirring rate and the mass ratio of paraffin to styrene on the thermal properties of MPCM. More studies using this process were done by Borreguero et al.,[89] who considered two main steps: a continuous one with deionized water and the stabilizer (polyvinyl-pyrrolidone, PVP), and a discontinuous one containing the styrene monomer, the paraffin wax, and the benzoyl peroxide.

62.5.1.4 Interfacial Polycondensation

It involves an organic phase (containing poly-functional monomers and/or oligomers) into an aqueous phase (containing a mixture of emulsifiers and protective colloid stabilizers) along with the material to encapsulate. The presence of cross-linked materials influences the morphology of the external microcapsule surface.[90] The major advantages of this technique are the high reaction speed and also its products have low penetrability. As an example of this procedure usage, Liang et al.[75] had used the interfacial polymerization to prepare MPCM of butyl stearate as a PCM in a polyurea system. The nanoparticles SiO_2/paraffin study by Li et al.[67] using polycondensation technique led to conclude that this methodology can be also used to fabricate other organic and inorganic PCM with different core/shell compositions.

62.5.1.5 Dispersion Polymerization

In this type of polymerization, it is important to study the parameters as initiator, monomer, stabilizer concentration, and the reaction time on the characteristics of the final micrcapsules. In this methodology, the inherent simplicity of the single-step process makes the procedure useful. Typical examples are alcohols, alcohol-ether, and alcohol-water mixtures. By UV photoinitiated dispersion polymerization, Wang et al.[91] prepared a composite composed by stearic acid PCM in PMMA shell.

62.5.2 PHYSICO-CHEMICAL METHODOLOGY

This technique encloses the sol-gel encapsulation, the coacervation, and the supercritical CO_2-assisted.

62.5.2.1 Sol-Gel Encapsulation or Core Templating

Fang et al.[92] described the preparation of form stable lauric acid/silicon dioxide composite PCM by sol-gel encapsulation. Furthermore, Li et al.[93] prepared a shape/stabilized paraffin/silicon dioxide composite PCM with the same procedure. Also, Tang et al.[94] enhance the thermal conductivity of PEG (polyethylene glycol)/SiO_2 via in situ chemical reduction of $CuSO_4$ through ultrasound-assisted sol-gel process, concluding that the Cu/PEG/SiO_2 hybrid material have an excellent thermal stability and a good-stable performance. However, this process technology does not allow obtaining a polymeric shell sufficiently tightened to prevent the small water molecules diffusion during the phase changes of a salt hydrate used as PCM.[95]

62.5.2.2 Coacervation

It is a phenomenon arranged in colloid systems, where macromolecular colloid rich coacervate droplets surround dispersed microcapsule cores, and generate viscous microcapsule wall, solidifying with cross-linking agents. Hence, the polymeric solute is separated in the form of small liquid droplets, forming the coacervate. Then, it is put around the insoluble particles dispersed into a liquid. These droplets slowly unite and form a continuous cover around the core. For obtaining longer lifetime microcapsules, this procedure has to be finished adding the polymer twice. In the course of this mechanism, a thin microcapsule shell thickness is obtained. Besides, the compatibility and impermeability are enhanced (a lower speed of polymer deposition increases impermeability), providing more stability to the microcapsules, preserving the size and the spherical shape. The texture is smoother and the form is more regular, compared to one-step coacervation, where microcapsules with many protrusions, rougher, coarser, and more porous are obtained. Therefore, it is a fact that takes place in colloid systems, where macromolecular colloid rich coacervate droplets surround dispersed microcapsule cores, and form a viscous microcapsule wall, which is solidified with cross-linking agents. This methodology is useful when hydrosoluble polymers are needed to create the MPCM. Some researchers use this technique such as Özonur et al.[96] concluding that gelatin + gum arabic mixture was the best wall material for microencapsulating coco fatty

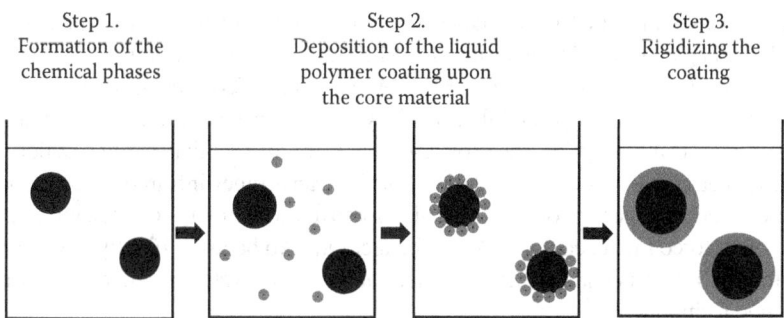

FIGURE 62.4 Steps for the coacervation technique.

acid mixtures compared with urea-formaldehyde resin, melamine-formaldehyde resin, and β-naphtol-formaldehyde. In Figure 62.4 is shown a scheme of the coacervation technique.

62.5.2.3 Supercritical CO_2-Assisted

This method is a good alternative to conventional processes because it is an effective synthetic method, which has gained interest in the polymeric composites synthesis. As an example, Haldorai et al.[97] overviewed on the synthesis of polymer-inorganic filler nanocomposites in supercritical CO_2.

62.5.3 Physico-Mechanical Methodology

This methodology includes the one-step method, and the spray-drying encapsulation. None of these techniques are capable to produce microcapsules smaller than 100 μm.[98]

62.5.3.1 One-Step Method

The advantages using this method are the easy scale-up and that there is no needed for a stabilizing agent due to self-stabilization. As an example, Jin et al.[99] used it without surfactants or dispersants nor acids/bases for stabilizing the capsules via oil/water emulsion. It allows tuning of the size and polydispersity of the capsules, and the use of nonadecane as core material.

62.5.3.2 Spray-Drying

This consisted on the preparation of an emulsion, dispersing the PCM material in a concentrated solution to form the capsule, until the preferred size of the MPCM is obtained. This emulsion is pulverized into droplets. Then it has to be dried to obtain the MPCM. Borreguero et al.[100] use this methodology mixing the two polymers LDPE and EVA with the purpose to create the polymeric shell. These two polymers have chemical similarity structure, besides the low density, their versatility, and the low cost. The technique involves the atomization of a homogeneous liquid stream in a drying chamber where the solvent is evaporated and solid particles are obtained, being suitable for heat-sensitive materials. Besides, Hawlader et al.[101] and Fei et al.[53] applied this technique to produce other type of microcapsules, using different polymeric shells and PCM cores. This methodology is also useful when hydrosoluble polymers are needed to create the MPCM.[95]

In Table 62.5, several references with the type of polymerization used to perform their experiments are summarized.

62.6 MPCM CHARACTERIZATION TECHNIQUES

There are several techniques described to characterize the microparticles. To analyze the shape and size of samples, particle size distribution (PSD), optical microscopy (OM), scanning electron microscopy (SEM), and transmission electron microscopy (TEM) are the most employed. When the crystallographic phases of MPCM are studied, x-ray diffraction (XRD), wide angle x-ray scattering (WAXS), and low angle laser light scattering (LLALS) are techniques used to characterize them. The way to describe the physical properties of an MPCM in a fluid (PCS) is using viscosimetry (η), density (ρ) technique. Also, conductivity (k) and flammability are parameters to take into account. Besides, to study the thermophysical properties, the most used ones are differential scanning calorimetry (DSC) and thermogravimetrical analysis (TGA). Using Fourier transformer infrared spectroscopy (FT-IR) is analyzed the chemical properties of the MPCM. Moreover, to extract the mechanical parameters of MPCM the atomic force microscopy (AFM) is the technique to be used. Finally, several MPCM studies describe the cycling tests and further characterization of the cycled microcapsules with any of the above mentioned technique.

Particle size distribution (PSD): This technique defines the relative amount of particles according to size. An exhaustive preparation of the sample is not necessary. A curve representing the % in number or the % in volume in front of the size is obtained. The results obtained must be estimated considering the size of the MPCM as well as the transparency of them. To calculate the approximation, the Mie and Fraunhofer models are available.[102] For this reason, to decide which of them fits better with the sample, it is necessary to evaluate the MPCM by SEM, verifying the results from the laser diffraction. The advantage on using PSD before SEM is that it is no need to recover the sample to be conductive. Mie model fits better for homogenous and spherical particles, opaque or transparent, and with diameters below 30 μm, and it is needed the refractive index of the substance and the absorption to calculate the size of the MPCM. To use the Fraunhofer model, opaque particles bigger than 30 μm are needed. A lot of MPCM studies include the use of this technique in their results. For instance, Yu et al.[74] used this technique to study the diameter distribution of MPCM prepared with different mass ratios of emulsifier to PCM.

Optical microscopy (OM): This technique supplies images of the microspheres in a high-quality resolution. Sánchez et al.[8] developed thermo-regulating textiles using MPCM with paraffin in the core, and evaluated their morphology, fixation and durability, using a transmitted light and reflection mode. Moreover, Bayés-García et al.[103] characterize gelatine/arabic gum (G/AG) MPCM by using a thermo-optical microscope at different temperatures.

Scanning electron microscopy (SEM): This technique is very useful to study the MPCM size and shape. Moreover, the preparation of the sample is very simple and fast. The SEM uses a focused beam of high-energy electrons to generate a variety of signals at the surface of solid specimens. The signals that derive from electron-sample interactions expose information about the sample including the texture, chemical composition, crystalline structures, and orientation of materials. Data is collected over a selected area of the surface of the sample, and a 2-D image is generated that displays spatial variations in these properties.

Transmission electron microscopy (TEM): This technique is used when the MPCM is in nanometer size range. The specimen must have a low density, allowing the electrons to travel through the sample. There are different ways to prepare the material: it can be cut in very thin slices either by fixing it in plastic or working with it as frozen material. Pan et al.[104] studied nanostructures that were prepared through the methodology in-situ interfacial polycondensation.

X-Ray diffraction (XRD): This technique is used to analyze the microencapsulated paraffin crystalloid phase, and it is used to guarantee the PCM encapsulation. Polystyrene shell nanocapsules were analyzed by Fang et al.[105] and Fang et al.[68] synthesized and characterized MPCMs paraffin composites with SiO_2 shells.

Wide angle x-ray scattering (WAXS) and low angle laser light scattering (LALLS): These techniques are used to determine the crystallinity degree of a PCM and the crystallographic forms of the MPCM. Zhang and Wang[106] analyzed the WAXS patterns of n-octadecane/resorcinol-modified melamine-formaldehyde MPCM. Besides, Sánchez-Silva et al.[86] have used LALLS technique to characterize the styrene-methyl methacrylate copolymer shell MPCM.

Viscosity measurement (η), density (ρ), and conductivity (k): The measurement of PCM viscosity is a key point in flowage, taking into account that the higher the temperature, the lower the viscosity. Also, it is important to measure it at different temperatures with a rotation viscosimeter. An increase of the viscosity will give rise to the increase of pump energy consumption, which will counteract the positive effect to some extent.

Another important property to consider in a PCS system is the density (ρ_{PCS}). The definition of ρ_{PCS} is expressed by the sum of the products of the weight fraction x_i and the density ρ_i of each component as Equation 62.4 define, and it can be obtained by the usage of a specific gravimeter or a pycnometer by determining the volume. Toppi and Mazzarella[107] proposed to calculate composite material properties as function of its composition instead of the necessity of determining the thermal properties on every occasion the composition changes.

$$\rho_{PCS} = \frac{1}{\sum_i \frac{x_i}{\rho_i}} \qquad (62.4)$$

An extra property very used to characterize the MPCM and PCS is the thermal conductivity. It is calculated in different manners, as Youssef et al.[108] reported in a review. One mode to calculate it is by Maxwell's[109] relation expressed in Equation 62.5, where k_{MPCM} is represented as the thermal conductivity of the microcapsule, k_{cont} is the thermal conductivity of the content, and c_{MPCM} is the volume fraction of the MPCM:

$$k_{MPCM} = k_{cont} \frac{2 + k_{MPCM}/k_{cont} + 2c_{MPCM}\left(k_{MPCM}/k_{cont} - 1\right)}{2 + k_{MPCM}/k_{cont} + c_{MPCM}\left(k_{MPCM}/k_{cont} - 1\right)} \qquad (62.5)$$

The study of these three explained properties (η, ρ, k) in a PCS was evaluated by Wang and Niu[110] where the density was calculated by weighted fraction of the densities of PCM, the coating material, and the water, based on the mass and energy balance, and the thermal conductivity of the PCS was calculated based on a composite sphere approach.

Flammability: As the lower thermal stability and inflammability properties have been severely restricted especially in building applications,[6] the flammability properties studies of PCM have high importance in numerous applications. To evaluate the combustion properties of polymer materials, the cone calorimeter is one of the most effective bench scale methods. One flammability composite studied by Cai et al.[111] evaluated PCM based on paraffin/HDPE. A proposed solution to reduce the flammability is introducing flame retardants, as Sittisart and Farid[112] studied. As microencapsulating PCM avoid leakage it will be expected that materials containing MPCM will have better fire resistant behavior than those with PCM without microencapsulation.

Differential scanning calorimeter (DSC): This is the most widely employed thermal analysis technique. It measures endothermic and exothermic transitions in function of the temperature, heat capacities and can detect glass, fusion, crystallization, and oxidation transitions. It provides the melting and solidifying enthalpy, as well as, the melting and solidifying temperature of a sample, giving an idea if the material has a good capacity to store energy. Zheng et al.[113] used this calorimetric technique to characterize two PCM suitable for applications in concentrating solar power systems.

Thermogravimetrical analysis (TGA): This technique is widely employed and it measures the amount and rate of change in the weight of a material as a function of temperature under a controlled atmosphere. The measurements are used primarily to evaluate the thermal stability. The technique can characterize materials that exhibit weight loss or gain due to decomposition, oxidation, or dehydration. As an example, Zhang et al.[114] evaluated the step thermal degradation and the thermal reliability of a silica/*n*-octadecane MPCM.

Fourier transformed infrared spectroscopy (FT-IR): The infrared segment of the electromagnetic spectrum is separated in three regions: the near (14000–4000 cm^{-1}), the mid (4000–400 cm^{-1}), and the far infrared (400–10 cm^{-1}). This FT-IR mid is used in MPCM studies to evaluate the shell material of the microcapsule, and also, to study the feasible degradation of it. It can be performed using attenuated total reflectance (ATR) or with KBr pellets,[115] without and with further preparation, respectively. Infrared spectroscopy uses the fact that the functional groups present in the molecules absorb some specific resonant frequencies depending on their structure. An example of a study using this technique is the ones of Zou et al.[116] to characterize the shell of the MPCM n-hexadecane in polyurea.

Atomic force microscopy (AFM): This is a surface technique that has the ability to obtain topographic images of surfaces with nanometric resolution and can also be used as a nanomanipulator in order to move and test the surface of the samples in a variety of ways, such as electrically, magnetically, and mechanically. AFM have been proposed to assess the mechanical properties of microcapsules. It is based on the measurement of deformation under well-defined stress. The microcapsule maximum force and the total deformation of an individual microcapsule[100] have to be measured increasing the applied force until obtaining the typical force displacement curves. It is necessary to repeat this procedure several times (minimum three times), because polymeric shells usually do not have repeatable results. A single microcapsule might not reflect the actual strength when microcapsules are piled together.[117] Moreover, AFM can extract the deformation and the Young's modulus histogram taking into account a selected small area on the top of the microcapsule. Giro-Paloma et al.[118] used it evaluating the maximum force that Micronal® DS 5001 MPCM can stand before their breakage at three different temperatures: 25°C, 45°C, and 80°C. The main conclusion is that the required applied load in order to break the sample was not constant, and it depends on the working temperature. In this research line, using atomic force microscopy (AFM) technique, Giro-Paloma et al.[119] estimated the highest force that the Micronal® DS 5007 PCS can hold on the top of the microsphere.

Cycling: The thermal cycling test exposes the thermal reliability and chemical stability of a microcapsule. This experiment is an important analysis to estimate the quality of the microcapsules to guarantee no alteration on its geometrical shape after numerous cycles.[120] Cycling can be performed on organic and inorganic PCM, as Shukla et al.[121] studied, concluding that although the studied inorganic PCMs were not found appropriate materials after some cycles, for the studied organic PCMs, after 1000 thermal cycles, have been observed gradual changes in the melting temperature as well as the latent heat of fusion. Fortuniak et al.[122] evaluated the 50 fusion-crystallization cycles tests in a DSC device. So this experiment can be performed in a DSC or using a thermocyclator. If the number of cycles increases until 5000 cycles, it is found the study of Sarı et al.[57] that tested the thermal cycling test of a PMMA shell and n-octacosane as PCM. Besides, when the cycles are done, the samples have to be evaluated by other techniques to estimate the possible PCM degradation, for instance by FT-IR device, as Sarı and Biçer[123] had evaluated. In most studies, the thermal reliability was studied following the thermal properties, and an example is the one made by Sarı and Biçer[123] with some fatty acid composites for building materials.

In Table 62.6, some studies using the different techniques exposed above are listed.

62.7 MPCM APPLICATIONS

There are several applications for MPCM,[3] reflected in a lot of studies taking into account these materials for storing energy. The first industrial application of microcapsules was introduced at the end of the 1950s in the production of pressure sensitive copying papers. Since then, this technology has been enhanced, modified, and adapted for different reasons and uses. Consequently, it has had a rapid patent applications growth, reflecting industrial research and development, as well as by an increase in the number of articles, and papers. Also, microcapsules have been used for graphic and

TABLE 62.6
Used Techniques

References	FTIR	TGA	SEM	TEM	PSD	DSC	Encapsulation Efficiency	OM	Washing/ Rubbing/Ironing Test	η	ρ	k	Cycling	WAXS	XRD	LALLS	Elasticity	AFM	UV	Microindentation
[8]	×		×			×		×	×											
[13]	×	×	×		×															
[14]	×	×	×		×	×														
[27]						×		×												
[30]					×	×				×	×									
[36]	×	×	×			×							×							
[40]					×	×							×							
[49]	×	×	×	×		×														
[51]	×		×		×	×							×		×					
[53]	×	×	×			×									×				×	
[54]	×	×	×		×	×							×							
[55]	×	×	×			×														
[56]	×	×	×		×	×							×							
[57]	×	×	×		×	×							×							
[58]	×	×	×		×	×							×							
[59]	×		×		×	×														
[60]	×							×					×				×			
[61]						×						×								
[62]				×						×										
[63]						×		×												
[65]				×																
[66]	×	×	×			×							×		×					
[67]	×	×	×			×									×					
[68]	×	×	×			×									×					
[69]		×	×			×														
[70]						×														
[71]	×		×			×							×							
[72]	×		×			×							×						×	
[73]	×	×	×		×	×														
[74]	×	×	×		×	×	×													

(Continued)

TABLE 62.6 (Continued)
Used Techniques

References	FTIR	TGA	SEM	TEM	PSD	DSC	Encapsulation Efficiency	OM	Washing/Rubbing/Ironing Test	η	ρ	k	Cycling	WAXS	XRD	LALLS	Elasticity	AFM	UV	Microindentation
[75]	×	—	—	—	×	×	—	×	—	—	—	—	—	—	—	—	—	—	—	—
[79]	—	×	×	—	×	—	—	—	—	—	—	—	—	—	—	—	—	—	—	—
[80]	—	—	—	—	—	—	—	—	—	—	—	—	—	—	—	—	—	—	—	—
[83]	×	—	—	×	—	×	—	—	—	×	—	—	—	—	—	—	—	—	—	—
[84]	×	×	×	—	×	×	×	—	—	—	—	—	—	—	—	—	—	—	—	—
[85]	×	×	×	—	—	×	—	—	—	—	—	—	—	—	—	—	—	—	—	—
[86]	×	×	×	—	×	×	—	×	—	—	—	—	—	—	—	×	—	—	—	—
[91]	×	×	×	—	—	×	—	—	—	—	—	—	—	—	—	—	—	—	—	—
[92]	×	×	×	—	—	×	×	—	—	—	—	—	—	—	×	—	—	—	—	—
[93]	—	—	×	—	×	×	—	—	—	—	—	—	×	—	×	—	—	—	—	—
[94]	—	—	—	—	×	×	—	×	—	×	—	—	×	—	—	—	—	—	—	—
[96]	—	×	×	—	×	×	—	—	—	—	×	×	×	—	—	—	—	—	—	—
[98]	×	×	×	×	—	×	×	×	—	—	×	×	×	—	×	—	—	—	—	—
[99]	×	×	×	×	—	×	—	—	—	—	—	—	×	—	—	—	—	—	—	—
[100]	—	—	—	—	×	—	—	—	—	×	×	—	—	—	—	—	—	—	—	—
[103]	—	—	×	—	—	×	—	—	—	—	—	—	—	—	×	—	—	—	—	—
[104]	×	×	×	—	×	×	—	×	—	—	—	—	×	—	—	—	—	—	—	—
[105]	—	—	—	—	×	×	—	—	—	—	—	—	—	—	—	—	—	—	—	—
[106]	—	—	×	—	×	×	—	—	—	—	—	—	×	—	—	—	—	—	—	—
[108]	×	—	×	—	—	×	—	×	—	×	×	×	×	—	—	—	—	×	—	—
[111]	—	—	—	—	—	×	—	—	—	—	—	—	—	—	—	—	—	—	—	—
[113]	—	—	×	—	×	×	—	×	—	×	—	×	×	—	—	—	—	—	—	—
[121]	—	—	—	—	×	×	—	—	—	×	×	×	—	—	—	—	—	—	—	—
[122]	—	—	×	—	×	×	—	—	—	×	×	—	—	—	—	—	—	—	—	—
[123]	—	—	×	—	—	×	—	—	—	—	×	×	—	—	—	—	—	—	—	×
[124]	—	—	×	—	×	×	—	—	—	—	—	—	—	—	—	—	—	—	—	×
[125]	—	—	—	—	—	×	—	—	—	—	—	—	—	—	—	—	—	—	×	—
[132]	—	—	×	—	×	×	—	—	—	—	—	—	—	—	—	—	—	—	—	—
[139]	—	—	—	—	—	—	—	—	—	—	—	—	—	—	—	—	—	—	—	—

TABLE 62.7
Properties Results of Some MPCM Studies

References	MPCM	Latent Heat (J/g)	Phase Change (°C) Peak Temperature	Diameter (μm)	Shape MPCM	Emulsifier	Cycles	PCM Content (%)
[8]	PS + paraffin wax	104.7	40 / 45	5.5	Tubular	—	—	—
[13]	Polyurea/polyurethane/butylstearate + paraffin	87.6	27.68	5–15	Spherical	—	500	63.7
[14]	PMMA-co-DVB	135	35 / 5–10	spherical	—	50	50–85	—
[25]	RT-20 emulsion	21.2	0.2–12.5	—	—	—	—	—
[49]	HDPE Wood + paraffin	45.9	16.7 / 7.9	—	Spherical	—	100	—
[51]	PMMA + n-octadecane	198.5	29.2 / 33.6	0.05–0.3	Spherical	89.5	—	—
[51]	PEMA + n-octadecane	208.7	31.1 / 32.3	0.06–0.36	Spherical	—	—	—
[53]	Titania + n-octadecane	92–97	—	—	—	—	—	—
[54]	PMMA + docosane	54.6	41 / 40.6	0.16	Spherical	—	—	—
[55]	BMA-MMA	173.7	— / —	—	Spherical	—	—	77.3
[56]	Acrylate/butyl stearate + paraffin	85	—	—	Spherical	—	500	48-68
[57]	PMMA + n-octacosane	86.4	50.6 / 53.2	10–30	Spherical	—	5000	—
[58]	PMMA + eicosane	35.2	54.2 / 87.5	0.25	Spherical	—	5000	—
[58]		34.9		0.7				
[59]	PMMA + n-hexadecane	68.89 and 145.61	—	0.22–1.050	Spherical	—	—	29.04–61.4
[60]	PMMA + paraffin	106.9	55.8 / 50.1	0.21	Spherical	66	3000	—
[60]		112.3						
[61]	MMA + n-pentadecane	107	10 / 9.5	650–760	Spherical	—	20	—
[62]	Micronal (BASF)	100	26	5	Spherical	—	—	—
[64]	PMMA + SA	187	67 / 66	—	—	—	—	80
[64]	PMMA + PA	173	60 / 59	—	—	—	—	80
[64]	PMMA + MA	166	51 / 50.7	—	—	—	—	80
[64]	PMMA + LA	149	41 / 41.5	—	—	—	—	80
[66]	SiO$_2$ + paraffin	156.86	57.96 / 55.78	40–60	Spherical	82.2	—	—
[67]	paraffin + SiO$_2$	45.5	56.5 / 45.5	0.2–0.5	Quasi-spherical	31.7	30	—
[68]	SiO$_2$ + paraffin	130.82	57.84 / 57.01	8–15	Spherical	—	—	—
[69]	PMMA+ paraffin	101	24–33	0.5–2	Spherical	—	—	61.2
[72]	Polycarbonate + Stearic Acid	91.4	60.2 / 51.2	0.5	Spherical	—	1000	—
[73]	SiO$_2$ + heptadecane	81.5	18.2 / 18.4	0.26	Spherical	—	5000	—
[74]	MF + n-dodecanol	187.5	21.5	30.6	Irregular spherical	93.1	—	—
[75]	Polyurea + butyl stearate	80	29	20–35	Spherical	—	—	—
[77]	PS + RT-31	75.7	31.56	4	Spherical	—	—	—

(Continued)

TABLE 62.7 (Continued)
Properties Results of Some MPCM Studies

References	MPCM	Latent Heat (J/g)	Phase Change (°C) Peak	Temperature	Diameter (μm)	Shape MPCM	Emulsifier	Cycles	PCM Content (%)
[78]	$C_{16}H_{33}Br$ + amino plastics	137		14.3	4.3	—	8.2	—	—
[83]	PMMA + PCM	66.26 60.62	2.95 5.97		5 to 30	Spherical	—	—	—
	PEMA + PCM	80.62 65.35	3.19 5.68		6 to 30	Spherical	84.6	—	—
[84]	Acrylonitrile-styrene copolymer + n-tetradecane	142.3		10	<1	Spherical	66.3	—	—
	ABS + n-tetradecane	107.1		10	<1	Spherical	30.6	—	—
	PC + n-tetradecane	49.5		10	<1	Spherical	—	—	—
[85]	Urea/formaldehyde + n-tetradecane	134.16	—	—	0.1	—	—	—	60
[86]	Melamine-formaldehyde + paraffin	87.5		45	380	Spherical	—	—	—
[91]	Acrylonitrile-styrene + n-tetradecane	60.4 50.6	92.1 95.9		2–3	Spherical	—	500	—
[92]	SiO_2 + polyethylene glycol	117.21 90	44.78 40.33		—	—	—	—	—
[93]	SiO_2 + paraffin	35.8 32.51	57.8 56.85		—	Spherical	—	—	—
[94]	$Cu/PEG/SiO_2$	110.2	—	—	—	—	—	—	—
[99]	Silica + nonadecane-	124.7	—	—	27	Spherical	—	—	—
[100]	LDPE/EVA + RT27	98.14	25.42 31.15		3.5	Spherical	63	3000	49.32
	Polystyrene + RT27	96.14	26.12 30.28		360	Spherical	—	—	48.61
[103]	Gelatine/Arabic Gum + RT-27	—	—	—	9	Spherical	49	—	—
	Sterilized Gelatine/Arabic Gum + RT-27	79	25.15 28.15		12	Spherical	—	—	—
[104]	Agar-Agar/Arabic Gum + RT-27	78	26.35 29.35		4.3	Spherical	48	—	—
	AlOOH + Palmitic Acid	19		12.7	0.2	Spherical	—	—	—
[105]	Polystyrene + n-octadecane	124.4	—	—	0.100–0.123	Spherical	—	—	—
[106]	MF + n-octadecane	122.5 132.2	26.75 20.33		12.45	Spherical	88	—	—
[111]	HDPE + paraffin	96.79	37	55	300	Lamellar	—	—	—
[113]	Silica + n-octadecane	—	—	—	18.72	Spherical	—	—	—
[121]	Polysiloxane + n-eicosane	139	37.4	30	22.9	Spherical	—	80	—
[122]	galactitol hexa myristate (GHM)	61	39	46	—	—	—	1000	—
	galactitol hexa laurate (GHL) esters	121							
[124]	PCS	90–100		28	11.2	Spherical	—	—	—
[125]	Micronal® DS 5008	135		23	2–8	Spherical	—	—	—
[133]	BASF slurries	—	up to 70°C			—	—	—	—

printing industries, in food and cosmetic products, for pharmaceutical and medical purposes, in agricultural formulations. Besides, they are used in the chemical, textile and construction materials industries, biotechnology, photography, electronics, and waste treatment.

The applications of MPCM can be divided into two major groups: thermal protection or inertia and storage. The main difference between them reports to the thermal conductivity of the substance. Zalba et al.[3] describe that storage systems with low thermal conductivity can generate a true problem as there can be enough energy stored but insufficient capacity to dispose of this energy quick enough.

One of the applications for MPCM is in *refrigeration systems*.[128] PCS are another important application to take into account.[124–126] When MPCM is suspended in a carrier fluid, mainly water[129] or with other substances such as glycerol[130] a PCS is created, enhancing the heat transfer to the MPCM. These substances are extensively employed in active pumping systems,[131] and they can be used for refrigeration,[110] heat exchangers,[132] heating,[133] ventilation, air-conditioning (HVAC),[134] and solar energy[135] applications. PCS not only act as an energy storage device but also as a heat transport system, as Salunkhe and Shembekar[136] explained in their review. The way of pumping the slurries in an active system is very important,[131] because it is known that the microcapsules can be broken when they collide among them. Moreover, Zhang and Niu[137] studied numerically the influences of microparticles and phase change in fluid-pure water, PCS, and MPCM.

Besides the *air conditioning*[128] applications, *heat exchangers* are also a possible application. Furthermore, as it was mentioned before, the introduction of PCM in *building constructive solutions* such as in floor,[138,139] walls,[140] and ceiling[141] is very common, because of the interest on evaluating the consequences when PCM are incorporated in passive systems.

62.8 DATA

In Table 62.7, a summary of some data reported in various studies is found. As a main conclusion, it is important to remark that it is very interesting to measure several times the same sample, due to the repeatability. As it can be seen in Table 62.7, there is a big dispersion of values. For this reason, to assure the reproducibility in results, it is essential to specify the experimental procedure, the instruments used, and the conditions applied in the measurement.

62.9 CONCLUSIONS

Thermal energy storage is studied today as a very good technology for energy efficiency in many applications. Latent heat storage is one of the technologies used, where PCMs are employed. Since PCM can leak when included in a system, they are always encapsulated, and microencapsulation is sometimes considered, producing the called MPCM.

The most used MPCMs have paraffin wax as core material and PMMA as a shell. Commercially, this type is the one MPCM available. Nowadays, there are several studies with organic, eutectic, and inorganic salts as PCM in MPCM. Emulsion polymerization and in situ polymerization method are the two more employed to create MPCM. Moreover, DSC is the most employed technique to measure MPCM main thermophysical properties.

ACKNOWLEDGMENTS

The work is partially funded by the Spanish government (ENE2011-28269-C03-02 and ENE2011-22722). The authors would like to thank the Catalan Government for the quality accreditation given to their research group GREA (2009 SGR 534) and research group DIOPMA (2009 SGR 645). The research leading to these results has received funding from the European Union's Seventh Framework Programme (FP7/2007-2013) under grant agreement n° PIRSES-GA-2013-610692 (INNOSTORAGE).

REFERENCES

1. Sharma, A.; Tyagi, V.V.; Chen, C.R.; Buddhi, D. Review on thermal energy storage with phase change materials and applications. *Renew Sust Energ Rev* 13 (2009): 318–345.
2. McCormack, S.; Griffiths, P. Phase Change Materials. A primer for Architects and Engineers. TU0802 next generation cost effective phase change materials for increased energy efficiency in renewable energy systems in buildings (NeCoEPCM).
3. Zalba, B.; Marín, J.M.; Cabeza, L.F.; Mehling, H. Review on thermal energy storage with phase change: Materials, heat transfer analysis and applications. *Appl Therm Eng* 23 (2003): 251–283.
4. Hamdan, M.A.; Elwerr, F.A. Thermal energy storage using phase change material. *Sol Energ* 56(2) (1996): 183–189.
5. Kabbaraa, M.J.; Abdallaha, N.B. Experimental investigation on phase change material based thermal energy storage unit. *Procedia Comput Sci* 19 (2013): 694–701.
6. Khudhair, A.M.; Farid, M.M. A review on energy conservation in building applications with thermal storage by latent heat using phase change materials. *Energ Convers Manage* 45 (2004): 263–275.
7. Pasupathy, A.; Velraj, R.; Seeniraj, R.V. Phase change material-based building architecture for thermal management in residential and commercial establishments. *Renew Sust Energ Rev* 12 (2008): 39–64.
8. Sánchez, P.; Sánchez-Fernández, M.V.; Romero, A.; Rodríguez, J.F.; Sánchez-Silva, L. Development of thermo-regulating textiles using paraffin wax microcapsules. *Thermochim Acta* 498 (2010): 16–21.
9. Baetens, R.; Petter Jelle, B.; Gustavsen, A. Phase change materials for building applications: A state-of-the-art review. *Energ Buildings* 42 (2010): 1361–1368.
10. Tyagi, V.V.; Buddhi, D. PCM thermal storage in buildings: A state of art. *Renew Sust Energ Rev* 11 (2007): 1146–1166.
11. Ling, T.-C.; Poon, C.-S. Use of phase change materials for thermal energy storage in concrete: An overview. *Constr Build Mater* 46 (2013): 55–62.
12. Zhang, D.; Li, Z.; Zhou, J.; Wu, K. Development of thermal energy storage concrete. *Cement Concrete Res* 34 (2004): 927–934.
13. Ma, Y.; Chu, X.; Tang, G.; Yao, Y. Adjusting phase change temperature of microcapsules by regulating their core compositions. *Mater Lett* 82 (2012): 39–41.
14. Ma, Y.; Chu, X.; Li, W.; Tang, G. Preparation and characterization of poly(methyl methacrylate-co-divinylbenzene) microcapsules containing phase change temperature adjustable binary core materials. *Sol Energ* 86 (2012): 2056–2066.
15. Lu, W.; Tassou, S.A. Characterization and experimental investigation of phase change materials for chilled food refrigerated cabinet applications. *Appl Energ* 112 (2013): 1376–1382.
16. Oró, E.; Miró, L.; Farid, M.M.; Cabeza, L.F. Thermal analysis of a low temperature storage unit using phase change materials without refrigeration system. *Int J Refrig* 35 (2012): 1709–1714.
17. Oró, E.; de Gracia, A.; Cabeza, L.F. Active phase change material package for thermal protection of ice cream containers. *Int J Refrig* 36 (2013): 102–109.
18. Oró, E.; Cabeza, L.F.; Farid, M.M. Experimental and numerical analysis of a chilly bin incorporating phase change material. *Appl Therm Eng* 58 (2013): 61–67.
19. Oró, E.; Gil, A.; Miró, L.; Peiró, G.; Álvarez, S.; Cabeza, L.F. Thermal energy storage implementation using phase change materials for solar cooling and refrigeration applications. *Energ Procedia* 30 (2012): 947–956.
20. Gil, A.; Oró, E.; Miró, L.; Peiró, G.; Ruiz, A.; Salmerón, J.M.; Cabeza, L.F. Experimental analysis of hydroquinone used as phase change material (PCM) to be applied in solar cooling refrigeration. *Int J Refrig* 39 (2014): 95–103.
21. Oró, E.; Miró, L.; Barreneche, C.; Martorell, I.; Farid, M.M.; Cabeza, L.F. Corrosion of metal and polymer containers for use in PCM cold storage. *Appl Energ* 109 (2013): 449–453. http://dx.doi.org/10.1016/j.apenergy.2012.10.049.
22. Castellón, C.; Martorell, I.; Cabeza, L.F.; Fernández, A.I.; Manich, A.M. Compatibility of plastic with phase change materials (PCM). *Int. J. Energy Res* 35 (2011): 765–771.
23. Giro-Paloma, J.; Roa, J.J.; Diez-Pascual, A.M.; Rayón, E.; Flores, A.; Martínez, M.; Chimenos, J.M.; Fernández, A.I. Depth-sensing indentation applied to polymers: A comparison between standard methods of analysis in relation to the nature of the materials. *Eur Polym J* 49 (2013): 4047–4053.
24. Günther, E.; Schmid, T.; Mehling, H.; Hiebler, S.; Huang, L. Subcooling in hexadecane emulsions. *Int J Refrig* 33 (2010): 1605–1611.
25. Huang, L.; Günther, E.; Doetsch, C.; Mehling, H. Subcooling in PCM emulsions—Part 1: Experimental. *Thermochim Acta* 509 (2010): 93–99.

26. Günther, E.; Huang, L.; Mehling, H.; Dötsch, C. Subcooling in PCM emulsions—Part 2: Interpretation in terms of nucleation theory. *Thermochim Acta* 522 (2011): 199–204.
27. Huang, L.; Petermann, M.; Doetsch, C. Evaluation of paraffin/water emulsion as a phase change slurry for cooling applications. *Energy* 34 (2009): 1145–1155.
28. Abduljalil, A.A.; Mat, S.B.; Sopian, K.; Sulaiman, M.Y.; Lim, C.H.; Abdulrahman, T. Review of thermal energy storage for air conditioning systems. *Renew Sust Energ Rev* 16 (2012): 5802–5819.
29. Domínguez, M.; García, C. Aprovechamiento de los Materiales de Cambio de Fase (PCM) en la Climatización. *Información Tecnológica* 20(4) (2009): 107–115.
30. Huang, L.; Doetsch, C.; Pollerberg, C. Low temperature paraffin phase change emulsions. *Int J Refrig* 33 (2010): 1583–1589.
31. Augood, P.C.; Newborough, M.; Highgate, D.J. Thermal behaviour of phase-change slurries incorporating hydrated hydrophilic polymeric particles. *Exp Therm Fluid Sci* 25 (2001): 457–468.
32. Cabeza, L.F.; Ibáñez, M.; Solé, C.; Roca, J.; Nogués, M. Experimentation with a water tank including a PCM module. *Sol Energ Mat Sol C* 90 (2006): 1273–1282.
33. Jordan, U.; Vajen, K. Realistic domestic hot-water profiles in different time scales, IEA SHC. Task 26: Solar combisystems (2001).
34. Mehling, H.; Cabeza, L.F.; Hippeli, S.; Hiebler, S. PCM-module to improve hot water heat stores with stratification. *Renew Energ* 28 (2003): 699–6711.
35. Haillot, D.; Nepveu, F.; Goetz, V.; Py, X.; Benabdelkarim, M. High performance storage composite for the enhancement of solar domestic hot water systems Part 2: Numerical system analysis. *Sol Energ* 86 (2012): 64–77.
36. Zhao, W.; Neti, S.; Oztekin, A. Heat transfer analysis of encapsulated phase change materials. *Appl Therm Eng* 50 (2013): 143–151.
37. Joulin, A.; Younsic, Z.; Zalewski, L.; Lassue, S.; Rousse, D.R.; Cavrot, J.-P. Experimental and numerical investigation of a phase change material: Thermal-energy storage and release. *Appl Energ* 88 (2011): 2454–2462.
38. Farid, M.M.; Khudhair, A.M.; Razack, S.A.K.; Al-Hallaj, S. A review on phase change energy storage: Materials and applications. *Energ Convers Manage* 45 (2004): 1597–1615.
39. Soares, N.; Costa, J.J.; Gaspar, A.R.; Santos, P. Review of passive PCM latent heat thermal energy storage systems towards buildings' energy efficiency. *Energ Buildings* 59 (2013): 82–103.
40. Tyagi, V.V.; Kaushik, S.C.; Tyagi, S.K.; Akiyama, T. Development of phase change materials based microencapsulated technology for buildings: A review. *Renew Sust Energ Rev* 15 (2011): 1373–1391.
41. Zheng, Y.; Zhao, W.; Sabol, J.C.; Tuzla, K.; Neti, S.; Oztekin, A.; Chen, J.C. Encapsulated phase change materials for energy storage—Characterization by calorimetry. *Sol Energ* 87 (2013): 117–126.
42. Kuznik, F.; David, D.; Johannes, K.; Roux, J.-J. A review on phase change materials integrated in building walls. *Renew Sust Energ Rev* 15 (2011): 379–391.
43. Dutil, Y.; Rousse, D.; Lassue, S.; Zalewski, L.; Joulin, J.; Virgone, A.; Kuznik, F. et al. Modeling phase change materials behavior in building applications: Comments on material characterization and model validation. *Renew Energ* 61 (2014): 132–135.
44. Cabeza, L.F.; Castell, A.; Barreneche, C.; de Gracia, A.; Fernández, A.I. Materials used as PCM in thermal energy storage in buildings: A review. *Renew Sust Energ Rev* 15 (2011): 1675–1695.
45. Barreneche, C.; Navarro, H.; Serrano, S.; Cabeza, L.F.; Fernández, A.I. New database on phase change materials for thermal energy storage in buildings to help PCM selection. *ISES Solar World Congress*, Cancun, Mexico, 2013.
46. Mochane, M.J. Polymer encapsulated paraffin wax to be used as phase change material for energy storage. Thesis, Master of Science, 2011.
47. Sharma, S.D.; Buddhi, D.; Sawhney, R.L. Accelerated thermal cycle test of latent heat-storage materials. *Sol Energ* 66 (1999): 483–490.
48. Silakhori, M.; Naghavi, M.S.; Metselaar, H.S.C.; Mahlia, T.M.I.; Mehrali, H.; Fauzi, M. Accelerated thermal cycling test of microencapsulated paraffin wax/polyaniline made by simple preparation method for solar thermal energy storage. *Materials* 6 (2013): 1608–1620.
49. Li, J.; Xue, P.; Ding, W.; Han, J.; Sun, G. Micro-encapsulated paraffin/high-density polyethylene/wood flour composite as form-stable phase change material for thermal energy storage. *Sol Energ Mat Sol C* 93 (2009): 1761–1767.
50. Tan, F.L.; Hosseinizadeh, S.F.; Khodadadi, J.M.; Fan, L. Experimental and computational study of constrained melting of phase change materials (PCM) inside a spherical capsule. *Int J Heat Mass Tran* 52 (2009): 3464–3472.

51. Zhang, G.H.; Bon, S.A.F; Zhao, C.Y. Synthesis, characterization and thermal properties of novel nanoencapsulated phase change materials for thermal energy storage. *Sol Energ* 86 (2012): 1149–1154.
52. Zhang, X.X.; Tao, X.M.; Yick, K.L.; Wang, X.C. Structure and thermal stability of microencapsulated phase-change materials. *Colloid Poly Sci* 282(4) (2004): 330–336.
53. Fei, B.; Lu, H.; Qi, K.; Shi, H.; Liu, T.; Li, X.; Xin, J.H. Multi-functional microcapsules produced by aerosol reaction. *J Aerosol Sci* 39 (2008): 1089–1098.
54. Alkan, C.; Sarı, A.; Karaipekli, A.; Uzun, O. Preparation, characterization, and thermal properties of microencapsulated phase change material for thermal energy storage. *Sol Energ Mat Sol C* 93 (2009): 143–147.
55. Qiu, X.; Li, W.; Song, G.; Chu, X.; Tang, G. Microencapsulated *n*-octadecane with different methyl-methacrylate-based copolymer shells as phase change materials for thermal energy storage. *Energy* 46 (2012): 188–199.
56. Ma, Y.; Chu, X.; Tang, G.; Yao, Y. Synthesis and thermal properties of acrylate-based polymer shell microcapsules with binary core as phase change materials. *Mater Lett* 91 (2013): 133–135.
57. Sarı, A.; Alkan, C.; Karaipekli, A.; Uzun, O. Microencapsulated *n*-octacosane as phase change material for thermal energy storage. *Sol Energ* 83 (2009): 1757–1763.
58. Alkan, C.; Sarı, A.; Karaipekli, A. Preparation, thermal properties and thermal reliability of microencapsulated *n*-eicosane as novel phase change material for thermal energy storage. *Energ Convers Manage* 52 (2011): 687–692.
59. Alay, S.; Alkan, C.; Göde, F. Synthesis and characterization of poly(methyl methacrylate)/*n*-hexadecane microcapsules using different cross-linkers and their application to some fabrics. *Thermochim Acta* 518 (2011): 1–8.
60. Wang, Y.; Shi, H.; Xia, T.D.; Zhang, T.; Feng, H.X. Fabrication and performances of microencapsulated paraffin composites with polymethylmethacrylate shell based on ultraviolet irradiation-initiated. *Mater Chem Phys* 135 (2012): 181–187.
61. Taguchi, Y.; Yokoyama, H.; Kado, H.; Tanaka, M. Preparation of PCM microcapsules by using oil absorbable polymer particles. *Colloid Surf A* 301 (2007): 41–47.
62. Castellón, C.; Medrano, M.; Roca, J.; Cabeza, L.F.; Navarro, M.E.; Fernández, A.I.; Lázaro, A.; Zalba, B. Effect of microencapsulated phase change material in sandwich panels. *Renew Energ* 35 (2010): 2370–2374.
63. Tzvetkov, G.; Graf, B.; Wiegner, R.; Raabe, J.; Quitmann, C.; Fink, R. Soft X-ray spectromicroscopy of phase-change microcapsules. *Micron* 39 (2008): 275–279.
64. Alkan, C.; Sarı, A. Fatty acid/poly(methyl methacrylate) (PMMA) blends as form-stable phase change materials for latent heat thermal energy storage. *Sol Energ* 82 (2008): 118–124.
65. Su, J.-F.; Wang, X.-Y.; Dong, H. Influence of temperature on the deformation behaviours of melamine-formaldehyde microcapsules containing phase change material. *Mater Lett* 84 (2012): 158–161.
66. Chen, Z.; Cao, L.; Fang, G.; Shan, F. Synthesis and characterization of microencapsulated paraffin microcapsules as shape-stabilized thermal energy storage materials. *Nanosc Microsce Therm* 17(2) (2013): 112–123.
67. Li, B.; Liu, T.; Hu, L.; Wang, Y.; Gao, L. Fabrication and properties of microencapsulated paraffin@SiO_2 phase change composite for thermal energy storage. *ACS Sustainable Chem Eng* 1 (2013): 374–380.
68. Fang, G.; Chen, Z.; Li, H. Synthesis and properties of microencapsulated paraffin composites with SiO_2 shell as thermal energy storage materials. *Chem Eng J* 163 (2010): 154–159.
69. Ma, S.; Song, G.; Li, W.; Fan, P.; Tang, G. UV irradiation-initiated MMA polymerization to prepare microcapsules containing phase change paraffin. *Sol Energ Mater Sol C* 94 (2010): 1643–1647.
70. Wu, X.; Wang, Y.; Zhu, P.; Sun, R.; Yu, S.; Du, R. Using UV-vis spectrum to investigate the phase transition process of PMMA–SiO_2@paraffin microcapsules with copper-chelating as the ion probe. *Mater Lett* 65 (2011): 705–707.
71. Kenisarin, M.M.; Kenisarina, K.M. Form-stable phase change materials for thermal energy storage. *Renew Sust Energ Rev* 16 (2012): 1999–2040.
72. Zhang, T.; Wang, Y.; Shi, H.; Yang, W. Fabrication and performances of new kind microencapsulated phase change material based on stearic acid core and polycarbonate shell. *Energ Convers Manage* 64 (2012): 1–7.
73. Sarı, A.; Alkan, C.; Karaipekli, A. Preparation, characterization and thermal properties of PMMA/*n*-heptadecane microcapsules as novel solid–liquid microPCM for thermal energy storage. *Appl Energ* 87 (2010): 1529–1534.

74. Yu, F.; Chen, Z.-H.; Zeng, X.-R. Preparation, characterization and thermal properties of microPCMs containing n-dodecanol by using different types of styrene-maleic anhydride as emulsifier. *Colloid Polym Sci* 287 (2009): 549–560.
75. Liang, C.; Lingling, X.; Honbo, S.; Zhibin, Z. Microencapsulation of butyl stearate as a phase change material by interfacial polycondensation in a polyurea system. *Energ Conver Manag* 50 (2009): 723–729.
76. Zuidam, N.J.; Shimoni, E. Overview of microencapsulates for use in food products or processes and methods to make them. In: Zuidam, N.J., Nedovic, V.A., eds. *Encapsulation Technologies for Food Active Ingredients and Food Processing.* Springer: Dordrecht, the Netherlands; 2009, pp. 3–31.
77. Sánchez-Silva, L.; Rodríguez, J.F.; Sánchez, P. Influence of different suspension stabilizers on the preparation of Rubitherm RT31 microcapsules. *Colloid Surf A* 390 (2011): 62–66.
78. Zeng, R.; Wang, X.; Chen, B.; Zhang, Y.; Niu, J.; Wang, X.; Hongfa D. Heat transfer characteristics of microencapsulated phase change material slurry in laminar flow under constant heat flux. *Appl Energ* 86 (2009): 2661–2670.
79. Choi, J.K.; Lee, J.G.; Kim, J.H.; Yang, H.S. Preparation of microcapsules containing phase change materials as heat transfer media by in-situ polymerization. *J Ind Eng Chem* 7 (2001): 358–362.
80. Šumiga, B.; Knez, E.; Vrtacnik, M.; Ferk-Savec, V.; Staresinic, M.; Boh, B. Production of melamine-formaldehyde PCM microcapsules with ammonia scavenger used for residual formaldehyde reduction. *Acta Chim Slov* 58 (2011): 14–25.
81. Boh, B.; Knez, E.; Staresinic, M. Microencapsulation of higher hydrocarbon phase change materials by *in situ* polymerization. *J Microencapsul* 22(7) (2005): 715–735.
82. Boh, B.; Šumiga, B. Microencapsulation technology and its applications in building construction materials. *RMZ—Mater Geoenviron* 55(3) (2008): 329–344.
83. Yang, R.; Xu, H.; Zhang, Y. Preparation, physical property and thermal physical property of phase change microcapsule slurry and phase change emulsion. *Sol Energ Mat Sol C* 80 (2003): 405–416.
84. Yang, R.; Zhang, Y.; Wang, X.; Zhang, Y.; Zhang, Q. Preparation of *n*-tetradecane-containing microcapsules with different shell materials by phase separation method. *Sol Energ Mat Sol C* 93 (2009): 1817–1822.
85. Fang, G.; Li, H.; Yang, F.; Liua, X.; Wu, S. Preparation and characterization of nano-encapsulated *n*-tetradecane as phase change material for thermal energy storage. *Chem Eng J* 153 (2009): 217–221.
86. Sánchez-Silva, L.; Rodríguez, J.F.; Romero, A.; Borreguero, A.M.; Carmona, M.; Sánchez, P. Microencapsulation of PCMs with a styrene-methyl methacrylate copolymer shell by suspension-like polymerization. *Chem Eng J* 157 (2010): 216–222.
87. Sánchez, L.; Sánchez, P.; de Lucas, A.; Carmona, M.; Rodríguez, J. Microencapsulation of PCMs with a polystyrene shell. *Colloid Polym Sci* 285 (2007): 1377–1385.
88. Sánchez, L.; Sánchez, P.; Carmona, M.; de Lucas, A.; Rodríguez, J.F. Influence of operation conditions on the microencapsulation of PCMs by means of suspension-like polymerization. *Colloid Polym Sci* 286 (2008): 1019–1027.
89. Borreguero, A.M.; Carmona, M.; Sánchez, M.L.; Valverde, J.L.; Rodríguez, J.F. Improvement of the thermal behavior of gypsum blocks by the incorporation of microcapsules containing PCMs obtained by suspension polymerization with an optimal core/coating mass ratio. *Appl Therm Eng* 30 (2010): 1164–1169.
90. Pascu, O.; Garcia-Valls, R.; Giamberini, M. Interfacial polymerization of an epoxy resin and carboxylic acids for the synthesis of microcapsules. *Poly Int* 57 (2008): 995–1006.
91. Wang, Y.; Dong X.T.; Xia Feng, H.; Zhang, H. Stearic acid/polymethylmethacrylate composite as formstable phase change materials for latent heat thermal energy storage. *Renew Energ* 36 (2011): 1814–1820.
92. Fang, G.; Li, H.; Liu, X. Preparation and properties of lauric acid/silicon dioxide composites as formstable phase change materials for thermal energy storage. *Mater Chem Phys* 122 (2010): 533–536.
93. Li, H.; Fang, G.; Liu, X. Synthesis of shape-stabilized paraffin/silicon dioxide composites as phase change material for thermal energy storage. *J Mater Sci* 45 (2010): 1672–1676.
94. Tang, B.; Qiu, M.; Zhang, S. Thermal conductivity enhancement of PEG/SiO$_2$ composite PCM by *in situ* Cu doping. *Sol Energ Mat Sol C* 105 (2012): 242–248.
95. Fabien, S. The manufacture of microencapsulated thermal energy storage compounds suitable for smart textile. Univ Lille Nord de France, Ensait, France, September 15, 2011. doi: 10.5772/17221.
96. Özonur, Y.; Mazman, M.; Paksoy, H.Ö.; Evliya, H. Microencapsulation of coco fatty acid mixture for thermal energy storage with phase change material. *Intl J Energ Res* 30(10) (2006): 741–749.

97. Haldorai, Y.; Shim, J.-J.; Lim, K.T. Synthesis of polymer-inorganic filler nanocomposites in supercritical CO_2. *J Supercrit Fluid* 71 (2012): 45–63.
98. Zhao, C.Y.; Zhang, G.H. Review on microencapsulated phase change materials (MEPCMs): Fabrication, characterization and applications. *Renew Sust Energ Rev* 15 (2011): 3813–3832.
99. Jin, Y.; Lee, W.; Musina, Z.; Ding, Y. A one-step method for producing microencapsulated phase change materials. *Particuology* 8(6) (2010): 588–590.
100. Borreguero, A.M.; Valverde, J.L.; Rodríguez, J.F.; Barber, A.H.; Cubillo, J.J.; Carmona, M. Synthesis and characterization of microcapsules containing Rubitherm® RT27 obtained by spray drying. *Chem Eng J* 166 (2011): 384–390.
101. Hawlader, M.N.A.; Uddin, M.S.; Kihn, M.M. Microencapsulated PCM thermal energy storage system. *Appl Energ* 74 (2003): 195–202.
102. de Boer, G.B.J.; de Weerd, C.; Thoenes, D.; Goossens, H.W.J. Laser diffraction spectrometry: Fraunhofer diffraction versus Mie scattering. *Part Syst Charact* 4 (1987): 9–14.
103. Bayés-García, L.; Ventolà, L.; Cordobilla, R.; Benages, R.; Calvet, T.; Cuevas-Diarte, M.A. Phase change materials (PCM) microcapsules with different shell compositions: Preparation, characterization and thermal stability. *Sol Energ Mat Sol C* 94 (2010): 1235–1240.
104. Pan, L.; Tao, Q.; Zhang, S.; Wang, S.; Zhang, J.; Wang, S.; Wang, Z.; Zhang, Z. Preparation, characterization and thermal properties of micro-encapsulated phase change materials. *Sol Energ Mat Sol C* 98 (2012): 66–70.
105. Fang, Y.; Kuang, S.; Gao, X.; Zhang, Z. Preparation and characterization of novel nanoencapsulated phase change materials. *Energ Convers Manage* 49 (2008): 3704–3707.
106. Zhang, H.; Wang, X. Fabrication and performances of microencapsulated phase change materials based on *n*-octadecane core and resorcinol-modified melamine-formaldehyde shell. *Colloid Surf A* 332 (2009): 129–138.
107. Toppi, T.; Mazzarella, L. Gypsum based composite materials with micro-encapsulated PCM: Experimental correlations for thermal properties estimation on the basis of the composition. *Energ Buildings* 57 (2013): 227–236.
108. Youssef, Z.; Delahaye, A.; Huang, L.; Trinquet, F.; Fournaison, L.; Pollerberg, C.; Doetsch, C. State of the art on phase change material slurries. *Energ Convers Manage* 65 (2013): 120–132.
109. Maxwell, J.C. *A Treatise on Electricity and Magnetism*. Dover: New York; 1954, pp. 1–440.
110. Wang, X.; Niu, J. Heat transfer of microencapsulated PCM slurry flow in a circular tube. *AIChE J* 54(4) (2008): 1110–1120.
111. Cai, Y.; Wei, Q.; Huang, F.; Lin, S.; Chen, F.; Gao, W. Thermal stability, latent heat and flame retardant properties of the thermal energy storage phase change materials based on paraffin/high density polyethylene composites. *Renew Energ* 34 (2009): 2117–2123.
112. Sittisart, P.; Farid, M.M. Fire retardants for phase change materials. *Appl Energ* 88 (2011): 3140–3145.
113. Zhang, H.; Sun, S.; Wang, X.; Wu, D. Fabrication of microencapsulated phase change materials based on *n*-octadecane core and silica shell through interfacial polycondensation. *Colloid Surf A* 389 (2011): 104–117.
114. Alkan, C. Enthalpy of melting and solidification of sulfonated paraffins as phase change materials for termal energy storage. *Thermochim Acta* 451 (2006): 126–130.
115. Zou, G.-L.; Lan, X.-Z.; Tan, Z.-C.; Sun, L.-Z.; Zhang, T. Microencapsulation of *n*-hexadecane as a phase change material in polyurea. *Acta Phys Chim Sin* 20(1) (2004): 90–93.
116. Su, J.F.;Wang, S.B.; Zhang, Y.Y.; Huang, Z. Physicochemical properties and mechanical characters of methanol modified melamine–formaldehyde (MMF) shell microPCMs containing paraffin. *Colloid Polym Sci* 289 (2011): 111–119.
117. Giro-Paloma, J.; Oncins, G.; Barreneche, C.; Martínez, M.; Fernández, A.I.; Cabeza, L.F. Physicochemical and mechanical properties of microencapsulated phase change material. *Appl Energ* 109 (2013): 441–448.
118. Giro-Paloma, J.; Barreneche, C.; Delgado, M.; Martínez, M.; Fernández, A.I.; Cabeza, L.F. Physicochemical and thermal study of a MPCM of PMMA shell and paraffin wax as a core. *Energy Procedia, 2014. International Conference on Solar Heating and Cooling for Buildings and Industry* 48 (2014): 347–354.
119. Hawlader, M.N.A.; Uddin, M.S.; Zhu, H.J. Encapsulated phase change materials for thermal energy storage: Experiments and simulation. *Int J Energ Res* 26 (2002): 159–171.
120. Shukla, A.; Buddhi, D.; Sawhney, R.L. Thermal cycling test of new selected inorganic and organic phase change materials. *Renew Energ* 33 (2008): 2606–2614.

121. Fortuniak, W.; Slomkowski, S.; Chojnowski, J.; Kurjata, J.; Tracz, A.; Mizerska, U. Synthesis of a paraffin phase change material microencapsulated in a siloxane polymer. *Colloid Polym Sci* 291 (2013): 725–733.
122. Sarı, A.; Biçer, A. Thermal energy storage properties and thermal reliability of some fatty acid esters/building material composites as novel form-stable PCMs. *Sol Energ Mat Sol C* 101 (2012): 114–122.
123. Vorbeck, L.; Gschwander, S.; Thiel, P.; Lüdemann, B.; Schossig, P. Pilot application of phase change slurry in a 5 m^3 storage. *Appl Energ* 109 (2013): 538–543.
124. Gschwander, S.; Schossig, P.; Henning, H.M. Micro-encapsulated paraffin in phase-change slurries. *Sol Energ Mat Sol C* 89 (2005): 307–315.
125. Rahman, A.; Dickinson, M.; Farid, M.M. Microindentation of microencapsulated phase change materials. *Adv Mat Res* 275 (2011): 85–88.
126. Zhang, P.; Ma, Z.W. An overview of fundamental studies and applications of phase change material slurries to secondary loop refrigeration and air conditioning systems. *Renew Sust Energ Rev* 16 (2012): 5021–5058.
127. Baronetto, S.; Serale, G.; Goia, F.; Perino, M. Numerical model of a slurry PCM-based solar thermal collector. *Proceedings of the 8th International Symposium on Heating, Ventilation and Air Conditioning*. Lecture Notes in Electrical Engineering 263. Springer: Heidelberg, Germany; 2014, pp. 13–20.
128. Hideo, I.; Yanlai, Z.; Akihiko, H.; Naoto, H. Numerical simulation of natural convection of latent heat phase-change-material microcapsulate slurry packed in a horizontal rectangular enclosure heated from below and cooled from above. *Heat Mass Transfer* 43 (2007): 459–470.
129. Lu, W.; Tassou, S.A. Experimental study of the thermal characteristics of phase change slurries for active cooling. *Appl Energ* 91(1) (2012): 366–374.
130. Delgado, M.; Lázaro, A.; Mazo, J.; Zalba, B. Review on phase change material emulsions and microencapsulated phase change material slurries: Materials, heat transfer studies and applications. *Renew Sust Energ Rev* 16 (2012): 253–273.
131. Alvarado, J.L.; Marsh, C.; Sohn, C.; Phetteplace, G.; Newell, T. Thermal performance of microencapsulated phase change material slurry in turbulent flow under constant heat flux. *Int J Heat Mass Tran* 50 (2007): 1938–1952.
132. Zhang, P.; Ma, Z.W.; Wang, R.Z. An overview of phase change material slurries: MPCS and CHS. *Renew Sust Energ Rev* 14 (2010): 598–614.
133. Huang, M.J.; Eames, P.C.; McCormack, S.; Griffiths, P.; Hewitt, N.J. Microencapsulated phase change slurries for thermal energy storage in a residential solar energy system. *Renew Energ* 36 (2011): 2932–2939.
134. Salunkhe, P.B.; Shembekar, P.S. A review on effect of phase change material encapsulation on the thermal performance of a system. *Renew Sust Energ Rev* 16 (2012): 5603–5616.
135. Zhang, S; Niu, J. Experimental investigation of effects of super-cooling on microencapsulated phase-change material (MPCM) slurry thermal storage capacities. *Sol Energ Mat Sol C* 94 (6) (2010): 1038–1048.
136. Lin, C.C.; Yu, K.P.; Zhao, P.; Lee, G.W.M. Evaluation of impact factors on VOC emissions and concentrations from wooden flooring based on chamber tests. *Build Environ* 44 (2009): 525–533.
137. Schossig, P.; Henning, H.M.; Gschwander, S.; Haussmann, T. Micro-encapsulated phase change materials integrated into construction materials. *Sol Energ Mat Sol C* 89 (2005): 297–306.
138. Kuznik, F.; Virgone, J.; Roux, J.J. Energetic efficiency of room wall containing PCM wallboard: A full-scale experimental investigation. *Energ Buildings* 40 (2008): 148–156.
139. Griffiths, P.W.; Eames, P.C. Performance of chilled ceiling panels using phase change materials slurries as the heat transport medium. *Appl Therm Eng* 27 (2007): 1756–1760.

Index

A

2-Acetamido-2-deoxy-D-glucose, 1261
Acetone, 135–136, 460–461, 519
Acrylate capsules, 1423–1424
Acrylic acid derivative grafting, 579
Acrylic copolymer urea formaldehyde (ACUF) capsule, 1418
Active pharmaceutical ingredients (APIs)
    hard-shell capsules, 521–522
    hot-melt extrusion, 213, 218
    immediate-release (IR) application, 520–521
    modified-release (MR) oral dosage, 520, 524–526
    solubility enhancement, 522–523
AFM, *see* Atomic force microscopy (AFM)
Agar, 277, 503, 701–702
Air suspension technique, *see* Fluid bed granulation
Alginate-poly-L-lysine system, 288
Alginates, *see* Alginic acid
Alginate-WGA microparticles, 1371
Alginic acid, 1258–1259
Allyl isothiocyanate (AITC), 1448
Aminoplastic microcapsules, 1423–1424
Amino resins
    aldehyde, 307
    emulsifiers, 310
    pH value, 310
    wall formation, 309
5-Aminosalicylicacid (ASA), 229
Amino-terminal domain, 1244
Amorphous solid dispersions (ASDs), 522, 1138
Anhydro-D-glucopyranose units (AGU), 537, 546
Anionic surfactants, 253–254
Annular jet process
    centrifugal extrusion, 208–209
    electrostatic nozzle, 206–207
    flow focusing technology, 208–209
    laminar flow breakup
        Ohnsorge–Reynolds plot, 203
        satellite droplets, 203
        surface tension, 202
        vibrating nozzle, 203–204
    spinning disk device, 208–209
    submerged nozzle systems, 207
    vibrational drip casting, 204–206
    visible/invisible capsules, 201–202
Annular Jet technology, 7
Antimicrobial packaging
    AITC, 1448
    BioSwitch, 1446
    BOPP film, 1447
    edible coatings/films, 1447
    encapsulated carvacrol and thymol, 1447
    liposome encapsulation, 1447
    micro-organisms hydrolyze starch-based particles, 1446
    PDA-NHS nanoliposomes, 1447
    trilayer films, 1447
    types, 1446
APIs, *see* Active pharmaceutical ingredients (APIs)
Aprotinin (AP), 563, 1369
Aromas
    applications, 834
    characteristics, 834
    coacervation, 853–854
    coating, 856–858
    cocrystallization, 853
    complementary processes
        chewing gum encapsulation, 859–861
        flavor material delayed release, 861
        soluble beverage powder, 861
        vegetable oil encapsulation, 858–859
    emulsification
        concentrated emulsion, 841–842
        emulsifiers, 840
        equipments, 841
        liquid encapsulation, 840
        nanoemulsion, 841
        stabilizers, 840
        thickeners, 841
    extrusion, 854–855
    freeze-drying, 852–853
    impregnation, 856
    microcapsules
        encapsulating supports, 837
        encapsulation process, 835
        gums, 838
        lipids, 838
        polysaccharides, 837–838
        powder particle encapsulation, 838–839
        properties, 835–837
        proteins, 838
    molecular inclusion, 856
    spray cooling, 849, 852
    spray-drying
        atomizers, 845–847
        emulsion, 847–849
        fine emulsion, 843
        operating conditions, 843–844
        parameters, 844–845
        powder characteristics, 844
        principle, 842–843
        retention efficiency, 849–851
        safety considerations, 841
        schematic diagram, 844
        spraying mode selection, 845–847
Artificial cells
    bioactive sorbents, 909–910
    biodegradable artificial cells, 912–913
    in biotechnology, 913
    cells/microorganisms
        cholesterol-removing microorganisms, 911–912
        diabetes mellitus, 911
        drop method, 910

encapsulation procedure, 910
genetically engineered microorganisms, 911
liver failure, 911
smaller cells concentrations, 911
enzyme therapy, 912
immunosorbents, 910
materials variation, 907–908
in medicine, 913
multienzyme system, 912
preparation methods
contents variations, 909
membranes, 909
principles, 908–909
red blood cell substitutes, 912
urea removal, 912
Association of Official Analytical Chemists (AOACs), 527
Atomic force microscopy (AFM), 222, 1470
Atom-transfer radical polymerization (ATRP), 1282, 1290–1292
Avrami–Erofe'ev kinetic model A3, 485
2,2′-Azo-bis-isobutyronitrile (AIBN), 319

B

BAB, see Blood-aqueous barrier (BAB)
Bacterial adhesion, 1010, 1246–1247
Bangham method, 663
BBB, see Blood–brain barrier (BBB)
Beeswax, 482, 507, 560, 811–812
Benzoyl peroxide (BPO), 335
Beta-lactoglobulin (β-LG)
advantages and disadvantages, 736
bioactive molecules, 734–735
biological activity, 736
BSA, 735
EGCG, 734
enzymatic hydrolysis, 736
hydrophobic molecules
cholesterol, 732–733
fatty acids, 732
protoporphyrin IX, 732–733
retinol, 733
vitamins, 732–733
α-lactalbumin, 735
ligand-binding site, 731
p-nitrophenyl phosphate, 734
polar compounds, 734
pyridoxal phosphate, 734
structure, 730–731
whey protein, 729
Bi-axially oriented polypropylene (BOPP) film, 1447
Bicontinuous microemulsions, 1389
diffusion-ordered NMR spectroscopy measurements, 263
drug delivery system, 263–269
ME-30, 263
naproxen, 263
physicochemical and spectroscopic methods, 263
structure and composition, 249–250
Bigels, 1395–1396, 1398
Bioactive compounds, food
in body behavior, 766
coacervation, 785
coextrusion, 784

colloid milling, 784
core-shell systems
classification, 782–783
enriched shell, 779
liposomes, 778–779
microscale, 780
reservoir type, 776
submicrometric scale, 780–782
dietary fibers, 771–772
extranutritional, 767
fluid-bed coating, 783
food industry, 766
high pressure homogenization, 784
layer-by-layer method, 785–786
matrix type systems
biopolymeric particles and nanoparticles, 777
classification, 782–783
coated-matrix type, 779
emulsions and nanoemulsions, 778
inclusion complexes, 777
lipid particles, 778
liposomes, 778–779
micelles, 776–777
microemulsions, 777
microscale, 780
submicrometric scale, 780–782
mechanical processes, 769–771
media milling, 784
melt extrusion, 783–784
micronutrients, 771
microscale, 780
nutrition and chronic diseases, 766
phase inversion methods, 785
phase separation method, 785
phytochemicals
bioactive lipids, 769–771
carotenoids, 768, 770
health-promoting functions, 767
medicinal/pesticidal properties, 767
organoleptic properties, 767
polyphenols, 768, 770
prebiotics, 772–773
probiotics
calcium/sodium alginates, 786
cellulose acetate phthalate, 786
chitosan, 786
dairy products, 775
definition, 773
encapsulation materials and process, 787–788
gellan/xanthan gums, 786
K-carrageenan, 786
LAB, 773
maltodextrins, 786
microbial cells, 787–788
microorganisms, 773–775
prevention and treatment, 774
starch and derivatives, 786
vegetable and animal proteins, 786
yogurt, 774
in product behavior, 765
spray chilling, 783
spray drying, 783
submicrometric scale, 780–782

Index

Bioactive nutrients delivery, *see* Beta-lactoglobulin (β-LG)
Bioadhesion, 542–543, 1099, 1212, 1367
Biodegradable nanoparticles, 1151
Biodegradable polymers, 1256
    aerobic biodegradation, 584
    anaerobic biodegradation, 584
    block copolymers, 586
    branched polymers, 586–587
    cross-linked polymers, 587
    drug delivery system
        biodegradable aliphatic polyesters, 587
        buccal, 1266
        chemical composition, 586
        colon specific drug delivery, 1267
        controlled and targeted drug delivery, 1257
        crystallinity, 586
        dendrimers, 601–603
        ELPs, 1262
        FDDS, 606
        glass transition temperature, 585
        homopolymers, 586
        hydrogels, 603–605
        hydrophilic blocks, 587–588
        hydrophilicity, 586
        implants/depots, 587–588
        insoluble matrix, 1257
        intra-articular drug delivery, 1267
        lactide-based polymers, 586
        liposomal drug delivery system, 1263–1265
        liposomes, 1257–1258
        melting temperature, 585
        micelles, 607–608
        microspheres, 588–589, 600–601
        mucosal vaccination, 1267
        nanoparticles, 1265–1266
        nanospheres, 589–601
        oral/intravenous route, 1257
        physicochemical properties, 585
        polyethylene glycol (PEG), 1262–1263
        polyethylene oxide, 1262
        polymersomes, 607
        polysaccharides, *see* Polysaccharides
        polyurethane, 1262
        transdermal, 1267
    environmental degradation, 585
    linear polymers, 586
    metabolic processes, 585
    microorganisms, 584
    natural polymers, 1256, 1258
        albumin, 588
        alginate, 588–589
        chitosan, 589–590
        collagen, 588
        cyclodextrins, 589, 591
        dextran, 589–590
        fibrin, 590
        gelatin, 588–589
        hyaluronic acid, 589, 591
        polysaccharides and proteins, 587
        starch, 589, 591
    non-biotic effects, 585
    surface erosion process, 585
    synthetic, *see* Synthetic biodegradable polymers
    synthetic polymers, 1256

Bioencapsulation
    bioactive substance, 618
    carbohydrates/glycans
        dextrin, 620–621
        hydroxypropyl methylcellulose, 621–622
        microcrystalline cellulose, 621
        pullulan, 622–625
    classification, 619–620
    controlled release, 618
    microcapsules, 618
    proteins, 626, 628
Bionanoparticles
    dendrimers, *see* Dendrimers
    drug–polymer conjugates, 1152–1154
    lipid–polymer hybrids, 1160–1161
    liposomes, 1156–1158
    niosomes, 1159–1160
    NLCs, 1156
    particle size, 1151–1152
    polymerosomes, 1158–1159
    SLNs, *see* Solid lipid nanoparticles
Biopharmaceutic Classification System (BCS) III drugs, 262
Biopolymers
    biochemical and biophysical function,
        living cells, 1255
    biocompatibility, 1256
    cooperative interactions, 1255
Biotech-derived pharmaceutical products, 1359–1360
β-LG, *see* Beta-lactoglobulin (β-LG)
Blood-aqueous barrier (BAB), 1173–1174
Blood–brain barrier (BBB), 1155–1156
Blood-retinal barrier (BRB), 1174
Bovine serum albumin (BSA), 735, 992, 1374
Bovine Spongiform Encephalopathy (BSE), 241
Bowman-Birk inhibitor (BBI), 1369
Bowman's membrane, 1169
Bragg's law of diffraction, 221
BRB, *see* Blood-retinal barrier (BRB)
Brick dust, 1127
Brij surfactant-laden p-HEMA hydrogels, 1207
BSA, *see* Bovine serum albumin (BSA)
Buccal absorption test, 1093
Buccal drug delivery, biodegradable polymers, 1266
Buccal mucosa, 1226–1228
Büchi Encapsulator, 206

C

Caco-2 cell models, 1369
Calcium chelators, 1361, 1369
Calcium phosphate–PEG–insulin–casein (CAPIC), 1371
Candelilla wax (CLW), 482, 507, 713
Candida albicans, 260, 1321
Captopril, 1320
Carbohydrate-based polymeric systems, 575
Carbohydrate polymers
    alginate polymer structure, 493–494
    cellulose and cellulose derivatives
        CMC, 498
        ethylcellulose and ethyl methylcellulose, 497
        β-(1→4)-glycosidic bonds, 496
        HPMC, 497
        methylcellulose, 496–497
        microcrystalline, 495

gum/hydrocolloid, 493
marine extracts
  alginates, 501–502
  carrageenans, 501
plant exudates
  galactomannans, 499–500
  gum arabic, 498
  gum karaya, 499
  gum tragacanth, 499
  mesquite gum, 499
  pectins, 500
  SSP, 500
polysaccharides via microbial process
  chitosan, 503
  dextran, 503
  gellan, 502–503
  xanthan, 502
starch and starch derivatives, 495–496
structural characteristics, 493, 495
Carbopols, 555, 1236, 1243, 1369
Carboxyl-terminal domain, 1244
Carboxymethyl cellulose (CMC), 280
  carbohydrate polymers, 498
  chalk suspensions, 552, 554, 556, 562
  commercial microencapsulation, 1417
  drug-loaded hydrogel beads, 280
  ionotropic gelation technique, 280
  microemulsion, 252
  nanoparticle drug-delivery systems, 561–562
Carnauba wax, 482, 484–485, 487, 507
Catastrophic phase inversion (PIC), 257–258
Cationic surfactants, 254–255
CDs, *see* Cyclodextrins (CDs)
Cell encapsulation
  alginates, 920–922
  chitosan, 923–925
  definition, 917
  material properties
    cell death, 919
    degradation rate, 917–918
    diffusivity, 917–918
    immunogenicity, 917–919
    MWCO, 917–918
    toxicity, 917–919
  poly(ethylene glycol), 922–923
  poly(lactic-co-glycolic acid), 925
Cell immobilization
  advantages, 933
  brewing
    alginate gels, 945
    cider production, 947–948
    continuous beer fermentation system, 944
    ethanol, 946–947
    polysaccharide hydrogels, 945
    rapid maturation, 943
    two-stage reactor system, 945
    whey cheese production, 948
  carrier-free immobilization, 938
  definition, 933
  fermentation, 933–934
  low-temperature fermentation, 940–941
  mechanical containment, 938–939
  porous matrix entrapment, 936–937
  product flavor, 939–940
  solid carrier surface
    cellulosic materials, 935
    cold pasteurization, 936
    covalent binding, 934–935
    DCMs, 935
    natural entrapment, 934–935
    physical adsorption, 934–935
  wine making
    applications, 941–942
    disadvantages, 941
    malolactic fermentation, 942–943
    yeast immobilization, 941–942
Cellulose-based biopolymers
  bacterial modification, 545
  cellulose esters, 540
  cellulose ethers
    alkaline-catalyzed oxalkylation, 547
    ethylcellulose, 539
    HPMC, 539
    methylcellulose, 538–539
    SCMC, 538
    William etherification, 547
  chemical modification, 544–546
    cellobiohydrolase, 548
    cellulose-hydrolyzing enzymes, 547
    endo-1,4-β-glucanase, 547
    β-glucanase, 548
  cross-linking sites, 548–549
  dissolving and regenerating cellulose materials, 549
  drug delivery application
    controlled-release. *see* Controlled-release drug delivery systems
    hydrogel-based devices, 557–558
    immediate-release dosage forms, 552–553
    immediate-release drug, 559
    MCC, 559–560
    modified-release dosage, 553
    mucoadhesive, 542–543
    nanoparticles, *see* Nanoparticle drug-delivery systems
    osmotic pump, 558
    particle properties, 541–542
    polymer crystallinity, 543
    prolonged-release drugs, 559
    protein and gene delivery, 563–564
    release rates, 559
    semi-crystalline polymers, 543
    sustained-release, 554–556
    swelling, 542
    wound healing, 564
  hemicelluloses, 540–541
  hydrogels, 548–549
  ionic liquids
    acetylation, 551
    blends, 551
    cellulose dissolution, 549–550
    grafting copolymerization, 551
    modifying cellulose, 549–550
  mercerization, 544–545
  microcrystalline cellulose, 538
  oxycellulose, 538
  physical modifications, 544
  silylation, 544

Index

Cellulose derivatives, mucoadhesive polymers, 1243
Cellulose ether and ether ester
  biomedical and pharmaceutical applications, 517
  blood cholesterol, 528
  celiac disease, 527
  dietary fibers, 526–527
  ethylcellulose, 519
  fat content, fried food, 527
  gelation, 518
  glycemic response, 528
  health enhancement, 526
  HPMC, 519
  HPMCAS, 519–520
  hydroxyalkyl substituents, 518
  lipid metabolism, 528–529
  methylcellulose, 519
  molecular weight, 518
  oral drug delivery
    hard-shell capsules, 521–522
    immediate-release (IR) application, 520–521
    modified-release (MR) oral dosage, 520, 524–526
    solubility enhancement, 522–523
  solubility, 518
  structural element, 518
  trans- and saturated fats, 527–528
  water retention, 518
Centrifugal extrusion, 8, 208–209, 1086–1087
Chemical immobilization, 1318–1319
Chitosan (CS) nanofibers
  antibacterial measurement, 1327
  biological properties, 1325–1326
  lysozyme-CLEA-immobilization, 1326–1327
3-[(3-Cholamidopropyl)dimethylammonio]-1-propanesulfonate (CHAPS), 254
Ciprofloxacin HCl (CipHCl), 1323–1324
CLEAs, *see* Cross-linked enzyme aggregates (CLEAs)
Clobetasol propionate niosomes, 1159
CMC, *see* Carboxymethylcellulose (CMC)
CNTs-liposomes conjugates (CLC), 1158
Coacervation, 1466–1467
  aqueous/organic liquids, 235
  aromas, 853–854
  bioactive compounds, food, 785
  complex coacervation, *see* Complex coacervation
  definition, 235
  edible films and coatings, 805
  essential oils, 871–872
  hydrophilic polymers, 236
  proteins, 504
  simple coacervation, 236–267
Coaxial electrospinning, 1312–1313
Cocrystallization encapsulation process, 9
Coextrusion, 23–24, 784
Colloid drug delivery systems
  aerosols, 1121–1122
  cubosomes, 1115–1116
  dendrimers, 1114–1115
  emulsions, 1116
  foams, 1122
  gels, 1116
  hexosomes, 1115–1116
  liposomes, 1118–1119
  liquid crystals, 1114–1115
  micellar cubosomes, 1115–1116
  micelles, 1112–1114
  microemulsions, 1117–1118
  microspheres, 1121
  multiple emulsions, 1117
  nanocapsules, 1120–1121
  nanoemulsions, 1117
  niosomes, 1119–1120
  polymersomes, 1120
Colloid milling, 784
Colon specific drug delivery, 1267
Coloring agents and pigments
  color control indicators, 1443–1444
  color migration, 1443
  efficiency and stability, 1442
  fresh indicator, 1444
  gas and oxygen indicators, 1444
  gelatin/alginate matrix, 1443
  mechanical damage detection, 1446
  modified paper, microcapsule layer, 1443
  natural color agents, 1442
  optical indicator, 1444–1445
  RH indicator, 1444, 1446
  self-healing packaging, 1446
  spray-drying, 1443
  time-temperature indicators, 1444
  wall material, 1442
Commercial microencapsulation
  ACUF capsule, 1418
  carbonless paper
    Camp Hill microencapsulation plant, 1416
    Colorformers and oil-filled microcapsules, 1414
    cross-section, 1414–1415
    NCR PAPER* brand, 1413–1415
    Original Portage microencapsulation plant, 1414, 1416
    polyacrylic capsule, 1413–1414
  gelatin capsule
    aggregation on coated paper, 1418
    CMC, 1417
    coacervation tank, 1417
    elasticity, 1418
    glutaraldehyde, 1417
    manual and labor intensive, 1416
    milling/emulsification, 1417
    premix tank, 1417
    stock pot and stand pipe, 1416
    wall chemistry, 1418
  PAC process, *see* Polyacrylic copolymer (PAC) process
Commercial scale encapsulation, 1413
Complex coacervation, 9, 235, 1374–1375
  definition, 237
  gelatin, *see* Gelatin-based complex coacervation systems
  gum arabic
    flow diagram, 238, 240
    negatively charged, 237
    pH, 237
    supernatant, 238
  non-gelatin based systems, 241–242
  vs. polymers, 238
  single core/multicore morphologies, 238–239
Conductivity, PECs, 1341–1342
Constant drying rate period (CDRP), 60

Contact lens-based ophthalmic drug delivery
  categories, 1183
  corneal bioavailability, 1182
  cyclodextrins, 1188–1189
  drug-eluting conventional contact lenses, 1183–1185
  hydrophilic and hydrophobic, 1183
  nanoparticle-loaded contact lenses
    conventional CLs, 1205–1206
    drug-eluting CLs, 1206
    lipophilic drug encapsulation, 1206–1207
    particle-laden lens, 1207
    PLGA particles, 1206–1207
    surfactant-laden gels, 1207–1208
  pendant cyclodextrins, 1189–1191
  perfect sink conditions, 1183
  polymeric hydrogels
    molecularly imprinted, 1191–1193
    Piggyback contact lens, 1186–1187
  surface-immobilized liposomes, 1188–1189
Controlled drug delivery
  advantages, 1381
  diffusion-mediated, 1381
  disadvantages, 1381–1382
  vegetable oil–based formulations, *see* Vegetable oil–based formulations
Controlled-release drug delivery systems
  ciprofloxacin hydrochloride, 556
  CLA, 555
  HPC, 554
  HPMC, 554
  indomethacin, 554
  metronidazole, 555
  modified-release, 553
  porcine mucin and MCC, 555
  pulsatile release, 553
  SCMC, 554
Conventional pan-coating
  applications, 172–173
  charging and discharging, 153
  coating delivery, 152
  inclined axis design, 149, 152
  Pellegrini design, 149, 152
  ventilation, 152
Core-shell encapsulation, 7, 399
Critical micelle concentration (CMC), 253, 608, 776
Cross-linked enzyme aggregates (CLEAs), 1325–1327
Cross-linked high amylose starch (CLA), 555
CS nanofibers, *see* Chitosan (CS) nanofibers
Cyclodextrin-complexed insulin, 1371
Cyclodextrin molecules, 11
Cyclodextrins (CDs), 10, 496
  biodegradable polymers, 589, 591
  contact lens-based ophthalmic drug delivery, 1188–1189
  essential oils
    complexation process, 876–877
    drug interaction, 875
    phase solubility, 875–876
    physicochemical properties, 876–877
    structure, 874–875
  SCLs, 1189–1191

D

Deacetylated gelan gum, 279
Debye–Huckel theory, 1336
Degradation-controlled monolithic system, 14
Delignified cellulosic materials (DCMs), 935
Dendrimer hydrogel (DH), 1202
Dendrimers
  advantages, 1162
  anionic and neutral dendrimers, 1162
  antiretroviral drug delivery, 1163
  bioadhesive polymers, 1201
  convergent synthesis, 1161–1162
  covalent dendrimer–drug conjugation, 1163
  DH, 1202
  divergent synthesis, 1161–1162
  lymphatic drug delivery, 1163
  molecular simulations, 1162
  non-covalent drug encapsulation, 1163
  PAMAM, 1201–1202
  Penicillin V delivery, 1163
  photoreactive G3.0-PEG acrylate conjugates, 1202–1203
  potential applications, 1204
  structures, 1162
  surface functionalization, 1204
  tissue localization, 1163
Descemet's membrane, 1169–1170, 1172
Dextrose equivalent (DE) value, 495, 837
Differential scanning calorimetry (DSC), 221
  melt extrusion, 1132–1133
  MPCM, 1469
  organogels, 1053–1054
  PECs, 1343
  spray chilling, 80–82
Diffusion-controlled monolithic system, 14
Diffusion-controlled mononuclear, 14
Dimyristoylphosphatidylcholine (DMPC) liposomes, 1188
Dispersion polymerization
  nanogel synthesis, 1299
  organic PCM, 1466
  RAFT, 1294–1295
DNPs, *see* Drug-loaded nanoparticles (DNPs)
Double emulsions, 1389–1390
Dripping and jet break-up techniques
  advantages and disadvantages, 195–196
  application, 192–193
  beads production
    electrostatic extrusion, 180, 183–185, 193
    flow focusing, 180, 182–183, 193
    jet cutter, 180, 187–188
    monodisperse drop, 179
    Ohnesorge number, 180
    Reynolds number, 180
    simple dripping, 180–182, 193
    spinning disk atomization, 180, 188
    vibrating nozzle, 180, 185–187
  collecting distance, 188–191
  droplet formation, 179
  equipment, 194–195
  gelation bath, 191–192
  mechanical techniques, 178–179

Index

Drug delivery systems
  bicontinuous microemulsions, 263–269
  biodegradable polymers, *see* Biodegradable polymers, drug delivery system
  controlled-release, *see* Controlled-release drug delivery systems
  liposomes, 660
  microparticulate, *see* Microparticulate drug delivery systems
  nanoemulsions/miniemulsions, 669
  nanoparticles, *see* Nanoparticle drug-delivery systems
  nanotechnology, 1193
  ocular, *see* Ocular drug delivery
  oral mucosal, *see* Oral mucosal drug delivery
  starches, *see* Starches

Drug-loaded hydrogel beads
  biomolecules, 276
  CMC, 280
  coacervation, 282
  cross-linking solution and pH, 281
  emulsification-internal gelation, 282
  encapsulating agent and polymer, 281
  gas-generating agent, 281
  high-voltage electrostatic field, 282
  millimeter size range, 276
  PEC, 283
  PEG-VS, 282–283
  polymer and cross-linking electrolyte concentration, 281
  pregelation/polyelectrolyte complexation, 282
  temperature, 281

Drug-loaded nanoparticles (DNPs), 1208, 1211–1212

Drug loading method, nanofiber surface
  chemical immobilization, 1318–1319
  physical adsorption, 1317–1318

Drug–polymer conjugates
  advantages, 1152–1153
  cellulose-based polymers, 1154
  chemical conjugation, 1152
  composition and structure, 1153
  daunorubicin and doxorubicin, 1153
  hydrogen bonds, 1152
  lipophilicity and hydrophilicity, 1154
  NHS ester, 1154
  passive/active targeting, 1153
  polyarabogalactan, 1154
  polymeric functional groups, 1152
  polysaccharides, 1154
  Ringsdorf model, synthetic polymer drugs, 1152
  star-shaped drug delivery systems, 1154

DSC, *see* Differential scanning calorimetry (DSC)

E

EC, *see* Ethylcellulose (EC)

Edible films and coatings
  nonvolatile molecules
    antibacterials, 824–825
    chemical structure, 827
    HPMC, 825
    molecular diffusion, 826–827
    nutraceuticals, 823–824
    plasticizers, 826
    polyacetic acid, 825–826
  volatile molecules
    advantages, 810
    appearance and transparency, 818
    barrier performances, 814–815
    barrier properties, 821–822
    carbohydrates, 809
    carrageenan films, 812–813
    carvacrol retention, 812
    chitosan films, 813
    deposition, 810
    flavor binding mechanism, 809
    flavor compounds, 807–809
    flavor storage, 815, 818
    food coating, 810
    lipid droplets, 820
    lipids, 809–810
    liquid media, 813
    matrices, 807–809
    mechanical properties, 820–821
    mechanism, 814
    n-hexanal release, 811
    partition coefficient, 812
    permeability, 816–817, 819

EGCG, *see* Epigallocatechin gallate (EGCG)

EHDA spray technology, *see* Electrohydrodynamic atomization (EHDA) spray technology

Elastin-like polypeptides (ELPs), 1263, 1267

Electrical storage, 1456

Electrohydrodynamic atomization (EHDA) spray technology
  advantage, 420, 422
  biomaterials, 423
  bipolar configuration, 420
  coating configuration, 420–421
  conical jets, 413–414
  Coulomb fission, 415
  cylindrical chamber, 418, 420
  disadvantage, 422
  drug release, 431–433
  electrical potential, 412
  electrospinning/ spraying configuration, 420
  fine particles, 422
  flow rate
    chitosan particles, 426
    droplets size, 426, 428
    fabricated particles, 426, 428
    Ganan-Calvo, 424
    heptane, 426
    liquid, 426
    paricles size and morphology, 424
    particles mean size, 426, 429
    PLGA microspheres, 424, 426–427
    stability map, 426–427, 430
    voltages, 426, 428
  high-voltage electrical current, 411–412
  jet shape, 413
  morphology, 415–416
  nanotechnology, 423
  nanoxerography, 423
  PLGA, 424–425
  polymer concentration, 430–431
  Rayleigh limit, 415
  schematic diagram, 420–421
  spray head configuration, 417–418

spray system, 413
    Taylor cone-jet mode, 413–414
    thin film formation, 424
    Zeleny, 412
Electrospinning technique
    advantages, 1312
    coaxial electrospinning, 1312–1313
    electrospun nanofiber, *see* Electrospun nanofibers
    emulsion electrospinning, 1312–1314
    polymer nanofiber drug delivery, *see* Polymer nanofiber drug delivery
Electrospray, *see* Electrohydrodynamic atomization (EHDA) spray technology
Electrospraying process, 411, 422–424, 1089, 1318
Electrospun nanofibers
    drug loading method
        chemical immobilization, 1318–1319
        physical adsorption, 1317–1318
    features, 1312
    surface modification techniques
        co-electrospinning process, 1316
        functionalization strategies, 1314
        plasma treatment, 1314–1315
        surface graft polymerization, 1316
        synthetic polymers, 1314
        wet chemical etching methods, 1315–1316
Electrospun PU–dextran nanofiber
    antimicrobial test, 1323–1325
    CipHCl, 1323
    morphology, 1323–1324
ELPs, *see* Elastin-like polypeptides (ELPs)
Emulgels
    emulsion hydrogel, 1393–1394
    emulsion organogel, 1393–1394
    fluid-filled structure mechanism, 1393–1394
    liquid gelators, 1394
    oil arrangement, 1394–1395
    solid gelators, 1394
Emulsification
    aromas
        concentrated emulsion, 841–843
        emulsifiers, 840
        equipments, 841
        liquid encapsulation, 840
        nanoemulsion, 841
        stabilizers, 840
        thickeners, 841
    essential oils, 871
Emulsion electrospinning, 1312–1314
Emulsion polymerization, 9–10, 315–316, 1372–1373, 1465
Emulsion solvent removal system
    agitation intensity, 1001
    biodegradable polymers, 997–998
    combined evaporation and extraction method, 995–996
    drug absorption process, 983
    drug encapsulation, 983
    emulsion formation, 986–988
    evaporation method, 993–994
    extraction method, 994–995
    microencapsulation products, 982
    microspheres characterization
        drug loading capacity, 1003
        encapsulation efficiency, 1003
        heterogeneous matrix, 1007–1009
        homogeneous matrix, 1005–1007
        particle size, 1001–1002
    vs. multiparticulate delivery systems, 983
    multiple purpose systems, 984–985
    nonbiodegradable polymers
        acrylics, 998–999
        celluloses, 998
        emulsifier, 999–1001
        lipids, 999
        polysaccharides, 999
        proteins, 999
    oral delivery
        bitter taste formulation, 1011–1012
        enteric microspheres, 1010–1011
        gastric retentive dosage, 1010
        hollow-bioadhesive microspheres, 1010
        mucoadhesive microspheres, 1010
        nondisintegrating multiparticulate systems, 1009
    O/W emulsion technique
        cellulose esters, 990
        hydrophilic encapsulation, 988
        hydrophobic encapsulation, 988
        neurotensin, 988–989
        schematic diagram, 988–989
    parenteral delivery, 1012–1013
    scanning electron micrograph, 984–985
    solvent selection, 996–997
    $W_I/O/W_{II}$ emulsions technique
        advantages, 992
        hydrophilic encapsulation, 990
        schematic diagram, 990–991
Emulsion technique, food
    gel microparticles/gelation
        Ca-alginate gel particles, 656–658
        membrane emulsification, 655–656
        microfluidic methods, 655–656
        static mixers, 655–656
        stirring, 655–656
    microemulsions
        characterization, 667–668
        food applications, 665
        PIT method, 667
        single-phase system, 665–666
        surfactants, 665
        Winsor classification system, 665
    multiple emulsion, 664
    nanoemulsions/miniemulsions
        advantages and disadvantages, 666
        definition, 666
        drug delivery systems, 669
        metastable structures, 666
        morphological parameters, 668
        PIT method, 668
    O/W and W/O emulsions, 663–664
    thermodynamic stability, 664–665
    W/O/W emulsion, 664
Encapsulation; *see also* Microfluidic system
    active compound stability, 802
    active ingredients, 4
    Annular Jet technology, 7
    anticaking agents, 30
    atomization, 23–24
    bioactive compounds, *see* Bioactive compounds, food
    capsules characteristics, 177–178

Index

capsule size, 24–26
capsule size reduction, 802
capsules matrices
    carbohydrates, 804
    core, 803
    film/capsules systems, 803
    films/coatings, 803
    polymeric matrix, 803
    proteins, 804
carbohydrate polymers, *see* Carbohydrate polymers
centrifugal suspension separation, 8
chemical processes, 7, 178
coacervation, 9, 805
cocrystallization, 9
coextrusion, 23–24
controlled release, 14
core material, 24, 28–29
definitions, 4
designing, 6
drug delivery systems, 274
edible films and coatings, *see* Edible films and coatings
emulsification/emulsion polymerization, 9–10
emulsion-based processes, 23–24
equipment, 30–31
evaluation process, 6–7
extrusion, 805
fluid-bed coating, 10
fluidized bed coating technology, *see* Fluidized bed coating technology
freeze-drying, 805
FTIR, *see* Fourier transform infrared (FTIR) spectroscopy
granulation, *see* Granulation technology
inclusion complexation, 10–11
interfacial polymerization, 10
ionic gelation, 11
jet cutting, 89–90
lipids, 506–508
liposome, 11
lyophilization/freeze-drying, 11
matrix encapsulation, 5
mechanical/physical processes, 7
mechanical processes, 178
melt-dispersion technique, *see* Melt-dispersion technique
melt extrusion, 12
microbeads/capsules production, *see* Dripping and jet break-up techniques
microemulsion, *see* Microemulsion
mononuclear, 5
morphology
    complex, 27
    core-shell morphology, 27
    microsphere, 26–27
organic solvents, 30
pan-coating, *see* Pan coating process
particle size ranges, 7–8
payload, 24, 27–28
phase separation process, 12
physical and chemical properties, 6
physicochemical processes, 178
polynuclear capsules, 5
polyphenolics, *see* Polyphenolics
preprocessing
    batch *vs.* continuous production, 32
    capsules, collection, 31
probiotics, *see* Probiotics
process, 6
proteins, *see* Proteins
pulsed combustion spray drying, *see* Pulsed combustion spray drying
release mechanisms, 14
release rates, 14–15
SCF, *see* Supercritical fluid (SCF)
selection process, 32–33
shell material, 24, 28
shell/wall/coating material, 508
size/morphology, 5
solvent evaporation, 13
spinning disk, 13
    advantages and disadvantages, 107
    applications, 102
    atomization, 90–92
    droplet collection and bead solidification, 98–102
    droplet formation, 95–97
    equipment, 102–106
    industrial applications, 89–90
    liquid distribution, 92–95
spray chilling, *see* Spray chilling
spray coating, 23–24
spray congealing, 13
spray-drying, *see* Spray drying (SD)
supercritical fluids, 13
surfactants, 30
types, 177–178
vibrating-jet (nozzle) method, 90
Encapsys, 440–441, 1422
Energy storage
    electrical, 1456
    mechanical, 1456
    thermal, 1456
    thermochemical, 1456
Enzyme immobilization
    advantages, 973–974
    biocatalysts, 961
    biomedical applications
        L-asparaginase, 972
        asparagine, 970
        bilirubin, 972
        biodegradable polymers, 969
        biomaterials, 967–968
        cancer therapy, 970
        diagnostic assays and biosensors, 972
        glucose-6-phosphate dehydrogenase, 971
        heparinase, 972
        lysozyme, 970–971
        trypsin, 970
        urease, 971–972
        urokinase, 969
    clinical diagnosis, 958
    disadvantages, 973–974
    enzyme therapy
        enzyme deficiencies, 958–959
        inborn metabolic disorders, 958–959
        intracellular application, 960
        topical application, 960

polymeric carriers
    adsorption, 961–962
    covalent binding, 962–964
    crosslinking method, 964–965
    entrapment method, 965
    ionic binding, 962
    protein fusion, 965–967
Enzyme-inhibiting polymers, 1244
EO, *see* Essential oils (EO)
Epigallocatechin gallate (EGCG), 734
    Alzheimer's disease, 748
    cellular uptake, 750
    layer-by-layer process, 751
    lipid-coated nanoparticles, 748
    liposomes, 753
    PEG-PLA nanoparticles, 749
    solvent evaporation method, 749
    spray drying, 747
Erodible ophthalmic inserts
    drug release *vs.* insert erosion kinetics, 1178–1179
    Lacisert, 1179
    minidisc, 1179
    PEO matrices, 1178
    SODI, 1179
Essential oils (EO)
    applications, 868
    biological properties, 868
    coacervation, 871–872
    cocrystallization, 874
    cyclodextrins molecular encapsulation
        complexation process, 876–877
        drug interaction, 875
        phase solubility, 875–876
        physicochemical properties, 876–877
        structure, 874–875
    definition, 867
    emulsification, 871
    extrusion, 869
    flavors/fragrances
        aromatherapy effect, 878
        βCD complexes, 877–878
        B. subtilis cells images, 895, 897
        CG/MS chromatogram, 891
        cinnamaldehyde release, 879–880
        complex powder recoveries, 880–881
        E. coli cells images, 895–896
        evolved gas detection curves, 891
        flavor constituents retention, 884, 886
        flavor oil proportion, 880–881
        fluorescence intensity, 892
        food applications, 884, 886
        formation constant, 892–893
        fractional inhibitory concentration indices, 894
        GAB model, 879
        LCEO, 880
        lemongrass oleoresin, 898
        locus ceruleus, 880, 882
        model flavor composition, 883, 885
        motor cortex, 880, 882
        multiple ion detection, 889–890
        nucleus raphe magnus, 880, 882
        particle size distribution, 887–888
        periaqueductal gray, 880, 882
        pharmaceutical effects, 878
        phase solubility diagram, 881–883
        pure eugenol, 881
        sedative effects, 878
        thermodynamic parameters, 893
        thymol release, 879
        unsaturated fatty acids, 894–895
        x-ray powder diffraction, 886–887
    fluid bed coating, 869
    interfacial polymerization, 874
    liposomes, 873–874
    microcapsules, 868
    microspheres, 868
    multiple emulsions, 871
    spray-chilling, 870
    spray-cooling, 870
    spray-drying, 871
Ethylcellulose (EC), 497, 519, 524–526, 1140
Ethylene glycol dimethacrylate (EGDMA), 319–320
Ethyl methylcellulose (EMC), 497
Eudragit organogel, 711
Euler–Lagrangian approach, 59
Eutectics, 1461
Extruders
    classes and subclasses, 1128
    classification, 1127–1128
Eye anatomy
    conjunctiva, 1170
    cornea, 1169–1170
    nasolacrimal drainage system, 1170
    structure, 1169
    tear film, 1170–1171

F

Falling drying rate period (FDRP), 60
Fenton reaction, 742
Fiber scaffolds, 1322–1323
Fimbriae, 1246
Fish gelatin, 241, 506
Flavor releasing systems
    ι-carrageenans, 1442
    edible films, 1442
    scented microencapsulated coatings, 1440–1441
    ScentSational Technologies, 1440
    scratch and sniff, 1440
    U.S. Patent Application No. 09/828807, 1441
    U.S. Patent Application No. 13/287122, 1441
    U.S. Patent Application No. 13/861936, 1441
Floating drug delivery systems (FDDS), 606
Flory–Huggins theory, 349, 1162
Fluid bed granulation
    advantages, 396
    components, 395–396
    definition, 395
    disadvantages, 397
    starch and starch derivatives, 495–496
Fluid bed rotor granulation, 397–398
Fluidized bed coating technology
    application
        coat functions, 142–144
        coating ingredients, 142–143
        coat thickness and coating percentage, 141–142
        processing obstacles, 140
        release mechanisms, 142

Index

bioactive compounds, food, 783
drying capacity
    organic solvent vehicles, 136
    safety precautions, 136
    solvent vapor capacity, 135
    spray rates, 135
    upper and lower explosive limits, 136
    volatile organic materials, 136
    water, 136–140
essential oils, 869
fluidization, 113–115
history, 112–113
hot melt temperatures, 134
nozzles and atomization, 131–133
process air, *see* Process air
process temperatures, 133–134
rotor rotation, 140
spouted bed process, 118–119
tangential spray configuration, 119–120
top spray process, 119
Wurster process
    coating chamber, 115–116
    cylindrical design, 115
    down-bed region, 115–116
    particle size distributions, 114, 118
    up-bed region, 115–116
Fluidized bed spray drying, 495–496
Foamed binder technology (FBT), 402–403
Foam extrusion
    application, 1143–1144
    cell growth and expansion, 1141, 1143
    cell nucleation, 1141, 1143
    gas dissolution, 1141, 1143
    HPMCAS, 1141, 1143
Follicle-associated epithelium (FAE), 1364
Food industry
    active ingredient, 649
        antioxidants, 670–671
        flavors, 669–670
        omega-3 acids, 671–672
        probiotics, *see* Probiotics
        sweeteners, 669
        vitamins and minerals, 672–673
    applications, 645
    diffusion, 648
    food ingredients
        microencapsulation, 644–645
        physical, physicochemical/chemical techniques, 648
    functional foods, 643–644
    gel microparticles/gelation
        droplet extrusion, 654–655
        emulsification, *see* Emulsion technique, food
        hydrophilic polymer, 653
        ionotropic gelation, 653
        nebulization, 655
    GRAS, 645
    ingredients, 645–646
    liposomes
        advantage, 662
        antimicrobial peptides, 661
        articles and patents publish, 660–661
        biomimetic model systems, 657
        cheese ripening, 661
        CoEnzyme Q10, 662
        disadvantage, 663
        drug-delivery system, 660
        electron microscopic studies, 661
        GI system, 662
        hydrophilic and hydrophobic compounds, 663
        instability, 660
        lecithin, 662
        microfluidization techniques, 662
        MLVs, 657
        multitubular systems, 663
        nutrient benefits, 661
        OLVs, 657
        phase transition, 660
        phosphatidylcholine, 662
        phospholipid aggregation, 658–659
        physicochemical stability, 660
        PLVs, 657
        precursor phase, 658
        protects labile compounds, 661
        rigorous analysis, 659
        supercritical fluid, 663
        surface modifications, 658–659
    matrix delivery systems, 646–647
    reservoir system, 646–647
    spray-dried microparticles, *see* Spray drying (SD), food industry
    thermal release, 648
Food packaging
    antimicrobial packaging
        AITC, 1448
        BioSwitch, 1446
        BOPP film, 1447
        edible coatings/films, 1447
        encapsulated carvacrol and thymol, 1447
        liposome encapsulation, 1447
        micro-organisms hydrolyze starch-based particles, 1446
        PDA-NHS nanoliposomes, 1447
        trilayer films, 1447
        types, 1446
    coloring agents and pigments
        color control indicators, 1443–1444
        color migration, 1443
        efficiency and stability, 1442
        fresh indicator, 1444
        gas and oxygen indicators, 1444
        gelatin/alginate matrix, 1443
        mechanical damage detection, 1446
        modified paper, microcapsule layer, 1443
        natural color agents, 1442
        optical indicator, 1444–1445
        RH indicator, 1444, 1446
        self-healing packaging, 1446
        spray-drying, 1443
        time-temperature indicators, 1444
        wall material, 1442
    flavor releasing systems
        ι-carrageenans, 1442
        edible films, 1442
        scented microencapsulated coatings, 1440–1441
        ScentSational Technologies, 1440
        scratch and sniff, 1440
        U.S. Patent Application No. 09/828807, 1441

U.S. Patent Application No. 13/287122, 1441
U.S. Patent Application No. 13/861936, 1441
functions, 1439
future perspectives, 1450
insect/rodent-resistant food packaging
   commercial pesticides, 1450
   essential oils, 1449
   eucalyptus oil, 1450
   LDPE films, 1449
   natural polymers, 1449
   penetrators, 1448
   rodent repellents, 1449
   synthetic and natural insect repellents, 1448
   synthetic polymers, 1449
   terpenes, 1450
tailored-made packaging materials, 1439
Fourier Transform Infrared–Mid Infrared–Attenuated Total Reflectance (FT–MIR–ATR) fingerprint
   biopolymers, 626, 628
   spectra
      alginate, 630, 632
      κ-carrageenan, 630, 632–633
      chitosan, 630, 633–634
      dextrin, 628–630
      gelatin, 630, 634–635
      guar gum, 630, 634
      heterogeneous carbohydrates, 629
      hydroxypropyl methylcellulose, 629–631
      microcrystalline cellulose, 628, 630–631
      pullulan, 629–630, 632
Fourier transform infrared (FTIR) spectroscopy
   metformin, 1344–1345
   polyacrylic acid, 1344–1345
   PVP, 1344–1345
      absorption, emission/scattering, 618–619
      advantages, 619
      disadvantages, 619
      FT–MIR–ATR fingerprint, *see* Fourier Transform Infrared–Mid Infrared–Attenuated Total Reflectance (FT–MIR–ATR) fingerprint
      MPCM, 1470
      spray chilling, 81, 82
Freeze-drying, 805
   advantages, 686–687
   characteristics, 852
   liposome suspension, 663
   principle, 852
Freeze granulation (FG) technology, 401–402
Fresh indicator, 1444
FTIR spectroscopy, *see* Fourier transform infrared (FTIR) spectroscopy
FT–MIR–ATR fingerprint, *see* Fourier Transform Infrared–Mid Infrared–Attenuated Total Reflectance (FT–MIR–ATR) fingerprint

G

Gas antisolvent (GAS) process, 449–450, 1085
Gas dynamic atomization, 439
Gas indicators, 1444
Gelatin-based complex coacervation systems
   amino groups, 237, 239–240
   applications, 240–241
   BSE, 241
   carboxylic groups, 237, 239–240
   chemicals, 239
   fish gelatin, 241
   flow diagram, 238, 240
   IEP values, Type A and B, 240–241
   pH, 237
   plant/milk-based proteins, 242
   supernatant, 238
Gelatin capsules, 1418, 1423–1424
Generally recognized as safe (GRAS), 72, 251, 645
Glucomannan, 540–541
D-Glucuronic acid, 1261
Granulation technology
   extrusion–spheronization
      advantages, 398
      disadvantages, 398
      plastics processing area, 397
      uniform-sized spherical particles, 397
   FBT, 402–403
   fluid bed rotor, 397–398
   fluid bed spray granulation, 397
   freeze granulation, 401–402
   Granurex® technology, 408
   MADG process, 406–408
   melt granulation, *see* Melt granulation technology
   objective, 386
   PDG, 400–401
   schematic representation, 386–387
   solid/liquid active ingredients, 386–388
   spray drying, *see* Spray drying, granulation technology
   steam granulation, 406
   TAG, 408
   vibrational nozzle technology, 399
   wet granulation, 388–389
   Wurster process
      advantages, 396
      components, 396
      disadvantages, 397
GRAS, *see* Generally recognized as safe (GRAS)
Guggenheim–Anderson–de Boer (GAB) model, 879
Gum arabic complex coacervation system
   flow diagram, 238, 240
   negatively charged, 237
   pH, 237
   supernatant, 238
Gums, mucoadhesive polymers, 1243
Gut-associated lymphoid tissue (GALT), 1364

H

Hen Egg Test-Chorioallantoic Membrane (HET-CAM) method, 1211
Hioxifilcon B, 1188
HLB, *see* Hydrophilic–lipophilic balance (HLB)
HME, *see* Hot-melt extrusion (HME)
Hofmeister/lyotropic series, 236
Host-guest complexation, 10
Hot-melt extrusion (HME), 556
   advantages, 217–218, 1138
   AFM, 222
   amorphous solid dispersions, 1139
   API, 213, 218
   carrier, dosage forms, 218–219

conveying of mass, 216
crystalline products, 218
crystalline solid dispersions, 1139
dexamethasone-loaded implants, 226
directly shaped products, 1145
disadvantages, 218
downstream processing, 216–217
drug substances, 218, 220
encapsulation
    advantages, 227
    bitter APIs, taste masking of, 227
    core-shell microcapsule/reservoir, 226
    limitations, 227
    lipid extrudates, 227
    matrix type microcapsule, 226
    rationale for, 226–227
    residual dry polymer, 227–228
    TEC concentration and pre-plasticization, 229
    water uptake, 227–228
flow through die, 216
guaifenesin release, 1140
HPC and HPC:HPMC films, 225
HSM and SEM, 222
marketed products, 229–230
material feeding, 216
meltable substances, 218
modular screws design, 216–217
oral drug delivery, 224
patents, 214
pharmaceutical-class extruders, 214
plasticizers, 220–221
platform technology, 213
poly(lactic acid) implants, 226
polymers, 1139
schematic presentation of, 217
solid dispersions, 213–214, 222–224
spectroscopic techniques, 222
SSE, *see* Single screw extruders (SSE)
thermal analysis, 221
TSE, *see* Twin-screw extruder (TSE)
WVS, 222
XRD, 221
Hot melt microencapsulation, 1374
Hot-stage microscopy (HSM), 222
HPC, *see* Hydroxypropyl cellulose (HPC)
HPDs, 564
HPMC, *see* Hydroxypropyl methyl cellulose (HPMC)
HPMCAS, *see* Hydroxypropyl methyl cellulose acetate succinate (HPMCAS)
Hydrogel microspheres, 11
Hydrophilic cellulose-based biopolymer matrix systems, 553
Hydrophilic–lipophilic balance (HLB), 253, 255, 257, 334, 473
Hydroxy ethyl methyl cellulose (HEMC), 529
Hydroxypropyl cellulose (HPC), 222, 225, 227–228, 497, 524, 539, 552, 554, 556, 564, 1179
Hydroxypropyl methyl cellulose (HPMC), 225, 497
    blood cholesterol, 528
    body weight gain, 528–529
    CLA, 555
    controlled-release tablets, 539, 554
    dietary fibers, 527
    fat content, fried food, 527
    gelatin cross-links, 522
    gluten replacement, 527
    glycemic response, 528
    hydration and gel-forming properties, 554
    hydroxypropyl and methyl substitution, 519
    immediate-release (IR) dosage, 521
    modified release oral drug delivery, 524
    molecular weight, 519
    PVP, 555
    renewable cellulose resources, 521
    solubilization enhancing polymer, 522–523
Hydroxypropyl methyl cellulose acetate succinate (HPMCAS)
    amphiphilic properties, 520
    foam extrusion, 1141, 1143
    pH, 519–520
    solubilization enhancing polymer, 523
    THF, 519
Hypromellose, 403, 522–523, 1140; *see also* Hydroxypropyl methyl cellulose (HPMC)

I

Iminothiolane-modified poly(ethylene glycol)-*block*-poly(L-lysine) (PEG-*b*-PLL(IM)), 1274–1275
Inorganic PCM
    advantages and disadvantages, 1459
    metals, 1460
    salt hydrates, 1460
    salts, 1460
Insect/rodent-resistant food packaging
    commercial pesticides, 1450
    essential oils, 1449
    eucalyptus oil, 1450
    LDPE films, 1449
    natural polymers, 1449
    penetrators, 1448
    rodent repellents, 1449
    synthetic and natural insect repellents, 1448
    synthetic polymers, 1449
    terpenes, 1450
In situ polymerization, 297
    advantages, 307
    amino resins, *see* Amino resins
    disadvantages, 307
    formaldehyde scavengers, 310–311
    MF, *see* Melamineformaldehyde (MF)
    organic PCM, 1464–1465
    related chemistry, 309
    salts, 310
    UF systems, *see* Urea-formaldehyde (UF) systems
Interfacial polycondensation of isophorone diisocyanate (IPDI), 320
Interfacial polycondensation/polyaddition
    applications, 298
    chemical structure, reactants, 299
    compact capsules, 298–299
    concentration and temperature, 299
    core/wall ratio, 299–300
    microcapsule wall formation, 298
    monodispersed microcapsules, 300
    oil phase, 297
    organic PCM, 1466

polymers and reactant systems, 297–298
polyurea microcapsules, 300–301
structure and fabrication method, 300
Interfacial polymerization
dispersion polymerization, 297
emulsion polymerization, 297
essential oils, 874
in situ polymerization, *see* In situ polymerization
polycondensation/polyaddition, *see* Interfacial polycondensation/polyaddition
suspension polymerization, 297
Internal gelation, 276–277, 653, 657, 1084
International Sol-Gel Workshop, 340
International Union of Pure and Applied Chemistry (IUPAC), 235, 248
Intra-articular drug delivery, 1267
Ionic gelation, 11, 750–751
Ionic liquids (ILs)
acetylation, 551
blends, 551
cellulose dissolution, 549–550
grafting copolymerization, 551
modifying cellulose, 549–550
Ionotropic gelation technique
biopolymers
alginate, 277–278
chitosan, 278–279
CMC, 280
gelan gum, 279–280
K-carrageenan, 280
divalent cation characteristics, 280–281
emulsion/gelation/polymerization methods, 276
extrusion method, 276
hydrogel beads, *see* Drug-loaded hydrogel beads
hydroxyethylcellulose, 283
internal gelation, 276–277
laboratory method, 288–290
multipolyelectrolyte gelispheres, 283
natural, chemically modified, and synthetic polymers, 288–289
organic solvents, 276
polyelectrolytes, 275
Iontophoresis, 1181–1182, 1235
Isoelectric point (IEP), 240, 563

K

Kernel lipocalins, 733
Ketorolac tromethamine, 554
Konjac, 502–503

L

Lacisert insert, 1179
Lactic acid bacteria (LAB), 773, 825
Langmuir model, 363
Large unilamellar vesicles (LUVs), 657, 1385–1386
Latent heat storage (LHS), 1456; *see also* Phase change materials (PCM)
Layer-by-layer process (LbL), 751–752
LBG, *see* Locust bean gum (LBG)
Lecithin organogels, 712, 1059
Lectin-mediated polymers, 1246
Lidocaine hydrochloride (LH), 1321, 1323

Light scattering, PECs, 1345
Lining mucosa, 1226
Lipid–polymer hybrids, 1160–1161
Lipids
fatty acids and fatty alcohols, 507
glycerides, 507
liposomes, 507
paraffin, 508
phospholipids, 507
waxes, 507
Lipophilicity, 1127, 1156
Liposomal drug delivery system, 1195, 1263–1265
Liposomes, 11, 1385–1386, 1397
circulation, 1157
CLC system, 1158
disadvantage, 1157
drug-loaded liposome producing methods, 1157
echogenic liposomes, 1158
multilamellar and unilamellar vesicles structure, 1156–1157
opsonization, 1157
phase transitions, 1158
physico-chemical properties, 1156–1157
proteins and peptides, 1158
temperature sensitization, 1158
water soluble drugs, 1156
Litsea cubeba EO (LCEO), 880
Locust bean gum (LBG), 497, 499–500
Low angle laser light scattering (LALLS), 1468
Low critical swelling temperature (LCST), 714
Low-density polyethylene (LDPE), 1449, 1457, 1467
Lyophilization, 11, 994
Lyotropic gels
lyotropic liquid crystals, 1395
amphiphilic molecule, 1396
disadvantages, 1397
hydrophilic solvent, 1396–1397
thermotropic liquid crystals, 1395–1396
Lyotropic liquid crystals, 1395
amphiphilic molecule, 1396
disadvantages, 1397
hydrophilic solvent, 1396–1397
Lysozyme-CLEA-immobilized CS nanofibers, 1326–1327

M

Macrocapsules, 1423
Macroemulsion, 247–248, 664, 1390, 1397–1398
MADG process, *see* Moisture-activated dry granulation (MADG) process
Malolactic fermentation (MLF), 936, 942–943, 947
Marangoni flow, 361
Masticatory mucosa, 1226
MC, *see* Methylcellulose (MC)
MCC, *see* Microcrystalline cellulose (MCC)
M-cells, 1362, 1364–1365
Mechanical energy storage, 1456
Media milling, 784
Melamineformaldehyde (MF)
core materials, 307–308
microscope and SEM images, 308–309
perfume microcapsule, 308–309
pH value, 310

Index

Melt agglomeration, *see* Melt granulation technology
Melt-dispersion technique
    disadvantages, 485–486
    emulsification/homogenization
        applications, 474–479
        continuous phase, 472
        critical points, 472–473
        disadvantages, 483–484
        dispersed phase, 472
        high-energy inputs, 474
        HLB, 473
        postloading methodology, 473
        SHM, 474
        surfactant, 484
        system matrix/active compound, 473
    lipid-based formulations, 482–483
    spray-congealing
        chemical compounds, 475–479, 483
        erythritol, 482
        fatty acids and fatty alcohols, 480–481
        glycerides and polyglycolyzed glycerides, 480–481
        lipophilic materials, 482
        natural and synthetic waxes, 480–482
        principle, 470
        waxy particles surface morphology, 483–484
    spray cooling, 470–472
    thermal characteristics, 485
Melt extrusion
    bioactive compounds, food, 783–784
    controlled released products
        abuse deterrence, 1138, 1141–1142
        delayed release, 1138
        excipients and applications, 1140–1141
        extended release, 1138
        fugitive plasticizers, 1140
        HME, 1138–1139
        innovators, 1138
        non-fugitive plasticizers, 1140
        polymers, 1139
    dissolution-enhanced products
        amorphous dispersion, 1135
        crystalline solid dispersion, 1136
        dalcetrapib, 1137
        embedded engineered particles, 1137–1138
        griseofulvin, 1136–1137
        lopinavir and ritonavir, 1136
        miscibility regime, 1138
        screw elements and designs, 1136
        solubilization regime, 1138
    foamed products, *see* Foam extrusion
    formulation design and characterization
        carrier selection, 1134
        controlled release characteristics, 1135
        crystallinity, 1132–1134
        disadvantages, 1135
        DSC, 1132–1133
        functional excipient roles, 1130–1131
        hot stage microscopy, 1132–1133
        miscibility regime, 1132
        multicomponent dispersions, 1130
        polymer selection, 1134
        solid solution, 1129–1130
        solubilization regime, 1132
        spring and parachute effect, 1134–1135
        types of solid dispersions, 1129–1130
        wettability, 1130
    marketed pharmaceutical products, 1128–1129
    shaped delivery systems, 1144–1145
Melt granulation technology
    advantage, 405
    attrition and breakage, 404
    coalescence, 404
    definition, 403
    disadvantages, 405–406
    meltable binders, 404
    requirements, 404–405
    wetting and nucleation, 403–404
Membrane-coating granules (MCGs), 1227–1228
Mercerization, 518, 544–545
Methylcellulose (MC)
    bulk laxative, 538
    carbohydrate polymers, 496–497
    fiber content, 527
    microgel system, 563
    mono methyl-substituted cellulose ether, 519
    satiety, 529
    solubilization enhancing polymer, 523
    thermoreversible gels, 548
Methyl methacrylate (MMA), 317, 319, 755, 1183
MF, *see* Melamineformaldehyde (MF)
Micelles
    bioactive compounds, food, 776–777
    biodegradable polymers, 607–608
    PIC, *see* Polyion complex (PIC) micelle
    vegetable oil–based formulations, 1384–1385, 1397
Microcapsule Processing and Technology, 676
Microcapsules
    capsule size, 1423–1424
    industrial application history, 1425–1426
    manufacture
        aminoplastic resin microcapsules, 1426–1427
        etherified melamine derivatives, 1428
        formaldehyde equilibrium, 1428–1429
        laundry microcapsules, 1428
        melamine–formaldehyde systems, 1426
        oil-in-water encapsulation, 1426
        particle size distribution, 1426–1427
        precondensate, 1428–1429
    stable core-shell, *see* Stable core-shell microcapsules
Microcrystalline cellulose (MCC)
    advantages, 538
    Avicel PH-200 LM, 407
    drug delivery application, 555, 559–560
    moisture-absorbing materials, 406
    protein retention, 563
Microemulsion, 247
    advantages, 248–249
    bicontinuous, *see* Bicontinuous microemulsions
    cosolvents, 256
    cosurfactants, 255–256
    GRAS, 251
    IUPAC definition, 248
    vs. macroemulsion, 248
    oil-in-water microemulsions, *see* Oil-in-water microemulsions
    oil phase, 252
    pharmaceutical industry, 248

phase diagrams, 250–251
phase inversion method
    PIC, 257–258
    PIT, 257
phase titration method, 256
production, 248
surfactants
    amphiphilic compounds, 252
    anionic surfactants, 253–254
    cationic, 254–255
    CMC value, 252–253
    HLB value, 253
    nonionic surfactants, 253–254
    Zwitterionic, 253–254
surfactants and cosurfactants, 249
vegetable oil-based formulations, 1388–1389, 1397
water-in-oil microemulsion, 261–262
Winsor phases, 250
Microencapsulated phase change material (MPCM), 1458
    applications, 1470, 1475
    characterization techniques, 1470–1472
        AFM, 1470
        cycling, 1470
        DSC, 1469
        flammability, 1469
        FTIR, 1470
        LALLS, 1468
        optical microscopy, 1468
        PSD, 1468
        SEM, 1468
        TEM, 1468
        TGA, 1469
        viscosity measurement, density, and conductivity, 1469
        WAXS, 1468
        XRD, 1468
    materials used, 1461–1462
    properties results, 1473–1475
    types
        mononuclear, polynuclear, and matrix encapsulation, 1462–1463
        shell and PCM combinations, 1462–1464
Microencapsulation; *see also* Encapsulation
    annular jet process, *see* Annular jet process
    artificial cells, *see* Artificial cells
    cell entrapment, 938–939
    coacervation, *see* Coacervation
    interfacial polymerization, *see* Interfacial polymerization
    microparticulate drug delivery systems, *see* Microparticulate drug delivery systems
    miniemulsion technology, *see* Miniemulsion technology
    phase inversion precipitation, *see* Phase inversion precipitation
    polymeric material, 274
    probiotics, *see* Probiotics
    sol-gel microencapsulation, *see* Sol-gel microencapsulation
Microfluidic system
    coflow-focusing devices, 370
    droplet formation and breakup, 363–364
    efficiency of mixing, 371
    flow-focusing systems
        coaxial capillary, 368
        configurations, 370
        cylindrical jet, 369
        gas and a liquid phases, 366
        quasistatic model, 367
        Rayleigh–Plateau instability, 369
        squeezing, 366
    interfacial tension, 361
    Langmuir model, 363
    Laplace pressure, 361–362
    loading efficiency control, 371–372
    low-Reynolds-number effects, 360
    Marangoni flow, 361
    micro- and nanodevices fabrication techniques, 360
    multiple emulsions/Janus particles, 377–378
    multiscale and multiphysics effects, 360
    Navier–Stokes equations, 361
    noncontinuum effects, 360
    photolithography-based microfabrication, 371–373
    Poisson–Boltzmann equation, 360–361
    replication-based methods
        advantages, 374
        hot embossing technique, 375
        injection molding, 376, 378
        soft lithography process, 374–375
    shape control, 371
    shell thickness control, 371–372
    size distribution, 370
    square cube law, 360
    surface-dominated effects, 360
    surfactant transport, 363
    tangential velocity, 362
    T-junctions, 364–366
    Weber number, 363
Microparticulate drug delivery systems
    advantages, 1070
    air suspension technique, 1086–1087
    applications, 1069
        agriculture, 1102
        beverage production, 1101
        building construction, 1102
        cell immobilization, 1101
        defense, 1102–1103
        drug delivery, 1101–1102
        food safety, 1102
        molecules protection, 1101
        soil inoculation, 1102
    bioadhesive microspheres, 1099–1101
    centrifugal extrusion, 1086–1087
    characterization
        compressibility index, 1091
        drug content, 1092
        drug entrapment, 1092
        particle size, 1091
        powder x-ray diffraction, 1091
        surface morphology, 1091
        swelling measurements, 1092
        tapped density, 1091
        yield, 1091
    coacervation
        complex coacervation, 1082
        cooling and hardening phase, 1082
        dispersion, 1081

drying phase, 1082
phase separation, 1082–1083
simple coacervation, 1082
coating material, 1071–1072
commercial products, 1103
core material, 1071
double-walled microspheres, 1095–1096
drug release
bioadhesivity testing, 1094
floating behavior, 1094–1095
mucoadhesive property, 1093
in vitro methods, 1092–1093
in vivo methods, 1093
electrospraying, 1089
floating microspheres, 1096–1097
fluidized-bed technology, 1086
in situ polymerization, 1073
interfacial polymerization, 1073
internal gelation, 1084
jet excitation, 1088
magnetic microspheres, 1099
matrix type microcapsules, 1069
modified melt-dispersion method, 1084
mononuclear microcapsules, 1069
O/O/O extraction method, 1081
physicochemical evaluation, 1090–1091
polynuclear microcapsules, 1069
radioactive microspheres, 1101
single emulsion method, 1078–1079
solid lipid microparticles, 1097–1098
solvent evaporation, *see* Solvent evaporation
S/O/O double emulsion technique, 1081
S/O/W double emulsion technique, 1080–1081
spray-congealing, 1088–1089
spray-drying, 1088–1089
targeted microsheres, 1098–1099
ultrasonic atomization, 1090
W/O/O double emulsion technique, 1081
W/O/W double emulsion solvent evaporation method, 1080–1081
Miniemulsion technology
conventional emulsion polymerization, 315–316
costabilizers, 317
hydrophilic core
interfacial polymerization, 321
nanoprecipitation, 321
hydrophobic oil
interfacial polymerization, 320
radical polymerization, 319–320
inorganic nanoparticles
direct miniemulsion, 323–324
inverse miniemulsion process, 323–325
monomers, 316–317
oil-containing nanocapsules, 318
oil-soluble initiators, 317–318
organic pigments, 321–323
polymer materials, 318–319
surfactants, 317
water-soluble initiator, 317–318
MLF, *see* Malolactic fermentation (MLF)
MLVs, *see* Multilamellar vesicles (MLVs)
MMA, *see* Methyl methacrylate (MMA)

Modified-release dosage, 553
Moisture-activated dry granulation (MADG) process
advantages, 407–408
Avicel HFE-102, 407
Avicel PH-200 LM, 407
FMC Biopolymer, 407
we agglomeration, 406–407
Molecular encapsulation process, 10
Molecular weight cutoff (MWCO), 917–918
Monomer miniemulsion, 316, 322
Monomer polymerization
emulsion polymerization, 1372–1373
nanogels
ATRP, 1290–1292
free-radical polymerization, 1287
heterogeneous controlled/living radical polymerization, 1290
heterogeneous free radical polymerization, 1288
inverse (mini) emulsion polymerization, 1288–1289
inverse microemulsion polymerization, 1289–1290
one-pot protocol, 1287
precipitation polymerization, 1288
RAFT, *see* Reversible addition–fragmentation chain transfer (RAFT) polymerization
precipitation and dispersion polymerization, 1373
suspension polymerization, 1373–1374
MPCM, *see* Microencapsulated phase change material (MPCM)
Mucoadhesion, 1367–1368
adsorption theory, 1230
chewing gums, 1234–1235
diffusion theory, 1231
electronic theory, 1230
fracture theory, 1231
liquids, gels, and ointments, 1235
lozenges, 1235
mucins role, 1229–1230
multidirectional and unidirectional release, 1232
patches/films, 1234
polymers, *see* Mucoadhesive polymers
products list, 1232–1233
sprays, 1235
tablets, 1232–1234
wetting theory, 1230
Mucoadhesive polymers
bacterial adhesion, 1246–1247
cellulose derivatives, 1243
chitosan, 1243–1244
enzyme-inhibiting polymers, 1244
formation and persistence, 1236
gums, 1243
lectin-mediated polymers, 1246
permeation-enhancing polymers, 1244
pH, 1236
polyacrylic acid derivatives, 1236, 1243
properties and characteristics, 1236–1242
semi-natural polymers, 1243
thiolated polymers, 1244–1246
Mucosal vaccination, 580, 1267
Multilamellar vesicles (MLVs), 657, 779, 873, 1385–1386
Multiple emulsions, 871, 1388
Multiple screw extruders, 1127
Mupirocin, 1321, 1323

N

Nanocrystalline cellulose (NCC), 561, 563
Nanoemulsion
   aromas, 841
   bioactive compounds, food, 778
   colloid drug delivery systems, 1117
   PIT method, 668
Nanogels
   amine-based cross-linking
      advantage, 1275
      amphiphilic block copolymers, 1275–1276
      core cross-linked block copolymer micelles, 1279
      design and synthesis, 1277–1278
      PAPMA shell, 1277
      PFPA, 1277
      pH-responsive features, 1279
      SCKs, 1275–1276
      SCL micelle, 1276–1277
   applications
      inflammatory disorder treatment, 1295–1296
      local anesthetic drugs, 1296–1297
      N/P ratio, 1297
      nucleic acid delivery, 1297
      pDNA, 1297
      pH-responsive charge-conversional nanogel, 1295
      siRNA delivery, 1297
   click chemistry-based cross-linking
      alkynyl core-functionalized block copolymer micelles, 1281
      alkynyl shell functionalized block copolymer micelles, 1279
      azido terminated dendrimers, 1279, 1281
      click-readied micelles and dendrimers, 1279–1280
      PNIPAM nanocapsules, 1282
      P(PAMA)-*b*-P(PEGMA) structure, 1281–1282
      schizophrenic micellization behavior, 1282–1283
      SCL micelles, 1282
      thermosensitive PIC micelles, 1281–1282
   disulfide-based cross-linking
      core–shell-type PIC micelle, 1274
      environmental thiol concentration, 1272–1273
      PDS groups, 1273
      PEG-*b*-PLL(IM), 1274–1275
      PEG-PLL, 1274–1275
      polymer nanoparticles design and synthesis, 1273–1274
      RAFT-synthesized copolymers, 1273
      siRNA, 1275–1276
   functionalization and bioconjugation, 1272
   hydrogel networks, 1271–1272
   imine bonds-induced cross-linking, 1283–1284
   monomer polymerization, *see* Monomer polymerization, nanogels
   photo-induced cross-linking
      core and shell-cross-linked nanogels, 1285
      NIPAAm-DMIAAm copolymer structure, 1285–1286
      PEOVE-*b*-P(HOVE/VEM) structure, 1286
      pH-dependent micellization, 1285–1286
      PNIPAAm graft terpolymer structure, 1287
      polymer precursor with photopolymerizable functionality, 1284
      preparation and photocontrolled volume change, 1284–1285
      temperature-responsive colloidal nanogels, 1287
      thermoresponsive SCL micelles, 1286
      thermosensitive nanogels, 1285
   physical cross-linking, 1287
   release mechanisms, 1297–1298
   synthesis methods, 1272–1273
Nanoparticle drug-delivery systems
   advantages, 560–561
   applications, 560
   biodegradable polymers, 1265–1266
   cellulose nanocrystals, 561–562
   chitosan, 561
   CMC, 561–562
   curcumin, 562
   definition, 560
   microgel systems, 563
   NCC, 563
   TTAB, 563
   uses, 560
   WSMM, 561
Nanoparticulate-based ophthalmic drug delivery systems
   dendrimers
      bioadhesive polymers, 1201
      DH, 1202
      PAMAM, 1201–1202
      photoreactive G3.0–PEG acrylate conjugates, 1202–1203
      potential applications, 1204
      surface functionalization, 1204
   discomes, 1205
   liposomes, 1195–1196
   microemulsions, 1197–1198
   nanoparticle-loaded contact lenses
      conventional CLs, 1205–1206
      drug-eluting CLs, 1206
      lipophilic drug encapsulation, 1206–1207
      particle-laden lens, 1207
      PLGA particles, 1206–1207
      surfactant-laden gels, 1207–1208
   nanosuspensions
      drug micro-particle suspensions, 1199
      Eudragit RS100 (RS) and RL100 (RL) polymer, 1199–1200
      high pressure homogenization method, 1199
      IBU-loaded polymeric nanoparticle suspensions, 1199
      modified Draize test, 1200
      nanoparticle suspensions (NS), 1199–1200
      poorly water-soluble drug, 1198
      stability tests, 1200
      sub-micron colloidal dispersion, 1198
   niosomes, 1204–1206
   polymeric nanoparticles
      advantages, 1211
      biodegradable polymers, 1209, 1213–1214
      colloidal particles stability, 1208
      DNPs interaction, 1208–1209
      drug-loaded polymeric micelles, 1208, 1210
      fluconazole, 1209–1210
      gum cordia, 1209
      HET-CAM method, 1211
      mucoadhesive polymers, 1212–1213

Index

PLA nanoparticle localization, 1210
PLGA nanoparticles, 1211
precorneal residence time, 1211
potential advantages, 1194
Nanostructured lipid carriers (NLCs), 748, 1156
Nasolacrimal drainage system, 1170, 1211
Natural polymers
albumin, 588
alginate, 588–589
chitosan, 589–590
collagen, 588
cyclodextrins, 589, 591
dextran, 589–590
fibrin, 590
gelatin, 588–589
hyaluronic acid, 589, 591
polysaccharides and proteins, 587
starch, 589, 591
Navier–Stokes equations, 60, 361
NCR PAPER* brand of carbonless paper, 1413
NeutrAvidin, 1188
NHS-PEG-biotin, 1188
N-hydroxysuccinimide (NHS) ester, 1154
Niosomes
bilayer structure, 1119, 1204–1205
hydrophilic and lipophilic drug molecules, 1159
ocular drug delivery, 1160
PEGylation, 1119–1120
NMR spectroscopy, see Nuclear magnetic resonance (NMR) spectroscopy
Nonbiodegradable synthetic polymers
bioadhesion, 1367
caco-2 cell models, 1369
calcium chelators, 1369
encapsulated bioactive agent, 1367
Eudragit L100 microspheres, 1369
Eudragit S100 microspheres, 1369–1370
mucoadhesion, 1367–1368
occludin and claudin, 1368
PAA-based systems, 1368–1369
paracellular transport, 1368
passive diffusion processes, 1368
tight junction, 1368
Non-gelatin-based complex coacervation systems, 241–242
Non-ionic surfactant-based vesicles, see Niosomes
Nonionic surfactants, 253–254
Nuclear magnetic resonance (NMR) spectroscopy
diffusion-ordered, 263
PECs, 1345
solid state, 1134
Nucleoprotein complex, 1335
Nurofen®, 224

O

Ocular drug delivery
bioavailability and systemic side-effects, 1171
blood-aqueous barrier, 1173–1174
blood-retinal barrier, 1174
contact lens, see Contact lens-based ophthalmic drug delivery
corneal barrier, 1172–1173
drug absorption, 1168
implants, 1181
intravitreal administration, 1180–1181
intrinsic barriers, 1171
iontophoresis, 1181–1182
nanoparticulate-based ophthalmic, see Nanoparticulate-based ophthalmic drug delivery systems
nanotechnology, 1193–1194
ocular inserts, see Ocular inserts
pulsed administration and patient noncompliance, 1171
spray, 1182
subconjunctival administration, 1180
systemic absorption, 1171
tear film barrier, 1171, 1173
therapeutic active agent absorption and elimination, 1168
therapeutic significance, 1194
topical ocular delivery
aqueous solutions, 1174–1175
corneal absorption enhancement, 1174
disadvantage of eye drops, 1174
emulsions, 1175
gels, 1176
ointments, 1175–1176
sol to gel systems, 1176
suspensions, 1175
Ocular inserts
drug release procedures, 1177
erodible ophthalmic inserts, 1178–1179
insoluble inserts, 1177
nonerodible inserts, 1177–1178
soluble inserts, 1177
Ocusert inserts, 1177–1178
Ohnesorge number (Oh), 180
Oil-in-water microemulsions, 1389
advantages, 258
drug penetration, 259
fluconazole, 260
pseudoternary phase diagrams, 259–260
skin permeation enhancement, 258
structure and composition, 249–250
ternary diagram, 259
transdermal delivery system, 259
Oleic acid, 259, 323–324, 1383
Oleogels, see Organogels
Oligolamellar vesicles (OLVs), 657
Open-cycle process, 48
Optical indicator, 1444–1445
Optical microscopy (OM), 257, 551, 1468
Orally disintegrating tablets (ODT), 224
Oral mucosal drug delivery
buccal mucosa, 1226–1228
lining mucosa, 1226
masticatory mucosa, 1226
mucoadhesion, see Mucoadhesion
specialized mucosa, 1226
sublingual mucosa, 1228–1229
Organic PCM
advantages and disadvantages, 1458
fatty acids, 1459
microencapsulation process
chemical methodology, 1464–1466
emulsification step, 1463
formation of microcapsules, 1463

physico-chemical methodology, 1466–1467
physico-mechanical methodology, 1467
paraffin compounds
    hydrocarbons, 1458
    melting point *vs.* latent heat of fusion, 1459, 1461
    source of, 1458
    in TES systems, 1459–1460
Organogels
    controlled drug delivery
        accelerated stability method, 1040–1041
        advantages, 1036
        atomic force microscopy, 1044–1046
        backward extrusion, 1051–1052
        biocompatibility, 1055–1056
        bright field microscopy, 1041, 1043
        controlled delivery systems, 1056–1057
        definition, 1035
        dermal delivery, 1059
        differential scanning calorimetry, 1053–1054
        drop-ball melting method, 1053
        drug administration, 1056, 1058
        dynamic viscosity, 1047–1048
        environmental scanning electron microscopy, 1044–1046
        fluid filled structure mechanism, 1038–1039
        fluorescent microscopy, 1043–1044
        forward extrusion, 1051–1052
        hemocompatibility, 1055–1057
        impedance spectroscopy, 1054–1055
        infrared spectroscopic technique, 1046–1047
        kinematic viscosity, 1047–1048
        microemulsion-based mechanism, 1040
        nasal drug delivery, 1059
        Newtonian fluids, 1047–1048
        oral delivery, 1060
        parenteral delivery, 1059
        phase contrast microscopy, 1042–1043
        polymeric matrix mechanism, 1039–1040
        rheology, 1049
        scanning electron microscopy, 1044–1046
        shear thinning, 1048–1049
        simple image processing, 1042
        solid fiber mechanism, 1036–1038
        spreadability, 1051
        stress relaxation, 1050
        time dependent non-Newtonian fluids, 1047–1048
        time independent non-Newtonian fluids, 1047–1048
        topical delivery systems, 1056, 1058–1059
        transdermal delivery, 1059
    food delivery systems
        active constituent content, 715–716
        active ingredients, 721–722
        advantages, 700–701
        L-alanine derivative, 711
        antimicrobial testing, 716
        ball indentation, 715
        biocompatibility, 703–704
        biodegradable, 704
        chirality effects, 705
        chocolates, 719–721
        classification of, 699–700
        differential scanning calorimetry, 713
        disadvantages, 701–702
        edible agents, 716–717
        essential components, 699
        Eudragit organogel, 711
        fluid-filled fiber, 702, 706–708
        FTIR spectroscopy, 714
        gel structures, 713
        homogeneity test, 714
        hydration method, 708
        kinetic approach, 703
        LCST, 714
        lecithin, 712
        loading capacity release kinetics, 705
        mechanical properties, 705
        microemulsion-based organogel, 711
        nonbirefringence, 705
        optical clarity, 705
        organogelators, 709–710
        pH determination, 715
        physicochemical parameters, 706
        PLO, 712
        processability, 705
        PTT, 713
        SARS/SAXS, 713–714
        saturated animal fat, 717–718
        solid fat crystal network, 718–719
        solid fiber mechanism, 702
        solid fiber method, 708
        solvent absorbency determination, 714
        sorbitan monostearate, 709–711
        spreadability, 715
        stability and safety, 706
        stability testing, 714
        statistical approach, 703
        structure of, 699
        thermoreversibility, 704
        thermostability, 704
        uniaxial tensile testing, 715
        viscoelasticity, 704, 715
        in vitro permeation study, 716
        in vitro release studies, 716
    vegetable oil–based formulations
        advantages, 1391
        apolar phase polymerization mechanism, 1392–1393
        biphasic organogels, 1391
        fluid-filled structure gelation, 1392
        organogelators, 1391
        physical and chemical organogels, 1391
        polymeric organogels, 1392–1393
        solid fiber gelation, 1392–1393
        supramolecular organogels, 1392–1393
Ostwald ripening, 315–318, 324, 666
Overbeek theory, 1336
Oxygen indicators, 1444
Ozurdex®, 1145

P

PAC process, *see* Polyacrylic copolymer (PAC) process
PAMAM dendrimers, *see* Poly(amidoamine) (PAMAM) dendrimers
Pan coating process, 12
    atomizing and pattern air
        gun spray techniques, 164–165
        parameter, 169–171

Index

characteristics, 173
charge range, 156
coating materials, 172
coating weight gain, 174
conventional, *see* Conventional pan-coating
definition, 148
description, 150–152
drying capacity
    air capacity, water vapor, 164
    batch time concern/coating gain rate requirement, 159
    dew point and exhaust temperature, 162–163
    lower explosive limit, 160–161
    solvent vapors *vs.* temperature, 159–160
    upper explosive limit, 160
    water, 160, 162
gun position, 158–159
history, 149–150
pan dimensions, 167–168
pan speed, 168
particles/tablets, 174–175
plasticizer content, 174
process air, 158, 168–169
process endpoint, 165, 167
speed, 157–158
spray rate scale-up, 169
temperature, 158
vented, *see* Vented pan-coating
Particles from gas saturated solutions (PGSS) process, 455, 457–459, 871, 1085
Particle size distribution (PSD)
    essential oils, 887–888
    fluidized bed coating technology, 114, 118
    microcapsules, 1426–1427
    MPCM, 1468, 1471–1472
    RESS process, 450
Particle-source-in-cell model (PSI-Cell model), 57–58
PCM, *see* Phase change materials (PCM)
PCS, *see* Pulse Combustion Systems (PCS)
PDG technology, *see* Pneumatic dry granulation (PDG) technology
PDT, *see* Photodynamic therapy (PDT)
PECs, *see* Polyelectrolyte complexes (PECs)
PEG-biotinylated lipids, 1188
Pentafluorophenyl acrylate (PFPA), 1277
PEO, *see* Poly(ethylene oxide) (PEO)
Peptide-based therapeutics, 1359
Perfume microcapsule (PMC), 1420–1421
Permeation-enhancing polymers, 1244
PEs, *see* Polyelectrolytes (PEs)
PGSS process, *see* Particles from gas saturated solutions (PGSS) process
Pharmacokinetic/pharmacodynamics (PK/PD) analysis, 1367
Phase change materials (PCM), 340, 1435–1436
applications
    building materials, 1457
    disadvantages, 1457–1458
    domestic heating applications, 1457
    domestic hot water, 1457
    food conservation, 1457
    PCS, 1457
    refrigeration, 1457
    subcooling, 1457
chemical properties, 1457
economical, 1457
kinetic properties, 1456
latent heat of fusion *vs.* melting point, 1459
microencapsulate, *see* Microencapsulated phase change material (MPCM)
MPCM, 1458
thermophysical properties, 1456
types
    eutectics, 1461
    inorganic, 1459–1460
    organic, *see* Organic PCM
Phase change slurries (PCS), 1457
Phase inversion method
    PIC method, 257–258
    PIT, 257
Phase inversion precipitation
    binodal changes, 350–351
    flat membranes and microcapsules, 347–348
    Flory–Huggins theory, 349
    Gibbs free energy, 349
    immersion in nonsolvent, 352–353
    liquid–liquid demixing process
        instantaneous/delayed, 349, 351–352
        polymer concentration, 348
    polysulfone/vanillin microcapsules
        air atomizing nozzle, 354
        materials, 353
        mean size and size distribution, 354–355
        morphology, 355
        release experiments, 354
        SEM, 354
        vanillin release, 356
    solvent and nonsolvent, 347, 351
    ternary system, 348–349
    vapor nonsolvent, 353
Phase inversion temperature (PIT) method
    microemulsions, 667
    nanoemulsions, 668
Phase transition temperatures (PTT), 713
pHEMAco-β-CD networks, 1189, 1191
Phosphatidylcholine, 662, 873, 1385
Photodynamic therapy (PDT), 260, 1272, 1290, 1295–1296
Physical adsorption
    Biteral®, 1318
    heparin, 1317
    layer-by-layer multilayer assembly, 1318
    modes of drug loading, 1317
    nanoparticles assembly, 1318
Phytoestrogens, 744
Pickering emulsions, 1388
PIC micelle, *see* Polyion complex (PIC) micelle
Piloplex, 1175
PIT, *see* Phase inversion temperature (PIT) method; Transitional phase inversion (PIT)
Pituitary adenylate cyclase-activating polypeptide (PACAP), 1246
PLGA, *see* Poly(lactic-co-glycolic acid) (PLGA)
PLLA, *see* Poly(L-lactic acid) (PLLA)
Plurilamellar vesicles (PLVs), 657
Pluronic lecithin organogels (PLO), 708, 712, 716
Pluronics-based polymers, 1373
PMAA, *see* Polymethacrylic acid (PMAA)
PMMA, *see* Poly(methyl methacrylate) (PMMA)

Pneumatic dry granulation (PDG) technology
   advantages, 400–401
   pharmaceutical companies, benefits, 401
   solid dosage ingredient, 400
   wet granulation, 400
PNIPAM nanocapsules, *see* Poly(*N*-isopropylacrylamide) (PNIPAM) nanocapsules
Poisson–Boltzmann equation, 360–361
Polyacrylic acid derivatives, 1236, 1243
Polyacrylic copolymer (PAC) process
   capacity expansion, Portage plant, 1420–1421
   cross-section, 1419–1420
   cycle times, 1420
   Encapsys encapsulation plant, 1422
   on fabric, 1422
   on human hair, 1420–1421
   melamine cross-linked structure, 1419
   melamine formaldehyde cross-linker, 1418
   phases, 1419
   PMC, 1420–1421
   proprietary emulsifiers, 1418
Poly(alkyl cyanoacrylate) nanocapsules, 1236, 1366
Polyamides, 595
Poly(amidoamine) (PAMAM) dendrimers, 602, 1163, 1201–1202
Polyanhydride microspheres, 1366
Polyanhydrides, 595, 1366
Polycaprolactone (PCL), 594
Polydiacetylene-N-hydroxysuccinimide (PDA-NHS) nanoliposomes, 1447
Poly[(2-dimethyl amino) ethyl methacrylate [P(DMAEMA)] side chains, 564
Polydioxanone (PDS), 594–595
Polyelectrolyte complexation (PEC) technique
   biopolymers, 285
   chemical gelation processes, 285
   chemical gelling agents, 286–287
   chemical/irreversible gelation, 286
   chitosan applications, 284
   chitosan-reinforced nanoparticles, 284
   complex coacervation, 286
   complex formation, 287
   demerits, 288
   freeze–thawing, 286
   gel/network, 285
   grafting, 287
   heat-induced aggregation/maturation gelation, 286
   heating/cooling, polymer solution, 285–286
   hydrogen bonding, 286
   ionic interaction, 286
   laboratory method, 286, 288–290
   merits, 287–288
   natural and synthetic polymer, 285
   physical/reversible gelation, 285
   polyanionic molecules, 283
   properties, 283–284
   sol, 285
Polyelectrolyte complexes (PECs)
   advantages, 1348
   applications in drug delivery, 1347–1348
      active substance incorporation, 1348
      biological stimuli, 1351
      electrochemical stimuli, 1349–1350
      ionic strength stimuli, 1349
      laser light stimuli, 1350
      magnetic stimuli, 1351
      mechanical stimuli, 1351
      microcapsule loading and release, 1348–1349
      pH stimuli, 1348–1349
      solvent stimuli, 1349
      temperature stimuli, 1350
      ultrasound stimuli, 1350–1351
   biocompatible polymer systems, 1335
   characterization
      conductivity, 1341–1342
      drug release, 1347
      DSC, 1343
      electron microscopy, 1347
      FTIR spectroscopy, 1344–1345
      light scattering, 1345
      NMR spectroscopy, 1345
      particle charge, 1345–1346
      particle size, 1346–1347
      potentiometry, 1340–1341
      turbidimetry, 1341
      viscosity, 1342–1343
      x-ray diffraction, 1345
   charge–charge interactions, 1334
   Debye–Huckel theory, 1336
   Flory interaction parameter, 1336
   formation
      charge density effect, 1340
      concentration effect, 1338–1339
      degree of neutralization, 1340
      between gum karaya and chitosan, 1336
      hydrophobicity effect, 1340
      within intracomplexes, 1337
      ionic group effect, 1339–1340
      ionic strength effect, 1339
      molecular structure effect, 1338
      molecular weight, 1340
      primary complex, 1337
      stoichiometric ratio effect, 1339
   intercomplex aggregation process, 1337
   ladder-like structure, 1337–1338
   layered structures, 1336
   oppositely charged polyions, 1334
   Overbeek and Voorn theory, 1336
   physicochemical properties, 1334
   scrambled egg model, 1337–1338
   soluble, colloidally stable, and coacervate complexes, 1337
Polyelectrolyte polymers, 1334
Polyelectrolytes (PEs)
   characterization, 1335
   classification, 1335
   definition, 1335
Poly(ethylene oxide) (PEO), 553, 556, 1140–1141, 1178, 1316
Polyhydroxybutyrate (PHB), 593–594
Polyion complex (PIC) micelle
   click chemistry-based cross-linking, 1281–1282
   disulfide-based cross-linking, 1274–1275
Poly(isobutyl cyanoacrylate) (PIBCA) nanocapsules, 1366
Poly(lactic-co-glycolic acid) (PLGA)
   AMX, 1320–1321
   biodegradable polymers, 997, 1319
   cellular uptake and cytotoxicity, 562

Index

curcumin, 749
dexamethasone, 1145
double-walled microspheres, 1096
drug-release rate control, 372
EGCG/ellagic acid, 749
electrospray, 424–425
electrospun nanofibers, 1314
extraction and evaporation, 995
HNTs, 1320
hydrolysis, 594
hydrolytically unstable, 925
implants, 1181
jet excitation, 1088
Lupron Depot®, 598
microparticles/microsphere, 431, 600–601
molecular structure of, 592–593
parenteral delivery, 1012
polymers, 587
scaffolds, 925
SEM images of, 427
spray drying, 783
synthetic polymers, 599
$W_I/O/W_{II}$ technique, 992
Polylactide (PLA), 592, 1210
Poly(L-lactic acid) (PLLA), 1096, 1320, 1323–1325
Polymer-based DDS, *see* Nanogels
Polymeric biomaterials
  biodegradable, *see* Biodegradable polymers
  classification, 619–620
  gelatin, 626, 628
  heterogeneous carbohydrates
    alginates, 623–624
    carrageenan, 624–625
    chitosan, 625
    guar gum, 626–627
  homogeneous carbohydrates (glycans)
    dextrin, 620–621
    hydroxypropyl methylcellulose, 621–622
    microcrystalline cellulose, 621
    pullulan, 622–625
  in vivo applications, 1256
Polymeric microparticles, 274–275
Polymeric nano/microparticles
  advantages, 1361
  hydrophobic biodegradable polymers, 1362
  M-cells, 1362
  monomers polymerization
    emulsion polymerization, 1372–1373
    precipitation and dispersion polymerization, 1373
    suspension polymerization, 1373–1374
  net surface charge, 1362
  oral peptide-delivery system
    natural and protein-based polymers, 1370–1371
    nonbiodegradable synthetic polymers, *see*
      Nonbiodegradable synthetic polymers
    synthetic biodegradable polymers, *see* Synthetic
      biodegradable polymeric nano/microparticles
  particle manufacturing techniques, 1372
  preformed polymers
    complex coacervation, 1374–1375
    hot melt microencapsulation, 1374
    solvent evaporation, 1374
    spray drying, 1374
    thermal denaturation, 1374–1375

preparation methods, 1372–1373
  transportation of nanoparticles, 1362
Polymeric organogels, 1040, 1392–1393
Polymer nanofiber drug delivery
  applications, 1319
    amoxicillin-loaded electrospun nano-HAp/PLA,
      1320–1321
    chitosan nanofibers, 1325–1327
    electrospun PU–dextran nanofiber, 1323–1325
    PLLA/Captopril, 1320
    rifampin PLLA electrospun fibers, 1324, 1326
    wound dressing, 1321–1323
  principle of, 1319
Polymersomes, 607, 1120, 1158–1159
Polymethacrylic acid (PMAA), 1338, 1368, 1374
Poly(methyl methacrylate) (PMMA), 319–320, 1140, 1318, 1461–1462
Poly(*N*-isopropylacrylamide) (PNIPAM) nanocapsules, 1282, 1288, 1294, 1296
Polyorthoesters (POEs), 596
Polyoxymethylene (POM), 595
Polyphenolics
  bifunctional cross-linkers, 755
  CDs, 755–757
  cocrystallization method, 757
  complex coacervation, 751–752
  cooling, emulsions, 748
  emulsification-solvent removal methods, 748–750
  food/cosmetics, 744
  in situ polymerization reactions, 755
  ionic gelation methods, 750–751
  LbL, 751–752
  liposomes, 753–755
  micelles, 751, 753
  natural polyphenols, 742, 744
  phytoestrogens, 744
  ROS-generating enzymes, 742
  shikimate pathway and polyacetate pathway, 742–743
  sodium caseinate/calcium caseinate, 750
  spray-drying, 746–747
  structures and biological properties, 744–746
  supercritical fluids, 747
  tanning effect, 744
  yeast cells, 757
Poly(propylene fumarate) (PPF), 594
Polysaccharide-derived polymers, 575
Polysaccharides, 536–537
  advantages, 1258
  alginic acid/alginates, 1258–1259
  chitin and chitosan, 1261–1262
  dextran, 1259–1260
  hyaluronic acid, 1261
  pullulan, 1260–1261
  starch, 1259
Polyurethane capsules, 1423–1424
Polyvinyl pyrrolidone (PVP), 1344–1345
  absorption, emission/scattering, 618–619
  advantages, 619
  disadvantages, 619
  FT–MIR–ATR fingerprint, *see* Fourier Transform
    Infrared–Mid Infrared–Attenuated Total
    Reflectance (FT–MIR–ATR) fingerprint

HPMC, 555
MPCM, 1470
spray chilling, 80, 82
Posurdex1, 1181
Potentiometry, 1340–1341
Precipitation and dispersion polymerization, 1373
Precipitation polymerization technique, 1373
Pregelatinized starches, 576–578
Prilling, *see* Spray chilling
Probiotics
    active ingredient, food industry
        alginates, 675
        colonization, 673
        dietary supplements, 673
        emulsified systems, 676
        FAO/WHO Expert Consultation, 673
        gastric fluid, 675
        gel particles, 674
        non-dairy-based probiotic products, 675
        spray drying, 674
    advantages, 75
    alginate microcapsules
        alginate–chitosan, 693
        alginate–polylysine, 692–693
        brown seaweed, 687
        chitosan-coated alginate microcapsules, 690–692
        L. plantarum ATCC 8014, 687–689
        palm oil, 693
        polylysine, 692
        polylysine-coated, 692
        whey protein, 693
    cytotoxic/antimicrobial, 74
    emulsion and extrusion methods, 686
    freeze drying, 686
    gastrointestinal passage, 687
    health benefits, 686
    ileum, 73
    immobilization of bifidobacteria, 74
    Lactobacillus paracasei IMC 502, 74
    Lactobacillus rhamnosus IMC 501, 74
    lipid matrices, 75–76
    microorganisms, 73
    solid lipid microparticles, 75
    spray chilling, 75
    symbiotic microparticles, 75
    vegetable fats, 74
Process air
    material fluidization properties, 120
    vs. process air velocity, 121–122
    slit air velocity and slit area, 122–124
    spouted bed and top stray processes, 121
    tangential spray systems, 121, 124
    temperature and volume changes, 129–131
    Wurster process
        air distribution plates, 124–125
        air feeding, 126
        air velocity, 126–127
        down bed air flow, 124
        multiple nozzle configurations, 128–129
        partition parameters, 127
        total process air, 125–126
Proteins
    advantages, 503
    animal derived products, 503
        casein, 505–506
        gelatin, 506
        whey proteins, 506
    coacervation, 504
    gelation, 504
    phase separation, 504
    precipitation, 504
    solvent evaporation, 504
    spray drying, 504
    vegetable sourced protein, 503
        gluten, 505
        pea, 505
        rice, 505
        soybean, 504–505
        wheat, 505
Proteins/peptides oral delivery
    barriers, 1360
    PEG and fatty acids, chemical modification, 1361
    permeation enhancers, 1361
    polymeric nano/microparticles, *see* Polymeric nano/microparticles
    protease inhibitors, coadministration of, 1361
PSD, *see* Particle size distribution (PSD)
Pulse Combustion Systems (PCS), 440
    Arizona plant, 441
    data, chart, and discussion points, 442
    ingredients/materials, 442
Pulsed combustion spray drying
    advantages, 441
    disadvantage, 441
    microcapsules, 442
    PCS
        Arizona plant, 441
        data, chart, and discussion points, 442
        ingredients/materials, 442
        percent free core
            vs. percent feed solid, 442–444
            vs. percent flow aid, 442–443
    percent moisture *vs.* exit temperature, 443–444
    principles, 439–440
    slurry properties, 442
    thermal efficiency, 443, 445

Q

Quadruple emulsions, 1390
Quaternary ammonium HPDs (QHPDs), 564

R

Radioimmunoassay (RIA), 1371
RAFT polymerization, *see* Reversible addition–fragmentation chain transfer (RAFT) polymerization
Rapid expansion of supercritical solutions (RESS) process, 449, 1085
Rayleigh–Plateau instability, 364, 369
Relative humidity (RH) indicator, 1094, 1444, 1446
Resveratrol, 732–735, 749, 753
Reticuloendothelial system (RES), 1361, 1365
Reversible addition–fragmentation chain transfer (RAFT) polymerization
    direct RAFT polymerization, 1292–1293
    inverse RAFT miniemulsion, 1295–1296

Index

strengths of, 1292
in water, 1293–1295
Reynolds stress model (RSM), 60
Rifampin PLLA electrospun fibers, 1324, 1326
Ringsdorf model, synthetic polymer drugs, 1152
RNG k–ε model, 60
Rotary atomizer, 38, 65, 102, 208, 391–392

S

Saccharomyces cerevisiae, 757, 936
SAS, *see* Supercritical antisolvent (SAS) process
Satellite droplets, 96, 203
Scanning electron microscopy (SEM)
    hot-melt extrusion, 222
    melamineformaldehyde, 308–309
    MPCM, 1468
    phase inversion precipitation, 354
    spray chilling, 80–81
    supercritical antisolvent process, 461–462
SC-CO_2, *see* Supercritical carbon dioxide (SC-CO_2)
SCF, *see* Supercritical fluid (SCF)
SCLs, *see* Soft contact lenses (SCLs)
SCMC, *see* Sodium carboxymethyl cellulose (SCMC)
SD, *see* Spray drying (SD)
SEM, *see* Scanning electron microscopy (SEM)
Semi-natural polymers, 1243
Sensible heat storage (SHS), 1456
SFEE, *see* Supercritical fluid extraction of emulsions (SFEE)
Shell-cross-linked knedel-like structures (SCKs), 1275–1276
Shell cross-linked (SCL) micelle, 1276–1277, 1282
Simple coacervation, 9, 236–267, 854, 872, 1082
Simple emulsions, 1388
Simultaneous Homogenizing and Mixing (SHM), 474
Single screw extruders (SSE), 214, 1127
    cross-section of, 215–216
    functions of, 214
    limitations, 214
    vs. TSE, 215
    zones, 214
SLM, *see* Solid lipid microparticles (SLM)
SLNs, *see* Solid lipid nanoparticles (SLNs)
Small angle x-ray scattering (SARS/SAXS), 713–714
Small unilamellar vesicles (SUVs), 657
Sodium alginate, 1375
Sodium carboxymethyl cellulose (SCMC), 498, 523, 536, 538, 548, 552–555, 558, 1243
Sodium dodecyl sulfate (SDS), 317, 753, 1077
Sodium starch glycolate (SSG), 577
Soft contact lenses (SCLs)
    cyclodextrins, 1188–1189
    drug-eluting conventional contact lenses, 1183–1185
    hydrophilic and non-hydrophilic, 1186–1187
    imprinted hydrogel, 1191–1192
    pendant cyclodextrins, 1189–1191
    surface-immobilized liposomes, 1188–1189
Sol-gel encapsulation, 330, 340–341, 1466–1467
Sol-gel microencapsulation
    adsorption isotherms, 332–333
    advantages, 335–336
    capsule burst, 339
    chemistry, 330
    commercial cosmetic preparations, 344
    controlled release, 337–339
    curing agents, 342–343
    flavor and fragrance, 340–342
    gelation, 332, 334
    hydrolysis, 332
    hydrophobicity, 338–339
    hydrophobic molecules, 331
    interfacial polymerization process, 330
    ionic detergents, 332, 334
    Laplace pressure, 338–339
    lyotropic mesophases, foams and emulsion, 330
    oil-in-water emulsions, 331
    particle size and release rate, 330
    patents, 336–337
    PCM, 340–341
    pore size distribution, 332–333
    Si alkoxides, 339
    sorts, molecules, 330–331
    sunscreens, 334–335, 337–338
    system materials, 331
    TEOS, 333, 344
    1000 L pilot plant built at Ceramisphere, 344
    UVA and UVB filters, 344
Solid lipid microparticles (SLM), 999, 1387
    microparticulate drug delivery systems, 1097–1098
    spray chilling, 74–75, 80–84
Solid lipid nanoparticles (SLNs)
    advantages, 1154
    BBB, 1155–1156
    excipients, lipids as, 1154
    high pressure homogenization, 1154–1155
    microemulsion method, 1154–1155
    proteins and peptides oral delivery, 1155
    surfactant and cosurfactant, 1155
Soluble ocular drug insert (SODI), 1179
Soluble Soybean Polysaccharide (SSP), 500
Soluplus®, 223
Solvent evaporation
    advantages, 1073
    agitation rate, 1077
    aqueous phase volumes, 1078
    baffles, 1077
    bioactive compounds incorporation, 1074
    diffusion rate, 1078
    dispersed phase volume, 1077–1078
    droplet formation, 1074–1075
    harvest and drying, 1075
    loading extent, 1077
    material selection
        antifoam, 1077
        cosolvent, 1076
        polymer, 1075–1076
        solvents, 1075–1076
        surfactant, 1076–1077
    polymer concentration, 1077
    preformed polymers, 1374
    solvent removal, 1075
    stabilizer concentration, 1077
    temperature, 1077
Soy protein isolate (SPI), 505, 804, 818, 820, 822
Span 80, 708, 1039, 1394
Sphingomones elodea, 279

SPI, *see* Soy protein isolate (SPI)
Spray chilling, 13, 805
    active ingredient, 83–84
    bioactive components, 73–74
    bioactive compounds, food, 783
    chemical industries, 72
    cooling chamber and atomizer, 72
    disadvantage, 77
    drugs/cosmetics, 76
    DSC, 80–82
    essential oils, 870
    evaluation and characterization, 77–78
    feedstock, 72
    food industries, 71–72
    FTIR, 81, 82
    hydrogenated palm oil, 84
    melting conditions, 72–73
    microparticles, 71
    microspheres, 71–72
    particle engineering technologies, 72
    particle size and morphology, 78–79
    phosphate buffer pH, 84
    probiotic, 74–75
    SEM, 80–81
    solid-lipid particles, 73
    uses, 73
    vitamins, 76–77
    x-ray powder diffraction, 82–83
Spray congealing, *see* Spray chilling
Spray cooling, *see* Spray chilling
Spray drying (SD), 804–805
    advantages, 47–48
    air-heating system, 56
    aromas, 841–843
        atomizers, 845–847
        emulsion, 847–849
        fine emulsion, 843
        operating conditions, 843–844
        parameters, 844–845
        powder characteristics, 844
        principle, 842–843
        retention efficiency, 849–851
        safety considerations, 841
        schematic diagram, 844
        spraying mode selection, 845–847
    atomization
        centrifugal atomizer, 51–52
        droplet size reduction, 35–36
        mechanical atomizer, 38
        pneumatic nozzle, 52–53
        pressure nozzles, 37–38, 52
        two-fluid nozzles/pneumatic nozzles, 37–38
        ultrasonic nozzle, 53
    bioactive compounds, food, 783
    carbohydrate-based matrix capsules, 35
    computational fluid dynamics
        agglomeration model, 58
        centrifugal atomizer, 58
        continuous phase, 58–59
        discretization, 62
        drying chamber dimension, 62
        heat transfer model, 60–61
        k–ε turbulence model, 58, 60
        linearization, 62
        main operating parameters, 62
        mass transfer models, 61
        NIZO-DrySim, 57
        one-way coupling, 57
        particle trajectory, 62–63
        PSI-Cell model, 57
        streamline, 62–63
        temperature contours, 62–63
        3D simulation, 58
        two-way coupling, 57
        velocity vector, 62–63
    core–shell capsules, 27
    dairy industry, 48
    disadvantages, 48
    dried-particle-collection system, 56
    essential oils, 870–871
    evaporation and particle formation, 39–41
    food industry, 12
        activation energy, 652
        carbohydrates, 651
        coffee, 66–67
        dairy products, 649
        eggs, 49, 67
        isothermal drying curve, 651–652
        liquid atomization, 649
        liquid nebulization, 650
        maltodextrin and gum arabic, 651
        microcapsules, 650–651
        milk, 63–65
        spray-dried particles, 650–651
        stages, 649–650
        tea extracts, 66
        tomato juice, 65
        wall materials, 652
    granulation technology
        advantages, 395
        air disperser, 393–394
        atomization, 390–391
        disadvantages, 395
        drying chamber, 393–394
        pharmaceutical spray-dried products, 394–395
        powder separation, 394
        pressure nozzles, 392
        principle, 389–390
        rotary atomizers, 391–392
        two-fluid/pneumatic nozzles, 392–393
        ultrasonic atomization, 393
MPCM, 1467
nanoparticles, 25
nutritional oils
    confocal laser scanning microscopy, 43
    encapsulated oils, 43
    interfacial dilatational rheology, 44
    octenylsuccinate-derivatized starch, 42–43
    oxidative stability, 42
    particle ballooning, 42
    polyunsaturated fatty acids, 42
    spray granulation, 42
one-stage spray drier, 36–37
organic solvents/water, 30
polyphenolics, 746–747
preformed polymers, 1374

Index

pregelatinized starch, 578
principles
  closed-cycle spray-drying plants, 48–49
  cocurrent flow configuration, 50
  flow diagram, 48–49
  foodstuffs, 48–49
  industrial spray drying, 48–49
  three-stage system, 50–51
  two-stage spray dryer, 50
  VFB, 50–51
proteins, 504
pulsed combustion spray drying, *see* Pulsed combustion spray drying
spray-air contact, 36, 38–39
  drying chamber designs, 55
  hot air distribution, 54–55
starch and starch derivatives, 495–496
system control, 56
Spray-fluidizers, 50
Square cube law, 360
SSE, *see* Single screw extruders (SSE)
Stable core-shell microcapsules
  formaldehyde-free systems, 1429
    acid catalyzed methylene bridge formation, 1430, 1432
    acidic precondensate formation, 1430, 1432
    alkaline precondensate formation, 1430, 1433
    in aqueous dispersion, 1433
    aromatic/heteroaromatic alcohol, 1430
    carbocations and carboanions, 1430
    monodisperse particles, 1433
    novolac- and resole-formation, 1430–1431
    phloroglucinol, 1430
    Pi–Pi interactions, 1430
    structural formula selection, 1430
  self-healing system
    anti-friction coatings, 1435–1436
    core material oxidation, 1434–1435
    PCM technology, 1435–1436
    principle of, 1434
    protective coating, 1435
  shell materials, 1423–1424
Starburst® PAMAM dendrimers, 1163
Starches
  amylopectin, 575–576
  amylose, 575–576
  dried starch utilization, 575
  maltodextrin/dextrin, 576
  modified starches
    chemical modification, 576, 578
    enzymatic modification, 576, 578
    gene therapy, 579–580
    nanoparticles, 579
    pharmaceutical industry, 576–577
    physical modification, 576, 578
    plasma volume expanders, 580
    solid dosage form, 577–578
    structure, 576
    sustained release system, 578–579
    vaccine, 580
  native starches
    drug development and drug delivery, 576–577
    structure, 576
  sources, 575

Stealth character, 588
Steam granulation, 406
Stokes law, 316, 840
Sublingual mucosa, 1228–1229
Supercritical antisolvent (SAS) process, 750
  applications, 452–454
  disadvantage, 452
  organic solvents, 451–452
  quercetin in Pluronic® F127 poloxamer
    amorphous polymer matrix, 463
    intestinal fluid, 463
    Pluronic-CO_2 mixtures, 459, 462
    pressure and temperature conditions, 460
    schematic representation, 460
    SEM micrographs, 461–462
    x-ray diffractograms, 462–463
  schematic diagram, 451
Supercritical carbon dioxide (SC-CO_2)
  batch GAS process, 449–450
  PGSS, 455, 457–459
  polymeric carrier materials, 457, 459
  RESS, 449
  SAS, *see* Supercritical antisolvent (SAS) process
  SFEE, 452, 455–456
Supercritical CO_2-assisted method, 1467
Supercritical fluid (SCF)
  precipitation processes, *see* Supercritical carbon dioxide (SC-CO_2)
  properties, 447–448
Supercritical fluid extraction of emulsions (SFEE), 452, 455–456
Super-hydrophilic hydrogels (SHHs), 1189, 1191
Supernatant, 238
Supramolecular organogels, 1040, 1392–1393
Surodex1, 1181
Suspension polymerization, 1373–1374, 1465
Synthetic biodegradable polymeric nano/microparticles
  disadvantages, 1364–1365
  drug absorption, lymphatic system, 1363
  FAE, 1364
  first-pass metabolism, 1363
  GALT, 1364
  insulin-loaded poly(lactic-*co*-glycolic acid) nanoparticles, 1366
  lectins, 1365
  linear polyesters, 1366
  M-cells, 1364
  Peyer's patches uptake, 1363–1364
  pharmacokinetic/pharmacodynamics analysis, 1367
  PIBCA nanocapsules, 1366
  PLA and PLGA, 1366
  PLGA nanoparticles, 1364
  poly(alkyl cyanoacrylate) nanocapsules, 1366
  RES clearance, 1365
Synthetic biodegradable polymers
  advantages, 584
  PCL, 592, 594
  PDS, 592, 594–595
  PHB, 592–594
  PLA, 592–593
  POEs, 592, 596
  polyamides, 595
  polyanhydrides, 595
  polycarbonates, 592, 596

polyglycolide, 592–593
poly(lactide-*co*-glycolide), 592–593
polyphosphazenes, 592, 596–597
polyphosphoesters, 592, 597
polyurethanes, 592, 596
POM, 592, 595
PPF, 592, 594

T

Tandem repeat array, 1244
TEM, *see* Transmission electron microscopy (TEM)
Tetradecyl trimethyl ammonium bromide (TTAB), 563
Tetraethoxysilane (TEOS), 332–334, 339–342, 344
Tetrahydrofuran (THF), 430–431, 519
Thermal adhesion granulation (TAG) process, 408
Thermal denaturation, 1343, 1374–1375
Thermal energy storage, 1456–1457, 1465
Thermochemical energy storage, 1456
Thermo gravimetric analysis (TGA), 221, 224, 1469
Thermoplastic granulation, *see* Melt granulation technology
Thermoreversible hydrogels, 279
Thiolated polymers
 advantages, 1245
 chitosan–thiobutylamidine conjugate, 1245
 disulfide bond, 1244
 in situ cross-linking effect, 1245
 PACAP, 1246
 polycarbophil–cysteine conjugates, 1245–1246
 polypeptide backbone of mucin, 1244
Thiomers, *see* Thiolated polymers
Time–temperature indicators, 1444
Transdermal drug delivery, 1059, 1267, 1383, 1397–1398
Transdermal drug delivery system (TDDS), 259, 556
Transitional phase inversion (PIT), 257–258
Transmission electron microscopy (TEM), 325, 1347, 1468
Triple emulsions, 1390
Tryptophan at position 19 (Trp19), 731
TSE, *see* Twin-screw extruder (TSE)
Turbidimetry, PECs, 1341
Tween 80, 1394
Twin-screw extruder (TSE), 1127–1128
 characteristics, 214–215
 corotating/counte-rotating, 214
 cross-section of, 215–216
 schematic presentation of, 214–215
 vs. SSE, 215

U

Urea-formaldehyde (UF) systems
 functionality/application, 307–308
 oil-in-water phase, 308
 temperature and stirring, 310

V

Vegetable oil–based formulations
 classification, 1384–1385
 commercially available vegetable oil–based products, 1398–1400
 composition, 1382
 emulsion
  immiscible liquids, 1388
  macroemulsion, 1390, 1397–1398
  microemulsions, 1388–1389, 1397
  multiple emulsions, 1388, 1390
  pickering emulsions, 1388
  stability, 1388
  types, 1388–1389
 free fatty acids, 1383–1384
 gels
  application, 1398
  bigels, 1395–1396
  definition, 1391
  emulgels, 1393–1395
  lyotropic gels, 1395–1397
  organogels, 1391–1393
 hydrophilic and lipophilic drug delivery, 1384
 ionic and nonionic surfactants, 1383
 liposomes, 1385–1386, 1397
 micelles, 1384–1385, 1397
 phospholipids, 1383
 pH-responsive food-grade lyotropic liquid crystals, 1398
 raw materials, 1382–1383
 solid lipid nanoparticles, 1387–1388
 sustained-release liquid crystal, 1400
 triacylglycerides, 1383
 U-type microemulsions, 1400
Vented pan-coating
 application, 173
 charging and discharging, 156
 design, 148, 150, 153–154
 exhaust filter, 156
 spray guns, 155
 ventilation, 153–155
Vibrated fluid bed (VFB), 50–51, 65
Vibrating-jet technique, 180, 185–187
Vibrational nozzle technology, 7, 399
Vinyl sulfone-terminated poly(ethylene glycol) (PEG-VS), 282–283
Viscosity measurement, density, and conductivity, 1469
Viscosity, PECs, 1342–1343
Voorn theory, 1336

W

Water drying capacity
 air capacity *vs.* temperature, 135–136
 dew points, 136–138
 saturation value, temperature, 138–139
Water-in-oil microemulsion
 Capmul MCM, Tween 80 and water, 261
 Miglyol 812, 261
 nasal route, 262
 oral bioavailability, 261–262
 skin permeation, diclofenac, 261
 structure and composition, 249–250
 TAMRA-TAT, 261
Water-in-oil microemulsions, 1389
Water vapor sorption (WVS), 222
Waxy starch, 575–576, 580
Weber number, 95, 186, 363, 1001
Web of Science database, 644–645

Index

Wet granulation, 388–389, 397, 400–402, 406–408, 538, 554, 577
Wet stirred media milling (WSMM), 561
Wheat germ agglutinin (WGA), 1371
Wide angle x-ray scattering (WAXS), 1468
Wide pneumatic nozzle, 486
William etherification, 547
Wound dressing
    cell viability, attachment, and proliferation, 1323
    dual drug scaffold, 1321
    dual spinneret electrospinning apparatus, 1322
    fiber scaffolds, 1322–1323
    lidocaine, 1321, 1323
    mupirocin, 1321
WSMM, *see* Wet stirred media milling (WSMM)
Wurster coating, 10, 32, 116, 127, 1322
Wurster process
    fluidized bed coating technology
        coating chamber, 115–116
        cylindrical design, 115
        down-bed region, 115–116
        particle size distributions, 114, 118
        up-bed region, 115–116
    granulation technology, 396–397
    process air
        air distribution plates, 124–125
        air feeding, 126
        air velocity, 126–127
        down bed air flow, 124
        multiple nozzle configurations, 128–129
        partition parameters, 127
        total process air, 125–126

X

X-ray diffraction (XRD)
    HME, 221
    microparticulate drug delivery systems, 1091
    MPCM, 1468
    PECs, 1345
    schematic diagram, 1047

Z

Zidovudine (AZT), 556
Zwitterionic surfactants, 253–254